Standard Atomic Weights of the Elements 2009, IUPAC

Based on Relative Atomic Mass of $^{12}C = 12$, where ^{12}C is a neutral atom in its nuclear and electronic ground state.[1]

Name	Symbol	Atomic Number	Atomic Weight
Actinium[2]	Ac	89	(227)
Aluminum	Al	13	26.981 5386(8)
Americium[2]	Am	95	(243)
Antimony	Sb	51	121.760(1)
Argon	Ar	18	39.948(1)
Arsenic	As	33	74.921 60(2)
Astatine[2]	At	85	(210)
Barium	Ba	56	137.327(7)
Berkelium[2]	Bk	97	(247)
Beryllium	Be	4	9.012 182(3)
Bismuth	Bi	83	208.980 40(1)
Bohrium[2]	Bh	107	(272)
Boron	B	5	10.811(7)
Bromine	Br	35	79.904(1)
Cadmium	Cd	48	112.411(8)
Calcium	Ca	20	40.078(4)
Californium[2]	Cf	98	(251)
Carbon	C	6	12.0107(8)
Cerium	Ce	58	140.116(1)
Cesium	Cs	55	132.905 4519(2)
Chlorine	Cl	17	35.453(2)
Chromium	Cr	24	51.9961(6)
Cobalt	Co	27	58.933 195(5)
Copper	Cu	29	63.546(3)
Curium[2]	Cm	96	(247)
Darmstadtium[2]	Ds	110	(281)
Dubnium[2]	Db	105	(268)
Dysprosium	Dy	66	162.500(1)
Einsteinium[2]	Es	99	(252)
Erbium	Er	68	167.259(3)
Europium	Eu	63	151.964(1)
Fermium[2]	Fm	100	(257)
Fluorine	F	9	18.998 4032(5)
Francium[2]	Fr	87	(223)
Gadolinium	Gd	64	157.25(3)
Gallium	Ga	31	69.723(1)
Germanium	Ge	32	72.64(1)
Gold	Au	79	196.966 569(4)
Hafnium	Hf	72	178.49(2)
Hassium[2]	Hs	108	(277)
Helium	He	2	4.002 602(2)
Holmium	Ho	67	164.930 32(2)
Hydrogen	H	1	1.00794(7)
Indium	In	49	114.818(3)
Iodine	I	53	126.904 47(3)
Iridium	Ir	77	192.217(3)
Iron	Fe	26	55.845(2)
Krypton	Kr	36	83.798(2)
Lanthanum	La	57	138.905 47(7)
Lawrencium[2]	Lr	103	(262)
Lead	Pb	82	207.2(1)
Lithium	Li	3	[6.941(2)]†
Lutetium	Lu	71	174.9668(1)
Magnesium	Mg	12	24.3050(6)
Manganese	Mn	25	54.938 045(5)
Meitnerium[2]	Mt	109	(276)
Mendelevium[2]	Md	101	(258)
Mercury	Hg	80	200.59(2)
Molybdenum	Mo	42	95.96(2)
Neodymium	Nd	60	144.242(3)
Neon	Ne	10	20.1797(6)
Neptunium[2]	Np	93	(237)
Nickel	Ni	28	58.6934(4)
Niobium	Nb	41	92.906 38(2)
Nitrogen	N	7	14.0067(2)
Nobelium[2]	No	102	(259)
Osmium	Os	76	190.23(3)
Oxygen	O	8	15.9994(3)
Palladium	Pd	46	106.42(1)
Phosphorus	P	15	30.973 762(2)
Platinum	Pt	78	195.084(9)
Plutonium[2]	Pu	94	(244)
Polonium[2]	Po	84	(209)
Potassium	K	19	39.0983(1)
Praseodymium	Pr	59	140.907 65(2)
Promethium[2]	Pm	61	(145)
Protactinium[2]	Pa	91	231.035 88(2)
Radium[2]	Ra	88	(226)
Radon[2]	Rn	86	(222)
Rhenium	Re	75	186.207(1)
Rhodium	Rh	45	102.905 50(2)
Roentgenium[2]	Rg	111	(280)
Rubidium	Rb	37	85.4678(3)
Ruthenium	Ru	44	101.07(2)
Rutherfordium[2]	Rf	104	(267)
Samarium	Sm	62	150.36(2)
Scandium	Sc	21	44.955 912(6)
Seaborgium[2]	Sg	106	(271)
Selenium	Se	34	78.96(3)
Silicon	Si	14	28.0855(3)
Silver	Ag	47	107.8682(2)
Sodium	Na	11	22.989 769 28(2)
Strontium	Sr	38	87.62(1)
Sulfur	S	16	32.065(5)
Tantalum	Ta	73	180.947 88(2)
Technetium[2]	Tc	43	(98)
Tellurium	Te	52	127.60(3)
Terbium	Tb	65	158.925 35(2)
Thallium	Tl	81	204.3833(2)
Thorium[2]	Th	90	232.038 06(2)
Thulium	Tm	69	168.934 21(2)
Tin	Sn	50	118.710(7)
Titanium	Ti	22	47.867(1)
Tungsten	W	74	183.84(1)
Uranium[2]	U	92	238.028 91(3)
Vanadium	V	23	50.9415(1)
Xenon	Xe	54	131.293(6)
Ytterbium	Yb	70	173.054(5)
Yttrium	Y	39	88.905 85(2)
Zinc	Zn	30	65.38(2)
Zirconium	Zr	40	91.224(2)
—[2,3,4]		112	(285)
—[2,3]		113	(284)
—[2,3]		114	(287)
—[2,3]		115	(288)
—[2,3]		116	(293)
—[2,3]		118	(294)

1. The atomic weights of many elements vary depending on the origin and treatment of the sample. This is particularly true for Li; commercially available lithium-containing materials have Li atomic weights in the range of 6.939 and 6.996. Uncertainties are given in parentheses following the last significant figure to which they are attributed.
2. Elements with no stable nuclide; the value given in parentheses is the atomic mass number of the isotope of longest known half-life. However, three such elements (Th, Pa, and U) have a characteristic terrestrial isotopic composition, and the atomic weight is tabulated for these.
3. Not yet named.
4. The name copernicium was proposed for element 112 in July 2009, but at that time this name had not been officially accepted by IUPAC.

FOURTH EDITION

Chemistry

THE MOLECULAR SCIENCE

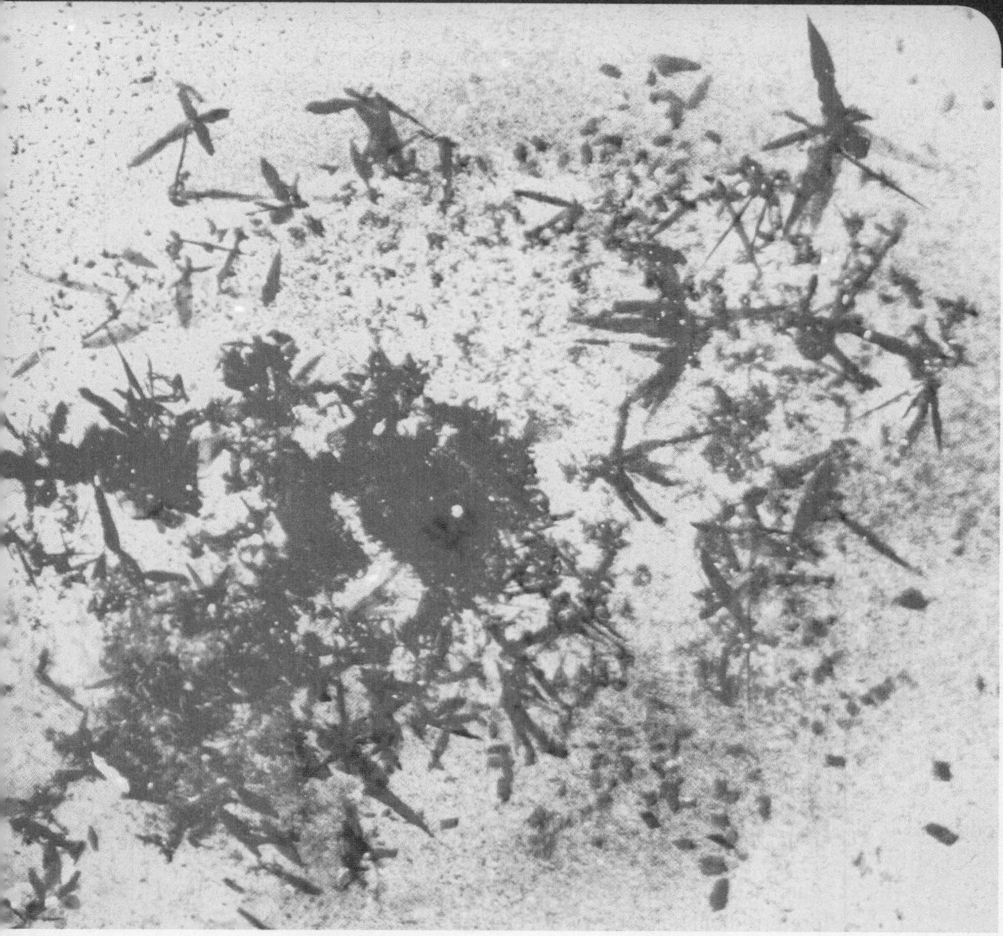

John W. Moore
University of Wisconsin–Madison

Conrad L. Stanitski
Franklin and Marshall College

Peter C. Jurs
Pennsylvania State University

BROOKS/COLE
CENGAGE Learning™

Australia • Brazil • Japan • Korea • Mexico • Singapore • Spain • United Kingdom • United States

BROOKS/COLE
CENGAGE Learning™

***Chemistry: The Molecular Science*, Fourth Edition**
John W. Moore, Conrad L. Stanitski, Peter C. Jurs

Publisher: Mary Finch
Executive Editor: Lisa M. Lockwood
Acquisitions Editor: Kilean Kennedy
Senior Developmental Editor: Peter McGahey
Assistant Editors: Ashley Summers, Liz Woods
Editorial Assistant: Laura Bowen
Media Editors: Lisa Weber, Stephanie VanCamp
Marketing Manager: Nicole Hamm
Marketing Assistant: Kevin Carroll
Marketing Communications Manager: Linda Yip
Content Project Manager: Teresa L. Trego
Creative Director: Rob Hugel
Art Director: John Walker
Print Buyer: Judy Inouye
Rights Acquisitions Account Manager, Text: Timothy Sisler
Rights Acquisitions Account Manager, Image: Don Schlotman
Production Service: Graphic World Inc.
Text Designer: tani hasegawa
Photo Researcher: Emma Hopson/Bill Smith Group
Copy Editor: Graphic World Inc.
OWL Producers: Stephen Battisti, Cindy Stein, and David Hart in the Center for Educational Software Development at the University of Massachusetts, Amherst, and Cow Town Productions
Cover Designer: John Walker
Cover Image: All Images © John W. Moore and James H. Maynard, University of Wisconsin–Madison
Compositor: Graphic World Inc.

© 2011, 2008 Brooks/Cole, Cengage Learning

ALL RIGHTS RESERVED. No part of this work covered by the copyright herein may be reproduced, transmitted, stored, or used in any form or by any means, graphic, electronic, or mechanical, including but not limited to photocopying, recording, scanning, digitizing, taping, Web distribution, information networks, or information storage and retrieval systems, except as permitted under Section 107 or 108 of the 1976 United States Copyright Act, without the prior written permission of the publisher.

For product information and technology assistance, contact us at **Cengage Learning Customer & Sales Support, 1-800-354-9706.**

For permission to use material from this text or product, submit all requests online at **www.cengage.com/permissions.**
Further permissions questions can be emailed to **permissionrequest@cengage.com.**

Library of Congress Control Number: 2009939226
ISBN-13: 978-1-4390-4930-3
ISBN-10: 1-4390-4930-0

Brooks/Cole
20 Davis Drive
Belmont, CA 94002
USA

Cengage Learning is a leading provider of customized learning solutions with office locations around the globe, including Singapore, the United Kingdom, Australia, Mexico, Brazil, and Japan. Locate your local office at **www.cengage.com/global.**

Cengage Learning products are represented in Canada by Nelson Education, Ltd.

To learn more about Brooks/Cole, visit **www.cengage.com/brookscole.**

Purchase any of our products at your local college store or at our preferred online store **www.CengageBrain.com**

Printed in the United States of America
2 3 4 5 14 13 12 11

To All Students of Chemistry

We intend that this book will help you to discover that chemistry is relevant to your lives and careers, full of beautiful ideas and phenomena, and of great benefit to society. May your study of this fascinating subject be exciting, successful, and fun!

We thank our wives—Betty (JWM), Barbara (CLS), and Elaine (PCJ)—for their patience, support, understanding, and love.

It does not do harm to the mystery
to know a little more about it.

Richard Feynman

About the Authors

John Moore, Conrad Stanitski, and Peter Jurs

John W. Moore received an A.B. magna cum laude from Franklin and Marshall College and a Ph.D. from Northwestern University. He held a National Science Foundation (NSF) postdoctoral fellowship at the University of Copenhagen and taught at Indiana University and Eastern Michigan University before joining the faculty of the University of Wisconsin–Madison in 1989. At the University of Wisconsin, Dr. Moore is W. T. Lippincott Professor of Chemistry and Director of the Institute for Chemical Education. He was Editor of the *Journal of Chemical Education (JCE)* from 1996 to 2009. Among his many awards are the American Chemical Society (ACS) George C. Pimentel Award in Chemical Education and the James Flack Norris Award for Excellence in Teaching Chemistry. He is a Fellow of the ACS and of the American Association for the Advancement of Science (AAAS). In 2003 he won the Benjamin Smith Reynolds Award at the University of Wisconsin–Madison in recognition of his excellence in teaching chemistry to engineering students. Dr. Moore has recently received the third in a series of major grants from the NSF to support development of online chemistry learning materials for the NSF-sponsored National Science Distributed Learning (NSDL) initiative.

Conrad L. Stanitski is Distinguished Emeritus Professor of Chemistry at the University of Central Arkansas and is currently Visiting Professor at Franklin and Marshall College. He received his B.S. in Science Education from Bloomsburg State College, M.A. in Chemical Education from the University of Northern Iowa, and Ph.D. in Inorganic Chemistry from the University of Connecticut. He has co-authored chemistry textbooks for science majors, allied health science students, nonscience majors, and high school chemistry students. Dr. Stanitski has won many teaching awards, including the CMA CATALYST National Award for Excellence in Chemistry Teaching, the Gustav Ohaus–National Science Teachers Association Award for Creative Innovations in College Science Teaching, the Thomas R. Branch Award for Teaching Excellence and the Samuel Nelson Gray Distinguished Professor Award from Randolph-Macon College, and the 2002 Western Connecticut ACS Section Visiting Scientist Award. He was Chair of the American Chemical Society Division of Chemical Education (2001) and has been an elected Councilor for that division. He is a Fellow of the American Association for the Advancement of Science (AAAS). An instrumental and vocal performer, he also enjoys jogging, tennis, rowing, and reading.

Peter C. Jurs is Professor Emeritus of Chemistry at the Pennsylvania State University. Dr. Jurs earned his B.S. in Chemistry from Stanford University and his Ph.D. in Chemistry from the University of Washington. He then joined the faculty of Pennsylvania State University, where he has been Professor of Chemistry since 1978. Jurs's research interests have focused on the application of computational methods to chemical and biological problems, including the development of models linking molecular structure to chemical or biological properties (drug design). For this work he was awarded the ACS Award for Computers in Chemistry in 1990. Dr. Jurs has been Assistant Head for Undergraduate Education at Penn State, and he works with the Chemical Education Interest Group to enhance and improve the undergraduate program. In 1995 he was awarded the C. I. Noll Award for Outstanding Undergraduate Teaching. Dr. Jurs serves as an elected Councilor for the American Chemical Society Computer Division, and he was recently selected as a Fellow of the ACS.

Contents Overview

Detailed Contents

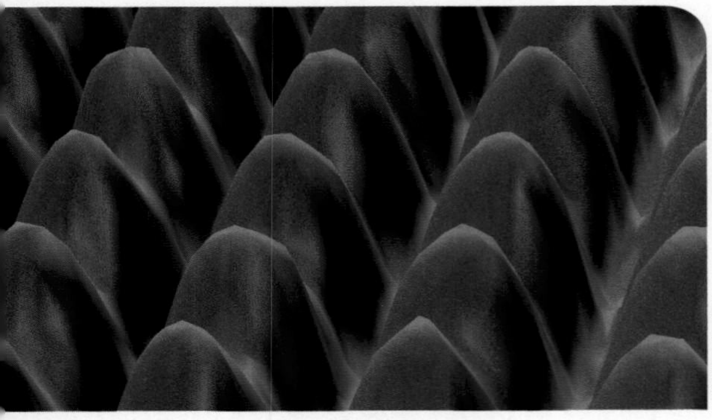
IBM Almaden Labs

© Cengage Learning/Charles D. Winters

© Breitling

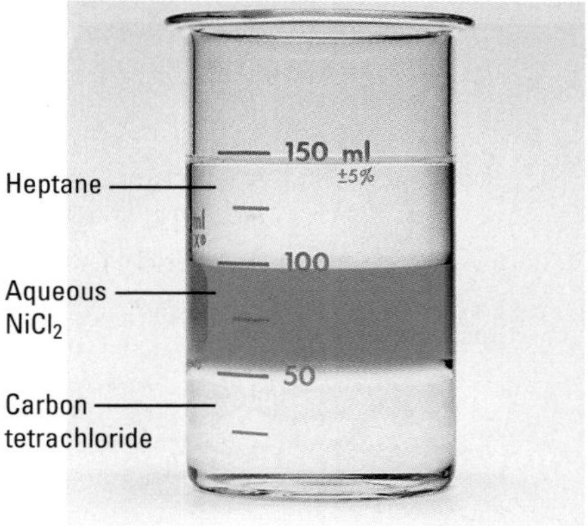
150 ml
±5%
100
50
Heptane
Aqueous $NiCl_2$
Carbon tetrachloride

© Cengage Learning/Charles D. Winters

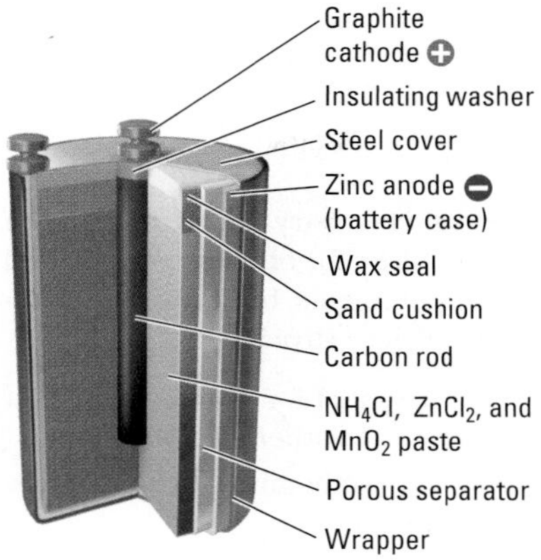
Graphite
cathode ⊕
Insulating washer
Steel cover
Zinc anode ⊖
(battery case)
Wax seal
Sand cushion
Carbon rod
NH_4Cl, $ZnCl_2$, and
MnO_2 paste
Porous separator
Wrapper

Preface

Students have many reasons for taking a two-semester general chemistry course for science majors, but the most likely is that the course is a pre- or co-requisite for other science-related courses or careers. There are important reasons for such requirements, but they are not always obvious to students. The authors of this textbook believe very strongly that

- Students need to recognize that chemical knowledge is essential for solving important problems and that chemistry makes important contributions to other disciplines; and
- It is essential that students gain a working knowledge of how chemistry principles are applied to solve problems in a broad spectrum of applications.

Examples of such applications are creating new and improving existing chemical pathways that lead to the more efficient synthesis of new pharmaceuticals; developing a deeper understanding of alternative energy sources to mitigate global warming; and understanding how new, more efficient catalysts could help to decrease air pollution and to minimize production of chemical waste from industrial processes. Knowledge of chemistry provides a way of interpreting macroscale phenomena at the molecular level that can be applied to many critical 21st century problems, including those just given. This fourth edition of *Chemistry: The Molecular Science* continues our tradition of integrating other sciences with chemistry and has been updated to include a broad range of recent chemical innovations that illustrate the importance of multidisciplinary science.

Goals

Our overarching goal is to involve science and engineering students in active study of what modern chemistry is, how it applies to a broad range of disciplines, and what effects it has on their own lives. We maintain a high level of rigor so that students in mainstream general chemistry courses for science majors and engineers will learn the concepts and develop the problem-solving skills essential to their future ability to use chemical ideas effectively. We have selected and carefully refined the book's many unique features in support of this goal.

More specifically, we intend that this textbook will help students develop:

- A broad overview of chemistry and chemical reactions,
- An understanding of the most important concepts and models used by chemists and scientists in chemistry-related fields,
- The ability to apply the facts, concepts, and models of chemistry appropriately to new situations in chemistry, to other sciences and engineering, and to other disciplines,
- Knowledge of the many practical applications of chemistry in other sciences, in engineering, and in other fields,
- An appreciation of the many ways that chemistry affects the daily lives of all people, students included, and
- Motivation to study in ways that help all students achieve real learning that results in long-term retention of facts and concepts and how to apply them.

Because modern chemistry is inextricably entwined with so many other disciplines, we have integrated organic chemistry, biochemistry, environmental chemistry, industrial chemistry, and materials chemistry into the discussions of chemical principles and facts. Applications in these areas are discussed together with the principles on which they are based. This approach serves to motivate students whose interests lie in related disciplines and also gives a more accurate picture of the multidisciplinary collaborations so prevalent in contemporary chemical research and modern industrial chemistry.

Audience

Chemistry: The Molecular Science is intended for mainstream general chemistry courses for students who expect to pursue further study in science, engineering, or science-related disciplines. Those planning to major in chemistry, biochemistry, biological sciences, engineering, geological sciences, agricultural sciences, materials science, physics, and many related areas will benefit from this book and its approach. The book has an extensive glossary and an excellent index, making it especially useful as a reference for study or review for standardized examinations, such as the MCAT.

We assume that the students who use this book have a basic foundation in mathematics (algebra and geometry) and in general science. Almost all will also have had a chemistry course before coming to college. The book is suitable for the typical two-semester sequence of general chemistry, and it has also been used quite successfully in a one-semester accelerated course that presumes students have a strong background in chemistry and mathematics.

New in This Edition

Users of the first three editions of this book have been most enthusiastic about its many features and as a result have provided superb feedback that we have taken into account to enhance its usefulness to students. Reviewers have also been helpful in pointing out things we could improve. Like the third edition, this fourth edition is a thorough revision. Although the art program in the first edition won an award for visual excellence, in preparation for this fourth edition we have had every figure critically reviewed. Based on those reviews we have updated nearly all of the art to further enhance a student reader's ability to visualize molecular-scale processes and to connect these processes with real-world, macroscale phenomena. We have also enhanced popular, pedagogically sound features, such as *Chemistry in the News, Chemistry You Can Do, Estimation, Portrait of a Scientist,* and *Tools of Chemistry.* Most of these features have been updated; nearly every *Chemistry in the News* is entirely new.

Our emphasis on conceptual understanding continues. We have revised the text and created additional conceptual questions at the ends of the chapters to help students gain a thorough mastery of important chemical principles. We have moved some sections from one chapter to another and reorganized content to present the material in the most logical way possible. We continue to use pedagogical research reported in recent articles in the *Journal of Chemical Education* that points the way toward teaching methods and writing characteristics that are most effective in helping students learn chemistry and retain their knowledge over the long term.

To support our emphasis on developing students' ability to approach problems systematically and logically, we have placed additional emphasis on the approach to problem solving that we have used in all three previous editions. In each chapter we have added text in the margin to remind students that in solving problems they should analyze the problem, plan a solution, execute the plan, and check that the result is reasonable. We have also more directly called to students' attention how to use the Exercises, Conceptual Exercises, Problem-Solving Examples, and Problem-Solving Practice Problems in each chapter, and the Questions for Review and Thought at the end of each chapter. We have added 226 new questions at the ends of the chapters, and a much larger fraction of the Questions for Review and Thought are accompanied by OWL assignments that will help students learn appropriate problem-solving techniques. In this new edition, solving real problems has been a major focus of the revision.

The **PROBLEM-SOLVING STRATEGY** in this book is

- Analyze the problem
- Plan a solution
- Execute the plan
- Check that the result is reasonable

Appendix A.1 explains this in detail.

Specifically, we have made these global changes from the third edition:

- Carefully examined each piece of art with respect to scientific accuracy and pedagogical efficacy, modifying or replacing figures whenever doing so would improve students' ability to understand the point being made;
- Re-emphasized our problem-solving approach to make it easier for students to remember and follow;
- Revised many Problem-Solving Examples, introducing a bullet style to the Strategy and Explanation section so that students can more easily see a step-by-step approach to the problem;
- Reworked text in many places into bullet format to make it easier for students to identify the most important ideas and to return to them for review and further study;
- Updated existing and added new pedagogically sound features: *Chemistry in the News, Chemistry You Can Do, Estimation, Portrait of a Scientist,* and *Tools of Chemistry;*
- Revised the end-of-chapter questions to provide better organization and increased the number of questions by 226;
- Added at the ends of many chapters new and unique questions, grid questions, that are based on cognition research results;
- Greatly increased the number of end-of-chapter questions that are associated with parameterized assignments in OWL;

- Correlated Go Chemistry mini-lecture videos for iPods and other mobile devices to book sections;
- Made use of the most accurate and up-to-date sources for data such as atomic weights, electronegativities, ionization energies, atomic and ionic radii, acid ionization constants, solubility product constants, and standard reduction potentials, and updated all tables, problem-solving examples, exercises, and appendixes to reflect the best data;
- Added newly discovered elements and updated atomic weight values (IUPAC) to periodic tables and data tables throughout the book;
- Updated the definitions in the extensive glossary and improved the index.

Revisions to each chapter include

Chapter 1

- Revised or replaced 20 figures and added a new figure;
- Added new questions about real-world situations that are answered later in the book;
- Emphasized a general approach to solving problems and demonstrated how to apply it to a specific problem;
- Replaced *Chemistry in the News*;
- Added 16 end-of-chapter questions, six of which are More Challenging Questions.

Chapter 2

- Revised most figures and made major changes in six figures;
- Added discussion of atomic force microscopy to *Tools of Chemistry* feature;
- Replaced one Problem-Solving Example;
- Replaced one *Chemistry in the News* and added a second;
- Added two end-of-chapter questions and renumbered questions for a more logical order.

Chapter 3

- Revised or replaced 12 figures and added a new figure;
- Reworked text into bullet format in several places to make it easier for students to identify important points;
- Added a new *Estimation* box;
- Added a new *Chemistry in the News* and updated the existing one;
- Revised four Problem-Solving Examples;
- Added two new Key Terms;
- Added 15 end-of-chapter questions, several of which involve atomic-scale interpretations.

Chapter 4

- Revised or replaced 11 figures;
- Revised six Problem-Solving Examples to make the explanations more vivid to students;
- Updated *Chemistry in the News* feature;
- Added seven new end-of-chapter questions, six with graphics that require students to apply atomic/molecular-scale thinking.

Chapter 5

- Revised or replaced eight figures;
- Reworked text to bullet format in several places to make it easier for students to identify important ideas;
- Revised or replaced eight Problem-Solving Examples;
- Replaced *Chemistry in the News*;
- Added a new Key Term;
- Added seven new end-of-chapter questions, four with graphics that require students to apply atomic/molecular-scale thinking.

Chapter 6

- Revised or replaced 16 figures;
- Reworked text to bullet format in several places to make it easier for students to identify important ideas;
- Added a new Problem-Solving Example;
- Replaced *Chemistry in the News*;
- Reworked material formerly in Chapter 12 to consolidate information on fuels and their importance to society;
- Added 28 new end-of-chapter questions, four with graphics that require students to apply atomic/molecular-scale thinking.

Chapter 7

- Revised or replaced more than 20 figures;
- Completely rewrote five pages to improve clarity;
- Revised and updated data for ionic radii, ionization energies, and electron affinities;
- Added three new Problem-Solving Examples and modified two;
- Added four new Exercises and modified one;
- Added a new *Chemistry in the News*;
- Reworked Sections 7.13 and 7.14 into a single section on bonding in ionic compounds;
- Added six new end-of-chapter questions, two of which are a new type (grid questions) unique to this book.

Chapter 8

- Revised or replaced nine figures;
- Reworked text to bullet format in several places to make it easier for students to identify important ideas;
- Revised or replaced two Problem-Solving Examples;
- Added a new *Chemistry in the News*;
- Completely reworked two subsections on *cis/trans* isomers and resonance in benzene;
- Added eight new end-of-chapter questions, two of which are a new type (grid questions) unique to this book;
- Revised and updated electronegativity data.

Chapter 9

- Revised or replaced 11 figures;
- Reworked text to bullet format in several places to make it easier for students to identify important ideas;
- Added three new Problem-Solving Examples and modified two;
- Replaced *Chemistry in the News*;
- Completely reworked section on Expanded Octets and Hybridization
- Revised the Summary Problem;
- Added six new end-of-chapter questions, two of which are a new type (grid questions) unique to this book.

Chapter 10

- Revised or replaced 11 figures;
- Reworked text and Problem-Solving Examples to bullet format in several places to make it easier for students to identify important ideas;
- Merged Sections 10.4 and 10.5 into a single, more coherent section;
- Replaced or revised three Problem-Solving Examples;
- Replaced one *Chemistry in the News*;
- Added new *Chemistry You Can Do*;
- Added three new end-of-chapter questions with graphics that require students to apply atomic/molecular-scale thinking.

Chapter 11

- Revised or replaced 10 figures;
- Reworked text to bullet format in several places to make it easier for students to identify important ideas;
- Revised and updated treatment of solid-state structure and close-packing of spheres;
- Replaced or edited two Problem-Solving Examples; added a new problem-solving practice;
- Added three new *Chemistry in the News* and deleted two existing ones;
- Moved and edited one subsection to make the presentation clearer;
- Added six new end-of-chapter questions.

Chapter 12

- Revised or replaced six figures;
- Added new material to Section 12.1, Petroleum;
- Completely revised Section 12.2, adding material on U.S. Energy Sources and Consumption;
- Updated and expanded discussion of plastics recycling;
- Reworked and switched order of main topics in Section 12.7, Biopolymers;
- Added new *Estimation* box;
- Added new *Chemistry in the News*;
- Revised *Tools of Chemistry* on MRI;
- Added three new end-of-chapter questions, two of which are a new type (grid questions) unique to this book.

Chapter 13

- Revised or replaced 15 figures;
- Reworked text to bullet format in several places to make it easier for students to identify important ideas;
- Revised three Problem-Solving Practice problems and two exercises;
- Replaced *Chemistry in the News*;
- Reworked the section on catalysis;
- Added 27 new end-of-chapter questions.

Chapter 14

- Revised or replaced 12 figures;
- Reworked text to bullet format in several places to make it easier for students to identify important ideas;
- To reinforce pedagogy, added color coding to section teaching how to solve equilibrium problems;
- Added new section Changing Volume by Adding Solvent to material on LeChatelier's principle;
- Replaced one Problem-Solving Practice;
- Updated *Chemistry in the News*;
- Added 47 new end-of-chapter questions.

Chapter 15

- Revised or replaced eight figures;
- Reworked text to bullet format in several places to make it easier for students to identify important ideas;
- Added one new Problem-Solving Practice problem and one exercise;
- Replaced *Chemistry in the News*;
- Added six new end-of-chapter questions including macro/nano modeling and interpretation of graphical data.

Chapter 16

- Revised or replaced 16 figures;
- Reworked text and Problem-Solving Examples to bullet format in several places to make it easier for students to identify important ideas;
- Updated table of acid ionization constants with the latest data and revised examples that use the new data;
- Revised section on Metal Ions as Acids;
- Revised three Exercises;
- Replaced *Chemistry in the News* with a new one;
- Reworked the section on Lewis acids and bases;
- Added seven new end-of-chapter questions, two of which are a new type (grid questions) unique to this book, and some of which are macro/nano modeling questions.

Chapter 17

- Revised or replaced six figures;
- Reworked text and Problem-Solving Examples to bullet format in several places to make it easier for students to identify important ideas;
- Updated table of solubility product constants with the latest data and revised examples that use the new data;
- Revised coverage of acid rain;
- Revised three Problem-Solving Practice problems and added one new one;
- Replaced *Chemistry in the News*;
- Added four new end-of-chapter questions, two of which are a new type (grid questions) unique to this book, and some of which are macro/nano modeling questions.

Chapter 18

- Revised or replaced 13 figures;
- Reworked text to bullet format in several places to make it easier for students to identify important ideas;
- Added new *Portrait of a Scientist*;
- Updated *Chemistry in the News*;
- Added four new end-of-chapter questions, including two macro/nano modeling questions.

Chapter 19

- Revised or replaced 12 figures;
- Reworked text to bullet format in several places to make it easier for students to identify important ideas;
- Replaced one Problem-Solving Example;
- Added new *Chemistry in the News*.

Chapter 20

- Revised or replaced one figure;
- Added new *Portrait of a Scientist*;
- Added new *Chemistry in the News*.

Chapter 21

- Added two new figures;
- Updated data to latest, best values for all elemental groups in the periodic table;
- Added new *Chemistry in the News*;
- Added new *Portrait of a Scientist*.

Chapter 22

- Updated *Estimation* box.

Appendixes

- Expanded Appendix A coverage of problem solving;
- Updated Appendix C to include latest values of physical constants and references to sources of data;
- Updated Appendix D with most recent references on electron configurations of the elements;
- Updated Appendix F with consistent values from a standard compilation of data;
- Updated Appendix G with consistent values from a standard compilation of data;
- Created a new Appendix H with solubility product data from a standard reference source;
- Updated Appendix I with consistent values from a standard compilation of data;
- Completely revised atomic weights in data table and periodic table on endpapers to latest values from IUPAC.

Features

We strongly encourage students to understand concepts and to learn to apply those concepts to problem solving. We believe that such understanding is essential if students are to be able to use what they learn in subsequent courses and in their future careers. All too often we hear professors in courses for which general chemistry is a prerequisite complain that students have not retained what they were taught in general chemistry. This book is unique in its thoughtful choice of features that address this issue and help students achieve long-term retention of the material.

Problem Solving

This book places major emphasis on helping students learn to approach and solve real problems. Problem solving is introduced in Chapter 1, and a framework is built there that is followed throughout the book. Four important components of our strategy for teaching problem solving are

- *Problem-Solving Example/Problem-Solving Practice* problems that outline how to approach and solve a specific problem, check the answer, and practice a similar problem;
- *Estimation* boxes that help students learn how to do back-of-the-envelope calculations and apply concepts to new situations;
- *Exercises,* many of which deal with conceptual learning and are identified as *Conceptual Exercises,* that follow introduction of new material and for which answers are not immediately available, forcing students to work out the Exercise before seeing the answer;
- *General Questions, Applying Concepts, More Challenging Questions,* and *Conceptual Challenge Problems* at the end of each chapter that are not keyed to specific textual material and require integration of concepts and out-of-the-box thinking to solve.

Problem-Solving Example/Problem-Solving Practice Each chapter contains many worked-out *Problem-Solving Examples*—a total of 257 in the book as a whole. Most consist of five parts:

- a *Question* (problem);
- an *Answer,* stated briefly;
- a *Strategy and Explanation* section that outlines one approach to analyzing the problem, planning a solution, and executing the plan, thereby providing significant help for students whose answer did not agree with ours;
- a *Reasonable Answer Check* section marked with a ☑ that indicates how a student could check whether a result is reasonable; and
- a companion *Problem-Solving Practice* that provides a similar question or questions, with answers appearing only in an Appendix.

We explicitly encourage students first to analyze the problem, plan a solution, and work out an answer without looking at either the Answer or the Explanation, and only then to compare their answer with ours. If their answer did not agree with ours, students are asked to repeat their work. Only then do we suggest that they look at the Strategy and Explanation, which is couched in conceptual as well as numeric terms so that it will improve students' understanding, not just their ability to answer an identical question on an exam. The Reasonable Answer Check section helps students learn how to use estimated results and other criteria to decide whether an answer is reasonable, an ability that will serve them well in the future. By providing related *Problem-Solving Practice* problems that are answered only in the back of the book, we encourage students to immediately consolidate their thinking and improve their ability to apply their new understanding to other problems based on the same concept.

An example Problem-Solving Example and Problem-Solving Practice taken from Chapter 1 is shown below. It explicitly describes the strategy of analyzing the problem, planning a solution, executing the plan, checking that the answer is reasonable, and solving another similar problem.

PROBLEM-SOLVING EXAMPLE 1.1 Density

In an old movie thieves are shown running off with pieces of gold bullion that are about a foot long and have a square cross section of about six inches. The volume of each piece of gold is 7000 mL. Calculate the mass of gold and express the result in pounds (lb). Based on your result, is what the movie shows physically possible? (1 lb = 454 g)

Answer 1.4×10^5 g; 300 lb; probably not

Strategy and Explanation A good approach to problem solving is to (1) analyze the problem, (2) plan a solution, (3) execute the plan, and (4) check your result to see whether it is reasonable. (These four steps are described in more detail in Appendix A.1.)

Step 1: *Analyze the problem.* You are asked to calculate the mass of the gold, and you know the volume.

Analyze the problem.

Step 2: *Plan a solution.* Density relates mass and volume and is the appropriate proportionality factor, so look up the density in a table. Mass is proportional to volume, so the volume either has to be multiplied by the density or divided by the density. Use the units to decide which.

Plan a solution.

Step 3: *Execute the plan.* According to Table 1.1, the density of gold is 19.32 g/mL. Setting up the calculation so that the unit (milliliter) cancels gives

Execute the plan.

$$7000 \cancel{\text{mL}} \times \frac{19.32 \text{ g}}{1 \cancel{\text{mL}}} = 1.35 \times 10^5 \text{ g}$$

This can be converted to pounds

$$1.35 \times 10^5 \cancel{\text{g}} \times \frac{1 \text{ lb}}{454 \cancel{\text{g}}} = 300 \text{ lb}$$

Notice that the result is expressed to one significant figure, because the volume was given to only one significant figure and only multiplications and divisions were done.

☑ **Reasonable Answer Check** Gold is nearly 20 times denser than water. A liter (1000 mL) of water is about a quart and a quart of water (2 pints) weighs about two pounds. Seven liters (7000 mL) of water should weigh 14 lb, and 20 times 14 gives 280 lb, so the answer is reasonable. The movie is not—few people could run while carrying a 300-lb object!

Check that the result is reasonable.

PROBLEM-SOLVING PRACTICE 1.1

Find the volume occupied by a 4.33-g sample of benzene.

Solve another related problem.

Estimation Enhancing students' abilities to estimate results is the goal of the *Estimation* boxes found in most chapters. These are a unique feature of this book. Each Estimation poses a problem that relates to the content of the chapter in which it appears and for which an approximate solution suffices. Students gain knowledge of various means of approximation, such as back-of-the-envelope calculations and graphing, and are encouraged to use diverse sources of information, such as encyclopedias, handbooks, and the Internet.

Exercises To further ensure that students do not merely memorize algorithmic solutions to specific problems, we provide 338 *Exercises,* which immediately follow introduction of new concepts within each chapter. Often the results that students obtain from a numeric Exercise provide insights into the concepts. Most Exercises are thought provoking and require that students apply conceptual thinking. Exercises that are conceptual rather than mathematical are clearly designated as shown below.

Exercises that are designed to test understanding of a concept are identified as conceptual.

CONCEPTUAL EXERCISE 7.13 *g* Atomic Orbitals

Using the same reasoning as was developed for *s, p, d,* and *f* atomic orbitals, what should be the *n* value of the first shell that could contain *g* atomic orbitals, and how many *g* atomic orbitals would be in that shell?

End-of-Chapter Questions At the end of each chapter we provide *General Questions, Applying Concepts, More Challenging Questions,* and *Conceptual Challenge Problems* in addition to the traditional *Review Questions* and *Topical Questions* keyed to the sections in the chapter. General Questions typically involve only one concept or topic, but students are required to think about which concept is needed to answer the question; no immediate indication is given regarding where to look in the chapter for the concept. Applying Concepts questions explicitly require conceptual thinking instead of numerical calculations and are designed to test students' understanding of concepts. It has been clearly established by research on cognition in both chemistry and physics that many students can correctly answer numerical-calculation questions yet not understand concepts well enough to answer simple conceptual questions. Applying Concepts questions have been designed to address this issue. More Challenging Questions are provided so that students' minds can be stretched to link two or more concepts and apply them to a problem. Conceptual Challenge Problems require out-of-the-box thinking and are suitable for group work by students.

Examples, Practice problems, Estimation boxes, Exercises, and End-of-Chapter Questions are all designed to stimulate active thinking and participation by students as they read the text and to help them hone their understanding of concepts. The grand total of more than 600 of these **active-learning items** exceeds the number found in any similar textbook.

Conceptual Understanding

We believe that a sound conceptual foundation is the best means by which students can approach and solve a wide variety of real-world problems. This approach is supported by considerable evidence in the literature: Students learn better and retain what they learn longer when they have mastered fundamental concepts. Chemistry requires familiarity with at least three conceptual levels:

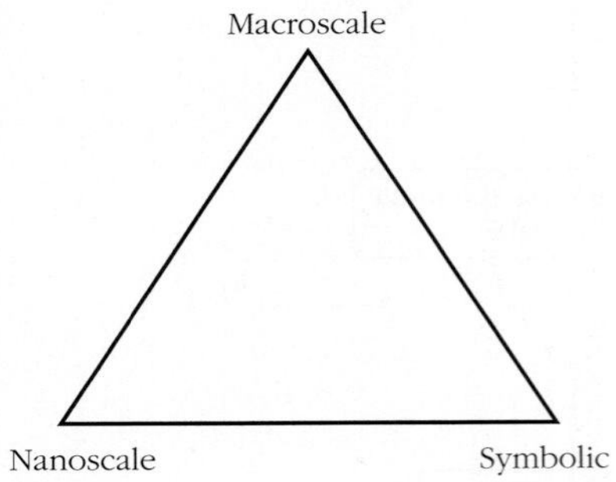

- **Macroscale** (laboratory and real-world phenomena)
- **Nanoscale** (models involving particles: atoms, molecules, and ions)
- **Symbolic** (chemical formulas and equations, as well as mathematical equations)

These three conceptual levels are explicitly defined in Chapter 1. This chapter emphasizes the value of the chemist's unique nanoscale perspective on science and the world with a specific example of how chemical thinking can help solve a real-world problem—how the anticancer agent paclitaxel (Taxol®) was discovered and synthesized in large quantities for use as a drug. This theme of conceptual understanding and its application to problems continues throughout the book. Many of the problem-solving features already mentioned have been specifically designed to support conceptual understanding.

Units are introduced on a need-to-know basis at the first point in the book where they contribute to the discussion. Units for length and mass are defined in Chapter 2, in conjunction with the discussion of the sizes and masses of atoms and subatomic particles. Energy units are defined in Chapter 6, where they are first needed to deal with kinetic and potential energy, work, and heat. In each case, defining units at the time when the need for them can be made clear allows definitions that would otherwise appear pointless and arbitrary to support the development of closely related concepts.

We use real chemical systems in examples and problems whenever possible, both in the text and in the end-of-chapter questions. In the kinetics chapter, for example, the text and problems utilize real reactions and real data from which to determine reaction rates or orders. Instead of $A + B \longrightarrow C + D$, students will find $I^- + CH_3Br \longrightarrow CH_3I + Br^-$. Data have been taken from the recent research literature. The same approach is employed in many other chapters, where real chemical systems are used as examples.

Most important, we provide **clear, direct, thorough, and understandable explanations** of all topics, including those such as stoichiometry, chemical kinetics, chemical thermodynamics, and electrochemistry that many students find daunting. The methods of science and concepts such as chemical and physical properties; purification and separation; the relation of macroscale, nanoscale, and symbolic representations; elements and compounds; and kinetic-molecular theory are introduced in Chapter 1 so that they can be used throughout the later discussion. Rather than being bogged down with discussions of units and nomenclature, students begin this book with an overview of what real chemistry is about—together with fundamental ideas that they will need to understand it.

Visualization for Understanding

The **illustrations** in *Chemistry: The Molecular Science* have been designed to engage today's visually oriented students. The success of the illustration program is exemplified by the fact that the first edition was awarded a national prize for visual excellence. Nevertheless, for this edition a special reviewer, Kathy Thrush Shaginaw, has examined carefully each piece of art and recommended revisions. Based on her suggestions we have revised, replaced, or added art in every chapter. Illustrations help students to visualize atoms and molecules and to make connections among macroscale observations, nanoscale models, and symbolic representations of chemistry. Excellent color photographs of substances and reactions, many by Charles D. Winters, are presented together with greatly magnified illustrations of the atoms, molecules, and/or ions involved that have been created by J/B Woolsey Associates LLC. New drawings for this edition have been created by Graphic World Inc. Often these are accompanied by the symbolic formula for a substance or equation for a reaction, as in the example shown below. These **nanoscale views of atoms, molecules, and ions** have been generated with molecular modeling software and then combined by a skilled artist with the photographs and formulas or equations. Similar illustrations appear in exercises, examples, and end-of-chapter problems, thereby ensuring that students are tested on the ideas the illustrations represent. This provides an exceptionally effective way for students to learn how chemists think about the nanoscale world of atoms, molecules, and ions.

Often the story is carried solely by an illustration and accompanying text that points out the most important parts of the figure. An example is the visual story of molecular structure on p. xxii. In other cases, text in balloons explains the operation of instruments, apparatus, and experiments; clarifies the development of a mathematical derivation; or points out salient features of graphs or nanoscale pictures. Throughout the book visual interest is high, and visualizations of many kinds are used to support conceptual development.

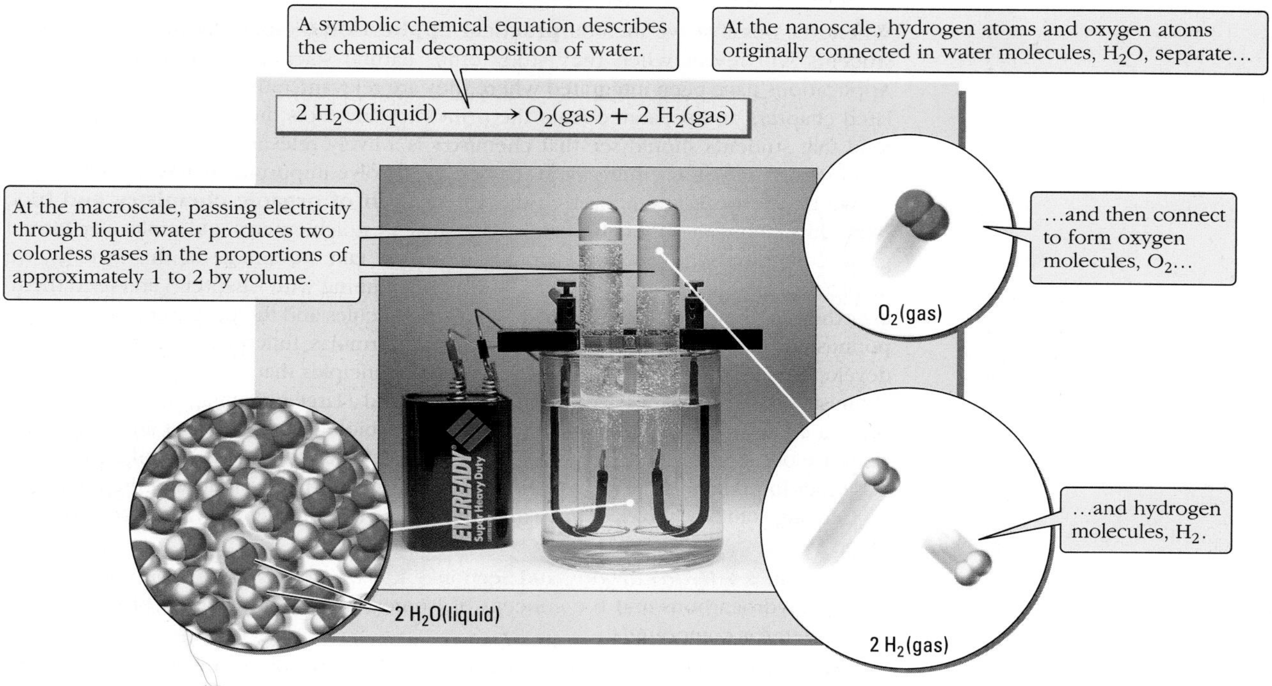
A symbolic chemical equation describes the chemical decomposition of water.
At the nanoscale, hydrogen atoms and oxygen atoms originally connected in water molecules, H_2O, separate...
$2\ H_2O(liquid) \longrightarrow O_2(gas) + 2\ H_2(gas)$
At the macroscale, passing electricity through liquid water produces two colorless gases in the proportions of approximately 1 to 2 by volume.
...and then connect to form oxygen molecules, O_2...
$O_2(gas)$
EVEREADY Super Heavy Duty
...and hydrogen molecules, H_2.
$2\ H_2O(liquid)$
$2\ H_2(gas)$

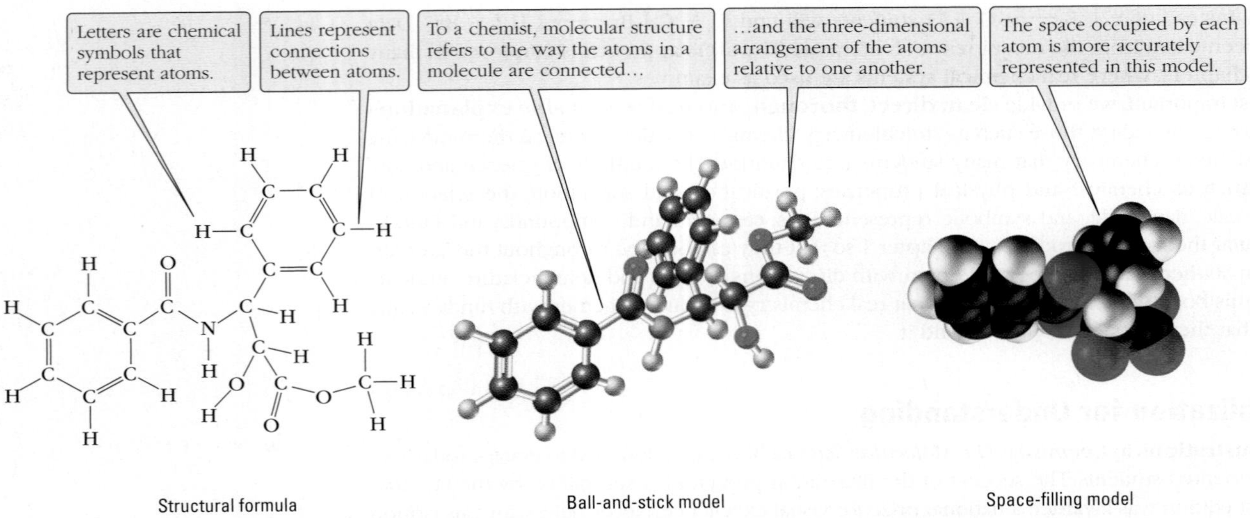
Letters are chemical symbols that represent atoms.
Lines represent connections between atoms.
To a chemist, molecular structure refers to the way the atoms in a molecule are connected…
…and the three-dimensional arrangement of the atoms relative to one another.
The space occupied by each atom is more accurately represented in this model.
Structural formula
Ball-and-stick model
Space-filling model

Integrated Media

Web-based tools have proven effective in helping general chemistry students with conceptual understanding. We have integrated these web-based study tools into this textbook.

- **Active Figures** are available in each chapter designated by an annotation in the caption of the figure. On the companion website, Active Figures provide an animated version of the text figure accompanied by an exercise, further illustrating the important concept embodied in the figure.
- **Estimation Boxes** from the text are expanded into modules on the companion web site to allow for continued development of students' skills in making approximations and estimations.
- **Go Chemistry** mini lecture videos are available for students to use on their portable electronic devices; a note in the margin indicates every topic where a Go Chemistry video is available.
- **OWL** Online Web Learning for this text has been correlated with many more end-of-chapter questions than in previous editions; OWL provides parameterized versions of such questions so that students can further develop their problem-solving skills.

Interdisciplinary Applications

Whenever possible we include **practical applications,** especially those applications that students will revisit when they study other natural science and engineering disciplines. Applications have been integrated where they are relevant, rather than being relegated to isolated chapters and separated from the principles and facts on which they are based. We intend that students should see that chemistry is a lively, relevant subject that is fundamental to a broad range of disciplines and that can help solve important, real-world problems.

We have especially emphasized the **integration of organic chemistry and biochemistry** throughout the book. In many areas, such as stoichiometry and molecular formulas, organic compounds provide excellent examples. To take advantage of this synergy, we have incorporated basic organic topics into the text beginning with Chapter 3 and used them wherever they are appropriate. In the discussion of molecules and the properties of molecular compounds, for example, the concepts of structural formulas, functional groups, and isomers are developed naturally and effectively. Many of the principles that students encounter in general chemistry are directly applicable to biochemistry, and a large percentage of the students in most general chemistry courses are planning careers in biological or medical areas that make constant use of biochemistry. For this reason, we have chosen to deal with fundamental biochemical topics in juxtaposition with the general chemistry principles that underlie them.

Here are some examples of integration of organic and biochemistry; the book contains many more:

- Section 3.3, *Hydrocarbons,* and Section 3.4, *Alkanes and Their Isomers,* introduce simple hydrocarbons and the concept of isomerism as a natural part of the discussion of molecular compounds.
- Section 3.11, *The Biological Periodic Table,* describes the many elements that are essential to living systems and why they are important.

STYLE KEY

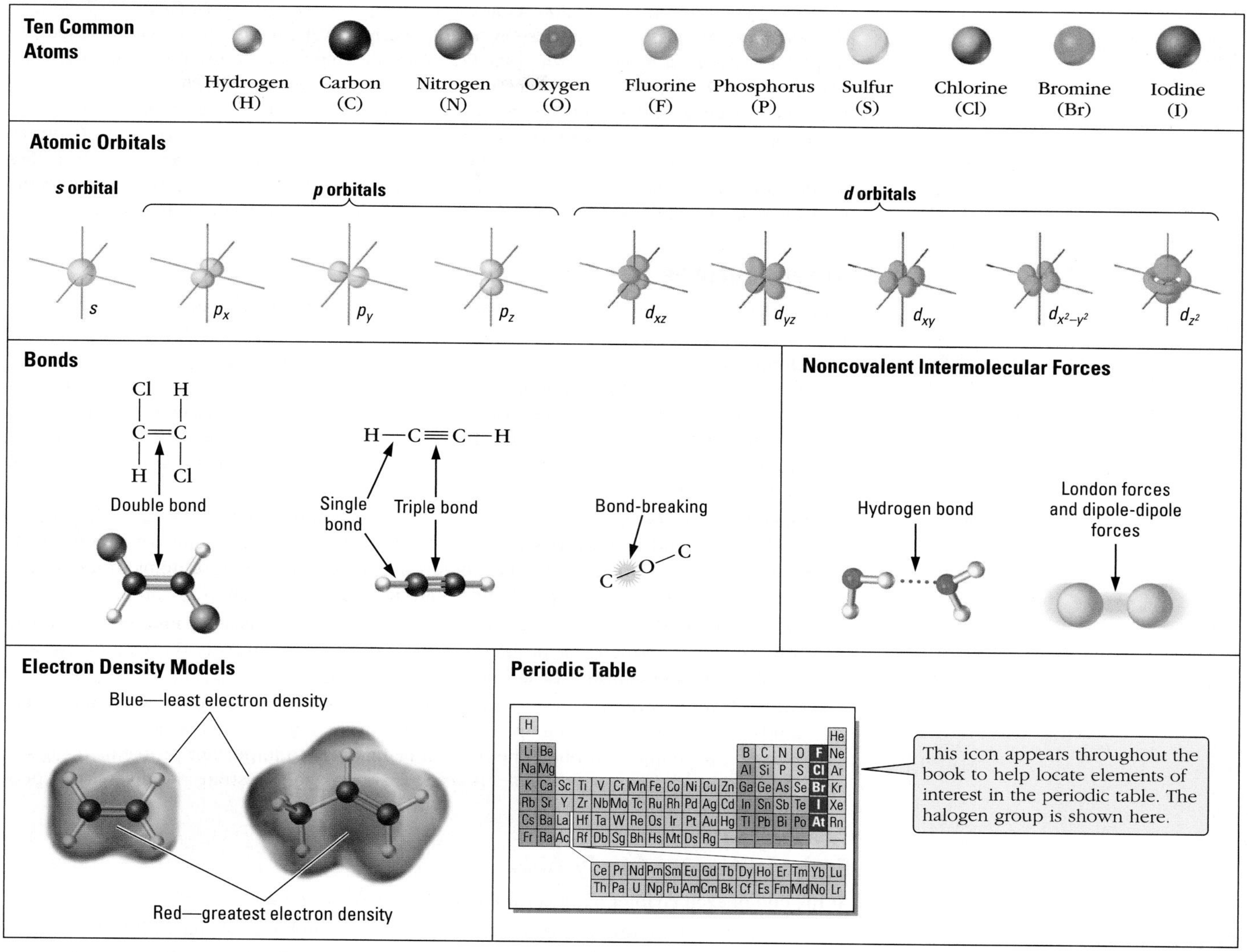

- Section 6.12, *Foods: Fuels for Our Bodies,* applies thermochemical and calorimetric principles learned earlier in the chapter to the caloric values of proteins, fats, and carbohydrates in food.
- Section 9.7, *Biomolecules: DNA and the Importance of Molecular Structure,* extends the concepts of molecular shapes and noncovalent intermolecular forces developed earlier in Chapter 9 to the structure and function of DNA, explaining how chemical principles can be applied to the storage and transmission of genetic information.
- Chapter 12, *Fuels, Organic Chemicals, and Polymers,* builds on principles and facts introduced earlier, applying them to organic molecules and functional groups selected for their relevance to synthetic and natural polymers. Proteins and polysaccharides illustrate the importance of biopolymers.
- Section 13.9, *Enzymes: Biological Catalysts,* applies kinetics principles developed earlier in the chapter and ideas about molecular structure from earlier chapters to enzyme catalysis and the way in which it is influenced by protein structure.
- Section 18.9, *Gibbs Free Energy and Biological Systems,* discusses the role of Gibbs free energy and coupling of thermodynamic systems in metabolism, making clear the fact that metabolic pathways are governed by the rules of thermodynamics.
- Section 19.8, *Neuron Cells,* applies electrochemical principles to the transmission of nerve impulses from one neuron to another, showing that changes in concentrations of ions result in changes in voltage and hence electrical signals.

Environmental and industrial chemistry are also integrated. In Chapter 6, *Energy and Chemical Reactions,* thermochemical principles are used to evaluate the energy densities of fuels. In Chapter 10, *Gases and the Atmosphere,* a discussion of gas-phase chemical reactions leads into the stories of stratospheric ozone depletion and air pollution. Chapter 10 also deals with the consequences of combustion in a section on global warming. Chapter 12, *Fuels, Organic Chemicals, and Polymers,* discusses energy resources and recent developments in recycling plastics. In Chapter 13, *Chemical Kinetics: Rates and Reactions,* the importance of catalysts is illustrated by several industrial processes and exhaust-emission control on automobiles. In Chapter 16, *Acids and Bases,* practical acid-base chemistry illustrates many of the principles developed in the same chapter. In Chapter 21, *The Chemistry of the Main Group Elements,* and Chapter 22, *Chemistry of Selected Transition Metals and Coordination Compounds,* principles developed in earlier chapters are applied to uses of the elements and to extraction of elements from their ores. Students in a variety of disciplines will discover that chemistry is fundamental to their other studies.

Other Features

Additional features of the book that we have designed specifically to address the needs of students are

- ***Chemistry You Can Do.*** Most chapters include a *Chemistry You Can Do* experiment that requires only simple equipment and familiar chemicals available at home or on a college campus, can be performed in a kitchen or residence hall room, and illustrates a topic included in the chapter. Including these experiments reflects our beliefs that students should be involved in doing chemistry, and they ought to learn that common household materials are also chemicals.
- ***Chemistry in the News*** boxes bring to the attention of students the latest discoveries in chemistry and applications of chemistry, making clear that chemistry is continually changing and developing—it is not merely a static compendium of items to memorize. These boxes have been updated, and 23 are new to this edition.
- ***Tools of Chemistry*** boxes provide examples of how chemists use modern instrumentation to solve challenging problems. They introduce to students the excitement and broad range of chemical measurements.
- ***Portrait of a Scientist*** items show that, like any other human pursuit, chemistry depends on people. These biographical sketches of men and women who have advanced our understanding or applied chemistry imaginatively to important problems bring the human side of chemistry to students using this book and illustrate the diversity of people who do science.

End-of-Chapter Study Aids

At the end of each chapter, students will find many ways to test and consolidate their learning.

- A **Summary Problem** brings together concepts and problem-solving skills from throughout the chapter. Students are challenged to answer a multifaceted question that builds on and integrates the chapter's content.
- **In Closing** highlights the learning goals for the chapter, provides references to the sections in the chapter that address each goal, and identifies end-of-chapter questions appropriate to test each goal.
- **Key Terms** are listed, with references to the sections where they are defined.

A broad range of chapter-end **Questions for Review and Thought** are provided to serve as a basis for homework or in-class problem solving.

- **Review Questions,** which are not answered in the back of the book, test vocabulary and simple concepts.
- **Topical Questions** are keyed to the major topics in the chapter and listed under headings that correspond with each section in the chapter. Questions are often accompanied by photographs, graphs, and diagrams that make the situations described more concrete and realistic. Usually a question that is answered at the end of the book is paired with a similar one that is not.
- **General Questions** are not explicitly keyed to chapter topics. These questions are designed to help students analyze questions and learn to apply appropriate ideas to solving problems.
- **Applying Concepts** includes questions specifically designed to test conceptual learning. Many of these questions include diagrams of atoms, molecules, or ions and require students to relate macroscopic observations, atomic-scale models, and symbolic formulas and equations.
- **More Challenging Questions** require students to apply more thought and to better integrate multiple concepts than do typical end-of chapter questions.

- **Conceptual Challenge Problems,** most of which were written by H. Graden Kirksey, emeritus faculty member of the University of Memphis, are especially important in helping students assess and improve their conceptual thinking ability. Designed for group work, the Conceptual Challenge Problems are rigorous and thought provoking. Much effective learning can be induced by dividing a class into groups of three or four students and then assigning these groups to work collaboratively on these problems.

Organization

The order of chapters reflects the most common division of content between the first and second semesters of a typical general chemistry course. The first few chapters briefly review basic material that most students should have encountered in high school. Next, the book develops the ideas of chemical reactions, stoichiometry, and energy transfers during reactions. Throughout these early chapters, organic chemistry, biochemistry, and applications of chemistry are integrated. We then deal with the electronic structure of atoms, bonding and molecular structures, and the way in which structure affects properties. To finish up a first-semester course, there are adjacent chapters on gases and on liquids and solids.

Next, we extend our integration of organic chemistry in a chapter that describes the role of organic chemicals in fuels, polymers, and biopolymers. Chapters on kinetics and equilibrium establish fundamental understanding of how fast reactions will go and what concentrations of reactants and products will remain when equilibrium is reached. These ideas are then applied to solutions, as well as to acid-base and solubility equilibria in aqueous solutions. A chapter on thermodynamics and Gibbs free energy is followed by one on electrochemistry, which makes use of thermodynamic ideas. Finally, the book focuses on nuclear chemistry and the descriptive chemistry of main group and transition elements.

To help students connect chemical ideas that are closely related but are presented in different chapters, we have included **numerous cross references (indicated by the ← symbol).** These cross references will help students link a concept being developed in the chapter they are currently reading with an earlier, related principle or fact. They also provide many opportunities for students to review material encountered earlier.

Varying Chapter Order

A number of variations in the order of presentation are possible. For example, in the classes of one of the authors, the first six sections of Chapter 18 on thermodynamics follow Chapter 13 on chemical kinetics and precede Chapter 14 on equilibrium. Section 14.7 is omitted, and the last five sections of Chapter 18 follow Chapter 14. The material on thermochemistry in Chapter 6 could be postponed and combined with Chapter 18 on thermodynamics with only minor adjustments in the teaching of other chapters, so long as the treatment of thermochemistry precedes the material in Chapter 13, which uses thermochemical concepts in the discussion of activation energy. Many other reorderings of chapters or sections within chapters are possible. The numerous cross references will aid students in picking up concepts that they would be assumed to know, had the chapters been taught consecutively.

At the University of Wisconsin–Madison, this textbook is used in a one-semester accelerated course that is required for most engineering students. We assume substantial high school background in both chemistry and mathematics, and the syllabus includes Chapters 1, 7, 8, 9, 12, 13, 14, 16, 17, 18, and 19. This presentation strategy works quite well, and some engineering students have commented favorably on the inclusion of practical applications of chemistry, such as octane rating and catalysis, in which they were interested.

Chemistry: The Molecular Science, Fourth Edition, can be divided into a number of sections, each of which treats an important aspect of chemistry:

Fundamental Ideas of Chemistry

Chapter 1, The Nature of Chemistry, is designed to capture students' interest from the start by concentrating on chemistry (not on math, units, and significant figures, which are treated comprehensively in an appendix). It asks Why Care About Chemistry? and then tells a story of modern drug discovery and development that illustrates interdisciplinary chemical research. It also introduces major concepts that bear on all of chemistry, emphasizing the three conceptual levels with which students must be familiar—macroscale, nanoscale, and symbolic.

Chapter 2, Atoms and Elements, introduces units and dimensional analysis on a need-to-know basis in the context of the sizes of atoms. It concentrates on thorough, understandable treatment of the concepts of atomic structure, atomic weight, and moles of elements, making the connections among them clear. It concludes by introducing the periodic table and highlighting the periodicity of properties of elements.

Chapter 3, Chemical Compounds, distinguishes ionic compounds from molecular compounds and illustrates molecular compounds with the simplest alkanes. The important theme of structure is reinforced by showing several ways that organic structures can be written. Charges of monatomic ions are related to the periodic table, which is also used to show elements that are important in living systems. Molar masses of compounds and determining formulas fit logically into the chapter's structure.

Chemical Reactions

Chapter 4, Quantities of Reactants and Products, begins a three-chapter sequence that treats chemical reactions qualitatively and quantitatively. Students learn how to balance equations and to use typical inorganic reaction patterns to predict products. A single stepwise method is provided for solving all stoichiometry problems, and 11 examples demonstrate a broad range of stoichiometry calculations.

Chapter 5, Chemical Reactions, has a strong descriptive chemistry focus, dealing with exchange reactions, acid-base reactions, and oxidation-reduction reactions in aqueous solutions. It includes real-world occurrences of each type of reaction. Students learn how to recognize a redox reaction from the chemical nature of the reactants (not just by using oxidation numbers) and how to do titration calculations.

Chapter 6, Energy and Chemical Reactions, begins with a thorough and straightforward introduction to forms of energy, conservation of energy, heat and work, system and surroundings, and exothermic and endothermic processes. Carefully designed figures help students to understand thermodynamic principles. Heat capacity, heats of changes of state, and heats of reactions are clearly explained, as are calorimetry and standard enthalpy changes. These ideas are then applied to fossil fuel combustion and to metabolism of biochemical fuels (proteins, carbohydrates, and fats).

Electrons, Bonding, and Structure

Chapter 7, Electron Configurations and the Periodic Table, introduces spectra, quantum theory, and quantum numbers, using color-coded illustrations to visualize the different energy levels of *s, p, d,* and *f* orbitals. The *s*-, *p*-, *d*-, and *f*-block locations in the periodic table are used to predict electron configurations.

Chapter 8, Covalent Bonding, provides simple stepwise guidelines for writing Lewis structures, with many examples of how to use them. The role of single and multiple bonds in hydrocarbons is smoothly integrated with the introduction to covalent bonding. The discussion of polar bonds is enhanced by molecular models that show variations in electron density. Molecular orbital theory is introduced as well.

Chapter 9, Molecular Structures, provides a thorough presentation of valence-shell electron-pair repulsion (VSEPR) theory and orbital hybridization. Molecular geometry and polarity are extensively illustrated with computer-generated models, and the relation of structure, polarity, and hydrogen bonding to attractions among molecules is clearly developed and illustrated in solved problems. The importance of noncovalent interactions is emphasized early and then reinforced by describing how noncovalent attractions determine the structure of DNA.

States of Matter

Chapter 10, Gases and the Atmosphere, uses kinetic-molecular theory to interpret the behavior of gases and then describes each of the individual gas laws. Mathematical problem solving focuses on the ideal gas law or the combined gas law, and many conceptual Exercises through-out the chapter emphasize qualitative understanding of gas properties. Gas stoichiometry is presented in a uniquely concise and clear manner. Then the properties of gases are applied to chemical reactions in the atmosphere, the role of ozone in both the troposphere and the stratosphere, industrial and photochemical smog, and global warming.

Chapter 11, Liquids, Solids, and Materials, begins by discussing the properties of liquids and the nature of phase changes. The unique and vitally important properties of water are covered thoroughly. The principles of crystal structure are introduced using cubic unit cells only. The fact that much current chemical research involves materials is illustrated by the discussions of metals, *n*- and *p*-type semiconductors, insulators, superconductors, network solids, carbon nanotubes, cement, ceramics and ceramic composites, and glasses, including optical fibers.

Important Industrial, Environmental, and Biological Molecules

Chapter 12, Fuels, Organic Chemicals, and Polymers, offers a distinctive combination of topics of major relevance to industrial, energy, and environmental concerns. Petroleum, natural gas, and coal are discussed as resources for energy and chemical materials. Enough organic functional groups are introduced so that students can understand polymer formation, and the idea of condensation polymerization is extended to carbohydrates and proteins, which are compared with synthetic polymers.

Reactions: How Fast and How Far?

Chapter 13, Chemical Kinetics: Rates of Reactions, presents one of the most difficult topics in the course with extraordinary clarity. Defining reaction rate, finding rate laws from initial rates and integrated rate laws, and using the Arrhenius equation are thoroughly developed. How molecular changes during unimolecular and bimolecular elementary reactions relate to activation energy initiates the treatment of reaction mechanisms (including those with an initial fast equilibrium). Catalysis is shown to involve changing a reaction mechanism. Both enzymes and industrial catalysts are described using concepts developed earlier in the chapter.

Chapter 14, Chemical Equilibrium, emphasizes equally a qualitative understanding of the nature of equilibrium and the solving of mathematical problems. That equilibrium results from equal but opposite reaction rates is fully explained. Both Le Chatelier's principle and the reaction quotient, Q, are used to predict shifts in equilibria. A unique section on equilibrium at the nanoscale introduces briefly and qualitatively how enthalpy changes and entropy changes affect equilibria. Optimizing the yield of the Haber-Bosch ammonia synthesis elegantly illustrates how kinetics, equilibrium, and enthalpy and entropy changes control the outcome of a chemical reaction.

Reactions in Aqueous Solution

Chapter 15, The Chemistry of Solutes and Solutions, builds on principles previously introduced, showing the influence of enthalpy and entropy on solution properties. Understanding of solubility, Henry's law, concentration units (including ppm and ppb), and colligative properties (including osmosis) is reinforced by applying these ideas to water as a resource, hard water, and municipal water treatment.

Chapter 16, Acids and Bases, concentrates initially on the Brønsted-Lowry acid-base concept, clearly delineating proton transfers using color coding and molecular models. In addition to a full exploration of pH and the meaning and use of K_a and K_b, acid strength is related to molecular structure, and the acid-base properties of carboxylic acids, amines, and amino acids are introduced. Lewis acids and bases are defined and illustrated using examples. Student interest is enhanced by a discussion of everyday uses of acids and bases.

Chapter 17, Additional Aqueous Equilibria, extends the treatment of acid-base and solubility equilibria to buffers, titration, and precipitation. The Henderson-Hasselbalch equation, which is widely used in biochemistry, is applied to buffer pH. Calculations of points on titration curves are shown, and the interpretation of several types of titration curves provides conceptual understanding. Acid-base concepts are applied to the formation of acid rain. The final section deals with the various factors that affect solubility (pH, common ions, complex ions, and amphoterism) and with selective precipitation.

Thermodynamics and Electrochemistry

Chapter 18, Thermodynamics: Directionality of Chemical Reactions, explores the nature and significance of entropy, both qualitatively and quantitatively. The signs of Gibbs free energy changes are related to the easily understood classification of reactions as reactant- or product-favored, with the discussion deliberately avoiding the often-misinterpreted term "spontaneous." The thermodynamic significance of coupling one reaction with another is illustrated using industrial, metabolic, and photosynthetic examples. Energy conservation is defined thermodynamically. A closing section reinforces the important distinction between thermodynamic and kinetic stability.

Chapter 19, Electrochemistry and Its Applications, defines redox reactions and uses half-reactions to balance redox equations. Electrochemical cells, cell voltage, standard cell potentials, the relation of cell potential to Gibbs free energy, and the effect of concentrations on cell potential are all explored. These ideas are then applied to the transmission of nerve impulses. Practical applications include batteries, fuel cells, electrolysis, and corrosion.

Nuclear Chemistry

Chapter 20, Nuclear Chemistry, deals with radioactivity, nuclear reactions, nuclear stability, and rates of disintegration reactions. Also provided are a thorough description of nuclear fission and nuclear fusion, and a thorough discussion of nuclear radiation, background radiation, and applications of radioisotopes.

More Descriptive Chemistry

Chapter 21, The Chemistry of the Main Group Elements, consists of two main parts. The first part tells the interesting story of how the elements were formed and which are most important on Earth. The physical separation of nitrogen, oxygen, and sulfur from natural sources, and the extraction of sodium, chlorine, magnesium, and aluminum by electrolysis, provide important industrial examples as well as an opportunity for students to apply principles learned earlier in the book. The second part (Section 21.6) discusses the properties, chemistry, and uses of the elements of Groups 1A–7A and their compounds in a systematic way, based on groups of the periodic table. Trends in atomic and ionic radii, melting points and boiling points, and densities of each group's elements are summarized. Group 8A is covered briefly.

Chapter 22, Chemistry of Selected Transition Elements and Coordination Compounds, treats a few important elements in depth and integrates the review of principles learned earlier. Iron, copper, chromium, silver, and gold provide an interesting, motivating collection of elements from which students can learn the principles of transition metal chemistry. In addition to the treatment of complex ions and coordination compounds, this chapter includes an extensive section on crystal-field theory, electron configurations, color, and magnetism in coordination complexes.

Supporting Materials

For the Instructor

Supporting instructor materials are available to qualified adopters. Please consult your local Cengage Learning Brooks/Cole representative for details. Go to **www.cengage.com/chemistry/moore** and click this textbook's Faculty Companion Site to

- See samples of materials
- Request a desk copy
- Locate your local representative
- Download digital files of the *Test Bank* and other helpful materials for instructors and students

PowerLecture with JoinIn™ and ExamView® Instructor's CD/DVD.
ISBN-10: 1-4390-4953-X, ISBN-13: 978-1-439-04953-2
PowerLecture is a one-stop digital library and presentation tool that includes

- Prepared **Microsoft® PowerPoint® Lecture Slides** authored by Stephen C. Foster of Mississippi State University that cover all key points from the text in a convenient format that you can enhance with your own materials or with the supplied interactive videos and animations for personalized, media-enhanced lectures.
- Image libraries in PowerPoint and JPEG formats that contain **digital files for all text art, most photographs, and all numbered tables** in the text. These files can be used to create your own transparencies or PowerPoint lectures.
- Digital files for the **complete *Instructor's Solutions Manual* and *Test Bank*.**
- **Sample chapters from the *Student Solutions Manual* and *Study Guide*.**
- **ExamView Computerized Testing** by David Treichel, Nebraska Wesleyan University, enables you to create customized tests of up to 250 items in print or online using more than 1000 questions carefully matched to the corresponding text sections. Tests can be taken electronically or printed for class distribution.
- **JoinIn™ clicker questions** specifically for this text, for use with the classroom response system of your choice. Assess student progress with instant quizzes and polls, and display student answers seamlessly within the Microsoft PowerPoint slides of your own lecture questions. Please consult your Cengage Learning Brooks/Cole representative for more details.

- **ChemQuiz and Lecture Slide** collections, developed by Mark Kubinec for teaching general chemistry at the University of California, Berkeley, correlated to this book. These thought-provoking conceptual questions require students to recognize trends or general concepts and stimulate discussion during lecture. Each question is accompanied by several lecture slides providing the background needed to answer the question.

Instructor's Solutions Manual by Judy L. Ozment, Pennsylvania State University. Contains fully worked-out solutions to all end-of-chapter questions, Summary Problems, and Conceptual Challenge Problems. Solutions match the problem-solving strategies used in the text. Available on the instructor's PowerLecture CD/DVD.

OWL: Online Web by Roberta Day and Beatrice Botch, University of Massachusetts, Amherst and William Vining, State University of New York at Oneonta.
Instant Access to OWL (four semesters) ISBN-10: 0-495-05099-7, ISBN-13: 978-0-495-05099-5
Instant Access to OWL (one semester) ISBN-10: 0-495-38442-9, ISBN-13: 978-0-495-38442-7
Instant Access to OWL e-Book (four semesters)
ISBN-10: 0-538-73817-0, ISBN-13: 978-0-538-73817-0
Instant Access to OWL e-Book (one semester)
ISBN-10: 0-538-73815-4, ISBN-13: 978-0-538-73815-6
Featuring an updated and more intuitive instructor interface, OWL offers more assignable, gradable content (including end-of-chapter questions specific to each textbook), more reliability, and more flexibility than any other system. Developed by chemistry instructors for teaching chemistry, OWL makes homework management a breeze and has already helped hundreds of thousands of students master chemistry through tutorials, interactive simulations, and algorithmically generated homework questions that provide instant, answer-specific feedback. In addition, when you become an OWL user, you can expect service that goes far beyond the ordinary.

OWL is continually enhanced with online learning tools to address the various learning styles of today's students such as:

- e-Books, which offer a fully integrated electronic textbook correlated to OWL questions;
- Video examples utilizing the Problem-Solving Examples from the textbook;
- Go Chemistry® mini video lectures on key concepts that can be viewed onscreen or downloaded to students' video iPods, iPhones, or personal video players;
- Quick Prep review courses that help students learn essential skills to succeed in General and Organic Chemistry;
- Thinkwell Video Lessons that teach key concepts through video, audio, and whiteboard examples;
- Jmol molecular visualization program for rotating molecules and measuring bond distances and angles.

For *Chemistry: The Molecular Science,* Fourth Edition, OWL includes parameterized end-of-chapter questions from the text and tutorials based on the Estimation boxes in the text. To view an OWL demo, and for more information, visit **www.cengage.com/owl** or contact your Cengage Learning Brooks/Cole representative.

ExamView Computerized Testing by David Treichel, Nebraska Wesleyan University. Containing more than 1000 questions carefully matched to the corresponding text sections, the *Test Bank* is available as PDF files on the instructor's PowerLecture CD/DVD and in the ExamView Computerized Testing.

Faculty Companion Website. Go to **www.cengage.com/chemistry/moore** and click this book's Faculty Companion Site to access resources such as Blackboard and WebCT versions of ExamView.

Cengage Learning Custom Solutions develops personalized text solutions to meet your course needs. Match your learning materials to your syllabus and create the perfect learning solution—your customized text will contain the same thought-provoking, scientifically sound content, superior authorship, and stunning art that you've come to expect from Cengage Learning Brooks/Cole texts, yet in a more flexible format. Visit **www.cengage.com/custom** to start building your book today.

Chemistry Comes Alive! (available separately by subscription from *JCE Web Software*; see **http://www.jce.divched.org/JCESoft/jcesoftSubscriber.html**). Chemistry Comes Alive! provides online, HTML-format access to a broad range of videos and animations suitable for use in lecture presentations, for independent study, or for incorporation into the instructor's own tutorials.

***JCE* QBank** (Available separately from the *Journal of Chemical Education*; see **http://www.jce.divched.org/JCEDLib/QBank/index.html**). Contains more than 3500 homework and quiz questions suitable for delivery via WebCT, Desire2Learn, or Moodle course management systems, hundreds of ConcepTest questions that can be used with "clickers" to make lectures more interactive, and a collection of conceptual questions together with a discussion of how to write conceptual questions. Available to all *JCE* subscribers.

For the Student

Go to **www.cengage.com/chemistry/moore** and click this textbook's Student Companion Site for study resources and samples of select student supplements. You can purchase any Cengage Learning Brooks/Cole product at your local college store or at **www.CengageBrain.com.**

OWL for General Chemistry.
See the above description in the instructor support materials section.

OWL Quick Prep for General Chemistry by Beatrice Botch and Roberta Day, University of Massachusetts, Amherst.
Instant Access (90 days): ISBN-10: 0-495-11042-6, ISBN-13: 978-0-495-11042-2
Quick Prep is a self-paced online short course that helps students succeed in general chemistry. Students who completed Quick Prep through an organized class or self-study averaged almost a full letter grade higher in their subsequent general chemistry courses than those who did not. Intended to be taken prior to the start of the semester, Quick Prep is appropriate for both underprepared students and for students who seek a review of basic skills and concepts. Quick Prep is an approximately 20-hour commitment delivered through the online learning system OWL with no textbook required and can be completed at any time on the student's schedule. To view an OWL Quick Prep demonstration and for more information, visit **www.cengage.com/chemistry/quickprep** or contact your Cengage Learning Brooks/Cole representative. Search by ISBN to purchase Instant Access Codes from **www.CengageBrain.com.**

Go Chemistry® for General Chemistry.
Instant Access to the 27-Video Set: ISBN-10: 1-4390-4700-6, ISBN-13: 978-1-4390-4700-2
Instant Access to Individual Videos: ISBN-10: 0-495-38228-0, ISBN-13: 978-0-495-38228-7
Go Chemistry is a set of 27 easy-to-use videos of essential general chemistry topics that can be downloaded to your video iPod, iPhone, or portable video player—ideal for the student on the go! Developed by chemistry textbook author John Kotz, these new electronic tools are designed to help students quickly review essential chemistry topics. Mini video lectures include animations and problems for a quick summary of key concepts. Selected Go Chemistry modules have flashcards to briefly introduce a key concept and then test student understanding of the basics with a series of questions. Go Chemistry also plays on iTunes, Windows Media Player, and QuickTime. To purchase Go Chemistry, search by ISBN at **www.CengageBrain.com.**

Student Solutions Manual by Judy L. Ozment, Pennsylvania State University.
ISBN-10: 1-4390-4963-7, ISBN-13: 978-1-4390-4963-1
Contains fully worked-out solutions to end-of-chapter questions that have blue, boldfaced numbers. Solutions match the problem-solving strategies used in the main text. Download a sample chapter from the Student Companion Website, which is accessible from **www.cengage.com/chemistry/moore.**

Study Guide by Michael J. Sanger, Middle Tennessee State University.
ISBN-10: 1-4390-4964-5, ISBN-13: 978-1-4390-4964-8
Contains learning tools such as brief notes on chapter sections with examples, reviews of key terms, and practice tests with answers. This new edition has been carefully revised to complement and match the changes to the core text. This revision includes revised objectives and

new and modified study questions. A sample is available on the Student Companion Website, which is accessible from **www.cengage.com/chemistry/moore.**

Student Companion Website.
Accessible from **www.cengage.com/chemistry/moore,** this site provides online study tools including an online glossary and flashcards, interactive versions of Active Figures and Estimation boxes from the text, and samples of the *Study Guide* and *Student Solutions Manual.*

General Chemistry: Guided Explorations, Fourth Edition
by David Hanson, Stony Brook University.
ISBN-10: 1-4390-4965-3, ISBN-13: 978-1-4390-4965-5
This student workbook is designed to support Process Oriented Guided Inquiry Learning (POGIL) with activities that promote a student-focused active classroom. It is an excellent ancillary to *Chemistry: The Molecular Science* or any other general chemistry text.

Survival Guide for General Chemistry with Math Review and Proficiency Questions, Second Edition by Charles H. Atwood, University of Georgia.
ISBN-10: 0-495-38751-7, ISBN-13 978-0-495-38751-0
Intended to help you practice for exams, this survival guide shows you how to solve difficult problems by dissecting them into manageable chunks. The guide includes three levels of proficiency questions—A, B, and minimal—to quickly build confidence as you master the knowledge you need to succeed in your course.

Essential Algebra for Chemistry Students, Second Edition
by David W. Ball, Cleveland State University.
ISBN-10: 0-495-01327-7, ISBN-13 978-0-495-01327-3
This short book is intended for students who lack confidence and/or competency in their essential mathematics skills necessary to succeed in general chemistry. Each chapter focuses on a specific type of skill and has worked-out examples to show how these skills translate to chemical problem solving. Includes references to OWL, our web-based tutorial program, offering students access to online algebra skills exercises.

ChemPages Laboratory (available separately by subscription from *JCE Web Software*; see **http://www.jce.divched.org/JCESoft/jcesoftSubscriber.html**). A collection of videos with voiceover and text showing how to perform the most common laboratory techniques used by students in first-year chemistry courses.

Netorials (available separately by subscription from *JCE Web Software*; see **http://www.jce.divched.org/JCESoft/jcesoftSubscriber.html**). The Netorials are online tutorials that cover selected topics in first-year chemistry including: Chemical Reactions, Stoichiometry, Thermodynamics, Intermolecular Forces, Acids & Bases, Biomolecules, and Electrochemistry.

Periodic Table Live! (available separately by subscription from *JCE Web Software*; see **http://www.jce.divched.org/JCESoft/jcesoftSubscriber.html**). Periodic Table Live! may not include *everything* you ever wanted to know about the elements, but it will probably answer any question you aren't afraid to ask. It includes a new interactive graphing and sorting capability.

Window on the Solid State (available separately by subscription from *JCE Web Software*; see **http://www.jce.divched.org/JCESoft/jcesoftSubscriber.html**). Four tutorials on solid-state structures that nicely complement the content in Chapter 11, helping students understand and instructors present the structural features of solids.

For the Laboratory

Laboratory Handbook for General Chemistry, Third Edition by Conrad L. Stanitski, Franklin and Marshall College; Norman E. Griswold, Nebraska Wesleyan College; H. A. Neidig, Lebanon Valley College; and James N. Spencer, Franklin and Marshall College.
ISBN-10: 0-495-01890-2, ISBN-13: 978-0-495-01890-2
This "how-to" guide containing specific information about the basic equipment, techniques, and operations necessary for successful laboratory experiments helps students perform their laboratory work more effectively, efficiently, and safely. The third edition includes video demonstrations of a number of common laboratory techniques.

Cengage Learning Brooks/Cole Lab Manuals.
We offer a variety of printed manuals to meet all your general chemistry laboratory needs. Instructors can visit the chemistry site at **www.cengage.com/chemistry** for a full listing and description of these laboratory manuals, laboratory notebooks, and laboratory handbooks. All Cengage Learning laboratory manuals can be customized for your specific needs.

Signature Labs . . . for the customized laboratory.
Signature Labs is Cengage Learning's digital library of tried-and-true labs that help you take the guesswork out of running your chemistry laboratory. Select just the experiments you want from hundreds of options and approaches. Provide your students with only the experiments they will conduct and know you will get the results you seek. Visit **www.signaturelabs.com** to begin building your manual today.

Note: Unless otherwise noted, the website domain names (URLs) provided here are not published by Cengage Learning Brooks/Cole and the Publisher can accept no responsibility or liability for these sites' content. Because of the dynamic nature of the Internet, Cengage Learning Brooks/Cole cannot in any case guarantee the continued availability of third-party websites.

Reviewers

Reviewers have played a critical role in the preparation of this textbook. The individuals listed below helped to shape this text into one that is not merely accurate and up to date, but a valuable practical resource for teaching and testing students.

Reviewers of the Fourth Edition
Margaret Czerw, *Raritan Valley Community College*
Michelle Driessen, *University of Minnesota*
Harold Goldwhite, *California State University, Los Angeles*
Steven C. Haefner, *Bridgewater State College*
David M. Hanson, *Stony Brook University*
Andy Jorgensen, *University of Toledo*
Roy Kennedy, *Massachusetts Bay Community College*
Mahesh Mahanthappa, *University of Wisconsin–Madison*
Joe L. March, *University of Alabama at Birmingham*
Wyatt R. Murphy, Jr., *Seton Hall University*
Jeff R. Schoonover, *St. Mary's University*
Clarissa Sorensen-Unruh, *Central New Mexico Community College*
Anton Wallner, *Barry University*

Kathy Thrush Shaginaw meticulously evaluated all art in this fourth edition. She provided a detailed review of each figure with suggestions for improving an already excellent illustration program. Her work in this regard was outstanding and has resulted in figures that will help students learn more effectively.

Editorial Advisory Board for the Third Edition
David Grainger, *University of Utah*
Benjamin R. Martin, *Texas State University, San Marcos*
David Miller, *California State University, Northridge*
Michael J. Sanger, *Middle Tennessee State University*
Sherril Soman, *Grand Valley State University*
Richard T. Toomey, *Northwest Missouri State University*
Katherine R. Miller, *Salisbury University*
Robert Milofsky, *Fort Lewis College*
Mark E. Ott, *Jackson Community College*
Philip J. Reid, *University of Washington*
Joel Tellinghuisen, *Vanderbilt University*
Richard T. Toomey, *Northwest Missouri State University*
Peter A. Wade, *Drexel University*
Keith A. Walters, *Northern Kentucky University*

Reviewers of the Third Edition
Patricia Amateis, *Virginia Tech*
Debra Boehmler, *University of Maryland*
Norman C. Craig, *Oberlin College*
Michael G. Finnegan, *Washington State University*
Milton D. Johnson, *University of South Florida*

Reviewers of the Second Edition
Ruth Ann Armitage, *Eastern Michigan University*
Margaret Asirvatham, *University of Colorado*
David Ball, *Cleveland State University*
Debbie J. Beard, *Mississippi State University*
Mary Jo Bojan, *Pennsylvania State University*
Simon Bott, *University of Houston*

Judith N. Burstyn, *University of Wisconsin-Madison*
Kathy Carrigan, *Portland Community College*
James A. Collier, *Truckee Meadows Community College*
Susan Collins, *California State University, Northridge*
Roberta Day, *University of Massachusetts-Amherst*
Norman Dean, *California State University, Northridge*
Barbara L. Edgar, *University of Minnesota, Twin Cities*
Paul Edwards, *Edinboro University of Pennsylvania*
Amina K. El-Ashmawy, *Collin County Community College*
Thomas P. Fehlner, *University of Notre Dame*
Daniel Fraser, *University of Toledo*
Mark B. Freilich, *The University of Memphis*
Noel George, *Ryerson University*
Stephen Z. Goldberg, *Adelphi University*
Gregory V. Hartland, *University of Notre Dame*
Ronald C. Johnson, *Emory University*
Jeffrey Kovac, *University of Tennessee*
John Z. Larese, *University of Tennessee*
Joe March, *University of Alabama at Birmingham*
Lyle V. McAfee, *The Citadel*
David Miller, *California State University, Northridge*
Wyatt R. Murphy, Jr., *Seton Hall University*
Mary-Ann Pearsall, *Drew University*
Vicente Talanquer, *University of Arizona*
Wayne Tikkanen, *California State University, Los Angeles*
Patricia Metthe Todebush, *Northwestern University*
Andrew V. Wells, *Chabot Community College*
Steven M. Wietstock, *Indiana University*
Martel Zeldin, *Hobart & William Smith Colleges*
William H. Zoller, *University of Washington*

Reviewers of the First Edition

Margaret Asirvatham, *University of Colorado-Boulder*
Donald Berry, *University of Pennsylvania*
Barbara Burke, *California State Polytechnic University, Pomona*
Dana Chatellier, *University of Delaware*
Mapi Cuevas, *Santa Fe Community College*
Cheryl Dammann, *University of North Carolina-Charlotte*
John DeKorte, *Glendale Community College*
Russ Geanangel, *University of Houston*
Peter Gold, *Pennsylvania State University*
Albert Martin, *Moravian College*
Marcy McDonald, *University of Alabama-Tuscaloosa*
Charles W. McLaughlin, *University of Nebraska*
David Metcalf, *University of Virginia*
David Miller, *California State University, Northridge*
Kathleen Murphy, *Daemen College*
William Reinhardt, *University of Washington*
Eugene Rochow, *Fort Myers, Florida*
Steven Socol, *McHenry County College*
Richard Thompson, *University of Missouri-Columbia*
Sheryl Tucker, *University of Missouri-Columbia*
Jose Vites, *Eastern Michigan University*
Sarah West, *University of Notre Dame*
Rick White, *Sam Houston State University*

We also thank the following people who were dedicated to checking the accuracy of the text and art.

Accuracy Reviewers of the Fourth Edition

Patrick J. Desrochers, *University of Central Arkansas*
Paul T. Kaiser, *United States Naval Academy*
Karen Pesis, *American River College*

Accuracy Reviewers of the Third Edition

Julie B. Ealy, *Pennsylvania State University*
Stephen Z. Goldberg, *Adelphi University*
Barbara Mowery, *York College of Pennsylvania*
David Shinn, *University of Hawaii at Manoa*

Accuracy Reviewers of the Second Edition

Larry Fishel, *East Lansing, Michigan*
Stephen Z. Goldberg, *Adelphi University*
Robert Milofsky, *Fort Lewis College*
Barbara D. Mowery, *Thomas Nelson Community College*

Accuracy Reviewers of the First Edition

John DeKorte, *Glendale Community College*
Larry Fishel, *East Lansing, Michigan*
Leslie Kinsland, *Cornell University*
Judy L. Ozment, *Pennsylvania State University-Abington*
Gary Riley, *St. Louis School of Pharmacy*

Acknowledgments

No project on the scale of a textbook revision is accomplished solely by the authors. We have had assistance of the very highest quality in all aspects of production of this book, and we extend hearty thanks to everyone who contributed to the project.

Lisa Lockwood, chemistry executive editor, and Kilean Kennedy, acquiring sponsoring editor, have overseen the entire project and have collaborated effectively with the author team on decisions and initiatives that have greatly improved what was already an excellent, rigorous, mainstream general chemistry textbook. They are also responsible for assembling the excellent editorial team that provided strong support for the authors.

Peter McGahey, development editor, provided advice and active support throughout the revision and was always available when things needed to be done or authors needed to be prompted to provide copy. He assembled an excellent group of expert reviewers, obtained reviews from them in timely fashion, and provided feedback based on their comments that was invaluable. He has also served as a calm, conscientious, and caring interface between the authors and the many other members of the production staff. During Peter's paternity leave, Rebecca Heider, a freelance developmental editor, ably replaced Peter and helped the authors to keep on schedule. Thanks, Peter and Rebecca!

Teresa Trego, content project manager, helped keep the authors on track and provided timely queries and suggestions regarding editing, layout, and appearance of the book. We thank her for her invaluable contribution. Lisa Weber and Stephanie VanCamp served as media editors and their ability to organize all the multimedia elements and the references to them in the printed book is much appreciated. Ashley Summers and Elizabeth Woods, both assistant editors, have ably handled all of the ancillary print materials. We also thank Laura Bowen, editorial assistant, for handling many tedious tasks.

The success of a book such as this one depends also on its being adopted and read. Nicole Hamm, marketing manager, directs the marketing and sales programs, and many local representatives throughout the country have helped and will help get this book to students who can benefit from it.

This book is beautiful to look at, and its beauty is more than skin deep. The illustration program has been carefully designed to support student learning in every possible way. The many photographs of Charles D. Winters of Oneonta, New York, provide students with close-up views of chemistry in action. We thank Charlie for doing many new shoots for this new edition. Jennifer Lim, Chris Althof, and Emma Hopson, photo researchers with the Bill Smith Group, carried out photo research in a most effective and friendly fashion, and we thank them for helping to improve the illustration program.

Julie Ninnis and Dan Fitzgerald, together with the staff at Graphic World Publishing Services, have handled copy editing, layout, and production of the book. Julie and Dan worked calmly and effectively with the authors to make certain that this book is of the highest possible quality. Thanks go to copy editor Maryalice Ditzler, who removed infelicities, made the entire book consistent, and even discovered typos that had made it through three previous editions. We thank all of the staff at Graphic World who contributed to this edition.

Elizabeth Moore has carefully read three rounds of page proofs, making certain that the changes requested by the authors were made and helping to improve clarity and layout.

Judy Ozment has solved all of the end-of-chapter questions in this book for all four editions. She has produced excellent student solution manuals and answers to selected questions at the end of the book. Judy is diligent in finding ways in which questions can be stated more clearly, cases where data used in a question are inconsistent with other material in the book, and situations where authors may not have asked what they wanted to ask. For all of her work and help we thank her profusely. Karen Pesis provided a second set of eyes for Judy and we thank her for excellent work.

Many of the take-home *Chemistry You Can Do* experiments in this book were adapted from activities published by the Institute for Chemical Education as *Fun with Chemistry: Volumes I and II,* by Mickey and Jerry Sarquis of Miami University (Ohio). Some were adapted from Classroom Activities published in the *Journal of Chemical Education.* Conceptual Challenge Problems at the end of most chapters were written by H. Graden Kirksey, emeritus faculty of the University of Memphis, and we very much appreciate his contribution. The active-learning, conceptual approach of this book has been greatly influenced by the systemic curriculum enhancement project, *Establishing New Traditions: Revitalizing the Curriculum,* funded by the National Science Foundation, Directorate for Education and Human Resources, Division of Undergraduate Education, grant DUE-9455928.

We also thank the many teachers, colleagues, students, and others who have contributed to our knowledge of chemistry and helped us devise better ways to help others learn it. Collectively, the authors of this book have many years of experience teaching and learning, and we have tried to incorporate as much of that as possible into our presentation of chemistry.

Finally, we thank our families and friends who have supported all of our efforts—and who can reasonably expect more of our time and attention now that this new edition is complete.

We hope that using this book results in a lively and productive experience for both faculty and students.

John W. Moore
Madison, Wisconsin

Conrad L. Stanitski
Lancaster, Pennsylvania

Peter C. Jurs
State College, Pennsylvania

Special Features

CHEMISTRY IN THE NEWS

CHEMISTRY YOU CAN DO

ESTIMATION

TOOLS OF CHEMISTRY

PORTRAIT OF A SCIENTIST

1 The Nature of Chemistry

Tom & Pat Leeson/Photo Researchers, Dennis Flaherty, Mike Trumball/Hauser Northwest/NIH

Chemistry, in collaboration with many other sciences, can produce spectacular advances in dealing with human suffering and pain. The Pacific yew tree, shown above, harbors in its bark a substance that has been amazingly successful in treating cancer. Chemists first separated the active ingredient from the bark in the late 1960s. A chemical pharmacologist discovered how it works in the body, and other chemists found ways to manufacture it without destroying the trees in which it was discovered. Chemical science involves a unique atomic and molecular perspective that enables us to find useful substances, separate them or synthesize them, and figure out how they work by imagining the behavior of particles that are too small to see.

Welcome to the world of chemical science! This chapter describes how modern chemical research is done and how it can be applied to questions and problems that affect our daily lives. It also provides an overview of the methods of science and the fundamental ideas of chemistry. These ideas are extremely important and very powerful. They will be applied over and over throughout your study of chemistry and of many other sciences.

OWL Sign in to OWL at **www.cengage.com/owl** to view tutorials and simulations, develop problem-solving skills, and complete online homework assigned by your professor.

go Chemistry Download mini lecture videos for key concept review and exam prep from OWL or purchase them from **www.CengageBrain.com.**

Companion Website

Visit this book's companion website at **www.cengage.com/chemistry/moore** to work interactive modules for the Estimation boxes and Active Figures in this text.

Very human accounts of how fascinating—even romantic—chemistry can be are provided by Primo Levi in his autobiography, *The Periodic Table* (New York: Schocken Books, 1984), and by Oliver Sacks in *Uncle Tungsten: Memories of a Chemical Boyhood* (New York: Knopf, 2001). Levi was sentenced to a death camp during World War II but survived because the Nazis found his chemistry skills useful; those same skills made him a special kind of writer. Sacks describes how his mother and other relatives encouraged his interest in metals, diamonds, magnets, medicines, and other chemicals and how he learned that "science is a territory of freedom and friendship in the midst of tyranny and hatred."

Atoms are the extremely small particles that are the building blocks of all matter (Section 1.9). In molecules, atoms combine to give the smallest particles with the properties of a particular substance (Section 1.10).

Chemistry in daily life. Chemists transform matter to make these and many other items we use every day.

1.1 Why Care About Chemistry?

Why study chemistry? There are many good reasons. **Chemistry** is the science of matter and its transformations from one form to another. **Matter** is anything that has mass and occupies space. Consequently, chemistry has enormous impact on our daily lives, on other sciences, and even on areas as diverse as art, music, cooking, and recreation. Chemical transformations happen all the time, everywhere. Chemistry is intimately involved in the air we breathe and the reasons we need to breathe it; in purifying the water we drink; in growing, cooking, and digesting the food we eat; and in the discovery and production of medicines to help maintain health. Chemists continually provide new ways of transforming matter into different forms with useful properties. Examples include the plastic disks used in CD and DVD players; the microchips and batteries in cell phones or laptop computers; and the steel, aluminum, rubber, plastic, and other components of automobiles.

Chemists are people who are fascinated by matter and its transformations—as you are likely to be after seeing and experiencing chemistry in action. Chemists have a unique and spectacularly successful way of thinking about and interpreting the material world around them—an atomic and molecular perspective. Knowledge and understanding of chemistry are crucial in biology, pharmacology, medicine, geology, materials science, many branches of engineering, and other sciences. Modern research is often done by teams of scientists whose members represent several of these different disciplines. In such teams, ability to communicate and collaborate is just as important as knowledge in a single field. Studying chemistry can help you learn how chemists think about the world and solve problems, which in turn can lead to effective collaborations. Such knowledge will be useful in many career paths and will help you become a better-informed citizen in a world that is becoming technologically more and more complex—and interesting.

Chemistry, and the chemist's way of thinking, can help answer a broad range of questions—questions that might arise in your mind as you carefully observe the world around you. Here are some that have occurred to us and are answered later in this book:

- How can a disease be caused or cured by a tiny change in a molecule? (Section 12.7)
- Why does a metal get red hot and then white hot when it is heated? (Section 7.2)
- Why does rain fall as drops instead of cubes or cylinders? (Section 11.1)
- Why does salt help to clear snow and ice from roads? (Section 15.7)
- What is the difference between a saturated fat, an unsaturated fat, and a polyunsaturated fat? (Section 12.6)
- Where does the energy come from to make my muscles work? (Sections 6.12 and 18.9)
- What are the molecules in my eyes doing when I watch a movie? (Section 13.4)
- Why does frost form on top of a parked car in winter, but not on the sides? (Section 10.11)
- Why is the sky blue? (Section 15.8)
- Why can some insects walk on water? (Section 11.1)
- Why is stratospheric ozone depletion harmful? I thought too much ozone was bad for your lungs. (Section 10.9)
- How does soap help to get clothes clean? (Section 15.9)
- What happens to an egg when I cook it? (Section 13.9)
- How are plastics made, and why are there so many different kinds? (Section 12.6)
- Why is there a warning on a container of household bleach that says not to mix the bleach with other cleaners, such as toilet-bowl cleaner? (Section 16.10)

- I've heard that most homes in the United States contain small quantities of a radioactive gas. How can I find out whether my home is safe? (Section 20.8)
- Why is iron strongly attracted to a magnet, but most substances are not? (Section 7.8)
- Why do some antacids fizz when added to vinegar? (Section 5.2)
- Why is Dry Ice always surrounded by fog? (Section 11.3)

The section number following each question indicates where you can find the answer. There are probably many more questions like these that have occurred to you. We encourage you to add them to the list and think about them as you study chemistry.

Dry Ice is surrounded by fog.

1.2 Molecular Medicine

How modern science works, and why the chemist's unique perspective is so valuable, can be seen through an example. (At this point you need not fully understand the science, so don't worry if some words or ideas are unfamiliar.) From many possibilities, we have chosen the anticancer agent paclitaxel. Paclitaxel was brought to market in 1993 by Bristol-Myers Squibb under the trade name Taxol®. It is recognized as an effective drug for treating ovarian cancer, breast cancer, certain forms of lung cancer, and some other cancers. Total sales of this drug reached $10 billion shortly after 2000, and at present it is the all-time best-selling anticancer drug.

"The whole of science is nothing more than a refinement of everyday thinking."—Albert Einstein

The story of paclitaxel began about 50 years ago when Jonathan Hartwell of the National Cancer Institute initiated a program to collect samples of 35,000 kinds of plants from within the United States and examine them as potential sources of anticancer drugs. On August 21, 1962, in a forest near Mount St. Helens in Washington, a group of three graduate students led by U.S. Department of Agriculture botanist Arthur S. Barclay collected three quarters of a pound of bark from *Taxus brevifolia*, commonly called the Pacific yew tree. Preliminary tests at a research center in Wisconsin showed that extracts from the bark were active against cancer, so a larger sample of bark was collected and sent to chemists Monroe Wall and Mansukh Wani at the Research Triangle Institute in North Carolina.

Wall and Wani carried out a painstaking chemical analysis in which they separated and purified several substances found in the bark. By 1967 they had isolated the active ingredient, which they named "taxol." The name was based on the botanic name of the yew (*Taxus*, giving "tax") and the fact that the substance belongs to a class of compounds known as alcohols (giving "ol"). Later the name Taxol® was registered as a trademark by Bristol-Myers Squibb, so paclitaxel is now used to describe the drug generically.

In 1971 Wall and Wani determined the chemical formula of paclitaxel: $C_{47}H_{51}NO_{14}$, a molecule containing 113 atoms. This is small compared with giant biological molecules such as proteins and DNA, but the structure of paclitaxel is complicated enough that Wall and Wani had to chemically separate the molecule into two parts, determine the structure of each part using a technique called x-ray crystallography, and then figure out how the two parts were connected. The structure of the smaller of those parts is shown in the figure on p. 4.

Throughout this book, computer-generated models of molecular structures will be used to help you visualize chemistry at the atomic and molecular levels.

Even after its structure was known, there was not a great deal of interest in paclitaxel as a drug because the compound was so hard to get. Removing the bark from a Pacific yew kills the tree, the bark from a 40-foot tree yields a very small quantity of the drug (less than half a gram), and it takes more than 100 years for a tree to grow to 40 feet. Had it not been for the discovery that paclitaxel's method for killing cancer cells was unique, which opened new avenues for research, the drug probably would not have been commercialized.

You are probably curious about why this substance can kill cancer cells when thousands of other substances collected by botanists do not. The answer to this question was found in 1979 by Susan Band Horwitz, who was interested in how

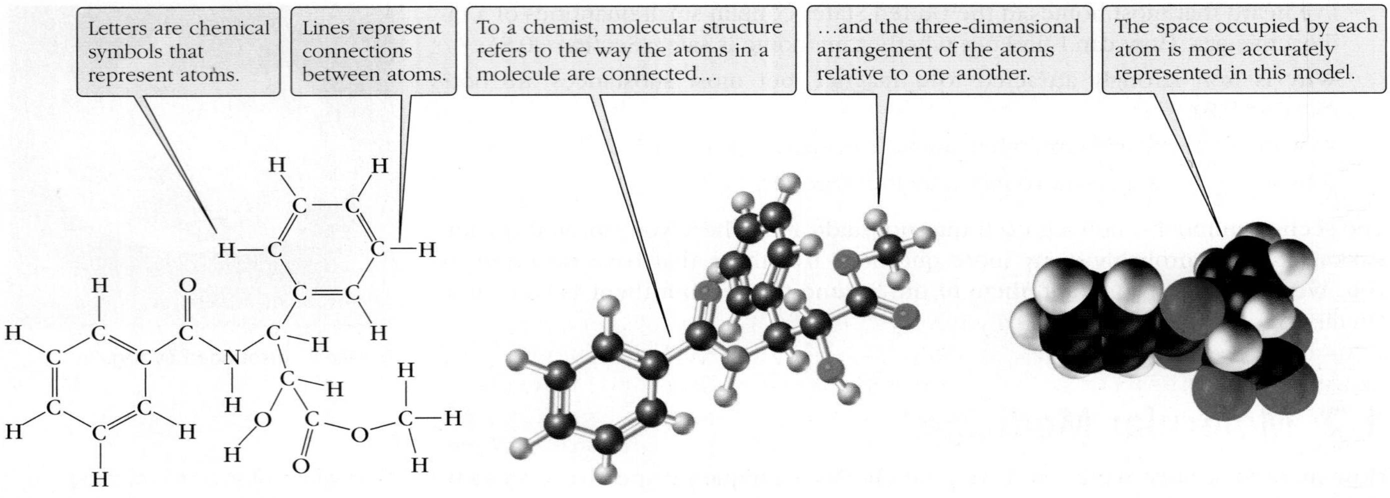
Letters are chemical symbols that represent atoms.
Lines represent connections between atoms.
To a chemist, molecular structure refers to the way the atoms in a molecule are connected…
…and the three-dimensional arrangement of the atoms relative to one another.
The space occupied by each atom is more accurately represented in this model.

Structural formula Ball-and-stick model Space-filling model

The molecular structure of one of the two molecular parts of paclitaxel used by Wall and Wani to determine the structure of the drug.

Courtesy of Dr. Susan Band Horwitz/ Albert Einstein School of Medicine

Susan Band Horwitz

1937–

Susan Band Horwitz is Falkenstein Professor of Cancer Research and co-Chair of the Department of Molecular Pharmacology at the Albert Einstein College of Medicine at Yeshiva University. She is past president of the American Association for Cancer Research and has won the Warren Alpert Foundation Prize for her work in developing Taxol®. She is currently studying other molecules that are even more effective than Taxol® for treating cancer.

small molecules, particularly those from natural sources, could be used to treat disease. She obtained a sample of paclitaxel from Wall and Wani and experimented to find out how it worked within biological cells. What she discovered was a unique atomic-scale mechanism of action: Paclitaxel aids the formation of structures known as microtubules and prevents their breakdown once they form. Microtubules help to maintain cell structure and shape. They also serve as conveyor belts, transporting other cell components from place to place. Microtubules form when many molecules of the protein tubulin assemble into long, hollow, cylindrical fibers. Tubulin is an extremely large molecule that consists of two very similar parts, one called alpha and the other beta.

During cell division, microtubules move chromosomes to opposite sides of the cell so that the chromosomes can be incorporated into two new cell nuclei as the cell divides. To carry out this process, the microtubules must grow and shrink by either adding or losing tubulin molecules. Because paclitaxel prevents the microtubules from shrinking, it prevents new cell nuclei from forming. The cells cannot divide and eventually die. An important characteristic of cancer is rapid, uncontrolled division of cells. Because cancer cells divide much faster than most other cells, substances that adversely affect cell division affect cancer cells more than normal cells. This provides an effective treatment, especially for cancers that are particularly virulent.

Horwitz's discovery of a new molecular mode of action, together with dramatic improvement in some patients for whom no other treatment had been successful, generated a great deal of interest in paclitaxel, but the drug faced two additional barriers to its widespread use. First, paclitaxel is almost completely insoluble in water, which makes it extremely difficult to deliver into the human body. Researchers at the National Cancer Institute found that castor oil could dissolve the drug, which allowed its use in clinical trials. For patients who were allergic to the castor oil, special medication was developed to alleviate the allergic reaction. These clinical trials showed paclitaxel was extremely effective against ovarian and breast cancers.

But there was another problem. A complete course of treatment for cancer required about two grams of paclitaxel—a mass obtained from six trees that had grown for 100 years. A simple calculation showed that harvesting enough bark to treat all cancer patients would soon cause extinction of the Pacific yew. To find a better way to obtain paclitaxel, chemists tried to synthesize it from simple ingredients. In 1994 they succeeded; paclitaxel was made by two independent groups of chemists without help from *Taxus brevifolia* or any other plant. To date, however, complete chemical synthesis produces only a tiny portion of product from large quantities of starting materials and has not been developed into a cost-effective industrial-scale process. Instead, paclitaxel is produced by the pharmaceutical industry in a four-step synthesis process that begins with the compound 10-deacetylbaccatin, which is isolated from the English yew, *Taxus baccata*. This species is more common than the Pacific yew and grows much faster. Thus, plenty of the drug can be obtained without concerns about extinction of the plant that produces its precursor.

The success of paclitaxel as an anticancer agent led to many attempts to discover even more about how it works. In 1998 Eva Nogales, Sharon Wolf, and Kenneth Downing at the Lawrence Berkeley National Laboratory created the first picture showing how all of the atoms are arranged in three-dimensional space when paclitaxel interacts with tubulin. This provided further insight into how paclitaxel prevents tubulin molecules from leaving a microtubule, thereby causing cell death. In 2001 Nogales, Downing, and co-workers confirmed experimentally a proposal by James P. Snyder of Emory University that the paclitaxel molecule assumes a T-shape when attached to tubulin. Figure 1.1 shows the huge tubulin molecule with a paclitaxel molecule neatly fitting into a "pocket" at the lower right. By attaching to tubulin in this way, paclitaxel makes the tubulin less flexible and prevents tubulin molecules from leaving microtubules. In 2004 James P. Snyder, David G. I. Kingston (Virginia Tech University), Susan Bane (SUNY Binghamton), and five co-workers synthesized a new molecule, similar to paclitaxel but held by chemical bonds in the T-shape required for binding to tubulin. The new compound, designated "**13a**" because it was the 13th mentioned in their article, showed increased activity against cancer of up to 20 times that of paclitaxel. Now that the required structure is known, it will be easier for chemists to devise other molecular structures that are even more effective.

© Michael W. Davidson and The Florida State University

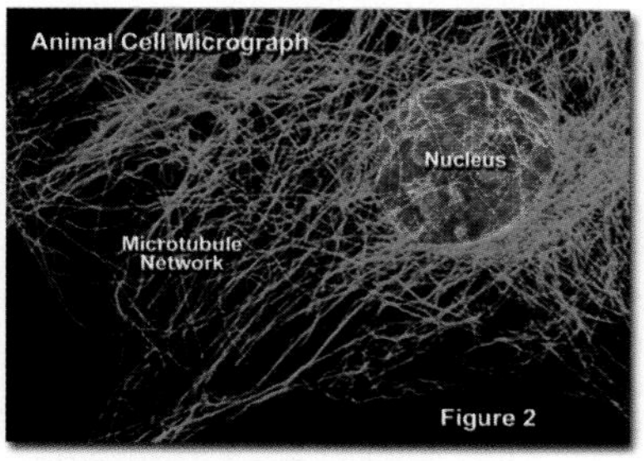

Microtubules as seen through a fluorescence optical microscope. These are from an embryonic mouse cell.

© Michael W. Davidson and The Florida State University

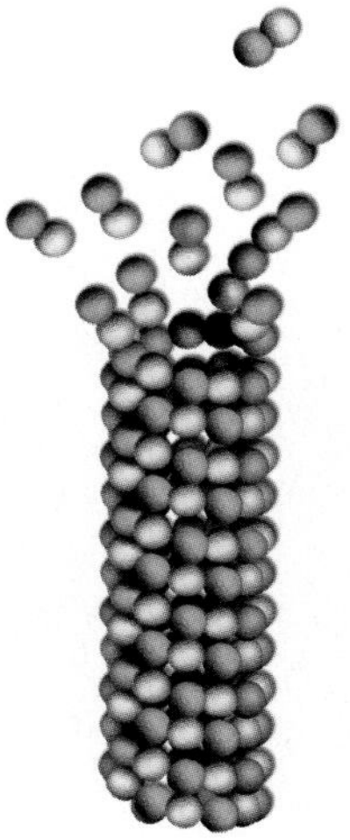

Assembly of a microtubule by addition of tubulin molecules. (Each two-part tubulin molecule is shown in yellow and green.)

From **http://www.lbl.gov/Science-Articles/Archive/3D-tubulin.html**

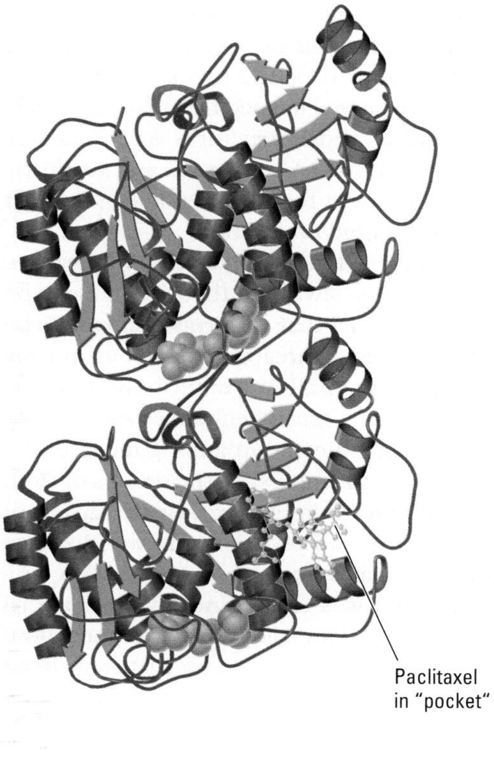

Figure 1.1 **Tubulin with paclitaxel attached.** The blue and green ribbon-like structures and the string-like structures represent the "backbone" of tubulin, giving a rough indication of where its atoms are located. Paclitaxel is the light tan group of atoms at the lower right.

1.3 How Science Is Done

How science is done is dealt with in *Oxygen*, a play written by chemists Carl Djerassi and Roald Hoffmann that premiered in 2001. By revisiting the discovery of oxygen, the play provides many insights regarding the process of science and the people who make science their life's work.

The story of paclitaxel illustrates many aspects of how people do science and how scientific knowledge changes and improves over time. In antiquity it was known that extracts of many plants had medicinal properties. For example, Native Americans in the Pacific Northwest made tonics from the bark of the Pacific yew. Probably this involved a chance observation that led to the hypothesis that yew bark was beneficial. A **hypothesis** is an idea that is tentatively proposed as an explanation for some observation and provides a basis for experimentation. The hypothesis that extracts of some plants might be effective against cancer led Jonathan Hartwell to initiate his program of collecting plant material. Tests of the extract from Pacific yew bark verified that this hypothesis was correct.

Testing a hypothesis may involve collecting qualitative or quantitative data. The observation that extracts from the bark of the Pacific yew killed cancer cells was **qualitative:** It did not involve numeric data. It was clear that there was an effect, but further studies were needed to find out how significant the effect was. **Quantitative** information is obtained from measurements that produce numeric data. Studies of the paclitaxel analog "**13a**" discovered in 2004 showed that it was 20 times more active than paclitaxel against ovarian cancer cells. These quantitative data were important in confirming that a T-shaped molecular structure was important.

A scientific **law** is a statement that summarizes and explains a wide range of experimental results and has not been contradicted by experiments. *A law can predict unknown results and also can be disproved or falsified by new experiments.* When the results of a new experiment contradict a law, that's exciting to a scientist. If several scientists repeat a contradictory experiment and get the same result, then the law must be modified to account for the new results—or even discarded altogether.

A successful hypothesis is often designated as a **theory**—a unifying principle that explains a body of facts and the laws based on them. A theory usually suggests new hypotheses and experiments, and, like a law, it may have to be modified or even discarded if contradicted by new experimental results. A **model** makes a theory more concrete, often in a physical or a mathematical form. Models of molecules, for example, were important in determining how paclitaxel binds to tubulin and kills cells. Molecular models can be constructed by using spheres to represent atoms and sticks to represent the connections between the atoms. Or a computer can be used to calculate the locations of the atoms and display model molecular structures on a screen (as was done to create Figure 1.1). The theories that matter is made of atoms and molecules, that atoms are arranged in specific molecular structures, and that the properties of matter depend on those structures are fundamental to chemists' unique atomic/molecular perspective on the world and to nearly everything modern chemists do. Clearly it is important that you become as familiar as you can with these theories and with models based on them.

Another important aspect of the way science is done involves communication. Science is based on experiments and on hypotheses, laws, and theories that can be contradicted by experiments. Therefore, it is essential that experimental results be communicated to all scientists working in any specific area of research as quickly and accurately as possible. Scientific communication allows contributions to be made by scientists in different parts of the world and greatly enhances the rapidity with which science can develop. In addition, communication among members of scientific research teams is crucial to their success. Examples are the groups of from five to twenty chemists who synthesized paclitaxel from scratch, or the group of eight scientists from three universities that made and tested the new substance "**13a**" with even greater anticancer activity. The importance of scientific communication is emphasized by the fact that the Internet was created not by commercial interests, but by scientists who saw its great potential for scientific communication.

In this chapter we discuss fundamental concepts of chemistry that have been revealed by applying the processes of science to the study of matter. We begin by considering how matter can be classified according to characteristic properties.

1.4 Identifying Matter: Physical Properties

One type of matter can be distinguished from another by observing the properties of samples of matter and classifying the matter according to those properties. A **substance** is a type of matter that has the same properties and the same composition throughout a sample. Each substance has characteristic properties that are different from the properties of any other substance (Figure 1.2). In addition, one sample of a substance has the same composition as every other sample of that substance—it consists of the same stuff in the same proportions.

You can distinguish sugar from water because you know that sugar consists of small, white particles of solid, while water is a colorless liquid. Metals can be recognized as a class of substances because they usually are solids, have high densities, feel cold to the touch, and have shiny surfaces. These properties can be observed and measured without changing the composition of a substance. They are called **physical properties.**

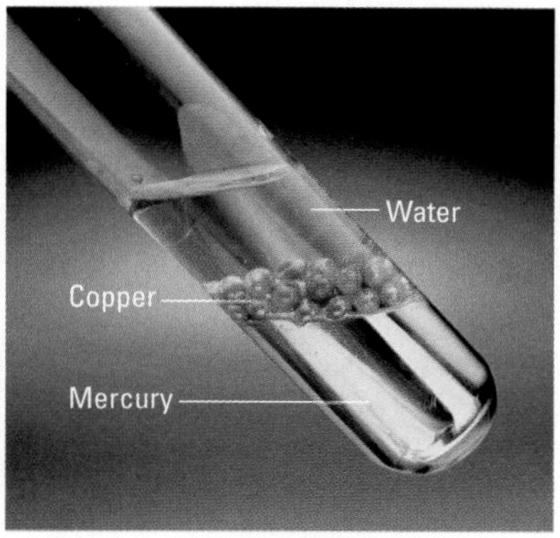

Figure 1.2 Substances have characteristic physical properties. The test tube contains silvery liquid mercury, small spheres of solid orange copper, and colorless liquid water. Each of these substances has characteristic properties that differentiate it from the others.

Physical Change

As a substance's temperature or pressure changes, or if it is mechanically manipulated, some of its physical properties may change. Changes in the physical properties of a substance are called **physical changes.** The same substance is present before and after a physical change, but the substance's physical state or the gross size and shape of its pieces may have changed. Examples are melting a solid (Figure 1.3), boiling a liquid, hammering a copper wire into a flat shape, and grinding sugar into a fine powder.

Some Physical Properties

Temperature	Shape of solid crystals
Pressure	Hardness, brittleness
Mass	Heat capacity
Volume	Thermal conductivity
State (solid, liquid, gas)	Electrical conductivity
Melting point	
Boiling point	
Density	
Color	

Melting and Boiling Point

An important way to help identify a substance is to measure the substance's **melting point,** the temperature at which the solid melts (or the temperature at which the liquid freezes, the **freezing point,** which is the same thing). Also characteristic is the substance's **boiling point,** the temperature at which the liquid boils. If two or more substances are in a mixture, the melting point depends on how much of each is present, but for a single substance the melting point is always the same. This is also true of the boiling point (as long as the pressure on the boiling liquid is the same). In addition, the melting point of a pure crystalline sample of a substance is sharp—there is almost no change in temperature as the sample melts. When a mixture of two or more substances melts, the temperature when liquid first appears can be quite different from the temperature when the last of the solid is gone.

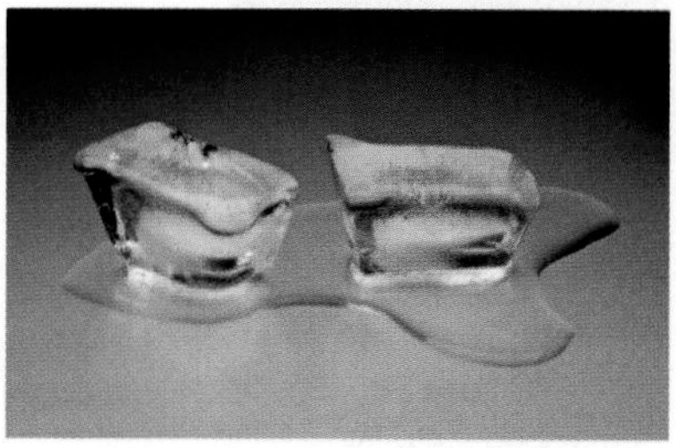

Figure 1.3 Physical change. When ice melts it changes—physically—from a solid to a liquid, but it is still water.

Temperature is the property of matter that determines whether there can be heat energy transfer from one object to another. It is represented by the symbol *T.* Energy transfers of its own accord from an object at a higher temperature to a cooler object. In the United States, everyday temperatures are reported using the Fahrenheit temperature scale. On this scale the freezing point of water is by definition 32 °F and the boiling point is 212 °F. The **Celsius temperature scale** is used in most countries of the world and in science. On this scale, 0 °C is the freezing point and 100 °C is the boiling point of pure water at a pressure of one atmosphere. The number of units between the freezing and boiling points of water is 180 Fahrenheit degrees and 100 Celsius degrees. This means that the Celsius degree is almost twice as large as the Fahrenheit degree. It takes only 5 Celsius degrees to cover the same temperature range as 9 Fahrenheit degrees, and this relationship can be used to calculate a temperature on one scale from a temperature on the other (see Appendix B.2).

If you need to convert from the Fahrenheit to the Celsius scale or from Celsius to Fahrenheit, an explanation of how to do so is in Appendix B.2.

Because temperatures in scientific studies are usually measured in Celsius units, there is little need to make conversions to and from the Fahrenheit scale, but it is quite useful to be familiar with how large various Celsius temperatures are. For example, it is useful to know that water freezes at 0 °C and boils at 100 °C, a comfortable room temperature is about 22 °C, your body temperature is 37 °C, and the hottest water you could put your hand into without serious burns is about 60 °C.

Answers to **EXERCISES** are provided at the back of this book in Appendix L.

EXERCISES that are labeled **CONCEPTUAL** are designed to test your understanding of one or more concepts; they usually involve qualitative rather than quantitative thinking.

CONCEPTUAL EXERCISE 1.1 Temperature

(a) Which is the higher temperature, 110 °C or 180 °F?
(b) Which is the lower temperature, 36 °C or 100 °F?
(c) The melting point of gallium is 29.8 °C. If you hold a sample of gallium in your hand, will it melt?

Density

Another property that is often used to help identify a substance is **density,** the ratio of the mass of a sample to its volume. If you have ten pounds of sugar, it occupies ten times the volume that one pound of sugar does. In mathematical terms, a substance's volume is directly proportional to its mass. This means that a substance's density has the same value regardless of how big the sample is.

The density of a substance varies depending on the temperature and the pressure. Densities of liquids and solids change very little as pressure changes, and they change less with temperature than do densities of gases. Because the volume of a gas varies significantly with temperature and pressure, the density of a gas can help identify the gas only if the temperature and pressure are specified.

$$\text{Density} = \frac{\text{mass}}{\text{volume}} \qquad d = \frac{m}{V}$$

Even if they look similar, you can tell a sample of aluminum from a sample of lead by picking each up. Your brain will automatically estimate which sample has greater mass for the same volume, telling you which is the lead. Aluminum has a density of 2.70 g/mL, placing it among the least dense metals. Lead's density is 11.34 g/mL, so a sample of lead is much heavier than a sample of aluminum of the same size.

Suppose that you are trying to identify a liquid that you think might be ethanol (ethyl alcohol), and you want to determine its density. You could weigh a clean, dry graduated cylinder and then add some of the liquid to it. Suppose that, from the markings on the cylinder, you read the volume of liquid to be 8.30 mL (at 20 °C). You could then weigh the cylinder with the liquid and subtract the mass of the empty cylinder to obtain the mass of liquid. Suppose the liquid mass is 6.544 g. The density can then be calculated as

$$d = \frac{m}{V} = \frac{6.544\ \text{g}}{8.30\ \text{mL}} = 0.788\ \text{g/mL}$$

From a table of physical properties of various substances you find that the density of ethanol is 0.789 g/mL, which helps confirm your suspicion that the substance is ethanol.

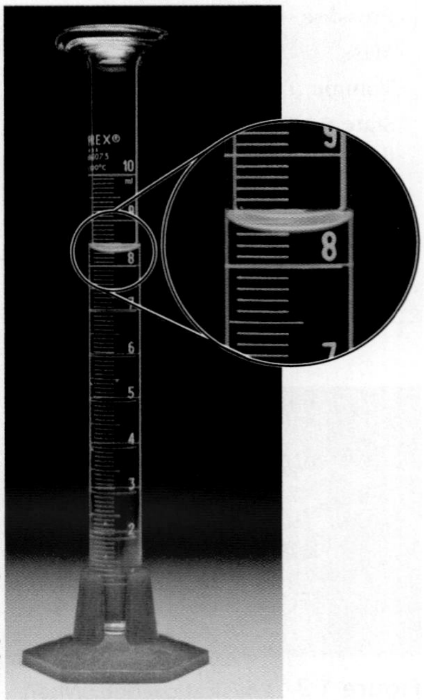

© Cengage Learning/Charles D. Winters

Graduated cylinder containing 8.30 mL of liquid.

If you divide 6.544 by 8.30 on a scientific calculator, the answer might come up as 0.788433735. This displays more digits than are meaningful, and we have rounded the result to only three significant digits. Rules for deciding how many digits should be reported in the result of a calculation and procedures for rounding numbers are introduced in Section 2.4 and Appendix A.3.

CONCEPTUAL EXERCISE 1.2 Density of Liquids

When 5.0 mL each of vegetable oil, water, and kerosene are put into a large test tube, they form three layers, as shown in the photo on p. 9.

(a) List the three liquids in order of increasing density (smallest density first, largest density last).
(b) If an additional 5.0 mL of vegetable oil is poured into the test tube, what will happen? Describe the appearance of the tube.
(c) If 5.0 mL of kerosene is added to the test tube with the 5.0 mL of vegetable oil in part (b), will there be a permanent change in the order of liquids from top to bottom of the tube? Why or why not?

EXERCISE 1.3 Physical Properties and Changes

Identify each physical property and physical change mentioned in each of these statements. Also identify the qualitative and the quantitative information given in each statement.

(a) The blue chemical compound azulene melts at 99 °C.
(b) The white crystals of table salt are cubic.
(c) A sample of lead has a mass of 0.123 g and melts at 327 °C.
(d) Ethanol is a colorless liquid that vaporizes easily; it boils at 78 °C and its density is 0.789 g/mL.

© Cengage Learning/Charles D. Winters

Liquid densities. Kerosene, vegetable oil, and water have different densities.

Measurements and Calculations: Dimensional Analysis

Determining a property such as density requires scientific measurements and calculations. The result of a measurement, such as 6.544 g or 8.30 mL, usually consists of a number and a unit. Both the number and the unit should be included in calculations. For example, the densities in Table 1.1 have units of grams per milliliter, g/mL, because density is defined as the mass of a sample divided by its volume. When a mass is divided by a volume, the units (g for the mass and mL for the volume) are also divided. The result is grams divided by milliliters, g/mL. That is, both numbers *and units* follow the rules of algebra. This is an example of **dimensional analysis,** a method of using units in calculations to check for correctness. More detailed descriptions of dimensional analysis are given in Section 2.3 and Appendix A.2. We will use this technique for problem solving throughout the book.

Suppose that you want to know whether you could lift a gallon (3784 mL) of the liquid metal mercury. To answer the question, calculate the mass of the mercury using the density, 13.55 g/mL, obtained from Table 1.1. One way to do this is to use the equation that defines density, $d = m/V$. Then solve algebraically for m, and calculate the result:

$$m = V \times d = 3784 \cancel{\text{mL}} \times \frac{13.55 \text{ g}}{1 \cancel{\text{mL}}} = 51{,}270 \text{ g} \qquad [1.1]$$

Because mercury and mercury vapor are poisonous, carrying a gallon of it around is not a good idea unless it is in a sealed container.

This equation emphasizes the fact that mass is proportional to volume, because the volume is multiplied by a proportionality constant, the density. Notice also that the units of volume (mL) appeared once in the denominator of a fraction and once in the numerator, thereby dividing out (canceling) and leaving only mass units (g). The result, 51,270 g, is more than 100 pounds, so you could probably lift the mercury, but not easily.

In Equation 1.1, a known quantity (the volume) was multiplied by a proportionality factor (the density), and the units canceled, giving an answer (the mass) with

Table 1.1 Densities of Some Substances at 20 °C

Substance	Density (g/mL)	Substance	Density (g/mL)
Butane	0.579	Titanium	4.50
Ethanol	0.789	Zinc	7.14
Benzene	0.880	Iron	7.86
Water	0.998	Nickel	8.90
Bromobenzene	1.49	Copper	8.93
Magnesium	1.74	Lead	11.34
Sodium chloride	2.16	Mercury	13.55
Aluminum	2.70	Gold	19.32

A useful source of data on densities and other physical properties of substances is the *CRC Handbook of Chemistry and Physics,* published by the CRC Press. Information is also available via the Internet—for example, the National Institute for Standards and Technology's Webbook at **http://webbook.nist.gov**.

In this book, units and dimensional analysis techniques are introduced at the first point where you need to know them. Appendices A and B provide all of this information in one place.

appropriate units. A general approach to this kind of problem is to recognize that the quantity you want to calculate (the mass) is proportional to a quantity whose value you know (the volume). Then use a proportionality factor that relates the two quantities, setting things up so that the units cancel.

$$\cancel{\text{known quantity units}} \times \frac{\text{desired quantity units}}{\cancel{\text{known quantity units}}} = \text{desired quantity units}$$

proportionality (conversion) factor

$$3784\ \cancel{\text{mL}} \times \frac{13.55\text{ g}}{1\ \cancel{\text{mL}}} = 51{,}270\text{ g}$$

A **proportionality factor** is a ratio (fraction) whose numerator and denominator have different units but refer to the same thing. In the preceding example, the proportionality factor is the density, which relates the mass and volume of the same sample of mercury. A proportionality factor is often called a **conversion factor** because it enables us to convert from one kind of unit to a different kind of unit.

Because a conversion factor is a fraction, every conversion factor can be expressed in two ways. The conversion factor in the example just given could be expressed either as the density or as its reciprocal:

$$\frac{13.55\text{ g}}{1\text{ mL}} \quad \text{or} \quad \frac{1\text{ mL}}{13.55\text{ g}}$$

The first fraction enables conversion from volume units (mL) to mass units (g). The second allows mass units to be converted to volume units. Which conversion factor to use depends on which units are in the known quantity and which units are in the quantity that we want to calculate. Setting up the calculation so that the units cancel ensures that we are using the appropriate conversion factor. (See Appendix A.2 for more examples.)

The **PROBLEM-SOLVING STRATEGY** in this book is

- Analyze the problem
- Plan a solution
- Execute the plan
- Check that the result is reasonable

Appendix A.1 explains this in detail.

PROBLEM-SOLVING EXAMPLE 1.1 Density

In an old movie thieves are shown running off with pieces of gold bullion that are about a foot long and have a square cross section of about six inches. The volume of each piece of gold is 7000 mL. Calculate the mass of gold and express the result in pounds (lb). Based on your result, is what the movie shows physically possible? (1 lb = 454 g)

Answer 1.4×10^5 g; 300 lb; probably not

Strategy and Explanation A good approach to problem solving is to (1) analyze the problem, (2) plan a solution, (3) execute the plan, and (4) check your result to see whether it is reasonable. (These four steps are described in more detail in Appendix A.1.)

Step 1: *Analyze the problem.* You are asked to calculate the mass of the gold, and you know the volume.

Step 2: *Plan a solution.* Density relates mass and volume and is the appropriate proportionality factor, so look up the density in a table. Mass is proportional to volume, so the volume either has to be multiplied by the density or divided by the density. Use the units to decide which.

Step 3: *Execute the plan.* According to Table 1.1, the density of gold is 19.32 g/mL. Setting up the calculation so that the unit (milliliter) cancels gives

$$7000\ \cancel{\text{mL}} \times \frac{19.32\text{ g}}{1\ \cancel{\text{mL}}} = 1.35 \times 10^5\text{ g}$$

This can be converted to pounds

$$1.35 \times 10^5 \text{ g} \times \frac{1 \text{ lb}}{454 \text{ g}} = 300 \text{ lb}$$

Notice that the result is expressed to one significant figure, because the volume was given to only one significant figure and only multiplications and divisions were done.

☑ **Reasonable Answer Check** Gold is nearly 20 times denser than water. A liter (1000 mL) of water is about a quart and a quart of water (2 pints) weighs about two pounds. Seven liters (7000 mL) of water should weigh 14 lb, and 20 times 14 gives 280 lb, so the answer is reasonable. The movie is not—few people could run while carrying a 300-lb object!

Rules for assigning the appropriate number of significant figures to a result are given in Appendix A.3.

The checkmark symbol accompanied by the words "Reasonable Answer Check" will be used throughout this book to indicate how to check the answer to a problem to make certain a reasonable result has been obtained.

PROBLEM-SOLVING PRACTICE 1.1

Find the volume occupied by a 4.33-g sample of benzene.

PROBLEM-SOLVING PRACTICE answers are provided at the back of this book in Appendix K.

This book includes many examples, like Problem-Solving Example 1.1, that illustrate general problem-solving techniques and ways to approach specific types of problems. Usually, each of these examples states a problem; gives the answer; explains one way to analyze the problem, plan a solution, and execute the plan; and describes a way to check that the result is reasonable. We urge you to first try to solve the problem on your own. Then check to see whether your answer matches the one given. If it does not match, try again before reading the explanation. After you have tried twice, read the explanation to find out why your reasoning differs from that given. If your answer is correct, but your reasoning differs from the explanation, you may have discovered an alternative solution to the problem. Finally, work out the Problem-Solving Practice that accompanies the example. It relates to the same concept and allows you to improve your problem-solving skills.

1.5 Chemical Changes and Chemical Properties

Another way to identify a substance is to observe how it reacts chemically. For example, if you heat a white, granular solid carefully and it caramelizes (turns brown and becomes a syrupy liquid—see Figure 1.4), it is a good bet that the white solid is ordinary table sugar (sucrose). When heated gently, sucrose decomposes to give water and other new substances. If you heat sucrose very hot, it will char, leaving be-

Sucrose (Reactant)
changes to
Carbon + Water (Products)
Water
Carbon
© Cengage Learning/Charles D. Winters
1 When table sugar (sucrose) is heated . . .
2 . . . it caramelizes, turning brown.
3 Heating to a higher temperature causes further decomposition (charring) to carbon and water vapor.

Figure 1.4 **Chemical change.** Heat can caramelize or char sugar.

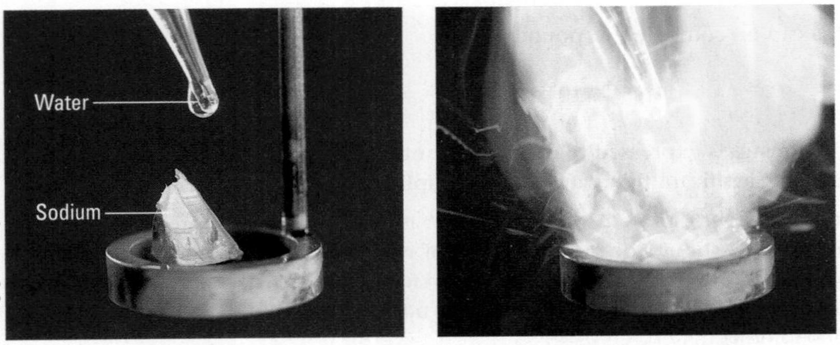
Water
Sodium
© Cengage Learning/Charles D. Winters

Figure 1.5 Chemical change. When a drop of water contacts sodium, a violent reaction produces flammable hydrogen gas and a solution of sodium hydroxide (lye). Production of motion, heat, and light when substances are mixed is evidence that a chemical reaction is occurring.

hind a black residue that is mainly carbon (and is hard to clean up). If you drip some water onto a sample of sodium metal, the sodium will react violently with the water, producing an aqueous solution of lye (sodium hydroxide) and a flammable gas, hydrogen (Figure 1.5). These are examples of **chemical changes** or **chemical reactions.** In a chemical reaction, one or more substances (the **reactants**) are transformed into one or more different substances (the **products**). Reactant substances are replaced by product substances as the reaction occurs. This process is indicated by writing the reactants, an arrow, and then the products:

Sucrose	$\longrightarrow$	Carbon + Water
Reactant	changes to	Products

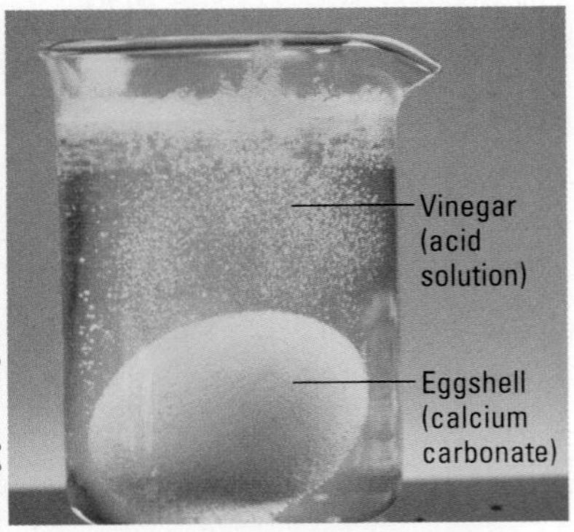
Vinegar (acid solution)
Eggshell (calcium carbonate)
© Cengage Learning/Charles D. Winters

Figure 1.6 Chemical change. Vinegar, which is an acid, reacts with an eggshell, which is mainly calcium carbonate, causing colorless carbon dioxide gas to bubble away. Production of gas bubbles when substances come into contact is one kind of evidence that a chemical reaction is occurring.

Chemical reactions make chemistry interesting, exciting, and valuable. If you know how, you can make a medicine from the bark of a tree, clothing from crude petroleum, or even a silk purse from a sow's ear (it has been done). This is a very empowering idea, and human society has gained a great deal from it. Our way of life is greatly enhanced by our ability to use and control chemical reactions. And life itself is based on chemical reactions. Biological cells are filled with water-based solutions in which thousands of chemical reactions are happening all the time.

Chemical Properties

A substance's **chemical properties** describe the kinds of chemical reactions the substance can undergo. One chemical property of metallic sodium is that it reacts rapidly with water to produce hydrogen and a solution of sodium hydroxide (Figure 1.5). Because it also reacts rapidly with air and a number of other substances, sodium is also said to have a more general chemical property: It is highly reactive. A chemical property of substances known as metal carbonates is that they produce carbon dioxide when treated with an acid (Figure 1.6). Fuels are substances that have the chemical property of reacting with oxygen or air and at the same time transferring large quantities of energy to their surroundings. An example is natural gas (mainly methane), which is shown reacting with oxygen from the air in a gas stove in Figure 1.7. A substance's chemical properties tell us how it will behave when it contacts air or water, when it is heated or cooled, when it is exposed to sunlight, or when it is mixed with another substance. Such knowledge is very useful to chemists, biochemists, geologists, chemical engineers, and many other kinds of scientists.

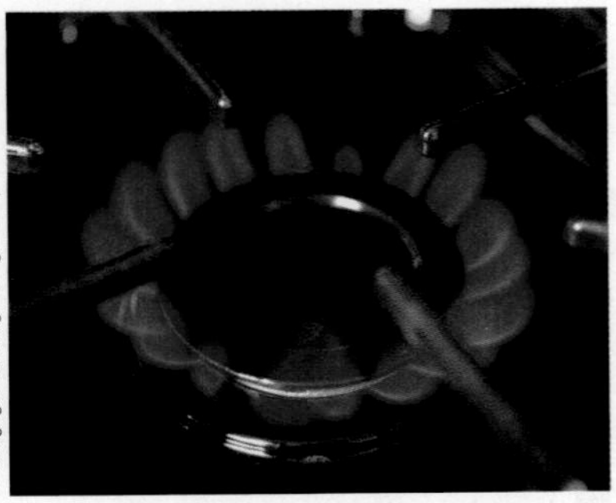
© Cengage Learning/George Semple

Figure 1.7 Combustion of natural gas. Natural gas, which in the United States consists mostly of methane, burns in air, transferring energy that raises the temperature of its surroundings.

Energy

Chemical reactions are usually accompanied by transfers of energy. (Physical changes also involve energy transfers, but usually they are smaller than those for chemical changes.) **Energy** is defined as the capacity to do work—that is, to make something happen. Combustion of a fuel, as in Figure 1.7, transforms energy stored in chemical bonds in the fuel molecules and oxygen molecules into motion of the product molecules and of other nearby molecules. This corresponds to a higher temperature in the vicinity of the flame. The chemical reaction in a light stick transforms energy stored in molecules into light energy, with only a little heat transfer (Figure 1.8). A chemical reaction in a battery makes a calculator work by forcing electrons to flow through an electric circuit.

Energy supplied from somewhere else can cause chemical reactions to occur. For example, photosynthesis takes place when sunlight illuminates green plants. Some of the sunlight's energy is stored in carbohydrate molecules and oxygen molecules that are produced from carbon dioxide and water by photosynthesis. Aluminum, which you may have used as foil to wrap and store food, is produced by passing electricity through a molten, aluminum-containing ore (Section 21.4). You consume and metabolize food, using the energy stored in food molecules to cause chemical reactions to occur in the cells of your body. The relation between chemical changes and energy is an important theme of chemistry.

Figure 1.8 Transforming energy. In each of these light sticks a chemical reaction transforms energy stored in molecules into light.

CONCEPTUAL EXERCISE 1.4 Chemical and Physical Changes

Identify the chemical and physical changes that are described in this statement: Propane gas burns, and the heat of the combustion reaction is used to hard-boil an egg.

1.6 Classifying Matter: Substances and Mixtures

Once its chemical and physical properties are known, a sample of matter can be classified on the basis of those properties. Most of the matter we encounter every day is like the bark of a yew tree, concrete, or the carbon fiber composite frame of a high-tech bicycle—not uniform throughout. There are variations in color, hardness, and other properties from one part of a sample to another. This makes these materials complicated, but also interesting. A major advance in chemistry occurred when it was realized that it was possible to separate several component substances from such nonuniform samples. For example, in 1967 appropriate treatment of yew bark produced a substance, paclitaxel, that was shown to be active against cancer.

Often, as in the case of the bark of a tree, we can easily see that one part of a sample is different from another part. In other cases a sample may appear completely uniform to the unaided eye, but a microscope can reveal that it is not. For example, blood appears smooth in texture, but magnification reveals red and white cells within the liquid (Figure 1.9). The same is true of milk. A mixture in which the uneven texture of the material can be seen with the naked eye or with a microscope is classified as a **heterogeneous mixture.** Properties in one region are different from the properties in another region.

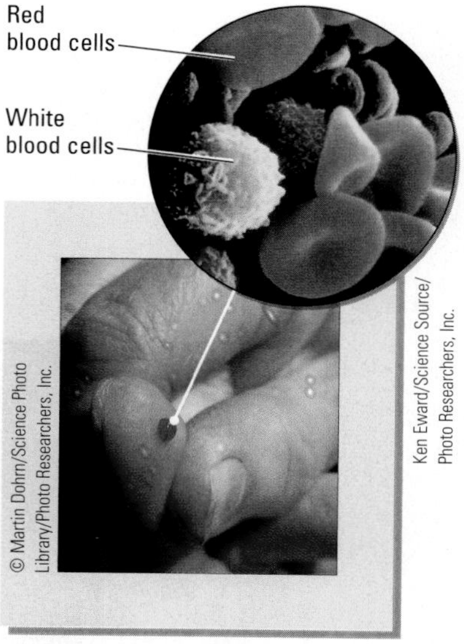

Figure 1.9 A heterogeneous mixture. Blood appears to be uniform to the unaided eye, but a microscope reveals that it is not homogeneous. The properties of red blood cells differ from the properties of the surrounding blood plasma, for example.

A **homogeneous mixture,** or **solution,** is completely uniform and consists of two or more substances in the same phase—solid, liquid, or gas (Figure 1.10). No amount of optical magnification will reveal different properties in one region of a solution compared with those in another. Heterogeneity exists in a solution only at the scale of atoms and molecules, which are too small to be seen with visible light. Examples of solutions are clear air (mostly a mixture of nitrogen and oxygen gases), sugar water, and some brass alloys (which are homogeneous mixtures of copper and zinc). The properties of a homogeneous mixture are the same

Figure 1.10 **A solution.** When solid salt (sodium chloride) is stirred into liquid water, it dissolves to form a homogeneous liquid mixture. Each portion of the solution has exactly the same saltiness as every other portion, and other properties are also the same throughout the solution.

everywhere in any particular sample, but they can vary from one sample to another depending on how much of one component is present relative to another component.

Separation and Purification

Earlier in this chapter we stated that a substance has characteristic properties that distinguish it from all other substances. However, for those characteristic properties to be observed, the substance must be separated from all other substances; that is, it must be purified. The melting point of an impure substance is different from that of the purified substance. The color and appearance of a mixture may also differ from those of a pure substance. Therefore, when we talk about the properties of a substance, it is assumed that we are referring to a pure substance—one from which all other substances have been separated.

Purification usually has to be done in several repeated steps and monitored by observing some property of the substance being purified. For example, iron can be separated from a heterogeneous mixture of iron and sulfur with a magnet, as shown in Figure 1.11. In this example, color, which depends on the relative quantities of iron and sulfur, indicates purity. The bright yellow color of pure sulfur is assumed to indicate that all the iron has been removed.

Concluding that a substance is pure on the basis of a single property of the mixture could be misleading because other methods of purification might change some other properties of the sample. It is safe to call sulfur pure when a variety of methods of purification fail to change its physical and chemical properties. Purification is important because it allows us to attribute properties (such as activity against cancer) to specific substances and then to study systematically which kinds of substances have properties that we find useful. In some cases, insufficient purification of a substance has led scientists to attribute to that substance properties that were actually due to a tiny trace of another substance.

Only a few substances occur in nature in pure form. Gold, diamonds, and silicon dioxide (quartz) are examples. We live in a world of mixtures; all living things, the air and food on which we depend, and many products of technology are mixtures. Much of what we know about chemistry, however, is based on separating and purifying the components of those mixtures and then determining their properties. To date, more than 50 million substances have been reported, and many more are being discovered or synthesized by chemists every year. When pure, each of these substances has its own particular composition and its own characteristic properties.

1 Iron and sulfur can be separated by stirring with a magnet.
2 The first time that the magnet is removed, much of the iron is removed with it.
3 The sulfur still looks dirty because a small quantity of iron remains.
4 Repeated extractions of iron with the magnet leave a bright yellow sample of sulfur that cannot be purified further by this technique.

Figure 1.11 **Separating a mixture:** iron and sulfur.

A good example of the importance of purification is the high-purity silicon needed to produce transistors and computer chips. In one billion grams (about 1000 tons) of highly pure silicon there has to be less than one gram of impurity. Once the silicon has been purified, small but accurately known quantities of specific substances, such as boron or arsenic, can be introduced to give the electronic chip the desired properties. (See Sections 11.8 and 11.9.)

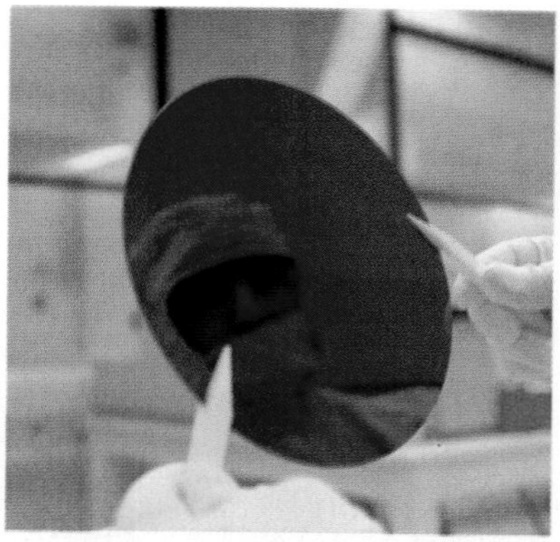
press photo BASF

High-purity silicon. A worker in a "clean room" holds a sample of high-purity silicon.

Detection and Analysis

Once we know that a given substance (such as an anticancer drug) is either valuable or harmful, it becomes important to know whether that substance is present in a sample and to be able to find out how much of it is there. Does an ore contain enough of a valuable metal to make it worthwhile to mine the ore? Is there enough mercury in a sample of fish to make it unsafe for humans to eat the fish?

Answering questions like these is the job of *analytical chemists,* and they improve their methods every year. For example, in 1960 mercury could be detected at a concentration of one part per million, in 1970 the detection limit was one part per billion, and by 1980 the limit had dropped to one part per trillion. Thus, in 20 years the ability to detect small concentrations of mercury had increased by a factor of one million. This improvement has an important effect. Because we can detect smaller and smaller concentrations of contaminants, such contaminants can be found in many more samples. A few decades ago, toxic substances were usually not found when food, air, or water was tested, but that did not mean they were not there. It just meant that our analytical methods were unable to detect them. Today, with much better methods, toxic substances can be detected in most samples, which prompts demands that concentrations of such substances should be reduced to zero.

One part per million (ppm) means we can find one gram of a substance in one million grams of total sample. That corresponds to one-tenth of a drop of water in a bucket of water. One part per billion corresponds to a drop in a swimming pool, and one part per trillion corresponds to a drop in a large supermarket.

Absence of evidence is not evidence of absence.

Although we expect that chemistry will push detection limits lower and lower, there will always be a limit below which an impurity will be undetectable. Proving that there are no contaminants in a sample will never be possible. This is a specific instance of the general rule that it is impossible to prove a negative. To put this idea another way, it will never be possible to prove that we have produced a completely pure sample of a substance, and therefore it is unproductive to legislate that there should be zero contamination in food or other substances. It is more important to use chemical analysis to determine a safe level of a toxin than to try to prove that the toxin is completely absent. In some cases, very small concentrations of a substance are beneficial but larger concentrations are toxic. (An example of this is selenium in the human diet.) Analytical chemistry can help us to determine the optimal ranges of concentration.

1.7 Classifying Matter: Elements and Compounds

Most of the substances separated from mixtures can be converted to two or more simpler substances by chemical reactions—a process called *decomposition.* Substances are often decomposed by heating them, illuminating them with sunlight, or passing electricity through them. For example, table sugar (sucrose) can be separated from sugarcane and purified. When heated it decomposes via a complex series of chemical changes (caramelization—shown earlier in Figure 1.4) that produces the brown color and flavor of caramel candy. If heated for a longer time at a high enough temperature, sucrose is converted completely to two other substances, carbon and water. Furthermore, if the water is collected, it can be decomposed still further to pure hydrogen and oxygen by passing a direct electric current through it. However, nobody has found a way to decompose carbon, hydrogen, or oxygen.

In 1661 Robert Boyle was the first to propose that elements could be defined by the fact that they could not be decomposed into two or more simpler substances.

Substances like carbon, hydrogen, and oxygen that cannot be changed by chemical reactions into two or more new substances are called **chemical elements** (or just elements). Substances that can be decomposed, like sucrose and water, are **chemical compounds** (or just compounds). When elements are chemically combined in a compound, their original characteristic properties—such as color, hardness, and melting point—are replaced by the characteristic properties of the compound. For example, in sucrose these three elements are chemically combined:

- Carbon, which is usually a black powder, but is also commonly seen in the form of diamonds
- Hydrogen, a colorless, flammable gas with the lowest density known
- Oxygen, a colorless gas necessary for human respiration

Figure 1.12 A compound and its elements. Table sugar, sucrose, is composed of the elements carbon, oxygen, and hydrogen. When elements are combined in a compound, the properties of the elements are no longer evident. Only the properties of the compound can be observed.

As you know from experience, sucrose is a white, crystalline powder that is completely unlike any of these three elements (Figure 1.12).

If a compound consists of two or more different elements, how is it different from a mixture? There are two ways: (1) A compound has specific composition and (2) a compound has specific properties. Both the composition and the properties of a mixture can vary. A solution of sugar in water can be very sweet or only a little sweet, depending on how much sugar has been dissolved. There is no particular composition of a sugar solution that is favored over any other, and each different composition has its own set of properties. On the other hand, 100.0 g pure water always contains 11.2 g hydrogen and 88.8 g oxygen. Pure water always melts at 0.0 °C and boils at 100 °C (at one atmosphere pressure), and it is always a colorless liquid at room temperature.

PROBLEM-SOLVING EXAMPLE 1.2 Elements and Compounds

A shiny, hard solid (substance A) is heated in the presence of carbon dioxide gas. After a few minutes, a white solid (substance B) and a black solid (substance C) are formed. No other substances are found. When the black solid is heated in the presence of pure oxygen, carbon dioxide is formed. Decide whether each substance (A, B, and C) is an element or a compound, and give a reason for your choice in each case. If there is insufficient evidence to decide, say so.

Answer A, insufficient evidence; B, compound; C, element

Explanation Substance C must be an element, because it combines with oxygen to form a compound, carbon dioxide, that contains only two elements; in fact, substance C must be carbon. Substance B must be a compound, because it must contain oxygen (from the carbon dioxide) and at least one other element (from substance A). There is not enough evidence to decide whether substance A is an element or a compound. If substance A is an element, then substance B must be an oxide of that element. However, there could be two or more elements in substance A (that is, it could be a compound), and the compound could still combine with oxygen from carbon dioxide to form a new compound.

☑ **Reasonable Answer Check** Substance C is black, and carbon (graphite) is black. You could test experimentally to see whether substance A could be decomposed by heating it in a vacuum; if two or more new substances were formed, then substance A would have to be a compound. If there was no change, substance A could be assumed to be an element.

PROBLEM-SOLVING PRACTICE 1.2

A student grinds an unknown sample (A) to a fine powder and attempts to dissolve the sample in 100 mL pure water. Some solid (B) remains undissolved. When the water is separated from the solid and allowed to evaporate, a white powder (C) forms. The dry white powder (C) is found to weigh 0.034 g. All of sample C can be dissolved in 25 mL pure water. Can you say whether each sample A, B, and C is an element, a compound, or a mixture? Explain briefly.

Types of Matter

What we have just said about separating mixtures to obtain elements or compounds and decomposing compounds to obtain elements leads to a useful way to classify matter (Figure 1.13). Heterogeneous mixtures such as iron with sulfur can be separated using simple manipulation—such as a magnet. Homogeneous mixtures are somewhat more difficult to separate, but physical processes will serve. For example, salt water can be purified for drinking by distilling: heating to evaporate the water and cooling to condense the water vapor back to liquid. When enough water has evaporated, salt crystals will form and they can be separated from the solution. Most difficult of all is separation of the elements that are combined in a compound. Such a separation requires a chemical change, which may involve reactions with other substances or sizable inputs of energy.

Desalinization of water (removing salt) could provide drinking water for large numbers of people who live in dry climates near the ocean. However, distillation requires a lot of energy resources and, therefore, is expensive. When solar energy can be used to evaporate the water, desalinization is less costly. For more information about desalinization of water, see Section 15.8.

EXERCISE 1.5 Classifying Matter

Classify each of these with regard to the type of matter described:

(a) Sugar dissolved in water
(b) The soda pop in a can of carbonated beverage
(c) Used motor oil freshly drained from a car
(d) The diamond in a piece of jewelry
(e) A 25-cent coin
(f) A single crystal of sugar

1.8 Nanoscale Theories and Models

To further illustrate how the methods of science are applied to matter, we now consider how a theory based on atoms and molecules can account for the physical properties, chemical properties, and classification scheme that we have just described. Physical and chemical properties can be observed by the unaided human senses and refer to samples of matter large enough to be seen, measured, and

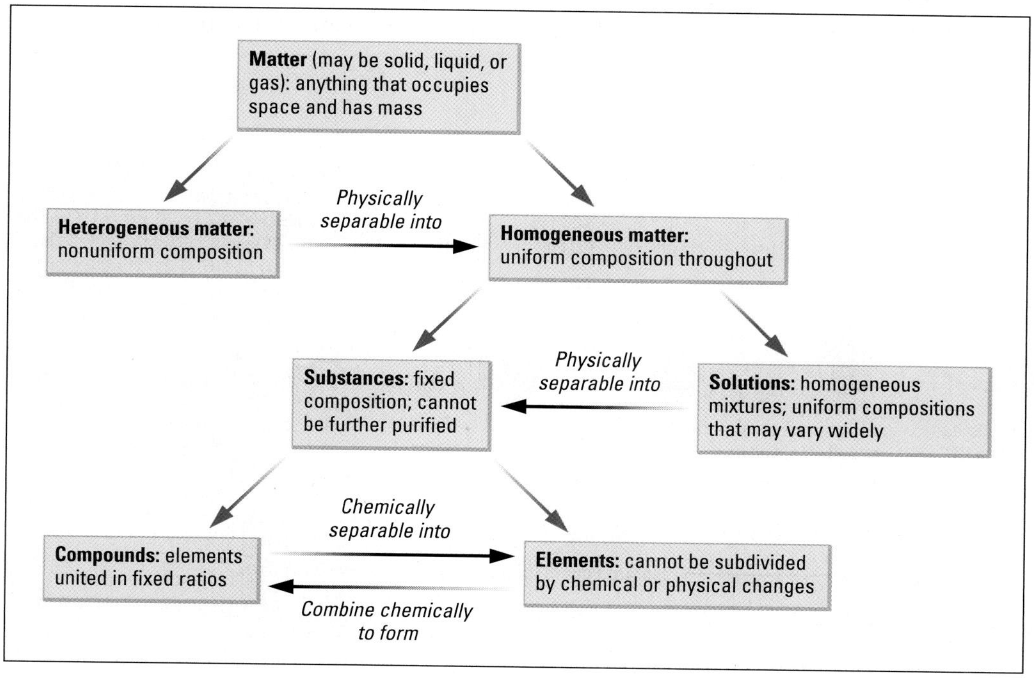

Figure 1.13 A scheme for classifying matter.

handled. Such samples are macroscopic; their size places them at the **macroscale.** By contrast, samples of matter so small that they have to be viewed with a microscope are **microscale** samples. Blood cells and bacteria, for example, are matter at the microscale. The matter that really interests chemists, however, is at the **nanoscale.** The term is based on the prefix "nano," which comes from the International System of Units (SI units) and indicates something one billion times smaller than something else. (See Table 1.2 for some important SI prefixes and length units.) For example, a line that is one billion (10^9) times shorter than 1 meter is 1 *nano*meter (1×10^{-9} m) long. The sizes of atoms and molecules are at the nanoscale. An average-sized atom such as a sulfur atom has a diameter of two tenths of a nanometer (0.2 nm = 2×10^{-10} m), a water molecule is about the same size, and an aspirin molecule is about three quarters of a nanometer (0.75 nm = 7.5×10^{-10} m) across. Figure 1.14 indicates the relative sizes of various objects at the macroscale, microscale, and nanoscale.

The International System of Units is the modern version of the metric system. It is described in more detail in Appendix B.

Using 10^9 to represent 1,000,000,000 or one billion is called scientific notation. It is reviewed in Appendix A.5.

Earlier, we described the chemist's unique atomic and molecular perspective. It is a fundamental idea of chemistry that matter is the way it is because of the nature of its constituent atoms and molecules. Those atoms and molecules are very, very tiny. Therefore, we need to use imagination creatively to discover useful theories that connect the behavior of tiny nanoscale constituents to the observed behavior of chemical substances at the macroscale. Learning chemistry enables you to "see" in the things all around you nanoscale structure that cannot be seen with your eyes.

Jacob Bronowski, in a television series and book titled *The Ascent of Man,* had this to say about the importance of imagination: "There are many gifts that are unique in man; but at the center of them all, the root from which all knowledge grows, lies the ability to draw conclusions from what we see to what we do not see."

Table 1.2 Some SI (Metric) Prefixes and Units for Length

Prefix	Abbreviation	Meaning	Example
kilo	k	10^3	1 kilometer (km) = 1×10^3 meter (m)
deci	d	10^{-1}	1 decimeter (dm) = 1×10^{-1} m = 0.1 m
centi	c	10^{-2}	1 centimeter (cm) = 1×10^{-2} m = 0.01 m
milli	m	10^{-3}	1 millimeter (mm) = 1×10^{-3} m = 0.001 m
micro	μ	10^{-6}	1 micrometer (μm) = 1×10^{-6} m
nano	n	10^{-9}	1 nanometer (nm) = 1×10^{-9} m
pico	p	10^{-12}	1 picometer (pm) = 1×10^{-12} m

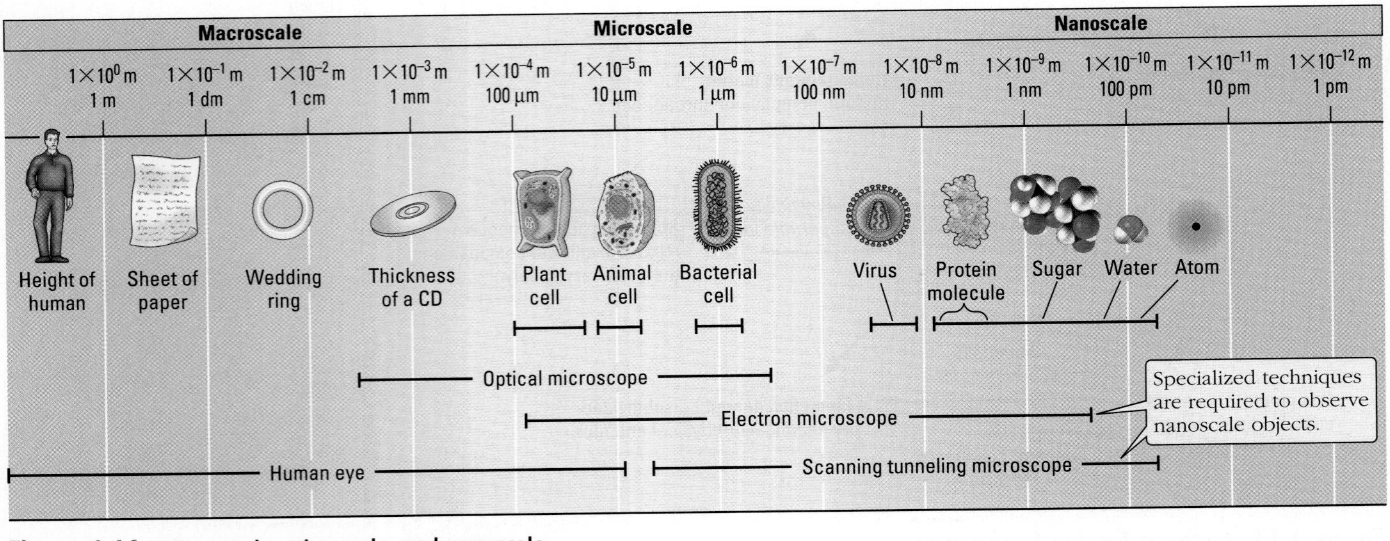

Figure 1.14 **Macroscale, microscale, and nanoscale.**

States of Matter: Solids, Liquids, and Gases

An easily observed and very useful property of matter is its physical state. Is it a solid, liquid, or gas? A **solid** can be recognized because it has a rigid shape and a fixed volume that changes very little as temperature and pressure change (Figure 1.15). Like a solid, a **liquid** has a fixed volume, but a liquid is fluid—it takes on the shape of its container and has no definite form of its own. **Gases** are also fluid, but gases expand to fill whatever containers they occupy and their volumes vary considerably with temperature and pressure. For most substances, when compared at the same conditions, the volume of the solid is slightly less than the volume of the same mass of liquid, but the volume of the same mass of gas is much, much larger. As the temperature is raised, most solids melt to form liquids; eventually, if the temperature is raised enough, most liquids boil to form gases.

Figure 1.15 **Quartz crystal.** Quartz, like any solid, has a rigid shape. Its volume changes very little with changes in temperature or pressure.

A theory that deals with matter at the nanoscale, the **kinetic-molecular theory,** states that all matter consists of extremely tiny particles (atoms or molecules) that are in constant motion. In a solid these particles are packed closely together in a regular array, as shown in Figure 1.16. The particles vibrate back and forth about their average positions, but seldom does a particle in a solid squeeze past its immediate neighbors to come into contact with a new set of particles. Because the particles are packed so tightly and in such a regular arrangement, a solid is rigid, its volume is fixed, and the volume of a given mass is small. The external shape of a solid often reflects the internal arrangement of its particles. This relation between the observable structure of the solid and the arrangement of the particles from which it is made is one reason that scientists have long been fascinated by the shapes of crystals and minerals (Figure 1.17).

The late Richard Feynmann, a Nobel laureate in physics, said, "If in some cataclysm all of scientific knowledge were to be destroyed, and only one sentence passed on to the next generation of creatures, what statement would contain the most information in the fewest words? I believe it is the atomic hypothesis, that all things are made of atoms, little particles that move around in perpetual motion."

The kinetic-molecular theory of matter can also be used to interpret the properties of liquids, as shown in Figure 1.16. Liquids are fluid because the atoms or molecules are arranged more haphazardly than in solids. Particles are not confined to specific locations but rather can move past one another. No particle goes very far without bumping into another—the particles in a liquid interact with their neighbors continually. Because the particles are usually a little farther apart in a liquid than in the corresponding solid, the volume is usually a little bigger. (Ice and liquid water, which are shown in Figure 1.16, are an important exception to this last

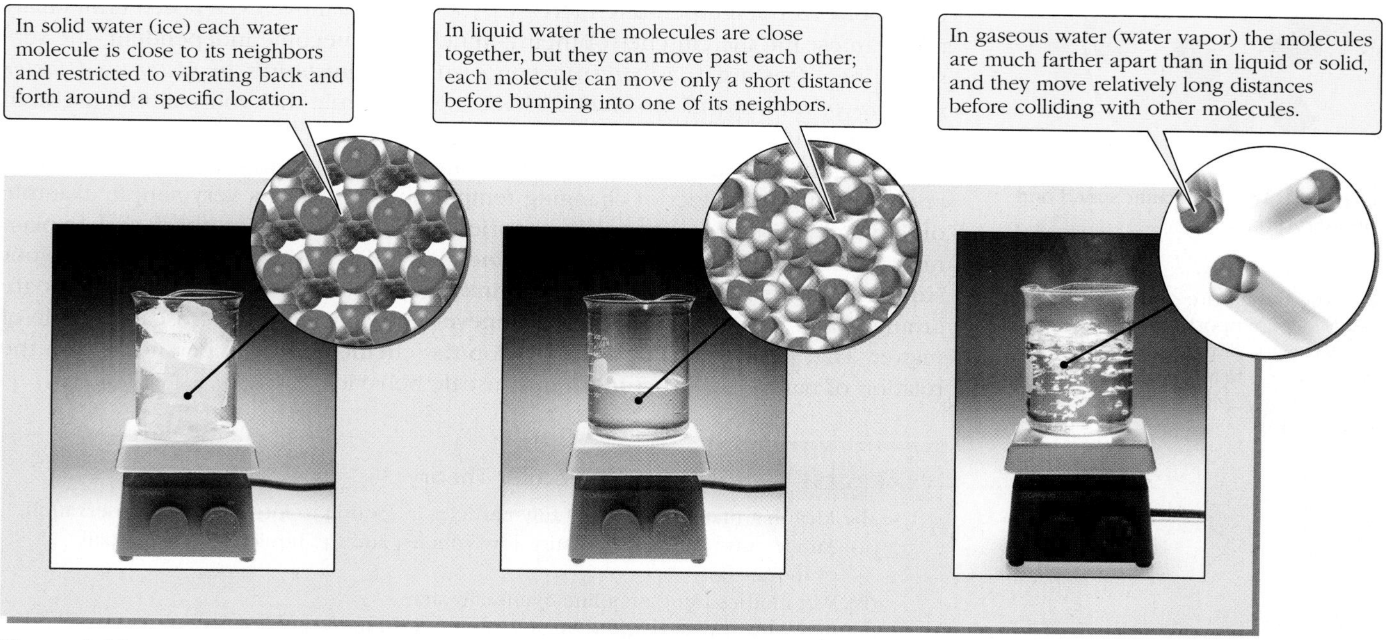
In solid water (ice) each water molecule is close to its neighbors and restricted to vibrating back and forth around a specific location.
In liquid water the molecules are close together, but they can move past each other; each molecule can move only a short distance before bumping into one of its neighbors.
In gaseous water (water vapor) the molecules are much farther apart than in liquid or solid, and they move relatively long distances before colliding with other molecules.

Figure 1.16 **Nanoscale representation of three states of matter.** Water is transformed from solid (ice) to liquid to gas as it is heated.

Figure 1.17 Structure and form. (a) In the nanoscale structure of ice, each water molecule occupies a position in a regular array or lattice that includes hexagonal units (outlined). (b) The form of a snowflake reflects the hexagonal symmetry of the nanoscale structure of ice.

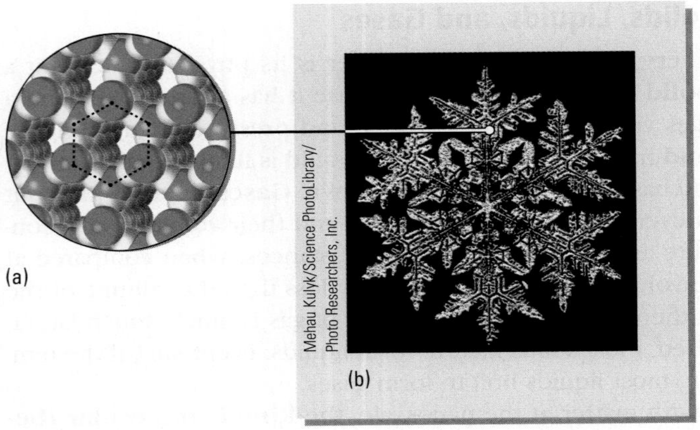
(a)
Mehau Kulyk/Science PhotoLibrary/
Photo Researchers, Inc.
(b)

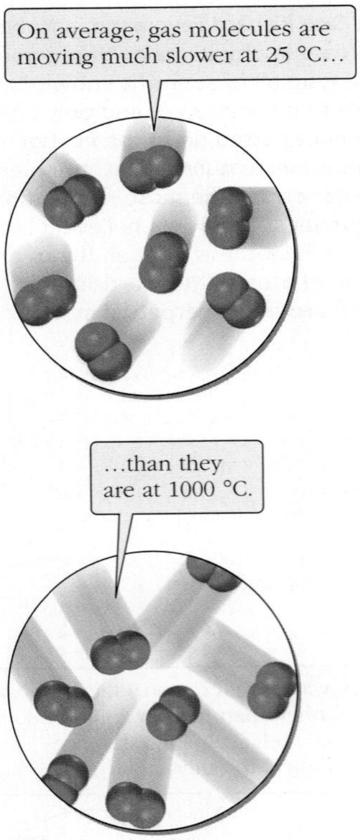
On average, gas molecules are moving much slower at 25 °C…
…than they are at 1000 °C.

Figure 1.18 Molecular speed and temperature.

Because the nature of the particles is relatively unimportant in determining the behavior of gases, all gases can be described fairly accurately by the ideal gas law, which is introduced in Chapter 10.

generality. As you can see from the figure, the water molecules in ice are arranged so that there are empty hexagonal channels. When ice melts, these channels become partially filled by water molecules, accounting for the slightly smaller volume of the same mass of liquid water.)

Like liquids, gases are fluid because their nanoscale particles can easily move past one another. As shown in Figure 1.16, the particles fly about to fill any container they are in; hence, a gas has no fixed shape or volume. In a gas the particles are much farther apart than in a solid or a liquid. They move significant distances before hitting other particles or the walls of the container. The particles also move quite rapidly. In air at room temperature, for example, the average molecule is going faster than 1000 miles per hour. A particle hits another particle every so often, but most of the time each is quite far away from all the others. Consequently, the nature of the particles is much less important in determining the properties of a gas.

Temperature can also be interpreted using the kinetic-molecular theory. The higher the temperature is, the more active the nanoscale particles are. A solid melts when its temperature is raised to the point where the particles vibrate fast enough and far enough to push each other out of the way and move out of their regularly spaced positions. The substance becomes a liquid because the particles are now behaving as they do in a liquid, bumping into one another and pushing past their neighbors. As the temperature increases, the particles move even faster, until finally they can escape the clutches of their comrades and become independent; the substance becomes a gas. *Increasing temperature corresponds to faster and faster motions of atoms and molecules.* This is a general rule that you will find useful in many future discussions of chemistry (Figure 1.18).

Using the kinetic-molecular theory to interpret the properties of solids, liquids, and gases and the effect of changing temperature provides a very simple example of how chemists use nanoscale theories and models to interpret and explain macroscale observations. In the remainder of this chapter and throughout your study of chemistry, you should try to imagine how the atoms and molecules are arranged and what they are doing whenever you consider a macroscale sample of matter. That is, you should try to develop the chemist's special perspective on the relation of nanoscale structure to macroscale behavior.

CONCEPTUAL EXERCISE 1.6 Kinetic-Molecular Theory

Use the idea that matter consists of tiny particles in motion to interpret each observation.

(a) An ice cube sitting in the sun slowly melts, and the liquid water eventually evaporates.

(b) Wet clothes hung on a line eventually dry.

(c) Moisture appears on the outside of a glass of ice water.

(d) Evaporation of a solution of sugar in water forms crystals.

CHEMISTRY IN THE NEWS Atomic Scale Electric Switches

Modern computers have become faster and faster, disk drives can store greater quantities of information, and cell phones are possible because the sizes of electronic circuits have steadily decreased during the past half century. More and more circuits can be squeezed into less and less volume, allowing greater computing power in smaller devices. However, electronic components, like everything else, must consist of atoms, so decreasing the size of electronic components will ultimately be limited by the sizes of atoms.

That limit has nearly been reached in practical devices, where electronic components have sizes as small as 30–40 nanometers. One approach to even smaller circuits involves *quantum dots,* nanoparticles that consist of fewer than 1000 atoms. For quantum dots to exhibit properties needed for electronic computing, the quantum dots usually need to be at very low temperatures—a few thousandths of a degree above the lowest possible temperature, absolute zero. Such low temperatures are very difficult to maintain, and this is a major limiting factor in the use of quantum dots in electronic circuits.

Now it has been found that properties characteristic of quantum dots can be obtained in much smaller arrays of atoms. Even a single silicon atom can behave as a quantum dot under the right circumstances. An array of four silicon atoms can become an electronic switch that does not require refrigeration and can retain its setting at room temperature.

Robert A. Wolkow and co-workers at the Canadian National Institute for Nanotechnology at the University of Alberta have developed a technique for coating a silicon surface with a single-atom-thick layer of hydrogen atoms. They can then selectively remove one or more hydrogen atoms, leaving negatively charged silicon atoms at the surface. When one hydrogen atom is removed, the single, negatively charged silicon atom at the surface behaves as a quantum dot. If four hydrogen atoms are removed from the surface, a four-silicon group of atoms can be made that is fixed in a specific position (blue and pink in the center of the figure). Two electrons in this four-atom group can be manipulated by removing other hydrogen atoms (green circles at upper left and lower right in the figure). One arrangement of the two electrons can be considered "switch off" and the other (diagonal to the first) can be considered "switch on." Techniques exist by which such switches could be made to behave as a circuit for a computer, thereby allowing much smaller computers and thinner cell phones than ever before.

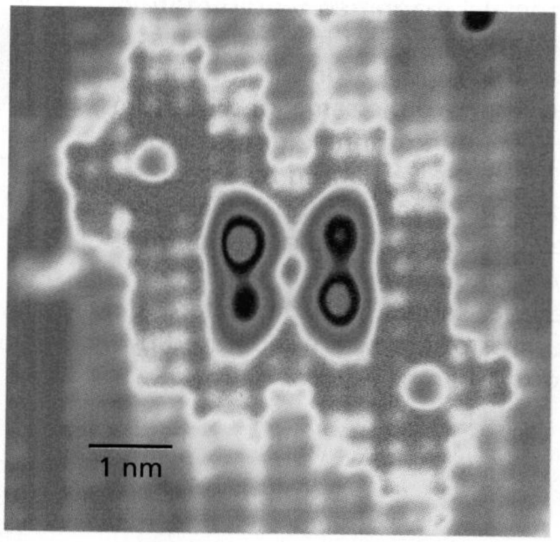

Two electrons occupy diagonal positions (dark blue) in a four-atom group of silicon atoms (two dark blue and two pink). The two electrons are locked into these positions by two additional negative silicon atoms that are seen as green circles within the red area.

Sources: Jacoby, M. *Chemical and Engineering News,* February 9, 2009, p. 10; the online supplement at **http://pubs.acs.org/cen/news/87/i06/8706notw8.html** includes a video showing how a computer based on this technology might work.

"Controlled Coupling and Occupation of Silicon Atomic Quantum Dots at Room Temperature." Haider, M. B., Pitters, J. L., DiLabio, G. A., Livadaru, L., Mutus, J. Y., and Wolkow, R. A. *Physical Review Letters,* Vol. 102, 2009; p. 046805.

"After the Transistor, A Leap Into the Microcosm." Markoff, J. *New York Times* September 1, 2009, p. D1.

1.9 The Atomic Theory

The existence of elements can be explained by a nanoscale model involving particles, just as the properties of solids, liquids, and gases can be. This model, which is closely related to the kinetic-molecular theory, is called the **atomic theory.** It was proposed in 1803 by John Dalton. According to Dalton's theory, an element cannot be decomposed into two or more new substances because at the nanoscale it consists of one and only one kind of atom and because atoms are indivisible under the conditions of chemical reactions. An **atom** is the smallest particle of an element that embodies the chemical properties of that element. An element, such as the sample of copper in Figure 1.19, is made up entirely of atoms of the same kind.

The fact that a compound can be decomposed into two or more different substances can be explained by saying that each compound must contain two or more different kinds of atoms. The process of decomposition involves separating at least one type of atom from atoms of the other kind(s). For example, charring of sugar corresponds to separating atoms of carbon from atoms of oxygen and atoms of hydrogen.

Dalton also said that each kind of atom must have its own properties—in particular, a characteristic mass. This idea allowed his theory to account for the masses

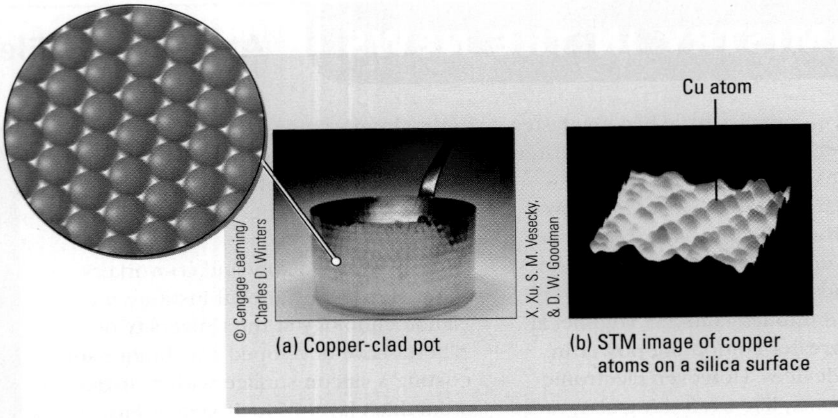
Cu atom
© Cengage Learning/ Charles D. Winters
X. Xu, S. M. Vesecky, & D. W. Goodman
(a) Copper-clad pot

(b) STM image of copper atoms on a silica surface

Figure 1.19 Elements, atoms, and the nanoscale world of chemistry. (a) A macroscopic sample of copper metal on a copper-clad pot with a nanoscale, magnified representation of a tiny portion of its surface. It is clear that all the atoms in the sample of copper are the same kind of atoms. (b) A scanning tunneling microscopy (STM) image, enhanced by a computer, of a layer of copper atoms on the surface of silica (a compound of silicon and oxygen). The section of the layer shown is 1.7 nm square and the rows of atoms are separated by about 0.44 nm.

of different elements that combine in chemical reactions to form compounds. An important success of Dalton's ideas was that they could be used to interpret known chemical facts quantitatively.

Two laws known in Dalton's time could be explained by the atomic theory. One was based on experiments in which the reactants were carefully weighed before a chemical reaction, and the reaction products were carefully collected and weighed afterward. The results led to the **law of conservation of mass** (also called the law of conservation of matter): *There is no detectable change in mass during an ordinary chemical reaction.* The atomic theory says that mass is conserved because the same number of atoms of each kind is present before and after a reaction, and each of those kinds of atoms has its same characteristic mass before and after the reaction.

Our bodies are made up of atoms from the distant past—atoms from other people and other things. Some of the carbon, hydrogen, and oxygen atoms in our carbohydrates have come from the breaths (first and last) of both famous and ordinary persons of the past.

The other law was based on the observation that in a chemical compound the proportions of the elements by mass are always the same. Water always contains 1 g hydrogen for every 8 g oxygen, and carbon monoxide always contains 4 g oxygen for every 3 g carbon. The **law of constant composition** summarizes such observations: *A chemical compound always contains the same elements in the same proportions by mass.* The atomic theory explains this observation by saying that atoms of different elements always combine in the same ratio in a compound. For example, in carbon monoxide there is always one carbon atom for each oxygen atom. If the mass of an oxygen atom is $\frac{4}{3}$ times the mass of a carbon atom, then the ratio of mass of oxygen to mass of carbon in carbon monoxide will always be 4:3.

Dalton's theory has been modified to account for discoveries since his time. The *modern* atomic theory is based on these assumptions:

All matter is composed of atoms, which are extremely tiny. Interactions among atoms account for the properties of matter.

All atoms of a given element have the same chemical properties. Atoms of different elements have different chemical properties.

According to the modern atomic theory, atoms of the same element have the same chemical properties but are not necessarily identical in all respects. The discussion of isotopes in Chapter 2 shows how atoms of the same element can differ in mass.

Compounds are formed by the chemical combination of two or more different kinds of atoms. Atoms usually combine in the ratio of small whole numbers. For example, in a carbon monoxide molecule there is one carbon atom and one oxygen atom; a carbon dioxide molecule consists of one carbon atom and two oxygen atoms.

A chemical reaction involves joining, separating, or rearranging atoms. Atoms in the reactant substances form new combinations in the product substances. Atoms are not created, destroyed, or converted into other kinds of atoms during a chemical reaction.

The hallmark of a good theory is that it suggests new experiments, and this was true of the atomic theory. Dalton realized that it predicted a law that had not yet been discovered. If compounds are formed by combining atoms of different

ESTIMATION How Tiny Are Atoms and Molecules?

It is often useful to estimate an approximate value for something. Usually this estimation can be done quickly, and often it can be done without a calculator. The idea is to pick round numbers that you can work with in your head, or to use some other method that allows a quick estimate. If you really need an accurate value, an estimate is still useful to check whether the accurate value is in the right ballpark. Often an estimate is referred to as a "back-of-the-envelope" calculation, because estimates might be done over lunch on any piece of paper that is at hand. Some estimates are referred to as "order-of-magnitude calculations" because only the power of ten (the order of magnitude) in the answer is obtained. To help you develop estimation skills, most chapters in this book will provide you with an example of estimating something.

To get a more intuitive feeling for how small atoms and molecules are, estimate how many hydrogen atoms could fit inside a 12-oz (355-mL) soft-drink can. Make the same estimate for protein molecules. Use the approximate sizes given in Figure 1.14.

Because 1 mL is the same volume as a cube 1 cm on each side (1 cm^3), the volume of the can is the same as the volume of 355 cubes 1 cm on each side. Therefore we can first estimate how many atoms would fit into a 1-cm cube and then multiply that number by 355.

According to Figure 1.14, a typical atom has a diameter slightly less than 100 pm. Because this is an estimate, and to make the numbers easy to handle, assume that we are dealing with an atom that is 100 pm in diameter. Then the atom's diameter is 100×10^{-12} m $= 1 \times 10^{-10}$ m, and it will require 10^{10} of these atoms lined up in a row to make a length of 1 m. Since 1 cm is $\frac{1}{100}$, that is, 10^{-2}, of a meter, only $10^{-2} \times 10^{10} = 10^8$ atoms would fit in 1 cm.

In three dimensions, there could be 10^8 atoms along each of the three perpendicular edges of a 1-cm cube (the *x*, *y*, and *z* directions). The one row along the *x*-axis could be repeated 10^8 times along the *y*-axis, and then that layer of atoms could be repeated 10^8 times along the *z*-axis. Therefore, the number of atoms that we estimate would fit inside the cube is $10^8 \times 10^8 \times 10^8 = 10^{24}$ atoms. Multiplying this by 355 gives $355 \times 10^{24} = 3.6 \times 10^{26}$ atoms in the soft-drink can.

This estimate is a bit low. A hydrogen atom's diameter is less than 100 pm, so more hydrogen atoms would fit inside the can. Also, atoms are usually thought of as spheres, so they could pack together more closely than they would if just lined up in rows. Therefore, an even larger number of atoms than 3.6×10^{26} could fit inside the can.

For a typical protein molecule, Figure 1.14 indicates a diameter on the order of 5 nm = 5000 pm. That is 50 times bigger than the 100-pm diameter we used for the hydrogen atom. Thus, there would be 50 times fewer protein molecules in the *x* direction, 50 times fewer in the *y* direction, and 50 times fewer in the *z* direction. Therefore, the number of protein molecules would be fewer by $50 \times 50 \times 50 = 125{,}000$. The number of protein molecules can thus be estimated as $(3.6 \times 10^{26})/(1.25 \times 10^5)$. Because 3.6 is roughly three times 1.25, and because we are estimating, not calculating accurately, we can take the result to be 3×10^{21} protein molecules. That's still a whole lot of molecules!

Visit this book's companion website at **www.cengage.com/chemistry/moore** to work an interactive module based on this material.

elements on the nanoscale, then in some cases there might be more than a single combination. An example is carbon monoxide and carbon dioxide. In carbon monoxide there is one oxygen atom for each carbon atom, while in carbon dioxide there are two oxygen atoms per carbon atom. Therefore, in carbon dioxide the mass of oxygen per gram of carbon ought to be twice as great as it is in carbon monoxide (because twice as many oxygen atoms will weigh twice as much). Dalton called this the **law of multiple proportions,** and he carried out quantitative experiments seeking data to confirm or deny it. Dalton and others obtained data consistent with the law of multiple proportions, thereby enhancing acceptance of the atomic theory.

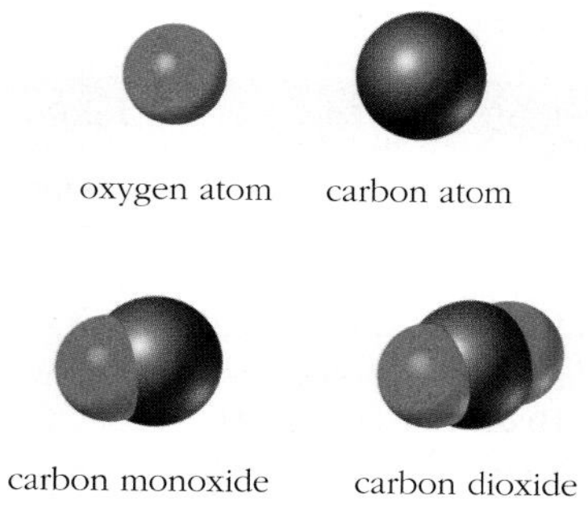
oxygen atom
carbon atom
carbon monoxide
carbon dioxide

1.10 The Chemical Elements

Every element has been given a unique *name* and a *symbol* derived from the name. These names and symbols are listed in the periodic table inside the front cover of the book. The first letter of each symbol is capitalized; the second letter, if there is one, is lowercase, as in He, the symbol for helium. Elements discovered a long time ago have names and symbols with Latin or other origins, such as Au for gold (from *aurum,* meaning "bright dawn") and Fe for iron (from *ferrum*). The names of more recently discovered elements are derived from their place of discovery or from a person or place of significance (Table 1.3).

Elements are being synthesized even now. Element 115 was made in 2004, but only a few atoms were produced.

Ancient people knew of nine elements—gold (Au), silver (Ag), copper (Cu), tin (Sn), lead (Pb), mercury (Hg), iron (Fe), sulfur (S), and carbon (C). Most of the other naturally occurring elements were discovered during the 1800s, as one by one they were separated from minerals in Earth's crust or from Earth's oceans or atmosphere. Currently, more than 110 elements are known, but only 90 occur in nature. Elements such as technetium (Tc), neptunium (Np), mendelevium (Md), seaborgium (Sg), and meitnerium (Mt) have been made using nuclear reactions (see Chapter 20), beginning in the 1930s.

Types of Elements

Figure 1.20 **Some metallic elements**—iron, aluminum, copper, and gold. Metals are malleable, ductile, and conduct electricity.

The vast majority of the elements are **metals**—only 24 are not. You are probably familiar with many properties of metals. At room temperature they are solids (except for mercury, which is a liquid), they conduct electricity (and conduct better as the temperature decreases), they are ductile (can be drawn into wires), they are malleable (can be rolled into sheets), and they can form alloys (solutions of one or more metals in another metal). In a solid metal, individual metal atoms are packed close to each other, so metals usually have fairly high densities. Figure 1.20 shows some common metals. Iron (Fe) and aluminum (Al) are used in automobile parts because of their ductility, malleability, and relatively low cost. Copper (Cu) is used in electrical wiring because it conducts electricity better than most metals. Gold (Au) is used for the vital electrical contacts in automobile air bags and in some computers because it does not corrode and is an excellent electrical conductor.

In contrast, **nonmetals** do not conduct electricity (with a few exceptions, such as graphite, one form of carbon). Nonmetals are more diverse in their physical properties than are metals (Figure 1.21). At room temperature some nonmetals are solids (such as phosphorus, sulfur, and iodine), bromine is a liquid, and others are gases (such as hydrogen, nitrogen, and chlorine). The nonmetals helium (He), neon (Ne), argon (Ar), krypton (Kr), xenon (Xe), and radon (Rn) are gases that consist of individual atoms.

On the periodic table inside the front cover of this book, the metals, nonmetals, and metalloids are color-coded: gray and blue for metals, lavender for nonmetals, and orange for metalloids.

A few elements—boron, silicon, germanium, arsenic, antimony, and tellurium—are classified as **metalloids.** Some properties of metalloids are typical of metals and other properties are characteristic of nonmetals. For example, some metalloids are shiny like metals, but they do not conduct electricity as well as metals. Many of them

Table 1.3 The Names of Some Chemical Elements

Element	Symbol	Date Discovered	Discoverer	Derivation of Name/Symbol
Carbon	C	Ancient	Ancient	Latin, *carbo* (charcoal)
Curium	Cm	1944	G. Seaborg, et al.	Honoring Marie and Pierre Curie, Nobel Prize winners for discovery of radioactive elements
Hydrogen	H	1766	H. Cavendish	Greek, *hydro* (water) and *genes* (generator)
Meitnerium	Mt	1982	P. Armbruster, et al.	Honoring Lise Meitner, codiscoverer of nuclear fission
Mendelevium	Md	1955	G. Seaborg, et al.	Honoring Dmitri Mendeleev, who devised the periodic table
Mercury	Hg	Ancient	Ancient	For Mercury, messenger of the gods, because it flows quickly; symbol from Greek *hydrargyrum,* liquid silver
Polonium	Po	1898	M. Curie and P. Curie	In honor of Poland, Marie Curie's native country
Seaborgium	Sg	1974	G. Seaborg, et al.	Honoring Glenn Seaborg, Nobel Prize winner for synthesis of new elements
Sodium	Na	1807	H. Davy	Latin, *soda* (sodium carbonate); symbol from Latin *natrium*
Tin	Sn	Ancient	Ancient	German, *Zinn;* symbol from Latin, *stannum*

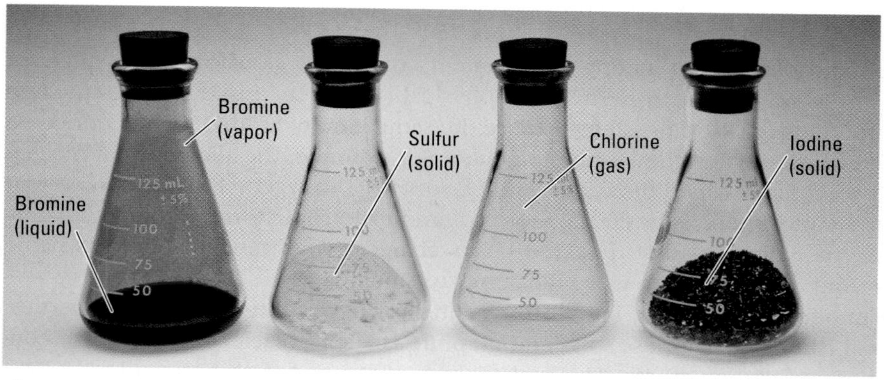

Figure 1.21 **Some nonmetallic elements**—(a) bromine, (b) sulfur, (c) chlorine, (d) iodine. Nonmetals occur as solids, liquids, and gases and have very low electrical conductivities. Bromine is the only nonmetal that is a liquid at room temperature.

are semiconductors and are essential for the electronics industry. (See Sections 11.8 and 11.9.)

EXERCISE 1.7 Elements

Use Table 1.3, the periodic table inside the front cover, and/or the list of elements inside the front cover to answer these questions.

(a) Four elements are named for planets in our solar system (including the ex-planet Pluto). Give their names and symbols.
(b) One element is named for a state in the United States. Name the element and give its symbol.
(c) Two elements are named in honor of women. What are their names and symbols?
(d) Several elements are named for countries or regions of the world. Find at least four of these and give names and symbols.
(e) List the symbols of all elements that are nonmetals.

Elements That Consist of Molecules

Most elements that are nonmetals consist of molecules on the nanoscale. A **molecule** is a unit of matter in which two or more atoms are held together by chemical bonds. For example, a chlorine molecule contains two chlorine atoms and can be represented by the chemical formula Cl_2. A **chemical formula** uses the symbols for the elements to represent the atomic composition of a substance. In gaseous chlorine, Cl_2 molecules are the particles that fly about and collide with each other and the container walls. Molecules, like Cl_2, that consist of two atoms are called **diatomic molecules.** Oxygen, O_2, and nitrogen, N_2, also exist as diatomic molecules, as do hydrogen, H_2; fluorine, F_2; bromine, Br_2; and iodine, I_2.

You have already seen (in Figure 1.1) that really big molecules, such as tubulin, may be represented by ribbons or sticks, without showing individual atoms at all. Often such representations are drawn by computers, which help chemists to visualize, manipulate, and understand the molecular structures.

Chemical bonds are strong attractions that hold atoms together. Bonding is discussed in detail in Chapter 8.

Space-filling model of Cl_2

Elements that consist of diatomic molecules are H_2, N_2, O_2, F_2, Cl_2, Br_2, and I_2. You need to remember that these elements consist of diatomic molecules, because they will be encountered frequently. Most of these elements are close together in the periodic table, which makes it easier to remember them.

EXERCISE 1.8 Elements That Consist of Diatomic Molecules or Are Metalloids

On a copy of the periodic table, circle the symbols of the elements that

(a) Consist of diatomic molecules; (b) Are metalloids.

Devise rules related to the periodic table that will help you to remember which elements these are.

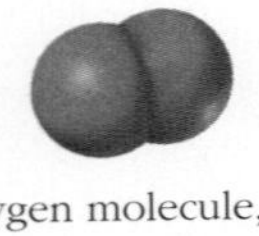

oxygen molecule, O_2

ozone molecule, O_3

Paul Seheult/Corbis

Sir Harold Kroto

1939–

Along with the late Richard E. Smalley and Robert Curl, Harold Kroto received the Nobel Prize in Chemistry in 1996 for discovering fullerenes. In the same year Kroto, who is British, received a knighthood for his work. At the time of the discovery, Smalley assembled paper hexagons and pentagons to make a model of the C_{60} molecule. He asked Kroto, "Who was the architect who worked with big domes?" When Kroto replied, "Buckminster Fuller," the two shouted with glee, "It's Buckminster Fuller—ene!" The structure's name had been coined.

Allotropes

Oxygen and carbon are among the elements that exist as **allotropes,** different forms of the same element in the same physical state at the same temperature and pressure. Allotropes are possible because the same kind of atoms can be connected in different ways when they form molecules. For example, the allotropes of oxygen are O_2, sometimes called dioxygen, and O_3, ozone. Dioxygen, a major component of Earth's atmosphere, is by far the more common allotropic form. Ozone is a highly reactive, pale blue gas first detected by its characteristic pungent odor. Its name comes from *ozein,* a Greek word meaning "to smell."

Diamond and graphite, known for centuries, have quite different properties. Diamond is a hard, colorless solid and graphite is a soft, black solid, but both consist entirely of carbon atoms. This makes them allotropes of carbon, and for a long time they were thought to be the only allotropes of carbon with well-defined structures. Therefore, it was a surprise in the 1980s when another carbon allotrope was discovered in soot produced when carbon-containing materials are burned with very little oxygen. The new allotrope consists of 60-carbon atom cages and represents a new class of molecules. The C_{60} molecule resembles a soccer ball with a carbon atom at each corner of each of the black pentagons in Figure 1.22b. Each five-membered ring of carbon atoms is surrounded by five six-membered rings. This molecular structure of carbon pentagons and hexagons reminded its discoverers of a geodesic dome (Figure 1.22a), a structure popularized years ago by the innovative American philosopher and engineer R. Buckminster Fuller. Therefore, the official name of the C_{60} allotrope is buckminsterfullerene. Chemists often call it simply a "buckyball." C_{60} buckyballs belong to a larger molecular family of even-numbered carbon cages that is collectively called fullerenes.

Carbon atoms can also form concentric tubes that resemble rolled-up chicken wire (see Section 11.7). These single- and multi-walled *nanotubes* of only carbon atoms are excellent electrical conductors and extremely strong. Imagine the exciting applications for such properties, including making buckyfibers that could substitute for the metal wires now used to transmit electrical power. Dozens of uses have been proposed for fullerenes, buckytubes, and buckyfibers, among them microscopic ball bearings, lightweight batteries, new lubricants, nanoscale electric switches, new plastics, and antitumor therapy for cancer patients (by enclosing a radioactive atom within the cage). All these applications await an inexpensive way of making buckyballs and other fullerenes. Currently buckyballs, the cheapest fullerene, are more expensive than gold.

EXERCISE 1.9 Allotropes

A student says that tin and lead are allotropes because they are both dull gray metals. Why is the statement wrong?

Norman Pogson/iStockphoto; © Cengage Learning/Charles D. Winters; © Cengage Learning/Charles D. Winters

(a) (b) (c)

Figure 1.22 Models for fullerenes. (a) A geodesic dome in Montreal, Canada. Geodesic domes, such as those designed originally by R. Buckminster Fuller, contain linked hexagons and pentagons. (b) A soccer ball is a model for the C_{60} structure. (c) The C_{60} fullerene molecule, which is made up of five-membered rings (black rings on the soccer ball) and six-membered rings (white rings on the ball). A C_{60} molecule's size compared with a soccer ball is almost the same as a soccer ball's size compared with planet Earth.

1.11 Communicating Chemistry: Symbolism

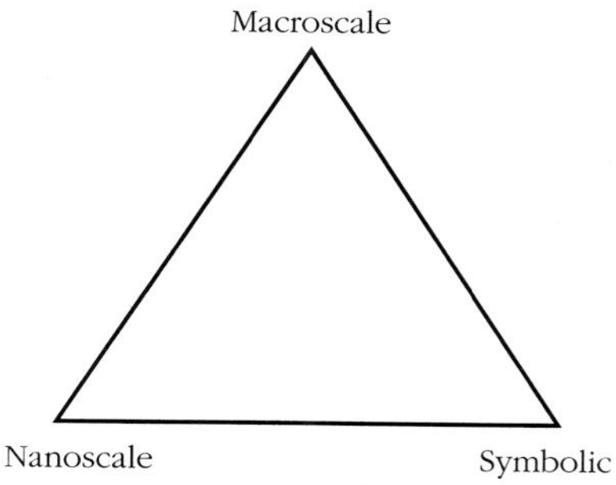
Macroscale
Nanoscale
Symbolic

Chemists' three ways of representing the natural world.

Chemical symbols—such as Na, I, or Mt—are a shorthand way of indicating what kind of atoms we are talking about. Chemical formulas tell us how many atoms of an element are combined in a molecule and in what ratios atoms are combined in compounds. For example, the formula Cl_2 tells us that there are two chlorine atoms in a chlorine molecule. The formulas CO and CO_2 tell us that carbon and oxygen form two different compounds—one that has equal numbers of C and O atoms and one that has twice as many O atoms as C atoms. In other words, chemical symbols and formulas symbolize the nanoscale composition of each substance.

Chemical symbols and formulas also represent the macroscale properties of elements and compounds. That is, the symbol Na brings to mind a highly reactive metal, and the formula H_2O represents a colorless liquid that freezes at 0 °C, boils at 100 °C, and reacts violently with Na. Because chemists are familiar with both the nanoscale and macroscale characteristics of substances, they usually use symbols to abbreviate their representations of both. Symbols are also useful for representing chemical reactions. For example, the charring of sucrose mentioned earlier is represented by

$$\begin{array}{ccccc} \text{Sucrose} & \longrightarrow & \text{Carbon} & + & \text{Water} \\ C_{12}H_{22}O_{11} & \longrightarrow & 12\ C & + & 11\ H_2O \\ \text{Reactant} & \text{changes to} & & \text{Products} & \end{array}$$

The symbolic aspect of chemistry is the third part of the chemist's special view of the natural world. It is important that you become familiar and comfortable with using chemical symbols and formulas to represent chemical substances and their reactions. Figure 1.23 shows how chemical symbolism can be applied to the process of decomposing water with electricity (electrolysis of water).

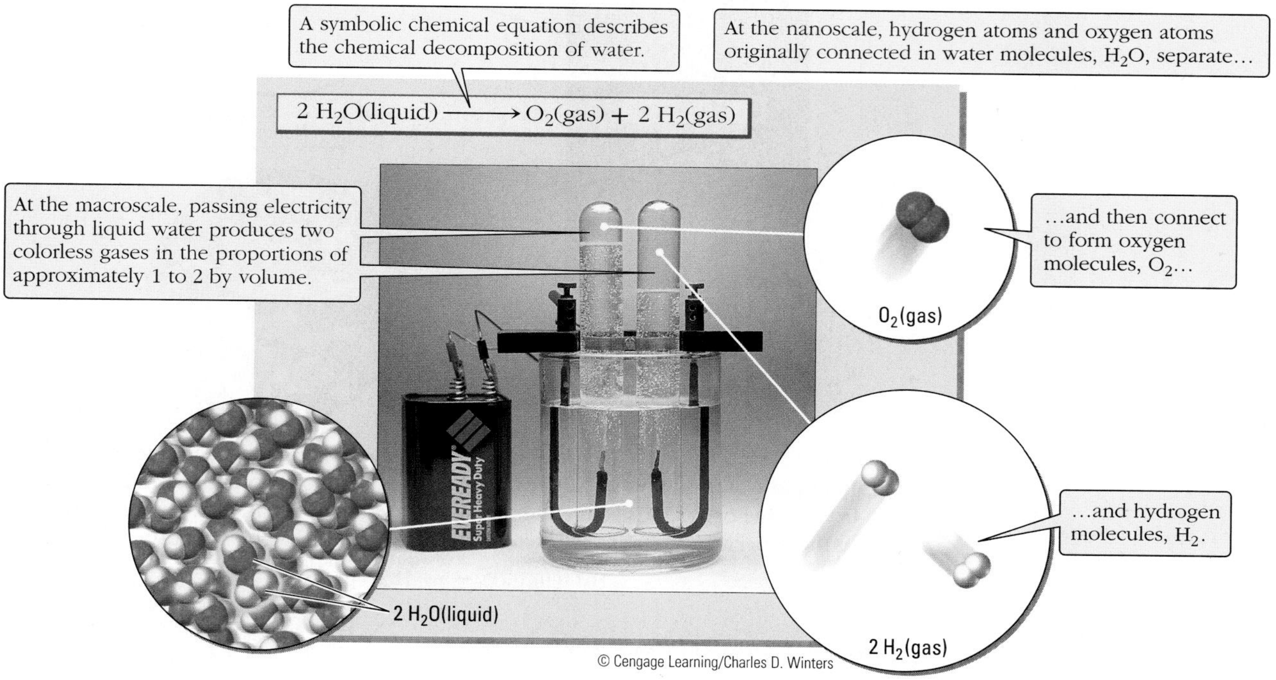
A symbolic chemical equation describes the chemical decomposition of water.
At the nanoscale, hydrogen atoms and oxygen atoms originally connected in water molecules, H_2O, separate...
2 H_2O(liquid) ⟶ O_2(gas) + 2 H_2(gas)
At the macroscale, passing electricity through liquid water produces two colorless gases in the proportions of approximately 1 to 2 by volume.
...and then connect to form oxygen molecules, O_2...
O_2(gas)
EVEREADY
Super Heavy Duty
...and hydrogen molecules, H_2.
2 H_2O(liquid)
2 H_2(gas)

Active Figure 1.23 Symbolic, macroscale, and nanoscale representations of a chemical reaction. Visit this book's companion website at **www.cengage.com/chemistry/moore** to test your understanding of the concepts in this figure.

PROBLEM-SOLVING EXAMPLE 1.3 Macroscale, Nanoscale, and Symbolic Representations

The figure shows a sample of water boiling. In spaces labeled C, indicate whether the macroscale or the nanoscale is represented. In spaces labeled B, draw the molecules that would be present with appropriate distances between them. One of the circles represents a bubble of gas within the liquid. The other represents the liquid. In space A, write a symbolic representation of the boiling process.

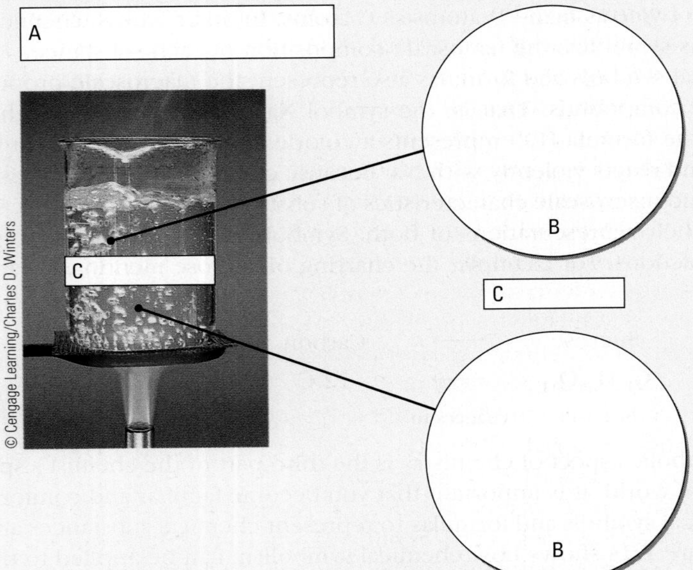
A
B
C
C
B
© Cengage Learning/Charles D. Winters

Answer

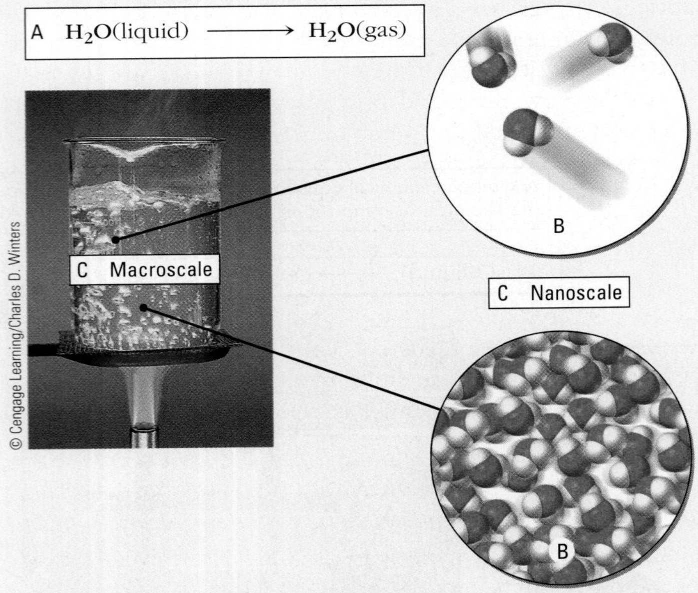
A H_2O(liquid) ⟶ H_2O(gas)
B
C Macroscale
C Nanoscale
B
© Cengage Learning/Charles D. Winters

Explanation Each water molecule consists of two hydrogen atoms and one oxygen atom. In liquid water the molecules are close together and oriented in various directions. In a bubble of gaseous water the molecules are much farther apart—there are fewer of them per unit volume. The symbolic representation is the equation

$$H_2O(\text{liquid}) \longrightarrow H_2O(\text{gas})$$

PROBLEM-SOLVING PRACTICE 1.3

Draw a nanoscale representation and a symbolic representation for both allotropes of oxygen. Describe the properties of each allotrope at the macroscale.

1.12 Modern Chemical Sciences

Our goal in this chapter has been to make clear many of the reasons you should care about chemistry. Chemistry is fundamental to understanding many other sciences and to understanding how the material world around us works. Chemistry provides a unique, nanoscale perspective that has been highly successful in stimulating scientific inquiry and in the development of high-tech materials, modern medicines, and many other advances that benefit us every day. Chemistry is happening all around us and within us all the time. Knowledge of chemistry is a key to understanding and making the most of our internal and external environments and to making better decisions about how we live our lives and structure our economy and society. Finally, chemistry—the properties of elements and compounds, the nanoscale theories and models that interpret those properties, and the changes of one kind of substance into another—is just plain interesting and fun. Chemistry presents an intellectual challenge and provides ways to satisfy intellectual curiosity while helping us to better understand the world in which we live.

Modern chemistry overlaps more and more with biology, medicine, engineering, and other sciences. Because it is central to understanding matter and its transformations, chemistry becomes continually more important in a world that relies on chemical knowledge to produce the materials and energy required for a comfortable and productive way of life. The breadth of chemistry is recognized by the term *chemical sciences*, which includes chemistry, chemical engineering, chemical biology, materials chemistry, industrial chemistry, environmental chemistry, medicinal chemistry, and many other fields. Practitioners of the chemical sciences produce new plastics, medicines, superconductors, composite materials, and electronic devices. The chemical sciences enable better understanding of the mechanisms of biological processes, how proteins function, and how to imitate the action of organisms that carry out important functions. The chemical sciences enable us to measure tiny concentrations of substances, separate one substance from another, and determine whether a given substance is helpful or harmful to humans. Practitioners of the chemical sciences are involved in "green chemistry"—creating new industrial processes and products that are less hazardous, produce less pollution, and generate smaller quantities of wastes. The enthusiasm of chemists for research in all of these areas and the many discoveries that are made every day offer ample evidence that chemistry is an energetic and exciting science. We hope that this excitement is evident in this chapter and in the rest of the book.

Near the beginning of this chapter we listed questions you may have wondered about that will be answered in this book. More important by far, however, are questions that chemists or other scientists cannot yet answer. Here are some big challenges for chemists of the future, as envisioned by a blue-ribbon panel of experts convened by the U.S. National Academies of Science, Engineering, and Medicine.

- How can chemists design and synthesize new substances with well-defined properties that can be predicted ahead of time? Example substances are medicines, electronic devices, composite materials, and polymeric plastics.
- How can chemists learn from nature to produce new substances efficiently and use biological processes as models for industrial production? An example is extracting a compound from the English yew and then processing it chemically to produce paclitaxel.
- How can chemists design mixtures of molecules that will assemble themselves into useful, more complex structures? Such self-assembly processes, in which each different molecule falls of its own accord into the right place, could be used to create a variety of new nanostructures.
- How can chemists learn to measure more accurately how much of a substance is present, determine the substance's properties, and predict how long it will last? Sensors are now being invented that combine chemical and biological processes to allow rapid, accurate determination of composition and structure.

The report "Beyond the Molecular Frontier: Challenges for Chemistry and Chemical Engineering" is available from the National Academies Press, **http://www.nap.edu/catalog/10633.html**.

- How can chemists devise better theories that will predict more accurately the behavior of molecules and of large collections of molecules?
- How can chemists devise new ways of making large quantities of materials without significant negative consequences for Earth's environment and the many species that inhabit our planet?
- How can chemists use improved knowledge of the molecular structures of genes and proteins to create new medicines and therapies to deal with viral diseases such as AIDS; major killers such as cancer, stroke, and heart disease; and psychological problems?
- How can chemists find alternatives to fossil fuels, avoid global warming, enable mobility for all without polluting the planet, and make use of the huge quantities of solar energy that are available?
- How can chemists contribute to national and global security?
- How can chemists improve the effectiveness of education, conveying chemical knowledge to the many students and others who can make use of it in their chosen fields?

A major goal of the authors of this book is to help you along the pathway to becoming a scientist. We encourage you to choose one of the preceding problems or another problem of similar importance and devote your life's work to finding new approaches and useful solutions to it. The future of our society depends on it!

IN CLOSING

OWL and Go Chemistry

Sign in at **www.cengage.com/owl** to:

- View tutorials and simulations, develop problem-solving skills, and complete online homework assigned by your professor.
- For quick review and exam prep, download Go Chemistry mini lecture modules from OWL (or purchase them at **www.CengageBrain.com**).

Having studied this chapter, you should be able to . . .

- Appreciate the power of chemistry to answer intriguing questions (Section 1.1).
- Describe the approach used by scientists in solving problems (Sections 1.2, 1.3).
- Understand the differences among a hypothesis, a theory, and a law (Section 1.3).
- Define quantitative and qualitative observations (Section 1.3). End-of-chapter questions: 11, 13
- Identify the physical properties of matter or physical changes occurring in a sample of matter (Section 1.4). Questions 15, 17
- Estimate Celsius temperatures for commonly encountered situations (Section 1.4). Question 19
- Calculate mass, volume, or density, given any two of the three (Section 1.4). Questions 21, 23, 25
- Identify the chemical properties of matter or chemical changes occurring in a sample of matter (Section 1.5). Questions 27, 29, 31
- Explain the difference between homogeneous and heterogeneous mixtures (Section 1.6). Questions 33, 35
- Describe the importance of separation, purification, and analysis (Section 1.6). Question 37
- Understand the difference between a chemical element and a chemical compound (Sections 1.7, 1.9). Questions 39, 45
- Classify matter (Section 1.7, Figure 1.13). Questions 41, 43
- Describe characteristic properties of the three states of matter—gases, liquids, and solids (Section 1.8).
- Identify relative sizes at the macroscale, microscale, and nanoscale levels (Section 1.8). Questions 47, 49, 51
- Describe the kinetic-molecular theory at the nanoscale level (Section 1.8). Questions 53, 55

- Use the postulates of modern atomic theory to explain macroscale observations about elements, compounds, conservation of mass, constant composition, and multiple proportions (Section 1.9). Questions 59, 61, 63
- Distinguish metals, nonmetals, and metalloids according to their properties (Section 1.10). Question 65
- Identify elements that consist of molecules, and define allotropes (Section 1.10). Question 65
- Distinguish among macroscale, nanoscale, and symbolic representations of substances and chemical processes (Section 1.11). Questions 67, 69, 71

KEY TERMS

These terms were defined and given in boldface type in this chapter. Be sure to understand each of these terms and the concepts with which they are associated.

allotrope *(Section 1.10)*
atom *(1.9)*
atomic theory *(1.9)*
boiling point *(1.4)*
Celsius temperature scale *(1.4)*
chemical change *(1.5)*
chemical compound *(1.7)*
chemical element *(1.7)*
chemical formula *(1.10)*
chemical property *(1.5)*
chemical reaction *(1.5)*
chemistry *(1.1)*
conservation of mass, law of *(1.9)*
constant composition, law of *(1.9)*
conversion factor *(1.4)*
density *(1.4)*
diatomic molecule *(1.10)*
dimensional analysis *(1.4)*
energy *(1.5)*
freezing point *(1.4)*
gas *(1.8)*
heterogeneous mixture *(1.6)*
homogeneous mixture *(1.6)*
hypothesis *(1.3)*
kinetic-molecular theory *(1.8)*
law *(1.3)*
liquid *(1.8)*
macroscale *(1.8)*
matter *(1.1)*
melting point *(1.4)*
metal *(1.10)*
metalloid *(1.10)*
microscale *(1.8)*
model *(1.3)*
molecule *(1.10)*
multiple proportions, law of *(1.9)*
nanoscale *(1.8)*
nonmetal *(1.10)*
physical changes *(1.4)*
physical properties *(1.4)*
product *(1.5)*
proportionality factor *(1.4)*
qualitative *(1.3)*
quantitative *(1.3)*
reactant *(1.5)*
solid *(1.8)*
solution *(1.6)*
substance *(1.4)*
temperature *(1.4)*
theory *(1.3)*

QUESTIONS FOR REVIEW AND THOUGHT

OWL Interactive versions of these problems are assignable in OWL.

Blue-numbered questions have short answers at the back of this book in Appendix M and fully worked solutions in the *Student Solutions Manual.*

Review Questions

These questions test vocabulary and simple concepts.

1. What is meant by the structure of a molecule? Why are molecular structures important?
2. Why is it often important to know the structure of an enzyme? How can knowledge of enzyme structures be useful in medicine?
3. Choose an object in your room, such as a CD player or television set. Write down five qualitative observations and five quantitative observations regarding the object you chose.
4. What are three important characteristics of a scientific law? Name two laws that were mentioned in this chapter. State each of the laws that you named.
5. How does a scientific theory differ from a law? How are theories and models related?

6. When James Snyder proposed on the basis of molecular models that paclitaxel assumes the shape of a letter T when attached to tubulin, was this a theory or a hypothesis?
7. What is the unique perspective that chemists use to make sense out of the material world? Give at least one example of how that perspective can be applied to a significant problem.
8. Give two examples of situations in which purity of a chemical substance is important.

Topical Questions

These questions are keyed to the major topics in the chapter. Usually a question that is answered at the back of this book is paired with a similar one that is not.

Why Care About Chemistry? (Section 1.1)

9. Make a list of at least four issues faced by our society that require scientific studies and scientific data before a democratic society can make informed, rational decisions. Exchange lists with another student and evaluate the quality of each other's choices.
10. Make a list of at least four questions you have wondered about that may involve chemistry. Compare your list with a list from another student taking the same chemistry course. Evaluate the quality of each other's questions and decide how "chemical" they are.

How Science Is Done (Section 1.3)

11. Identify the information in each sentence as qualitative or quantitative.
    (a) The element gallium melts at 29.8 °C.
    (b) A chemical compound containing cobalt and chlorine is blue.
    (c) Aluminum metal is a conductor of electricity.
    (d) The chemical compound ethanol boils at 79 °C.
    (e) A chemical compound containing lead and sulfur forms shiny, plate-like, yellow crystals.
12. Make as many qualitative and quantitative observations as you can regarding what is shown in the photograph.

© Cengage Learning/Charles D. Winters

13. Which of these statements are qualitative? Which are quantitative? Explain your choice in each case.
    (a) Sodium is a silvery-white metal.
    (b) Aluminum melts at 660 °C.
    (c) Carbon makes up about 23% of the human body by mass.
    (d) Pure carbon occurs in different forms: graphite, diamond, and fullerenes.
14. Which of the these statements are qualitative? Which are quantitative? Explain your choice in each case.
    (a) The atomic mass of carbon is 12.011 amu (atomic mass units).
    (b) Pure aluminum is a silvery-white metal that is nonmagnetic, has a low density, and does not produce sparks when struck.
    (c) Sodium has a density of 0.968 g/mL.
    (d) In animals the sodium cation, Na^+, is the main extracellular cation and is important for nerve function.

Identifying Matter: Physical Properties (Section 1.4)

15. The elements sulfur and bromine are shown in the photograph. Based on the photograph, describe as many properties of each sample as you can. Are any properties the same? Which properties are different?

© Cengage Learning/Charles D. Winters

Sulfur and bromine. The sulfur is on the flat dish; the bromine is in a closed flask.

16. In the accompanying photo, you see a crystal of the mineral calcite surrounded by piles of calcium and carbon, two of the elements that combine to make the mineral. (The other element combined in calcite is oxygen.) Based on the photo, describe some of the physical properties of the elements and the mineral. Are any properties the same? Are any properties different?

© Cengage Learning/Charles D. Winters

Calcite (the transparent, cube-like crystal) and two of its constituent elements, calcium (chips) and carbon (black grains). The calcium chips are covered with a thin film of calcium oxide.

17. The boiling point of a liquid is 20 °C. If you hold a sample of the substance in your hand, will it boil? Explain briefly.
18. Dry Ice (solid carbon dioxide) sublimes (changes from solid to gas without forming liquid) at −78.6 °C. Suppose you had a sample of gaseous carbon dioxide and the tem-

perature was 30 degrees Fahrenheit below zero (−30 °F, a very cold day). Would solid carbon dioxide form? Explain briefly how you answered the question.

19. Which temperature is higher?
    (a) 20 °C or 20 °F (b) 100 °C or 180 °F
    (c) 60 °C or 100 °F (d) −12 °C or 20 °F
20. These temperatures are measured at various locations during the same day in the winter in North America: −10 °C at Montreal, 28 °F at Chicago, 20 °C at Charlotte, and 40 °F at Philadelphia. Which city is the warmest? Which city is the coldest?
21. A 105.5-g sample of a metal was placed into water in a graduated cylinder, and it completely submerged. The water level rose from 25.4 mL to 37.2 mL. Use data in Table 1.1 to identify the metal.
22. An irregularly shaped piece of lead weighs 10.0 g. It is carefully lowered into a graduated cylinder containing 30.0 mL ethanol, and it sinks to the bottom of the cylinder. To what volume reading does the ethanol rise?
23. An unknown sample of a metal is 1.0 cm thick, 2.0 cm wide, and 10.0 cm long. Its mass is 54.0 g. Use data in Table 1.1 to identify the metal. (Remember that 1 cm^3 = 1 mL.)
24. Calculate the volume of a 23.4-g sample of bromobenzene, density = 1.49 g/mL.
25. Calculate the mass of the sodium chloride crystal in the photo that accompanies Question 47 if the dimensions of the crystal are 10 cm thick by 12 cm long by 15 cm wide. (Remember that 1 cm^3 = 1 mL.)
26. Find the volume occupied by a 4.33-g sample of iron.

Chemical Changes and Chemical Properties (Section 1.5)

27. In each case, identify the underlined property as a physical or chemical property. Give a reason for your choice.
    (a) The normal <u>color</u> of the element bromine is red-orange.
    (b) Iron is <u>transformed into rust</u> in the presence of air and water.
    (c) Dynamite can <u>explode</u>.
    (d) Aluminum metal, the <u>shiny</u> "foil" you use in the kitchen, <u>melts</u> at 660 °C.
28. In each case, identify the underlined property as a physical or a chemical property. Give a reason for your choice.
    (a) Dry Ice <u>sublimes</u> (changes directly from a solid to a gas) at −78.6 °C.
    (b) Methanol (methyl alcohol) <u>burns in air</u> with a colorless flame.
    (c) Sugar is <u>soluble in water</u>.
    (d) Hydrogen peroxide, H_2O_2, <u>decomposes to form oxygen, O_2, and water, H_2O</u>.
29. In each case, describe the change as a chemical or physical change. Give a reason for your choice.
    (a) A cup of household bleach changes the color of your favorite T-shirt from purple to pink.
    (b) The fuels in the space shuttle (hydrogen and oxygen) combine to give water and provide the energy to lift the shuttle into space.
    (c) An ice cube in your glass of lemonade melts.
30. In each case, describe the change as a chemical or physical change. Give a reason for your choice.
    (a) Salt dissolves when you add it to water.
    (b) Food is digested and metabolized in your body.
    (c) Crystalline sugar is ground into a fine powder.
    (d) When potassium is added to water there is a purplish-pink flame and the water becomes basic (alkaline).
31. In each situation, decide whether a chemical reaction is releasing energy and causing work to be done, or whether an outside source of energy is forcing a chemical reaction to occur.
    (a) Your body converts excess intake of food into fat molecules.
    (b) Sodium reacts with water as shown in Figure 1.5.
    (c) Sodium azide in an automobile air bag decomposes, causing the bag to inflate.
    (d) An egg is hard-boiled in a pan on your kitchen stove.
32. While camping in the mountains you build a small fire out of tree limbs you find on the ground near your campsite. The dry wood crackles and burns brightly and warms you. Before slipping into your sleeping bag for the night, you put the fire out by dousing it with cold water from a nearby stream. Steam rises when the water hits the hot coals. Describe the physical and chemical changes in this scene.

Classifying Matter: Substances and Mixtures (Section 1.6)

33. Small chips of iron are mixed with sand (see photo). Is this a homogeneous or heterogeneous mixture? Suggest a way to separate the iron and sand from each other.

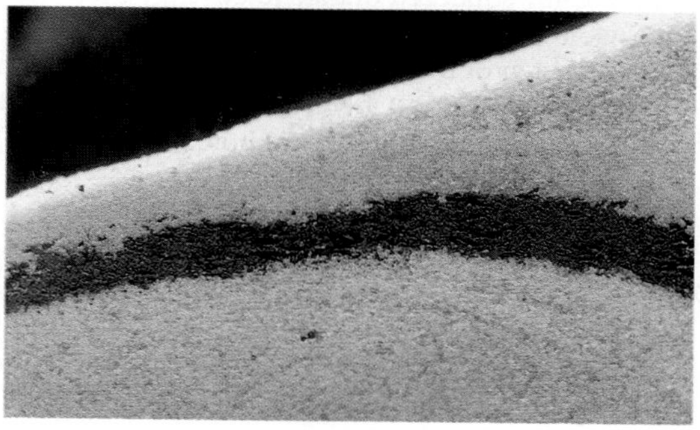

Layers of sand, iron, and sand.

34. Suppose that you have a solution of sugar in water. Is this a homogeneous or heterogeneous mixture? Describe an experimental procedure by which you can separate the two substances.
35. Identify each of these as a homogeneous or a heterogeneous mixture.
    (a) Vodka (b) Blood
    (c) Cowhide (d) Bread
36. Identify each of these as a homogeneous or a heterogeneous mixture.
    (a) An asphalt (blacktop) road (b) Clear ocean water
    (c) Iced tea with ice cubes (d) Filtered apple cider
37. Devise and describe an experiment to
    (a) Separate table salt (sodium chloride) from water.
    (b) Separate iron filings from small pieces of magnesium.
    (c) Separate the element zinc from sugar (sucrose).
38. Devise and describe an experiment to
    (a) Separate sucrose (table sugar) from water.
    (b) Separate the element sulfur from table salt (sodium chloride).
    (c) Separate iron filings from granular zinc.

Classifying Matter: Elements and Compounds (Section 1.7)

39. For each of the changes described, decide whether two or more elements formed a compound or if a compound decomposed (to form elements or other compounds). Explain your reasoning in each case.
    (a) Upon heating, a blue powder turned white and lost mass.
    (b) A white solid forms three different gases when heated. The total mass of the gases is the same as that of the solid.
40. For each of the changes described, decide whether two or more elements formed a compound or if a compound decomposed (to form elements or other compounds). Explain your reasoning in each case.
    (a) After a reddish-colored metal is placed in a flame, it turns black and has a higher mass.
    (b) A white solid is heated in oxygen and forms two gases. The mass of the gases is the same as the masses of the solid and the oxygen.
41. Classify each of these with regard to the type of matter (element, compound, heterogeneous mixture, or homogeneous mixture). Explain your choice in each case.
    (a) A piece of newspaper
    (b) Solid, granulated sugar
    (c) Freshly squeezed orange juice
    (d) Gold jewelry
42. Classify each of these with regard to the type of matter (element, compound, heterogeneous mixture, or homogeneous mixture). Explain your choice in each case.
    (a) A cup of coffee
    (b) A soft drink such as a Coke or Pepsi
    (c) A piece of Dry Ice (a solid form of carbon dioxide)
43. Classify each of these as an element, a compound, a heterogeneous mixture, or a homogeneous mixture. Explain your choice in each case.
    (a) Chunky peanut butter (b) Distilled water
    (c) Platinum (d) Air
44. Classify each of these as an element, a compound, a heterogeneous mixture, or a homogeneous mixture. Explain your choice in each case.
    (a) Table salt (sodium chloride)
    (b) Methane (which burns in pure oxygen to form only carbon dioxide and water)
    (c) Chocolate chip cookie
    (d) Silicon
45. A black powder is placed in a long glass tube. Hydrogen gas is passed into the tube so that the hydrogen sweeps out all other gases. The powder is then heated with a Bunsen burner. The powder turns red-orange, and water vapor can be seen condensing at the unheated far end of the tube. The red-orange color remains after the tube cools.
    (a) Was the original black substance an element? Explain briefly.
    (b) Is the new red-orange substance an element? Explain briefly.
46. A finely divided black substance is placed in a glass tube filled with air. When the tube is heated with a Bunsen burner, the black substance turns red-orange. The total mass of the red-orange substance is greater than that of the black substance.
    (a) Can you conclude that the black substance is an element? Explain briefly.
    (b) Can you conclude that the red-orange substance is a compound? Explain briefly.

Nanoscale Theories and Models (Section 1.8)

47. The accompanying photo shows a crystal of the mineral halite, a form of ordinary salt. Are these crystals at the macroscale, microscale, or nanoscale? How would you describe the shape of these crystals? What might this tell you about the arrangement of the atoms deep inside the crystal?

A halite (sodium chloride) crystal.

48. Galena, shown in the photo, is a black mineral that contains lead and sulfur. It shares its name with a number of towns in the United States; they are located in Alaska, Illinois, Kansas, Maryland, Missouri, and Ohio. How would you describe the shape of the galena crystals? What might this tell you about the arrangement of the atoms deep inside the crystal?

Black crystals of galena (lead sulfide).

49. The photograph on p. 35 on the left shows the four-celled alga *Scenedesmus.* This image has been enlarged by a factor of 400. Are algae such as *Scenedesmus* at the macroscale, microscale, or nanoscale?
50. The photograph on p. 35 on the right shows an end-on view of tiny wires made from nickel metal by special processing. The scale bar is 1 μm long. Are these wires at the macroscale, microscale, or nanoscale?

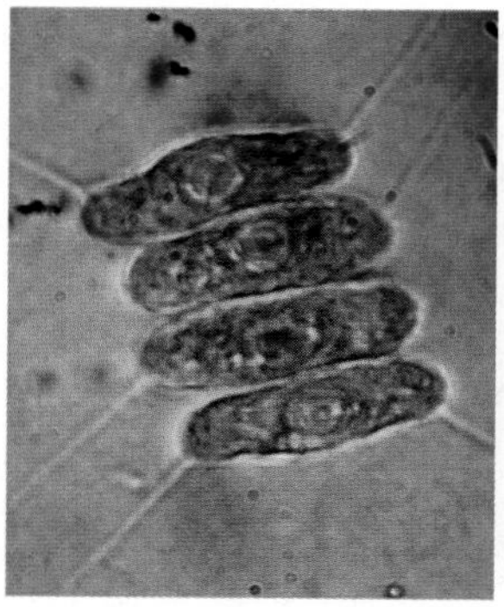

NOAA

Scenedesmus.

© Science VU/NASA/ARC/Visuals Unlimited

Nickel wires. (The scale bar is 1 μm long.)

51. The Chemistry in the News box on p. 21 shows an image obtained with a scanning tunneling microscope. Is this image at the macroscale, the microscale, or the nanoscale?
52. The image below shows several examples of tobacco mosaic virus. Is this virus at the macroscale, the microscale, or the nanoscale? (The scale bar at the bottom of the image is 100 nm long.)

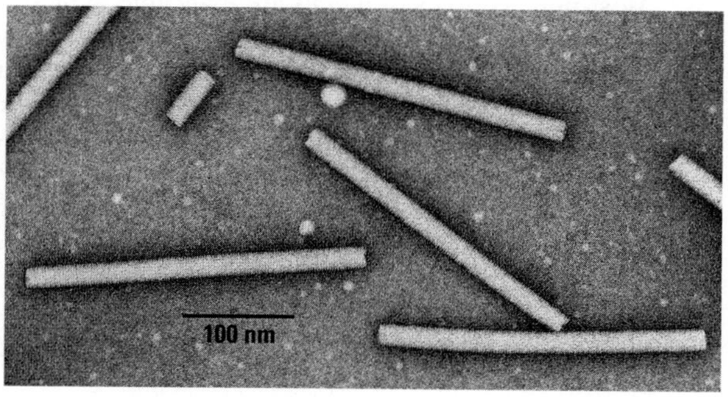

Robert G. Milne, Plant Virus Institute, National Research Council, Turin, Italy

53. When you open a can of a carbonated drink, the carbon dioxide gas inside expands rapidly as it rushes from the can. Describe this process in terms of the kinetic-molecular theory.
54. After you wash your clothes, you hang them on a line in the sun to dry. Describe the change or changes that occur in terms of the kinetic-molecular theory. Are the changes that occur physical or chemical changes?
55. Sucrose has to be heated to a high temperature before it caramelizes. Use the kinetic-molecular theory to explain why sugar caramelizes only at high temperatures.
56. Give a nanoscale interpretation of the fact that at the melting point the density of solid mercury is greater than the density of liquid mercury, and at the boiling point the density of liquid mercury is greater than the density of gaseous mercury.
57. Make these unit conversions, using the prefixes in Table 1.2.
    (a) 32.75 km to meters
    (b) 0.0342 mm to nanometers
    (c) 1.21×10^{-12} km to micrometers
58. Make these unit conversions, using the prefixes in Table 1.2.
    (a) 0.00572 kg to grams
    (b) 8.347×10^{7} nL to liters
    (c) 423.7 g to kilograms

The Atomic Theory (Section 1.9)

59. Explain in your own words, by writing a short paragraph, how the atomic theory explains conservation of mass during a chemical reaction and during a physical change.
60. Explain in your own words, by writing a short paragraph, how the atomic theory explains constant composition of chemical compounds.
61. Explain in your own words, by writing a short paragraph, how the atomic theory predicts the law of multiple proportions.
62. State the four postulates of the modern atomic theory.
63. State the law of multiple proportions in your own words.
64. The element chromium forms three different oxides (that contain only chromium and oxygen). The percentage of chromium (number of grams of chromium in 100 g oxide) in these compounds is 52.0%, 68.4%, and 76.5%. Do these data conform to the law of multiple proportions? Explain why or why not.

The Chemical Elements (Section 1.10)

65. Name and give the symbols for two elements that
    (a) Are metals
    (b) Are nonmetals
    (c) Are metalloids
    (d) Consist of diatomic molecules
66. Name and give the symbols for two elements that
    (a) Are gases at room temperature
    (b) Are solids at room temperature
    (c) Do not consist of molecules
    (d) Have different allotropic forms

Communicating Chemistry: Symbolism (Section 1.11)

67. Write a chemical formula for each substance, and draw a nanoscale picture of how the molecules are arranged at room temperature.
    (a) Water, a liquid whose molecules contain two hydrogen atoms and one oxygen atom each
    (b) Nitrogen, a gas that consists of diatomic molecules
    (c) Neon
    (d) Chlorine
68. Write a chemical formula for each substance and draw a nanoscale picture of how the molecules are arranged at room temperature.
    (a) Iodine, a solid that consists of diatomic molecules
    (b) Ozone
    (c) Helium
    (d) Carbon dioxide
69. Write a nanoscale representation and a symbolic representation and describe what happens at the macroscale when hydrogen reacts chemically with oxygen to form water vapor.
70. Write a nanoscale representation and a symbolic representation and describe what happens at the macroscale when carbon monoxide reacts with oxygen to form carbon dioxide.
71. Write a nanoscale representation and a symbolic representation and describe what happens at the macroscale when iodine sublimes (passes directly from solid to gas with no liquid formation) to form iodine vapor.

72. Write a nanoscale representation and a symbolic representation and describe what happens at the macroscale when bromine evaporates to form bromine vapor.

General Questions

These questions are not explicitly keyed to chapter topics; many require integration of several concepts.

73. Classify the information in each of these statements as quantitative or qualitative and as relating to a physical or chemical property.
    (a) A white chemical compound has a mass of 1.456 g. When placed in water containing a dye, it causes the red color of the dye to fade to colorless.
    (b) A sample of lithium metal, with a mass of 0.6 g, was placed in water. The metal reacted with the water to produce the compound lithium hydroxide and the element hydrogen.
74. Classify the information in each of these statements as quantitative or qualitative and as relating to a physical or chemical property.
    (a) A liter of water, colored with a purple dye, was passed through a charcoal filter. The charcoal adsorbed the dye, and colorless water came through. Later, the purple dye was removed from the charcoal and retained its color.
    (b) When a white powder dissolved in a test tube of water, the test tube felt cold. Hydrochloric acid was then added, and a white solid formed.
75. The density of solid potassium is 0.86 g/mL. The density of solid calcium is 1.55 g/mL, almost twice as great. However, the mass of a potassium atom is only slightly less than the mass of a calcium atom. Provide a nanoscale explanation of these facts.
76. The density of gaseous helium at 25 °C and normal atmospheric pressure is 1.64×10^{-4} g/mL. At the same temperature and pressure the density of argon gas is 1.63×10^{-3} g/mL. The mass of an atom of argon is almost exactly ten times the mass of an atom of helium. Provide a nanoscale explanation of why the densities differ as they do.
77. The gauge number of a wire is related to the diameter of the wire. Suppose you have a 10.0-lb spool of 12-gauge aluminum wire (diameter 0.0808 in). What length of wire (in meters) is on the spool? Assume that the wire is a cylinder ($V = \pi r^2 \ell$, where V is the volume, r is the radius, and ℓ is the length) and obtain the density of aluminum from Table 1.1. (1 in = 2.54 cm; 1 lb = 453.59 g)
78. The dimensions of aluminum foil in a box for sale at a supermarket are $66\frac{2}{3}$ yards by 12 inches. The mass of the foil is 0.83 kg. What is the thickness of the foil (in cm)? (1 in = 2.54 cm)
79. Hexane (density = 0.766 g/cm^3), perfluorohexane (density = 1.669 g/cm^3), and water are immiscible liquids; that is, they do not dissolve in one another. You place 10 mL of each in a graduated cylinder, along with pieces of high-density polyethylene (HDPE, density 0.97 g/mL), polyvinyl chloride (PVC, density = 1.36 g/cm^3), and Teflon (density = 2.3 g/cm^3). None of these common plastics dissolves in these liquids. Describe what you expect to see.
80. You can figure out whether a substance floats or sinks if you know its density and the density of the liquid. In which of the liquids listed below will high-density polyethylene (HDPE) float? HDPE, a common plastic, has a density of 0.97 g/mL. It does not dissolve in any of these liquids.

Substance	Density (g/mL)	Properties, Uses
Ethylene glycol	1.113	Toxic; the major component of automobile antifreeze
Water	0.9982	
Ethanol	0.7893	The alcohol in alcoholic beverages
Methanol	0.7917	Toxic; gasoline additive to prevent gas line freezing
Acetic acid	1.0498	Component of vinegar
Glycerol	1.2611	Solvent used in home care products

81. Describe in your own words how different allotropic forms of an element are different at the nanoscale.
82. Most pure samples of metals are malleable, which means that if you try to grind up a sample of a metal by pounding on it with a hard object, the pieces of metal change shape but do not break apart. Solid samples of nonmetallic elements, such as sulfur or graphite, are often brittle and break into smaller particles when hit by a hard object. Devise a nanoscale theory about the structures of metals and nonmetals that can account for this difference in macroscale properties.

Applying Concepts

These questions test conceptual learning.

83. Using Table 1.1, but without using your calculator, decide which has the larger mass:
    (a) 20. mL butane or 20. mL bromobenzene
    (b) 10. mL benzene or 1.0 mL gold
    (c) 0.732 mL copper or 0.732 mL lead
84. Using Table 1.1, but without using your calculator, decide which has the larger volume:
    (a) 1.0 g ethanol or 1.0 g bromobenzene
    (b) 10. g aluminum or 12 g water
    (c) 20 g gold or 40 g magnesium
85. At 25 °C the density of water is 0.997 g/mL, whereas the density of ice at −10 °C is 0.917 g/mL.
    (a) If a plastic soft-drink bottle (volume = 250 mL) is completely filled with pure water, capped, and then frozen at −10 °C, what volume will the solid occupy?
    (b) What will the bottle look like when you take it out of the freezer?
86. When water alone (instead of engine coolant, which contains water and other substances) was used in automobile radiators to cool cast-iron engine blocks, it sometimes happened in winter that the engine block would crack, ruining the engine. Cast iron is not pure iron and is relatively hard and brittle. Explain in your own words how the engine block in a car might crack in cold weather.

87. Of the substances listed in Table 1.1, which would not float on liquid mercury? (Assume that none of the substances would dissolve in the mercury.)
88. Which of the substances in Figure 1.2 has the greatest density? Which has the lowest density?
89. Water does not dissolve in bromobenzene.
    (a) If you pour 2 mL water into a test tube that contains 2 mL bromobenzene, which liquid will be on top?
    (b) If you pour 2 mL ethanol carefully into the test tube containing bromobenzene and water described in part (a) without shaking or mixing the liquids, what will happen?
    (c) What will happen if you thoroughly stir the mixture in part (b)?
90. Water does not mix with either benzene or bromobenzene when it is stirred together with either of them, but benzene and bromobenzene do mix.
    (a) If you pour 2 mL bromobenzene into a test tube, then add 2 mL water and stir, what would the test tube look like a few minutes later?
    (b) Suppose you add 2 mL benzene to the test tube in part (a), pouring the benzene carefully down the side of the tube so that the liquids do not mix. Describe the appearance of the test tube now.
    (c) If the test tube containing all three liquids is thoroughly shaken and then allowed to stand for five minutes, what will the tube look like?
91. The figure shows a nanoscale view of the atoms of mercury in a thermometer registering 10 °C.

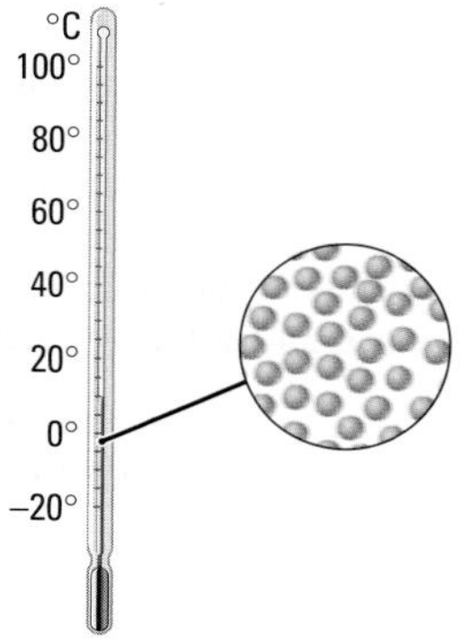

Which nanoscale drawing best represents the atoms in the liquid in this same thermometer at 90 °C? (Assume that the same volume of liquid is shown in each nanoscale drawing.)

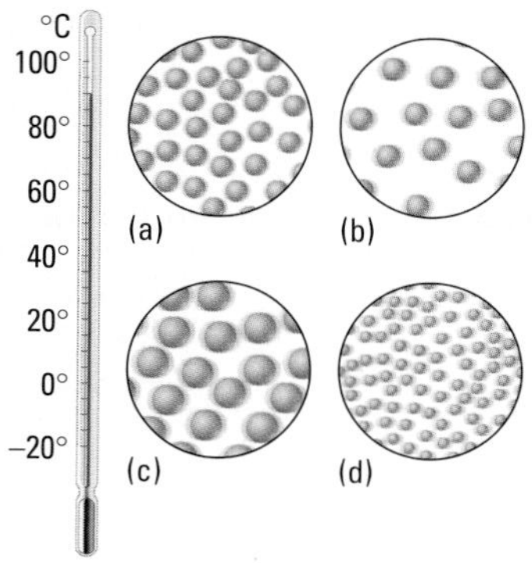

92. Answer these questions using figures (a) through (i). (Each question may have more than one answer.)

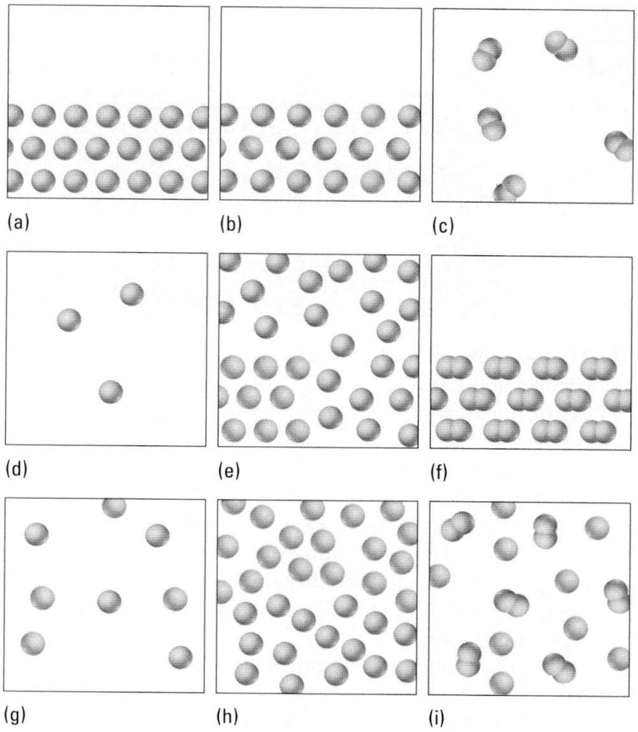

(a) Which represents nanoscale particles in a sample of solid?
(b) Which represents nanoscale particles in a sample of liquid?
(c) Which represents nanoscale particles in a sample of gas?
(d) Which represents nanoscale particles in a sample of an element?
(e) Which represents nanoscale particles in a sample of a compound?
(f) Which represents nanoscale particles in a sample of a pure substance?
(g) Which represents nanoscale particles in a sample of a mixture?

More Challenging Questions

These questions require more thought and integrate several concepts.

93. The element platinum has a solid-state structure in which platinum atoms are arranged in a cubic shape that repeats throughout the solid. The length of an edge of the cube is 392 pm (1 pm = 1×10^{-12} m). Calculate the volume of the cube in cubic meters.
94. The compound sodium chloride has a solid-state structure in which there is a repeating cubic arrangement of sodium ions and chloride ions. The volume of the cube is 1.81×10^{-22} cm^3. Calculate the length of an edge of the cube in pm (1 pm = 1×10^{-12} m).

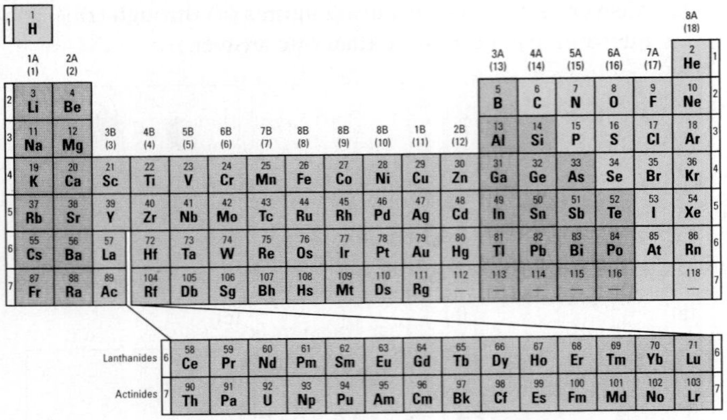

95. The periodic table shown here is color coded gray, blue, orange, and lavender. Identify the color of the area (or colors of the areas) in which you would expect to find each type of element.
    (a) A colorless gas
    (b) A solid that is ductile and malleable
    (c) A solid that has poor electrical conductivity
96. The periodic table shown here is color coded gray, blue, orange, and lavender. Identify the color of the area (or colors of the areas) in which you would expect to find each type of element.
    (a) A shiny solid that conducts electricity
    (b) A gas whose molecules consist of single atoms
    (c) An element that is a semiconductor
    (d) A yellow solid that has very low electrical conductivity
97. Which two elements from this list exhibit the greatest similarity in physical and chemical properties? Explain your choice. S, Ga, Se, Ti.
98. Which two elements from this list exhibit the greatest similarity in physical and chemical properties? Explain your choice. Mg, Br, Si, Sr.
99. When someone discovers a new substance, it is relatively easy to show that the substance is not an element, but it is quite difficult to prove that the substance is an element. Explain why this is so, and relate your explanation to the discussion of scientific laws and theories in Section 1.3.
100. Soap can be made by mixing animal or vegetable fat with a concentrated solution of lye and heating it in a large vat. Suppose that 3.24 kg vegetable fat is placed in a large iron vat and then 50.0 L water and 5.0 kg lye (sodium hydroxide, NaOH) are added. The vat is placed over a fire and heated for two hours, and soap forms.
    (a) Classify each of the materials identified in the soap-making process as a substance or a mixture. For each substance, indicate whether it is an element or a compound. For each mixture, indicate whether it is homogeneous or heterogeneous.
    (b) Assuming that the fat and lye are completely converted into soap, what mass of soap is produced?
    (c) What physical and chemical processes occur as the soap is made?
101. The densities of several elements are given in Table 1.1.
    (a) Of the elements nickel, gold, lead, and magnesium, which will float on liquid mercury at 20 °C?
    (b) Of the elements titanium, copper, iron, and gold, which will float highest on the mercury? That is, which element will have the smallest fraction of its volume below the surface of the liquid?
102. You have some metal shot (small spheres like BBs), and you want to identify the metal. You have a flask that is known to contain exactly 100.0 mL when filled with liquid to a mark in the flask's neck. When the flask is filled with water at 20 °C, the mass of flask and water is 122.3 g. The water is emptied from the flask and 20 of the small spheres of metal are carefully placed in the flask. The 20 small spheres had a mass of 42.3 g. The flask is again filled to the mark with water at 20 °C and weighed. This time the mass is 159.9 g.
    (a) What metal is in the spheres? (Assume that the spheres are all the same and consist of pure metal.)
    (b) What volume would 500 spheres occupy?
103. The element zinc reacts with the element sulfur to form a white solid compound, zinc sulfide. When a sample of zinc that weighs 65.4 g reacts with sulfur, it is found that the zinc sulfide produced weighs exactly 97.5 g.
    (a) What mass of sulfur is present in the zinc sulfide?
    (b) What mass of zinc sulfide could be produced from 20.0 g zinc?
104. The element magnesium reacts with the element oxygen to form a white solid compound, magnesium oxide. When a sample of magnesium that weighs 24.30 g reacts with oxygen, it is found that the magnesium oxide produced weighs exactly 40.30 g.
    (a) What mass of oxygen is present in the magnesium oxide?
    (b) What mass of magnesium oxide could be produced from 40.0 g magnesium?
105. A chemist analyzed several portions taken from different parts of a sample that contained only iron and sulfur. She reported the results in the table. Could this sample be a compound of iron and sulfur? Explain why or why not.

Portion Number	Mass of Portion (g)	Mass of Iron (g)
1	1.518	0.964
2	2.056	1.203
3	1.873	1.290

106. A chemist analyzed several portions taken from different parts of a sample that contained only selenium and oxygen. She reported the results in the table. Could this sample be a compound of selenium and oxygen? Explain why or why not.

Portion Number	Mass of Portion (g)	Mass of Selenium (g)
1	1.518	1.08
2	2.056	1.46
3	1.873	1.33

107. When 12.6 g calcium carbonate (the principal component of chalk) is treated with 63.0 mL hydrochloric acid (muriatic acid sold in hardware stores; density = 1.096 g/mL), the calcium carbonate reacts with the acid, goes into solution, and carbon dioxide gas bubbles out of the solution. After all of the carbon dioxide has escaped, the solution weighs 76.1 g. Calculate the volume (in liters) of carbon dioxide gas that was produced. (The density of the carbon dioxide gas is 1.798 g/L.)

108. When 15.6 g sodium carbonate (washing soda used in tie dies) is treated with 63.0 mL hydrochloric acid (muriatic acid sold in hardware stores; density = 1.096 g/mL), the sodium carbonate reacts with the acid, goes into solution, and carbon dioxide gas bubbles out of the solution. All of the carbon dioxide is collected and its volume is found to be 3.29 L. Calculate the mass of solution that remains. (The density of the carbon dioxide gas is 1.798 g/L.)

109. Suppose you are trying to get lemon juice and you have no juicer. Some people say that you can get more juice from a lemon if you roll it on a hard surface, applying pressure with the palm of your hand before you cut it and squeeze out the juice. Others claim that you will get more juice if you first heat the lemon in a microwave and then cut and squeeze it. Apply the methods of science to arrive at a technique that will give the most juice from a lemon. Carry out experiments and draw conclusions based on them. Try to generate a hypothesis to explain your results.

110. If you drink orange juice soon after you brush your teeth, the orange juice tastes quite different. Apply the methods of science to find what causes this effect. Carry out experiments and draw conclusions based on them.

Conceptual Challenge Problems

These rigorous, thought-provoking problems integrate conceptual learning with problem solving and are suitable for group work.

CP1.A (Section 1.3) Some people use expressions such as "a rolling stone gathers no moss" and "where there is no light there is no life." Why do you believe these are "laws of nature"?

CP1.B (Section 1.3) Parents teach their children to wash their hands before eating. (a) Do all parents accept the germ theory of disease? (b) Are all diseases caused by germs?

CP1.C (Section 1.8) In Section 1.8 you read that, on an atomic scale, all matter is in constant motion. (For example, the average speed of a molecule of nitrogen or oxygen in the air is greater than 1000 miles per hour at room temperature.) (a) What evidence can you put forward that supports the kinetic-molecular theory? (b) Suppose you accept the notion that molecules of air are moving at speeds near 1000 miles per hour. What can you then reason about the paths that these molecules take when moving at this speed?

CP1.D (Section 1.8) Some scientists think there are living things smaller than bacteria (*New York Times,* January 18, 2000, p. D1). Called "nanobes," they are roughly cylindrical and range from 20 to 150 nm long and about 10 nm in diameter. One approach to determining whether nanobes are living is to estimate how many atoms and molecules could make up a nanobe. If the number is too small, then there would not be enough DNA, protein, and other biological molecules to carry out life processes. To test this method, estimate an upper limit for the number of atoms that could be in a nanobe. (Use a small atom, such as hydrogen.) Also estimate how many protein molecules could fit inside a nanobe. Do your estimates rule out the possibility that a nanobe could be living? Explain why or why not.

CP1.E (Section 1.12) The life expectancy of U.S. citizens in 1992 was 76 years. In 1916 the life expectancy was only 52 years. This is an increase of 46% in a lifetime. (a) Could this astonishing increase occur again? (b) To what single source would you attribute this noteworthy increase in life expectancy? Why did you identify this one source as being most influential?

CP1.F Helium-filled balloons rise and will fly away unless tethered by a string. Use the kinetic-molecular theory to explain why a helium-filled balloon is "lighter than air."

2 Atoms and Elements

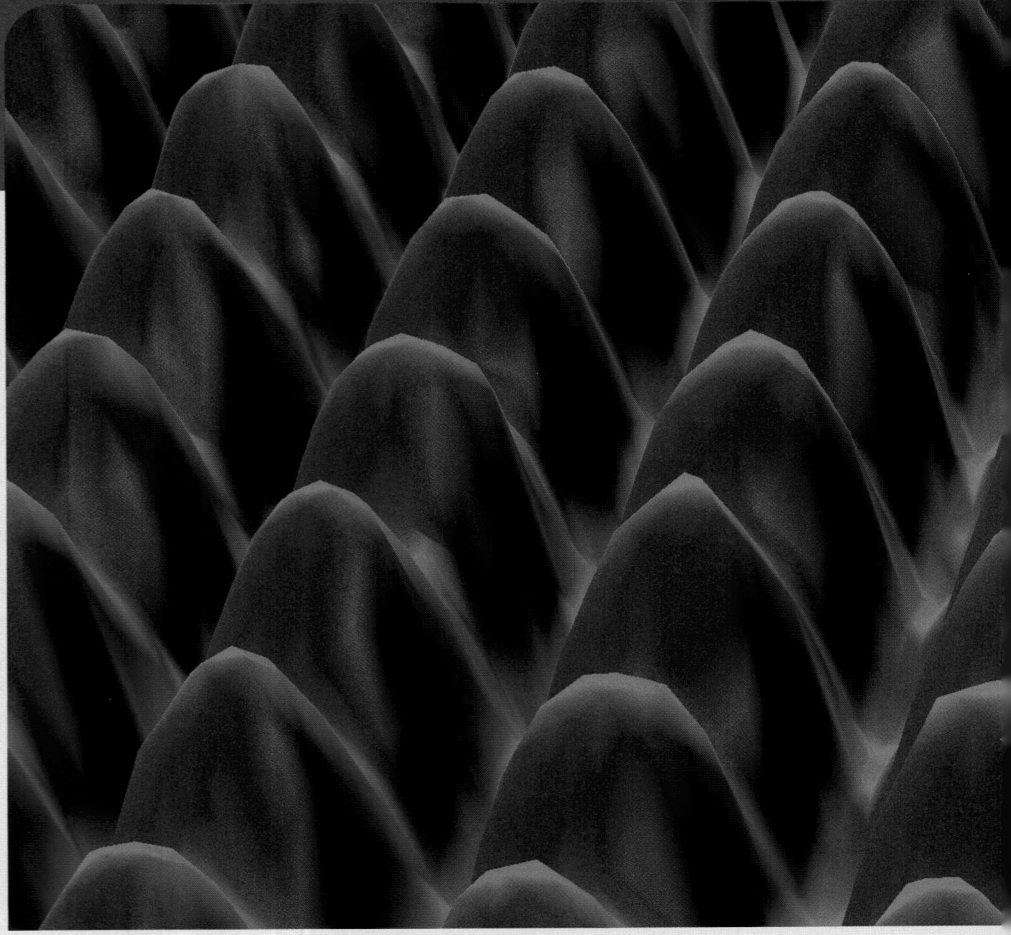

Image reproduced by permission of IBM Research, Almaden Research

Each rounded peak in this image represents one nickel atom on a uniform surface of metallic nickel. The atoms are spaced about 0.25 nm (2.5×10^{-10} m) apart, so the entire image covers an area of about one square nanometer. The image was generated with a scanning tunneling microscope (STM) that can detect individual atoms or molecules, allowing scientists to make images of nanoscale atomic arrangements. (See *Tools of Chemistry,* p. 46.)

To study chemistry, we need to start with atoms—the basic building blocks of matter. Early theories of the atom considered atoms to be indivisible, but we know now that atoms are more complicated than that. Elements differ from one another because of differences in the internal structure of their atoms. Under the right conditions, smaller particles within atoms—known as *subatomic particles*—can be removed or rearranged. The term **atomic structure** refers to the identity and arrangement of these subatomic particles in the atom. An understanding of the details of atomic structure aids in the understanding of how elements combine to form compounds and how atoms are rearranged in chemical reactions. Atomic structure also accounts for the properties of materials. The next few sections describe how experiments support the idea that atoms are composed of smaller (subatomic) particles.

OWL Sign in to OWL at **www.cengage.com/owl** to view tutorials and simulations, develop problem-solving skills, and complete online homework assigned by your professor.

go Chemistry Download mini lecture videos for key concept review and exam prep from OWL or purchase them from **www.CengageBrain.com.**

Companion Website

Visit this book's companion website at **www.cengage.com/chemistry/moore** to work interactive modules for the Estimation boxes and Active Figures in this text.

2.1 Atomic Structure and Subatomic Particles

Electrical charges played an important role in many of the experiments from which the theory of atomic structure was derived. Two types of electrical charge exist: positive and negative. ***Electrical charges of the same type repel one another, and charges of the opposite type attract one another.*** A positively charged particle repels another positively charged particle. Likewise, two negatively charged particles repel each other. In contrast, two particles with opposite signs attract each other.

CONCEPTUAL EXERCISE 2.1 Electrical Charge

When you comb your hair on a dry day, your hair sticks to the comb. How could you explain this behavior in terms of a nanoscale model in which atoms contain positive and negative charges?

Answers to **EXERCISES** are provided at the back of this book in Appendix L.

EXERCISES that are labeled **CONCEPTUAL** are designed to test your understanding of one or more concepts; they usually involve qualitative rather than quantitative thinking.

Radioactivity

In 1896 Henri Becquerel discovered that a sample of a uranium ore emitted rays that darkened a photographic plate, even though the plate was covered by a protective black paper. In 1898 Marie and Pierre Curie isolated the new elements polonium and radium, which emitted the same kind of rays. Marie suggested that atoms of such elements spontaneously emit these rays and named the phenomenon **radioactivity.**

Atoms of radioactive elements can emit three types of radiation: alpha (α), beta (β), and gamma (γ) rays. These radiations behave differently when passed between electrically charged plates (Figure 2.1). Alpha and beta rays are deflected, but gamma rays are not. These events occur because alpha rays and beta rays are composed of charged particles that come from within the radioactive atom. Alpha rays have a 2+ charge, and beta rays have a 1− charge. Alpha rays and beta rays are par-

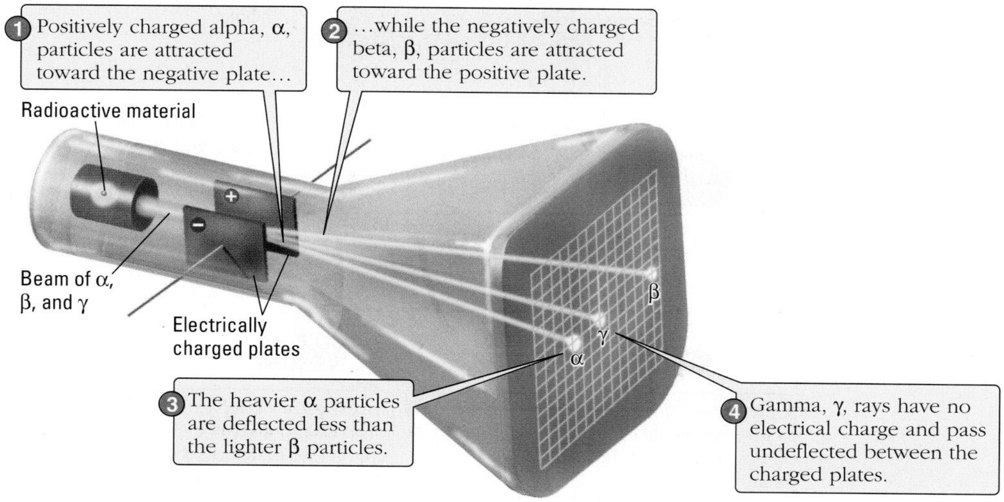
1 Positively charged alpha, α, particles are attracted toward the negative plate...
2 ...while the negatively charged beta, β, particles are attracted toward the positive plate.
Radioactive material
+
−
Beam of α, β, and γ
Electrically charged plates
β
γ
α
3 The heavier α particles are deflected less than the lighter β particles.
4 Gamma, γ, rays have no electrical charge and pass undeflected between the charged plates.

Figure 2.1 The alpha (α), beta (β), and gamma (γ) rays from a radioactive sample can be separated by an electrical field.

ticles because they have mass—they are matter. In the experiment shown in Figure 2.1, alpha particles are deflected less and so must be heavier than beta particles. Gamma rays have no detectable charge or mass; they behave like light rays. If radioactive atoms can break apart to produce subatomic alpha and beta particles, then there must be something smaller inside the atoms.

Electrons

Further evidence that atoms are composed of subatomic particles came from experiments with specially constructed glass tubes called cathode-ray tubes. Most of the air has been removed from these tubes and a metal electrode sealed into each end. When a sufficiently high voltage is applied to the electrodes, a beam of rays known as cathode rays flows from the metal atoms of the negatively charged electrode (the *cathode*) to the positively charged electrode (the *anode*). The cathode rays travel in straight lines, are attracted toward positively charged plates, can be deflected by a magnetic field, and can cause gases and fluorescent materials to glow. Thus, the properties of a cathode ray are those of a beam of negatively charged particles, each of which produces a light flash when it hits a fluorescent screen. Sir Joseph John Thomson suggested that cathode rays consist of the same particles that had earlier been named **electrons** and had been suggested to be the carriers of electricity. He also observed that cathode rays were produced from cathodes made of different metals, which implied that electrons are constituents of the atoms of each of those different metallic elements.

In 1897 Thomson used a specially designed cathode-ray tube to simultaneously apply electric and magnetic fields to a beam of cathode rays. By balancing the electric field against the magnetic field and using the basic laws of electricity and magnetism, Thomson calculated the *ratio* of mass to charge for the electrons in the cathode-ray beam: 5.60×10^{-9} grams per coulomb (g/C). (The coulomb, C, is a fundamental unit of electrical charge.)

See Appendix A.2 for a review of scientific notation, which is used to represent very small or very large numbers as powers of 10. For example, 0.000001 is 1×10^{-6} (the decimal point moves six places to the right to give the -6 exponent) and 2,000,000 is 2×10^{6} (the decimal point moves six places to the left to give the $+6$ exponent).

Fourteen years later Robert Millikan used a cleverly devised experiment to measure the charge of an electron (Figure 2.2). Tiny oil droplets were sprayed into

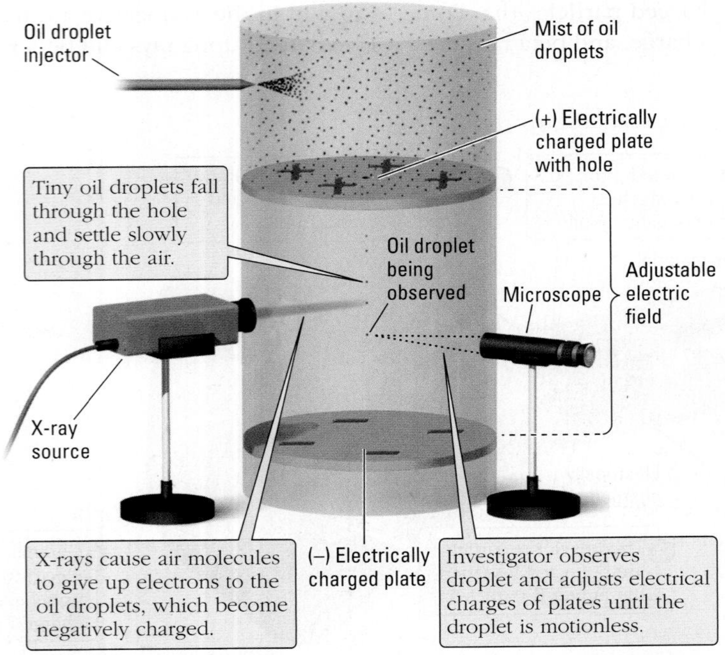
Oil droplet injector
Mist of oil droplets
(+) Electrically charged plate with hole
Tiny oil droplets fall through the hole and settle slowly through the air.
Oil droplet being observed
Microscope
Adjustable electric field
X-ray source
X-rays cause air molecules to give up electrons to the oil droplets, which become negatively charged.
(–) Electrically charged plate
Investigator observes droplet and adjusts electrical charges of plates until the droplet is motionless.

Figure 2.2 Millikan oil-drop experiment. From the known mass of the droplets and the applied voltage at which the charged droplets were held stationary, Millikan could calculate the charges on the droplets.

a chamber. As they settled slowly through the air, the droplets were exposed to x-rays, which caused electrons to be transferred from gas molecules in the air to the droplets. Using a small microscope to observe individual droplets, Millikan adjusted the electrical charge of plates above and below the droplets so that the electrostatic attraction just balanced the gravitational attraction. In this way he could suspend a single droplet motionless. From equations describing these forces, Millikan calculated the charge on the suspended droplets. Different droplets had different amounts of charge, but Millikan found that each was an integer multiple of the smallest charge. The smallest charge was 1.60×10^{-19} C. Millikan assumed this to be the fundamental quantity of charge, the charge on an electron. Given this value and the mass-to-charge ratio determined by Thomson, the mass of an electron could be computed: $(1.60 \times 10^{-19}\ \text{C})(5.60 \times 10^{-9}\ \text{g/C}) = 8.96 \times 10^{-28}$ g. Currently the most accurate value for the electron's mass is $9.10938215 \times 10^{-28}$ g, and the currently accepted most accurate value for the electron's charge is $-1.602176487 \times 10^{-19}$ C. This latter quantity is called the *electronic charge.* For convenience, the charges on subatomic particles are given in multiples of electronic charge rather than in coulombs. Using this convention, the charge on an electron is $1-$.

Electron
Charge = $1-$
Mass = 9.10938×10^{-28} g

Other experiments provided further evidence that the electron is a *fundamental* particle of matter; that is, it is present in *all* matter. The beta particles emitted by radioactive elements were found to have the same properties as cathode rays, which are streams of electrons.

Protons

When atoms *lose* electrons, the atoms become positively charged. When atoms *gain* electrons, the atoms become negatively charged. Such charged atoms, or similarly charged groups of atoms, are known as **ions.** From experiments with positive ions, formed by knocking electrons out of atoms, the existence of a positively charged, fundamental particle was deduced. Positively charged particles with different mass-to-charge ratios were formed by atoms of different elements. The variation in masses showed that atoms of different elements must contain different numbers of positive particles. Those from hydrogen atoms had the smallest mass-to-charge ratio, indicating that they are the fundamental positively charged particles of atomic structure. Such particles are called **protons.** The mass of a proton is known from experiment to be $1.672621637 \times 10^{-24}$ g, which is about 1800 times the mass of an electron. The charge on a proton is $1.602176487 \times 10^{-19}$ C, equal in size, but opposite in sign, to the charge on an electron. The proton's charge is represented by $1+$. Thus, an atom that has lost two electrons has a charge of $2+$.

As mass increases, mass-to-charge ratio increases for a given amount of charge. For a fixed charge, doubling the mass will double the mass-to-charge ratio. For a fixed mass, doubling the charge will halve the mass-to-charge ratio.

Proton
Charge = $1+$
Mass = 1.6726×10^{-24} g

2.2 The Nuclear Atom

The Nucleus

Once it was known that there were subatomic particles, the next question scientists wanted to answer was, How are these particles arranged in an atom? During 1910 and 1911 Ernest Rutherford reported experiments (Figure 2.3) that led to a better understanding of atomic structure. Alpha particles (which have the same mass as helium atoms and a $2+$ charge) were allowed to hit a very thin sheet of gold foil. Almost all of the alpha particles passed through undeflected. However, a very few alpha particles were deflected through large angles, and some came almost straight back toward the source. Rutherford described this unexpected result by saying, "It was about as credible as if you had fired a 15-inch [artillery] shell at a piece of paper and it came back and hit you."

The only way to account for the observations was to conclude that all of the positive charge and most of the mass of the atom are concentrated in a very small region (Figure 2.3). Rutherford called this tiny atomic core the **nucleus.** Only such a region could be sufficiently dense and highly charged to repel an alpha particle.

Alpha particles are four times heavier than the lightest atoms, which are hydrogen atoms.

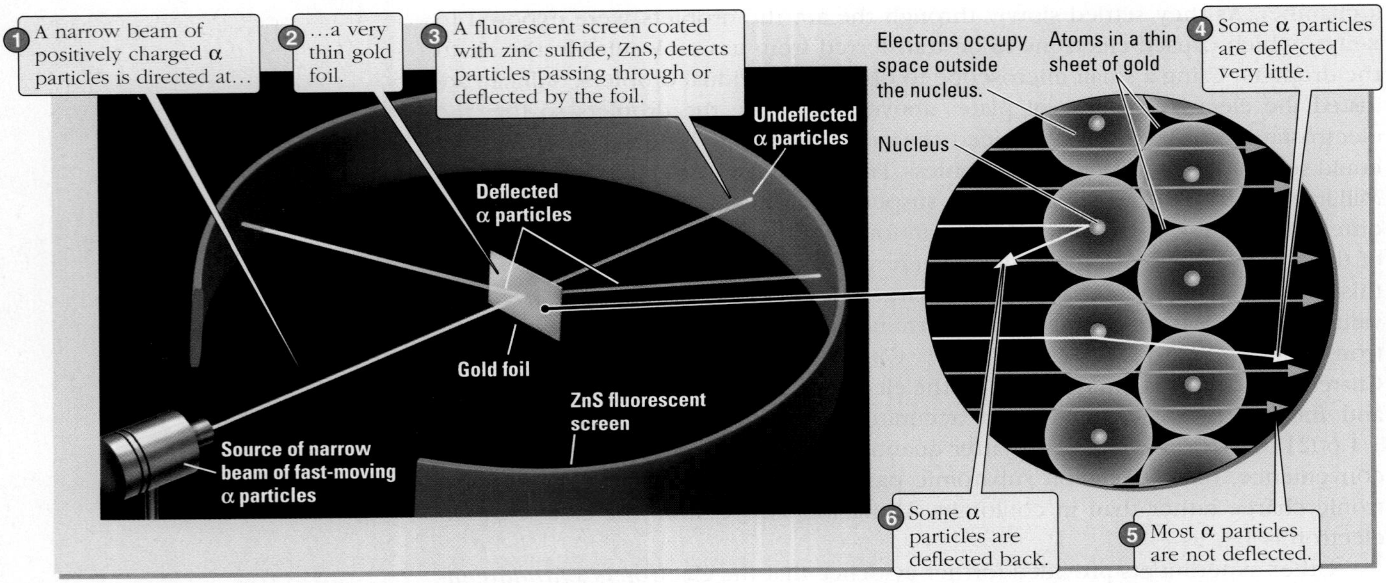

Figure 2.3 **The Rutherford experiment and its interpretation.**

From their results, Rutherford and his associates calculated values for the charge and radius of the gold nucleus. The currently accepted values are a charge of 79+ and a radius of 5.1×10^{-15} m. This makes the nucleus about 10,000 times *smaller* than the atom. Most of the volume of the atom is occupied by the electrons. Somehow the space outside the nucleus is occupied by the negatively charged electrons, but their arrangement was unknown to Rutherford and other scientists of the time. The arrangement of electrons in atoms is now well understood and is described in Chapter 7.

Neutrons

Neutron
Charge = 0
Mass = 1.6749×10^{-24} g

In 1920 Ernest Rutherford proposed that the nucleus might contain an uncharged particle whose mass approximated that of a proton.

Atoms are electrically neutral (no net charge), so they must contain equal numbers of protons and electrons. However, most neutral atoms have masses greater than the sum of the masses of their protons and electrons. The additional mass indicates that subatomic particles with mass but no charge must also be present. Because they have no charge, these particles are more difficult to detect experimentally. In 1932 James Chadwick devised a clever experiment that detected the neutral particles by having them knock protons out of atoms and then detecting the protons. The neutral subatomic particles called **neutrons** have no electrical charge and a mass of $1.674927211 \times 10^{-24}$ g, nearly the same as the mass of a proton.

In summary, there are three primary subatomic particles: protons, neutrons, and electrons.

- Protons and neutrons make up the nucleus, providing most of the atom's mass; the protons provide all of its positive charge.
- The nuclear radius is approximately 10,000 times smaller than the radius of the atom.
- Negatively charged electrons outside the nucleus occupy most of the volume of the atom, but contribute very little mass.
- A neutral atom has no net electrical charge because the number of electrons outside the nucleus equals the number of protons inside the nucleus.

To chemists, the electrons are the most important subatomic particles because they are the first part of the atom to contact another atom. The electrons in atoms largely determine how elements combine to form chemical compounds.

CONCEPTUAL EXERCISE 2.2 Describing Atoms

If an atom had a radius of 100 m, it would approximately fill a football stadium.

(a) What would the approximate radius of the nucleus of such an atom be?

(b) What common object is about that size?

2.3 The Sizes of Atoms and the Units Used to Represent Them

Atoms are extremely small. One teaspoon of water contains about three times as many atoms as the Atlantic Ocean contains teaspoons of water. To do quantitative calculations in chemistry, it is important to understand the units used to express the sizes of very large and very small quantities.

To state the size of an object on the macroscale in the United States (for example, yourself), we would give your weight in pounds and your height in feet and inches. Pounds, feet, and inches are part of the measurement system used in the United States, but almost nowhere else in the world. Most of the world uses the **metric system** of units for recording and reporting measurements. The metric system is a decimal system that adjusts the size of its basic units by multiplying or dividing them by multiples of 10.

The International System of Units, or *Système International* (**SI units**), is the officially recognized measurement system of science. It is derived from the metric system and is constructed from seven base units. The SI system is described in detail in Appendix B. The units for mass (kilogram) and length (meter), which are base units in the SI system, are introduced here, along with the unit of volume (liter). Other units are introduced as they are needed in later chapters.

In the metric system, your weight (really, your mass) would be given in kilograms. The **mass** of an object is a fundamental measure of the quantity of matter in that object. The metric units for mass are *grams* or multiples or fractions of grams. The prefixes listed in Table 2.1 are used with all metric units. A *kilo*gram, for example, is equal to 1000 grams and is a convenient size for measuring the mass of a person.

For objects much smaller than people, prefixes that represent negative powers of 10 are used. For example, 1 *milli*gram equals 1×10^{-3} g.

$$1 \text{ milligram (mg)} = \frac{1}{1000} \times 1 \text{ g} = 0.001 \text{ g} = 1 \times 10^{-3} \text{ g}$$

Bettmann/Corbis

Ernest Rutherford

1871–1937

Born on a farm in New Zealand, Rutherford earned his Ph.D. in physics from Cambridge University in 1895. He discovered alpha and beta radiation and coined the term "half-life." For proving that alpha radiation is composed of helium nuclei and that beta radiation consists of electrons, Rutherford received the Nobel Prize in Chemistry in 1908. As a professor at Cambridge University, he guided the work of no fewer than ten future Nobel Prize recipients. Element 104 is named in Rutherford's honor.

Strictly speaking, the pound is a unit of weight rather than mass. The weight of an object depends on the local force of gravity. For measurements made at Earth's surface, the distinction between mass and weight is not generally useful.

Table 2.1 Some Prefixes Used in the SI and Metric Systems

Prefix	Abbreviation	Meaning	Example
mega	M	10^6	1 megaton = 1×10^6 tons
kilo	k	10^3	1 kilometer (km) = 1×10^3 meter (m) 1 kilogram (kg) = 1×10^3 gram (g)
deci	d	10^{-1}	1 decimeter (dm) = 1×10^{-1} m 1 deciliter (dL) = 1×10^{-1} liter (L)
centi	c	10^{-2}	1 centimeter (cm) = 1×10^{-2} m
milli	m	10^{-3}	1 milligram (mg) = 1×10^{-3} g
micro	μ	10^{-6}	1 micrometer (μm) = 1×10^{-6} m
nano	n	10^{-9}	1 nanometer (nm) = 1×10^{-9} m 1 nanogram (ng) = 1×10^{-9} g
pico	p	10^{-12}	1 picometer (pm) = 1×10^{-12} m
femto	f	10^{-15}	1 femtogram (fg) = 1×10^{-15} g

TOOLS OF CHEMISTRY

Scanning Tunneling Microscopy and Atomic Force Microscopy

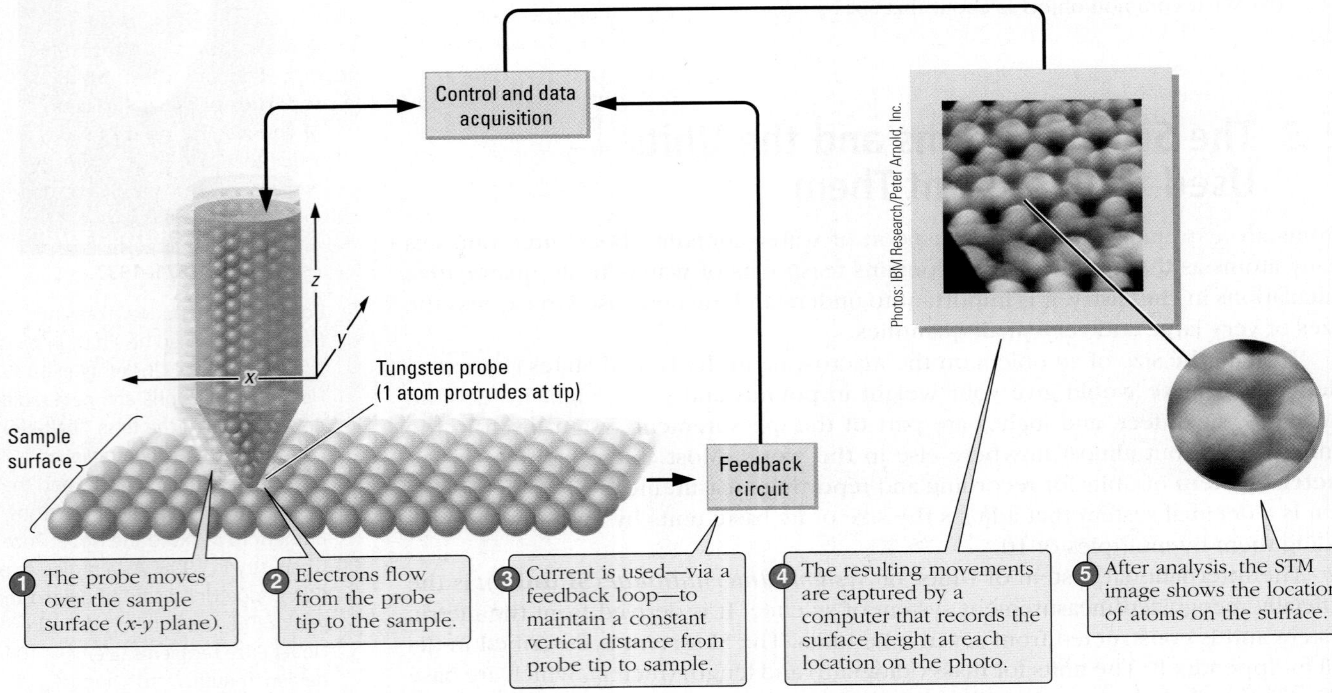
Control and data acquisition
Photos: IBM Research/Peter Arnold, Inc.
z
y
x
Tungsten probe (1 atom protrudes at tip)
Sample surface
Feedback circuit
1 The probe moves over the sample surface (x-y plane).
2 Electrons flow from the probe tip to the sample.
3 Current is used—via a feedback loop—to maintain a constant vertical distance from probe tip to sample.
4 The resulting movements are captured by a computer that records the surface height at each location on the photo.
5 After analysis, the STM image shows the location of atoms on the surface.

Schematic diagram of scanning tunneling microscopy.

The **scanning tunneling microscope** (STM) is an analytical instrument that provides images of individual atoms or molecules on a surface. To do this, a metal probe in the shape of a needle with an extremely fine point (a nanoscale tip) is brought extremely close (within one or two atomic diameters, a few tenths of a nanometer) to the sample surface being examined. When the tip is close enough to the sample, electrons jump between the probe and the sample. The size and

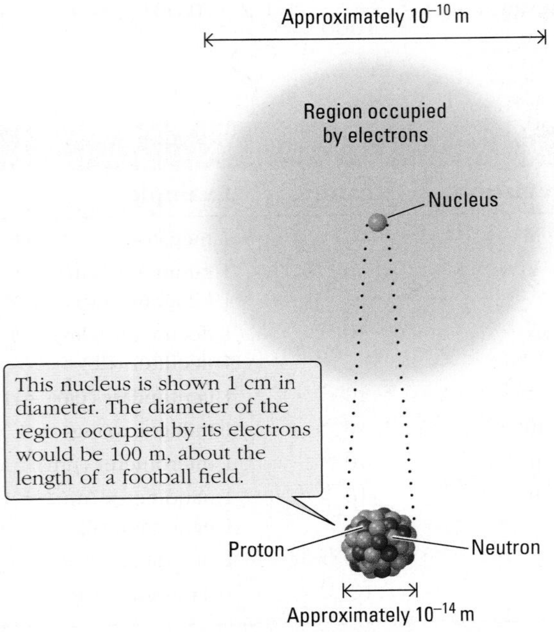
Approximately 10^{-10} m
Region occupied by electrons
Nucleus
This nucleus is shown 1 cm in diameter. The diameter of the region occupied by its electrons would be 100 m, about the length of a football field.
Proton
Neutron
Approximately 10^{-14} m

Relative sizes of the atomic nucleus and an atom (not to scale).

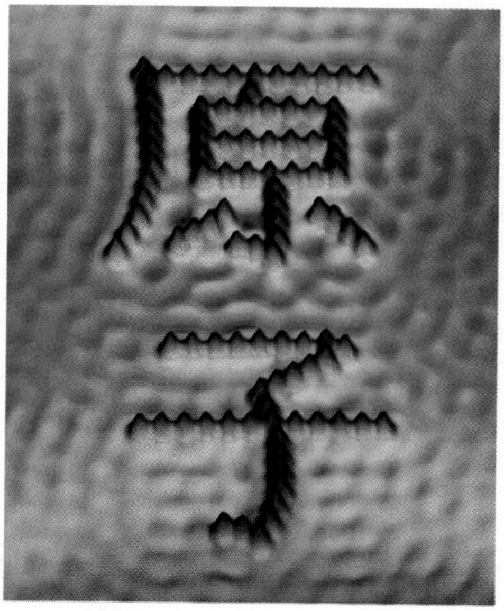

Image reproduced by permission of IBM Research, Almaden Research

An STM image of individual iron atoms arranged on a copper surface. The iron atoms form the Chinese characters for "atom."

direction of this electron flow (the current) depend on the applied voltage, the distance between probe tip and sample, and the identity and location of the nearest sample atom on the surface and its closest neighboring atoms.

The probe tip is attached to a control mechanism that maintains a constant distance between the tip and the sample. The current provides an extremely sensitive measure of the interatomic separation between probe tip and sample. The probe tip is scanned across the surface to form a topographic map of that part of the surface. The STM image, which appears much like a photographic image, shows the locations of atoms on the surface being investigated. The figure on this page shows an STM image of a copper surface with a well-ordered array of iron atoms on it.

The STM has been applied to a wide variety of problems throughout science and engineering. The greatest number of studies have focused on the properties of clean surfaces that have been modified. The STM can also be used to study electrode surfaces in a liquid. Applications of STM to biological molecules on surfaces represent another growing area of scientific research. The STM can be used to move atoms on a surface, and researchers have taken advantage of this capability to generate spectacular images such as the characters in the figure.

The scanning tunneling microscope was invented by Gerd Binnig and Heinrich Rohrer at IBM, Zurich, in 1981, and they shared a Nobel Prize in Physics in 1986 for this work.

The **atomic force microscope** (AFM) is a close relative of the STM. AFM imaging is done by bringing a very small ceramic or semiconductor tip into contact with the surface being studied. The tip, mounted at one end of a tiny, flexible cantilever, is deflected up or down by forces between the tip and the surface that depend on the identities of the atoms or molecules at the surface. The extent of deflection is measured with a laser beam. To generate a topographical map of the bumps and grooves of the surface, the tip is moved systematically over the surface while the tip deflection is measured and graphed versus the position of the tip.

AFM can be used to generate an image of atoms or molecules on any surface, whereas STM requires a conducting surface. Like STM, AFM allows scientists to study atoms and molecules on surfaces in areas that include life sciences, materials science, electrochemistry, polymer science, nanotechnology, and biotechnology.

Individual atoms are too small to be weighed directly; their masses can be measured only by indirect experiments. An atom's mass is on the order of 1×10^{-22} g. For example, a sample of copper that weighs one *nano*gram (1 ng = 1×10^{-9} g) contains about 9×10^{12} copper atoms. High-quality laboratory microbalances can weigh samples of about 0.0000001 g (1×10^{-7} g = 0.1 microgram, μg).

Your height in metric units would be given in *meters,* the metric unit for length. Six feet is equivalent to 1.83 m. Atoms aren't nearly this big. The sizes of atoms are reported in *pico*meters (1 pm = 1×10^{-12} m), and the radius of a typical atom is very small—between 30 and 300 pm. For example, the radius of a copper atom is 126 pm (126×10^{-12} m).

To get a feeling for these dimensions, consider how many copper atoms it would take to form a single file of copper atoms across a U.S. penny with a diameter of 1.90×10^{-2} m. This distance can be expressed in picometers by using a conversion factor **(← *p. 10*)** based on 1 pm = 1×10^{-12} m.

$$1.90 \times 10^{-2}\,\cancel{\text{m}} \times \underbrace{\frac{1\text{ pm}}{1 \times 10^{-12}\,\cancel{\text{m}}}}_{\text{conversion factor}} = 1.90 \times 10^{10}\text{ pm}$$

Conversion factors are the basis for dimensional analysis, a commonly used problem-solving technique. It is described in detail in Appendix A.2.

Note that the units m (for meters) cancel, leaving the answer in pm, the units we want. A penny is 1.90×10^{10} pm in diameter.

Table 2.2 Some Common Unit Equalities*

Length		Volume		Mass	
1 kilometer	= 0.62137 mile	1 liter (L)	$= 1 \times 10^{-3}$ m^3	1 kilogram	= 1000 gram
1 meter	= 100 centimeter		= 1000 cm^3	1 gram	= 1000 milligram
1 centimeter	= 10 millimeter		= 1000 mL	1 amu	$= 1.66054 \times 10^{-24}$ gram
1 nanometer	$= 1 \times 10^{-9}$ meter		= 1.056710 quart	1 pound	= 453.59237 gram
1 picometer	$= 1 \times 10^{-12}$ meter	1 gallon	= 4 quart		= 16 ounce
1 inch	= 2.54 centimeters (exactly†)	1 quart	= 2 pint	1 tonne (metric ton)	= 1000 kilogram
		1 cubic meter (m^3)	$= 1 \times 10^{3}$ liter		
1 angstrom (Å)	$= 1 \times 10^{-10}$ meter			1 short ton (American)	= 2000 pound

*See Appendix B for other unit equalities.
†This exact equality is important for length conversions.

Every conversion factor can be used in two ways. We just converted meters to picometers by using

$$\frac{1 \text{ pm}}{1 \times 10^{-12} \text{ m}}$$

Picometers can be converted to meters by inverting this conversion factor:

$$8.70 \times 10^{10} \cancel{\text{pm}} \times \frac{1 \times 10^{-12} \text{ m}}{1 \cancel{\text{pm}}} = 8.70 \times 10^{-2} \text{ m} = 0.0870 \text{ m}$$

The number of copper atoms needed to stretch across a penny can be calculated by using a conversion factor linking the penny's diameter in picometers with the diameter of a single copper atom. The diameter of a Cu atom is twice the radius, 2×126 pm = 252 pm. Therefore, the conversion factor is 1 Cu atom per 252 pm, and

Notice how "Cu atom" is included in the conversion to keep track of what kind of atom we are interested in.

$$1.90 \times 10^{10} \cancel{\text{pm}} \times \frac{1 \text{ Cu atom}}{252 \cancel{\text{pm}}} = 7.54 \times 10^{7} \text{ Cu atoms, or } 75{,}400{,}000 \text{ Cu atoms}$$

Thus, it takes more than 75 million copper atoms to reach across the penny's diameter. Atoms are indeed tiny.

In chemistry, the most commonly used length units are the centimeter, the millimeter, the nanometer, and the picometer. The most commonly used mass units are the kilogram, gram, and milligram. The relationships among these units and some other units are given in Table 2.2.

Problem-Solving Examples 2.1 and 2.2 illustrate the use of dimensional analysis in unit conversion problems. Notice that in these examples, and throughout the book, the answers are given before the strategy and explanation of how the answers are found. We urge you to first try to answer the problem on your own. Then, check to see whether your answer is correct. If it does not match, try again. Finally, read the explanation, which usually also includes the strategy for solving the problem. If your answer is correct, but your reasoning differs from the explanation, you might have discovered an alternative way to solve the problem.

The **PROBLEM-SOLVING STRATEGY** in this book is

- Analyze the problem
- Plan a solution
- Execute the plan
- Check that the result is reasonable

Appendix A.1 explains this in detail.

PROBLEM-SOLVING EXAMPLE 2.1 Conversion of Units

A medium-sized paperback book has a mass of 530. g. What is the book's mass in kilograms and in pounds?

Answer 0.530 kg and 1.17 lb

Strategy and Explanation Our strategy is to use conversion factors that relate what we know to what we are trying to calculate. For this problem, we use conversion factors derived from Table 2.2 and metric prefixes from Table 2.1 to calculate the answer, being sure to set up the calculation so that only the desired unit remains. The relationship between grams and kilograms is 1 kg = 10^3 g, so 530. g is

$$530.\ \cancel{g} \times \frac{1\ \text{kg}}{10^3\ \cancel{g}} = 530. \times 10^{-3}\ \text{kg} = 0.530\ \text{kg}$$

One pound equals 453.6 g, so we convert grams to pounds as follows:

$$530.\ \cancel{g} \times \frac{1\ \text{lb}}{453.6\ \cancel{g}} = 1.17\ \text{lb}$$

☑ **Reasonable Answer Check** There are about 2.2 lb per kg, and the book's mass is about ½ kg. Thus its mass in pounds should be about 2.2/2 = 1.1, which is close to our more accurate answer.

On September 23, 1999, the NASA Mars Climate Orbiter spacecraft approached too close and burned up in Mars's atmosphere because of a navigational error due to a failed translation of English units into metric units by the spacecraft's software program.

PROBLEM-SOLVING PRACTICE 2.1

(a) A car requires 10. gallons to fill its gas tank. How many liters is this?
(b) An American football field is 100. yards long. How many meters is this?

PROBLEM-SOLVING PRACTICE answers are provided at the back of this book in Appendix K.

PROBLEM-SOLVING EXAMPLE 2.2 Nanoscale Distances

Consider an STM image, similar to the one on p. 47, that contains an image of an object that is 3.0 nm wide. How many such objects could fit in a distance of 1.00 mm (about the width of the head of a pin)?

Answer 3.3×10^5 objects

Strategy and Explanation Our strategy uses conversion factors to relate what we know to what we want to calculate. A nanometer is 1×10^{-9} m and a millimeter is 1×10^{-3} m. The width of the object in meters is

$$3.0\ \cancel{\text{nm}} \times \frac{1 \times 10^{-9}\ \text{m}}{1\ \cancel{\text{nm}}} = 3.0 \times 10^{-9}\ \text{m}$$

Therefore, the number of 3.0-nm-wide objects that could fit into 1.00 mm is

$$1.0 \times 10^{-3}\ \cancel{\text{m}} \times \frac{1\ \text{object}}{3.0 \times 10^{-9}\ \cancel{\text{m}}} = 3.3 \times 10^5\ \text{objects}$$

☑ **Reasonable Answer Check** If the objects are 3.0 nm wide, then about one third of 10^9 could fit into 1 m. One mm is 1/1000 of a meter, and multiplying these two estimates gives $(0.33 \times 10^9)(1/1000) = 0.33 \times 10^6$, which is another way of writing the answer we calculated.

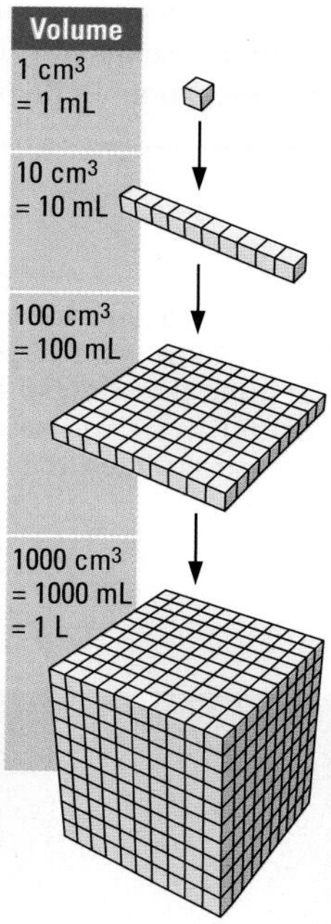

Relative sizes of volume units.

PROBLEM-SOLVING PRACTICE 2.2

Do these conversions using factors based on the equalities in Table 2.2.

(a) How many grams of sugar are in a 5-lb bag of sugar?
(b) Over a period of time, a donor gives 3 pints of blood. How many milliliters (mL) has the donor given?
(c) The same donor's 160-lb body contains approximately 5 L of blood. Considering that 1 L is nearly equal to 1 quart, estimate the percentage of the donor's blood that has been donated in all.

Table 2.2 also lists the liter (L) and milliliter (mL), which are the most common volume units of chemistry. There are 1000 mL in 1 L. One liter is a bit larger than a quart, and a teaspoon of water has a volume of about 5 mL. Chemists often use the terms *milliliter* and *cubic centimeter* (cm^3, or sometimes cc) interchangeably because they are equivalent (1 mL = 1 cm^3).

CHEMISTRY IN THE NEWS The Kilogram Redefined

Six International System of Units (SI) base units—meter (length), second (time), ampere (electric current), kelvin (temperature), mole (amount of substance), candela (light intensity)—are defined by an unvarying physical property of nature. The kilogram (mass) is the only SI base unit still defined by a physical object. A small cylinder of platinum-iridium alloy was declared to be the International Prototype Kilogram (IPK) in 1889. Very accurate copies of this standard kilogram are distributed around the world, including the official U.S. kilogram, K20, which is in a vault at the National Institute of Standards and Technology (NIST), Gaithersburg, MD. In turn, NIST has multiple stainless steel copies of K20 that are used to calibrate laboratory weight sets, balances, etc.

Several methods for defining the kilogram without using a physical object are under investigation. Defining an exact value for Planck's constant (h) or Avogadro's constant would suffice. Currently, these two constants are experimentally measured quantities that contain some very slight experimental uncertainty. One proposal is to set Planck's constant at exactly $6.6260693 \times 10^{-34}$ J s (joule-seconds), which would also set the definition of the kilogram. A second proposal is to set Avogadro's constant at exactly 6.0221418×10^{23}, which would establish the kilogram as the mass of 5.0184515×10^{25} carbon atoms, each containing six protons and six neutrons. Either way, the kilogram would no longer be based on a physical object.

Sources:

Ritter, S. K. "Redefining the Kilogram." *Chemical & Engineering News,* May 26, 2008; p. 43.

Sobel, D. "Within a Secure, Climate-Controlled Vault in France, the Perfect Kilogram Watches Over Every Weight Measurement in the World." *Discovery,* March 2009; p. 28.

Courtesy NIST

Standard kilogram replica. A replica of the standard kilogram is stored in an inert atmosphere, protected by two bell jars.

As illustrated in Problem-Solving Example 2.3, two or more steps of a calculation using dimensional analysis are best written in a single setup and entered into a calculator as a single calculation.

PROBLEM-SOLVING EXAMPLE 2.3 Volume Units

A chemist uses 50. μL (microliters) of a sample for her analysis. What is the volume in mL? In cm^3? In L?

Answer 5.0×10^{-2} mL; 5.0×10^{-2} cm^3; 5.0×10^{-5} L

Strategy and Explanation Use the conversion factors in Table 2.1 that involve micro and milli: 1 μL = 1×10^{-6} L and 1 L = 1000 mL. Multiply the conversion factors to cancel μL and L, leaving only mL.

$$50.\ \cancel{\mu L} \times \frac{1 \times 10^{-6}\ \cancel{L}}{1\ \cancel{\mu L}} \times \frac{10^3\ mL}{1\ \cancel{L}} = 5.0 \times 10^{-2}\ mL$$

Because 1 mL and 1 cm^3 are equivalent, the sample volume can also be expressed as 5.0×10^{-2} cm^3. Since there are 1000 mL in 1 L, the sample volume expressed in liters is 5.0×10^{-5} L.

☑ **Reasonable Answer Check** There is a factor of 1000 between μL and mL, and the final number is a factor of 1000 smaller than the original volume, so the answer is reasonable.

PROBLEM-SOLVING PRACTICE 2.3

A patient's blood cholesterol level measures 165 mg/dL. Express this value in g/L.

© Cengage Learning/Charles D. Winters

Glassware for measuring the volume of liquids.

2.4 Uncertainty and Significant Figures

Measurements always include some degree of uncertainty, because we can never measure quantities exactly or know the value of a quantity with absolute certainty. Scientists have adopted standardized ways of expressing uncertainty in numerical results of measurements.

When the result of a measurement is expressed numerically, the final digit reported is uncertain. The digits we write down from a measurement—both the certain ones and the final, uncertain one—are called **significant figures.** For example, the number 5.025 has four significant figures and the number 4.0 has two significant figures.

To determine the number of significant figures in a measurement, read the number from left to right and count all digits, starting with the first digit that is not zero. All the digits counted are significant except any zeros that are used only to position the decimal point.

Appendix A.3 discusses precision, accuracy, and significant figures in detail.

Example	Number of Significant Figures
1.23 g	3
0.00123 g	3; the zeros to the left of the 1 simply locate the decimal point. The number of significant figures is more obvious if you write numbers in scientific notation. Thus, 0.00123 = 1.23×10^{-3}.
2.0 g and 0.020 g	2; both have two significant figures. When a number is greater than 1, *all zeros to the right of the decimal point are significant.* For a number less than 1, only zeros to the right of the first significant figure are significant.
100 g	1; in numbers that do not contain a decimal point, trailing zeros may or may not be significant. To eliminate possible confusion, the practice followed in this book is to include a decimal point if the zeros are significant. Thus, 100. has three significant figures, while 100 has only one. Alternatively, we can write in scientific notation 1.00×10^2 (three significant figures) or 1×10^2 (one significant figure). For a number written in scientific notation, all digits are significant.
100 cm/m	Infinite number of significant figures, because this is a defined quantity. There are *exactly* 100 centimeters in one meter.
π = 3.1415926 . . .	The value of π is known to a greater number of significant figures than any data you will ever use in a calculation.

In Section 2.5, we discuss the masses of atoms and begin calculating with numbers that involve uncertainty. Significant figures provide a simple means for keeping track of these uncertainties.

PROBLEM-SOLVING EXAMPLE 2.4 Significant Figures

How many significant figures are present in each of these numbers?

(a) 0.0001171 m (b) 26.94 mL (c) 207 cm
(d) 0.7011 g (e) 0.0010 L (f) 12,400. s

Answer

(a) Four (b) Four (c) Three
(d) Four (e) Two (f) Five

Strategy and Explanation Apply the principles of significant figures given in the preceding examples.

(a) The leading zeros do not count, so there are four significant figures.
(b) All four of the digits are significant.
(c) All three digits are significant figures.
(d) Four digits follow the decimal, and all are significant figures.
(e) The leading zeros do not count, so there are two significant figures.
(f) Since there is a decimal point, all five of the digits are significant.

PROBLEM-SOLVING PRACTICE 2.4

Determine the number of significant figures in these numbers: (a) 0.00602 g; (b) 22.871 mg; (c) 344 °C; (d) 100.0 mL; (e) 0.00042 m; (f) 0.002001 L.

Significant Figures in Calculations

When numbers are combined in a calculation, the number of significant figures in the result is determined by the number of significant figures in the starting numbers and the nature of the arithmetic operation being performed.

Addition and Subtraction: *In addition or subtraction, the number of decimal places in the answer equals the number of decimal places in the number with the fewest decimal places.* Suppose you add these three numbers:

0.12	2 significant figures	2 decimal places
1.6	2 significant figures	1 decimal place
10.976	5 significant figures	3 decimal places
12.696 rounds to 12.7		

This sum should be reported as 12.7, a number with one decimal place, because 1.6 has only one decimal place.

Multiplication and Division: *In multiplication or division, the number of significant figures in the answer is the same as that in the quantity with the fewest significant figures.*

$$\frac{0.7608}{0.0546} = 13.9 \text{ or, in scientific notation, } 1.39 \times 10^1$$

The numerator, 0.7608, has four significant figures, but the denominator has only three, so the result must be reported with three significant figures.

Rules for Rounding

The numerical result obtained in many calculations must be rounded to retain the proper number of significant figures. When you round a number to reduce the number of digits, follow these rules:

- The last digit retained is left unchanged if the following digit is less than 5 (4.327 rounded to two significant digits is 4.3)
- The last digit retained is increased by 1 if the following digit is greater than 5 or is a 5 followed by other non-zero digits (4.573 rounded to two significant figures is 4.6)
- If the last digit retained is followed by a single digit 5 only or by a 5 followed by zeroes, then the last digit retained is increased by 1 if it is odd and remains the same if it is even (4.75 rounded to two significant digits is 4.8; 4.850 rounded to two significant digits is 4.8)

Full Number	Number Rounded to Three Significant Digits
12.696	12.7
16.249	16.2
18.35	18.4
18.45	18.4
24.750	24.8
18.351	18.4

One last word regarding significant figures, rounding, and calculations. In working problems on a calculator, you should do the calculation using all the digits allowed by the calculator and round only at the end of the problem. Rounding in the middle of a calculation sequence can introduce small errors that can accumulate later in the calculation. If your answers do not quite agree with those in the appendices of this book, this practice may be the source of the disagreement.

PROBLEM-SOLVING EXAMPLE 2.5 Rounding Significant Figures

Do these calculations and round the result to the proper number of significant figures.

(a) $15.80 + 0.0060 - 2.0 + 0.081$ (b) $\frac{55.0}{12.34}$

(c) $\frac{12.7732 - 2.3317}{5.007}$ (d) $2.16 \times 10^3 + 4.01 \times 10^2$

Answer

(a) 13.9 (b) 4.46 (c) 2.085 (d) 2.56×10^3

Strategy and Explanation In each case, do the calculation with no rounding. Then, apply the rules for rounding to the answer.

(a) 13.887 is rounded to 13.9 with one decimal place because 2.0 has one decimal place.

(b) 4.457 is rounded to 4.46 with three significant figures just as in 55.0, which also has three significant figures.

(c) The number of significant figures in the result is governed by the least number of significant figures in the quotient. The denominator, 5.007, has four significant figures, so the calculator result, 2.08538 . . . , is properly expressed with four significant figures as 2.085.

(d) We change 4.01×10^2 to 0.401×10^3 and then sum 2.16×10^3 plus 0.401×10^3 and round to three significant figures to get 2.56×10^3.

PROBLEM-SOLVING PRACTICE 2.5

Do these calculations and round the result to the proper number of significant figures.

(a) $244.2 + 0.1732$ (b) 6.19×5.2222 (c) $\dfrac{7.2234 - 11.3851}{4.22}$

2.5 Atomic Numbers and Mass Numbers

Experiments done early in the 20th century found that *all atoms of the same element have the same number of protons in their nuclei.* This number is called the **atomic number** of the element and is given the symbol *Z*. In the periodic table on the inside front cover of this book, the atomic number for each element is written above the element's symbol. For example, a copper atom has a nucleus containing 29 protons, so its atomic number is 29 ($Z = 29$). A lead atom (Pb) has 82 protons in its nucleus, so the atomic number for lead is 82.

The periodic table is organized by atomic number; it is discussed in Section 2.9.

The atomic number of each element is unique.

Atomic masses are established relative to a standard, the mass of a carbon atom that has six protons and six neutrons in its nucleus. The masses of atoms of every other element are compared to the mass of this carbon atom, carbon-12, which is defined as having a mass of exactly 12 atomic mass units. In terms of macroscale mass units, 1 **atomic mass unit (amu)** = 1.66054×10^{-24} g. For example, when an experiment shows that a gold atom, on average, is 16.4 times as massive as the standard carbon atom, we then know the mass of the gold atom in amu and grams.

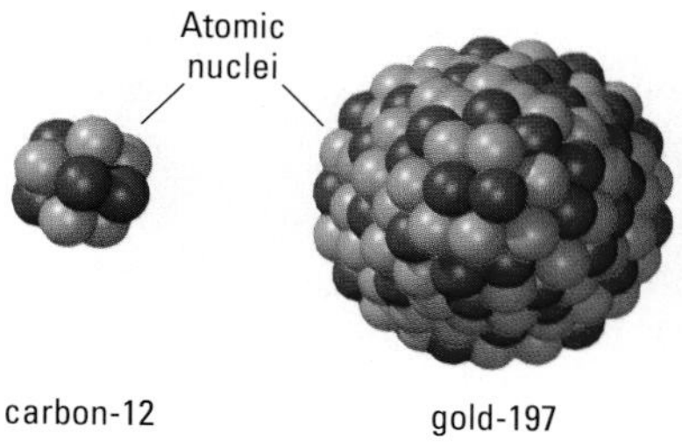

$$16.4 \times 12 \text{ amu} = 197 \text{ amu}$$

$$197 \cancel{\text{amu}} \times \frac{1.66054 \times 10^{-24} \text{ g}}{1 \cancel{\text{amu}}} = 3.27 \times 10^{-22} \text{ g}$$

The masses of the fundamental subatomic particles in atomic mass units have been determined experimentally. The proton and the neutron have masses very close to 1 amu, whereas the electron's mass is approximately 1800 times smaller.

1 atomic mass unit (amu) = $\frac{1}{12}$ the mass of a carbon atom having six protons and six neutrons in the nucleus.

Particle	Mass (grams)	Mass (atomic mass units)	Charge
Electron	$9.10938215 \times 10^{-28}$	0.000548579	1−
Proton	$1.672621637 \times 10^{-24}$	1.00728	1+
Neutron	$1.674927211 \times 10^{-24}$	1.00866	0

A relative scale of atomic masses allows us to estimate the mass of any atom whose nuclear composition is known. The proton and neutron have masses so close to 1 amu that the difference can be ignored in an estimate. Electrons have much less mass than protons or neutrons. Even though the number of electrons in an atom must equal the number of protons, the mass of the electrons is so small that it never affects the atomic mass by more than 0.1%, so the electrons' mass need not be considered. To estimate an atom's mass, we add up its number of protons and neutrons. This sum, called the **mass number** *of that particular atom,* is given the symbol *A*. For example, a copper atom that has 29 protons and 34 neutrons in its nucleus has a mass number, *A*, of 63. A lead atom that has 82 protons and 126 neutrons has $A = 208$.

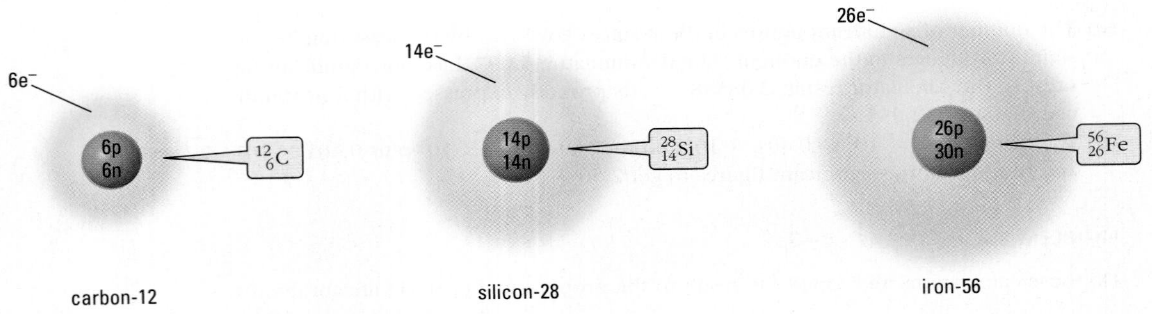

With this information, an atom of known composition, such as a lead-208 atom, can be represented as follows:

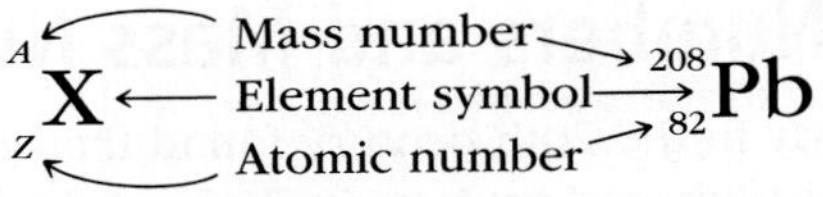

Each element has its own unique one- or two-letter symbol. Because each element is defined by the number of protons each of its atoms contains, knowing which element you are dealing with means you automatically know the number of protons its atoms have. Thus, the *Z* part of the notation is redundant. For example, the lead atom might be represented by the symbol ^{208}Pb because the Pb tells us the element is lead, and lead atoms by definition always contain 82 protons. Whether we use the symbol or the alternative notation lead-208, we simply say "lead-208."

PROBLEM-SOLVING EXAMPLE 2.6 Atomic Nuclei

Iodine-131 is used in medicine to assess thyroid gland function. How many protons and neutrons are present in an iodine-131 atom?

Answer 53 protons and 78 neutrons

Strategy and Explanation The periodic table inside the front cover of this book shows that the atomic number of iodine (I) is 53. Therefore, the atom has 53 protons in its nucleus. Because the mass number of the atom is the sum of the number of protons and neutrons in the nucleus,

$$\text{Mass number} = \text{number of protons} + \text{number of neutrons}$$

$$131 = 53 + \text{number of neutrons}$$

$$\text{Number of neutrons} = 131 - 53 = 78$$

PROBLEM-SOLVING PRACTICE 2.6

(a) What is the mass number of a phosphorus atom with 16 neutrons?
(b) How many protons, neutrons, and electrons are there in a neon-22 atom?
(c) Write the symbol for the atom with 82 protons and 125 neutrons.

The actual mass of an atom is slightly less than the sum of the masses of its protons, neutrons, and electrons. The difference, known as the mass defect, is related to the energy that binds nuclear particles together, a topic discussed in Chapter 20.

Although an atom's mass approximately equals its mass number, the actual mass is not an integral number. For example, the actual mass of a gold-196 atom is 195.96655 amu, slightly less than the mass number 196. The masses of atoms are determined experimentally using mass spectrometers (see Figure 2.4). The *Tools of Chemistry* discussion illustrates the use of a mass spectrometer to determine the atomic masses of neon atoms.

Mass spectrometric analysis of most naturally occurring elements reveals that not all atoms of an element have the same mass. For example, all silicon atoms have 14 protons, but some silicon nuclei have 14 neutrons, others have 15, and others have 16. Thus, naturally occurring silicon (atomic number 14) is always a mixture

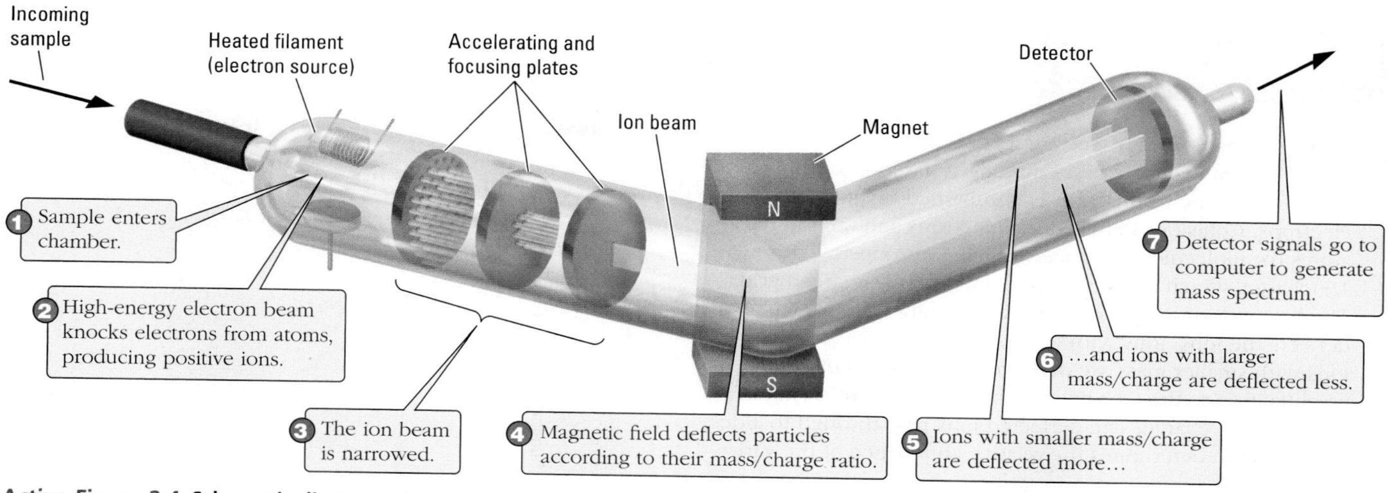
Incoming sample
Heated filament (electron source)
Accelerating and focusing plates
Ion beam
Magnet
Detector
N
S
1 Sample enters chamber.
2 High-energy electron beam knocks electrons from atoms, producing positive ions.
3 The ion beam is narrowed.
4 Magnetic field deflects particles according to their mass/charge ratio.
5 Ions with smaller mass/charge are deflected more…
6 …and ions with larger mass/charge are deflected less.
7 Detector signals go to computer to generate mass spectrum.

Active Figure 2.4 Schematic diagram of a mass spectrometer. This analytical instrument uses a magnetic field to measure atomic and molecular masses directly. Visit this book's companion website at **www.cengage.com/chemistry/moore** to test your understanding of the concepts in this figure.

of silicon-28, silicon-29, and silicon-30 atoms. Such different atoms of the *same* element are called **isotopes**—atoms with the *same* atomic number (*Z*) but different mass numbers (*A*). Naturally occurring silicon has three isotopes:

Isotope	^{28}Si	^{29}Si	^{30}Si
Atomic Number (*Z*), Protons	14	14	14
Neutrons	14	15	16
Mass Number (*A*), Protons + Neutrons	28	29	30

Two elements can't have the same atomic number. If two atoms differ in their number of protons, they are atoms of different elements. If only their number of neutrons differs, they are isotopes of a single element, such as neon-20, neon-21, and neon-22.

PROBLEM-SOLVING EXAMPLE 2.7 Isotopes

Boron has two isotopes, one with five neutrons and the other with six neutrons. What are the mass numbers and symbols of these isotopes?

Answer The mass numbers are 10 and 11. The symbols are $^{10}_{5}B$ and $^{11}_{5}B$.

Strategy and Explanation We use the entry for boron in the periodic table to find the answer. Boron has an atomic number of 5, so it has five protons in its nucleus. Therefore, the mass numbers of the two isotopes are given by the sum of their numbers of protons and neutrons:

$$\text{Isotope 1: B} = 5 \text{ protons} + 5 \text{ neutrons} = 10 \text{ (boron-10)}$$

$$\text{Isotope 2: B} = 5 \text{ protons} + 6 \text{ neutrons} = 11 \text{ (boron-11)}$$

Placing the atomic number at the bottom left and the mass number at the top left gives the symbols $^{10}_{5}B$ and $^{11}_{5}B$.

PROBLEM-SOLVING PRACTICE 2.7

Naturally occurring magnesium has three isotopes with 12, 13, and 14 neutrons. What are the mass numbers and symbols of these three isotopes?

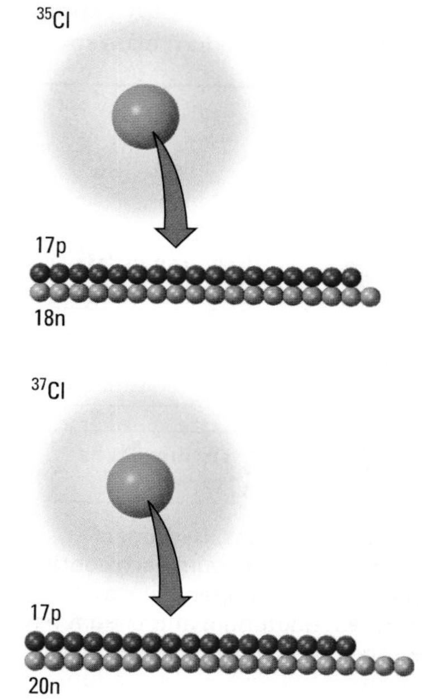
35Cl
17p
18n
37Cl
17p
20n

Chlorine isotopes. Chlorine-35 and chlorine-37 atoms each contain 17 protons; chlorine-35 atoms have 18 neutrons, and chlorine-37 atoms contain 20 neutrons.

We usually refer to a particular isotope by giving its mass number. For example, $^{238}_{92}U$ is referred to as uranium-238. But a few isotopes have distinctive names and symbols because of their importance, such as the isotopes of hydrogen. All hydrogen isotopes have just one proton. When the single proton is the only nuclear particle, the element is simply called hydrogen. With one neutron as well as one proton

TOOLS OF CHEMISTRY Mass Spectrometer

A **mass spectrometer** (see Figure 2.4) is used to measure atomic and molecular masses directly. A gaseous sample of the substance being analyzed is bombarded by high-energy electrons. Collisions between the electrons and the sample's atoms (or molecules) produce positive ions, mostly with 1+ charge, which are attracted to a negatively charged grid. The beam of ions is passed through a magnetic field, which deflects the ions. Ions with larger mass are deflected less; ions with smaller mass are deflected more. This deflection essentially sorts the ions by mass because most of them have the same 1+ charge. The deflected ions pass through to a detector, which determines the relative abundance of the various ions in the sample.

In practice, the mass spectrometer settings are varied to focus ions of different masses on the stationary detector at different times. The mass spectrometer records the current of ions (the ion abundance) as the magnetic field is varied systematically. After processing by software, the data are plotted as a graph of the ion abundance versus the mass number of the ions, which is a **mass spectrum.**

The means of measuring the mass spectrum of neon is shown in Figure 2.4 and the resulting mass spectrum is shown in the figure here. The beam of Ne^+ ions passing through the mass spectrometer is divided into three segments because three isotopes are present: ^{20}Ne, with an atomic mass of 19.9924 amu and an abundance of 90.92%; ^{21}Ne, with an atomic mass of 20.9940 and an abundance of only 0.26%; and ^{22}Ne, with an atomic mass of 21.9914 amu and an abundance of 8.82%.

The mass spectrometer described here is a simple one based on magnetic field deflection of the ions, and it is similar to those used in early experiments to determine isotopic abundances. Modern mass spectrometers operate on quite different principles, although they generate similar mass spectra. These instruments are used to measure the masses of molecules as well as atoms. In addition, mass spectrometers are used to investigate details of molecular structure of compounds ranging in complexity from simple organic and inorganic compounds to biomolecules such as proteins.

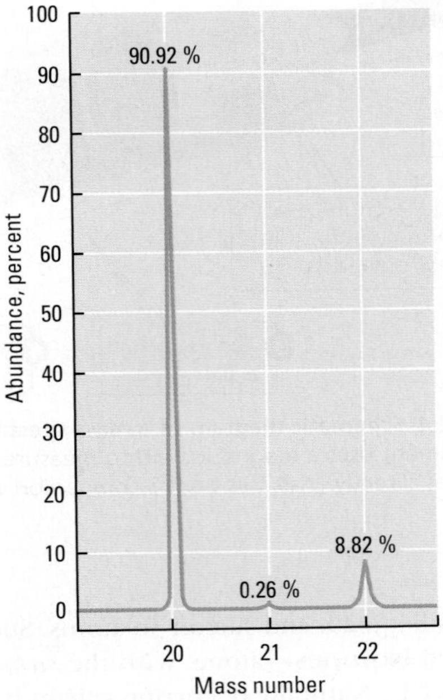

Mass spectrum of neon. The principal peak corresponds to the most abundant isotope, neon-20. The height of each peak indicates the percent relative abundance of each isotope.

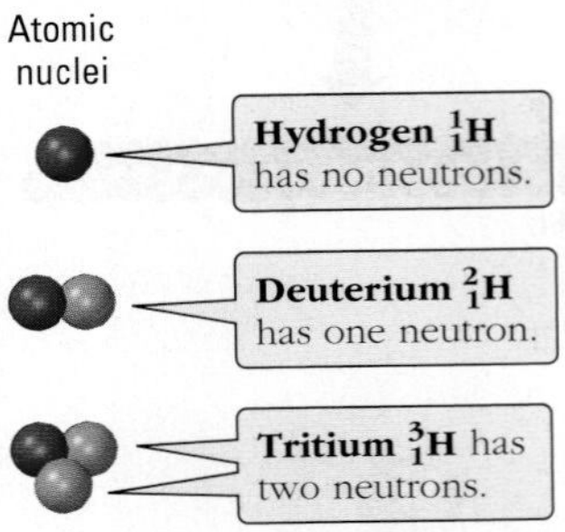

Hydrogen isotopes. Hydrogen, deuterium, and tritium each contain one proton. Hydrogen has no neutrons; deuterium and tritium have one and two neutrons, respectively.

present, the isotope ^2_1H is called either deuterium or heavy hydrogen (symbol D). When two neutrons are present, the isotope ^3_1H is called tritium (symbol T).

CONCEPTUAL EXERCISE 2.3 Isotopes

A student in your chemistry class tells you that nitrogen-14 and nitrogen-15 are not isotopes because they have the same number of protons. How would you refute this statement?

2.6 Isotopes and Atomic Weight

Copper has two naturally occurring isotopes, copper-63 and copper-65, that differ by two neutrons. These isotopes have atomic masses of 62.9296 amu and 64.9278 amu, respectively. In a macroscopic collection of naturally occurring copper atoms, the average mass of the atoms is neither 63 (all copper-63) nor 65 (all copper-65). Rather, the average atomic mass will fall between 63 and 65, with its exact value depending on the proportion of each isotope in the mixture. The proportion of atoms of each isotope in a natural sample of an element is called the **percent abundance,** the percentage of atoms of a *particular* isotope.

The concept of percent is widely used in chemistry, and it is worth briefly reviewing here. For example, Earth's atmosphere contains approximately 78% nitrogen, 21% oxygen, and 1% argon. U.S. pennies minted after 1982 contain 2.4% copper; the remainder is zinc.

PROBLEM-SOLVING EXAMPLE 2.8 Applying Percent

The U.S. Mint issued state quarters over the ten-year period 1999–2008. The quarters each weigh 5.670 g and contain 8.33% nickel and the remainder copper. What mass of each element is contained in each quarter?

Answer 0.472 g nickel and 5.20 g copper

Strategy and Explanation We need the mass of each element in each quarter. We start by calculating the mass of nickel. Its percentage, 8.33%, means that every 100. g of coin contains 8.33 g nickel.

$$5.670 \text{ g } \cancel{\text{quarter}} \times \frac{8.33 \text{ g nickel}}{100. \text{ g } \cancel{\text{quarter}}} = 0.472 \text{ g nickel}$$

The mass of copper is found the same way, using the conversion factor 91.67 g copper per 100. g of quarter.

$$5.670 \text{ g } \cancel{\text{quarter}} \times \frac{91.67 \text{ g copper}}{100. \text{ g } \cancel{\text{quarter}}} = 5.198 \text{ g copper}$$

We could have obtained this value directly by recognizing that the masses of nickel and copper must sum to the mass of the quarter. Therefore,

$$5.670 \text{ g quarter} = 0.472 \text{ g nickel} + x \text{ g copper}$$

Solving for x, the mass of copper, gives

$$5.670 \text{ g} - 0.472 \text{ g} = 5.198 \text{ g copper}$$

☑ **Reasonable Answer Check** The ratio of nickel to copper is about 11:1 (91.67/8.33 = 11), and the ratio of the masses calculated is also about 11:1 (5.198/0.472 = 11), so the answer is reasonable.

PROBLEM-SOLVING PRACTICE 2.8

Many heating devices such as hair dryers contain nichrome wire, an alloy containing 80.% nickel and 20.% chromium, which gets hot when an electric current passes through it. If a heating device contains 75 g nichrome wire, how many grams of nickel and how many grams of chromium does the wire contain?

The Pennsylvania quarter. It shows the statue "Commonwealth," an outline of the state, the state motto, and a keystone.

The sum of the percentages for the composition of a sample must be 100%.

The percent abundance of each isotope in a sample of an element is given as follows:

$$\text{Percent abundance} = \frac{\text{number of atoms of a given isotope}}{\text{total number of atoms of all isotopes of that element}} \times 100\%$$

Table 2.3 gives information about the percent abundance for naturally occurring isotopes of hydrogen, boron, and bromine. The percent abundance and isotopic mass of each isotope can be used to find the average mass of atoms of that element, and this average mass is called the atomic weight of the element. The **atomic weight** of an element is the average mass of a representative sample of atoms of the element, expressed in atomic mass units.

The mass number and atomic weight of an element are *not* the same. Mass number is the sum of the number of protons plus neutrons for a particular isotope. The atomic weight of an element depends on the masses and relative abundances of all naturally occurring isotopes of that element.

Boron, for example, is a relatively rare element present in compounds used in laundry detergents, mild antiseptics, and Pyrex cookware. It has two naturally occurring isotopes: boron-10, with a mass of 10.0129 amu and 19.91% abundance, and

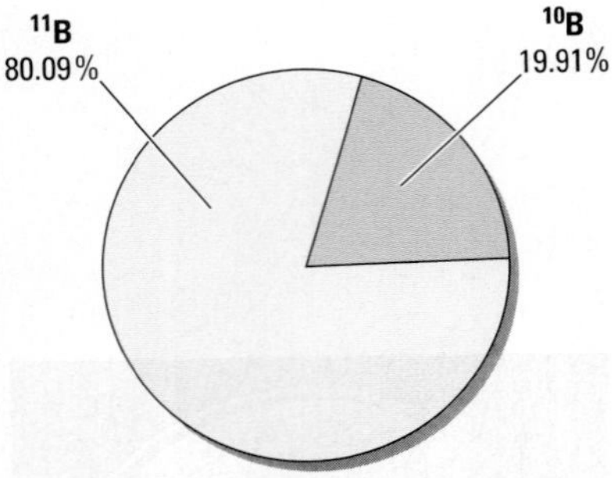

^{11}B
80.09%
^{10}B
19.91%

Percent abundance of boron-10 and boron-11.

Table 2.3 Isotopic Masses of the Stable Isotopes of Hydrogen, Boron, and Bromine

Element	Symbol	Atomic Weight (amu)	Mass Number	Isotopic Mass (amu)	Percent Abundance
Hydrogen	H	1.00794	1	1.007825	99.9855
	D		2	2.0141022	0.0145
Boron	B	10.811	10	10.012937	19.91
			11	11.009305	80.09
Bromine	Br	79.904	79	78.918336	50.69
			81	80.916289	49.31

boron-11, with a mass of 11.0093 amu and 80.09% abundance. Since the abundances are approximately 20% and 80%, respectively, you can estimate the atomic weight of boron: 20 atoms out of every 100, or 2 atoms out of every 10, are boron-10. If you then add up the masses of 10 atoms, you have 2 atoms with a mass of about 10 amu and 8 atoms with a mass of about 11 amu, so the sum is 108 amu, and the average is 108 amu/10 = 10.8 amu. This approximation is about right when you consider that the mass numbers of the boron isotopes are 10 and 11 and that boron is 80% boron-11 and only 20% boron-10. Therefore, the atomic weight should be about two tenths of the way down from 11 to 10, or 10.8. Thus, each atomic weight is a *weighted* average that accounts for the proportion of each isotope, not just the usual arithmetic average in which the values are simply summed and divided by the number of values.

In general, the atomic weight of an element is found from the percent abundance data as shown by this more exact calculation for boron. The mass of each isotope is multiplied by its fractional abundance, the percent abundance expressed as a decimal, to calculate the weighted average, the atomic weight.

1 amu = $\frac{1}{12}$ mass of carbon-12 atom

$$\begin{aligned}\text{Atomic weight} &= [(\text{fractional abundance } ^{10}\text{B})(\text{isotopic mass } ^{10}\text{B}) \\ &\quad + (\text{fractional abundance } ^{11}\text{B})(\text{isotopic mass } ^{11}\text{B})] \\ &= (0.1991)(10.0129 \text{ amu}) + (0.8009)(11.0093 \text{ amu}) \\ &= 10.81 \text{ amu}\end{aligned}$$

Our earlier estimate was quite close to the more exact result. The arithmetic average of the isotopic masses of boron is (10.0129 + 11.093)/2 = 10.55, which is quite different from the actual atomic weight.

The term "atomic weight" is so commonly used that it has become accepted, even though it is really a mass rather than a weight.

The atomic weight of each stable (nonradioactive) element has been determined; these values appear in the periodic table in the inside front cover of this book. For most elements, the abundances of the isotopes are the same no matter where a sample is collected. Therefore, the atomic weights in the periodic table are used whenever an atomic weight is needed. In the periodic table, each element's box contains the atomic number, the symbol, and the weighted average atomic weight. For example, the periodic table entry for zinc is

Entry	
30	← Atomic number
Zn	← Symbol
65.38	← Atomic weight

Astrid & Hamms-Frieder Michler/ Photo Researchers, Inc.

Elemental zinc.

EXERCISE **2.4 Atomic Weight**

Verify that the atomic weight of lithium is 6.941 amu, given this information:

$^{6}_{3}$Li mass = 6.015121 amu and percent abundance = 7.500%

$^{7}_{3}$Li mass = 7.016003 amu and percent abundance = 92.50%

CONCEPTUAL EXERCISE **2.5 Isotopic Abundance**

Naturally occurring magnesium contains three isotopes: ^{24}Mg (78.70%), ^{25}Mg (10.13%), and ^{26}Mg (11.17%). Estimate the atomic weight of Mg and compare your estimate with the atomic weight calculated by finding the arithmetic average of the atomic masses. Which value is larger? Why is it larger?

CONCEPTUAL EXERCISE **2.6 Percent Abundance**

Gallium has two abundant isotopes, and its atomic weight is 69.72 amu. If you knew only this value and not the percent abundance of the isotopes, make the case that the percent abundance of each of the two gallium isotopes cannot be 50%.

2.7 Amounts of Substances: The Mole

go Chemistry **Module 4: The Mole** covers concepts in this section.

As noted earlier, atoms are much too small to be seen directly or weighed individually on the most sensitive laboratory balance. However, when working with chemical reagents, it is essential to know how many atoms, molecules, or other nanoscale units of an element or compound you have. To connect the macroscale world, where chemicals can be manipulated and weighed, to the nanoscale world of individual atoms or molecules, chemists have defined a convenient unit of matter that contains a known number of particles. This chemical counting unit is the **mole (mol),** defined as the amount of substance that contains as many atoms, molecules, ions, or other nanoscale units as there are atoms in *exactly* 12 g of carbon-12.

One mole of carbon has a mass of 12.01 g, not exactly 12 g, because naturally occurring carbon contains both carbon-12 (98.89%) and carbon-13 (1.11%). By definition, one mole of carbon-12 has a mass of exactly 12 g.

The term "mole" is derived from the Latin word *moles* meaning a "heap" or "pile."

The mole is the connection between the macroscale and nanoscale worlds, the visible and the not directly visible.

The essential point to understand about moles is that ***one mole always contains the same number of particles,*** no matter what substance or what kind of particles we are talking about. The number of particles in a mole is

$$1 \text{ mol} = 6.02214179 \times 10^{23} \text{ particles}$$

When used with a number, mole is abbreviated mol, for example, 0.5 mol.

The number of particles in a mole is known as **Avogadro's number** after Amadeo Avogadro (1776–1856), an Italian physicist who conceived the basic idea but never experimentally determined the number, which came later. It is important to realize that the value of Avogadro's number is a definition tied to the number of atoms in 12 g of carbon-12.

$$\text{Avogadro's number} = 6.02214179 \times 10^{23} \text{ per mole} = 6.02214179 \times 10^{23} \text{ mol}^{-1}$$

One difficulty in comprehending Avogadro's number is its sheer size. Writing it out fully yields

$$6.02214179 \times 10^{23} = 602{,}214{,}179{,}000{,}000{,}000{,}000{,}000$$

or 602,214.179 × 1 million × 1 million × 1 million. Although Avogadro's number is known to nine significant figures, we will most often use it rounded to 6.022×10^{23}. There are many analogies used to try to give a feeling for the size of this number. If you poured Avogadro's number of marshmallows over the continental United States, the marshmallows would cover the country to a depth of approximately 650 miles. Or, if one mole of pennies were divided evenly among every man, woman, and child in the United States, your share alone would pay off the national debt (about \$11 trillion, or $\$11 \times 10^{12}$) about twice.

You can think of the mole simply as a counting unit, analogous to the counting units we use for ordinary items such as doughnuts or bagels by the dozen, shoes by the pair, or sheets of paper by the ream (500 sheets). Atoms, molecules, and other particles in chemistry are counted by the mole. The different masses of the ele-

Atomic weights are given in the periodic table of elements in the inside front cover of this book.

ments shown in Figure 2.5 each contain one mole of atoms. For each element in the figure, the mass in grams (the macroscale) is numerically equal to the atomic weight in atomic mass units (the nanoscale). *The* **molar mass** *of any substance is the mass, in grams, of one mole of that substance.* Molar mass has the units of grams per mole (g/mol).

For example,

$$\begin{aligned} \text{molar mass of copper (Cu)} &= \text{mass of 1 mol Cu atoms} \\ &= \text{mass of } 6.022 \times 10^{23} \text{ Cu atoms} \\ &= 63.546 \text{ g/mol} \\ \text{molar mass of aluminum (Al)} &= \text{mass of 1 mol Al atoms} \\ &= \text{mass of } 6.022 \times 10^{23} \text{ Al atoms} \\ &= 26.9815 \text{ g/mol} \end{aligned}$$

Cu 63.55 g
Al 26.98 g
Pb 207.2 g
S 32.07 g
Mg 24.31 g
Cr 52.00 g

© Cengage Learning/Charles D. Winters

Figure 2.5 **One-mole quantities of six elements.**

Each molar mass of copper or aluminum contains Avogadro's number of atoms. *Molar mass differs from one element to the next because the atoms of different elements have different masses.* Think of a mole as analogous to a dozen. We could have a dozen golf balls, a dozen baseballs, or a dozen bowling balls, 12 items in each case. The dozen items do not weigh the same, however, because the individual items do not weigh the same: 45 g per golf ball, 134 g per baseball, and 7200 g per bowling ball. In a similar way, the mass of Avogadro's number of atoms of one element is different from the mass of Avogadro's number of atoms of another element because the atoms of different elements differ in mass.

ESTIMATION The Size of Avogadro's Number

Chemists and other scientists often use estimates in place of exact calculations when they want to know the approximate value of a quantity. Analogies to help us understand the extremely large value of Avogadro's number are an example.

If 1 mol green peas were spread evenly over the continental United States, how deep would the layer of peas be? The surface area of the continental United States is about 3.0×10^6 square miles (mi^2). There are 5280 feet per mile.

Let's start with an estimate of a green pea's size: $\frac{1}{4}$-inch diameter. Then 4 peas would fit along a 1-inch line, and 48 would fit along a 1-foot line, and $48^3 = 110{,}592$ would fit into 1 cubic foot (ft^3). Since we are estimating, we will approximate by saying 1×10^5 peas per cubic foot.

Now, let's estimate how many cubic feet of peas are in 1 mol peas:

$$\frac{6.022 \times 10^{23} \text{ peas}}{1 \text{ mol peas}} \times \frac{1 \text{ ft}^3}{1 \times 10^5 \text{ peas}} \cong \frac{6.0 \times 10^{18} \text{ ft}^3}{1 \text{ mol peas}}$$

(The "approximately equal" sign, $\cong$, is an indicator of these approximations.) The surface area of the continental United States is 3.0×10^6 square miles (mi^2), which is about

$$3.0 \times 10^6 \text{ mi}^2 \times \left(\frac{5280 \text{ ft}}{1 \text{ mi}}\right)^2 \cong 8.4 \times 10^{13} \text{ ft}^2$$

so 1 mol peas spread evenly over this area has a depth of

$$\frac{6.0 \times 10^{18} \text{ ft}^3}{1 \text{ mol peas}} \times \frac{1}{8.4 \times 10^{13} \text{ ft}^2} \cong \frac{7.1 \times 10^4 \text{ ft}}{1 \text{ mol peas}}$$

or

$$\frac{7.1 \times 10^4 \text{ ft}}{1 \text{ mol peas}} \times \frac{1 \text{ mi}}{5280 \text{ ft}} \cong \frac{14 \text{ mi}}{1 \text{ mol peas}}$$

Note that in many parts of the estimate, we rounded or used fewer significant figures than we could have used. Our purpose was to *estimate* the final answer, not to compute it exactly. The final answer, 14 miles, is not particularly accurate, but it is a valid estimate. The depth would be more than 10 miles but less than 20 miles. It would not be 6 inches or even 6 feet. Estimating served the overall purpose of developing the analogy for understanding the size of Avogadro's number.

Visit this book's companion website at **www.cengage.com/chemistry/moore** to work an interactive module based on this material.

2.8 Molar Mass and Problem Solving

go Chemistry **Module 4: The Mole** covers concepts in this section.

Understanding the idea of a mole and applying it properly are *essential* to doing quantitative chemistry. In particular, it is *absolutely necessary* to be able to make two basic conversions: *moles* → *mass* and *mass* → *moles*. To do these and many other calculations in chemistry, it is most helpful to use dimensional analysis in the same way it is used in unit conversions. Along with calculating the final answer, write the units with all quantities in a calculation and cancel the units. If the problem is set up properly, the answer will have the desired units.

Let's see how these concepts apply to converting mass to moles or moles to mass. *In either case, the conversion factor is provided by the molar mass of the substance, the number of grams in one mole, that is, grams per mole (g/mol).*

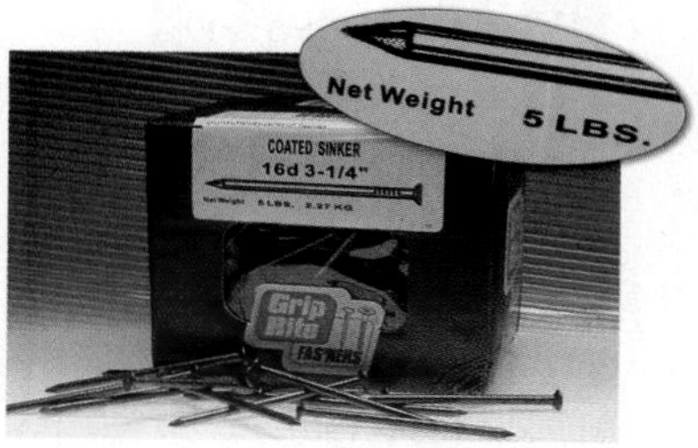

© Cengage Learning/Charles D. Winters

Items can be counted by weighing. Knowing the mass of one nail, we can estimate the number of nails in this 5-lb box.

Mass ⇌ Moles conversions for substance A

Mass A ⟶ moles A

$$\text{Grams A} \times \frac{1\ \text{mol A}}{\text{grams A}} = \text{moles A}$$

where $\frac{1\ \text{mol A}}{\text{grams A}} = \frac{1}{\text{molar mass}}$

Moles A ⟶ mass A

$$\text{Moles A} \times \frac{\text{grams A}}{1\ \text{mol A}} = \text{grams A}$$

where $\frac{\text{grams A}}{1\ \text{mol A}}$ = molar mass

Suppose you need 0.250 mol Cu for an experiment. How many grams of Cu should you use? The atomic weight of Cu is 63.546 amu, so the molar mass of Cu is 63.546 g/mol. To calculate the mass of 0.250 mol Cu, you need the conversion factor 63.546 g Cu/1 mol Cu.

$$0.250\ \cancel{\text{mol Cu}} \times \frac{63.55\ \text{g Cu}}{1\ \cancel{\text{mol Cu}}} = 15.9\ \text{g Cu}$$

In this book we will, when possible, *use one more significant figure in the molar mass than in any of the other data in the problem.* In the problem just completed, note that we used four significant figures in the molar mass of Cu when three were given in the number of moles. Using one more significant figure in the molar mass guarantees that its precision is greater than that of the other numbers and does not limit the precision of the result of the computation.

Frequently, a problem requires converting a mass to the equivalent number of moles, such as calculating the number of moles of bromine in 10.00 g of bromine. Because bromine is a diatomic element, it consists of Br_2 molecules. Therefore, there are 2 mol Br atoms in 1 mol Br_2 molecules. The molar mass of Br_2 is twice its atomic mass, 2 × 79.904 g/mol = 159.81 g/mol. To calculate the moles of bromine in 10.00 g of Br_2, use the molar mass of Br_2 as the conversion factor, 1 mol Br_2/159.81 g Br_2.

$$10.00\ \cancel{\text{g Br}_2} \times \frac{1\ \text{mol Br}_2}{159.81\ \cancel{\text{g Br}_2}} = 6.257 \times 10^{-2}\ \text{mol Br}_2$$

PROBLEM-SOLVING EXAMPLE 2.9 Mass and Moles

(a) Copper is an important metal in the world economy. How many moles of Cu are in a 500.-g sample of the pure metal?

(b) Zinc is also an important metal commercially. A sample of Zn contains 6.00 mol Zn. Is the mass of Zn greater or less than the mass of Cu in part (a)?

Answer (a) 7.87 mol Cu (b) 392 g Zn, which is less than the mass of Cu

Strategy and Explanation

(a) Convert the mass of Cu to moles using the molar mass of Cu as the conversion factor.

$$500.\ \cancel{\text{g Cu}} \times \frac{1\ \text{mol Cu}}{63.546\ \cancel{\text{g Cu}}} = 7.87\ \text{mol Cu}$$

Courtesy of Serge Lachinov

1834–1907

Originally from Siberia, Mendeleev spent most of his life in St. Petersburg. He taught at the University of St. Petersburg, where he wrote books and published his concept of chemical periodicity, which helped systematize inorganic chemistry. Later in life he moved on to other interests, including studying the natural resources of Russia and their commercial applications.

(b) Convert moles of Zn to mass using the molar mass of Zn as the conversion factor.

$$6.00 \text{ mol Zn} \times \frac{65.38 \text{ g Zn}}{1 \text{ mol Zn}} = 392 \text{ g Zn}$$

The mass of Zn is less than the 500.-g mass of Cu.

☑ **Reasonable Answer Check** The molar masses of Cu and Zn are both approximately 64 g/mol. The mass of Cu and the molar amount of Cu are both larger than those for Zn. The answer is reasonable.

PROBLEM-SOLVING PRACTICE **2.9**

Calculate (a) the number of moles in 4.00 g titanium (Ti) and (b) the number of grams in 3.00×10^{-2} mol silver (Ag).

EXERCISE **2.7 Grams, Moles, and Avogadro's Number**

You have a 10.00-g sample of lithium and a 10.00-g sample of iridium. How many atoms are in each sample, and how many more atoms are in the lithium sample than in the iridium sample?

2.9 The Periodic Table

go Chemistry **Module 1: The Periodic Table** covers concepts in this section.

It gives a useful perspective to realize that Mendeleev developed the periodic table nearly a half-century before electrons, protons, and neutrons were known.

Periodicity of the elements means a recurrence of similar properties at regular intervals when the elements are arranged in the correct order.

You have already used the periodic table inside the front cover of this book to obtain atomic numbers and atomic weights of elements. But it is much more valuable than this. The periodic table is an exceptionally useful tool in chemistry. It allows us to organize and interrelate the chemical and physical properties of the elements. For example, elements can be classified as metals, nonmetals, or metalloids by their positions in the periodic table. You should become familiar with the periodic table's main features and terminology.

Dmitri Mendeleev (1834–1907), while a professor at the University of St. Petersburg, realized that listing the elements in order of increasing atomic weight revealed a periodic repetition of their properties. He summarized his findings in the table that has come to be called the periodic table. By lining up the elements in horizontal rows in order of increasing atomic weight and starting a new row when he came to an element with properties similar to one already in the previous row, he saw that the resulting columns contained elements with similar properties. Mendeleev found that some positions in his table were not filled. He predicted that new elements would be found that filled the gaps, and he predicted properties of the undiscovered elements. Two of the missing elements—gallium (Ga) and germanium (Ge)—were soon discovered, with properties very close to those Mendeleev had predicted.

Later experiments by H. G. J. Moseley demonstrated that elements in the periodic table should be ordered by atomic numbers rather than atomic weights. Arranging the elements in order of increasing atomic number gives the **law of chemical periodicity:** ***The properties of the elements are periodic functions of their atomic numbers (numbers of protons).***

© Cengage Learning/George Semple

Periodicity of piano keys.

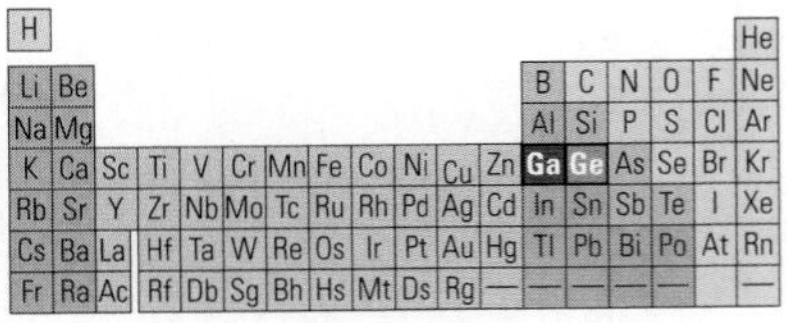

An alternative convention for numbering the groups in the periodic table uses the numbers 1 through 18, with no letters.

Periodic Table Features

Elements in the **periodic table** are arranged according to atomic number so that *elements with similar chemical properties occur in vertical columns* called **groups.** The table commonly used in the United States has groups numbered 1 through 8 (Figure 2.6), with each number followed by either an A or a B. The A groups (Groups 1A and 2A on the left of the table and Groups 3A through 8A at the right) are col-

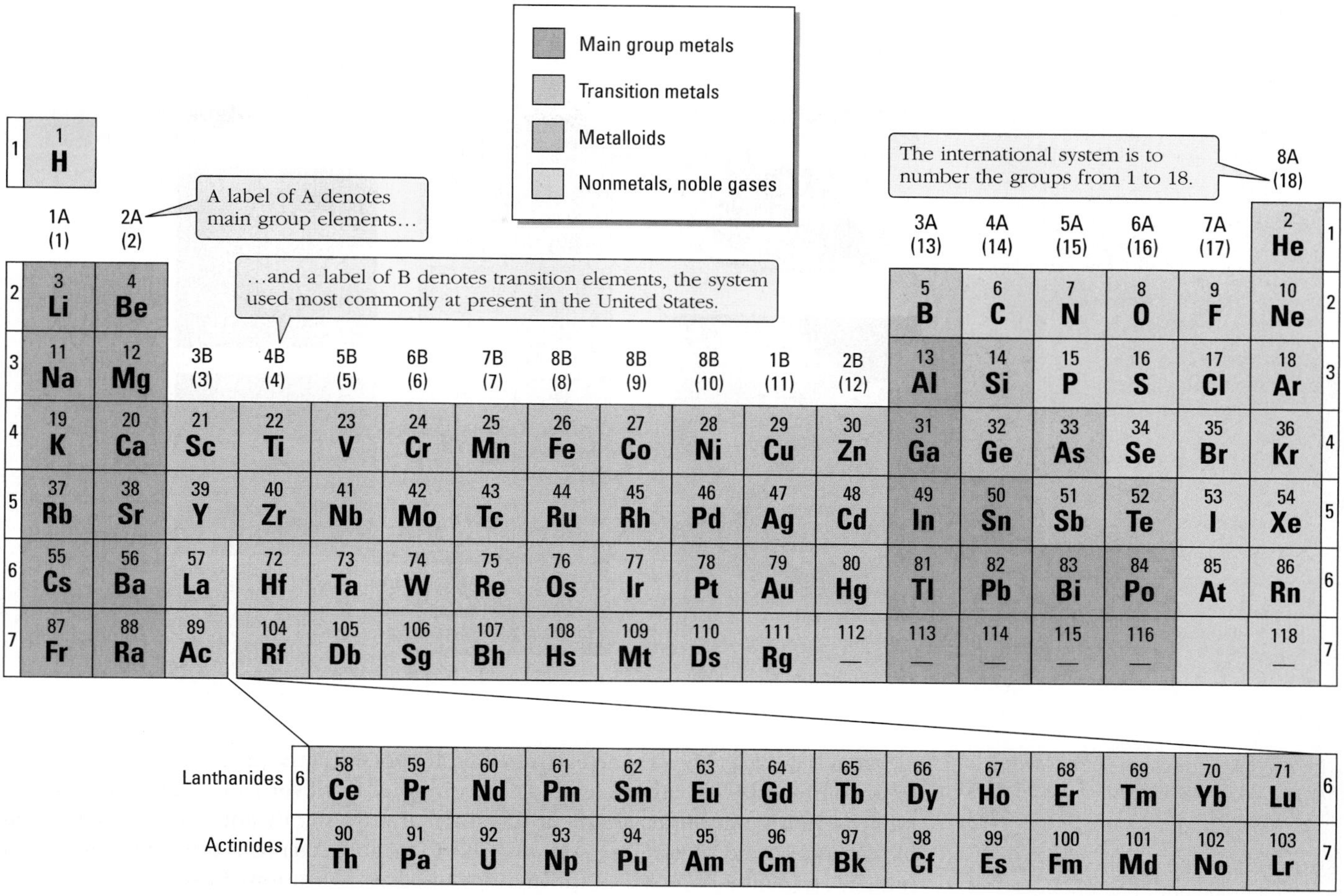

Figure 2.6 Modern periodic table of the elements. Elements are listed in order of increasing atomic number across horizontal rows called periods. Groups are vertical columns of elements. Some groups have common names: Group 1A, alkali metals; Group 2A, alkaline-earth metals; Group 7A, halogens; Group 8A, noble gases.

lectively known as **main group elements.** The B groups (in the middle of the table) are called **transition elements.**

The horizontal rows of the table are called **periods,** and they are numbered beginning with 1 for the period containing only H and He. Sodium (Na) is, for example, in Group 1A and is the first element in the third period. Silver (Ag) is in Group 1B and is in the fifth period.

Many interactive periodic tables are available on the Internet. Two to explore are at **http://www.chemeddl.org/collections/ptl** and **http://periodictable.com**.

The table in Figure 2.6 and inside the front cover helps us to recognize that most elements are metals (gray and blue), far fewer elements are nonmetals (lavender), and even fewer are metalloids (orange). Elements become less metallic from left to right across a period, and eventually one or more nonmetals are found in each period. The six metalloids (B, Si, Ge, As, Sb, Te) fall along a zigzag line passing between Al and Si, Ge and As, and Sb and Te.

The Alkali Metals (Group 1A) and Alkaline-Earth Metals (Group 2A)

The elements (except hydrogen) in the leftmost column (Group 1A) are called **alkali metals** because their aqueous solutions are alkaline (basic). Elements in Group 2A, known as **alkaline-earth metals,** are extracted from minerals (earths) and also produce alkaline aqueous solutions (except beryllium).

"Alkali" comes from the Arabic language. Ancient Arabian chemists discovered that ashes of certain plants, which they called *al-qali,* produced water solutions that felt slippery and burned the skin.

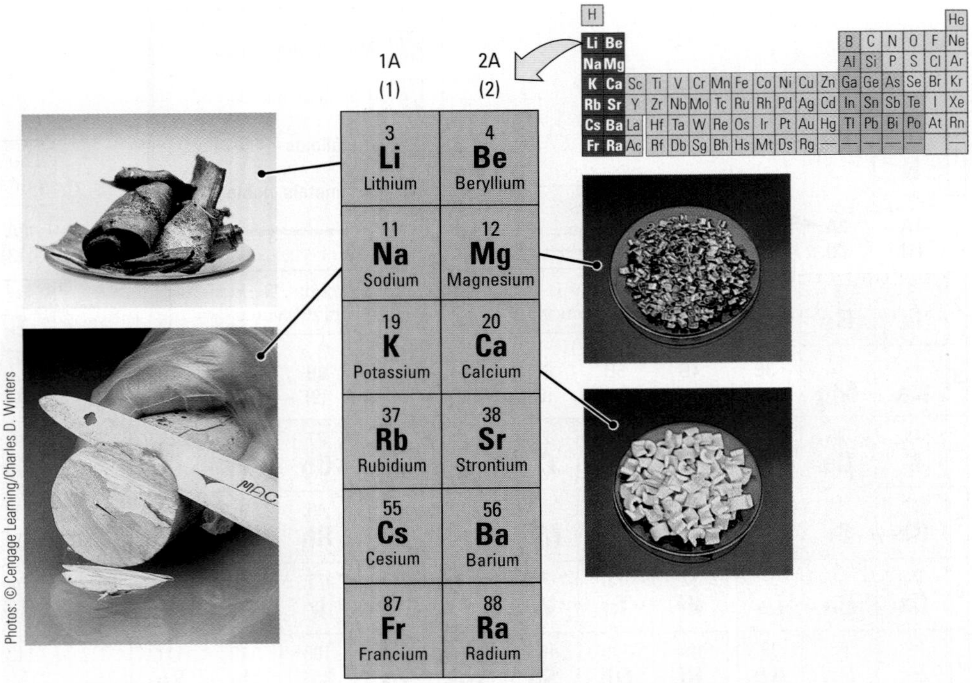

The chemistry of elements in Groups 1A and 2A as well as other elements is discussed in Chapters 21 and 22.

Elements found uncombined with any other element in nature are sometimes called "free" elements.

Gold and silver as free metals in nature triggered the great gold and silver rushes of the 1800s in the United States.

Alkali metals and alkaline-earth metals are very reactive and are found in nature only combined with other elements in compounds, never as the free metallic elements. Their compounds are plentiful, and many are significant to human and plant life. A compound of sodium, sodium chloride (table salt), is a fundamental part of human and animal diets, and throughout history civilizations have sought salt as a dietary necessity and a commercial commodity. Today, two of the most important industrial chemicals—sodium hydroxide and chlorine—are produced commercially from sodium chloride. Magnesium (Mg) and calcium (Ca), the sixth and fifth most abundant elements in Earth's crust, respectively, are present as ions (Mg^{2+}, Ca^{2+}) in a vast array of chemical compounds.

Sample of manmade transuranium elements coating the tip of a microscopic spatula. The sample was produced by neutron irradiation of uranium in a nuclear reactor in the 1940s.

The Transition Elements, Lanthanides, and Actinides

The transition elements (also known as the *transition metals*) fill the middle of the periodic table in Periods 4 through 7, and most are found in nature only in compounds. The notable exceptions are gold, silver, platinum, copper, and liquid mercury, which can be found in elemental form. Iron, zinc, copper, and chromium are among the most important commercial metals. Because of their vivid colors, transition metal compounds are used for pigments (Section 22.7).

The **lanthanides** and **actinides** are metallic elements listed separately in two rows at the bottom of the periodic table. Using the extra, separate rows keeps the periodic table from becoming too wide and too cumbersome. These elements are relatively rare and not as commercially important as the transition elements. The elements beyond uranium are synthesized by special nuclear techniques.

Groups 3A to 6A

These four groups contain the most abundant elements in Earth's crust and atmosphere (Table 2.4). They also contain the elements—carbon (C), nitrogen (N), and oxygen (O)—present in most of the important molecules in our bodies. Because of the ability of carbon atoms to bond extensively with each other, huge numbers of carbon compounds exist. Organic chemistry is the branch of chemistry devoted to

Table 2.4 Selected Group 3A–6A Elements

Group 3A

Aluminum: Most abundant *metal* in Earth's crust (7%). In nature, always found in compounds, especially with silicon and oxygen in clay minerals.

Group 4A

Carbon: Second most abundant element in living things. Provides the framework for organic and biochemical molecules.

Silicon: Second most abundant element in Earth's crust (25%). Always found combined naturally, usually with oxygen in quartz and silicate minerals.

Group 5A

Nitrogen: Most abundant element in Earth's atmosphere (78%) but not abundant in Earth's crust because of the relatively low chemical reactivity of N_2.

Group 6A

Oxygen: Most abundant element in Earth's crust (47%) because of its high chemical reactivity. Second most abundant element in Earth's atmosphere (21%).

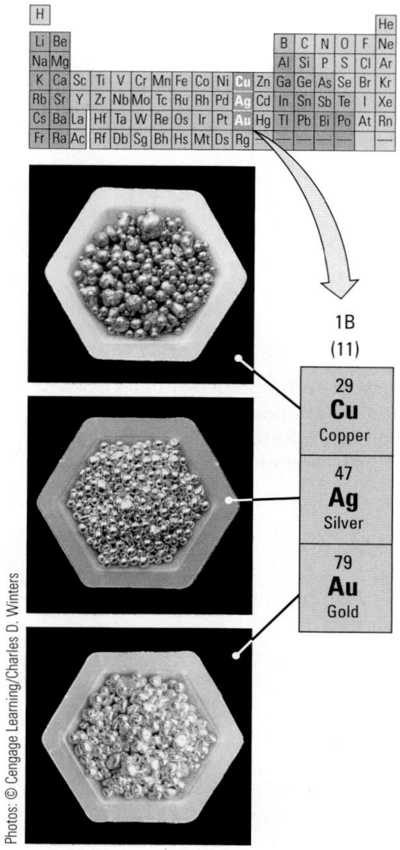
1B
(11)
29
Cu
Copper
47
Ag
Silver
79
Au
Gold
Photos: © Cengage Learning/Charles D. Winters

the study of carbon compounds. Carbon atoms also provide the framework for the molecules essential to living things, which are the subject of the branch of chemistry known as biochemistry.

Groups 4A to 6A each begin with one or more nonmetals, include one or more metalloids, and end with a metal. Group 4A, for example, contains carbon, a nonmetal, includes two metalloids (Si and Ge), and finishes with two metals (Sn and Pb). Group 3A starts with boron (B), a metalloid (Section 21.6).

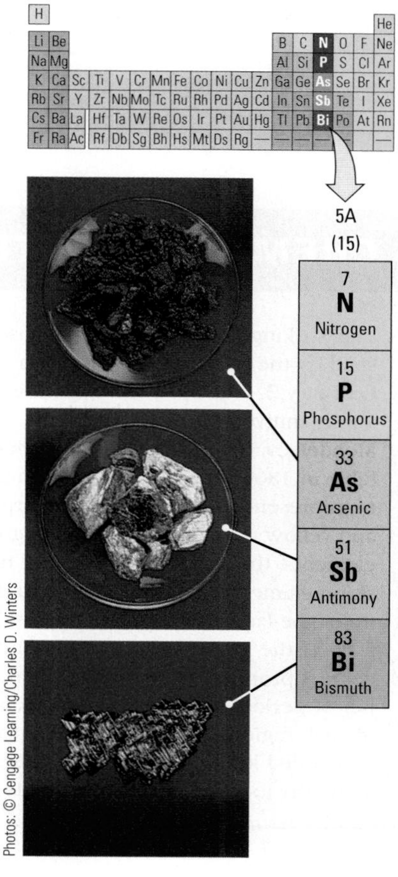
5A
(15)
7
N
Nitrogen
15
P
Phosphorus
33
As
Arsenic
51
Sb
Antimony
83
Bi
Bismuth
Photos: © Cengage Learning/Charles D. Winters

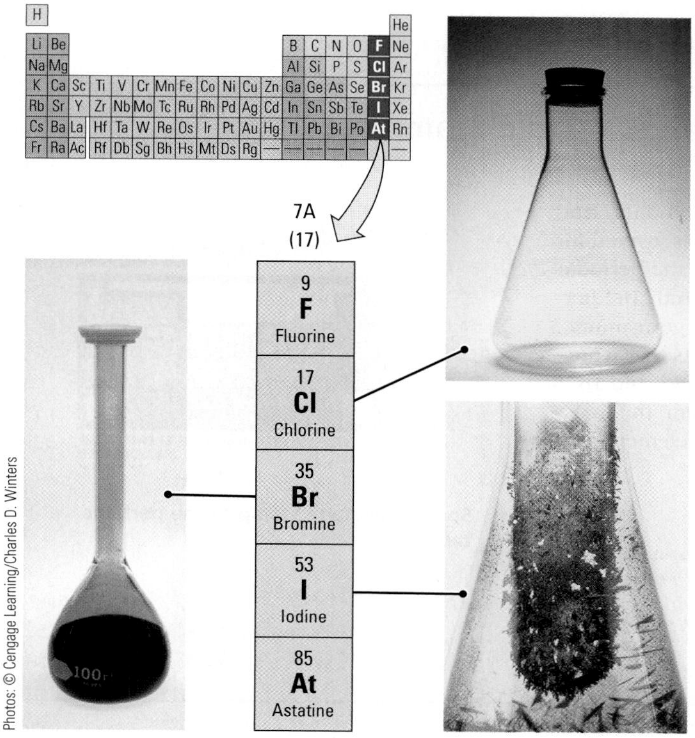
7A
(17)
9
F
Fluorine
17
Cl
Chlorine
35
Br
Bromine
53
I
Iodine
85
At
Astatine
Photos: © Cengage Learning/Charles D. Winters

The Halogens (Group 7A)

The elements in this group consist of diatomic molecules and are highly reactive. The group name, **halogens,** comes from the Greek words *hals,* meaning "salt," and *genes,* meaning "forming." The halogens all form salts—compounds similar to sodium chloride, NaCl—by reacting vigorously with alkali metals and with other metals as well. Halogens also react with most nonmetals. Small carbon compounds containing chlorine and fluorine are relatively unreactive but are involved in seasonal ozone depletion in the upper atmosphere (Section 10.9).

The Noble Gases (Group 8A)

Once the arrangement of electrons in atoms was understood (Section 7.6), the place where the noble gases fit into the periodic table was obvious.

The **noble gases** at the far right of the periodic table are the least reactive elements. They were not discovered on Earth until late in the nineteenth century, although helium was detected in the sun by analysis of solar radiation in 1868. Mendeleev did not know about the noble gases when he developed his periodic table (1869). For many years the noble gases were called the *inert gases,* because they were thought not to combine with any element to form compounds. In 1962 this basic canon of chemistry was overturned when compounds of xenon with fluorine and with oxygen were synthesized. Since then other xenon compounds have been made, as well as compounds of fluorine with krypton and with radon.

EXERCISE 2.8 The Periodic Table

1. How many (a) metals, (b) nonmetals, and (c) metalloids are in the fourth period of the periodic table? Give the name and symbol for each element.
2. Which groups of the periodic table contain (a) only metals, (b) only nonmetals, (c) only metalloids?
3. Which period of the periodic table contains the most metals?

CHEMISTRY IN THE NEWS Periodic Table Stamp

This striking 0.30-euro stamp was issued by the Spanish Post Office on February 2, 2007, to commemorate the 100th anniversary of the death of Mendeleev, who proposed the periodic table in 1869. The stamp shows an artistic representation in which the purple and yellow areas denote the main group elements, the red area denotes the transition elements, and the green area denotes the lanthanides and actinides. The sizes of the four colored areas are in the correct proportions regarding the number of periods and groups of elements in each region. The small white squares embedded in the red and yellow regions show the locations of four elements—gallium, germanium, scandium, and technetium—not yet discovered in 1869. Mendeleev used the periodicity of the table to correctly predict properties for gallium, germanium, and scandium, elements that were discovered in 1875, 1886, and 1878, respectively. Technetium, the first artificially synthesized element, was produced in 1937.

Sources:

Pinto, G. "A Postage Stamp About the Periodic Table." *Journal of Chemical Education,* Vol. 84, 2007; p. 1919.

http://www.cpossu.org;
http://www.correos.es/

Photo courtesy of Gabriel Pinto/Stamp design by Dr. Javier Garcia Martinez

Spanish postage stamp of the periodic table.

CHEMISTRY YOU CAN DO Preparing a Pure Sample of an Element

You will need these items to do this experiment:

- Two glasses or plastic cups that will each hold about 250 mL of liquid
- Approximately 100 mL (about 3.5 oz) of vinegar
- Soap
- An iron nail, paper clip, or other similar-sized piece of iron
- Something abrasive, such as a piece of steel wool, Brillo, sandpaper, or nail file
- About 40 to 50 cm of thin string or thread
- Some table salt
- A magnifying glass (optional)
- 15 to 20 dull pennies (shiny pennies will not work)

Wash the piece of iron with soap, dry it, and clean the surface further with the steel wool or other abrasive until the iron is shiny. Tie one end of the string around one end of the piece of iron.

Place the pennies in one cup (labeled A) and pour in enough vinegar to cover them. Sprinkle on a little salt, swirl the liquid around so it contacts all the pennies, and observe what happens. When nothing more seems to be happening, pour the liquid into the second cup (labeled B), leaving the pennies in cup A (that is, pour off the liquid). Suspend the piece of iron from the thread so that it is half-submerged in the liquid in cup B.

Observe the piece of iron over a period of 10 minutes or so, and then use the thread to pull it out of the liquid. Observe it carefully, using a magnifying glass if you have one. Compare the part that was submerged with the part that remained above the surface of the liquid.

Think about these questions:

1. What did you observe happening to the pennies?
2. How could you account for what happened to the pennies in terms of a nanoscale model? Cite observations that support your conclusion.
3. What did you observe happening to the piece of iron?
4. Interpret the experiment in terms of a nanoscale model, citing observations that support your conclusions.
5. Would this method be of use in purifying copper? If so, can you suggest ways that it could be used effectively to obtain copper from ores?

EXERCISE 2.9 Element Names

On June 14, 2000, a major daily American newspaper published this paragraph:

> ABC's *Who Wants to Be a Millionaire* crowned its fourth million-dollar winner Tuesday night. Bob House . . . [answered] the final question: Which of these men does not have a chemical compound named after him? (a) Enrico Fermi, (b) Albert Einstein, (c) Niels Bohr, (d) Isaac Newton

What is wrong with the question? What is the correct answer to the question after it is properly posed? (The question was properly posed and correctly answered on the TV show.)

SUMMARY PROBLEM

The atoms of one of the elements contain 47 protons and 62 neutrons.

(a) Identify the element and give its symbol.

(b) What is this atom's atomic number? Mass number?

(c) This element has two naturally occurring isotopes. Calculate the atomic weight of the element.

Isotope	Mass Number	Percent Abundance	Isotopic Mass (amu)
1	107	51.84	106.905
2	109	48.16	108.905

(d) This element is a member of which group in the periodic table? Is this element a metal, nonmetal, or metalloid? Explain your answer.

(e) Consider a piece of jewelry that contains 1.00 g of the element. (i) How many moles of the element are in this mass? (ii) How many atoms of the element are in this mass? (iii) Atoms of this element have an atomic diameter of 288 pm. If all the atoms of this element in the sample were put into a row, how many meters long would the chain of atoms be?

IN CLOSING

OWL and go Chemistry

Sign in at **www.cengage.com/owl** to:

- View tutorials and simulations, develop problem-solving skills, and complete online homework assigned by your professor.
- For quick review and exam prep, download Go Chemistry mini lecture modules from OWL (or purchase them at **www.CengageBrain.com**).

Having studied this chapter, you should be able to . . .

- Describe radioactivity, electrons, protons, and neutrons and the general structure of the atom (Sections 2.1, 2.2).
- Use conversion factors for the units for mass, volume, and length common in chemistry (Section 2.3). End-of-chapter questions: 9, 11, 13, 15, 113
- Identify the correct number of significant figures in a number and carry significant figures through calculations (Section 2.4). Questions 22, 26
- Define isotope and give the mass number and number of neutrons for a specific isotope (Section 2.5). Questions 37, 41
- Calculate the atomic weight of an element from isotopic abundances (Section 2.6). Questions 57, 59, 106
- Explain the difference between the atomic number and the atomic weight of an element and find this information for any element (Sections 2.5, 2.6).
- Relate masses of elements to the mole, Avogadro's number, and molar mass (Section 2.7). Questions 76, 118
- Do gram–mole and mole–gram conversions for elements (Section 2.8). Questions 66, 68
- Identify the periodic table location of groups, periods, alkali metals, alkaline-earth metals, halogens, noble gases, transition elements, lanthanides, and actinides (Section 2.9). Questions 83, 85, 87, 89, 91

KEY TERMS

actinides *(Section 2.9)*

alkali metals *(2.9)*

alkaline-earth metals *(2.9)*

atomic force microscope *(p. 47)*

atomic mass unit (amu) *(2.5)*

atomic number *(2.5)*

atomic structure *(Introduction)*

atomic weight *(2.6)*

Avogadro's number *(2.7)*

chemical periodicity, law of *(2.9)*

electron *(2.1)*

group *(2.9)*

halogens *(2.9)*

ion *(2.1)*

isotope *(2.5)*

lanthanides *(2.9)*

main group elements *(2.9)*

mass *(2.3)*

mass number *(2.5)*

mass spectrometer *(p. 56)*

mass spectrum *(p. 56)*

metric system *(2.3)*

molar mass *(2.7)*

mole (mol) *(2.7)*

neutron *(2.2)*

noble gases *(2.9)*

nucleus *(2.2)*

percent abundance *(2.6)*

period *(2.9)*

periodic table *(2.9)*

proton *(2.1)*

radioactivity *(2.1)*

scanning tunneling microscope *(p. 46)*

significant figures *(2.4)*

SI units *(2.3)*

transition elements *(2.9)*

QUESTIONS FOR REVIEW AND THOUGHT

OWL Interactive versions of these problems are assignable in OWL.

Blue-numbered questions have short answers at the back of this book in Appendix M and fully worked solutions in the *Student Solutions Manual.*

Review Questions

These questions test vocabulary and simple concepts.

1. What is the fundamental unit of electrical charge?
2. Millikan was able to determine the charge on an electron using his famous oil-drop experiment. Describe the experiment and explain how Millikan was able to calculate the mass of an electron using his results and the ratio discovered earlier by Thomson.
3. The positively charged particle in an atom is called the proton.
   (a) How much heavier is a proton than an electron?
   (b) What is the difference in the charge on a proton and an electron?
4. Ernest Rutherford's famous gold-foil experiment examined the structure of atoms.
   (a) What surprising result was observed?
   (b) The results of the gold-foil experiment enabled Rutherford to calculate that the nucleus is much smaller than the atom. How much smaller?
5. In any given *neutral* atom, how many protons are there compared with the number of electrons?
6. Atoms of elements can have varying numbers of neutrons in their nuclei.
   (a) What are species called that have varying numbers of neutrons for the same element?
   (b) How do the mass numbers vary for these species?
   (c) What are two common elements that exemplify this property?

Topical Questions

These questions are keyed to the major topics in the chapter. Usually a question that is answered at the back of the book is paired with a similar one that is not.

Units and Unit Conversions (Section 2.3)

7. If the nucleus of an atom were the size of a golf ball (4 cm diameter), what would be the diameter of the atom?
8. If a sheet of business paper is exactly 11 inches high, what is its height in centimeters? Millimeters? Meters?
9. The pole vault world record is 6.14 m. What is this in centimeters? In feet and inches?
10. The maximum speed limit in many states is 65 miles per hour. What is this speed in kilometers per hour?
11. A student weighs 168 lb. What is the student's weight in kilograms?
12. Basketball hoops are exactly 10 ft off the floor. How far is this in meters? Centimeters?
13. A Volkswagen engine has a displacement of 120. in.^{3}. What is this volume in cubic centimeters? In liters?
14. An automobile engine has a displacement of 250. in.^{3}. What is this volume in cubic centimeters? In liters?
15. Calculate how many square inches there are in one square meter.
16. One square mile contains exactly 640 acres. How many square meters are in one acre?
17. On May 18, 1980, Mt. St. Helens in Washington erupted. The 9677-ft high summit was lowered by 1314 ft by the eruption. Approximately 0.67 cubic miles of debris was released into the atmosphere. How many cubic meters of debris was released?
18. Suppose a room is 18 ft long, 15 ft wide, and the distance from floor to ceiling is 8 ft, 6 in. You need to know the volume of the room in metric units for some scientific calculations. What is the room's volume in cubic meters? In liters?
19. A crystal of fluorite (a mineral that contains calcium and fluorine) has a mass of 2.83 g. What is this mass in kilograms? In pounds? Give the symbols for the elements in this crystal.

Scanning Tunneling Microscopy (Section 2.3)

20. Comment on this statement: The scanning tunneling microscope enables scientists to image individual atoms on surfaces directly.
21. The scanning tunneling microscope is based on the flow of electrons from the instrument to the sample surface being investigated. How is this flow converted into an image of the surface?

Significant Figures (Section 2.4)

22. How many significant figures are present in these measured quantities?
   (a) 1374 kg (b) 0.00348 s
   (c) 5.619 mm (d) 2.475×10^{-3} cm
   (e) 33.1 mL
23. How many significant figures are present in these measured quantities?
   (a) 1.022×10^2 km (b) 34 m^2
   (c) 0.042 L (d) 28.2 °C
   (e) 323. mg
24. For each of these numbers, round to three significant digits and write the result in scientific notation.
   (a) 0.0004332 (b) 44.7337
   (c) 22.4555 (d) 0.0088418
25. For each of these numbers, round to four significant digits and write the result in scientific notation.
   (a) 247.583 (b) 100,578
   (c) 0.0000348719 (d) 0.004003881
26. Perform these calculations and express the result with the proper number of significant figures.
   (a) $\frac{4.850 \text{ g} - 2.34 \text{ g}}{1.3 \text{ mL}}$
   (b) $V = \pi r^3$ where $r = 4.112$ cm

(c) $(4.66 \times 10^{-3}) \times 4.666$

(d) $\dfrac{0.003400}{65.2}$

27. Perform these calculations and express the result with the proper number of significant figures.

(a) $2221.05 - \dfrac{3256.5}{3.20}$

(b) $343.2 \times (2.01 \times 10^{-3})$

(c) $S = 4\pi r^2$ where $r = 2.55$ cm

(d) $\dfrac{2802}{15} - (0.0025 \times 10{,}000.)$

Mass Spectrometry (Section 2.5)

28. What nanoscale species are moving through a mass spectrometer during its operation?
29. What is plotted on the *x*-axis and on the *y*-axis in a mass spectrum? What information does a mass spectrum convey?
30. How are the ions in the mass spectrometer separated from one another?
31. How can a mass spectrometer be used to measure the masses of individual isotopes of an element?

Isotopes (Sections 2.5 and 2.6)

32. Are these statements true or false? Explain why in each case.
    (a) Atoms of the same element always have the same mass number.
    (b) Atoms of the same element can have different atomic numbers.
33. What is the definition of the atomic mass unit?
34. What is the difference between the mass number and the atomic number of an atom?
35. Uranium-235 and uranium-238 differ in terms of the number of which subatomic particle?
36. When you subtract the atomic number from the mass number for an atom, what do you obtain?
37. How many electrons, protons, and neutrons are present in an atom of cobalt-60?
38. The artificial radioactive element technetium is used in many medical studies. Give the number of electrons, protons, and neutrons in an atom of technetium-99.
39. The atomic weight of bromine is 79.904. The natural abundance of ^{81}Br is 49.31%. What is the atomic weight of the only other natural isotope of bromine?
40. The atomic weight of boron is 10.811. The natural abundance of ^{10}B is 19.91%. What is the atomic weight of the only other natural isotope of boron?
41. Give the mass number of each of these atoms: (a) beryllium with 5 neutrons, (b) titanium with 26 neutrons, and (c) gallium with 39 neutrons.
42. Give the mass number of (a) an iron atom with 30 neutrons, (b) an americium atom with 148 neutrons, and (c) a tungsten atom with 110 neutrons.
43. Give the complete symbol $^{A}_{Z}X$ for each of these atoms: (a) sodium with 12 neutrons, (b) argon with 21 neutrons, and (c) gallium with 38 neutrons.
44. Give the complete symbol $^{A}_{Z}X$ for each of these atoms: (a) nitrogen with 8 neutrons, (b) zinc with 34 neutrons, and (c) xenon with 75 neutrons.
45. How many electrons, protons, and neutrons are there in an atom of (a) calcium-40, $^{40}_{20}$Ca, (b) tin-119, $^{119}_{50}$Sn, and (c) plutonium-244, $^{244}_{94}$Pu?
46. How many electrons, protons, and neutrons are there in an atom of (a) carbon-13, $^{13}_{6}$C, (b) chromium-50, $^{50}_{24}$Cr, and (c) bismuth-205, $^{205}_{83}$Bi?
47. Fill in this table:

Z	*A*	Number of Neutrons	Element
35	81	______	______
______	______	62	Pd
77	______	115	______
______	151	______	Eu

48. Fill in this table:

Z	*A*	Number of Neutrons	Element
60	144	______	______
______	______	12	Mg
64	______	94	______
______	37	______	Cl

49. Which of these are isotopes of element X, whose atomic number is 9: $^{18}_{9}$X, $^{20}_{9}$X, $^{9}_{4}$X, $^{15}_{9}$X?
50. Which of these species are isotopes of the same element: $^{20}_{10}$X, $^{20}_{11}$X, $^{21}_{10}$X, $^{20}_{12}$X? Explain.

Percent (Section 2.6)

51. Silver jewelry is actually a mixture of silver and copper. If a bracelet with a mass of 17.6 g contains 14.1 g silver, what is the percentage of silver? Of copper?
52. The solder once used by plumbers to fasten copper pipes together consists of 67% lead and 33% tin. What is the mass of lead (in grams) in a 1.00-lb block of solder? What is the mass of tin?
53. Many automobile batteries are filled with sulfuric acid. What is the mass of the acid (in grams) in 500. mL of the battery acid solution if the density of the solution is 1.285 g/cm^3 and the solution is 38.08% sulfuric acid by mass?
54. When popcorn pops, it loses water explosively. If a kernel of corn weighing 0.125 g before popping weighs 0.106 g afterward, what percentage of its mass did it lose on popping?
55. A well-known breakfast cereal contains 280. mg sodium per 30.-g serving. What percentage of the cereal is sodium?
56. If a 6.0-oz cup of regular coffee contains 100. mg caffeine, what is the percentage of caffeine in the coffee?

Atomic Weight (Section 2.6)

57. Verify that the atomic weight of lithium is 6.941 amu, given this information:

    ^{6}Li, exact mass = 6.015121 amu

    percent abundance = 7.500%

    ^{7}Li, exact mass = 7.016003 amu

    percent abundance = 92.50%

58. Verify that the atomic weight of magnesium is 24.3050 amu, given this information:

^{24}Mg, exact mass = 23.985042 amu

percent abundance = 78.99%

^{25}Mg, exact mass = 24.98537 amu

percent abundance = 10.00%

^{26}Mg, exact mass = 25.982593 amu

percent abundance = 11.01%

59. Gallium has two naturally occurring isotopes, ^{69}Ga and ^{71}Ga, with masses of 68.9257 amu and 70.9249 amu, respectively. Calculate the abundances of these isotopes of gallium.
60. Silver has two stable isotopes, ^{107}Ag and ^{109}Ag, with masses of 106.90509 amu and 108.90476 amu, respectively. Calculate the abundances of these isotopes of silver.
61. Lithium has two stable isotopes, ^{6}Li and ^{7}Li. Since the atomic weight of lithium is 6.941, which is the more abundant isotope?
62. Argon has three naturally occurring isotopes: 0.337% ^{36}Ar, 0.063% ^{38}Ar, and 99.60% ^{40}Ar. Estimate the atomic weight of argon. If the masses of the isotopes are 35.968, 37.963, and 39.962, respectively, what is the atomic weight of natural argon?

The Mole, Molar Mass, and Problem Solving (Sections 2.7 and 2.8)

63. The mole is simply a convenient unit for counting molecules and atoms. Name four "counting units" (such as a dozen for eggs and cookies) that you commonly encounter.
64. If you divide Avogadro's number of pennies among the nearly 300 million people in the United States, and if each person could count one penny each second every day of the year for eight hours per day, how long would it take to count the pennies?
65. Why do you think it is more convenient to use some chemical counting unit when doing calculations (chemists have adopted the unit of the mole, but it could have been something different) rather than using individual molecules?
66. Calculate the number of grams in
    (a) 2.5 mol boron
    (b) 0.015 mol O_2
    (c) 1.25×10^{-3} mol iron
    (d) 653 mol helium
67. Calculate the number of grams in
    (a) 6.03 mol gold
    (b) 0.045 mol uranium
    (c) 15.6 mol Ne
    (d) 3.63×10^{-4} mol plutonium
68. Calculate the number of moles represented by each of these:
    (a) 127.08 g Cu
    (b) 20.0 g calcium
    (c) 16.75 g Al
    (d) 0.012 g potassium
    (e) 5.0 mg americium
69. Calculate the number of moles represented by each of these:
    (a) 16.0 g Na
    (b) 0.0034 g platinum
    (c) 1.54 g P
    (d) 0.876 g arsenic
    (e) 0.983 g Xe
70. How many moles of Na are in 50.4 g sodium?
71. How many moles of zinc are in 79.3 g Zn?
72. If you have 0.00789 g of the gaseous element krypton, how many moles does this mass represent?
73. If you have 4.6×10^{-3} g gaseous helium, how many moles of helium do you have?
74. If you have a 35.67-g piece of chromium metal on your car, how many atoms of chromium do you have?
75. If you have a ring that contains 1.94 g gold, how many atoms of gold are in the ring?
76. What is the average mass in grams of one copper atom?
77. What is the average mass in grams of one atom of titanium?

The Periodic Table (Section 2.9)

78. What is the difference between a group and a period in the periodic table?
79. Name and give symbols for (a) three elements that are metals; (b) four elements that are nonmetals; and (c) two elements that are metalloids. In each case, also locate the element in the periodic table by giving the group and period in which the element is found.
80. Name and give symbols for three transition metals in the fourth period. Look up each of your choices in a dictionary, a book such as *The Handbook of Chemistry and Physics,* or on the Internet, and make a list of their properties. Also list the uses of each element.
81. Name an element discovered by Madame Curie. Give its name, symbol, and atomic number. Use a dictionary, a book such as *The Handbook of Chemistry and Physics,* or the Internet to find the origin of the name of this element.
82. Name two halogens. Look up each of your choices in a dictionary, in a book such as *The Handbook of Chemistry and Physics,* or on the Internet, and make a list of their properties. Also list any uses of each element that are given by the source.
83. Name three transition elements, two halogens, and one alkali metal.
84. Name an alkali metal, an alkaline-earth metal, and a halogen.
85. How many elements are there in Group 4A of the periodic table? Give the name and symbol of each of these elements. Tell whether each is a metal, nonmetal, or metalloid.
86. How many elements are there in the fourth period of the periodic table? Give the name and symbol of each of these elements. Tell whether each is a metal, metalloid, or nonmetal.
87. The symbols for the four elements whose names begin with the letter I are In, I, Ir, and Fe. Match each symbol with one of the statements below.
    (a) a halogen
    (b) a main group metal
    (c) a transition metal in Period 6
    (d) a transition metal in Period 4
88. The symbols for four of the eight elements whose names begin with the letter S are Si, Ag, Na, and S. Match each symbol with one of the statements below.
    (a) a solid nonmetal
    (b) an alkali metal
    (c) a transition metal
    (d) a metalloid
89. Which single period in the periodic table contains the most (a) metals, (b) metalloids, and (c) nonmetals?
90. How many periods of the periodic table have 8 elements, how many have 18 elements, and how many have 32 elements?

91. Use the periodic table to identify these elements:
   (a) Name an element in Group 2A.
   (b) Name an element in the third period.
   (c) What element is in the second period in Group 4A?
   (d) What element is in the third period in Group 6A?
   (e) What halogen is in the fifth period?
   (f) What alkaline-earth element is in the third period?
   (g) What noble gas element is in the fourth period?
   (h) What nonmetal is in Group 6A and the second period?
   (i) Name a metalloid in the fourth period.

92. Use the periodic table to identify these elements:
   (a) Name an element in Group 2B.
   (b) Name an element in the fifth period.
   (c) What element is in the sixth period in Group 4A?
   (d) What element is in the third period in Group 5A?
   (e) What alkali metal is in the third period?
   (f) What noble gas is in the fifth period?
   (g) Name the element in Group 6A and the fourth period. Is it a metal, nonmetal, or metalloid?
   (h) Name a metalloid in Group 5A.

93. This chart is a plot of the logarithm of the relative abundances in the solar system of elements 1 through 36. The abundances are given on a scale that assigns silicon a relative value of 1.00×10^6 (the logarithm of which is 6).

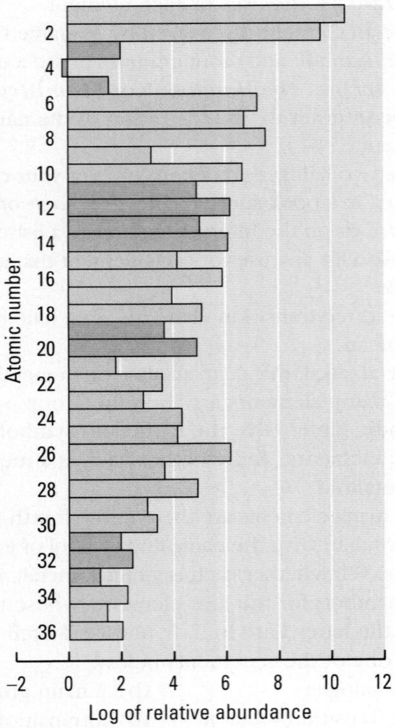

   (a) What is the most abundant metal?
   (b) What is the most abundant nonmetal?
   (c) What is the most abundant metalloid?
   (d) Which of the transition elements is most abundant?
   (e) How many halogens are considered on this plot, and which is the most abundant?

94. Consider the plot of relative abundance versus atomic number once again (Question 93). Uncover any relation between abundance and atomic number. Is there any difference between elements of even atomic number and those of odd atomic number?

General Questions

These questions are not explicitly keyed to chapter topics; many require integration of several concepts.

95. In his beautifully written autobiography, *The Periodic Table,* Primo Levi says of zinc that "it is not an element which says much to the imagination; it is gray and its salts are colorless; it is not toxic, nor does it produce striking chromatic reactions; in short, it is a boring metal. It has been known to humanity for two or three centuries, so it is not a veteran covered with glory like copper, nor even one of these newly minted elements which are still surrounded with the glamour of their discovery." From this description, and from reading this chapter, make a list of the properties of zinc. For example, include in your list the position of the element in the periodic table, and tell how many electrons and protons an atom of zinc has. What are its atomic number and atomic weight? Zinc is important in our economy. Check in your dictionary, in a book such as *The Handbook of Chemistry and Physics,* or on the Internet, and make a list of the uses of the element.

96. The density of a solution of sulfuric acid is 1.285 g/cm^3, and it is 38.08% acid by mass. What volume of the acid solution (in mL) do you need to supply 125 g of sulfuric acid?

97. In addition to the metric units of nm and pm, a commonly used unit is the angstrom, where $1\ \text{Å} = 1 \times 10^{-10}$ m. If the distance between the Pt atom and the N atom in a compound is 1.97 Å, what is the distance in nm? In pm?

98. The separation between carbon atoms in diamond is 0.154 nm. (a) What is their separation in meters? (b) What is the carbon atom separation in angstroms (where $1\ \text{Å} = 1 \times 10^{-10}$ m)?

99. The smallest repeating unit of a crystal of common salt is a cube with an edge length of 0.563 nm. What is the volume of this cube in nm^3? In cm^3?

100. The cancer drug cisplatin contains 65.0% platinum. If you have 1.53 g of the compound, how many grams of platinum does this sample contain?

101. Ethyl alcohol, C_2H_5OH, has a density of 0.789 g/mL at 25 °C. Water weighs 1.00 kg per liter at 25 °C. What volume of ethanol contains the same number of molecules as the liter of water?

102. One drop of water has a volume of one-twentieth (0.0500) of a milliliter. (a) How many water molecules are in this amount of water? (Water density = 1.00 g/mL) (b) Water molecules are about 100. pm in size. If the number of water molecules in the drop of water were laid end to end, how far would they reach?

103. A common fertilizer used on lawns is designated as "16-4-8." These numbers mean that the fertilizer contains 16% nitrogen-containing compounds, 4.0% phosphorus-containing compounds, and 8.0% potassium-containing compounds. You buy a 40.0-lb bag of this fertilizer and use all of it on your lawn. How many grams of the phosphorus-containing compound are you putting on your lawn? If the phosphorus-containing compound consists of 43.64% phosphorus (the rest is oxygen), how many grams of phosphorus are there in 40.0 lb of fertilizer?

104. The fluoridation of city water supplies has been practiced in the United States for several decades because it is believed that fluoride prevents tooth decay, especially in

young children. This is done by continuously adding sodium fluoride to water as it comes from a reservoir. Assume you live in a medium-sized city of 150,000 people and that each person uses 175 gal water per day. How many tons of sodium fluoride must you add to the water supply each year (365 days) to have the required fluoride concentration of 1 part per million (that is, 1 ton of fluoride per million tons of water)? (Sodium fluoride is 45.0% fluoride, and 1 U.S. gallon of water has a mass of 8.34 lb.)

105. Name three elements that you have encountered today. (Name only those that you have seen as elements, not those combined into compounds.) Give the location of each of these elements in the periodic table by specifying the group and period in which it is found.

106. Potassium has three stable isotopes, ^{39}K, ^{40}K, and ^{41}K, but ^{40}K has a very low natural abundance. Which of the other two is the more abundant?

107. Which one of these symbols conveys more information about the atom: ^{37}Cl or $_{17}Cl$? Explain.

108. The figure in the *Tools of Chemistry* box (p. 56) shows the mass spectrum of neon isotopes. What are the symbols of the isotopes? Which is the most abundant isotope? How many protons, neutrons, and electrons does this isotope have? Without looking at a periodic table, give the approximate atomic weight of neon.

109. When an athlete tears ligaments and tendons, they can be surgically attached to bone to keep them in place until they reattach themselves. A problem with current techniques, though, is that the screws and washers used are often too big to be positioned accurately or properly. Therefore, a titanium-containing device is used.
(a) What are the symbol, atomic number, and atomic weight of titanium?
(b) In what group and period is it found? Name the other elements of its group.
(c) What chemical properties do you suppose make titanium an excellent choice for this and other surgical applications?
(d) Use a dictionary, a book such as *The Handbook of Chemistry and Physics,* or the Internet to make a list of the properties of the element and its uses.

110. Draw a picture showing the approximate positions of all protons, electrons, and neutrons in an atom of helium-4. Make certain that your diagram indicates both the number and position of each type of particle.

111. Gems and precious stones are measured in carats, a weight unit equivalent to 200. mg. If you have a 2.3-carat diamond in a ring, how many moles of carbon do you have?

112. The international markets in precious metals operate in the weight unit "troy ounce" (where 1 troy ounce is equivalent to 31.1 g). Platinum sells for $1100 per troy ounce.
(a) How many moles of Pt are there in 1 troy ounce of the metal?
(b) If you have $5000 to spend, how many grams and how many moles of platinum can you purchase?

113. Gold prices fluctuate, depending on the international situation. If gold currently sells for $900 per troy ounce, how much must you spend to purchase 1.00 mol gold (1 troy ounce is equivalent to 31.1 g)?

114. The Statue of Liberty in New York harbor is made of 2.00×10^5 lb copper sheets bolted to an iron framework. How many grams and how many moles of copper does this represent (1 lb = 454 g)?

115. A piece of copper wire is 25 ft long and has a diameter of 2.0 mm. Copper has a density of 8.92 g/cm^3. How many moles of copper and how many atoms of copper are there in the piece of wire?

Applying Concepts

These questions test conceptual learning.

116. Which sets of values are possible? Why are the others not possible? Explain your reasoning.

	Mass Number	Atomic Number	Number of Protons	Number of Neutrons
(a)	19	42	19	23
(b)	235	92	92	143
(c)	53	131	131	79
(d)	32	15	15	15
(e)	14	7	7	7
(f)	40	18	18	40

117. Which sets of values are possible? Why are the others not possible? Explain your reasoning.

	Mass Number	Atomic Number	Number of Protons	Number of Neutrons
(a)	53	25	25	29
(b)	195	78	195	117
(c)	33	16	16	16
(d)	52	24	24	28
(e)	35	17	18	17

118. Which member of each pair has the greater number of particles? Explain why.
(a) 1 mol Cl or 1 mol Cl_2
(b) 1 molecule O_2 or 1 mol O_2
(c) 1 nitrogen atom or 1 nitrogen molecule
(d) 6.022×10^{23} fluorine molecules or 1 mol fluorine molecules
(e) 20.2 g Ne or 1 mol Ne
(f) 1 molecule Br_2 or 159.8 g Br_2
(g) 107.9 g Ag or 6.9 g Li
(h) 58.9 g Co or 58.9 g Cu
(i) 1 g calcium or 6.022×10^{23} calcium atoms
(j) 1 g chlorine atoms or 1 g chlorine molecules

119. Which member of each pair has the greater mass? Explain why.
(a) 1 mol iron or 1 mol aluminum
(b) 6.022×10^{23} lead atoms or 1 mol lead
(c) 1 copper atom or 1 mol copper
(d) 1 mol Cl or 1 mol Cl_2
(e) 1 g oxygen atoms or 1 g oxygen molecules
(f) 24.3 g Mg or 1 mol Mg
(g) 1 mol Na or 1 g Na
(h) 4.0 g He or 6.022×10^{23} He atoms
(i) 1 molecule I_2 or 1 mol I_2
(j) 1 oxygen molecule or 1 oxygen atom

120. Eleven of the elements in the periodic table are found in nature as gases at room temperature. List them. Where are they located in the periodic table?

121. Ten of the elements are O, H, Ar, Al, Ca, Br, Ge, K, Cu, and P. Pick the one that best fits each description: (a) an alkali metal; (b) a noble gas; (c) a transition metal; (d) a metalloid; (e) a Group 1 nonmetal; (f) a metal that forms a 3+ ion; (g) a nonmetal that forms a 2− ion; (h) an alkaline-earth metal; (i) a halogen; (j) a nonmetal that is a solid.

122. Air mostly consists of diatomic molecules of nitrogen (about 80%) and oxygen (about 20%). Draw a nanoscale picture of a sample of air that contains a total of 10 molecules.

123. Identify the element that satisfies each of these descriptions:
(a) A member of the same group as oxygen whose atoms contain 34 electrons
(b) A member of the alkali metal group whose atoms contain 20 neutrons
(c) A halogen whose atoms contain 35 protons and 44 neutrons
(d) A noble gas whose atoms contain 10 protons and 10 neutrons

More Challenging Questions

These questions require more thought and integrate several concepts.

124. At 25 °C, the density of water is 0.997 g/cm³, whereas the density of ice at −10 °C is 0.917 g/cm³.
(a) If a plastic soft-drink bottle (volume = 250 mL) is filled with pure water, capped, and then frozen at −10 °C, what volume will the solid occupy?
(b) Could the ice be contained within the bottle?

125. A high-quality analytical balance can weigh accurately to the nearest 1.0×10^{-4} g. How many carbon atoms are present in 1.000 mg carbon, that could be weighed by such a balance? Given the precision of the balance, what are the high and low limits on the number of atoms present in the 1.000-mg sample?

126. A group of astronauts in a spaceship accidentally encounters a space warp that traps them in an alternative universe where the chemical elements are quite different from the ones they are used to. The astronauts find these properties for the elements that they have discovered:

Atomic Symbol	Atomic Weight	State	Color	Electrical Conductivity	Electrical Reactivity
A	3.2	Solid	Silvery	High	Medium
D	13.5	Gas	Colorless	Very low	Very high
E	5.31	Solid	Golden	Very high	Medium
G	15.43	Solid	Silvery	High	Medium
J	27.89	Solid	Silvery	High	Medium
L	21.57	Liquid	Colorless	Very low	Medium
M	11.23	Gas	Colorless	Very low	Very low
Q	8.97	Liquid	Colorless	Very low	Medium
R	1.02	Gas	Colorless	Very low	Very high
T	33.85	Solid	Colorless	Very low	Medium
X	23.68	Gas	Colorless	Very low	Very low
Z	36.2	Gas	Colorless	Very low	Medium
Ab	29.85	Solid	Golden	Very high	Medium

(a) Arrange these elements into a periodic table.
(b) If a new element, X, with atomic weight 25.84 is discovered, what would its properties be? Where would it fit in the periodic table you constructed?
(c) Are there any elements that have not yet been discovered? If so, what would their properties be?

127. The element bromine is Br_2, so the mass of a Br_2 molecule is the sum of the mass of its two atoms. Bromine has two different isotopes. The mass spectrum of Br_2 produces three peaks with masses of 157.836, 159.834, and 161.832 amu, and relative heights of 25.54%, 49.99%, and 24.46%, respectively.
(a) What isotopes of bromine are present in each of the three peaks?
(b) What is the mass of each bromine isotope?
(c) What is the average atomic mass of bromine?
(d) What is the abundance of each of the two bromine isotopes?

Conceptual Challenge Problems

These rigorous, thought-provoking problems integrate conceptual learning with problem solving and are suitable for group work.

CP2.A (Section 2.1) Suppose you are faced with a problem similar to the one faced by Robert Millikan when he analyzed data from his oil-drop experiment. Below are the masses of three stacks of dimes. What do you conclude to be the mass of a dime, and what is your argument?

Stack 1 = 9.12 g   Stack 2 = 15.96 g   Stack 3 = 27.36 g

CP2.B (Section 2.3) The age of the universe is unknown, but some conclude from measuring Hubble's constant that it is about 18 billion years old, which is about four times the age of Earth. If so, what is the age of the universe in seconds? If you had a sample of carbon with the same number of carbon atoms as there have been seconds since the universe began, could you measure this sample on a laboratory balance that can detect masses as small as 0.1 mg?

Chemical Compounds

3

Javier Trueba/MSF/Photo Researchers, Inc.

These enormous selenite gypsum crystals, $CaSO_4 \cdot 2H_2O$, in Cueva de los Cristales (Cave of Crystals) were recently discovered hundreds of feet beneath the Chihuahuan desert of northern Mexico. The ionic hydrate crystals, up to 12 m (36 feet) long, were precipitated by natural processes over hundreds of thousands of years from a hot (58 °C) aqueous solution of calcium and sulfate ions. Pure selenite gypsum is transparent and colorless, but these giant crystals are translucent due to impurities.

One of the most important things chemists do is synthesize new chemical compounds, substances that on the nanoscale consist of new, unique combinations of atoms. These compounds may have properties similar to those of existing compounds, or they may be very different. Often chemists can custom-design a new compound to have desirable properties. All compounds contain at least two elements, and most compounds contain more than two elements. This chapter deals with two major, general types of chemical compounds—those consisting of individual

OWL Sign in to OWL at **www.cengage.com/owl** to view tutorials and simulations, develop problem-solving skills, and complete online homework assigned by your professor.

go Chemistry Download mini lecture videos for key concept review and exam prep from OWL or purchase them from **www.CengageBrain.com.**

Companion Website

Visit this book's companion website at **www.cengage.com/chemistry/moore** to work interactive modules for the Estimation boxes and Active Figures in this text.

Metabolism is a general term for all of the chemical reactions that act to keep a living thing functioning. We *metabolize* food molecules to extract energy and produce other molecules needed by our bodies. Our *metabolic* reactions are controlled by enzymes (discussed in Section 13.9) and other kinds of molecules.

H_2O H—O—H

Space-filling model | Ball-and-stick model

Some elements are also composed of molecules. In oxygen, for example, two oxygen atoms are joined in an O_2 molecule *(← p. 25)*.

molecules, and those made of the positively and negatively charged atoms or charged groups of atoms called ions. We will now examine how compounds are represented by symbols, formulas, and names and how formulas represent the macroscale masses and compositions of compounds.

3.1 Molecular Compounds

In a **molecular compound** at the nanoscale level, atoms of two or more different elements are combined into the independent units known as molecules ***(← p. 25)***. Every day we inhale, exhale, metabolize, and in other ways use thousands of molecular compounds. Water, carbon dioxide, sucrose (table sugar), and caffeine, as well as carbohydrates, proteins, and fats, are among the many common molecular compounds in our bodies.

Molecular Formulas

The composition of a molecular compound is represented in writing by its **molecular formula,** in which the number and kinds of atoms combined to make one molecule of the compound are indicated by subscripts and elemental symbols. For example, the molecular formula for water, H_2O, shows that there are three atoms per molecule—two hydrogen atoms and one oxygen atom. The subscript to the right of each element's symbol indicates the number of atoms of that element present in the molecule. If the subscript is omitted, it is understood to be 1, as for the O in H_2O. These same principles apply to the molecular formulas of all molecules.

Some molecules are classified as **inorganic compounds** because they do not contain carbon—for example, sulfur dioxide, SO_2, an air pollutant, or ammonia, NH_3, which, dissolved in water, is used as a household cleaning agent. Many inorganic compounds are ionic compounds, which are described in Section 3.5 of this chapter. The majority of **organic compounds** are composed of molecules. Organic compounds invariably contain carbon, usually contain hydrogen, and may also contain oxygen, nitrogen, sulfur, phosphorus, or halogens. Such compounds are of great interest because they are the basis for the clothes we wear, the food we eat, the fuels we burn, and the living organisms in our environment. For example, ethanol, C_2H_6O, is the organic compound familiar as a component of "alcoholic" beverages, and methane, CH_4, is the organic compound that is the major component of natural gas.

The formula of a molecular compound, especially an organic compound, can be written in several different ways. The molecular formula given previously for ethanol, C_2H_6O, is one example. For an organic compound, the symbols of the elements other than carbon are frequently written in alphabetical order, and each has a subscript indicating the total number of atoms of that type in the molecule, as illustrated by C_2H_6O. Because of the huge number of organic compounds, this formula may not give sufficient information to indicate what compound is represented. Such identification requires more information about how the atoms are connected to each other. A **structural formula** shows exactly how atoms are connected. In ethanol, for example, the first carbon atom is connected to three hydrogen atoms, and the second carbon atom is connected to two hydrogen atoms and an —OH group.

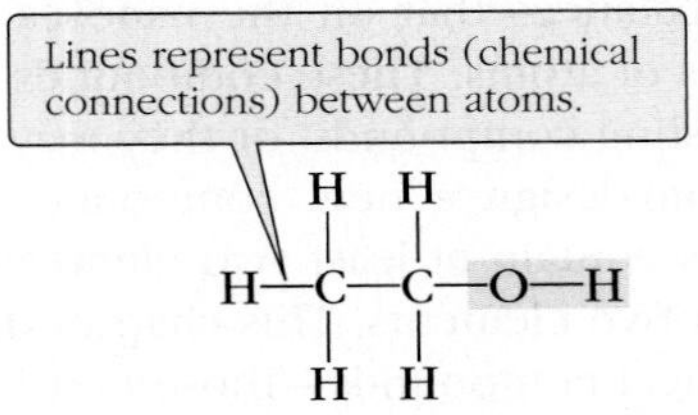
Lines represent bonds (chemical connections) between atoms.
H H
H—C—C—O—H
H H

The formula can also be written in a modified form to show how the atoms are grouped together in the molecule. Such formulas, called **condensed formulas,** emphasize the atoms or groups of atoms connected to each carbon atom. For ethanol, the condensed formula is CH_3CH_2OH. If you compare this to the structural formula for ethanol, you can easily see that they represent the same structure.

To summarize, three different ways of writing formulas are shown here for ethanol:

Molecular formula	Condensed formula	Structural formula
C_2H_6O	CH_3CH_2OH	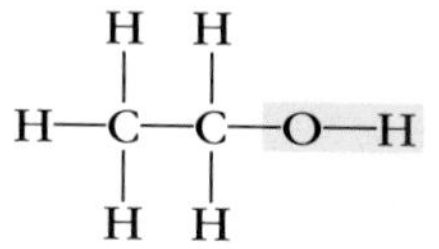

The —OH group of atoms is called a functional group.

The —OH attached to the C atom is a distinctive grouping of atoms that characterizes the group of organic compounds known as alcohols. Such groups distinctive to the various classes of organic compounds are known as **functional groups.**

As illustrated earlier for molecular elements ***(← p. 26),*** molecular compounds can also be represented by ball-and-stick and space-filling models.

In this figure and throughout the book, atoms in molecular models are color-coded: H, light gray; C, dark gray; O, red. A chart showing the full color key is inside the back cover.

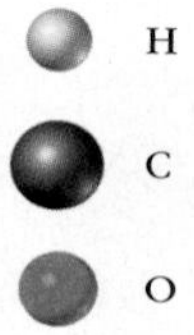
H
C
O

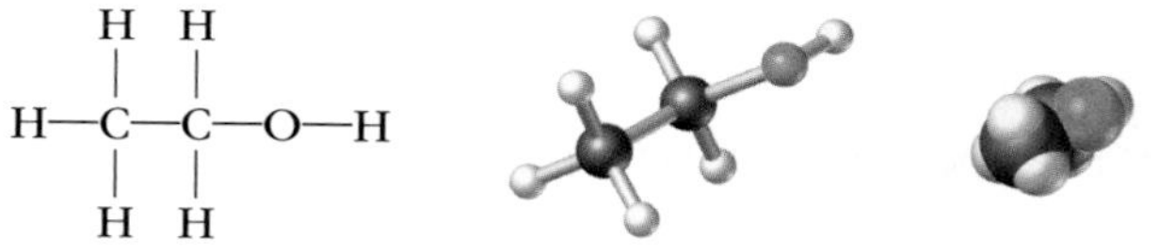

Structural formula   Ball-and-stick model   Space-filling model

Some additional examples of ball-and-stick molecular models are given below in Table 3.1.

Table 3.1 Examples of Simple Molecular Compounds

Name	Molecular Formula	Number and Kind of Atoms	Molecular Model
Carbon dioxide	CO_2	3 total: 1 carbon, 2 oxygen	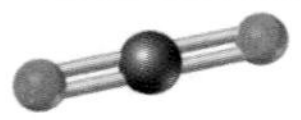
Ammonia	NH_3	4 total: 1 nitrogen, 3 hydrogen	
Nitrogen dioxide	NO_2	3 total: 1 nitrogen, 2 oxygen	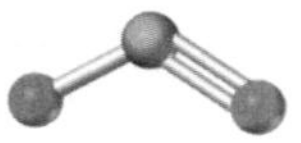
Carbon tetrachloride	CCl_4	5 total: 1 carbon, 4 chlorine	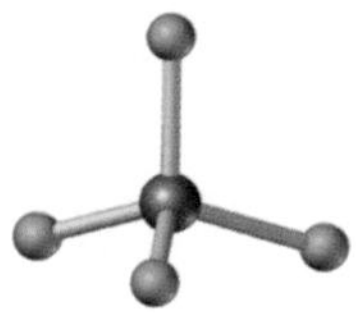
Octane	C_8H_{18}	26 total: 8 carbon, 18 hydrogen	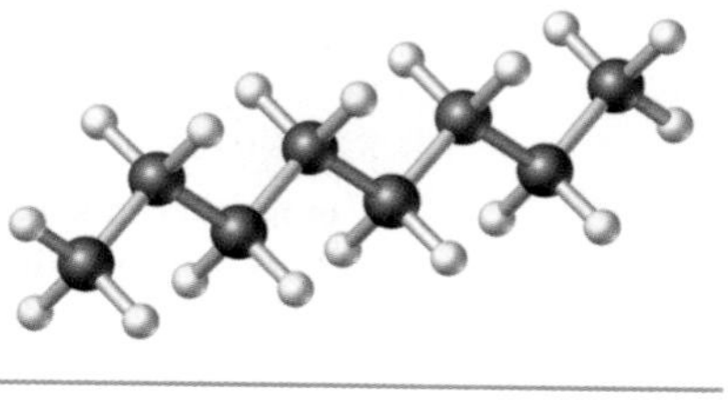

The **PROBLEM-SOLVING STRATEGY** in this book is
- Analyze the problem
- Plan a solution
- Execute the plan
- Check that the result is reasonable

Appendix A.1 explains this in detail.

An automotive antifreeze that contains propylene glycol.

PROBLEM-SOLVING PRACTICE answers are provided at the back of this book in Appendix K.

Answers to **EXERCISES** are provided at the back of this book in Appendix L.

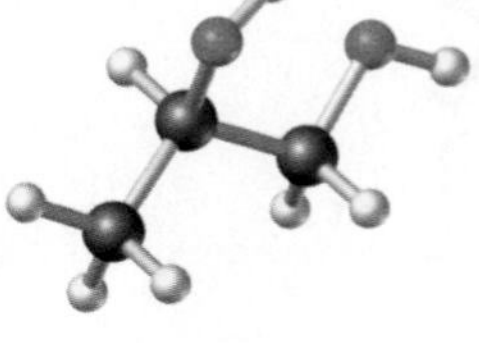

propylene glycol

PROBLEM-SOLVING EXAMPLE 3.1 Condensed and Molecular Formulas

(a) Write the molecular formulas for these molecules:

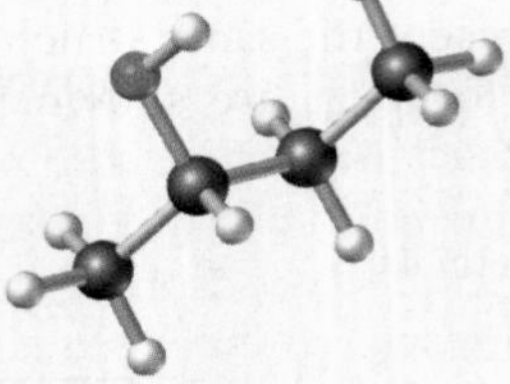

2-butanol

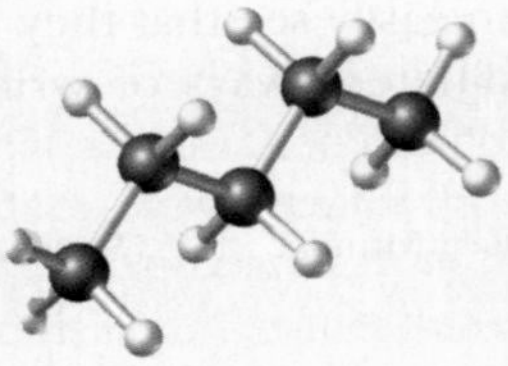

pentane

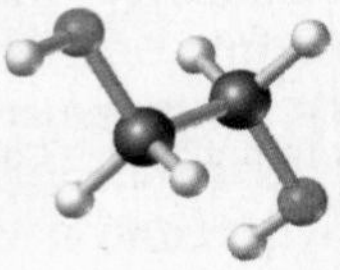

ethylene glycol

(b) Write the condensed formulas for 2-butanol, pentane, and ethylene glycol.

(c) Write the molecular formula for this molecule:

butanol

Answer

(a) 2-butanol, $C_4H_{10}O$; pentane, C_5H_{12}; ethylene glycol, $C_2H_6O_2$

(b)

```
              OH
              |
CH3—CH2—CH—CH3
```

$CH_3—CH_2—CH(OH)—CH_3$ 2-butanol

$CH_3CH_2CH_2CH_2CH_3$ pentane

$HOCH_2CH_2OH$ ethylene glycol

(c) $C_4H_{10}O$

Strategy and Explanation

(a) Count the atoms of each type in each molecule to obtain the molecular formulas. Then write the symbols with their subscripts, putting C first and the others in alphabetical order.

(b) In condensed formulas, each carbon atom and its hydrogen atoms are written without connecting lines (CH_3, CH_2, or CH). Other groups are usually written on the same line with the carbon and hydrogen atoms if the groups are at the beginning or end of the molecule. Otherwise, they are connected above or below the line by straight lines to the respective carbon atoms. Condensed formulas emphasize important groups in molecules, such as the —OH groups in 2-butanol and ethylene glycol.

(c) Count the atoms of each type in the molecule to obtain the molecular formula.

PROBLEM-SOLVING PRACTICE 3.1

Write the molecular formulas for these compounds.

(a) Adenosine triphosphate (ATP), an energy source in biochemical reactions, contains 10 carbon, 11 hydrogen, 13 oxygen, 5 nitrogen, and 3 phosphorus atoms per molecule

(b) Capsaicin, the active ingredient in chili peppers, has 18 carbon, 27 hydrogen, 3 oxygen, and 1 nitrogen atoms per molecule

(c) Oxalic acid has the condensed formula HOOCCOOH and is found in rhubarb.

EXERCISE 3.1 Structural, Condensed, and Molecular Formulas

A molecular model of propylene glycol, used in some "environmentally friendly" antifreezes, is shown in the margin. Write the molecular formula, the structural formula, and the condensed formula for propylene glycol.

3.2 Naming Binary Inorganic Compounds

Each chemical compound has a unique name that is assigned in a systematic way based on well-established rules. We describe some naming rules here and introduce other naming rules as we need them.

Binary molecular compounds consist of molecules that contain atoms of only two elements. There is a binary compound of hydrogen with every nonmetal except the noble gases. For hydrogen compounds of most nonmetals, particularly those in Groups 6A and 7A, the hydrogen is written first in the formula and named first. The other nonmetal is then named, with the nonmetal's name changed to end in *-ide*. For example, HCl is named hydrogen chloride.

Hydrogen compounds with carbon are discussed in the next section.

H_2O is written with H before O, as are the hydrogen compounds of Groups 6A and 7A: H_2S, H_2Se, and HF, HCl, HBr, and HI. Other H-containing compounds are usually written with the H atom after the other atom.

Formula	Name
HCl	Hydrogen chloride
HBr	Hydrogen bromide
HI	Hydrogen iodide
H_2Se	Hydrogen selenide

hydrogen chloride   hydrogen bromide   hydrogen iodide

Many binary molecular compounds contain nonmetallic elements from Groups 4A, 5A, 6A, and 7A of the periodic table. In these compounds the elements are listed in formulas and names in the order of the group numbers, and prefixes are used to designate the number of a particular kind of atom. The prefixes are listed in Table 3.2. Table 3.3 illustrates how these prefixes are applied.

A number of binary nonmetal compounds were discovered and named before systematic naming rules were developed. Their *common names* are still used today and must simply be learned.

Formula	Common Name	Formula	Common Name
H_2O	Water	NO	Nitric oxide
NH_3	Ammonia	N_2O	Nitrous oxide ("laughing gas")
N_2H_4	Hydrazine	PH_3	Phosphine

Table 3.2 Prefixes Used in Naming Chemical Compounds

Prefix	Number
Mono-	1
Di-	2
Tri-	3
Tetra-	4
Penta-	5
Hexa-	6
Hepta-	7
Octa-	8
Nona-	9
Deca-	10

Table 3.3 Examples of Binary Compounds

Molecular Formula	Name	Use
CO	Carbon monoxide	Steel manufacturing
NO_2	Nitrogen dioxide	Preparation of nitric acid
N_2O	Dinitrogen oxide	Anesthetic; spray can propellant
N_2O_5	Dinitrogen pentaoxide	Forms nitric acid
PBr_3	Phosphorus tribromide	Forms phosphorous acid
PBr_5	Phosphorus pentabromide	Forms phosphoric acid
SF_6	Sulfur hexafluoride	Transformer insulator
P_4O_{10}	Tetraphosphorus decaoxide	Drying agent

PROBLEM-SOLVING EXAMPLE 3.2 Naming Binary Inorganic Compounds

Name these compounds: (a) CO_2, (b) SiO_2, (c) SO_3, (d) N_2O_4, (e) PCl_5.

Answer

(a) Carbon dioxide (b) Silicon dioxide (c) Sulfur trioxide
(d) Dinitrogen tetraoxide (e) Phosphorus pentachloride

Strategy and Explanation These compounds consist entirely of nonmetals, so they are all molecular compounds. The prefixes in Table 3.2 are used as necessary.
(a) Use *di-* to represent the two oxygen atoms.
(b) Use *di-* for the two oxygen atoms.
(c) Use *tri-* for the three oxygen atoms.
(d) Use *di-* for the two nitrogen atoms and *tetra-* for the four oxygen atoms.
(e) Use *penta-* for the five chlorine atoms.

PROBLEM-SOLVING PRACTICE 3.2

Name these compounds: (a) SO_2, (b) BF_3, (c) CCl_4.

EXERCISE 3.2 Names and Formulas of Compounds

Give the formula for each of these binary nonmetal compounds:

(a) Carbon disulfide (b) Phosphorus trichloride (c) Sulfur dibromide
(d) Selenium dioxide (e) Oxygen difluoride (f) Xenon trioxide

3.3 Hydrocarbons

go Chemistry **Module 15: Naming Organic Compounds** covers concepts in this section.

Millions of organic compounds, including hydrocarbons, are known. They vary enormously in structure and function, ranging from the simple molecule methane (CH_4, the major constituent of natural gas) to large, complex biochemical molecules such as proteins, which often contain hundreds or thousands of atoms. Organic compounds are the main constituents of living matter. In organic compounds the carbon atoms are nearly always bonded to other carbon atoms and to hydrogen atoms. Among the reasons for the enormous variety of organic compounds is the characteristic property of carbon atoms to form strong, stable bonds with up to four other carbon atoms. A **chemical bond** is an attractive force between two atoms holding them together. Through their carbon-carbon bonds, carbon atoms can form chains, branched chains, rings, and other more complicated structures. With such a large number of compounds, dividing them into classes is necessary to make organic chemistry manageable.

Hydrocarbons, the simplest of the organic compounds, are composed of only carbon and hydrogen. A major class of hydrocarbons is the **alkanes,** which are economically important fuels and lubricants. The simplest alkane is methane, CH_4, which has a central carbon atom with four bonds joining it to four H atoms (p. 81). The general formula for noncyclic alkanes is C_nH_{2n+2}, where n is an integer. Table 3.4 provides some information about the first ten such alkanes. The first four (methane, ethane, propane, butane) have common names that must be memorized. For $n = 5$ or greater, the names are systematic. The prefixes of Table 3.2 indicate the number of carbon atoms in the molecule, and the ending *-ane* indicates that the compound is an alk*ane*. For example, the five-carbon alkane is *pentane*.

Methane, the simplest alkane, makes up about 85% of natural gas in the United States. Methane is also known to be one of the greenhouse gases (Section 10.11), meaning that it is one of the chemicals implicated in the problem of global warming. Ethane, propane, and butane are used as heating fuel for homes and in industry. In these simple alkanes, the carbon atoms are connected in unbranched (straight) chains, and each carbon atom is connected to either two or three hydrogen atoms.

In unbranched alkanes, all but two of the carbon atoms are each bonded to two hydrogen atoms. Two carbon atoms (one at each end of the molecule) are bonded to three hydrogen atoms each.

Table 3.4 The First Ten Alkane Hydrocarbons, C_nH_{2n+2}

Molecular Formula	Name	Boiling Point (°C)	Physical State at Room Temperature
CH_4	Methane	−161.6	Gas
C_2H_6	Ethane	−88.6	Gas
C_3H_8	Propane	−42.1	Gas
C_4H_{10}	Butane	−0.5	Gas
C_5H_{12}	Pentane	36.1	Liquid
C_6H_{14}	Hexane	68.7	Liquid
C_7H_{16}	Heptane	98.4	Liquid
C_8H_{18}	Octane	125.7	Liquid
C_9H_{20}	Nonane	150.8	Liquid
$C_{10}H_{22}$	Decane	174.0	Liquid

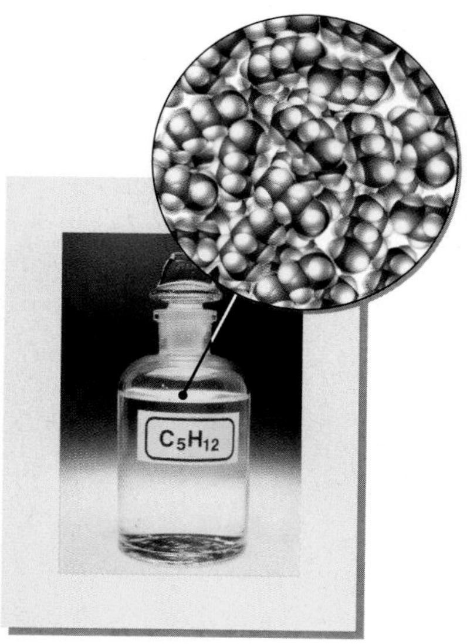

© Cengage Learning/Charles D. Winters

Pentane is an alkane used as a solvent.

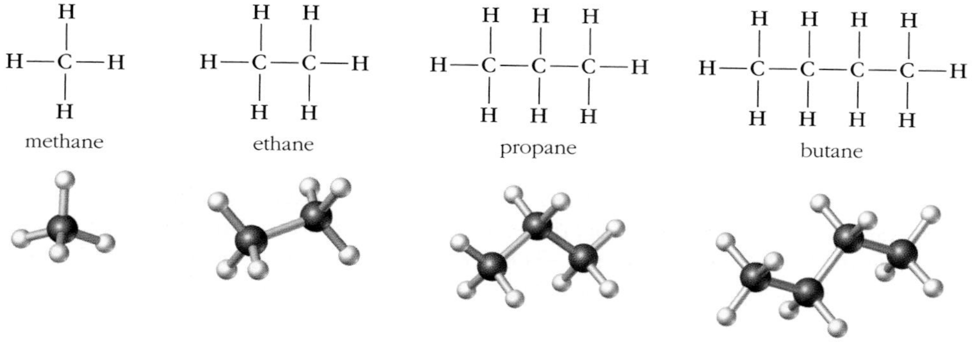

Larger alkanes have longer chains of carbon atoms with hydrogens attached to each carbon. For example, heptane, C_7H_{16}, is found in gasoline, and eicosane, $C_{20}H_{42}$, is found in paraffin wax.

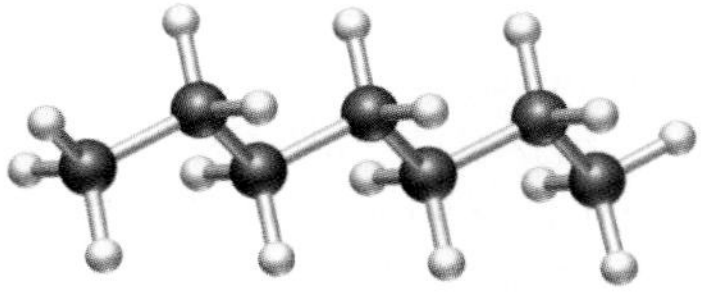

$CH_3(CH_2)_5CH_3$
heptane

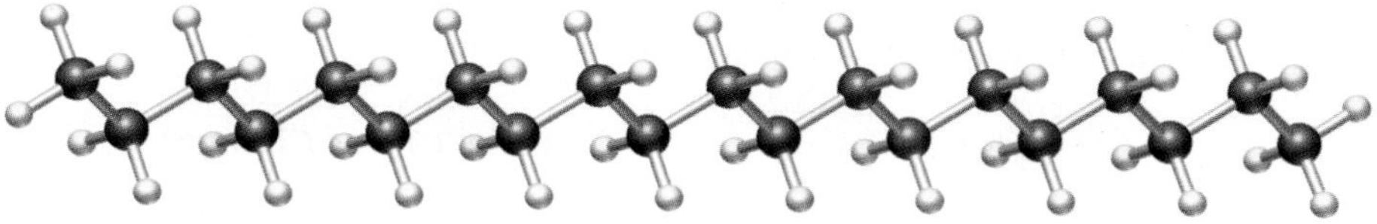

$CH_3(CH_2)_{18}CH_3$
eicosane

There are also cyclic hydrocarbons in which the carbon atoms are connected in rings, for example, cyclopentane. These cyclic alkanes have the general formula C_nH_{2n}.

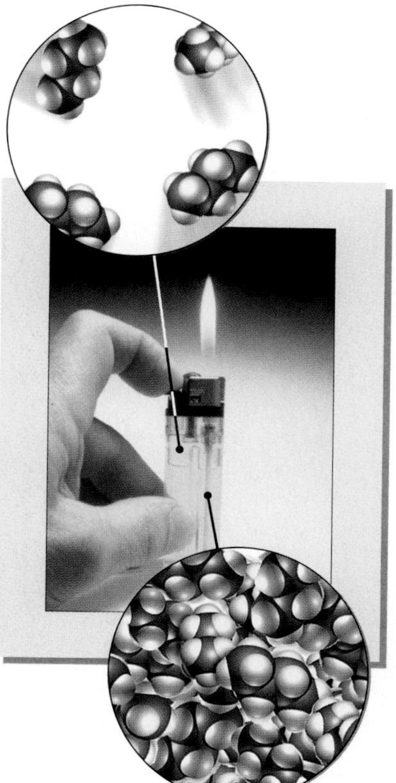

© Cengage Learning/Charles D. Winters

Butane, $CH_3CH_2CH_2CH_3$, is the fuel in this lighter. Butane molecules are present in the liquid and gaseous states in the lighter.

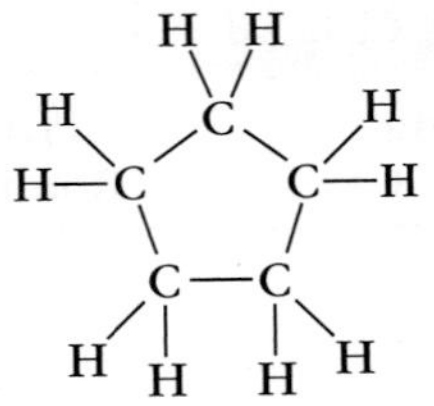

cyclopentane, C_5H_{10}

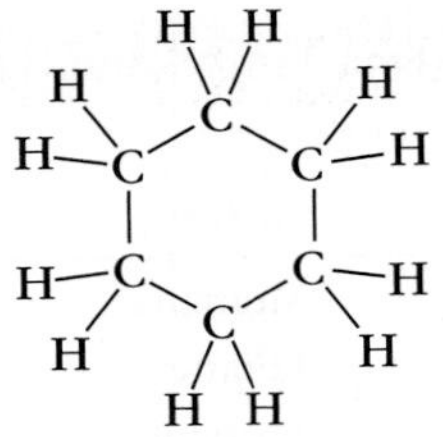

cyclohexane, C_6H_{12}

EXERCISE 3.3 Alkane Molecular Formulas

(a) Using the general formula for noncyclic alkanes, C_nH_{2n+2}, write the molecular formulas for the alkanes containing 16 and 28 carbon atoms.

(b) How many hydrogen atoms are present in tetradecane, which has 14 carbon atoms?

(c) Verify that each of these formulas corresponds to the general formula for noncyclic alkanes.

The molecular structures of hydrocarbons provide the framework for the discussion of the structures of all other organic compounds. If a different atom or combination of atoms replaces one or more of the hydrogens in the molecular structure of an alkane, a compound with different properties results. A hydrogen atom in an alkane can be replaced by a single atom such as a halogen, for example. In this way ethane, CH_3CH_3, becomes chloroethane, CH_3CH_2Cl. The replacement can also be a combination of atoms such as an oxygen bonded to a hydrogen, —OH, so ethane, CH_3CH_3, can be changed to ethanol, CH_3CH_2OH.

The molecular structures of organic compounds determine their properties. For example, the boiling point of ethane, CH_3CH_3, is −88.6 °C, but the boiling point of ethanol, CH_3CH_2OH, where the —OH group is substituted for one of the hydrogens, is 78.5 °C—quite a difference. This is due to the types of intermolecular interactions that are present; these effects will be fully explained in Section 9.6.

PROBLEM-SOLVING EXAMPLE 3.3 Alkanes

Table 3.4 gives the boiling points for the first ten noncyclic alkane hydrocarbons.

(a) Is the change in boiling point constant from one noncyclic alkane to the next in the series?

(b) Propose an explanation for the manner in which the boiling point changes from one noncyclic alkane to the next.

Answer

(a) No. The increment is large between methane and ethane and gets progressively smaller as the molecules get larger. It is only 23 °C between nonane and decane.

(b) Larger molecules interact more strongly and therefore require a higher temperature to move them apart to convert them from liquid to gas.

Strategy and Explanation We analyze the data given in Table 3.4.

(a) The boiling point differences between successive alkanes are

	B.P. (°C)		B.P. (°C)	Change (°C)
Methane	−162	to ethane	−89	73
Ethane	−89	to propane	−42	47
Propane	−42	to butane	0	42
Butane	0	to pentane	36	36

(continued on next page)

	B.P. (°C)		B.P. (°C)	Change (°C)
Pentane	36	to hexane	69	33
Hexane	69	to heptane	98	29
Heptane	98	to octane	126	28
Octane	126	to nonane	151	25
Nonane	151	to decane	174	23

The increments get smaller as the alkanes get larger.

(b) The larger molecules require a higher temperature to overcome their attraction to one another and to convert from liquid to gas *(← p. 20)*.

PROBLEM-SOLVING PRACTICE 3.3

Consider a series of molecules formed from the alkanes by substituting one of the hydrogen atoms with a chlorine atom. Would you expect a similar trend in changes in boiling points among this set of compounds as you observed with the alkanes themselves?

3.4 Alkanes and Their Isomers

Two or more compounds that have the same molecular formula but different arrangements of their atoms are called **isomers.** Because of the different arrangement of atoms in their molecules, isomers differ from one another in one or more physical properties, such as boiling point, color, and solubility; chemical reactivity differs as well. Several types of isomerism are possible, particularly in organic compounds. **Constitutional isomers** (also called structural isomers) are compounds with the same molecular formula that differ in the order in which their atoms are bonded.

Module 15: Naming Organic Compounds covers concepts in this section.

Straight-Chain and Branched-Chain Isomers of Alkanes

The first three alkanes—methane, ethane, and propane—have only one possible structural arrangement. When we come to an alkane with four carbon atoms, C_4H_{10}, there are two possible arrangements—a *straight* chain of four carbons (butane) or a *branched* chain of three carbons with the fourth carbon attached to the central atom of the chain of three (methylpropane), as shown in the table.

In this context, "straight chain" means a chain of carbon atoms with no branches to other carbon atoms; the carbon atoms are in an unbranched sequence. As you can see from the molecular model of butane, the chain is not actually straight, but rather a zigzag.

Molecular Formula	Condensed Formula	Structural Formula	Molecular Model	Melting Point (°C) Boiling Point (°C)
Butane C_4H_{10}	$CH_3CH_2CH_2CH_3$	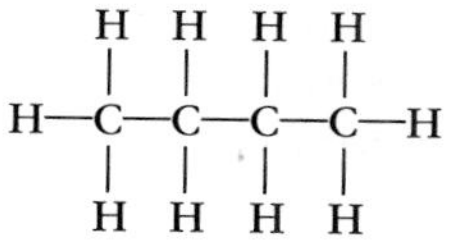	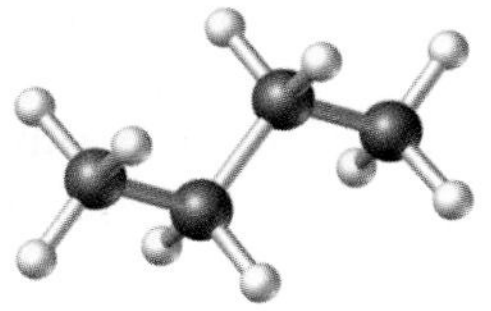	Melting point −138 Boiling point −0.5
Methylpropane C_4H_{10}	CH_3 | CH_3—CH—CH_3	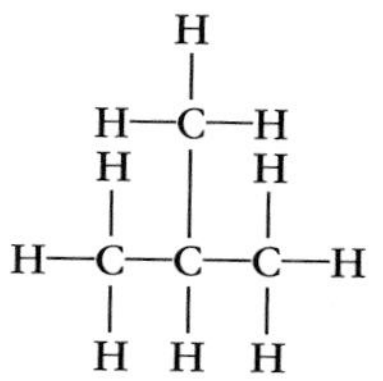	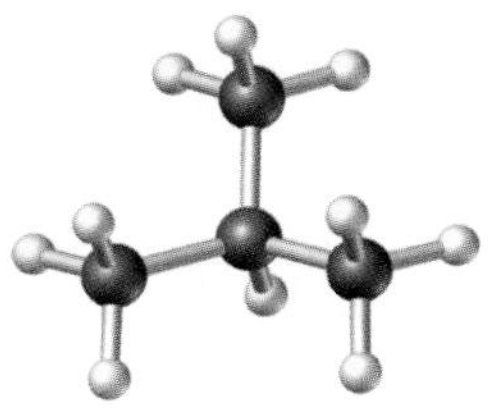	Melting point −145 Boiling point −11.6

Historically, straight-chain hydrocarbons were referred to as *normal* hydrocarbons, and *n*- was used as a prefix in their names. The current practice is not to use *n*-. If a hydrocarbon's name is given without indication that it is a branched-chain molecule, assume it is a straight-chain hydrocarbon.

Butane and methylpropane are constitutional isomers because they have the same molecular formula, but they are different compounds with different properties. Two constitutional isomers are different from each other in the same sense that two different structures built with identical Lego blocks are different from each other.

Methylpropane, the branched isomer of butane, has a *methyl* group, $—CH_3$, bonded to the central carbon atom. A methyl group is the simplest example of an **alkyl group,** the fragment of the molecule that remains when a hydrogen atom is removed from an alkane. Removal of one hydrogen atom from methane gives a methyl group:

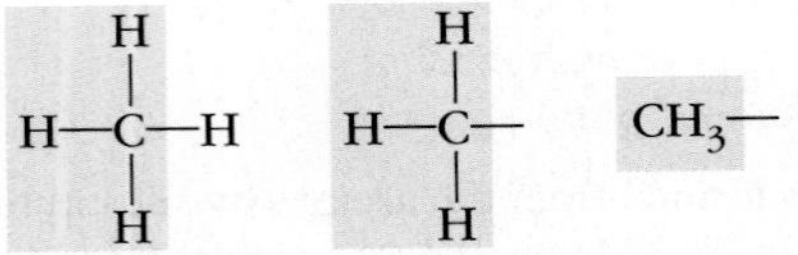

methyl group

Removal of one hydrogen atom from ethane gives an ethyl group:

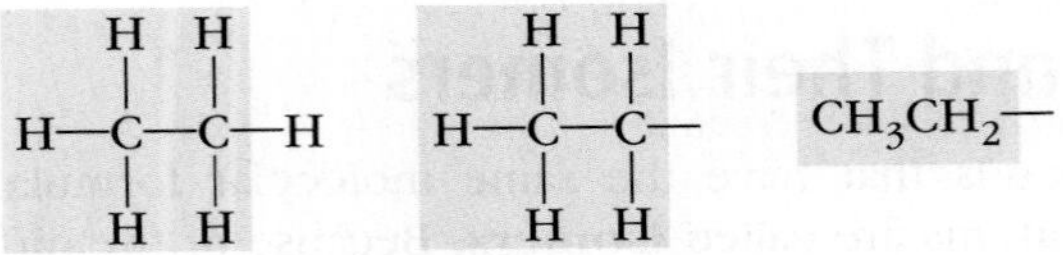

ethyl group

Alkyl groups are named by dropping *-ane* from the parent alkane name and adding *-yl.*

When we consider an alkyl group with three carbon atoms, there are two possibilities:

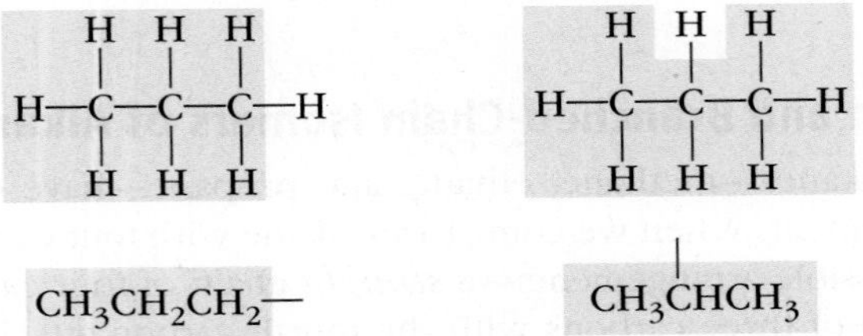

propyl group

isopropyl group

Theoretically, an alkyl group can be derived from any alkane. Some of the more common examples of alkyl groups are given in Table 3.5.

The number of alkane constitutional isomers grows rapidly as the number of carbon atoms increases because of the possibility of chain branching (Estimation box, p. 85). Table 3.6 shows the number of isomers for some alkanes. Chain branching is another reason for the enormous number of organic compounds.

Table 3.5 Some Common Alkyl Groups

Name	Condensed Structural Representation
Methyl	$CH_3—$
Ethyl	$CH_3CH_2—$
Propyl	$CH_3CH_2CH_2—$
Isopropyl	$CH_3CH—$ (with CH_3 attached below the CH) or $(CH_3)_2CH—$

Table 3.6 Alkane Isomers

Molecular Formula	Number of Isomers
C_4H_{10}	2
C_5H_{12}	3
C_6H_{14}	5
C_7H_{16}	9
C_8H_{18}	18
C_9H_{20}	35
$C_{10}H_{22}$	75
$C_{12}H_{26}$	355
$C_{15}H_{32}$	4347
$C_{20}H_{42}$	366,319

EXERCISES that are labeled **CONCEPTUAL** are designed to test your understanding of one or more concepts; they usually involve qualitative rather than quantitative thinking.

CONCEPTUAL EXERCISE 3.4 Straight-Chain and Branched-Chain Isomers

Three constitutional isomers are possible for pentane. Write structural and condensed formulas for these isomers.

ESTIMATION Number of Alkane Isomers

The number of possible carbon compounds is truly enormous. Table 3.6 (p. 84) shows how the number of isomers of the simplest hydrocarbon compounds, alkanes, increases as the number of carbon atoms increases. How could we use these data to estimate the number of alkane isomers for a much larger number of carbon atoms? More specifically, let's estimate the number of alkane isomers for C_{40} and check the result against the last entry in the table.

To picture the growth rate, we could plot the number of alkane isomers versus the number of carbon atoms. If we made such a linear plot—that is, with the *x*-axis as the number of carbon atoms and the *y*-axis as the number of alkane isomers—we would see very little, because the plot would be rising so fast. To keep the final point on the plot, the *y*-axis would be so expanded that all the other points would be squashed toward the bottom of the plot. Therefore, to make these data easier to view, we plot the logarithm of the number of isomers, $\log(N_i)$, versus the number of carbon atoms (see the figure). The points lie on a slightly concave-upward curve, but a line fitted through them would be reasonably close to a straight line.

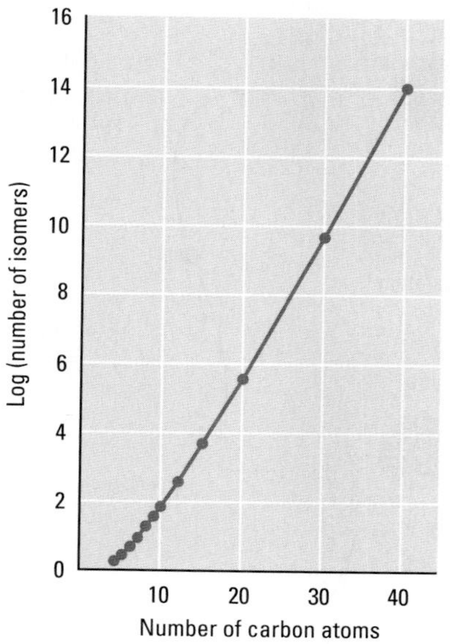

Semilog plot of the number of isomers versus the number of carbon atoms for alkanes.

Now we are ready to make our estimate. To estimate how many isomers there are for C_{40}, we will extrapolate from the C_{20} and C_{30} points. The $\log(N_i)$ for C_{20} is 5.56, and the $\log(N_i)$ for C_{30} is 9.61; the difference is $9.61 - 5.56 = 4.05$. Our estimate of the $\log(N_i)$ at C_{40} will be this increment added to the value of the $\log(N_i)$ for C_{30}:

$$\log(N_i) \text{ for } C_{40} = [\log(N_i) \text{ for } C_{30}] + \text{increment}$$
$$= 9.61 + 4.05 = 13.66$$

To calculate the number of isomers at C_{40} we take the antilog$(13.66) = 4.57 \times 10^{13}$. Since the curve on the plot is concave upward, we know that our estimate will be a little too low, but it is still reasonable. The actual number of alkane isomers for C_{40} is 6.25×10^{13}. Our estimate is only 27% off.

$$\frac{(6.25 - 4.57)}{6.25} \times 100\% = 27\%$$

Visit this book's companion website at **www.cengage.com/chemistry/moore** to work an interactive module based on this material.

3.5 Ions and Ionic Compounds

go Chemistry Module 2: Predicting Ion Charges covers concepts in this section.

Not all compounds are molecular. A compound whose nanoscale composition consists of positive and negative ions is classified as an **ionic compound.** Many common substances, such as table salt, NaCl; lime, CaO; and lye, NaOH, are ionic.

When metals react with nonmetals:

- **Metal atoms typically lose electrons to form positive ions, cations** (pronounced CAT-ions).
- **The quantity of positive charge on the positive ion equals the number of electrons lost by the neutral metal atom.**

The terms "cation" and "anion" are derived from the Greek words *ion* (traveling), *cat* (down), and *an* (up).

Cations *always* have *fewer* electrons than protons. For example, Figure 3.1 shows that an electrically neutral sodium atom, which has 11 protons and 11 electrons, loses one electron to become a sodium cation, which has 11 protons but only 10 electrons, and thus a *net* 1+ charge, symbolized as Na^+. When a neutral magnesium atom loses two electrons, it forms a 2+ magnesium ion, Mg^{2+}.

Conversely, when nonmetals react with metals:

- **Nonmetal atoms typically gain electrons to form negatively charged ions, anions** (pronounced ANN-ions).
- **The quantity of negative charge on the negative ion equals the number of electrons gained by the neutral nonmetal atom.**

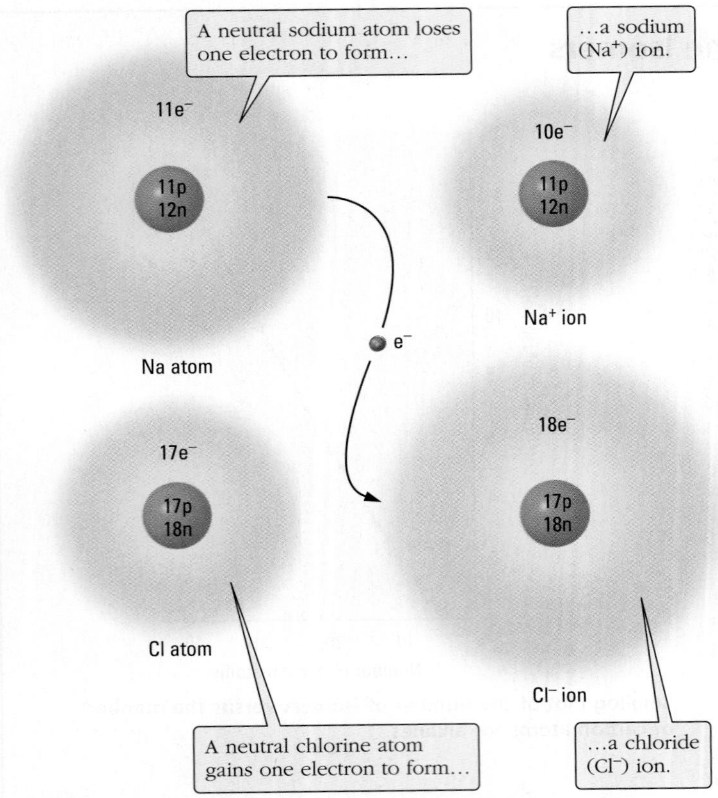

Active Figure 3.1 Formation of the ionic compound NaCl. Visit this book's companion website at **www.cengage.com/chemistry/moore** to test your understanding of the concepts in this figure.

Ionic compounds. Red iron(III) oxide, black copper(II) bromide, CaF_2 (front crystal), and NaCl (rear crystal).

Anions *always* have *more* electrons than protons. Figure 3.1 shows that a neutral chlorine atom (17 protons, 17 electrons) gains an electron to form a chlor*ide* ion, Cl^-. With 17 protons and 18 electrons, the chloride ion has a *net* 1− charge. A sulfur atom that gains two electrons forms a sulf*ide* ion, S^{2-}.

Monatomic Ions

A **monatomic ion** is a single atom that has lost or gained electrons. The charges of the common monatomic ions are given in Figure 3.2.

- **Metals of Groups 1A, 2A, and 3A form monatomic ions with charges *equal to the A-group number.***

Group	Neutral Metal Atom	Electrons Lost = (A-Group Number)	Metal Ion
1A	K (19 protons, 19 electrons)	1	K^+ (19 protons, 18 electrons)
2A	Mg (12 protons, 12 electrons)	2	Mg^{2+} (12 protons, 10 electrons)
3A	Al (13 protons, 13 electrons)	3	Al^{3+} (13 protons, 10 electrons)

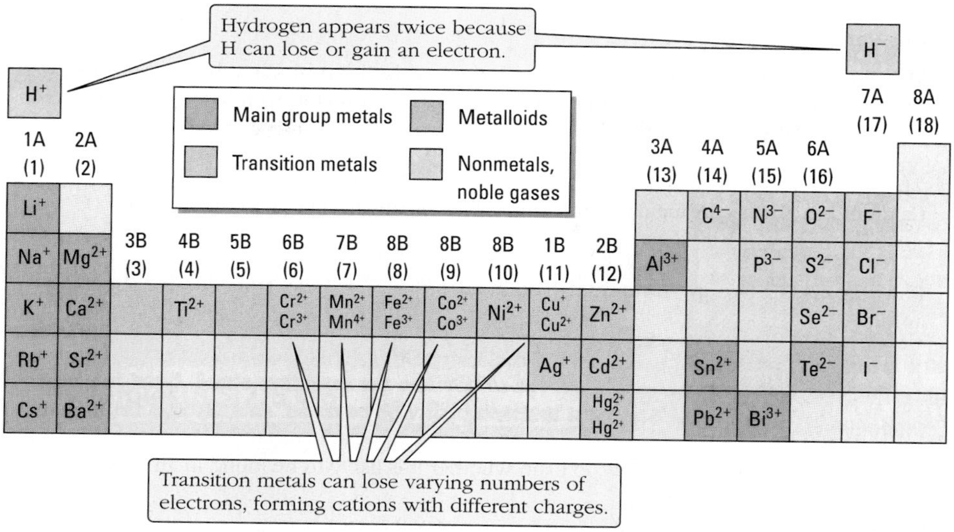

Figure 3.2 Charges on some common monatomic cations and anions. Note that metals generally form cations. The cation charge is given by the group number in the case of the main group elements of Groups 1A, 2A, and 3A *(gray)*. For transition elements *(blue)*, the positive charge is variable, and other ions in addition to those illustrated are possible. Nonmetals *(lavender)* generally form anions that have a negative charge equal to 8 minus the A-group number.

It is extremely important that you know the ions commonly formed by the elements shown in Figure 3.2 so that you can recognize ionic compounds and their formulas and write their formulas as reaction products (Section 5.1).

- **Nonmetals of Groups 5A, 6A, and 7A form monatomic ions that have a negative charge *usually equal to 8 minus the A-group number.***

In Section 8.2 we explain the basis for the (8 − A-group number) relationship for nonmetals.

Group	Neutral Nonmetal Atom	Electrons Gained = 8 − (A-Group Number)	Nonmetal Ion
5A	N (7 protons, 7 electrons)	3 = (8 − 5)	N^{3-} (7 protons, 10 electrons)
6A	S (16 protons, 16 electrons)	2 = (8 − 6)	S^{2-} (16 protons, 18 electrons)
7A	F (9 protons, 9 electrons)	1 = (8 − 7)	F^{-} (9 protons, 10 electrons)

You might have noticed in Figure 3.2 that hydrogen appears at two locations in the periodic table. This is because a hydrogen atom can either lose or gain an electron. When it loses an electron, it forms a hydrogen ion, H^+ (1 proton, 0 electrons). When it gains an electron, it forms a hydride ion, H^- (1 proton, 2 electrons).

Noble gas atoms do not easily lose or gain electrons and have no common ions to list in Figure 3.2.

Transition metals form cations, but these metal atoms can lose varying numbers of electrons, thus forming ions of different charges (Figure 3.2). Therefore, the group number is not an accurate guide to charges in these cases. It is important to learn which ions are formed most frequently by these transition metals. Many transition metals form 2+ and 3+ ions. For example, iron atoms can lose two or three electrons to form Fe^{2+} (26 protons, 24 electrons) or Fe^{3+} (26 protons, 23 electrons), respectively.

An older naming system for distinguishing between metal ions of different charges uses the ending *-ic* for the ion of higher charge and *-ous* for the ion of lower charge. These endings are combined with the element's name—for example, Fe^{2+} (ferrous) and Fe^{3+} (ferric) or Cu^+ (cuprous) and Cu^{2+} (cupric). We will not use these names in this book, but you might encounter them elsewhere.

PROBLEM-SOLVING EXAMPLE **3.4** **Predicting Ion Charges**

Using a periodic table, predict the charges on ions of aluminum, calcium, and phosphorus, and write symbols for these ions.

Answer Al^{3+}, Ca^{2+}, P^{3-}

Strategy and Explanation We find each element in the periodic table and use its position to answer the question. Aluminum is a Group 3A metal, so it loses three electrons to give the Al^{3+} cation.

$$Al \longrightarrow Al^{3+} + 3\,e^-$$

Calcium is a Group 2A metal, so it loses two electrons to give the Ca^{2+} cation.

$$Ca \longrightarrow Ca^{2+} + 2\,e^-$$

Phosphorus is a Group 5A nonmetal, so it gains $8 - 5 = 3$ electrons to give the P^{3-} anion.

$$P + 3\,e^- \longrightarrow P^{3-}$$

PROBLEM-SOLVING PRACTICE 3.4

For each of the ions listed below, explain whether it is likely to be found in an ionic compound.

(a) Ca^{4+} (b) Cr^{2+} (c) Sr^-

Polyatomic Ions

A **polyatomic ion** is a *unit* of two or more atoms that bears a net electrical charge. Table 3.7 lists some common polyatomic ions. Polyatomic ions are found in many places—oceans, minerals, living cells, and foods. For example, hydrogen carbonate (bicarbonate) ion, HCO_3^-, is present in rain water, sea water, blood, and baking soda. This polyatomic ion consists of one carbon atom, three oxygen atoms, and one hydrogen atom, with one unit of negative charge spread over the group of five atoms. The polyatomic sulfate ion, SO_4^{2-}, consists of one sulfur atom and four oxygen atoms and has an overall charge of 2−. One of the most common polyatomic cations is

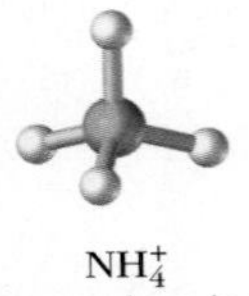

NH_4^+
ammonium ion

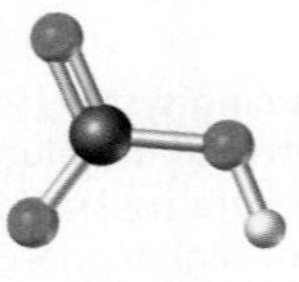

HCO_3^-
hydrogen carbonate ion

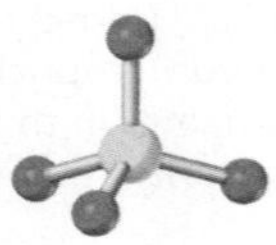

SO_4^{2-}
sulfate ion

Table 3.7 Common Polyatomic Ions

Cation (1+)			
NH_4^+	Ammonium	Hg_2^{2+}	Mercury(I)
Anions (1−)			
OH^-	Hydroxide	NO_2^-	Nitrite
HSO_4^-	Hydrogen sulfate	NO_3^-	Nitrate
CH_3COO^-	Acetate	MnO_4^-	Permanganate
ClO^-	Hypochlorite	$H_2PO_4^-$	Dihydrogen phosphate
ClO_2^-	Chlorite	CN^-	Cyanide
ClO_3^-	Chlorate	HCO_3^-	Hydrogen carbonate (bicarbonate)
ClO_4^-	Perchlorate		
Anions (2−)			
CO_3^{2-}	Carbonate	SO_3^{2-}	Sulfite
HPO_4^{2-}	Monohydrogen phosphate	SO_4^{2-}	Sulfate
$Cr_2O_7^{2-}$	Dichromate	$C_2O_4^{2-}$	Oxalate
$S_2O_3^{2-}$	Thiosulfate		
Anion (3−)			
PO_4^{3-}	Phosphate		

NH_4^+, the ammonium ion. In this case, four hydrogen atoms are connected to a nitrogen atom, and the group bears a net 1+ charge. (We discuss the naming of polyatomic ions containing oxygen atoms in Section 3.6.)

In many chemical reactions the polyatomic ion unit remains intact. *It is important to know the names, formulas, and charges of the common polyatomic ions listed in Table 3.7.*

Writing Formulas for Ionic Compounds

All compounds are electrically neutral. Therefore, when cations and anions combine to form an ionic compound, there must be zero net charge. The *total* positive charge of all the cations must equal the *total* negative charge of all the anions. For example, consider the ionic compound formed when potassium reacts with sulfur. Potassium is a Group 1A metal, so a potassium atom loses one electron to become a K^+ ion. Sulfur is a Group 6A nonmetal, so a sulfur atom gains two electrons to become an S^{2-} ion. To make the compound electrically neutral, two K^+ ions (total charge 2+) are needed for each S^{2-} ion. Consequently, the compound has the formula K_2S. The subscripts in an ionic compound formula show the numbers of ions included in the simplest formula unit. In this case, the subscript 2 indicates two K^+ ions for every S^{2-} ion.

As with the formulas for molecular compounds, a subscript of 1 in formulas of ionic compounds is understood to be there and is not written.

Similarly, aluminum oxide, a combination of Al^{3+} and O^{2-} ions, has the formula Al_2O_3: 2 Al^{3+} gives 6+ charge; 3 O^{2-} gives 6− charge; total charge = 0.

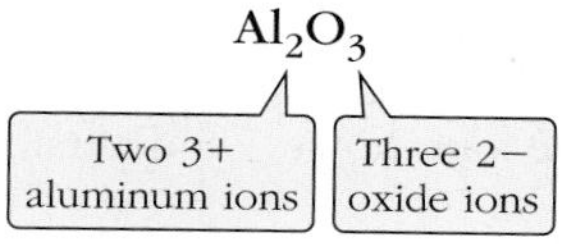
Al_2O_3
Two 3+ aluminum ions
Three 2− oxide ions

Notice that in writing the formulas for ionic compounds, *the cation symbol is written first, followed by the anion symbol.* The charges of the ions are *not* included in the formulas of ionic compounds.

Let's now consider several ionic compounds of magnesium, a Group 2A metal that forms Mg^{2+} ions.

Combining Ions	Overall Charge	Formula
Mg^{2+} and Br^-	$(2+) + 2(1-) = 0$	$MgBr_2$
Mg^{2+} and SO_4^{2-}	$(2+) + (2-) = 0$	$MgSO_4$
Mg^{2+} and OH^-	$(2+) + 2(1-) = 0$	$Mg(OH)_2$
Mg^{2+} and PO_4^{3-}	$3(2+) + 2(3-) = 0$	$Mg_3(PO_4)_2$

Notice in the latter two cases that when a polyatomic ion occurs more than once in a formula, the polyatomic ion's formula is put in parentheses followed by the necessary subscript.

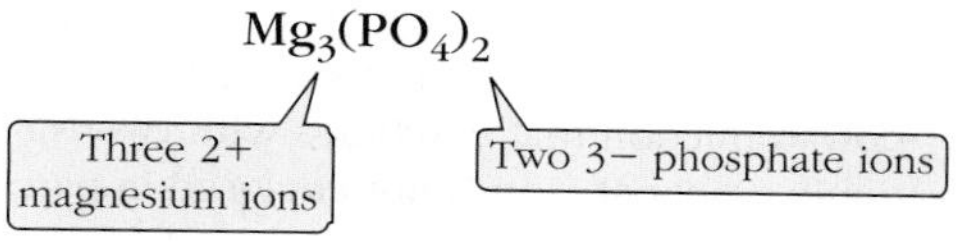
$Mg_3(PO_4)_2$
Three 2+ magnesium ions
Two 3− phosphate ions

PROBLEM-SOLVING EXAMPLE 3.5 Ions in Ionic Compounds

For each of these compounds, give the symbol or formula of each ion present and indicate how many of each ion are represented in the formula.

(a) K_2SO_4 (b) Na_2S (c) $Mg(CH_3COO)_2$ (d) $(NH_4)_2CO_3$

Answer

(a) Two K^+, one SO_4^{2-} (b) Two Na^+, one S^{2-}
(c) One Mg^{2+}, two CH_3COO^- (d) Two NH_4^+, one CO_3^{2-}

Strategy and Explanation

(a) Potassium is a Group 1A element and therefore forms a 1+ ion. Two K+ ions are necessary to balance the 2− charge of the sulfate ion.
(b) Sodium is a Group 1A element and therefore forms Na^+. The S^{2-} ion is formed from sulfur, which is a Group 6A element, by gaining two electrons (8 − 6 = 2). Two Na^+ ions and one S^{2-} ion maintain electrical neutrality.
(c) Magnesium is a Group 2A element and therefore forms Mg^{2+} ions. To form an electrically neutral compound, two acetate ions, CH_3COO^-, each with a 1− charge, are necessary to offset the 2+ charge on the Mg^{2+}.
(d) Ammonium ions each have a net 1+ charge. Each carbonate ion has a net 2− charge. Therefore, to maintain electrical neutrality, two ammonium ions are needed to offset the carbonate ion's 2− charge.

PROBLEM-SOLVING PRACTICE 3.5

Determine how many ions are present in each of these formulas.
(a) $CaSO_3$ (b) $Mg_3(PO_4)_2$

PROBLEM-SOLVING EXAMPLE 3.6 Formulas of Ionic Compounds

Write the correct formulas for ionic compounds composed of (a) calcium and fluoride ions, (b) barium and phosphate ions, (c) Fe^{3+} and nitrate ions, and (d) sodium and carbonate ions.

Answer

(a) CaF_2 (b) $Ba_3(PO_4)_2$ (c) $Fe(NO_3)_3$ (d) Na_2CO_3

Strategy and Explanation We solve the problem by applying these rules:

- **Locate each element in the periodic table.**
- **Use these locations (Groups) to determine the cation (metal) and anion (nonmetal) charges. Metal atoms form cations, and nonmetal atoms form anions.**
- **Identify any polyatomic anions and their charges.**
- **An ionic compound is electrically neutral; the sum of the positive charges of the cations must equal the sum of the negative charges of the anions.**

(a) Calcium is a Group 2A metal, so it forms 2+ ions. Fluorine is a Group 7A nonmetal, so it forms 1− ions. We need two F^- ions for every Ca^{2+} ion to make CaF_2 electrically neutral.
(b) Barium is a Group 2A metal, so it forms 2+ ions. Phosphate is a 3− polyatomic ion. For electrical neutrality, we need three Ba^{2+} ions and two PO_4^{3-} ions to form $Ba_3(PO_4)_2$. Because there is more than one polyatomic phosphate ion, its formula is enclosed in parentheses followed by the proper subscript.
(c) Iron is in its Fe^{3+} state. Nitrate is a 1− polyatomic ion. Therefore, we need three nitrate ions for each Fe^{3+} ion to form $Fe(NO_3)_3$. The polyatomic nitrate ion is enclosed in parentheses followed by the proper subscript.
(d) Carbonate is a 2− polyatomic ion that combines with two Na^+ ions to form Na_2CO_3. The polyatomic carbonate ion is not enclosed in parentheses since the formula contains only one carbonate.

PROBLEM-SOLVING PRACTICE 3.6

For each of these ionic compounds, write a list of which ions and how many of each are present.
(a) $MgBr_2$ (b) Li_2CO_3 (c) NH_4Cl (d) $Fe_2(SO_4)_3$
(e) Copper is a transition element that can form two compounds with bromine containing either Cu^+ or Cu^{2+}. Write the formulas for these compounds.

3.6 Naming Ions and Ionic Compounds

Ionic compounds can be named unambiguously by using the rules given in this section. You should learn these rules thoroughly.

go Chemistry **Module 3: Names to Formulas of Ionic Compounds** covers concepts in this section.

Naming Positive Ions

Most cations used in this book are metal ions that can be named by the rules given below. The ammonium ion, NH_4^+, is the major exception; it is a polyatomic ion composed of nonmetal atoms.

1a. ***For metals that form only one kind of cation, the name is simply the name of the metal plus the word "ion."*** For example, Mg^{2+} is the magnesium ion.

1b. ***For metals that can form more than one kind of cation, the name of each ion must indicate its charge. The charge is indicated by a Roman numeral in parentheses immediately following the ion's name (the Stock system).*** For example, Cu^{2+} is the copper(II) ion and Cu^+ is the copper(I) ion.

The Stock system is named after Alfred Stock (1876–1946), a German chemist famous for his work on the hydrogen compounds of boron and silicon.

Naming Negative Ions

These rules apply to naming anions.

2a. ***A monatomic anion is named by adding -ide to the stem of the name of the nonmetal element from which the ion is derived.*** For example, a *phosphorus* atom gives a *phosphide* ion, and a *chlorine* atom forms a *chloride* ion. Anions of Group 7A elements, the halogens, are collectively called **halide ions.**

2b. ***The names of the most common polyatomic ions are given in Table 3.7*** (p. 88). Most must simply be memorized. However, some guidelines can help, especially for **oxoanions,** which are polyatomic ions containing oxygen.

© Cengage Learning/Charles D. Winters

Copper(I) oxide *(left)* and copper(II) oxide. The different copper ion charges result in different colors.

For oxoanions with a nonmetal in addition to oxygen, the oxoanion with the greater number of oxygen atoms is given the suffix **-ate**.

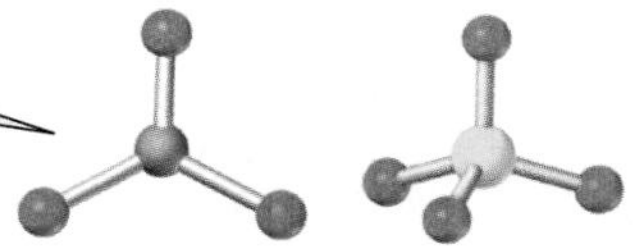

NO_3^- nitr*ate* ion $\quad$ SO_4^{2-} sulf*ate* ion

The oxoanion with the smaller number of oxygen atoms is given the suffix **-ite**.

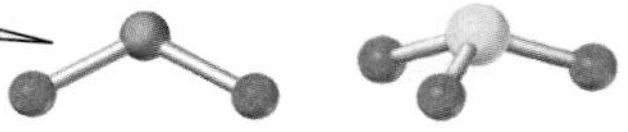

NO_2^- nitr*ite* ion $\quad$ SO_3^{2-} sulf*ite* ion

When more than two different oxoanions of a given nonmetal exist, a more extended naming scheme must be used. When there are four different oxoanions, the two middle ones are named according to the *-ate* and *-ite* endings; the oxoanion containing the largest number of oxygen atoms is given the prefix *per-* and the suffix *-ate,* and the oxoanion containing the smallest number is given the prefix *hypo-* and the suffix *-ite.* The oxoanions of chlorine are good examples:

Oxoanions having one more oxygen atom than the **-ate** ion are named using the prefix **per-**.

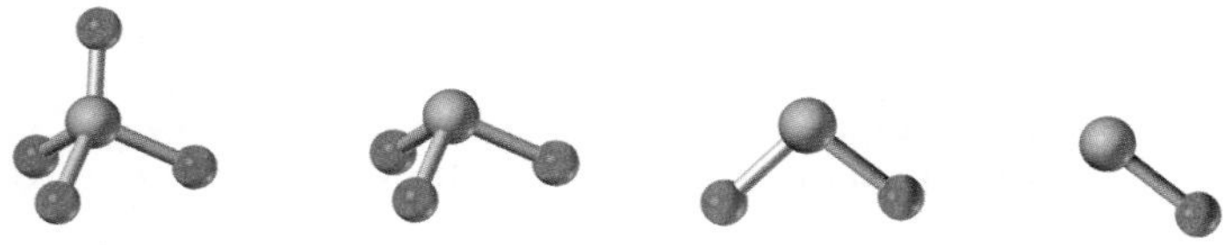

ClO_4^- *per*chlor*ate* ion $\quad$ ClO_3^- chlor*ate* ion $\quad$ ClO_2^- chlor*ite* ion $\quad$ ClO^- *hypo*chlor*ite* ion

Oxoanions having one fewer oxygen atom than the **-ite** ion are named using the prefix **hypo-**.

The same naming rules also apply to the oxoanions of bromine.

Note that the negative charge of the polyatomic ion decreases by one for each hydrogen added.

Oxoanions containing hydrogen are named simply by adding the word "hydrogen" before the name of the oxoanion, for example, hydrogen sulfate ion, HSO_4^-. When an oxoanion of a given nonmetal can combine with different numbers of hydrogen atoms, we must use prefixes to indicate which ion we are talking about: *di*hydrogen phosphate for $H_2PO_4^-$ and *mono*hydrogen phosphate for HPO_4^{2-}. Because many hydrogen-containing oxoanions have common names that are used often, you should know them. For example, the hydrogen carbonate ion, HCO_3^-, is often called the bicarbonate ion.

Naming Ionic Compounds

Table 3.8 lists common names and systematic names of several useful ionic compounds. In the systematic name, ***the name of the cation comes first, then the name of the anion.*** The cation is named using the rules for naming positive ions on p. 91, and the anion is named using the rules for negative ions. Also, the word "ion" is not used with the cation name.

Notice these examples from Table 3.8:

- Calcium oxide, CaO, is named from calcium for Ca^{2+} (Rule 1a) and oxide for O^{2-} (Rule 2a). Likewise, sodium chloride is derived from sodium (Na^+, Rule 1a) and chloride (Cl^-, Rule 2a).
- Ammonium carbonate, $(NH_4)_2CO_3$, contains two polyatomic ions named in Table 3.7 (p. 88).
- In the name copper(II) sulfate, the (II) indicates that Cu^{2+} is present, not Cu^+, the other possibility.

PROBLEM-SOLVING EXAMPLE 3.7 Using Formulas to Name Ionic Compounds

Name each of these ionic compounds.

(a) KCl (b) $Ca(OH)_2$ (c) $Fe_3(PO_4)_2$
(d) $Al(NO_3)_3$ (e) $(NH_4)_2SO_4$

Answer

(a) Potassium chloride (b) Calcium hydroxide (c) Iron(II) phosphate
(d) Aluminum nitrate (e) Ammonium sulfate

Strategy and Explanation

(a) The potassium ion, K^+, and the chloride ion, Cl^-, combine to form potassium chloride.
(b) The calcium ion, Ca^{2+}, and the hydroxide ion, OH^-, combine to form calcium hydroxide.
(c) The iron(II) ion, Fe^{2+}, and the phosphate ion, PO_4^{3-}, combine to give iron(II) phosphate.
(d) The aluminum ion, Al^{3+}, combines with the nitrate ion, NO_3^-, to form aluminum nitrate.
(e) The ammonium ion, NH_4^+, and the sulfate ion, SO_4^{2-}, combine to form ammonium sulfate.

PROBLEM-SOLVING PRACTICE 3.7

Name each of these ionic compounds:

(a) KNO_2 (b) $NaHSO_3$ (c) $Mn(OH)_2$
(d) $Mn_2(SO_4)_3$ (e) Ba_3N_2 (f) LiH

Table 3.8 Names of Some Useful Ionic Compounds

Common Name	Systematic Name	Formula
Baking soda	Sodium hydrogen carbonate	$NaHCO_3$
Lime	Calcium oxide	CaO
Milk of magnesia	Magnesium hydroxide	$Mg(OH)_2$
Table salt	Sodium chloride	$NaCl$
Smelling salts	Ammonium carbonate	$(NH_4)_2CO_3$
Lye	Sodium hydroxide	$NaOH$

© Cengage Learning/Charles D. Winters

Sodium chloride, NaCl. This common ionic compound contains sodium ions (Na^+) and chloride ions (Cl^-).

PROBLEM-SOLVING EXAMPLE 3.8 Using Names to Write Formulas of Ionic Compounds

Write the correct formula for each of these ionic compounds:

(a) Ammonium sulfide (b) Potassium sulfate
(c) Copper(II) nitrate (d) Iron(II) chloride

Answer

(a) $(NH_4)_2S$ (b) K_2SO_4 (c) $Cu(NO_3)_2$ (d) $FeCl_2$

Strategy and Explanation Determine the charge of each ion and then make certain that the total charge for the formula is zero.

(a) The ammonium cation is NH_4^+ and the sulfide ion is S^{2-}, so two ammonium ions are needed for one sulfide ion to make the electrically neutral compound.

(b) The potassium cation is K^+ and the sulfate anion is SO_4^{2-}, so two potassium ions are needed for each sulfate ion.

(c) The copper(II) cation is Cu^{2+} and the anion is NO_3^-, so two nitrate ions are needed.

(d) The iron(II) ion is Fe^{2+} and the chloride ion is Cl^-, so two chloride ions are needed.

PROBLEM-SOLVING PRACTICE 3.8

Write the correct formula for each of these ionic compounds:

(a) Potassium dihydrogen phosphate (b) Copper(I) hydroxide
(c) Sodium hypochlorite (d) Ammonium perchlorate
(e) Chromium(III) chloride (f) Iron(II) sulfite

CHEMISTRY IN THE NEWS Airport Runway Deicer Shortage

U.S. metropolitan airports with the largest average annual snowfall are in Cleveland, Denver, Salt Lake City, and Minneapolis-St. Paul, each with more than 50 inches of snow per year. For safety, airport runways must be treated when snow- or ice-covered. One of the primary deicing agents used for this purpose is Cryotech E36, an aqueous solution of potassium acetate, KCH_3COO. The deicer is a 50% by weight aqueous solution of potassium acetate; it also contains small amounts of corrosion inhibitors. Millions of gallons of this product are used annually. Cryotech E36 is biodegradable and nonpersistent, and it was approved by the Federal Aviation Administration (FAA) in 1992. This deicer replaced other deicers that used urea, a compound that releases ammonia into nearby streams. The FAA warned airports of a potential shortage of Cryotech E36 during the winter of 2009 due to a drop in production at a potash, K_2CO_3, mine in Canada. The shortage caused a spike in prices for the deicer during the winter of 2009.

Alternative deicers are available, but they are more expensive and have some less desirable characteristics. A recent study reported that potassium acetate may be more harmful to aquatic life near airports than was previously thought,[1] so further examination of alternatives may occur in the coming years.

Source: U.S.A. Today, Dec. 4, 2008.

[1]U.S. Geological Survey report: **http://www.usgs.gov/newsroom/article.asp?ID=2107**

3.7 Ionic Compounds: Bonding and Properties

The book *Salt: A World History,* by Mark Kurlansky, is a compelling account of table salt's importance through the ages.

The properties of an ionic compound differ significantly from those of its component elements. Consider the familiar ionic compound, table salt (sodium chloride, NaCl), composed of Na^+ and Cl^- ions. Sodium chloride is a white, crystalline, water-soluble solid, very different from its component elements, metallic sodium and gaseous chlorine. Sodium is an extremely reactive metal that reacts violently with water. Chlorine is a diatomic, toxic gas that reacts with water. Sodium *ions* and chloride *ions* do not undergo such reactions, and NaCl dissolves uneventfully in water.

Ionic Compounds

Ionic compounds contain cations and anions that are attracted to each other by electrostatic forces, the forces of attraction between positive and negative charges. This attraction is called **ionic bonding.** The strength of the electrostatic force dictates many of the properties of ionic compounds.

The attraction between oppositely charged ions increases with charge and decreases with the square of the distance between the ions. The force between two charged particles is given quantitatively by **Coulomb's law:**

$$F = k\frac{Q_1Q_2}{d^2}$$

where Q_1 and Q_2 are the magnitudes of the charges on the two interacting particles, d is the distance between the two particles, and k is a constant. For ions separated by the same distance, the attractive force between 2+ and 2− ions is four times greater than that between 1+ and 1− ions. The attractive force also increases as the distance between the centers of the ions decreases. Thus, a small cation and a small anion will attract each other more strongly than will larger ions. (We discuss the sizes of ions in Section 7.10.)

Classifying compounds as ionic or molecular is very useful because these two types of compounds have quite different properties and thus different uses. The following generalizations will help you to predict whether a compound is ionic.

A compound is ionic if it satisfies either of these conditions:

- **It is composed of a metal cation** (ions in the gray and blue areas in Figure 3.2) **and a nonmetal anion** (ions in the lavender area of Figure 3.2). Examples of such compounds include NaCl, $CaCl_2$, and KI.
- **It includes a polyatomic ion** (see Table 3.7). Examples include $CaSO_4$, $NaNO_3$, $Sr(OH)_2$, NH_4Cl, $(NH_4)_2SO_4$, $KMnO_4$, and $(NH_4)_2Cr_2O_7$.

Notice that ammonium ion, NH_4^+, is a polyatomic ion. Do not confuse it with ammonia, NH_3, a molecular compound.

A compound is likely to be molecular if it is composed of two or more nonmetals and does not contain ions. Examples include acetic acid, CH_3COOH, and urea, CH_4N_2O.

Metalloids (elements in the orange area of Figure 3.2) are present in either ionic or molecular compounds. For example, the metalloid boron can combine with a nonmetal chlorine to form the molecular compound BCl_3. The metalloid arsenic is found in the polyatomic arsenate ion, AsO_4^{3-}, in the ionic compound K_3AsO_4.

Potassium dichromate, $K_2Cr_2O_7$. This beautiful orange-red compound contains potassium ions, K^+, and dichromate ions, $Cr_2O_7^{2-}$.

PROBLEM-SOLVING EXAMPLE 3.9 Ionic and Molecular Compounds

Predict whether each compound is likely to be ionic or molecular. Explain your answers.

(a) Li_2CO_3 (b) $C_{10}H_{22}$ (c) $(NH_4)_2SO_3$
(d) N_2H_4 (e) Na_2S (f) P_4S_3

Answer

(a) Ionic (b) Molecular (c) Ionic
(d) Molecular (e) Ionic (f) Molecular

Strategy and Explanation For each compound we use the location of its elements in the periodic table and our knowledge of polyatomic ions to answer the question.

(a) Lithium is a Group 1A metal, so it always forms lithium ions, Li^+. The formula also contains the polyatomic carbonate ion, CO_3^{2-}, so the compound is ionic.
(b) This compound is composed entirely of nonmetal atoms, so it is molecular. It is a noncyclic alkane.
(c) This ionic compound contains the polyatomic ammonium cation, NH_4^+, and the polyatomic sulfite ion, SO_3^{2-}.
(d) This compound is composed entirely of nonmetal atoms, so it is molecular.
(e) Sodium is a Group 1A metal and sulfur is a Group 6A nonmetal, so they react to form an ionic compound of Na^+ and S^{2-} ions.
(f) This is a molecular compound composed entirely of two nonmetal atoms, phosphorus and sulfur.

PROBLEM-SOLVING PRACTICE 3.9

Predict whether each of these compounds is likely to be ionic or molecular:
(a) CH_4 (b) $CaBr_2$ (c) $MgCl_2$ (d) PCl_3 (e) KCl

In solid ionic compounds, cations and anions are held by ionic bonding in an orderly array called a **crystal lattice,** in which each cation is surrounded by anions and each anion is surrounded by cations. Such an arrangement maximizes the attraction between cations and anions and minimizes the repulsion between ions of like charge. In sodium chloride, as shown in Figure 3.3, six chloride ions surround each sodium ion, and six sodium ions surround each chloride ion. As indicated in the formula, there is one sodium ion for each chloride ion.

The formula of an ionic compound indicates only the smallest whole-number ratio of the number of cations to the number of anions in the compound. In NaCl that ratio is 1:1. An Na^+Cl^- pair is referred to as a **formula unit** of sodium chloride. Note that the formula unit of an ionic compound has no independent exis-

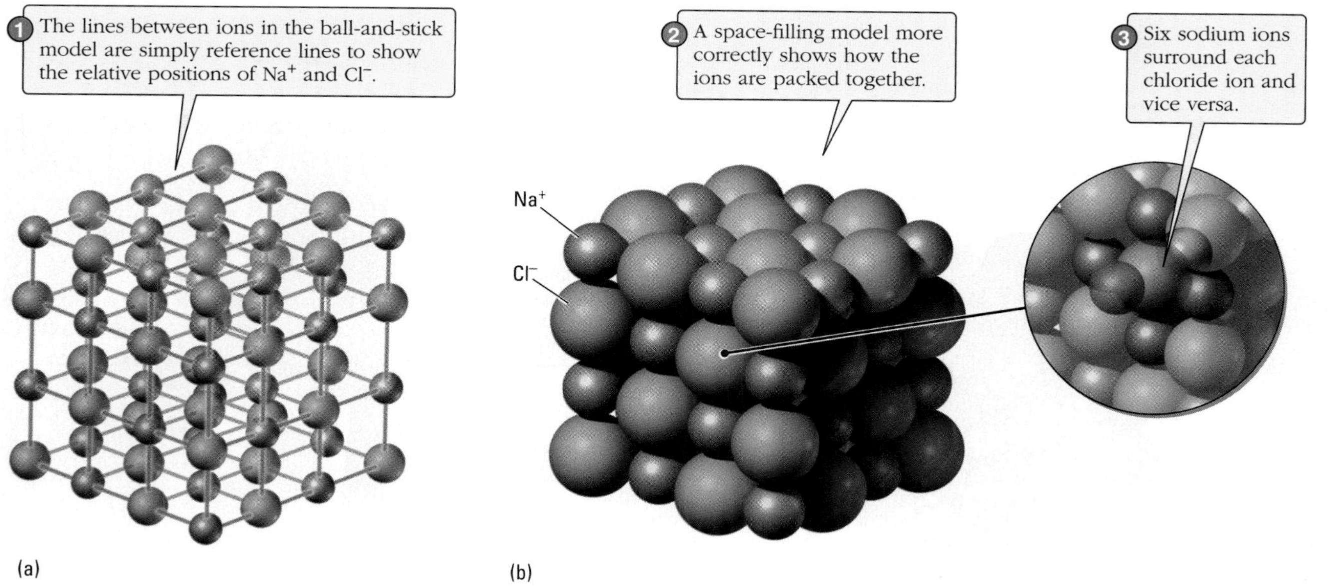
1 The lines between ions in the ball-and-stick model are simply reference lines to show the relative positions of Na+ and Cl−.
2 A space-filling model more correctly shows how the ions are packed together.
3 Six sodium ions surround each chloride ion and vice versa.
Na+
Cl−
(a)
(b)

Figure 3.3 Two models of a sodium chloride crystal lattice. (a) This ball-and-stick model illustrates clearly how the ions are arranged, although it shows the ions too far apart. (b) Although a space-filling model shows how the ions are packed and their relative sizes, it is difficult to see the locations of ions other than those on the faces of the crystal lattice.

tence outside of the crystal, which is different from the individual molecules of a molecular compound such as H_2O.

The regular array of cations and anions in a crystal lattice and the strong electrostatic attractions that hold the ions rather rigidly in position help us understand **three characteristic properties of ionic compounds:**

- **Ionic compounds have distinctive crystalline shapes and are easily cleaved.**
- **Ionic compounds have high melting points and are solids at room temperature.**
- **Ionic compounds do not conduct electricity when solid but do conduct when molten.**

Crystalline shapes are distinctive because the cations and anions are held rather rigidly in position. Such alignment creates planes of ions within the crystals and the angles between those planes determine the angles between sides of macroscopic crystals. For example, if you look closely (perhaps with a magnifying glass) at table salt, you will see that many of the salt crystals have 90° angles between sides. This is consistent with the 90° angles between layers of ions shown in Figure 3.3. Ionic crystals also can be split parallel to the planes of ions (Figure 3.4). When an outside force causes one plane to shift slightly relative to the next, ions of like charge are brought close together and repel strongly. The repulsion causes the layers on opposite sides of the cleavage plane to separate, and the crystal splits.

Melting points are high because, according to the kinetic-molecular theory (← *p. 19*), melting requires that ions break out of the rigid array in the solid and move about independently in the liquid. Because of the strong attractions among the cations and anions, a high temperature is required before molecular motion is great enough to overcome these attractions. Melting points are also related to the charges and sizes of the ions. For ions of similar size, such as the cations Na^+ and Ca^{2+} and the anions O^{2-} and F^-, Coulomb's law predicts that *the larger the charges, the greater the attraction and the higher the melting point.* For example, CaO (composed of doubly charged Ca^{2+} and O^{2-} ions) melts at 2572 °C, whereas NaF (composed of singly charged Na^+ and F^- ions) melts at 993 °C. For ions of similar charge but different sizes, such as F^- and the much larger I^-, Coulomb's law predicts that *the smaller the ion, the higher the melting point.* For example, NaF melts at 990 °C, whereas NaI melts at 651 °C.

Electric current involves movement of charged nanoscale particles such as electrons or ions. Because the ions in a crystal can only vibrate about fixed positions,

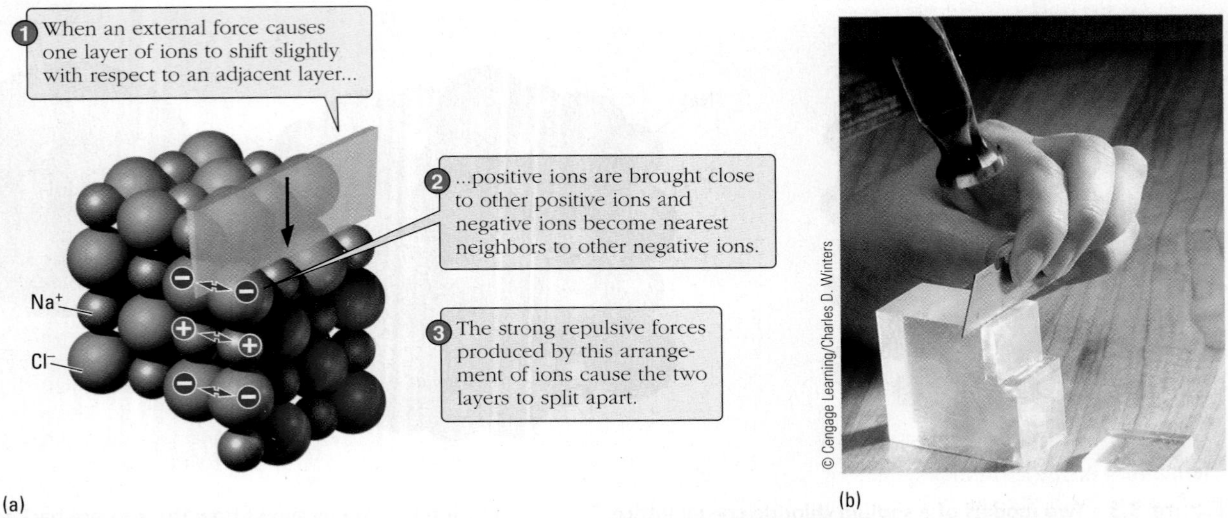
1 When an external force causes one layer of ions to shift slightly with respect to an adjacent layer...
2 ...positive ions are brought close to other positive ions and negative ions become nearest neighbors to other negative ions.
3 The strong repulsive forces produced by this arrangement of ions cause the two layers to split apart.
Na+
Cl−
© Cengage Learning/Charles D. Winters
(a)
(b)

Figure 3.4 Cleavage of an ionic crystal. (a) Diagram of the forces involved in cleaving an ionic crystal. (b) A sharp blow on a knife edge lying along a plane of a salt crystal causes the crystal to split.

© Cengage Learning/Charles D. Winters

Figure 3.5 A molten ionic compound conducts an electric current. A battery is connected by a wire to one metal electrode that dips into an ionic compound. A second electrode is connected to a light bulb and then back to the battery. When the ionic compound melts, the bulb lights, showing that the electrical circuit is complete and current is flowing. In the molten ionic compound, ions are freed from the crystal lattice of the solid. Cations move in one direction and anions in the opposite direction; this movement of electric charges constitutes the electric current that lights the bulb.

ionic solids do not conduct electricity. However, when an ionic solid melts, the liquid conducts electricity as shown in Figure 3.5. This is because the ions are now free to move relative to each other. Cations moving in one direction and anions moving in the opposite direction carry an electric current in the molten ionic compound, just as electrons carry current in a copper wire.

The general properties of molecular and ionic compounds are summarized in Table 3.9. In particular, note the differences in physical state, electrical conductivity, melting point, and water solubility.

Ionic Compounds in Aqueous Solution: Electrolytes

Many ionic compounds are soluble in water. As a result, the oceans, rivers, lakes, and even the tap water in our residences contain many kinds of ions in solution. This makes the solubilities of ionic compounds and the properties of ions in solution of great practical interest.

When dissolved in water, an ionic compound **dissociates**—the oppositely charged ions separate from one another. For example, when solid NaCl dissolves in water, it dissociates into Na^+ and Cl^- ions that become uniformly mixed with water molecules and dispersed throughout the solution.

Aqueous solutions of ionic compounds conduct electricity because the ions are free to move about (Figure 3.6). (This is the same mechanism of conductivity as for

© Cengage Learning/Charles D. Winters
K⁺ ion
H₂O
Cl⁻ ion

Figure 3.6 Electrical conductivity of an ionic compound solution. When an electrolyte, such as KCl, is dissolved in water and provides ions that move about, the electrical circuit is completed and the light bulb in the circuit glows. The ions of every KCl unit have dissociated: K^+ and Cl^-. The Cl^- ions move toward the positive electrode, and the K^+ ions move toward the negative electrode, transporting electrical charge through the solution.

Table 3.9 Properties of Molecular and Ionic Compounds

Molecular Compounds	Ionic Compounds
Many are formed by combination of nonmetals with other nonmetals or with some metals	Typically formed by combination of reactive metals with reactive nonmetals
Gases, liquids, solids	Crystalline solids
Brittle and weak or soft and waxy solids	Hard and brittle solids
Low melting points	High melting points
Low boiling points (−250 to 600 °C)	High boiling points (700 to 3500 °C)
Poor conductors of electricity	Good conductors of electricity when molten; poor conductors of electricity when solid
Poor conductors of heat	Poor conductors of heat
Solubility depends on molecular structure, often soluble in organic solvents	Often soluble in water
Examples: hydrocarbons, H_2O, CO_2, sugar	Examples: NaCl, CaF_2, NH_4NO_3

We use the term "dissociate" for ionic compounds that separate into their constituent ions in water. The term "ionize" is used for molecular compounds whose molecules react with water to form ions.

molten ionic compounds.) **Electrolytes** are substances that conduct electricity when dissolved in water.

Most molecular compounds that are water-soluble continue to exist as molecules in solution; table sugar (chemical name, sucrose) is an example. Substances such as sucrose whose solutions do not conduct electricity are called **nonelectrolytes.** Be sure you understand the difference between these two important properties of a compound: (a) its *solubility,* and (b) its ability to *dissociate,* releasing ions into solution. In Section 5.1 we will provide a much more detailed discussion of solubility and the dissociation of ions from electrolytes in solution.

CONCEPTUAL EXERCISE 3.5 Properties of Molecular and Ionic Compounds

Is a compound that is solid at room temperature and conducts electricity when in solution likely to be a molecular or ionic compound? Why?

go Chemistry **Module 4: The Mole** covers concepts in this section.

3.8 Moles of Compounds

In talking about compounds in quantities large enough to manipulate conveniently, we deal with moles of these compounds.

Molar Mass of Molecular Compounds

The most recognizable molecular formula, H_2O, shows us that there are two H atoms for every O atom in a water molecule. In two water molecules, therefore, there are four H atoms and two O atoms; in a dozen water molecules, there are two dozen H atoms and one dozen O atoms. We can extend this until we have one mole of water molecules (Avogadro's number of molecules, 6.022×10^{23}), each containing two moles of hydrogen atoms and one mole of oxygen atoms ***(← p. 59).*** We can also say that in 1.000 mol water there are 2.000 mol H atoms and 1.000 mol O atoms:

One mole of a molecular compound means 6.022×10^{23} molecules, not one molecule.

H_2O	H	O
6.022×10^{23} water molecules	$2\,(6.022 \times 10^{23}$ H atoms)	6.022×10^{23} O atoms
1.000 mol H_2O molecules	2.000 mol H atoms	1.000 mol O atoms
18.0153 g H_2O	2 (1.0079 g H) = 2.0158 g H	15.9994 g O

The mass of one mole of water molecules—its *molar mass*—is the sum of the masses of two moles of H atoms and one mole of O atoms: 2.0158 g H + 15.9994 g O = 18.0152 g water in a mole of water. For chemical compounds, the *molar mass,* in grams per mole, is *numerically the same* as the **molecular weight,** the sum of the atomic weights (in amu) of *all* the atoms in the compound's formula. The molar masses of several molecular compounds are shown in this table.

Compound	Structural Formula	Molecular Weight	Molar Mass
Ammonia NH_3	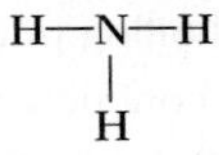	14.01 amu, N + 3 (1.01 amu, H) = 17.04 amu	17.04 g/mol
Trifluoromethane CHF_3	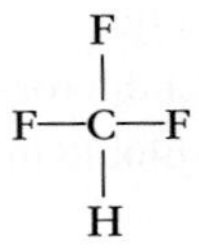	12.01 amu, C + 1.01 amu, H + 3 (19.00 amu, F) = 70.02 amu	70.02 g/mol

(continued on next page)

Compound	Structural Formula	Molecular Weight	Molar Mass
Sulfur dioxide SO_2	O=S—O	32.07 amu, S + 2 (16.00 amu, O) = 64.07 amu	64.07 g/mol
Glycerol $C_3H_8O_3$	CH_2OH | CHOH | CH_2OH	3 (12.01 amu, C) + 8 (1.01 amu, H) + 3 (16.00 amu, O) = 92.11 amu	92.11 g/mol

Molar Mass of Ionic Compounds

Because ionic compounds do not contain individual molecules, the term "formula weight" is sometimes used for ionic compounds instead of "molecular weight." As with molecular weight, an ionic compound's **formula weight** is the sum of the atomic weights of all the atoms in the compound's formula. The molar mass of an ionic compound, expressed in grams per mole (g/mol), is numerically equivalent to its formula weight. The term "molar mass" is used for both molecular and ionic compounds.

ESTIMATION Is Each Snowflake Unique?

An old adage says that no two snowflakes are alike. To see whether this assertion is plausible, let's estimate how many water molecules are contained in one snowflake. Our determination hinges on the enormous size of Avogadro's number.

The molar mass of H_2O is approximately 18 g/mol, and 1 mol H_2O contains Avogadro's number of molecules. How many molecules are in 1 g water?

$$\frac{6.022 \times 10^{23} \text{ molecules}}{1 \text{ mol}} \times \frac{1 \text{ mol } H_2O}{18 \text{ g } H_2O} \approx \frac{3 \times 10^{22} \text{ } H_2O \text{ molecules}}{1 \text{ g } H_2O}$$

While snowflakes range in size from too small to see to up to an inch across, we estimate the mass of a "typical" snowflake to be 1 mg (1×10^{-3} g). If so, one snowflake consists of approximately

$$\frac{1 \times 10^{-3} \text{ g}}{1 \text{ mg}} \times \frac{3 \times 10^{22} \text{ } H_2O \text{ molecules}}{1 \text{ g } H_2O} = 3 \times 10^{19} \text{ } H_2O \text{ molecules}$$

As a snowflake grows, it has a large number of possible sites for the next water molecule. Because of the enormous number of water molecules, it is plausible that each snowflake can be different because it would be highly unlikely that all 10^{19} H_2O molecules would be aligned in exactly the same way in two snowflakes.

However, a very large number of snowflakes has fallen on Earth throughout its history. This may be a large enough number that over time even the highly improbable event that all 3×10^{19} H_2O molecules would align exactly the same way might occur. But the probability that in your lifetime you would find two snowflakes that are alike is vanishingly small. The old adage is correct for all practical purposes.

Mehau Kulyk/Photo Researchers, Inc.

Visit this book's companion website at www.cengage.com/chemistry/moore to work an interactive module based on this material.

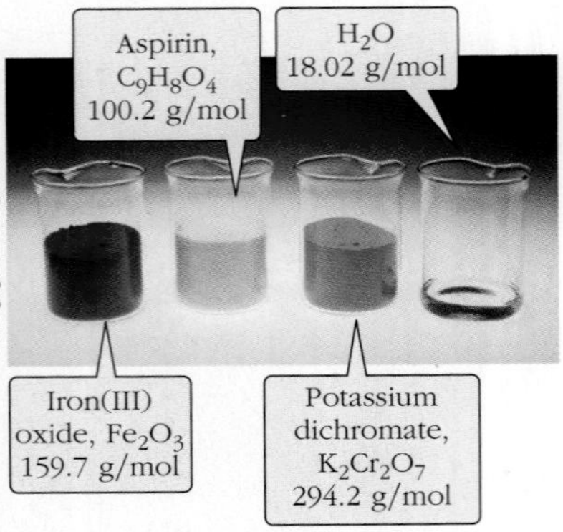

One-mole quantities of four compounds.

Compound	Formula Weight	Molar Mass
Sodium chloride $NaCl$	22.99 amu, Na + 35.45 amu, Cl = 58.44 amu	58.44 g/mol
Magnesium oxide MgO	24.31 amu, Mg + 16.00 amu, O = 40.31 amu	40.31 g/mol
Potassium sulfide K_2S	2 (39.10 amu, K) + 32.07 amu, S = 110.27 amu	110.27 g/mol
Calcium nitrate $Ca(NO_3)_2$	40.08 amu, Ca + 2 (14.01 amu, N) + 6 (16.00 amu, O) = 164.10 amu	164.10 g/mol
Magnesium phosphate $Mg_3(PO_4)_2$	3 (24.31 amu, Mg) + 2 (30.97 amu, P) + 8 (16.00 amu, O) = 262.87 amu	262.87 g/mol

Notice that $Mg_3(PO_4)_2$ has 2 P atoms and $2 \times 4 = 8$ O atoms because there are two PO_4^{3-} ions in the formula.

EXERCISE 3.6 Molar Masses

Calculate the molar mass of each of these compounds:

(a) K_2HPO_4 (b) $C_{27}H_{46}O$ (cholesterol)
(c) $Mn_2(SO_4)_3$ (d) $C_8H_{10}N_4O_2$ (caffeine)

Gram-Mole Conversions

As you might expect, it is essential to be able to do gram-mole conversions for compounds, just as we did for elements (**← *p. 61***). Here also, the key to such conversions is using molar mass as a conversion factor.

PROBLEM-SOLVING EXAMPLE 3.10 Grams to Moles

Ammonium carbonate, $(NH_4)_2CO_3$, is a component of baking powder. How many moles of ammonium carbonate are in 20.0 g of this ionic compound?

Answer 0.208 mol $(NH_4)_2CO_3$

Strategy and Explanation To convert grams of ammonium carbonate to moles, we follow these steps.

- **Molar mass relates grams and moles, so use the compound's formula to calculate molar mass.**

$$[2 \text{ mol N} \times (14.0067 \text{ g N/mol N})] + [8 \text{ mol H} \times (1.0079 \text{ g H/mol H})] + [1 \text{ mol C} \times (12.011 \text{ g C/mol C})] + [3 \text{ mol O} \times (15.9994 \text{ g O/mol O})] = 96.09 \text{ g/mol } (NH_4)_2CO_3$$

- **Use the molar mass to convert mass to moles.**

$$20.0 \text{ g } (NH_4)_2CO_3 \times \frac{1 \text{ mol } (NH_4)_2CO_3}{96.09 \text{ mol } (NH_4)_2CO_3} = 0.208 \text{ mol } (NH_4)_2CO_3$$

☑ **Reasonable Answer Check** The molar mass of $(NH_4)_2CO_3$ is about 100 g/mol, so 20 grams is about 0.2 mol, which is close to the result of our exact calculation.

PROBLEM-SOLVING PRACTICE 3.10

Calculate the number of moles in 12.0 g of each of these ionic compounds:

(a) $Ca(NO_3)_2$ (b) $KMnO_4$ (c) Na_2SO_4

PROBLEM-SOLVING EXAMPLE 3.11 Moles to Grams

Cortisone, $C_{21}H_{28}O_5$, is an anti-inflammatory steroid. How many grams of cortisone are in 5.00×10^{-3} mol cortisone?

Answer 1.80 g

Strategy and Explanation Molar mass relates mass to molar amount. Therefore, first calculate the molar mass and then use it to calculate the mass from the amount.

- **Calculate the molar mass of cortisone.**

$$[21 \text{ mol C} \times (12.011 \text{ g C/mol C})] + [28 \text{ mol H} \times (1.0079 \text{ g H/mol H})] + [5 \text{ mol O} \times 15.9994 \text{ g O/mol O})] = 360.45 \text{ g/mol cortisone}$$

- **Use the molar mass to convert from moles to grams.**

$$5.00 \times 10^{-3} \text{ mol cortisone} \times \frac{360.45 \text{ g cortisone}}{1 \text{ mol cortisone}} = 1.80 \text{ g cortisone}$$

☑ **Reasonable Answer Check** The molar amount of cortisone is $\frac{1}{200}$ mol, and this fraction of the molar mass is about (360 g/mol)/200 = 1.8 g. The answer is reasonable.

PROBLEM-SOLVING PRACTICE 3.11

(a) How many grams are in 5.00×10^{-3} mol sucrose, $C_{12}H_{22}O_{11}$?
(b) How many grams are in 3.00×10^{-6} mol adrenocorticotropic hormone (ACTH), which has a molar mass of approximately 4600 g/mol?

PROBLEM-SOLVING EXAMPLE 3.12 Grams and Moles

Sucralose, an artificial sweetner (Splenda®), $C_{12}H_{19}Cl_3O_8$, is about 600 times sweeter than sucrose. You have one sample of sucralose with a mass of 2.00 g and a second sample containing 0.0350 mol of the sweetener. Answer these questions about the samples.
(a) What is the molar mass of sucralose?
(b) How many moles of sucralose are in the 2.00-g sample?
(c) How many grams of sucralose are in the 0.0350-mol sample?
(d) Which sample has the larger mass?

Answer
(a) 397.6 g/mol (b) 5.03×10^{-3} mol
(c) 13.9 g (d) the 0.0350-mol sample has the larger mass.

Strategy and Explanation
(a) Calculate the molar mass of sucralose from its molecular formula.

$$[12 \text{ mol C} \times (12.011 \text{ g C/mol C})] + [19 \text{ mol H} \times (1.008 \text{ g H/mol H})] + [3 \text{ mol Cl} \times (35.453 \text{ g Cl/mol Cl})] + [8 \text{ mol O} \times (15.999 \text{ g O/mol O})] = 397.64 \text{ g/mol}$$

(b) Use the molar mass to convert from grams to moles.

$$2.00 \text{ g} \times \frac{1 \text{ mol sucralose}}{397.64 \text{ g sucralose}} = 5.03 \times 10^{-3} \text{ mol sucralose}$$

(c) Use the molar mass to calculate grams of sucralose from moles.

$$0.0350 \text{ mol sucralose} \times \frac{397.64 \text{ g sucralose}}{1 \text{ mol sucralose}} = 13.9 \text{ g sucralose}$$

(d) The 0.0350-mol sample of sucralose has the larger mass.

☑ **Reasonable Answer Check** Because the molar mass of sucralose is nearly 400 g/mol, 2.0 g sucralose is about 2/400 = 1/200 or 0.005 mol, much less than the 0.0350-mol sample.

PROBLEM-SOLVING PRACTICE 3.12

(a) Calculate the molar amount of each compound in 12.0 g aspirin, $C_9H_8O_4$, and 12.0 g $Ca_3(PO_4)_2$.

(b) Calculate the number of grams in 0.350 mol $Mn_2(SO_4)_3$ and in 0.350 mol caffeine, $C_8H_{10}N_4O_2$.

CONCEPTUAL EXERCISE 3.7 Moles and Formulas

Is this statement true? "Two different compounds have the same formula. Therefore, 100 g of each compound contains the same number of moles." Justify your answer.

Moles of Ionic Hydrates

$CuSO_4 \cdot 5\ H_2O$.

Many ionic compounds, known as **ionic hydrates** or hydrated compounds, have water molecules trapped within the crystal lattice. The associated water is called the **water of hydration.** For example, the formula for a beautiful deep-blue compound named copper(II) sulfate pentahydrate is $CuSO_4 \cdot 5\ H_2O$. The $\cdot 5\ H_2O$ and the term "pentahydrate" indicate five moles of water associated with every mole of copper(II) sulfate. *The molar mass of a hydrate includes the mass of the water of hydration.* Thus, the molar mass of $CuSO_4 \cdot 5\ H_2O$ is 249.7 g: 159.6 g $CuSO_4$ + 90.1 g (for 5 mol H_2O) = 249.7 g. There are many ionic hydrates, including the frequently encountered ones listed in Table 3.10.

One commonly used hydrate may well be in the walls of your room. Plasterboard (sometimes called wallboard or gypsum board) contains hydrated calcium sulfate, or gypsum, $CaSO_4 \cdot 2\ H_2O$, as well as unhydrated $CaSO_4$, sandwiched between two thicknesses of paper. Gypsum is a natural mineral that can be mined. It is also formed when sulfur dioxide is removed from electric power plant exhaust gases by reacting the SO_2 with an aqueous slurry of lime, calcium oxide.

Calcium sulfate *hemi*hydrate contains one water molecule per two $CaSO_4$ units. The prefix *hemi-* refers to $\frac{1}{2}$ just as in the familiar word "hemisphere."

Heating gypsum to 180 °C drives off some of the water of hydration to form calcium sulfate hemihydrate, $CaSO_4 \cdot \frac{1}{2} H_2O$, commonly called Plaster of Paris. This compound is widely used in casts for broken limbs. When water is added to it, it forms a thick slurry that can be poured into a mold or spread over a part of the body. As the slurry hardens, it takes on additional water of hydration and its volume increases, forming a rigid protective cast.

EXERCISE 3.8 Moles of an Ionic Hydrate

A home remedy calls for 2 teaspoons (20 g) Epsom salt (see Table 3.10). Calculate the number of moles of the hydrate represented by this mass.

Gypsum in its crystalline form. Gypsum is hydrated calcium sulfate, $CaSO_4 \cdot 2\ H_2O$.

Table 3.10 Some Common Hydrated Ionic Compounds

Formula	Systematic Name	Common Name	Uses
$Na_2CO_3 \cdot 10\ H_2O$	Sodium carbonate decahydrate	Washing soda	Water softener
$MgSO_4 \cdot 7\ H_2O$	Magnesium sulfate heptahydrate	Epsom salt	Dyeing and tanning
$CaSO_4 \cdot 2\ H_2O$	Calcium sulfate dihydrate	Gypsum	Wallboard
$CaSO_4 \cdot \frac{1}{2} H_2O$	Calcium sulfate hemihydrate	Plaster of Paris	Casts, molds

3.9 Percent Composition

You saw in the previous section that the composition of any compound can be expressed as either *(1) the number of atoms of each type per molecule or formula unit* or *(2) the mass of each element in a mole of the compound.* The latter relationship provides the information needed to find the **percent composition by mass** (also called the *mass percent*) of the compound.

It is important to recognize that the percent composition of a compound by mass is independent of the quantity of the compound. The percent composition by mass remains the same whether a sample contains 1 mg, 1 g, or 1 kg of the compound.

PROBLEM-SOLVING EXAMPLE 3.13 Percent Composition by Mass

Propane, C_3H_8, is the fuel used in gas grills. Calculate the percentages of carbon and hydrogen in propane.

Answer 81.72% carbon and 18.28% hydrogen

Strategy and Explanation We must relate grams to moles, so molar mass is needed. We then use the molar mass and the mass of each element to calculate the mass percentages.

- **Calculate the molar mass of propane.**

$$[3 \text{ mol C} \times (12.011 \text{ g C/mol C})] + [8 \text{ mol H} \times (1.0079 \text{ g H/mol H})] = 44.09 \text{ g/mol } C_3H_8$$

- **Calculate the percentages of each element from the mass of each element in 1 mol C_3H_8.**

$$\% \text{ C} = \frac{\text{mass of C in 1 mol } C_3H_8}{\text{mass of } C_3H_8 \text{ in 1 mol } C_3H_8} \times 100\%$$

$$= \frac{3 \times 12.01 \text{ g C}}{44.09 \text{ g } C_3H_8} \times 100\% = 81.72\% \text{ C}$$

$$\% \text{ H} = \frac{\text{mass of H in 1 mol } C_3H_8}{\text{mass of } C_3H_8 \text{ in 1 mol } C_3H_8} \times 100\%$$

$$= \frac{8 \times 1.0079 \text{ g H}}{44.09 \text{ g } C_3H_8} \times 100\% = 18.28\% \text{ H}$$

These answers can also be expressed as 81.72 g C per 100.0 g C_3H_8 and 18.28 g H per 100.0 g C_3H_8.

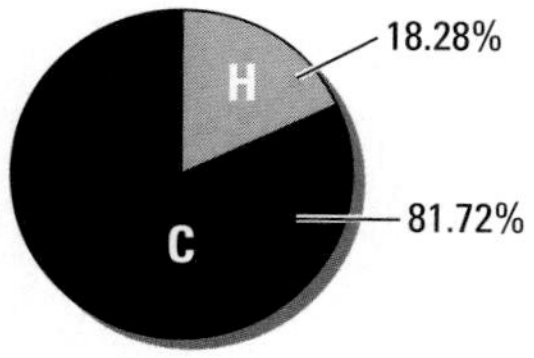

Mass percent carbon and hydrogen in propane, C_3H_8.

☑ **Reasonable Answer Check** Each carbon atom has 12 times the mass of a hydrogen atom. Propane has approximately 12 × 3 = 36 amu of carbon and approximately 1 × 8 = 8 amu of hydrogen. So the percentage of carbon should be about 36/8 = 4.5 times larger than the percentage of hydrogen. This agrees with our more carefully calculated answer.

PROBLEM-SOLVING PRACTICE 3.13

Calculate the percentage of each element in silicon dioxide, SiO_2.

Note that the percentages calculated in Problem-Solving Example 3.13 add up to 100%. Therefore, once we calculated the percentage of carbon, we also could have determined the percentage of hydrogen simply by subtracting: 100% − 81.72% C = 18.28% H. Calculating all percentages and adding them to confirm that they give 100% is a good way to check for errors.

PROBLEM-SOLVING EXAMPLE 3.14 Percent Composition of Hydrated Salt

Epsom salt is $MgSO_4 \cdot 7\ H_2O$. (a) What is the percent by mass of water in Epsom salt? (b) What are the percentages of each element in Epsom salt?

Epsom salt, $MgSO_4 \cdot 7\ H_2O$.

Answer

(a) 51.15% water (b) 9.86% Mg; 13.01% S; 71.40% O; 5.72% H

Strategy and Explanation

(a) We first find the molar mass of Epsom salt, which is the sum of the molar masses of the atoms in the chemical formula:

$$[1 \text{ mol Mg} \times (24.31 \text{ g Mg/mol Mg})] + [1 \text{ mol S} \times (32.07 \text{ g S/mol S})]$$
$$+ [4 \text{ mol O} \times (15.9994 \text{ g O/mol O})] + [14 \text{ mol H} \times (1.0079 \text{ g H/mol H})]$$
$$+ [7 \text{ mol O} \times (15.9994 \text{ g O/mol O})] = 246.49 \text{ g/mol Epsom salt}$$

Because 1 mol Epsom salt contains 7 mol H_2O, the mass of water in 1 mol Epsom salt is

$$7 \text{ mol } H_2O \times (18.012 \text{ g } H_2O/\text{mol } H_2O) = 126.08 \text{ g } H_2O$$

$$\% H_2O = \frac{126.08 \text{ g water per mol Epsom salt}}{246.49 \text{ g/mol Epsom salt}} \times 100\% = 51.15\% \ H_2O$$

(b) We calculate the percentage of magnesium from the ratio of the mass of magnesium in 1 mol Epsom salt to the mass of Epsom salt in 1 mol:

$$\% \text{ Mg} = \frac{\text{mass of Mg in 1 mol Epsom salt}}{\text{mass of Epsom salt in 1 mol}}$$
$$= \frac{24.31 \text{ g Mg}}{246.49 \text{ g/mol Epsom salt}} \times 100\% = 9.862\% \text{ Mg}$$

We calculate the percentages for the remaining elements in the same way:

$$\% \text{ S} = \frac{32.07 \text{ g S}}{246.49 \text{ g/mol Epsom salt}} \times 100\% = 13.01\% \text{ S}$$

$$\% \text{ O} = \frac{(64.00 + 112.00) \text{ g O}}{246.49 \text{ g/mol Epsom salt}} \times 100\% = 71.40\% \text{ O}$$

$$\% \text{ H} = \frac{14.11 \text{ g H}}{246.49 \text{ g/mol Epsom salt}} \times 100\% = 5.724\% \text{ H}$$

☑ **Reasonable Answer Check** In the formula of the hydrated salt, there are seven waters with a combined mass of $7 \times 18 = 126$ g, and there are six other atoms with molar masses ranging between 16 and 32 that total to 120 g. Thus, the hydrated salt should be about 50% water by weight, and it is. There are 11 oxygen atoms in the formula, so oxygen should have the largest percent by weight, and it does. The percentages sum to 99.996% due to rounding.

PROBLEM-SOLVING PRACTICE 3.14

What is the mass percent of each element in hydrated nickel(II) chloride, $NiCl_2 \cdot 6\ H_2O$?

EXERCISE 3.9 Percent Composition

Express the composition of each compound first as the mass of each element in 1.000 mol of the compound and then as the mass percent of each element:

(a) SF_6 (b) $C_{12}H_{22}O_{11}$ (c) $Al_2(SO_4)_3$ (d) $U(OTeF_5)_6$

3.10 Determining Empirical and Molecular Formulas

A formula can be used to derive the percent composition by mass of a compound, and the reverse process also works—we can determine the formula of a compound from mass percent data. In doing so, keep in mind that the subscripts in a formula indicate the relative numbers of moles of each element in one mole of that compound.

We can apply this method to finding the formula of diborane, a compound consisting of boron and hydrogen. Experiments show that diborane is 78.13% B and 21.87% H. Based on these percentages, a 100.0-g diborane sample contains 78.13 g B and 21.87 g H. From this information we can calculate the number of moles of each element in the sample:

$$78.13 \text{ g B} \times \frac{1 \text{ mol B}}{10.811 \text{ g B}} = 7.227 \text{ mol B}$$

$$21.87 \text{ g H} \times \frac{1 \text{ mol H}}{1.0079 \text{ g H}} = 21.70 \text{ mol H}$$

To determine the formula from these data, we next need to find the number of moles of each element *relative to the other element*—in this case, the ratio of moles of hydrogen to moles of boron. Looking at the numbers reveals that there are about three times as many moles of H atoms as there are moles of B atoms. To calculate the ratio exactly, we divide the larger number of moles by the smaller number of moles. For diborane that ratio is

$$\frac{21.70 \text{ mol H}}{7.227 \text{ mol B}} = \frac{3.003 \text{ mol H}}{1.000 \text{ mol B}}$$

This ratio confirms that there are three moles of H atoms for every one mole of B atoms and that there are three hydrogen atoms for each boron atom. This information gives the formula BH_3, which may or may not be the molecular formula of diborane.

For a molecular compound such as diborane, the molecular formula must also accurately reflect the *total number of atoms in a molecule of the compound.* The calculation we have done gives the *simplest possible ratio of atoms in the molecule,* and BH_3 is the simplest formula for diborane. A formula that reports the simplest possible whole number ratio of atoms in the molecule is called an **empirical formula.** Multiples of the simplest formula are possible, such as B_2H_6, B_3H_9, and so on.

To determine the actual molecular formula from the empirical formula requires that we *experimentally determine* the molar mass of the compound and then compare that result with the molar mass of the empirical formula. If the two molar masses are the same, the empirical and molecular formulas are the same. However, if the experimentally determined molar mass is some multiple of the empirical formula value, the molecular formula is that multiple of the empirical formula. In the case of diborane, experiments indicate that the molar mass is 27.67 g/mol. This compares with the molar mass of 13.84 g/mol for BH_3, and so the molecular formula is a multiple of the empirical formula. That multiple is 27.67/13.84 = 2.00. Thus, the molecular formula of diborane is B_2H_6, two times BH_3.

$$\frac{\text{molecular formula molar mass}}{\text{empirical formula molar mass}} = n, \text{ an integer}$$

If $n = 1$, the molecular formula and the empirical formula are the same. When $n > 1$, the subscripts in the molecular formula are all n times the subscripts in the empirical formula.

PROBLEM-SOLVING EXAMPLE 3.15 Molecular Formula from Percent Composition by Mass Data

Hydrazine is a compound composed of 87.42% nitrogen and 12.58% hydrogen by mass. Its experimentally determined molar mass is 32.05 g/mol. Determine the empirical and molecular formulas of hydrazine.

Answer Empirical formula: NH_2; molecular formula: N_2H_4

Strategy and Explanation Use the number of moles of each element to determine the empirical formula using mole ratios; then compare the experimental molar mass with the molar mass from the empirical formula.

- **Determine the number of moles of each element.**

We know from the mass percentage data that a 100.-g hydrazine sample contains 87.42 g N and 12.58 g H, so the numbers of moles of each element are

$$87.42 \text{ g N} \times \frac{1 \text{ mol N}}{14.0067 \text{ g N}} = 6.241 \text{ mol N}$$

$$12.58 \text{ g H} \times \frac{1 \text{ mol H}}{1.0079 \text{ g H}} = 12.48 \text{ mol H}$$

- **Calculate the mole ratio of elements in the compound.**

  From the number of moles of each element, we calculate the mole ratio (and atom ratio) as

$$\frac{12.48 \text{ mol H}}{6.241 \text{ mol N}} = \frac{2.00 \text{ mol H}}{1.00 \text{ mol N}}$$

  Therefore, the H : N mole ratio (and atom ratio) is 2 : 1, and the empirical (simplest) formula of hydrazine is NH_2.

- **Calculate the molar mass corresponding to the empirical formula.**

$$[1 \text{ mol N} \times (14.0067 \text{ g N/mol N})] + 2 \text{ mol H} \times (1.0079 \text{ g H/mol H})]$$
$$= 16.02 \text{ g/mol hydrazine}$$

- **Compare the molar mass of the empirical formula with the experimental molar mass to determine the molecular formula of the compound.**

  The molar mass of the empirical formula is 16.02 g/mol, and the experimental molar mass is 32.05 g/mol. The experimental molar mass is $\frac{32.05}{16.02} = 2$ times the empirical molar mass, so the molecular formula is N_2H_4, twice the empirical formula.

☑ **Reasonable Answer Check** The molar mass of N is about 14 g/mol, so we should have 14 g/mol × 2 = 28 g/mol N in 1 mol hydrazine. This would give an N percent by mass of 28/32 = 87.5%, which is just about the right value. Our answer is reasonable.

PROBLEM-SOLVING PRACTICE 3.15

An oxide of phosphorus contains 56.34% P and 43.66% O, and its experimentally determined molar mass is 219.90 g/mol. Determine the empirical and molecular formulas of this compound.

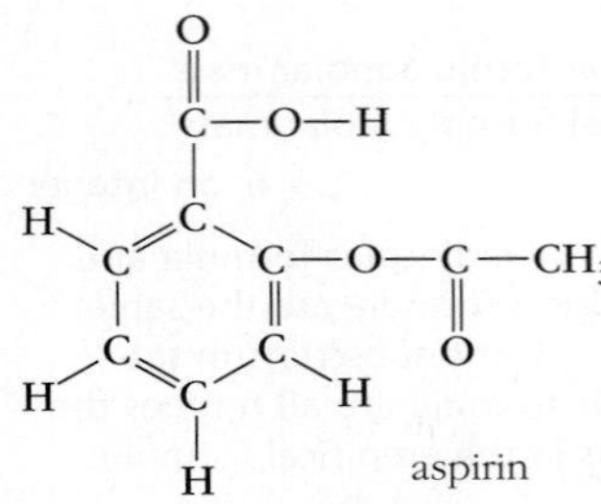

aspirin

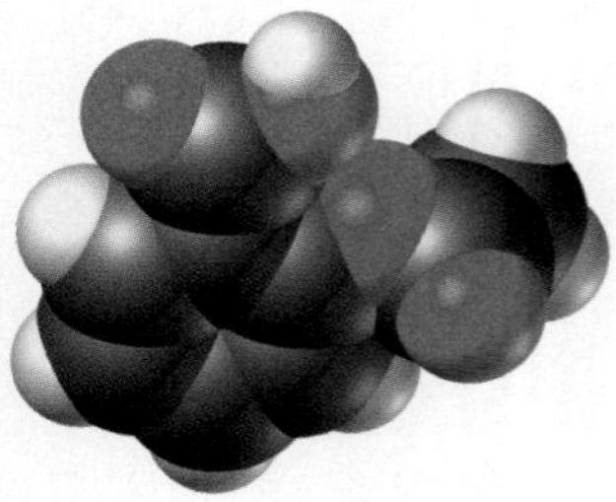

PROBLEM-SOLVING EXAMPLE 3.16 Molecular Formula from Percent Composition by Mass Data

Aspirin, a commonly used analgesic, has a molar mass of 180.16 g/mol. It contains 60.00% C, 4.4756% H, and the rest is oxygen. What are its empirical and molecular formulas?

Answer The empirical and molecular formulas are both $C_9H_8O_4$.

Strategy and Explanation Determine the empirical formula from the relative number of moles of each element in the compound. Then compare the empirical-formula molar mass and the experimental molar mass to find the molecular formula.

- **Use the mass percentages to find the number of moles of each element in the compound.**

  Find the number of moles of each element in 100.0 g of the compound using the mass percent data.

$$60.00 \text{ g C} \times \frac{1 \text{ mol C}}{12.011 \text{ g C}} = 4.995 \text{ mol C}$$

$$4.4756 \text{ g H} \times \frac{1 \text{ mol H}}{1.0079 \text{ g H}} = 4.441 \text{ mol H}$$

  The mass of oxygen in the 100.0-g sample is

$$100.0 \text{ g sample} - 60.00 \text{ g C} - 4.4756 \text{ g H} = 35.52 \text{ g O}$$

  Converting this to moles of oxygen gives

$$35.52 \text{ g O} \times \frac{1 \text{ mol O}}{15.9994 \text{ g O}} = 2.220 \text{ mol O}$$

- **Find the mole ratios of the elements in the compound.**

  Base the mole ratios on the smallest number of moles present, in this case, moles of oxygen. This relates the mol C and mol H to 1.00 mol O.

$$\frac{4.495 \text{ mol C}}{2.220 \text{ mol O}} = \frac{2.25 \text{ mol C}}{1.00 \text{ mol O}}$$

$$\frac{4.441 \text{ mol H}}{2.220 \text{ mol O}} = \frac{2.00 \text{ mol H}}{1.00 \text{ mol O}}$$

  Therefore, the empirical formula has a mole (atom) ratio of 2.25 C to 1.00 O and 2.00 H to 1.00 O. This gives a formula with subscripts that are not whole numbers, $C_{2.25}H_{2.00}O_{1.00}$. To convert these values to whole numbers, multiply by 4, so the empirical formula is $C_9H_8O_4$.

- **Calculate the molar mass corresponding to the empirical formula.**

$$\left(9 \text{ mol C} \times \frac{12.011 \text{ g C}}{1 \text{ mol C}}\right) + \left(8 \text{ mol H} \times \frac{1.0079 \text{ g H}}{1 \text{ mol H}}\right) +$$

$$\left(4 \text{ mol O} \times \frac{15.9994 \text{ g O}}{1 \text{ mol O}}\right) = 180.16 \text{ g } C_9H_8O_4 \text{ per mol of } C_9H_8O_4$$

- **Compare molar masses to find the molecular formula of the compound.**

  In this case, the molar mass of the empirical formula and the experimentally determined molar mass of the molecular formula are the same ($n = 1$), so the molecular formula is the same as the empirical formula, $C_9H_8O_4$.

☑ **Reasonable Answer Check** The molar mass of C is 12 g/mol, so we should have 12 g/mol × 9 mol = 108 g C in 1 mol $C_9H_8O_4$. This would give a mass percent C of 108/180 = 60%, which is the percent given in the problem. The answer is reasonable.

PROBLEM-SOLVING PRACTICE 3.16

Vitamin C (ascorbic acid) contains 40.9% C, 4.58% H, and 54.5% O and has an experimentally determined molar mass of 176.13 g/mol. Determine its empirical and molecular formulas.

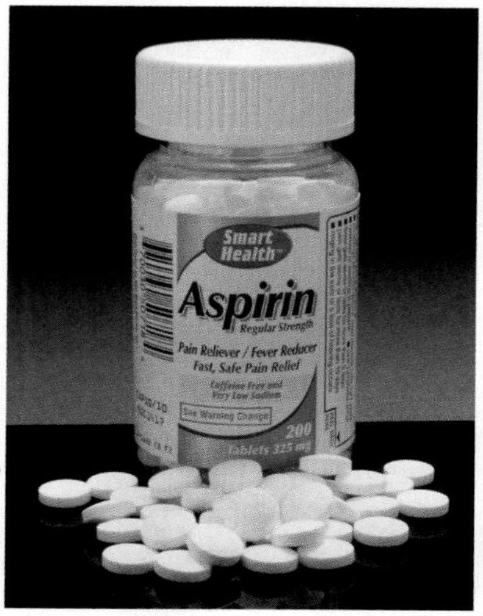

© Cengage Learning/Charles D. Winters

Aspirin tablets.

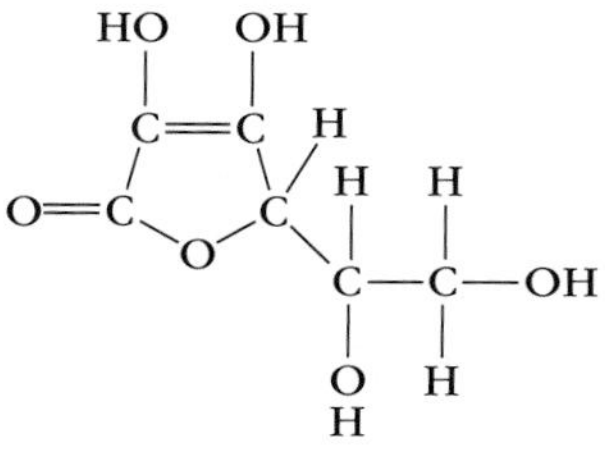

vitamin C

3.11 The Biological Periodic Table

Most of the more than 100 known elements are not directly involved with our personal health and well-being. However, more than 25 of the elements, listed in Figure 3.7, are absolutely essential to human life. Among these essential elements are metals, nonmetals, and metalloids from across the periodic table. All are necessary as part of a well-balanced diet.

Table 3.11 lists the building block elements and major minerals in order of their relative abundances per million atoms in the body, showing the preeminence of

If you weigh 150 lb, about 90 lb (60%) is water, 30 lb is fat, and the remaining 30 lb is a combination of proteins, carbohydrates, and calcium, phosphorus, and other dietary minerals.

Table 3.11 Major Elements of the Human Body

Element	Symbol	Relative Abundance in Atoms/Million Atoms in the Body	Element	Symbol	Relative Abundance in Atoms/Million Atoms in the Body
Hydrogen	H	628,000	Phosphorus	P	2100
Oxygen	O	257,000	Sulfur	S	500
Carbon	C	95,000	Sodium	Na	420
Nitrogen	N	13,600	Potassium	K	330
Calcium	Ca	2400	Chlorine	Cl	270
			Magnesium	Mg	130

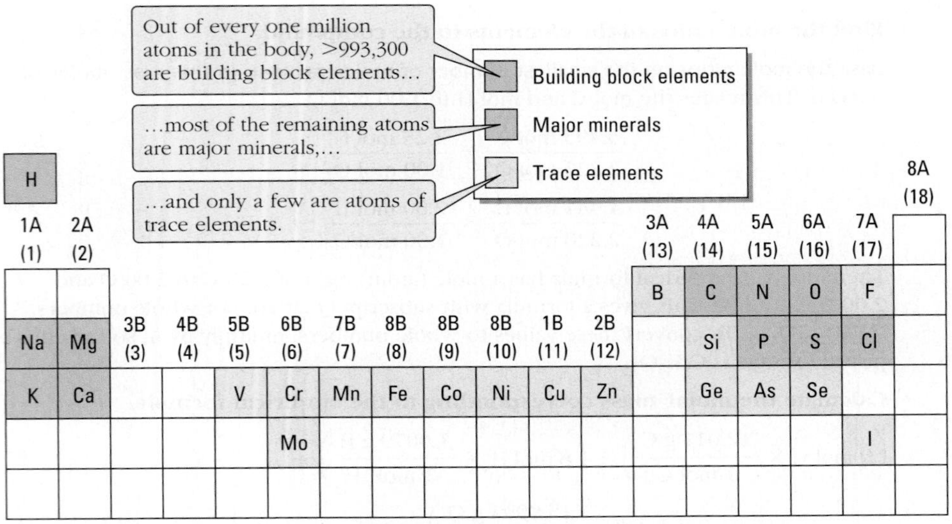

Figure 3.7 Elements essential to human health. Four elements—C, H, N, and O—form the many organic compounds that make up living organisms. The major minerals are required in relatively large amounts; trace elements are required in lesser amounts.

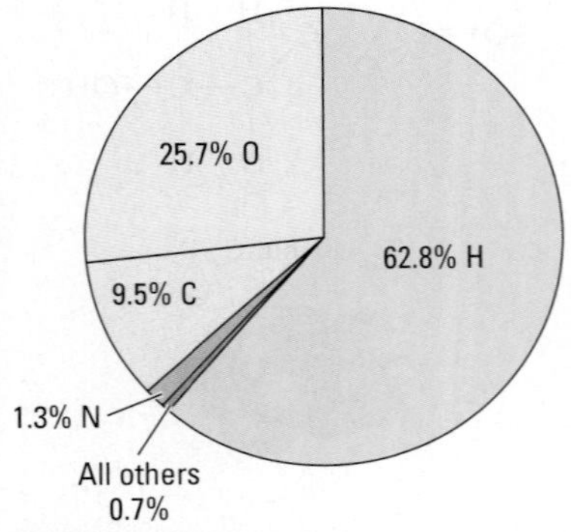

The four building block elements are more than 99% of the atoms in the human body.

four of the nonmetals—hydrogen, oxygen, carbon, and nitrogen. These four building block elements contribute most of the atoms in the biologically significant chemicals—the biochemicals—composing all plants and animals. With few exceptions, a major one being water, the biochemicals that these nonmetals form are organic compounds.

Nonmetals are also present in anions in body fluids. Oxygen and hydrogen are found in oxyanions such as PO_4^{3-} and HCO_3^-; chlorine is present as chloride ion Cl^-; phosphorus is found in three forms, PO_4^{3-}, HPO_4^{2-}, and $H_2PO_4^-$; and carbon is present as hydrogen carbonate HCO_3^- and carbonate CO_3^{2-} ions. Metals are present in the body as cations in solution (for example, Na^+, K^+) and in solids (Ca^{2+} in bones and teeth). Metals are also incorporated into large biomolecules (for example, Fe^{2+} in hemoglobin and Co^{3+} in vitamin B-12).

The Dietary Minerals

The general term **dietary minerals** refers to the essential elements other than carbon, hydrogen, oxygen, or nitrogen. The dietary necessity and effects of these elements go far beyond those implied by their collective presence as only about 4% of our body weight. They exemplify the old saying, "Good things come in small packages." Because the body uses them efficiently, recycling them through many reactions, dietary minerals are required in only small amounts, but their absence from your diet can cause significant health problems.

The dietary minerals indicated in Figure 3.7 are classified into the relatively more abundant **major minerals** and the less plentiful **trace elements.** Major minerals are those present in quantities greater than 0.01% of body mass (100 mg per kg)—for example, more than 6 g for a 60-kg (132-lb) individual. Trace elements are present in smaller (sometimes far smaller) amounts. For example, the necessary daily total intake of iodine is only 150 μg. In the context of nutrition, major minerals and trace elements usually refer to ions in soluble ionic compounds in the diet.

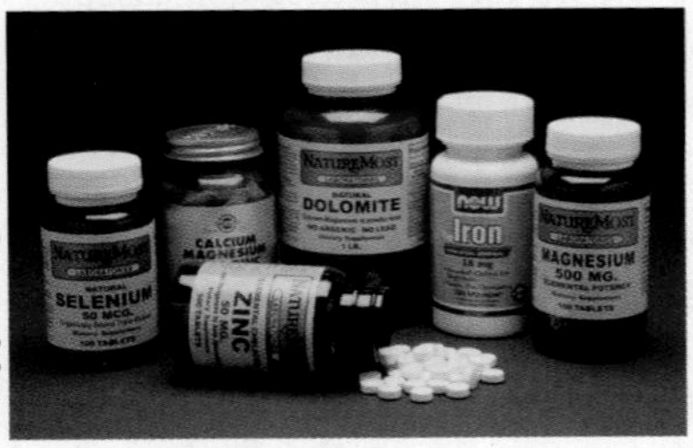

© Cengage Learning/Charles D. Winters

Vitamin and mineral supplements.

EXERCISE 3.10 Essential Elements

Using Figure 3.7, identify (a) the essential nonmetals, (b) the essential alkaline-earth metals, (c) the essential halide ions, and (d) four essential transition metals.

CHEMISTRY YOU CAN DO

Pumping Iron: How Strong Is Your Breakfast Cereal?

Iron is an essential dietary mineral that enters our diets in many ways. To prevent dietary iron deficiency, the U.S. Food and Drug Administration (FDA) allows food manufacturers to "fortify" (add iron to) their products. One way of fortifying a cereal is to add an iron compound, such as iron(III) phosphate, that dissolves in the stomach acid (HCl). Another is the iron in Special K cereal, which consists of particles of pure iron metal baked into the flakes. It is listed on the ingredients label as "reduced iron."

Can you detect metallic iron in a cereal? You can find out with this experiment, for which you will need these items:

- A reasonably strong magnet. You might find a good one holding things to your refrigerator door or in a toy store or hobby shop. Agricultural supply stores have "cow magnets," which will work well.
- A plastic freezer bag (quart size) and a rolling pin (or something else to crush the cereal).
- One serving of Special K (1 cup = about 30 g).

Put the Special K into the plastic bag and crush the cereal into small particles. Place the magnet into the bag and mix it well with the cereal for several minutes. Carefully remove the magnet from the bag and examine it closely. A magnifying glass is helpful. Is anything clinging to the magnet that was not there before?

Think about these questions:

1. Based on this experiment, is there metallic iron in the cereal?
2. What happens to the metallic iron after you swallow it?
3. Does this iron contribute to your daily iron requirement?

CHEMISTRY IN THE NEWS

Removing Arsenic from Drinking Water

Who wants to be a millionaire? In February 2005, the U.S. National Academy of Engineering announced the establishment of a $1 million Grainger Challenge Prize for Sustainable Development for development of technologies to help improve the quality of living throughout the world. Sponsored by the Grainger Foundation, a private philanthropy, the first prize was awarded for designing an inexpensive system to reduce arsenic levels in drinking water in developing countries.

Arsenic due to natural sources is found in low levels in the drinking water of tens of millions of people around the world. Arsenic is, of course, a poison when taken in sufficient quantity, and consumption of even small amounts of arsenic over a long period of time can lead to serious health problems. The current maximum contaminant level (MCL) in the United States is 10 parts per billion (ppb), which is the same as the international standard set by the World Health Organization. All water systems in the United States had to meet this standard by January 2006.

Arsenic in drinking water is a worldwide problem, but it is an acute public health issue in Bangladesh. In the 1980s many shallow wells were dug to supply better quality drinking water than the bacteria-laden groundwater then being used. Recent testing of the well water has shown that tens of millions of people (estimates range from 35 to 77 million people of the total population of 135 million) are now drinking water with levels of arsenic well above the 10 ppb MCL. The situation has been called the largest mass poisoning in history.

In February 2007 the $1 million Gold Award was presented to analytical chemistry professor Abul Hussam of George Mason University. His simple and inexpensive SONO filter for removing arsenic from drinking water uses sand, charcoal, brick chips, and a composite iron matrix made from iron pieces. Thousands of these filters are now being manufactured and used to provide drinking water with less than 10 ppb arsenic in Bangladesh, where arsenic levels as high as 14,000 ppb have been found in some wells.

Reuters/Rafiqur Rahman/Landov

Bangladeshi boy collecting safe drinking water from a well. The green paint indicates that the water from this well has been tested, shown to be free of arsenic contamination, and is therefore safe to drink.

Sources: Wang, J.S., and Wai, C.M. "Arsenic in Drinking Water—A Global Environmental Problem." *Journal of Chemical Education,* Vol. 81, 2004; pp. 207–213.

Ritter, S. "Million-Dollar Challenge." *Chemical & Engineering News,* Feb. 7, 2005; p. 10.

Ritter, S. "Chemist Wins Arsenic Challenge." *Chemical & Engineering News,* Feb. 12, 2007; p. 19.

SUMMARY PROBLEM

Part I

During each launch of the Space Shuttle, the booster rocket uses about 1.5×10^6 lb of ammonium perchlorate as fuel.

1. Write the chemical formulas for (a) ammonium perchlorate, (b) ammonium chlorate, and (c) ammonium chlorite.
2. (a) When ammonium perchlorate dissociates in water, what ions are dispersed in the solution?
   (b) Would this aqueous solution conduct an electric current? Explain your answer.
3. How many moles of ammonium perchlorate are used in Space Shuttle booster rockets during a launch?

Part II

Chemical analysis of ibuprofen (Advil) indicates that it contains 75.69% carbon, 8.80% hydrogen, and the remainder oxygen. The empirical formula is also the molecular formula.

1. Determine the molecular formula of ibuprofen.
2. Two 200-mg ibuprofen tablets were taken by a patient to relieve pain. Calculate the number of moles of ibuprofen contained in the two tablets.

IN CLOSING

OWL and go Chemistry

Sign in at **www.cengage.com/owl** to:

- View tutorials and simulations, develop problem-solving skills, and complete online homework assigned by your professor.
- For quick review and exam prep, download Go Chemistry mini lecture modules from OWL (or purchase them at **www.CengageBrain.com**).

Having studied this chapter, you should be able to . . .

- Interpret the meaning of molecular formulas, condensed formulas, and structural formulas (Section 3.1).
- Name binary molecular compounds, including straight-chain alkanes (Sections 3.2, 3.3). End-of-chapter question: 9
- Write structural formulas for and identify straight- and branched-chain alkane constitutional isomers (Section 3.4). Question 11
- Predict the charges on monatomic ions of metals and nonmetals (Section 3.5). Question 28
- Know the names and formulas of polyatomic ions (Section 3.5).
- Describe the properties of ionic compounds and compare them with the properties of molecular compounds (Sections 3.5, 3.7). Question 54
- Given their names, write the formulas of ionic compounds (Section 3.5). Question 42
- Given their formulas, name ionic compounds (Section 3.6). Question 44
- Describe electrolytes in aqueous solution and summarize the differences between electrolytes and nonelectrolytes (Section 3.7). Question 60
- Thoroughly explain the use of the mole concept for chemical compounds (Section 3.8).
- Calculate the molar mass of a compound (Section 3.8). Question 68
- Calculate the number of moles of a compound given the mass, and vice versa (Section 3.8). Questions 70, 74, 78
- Explain the formula of a hydrated ionic compound and calculate its molar mass (Section 3.8).
- Express molecular composition in terms of percent composition (Section 3.9). Questions 87, 89

- Use percent composition and molar mass to determine the empirical and molecular formulas of a compound (Section 3.10). Questions 102, 106, 108
- Identify biologically important elements (Section 3.11).

KEY TERMS

alkane *(Section 3.3)*
alkyl group *(3.4)*
anion *(3.5)*
binary molecular compound *(3.2)*
cation *(3.5)*
chemical bond *(3.3)*
condensed formula *(3.1)*
constitutional isomer *(3.4)*
Coulomb's law *(3.7)*
crystal lattice *(3.7)*
dietary mineral *(3.11)*
dissociation *(3.7)*
electrolyte *(3.7)*
empirical formula *(3.10)*
formula unit *(3.7)*
formula weight *(3.8)*
functional groups *(3.1)*
halide ion *(3.6)*
hydrocarbon *(3.3)*
inorganic compound *(3.1)*
ionic bonding *(3.7)*
ionic compound *(3.5)*
ionic hydrate *(3.8)*
isomer *(3.4)*
major mineral *(3.11)*
molecular compound *(3.1)*
molecular formula *(3.1)*
molecular weight *(3.8)*
monatomic ion *(3.5)*
nonelectrolyte *(3.7)*
organic compound *(3.1)*
oxoanion *(3.6)*
percent composition by mass *(3.9)*
polyatomic ion *(3.5)*
structural formula *(3.1)*
trace element *(3.11)*
water of hydration *(3.8)*

QUESTIONS FOR REVIEW AND THOUGHT

OWL Interactive versions of these problems are assignable in OWL.

Blue-numbered questions have short answers at the back of this book in Appendix M and fully worked solutions in the *Student Solutions Manual.*

Review Questions

These questions test vocabulary and simple concepts.

1. A dictionary defines the word "compound" as a "combination of two or more parts." What are the "parts" of a chemical compound? Identify three pure (or nearly pure) compounds you have encountered today. What is the difference between a compound and a mixture?
2. For each of these structural formulas, write the molecular formula and condensed formula.

(a) 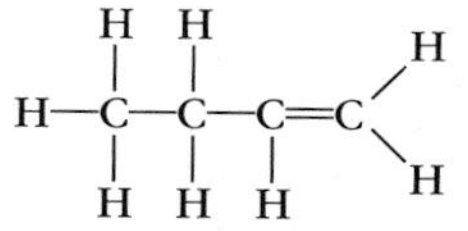

(b) 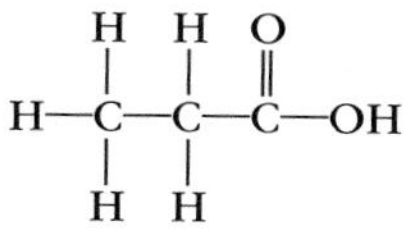

(c) 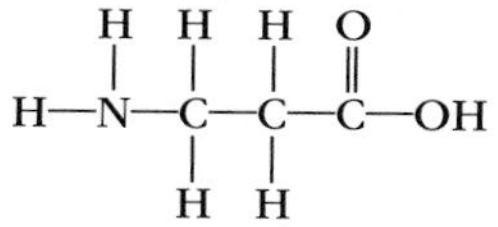

3. Given these condensed formulas, write the structural and molecular formulas.
(a) CH_3OH (b) $CH_3CH_2NH_2$
(c) $CH_3CH_2SCH_2CH_3$
4. Give a molecular formula for each of these organic acids.

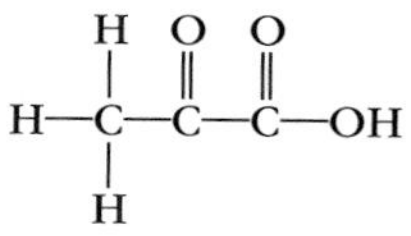

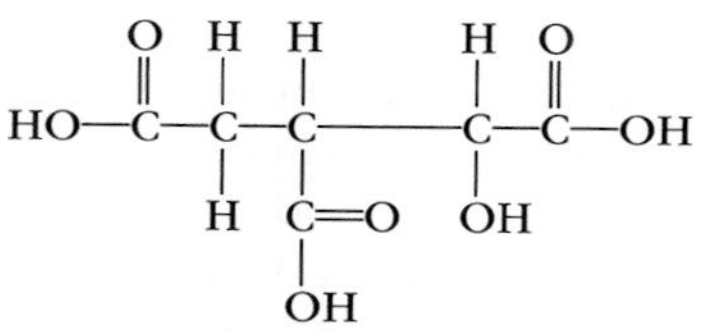

(a) pyruvic acid (b) isocitric acid
5. Give a molecular formula for each of these molecules.

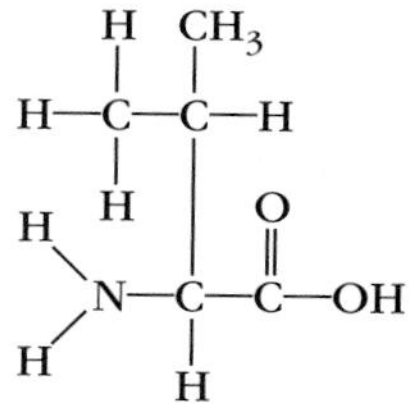

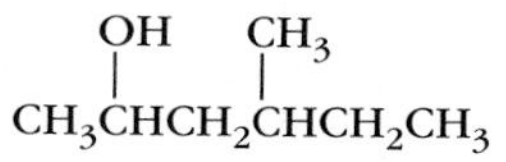

(a) valine (b) 4-methyl-2-hexanol
6. Give the name for each of these binary nonmetal compounds.
(a) NF_3 (b) HI
(c) BBr_3 (d) C_6H_{14}
7. Give the name for each of these binary nonmetal compounds.
(a) C_8H_{18} (b) P_2S_3
(c) OF_2 (d) XeF_4

8. Give the formula for each of these binary nonmetal compounds.
   (a) Sulfur trioxide
   (b) Dinitrogen pentaoxide
   (c) Phosphorous pentachloride
   (d) Silicon tetrachloride
   (e) Diboron trioxide (commonly called boric oxide)

Topical Questions

These questions are keyed to the major topics in the chapter. Usually a question that is answered at the back of this book is paired with a similar one that is not.

Molecular and Structural Formulas (Section 3.1)

9. Give the formula for each of these nonmetal compounds.
   (a) Bromine trifluoride
   (b) Xenon difluoride
   (c) Diphosphorus tetrafluoride
   (d) Pentadecane
   (e) Hydrazine
10. Write structural formulas for these alkanes.
   (a) Butane (b) Nonane
   (c) Hexane (d) Octane
   (e) Octadecane
11. Write the molecular, condensed, and structural formulas for the simplest alcohols derived from butane and pentane.
12. Octane is an alkane (Table 3.4). For the sake of this problem, we will assume that gasoline, a complex mixture of hydrocarbons, is represented by octane. If you fill the tank of your car with 18 gal gasoline, how many grams and how many pounds of gasoline have you put into the car? Information you may need is (a) the density of octane, 0.692 g/cm^3, and (b) the volume of 1 gal in milliliters, 3790 mL.
13. Which of these molecules contains more O atoms, and which contains more atoms of all kinds?
   (a) Sucrose, $C_{12}H_{22}O_{11}$
   (b) Glutathione, $C_{10}H_{17}N_3O_6S$ (the major low-molecular-weight sulfur-containing compound in plant or animal cells)
14. Write the molecular formula of each of these compounds.
   (a) Benzene (a liquid hydrocarbon), with six carbon atoms and six hydrogen atoms per molecule
   (b) Vitamin C, with six carbon atoms, eight hydrogen atoms, and six oxygen atoms per molecule
15. Write the molecular formula for each molecule.
   (a) A molecule of the hydrocarbon heptane, which has seven carbon atoms and 16 hydrogen atoms
   (b) A molecule of acrylonitrile (the basis of Orlon and Acrilan fibers), which has three carbon atoms, three hydrogen atoms, and one nitrogen atom
   (c) A molecule of Fenclorac (an anti-inflammatory drug), which has 14 carbon atoms, 16 hydrogen atoms, two chlorine atoms, and two oxygen atoms
16. Give the total number of atoms of each element in one formula unit of each of these compounds.
   (a) CaC_2O_4 (b) $C_6H_5CHCH_2$
   (c) $(NH_4)_2SO_4$ (d) $Pt(NH_3)_2Cl_2$
   (e) $K_4Fe(CN)_6$
17. Give the total number of atoms of each element in each of these molecules.
   (a) $C_6H_5COOC_2H_5$ (b) $HOOCCH_2CH_2COOH$
   (c) $NH_2CH_2CH_2COOH$ (d) $C_{10}H_9NH_2Fe$
   (e) $C_6H_2CH_3(NO_2)_3$

Binary Inorganic Compounds (Section 3.2)

18. Give the correct name for each of these binary nonmetal compounds.
   (a) SO_2 (b) CCl_4
   (c) P_4S_{10} (d) SF_4
19. Write the correct formula for each of these nonmetal compounds.
   (a) nitrogen triiodide (b) carbon disulfide
   (c) dinitrogen tetraoxide (d) selenium hexafluoride
20. Give the correct name for each of these binary nonmetal compounds.
   (a) HBr (b) ClF_3
   (c) Cl_2O_7 (d) BI_3
21. Write the correct formula for each of these nonmetal compounds.
   (a) bromine trichloride
   (b) xenon trioxide
   (c) diphosphorus tetrafluoride
   (d) oxygen difluoride

Hydrocarbons (Section 3.3)

22. In a noncyclic alkane other than methane, what is the maximum number of hydrogen atoms that can be bonded to one carbon atom?
23. In a noncyclic alkane, what is the maximum number of carbon atoms that can be bonded to one carbon atom?
24. How many hydrogen atoms are contained in one molecule of heptane?
25. Write the molecular formula for the molecule that results from substituting one chlorine atom for a hydrogen atom in butane.

Constitutional Isomers (Section 3.4)

26. Consider two molecules that are constitutional isomers.
   (a) What is the same on the molecular level between these two molecules?
   (b) What is different on the molecular level between these two molecules?
27. Draw structural formulas for the five constitutional isomers of C_6H_{14}.

Predicting Ion Charges (Section 3.5)

28. For each of these metals, write the chemical symbol for the corresponding ion (with charge).
   (a) Lithium (b) Strontium
   (c) Aluminum (d) Calcium
   (e) Zinc
29. For each of these nonmetals, write the chemical symbol for the corresponding ion (with charge).
   (a) Nitrogen (b) Sulfur
   (c) Chlorine (d) Iodine
   (e) Phosphorus

30. Predict the charges of the ions in an ionic compound composed of barium and bromine.
31. Predict the charges of the ions in calcium chloride, an ionic compound.
32. Predict the charges for ions of these elements.
(a) Magnesium (b) Zinc
(c) Iron (d) Gallium
33. Predict the charges for ions of these elements.
(a) Selenium (b) Fluorine
(c) Silver (d) Nitrogen
34. Cobalt is a transition metal and so can form ions with at least two different charges. Write the formulas for the compounds formed between cobalt ions and the oxide ion.
35. Although not a transition element, lead can form two cations: Pb^{2+} and Pb^{4+}. Write the formulas for the compounds of these ions with the chloride ion.
36. Which of these are the correct formulas of compounds? For those that are not, give the correct formula.
(a) AlCl (b) NaF_2
(c) Ga_2O_3 (d) MgS
37. Which of these are the correct formulas of compounds? For those that are not, give the correct formula.
(a) Ca_2O (b) $SrCl_2$
(c) Fe_2O_5 (d) K_2O
38. A monatomic ion X^{2+} has 23 electrons and 27 neutrons. Identify element X.
39. A monatomic ion X^{2+} has 22 electrons and 28 neutrons. Identify element X.

Polyatomic Ions (Section 3.5)

40. For each of these compounds, tell what ions are present and how many there are per formula unit.
(a) $Pb(NO_3)_2$ (b) $NiCO_3$
(c) $(NH_4)_3PO_4$ (d) K_2SO_4
41. For each of these compounds, tell what ions are present and how many there are per formula unit.
(a) $Ca(CH_3CO_2)_2$ (b) $Co_2(SO_4)_3$
(c) $Al(OH)_3$ (d) $(NH_4)_2CO_3$
42. Determine the chemical formulas for barium sulfate, magnesium nitrate, and sodium acetate. Each compound contains a monatomic cation and a polyatomic anion. What are the names and electrical charges of these ions?
43. Write the chemical formula for calcium nitrate, potassium chloride, and barium phosphate. What are the names and charges of all the ions in these three compounds?
44. Write the chemical formulas for these compounds.
(a) Nickel(II) nitrate (b) Sodium bicarbonate
(c) Lithium hypochlorite (d) Magnesium chlorate
(e) Calcium sulfite
45. Write the chemical formulas for these compounds.
(a) Iron(III) nitrate (b) Potassium carbonate
(c) Sodium phosphate (d) Calcium chlorite
(e) Sodium sulfate

Ionic Compounds (Sections 3.5, 3.6, 3.7)

46. Which of these substances are ionic?
(a) CF_4 (b) $SrBr_2$
(c) $Co(NO_3)_3$ (d) SiO_2
(e) KCN (f) SCl_2
47. Write the formula for each substance. Which substances are ionic?
(a) Methane (b) Dinitrogen pentaoxide
(c) Ammonium sulfide (d) Hydrogen selenide
(e) Sodium perchlorate
48. Which of these compounds would you expect to be ionic? Explain your answers.
(a) SF_6 (b) CH_4
(c) H_2O_2 (d) NH_3
(e) CaO
49. Which of these compounds would you expect to be ionic? Explain your answers.
(a) NaH (b) HCl
(c) NH_3 (d) CH_4
(e) HI
50. Give the correct formula for each of these ionic compounds.
(a) Ammonium carbonate (b) Calcium iodide
(c) Copper(II) bromide (d) Aluminum phosphate
51. Give the correct formula for each of these ionic compounds.
(a) Calcium hydrogen carbonate
(b) Potassium permanganate
(c) Magnesium perchlorate
(d) Ammonium monohydrogen phosphate
52. Correctly name each of these ionic compounds.
(a) K_2S (b) $NiSO_4$
(c) $(NH_4)_3PO_4$ (d) $Al(OH)_3$
(e) $Co_2(SO_4)_3$
53. Correctly name each of these ionic compounds.
(a) KH_2PO_4 (b) $CuSO_4$
(c) $CrCl_3$ (d) $Ca(CH_3COO)_2$
(e) $Fe_2(SO_4)_3$
54. Solid magnesium oxide melts at 2800 °C. This property, combined with the fact that magnesium oxide is not an electrical conductor, makes it an ideal heat insulator for electric wires in cooking ovens and toasters. In contrast, solid NaCl melts at the relatively low temperature of 801 °C. What is the formula of magnesium oxide? Suggest a reason that it has a melting temperature so much higher than that of NaCl.
55. Assume you have an unlabeled bottle containing a white, crystalline powder. The powder melts at 310 °C. You are told that it could be NH_3, NO_2, or $NaNO_3$. What do you think it is and why?

Electrolytes (Section 3.7)

56. What is an electrolyte? How can we differentiate between a strong electrolyte and a nonelectrolyte? Give an example of each.
57. Epsom salt, $MgSO_4 \cdot 7\ H_2O$, is sold for various purposes over the counter in drug stores. Methanol, CH_3OH, is a small organic molecule that is readily soluble in either water or gasoline. Which of these two compounds is an electrolyte and which is a nonelectrolyte?
58. Comment on this statement: "Molecular compounds are generally nonelectrolytes."
59. Comment on this statement: "Ionic compounds are generally electrolytes."
60. For each of these electrolytes, what ions will be present in an aqueous solution?
(a) KOH (b) K_2SO_4
(c) $NaNO_3$ (d) NH_4Cl

61. For each of these electrolytes, what ions will be present in an aqueous solution?
(a) CaI_2 (b) $Mg_3(PO_4)_2$
(c) NiS (d) $MgBr_2$

62. Each of these three boxes shows a representation of a tiny volume in an aqueous solution of an ionic compound. (For simplicity, the water molecules are not shown.) Which box represents: (a) $MgCl_2$, (b) K_2SO_4, (c) NH_4Cl? Explain your reasoning.

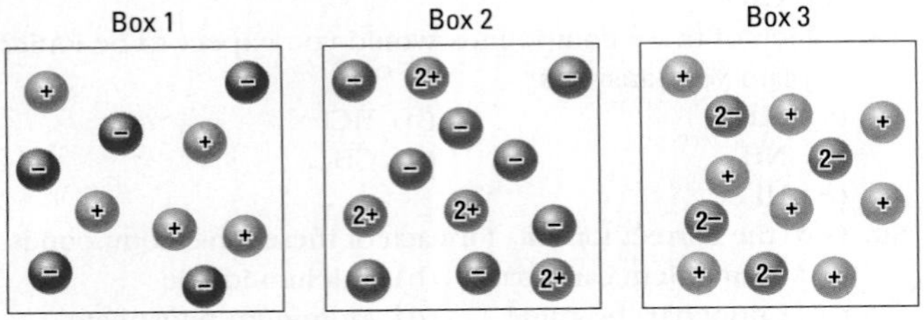
Box 1
Box 2
Box 3

63. Each of these three boxes shows a representation of a tiny volume in an aqueous solution of an ionic compound. (For simplicity, the water molecules are not shown.) Which box represents: (a) Na_2CO_3, (b) NH_4NO_3, (c) $CaBr_2$? Explain your reasoning.

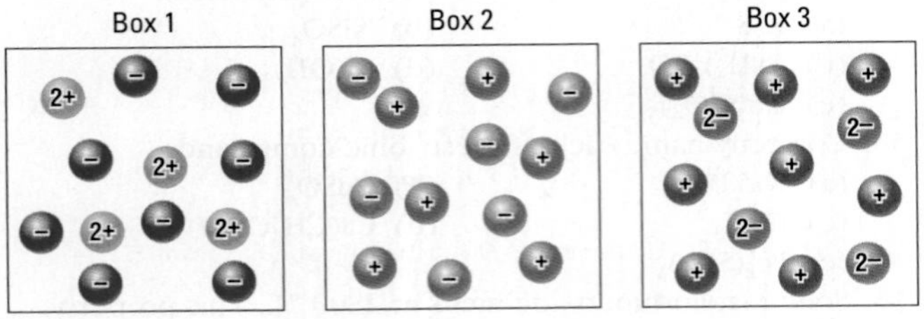
Box 1
Box 2
Box 3

64. Which of these substances would conduct electricity when dissolved in water?
(a) NaCl (b) $CH_3CH_2CH_3$ (propane)
(c) CH_3OH (methanol) (d) $Ca(NO_3)_2$

65. Which of these substances would conduct electricity when dissolved in water?
(a) NH_4Cl (b) $CH_3CH_2CH_2CH_3$ (butane)
(c) $C_{12}H_{22}O_{11}$ (table sugar) (d) $Ba(NO_3)_2$

Moles of Compounds (Section 3.8)

66. Fill in this table for 1 mol methanol, CH_3OH.

	CH_3OH	Carbon	Hydrogen	Oxygen
Number of moles				
Number of molecules or atoms				
Molar mass				

67. Fill in this table for 1 mol glucose, $C_6H_{12}O_6$.

	$C_6H_{12}O_6$	Carbon	Hydrogen	Oxygen
Number of moles				
Number of molecules or atoms				
Molar mass				

68. Calculate the molar mass of each of these compounds.
(a) Iron(III) oxide
(b) Boron trifluoride
(c) Dinitrogen oxide (laughing gas)
(d) Manganese(II) chloride tetrahydrate
(e) $C_6H_8O_6$, ascorbic acid

69. Calculate the molar mass of each of these compounds.
(a) $B_{10}H_{14}$, a boron hydride once considered as a rocket fuel
(b) $C_6H_2(CH_3)(NO_2)_3$, TNT, an explosive
(c) $PtCl_2(NH_3)_2$, a cancer chemotherapy agent called cisplatin
(d) $CH_3(CH_2)_3SH$, butyl mercaptan, a compound with a skunk-like odor
(e) $C_{20}H_{24}N_2O_2$, quinine, used as an antimalarial drug

70. How many moles are represented by 1.00 g of each of these compounds?
(a) CH_3OH, methanol
(b) Cl_2CO, phosgene, a poisonous gas
(c) Ammonium nitrate
(d) Magnesium sulfate heptahydrate (Epsom salt)
(e) Silver acetate

71. How many moles are present in 0.250 g of each of these compounds?
(a) $C_7H_5NO_3S$, saccharin, an artificial sweetener
(b) $C_{13}H_{20}N_2O_2$, procaine, a painkiller used by dentists
(c) $C_{20}H_{14}O_4$, phenolphthalein, a dye

72. (a) What is the molar mass of iron(II) nitrate, $Fe(NO_3)_2$?
(b) What is the mass, in grams, of 0.200 mol $Fe(NO_3)_2$?
(c) How many moles of $Fe(NO_3)_2$ are present in 4.66 g?

73. Calcium carbonate, found in marine fossils, has the molecular formula $CaCO_3$.
(a) What is the molar mass of $CaCO_3$?
(b) What is the mass, in grams, of 0.400 mol $CaCO_3$?
(c) How many moles of $CaCO_3$ are present in 7.63 g?

74. Acetaminophen, an analgesic, has the molecular formula $C_8H_9O_2N$.
(a) What is the molar mass of acetaminophen?
(b) How many moles are present in 5.32 g acetaminophen?
(c) How many grams are present in 0.166 mol acetaminophen?

75. A tablet contains 500. mg vitamin C, $C_6H_8O_6$. How many moles and molecules of vitamin C does the tablet contain?

76. An adult-strength tablet contains 325 mg aspirin, $C_9H_8O_4$. How many moles and molecules of aspirin does the tablet contain?

77. An Alka-Seltzer tablet contains 324 mg aspirin, $C_9H_8O_4$; 1904 mg $NaHCO_3$; and 1000 mg citric acid, $C_6H_8O_7$. (The

last two compounds react with each other to provide the "fizz," bubbles of CO_2, when the tablet is put into water.)

(a) Calculate the number of moles of each substance in the tablet.

(b) If you take one tablet, how many molecules of aspirin are you consuming?

78. How many moles of compound are in
(a) 39.2 g H_2SO_4? (b) 8.00 g O_2?
(c) 10.7 g NH_3?

79. How many moles of compound are in
(a) 46.1 g NH_4Cl? (b) 22.8 g CH_4?
(c) 9.63 g $CaCO_3$?

80. How many oxygen atoms are present in a 14.0-g sample of copper(II) nitrate?

81. How many sulfur atoms are present in a 21.0-g sample of iron(III) sulfate?

82. The use of CFCs (chlorofluorocarbons) has been curtailed because there is strong evidence that they cause environmental damage. If a spray can contains 250 g of one of these compounds, CCl_2F_2, how many molecules of this CFC are you releasing to the air when you empty the can?

83. Sulfur trioxide, SO_3, is made in enormous quantities by combining oxygen and sulfur dioxide, SO_2. The trioxide is not usually isolated but is converted to sulfuric acid.
(a) If you have 1.00 lb (454 g) sulfur trioxide, how many moles does this represent?
(b) How many molecules?
(c) How many sulfur atoms?
(d) How many oxygen atoms?

84. CFCs (chlorofluorocarbons) are implicated in decreasing the ozone concentration in the stratosphere. A CFC substitute is CF_3CH_2F.
(a) If you have 25.5 g of this compound, how many moles does this represent?
(b) How many atoms of fluorine are contained in 25.5 g of the compound?

85. How many water molecules are in one drop of water? (One drop of water is $\frac{1}{20}$ mL, and the density of water is 1.0 g/mL.)

86. If the water from a well contains 0.10 ppb (parts per billion) chloroform, $CHCl_3$, how many molecules of chloroform are present in one drop of the water? (One drop of water is $\frac{1}{20}$ mL, and the density of water is 1.0 g/mL.)

Percent Composition (Section 3.9)

87. Calculate the molar mass of each of these compounds and the mass percent of each element.
(a) PbS, lead(II) sulfide, galena
(b) C_2H_6, ethane, a hydrocarbon fuel
(c) CH_3COOH, acetic acid, an important ingredient in vinegar
(d) NH_4NO_3, ammonium nitrate, a fertilizer

88. Calculate the molar mass of each of these compounds and the mass percent of each element.
(a) $MgCO_3$, magnesium carbonate
(b) C_6H_5OH, phenol, an organic compound used in some cleaners
(c) $C_2H_3O_5N$, peroxyacetyl nitrate, an objectionable compound in photochemical smog
(d) $C_4H_{10}O_3NPS$, acephate, an insecticide

89. A certain metal, M, forms two oxides, M_2O and MO. If the percent by mass of M in M_2O is 73.4%, what is its percent by mass in MO?

90. Three oxygen-containing compounds of iron are $FeCO_3$, Fe_2O_3, and Fe_3O_4. What are the percentages of iron in each of these iron compounds?

91. The copper-containing compound $Cu(NH_3)_4SO_4 \cdot H_2O$ is a beautiful blue solid. Calculate the molar mass of the compound and the mass percent of each element.

92. Sucrose, table sugar, is $C_{12}H_{22}O_{11}$. When sucrose is heated, water is driven off. How many grams of pure carbon can be obtained from exactly one pound of sugar?

93. Carbonic anhydrase, an important enzyme in mammalian respiration, is a large zinc-containing protein with a molar mass of 3.00×10^4 g/mol. The zinc is 0.218% by mass of the protein. How many zinc atoms does each carbonic anhydrase molecule contain?

94. Nitrogen fixation in the root nodules of peas and other legumes occurs with a reaction involving a molybdenum-containing enzyme named *nitrogenase.* This enzyme contains two Mo atoms per molecule and is 0.0872% Mo by mass. What is the molar mass of the enzyme?

95. If you heat Al with an element from Group 6A, an ionic compound is formed that contains 18.55% Al by mass.
(a) What is the likely charge on the nonmetal in the compound formed?
(b) Using X to represent the nonmetal, what is the empirical formula for this ionic compound?
(c) Which element in Group 6A has been combined with Al?

96. Disilane, Si_2H_x, contains 90.28% silicon by mass. What is the value of x in this compound?

97. Chalky, white crystals in mineral collections are often labeled borax, which has the molecular formula $Na_2B_4O_7 \cdot 10\ H_2O$, when actually they are partially dehydrated samples with the molecular formula $Na_2B_4O_7 \cdot 5\ H_2O$, which is more stable under the storage conditions. Real crystals of borax are colorless, transparent crystals.
(a) What percent of the mass has the mineral lost when it partially dehydrates?
(b) Will the percent boron by mass be the same in the two compounds?

Empirical and Molecular Formulas (Section 3.10)

98. What is the difference between an empirical formula and a molecular formula? Use the compound ethane, C_2H_6, to illustrate your answer.

99. The molecular formula of ascorbic acid (vitamin C) is $C_6H_8O_6$. What is its empirical formula?

100. The empirical formula of maleic acid is CHO. Its molar mass is 116.1 g/mol. What is its molecular formula?

101. A well-known reagent in analytical chemistry, dimethylglyoxime, has the empirical formula C_2H_4NO. If its molar mass is 116.1 g/mol, what is the molecular formula of the compound?

102. A compound with a molar mass of 100.0 g/mol has an elemental composition of 24.0% C, 3.0% H, 16.0% O, and 57.0% F. What is the molecular formula of the compound?

103. Acetylene is a colorless gas that is used as a fuel in welding torches, among other things. It is 92.26% C and

7.74% H. Its molar mass is 26.02 g/mol. Calculate the empirical and molecular formulas.

104. A compound contains 38.67% K, 13.85% N, and 47.47% O by mass. What is the empirical formula of the compound?

105. A compound contains 36.76% Fe, 21.11% S, and 42.13% O by mass. What is the empirical formula of the compound?

106. There is a large family of boron-hydrogen compounds called boron hydrides. All have the formula B_xH_y and almost all react with air and burn or explode. One member of this family contains 88.5% B; the remainder is hydrogen. Which of these is its empirical formula: BH_2, BH_3, B_2H_5, B_5H_7, B_5H_{11}?

107. Nitrogen and oxygen form an extensive series of at least seven oxides of general formula N_xO_y. One of them is a blue solid that comes apart, or "dissociates," reversibly, in the gas phase. It contains 36.84% N. What is the empirical formula of this oxide?

108. Cumene, a hydrocarbon, is 89.94% carbon, and its molar mass is 120.2 g/mol. What are the empirical and molecular formulas of cumene?

109. Acetic acid is the important ingredient in vinegar. It is composed of carbon (40.0%), hydrogen (6.71%), and oxygen (53.29%). Its molar mass is 60.0 g/mol. Determine the empirical and molecular formulas of the acid.

110. An analysis of nicotine, a poisonous compound found in tobacco leaves, shows that it is 74.0% C, 8.65% H, and 17.35% N. Its molar mass is 162 g/mol. What are the empirical and molecular formulas of nicotine?

111. Cacodyl, a compound containing arsenic, was reported in 1842 by the German chemist Bunsen. It has an almost intolerable garlic-like odor. Its molar mass is 210. g/mol, and it is 22.88% C, 5.76% H, and 71.36% As. Determine its empirical and molecular formulas.

112. The action of bacteria on meat and fish produces a poisonous and stinky compound called cadaverine. It is 58.77% C, 13.81% H, and 27.42% N. Its molar mass is 102.2 g/mol. Determine the molecular formula of cadaverine.

113. DDT is an insecticide with this percent composition: 47.5% C, 2.54% H, and the remainder chlorine. What is the empirical formula of DDT?

114. If blue vitriol, $CuSO_4 \cdot x\, H_2O$, is heated, all of the water of hydration is lost. When a 5.172-g sample of the hydrate is heated, 3.306 g $CuSO_4$ remains. How many molecules of water are there per formula unit of blue vitriol?

115. The alum used in cooking is potassium aluminum sulfate hydrate, $KAl(SO_4)_2 \cdot x\, H_2O$. To find the value of x, you can heat a sample of the compound to drive off all the water and leave only $KAl(SO_4)_2$. Assume that you heat 4.74 g of the hydrated compound and that it loses 2.16 g water. What is the value of x?

Biological Periodic Table (Section 3.11)

116. Make a list of the top ten most abundant essential elements needed by the human body.

117. Which types of compounds contain the majority of the oxygen found in the human body?

118. (a) In what form are metals found in the body, as atoms or as ions?
(b) What are two uses for metals in the human body?

119. Distinguish between macrominerals and microminerals.

120. Which minerals are essential at smaller concentrations but toxic at higher concentrations?

General Questions

These questions are not explicitly keyed to chapter topics; many require integration of several concepts.

121. (a) Draw a diagram showing the crystal lattice of sodium chloride, NaCl. Show clearly why such a crystal can be cleaved easily by tapping on a knife blade properly aligned along the crystal.
(b) Describe in words why the cleavage occurs as it does.

122. Give the molecular formula for each of these molecules.

(a)

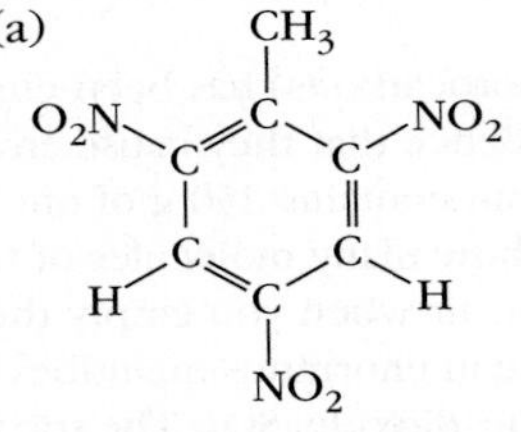

trinitrotoluene, TNT

(b)

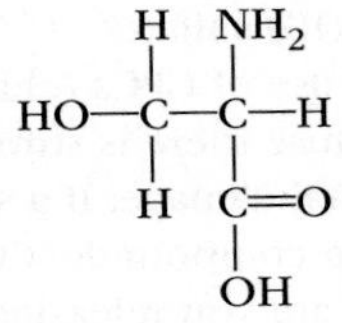

serine, an essential amino acid

123. (a) Calculate the mass of one molecule of nitrogen.
(b) Calculate the mass of one molecule of oxygen.
(c) What is the ratio of masses of these two molecules? How does it compare to the ratio of the atomic weights of N and O?

124. (a) Which of these pairs of elements are likely to form ionic compounds?
(b) Write appropriate formulas for the compounds you expect to form, and name each.
(i) Chlorine and bromine
(ii) Lithium and tellurium
(iii) Sodium and argon
(iv) Magnesium and fluorine
(v) Nitrogen and bromine
(vi) Indium and sulfur
(vii) Selenium and bromine

125. (a) Name each of these compounds.
(b) Tell which ones are best described as ionic.

(i) $ClBr_3$	(ii) NCl_3	(iii) $CaSO_4$
(iv) C_7H_{16}	(v) XeF_4	(vi) OF_2
(vii) NaI	(viii) Al_2S_3	(ix) PCl_5
(x) K_3PO_4		

126. (a) Write the formula for each of these compounds.
(b) Tell which ones are best described as ionic.
(i) Sodium hypochlorite
(ii) Aluminum perchlorate
(iii) Potassium permanganate
(iv) Potassium dihydrogen phosphate
(v) Chlorine trifluoride
(vi) Boron tribromide
(vii) Calcium acetate
(viii) Sodium sulfite
(ix) Disulfur tetrachloride
(x) Phosphorus trifluoride

127. Precious metals such as gold and platinum are sold in units of "troy ounces," where 1 troy ounce is equivalent to 31.1 g.
   (a) If you have a block of platinum with a mass of 15.0 troy ounces, how many moles of the metal do you have?
   (b) What is the size of the block in cubic centimeters? (The density of platinum is 21.45 g/cm^3 at 20 °C.)

128. "Dilithium" is the fuel for the Starship *Enterprise.* However, because its density is quite low, you will need a large space to store a large mass. As an estimate for the volume required, we shall use the element lithium.
   (a) If you want to have 256 mol for an interplanetary trip, what must the volume of a piece of lithium be?
   (b) If the piece of lithium is a cube, what is the dimension of an edge of the cube? (The density of lithium is 0.534 g/cm^3 at 20 °C.)

129. Elemental analysis of fluorocarbonyl hypofluorite gave 14.6% C, 39.0% O, and 46.3% F. If the molar mass of the compound is 82.0 g/mol, determine the (a) empirical and (b) molecular formulas of the compound.

130. Azulene, a beautiful blue hydrocarbon, is 93.71% C and has a molar mass of 128.16 g/mol. What are the (a) empirical and (b) molecular formulas of azulene?

131. A major oil company has used an additive called MMT to boost the octane rating of its gasoline. What is the empirical formula of MMT if it is 49.5% C, 3.2% H, 22.0% O, and 25.2% Mn?

132. Direct reaction of iodine, I_2, and chlorine, Cl_2, produces an iodine chloride, I_xCl_y, a bright yellow solid.
   (a) If you completely react 0.678 g iodine to produce 1.246 g I_xCl_y, what is the empirical formula of the compound?
   (b) A later experiment shows that the molar mass of I_xCl_y is 467 g/mol. What is the molecular formula of the compound?

133. Pepto-Bismol, which helps provide relief for an upset stomach, contains 300 mg bismuth subsalicylate, $C_7H_5BiO_4$, per tablet.
   (a) If you take two tablets for your stomach distress, how many moles of the "active ingredient" are you taking?
   (b) How many grams of Bi are you consuming in two tablets?

134. Iron pyrite, often called "fool's gold," has the formula FeS_2. If you could convert 15.8 kg iron pyrite to iron metal and remove the sulfur, how many kilograms of the metal could you obtain?

135. Ilmenite is a mineral that is an oxide of iron and titanium, $FeTiO_3$. If an ore that contains ilmenite is 6.75% titanium, what is the mass (in grams) of ilmenite in 1.00 metric ton (exactly 1000 kg) of the ore?

136. Stibnite, Sb_2S_3, is a dark gray mineral from which antimony metal is obtained. If you have one pound of an ore that contains 10.6% antimony, what mass of Sb_2S_3 (in grams) is there in the ore?

137. Draw a diagram to indicate the arrangement of nanoscale particles of each substance. Consider each drawing to hold a very tiny portion of each substance. Each drawing should contain at least 16 particles, and it need not be three-dimensional.

$Br_2(\ell)$ | LiF(s)

138. Draw diagrams of each nanoscale situation below. Represent atoms or monatomic ions as circles; represent molecules or polyatomic ions by overlapping circles for the atoms that make up the molecule or ion; and distinguish among different kinds of atoms by labeling or shading the circles. In each case draw representations of at least five nanoscale particles. Your diagrams can be two-dimensional.
   (a) A crystal of solid sodium chloride
   (b) The sodium chloride from part (a) after it has been melted

139. Draw diagrams of each nanoscale situation below. Represent atoms or monatomic ions as circles; represent molecules or polyatomic ions by overlapping circles for the atoms that make up the molecule or ion; and distinguish among different kinds of atoms by labeling or shading the circles. In each case draw representations of at least five nanoscale particles. Your diagrams can be two-dimensional.
   (a) A sample of solid lithium nitrate, $LiNO_3$
   (b) A sample of molten lithium nitrate
   (c) A molten sample of lithium nitrate after electrodes have been placed into it and a direct current applied to the electrodes

Applying Concepts

These questions test conceptual learning.

140. When asked to draw all the possible constitutional isomers for C_3H_8O, a student drew these structures. The student's instructor said some of the structures were identical.
   (a) How many actual isomers are there?
   (b) Which structures are identical?

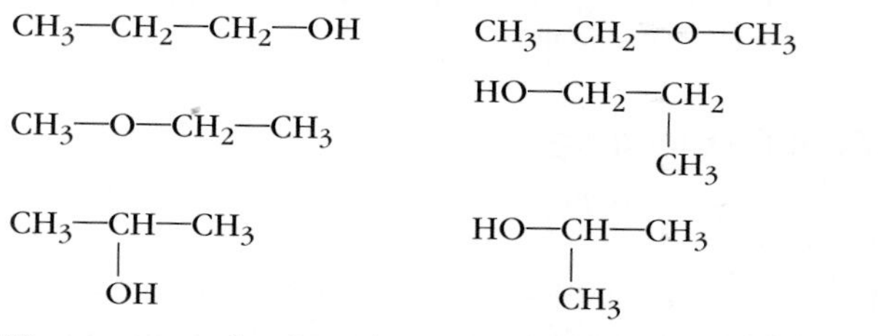
CH3—CH2—CH2—OH
CH3—CH2—O—CH3
HO—CH2—CH2
CH3
CH3—O—CH2—CH3
CH3—CH—CH3
OH
HO—CH—CH3
CH3

141. The statement that "metals form positive ions by losing electrons" is difficult to grasp for some students because positive signs indicate a gain and negative signs indicate a loss. How would you explain this contradiction to a classmate?

142. The formula for thallium nitrate is $TlNO_3$. Based on this information, what would be the formulas for thallium carbonate and thallium sulfate?

143. The name given with each of these formulas is incorrect. What are the correct names?
(a) CaF_2, calcium difluoride
(b) CuO, copper oxide
(c) $NaNO_3$, sodium nitroxide
(d) NI_3, nitrogen iodide
(e) $FeCl_3$, iron(I) chloride
(f) Li_2SO_4, dilithium sulfate

144. Based on the guidelines for naming oxoanions in a series, how would you name these species?
(a) BrO_4^-, BrO_3^-, BrO_2^-, BrO^-
(b) SeO_4^{2-}, SeO_3^{2-}

145. Which illustration best represents $CaCl_2$ dissolved in water?

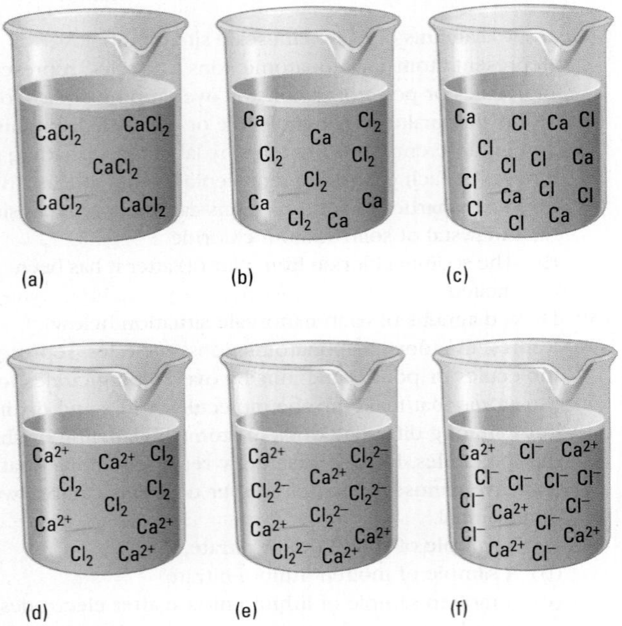

146. Which sample has the largest amount of NH_3?
(a) 6.022×10^{24} molecules of NH_3
(b) 0.1 mol NH_3
(c) 17.03 g NH_3

147. One molecule of an unknown compound has a mass of 7.308×10^{-23} g, and 27.3% of that mass is due to carbon; the rest is oxygen. What is the compound?

148. What is the empirical formula of a compound that contains 23.3% Co, 25.3% Mo, and the remainder Cl?

149. What is the empirical formula of a compound that contains 23.3% Mg, 46.0% O, and the remainder S? What is the compound's name?

More Challenging Questions

These questions require more thought and integrate several concepts.

150. A piece of nickel foil, 0.550 mm thick and 1.25 cm square, was allowed to react with fluorine, F_2, to give a nickel fluoride. (The density of nickel is 8.908 g/cm^3.)
(a) How many moles of nickel foil were used?
(b) If you isolate 1.261 g nickel fluoride, what is its formula?
(c) What is its name?

151. Uranium is used as a fuel, primarily in the form of uranium(IV) oxide, in nuclear power plants. This question considers some uranium chemistry.
(a) A small sample of uranium metal (0.169 g) is heated to 800 to 900 °C in air to give 0.199 g of a dark green oxide, U_xO_y. How many moles of uranium metal were used? What is the empirical formula of the oxide U_xO_y? What is the name of the oxide? How many moles of U_xO_y must have been obtained?
(b) The oxide U_xO_y is obtained if $UO_2NO_3 \cdot n\,H_2O$ is heated to temperatures greater than 800 °C in air. However, if you heat it gently, only the water of hydration is lost. If you have 0.865 g $UO_2NO_3 \cdot n\,H_2O$ and obtain 0.679 g UO_2NO_3 on heating, how many molecules of water of hydration were there in each formula unit of the original compound?

152. What are the empirical formulas of these compounds?
(a) A hydrate of Fe(III) thiocyanate, $Fe(SCN)_3$, that contains 19.0% water
(b) A mineral hydrate of zinc sulfate that contains 43.86% water
(c) A sodium aluminum silicate hydrate that contains 12.10% Na, 14.19% Al, 22.14% Si, 42.09% O, and 9.48% H_2O

153. The active ingredients in lawn and garden fertilizer are nitrogen, phosphorus, and potassium. Bags of fertilizer usually carry three numbers, as in 5-10-5 fertilizer (a typical fertilizer for flowers). The first number is the mass percent N; the second is the mass percent P_2O_5; the third is the mass percent K_2O. Thus, the active ingredients in a 5-10-5 product are equivalent to 5% N, 10% P_2O_5, and 5% K_2O, by weight. (The reporting of fertilizer ingredients in this way is a convention agreed on by fertilizer manufacturers.) What is the mass in pounds of each of these three elements (N, P, K) in a 100-lb bag of fertilizer?

© Cengage Learning/George Semple

154. Four common ingredients in fertilizers are ammonium nitrate, NH_4NO_3; ammonium sulfate, $(NH_4)_2SO_4$; urea, $(NH_2)_2CO$; and ammonium hydrogen phosphate, $(NH_4)_2HPO_4$. On the basis of mass, which of these compounds has the largest mass percent nitrogen? What is the mass percent of nitrogen in this compound?

155. A mixture contains only $MgSO_4$ and $(NH_4)_2SO_4$. If the mass percent of $MgSO_4$ in the mixture is 32.0%, what is the mass percent of sulfate in the mixture?

156. Tyrosine is an amino acid with this molecular structure. Calculate the mass percent of each element in tyrosine.

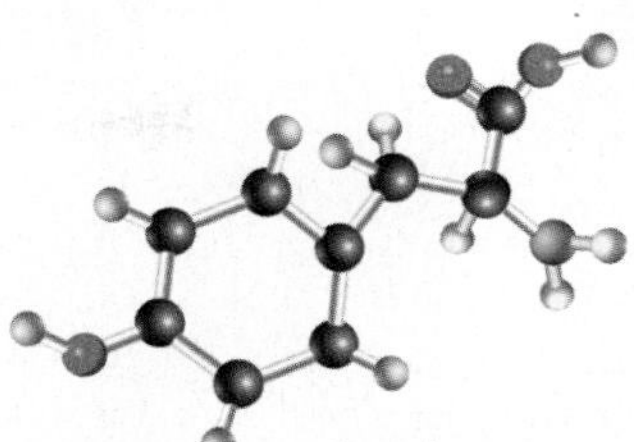

tyrosine

157. Hemoglobin is an iron-containing protein (molar mass 64,458 g/mol) that is responsible for oxygen transport in our blood. Hemoglobin is 0.35% iron by mass. Calculate how many iron atoms are in each hemoglobin molecule.

Conceptual Challenge Problems

These rigorous, thought-provoking problems integrate conceptual learning with problem solving and are suitable for group work.

CP3.A (Section 3.9) A chemist analyzes three compounds and reports these data for the percent by mass of the elements Ex, Ey, and Ez in each compound.

Compound	% Ex	% Ey	% Ez
A	37.485	12.583	49.931
B	40.002	6.7142	53.284
C	40.685	5.1216	54.193

Assume that you accept the notion that the numbers of atoms of the elements in compounds are in small whole-number ratios and that the number of atoms in a sample of any element is directly proportional to that sample's mass. What is possible for you to know about the empirical formulas for these three compounds?

CP3.B (Section 3.10) The following table displays on each horizontal row an empirical formula for one of the three compounds noted in CP3.A.

Compound A	Compound B	Compound C
______	$ExEy_2Ez$	______
$Ex_6Ey_8Ez_3$	______	______
______	______	Ex_3Ey_2Ez
______	$Ex_9Ey_2Ez_6$	______
______	______	$ExEy_2Ez_3$
$Ex_3Ey_8Ez_3$	______	______

Based only on what was learned in that problem, what is the empirical formula for the other two compounds in that row?

CP3.C (Section 3.10)

(a) Suppose that a chemist now determines that the ratio of the masses of equal numbers of atoms of Ez and Ex atoms is 1.3320 g Ez/1 g Ex. With this added information, what can now be known about the formulas for compounds A, B, and C in Problem CP3.A?

(b) Suppose that this chemist further determines that the ratio of the masses of equal numbers of atoms of Ex and Ey is 11.916 g Ex/1 g Ey. What is the ratio of the masses of equal numbers of Ez and Ey atoms?

(c) If the mass ratios of equal numbers of atoms of Ex, Ey, and Ez are known, what can be known about the formulas of the three compounds A, B, and C?

4 Quantities of Reactants and Products

© Cengage Learning/Charles D. Winters

When an Alka-Seltzer tablet dissolves in water, carbon dioxide gas is produced by the reaction of citric acid, $C_6H_8O_7$, and sodium hydrogen carbonate, $NaHCO_3$ (baking soda), contained in the tablet. The reaction of 1 mol $C_6H_8O_7$ with 3 mol $NaHCO_3$ produces 3 mol H_2O, 3 mol CO_2, and 1 mol sodium citrate, $Na_3C_6H_5O_7$, as shown by the balanced chemical equation:

$$C_6H_8O_7\,(aq) + 3\ NaHCO_3(aq) \longrightarrow 3\ H_2O(\ell) + 3\ CO_2(g) + Na_3C_6H_5O_7(aq)$$

where (aq), (ℓ), and (g) indicate aqueous solution, liquid, and gas, respectively. The coefficients in a balanced chemical equation provide important quantitative relations between amounts (moles) of reactants and products. Such relations are a major focus of this chapter.

A major emphasis of chemistry is understanding chemical reactions. To work with chemical reactions requires knowing the correct formulas and the relative molar amounts of all reactants and products involved in the reaction. This information is contained in balanced chemical equations—equations that are consistent with the law of conservation of mass and the existence of atoms. This chapter begins with a discussion of the nature of chemical equations. It is followed by a brief introduction to some general types of chemical reactions. Many of the very large number of known chemical reactions can be assigned to a few categories: combination, decomposition, displacement, and exchange. Next comes a description of how to write balanced chemical equations. After that, we will look at how the balanced chemical equation can be used to move from an understanding of what the reactants and products are to how much of each is involved under various conditions. Finally, we will introduce methods used to find the formulas of chemical compounds.

While the fundamentals of chemical reactions must be understood at the atomic and molecular levels (the nanoscale), the chemical reactions that we will describe are observable at the macroscale in the everyday world around us. Understanding and applying the quantitative relationships in chemical reactions are essential skills. You should know how to calculate the quantity of product that a reaction will generate from a given quantity of reactants. Facility with these quantitative calculations connects the nanoscale world of the chemical reaction to the macroscale world of laboratory and industrial manipulations of measurable quantities of chemicals.

OWL Sign in to OWL at **www.cengage.com/owl** to view tutorials and simulations, develop problem-solving skills, and complete online homework assigned by your professor.

go Chemistry Download mini lecture videos for key concept review and exam prep from OWL or purchase them from **www.CengageBrain.com.**

Companion Website

Visit this book's companion website at **www.cengage.com/chemistry/moore** to work interactive modules for the Estimation boxes and Active Figures in this text.

4.1 Chemical Equations

A candle flame can create a mood as well as provide light. It also is the result of a chemical reaction, a process in which reactants are converted into products ***(← p. 12)***. The reactants and products can be elements, compounds, or both. In equation form we write

$$\text{Reactant(s)} \longrightarrow \text{product(s)}$$

where the arrow means "forms" or "yields" or "changes to."

In a burning candle, the reactants are hydrocarbons from the candle wax and oxygen from the air. Such reactions, in which an element or compound burns in air or oxygen, are called **combustion reactions.** The products of the complete combustion of hydrocarbons ***(← p. 80)*** are always carbon dioxide and water.

$$\underset{\text{a hydrocarbon in candle wax}}{C_{25}H_{52}(s)} + \underset{\text{oxygen}}{38\ O_2(g)} \longrightarrow \underset{\text{carbon dioxide}}{25\ CO_2(g)} + \underset{\text{water}}{26\ H_2O(g)}$$

© PhotoLink/PhotoDisc Green/Getty Images

Combustion of hydrocarbons from candle wax produces a candle flame.

This **balanced chemical equation** indicates the relative amounts of reactants and products required so that the number of atoms of each element in the reactants equals the number of atoms of the same element in the products. In the next section we discuss how to write balanced equations.

Usually the physical states of the reactants and products are indicated in a chemical equation by placing one of these symbols after each reactant and product: (s) for solid, (ℓ) for liquid, and (g) for gas. The symbol (aq) is used to represent an **aqueous solution,** a substance dissolved in water. This is illustrated by the equation for the reaction of zinc metal with hydrochloric acid, an aqueous solution of hydrogen chloride, HCl. The products are hydrogen gas and an aqueous solution of zinc chloride, a soluble ionic compound ***(← p. 85)***.

$$\underbrace{Zn(s) + 2\ HCl(aq)}_{\text{Reactants}} \longrightarrow \underbrace{H_2(g) + ZnCl_2(aq)}_{\text{Products}}$$

In the 18th century, the great French scientist Antoine Lavoisier introduced the **law of conservation of mass,** which later became part of Dalton's atomic theory. Lavoisier showed that **mass is neither created nor destroyed in chemical reactions.** Therefore, if you use 5 g of reactants they will form 5 g of products when the reaction is complete; if you use 500 mg of reactants they will form 500 mg of products; and so on. Combined with Dalton's atomic theory ***(← p. 21)***, this also means that if there are 1000 atoms of a particular element in the reactants, then those 1000 atoms must appear in the products.

There must be an accounting for *all* atoms in a chemical reaction.

Recall that there are 6.022×10^{23} atoms in a mole of any atomic element ***(← p. 59)*** or 6.022×10^{23} molecules in a mole of any molecular element or compound ***(← p. 61)***.

1743–1794

One of the first to recognize the importance of exact scientific measurements and of carefully planned experiments, Lavoisier introduced principles for naming chemical substances that are still in use today. Further, he wrote a textbook, *Elements of Chemistry* (1789), in which he applied for the first time the principle of conservation of mass to chemistry and used the idea to write early versions of chemical equations. His life was cut short during the Reign of Terror of the French Revolution.

Consider, for example, the reaction between gaseous hydrogen and chlorine to produce hydrogen chloride gas:

$$H_2(g) + Cl_2(g) \longrightarrow 2\ HCl(g)$$

When applied to this reaction, the law of conservation of mass means that one diatomic molecule of H_2 (two atoms of hydrogen) and one diatomic molecule of Cl_2 (two atoms of Cl) must produce two molecules of HCl. The numbers in front of the formulas—the **coefficients**—in balanced equations show how matter is conserved. The 2 HCl indicates that two HCl molecules are formed, each containing one hydrogen atom and one chlorine atom. Note how the symbol of an element or the formula of a compound is multiplied through by the coefficient that precedes it. The equality of the number of atoms of each kind in the reactants and in the products is what makes the equation "balanced."

Multiplying all the coefficients by the same factor gives the relative amounts of reactants and products at any scale. For example, 4 H_2 molecules will react with 4 Cl_2 molecules to produce 8 HCl molecules (Figure 4.1). If we continue to scale up the reaction, we can use Avogadro's number as the common factor. Thus, 1 mol H_2 molecules reacting with 1 mol Cl_2 molecules will produce 2 mol HCl molecules. As demanded by the conservation of mass, the number of atoms of each type in the reactants and the products is the same.

With the numbers of atoms balanced, the masses represented by the equation are also balanced: 1.000 mol H_2 is equivalent to 2.016 g H_2 and 1.000 mol Cl_2 is equivalent to 70.90 g Cl_2, so the total mass of reactants must be 2.016 g + 70.90 g = 72.92 g when 1.000 mol each of H_2 and Cl_2 are used. Conservation of mass demands that the same mass, 72.92 g HCl, must result from the reaction, and it does.

$$2.000\ \text{mol HCl} \times \frac{36.45\ \text{g HCl}}{1\ \text{mol HCl}} = 72.90\ \text{g HCl}$$

Relations among the masses of chemical reactants and products are called **stoichiometry** (stoy-key-AHM-uh-tree), and the coefficients (the multiplying numbers) in a balanced equation are called **stoichiometric coefficients** (or just coefficients).

Answers to **EXERCISES** are provided at the back of this book in Appendix L.

EXERCISES that are labeled **CONCEPTUAL** are designed to test your understanding of one or more concepts; they usually involve qualitative rather than quantitative thinking.

CONCEPTUAL EXERCISE 4.1 Chemical Equations

When methane burns this reaction is occurring:

$$CH_4(g) + 2\ O_2(g) \longrightarrow CO_2(g) + 2\ H_2O(g)$$

Write out in words the meaning of this chemical equation.

EXERCISE 4.2 Stoichiometric Coefficients

Heating iron metal in oxygen forms iron(III) oxide, Fe_2O_3 (Figure 4.2):

$$4\ Fe(s) + 3\ O_2(g) \longrightarrow 2\ Fe_2O_3(s)$$

(a) If 2.50 g Fe_2O_3 is formed by this reaction, what is the maximum *total* mass of iron metal and oxygen that reacted?

(b) Identify the stoichiometric coefficients in this equation.

(c) If 10,000 oxygen atoms reacted, how many Fe atoms were needed to react with this amount of oxygen?

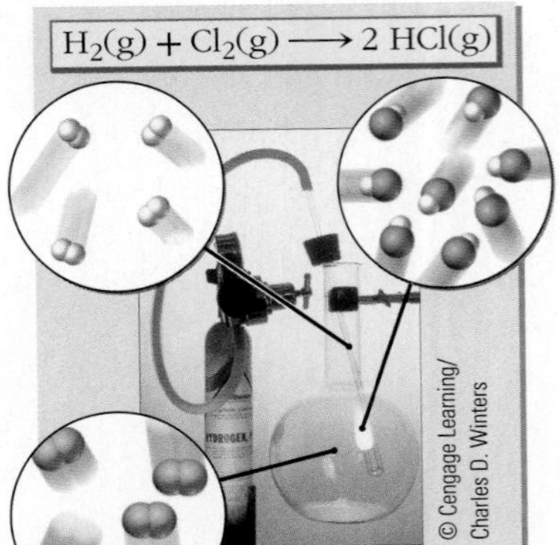

Figure 4.1 Hydrogen, H_2, and chlorine, Cl_2, react to form hydrogen chloride, HCl. Two molecules of HCl are formed when one H_2 molecule reacts with one Cl_2 molecule. This ratio is maintained when the reaction is carried out on a larger scale.

4.2 Patterns of Chemical Reactions

Many simple inorganic chemical reactions can be characterized as one of these reaction patterns: combination, decomposition, displacement, or exchange. Learning to recognize these reaction patterns is useful because they serve as a guide to predict what might happen when chemicals are mixed or heated.

Take a look at the equation below. What does it mean to you at this stage in your study of chemistry?

$$Cl_2(g) + 2\ KBr(aq) \longrightarrow 2\ KCl(aq) + Br_2(\ell)$$

It's easy to let your eye slide by an equation on the printed page, but *don't* do that; ***read the equation.*** In this case it shows you that gaseous diatomic chlorine mixed with an aqueous solution of potassium bromide reacts to produce an aqueous solution of potassium chloride plus liquid diatomic bromine. After you learn to recognize reaction patterns, you will see that this is a *displacement* reaction—chlorine has displaced bromine from KBr so that the resulting compound in solution is KCl instead of the KBr originally present.

Throughout the rest of this book, you should read and interpret chemical equations as we have just illustrated. Note the physical states of reactants and products. Mentally classify the reactions as described in the next few sections and look for what can be learned from each example. Most importantly, *don't* think that the equations must be memorized. There are far too many chemical reactions for that. Instead, look for patterns, classes of reactions and the kinds of substances that undergo them, and information that can be applied in other situations. Doing so will give you insight into how chemistry is used every day in a wide variety of applications.

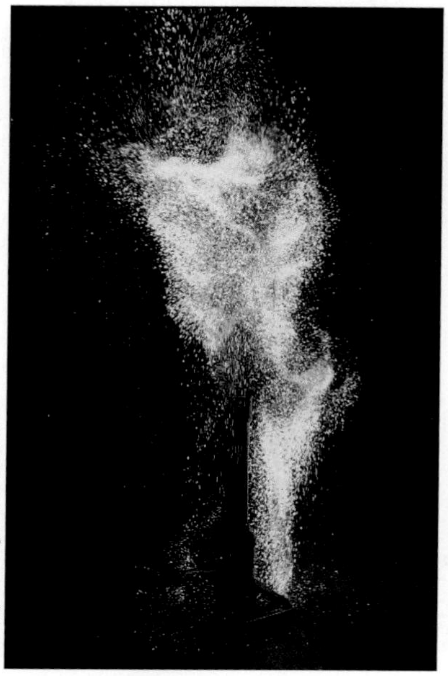

© Cengage Learning/Charles D. Winters

Figure 4.2 **Powdered iron burns in air to form the iron oxide Fe_2O_3.** The energy released during the reaction heats the particles to incandescence, seen here as tiny bright spots.

Combination Reactions

In a **combination reaction,** two or more substances react to form a *single* product.

The halogens (Group 7A) and oxygen are such reactive elements that they undergo combination reactions with most other elements. Thus, if one of two possible reactants is oxygen or a halogen and the other reactant is another element, it is reasonable to expect that a combination reaction will occur.

Recall that the halogens are F_2, Cl_2, Br_2, and I_2.

The combination reaction of a metal with oxygen produces an *ionic compound,* a metal oxide. Like any ionic compound, the metal oxide must be electrically neutral. Because you can predict a reasonable positive charge for the metal ion by knowing its position in the periodic table and by using the guidelines in Section 3.5 ***(← p. 85)*** you can determine the formula of the metal oxide. For example, when aluminum (Al, Group 3A), which forms Al^{3+} ions, reacts with O_2, which forms O^{2-} ions, the product must be aluminum oxide Al_2O_3 (2 Al^{3+} and 3 O^{2-} ions).

Aluminum oxide is also known as alumina or corundum.

$$4\ Al(s) + 3\ O_2(g) \longrightarrow 2\ Al_2O_3(s)$$

aluminum oxide

The halogens also combine with metals to form ionic compounds with formulas that are predictable based on the charges of the ions formed. The halogens form 1− ions in simple ionic compounds. For example, sodium combines with chlorine, and zinc combines with iodine, to form sodium chloride and zinc iodide, respectively (Figure 4.3, p. 124).

$$2\ Na(s) + Cl_2(g) \longrightarrow 2\ NaCl(s)$$

$$Zn(s) + I_2(s) \longrightarrow ZnI_2(s)$$

When nonmetals combine with oxygen or chlorine, the compounds formed are molecular, composed of *molecules,* not ions. For example, sulfur, the Group 6A neighbor of oxygen, combines with oxygen to form two oxides, SO_2 and SO_3, in reactions of great environmental and industrial significance.

$$S_8(s) + 8\ O_2(g) \longrightarrow 8\ SO_2(g)$$

sulfur dioxide
(colorless; choking odor)

$$2\ SO_2(g) + O_2(g) \longrightarrow 2\ SO_3(g)$$

sulfur trioxide
(colorless; even more choking odor)

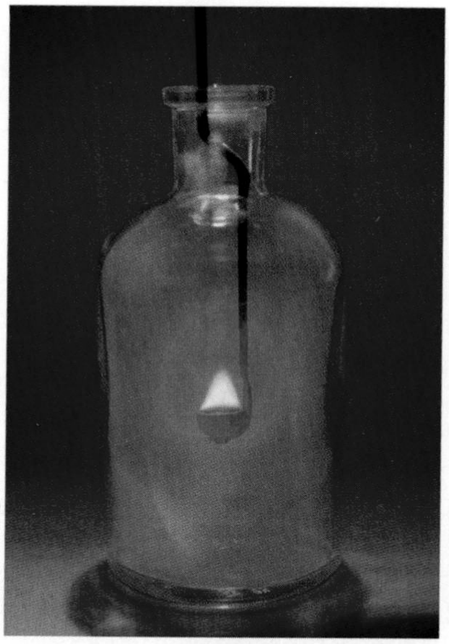

© Cengage Learning/Charles D. Winters

Sulfur burns in oxygen to produce a bright blue flame and SO_2(g).

At room temperature, sulfur is a bright yellow solid. At the nanoscale it consists of molecules with eight-membered rings, S_8.

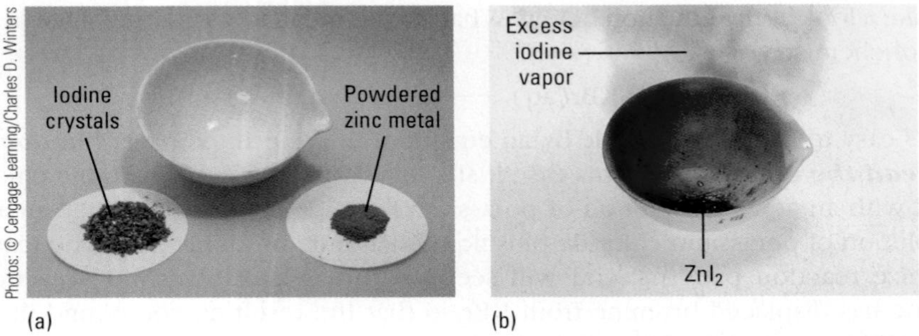
Photos: © Cengage Learning/Charles D. Winters
Iodine crystals
Powdered zinc metal
Excess iodine vapor
ZnI_2
(a)
(b)

Figure 4.3 **Combination reaction of zinc and iodine.** (a) The reactants. (b) The reaction. In a vigorous combination reaction, zinc atoms react with diatomic iodine molecules to form zinc iodide, an ionic compound, and the heat of the reaction is great enough that excess iodine forms a purple vapor.

Sulfur dioxide enters the atmosphere both from natural sources and from human activities. About 75% of the sulfur oxides in the atmosphere come from human activities, such as burning coal in electrical power plants. All coal contains sulfur, usually from 1% to 4% by weight.

Yet another example of a combination reaction is that between an organic molecule such as ethylene, C_2H_2, and bromine, Br_2, to form dibromoethane:

The reaction between C_2H_2 and Br_2 is also called an *addition reaction.*

$$C_2H_2(g) + Br_2(\ell) \longrightarrow C_2H_2Br_2(g)$$

EXERCISE 4.3 Combination Reactions

Indicate whether each equation for a combination reaction is balanced, and if it is not, why not.

(a) $Cu + O_2 \rightarrow CuO$ (b) $Cr + Br_2 \rightarrow CrBr_3$ (c) $S_8 + 3\ F_2 \rightarrow SF_6$

CONCEPTUAL EXERCISE 4.4 Combination Reactions

(a) What information is needed to predict the product of a combination reaction between two elements? (b) What specific information is needed to predict the product of a combination reaction between calcium and fluorine? (c) What is the product formed by this reaction?

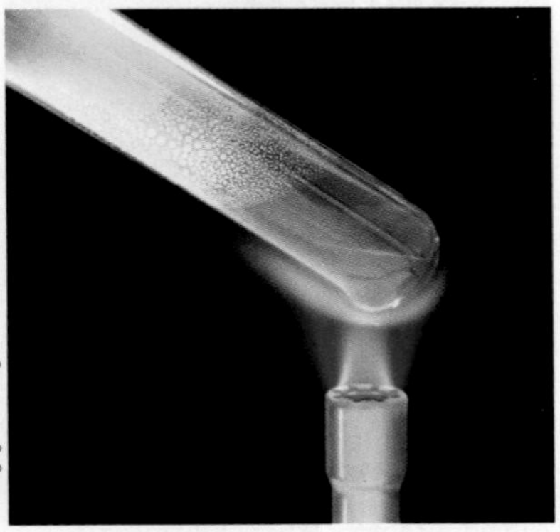
© Cengage Learning/Charles D. Winters

Decomposition of HgO. When heated, solid red mercury(II) oxide decomposes into liquid mercury and oxygen gas.

Decomposition Reactions

Decomposition reactions can be considered the opposite of combination reactions. In a **decomposition reaction,** one substance decomposes to form two or more products. The general reaction is

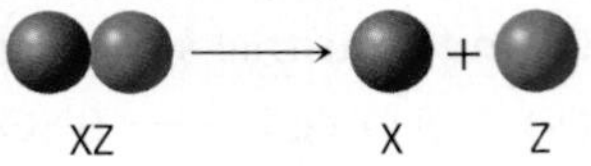
XZ
X
Z

Many compounds that we would describe as "stable" because they exist without change under normal conditions of temperature and pressure undergo decomposition when the temperature is raised, a process known as *thermal decomposition.* For example, a few metal oxides decompose upon heating to give the metal and oxygen gas, the reverse of combination reactions. One of the best-known metal oxide decomposition reactions is the reaction by which Joseph Priestley discovered oxygen in 1775:

$$2\ HgO(s) \xrightarrow{\text{heat}} 2\ Hg(\ell) + O_2(g)$$

A very common and commercially important type of decomposition reaction is illustrated by the chemistry of *metal carbonates,* and calcium carbonate in particular. Many metal carbonates decompose when heated to give metal oxides plus carbon dioxide:

$$\underset{\text{calcium carbonate}}{CaCO_3(s)} \xrightarrow{800\text{–}1000\ ^\circ C} \underset{\text{calcium oxide}}{CaO(s)} + CO_2(g)$$

Some compounds, such as nitroglycerin, $C_3H_5(NO_3)_3$, are sufficiently unstable that their decomposition reactions are explosive. The formula for nitroglycerin contains parentheses around the NO_3 groups because more than one is present; all three must be accounted for when balancing chemical equations. Nitroglycerin is a molecular organic compound with this structure:

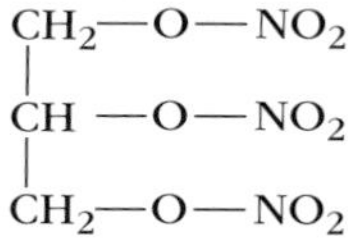

nitroglycerin

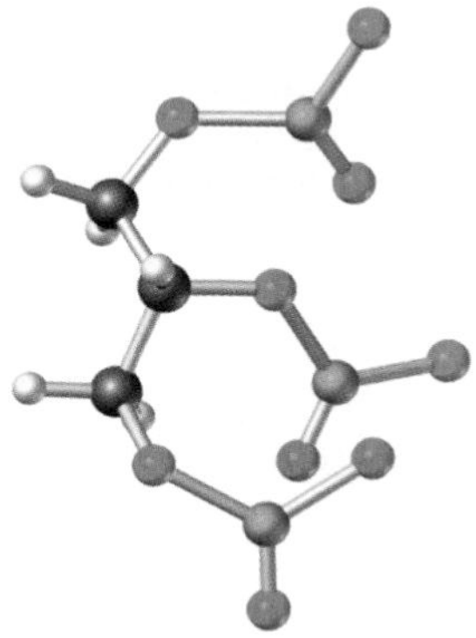

Nitroglycerin is very sensitive to vibrations and jostling, which can cause it to decompose violently.

$$4\ C_3H_5(NO_3)_3(\ell) \longrightarrow 12\ CO_2(g) + 10\ H_2O(g) + 6\ N_2(g) + O_2(g)$$

Water, by contrast, is such a stable compound that it can be decomposed to hydrogen and oxygen only at a very high temperature or by using a direct electric current, a process called electrolysis (Figure 4.4).

$$2\ H_2O(\ell) \xrightarrow{\text{direct current}} 2\ H_2(g) + O_2(g)$$

PROBLEM-SOLVING EXAMPLE 4.1 Combination and Decomposition Reactions

Predict the reaction type and the formula of the missing species for each of these reactions:

(a) $2\ Fe(s) + 3$ ________ $(g) \longrightarrow 2\ FeCl_3(s)$

(b) $Cu(OH)_2(s) \longrightarrow CuO(s) +$ ________ (ℓ)

(c) $P_4(s) + 5\ O_2(g) \longrightarrow$ ________ (s)

(d) $CaSO_3(s) \longrightarrow$ ________ $(s) + SO_2(g)$

Answer

(a) Combination: Cl_2
(b) Decomposition: H_2O
(c) Combination: P_4O_{10}
(d) Decomposition: CaO

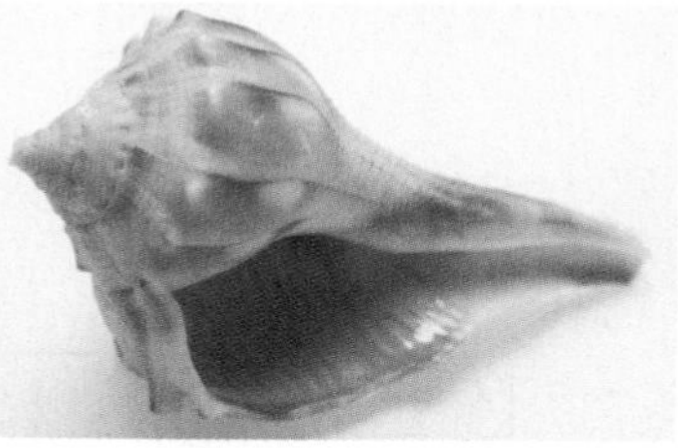

© Cengage Learning/Charles D. Winters

Sea shells are composed largely of calcium carbonate.

The Granger Collection

1833–1896

A Swedish chemist and engineer, Nobel discovered how to mix nitroglycerin (a liquid explosive that is extremely sensitive to light and heat) with diatomaceous earth to make dynamite, which could be handled and shipped safely. Nobel's talent as an entrepreneur combined with his many inventions (he held 355 patents) made him a very rich man. He never married and left his fortune to establish the Nobel Prizes, awarded annually to individuals who "have conferred the greatest benefits on mankind in the fields of physics, chemistry, physiology or medicine, literature and peace."

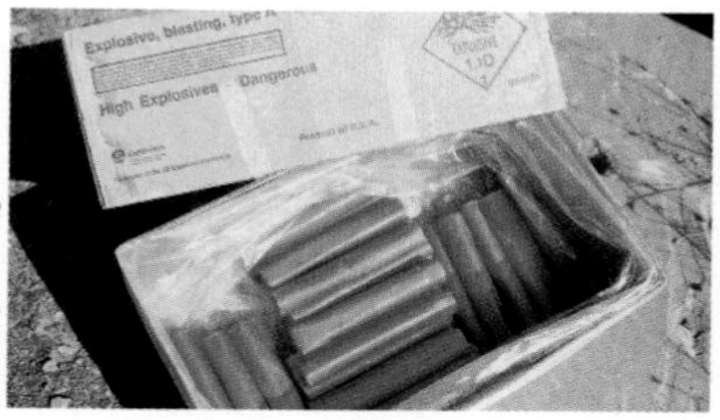

© Cengage Learning/Charles D. Winters

Dynamite contains nitroglycerin.

The **PROBLEM-SOLVING STRATEGY** in this book is

- Analyze the problem
- Plan a solution
- Execute the plan
- Check that the result is reasonable

Appendix A.1 explains this in detail.

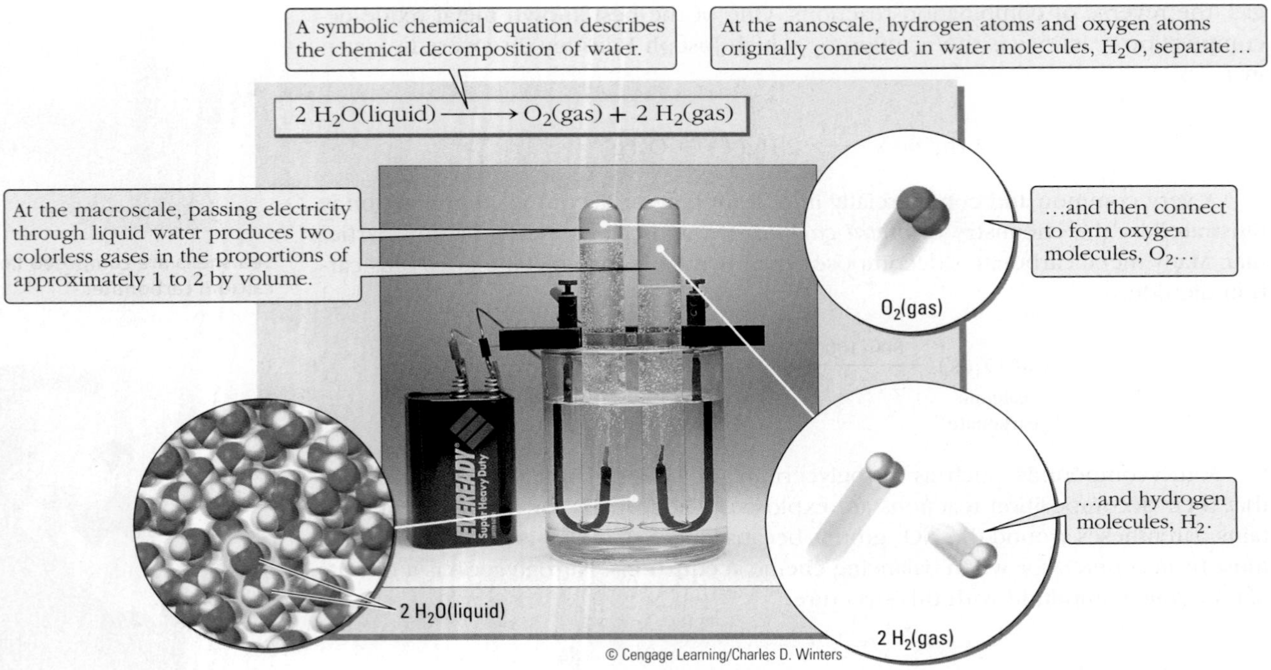
A symbolic chemical equation describes the chemical decomposition of water.
At the nanoscale, hydrogen atoms and oxygen atoms originally connected in water molecules, H_2O, separate…
2 H_2O(liquid) ⟶ O_2(gas) + 2 H_2(gas)
At the macroscale, passing electricity through liquid water produces two colorless gases in the proportions of approximately 1 to 2 by volume.
…and then connect to form oxygen molecules, O_2…
O_2(gas)
…and hydrogen molecules, H_2.
EVEREADY Super Heavy Duty
2 H_2O(liquid)
2 H_2(gas)
© Cengage Learning/Charles D. Winters

Active Figure 4.4 Decomposition of water. A direct electric current from a battery decomposes water into gaseous hydrogen, H_2, and oxygen, O_2. Visit this book's companion website at **www.cengage.com/chemistry/moore** to test your understanding of the concepts in this figure.

Strategy and Explanation Use the fact that decomposition reactions have one reactant and combination reactions have one product. (a) Combination of Fe(s) and Cl_2(g) produces $FeCl_3$(s). (b) Decomposition of $Cu(OH)_2$(s) produces CuO(s) and $H_2O(\ell)$. (c) Combination of P_4(s) and O_2(g) produces P_4O_{10}(s). (d) Decomposition of $CaSO_3$(s) produces CaO(s) and SO_2(g).

PROBLEM-SOLVING PRACTICE answers are provided at the back of this book in Appendix K.

PROBLEM-SOLVING PRACTICE 4.1

Predict the reaction type and the missing substance for each of these reactions:

(a) ________ (g) + 2 O_2(g) ⟶ 2 NO_2(g)

(b) 4 Fe(s) + 3 ________ (g) ⟶ 2 Fe_2O_3(s)

(c) 2 NaN_3(s) ⟶ 2 Na(s) + 3 ________ (g)

CONCEPTUAL EXERCISE 4.5 Combination and Decomposition Reactions

Predict the products formed by these reactions:

(a) Magnesium with chlorine

(b) The thermal decomposition of magnesium carbonate

Displacement Reactions

Displacement reactions are those in which one element reacts with a compound to form a new compound and release a different element. The element released is said to have been displaced. The general equation for a **displacement reaction** is

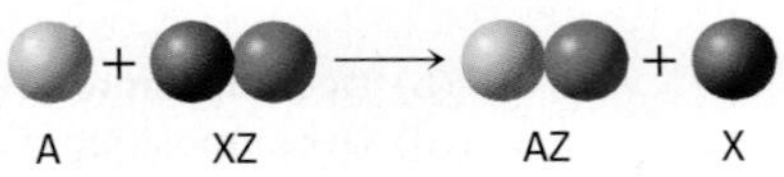
A
XZ
AZ
X

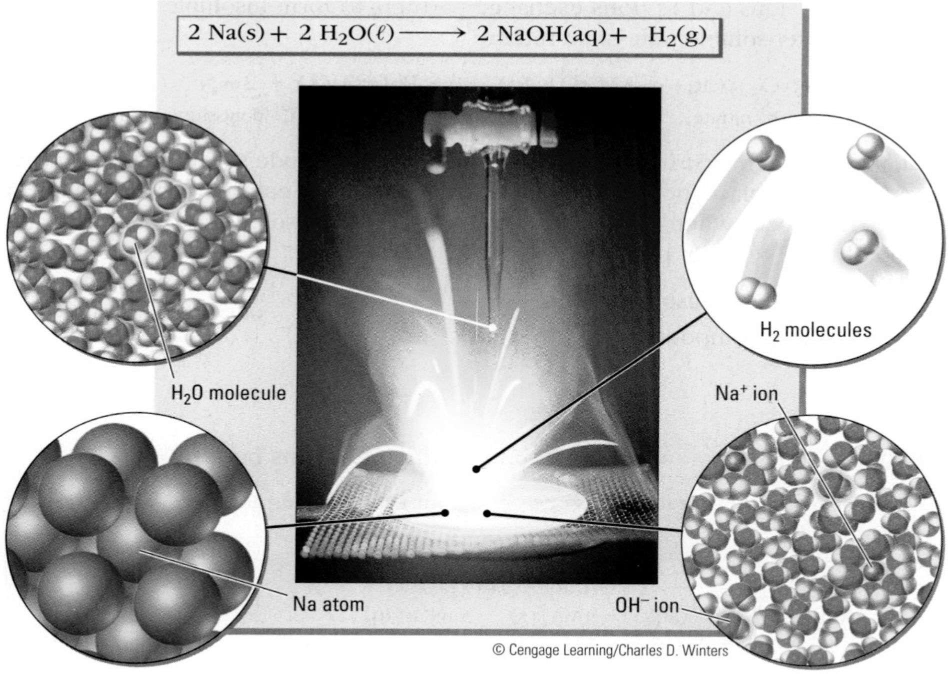
2 Na(s) + 2 H2O(ℓ) ⟶ 2 NaOH(aq) + H2(g)
H2 molecules
H2O molecule
Na+ ion
Na atom
OH− ion
© Cengage Learning/Charles D. Winters

Figure 4.5 A displacement reaction. When liquid water drips from a buret onto a sample of solid sodium metal, the sodium displaces hydrogen gas from the water, and an aqueous solution of sodium hydroxide is formed. The hydrogen gas burns, producing the flame shown in the photograph. In the nanoscale illustrations, the numbers of atoms, molecules, and ions that appear in the balanced equation are shown with yellow highlights.

The reaction of metallic sodium with water is such a reaction.

$$2\ Na(s) + 2\ H_2O(\ell) \longrightarrow 2\ NaOH(aq) + H_2(g)$$

If you think of water as H—O—H, it is easier to see that the Na displaces H from H_2O.

Here sodium displaces hydrogen from water (Figure 4.5). All the alkali metals (Group 1A), which are very reactive elements, react in this way when exposed to water.

Another example is the displacement reaction that occurs between metallic copper and an aqueous solution of silver nitrate.

$$Cu(s) + 2\ AgNO_3(aq) \longrightarrow Cu(NO_3)_2(aq) + 2\ Ag(s)$$

In this case, copper metal displaces silver ions, Ag^+, from its compound. As you will see in Chapter 5, the metals can be arranged in a series from most reactive to least reactive (Section 5.5). This activity series can be used to predict the outcome of displacement reactions.

Exchange Reactions

In an **exchange reaction,** there is an interchange of partners between two compounds. In general:

Exchange reactions are also called metathesis or double-displacement reactions.

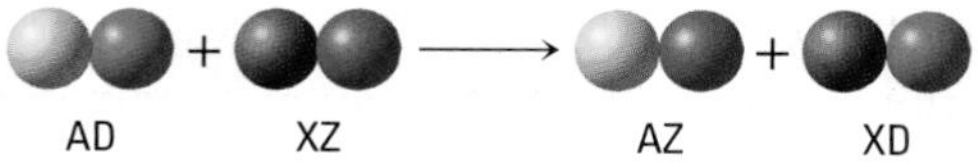
AD
XZ
AZ
XD

Mixing aqueous solutions of lead(II) nitrate and potassium chromate, for example, illustrates an exchange reaction in which an insoluble product is formed. The

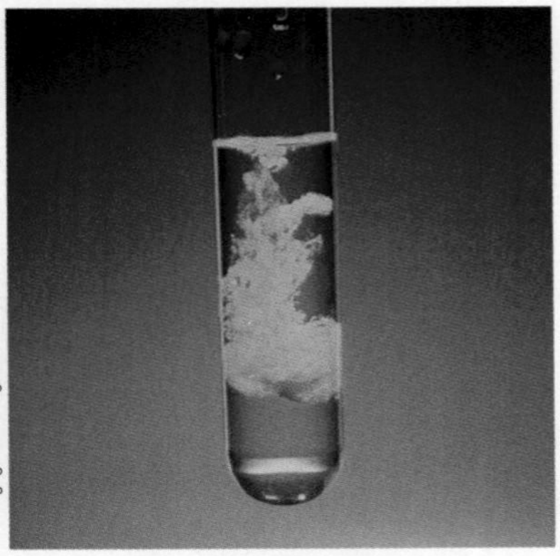
© Cengage Learning/Charles D. Winters

When aqueous solutions of lead(II) nitrate and potassium chromate are mixed, a brilliant yellow precipitate of lead(II) chromate is formed.

aqueous Pb^{2+} ions and K^+ ions exchange partners to form insoluble lead(II) chromate and water-soluble potassium nitrate:

$$\underset{\text{lead(II) nitrate}}{Pb(NO_3)_2(aq)} + \underset{\text{potassium chromate}}{K_2CrO_4(aq)} \longrightarrow \underset{\text{lead(II) chromate}}{PbCrO_4(s)} + \underset{\text{potassium nitrate}}{2\ KNO_3(aq)}$$

Exchange reactions (further discussed in Chapter 5) include several kinds of reactions that take place between reactants that are ionic compounds dissolved in water. They occur when reactant ions are removed from solution by the formation of one of three types of product:

- a precipitate, an insoluble solid
- a molecular compound
- a gas

PROBLEM-SOLVING EXAMPLE **4.2** **Classifying Reactions by Type**

Classify each of these reactions as one of the four general types discussed in this section.

(a) $2\ Al(s) + 3\ Br_2(\ell) \longrightarrow Al_2Br_6(s)$
(b) $2\ K(s) + H_2O(\ell) \longrightarrow 2\ KOH(aq) + H_2(g)$
(c) $AgNO_3(aq) + NaCl(aq) \longrightarrow AgCl(s) + NaNO_3(aq)$
(d) $NH_4NO_3(s) \longrightarrow N_2O(g) + 2\ H_2O(g)$

Answer

(a) Combination (b) Displacement (c) Exchange (d) Decomposition

Strategy and Explanation Use the fact that in displacement reactions, one reactant is an element, and in exchange reactions, both reactants are compounds.

(a) With two reactants and a single product, this must be a combination reaction.
(b) The general equation for a displacement reaction, $A + XZ \longrightarrow AZ + X$, matches what occurs in the given reaction. Potassium (A) displaces hydrogen (X) from water (XZ) to form KOH (AZ) plus H_2 (X).
(c) The reactants exchange partners in this exchange reaction. Applying the general equation for an exchange reaction, $AD + XZ \longrightarrow AZ + XD$, to this case, we find A is Ag^+, D is NO_3^-, X is Na^+, and Z is Cl^-.
(d) In this reaction, a single substance, NH_4NO_3, decomposes to form two products, N_2O and H_2O.

PROBLEM-SOLVING PRACTICE **4.2**

Classify each of these reactions as one of the four general reaction types described in this section.

(a) $2\ Al(OH)_3(s) \longrightarrow Al_2O_3(s) + 3\ H_2O(g)$
(b) $Na_2O(s) + H_2O(\ell) \longrightarrow 2\ NaOH(aq)$
(c) $S_8(s) + 24\ F_2(g) \longrightarrow 8\ SF_6(g)$
(d) $3\ NaOH(aq) + H_3PO_4(aq) \longrightarrow Na_3PO_4(aq) + 3\ H_2O(\ell)$
(e) $3\ C(s) + Fe_2O_3(s) \longrightarrow 3\ CO(g) + 2\ Fe(\ell)$

go Chemistry **Module 7: Simple Stoichiometry** covers concepts in this section.

4.3 Balancing Chemical Equations

Balancing a chemical equation means using correct coefficients so that the same number of atoms of each element appears on each side of the equation. We will begin with one of the general classes of reactions, the combination of reactants to produce a single product, to illustrate how to balance chemical equations by a largely trial-and-error process.

We will balance the equation for the formation of ammonia from nitrogen and hydrogen. Millions of tons of ammonia, NH_3, are manufactured worldwide annually by this reaction, using nitrogen extracted from air and hydrogen obtained from natural gas.

- **Write an unbalanced equation containing the correct formulas of all reactants and products.**

$$\text{(unbalanced equation)} \qquad N_2 + H_2 \longrightarrow NH_3$$

Clearly, both nitrogen and hydrogen are unbalanced. There are two nitrogen atoms on the left and only one on the right, and two hydrogen atoms on the left and three on the right.

- **Balance atoms of one of the elements.** Start by using a coefficient of 2 on the right to balance the nitrogen atoms: 2 NH_3 indicates two ammonia molecules, each containing a nitrogen atom and three hydrogen atoms. On the right we now have two nitrogen atoms and six hydrogen atoms.

Balancing an equation involves changing the coefficients, but the subscripts in the formulas *cannot* be changed.

$$\text{(unbalanced equation)} \qquad N_2 + H_2 \longrightarrow 2\ NH_3$$

- **Balance atoms of the remaining elements.** Balance the six hydrogen atoms on the right using a coefficient of 3 for the H_2 on the left to furnish six hydrogen atoms.

$$\text{(balanced equation)} \qquad N_2 + 3\ H_2 \longrightarrow 2\ NH_3$$

- **Verify that the number of atoms of each element is balanced.** Do an atom count to check that the numbers of nitrogen and hydrogen atoms are the same on each side of the equation.

(balanced equation)	N_2	+	$3\ H_2$	$\longrightarrow$	$2\ NH_3$
atom count:	2 N	+	(3×2) H	=	2 N + (2×3) H
	2 N	+	6 H	=	2 N + 6 H

The physical states of the reactants and products are frequently included in the balanced equation. Thus, the final equation for ammonia formation is

$$N_2(g) + 3\ H_2(g) \longrightarrow 2\ NH_3(g)$$

PROBLEM-SOLVING EXAMPLE 4.3 Balancing a Chemical Equation

Ammonia gas reacts with oxygen gas to form gaseous nitrogen monoxide, NO, and water vapor at 1000 °C. Write the balanced equation for this reaction.

Answer $4\ NH_3(g) + 5\ O_2(g) \longrightarrow 4\ NO(g) + 6\ H_2O(g)$

Strategy and Explanation Use the stepwise approach to balancing chemical equations. Note that oxygen is in both products.

- **Write an unbalanced equation containing the correct formulas of all reactants and products.** The formula for ammonia is NH_3. The unbalanced equation is

$$\text{(unbalanced equation)} \qquad NH_3(g) + O_2(g) \longrightarrow NO(g) + H_2O(g)$$

- **Balance atoms of one of the elements.** Hydrogen is unbalanced since there are three hydrogen atoms on the left and two on the right. Whenever three and two atoms must be balanced, use coefficients to give six atoms on both sides of the equation. To do so, use a coefficient of 2 on the left and 3 on the right to have six hydrogens on each side.

$$\text{(unbalanced equation)} \qquad 2\ NH_3(g) + O_2(g) \longrightarrow NO(g) + 3\ H_2O(g)$$

- **Balance atoms of the remaining elements.** There are now two nitrogen atoms on the left and one on the right, so we balance nitrogen by using the coefficient 2 for the NO molecule.

$$\text{(unbalanced equation)} \qquad 2\ NH_3(g) + O_2(g) \longrightarrow 2\ NO(g) + 3\ H_2O(g)$$

Now there are two oxygen atoms on the left and five on the right. We use a coefficient of $\frac{5}{2}$ to balance the atoms of O_2.

(balanced equation) $2\ NH_3(g) + \frac{5}{2}O_2(g) \longrightarrow 2\ NO(g) + 3\ H_2O(g)$

The equation is now balanced, but it is customary to use whole-number coefficients. Therefore, we multiply all the coefficients by 2 to get the final balanced equation with whole-number coefficients.

(balanced equation) $4\ NH_3(g) + 5\ O_2(g) \longrightarrow 4\ NO(g) + 6\ H_2O(g)$

- **Verify that the number of atoms of each element is balanced.**

(balanced equation) $4\ NH_3(g) + 5\ O_2(g) \longrightarrow 4\ NO(g) + 6\ H_2O(g)$

4 N		=	4 N	
12 H		=		12 H
	10 O	=	4 O	6 O

PROBLEM-SOLVING PRACTICE **4.3**

Balance these equations:

(a) $Cr(s) + Cl_2(g) \longrightarrow CrCl_3(s)$ (b) $As_2O_3(s) + H_2(g) \longrightarrow As(s) + H_2O(\ell)$

The combustion of propane, C_3H_8, illustrates balancing a somewhat more complex chemical equation. We will assume that the propane reacts completely with O_2, so the only products are carbon dioxide and water.

PROBLEM-SOLVING EXAMPLE **4.4** **Balancing a Combustion Reaction Equation**

Write a balanced equation for the complete combustion of propane, C_3H_8, the fuel used in gas grills.

Answer $C_3H_8(g) + 5\ O_2(g) \longrightarrow 3\ CO_2(g) + 4\ H_2O(\ell)$

Strategy and Explanation Recall that complete combustion of a hydrocarbon produces carbon dioxide and water.

- **Write an unbalanced equation containing the correct formulas of all reactants and products.** The initial equation is

(unbalanced equation) $C_3H_8 + O_2 \longrightarrow CO_2 + H_2O$

It usually works best to first balance the element that appears in the fewest formulas; balance the element that appears in the most formulas last.

- **Balance the atoms of one of the elements.** None of the elements are balanced, so we could start with C, H, or O. We will start with C because it appears in only one reactant and one product. The three carbon atoms in C_3H_8 will produce three CO_2 molecules.

(unbalanced equation) $C_3H_8 + O_2 \longrightarrow 3\ CO_2 + H_2O$

- **Balance atoms of the remaining elements.** We next balance the H atoms. The eight H atoms in the reactants will combine with oxygen to produce four water molecules, each containing two H atoms.

(unbalanced equation) $C_3H_8 + O_2 \longrightarrow 3\ CO_2 + 4\ H_2O$

Oxygen is the remaining element to balance. At this point, there are ten oxygen atoms in the products (3×2 in three CO_2 molecules, and 4×1 in four water molecules), but only two in O_2, a reactant. Therefore, O_2 in the reactants needs a coefficient of 5 to have ten oxygen atoms in the reactants.

$$C_3H_8(g) + 5\ O_2(g) \longrightarrow 3\ CO_2(g) + 4\ H_2O(\ell)$$

This combustion equation is now balanced.

- **Verify that the number of atoms of each element is balanced.**

$$C_3H_8(g) + 5\,O_2(g) \longrightarrow 3\,CO_2(g) + 4\,H_2O(\ell)$$

3 C		=	3 C		
8 H		=			8 H
	10 O	=	6 O	+	4 O

PROBLEM-SOLVING PRACTICE 4.4

Ethyl alcohol, C_2H_5OH, can be added to gasoline to create a cleaner-burning fuel. Write the balanced equation for:

(a) Complete combustion of ethyl alcohol to produce carbon dioxide and water

(b) Incomplete combustion of ethyl alcohol to produce carbon monoxide and water

When one or more polyatomic ions appear on both sides of a chemical equation, each one is treated as a whole during the balancing steps. When such an ion must have a subscript in the chemical formula, the polyatomic ion is enclosed in parentheses. For example, in the equation for the reaction between sodium phosphate and barium nitrate to produce barium phosphate and sodium nitrate,

$$2\,Na_3PO_4(aq) + 3\,Ba(NO_3)_2(aq) \longrightarrow Ba_3(PO_4)_2(s) + 6\,NaNO_3(aq)$$

the nitrate ions, NO_3^-, and phosphate ions, PO_4^{3-}, are kept together as units and are enclosed in parentheses when the polyatomic ion occurs more than once in a chemical formula.

4.4 The Mole and Chemical Reactions: The Macro-Nano Connection

go Chemistry **Module 7: Simple Stoichiometry** covers concepts in this section.

Molar mass links the number of atoms, molecules, or formula units with the mass of atoms, molecules, or ionic compounds. When molar mass is combined with a balanced chemical equation, the masses of the reactants and products can be calculated. In this way the nanoscale of chemical reactions is linked with the macroscale, at which we can measure masses of reactants and products by weighing.

We will explore these relationships using the combustion reaction between methane, CH_4, and oxygen, O_2, as an example (Figure 4.6). The balanced equation that follows shows the number of molecules of the reactants, which are methane and oxygen, and of the products, which are carbon dioxide and water. The coefficients on each formula can also be interpreted as the numbers of moles of each compound. We can use the molar mass of each compound to calculate the mass of each reactant and product represented in the balanced equation.

$$CH_4(g) \quad + \quad 2\,O_2(g) \longrightarrow CO_2(g) \quad + \quad 2\,H_2O(g)$$

$CH_4(g)$	$2\,O_2(g)$	$CO_2(g)$	$2\,H_2O(g)$
1 CH_4 molecule	2 O_2 molecules	1 CO_2 molecule	2 H_2O molecules
1 mol CH_4	2 mol O_2	1 mol CO_2	2 mol H_2O
16.0 g CH_4	64.0 g O_2	44.0 g CO_2	36.0 g H_2O
80.0 g total		80.0 g total	

Figure 4.6 Combustion of methane with oxygen. Methane is the main component of natural gas, a primary fuel for industrial economies and the gas commonly used in laboratory Bunsen burners.

Notice that the total mass of reactants (16.0 g CH_4 + 64.0 g O_2 = 80.0 g reactants) equals the total mass of products (44.0 g CO_2 + 36.0 g H_2O = 80.0 g products), as must always be the case for a balanced equation.

The stoichiometric coefficients in a balanced chemical equation provide the **mole ratios** that relate the numbers of moles of reactants and products to each

The mole ratio is also known as the stoichiometric factor.

other. These mole ratios are used in all quantitative calculations involving chemical reactions. Several of the mole ratios for the example equation are

$$\frac{2 \text{ mol } O_2}{1 \text{ mol } CH_4} \qquad \frac{1 \text{ mol } CO_2}{1 \text{ mol } CH_4} \qquad \frac{2 \text{ mol } H_2O}{2 \text{ mol } O_2}$$

These mole ratios tell us that 2 mol O_2 react with every 1 mol CH_4, or 1 mol CO_2 is formed for each 1 mol CH_4 that reacts, or 2 mol H_2O is formed for each 2 mol O_2 that reacts.

We can use the mole ratios in the balanced equation to calculate the molar amount of one reactant or product from the molar amount of another reactant or product. For example, we can calculate the number of moles of H_2O produced when 0.40 mol CH_4 is reacted fully with oxygen.

$$0.40 \text{ mol } CH_4 \times \frac{2 \text{ mol } H_2O}{1 \text{ mol } CH_4} = 0.80 \text{ mol } H_2O$$

CONCEPTUAL EXERCISE 4.6 Mole Ratios

Write all the possible mole ratios that can be obtained from the balanced equation for the reaction between Al and Br_2 to form Al_2Br_6.

The molar mass and the mole ratio, as illustrated in Figure 4.7, provide the links between masses and molar amounts of reactants and products. When the quantity of a reactant is given in grams, we use its molar mass to convert to moles of reactant as a first step in using the balanced chemical equation. Then we use the balanced chemical equation to convert from moles of reactant to moles of product. Finally, we convert to grams of product if necessary by using the product's molar mass. Figure 4.7 illustrates this sequence of calculations.

We will illustrate the stoichiometric relationships of Figure 4.7 and a stepwise method for solving problems involving mass relations in chemical reactions to answer gram-to-gram conversion questions such as: How many grams of O_2 are needed to react completely with 5.0 g CH_4?

Step 1: ***Write the correct formulas for reactants and products and balance the chemical equation.*** The balanced equation is

$$CH_4(g) + 2\,O_2(g) \longrightarrow CO_2(g) + 2\,H_2O(g)$$

Step 2: ***Decide what information about the problem is known and what is unknown. Map out a strategy for answering the question.*** In this example, you know the mass of CH_4 and you want to calculate the mass of O_2. You also know that you can use molar mass to convert mass of CH_4 to

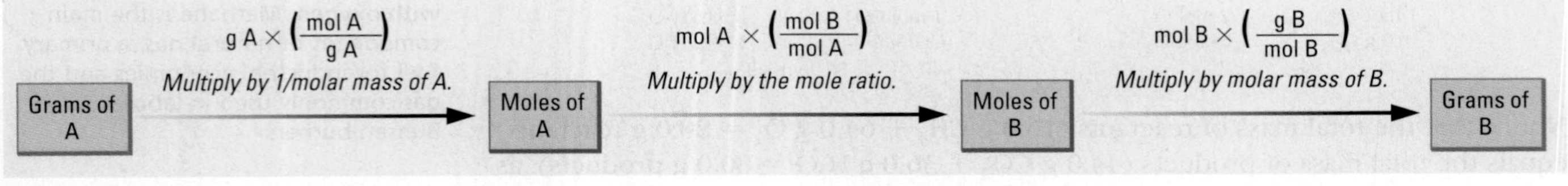

Figure 4.7 Stoichiometric relationships in a chemical reaction. The mass or molar amount of one reactant or product (A) is related to the mass or molar amount of another reactant or product (B) by the series of calculations shown.

molar amount of CH_4. Then you can use the mole ratio from the balanced equation (2 mol O_2/1 mol CH_4) to calculate the moles of O_2 needed. Finally, you can use the molar mass of O_2 to convert the moles of O_2 to grams of O_2.

Grams of CH_4 → Moles of CH_4 → Moles of O_2 → Grams of O_2

Step 3: Calculate moles from grams (if necessary). The known mass of CH_4 must be converted to molar amount because the coefficients of the balanced equation express mole relationships.

$$5.0 \text{ g } CH_4 \times \frac{1 \text{ mol } CH_4}{16.0 \text{ g } CH_4} = 0.313 \text{ mol } CH_4$$

In multistep calculations, remember to carry one additional significant figure in intermediate steps before rounding to the final value.

Step 4: Use the mole ratio to calculate the unknown number of moles, and then convert the number of moles to number of grams (if necessary). Calculate the number of moles and then grams of O_2.

$$0.313 \text{ mol } CH_4 \times \frac{2 \text{ mol } O_2}{1 \text{ mol } CH_4} = 0.626 \text{ mol } O_2$$

$$0.626 \text{ mol } O_2 \times \frac{32.0 \text{ g } O_2}{1 \text{ mol } O_2} = 20. \text{ g } O_2$$

Step 5: Check the answer to see whether it is reasonable. The starting mass of CH_4 of 5.0 g is about one third of a mole of CH_4. One third of a mole of CH_4 should react with two thirds of a mole of O_2 because the mole ratio is 1:2. The molar mass of O_2 is 32.0 g/mol, so two thirds of this amount is about 20. Therefore, the answer of 20. g O_2 is reasonable.

PROBLEM-SOLVING EXAMPLE 4.5 Moles and Grams in Chemical Reactions

An iron ore named hematite, Fe_2O_3, can be reacted with carbon monoxide, CO, to form iron and carbon dioxide. How many moles and grams of iron are produced when 45.0 g hematite is reacted with sufficient CO?

Answer 0.564 mol and 31.5 g Fe

Strategy and Explanation Solving stoichiometric problems relies on the relationships illustrated in Figure 4.7 to connect masses and molar amounts of reactants or products.

- **Write the balanced chemical equation.**

$$Fe_2O_3(s) + 3\ CO(g) \longrightarrow 2\ Fe(s) + 3\ CO_2(g)$$

- **Use the relationships connecting masses and moles of reactants and products as illustrated in Figure 4.7.**

We know the mass of hematite that reacted, but to proceed we need to know the molar amount of hematite. The molar mass of hematite is 159.69 g/mol.

Grams of hematite ⟶ moles of hematite

The amount (moles) of hematite reacted is

$$45.0 \text{ g hematite} \times \frac{1 \text{ mol hematite}}{159.69 \text{ g hematite}} = 0.282 \text{ mol hematite}$$

- **Use the mole ratio (stoichiometric factor) derived from the balanced reaction to convert moles of reactant to moles of product.**

Moles of hematite ⟶ moles of iron

The amount of iron formed is

$$0.282 \text{ mol hematite} \times \frac{2 \text{ mol Fe}}{1 \text{ mol hematite}} = 0.564 \text{ mol Fe}$$

- **Use the molar mass to convert from moles to grams of iron.**
  Multiply the molar amount of Fe formed by the molar mass of iron to obtain the mass of iron produced.

$$0.564 \text{ mol Fe} \times \frac{55.845 \text{ g Fe}}{1 \text{ mol Fe}} = 31.5 \text{ g Fe}$$

☑ **Reasonable Answer Check** The balanced equation shows that for every one mole of hematite reacted, two moles of iron are produced. Therefore, approximately 0.3 mol hematite should produce twice that molar amount, or approximately 0.6 mol iron. The answer is reasonable.

PROBLEM-SOLVING PRACTICE 4.5

How many grams of carbon monoxide are required to react completely with 0.433 mol hematite?

EXERCISE 4.7 Moles and Grams in Chemical Reactions

Verify that 10.8 g water is produced by the reaction of sufficient O_2 with 0.300 mol CH_4.

Problem-Solving Examples 4.6 and 4.7 illustrate further the application of the steps for solving problems involving mass relations in chemical reactions.

PROBLEM-SOLVING EXAMPLE 4.6 Moles and Grams in Chemical Reactions

When silicon dioxide, SiO_2, and carbon are reacted at high temperature, silicon carbide, SiC (also known as carborundum, an important industrial abrasive), and carbon monoxide, CO, are produced. Calculate the mass in grams of silicon carbide that will be formed by the complete reaction of 0.400 mol SiO_2 with sufficient carbon.

Answer 16.0 g SiC

Strategy and Explanation Use the stepwise approach to solve mass relations problems.

Step 1: Write the correct formulas for reactants and products and balance the chemical equation. SiO_2 and C are the reactants; SiC and CO are the products.

$$SiO_2(s) + 3\,C(s) \longrightarrow SiC(s) + 2\,CO(g)$$

Step 2: Decide what information about the problem is known and what is unknown. Map out a strategy for answering the question. We know how many moles of SiO_2 are available. Sufficient C is present, so there is enough C to react with all of the SiO_2. If we can calculate how many moles of SiC are formed, then the mass of SiC can be calculated.

Step 3: Calculate moles from grams (if necessary). We know there is 0.400 mol SiO_2.

Step 4: Use the mole ratio to calculate the unknown number of moles, and then convert the number of moles to number of grams (if necessary). Both steps are needed here to convert moles of SiO_2 to moles of SiC and then to grams of SiC. In a single setup the calculation is

$$0.400 \text{ mol } SiO_2 \times \frac{1 \text{ mol SiC}}{1 \text{ mol } SiO_2} \times \frac{40.10 \text{ g SiC}}{1 \text{ mol SiC}} = 16.0 \text{ g SiC}$$

Step 5: Check the answer to see whether it is reasonable. The answer, 16.0 g SiC, is reasonable because 1 mol SiO_2 would produce 1 mol SiC (40.10 g); therefore, four tenths of a mole of SiO_2 would produce four tenths of a mole of SiC (16.0 g).

☑ **Reasonable Answer Check** Step 5 verified that the answer is reasonable.

Silicon carbide, SiC. The grinding wheel *(left)* is coated with SiC. Naturally occurring silicon carbide *(right)* is also known as carborundum. It is one of the hardest substances known, making it valuable as an abrasive.

PROBLEM-SOLVING PRACTICE 4.6

Tin is extracted from its ore cassiterite, SnO_2, by reaction with carbon from coal.

$$SnO_2(s) + 2\,C(s) \rightarrow Sn(\ell) + 2\,CO(g)$$

(a) What mass of tin can be produced from 0.300 mol cassiterite?
(b) How many grams of carbon are required to produce this much tin?

PROBLEM-SOLVING EXAMPLE 4.7 Grams, Moles, and Grams

A 12-fluid ounce can of a soft drink (355 mL) contains 35.0 g sugar, which can be considered to be sucrose (table sugar), $C_{12}H_{22}O_{11}$. When you drink the soda, the sucrose is metabolized. Metabolism involves reaction with oxygen to produce carbon dioxide and water. (a) Balance the chemical equation for this reaction. (b) Calculate the mass of O_2 consumed and the masses of CO_2 and H_2O produced.

Answer
(a) $C_{12}H_{22}O_{11}(s) + 12\,O_2(g) \rightarrow 12\,CO_2(g) + 11\,H_2O(\ell)$
(b) 39.3 g O_2 is consumed; 54.0 g CO_2 and 20.3 g H_2O are produced.

Strategy and Explanation Write the correct formulas for all the reactants and products. Then balance the chemical equation. Use the equation's mole ratios to calculate the masses required.

- **Write the correct formulas for the reactants and products and balance the chemical equation.**

  The formula for sucrose is given in the statement of the problem, and the formulas of the other species are O_2, CO_2, and H_2O. The equation is

  $$\text{(unbalanced equation)} \qquad C_{12}H_{22}O_{11}(s) + O_2(g) \longrightarrow CO_2(g) + H_2O(\ell)$$

  Coefficients of 11 and 12 for the CO_2 and H_2O balance the C and H, giving 35 oxygen atoms in the products. In the reactants, these oxygen atoms are balanced by the 12 O_2 molecules plus the 11 oxygen atoms in sucrose. The balanced equation is

  $$C_{12}H_{22}O_{11}(s) + 12\,O_2(g) \rightarrow 12\,CO_2(g) + 11\,H_2O(\ell)$$

- **Use the molar mass of the reactant to find the moles of reactant.**

  Using atomic molar masses, we calculate the molar mass of sucrose to be 342.3 g/mol and use it to convert grams of sucrose to moles of sucrose.

  $$35.0 \text{ g sucrose} \times \frac{1 \text{ mol sucrose}}{342.3 \text{ g sucrose}} = 0.1022 \text{ mol sucrose}$$

- **Use the mole ratios (stoichiometric factors) from the balanced reaction to convert moles of one reactant to moles of other reactants or products. Use molar masses to convert moles of reactants or products to grams of reactants or products.**

  $$0.1022 \text{ mol sucrose} \times \frac{12 \text{ mol } O_2}{1 \text{ mol sucrose}} \times \frac{31.99 \text{ g } O_2}{1 \text{ mol } O_2} = 39.3 \text{ g } O_2$$

  $$0.1022 \text{ mol sucrose} \times \frac{12 \text{ mol } CO_2}{1 \text{ mol sucrose}} \times \frac{44.01 \text{ g } CO_2}{1 \text{ mol } CO_2} = 54.0 \text{ g } CO_2$$

  $$0.1022 \text{ mol sucrose} \times \frac{11 \text{ mol } H_2O}{1 \text{ mol sucrose}} \times \frac{18.02 \text{ g } H_2O}{1 \text{ mol } H_2O} = 20.3 \text{ g } H_2O$$

  The mass of water could have been found by using the conservation of mass.

  $$\text{Total mass of reactants} = \text{total mass of products}$$

  $$\text{Total mass of reactants} = 35.0 \text{ g sucrose} + 39.3 \text{ g } O_2 = 74.3 \text{ g}$$

  $$\text{Total mass of products} = 54.0 \text{ g } CO_2 + ? \text{ g } H_2O = 74.3 \text{ g}$$

  $$\text{Mass of } H_2O = 74.3 \text{ g} - 54.0 \text{ g} = 20.3 \text{ g } H_2O$$

☑ **Reasonable Answer Check** Approximately 0.1 mol sucrose would require approximately 1.2 mol O_2 and would produce approximately 1.2 mol CO_2 and 1.1 mol H_2O.

Therefore, the calculated masses of the products should be somewhat larger than the molar masses, and they are. The answers are reasonable.

PROBLEM-SOLVING PRACTICE 4.7

A lump of coke (carbon) weighs 57 g.
(a) What mass of oxygen is required to burn the coke to carbon monoxide?
(b) How many grams of CO are produced?

To this point we have used the methods of stoichiometry to compute the quantity of products given the quantity of reactants. Now we turn to the reverse problem: Given the quantity of products, what quantitative information can we deduce about the reactants? Questions such as these are often confronted by *analytical chemistry,* a field in which chemists creatively identify pure substances and measure the quantities of components of mixtures. The analysis of mixtures is often challenging. It can take a great deal of imagination to figure out how to use chemistry to determine what, and how much, is present in a mixture such as an environmental sample containing air or water pollutants.

PROBLEM-SOLVING EXAMPLE 4.8 Evaluating a Metal-Containing Sample

Chromium metal is obtained from chromium(III) oxide, Cr_2O_3, by reacting the oxide with aluminum metal.

$$Cr_2O_3(s) + 2\,Al(s) \longrightarrow 2\,Cr(s) + Al_2O_3(s)$$

If 10.0 g of a Cr_2O_3-containing sample yields 0.821 g chromium metal, what is the mass percent of Cr_2O_3 in the sample?

Answer 12.0%

Strategy and Explanation The mass percent of Cr_2O_3 is

$$\text{Mass percent } Cr_2O_3 = \frac{\text{mass of } Cr_2O_3}{\text{mass of sample}} \times 100\%$$

The mass of the sample is given. The mass of Cr_2O_3 can be determined from the known mass of chromium metal produced and the balanced equation. The molar mass of Cr_2O_3 is 151.9961 g/mol, and the molar mass of Cr is 51.9961 g/mol.

$$0.821 \text{ g Cr} \times \frac{1 \text{ mol Cr}}{51.9961 \text{ g Cr}} \times \frac{1 \text{ mol } Cr_2O_3}{2 \text{ mol Cr}} \times \frac{151.9904 \text{ g } Cr_2O_3}{1 \text{ mol } Cr_2O_3} = 1.20 \text{ g } Cr_2O_3$$

The mass percent of Cr_2O_3 is

$$\frac{1.20 \text{ g } Cr_2O_3}{10.0 \text{ g sample}} \times 100\% = 12.0\%$$

☑ **Reasonable Answer Check** Based on molar masses and a mole ratio from the balanced equation, 10.0 g Cr_2O_3 would ideally produce

$$10 \text{ g } Cr_2O_3 \times \frac{1 \text{ mol } Cr_2O_3}{152 \text{ g } Cr_2O_3} \times \frac{2 \text{ mol Cr}}{1 \text{ mol } Cr_2O_3} \times \frac{52 \text{ g Cr}}{1 \text{ mol Cr}} \approx 6.8 \text{ g Cr}$$

This reaction actually produced only 0.82 g Cr. Since 0.82 is 12% of 6.8, the more precisely calculated answer is reasonable.

PROBLEM-SOLVING PRACTICE 4.8

The purity of magnesium metal can be determined by reacting the metal with sufficient hydrochloric acid to form $MgCl_2$, evaporating the water from the resulting solution, and weighing the solid $MgCl_2$ formed.

$$Mg(s) + 2\,HCl(aq) \longrightarrow MgCl_2(aq) + H_2(g)$$

Calculate the percentage of magnesium in a 1.72-g sample that produced 6.46 g $MgCl_2$ when reacted with sufficient HCl.

4.5 Reactions with One Reactant in Limited Supply

go Chemistry Modules 8a and b: Stoichiometry and Limiting Reactants (Parts 1 and 2) cover concepts in this section.

In the previous section, we assumed that exactly stoichiometric amounts of reactants were present; each reactant was entirely converted to products when the reaction was over. However, this is rarely the case when chemists carry out an actual synthesis, whether for small quantities in a laboratory or on a large scale in an industrial process. Usually, one reactant is more expensive or less readily available than others. The cheaper or more available reactants are used in excess to ensure that the more expensive material is completely converted to product.

The industrial production of methanol, CH_3OH, is such a case. Methanol, an important industrial product, is manufactured by the reaction of carbon monoxide and hydrogen.

$$CO(g) + 2\ H_2(g) \longrightarrow CH_3OH(\ell)$$

Carbon monoxide is manufactured cheaply by burning coke, which is mostly carbon in a limited supply of air so that there is insufficient oxygen to form carbon dioxide. Hydrogen is more expensive to manufacture. Therefore, industrial methanol synthesis uses an excess of carbon monoxide, and the quantity of methanol produced is dictated by the quantity of hydrogen available. In this case, hydrogen acts as the limiting reactant.

A **limiting reactant** is the reactant that is completely converted to products during a reaction. Once the limiting reactant has been used up, no more product can

The limiting reactant is always a reactant, never a product.

ESTIMATION How Much CO_2 Is Produced by Your Car?

Your car burns gasoline in a combustion reaction and produces water and carbon dioxide, CO_2, one of the major greenhouse gases, which is involved in global warming. For each gallon of gasoline that you burn in your car, how much CO_2 is produced? How much CO_2 is produced by your car per year?

To proceed with the estimation, we need to write a balanced chemical equation with the stoichiometric relationship between the reactant, gasoline, and the product of interest, CO_2. To write the chemical equation we need to make an assumption about the composition of gasoline. We will assume that the gasoline is octane, C_8H_{18}, so the reaction of interest is

$$2\ C_8H_{18} + 25\ O_2 \longrightarrow 16\ CO_2 + 18\ H_2O$$

One gallon equals 4 quarts, which equals

$$4\ \text{qt} \times \frac{1\ \text{L}}{1.057\ \text{qt}} = 3.78\ \text{L}$$

Gasoline floats on water, so its density must be less than that of water. Assume it is 0.80 g/mL, so

$$3.78\ \text{L} \times 0.80\ \text{g/mL} \times 10^3\ \text{mL/L} = 3.02 \times 10^3\ \text{g}$$

We now convert the grams of octane to moles using the molar mass of octane.

$$3020\ \text{g octane} \times \frac{1\ \text{mol}}{114.2\ \text{g}} = 26.4\ \text{mol octane}$$

To convert to moles of CO_2 we use the balanced equation, which shows that for every mole of octane consumed, eight moles of CO_2 are produced; thus, we have

$$26.4\ \text{mol}\ C_8H_{18} \times \frac{8\ \text{mol}\ CO_2}{1\ \text{mol}\ C_8H_{18}} = 211\ \text{mol}\ CO_2$$

The molar mass of CO_2 is 44 g/mol, so 211 mol × 44 g/mol = 9280 g CO_2. Thus, for every gallon of gasoline burned, 9.3 kg (20.5 lb) CO_2 is produced.

If you drive your car 10,000 miles per year and get an average of 25 miles per gallon, you use about 400 gallons of gasoline per year. Burning this quantity of gasoline produces 400 × 9.3 kg = 3720 kg CO_2. That's 3720 kg × 2.2 lb/kg = 8200 lb, or more than 4 tons CO_2.

How much CO_2 is that? The 3720 kg CO_2 is about 85,000 mol CO_2. At room temperature and atmospheric pressure, that's about 2,080,000 L CO_2 or 2080 m^3 CO_2—enough to fill about 4000 1-m diameter balloons, or 11 such balloons each day of the year. In Section 10.11 we will discuss the effect that CO_2 is having on Earth's atmosphere and its link to global warming.

Visit this book's companion website at **www.cengage.com/chemistry/moore** to work an interactive module based on this material.

form. *The limiting reactant must be used as the basis for calculating the maximum possible amount of product(s)* because the limiting reactant limits the amount of product(s) that can be formed. ***The moles of product(s) formed are always determined by the starting number of moles of the limiting reactant.***

An analogy to a chemistry limiting reactant is the assembly of grilled cheese sandwiches. Each sandwich must have two slices of bread and one slice of cheese. Suppose we have eight slices of bread and six slices of cheese. How many complete grilled cheese sandwiches can be made?

Each sandwich must have the ratio of 2 bread : 1 cheese (analogous to coefficients in a balanced chemical equation):

$$2 \text{ bread slices} + 1 \text{ cheese slice} \longrightarrow 1 \text{ grilled cheese sandwich}$$

There are enough bread slices (eight) to make four sandwiches and enough cheese slices (six) to make six sandwiches. Thus, using the quantities of ingredients on hand and the 2:1 "stoichiometric ratio" requirement for bread to cheese, only four complete sandwiches can be made. At that point there is no more bread even though there are two unused cheese slices; cheese is in excess. Bread slices are the "limiting reactant" because they limit the number of complete sandwiches that can be made.

The "reactant" in excess, cheese in this case, cannot be the limiting reactant.

In determining the maximum number of grilled cheese sandwiches that could be made, the "limiting reactant" was bread slices. We ran out of bread slices before using all the available cheese. Similarly, the limiting reactant must be identified in a chemical reaction to determine how much product(s) will be produced when all the limiting reactant is converted to the desired product(s).

If we know which one of a set of reactants is the limiting reactant, we can use that information to solve a quantitative problem directly, as illustrated in Problem-Solving Example 4.9.

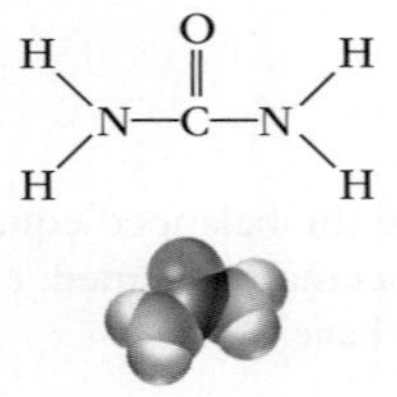

urea

You should be able to explain why NH_3 is the limiting reactant.

PROBLEM-SOLVING EXAMPLE 4.9 Moles of Product from Limiting Reactant

The organic compound urea, $(NH_2)_2CO$, can be prepared by reacting ammonia and carbon dioxide:

$$2\,NH_3(g) + CO_2(g) \longrightarrow (NH_2)_2CO(aq) + H_2O(\ell)$$

If 2.0 mol ammonia and 2.0 mol carbon dioxide are mixed, how many moles of urea are produced? Ammonia is the limiting reactant.

Answer 1.0 mol $(NH_2)_2CO$

Strategy and Explanation Start with the balanced equation and consider the stoichiometric coefficients. Concentrate on the NH_3 since it is given as the limiting reactant. Thus, the amount of urea produced must be based on the amount of NH_3 that is available. The coefficients show that for every 2 mol NH_3 reacted, 1 mol $(NH_2)_2CO$ will be produced. Thus, we use this information to answer the question

$$2.0 \text{ mol } NH_3 \times \frac{1 \text{ mol } (NH_2)_2CO}{2 \text{ mol } NH_3} = 1.0 \text{ mol } (NH_2)_2CO$$

☑ **Reasonable Answer Check** The balanced equation shows that the number of moles of $(NH_2)_2CO$ produced must be one half the number of moles of NH_3 that reacted, and it is.

PROBLEM-SOLVING PRACTICE 4.9

If we reacted 2.0 mol NH_3 with 0.75 mol CO_2 (and CO_2 is now the limiting reactant), how many moles of $(NH_2)_2CO$ would be produced?

Next, we consider the case where the quantities of reactants are given, but the limiting reactant is not identified and therefore must be determined. There are two approaches to identifying the limiting reactant—the mole ratio method and the mass method. You can rely on whichever method works better for you.

Mole Ratio Method

Calculate the number of moles of each reactant available and use the results to calculate the actual mole ratio. Compare the actual mole ratio with the stoichiometric mole ratio from the stoichiometric coefficients of the balanced equation. If the actual mole ratio is less than the theoretical mole ratio, the substance in the *numerator* is the limiting reactant.

Mass Method

Calculate the mass of product that would be produced from the available quantity of each reactant, assuming that an unlimited quantity of the other reactant were available. The limiting reactant is the one that produces the smaller mass of product.

Problem-Solving Example 4.10 illustrates these two methods for identifying a limiting reactant.

PROBLEM-SOLVING EXAMPLE 4.10 Limiting Reactant

Cisplatin is an anticancer drug used for treatment of solid tumors. It can be produced by reacting ammonia with potassium tetrachloroplatinate (ktcp).

$$\underset{\text{potassium tetrachloroplatinate}}{K_2PtCl_4(s)} + \underset{\text{ammonia}}{2\ NH_3(aq)} \longrightarrow \underset{\text{cisplatin}}{Pt(NH_3)_2Cl_2(s)} + 2\ KCl(aq)$$

If the reaction starts with 5.00 g ammonia and 50.0 g ktcp: (a) Which is the limiting reactant? (b) How many grams of cisplatin are produced? Assume that all the limiting reactant is converted to cisplatin.

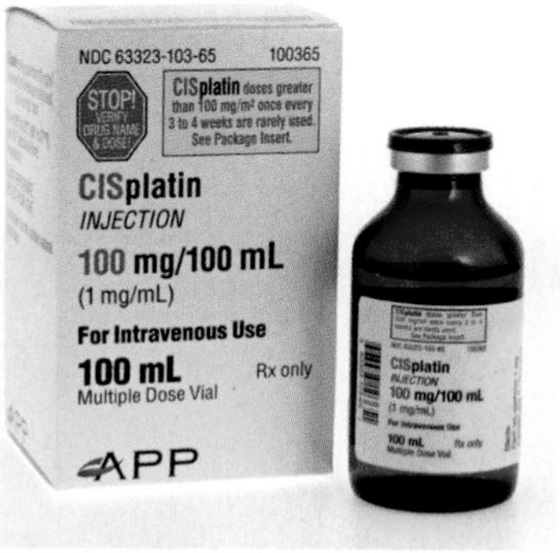

Courtesy of APP Pharmaceuticals

Vial of cisplatin, a drug used for cancer treatment.

Answer

(a) Potassium tetrachloroplatinate (b) 36.0 g cisplatin

Strategy and Explanation The problem can be solved for the limiting reactant by either of two methods: (1) mole ratio method or (2) mass method.

Mole Ratio Method

- **Calculate the molar mass and number of moles of each reactant available.**

  The molar mass of ktcp is 2(39.0983) + (195.078) + 4(35.453) = 415.09 g/mol.
  The molar mass of NH_3 is (14.007) + 3(1.0079) = 17.03 g/mol.

  The moles of each reactant available are given by

$$50.0\ \text{g ktcp} \times \frac{1\ \text{mol ktcp}}{415.09\ \text{g ktcp}} = 0.120\ \text{mol ktcp}$$

$$5.00\ \text{g } NH_3 \times \frac{1\ \text{mol } NH_3}{17.03\ \text{g } NH_3} = 0.294\ \text{mol } NH_3$$

- **Use the number of moles of each reactant to calculate the actual mole ratio. Compare this with the stoichiometric mole ratio.**

  The balanced equation shows that every 2 mol NH_3 reacted requires 1 mol ktcp.

$$\text{Actual mole ratio} = \frac{0.294\ \text{mol } NH_3}{0.120\ \text{mol ktcp}} = \frac{2.45\ \text{mol } NH_3}{1\ \text{mol ktcp}}$$

$$\text{Stoichiometric mole ratio} = \frac{2\ \text{mol } NH_3}{1\ \text{mol ktcp}}$$

  The actual mole ratio is greater than the stoichiometric mole ratio. This means that more NH_3 is present than is needed (NH_3 is in excess). Therefore, ktcp is the limiting reactant, and its amount must be used to calculate the quantities of products.

- **Use the amount of limiting reactant to determine the mass of product produced.**

  The mass of cisplatin produced by the reaction must be based on 0.120 mol ktcp, the amount of limiting reactant available. Calculate the mass of cisplatin that is produced using a mole ratio (1 mol cisplatin/1 mol ktcp) and the molar mass of cisplatin.

$$195.078 \text{ g Pt} + [2 \text{ mol N} \times (14.0067 \text{ g N/mol N})] + [6 \text{ mol H} \times (1.0079 \text{ g H/mol H})] + [2 \text{ mol Cl} \times 35.453 \text{ g Cl/mol Cl})] = 300.04 \text{ g/mol cisplatin}$$

$$0.120 \text{ mol ktcp} \times \frac{1 \text{ mol cisplatin}}{1 \text{ mol ktcp}} \times \frac{300.04 \text{ g cisplatin}}{1 \text{ mol cisplatin}} = 36.0 \text{ g cisplatin}$$

Mass Method

Calculate the mass of cisplatin that would be produced from 0.120 mol ktcp and sufficient ammonia. Then calculate the mass of cisplatin that would be produced from 0.294 mol NH_3 and sufficient ktcp. Compare the results to determine which reactant produces the smaller mass of product. Therefore, that reactant is the limiting reactant.

- **Calculate the mass of product using the mass of each reactant.**
  The mass of cisplatin produced from 0.120 mol ktcp and sufficient NH_3 is

The mass method directly gives the maximum mass of product.

$$0.120 \text{ mol ktcp} \times \frac{1 \text{ mol cisplatin}}{1 \text{ mol ktcp}} \times \frac{300.04 \text{ g cisplatin}}{1 \text{ mol cisplatin}} = 36.0 \text{ g cisplatin}$$

The mass of cisplatin produced from 0.294 mol NH_3 and sufficient ktcp is

$$0.294 \text{ mol } NH_3 \times \frac{1 \text{ mol cisplatin}}{2 \text{ mol } NH_3} \times \frac{300.04 \text{ g cisplatin}}{1 \text{ mol cisplatin}} = 44.1 \text{ g cisplatin}$$

- **Compare the masses of product to determine which reactant is limiting.**
  The amount of ktcp available would produce a smaller mass of cisplatin than the available amount of NH_3 would produce. Therefore, ktcp is the limiting reactant.

☑ **Reasonable Answer Check** The ratio of molar masses of cisplatin and ktcp is about three quarters (300/415), so we should have approximately three quarters as much cisplatin product as ktcp reactant (36/50), and we do.

PROBLEM-SOLVING PRACTICE 4.10

Carbon disulfide reacts with oxygen to form carbon dioxide and sulfur dioxide.

$$CS_2(\ell) + O_2(g) \longrightarrow CO_2(g) + SO_2(g)$$

A mixture of 3.5 g CS_2 and 17.5 g O_2 is reacted.
(a) Balance the equation.
(b) What is the limiting reactant?
(c) What is the maximum mass of sulfur dioxide that can be formed?

Powdered aluminum reacts with iron(III) oxide extremely vigorously in the thermite reaction to form molten iron and aluminum oxide.

PROBLEM-SOLVING EXAMPLE 4.11 Limiting Reactant

Powdered aluminum reacts with iron(III) oxide in the *thermite reaction* to form molten iron and aluminum oxide:

$$2\,Al(s) + Fe_2O_3(s) \longrightarrow 2\,Fe(\ell) + Al_2O_3(s)$$

Liquid iron is produced because the reaction releases enough energy to melt the iron. This liquid iron can be used to weld steel railroad rails. (a) Determine the limiting reactant when a mixture of 100. g Al and 100. g Fe_2O_3 react. (b) How many grams of liquid iron are formed? (c) How many grams of the excess reactant remain after the reaction is complete?

Answer

(a) Fe_2O_3 (b) 69.9 g Fe (c) 66.4 g Al

Strategy and Explanation

(a) We begin by determining how many moles of each reactant are available.

$$100.\text{ g Al} \times \frac{1 \text{ mol Al}}{26.98 \text{ g Al}} = 3.71 \text{ mol Al}$$

$$100.\text{ g } Fe_2O_3 \times \frac{1 \text{ mol } Fe_2O_3}{159.69 \text{ g } Fe_2O_3} = 0.626 \text{ mol } Fe_2O_3$$

Next, using the mass method for this limiting reactant problem, we find the masses of iron produced, based on the available masses of each reactant.

$$3.71 \text{ mol Al} \times \frac{2 \text{ mol Fe}}{2 \text{ mol Al}} \times \frac{55.845 \text{ g Fe}}{1 \text{ mol Fe}} = 207 \text{ g Fe}$$

$$0.626 \text{ mol } Fe_2O_3 \times \frac{2 \text{ mol Fe}}{1 \text{ mol } Fe_2O_3} \times \frac{55.845 \text{ g Fe}}{1 \text{ mol Fe}} = 69.9 \text{ g Fe}$$

Clearly, the mass of iron that can be formed using the given masses of aluminum metal and iron(III) oxide is controlled by the quantity of the iron(III) oxide. The iron(III) oxide is the limiting reactant because it produces less iron. Aluminum is in excess.

(b) The mass of iron formed is 69.9 g Fe.

(c) We can find the number of moles of Al that reacted from the moles of Fe_2O_3 used and the mole ratio of Al to Fe_2O_3.

$$0.626 \text{ mol } Fe_2O_3 \times \frac{2 \text{ mol Al}}{1 \text{ mol } Fe_2O_3} = 1.25 \text{ mol Al}$$

By subtracting the moles of Al reacted from the initial amount of Al, we find that 2.46 mol Al remains unreacted (3.71 mol Al − 1.25 mol Al = 2.46 mol Al). Therefore, the mass of unreacted Al is

$$2.46 \text{ mol Al} \times \frac{26.98 \text{ g Al}}{1 \text{ mol Al}} = 66.4 \text{ g Al}$$

It is useful, although not necessary, to calculate the quantity of excess reactant remaining to verify that the reactant in excess is not the limiting reactant.

☑ **Reasonable Answer Check** The molar masses of Fe_2O_3 (160 g/mol) and Fe (56 g/mol) are in the ratio of approximately 3:1. One mole of Fe_2O_3 produces 2 mol Fe according to the balanced equation. So a given mass of Fe_2O_3 should produce about two thirds as much Fe, and this agrees with our more exact calculation.

PROBLEM-SOLVING PRACTICE 4.11

Preparation of the pure silicon used in silicon chips involves the reaction between purified liquid silicon tetrachloride and magnesium.

$$SiCl_4(\ell) + 2\,Mg(s) \longrightarrow Si(s) + 2\,MgCl_2(s)$$

If the reaction were run with 100. g each of $SiCl_4$ and Mg, which reactant would be limiting, and what mass of Si would be produced?

CHEMISTRY IN THE NEWS Smothering Fire—Water That Isn't Wet

Fire is a combustion reaction in which fuel and oxygen, O_2, combine, usually at high temperatures, to form water and carbon dioxide. Three factors are necessary for a fire: combustible fuel, oxygen, and a temperature above the ignition temperature of the fuel. Once the fire has started, it is self-supporting because it supplies the heat necessary to keep the temperature high (if sufficient fuel and oxygen are available). Quenching a fire requires removing the fuel, lowering the oxygen level, cooling the reaction mixture below the ignition temperature or some combination of these.

An effective way to quench a fire is smothering, which reduces the amount of available oxygen below the level needed to support combustion. In other words, smothering decreases the amount of the limiting reactant. Foams, inert gas, and CO_2 are effective substances for smothering.

Developed by 3M, Novec 1230 is a new compound with very desirable fire suppression properties. An organic compound with many carbon-fluorine bonds, it is a colorless liquid at room temperature (B.P. 49.2 °C) that feels like water. When placed on a fire, Novec vaporizes and smothers the fire. Novec 1230 is not an electrical conductor, so it can be used for electrical fires. In fact, in a televised demonstration, a laptop computer was immersed in Novec 1230 and continued to work. The compound is used in situations where water cannot be used, for example, clean rooms, hospitals, or museums. The new compound also does not affect the stratospheric ozone layer because its lifetime in the lower atmosphere is short, making it environmentally friendly.

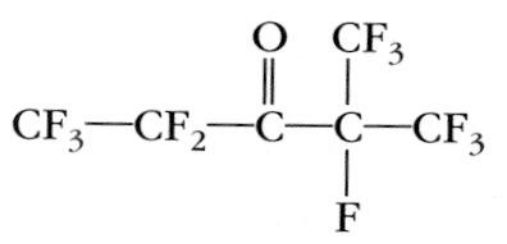

Structural formula for Novec 1230.

Sources: New York Times, Dec. 12, 2004; p. 103.
http://solutions.3m.com/wps/portal/3M/en_US/Novec/Home/Product_Information/Fire_Protection/
http://en.wikipedia.org/wiki/Novec_1230

EXERCISE 4.8 Limiting Reactant

Urea is used as a fertilizer because it can react with water to release ammonia, which provides nitrogen to plants.

$$(NH_2)_2CO(s) + H_2O(\ell) \longrightarrow 2\ NH_3(aq) + CO_2(g)$$

(a) Determine the limiting reactant when 300. g urea and 100. g water are combined.
(b) How many grams of ammonia and how many grams of carbon dioxide form?
(c) What mass of the excess reactant remains after reaction?

4.6 Evaluating the Success of a Synthesis: Percent Yield

A reaction that converts the limiting reactant into the maximum possible quantity of product is said to have a 100% yield. This maximum possible quantity of product is called the **theoretical yield.** Often the **actual yield,** the quantity of desired product actually obtained from a synthesis in a laboratory or industrial chemical plant, is less than the theoretical yield.

The efficiency of a particular synthesis method is evaluated by calculating the **percent yield,** which is defined as

$$\text{Percent yield} = \frac{\text{actual yield}}{\text{theoretical yield}} \times 100\%$$

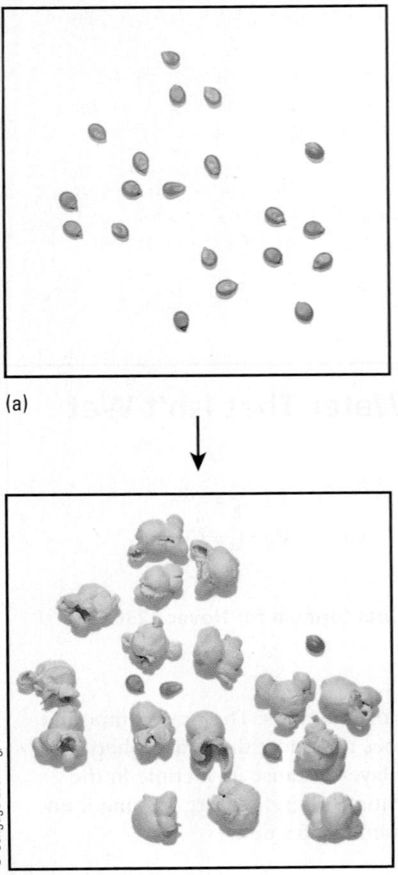
(a)

(b)

Popcorn yield. We began with 20 popcorn kernels, but only 16 of them popped. The percent yield of popcorn was (16/20) × 100% = 80%.

Percent yield can be applied, for example, to the synthesis of aspirin. Suppose a student carried out the synthesis and obtained 2.2 g aspirin rather than the calculated theoretical yield of 2.6 g. What is the percent yield of this reaction?

$$\text{Percent yield} = \frac{\text{actual yield of product}}{\text{theoretical yield of product}} \times 100\% = \frac{2.2\text{ g}}{2.6\text{ g}} \times 100\% = 85\%$$

Although we hope to obtain as close to the theoretical yield as possible when carrying out a reaction, few reactions or experimental manipulations are 100% efficient, despite controlled conditions and careful laboratory techniques. Side reactions can occur that form products other than the desired one, and during the isolation and purification of the desired product, some of it may be lost. When chemists report the synthesis of a new compound or the development of a new synthesis, they also report the percent yield of the reaction or the overall series of reactions. Other chemists who wish to repeat the synthesis then have an idea of how much product can be expected from a certain amount of reactants.

PROBLEM-SOLVING EXAMPLE 4.12 Calculating Percent Yield

Methanol, CH_3OH, is an excellent fuel, and it can be produced from carbon monoxide and hydrogen.

$$CO(g) + 2\ H_2(g) \longrightarrow CH_3OH(\ell)$$

If 500. g CO reacts with sufficient H_2 and 485 g CH_3OH is produced, what is the percent yield of the reaction?

Answer 85.0%

Strategy and Explanation Calculate the theoretical yield of CH_3OH and compare it to the mass actually produced.

- **Calculate the number of moles of the limiting reactant.**
  Calculate the moles of the limiting reactant, CO. Hydrogen is present in sufficient amount, so it is not the limiting reactant.

CHEMISTRY YOU CAN DO

Vinegar and Baking Soda: A Stoichiometry Experiment

This experiment focuses on the reactions of metal carbonates with acid. For example, limestone reacts with hydrochloric acid to give calcium chloride, carbon dioxide, and water:

$$\underset{\text{limestone}}{CaCO_3(s)} + 2\,HCl(aq) \longrightarrow CaCl_2(aq) + CO_2(g) + H_2O(\ell)$$

In a similar way, baking soda (sodium hydrogen carbonate) and vinegar (aqueous acetic acid) react to give sodium acetate, carbon dioxide, and water:

$$\underset{\text{baking soda}}{NaHCO_3(s)} + \underset{\text{acetic acid}}{CH_3COOH(aq)} \longrightarrow \underset{\text{sodium acetate}}{NaCH_3COO(aq)} + CO_2(g) + H_2O(\ell)$$

In this experiment we want to explore the relationship between the quantity of acid or hydrogen carbonate used and the quantity of carbon dioxide evolved. To do the experiment you need some baking soda, vinegar, small balloons, and a bottle with a narrow neck and a volume of about 100 mL. The balloon should fit tightly but easily over the top of the bottle. (It may slip on more easily if the bottle and balloon are wet.)

Place 1 level teaspoon of baking soda in the balloon. (You can make a funnel out of a rolled-up piece of paper to help get the baking soda into the balloon.) Add 3 teaspoons of vinegar to the bottle, and then slip the lip of the balloon over the neck of the bottle. Turn the balloon over so that the baking soda runs into the bottle, and then shake the bottle to make sure that the vinegar and baking soda are well mixed. What do you see? Does the balloon inflate? If so, why?

Now repeat the experiment several times using these quantities of vinegar and baking soda:

The setup for the study of the reaction of baking soda with vinegar (acetic acid).

Baking Soda	Vinegar
1 tsp	1 tsp
1 tsp	4 tsp
1 tsp	7 tsp
1 tsp	10 tsp

Be sure to use a new balloon each time and rinse out the bottle between tests. In each test, record how much the balloon inflates.

Think about these questions: Is there a relationship between the quantity of vinegar and baking soda used and the extent to which the balloon inflates? If so, how can you explain this connection?

At which point does an increase in the quantity of vinegar not increase the volume of the balloon? Based on what we know about chemical reactions, how could increasing the amount of one reactant not have an effect on the balloon's size?

$$500.\text{ g CO} \times \frac{1\text{ mol CO}}{28.0\text{ g CO}} = 17.86\text{ mol CO}$$

- **Use the balanced chemical equation to determine the number of moles of product formed.**

  The coefficients of the balanced chemical equation show that when 1 mol CO reacts, 1 mol CH_3OH forms. Therefore, 17.86 mol is the maximum amount of CH_3OH that can be produced.

- **Convert the amount of product to mass of product, the theoretical yield.**

  The molar mass of CH_3OH is 32.0 g/mol.

$$17.86\text{ mol }CH_3OH \times \frac{32.0\text{ g }CH_3OH}{1\text{ mol }CH_3OH} = 571\text{ g }CH_3OH$$

  Thus, the theoretical yield is 571 g CH_3OH.

- **Calculate the ratio of the actual yield to the theoretical yield to get the percent yield.**

  The problem states that 485 g CH_3OH was produced.

$$\frac{485\text{ g }CH_3OH\text{ (actual yield)}}{571\text{ g }CH_3OH\text{ (theoretical yield)}} \times 100\% = 85.0\%$$

☑ **Reasonable Answer Check** The molar mass of CH_3OH is slightly greater than that of CO, and an equal number of moles of CO and CH_3OH are in the balanced equation. So *x* g CO should produce somewhat more than *x* g CH_3OH. Slightly less is actually produced, however, so a percent yield slightly less than 100% is about right.

PROBLEM-SOLVING PRACTICE 4.12

If the methanol synthesis reaction is run with an 85% yield, and you want to make 1.0 kg CH_3OH, how many grams of H_2 should you use if you have sufficient CO?

PROBLEM-SOLVING EXAMPLE 4.13 Percent Yield

Ammonia can be produced from the reaction of a metal oxide such as calcium oxide with ammonium chloride:

$$CaO(s) + 2\ NH_4Cl(s) \rightarrow 2\ NH_3(g) + H_2O(g) + CaCl_2(s)$$

How many grams of calcium oxide would be needed to react with excess ammonium chloride to produce 1.00 g ammonia if the expected percent yield were 25%?

Answer 6.6 g CaO

Strategy and Explanation The expected percent yield expressed as a decimal is 0.25. So the theoretical yield is

$$\text{Theoretical yield} = \frac{\text{actual yield}}{\text{percent yield}} = \frac{1.00\ \text{g NH}_3}{0.25} = 4.00\ \text{g NH}_3$$

Ammonium chloride is in excess, so CaO is the limiting reactant and determines the amount of ammonia that will be produced. The mass of CaO needed can be calculated from the theoretical yield of ammonia and the 1:2 mole ratio for CaO and ammonia as given in the balanced equation.

$$4.00\ \text{g NH}_3 \times \frac{1\ \text{mol NH}_3}{17.03\ \text{g NH}_3} \times \frac{1\ \text{mol CaO}}{2\ \text{mol NH}_3} \times \frac{56.077\ \text{g CaO}}{1\ \text{mol CaO}} = 6.6\ \text{g CaO}$$

☑ **Reasonable Answer Check** To check the answer, we solve the problem a different way. The molar mass of CaO is approximately 56 g/mol, so 6.6 g CaO is approximately 0.12 mol CaO. The coefficients in the balanced equation tell us that this would produce twice as many moles of NH_3 or 0.24 mol NH_3, which is 0.24×17 g/mol = 4 g NH_3, if the yield were 100%. The actual yield is 25%, so the amount of ammonia expected is reduced by one fourth to approximately 1 g. This approximate calculation is consistent with our more accurate calculation, and the answer is reasonable.

PROBLEM-SOLVING PRACTICE 4.13

You heat 2.50 g copper with an excess of sulfur and synthesize 2.53 g copper(I) sulfide, Cu_2S:

$$16\ Cu(s) + S_8(s) \longrightarrow 8\ Cu_2S(s)$$

Your laboratory instructor expects students to have at least a 60% yield for this reaction. Did your synthesis meet this standard?

CONCEPTUAL EXERCISE 4.9 Percent Yield

Percent yield can be reduced by side reactions that produce undesired product(s) and by poor laboratory technique in isolating and purifying the desired product. Identify two other factors that could lead to a low percent yield.

Atom Economy—Another Approach to Tracing Starting Materials

Rather than concentrating simply on percent yield, the concept of **atom economy** focuses on the amounts of starting materials that are incorporated into the desired final product. The greater the fraction of starting atoms incorporated into the desired final product, the fewer waste by-products created. Of course, the objective is to devise syntheses that are as efficient as possible.

Whereas a high percent yield has often been the major goal of chemical synthesis, the concept of atom economy, quantified by the definition of percent atom economy, is becoming important.

$$\text{Percent atom economy} = \frac{\text{sum of atomic weight of atoms in the useful product}}{\text{sum of atomic weight of all atoms in reactants}} \times 100\%$$

Reactions for which all atoms in the reactants are found in the desired product have a percent atom economy of 100%. As an example, consider this combination reaction:

$$CO(g) + 2\,H_2(g) \longrightarrow CH_3OH(\ell)$$

The sum of atomic weights of all atoms in the reactants is $12.011 + 15.9994 + \{2 \times (2 \times 1.0079)\} = 32.042$ amu. The atomic weight of all the atoms in the product is $12.011 + \{3 \times (1.0079)\} + 15.9994 + 1.0079 = 32.042$ amu. The percent atom economy for this reaction is 100%.

Many other reactions in organic synthesis, however, generate other products in addition to the desired product. In such cases, the percent atom economy is far less than 100%. Devising strategies for synthesis of desired compounds with the least waste is a major goal of the current push toward "green chemistry."

Green chemistry aims to eliminate pollution by making chemical products that do not harm health or the environment. It encourages the development of production processes that reduce or eliminate hazardous chemicals. Green chemistry also aims to prevent pollution at its source rather than having to clean up problems after they occur. Each year since 1996 the U.S. Environmental Protection Agency has given Presidential Green Chemistry Challenge Awards for noteworthy green chemistry advances.

One of the 2008 Green Chemistry Challenge Awards was given for developing a soy-based laser printer cartridge toner that is much easier to remove from paper than the traditional toners. The new toner will allow greater recycling of waste paper from printers and copiers, and it is made via greener manufacturing processes. **http://www.epa.gov/greenchemistry/pubs/pgcc/presgcc.html**

4.7 Percent Composition and Empirical Formulas

In Section 3.10, percent composition data were used to derive empirical and molecular formulas, but nothing was mentioned about how such data are obtained. One way to obtain such data is combustion analysis, which is often employed with organic compounds, most of which contain carbon and hydrogen. In **combustion analysis** a compound is burned in excess oxygen, which converts the carbon to carbon dioxide and the hydrogen to water. These combustion products are collected and weighed, and the masses are used to calculate the quantities of carbon and hydrogen in the original substance using the balanced combustion equation. A schematic diagram of the apparatus is shown in Figure 4.8. Many organic compounds also contain oxygen. In such cases, the mass of oxygen in the sample can be determined simply by difference.

$$\text{Mass of oxygen} = \text{mass of sample} - (\text{mass of C} + \text{mass of H})$$

As an example, consider this problem. An analytical chemist used combustion analysis to determine the empirical formula of vitamin C, an organic compound containing only carbon, hydrogen, and oxygen. Combustion of 1.0000 g pure vitamin C produced 1.502 g CO_2 and 0.409 g H_2O. A different experiment determined that the molar mass of vitamin C is 176.12 g/mol.

The task is to determine the subscripts on C, H, and O in the empirical formula of vitamin C. Recall from Chapter 3 that the subscripts in a chemical formula tell how many moles of atoms of each element are in 1 mol of the compound. *All* of the carbon in the CO_2 and *all* of the hydrogen in the H_2O came from the vitamin C sample that was burned, so we can work backward to assess the composition of vitamin C.

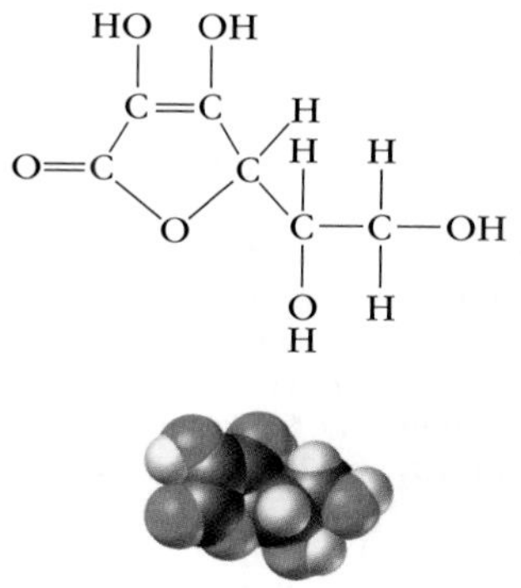

vitamin C

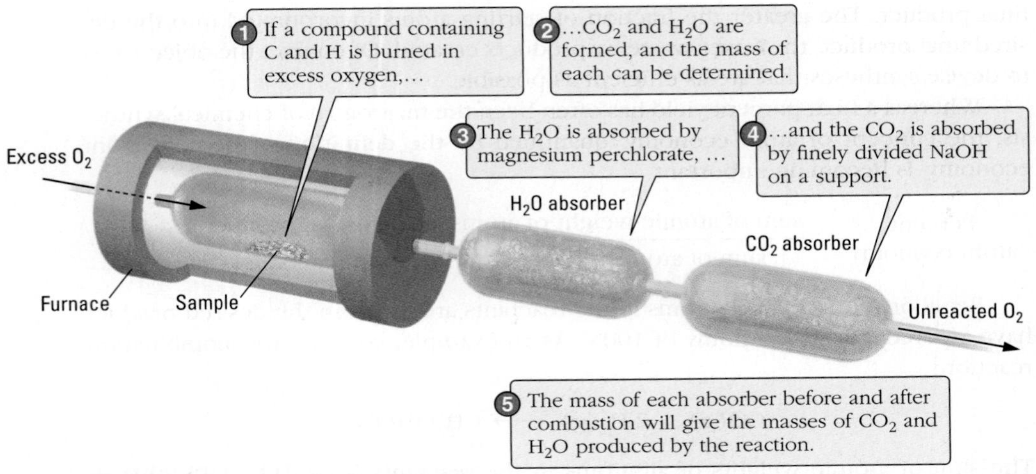

Figure 4.8 Combustion analysis. Schematic diagram of an apparatus for determining the empirical formula of an organic compound. Only a few milligrams of a combustible compound are needed for analysis.

First, we determine the masses of carbon and hydrogen in the original sample.

$$1.502 \text{ g } CO_2 \times \frac{1 \text{ mol } CO_2}{44.009 \text{ g } CO_2} \times \frac{1 \text{ mol C}}{1 \text{ mol } CO_2} \times \frac{12.011 \text{ g C}}{1 \text{ mol C}} = 0.4100 \text{ g C}$$

$$0.409 \text{ g } H_2O \times \frac{1 \text{ mol } H_2O}{18.015 \text{ g } H_2O} \times \frac{2 \text{ mol H}}{1 \text{ mol } H_2O} \times \frac{1.0079 \text{ g H}}{1 \text{ mol H}} = 0.04577 \text{ g H}$$

The mass of oxygen in the original sample can be calculated by difference.

$$1.0000 \text{ g sample} - (0.4100 \text{ g C in sample} + 0.04577 \text{ g H in sample})$$
$$= 0.5442 \text{ g O in the sample}$$

From the mass data, we can now calculate how many moles of each element were in the sample.

Carrying an extra digit during the intermediate parts of a multistep problem and rounding at the end is good practice.

$$0.4100 \text{ g C} \times \frac{1 \text{ mol C}}{12.011 \text{ g C}} = 0.03414 \text{ mol C}$$

$$0.04577 \text{ g H} \times \frac{1 \text{ mol H}}{1.0079 \text{ g H}} = 0.04541 \text{ mol H}$$

$$0.5442 \text{ g O} \times \frac{1 \text{ mol O}}{15.999 \text{ g O}} = 0.03401 \text{ mol O}$$

Next, we find the mole ratios of the elements in the compound by dividing by the smallest number of moles.

$$\frac{0.04541 \text{ mol H}}{0.03401 \text{ mol O}} = \frac{1.335 \text{ mol H}}{1.000 \text{ mol O}}$$

$$\frac{0.03414 \text{ mol C}}{0.03401 \text{ mol O}} = \frac{1.004 \text{ mol C}}{1.000 \text{ mol O}}$$

If after dividing by the smallest number of moles, the ratios are not whole numbers, multiply each subscript by a number that converts the fractions to whole numbers. For example, multiplying $NO_{2.5}$ by 2 changes it to N_2O_5.

The ratios are very close to 1.33 mol H to 1.00 mol O and a one-to-one ratio of C to O. Multiplying by 3 to get whole numbers gives the empirical formula of vitamin C as $C_3H_4O_3$. From this we can calculate an empirical formula mass of 88.06 g. Because the experimental molar mass is twice the empirical formula mass, the molecular formula of vitamin C is $C_6H_8O_6$, twice the empirical formula.

For many organic compounds, the empirical and molecular formulas are the same. In addition, several different organic compounds can have the identical empirical and molecular formulas—they are isomers (← *p. 83*). In such cases, you must know the structural formula to fully describe the compound. Ethanol, CH_3CH_2OH, and dimethyl ether, CH_3OCH_3, for example, each have the same empirical formula and molecular formula, C_2H_6O, but they are different compounds with different properties.

CH_3CH_2OH
ethanol

CH_3OCH_3
dimethyl ether

PROBLEM-SOLVING EXAMPLE 4.14 Empirical Formula from Combustion Analysis

Butyric acid, an organic compound with an extremely unpleasant odor, contains only carbon, hydrogen, and oxygen. When 1.20 g butyric acid was burned, 2.41 g CO_2 and 0.982 g H_2O were produced. Calculate the empirical formula of butyric acid. In a separate experiment, the molar mass of butyric acid was determined to be 88.1 g/mol. Determine butyric acid's molecular formula.

Answer The empirical formula is C_2H_4O. The molecular formula is $C_4H_8O_2$.

Strategy and Explanation All of the carbon and hydrogen in the butyric acid are burned to form CO_2 and H_2O, respectively. Therefore, use the masses of CO_2 and H_2O formed to calculate how many grams of C and H, respectively, were in the original butyric acid sample. Then, find the number of moles of each element in the sample, the mole ratios, and the empirical formula. To determine the molecular formula, compare the formula mass of the empirical formula with the known molar mass.

- **Calculate the number of grams of C, H, and O in the sample.**

  Start with the mass of CO_2 and H_2O and use their molar masses and the mole ratios from the balanced equation to complete the calculation.

$$2.41 \text{ g } CO_2 \times \frac{1 \text{ mol } CO_2}{44.01 \text{ g } CO_2} \times \frac{1 \text{ mol C}}{1 \text{ mol } CO_2} \times \frac{12.01 \text{ g C}}{1 \text{ mol C}} = 0.658 \text{ g C}$$

$$0.982 \text{ g } H_2O \times \frac{1 \text{ mol } H_2O}{18.02 \text{ g } H_2O} \times \frac{2 \text{ mol H}}{1 \text{ mol } H_2O} \times \frac{1.0079 \text{ g H}}{1 \text{ mol H}} = 0.110 \text{ g H}$$

  The remaining mass of the sample must be oxygen:

$$1.200 \text{ g} - 0.658 \text{ g C} - 0.110 \text{ g H} = 0.432 \text{ g O}$$

- **Calculate the number of moles of C, H, and O in the sample.**

$$0.658 \text{ g C} \times \frac{1 \text{ mol C}}{12.01 \text{ g C}} = 0.0548 \text{ mol C}$$

$$0.110 \text{ g H} \times \frac{1 \text{ mol H}}{1.0079 \text{ g H}} = 0.109 \text{ mol H}$$

$$0.432 \text{ g O} \times \frac{1 \text{ mol O}}{16.00 \text{ g O}} = 0.0270 \text{ mol O}$$

- **To find the empirical formula, calculate the mole ratios of C, H, and O in the sample.**

  Divide the molar amount of each element by the smallest number of moles

$$\frac{0.0548 \text{ mol C}}{0.0270 \text{ mol O}} = \frac{2.03 \text{ mol C}}{1.00 \text{ mol O}} \quad \text{and} \quad \frac{0.109 \text{ mol H}}{0.0270 \text{ mol O}} = \frac{4.03 \text{ mol H}}{1.00 \text{ mol O}}$$

  The mole ratios show that for every oxygen atom in the molecule, there are two carbon atoms and four hydrogen atoms. Therefore, the empirical formula of butyric acid is C_2H_4O, which has an empirical formula mass of 44.05 g/mol.

- **Compare the experimental molar mass to the empirical formula mass to determine the molecular formula.**

  The experimental molar mass is known to be 88.10 g/mol, twice the empirical formula mass. Therefore, the molecular formula of butyric acid is $C_4H_8O_2$, twice the empirical formula.

☑ **Reasonable Answer Check** The molar mass of butyric acid, $C_4H_8O_2$, is 4(12.01) + 8(1.0079) + 2(16.00) = 88.10 g/mol, so the answer is reasonable.

PROBLEM-SOLVING PRACTICE 4.14

Phenol is a compound of carbon, hydrogen, and oxygen that is used commonly as a disinfectant. Combustion analysis of a 175-mg sample of phenol yielded 491. mg CO_2 and 100. mg H_2O.

(a) Calculate the empirical formula of phenol.

(b) What other information is necessary to determine whether the empirical formula is the actual molecular formula?

CONCEPTUAL EXERCISE 4.10 Formula from Combustion Analysis

Nicotine, a compound found in cigarettes, contains C, H, and N. Outline a method by which you could use combustion analysis to determine the empirical formula for nicotine.

Determining Formulas from Experimental Data

One technique to determine the formula of a binary compound formed by direct combination of its two elements is to measure the mass of reactants that is converted to the product compound.

PROBLEM-SOLVING EXAMPLE 4.15 Empirical Formula from Experimental Data

Solid red phosphorus reacts with liquid bromine to produce a phosphorus bromide.

$$P_4(s) + Br_2(\ell) \longrightarrow P_xBr_y(\ell)$$

If 0.347 g P_4 reacts with 0.860 mL Br_2, what is the empirical formula of the product? The density of bromine is 3.12 g/mL.

Answer PBr_3

Strategy and Explanation Calculate the molar amount of each reactant and then the mole ratio of the product molecule to determine the empirical formula.

- **Calculate the molar amount of each reactant.**

$$0.347 \text{ g } P_4 \times \frac{1 \text{ mol } P_4}{123.90 \text{ g } P_4} = 2.801 \times 10^{-3} \text{ mol } P_4$$

To determine the mol Br_2, use the density of Br_2 to convert milliliters of Br_2 to grams:

$$0.860 \text{ mL } Br_2 \times \frac{3.12 \text{ g } Br_2}{1 \text{ mL } Br_2} = 2.68 \text{ g } Br_2$$

Then convert the mass of Br_2 to mol Br_2

$$2.68 \text{ g } Br_2 \times \frac{1 \text{ mol } Br_2}{159.8 \text{ g } Br_2} = 1.677 \times 10^{-2} \text{ mol } Br_2$$

- **To find the empirical formula, calculate the mole ratio of the atoms in the product molecule.**

$$1.677 \times 10^{-2} \text{ mol } Br_2 \text{ molecules} \times \frac{2 \text{ mol Br atoms}}{1 \text{ mol } Br_2 \text{ molecules}} = 3.35 \times 10^{-2} \text{ mol Br atoms}$$

$$2.801 \times 10^{-3} \text{ mol } P_4 \text{ molecules} \times \frac{4 \text{ mol P atoms}}{1 \text{ mol } P_4 \text{ molecules}} = 1.12 \times 10^{-2} \text{ mol P atoms}$$

$$\frac{3.35 \times 10^{-2} \text{ mol Br atoms}}{1.12 \times 10^{-2} \text{ mol P atoms}} = \frac{2.99 \text{ mol Br atoms}}{1.00 \text{ mol P atoms}}$$

The mole ratio in the compound is 3.00 mol Br atoms for 1.00 mol P atoms. The empirical formula is PBr_3.

☑ **Reasonable Answer Check** Phosphorus is a Group 5A element, so combining it with three bromines results in a reasonable molecular formula for such a combination of elements.

PROBLEM-SOLVING PRACTICE **4.15**

The complete reaction of 0.569 g tin with 2.434 g iodine formed Sn_xI_y. What is the empirical formula of this tin iodide?

SUMMARY PROBLEM

Iron can be smelted from iron(III) oxide in ore via this high-temperature reaction in a blast furnace:

$$Fe_2O_3(s) + 3\ CO(g) \longrightarrow 2\ Fe(\ell) + 3\ CO_2(g)$$

The liquid iron produced is cooled and weighed.

(a) For 19.0 g Fe_2O_3, what mass of CO is required to react completely?

(b) What mass of CO_2 is produced when the reaction runs to completion with 10.0 g Fe_2O_3 as starting material?

When the reaction was run repeatedly with the same mass of iron oxide, 19.0 g Fe_2O_3, but differing masses of carbon monoxide, this graph was obtained.

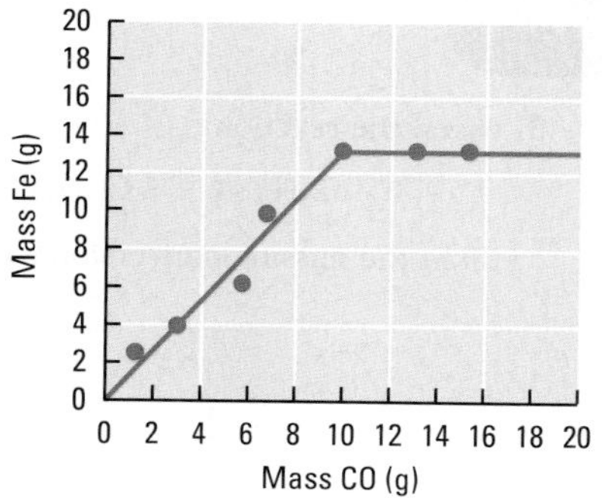

(c) Which reactant is limiting in the part of the graph where there is less than 10.0 g CO available to react with 19.0 g Fe_2O_3?

(d) Which reactant is limiting when more than 10.0 g CO is available to react with 19.0 g Fe_2O_3?

(e) If 24.0 g Fe_2O_3 reacted with 20.0 g CO and 15.9 g Fe was produced, what was the percent yield of the reaction?

IN CLOSING

Having studied this chapter, you should be able to . . .

- Interpret the information conveyed by a balanced chemical equation (Section 4.1). End-of-chapter questions: 10, 12, 18
- Recognize the general reaction types: combination, decomposition, displacement, and exchange (Section 4.2). Questions 24, 26
- Balance simple chemical equations (Section 4.3). Questions 34, 36

OWL and **go Chemistry**

Sign in at **www.cengage.com/owl** to:

- View tutorials and simulations, develop problem-solving skills, and complete online homework assigned by your professor.
- For quick review and exam prep, download Go Chemistry mini lecture modules from OWL (or purchase them at **www.CengageBrain.com**).

- Use mole ratios to calculate the number of moles or number of grams of one reactant or product from the number of moles or number of grams of another reactant or product by using the balanced chemical equation (Section 4.4). Questions 42, 44, 48, 95
- Use principles of stoichiometry in the chemical analysis of a mixture (Section 4.4). Question 118
- Determine which of the reactants is the limiting reactant (Section 4.5). Questions 66, 70, 72
- Explain the differences among actual yield, theoretical yield, and percent yield, and calculate theoretical and percent yields (Section 4.6). Questions 77, 81, 83
- Use principles of stoichiometry to find the empirical formula of an unknown compound using combustion analysis or other mass data (Section 4.7). Questions 88, 97

KEY TERMS

actual yield *(Section 4.6)*
aqueous solution *(4.1)*
atom economy *(4.6)*
balanced chemical equation *(4.1)*
coefficient *(4.1)*
combination reaction *(4.2)*
combustion analysis *(4.7)*
combustion reaction *(4.1)*
decomposition reaction *(4.2)*
displacement reaction *(4.2)*
exchange reaction *(4.2)*
limiting reactant *(4.5)*
mole ratio *(4.4)*
percent yield *(4.6)*
stoichiometric coefficient *(4.1)*
stoichiometry *(4.1)*
theoretical yield *(4.6)*

QUESTIONS FOR REVIEW AND THOUGHT

OWL Interactive versions of these problems are assignable in OWL.

Blue-numbered questions have short answers at the back of this book in Appendix M and fully worked solutions in the *Student Solutions Manual.*

Review Questions

These questions test vocabulary and simple concepts.

1. What information is provided by a balanced chemical equation?
2. Complete the table for the reaction

$$3\ H_2(g) + N_2(g) \longrightarrow 2\ NH_3(g)$$

H_2	N_2	NH_3
______ mol	1 mol	______ mol
3 molecules	______ molecules	______ molecules
______ g	______ g	34.08 g

3. What is meant by the statement, "The reactants were present in stoichiometric amounts"?
4. Write all the possible mole ratios for the reaction

$$3\ MgO(s) + 2\ Fe(s) \longrightarrow Fe_2O_3(s) + 3\ Mg(s)$$

5. If a 10.0-g mass of carbon is combined with an exact stoichiometric amount of oxygen (26.6 g) to make carbon dioxide, how many grams of CO_2 can be isolated?
6. Given the reaction

$$2\ Fe(s) + 3\ Cl_2(g) \longrightarrow 2\ FeCl_3(s)$$

fill in the missing conversion factors for the scheme

g Cl_2 —?→ g $FeCl_3$
|? ↓ (both sides)
mol Cl_2 —?→ mol $FeCl_3$

7. When an exam question asks, "What is the limiting reactant?" students may be tempted to guess the reactant with the smallest mass. Why is this not a good strategy?
8. Why can't the product of a reaction ever be the limiting reactant?
9. Does the limiting reactant determine the theoretical yield, actual yield, or both? Explain.

Topical Questions

These questions are keyed to the major topics in the chapter. Usually a question that is answered at the back of this book is paired with a similar one that is not.

Stoichiometry (Section 4.1)

10. For this reaction, fill in the table on the next page with the indicated quantities for the balanced equation.

$$4\ NH_3(g) + 5\ O_2(g) \longrightarrow 4\ NO(g) + 6\ H_2O(g)$$

	NH_3	O_2	NO	H_2O
No. of molecules				
No. of atoms				
No. of moles of molecules				
Mass				
Total mass of reactants				
Total mass of products				

11. For this reaction, fill in the table with the indicated quantities for the balanced equation.

$$2\ C_2H_6(g) + 7\ O_2(g) \longrightarrow 4\ CO_2(g) + 6\ H_2O(g)$$

	C_2H_6	O_2	CO_2	H_2O
No. of molecules				
No. of atoms				
No. of moles of molecules				
Mass				
Total mass of reactants				
Total mass of products				

12. Magnesium metal burns brightly in the presence of oxygen to produce a white powdery substance, MgO.

$$Mg(s) + O_2(g) \longrightarrow MgO(s) \quad \text{(unbalanced)}$$

(a) If 1.00 g MgO(s) is formed by this reaction, what is the *total* mass of magnesium metal and oxygen that reacted?

(b) Identify the stoichiometric coefficients in this equation.

(c) If 50 *atoms* of oxygen reacted, how many magnesium *atoms* were needed to react with this much oxygen?

13. Sucrose (table sugar) reacts with oxygen as follows:

$$C_{12}H_{22}O_{11} + 12\ O_2 \longrightarrow 12\ CO_2 + 11\ H_2O$$

When 1.0 g sucrose is reacted, how many grams of CO_2 are produced? How many grams of O_2 are required to react with 1.0 g sucrose?

14. Balance this *combination reaction* by adding coefficients as needed.

$$Fe(s) + O_2(g) \longrightarrow Fe_2O_3(s)$$

15. Balance this *decomposition reaction* by adding coefficients as needed.

$$KClO_3(s) \longrightarrow KCl(s) + O_2(g)$$

16. This diagram shows A (blue spheres) reacting with B (tan spheres). Which equation best describes the stoichiometry of the reaction depicted in this diagram?

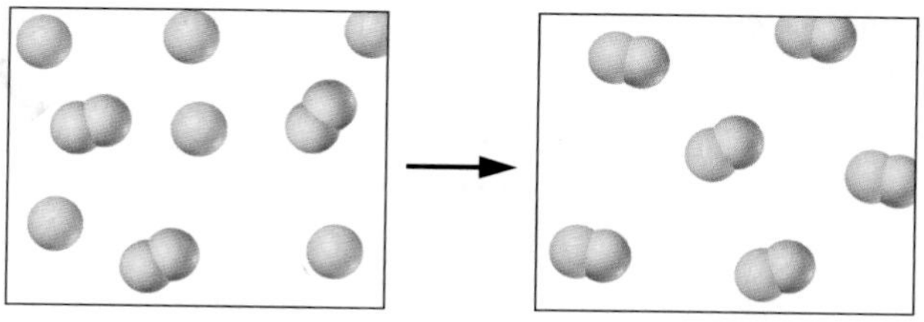

(a) $3\ A_2 + 6\ B \rightarrow 6\ AB$ (b) $A_2 + 2\ B \rightarrow 2\ AB$
(c) $2\ A + B \rightarrow AB$ (d) $3\ A + 6\ B \rightarrow 6\ AB$

17. This diagram shows A (blue spheres) reacting with B (tan spheres). Write a balanced equation that describes the stoichiometry of the reaction shown in the diagram.

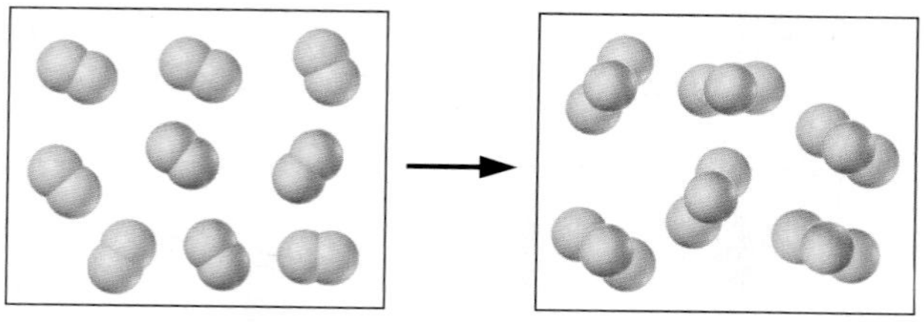

18. Given this equation,

$$4\ A_2 + 3\ B \longrightarrow B_3A_8$$

use a diagram to illustrate the amount of reactant A and product, B_3A_8, that would be needed/produced from the reaction of six atoms of B.

19. Balance this equation and determine which box represents reactants and which box represents products.

$$Sb(g) + Cl_2(g) \longrightarrow SbCl_3(g)$$

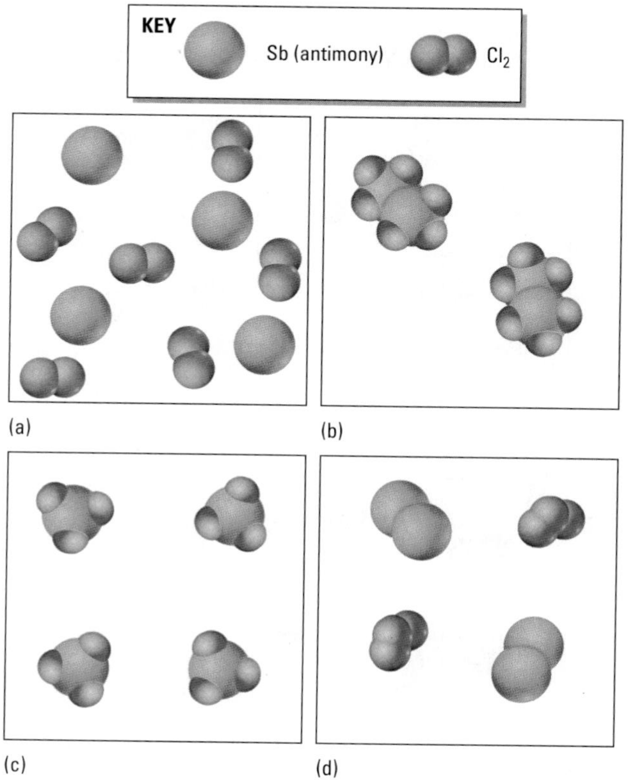
KEY
Sb (antimony)
Cl_2
(a)
(b)
(c)
(d)

20. The elements X (blue) and Y (pink) react by this equation

$$X_2 + 2\,Y_2 \longrightarrow 2\,XY_2$$

(a) Which drawing represents the reactants?
(b) Which drawing represents the products?

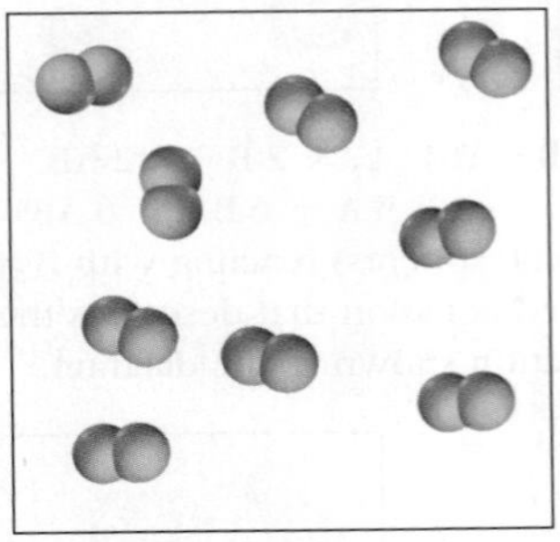

(a)

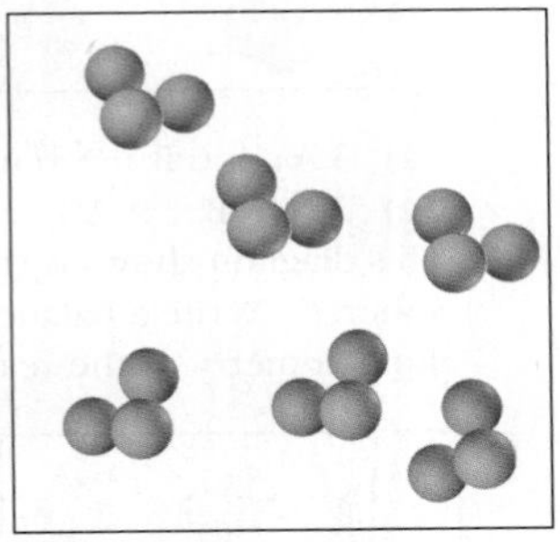

(b)

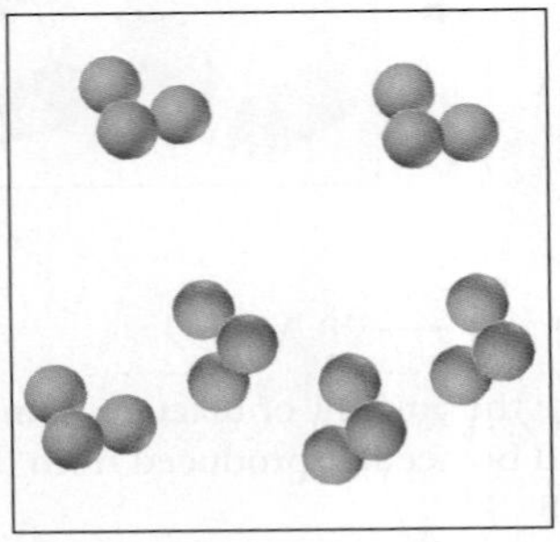

(c)

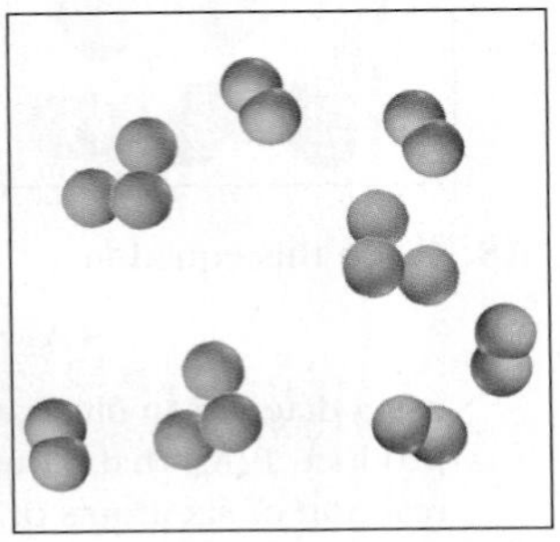

(d)

21. The elements X (blue) and Y (pink) react by this equation

$$2\,X_2 + Y_2 \longrightarrow 2\,X_2Y$$

(a) Which drawing represents the reactants?
(b) Which drawing represents the products?

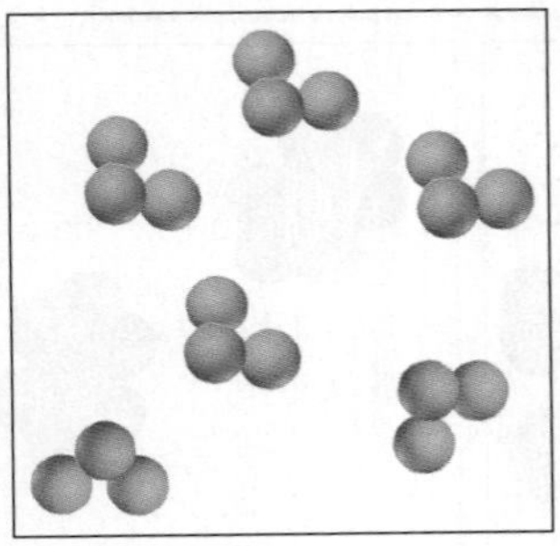

(a)

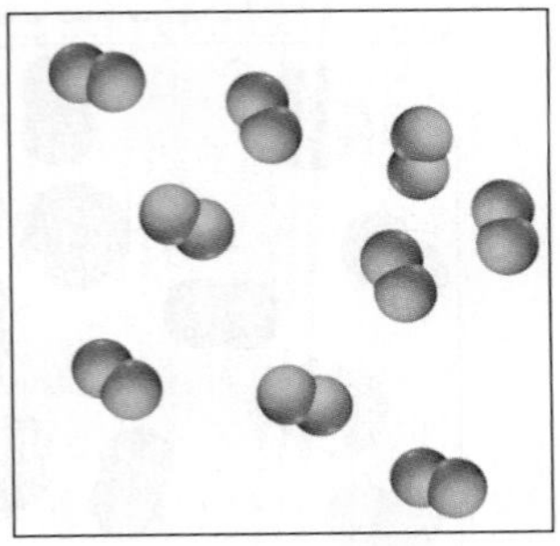

(b)

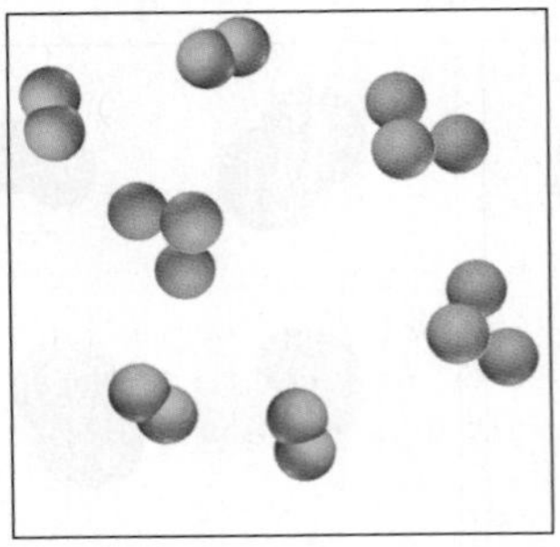

(c)

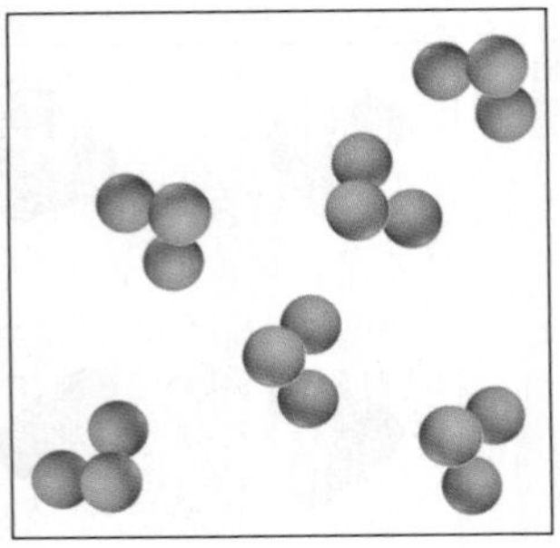

(d)

22. Write a balanced chemical equation that represents the reaction shown in the two drawings.

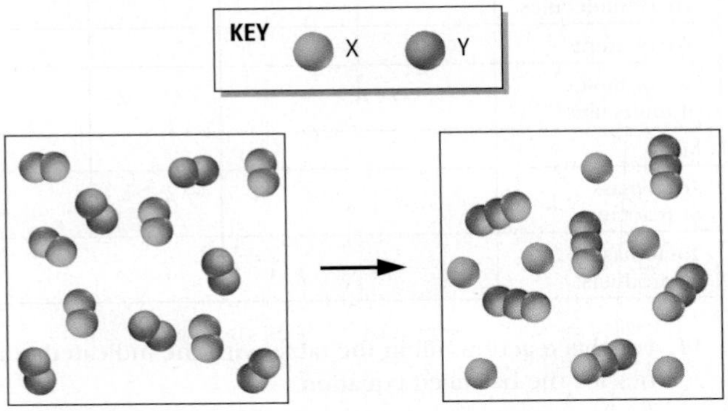

23. Write a balanced chemical equation that represents the reaction shown in the two drawings.

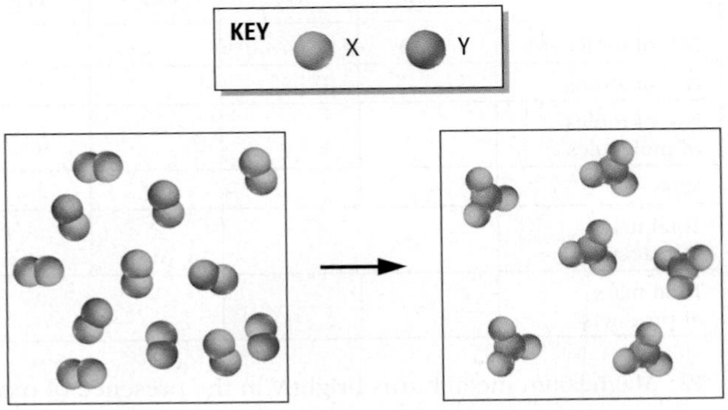

Classification of Chemical Reactions (Section 4.2)

24. Indicate whether each of these equations represents a combination, decomposition, displacement, or exchange reaction.
(a) $Cu(s) + O_2(g) \longrightarrow 2\,CuO(s)$
(b) $NH_4NO_3(s) \longrightarrow N_2O(g) + 2\,H_2O(\ell)$
(c) $AgNO_3(aq) + KCl(aq) \longrightarrow AgCl(s) + KNO_3(aq)$
(d) $Mg(s) + 2\,HCl(aq) \longrightarrow MgCl_2(aq) + H_2(g)$

25. Indicate whether each of these equations represents a combination, decomposition, displacement, or exchange reaction.
(a) $C(s) + O_2(g) \longrightarrow CO_2(g)$
(b) $2\,KClO_3(s) \longrightarrow 2\,KCl(s) + 3\,O_2(g)$
(c) $BaCl_2(aq) + K_2SO_4(aq) \longrightarrow BaSO_4(s) + 2\,KCl(aq)$
(d) $Mg(s) + CoSO_4(aq) \longrightarrow MgSO_4(aq) + Co(s)$

26. Indicate whether each of these equations represents a combination, decomposition, displacement, or exchange reaction.
(a) $PbCO_3(s) \longrightarrow PbO(s) + CO_2(g)$
(b) $Cu(s) + 4\,HNO_3(aq) \longrightarrow Cu(NO_3)_2(aq) + 2\,H_2O(\ell) + 2\,NO_2(g)$
(c) $2\,Zn(s) + O_2(g) \longrightarrow 2\,ZnO(s)$
(d) $Pb(NO_3)_2(aq) + 2\,KI(aq) \longrightarrow PbI_2(s) + 2\,KNO_3(aq)$

27. Indicate whether each of these equations represents a combination, decomposition, displacement, or exchange reaction.
(a) $Mg(s) + FeCl_2(aq) \longrightarrow MgCl_2(aq) + Fe(s)$
(b) $ZnCO_3(s) \longrightarrow ZnO(s) + CO_2(g)$
(c) $2\,C(s) + O_2(g) \longrightarrow 2\,CO(g)$
(d) $CaCl_2(aq) + Na_2CO_3(aq) \longrightarrow CaCO_3(s) + 2\,NaCl(aq)$

Balancing Equations (Section 4.3)

28. Write a balanced equation for each of these combustion reactions.
    (a) $C_4H_{10}(g) + O_2(g) \longrightarrow$
    (b) $C_6H_{12}O_6(s) + O_2(g) \longrightarrow$
    (c) $C_4H_8O(\ell) + O_2(g) \longrightarrow$
29. Write a balanced equation for each of these combustion reactions.
    (a) $C_3H_8O(g) + O_2(g) \longrightarrow$
    (b) $C_5H_{12}(\ell) + O_2(g) \longrightarrow$
    (c) $C_{12}H_{22}O_{11}(s) + O_2(g) \longrightarrow$
30. Complete and balance these equations involving oxygen reacting with an element. Name the product in each case.
    (a) $Mg(s) + O_2(g) \longrightarrow$
    (b) $Ca(s) + O_2(g) \longrightarrow$
    (c) $In(s) + O_2(g) \longrightarrow$
31. Complete and balance these equations involving oxygen reacting with an element.
    (a) $Ti(s) + O_2(g) \longrightarrow$ titanium(IV) oxide
    (b) $S_8(s) + O_2(g) \longrightarrow$ sulfur dioxide
    (c) $Se(s) + O_2(g) \longrightarrow$ selenium dioxide
32. Complete and balance these equations involving the reaction of a halogen with a metal. Name the product in each case.
    (a) $K(s) + Cl_2(g) \longrightarrow$
    (b) $Mg(s) + Br_2(\ell) \longrightarrow$
    (c) $Al(s) + F_2(g) \longrightarrow$
33. Complete and balance these equations involving the reaction of a halogen with a metal.
    (a) $Cr(s) + Cl_2(g) \longrightarrow$ chromium(III) chloride
    (b) $Cu(s) + Br_2(\ell) \longrightarrow$ copper(II) bromide
    (c) $Pt(s) + F_2(g) \longrightarrow$ platinum(IV) fluoride
34. Balance these equations.
    (a) $Al(s) + O_2(g) \longrightarrow Al_2O_3(s)$
    (b) $N_2(g) + H_2(g) \longrightarrow NH_3(g)$
    (c) $C_6H_6(\ell) + O_2(g) \longrightarrow H_2O(\ell) + CO_2(g)$
35. Balance these equations.
    (a) $Fe(s) + Cl_2(g) \longrightarrow FeCl_3(s)$
    (b) $SiO_2(s) + C(s) \longrightarrow Si(s) + CO(g)$
    (c) $Fe(s) + H_2O(g) \longrightarrow Fe_3O_4(s) + H_2(g)$
36. Balance these equations.
    (a) $UO_2(s) + HF(\ell) \longrightarrow UF_4(s) + H_2O(\ell)$
    (b) $B_2O_3(s) + HF(\ell) \longrightarrow BF_3(g) + H_2O(\ell)$
    (c) $BF_3(g) + H_2O(\ell) \longrightarrow HF(\ell) + H_3BO_3(s)$
37. Balance these equations.
    (a) $MgO(s) + Fe(s) \longrightarrow Fe_2O_3(s) + Mg(s)$
    (b) $H_3BO_3(s) \longrightarrow B_2O_3(s) + H_2O(\ell)$
    (c) $NaNO_3(s) + H_2SO_4(aq) \longrightarrow Na_2SO_4(aq) + HNO_3(g)$
38. Balance these equations.
    (a) Reaction to produce hydrazine, N_2H_4:

    $$H_2NCl(aq) + NH_3(g) \longrightarrow NH_4Cl(aq) + N_2H_4(aq)$$

    (b) Reaction of the fuels (dimethylhydrazine and dinitrogen tetraoxide) used in the Moon Lander and Space Shuttle:

    $$(CH_3)_2N_2H_2(\ell) + N_2O_4(g) \longrightarrow N_2(g) + H_2O(g) + CO_2(g)$$

    (c) Reaction of calcium carbide with water to produce acetylene, C_2H_2:

    $$CaC_2(s) + H_2O(\ell) \longrightarrow Ca(OH)_2(s) + C_2H_2(g)$$

39. Balance these equations.
    (a) Reaction of calcium cyanamide to produce ammonia:

    $$CaNCN(s) + H_2O(\ell) \longrightarrow CaCO_3(s) + NH_3(g)$$

    (b) Reaction to produce diborane, B_2H_6:

    $$NaBH_4(s) + H_2SO_4(aq) \longrightarrow B_2H_6(g) + H_2(g) + Na_2SO_4(aq)$$

    (c) Reaction to rid water of hydrogen sulfide, H_2S, a foul-smelling compound:

    $$H_2S(aq) + Cl_2(aq) \longrightarrow S_8(s) + HCl(aq)$$

40. Balance these combustion reactions.
    (a) $C_6H_{12}O_6 + O_2 \longrightarrow CO_2 + H_2O$
    (b) $C_5H_{12} + O_2 \longrightarrow CO_2 + H_2O$
    (c) $C_7H_{14}O_2 + O_2 \longrightarrow CO_2 + H_2O$
    (d) $C_2H_4O_2 + O_2 \longrightarrow CO_2 + H_2O$
41. Balance these equations.
    (a) $Mg + HNO_3 \longrightarrow H_2 + Mg(NO_3)_2$
    (b) $Al + Fe_2O_3 \longrightarrow Al_2O_3 + Fe$
    (c) $S_8 + O_2 \longrightarrow SO_3$
    (d) $SO_3 + H_2O \longrightarrow H_2SO_4$

The Mole and Chemical Reactions (Section 4.4)

42. Chlorine can be produced in the laboratory by the reaction of hydrochloric acid with excess manganese(IV) oxide.

    $$4\ HCl(aq) + MnO_2(s) \longrightarrow Cl_2(g) + 2\ H_2O(\ell) + MnCl_2(aq)$$

    How many moles of HCl are needed to form 12.5 mol Cl_2?
43. Methane, CH_4, is the major component of natural gas. How many moles of oxygen are needed to burn 16.5 mol CH_4?

    $$CH_4(g) + 2\ O_2(g) \longrightarrow CO_2(g) + 2\ H_2O(\ell)$$

44. In the laboratory, the salt potassium chlorate, $KClO_3$, can be decomposed thermally to generate small amounts of oxygen gas.

    $$2\ KClO_3(s) \longrightarrow 2\ KCl(s) + 3\ O_2(g)$$

    How many grams of potassium chlorate must be decomposed to produce 5.00 g O_2?
45. An ingredient in many baking recipes is baking soda, $NaHCO_3$, which decomposes when heated to produce carbon dioxide, causing the baked goods to rise.

    $$2\ NaHCO_3(s) \longrightarrow Na_2CO_3(s) + CO_2(g) + H_2O(g)$$

    How many grams of carbon dioxide are produced per gram of baking soda?
46. Nitrogen monoxide is oxidized in air to give brown nitrogen dioxide.

    $$2\ NO(g) + O_2(g) \longrightarrow 2\ NO_2(g)$$

    Starting with 2.2 mol NO, how many moles and how many grams of O_2 are required for complete reaction? What mass of NO_2, in grams, is produced?
47. Aluminum reacts with oxygen to give aluminum oxide.

    $$4\ Al(s) + 3\ O_2(g) \longrightarrow 2\ Al_2O_3(s)$$

    If you have 6.0 mol Al, how many moles and how many grams of O_2 are needed for complete reaction? What mass of Al_2O_3, in grams, is produced?

48. Many metals react with halogens to give metal halides. For example, iron reacts with chlorine to give iron(II) chloride, $FeCl_2$.

$$Fe(s) + Cl_2(g) \longrightarrow FeCl_2(s)$$

(a) Beginning with 10.0 g iron, what mass of Cl_2, in grams, is required for complete reaction?
(b) What quantity of $FeCl_2$, in moles and in grams, is expected?

49. Like many metals, manganese reacts with a halogen to give a metal halide.

$$2\ Mn(s) + 3\ F_2(g) \longrightarrow 2\ MnF_3(s)$$

(a) If you begin with 5.12 g Mn, what mass in grams of F_2 is required for complete reaction?
(b) What quantity in moles and in grams of the red solid MnF_3 is expected?

50. The final step in the manufacture of platinum metal (for use in automotive catalytic converters and other products) is the reaction

$$3\ (NH_4)_2PtCl_6(s) \longrightarrow 3\ Pt(s) + 2\ NH_4Cl(s) + 2\ N_2(g) + 16\ HCl(g)$$

Complete this table of reaction quantities for the reaction of 12.35 g $(NH_4)_2PtCl_6$.

$(NH_4)_2PtCl_6$	Pt	HCl
12.35 g	________ g	________ g
________ mol	________ mol	________ mol

51. Disulfur dichloride, S_2Cl_2, is used to vulcanize rubber. It can be made by treating molten sulfur with gaseous chlorine.

$$S_8(\ell) + 4\ Cl_2(g) \longrightarrow 4\ S_2Cl_2(g)$$

Complete this table of reaction quantities for the production of 103.5 g S_2Cl_2.

S_8	Cl_2	S_2Cl_2
________ g	________ g	103.5 g
________ mol	________ mol	________ mol

52. Many metal halides react with water to produce the metal oxide (or hydroxide) and the appropriate hydrogen halide. For example,

$$TiCl_4(\ell) + 2\ H_2O(g) \longrightarrow TiO_2(s) + 4\ HCl(g)$$

(a) If you begin with 14.0 g $TiCl_4$, how many moles of water are required for complete reaction?
(b) How many grams of each product are expected?

Liquid titanium tetrachloride, $TiCl_4$. When exposed to moist air, the reaction forms a dense fog of titanium(IV) oxide, TiO_2.

53. Gaseous sulfur dioxide, SO_2, can be removed from smokestacks by treatment with limestone and oxygen.

$$2\ SO_2(g) + 2\ CaCO_3(s) + O_2(g) \longrightarrow 2\ CaSO_4(s) + 2\ CO_2(g)$$

(a) How many moles each of $CaCO_3$ and O_2 are required to remove 150. g SO_2?
(b) What mass of $CaSO_4$ is formed when 150. g SO_2 is consumed completely?

54. If 2.5 mol O_2 reacts with propane, C_3H_8, by combustion, how many moles of H_2O will be produced? How many grams of H_2O will be produced?

55. Tungsten(VI) oxide can be reduced to tungsten metal.

$$WO_3(s) + 3\ H_2(g) \longrightarrow W(s) + 3\ H_2O(\ell)$$

How many grams of tungsten are formed from 1.00 kg WO_3?

56. If you want to synthesize 1.45 g of the semiconducting material GaAs, what masses of Ga and of As, in grams, are required?

57. Ammonium nitrate, NH_4NO_3, is a common fertilizer and explosive. When heated, it decomposes into gaseous products.

$$2\ NH_4NO_3(s) \longrightarrow 2\ N_2(g) + 4\ H_2O(g) + O_2(g)$$

How many grams of each product are formed from 1.0 kg NH_4NO_3?

58. Iron reacts with oxygen to give iron(III) oxide, Fe_2O_3.
(a) Write a balanced equation for this reaction.
(b) If an ordinary iron nail (assumed to be pure iron) has a mass of 5.58 g, what mass in grams of Fe_2O_3 would be produced if the nail is converted completely to this oxide?
(c) What mass of O_2 (in grams) is required for the reaction?

59. Nitroglycerin decomposes violently according to the equation

$$4\ C_3H_5(NO_3)_3(\ell) \longrightarrow 12\ CO_2(g) + 10\ H_2O(\ell) + 6\ N_2(g) + O_2(g)$$

How many grams of each gaseous product are produced from 1.00 g nitroglycerin?

60. Chlorinated fluorocarbons, such as CCl_2F_2, have been banned from use in automobile air conditioners because the compounds are destructive to the stratospheric ozone layer. Researchers at MIT have found an environmentally safe way to decompose these compounds by treating them with sodium oxalate, $Na_2C_2O_4$. The products of the reaction are carbon, carbon dioxide, sodium chloride, and sodium fluoride.
(a) Write a balanced equation for this reaction of CCl_2F_2.
(b) What mass of $Na_2C_2O_4$ is needed to remove 76.8 g CCl_2F_2?
(c) What mass of CO_2 is produced?

61. Careful decomposition of ammonium nitrate, NH_4NO_3, gives laughing gas (dinitrogen monoxide, N_2O) and water.
(a) Write a balanced equation for this reaction.
(b) Beginning with 10.0 g NH_4NO_3, what masses of N_2O and water are expected?

62. In making iron from iron ore, this reaction occurs.

$$Fe_2O_3(s) + 3\ CO(g) \longrightarrow 2\ Fe(s) + 3\ CO_2(g)$$

(a) How many grams of iron can be obtained from 1.00 kg iron(III) oxide?
(b) How many grams of CO are required?

63. Cisplatin, $Pt(NH_3)_2Cl_2$, a drug used in the treatment of cancer, can be made by the reaction of K_2PtCl_4 with ammonia, NH_3. Besides cisplatin, the other product is KCl.
(a) Write a balanced equation for this reaction.
(b) To obtain 2.50 g cisplatin, what masses in grams of K_2PtCl_4 and ammonia do you need?

Limiting Reactant (Section 4.5)

64. The reaction of Na_2SO_4 with $BaCl_2$ is

$$Na_2SO_4(aq) + BaCl_2(aq) \longrightarrow BaSO_4(s) + 2\ NaCl(aq)$$

If solutions containing exactly one gram of each reactant are mixed, which reactant is the limiting reactant, and how many grams of $BaSO_4$ are produced?

65. If a mixture of 100. g Al and 200. g MnO is reacted according to the reaction

$$2\ Al(s) + 3\ MnO(s) \longrightarrow Al_2O_3(s) + 3\ Mn(s)$$

which reactant is in excess and how many grams of it remain when the reaction is complete?

66. Aluminum chloride, Al_2Cl_6, is an inexpensive reagent used in many industrial processes. It is made by treating scrap aluminum with chlorine according to the balanced equation

$$2\ Al(s) + 3\ Cl_2(g) \longrightarrow Al_2Cl_6(s)$$

(a) Which reactant is limiting if 2.70 g Al and 4.05 g Cl_2 are mixed?
(b) What mass of Al_2Cl_6 can be produced?
(c) What mass of the excess reactant will remain when the reaction is complete?

67. Hydrogen and oxygen react to form water by combustion.

$$2\ H_2(g) + O_2(g) \longrightarrow 2\ H_2O(\ell)$$

(a) If a mixture containing 100. g of each reactant is ignited, what is the limiting reactant?
(b) How many moles and grams of water are produced?

68. Methanol, CH_3OH, is a clean-burning, easily handled fuel. It can be made by the direct reaction of CO and H_2.

$$CO(g) + 2\ H_2(g) \longrightarrow CH_3OH(\ell)$$

(a) Starting with a mixture of 12.0 g H_2 and 74.5 g CO, which is the limiting reactant?
(b) What mass of the excess reactant, in grams, is left after reaction is complete?
(c) What mass of methanol can be obtained, in theory?

69. The reaction of methane and water is one way to prepare hydrogen.

$$CH_4(g) + 2\ H_2O(g) \longrightarrow CO_2(g) + 4\ H_2(g)$$

If 995 g CH_4 reacts with 2510 g water, how many moles of reactants and products are there when the reaction is complete?

70. Ammonia gas can be prepared by the reaction

$$CaO(s) + 2\ NH_4Cl(s) \longrightarrow 2\ NH_3(g) + H_2O(g) + CaCl_2(s)$$

If 112 g CaO reacts with 224 g NH_4Cl, how many moles of reactants and products are there when the reaction is complete?

71. This reaction between lithium hydroxide and carbon dioxide has been used to remove CO_2 from spacecraft atmospheres:

$$2\ LiOH + CO_2 \longrightarrow Li_2CO_3 + H_2O$$

(a) If 0.500 kg LiOH were available, how many grams of CO_2 could be consumed?
(b) How many grams of water would be produced?

72. The equation for one of the reactions in the process of turning iron ore into the metal is

$$Fe_2O_3(s) + 3\ CO(g) \longrightarrow 2\ Fe(s) + 3\ CO_2(g)$$

If you start with 2.00 kg of each reactant, what is the maximum mass of iron you can produce?

73. Aspirin is produced by the reaction of salicylic acid and acetic anhydride.

$$\underset{\text{salicylic acid}}{2\ C_7H_6O_3(s)} + \underset{\text{acetic anhydride}}{C_4H_6O_3(\ell)} \longrightarrow \underset{\text{aspirin}}{2\ C_9H_8O_4(s)} + H_2O(\ell)$$

If you mix 100. g of each of the reactants, what is the maximum mass of aspirin that can be obtained?

74. Consider the chemical reaction $2\ S + 3\ O_2 \longrightarrow 2\ SO_3$. If the reaction is run by adding S indefinitely to a fixed amount of O_2, which of these graphs best represents the formation of SO_3? Explain your choice.

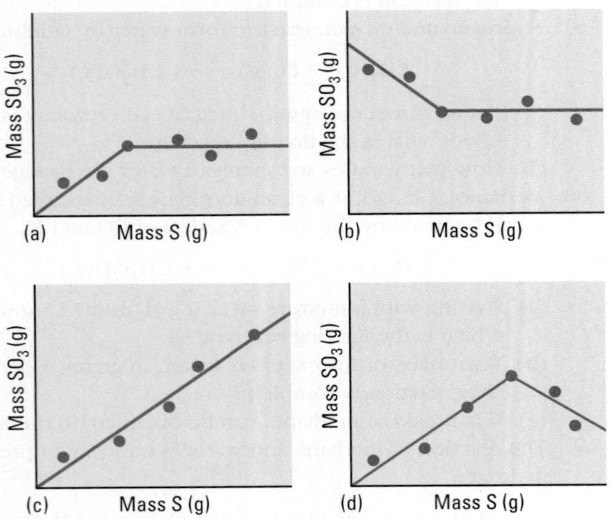

75. This reaction can be used to generate hydrogen gas from methane:

$$CH_4(g) + H_2O(g) \longrightarrow CO(g) + 3\ H_2(g)$$

If you use 500. g CH_4 and 1300. g water:
(a) Which reactant is the limiting reactant?
(b) How many grams H_2 can be produced?
(c) How many grams of the excess reactant remain when the reaction is complete?

Percent Yield (Section 4.6)

76. Iron oxide can be reduced to the metal as follows:

$$Fe_2O_3(s) + 3\ CO(g) \longrightarrow 2\ Fe(s) + 3\ CO_2(g)$$

How many grams of iron can be obtained from 1.00 kg of the iron oxide? If 654 g Fe was obtained from the reaction, what was the percent yield?

77. Aluminum bromide is a valuable laboratory chemical. What is the theoretical yield, in grams, of Al_2Br_6 if 25.0 mL liquid bromine (density = 3.12 g/mL) and excess aluminum metal are reacted?

$$2\ Al(s) + 3\ Br_2(\ell) \longrightarrow Al_2Br_6(s)$$

78. Ammonia gas can be prepared by the reaction of calcium oxide with ammonium chloride.

$$CaO(s) + 2\ NH_4Cl(s) \longrightarrow 2\ NH_3(g) + H_2O(g) + CaCl_2(s)$$

If exactly 100 g ammonia is isolated but the theoretical yield is 136 g, what is the percent yield of this gas?

79. Quicklime, CaO, is formed when calcium hydroxide is heated.

$$Ca(OH)_2(s) \longrightarrow CaO(s) + H_2O(\ell)$$

If the theoretical yield is 65.5 g but only 36.7 g quicklime is produced, what is the percent yield?

80. Diborane, B_2H_6, is valuable for the synthesis of new organic compounds. The boron compound can be made by the reaction

$$2\ NaBH_4(s) + I_2(s) \longrightarrow B_2H_6(g) + 2\ NaI(s) + H_2(g)$$

Suppose you use 1.203 g $NaBH_4$ and excess iodine, and you isolate 0.295 g B_2H_6. What is the percent yield of B_2H_6?

81. Methanol, CH_3OH, is used in racing cars because it is a clean-burning fuel. It can be made by this reaction:

$$CO(g) + 2\ H_2(g) \longrightarrow CH_3OH(\ell)$$

What is the percent yield if 5.0×10^3 g H_2 reacts with excess CO to form 3.5×10^3 g CH_3OH?

82. If 3.7 g sodium metal and 4.3 g chlorine gas react to form NaCl, what is the theoretical yield? If 5.5 g NaCl was formed, what is the percent yield?

83. Disulfur dichloride, which has a revolting smell, can be prepared by directly combining S_8 and Cl_2, but it can also be made by this reaction:

$$3\ SCl_2(\ell) + 4\ NaF(s) \longrightarrow SF_4(g) + S_2Cl_2(\ell) + 4\ NaCl(s)$$

What mass of SCl_2 is needed to react with excess NaF to prepare 1.19 g S_2Cl_2, if the expected yield is 51%?

84. The ceramic silicon nitride, Si_3N_4, is made by heating silicon and nitrogen at an elevated temperature.

$$3\ Si(s) + 2\ N_2(g) \longrightarrow Si_3N_4(s)$$

How many grams of silicon must combine with excess N_2 to produce 1.0 kg Si_3N_4 if this process is 92% efficient?

85. Disulfur dichloride can be prepared by

$$3\ SCl_2 + 4\ NaF \longrightarrow SF_4 + S_2Cl_2 + 4\ NaCl$$

What is the percent yield of the reaction if 5.00 g SCl_2 reacts with excess NaF to produce 1.262 g S_2Cl_2?

Empirical Formulas (Section 4.7)

86. What is the empirical formula of a compound that contains 60.0% oxygen and 40.0% sulfur by mass?

87. A potassium salt was analyzed to have this percent composition: 26.57% K, 35.36% Cr, and 38.07% O. What is its empirical formula?

88. Styrene, the building block of polystyrene, is a hydrocarbon. If 0.438 g of the compound is burned and produces 1.481 g CO_2 and 0.303 g H_2O, what is the empirical formula of the compound?

89. Mesitylene is a liquid hydrocarbon. If 0.115 g of the compound is burned in pure O_2 to give 0.379 g CO_2 and 0.1035 g H_2O, what is the empirical formula of the compound?

90. Propionic acid, an organic acid, contains only C, H, and O. If 0.236 g of the acid burns completely in O_2 and gives 0.421 g CO_2 and 0.172 g H_2O, what is the empirical formula of the acid?

91. Quinone, which is used in the dye industry and in photography, is an organic compound containing only C, H, and O. What is the empirical formula of the compound if 0.105 g of the compound gives 0.257 g CO_2 and 0.0350 g H_2O when burned completely?

92. Combustion of a 0.2500-g sample of a compound containing only C, H, and O shows that the compound contains 0.1614 g C, 0.02715 g H, and 0.06145 g O. What is the empirical formula of the compound?

93. Write the balanced chemical equation for the complete combustion of adipic acid, an organic acid containing 49.31% C, 6.90% H, and the remainder O, by mass.

94. L-Dopa is a drug used for the treatment of Parkinson's disease. Elemental analysis shows it to be 54.82% carbon, 7.10% nitrogen, 32.46% oxygen, and the remainder hydrogen.
   (a) What is L-dopa's empirical formula?
   (b) The molar mass of L-dopa is 197.19 g/mol; what is its molecular formula?

General Questions

These questions are not explicitly keyed to chapter topics; many require integration of several concepts.

95. Nitrogen gas can be prepared in the laboratory by the reaction of ammonia with copper(II) oxide according to this unbalanced equation:

$$NH_3(g) + CuO(s) \longrightarrow N_2(g) + Cu(s) + H_2O(g)$$

If 26.3 g of gaseous NH_3 is passed over a bed of solid CuO in stoichiometric excess, what mass, in grams, of N_2 can be isolated?

96. The overall chemical equation for the photosynthesis reaction in green plants is

$$6\ CO_2(g) + 6\ H_2O(\ell) \longrightarrow C_6H_{12}O_6(aq) + 6\ O_2(g)$$

How many grams of oxygen are produced by a plant when 50.0 g CO_2 is consumed?

97. In an experiment, 1.056 g of a metal carbonate containing an unknown metal M was heated to give the metal oxide and 0.376 g CO_2.

$$MCO_3(s) \xrightarrow{\text{heat}} MO(s) + CO_2(g)$$

What is the identity of the metal M?
   (a) Ni   (b) Cu
   (c) Zn   (d) Ba

98. Uranium(VI) oxide reacts with bromine trifluoride to give uranium(IV) fluoride, an important step in the purification of uranium ore.

$$6\ UO_3(s) + 8\ BrF_3(\ell) \longrightarrow 6\ UF_4(s) + 4\ Br_2(\ell) + 9\ O_2(g)$$

If you begin with 365 g each of UO_3 and BrF_3, what is the maximum yield, in grams, of UF_4?

99. The cancer chemotherapy agent cisplatin is made by the reaction

$$(NH_4)_2PtCl_4(s) + 2\ NH_3(aq) \longrightarrow 2\ NH_4Cl(aq) + Pt(NH_3)_2Cl_2(s)$$

Assume that 15.5 g $(NH_4)_2PtCl_4$ is combined with 0.15 mol aqueous NH_3 to make cisplatin. What is the theoretical mass, in grams, of cisplatin that can be formed?

100. Diborane, B_2H_6, can be produced by the reaction

$$2\ NaBH_4(aq) + H_2SO_4(aq) \longrightarrow 2\ H_2(g) + Na_2SO_4(aq) + B_2H_6(g)$$

What is the maximum yield, in grams, of B_2H_6 that can be prepared starting with 2.19×10^{-2} mol H_2SO_4 and 1.55 g $NaBH_4$?

101. Silicon and hydrogen form a series of interesting compounds, Si_xH_y. To find the formula of one of them, a 6.22-g sample of the compound is burned in oxygen. All of the Si is converted to 11.64 g SiO_2 and all of the H to 6.980 g H_2O. What is the empirical formula of the silicon compound?

102. Boron forms an extensive series of compounds with hydrogen, all with the general formula B_xH_y. To analyze one of these compounds, you burn it in air and isolate the boron in the form of B_2O_3 and the hydrogen in the form of water. If 0.148 g B_xH_y gives 0.422 g B_2O_3 when burned in excess O_2, what is the empirical formula of B_xH_y?

103. What is the limiting reactant for the reaction

$$4\ KOH + 2\ MnO_2 + O_2 + Cl_2 \longrightarrow 2\ KMnO_4 + 2\ KCl + 2\ H_2O$$

if 5 mol of each reactant is present? What is the limiting reactant when 5 g of each reactant is present?

104. The Hargraves process is an industrial method for making sodium sulfate for use in papermaking.

$$4\ NaCl + 2\ SO_2 + 2\ H_2O + O_2 \longrightarrow 2\ Na_2SO_4 + 4\ HCl$$

   (a) If you start with 10. mol of each reactant, which one will determine the amount of Na_2SO_4 produced?
   (b) What if you start with 100. g of each reactant?

Applying Concepts

These questions test conceptual learning.

105. Chemical equations can be interpreted on either a nanoscale level (atoms, molecules, ions) or a mole level (moles of reactants and products). Write word statements to describe the combustion of butane on a nanoscale level and a mole level.

$$2\ C_4H_{10}(g) + 13\ O_2(g) \longrightarrow 8\ CO_2(g) + 10\ H_2O(\ell)$$

106. Write word statements to describe this reaction on a nanoscale level and a mole level.

$$P_4(s) + 6\ Cl_2(g) \longrightarrow 4\ PCl_3(\ell)$$

107. What is the single product of this hypothetical reaction?

$$4\ A_2 + AB_3 \longrightarrow 3\_\_\_\_\_\_\_\_$$

108. What is the single product of this hypothetical reaction?

$$3\ A_2B_3 + B_3 \longrightarrow 6\_\_\_\_\_\_\_\_$$

109. If 1.5 mol Cu reacts with a solution containing 4.0 mol $AgNO_3$, what ions will be present in the solution at the end of the reaction?

$$Cu(s) + 2\ AgNO_3(aq) \longrightarrow Cu(NO_3)_2(aq) + 2\ Ag(s)$$

110. Ammonia can be formed by a direct reaction of nitrogen and hydrogen.

$$N_2(g) + 3\,H_2(g) \longrightarrow 2\,NH_3(g)$$

A tiny portion of the starting mixture is represented by the diagram below, where the blue circles represent N and the white circles represent H.

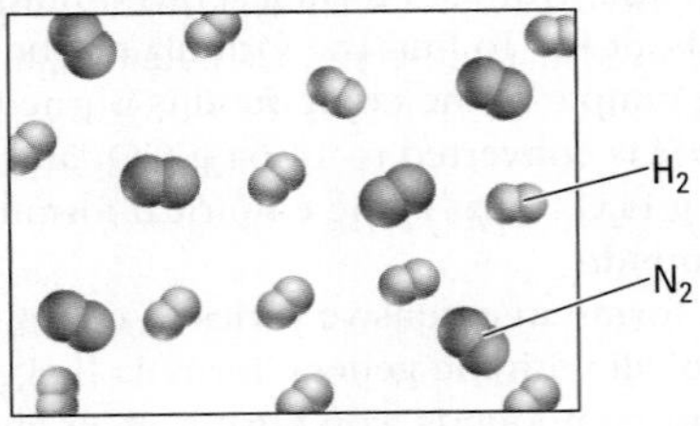

Which of these represents the product mixture?

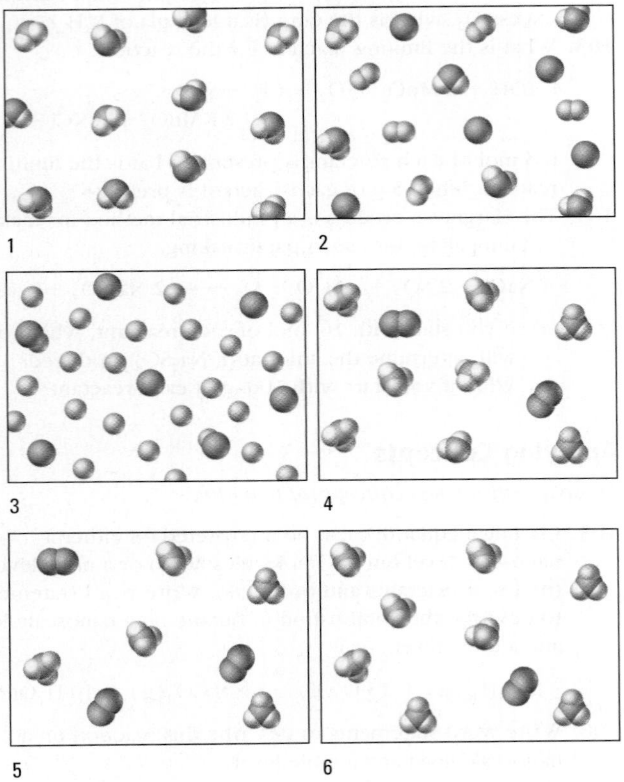

For the reaction of the given sample, which of these statements is true?

(a) N_2 is the limiting reactant.
(b) H_2 is the limiting reactant.
(c) NH_3 is the limiting reactant.
(d) No reactant is limiting; they are present in the correct stoichiometric ratio.

111. Carbon monoxide burns readily in oxygen to form carbon dioxide.

$$2\,CO(g) + O_2(g) \longrightarrow 2\,CO_2(g)$$

The box on the left represents a tiny portion of a mixture of CO and O_2. If these molecules react to form CO_2, what should the contents of the box on the right look like?

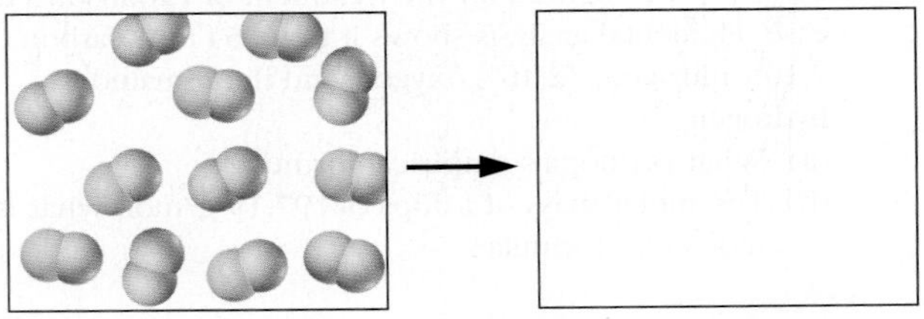

112. Which chemical equation best represents the reaction taking place in the illustration below?
(a) $X_2 + Y_2 \longrightarrow n\,XY_3$
(b) $X_2 + 3\,Y_2 \longrightarrow 2\,XY_3$
(c) $6\,X_2 + 6\,Y_2 \longrightarrow 4\,XY_3 + 4\,X_2$
(d) $6\,X_2 + 6\,Y_2 \longrightarrow 4\,X_3Y + 4\,Y_2$

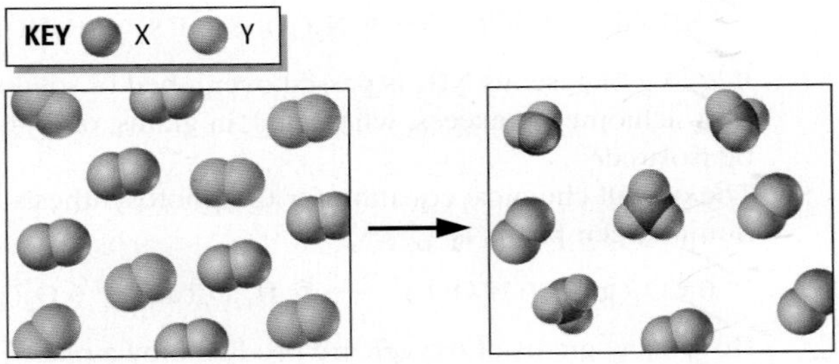

113. Write a balanced chemical equation that represents the reaction shown in the two drawings below.

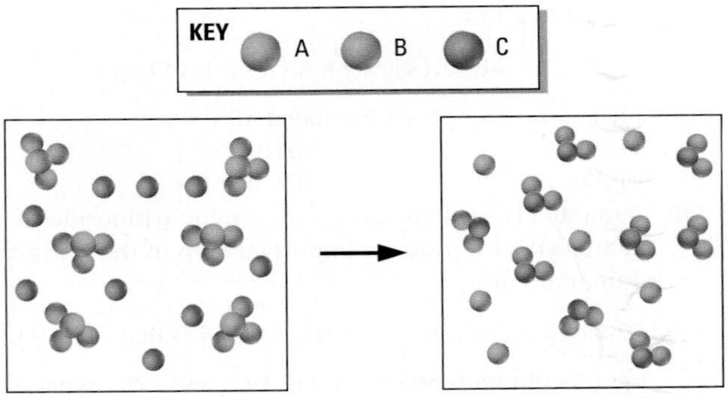

114. Write a balanced chemical equation that represents the reaction shown in the two drawings.

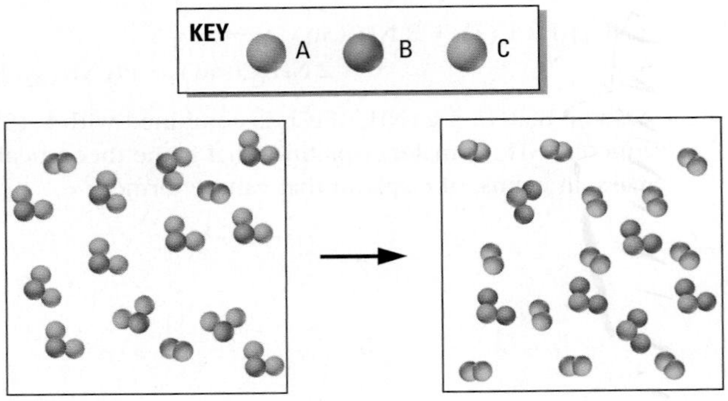

115. A student set up an experiment, like the one described in Chemistry You Can Do (p. 143), for six different trials between acetic acid, CH_3COOH, and sodium hydrogen carbonate, $NaHCO_3$.

$$CH_3COOH(aq) + NaHCO_3(s) \longrightarrow NaCH_3CO_2(aq) + CO_2(g) + H_2O(\ell)$$

The volume of acetic acid is kept constant, but the mass of sodium bicarbonate increased with each trial. The results of the tests are shown in the figure.

(a) In which trial(s) is the acetic acid the limiting reactant?

(b) In which trial(s) is sodium bicarbonate the limiting reactant?

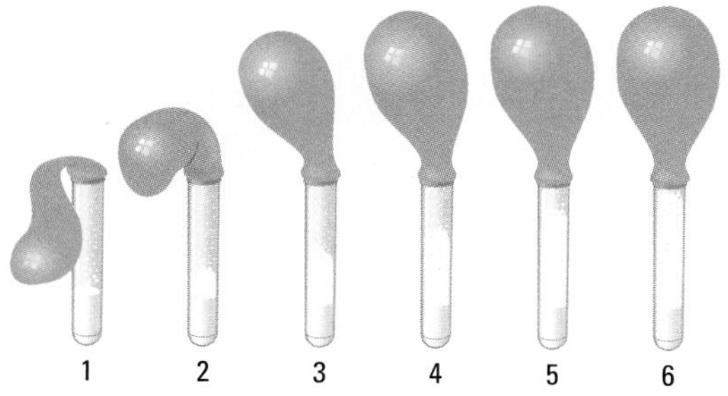

116. A weighed sample of a metal is added to liquid bromine and allowed to react completely. The product substance is then separated from any leftover reactants and weighed. This experiment is repeated with several masses of the metal but with the same volume of bromine. This graph indicates the results. Explain why the graph has the shape that it does.

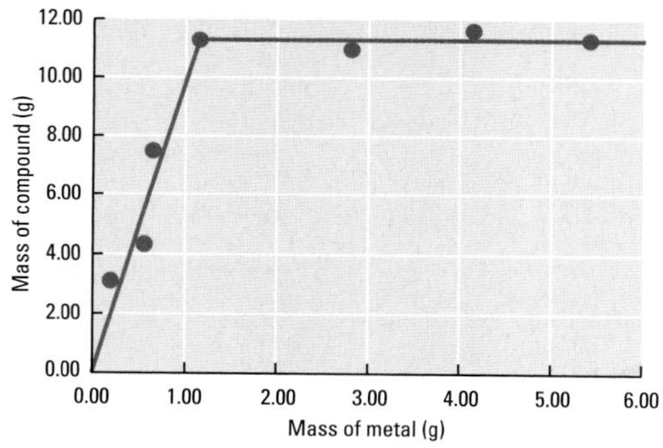

117. A series of experimental measurements like the ones described in Question 116 is carried out for iron reacting with bromine. This graph is obtained. What is the empirical formula of the compound formed by iron and bromine? Write a balanced equation for the reaction between iron and bromine. Name the product.

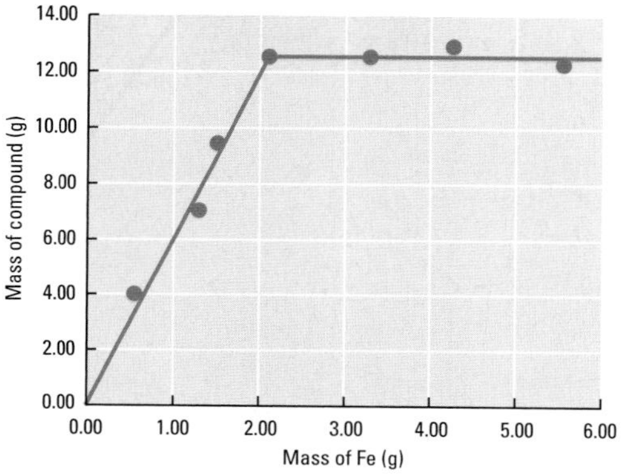

More Challenging Questions

These questions require more thought and integrate several concepts.

118. Hydrogen gas $H_2(g)$ is reacted with a sample of $Fe_2O_3(s)$ at 400 °C. Two products are formed: water vapor and a black solid compound that is 72.3% Fe and 27.7% O by mass. Write the balanced chemical equation for the reaction.

119. Write the balanced chemical equation for the complete combustion of malonic acid, an organic acid containing 34.62% C, 3.88% H, and the remainder O, by mass.

120. In a reaction, 1.2 g element A reacts with exactly 3.2 g oxygen to form an oxide, AO_x; 2.4 g element A reacts with exactly 3.2 g oxygen to form a second oxide, AO_y.

(a) What is the ratio x/y?

(b) If $x = 2$, what might be the identity of element A?

121. A copper ore contained Cu_2S and CuS plus 10% inert impurities. When 200.0 g of the ore was "roasted," it yielded 150.8 g of 90.0% pure copper and sulfur dioxide gas. What is the percentage of Cu_2S in the ore?

$$Cu_2S + O_2 \longrightarrow 2\ Cu + SO_2 \qquad CuS + O_2 \longrightarrow Cu + SO_2$$

122. A sample of a compound with the formula X_2S_3 has a mass of 10.00 g. It is then roasted (reacted with oxygen) to convert it to X_2O_3. After roasting, it weighs 7.410 g. What is the atomic mass of element X?

123. A metal carbonate decomposed to form its metal oxide and CO_2 when it was heated:

$$MCO_3(s) \longrightarrow MO(s) + CO_2(g) \qquad \text{(balanced)}$$

After the reaction, the metal oxide was found to have a mass 56.0% as large as the starting MCO_3. What metal was in the carbonate?

124. When solutions of silver nitrate and sodium carbonate are mixed, solid silver carbonate is formed and sodium nitrate remains in solution. If a solution containing 12.43 g sodium carbonate is mixed with a solution containing 8.37 g silver nitrate, how many grams of the four species are present after the reaction is complete?

125. This reaction produces sulfuric acid:

$$2\ SO_2 + O_2 + 2\ H_2O \longrightarrow 2\ H_2SO_4$$

If 200. g SO_2, 85 g O_2, and 66 g H_2O are mixed and the reaction proceeds to completion, which reactant is limiting, how many grams of H_2SO_4 are produced, and how many grams of the other two reactants are left over?

126. You have an organic liquid that contains either ethyl alcohol, C_2H_5OH, or methyl alcohol, CH_3OH, or both. You burned a sample of the liquid weighing 0.280 g to form 0.385 g $CO_2(g)$. What was the composition of the sample of liquid?

127. Nickel metal reacts with aqueous silver nitrate in a displacement reaction to produce silver metal and aqueous nickel nitrate. Consider an experiment in which the reaction starts with 12.0 g nickel metal and stops before all the nickel reacts. A total of 24.0 g of metal is present when the reaction stops. Calculate how many grams of each metal are present in the 24.0-g mixture of metals.

Conceptual Challenge Problems

These rigorous, thought-provoking problems integrate conceptual learning with problem solving and are suitable for group work.

CP4.A (Section 4.4) In Example 4.7 it was not possible to find the mass of O_2 directly from a knowledge of the mass of sucrose. Are there chemical reactions in which the mass of a product or another reactant can be known directly if you know the mass of a reactant? Cite a few of these reactions.

CP4.B (Section 4.4) Glucose, $C_6H_{12}O_6$, a monosaccharide, and sucrose, $C_{12}H_{22}O_{11}$, a disaccharide, undergo complete combustion with O_2 (metabolic conversion) to produce H_2O and CO_2.

(a) How many moles of O_2 are needed per mole of each sugar for the reaction to proceed?

(b) How many grams of O_2 are needed per mole of each sugar for the reaction to proceed?

(c) Which combustion reaction produces more H_2O per gram of sugar? How many grams of H_2O are produced per gram of each sugar?

Chemical Reactions

5

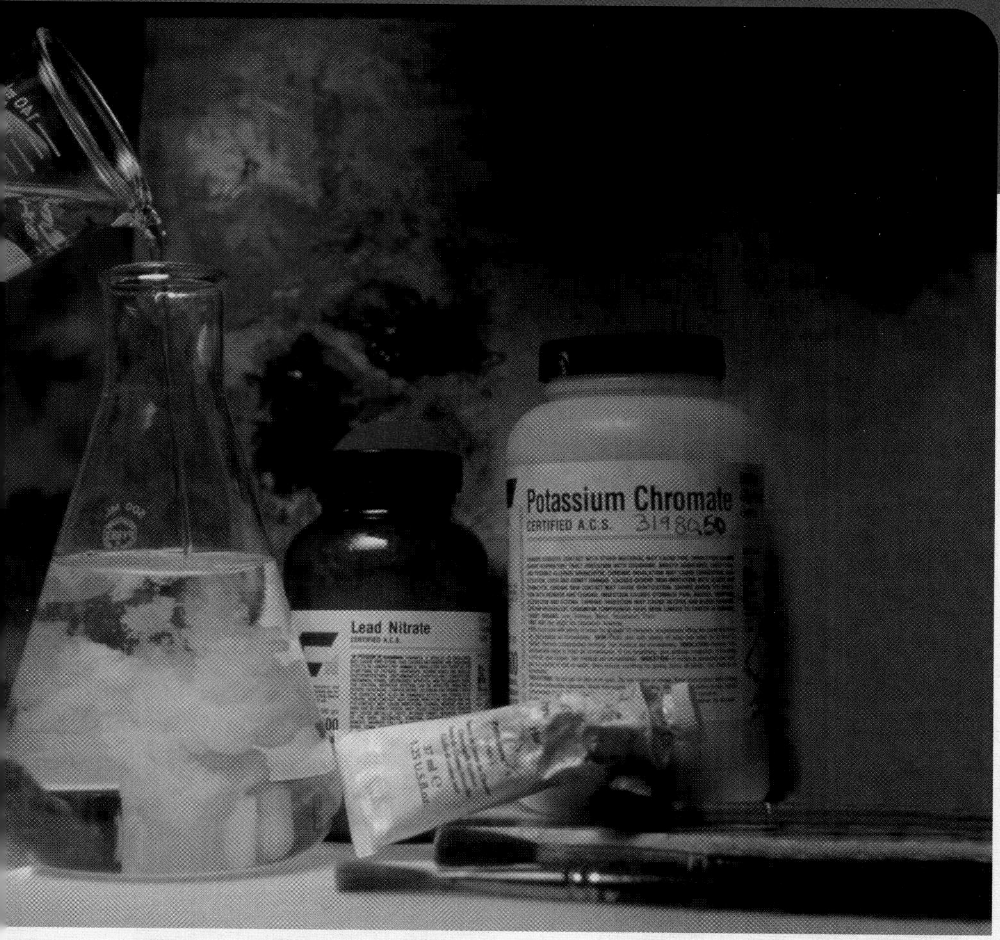

© Cengage Learning/Charles D. Winters

The brilliant yellow precipitate lead(II) chromate, $Pb(CrO)_4$, is formed when lead(II) ions, Pb^{2+}, and chromate ions, CrO_4^{2-}, react in an aqueous solution. The precipitation reaction occurs because the lead(II) chromate product is insoluble. Although the color of lead(II) chromate is so striking that it is used as a pigment in yellow chrome paint, lead and chromate ions are both toxic, and the paint must be handled carefully.

Chemistry is concerned with how substances react and what products are formed when they react. A chemical compound can consist of molecules or oppositely charged ions, and often the compound's properties can be deduced from the behavior of these molecules or ions. The *chemical* properties of a compound are the transformations that the molecules or ions can undergo when the substance reacts. A central focus of chemistry is providing answers to questions such as these: When two

OWL Sign in to OWL at **www.cengage.com/owl** to view tutorials and simulations, develop problem-solving skills, and complete online homework assigned by your professor.

go Chemistry Download mini lecture videos for key concept review and exam prep from OWL or purchase them from **www.CengageBrain.com.**

Companion Website

Visit this book's companion website at **www.cengage.com/chemistry/moore** to work interactive modules for the Estimation boxes and Active Figures in this text.

substances are mixed, will a chemical reaction occur? If a chemical reaction occurs, what will the products be?

As you saw in Chapter 4 ***(← p. 122)***, most reactions of simple ionic and molecular compounds can be assigned to a few general categories: combination, decomposition, displacement, and exchange. In this chapter we discuss chemical reactions in more detail, including oxidation-reduction reactions. The ability to recognize which type of reaction occurs for a particular set of reactants will allow you to predict the products.

Chemical reactions involving exchange of ions to form precipitates are discussed first, followed by net ionic equations, which focus on the active participants in such reactions. We then consider acid-base reactions, neutralization reactions, and reactions that form gases as products. Next comes a discussion of oxidation-reduction (redox) reactions, oxidation numbers as a means to organize our understanding of redox reactions, and the activity series of metals.

A great deal of chemistry—perhaps most—occurs in solution, and we introduce the means for quantitatively describing the concentrations of solutes in solutions. This discussion is followed by explorations of solution stoichiometry and finally aqueous titration, an analytical technique that is used to measure solute concentrations.

go Chemistry **Module 5: Predicting the Water Solubility of Common Ionic Compounds** and **Module 6: Writing Net Ionic Equations** cover concepts in this section.

5.1 Exchange Reactions: Precipitation and Net Ionic Equations

Aqueous Solubility of Ionic Compounds

Many of the ionic compounds that you frequently encounter, such as table salt, baking soda, and household plant fertilizers, are soluble in water. It is therefore tempting to conclude that all ionic compounds are soluble in water, but such is not the case. Although many ionic compounds are water-soluble, some are only slightly soluble, and others dissolve hardly at all.

When an ionic compound dissolves in water, its ions separate and become surrounded by water molecules, as illustrated in Figure 5.1a. The process in which ions

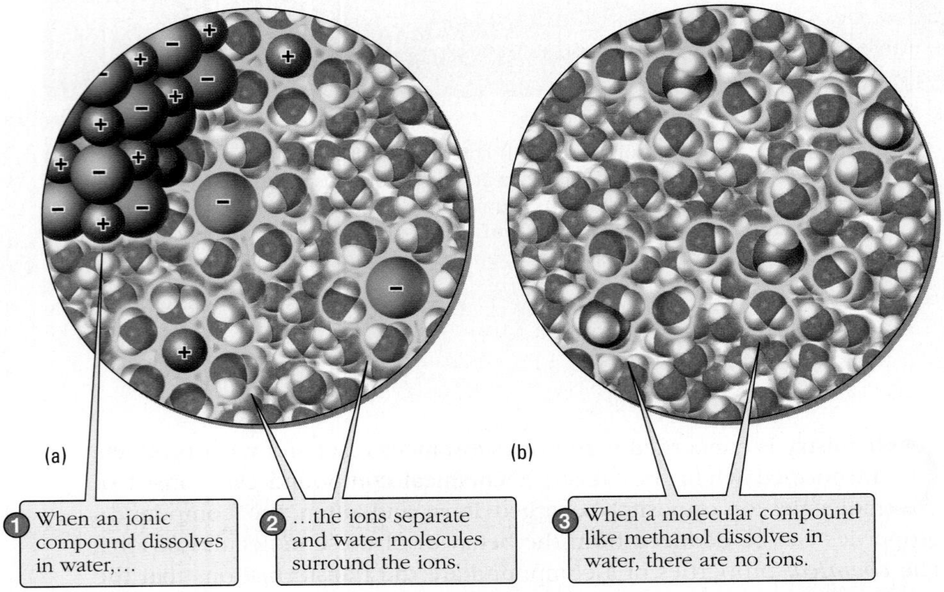
(a)
(b)
1 When an ionic compound dissolves in water,...
2 ...the ions separate and water molecules surround the ions.
3 When a molecular compound like methanol dissolves in water, there are no ions.

Figure 5.1 Dissolution of (a) an ionic compound (sodium chloride, NaCl) and (b) a molecular compound (methanol, CH_3OH) in water.

Table 5.1 Solubility Rules for Ionic Compounds

Usually Soluble	
Group 1A: Li^+, Na^+, K^+, Rb^+, Cs^+, ammonium, NH_4^+	All Group 1A (alkali metal) and ammonium salts are soluble.
Nitrates: NO_3^-	All nitrates are soluble.
Chlorides, bromides, iodides: Cl^-, Br^-, I^-	All common chlorides, bromides, and iodides are soluble except AgCl, Hg_2Cl_2, $PbCl_2$; AgBr, Hg_2Br_2, $PbBr_2$; AgI, Hg_2I_2, PbI_2.
Sulfates: SO_4^{2-}	Most sulfates are soluble; exceptions include $CaSO_4$, $SrSO_4$, $BaSO_4$, and $PbSO_4$.
Chlorates: ClO_3^-	All chlorates are soluble.
Perchlorates: ClO_4^-	All perchlorates are soluble.
Acetates: CH_3COO^-	All acetates are soluble.
Usually Insoluble	
Phosphates: PO_4^{3-}	All phosphates are insoluble except those of NH_4^+ and Group 1A ions (alkali metal cations).
Carbonates: CO_3^{2-}	All carbonates are insoluble except those of NH_4^+ and Group 1A ions (alkali metal cations).
Hydroxides: OH^-	All hydroxides are insoluble except those of NH_4^+ and Group 1A (alkali metal cations). $Sr(OH)_2$, $Ba(OH)_2$, and $Ca(OH)_2$ are slightly soluble.
Oxalates: $C_2O_4^{2-}$	All oxalates are insoluble except those of NH_4^+ and Group 1A (alkali metal cations).
Sulfides: S^{2-}	All sulfides are insoluble except those of NH_4^+, Group 1A (alkali metal cations) and Group 2A (MgS, CaS, and BaS are sparingly soluble).

separate is called *dissociation*. Soluble ionic compounds are one type of strong electrolyte. Recall that an electrolyte is a substance whose aqueous solution contains ions and therefore conducts electricity. A **strong electrolyte** is completely converted to ions when it forms an aqueous solution. By contrast, most water-soluble molecular compounds do not ionize when they dissolve. This is illustrated in Figure 5.1b.

The solubility rules given in Table 5.1 are *general guidelines* for predicting the water solubilities of ionic compounds based on the ions they contain. *If a compound contains at least one of the ions indicated for soluble compounds in Table 5.1, then the compound is at least moderately soluble.*

Suppose you want to know whether $NiSO_4$ is soluble in water. $NiSO_4$ contains Ni^{2+} and SO_4^{2-} ions. Although Ni^{2+} is not mentioned in Table 5.1, substances containing SO_4^{2-} are described as soluble (except for $SrSO_4$, $BaSO_4$, and $PbSO_4$). Because $NiSO_4$ contains an ion, SO_4^{2-}, that indicates solubility and $NiSO_4$ is not one of the sulfate exceptions, it is predicted to be soluble. Figure 5.2 shows examples illustrating the solubility rules for a few nitrates, hydroxides, and sulfides.

The **PROBLEM-SOLVING STRATEGY** in this book is

- Analyze the problem
- Plan a solution
- Execute the plan
- Check that the result is reasonable

Appendix A.1 explains this in detail.

PROBLEM-SOLVING EXAMPLE 5.1 Using Solubility Rules

Indicate what ions are present in each of these compounds, and then predict whether each compound is water-soluble.

(a) $CaCl_2$ (b) $Fe(OH)_3$ (c) NH_4NO_3

(d) $CuCO_3$ (e) $Ni(ClO_3)_2$

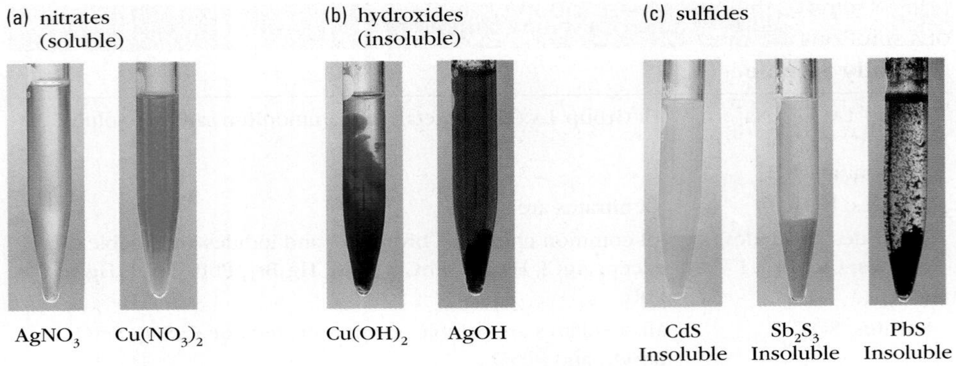
(a) nitrates
(soluble)
(b) hydroxides
(insoluble)
(c) sulfides
AgNO3
Cu(NO3)2
Cu(OH)2
AgOH
CdS
Insoluble
Sb2S3
Insoluble
PbS
Insoluble

Photos: © Cengage Learning/ Charles D. Winters

Figure 5.2 **Illustration of some of the solubility guidelines in Table 5.1.**

Answer (a) Ca^{2+} and Cl^-, soluble. (b) Fe^{3+} and OH^-, insoluble. (c) NH_4^+ and NO_3^-, soluble. (d) Cu^{2+} and CO_3^{2-}, insoluble. (e) Ni^{2+} and ClO_3^-, soluble.

Strategy and Explanation Use the solubility rules, which require identifying the ions present and checking their aqueous solubility (Table 5.1).

(a) $CaCl_2$ contains Ca^{2+} and Cl^- ions. All chlorides are soluble, with a few exceptions for transition metals, so calcium chloride is soluble.

(b) $Fe(OH)_3$ contains Fe^{3+} and OH^- ions. As indicated in Table 5.1, all hydroxides are insoluble except those of alkali metals and a few other exceptions, so iron(III) hydroxide is insoluble.

(c) NH_4NO_3 contains NH_4^+ and NO_3^- ions. All ammonium salts are soluble, and all nitrates are soluble, so NH_4NO_3 is soluble.

(d) $CuCO_3$ contains Cu^{2+} and CO_3^{2-} ions. All carbonates are insoluble except those of ammonium and alkali metals, and copper is not an alkali metal, so $CuCO_3$ is insoluble.

(e) $Ni(ClO_3)_2$ contains Ni^{2+} and ClO_3^- ions. All chlorates are soluble, so $Ni(ClO_3)_2$ is soluble.

PROBLEM-SOLVING PRACTICE answers are provided at the back of this book in Appendix K.

PROBLEM-SOLVING PRACTICE 5.1

Predict whether each of these compounds is likely to be water-soluble.

(a) NaF (b) $Ca(CH_3COO)_2$ (c) $SrCl_2$
(d) MgO (e) $PbCl_2$ (f) HgS

Recall that exchange reactions (*← p. 127*) have this reaction pattern:

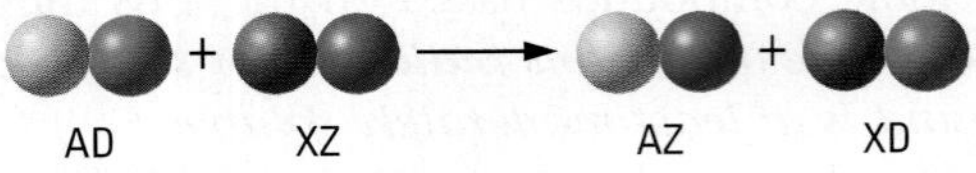
AD
XZ
AZ
XD

If all reactants and all products of such a reaction are water-soluble ionic compounds, no overall reaction takes place. In such cases, mixing the solutions of AD and XZ just results in an aqueous solution containing the A, D, X, and Z ions.

What will happen when two aqueous solutions are mixed, one containing dissolved calcium nitrate, $Ca(NO_3)_2$, and the other containing dissolved sodium chloride, NaCl? Both are soluble ionic compounds (Table 5.1), so the resulting solution contains Ca^{2+}, NO_3^-, Na^+, and Cl^- ions. To decide whether a reaction will occur requires determining whether any two of these ions can react with each other to form a *new* compound. For an exchange reaction to occur, the calcium ion and the chloride ion would have to form calcium chloride, $CaCl_2$, and the sodium ion and the nitrate ion would have to form sodium nitrate, $NaNO_3$. Is either a possible chemical reaction? When either product is insoluble, the answer is yes. Checking the solubility rules shows that both of these compounds are water-soluble. No reaction to remove the ions from solution is possible; therefore, when these two aqueous solutions are mixed, no reaction occurs.

If, however, one or both of the potential products of the reaction remove ions from the solution, a reaction will occur. Three different kinds of products can cause an exchange reaction to occur in aqueous solution:

1. Formation of an *insoluble ionic compound:*

$$AgNO_3(aq) + KCl(aq) \longrightarrow KNO_3(aq) + AgCl(s)$$

2. Formation of a *molecular compound* that remains in solution. Most commonly this happens when water is produced in acid-base neutralization reactions:

$$H_2SO_4(aq) + 2\ NaOH(aq) \longrightarrow Na_2SO_4(aq) + 2\ H_2O(\ell)$$

3. Formation of a *gaseous molecular compound* that escapes from the solution:

$$2\ HCl(aq) + Na_2S(aq) \longrightarrow 2\ NaCl(aq) + H_2S(g)$$

Precipitation Reactions

Consider the possibility of an exchange reaction when aqueous solutions of barium chloride and sodium sulfate are mixed:

$$BaCl_2(aq) + Na_2SO_4(aq) \longrightarrow ? + ?$$

If the barium ions and sodium ions exchange partners to form $BaSO_4$ and NaCl, the equation will be

$$\underset{\text{barium chloride}}{BaCl_2(aq)} + \underset{\text{sodium sulfate}}{Na_2SO_4(aq)} \longrightarrow \underset{\text{barium sulfate}}{BaSO_4} + \underset{\text{sodium chloride}}{2\ NaCl}$$

Will a reaction occur? The answer is yes when an insoluble solid product—a **precipitate**—forms. Checking Table 5.1, we find that NaCl is soluble, but $BaSO_4$ is not soluble (sulfates of Ca^{2+}, Sr^{2+}, Ba^{2+}, and Pb^{2+} are insoluble). Therefore, an exchange reaction will occur, and solid barium sulfate will precipitate from the solution (Figure 5.3). Precipitate formation is indicated by an (s) next to the formula of the precipitate, a solid, in the overall equation. Because it is soluble, NaCl remains dissolved in solution, and we put (aq) next to NaCl in the equation.

$$BaCl_2(aq) + Na_2SO_4(aq) \longrightarrow BaSO_4(s) + 2\ NaCl(aq)$$

© Cengage Learning/Charles D. Winters

Figure 5.3 Precipitation of barium sulfate. Mixing aqueous solutions of barium chloride, $BaCl_2$, and sodium sulfate, Na_2SO_4, forms a precipitate of barium sulfate, $BaSO_4$. Sodium chloride, NaCl, the other product of this exchange reaction, is water soluble, and Na^+ and Cl^- ions remain in solution.

Ba^{2+} and Na^+ do not react with each other, and neither do Cl^- and SO_4^{2-}.

PROBLEM-SOLVING EXAMPLE 5.2 Exchange Reactions

For each of these pairs of ionic compounds, decide whether an exchange reaction will occur when their aqueous solutions are mixed, and write a balanced chemical equation for those reactions that will occur.

(a) $(NH_4)_2S$ and $Cu(NO_3)_2$ (b) $ZnCl_2$ and Na_2CO_3 (c) $CaCl_2$ and KNO_3

Answer

(a) CuS precipitates. $(NH_4)_2S(aq) + Cu(NO_3)_2(aq) \longrightarrow CuS(s) + 2\ NH_4NO_3(aq)$

(b) $ZnCO_3$ precipitates. $ZnCl_2(aq) + Na_2CO_3(aq) \longrightarrow ZnCO_3(s) + 2\ NaCl(aq)$

(c) No reaction occurs. Both possible products, $Ca(NO_3)_2$ and KCl, are soluble.

Strategy and Explanation In each case, consider which cation-anion combinations can form, and then decide whether the possible compounds will precipitate. You have to be careful to take into account any polyatomic ions, which must be kept together as a unit during the balancing of the chemical reaction.

(a) An exchange reaction between $(NH_4)_2S$ and $Cu(NO_3)_2$ forms CuS(s) and NH_4NO_3. Table 5.1 shows that all nitrates are soluble, so NH_4NO_3 remains in solution. CuS is not soluble and therefore precipitates.

(b) The exchange reaction between $ZnCl_2$ and Na_2CO_3 forms the insoluble product $ZnCO_3(s)$ and leaves soluble NaCl in solution.

(c) No precipitate forms when $CaCl_2$ and KNO_3 are mixed because each product, $Ca(NO_3)_2$ and KCl, is soluble. All four of the ions (Ca^{2+}, Cl^-, K^+, NO_3^-) remain in solution. No exchange reaction occurs because no product is formed that removes ions from the solution.

PROBLEM-SOLVING PRACTICE 5.2

Predict the products and write a balanced chemical equation for the exchange reaction in aqueous solution between each pair of ionic compounds. Use Table 5.1 to determine solubilities and indicate in the equation whether a precipitate forms.

(a) $NiCl_2$ and NaOH (b) K_2CO_3 and $CaBr_2$

go Chemistry **Module 6: Writing Net Ionic Equations** covers concepts in this section.

Net Ionic Equations

In writing equations for exchange reactions in the preceding section, we used overall equations. There is another way to represent what happens. In each case in which a precipitate forms, the product that does not precipitate remains in solution. Therefore, its ions are in solution as reactants and remain there after the reaction. Such ions are commonly called **spectator ions** because, like the spectators at a play or game, they are present but are not involved directly in the real action. Consequently, the spectator ions can be left out of the equation that represents the chemical change that occurs. An equation that includes only the symbols or formulas of ions in solution or compounds that undergo change is called a **net ionic equation.** We will use the reaction of aqueous NaCl with $AgNO_3$ to form AgCl and $NaNO_3$ (Figure 5.4) to illustrate the general steps for writing a net ionic equation.

Step 1: ***Write the overall balanced equation using the correct formulas for the reactants and products.***

Overall chemical reaction:

$$\underset{\text{silver nitrate}}{AgNO_3} + \underset{\text{sodium chloride}}{NaCl} \longrightarrow \underset{\text{silver chloride}}{AgCl} + \underset{\text{sodium nitrate}}{NaNO_3}$$

Step 1 actually consists of two parts: first, write the unbalanced equation with the correct formulas for reactants and products; second, balance the equation.

Step 2: ***Use the general guidelines in Table 5.1 to determine the solubilities of reactants and products.*** In this case, the guidelines indicate that nitrates are soluble, so $AgNO_3$ and $NaNO_3$ are soluble. NaCl is water-soluble because almost all chlorides are soluble. However, AgCl is one of the insoluble chlorides (AgCl, Hg_2Cl_2, and $PbCl_2$). Using this information we can write

$$AgNO_3(aq) + NaCl(aq) \longrightarrow AgCl(s) + NaNO_3(aq)$$

Step 3: ***Recognize that all soluble ionic compounds dissociate into their component ions in aqueous solution.*** Therefore we have

$AgNO_3(aq)$ consists of $Ag^+(aq) + NO_3^-(aq)$.

$NaCl(aq)$ consists of $Na^+(aq) + Cl^-(aq)$.

$NaNO_3(aq)$ consists of $Na^+(aq) + NO_3^-(aq)$.

Step 4: ***Use the ions from Step 3 to write a complete ionic equation with the ions in solution from each soluble compound shown separately.***

Complete ionic equation:

$$Ag^+(aq) + NO_3^-(aq) + Na^+(aq) + Cl^-(aq) \longrightarrow AgCl(s) + Na^+(aq) + NO_3^-(aq)$$

Note that the precipitate is represented by its complete formula.

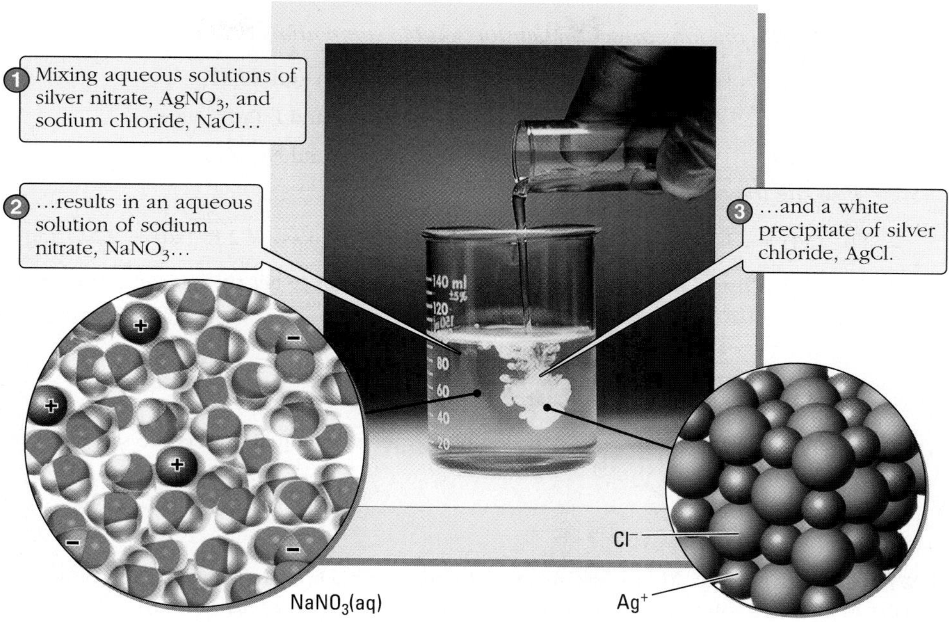

Photo: © Cengage Learning/Charles D. Winters

Active Figure 5.4 **Precipitation of silver chloride.** Visit this book's companion website at **www.cengage.com/chemistry/moore** to test your understanding of the concepts in this figure.

Step 5: ***To obtain the net ionic equation, cancel the spectator ions from each side of the complete ionic equation.*** Sodium ions and nitrate ions are the spectator ions in this example, and we cancel them from the complete ionic equation to give the net ionic equation.

Complete ionic equation:

$$Ag^+(aq) + \cancel{NO_3^-(aq)} + \cancel{Na^+(aq)} + Cl^-(aq) \longrightarrow AgCl(s) + \cancel{Na^+(aq)} + \cancel{NO_3^-(aq)}$$

Net ionic equation:

$$Ag^+(aq) + Cl^-(aq) \longrightarrow AgCl(s)$$

Step 6: ***Check that the sum of the charges is the same on each side of the net ionic equation.*** For the equation in Step 5 the sum of charges is zero on each side: (1+) + (1−) = 0 on the left; AgCl is an ionic compound with zero net charge on the right.

The charge must be the same on both sides of a balanced equation since electrons are neither created nor destroyed.

PROBLEM-SOLVING EXAMPLE 5.3 Net Ionic Equations

Write the net ionic equation that occurs when aqueous solutions of lead nitrate, $Pb(NO_3)_2$, and potassium iodide, KI, are mixed.

Answer $Pb^{2+}(aq) + 2\,I^-(aq) \longrightarrow PbI_2(s)$

Strategy and Explanation Use the stepwise procedure presented above.

Step 1: *Write the overall balanced equation using the correct formulas for the reactants and products.* This is an exchange reaction (← *p. 127*).

$$Pb(NO_3)_2 + 2\,KI \longrightarrow PbI_2 + 2\,KNO_3$$

Step 2: *Determine the solubilities of reactants and products.* The solubility rules in Table 5.1 predict that all these reactants and products are soluble except PbI_2.

$$Pb(NO_3)_2(aq) + 2\,KI(aq) \longrightarrow PbI_2(s) + 2\,KNO_3(aq)$$

1 mol $Pb(NO_3)_2$ contains 2 mol NO_3^- ions along with 1 mol Pb^{2+} ions.

Step 3: *Identify the ions present when the soluble compounds dissociate in solution.*

$Pb(NO_3)_2(aq)$ consists of $Pb^{2+}(aq)$ and 2 NO_3^-.

$KI(aq)$ consists of $K^+(aq)$ and $I^-(aq)$.

$KNO_3(aq)$ consists of $K^+(aq)$ and $NO_3^-(aq)$.

Step 4: *Write the complete ionic equation.*

$$Pb^{2+}(aq) + 2\,NO_3^-(aq) + 2\,K^+(aq) + 2\,I^-(aq) \longrightarrow PbI_2(s) + 2\,K^+(aq) + 2\,NO_3^-(aq)$$

Step 5 and 6: *Cancel spectator ions* [$K^+(aq)$ and $NO_3^-(aq)$] *to get the net ionic equation; check that charge is balanced.*

$$\text{Net ionic equation: } Pb^{2+}(aq) + 2\,I^-(aq) \longrightarrow PbI_2(s)$$

$$\text{Net charge} = (2+) + 2 \times (1-) = 0$$

☑ **Reasonable Answer Check** Each side of the net ionic equation has the same charge and types of ions.

PROBLEM-SOLVING PRACTICE 5.3

Write a balanced equation for the reaction (if any) for each of these ionic compound pairs in aqueous solution. Then use the complete ionic equation to write their balanced net ionic equations.

(a) $BaCl_2$ and Na_2SO_4 (b) $(NH_4)_2S$ and $FeCl_2$

Answers to **EXERCISES** are provided at the back of this book in Appendix L.

EXERCISES that are labeled **CONCEPTUAL** are designed to test your understanding of one or more concepts; they usually involve qualitative rather than quantitative thinking.

CONCEPTUAL EXERCISE 5.1 Net Ionic Equations

There are some exchange reactions where both products are insoluble and precipitate from aqueous solution. Use Table 5.1 to find an example of such a reaction. Write a balanced net ionic equation for the reaction.

If you live in an area with "hard water," you have probably noticed the scale that forms inside your teakettle or saucepans when you boil water in them. Hard water is mostly caused by the presence of the cations Ca^{2+}, Mg^{2+}, and also Fe^{2+} or Fe^{3+}. When the water also contains hydrogen carbonate ion, HCO_3^-, this reaction occurs when the water is heated:

$$2\,HCO_3^-(aq) \longrightarrow H_2O(\ell) + CO_2(g) + CO_3^{2-}(aq)$$

The carbon dioxide escapes from the hot water, and the hydrogen carbonate ions are slowly converted to carbonate ions. The carbonate ions can form precipitates with calcium, magnesium, or iron ions to produce a solid that sticks to metal surfaces. In hot-water heating systems in areas with high calcium ion concentrations, the buildup of such *boiler scale* can plug the pipes.

$$Ca^{2+}(aq) + 2\,HCO_3^-(aq) \longrightarrow CaCO_3(s) + H_2O(\ell) + CO_2(g)$$

© Cengage Learning/Charles D. Winters

A solid forms on the inside of saucepans and teakettles when hard water is boiled in them.

5.2 Acids, Bases, and Acid-Base Exchange Reactions

Acids and bases are two extremely important classes of compounds—so important that this book devotes two chapters to them later (Chapters 16 and 17). Here, we focus on a few general properties and consider how acids and bases react with each other. Acids have a number of properties in common, and so do bases. Some properties of acids are related to properties of bases.

- Acidic solutions change the color of litmus from blue to red, and basic solutions change the color of litmus from red to blue.
- Acidic solutions cause the dye phenolphthalein to be colorless, but basic solutions make phenolphthalein pink.
- If an acid has made litmus red, adding a base will reverse the effect, making the litmus blue.
- A base can *neutralize* the effect of an acid, and an acid can *neutralize* the effect of a base. Acids and bases are chemical opposites.

Litmus is a dye derived from lichens. Phenolphthalein is a synthetic dye.

Acids have other characteristic properties. They taste sour, they produce bubbles of gas when reacting with limestone, and they dissolve many metals while producing a flammable gas. Although you should never taste substances in a chemistry laboratory, you have probably experienced the sour taste of at least one acid—vinegar, which is a dilute solution of acetic acid in water. Bases, in contrast, have a bitter taste. Soap, for example, contains a base. Rather than dissolving metals, bases often cause metal ions to form insoluble compounds that precipitate from solution as metal hydroxides. Such precipitates can be made to dissolve by adding an acid, another case in which an acid counteracts a property of a base.

Acids

The properties of acids in aqueous solutions can be explained by defining an **acid** as any substance that increases the concentration of aqueous hydrogen ions, $H^+(aq)$, when dissolved in pure water. ***The properties acidic solutions have in common are the properties of $H^+(aq)$.*** The "(aq)" is important here, because a hydrogen ion, H^+, is a hydrogen atom that has lost its single electron; that is, H^+ is a proton—the nucleus of a hydrogen atom. Remember that the diameter of the nucleus of an atom is about 1/10,000 the diameter of the atom ***(← p. 43)***. Consequently a proton, H^+, is much smaller than a hydrogen atom and has a very high ratio of positive charge to size. This very concentrated positive charge interacts strongly with electrons in oxygen atoms of water molecules and H^+ combines with H_2O to form H_3O^+, known as the **hydronium ion**. Other water molecules are attracted to each H_3O^+ ion, forming even larger clusters. $H^+(aq)$ is used to represent H_3O^+ and these larger clusters of water molecules.

Chapter 16 explores the importance of the hydronium ion to acid-base chemistry.

$$H^+ + H_2O \longrightarrow H_3O^+$$

Acids that are entirely converted to ions (completely ionized) when dissolved in water are **strong electrolytes** and are called **strong acids.** One of the most common strong acids is hydrochloric acid, which ionizes completely in aqueous solution to form $H^+(aq)$ and chloride ions (Figure 5.5a).

$$HCl(aq) \longrightarrow H^+(aq) + Cl^-(aq)$$

The more complete, and proper, way to write an equation for the reaction of a strong acid like HCl with water is

$$HCl(aq) + H_2O(\ell) \longrightarrow H_3O^+(aq) + Cl^-(aq)$$

which explicitly shows the hydronium ion, H_3O^+. Table 5.2 lists some other common strong acids.

In contrast, acids and other substances that ionize only slightly are termed **weak electrolytes.** Acids that are only partially ionized in aqueous solution are termed **weak acids** (Table 5.2). For example, when acetic acid, CH_3COOH, dissolves in water, usually fewer than 5% of the acetic acid molecules are ionized at any time. The remainder of the acetic acid exists as nonionized molecules. Thus,

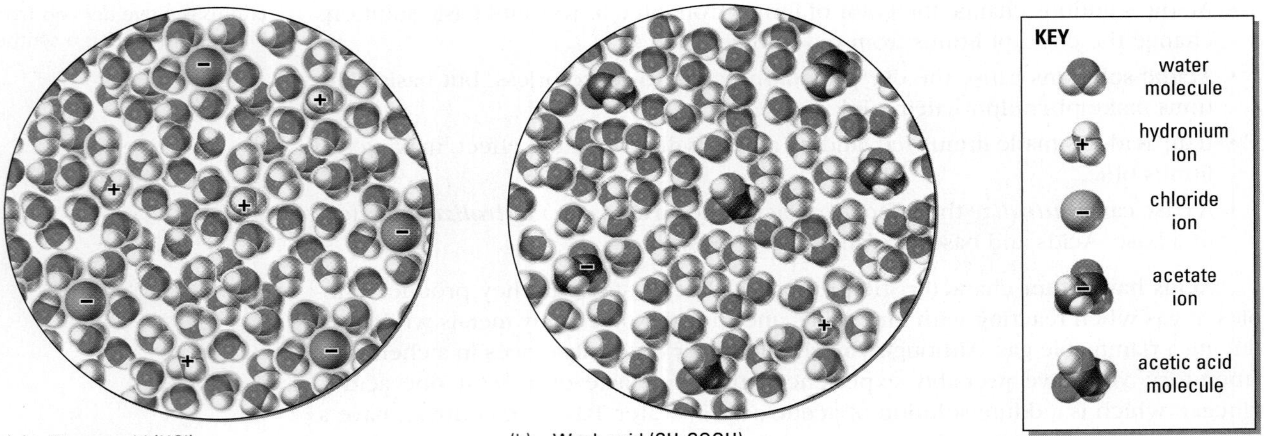

(a) Strong acid (HCl) (b) Weak acid (CH_3COOH)

Figure 5.5 The ionization of acids in water. (a) A strong acid such as hydrochloric acid, HCl, is completely ionized in water; all the HCl molecules ionize to form $H_3O^+(aq)$ and $Cl^-(aq)$ ions. (b) Weak acids such as acetic acid, CH_3COOH, are only slightly ionized in water. Nonionized acetic acid molecules far outnumber aqueous H_3O^+ and CH_3COO^- ions formed by the ionization of acetic acid molecules.

because acetic acid is only slightly ionized in aqueous solution, it is a weak electrolyte and classified as a weak acid (Figure 5.5b).

The organic functional group —COOH is present in all organic carboxylic acids (Section 12.6).

$$\underset{\text{(molecular) form}}{\underset{\text{Nonionized}}{CH_3COOH(aq)}} \rightleftharpoons H^+(aq) + CH_3COO^-(aq)$$

The double arrow in this equation for the ionization of acetic acid signifies a characteristic property of the reaction of a weak electrolyte with water: there is a *dynamic equilibrium* in which the nonionized, molecular form, CH_3COOH, is ionizing at

Table 5.2 Common Acids and Bases

Strong Acids (Strong Electrolytes)		Strong Bases (Strong Electrolytes)	
HCl	Hydrochloric acid	LiOH	Lithium hydroxide
HNO_3	Nitric acid	NaOH	Sodium hydroxide
H_2SO_4	Sulfuric acid	KOH	Potassium hydroxide
$HClO_4$	Perchloric acid	$Ca(OH)_2$	Calcium hydroxide‡
HBr	Hydrobromic acid	$Ba(OH)_2$	Barium hydroxide‡
HI	Hydroiodic acid	$Sr(OH)_2$	Strontium hydroxide‡
Weak Acids* (Weak Electrolytes)		**Weak Bases† (Weak Electrolytes)**	
H_3PO_4	Phosphoric acid	NH_3	Ammonia
CH_3COOH	Acetic acid	CH_3NH_2	Methylamine
H_2CO_3	Carbonic acid		
HCN	Hydrocyanic acid		
HCOOH	Formic acid		
C_6H_5COOH	Benzoic acid		

*Many organic acids are weak acids.
†Many organic amines (related to ammonia) are weak bases.
‡The hydroxides of calcium, barium, and strontium are only slightly soluble, but all of the solute that dissolves is dissociated into ions.

exactly the same rate that the ions, $H^+(aq)$ and $CH_3COO^-(aq)$, are reacting to form nonionized acetic acid molecules. Dynamic equilibrium will be described in more detail in Chapter 14.

Some common acids, such as sulfuric acid, can provide more than 1 mol H^+ ions per mole of acid:

$$\underset{\text{sulfuric acid}}{H_2SO_4(aq)} \longrightarrow H^+(aq) + \underset{\text{hydrogen sulfate ion}}{HSO_4^-(aq)}$$

$$\underset{\text{hydrogen sulfate ion}}{HSO_4^-(aq)} \rightleftharpoons H^+(aq) + \underset{\text{sulfate ion}}{SO_4^{2-}(aq)}$$

The first ionization reaction is essentially complete, so sulfuric acid is considered a strong electrolyte (and a strong acid as well). However, the hydrogen sulfate ion, like acetic acid, is only partially ionized, so it is a weak electrolyte and also a weak acid.

CONCEPTUAL EXERCISE 5.2 Dissociation of Acids

Phosphoric acid, H_3PO_4, has three protons that can ionize. Write the equations for its three ionization reactions, each of which is a dynamic equilibrium.

Bases

A **base** is a substance that increases the concentration of the aqueous **hydroxide ions**, $OH^-(aq)$, when dissolved in pure water. ***The properties that basic solutions have in common are properties attributable to the aqueous hydroxide ion, $OH^-(aq)$.*** Compounds that contain hydroxide ions, such as sodium hydroxide or potassium hydroxide, are obvious bases. As ionic compounds they are strong electrolytes and **strong bases** (Table 5.2).

$$NaOH(s) \xrightarrow{H_2O} Na^+(aq) + OH^-(aq)$$

A base that is slightly water-soluble, such as $Ca(OH)_2$, can still be a strong electrolyte if the amount of the compound that dissolves completely dissociates into ions.

Acids and bases that are strong electrolytes are strong acids and bases. Acids and bases that are weak electrolytes are weak acids and bases.

Ammonia, NH_3, is another very common base. Although the compound does not have an OH^- ion as part of its formula, it produces the ion by reaction with water.

$$NH_3(aq) + H_2O(\ell) \rightleftharpoons NH_4^+(aq) + OH^-(aq)$$

In the equilibrium between NH_3 and the NH_4^+ and OH^- ions, only a small concentration of the ions is present, so ammonia is a weak electrolyte (<5% ionized), and it is a **weak base** (Table 5.2).

To summarize:

- *Strong electrolytes* are compounds that exist *completely* as ions in aqueous solutions. They can be ionic compounds (salts or strong bases) or molecular compounds that are strong acids and ionize completely.
- *Weak electrolytes* are molecular compounds that are weak acids or bases.
- *Nonelectrolytes* are molecular compounds that do not ionize in aqueous solution.

CONCEPTUAL EXERCISE 5.3 Acids and Bases

(a) What ions are produced when perchloric acid, $HClO_4$, dissolves in water?
(b) Calcium hydroxide is only slightly soluble in water. What little does dissolve, however, is dissociated. What ions are produced? Write an equation for the dissociation of calcium hydroxide.

(a) Many common foods and household products are acidic or basic. Citrus fruits contain citric acid; household ammonia and oven cleaner are both basic.

(b) The acid in lemon juice turns blue litmus paper red.

(c) Household ammonia turns red litmus paper blue.

Acids and bases.

PROBLEM-SOLVING EXAMPLE 5.4 Strong Electrolytes, Weak Electrolytes, and Nonelectrolytes

Identify whether each of these substances in an aqueous solution is a strong electrolyte, a weak electrolyte, or a nonelectrolyte: HBr, hydrogen bromide; LiOH, lithium hydroxide; HCOOH, formic acid; CH_3CH_2OH, ethanol.

Answer HBr is a strong electrolyte; LiOH is a strong electrolyte; HCOOH is a weak electrolyte; CH_3CH_2OH is a nonelectrolyte.

Strategy and Explanation For the acids and bases we refer to Table 5.2. Hydrogen bromide is a common strong acid and, therefore, is a strong electrolyte. Lithium hydroxide is a strong base that completely dissociates into ions in aqueous solution, so it is a strong electrolyte. Formic acid is a weak acid because it only partially ionizes in aqueous solution, so it is a weak electrolyte. Ethanol is a molecular compound that does not dissociate into ions in aqueous solution, so it is a nonelectrolye.

PROBLEM-SOLVING PRACTICE 5.4

Look back through the discussion of electrolytes and Table 5.2 and identify at least one additional strong electrolyte, one additional weak electrolyte, and one additional nonelectrolyte different from those discussed in Problem-Solving Example 5.4.

Neutralization Reactions

When aqueous solutions of a strong acid (such as HCl) and a strong base (such as NaOH) are mixed, the ions in solution are $H^+(aq)$ and the anion from the acid, the metal cation, and the hydroxide ion from the base:

From hydrochloric acid: $H^+(aq)$, $Cl^-(aq)$
From sodium hydroxide: $Na^+(aq)$, $OH^-(aq)$

As in precipitation reactions, an exchange reaction will occur *whenever two of these ions can react with each other to form a compound that removes ions from solution*. In an acid-base reaction, that compound is water, formed by the combination of $H^+(aq)$ with $OH^-(aq)$.

When a strong acid and a strong base react, they neutralize each other. This happens because the hydrogen ions from the acid react with hydroxide ions from the base to form water, a molecular compound. The other ions form a **salt,** an ionic compound whose cation comes from a base and whose anion comes from an acid. When the water is evaporated, the solid salt remains.

A **general equation for a neutralization reaction** can be written as

$$\underset{\text{acid}}{HX(aq)} + \underset{\text{base}}{MOH(aq)} \longrightarrow \underset{\text{water}}{H\,OH(\ell)} + \underset{\text{salt}}{M\,X(aq)}$$

Note that the cation of the salt comes from the base, and the anion of the salt comes from the acid.

You should recognize this as an exchange reaction in which the $H^+(aq)$ ions from the aqueous acid and the $M^+(aq)$ ions from the metal hydroxide exchange partners.

In the case of HCl plus NaOH, the salt is sodium chloride, NaCl.

$$\underset{\text{acid}}{HCl(aq)} + \underset{\text{base}}{NaOH(aq)} \longrightarrow \underset{\text{water}}{H\,OH(\ell)} + \underset{\text{salt}}{Na\,Cl(aq)}$$

The salt that forms depends on the acid and base that react. Magnesium chloride, another salt, is formed when a commercial antacid containing magnesium hydroxide is swallowed to neutralize excess hydrochloric acid in the stomach.

Milk of magnesia consists of a suspension of finely divided particles of $Mg(OH)_2(s)$ in water.

$$\underset{\substack{\text{hydrochloric}\\\text{acid}}}{2\,HCl(aq)} + \underset{\substack{\text{magnesium}\\\text{hydroxide}}}{Mg(OH)_2(s)} \longrightarrow 2\,H_2O(\ell) + \underset{\substack{\text{magnesium}\\\text{chloride}}}{MgCl_2(aq)}$$

The —COOH structure, called the *acid functional group*, is present in all organic acids and imparts acidic properties to compounds containing it.

Organic acids, such as acetic acid and propanoic acid, which contain the acid functional group —COOH, also neutralize bases to form salts. The H in the —COOH functional group is the acidic proton. Its removal generates the —COO$^-$ anion. The reaction of propanoic acid, CH_3CH_2COOH, and sodium hydroxide produces the salt sodium propanoate, $NaCH_3CH_2COO$, containing sodium ions, Na^+, and propanoate ions, $CH_3CH_2COO^-$. Sodium propanoate is commonly used as a food preservative.

$$\underset{\text{propanoic acid}}{CH_3CH_2COOH(aq)} + NaOH(aq) \longrightarrow H_2O(\ell) + \underset{\text{sodium propanoate}}{NaCH_3CH_2COO(aq)}$$

Note that this chemical reaction is an example of the general equation for a neutralization reaction on p. 172.

Although the propanoic acid molecule contains six H atoms, it is only the H atom that is part of the acid functional group —COOH that is involved in this neutralization reaction.

PROBLEM-SOLVING EXAMPLE 5.5 Balancing Neutralization Equations

Write a balanced chemical equation for the reaction of nitric acid, HNO_3, with calcium hydroxide, $Ca(OH)_2$, in aqueous solution.

Answer $2\ HNO_3(aq) + Ca(OH)_2(aq) \longrightarrow Ca(NO_3)_2(aq) + H_2O\ (\ell)$

Strategy and Explanation This is a neutralization reaction between an acid and a base, so the products are a salt and water.

- **First, write the unbalanced equation with all the reactants and products.**

(unbalanced equation) $HNO_3(aq) + Ca(OH)_2(aq) \longrightarrow Ca(NO_3)_2(aq) + H_2O(\ell)$

- **Balance the ions first.**

It is usually a good idea to start with the ions and balance the hydrogen and oxygen atoms later. The calcium ions are in balance, but we need to add a coefficient of 2 to the nitric acid since two nitrate ions appear in the products.

(unbalanced equation) $2\ HNO_3(aq) + Ca(OH)_2(aq) \longrightarrow Ca(NO_3)_2(aq) + H_2O(\ell)$

- **Balance the hydrogen and oxygen atoms.**

All the coefficients are correct except for that of water. We count four hydrogen atoms in the reactants (two from nitric acid and two from calcium hydroxide), so we must put a coefficient of 2 in front of the water to balance the equation.

(balanced equation) $2\ HNO_3(aq) + Ca(OH)_2(aq) \longrightarrow Ca(NO_3)_2(aq) + 2\ H_2O(\ell)$

☑ **Reasonable Answer Check** There are eight oxygen atoms in the reactants (six from nitric acid and two from calcium hydroxide), and there are eight oxygen atoms in the products (six from calcium nitrate and two from water).

PROBLEM-SOLVING PRACTICE 5.5

Write a balanced equation for the reaction of phosphoric acid, H_3PO_4, with sodium hydroxide, NaOH.

PROBLEM-SOLVING EXAMPLE 5.6 Acids, Bases, and Salts

Identify the acid and base used to form each of these salts: (a) $CaSO_4$, (b) $KClO_4$. Write balanced equations for the formation of these compounds.

Answer (a) Calcium hydroxide, $Ca(OH)_2$, and sulfuric acid, H_2SO_4. (b) Potassium hydroxide, KOH, and perchloric acid, $HClO_4$.

Strategy and Explanation A salt is formed from the cation of a base and the anion of an acid.

(a) $CaSO_4$ contains calcium and sulfate ions. Ca^{2+} ions come from $Ca(OH)_2$, calcium hydroxide, and SO_4^{2-} ions come from H_2SO_4, sulfuric acid. The neutralization reaction between $Ca(OH)_2$ and H_2SO_4 produces $CaSO_4$ and water.

$$Ca(OH)_2(aq) + H_2SO_4(aq) \longrightarrow CaSO_4(s) + 2\ H_2O(\ell)$$

(b) Potassium perchlorate contains potassium ions and perchlorate ions, K^+ and ClO_4^-. K^+ ions come from KOH, potassium hydroxide, and ClO_4^- ions come from $HClO_4$, perchloric acid. The neutralization reaction between KOH and $HClO_4$ produces $KClO_4$ and water.

$$KOH(aq) + HClO_4(aq) \longrightarrow KClO_4(aq) + H_2O(\ell)$$

☑ **Reasonable Answer Check** The final neutralization equations each have the same types and numbers of atoms on each side.

PROBLEM-SOLVING PRACTICE 5.6

Identify the acid and the base that can react to form (a) $MgSO_4$ and (b) $SrCO_3$.

Net Ionic Equations for Acid-Base Reactions

Net ionic equations can be written for acid-base reactions as well as for precipitation reactions. This should not be surprising because precipitation and acid-base neutralization reactions are both exchange reactions.

Consider the reaction given earlier of magnesium hydroxide with hydrochloric acid to relieve excess stomach acid, HCl. The overall balanced equation is

$$2\,HCl(aq) + Mg(OH)_2(s) \longrightarrow 2\,H_2O(\ell) + MgCl_2(aq)$$

The acid and base furnish aqueous hydrogen ions and hydroxide ions, respectively.

Although magnesium hydroxide is not very soluble, the little that dissolves is completely dissociated.

$$2\,HCl(aq) \longrightarrow 2\,H^+(aq) + 2\,Cl^-(aq)$$

$$Mg(OH)_2(s) \rightleftharpoons Mg^{2+}(aq) + 2\,OH^-(aq)$$

Note that we retain the coefficients from the balanced overall equation (first step).

We now use this information to write a complete ionic equation. We use Table 5.1 to check the solubility of the product salt, $MgCl_2$. Magnesium chloride is soluble, so the Mg^{2+} and Cl^- ions remain in solution. The complete ionic equation is

$$\cancel{Mg^{2+}(aq)} + 2\,OH^-(aq) + 2\,H^+(aq) + \cancel{2\,Cl^-(aq)} \longrightarrow \cancel{Mg^{2+}(aq)} + \cancel{2\,Cl^-(aq)} + 2\,H_2O(\ell)$$

Canceling spectator ions from each side of the complete ionic equation yields the net ionic equation. In this case, magnesium ions and chloride ions are the spectator ions. Canceling them leaves us with this net ionic equation:

$$2\,H^+(aq) + 2\,OH^-(aq) \longrightarrow 2\,H_2O(\ell)$$

or simply

$$H^+(aq) + OH^-(aq) \longrightarrow H_2O(\ell)$$

This is the net ionic equation for the neutralization reaction between a strong acid and a strong base that yields a soluble salt. Note that, as always, there is *conservation of charge* in the net ionic equation. On the left, $(1+) + (1-) = 0$; on the right, water has zero net charge.

Next, consider a neutralization reaction between a weak acid, HCN, and a strong base, KOH.

$$HCN(aq) + KOH(aq) \longrightarrow KCN(aq) + H_2O(\ell)$$

The weak acid HCN is not completely ionized, so we leave it in the molecular form, but KOH and KCN are strong electrolytes. The complete ionic equation is

$$HCN(aq) + \cancel{K^+(aq)} + OH^-(aq) \longrightarrow \cancel{K^+(aq)} + CN^-(aq) + H_2O(\ell)$$

Canceling spectator ions yields

$$HCN(aq) + OH^-(aq) \longrightarrow CN^-(aq) + H_2O(\ell)$$

The net ionic equation for the neutralization of a weak acid by a strong base contains the molecular form of the acid and the anion of the salt. The net ionic equation shows that charge is conserved.

PROBLEM-SOLVING EXAMPLE 5.7 Neutralization Reaction with a Weak Acid

Write a balanced equation for the reaction of acetic acid, CH_3COOH, with calcium hydroxide, $Ca(OH)_2$. Then write the net ionic equation for this neutralization reaction.

Answer

Equation: $2\ CH_3COOH(aq) + Ca(OH)_2(aq) \longrightarrow Ca(CH_3COO)_2(aq) + 2\ H_2O\ (\ell)$

Net ionic equation: $CH_3COOH(aq) + OH^-(aq) \longrightarrow CH_3COO^-(aq) + H_2O\ (\ell)$

Strategy and Explanation We have been given the formula of an acid and a base that will react. The two products of the neutralization reaction are water and the salt calcium acetate, $Ca(CH_3COO)_2$, formed from the base's cation, Ca^{2+}, and the acid's anion, CH_3COO^-. Table 5.1 shows that calcium acetate is soluble.

- **Write the unbalanced equation with all reactants and products.**

Write the four species involved in the reaction, not worrying for the moment about balancing the equation.

(unbalanced equation) $CH_3COOH(aq) + Ca(OH)_2(aq) \longrightarrow Ca(CH_3COO)_2(aq) + H_2O(\ell)$

- **Balance the ions first, then the hydrogen and oxygen atoms.**

Add a coefficient of 2 to the acetic acid since two acetate ions appear in the products.

(unbalanced equation) $2\ CH_3COOH(aq) + Ca(OH)_2(aq) \longrightarrow Ca(CH_3COO)_2(aq) + H_2O(\ell)$

Two hydrogen ions, $H^+(aq)$, from the acetic acid react with two hydroxide ions, $OH^-(aq)$, from the calcium hydroxide, producing two water molecules. To balance the equation, we must put a coefficient of 2 in front of the water.

(balanced equation) $2\ CH_3COOH(aq) + Ca(OH)_2(aq) \longrightarrow Ca(CH_3COO)_2(aq) + 2\ H_2O(\ell)$

- **Write the complete ionic equation.**

To write a complete ionic equation, we determine whether the four substances involved in the reaction are strong or weak electrolytes. Acetic acid is a weak electrolyte (it is a weak acid). Calcium hydroxide is a strong electrolyte (strong base). Calcium acetate is a strong electrolyte (Table 5.1). Water is a molecular compound and a nonelectrolyte. The complete ionic equation is

$$2\ CH_3COOH(aq) + Ca^{2+}(aq) + 2\ OH^-(aq) \longrightarrow Ca^{2+}(aq) + 2\ CH_3COO^-(aq) + 2\ H_2O(\ell)$$

- **Write the net ionic equation by canceling spectator ions from the complete ionic equation.**

The calcium ions are spectator ions and are canceled to give the net ionic equation.

$$CH_3COOH(aq) + OH^-(aq) \longrightarrow CH_3COO^-(aq) + H_2O(\ell)$$

☑ **Reasonable Answer Check** Each side of the net ionic equation has the same charge (1−) and the same number and types of atoms. The answer is reasonable.

PROBLEM-SOLVING PRACTICE 5.7

Write a balanced equation for the reaction of hydroiodic acid, HI, with calcium hydroxide, $Ca(OH)_2$. Then write the balanced complete ionic equation and the net ionic equation for this neutralization reaction.

CONCEPTUAL EXERCISE 5.4 Neutralizations and Net Ionic Equations

Write balanced complete ionic equations and net ionic equations for the neutralization reactions of these acids and bases:

(a) HCl and KOH

(b) H_2SO_4 and $Ba(OH)_2$ (Remember that sulfuric acid can provide 2 mol $H^+(aq)$ per 1 mol sulfuric acid.)

(c) CH_3COOH and NaOH

CONCEPTUAL EXERCISE 5.5 Net Ionic Equations and Antacids

The commercial antacids Maalox, Di-Gel tablets, and Mylanta contain aluminum hydroxide or magnesium hydroxide that reacts with excess hydrochloric acid in the stomach. Write the balanced complete ionic equation and net ionic equation for the soothing neutralization reaction of aluminum hydroxide with HCl. Assume that dissolved aluminum hydroxide is completely dissociated.

Figure 5.6 Reaction of coral, calcium carbonate, with an acid. A piece of coral that is largely calcium carbonate, $CaCO_3$, reacts readily with hydrochloric acid to give CO_2 gas and aqueous calcium chloride.

This answers the question posed in Chapter 1 (← *p. 3*), "Why do some antacids fizz when added to vinegar?"

Gas-Forming Exchange Reactions

The formation of a gas is the third way that exchange reactions can occur, since formation of the gas removes the molecular product from the solution. Escape of the gas from the solution removes ions from the solution. Acids are involved in many gas-forming exchange reactions.

The reaction of a metal carbonate with an acid is an excellent example of a gas-forming exchange reaction. Figure 5.6 shows that coral, which is mainly calcium carbonate, reacts with acid. As the world's oceans become more acidic due to industrial pollutants and increasing levels of CO_2, dissolution of coral is becoming a significant environmental issue.

$$CaCO_3(s) + 2\,HCl(aq) \longrightarrow CaCl_2(aq) + \cancel{H_2CO_3(aq)}$$
$$\cancel{H_2CO_3(aq)} \longrightarrow H_2O(\ell) + CO_2(g)$$
$$\text{Overall reaction:}\quad CaCO_3(s) + 2\,HCl(aq) \longrightarrow CaCl_2(aq) + H_2O(\ell) + CO_2(g)$$

A salt and H_2CO_3 (carbonic acid) are always the products from an acid reacting with a metal carbonate, and their formation illustrates the exchange reaction pattern. Carbonic acid is unstable, however, and much of it is rapidly converted to water and CO_2 gas. If the reaction is done in an open container, most of the gas bubbles out of the solution.

Carbon dioxide is always released when acids react with a metal carbonate or a metal hydrogen carbonate. For example, excess hydrochloric acid in the stomach is neutralized by ingesting commercial antacids such as Alka-Seltzer ($NaHCO_3$), Tums ($CaCO_3$), or Di-Gel liquid ($MgCO_3$). Taking an Alka-Seltzer or a Tums to relieve excess stomach acid produces these helpful reactions:

$$\text{Alka-Seltzer:}\quad NaHCO_3(aq) + HCl(aq) \longrightarrow NaCl(aq) + H_2O(\ell) + CO_2(g)$$
$$\text{Tums:}\quad CaCO_3(aq) + 2\,HCl(aq) \longrightarrow CaCl_2(aq) + H_2O(\ell) + CO_2(g)$$

The net ionic equations for these two reactions are

$$HCO_3^-(aq) + H^+(aq) \longrightarrow H_2O(\ell) + CO_2(g)$$
$$CO_3^{2-}(aq) + 2\,H^+(aq) \longrightarrow H_2O(\ell) + CO_2(g)$$

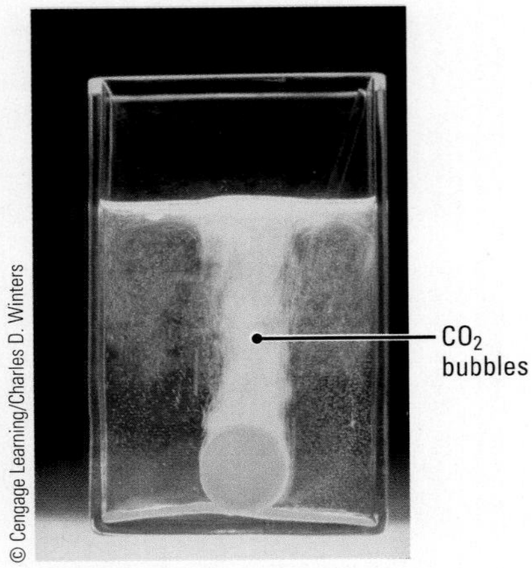

Antacid reacting with HCl.

Acids also react by exchange reactions with metal sulfites or sulfides to produce foul-smelling gaseous SO_2 or H_2S, respectively. With sulfites, the initial product is sulfurous acid, H_2SO_3, which, like carbonic acid, quickly decomposes.

$$CaSO_3(aq) + 2\,HCl(aq) \longrightarrow CaCl_2(aq) + \cancel{H_2SO_3(aq)}$$
$$\cancel{H_2SO_3(aq)} \longrightarrow H_2O(\ell) + SO_2(g)$$
$$\text{Overall reaction:}\quad CaSO_3(aq) + 2\,HCl(aq) \longrightarrow CaCl_2(aq) + H_2O(\ell) + SO_2(g)$$

With sulfides, the gaseous product H_2S is formed directly.

$$Na_2S(aq) + 2\,HCl(aq) \longrightarrow 2\,NaCl(aq) + H_2S(g)$$

CHEMISTRY IN THE NEWS Stream Cleaning with Chemistry

One hundred and fifty years of anthracite coal mining in northeastern Pennsylvania has left more than 2000 miles of streams degraded. Two common types of such degradation are due to acid runoff and to deposits of iron(III) hydroxide, $Fe(OH)_3$. Coal contains Fe^{2+} in pyrite, FeS_2, and the Fe^{2+} is oxidized to Fe^{+3} by exposure to air and water. When Fe^{3+} forms, it precipitates as red-orange $Fe(OH)_3$. Iron(III) hydroxide coats banks and rocks of streambeds in the coal regions of northeastern Pennsylvania, an unmistakable sign of water pollution.

One method for remediating this damage involves passing acidic stream water over limestone, $CaCO_3$. The pulverized limestone is dispersed in the stream water, releasing Ca^{2+} and CO_3^{2-} ions. The acidity of the stream is neutralized by these reactions.

$$CaCO_3(aq) + H^+(aq) \longrightarrow HCO_3^-(aq) + Ca^{2+}(aq)$$

$$HCO_3^-(aq) + H^+(aq) \longrightarrow H_2CO_3(aq) \longrightarrow H_2O(\ell) + CO_2(g)$$

Through this process, the stream water's acidity is almost neutralized, creating a condition much more favorable for aquatic life.

Source: Ritter, S.K. "Chemistry Unexpected," *Chemical & Engineering News*, Aug. 27, 2007; p. 40.

Carl S. Kirby

Acid coal mine drainage produces red iron(III) hydroxide. Here, acid runoff is being investigated by a Bucknell student and Harriet, the Geochemistry Field Dog.

CONCEPTUAL EXERCISE 5.6 Gas-Forming Reactions

Predict the products and write the balanced overall equation and the net ionic equation for each of these gas-generating reactions.

(a) $Na_2CO_3(aq) + H_2SO_4(aq) \longrightarrow$

(b) $FeS(s) + HCl(aq) \longrightarrow$

(c) $K_2SO_3(aq) + HCl(aq) \longrightarrow$

CONCEPTUAL EXERCISE 5.7 Exchange Reaction Classification

Identify each of these exchange reactions as a precipitation reaction, an acid-base reaction, or a gas-forming reaction. Predict the products of each reaction and write an overall balanced equation and net ionic equation for the reaction.

(a) $NiCO_3(s) + H_2SO_4(aq) \longrightarrow$ (b) $Sr(OH)_2(s) + HNO_3(aq) \longrightarrow$

(c) $BaCl_2(aq) + Na_2C_2O_4(aq) \longrightarrow$ (d) $PbCO_3(s) + H_2SO_4(aq) \longrightarrow$

5.3 Oxidation-Reduction Reactions

Now we turn to oxidation-reduction reactions, which are classified by what happens with electrons at the nanoscale level as a result of the reaction.

The terms "oxidation" and "reduction" come from reactions that have been known for centuries. Ancient civilizations learned how to change metal oxides and sulfides to the metal—that is, how to *reduce* ore to the metal. For example, cassiterite or tin(IV) oxide, SnO_2, is a tin ore discovered in Britain centuries ago. It is very easily reduced to tin by heating with carbon. In this reaction, tin is reduced from tin(IV) in the ore to tin metal.

SnO_2 loses oxygen and is reduced.

$$SnO_2(s) + 2\ C(s) \longrightarrow Sn(s) + 2\ CO(g)$$

When SnO_2 is reduced by carbon, oxygen is removed from the tin and added to the carbon, which is "oxidized" by the addition of oxygen. In fact, *any process in which oxygen is added to another substance is an oxidation.*

When magnesium burns in air, the magnesium is oxidized to magnesium oxide, MgO.

Mg combines with oxygen and is oxidized.

$$2\ Mg(s) + O_2(g) \longrightarrow 2\ MgO(s)$$

The experimental observations we have just outlined point to several fundamental conclusions:

- **If one reactant is oxidized, another reactant in the same reaction must simultaneously be reduced.** For this reason, we refer to such reactions as **oxidation-reduction reactions**, or **redox reactions** for short.
- **Oxidation is the reverse of reduction.**

For example, the reactions we have just described show that addition of oxygen is oxidation and removal of oxygen is reduction. But oxidation and reduction are more than that, as we see next.

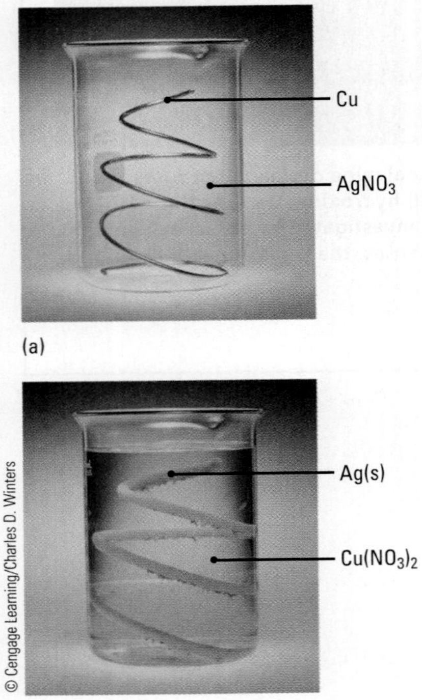

(b)

Figure 5.7 Oxidation of copper metal by silver ion. (a) A spiral of copper wire was immersed in an aqueous solution of silver nitrate, $AgNO_3$. (b) Copper metal reduces Ag^+ ions to silver metal crystals, and copper metal is oxidized to Cu^{2+} ions. The blue color of the solution is due to the presence of aqueous copper(II) ions formed by this oxidation.

Redox Reactions and Electron Transfer

Oxidation and reduction reactions involve transfer of electrons from one reactant atom, molecule, or ion to another.

- **When a species *accepts electrons*, it is *reduced*.**

The language is descriptive because in a **reduction** there is a decrease (reduction) in the real or apparent electric charge on an atom, molecule, or ion. For example, in this net ionic equation, Ag^+ ions are reduced to uncharged Ag atoms by accepting electrons from copper atoms (Figure 5.7).

Each Ag^+ accepts an electron and is reduced to Ag. ($+2e^-$)

$$2\ Ag^+(aq) + Cu(s) \longrightarrow 2\ Ag(s) + Cu^{2+}(aq)$$

($-2e^-$) Each Cu donates two electrons and is oxidized to Cu^{2+}.

- **When a species *loses electrons*, it is *oxidized*.**

In **oxidation**, the real or apparent electrical charge on an atom, molecule, or ion of the substance increases when it gives up electrons. In our example, a copper metal atom releases two electrons, forming a Cu^{2+} ion; the electric charge has increased from 0 to 2+, and the copper atom has been oxidized. For this to happen, another substance must be available to take the electrons donated by the copper. In this case, Ag^+ is the electron acceptor.

- **In every redox reaction, a reactant is reduced and a reactant is oxidized simultaneously.**

In the reaction of magnesium with oxygen (Figure 5.8), each oxygen atom gains electrons when converted to an oxide ion. The charge of each O atom changes from 0 to 2− as it is reduced.

Oxidation is the loss of electrons.

$$X \rightarrow X^+ + e^-$$

X loses one or more electrons and is oxidized.

Reduction is the gain of electrons.

$$Y + e^- \rightarrow Y^-$$

Y gains one or more electrons and is reduced.

Mg loses $2e^-$ per atom.
Mg is oxidized.

$$2\,Mg(s) + O_2(g) \longrightarrow 2\,[Mg^{2+} + O^{2-}]$$

O_2 gains $4e^-$ per molecule, 2 for each O.
O_2 is reduced.

Each magnesium atom changes from 0 to 2+ as it is oxidized. All redox reactions can be analyzed in a similar manner.

© Cengage Learning/Charles D. Winters

Figure 5.8 **$Mg(s) + O_2(g)$.** A piece of magnesium ribbon burns in air, oxidizing the metal to the white solid magnesium oxide, MgO.

Common Oxidizing and Reducing Agents

As stated previously, in every redox reaction, a reactant species is oxidized and a reactant species is reduced. The species causing the oxidation (electron loss) is the **oxidizing agent,** and the species causing the reduction (electron gain) is the **reducing agent.** As a redox reaction proceeds, *the oxidizing agent is reduced and the reducing agent is oxidized.* In the reaction just described between Mg and O_2, the reactant species Mg atoms are oxidized, and Mg is the reducing agent; the reactant species O_2 molecules are reduced, and O_2 is the oxidizing agent. *Note that the oxidizing agent and the reducing agent are always reactants, not products.* Figure 5.9 provides some guidelines to determine which species involved in a redox reaction is the oxidizing agent and which is the reducing agent.

Like oxygen, the halogens (F_2, Cl_2, Br_2, and I_2) are always oxidizing agents in their reactions with metals and most nonmetals. For example, consider the combination reaction of sodium metal with chlorine:

A useful memory aid for keeping the oxidation and reduction definitions straight is **OIL RIG (Oxidation Is Loss; Reduction Is Gain).**

Na metal loses $1e^-$ per atom to form a Na^+ ion.
Na metal is oxidized and is the reducing agent.

$$2\,Na(s) + Cl_2(g) \longrightarrow 2\,[Na^+ + Cl^-]$$

Cl_2 gains $2e^-$ per molecule to form 2 Cl^-.
Cl_2 is reduced and is the oxidizing agent.

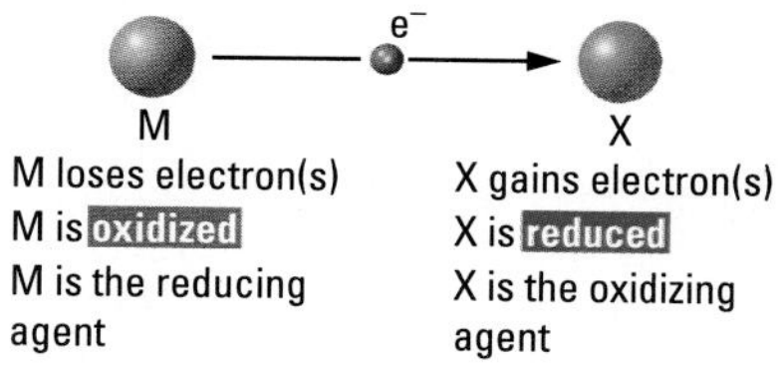

Figure 5.9 **Oxidation-reduction relationships and electron transfer.**

Here, sodium begins as atoms in the metallic element, but it ends up as Na^+ ions in NaCl after combining with chlorine. Thus, sodium metal is oxidized (loses electrons) and is the reducing agent. Chlorine ends up as Cl^-; Cl_2 has been reduced (gains electrons) and therefore is the oxidizing agent.

The general reaction for halogen, X_2, reduction is

$$\text{Reduction reaction:} \quad \underset{\text{oxidizing agent}}{X_2} + 2e^- \longrightarrow 2\,X^-$$

That is, a halogen will always oxidize a metal to give a metal halide, and the formula of the product can be predicted from the charge on the metal ion and the charge of the halide. The halogens in decreasing order of oxidizing ability are as follows:

Note that the oxidizing agent is reduced, and the reducing agent is oxidized.

Oxidizing Agent	Usual Reduction Product
F_2 (strongest)	F^-
Cl_2	Cl^-
Br_2	Br^-
I_2 (weakest)	I^-

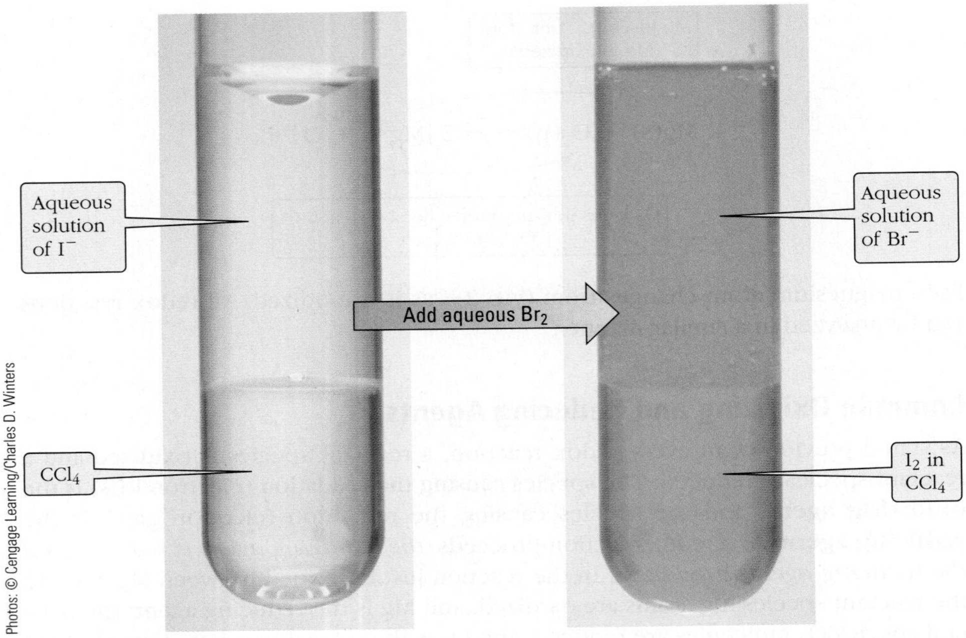
Aqueous solution of I⁻
CCl4
Add aqueous Br2
Aqueous solution of Br⁻
I2 in CCl4
Photos: © Cengage Learning/Charles D. Winters

Figure 5.10 **Oxidation of I^- by Br_2.** This demonstration shows that Br_2 is a better oxidizing agent than I_2. Iodide ions, I^-, in the colorless NaI solution are oxidized by the Br_2 to form I_2, which produces the violet color in the right-hand photo.

A consequence of the relative halogen oxidizing agent strength is that a halogen higher in the list will oxidize the halide ion of another halogen lower in the list. For example, as illustrated in Figure 5.10, Br_2 oxidizes $I^-(aq)$ to form I_2 in this redox reaction

$$Br_2(aq) + 2\,I^-(aq) \longrightarrow 2\,Br^-(aq) + I_2(aq)$$

CONCEPTUAL EXERCISE 5.8 Oxidizing and Reducing Agents

Identify which species is losing electrons and which is gaining electrons, which is oxidized and which is reduced, and which is the oxidizing agent and which is the reducing agent in this reaction:

$$2\,Ca(s) + O_2(g) \longrightarrow 2\,CaO(s)$$

CONCEPTUAL EXERCISE 5.9 Redox Reactions

Write the balanced chemical equation for chlorine gas undergoing a redox reaction with calcium metal. Which species is the oxidizing agent?

Chlorine is widely used as an oxidizing agent in water and sewage treatment. A common contaminant of water is hydrogen sulfide, H_2S, which gives a thoroughly unpleasant "rotten egg" odor to the water and may come from the decay of organic matter or from underground mineral deposits. Chlorine oxidizes H_2S to insoluble elemental sulfur, which is easily removed.

$$8\,Cl_2(g) + 8\,H_2S(aq) \longrightarrow S_8(s) + 16\,HCl(aq)$$

Oxidation and reduction occur readily when a strong oxidizing agent comes into contact with a strong reducing agent. Knowing the easily recognized oxidizing and reducing agents enables you to predict that a reaction will take place when they are combined and in some cases to predict what the products will be. Table 5.3 and the following points provide some guidelines.

- **An element that has combined with oxygen has been oxidized**. In the process each oxygen atom in oxygen, O_2, gains two electrons from the other element and becomes the oxide ion, O^{2-} (as in a metal oxide). When reacting, oxygen can also be combined in a molecule such as CO_2 or H_2O (as occurs in the combustion reaction of a hydrocarbon). In such cases, oxygen is reduced, and since it has accepted electrons, oxygen is the oxidizing agent.
- **An element that has combined with a halogen has been oxidized**. In the process the halogen, X_2, is changed to halide ions, X^-, by adding an electron to each halogen atom. Therefore, the halogen atom has been reduced to the halide ion, and the halogen is the oxidizing agent. A halogen can also be combined in a molecule such as HCl. Among the halogens, fluorine and chlorine are particularly strong oxidizing agents.
- **When an elemental metal combines with something to form an ionic compound, the metal has been oxidized**. In the process, it has lost electrons, usually to form a positive ion.

$$\text{Oxidation reaction:} \quad \underset{\text{reducing agent}}{\text{M}} \longrightarrow \text{M}^{n+} + n\text{e}^-$$

  Therefore, the metal (an electron donor) has been oxidized and has functioned as a reducing agent. Most metals are reasonably good reducing agents, and metals from Groups 1A, 2A, and 3A such as sodium, magnesium, and aluminum are particularly good ones.
- **Other common oxidizing and reducing agents are listed in Table 5.3, and some are described below**. When one of these agents takes part in a reaction, it is reasonably certain that it is a redox reaction. (Nitric acid can be an exception. In addition to being a good oxidizing agent, it is an acid and functions only as an acid in reactions such as the decomposition of a metal carbonate, a non-redox reaction.)

There are exceptions to the guideline that metals are always positively charged in compounds. However, you probably will not encounter these exceptions in introductory chemistry.

© Cengage Learning/Charles D. Winters

Figure 5.11 **Cu(s) + HNO_3(aq).** Copper reacts vigorously with concentrated nitric acid to give brown NO_2 gas.

Figure 5.11 illustrates the action of concentrated nitric acid, HNO_3, as an oxidizing agent. Nitric acid oxidizes copper metal to give copper(II) nitrate, and copper metal reduces nitric acid to the brown gas NO_2. The net ionic equation is

$$\underset{\text{reducing agent}}{\text{Cu(s)}} + \underset{\text{oxidizing agent}}{4\,\text{H}^+\text{(aq)} + 2\,\text{NO}_3^-\text{(aq)}} \longrightarrow \text{Cu}^{2+}\text{(aq)} + 2\,\text{NO}_2\text{(g)} + 2\,\text{H}_2\text{O}(\ell)$$

Table 5.3 Common Oxidizing and Reducing Agents

Oxidizing Agent (oxidizing agents are reduced)	**Reaction Product** (reduced form)	**Reducing Agent** (reducing agents are oxidized)	**Reaction Product** (oxidized form)
O_2	O^{2-} or an oxygen-containing molecular compound	H_2 or hydrogen-containing molecular compound	H^+ or H combined in H_2O
H_2O_2 (hydrogen peroxide)	$H_2O(\ell)$	C used to reduce metal oxides	CO and CO_2
F_2, Cl_2, Br_2, or I_2 (halogens)	F^-, Cl^-, Br^-, or I^- (halide ions)	M, metals such as Na, K, Fe, or Al	M^{n+}, metal ions such as Na^+, K^+, Fe^{3+}, or Al^{3+}
HNO_3 (nitric acid)	Nitrogen oxides such as NO and NO_2		
$Cr_2O_7^{2-}$ (dichromate ion)	Cr^{3+} (chromium(III) ion), in acid solution		
MnO_4^- (permanganate ion)	Mn^{2+} (manganese(II) ion), in acid solution		

The metal is the reducing agent, since it is the substance oxidized. In fact, metals are the most common reducing agents.

Some metal ions such as Fe^{2+} can also be reducing agents because they can be oxidized to ions of higher charge. Aqueous Fe^{2+} ion reacts readily with the strong oxidizing agent MnO_4^-, the permanganate ion. The Fe^{2+} ion is oxidized to Fe^{3+}, and the MnO_4^- ion is reduced to the Mn^{2+} ion.

$$5\ Fe^{2+}(aq) + MnO_4^-(aq) + 8\ H^+(aq) \longrightarrow 5\ Fe^{3+}(aq) + Mn^{2+}(aq) + 4\ H_2O(\ell)$$

Carbon can reduce many metal oxides to metals, and it is widely used in the metals industry to obtain metals from their compounds in ores. For example, titanium is produced by treating a mineral containing titanium(IV) oxide with carbon and chlorine.

$$TiO_2(s) + C(s) + 2\ Cl_2(g) \longrightarrow TiCl_4(\ell) + CO_2(g)$$

In effect, the carbon reduces the metal oxide to titanium metal, and the chlorine then oxidizes it to titanium(IV) chloride. Because $TiCl_4$ is easily converted to a gas, it can be removed from the reaction mixture and purified. The $TiCl_4$ is then reduced with another metal, such as magnesium, to give titanium metal.

$$TiCl_4(\ell) + 2\ Mg(s) \longrightarrow Ti(s) + 2\ MgCl_2(s)$$

Finally, H_2 gas is a common reducing agent, widely used in the laboratory and in industry. For example, it readily reduces copper(II) oxide to copper metal (Figure 5.12).

$$\underset{\text{reducing agent}}{H_2(g)} + \underset{\text{oxidizing agent}}{CuO(s)} \longrightarrow Cu(s) + H_2O(g)$$

It is important to be aware that it can be dangerous to mix a strong oxidizing agent with a strong reducing agent. A violent reaction, even an explosion, may take place. Chemicals should not be stored on laboratory shelves in alphabetical order, because such an ordering may place a strong oxidizing agent next to a strong reducing agent. In particular, swimming pool chemicals that contain chlorine and are strong oxidizing agents should not be stored in the hardware store or the garage next to easily oxidized materials such as ammonia.

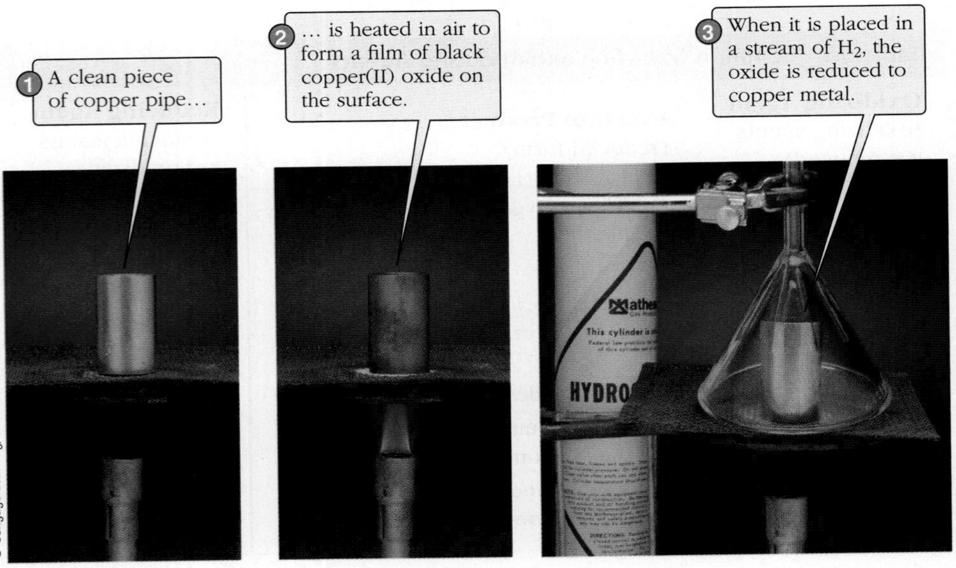

© Cengage Learning/Charles D. Winters

Figure 5.12 Reduction of copper oxide with hydrogen. A clean piece of copper pipe is heated in air to form a film of black copper(II) oxide on the surface. When the hot copper metal, with its film of CuO, is placed in a stream of hydrogen gas (from the yellow tank at the rear), the oxide is reduced to copper metal, and water forms as the by-product.

CONCEPTUAL EXERCISE 5.10 Oxidation-Reduction Reactions

Decide which of these reactions are oxidation-reduction reactions. In each case explain your choice and identify the oxidizing and reducing agents in the redox reactions.

(a) $NaOH(aq) + HNO_3(aq) \longrightarrow NaNO_3(aq) + H_2O(\ell)$

(b) $4\ Cr(s) + 3\ O_2(g) \longrightarrow 2\ Cr_2O_3(s)$

(c) $NiCO_3(s) + 2\ HCl(aq) \longrightarrow NiCl_2(aq) + H_2O(\ell) + CO_2(g)$

(d) $Cu(s) + Cl_2(g) \longrightarrow CuCl_2(s)$

5.4 Oxidation Numbers and Redox Reactions

An arbitrary bookkeeping system has been devised for keeping track of electrons in redox reactions. It extends the obvious oxidation and reduction case when neutral atoms become ions to reactions in which the changes are less obvious. The system is set up so that ***oxidation numbers always change in redox reactions.*** As a result, oxidation and reduction can be determined in the ways shown in Table 5.4.

How electrons participate in bonding atoms in molecules is the subject of Chapter 8.

An **oxidation number** compares the charge of an uncombined atom with its actual charge or its relative charge in a compound. All neutral atoms have an equal number of protons and electrons and thus have no net charge.

Oxidation numbers of atoms in molecular compounds are assigned as though electrons were completely transferred to form ions. In the molecular compound phosphorus trichloride (PCl_3), for example, chlorine is assigned an oxidation number of -1 even though it is not a Cl^- ion; the chlorine is directly bonded to the phosphorus. Each chlorine atom in PCl_3 is thought of as "possessing" more electrons than each chlorine has in Cl_2. This parallels the fact that a Cl^- ion has one more electron than a Cl atom.

You can use this set of rules to determine oxidation numbers.

Oxidation numbers are also called oxidation states.

Rule 1: ***The oxidation number of an atom of a pure element is 0.*** When the atoms are not combined with those of any other element (for example, oxygen in O_2, sulfur in S_8, iron in metallic Fe, or chlorine in Cl_2), the oxidation number is 0.

Rule 2: ***The oxidation number of a monatomic ion equals its charge.*** Thus, the oxidation number of Cu^{2+} is $+2$; that of S^{2-} is -2.

In this book, oxidation numbers are written as $+1$, $+2$, etc., whereas charges on ions are written as $1+$, $2+$, etc.

Rule 3: ***Atoms of some elements have the same oxidation number in almost all their compounds and can be used as references for oxidation numbers of atoms of other elements in compounds.***

(a) Hydrogen has an oxidation number of $+1$ unless it is combined with a metal, in which case its oxidation number is -1.

(b) Fluorine has an oxidation number of -1 in all its compounds. Halogens other than fluorine have an oxidation number of -1 except when combined with a halogen above them in the periodic table or with oxygen.

Table 5.4 Recognizing Oxidation-Reduction Reactions

	Oxidation	Reduction
In terms of oxygen	Gain of oxygen	Loss of oxygen
In terms of halogen	Gain of halogen	Loss of halogen
In terms of hydrogen	Loss of hydrogen	Gain of hydrogen
In terms of electrons	Loss of electrons	Gain of electrons
In terms of oxidation numbers	Increase of oxidation number	Decrease of oxidation number

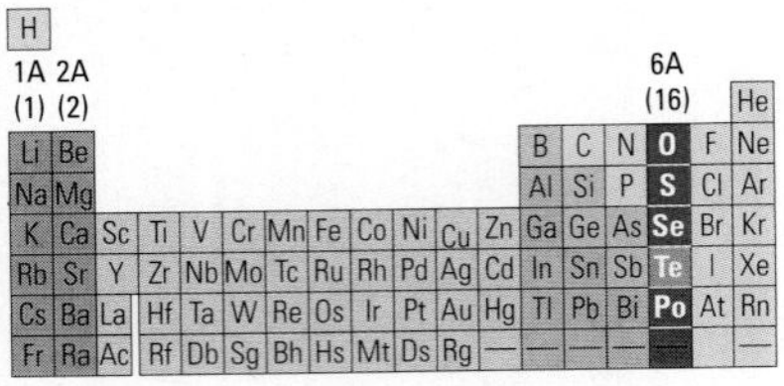

(c) Oxygen has an oxidation number of −2 except in peroxides, such as hydrogen peroxide, H_2O_2, in which oxygen has an oxidation number of −1 (and hydrogen is +1).

(d) In binary compounds (compounds of two elements), atoms of the alkali metals of Group 1A elements (Li, Na, K, Rb, Cs) always have an oxidation number of +1.

(e) In binary compounds, atoms of the alkaline-earth metals of Group 2A elements (Be, Mg, Ca, Sr, Ba) always have an oxidation number of +2.

(f) In binary compounds, atoms of Group 6A elements (O, S, Se, Te) have an oxidation number of −2 except when combined with oxygen or halogens, in which case the Group 6A elements have positive oxidation numbers.

Rule 4: ***The sum of the oxidation numbers of the atoms in an uncharged compound is 0; the sum of the oxidation numbers of the atoms in a polyatomic ion equals the charge on the ion.*** For example, in SO_2, the oxidation number of oxygen is −2, and with two O atoms, the total for oxygen is −4. Because the sum of the oxidation numbers must equal zero, the oxidation number of sulfur must be +4: (+4) + 2(−2) = 0. In the sulfite ion, SO_3^{2-}, the net charge is 2−. Because each oxygen is −2, the oxidation number of sulfur in sulfite must be +4: (+4) + 3(−2) = 2−.

Oxidation numbers are written above the chemical symbols for the atoms to which they apply.

$$\overset{+4-2}{SO_3^{2-}}$$

Now, let's apply these rules to the equations for simple combination and displacement reactions involving sulfur and oxygen.

$$\text{Combination:}\quad \overset{0}{S_8}(s) + 8\,\overset{0}{O_2}(g) \longrightarrow 8\,\overset{+4-2}{SO_2}(g)$$

$$\text{Combination:}\quad \overset{+2-2}{ZnS}(s) + 2\,\overset{0}{O_2}(aq) \longrightarrow \overset{+2+6-2}{ZnSO_4}(aq)$$

$$\text{Displacement:}\quad \overset{+1\ -2}{Cu_2S}(s) + \overset{0}{O_2}(g) \longrightarrow 2\,\overset{0}{Cu}(s) + \overset{+4-2}{SO_2}(g)$$

These are all oxidation-reduction reactions, as shown by the fact that there has been a change in the oxidation numbers of atoms from reactants to products.

Every reaction in which a free (uncombined) element becomes combined in a compound is a redox reaction. The oxidation number of the free element must increase or decrease from its original value of zero. Combination reactions and displacement reactions in which one element displaces another are all redox reactions.

Those decomposition reactions in which elemental gases are produced are also redox reactions. Millions of tons of ammonium nitrate, NH_4NO_3, are used as fertilizer to supply nitrogen to crops. Ammonium nitrate is also used as an explosive that is decomposed by heating.

Ammonium nitrate was used in the 1995 bombing of the Federal Building in Oklahoma City, Oklahoma.

$$2\,NH_4NO_3(s) \longrightarrow 2\,N_2(g) + 4\,H_2O(g) + O_2(g)$$

Like a number of other explosives, ammonium nitrate contains an element with two different oxidation numbers, in effect having an oxidizing and reducing agent in the same compound.

$$\overset{-3+1}{NH_4^+} \qquad \overset{+5-2}{NO_3^-}$$

Note that nitrogen's oxidation number is −3 in the ammonium ion and +5 in the nitrate ion. Therefore, when ammonium nitrate decomposes and forms N_2, the N in the ammonium ion is oxidized from −3 to 0, and the ammonium ion is the reducing agent. The N in the nitrate ion is reduced from +5 to 0, and the nitrate ion is the oxidizing agent.

PROBLEM-SOLVING EXAMPLE 5.8 Applying Oxidation Numbers

Metallic copper and dilute nitric acid react according to this redox equation:

$$3\ Cu(s) + 8\ HNO_3(aq) \longrightarrow 3\ Cu(NO_3)_2(aq) + 2\ NO(g) + 4\ H_2O(\ell)$$

Assign oxidation numbers for each atom in the equation. Identify which element has been oxidized and which has been reduced.

Answer

$$\overset{0}{3\ Cu(s)} + 8\ \overset{+1+5-2}{HNO_3(aq)} \longrightarrow 3\ \overset{+2+5-2}{Cu(NO_3)_2(aq)} + 2\ \overset{+2-2}{NO(g)} + 4\ \overset{+1\ -2}{H_2O(\ell)}$$

Copper metal is oxidized. Nitrogen (in HNO_3) is reduced.

Strategy and Explanation Use the four rules introduced earlier and knowledge of the formulas of polyatomic ions to assign oxidation numbers.

- **The first reactant, copper, is in its elemental state, so its oxidation number is 0** (Rule 1).
- **Consider the second reactant, HNO_3.**

Nitric acid is a strong acid and therefore consists of $H^+(aq)$ and $NO_3^-(aq)$. $H^+(aq)$ has oxidation number +1. Because the oxidation number of each oxygen is −2 (Rule 3c), for a total of −6, and the charge on a nitrate ion is −1, the oxidation number of nitrogen in nitrate ion must be +5: $-1 = 3(-2) + (+5)$.

- **Consider the product, $Cu(NO_3)_2$.**

Assign the oxidation number +2 to copper to balance the charges of the two nitrate ions. The oxidation numbers of the oxygen and nitrogen atoms within the nitrate anion are the same as in the reactant nitric acid.

- **Consider the product NO.**

Rule 3c assigns an oxidation number of −2 to oxygen, so the oxidation number of the nitrogen in NO must be +2 (Rule 4).

- **Consider water.**

The oxygen in water has the oxidation number −2 (Rule 3c), and the hydrogen is +1 (Rule 3a).

- **Identify oxidation number changes to determine which atoms, if any, have been oxidized or reduced.**

Copper has changed from an oxidation number of 0 in copper metal to +2 in Cu^{2+}; copper metal has been oxidized. Nitrogen has changed from an oxidation number of +5 in NO_3^- to +2 in NO; the N in NO_3^- has been reduced.

PROBLEM-SOLVING PRACTICE 5.8

Determine the oxidation number for each atom in this equation:

$$Sb_2S_3(s) + 3\ Fe(s) \longrightarrow 3\ FeS(s) + 2\ Sb(s)$$

Cite the oxidation number rule(s) you used to obtain your answers.

PROBLEM-SOLVING EXAMPLE 5.9 Oxidation-Reduction Reaction

Most metals we use are found in nature as cations in ores. The metal ion must be reduced to its elemental form, which is done with an appropriate oxidation-reduction reaction. The copper ore chalcocite, Cu_2S, is reacted with oxygen to form metallic copper.

$$Cu_2S(s) + O_2(g) \longrightarrow 2\ Cu(s) + SO_2(g)$$

Identify the atoms that are oxidized and reduced, and name the oxidizing and reducing agents.

Answer Cu^+ ions and O_2 are reduced; S^{2-} ions are oxidized. O_2 and Cu^+ are the oxidizing agents. S^{2-} ion in Cu_2S is the reducing agent.

Strategy and Explanation We first assign the oxidation numbers for all the atoms in the reaction according to Rules 1 through 4.

$$\overset{+1\ -2}{Cu_2S(s)} + \overset{0}{O_2(g)} \longrightarrow 2\,\overset{0}{Cu(s)} + \overset{+4-2}{SO_2(g)}$$

The oxidation number of Cu^+ decreases from +1 to 0, so Cu^+ is reduced. The oxidation number of S increases from −2 to +4, so S^{2-} is oxidized. The oxidation number of oxygen decreases from 0 to −2, so oxygen is reduced. The oxidizing agents are Cu^+ and O_2, which accept electrons. The reducing agent is S^{2-} in Cu_2S, which donates electrons.

PROBLEM-SOLVING PRACTICE 5.9

Which atoms are oxidized and reduced in this reaction? Which are the oxidizing and reducing agents?

$$PbO(s) + CO(g) \longrightarrow Pb(s) + CO_2(g)$$

Exchange reactions of ionic compounds in aqueous solutions are not redox reactions because no change of oxidation numbers occurs. Consider, for example, the precipitation of barium sulfate when aqueous solutions of barium chloride and sulfuric acid are mixed.

$$Ba^{2+}(aq) + 2\,Cl^-(aq) + 2\,H^+(aq) + SO_4^{2-}(aq) \longrightarrow BaSO_4(s) + 2\,H^+(aq) + 2\,Cl^-(aq)$$

Net ionic equation: $Ba^{2+}(aq) + SO_4^{2-}(aq) \longrightarrow BaSO_4(s)$

The oxidation numbers of all atoms remain unchanged from the reactants to products, so this is not a redox reaction.

CONCEPTUAL EXERCISE **5.11 Redox in CFC Disposal**

This redox reaction is used for the disposal of chlorofluorocarbons (CFCs) by their reaction with sodium oxalate, $Na_2C_2O_4$:

$$CF_2Cl_2(g) + 2\,Na_2C_2O_4(s) \longrightarrow 2\,NaF(s) + 2\,NaCl(s) + C(s) + 4\,CO_2(g)$$

(a) What is oxidized in this reaction?
(b) What is reduced?

5.5 Displacement Reactions, Redox, and the Activity Series

Recall that displacement reactions (← p. 126) have this reaction pattern:

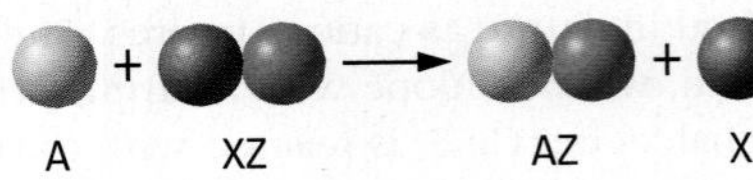

Displacement reactions, like combination reactions, are oxidation-reduction reactions. For example, in the reaction of hydrochloric acid with iron (Figure 5.13),

$$Fe(s) + 2\,HCl(aq) \longrightarrow FeCl_2(aq) + H_2(g)$$

metallic iron is the reducing agent; it is oxidized from an oxidation number of 0 in Fe(s) to +2 in $FeCl_2$. Hydrogen ions, H^+, in hydrochloric acid are reduced to hydrogen gas (H_2), in which hydrogen has an oxidation number of 0.

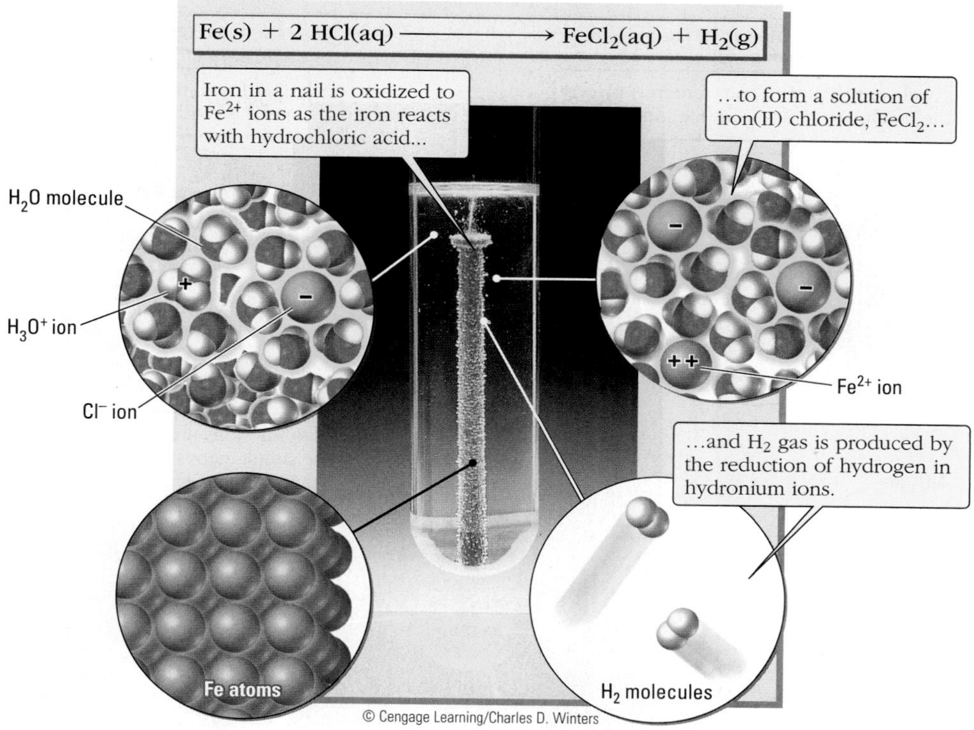

Fe(s) + 2 HCl(aq) ⟶ FeCl2(aq) + H2(g)
Iron in a nail is oxidized to Fe2+ ions as the iron reacts with hydrochloric acid...
...to form a solution of iron(II) chloride, FeCl2...
...and H2 gas is produced by the reduction of hydrogen in hydronium ions.
H2O molecule
H3O+ ion
Cl− ion
Fe2+ ion
Fe atoms
H2 molecules
© Cengage Learning/Charles D. Winters

Figure 5.13 **Metal + acid displacement reaction.**

Extensive studies with many metals have led to the development of a **metal activity series,** a ranking of relative reactivity of metals in displacement and other kinds of reactions (Table 5.5). The most reactive metals appear at the top of the series, and activity decreases going down the series. Metals at the top are powerful reducing agents and readily lose electrons to form cations. The metals at the lower end of the series are poor reducing agents. However, their cations (Au^+, Ag^+) are powerful oxidizing agents that readily gain an electron to form the free metal.

- **An element higher in the activity series (Table 5.5) will displace from its compounds an element below it in the series.**

For example, zinc displaces copper ions from copper(II) sulfate, and copper metal displaces silver ions from silver nitrate solution (Figure 5.14).

$$Zn(s) + CuSO_4(aq) \longrightarrow ZnSO_4(aq) + Cu(s)$$

$$Cu(s) + 2\,AgNO_3(aq) \longrightarrow Cu(NO_3)_2(aq) + 2\,Ag(s)$$

In each case, the elemental metal (Zn, Cu) is the reducing agent and is oxidized; Cu^{2+} ions and Ag^+ ions are oxidizing agents and are reduced to Cu(s) and Ag(s), respectively.

- **Metals above hydrogen in the activity series react with acids whose anions are not oxidizing agents to form hydrogen, H_2, and the metal salt containing the cation of the metal and the anion of the acid.**

HCl and HBr are examples of such acids. HCl reacts with metallic iron to form $FeCl_2$, and HBr reacts with zinc to form $ZnBr_2$.

$$Fe(s) + 2\,HCl(aq) \longrightarrow FeCl_2(aq) + H_2(g)$$

$$Zn(s) + 2\,HBr(aq) \longrightarrow ZnBr_2(aq) + H_2(g)$$

Table 5.5 Activity Series of Metals

Displace H_2 from $H_2O(\ell)$, steam, or acid	Li, K, Ba, Sr, Ca, Na	Ease of oxidation decreases (top to bottom)
Displace H_2 from steam or acid	Mg, Al, Mn, Zn, Cr	
Displace H_2 from acid	Fe, Ni, Sn, Pb	
	H_2	
Do not displace H_2 from $H_2O(\ell)$, steam, or acid	Sb, Cu, Hg, Ag, Pd, Pt, Au	

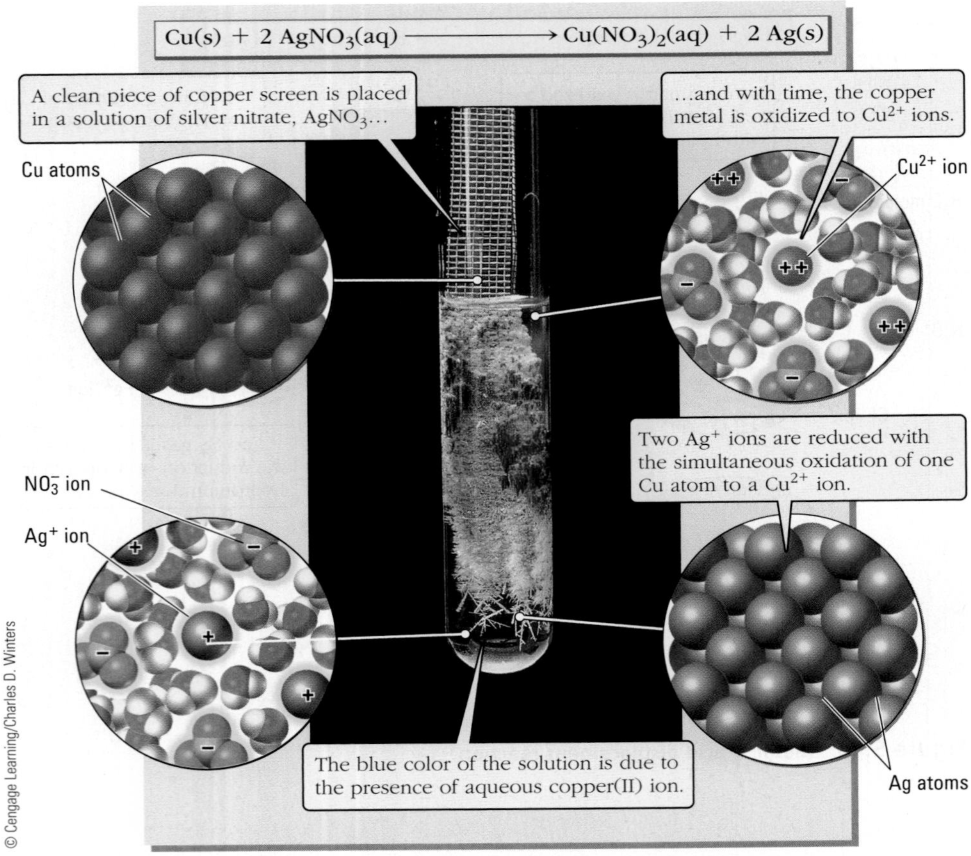

$Cu(s) + 2\ AgNO_3(aq) \longrightarrow Cu(NO_3)_2(aq) + 2\ Ag(s)$
A clean piece of copper screen is placed in a solution of silver nitrate, $AgNO_3$...
...and with time, the copper metal is oxidized to Cu^{2+} ions.
Cu atoms
Cu^{2+} ion
Two Ag^+ ions are reduced with the simultaneous oxidation of one Cu atom to a Cu^{2+} ion.
NO_3^- ion
Ag^+ ion
The blue color of the solution is due to the presence of aqueous copper(II) ion.
Ag atoms

© Cengage Learning/Charles D. Winters

Active Figure 5.14 Metal + aqueous metal salt displacement reaction. The oxidation of copper metal by silver ion. (Atoms or ions that take part in the reaction have been highlighted in the nanoscale pictures.) Visit this book's companion website at **www.cengage.com/chemistry/moore** to test your understanding of the concepts in this figure.

© Cengage Learning/Charles D. Winters

Figure 5.15 Potassium, an active metal. When a drop of water falls onto a sample of potassium metal, a vigorous reaction gives hydrogen gas and a solution of potassium hydroxide.

- **Metals below hydrogen in the activity series do not displace hydrogen from acids.**
- **Very reactive metals displace hydrogen from water.** These metals are at the top of the activity series, from lithium (Li) through sodium (Na). Some do this displacement violently (Figure 5.15).
- **Metals of intermediate reactivity displace hydrogen from steam, but not from liquid water at room temperature.** Intermediate metals include magnesium (Mg) through chromium (Cr).
- **Elements very low in the activity series are unreactive.** Sometimes called noble metals (Au, Ag, Pt, Pd), they are prized for their nonreactivity.

It is no accident that gold and silver have been used extensively for coinage since antiquity. These metals do not react with air, water, or even common acids, thus maintaining their luster (and value) for many years. Their low reactivity explains why they occur naturally as free metals and have been known as elements since antiquity. These metals are discussed in Chapter 22.

PROBLEM-SOLVING EXAMPLE 5.10 Activity Series of Metals

Use the activity series found in Table 5.5 to predict which of these reactions will occur. Complete and balance the equations for those reactions that will occur.

(a) $Cr(s) + MgCl_2(aq) \longrightarrow$

(b) $Al(s) + Pb(NO_3)_2(aq) \longrightarrow$

(c) $Mg(s) +$ hydrochloric acid $\longrightarrow$

Answer

(a) No reaction

(b) $2\ Al(s) + 3\ Pb(NO_3)_2(aq) \longrightarrow 2\ Al(NO_3)_3(aq) + 3\ Pb(s)$

(c) $Mg(s) + 2\ HCl(aq) \longrightarrow MgCl_2(aq) + H_2(g)$

Strategy and Explanation

(a) Chromium is less active than magnesium, so it will not displace magnesium ions from magnesium chloride. Therefore, no reaction occurs.

(b) Aluminum is above lead in the activity series, so aluminum will displace lead ions from a lead(II) nitrate solution to form metallic lead and Al^{3+} ions.

$$2\ Al(s) + 3\ Pb(NO_3)_2(aq) \longrightarrow 2\ Al(NO_3)_3(aq) + 3\ Pb(s)$$

(c) Magnesium is above hydrogen in the activity series, so magnesium will displace hydrogen ions from HCl to form the metal salt, $MgCl_2$, plus hydrogen gas.

$$Mg(s) + 2\ HCl(aq) \longrightarrow MgCl_2(aq) + H_2(g)$$

PROBLEM-SOLVING PRACTICE 5.10

Use Table 5.5 to predict whether each of these reactions will occur. If a reaction occurs, identify what has been oxidized or reduced and what the oxidizing agent and the reducing agent are.

(a) $2\ Al(s) + 3\ CuSO_4(aq) \longrightarrow Al_2(SO_4)_3(aq) + 3\ Cu(s)$

(b) $2\ Al(s) + Cr_2O_3(s) \longrightarrow Al_2O_3(s) + 2\ Cr(s)$

(c) $Pt(s) + 4\ HCl(aq) \longrightarrow PtCl_4(aq) + 2\ H_2(g)$

(d) $Au(s) + 3\ AgNO_3(aq) \longrightarrow Au(NO_3)_3(aq) + 3\ Ag(s)$

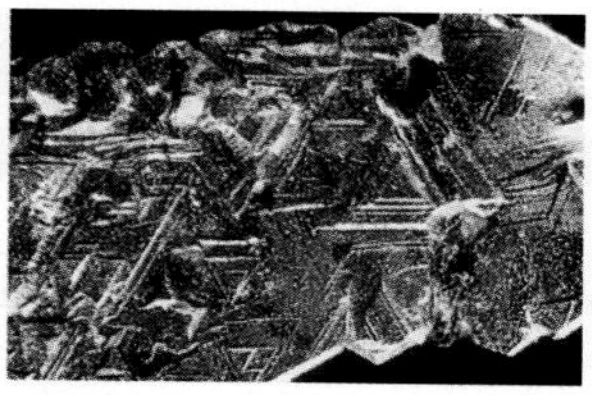

Gold

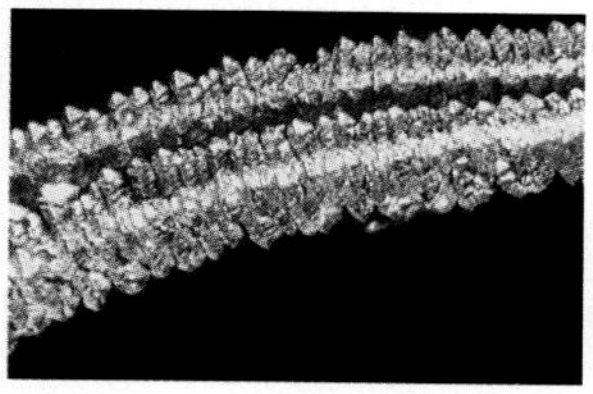

Silver

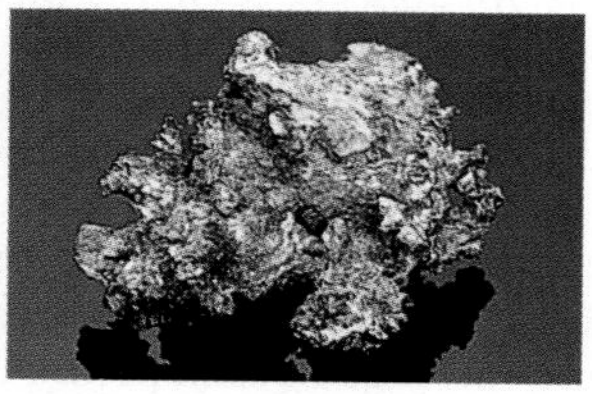

Copper

© Cengage Learning/Charles D. Winters

CONCEPTUAL EXERCISE 5.12 Reaction Product Prediction

For these pairs of reactants, predict what kind of reaction would occur and what the products might be. Which reactions are redox reactions?

(a) Combustion of ethanol: $CH_3CH_2OH(\ell) + O_2(g) \longrightarrow ?$

(b) $Fe(s) + HNO_3(aq) \longrightarrow ?$

(c) $AgNO_3(aq) + KBr(aq) \longrightarrow ?$

Other elements can also be ranked according to their oxidizing strength. Consider the halogens. As was shown previously (p. 179), the halogens have oxidizing strength in this order: F_2, Cl_2, Br_2, I_2. As an example of the relative reactivity of the halogens, bromine, Br_2, will oxidize iodide ions, I^-, to molecular iodine:

$$Br_2(\ell) + 2\ KI(aq) \longrightarrow 2\ KBr(aq) + I_2(s)$$

5.6 Solution Concentration

Many of the substances in your body or in other living systems are dissolved in water—that is, they are in an aqueous solution. Like chemical reactions in living systems, many reactions studied in the chemical laboratory are carried out in solution. Frequently, this chemistry must be done quantitatively. For example, intravenous fluids administered to patients contain many compounds (salts, nutrients, drugs, and so on), and the concentration of each must be known accurately. To accomplish this task, we continue to use balanced equations and moles, but we measure volumes of solution rather than masses of solids, liquids, and gases.

CHEMISTRY YOU CAN DO

Pennies, Redox, and the Activity Series of Metals

In this experiment, you will use pennies to test the reactivity of copper and zinc with acid. Post-1982 pennies are a copper and zinc "sandwich" with zinc in the middle covered by a layer of copper. Pre-1982 pennies do not have this composition.

To do this experiment you will need:

- Two glasses or plastic cups that will each hold 50 mL (about 1.5 oz) of liquid
- About 100 mL "pickling" vinegar, such as Heinz Ultrastrength brand (Regular vinegar is only about 4–5% acetic acid.)
- An abrasive such as a piece of sandpaper, steel wool, or a Brillo pad
- A small file such as a nail file
- Four pennies—two pre-1982 and two post-1982

Clean the pennies with the abrasive until all the surfaces (including the edges) are shiny. Use the file to make two cuts into the edge of each penny, one across from the other. If you look carefully, you might observe a shiny metal where you cut into the post-1982 pennies.

Caution: Keep the vinegar away from your skin and clothes and especially your eyes. If vinegar spills on you, rinse it off with flowing water.

Place the two pre-1982 pennies into one of the cups and the post-1982 pennies into the other cup. Add the same volume of vinegar to each cup, making sure that the pennies are completely covered by the liquid. Let the pennies remain in the liquid for several hours (even overnight), and periodically observe any changes in them. After several hours, pour off the vinegar and remove the pennies. Dry them carefully and observe any changes that have occurred.

Think about these questions:

1. What difference did you observe between the pre-1982 pennies and the post-1982 ones?
2. Which is the more reactive element—copper or zinc?
3. What happened to the zinc in the post-1982 pennies? Interpret the change in redox terms, and write a chemical equation to represent the reaction.
4. How could this experiment be modified to determine the percent zinc and percent copper in post-1982 pennies?

A solution is a homogeneous mixture (← *p. 13*) of a **solute,** the substance that has been dissolved, and the **solvent,** the substance in which the solute has been dissolved. To know the quantity of solute in a given volume of a liquid solution requires knowing the **concentration** of the solution—the relative quantities of solute and solvent. Molarity, which relates the amount of solute expressed in moles to the solution volume expressed in liters, is the most useful of the many ways of expressing solution concentration for studying chemical reactions in solution.

Molarity

The **molarity** of a solution is defined as the amount of solute expressed in moles per unit volume of solution, expressed in liters (mol/L).

$$\text{Molarity} = \frac{\text{moles of solute}}{\text{liters of solution}}$$

Note that the volume term in the denominator is liters of *solution,* not liters of solvent.

If, for example, 40.0 g (1.00 mol) NaOH is dissolved in sufficient water to produce a solution with a total volume of 1.00 L, the solution has a concentration of 1.00 mol NaOH/1.00 L of solution, which is a 1.00 molar solution. The molarity of this solution is reported as 1.00 M, where the capital M stands for moles/liter. Molarity is also represented by square brackets around the formula of a compound or ion, such as [NaOH] or $[OH^-]$. The brackets indicate amount (in moles) of the species (compound or ion) per unit volume (in liters) of solution.

$$\underbrace{[\text{NaOH}]}_{\text{molarity of NaOH solution}} = \frac{2\text{ mol NaOH}}{1\text{ L solution}} = \underbrace{2\text{ M}}_{\text{2 mol NaOH per liter of solution}}$$

A solution of known molarity can be made by adding the required amount of solute to a volumetric flask, adding some solvent to dissolve all the solute, and then

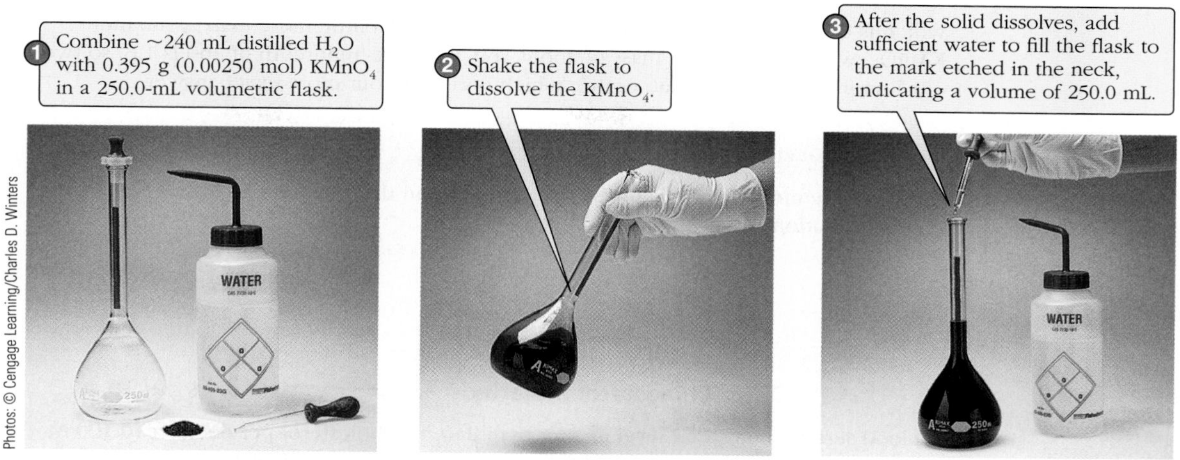
1 Combine ~240 mL distilled H_2O with 0.395 g (0.00250 mol) $KMnO_4$ in a 250.0-mL volumetric flask.
2 Shake the flask to dissolve the $KMnO_4$.
3 After the solid dissolves, add sufficient water to fill the flask to the mark etched in the neck, indicating a volume of 250.0 mL.
WATER
WATER
Photos: © Cengage Learning/Charles D. Winters

Figure 5.16 Solution preparation from a solid solute. Making 250.0 mL of a 0.0100-M aqueous solution of $KMnO_4$.

adding sufficient solvent with continual mixing to fill the flask to the mark etched on the flask's neck, as shown in Figure 5.16. The etched marking indicates the liquid level equal to the specified volume of the flask.

PROBLEM-SOLVING EXAMPLE 5.11 Molarity

Potassium permanganate, $KMnO_4$, is a strong oxidizing agent whose solutions are often used in laboratory experiments.

(a) If 0.433 g $KMnO_4$ is added to a 500.0-mL volumetric flask and water is added until the solution volume is exactly 500.0 mL, what is the molarity of the resulting solution?

(b) You need to prepare a 0.0250-M solution of $KMnO_4$ for an experiment. How many grams of $KMnO_4$ should be added with sufficient water to a 1.00-L volumetric flask to give the desired solution?

Answer

(a) 0.00548 M (b) 3.95 g

Strategy and Explanation

(a) To calculate the molarity of the solution, we need to calculate the amount of solute in moles and the solution volume in liters. The volume was given as 500.0 mL, which is 0.5000 L.

- **Use the molar mass of $KMnO_4$ to calculate the amount (moles) of solute.**

The molar mass of $KMnO_4$ is 158.03 g/mol.

$$0.433 \text{ g } KMnO_4 \times \frac{1 \text{ mol } KMnO_4}{158.03 \text{ g } KMnO_4} = 2.74 \times 10^{-3} \text{ mol } KMnO_4$$

- **Use the moles of solute and volume of solution to calculate the molarity.**

$$\text{Molarity of } KMnO_4 = \frac{2.74 \times 10^{-3} \text{ mol } KMnO_4}{0.500 \text{ L solution}} = 5.48 \times 10^{-3} \text{ mol/L}$$

This can be expressed as 0.00548 M or in the notation $[KMnO_4] = 0.00548$ M.

(b) To make a 0.0250-M solution in a 1.00-L volumetric flask requires 0.0250 mol $KMnO_4$.

- **Convert moles to grams using the molar mass of $KMnO_4$.**

$$0.0250 \text{ mol } KMnO_4 \times \frac{158.03 \text{ g } KMnO_4}{1 \text{ mol } KMnO_4} = 3.95 \text{ g } KMnO_4$$

☑ **Reasonable Answer Check** (a) We have about a half gram of solute with a molar mass of about 160. We will put this solute into a half-liter flask, so it is as if we put about one gram into a one-liter flask. The molarity should be about 1/160 = 0.00625, which is close to our more exact answer. (b) We need a little more than 2/100 mol $KMnO_4$, which has a molar mass of about 160. One one-hundredth of 160 is 1.6, so two one-hundredths is twice that, or 3.2, which is close to our more exact answer.

PROBLEM-SOLVING PRACTICE 5.11

Calculate the molarity of sodium sulfate in a solution that contains 36.0 g Na_2SO_4 in 750. mL solution.

EXERCISE 5.13 Cholesterol Molarity

A blood serum cholesterol level greater than 240 mg cholesterol per deciliter (0.100 L) of blood usually indicates the need for medical intervention. Calculate this serum cholesterol level in molarity. Cholesterol's molecular formula is $C_{27}H_{46}O$.

Sometimes the molarity of a particular ion in a solution is required, a value that depends on the nature of the solute. For example, potassium chromate is a soluble ionic compound and a strong electrolyte that completely dissociates in solution to form 2 mol K^+ ions and 1 mol CrO_4^{2-} ions for each mole of K_2CrO_4 that dissolves:

$$\underset{\substack{1\text{ mol}\\100\%\text{ dissociation}}}{K_2CrO_4(aq)} \longrightarrow \underset{2\text{ mol}}{2\,K^+(aq)} + \underset{1\text{ mol}}{CrO_4^{2-}(aq)}$$

The K^+ concentration is twice the K_2CrO_4 concentration because each mole of K_2CrO_4 contains 2 mol K^+. Therefore, a 0.00283-M K_2CrO_4 solution has a K^+ concentration of $2 \times 0.00283\text{ M} = 0.00566\text{ M}$ and a CrO_4^{2-} concentration of 0.00283 M.

EXERCISE 5.14 Molarity

A student dissolves 6.37 g aluminum nitrate in sufficient water to make 250.0 mL of solution. Calculate (a) the molarity of aluminum nitrate in this solution and (b) the molarity of aluminum ions and of nitrate ions in this solution.

CONCEPTUAL EXERCISE 5.15 Molarity

When solutions are prepared, the final volume of solution can be different from the sum of the volumes of the solute and solvent because some expansion or contraction can occur. Why is it always better to describe solution preparation as "adding enough solvent" to make a certain volume of solution?

Preparing a Solution of Known Molarity by Diluting a More Concentrated One

Frequently, solutions of the same solute need to be available at several different molarities. For example, hydrochloric acid is often used at concentrations of 6.0 M, 1.0 M, and 0.050 M. To make these solutions, chemists often use a concentrated solution of known molarity and through **dilution** with water make solutions of lesser molarity. *The number of moles of solute in the sample that is diluted remains constant throughout the dilution operation.* Therefore, the number of moles of solute in the dilute solution must be the same as the number of moles of solute in the sample of the more concentrated solution. Diluting a solution does increase the volume, so the molarity of the solution is lowered by the dilution operation, even though the number of moles of solute remains unchanged.

The number of moles in each case is the same and a simple relationship applies:

$$Molarity(\text{conc.}) \times V(\text{conc.}) = Molarity(\text{dil}) \times V(\text{dil})$$

Molarity(conc.) and *V*(conc.) represent the molarity and the volume (in liters) of the concentrated solution. *Molarity*(dil) and *V*(dil) represent the molarity and volume of the dilute solution. *Multiplying a volume in liters by a solution's molarity (moles/liter) yields the number of moles of solute.*

We can calculate, for example, the concentration of a hydrochloric acid solution made by diluting 25.0 mL of 6.0-M HCl to 500. mL. In this case, we want to determine *Molarity*(dil) when *Molarity*(conc.) = 6.0 M, *V*(conc.) = 0.0250 L, and *V*(dil) = 0.500 L. We algebraically rearrange the relationship to get the concentration of the diluted HCl.

$$Molarity(\text{dil}) = \frac{Molarity(\text{conc.}) \times V(\text{conc.})}{V(\text{dil})}$$

$$= \frac{6.0\ \text{mol/L} \times 0.0250\ \text{L}}{0.500\ \text{L}} = 0.30\ \text{mol/L}$$

A diluted solution will always be less concentrated (lower molarity) than the more concentrated solution (Figure 5.17).

Use caution when diluting a concentrated acid. The more concentrated acid should be added slowly to the solvent (water) so that the heat released during the dilution is rapidly dissipated into a large volume of water. If water is added to the acid, the heat released by the dissolving could be sufficient to vaporize the solution, spraying the acid over you and anyone nearby.

Consider two cases: A teaspoonful of sugar, $C_{12}H_{22}O_{11}$, is dissolved in a glass of water and a teaspoonful of sugar is dissolved in a swimming pool full of water. The swimming pool and the glass contain the same number of moles of sugar, but the concentration of sugar in the swimming pool is far less because the volume of solution in the pool is much greater than that in the glass.

A quick and useful check on a dilution calculation is to make certain that the molarity of the diluted solution is lower than that of the concentrated solution.

EXERCISE 5.16 Moles of Solute in Solutions

Consider 100. mL of 6.0-M HCl solution, which is diluted with water to yield 500. mL of 1.20-M HCl. Show that 100. mL of the more concentrated solution contains the same number of moles of HCl as 500. mL of the more dilute solution.

PROBLEM-SOLVING EXAMPLE 5.12 Solution Concentration and Dilution

Describe how to prepare 500.0 mL of 1.00-M H_2SO_4 solution from a concentrated sulfuric acid solution that is 18.0 M.

Answer Add 27.8 mL of the concentrated sulfuric acid slowly, carefully, and with stirring to about 450 mL water. After the sulfuric acid is thoroughly mixed with the water, add water to make up a total volume of 500.0 mL solution.

Strategy and Explanation In this dilution problem, the concentrations of the concentrated (18.0 M) and less concentrated (1.00 M) solutions are given, as well as the volume

1 A 100.0-mL volumetric flask has been filled to the mark with a 0.100-M $K_2Cr_2O_7$ solution.
2 This is transferred to a 1.000-L volumetric flask.
3 All of the initial solution is rinsed out of the 100.0-mL flask.
4 The 1.000-L flask is filled with distilled water until almost full, and then its contents are mixed by shaking the flask. Finally, distilled water is added to reach the mark on the neck of the flask. The concentration of the $K_2Cr_2O_7$ in the diluted solution is 0.0100 M.
WATER

Photos: © Cengage Learning/Charles D. Winters

Figure 5.17 Solution preparation by dilution.

of the diluted solution (500.0 mL). The volume of the concentrated sulfuric acid, *V*(conc.), to be diluted is needed, and can be calculated from this relationship:

$$Molarity(\text{conc.}) \times V(\text{conc.}) = Molarity(\text{dil}) \times V(\text{dil})$$

$$V(\text{conc.}) = \frac{Molarity(\text{dil}) \times V(\text{dil})}{Molarity(\text{conc.})}$$

$$= \frac{1.00\ \text{mol/L} \times 0.500\ \text{L}}{18.0\ \text{mol/L}} = 0.0278\ \text{L} = 27.8\ \text{mL}$$

Thus, 27.8 mL of concentrated sulfuric acid is added slowly, with stirring, to about 450 mL of distilled water. When the solution has cooled to room temperature, sufficient water is added to bring the final volume to 500.0 mL, resulting in a 1.00-M sulfuric acid solution.

☑ **Reasonable Answer Check** The ratio of molarities is 18:1, so the ratio of volumes should be 1:18, and it is.

PROBLEM-SOLVING PRACTICE **5.12**

A laboratory procedure calls for 50.0 mL of 0.150-M NaOH. You have available 100. mL of 0.500-M NaOH. What volume of the more concentrated solution should be diluted to make the desired solution?

CONCEPTUAL EXERCISE **5.17 Solution Concentration**

The molarity of a solution can be decreased by dilution. How could the molarity of a solution be increased without adding additional solute?

Preparing a Solution of Known Molarity from a Pure Solute

In Problem-Solving Example 5.11, we described finding the molarity of a $KMnO_4$ solution that was prepared from known quantities of solute and solution. More frequently, a solid or liquid solute (sometimes even a gas) must be used to make up a solution of known molarity. The problem becomes one of calculating what mass of solute to use to provide the proper number of moles.

Consider a laboratory experiment that requires 2.00 L of 0.750-M NH_4Cl solution. What mass of NH_4Cl must be dissolved in water to make 2.00 L of solution? The number of moles of NH_4Cl required can be calculated from the molarity.

$$0.750 \text{ mol/L } NH_4Cl \text{ solution} \times 2.00 \text{ L solution} = 1.500 \text{ mol } NH_4Cl$$

$M \times L = mol/L \times L = mol$

Then the molar mass can be used to calculate the number of grams of NH_4Cl needed.

$$1.500 \text{ mol } NH_4Cl \times 53.49 \text{ g/mol } NH_4Cl = 80.2 \text{ g } NH_4Cl$$

The solution is prepared by putting 80.2 g NH_4Cl into a beaker, dissolving it in pure water, rinsing all of the solution into a volumetric flask, and adding distilled water until the solution volume is 2.00 L, which results in a 0.750-M NH_4Cl solution.

PROBLEM-SOLVING EXAMPLE 5.13 Solute Mass and Molarity

Describe how to prepare 500.0 mL of 0.0250-M $K_2Cr_2O_7$ solution starting with solid potassium dichromate.

Answer Dissolve 3.68 g $K_2Cr_2O_7$ in water and add enough water to make 500.0 mL of solution.

Strategy and Explanation Use the definition of molarity and the molar mass of potassium dichromate the solve the problem.

- **Calculate the number of moles of solute.**

Find the number of moles of solute, $K_2Cr_2O_7$, in 500.0 mL of 0.0250-M $K_2Cr_2O_7$ solution by multiplying the volume in liters times the molarity of the solution.

$$0.0500 \text{ L solution} \times \frac{0.0250 \text{ mol } K_2Cr_2O_7}{1 \text{ L solution}} = 1.25 \times 10^{-2} \text{ mol } K_2Cr_2O_7$$

- **Calculate the number of grams of solute required.**

The molar mass of potassium dichromate is 294.2 g/mol.

$$1.25 \times 10^{-2} \text{ mol } K_2Cr_2O_7 \times \frac{294.2 \text{ g } K_2Cr_2O_7}{1 \text{ mol } K_2Cr_2O_7} = 3.68 \text{ g } K_2Cr_2O_7$$

The solution is prepared by putting 3.68 g $K_2Cr_2O_7$ into a 500-mL volumetric flask and adding enough distilled water to dissolve the solute and then additional water sufficient to bring the solution volume up to the mark on the flask. This results in 500.0 mL of 0.0250-M $K_2Cr_2O_7$ solution.

☑ **Reasonable Answer Check** The molar mass of $K_2Cr_2O_7$ is about 300 g/mol, and we want a 0.025-M solution, so we need about 300 g/mol × 0.025 mol/L = 7.5 g/L. But only one-half liter is required, so 0.5 L × 7.5 g/L = 3.75 g is needed, which agrees with our more accurate answer.

PROBLEM-SOLVING PRACTICE 5.13

Describe how you would prepare these solutions:
(a) 1.00 L of 0.125-M Na_2CO_3 from solid Na_2CO_3
(b) 100. mL of 0.0500-M Na_2CO_3 from a 0.125 M Na_2CO_3 solution
(c) 500. mL of 0.0215-M $KMnO_4$ from solid $KMnO_4$
(d) 250. mL of 0.00450-M $KMnO_4$ from 0.0215 M $KMnO_4$

5.7 Molarity and Reactions in Aqueous Solutions

Many kinds of reactions—acid-base neutralization (p. 172), precipitation (p. 165), and redox (p. 177)—occur in aqueous solutions. In such reactions, molarity is the concentration unit of choice because it allows us to make conversions between volumes of solutions and moles of reactants and products as given by the stoichiometric coefficients. Molarity is used to link mass, amount (moles), and volume of solution (Figure 5.18).

PROBLEM-SOLVING EXAMPLE 5.14 Solution Reaction Stoichiometry

Limestone, $CaCO_3$, reacts with HCl to produce the salt $CaCl_2$, carbon dioxide, and water:

$$CaCO_3(s) + 2\ HCl(aq) \longrightarrow CaCl_2(aq) + CO_2(g) + H_2O(\ell)$$

How many grams of calcium carbonate will react completely with 10.0 mL of 3.00-M HCl?

Answer 1.50 g $CaCO_3$

Strategy and Explanation Use the stoichiometry relationships in Figure 5.18. First, calculate the number of moles of HCl, then the number of moles of $CaCO_3$, and finally the mass of $CaCO_3$.

- **Calculate the number of moles of HCl.**

$$10.0\ \text{mL HCl} \times \frac{1\ \text{L}}{1000\ \text{mL}} \times \frac{3.00\ \text{mol HCl}}{\text{L HCl}} = 0.0300\ \text{mol HCl}$$

- **Calculate the number of moles of $CaCO_3$ using moles of HCl and the stoichiometric 1:2 ratio.**

$$0.0300\ \text{mol HCl} \times \frac{1\ \text{mol CaCO}_3}{2\ \text{mol HCl}} = 0.0150\ \text{mol CaCO}_3$$

- **Calculate the mass of $CaCO_3$ using its molar mass.**

The molar mass of $CaCO_3$ is 100.07 g/mol.

$$0.0150\ \text{mol CaCO}_3 \times \frac{100.07\ \text{g CaCO}_3}{1\ \text{mol CaCO}_3} = 1.50\ \text{g CaCO}_3$$

☑ **Reasonable Answer Check** The HCl concentration is 3 M; we have 0.0100 L of solution, so we have 0.03 mol HCl. Because of the 1:2 stoichiometry, HCl will react with half as many moles of $CaCO_3$, or 0.015 mol $CaCO_3$. The molar mass of $CaCO_3$ is about 100 g/mol, so 0.15 mol is about 0.15 mol × 100 g/mol = 1.5 g, which agrees with the more exact answer.

PROBLEM-SOLVING PRACTICE 5.14

In a recent year, 1.2×10^{10} kg sodium hydroxide (NaOH) was produced in the United States by passing an electric current through brine, an aqueous solution of sodium chloride.

$$2\ NaCl(aq) + 2\ H_2O(\ell) \longrightarrow 2\ NaOH(aq) + Cl_2(g) + H_2(g)$$

What volume of brine is needed to produce this mass of NaOH? (*Note*: 1.0 L brine contains 360 g dissolved NaCl.)

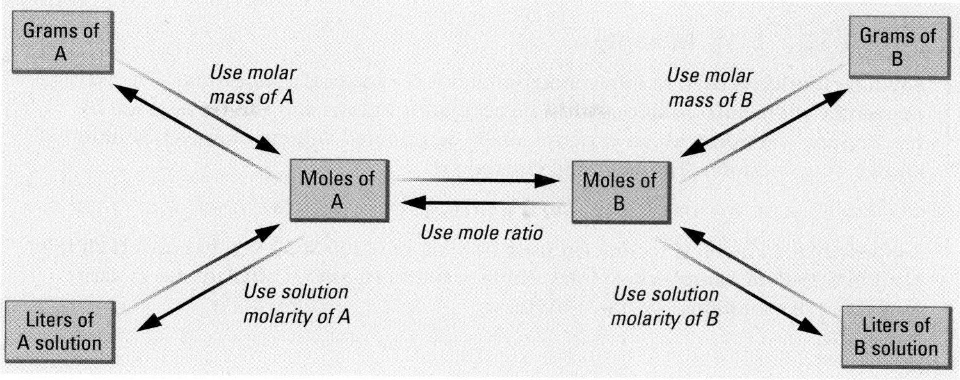

Figure 5.18 **Stoichiometric relationships for a chemical reaction in aqueous solution.** A mole ratio provides the connection between moles of a reactant or product and moles of another reactant or product.

PROBLEM-SOLVING EXAMPLE 5.15 Solution Reaction Stoichiometry

When aqueous potassium iodide, KI, is added to aqueous lead(II) nitrate, $Pb(NO_3)_2$, a brilliant yellow precipitate of PbI_2 is produced.

$$Pb(NO_3)_2(aq) + 2\ KI(aq) \longrightarrow PbI_2(s) + 2\ KNO_3(aq)$$

Calculate the minimum volume (mL) of 3.00-M KI required to react completely with 55.0 mL of 0.740-M $Pb(NO_3)_2$.

Answer 27.1 mL of 3.00-M KI solution

Strategy and Explanation Use the diagram in Figure 5.18 to decide on the sequence of calculations. Find the number of moles of $Pb(NO_3)_2$ and then use the balanced chemical equation to deduce how many moles of KI are required. From that, calculate the volume of 3.00-M KI solution required.

- **Calculate the number of moles of $Pb(NO_3)_2$ in the solution.**

$$0.0550\ L \times \frac{0.740\ mol\ Pb(NO_3)_2}{1\ L} = 4.07 \times 10^{-2}\ mol\ Pb(NO_3)_3$$

- **Calculate the number of moles of KI required to react completely.**

$$4.07 \times 10^{-2}\ mol\ Pb(NO_3)_2 \times \frac{2\ mol\ KI}{1\ mol\ Pb(NO_3)_2} = 8.14 \times 10^{-2}\ mol\ KI$$

- **Calculate the minimum volume of KI required.**

$$8.14 \times 10^{-2}\ mol\ KI \times \frac{1\ L\ KI}{3.00\ mol\ KI} = 2.71 \times 10^{-2}\ L\ KI\ or\ 27.1\ mL\ KI$$

☑ **Reasonable Answer Check** We have 0.055 L × 0.74 mol/L ≈ 0.041 mol $Pb(NO_3)_2$. We have 0.0271 L × 3 mol/L ≈ 0.081 mol KI. This is the 2:1 ratio of reactants needed according to the balanced equation. The answer is reasonable.

PROBLEM-SOLVING PRACTICE 5.15

Insoluble solid silver bromide, AgBr, can be dissolved by aqueous sodium thiosulfate, $Na_2S_2O_3$, as described by this net ionic equation

$$AgBr(s) + 2\ S_2O_3^{2-}(aq) \longrightarrow Ag(S_2O_3)_2^{3-}(aq) + Br^-(aq)$$

If you want to dissolve 50.0 mg AgBr, what is the minimum number of mL of 0.0150-M $Na_2S_2O_3$ you must use?

© Cengage Learning/Charles D. Winters

Precipitation of lead(II) iodide. Adding a drop of aqueous KI to an aqueous solution of lead(II) nitrate precipitates yellow lead(II) iodide. $KNO_3(aq)$ remains dissolved in the solution.

EXERCISE 5.18 Molarity

Sodium chloride is used in intravenous solutions for medical applications. The NaCl concentration in such solutions must be accurately known and can be assessed by reacting the solution with an experimentally determined volume of $AgNO_3$ solution of known concentration. The net ionic equation is

$$Ag^+(aq) + Cl^-(aq) \longrightarrow AgCl(s)$$

Suppose that a chemical technician uses 19.3 mL of 0.200-M $AgNO_3$ to convert all the NaCl in a 25.0-mL sample of an intravenous solution to AgCl. Calculate the molarity of NaCl in the solution.

5.8 Aqueous Solution Titrations

One important quantitative use of aqueous solution reactions is to determine the unknown concentration of a reactant in a solution, such as the concentration of HCl in a solution of HCl. This is done with a titration using a **standard solution,** a solution whose concentration is known accurately. In a **titration,** a substance in the standard solution reacts with a known stoichiometry with the substance whose concentration is to be determined. When the stoichiometrically equivalent amount of standard solution has been added, the **equivalence point** is reached. At that point, the molar amount of reactant that has been added from the standard solution is exactly what is needed to react completely with the substance whose concentration is to be determined. The progress of the reaction is monitored by an indicator, a dye that changes color at the equivalence point, or through some other means with appropriate instruments. Phenolphthalein, for example, is commonly used as the indicator in strong acid-strong base titrations because it is colorless in acidic solutions and pink in basic solutions. The point at which the indicator is seen to change color is called the end point.

Acid-base titrations are described more extensively in Chapter 17.

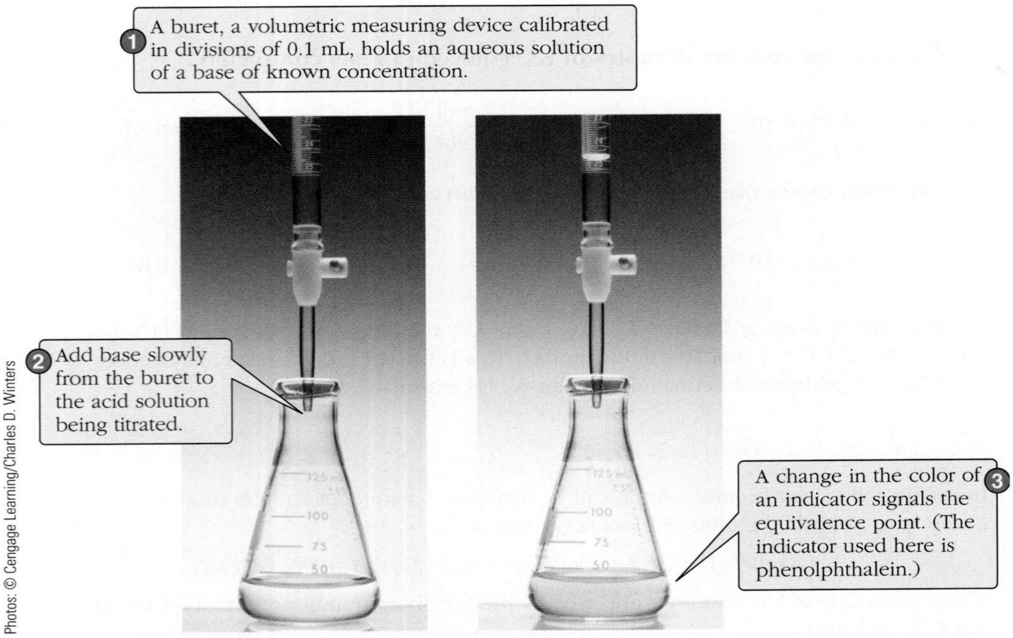
1 A buret, a volumetric measuring device calibrated in divisions of 0.1 mL, holds an aqueous solution of a base of known concentration.
2 Add base slowly from the buret to the acid solution being titrated.
3 A change in the color of an indicator signals the equivalence point. (The indicator used here is phenolphthalein.)
Photos: © Cengage Learning/Charles D. Winters

Figure 5.19 Titration of an acid in aqueous solution with a standard solution of base.

A common example of a titration is the determination of the molarity of an acid by titration of the acid with a standard solution of a base. For example, we can use a standard solution of 0.100-M KOH to determine the concentration of an HCl solution. To carry out this titration, we use a carefully measured volume of the HCl solution and slowly add the standardized KOH solution until the equivalence point is reached (Figure 5.19). At that point, the number of moles of OH^- added to the HCl solution exactly matches the number of moles of H^+ that were in the original acid sample.

PROBLEM-SOLVING EXAMPLE 5.16 Acid-Base Titration

A student has an aqueous solution of calcium hydroxide that is approximately 0.10 M. She titrated a 50.0-mL sample of the calcium hydroxide solution with a standardized solution of 0.300-M HNO_3(aq). To reach the end point, 41.4 mL of the HNO_3 solution was needed. What is the molarity of the calcium hydroxide solution?

Answer 0.124-M $Ca(OH)_2$

Strategy and Explanation Calculate the number of moles of acid titrant used, and then calculate the number of moles and the concentration of the base being titrated.

- **Write the balanced equation for the reaction.**

$$2\ HNO_3(aq) + Ca(OH)_2(aq) \longrightarrow Ca(NO_3)_2(aq) + 2\ H_2O(\ell)$$

The net ionic equation is

$$H^+(aq) + OH^-(aq) \longrightarrow H_2O(\ell)$$

- **Calculate the number of moles of acid titrant reacted.**

The number of moles of HNO_3 reacted is

$$0.0414\ L\ HNO_3 \times \frac{0.300\ mol\ HNO_3}{1.00\ L\ HNO_3\ solution} = 0.0124\ mol\ HNO_3$$

- **Use the balanced chemical equation to find the number of moles of base used.**

The balanced equation shows that for every 2 mol HNO_3 reacted, 1 mol $Ca(OH)_2$ is used. Therefore, since 1.24×10^{-2} mol HNO_3 reacted, then

$$1.24 \times 10^{-2}\ mol\ HNO_3 \times \frac{1\ mol\ Ca(OH)_2}{2\ mol\ HNO_3} = 6.21 \times 10^{-3}\ mol\ Ca(OH)_2$$

must have reacted.

- **Calculate the molarity of the $Ca(OH)_2$ solution.**

From the number of moles of $Ca(OH)_2$ and the volume of the $Ca(OH)_2$ solution, calculate the molarity of the solution

$$\frac{6.21 \times 10^{-3}\ mol\ Ca(OH)_2}{0.0500\ L\ Ca(OH)_2\ solution} = 0.124\text{-}M\ Ca(OH)_2$$

☑ **Reasonable Answer Check** At the equivalence point, the number of moles of H^+(aq) added and OH^-(aq) in the initial sample must be equal. The number of moles of each reactant is its volume multiplied by its molarity. For the HNO_3 we have $0.0414\ L \times 0.300\ M = 0.0124$ mol HNO_3. For the OH^- (from $Ca(OH)_2$) we have $0.050\ L \times 0.124\ M \times 2 = 0.0124$ mol OH^-. The answer is reasonable.

PROBLEM-SOLVING PRACTICE 5.16

In a titration, a 20.0-mL sample of sulfuric acid, H_2SO_4, was titrated to the end point with 41.3 mL of 0.100-M NaOH. What is the molarity of the H_2SO_4 solution?

SUMMARY PROBLEM

Gold in its elemental state can be separated from gold-bearing rock by treating the ore with cyanide, CN^-, in the presence of oxygen via this reaction:

$$4\ Au(s) + 8\ CN^-(aq) + O_2(g) + 2\ H_2O(\ell) \longrightarrow 4\ Au(CN)_2^-(aq) + 4\ OH^-(aq)$$

The CN^- is supplied by NaCN, but Na^+ is a spectator ion and is left out of the net ionic equation.

(a) Which reactant is oxidized? What are the oxidation numbers of this species as a reactant and as a product?

(b) Which reactant is reduced? What are the oxidation numbers of this species as a reactant and as a product?

(c) What is the oxidizing agent?

(d) What is the reducing agent?

(e) What mass of NaCN would it take to prepare 1.0 L of 0.075-M NaCN?

(f) If the ore contains 0.019% gold by weight, what mass of gold is found in one metric ton (exactly 1000 kg) of the ore?

(g) How many grams of NaCN would you require to extract the gold in one metric ton of this ore?

(h) How many liters of the 0.075-M NaCN solution would you require to extract the gold in one metric ton of this ore?

IN CLOSING

OWL and go Chemistry

Sign in at **www.cengage.com/owl** to:

- View tutorials and simulations, develop problem-solving skills, and complete online homework assigned by your professor.
- For quick review and exam prep, download Go Chemistry mini lecture modules from OWL (or purchase them at **www.CengageBrain.com**).

Having studied this chapter, you should be able to . . .

- Predict products of common types of exchange reactions: precipitation, acid-base, and gas-forming (Sections 5.1–5.3). End-of-chapter question 42
- Write a net ionic equation for a given reaction in aqueous solution (Section 5.1). Questions 23, 27, 29
- Recognize common acids and bases and predict when neutralization reactions will occur (Section 5.2). Question 38
- Identify the acid and base used to form a specific salt (Section 5.2). Question 40
- Recognize oxidation-reduction reactions and common oxidizing and reducing agents (Section 5.3). Questions 52, 55
- Assign oxidation numbers to reactants and products in a redox reaction, identify what has been oxidized or reduced, and identify oxidizing agents and reducing agents (Section 5.4). Questions 44, 61
- Use the activity series to predict products of displacement redox reactions (Section 5.5). Questions 60, 66
- Define molarity and calculate molar concentrations (Section 5.6). Questions 70, 72, 74
- Determine how to prepare a solution of a given molarity from the solute and water or by dilution of a more concentrated solution (Section 5.6). Questions 76, 78, 80
- Solve stoichiometry problems by using solution molarities (Section 5.7). Questions 83, 85
- Understand how aqueous solution titrations can be used to determine the concentration of an unknown solution (Section 5.8). Questions 87, 93

KEY TERMS

acid *(Section 5.2)*
base *(5.2)*
concentration *(5.6)*
dilution *(5.6)*
equivalence point *(5.8)*
hydronium ion *(5.2)*
hydroxide ion *(5.2)*
metal activity series *(5.5)*
molarity *(5.6)*
net ionic equation *(5.1)*
oxidation *(5.3)*
oxidation number *(5.4)*
oxidation-reduction reaction *(5.3)*
oxidized *(5.3)*
oxidizing agent *(5.3)*
precipitate *(5.1)*
redox reactions *(5.3)*
reduced *(5.3)*
reducing agent *(5.3)*
reduction *(5.3)*
salt *(5.2)*
solute *(5.6)*
solvent *(5.6)*
spectator ion *(5.1)*
standard solution *(5.8)*
strong acid *(5.2)*
strong base *(5.2)*
strong electrolyte *(5.1)*
titration *(5.8)*
weak acid *(5.2)*
weak base *(5.2)*
weak electrolyte *(5.2)*

QUESTIONS FOR REVIEW AND THOUGHT

OWL Interactive versions of these problems are assignable in OWL.

Blue-numbered questions have short answers at the back of this book in Appendix M and fully worked solutions in the *Student Solutions Manual.*

Review Questions

These questions test vocabulary and simple concepts.

1. Find in this chapter one example of each of these reaction types, and write the balanced equation for the reaction: (a) combustion, (b) combination, (c) exchange, (d) decomposition, and (e) oxidation-reduction. Name the products of each reaction.
2. Classify each of these reactions as a combination, decomposition, exchange, acid-base, or oxidation-reduction reaction.
   (a) $MgO(s) + 2\,HCl(aq) \longrightarrow MgCl_2(aq) + H_2O(\ell)$
   (b) $2\,NaHCO_3(s) \xrightarrow{\text{heat}} Na_2CO_3(s) + CO_2(g) + H_2O(g)$
   (c) $CaO(s) + SO_2(g) \longrightarrow CaSO_3(s)$
   (d) $3\,Cu(s) + 8\,HNO_3(aq) \longrightarrow$
   $3\,Cu(NO_3)_2(aq) + 2\,NO(g) + 4\,H_2O(\ell)$
   (e) $2\,NO(g) + O_2(g) \longrightarrow 2\,NO_2(g)$
3. Find two examples in this chapter of the reaction of a metal with a halogen. Write a balanced equation for each example, and name the product.
4. Find two examples of acid-base reactions in this chapter. Write balanced equations for these reactions, and name the reactants and products.
5. Find two examples of precipitation reactions in this chapter. Write balanced equations for these reactions, and name the reactants and products.
6. Find an example of a gas-forming reaction in this chapter. Write a balanced equation for the reaction, and name the reactants and products.
7. Explain the difference between oxidation and reduction. Give an example of each.
8. For each of the following, does the oxidation number increase or decrease in the course of a redox reaction?
   (a) An oxidizing agent
   (b) A reducing agent
   (c) A substance undergoing oxidation
   (d) A substance undergoing reduction
9. Explain the difference between an oxidizing agent and a reducing agent. Give an example of each.

Topical Questions

These questions are keyed to the major topics in the chapter. Usually a question that is answered at the back of this book is paired with a similar one that is not.

Solubility (Section 5.1)

10. Tell how solubility rules predict that $Ni(NO_3)_2$ is soluble in water, whereas $NiCO_3$ is not soluble in water.
11. Predict whether each of these compounds is likely to be water-soluble. Indicate which ions are present in solution for the water-soluble compounds.
    (a) $Fe(ClO_4)_2$ (b) Na_2SO_4
    (c) KBr (d) Na_2CO_3
12. Predict whether each of these compounds is likely to be water-soluble. For those compounds which are soluble, indicate which ions are present in solution.
    (a) $Ca(NO_3)_2$ (b) KCl
    (c) $CuSO_4$ (d) $FeCl_3$

13. Predict whether each of these compounds is likely to be water-soluble. Indicate which ions are present in solution for the water-soluble compounds.
    (a) Potassium monohydrogen phosphate
    (b) Sodium hypochlorite (c) Magnesium chloride
    (d) Calcium hydroxide (e) Aluminum bromide
14. Predict whether each of these compounds is likely to be water-soluble. Indicate which ions are present in solution for the water-soluble compounds.
    (a) Ammonium nitrate (b) Barium sulfate
    (c) Potassium acetate (d) Calcium carbonate
    (e) Sodium perchlorate
15. Which of these drawings is the best nanoscale representation of an aqueous solution of calcium chloride? (Water molecules are not shown for simplicity.)

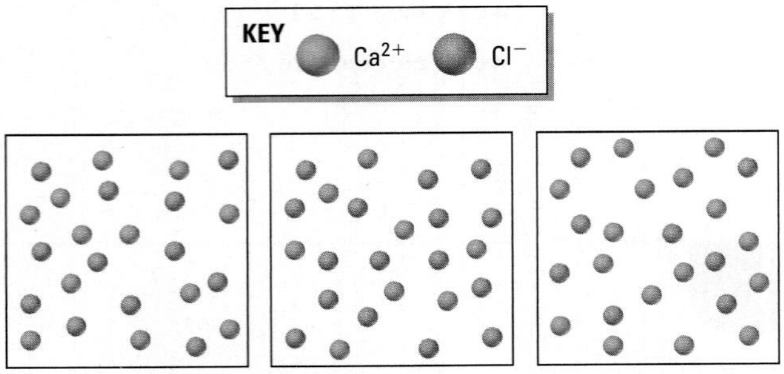

(a) (b) (c)

16. Which of these drawings is the best nanoscale representation of an aqueous solution of magnesium nitrate? (Water molecules are not shown for simplicity.)

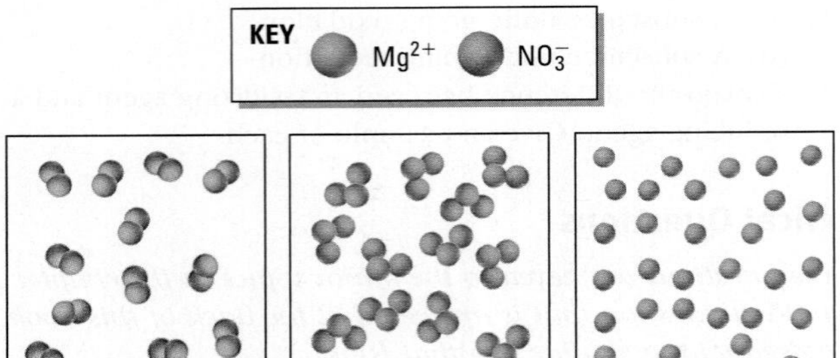

(a) (b) (c)

17. Which of these drawings is the best nanoscale representation of an aqueous solution of calcium phosphate? (Water molecules are not shown for simplicity.)

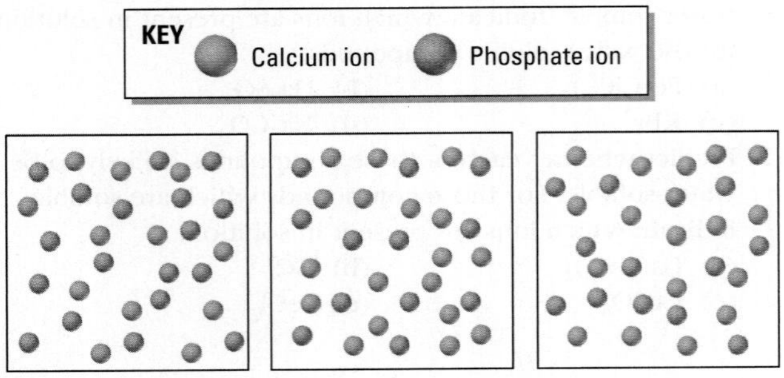

(a) (b) (c)

18. Which of these drawings is the best nanoscale representation of an aqueous solution of potassium sulfate? (Water molecules are not shown for simplicity.)

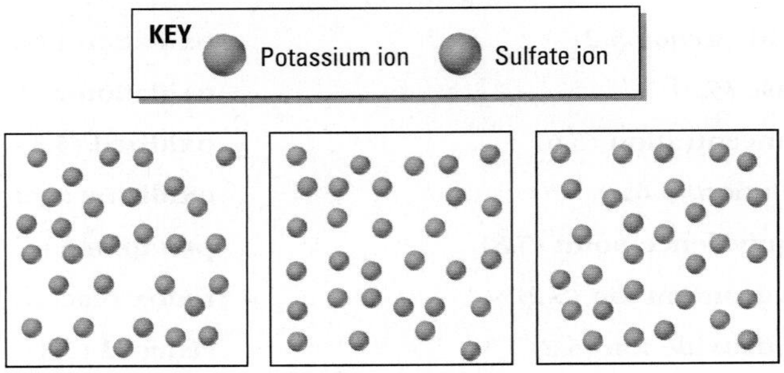

(a) (b) (c)

Exchange Reactions (Sections 5.1 & 5.2)

19. Write a balanced equation for the reaction of nitric acid with calcium hydroxide.
20. Write a balanced equation for the reaction of hydrochloric acid with magnesium hydroxide.
21. For each of these pairs of ionic compounds, write a balanced equation reflecting whether precipitation will occur in aqueous solution. For those combinations that do not produce a precipitate, write "NP."
    (a) $MnCl_2 + Na_2S$ (b) $HNO_3 + CuSO_4$
    (c) $NaOH + HClO_4$ (d) $Hg(NO_3)_2 + Na_2S$
    (e) $Pb(NO_3)_2 + HCl$ (f) $BaCl_2 + H_2SO_4$
22. For each of these pairs of ionic compounds, write a balanced equation reflecting whether precipitation will occur in aqueous solution. For those combinations that do not produce a precipitate, write "NP."
    (a) $HNO_3 + Na_3PO_4$ (b) $NaCl + Pb(CH_3COO)_2$
    (c) $(NH_4)_2S + NiCl_2$ (d) $K_2SO_4 + Cu(NO_3)_2$
    (e) $FeCl_3 + NaOH$ (f) $AgNO_3 + KCl$
23. Identify the water-insoluble product in each of these reactions. Write the net ionic equations for these reactions. Identify the spectator ions.
    (a) $CuCl_2(aq) + H_2S(aq) \longrightarrow CuS + 2\ HCl$
    (b) $CaCl_2(aq) + K_2CO_3(aq) \longrightarrow 2\ KCl + CaCO_3$
    (c) $AgNO_3(aq) + NaI(aq) \longrightarrow AgI + NaNO_3$
24. Identify the water-insoluble product in each of these reactions. Write the net ionic equations for these reactions. Identify the spectator ions.
    (a) $Pb(NO_3)_2(aq) + Na_2SO_4(aq) \longrightarrow PbSO_4 + 2\ NaNO_3$
    (b) $2\ K_3PO_4(aq) + 3\ Mg(NO_3)_2(aq) \longrightarrow$
    $Mg_3(PO_4)_2 + 6\ KNO_3$
    (c) $(NH_4)_2SO_4(aq) + BaBr_2(aq) \longrightarrow BaSO_4 + 2\ NH_4Br$
25. If aqueous solutions of potassium carbonate and copper(II) nitrate are mixed, a precipitate is formed. Write the complete and net ionic equations for this reaction, and name the precipitate.
26. If aqueous solutions of potassium sulfide and iron(III) chloride are mixed, a precipitate is formed. Write the complete and net ionic equations for this reaction, and name the precipitate.
27. Balance each of these equations, and then write the complete ionic and net ionic equations.
    (a) $Zn(s) + HCl(aq) \longrightarrow H_2(g) + ZnCl_2(aq)$
    (b) $Mg(OH)_2(s) + HCl(aq) \longrightarrow MgCl_2(aq) + H_2O(\ell)$
    (c) $HNO_3(aq) + CaCO_3(s) \longrightarrow$
    $Ca(NO_3)_2(aq) + H_2O(\ell) + CO_2(g)$
    (d) $HCl(aq) + MnO_2(s) \longrightarrow MnCl_2(aq) + Cl_2(g) + H_2O(\ell)$

28. Balance each of these equations, and then write the complete ionic and net ionic equations.
(a) $(NH_4)_2CO_3(aq) + Cu(NO_3)_2(aq) \longrightarrow CuCO_3(s) + NH_4NO_3(aq)$
(b) $Pb(NO_3)_2(aq) + HCl(aq) \longrightarrow PbCl_2(s) + HNO_3(aq)$
(c) $BaCO_3(s) + HCl(aq) \longrightarrow BaCl_2(aq) + H_2O(\ell) + CO_2(g)$

29. Balance each of these equations, and then write the complete ionic and net ionic equations. Refer to Tables 5.1 and 5.2 for information on solubility and on acids and bases. Show states (s, ℓ, g, aq) for all reactants and products.
(a) $Ca(OH)_2 + HNO_3 \longrightarrow Ca(NO_3)_2 + H_2O$
(b) $BaCl_2 + Na_2CO_3 \longrightarrow BaCO_3 + NaCl$
(c) $Na_3PO_4 + Ni(NO_3)_2 \longrightarrow Ni_3(PO_4)_2 + NaNO_3$

30. Balance each of these equations, and then write the complete ionic and net ionic equations. Refer to Tables 5.1 and 5.2 for information on solubility and on acids and bases. Show states (s, ℓ, g, aq) for all reactants and products.
(a) $ZnCl_2 + KOH \longrightarrow KCl + Zn(OH)_2$
(b) $AgNO_3 + KI \longrightarrow AgI + KNO_3$
(c) $NaOH + FeCl_2 \longrightarrow Fe(OH)_2 + NaCl$

31. Barium hydroxide is used in lubricating oils and greases. Write a balanced equation for the reaction of this hydroxide with nitric acid to give barium nitrate, a compound used in pyrotechnics devices such as green flares.

32. Aluminum is obtained from bauxite, which is not a specific mineral but a name applied to a mixture of minerals. One of those minerals, which can dissolve in acids, is gibbsite, $Al(OH)_3$. Write a balanced equation for the reaction of gibbsite with sulfuric acid.

33. Balance the equation for this precipitation reaction, and then write the complete ionic and net ionic equations.

$$CdCl_2 + NaOH \longrightarrow Cd(OH)_2 + NaCl$$

34. Balance the equation for this precipitation reaction, and then write the complete ionic and net ionic equations.

$$Ni(NO_3)_2 + Na_2CO_3 \longrightarrow NiCO_3 + NaNO_3$$

35. Write an overall balanced equation for the precipitation reaction that occurs when aqueous lead(II) nitrate is mixed with an aqueous solution of potassium chloride. Name each reactant and product. Indicate the state of each substance (s, ℓ, g, or aq).

36. Write an overall balanced equation for the precipitation reaction that occurs when aqueous copper(II) nitrate is mixed with an aqueous solution of sodium carbonate. Name each reactant and product. Indicate the state of each substance (s, ℓ, g, or aq).

37. The beautiful mineral rhodochrosite is manganese(II) carbonate. Write an overall balanced equation for the reaction of the mineral with hydrochloric acid. Name each reactant and product.

38. Classify each of these as an acid or a base. Which are strong and which are weak? What ions are produced when each is dissolved in water?
(a) KOH (b) $Mg(OH)_2$
(c) HClO (d) HBr
(e) LiOH (f) H_2SO_3

39. Classify each of these as an acid or a base. Which are strong and which are weak? What ions are produced when each is dissolved in water?
(a) HNO_3 (b) $Ca(OH)_2$
(c) NH_3 (d) H_3PO_4
(e) KOH (f) CH_3COOH

40. Identify the acid and base used to form these salts, and write the overall neutralization reaction in both complete and net ionic form.
(a) $NaNO_2$ (b) $CaSO_4$
(c) NaI (d) $Mg_3(PO_4)_2$

41. Identify the acid and base used to form these salts, and write the overall neutralization reaction in both complete and net ionic form.
(a) $NaCH_3COO$ (b) $CaCl_2$
(c) LiBr (d) $Ba(NO_3)_2$

42. Classify each of these exchange reactions as an acid-base reaction, a precipitation reaction, or a gas-forming reaction. Predict the products of the reaction, and then balance the completed equation.
(a) $MnCl_2(aq) + Na_2S(aq) \longrightarrow$
(b) $Na_2CO_3(aq) + ZnCl_2(aq) \longrightarrow$
(c) $K_2CO_3(aq) + HClO_4(aq) \longrightarrow$

43. Classify each of these exchange reactions as an acid-base reaction, a precipitation reaction, or a gas-forming reaction. Predict the products of the reaction, and then balance the completed equation.
(a) $Fe(OH)_3(s) + HNO_3(aq) \longrightarrow$
(b) $FeCO_3(s) + H_2SO_4(aq) \longrightarrow$
(c) $FeCl_2(aq) + (NH_4)_2S(aq) \longrightarrow$
(d) $Fe(NO_3)_2(aq) + Na_2CO_3(aq) \longrightarrow$

Oxidation-Reduction Reactions (Sections 5.3 & 5.4)

44. Assign oxidation numbers to each atom in these compounds.
(a) SO_3 (b) HNO_3
(c) $KMnO_4$ (d) H_2O
(e) LiOH (f) CH_2Cl_2

45. Assign oxidation numbers to each atom in these compounds.
(a) $Fe(OH)_3$ (b) $HClO_3$
(c) $CuCl_2$ (d) K_2CrO_4
(e) $Ni(OH)_2$ (f) N_2H_4

46. Assign oxidation numbers to each atom in these ions.
(a) SO_4^{2-} (b) NO_3^-
(c) MnO_4^- (d) $Cr(OH)_4^-$
(e) $H_2PO_4^-$ (f) $S_2O_3^{2-}$

47. What is the oxidation number of Mn in each of these species?
(a) $(MnF_6)^{3-}$ (b) Mn_2O_7
(c) MnO_4^- (d) $Mn(CN)_6^-$
(e) MnO_2

48. What is the oxidation number of Cl in each of these species?
(a) HCl (b) HClO
(c) $HClO_2$ (d) $HClO_3$
(e) $HClO_4$

49. What is the oxidation number of S in each of these species?
(a) H_2SO_4 (b) H_2SO_3
(c) SO_2 (d) SO_3
(e) $H_2S_2O_7$ (f) $Na_2S_2O_3$

50. Sulfur can exist in many oxidation states. What is the oxidation state of S in each of these species?
(a) H_2S (b) S_8
(c) SCl_2 (d) SO_3^{2-}
(e) K_2SO_4

51. What is the oxidation state of Cr in each of these species?
(a) $CrCl_3$ (b) Na_2CrO_4
(c) $K_2Cr_2O_7$

52. Which of these reactions are oxidation-reduction reactions? Explain your answer briefly. Classify the remaining reactions.
(a) $CdCl_2(aq) + Na_2S(aq) \longrightarrow CdS(s) + 2\ NaCl(aq)$
(b) $2\ Ca(s) + O_2(g) \longrightarrow 2\ CaO(s)$
(c) $Ca(OH)_2(s) + 2\ HCl(aq) \longrightarrow CaCl_2(aq) + 2\ H_2O(\ell)$

53. Which of these reactions are oxidation-reduction reactions? Explain your answer briefly. Classify the remaining reactions.
(a) $Zn(s) + 2\ NO_3^-(aq) + 4\ H_3O^+(aq) \longrightarrow Zn^{2+}(aq) + 2\ NO_2(g) + 6\ H_2O(\ell)$
(b) $Zn(OH)_2(s) + H_2SO_4(aq) \longrightarrow ZnSO_4(aq) + 2\ H_2O(\ell)$
(c) $Ca(s) + 2\ H_2O(\ell) \longrightarrow Ca(OH)_2(s) + H_2(g)$

54. Which region of the periodic table has the best reducing agents? The best oxidizing agents?

55. Which of these substances are oxidizing agents?
(a) Zn (b) O_2
(c) HNO_3 (d) MnO_4^-
(e) H_2 (f) H^+

56. Which of these substances are reducing agents?
(a) Ca (b) Ca^{2+}
(c) $Cr_2O_7^{2-}$ (d) Al
(e) Br_2 (f) H_2

57. Identify the products of these redox combination reactions.
(a) $C(s) + O_2(g) \longrightarrow$
(b) $P_4(s) + Cl_2(g) \longrightarrow$
(c) $Ti(s) + Cl_2(g) \longrightarrow$
(d) $Mg(s) + N_2(g) \longrightarrow$
(e) $FeO(s) + O_2(g) \longrightarrow$
(f) $NO(g) + O_2(g) \longrightarrow$

58. Complete and balance these equations for redox displacement reactions.
(a) $K(s) + H_2O(\ell) \longrightarrow$
(b) $Mg(s) + HBr(aq) \longrightarrow$
(c) $NaBr(aq) + Cl_2(aq) \longrightarrow$
(d) $WO_3(s) + H_2(g) \longrightarrow$
(e) $H_2S(aq) + Cl_2(aq) \longrightarrow$

59. Which halogen is the strongest oxidizing agent? Which is the strongest reducing agent?

60. Predict the products of these halogen displacement reactions. If no reaction occurs, write "NR."
(a) $I_2(s) + NaBr(aq) \longrightarrow$
(b) $Br_2(\ell) + NaI(aq) \longrightarrow$
(c) $F_2(g) + NaCl(aq) \longrightarrow$
(d) $Cl_2(g) + NaBr(aq) \longrightarrow$
(e) $Br_2(\ell) + NaCl(aq) \longrightarrow$
(f) $Cl_2(g) + NaF(aq) \longrightarrow$

61. For the reactions in Question 60 that occur, identify the species oxidized or reduced as well as the oxidizing and reducing agents.

62. For the reactions in Question 60 that do not occur, rewrite the equation so that a reaction does occur (consider the halogen activity series).

63. The drug methamphetamine, also called "speed," has the molecular formula $C_{10}H_{15}N$. In the body it undergoes a series of metabolic reactions by which it is ultimately oxidized to produce carbon dioxide, nitrogen, and water. Write a balanced equation for this overall reaction.

Activity Series (Section 5.5)

64. Give an example of a displacement reaction that is also a redox reaction and identify which species is (a) oxidized, (b) reduced, (c) the reducing agent, and (d) the oxidizing agent.

65. (a) In what groups of the periodic table are the most reactive metals found? Where do we find the least reactive metals?
(b) Silver (Ag) does not react with 1-M HCl solution. Will Ag react with a solution of aluminum nitrate, $Al(NO_3)_3$? If so, write a chemical equation for the reaction.
(c) Lead (Pb) will react very slowly with 1-M HCl solution. Aluminum will react with lead(II) sulfate solution, $PbSO_4$. Will Pb react with an $AgNO_3$ solution? If so, write a chemical equation for the reaction.
(d) On the basis of the information obtained in answering parts (a), (b), and (c), arrange Ag, Al, and Pb in decreasing order of reactivity.

66. Use the activity series of metals (Table 5.5) to predict the outcome of each of these reactions. If no reaction occurs, write "NR."
(a) $Na^+(aq) + Zn(s) \longrightarrow$
(b) $HCl(aq) + Pt(s) \longrightarrow$
(c) $Ag^+(aq) + Au(s) \longrightarrow$
(d) $Au^{3+}(aq) + Ag(s) \longrightarrow$

67. Using the activity series of metals (Table 5.5), predict whether these reactions will occur in aqueous solution.
(a) $Mg(s) + Ca(s) \longrightarrow Mg^{2+}(aq) + Ca^{2+}(aq)$
(b) $2\ Al^{3+}(aq) + 3\ Pb^{2+}(aq) \longrightarrow 2\ Al(s) + 3\ Pb(s)$
(c) $H_2(g) + Zn^{2+}(aq) \longrightarrow 2\ H^+(aq) + Zn(s)$
(d) $Mg(s) + Cu^{2+}(aq) \longrightarrow Mg^{2+}(aq) + Cu(s)$
(e) $Pb(s) + 2\ H^+(aq) \longrightarrow H_2(g) + Pb^{2+}(aq)$
(f) $2\ Ag^+(aq) + Cu(s) \longrightarrow 2\ Ag(s) + Cu^{2+}(aq)$
(g) $2\ Al^{3+}(aq) + 3\ Zn(s) \longrightarrow 3\ Zn^{2+}(aq) + 2\ Al(s)$

Solution Concentrations (Section 5.6)

68. You have a 0.12-M solution of $BaCl_2$. What ions exist in the solution, and what are their concentrations?

69. A flask contains 0.25-M $(NH_4)_2SO_4$. What ions exist in the solution, and what are their concentrations?

70. Assume that 6.73 g Na_2CO_3 is dissolved in enough water to make 250. mL of solution.
(a) What is the molarity of the sodium carbonate?
(b) What are the concentrations of the Na^+ and CO_3^{2-} ions?

71. Some $K_2Cr_2O_7$, with a mass of 2.335 g, is dissolved in enough water to make 500. mL of solution.
(a) What is the molarity of the potassium dichromate?
(b) What are the concentrations of the K^+ and $Cr_2O_7^{2-}$ ions?

72. What is the mass, in grams, of solute in 250. mL of a 0.0125-M solution of $KMnO_4$?

73. What is the mass, in grams, of solute in 100. mL of a 1.023×10^{-3}-M solution of Na_3PO_4?

74. What volume of 0.123-M NaOH, in milliliters, contains 25.0 g NaOH?

75. What volume of 2.06-M $KMnO_4$, in liters, contains 322 g solute?

76. If 6.00 mL of 0.0250-M $CuSO_4$ is diluted to 10.0 mL with pure water, what is the concentration of copper(II) sulfate in the diluted solution?

77. If you dilute 25.0 mL of 1.50-M HCl to 500. mL, what is the molar concentration of the diluted HCl?

78. If you need 1.00 L of 0.125-M H_2SO_4, which method would you use to prepare this solution?
(a) Dilute 36.0 mL of 1.25-M H_2SO_4 to a volume of 1.00 L.
(b) Dilute 20.8 mL of 6.00-M H_2SO_4 to a volume of 1.00 L.
(c) Add 50.0 mL of 3.00-M H_2SO_4 to 950. mL water.
(d) Add 500. mL of 0.500-M H_2SO_4 to 500. mL water.

79. If you need 300. mL of 0.500-M $K_2Cr_2O_7$, which method would you use to prepare this solution?
(a) Dilute 250. mL of 0.600-M $K_2Cr_2O_7$ to 300. mL.
(b) Add 50.0 mL of water to 250. mL of 0.250-M $K_2Cr_2O_7$.
(c) Dilute 125 mL of 1.00-M $K_2Cr_2O_7$ to 300. mL.
(d) Add 30.0 mL of 1.50-M $K_2Cr_2O_7$ to 270. mL of water.

80. You need to make a 0.300-M solution of $NiSO_4$(aq). How many grams of $NiSO_4 \cdot 6\ H_2O$ should you put into a 0.500-L volumetric flask?

81. You wish to make a 0.200-M solution of $NiSO_4$(aq). How many grams of $NiSO_4 \cdot 6\ H_2O$ should you put in a 0.500-L volumetric flask?

82. A typical mug (250 mL) of coffee contains 125 mg caffeine, $C_8H_{10}N_4O_2$. What is the molarity of the caffeine?

Calculations for Reactions in Solution (Sections 5.7 & 5.8)

83. What mass, in grams, of Na_2CO_3 is required for complete reaction with 25.0 mL of 0.155-M HNO_3?

$$Na_2CO_3(aq) + 2\ HNO_3(aq) \longrightarrow 2\ NaNO_3(aq) + CO_2(g) + H_2O(\ell)$$

84. Hydrazine, N_2H_4, a base like ammonia, can react with an acid such as sulfuric acid.

$$2\ N_2H_4(aq) + H_2SO_4(aq) \longrightarrow 2\ N_2H_5^+(aq) + SO_4^{-2}(aq)$$

What mass of hydrazine can react with 250. mL of 0.225-M H_2SO_4?

85. What volume, in milliliters, of 0.125-M HNO_3 is required to react completely with 1.30 g $Ba(OH)_2$?

$$2\ HNO_3(aq) + Ba(OH)_2(s) \longrightarrow Ba(NO_3)_2(aq) + 2\ H_2O(\ell)$$

86. Diborane, B_2H_6, can be produced by this reaction:

$$2\ NaBH_4(s) + H_2SO_4(aq) \longrightarrow 2\ H_2(g) + Na_2SO_4(aq) + B_2H_6(g)$$

What volume, in milliliters, of 0.0875-M H_2SO_4 should be used to completely react with 1.35 g $NaBH_4$?

87. What volume, in milliliters, of 0.512-M NaOH is required to react completely with 25.0 mL of 0.234-M H_2SO_4?

88. What volume, in milliliters, of 0.812-M HCl would be required to neutralize 15.0 mL of 0.635-M NaOH?

89. What is the maximum mass, in grams, of AgCl that can be precipitated by mixing 50.0 mL of 0.025-M $AgNO_3$ solution with 100.0 mL of 0.025-M NaCl solution? Which reactant is in excess? What is the concentration of the excess reactant remaining in solution after the AgCl has precipitated?

90. Suppose you mix 25.0 mL of 0.234-M $FeCl_3$ solution with 42.5 mL of 0.453-M NaOH.
(a) What is the maximum mass, in grams, of $Fe(OH)_3$ that will precipitate?
(b) Which reactant is in excess?
(c) What is the concentration of the excess reactant remaining in solution after the maximum mass of $Fe(OH)_3$ has precipitated?

91. A soft drink contains an unknown amount of citric acid, $C_3H_5O(COOH)_3$. A volume of 10.0 mL of the soft drink requires 6.42 mL of 9.580×10^{-2}-M NaOH to neutralize the citric acid.

$$C_3H_5O(COOH)_3(aq) + 3\ NaOH(aq) \longrightarrow Na_3C_3H_5O(COO)_3(aq) + 3\ H_2O(\ell)$$

(a) Which step in these calculations for the mass of citric acid in 1 mL of soft drink is not correct?
(b) What is the correct answer?
(i) Moles NaOH = $(6.42\ mL)(1L/1000\ mL)(9.580 \times 10^{-2}\ mol/L)$
(ii) Moles citric acid = $(6.15 \times 10^{-4}\ mol\ NaOH) \times (3\ mol\ citric\ acid/1\ mol\ NaOH)$
(iii) Mass citric acid in sample = $(1.85 \times 10^{-3}\ mol\ citric\ acid) \times (192.12\ g/mol\ citric\ acid)$
(iv) Mass citric acid in 1 mL soft drink = (0.354 g citric acid)/(10 mL soft drink)

92. Vitamin C is the compound $C_6H_8O_6$. Besides being an acid, it is a reducing agent that reacts readily with bromine, Br_2 a good oxidizing agent.

$$C_6H_8O_6(aq) + Br_2(aq) \longrightarrow 2\ HBr(aq) + C_6H_6O_6(aq)$$

Suppose a 1.00-g chewable vitamin C tablet requires 27.85 mL of 0.102-M Br_2 to react completely.
(a) Which step in these calculations for the mass, in grams, of vitamin C in the tablet is incorrect?
(b) What is the correct answer?
(i) Mole Br_2 = (27.85 mL)(0.102 mol/L)
(ii) Moles $C_6H_8O_6$ = (2.84 mol Br_2)(1 mol $C_6H_8O_6$/1 mol Br_2)
(iii) Mass $C_6H_8O_6$ = (2.84 mol $C_6H_8O_6$)(176 g/mol $C_6H_8O_6$)
(iv) Mass $C_6H_8O_6$ = (500 g $C_6H_8O_6$)/(1 g tablet)

93. If a volume of 32.45 mL HCl is used to completely neutralize 2.050 g Na_2CO_3 according to this equation, what is the molarity of the HCl?

$$Na_2CO_3(aq) + 2\ HCl(aq) \longrightarrow 2\ NaCl(aq) + CO_2(g) + H_2O(\ell)$$

94. Potassium acid phthalate, $KHC_8H_4O_4$, is used to standardize solutions of bases. The acidic anion reacts with bases according to this net ionic equation:

$$HC_8H_4O_4^-(aq) + OH^-(aq) \longrightarrow H_2O(\ell) + C_8H_4O_4^{2-}(aq)$$

If a 0.902-g sample of potassium acid phthalate requires 26.45 mL NaOH to react, what is the molarity of the NaOH?

95. Sodium thiosulfate, $Na_2S_2O_3$, is used as a “fixer” in black-and-white photography. Assume you have a bottle of sodium thiosulfate and want to determine its purity. The

thiosulfate ion can be oxidized with I_2 according to this equation:

$$I_2(aq) + 2\,S_2O_3^{2-}(aq) \longrightarrow 2\,I^-(aq) + S_4O_6^{2-}(aq)$$

If you use 40.21 mL of 0.246-M I_2 to completely react a 3.232-g sample of impure $Na_2S_2O_3$, what is the percent purity of the $Na_2S_2O_3$?

96. A sample of a mixture of oxalic acid, $H_2C_2O_4$, and sodium chloride, NaCl, has a mass of 4.554 g. If a volume of 29.58 mL of 0.550-M NaOH is required to neutralize all the $H_2C_2O_4$, what is the weight percent of oxalic acid in the mixture? Oxalic acid and NaOH react according to this equation:

$$H_2C_2O_4(aq) + 2\,NaOH(aq) \longrightarrow Na_2C_2O_4(aq) + 2\,H_2O(\ell)$$

97. You are given 0.954 g of an unknown acid, H_2A, which reacts with NaOH according to the balanced equation

$$H_2A(aq) + 2\,NaOH(aq) \longrightarrow Na_2A(aq) + 2\,H_2O(\ell)$$

If 36.04 mL of 0.509-M NaOH is required to react with all of the acid, what is the molar mass of the acid?

98. (a) How many mL of 0.050-M HCl should be added to 50.0 mL of 0.40-M HCl to have a final solution with a molarity of 0.30 M?
(b) What volume of 0.154-M NaCl, "physiological saline solution," can you prepare by diluting 100 mL of 6.0-M NaCl solution?

General Questions

These questions are not explicitly keyed to chapter topics; many require integration of several concepts.

99. Name the spectator ions in the reaction of calcium carbonate and hydrochloric acid, and write the net ionic equation.

$$CaCO_3(s) + 2\,H^+(aq) + 2\,Cl^-(aq) \longrightarrow CO_2(g) + Ca^{2+}(aq) + 2\,Cl^-(aq) + H_2O(\ell)$$

What type of reaction is this?

100. Magnesium metal reacts readily with HNO_3, as shown in this equation:

$$Mg(s) + HNO_3(aq) \longrightarrow Mg(NO_3)_2(aq) + NO_2(g) + H_2O(\ell)$$

(a) Balance the equation.
(b) Name each reactant and product.
(c) Write the net ionic equation.
(d) What type of reaction is this?

101. Aqueous solutions of $(NH_4)_2S$ and $Hg(NO_3)_2$ react to give HgS and NH_4NO_3.
(a) Write the overall balanced equation. Indicate the state (s or aq) for each compound.
(b) Name each compound.
(c) Write the net ionic equation.
(d) What type of reaction does this appear to be?

102. Classify these reactions and predict the products formed.
(a) $Li(s) + H_2O(\ell) \longrightarrow$
(b) $Ag_2O(s) \xrightarrow{\text{heat}}$
(c) $Li_2O(s) + H_2O(\ell) \longrightarrow$
(d) $I_2(s) + Cl^-(aq) \longrightarrow$
(e) $Cu(s) + HCl(aq) \longrightarrow$
(f) $BaCO_3(s) \xrightarrow{\text{heat}}$

103. Classify these reactions and predict the products formed.
(a) $SO_3(g) + H_2O(\ell) \longrightarrow$
(b) $Sr(s) + H_2(g) \longrightarrow$
(c) $Mg(s) + H_2SO_4(aq, \text{dilute}) \longrightarrow$
(d) $Na_3PO_4(aq) + AgNO_3(aq) \longrightarrow$
(e) $Ca(HCO_3)_2(s) \xrightarrow{\text{heat}}$
(f) $Fe^{3+}(aq) + Sn^{2+}(aq) \longrightarrow$

104. Azurite is a copper-containing mineral that often forms beautiful crystals. Its formula is $Cu_3(CO_3)_2(OH)_2$. Write a balanced equation for the reaction of this mineral with hydrochloric acid.

105. What species (atoms, molecules, ions) are present in an aqueous solution of each of these compounds?
(a) NH_3 (b) CH_3COOH
(c) NaOH (d) HBr

106. Use the activity series to predict whether these reactions will occur.
(a) $Fe(s) + Mg^{2+}(aq) \longrightarrow Mg(s) + Fe^{2+}(aq)$
(b) $Ni(s) + Cu^{2+}(aq) \longrightarrow Ni^{2+}(aq) + Cu(s)$
(c) $Cu(s) + 2\,H^+(aq) \longrightarrow Cu^{2+}(aq) + H_2(g)$
(d) $Mg(s) + H_2O(g) \longrightarrow MgO(s) + H_2(g)$

107. Determine which of these are redox reactions. Identify the oxidizing and reducing agents in each of the redox reactions.
(a) $NaOH(aq) + H_3PO_4(aq) \rightarrow NaH_2PO_4(aq) + H_2O(\ell)$
(b) $NH_3(g) + CO_2(g) + H_2O(\ell) \rightarrow NH_4HCO_3(aq)$
(c) $TiCl_4(g) + 2\,Mg(\ell) \longrightarrow Ti(s) + 2\,MgCl_2(\ell)$
(d) $NaCl(s) + NaHSO_4(aq) \longrightarrow HCl(g) + Na_2SO_4(aq)$

108. Identify the substance oxidized, the substance reduced, the reducing agent, and the oxidizing agent in the equations in Question 107. For each oxidized or reduced substance, identify the change in its oxidation number.

109. Much has been written about chlorofluorocarbons and their effects on our environment. Their manufacture begins with the preparation of HF from the mineral fluorspar, CaF_2, according to this unbalanced equation:

$$CaF_2(s) + H_2SO_4(aq) \longrightarrow HF(g) + CaSO_4(s)$$

HF is combined with, for example, CCl_4 in the presence of $SbCl_5$ to make CCl_2F_2, called dichlorodifluoromethane or CFC-12, and other chlorofluorocarbons.

$$2\,HF(g) + CCl_4(\ell) \longrightarrow CCl_2F_2(g) + 2\,HCl(g)$$

(a) Balance the first equation above and name each substance.
(b) Is the first reaction best classified as an acid-base reaction, an oxidation-reduction reaction, or a precipitation reaction?
(c) Give the names of the compounds CCl_4, $SbCl_5$, and HCl.
(d) Another chlorofluorocarbon produced in the reaction is composed of 8.74% C, 77.43% Cl, and 13.83% F. What is the empirical formula of the compound?

110. How much salt is in your chicken soup? A student added excess $AgNO_3(aq)$ to a 1-cup serving of regular chicken soup (240 mL) and got 5.55 g AgCl precipitate. How many grams of NaCl were in the regular chicken soup? Assume that all the chloride ions in the soup were from NaCl. In a second experiment, the same procedure was done with chicken soup advertised to have "less salt," and the student got 3.55 g AgCl precipitate. How many grams of NaCl are in the "less salt" version?

111. Vitamin C is ascorbic acid, $HC_6H_7O_6$, which can be titrated with a strong base.

$$HC_6H_7O_6(aq) + NaOH(aq) \longrightarrow NaC_6H_7O_6(aq) + H_2O(\ell)$$

In a laboratory experiment, a student dissolved a 500.0-mg vitamin C tablet in 200.0 mL water and then titrated it with 0.1250-M NaOH. It required 21.30 mL of the base to reach the equivalence point. What percentage of the mass of the tablet is impurity?

Applying Concepts

These questions test conceptual learning.

112. When these pairs of reactants are combined in a beaker, (a) describe in words what the contents of the beaker would look like before and after any reaction occurs, (b) use different circles for atoms, molecules, and ions to draw a nanoscale (particulate-level) diagram of what the contents would look like, and (c) write a chemical equation to represent symbolically what the contents would look like.

$LiCl(aq)$ and $AgNO_3(aq)$

$NaOH(aq)$ and $HCl(aq)$

113. When these pairs of reactants are combined in a beaker, (a) describe in words what the contents of the beaker would look like before and after any reaction occurs, (b) use different circles for atoms, molecules, and ions to draw a particulate-level diagram of what the contents would look like, and (c) write a chemical equation to represent symbolically what the contents would look like.

$CaCO_3(s)$ and $HCl(aq)$

$NH_4NO_3(aq)$ and $KOH(aq)$

114. Explain how you could prepare barium sulfate by (a) an acid-base reaction, (b) a precipitation reaction, and (c) a gas-forming reaction. The materials you have to start with are $BaCO_3$, $Ba(OH)_2$, Na_2SO_4, and H_2SO_4.

115. Students were asked to prepare nickel sulfate by reacting a nickel compound with a sulfate compound in water and then evaporating the water. Three students chose these pairs of reactants:

Student 1	$Ni(OH)_2$ and H_2SO_4
Student 2	$Ni(NO_3)_2$ and Na_2SO_4
Student 3	$NiCO_3$ and H_2SO_4

Comment on each student's choice of reactants and how successful you think each student will be at preparing nickel sulfate by the procedure indicated.

116. An unknown solution contains either lead ions or barium ions, but not both. Which one of these solutions could you use to tell whether the ions present are Pb^{2+} or Ba^{2+}? Explain the reasoning behind your choice.

$HCl(aq)$, $H_2SO_4(aq)$, $H_3PO_4(aq)$

117. An unknown solution contains either calcium ions or strontium ions, but not both. Which one of these solutions could you use to tell whether the ions present are Ca^{2+} or Sr^{2+}? Explain the reasoning behind your choice.

$NaOH(aq)$, $H_2SO_4(aq)$, $H_2S(aq)$

118. When you are given an oxidation-reduction reaction and asked what is oxidized or what is reduced, why should you never choose one of the products for your answer?

119. When you are given an oxidation-reduction reaction and asked what is the oxidizing agent or what is the reducing agent, why should you never choose one of the products for your answer?

120. You prepared a NaCl solution by adding 58.44 g NaCl to a 1-L volumetric flask and then adding water to dissolve it. When you were finished, the final volume in your flask looked like this:

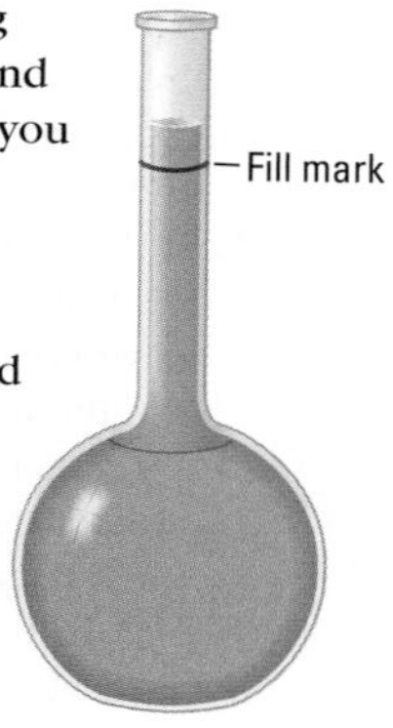

1.00-L flask

The solution you prepared is

(a) Greater than 1 M because you added more solvent than necessary.
(b) Less than 1 M because you added less solvent than necessary.
(c) Greater than 1 M because you added less solvent than necessary.
(d) Less than 1 M because you added more solvent than necessary.
(e) 1 M because the amount of solute, not solvent, determines the concentration.

121. These drawings represent beakers of aqueous solutions. Each orange circle represents a dissolved solute particle.

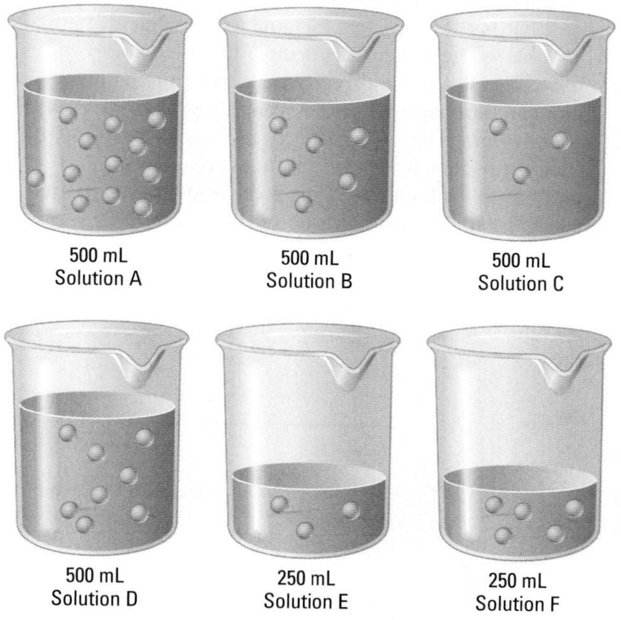

(a) Which solution is most concentrated?
(b) Which solution is least concentrated?
(c) Which two solutions have the same concentration?
(d) When solutions E and F are combined, the resulting solution has the same concentration as solution ______.
(e) When solutions B and E are combined, the resulting solution has the same concentration as solution ______.
(f) If you evaporate half of the water from solution B, the resulting solution will have the same concentration as solution ________.
(g) If you place half of solution A in another beaker and then add 250 mL water, the resulting solution will have the same concentration as solution ________.

122. Ten milliliters of a solution of an acid is mixed with 10 mL of a solution of a base. When the mixture was tested with litmus paper, the blue litmus turned red, and the red litmus remained red. Which of these interpretations is (are) correct?
   (a) The mixture contains more hydrogen ions than hydroxide ions.
   (b) The mixture contains more hydroxide ions than hydrogen ions.
   (c) When an acid and a base react, water is formed, so the mixture cannot be acidic or basic.
   (d) If the acid was HCl and the base was NaOH, the concentration of HCl in the initial acidic solution must have been greater than the concentration of NaOH in the initial basic solution.
   (e) If the acid was H_2SO_4 and the base was NaOH, the concentration of H_2SO_4 in the initial acidic solution must have been greater than the concentration of NaOH in the initial basic solution.

123. A chemical company was interested in characterizing a competitor's organic acid (it consists of C, H, and O). After determining that it was a diacid, H_2X, a 0.1235-g sample was neutralized with 15.55 mL of 0.1087-M NaOH. Next, a 0.3469-g sample was burned completely in pure oxygen, producing 0.6268 g CO_2 and 0.2138 g H_2O.
   (a) What is the molar mass of H_2X?
   (b) What is the empirical formula for the diacid?
   (c) What is the molecular formula for the diacid?

124. Various masses of the three Group 2A elements magnesium, calcium, and strontium were allowed to react with liquid bromine, Br_2. After the reaction was complete, the reaction product was freed of excess reactant(s) and weighed. In each case, the mass of product was plotted against the mass of metal used in the reaction (as shown below).

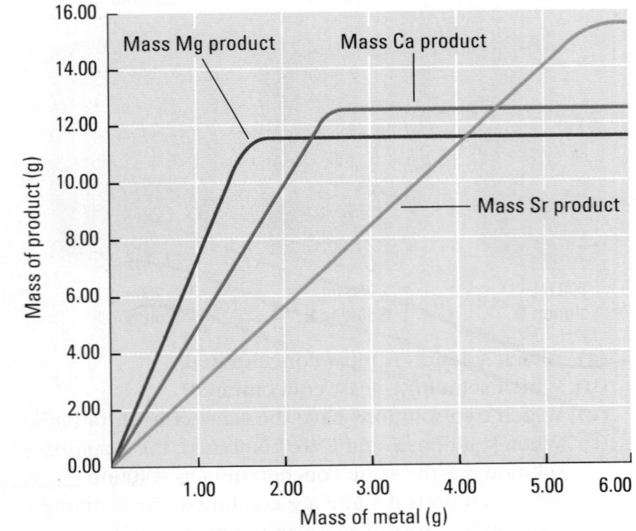

   (a) Based on your knowledge of the reactions of metals with halogens, what product is predicted for each reaction? What are the name and formula for the reaction product in each case?
   (b) Write a balanced equation for the reaction occurring in each case.
   (c) What kind of reaction occurs between the metals and bromine—that is, is the reaction a gas-forming reaction, a precipitation reaction, or an oxidation-reduction reaction?
   (d) Each plot shows that the mass of product increases with increasing mass of metal used, but the plot levels out at some point. Use these plots to verify your prediction of the formula of each product, and explain why the plots become level at different masses of metal and different masses of product.

125. Gold can be dissolved from gold-bearing rock by treating the rock with sodium cyanide in the presence of the oxygen in air.

$$4\ Au(s) + 8\ NaCN(aq) + O_2(g) + 2\ H_2O(\ell) \longrightarrow 4\ NaAu(CN)_2(aq) + 4\ NaOH(aq)$$

Once the gold is in solution in the form of the $[Au(CN)_2]^-$ ion, it can be precipitated as the metal according to this unbalanced equation:

$$[Au(CN)_2]^-(aq) + Zn(s) \longrightarrow Zn^{2+}(aq) + Au(s) + CN^-(aq)$$

   (a) Are the two reactions above acid-base or oxidation-reduction reactions? Briefly describe your reasoning.
   (b) How many liters of 0.075-M NaCN will you need to extract the gold from 1000 kg of rock if the rock is 0.019% gold?
   (c) How many kilograms of metallic zinc will you need to recover the gold from the $[Au(CN)_2]^-$ obtained from the gold in the rock?
   (d) If the gold is recovered completely from the rock and the metal is made into a cylindrical rod 15.0 cm long, what is the diameter of the rod? (The density of gold is 19.3 g/cm^3.)

126. Four groups of students from an introductory chemistry laboratory are studying the reactions of solutions of alkali metal halides with aqueous silver nitrate, $AgNO_3$. They use these salts.
Group A: NaCl
Group B: KCl
Group C: NaBr
Group D: KBr
Each of the four groups dissolves 0.004 mol of their salt in some water. Each then adds various masses of silver nitrate, $AgNO_3$, to their solutions. After each group collects the precipitated silver halide, the mass of this product is plotted versus the mass of $AgNO_3$ added. The results are given on this graph.

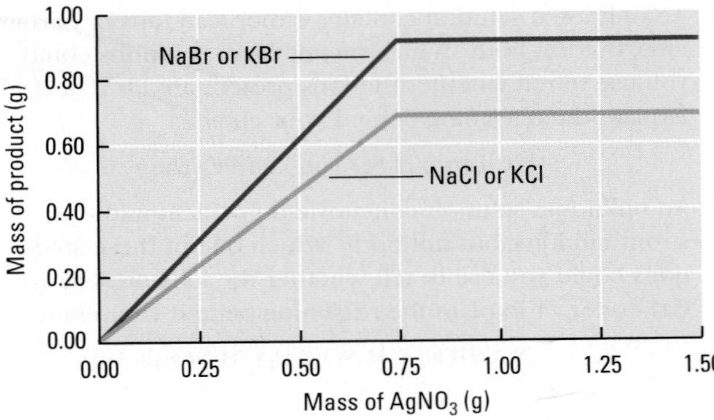

(a) Write the balanced net ionic equation for the reaction observed by each group.
(b) Explain why the data for groups A and B lie on the same line, whereas those for groups C and D lie on a different line.
(c) Explain the shape of the plot observed by each group. Why do the plots level off at the same mass of added $AgNO_3$ (0.75 g) but give different masses of product?

127. One way to determine the stoichiometric relationships among reactants is continuous variations. In this process, a series of reactions is carried out in which the reactants are varied systematically, while keeping the total volume of each reaction mixture constant. When the reactants combine stoichiometrically, they react completely; none is in excess. These data were collected to determine the stoichiometric relationship for the reaction.

$$mX^{n+} + nY^{m-} \longrightarrow X_mY_n$$

Trial	A	B	C	D	E
0.10 M X^{n+}	7 mL	6 mL	5 mL	4 mL	3 mL
0.20 M Y^{m-}	3 mL	4 mL	5 mL	6 mL	7 mL
Excess X^{n+} present?	Yes	Yes	Yes	No	No
Excess Y^{m-} present?	No	No	No	No	Yes

(a) In which trial are the reactants present in stoichiometric amounts?
(b) How many moles of X^{n+} reacted in that trial?
(c) How many moles of Y^{m-} reacted in that trial?
(d) What is the whole-number mole ratio of X^{n+} to Y^{m-}?
(e) What is the chemical formula for the product X_mY_n in terms of x and y?

More Challenging Questions

These questions require more thought and integrate several concepts.

128. You are given an acid and told only that it could be citric acid (molar mass = 192.1 g/mol) or tartaric acid (molar mass = 150.1 g/mol). To determine which acid you have, you react it with NaOH. The appropriate reactions are

Citric acid:

$$C_6H_8O_7(aq) + 3\ NaOH(aq) \longrightarrow Na_3C_6H_5O_7(aq) + 3\ H_2O(\ell)$$

Tartaric acid:

$$C_4H_6O_6(aq) + 2\ NaOH(aq) \longrightarrow Na_2C_4H_4O_6(aq) + 2\ H_2O(\ell)$$

You find that a 0.956-g sample requires 29.1 mL 0.513-M NaOH to reach the equivalence point. What is the unknown acid?

129. In the past, devices for testing a driver's breath for alcohol depended on this reaction:

$$\underset{\text{ethanol}}{3\ C_2H_5OH(aq)} + 2\ K_2Cr_2O_7(aq) + 8\ H_2SO_4(aq) \longrightarrow$$
$$\underset{\text{acetic acid}}{3\ CH_3COOH(aq)} + 2\ Cr_2(SO_4)_3(aq) + 2\ K_2SO_4(aq) + 11\ H_2O(\ell)$$

Write the net ionic equation for this reaction. What oxidation numbers are changing in the course of this reaction? Which substances are being oxidized and reduced? Which substance is the oxidizing agent and which is the reducing agent?

130. The salt calcium sulfate is sparingly soluble in water with a solubility of 0.209 g/100 mL of water at 30° C. If you stirred 0.550 g $CaSO_4$ into 100.0 mL water at 30° C, what would the molarity of the resulting solution be? How many grams of $CaSO_4$ would remain undissolved?

131. What is the molarity of water in pure water?

132. The balanced equation for the oxidation of ethanol to acetic acid by potassium dichromate in an acidic aqueous HCl solution is

$$3\ C_2H_5OH(aq) + 2\ K_2Cr_2O_7(aq) + 16\ HCl(aq) \longrightarrow$$
$$3\ CH_3COOH(aq) + 4\ CrCl_3(aq) + 4\ KCl(aq) + 11\ H_2O(\ell)$$

What volume of a 0.600-M potassium dichromate solution is needed to generate 0.166 mol acetic acid, CH_3COOH, from a solution containing excess ethanol and HCl?

133. Dolomite, found in soil, is $CaMg(CO_3)_2$. If a 20.0-g sample of soil is titrated with 65.25 mL of 0.2500-M HCl, what is the mass percent of dolomite in the soil sample?

134. In this redox reaction methanol reduces chlorate ion to chlorine dioxide in the presence of acid, and the methanol is oxidized to CO_2:

$$CH_3OH(\ell) + 6\ HClO_3(aq) \longrightarrow$$
$$6\ ClO_2(g) + CO_2(g) + 5\ H_2O(\ell)$$

How many mL of methanol would be needed to produce 100.0 kg of chlorine dioxide? The density of methanol is 0.791 g/mL.

135. Citric acid, $C_6H_8O_7$, is found in citrus fruit. An aqueous solution consists of 0.400 g citric acid dissolved in 25.0 mL of water. To completely neutralize this citric acid solution required 31.2 mL of 0.200-M NaOH. How many acidic hydrogen atoms does citric acid have per molecule?

136. A 60.0-mL sample of 2.00-M NaCl and a 40.0-mL sample of 0.500-M NaCl are mixed. Then, additional distilled water is added until the total volume is 500. mL. What is the molarity of the NaCl in the final solution?

137. Hard water contains Ca^{2+} and Mg^{2+} ions that interfere with the cleaning action of soap. Hard water can be softened by replacing these ions with Na^+ ions by reacting the hard water with sodium carbonate. Consider a sample of moderately hard water containing 250 ppm calcium and magnesium ions: 1.50×10^{-3}-M Ca^{2+} and 1.00×10^{-3}-M Mg^{2+}. How many moles and grams of sodium carbonate are needed to replace the calcium and magnesium ions in 1000. L of this hard water?

Conceptual Challenge Problems

These rigorous, thought-provoking problems integrate conceptual learning with problem solving and are suitable for group work.

CP5.A (Section 5.3) There is a conservation of the number of electrons exchanged during redox reactions, which is tantamount to stating that electrical charge is conserved during chemical reactions. The assignment of oxidation numbers is an arbitrary yet clever way to do the bookkeeping for these electrons. What makes it possible to assign the same oxidation number to all elements that are not bound to other elements not in chemical compounds?

CP5.B (Section 5.4) Consider these redox reactions:

$$HIO_3 + FeI_2 + HCl \longrightarrow FeCl_3 + ICl + H_2O$$

$$CuSCN + KIO_3 + HCl \longrightarrow CuSO_4 + KCl + HCN + ICl + H_2O$$

(a) Identify the species that have been oxidized or reduced in each of the reactions.

(b) After you have correctly identified the species that have been oxidized or reduced in each equation, you might like to try using oxidation numbers to balance each equation. This will be a challenge because, as you have discovered, more than one kind of atom is oxidized or reduced, although in all cases the product of the oxidation and reduction is unambiguous. Record the initial and final oxidation states of each kind of atom that is oxidized or reduced in each equation. Then decide on the coefficients that will equalize the oxidation number changes and satisfy any other atom balancing needed. Finally, balance the equation by adding the correct coefficients to it.

CP5.C (Section 5.5) A student was given four metals (A, B, C, and D) and solutions of their corresponding salts (AZ, BZ, CZ, and DZ). The student was asked to determine the relative reactivity of the four metals by reacting the metals with the solutions. The student's laboratory observations are indicated in the table. Arrange the four metals in order of decreasing activity.

Metal	AZ(aq)	BZ(aq)	CZ(aq)	DZ(aq)
A	No reaction	No reaction	No reaction	No reaction
B	Reaction	No reaction	Reaction	No reaction
C	Reaction	No reaction	No reaction	No reaction
D	Reaction	Reaction	Reaction	No reaction

CP5.D (Section 5.6) How would you prepare 1 L of 1.00×10^{-6}-M NaCl (molar mass = 58.44 g/mol) solution by using a balance that can measure mass only to 0.01 g?

CP5.E (Section 5.8) How could you show that when baking soda reacts with the acetic acid, CH_3COOH, in vinegar, all of the carbon and oxygen atoms in the carbon dioxide produced come from the baking soda alone and none comes from the acetic acid in vinegar?

6

Energy and Chemical Reactions

© Cengage Learning/Charles D. Winters

Combustion of a fuel. Burning charcoal, which is mostly carbon, releases a great deal of energy to anything in contact with the reactant and product molecules. In this case a metal grill, four hamburgers, and the air above the fire are all heated. The energy released when a fuel such as charcoal burns can be transformed to provide many of the benefits of our technology-intensive society—and a good hot meal!

In our industrialized, high-technology, appliance-oriented society, the average use of energy per person is at nearly its highest point in history. The United States, with only 5% of the world's population, consumes about 30% of the world's energy resources. In every year since 1958 we have consumed more energy resources than have been produced within our borders. Most of the energy we use comes from chemical reactions: combustion of the fossil fuels coal, petroleum, and natural gas. The rest

OWL Sign in to OWL at **www.cengage.com/owl** to view tutorials and simulations, develop problem-solving skills, and complete online homework assigned by your professor.

go Chemistry Download mini lecture videos for key concept review and exam prep from OWL or purchase them from **www.CengageBrain.com.**

Companion Website

Visit this book's companion website at **www.cengage.com/chemistry/moore** to work interactive modules for the Estimation boxes and Active Figures in this text.

"A theory is the more impressive the greater the simplicity of its premises is, the more different kinds of things it relates, and the more extended is its area of applicability. Therefore, the deep impression which classical thermodynamics made upon me. It is the only physical theory of universal content concerning which I am convinced that, within the framework of the applicability of its basic concepts, it will never be overthrown." (Albert Einstein, quoted in Schlipp, P. A. [ed.] "Albert Einstein: Philosopher-Scientist." In *The Library of Living Philosophers,* Vol. VII. Autobiographical notes, 3rd ed. LaSalle, IL: Open Court Publishing, 1969; p. 33.)

comes from hydroelectric power plants, nuclear power plants, solar energy and wind collectors, and burning wood and other plant material. Both U.S. and world energy use are growing rapidly.

Chemical reactions involve transfers of energy. When a fuel burns, the energy of the products is less than the energy of the reactants. The leftover energy shows up in anything that is in contact with the reactants and products. For example, when propane burns in air, the carbon, hydrogen, and oxygen atoms in the C_3H_8 and O_2 reactant molecules rearrange to form CO_2 and H_2O product molecules.

$$C_3H_8(g) + 5\,O_2(g) \longrightarrow 3\,CO_2(g) + 4\,H_2O(g)$$

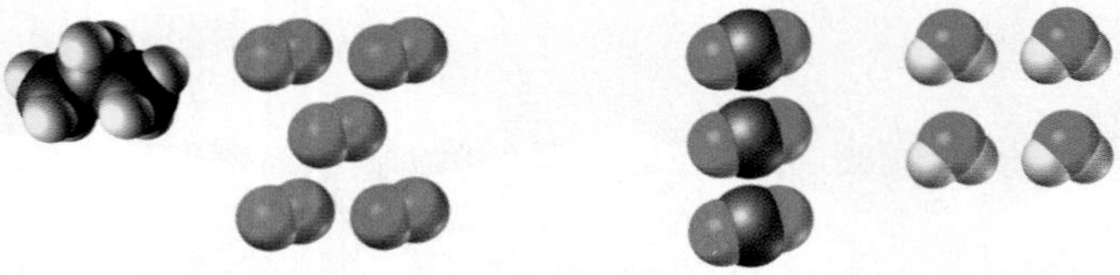

Because of the way their atoms are connected, the CO_2 and H_2O molecules have less total energy than the C_3H_8 and O_2 reactant molecules did. After the reaction, some energy that was in the reactants is not contained in the product molecules. That energy heats everything that is close to where the reaction takes place. The reaction transfers energy to its surroundings.

For the past hundred years or so, most of the energy society has used has come from combustion of fossil fuels, and this will continue to be true well into the future. Consequently, it is very important to understand how energy and chemical reactions are related and how chemistry might be used to alter our dependence on fossil fuels. This requires knowledge of **thermodynamics,** the science of heat, work, and transformations of one to the other. The fastest-growing new industries in the twenty-first century may well be those that capitalize on such knowledge and the new chemistry and chemical industries it spawns.

6.1 The Nature of Energy

What is energy? Where does the energy we use come from? And how can chemical reactions result in the transfer of energy to or from their surroundings? *Energy,* represented by E, was defined in Section 1.5 **(← p. 13)** as the capacity to do work. If you climb a mountain or a staircase, you work against the force of gravity as you move upward, and your gravitational energy increases. The energy you use to do this work is released when food you have eaten is metabolized (undergoes chemical reactions) within your body. Energy from food enables you to work against the force of gravity as you climb, and it warms your body (climbing makes you hotter as well as higher). Therefore our study of the relations between energy and chemistry also needs to consider processes that involve work and processes that involve heat.

Energy can be classified as kinetic or potential. **Kinetic energy** is energy that something has because it is moving (Figure 6.1). Examples of kinetic energy are

- Energy of motion of a macroscale object, such as a moving baseball or automobile; this is often called *mechanical energy.*
- Energy of motion of nanoscale objects such as atoms, molecules, or ions; this is often called *thermal energy.*
- Energy of motion of electrons through an electrical conductor; this is often called *electrical energy.*
- Energy of periodic motion of nanoscale particles when a macroscale sample is alternately compressed and expanded (as when a sound wave passes through air).

Kinetic energy, E_k, can be calculated as $E_k = \frac{1}{2}mv^2$, where m represents the mass and v represents the velocity of a moving object.

Riddle Photography/Shutterstock

Figure 6.1 **Kinetic energy.** As it speeds away from the club, the golf ball has kinetic energy that depends on its mass and velocity.

PhotoGorfer/iStockphoto

(a)

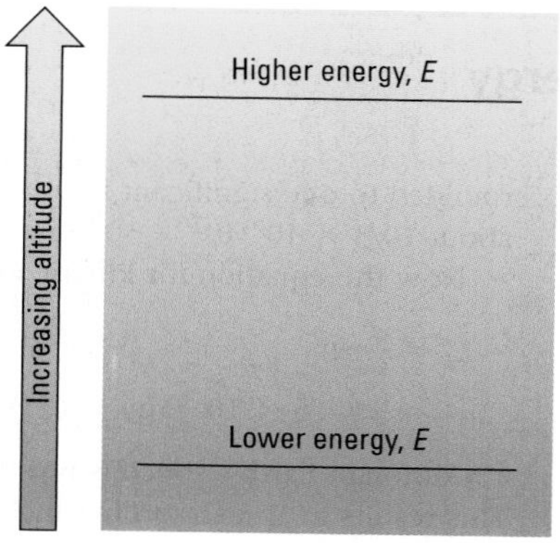

(b)

Rock climbing. (a) Climbing requires energy. (b) The higher the altitude, the greater the climber's gravitational energy.

Gary Muth/iStockphoto

Figure 6.2 Gravitational potential energy. Water on the brink of a waterfall has potential energy (stored energy that could be used to do work) because of its position relative to Earth; that energy could be used to generate electricity, for example, as in a hydroelectric power plant.

Potential energy is energy that something has as a result of its position and some force that is capable of changing that position. Examples include

- Energy that a ball held in your hand has because the force of gravity attracts it toward the floor; this is often called *gravitational energy.*
- Energy that charged particles have because they attract or repel each other; this is often called *electrostatic energy.* An example is the potential energy of positive and negative ions close together.
- Energy resulting from attractions and repulsions among electrons and atomic nuclei in molecules; this is often called *chemical potential energy* and is the kind of energy stored in foods and fuels.

Potential energy can be calculated in different ways, depending on the type of force that is involved. For example, near the surface of Earth, gravitational potential energy, E_p, can be calculated as $E_p = mgh$, where m is mass, g is the gravitational constant ($g = 9.8\ \text{m/s}^2$), and h is the height above the surface.

Potential energy can be converted to kinetic energy, and vice versa. As droplets of water fall over a waterfall (Figure 6.2), the potential energy they had at the top is converted to kinetic energy—they move faster and faster. Conversely, the kinetic energy of falling water could drive a water wheel to pump water to an elevated reservoir, where its potential energy would be higher.

Oesper Collection in the History of Chemistry, University of Cincinnati

1818–1889

The energy unit joule is named for James P. Joule. The son of a brewer in Manchester, England, Joule was a student of John Dalton (***← p. 21***). Joule established the idea that working and heating are both processes by which energy can be transferred from one sample of matter to another.

Energy Units

The SI unit of energy is the *joule* (rhymes with rule), symbol J. The joule is a derived unit, which means that it can be expressed as a combination of other more fundamental units: $1\ \text{J} = 1\ \text{kg m}^2/\text{s}^2$. If a 2.0-kg object (which weighs about $4\frac{1}{2}$ pounds) is moving with a velocity of 1.0 meter per second (roughly 2 miles per hour), its kinetic energy is

$$E_k = \tfrac{1}{2}mv^2 = \tfrac{1}{2} \times (2.0\ \text{kg})(1.0\ \text{m/s})^2 = 1.0\ \text{kg m}^2/\text{s}^2 = 1.0\ \text{J}$$

This is a relatively small quantity of energy. Because the joule is so small, we often use the kilojoule (1 kilojoule = 1 kJ = 1000 J) as a unit of energy.

Another energy unit is the *calorie*, symbol cal. By definition 1 cal = 4.184 J exactly. A calorie is very close to the quantity of energy required to raise the temperature of one gram of water by one degree Celsius. (The calorie was originally defined as the quantity of energy required to raise the temperature of 1 g $H_2O(\ell)$ from 14.5 °C to 15.5 °C.) The "calorie" that you hear about in connection with nutrition and dieting is actually a kilocalorie (kcal) and is usually represented with a capital C.

The joule is the unit of energy in the International System of units (SI units). SI units are described in Appendix B. A joule is approximately the quantity of energy required for one human heartbeat.

ESTIMATION Earth's Kinetic Energy

Estimate Earth's kinetic energy as it moves through space in orbit around the sun.

From an encyclopedia, a dictionary, or the Internet, you can obtain the facts that Earth's mass is about 6×10^{24} kg and its distance from the sun is about 150,000,000 km. Assume Earth's orbit is a circle and calculate the distance traveled in a year as the circumference of this circle, $\pi d = 2\pi r = 2 \times 3.14 \times 1.5 \times 10^8$ km. Since $2 \times 3 = 6$, $1.5 \times 6 = 9$, and 3.14 is a bit more than 3, estimate the distance as 10×10^8 km. Earth's speed, then, is a bit less than 10×10^8 km/yr.

Because $1\ \text{J} = 1\ \text{kg m}^2/\text{s}^2$, convert the time unit from years to seconds. Estimate the number of seconds in one year as $60\ \text{s/min} \times 60\ \text{min/h} \times 24\ \text{h/d} \times 365\ \text{d/yr} = 60 \times 60 \times 24 \times 365$ s/yr. To make the arithmetic easy, round 24 to 20 and 365 to 400, giving $60 \times 60 \times 20 \times 400\ \text{s/yr} = 6 \times 6 \times 2 \times 4 \times 10^5$ s/yr. This gives 288×10^5 s/yr, or 3×10^7 s/yr rounded to one significant figure. Therefore Earth's speed is about $10/3 \times 10^8/10^7 \cong 30$ km/s or 3×10^4 m/s.

Now the equation for kinetic energy can be used.

$$E_k = \tfrac{1}{2}mv^2$$
$$= \tfrac{1}{2} \times (6 \times 10^{24}\ \text{kg}) \times (3 \times 10^4\ \text{m/s})^2 \cong 2 \times 10^{33}\ \text{J}$$

Although Earth's speed is not high, its mass is very large. This results in an extraordinarily large kinetic energy—far more energy than has been involved in all of the hurricanes and typhoons that Earth has ever experienced.

Visit this book's companion website at **www.cengage.com/chemistry/moore** to work an interactive module based on this material.

1 cal = 4.184 J exactly

1 Cal = 1 kcal = 1000 cal = 4.184 kJ = 4184 J

The food Calorie measures how much energy is released when a given quantity of food undergoes combustion with oxygen.

The **PROBLEM-SOLVING STRATEGY** in this book is

- Analyze the problem
- Plan a solution
- Execute the plan
- Check that the result is reasonable

Appendix A.1 explains this in detail.

Figure 6.3 Food energy. A packet of artificial sweetener from Australia. As its label shows, the sweetener in the packet supplies 16 kJ of nutritional energy. It is equivalent in sweetness to 2 level teaspoonfuls of sugar, which would supply 140 kJ of nutritional energy.

Thus, a breakfast cereal that gives you 100 Calories of nutritional energy actually provides 100 kcal = 100×10^3 cal. In many countries food energy is reported in kilojoules rather than in Calories. For example, the label on the packet of nonsugar sweetener shown in Figure 6.3 indicates that it provides 16 kJ of nutritional energy.

PROBLEM-SOLVING EXAMPLE 6.1 Energy Units

A single Fritos snack chip has a food energy of 5.0 Cal. What is this energy in joules?

Answer 2.1×10^4 J

Strategy and Explanation To find the energy in joules, we use the fact that 1 Cal = 1 kcal, the definition of the prefix *kilo-* (= 1000), and the definition 1 cal = 4.184 J to generate appropriate proportionality factors (conversion factors).

$$E = 5.0\ \text{Cal} \times \frac{1\ \text{kcal}}{1\ \text{Cal}} \times \frac{1000\ \text{cal}}{1\ \text{kcal}} \times \frac{4.184\ \text{J}}{1\ \text{cal}} = 2.1 \times 10^4\ \text{J}$$

☑ **Reasonable Answer Check** 2.1×10^4 J is 21 kJ. Because 1 Cal = 1 kcal = 4.184 kJ, the result in kJ should be about four times the original 5 Cal (that is, about 20 kJ), which it is.

(a) (b)

Food energy. When a Fritos chip (a) is dropped into molten potassium chlorate, the chip burns (b) in oxygen generated by thermal decomposition of the potassium chlorate, and about 20 kJ is transferred to the surroundings.

PROBLEM-SOLVING PRACTICE 6.1

(a) If you eat a hot dog, it will provide 160 Calories of energy. Express this energy in joules.

(b) A watt (W) is a unit of power that corresponds to the transfer of one joule of energy in one second. The energy used by an x-watt light bulb operating for y seconds is $x \times y$ joules. If you turn on a 75-watt bulb for 3.0 hours, how many joules of electrical energy will be transformed into light and heat?

(c) The packet of nonsugar sweetener in Figure 6.3 provides 16 kJ of nutritional energy. Express this energy in kilocalories.

PROBLEM-SOLVING PRACTICE answers are provided at the back of this book in Appendix K.

6.2 Conservation of Energy

When you dive from a diving board into a pool of water, several transformations of energy occur (Figure 6.4). Eventually, you float on the surface and the water becomes still. However, on average, the water molecules are moving a little faster in the vicinity of your point of impact; that is, the temperature of the water is now a little higher. Energy has been transformed from potential to kinetic and from macroscale kinetic to nanoscale kinetic (that is, thermal). Nevertheless, the total quantity of energy, kinetic plus potential, is the same before and after the dive. In many, many experiments, the total energy has always been found to be the same before and after an event. These experiments are summarized by the **law of conservation of energy,** which states that *energy can neither be created nor destroyed—the total energy of the universe is constant.* This is also called the **first law of thermodynamics.**

In Section 1.8 (← *p. 19*) the kinetic-molecular theory was described qualitatively. A corollary to this theory is that *molecules move faster, on average, as the temperature increases.*

The nature of scientific laws is discussed in Chapter 1 (← *p. 6*).

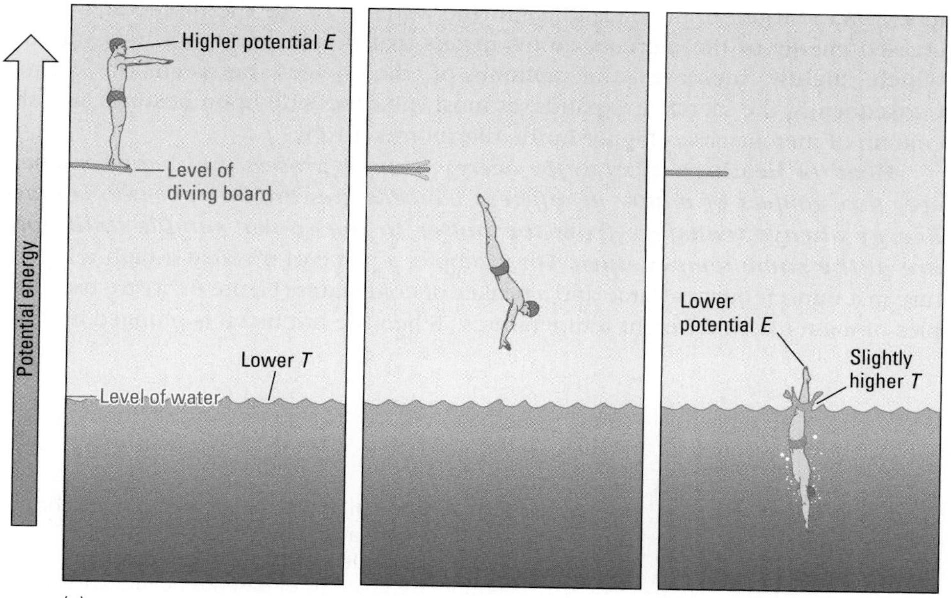

Figure 6.4 Energy transformations. Potential and kinetic energy are interconverted when someone dives into water. These interconversions are governed by the law of conservation of energy. (a) The diver has greater gravitational potential energy on the diving board than at the surface of the water, because the platform is higher above the earth. (b) Some of the potential energy has been converted into kinetic energy as the diver's altitude above the water decreases and velocity increases; maximum kinetic energy occurs just before impact with the water. (c) Upon impact, the diver works on the water, splashing it aside; eventually, the initial potential energy difference is converted into motion on the nanoscale—the temperature of the water has become slightly higher.

Answers to **EXERCISES** are provided at the back of this book in Appendix L.

EXERCISES that are labeled **CONCEPTUAL** are designed to test your understanding of one or more concepts; they usually involve qualitative rather than quantitative thinking.

CONCEPTUAL EXERCISE 6.1 Energy Transfers

You toss a rubber ball up into the air. It falls to the floor, bounces for a while, and eventually comes to rest. Several energy transfers are involved. Describe them and the changes they cause.

Energy and Working

When a force acts on an object and moves the object, the change in the object's kinetic energy is equal to the work done on the object. Work has to be done, for example, to accelerate a car from 0 to 60 miles per hour or to hit a baseball out of a stadium. Work is also required to increase the potential energy of an object. Thus, work has to be done to raise an object against the force of gravity (as in an elevator), to separate a sodium ion, Na^+, from a chloride ion, Cl^-, or to move an electron away from an atomic nucleus. The work done on an object corresponds to the quantity of energy transferred to that object; that is, doing **work** (or **working**) on an object is *a process that transfers energy to an object.* Conversely, if an object does work on something else, the quantity of energy associated with the object must decrease. In the rest of this chapter (and book), we will refer to a transfer of energy by doing work as a "work transfer."

Work and *heat* refer to the quantity of energy transferred from one object or sample to another by working or heating *processes.* However, we often talk about work and heat as if they were forms of energy. Working and heating processes transfer energy from one form or one place to another. To emphasize this, we often will use the words *working* and *heating* where many people would use *work* and *heat.*

Work is required to cause some chemical and biochemical processes to occur. Examples are moving ions across a cell membrane and synthesizing adenosine triphosphate (ATP) from adenosine diphosphate (ADP).

Energy, Temperature, and Heating

According to the kinetic-molecular theory (← *p. 19*), all matter consists of nanoscale particles that are in constant motion (Figure 6.5). Therefore, all matter has thermal energy. For a given sample, the quantity of thermal energy is greater the higher the temperature is. Transferring energy to a sample of matter usually results in a temperature increase that can be measured with a thermometer. For example, when a mercury thermometer is placed into warm water (Figure 6.6), energy transfers from the water to the thermometer (the water heats the thermometer). The increased energy of the mercury atoms means that they move about more rapidly, which slightly increases the volume of the spaces between the atoms. Consequently, the mercury expands (as most substances do upon heating), and the column of mercury rises higher in the thermometer tube.

Heat (or **heating**) refers to *the energy transfer process that happens whenever two samples of matter at different temperatures are brought into contact.* ***Energy always transfers from the hotter to the cooler sample until both are at the same temperature.*** For example, a piece of metal at a high temperature in a Bunsen burner flame and a beaker of cold water (Figure 6.7a) are two samples of matter with different temperatures. When the hot metal is plunged into the

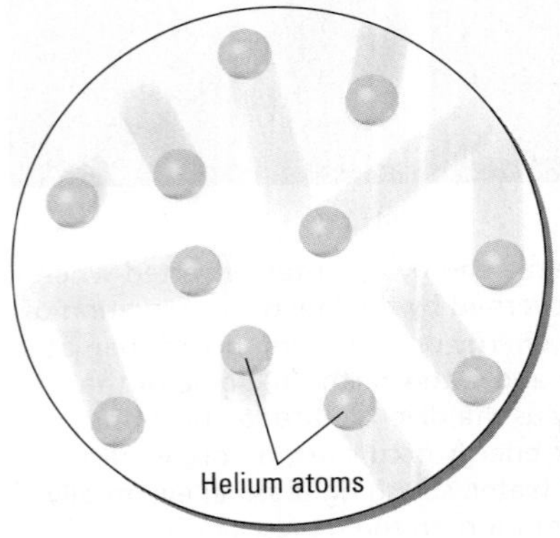

Figure 6.5 Thermal energy. According to the kinetic-molecular theory, nanoscale particles (atoms, molecules, and ions) are in constant motion. Here, atoms of gaseous helium are shown. Each atom has kinetic energy that depends on its mass and how fast it is moving (as indicated by the length of the "tail," which shows how far each atom travels per unit time). The thermal energy of the sample is the sum of the kinetic energies of all the helium atoms. The higher the temperature of the helium, the faster the average speed of the molecules, and therefore the greater the thermal energy.

Figure 6.6 Measuring temperature. The mercury in a thermometer expands because the mercury atoms are moving faster (have more energy) after the boiling water transfers energy to (heats) the mercury; the temperature and the volume of the mercury have both increased.

cold water (Figure 6.7b), energy transfers from the metal to the water until the two samples reach the same temperature. Once that happens, the metal and water are said to be in **thermal equilibrium.** When thermal equilibrium is reached, the metal has heated the water (and the water has cooled the metal) to a common temperature. In the rest of this chapter (and book), we will refer to a transfer of energy by heating and cooling as a "heat transfer."

Usually most objects in a given region, such as your room, are at about the same temperature—at thermal equilibrium. A fresh cup of coffee, which is hotter than room temperature, transfers energy by heating the rest of the room until the coffee cools off (and the rest of the room warms up a bit). A can of cold soft drink, which is much cooler than its surroundings, receives energy from everything else until it warms up (and your room cools off a little). Because the total quantity of material in your room is very much greater than that in a cup of coffee or a can of soda, the

Transferring energy by heating is a process, but it is common to talk about that process as if heat were a form of energy. It is often said that one sample transfers heat to another, when what is meant is that one sample transfers energy by heating the other.

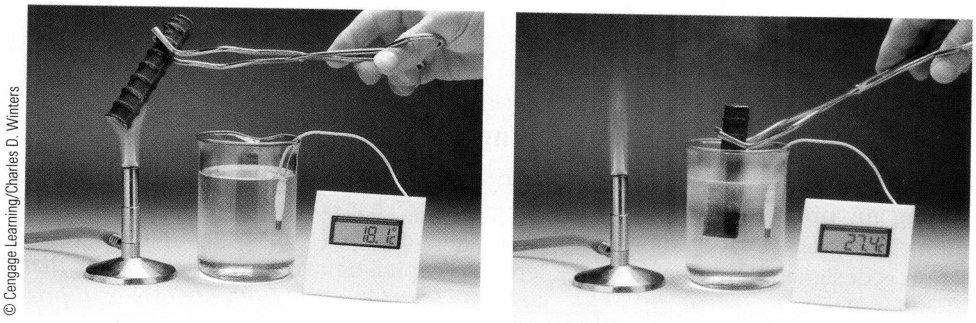

Figure 6.7 Energy transfer by heating. Water in a beaker with a probe that measures temperature in °C is heated when a hotter sample (a steel bar) is plunged into the water. There is a transfer of energy from the hotter metal bar to the cooler water. Eventually, enough energy is transferred so that the bar and the water reach the same temperature—that is, thermal equilibrium is achieved.

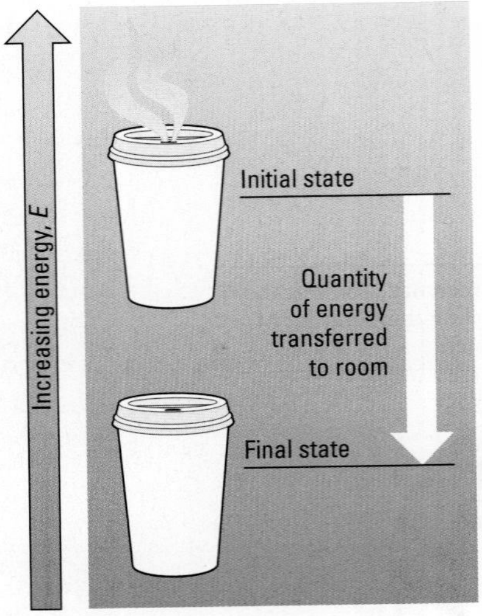

Figure 6.8 Energy diagram for a cup of hot coffee. The diagram compares the energy of a cup of hot coffee with the energy after the coffee has cooled to room temperature. The higher something is in the diagram, the more energy it has. As the coffee and cup cool to room temperature, energy is transferred to the surrounding matter in the room. According to the law of conservation of energy, the energy remaining in the coffee must be less after the change (in the final state) than it was before the change (in the initial state). The quantity of energy transferred is represented by the arrow from the initial to the final state.

room temperature changes only a tiny bit to reach thermal equilibrium, whereas the temperature of the coffee or the soda changes a lot.

A diagram such as Figure 6.8 can be used to show the energy transfer from a cup of hot coffee to your room. The upper horizontal line represents the energy of the hot coffee and the lower line represents the energy of the room-temperature coffee. Because the coffee started at a higher temperature (higher energy), the upper line is labeled the initial state. The lower line is the final state. During the change from initial to final state, energy transfers from the coffee to your room. Therefore, the energy of the coffee is lower in the final state than it was in the initial state.

CONCEPTUAL EXERCISE 6.2 Energy Diagrams

(a) Draw an energy diagram like the one in Figure 6.8 for warming a can of cold soft drink to room temperature. Label the initial and final states and use an arrow to represent the change in energy of the can of soda.

(b) Draw a second energy diagram, to the same scale, to show the change in energy of the room as the can of cold soda warms to room temperature.

System, Surroundings, and Internal Energy

In thermodynamics it is useful to define a *region of primary concern* as the **system.** Then we can decide whether energy transfers into or out of the system and keep an accounting of how much energy transfers in each direction. *Everything that can exchange energy with the system* is defined as the **surroundings.** A system may be delineated by an actual physical boundary, such as the inside surface of a flask or the membrane of a cell in your body. Or the boundary may be indistinct, as in the case of the solar system within its surroundings, the rest of the galaxy. In the case of a hot cup of coffee in your room, the cup and the coffee might be the system, and your room would be the surroundings. For a chemical reaction, the system is usually defined to be all of the atoms that make up the reactants. These same atoms will be bonded in a different way in the products after the reaction, and it is their energy before and after reaction that interests us most.

The **internal energy** of a system is the *sum of the individual energies (kinetic and potential) of all nanoscale particles (atoms, molecules, or ions)* in that system. Increasing the temperature increases the internal energy because it increases the average speed of motion of nanoscale particles. ***The total internal energy of a sample of matter depends on temperature, the type of particles, and the number of particles in the sample.*** For a given substance, internal energy depends on temperature and the size of the sample. Thus, despite being at a higher temperature, a cupful of boiling water contains less energy than a bathtub full of warm water.

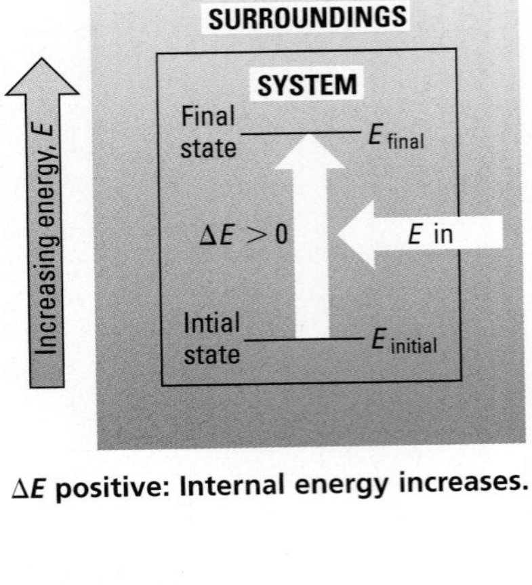

ΔE positive: Internal energy increases.

Calculating Thermodynamic Changes

If we represent a system's internal energy by E, then the change in internal energy during any process is calculated as $E_{final} - E_{initial}$. That is, from the internal energy after the process is over, subtract the internal energy before it began. Such a calculation is designated by using a Greek letter Δ (capital delta) before the quantity that changes. Thus, $E_{final} - E_{initial} = \Delta E$. *Whenever a change is indicated by* Δ, *a positive value indicates an increase and a negative value indicates a decrease.* Therefore, if the internal energy increases during a process, ΔE has a positive value ($\Delta E > 0$); if the internal energy decreases, ΔE is negative ($\Delta E < 0$).

A good analogy to this thermodynamic calculation is your bank account. Assume that in your account (the system) you have a balance B of \$260 ($B_{initial}$), and you withdraw \$60 in spending money. After the withdrawal the balance is \$200 (B_{final}). The change in your balance is

$$\text{Change in balance} = \Delta B = B_{final} - B_{initial} = \$200 - \$260 = -\$60$$

The negative sign on the \$60 indicates that money has been withdrawn from the account (system) and transferred to you (the surroundings). The cash itself is not negative, but during the process of withdrawing your money the balance in the bank went down, so ΔB was negative. Similarly, the *magnitude* of change of a thermodynamic quantity is a number with no sign. To indicate the *direction* of a change, we attach a negative sign (transferred out of the system) or a positive sign (transferred into the system).

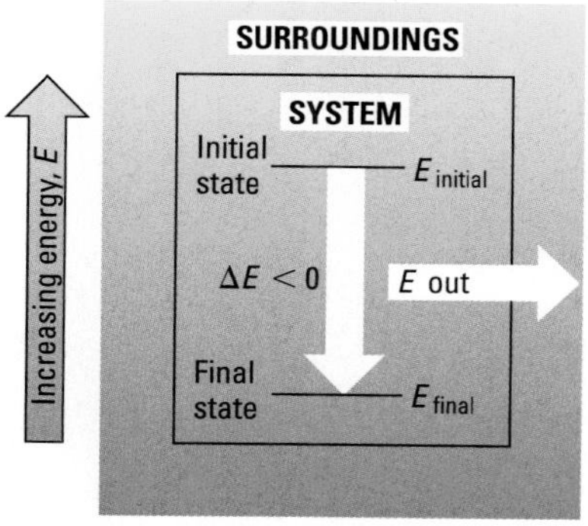

ΔE negative: Internal energy decreases.

CONCEPTUAL EXERCISE 6.3 Direction of Energy Transfer

It takes about 1.5 kJ to raise the temperature of a can of beverage from 25.0 °C to 26.0 °C. You put the can of beverage into a refrigerator to cool it from room temperature (25.0 °C) to 1.0 °C.

(a) What quantity of heat transfer is required? Express your answer in kilojoules.
(b) What is a reasonable choice of system for this situation?
(c) What constitutes the surroundings?
(d) What is the sign of ΔE for this situation? What is the calculated value of ΔE?
(e) Draw an energy diagram showing the system, the surroundings, the change in energy of the system, and the energy transfer between the system and the surroundings.

Conservation of Energy and Chemical Reactions

For many chemical reactions the only energy transfer processes are heating and doing work. If no other energy transfers (such as emitting light) take place, the law of conservation of energy for any system can be written as

$$\Delta E = q + w \qquad [6.1]$$

where q represents the quantity of energy transferred by heating the system, and w represents the quantity of energy transferred by doing work on the system. If energy is transferred into the system from the surroundings by heating, then q is positive; if energy is transferred into the system because the surroundings do work on the system, then w is positive. If energy is transferred out of the system by heating the surroundings, then q has a negative value; if energy is transferred out of the system because work is done on the surroundings, then w has a negative value. The *magnitudes* of q and w indicate the *quantities* of energy transferred, and the *signs* of q and w indicate the *direction* in which the energy is transferred. The relationships among ΔE, q, and w for a system are shown in Figure 6.9.

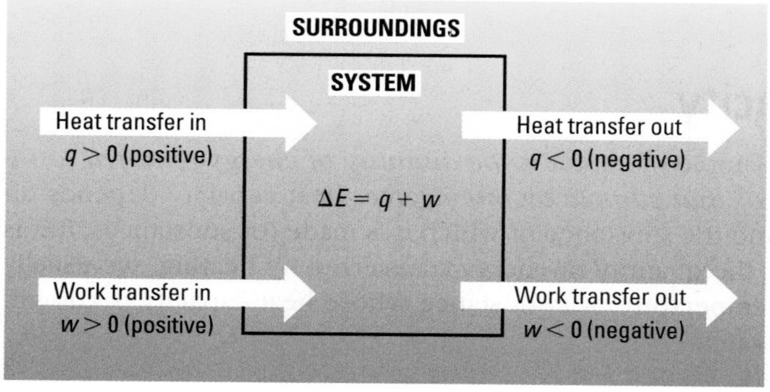

Figure 6.9 Internal energy, heat, and work. Schematic diagram showing energy transfers between a thermodynamic system and its surroundings.

PROBLEM-SOLVING EXAMPLE 6.2 Internal Energy, Heat, and Work

A fuel cell based on the reaction of hydrogen with oxygen powers a small automobile by running an electric motor. The motor draws 75.0 kilowatts (75.0 kJ/s) and runs for 2 minutes and 20 seconds. During this period, 5.0×10^3 kJ must be carried away from the fuel cell to prevent it from overheating. If the system is defined to be the hydrogen and oxygen that react, what is the change in the system's internal energy?

Answer -1.55×10^4 kJ

Strategy and Explanation Because the system is the reactants and products, the rest of the fuel cell is part of the surroundings. The problem states that the motor powers a car, which means that the system is doing work (to cause electric current to flow, which then makes the car move). Therefore energy is transferred out of the system and w must be negative. The work is

$$w = -75.0 \frac{\text{kJ}}{\text{s}} \times 140.\text{ s} = -10.5 \times 10^3 \text{ kJ}$$

Energy is also transferred out of the system when the system heats its surroundings, which means that q must be negative, and $q = -5.0 \times 10^3$ kJ. Using Equation 6.1

$$\Delta E = q + w = (-5.0 \times 10^3 \text{ kJ}) + (-10.5 \times 10^3 \text{ kJ}) = -1.55 \times 10^4 \text{ kJ}$$

Thus the internal energy of the reaction product (water) is 1.55×10^4 kJ lower than the internal energy of the reactants (hydrogen and oxygen).

☑ **Reasonable Answer Check** ΔE is negative, which is reasonable. The internal energy of the reaction products should be lower than that of the reactants, because energy transferred from the reactions heats the surroundings and does work on the surroundings.

PROBLEM-SOLVING PRACTICE 6.2

Suppose that the internal energy decreases by 2400 J when a mixture of natural gas (methane) and oxygen is ignited and burns. If the surroundings are heated by 1.89 kJ, how much work was done by this system on the surroundings?

So far we have seen that

- energy transfers can occur either by heating or by working;
- it is convenient to define a system so that energy transfers into a system (positive) and out of a system (negative) can be accounted for; and
- the internal energy of a system can change as a result of heating or doing work on the system.

Our primary interest in this chapter is heat transfers (the "thermo" in thermodynamics). Heat transfers can take place between two objects at different temperatures. Heat transfers also accompany physical changes and chemical changes. The next three sections (6.3 to 6.5) show how quantitative measurements of heat transfers can be made, first for heating that results from a temperature difference and then for heating that accompanies a physical change.

6.3 Heat Capacity

The **heat capacity** of a sample of matter is *the quantity of energy required to increase the temperature of that sample by one degree.* Heat capacity depends on the mass of the sample and the substance of which it is made (or substances, if it is not pure). To determine the quantity of energy transferred by heating, we usually measure the change in temperature of a substance whose heat capacity is known. Often that substance is water.

Specific Heat Capacity

To make useful comparisons among samples of different substances with different masses, the **specific heat capacity** (which is sometimes just called *specific heat*) is defined as the quantity of energy needed to increase the temperature of one gram of a substance by one degree Celsius. For water at 15 °C, the specific heat capacity is 1.00 cal g^{-1} $°C^{-1}$ or 4.184 J g^{-1} $°C^{-1}$; for common window glass, it is only about 0.8 J g^{-1} $°C^{-1}$. That is, it takes about five times as much heat transfer of energy to raise the temperature of a gram of water by 1 °C as it does for a gram of glass. Like density (← *p. 8*), specific heat capacity is a property that can be used to distinguish one substance from another. It can also be used to distinguish a pure substance from a solution or mixture, because the specific heat capacity of a mixture will vary with the proportions of the mixture's components.

The notation J g^{-1} $°C^{-1}$ means that the units are joules divided by grams and divided by degrees Celsius; that is, J/(g °C). We will use negative exponents to show unambiguously which units are in the denominator whenever the denominator includes two or more units.

The specific heat capacity, *c*, of a substance can be determined experimentally by measuring the quantity of energy transferred to or from a known mass of the substance as its temperature rises or falls. We assume that there is no work transfer of energy to or from the sample and we treat the sample as a thermodynamic system, so $\Delta E = q$.

$$\text{Specific heat capacity} = \frac{\text{quantity of energy transferred by heating}}{\text{sample mass} \times \text{temperature change}}$$

or

$$c = \frac{q}{m \times \Delta T} \qquad [6.2]$$

Suppose that for a 25.0-g sample of ethylene glycol, 90.7 J is required to change the temperature from 22.4 °C to 23.9 °C. (Ethylene glycol is used as a coolant in automobile engines.) Thus,

$$\Delta T = (23.9\ °C - 22.4\ °C) = 1.5\ °C$$

From Equation 6.2, the specific heat capacity of ethylene glycol is

$$c = \frac{q}{m \times \Delta T} = \frac{90.7\ \text{J}}{25.0\ \text{g} \times 1.5\ °C} = 2.4\ \text{J g}^{-1}\ °C^{-1}$$

Akropot/iStockphoto

Moderation of microclimate by water. In cities near bodies of water (such as San Francisco, shown here), summertime temperatures are lower within a few hundred meters of the waterfront than they are a few kilometers away from the water. Wintertime temperatures are higher, unless the water freezes, in which case the moderating effect is less, because ice on the surface insulates the rest of the water from the air.

The specific heat capacities of many substances have been determined. A few values are listed in Table 6.1. Notice that water has one of the highest values. This is important because a high specific heat capacity means that a great deal of energy must be transferred to a large body of water to raise its temperature by just one degree. Conversely, a lot of energy must be transferred away from the water before its temperature falls by one degree. Thus, a lake or ocean can store an enormous quantity of energy and thereby moderate local temperatures. This has a profound influence on weather near lakes or oceans.

When the specific heat capacity of a substance is known, you can calculate the temperature change that should occur when a given quantity of energy is transferred to or from a sample of known mass. More important, by measuring the temperature change and the mass of a substance, you can calculate *q*, the quantity of energy transferred to or from it by heating. For these calculations it is convenient to rearrange Equation 6.2 algebraically as

$$\Delta T = \frac{q}{c \times m} \quad \text{or} \quad q = c \times m \times \Delta T \qquad [6.2']$$

The high specific heat capacity of water helps to keep your body temperature relatively constant. Water accounts for a large fraction of your body mass, and warming or cooling that water requires a lot of energy transfer.

PROBLEM-SOLVING EXAMPLE **6.3** **Using Specific Heat Capacity**

If 100.0 g water is cooled from 25.3 °C to 16.9 °C, what quantity of energy has been transferred from the water?

Answer 3.5 kJ transferred from the water

Table 6.1 Specific Heat Capacities for Some Elements, Compounds, and Common Solids

Substance	Specific Heat Capacity ($J\ g^{-1}\ °C^{-1}$)
Elements	
Aluminum, Al	0.902
Carbon (graphite), C	0.720
Iron, Fe	0.451
Copper, Cu	0.385
Gold, Au	0.128
Compounds	
Ammonia, $NH_3(\ell)$	4.70
Water, $H_2O(\ell)$	4.184
Ethanol, $C_2H_5OH(\ell)$	2.46
Ethylene glycol (antifreeze), $HOCH_2CH_2OH(\ell)$	2.42
Water, $H_2O(s)$	2.06
Carbon tetrachloride, $CCl_4(\ell)$	0.861
A chlorofluorocarbon (CFC), $CCl_2F_2(\ell)$	0.598
Common solids	
Wood	1.76
Concrete	0.88
Glass	0.84
Granite	0.79

© Cengage Learning/Charles D. Winters

Samples of substances listed in Table 6.1: glass beaker containing water, copper wire, aluminum rod, powdered graphite, and iron bar.

Strategy and Explanation Treat the water as a system. The quantity of energy is proportional to the specific heat capacity of water (Table 6.1), the mass of water, and the change in temperature. This is summarized in Equation 6.2′ as

$$\Delta E = q = c \times m \times \Delta T$$

$$= 4.184\ J\ g^{-1}\ °C^{-1} \times 100.0\ g \times (16.9\ °C - 25.3\ °C) = -3.5 \times 10^3\ J = -3.5\ kJ$$

☑ **Reasonable Answer Check** It requires about 4 J to heat 1 g water by 1 °C. In this case the temperature change is not quite 10 °C and we have 100 g water, so q should be about $4\ J\ g^{-1}\ °C^{-1} \times (-10\ °C) \times 100\ g = -4000\ J = -4\ kJ$, which it is. The sign is negative because energy is transferred out of the water as it cools.

PROBLEM-SOLVING PRACTICE **6.3**

A piece of aluminum with a mass of 250. g is at an initial temperature of 5.0 °C. If 24.1 kJ is supplied to warm the Al, what is its final temperature? Obtain the specific heat capacity of Al from Table 6.1.

CONCEPTUAL EXERCISE **6.4 Specific Heat Capacity and Temperature Change**

Suppose you put two 50-mL beakers in a refrigerator so energy is transferred out of each sample at the same constant rate. If one beaker contains 10. g pulverized glass and one contains 10. g carbon (graphite), which beaker has the lower temperature after 3 min in the refrigerator?

Molar Heat Capacity

It is often useful to know the heat capacity of a sample in terms of the same number of particles instead of the same mass. For this purpose we use the **molar heat capacity,** symbol c_m. This is *the quantity of energy that must be transferred to increase the temperature of one mole of a substance by 1 °C.* The molar heat capacity is easily calculated from the specific heat capacity by using the molar mass of the substance. For example, the specific heat capacity of liquid ethanol is given in Table 6.1 as $2.46\ J\ g^{-1}\ °C^{-1}$. The molecular formula of ethanol is CH_3CH_2OH, so its molar mass is 46.07 g/mol. The molar heat capacity is

$$c_m = \frac{2.46\ J}{g\ °C} \times \frac{46.07\ g}{mol} = 113\ J\ mol^{-1}\ °C^{-1}$$

CONCEPTUAL EXERCISE **6.5 Molar Heat Capacity**

Calculate the molar heat capacities of all the metals listed in Table 6.1. Compare these with the value just calculated for ethanol. Based on your results, suggest a way to predict the molar heat capacity of a metal. Can this same rule be applied to other kinds of substances?

As you should have found in Conceptual Exercise 6.5, molar heat capacities of metals are very similar. This can be explained if we consider what happens on the nanoscale when a metal is heated. The energy transferred by heating a solid makes the atoms vibrate more extensively about their average positions in the solid crystal lattice. Every metal consists of many, many atoms, all of the same kind and packed closely together; that is, the structures of all metals are very similar. As a consequence, the ways that the metal atoms can vibrate (and therefore the ways that their energies can be increased) are very similar. Thus, no matter what the metal, nearly the same quantity of energy must be transferred per metal atom to increase the temperature by one degree. The quantity of energy per mole is therefore very similar for all metals.

PROBLEM-SOLVING EXAMPLE 6.4 Direction of Energy Transfer

People sometimes drink hot tea to keep warm. Suppose that you drink a 250.-mL cup of tea that is at 65.0 °C. Calculate the quantity of energy transferred to your body and the surrounding air when the temperature of the tea drops to 37.0 °C (normal body temperature). Make reasonable assumptions to obtain the mass and specific heat capacity of the liquid.

Answer 29.3 kJ transferred out of the tea

Strategy and Explanation Treat the tea as the system. Assume the density of tea, which is mostly water, is 1.00 g/mL, so the tea has a mass of 250. g; also assume the specific heat capacity of tea is the same as that of water, 4.184 J g^{-1} $°C^{-1}$. The initial temperature is 63.0 °C and the final temperature is 37.0 °C. Thus the quantity of energy transferred is

$$\Delta E = q = c \times m \times \Delta T$$

$$= 4.184 \text{ J g}^{-1} \text{ °C}^{-1} \times 250. \text{ g} \times (37.0 - 65.0) \text{ °C} = -29.3 \times 10^3 \text{ J} = -29.3 \text{ kJ}$$

The negative sign of the result indicates that 29.3 kJ is transferred from the tea (system) to the surroundings (you) as the temperature of the tea decreases.

☑ **Reasonable Answer Check** Estimate the heat transfer as a bit more than $(4 \times 25 \times 250)$ J = (100×250) J = 25,000 J = 25 kJ, which it is. The transfer is from the tea so q should be negative, which it is.

Usually the surroundings contain a great deal more matter than the system and hence have a much greater heat capacity. Consequently the change in temperature of the surroundings is often so small that it cannot be measured.

PROBLEM-SOLVING PRACTICE 6.4

Assume that the same cup of tea described in Problem-Solving Example 6.4 is warmed from 37 °C to 65 °C and there is no work done by the heating process. What is ΔE for this process?

PROBLEM-SOLVING EXAMPLE 6.5 Transfer of Energy Between Samples by Heating

Suppose that you have 100. mL H_2O at 20.0 °C and you add to the water 55.0 g iron pellets that had been heated to 425 °C. What is the temperature of both the water and the iron when thermal equilibrium is reached? (Assume that there is no energy transfer to the glass beaker or to the air or to anything else but the water. Assume also that no work is done, that no liquid water vaporizes, and that the density of water is 1.00 g/mL.)

Answer T_{final} = 42.7 °C

Strategy and Explanation Thermal equilibrium means that the water and the iron pellets will have the same final temperature, which is what we want to calculate. Consider the iron to be the system and the water to be the surroundings. The energy transferred from the iron is the same energy that is transferred to the water. None of this energy goes anywhere other than to the water. Therefore $\Delta E_{water} = -\Delta E_{iron}$ and $q_{water} = -q_{iron}$. The quantity of energy transferred to the water and the quantity transferred from the iron are equal. They are opposite in algebraic sign because energy was transferred *from* the iron as its temperature dropped, and energy was transferred *to* the water to raise its temperature.

The quantities of energy transferred have opposite signs because they take place from the iron (negative) to the water (positive).

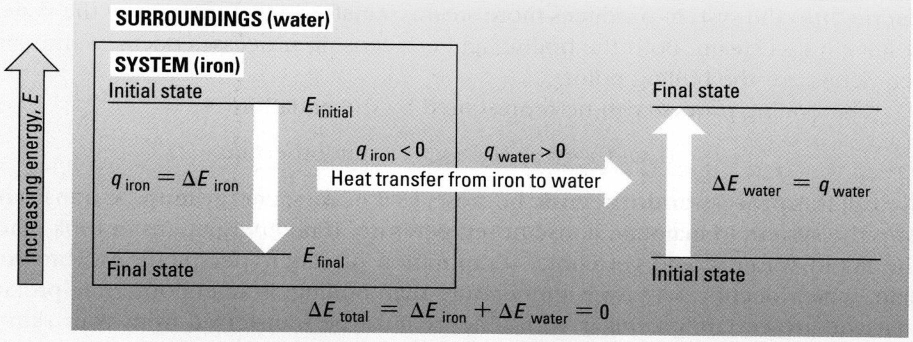

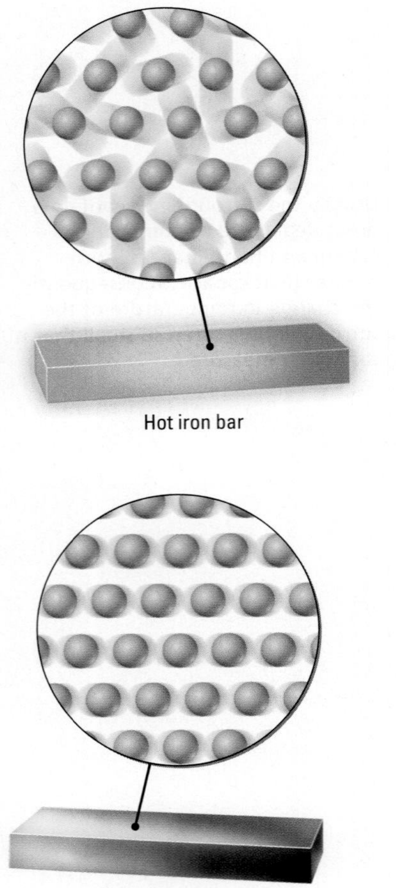

Hot and cold iron. On the nanoscale the atoms in the sample of hot iron are vibrating much farther from their average positions than those in the sample of room-temperature iron. The greater vibration of atoms in hot iron means harder collisions of iron atoms with water molecules. Such collisions transfer energy to the water molecules, heating the water.

Specific heat capacities for iron and water are listed in Table 6.1. The mass of water is 100. mL × 1.00 g/mL = 100. g. $T_{initial}$ for the iron is 425 °C and $T_{initial}$ for the water is 20.0 °C.

$$q_{water} = -q_{iron}$$

$$c_{water} \times m_{water} \times \Delta T_{water} = -c_{iron} \times m_{iron} \times \Delta T_{iron}$$

$$(4.184\ \text{J g}^{-1}\ {}^\circ\text{C}^{-1})(100.\ \text{g})(T_{final} - 20.0\ {}^\circ\text{C}) = -(0.451\ \text{J g}^{-1}\ {}^\circ\text{C}^{-1})(55.0\ \text{g})(T_{final} - 425\ {}^\circ\text{C})$$

$$(418.4\ \text{J}\ {}^\circ\text{C}^{-1})T_{final} - (8.368 \times 10^3\ \text{J}) = -(24.80\ \text{J}\ {}^\circ\text{C}^{-1})T_{final} + (1.054 \times 10^4\ \text{J})$$

$$(443.2\ \text{J}\ {}^\circ\text{C}^{-1})T_{final} = 1.891 \times 10^4\ \text{J}$$

Solving, we find T_{final} = 42.7 °C. The iron has cooled a lot (ΔT_{iron} = −382 °C) and the water has warmed a little (ΔT_{water} = 22.7 °C).

☑ **Reasonable Answer Check** As a check, note that the final temperature must be between the two initial values, which it is. Also, don't be concerned by the fact that transferring the same quantity of energy resulted in two very different values of ΔT; this difference arises because the specific heat capacities and masses of iron and water are different. There is much less iron and its specific heat capacity is smaller, so its temperature changes much more than the temperature of the water.

PROBLEM-SOLVING PRACTICE 6.5

A 400.-g iron bar is heated in a flame and then immersed in 1000. g water in a beaker. The initial temperature of the water was 20.0 °C, and both the iron and the water are at 32.8 °C at the end of the experiment. What was the original temperature of the hot iron bar? (Assume that all energy transfer is between the water and the iron.)

6.4 Energy and Enthalpy

Using heat capacity we can account for transfers of energy between samples of matter as a result of temperature differences. But energy transfers also accompany physical or chemical changes, *even though there may be no change in temperature.* We will first consider the simpler case of physical change and then apply the same ideas to chemical changes.

Conservation of Energy and Changes of State

Consider a system that consists of water at its boiling temperature in a container with a balloon attached (Figure 6.10). The system is under a constant atmospheric pressure. If the water is heated, it will boil, the temperature will remain at 100 °C, and the steam produced by boiling the water will inflate the balloon (Figure 6.10b). If the heating stops, then the water will stop boiling, some of the steam will condense to liquid, and the volume of steam will decrease (Figure 6.10c). There will be heat transfer of energy to the surroundings. However, as long as steam is condensing to liquid water, the temperature will remain at 100 °C. In summary, transferring energy *into* the system produces more steam; transferring energy *out of* the system results in less steam. Both the boiling and condensing processes occur at the same temperature—the boiling point.

Changes of state (between solid and liquid, liquid and gas, or solid and gas) are described in more detail in Section 11.3. Because the temperature remains constant during a change of state, melting points and boiling points can be measured relatively easily and used to identify substances **(← *p. 7*).**

The boiling process can be represented by the equation

$$H_2O(\ell) \longrightarrow H_2O(g) \qquad \text{endothermic}$$

We call this process **endothermic** because, as it occurs, energy must be transferred *into* the system to maintain constant temperature. If no energy transfer took place, the liquid water would get cooler. Evaporation of water (perspiration) from your skin, which occurs at a lower temperature than boiling, is an endothermic process that you are certainly familiar with. Energy must be transferred from your skin to the evaporating water, and this energy transfer cools your skin.

Thermic or *thermo* comes from the Greek word *thermé*, meaning "heat." *Endo* comes from the Greek word *endon*, meaning "within or inside." *Endothermic* therefore indicates transfer of energy *into* the system.

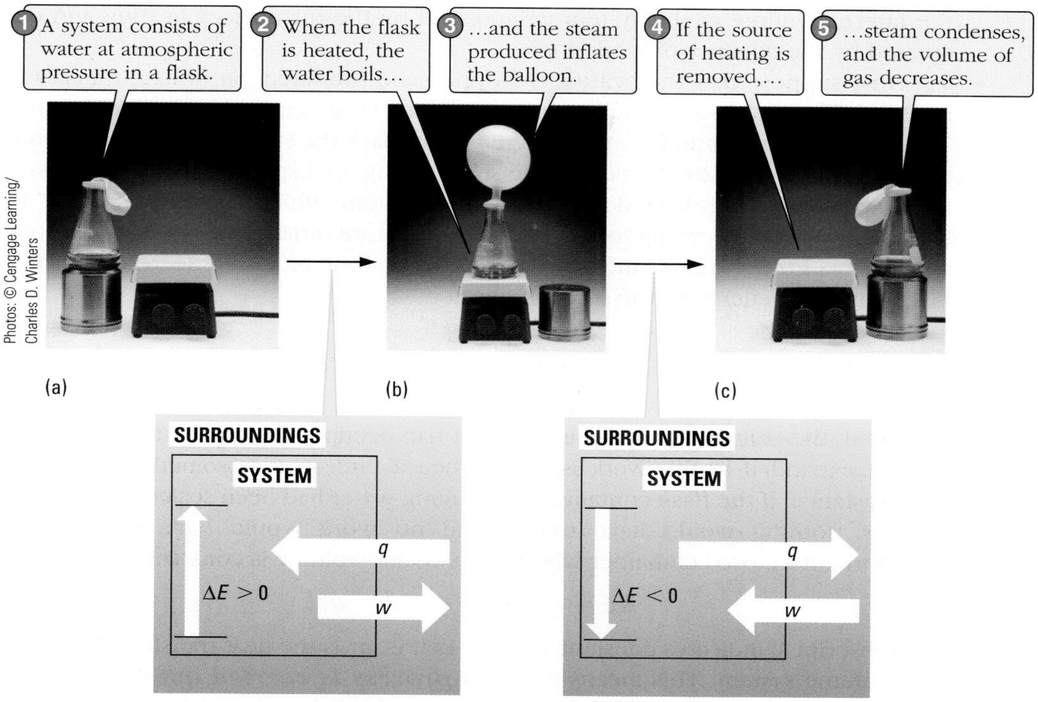

Active Figure 6.10 Boiling water at constant pressure. When water boils, the steam pushes against atmospheric pressure and does work on the atmosphere (which is part of the surroundings). The balloon allows the expansion of the steam to be seen; even if the balloon were not there, the steam would push back the surrounding air. In general, for any constant-pressure process, if a change in volume occurs, some work is done, either on the surroundings or on the system. Visit this book's companion website at **www.cengage.com/chemistry/moore** to test your understanding of the concepts in this figure.

The opposite of boiling is condensation. It can be represented by the opposite equation,

$$H_2O(g) \longrightarrow H_2O(\ell) \qquad \text{exothermic}$$

This process is said to be **exothermic** because energy must be transferred *out of* the system to maintain constant temperature. Because condensation of $H_2O(g)$ (steam) is exothermic, a burn from steam at 100 °C is much worse than a burn from liquid water at 100 °C. The steam heats the skin a lot more because there is a heat transfer due to the condensation as well as the difference in temperature between the water and your skin.

Exo comes from the Greek word *exō,* meaning "out of." *Exothermic* indicates transfer of energy *out of* the system.

Phase Change	Direction of Energy Transfer	Sign of q	Type of Change
$H_2O(\ell) \rightarrow H_2O(g)$	Surroundings → system	Positive ($q > 0$)	*Endo*thermic
$H_2O(g) \rightarrow H_2O(\ell)$	System → surroundings	Negative ($q < 0$)	*Exo*thermic

The system in Figure 6.10 can be analyzed by using the law of conservation of energy, $\Delta E = q + w$. Vaporizing 1.0 g water requires heat transfer of 2260 J, so $q =$ 2260 J (a positive value because the transfer is from the surroundings to the system). At the same time, the expansion of the steam pushes back the atmosphere, doing work. The quantity of work is more difficult to calculate, but it is clear that w must be negative, because the system does work on the surroundings. Therefore,

The device described here is a crude example of a steam engine. Burning fuel boils water, and the steam does work. In a real steam engine the steam would drive a piston and then be allowed to escape, providing a means for the system to continually do work on its surroundings. Systems that convert heat into work are called heat engines. Another example is an internal combustion engine in an automobile, which converts heat from the combustion of fuel into work to move the car.

the internal energy of the system is increased by the quantity of heating and decreased by the quantity of work done.

Now suppose that the heating is stopped and the direction of heat transfer is reversed. The water stops boiling and some of the steam condenses to liquid water. The balloon deflates and the atmosphere pushes back the steam. If 1.0 g steam condenses, then $q = -2260$ J. Because the surrounding atmosphere pushes on the system, the surroundings have done work on the system, which makes w positive. As long as steam is condensing to liquid, the temperature remains at 100 °C. The internal energy of the system is increased by the work done on it and decreased by the heat transfer of energy to the surroundings.

Enthalpy: Heat Transfer at Constant Pressure

In the previous section it was clear that work was done. When the balloon's volume increased, the balloon pushed aside air that had occupied the space the gas in the balloon expanded to fill. Work is done when a force moves something through some distance. If the flask containing the boiling water had been sealed with a solid stopper, nothing would have moved and no work would have been done. Therefore, in a closed container where the system's volume is constant, $w = 0$, and

$$\Delta E = q + w = q + 0 = q_V$$

The subscript V indicates constant volume; that is, q_V is the heat transfer into a constant-volume system. This means that ***if a process is carried out in a closed container and the heat transfer is measured, ΔE has been determined***.

In plants, animals, laboratories, and the environment, physical processes and chemical reactions seldom take place in closed containers. Instead they are carried out in contact with the atmosphere. For example, the vaporization of water shown in Figure 6.10 took place under conditions of constant atmospheric pressure, and the expanding steam had to push back the atmosphere. In such a case,

$$\Delta E = q_P + w_{atm}$$

That is, ΔE differs from the heat transfer at constant pressure, q_P, by the work done to push back the atmosphere, w_{atm}.

Steam
h_1
Liquid water

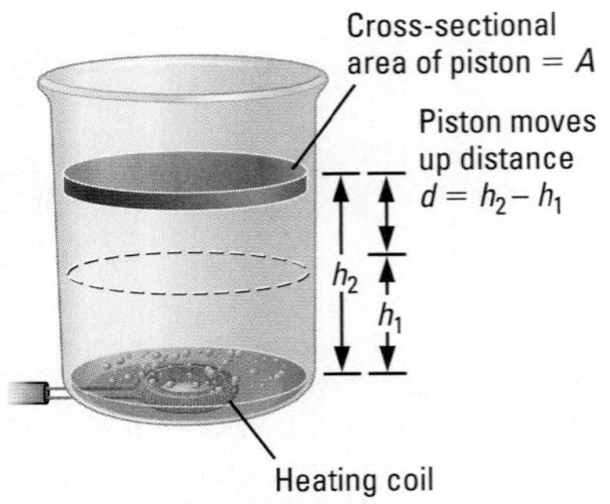
Cross-sectional area of piston = A
Piston moves up distance $d = h_2 - h_1$
h_2
h_1
Heating coil

Vaporization of water. When a sample of water boils at constant pressure, energy must be supplied to expand the steam against atmospheric pressure.

To see how much work is required, consider the diagram of the idealized system shown in the margin. There is a cylinder with a weightless piston. (The purpose of the piston is to distinguish the water system from the surrounding air.) When some of the water in the bottom of the cylinder boils, the volume of the system increases. The piston and the atmosphere are forced upward. The system (water and steam) does work to raise the surrounding air. The work can be calculated as $w_{atm} = -(\text{force} \times \text{distance}) = -(F \times d)$. (The negative sign indicates that the system is doing work on the surroundings.) The distance the piston moves is $h_2 - h_1$. The force can be calculated from the pressure (P), which is defined as force per unit area (A). Since $P = F/A$, the force is $F = P \times A$, and the work is $w_{atm} = -F \times d = -P \times A \times d = -P\Delta V$. The change in the volume of the system, ΔV, is the volume of the cylinder through which the piston moves. This is calculated as the area of the base times the height, or $\Delta V = A \times d$. Thus the work is $w_{atm} = -P \times A \times d = -P\Delta V$. This means that the work of pushing back the atmosphere is always equal to the atmospheric pressure times the change in volume of the system.

The law of conservation of energy for a constant-pressure process can now be written as

$$\Delta E = q_P + w_{atm} \quad \text{or} \quad q_P = \Delta E - w_{atm} = \Delta E + P\Delta V$$

This equation says that when we carry out reactions in beakers or other containers open to the atmosphere, the heat transfer differs from the change in energy by an easily calculated term, $P\Delta V$. Therefore it is convenient to use q_P to characterize energy transfers in typical chemical and physical processes. *The quantity of thermal energy transferred into a system at constant pressure*, q_P, is called the **enthalpy change** of the system, symbolized by ΔH. Thus, $\Delta H = q_P$, and

$$\Delta H = \Delta E + P\Delta V$$

ΔH accounts for all the energy transferred except the quantity that does the work of pushing back the atmosphere. For processes that do not involve gases, w_{atm} is very small. Even when gases are involved, w_{atm} is usually much smaller than q_P. That is, ΔH is closely related to the change in the internal energy of the system but is slightly different in magnitude. ***Whenever heat transfer is measured at constant pressure, it is ΔH that is determined.***

Because it is equal to the quantity of thermal energy transferred at constant pressure, and because most chemical reactions are carried out at atmospheric (constant) pressure, the enthalpy change for a process is often called the *heat of that process*. For example, the enthalpy change for melting (fusion) is also called the heat of fusion.

PROBLEM-SOLVING EXAMPLE **6.6** **Changes of State, ΔH, and ΔE**

Methanol, CH_3OH, boils at 65.0 °C. When 5.0 g methanol boils at 1 atm, the volume of $CH_3OH(g)$ is 4.32 L greater than the volume of the liquid. The heat transfer is 5865 J, and the process is endothermic. Calculate ΔH and ΔE. (The units 1 L × 1 atm = 101.3 J.)

Answer $\Delta H = 5865$ J; $\Delta E = 5427$ J

Strategy and Explanation The process takes place at constant pressure. By definition, $\Delta H = q_P$. Because thermal energy is transferred to the system, the sign of ΔH must be positive. Therefore $\Delta H = 5865$ J. Because the system expands, ΔV is positive. This makes the sign of w_{atm} negative, and

$$w_{atm} = -P\Delta V = -1\text{ atm} \times 4.32\text{ L} = -4.32\text{ L atm} = -4.32 \times 101.3\text{ J} = -438\text{ J}$$

To calculate ΔE, add the expansion work to the enthalpy change.

$$\Delta E = q_P + w_{atm} = \Delta H + w_{atm} = 5865\text{ J} - 438\text{ J} = 5427\text{ J}$$

☑ **Reasonable Answer Check** Boiling is an endothermic process, so ΔH must be positive. Because the system did work on the surroundings, the change in internal energy must be less than the enthalpy change, and it is.

The results of this example show that ΔE differs by less than 10% from ΔH—that is, by 440 J out of 5865 J, which is 7.5%. It is true for most physical and chemical processes that the work of pushing back the atmosphere is only a small fraction of the heat transfer of energy. Because ΔH is so close to ΔE, chemists often refer to enthalpy changes as energy changes.

PROBLEM-SOLVING PRACTICE **6.6**

When potassium melts at atmospheric pressure, the heat transfer is 14.6 cal/g. The density of liquid potassium at its melting point is 0.82 g/mL, and that of solid potassium is 0.86 g/mL. Given that a volume change of 1.00 mL at atmospheric pressure corresponds to 0.10 J, calculate ΔH and ΔE for melting 1.00 g potassium.

Freezing and Melting (Fusion)

Consider what happens when ice is heated at a slow, constant rate from −50 °C to +50 °C. A graph of temperature as a function of quantity of transferred energy is shown in Figure 6.11. When the temperature reaches 0 °C, it remains constant, despite the fact that energy is still being transferred to the sample. As long as ice is melting, thermal energy must be continually supplied to overcome forces that hold the water molecules in their regularly spaced positions in the nanoscale structure of solid ice. Overcoming these forces raises the potential energy of the water molecules and therefore requires a transfer of energy into the system.

Melting a solid is an example of a **change of state** or **phase change,** a physical process in which one state of matter is transformed into another. During a phase change, *the temperature remains constant, but energy must be continually transferred* into the system (melting, boiling) or out of the system (condensing, freezing) because the nanoscale particles have higher or lower potential energy after the phase change than they did before it. As shown in Figure 6.11, the quantity of energy transferred during a phase change is significant.

The quantity of thermal energy that must be transferred to a solid as it melts at constant pressure is called the **enthalpy of fusion,** ΔH_{fusion}. For ice the enthalpy of fusion

Tony Ranze/AFP/Getty Images

Protecting crops from freezing. Because heat transfer to the surroundings occurs as water freezes, one way to protect plants from freezing if the temperature drops just below the freezing point is to spray water on them. As the water freezes, energy transfer to the leaves, stems, and fruits keeps the plants themselves from freezing.

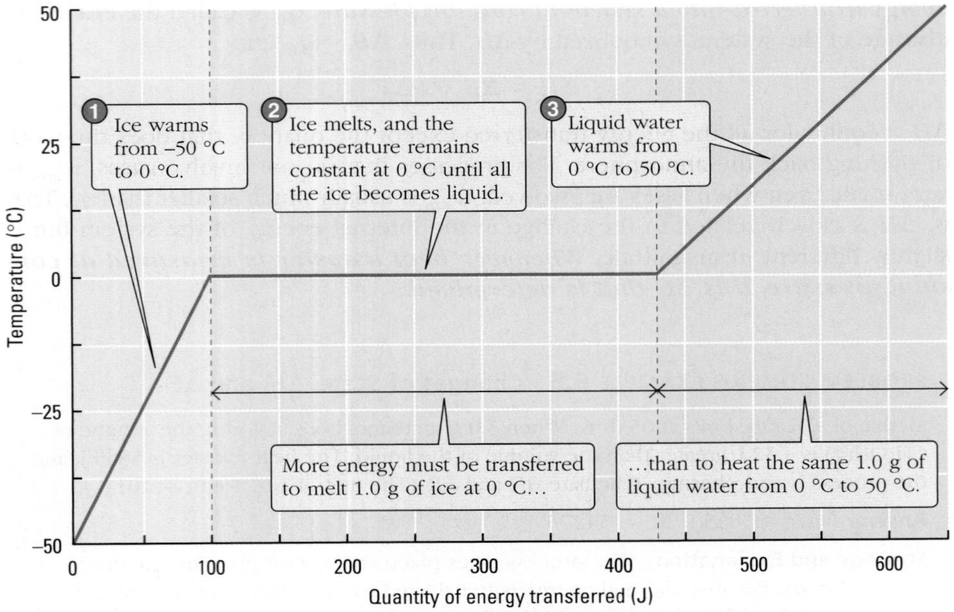

Figure 6.11 Heating graph. When a 1.0-g sample of ice is heated at a constant rate, the temperature does not always increase at a constant rate.

is 333 J/g at 0 °C. This same quantity of energy could raise the temperature of a 1.00-g block of iron from 0 °C to 738 °C (red hot), or it could melt 0.50 g ice and heat the liquid water from 0 °C to 80 °C. This is illustrated schematically in Figure 6.12.

The opposite of melting is freezing. When water freezes, the quantity of energy transferred is the same as when the same mass of water melts, but energy transfers in the opposite direction—from the system to the surroundings. Thus, under the same conditions of temperature and pressure, $\Delta H_{fusion} = -\Delta H_{freezing}$.

Vaporization and Condensation

The quantity of energy that must be transferred at constant pressure to convert a liquid to vapor (gas) is called the **enthalpy of vaporization,** $\Delta H_{vaporization}$. For water it is 2260. J/g at 100 °C. This is considerably larger than the enthalpy of fusion, be-

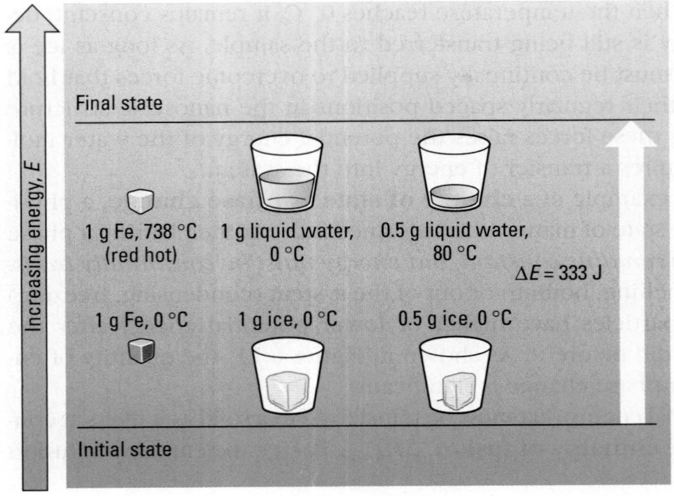

Figure 6.12 Heating, temperature change, and phase change. Heating a substance can cause a temperature change, a phase change, or both. Here, 333 J has been transferred to each of three samples: a 1-g block of iron at 0 °C; a 1-g block of ice at 0 °C; and a 0.5-g block of ice at 0 °C. The iron block becomes red hot; its temperature increases to 738 °C. The 1-g block of ice melts, resulting in 1 g of liquid water at 0 °C. The 0.5-g block of ice melts, and there is enough energy to heat the liquid water to 80 °C.

cause the water molecules become completely separated during the transition from liquid to vapor. As they separate, a great deal of energy is required to overcome the attractions among the water molecules. Therefore, the potential energy of the vapor is considerably higher than that of the liquid. Although 333 J can melt 1.00 g ice at 0 °C, it will boil only 0.147 g water at 100 °C.

$$333 \text{ J} \times \frac{1.00 \text{ g water vaporized}}{2260 \text{ J}}$$
$$= 0.147 \text{ g water vaporized}$$

The opposite of vaporization is condensation. Therefore, under the same conditions of temperature and pressure, $\Delta H_{\text{vaporization}} = -\Delta H_{\text{condensation}}$.

You experience cooling due to evaporation of water when you perspire. If you work up a real sweat, then lots of water evaporates from your skin, producing a much greater cooling effect. People who exercise in cool weather need to carry a sweatshirt or jacket. When they stop exercising they generate less body heat, but lots of perspiration remains on their skin. Its evaporation can cool the body enough to cause a chill.

CONCEPTUAL EXERCISE **6.6 Heating and Cooling Graphs**

(a) Assume that a 1.0-g sample of ice at −5 °C is heated at a uniform rate until the temperature is 105 °C. Draw a graph like the one in Figure 6.10 to show how temperature varies with energy transferred. Your graph should be drawn to approximately the correct scale.

(b) Assume that a 0.50-g sample of water is cooled [at the same uniform rate as the heating in part (a)] from 105 °C to −5 °C. Draw a cooling curve to show how temperature varies with energy transferred. Your graph should be drawn to the same scale as in part (a).

EXERCISE **6.7 Changes of State**

Assume you have 1 cup of ice (237 g) at 0.0 °C. How much heating is required to melt the ice, warm the resulting water to 100.0 °C, and then boil the water to vapor at 100.0 °C? (*Hint:* Do three separate calculations and then add the results.)

State Functions and Path Independence

Both energy and enthalpy are **state functions,** *properties whose values are invariably the same if a system is in the same state.* A system's state is defined by its temperature, pressure, volume, mass, and composition. For the same initial and final states, *a change in a state function does not depend on the path by which the system changes from one state to another* (Figure 6.13). Returning to the bank account

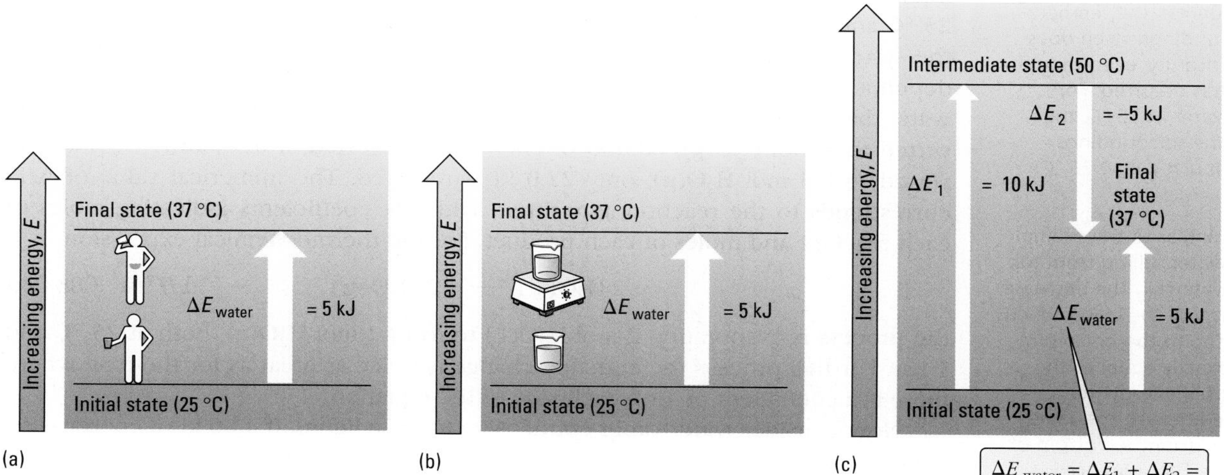

Figure 6.13 Energy change is independent of path. If 100. g water at 25 °C is warmed to 37 °C (body temperature) at atmospheric pressure, the change in energy of the water is the same whether (a) you drank the water and your body warmed it to 37 °C, (b) you put the water in a beaker and heated it with a hot plate, or (c) you heated the water to 50 °C and then cooled it to 37 °C.

analogy (p. 218), your bank balance is independent of the path by which you change it. If you have \$1000 in the bank (initial state) and withdraw \$100, your balance will go down to \$900 (final state) and $\Delta B = -\$100$. If instead you had deposited \$500 and withdrawn \$600 you would have achieved the same change of $\Delta B = -\$100$ by a different pathway, and your final balance would still be \$900.

The fact that changes in a state function are independent of the sequence of events by which change occurs is important, because it allows us to apply laboratory measurements to real-life situations. For example, if you measure in the lab the heat transfer when 1.0 g glucose (dextrose sugar) burns in exactly the amount of oxygen required to convert it to carbon dioxide and water, you will find that $\Delta H = -15.5$ kJ. When you eat something that contains 1.0 g glucose and your body metabolizes the glucose (producing the same products at the same temperature and pressure), there is the same change in enthalpy. Thus, laboratory measurements can be used to determine how much energy you can get from a given quantity of food, which is the basis for the caloric values listed on labels.

6.5 Thermochemical Expressions

To indicate the heat transfer that occurs when either a physical or chemical process takes place, we write a **thermochemical expression,** *a balanced chemical equation together with the corresponding value of the enthalpy change.* For evaporation of water near room temperature and at typical atmospheric pressure, this thermochemical expression can be written:

$$H_2O(\ell) \longrightarrow H_2O(g) \qquad \Delta H° = +44.0 \text{ kJ} \qquad (25\ °C, 1\ bar)$$

In 1982 the International Union of Pure and Applied Chemistry chose a pressure of 1 bar as the standard for tabulating information for thermochemical expressions. This pressure is very close to the standard atmosphere: 1 bar = 0.98692 atm = 1×10^5 kg m^{-1} s^{-2}. (Pressure units are discussed further in Section 10.2.)

The symbol $\Delta H°$ (pronounced "delta-aitch-standard") represents the **standard enthalpy change,** which is defined as the *enthalpy change at the standard pressure of 1 bar and a specified temperature.* Because the value of the enthalpy change depends on the pressure at which the process is carried out, all enthalpy changes are reported at the same standard pressure, 1 bar. (The bar is a unit of pressure that is very close to the pressure of Earth's atmosphere at sea level; you may have heard this unit used in a weather report.) The value of the enthalpy change also varies slightly with temperature. For thermochemical expressions in this book, the temperature can be assumed to be 25 °C, unless some other temperature is specified.

Usually the surroundings contain far more matter than the system and hence have a much greater heat capacity. Consequently, the temperature of the surroundings often does not change significantly, even though energy transfer has occurred. For evaporation of water at 25 °C, the temperature of the surroundings would not drop much below 25 °C.

The thermochemical expression given above indicates that when *one mole* of liquid water (at 25 °C and 1 bar) evaporates to form *one mole* of water vapor (at 25 °C and 1 bar), 44.0 kJ of energy must be transferred from the surroundings to the system to maintain the temperature at 25 °C. The size of the enthalpy change depends on how much process (in this case evaporation) takes place. The more water that evaporates, the more the surroundings are cooled. If 2 mol $H_2O(\ell)$ is converted to 2 mol $H_2O(g)$, 88.0 kJ of energy is transferred; if 0.5 mol $H_2O(\ell)$ is converted to 0.5 mol $H_2O(g)$, only 22.0 kJ is required. The numerical value of $\Delta H°$ corresponds to the reaction as written, with the coefficients indicating moles of each reactant and moles of each product. For the thermochemical expression

$$2\,H_2O(\ell) \longrightarrow 2\,H_2O(g) \qquad \Delta H° = +88.0 \text{ kJ}$$

The idea here is similar to the example given earlier of water falling from top to bottom of a waterfall. The decrease in potential energy of the water when it falls from the top to the bottom of the waterfall is exactly equal to the increase in potential energy that would be required to take the same quantity of water from the bottom of the fall to the top. The signs are opposite because in one case potential energy is transferred *from* the water and in the other case it is transferred *to* the water.

the process is evaporating 2 mol $H_2O(\ell)$ to form 2 mol $H_2O(g)$, both at 25 °C and 1 bar. For this process the enthalpy change is twice as great as for the case where there is a coefficient of 1 on each side of the equation.

Now consider water vapor condensing to form liquid. If 44.0 kJ of energy is required to do the work of separating the water molecules in 1 mol of the liquid as it vaporizes, the same quantity of energy will be released when the molecules move closer together as the vapor condenses to form liquid.

$$H_2O(g) \longrightarrow H_2O(\ell) \qquad \Delta H° = -44.0 \text{ kJ}$$

CHEMISTRY YOU CAN DO Work and Volume Change

Obtain an empty aluminum soft-drink can; a hot plate or electric stove that can boil water; tongs, a glove, or a potholder you can use to pick up the can when it is hot; and a container of cold water large enough that you can immerse the soft-drink can in the water. Rinse out the can with clean water and then pour water into the can until it is about 1 cm deep. Put the can on the hot plate and heat it until the water starts to boil. Let the water boil until steam has been coming out of the opening for at least 1 min. (**Caution:** Watch the can carefully while it is being heated. If it boils dry, the temperature will go way above 100 °C, the aluminum can will melt, your hot plate will be messed up, and you might start a fire.) While a steady stream of steam continues to come out of the can, pick it up with the tongs and in one smooth, quick motion turn it upside down and immerse the opening in the cold water. Be prepared for a surprise. What happens?

Now analyze what happened thermodynamically. Think about these questions:

1. Write a thermochemical expression for the process of boiling the water.
2. What energy transfers occur between the system and the surroundings as the water boils?
3. What was in the can after the water had boiled for a minute or two? What happened to the air that was originally in the can?
4. What happened to the contents of the can as soon as it was immersed in the cold water?
5. Write a thermochemical expression for the process in Question 4.
6. Did the atmosphere do work on the can and its contents after the can was immersed in the water? Cite observations to support your answer.

This thermochemical expression indicates that 44.0 kJ of energy is transferred to the surroundings *from* the system when 1 mol of water vapor condenses to liquid at 25 °C and 1 bar.

CONCEPTUAL EXERCISE 6.8 Interpreting Thermochemical Expressions

What part of the thermochemical expression for vaporization of water indicates that energy is transferred from the surroundings to the system when the evaporation process occurs?

CONCEPTUAL EXERCISE 6.9 Thermochemical Expressions

Why is it essential to specify the state (s, ℓ, or g) of each reactant and each product in a thermochemical expression?

PROBLEM-SOLVING EXAMPLE 6.7 Changes of State and $\Delta H°$

Calculate the energy transferred to the surroundings when water vapor in the air condenses at 25 °C to give rain in a thunderstorm. Suppose that one inch of rain falls over one square mile of ground, so that 6.6×10^{10} mL has fallen. (Assume $d_{H_2O(\ell)} = 1.0$ g/mL.)

Answer 1.6×10^{11} kJ

Agronomists and meteorologists measure quantities of rainwater in units of acre-feet; an acre-foot is enough water to cover an acre of land to a depth of one foot.

Since the explosion of 1000 tons of dynamite is equivalent to 4.2×10^9 kJ, the energy transferred by our hypothetical thunderstorm is about the same as that released when 38,000 tons of dynamite explodes! A great deal of energy can be stored in water vapor, which is one reason why storms can cause so much damage.

Like all examples in this chapter, this one assumes that the temperature of the system remains constant, so that all the energy transfer associated with the phase change goes to or from the surroundings.

Richard Ramette

Iodine "thermometer." A glass sphere containing a few iodine crystals rests on the ground in desert sunshine. The higher the temperature, the more iodine sublimes inside the sphere, and the darker the beautiful violet color becomes.

Strategy and Explanation The thermochemical expression for condensation of 1 mol water at 25 °C is

$$H_2O(g) \longrightarrow H_2O(\ell) \qquad \Delta H° = -44.0 \text{ kJ}$$

The standard enthalpy change tells how much heat transfer is required when 1 mol water condenses at constant pressure, so we first calculate how many moles of water condensed.

$$\text{Amount of water condensed} = 6.6 \times 10^{10} \text{ g water} \times \frac{1 \text{ mol}}{18.0 \text{ g}} = 3.66 \times 10^9 \text{ mol water}$$

Next, calculate the quantity of energy transferred from the fact that 44.0 kJ is transferred per mole of water.

$$\text{Quantity of energy transferred} = 3.66 \times 10^9 \text{ mol water} \times \frac{44.0 \text{ kJ}}{1 \text{ mol}} = 1.6 \times 10^{11} \text{ kJ}$$

The negative sign of $\Delta H°$ in the thermochemical expression indicates transfer of the 1.6×10^{11} kJ *from the water* (system) *to the surroundings.*

☑ **Reasonable Answer Check** The quantity of water is about 10^{11} g. The energy transfer is 44 kJ for 1 mol (18 g) water. Since 44 is about twice 18, this is about 2 kJ/g. Therefore, the number of kJ transferred should be about twice the number of grams, or about 2×10^{11} kJ, and it is.

PROBLEM-SOLVING PRACTICE 6.7

The enthalpy change for sublimation of 1 mol solid iodine at 25 °C and 1 bar is 62.4 kJ. (Sublimation means changing directly from solid to gas.)

$$I_2(s) \longrightarrow I_2(g) \qquad \Delta H° = +62.4 \text{ kJ}$$

(a) What quantity of energy must be transferred to vaporize 10.0 g solid iodine?
(b) If 3.42 g iodine vapor changes to solid iodine, what quantity of energy is transferred?
(c) Is the process in part (b) exothermic or endothermic?

6.6 Enthalpy Changes for Chemical Reactions

Having developed methods for quantitative treatment of energy transfers as a result of temperature differences and as a result of phase changes, we are now ready to apply these ideas to energy transfers that accompany chemical reactions.

Like phase changes, chemical reactions can be exothermic or endothermic, but reactions usually involve much larger energy transfers than do phase changes. Indeed, a significant temperature change is one piece of evidence that a chemical reaction has taken place. The large energy transfers that occur during chemical reactions result from breaking and forming chemical bonds as reactants are converted into products. These energy transfers have important applications in living systems, in industrial processes, in heating or cooling your home, and in many other situations.

Hydrogen is an excellent fuel. It produces very little pollution when it burns in air, and its reaction with oxygen to form water is highly exothermic. It is used as a fuel in the Space Shuttle, for example. The thermochemical expression for formation of 1 mol water vapor from hydrogen and oxygen is

$$H_2(g) + \tfrac{1}{2}O_2(g) \longrightarrow H_2O(g) \qquad \Delta H° = -241.8 \text{ kJ} \quad [6.3]$$

The relationship of reaction heat transfer and enthalpy change:

Reactant → product with transfer of thermal energy *from* system to surroundings. ΔH is negative; reaction is *exo*thermic.

Reactant → product with transfer of thermal energy *into* system from surroundings. ΔH is positive; reaction is *endo*thermic.

Like all thermochemical expressions, this one has four important characteristics:

- The *sign of* $\Delta H°$ indicates the direction of energy transfer.
- The magnitude of $\Delta H°$ depends on the *states of matter* of the reactants and products.

- The *balanced equation represents moles* of reactants and of products.
- The *quantity of energy transferred is proportional to the quantity of reaction* that occurs.

Sign of $\Delta H°$ Thermochemical Expression 6.3 tells us that this process is exothermic, because $\Delta H°$ is negative. Formation of 1 mol water vapor transfers 241.8 kJ of energy from the reacting chemicals to the surroundings. If 1 mol water vapor is decomposed to hydrogen and oxygen (the reverse process), the magnitude of $\Delta H°$ is the same, but the sign is opposite, indicating transfer of energy from the surroundings to the system:

$$H_2O(g) \longrightarrow H_2(g) + \tfrac{1}{2}O_2(g) \qquad \Delta H° = +241.8 \text{ kJ} \quad [6.4]$$

The reverse of an exothermic process is endothermic. The magnitude of the energy transfer is the same, but the direction of transfer is opposite.

States of Matter If liquid water is involved instead of water vapor, the magnitude of $\Delta H°$ is different from that in Thermochemical Expression 6.3:

$$H_2(g) + \tfrac{1}{2}O_2(g) \longrightarrow H_2O(\ell) \qquad \Delta H° = -285.8 \text{ kJ} \quad [6.5]$$

Our discussion of phase changes (p. 224) showed that an enthalpy change occurs when a substance changes state. Vaporizing 1 mol $H_2O(\ell)$ requires 44.0 kJ. Forming 1 mol $H_2O(\ell)$ from $H_2(g)$ and $O_2(g)$ is 285.8 kJ − 241.8 kJ = 44.0 kJ more exothermic than is forming 1 mol $H_2O(g)$. Figure 6.14 shows the relationships among these quantities. The enthalpy of the reactants [$H_2(g)$ and $\frac{1}{2}$ $O_2(g)$] is greater than that of the product [$H_2O(g)$]. Because the system has less enthalpy after the reaction than before, the law of conservation of energy requires that 241.8 kJ must be transferred *to* the surroundings as the reaction takes place. $H_2O(\ell)$ has even less enthalpy than $H_2O(g)$, so when $H_2O(\ell)$ is formed, even more energy, 285.8 kJ, must be transferred to the surroundings.

Balanced Equation Represents Moles To write an equation for the formation of 1 mol H_2O it is necessary to use a fractional coefficient for O_2. This is acceptable in a thermochemical expression, because the coefficients mean moles, not molecules, and half a mole of O_2 is a perfectly reasonable quantity.

Quantity of Energy Is Proportional to Quantity of Reaction Thermochemical expressions obey the rules of stoichiometry ***(← p. 128).*** The more reaction there is, the more energy is transferred. Because the balanced equation represents moles, we

(a)

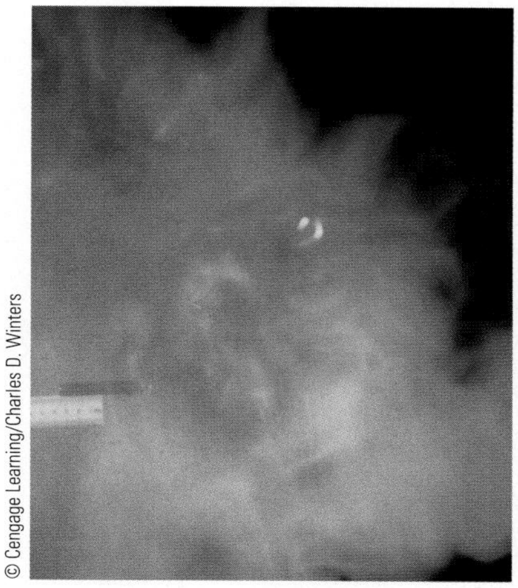

(b)

Combustion of hydrogen is exothermic. (a) When hydrogen gas in a balloon is ignited by a candle flame, (b) the hydrogen combines with oxygen from the air to form water vapor, $2\,H_2(g) + O_2(g) \rightarrow 2\,H_2O(g)$. The flame indicates that the reaction is highly exothermic; it transfers to the surroundings 241.8 kJ per mole of hydrogen burned.

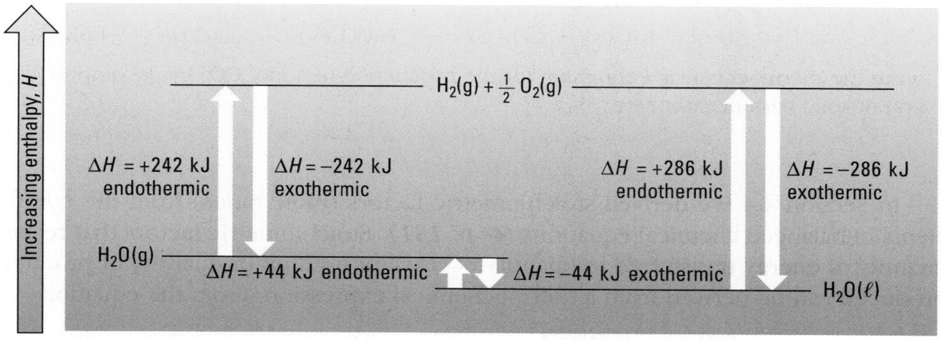

Figure 6.14 Enthalpy diagram. Water vapor [1 mol $H_2O(g)$], liquid water [1 mol $H_2O(\ell)$], and a stoichiometric mixture of hydrogen and oxygen gases [1 mol $H_2(g)$ and $\frac{1}{2}$ mol $O_2(g)$] all have different enthalpy values. The figure shows how these are related, with the highest enthalpy at the top.

The direct proportionality between quantity of reaction and quantity of heat transfer is in line with your everyday experience. Burning twice as much natural gas produces twice as much heating.

can calculate how much heat transfer occurs from the number of moles of a reactant that is consumed or the number of moles of a product that is formed. If the thermochemical expression for combination of gaseous hydrogen and oxygen is written without the fractional coefficient, so that 2 mol $H_2O(g)$ is produced, then the energy transfer is twice as great; that is, 2(−241.8 kJ) = −483.6 kJ.

$$2\ H_2(g) + O_2(g) \longrightarrow 2\ H_2O(g) \qquad \Delta H° = -483.6\ kJ \quad [6.6]$$

EXERCISE 6.10 Enthalpy Change and Stoichiometry

Calculate the change in enthalpy if 0.5000 mol $H_2(g)$ reacts with an excess of $O_2(g)$ to form water vapor at 25 °C.

PROBLEM-SOLVING EXAMPLE 6.8 Thermochemical Expressions

Given the thermochemical expression

$$2\ C_2H_6(g) + 7\ O_2(g) \longrightarrow 4\ CO_2(g) + 6\ H_2O(g) \qquad \Delta H° = -2856\ kJ$$

write a thermochemical expression for
(a) Formation of 1 mol $CO_2(g)$ by burning $C_2H_6(g)$
(b) Formation of 1 mol $C_2H_6(g)$ by reacting $CO_2(g)$ with $H_2O(g)$
(c) Combination of 1 mol $O_2(g)$ with a stoichiometric quantity of $C_2H_6(g)$

Answer

(a) $\frac{1}{2}\ C_2H_6(g) + \frac{7}{4}\ O_2(g) \rightarrow CO_2(g) + \frac{3}{2}\ H_2(g) \qquad \Delta H° = -714.0\ kJ$

(b) $2\ CO_2(g) + 3\ H_2O(g) \rightarrow C_2H_6(g) + \frac{7}{2}\ O_2(g) \qquad \Delta H° = +1428\ kJ$

(c) $\frac{2}{7}\ C_2H_6(g) + O_2(g) \rightarrow \frac{4}{7}\ CO_2(g) + \frac{6}{7}\ H_2O(g) \qquad \Delta H° = -408.0\ kJ$

Strategy and Explanation

(a) Producing 1 mol $CO_2(g)$ requires that one quarter the molar amount of each reactant and product be used and also makes the $\Delta H°$ value one quarter as big.
(b) Forming $C_2H_6(g)$ means that $C_2H_6(g)$ must be a product. This changes the direction of the reaction and the sign of $\Delta H°$; forming 1 mol $C_2H_6(g)$ requires that each coefficient be halved and this halves the size of $\Delta H°$.
(c) If 1 mol $O_2(g)$ reacts, only $\frac{2}{7}$ mol $C_2H_6(g)$ is required; each coefficient is one seventh its original value, and $\Delta H°$ is also one seventh the original value.

☑ **Reasonable Answer Check** In each case examine the coefficients, the direction of the chemical equation, and the sign of $\Delta H°$ to make certain that the appropriate quantity of reactant or product and the appropriate sign have been written.

PROBLEM-SOLVING PRACTICE 6.8

Given the thermochemical expression

$$BaO(s) + CO_2(g) \longrightarrow BaCO_3(s) \qquad \Delta H° = -662.8\ kJ$$

write the thermochemical expression for the production of 4 mol CO_2 by decomposition of solid barium carbonate.

In Section 4.4 we derived stoichiometric factors (mole ratios) from the coefficients in balanced chemical equations (← *p. 131*). Stoichiometric factors that relate quantity of energy transferred to quantity of reactant used up or quantity of product produced can be derived from a thermochemical expression. From the equation

$$2\ H_2(g) + O_2(g) \longrightarrow 2\ H_2O(g) \qquad \Delta H° = -483.6\ kJ$$

these factors (and their reciprocals) can be derived:

$$\frac{-483.6\ kJ}{2\ mol\ H_2\ reacted} \qquad \frac{-483.6\ kJ}{1\ mol\ O_2\ reacted} \qquad \frac{-483.6\ kJ}{2\ mol\ H_2O\ produced}$$

The first factor says that 483.6 kJ of energy will transfer from the system to the surroundings whenever 2 mol H_2 is consumed in this reaction. The reciprocal of the second factor says that if the reaction transfers 483.6 kJ to the surroundings, then 1 mol O_2 must have been used up. We shall refer to stoichiometric factors that include thermochemical information as *thermostoichiometric factors*.

EXERCISE 6.11 Thermostoichiometric Factors from Thermochemical Expressions

Write all of the thermostoichiometric factors (including their reciprocals) that can be derived from this expression:

$$N_2(g) + 3\,H_2(g) \longrightarrow 2\,NH_3(g) \qquad \Delta H^\circ = -92.22\text{ kJ}$$

CONCEPTUAL EXERCISE 6.12 Hand Warmer

When the tightly sealed outer package is opened, the portable hand warmer shown in the margin transfers energy to its surroundings. In cold weather it can keep fingers or toes warm for several hours. Suggest a way that such a hand warmer could be designed. What chemicals might be used? Why is the tightly sealed package needed?

Portable hand warmer. When the sealed package is opened, the temperature rises as shown by the thermometer.

Enthalpy changes for reactions have many practical applications. For instance, when enthalpies of combustion are known, the quantity of energy transferred by the combustion of a given mass of fuel can be calculated. Suppose you are designing a heating system, and you want to know how much heating can be provided per pound (454 g) of propane, C_3H_8, burned in a furnace. The reaction that occurs is *exothermic* (which is not surprising, given that it is a combustion reaction).

$$C_3H_8(g) + 5\,O_2(g) \longrightarrow 3\,CO_2(g) + 4\,H_2O(\ell) \qquad \Delta H^\circ = -2220\text{ kJ}$$

According to this thermochemical expression, 2220 kJ of energy transfers to the surroundings for every 1 mol $C_3H_8(g)$ burned, for every 5 mol $O_2(g)$ consumed, for every 3 mol $CO_2(g)$ formed, and for every 4 mol $H_2O(\ell)$ produced. We know that

CHEMISTRY YOU CAN DO Rusting and Heating

Chemical reactions can heat their surroundings, and a simple experiment demonstrates this fact very well. To perform the experiment you will need a steel wool pad (without soap), $\frac{1}{4}$ cup of vinegar, a cooking or outdoor thermometer, and a large jar with a lid. (The thermometer must fit inside the jar.)

Soak the steel wool pad in vinegar for several minutes. While doing so, place the thermometer in the jar, close the lid, and let it stand for several minutes. Read the temperature.

Squeeze the excess vinegar out of the steel wool pad, wrap the pad around the bulb of the thermometer, and place both in the jar. Close the lid but do not seal it tightly. After about 5 min, read the temperature again. What has happened?

Repeat the experiment with another steel wool pad, but wash it with water instead of vinegar. Try a third pad that is not washed at all. Allow each pad to stand in air for a few hours or for a day and observe the pad carefully. Do you see any change in the metal? Suggest an explanation for your observations of temperature changes and appearance of the steel wool.

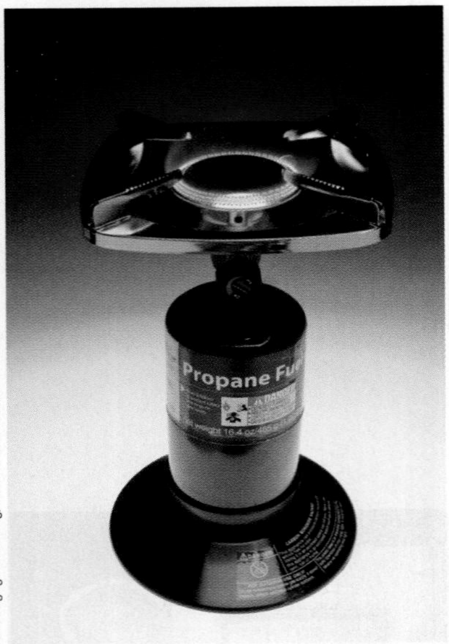

Propane burning. This portable camp stove burns propane fuel. Propane is a major component of liquified petroleum (LP) gas, which is used for heating some houses.

454 g $C_3H_8(g)$ has been burned, so we can calculate how many moles of propane that is.

$$\text{Amount of propane} = 454 \text{ g} \times \frac{1 \text{ mol } C_3H_8}{44.10 \text{ g}} = 10.29 \text{ mol } C_3H_8$$

Then we multiply by the appropriate thermostoichiometric factor to find the total energy transferred.

$$\text{Energy transferred} = 10.29 \text{ mol } C_3H_8 \times \frac{-2220 \text{ kJ}}{1 \text{ mol } C_3H_8} = -22{,}900 \text{ kJ}$$

Burning a pound of fuel such as propane releases a substantial quantity of energy.

PROBLEM-SOLVING EXAMPLE 6.9 Calculating Energy Transferred

The reaction of iron with oxygen from the air provides the energy transferred by the hot pack described in Conceptual Exercise 6.12. Assuming that the iron is converted to iron(III) oxide, how much heating can be provided by a hot pack that contains 0.100 pound of iron? The thermochemical expression is

$$2\ Fe(s) + \tfrac{3}{2}\ O_2(g) \longrightarrow Fe_2O_3(s) \qquad \Delta H° = -824.2 \text{ kJ}$$

Answer −335 kJ

Strategy and Explanation Begin by calculating how many moles of iron are present. A pound is 454 g, so

$$\text{Amount of iron} = 0.100 \text{ lb} \times \frac{454 \text{ g}}{1 \text{ lb}} \times \frac{1 \text{ mol Fe}}{55.84 \text{ g}} = 0.8130 \text{ mol Fe}$$

Then use a thermostoichiometric factor to calculate the energy transferred. The appropriate factor is 824.2 kJ transferred to the surroundings per 2 mol Fe, so

$$\text{Energy transferred} = 0.8130 \text{ mol Fe} \times \frac{-824.2 \text{ kJ}}{2 \text{ mol Fe}} = -335 \text{ kJ}$$

Thus, 335 kJ is transferred by the reaction to heat your hand.

☑ **Reasonable Answer Check** A tenth of a pound is about 45 g, which is a bit less than the molar mass of iron, so we are oxidizing less than a mole of iron. Two moles of iron gives about 800 kJ, so less than a mole should give less than 400 kJ, which makes 335 kJ a reasonable value. The sign should be negative because the enthalpy of the hand warmer (system) should go down when it transfers energy to your hand.

PROBLEM-SOLVING PRACTICE 6.9

How much thermal energy transfer is required to maintain constant temperature during decomposition of 12.6 g liquid water to the elements hydrogen and oxygen at 25.0 °C? In what direction does the energy transfer?

$$H_2O(\ell) \longrightarrow H_2(g) + \tfrac{1}{2}\ O_2(g) \qquad \Delta H° = 285.8 \text{ kJ}$$

6.7 Where Does the Energy Come From?

During melting or boiling, nanoscale particles (atoms, molecules, or ions) that attract each other are separated, which increases their potential energy. This requires transfer of energy from the surroundings to enable the particles to overcome their mutual attractions. During a chemical reaction, chemical compounds are created or broken down; that is, reactant molecules are converted into product molecules. Atoms in molecules are held together by chemical bonds. When existing chemical bonds are broken and new chemical bonds are formed, atomic nuclei and electrons move farther apart or closer together, and their energy increases or decreases. These energy differences are usually much greater than those for phase changes.

Consider the reaction of hydrogen gas with chlorine gas to form hydrogen chloride gas.

$$H_2(g) + Cl_2(g) \longrightarrow 2\ HCl(g) \qquad [6.7]$$

When this reaction occurs, the two hydrogen atoms in a H_2 molecule separate, as do the two chlorine atoms in a Cl_2 molecule. In the product the atoms are combined in a different way—as two HCl molecules. We can think of this change as involving two steps:

$$H_2(g) + Cl_2(g) \longrightarrow 2\ H(g) + 2\ Cl(g) \longrightarrow 2\ HCl(g)$$

The first step is to break all bonds in the reactant H_2 and Cl_2 molecules. The second step is to form the bonds in the two product HCl molecules. The net effect of these two steps is the same as for Equation 6.7: One hydrogen molecule and one chlorine molecule change into two hydrogen chloride molecules. The enthalpy changes for these two processes are shown in Figure 6.15.

The reaction of hydrogen with chlorine actually occurs by a complicated series of steps, but the details of how the atoms rearrange do not matter, because enthalpy is a state function and the initial and final states are the same. This means that we can concentrate on products and reactants and not worry about exactly what happens in between.

Another analogy for the enthalpy change for a reaction is the change in altitude when you climb a mountain. No matter which route you take to the summit (which atoms you separate or combine first), the difference in altitude between the summit and where you started to climb (the enthalpy difference between products and reactants) is the same.

CONCEPTUAL EXERCISE 6.13 Reaction Pathways and Enthalpy Change

Suppose that the enthalpy change differed depending on the pathway a reaction took from reactants to products. For example, suppose that 190. kJ was released when a mole of hydrogen gas and a mole of chlorine gas combined to form two moles of hydrogen chloride (Equation 6.7), but that only 185 kJ was released when the same reactant molecules were broken into atoms and the atoms then recombined to form hydrogen chloride (Equation 6.8). Would this violate the first law of thermodynamics? Explain why or why not.

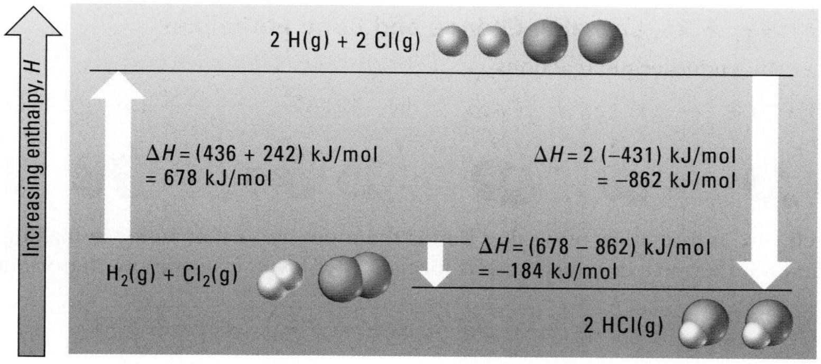

Figure 6.15 Stepwise energy changes in a reaction. Breaking a mole of H_2 molecules into H atoms requires 436 kJ. Breaking a mole of Cl_2 molecules into Cl atoms requires 242 kJ. Putting 2 mol H atoms together with 2 mol Cl atoms to form 2 mol HCl provides $2 \times (-431\ kJ) = -862\ kJ$, so the reaction is exothermic. $\Delta H° = 436\ kJ + 242\ kJ - 862\ kJ = -184\ kJ$. The relatively weak Cl—Cl bond in the reactants accounts for the fact that this reaction is exothermic.

Bond Enthalpies

Bond enthalpy and bond energy differ because a volume change occurs when one molecule changes to two atoms at constant pressure. Therefore work is done on the surroundings (p. 226) and $\Delta E \neq \Delta H$. For a more detailed discussion, see Treptow, R. S., *Journal of Chemical Education*, Vol. 72, 1995; p. 497.

Separating two atoms that are bonded together requires a transfer of energy into the system, because work must be done against the force holding the pair of atoms together. *The enthalpy change that occurs when two bonded atoms in a gas-phase molecule are separated completely at constant pressure* is called the **bond enthalpy** (or the **bond energy**—the two terms are often used interchangeably). The bond enthalpy is usually expressed per mole of bonds. For example, the bond enthalpy for a Cl_2 molecule is 242 kJ/mol, so we can write

$$Cl_2(g) \longrightarrow 2\ Cl(g) \qquad \Delta H° = 242\ kJ \qquad [6.8]$$

Bond *breaking* is *endo*thermic.

Bond *making* is *exo*thermic.

Bond enthalpies are always positive, and they range in magnitude from about 150 kJ/mol to a little more than 1000 kJ/mol. ***Bond breaking is always endothermic,*** because there is always a transfer of energy into the system (in this case, the mole of Cl_2 molecules) to separate pairs of bonded atoms. Conversely, when atoms come together to form a bond, energy will invariably be transferred to the surroundings because the potential energy of the atoms is lower when they are bonded together. Conservation of energy requires that if the system's energy goes down, the energy of the surroundings must go up. Thus, ***formation of bonds from separated atoms is always exothermic.*** How these generalizations apply to the reaction of hydrogen with chlorine to form hydrogen chloride is shown in Figure 6.15.

Bond enthalpies provide a way to see what makes a process exothermic or endothermic. If, as in Figure 6.15, the total energy transferred out of the system when new bonds form is greater than the total energy transferred in to break all of the bonds in the reactants, then the reaction is exothermic. In terms of bond enthalpies *there are two ways for an exothermic reaction to happen:*

- Weaker bonds are broken, stronger bonds are formed, and the number of bonds is the same.
- Bonds in reactants and products are of about the same strength, but more bonds are formed than are broken.

An endothermic reaction involves breaking stronger bonds than are formed, breaking more bonds than are formed, or both.

CONCEPTUAL EXERCISE 6.14 Enthalpy Change and Bond Enthalpies

Consider the endothermic reactions

(a) $2\ HF(g) \longrightarrow H_2(g) + F_2(g)$ (b) $2\ H_2O(g) \longrightarrow 2\ H_2(g) + O_2(g)$

In which case is formation of weaker bonds the more important factor in making the reaction endothermic? In which case is formation of fewer bonds more important?

6.8 Measuring Enthalpy Changes: Calorimetry

A thermochemical expression tells us how much energy is transferred as a chemical process occurs. This knowledge enables us to calculate the heat obtainable when a fuel is burned, as was done in the preceding section. Also, when reactions are carried out on a larger scale—say, in a chemical plant that manufactures sulfuric acid—

the surroundings must have enough cooling capacity to prevent an exothermic reaction from overheating, speeding up, running out of control, and possibly damaging the plant. For these and many other reasons it is useful to know as many $\Delta H°$ values as possible.

For many reactions, direct experimental measurements can be made by using a **calorimeter,** *a device that measures heat transfers.* Calorimetric measurements can be made at constant volume or at constant pressure. Often, in finding heats of combustion or the caloric value of foods, where at least one of the reactants is a gas, the measurement is done at a constant volume in a *bomb calorimeter* (Figure 6.16). The "bomb" is a capped cylinder about the size of a large fruit juice can with heavy steel walls so that it can contain high pressures. A weighed sample of a combustible solid or liquid is placed in a dish inside the bomb. The bomb is then filled with pure $O_2(g)$ and placed in a water-filled container with well-insulated walls. The sample is ignited, usually by an electrical spark. When the sample burns, it warms the bomb and the water around it to the same temperature.

In this configuration, the oxygen and the compound represent the *system* and the bomb and the water around it are the *surroundings*. For the system, $\Delta E = q + w$. Because there is no change in volume of the sealed, rigid bomb, $w = 0$. Therefore, $\Delta E = q_V$. To calculate q_V and ΔE, we can sum the energy transfers from the reaction to the bomb and to the water. Because each of these is a transfer out of the system, each will be negative. For example, the energy transfer to heat the water can be calculated as $c_{water} \times m_{water} \times \Delta T_{water}$, where c_{water} is the specific heat capacity of water, m_{water} is the mass of water, and ΔT_{water} is the change in temperature of the water. Because this energy transfer is out of the system, the energy transfer from the system to the water is negative, that is, $-(c_{water} \times m_{water} \times \Delta T_{water})$. Problem-Solving Example 6.10 illustrates how this works.

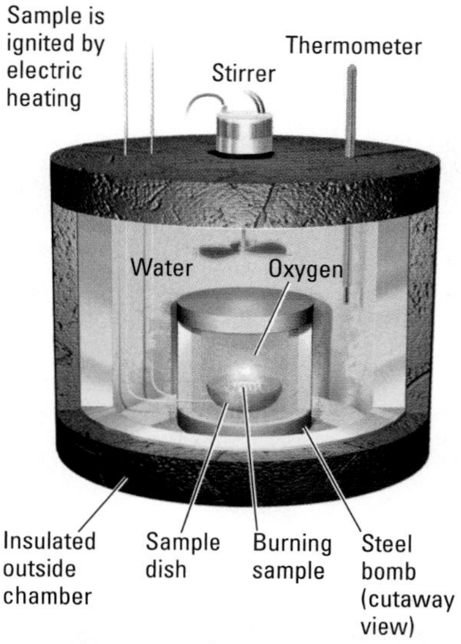

Figure 6.16 Combustion (bomb) calorimeter. A sample in a strong container (bomb) full of oxygen is ignited by electric heating. The heat transfer from combustion of the sample raises the temperature of the bomb and the surrounding water.

PROBLEM-SOLVING EXAMPLE 6.10 Measuring Energy Change with a Bomb Calorimeter

A 3.30-g sample of the sugar glucose, $C_6H_{12}O_6(s)$, was placed in a bomb calorimeter, ignited, and burned to form carbon dioxide and water. The temperature of the water and the bomb changed from 22.4 °C to 34.1 °C. If the calorimeter contained 850. g water and had a heat capacity of 847 J/°C, what is ΔE for combustion of 1 mol glucose? (The heat capacity of the bomb is the energy transfer required to raise the bomb's temperature by 1 °C.)

Answer −2810 kJ

Strategy and Explanation When the glucose burns, it heats the calorimeter and the water. Calculate the heat transfer from the reaction to the calorimeter and the water from the temperature change and their heat capacities. (Look up the heat capacity of water in Table 6.1.) Use this result to calculate the heat transfer from the reaction. Then use a proportion to find the heat transfer for 1 mol glucose.

$$\Delta T = (34.1 - 22.4)\text{°C} = 11.7\ \text{°C}$$

$$\text{Energy transferred from system to bomb} = -(\text{heat capacity of bomb} \times \Delta T)$$

$$= -\left(\frac{847\ \text{J}}{\text{°C}} \times 11.7\ \text{°C}\right) = -9910\ \text{J} = -9.910\ \text{kJ}$$

$$\text{Energy transferred from system to water} = -(c \times m \times \Delta T)$$

$$= -\left(\frac{4.184\ \text{J}}{\text{g °C}} \times 850.\ \text{g} \times 11.7\ \text{°C}\right) = -41{,}610\ \text{J} = -41.61\ \text{kJ}$$

$$\Delta E = q_V = -9.910\ \text{kJ} - 41.61\ \text{kJ} = -51.52\ \text{kJ}$$

This quantity of energy transfer corresponds to burning 3.30 g glucose. To scale to 1 mol glucose, first calculate how many moles of glucose were burned.

$$3.30 \text{ g } C_6H_{12}O_6 \times \frac{1 \text{ mol}}{180.16 \text{ g}} = 1.832 \times 10^{-2} \text{ mol } C_6H_{12}O_6$$

Then set up this proportion.

$$\frac{-51.52 \text{ kJ}}{1.832 \times 10^{-2} \text{ mol}} = \frac{\Delta E}{1 \text{ mol}} \qquad \Delta E = 1 \text{ mol} \times \frac{-51.52 \text{ kJ}}{1.832 \times 10^{-2} \text{ mol}} = -2.81 \times 10^3 \text{ kJ}$$

☑ **Reasonable Answer Check** The result is negative, which correctly reflects the fact that burning sugar is exothermic. A mole of glucose (180 g) is more than a third of a pound, and a third of a pound of sugar contains quite a bit of energy (many Calories), so it is reasonable that the magnitude of the answer is in the thousands of kilojoules.

PROBLEM-SOLVING PRACTICE 6.10

In Problem-Solving Example 6.1, a single Fritos chip was oxidized by potassium chlorate. Suppose that a single chip weighing 1.0 g is placed in a bomb calorimeter that has a heat capacity of 877 J/°C. The calorimeter contains 832 g water. When the bomb is filled with excess oxygen and the chip is ignited, the temperature rises from 20.64 °C to 25.43 °C. Use these data to verify the statement that the chip provides 5 Cal when metabolized.

CONCEPTUAL EXERCISE 6.15 Comparing Enthalpy Change and Energy Change

Write a balanced equation for the combustion of glucose to form CO_2(g) and H_2O(ℓ). Use what you already know about the volume of a mole of any gas at a given temperature and pressure (or look in Section 10.4) to predict whether ΔH would differ significantly from ΔE for the reaction in Problem-Solving Example 6.10.

When reactions take place in solution, it is much easier to use a calorimeter that is open to the atmosphere. An example, often encountered in introductory chemistry courses, is the *coffee cup calorimeter* shown in Figure 6.17. The nested coffee cups (which are made of expanded polystyrene) provide good thermal insulation; reactions can occur when solutions are poured together in the inner cup. Because a coffee cup calorimeter is a constant-pressure device, the measured heat transfer is q_P, which can be used to calculate ΔH as shown in Problem-Solving Example 6.11.

(a)

(b)

Photos: © Cengage Learning/Charles D. Winters

Figure 6.17 Coffee cup calorimeter. (a) A simple constant-pressure calorimeter can be made from two coffee cups that are good thermal insulators, a cork or other insulating lid, a temperature probe, and a stirrer. (b) Close-up of the nested cups that make up the calorimeter. A reaction carried out in an aqueous solution within the calorimeter will change the temperature of the solution. Because the thermal insulation is extremely good, essentially no energy transfer can occur to or from anything outside the calorimeter. Therefore, the heat capacity of the solution and its change in temperature can be used to calculate q_P and ΔH.

PROBLEM-SOLVING EXAMPLE 6.11 Measuring Enthalpy Change with a Coffee Cup Calorimeter

A coffee cup calorimeter is used to determine ΔH for the reaction

$$NaOH(aq) + HCl(aq) \longrightarrow H_2O(\ell) + NaCl(aq) \qquad \Delta H = ?$$

When 250. mL of 1.00-M NaOH was added to 250. mL of 1.00-M HCl at 1 bar, the temperature of the solution increased from 23.4 °C to 30.4 °C. Use this information to determine ΔH and complete the thermochemical expression. Assume that the heat capacities of the coffee cups, the temperature probe, and the stirrer are negligible, that the solution has the same density and the same specific heat capacity as water, and that there is no change in volume of the solutions upon mixing.

Answer $\Delta H = -58.7$ kJ

Strategy and Explanation Use the definition of specific heat capacity [Equation 6.2′ (p. 221)] to calculate q_P, the heat transfer for the constant-pressure conditions

(1 bar). Because the density of the solution is assumed to be the same as for water and the total volume is 500. mL, the mass of solution is 500. g. Because the reaction system heats the solution, q_P is negative and

$$q_P = -c \times m \times \Delta T$$
$$= -(4.184\ \text{J g}^{-1}\ {}^\circ\text{C}^{-1})(500.\ \text{g})(30.4\ {}^\circ\text{C} - 23.4\ {}^\circ\text{C}) = -1.46 \times 10^4\ \text{J} = -14.6\ \text{kJ}$$

This quantity of heat transfer does not correspond to the equation as written, however; instead it corresponds to consumption of

$$250.\ \text{mL} \times \frac{1.00\ \text{mol}}{1000\ \text{mL}} = 0.250\ \text{mol HCl} \quad \text{and} \quad 250.\ \text{mL} \times \frac{1.00\ \text{mol}}{1000\ \text{mL}} = 0.250\ \text{mol NaOH}$$

From the balanced equation, 1 mol HCl is required for each 1 mol NaOH. Therefore the reactants are in the stoichiometric ratio and neither reactant is a limiting reactant. Because the chemical equation involves 1 mol HCl, the heat transfer must be scaled in proportion to this quantity of HCl.

$$\frac{\Delta H}{1\ \text{mol HCl}} = \frac{-14.6\ \text{kJ}}{0.250\ \text{mol HCl}}$$

$$\Delta H = 1\ \text{mol HCl} \times \frac{-14.6\ \text{kJ}}{0.250\ \text{mol HCl}} = -58.6\ \text{kJ}$$

(Note that because the reactants were in the stoichiometric ratio, the NaOH could also have been used in the preceding calculation.)

☑ **Reasonable Answer Check** The temperature of the surroundings increased, so the reaction is exothermic and ΔH° must be negative. The temperature of 500. g solution went up 7.0 °C, so the heat transfer was about $(500 \times 7 \times 4)$ J = 14,000 J = 14 kJ. This corresponded to one-quarter mole of each reactant, so the heat transfer per mole must be about 4×14 kJ = 56 kJ. Therefore ΔH should be about −56 kJ, which it is.

When expanded polystyrene coffee cups are used to make a calorimeter, the masses of substances other than the solvent water are often so small that their heat capacities can be ignored; all of the energy of a reacton can be assumed to be transferred to the water.

PROBLEM-SOLVING PRACTICE 6.11

Suppose that 100. mL of 1.00-M HCl and 100. mL of 0.50-M NaOH, both at 20.4 °C, are mixed in a coffee cup calorimeter. Use the result from Problem-Solving Example 6.11 to predict what will be the highest temperature reached in the calorimeter after mixing the solutions. Make assumptions similar to those made in Problem-Solving Example 6.11.

CONCEPTUAL EXERCISE 6.16 Calorimetry

In Problem-Solving Example 6.11, ΔT was observed to be 7.0 °C for mixing 250. mL of 1.00-M HCl and 250. mL of 1.00-M NaOH in a coffee cup calorimeter. Predict ΔT for mixing

(a) 200. mL of 1.0-M HCl and 200. mL 1.0-M NaOH.
(b) 100. mL of 1.0-M H_2SO_4 and 100. mL 1.0-M NaOH.

PROBLEM-SOLVING EXAMPLE 6.12 Measuring Enthalpy Change for Dissolving

The process of dissolving solid ammonium nitrate is often used in cold packs. When 7.07 g NH_4NO_3 is added to 150. mL H_2O in a coffee cup calorimeter and the mixture is stirred to dissolve all of the NH_4NO_3, the temperature falls from 22.3 °C to 19.2 °C. Write a balanced equation for dissolving 1 mol NH_4NO_3 in water. Assuming that the solution has the same specific heat capacity as water and the density of water is 1.00 g/mL, what is the enthalpy change for dissolving NH_4NO_3? (Express your result in kJ.)

Answer $NH_4NO_3(s) \longrightarrow NH_4^+(aq) + NO_3^-(aq)$; $\Delta H = 23$ kJ

Strategy and Explanation

- **Write a chemical equation for the process.**
- **Use the definition of specific heat capacity to calculate q_P, the heat transfer.**
- **Use q_P to calculate ΔH for the reaction as written.**

Solid NH_4NO_3 dissolves in water to form aqueous ammonium ions and nitrate ions (← p. 88), so the equation is

$$NH_4NO_3(s) \longrightarrow NH_4^+(aq) + NO_3^-(aq)$$

Because the density of water is 1.00 g/mL, the mass of water is 150. g. The total mass of solution is 150. g + 7.07 g = 157.1 g, so

$$q_P = -c \times m \times \Delta T = -(4.184 \text{ J g}^{-1} \text{ °C}^{-1})(157.1 \text{ g})(19.2 \text{ °C} - 22.3 \text{ °C}) = 2.04 \times 10^3 \text{ J}$$

Note that because q_P is for the system but is being calculated from data for the surroundings, a minus sign is needed; thus, $q_P = -c \times m \times \Delta T$.

This quantity of heat transfer does not correspond with the chemical equation as written for 1 mol NH_4NO_3. The 2.04×10^3 J is transferred when 7.07 g NH_4NO_3 dissolves. This mass of NH_4NO_3 corresponds to

$$7.07 \text{ g } NH_4NO_3 \times \frac{1 \text{ mol } NH_4NO_3}{80.05 \text{ g } NH_4NO_3} = 8.83 \times 10^{-2} \text{ mol } NH_4NO_3$$

For the balanced chemical equation, in which 1 mol NH_4NO_3 dissolves,

$$\frac{\Delta H}{1 \text{ mol } NH_4NO_3} = \frac{2.04 \times 10^3 \text{ J}}{8.83 \times 10^{-2} \text{ mol } NH_4NO_3}$$

$$\Delta H = 1 \text{ mol } NH_4NO_3 \times \frac{2.04 \times 10^3 \text{ J}}{8.83 \times 10^{-2} \text{ mol } NH_4NO_3} = 2.3 \times 10^4 \text{ J} = 23 \text{ kJ}$$

☑ **Reasonable Answer Check** The temperature of the surroundings decreased, so the reaction is endothermic and ΔH should therefore be positive. The temperature changed by about 3 °C, and the specific heat capacity is about 4 J g^{-1} °C^{-1}. This corresponds to about 12 J/g. Since a little more than 150 g solution changed temperature, the total heat transfer should be a little more than 12 × 150 = 1800 J, and it is. This change was for a little less than 0.1 mol NH_4NO_3, so the change for 1 mol should be more than 10 times as great, and it is.

PROBLEM-SOLVING PRACTICE 6.12

When pure sulfuric acid, a liquid, dissolves in water, the process is highly exothermic. (Sulfuric acid should always be added to water, not water to sulfuric acid, because if there is only a little water, the water will boil and spatter.) When 10.4 g $H_2SO_4(\ell)$ is added to 270. mL water in a coffee cup calorimeter, with stirring, the temperature rises 8.6 °C. Make assumptions similar to those in Problem-Solving Example 6.12 and calculate the enthalpy change (in kJ) for the process $H_2SO_4(\ell) \longrightarrow H_2SO_4(aq)$.

6.9 Hess's Law

go Chemistry **Module 10: Thermochemistry and Hess's Law** covers concepts in this section.

Calorimetry works well for some reactions, but for many others it is difficult to use. Besides, it would be very time-consuming to measure values for every conceivable reaction, and it would take a great deal of space to tabulate so many values. Fortunately, there is a better way. It is based on **Hess's law,** which states that, *if the equation for a reaction is the sum of the equations for two or more other reactions, then* $\Delta H°$ *for the first reaction must be the sum of* $\Delta H°$ *values of the other reactions.* Hess's law is a corollary of the law of conservation of energy. It works even if the overall reaction does not actually occur by way of the separate equations that are summed.

For example, in Figure 6.14 (p. 233) we noted that the formation of liquid water from its elements $H_2(g)$ and $O_2(g)$ could be thought of as two successive changes: (a) formation of water vapor from the elements and (b) condensation of water vapor to liquid water. As shown below, the equation for formation of liquid water can be obtained by adding algebraically the chemical equations for these two steps.

Therefore, according to Hess's law, the $\Delta H°$ value can be found by adding the $\Delta H°$ values for the two steps.

$$\text{(a)} \quad H_2(g) + \tfrac{1}{2}O_2(g) \longrightarrow \cancel{H_2O(g)} \qquad \Delta H_1° = -241.8 \text{ kJ}$$

$$\text{(b)} \quad \cancel{H_2O(g)} \longrightarrow H_2O(\ell) \qquad \Delta H_2° = -44.0 \text{ kJ}$$

$$\text{(a) + (b)} \quad H_2(g) + \tfrac{1}{2}O_2(g) \longrightarrow H_2O(\ell) \qquad \Delta H° = \Delta H_1° + \Delta H_2° = -285.8 \text{ kJ}$$

Here, 1 mol $H_2O(g)$ is a product of the first reaction and a reactant in the second. Thus, $H_2O(g)$ can be canceled out. This is similar to adding two algebraic equations: If the same quantity or term appears on both sides of the equation, it cancels. The net result is an equation for the overall reaction and its associated enthalpy change. This overall enthalpy change applies even if the liquid water is formed directly from hydrogen and oxygen.

Hess's law is based on a fact we mentioned earlier (p. 229). A system's enthalpy will be the same no matter how the system is prepared. Therefore, at 25 °C and 1 bar, the initial system, $H_2(g) + \frac{1}{2}O_2(g)$, has a particular enthalpy value. The final system, $H_2O(\ell)$, also has a characteristic (but different) enthalpy. Whether we get from initial system to final system by a single step or by the two-step process of the chemical equations (a) and (b), the enthalpy change will be the same.

A useful approach to Hess's law is

- **begin with the thermochemical expression for which you want to calculate $\Delta H°$; call this the target expression;**
- **in the target expression, identify which reactants are desired in what quantities and which products are desired in what quantities;**
- **look at the known thermochemical expressions and decide how each needs to be changed to give reactants and products in the quantities that are in the target expression.**

Note that it takes 1 mol $H_2O(g)$ to cancel 1 mol $H_2O(g)$. If the coefficient of $H_2O(g)$ had been different on one side of the chemical from the coefficient on the other side, $H_2O(g)$ could not have been completely canceled.

For example, suppose you want the thermochemical expression for the reaction

$$\tfrac{1}{2}CH_4(g) + O_2(g) \longrightarrow \tfrac{1}{2}CO_2(g) + H_2O(\ell) \qquad \Delta H° = ?$$

and you already know the thermochemical expressions

$$\text{(a)}\ CH_4(g) + 2\,O_2(g) \longrightarrow CO_2(g) + 2\,H_2O(g) \qquad \Delta H_a° = -802.34 \text{ kJ}$$

and

$$\text{(b)}\ H_2O(\ell) \longrightarrow H_2O(g) \qquad \Delta H_b° = 44.01 \text{ kJ}$$

The target expression has only $\frac{1}{2}$ mol $CH_4(g)$ as a reactant; it also has $\frac{1}{2}$ mol $CO_2(g)$ and 1 mol $H_2O(\ell)$ as products. Expression (a) has the same reactants and products, but twice as many moles of each; also, water is in the gaseous state in expression (a). If we change each coefficient and the $\Delta H°$ value of expression (a) to one half their original values, we have the thermochemical expression

$$\text{(a}'\text{)}\ \tfrac{1}{2}CH_4(g) + O_2(g) \longrightarrow \tfrac{1}{2}CO_2(g) + H_2O(g) \qquad \Delta H_{a'}° = -401.17 \text{ kJ}$$

which differs from the target expression only in the phase of water. Expression (b) has liquid water on the left and gaseous water on the right, but our target expression has liquid water on the right. If the equation in (b) is reversed (which changes the sign of $\Delta H°$), the thermochemical expression becomes

$$\text{(b}'\text{)}\ H_2O(g) \longrightarrow H_2O(\ell) \qquad \Delta H_{b'}° = -44.01 \text{ kJ}$$

Summing the expressions (a′) and (b′) gives the target expression, from which $H_2O(g)$ has been canceled.

$$\text{(a}' + \text{b}'\text{)}\ \tfrac{1}{2}CH_4(g) + O_2(g) \longrightarrow \tfrac{1}{2}CO_2(g) + H_2O(\ell) \qquad \Delta H° = \Delta H_{a'}° + \Delta H_{b'}°$$

$$\Delta H° = (-401.17 \text{ kJ}) + (-44.01 \text{ kJ}) = -445.18 \text{ kJ}$$

© Cengage Learning/Charles D. Winters

Polyethylene is a common plastic. Many products are packaged in polyethylene bottles.

PROBLEM-SOLVING EXAMPLE 6.13 **Using Hess's Law**

In designing a chemical plant for manufacturing the plastic polyethylene, you need to know the enthalpy change for the removal of H_2 from C_2H_6 (ethane) to give C_2H_4 (ethylene), a key step in the process.

$$C_2H_6(g) \longrightarrow C_2H_4(g) + H_2(g) \qquad \Delta H° = ?$$

From experiments you know these thermochemical expressions:

(a) $2\,C_2H_6(g) + 7\,O_2(g) \longrightarrow 4\,CO_2(g) + 6\,H_2O(\ell)$ $\Delta H_a^\circ = -3119.4$ kJ

(b) $C_2H_4(g) + 3\,O_2(g) \longrightarrow 2\,CO_2(g) + 2\,H_2O(\ell)$ $\Delta H_b^\circ = -1410.9$ kJ

(c) $2\,H_2(g) + O_2(g) \longrightarrow 2\,H_2O(\ell)$ $\Delta H_c^\circ = -571.66$ kJ

Use this information to find the value of ΔH° for the formation of ethylene from ethane.

Answer $\Delta H^\circ = 137.0$ kJ

Strategy and Explanation Compare reactions (a), (b), and (c) with the target expression and decide how to change each to match the target expression.

- **Reaction (a) involves 2 mol ethane on the reactant side, but the target expression requires only 1 mol ethane.**
- **Reaction (b) has 1 mol C_2H_4 as a reactant, but 1 mol C_2H_4 is a product in the target expression.**
- **Reaction (c) has 2 mol H_2 as a reactant, but 1 mol H_2 is a product in the target expression.**

First, since the desired expression has only 1 mol ethane on the reactant side, we multiply expression (a) by $\frac{1}{2}$ to give an expression (a′) that also has 1 mol ethane on the reactant side. Halving the coefficients in the equation also halves the enthalpy change.

(a′) $= \frac{1}{2}$(a) $C_2H_6(g) + \frac{7}{2}\,O_2(g) \longrightarrow 2\,CO_2(g) + 3\,H_2O(\ell)$ $\Delta H_{a'}^\circ = -1559.7$ kJ

Next, we reverse expression (b) so that C_2H_4 is on the product side, giving expression (b′). This also reverses the sign of the enthalpy change.

(b′) $= -$(b) $2\,CO_2(g) + 2\,H_2O(\ell) \longrightarrow C_2H_4(g) + 3\,O_2(g)$

$\Delta H_{b'}^\circ = -\Delta H_b^\circ = +1410.9$ kJ

To get 1 mol $H_2(g)$ on the product side, we reverse expression (c) and multiply all coefficients by $\frac{1}{2}$. This changes the sign and halves the enthalpy change.

(c′) $= -\frac{1}{2}$(c) $H_2O(\ell) \longrightarrow H_2(g) + \frac{1}{2}\,O_2(g)$ $\Delta H_{c'}^\circ = -\frac{1}{2}\Delta H_c^\circ = +285.83$ kJ

Now it is possible to add expressions (a′), (b′), and (c′) to give the desired expression.

(a′) $C_2H_6(g) + \cancel{\frac{7}{2}\,O_2(g)} \longrightarrow \cancel{2\,CO_2(g)} + \cancel{3\,H_2O(\ell)}$ $\Delta H_{a'}^\circ = -1559.7$ kJ

(b′) $\cancel{2\,CO_2(g)} + \cancel{2\,H_2O(\ell)} \longrightarrow C_2H_4(g) + \cancel{3\,O_2(g)}$ $\Delta H_{b'}^\circ = +1410.9$ kJ

(c′) $\cancel{H_2O(\ell)} \longrightarrow H_2(g) + \cancel{\frac{1}{2}\,O_2(g)}$ $\Delta H_{c'}^\circ = +285.83$ kJ

Net expression: $C_2H_6(g) \longrightarrow C_2H_4(g) + H_2(g)$ $\Delta H_{net}^\circ = 137.0$ kJ

When the chemical equations are added, there is $\frac{7}{2}$ mol $O_2(g)$ on the reactant side and $(3 + \frac{1}{2}) = \frac{7}{2}$ mol $O_2(g)$ on the product side. There is 3 mol $H_2O(\ell)$ on each side and 2 mol $CO_2(g)$ on each side. Therefore, $O_2(g)$, $CO_2(g)$, and $H_2O(\ell)$ all cancel, and the chemical equation for the conversion of ethane to ethylene and hydrogen remains.

☑ **Reasonable Answer Check** The overall process involves breaking a molecule apart into simpler molecules, which is likely to involve breaking bonds. Therefore it should be endothermic, and ΔH° should be positive.

PROBLEM-SOLVING PRACTICE **6.13**

When iron is obtained from iron ore, an important reaction is conversion of $Fe_3O_4(s)$ to $FeO(s)$. Write a balanced equation for this reaction. Then use these thermochemical expressions to calculate ΔH° for the reaction.

$3\,Fe(s) + 2\,O_2(g) \longrightarrow Fe_3O_4(s)$ $\Delta H_1^\circ = -1118.4$ kJ

$Fe(s) + \frac{1}{2}\,O_2(g) \longrightarrow FeO(s)$ $\Delta H_2^\circ = -272.0$ kJ

6.10 Standard Molar Enthalpies of Formation

Hess's law makes it possible to tabulate ΔH° values for a relatively few reactions and, by suitable combinations of these few reactions, to calculate ΔH° values for a great many other reactions. To make such a tabulation we use standard molar enthalpies of formation. The **standard molar enthalpy of formation, ΔH_f°,** is the *standard enthalpy change for formation of one mole of a compound from its elements in their standard states.* The subscript *f* indicates *f*ormation of the compound. The **standard state** of an element or compound is the physical state in which it exists at 1 bar and a specified temperature. At 25 °C the standard state for hydrogen is $H_2(g)$ and for sodium chloride is NaCl(s). For an element that can exist in several different allotropic forms (← p. 26) at 1 bar and 25 °C, the most stable form is usually selected as the standard state. For example, graphite, not diamond or buckminsterfullerene, is the standard state for carbon; $O_2(g)$, not $O_3(g)$, is the standard state for oxygen.

Some examples of thermochemical expressions involving standard molar enthalpies of formation are

$$H_2(g) + \tfrac{1}{2}O_2(g) \longrightarrow H_2O(\ell) \qquad \Delta H^\circ = \Delta H_f^\circ\{H_2O(\ell)\} = -285.8 \text{ kJ/mol}$$

$$2\ C(\text{graphite}) + 2\ H_2(g) \longrightarrow C_2H_4(g) \qquad \Delta H^\circ = \Delta H_f^\circ\{C_2H_4(g)\} = 52.26 \text{ kJ/mol}$$

$$2\ C(\text{graphite}) + 3\ H_2(g) + \tfrac{1}{2}O_2 \longrightarrow C_2H_5OH(\ell)$$

$$\Delta H^\circ = \Delta H_f^\circ\{C_2H_5OH(\ell)\} = -277.69 \text{ kJ/mol}$$

Notice that in each case *1 mol of a compound in its standard state is formed directly from appropriate amounts of elements in their standard states.*

Some examples of thermochemical expressions at 25 °C and 1 bar where ΔH° is *not* a standard molar enthalpy of formation (and the reason why it is not) are

$$MgO(s) + SO_3(g) \longrightarrow MgSO_4(s) \qquad \Delta H^\circ = -287.5 \text{ kJ} \quad [6.9]$$

(reactants are not elements)

and

$$P_4(s) + 6\ Cl_2(g) \longrightarrow 4\ PCl_3(\ell) \qquad \Delta H^\circ = -1278.8 \text{ kJ} \quad [6.10]$$

(4 mol product formed instead of 1 mol)

The word *molar* means "per mole." Thus, the standard molar enthalpy of formation is the standard enthalpy of formation per mole of compound formed.

It is common to use the term "heat of formation" interchangeably with "enthalpy of formation." It is only the heat of reaction at constant pressure that is equivalent to the enthalpy change. If heat of reaction is measured under other conditions, it may not equal the enthalpy change. For example, when measured at constant volume in a bomb calorimeter, heat of reaction corresponds to the change of internal energy, not enthalpy.

PROBLEM-SOLVING EXAMPLE 6.14 Thermochemical Expressions for Standard Molar Enthalpies of Formation

Rewrite Thermochemical Expressions 6.9 and 6.10 so that they represent standard molar enthalpies of formation of their products. (The standard molar enthalpy of formation value for $MgSO_4(s)$ is −1284.9 kJ/mol at 25 °C.)

Answer

$$Mg(s) + \tfrac{1}{8}S_8(s) + 2\ O_2(g) \longrightarrow MgSO_4(s) \qquad \Delta H_f^\circ\{MgSO_4(s)\} = -1284.9 \text{ kJ/mol}$$

and

$$\tfrac{1}{4}P_4(s) + \tfrac{3}{2}Cl_2(g) \longrightarrow PCl_3(\ell) \qquad \Delta H_f^\circ\{PCl_3(\ell)\} = -319.7 \text{ kJ/mol}$$

Strategy and Explanation Thermochemical Expression 6.9 has 1 mol $MgSO_4(s)$ on the right side, but the reactants are not elements in their standard states. Write a new thermochemical expression so that the left side contains the elements Mg(s), $S_8(s)$, and $O_2(g)$. For this thermochemical expression ΔH° is the standard molar enthalpy of formation, −1284.9 kJ/mol. The new thermochemical expression is given in the Answer section above.

Thermochemical Expression 6.10 has elements in their standard states on the left side, but more than 1 mol of product is formed. Rewrite the thermochemical expression so the right side involves only 1 mol $PCl_3(\ell)$, and reduce the coefficients of the elements on the left side in proportion—that is, divide all coefficients by 4. Then ΔH° must also be divided by 4 to obtain the second thermochemical expression in the Answer section.

☑ **Reasonable Answer Check** Check each thermochemical expression carefully to make certain the substance whose standard enthalpy of formation you want is on the right side and has a coefficient of 1. For $PCl_3(\ell)$, ΔH_f° should be one fourth of about −1300 kJ, and it is.

Burning charcoal. Charcoal is mainly carbon, and it burns to form mainly carbon dioxide gas. The energy transfer from a charcoal grill could be estimated from the mass of charcoal and the standard molar enthalpy of formation of $CO_2(g)$. The thermochemical expression is

$$C(s) + O_2(g) \longrightarrow CO_2(g) \qquad \Delta H_f^\circ = -393.509 \text{ kJ/mol.}$$

PROBLEM-SOLVING PRACTICE 6.14

Write an appropriate thermochemical expression in each case. (You may need to use fractional coefficients.)

(a) The standard molar enthalpy of formation of $NH_3(g)$ at 25 °C is −46.11 kJ/mol.

(b) The standard molar enthalpy of formation of CO(g) at 25 °C is −110.525 kJ/mol.

CONCEPTUAL EXERCISE 6.17 Standard Molar Enthalpies of Formation of Elements

Write the thermochemical expression that corresponds to the standard molar enthalpy of formation of $N_2(g)$.

(a) What process, if any, takes place in the chemical equation?

(b) What does this imply about the enthalpy change?

Table 6.2 and Appendix J list values of ΔH_f°, obtained from the National Institute for Standards and Technology (NIST), for many compounds. Notice that no values are listed in these tables for elements in their most stable forms, such as C(graphite) or $O_2(g)$. As you probably realized from Conceptual Exercise 6.17,

Table 6.2 Selected Standard Molar Enthalpies of Formation at 25 °C*

Formula	Name	Standard Molar Enthalpy of Formation (kJ/mol)	Formula	Name	Standard Molar Enthalpy of Formation (kJ/mol)
$Al_2O_3(s)$	Aluminum oxide	−1675.7	HI(g)	Hydrogen iodide	26.48
$BaCO_3(s)$	Barium carbonate	−1216.3	KF(s)	Potassium fluoride	−567.27
$CaCO_3(s)$	Calcium carbonate	−1206.92	KCl(s)	Potassium chloride	−436.747
CaO(s)	Calcium oxide	−635.09	KBr(s)	Potassium bromide	−393.8
C(s, diamond)	Diamond	1.895	MgO(s)	Magnesium oxide	−601.70
$CCl_4(\ell)$	Carbon tetrachloride	−135.44	$MgSO_4(s)$	Magnesium sulfate	−1284.9
$CH_4(g)$	Methane	−74.81	$Mg(OH)_2(s)$	Magnesium hydroxide	−924.54
$C_2H_5OH(\ell)$	Ethyl alcohol	−277.69	NaF(s)	Sodium fluoride	−573.647
CO(g)	Carbon monoxide	−110.525	NaCl(s)	Sodium chloride	−411.153
$CO_2(g)$	Carbon dioxide	−393.509	NaBr(s)	Sodium bromide	−361.062
$C_2H_2(g)$	Acetylene (ethyne)	226.73	NaI(s)	Sodium iodide	−287.78
$C_2H_4(g)$	Ethylene (ethene)	52.26	$NH_3(g)$	Ammonia	−46.11
$C_2H_6(g)$	Ethane	−84.68	NO(g)	Nitrogen monoxide	90.25
$C_3H_8(g)$	Propane	−103.8	$NO_2(g)$	Nitrogen dioxide	33.18
$C_4H_{10}(g)$	Butane	−126.148	$O_3(g)$	Ozone	142.7
$C_6H_{12}O_6(s)$	α-D-Glucose	−1274.4	$PCl_3(\ell)$	Phosphorus trichloride	−319.7
$CuSO_4(s)$	Copper(II) sulfate	−771.36	$PCl_5(s)$	Phosphorus pentachloride	−443.5
$H_2O(g)$	Water vapor	−241.818	$SiO_2(s)$	Silicon dioxide (quartz)	−910.94
$H_2O(\ell)$	Liquid water	−285.830	$SnCl_2(s)$	Tin(II) chloride	−325.1
HF(g)	Hydrogen fluoride	−271.1	$SnCl_4(\ell)$	Tin(IV) chloride	−511.3
HCl(g)	Hydrogen chloride	−92.307	$SO_2(g)$	Sulfur dioxide	−296.830
HBr(g)	Hydrogen bromide	−36.40	$SO_3(g)$	Sulfur trioxide	−395.72

*From Wagman, D. D., Evans, W. H., Parker, V. B., Schumm, R. H., Halow, I., Bailey, S. M., Churney, K. L., and Nuttall, R. The NBS Tables of Chemical Thermodynamic Properties. *Journal of Physical and Chemical Reference Data*, Vol. 11, Suppl. 2, 1982. (NBS, the National Bureau of Standards, is now NIST, the National Institute for Standards and Technology.)

standard enthalpies of formation for the elements in their standard states are zero, because forming an element in its standard state from the same element in its standard state involves no chemical or physical change.

Hess's law can be used to find the standard enthalpy change for any reaction if there is a set of reactions whose enthalpy changes are known and whose chemical equations, when added together, will give the equation for the desired reaction. For example, suppose you are a chemical engineer and want to know how much heating is required to decompose limestone (calcium carbonate) to lime (calcium oxide) and carbon dioxide.

$$\mathrm{CaCO_3(s)} \longrightarrow \mathrm{CaO(s)} + \mathrm{CO_2(g)} \qquad \Delta H^\circ = ?$$

As a first approximation you can assume that all substances are in their standard states at 25 °C and look up the standard molar enthalpy of formation of each substance in a table such as Table 6.2 or Appendix J. This gives the thermochemical expressions,

$$\text{(a)} \quad \mathrm{Ca(s)} + \mathrm{C(graphite)} + \tfrac{3}{2}\mathrm{O_2(g)} \longrightarrow \mathrm{CaCO_3(s)} \qquad \Delta H_a^\circ = -1206.9\ \mathrm{kJ}$$

$$\text{(b)} \quad \mathrm{Ca(s)} + \tfrac{1}{2}\mathrm{O_2(g)} \longrightarrow \mathrm{CaO(s)} \qquad \Delta H_b^\circ = -635.1\ \mathrm{kJ}$$

$$\text{(c)} \quad \mathrm{C(graphite)} + \mathrm{O_2(g)} \longrightarrow \mathrm{CO_2(g)} \qquad \Delta H_c^\circ = -393.5\ \mathrm{kJ}$$

Now add the three chemical equations in such a way that the resulting equation is the one given above for the decomposition of limestone. In expression (a), $CaCO_3$(s) is a product, but it must appear in the desired expression as a reactant. Therefore, the equation in (a) must be reversed, and the sign of ΔH_a° must also be reversed. On the other hand, CaO(s) and CO_2(g) are products in the desired expression, so expressions (b) and (c) can be added with the same direction and sign of ΔH° as they have in the ΔH_f° equations:

$$\text{(a}'\text{)} = -\text{(a)} \quad \mathrm{CaCO_3(s)} \longrightarrow \cancel{\mathrm{Ca(s)}} + \cancel{\mathrm{C(graphite)}} + \cancel{\tfrac{3}{2}\mathrm{O_2(g)}}$$

$$\Delta H_{a'}^\circ = +1206.9\ \mathrm{kJ}$$

$$\text{(b)} \quad \cancel{\mathrm{Ca(s)}} + \cancel{\tfrac{1}{2}\mathrm{O_2(g)}} \longrightarrow \mathrm{CaO(s)} \qquad \Delta H_b^\circ = -635.1\ \mathrm{kJ}$$

$$\text{(c)} \quad \cancel{\mathrm{C(graphite)}} + \cancel{\mathrm{O_2(g)}} \longrightarrow \mathrm{CO_2(g)} \qquad \Delta H_c^\circ = -393.5\ \mathrm{kJ}$$

$$\mathrm{CaCO_3(s)} \longrightarrow \mathrm{CaO(s)} + \mathrm{CO_2(g)} \qquad \Delta H^\circ = +178.3\ \mathrm{kJ}$$

When the expressions are added in this fashion, 1 mol each of C(graphite) and Ca(s) and $\frac{3}{2}$ mol O_2(g) appear on opposite sides and so are canceled out. Thus, the sum of these chemical equations is the desired one for the decomposition of calcium carbonate, and the sum of the enthalpy changes of the three expressions gives that for the desired expression.

Another very useful conclusion can be drawn from this example. The calculation can be written mathematically as

$$\Delta H^\circ = \Delta H_f^\circ\{\mathrm{CaO(s)}\} + \Delta H_f^\circ\{\mathrm{CO_2(g)}\} - \Delta H_f^\circ\{\mathrm{CaCO_3(s)}\}$$
$$= (-635.1\ \mathrm{kJ}) + (-393.5\ \mathrm{kJ}) - (-1206.9\ \mathrm{kJ}) = 178.3\ \mathrm{kJ}$$

which involves adding the ΔH_f° values for the products of the reaction, CaO(s) and CO_2(g), and subtracting the ΔH_f° value for the reactant, $CaCO_3$(s). The mathematics of the problem can be summarized by the equation

$$\Delta H^\circ = \sum\{(\text{moles of product}) \times \Delta H_f^\circ(\text{product})\}$$
$$-\sum\{(\text{moles of reactant})\} \times \Delta H_f^\circ(\text{reactant})\} \qquad [6.11]$$

According to this equation you should **follow these steps to calculate the standard enthalpy of a reaction from standard molar enthalpies of formation:**

- **Multiply the standard molar enthalpy of formation of each product by the number of moles of that product and then sum over all products;**

Courtesy of João Paiva

Lime production. At high temperature in a lime kiln, calcium carbonate (limestone, $CaCO_3$) decomposes to calcium oxide (lime, CaO) and carbon dioxide (CO_2).

Courtesy Reatha Clark King

1938–

Reatha Clark King was born in Georgia. She obtained degrees from Clark Atlanta University and the University of Chicago and began her career with the National Bureau of Standards (now the National Institute for Standards and Technology, NIST), where she determined enthalpies of formation of fluorine compounds that were important to the U.S. space program and NASA. She became a dean at York College, president of Metropolitan State University (Minneapolis), and served as president of the General Mills Foundation.

- **Multiply the standard molar enthalpy of formation of each reactant by the number of moles of that reactant and then sum over all reactants;**
- **Subtract the sum for the reactants from the sum for the products.**

This is a useful shortcut for writing the thermochemical expressions for all appropriate formation reactions and applying Hess's law.

PROBLEM-SOLVING EXAMPLE 6.15 Using Standard Molar Enthalpies of Formation

Benzene, C_6H_6, is a commercially important hydrocarbon that is present in gasoline, where it enhances the octane rating. Calculate its enthalpy of combustion per mole; that is, find the value of $\Delta H°$ for the reaction

$$C_6H_6(\ell) + \tfrac{15}{2}\,O_2(g) \longrightarrow 6\,CO_2(g) + 3\,H_2O(\ell)$$

For benzene, $\Delta H_f°\{C_6H_6(\ell)\} = 49.0$ kJ/mol. Use Table 6.2 for any other values you may need.

Answer $\Delta H° = -3267.5$ kJ

Strategy and Explanation To calculate $\Delta H°$ you need standard molar enthalpies of formation for all compounds (and elements, if they are not in their standard states) involved in the reaction. (Since $O_2(g)$ is in its standard state, it is not included.) From Table 6.2,

$$C(\text{graphite}) + O_2(g) \longrightarrow CO_2(g) \qquad \Delta H_f° = -393.509 \text{ kJ/mol}$$

$$H_2(g) + \tfrac{1}{2}\,O_2(g) \longrightarrow H_2O(\ell) \qquad \Delta H_f° = -285.830 \text{ kJ/mol}$$

Using Equation 6.11,

$$\Delta H° = [6 \text{ mol} \times \Delta H_f°\{CO_2(g)\} + 3 \text{ mol} \times \Delta H_f°\{H_2O(\ell)\}] - [1 \text{ mol } C_6H_6(\ell) \times \Delta H_f°\{C_6H_6(\ell)\}]$$
$$= [6 \text{ mol} \times (-393.509 \text{ kJ/mol}) + 3 \text{ mol} \times (-285.830 \text{ kJ/mol})]$$
$$- [1 \text{ mol} \times (49.0 \text{ kJ/mol})] = -3267.5 \text{ kJ}$$

☑ **Reasonable Answer Check** As expected, the enthalpy change for combustion of a fuel is negative and large.

PROBLEM-SOLVING PRACTICE 6.15

Nitroglycerin is a powerful explosive because it decomposes exothermically and four different gases are formed.

$$2\,C_3H_5(NO_3)_3(\ell) \longrightarrow 3\,N_2(g) + \tfrac{1}{2}\,O_2(g) + 6\,CO_2(g) + 5\,H_2O(g)$$

For nitroglycerin, $\Delta H_f°\{C_3H_5(NO_3)_3(\ell)\} = -364$ kJ/mol. Using data from Table 6.2, calculate the energy transfer when 10.0 g nitroglycerin explodes.

When the enthalpy change for a reaction is known, it is possible to use that information to calculate $\Delta H_f°$ for one substance in the reaction provided that $\Delta H_f°$ values are known for all of the rest of the substances. Problem-Solving Example 6.16 indicates how to do this.

PROBLEM-SOLVING EXAMPLE 6.16 Standard Molar Enthalpy of Formation from Enthalpy of Combustion

Octane, C_8H_{18}, is a hydrocarbon that is present in gasoline. At 25 °C the enthalpy of combustion per mole for octane is −5116.0 kJ/mol. Use data from Table 6.2 to calculate the standard molar enthalpy of formation of octane. (Assume that water vapor is produced by the combustion reaction.)

Answer $\Delta H_f° = -208.4$ kJ

Strategy and Explanation Write a balanced equation for the target reaction whose $\Delta H°$ you want to calculate. Also write a balanced equation for combustion of octane, for which you know the standard enthalpy change. By studying these two equations, decide what additional information is needed to set up a Hess's law calculation that will yield the standard molar enthalpy of formation of octane.

(target reaction) $8\,C(s) + 9\,H_2(g) \longrightarrow C_8H_{18}(\ell)$ $\Delta H_f^\circ = ??$ kJ/mol

(a) $C_8H_{18}(\ell) + \frac{25}{2}\,O_2(g) \longrightarrow 8\,CO_2(g) + 9\,H_2O(g)$ $\Delta H^\circ_{combustion} = -5116.0$ kJ

Notice that the combustion equation involves 1 mol $C_8H_{18}(\ell)$ as a reactant and the target equation (for enthalpy of formation) involves 1 mol $C_8H_{18}(\ell)$ as a product. Therefore, it seems reasonable to reverse the combustion equation and see where that leads.

(a′) $8\,CO_2(g) + 9\,H_2O(g) \longrightarrow C_8H_{18}(\ell) + \frac{25}{2}\,O_2(g)$ $\Delta H^\circ = +5116.0$ kJ

On the reactant side of the target equation we have 8 C(s) and 9 $H_2(g)$. These elements, combined with $O_2(g)$, are on the left side of equation (a′), so perhaps it would be reasonable to use the equations corresponding to standard molar enthalpies of formation of carbon dioxide and water. From Table 6.2, we have

(b) $C(s) + O_2(g) \longrightarrow CO_2(g)$ $\Delta H_f^\circ = -393.509$ kJ/mol

(c) $H_2(g) + \frac{1}{2}\,O_2(g) \longrightarrow H_2O(g)$ $\Delta H_f^\circ = -241.818$ kJ/mol

Multiplying equation (b) by 8 and equation (c) by 9 gives the correct number of moles of C(s) and of $H_2(g)$ on the reactant side of the target equation. This gives

(a′) $8\,CO_2(g) + 9\,H_2O(g) \longrightarrow C_8H_{18}(\ell) + \frac{25}{2}\,O_2(g)$ $\Delta H^\circ = +5116.0$ kJ

(b′) $8\,C(s) + 8\,O_2(g) \longrightarrow 8\,CO_2(g)$ $\Delta H^\circ = -3148.072$ kJ

(c′) $9\,H_2(g) + \frac{9}{2}\,O_2(g) \longrightarrow 9\,H_2O(g)$ $\Delta H^\circ = -2176.362$ kJ/mol

$8\,C(s) + 9\,H_2(g) \longrightarrow C_8H_{18}(\ell)$ $\Delta H_f^\circ = \Delta H^\circ_{a'} + \Delta H^\circ_{b'} + \Delta H^\circ_{c'} = -208.4$ kJ/mol

PROBLEM-SOLVING PRACTICE 6.16

Use data from Table 6.2 to calculate the molar heat of combustion of sulfur dioxide, $SO_2(g)$, to form sulfur trioxide, $SO_3(g)$.

6.11 Chemical Fuels for Home and Industry

As you probably realize, energy used by our society comes from many sources. When you drive a car, the gasoline that provides the energy may have come from the United States or another country. Some vehicles are powered by natural gas (buses), electricity (golf carts), or a combination of electricity and gasoline (hybrid cars). Electricity used in your home may have come from a power plant that burns coal, natural gas, or fuel oil—or it might come from a nuclear power plant.

A **chemical fuel** is any substance that will react exothermically with atmospheric oxygen and is available at reasonable cost and in reasonable quantity. It is desirable that when a fuel burns, the products create as little environmental damage as possible. As indicated in Figure 6.18, most of the fuels that supply us with thermal energy are fossil fuels: coal, petroleum, and natural gas. Biomass fuels consisting of wood, peat, and other plant matter are a distant second among chemical fuels. A significant quantity of energy comes from nuclear reactors and hydroelectric power plants, and a much smaller quantity from solar, wind, and geothermal (hot springs or other sources of heat within Earth) sources. For specific applications, other fuels are sometimes chosen because of their special properties. For example, hydrogen is used as the fuel for the Space Shuttle, and hydrazine, N_2H_4, is used as a rocket fuel in some applications.

What nanoscale characteristics make for a good fuel? If some or all of the bonds in the molecules of a fuel are weak, or if the bonds in the products of its combustion are strong, then the combustion reaction will be exothermic. An example of a

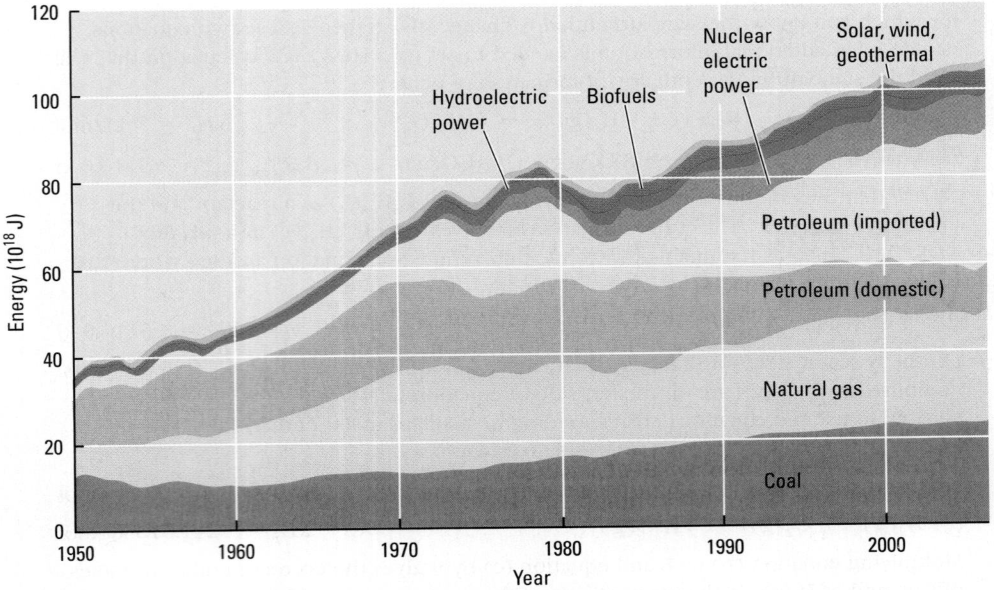

Figure 6.18 **Use of energy resources in the United States.** Use of energy resources in the United States is plotted from 1950 to 2004. (An energy resource is a naturally occurring fuel, such as petroleum, or a continuous supply, such as sunlight.) At the midpoint of the twentieth century, coal and petroleum were almost equally important, with natural gas coming in third. Today, petroleum and natural gas are used in greater quantities than coal, and more than half of U.S. petroleum is imported. Nuclear electric power did not exist in 1949 but contributes significantly to energy resources today, whereas hydroelectric electricity generation has grown only slightly since 1950.

molecule with a weak bond is hydrazine, $N_2H_4(g)$, which burns according to the equation

$$N_2H_4(g) + O_2(g) \longrightarrow N_2(g) + 2\ H_2O(g)$$

In N_2H_4 the N—N bond enthalpy is only 160 kJ/mol, although its four N—H bonds are reasonably strong at 391 kJ/mol. The O_2 bond enthalpy is 498 kJ/mol. The reaction products are N_2, which has a very strong bond at 946 kJ/mol, and H_2O, which also has strong O—H bonds at 467 kJ/mol. In this case there are fewer bonds after the reaction than before, but the bonds are much stronger, so the reaction is exothermic.

EXERCISE 6.18 Using Bond Enthalpies to Evaluate a Fuel

Based on the molecular structures and bond enthalpies given above for combustion of hydrazine:

(a) Calculate $\Delta H°$ when all the bonds in the reactant molecules are broken.
(b) Calculate $\Delta H°$ when all the bonds in the product molecules are formed.
(c) Calculate $\Delta H°$ for the reaction and write the thermochemical expression.

Coal, petroleum, and natural gas consist mainly of hydrocarbon molecules. When these fuels burn completely in air, they produce water and carbon dioxide. A carbon dioxide molecule contains two very strong carbon-oxygen bonds (803 kJ/mol each), and a water molecule contains two very strong O—H bonds (467 kJ/mol each). As shown by the equation,

$$CH_4(g) + 2\ O_2(g) \longrightarrow CO_2(g) + 2\ H_2O(g)$$

when the hydrocarbon methane, CH_4, burns, the number of bonds in reactant molecules is the same as the number of bonds in product molecules. Because the bonds formed are stronger than the bonds broken, the reaction is exothermic.

Another very good fuel is hydrogen. It burns in air to produce only water, which is a big advantage from an environmental point of view. Burning hydrocarbon fuels increases CO_2 levels in the atmosphere and therefore is partly responsible for global warming (see Section 10.11). The thermochemical expression for combustion of hydrogen corresponds to the formation of water from its elements, so the standard enthalpy change is just $\Delta H_f^\circ\{H_2O(g)\}$:

$$H_2(g) + \tfrac{1}{2}\,O_2(g) \longrightarrow H_2O(g) \qquad \Delta H_f^\circ = -241.818 \text{ kJ}$$

On Earth, little or no hydrogen is available naturally as the element; it is always combined in compounds. Therefore hydrogen fails to meet the criterion of availability in reasonable quantity mentioned at the beginning of this section. It is manufactured as a by-product of petroleum refining at present, which makes it too expensive for most fuel applications. However, considerable research is aimed at finding ways to produce hydrogen either by electrolysis of water or photochemically. For example, electricity supplied by solar cells could be used to electrolyze water and produce hydrogen, which could then be used as fuel. Or solar energy might be used directly to cause a series of chemical reactions to occur in which hydrogen was produced from water. At present none of these ways of producing hydrogen is inexpensive enough to be competitive with fossil fuels. As supplies of fossil fuels become exhausted, however, hydrogen may become much more important.

Scientists at the University of California, Berkeley, and the National Renewable Energy Laboratory have found that when sulfur is removed from the growing environment of *Chlamydomonas reinhardtii,* the algae generate hydrogen gas (*Plant Physiology,* Vol. 122, 2000; p. 127). Optimizing hydrogen production from algae might provide an inexpensive source of fuel.

Before the Industrial Revolution the main fuel was wood. It is now referred to more generically as biomass, because plant matter other than wood can also be burned. Biomass is very important as a fuel in many less developed countries, and it is a renewable resource. Coal, petroleum, and natural gas will eventually be used up, but it is possible to continue growing plants to create biomass. Although biomass is a mixture of materials, it is primarily carbohydrate (cellulose in wood, for example) and can be represented by the empirical formula CH_2O. Combustion of biomass is highly exothermic:

$$CH_2O(s) + O_2(g) \longrightarrow CO_2(g) + H_2O(g) \qquad \Delta H^\circ = -425 \text{ kJ}$$

Two important criteria for a fuel are the **fuel value,** which is the quantity of energy released when 1 g of fuel is burned to form carbon dioxide and water, and the **energy density,** which is the quantity of energy released per unit volume of fuel. For gaseous fuels the fuel value may be high, but the energy density will be low because the density of a gas is low. Fuels with low energy density take a large volume for storage and therefore are not convenient to use unless, as in the case of natural gas, they can be supplied on demand via pipes (mains). Often gaseous fuels, such as propane and butane, are condensed and stored as liquids under pressure. Fuels such as methane and hydrogen cannot be liquefied by pressure alone. If they are to be stored as liquids, the temperature must be kept low and the pressure high. This makes them less convenient than liquid fuels for use in cars and airplanes, for example.

PROBLEM-SOLVING EXAMPLE 6.17 Comparing Fuels

Evaluate each of the fuels listed below on the basis of fuel value and energy density. Assume that $H_2O(g)$ is formed and use data from Table 6.2 or Appendix J. Which fuel provides the largest fuel value? Which provides the greatest energy density? Assume that when the fuels burn, carbon is converted to gaseous CO_2 and hydrogen to water vapor. The densities are given with the substances.

(a) Methane, $CH_4(g)$ (0.656 g/L)
(b) Ethanol, $C_8H_{18}(\ell)$ (0.789 g/mL)
(c) Hydrogen, $H_2(g)$ (0.082 g/L)

Answer

(a) 50.013 kJ/g CH_4, 3.28×10^1 kJ/L CH_4
(b) 26.807 kJ/g C_2H_5OH, 2.19×10^4 kJ/L C_2H_5OH
(c) 119.96 kJ/g H_2, 9.84 kJ/L H_2

Hydrogen has the highest fuel value; ethanol has the highest energy density.

Strategy and Explanation In each case, the balanced chemical equation must be written and then used in conjunction with the thermodynamic data to calculate the enthalpy change for the reaction. For methane:

$$CH_4(g) + 2\,O_2(g) \longrightarrow CO_2(g) + 2\,H_2O(g)$$

$$\Delta H° = [(-393.509 \text{ kJ}) + 2(-241.818 \text{ kJ}) - (-74.81 \text{ kJ})] = -802.34 \text{ kJ}$$

Therefore, the fuel value of methane is

$$\frac{802.34 \text{ kJ}}{1 \text{ mol } CH_4} \times \frac{1 \text{ mol } CH_4}{16.0426 \text{ g } CH_4} = 50.013 \text{ kJ/g } CH_4$$

and its energy density is

$$\frac{50.013 \text{ kJ}}{1 \text{ g } CH_4} \times \frac{0.656 \text{ g } CH_4}{1 \text{ L } CH_4} = 3.28 \times 10^1 \text{ kJ/L } CH_4$$

In a similar manner the fuel values and energy densities for ethanol and hydrogen were calculated.

PROBLEM-SOLVING PRACTICE 6.17

Evaluate each of the fuels listed below on the basis of fuel value and energy density. Use data from Appendix J. Which fuel provides the largest fuel value? Which provides the greatest energy density? Assume that when the fuels burn, carbon is converted to gaseous CO_2, hydrogen to water vapor, and nitrogen to N_2 gas. The densities are given with the substances.

(a) Octane, $C_8H_{18}(\ell)$ (0.703 g/mL) (b) Hydrazine, $N_2H_4(\ell)$ (1.004 g/mL)
(c) Glucose, $C_6H_{12}O_6(s)$ (1.56 g/mL)

All forms of energy are, in principle, interchangeable. The work a given quantity of energy can do is the same, no matter what the energy source. The energy contents of natural gas, fuel oil, and coal in terms of their equivalents in various units are compared in Table 6.3. As a consumer, it doesn't matter to you whether

Table 6.3 A Chart of Energy Equivalencies, Based on Fuel-Heating Values*

Cubic Feet of Natural Gas	Barrels of Oil	Tons of Bituminous Coal	Kilowatt Hours of Electricity	Joules	Btu
1	0.00018	0.00004	0.293	1.055×10^6	1.00×10^3
1000	0.18	0.04	293	1.055×10^9	1.00×10^6
5556	1	0.22	1628	5.9×10^9	5.6×10^6
25,000	4.50	1	7326	26.4×10^9	25.0×10^7
1×10^6	180	40	293,000	1.055×10^{12}	1.00×10^9
3.41×10^6	614	137	1×10^6	3.6×10^{12}	3.4×10^9
1×10^9	180,000	40,000	293×10^6	1.055×10^{15}	1.00×10^{12}
1×10^{12}	180×10^6	40×10^6	293×10^9	1.055×10^{18}	1.00×10^{15}

*The Btu (British thermal unit) is the quantity of energy required to raise the temperature of 1 lb water from 63 °F to 64 °F; 1 Btu = 252 cal = 1.055×10^3 J.

you travel using energy from a fossil fuel in the form of gasoline or whether you use electricity created by burning coal, although the energy is measured in different units. A trip might require 1 gal gasoline for the regular automobile. An electric automobile would require about 39 kilowatt-hours (kWh) of electricity to travel the same distance.

All fuel-burning engines, including automobile engines and electrical power plants, are less than 100% efficient; there is always thermal energy that cannot be used for the intended purpose and thus is wasted. For example, the generation and distribution of electricity is only about 33% efficient overall. This means that for every 100 J energy produced by burning a fossil fuel such as coal, only 33 J electrical energy reaches the consumer.

One watt (W) of power is 1 J/s, so a 100-W electric lamp uses energy at the rate of 100 J every second. Using data from Table 6.3 we find that one cubic foot of natural gas supplies sufficient energy (1.055×10^6 J) to light a 100-W bulb for about 3 hours; a barrel of oil has enough energy (5.9×10^9 J) to light 1000 100-W light bulbs for almost 16 hours.

PROBLEM-SOLVING EXAMPLE 6.18 Energy Conversions

A large ocean-going tanker holds 1.5 million barrels (bbl) of crude oil.
(a) What is the energy equivalent of this oil in joules?
(b) How many tons of coal is this oil equivalent to?

Answer
(a) 8.9×10^{15} J
(b) 3.3×10^5 ton coal

Strategy and Explanation
(a) Table 6.3 gives the thermal energy equivalent to 1 barrel of oil as 5.9×10^9 J. Use this information as a conversion factor to calculate the first answer.

$$1.5 \times 10^6 \text{ bbl} \times \frac{5.9 \times 10^9 \text{ J}}{1 \text{ bbl}} = 8.9 \times 10^{15} \text{ J}$$

(b) Look in Table 6.3 and find the conversion between barrels of oil and tons of coal. One barrel of oil is equivalent to 0.22 ton coal. Use this conversion factor to calculate the answer.

$$1.5 \times 10^6 \text{ bbl} \times \frac{0.22 \text{ ton coal}}{1 \text{ bbl}} = 3.3 \times 10^5 \text{ ton coal}$$

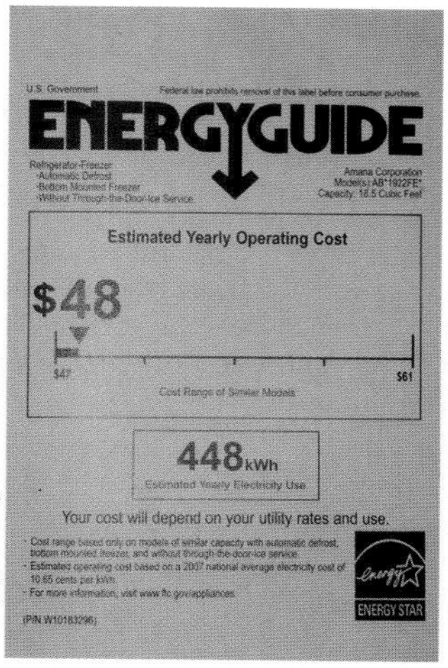

© Cengage Learning/John W. Moore

Energy efficiency comparison guide. Cost per kilowatt-hour of a fully self-defrosting refrigerator-freezer.

ESTIMATION Burning Coal

A coal-fired electric power plant burns about 1.5 million tons of coal a year. Coal has an approximate composition of $C_{135}H_{96}O_9NS$. When coal burns, it releases about 30 kJ/g. We can approximate how much energy (kJ) is released by this plant in a year and the mass of CO_2 released in that year by burning 1.5 million tons of coal.

The mass of coal is

$$1.5 \times 10^6 \text{ ton} \times \frac{2000 \text{ lb}}{1 \text{ ton}} \times \frac{1 \text{ kg}}{2.2 \text{ lb}} \approx 1 \times 10^9 \text{ kg}$$

This is approximately 1×10^9 kg, or 1×10^{12} g. Burning the coal at 30 kJ/g yields about 3×10^{13} kJ per year.

$$1 \times 10^{12} \text{ g} \times \frac{30 \text{ kJ}}{1 \text{ g}} = 3 \times 10^{13} \text{ kJ}$$

There are 135 mol carbon (1620 g C) in 1 mol coal (1906 g coal/mol). Thus, the ratio of grams of carbon to grams of coal is roughly 0.9 and so in 1×10^{12} g coal there is approximately $0.9 \times 1 \times 10^{12}$ g carbon, which is roughly 1×10^{12} g carbon. When burned, each gram of carbon produces about 4 g CO_2 because there are 44 g CO_2 formed per 12 g C.

$$\frac{44 \text{ g } CO_2}{12 \text{ g C}} \approx 4 \text{ g } CO_2/\text{g C}$$

Thus, in 1 year the plant generates

$$(1 \times 10^{12} \text{ g C})\left(\frac{4 \text{ g } CO_2}{1 \text{ g C}}\right) = 4 \times 10^{12} \text{ g } CO_2$$

This is more than 4 million tons, approximately the mass of 2 million SUVs.

Visit this book's companion website at **www.cengage.com/chemistry/moore** to work an interactive module based on this material.

☑ **Reasonable Answer Check** It takes only 0.22 ton coal (approximately one-fourth ton) to furnish the energy equivalent of a barrel of oil. Therefore, the coal equivalent of 1.5×10^6 barrels of oil should be about one fourth that number of barrels of oil, or about $(0.25)(1.5 \times 10^6) = 4 \times 10^5$, which is close to the calculated value.

PROBLEM-SOLVING PRACTICE **6.18**

How much energy, in joules, can be obtained by burning 4.2×10^9 ton coal? This is equivalent to how many cubic feet of natural gas?

CONCEPTUAL EXERCISE **6.19 The Energy Value of CO_2**

Explain why CO_2 has no fuel energy value.

6.12 Foods: Fuels for Our Bodies

This discussion answers the question "Where does the energy come from to make my muscles work?" that was posed in Chapter 1 (← p. 2).

Foods are similar to fuels, in that the **caloric value** of fats and carbohydrates corresponds to the standard enthalpy of combustion per gram of the food. Foods consist mainly of carbohydrate, fat, and protein. **Carbohydrates** have the general formula $C_x(H_2O)_y$. They are converted in the intestines to glucose, $C_6H_{12}O_6$, which is soluble in blood and thereby can be transported throughout the body. Glucose is metabolized in a complicated series of reactions that eventually produce $CO_2(g)$ and $H_2O(\ell)$, with release of energy. The net effect is equivalent to combustion of glucose,

$$C_6H_{12}O_6(s) + 6\,O_2(g) \longrightarrow 6\,CO_2(g) + 6\,H_2O(\ell) \qquad \Delta H° = -2801.6 \text{ kJ}$$

Because enthalpy is a state function and the initial and final states are the same, it is appropriate and convenient to measure the caloric values for carbohydrates using a bomb calorimeter (p. 239). A sample of glucose or other carbohydrate is ignited inside the bomb, and the heat transfer to the calorimeter and surrounding water is measured. The average caloric value of carbohydrates in food is 4 Cal/g (17 kJ/g).

EXERCISE **6.20 Caloric Value of Carbohydrate**

Use the thermochemical expression given above to verify that the caloric value of glucose corresponds to the average 4 Cal/g for carbohydrates.

Carbohydrates are metabolized quickly and large quantities are not usually stored in the body. The energy released by carbohydrates is used to power muscles, to transmit nerve impulses, and to cause chemical reactions to occur that construct and repair tissues. Energy is also required to maintain body temperature. Energy from carbohydrates that is not needed for these purposes is stored in fats.

Fats also compose a significant portion of most people's diets. Like carbohydrates, fats are metabolized to $CO_2(g)$ and $H_2O(\ell)$. For example, tristearin is oxidized according to the thermochemical expression

$$2\,C_{57}H_{110}O_6(s) + 163\,O_2(g) \longrightarrow 114\,CO_2(g) + 110\,H_2O(\ell) \quad \Delta H° = -75{,}520 \text{ kJ}$$

Again, bomb calorimetry simulates the metabolic process quite well. Fat molecules make excellent storehouses for energy. They are insoluble in water and therefore are not excreted easily. When metabolized they release 9 Cal/g (38 kJ/g)—more than twice the energy from the same mass of carbohydrate or protein.

The third dietary component, protein, contains nitrogen in addition to carbon, hydrogen, and oxygen. The nitrogen is metabolized in the body to produce new proteins or urea, $(NH_2)_2CO(aq)$, which is excreted. The carbon, hydrogen, and oxygen are metabolized in pathways that release energy. On average, protein metabolism releases 4 Cal/g (17 kJ/g), the same caloric value as for carbohydrate.

A few substances may be part of your diet that are not protein, carbohydrate, or fat. For example, ethanol in alcoholic beverages contributes calories because ethanol can be oxidized, or metabolized, exothermically. Most food labels provide information about the caloric value of a typical serving and about the percentages of carbohydrate, fat, and protein in the food. An example is shown in Figure 6.19. The caloric value of the food can be estimated from the quantities of protein, carbohydrate, and fat, as long as no other component (such as alcohol) is present. Table 6.4 gives the content of fat, carbohydrate, and protein along with caloric values for some representative foods.

The release of energy upon combustion or metabolism of carbohydrates, fats, proteins, and other substances can be understood in terms of bond enthalpies. For example, fats consist almost entirely of long chains of carbon atoms to which hydrogen atoms are attached. Therefore, fats contain mostly C—H and C—C bonds. When fats burn or are metabolized, each carbon atom becomes bonded to oxygen in CO_2 and each hydrogen becomes bonded to oxygen in H_2O. The bond in each O_2 molecule that reacts with the fat must be broken, as must the C—H and C—C bonds in the fat. The respective bond enthalpies are 498 kJ/mol, 416 kJ/mol, and 356 kJ/mol. The strengths of the bonds formed in the products are 803 kJ/mol for carbon-oxygen bonds and 467 kJ/mol for hydrogen-oxygen bonds. Because the number of bonds before and after reaction is nearly the same, and because the bonds formed are stronger than the bonds broken, metabolism of fats transfers energy to the surroundings.

As we implied earlier, there is a balance between the quantity of food energy that is taken into our bodies and the quantity that is used for body functions. If food intake exceeds consumption, the body stores energy in fat molecules. If consumption exceeds intake, some fat is burned to provide the needed energy. The **basal metabolic rate (BMR)** is the minimum energy intake required to maintain a body

Caloric Values (approximate)

Carbohydrate	4 Cal/g (17 kJ/g)
Fat	9 Cal/g (38 kJ/g)
Protein	4 Cal/g (17 kJ/g)

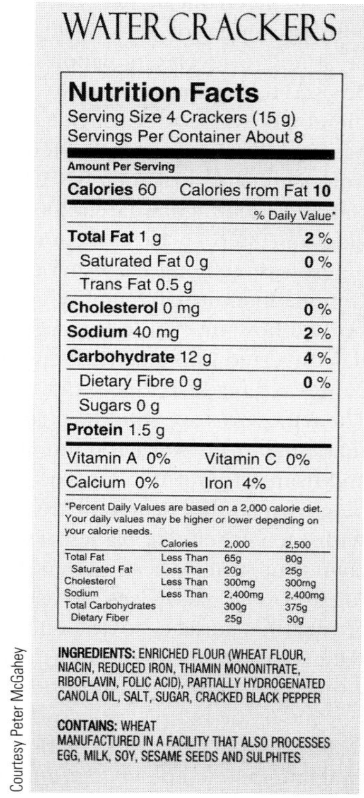

Figure 6.19 Nutrition facts from a package of snack crackers.

Table 6.4 Composition and Caloric Values of Some Foods

Food	Approximate Composition per 100. g			Caloric Value	
	Fat	Carbohydrate	Protein	Cal/g	kJ/g
All-purpose flour	0.0	73.3	13.3	3.33	13.95
Apple	0.5	13.0	0.4	0.59	2.47
Brownie with nuts	16.0	64.0	4.0	4.04	16.9
Cheese pizza	10.2	25.8	11.2	2.41	10.1
Egg	0.7	10.0	13.0	1.40	5.86
Egg noodle substitute	0.9	73.2	14.3	3.75	15.69
Grapes, white	0.6	17.5	0.6	0.7	2.9
Green beans	0.0	7.0	1.9	0.38	0.00
Hamburger	30.0	0.0	22.0	3.60	15.06
Microwave popcorn (popped)	7.1	11.4	2.9	1.00	4.18
Peanuts (unsalted)	50.0	21.4	28.6	5.71	23.91
Prunes (pitted)	0.0	65.0	2.5	2.75	11.51
Rice	1.0	77.6	8.2	3.47	14.52
Salad dressing (vinaigrette)	20.0	40.0	0.0	3.33	13.95
Tomato sauce	0.0	4.8	1.6	0.24	1.01
Wheat crackers	10.7	71.4	14.3	4.29	17.93

CHEMISTRY IN THE NEWS

Charge Your iPod with a Wave of Your Hand

Can the mechanical energy associated with movement of arms or fingers, breathing, or even circulation of blood be used to generate electricity and thereby power devices such as iPods or pacemakers? Research reported at the American Chemical Society's 237th National Meeting in March 2009 provides one step in that direction.

Professor Zhong Lin Wang of The Georgia Institute of Technology described how mechanical energy could be converted into electrical energy by nanowires made of zinc oxide, ZnO. The nanowires are solid hexagonal rods that range from 30 to 100 nm in width and from 1 to 3 μm in length. They can be grown so that nearly all are aligned with each other, like blades of grass in a lawn (but stiffer), as shown in the photograph. The nanowires are usually grown on a surface, so one end of each wire is firmly attached, just as grass would be rooted in the ground.

Zinc oxide is a piezoelectric material—squeezing or stretching it generates an electric voltage and that voltage can drive an electric current. A second surface can be placed onto the upper ends of the nanowires and, if that surface is rough or ridged (on the nanoscale), movement of the second surface will cause the wires to move along with the surface. When the nanowires are pushed back and forth in a direction perpendicular to their length, the nanowires bend in one direction and then the other like a waving field of grass in the wind. Each bend compresses one side of a nanowire and lengthens the other side, which generates a separation of positive and negative charges. When the nanowire bends back, the charges reverse position and an alternating electric current is generated. If both the surface on which the nanowires were grown and the second surface on top of the wires conduct electricity, the current can be conducted away from the nanowires. This current can be converted to direct current—the type generated by a battery—so the array of nanowires could replace a battery and would last for a much longer time.

Nanogenerators such as the one described here could have many uses. For example, biosensors or pacemakers could be implanted into a human and would not require periodic operations for replacement of batteries because the nanogenerator would be powered by normal body movements. Not having to replace batteries broadens the types of organs into which such sensors could be implanted. Nanogenerators would be ideal for powering nanorobots or nanomotors that would perform mechanical tasks at the nanoscale, and nanogenerators could also be used in tiny sensors that would sample air to detect drugs and substances used for bioterrorism.

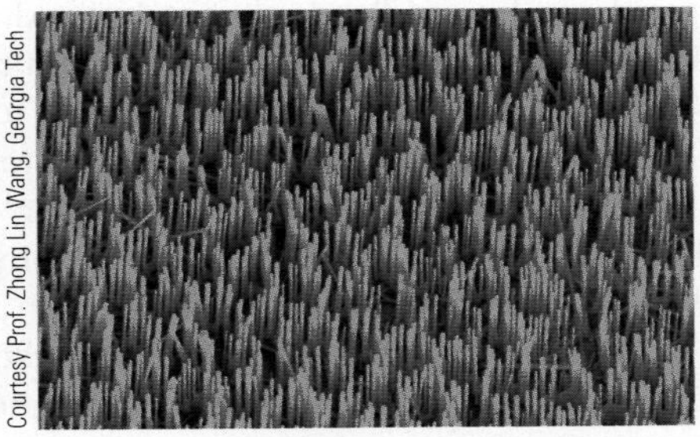

Zinc oxide nanowires are typically 30 to 100 nm across and from 1 to 3 μm long. The image was made with a scanning electron microscope.

Think about these questions:

What are some other ways in which such nanoscale power sources might be useful to society?

How might nanogenerators be used to address energy supply problems for the United States and other countries?

Sources: Wang, Z.L. Nanogenerators. Presented at the 237th American Chemical Society National Meeting, Mar. 26, 2009; Abstract I&EC 125. *Eureka! Science News*, Mar. 30, 2009. **http://esciencenews.com/articles/2009/03/26/new.nanogenerator.may.charge.ipods.and.cell.phones.with.a.wave.hand**

Wang, Z.L. "Self-Powered Nanotech," *Scientific American*, Jan. 2008; pp. 82-87.

that is awake and at rest, excluding the energy needed to digest, absorb, and metabolize the food, which is about 10% of the caloric intake. The BMR varies considerably depending on age, gender, and body mass. For a 70-kg (155-lb) human between 18 and 30 years old, the average BMR is 1750 Cal/day for a male and 1525 Cal/day for a female. The basal metabolic rate is approximately 1 Cal kg^{-1} h^{-1}; that is, about 1 Calorie is expended per hour for each kilogram of body mass. Thus, the average 60-kg (132-1b) person has a daily BMR of

$$\frac{1\ \text{Cal}}{\text{kg} \times \text{h}} \times 60\ \text{kg} \times \frac{24\ \text{h}}{\text{day}} = 1440\ \text{Cal/day}$$

This value is multiplied by factors of up to 7 depending on the level of muscular activity. For example, walking or other light work requires 2.5 times the BMR. Heavy work, such as playing basketball or soccer, requires 7 times the BMR. Therefore, the appropriate food energy intake varies greatly from one individual to another.

PROBLEM-SOLVING EXAMPLE 6.19 Energy Value of Food

A 70-kg, 22-year-old female eats 50. g unsalted peanuts and then exercises by playing basketball for 30 min.

(a) What fraction of the food energy of the peanuts comes from fat, carbohydrate, and protein?

(b) Is the exercise sufficient to use up the energy provided by the food?

Answer

(a) 69% from fat, 13% from carbohydrate, and 18% from protein

(b) No

Strategy and Explanation

(a) Data from Table 6.4 show that 100. g of unsalted peanuts contains 50.0 g fat, 21.4 g carbohydrate, and 28.6 g protein. The fat provides 38 kJ/g and the carbohydrate and protein provide 17 kJ/g each. Therefore the energy provided is

$$\text{Energy from fat} = 50.\text{ g peanuts} \times \frac{50.0\text{ g fat}}{100.\text{ g peanuts}} \times \frac{38\text{ kJ}}{1\text{ g fat}} = 950\text{ kJ}$$

$$\text{Energy from carbohydrate} = 50.\text{ g peanuts} \times \frac{21.4\text{ g carbo}}{100.\text{ g peanuts}} \times \frac{17\text{ kJ}}{1\text{ g carbo}} = 182\text{ kJ}$$

$$\text{Energy from protein} = 50.\text{ g peanuts} \times \frac{28.6\text{ g protein}}{100.\text{ g peanuts}} \times \frac{17\text{ kJ}}{1\text{ g protein}} = 243\text{ kJ}$$

The total caloric intake is 1375 kJ, and the fractions are $(950/1375) \times 100\% = 69\%$ from fat, $(182/1375) \times 100\% = 13\%$ from carbohydrate, and $(243/1375) \times 100\% = 18\%$ from protein.

(b) Playing basketball requires seven times the BMR—that is, 7×1525 Cal/day.

$$\text{Energy required} = 7 \times \frac{1525\text{ Cal}}{\text{day}} \times \frac{1\text{ day}}{24\text{ h}} \times \frac{1\text{ h}}{60\text{ min}} \times 30\text{ min} \times \frac{4.184\text{ kJ}}{1\text{ Cal}} = 930.\text{ kJ}$$

In addition, 10% of the caloric intake is required to digest, absorb, and metabolize the food. The quantity of energy required is thus

$$930.\text{ kJ} + (0.10 \times 1375\text{ kJ}) = 1068\text{ kJ}$$

which is less than the caloric value of the 50. g of peanuts.

☑ **Reasonable Answer Check** As you might have expected, peanuts contain fat (oil), protein, and some carbohydrate, so the quantities of energy from these sources seem reasonable. However, it takes a lot of exercise to work off what we eat, so you may have been surprised by the answer to part (b). A little less than 2 oz of peanuts corresponds to 50. g, so eating peanuts without exercising can easily increase your weight.

PROBLEM-SOLVING PRACTICE 6.19

Whole milk contains 5.0% carbohydrate, 4.0% fat, and 3.3% protein by mass.

(a) Estimate the caloric value of an 8-oz (227-g) glass of milk.

(b) For how long would this caloric intake support a 70-kg male who was taking a leisurely walk?

EXERCISE 6.21 Power of a Person

Use the average BMR values in the text to calculate the power (energy per unit time) required to sustain a 70-kg male (a) at rest and (b) playing basketball. Express your result in watts. (1 W = 1 J/s.) Compare your results with the power of a typical incandescent light bulb.

SUMMARY PROBLEM

Sulfur dioxide, SO_2, is a major pollutant emitted by coal-fired electric power generating plants. A large power plant can produce 8.64×10^{13} kJ electrical energy every day by burning 7000. ton coal (1 ton = 9.08×10^5 g).

(a) Assume that the fuel value of coal is approximately the same as for graphite. Calculate the quantity of energy transferred to the surroundings by the coal combustion reaction in a plant that burns 7000. ton coal.

(b) What is the efficiency of this power plant in converting chemical energy to electrical energy—that is, what percentage of the thermal energy transfer shows up as electrical energy?

(c) When SO_2 is given off by a power plant, it can be trapped by reaction with MgO in the smokestack to form $MgSO_4$.

$$MgO(s) + SO_2(g) + \frac{1}{2}O_2(g) \longrightarrow MgSO_4(s)$$

If 140. ton SO_2 is given off by a coal-burning power plant each day, how much MgO would you have to supply to remove all of this SO_2? How much $MgSO_4$ would be produced?

(d) How much heat transfer does the reaction in part (c) add or take away from the heat transfer of the coal combustion?

(e) Sulfuric acid comes from the oxidation of sulfur, first to SO_2 and then to SO_3. The SO_3 is then absorbed by water in 98% H_2SO_4 solution to make H_2SO_4.

$$S(s) + O_2(g) \longrightarrow SO_2(g)$$
$$SO_2(g) + \frac{1}{2}O_2(g) \longrightarrow SO_3(g)$$
$$SO_3(g) + H_2O\ (\text{in 98\% } H_2SO_4) \longrightarrow H_2SO_4(\ell)$$

For which of these reactions can you calculate $\Delta H°$ using data in Table 6.2 or Appendix J? Do the calculation for each case where data are available.

IN CLOSING

OWL and go Chemistry

Sign in at **www.cengage.com/owl** to:

- View tutorials and simulations, develop problem-solving skills, and complete online homework assigned by your professor.
- For quick review and exam prep, download Go Chemistry mini lecture modules from OWL (or purchase them at **www.CengageBrain.com**).

Having studied this chapter, you should be able to . . .

- Understand the difference between kinetic energy and potential energy (Section 6.1).
- Be familiar with typical energy units and be able to convert from one unit to another (Sections 6.1, 6.11). End-of-chapter questions: 13, 14, 16, 17
- Understand conservation of energy and energy transfer by heating and working (Section 6.2). Questions 25, 27
- Recognize and use thermodynamic terms: system, surroundings, heat, work, temperature, thermal equilibrium, exothermic, endothermic, and state function (Sections 6.2, 6.4). Questions 21, 23
- Use specific heat capacity and the sign conventions for transfer of energy (Section 6.3). Questions 33, 35, 37, 41, 43, 47
- Distinguish between the change in internal energy and the change in enthalpy for a system (Section 6.4). Questions 51, 53, 55, 59, 61, 63, 65
- Use thermochemical expressions and derive thermostoichiometric factors from them (Sections 6.5, 6.6). Questions 71, 75

- Use the fact that the standard enthalpy change for a reaction, ΔH°, is proportional to the quantity of reactants consumed or products produced when the reaction occurs (Section 6.6). Questions 77, 81, 83, 85, 87
- Understand the origin of the enthalpy change for a chemical reaction in terms of bond enthalpies (Section 6.7). Questions 89, 91
- Describe how calorimeters can measure the quantity of thermal energy transferred during a reaction (Section 6.8). Questions 94, 96, 98, 100, 102
- Apply Hess's law to find the enthalpy change for a reaction (Sections 6.9, 6.10). Questions 104, 106, 108
- Use standard molar enthalpies of formation to calculate the thermal energy transfer when a reaction takes place (Section 6.10). Questions 110, 114, 116, 118, 120, 122, 124, 126
- Define and give examples of some chemical fuels and evaluate their abilities to provide heating (Section 6.11). Question 125
- Describe the main components of food and evaluate their contributions to caloric intake (Section 6.12). Questions 132, 134

KEY TERMS

basal metabolic rate (BMR) *(Section 6.12)*

bond enthalpy (bond energy) *(6.7)*

caloric value *(6.12)*

calorimeter *(6.8)*

carbohydrate *(6.12)*

change of state *(6.4)*

chemical fuel *(6.11)*

conservation of energy, law of *(6.2)*

endothermic *(6.4)*

energy density *(6.11)*

enthalpy change *(6.4)*

enthalpy of fusion *(6.4)*

enthalpy of vaporization *(6.4)*

exothermic *(6.4)*

first law of thermodynamics *(6.2)*

fuel value *(6.11)*

heat/heating *(6.2)*

heat capacity *(6.3)*

Hess's law *(6.9)*

internal energy *(6.2)*

kinetic energy *(6.1)*

molar heat capacity *(6.3)*

phase change *(6.4)*

potential energy *(6.1)*

specific heat capacity *(6.3)*

standard enthalpy change *(6.5)*

standard molar enthalpy of formation *(6.10)*

standard state *(6.10)*

state function *(6.4)*

surroundings *(6.2)*

system *(6.2)*

thermal equilibrium *(6.2)*

thermochemical expression *(6.5)*

thermodynamics *(Introduction)*

work/working *(6.2)*

QUESTIONS FOR REVIEW AND THOUGHT

OWL Interactive versions of these problems are assignable in OWL.

Blue-numbered questions have short answers at the back of this book in Appendix M and fully worked solutions in the *Student Solutions Manual.*

Review Questions

These questions test vocabulary and simple concepts.

1. Name two laws stated in this chapter and explain each in your own words.
2. For each of the following, define a system and its surroundings and give the direction of heat transfer:
   (a) Propane is burning in a Bunsen burner in the laboratory.
   (b) After you have a swim, water droplets on your skin evaporate.
   (c) Water, originally at 25 °C, is placed in the freezing compartment of a refrigerator.
   (d) Two chemicals are mixed in a flask on a laboratory bench. A reaction occurs and heat is evolved.
3. What is the value of the standard enthalpy of formation for any element under standard conditions?
4. Criticize each of these statements:
   (a) Enthalpy of formation refers to a reaction in which 1 mol of one or more reactants produces some quantity of product.
   (b) The standard enthalpy of formation of O_2 as a gas at 25 °C and a pressure of 1 atm is 15.0 kJ/mol.

5. Explain how a coffee cup calorimeter may be used to measure the enthalpy change of (a) a change in state and (b) a chemical reaction.
6. What is required for heat transfer of energy from one sample of matter to another to occur?
7. Name two exothermic processes and two endothermic processes that you encountered recently and that were not associated with your chemistry course.
8. Explain what is meant by (a) energy density of a fuel and (b) caloric value of a food. Why is each of these terms important?
9. Explain in your own words why it is useful in thermodynamics to distinguish a system from its surroundings.
10. In a steelmaking plant, molten metal is poured from ladles into furnaces. Considering the molten iron to be the system, what is the sign of q in the process shown in the photograph?

© James Hardy/Photo Alto/Getty Images

Topical Questions

These questions are keyed to the major topics in this chapter. Usually a question that is answered at the back of this book is paired with a similar one that is not.

The Nature of Energy (Section 6.1)

11. (a) A 2-inch piece of two-layer chocolate cake with frosting provides 1670 kJ of energy. What is this in Cal?
    (b) If you were on a diet that calls for eating no more than 1200 Cal per day, how many joules would you consume per day?
12. Sulfur dioxide, SO_2, is found in wines and in polluted air. If a 32.1-g sample of sulfur is burned in the air to get 64.1 g SO_2, 297 kJ of energy is released. Express this energy in (a) joules, (b) calories, and (c) kilocalories.
13. Melting lead requires 5.91 cal/g. How many joules are required to melt 1.00 lb (454 g) lead?
14. On a sunny day, solar energy reaches Earth at a rate of 4.0 J min^{-1} cm^{-2}. Suppose a house has a square, flat roof of dimensions 12 m by 12 m. How much solar energy reaches this roof in 1.0 h? (*Note:* This is why roofs painted with light-reflecting paint keep buildings cooler than black, unpainted roofs. The painted roofs reflect most of this energy rather than absorb it.)
15. When an electrical appliance whose power usage is X watts is run for Y seconds, it uses $X \times Y$ J of energy. The energy unit used by electrical utilities in their monthly bills is the ***kilowatt hour*** (kWh; 1 kilowatt used for 1 hour). How many joules are there in a kilowatt hour? If electricity costs \$.09 per kilowatt hour, how much does it cost per megajoule?
16. A 100-W light bulb is left on for 14 h. How many joules of energy are used? With electricity at \$0.09 per kWh, how much does it cost to leave the light on for 14 h?
17. One food product has an energy content of 170 kcal per serving, and another has 280 kJ per serving. Which food provides the greater energy per serving?
18. The food Calorie is actually a kilocalorie, that is, 1000 cal. For a 70-kg male between the ages of 18 and 30 the basal metabolic rate is 1750 Calories per day. Express this in kJ/d.

Conservation of Energy (Section 6.2)

19. Describe how energy is changed from one form to another in these processes:
    (a) At a July 4th celebration, a match is lit and ignites the fuse of a rocket firecracker, which fires off and explodes at an altitude of 1000 ft.
    (b) A gallon of gasoline is pumped from an underground storage tank into the fuel tank of your car, and you use it up by driving 25 mi.
20. Analyze transfer of energy from one form to another in each situation below.
    (a) In a space shuttle, hydrogen and oxygen combine to form water, boosting the shuttle into orbit above Earth.
    (b) You eat a package of Fritos, go to class and listen to a lecture, walk back to your dorm, and climb the stairs to the fourth floor.
21. Suppose that you are studying the growth of a plant, and you want to apply thermodynamic ideas.
    (a) Make an appropriate choice of system and surroundings and describe it unambiguously.
    (b) Explain why you chose the system and surroundings you did.
    (c) Identify transfers of energy and material into and out of the system that would be important for you to monitor in your study.
22. Suppose that you are studying an ecosystem and want to apply thermodynamic ideas.
    (a) Make an appropriate choice of system and surroundings and describe it unambiguously.
    (b) Explain why you chose the system and surroundings you did.
    (c) Identify transfers of energy and material into and out of the system that would be important for you to monitor in your study.
23. Solid ammonium chloride is added to water in a beaker and dissolves. The beaker becomes cold to the touch.
    (a) Make an appropriate choice of system and surroundings and describe it unambiguously.
    (b) Explain why you chose the system and surroundings you did.

(c) Identify transfers of energy and material into and out of the system that would be important for you to monitor in your study.

(d) Is the process of dissolving $NH_4Cl(s)$ in water exothermic or endothermic?

24. A bar of Monel (an alloy of nickel, copper, iron, and manganese) is heated until it melts, poured into a mold, and solidifies.

(a) Make an appropriate choice of system and surroundings and describe it unambiguously.

(b) Explain why you chose the system and surroundings you did.

(c) Identify transfers of energy and material into and out of the system that would be important for you to monitor in your study.

25. If a system does 75.4 J of work on its surroundings and simultaneously there is 25.7 cal of heat transfer from the surroundings to the system, what is ΔE for the system?

26. A 20.0-g sample of water cools from 30 °C to 20.0 °C, which transfers 840 J to the surroundings. No work is done on the water. What is ΔE_{water}?

27. Make a diagram like the one in Figure 6.9 for a system in which 127.6 kJ of work is done on the surroundings and there is 843.2 kJ of heat transfer into the system. Use the diagram to help you determine ΔE_{system}.

28. Make a diagram like the one in Figure 6.9 for a system in which 876.3 J of work is done on the surroundings and there is 37.4 J of heat transfer into the system. Use the diagram to help you determine ΔE_{system}.

Heat Capacity (Section 6.3)

29. Which requires greater transfer of energy: (a) cooling 10.0 g water from 50 °C to 20 °C or (b) cooling 20.0 g Cu from 37 °C to 25 °C?

30. Which requires more energy: (a) warming 15.0 g water from 25 °C to 37 °C or (b) warming 60.0 g aluminum from 25 °C to 37 °C?

31. You hold a gram of copper in one hand and a gram of aluminum in the other. Each metal was originally at 0 °C. (Both metals are in the shape of a little ball that fits into your hand.) If energy is transferred to both at the same rate, which will warm to your body temperature first?

32. Ethylene glycol, $HOCH_2CH_2OH$, is often used as a coolant in cars. Which requires greater transfer of thermal energy to warm from 25.0 °C to 100.0 °C, pure water or an equal mass of pure ethylene glycol?

33. How much thermal energy is required to heat all of the water in a swimming pool by 1.0 °C if the dimensions are 4.0 ft deep by 20. ft wide by 75 ft long? Report your result in megajoules.

34. How much thermal energy is required to heat all the aluminum in a roll of aluminum foil (500. g) from room temperature (25 °C) to the temperature of a hot oven (250 °C)? Report your result in kilojoules.

35. The specific heat capacity of benzene, C_6H_6, is 1.74 J g^{-1} K^{-1}. What is its molar heat capacity (in J mol^{-1} K^{-1})?

36. The specific heat capacity of carbon tetrachloride, CCl_4, is 0.861 J g^{-1} K^{-1}. What is its molar heat capacity (in J mol^{-1} K^{-1})?

37. A 237-g piece of molybdenum, initially at 100.0 °C, is dropped into 244 g water at 10.0 °C. When the system comes to thermal equilibrium, the temperature is 15.3 °C. What is the specific heat capacity of molybdenum?

38. A sample of glass beads weighs 34.5 g. The beads are heated to 100.0 °C in a boiling water bath and poured into a beaker containing 100.0 g H_2O at 25.0 °C. When thermal equilibrium is reached, the temperature of the glass and the water is 29.9 °C. What is the specific heat capacity of the glass?

39. If the cooling system in an automobile has a capacity of 5.00 quarts of liquid, compare the quantity of thermal energy transferred to the liquid in the system when its temperature is raised from 25.0 °C to 100.0 °C for water and for ethylene glycol. The densities of water and ethylene glycol are 1.00 g/cm^3 and 1.113 g/cm^3, respectively. 1 quart = 0.946 L. Report your results in joules.

40. One way to cool a cup of coffee is to plunge an ice-cold piece of aluminum into it. Suppose a 20.0-g piece of aluminum is stored in the refrigerator at 32 °F (0.0 °C) and then put into a cup of coffee. The coffee's temperature drops from 90.0 °C to 75.0 °C. How much energy (in kilojoules) did the coffee transfer to the piece of aluminum?

41. A piece of iron (400. g) is heated in a flame and then plunged into a beaker containing 1.00 kg water. The original temperature of the water was 20.0 °C, but it is 32.8 °C after the iron bar is put in and thermal equilibrium is reached. What was the original temperature of the hot iron bar?

42. A 192-g piece of copper was heated to 100.0 °C in a boiling water bath and then put into a beaker containing 750. mL water (density = 1.00 g/cm^3) at 4.0 °C. What is the final temperature of the copper and water after they come to thermal equilibrium?

43. A 12.3-g sample of iron requires heat transfer of 41.0 J to raise its temperature from 17.3 °C to 24.7 °C.

(a) Calculate the specific heat capacity of iron.

(b) Calculate the molar heat capacity of iron.

44. A diamond weighing 310. mg requires 2.38 J to raise its temperature from 23.4 °C to 38.7 °C.

(a) Calculate the specific heat capacity of diamond.

(b) Calculate the molar heat capacity of diamond.

(c) Is the specific heat capacity of diamond the same as the specific heat capacity of carbon in the form of graphite? Give one reason why you might expect them to be the same and one reason why you might expect them to be different.

45. An unknown metal requires 34.7 J to heat a 23.4-g sample of it from 17.3°C to 28.9 °C. Which of the metals in Table 6.1 is most likely to be the unknown?

46. An unknown metal requires 336.9 J to heat a 46.3-g sample of it from 24.3 °C to 43.2 °C. Which of the metals in Table 6.1 is most likely to be the unknown?

47. A 200.-g sample of Al is heated in a flame and then immersed in 500. mL water in an insulated container. The initial temperature of the water was 22.0 °C. After the Al and water reached thermal equilibrium, the temperature of both was 33.6 °C. What was the temperature of the Al just before it was plunged into the water? (Assume that the density of water is 0.98 g/mL and that there is no heat transfer other than between the water and aluminum.)

48. A 200.-g sample of copper is heated to 500. °C in a flame and then plunged into 1000. g water in an insulated con-

tainer. If the water's initial temperature was 23.4 °C, what is the temperature of the water and Cu when thermal equilibrium has been reached? (Assume that the only energy transfer is between water and copper.)

49. A chemical reaction occurs, and 20.7 J is transferred from the chemical system to its surroundings. (Assume that no work is done.)
   (a) What is the algebraic sign of $\Delta T_{surroundings}$?
   (b) What is the algebraic sign of ΔE_{system}?

50. A physical process called a phase transition occurs in a sample of an alloy, and 437 kJ transfers from the surroundings to the alloy. (Assume that no work is done.)
   (a) What is the algebraic sign of ΔT_{alloy}?
   (b) What is the algebraic sign of ΔE_{alloy}?

Energy and Enthalpy (Section 6.4)

51. The energy required to melt 1.00 g ice at 0 °C is 333 J. If one ice cube has a mass of 62.0 g and a tray contains 16 ice cubes, what quantity of energy is required to melt a tray of ice cubes to form liquid water at 0 °C?

52. Calculate the quantity of heating required to convert the water in four ice cubes (60.1 g each) from $H_2O(s)$ at 0 °C to $H_2O(g)$ at 100. °C. The enthalpy of fusion of ice at 0 °C is 333 J/g and the enthalpy of vaporization of liquid water at 100. °C is 2260 J/g.

53. The heat of fusion of mercury is 2.82 cal/g. Calculate the quantity of energy transferred when 4.37 mol Hg freezes at a temperature of −39 °C.

54. Chloromethane, CH_3Cl, arises from microbial fermentation and is found throughout the environment. It is also produced industrially and is used in the manufacture of various chemicals and has been used as a topical anesthetic. How much energy is required to convert 92.5 g liquid to a vapor at its boiling point, −24.09 °C? (The heat of vaporization of CH_3Cl is 21.40 kJ/mol.)

55. The freezing point of mercury is −38.8 °C. What quantity of energy, in joules, is released to the surroundings if 1.00 mL mercury is cooled from 23.0 °C to −38.8 °C and then frozen to a solid? (The density of liquid mercury is 13.6 g/cm^3. Its specific heat capacity is 0.140 J g^{-1} K^{-1} and its heat of fusion is 11.4 J g^{-1}.)

56. On a cold day in winter, ice can sublime (go directly from solid to gas without melting). The heat of sublimation is approximately equal to the sum of the heat of fusion and the heat of vaporization (see Question 52). How much thermal energy in joules does it take to sublime 0.1 g of frost on a windowpane?

57. Draw a cooling graph for steam-to-water-to-ice.

58. Draw a heating graph for converting Dry Ice to carbon dioxide gas.

59. Based on the heating graph shown in the figure for a substance, *X*, at constant pressure,
   (a) Which has the largest specific heat capacity, *X*(s), *X*(ℓ), or *X*(g)?
   (b) Which is smaller, the heat of fusion or the heat of vaporization?
   (c) What is the algebraic sign of the enthalpy of vaporization at the boiling point?

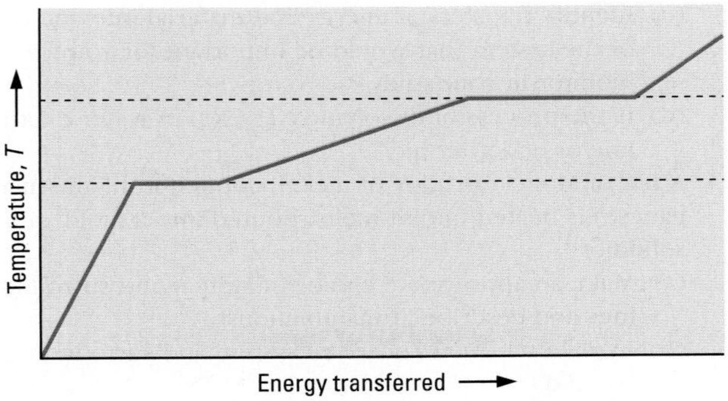

60. Based on the cooling graph shown in the figure for a substance, *Y*, at constant pressure,
   (a) Which has the smallest specific heat capacity, *Y*(s), *Y*(ℓ), or *Y*(g)?
   (b) Which is larger, the heat of fusion or the heat of vaporization?
   (c) What is the algebraic sign of the enthalpy of fusion at the melting point?

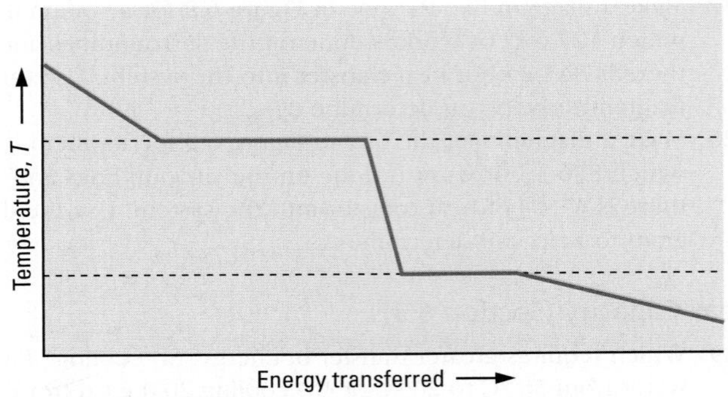

61. What is the sign of *w* for these processes if they occur at constant pressure? Consider only PV work from gases.
   (a) $Fe_2S_3(s) + 6\ HNO_3(aq) \longrightarrow 2\ Fe(NO_3)_3(aq) + 3\ H_2S(g)$
   (b) $C_3H_8(g) + 5\ O_2(g) \longrightarrow 3\ CO_2(g) + 4\ H_2O(\ell)$

62. Assume that these reactions occur under constant atmospheric pressure. What is the sign of *w* for each?
   (a) $CaO(s) + 3\ C(s) \longrightarrow CaC_2(s) + CO(g)$
   (b) $2\ C_6H_6(\ell) + 15\ O_2(g) \longrightarrow 12\ CO_2(g) + 6\ H_2O(\ell)$

63. Calculate *q* for a process in which $w = 0$ J and $\Delta E = 150$ J.

64. Calculate ΔE for a chemical reaction where there is no volume change and $q = -43.2$ kJ.

65. A chemical reaction occurs and 64.7 cal is transferred to the system. How many joules is this? Is the reaction endothermic or exothermic?

66. When 2 mol Fe reacts with air to form rust, 824.2 kJ is transferred to the surroundings. How many kilocalories is this? Is the reaction exothermic or endothermic?

Thermochemical Expressions (Section 6.5)

67. Energy is stored in the body in adenosine triphosphate, ATP, which is formed by the reaction between adenosine diphosphate, ADP, and dihydrogen phosphate ions.

$$ADP^{3-}(aq) + H_2PO_4^{2-}(aq) \longrightarrow ATP^{4-}(aq) + H_2O\ (\ell) \qquad \Delta H° = 38\ kJ$$

Is the reaction endothermic or exothermic?

68. Calcium carbide, CaC_2, is manufactured by reducing lime with carbon at high temperature. (The carbide is used in turn to make acetylene, an industrially important organic chemical.)

$$CaO(s) + 3\ C(s) \longrightarrow CaC_2(s) + CO(g) \qquad \Delta H° = 464.8\ kJ$$

Is the reaction endothermic or exothermic?

69. When calcium oxide, CaO(s), dissolves in water the water becomes hot. Write a chemical equation for this process and indicate whether it is exothermic or endothermic.

70. When table salt is dissolved in water, the temperature drops slightly. Write a chemical equation for this process, and indicate if it is exothermic or endothermic.

71. How much energy must be transferred to vaporize 125 g benzene, C_6H_6, at its boiling point, 80.1 °C? (The enthalpy of vaporization of benzene is 30.8 kJ/mol.)

72. The enthalpy of fusion (melting) of water is 6.0 kJ/mol. Calculate the quantity of energy that must be transferred to melt 25.0 g H_2O at 0 °C.

73. Write a word statement for the thermochemical expression

$$H_2O(s) \longrightarrow H_2O(\ell) \qquad \Delta H° = 6.0\ kJ$$

74. Write a word statement for the thermochemical expression

$$HI(\ell) \longrightarrow HI(s) \qquad \Delta H° = -2.87\ kJ$$

75. Given the thermochemical expression

$$H_2O(s) \longrightarrow H_2O(\ell) \qquad \Delta H° = 6.0\ kJ$$

what quantity of energy is transferred to the surroundings when
(a) 34.2 mol liquid water freezes?
(b) 100.0 g liquid water freezes?

76. Given the thermochemical expression

$$CaO(s) + 3\ C(s) \longrightarrow CaC_2(s) + CO(g) \qquad \Delta H° = 464.8\ kJ$$

what quantity of energy is transferred when
(a) 34.8 mol CO(g) is formed by this reaction?
(b) a metric ton (1000 kg) of $CaC_2(s)$ is manufactured?
(c) 0.432 mol carbon reacts with CaO(s)?

Enthalpy Changes for Chemical Reactions (Section 6.6)

77. Given the thermochemical expression for combustion of isooctane (a component of gasoline),

$$2\ C_8H_{18}(\ell) + 25\ O_2(g) \longrightarrow 16\ CO_2(g) + 18\ H_2O(\ell) \qquad \Delta H° = -10{,}992\ kJ$$

write a thermochemical expression for
(a) production of 4.00 mol $CO_2(g)$.
(b) combustion of 100. mol isooctane.
(c) combination of 1.00 mol isooctane with a stoichiometric quantity of oxygen from air.

78. Given the thermochemical expression for combustion of benzene,

$$2\ C_6H_6(\ell) + 15\ O_2(g) \longrightarrow 12\ CO_2(g) + 6\ H_2O(\ell) \qquad \Delta H° = -6534.8\ kJ$$

write a thermochemical expression for
(a) combustion of 0.50 mol benzene.
(b) consumption of 5 mol $O_2(g)$.
(c) production of 144 mol $CO_2(g)$.

79. Write all the thermostoichiometric factors that can be derived from the thermochemical expression

$$CaO(s) + 3\ C(s) \longrightarrow CaC_2(s) + CO(g) \qquad \Delta H° = 464.8\ kJ$$

80. Write all the thermostoichiometric factors that can be derived from the thermochemical expression

$$2\ CH_3OH(\ell) + 3\ O_2(g) \longrightarrow 2\ CO_2(g) + 4\ H_2O(\ell) \qquad \Delta H° = -1530\ kJ$$

81. Isooctane (2,2,4-trimethylpentane), one of the many hydrocarbons that make up gasoline, burns in air to give water and carbon dioxide.

$$2\ C_8H_{18}(\ell) + 25\ O_2(g) \longrightarrow 16\ CO_2(g) + 18\ H_2O(\ell) \qquad \Delta H° = -10{,}922\ kJ$$

What is the enthalpy change if you burn 1.00 L isooctane (density = 0.69 g/mL)?

82. When $KClO_3(s)$, potassium chlorate, is heated, it melts and decomposes to form oxygen gas. [Molten $KClO_3$ was shown reacting with a Fritos chip earlier in this chapter (p. 214).] The thermochemical expression for decomposition of potassium chlorate is

$$2\ KClO_3(s) \longrightarrow 2\ KCl(s) + 3\ O_2(g) \qquad \Delta H° = -89.4\ kJ$$

Calculate q at constant pressure for
(a) formation of 97.8 g KCl(s).
(b) production of 24.8 mol $O_2(g)$.
(c) decomposition of 35.2 g $KClO_3(s)$.

83. "Gasohol," a mixture of gasoline and ethanol, C_2H_5OH, is used as automobile fuel. The alcohol releases energy in a combustion reaction with O_2.

$$C_2H_5OH(\ell) + 3\ O_2(g) \longrightarrow 2\ CO_2(g) + 3\ H_2O(\ell)$$

If 0.115 g ethanol evolves 3.62 kJ when burned at constant pressure, what is the molar enthalpy of combustion for ethanol?

84. White phosphorus, P_4, ignites in air to produce P_4O_{10}.

$$P_4(s) + 5\ O_2(g) \longrightarrow P_4O_{10}(s)$$

When 3.56 g P_4 is burned, 85.8 kJ of thermal energy is evolved at constant pressure. What is the molar enthalpy of combustion of P_4?

85. Acetic acid, CH_3CO_2H, is made industrially by the reaction of methanol and carbon monoxide.

$$CH_3OH(\ell) + CO(g) \longrightarrow CH_3COOH(\ell) \qquad \Delta H° = -355.9\ kJ$$

If you produce 1.00 L of acetic acid (d = 1.044 g/mL) by this reaction, how much energy is transferred out of the system?

86. Rhombic sulfur, S_8, burns in air to give sulfur dioxide, SO_2. When 4.809 g S_8 is burned, 44.52 kJ is transferred to the surroundings under constant pressure conditions. Write a balanced chemical equation for formation of SO_2 from rhombic sulfur. What is the molar enthalpy of combustion of rhombic sulfur? Note: S(s, rhombic) is often used to represent $\frac{1}{8}S_8(s)$.

87. When wood is burned we may assume that the reaction is the combustion of cellulose (empirical formula, CH_2O).

$$CH_2O(s) + O_2(g) \longrightarrow CO_2(g) + H_2O(g) \qquad \Delta H° = -425\ kJ$$

How much energy is released when a 10.0-lb wood log burns completely? (Assume the wood is 100% dry and burns via the reaction above.)

88. A plant takes CO_2 and H_2O from its surroundings and makes cellulose by the reverse of the reaction in the preceding problem. The energy provided for this process comes from the sun via photosynthesis. How much energy does it take for a plant to make 100 g of cellulose?

Where Does the Energy Come From? (Section 6.7)

Use these bond enthalpy values to answer the questions below.

Bond	Bond Enthalpy (kJ/mol)	Bond	Bond Enthalpy (kJ/mol)
H—F	566	F—F	158
H—Cl	431	Cl—Cl	242
H—Br	366	Br—Br	193
H—I	299	I—I	151
H—H	436		

89. Which molecule, HF, HCl, HBr, or HI, has the strongest chemical bond?
90. Which molecule, F_2, Cl_2, Br_2, or I_2, has the weakest chemical bond?
91. For the reactions of molecular hydrogen with fluorine and with chlorine:
    (a) Calculate the enthalpy change for breaking all the bonds in the reactants.
    (b) Calculate the enthalpy change for forming all the bonds in the products.
    (c) From the results in parts (a) and (b), calculate the enthalpy change for the reaction.
    (d) Which reaction is most exothermic?
92. For the reactions of molecular hydrogen with bromine and with iodine:
    (a) Calculate the enthalpy change for breaking all the bonds in the reactants.
    (b) Calculate the enthalpy change for forming all the bonds in the products.
    (c) From the results in parts (a) and (b), calculate the enthalpy change for the reaction.
    (d) Which reaction is most exothermic?
93. A diamond can be considered a giant all-carbon supermolecule in which almost every carbon atom is bonded to four other carbons. When a diamond cutter cleaves (splits) a diamond, carbon-carbon bonds must be broken. Is the cleavage (splitting) of a diamond endothermic or exothermic? Explain.

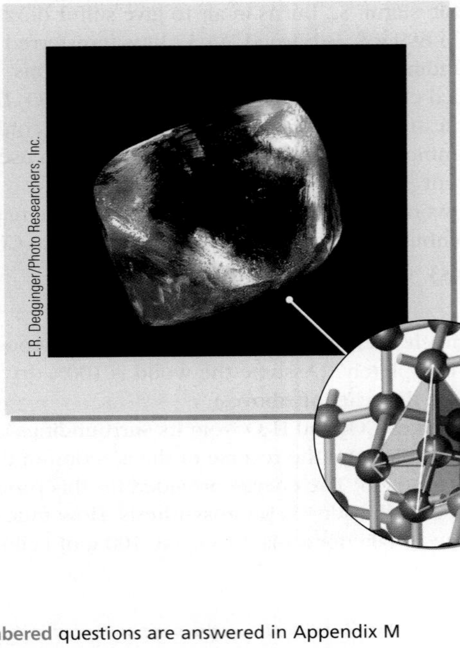

E.R. Degginger/Photo Researchers, Inc.

Measuring Enthalpy Changes: Calorimetry (Section 6.8)

94. A piece of titanium metal with a mass of 20.8 g is heated in boiling water to 99.5 °C and then dropped into a coffee cup calorimeter containing 75.0 g water at 21.7 °C. When thermal equilibrium is reached, the final temperature is 24.3 °C. Calculate the specific heat capacity of titanium.
95. Suppose you add a small ice cube to room-temperature water in a coffee cup calorimeter. What is the final temperature when all of the ice is melted? Assume that you have 200. mL water at 25 °C and that the ice cube weighs 15.0 g and is at 0 °C before being added to the water.
96. When 0.100 g CaO(s) is added to 125 g H_2O at 23.6 °C in a coffee cup calorimeter, the reaction below occurs. What will be the final temperature of the solution?

$$CaO(s) + H_2O(\ell) \longrightarrow Ca(OH)_2(aq) \qquad \Delta H° = -81.9 \text{ kJ}$$

97. A coffee cup calorimeter can be used to investigate the "cold pack reaction," the process that occurs when solid ammonium nitrate dissolves in water:

$$NH_4NO_3(s) \longrightarrow NH_4^+(aq) + NO_3^-(aq)$$

Suppose 25.0 g solid NH_4NO_3 at 23.0 °C is added to 250. mL H_2O at the same temperature. After all of the solid dissolves, the temperature is measured to be 15.6 °C. Calculate the enthalpy change for the cold pack reaction. (*Hint:* Calculate the energy transferred per mole of NH_4NO_3. Assume that the specific heat capacity of the solution is the same as for water.) Is the reaction endothermic or exothermic?

98. When a 13.0-g sample of NaOH(s) dissolves in 400.0 mL water in a coffee cup calorimeter, the temperature of the water changes from 22.6 °C to 30.7 °C. Assuming that the specific heat capacity of the solution is the same as for water, calculate
    (a) The heat transfer from system to surroundings.
    (b) ΔH for the reaction.

$$NaOH(s) \longrightarrow Na^+(aq) + OH^-(aq)$$

99. Suppose that you mix 200.0 mL of 0.200-M RbOH(aq) with 100. mL of 0.400-M HBr(aq) in a coffee cup calorimeter. If the temperature of each of the two solutions was 24.40 °C before mixing, and the temperature rises to 26.18 °C,
    (a) Calculate the heat transfer as a result of the reaction.
    (b) Write the thermochemical expression for the reaction.
100. A 0.692-g sample of glucose, $C_6H_{12}O_6$, is burned in a constant volume calorimeter. The temperature rises from 21.70 °C to 25.22 °C. The calorimeter contains 575 g water, and the bomb has a heat capacity of 650 J/K. What is ΔE per mole of glucose?
101. Sulfur (2.56 g) was burned in a bomb calorimeter with excess $O_2(g)$. The temperature increased from 21.25 °C to 26.72 °C. The bomb had a heat capacity of 923 J °C^{-1} and the calorimeter contained 815 g water. Calculate the heat transfer, per mole of SO_2 formed, in the course of the reaction

$$S(s) + O_2(g) \longrightarrow SO_2(g)$$

102. You can find the quantity of thermal energy transferred during combustion of carbon by carrying out the reaction in a combustion calorimeter. Suppose you burn 0.300 g C(graphite) in an excess of $O_2(g)$ to give $CO_2(g)$.

$$C(graphite) + O_2(g) \longrightarrow CO_2(g)$$

The temperature of the calorimeter, which contains 775 g water, increases from 25.00 °C to 27.38 °C. The heat capacity of the bomb is 893 J $°C^{-1}$. What quantity of thermal energy is transferred per mole of graphite?

103. Benzoic acid, $C_7H_6O_2$, occurs naturally in many berries. Suppose you burn 1.500 g of the compound in a combustion calorimeter and find that the temperature of the calorimeter increases from 22.50 °C to 31.69 °C. The calorimeter contains 775 g water, and the bomb has a heat capacity of 893 J $°C^{-1}$. How much thermal energy is transferred per mole of benzoic acid? What is the change in internal energy, ΔE, of the system?

Hess's Law (Section 6.9)

104. The enthalpy changes of these reactions can be measured:

$$C_2H_4(g) + 3\ O_2(g) \longrightarrow 2\ CO_2(g) + 2\ H_2O(\ell) \qquad \Delta H° = -1411.1\ kJ$$

$$C_2H_5OH(\ell) + 3\ O_2(g) \longrightarrow 2\ CO_2(g) + 3\ H_2O(\ell) \qquad \Delta H° = -1367.5\ kJ$$

Use these values and Hess's law to determine the enthalpy change for the reaction

$$C_2H_4(g) + H_2O(\ell) \longrightarrow C_2H_5OH(\ell)$$

105. Calculate the standard enthalpy change, $\Delta H°$, for the formation of 1 mol strontium carbonate (the material that gives the red color in fireworks) from its elements.

$$Sr(s) + C(graphite) + \tfrac{3}{2}\ O_2(g) \longrightarrow SrCO_3(s)$$

The information available is

$$Sr(s) + \tfrac{1}{2}\ O_2(g) \longrightarrow SrO(s) \qquad \Delta H° = -592\ kJ$$

$$SrO(s) + CO_2(g) \longrightarrow SrCO_3(s) \qquad \Delta H° = -234\ kJ$$

$$C(graphite) + O_2(g) \longrightarrow CO_2(g) \qquad \Delta H° = -394\ kJ$$

106. You wish to know the enthalpy change for the formation of liquid PCl_3 from the elements.

$$P_4(s) + 6\ Cl_2(g) \longrightarrow 4\ PCl_3(\ell) \qquad \Delta H° = ?$$

The enthalpy change for the formation of PCl_5 from the elements can be determined experimentally, as can the enthalpy change for the reaction of $PCl_3(\ell)$ with more chlorine to give $PCl_5(s)$:

$$P_4(s) + 10\ Cl_2(g) \longrightarrow 4\ PCl_5(s) \qquad \Delta H° = -1774.0\ kJ$$

$$PCl_3(\ell) + Cl_2(g) \longrightarrow PCl_5(s) \qquad \Delta H° = -123.8\ kJ$$

Use these data to calculate the enthalpy change for the formation of 1.00 mol of $PCl_3(\ell)$ from phosphorus and chlorine.

107. What is the standard enthalpy change for the reaction of lead(II) chloride with chlorine to give lead(IV) chloride?

$$PbCl_2(s) + Cl_2(g) \longrightarrow PbCl_4(\ell)$$

It is known that $PbCl_2(s)$ can be formed from the metal and $Cl_2(g)$,

$$Pb(s) + Cl_2(g) \longrightarrow PbCl_2(s) \qquad \Delta H° = -359.4\ kJ$$

and that $PbCl_4(\ell)$ can be formed directly from the elements.

$$Pb(s) + 2\ Cl_2(g) \longrightarrow PbCl_4(\ell) \qquad \Delta H° = -329.3\ kJ$$

108. Using these reactions, find the standard enthalpy change for the formation of 1 mol PbO(s) from lead metal and oxygen gas.

$$PbO(s) + C(graphite) \longrightarrow Pb(s) + CO(g) \qquad \Delta H° = 106.8\ kJ$$

$$2\ C(graphite) + O_2(g) \longrightarrow 2\ CO(g) \qquad \Delta H° = -221.0\ kJ$$

If 250. g lead reacts with oxygen to form lead(II) oxide, what quantity of thermal energy (in kJ) is absorbed or evolved?

109. Three reactions very important to the semiconductor industry are

(a) the reduction of silicon dioxide to crude silicon,

$$SiO_2(s) + 2\ C(s) \longrightarrow Si(s) + 2\ CO(g) \qquad \Delta H° = 689.9\ kJ$$

(b) the formation of silicon tetrachloride from crude silicon,

$$Si(s) + 2\ Cl_2(g) \longrightarrow SiCl_4(g) \qquad \Delta H° = -657.0\ kJ$$

(c) and the reduction of silicon tetrachloride to pure silicon with magnesium,

$$SiCl_4(g) + 2\ Mg(s) \longrightarrow 2\ MgCl_2(s) + Si(s) \qquad \Delta H° = -625.6\ kJ$$

What is the overall enthalpy change when 1.00 mol sand, SiO_2, changes into very pure silicon by this series of reactions?

Standard Molar Enthalpies of Formation (Section 6.10)

110. Write thermochemical expressions for the decomposition of one mole of the each compound into the elements in their stable states at 25 °C and 1 bar.

(a) ethanol, $C_2H_5OH(\ell)$
(b) sodium fluoride(s)
(c) magnesium sulfate(s)
(d) ammonium nitrate(s)

111. Write a thermochemical expression for the formation of each compound in the preceding question from its elements in their stable states at 25 °C and 1 bar.

112. For each compound below, write a balanced thermochemical expression depicting the formation of 1 mol of the compound. Standard molar enthalpies of formation are found in Appendix J.

(a) $Al_2O_3(s)$
(b) $TiCl_4(\ell)$
(c) $NH_4NO_3(s)$

113. The standard molar enthalpy of formation of glucose, $C_6H_{12}O_6(s)$, is −1274.4 kJ/mol.

(a) Is the formation of glucose from its elements exothermic or endothermic?
(b) Write a balanced equation depicting the formation of glucose from its elements and for which the enthalpy of reaction is −1274.4 kJ.

114. Given

$$2\ Al_2O_3(s) \longrightarrow 4\ Al(s) + 3\ O_2(g) \qquad \Delta H° = 3351.4\ kJ$$

(a) What is the standard molar enthalpy of formation of aluminum oxide?
(b) What is $\Delta H°$ for the formation of 12.50 g aluminum oxide?

115. Calorimetric measurements show that the reaction of magnesium with chlorine releases 26.4 kJ per gram of magnesium reacted.

(a) Write a balanced chemical equation for the reaction.
(b) Calculate the standard molar enthalpy of formation of magnesium chloride.

116. We burn 3.47 g lithium in excess oxygen at constant atmospheric pressure to form Li_2O. Then we bring the reaction mixture back to 25 °C. In this process 146 kJ of heat is given off. What is the standard molar enthalpy of formation of Li_2O?

117. When 43.2 g Mg is reacted with sufficient sulfur, 615 kJ is transferred to the surroundings. What is the standard molar enthalpy of formation for MgS?

118. An important reaction in the production of sulfuric acid is

$$SO_2(g) + \tfrac{1}{2}O_2(g) \longrightarrow SO_3(g)$$

It is also a key reaction in the formation of acid rain, beginning with the air pollutant SO_2. Using the data in Table 6.2, calculate the enthalpy change for the reaction.

119. The Romans used CaO as mortar in stone structures. The CaO was mixed with water to give $Ca(OH)_2$, which slowly reacted with CO_2 in the air to give limestone.

$$Ca(OH)_2(s) + CO_2(g) \longrightarrow CaCO_3(s) + H_2O(g)$$

Calculate the enthalpy change for this reaction.

120. Use data in Appendix J to find the enthalpy of reaction for
(a) $CaCO_3(s) \longrightarrow CaO(s) + CO_2(g)$
(b) $2\ HI(g) + F_2(g) \longrightarrow 2\ HF(g) + I_2(s)$
(c) $SF_6(g) + 3\ H_2O(\ell) \longrightarrow 6\ HF(g) + SO_3(g)$

121. In photosynthesis, the sun's energy brings about the combination of CO_2 and H_2O to form O_2 and a carbon-containing compound such as a sugar. In its simplest form, the reaction could be written

$$6\ CO_2(g) + 6\ H_2O(\ell) \longrightarrow 6\ O_2(g) + C_6H_{12}O_6(s)$$

Using the enthalpies of formation in Table 6.2, (a) calculate the enthalpy of reaction and (b) decide whether the reaction is exothermic or endothermic.

122. Use standard molar enthalpies of formation in Appendix J to calculate enthalpy change when
(a) 0.054 g sulfur burns, forming $SO_2(g)$
(b) 0.20 mol HgO(s) decomposes to $Hg(\ell)$ and $O_2(g)$
(c) 2.40 g $NH_3(g)$ is formed from $N_2(g)$ and excess $H_2(g)$
(d) 1.05×10^{-2} mol carbon is oxidized to $CO_2(g)$

123. The first step in the production of nitric acid from ammonia involves the oxidation of NH_3.

$$4\ NH_3(g) + 5\ O_2(g) \longrightarrow 4\ NO(g) + 6\ H_2O(g)$$

Use the information in Table 6.2 or Appendix J to calculate the enthalpy change for this reaction. Is the reaction exothermic or endothermic?

124. The enthalpy change for the oxidation of styrene, C_8H_8, is measured by calorimetry.

$$C_8H_8(\ell) + 10\ O_2(g) \longrightarrow 8\ CO_2(g) + 4\ H_2O(\ell) \qquad \Delta H^\circ = -4395.0\ kJ$$

Use this value, along with the standard heats of formation of $CO_2(g)$ and $H_2O(\ell)$, to calculate the standard molar enthalpy of formation of styrene, in kJ/mol.

125. Oxygen is not normally found in positive oxidation states, but when it is combined with fluorine in the compound oxygen difluoride, oxygen is in a +2 oxidation state. Oxygen difluoride is a strong oxidizing agent that oxidizes water to produce O_2 and HF.

$$OF_2(g) + H_2O(g) \longrightarrow 2\ HF(g) + O_2(g)$$

The standard molar enthalpy of formation of oxygen difluoride is 18 kJ/mol. Using this information and data from Appendix J, calculate the standard molar enthalpy change for the reaction of oxygen difluoride with water.

126. Iron can react with oxygen to give iron(III) oxide. If 5.58 g Fe is heated in pure O_2 to give $Fe_2O_3(s)$, how much thermal energy is transferred out of this system (at constant pressure)?

127. The formation of aluminum oxide from its elements is highly exothermic. If 2.70 g Al metal is burned in pure O_2 to give Al_2O_3, how much thermal energy is evolved in the process (at constant pressure)?

Chemical Fuels (Section 6.11)

128. You want to heat the air in your house with natural gas, CH_4. Assume your house has 275 m^2 (about 2800 ft^2) of floor area and that the ceilings are 2.50 m (about 8 ft) from the floors. The air in the house has a molar heat capacity of 29.1 J mol^{-1} K^{-1}. (The number of moles of air in the house can be found by assuming that the average molar mass of air is 28.9 g/mol and that the density of air at these temperatures is 1.22 g/L.) What mass of methane do you have to burn to heat the air from 15.0 °C to 22.0 °C?

129. If you want to convert 56.0 g ice (at 0 °C) to water at 75.0 °C, how many grams of propane, C_3H_8, would you have to burn to supply the energy to melt the ice and then warm it to the final temperature (at 1 bar)?

130. Companies around the world are constantly searching for compounds that can be used as substitutes for gasoline in automobiles. One possibility is methanol, CH_3OH, a compound that can be made relatively inexpensively from coal. The alcohol has a smaller energy per gram than gasoline, but with its higher octane rating it burns more efficiently than gasoline in internal combustion engines. (It also contributes smaller quantities of some air pollutants.) Compare the quantity of thermal energy produced per gram of CH_3OH and C_8H_{18} (octane), the latter being representative of the compounds in gasoline. The ΔH_f° for octane is −250.1 kJ/mol.

131. Hydrazine and 1,1-dimethylhydrazine both react with O_2 and can be used as rocket fuels.

$$\underset{\text{hydrazine}}{N_2H_4(\ell)} + O_2(g) \longrightarrow N_2(g) + 2\ H_2O(g)$$

$$\underset{\text{1,1-dimethylhydrazine}}{N_2H_2(CH_3)_2(\ell)} + 4\ O_2(g) \longrightarrow 2\ CO_2(g) + 4\ H_2O(g) + N_2(g)$$

The molar enthalpy of formation of liquid hydrazine is 50.6 kJ/mol, and that of liquid 1,1-dimethylhydrazine is 49.2 kJ/mol. By doing appropriate calculations, decide whether the reaction of hydrazine or 1,1-dimethylhydrazine with oxygen gives more heat per gram (at constant pressure). (Other enthalpy of formation data can be obtained from Table 6.2.)

Foods: Fuels for Our Bodies (Section 6.12)

132. M&M candies consist of 70.% carbohydrate, 21% fat, and 4.6% protein as well as other ingredients that do not have caloric value. What quantity of energy transfer would occur if 34.5 g M&Ms were burned in a bomb calorimeter?

133. A particular kind of cracker contains 71.4% carbohydrate, 10.7% fat, and 14.3% protein. Calculate the caloric value in Cal/g and in kJ/g.

134. Suppose you eat a quarter-pound hamburger (no bread, cheese, or other items—just meat) and then take a walk. For how long will you need to walk to use up the caloric value of the hamburger?

135. If you eat a quarter-pound of cheese pizza and then go out and play soccer vigorously, how long will it take before you have used up the calories in the pizza you ate?

General Questions

These questions are not explicitly keyed to chapter topics; many require integration of several concepts.

136. The specific heat capacity of copper is 0.385 $J\ g^{-1}\ °C^{-1}$, whereas it is 0.128 $J\ g^{-1}\ °C^{-1}$ for gold. Assume you place 100. g of each metal, originally at 25 °C, in a boiling water bath at 100 °C. If energy is transferred to each metal at the same rate, which piece of metal will reach 100 °C first?

137. Calculate the molar heat capacity, in $J\ mol^{-1}\ °C^{-1}$, for the four metals in Table 6.1. What observation can you make about these values? Are they very different or very similar? Using this information, can you estimate the heat capacity in the units $J\ g^{-1}\ °C^{-1}$ for silver? (The correct value for silver is 0.23 $J\ g^{-1}\ °C^{-1}$.)

138. What quantity of energy, in joules, is required to raise the temperature of 454 g tin from room temperature, 25.0 °C, to its melting point, 231.9 °C, and then melt the tin at that temperature? (The specific heat capacity of tin is 0.227 J/g · K, and the heat of fusion of this metal is 59.2 J/g.)

139. A 25.0-mL sample of benzene at 19.9 °C was cooled to its melting point, 5.5 °C, and then frozen. How much energy as heat was given off in this process? (The density of benzene is 0.80 g/mL; its specific heat capacity is 1.74 J/g · K, and its heat of fusion is 127 J/g.)

140. You add 100.0 g water at 60.0 °C to 100.0 g ice at 0.00 °C. Some of the ice melts and cools the water to 0.00 °C. When the ice and water mixture has come to a uniform temperature of 0 °C, what mass of ice has melted?

141. Suppose you have 40.0 g ice at −25.0 °C and you add to it 60.0 g liquid water at 10.0 °C. What will be the composition and state (s or ℓ) that results? What will be the temperature? (Assume that there is no heat transfer of energy except that between the water and the ice.)

142. The combustion of diborane, B_2H_6, proceeds according to the equation

$$B_2H_6(g) + 3\ O_2(g) \longrightarrow B_2O_3(s) + 3\ H_2O(\ell)$$

and 2166 kJ is liberated per mole of $B_2H_6(g)$ (at constant pressure). Calculate the molar enthalpy of formation of $B_2H_6(g)$ using this information, the data in Table 6.2, and the fact that the standard molar enthalpy of formation for $B_2O_3(s)$ is −1273 kJ/mol.

143. In principle, copper could be used to generate valuable hydrogen gas from water.

$$Cu(s) + H_2O(g) \longrightarrow CuO(s) + H_2(g)$$

(a) Is the reaction exothermic or endothermic?
(b) If 2.00 g copper metal reacts with excess water vapor at constant pressure, how much thermal energy transfer is involved (either into or out of the system) in the reaction?

144. From these enthalpies of reaction,

$$CaCO_3(s) \longrightarrow CaO(s) + CO_2(g) \qquad \Delta H° = 178.3\ kJ$$
$$CaO(s) + H_2O(\ell) \longrightarrow Ca(OH)_2(s) \qquad \Delta H° = -65.2\ kJ$$
$$Ca(OH)_2(s) \longrightarrow Ca^{2+}(aq) + 2\ OH^-(aq) \qquad \Delta H° = -16.7\ kJ$$

calculate $\Delta H°$ for

$$Ca^{2+}(aq) + 2\ OH^-(aq) + CO_2(g) \longrightarrow CaCO_3(s) + H_2O(\ell)$$

145. The combination of coke and steam produces a mixture called coal gas, which can be used as a fuel or as a starting material for other reactions. If we assume coke can be represented by graphite, the equation for the production of coal gas is

$$2\ C(s) + 2\ H_2O(g) \longrightarrow CH_4(g) + CO_2(g)$$

Determine the standard enthalpy change for this reaction from the following standard enthalpies of reaction:

$$C(s) + H_2O(g) \longrightarrow CO(g) + H_2(g) \qquad \Delta H° = 131.3\ kJ$$
$$CO(g) + H_2O(g) \longrightarrow CO_2(g) + H_2(g) \qquad \Delta H° = 41.2\ kJ$$
$$CH_4(g) + H_2O(g) \longrightarrow 3\ H_2(g) + CO(g) \qquad \Delta H° = 206.1\ kJ$$

146. The Romans used calcium oxide, CaO, to produce a strong mortar to build stone structures. The CaO was mixed with water to give $Ca(OH)_2$, which reacted slowly with CO_2 in the air to give $CaCO_3$.

$$Ca(OH)_2(s) + CO_2(g) \longrightarrow CaCO_3(s) + H_2O(g)$$

(a) Calculate the standard enthalpy change for this reaction.
(b) How much energy as heat is evolved or absorbed if 1.00 kg $Ca(OH)_2$ reacts with a stoichiometric amount of CO_2?

147. Given the following information and the data in Table 6.2, calculate the molar enthalpy of formation for liquid hydrazine, N_2H_4.

$$N_2H_4(\ell) + O_2(g) \longrightarrow N_2(g) + 2\ H_2O(g) \qquad \Delta H° = -534\ kJ$$

148. The standard molar enthalpy of combustion of methanol, $CH_3OH(\ell)$, to form carbon dioxide and water vapor is −676.485 kJ/mol. Use data from Table 6.2 to calculate the standard molar enthalpy of formation of methanol.

149. The standard molar enthalpy of combustion of formic acid, $CH_2O_2(\ell)$, to form carbon dioxide and water vapor is −210.607 kJ/mol. Use data from Table 6.2 to calculate the standard molar enthalpy of formation of formic acid.

150. How fast (in meters per second) must an iron ball with a mass of 56.6 g be traveling to have a kinetic energy of 15.75 J? The density of iron is 7.86 g/cm^3.

151. In 1947, Texas City, Texas, was devastated by the explosion of a shipload of ammonium nitrate, a compound intended to be used as a fertilizer. When heated, ammonium nitrate can decompose exothermically to N_2O and water.

$$NH_4NO_3(s) \longrightarrow N_2O(g) + 2\ H_2O(g)$$

If the heat from this exothermic reaction is contained, higher temperatures are generated, at which point ammonium nitrate can decompose explosively to N_2, H_2O, and O_2.

$$2\ NH_4NO_3(s) \longrightarrow 2\ N_2(g) + 4\ H_2O(g) + O_2(g)$$

If oxidizable materials are present, fires can break out, as was the case at Texas City. Using the information in Appendix J, answer the following questions.

(a) How much thermal energy is evolved (at constant pressure and under standard conditions) by the first reaction?

(b) If 8.00 kg ammonium nitrate explodes (the second reaction), how much thermal energy is evolved (at constant pressure and under standard conditions)?

152. Uranium-235 is used as a fuel in nuclear power plants. Since natural uranium contains only a small amount of this isotope, the uranium must be enriched in uranium-235 before it can be used. To do this, uranium(IV) oxide is first converted to a gaseous compound, UF_6, and the isotopes are separated by a gaseous diffusion technique. Some key reactions are

$$UO_2(s) + 4\ HF(g) \longrightarrow UF_4(s) + 2\ H_2O(g)$$

$$UF_4(s) + F_2(g) \longrightarrow UF_6(g)$$

How much thermal energy transfer (at constant pressure) would be involved in producing 225 ton $UF_6(g)$ from $UO_2(s)$ (1 ton $= 9.08 \times 10^5$ g)? Some necessary standard enthalpies of formation are

$$\Delta H_f^\circ\{UO_2(s)\} = -1085 \text{ kJ/mol}$$

$$\Delta H_f^\circ\{UF_4(s)\} = -1914 \text{ kJ/mol}$$

$$\Delta H_f^\circ\{UF_6(g)\} = -2147 \text{ kJ/mol}$$

153. One method of producing H_2 on a large scale is this chemical cycle.

Step 1: $SO_2(g) + 2\ H_2O(g) + Br_2(g) \longrightarrow H_2SO_4(\ell) + 2\ HBr(g)$

Step 2: $H_2SO_4(\ell) \longrightarrow H_2O(g) + SO_2(g) + \frac{1}{2}\ O_2(g)$

Step 3: $2\ HBr(g) \longrightarrow H_2(g) + Br_2(g)$

Using the table of standard enthalpies of formation in Appendix J, calculate ΔH° for each step. What is the equation for the overall process, and what is its enthalpy change? Is the overall process exothermic or endothermic?

154. One reaction involved in the conversion of iron ore to the metal is

$$FeO(s) + CO(g) \longrightarrow Fe(s) + CO_2(g)$$

Calculate the standard enthalpy change for this reaction from these reactions of iron oxides with CO:

$$3\ Fe_2O_3(s) + CO(g) \longrightarrow 2\ Fe_3O_4(s) + CO_2(g) \qquad \Delta H^\circ = -47 \text{ kJ}$$

$$Fe_2O_3(s) + 3\ CO(g) \longrightarrow 2\ Fe(s) + 3\ CO_2(g) \qquad \Delta H^\circ = -25 \text{ kJ}$$

$$Fe_3O_4(s) + CO(g) \longrightarrow 3\ FeO(s) + CO_2(g) \qquad \Delta H^\circ = 19 \text{ kJ}$$

Applying Concepts

These questions test conceptual learning.

155. Based on your experience, when ice melts to liquid water, is the process exothermic or endothermic? When liquid water freezes to ice at 0 °C, is this exothermic or endothermic? (Assume that the ice/water is the system in each case.)

156. You pick up a six-pack of soft drinks from the floor, but it slips from your hand and smashes onto your foot. Comment on the work and energy involved in this sequence. What forms of energy are involved and at what stages of the process?

157. Consider this graph, which presents data for a 1.0-g mass of each of four substances, A, B, C, and D. Which substance has the highest specific heat capacity?

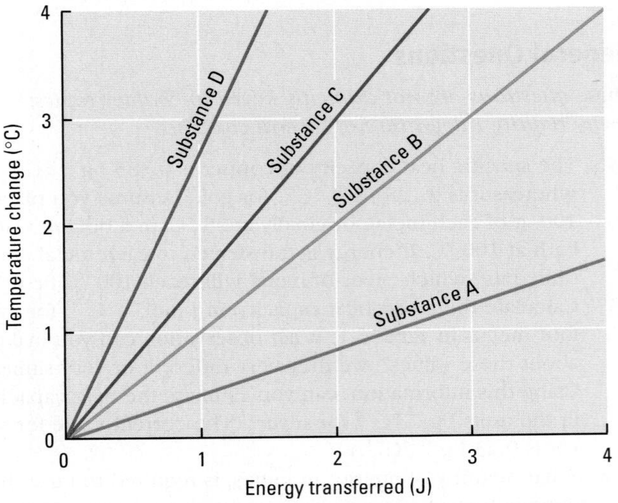

158. Based on the graph that accompanies Question 157, how much heat would need to be transferred to 10 g substance B to raise its temperature from 35 °C to 38 °C?

159. The sketch below shows two identical beakers with different volumes of water at the same temperature. Is the thermal energy content of beaker 1 greater than, less than, or equal to that of beaker 2? Explain your reasoning.

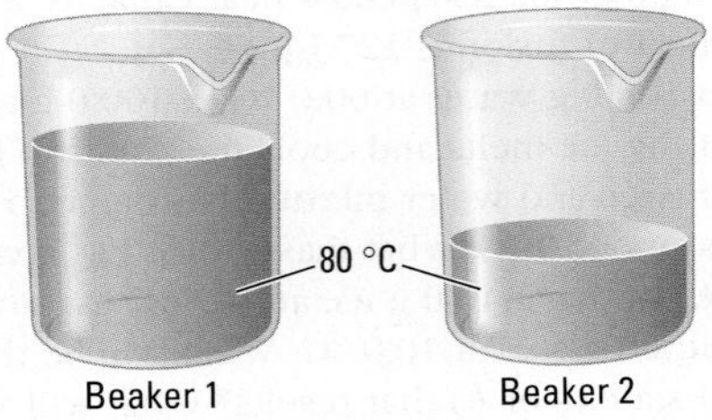

160. If the same quantity of thermal energy were transferred to each beaker in Question 159, would the temperature of beaker 1 be greater than, less than, or equal to that of beaker 2? Explain your reasoning.

161. Consider this thermochemical expression

$$2\ S(s) + 3\ O_2(g) \longrightarrow 2\ SO_3(g) \qquad \Delta H^\circ = -791 \text{ kJ}$$

and the standard molar enthalpy of formation for $SO_3(g)$ listed in Table 6.2. Why are the enthalpy values different?

162. In this chapter, the symbols ΔH_f° and ΔH° were used to denote a change in enthalpy. What is similar and what is different about the enthalpy changes they represent?

163. A fully inflated Mylar party balloon (the kind that does not expand once it is fully inflated) is heated, and 310 J transfers to the gas in the balloon. Calculate ΔE and w for the gas.

164. A sample of Dry Ice, $CO_2(s)$, is placed in a flask that is then tightly stoppered. Because the Dry Ice is very cold, 350 J transfers from the surroundings to the $CO_2(s)$. Calculate ΔE and w for the contents of the flask.

More Challenging Questions

These questions require more thought and integrate several concepts.

165. Oxygen difluoride, OF_2, is a colorless, very poisonous gas that reacts rapidly and exothermically with water vapor to produce O_2 and HF.

$$OF_2(g) + H_2O(g) \longrightarrow 2\ HF(g) + O_2(g) \qquad \Delta H° = -318\ kJ$$

Using this information and Table 6.2 or Appendix J, calculate the molar enthalpy of formation of $OF_2(g)$.

166. Suppose that 28.89 g $ClF_3(g)$ is reacted with 57.3 g Na(s) to form NaCl(s) and NaF(s). The reaction occurs at 1 bar and 25.0 °C. Calculate the quantity of energy transferred between the system and surroundings and describe in which direction the energy is transferred.

167. The four hydrocarbons of lowest molar mass are methane, ethane, propane, and butane. All are used extensively as fuels in our economy. Calculate the thermal energy transferred to the surroundings per gram (the fuel value) of each of these four fuels and rank them by this quantity.

168. For the four substances in the preceding question, calculate the energy density. Assume that substances that are gases at room temperature have been forced to condense to liquids by applying pressure and, if necessary, by lowering temperature. Obtain needed density information from the *CRC Handbook of Chemistry and Physics,* 90th ed. (London: Taylor & Francis CRC Press; 2009) **http://crcpress.com**.

169. You want to determine the value for the enthalpy of formation of $CaSO_4(s)$.

$$Ca(s) + S(s) + 2\ O_2(g) \longrightarrow CaSO_4(s)$$

This synthesis reaction cannot be done directly. You know, however, that both calcium and sulfur react with oxygen to produce oxides in reactions that can be studied calorimetrically. You also know that the basic oxide CaO reacts with the acidic oxide $SO_3(g)$ to produce $CaSO_4(s)$ with $\Delta H° = -403.3$ kJ. Outline a method for determining $\Delta H_f°$ for $CaSO_4(s)$, and identify the information that must be collected by experiment. Using information in Appendix J, confirm that $\Delta H_f°$ for $CaSO_4(s) = -1434.11$ kJ/mol.

170. When Tiger Woods drives a golf ball it is estimated that the ball is traveling about 180 miles per hour when it comes off the club. If the mass of a golf ball is 46 g, what is the quantity of energy that Tiger Woods has to supply to drive the ball? What is the minimum quantity of apples that Woods would have to eat to provide this energy?

171. When Roger Federer serves a tennis ball it is estimated that the ball is traveling about 135 miles per hour when it comes off the racquet. Find the mass of a tennis ball on the Internet and calculate the kinetic energy of the ball. What is the minimum quantity of peanuts that Federer would have to eat to provide the energy he imparted to the ball?

172. A student had five beakers, each containing 100. mL of 0.500-M NaOH(aq) and all at room temperature (20.0 °C). The student planned to add a carefully weighed quantity of solid ascorbic acid, $C_6H_8O_6$, to each beaker, stir until it dissolved, and measure the increase in temperature. After the fourth experiment, the student was interrupted and called away. The data table looked like this:

Experiment	Mass of Ascorbic Acid (g)	Final Temperature (°C)
1	2.20	21.7
2	4.40	23.3
3	8.81	26.7
4	13.22	26.6
5	17.62	—

(a) Predict the temperature the student would have observed in experiment 5. Explain why you predicted this temperature.
(b) For each experiment indicate which is the limiting reactant, sodium hydroxide or ascorbic acid.
(c) When ascorbic acid reacts with NaOH, how many hydrogen ions are involved? One, as in the case of HCl? Two, as in the case of H_2SO_4? Or three, as in the case of phosphoric acid, H_3PO_4? Explain clearly how you can tell, based on the student's calorimeter data.

173. In their home laboratory, two students do an experiment (a rather dangerous one—don't try it without proper safety precautions!) with drain cleaner (Drano, a solid) and toilet bowl cleaner (The Works, a liquid solution). The students measure 1 teaspoon (tsp) of Drano into each of four Styrofoam coffee cups and dissolve the solid in half a cup of water. Then they wash their hands and go have lunch. When they return, they measure the temperature of the solution in each of the four cups and find it to be 22.3 °C. Next they measure into separate small empty cups 1, 2, 3, and 4 tablespoons (Tbsp) of The Works. In each cup they add enough water to make the total volume 4 Tbsp. After a few minutes they measure the temperature of each cup and find it to be 22.3 °C. Finally the two students take each cup of The Works, pour it into a cup of Drano solution, and measure the temperature over a period of a few minutes. Their results are reported in the table below.

Experiment	Volume of The Works (Tbsp)	Highest Temperature (°C)
1	1	28.0
2	2	33.6
3	3	39.3
4	4	39.4

Discuss these results and interpret them in terms of the thermochemistry and stoichiometry of the reaction. Is the reaction exothermic or endothermic? Why is more energy transferred in some cases than others? For each experiment, which reactant, Drano or The Works, is limiting? Why are the final temperatures nearly the same in experiments 3 and 4? What can you conclude about the stoichiometric ratio between the two reactants?

174. On March 4, 1996, five railroad tank cars carrying liquid propane derailed in Weyauwega, Wisconsin, forcing evacuation of the town for more than a week. Residents who lived within the square mile centered on the accident

were unable to return to their homes for more than two weeks. Evaluate whether this evacuation was reasonable and necessary by considering these questions.

(a) Estimate the volume of a railroad tank car. Obtain the density of liquid propane, C_3H_8, at or near its boiling point and calculate the mass of propane in the five tank cars. Obtain the data you need to calculate the energy transfer if all the propane burned at once. (Assume that the reaction takes place at room temperature.)

(b) The enthalpy of decomposition of TNT, $C_7H_5N_3O_6$, to water, nitrogen, carbon monoxide, and carbon is -1066.1 kJ/mol. How many metric kilotons (1 tonne $= 1$ metric ton $= 1$ Mg $= 1 \times 10^6$ g) of TNT would provide energy transfer equivalent to that produced by combustion of propane in the five tank cars?

(c) Find the energy transfer (in kilotons of TNT) resulting from the nuclear fission bombs dropped on Hiroshima and Nagasaki, Japan, in 1945, and the energy transfer for modern fission weapons. What is the largest nuclear weapon thought to have been detonated to date? Compare the energy of the Hiroshima and Nagasaki bombs with the Weyauwega propane spill. What can you conclude about the wisdom of evacuating the town?

(d) Compare the energy that would have been released by burning the propane with the energy of a hurricane.

175. In some cities, taxicabs run on liquefied propane fuel instead of gasoline. This practice extends the lifetime of the vehicle and produces less pollution. Given that it costs about $2000 to modify the engine of a taxicab to run on propane and that the cost of gasoline and liquid propane are $3.00 per gallon and $2.00 per gallon, respectively, make reasonable assumptions and figure out how many miles a taxi would have to go so that the decreased fuel cost would balance the added cost of modifying the taxi's motor. [For enthalpy calculations gasoline can be approximated as octane, $C_8H_{18}(\ell)$.]

Conceptual Challenge Problems

These rigorous, thought-provoking problems integrate conceptual learning with problem solving and are suitable for group work.

CP6.A (Section 6.2) Suppose a scientist discovered that energy was not conserved, but rather that 1×10^{-7}% of the energy transferred from one system vanishes before it enters another system. How would this affect electric utilities, thermochemical experiments in scientific laboratories, and scientific thinking?

CP6.B (Section 6.2) Suppose that someone were to tell your teacher during class that energy is not always conserved. This person states that he or she had previously learned that in the case of nuclear reactions, mass is converted into energy according to Einstein's equation $E = mc^2$. Hence, energy is continuously produced as mass is changed into energy. Your teacher quickly responds by giving the following assignment to the class: "Please write a paragraph or two to refute or clarify this student's thesis." What would you say?

CP6.C (Section 6.3) The specific heat capacities at 25 °C for three metals with widely differing molar masses are 3.6 J g^{-1} $°C^{-1}$ for Li, 0.25 J g^{-1} $°C^{-1}$ for Ag, and 0.11 J g^{-1} $°C^{-1}$ for Th. Suppose that you have three samples, one of each metal and each containing the same number of atoms.

(a) Is the energy transfer required to increase the temperature of each sample by 1 °C significantly different from one sample to the next?

(b) What interpretation can you make about temperature based on the result you found in part (a)?

CP6.D (Section 6.4) During one of your chemistry classes a student asks the professor, "Why does hot water freeze more quickly than cold water?"

(a) What do you expect the professor to say in answer to the student's question?

(b) In one experiment, two 100.-g samples of water were placed in identical containers on the same surface 10. cm apart in a room at -25 °C. One sample had an initial temperature of 78 °C, while the second was at 24 °C. The second sample took 151 min to freeze, and the first took 166 min (only 10% longer) to freeze. Clearly the cooler sample froze more quickly, but not nearly as quickly as one might have expected. How can this be so?

CP6.E (Section 6.10) Assume that glass has the same properties as pure SiO_2. The thermal conductivity (the rate at which heat transfer occurs through a substance) for aluminum is eight times that for SiO_2.

(a) Is it more efficient in time and energy to bake brownies in an aluminum pan or a glass pan?

(b) It is said that things cook more evenly in a glass pan than in an aluminum pan. Are there scientific data that indicate that this statement is reasonable?

CP6.F (Section 6.11) Suppose that you are an athlete and exercise a lot. Consequently you need a large caloric intake each day. Choose at least ten foods that you normally eat and evaluate each one to find which provides the most calories per dollar.

7

Electron Configurations and the Periodic Table

Roman Krochuk/iStockphoto

The Northern Lights. In Earth's atmosphere, electrons and protons in the solar wind collide with nitrogen and oxygen molecules, breaking the molecules into their constituent atoms and exciting the atoms to higher energy states. The energy is subsequently released as a brilliant display of radiant energy called the Northern Lights. Excited nitrogen atoms produce blue, mauve, and pink light; excited oxygen atoms generate green and red light.

The periodic table was created by Dmitri Mendeleev to summarize experimental observations ***(← p. 62)***. He had no theory or model to explain why, for example, all alkaline-earth metals combine with oxygen in a 1:1 atom ratio—he just knew they did. In the early years of the twentieth century, however, it became evident that atoms contain electrons. As a result of these findings, explanations of periodic trends in physical and

OWL Sign in to OWL at **www.cengage.com/owl** to view tutorials and simulations, develop problem-solving skills, and complete online homework assigned by your professor.

go Chemistry Download mini lecture videos for key concept review and exam prep from OWL or purchase them from **www.CengageBrain.com.**

Companion Website

Visit this book's companion website at **www.cengage.com/chemistry/moore** to work interactive modules for the Estimation boxes and Active Figures in this text.

© Steve Taylor/Digital Vision/ Getty Images

The various colors of light emitted by neon-type signs are due to emissions from atoms of different gases.

The speed of light through a substance (air, glass, or water, for example) depends on the chemical constitution of the substance and the wavelength of the light. This is the basis for using a glass prism to disperse light and is the explanation for rainbows.

Wavelength units are summarized in Table 7.1.

chemical properties began to be based on an understanding of the arrangement of electrons within atoms—on what we now call *electron configurations.* Studies of the interaction of light with atoms and molecules revealed that electrons in atoms are arranged in energy levels or shells. Which shells have electrons in them is determined by the energies of the electrons. Electrons in the outermost shell are the *valence electrons; the number of valence electrons and the shell in which they occur are the chief factors that determine chemical reactivity.* This chapter describes the relationship of atomic electron configurations to atomic properties. Special emphasis is placed on the use of the periodic table to derive the electron configurations for atoms and ions.

7.1 Electromagnetic Radiation and Matter

Theories about the energy and arrangement of electrons in atoms are based on experimental studies of the interaction of matter with electromagnetic radiation. A good example is the interaction of visible light and matter. Interestingly, matter in some form is always associated with any color of light our eyes can see. The red glow of a neon sign comes from neon atoms losing energy they gained when electricity was passed through them; fireworks displays are visible because of light emitted by metal atoms that have extra energy from heating when the firework explodes. When an electron in an atom has gained energy, from electricity, heating, or some other way, the atom is said to be "excited." The extra energy can be released in the form of electromagnetic radiation, sometimes as visible light. Studying the quantity of energy gained or released when an atom is excited can help us understand the arrangement of electrons within the atom.

Electromagnetic radiation and its applications are familiar to all of us; sunlight, automobile headlights, dental X-rays, microwave ovens, and cell phones that we use for communications are a few examples (Figure 7.1). These kinds of radiation seem very different, but they are actually very similar. All **electromagnetic radiation** consists of oscillating, perpendicular electric and magnetic fields that travel through space at the same rate (the "speed of light": 186,000 m/s, or 2.998×10^8 m/s in a vacuum). Any of the various kinds of electromagnetic radiation can be described in terms of frequency (ν) and wavelength (λ). A **spectrum** is the distribution of intensities of wavelengths or frequencies of electromagnetic radiation emitted or absorbed by an object. Figure 7.1 gives wavelength and frequency values for several regions of the spectrum.

As illustrated in Figure 7.2, the **wavelength** is the distance between adjacent crests (or troughs) in a wave, and the **frequency** is the number of complete waves passing a point in a given period of time—that is, waves per second or simply per second, 1/s. Reciprocal units such as 1/s are often represented in the negative exponent form, s^{-1}, which means "per second." A frequency of 4.80×10^{14} s^{-1} means

Table 7.1 Useful Wavelength Units for Different Regions of the Electromagnetic Spectrum

Wavelength Unit	Unit Length (m)	Radiation Type
Picometer, pm	10^{-12}	γ (gamma)
Ångström, Å	10^{-10}	X-ray
Nanometer, nm	10^{-9}	Ultraviolet, visible
Micrometer, μm	10^{-6}	Infrared
Millimeter, mm	10^{-3}	Infrared
Centimeter, cm	10^{-2}	Microwave
Meter, m	1	TV, radio

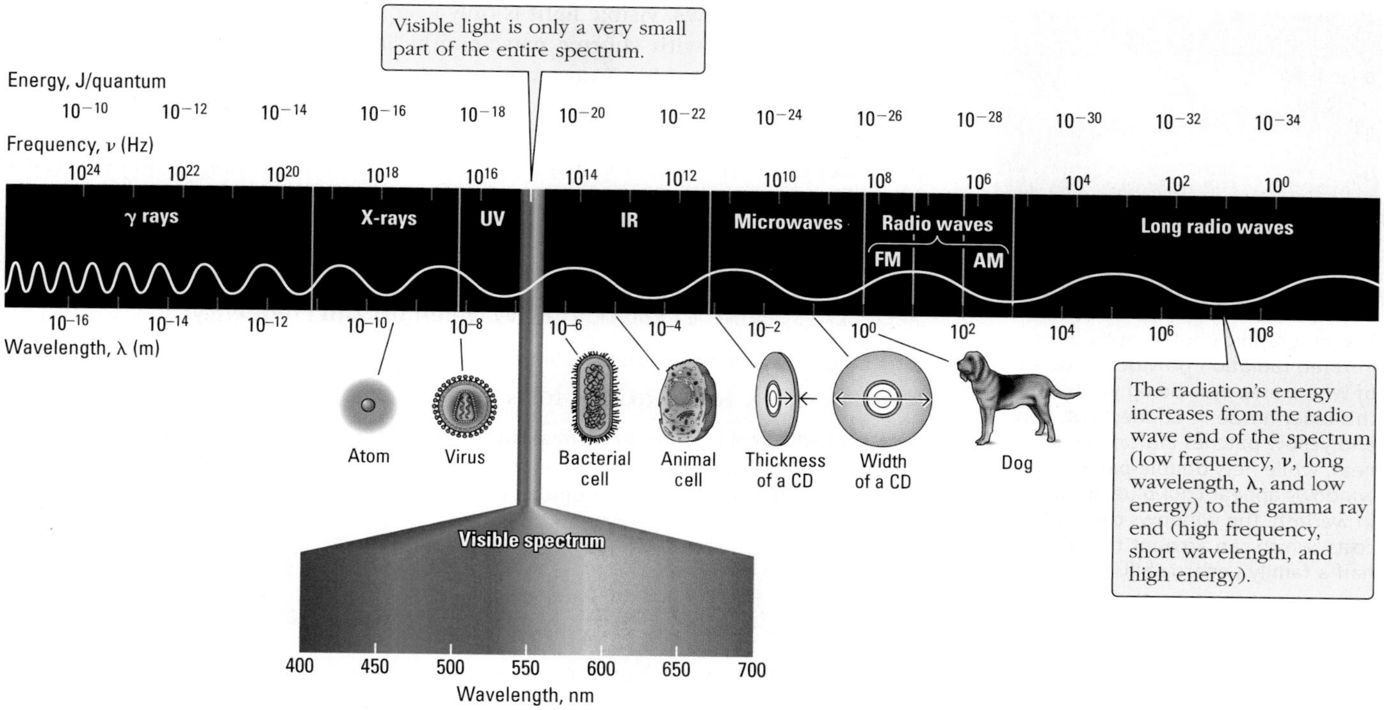

Figure 7.1 The electromagnetic spectrum. The size of wavelengths in the electromagnetic spectrum is compared with common objects. Visible light (enlarged section) is only a small part of the entire spectrum.

that 4.80×10^{14} waves pass a fixed point every second. The unit s^{-1} is given the name *hertz* (Hz).

The hertz unit (waves/s, or s^{-1}) was named in honor of Heinrich Hertz (1857–1894), a German physicist.

For wave phenomena such as sound, the intensity (loudness) is related to the *amplitude* (height of the wave crest) (Figure 7.2). The higher the amplitude, the more intense is the sound. For example, a middle C tone of a piano always has the same frequency and wavelength (pitch), but the sound can vary from loud (high amplitude) to soft (low amplitude).

The frequency of electromagnetic radiation is related to its wavelength by

$$\nu\lambda = c$$

where c is the speed of light, 2.998×10^8 m/s.

Note that frequency and wavelength are *inversely* related; that is, as one decreases, the other increases. *Short*-wavelength radiation has a *high* frequency.

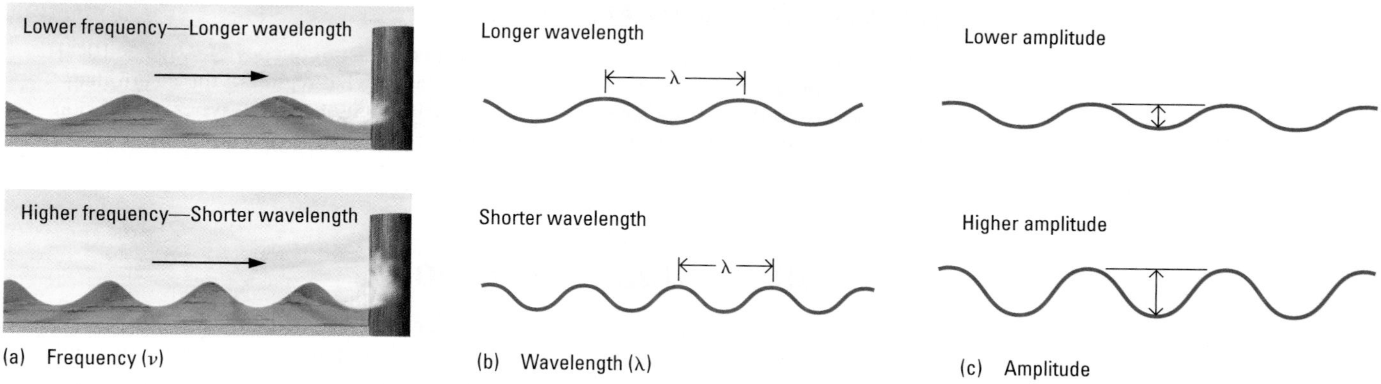

Figure 7.2 Wavelength, frequency, and amplitude of water waves. (a) The waves are moving from left to right toward the post at the same speed. The upper wave has a long wavelength (large λ) and low frequency (the number of times per second the peak of the wave hits the post). The lower wave has a shorter wavelength and a higher frequency (its peaks hit the post more often in a given time). (b) Variation in wavelength. (c) Variation in amplitude.

Tyrone Turner/National Geographic Image Collection

Thermographic photography using infrared radiation provides evidence of where energy is being wasted in this older house in Connecticut. Red and yellow patches indicate escaping heat, while new double-pane windows appear cool blue. By sealing in warmth, the windows cut heating costs, which can account for up to half a family's energy bill.

As Figure 7.1 shows, visible light is only a small portion of the electromagnetic spectrum. Radiation with shorter wavelengths includes ultraviolet radiation (the type that leads to sunburn), X-rays, and gamma (γ) rays (emitted in the process of radioactive disintegration of some atoms). Infrared radiation, the type that is sensed as heat from a fire, has wavelengths longer than those of visible light. Longer still is the wavelength of the radiation in a microwave oven or of television and radio transmissions, such as that used for cellular telephones.

The **PROBLEM-SOLVING STRATEGY** in this book is

- Analyze the problem
- Plan a solution
- Execute the plan
- Check that the result is reasonable

Appendix A.1 explains this in detail.

PROBLEM-SOLVING PRACTICE answers are provided at the back of this book in Appendix K.

Answers to **EXERCISES** are provided at the back of this book in Appendix L.

EXERCISES that are labeled **CONCEPTUAL** are designed to test your understanding of one or more concepts; they usually involve qualitative rather than quantitative thinking.

PROBLEM-SOLVING EXAMPLE 7.1 Wavelength and Frequency

Green light has a wavelength of 495 nm. Calculate the frequency of this light.

Answer 6.06×10^{14} Hz ($6.06 \times 10^{14}\ s^{-1}$)

Strategy and Explanation Rearrange the equation relating frequency, the speed of light, and wavelength to calculate the frequency of this radiation in Hz (s^{-1}). Recall that $1\ m = 1 \times 10^9$ nm. Convert nanometers to meters and substitute into the equation.

$$495\ \text{nm} \times \frac{1\ \text{m}}{1 \times 10^9\ \text{nm}} = 4.95 \times 10^{-7}\ \text{m}$$

$$\upsilon = \frac{c}{\lambda} = \frac{2.998 \times 10^8\ \text{m/s}}{4.95 \times 10^{-7}\ \text{m}} = 6.06 \times 10^{14}\ \text{s}^{-1}\ (\text{Hz})$$

☑ **Reasonable Answer Check** Compare the calculated frequency ($6.06 \times 10^{14}\ s^{-1}$) with frequency of the visible region of the electromagnetic spectrum illustrated in Figure 7.1. The wavelength and frequency fall into the visible region and thus the answer is reasonable.

PROBLEM-SOLVING PRACTICE 7.1

In the upper atmosphere there is solar radiation with a frequency of 9.99×10^{14} Hz. Calculate its wavelength in (a) meters; (b) nanometers.

CONCEPTUAL EXERCISE 7.1 Frequency and Wavelength

A fellow chemistry student says that low-frequency radiation is short-wavelength radiation. You disagree. Explain why the other student is wrong.

CONCEPTUAL EXERCISE 7.2 In the Atmosphere

One type of solar radiation in the upper atmosphere has a frequency of 7.898×10^{14} Hz; another type has a frequency of 1.20×10^{15} Hz. (a) In what region of the electromagnetic spectrum does this solar radiation occur? (b) Which of the two types of radiation has the shorter wavelength? Explain your answer.

7.2 Planck's Quantum Theory

Have you ever looked at the wires in a toaster as they heat up? Although you cannot see the metal atoms in the toaster wires, as electric energy flows through the wires, their atoms gain energy and then emit it as radiation. When heated, the wires first emit a slight amount of heat that you can feel (infrared radiation). As the wire gets hotter, it begins to glow, initially emitting red light, and then orange (Figure 7.3a). If a wire gets very hot, it appears almost white. Figure 7.3b shows the spectrum of a typical heated object.

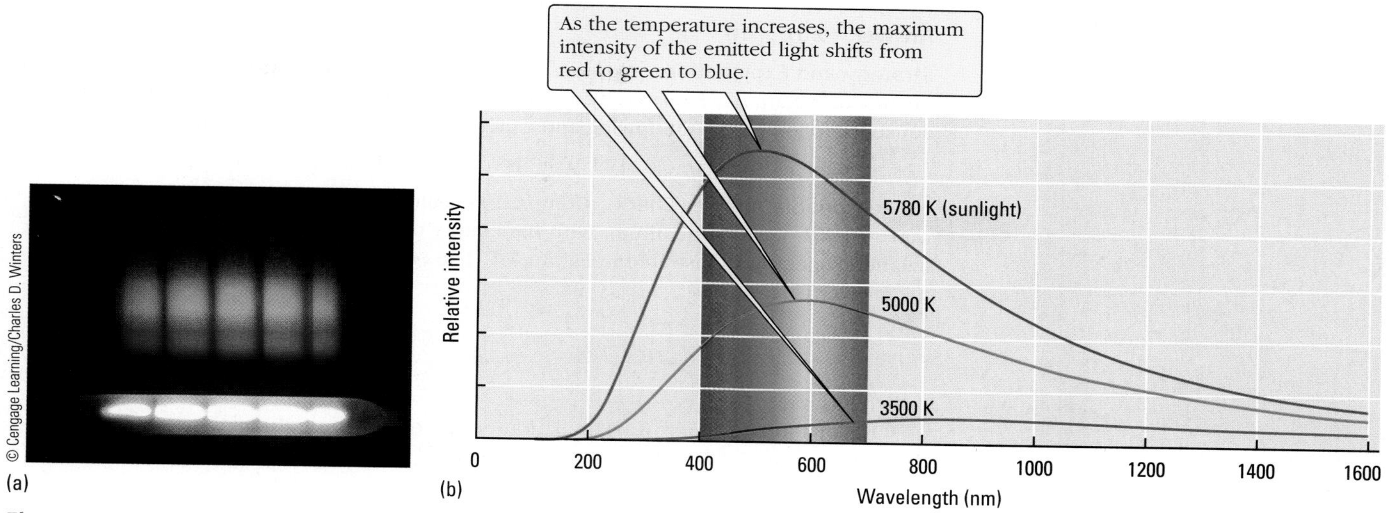

Figure 7.3 Heated objects and temperature. (a) A heated filament in a toaster. The color of the hot filament and its continuous spectrum change as the temperature is increased. (b) The spectrum of radiation given off by a heated object. At very high temperatures, the heated object becomes "white hot" as all wavelengths of visible light become equally intense.

In trying to explain the nature of these emissions from hot objects, late nineteenth-century scientists assumed that vibrating atoms in a hot wire caused the emission of electromagnetic vibrations (light waves). According to the laws of classical physics, the light waves could have any frequency along a continuously varying range. Using such laws, however, the scientists were unable to predict the experimentally observed spectrum.

In 1900 Max Planck (1858–1947) offered an explanation for the spectrum of a heated body. His ideas contained the seeds of a revolution in scientific thought. Planck made what was at that time an incredible assertion, one that departed dramatically from the laws of classical physics. He stated that when an atom in a hot object emits radiation, it does so only in packets having a minimum amount of energy. That is, there must be a small packet of energy such that no smaller quantity can be emitted, just as an atom is the smallest packet of an element. Planck called this packet of energy a **quantum.** He further asserted that the energy of a quantum is proportional to the frequency of the radiation according to the equation

Planck's explanation of the spectrum of a heated body answers the question, "Why does a metal get red hot and then white when it is heated?" that was posed in Chapter 1 (*← p. 2*).

$$E_{\text{quantum}} = h\nu_{\text{radiation}}$$

The proportionality constant h is called **Planck's constant** in his honor; it has a value of 6.626×10^{-34} Js and relates the frequency of radiation to its energy *per quantum.*

A quantum is the smallest possible unit of a distinct quantity—for example, the smallest possible unit of energy for electromagnetic radiation of a given frequency.

Orange light has a measured frequency of $4.80 \times 10^{14}\ \text{s}^{-1}$. The energy of *one* quantum of orange light is therefore

$$E = h\nu = (6.626 \times 10^{-34}\ \text{Js})(4.80 \times 10^{14}\ \text{s}^{-1}) = 3.18 \times 10^{-19}\ \text{J}$$

The theory based on Planck's work is called the **quantum theory.** By using his quantum theory, Planck was able to calculate results that agreed very well with the experimentally measured spectra of heated objects, spectra for which the laws of classical physics had no explanation.

Max Planck won the 1918 Nobel Prize in Physics for his quantum theory.

PROBLEM-SOLVING EXAMPLE 7.2 Calculating Quantum Energies

In the stratosphere, ultraviolet radiation with a frequency of $1.36 \times 10^{15}\ \text{s}^{-1}$ can break C—Cl bonds in chlorofluorocarbons (CFCs), which can lead to stratospheric ozone depletion. Calculate the energy per quantum of this radiation.

Answer 9.01×10^{-19} J

Strategy and Explanation According to Planck's quantum theory, the energy and frequency of radiation are related by $E = h\nu$. Substituting the values for h and ν into the equation, followed by multiplying and canceling the units, gives the correct answer.

$$E = h\nu = (6.626 \times 10^{-34}\ \text{Js})(1.36 \times 10^{15}\ \text{s}^{-1}) = 9.01 \times 10^{-19}\ \text{J}$$

☑ **Reasonable Answer Check** Compare the calculated energy, 9.01×10^{-19} J/quantum = 0.901×1^{-18} J/quantum, with the energy of the ultraviolet region of the electromagnetic spectrum illustrated in Figure 7.1. The energy is within the ultraviolet region; the answer is reasonable.

PROBLEM-SOLVING PRACTICE 7.2

Which has more energy,

(a) One quantum of microwave radiation or one quantum of ultraviolet radiation?

(b) One quantum of blue light or one quantum of green light?

A very important relationship links the energy and wavelength of a quantum of radiation. Since $E = h\nu$ and $\nu = c/\lambda$, then

$$E = \frac{hc}{\lambda}$$

where h is Planck's constant, c is the velocity of light (2.998×10^8 m/s), and λ is the wavelength of the radiation. Note that energy and wavelength are inversely proportional: ***The energy per quantum of radiation increases as the wavelength gets shorter.*** For example, red light and blue light have wavelengths of 656.3 nm and 434.1 nm, respectively. Because of its shorter wavelength, blue light has a higher energy per quantum than red light. We can calculate their different energies by applying the previous equation.

$$E_{\text{red}} = \frac{(6.626 \times 10^{-34}\ \text{Js})(2.998 \times 10^8\ \text{m/s})}{(656.3\ \text{nm})(1\ \text{m}/10^9\ \text{nm})} = 3.027 \times 10^{-19}\ \text{J}$$

$$E_{\text{blue}} = \frac{(6.626 \times 10^{-34}\ \text{Js})(2.998 \times 10^8\ \text{m/s})}{(434.1\ \text{nm})(1\ \text{m}/10^9\ \text{nm})} = 4.576 \times 10^{-19}\ \text{J}$$

EXERCISE 7.3 Carbon Dioxide, Energy, and Global Warming

Atmospheric carbon dioxide is a greenhouse gas associated with global warming. The carbon-to-oxygen bonds in CO_2 absorb electromagnetic radiation at 4.257 μm and at 15.00 μm (1 μm = 10^{-6} m). The 4.257 μm absorption increases the energy of stretching these bonds; the 15.00 μm absorption increases the energy of bending them. Calculate which absorption requires the greater amount of energy.

The Photoelectric Effect

When a theory can accurately predict experimental results, the theory is usually regarded as useful. Planck's quantum theory was not widely accepted at first because of its radical assertion that energy is quantized. The quantum theory of electromagnetic energy was firmly accepted only after Planck's quanta were used by Albert Einstein to explain another phenomenon, the photoelectric effect.

In the early 1900s it was known that certain metals exhibit a **photoelectric effect**: Such metals emit electrons when illuminated by light of certain wavelengths (Figure 7.4a). For each photosensitive metal there is a threshold wavelength below

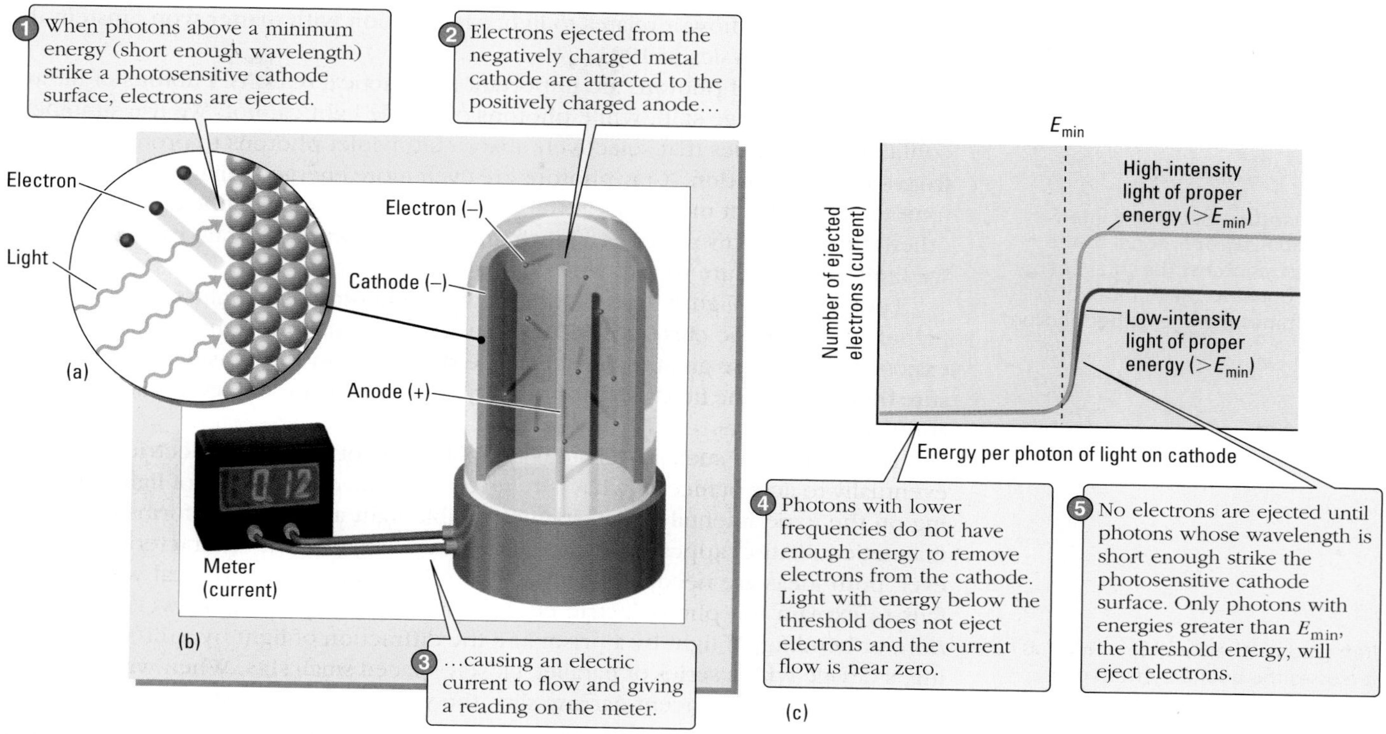

Active Figure 7.4 The photoelectric effect. (a) A photoelectric surface. (b) A photoelectric cell. Only photons with energies greater than the threshold energy will eject electrons from the photosensitive surface. (c) Ejected electrons and energy. Photons with lower energy (left side of Figure 7.4c) do not have enough energy to remove electrons. Note that the *number* of electrons ejected (the current) depends on the light's intensity, not its energy. Visit this book's companion website at **www.cengage.com/chemistry/moore** to test your understanding of the concepts in this figure.

which no photoelectric effect is observed. For example, cesium (Cs) metal emits electrons when illuminated by red light, whereas some metals require yellow light, and others require ultraviolet light to cause emission of electrons. The differences occur because each type of light has a different wavelength and thus a different energy per quantum. Red light, for example, has a lower energy (longer wavelength) than yellow light or ultraviolet light. Light with energy per quantum that is below the threshold needed to eject electrons from the metal will not cause an electric current to flow, *no matter how bright the light.* Figure 7.4b illustrates how the electrons ejected from the photosensitive cathode are attracted to the anode in the photoelectric cell, causing an electric current to flow, resulting in a display on the meter. Figure 7.4c shows how an electric current suddenly increases when light of sufficient energy per quantum (short enough wavelength) shines on a photosensitive metal.

Albert Einstein explained these experimental observations by assuming that Planck's quanta were *massless* "particles" of light. Einstein called them **photons** instead of quanta. That is, light could be described as a stream of photons that had particle-like properties as well as wave-like properties. To remove one electron from a photosensitive metal surface requires a certain minimum quantity of energy; we call it E_{min}. Since each photon has an energy given by $E = h\nu$, only photons whose E is greater than E_{min} will have enough energy to knock an electron loose. Thus, no electrons are ejected until photons with high enough frequencies (and short enough wavelengths) strike the photosensitive surface. Photons with lower frequencies (left side of Figure 7.4c) do not have enough energy to remove electrons. This means that if a photosensitive metal cathode requires photons of green light to eject electrons from its surface, then yellow light, red light, or light of any other lower frequency (longer wavelength) will not have sufficient energy to cause the photoelectric effect with that metal. This brilliant deduction about the quantized nature of light and how it relates to light's interaction with matter won Einstein the Nobel Prize for Physics in 1921.

A common misconception is that Einstein won the Nobel Prize for his theory of relativity.

The energies of photons are important for practical reasons. Photons of ultraviolet light can damage skin, while photons of visible light cannot. We use sunblocks containing molecules that selectively absorb ultraviolet photons to protect our skin from solar UV radiation. X-ray photons are even more energetic than ultraviolet photons and can disrupt molecules at the cellular level, causing genetic damage, among other effects. For this reason we try to limit our exposure to X-rays even more than we limit our exposure to ultraviolet light.

In the quantum theory, the intensity of radiation is proportional to the number of photons. The total energy is proportional to the number of photons times the energy per photon.

Look again at Figure 7.4c. Notice how a higher-intensity light source causes a higher photoelectric current. Higher intensities of ultraviolet light (or more time of exposure) can cause greater damage to the skin than lower intensities (or less exposure time). The same holds true for other high-energy forms of electromagnetic radiation, such as X-rays.

Developments such as Einstein's explanation of the photoelectric effect led eventually to acceptance of what is referred to as the *dual nature* of light. Depending on the experimental circumstances, visible light and all other forms of electromagnetic radiation appear to have either "wave" or "particle" characteristics. However, both ideas are needed to fully explain light's behavior. Classical wave theory fails to explain the photoelectric effect. But it does explain quite well the *refraction,* or bending, of light by a prism and the diffraction of light by a diffraction grating, a device with a series of parallel, closely spaced small slits. When waves of light pass through such adjacent narrow slits, the waves are scattered so that the emerging light waves spread out, a phenomenon called *diffraction.* The semicircular waves emerging from the narrow slits can either amplify or cancel each other. Such behavior creates a diffraction pattern of dark and bright spots, as seen in Figure 7.5.

Refraction is the bending of light as it crosses the boundary from one medium to another—for example, from air to water. Diffraction is the bending of light around the edges of objects, such as slits in a diffraction grating.

It is important to realize, however, that this "dual nature" description arises because of our attempts to explain observations by using inadequate models. Light is not changing back and forth from being a wave to being a particle, but has a single consistent nature that can be described by modern quantum theory. The dual-nature description arises when we try to explain our observations by using our familiarity with classical models for "wave" or "particle" behavior.

Prior to Einstein's explanation of the photoelectric effect, classical physics considered light as being only wave-like.

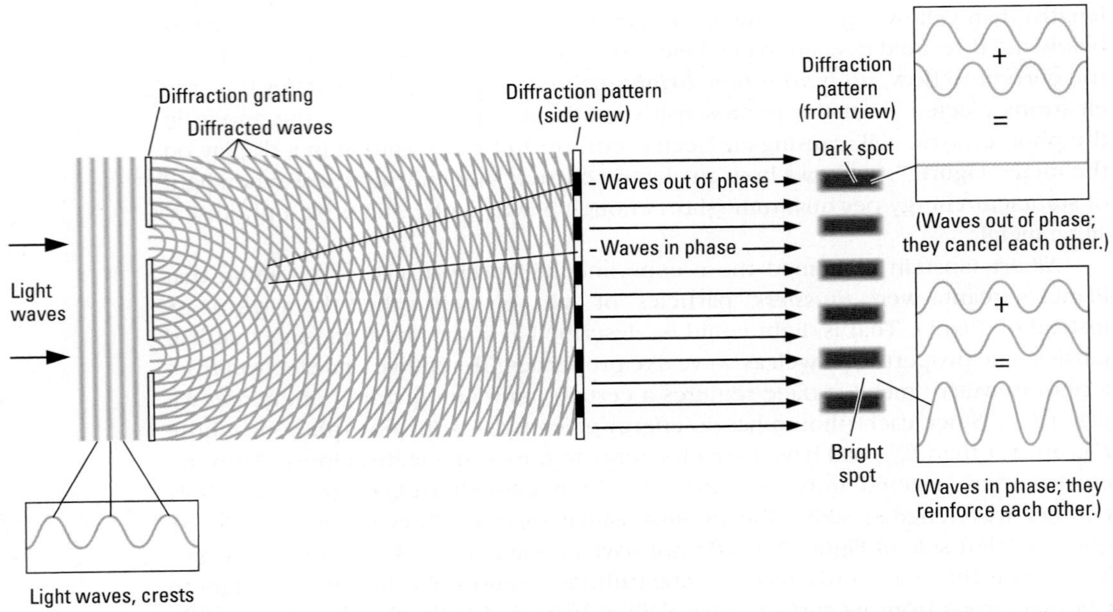

Figure 7.5 Diffraction pattern. Light waves passing through slits of a diffraction grating are scattered. When adjacent waves are in phase, they reinforce each other, producing a bright spot. When the waves are out of phase, they cancel each other and a dark spot occurs.

ESTIMATION Turning on the Light Bulb

You turn the switch and a living room lamp comes on. The lamp emits yellow light (585 nm), producing 20 J of energy in one second. Approximately how many quanta of yellow light are given off by the lamp in that time?

To find out, we first determine the energy of each 585-nm quantum by using the relationship $E = \frac{hc}{\lambda}$. We approximate by rounding the value of Planck's constant to 7×10^{-34} Js and that of *c*, the speed of light, to 3×10^{8} m/s. The wavelength of the quanta we round to approximately 600×10^{-9} m; remember we have to convert nanometers to meters: 1 nm = 1×10^{-9} m. Therefore, the energy, *E*, of each quantum is approximately

$$\frac{(7 \times 10^{-34}\ \text{Js})\ (3 \times 10^{8}\ \text{m/s})}{600 \times 10^{-9}\ \text{m}}$$

which is about 4×10^{-19} J per quantum. Using this value and the total energy generated in one second, we can calculate the number of quanta of yellow light emitted in that time.

$$\text{Number of quanta} = \frac{\text{total energy}}{\text{energy of each quantum}} =$$

$$20\ \text{J} \times \frac{1\ \text{quantum}}{4 \times 10^{-19}\ \text{J}} = 5 \times 10^{19}\ \text{quanta}$$

a considerable number, just slightly more than the number of milliliters of water (6×10^{18} mL) in the Atlantic Ocean.

Visit this book's companion website at www.cengage.com/chemistry/moore to work an interactive module based on this material.

CHEMISTRY IN THE NEWS Using an Ultra-Fast Laser to Make a More Efficient Incandescent Light Bulb

Researchers led by Dr. Chunlei Guo at the University of Rochester (NY) used an ultra-fast laser to improve the efficiency of the tungsten filament inside an incandescent light bulb. In a regular incandescent bulb, a tungsten filament gives off light when heated by an electric current. The enhanced filament is created by striking the tungsten surface with an extremely brief, ultra-intense laser light beam. Dr. Guo says, "We fired the laser beam right through the glass of the bulb and altered a small area of the filament. When we lit the bulb, we could actually see this one patch was clearly brighter than the rest of the filament, but there was no change in the bulb's energy usage." The laser pulse lasts only a femtosecond (1×10^{-15} s), releasing a significant amount of energy onto a spot the size of a needle point. (A femtosecond is to a second what a second is to 32 million years.)

Energy from the laser pulse creates nanostructures and microstructures on the tungsten surface that significantly increase how efficiently light radiates from the filament. Such a filament could make a light as bright as a 100-watt bulb use less energy than a 60-watt bulb and radiate a more pleasant light than that given off by a fluorescent bulb. Controlling the size and shape of the nanostructures formed on the filament's surface changes the overall spectrum radiated so that tungsten, which normally gives off a yellowish light, can radiate a purer white light.

Courtesy Dr. Guo

Dr. Chunlei Guo.

Source: **http://esciencenews.com/articles/2009/05/29/regular.light.bulbs.made.super.efficient.with.ultra.fast.laser**

7.3 The Bohr Model of the Hydrogen Atom

Within about a decade (1900–1911), three major discoveries were made about the atom and the nature of electromagnetic radiation. The first two, discussed in the previous section, were Max Planck's suggestion that energy was quantized and, five years later, Albert Einstein's application of the quantum idea to explain the photoelectric effect. The third came in 1911 when Ernest Rutherford demonstrated experimentally that atoms contain a dense positive core, the nucleus, surrounded by electrons. Niels Bohr linked these three powerful ideas when, in 1913, he used quantum theory to explain the behavior of the electron in a hydrogen atom. In doing so, Bohr developed a mathematical model to explain how excited hydrogen

Figure 7.6 A continuous spectrum from white light. When white light is passed through slits to produce a narrow beam and then refracted in a glass prism, the various colors blend smoothly into one another.

atoms emit or absorb only certain wavelengths of light, a phenomenon that had been unexplained for nearly a half century. To understand what Bohr did, we turn first to the visible spectrum.

The spectrum of white light, such as that from the sun or an incandescent light bulb, consists of a rainbow of colors, as shown in Figure 7.6. This **continuous spectrum** contains light of all wavelengths in the visible region.

If a high voltage is applied to a gaseous element at low pressure, the atoms absorb energy and are excited. The excited atoms then emit, as electromagnetic radiation, energy that was previously absorbed. A neon advertising sign, in which the neon atoms emit red-orange light, makes commercial use of the light from excited neon atoms. When light from such a source passes through a prism onto a white surface, only a few colored lines are seen in a **line emission spectrum,** characteristic of the element. The line spectra of the visible light emitted by excited atoms of hydrogen, mercury, and neon are shown in Figure 7.7.

Hydrogen has the simplest line emission spectrum, first studied in the 1880s. At that time, lines in the visible region of the spectrum and their wavelengths were accurately measured (Figure 7.7). Other series of lines in the ultraviolet region ($\lambda < 400$ nm) and in the infrared region ($\lambda > 700$ nm) were discovered subsequently.

Niels Bohr provided the first explanation of the line emission spectra of atoms by audaciously assuming that the single electron of a hydrogen atom moved in a circular orbit around its nucleus. He then related the energies of the electron to the radius of its orbit, but not by using the classic laws of physics. Such laws allowed the electron to have a wide range of energy. Instead, Bohr introduced quantum theory into his atomic model by invoking Planck's idea that energies are quantized. In the Bohr model, the electron could circle the nucleus in orbits of only certain radii, which correspond to specific energies. Thus, the energy of the electron is quantized, and the electron is restricted to certain energy levels unless it gains or loses a certain amount of energy. Bohr referred to these energy levels as *orbits* and represented the energy difference between any two adjacent orbits as a specific quantity of energy. Each allowed orbit was assigned a positive integer, n, known as the **principal quantum number.**

In the Bohr model, the value of n for the possible orbits can be any integer from 1 to infinity (∞). *The energy of the electron and the size of its orbit increase as the value of* n *increases.* The orbit of lowest energy, $n = 1$, is closest to the nucleus,

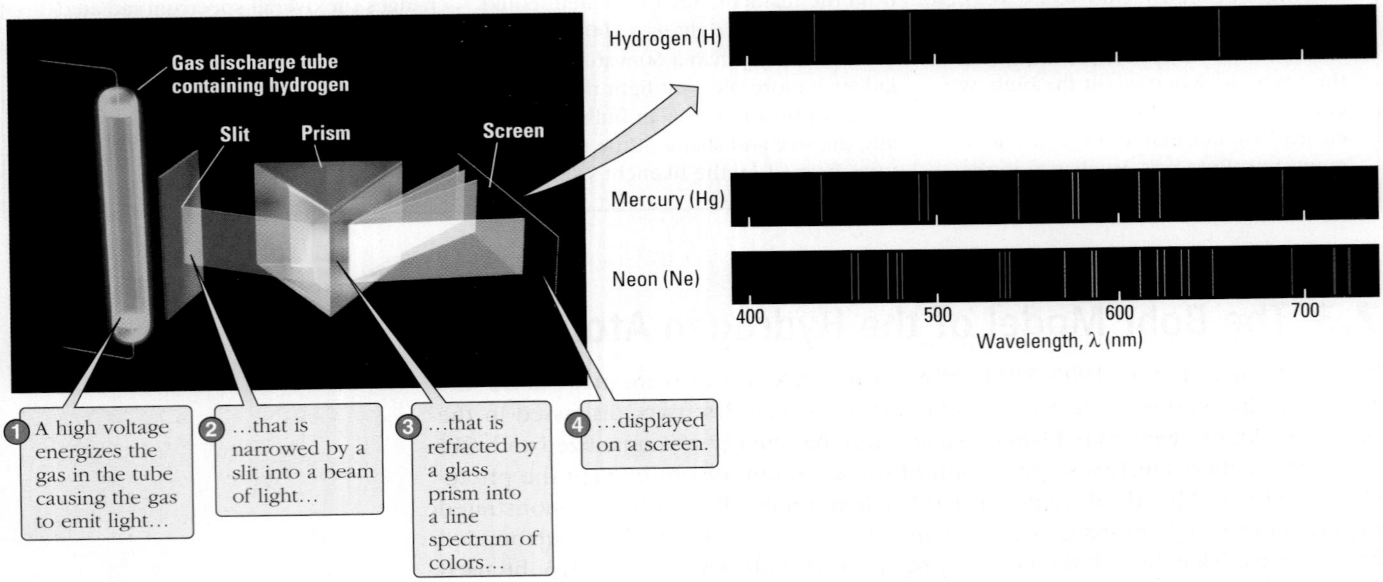

Figure 7.7 Line emission spectra of hydrogen, mercury, and neon. Excited gaseous elements produce characteristic line spectra that can be used to identify the elements as well as to determine how much of an element is present in a sample.

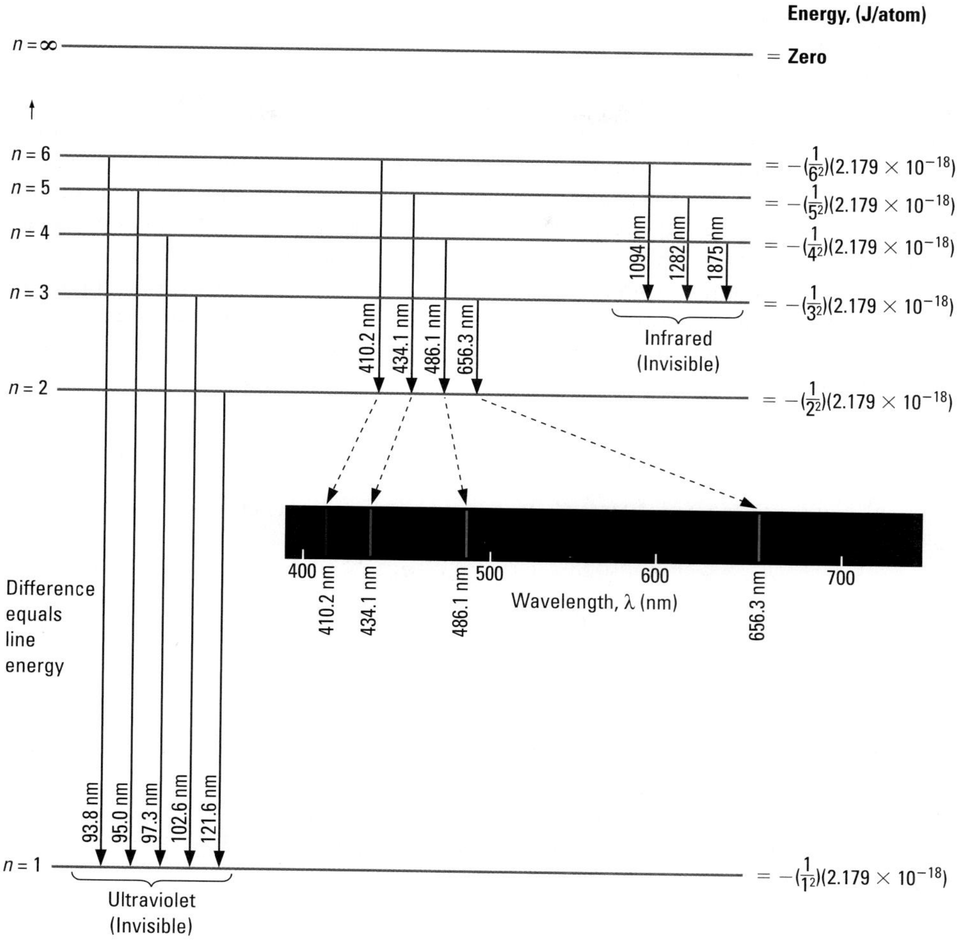

Figure 7.8 Electron transitions in an excited H atom. The lines in the ultraviolet region all result from transitions to the $n = 1$ level. Transitions from levels with values of $n > 2$ to the $n = 2$ level occur in the visible region. Lines in the infrared region result from the transitions from levels with values of $n > 3$ or 4 to the $n = 3$ or 4 levels (only the series for transitions to the $n = 3$ level is shown here).

and the electron of a hydrogen atom is normally in this energy level. Any atom with its electrons arranged to give the lowest total energy is said to be in its **ground state.** Because the positive nucleus attracts the negative electron, an electron must absorb energy to go from a lower energy state (lower n value) to a higher energy state (higher n value), an excited state. Conversely, energy is emitted when the reverse process occurs—an electron going from a higher energy state to a lower energy state (Figure 7.8). As seen in Figure 7.8 for the hydrogen atom, emission lines in the

- **ultraviolet** region are due to $n > 1$ to $n = 1$ transitions
- **visible** region are due to $n > 2$ to $n = 2$ transitions
- **infrared** region are due to $n > 3$ or $n > 4$ to $n = 3$ or 4 transitions

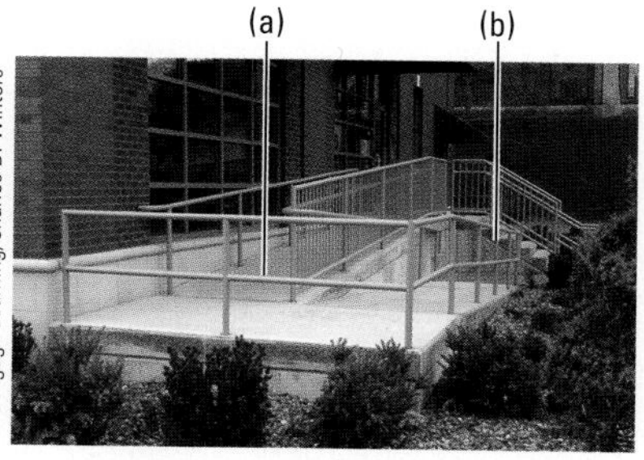

(a) A ramp (not quantized) (b) Stair steps (quantized)

The Bohr orbit model is analogous to a set of stairs in which the higher stairs are closer together. As you move up and down the stairs, you can stop at any step, *but not between steps* because the steps are quantized; only whole steps are allowed.

The highest potential energy the electron in a hydrogen atom can have is when sufficient energy has been added to separate the electron from the proton. Bohr designated this situation as zero energy when $n = \infty$ (Figure 7.8). Thus, the potential energy of the electron in $n < \infty$ is negative, as shown in Figure 7.8.

Consider this analogy to Bohr's arbitrarily selecting zero as the highest energy for the electron in a hydrogen atom. A book resting on a table is arbitrarily designated as having zero potential energy. As the book falls from the table to a chair, to a stool, and then to the floor, it loses energy so that when it is on the floor, its potential energy is negative with respect to when it was on the table or the chair or the stool. Such is also the case for an electron going from a higher energy level to a lower energy level. The electron must lose energy. Correspondingly, the electron's

Electron transitions in the visible region of the spectrum are responsible for the impressive displays of color in fireworks.

energy in all its allowed energy states within the atom must be less than zero; that is, the energy must be negative.

When the electron in a hydrogen atom absorbs a quantized amount of energy and moves to an orbit with $n > 1$, the atom is said to be in an **excited state** in which it has more energy than in its ground state. Any excited state of any atom is unstable. When the electron returns to the ground state from an excited state, the electron emits the energy gained when it moved from the ground state to the excited state. The emitted energy is given off as a photon of light having an energy of hc/λ. The energy of an emitted photon corresponds to the *difference* between two energy levels of the atom (Figure 7.8). According to Bohr, the light forming the lines in the emission spectrum of hydrogen (Figure 7.7) comes from electrons in hydrogen atoms moving from higher energy orbits to lower energy orbits closer to the nucleus (Figure 7.8).

PROBLEM-SOLVING EXAMPLE 7.3 Electron Transitions

Four possible electron transitions in a hydrogen atom are given below.

	$n_{initial}$	n_{final}
(1)	2	5
(2)	5	3
(3)	7	2
(4)	4	6

(a) Which transition(s) represent a loss of energy?
(b) For which transition does the atom gain the greatest quantity of energy?
(c) Which transition corresponds to emission of the greatest quantity of energy?

Answer
(a) (2) and (3) (b) (1) (c) (3)

Strategy and Explanation First, consider the initial and final energy levels, and then whether the initial level is greater or less than the final one.

(a) Energy is emitted when electrons fall from a higher n level to a lower n level. The $n_7 \rightarrow n_2$ and $n_5 \rightarrow n_3$ transitions release energy.
(b) Energy is gained when electrons go from a lower n level to a higher n level. The $n_2 \rightarrow n_5$ transition requires more energy than the $n_4 \rightarrow n_6$ one.
(c) The $n_7 \rightarrow n_2$ transition is the most energetic (Figure 7.8). It represents the greatest energy difference between n energy levels.

PROBLEM-SOLVING PRACTICE 7.3

Identify two electron transitions in a hydrogen atom that would be of greater energy than any of those listed in Problem-Solving Example 7.3.

CONCEPTUAL EXERCISE 7.4 Many Spectral Lines, but Only One Kind of Atom

The hydrogen atom contains only one electron, but there are several lines in its line emission spectrum (Figure 7.7). How does the Bohr theory explain this?

EXERCISE 7.5 Electron Transitions

In which of these electron transitions in the hydrogen atom is energy emitted and in which is it absorbed?

(a) $n = 3 \rightarrow n = 2$ (b) $n = 1 \rightarrow n = 4$
(c) $n = 5 \rightarrow n = 3$ (d) $n = 6 \rightarrow n = 2$

Through his calculations, Bohr found that the allowed energies of the electron in a hydrogen atom are restricted by n, the principal quantum number, according to the equation

$$E = -\left(\frac{1}{n^2}\right) 2.179 \times 10^{-18}\,\text{J} \qquad (n = 1, 2, 3, \ldots)$$

where the constant, 2.179×10^{-18} J, called the Rydberg constant, can be calculated from theory. The negative sign in the equation arises due to the choice Bohr made to select zero as the energy of a completely separated electron and proton ($n = \infty$). As a consequence of this choice, all energies for the electron in the atom are negative (Figure 7.8).

Division by infinity makes E equal to zero when $n = \infty$.

The energy of the electron in a ground-state hydrogen atom ($n = 1$) is

$$E = -\left(\frac{1}{n^2}\right) 2.179 \times 10^{-18}\,\text{J} = -\left(\frac{1}{1^2}\right) 2.179 \times 10^{-18}\,\text{J} = -2.179 \times 10^{-18}\,\text{J}$$

The difference in energy, ΔE, when the electron moves from its initial state, E_i, to its final state, E_f, can be calculated as $\Delta E = E_f - E_i$. When an electron moves from a higher energy state to a lower energy state, one with a lower n value, energy is emitted and ΔE is negative. When an electron moves from a lower energy state to a higher one, energy must be absorbed (ΔE is positive). These energy differences can be expressed in relation to the change in principal energy levels, where n_i and n_f are the initial and final energy states of the atom, respectively.

Think of the atom as the system and everything else as the surroundings *(← p. 218)*.

$$\Delta E = E_f - E_i = -2.179 \times 10^{-18}\,\text{J}\left(\frac{1}{n_f^2} - \frac{1}{n_i^2}\right)$$

If $n_f > n_i$, energy is absorbed; if $n_f < n_i$, energy is emitted.

The Bohr model asserts that only certain frequencies of light can be absorbed or emitted by an atom. This is because the energy of the photon absorbed or emitted must be the same as the difference in energy, ΔE, between the two energy levels involved. If ΔE is positive, then the photon must have transferred its energy *to* the atom (the photon must have been *absorbed*). If ΔE is negative, then the photon must have received energy *from* the atom (the photon must have been *emitted*). According to Planck's quantum theory, the energy of the photon equals Planck's constant times the frequency,

$$E = h\nu = \frac{hc}{\lambda}$$

This allows us to calculate the frequency and wavelength of the photon absorbed or emitted. Notice that the energy of the photon is always positive, but the sign of the energy change of the atom indicates whether the photon is absorbed (positive change in E_{atom}) or emitted (negative change in E_{atom}).

For example, the frequency of light absorbed in the $n = 2$ to $n = 3$ transition for a hydrogen atom electron can be determined by calculating the change in energy of the atom

$$\Delta E = E_f - E_i = -2.179 \times 10^{-18}\,\text{J}\left(\frac{1}{n_f^2} - \frac{1}{n_i^2}\right)$$

$$= -2.179 \times 10^{-18}\,\text{J}\left(\frac{1}{3^2} - \frac{1}{2^2}\right) = -2.179 \times 10^{-18}\,\text{J}\left(\frac{1}{9} - \frac{1}{4}\right)$$

$$= -2.179 \times 10^{-18}\,\text{J}(0.1111 - 0.2500) = 3.026 \times 10^{-19}\,\text{J}$$

The positive energy change indicates that the atom absorbed a photon whose energy is 3.026×10^{-19} J. Therefore the frequency can be calculated as

$$\nu = \frac{E_{photon}}{h} = \frac{3.026 \times 10^{-19}\,\text{J}}{6.626 \times 10^{-34}\,\text{Js}} = 4.567 \times 10^{14}\,\text{s}^{-1}$$

Niels Bohr

1885–1962

Born in Copenhagen, Denmark, Bohr worked with J. J. Thomson and Ernest Rutherford in England, where he began to develop the ideas that led to the publication of his explanation of atomic spectra. He received the Nobel Prize in Physics in 1922 for this work.

As the director of the Institute of Theoretical Physics in Copenhagen, Bohr was a mentor to many young physicists, seven of whom later received Nobel Prizes for their studies in physics or chemistry, including Werner Heisenberg, Wolfgang Pauli, and Linus Pauling.

The wavelength (λ) of the light absorbed can be obtained from its frequency by using the relationship $\lambda = c/\nu$, where c is the velocity of light (2.998×10^8 m/s). For the $n = 2$ to $n = 3$ transition,

$$\lambda = \frac{2.998 \times 10^8 \text{ m/s}}{4.567 \times 10^{14} \text{ s}^{-1}} = 6.565 \times 10^{-7} \text{ m} = 656.5 \text{ nm}$$

This absorption is in the red region of visible light.

As seen in Table 7.2, there is exceptional agreement between the experimentally measured wavelengths and those calculated by the Bohr theory. Thus, Niels Bohr had tied the unseen (the atom) to the seen (the observable lines of the hydrogen emission spectrum)—a fantastic achievement!

Table 7.2 Agreement Between Bohr's Theory and the Lines of the Hydrogen Emission Spectrum*

Changes in Energy Levels	Wavelength Predicted by Bohr's Theory (nm)	Wavelength Determined from Laboratory Measurement (nm)	Spectral Region
$2 \rightarrow 1$	121.6	121.7	Ultraviolet
$3 \rightarrow 1$	102.6	102.6	Ultraviolet
$4 \rightarrow 1$	97.28	97.32	Ultraviolet
$3 \rightarrow 2$	656.5	656.3	Visible red
$4 \rightarrow 2$	486.5	486.1	Visible blue-green
$5 \rightarrow 2$	434.3	434.1	Visible blue
$4 \rightarrow 3$	1876	1876	Infrared

*These lines are typical; other lines could be cited as well, with equally good agreement between theory and experiment. The unit of wavelength is the nanometer (nm), 10^{-9} m.

Spectroscopy is the science of measuring spectra. Many kinds of spectroscopy have emerged since the first studies of simple line spectra. Some spectral measurements are done for quantitative analytical purposes; others are done to determine molecular structure.

PROBLEM-SOLVING EXAMPLE 7.4 Electron Transitions

(a) Calculate the frequency and wavelength (nm) corresponding to the $n = 2$ to $n = 5$ transition in a hydrogen atom.

(b) In what region of the spectrum does this transition occur?

Answer

(a) 6.907×10^{14} s^{-1}; 434.1 nm (b) Visible region

Strategy and Explanation

(a) • First calculate the frequency of the transition from which the wavelength can then be determined. In this case, $n_i = 2$ and $n_f = 5$.

$$\nu = \frac{E_{\text{photon}}}{h} = \frac{2.179 \times 10^{-18}\text{ J}}{h}\left(\frac{1}{n_i^2} - \frac{1}{n_f^2}\right)$$

$$= \left(\frac{2.179 \times 10^{-18}\text{ J}}{6.626 \times 10^{-34}\text{ Js}}\right)\left(\frac{1}{2^2} - \frac{1}{5^2}\right)$$

$$= \left(\frac{2.179 \times 10^{-18}\text{ J}}{6.626 \times 10^{-34}\text{ Js}}\right)\left(\frac{1}{4} - \frac{1}{25}\right) = (3.289 \times 10^{15}\text{ s}^{-1})(0.2500 - 0.04000)$$

$$= (3.289 \times 10^{15}\text{ s}^{-1})(0.2100) = 6.907 \times 10^{14}\text{ s}^{-1}$$

• Convert the frequency to wavelength.

$$\lambda = \frac{c}{\nu} = \frac{2.998 \times 10^8\text{ m/s}}{6.907 \times 10^{14}\text{ s}^{-1}} = 4.341 \times 10^{-7}\text{ m} = 434.1\text{ nm}$$

(b) Checking Figure 7.8 shows that electron transitions between any higher n level and the $n = 2$ level emit radiation in the visible region ($\lambda > 400$ nm; $\nu < 7.5 \times 10^{14}$ s^{-1}).

PROBLEM-SOLVING PRACTICE 7.4

(a) Calculate the frequency and the wavelength of the line for the $n = 6$ to $n = 4$ transition.

(b) Is this wavelength longer or shorter than that of the $n = 7$ to $n = 4$ transition?

CONCEPTUAL EXERCISE 7.6 Conversions

Show that the value of the Rydberg constant per photon, 2.179×10^{-18} J, is equivalent to 1312 kJ/mol photons.

CHEMISTRY YOU CAN DO

Using a Compact Disc (CD) as a Diffraction Grating

Seeing the visible spectrum by using a prism to refract visible light is a familiar experiment (Figure 7.6, p. 280). Less familiar is the use of a diffraction grating for the same purpose. A diffraction grating consists of many equally spaced parallel lines—thousands of lines per centimeter. A grating that transmits diffracted light is made by cutting thousands of grooves on a piece of glass or clear plastic; a grating that reflects diffracted light is made by cutting the grooves on a piece of metal or opaque plastic.

You can get an idea of how diffraction gratings work by using a compact disc as a diffraction grating. Compare the spectra of light from two different sources—a mercury vapor lamp and a white incandescent light bulb. Stand about 20 to 60 m away from a mercury street lamp and hold the CD about waist high with the print side down. Tilt the CD down until you see the reflected image of the street lamp. Close one eye, and view along the line from the light source to your body as you slowly tilt the CD up toward you. Use the same procedure with an incandescent light bulb as the light source.

Think about these questions:

1. What colors do you see with the mercury vapor lamp?
2. Do you see the same colors with the incandescent light? Explain your results.

Source: Mebane, R. C., and Rybolt, T. R. "Atomic Spectroscopy with a Compact Disc." *Journal of Chemical Education,* Vol. 69, 1992; p. 401.

7.4 Beyond the Bohr Model: The Quantum Mechanical Model of the Atom

The Bohr atomic model was accepted almost immediately. Bohr's success with the hydrogen atom soon led to attempts by him and others to extend the same model to more complex atoms. Before long it became apparent, however, that line spectra for elements other than hydrogen had more lines than could be explained by the simple Bohr model. A totally different approach was needed to explain electron behavior in atoms or ions with more than one electron. The new approach again took a radical departure from classical physics.

In 1924 the young physicist Louis de Broglie posed the question: *If light can be viewed in terms of both wave and particle properties, why can't particles of matter, such as electrons, be treated the same way?* And so de Broglie proposed the revolutionary idea that electrons could have wave-like properties. According to de Broglie's proposal, the wavelength (λ) of the electron (or any other particle) would depend on its mass (m) and its velocity (v) according to the relationship

$$\lambda = \frac{h}{mv}$$

where h is Planck's constant. The product of $m \times v$ for any object, including electrons, is the **momentum** of the object. Note that particles of very small mass, such as the electron (mass = 9.11×10^{-28} g), have wavelengths that are measurable. Objects of ordinary size, such as a tennis ball (mass = 56.7 g), have wavelengths too short to be observed.

A served tennis ball. Even when the ball is served at 125 mi/h, the wavelength of the tennis ball is too short to be observed.

PROBLEM-SOLVING EXAMPLE 7.5 Tennis Balls and Electrons

At the U.S. Open tennis tournament, serves routinely reach more than 100 mi/h. Compare the de Broglie wavelength (nm) of an electron moving at a velocity of 5.0×10^6 m/s with that of a tennis ball traveling at 56.0 m/s (125 mi/h). Masses: electron = 9.11×10^{-31} kg; tennis ball = 0.0567 kg.

Answer The wavelength of the electron is *much* longer than that of the tennis ball: electron = 0.15 nm; tennis ball = 2.09×10^{-25} nm.

Strategy and Explanation We can substitute the mass and velocity into the de Broglie wave equation to calculate the corresponding wavelength. Planck's constant, *h*, is

$$6.626 \times 10^{-34}\ \text{Js, and } 1\ \text{J} = \frac{1\ \text{kg m}^2}{\text{s}^2}\ \text{so that } h = 6.626 \times 10^{-34}\ \text{kg m}^2\ \text{s}^{-1}.$$

For the electron:

$$\lambda = \frac{6.626 \times 10^{-34}\ \text{kg m}^2\ \text{s}^{-1}}{(9.11 \times 10^{-31}\ \text{kg})(5.0 \times 10^6\ \text{m/s})} = 1.5 \times 10^{-10}\ \text{m} \times \frac{1\ \text{nm}}{10^{-9}\ \text{m}} = 0.15\ \text{nm}$$

For the tennis ball:

$$\lambda = \frac{6.626 \times 10^{-34}\ \text{kg m}^2\ \text{s}^{-1}}{(0.0567\ \text{kg})(56.0\ \text{m/s})} = 2.09 \times 10^{-34}\ \text{m} \times \frac{1\ \text{nm}}{10^{-9}\ \text{m}} = 2.09 \times 10^{-25}\ \text{nm}$$

The wavelength of the electron is in the X-ray region of the electromagnetic spectrum (Figure 7.1, p. 273). The wavelength of the tennis ball is far too short to observe.

PROBLEM-SOLVING PRACTICE 7.5

Calculate the de Broglie wavelength of a neutron moving at 10% the velocity of light. The mass of a neutron is 1.67×10^{-24} g.

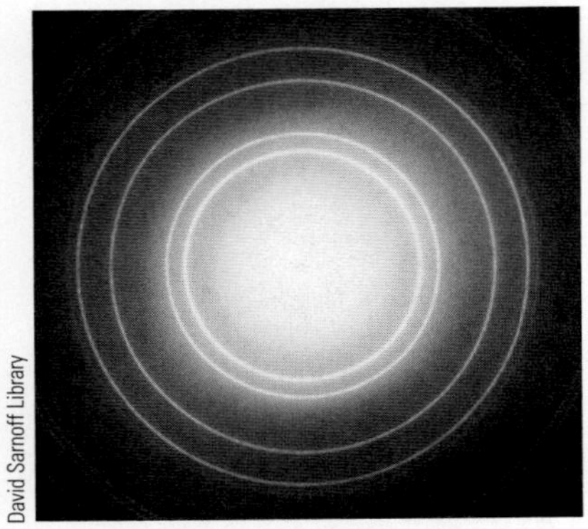

Figure 7.9 Electron diffraction pattern obtained from aluminum foil.

Many scientists found de Broglie's concept hard to accept, but experimental support for de Broglie's hypothesis was soon produced. In 1927 C. Davisson and L. H. Germer, researchers at the Bell Telephone Laboratories, found that a beam of electrons is diffracted by planes of atoms in a thin sheet of metal foil (Figure 7.9) in the same way that light waves are diffracted by a diffraction grating. Since diffraction is readily explained by the wave properties of light, it followed that electrons also can be described by the equations of waves under some circumstances.

A few years after de Broglie's hypothesis about the wave nature of the electron, Werner Heisenberg (1901–1976) proposed the **uncertainty principle,** which states that *it is impossible to simultaneously determine the exact position and the exact momentum of an electron.* This limitation is not a problem for a macroscopic object because the energy of photons used to locate such an object does not cause a measurable change in the position or momentum of that object. However, the very act of measurement would affect the position and momentum of the electron because of its very small size and mass. High-energy photons with short wavelengths would be required to locate the small electron; when such photons collide with the electron, the momentum of the electron would be changed. If lower-energy photons were used to avoid affecting the momentum, little information would be obtained about the location of the electron. Consider an analogy in photography. If you take a picture at a higher shutter speed setting of a car race, you get a clear picture of the cars but you can't tell how fast they are going or even whether they are moving. With a slow shutter speed, you can tell from the blur of the car images something about the speed and direction, but you have less information about where the car is.

The Heisenberg uncertainty principle illustrated another inadequacy in the Bohr model—its representation of the electron in the hydrogen atom in terms of well-defined orbits about the nucleus. In practical terms, the best we can do is to represent the *probability* of finding an electron of a given energy and momentum within a given space. This probability-based model of the atom is what chemists now use.

In 1926 Erwin Schrödinger (1877–1961) combined de Broglie's hypothesis with classic equations for wave motion. From these and other ideas he derived a new equation called the *wave equation* to describe the behavior of an electron in the hydrogen atom. Solutions to the wave equation are called *wave functions* and are represented by the Greek letter psi, ψ. Wave functions predict the allowed energy states of an electron and the probability of finding that electron in a given region of space. Only certain wave functions are satisfactory for an electron in an atom.

The wave function is a complex mathematical equation that has no direct physical meaning. However, the square of the wave function, ψ^2, can be represented in graphical form as a picture of the three-dimensional region of the atom where an electron with a given energy state is most likely to be found. That is, ψ^2 is related to the probability of finding the electron in a given region of space. This probability, known as the **electron density,** can be visualized as an array of dots, each dot representing the possible location of the electron with a given energy in relation to the nucleus. Figure 7.10a is an electron density distribution, sometimes called an electron cloud picture, for a ground state hydrogen atom. In such a plot, the density of the dots indicates the probability of the electron being there. The cloud of dots is denser nearer the nucleus than farther from it, indicating that the probability of finding the electron is higher nearer the nucleus than farther from it. The electron-cloud density decreases as the distance from the nucleus increases, illustrating the probability of finding the electron as distance from the nucleus increases.

Bryn Lennon/Getty Images, Inc.

Al Bello/Getty Images

A speeding race car. Top photo taken at high shutter speed. Bottom photo taken at low shutter speed.

Solving the Schrödinger wave equation results in a set of wave functions called atomic **orbitals**. Each atomic orbital contains information about the region of space where an electron of a given energy is most likely located because ψ^2 represents electron probability density. Another way to represent an electron is to draw a surface within which there is a 90% probability that the electron will be found. That is, nine times out of ten an electron will be somewhere inside such a **boundary surface**; there is one chance in ten that the electron will be outside of it (Figure 7.10b). A 100% probability is not chosen because such a surface would have no definite boundary. Consider that a typical dart board has a finite size, and, normally, players in a dart game hit the board more than 90% of the time. But if you wanted to be certain that any player, no matter how far away, would be able to hit the board on 100% of his or her throws, then the board would have to be considerably larger. By similar reasoning, a boundary surface in which there would be a 100% probability of finding a given kind of electron would have to be infinitely large. Atomic orbitals and the boundary-surface diagram such as given in Figure 7.11 will be discussed in detail in Sections 7.5 and 7.6.

Electron density is also called probability density.

The density of dots means the number of dots within a specified volume.

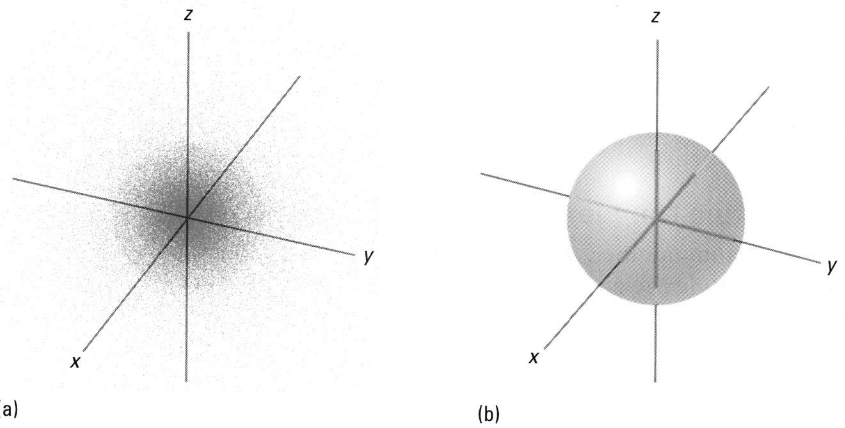

Figure 7.10 **(a) Electron density distribution in a ground-state hydrogen atom. (b) A boundary surface.**

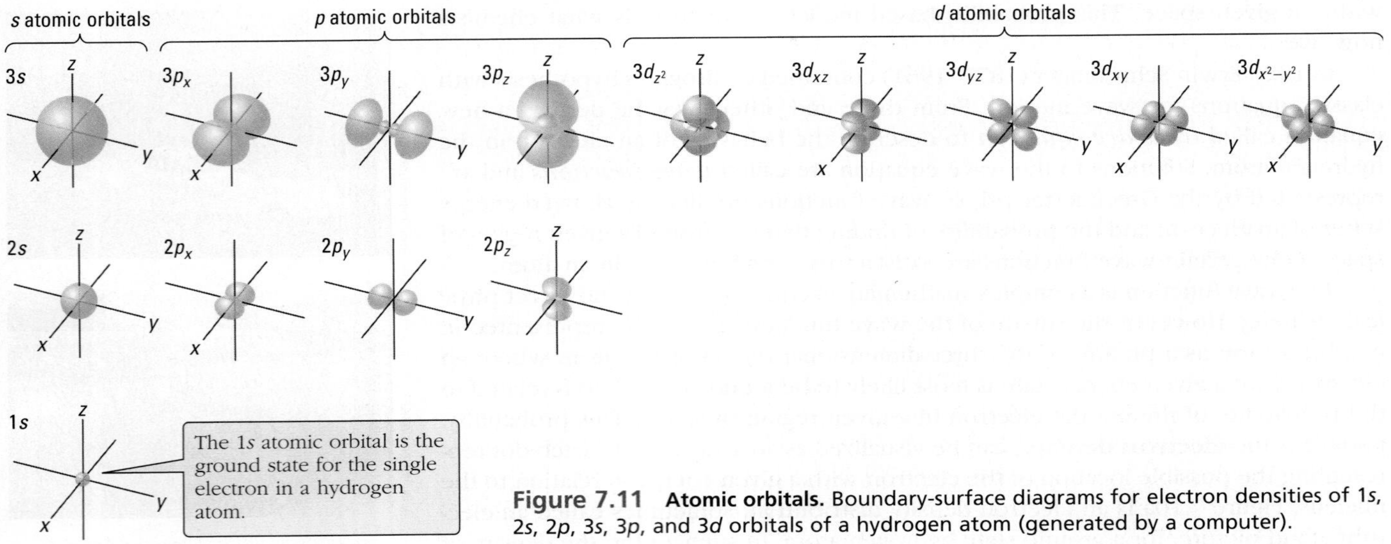

Figure 7.11 **Atomic orbitals.** Boundary-surface diagrams for electron densities of 1*s*, 2*s*, 2*p*, 3*s*, 3*p*, and 3*d* orbitals of a hydrogen atom (generated by a computer).

Note that an atomic *orbital* (quantum mechanical model) is not the same as an *orbit* (Bohr model). In the quantum mechanical model, the principal quantum number, n, is a measure of the most probable distance of the electron from the nucleus, not the radius of a well-defined orbit.

We turn next to describing quantum numbers, which define the orbitals and energy states available to an electron in an atom. Following that, we will relate quantum numbers to the shapes of atomic orbitals.

7.5 Quantum Numbers, Energy Levels, and Atomic Orbitals

In their quantum mechanical calculations, Schrödinger and others found that a set of three integers, called **quantum numbers**—the principal quantum number n and two others, ℓ and m_ℓ—are needed to describe the three-dimensional coordinates of an electron's motion in the atomic orbitals in a hydrogen atom. The need for a fourth quantum number, m_s, was identified in subsequent work by others. Thus, a set of four quantum numbers—n, ℓ, m_ℓ, and m_s—is used to denote the energy and the shape of the electron cloud for each electron.

We can apply quantum numbers to electrons in any atom, not just hydrogen. The quantum numbers, their meanings, and the quantum number notations used for atomic orbitals are given in the following subsections.

First Quantum Number, *n*: Principal Energy Levels

The principal quantum number, n, is the same n found from spectra (p. 281).

The first quantum number, n, the principal quantum number, is the most important one in determining the energy of an electron. The quantum number n has only integer values, starting with 1:

$$n = 1, 2, 3, 4, \ldots$$

The value of n corresponds to a **principal energy level** so that an electron is in the first principal energy level when $n = 1$, in the second principal level when

$n = 2$, and so on. As n increases, the energy of the electron increases as well, and the electron, on average, is farther away from the nucleus and is less tightly bound to it.

Second Quantum Number, ℓ: Atomic Orbital Shapes

Each principal energy level is also known as an electron **shell**, a collection of atomic orbitals with the *same* principal quantum number value, n. When $n = 1$, there is only one kind of atomic orbital possible. When $n = 2$, two kinds of atomic orbitals are possible; likewise, three kinds are possible when $n = 3$, four when $n = 4$, and so on.

Within each principal energy level (shell) are **subshells**. The $n = 1$ principal energy level has only one subshell; all other energy levels ($n > 1$) have more than one subshell. The subshells are designated by the second quantum number, ℓ (sometimes referred to as the *azimuthal quantum number*). *The* ***shape*** *of the electron cloud corresponding to an atomic orbital is determined by the value of* ℓ.

A subshell is one or more atomic orbitals with the same n and ℓ quantum numbers. Thus, the value of ℓ is associated with an n value. The value of ℓ is an integer that ranges from zero to a maximum of $n - 1$:

$$\ell = 0, 1, 2, 3, \ldots, (n - 1)$$

According to this relationship, when $n = 1$, then ℓ must be zero. This indicates that in the first principal energy level ($n = 1$) there is only one subshell, with an ℓ value of 0. The second principal energy level ($n = 2$) has two subshells—one with an ℓ value of 0 and a second with an ℓ value of 1. Continuing in this manner,

$$n = 3; \quad \ell = 0, 1, \text{ or } 2 \quad \text{(three subshells)}$$

$$n = 4; \quad \ell = 0, 1, 2 \text{ or } 3 \quad \text{(four subshells)}$$

As mentioned earlier, a boundary surface for an atomic orbital can be drawn within which there is a 90% probability that an electron can be found. Note in Figure 7.11 that as n increases, two things happen: (1) orbitals become larger and (2) there is an increase in the number and types of atomic orbitals within a given level. ***The number of atomic orbital types within a principal energy level equals* n.** Thus, in the $n = 3$ level, there are three different types of subshells—that is, s, p, and d orbitals. *Atomic orbitals with the same* n *and* ℓ *values are in the same subshell.*

Rather than using ℓ values, subshells are more commonly designated by letters: *s, p, d,* or *f*. The first four subshells are known as the *s* subshell, *p* subshell, *d* subshell, and *f* subshell.

The letters *s, p, d,* and *f* derived historically from spectral lines called *s*harp, *p*rincipal, *d*iffuse, and *f*undamental.

ℓ value	0	1	2	3
Subshell	*s*	*p*	*d*	*f*

A number (the n value) and a letter (*s, p, d,* or *f*) are used to designate a specific subshell.

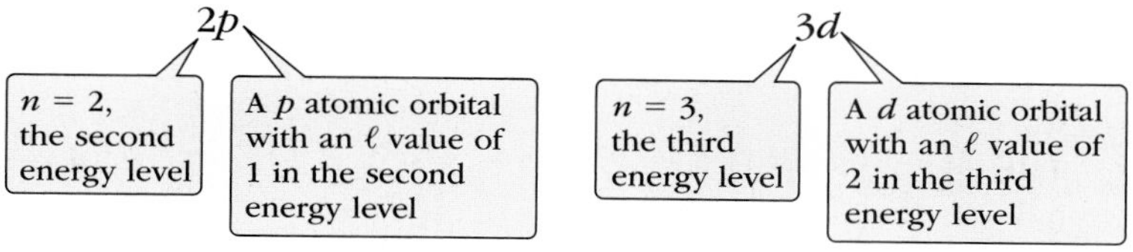
2p
n = 2, the second energy level
A p atomic orbital with an ℓ value of 1 in the second energy level
3d
n = 3, the third energy level
A d atomic orbital with an ℓ value of 2 in the third energy level

For the first four principal energy levels, these designations are as follows.

n	1	2		3			4			
ℓ	0	0	1	0	1	2	0	1	2	3
Level	1*s*	2*s*	2*p*	3*s*	3*p*	3*d*	4*s*	4*p*	4*d*	4*f*

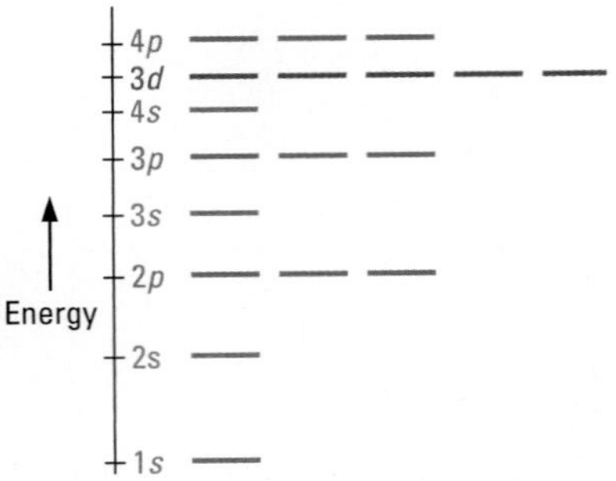

Figure 7.12 The relative ordering of atomic orbital energy levels in a many-electron atom from the 1*s* level through the 4*p* sublevels. Atomic orbitals in different sublevels of the same level have different energy. The distance between sublevels is nonlinear; it decreases as *n* increases.

For atoms with two or more electrons, atomic orbitals in different subshells but with the same n value have different energies. The energies of subshells within a given principal energy level always increase in the order $ns < np < nd < nf$. Consequently, a 3*p* subshell has a higher energy than a 3*s* subshell but less energy than a 3*d* subshell, which would have the highest energy in the $n = 3$ level. Figure 7.12 shows the relative energies of atomic orbitals in many-electron atoms through the 4*p* orbitals.

EXERCISE 7.7 Subshell Designations

Give the subshell designation for an electron with these quantum numbers.
(a) $n = 5$, $\ell = 2$ (b) $n = 4$, $\ell = 3$ (c) $n = 6$, $\ell = 1$

EXERCISE 7.8 Subshell Designations

Explain why (a) 3*f* is an incorrect subshell designation, (b) $n = 2$, $\ell = 2$ is incorrect for a subshell.

Third Quantum Number, m_ℓ: Orientation of Atomic Orbitals

The magnetic quantum number, m_ℓ, can have any integer value between ℓ and $-\ell$, including zero. Thus,

$$m_\ell = \ell, \ldots, +1, 0, -1, \ldots, -\ell$$

For an *s* subshell, which has an ℓ value of zero, m_ℓ has only one value—zero. Therefore, an *s* subshell, regardless of its n value—1*s*, 2*s*, and so on—contains only one orbital, whose m_ℓ value is zero. For a *p* subshell, ℓ equals 1, and so m_ℓ can be $+1$, 0, or -1. This means that within each *p* subshell there are three *different* atomic orbitals: one with $m_\ell = +1$, another with $m_\ell = 0$, and a third with $m_\ell = -1$. *In general, for a subshell of quantum number ℓ, there is a total of $2\ell + 1$ atomic orbitals within that subshell.* Atomic orbitals within the same *subshell* have essentially the same energy.

Table 7.3 summarizes the relationships among n, ℓ, and m_ℓ. ***The total number of atomic orbitals in a shell equals* n^2.** For example, an $n = 3$ shell has a total of nine atomic orbitals: one 3*s* + three 3*p* + five 3*d*.

Table 7.3 Relationships Among n, ℓ, and m_ℓ for the First Four Principal Energy Levels

n Value	ℓ Value	Subshell Designation	m_ℓ Values	Number of Atomic Orbitals in Subshell, $2\ell + 1$	Total Number of Atomic Orbitals in Shell, n^2
1	0	1*s*	0	1	1
2	0	2*s*	0	1	
2	1	2*p*	1, 0, −1	3	4
3	0	3*s*	0	1	
3	1	3*p*	1, 0, −1	3	
3	2	3*d*	2, 1, 0, −1, −2	5	9
4	0	4*s*	0	1	
4	1	4*p*	1, 0, −1	3	
4	2	4*d*	2, 1, 0, −1, −2	5	
4	3	4*f*	3, 2, 1, 0, −1, −2, −3	7	16

In summary, the relation between an atomic orbital and its first three quantum numbers is

- ***n* relates to the atomic orbital's size.**
- **ℓ relates to the atomic orbital's shape.**
- **m_ℓ relates to the atomic orbital's orientation.**

The m_ℓ value, in conjunction with the ℓ value, is related to the shape and orientation of an atomic orbital in space. The *s* atomic orbitals are spherical; their size enlarges as *n* increases (Figure 7.11, p. 288).

Beginning with the second principal energy level ($n = 2$) and at each higher *n* level, each energy level has three *p* atomic orbitals. The three *p* atomic orbitals are dumbbell-shaped and oriented at right angles to each other, with maximum electron density directed along either the *x*-, *y*-, or *z*-axis, as shown in Figure 7.11. They are sometimes designated as p_x, p_y, and p_z atomic orbitals. The five *d* atomic orbitals in the 3*d* sublevel ($\ell = 2$) are also illustrated in Figure 7.11. We will discuss *d* atomic orbitals further in Chapter 22 in connection with their role in the chemistry of transition metal ions.

We leave the discussion of *f* atomic orbitals, which are very complex, to subsequent courses.

PROBLEM-SOLVING EXAMPLE 7.6 Quantum Numbers, Subshells, and Atomic Orbitals

Consider the $n = 4$ principal energy level.

(a) Without referring to Table 7.3, predict the number of subshells in this level.
(b) Identify each of the subshells by its number and letter designation (as in 1*s*) and give its ℓ values.
(c) Use the $2\ell + 1$ rule to calculate how many atomic orbitals each subshell has and identify the m_ℓ value for each orbital.
(d) What is the total number of atomic orbitals in the $n = 4$ level?

Answer

(a) Four subshells
(b) 4*s*, 4*p*, 4*d*, and 4*f*; $\ell = 0$, 1, 2, and 3, respectively
(c) One 4*s* atomic orbital, three 4*p* atomic orbitals, five 4*d* atomic orbitals, and seven 4*f* atomic orbitals
(d) 16 orbitals

Strategy and Explanation

(a) There are *n* subshells in the *n*th level. Thus, the $n = 4$ level contains four subshells.
(b) The number refers to the principal quantum number, *n*; the letter is associated with the ℓ quantum number. The four sublevels correspond to the four possible ℓ values:

Sublevels	4*s*	4*p*	4*d*	4*f*
ℓ value	0	1	2	3

(c) There are $2\ell + 1$ atomic orbitals within a sublevel. Only one 4*s* atomic orbital is possible ($\ell = 0$, so m_ℓ must be zero). There are three 4*p* atomic orbitals ($\ell = 1$) with m_ℓ values of 1, 0, or -1. There are five 4*d* atomic orbitals ($\ell = 2$) corresponding to the five allowed values for m_ℓ: 2, 1, 0, -1, and -2. There are seven 4*f* atomic orbitals ($\ell = 3$), each with one of the seven permitted values of m_ℓ: 3, 2, 1, 0, -1, -2, and -3.
(d) The total number of atomic orbitals in a level is n^2. Therefore, the $n = 4$ level has a total of 16 atomic orbitals.

PROBLEM-SOLVING PRACTICE 7.6

(a) Identify the subshell with $n = 6$ and $\ell = 2$.
(b) How many atomic orbitals are in this subshell?
(c) What are the m_ℓ values for these atomic orbitals?

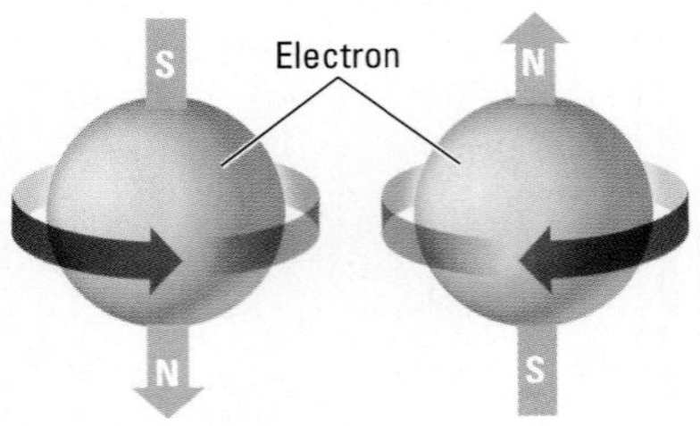

Figure 7.13 Spin directions. The electron can be pictured as though it were a charged sphere spinning about an axis through its center. The electron can have only two directions of spin; any other position is forbidden. Therefore, the spin of the electron is said to be quantized.

Fourth Quantum Number, m_s: Electron Spin

When spectroscopists more closely studied emission spectra of hydrogen and sodium atoms, they discovered that what were originally thought to be single lines were actually very closely spaced pairs of lines. In 1925 the Dutch physicists George Uhlenbeck and Samuel Goudsmit proposed that the line splitting could be explained by assuming that each electron in an atom can exist in one of two possible spin states. To visualize these states, consider an electron as a charged sphere rotating about an axis through its center (Figure 7.13). Such a spinning charge generates a magnetic field, so that each electron acts like a tiny bar magnet with north and south magnetic poles. Only two directions of spin are possible in relation to the direction of an external magnetic field—clockwise or counterclockwise. Spins in opposite directions produce oppositely directed magnetic fields, which result in two slightly different energies. This slight difference in energy splits the spectral lines into closely spaced pairs.

Thus, to describe an electron in an atom completely, a fourth quantum number, m_s, called the *spin quantum number,* is needed in conjunction with the other three quantum numbers.

The spin quantum number can have just one of two values: $+\frac{1}{2}$ or $-\frac{1}{2}$. Electrons are said to have *parallel* spins if they have the same m_s quantum number (both $+\frac{1}{2}$ or both $-\frac{1}{2}$). Electrons are said to be *paired* when they are in the same atomic orbital and have opposite spins—one has an m_s of $+\frac{1}{2}$ and the other has a $-\frac{1}{2}$ value.

To make quantum theory consistent with experiment, Wolfgang Pauli stated in 1925 what is now known as the **Pauli exclusion principle:** ***No more than two electrons can occupy the same atomic orbital in an atom, and these electrons must have opposite spins.*** This is equivalent to saying that no two electrons in the same atom can have the same set of four quantum numbers n, ℓ, m_ℓ, and m_s. For example, two electrons can occupy the same 3*p* atomic orbital only if their spins are paired ($+\frac{1}{2}$ and $-\frac{1}{2}$). In such a case, the electrons would have the same first three quantum numbers, for example, $n = 3$, $\ell = 1$, and $m_\ell = +1$, but would differ in their m_s values.

EXERCISE 7.9 Quantum Numbers

Give the set of four quantum numbers for each electron in a 3*s* atomic orbital in a sodium atom.

CONCEPTUAL EXERCISE 7.10 Orbitals and Quantum Numbers

Give sets of four quantum numbers for two electrons that are (a) in the same level and subshell, but in different atomic orbitals, and (b) in the same level, but in different subshells and in different atomic orbitals.

EXERCISE 7.11 Quantum Number Comparisons

Two electrons in the same atom have these sets of quantum numbers: electron (a) has 3, 1, 0, $+\frac{1}{2}$; electron (b) has 3, 1, -1, $+\frac{1}{2}$. Show that these two electrons are not in the same atomic orbital. Which subshell are these electrons in?

The restriction that only two electrons can occupy a single atomic orbital has the effect of establishing the maximum number of electrons for each principal energy level, *n,* as summarized in Table 7.4. Since each atomic orbital can hold a maximum of two paired electrons, the general rule is that ***each principal energy level,* n, *can accommodate a maximum number of* $2n^2$ *electrons.*** For exam-

Table 7.4 Number of Electrons Accommodated in Electron Shells and Subshells

Electron Shell (n)	Number of Subshells Available ($=n$)	Number of Atomic Orbitals Available ($=2\ell + 1$)	Number of Electrons Possible in Subshell	Maximum Electrons for nth Shell ($=2n^2$)
1	*s*	1	2	2
2	*s*	1	2	
	p	3	6	8
3	*s*	1	2	
	p	3	6	
	d	5	10	18
4	*s*	1	2	
	p	3	6	
	d	5	10	
	f	7	14	32
5	*s*	1	2	
	p	3	6	
	d	5	10	
	f	7	14	
	*g**	9	18	50
6	*s*	1	2	
	p	3	6	
	d	5	10	
	*f**	7	14	
	*g**	9	18	
	*h**	11	22	72
7	*s*	1	2	

*These atomic orbitals are not used in the ground state of any known element.

The results expressed in this table were predicted by the Schrödinger theory and have been confirmed by experiment.

ple, the $n = 2$ energy level has one *s* and three *p* atomic orbitals, each of which can accommodate two paired electrons. Therefore, this level can accommodate a total of $2 \times (2)^2 =$ eight electrons (two in the 2*s* orbital and three pairs in the three 2*p* orbitals).

In summary,

- ***The principal energy level (shell) has a quantum number*** n ***= 1, 2, 3, . . .***
- ***Within each principal energy level there are subshells equal in number to*** n ***and designated as*** s, p, d, ***and*** f ***subshells.***
- ***The number of atomic orbitals in each subshell is*** $2\ell + 1$***: one*** s ***atomic orbital (***$\ell = 0$***), three*** p ***atomic orbitals (***$\ell = 1$***), five*** d ***atomic orbitals (***$\ell = 2$***), and seven*** f ***atomic orbitals (***$\ell = 3$***).***
- ***Within a principal energy level*** n***, there are*** n^2 ***atomic orbitals.***
- ***Within a principal energy level*** n***, there is a maximum of*** $2n^2$ ***electrons.***

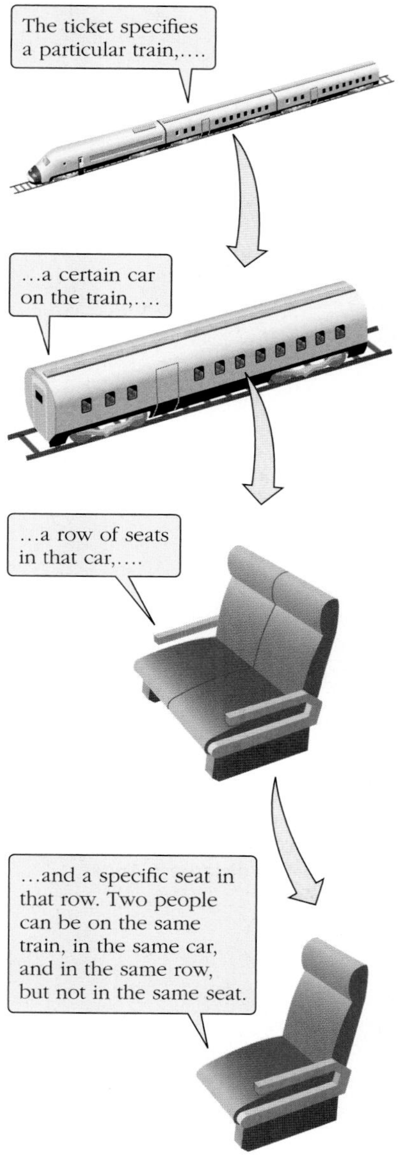

A quantum number analogy. A ticket for a reserved seat on a train is analogous to a set of four quantum numbers, n, ℓ, m_ℓ, and m_s. The ticket specifies a particular train, a certain car on the train, a row of seats in that car, and a specific seat in that row. The row is analogous to an atomic orbital, and the occupants of the row are like two electrons in the same orbital (same train, car, and row), but with opposite spins (different seats).

EXERCISE 7.12 Maximum Number of Electrons

(a) What is the maximum number of electrons in the $n = 3$ level? Identify the atomic orbital of each electron.

(b) What is the maximum number of electrons in the $n = 5$ level? Identify the atomic orbital of each electron. (*Hint:* *g* atomic orbitals follow *f* atomic orbitals.)

CONCEPTUAL EXERCISE 7.13 *g* Atomic Orbitals

Using the same reasoning as was developed for *s, p, d,* and *f* atomic orbitals, what should be the *n* value of the first shell that could contain *g* atomic orbitals, and how many *g* atomic orbitals would be in that shell?

7.6 Shapes of Atomic Orbitals

As noted in the previous section, the ℓ and m_ℓ quantum numbers relate to the shapes and spatial orientations of atomic orbitals. We now consider these in more detail.

s Atomic Orbitals ($\ell = 0$)

Electron density plots, such as those shown in Figure 7.10a, are one way to depict atomic orbitals. Figure 7.14a shows the dot-density diagram of the 1*s* atomic orbital of a hydrogen atom with its electron density decreasing rapidly as the distance from the nucleus increases. The value r_{90} in the figure is the radius of the sphere in which the electron is found 90% of the time. Note from Figure 7.14a that the probability of finding a 1*s* electron is the same *in any direction* at the same distance from the nucleus. Thus, the 1*s* atomic orbital is spherical.

In Figure 7.14a the electron density illustrates that the probability of finding the 1*s* electron in a hydrogen atom varies throughout space. This probability can be represented in another way known as a **radial distribution plot**, as shown in Figure 7.14b. In this case, the *y*-axis represents the probability of finding the electron at a given distance, *r*, from the nucleus (*x*-axis). The probability approaches zero as *r* approaches zero; that is, there is a very small probability of finding the electron at the nucleus. The greatest probability of finding the 1*s* electron in a ground-state hydrogen atom is 0.0529 nm (52.9 pm) from the nucleus. Note from the figure that at larger distances the probability drops very close to zero, but does not quite become zero. This indicates that there is a very small, but finite, probability of the electron being at a considerable distance from the nucleus.

In Figure 7.14b, the *y*-axis probability values equal the square of the wave function, ψ^2, times 4π times the square of the distance *r*: $4\pi r^2\psi^2$.

As the *n* value increases, the shape of *s* atomic orbitals remains the same; they are all spherical. Their sizes, however, increase as *n* increases, as shown in Figure 7.14c. For example, the boundary surface of a 3*s* orbital has a greater volume (and radius) than that of a 2*s* atomic orbital, which is greater than that of a 1*s* atomic orbital.

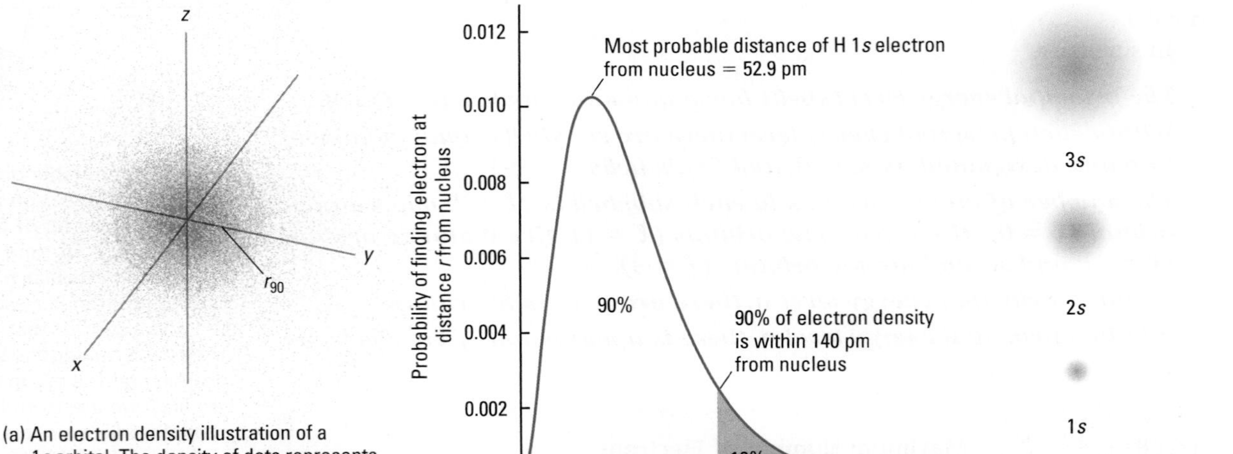
z
y
x
r_{90}
Probability of finding electron at distance *r* from nucleus
0.012
0.010
0.008
0.006
0.004
0.002
0.000
Most probable distance of H 1*s* electron from nucleus = 52.9 pm
90%
90% of electron density is within 140 pm from nucleus
10%
0
50
100
150
200
250
300
Distance from nucleus, *r* (pm)
3*s*
2*s*
1*s*

(a) An electron density illustration of a 1*s* orbital. The density of dots represents the probability of finding the electron in any region of the diagram.

(b) A radial distribution plot as a function of distance from the nucleus for a 1*s* atomic orbital of hydrogen.

(c) The relative sizes of *s* atomic orbitals. Only electron density within each boundary surface is shown.

Figure 7.14 Depicting *s* atomic orbitals.

p Atomic Orbitals ($\ell = 1$)

Unlike *s* atomic orbitals, which are all spherical, the *p* atomic orbitals, those for which $\ell = 1$, are all dumbbell-shaped (Figure 7.15). In a *p* atomic orbital, there are two lobes with electron density on either side of the nucleus. The *p* atomic orbitals within a given subshell (same n and ℓ value) differ from each other in their orientation in space; the three *p* atomic orbitals within the same subshell are mutually perpendicular and lie along the *x*-, *y*-, and *z*-axes. These orientations correspond to the three allowed m_ℓ values of +1, 0, and −1.

d Atomic Orbitals ($\ell = 2$)

Each energy level with $n \geq 3$ contains a subshell for which $\ell = 2$, consisting of five *d* atomic orbitals with corresponding m_ℓ values of +2, +1, 0, −1, and −2. The *d* atomic orbitals consist of two different types of shapes and spatial orientations. Three of them (d_{xz}, d_{yz}, and d_{xy}) each have four lobes that lie in the plane of but between the designated *x*-, *y*-, or *z*-axes; the other pair—the $d_{x^2-y^2}$ and d_{z^2} orbitals—have their principal electron density along the designated axes. The shapes of five *d* atomic orbitals, along with those of *s* and *p* atomic orbitals, are illustrated in Figure 7.16. We will discuss *d* atomic orbitals further in Chapter 22 in connection with their role in the chemistry of transition metal ions.

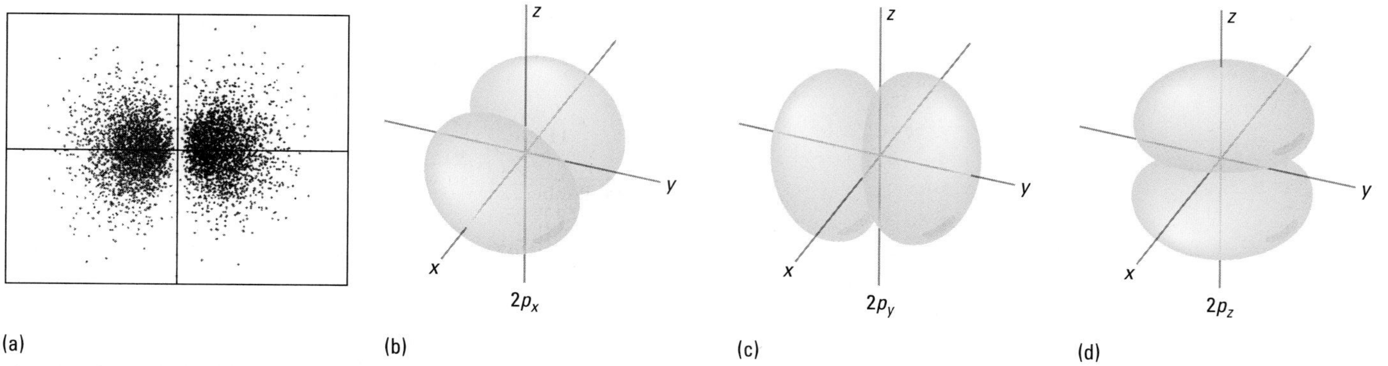

Figure 7.15 Depicting *p* atomic orbitals. (a) Electron density distribution of a 2*p* atomic orbital. (b), (c), and (d) Boundary surfaces for the set of three 2*p* atomic orbitals. (From Zumdahl and Zumdahl, *Chemistry,* 6th ed., Fig. 7.14, p. 310. Copyright © 2007 Brooks/Cole, a part of Cengage Learning, Inc. Reproduced by permission. www.cengage.com/permissions.)

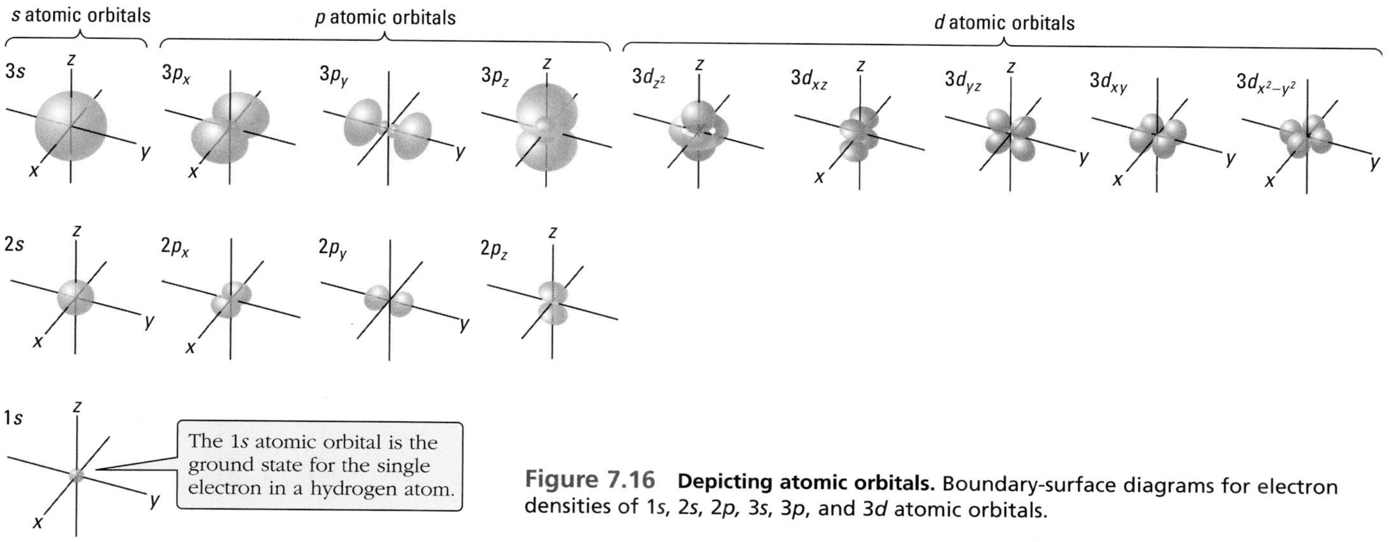

Figure 7.16 Depicting atomic orbitals. Boundary-surface diagrams for electron densities of 1*s*, 2*s*, 2*p*, 3*s*, 3*p*, and 3*d* atomic orbitals.

7.7 Atom Electron Configurations

The complete description of the atomic orbitals occupied by all the electrons in an atom or ion is called its **electron configuration.** The periodic table can serve as a guide to determine the electron configurations of atoms. As you will see, the chemical similarities of elements in the same periodic table group are explained by the similar electron configurations of their atoms.

Electron Configurations of Main Group Elements

The atomic numbers of the elements increase in numerical order throughout the periodic table. As a result, atoms of a particular element each contain one more electron (and proton) than atoms of the preceding element. How do we know which shell and atomic orbital each new electron occupies? An important principle for answering this question is this: *for an atom in its ground state, electrons are found in the energy shells, subshells, and atomic orbitals that produce the lowest energy for the atom.* In other words, electrons fill atomic orbitals starting with the 1*s* atomic orbital and work upward in the subshell energy order shown in Figure 7.12 (p. 290).

To better understand how this filling of atomic orbitals works, consider the experimentally determined electron configurations of the first ten elements, which are written in three different ways in Table 7.5—condensed, expanded, and atomic orbital box diagram. Since electrons assigned to the $n = 1$ shell are closest to the nucleus and therefore lowest in energy, electrons are assigned to it first (H and He). At the left in Table 7.5, the occupied atomic orbitals and the number of electrons in each atomic orbital are represented by this notation:

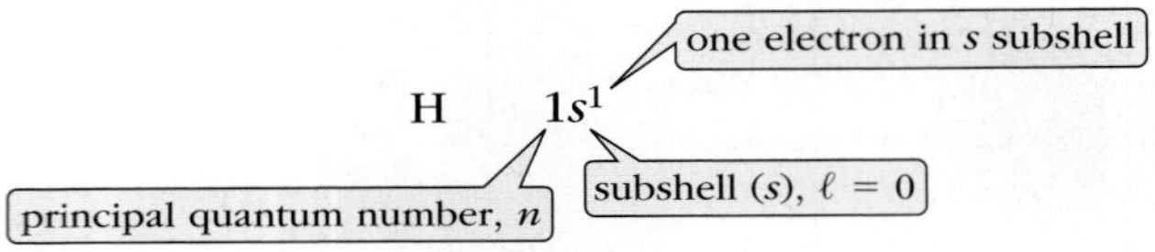

one electron in s subshell
H 1s¹
subshell (s), ℓ = 0
principal quantum number, n

At the right in the table, each occupied atomic orbital is represented by a box in which electrons are shown as arrows: ↑ for a single electron in an orbital and ↑↓ for paired electrons in an atomic orbital.

In helium the two electrons are paired in the 1*s* atomic orbital so that the lowest energy shell ($n = 1$) is filled. After the $n = 1$ shell is filled, electrons are assigned

Table 7.5 Electron Configurations of the First Ten Elements

	Electron Configurations		Atomic Orbital Box Diagrams				
	Condensed	*Expanded*	1*s*	2*s*	2*p*		
H	$1s^1$		↑				
He	$1s^2$		↑↓				
Li	$1s^2 2s^1$		↑↓	↑			
Be	$1s^2 2s^2$		↑↓	↑↓			
B	$1s^2 2s^2 2p^1$		↑↓	↑↓	↑		
C	$1s^2 2s^2 2p^2$	$1s^2 2s^2 2p^1 2p^1$	↑↓	↑↓	↑	↑	
N	$1s^2 2s^2 2p^3$	$1s^2 2s^2 2p^1 2p^1 2p^1$	↑↓	↑↓	↑	↑	↑
O	$1s^2 2s^2 2p^4$	$1s^2 2s^2 2p_x^2 2p^1 2p^1$	↑↓	↑↓	↑↓	↑	↑
F	$1s^2 2s^2 2p^5$	$1s^2 2s^2 2p^2 2p^2 2p^1$	↑↓	↑↓	↑↓	↑↓	↑
Ne	$1s^2 2s^2 2p^6$	$1s^2 2s^2 2p^2 2p^2 2p^2$	↑↓	↑↓	↑↓	↑↓	↑↓

to the next lowest unoccupied energy level (Table 7.5), the $n = 2$ shell, beginning with lithium. This second shell can hold eight electrons ($2n^2$), and its atomic orbitals are occupied sequentially in the eight elements from lithium to neon. Notice in the periodic table inside the front cover that these are the eight elements of the second period.

As happens in each principal energy level (and each period), the first two electrons fill the *s* atomic orbital. In the second period this occurs in Li ($1s^22s^1$) and in Be ($1s^22s^2$), which has the 2*s* atomic orbital filled. The next element is boron, B ($1s^22s^22p^1$), and the fifth electron goes into a 2*p* atomic orbital. The three 2*p* atomic orbitals are of equal energy, and it does not matter which 2*p* atomic orbital is occupied first. Adding a second *p* electron for the next element, carbon ($1s^22s^22p^2$), presents a choice. Does the second 2*p* electron in the carbon atom pair with the existing electron in a 2*p* atomic orbital, or does it occupy a 2*p* atomic orbital by itself? It has been shown experimentally that both 2*p* electrons have the same spin. (The electrons are said to have *parallel* spins.) Hence, they must occupy different 2*p* atomic orbitals; otherwise, they would violate the Pauli exclusion principle. The expanded electron configurations in the middle of Table 7.5 show the locations of the *p* electrons individually in the boron, carbon, and nitrogen atoms' 2*p* atomic orbitals.

Because electrons are negatively charged particles, electron configurations where electrons have parallel spins also minimize electron-electron repulsions, making the total energy of the set of electrons as low as possible. **Hund's rule** summarizes how subshells are filled: *The most stable arrangement of electrons in the same subshell has the maximum number of unpaired electrons, all with the same spin:* ***Electrons pair only after each atomic orbital in a subshell is occupied by a single electron.*** The general result of Hund's rule is that in *p, d,* or *f* subshells, each successive electron enters a different atomic orbital of the subshell until the subshell is half-full, after which electrons pair in the atomic orbitals one by one. We can see this in Table 7.5 for the elements that follow boron—carbon, nitrogen, oxygen, fluorine, and neon.

Suppose you need to predict the electron configuration of a ground state phosphorus atom. You can use a periodic table (inside the front cover) to do so. The expanded periodic table in Figure 7.17 shows how. First, locate phosphorus in the periodic table; it is in the third period, $n = 3$ and Group 5A. Then, start with H at the top of the periodic table and work your way along each row until you reach phosphorus, writing electron configurations as you go. In the first row (first

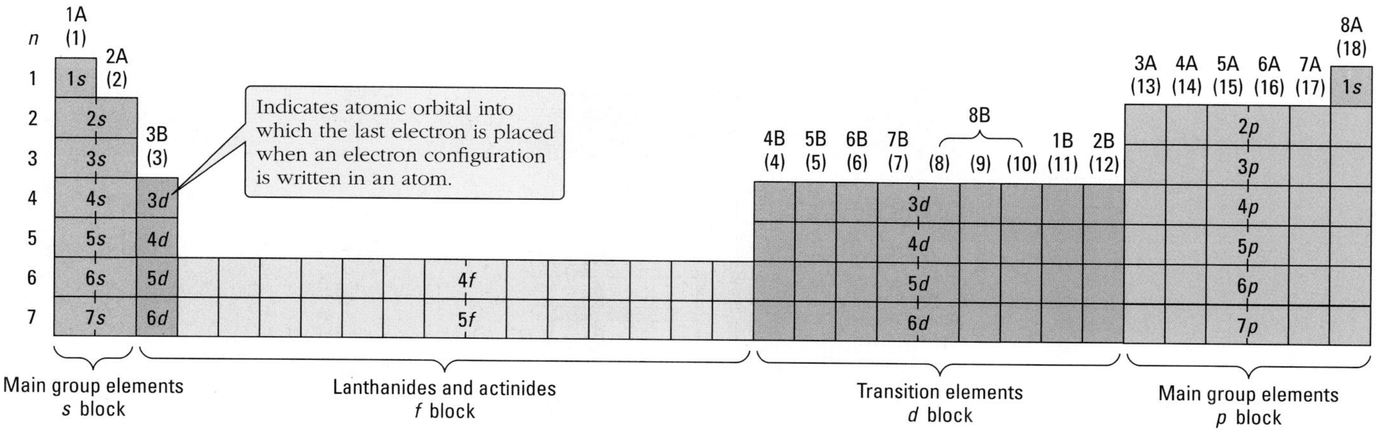

Figure 7.17 Electron configuration and the periodic table. Electron configurations can be generated by assigning electrons to the subshells shown here starting at H and moving through this table in atomic number order until the desired element is reached. The electron configurations in atomic number order are given in Appendix D. The *s* block refers to elements where the last electron in the electron configuration is in an *s* atomic orbital; for example, Mg is in the *s* block because its configuration is $1s^2\ 2s^2\ 2p^6\ 3s^2$. The *p*-block, *d*-block, and *f*-block elements are defined similarly.

period), Figure 7.17 indicates that the 1*s* atomic orbital is being filled, so write $1s^2$. By the end of the second row (second period) Figure 7.17 indicates that the 2*s* orbital and then the 2*p* atomic orbitals have been filled, so the electron configuration at that point is $1s^22s^22p^6$. In the third period, the 3*s* atomic orbital fills and the 3*p* atomic orbitals are only partly filled when you reach phosphorus in Group 5A, so the complete electron configuration is

$$\text{P} \quad 1s^22s^22p^63s^23p^3$$

You can check this by noting that phosphorus has atomic number 15 and therefore must have 15 electrons. Adding the superscripts in the electron configuration gives $2 + 2 + 6 + 2 + 3 = 15$.

Another way to write this electron configuration is called the **noble gas notation**. This uses brackets around the symbol of the noble gas at the end of the period *preceding* the element whose configuration is being written. In the case of phosphorus, the preceding noble gas is Ne, at the end of the second period, so [Ne] is used to represent the electron configuration up to the end of that period, $1s^22s^22p^6$. The electron configuration for phosphorus in noble gas notation is [Ne] $3s^23p^3$. Noble gas notation is used to present the ground state electron configurations for all elements in Appendix D.

According to Hund's rule, the three electrons in the 3*p* atomic orbitals of the phosphorus atom must be unpaired. To show this, you can write the *expanded* electron configuration

$$\text{P} \quad 1s^22s^22p^22p^22p^23s^23p^13p^13p^1$$

or the atomic orbital box diagram

P	1*s*	2*s*	2*p*	3*s*	3*p*
	↑↓	↑↓	↑↓ ↑↓ ↑↓	↑↓	↑ ↑ ↑

Thus, all the electrons are paired except for the three electrons in 3*p* atomic orbitals. These three electrons each occupy different 3*p* atomic orbitals, and they have parallel spins. The partial atomic orbital box diagram of elements in Period 3 is given in Figure 7.18.

At this point, you should be able to write electron configurations and atomic orbital box diagrams for main group elements through Ca, atomic number 20, using the periodic table to assist you.

Atomic number/ element	Partial atomic orbital box diagram (3*s* and 3*p* sublevels only) 3*s*	3*p*	Electron configuration	Noble gas notation
$_{11}$Na	[↑]	[ ] [ ] [ ]	$[1s^22s^22p^6]\,3s^1$	[Ne] $3s^1$
$_{12}$Mg	[↑↓]	[ ] [ ] [ ]	$[1s^22s^22p^6]\,3s^2$	[Ne] $3s^2$
$_{13}$Al	[↑↓]	[↑] [ ] [ ]	$[1s^22s^22p^6]\,3s^23p^1$	[Ne] $3s^23p^1$
$_{14}$Si	[↑↓]	[↑] [↑] [ ]	$[1s^22s^22p^6]\,3s^23p^2$	[Ne] $3s^23p^2$
$_{15}$P	[↑↓]	[↑] [↑] [↑]	$[1s^22s^22p^6]\,3s^23p^3$	[Ne] $3s^23p^3$
$_{16}$S	[↑↓]	[↑↓] [↑] [↑]	$[1s^22s^22p^6]\,3s^23p^4$	[Ne] $3s^23p^4$
$_{17}$Cl	[↑↓]	[↑↓] [↑↓] [↑]	$[1s^22s^22p^6]\,3s^23p^5$	[Ne] $3s^23p^5$
$_{18}$Ar	[↑↓]	[↑↓] [↑↓] [↑↓]	$[1s^22s^22p^6]\,3s^23p^6$	[Ne] $3s^23p^6$

Figure 7.18 **Partial atomic orbital diagrams for Period 3 elements.**

Electron Configurations of Transition Elements

The **transition elements** are the elements in B groups in Periods 4 through 7 in the middle of the periodic table. For these elements the last electron filled into the electron configuration is in a *d* subshell. For example, the electron configuration for the first transition element, scandium, is Sc [Ar] $4s^2 3d^1$. Notice that in each period in which they occur, the transition elements are preceded by two *s*-block elements (in the case of Sc, these are K and Ca). This might be a surprise, because $n = 3$ (3*d*) usually indicates lower energy than $n = 4$ (4*s*), but Figure 7.17 and the periodic table clearly predict that the electron configuration for Sc has only a single *d* electron; the experimental data in Appendix D confirm this prediction. ***In general,* (n − *1)*d *atomic orbitals are filled after* ns *atomic orbitals and before filling of* np *atomic orbitals begins.***

In all atoms except hydrogen, the subshell energies depend on both n and ℓ, not simply n. Subshells are filled sequentially in order of increasing $n + \ell$; when two subshells have the same $n + \ell$ value, the electrons are first assigned to the subshell with the lower n value (Figure 7.19). For example, the sequence for 3*d*, 4*s*, and 4*p* subshells is that the 4*s* subshell ($n + \ell = 4 + 0 = 4$) is filled before the 3*d* subshell ($n + \ell = 4 + 1 = 5$). The 3*d* and 4*p* subshells have the same $n + \ell$ value, but because of its lower n value, electrons fill the 3*d* subshell before the 4*p* subshell. As an example of how this works, compare Ca, [Ar] $4s^2$, Mn, [Ar] $3d^5 4s^2$, and Ge, [Ar] $3d^{10} 4s^2 4p^2$. This sequence illustrates that the 4*s* atomic orbitals are filled before 3*d* and 3*d* before 4*p*.

Occupancy of *d* atomic orbitals begins with the $n = 3$ level and therefore with the first transition element, scandium, which has the configuration [Ar] $3d^1 4s^2$. After scandium comes titanium with [Ar] $3d^2 4s^2$ and vanadium with [Ar] $3d^3 4s^2$. The next element, chromium, is predicted to be [Ar] $3d^4 4s^2$, but that is incorrect; based on spectroscopic and magnetic measurements, the correct configuration is [Ar] $3d^5 4s^1$. This illustrates one of several cases where predicted and experimental configurations differ. When half-filled or filled *d* subshells are possible, they sometimes occur, even though this forces an *s* subshell to also be half-filled. (Hund's rule indicates that half-filled subshells minimize electron repulsions.) As a result, *transition elements that have a half-filled* s *subshell and half-filled or filled* d *subshells have relatively stable electron configurations*. Chromium ([Ar] $4s^1 3d^5$) and copper ([Ar] $4s^1 3d^{10}$) are examples.

The number of unpaired electrons for atoms of most transition elements can be predicted according to Hund's rule by placing valence electrons in atomic orbital diagrams. For example, the electron configuration of Co ($Z = 27$) is [Ar] $3d^7 4s^2$, and the number of unpaired electrons is three, corroborated by experimental measurements, and seen from this atomic orbital box diagram.

		3*d*					4*s*
Co	[Ar]	↑↓	↑↓	↑	↑	↑	↑↓

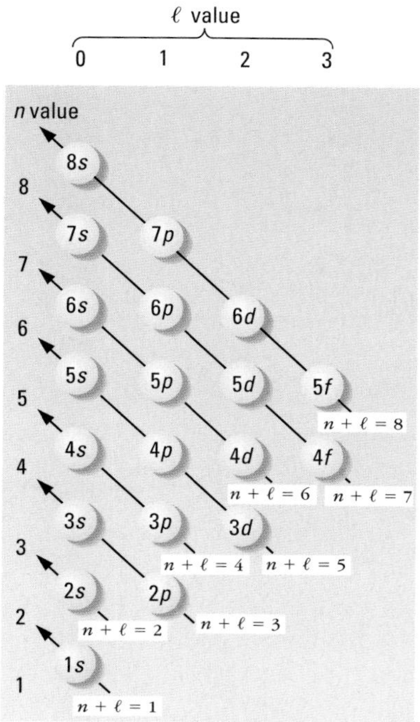

Figure 7.19 Order of subshell filling. As n increases, the energies of electron shells also increase. The energies of subshells within a shell increase with ℓ. Subshells are filled in increasing order of $n + \ell$. When two subshells have the same sum of $n + \ell$, the subshell with the lower n value is filled first.

Cr is [Ar] $4s^1 3d^5$ not [Ar] $4s^2 3d^4$; Cu is [Ar] $4s^1 3d^{10}$ not [Ar] $4s^2 3d^9$.

Unpredicted electron configurations also occur with silver and gold, elements that are in the same group as copper.

EXERCISE 7.14 Unpaired Electrons

Use atomic orbital box diagrams to determine which chromium ground-state configuration has the greater number of unpaired electrons: [Ar] $3d^4 4s^2$ or [Ar] $3d^5 4s^1$.

Elements with Incompletely Filled *f* Atomic Orbitals

In the elements of the sixth and seventh periods, *f* subshell atomic orbitals exist and can be filled (Figure 7.17). The elements (all metals) for which *f* subshells are filling are sometimes called the *inner transition* elements or, more usually, *lanthanides* (for lanthanum, the element just before those filling the 4*f* subshell) and *actinides* (for actinium, the element just before those filling the 5*f* subshell). The lanthanides start with lanthanum (La), which has the electron configuration [Xe] $5d^1 6s^2$. The next element, cerium (Ce), begins a separate row at the bottom of the periodic table, and it is with these elements that *f* atomic orbitals are filled (Appendix D). The electron configuration of Ce is [Xe] $4f^1 5d^1 6s^2$. Each of the lanthanide elements, from Ce to Lu, continues to add $4f$ electrons until the seven $4f$ atomic orbitals are filled by 14 electrons in lutetium (Lu, [Xe] $4f^{14} 5d^1 6s^2$). Note that both the $n = 5$ and $n = 6$ levels are partially filled before the $4f$ atomic orbitals start to be occupied.

PROBLEM-SOLVING EXAMPLE 7.7 Electron Configurations

Using only the periodic table as a guide, give the complete electron configuration, atomic orbital box diagram, and noble gas notation for vanadium, V.

Answer $1s^2 2s^2 2p^6 3s^2 3p^6 3d^3 4s^2$; [Ar] $3d^3 4s^2$

$1s$	$2s$	$2p$			$3s$	$3p$			$3d$					$4s$
↑↓	↑↓	↑↓	↑↓	↑↓	↑↓	↑↓	↑↓	↑↓	↑	↑	↑			↑↓

Strategy and Explanation The periodic table shows vanadium in the fourth period with atomic number 23. Therefore, vanadium has 23 electrons. The first 18 are represented by $1s^2 2s^2 2p^6 3s^2 3p^6$, the electron configuration of argon (Ar), the noble gas that precedes vanadium. The last five electrons in vanadium have the configuration $3d^3 4s^2$. The atomic orbital box diagram shown above follows from the total electron configuration.

PROBLEM-SOLVING PRACTICE 7.7

(a) Write the electron configuration of silicon in the noble gas notation.
(b) Determine how many unpaired electrons a silicon atom has by drawing the atomic orbital box diagram.

EXERCISE 7.15 Highest Energy Electrons in Ground-State Atoms

Give the electron configuration of electrons in the highest occupied principal energy level (highest n) in a ground-state chlorine atom. Do the same for a selenium atom.

We move now from considering the electron configuration of all electrons in an atom to those electrons of primary importance to the chemical reactivity of an element, its valence electrons in the outermost atomic orbitals of the atom.

Valence Electrons

Elements in the same group in the periodic table have similar chemical behavior. For example, the formulas of the oxides are similar; in Group 4A CO_2, SiO_2, GeO_2, SnO_2, and PbO_2 illustrate this. Elements in different periodic groups display different chemical behaviors. This can be explained if we assume that atoms of elements in the same group all have the same number and arrangement of electrons in the outermost region of their atoms. These are the electrons that can easily interact with electrons from another atom. This idea originated with Gilbert N. Lewis (1875–1946), who suggested as early as 1902 that electrons in atoms might be arranged in shells, starting close to the nucleus and building outward. The electrons in the outermost parts of the atom that determine chemical behavior are called **valence electrons**. Electrons in filled inner shells and subshells of an atom are referred to as **core electrons**.

For main group elements (those in groups designated by a number and the letter A), the valence electrons are always in the shell with the highest principal quantum number n. All of the other electrons, those with smaller values of n, are core electrons. For main group elements the number of valence electrons is given by the A-group number. Some examples of core and valence electrons for main group elements are

Element	Electron Configuration	Core Electrons	Valence Electrons	Periodic Group
Na	[Ne] $3s^1$	[Ne] $(1s^22s^22p^6)$	$3s^1$	1A
Si	[Ne] $3s^23p^2$	[Ne] $(1s^22s^22p^6)$	$3s^23p^2$	4A
As	[Ar] $3d^{10}4s^24p^3$	[Ar] $3d^{10}$	$4s^24p^3$	5A

Notice that the valence electrons all have the same principal quantum number and all are in the same, outermost, shell. In the case of As, the filled *d* subshell ($3d^{10}$) is part of the core even though it filled after the noble-gas configuration of Ar was complete.

For transition metals and inner transition metals (*d*-block and *f*-block elements), electrons in incompletely filled *d* and *f* subshells that are not part of the outermost shell may also be valence electrons. Examples are scandium, [Ar] $3d^14s^2$, which has three valence electrons, and cerium, [Xe] $4f^15d^16s^2$, which has four valence electrons. These numbers of valence electrons account for the formation of compounds that contain Sc^{3+} ions and Ce^{4+} ions in which three and four electrons have been lost, respectively, from the neutral atom. Examples of core and valence electrons for transition and inner transition elements are

Element	Electron Configuration	Core Electrons	Valence Electrons	Periodic Group
Mn	[Ar] $3d^54s^2$	[Ar]	$3d^54s^2$	7B
Ta	[Xe] $4f^{14}5d^36s^2$	[Xe] $4f^{14}$	$5d^36s^2$	5B
U	[Rn] $5f^36d^17s^2$	[Rn]	$5f^36d^17s^2$	Not numbered; forms UO_3; 6 valence electrons

While teaching his students about atomic structure, Lewis used the element's symbol to represent the atomic nucleus together with the core electrons. He introduced the practice of representing the valence electrons as dots. The dots are placed around the symbol one at a time until they are used up or until all four sides are occupied; any remaining electron dots are paired with ones already there. The result is a **Lewis dot symbol**. Table 7.6 shows Lewis dot symbols for the atoms of the elements in Periods 2 and 3.

Main group elements in Groups 1A and 2A are known as ***s*-block elements**, and their valence electrons are *s* electrons (ns^1 for Group 1A, ns^2 for Group 2A). Elements in the main groups at the right in the periodic table, Groups 3A through 8A, are known as ***p*-block elements**. Their valence electrons include the outermost *s* and *p* electrons. Notice in Table 7.6 how the Lewis dot symbols show that ***in each A group the number of valence electrons is equal to the group number.***

Figure 7.17, p. 297, defines and illustrates *s*-block, *p*-block, *d*-block, and *f*-block elements.

Table 7.6 Lewis Dot Symbols for Atoms

1A ns^1	2A ns^2	3A ns^2np^1	4A ns^2np^2	5A ns^2np^3	6A ns^2np^4	7A ns^2np^5	8A ns^2np^6	
Li·	·Be·	·Ḃ·	·Ċ·	·N̈·	:Ö·	:F̈·	:N̈e:	Period 2
Na·	·Mg·	·Ȧl·	·Ṡi·	·P̈·	:S̈·	:C̈l·	:Är:	Period 3

PROBLEM-SOLVING EXAMPLE 7.8 Valence Electrons

(a) Using the noble gas notation, write the electron configuration for bromine. Identify its core and valence electrons.

(b) Write the Lewis dot symbol for bromine.

Answer

(a) [Ar] $3d^{10}4s^24p^5$ Core electrons are [Ar] $3d^{10}$; valence electrons are $4s^24p^5$.

(b) $:\ddot{\underset{..}{Br}}\cdot$

Strategy and Explanation Bromine is element 35, a Group 7A element. Its first 18 electrons are represented by [Ar]. These, along with the ten 3*d* electrons, make up the core electrons. Because it is in Group 7A, bromine has seven valence electrons, as given by $4s^24p^5$ and shown in the Lewis dot symbol.

PROBLEM-SOLVING PRACTICE 7.8

Use the noble gas notation to write electron configurations and Lewis dot symbols for Se and Te. What do these configurations illustrate about elements in the same main group?

It is hard to overemphasize how useful the periodic table is as a guide to electron configurations. As another example, using Figure 7.17 as a guide we can find the configuration of Te. Because tellurium is in Group 6A, by now you should immediately recognize that it has *six* valence electrons with an outer electron configuration of ns^2np^4. And, because Te is in the fifth period, $n = 5$. Thus, the complete electron configuration is given by starting with the electron configuration of krypton [$1s^22s^22p^63s^23p^63d^{10}4s^24p^6$], the noble gas at the end of the fourth period. Then, add the filled $4d^{10}$ subshell and the six valence electrons of Te ($5s^25p^4$) to give $1s^22s^22p^63s^23p^63d^{10}4s^24p^64d^{10}5s^25p^4$, or [Kr] $4d^{10}5s^25p^4$. To predict the number of unpaired electrons, look at the outermost subshell's electron configuration, because the inner shells (represented by [Kr] in this case) are completely filled with paired electrons. For Te, [Kr] $4d^{10}5s^25p^25p^15p^1$ indicates two *p* atomic orbitals with unpaired electrons for a total of two unpaired electrons.

7.8 Ion Electron Configurations

In studying ionic compounds (← p. 85), you learned that atoms from Groups 1A through 3A form positive ions (cations) with charges equal to their group numbers—for example, Li^+, Mg^{2+}, and Al^{3+}. Nonmetals in Groups 5A through 7A that form ions do so by adding electrons to form negative ions (anions) with *charges equal to eight minus the A group number.* Examples of such anions are N^{3-}, O^{2-}, and F^-. Here's the explanation. When atoms from *s*- and *p*-block elements form ions, electrons are removed from or added to valence shells so that a noble gas configuration is achieved.

- Atoms from Groups 1A, 2A, and 3A lose 1, 2, or 3 electrons to form 1+, 2+, or 3+ ions, respectively; the positive ions have the same electron configuration as the ***preceding*** noble gas.
- Atoms from Groups 7A, 6A, and some in 5A gain 1, 2, or 3 electrons to form 1−, 2−, or 3− ions, respectively; the negative ions have the same electron configuration as the ***next*** noble gas.
- Metal atoms *lose* electrons to form cations with a positive charge equal to the A group number; nonmetals *gain* electrons to form anions with a negative charge equal to the A group number minus eight.

These relationships apply to Groups 1A to 3A and 5A to 7A. Atoms and ions that have the same electron configuration are said to be **isoelectronic**.

Figure 7.20 shows isoelectronic ions and the noble gas atom that has the same electron configuration, emphasizing that ***metal ions are isoelectronic with the preceding noble gas atom, while nonmetal ions have the electron configuration of the next noble gas atom.***

PROBLEM-SOLVING EXAMPLE 7.9 Atoms and Their Ions

Complete this table.

Neutral Atom	Neutral Atom Electron Configuration	Ion	Ion Electron Configuration
Se	________	________	[Kr]
Ba	________	Ba^{2+}	________
Br	________	Br^-	________
________	[Kr] $5s^1$	Rb^+	________
________	[Ne] $3s^2 3p^3$	________	[Ar]

Answer

Neutral Atom	Neutral Atom Electron Configuration	Ion	Ion Electron Configuration
Se	[Ar] $3d^{10}4s^2 4p^4$	Se^{2-}	[Kr]
Ba	[Xe] $6s^2$	Ba^{2+}	[Xe]
Br	[Ar] $3d^{10}4s^2 4p^5$	Br^-	[Kr]
Rb	[Kr] $5s^1$	Rb^+	[Kr]
P	[Ne] $3s^2 3p^3$	P^{3-}	[Ar]

Strategy and Explanation A neutral Se atom has the electron configuration [Ar] $3d^{10}4s^2 4p^4$; Se is in Group 6A, so it gains two electrons in the $4p$ subshell to form Se^{2-} and achieve the noble gas configuration of krypton, Kr (36 electrons). Barium is a Group 2A element and loses the two $6s$ electrons to acquire the electron configuration of xenon (54 electrons), the preceding noble gas. Therefore, Ba is [Xe] $6s^2$ and Ba^{2+} is [Xe]. Bromine is in Group 7A and will gain one electron to form Br^-, which has the electron configuration of krypton, the next noble gas. An electron configuration of [Kr] $5s^1$ indicates 37 electrons in the neutral atom, which is a rubidium atom, Rb. A neutral Rb atom loses the $5s^1$ electron to form an Rb^+ ion, which has a [Kr] configuration. The [Ne] $3s^2 3p^3$ configuration is for an element with 15 electrons, which is a phosphorus atom, P. By gaining three electrons, a neutral phosphorus atom becomes a P^{3-} ion with the [Ar] configuration.

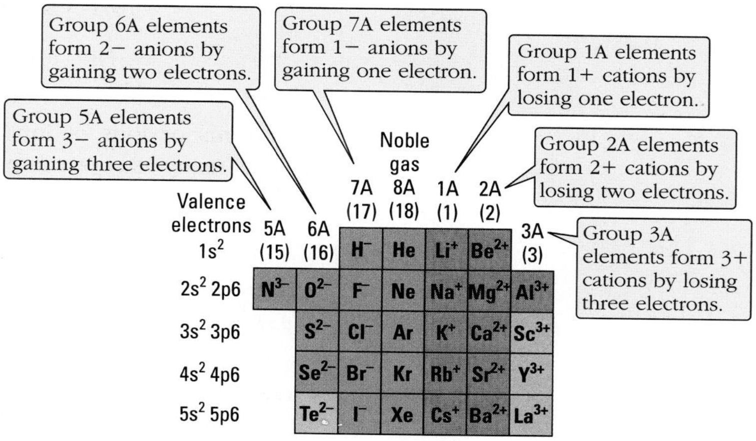

Figure 7.20 Cations, anions, and atoms with isoelectronic ground-state noble gas configurations. Atoms and ions in the same row are isoelectronic; they have the same electron configuration.

PROBLEM-SOLVING PRACTICE 7.9

(a) What Period 3 anion with a 3− charge has the [Ar] electron configuration?

(b) What Period 4 cation with a 2+ charge has the electron configuration of argon?

Transition Metal Ions

The closeness in energy of the 4*s* and 3*d* subshells was mentioned earlier in connection with the electron configurations of transition metal atoms. The 5*s* and 4*d* subshells are also close to each other in energy, as are the 6*s*, 4*f*, and 5*d* subshells and the 7*s*, 5*f*, and 6*d* subshells. The *ns* and $(n-1)d$ subshells are so close in energy that once *d* electrons are added, the $(n-1)d$ subshell becomes slightly lower in energy than the *ns* subshell. As a result, ***the* ns *electrons are at higher energy and are always removed before* (n−1)d *electrons when transition metals form cations***. A simple way to remember this is that when a transition metal atom loses electrons to form an ion, the electrons in the outermost shell (highest *n*) are lost first.

A neutral nickel atom, Ni^0, with an equal number of protons (28) and electrons (28), has no net charge. An Ni^{2+} ion has a net charge of 2+ because it has 28 protons, but only 26 electrons.

For example, a nickel atom, Ni, [Ar] $3d^84s^2$, will lose its two 4*s* electrons to form a Ni^{2+} ion.

$$Ni^0 \rightarrow Ni^{2+} + 2e^-$$

That two 4*s* electrons are lost, and not two electrons from the 3*d* subshell, is corroborated by the experimental evidence that Ni^{2+} has two unpaired electrons. This is shown by these atomic orbital box diagrams.

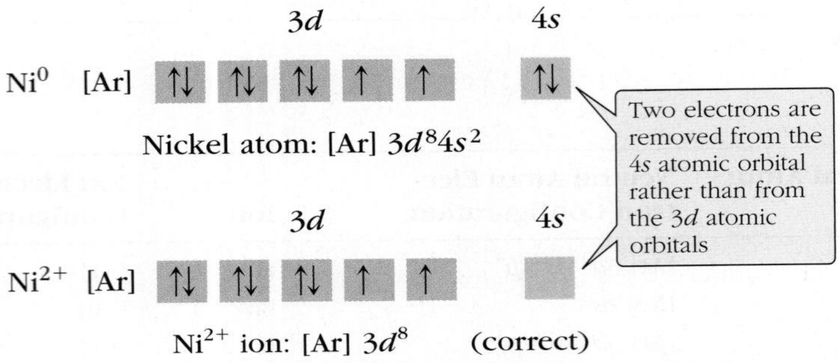
3d
4s
Ni0 [Ar]
Nickel atom: [Ar] 3d84s2
3d
4s
Ni2+ [Ar]
Ni2+ ion: [Ar] 3d8
(correct)
Two electrons are removed from the 4s atomic orbital rather than from the 3d atomic orbitals

If two electrons had been removed from the 3*d* subshell rather than the 4*s* subshell, a Ni^{2+} ion would have four unpaired electrons, which is inconsistent with experimental evidence.

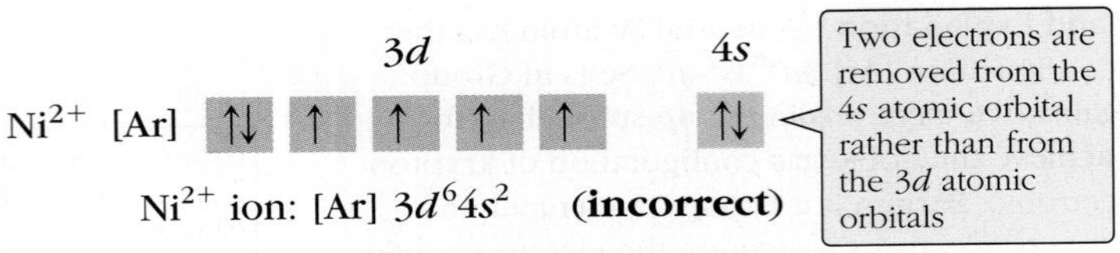
3d
4s
Ni2+ [Ar]
Ni2+ ion: [Ar] 3d64s2
(incorrect)
Two electrons are removed from the 4s atomic orbital rather than from the 3d atomic orbitals

Atoms or ions of inner transition elements (lanthanides and actinides) can have as many as seven unpaired electrons in the *f* subshell, as occurs in Eu^{2+} and Gd^{3+} ions.

PROBLEM-SOLVING EXAMPLE 7.10 Electron Configurations for Transition Elements and Ions

(a) Write the electron configuration for the Fe atom using the noble gas notation. Then draw the atomic orbital box diagram for the electrons beyond the preceding noble gas configuration.

(b) Iron commonly exists as 2+ and 3+ ions. How does the atomic orbital box diagram given in part (a) have to be changed to represent the outer electrons of Fe^{2+} and Fe^{3+}?

(c) How many unpaired electrons do Fe, Fe^{2+}, and Fe^{3+} each have?

Answer

(a)

		3d					4s
Fe [Ar] $3d^64s^2$;	[Ar]	↑↓	↑	↑	↑	↑	↑↓

(b) For Fe^{2+}, remove the two 4s electrons from Fe to give

		3d				
Fe^{2+}	[Ar]	↑↓	↑	↑	↑	↑

For Fe^{3+}, remove the two 4s electrons and one of the paired 3d electrons from Fe to give

Fe^{3+}	[Ar]	↑	↑	↑	↑	↑

(c) Fe has four unpaired electrons; Fe^{2+} has four unpaired electrons; Fe^{3+} has five unpaired electrons.

Strategy and Explanation

(a) Fe, with an atomic number of 26, is in the fourth period. It has eight more electrons than Ar, with two of the eight in the 4s subshell and six in the 3d subshell, so its electron configuration is [Ar] $3d^64s^2$. For the atomic orbital box diagram, all *d* atomic orbitals get one electron before pairing occurs (Hund's rule).

(b) To form Fe^{2+}, two electrons are removed from the outermost shell (4s subshell). To form Fe^{3+} from Fe^{2+}, one of the paired electrons is removed from a 3d atomic orbital.

(c) Looking at the atomic orbital box diagrams, you should see that the numbers of unpaired electrons are four for Fe, four for Fe^{2+}, and five for Fe^{3+}.

The fact that the two 4s electrons are removed to form Fe^{2+} shows that the 4s electrons are valence electrons. The fact that a 3d electron is removed when Fe^{3+} forms shows that the 3d electrons are also valence electrons.

PROBLEM-SOLVING PRACTICE 7.10

Use the electron configuration of a neutral ground state copper atom to explain why copper readily forms the Cu^+ ion.

Paramagnetism and Unpaired Electrons

The magnetic properties of a spinning electron were described in Section 7.5. In atoms and ions that have filled shells, all the electrons are paired (opposite spins) and their magnetic fields effectively cancel each other. Such substances are very weakly repelled by a magnetic field and are called **diamagnetic.** Atoms or ions with unpaired electrons are attracted to a magnetic field; the more unpaired electrons, the greater is the attraction. Such substances are called **paramagnetic.** For example, the greater paramagnetism of Fe^{3+} with five unpaired electrons rather than three is experimental evidence that Fe^{3+} has the electron configuration [Ar] $3d^5$, rather than [Ar] $3d^34s^2$.

Permanent magnets, such as those holding notes on a refrigerator, are made of **ferromagnetic** substances. The magnetic effect in ferromagnetic materials is much larger than that in paramagnetic materials. Ferromagnetism occurs when the spins of unpaired electrons in a cluster of atoms (called a domain) in a solid are all aligned in the same direction (Figure 7.21). Only the metals of the iron, cobalt, and nickel subgroups in the periodic table exhibit this property. They are also unique in that, once the domains are aligned in a magnetic field, the metal is permanently magnetized. In such a case, the magnetism can be eliminated only by heating or shaping the metal to rearrange the electron spin domains. Many alloys, such as alnico (an alloy of aluminum, nickel, and cobalt), exhibit greater ferromagnetism than do the

Paramagnetism: The atoms or ions with magnetic moments are not aligned. If the substance is in a magnetic field, they align with and against the field. Magnetism is weak.

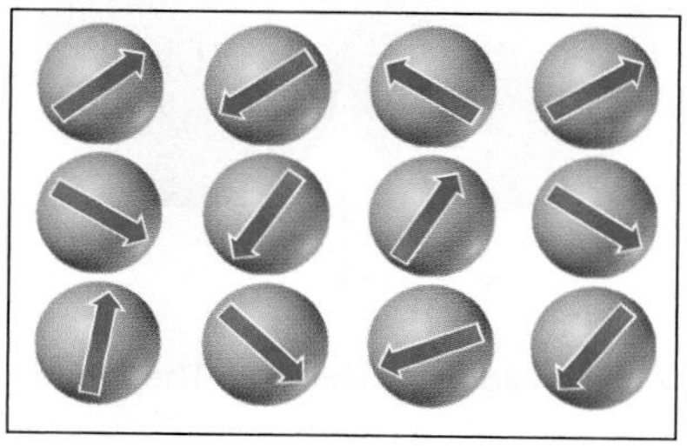

Ferromagnetism: The spins of unpaired electrons in clusters of atoms or ions are aligned in the same direction. In a magnetic field these domains all align and stay aligned when the field is removed.

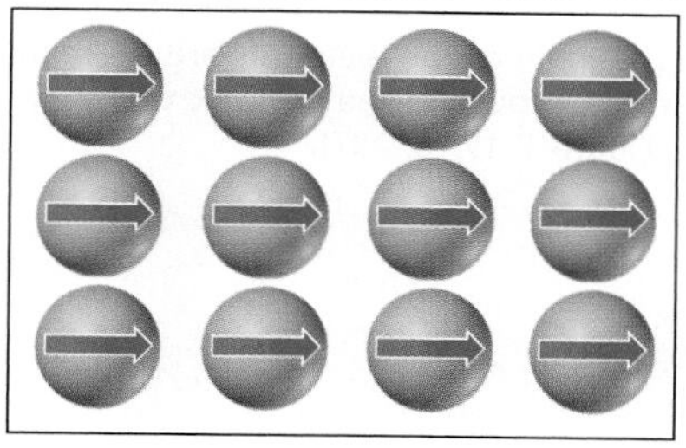

Figure 7.21 Types of magnetic behavior.

This answers the question posed in Chapter 1 (← p. 3), "Why is iron strongly attracted to a magnet, but most substances are not?"

pure metals themselves. Some metal oxides, such as CrO_2 and Fe_3O_4, are also ferromagnetic and are used in magnetic recording tape. Computer hard drives use ferromagnetic materials to store data in a binary code of zeroes and ones. Very small magnetic domains store bits of binary code as a magnetized region that represents a one and an unmagnetized region that represents a zero.

CONCEPTUAL EXERCISE 7.16 Unpaired Electrons

Fluoride ion, F^-, has no unpaired electrons. Vanadium forms four binary fluorides—VF_2, VF_3, VF_4, and VF_5. (a) Which fluoride is diamagnetic? (b) Which fluoride has the greatest attraction to a magnetic field? (c) Which fluoride has two unpaired electrons per vanadium?

7.9 Periodic Trends: Atomic Radii

go Chemistry **Module 11: Periodic Trends** covers concepts in this section.

Using knowledge of electron configurations, we can now answer fundamental questions about why atoms of different elements fit as they do in the periodic table, as well as explain the trends in the properties of the elements in the table.

For atoms that form simple diatomic molecules, such as Cl_2, the **atomic radius** can be defined experimentally by finding the distance between the centers (nuclei) of the two atoms in the molecule. Assuming the atoms to be spherical, one-half of this distance is a good estimate of the atom's radius. In the Cl_2 molecule, the internuclear distance (the distance from the center of one atom to the center of the other) is 198 pm. Dividing by 2 shows that the Cl radius is 99 pm. Similarly, the C—C distance in diamond is 154 pm, so the radius of the carbon atom is 77 pm. To test these estimates, we can add them together to estimate the distance between Cl and C in CCl_4. The estimated distance of 176 pm (99 pm + 77 pm) is in good agreement with the experimentally measured C—Cl distance of 176 pm.

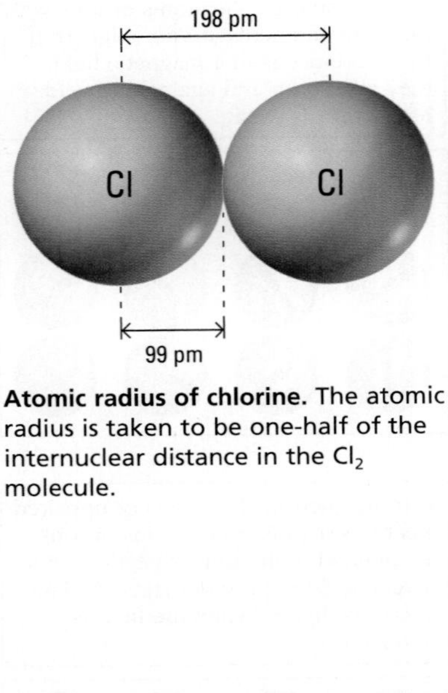

Atomic radius of chlorine. The atomic radius is taken to be one-half of the internuclear distance in the Cl_2 molecule.

This approach can be extended to other atomic radii. The radii of O, C, and S atoms can be estimated by measuring the O—H, C—Cl and H—S distances in H_2O, CCl_4, and H_2S, and then subtracting the H and Cl radii found from H_2 and Cl_2. By this and other techniques, a reasonable set of atomic radii for main group elements has been assembled (Figure 7.22).

Atomic Radii of the Main Group Elements

The main group elements are those in the A groups in the periodic table (Figure 7.17, p. 297).

For the main group elements, atomic radii increase going down a group in the periodic table and decrease going across a period (Figure 7.22). These trends reflect three important effects:

- *From the top to the bottom of a group in the periodic table, the atomic radii increase* because electrons occupy atomic orbitals that are successively larger as the value of *n*, the principal quantum number, increases.
- *The atomic radii decrease from left to right across a period.* The *n* value of the outermost atomic orbitals stays the same, so we might expect the radii of the occupied atomic orbitals to remain approximately constant. However, in crossing a period, as each successive electron is added, the nuclear charge also increases by the addition of one proton. The result is an increased attraction between the nucleus and electrons that is somewhat stronger than the increasing repulsion between electrons, causing atomic radii to decrease (Figure 7.22).

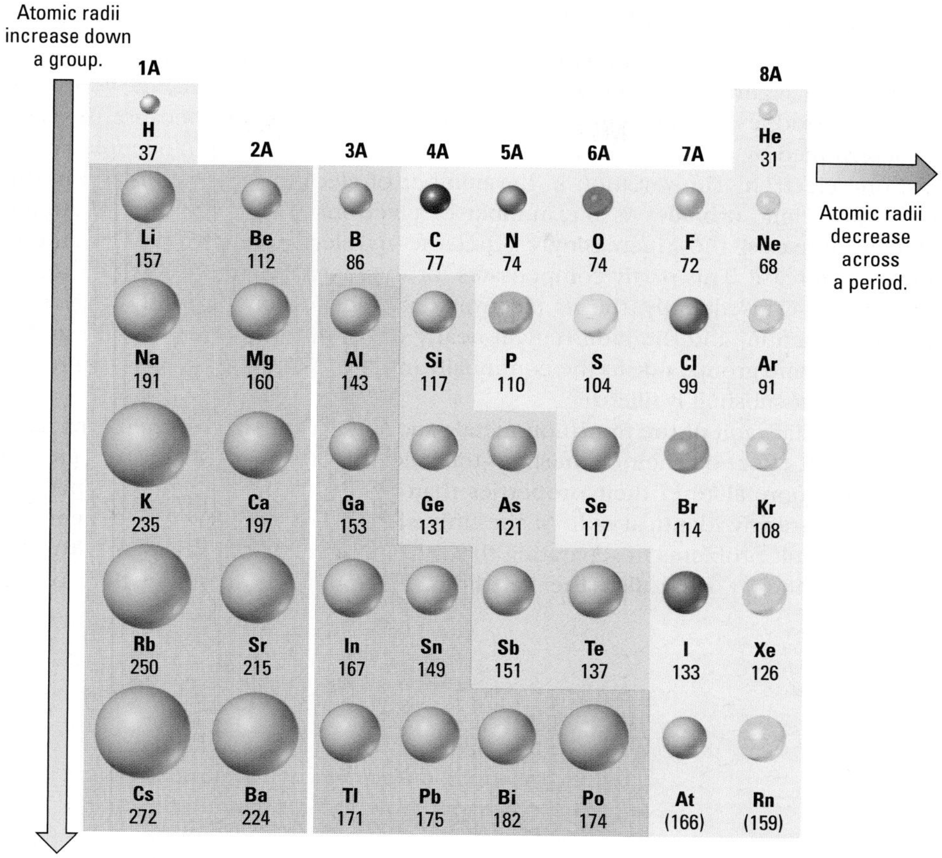

Figure 7.22 Atomic radii of the main group elements (in picometers; 1 pm = 10^{-12} m). The trends are for radii to decrease across a period and to increase down a group.

- *There is a large increase in atomic radius going from any noble gas atom to the following Group 1A atom,* where the outermost electron is assigned to the next higher energy level (the next larger shell, higher n value). For example, compare the atomic radii of Ne (68 pm) and Na (191 pm) or Ar (91 pm) and K (235 pm).

PROBLEM-SOLVING EXAMPLE 7.11 Atomic Radii and Periodic Trends

Using only a periodic table, list these atoms in order of decreasing size: Br, Cl, Ge, K, S.

Answer K > Ge > Br > S > Cl

Strategy and Explanation Based on the periodic trends in atomic size, we expect the radius of Cl to be smaller than that of Br, which is below it in Group 7A; S in Period 3 should be smaller than Br, which is in Period 4. Because size decreases across a period, we expect Cl to be smaller than S. Of the Period 4 atoms, Ge is earlier in the period than Br and therefore larger than it, but smaller than K, which starts the period. From these trends we can summarize the relative sizes of the radii, from largest to smallest, as K > Ge > Br > S > Cl.

PROBLEM-SOLVING PRACTICE 7.11

Using just a periodic table, arrange these atoms in order of increasing atomic radius: B, Mg, K, Na.

Atomic Radii of Transition Metals

The periodic trend in the atomic radii of main group and transition metal atoms is illustrated in Figure 7.23. The sizes of transition metal atoms change very little across a period, especially beginning at Group 5B (V, Nb, or Ta), because the sizes are determined by the radius of an *ns* atomic orbital (n = 4, 5, or 6) occupied by at least one electron. The variation in the number of electrons occurs instead in the $(n - 1)d$ atomic orbitals. As the number of electrons in these $(n - 1)d$ atomic orbitals increases, they increasingly repel the *ns* electrons, causing the atomic radius to increase. This partly compensates for the increased nuclear charge across the periods. Consequently, the *ns* electrons experience only a slightly increasing nuclear attraction, and the radii remain nearly constant until the slight rise at the copper and zinc groups due to the continually increasing electron-to-electron repulsions as the *d* subshell is filled.

The similar radii of the transition metals and the similar sizes of their ions have an important effect on their chemistry—transition elements in neighboring groups tend to be more alike in their properties than other elements in their respective groups. The nearly identical radii of the fifth- and sixth-period transition elements lead to difficult problems in separating them from one another. The metals Ru, Os, Rh, Ir, Pd, and Pt are called the "platinum group metals" because they occur

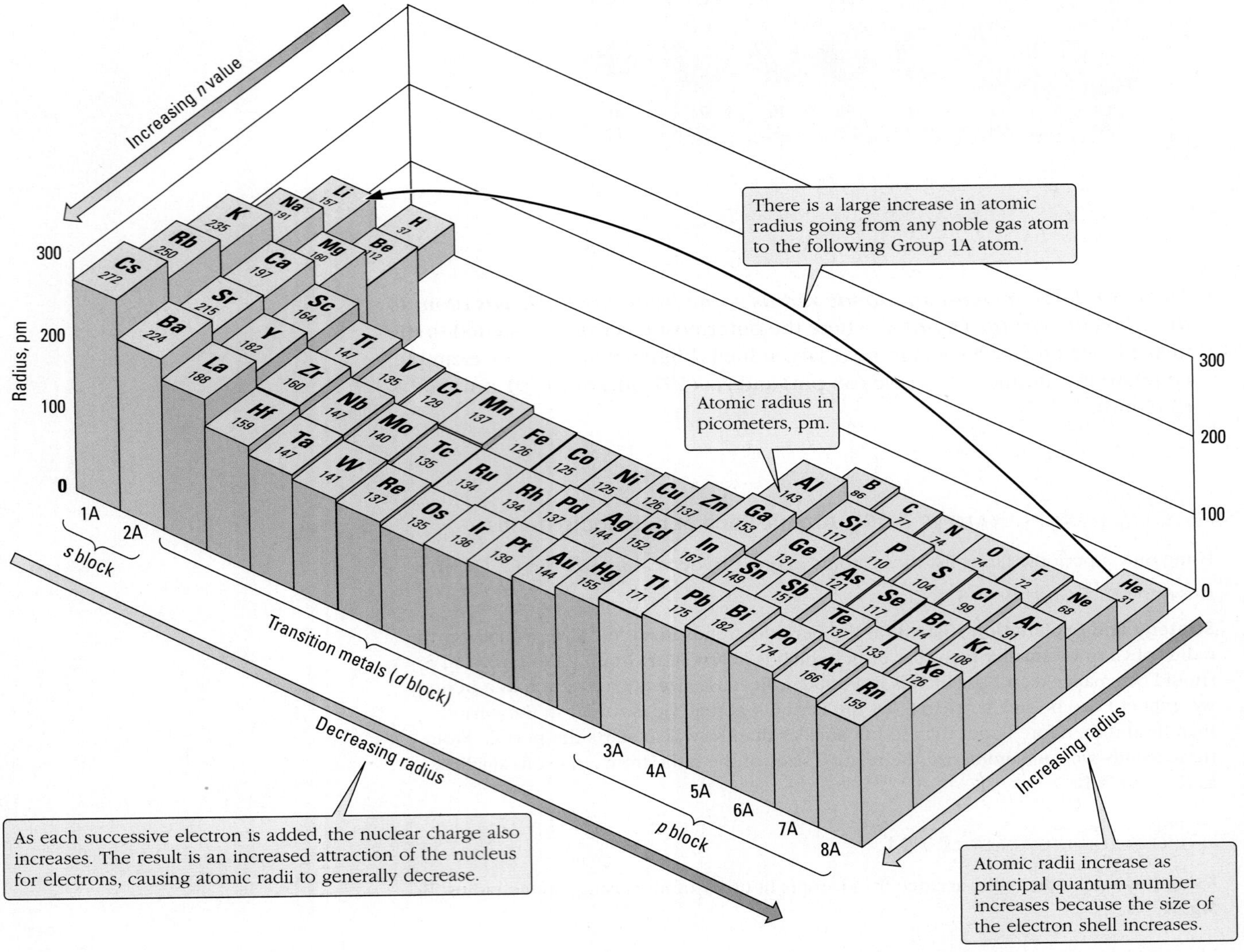

Figure 7.23 Atomic radii of *s*-, *p*-, and *d*-block elements (in picometers, pm; 1 pm = 10^{-12} m). Note the slight change in atomic radii for the transition elements (*d* block) across a period.

together in nature. Their radii and chemistry are so similar that their minerals are similar and are found in the same geologic zones.

Effective Nuclear Charge

We have seen that the atomic radii of main group elements decrease across a period as the number of protons and electrons increases with each element. In a many-electron atom, such as a phosphorus atom, $Z = 15$, the outermost electrons are attracted by the 15 protons in the nucleus, but they are also repelled by the other 14 electrons. The electrons in the first and second shells ($n = 1$ and $n = 2$) are closer to the nucleus most of the time than are the valence electrons, so those core electrons repel the valence electrons away from the nucleus. This repulsion is described by saying that there is a **screening effect**: the inner electrons shield or screen the valence electrons from the charge of 15+ due to the 15 protons in the nucleus. This qualitative discussion can be made quantitative by defining a screening constant, σ, for each electron. The screening constant measures how much the other electrons screen the electron in question from the nuclear charge. The **effective nuclear charge**, Z^*, is then calculated as $Z^* = Z - \sigma$. The effective nuclear charge is the positive charge on the nucleus, Z, corrected for the repulsions, σ, all other electrons exert on a given electron. Within a period, electrons are being added to the same shell and each added electron is not as effective in screening the nuclear charge as are the inner electrons. Thus the effective nuclear charge increases across a period. This change in effective nuclear charge predicts that the atomic radii should decrease and the ionization energies (p. 311) should increase across a period.

7.10 Periodic Trends: Ionic Radii

A consistent set of **ionic radii** can be determined from data showing the distances between adjacent positive and negative ions. For example, the distance between centers of an Na^+ ion and a Cl^- ion in a crystal of NaCl ***(← p. 95)*** can be measured and compared with distances between Na^+ ions and other anions to obtain a radius for Na^+. Ionic radii allow predictions of distances between ions in crystals for which no measurements have been made. Figure 7.24 shows both atomic and ionic radii for selected periodic groups. Important trends with respect to cations are

- **In the same group, cations increase in size down the group.**
- **The radius of the cation is always smaller than the radius of the neutral atom from which the cation is formed.**

Take time to compare these trends in Figure 7.24. For example, an Rb^+ ion is larger than an Na^+ or an Li^+ ion. When an electron (or electrons) is removed from an atom to form a cation, the size shrinks considerably such as in the case of lithium; the radius of a neutral Li atom is 157 pm, whereas that for an Li^+ ion is only 90 pm. This is understandable, because when an electron is removed from a lithium atom, the nuclear charge remains the same (3+), but there are fewer electrons repelling each other. Consequently, the positive nucleus can attract the two remaining electrons more strongly, causing the electrons to contract toward the nucleus. The decrease in ion size is especially great when the electron removed comes from a higher energy level larger shell than the new outer electron. This is the case for Li, for which the "old" outer electron was from a $2s$ atomic orbital and the "new" outer electron in Li^+ is in a $1s$ atomic orbital.

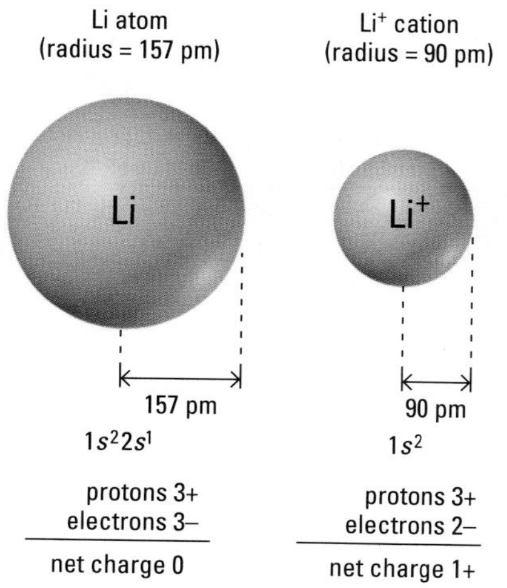

Li atom
(radius = 157 pm)
Li⁺ cation
(radius = 90 pm)
Li
Li⁺
157 pm
90 pm
1s²2s¹
1s²
protons 3+
electrons 3–
net charge 0
protons 3+
electrons 2–
net charge 1+

The shrinkage is also great when two or more electrons are removed—for example, Mg^{2+} and Al^{3+}.

Mg atom (radius = 160 pm) $1s^22s^22p^63s^2$	Mg^{2+} cation (radius = 86 pm) $1s^22s^22p^6$
Al atom (radius 143 pm) $1s^22s^22p^63s^23p^1$	Al^{3+} cation (radius = 68 pm) $1s^22s^22p^6$

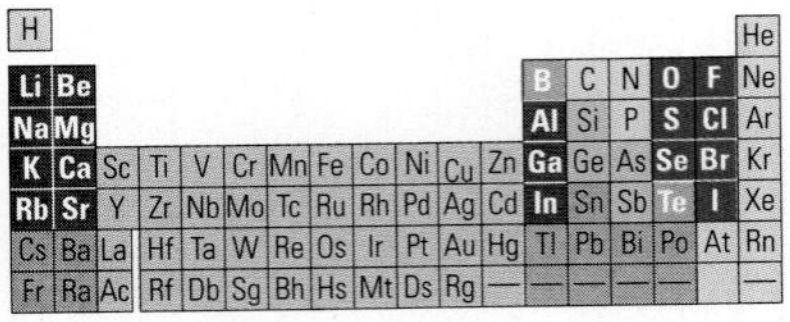

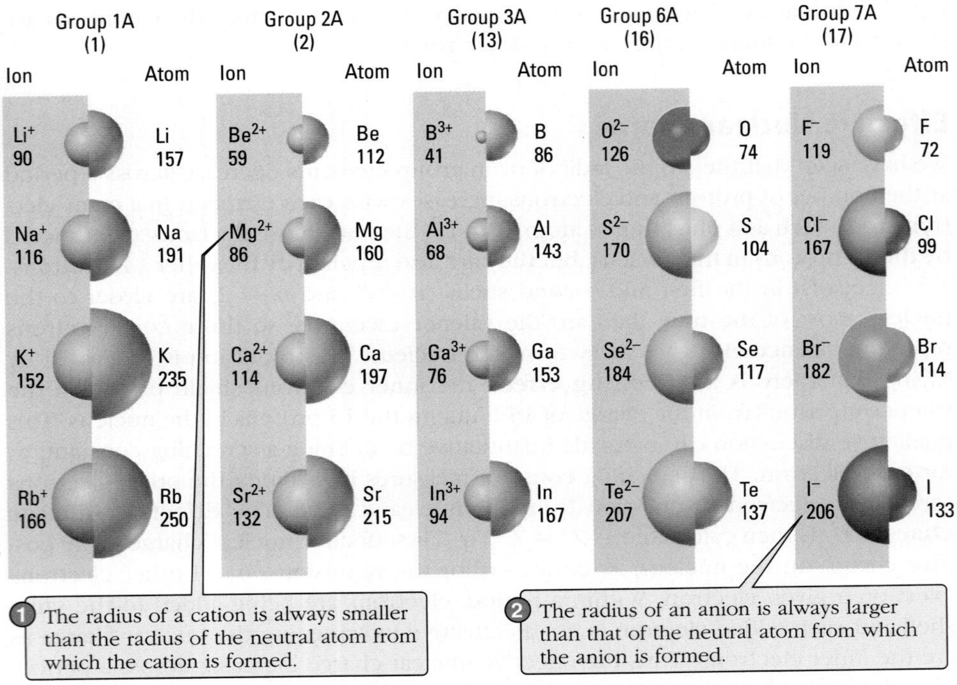

Figure 7.24 Sizes of ions and their neutral atoms. Radii are given in picometers (1 pm = 10^{-12} m).

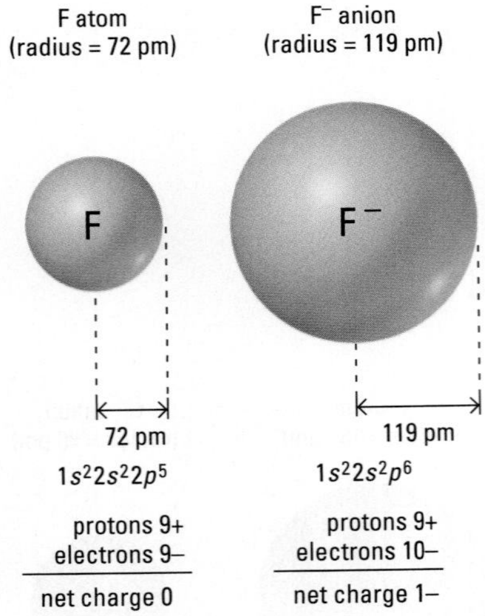

From Figure 7.24 you can see that

- **Anions in the same group increase in size down the group.**
- **The radius of the anion is always larger than the radius of the neutral atom from which the anion is formed.**

Here the argument is the opposite of that used to explain the radii of cations. For anions, the nuclear charge is also unchanged, but the added electron(s) introduce new repulsions and the electron clouds swell. The F atom has nine protons and nine electrons. When it forms the F^- anion, the nuclear charge is still 9+, but there are now ten electrons in the anion. The F^- ion (119 pm) is much larger than the F atom (72 pm) because of increased electron-to-electron repulsions.

The effect of nuclear charge is evident when the sizes of isoelectronic ions across the periodic table are compared. Consider the isoelectronic ions O^{2-}, F^-, Na^+, and Mg^{2+}.

Isoelectronic Ions	O^{2-}	F^-	Na^+	Mg^{2+}
Ionic radius (pm)	126	119	116	86
Number of protons	8	9	11	12
Number of electrons	10	10	10	10

Each ion contains ten electrons. However, the O^{2-} ion has only eight protons in its nucleus to attract these electrons, while F^- has nine, Na^+ has eleven, and Mg^{2+} has twelve. Thus, there is an increasing number of protons to attract the same number of electrons (10). As the proton-to-electron ratio increases in an isoelectronic series of ions, the overall electron cloud is attracted more tightly to the nucleus and the ion size shrinks, as seen for O^{2-} to Mg^{2+}.

PROBLEM-SOLVING EXAMPLE 7.12 Trends in Ion Sizes

Using only a periodic table, choose the smaller atom or ion for each of these pairs:
(a) An Fe atom or an Fe^{2+} ion (b) A P atom or a P^{3-} ion (c) An Fe^{2+} or an Fe^{3+} ion

Answer

(a) Fe^{2+} ion (b) P atom (c) Fe^{3+} ion

Strategy and Explanation Recall the differences in the sizes of atoms and their corresponding ions in terms of electron loss or gain.

(a) Due to loss of electrons from the parent atom to form the cation, a cation is smaller than its parent atom; therefore, a Fe^{2+} ion is smaller than a Fe atom.
(b) Anions are larger than their parent atom due to electron gain, so a phosphorus atom is smaller than a P^{3-} ion.
(c) Both iron ions contain 26 protons, but Fe^{3+} has one fewer electron than Fe^{2+}. Consequently, there is less electron-electron repulsion and the radius of Fe^{3+} is smaller than that of Fe^{2+}.

PROBLEM-SOLVING PRACTICE 7.12

Which of these isoelectronic ions, Ba^{2+}, Cs^{+}, or La^{3+}, will be (a) the largest? (b) the smallest?

7.11 Periodic Trends: Ionization Energies

go Chemistry **Module 11: Periodic Trends** covers concepts in this section.

An element's chemical reactivity is determined, in part, by how easily valence electrons are removed from its atoms, a process that requires energy. The **ionization energy (IE)** of an atom is the energy needed to remove an electron from that atom in the gas phase. For a gaseous sodium atom, the ionization process is

$$Na(g) \longrightarrow Na^{+}(g) + e^{-} \qquad \Delta E = \text{ionization energy (IE)}$$

Because energy is always required to remove an electron, the process is endothermic, and the sign of the ionization energy is always positive. The more difficult an electron is to remove, the greater the ionization energy. ***As illustrated in Figures 7.25 and 7.26, the first ionization energy (the energy needed to remove one electron from the neutral atom) generally decreases down a group and increases across a period for s- and p-block elements***. The decrease down a group reflects the increasing radii of the atoms—it is easier to remove a valence electron from a larger atom because the force of attraction between the electron and the nucleus is smaller due to the greater separation. Thus, for example, the ionization energy of Rb (403 kJ/mol) is less than that of K (419 kJ/mol) or Na (496 kJ/mol). Ionization energies increase across a period for the same reason that the radii decrease. Increasing nuclear charge attracts electrons in the same shell more tightly. However, as shown in Figure 7.26, the trend across a given period is not always smooth, as in the second, third, fourth, and fifth periods. The single *np* electron of the Group 3A element is more easily removed than one of the two *ns* electrons in the preceding Group 2A element. Another deviation occurs with Group 6A elements (ns^2np^4), which have smaller ionization energies than the Group 5A elements that precede them in the second, third, and fourth periods. Beginning in Group 6A, two electrons are assigned to the same *p* atomic orbital. Thus, greater electron repulsion is experienced by the fourth *p* electron, making it easier to remove.

Ionization energies increase much more gradually across the period for the transition and inner transition elements than for the main group elements (Figure 7.25). Just as the atomic radii of transition elements change very little across a period, the ionization energy for the removal of an *ns* electron also shows small changes, until the *d* orbitals are filled (Zn, Cd, Hg).

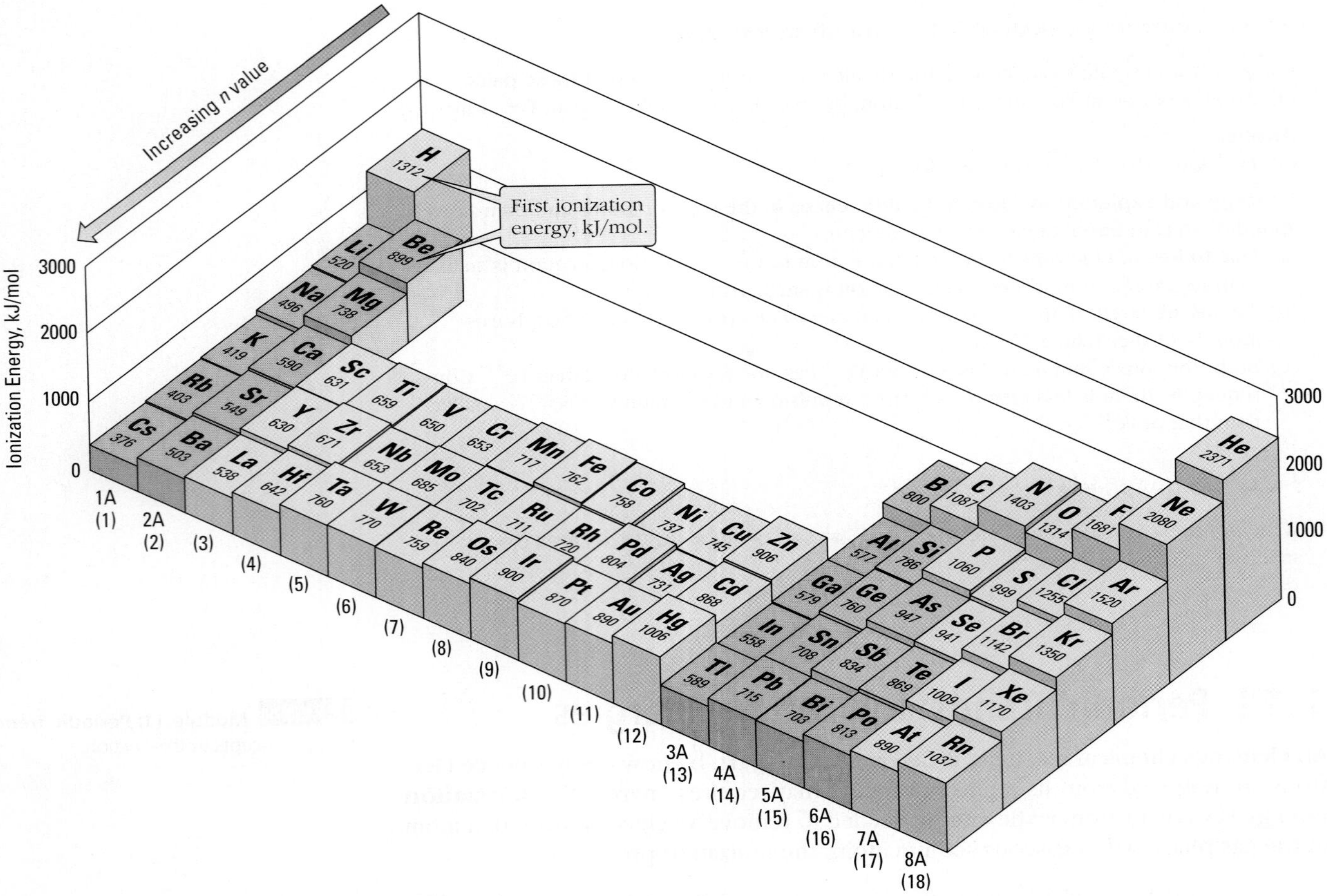

Figure 7.25 First ionization energies (kJ/mol) for elements in the first six periods.

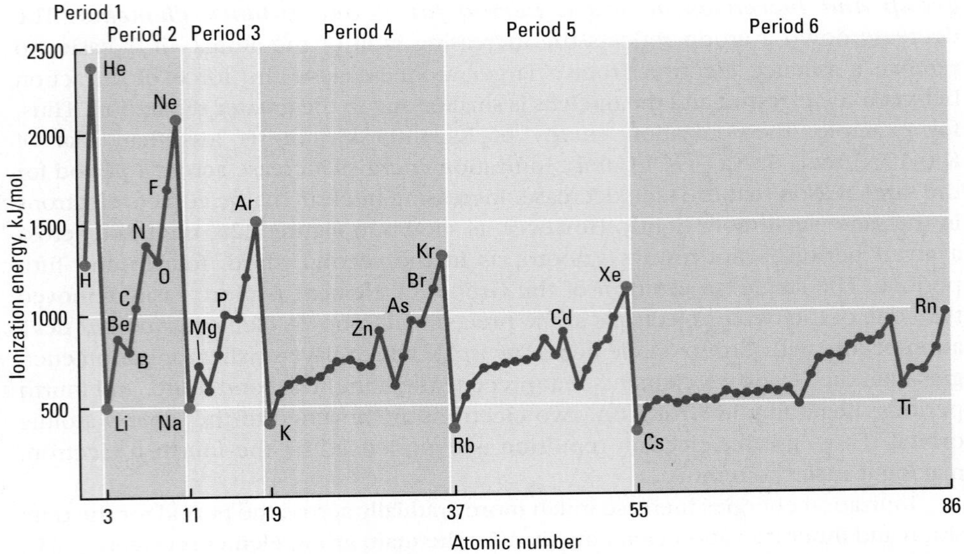

Figure 7.26 First ionization energies for elements in the first six periods plotted against atomic number. Note that in each period, the alkali metal has the lowest and the noble gas the highest ionization energy.

Every atom, except hydrogen, has a series of ionization energies because more than one electron can be removed. For example, the first three ionization energies (IEs) of Mg in the gaseous state are

$$Mg(g) \longrightarrow Mg^{+}(g) + e^{-} \qquad IE_1 = 738\ kJ/mol$$
$$1s^22s^22p^63s^2 \qquad 1s^22s^22p^63s^1$$

$$Mg^{+}(g) \longrightarrow Mg^{2+}(g) + e^{-} \qquad IE_2 = 1450\ kJ/mol$$
$$1s^22s^22p^63s^1 \qquad 1s^22s^22p^6$$

$$Mg^{2+}(g) \longrightarrow Mg^{3+}(g) + e^{-} \qquad IE_3 = 7734\ kJ/mol$$
$$1s^22s^22p^6 \qquad 1s^32s^32p^5$$

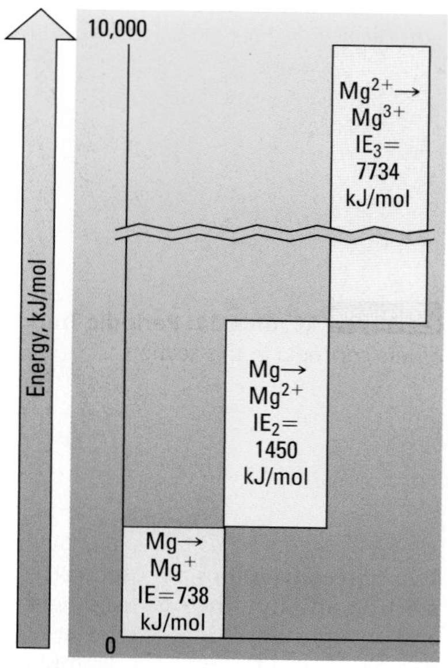

First three ionization energies for Mg.

Notice that removing each subsequent electron requires more energy, and the jump from the second (IE_2) to the third (IE_3) ionization energy of Mg is particularly great. The first electron removed from a magnesium atom comes from the 3*s* atomic orbital. The second ionization energy corresponds to removing a 3*s* electron from an Mg^{+} ion. As expected, the second ionization energy is higher than the first ionization energy because the electron is being removed from a positive ion, which strongly attracts the electron. The third ionization energy corresponds to removing a 2*p* electron, a core electron, from a filled *p* subshell in Mg^{2+}. The great difference between the second and third ionization energies for Mg is excellent experimental evidence for the existence of electron shells in atoms. Removal of the first core electron requires much more energy than removal of a valence electron. Table 7.7 gives the successive ionization energies of the second-period elements.

PROBLEM-SOLVING EXAMPLE 7.13 Ionization Energies

Using only a periodic table, arrange these atoms in order of increasing first ionization energy: Al, Ar, Cl, Na, K, Si.

Answer K < Na < Al < Si < Cl < Ar

Strategy and Explanation Since ionization energy increases across a period, we expect the ionization energies for Ar, Al, Cl, Na, and Si, which are all in Period 3, to be in the order Na < Al < Si < Cl < Ar. Because K is below Na in Group 1A, the first ionization energy of K is less than that of Na. Therefore, the final order is K < Na < Al < Si < Cl < Ar.

Table 7.7 Ionization Energies Required to Remove Successive Electrons from Second-Period Atoms

Ionization Energy (10^6 J/mol)	Li $1s^22s^1$	Be $1s^22s^2$	B $1s^22s^22p^1$	C $1s^22s^22p^2$	N $1s^22s^22p^3$	O $1s^22s^22p^4$	F $1s^22s^22p^5$	Ne $1s^22s^22p^6$
IE_1	0.52	0.90	0.80	1.09	1.40	1.31	1.68	2.08
IE_2	7.30	1.76	2.43	2.35	2.86	3.39	3.37	3.95
IE_3	11.81	14.85	3.66	4.62	4.58	5.30	6.05	6.12
IE_4		21.01	25.02	6.22	7.48	7.47	8.41	9.37
IE_5			32.82	37.83	9.44	10.98	11.02	12.18
IE_6				47.28	53.27	13.33	15.16	15.24
IE_7					64.37	71.33	17.87	20.00
IE_8		Core electrons				84.08	92.04	23.07
IE_9							106.43	115.38
IE_{10}								131.43

Note: In each of these elements the core electrons are those in the 1*s* orbital.

PROBLEM-SOLVING PRACTICE 7.13

Use only a periodic table to arrange these elements in decreasing order of their first ionization energy: Na, F, N, P.

go Chemistry **Module 11: Periodic Trends** covers concepts in this section.

7.12 Periodic Trends: Electron Affinities

The **electron affinity** of an element is the energy change when an electron is added to a gaseous atom to form a 1− ion. As the term implies, the electron affinity (EA) is a measure of the attraction an atom has for an additional electron. For example, the electron affinity of fluorine is −328 kJ/mol.

$$\underset{[\text{He}]\ 2s^2 2p^5}{\text{F(g)}} + e^- \longrightarrow \underset{[\text{He}]\ 2s^2 2p^6}{\text{F}^-\text{(g)}} \qquad \Delta E = \text{EA} = -328 \text{ kJ/mol}$$

Two sign conventions are used for electron affinity—the one used here and an alternative one. The latter uses the opposite sign—for example, +328 kJ/mol for the electron affinity of fluorine.

This large negative value indicates that fluorine atoms readily accept an electron. As expected, fluorine and the rest of the halogens have large negative electron affinities because, by acquiring an electron, the halogen atoms achieve a stable octet of valence electrons, ns^2np^6 (Table 7.8). Notice from Table 7.8 that electron affinities generally become more negative across a period toward the halogen, which is only one electron away from a noble gas configuration.

Some elements have a positive electron affinity, meaning that the negative ion is less stable than the neutral atom. Such electron affinity values are difficult to obtain experimentally, and the electron affinity is simply indicated as > 0, as in the case of neon.

$$\underset{[\text{He}]\ 2s^2 2p^6}{\text{Ne}} + e^- \longrightarrow \underset{[\text{He}]\ 2s^2 2p^6 3s^1}{\text{Ne}^-} \qquad \text{EA} > 0 \text{ kJ/mol}$$

The Ne^- ion would revert back to the neutral neon atom and an electron. From their electron configurations we can understand why the Ne^- ion would be less stable than the neutral neon atom; the anion would exceed the stable octet of valence electrons of Ne atoms.

Table 7.8 Electron Affinities (kJ/mol)

1A (1)	2A (2)	3A (13)	4A (14)	5A (15)	6A (16)	7A (17)	8A (18)
H −73							He +50
Li −60	Be +50	B −27	C −122	N +70	O −141	F −328	Ne +120
Na −53	Mg +40	Al −43	Si −134	P −72	S −200	Cl −349	Ar +96
K −48	Ca +96	Ga −30	Ge −120	As −78	Se −195	Br −325	Kr +96
Rb −47	Sr −5	In −30	Sn −120	Sb −103	Te −190	I −295	Xe +80

First electron affinity (kJ/mol)

Electron affinities tend to become more negative across a period with halogens having the most negative values.

7.13 Energy Considerations in Ionic Compound Formation

Energy is released when an alkali metal and a nonmetal react to form a salt. For example, the vigorous reaction of sodium metal and chlorine gas forms solid sodium chloride, NaCl, and also transfers energy to the surroundings (Figure 7.27).

$$Na(s) + \frac{1}{2}Cl_2(g) \longrightarrow NaCl(s) \qquad \Delta H_f^\circ = -410.9\ kJ$$

As seen in Figure 7.23, energy changes occur during the formation of an ionic compound such as NaCl. Let's consider the energy changes in the formation of solid potassium fluoride, formed by the reaction of potassium, a very reactive alkali metal, and fluorine, a highly reactive halogen. Potassium fluoride contains K^+ and F^- ions grouped in a crystalline lattice of alternating cations and anions (Figure 7.28). Potassium atoms have low ionization energy and readily lose their $4s^1$ valence electrons; fluorine atoms have a high electron affinity and accept the electrons into a $2p$ orbital. Thus, electron transfer occurs from potassium metal atoms to fluorine atoms during KF formation.

$$K\ 1s^22s^22p^63s^23p^64s^1 \longrightarrow K^+\ 1s^22s^22p^63s^23p^6 \qquad \text{Ionization energy} = 419\ kJ$$

$$F\ 1s^22s^22p^5 + e^- \longrightarrow F^-\ 1s^22s^22p^6 \qquad \text{Electron affinity} = -328\ kJ$$

The transfer of electrons converts potassium and fluorine atoms into K^+ and F^- ions, each of which has an octet of outer electrons, a very stable noble gas configuration.

Electron configurations: K:[Ar]$4s^1$; F:[He]$2s^22p^5$; K^+ [Ar]; F^- [Ne]

The standard molar enthalpy of formation, ΔH_f°, of KF is −567.27 kJ/mol.

$$K(s) + \frac{1}{2}F_2(g) \longrightarrow KF(s) \qquad \Delta H_f^\circ = -567.27\ kJ$$

We can evaluate the energy required to form the ions and the solid compound by applying Hess's law (← *p. 242*) to a series of five steps, known as a **Born-Haber cycle**, as outlined on the next page and illustrated in Figure 7.29. When combined with the thermochemical expression for standard molar enthalpy of formation, such a multistep sequence allows the calculation of the **lattice energy** of the crystal, the enthalpy of formation of 1 mol of an ionic solid from its separated gaseous ions.

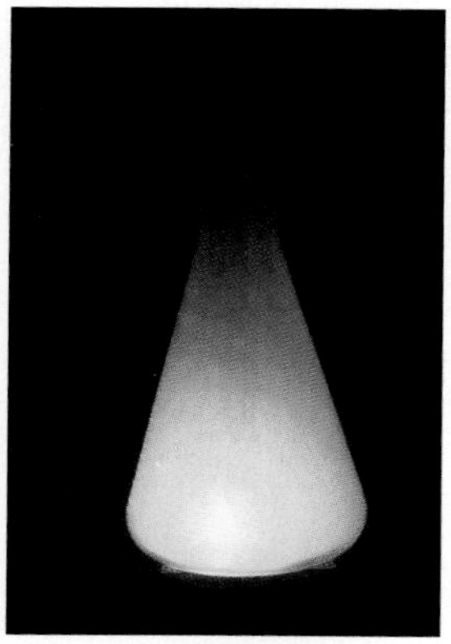

© Cengage Learning/Charles D. Winters

Figure 7.27 The reaction of sodium with chlorine. The reaction forms Na^+ and Cl^- ions plus liberates heat and light. Ionic compounds such as NaCl were first discussed in Section 3.5.

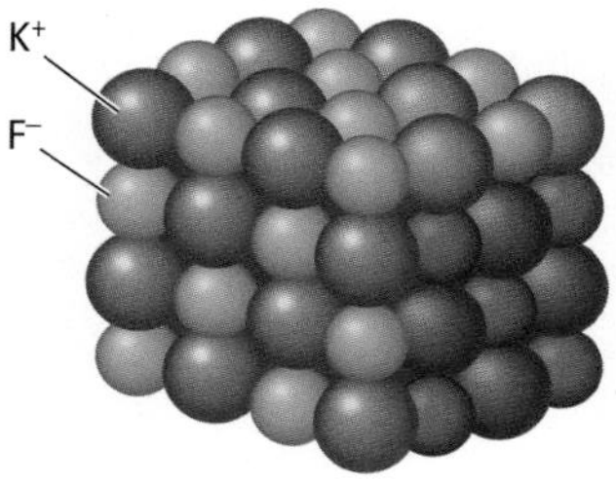

Figure 7.28 A potassium fluoride crystal illustrating the arrangement of K^+ and F^- ions in a lattice structure.

The five steps do not describe what actually happens when K(s) reacts with F_2(g). They do allow us to apply known thermochemical data to the problem.

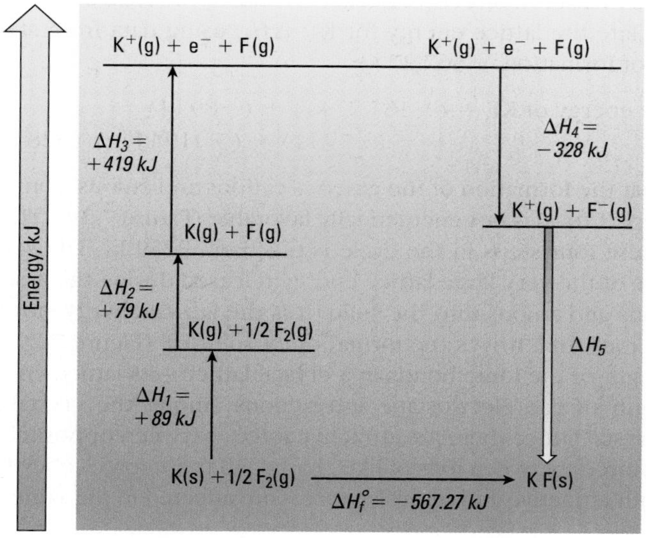

Figure 7.29 Born-Haber cycle for KF.

This cycle is named for Max Born and Fritz Haber, two Nobel Prize-winning German scientists.

Step 1: *Convert solid potassium to gaseous potassium atoms.* This is the enthalpy of sublimation, ΔH_1, which is known to be +89 kJ for 1 mol of potassium.

$$K(s) \longrightarrow K(g) \qquad \Delta H_1 = +89 \text{ kJ/mol}$$

Step 2: *Dissociate F_2 molecules into F atoms.* One mol of KF contains 1 mol fluoride ions, so $\frac{1}{2}$ mol F_2 is required. Breaking the bonds in diatomic fluorine molecules requires energy. This energy, ΔH_2, is +79 kJ, one-half of the bond energy of F_2, $\frac{1}{2}$(158 kJ/mol).

$$\tfrac{1}{2}\ F_2(g) \longrightarrow F(g) \quad \Delta H_2 = \tfrac{1}{2}\text{ BE of } F_2 = \tfrac{1}{2}\text{ mol } (+158 \text{ kJ/mol}) = +79 \text{ kJ}$$

Step 3: *Convert gaseous potassium atoms to gaseous potassium ions.* The energy, ΔH_3, required to remove the $4s^1$ electron is the first ionization energy, IE_1, of potassium, +419 kJ/mol, which can be obtained from a table of ionization energies.

$$K(g) \longrightarrow K^+(g) + e^- \qquad \Delta H_3 = IE_1 = +419 \text{ kJ}$$

Step 4: *Form F^- ions from F atoms.* The addition of a sixth electron to the 2*p* subshell completes the octet of valence electrons for F^-. For the formation of 1 mol F^- ions, this energy, ΔH_4, is the electron affinity (EA) of F, −328 kJ/mol, obtained from Table 7.8.

$$F(g) + e^- \longrightarrow F^-(g) \qquad \Delta H_4 = EA = -328 \text{ kJ}$$

Step 5: *Consolidate gaseous K^+ and F^- ions into a crystal.* The cations and anions assemble into a crystal, releasing a significant quantity of energy due to the electrostatic attraction between oppositely charged ions. This energy, ΔH_5, is the lattice energy, the enthalpy of formation of 1 mol of an ionic solid from its separated gaseous ions.

$$K^+(g) + F^-(g) \longrightarrow KF(s) \qquad \Delta H_5 = \text{lattice energy of KF}$$

The sum of the enthalpy changes in the five steps of the cycle represents the energy change for the overall reaction of solid potassium metal with gaseous fluorine to produce solid potassium fluoride (Figure 7.29). According to Hess's law, the sum of these enthalpy changes must equal the standard molar enthalpy of formation (ΔH_f° of KF, which is experimentally known to be −567.27 kJ.

$$\Delta H_f^\circ = -567.27 \text{ kJ} = \Delta H_1 + \Delta H_2 + \Delta H_3 + \Delta H_4 + \Delta H_5$$

Although data for Steps 1 to 4 can be determined experimentally, lattice energies cannot be measured directly. Instead, we can apply Hess's law in a Born-Haber cycle to calculate the lattice energy for KF, ΔH_5, using data from steps 1 to 4 and the enthalpy of formation, −567.27 kJ.

$$\Delta H_5 = \text{lattice energy of KF} = (-567.27 \text{ kJ}) + (-89 \text{ kJ}) + (-79 \text{ kJ}) + (-419 \text{ kJ}) + (328 \text{ kJ}) = -826 \text{ kJ}$$

Notice that the formation of the gaseous cations and anions from the respective elements (Steps 1 to 4) is not energetically favorable (Figure 7.29). The net enthalpy change for these four steps in the cycle is positive, +259 kJ. The solid compound forms because of the very large lattice energy released during the assembling of the gaseous cations and anions into the solid. It is the lattice energy, not the formation of K^+ and F^- ions, that drives the formation of solid KF (Figure 7.29).

The strength of the ionic bonds in a crystal lattice—its lattice energy—depends on the strength of the electrostatic interactions among the crystal's cations and anions. In a crystal lattice there are attractive forces between oppositely charged ions and repulsive forces between ions of like charge. All these forces obey Coulomb's law (← p. 94). Both attractive and repulsive forces are affected in the same proportion by two factors:

Table 7.9 Effect of Ion Size and Charge on Lattice Energy and Melting Point

Compound	Charges of Ions	$r_+ + r_-$	Lattice Energy, kJ/mol	Melting Point, °C
NaCl	+1, −1	102 pm + 181 pm = 283 pm	786	800
BaO	+2.−2	135 pm + 140 pm = 275 pm	3054	1920
MgO	+2. −2	66 pm + 140 pm = 206 pm	3791	2800

- *The sizes of the charges of the cations and anions.* As the charges get larger (from 1+ and 1−, to 2+ and 2−, etc.), both attractive and repulsive forces among the ions increase.
- The *distances among the ions.* The forces increase as the distances decrease. The shortest distance is $r_+ + r_-$, where r_+ is the cation radius and r_- is the anion radius. The distances among ions of like charge increase or decrease in proportion to this shortest distance. The smaller the ionic radii, the shorter the distances among the ions.

Ions are arranged in a crystal so that the attractive forces predominate over the repulsive forces. This makes the energy lower when the ions are in a crystal lattice than when they are completely separated from one another. Because the energy of the crystal lattice is lower than that for the separated ions (that is, a mole of KF(s) is lower in energy than a mole of separated K^+ and F^- ions), the lattice energy is negative. *The stronger the net ionic attractions are, the more negative the lattice energy is, and the more stable the crystal lattice is.* Thus, crystal lattices are more stable for small ions of large charge.

This is reflected in the fact that compounds with high lattice energies have high melting points. The entries in Table 7.9 exemplify this trend. Consider NaCl and BaO in Table 7.9. In both compounds the sum of the ionic radii of the cation and anion is about the same; it is 283 pm for NaCl and 275 pm for BaO. However, the product of the charges differs by a factor of four: [(1+) × (1−)] for NaCl versus [(2+) × (2−)] for BaO. This charge factor of four is reflected in a lattice energy of (−3054 kJ/mol) and a melting point (1920 °C) for BaO that are much larger than the lattice energy (−786 kJ/mol) and the melting point (800 °C) of NaCl.

The effect of cation size can be seen by comparing MgO and BaO as shown in Table 7.9. Because Mg^{2+} ($r_+ = 86$ pm) is much smaller than Ba^{2+} ($r_+ = 149$ pm), the 2+ charge on magnesium ion is more concentrated (higher charge density) than the 2+ charge on the barium ion. Consequently, the net attractive force between Mg^{2+} and O^{2-} is greater than the net attractive force between Ba^{2+} and O^{2-}, and the lattice energy of MgO is more negative. This results in a higher melting point for MgO (2800 °C) than for BaO (1920 °C).

SUMMARY PROBLEM

(a) Without looking back in the chapter, draw and label the first five energy levels of a hydrogen atom. Next, indicate the 2 → 1, 3 → 1, 5 → 2, and 4 → 3 transitions in a hydrogen atom. Look in Table 7.2 (p. 284) to get the measured wavelengths and spectral regions for these transitions. Now calculate the frequencies (ν) for these transitions. Next, calculate the energies of the photons that are produced in these transitions.

(b) The photoelectric threshold is the minimum wavelength a photon must have to produce a photoelectric effect for a metal. These three metals exhibit photoelectric effects when photons of sufficient energies strike their surfaces.

Element	Photoelectric Threshold (nm)
Lithium	540
Potassium	550
Cesium	660

Which photon energies calculated in part (a) would be sufficient to cause a photoelectric effect in lithium, in potassium, and in cesium?

(c) Niobium (Nb) and tantalum (Ta) are transition elements named after the mythical king Tantalus, son of Zeus, and his daughter Niobe. Niobium metal is a component of stainless steels fabricated for high temperature use. Tantalum metal is extremely corrosion resistant and thus is used in the manufacturing of surgical implants.

(i) Write the full electron configuration for niobium and for tantalum.

(ii) Using the noble gas notation, write the electron configuration for an Nb^{4+} ion.

(iii) Write the atomic orbital box diagram for a Ta^{5+} ion.

(iv) How many unpaired electrons are there in an Nb^{5+} ion? Explain your answer.

(v) Are either an Nb^{4+} ion or an Nb^{5+} ion paramagnetic? Explain your answer.

(iv) Explain why niobium and tantalum occur together in natural deposits of their compounds.

(d) An element is brittle with a steel-gray appearance. It is a relatively poor electrical conductor and forms a volatile molecular chloride and a molecular hydride that decomposes at room temperature. Is the element an alkaline-earth metal, a transition metal, a metalloid, or a halogen? Explain your answer.

(e) Give two ways in which *p* atomic orbitals and *d* atomic orbitals are (i) alike; (ii) different.

IN CLOSING

Having studied this chapter, you should be able to . . .

- Use the relationships among frequency, wavelength, and the speed of light for electromagnetic radiation (Section 7.1). End-of-chapter questions: 14, 16, 26
- Explain the relationship between Planck's quantum theory and the photoelectric effect (Section 7.2). Question 32
- Use the Bohr model of the atom to interpret line emission spectra and the energy absorbed or emitted when electrons in atoms change energy levels (Section 7.3). Questions 40, 44
- Calculate the frequency, energy, or wavelength of an electron transition in a hydrogen atom and determine in what region of the electromagnetic spectrum the emission would occur (Section 7.3). Question 46

OWL and go Chemistry

Sign in at **www.cengage.com/owl** to:

- View tutorials and simulations, develop problem-solving skills, and complete online homework assigned by your professor.
- For quick review and exam prep, download Go Chemistry mini lecture modules from OWL (or purchase them at **www.CengageBrain.com**).

- Explain the use of the quantum mechanical model of the atom to represent the energy and probable location of electrons (Section 7.4). Questions 58, 62
- Apply quantum numbers (Section 7.5). Questions 52, 54
- Understand the spin properties of electrons and how they affect electron configurations and the magnetic properties of atoms (Section 7.5). Question 87
- Describe and explain the relationships among shells, subshells, and atomic orbitals (Section 7.5). Questions 56, 74
- Describe how the shapes of *s*, *p*, and *d* atomic orbitals and their sets of the first three quantum numbers differ (Section 7.6).
- Use the periodic table to write the electron configurations of atoms and ions of main group and transition elements (Sections 7.7, 7.8). Questions 66, 69, 72
- Explain variations in valence electrons, electron configurations, ion formation, and paramagnetism of transition metals (Section 7.8). Question 76
- Describe trends in atomic radii, based on electron configurations (Section 7.9). Question 94
- Describe the role of effective nuclear charge in the trend of atomic radii across a period (Section 7.9).
- Describe trends in ionic radii and explain why ions differ in size from their atoms (Section 7.10). Questions 96, 106
- Use electron configurations to explain trends in the ionization energies of the elements (Section 7.11). Questions 98, 102
- Describe electron affinity (Section 7.12).
- Discuss the energy changes that occur during the formation of an ionic compound from its elements (Section 7.13).
- Use a Born-Haber cycle to determine the lattice energy of an ionic compound (Section 7.13). Question 110

KEY TERMS

atomic radius *(Section 7.9)*
Born-Haber cycle *(7.13)*
boundary surface *(7.4)*
continuous spectrum *(7.3)*
core electrons *(7.7)*
diamagnetic *(7.8)*
effective nuclear charge *(7.9)*
electromagnetic radiation *(7.1)*
electron affinity *(7.12)*
electron configuration *(7.7)*
electron density *(7.4)*
excited state *(7.3)*
ferromagnetic *(7.8)*
frequency *(7.1)*
ground state *(7.3)*
Hund's rule *(7.7)*
ionic radii *(7.10)*
ionization energy *(7.11)*
isoelectronic *(7.8)*
lattice energy *(7.13)*
Lewis dot symbol *(7.7)*
line emission spectrum *(7.3)*
momentum *(7.4)*
noble gas notation *(7.7)*
orbital *(7.4)*
orbital shape *(7.6)*
paramagnetic *(7.8)*
Pauli exclusion principle *(7.5)*
***p*-block elements** *(7.7)*
photoelectric effect *(7.2)*
photons *(7.2)*
Planck's constant *(7.2)*
principal energy level *(7.5)*
principal quantum number *(7.3)*
quantum *(7.2)*
quantum number *(7.5)*
quantum theory *(7.2)*
radial distribution plot *(7.6)*
***s*-block elements** *(7.7)*
screening effect *(7.9)*
shell *(7.5)*
spectrum *(7.1)*
subshell *(7.5)*
transition elements *(7.7)*
uncertainty principle *(7.4)*
valence electrons *(7.7)*
wavelength *(7.1)*

QUESTIONS FOR REVIEW AND THOUGHT

OWL Interactive versions of these problems are assignable in OWL.

Blue-numbered questions have short answers at the back of this book in Appendix M and fully worked solutions in the *Student Solutions Manual.*

Review Questions

These questions test vocabulary and simple concepts.

1. How is the frequency of electromagnetic radiation related to its wavelength?
2. Light is given off by a sodium- or mercury-containing streetlight when the atoms are excited in some way. The light you see arises for which of these reasons?
   (a) Electrons moving from a given quantum level to one of higher n.
   (b) Electrons being removed from the atom, thereby creating a metal cation.
   (c) Electrons moving from a given quantum level to one of lower n.
   (d) Electrons whizzing about the nucleus in an absolute frenzy.
3. What is Hund's rule? Give an example of the use of this rule.
4. Explain what it means when an electron occupies the $3p_x$ orbital.
5. Tell what happens to atomic size and ionization energy across a period and down a group.
6. Why is the radius of Na^+ much smaller than the radius of Na? Why is the radius of Cl^- much larger than the radius of Cl?
7. Write electron configurations to show the first two ionization steps for potassium. Explain why the second ionization energy is much larger than the first.
8. Explain how the sizes of atoms change and why they change across a period of the periodic table.
9. Write the electron configurations for the valence electrons of elements in the first three periods in Groups 1A through 8A.

Topical Questions

These questions are keyed to the major topics in the chapter. Usually a question that is answered at the back of this book is paired with a similar one that is not.

Electromagnetic Radiation (Sections 7.1, 7.2)

10. Electromagnetic radiation that is high in energy consists of waves that have __________ (long or short) wavelengths and __________ (high or low) frequencies. Give an example of radiation from the high-energy end of the electromagnetic spectrum.
11. Electromagnetic radiation that is low in energy consists of waves that have __________ (long or short) wavelengths and __________ (high or low) frequencies. Give an example of radiation from the low-energy end of the electromagnetic spectrum.
12. The regions of the electromagnetic spectrum are shown in Figure 7.1. Answer these questions on the basis of this figure.
    (a) Which type of radiation involves less energy, radio waves or infrared radiation?
    (b) Which radiation has the higher frequency, radio waves or microwaves?
13. The colors of the visible spectrum and the wavelengths corresponding to the colors are given in Figure 7.1.
    (a) Which colors of light involve photons with less energy than yellow light?
    (b) Which color of visible light has photons of greater energy, green or violet?
    (c) Which color of light has the greater frequency, blue or green?
14. Assume that a microwave oven operates at a frequency of $1.00 \times 10^{11}\ s^{-1}$.
    (a) What is the wavelength of this radiation in meters?
    (b) What is the energy in joules per photon?
    (c) What is the energy per mole of photons?
15. IBM scientists have developed a prototype computer chip that operates at 350 GHz (1 GHz = 10^9 Hz).
    (a) What is its wavelength in meters? In nanometers (nm)?
    (b) Calculate its energy in joules.
16. Place these types of radiation in order of increasing energy per photon.
    (a) Green light from a mercury lamp
    (b) X-rays from a dental X-ray
    (c) Microwaves in a microwave oven
    (d) An FM music station broadcasting at 89.1 MHz
17. Place these types of radiation in order of increasing energy per photon.
    (a) Radio signals
    (b) Radiation from a microwave oven
    (c) Gamma rays from a nuclear reaction
    (d) Red light from a neon sign
    (e) Ultraviolet radiation from a sun lamp
18. If green light has a wavelength of 495 nm, what is its frequency?
19. What kind of radiation has a frequency of 5×10^{12} Hz? What is its wavelength?
20. What is the energy of one photon of blue light that has a wavelength of 450 nm?
21. Which has more energy,
    (a) One photon of infrared radiation or one photon of microwave radiation?
    (b) One photon of yellow light or one photon of orange light?
22. Which kinds of electromagnetic radiation can interact with molecules at the cellular level?
23. What is the energy and wavelength associated with one quantum of laser light that has a frequency of $4.57 \times 10^{14}\ s^{-1}$?
24. Stratospheric ozone absorbs damaging UV-C radiation from the sun, preventing the radiation from reaching Earth's surface. Calculate the frequency and energy of UV-C radiation that has a wavelength of 270 nm.

25. When someone uses a sunscreen, which kind of radiation will be blocked? How does the sunscreen protect your skin from this type of radiation?
26. Calculate the energy of one photon of X-radiation having a wavelength of 2.36 nm, and compare it with the energy of one photon of orange light (3.18×10^{-19} J).
27. Green light of wavelength 516 nm is absorbed by an atomic gas. Calculate the energy difference between the two quantum states involved with this absorption.
28. Calculate the energy possessed by one mole of X-ray photons of wavelength 1.00×10^{-9} m.

Photoelectric Effect (Section 7.2)

29. Describe the role Einstein's explanation of the photoelectric effect played in the development of the quantum theory.
30. Light of very long wavelength strikes a photosensitive metallic surface and no electrons are ejected. Explain why increasing the intensity of this light on the metal still will not cause the photoelectric effect.
31. Photons of light with sufficient energy can eject electrons from a gold surface. To do so requires photons with a wavelength equal to or shorter than 257 nm. Will photons in the visible region of the spectrum dislodge electrons from a gold surface?
32. To eject electrons from the surface of potassium metal requires a minimum energy of 3.69×10^{-19} J. When 600.-nm photons shine on a potassium surface, will they cause the photoelectric effect? Explain your answer.
33. A bright red light strikes a photosensitive surface and no electrons are ejected, even though dim blue light ejects electrons from the surface. Explain.

Atomic Spectra and the Bohr Atom (Section 7.3)

34. How does a line emission spectrum differ from sunlight?
35. Any atom with its electrons in their lowest energy levels is said to be in a(n) __________ state.
36. Energy is emitted from an atom when an electron moves from the __________ state to the __________ state. The energy of the emitted radiation corresponds to the __________ between the two energy levels.
37. Which transition involves the emission of less energy in the H atom, an electron moving from $n = 4$ to $n = 3$ or an electron moving from $n = 3$ to $n = 1$? (See Figure 7.8.)
38. For which of these transitions in a hydrogen atom is energy absorbed? Emitted?
    (a) $n = 1$ to $n = 3$ (b) $n = 5$ to $n = 1$
    (c) $n = 2$ to $n = 4$ (d) $n = 5$ to $n = 4$
39. For the transitions in Question 38:
    (a) Which ones involve the ground state?
    (b) Which one involves the greatest energy change?
    (c) Which one absorbs the most energy?
40. If energy is absorbed by a hydrogen atom in its ground state, the atom is excited to a higher energy state. For example, the excitation of an electron from the energy level with $n = 1$ to a level with $n = 4$ requires radiation with a wavelength of 97.3 nm. Which of these transitions would require radiation of a wavelength longer than this? (See Figure 7.8.)
    (a) $n = 2$ to $n = 4$ (b) $n = 1$ to $n = 3$
    (c) $n = 1$ to $n = 5$ (d) $n = 3$ to $n = 5$
41. (a) Calculate the wavelength, in nanometers, of the emission line that results from the $n = 2$ to $n = 1$ transition in hydrogen.
    (b) The emitted radiation is in what region of the electromagnetic spectrum?
42. Spectroscopists have observed He^+ in outer space. This ion is a one-electron species like a neutral hydrogen atom. Calculate the energy emitted for the transition from the $n = 5$ to the $n = 3$ state in this ion using the equation $E_n = -\frac{Z^2}{n^2}(2.179 \times 10^{-18}\text{ J})$. Z is the positive charge of the nucleus and n is principal quantum number. In what part of the electromagnetic spectrum does this radiation lie?
43. The Brackett series of emissions has $n_f = 4$.
    (a) Calculate the wavelength, in nanometers, of the $n = 7$ to $n = 4$ electron transition.
    (b) The emitted radiation is in what region of the electromagnetic spectrum?
44. The Bohr equation for hydrogen can be modified to apply to one-electron species other than neutral hydrogen atoms, for example Li^{2+}, to calculate the energy of electron transitions in the ion. The modified equation is $E_n = -\frac{Z^2}{n^2}(2.179 \times 10^{-18}\text{ J})$. Z is the positive charge of the nucleus and n is the principal quantum number. Calculate the energy emitted for the transition from the $n = 4$ to the $n = 1$ state in this ion. In what region of the electromagnetic spectrum does it lie?
45. Calculate the energy and the wavelength of the electron transition from $n = 1$ to $n = 4$ in the hydrogen atom.
46. Calculate the energy and wavelength for the electron transition from $n = 2$ to $n = 5$ in the hydrogen atom.

de Broglie Wavelength (Section 7.4)

47. Calculate the de Broglie wavelength of a 4400-lb sport-utility vehicle moving at 75 miles per hour.
48. Calculate the de Broglie wavelength of an electron moving at 5% the speed of light.

Quantum Numbers (Section 7.5)

49. Assign a set of four quantum numbers for
    (a) *Each* electron in a nitrogen atom.
    (b) The valence electron in a sodium atom.
    (c) A 3*d* electron in a nickel atom.
50. Assign a correct set of four quantum numbers for
    (a) *Each* electron in a boron atom.
    (b) The 3*s* electrons in a magnesium atom.
    (c) A 3*d* electron in an iron atom.
51. One electron has the set of quantum numbers $n = 3$, $\ell = 1$, $m_\ell = -1$, and $m_s = +\frac{1}{2}$; another electron has the set $n = 3$, $\ell = 1$, $m_\ell = 1$, and $m_s = +\frac{1}{2}$.
    (a) Could the electrons be in the same atom? Explain.
    (b) Could they be in the same atomic orbital? Explain.
52. Some of these sets of quantum numbers (n, ℓ, m_ℓ, m_s) could not occur. Explain why.
    (a) 2, 1, 2, $+\frac{1}{2}$ (b) 3, 2, 0, $-\frac{1}{2}$
    (c) 1, 0, 0, 1 (d) 3, 3, 2, $-\frac{1}{2}$
    (e) 2, 0, 0, $+\frac{1}{2}$
53. Give the n, ℓ, and m_ℓ values for
    (a) Each atomic orbital in the 6*f* sublevel.
    (b) Each atomic orbital in the $n = 5$ level.

54. Assign a correct set of four quantum numbers for the circled electrons in these orbital diagrams.

(a) [Ar] 4s: (↑↓) — circled

(b) [Ne] 3s: ↑↓ 3p: (↑)↓ ↑ ↑ — circled

(c) [Ar] 4s: ↑↓ 3d: ↑↓ ↑↓ (↑) ↑ ↑ — circled

Quantum Mechanics (Section 7.6)

55. From memory, sketch the shape of the boundary surface for each of these atomic orbitals:
(a) $2p_z$ (b) 4s

56. How many subshells are there in the electron shell with the principal quantum number $n = 4$?

57. How many subshells are there in the electron shell with the principal quantum number $n = 5$?

58. Bohr pictured the electrons of the atom as being located in definite orbits about the nucleus, just as planets orbit the sun. Criticize this model in view of the quantum mechanical model.

59. Radiation is not the only thing that has both wave-like and particle-like characteristics. What component of matter can also be described in the same way?

60. How did the Heisenberg uncertainty principle illustrate the fundamental flaw in Bohr's model of the atom?

61. The three-dimensional boundary surfaces describing the probability of finding an electron and the energy of an electron can both be obtained from the __________ __________ of the electron.

62. Which types of atomic orbitals are found in the $n = 3$ shell? How many atomic orbitals altogether are found in this shell?

Electron Configurations (Section 7.7)

63. Write electron configurations for Mg and Cl atoms.

64. Write electron configurations for Al and S atoms.

65. Write electron configurations for these atoms.
(a) Strontium (Sr), named for a town in Scotland.
(b) Tin (Sn), a metal used in the ancient world. Alloys of tin (solder, bronze, and pewter) are important.

66. Germanium had not been discovered when Mendeleev formulated his ideas of chemical periodicity. He predicted its existence, however, and germanium was found in 1886 by Winkler. Write the electron configuration of germanium.

67. Name an element of Group 3A. What does the group designation tell you about the electron configuration of the element?

68. Name an element of Group 6A. What does the group designation tell you about the electron configuration of the element?

69. (a) Which ions in this list are likely to be formed: K^{2+}, Cs^+, Al^{4+}, F^{2-}, Se^{2-}?
(b) Which, if any, of these ions have a noble gas configuration?

70. These ground-state atomic orbital diagrams are incorrect. Explain why they are incorrect and how they should be corrected.

(a) Fe [Ar] 4s: ↑ 3d: ↑↓ ↑↓ ↑↓ ↑ ↑ ↑

(b) P [Ne] 2s: ↑↓ 2p: ↑↓ ↑ ☐

(c) Sn^{2+} [Kr] 4d: ↑↓ ↑↓ ↑↓ ↑↓ ↑ ↑ 5s: ↑↓ 5p: ↑ ↑ ☐

71. Write the atomic orbital diagrams for
(a) A nitrogen atom and a nitride, N^{3-}, ion.
(b) The 3p electrons of a sulfur atom and a sulfide, S^{2-}, ion.

72. Write the atomic orbital diagram for the 4s and 3d electrons in a
(a) Vanadium atom. (b) V^{2+} ion.
(c) V^{4+} ion.

73. Write the atomic orbital diagram for the 4s and 3d electrons in a
(a) Manganese atom. (b) Mn^+ ion.
(c) Mn^{5+} ion.

74. How many elements are there in the fourth period of the periodic table? Explain why it is not possible for there to be another element in this period.

75. When transition metals form ions, electrons are lost first from which type of atomic orbital? Why?

76. Give the electron configurations of Mn, Mn^{2+}, and Mn^{3+}. Use atomic orbital box diagrams to determine the number of unpaired electrons for each species.

77. Write the electron configurations of chromium: Cr, Cr^{2+}, and Cr^{3+}. Use atomic orbital box diagrams to determine the number of unpaired electrons for each species.

78. Write the electron configuration of sulfur, S. The element was known to the ancients as brimstone.

79. Write electron configurations for these elements.
(a) Zirconium (Zr). This metal is exceptionally resistant to corrosion and so has important industrial applications. Moon rocks show a surprisingly high zirconium content compared with rocks on Earth.
(b) Rhodium (Rh), used in jewelry and in industrial catalysts.

80. The lanthanides, or rare earths, are only "medium rare." All can be purchased for a reasonable price. Give electron configurations for atoms of these lanthanides.
(a) Europium (Eu), the most expensive of the rare earth elements; 1 g can be purchased for about $90.
(b) Ytterbium (Yb). Less expensive than Eu, Yb costs only about $12 per gram. It was named for the village of Ytterby in Sweden, where a mineral source of the element was found.

Valence Electrons (Sections 7.7, 7.8)

81. Locate these elements in the periodic table, and draw a Lewis dot symbol that represents the number of valence electrons for an atom of each element.
(a) F (b) In
(c) Te (d) Cs

82. Locate these elements in the periodic table, and draw a Lewis dot symbol that represents the number of valence electrons for an atom of each element.
(a) Sr (b) Br
(c) Ga (d) Sb

83. Give the electron configurations of these ions, and indicate which ones are isoelectronic.
(a) Na^+ (b) Al^{3+} (c) Cl^-

84. Give the electron configurations of these ions, and indicate which ones are isoelectronic.
(a) Ca^{2+} (b) K^+ (c) O^{2-}

85. What is the electron configuration for
(a) A bromine atom? (b) A bromide ion?

86. (a) What is the electron configuration for an atom of tin?
(b) What are the electron configurations for Sn^{2+} and Sn^{4+} ions?

Paramagnetism and Unpaired Electrons (Section 7.8)

87. (a) In the first transition series (in row four of the periodic table), which elements would you predict to be diamagnetic?
(b) Which element in this series has the greatest number of unpaired electrons?

88. What is ferromagnetism?

89. Which groups of elements in the periodic table are ferromagnetic?

90. How do the spins of unpaired electrons from paramagnetic and ferromagnetic materials differ in their behavior in a magnetic field?

91. The acetylacetonate ion, $(acac)^-$, has no unpaired electrons. It forms compounds with Fe^{2+} and with Fe^{3+} ions whose formulas are $Fe(acac)_2$ and $Fe(acac)_3$, respectively. Explain which compound will have the greater attraction to a magnetic field.

92. Consider titanium metal and its two oxides, TiO and TiO_2. The oxide ion has no unpaired electrons.
(a) Which of these titanium species is diamagnetic? Explain your answer.
(b) Which titanium species will be most attracted to a magnetic field? Explain your answer.

Periodic Trends (Sections 7.9–7.11)

93. Arrange these elements in order of increasing atomic size: Al, B, C, K, Na. (Try doing it without looking at Figure 7.22 and then check yourself by looking up the necessary atomic radii.)

94. Arrange these elements in order of increasing atomic size: Ca, Rb, P, Ge, Sr. (Try doing it without looking at Figure 7.22 and then check yourself by looking up the necessary atomic radii.)

95. Select the atom or ion in each pair that has the larger radius.
(a) Cl or Cl^- (b) Ca or Ca^{2+}
(c) Al or N (d) Cl^- or K^+
(e) In or Sn

96. Select the atom or ion in each pair that has the smaller radius.
(a) Cs or Rb (b) O^{2-} or O
(c) Br or As (d) Ba or Ba^{2+}
(e) Cl^- or Ca^{2+}

97. Write electron configurations to show the first two ionization steps for sodium. Explain why the second ionization energy is much larger than the first.

98. Arrange these atoms in order of increasing first ionization energy: F, Al, P, Mg.

99. Arrange these atoms in order of increasing first ionization energy: Li, K, C, N.

100. Which of these groups of elements is arranged correctly in order of increasing ionization energy?
(a) C, Si, Li, Ne (b) Ne, Si, C, Li
(c) Li, Si, C, Ne (d) Ne, C, Si, Li

101. Rank these ionization energies (IE) in order from the smallest value to the largest value. Briefly explain your answer.
(a) First IE of Be (b) First IE of Li
(c) Second IE of Be (d) Second IE of Na
(e) First IE of K

102. Predict which of these elements would have the greatest difference between the first and second ionization energies: Si, Na, P, Mg. Briefly explain your answer.

103. Compare the elements Li, K, C, N.
(a) Which has the largest atomic radius?
(b) Place the elements in order of increasing first ionization energy.

104. Compare the elements B, Al, C, Si.
(a) Which has the most metallic character?
(b) Which has the largest atomic radius?
(c) Place the three elements B, Al, and C in order of increasing first ionization energy.

105. Explain why the transition elements in a row are more alike in their properties than other elements in the same groups.

106. Select the atom or ion in each pair that has the larger radius. Explain your choice.
(a) H^+ or H^- (b) N or N^{3-} (c) F^- or Ne

107. Explain why nitrogen has a higher first ionization energy than does carbon, the preceding element in the periodic table.

108. The first electron affinity of oxygen is negative, the second is positive. Explain why this change in sign occurs.

109. Which group of the periodic table has elements with high first ionization energies *and* very negative electron affinities? Explain this behavior.

Born-Haber Cycle (Section 7.13)

110. Determine the lattice energy for LiCl(s) given these data (kJ/mol): Enthalpy of sublimation of Li, 161; IE_1 for Li, 520; BE of $Cl_2(g)$, 242; electron affinity of Cl, −349; enthalpy of formation of LiCl(s), −408.7.

111. The lattice energy of KCl(s) is −719 kJ/mol. Use data from the text to calculate the enthalpy of formation of KCl.

General Questions

These questions are not explicitly keyed to chapter topics; many require integration of several concepts.

112. Arrange these colors of the visible region of the electromagnetic spectrum in order of increasing energy per photon: green, red, yellow, and violet.

113. A neutral atom has two electrons with $n = 1$, eight electrons with $n = 2$, eight electrons with $n = 3$, and one electron with $n = 4$. Assuming this atom is in its ground state, supply its
(a) Atomic number and name.
(b) Total number of *s* electrons.
(c) Total number of *p* electrons.
(d) Total number of *d* electrons.

114. How many p atomic orbital electron pairs are there in an atom of selenium (Se) in its ground state?
115. Give the symbol of all the ground-state atoms that have
(a) No p electrons.
(b) From two to four d electrons.
(c) From two to four s electrons.
116. Give the symbol of the ground-state atom that
(a) Is in Group 8A but has no p electrons.
(b) Has a single electron in the $3d$ subshell.
(c) Forms a 1+ ion with a $1s^2 2s^2 2p^6$ electron configuration.
117. Answer these questions about the elements X and Z, which have the electron configurations shown.

$$X = [Kr]\ 4d^{10}5s^1 \qquad Z = [Ar]\ 3d^{10}4s^2 4p^4$$

(a) Is element X a metal or a nonmetal?
(b) Which element has the larger atomic radius?
(c) Which element would have the greater first ionization energy?
118. (a) Rank these in order of increasing atomic radius: O, S, F.
(b) Which has the largest first ionization energy: P, Si, S, Se?
119. (a) Place these in order of increasing radius: Ne, O^{2-}, N^{3-}, F^-.
(b) Place these in order of increasing first ionization energy: Cs, Sr, Ba.
120. Name the element corresponding to each of these characteristics.
(a) The element whose atoms have the electron configuration $1s^2 2s^2 2p^6 3s^2 3p^4$
(b) The element in the alkaline-earth group that has the largest atomic radius
(c) The element in Group 5A whose atoms have the largest first ionization energy
(d) The element whose 2^+ ion has the configuration [Kr] $4d^6$
(e) The element whose neutral atoms have the electron configuration [Ar] $3d^{10}4s^1$
121. The ionization energies for the removal of the first electron from atoms of Si, P, S, and Cl are listed below. Briefly rationalize this trend.

Element	First Ionization Energy (kJ/mol)
Si	780
P	1060
S	1005
Cl	1255

122. Answer these questions about the elements with the electron configurations shown.

$$X = [Ar]\ 3d^8 4s^2 \qquad Z = [Ar]\ 3d^{10}4s^2 4p^5$$

(a) An atom of which element is expected to have the larger first ionization energy?
(b) An atom of which element would be the smaller of the two?
123. Place these atoms and ions in order of decreasing size: Ar, K^+, Cl^-, S^{2-}, Ca^{2+}.
124. Which of these ions are unlikely, and why: Cs^+, In^{4+}, V^{6+}, Te^{2-}, Sn^{5+}, I^-?
125. Rank these in order of increasing first ionization energy: Zn, Ca, Ca^{2+}, Cl^-. Briefly explain your answer.
126. Classify these statements as being either true or false. If a statement is false, correct it to make it true.
(a) The wavelength of green light is longer than that of red light.
(b) Photons of green light have greater energy than those of red light.
(c) The frequency of green light is greater than that of red light.
(d) In the electromagnetic spectrum, frequency and wavelength of radiation are directly related.
127. Classify these statements as being either true or false. If a statement is false, correct it to make it true.
(a) A $3f$ atomic orbital can hold a maximum of 14 electrons.
(b) The ground-state electron configuration of a sulfur atom is $1s^2 2s^2 2p^6 3s^2 3p^4$.
(c) A ground-state sulfur atom has four unpaired electrons.
(d) An Mg^{2+} ion has an argon electron configuration.
(e) An N^{3-} ion and a P^{3-} ion have the same ground-state electron configuration.
128. Below is a grid with nine lettered boxes, each of which contains an item that is used to answer the questions that follow. Items may be used more than once and there may be more than one correct item in response to a question. The n_m are principal quantum numbers for H-atom energy levels.

A $n_4 \longrightarrow n_2$	**B** $n_3 \longrightarrow n_1$	**C** > 700 nm
D $n_4 \longrightarrow n_1$	**E** $n_2 \longrightarrow n_7$	**F** $n_3 \longrightarrow n_5$
G 486 nm	**H** $n_6 \longrightarrow n_3$	**I** < 400 nm

Place the letter(s) of the correct selection(s) on the appropriate line.
(a) Ultraviolet emission(s) ______________
(b) Highest energy absorption ______________
(c) Highest energy emission ______________
(d) Visible region emission(s) ______________
(e) Infrared region emission(s) ______________
129. Below is a grid with nine lettered boxes, each of which contains an item that is used to answer the questions that follow. Items may be used more than once and there may be more that one correct item in response to a question.

A Fe	**B** Br^-	**C** K^+
D Ne	**E** Fe^{3+}	**F** F
G K	**H** Sc	**I** Ar

Place the letter(s) of the correct selection(s) on the appropriate line.

(a) Neutral atom with a radius greater than the radius of Cl. ____________
(b) Radius is less than the radius of Si. ____________
(c) Isoelectronic species. ____________
(d) Radius is greater than that of Br ____________
(e) Neutral atom with highest first ionization energy. ____________
(f) Iron-containing species with radius less than the radius of Fe^{2+} ____________.

130. Criticize these statements.
(a) The energy of a photon is inversely related to its frequency.
(b) The energy of the hydrogen electron is inversely proportional to its principal quantum number *n*.
(c) Electrons start to enter the fourth energy level as soon as the third level is full.
(d) Light emitted by an $n = 4$ to $n = 2$ transition will have a longer frequency than that from an $n = 5$ to $n = 2$ transition.

131. A general chemistry student tells a chemistry classmate that when an electron goes from a 2*d* atomic orbital to a 1*s* atomic orbital, it emits more energy than that for a 2*p* to 1*s* transition. The other student is skeptical and says that such an energy change is not possible and explains why. What explanation was given?

132. Which of these types of radiation—infrared, visible, or ultraviolet—is required to ionize a hydrogen atom? Explain.

133. A certain minimum energy, E_{min}, is required to eject an electron from a photosensitive surface. Any energy absorbed beyond this minimum gives kinetic energy to the ejected electron. When 540-nm light falls on a cesium surface, an electron is ejected with a kinetic energy of 6.69×10^{-20} J. When the wavelength is 400 nm, the kinetic energy is 1.96×10^{-19} J.
(a) Calculate E_{min} for cesium, in joules.
(b) Calculate the longest wavelength, in nanometers, that will eject an electron from cesium.

134. Suppose a new element, extraterrestrium, tentatively given the symbol Et, has just been discovered. Its atomic number is 120.
(a) Write the electron configuration of the element.
(b) Name another element you would expect to find in the same group as Et.
(c) Give the formulas for the compounds of Et with O and Cl.

135. When sulfur dioxide reacts with chlorine, the products are thionyl chloride, $SOCl_2$, and dichlorine monoxide, OCl_2.

$$SO_2(g) + 2\,Cl_2(g) \longrightarrow SOCl_2(g) + OCl_2(g)$$

(a) In what period of the periodic table is S located?
(b) Give the complete electron configuration of S. Do *not* use the noble gas notation.
(c) An atom of which element involved in this reaction (O, S, or Cl) should have the smallest first ionization energy? The smallest radius?
(d) You want to make 675 g $SOCl_2$. Calculate what mass in grams of Cl_2 is required.
(e) If you use 10.0 g SO_2 and 20.0 g Cl_2, determine the theoretical yield of $SOCl_2$.

Applying Concepts

These questions test conceptual learning.

136. Write the electron configuration for the product of the first ionization of the smallest halogen.

137. Write the electron configuration for the product of the second ionization of the third largest alkaline-earth metal.

138. What compound will most likely form between chlorine and element X, if element X has the electronic configuration $1s^22s^22p^63s^1$?

139. Write the formula for the compound that most likely forms between potassium and element Z, if element Z has the electronic configuration $1s^22s^22p^63s^23p^4$.

140. Which of these electron configurations are for atoms in the ground state? In excited states? Which are impossible?
(a) $1s^22s^1$
(b) $1s^22s^22p^3$
(c) [Ne] $3s^23p^34s^1$
(d) [Ne] $3s^23p^64s^33d^2$
(e) [Ne] $3s^23p^64f^4$
(f) $1s^22s^22p^43s^2$

141. Which of these electron configurations are for atoms in the ground state? In excited states? Which are impossible?
(a) $1s^22s^2$
(b) $1s^22s^23s^1$
(c) [Ne] $3s^23p^84s^1$
(d) [He] $2s^22p^62d^2$
(e) [Ar] $4s^23d^3$
(f) [Ne] $3s^22p^54s^1$

142. Using the information in Appendix D and Figure 7.17, write the electron configuration for the undiscovered element with an atomic number of 164. Where would this element be located in the periodic table?

143. These questions refer to the graph below.

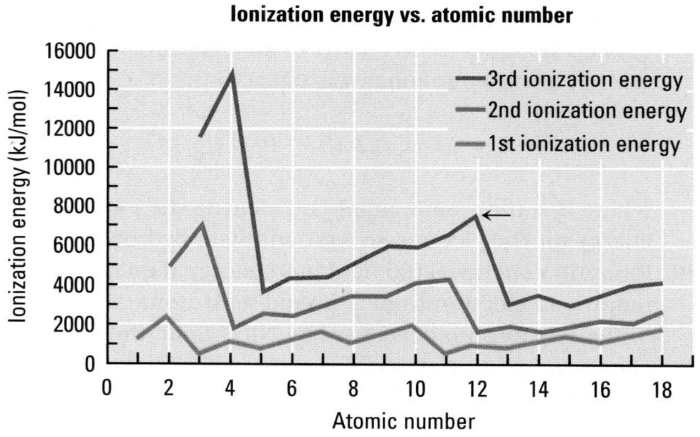

(a) Based on the graphic data, ionization energies __________ (decrease, increase) left to right and __________ (decrease, increase) top to bottom on the periodic table.
(b) Which element has the largest first ionization energy?
(c) A plot of the fourth ionization energy versus atomic number for elements 1 through 18 would have peaks at which atomic numbers?
(d) Why is there no third ionization energy for helium?
(e) What is the reason for the large second ionization energy for lithium?
(f) Find the arrow pointing to the third ionization energy curve. Write the equation for the process corresponding to this data point.

144. Use Coulomb's law to predict which substance in each of these pairs has the larger lattice energy.
(a) CaO or KI
(b) CaF_2 or BaF_2
(c) KCl or LiBr

More Challenging Questions

These questions require more thought and integrate several concepts.

145. An element has an ionization energy of 1.66×10^{-18} J. The three longest wavelengths in its absorption spectrum are 253.7 nm, 185.0 nm, and 158.5 nm.
    (a) Construct an energy-level diagram similar to Figure 7.8 for this element.
    (b) Indicate all possible emission lines. Start from the highest energy level on the diagram.
146. The energy of a photon needed to cause ejection of an electron from a photoemissive metal is expressed as the sum of the binding energy of the photon plus the kinetic energy of the emitted electron. When photons of 4.00×10^{-7} m light strike a calcium metal surface, electrons are ejected with a kinetic energy of 6.3×10^{-20} J.
    (a) Calculate the binding energy of the calcium electrons.
    (b) The minimum frequency at which the photoelectric effect occurs for a metal is that for which the emitted electron has no kinetic energy. Calculate the minimum frequency required for the photoelectric effect to occur with calcium metal.
147. Calculate the kinetic energy of an electron that is emitted from a strontium metal surface irradiated with photons of 4.20×10^{-7} m light. The binding energy of strontium is 4.39×10^{-19} J. See Question 146 for a statement about binding energy.
148. Ionization energy is the minimum energy required to remove an electron from a ground-state atom. According to Niels Bohr, the total energy of an electron in a stable orbit of quantum number n is equal to

$$E_n = -\frac{Z^2}{n^2}(2.18 \times 10^{-18}\text{ J})$$

where Z is the atomic number. Calculate the ionization energy for the electron in a ground-state hydrogen atom.
149. Ionization energy is the minimum energy required to remove an electron from a ground-state atom. According to the relationship developed by Niels Bohr, the total energy, E_n, of an electron in a stable orbit of quantum number n is equal to $E_n = -[Z^2/n^2]\ (2.18 \times 10^{-18}\text{ J})$ where Z is the atomic number. Calculate the ionization energy for the electron in a ground state He^+ ion.
150. Oxygen atoms are smaller than nitrogen atoms, yet oxygen has a lower first ionization energy than nitrogen. Explain.
151. Beryllium atoms are larger than boron atoms, yet boron has a lower first ionization energy than beryllium. Explain.
152. The element Meitnerium (Mt) honors Lise Meitner for her role in the discovery of nuclear fission.
    (a) Meitnerium atoms have the same outer electron configuration as what transition metal?
    (b) Using the noble gas notation, write the electron configuration for a ground state Mt atom.
153. Suppose two electrons in the same system each have $n = 3$, $\ell = 0$.
    (a) How many different electron arrangements would be possible if the Pauli exclusion principle did not apply in this case?
    (b) How many would apply if it is operative?
154. Although yet unnamed, element 112 is known and its existence verified.
    (a) The chemical properties of the element can be expected to be those of what kind of element—main group, transition metal, lanthanide, or actinide?
    (b) Using the noble gas notation, write the electron configuration for a ground-state atom of element 112 to corroborate your answer to part (a).
155. You are given the atomic radii of 110 pm, 118 pm, 120 pm, 122 pm, and 135 pm, but do not know to which element (As, Ga, Ge, P, Si) these values correspond. Which must be the value for Ge?
156. (a) Use a Born-Haber cycle to calculate ΔH_f° CaCl(s). Use these data (kJ/mol): $\Delta H_{sublimation}$ Ca(s) = 201; lattice energy CaCl(s) = −744 and obtain other necessary data from tables and figures in this book.
    (b) Use the answer from Part (a) and other ΔH_f° to calculate ΔH° for the reaction

$$2\ CaCl(s) \longrightarrow CaCl_2(s) + Ca(s)$$

    (c) Based on the answers to (a) and (b), is $CaCl_2$(s) more stable than CaCl(s)? Explain your answer.
157. Solid silver(I) chloride and sodium chloride have the same type of crystalline lattice, but differ in their ΔH_f°.
    (a) Use a Born-Haber cycle to calculate ΔH_f° AgCl(s). Use these data (kJ/mol): $\Delta H_{sublimation}$ Ag(s) = 284; lattice energy AgCl(s) = −910 and obtain other necessary data from tables and figures in this book.
    (b) Use a Born-Haber cycle to calculate ΔH_f° NaCl(s). Use these data (kJ/mol): $\Delta H_{sublimation}$ Na(s) = 108; lattice energy NaCl(s) = −787 and obtain other necessary data from tables and figures in this book.
    (c) Comment on the main factors for the differences in ΔH_f° of these two compounds.

Conceptual Challenge Problems

These rigorous, thought-provoking problems integrate conceptual learning with problem solving and are suitable for group work.

CP7.A (Section 7.2) Planck stated in 1900 that the energy of a single photon of electromagnetic radiation was directly proportional to the frequency of the radiation ($E = h\nu$). The constant h is known as Planck's constant and has a value of 6.626×10^{-34} Js. Soon after Planck's statement, Einstein proposed his famous equation ($E = mc^2$), which states that the total energy in any system is equal to its mass times the speed of light squared.

According to the de Broglie relation, what is the apparent mass of a photon emitted by an electron undergoing a change from the second to the first energy level in a hydrogen atom? How does the photon mass compare with the mass of the electron (9.109×10^{-31} kg)?

CP7.B (Section 7.7) When D. I. Mendeleev proposed a periodic law around 1870, he asserted that the properties of the elements are a periodic function of their atomic weights. Later, after H. G. J. Moseley measured the charge on the nuclei of atoms, the periodic law could be revised to state that the properties of the elements are a periodic function of their atomic numbers. What would be another way to define the periodic function that relates the properties of the elements?

CP7.C (Sections 7.3 and 7.11) Figure 7.8 shows a diagram of the energy states that an electron can occupy in a hydrogen atom. Use this diagram to show that the first ionization energy for hydrogen, given in Figure 7.25, is correct.

8

Covalent Bonding

Anton Gvozdikov/iStockphoto

Human body cells, such as those of these hikers, are made up of many thousands of compounds, most of which consist of atoms covalently bonded by sharing electron pairs. The chapter explains aspects of these kinds of bonds—their lengths, strengths, and polarities—and describes some molecular compounds and their covalent bonding.

Atoms of elements are rarely found uncombined in nature. Only the noble gases consist of individual uncombined atoms. Most nonmetallic elements consist of molecules, and in a solid metallic element each atom is closely surrounded by eight or twelve neighbors. What makes atoms stick to one another? Interactions among valence electrons of bonded atoms form the glue, but how? In ionic compounds, the transfer of one or more valence electrons from an atom of a metal to the valence shell of an atom of a nonmetal produces cations and anions of opposite charge. The electrostatic attractions among these oppositely charged ions hold them in a regular arrangement in a crystal lattice, such as in the ionic compounds NaCl (Na^+ and Cl^-) and CaO (Ca^{2+} and O^{2-}). Ionic compounds are

OWL Sign in to OWL at **www.cengage.com/owl** to view tutorials and simulations, develop problem-solving skills, and complete online homework assigned by your professor.

go Chemistry Download mini lecture videos for key concept review and exam prep from OWL or purchase them from **www.CengageBrain.com.**

Companion Website

Visit this book's companion website at **www.cengage.com/chemistry/moore** to work interactive modules for the Estimation boxes and Active Figures in this text.

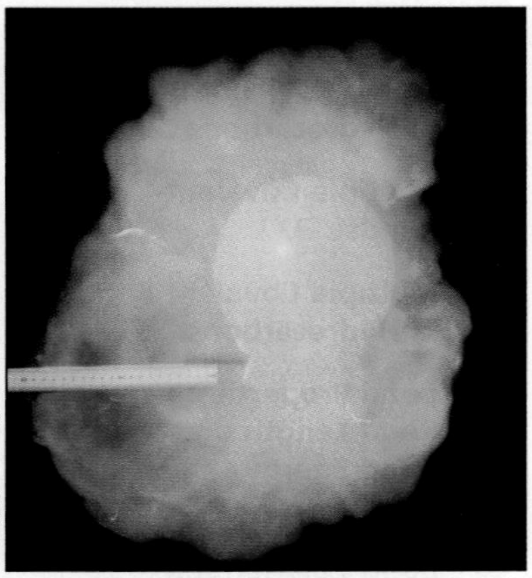

© Cengage Learning/Charles D. Winters

Hydrogen. When ignited, hydrogen in a balloon burns in air by reacting with oxygen to form water. Hydrogen and oxygen are molecular elements and water is a molecular compound.

solids at room temperature and conduct an electric current when melted due to movement of the mobile cations and anions ***(← p. 97).*** Many other substances, however, are composed of molecules, not ions. Examples are CO_2, H_2O, and $C_6H_{12}O_6$ (glucose). Molecular compounds have properties very different from those of ionic compounds. The melting points of molecular compounds are lower, and, when melted, these compounds do not conduct an electric current because they do not contain ions.

In these molecular compounds, atoms are held together by bonds consisting of one or more pairs of electrons *shared* between the bonded atoms. The attraction of positively charged nuclei for electrons between them pulls the nuclei together. This simple idea can account for the bonding in nearly all molecular compounds, allowing us to correlate their structures with their physical and chemical properties. This chapter describes the bonding found in molecules ranging from simple diatomic gases to hydrocarbons and more complex molecules.

8.1 Covalent Bonding

An explanation of bonding in the simplest stable molecule H_2 was proposed in 1916 by G. N. Lewis, who suggested that valence electrons rearrange to give noble gas electron configurations when atoms join together chemically. Lewis regarded the lack of reactivity of the noble gases as due to their stable electron configuration, an s^2 configuration for He or an s^2p^6 configuration for Ne, Ar, Kr, and Xe. To achieve these stable configurations, atoms share electrons. Lewis proposed that an attractive force called a **covalent bond** results when one or more pairs of electrons are shared between the bonded atoms. ***Covalent bonds, which are shared electron pairs, connect the atoms in molecular (covalent) compounds.*** For example, the two H atoms in H_2 and the H and Cl atoms in HCl are held together by covalent bonds.

But why does sharing electrons provide an attractive force between bonded atoms? Consider the formation of H_2. If the two hydrogen atoms are widely separated, there is little if any interaction between them. When the two atoms get close enough, however, their 1*s* electron clouds overlap. This overlap allows the electron from each atom to be attracted by the other atom's nucleus, causing a net attraction between the two atoms.

Experimental data and calculations indicate that an H_2 molecule has its lowest potential energy and is therefore most stable when the nuclei are 74 pm apart (Figure 8.1). At that distance—*the bond length*—the attractive and repulsive electrostatic forces are balanced. If the nuclei get closer than 74 pm, repulsion of each nucleus by the other raises the potential energy. When the H nuclei are 74 pm apart, it takes 436 kJ of energy to break the hydrogen-to-hydrogen covalent bonds when a mole of gaseous H_2 molecules is converted into isolated H atoms. This energy is the *bond energy* of H_2 ***(← p. 238).***

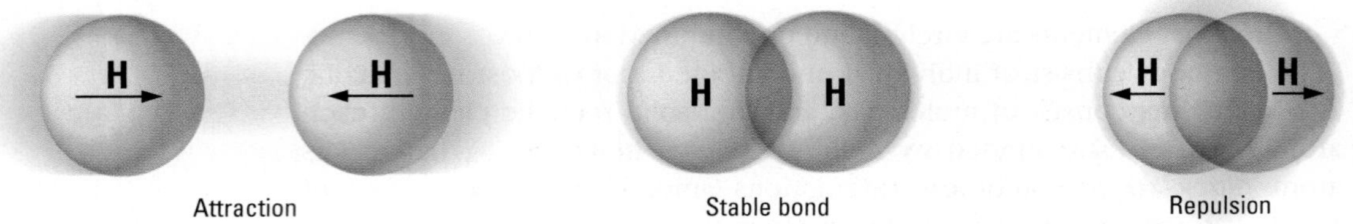

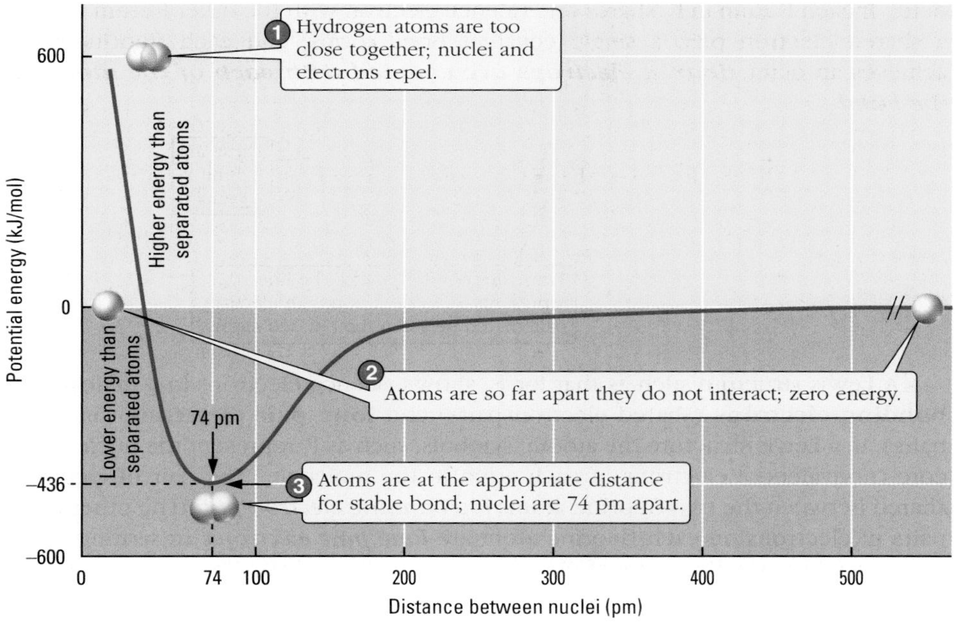

Figure 8.1 H—H bond formation from isolated H atoms. Energy is at a minimum at an internuclear distance of 74 pm, where there is a balance between electrostatic attractions and repulsions.

Lowering potential energy is favorable to bond formation.

8.2 Single Covalent Bonds and Lewis Structures

go Chemistry **Module 12: Drawing Lewis Electron Dot Structures** covers concepts in this section.

The shared electron pairs of covalent bonds occupy the same shell as the valence electrons of each atom. For main group elements, the electrons commonly contribute to a noble gas configuration on each atom. Lewis further proposed that by counting the valence electrons of an atom, it would be possible to predict how many bonds that atom can form. ***The number of covalent bonds an atom can form is determined by the number of electrons that the atom must share to achieve a noble gas configuration.***

A **single covalent bond** *is formed when two atoms share one pair of electrons.* The simplest examples are the bonds in diatomic molecules such as H_2, F_2, and Cl_2, which, like other simple molecules, can be represented by Lewis structures. A **Lewis structure** for a molecule shows all valence electrons as dots or lines that represent covalent bonds.

Lewis structures are drawn by starting with the Lewis dot symbols for the atoms ***(Table 7.6, ← p. 301)*** and arranging the valence electrons until each atom in the molecule has a noble gas configuration. For example, the Lewis structure for H_2 shows two bonding electrons shared between two hydrogen atoms.

$$H{:}H \qquad \text{or} \qquad H{-}H$$

The shared electron pair of a *single* covalent bond is often represented by a line instead of a pair of dots. Note that *each* hydrogen atom shares the pair of electrons, thereby achieving the same two-electron configuration as helium, the simplest noble gas.

Atoms with more than two valence electrons achieve a noble gas structure by sharing enough electrons to attain an octet of valence electrons (an s^2p^6 configuration.) This is known as the **octet rule:** ***To form bonds, main-group elements gain, lose, or share electrons to achieve a stable electron configuration characterized by eight valence electrons.*** To obtain the Lewis structure for F_2, for example, we start with the Lewis dot symbol for a fluorine atom. Fluorine, in Group 7A, has seven valence electrons, so there are seven dots, one less than an

Oesper Collection in the History of Chemistry, University of Cincinnati

1875–1946

In 1916, G. N. Lewis introduced the theory of the shared electron pair chemical bond in a paper published in the *Journal of the American Chemical Society.* His theory revolutionized chemistry, and it is in honor of this contribution that we refer to "electron dot" structures as Lewis structures. Of particular interest in this text is the extension of his theory of bonding to a generalized theory of acids and bases (Section 16.9).

Achieving a noble gas configuration is also referred to as "obeying the octet rule" because all noble gases except helium have eight valence electrons.

octet. If each F atom in F_2 shares one valence electron with the other F atom to form a shared electron pair, a single covalent bond forms, and each fluorine atom achieves an octet. ***Shared electrons are counted with each of the atoms in the bond.***

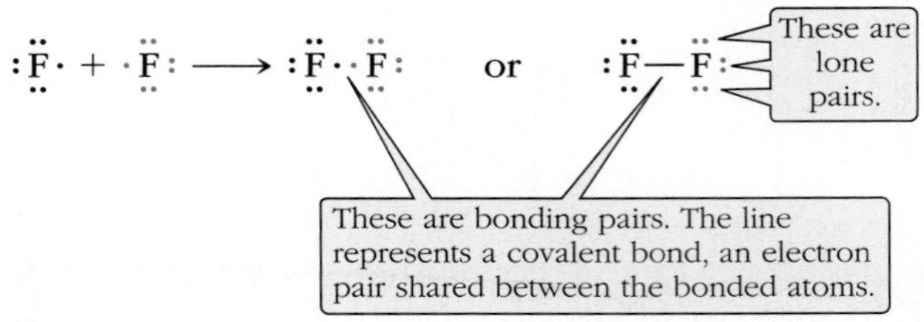

:F· + ·F: ⟶ :F··F: or :F—F:
These are lone pairs.
These are bonding pairs. The line represents a covalent bond, an electron pair shared between the bonded atoms.

The term "lone pairs" will be used in this text to refer to unshared electron pairs.

Magnetic measurements ***(Section 7.7,*** ***← p. 305)*** support the concept that each electron in a pair of electrons (bonding or lone pair) has its spin opposite that of the other electron.

A Lewis structure such as that for F_2 shows valence electrons in a molecule as **bonding electrons** (shared electron pairs) and **lone pair electrons** (unshared pairs). In a Lewis structure the atomic symbols, such as F, represent the nucleus and core (nonvalence) electrons of each atom in the molecule. The pair of electrons shared between the two fluorine atoms is a *bonding electron pair.* The other three pairs of electrons on each fluorine atom are *lone pair electrons.* In writing Lewis structures, the bonding pairs of electrons are usually indicated by lines connecting the atoms they hold together; lone pairs are usually represented by pairs of dots.

What about Lewis structures for molecules such as H_2O or NH_3? Oxygen (Group 6A) has six valence electrons and must share two electrons to satisfy the octet rule. This can be accomplished by forming covalent bonds with two hydrogen atoms.

H· + H· + ·Ö· forms H:Ö:H or H—Ö—H

Nitrogen (Group 5A) in NH_3 must share three electrons to achieve a noble gas configuration, which can be done by forming covalent bonds with three hydrogen atoms.

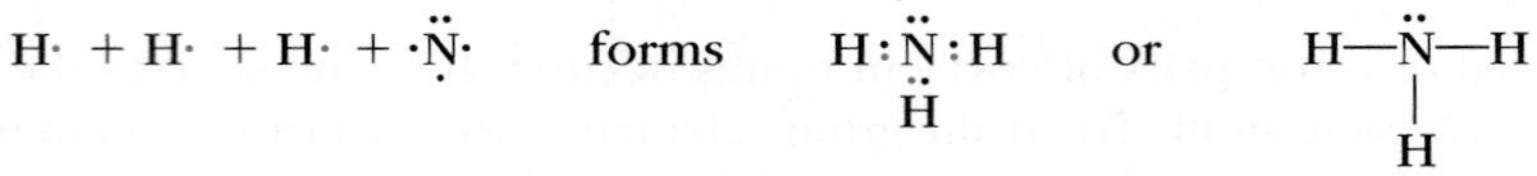

H· + H· + H· + ·N· forms H:N:H or H—N—H
H
H

Main-group elements are those in the groups labeled "A" in the periodic table inside the front cover of this book.

From the Lewis structures of F_2, H_2O, and NH_3, we can make an important generalization: ***The number of electrons that an atom of a main-group element must share to achieve an octet equals eight minus its A-group number.*** Carbon, for example, which is in Group 4A, needs to share four electrons to reach an octet.

Group Number	Number of Valence Electrons	Number of Electrons Shared to Complete an Octet	Example
4A	4	(8 − A-group number) = 8 − 4 = 4	C in CH_4 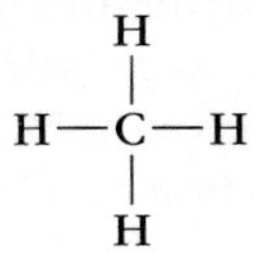
5A	5	3	N in NF_3 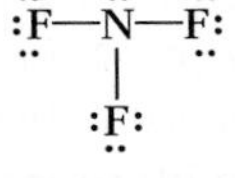
6A	6	2	O in H_2O 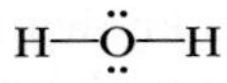
7A	7	1	F in HF 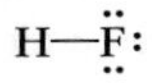

Many essential biochemical molecules contain carbon, hydrogen, oxygen, and nitrogen atoms. The structures of these molecules are dictated by the number of bonds C, O, and N need to complete their octets, as pointed out in the table. For example, glycerol, $C_3H_8O_3$,

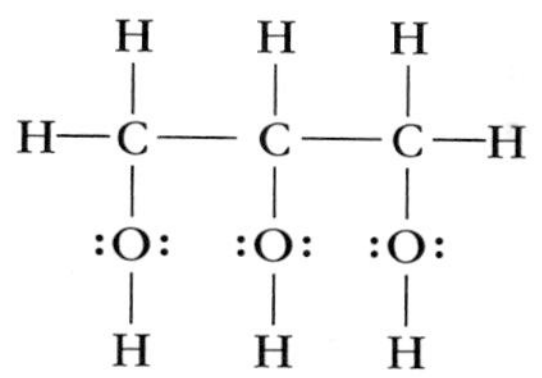

glycerol

is vital to the formation and metabolism of fats. In glycerol, each carbon atom has four shared pairs to complete an octet; each oxygen atom completes an octet by sharing two pairs and having two lone pairs; and each hydrogen atom achieves a $1s^2$ configuration by sharing a single pair of electrons. The same type of electron sharing occurs in deoxyribose, $C_5H_{10}O_4$, an essential component of DNA (deoxyribonucleic acid). Four of the carbon atoms are joined in a ring with an oxygen atom.

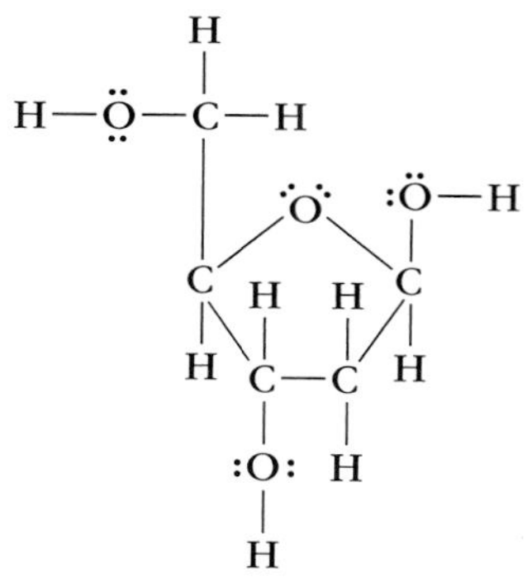

deoxyribose

Guidelines for Writing Lewis Structures

Guidelines have been developed for writing Lewis structures correctly, and we will illustrate using these guidelines to write the Lewis structure for PCl_3.

1. ***Count the total number of valence electrons in the molecule or ion.*** Use the A-group number in the periodic table as a guide to indicate the number of valence electrons in each atom. The total of the A-group numbers equals the total number of valence electrons of the atoms in a neutral molecule. For a negative ion, add electrons equal to the ion's charge. For a positive ion, subtract the number of electrons equal to the charge. For example, add one electron for the negative charge of OH^- (total of eight valence electrons); subtract one electron for the positive charge of NH_4^+, total of eight valence electrons: $5 + 4(1) - 1$.

   Because PCl_3 is a neutral molecule, its number of valence electrons is five for P (it is in Group 5A) and seven for *each* Cl (chlorine is in Group 7A): Total number of valence electrons $= 5 + (3 \times 7) = 26$.
2. ***Use atomic symbols to draw a skeleton structure by joining the atoms with shared pairs of electrons (a single line for each shared pair).*** A skeleton structure indicates the attachment of terminal atoms to a central atom. The central atom is usually the one written first in the molecular formula and is the one that can form the most bonds, such as Si in $SiCl_4$ and P in PO_4^{3-}. Hydrogen, oxygen, and the halogens are often terminal atoms.

   In PCl_3, the central atom is phosphorus, so we draw a skeleton structure with P as the central atom and the three terminal chlorine atoms arranged around it. The three bonding pairs account for six of the total of 26 valence electrons.

Although H is given first in the formulas of H_2O and H_2O_2, for example, it is not the central atom. H is never the central atom in a molecule or ion because a hydrogen atom forms only one covalent bond.

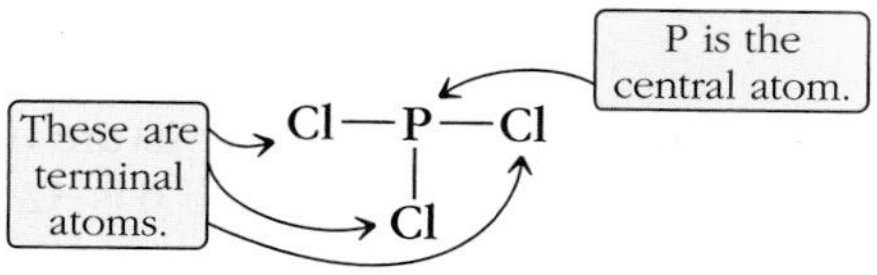
P is the central atom.
These are terminal atoms.
Cl—P—Cl
Cl

3. ***Place lone pairs of electrons around each atom (except H) to satisfy the octet rule, starting with the terminal atoms.*** Using lone pairs in this way on the Cl atoms accounts for 18 of the remaining 20 valence electrons, leaving two electrons for a lone pair on the P atom. When you check for octets, remember that shared electrons are counted as "belonging" to *each* of the

atoms bonded by the shared pair. Thus, each P—Cl bond has two shared electrons that count for phosphorus and also count for chlorine.

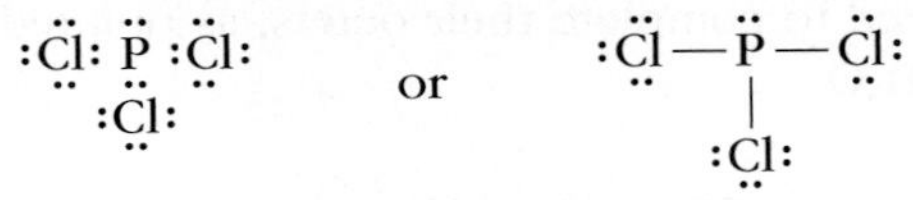

Counting dots and lines in the Lewis structure above shows 26 electrons, accounting for all valence electrons. This is the correct Lewis structure for PCl_3.

The next two steps apply when Steps 1 through 3 result in a structure that does not use all the valence electrons or fails to give an octet of electrons to each atom that should have an octet.

4. ***Place any remaining electrons on the central atom, even if it will give the central atom more than an octet.*** If the central atom is from the third or a higher period, it is large enough to accommodate more than an octet of electrons (Section 8.10). Sulfur tetrafluoride, SF_4, is such a molecule. It has a total of 34 valence electrons—six for S (Group 6A) plus seven for each F (Group 7A). In the Lewis structure, each F atom has an octet of electrons giving a total of 32. The remaining two valence electrons are on the central S atom, which, as a third-period element, can accommodate more than an octet of electrons.

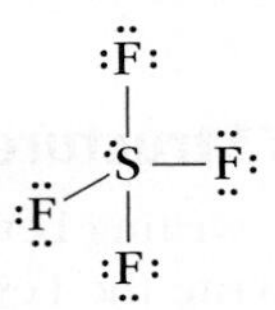

Double bond: two shared pairs of electrons, as in C=C.

Triple bond: three shared electron pairs, as in C≡N.

5. ***If the number of electrons around the central atom is less than eight, change single bonds to the central atom to multiple bonds.*** Some atoms can share more than one pair of electrons, resulting in a double covalent bond (two shared pairs) or a triple covalent bond (three shared pairs), known as *multiple bonds.* Where multiple bonds are needed to complete an octet, use one or more lone pairs of electrons from the terminal atoms to form *double* (two shared pairs) or *triple* (three shared pairs) covalent bonds until the central atom and all terminal atoms have octets. This guideline does not apply to PCl_3, but is illustrated here with $COCl_2$.

There are 24 valence electrons in $COCl_2$—four for carbon (Group 4A), six for oxygen (Group 6A), plus seven for each chlorine (Group 7A). However, these 24 valence electrons are not enough to give each atom an octet of electrons using just single bonds and lone pairs (Guideline 3) as seen from the skeleton structure for $COCl_2$.

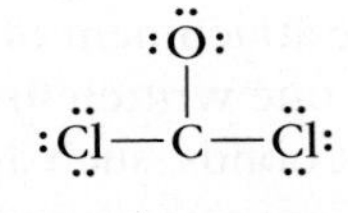

In this structure, carbon does not have an octet. We use one of the lone pairs on oxygen as a shared pair with carbon, which changes the C—O bond to a C=O double bond giving carbon an octet (Guideline 5).

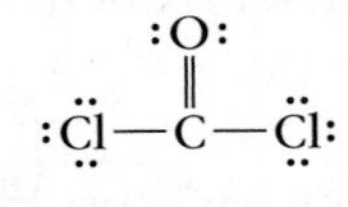

We apply these guidelines to write the Lewis structure of phosphate ion, PO_4^{3-}, one of the polyatomic ions listed in Table 3.7, which has 32 valence electrons (five from P, six from each O, and three for the 3− charge). Phosphorus is the central atom and oxygens are the terminal atoms, giving a skeleton structure of

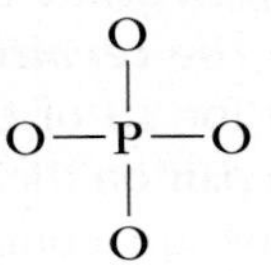

The four single P—O bonds account for eight of the valence electrons and provide phosphorus with an octet of electrons. The remaining 24 valence electrons are distributed as three lone pairs around each oxygen to complete their octets.

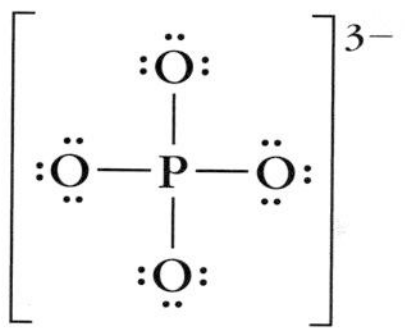

This is the correct Lewis structure for phosphate ion.

The brackets indicate that the polyatomic phosphate ion, consisting of one phosphorus atom and four oxygen atoms, collectively has a net charge of 3−. Such brackets are used with other polyatomic ions as well.

PROBLEM-SOLVING EXAMPLE 8.1 Lewis Structures

Write Lewis structures for these molecules or ions.

(a) HOCl (b) ClO_3^- (c) SF_6
(d) ClF_4^+ (e) HCN

Answer

(a) H—Ö—Cl: (b) 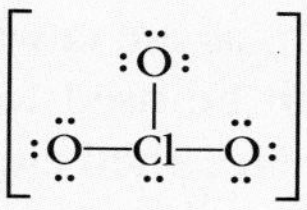

(c) 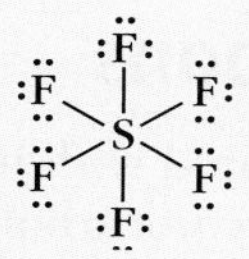 (d) 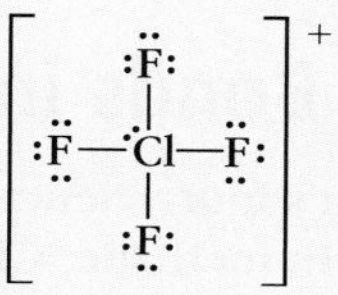 (e) H—C≡N:

Strategy and Explanation Apply the guidelines given previously. Count up the total valence electrons and distribute them in such a way that each atom (except hydrogen) has an octet of electrons; S in SF_6 is an exception (Guideline 4).

(a) HOCl: There are 14 valence electrons—one from hydrogen (Group 1A), six from oxygen (Group 6A), and seven from chlorine (Group 7A). The central atom is O and the skeleton structure is H—O—Cl. This arrangement uses four of the 14 valence electrons. The remaining ten valence electrons are used by placing three lone pairs on the chlorine and two lone pairs on the oxygen, thereby satisfying the octet rule for oxygen and chlorine; hydrogen needs just two electrons.

(b) ClO_3^-: There are 26 valence electrons—seven from chlorine and six from each oxygen, plus one for the 1− charge of the ion. Chlorine is the central atom with the three oxygen atoms each single bonded to it. The bonding uses six of the 26 valence electrons, leaving 20 for lone pairs. Placing three lone pairs on each oxygen and a lone pair on chlorine satisfies the octet rule for each atom and uses up the remaining valence electrons.

(c) SF_6: Sulfur forms more bonds than fluorine does, so sulfur is the central atom. There are 48 valence electrons: six from sulfur (Group 6A) and seven from each fluorine (Group 7A) for a total of 42 from fluorine. Because sulfur is a Period 3 nonmetal, it can expand its octet to accommodate six fluorine atoms around it, an exception to the octet rule. The 48 valence electrons are distributed as six bonding pairs in S—F bonds and the remaining 36 valence electrons as three lone pairs around each of the six fluorine atoms.

(d) ClF_4^+: Chlorine, a Period 3 element, can accommodate more than an octet of electrons and is the central atom. There are 34 valence electrons: seven from chlorine, seven for each fluorine, and one subtracted for the 1+ charge. The 34 valence electrons are distributed as four pairs in the four Cl—F bonds and the remaining 26 valence electrons as three lone pairs on each fluorine and one lone pair on the chlorine, making chlorine in this case an exception to the octet rule.

(e) HCN: There are 10 valence electrons—one from H (Group 1A), four from C (Group 4A), and five from N (Group 5A). Carbon is the central atom in the skeleton structure H—C—N, which uses four of the ten valence electrons. There are three lone

The **PROBLEM-SOLVING STRATEGY** in this book is

- Analyze the problem
- Plan a solution
- Execute the plan
- Check that the result is reasonable

Appendix A.1 explains this in detail.

pairs on the N. Use two of the lone pairs on N as shared pairs between nitrogen and carbon to complete a triple covalent bond, C≡N, giving carbon an octet (Guideline 5).

H—C—N: gives H—C≡N:

PROBLEM-SOLVING PRACTICE 8.1

Write Lewis structures for (a) NF_3, (b) N_2H_4, and (c) ClO_4^-.

PROBLEM-SOLVING PRACTICE answers are provided at the back of this book in Appendix K.

Although Lewis structures are useful for predicting the number of covalent bonds an atom will form, they do not give an accurate representation of where electrons are located in a molecule. Bonding electrons do not stay in fixed positions between nuclei, as Lewis's dots might imply. Instead, quantum mechanics tells us that there is a high probability of finding the bonding electrons between the nuclei. Also, Lewis structures do not convey the shapes of molecules. The angle between the two O—H bonds in a water molecule is not 180°, as the Lewis structure in the margin seems to imply. However, Lewis structures can be used to predict geometries by a method based on the repulsions between valence shell electron pairs (Section 9.1).

H—Ö—H

go Chemistry **Module 15: Naming Organic Compounds** covers concepts in this section.

8.3 Single Covalent Bonds in Hydrocarbons

In hydrocarbons ***(← p. 80)***, carbon's four valence electrons are shared with hydrogen atoms or other carbon atoms. In methane, CH_4, the simplest hydrocarbon, the four valence electrons are shared with electrons from four hydrogen atoms, forming four single covalent bonds. The bonding in methane can be represented by a Lewis structure and an *electron density model.* In an electron density model, a ball-and-stick model is surrounded by a space-filling model that represents the distribution of electron density on the surface of the molecule.

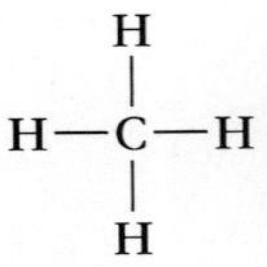

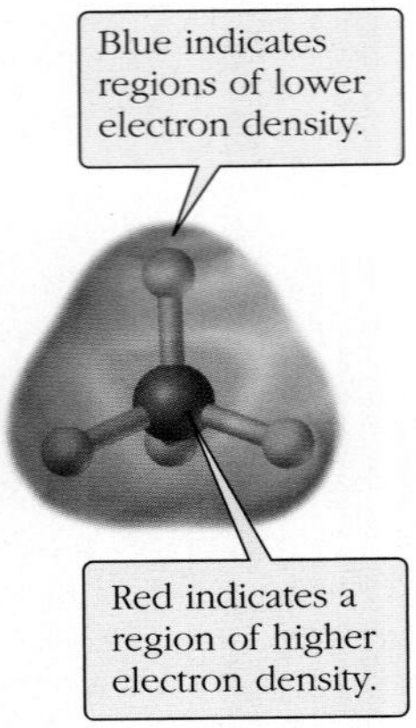
Blue indicates regions of lower electron density.
Red indicates a region of higher electron density.

An electron density model of methane.

Carbon is unique among the elements because of the ability of its atoms to form strong bonds with one another as well as with atoms of hydrogen, oxygen, nitrogen, sulfur, and the halogens. The strength of the carbon-carbon bond permits long chains to form:

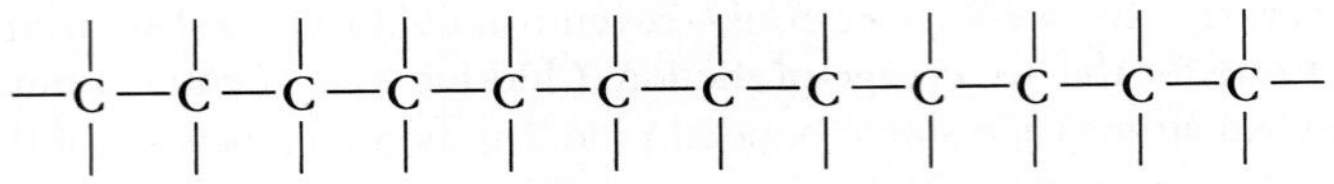

Because each carbon atom can form four covalent bonds, such chains contain numerous sites to which other atoms (including more carbon atoms) can bond, leading to isomeric structures ***(← p. 83)*** and the great variety of carbon compounds.

Hydrocarbons contain only carbon and hydrogen atoms. Alkanes, which contain only C—H and C—C single covalent bonds, are often referred to as **saturated hydrocarbons** because each carbon is bonded to a maximum number of hydrogen atoms ***(← p. 80)***. The carbon atoms in alkanes with four or more carbon atoms per molecule can be arranged in either a straight chain or a branched chain ***(← p. 83)***.

Review the discussion on alkanes in Section 3.4. See Table 3.4 for a list of selected alkanes.

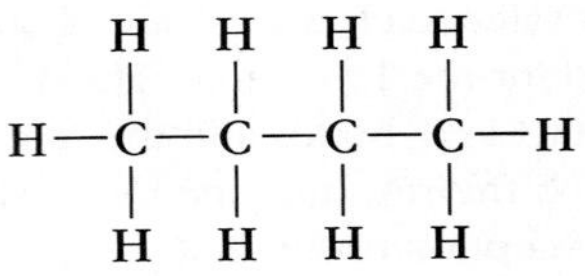

butane

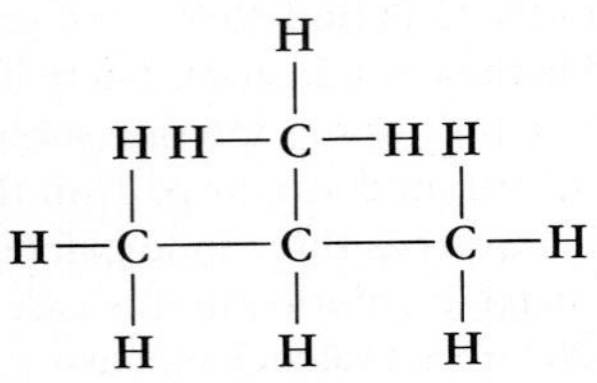

2-methylpropane

Rules for naming organic compounds are given in Appendix E.

In addition to straight-chain and branched-chain alkanes, there are *cycloalkanes,* saturated hydrocarbon compounds consisting of carbon atoms joined in rings of —CH_2— units. The simplest cycloalkane is cyclopropane; other common cycloalkanes include cyclobutane, cyclopentane, and cyclohexane.

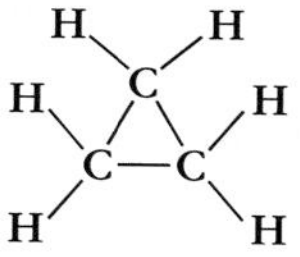

cyclopropane

or

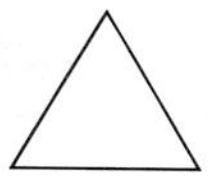

cyclopropane C_3H_6

cyclobutane C_4H_8

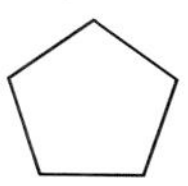

cyclopentane C_5H_{10}

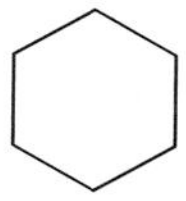

cyclohexane C_6H_{12}

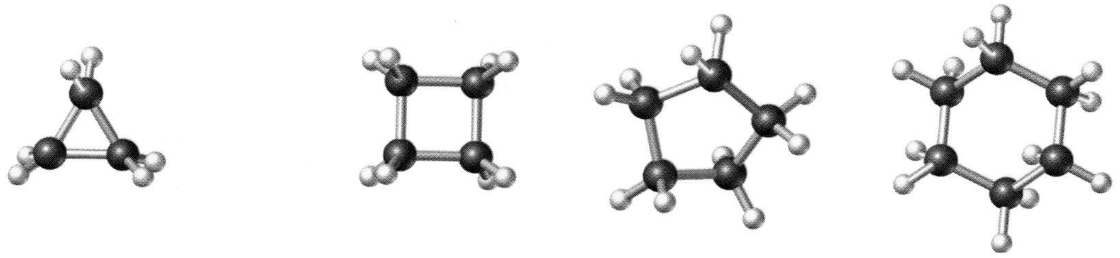

The cycloalkanes shown above are drawn using a common convention in which only the bonds between carbon atoms are shown. For example, each line in the drawing of cyclohexane represents a C—C bond. It is assumed that you understand that at the intersection of two lines there is a carbon atom and each carbon atom forms four bonds. Therefore, neither hydrogen atoms nor bonds from carbon to hydrogen atoms are shown. You have to fill in hydrogen atoms and C—H bonds for yourself. To show how this works, cyclopropane is represented both as a drawing with all bonds and atoms showing and as a drawing in which only C—C bonds are shown.

EXERCISE 8.1 Cyclic Hydrocarbons

Write the Lewis structure and the polygon that represents cyclooctane. Write the molecular formula for this compound.

Answers to **EXERCISES** are provided at the back of this book in Appendix L.

The great variety of organic compounds can also be accounted for by the fact that one or many carbon-hydrogen bonds in hydrocarbons can be replaced by bonds between carbon and other atoms. For example, the new bonds to carbon can connect to individual halogen atoms, thus creating entirely different compounds. Consider the new substances that result when a chlorine atom replaces a hydrogen atom in ethane, 2-methylbutane, and cyclopropane.

When we say that a chlorine atom replaces a hydrogen atom we are referring to the different bonding pattern in the molecule, not to a chemical reaction of the hydrocarbon molecule with a chlorine atom.

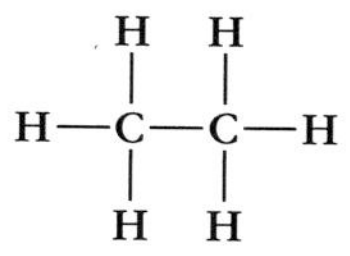

ethane

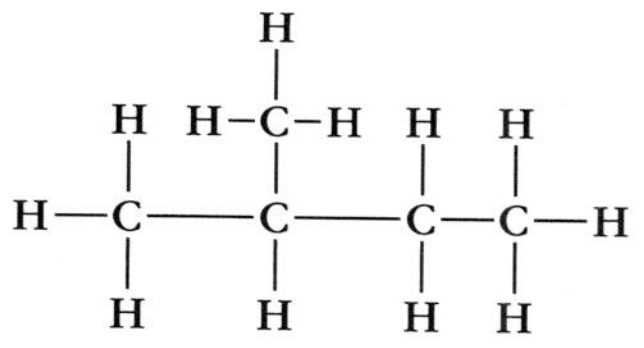

2-methylbutane

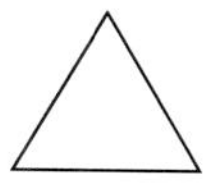

cyclopropane

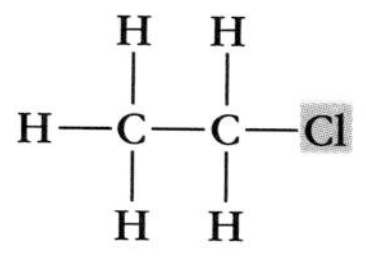

chloroethane

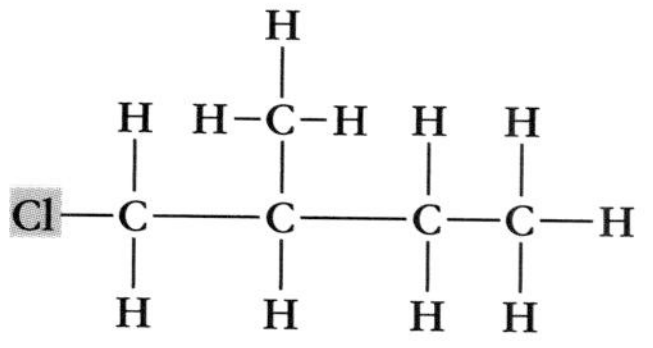

1-chloro-2-methylbutane

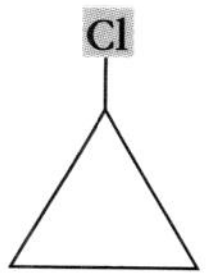

chlorocyclopropane

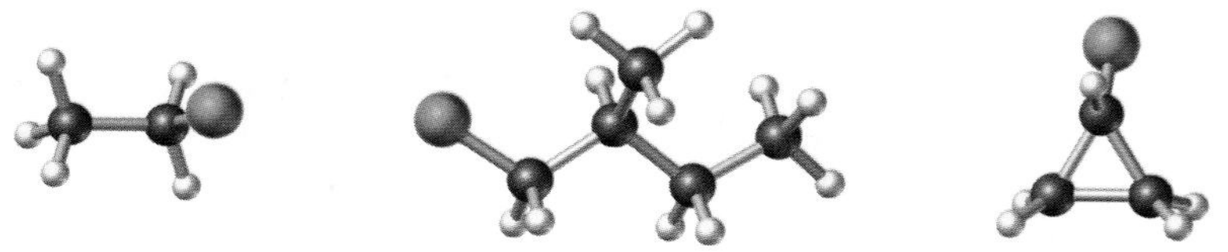

The —OH group in an alcohol is not an ion; it is different from the OH^- ion in a base.

Chapter 12 describes additional functional groups. Appendix E.2 includes a list of functional groups.

In another case, consider how an —OH functional group can replace one or more hydrogen atoms in an alkane. A **functional group** is a distinctive group of atoms in an organic molecule that imparts characteristic chemical properties to the molecule. The —OH functional group is characteristic of alcohols. A molecule of ethanol can be thought of as a molecule of ethane in which a hydrogen atom has been replaced by an —OH group. Replacing two hydrogen atoms on adjacent carbon atoms in ethane with —OH groups results in a molecule of ethylene glycol. In the structure of glycerol, three hydrogen atoms in propane are replaced by —OH groups.

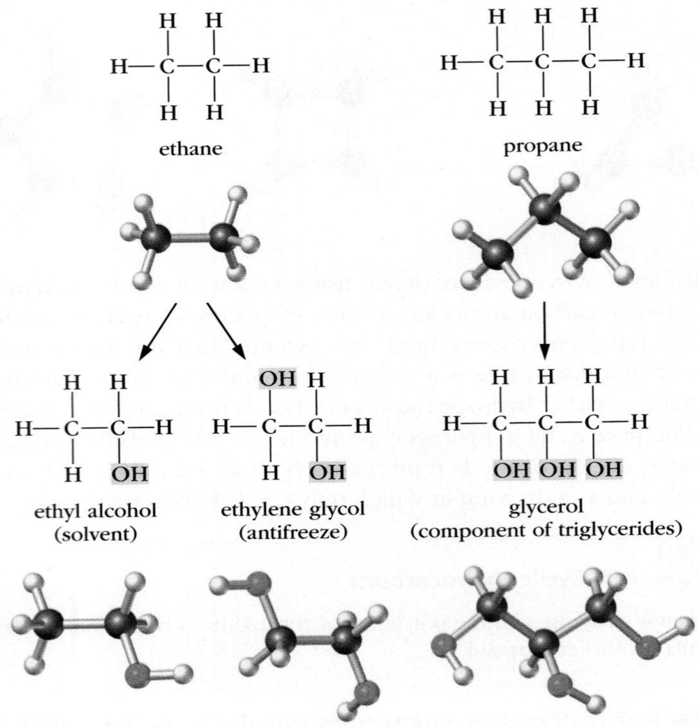
ethane
propane
ethyl alcohol
(solvent)
ethylene glycol
(antifreeze)
glycerol
(component of triglycerides)

PROBLEM-SOLVING EXAMPLE **8.2** **Structural Formulas and Cl Substitution**

Three hydrogens in propane can be replaced with three Cl atoms to form five different structures. Draw the structural formulas for two of five possible compounds.

Answer

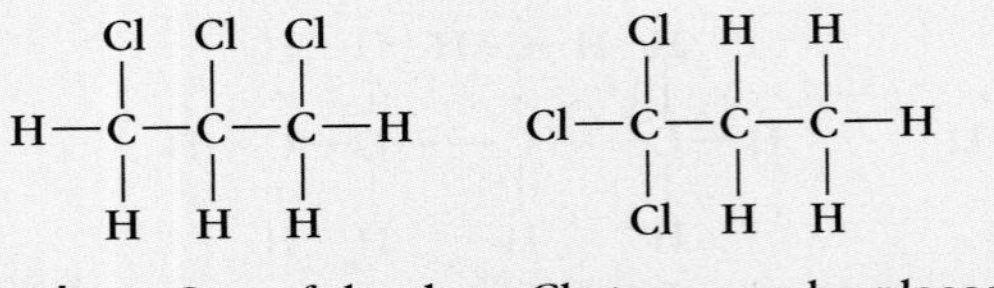

Strategy and Explanation One of the three Cl atoms can be placed on each of the carbons, or all three Cl atoms could be put on a terminal carbon atom, one at the end of the carbon chain. Two possibilities are

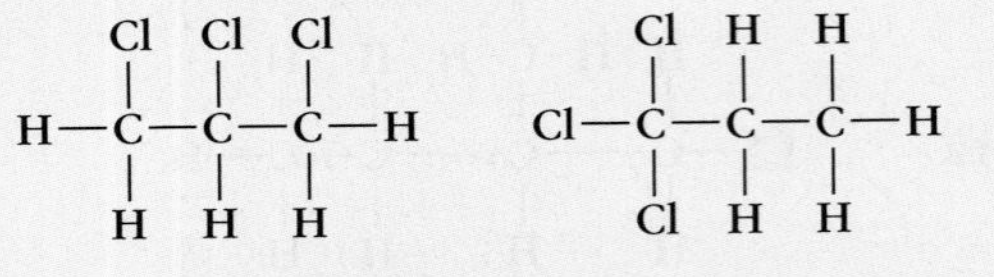

PROBLEM-SOLVING PRACTICE **8.2**

There are three other possible structures for three Cl atoms on a three-carbon chain. Draw these structures.

8.4 Multiple Covalent Bonds

A nonmetal atom with fewer than seven valence electrons can form covalent bonds in more than one way. The atom can share a single electron with another atom that also contributes a single electron; this process forms a shared electron pair—a *single* covalent bond. But some atoms can also share two or three pairs of electrons with another atom, in which case there will be two or three bonds, respectively, between the two atoms. When *two* shared pairs of electrons join the same pair of atoms, the bond is called a **double bond.** When *three* shared pairs are involved, the bond is called a **triple bond.** Double and triple bonds are referred to as **multiple covalent bonds.** Nitrous acid, HNO_2, contains an N=O double bond, and hydrogen cyanide, HCN, contains a C≡N triple bond.

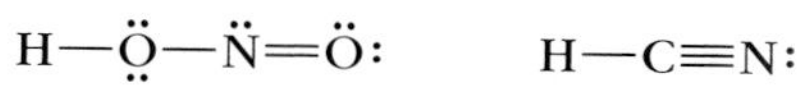

nitrous acid hydrogen cyanide

Guideline 5 (p. 332) says that, in molecules where there are not enough electrons to complete all octets using only single bonds, one or more lone pairs of electrons from the terminal atoms can be shared with the central atom to form double or triple bonds, so that all atoms have octets of electrons. Let's apply this guideline to the Lewis structure for formaldehyde, H_2CO. There are 12 valence electrons (Guideline 1): two from two H atoms (Group 1A), four from the C atom (Group 4A), and six from the O atom (Group 6A). To complete noble gas configurations, H should form one bond, C four bonds, and O two bonds. Because C forms the most bonds, it is the central atom, and we can write this skeleton structure (Guideline 2).

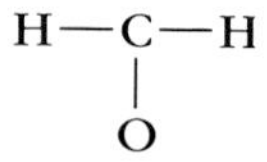

You might have written the skeleton structure O—C—H—H, but remember that H forms only one bond. Another possible skeleton is H—O—C—H, but with this skeleton it is impossible to achieve an octet around carbon without having more than two bonds to oxygen.

Putting bonding pairs and lone pairs in the skeleton structure according to Guideline 3 yields a structure using all 12 valence electrons in which oxygen has an octet, but carbon does not.

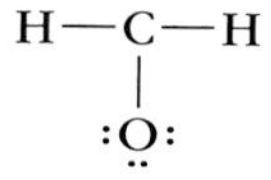

We use one of the lone pairs on oxygen as a shared pair with carbon to change the C—O single bond to a C=O double bond (Guideline 5).

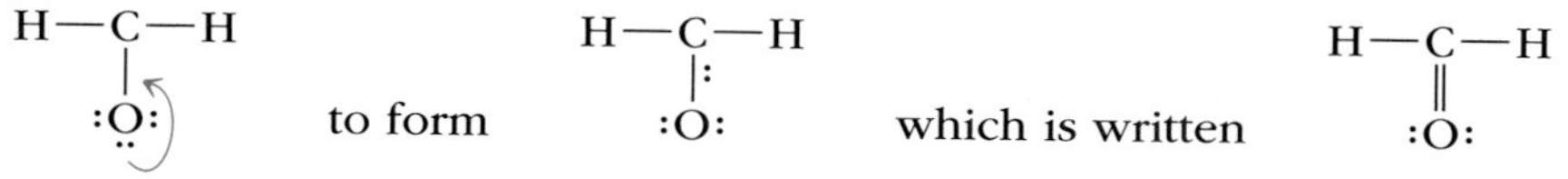

This gives carbon and oxygen a share in an octet of electrons, and each hydrogen has a share of two electrons, accounting for all 12 valence electrons and verifying that this is the correct Lewis structure for formaldehyde.

The >C=O combination, called the *carbonyl group,* is part of several functional groups that are very important in organic and biochemical molecules (Section 12.4). The carbonyl-containing —C(=O)—H group that appears in formaldehyde is known as the *aldehyde functional group.*

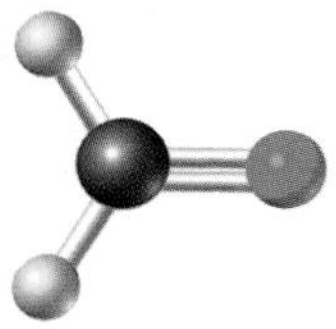

Formaldehyde

Formaldehyde, a gas at room temperature, is the simplest compound with an aldehyde functional group.

As another example of multiple bonds, let's write the Lewis structure for molecular nitrogen, N_2. There are ten valence electrons (five from each N). If two nonbonding pairs of electrons (one pair from each N) become bonding pairs to give a triple bond, the octet rule is satisfied. This is the correct Lewis structure of N_2.

N—N to form :N≡N:

EXERCISES that are labeled **CONCEPTUAL** are designed to test your understanding of one or more concepts; they usually involve qualitative rather than quantitative thinking.

CONCEPTUAL EXERCISE 8.2 Lewis Structures

Why is :N̈—N̈: an incorrect Lewis structure for N_2?

A molecule can have more than one multiple bond, as in carbon dioxide, where carbon is the central atom. There are 16 valence electrons in CO_2, and the skeleton structure uses four of them (two shared pairs):

O—C—O

Adding lone pairs to give each O an octet of electrons uses up the remaining 12 electrons, but leaves C needing four more valence electrons to complete an octet.

:Ö—C—Ö:

With no more valence electrons available, the only way that carbon can have four more valence electrons is to use one lone pair of electrons on each oxygen to form a covalent bond to carbon to give it an octet. In this way the 16 valence electrons are accounted for, and each atom has an electron octet.

:Ö—C—Ö: forms :O=C=O:

PROBLEM-SOLVING EXAMPLE 8.3 Lewis Structures

Write Lewis structures for (a) carbon monoxide, CO, an air pollutant; (b) nitrosyl chloride, ClNO, an unstable solid; (c) N_3^-, a polyatomic ion; and (d) acetylene, C_2H_2, a fuel.

Answer

(a) :C≡O: (b) :C̈l—N̈=Ö:

(c) [N̈=N=N̈]⁻ (d) H—C≡C—H

Strategy and Explanation Follow the steps given previously to write proper Lewis structures.

(a) The molecule CO contains ten valence electrons (four from C and six from O). We start with a single C—O bond as a skeleton structure.

C—O

Putting lone pairs around the carbon and oxygen atoms satsfies the octet rule for one of the atoms, but not both. Therefore, lone pairs must become bonding pairs to make up this deficiency and achieve an appropriate Lewis structure.

:C—Ö: to form :C≡O:

(b) In ClNO, nitrogen is the central atom (it can form more bonds than can oxygen or chlorine) and there are 18 valence electrons (five from N, six from O, and seven from Cl). The skeleton structure is

Cl—N—O

Placing lone pairs on the terminal atoms and a lone pair on the N atom uses the remaining valence electrons, but leaves the N atom without an octet.

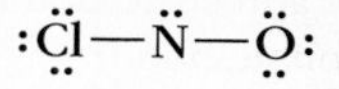

Converting a lone pair on the oxygen to a bonding pair results in the correct Lewis structure, with each atom having an octet of electrons.

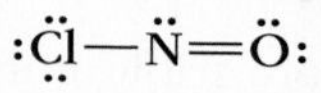

(c) There are 16 valence electrons in N_3^- (15 from three N atoms and one for the ion's 1− charge). The skeleton structure with lone pairs on the outer two nitrogen atoms uses all 16 valence electrons.

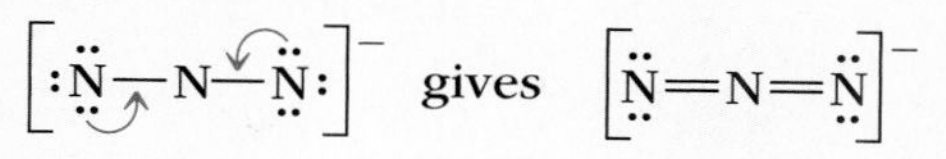

is a plausible Lewis structure.

(d) Acetylene has ten valence electrons (four from each C and one from each H.)

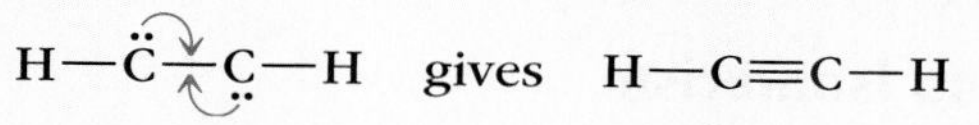

PROBLEM-SOLVING PRACTICE 8.3

Write Lewis structures for (a) nitrosyl ion, NO^+; (b) silicon sulfide dichloride, $SiSCl_2$.

CONCEPTUAL EXERCISE 8.3 Lewis Structures

Which of these are appropriate Lewis structures and which are not? Explain what is wrong with the incorrect ones.

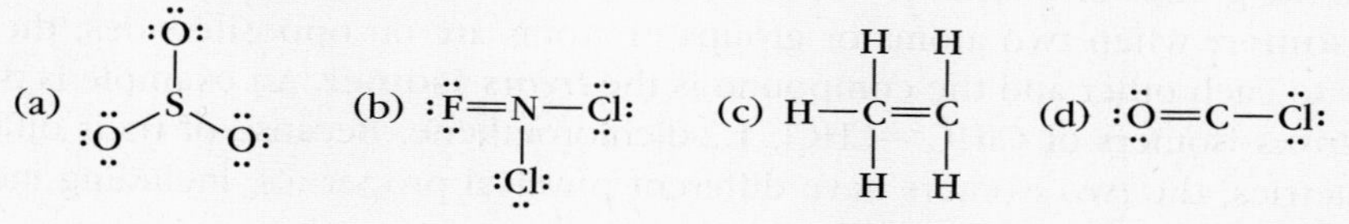
(a)
(b)
(c)
(d)

8.5 Multiple Covalent Bonds in Hydrocarbons

go Chemistry Module 15: Naming Organic Compounds covers concepts in this section.

Carbon atoms are connected by double bonds in some compounds and triple bonds in others as well as by single bonds. **Alkenes** *are hydrocarbons that have one or more carbon-carbon double bonds,* C=C. The general formula for alkenes with one double bond is C_nH_{2n}, where $n = 2, 3, 4$, and so on. The first two members of the alkene series are ethene, $CH_2{=}CH_2$, and propene, $CH_3CH{=}CH_2$; they are commonly called ethylene and propylene, particularly when referring to the polymers polyethylene and polypropylene, which will be discussed in Section 12.6.

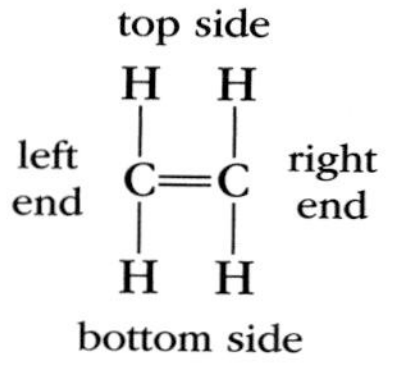
top side
left end
right end
bottom side
ethylene

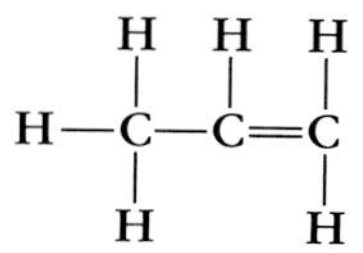
propylene

Alkenes are said to be **unsaturated hydrocarbons.** The carbon atoms connected by double bonds are the *unsaturated sites*—they contain fewer hydrogen atoms than the corresponding alkanes (ethene, $CH_2{=}CH_2$; ethane, $CH_3{-}CH_3$).

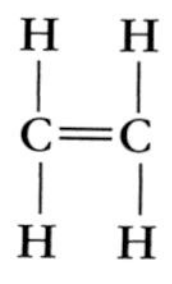
unsaturated

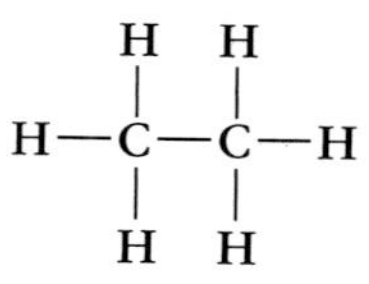
saturated

Alkenes are named by using the name of the corresponding alkane *(← p. 80)* to indicate the number of carbons and the suffix *-ene* to indicate one or more double bonds. The first member, ethene (ethylene), is the most important raw material in the organic chemical industry, where it is used in making polyethylene, antifreeze (ethylene glycol), ethanol, and other chemicals.

An oxyacetylene torch cutting steel. An oxyacetylene torch cuts through the steel door of a Titan II missile silo that is being retired as part of a disarmament treaty. A mixture of acetylene and oxygen burns with a flame hot enough (3000 °C) to cut steel.

CONCEPTUAL EXERCISE 8.4 Alkenes

(a) Write the molecular formula and structural formula of an alkene with five carbon atoms and one C=C double bond.

(b) How many different alkenes have five carbon atoms and one C=C double bond?

Hydrocarbons with one or more triple bonds, —C≡C—, per molecule are **alkynes.** The general formula for alkynes with one triple bond is C_nH_{2n-2}, where $n = 2, 3, 4$, and so on. The simplest one is ethyne, commonly called acetylene, C_2H_2.

H—C≡C—H

acetylene

Double Bonds and Isomerism

The C=C double bond creates an important difference between alkanes and alkenes—the degree of flexibility of the carbon-carbon bonds in the molecules. The C—C single bonds in alkanes allow the carbon atoms to rotate freely along the C—C bond axis (Figure 8.2). But in alkenes, the C=C double bond prevents such free rotation. This limitation is responsible for the *cis-trans* isomerism of alkenes.

Two or more compounds with the same molecular formula but different arrangements of atoms are known as isomers ***(← p. 83).*** *Cis-trans* isomerism occurs when molecules differ in their arrangement of atoms on either side of a C=C double bond because there is no rotation around the C=C double bond (Figure 8.2). When two atoms or groups of atoms are attached on the *same* side of the C=C bond, the groups or atoms are said to be *cis* to each other and the compound is the ***cis* isomer;** when two atoms or groups of atoms are on opposite sides, they are *trans* to each other and the compound is the ***trans* isomer.** An example is the *cis* and *trans* isomers of ClHC=CHCl, 1,2-dichloroethene. Because of their different geometries, the two isomers have different physical properties, including melting point, boiling point, and density as shown on the next page.

Cis-trans isomerism is also called geometric isomerism.

Cis-trans* isomerism** in alkenes is possible ***only when each of the carbon atoms connected by the double bond has two different groups attached. (For the sake of simplicity, the word "groups" refers to both atoms and groups of atoms.) For example, two chlorine atoms can also bond to the *first* carbon to give 1,1-dichloroethene, which does not have *cis* and *trans* isomers because each carbon atom is attached to two identical atoms (one carbon to two chlorines, the other carbon to two hydrogens).

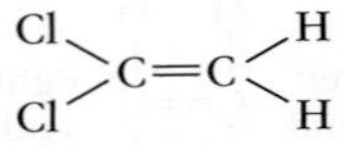
1,1-dichloroethene

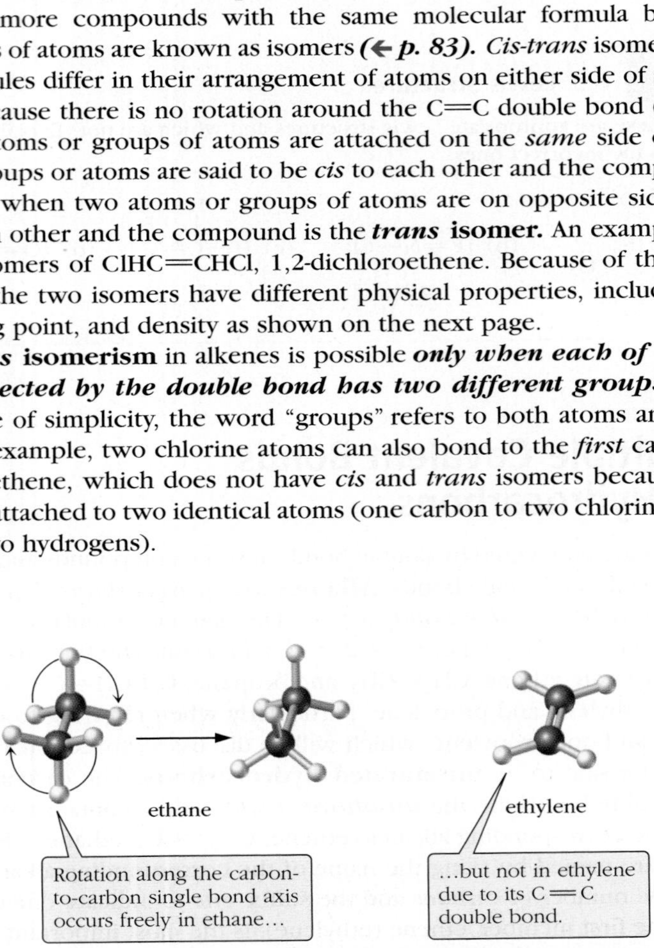

Figure 8.2 Nonrotation around C=C. At room temperature, rotation along the carbon-to-carbon axis occurs freely in ethane, but not in ethylene because of its double bond.

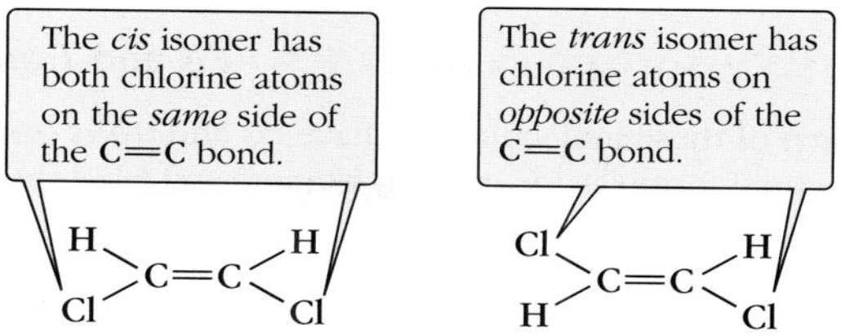

cis-1,2-dichloroethene   *trans*-1,2-dichloroethene

The 1 and 2 indicate that the two chlorine atoms are attached to the first and second carbon atoms, respectively.

Physical Property	*cis*-1,2-dichloroethene	*trans*-1,2-dichloroethene
Melting point	−80.0 °C	−49.8 °C
Boiling point (at 1 atm)	60.1 °C	48.7 °C
Density (at 20 °C)	1.284 g/mL	1.265 g/mL

If free rotation could occur around a carbon-carbon double bond, these two molecules would be the same. Therefore, there would only be a single compound.

CONCEPTUAL EXERCISE **8.5 *Cis-Trans* Isomerism and Biomolecules**

Maleic acid and fumaric acid are very important biomolecules that undergo different reactions in metabolism because they are *cis-trans* isomers.

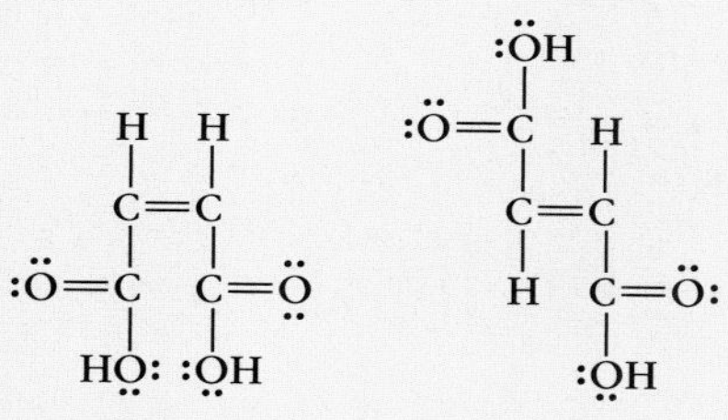

maleic acid   fumaric acid

Identify the *cis* isomer and the *trans* isomer.

The importance of *cis-trans* isomerism in biologically important molecules is discussed in Section 12.5.

When there are four or more carbon atoms in an alkene, the possibility exists for *cis* and *trans* isomers even when only carbon and hydrogen atoms are present. For example, 2-butene has both *cis* and *trans* isomers. (The 2 indicates that the double bond is at the second carbon atom, with the straight carbon chain beginning with carbon 1.)

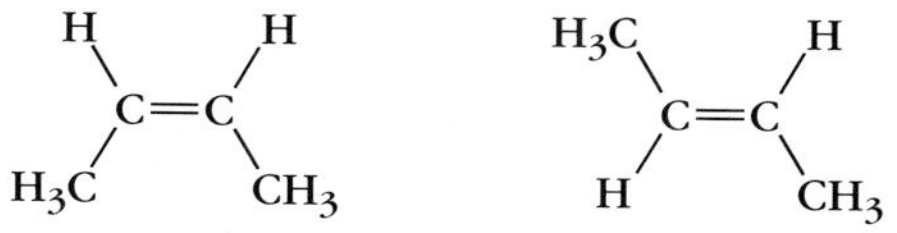

Physical Property	*cis*-2-butene	*trans*-2-butene
Melting point	−138.9 °C	−105.5 °C
Boiling point (at 1 atm)	3.7 °C	0.88 °C

Cis-trans isomerism plays a vital role in vision through the *cis-trans* conversion of retinal. A *Chemistry You Can Do* in Chapter 13 (Kinetics and Vision) explores this conversion.

PROBLEM-SOLVING EXAMPLE 8.4 *Cis* and *Trans* Isomers

Which of these molecules can have *cis* and *trans* isomers? For those that do, write the structural formulas for the two isomers and label them *cis* and *trans*.

(a) $(CH_3)_2C{=}CCl_2$
(b) $CH_3ClC{=}CClCH_3$
(c) $CH_3BrC{=}CClCH_3$
(d) $(CH_3)_2C{=}CBrCl$

Answer (b) and (c)

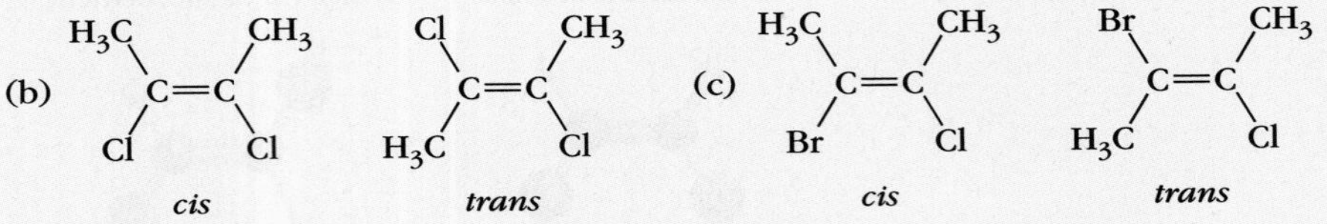

Strategy and Explanation Recall that molecules having two identical groups on a C=C carbon cannot be *cis* and *trans* isomers. Because both of the groups on each carbon are the same, (a) cannot have *cis* and *trans* isomers. In (b), the two —CH_3 groups and the two Cl atoms can both be on the same side of the C=C bond (the *cis* form) or on opposite sides (the *trans* form). The same holds true for the —CH_3 groups in (c). There are two CH_3 groups on the same carbon in (d), so *cis* and *trans* isomers are not possible.

PROBLEM-SOLVING PRACTICE 8.4

Which of these molecules can have *cis* and *trans* isomers? For those that do, write the structural formulas for the two isomers and label them *cis* and *trans*.

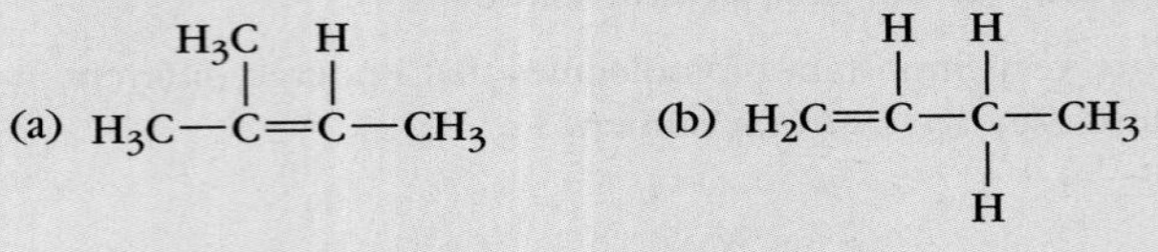

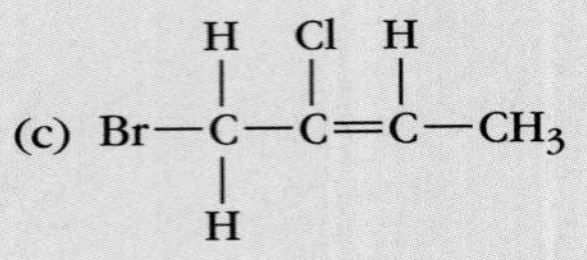

8.6 Bond Properties: Bond Length and Bond Energy

Bond Length

Bond length, the distance between nuclei of two bonded atoms, is determined primarily by the sizes of the atoms themselves (← p. 306). The bond length can be considered as the sum of the atomic radii of the two bonded atoms. Bond lengths are given in Table 8.1. As expected, the bond length is greater for larger atoms. Figure 8.3 illustrates the change in atomic size across Periods 2 and 3, and down Groups 4A–6A. Thus, single bonds with carbon increase in length along the series:

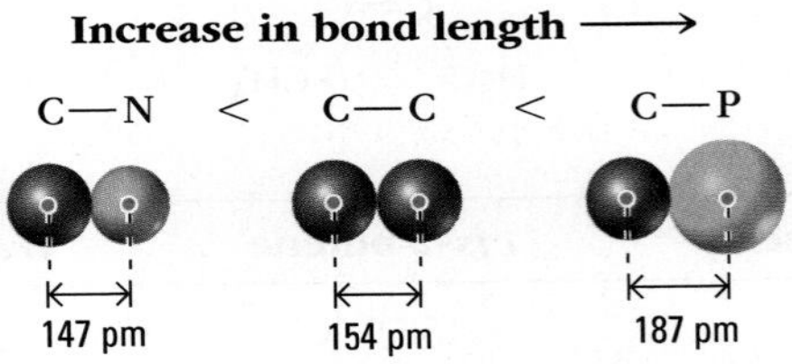
Increase in bond length
C—N < C—C < C—P
147 pm
154 pm
187 pm

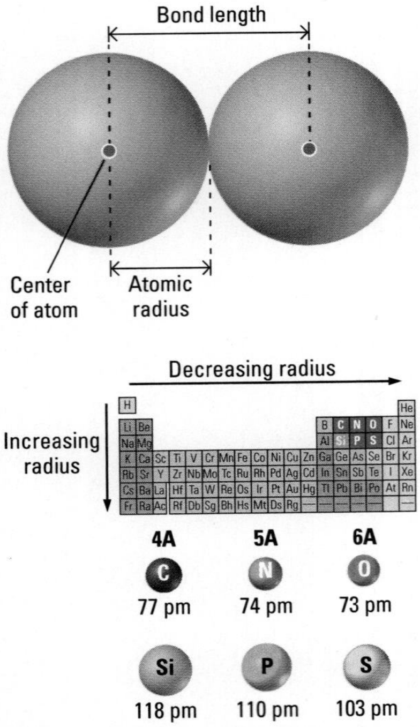
Bond length
Center of atom
Atomic radius
Decreasing radius
Increasing radius
4A
5A
6A
C
N
O
77 pm
74 pm
73 pm
Si
P
S
118 pm
110 pm
103 pm

Figure 8.3 Relative atom sizes (radii) for second- and third-period elements in Groups 4A, 5A, and 6A.

In the case of multiple bonds, a C=O bond is shorter than a C=S bond because O is a smaller atom than S. Likewise, a C≡N bond is shorter than a C≡C

Table 8.1 Some Average Single and Multiple Bond Lengths (in picometers, pm)*

Single Bonds

	I	Br	Cl	S	P	Si	F	O	N	C	H
H	161	142	127	132	138	145	92	94	98	110	74
C	210	191	176	181	187	194	141	143	147	154	
N	203	184	169	174	180	187	134	136	140		
O	199	180	165	170	176	183	130	132			
F	197	178	163	168	174	181	128				
Si	250	231	216	221	227	234					
P	243	224	209	214	220						
S	237	218	203	208							
Cl	232	213	200								
Br	247	228									
I	266										

Multiple Bonds

N=N	120	C=C	134
N≡N	110	C≡C	121
C=N	127	C=O	122
C≡N	115	C≡O	113
O=O (in O_2)	112	N≡O	108
N=O	115		

*1 pm = 10^{-12} m. Bond lengths are given in picometers (pm) in this table, but many scientists use nanometers (1 nm = 10^3 pm) or the older unit of Ångströms (Å). 1 Å equals 100 pm. A C—C single bond is 0.154 nm, 1.54 Å, or 154 pm in length.

bond because N is a smaller atom than C (Figure 8.3). Each of these trends can be predicted from the relative sizes shown in Figure 8.3 and is confirmed by the average bond lengths given in Table 8.1.

The effect of bond type—single, double, or triple—is evident when bonds between the same two atoms are compared. For example, structural data show that the bonds become shorter in the series C—O > C=O > C≡O. As the electron density between the atoms increases (number of bonds increases), the bond lengths decrease because the atoms are pulled together more strongly.

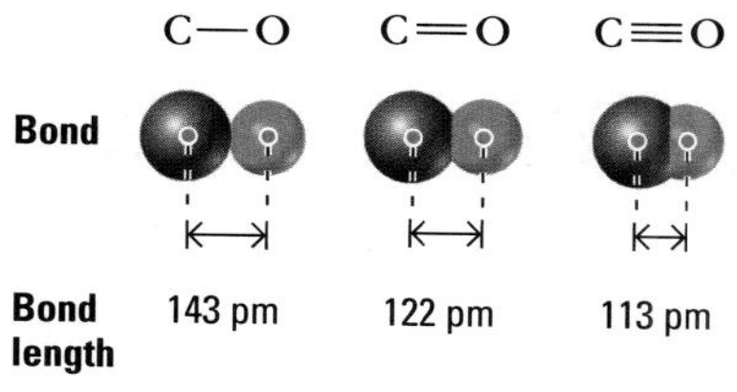

The bond lengths in Table 8.1 are average values, because variations in neighboring parts of a molecule can affect the length of a particular bond. For example, the C—H bond has a length of 105.9 pm in acetylene, HC≡CH, but a length of 109.3 pm in methane, CH_4. Although there can be a variation of as much as 10% from the average values listed in Table 8.1, the average bond lengths are useful for estimating bond lengths and building models of molecules.

PROBLEM-SOLVING EXAMPLE 8.5 Bond Lengths

In each pair of bonds, predict which will be shorter.

(a) P—O or S—O (b) C≡C or C=C (c) C—S or C—Cl

Answer The shorter bonds are (a) S—O, (b) C≡C, (c) C—Cl

Strategy and Explanation Apply the periodic trends in atomic radii to determine the relative sizes of the atoms in the question and consequently their relative bond distances.

(a) S—O is shorter than P—O because a P atom is larger than an S atom.

(b) C≡C is shorter than C=C because the more electrons that are shared by atoms, the more closely the atoms are pulled together.

(c) C—Cl is shorter than C—S because a Cl atom is smaller than an S atom.

PROBLEM-SOLVING PRACTICE **8.5**

Explain the increasing order of bond lengths in these pairs of bonds.

(a) C—S is shorter than C—Si. (b) C—Cl is shorter than C—Br.

(c) N≡O is shorter than N=O.

Bond Enthalpies

In any chemical reaction, bonds are broken and new bonds are formed.

- ***Bond breaking* is an *endothermic* process that requires transfer of energy into a system.**
- ***Bond making* is an *exothermic* process in which energy is transferred out of a system.**

The two terms "bond enthalpy" and "bond energy" are used interchangeably, although they are not quite equal.

These changes contribute to the enthalpy change for the overall reaction (← *p. 238*). **Bond enthalpy (bond energy)** *is the enthalpy change that occurs when the bond between two bonded atoms in the gas phase is broken and the atoms are separated completely at constant pressure.* Data on the strengths of bonds between atoms in gas phase molecules are summarized in Table 8.2. The quantity of energy released when 1 mol of a particular bond is made equals that needed when 1 mol of that bond is broken. For example, the H—Cl bond energy is 431 kJ/mol, indicating that 431 kJ must be supplied to break 1 mol H—Cl bonds ($\Delta H° = +431$ kJ). Conversely, when 1 mol H—Cl bonds is formed, 431 kJ is released ($\Delta H° = -431$ kJ).

You have seen that as the number of bonding electrons between a pair of atoms increases (single to double to triple bonds), the bond length decreases. It is therefore reasonable to expect that multiple bonds are stronger than single bonds. ***As the electron density between two atoms increases, the bond gets shorter and stronger.*** For example, the bond energy of C=O in CO_2 is 803 kJ/mol and that of C≡O is 1073 kJ/mol. In fact, the C≡O triple bond in carbon monoxide is the strongest known covalent bond.

The data in Table 8.2 can help us understand why an element such as nitrogen, which forms many compounds with oxygen, is unreactive enough to remain in Earth's atmosphere as N_2 molecules even though there is plenty of O_2 to react with. Reactions in which N_2 combines with other elements are less likely to occur, because they require breaking a very strong N≡N bond (946 kJ/mol). This allows us to inhale and exhale N_2 without its undergoing any chemical change. If this were not the case and N_2 reacted readily at body temperature (37 °C) to form oxides and other compounds, there would be severe consequences for us. In fact, life on Earth as we know it would not be possible.

We can use data from Table 8.2 to estimate $\Delta H°$ for reactions such as the reaction of hydrogen gas with chlorine gas: $H_2(g) + Cl_2(g) \rightarrow 2\ HCl(g)$.

- Breaking the covalent bond of H_2 requires an *input* to the system of 436 kJ/mol.
- Breaking the covalent bond of Cl_2 requires an *input* to the system of 242 kJ/mol. In each of these cases the sign of the enthalpy change is *positive*.
- Forming an H—Cl covalent bond in HCl transfers 431 kJ/mol *out of the system*. Because 2 mol HCl bonds are formed, there will be 2 mol × 431 kJ/mol = 862 kJ transferred *out*, making the sign of this enthalpy change *negative*.

Table 8.2 Average Bond Enthalpies (in kJ/mol)*

Single Bonds

	I	Br	Cl	S	P	Si	F	O	N	C	H
H	299	366	431	347	322	323	566	467	391	416	436
C	213	285	327	272	264	301	486	336	285	356	
N	—	—	193	—	~200	335	272	201	160		
O	201	—	205	—	~340	368	190	146			
F	—	—	255	326	490	582	158				
Si	234	310	391	226	—	226					
P	184	264	319	—	209						
S	—	213	255	226							
Cl	209	217	242								
Br	180	193									
I	151										

Multiple Bonds

N=N	418	C=C	598
N≡N	946	C≡C	813
C=N	616	C=O (as in CO_2, O=C=O)	803
C≡N	866	C=O (as in H_2C=O)	695
O=O (in O_2)	498	C≡O	1073

*Data from Cotton, F. A., Wilkinson, G., and Gaus, P. L. *Basic Inorganic Chemistry,* 3rd ed. New York: Wiley, 1995; p. 12.

The bond enthalpies in Table 8.2 are for gas phase reactions. If liquids or solids are involved, there are additional energy transfers for the phase changes needed to convert the liquids or solids to the gas phase. We shall restrict our use of bond enthalpies to gas phase reactions for that reason.

Bond enthalpy is represented by the letter *D* because it refers to *d*issociation of a bond.

If we represent bond enthalpy by the letter *D*, with a subscript to show which bond it refers to, then for the reaction $H_2(g) + Cl_2(g) \rightarrow 2\ HCl(g)$ the net energy transfer is

$$\begin{aligned}\Delta H^\circ &= \{[(1\ \text{mol H—H}) \times D_{\text{H—H}}] + [(1\ \text{mol Cl—Cl}) \times D_{\text{Cl—Cl}}]\}\\ &\quad -[(2\ \text{mol H—Cl}) \times D_{\text{H—Cl}}]\\ &= [(1\ \text{mol H—H})(436\ \text{kJ/mol})] + [(1\ \text{mol Cl—Cl})(242\ \text{kJ/mol})]\\ &\quad - (2\ \text{mol H—Cl})(431\ \text{kJ/mol})\\ &= -184\ \text{kJ}\end{aligned}$$

This differs only very slightly from the experimentally determined value of −184.614 kJ.

As illustrated in Figure 8.4, we can think of the process in the calculation in terms of breaking all the bonds in each reactant molecule and then forming all the bonds in the product molecules. Each bond enthalpy was multiplied by the number of moles of bonds that were broken or formed. For bonds in reactant molecules, we added the bond enthalpies because *breaking bonds is endothermic*. For products, we subtracted the bond enthalpies because *bond formation is exothermic.* This can be summarized in this equation:

$$\begin{aligned}\Delta H^\circ = &\sum[(\text{moles of bonds}) \times D(\text{bonds broken})]\\ &-\sum[(\text{moles of bonds}) \times D(\text{bonds formed})]\end{aligned} \qquad [8.1]$$

The Σ (Greek capital letter *sigma*) represents summation. We add the bond enthalpies for all bonds broken, and we subtract the bond enthalpies for all bonds formed. This equation and the values in Table 8.2 allow us to estimate enthalpy changes for a wide variety of gas-phase reactions.

There are several important points about the bond enthalpies in Table 8.2:

- **The enthalpies listed are often average bond enthalpies and may vary depending on the molecular structure.** For example, the enthalpy of a C—H bond is given as 413 kJ/mol, but C—H bond strengths are affected by other atoms and bonds in the same molecule. Depending on the structure of the molecule, the energy required to break a mole of C—H bonds may vary by 30 to 40 kJ/mol, so the values in Table 8.2 can be used only to estimate an enthalpy change, not to calculate it exactly.

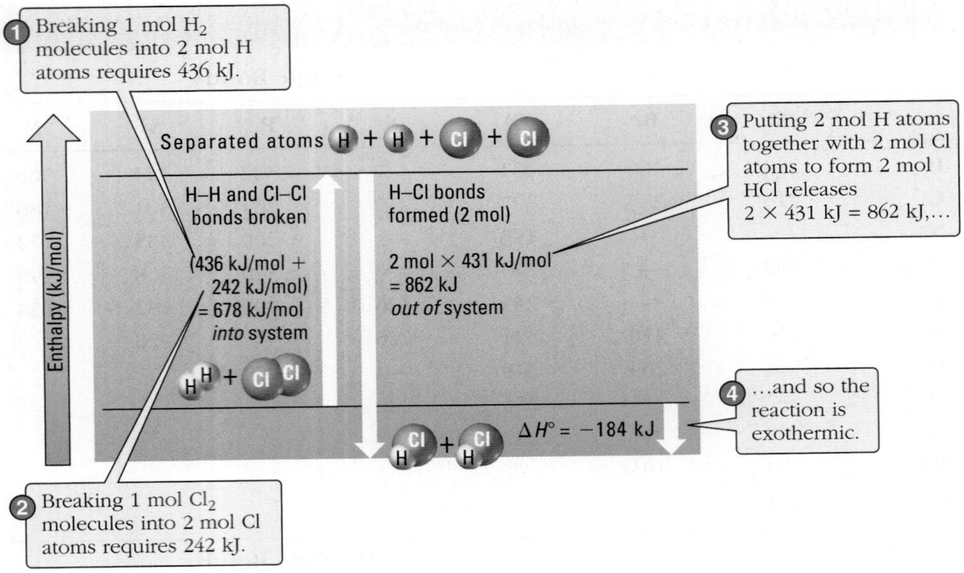

Active Figure 8.4 Stepwise energy changes in the reaction of hydrogen with chlorine. The enthalpy change for the reaction is −184 kJ; $\Delta H° = 436\ \text{kJ} + 242\ \text{kJ} - 862\ \text{kJ} = -184\ \text{kJ}$. Visit this book's companion website at **www.cengage.com/chemistry/moore** to test your understanding of the concepts in this figure.

- **The enthalpies in Table 8.2 are for breaking bonds in molecules in the gaseous state.** If a reactant or a product is in the liquid or solid state, the energy required to convert it to or from the gas phase will also contribute to the enthalpy change of a reaction and should be accounted for.
- **Multiple bonds, shown as double and triple lines between atoms, are listed at the bottom of Table 8.2.** In some cases, different enthalpies are given for multiple bonds in specific molecules such as $O_2(g)$ or $CO_2(g)$.

PROBLEM-SOLVING EXAMPLE 8.6 Estimating ΔH° from Bond Enthalpies

The conversion of diazomethane to ethene and nitrogen is given by this equation:

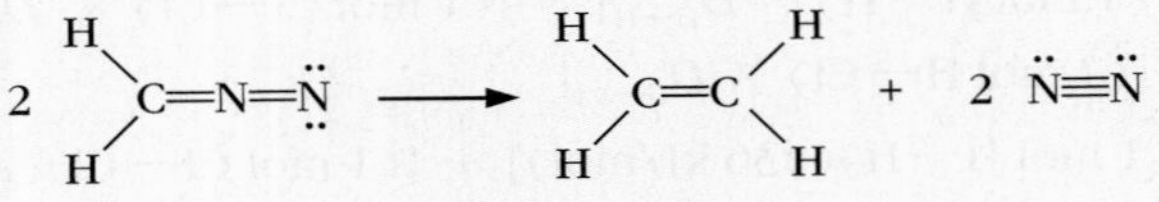

diazomethane ethene

Using the bond enthalpies in Table 8.2, calculate the $\Delta H°$ for this reaction.

Answer $\Delta H° = -422$ kJ

Strategy and Explanation Begin by using the given Lewis structures to compare bonds broken with those formed. These structures show that there are two C—H bonds, one C=N bond, and one N=N bond in each diazomethane molecule. In each ethene molecule there are four C—H bonds and one C=C bond; each nitrogen molecule contains one N≡N bond. Bond enthalpies for other than single bonds, such as C=N and C=C, are listed at the bottom of Table 8.2. We can use these data and the balanced chemical equation in the change-in-enthalpy Equation 8.1 for bond breaking and bond making.

$$\Delta H° = \Sigma[(\text{moles of bonds}) \times D(\text{bonds broken})] - \Sigma[(\text{moles of bonds}) \times D(\text{bonds formed})]$$

$$= \{[(4\text{ mol C—H}) \times D_{\text{C—H}}] + [(2\text{ mol C=N}) \times D_{\text{C=N}}] + [(2\text{ mol N=N}) \times D_{\text{N=N}}]\}$$
$$-\{[(4\text{ mol C—H}) \times D_{\text{C—H}}] + [(1\text{ mol C=C}) \times D_{\text{C=C}}] + [(2\text{ mol N≡N}) \times D_{\text{N≡N}}]\}$$

$$= \{[(4\text{ mol C—H}) \times (416\text{ kJ/mol})] + [(2\text{ mol C=N}) \times (616\text{ kJ/mol})] + [(2\text{ mol N=N}) \times (418\text{ kJ/mol})]\}$$
$$-\{[(4\text{ mol C—H}) \times (416\text{ kJ/mol}) + [(1\text{ mol C=C}) \times (598\text{ kJ/mol})] + [(2\text{ mol N≡N}) \times (946\text{ kJ/mol})]\}$$

$$= (3732\text{ kJ} - 4154\text{ kJ}) = -422\text{ kJ}$$

☑ **Reasonable Answer Check** Note that an N≡N triple bond, 946 kJ/mol, is more than 500 kJ/mol stronger than an N═N double bond, 418 kJ/mol. The C═C double bond, 598 kJ/mol, is about the same strength as a C═N double bond, 616 kJ/mol. Now analyze the bonds broken and bonds formed. There are 4 mol C—H bonds broken in the reactants and 4 mol C—H bonds formed in the products; this results in no overall change. There are 2 mol C═N bonds broken in the reactants and only 1 mol C═C bonds formed in the products; thus about 600 kJ must be transferred into the system to break the extra mole of double bonds in the reactants. There are 2 mol N═N double bonds in the reactants and 2 mol N≡N triple bonds in the products; thus about 2 × 500 kJ is released when the stronger bonds form. Overall there should be about 1000 kJ − 600 kJ = 400 kJ released, which agrees with the calculated −422 kJ.

PROBLEM-SOLVING PRACTICE 8.6

Use Equation 8.1 and values from Table 8.2 to estimate the enthalpy change when methane, CH_4, and oxygen combine according to the equation:

$$CH_4(g) + 2\,O_2(g) \longrightarrow CO_2(g) + 2\,H_2O(g)$$

Burning methane in a Bunsen burner. Combustion of methane is highly exothermic, releasing energy to the surroundings for every mole of CH_4 that burns. (See Problem-Solving Practice 8.6.)

CONCEPTUAL EXERCISE 8.6 Bond Length

Arrange C═N, C≡N, and C—N in order of decreasing bond length. Is the order for decreasing bond energy the same or the reverse order? Explain.

8.7 Bond Properties: Bond Polarity and Electronegativity

In a molecule such as H_2 or F_2, where both atoms are the same, there is *equal* sharing of the bonding electron pair and the bond is a **nonpolar covalent bond.** When two different atoms are bonded, however, the sharing of the bonding electrons is usually *unequal* and results in a displacement of the bonding electrons toward one of the atoms. If the displacement is complete, electron transfer occurs and the bond is ionic. If the displacement is less than complete, the bonding electrons are shared *unequally,* and the bond is a **polar covalent bond** (Figure 8.5). As you will see in Chapters 9 and 12, properties of molecules are dramatically affected by bond polarity.

Linus Pauling, in 1932, first proposed the concept of electronegativity based on an analysis of bond energies. **Electronegativity** represents the ability of an atom in a covalent bond to attract *shared* electrons to itself—creating a displacement of those bonding electrons toward the more electronegative atom.

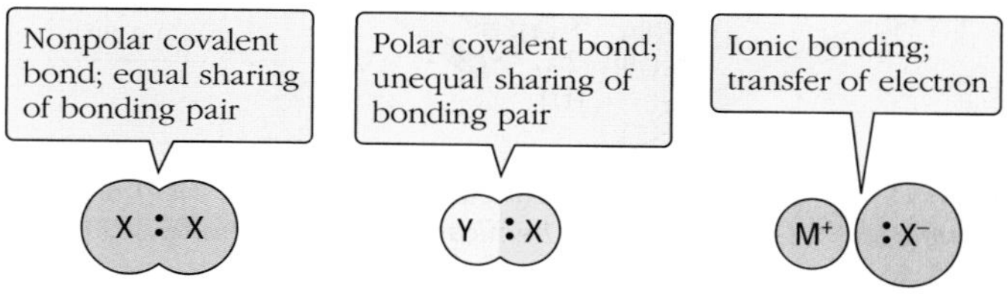

Figure 8.5 Nonpolar covalent, polar covalent, and ionic bonding.

1901–1994

The son of a druggist, Linus Pauling completed his Ph.D. in chemistry and then traveled to Europe, where he worked briefly with Erwin Schrödinger and Niels Bohr (see Chapter 7). For his bonding theories and his work with structural aspects of proteins, Pauling was awarded the Nobel Prize in Chemistry in 1954. Shortly after World War II, Pauling and his wife began a crusade to limit nuclear weapons, a crusade culminating in the limited test ban treaty of 1963. For this effort, Pauling was awarded the 1963 Nobel Peace Prize, the first time any person received two unshared Nobel Prizes.

Linus Pauling died in August 1994, leaving a remarkable legacy of breakthrough scientific research and a social consciousness to be remembered in guarding against the possible misapplications of technology.

go Chemistry **Module 13: Molecular Polarity** covers concepts in this section.

Pauling's electronegativity values are relative numbers with an arbitrary value of 4.0 assigned to fluorine, the most electronegative element. More recent experiments and theoretical calculations have modified many of his original values. The current values are given in Figure 8.6. The nonmetal with the next highest electronegativity is oxygen, with a value of 3.4, followed by nitrogen (3.0) and chlorine (2.7). Elements with electronegativities of 2.5 or more are all nonmetals in the upper-right corner of the periodic table. By contrast, elements with electronegativities of 1.1 or less are all metals on the left side of the periodic table. These elements are often referred to as the most electropositive elements; *they are the metals that invariably form ionic compounds.* Between these two extremes are most of the remaining metals (largely transition metals) with electronegativities between 1.1 and 1.7, the metalloids with electronegativities between 1.6 and 2.1, and some nonmetals (P, Se) with electronegativities between 2.0 and 2.4.

Electronegativities show periodic trends (Figure 8.6):

- **Electronegativity increases across a period.**
- **Electronegativity decreases down a group.**
- **In general, electronegativity increases diagonally upward and to the right in the periodic table.**

Because metals typically lose electrons, they are the least electronegative elements. Nonmetals, which have a tendency to gain electrons, are the most electronegative.

Electronegativity values are approximate and are primarily used to predict the polarity of covalent bonds. Bond polarity is indicated by writing $\delta+$ by the *less* electronegative atom and $\delta-$ by the *more* electronegative atom, where δ stands for par-

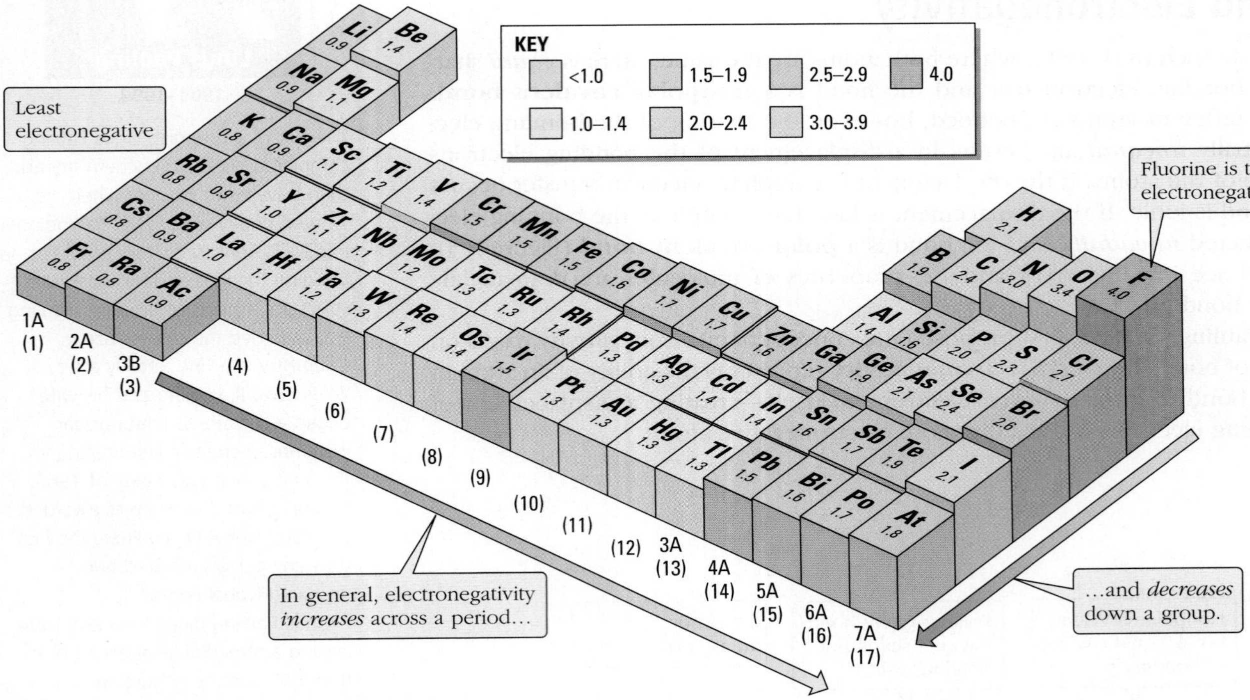

Figure 8.6 Periodic trends in electronegativities. Data from Allred, A. L., and Rochow, G. E. *Journal of Inorganic and Nuclear Chemistry*, Vol. 5, 1958; pp. 264, 269; Little, E. J., and Jones, M. M. *Journal of Chemical Education*, Vol. 37, 1960; p. 261; and Meek, T. L. *Journal of Chemical Education*, Vol. 83, 2005; p. 325.

tial charge. For example, the polar H—Cl bond in hydrogen chloride can be represented as shown to the right.

Except for bonds between identical atoms (equal electronegativity), all bonds are polar to some extent, and the difference in electronegativity values is a qualitative measure of the degree of polarity. ***The greater the difference in electronegativity between two atoms, the more polar will be the bond between them.*** The change from nonpolar covalent bonds to slightly polar covalent bonds to very polar covalent bonds to ionic bonds can be regarded as a continuum (Figure 8.7). Examples are F_2 (nonpolar), HBr (slightly polar), HF (very polar), and Na^+F^- (ionic).

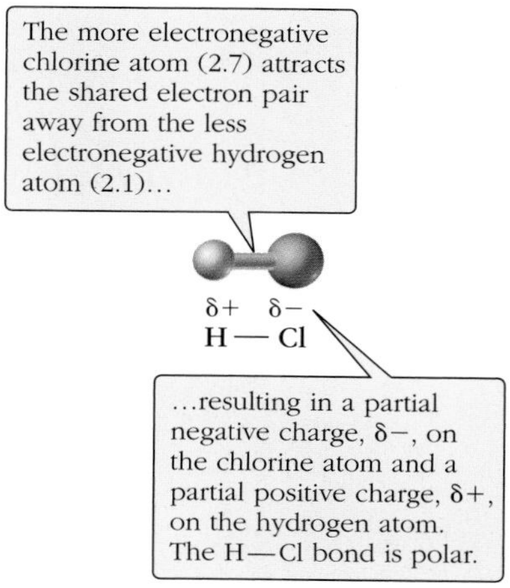

In HCl, the H and Cl atoms are joined by a polar covalent bond. Calculations indicate that the bond has about 18% ionic character; that is, δ = 0.18; the δ+ charge is 0.18 times the charge of a proton, and the δ− charge is 0.18 times the charge of an electron.

PROBLEM-SOLVING EXAMPLE 8.7 Bond Polarity

For each of these bond pairs, indicate the partial positive and negative atoms and tell which is the more polar bond.

(a) Cl—F and Br—F (b) N—Cl and P—Cl

Answer

(a) $\overset{\delta+}{\text{Br}}$—$\overset{\delta-}{\text{F}}$ (more polar); $\overset{\delta-}{\text{F}}$—$\overset{\delta+}{\text{Cl}}$ (b) $\overset{\delta+}{\text{P}}$—$\overset{\delta-}{\text{Cl}}$ (more polar); $\overset{\delta+}{\text{N}}$—$\overset{\delta-}{\text{Cl}}$

Strategy and Explanation Compare the electronegativities of the two bonded atoms by using Figure 8.6. The atom with the lesser electronegativity will have a partial positive charge and the one with the greater electronegativity will bear a partial negative charge.

(a) The Br—F bond is more polar because the difference in electronegativity is greater between Br and F (1.4) than between Cl and F (1.3). The F atom is the partial negative end in BrF; because F is more electronegative than Cl, the F atom is the partial negative end in ClF.

(b) P—Cl is more polar than N—Cl because P is less electronegative than N, so the electronegativity difference is greater between P and Cl (0.7) than between N and Cl (0.3).

PROBLEM-SOLVING PRACTICE 8.7

For each of these pairs of bonds, decide which is the more polar. For each polar bond, indicate the partial positive and partial negative atoms.

(a) B—C and B—Cl (b) N—H and O—H

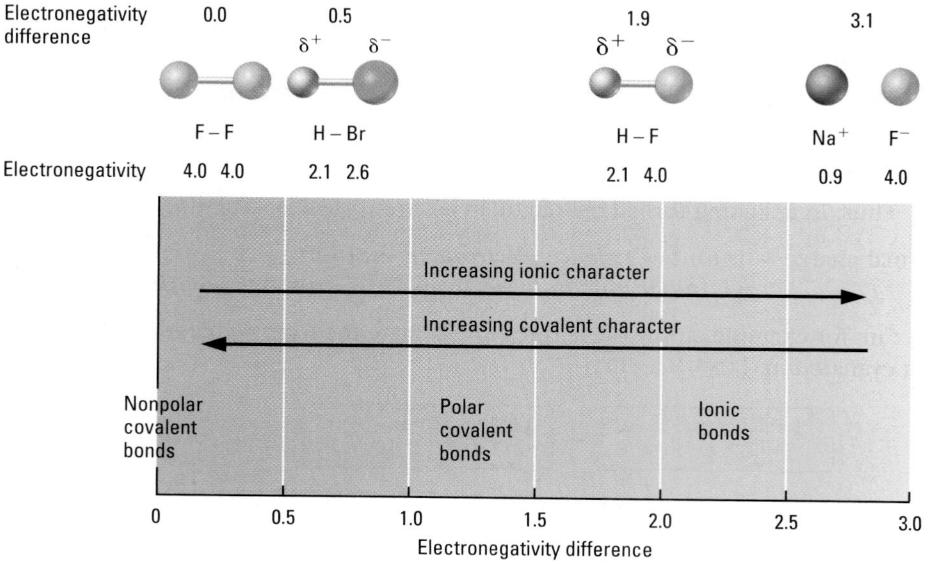

Figure 8.7 Bond character and electronegativity differences. This graph shows the relationship between electronegativity difference and the ionic character of a bond.

CONCEPTUAL EXERCISE 8.7 Bond Types

Use Figures 8.6 and 8.7 to explain why:

(a) NaCl is considered to be an ionic compound rather than a polar covalent compound.

(b) BrF is considered to be a polar covalent compound rather than an ionic compound.

In moving across the third period from sodium to sulfur, changes in electronegativity difference cause a significant shift in the properties of compounds of these elements with chlorine. The third period begins with the ionic compounds NaCl and $MgCl_2$; both are crystalline solids at room temperature. The electronegativity differences between the metal and chlorine are 1.8 and 1.6, respectively. Aluminum chloride is less ionic; with an aluminum-chlorine electronegativity difference of 1.3, its bonding is polar covalent. As electronegativity increases across the period, the other Period 3 chlorides—$SiCl_4$, PCl_3, and SCl_2—are molecular compounds, with decreasing electronegativity differences between the nonmetal and chlorine from Si to S. This results in a decrease in bond polarity from Si—Cl to P—Cl to S—Cl bonds, culminating in no electronegativity difference in Cl—Cl bonds.

Compound	$SiCl_4$	PCl_3	SCl_2	Cl_2
Bond	Si—Cl	P—Cl	S—Cl	Cl—Cl
Electronegativity difference	1.1	0.7	0.4	0.0

Electronegativities: Si = 1.6; P = 2.0; S = 2.3; Cl = 2.7.

8.8 Formal Charge

Lewis structures depict how valence electrons are distributed in a molecule or ion. For some molecules or ions, more than one Lewis structure can be written, each of which obeys the octet rule. Which of the structures is more correct? How do you decide? Using formal charge is one way to do so. **Formal charge *is the charge a bonded atom would have if its bonding electrons were shared equally.*** In calculating formal charges, these assignments are made:

- ***All of the lone pair electrons are assigned to the atom on which they are found.***
- ***Half of the bonding electrons are assigned to each atom in the bond.***
- ***The sum of the formal charges must equal the actual charge: zero for molecules and the ionic charge for an ion.***

Thus, in assigning formal charge to an atom in a Lewis structure,

Formal charge = [*number valence electrons in an atom*] − [(*number lone pair electrons*) + ($\frac{1}{2}$ *number bonding electrons*)]

We can, for example, apply these rules to calculate the formal charges of the atoms in a cyanate ion, $[:N\equiv C-\ddot{\underset{\cdot\cdot}{O}}:]^-$

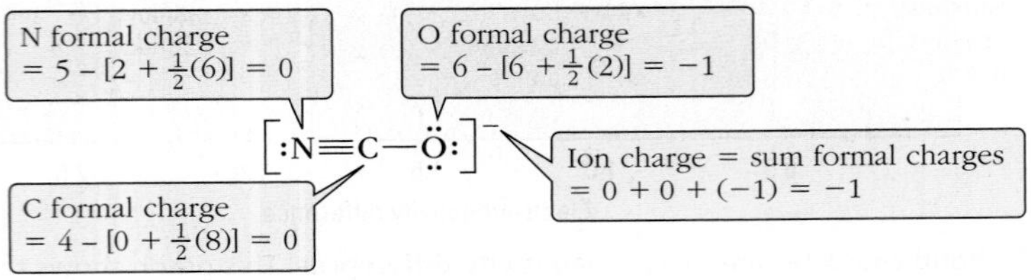
N formal charge
= 5 − [2 + ½(6)] = 0
O formal charge
= 6 − [6 + ½(2)] = −1
[:N≡C—O:]⁻
Ion charge = sum formal charges
= 0 + 0 + (−1) = −1
C formal charge
= 4 − [0 + ½(8)] = 0

$[:N\equiv C-\ddot{\underset{\cdot\cdot}{O}}:]^-$	N	C	O
Valence electrons	5	4	6
Lone pair electrons	2	0	6
$\frac{1}{2}$ shared electrons	3	4	1
Formal charge	0	0	−1

Note that the sum of the formal charges equals −1, the charge on the ion. The −1 formal charge is on oxygen, the most electronegative atom in the ion.

It is important to recognize that formal charges do *not* indicate actual charges on atoms. Formal charge is a useful way to determine the most likely structure from among several Lewis structures. In evaluating possible structures with different formal charge distributions, these principles apply:

- ***Smaller formal charges are more favorable than larger ones.***
- ***Negative formal charges should reside on the more electronegative atoms and positive formal charges on less electronegative atoms.***
- ***Like charges should not be on adjacent atoms.***

PROBLEM-SOLVING EXAMPLE 8.8 Formal Charges

(a) Two possible Lewis structures for N_2O are :Ö=N=N̈: and :Ö—N≡N:. Determine the formal charges on each atom and the preferred structure.
(b) Determine the formal charges on each atom of these two Lewis structures of dichlorine monoxide, Cl_2O. Which structure is preferred?

Structure 1 :Ö—C̈l—C̈l: Structure 2 :C̈l—Ö—C̈l:

Answer

(a) :Ö=N=N̈: (0, +1, −1)   :Ö—N≡N: (−1, +1, 0) (preferred)

(b) :Ö—C̈l—C̈l: (−1, +1, 0)   :C̈l—Ö—C̈l: (0, 0, 0) (preferred)

Strategy and Explanation Count the valence electrons, account for the number of lone pair electrons, and assign half the bonding electrons to each atom in the bond. The formal charges should add up to zero for N_2O and Cl_2O because they are both neutral molecules.

(a) In the first N_2O structure, :Ö=N=N̈:, the formal charges for O, N, and N are 0, +1, and −1, respectively. The sum of the formal charges is $0 + (+1) + (-1) = 0$. This is as it should be for a neutral molecule. Applying the rules for assigning electrons to the second structure, :Ö—N≡N:, the formal charges are (in order) −1, +1, and 0. The sum of the formal charges in this structure is also zero.

	Structure 1			Structure 2		
	O	**N**	**N**	**O**	**N**	**N**
Valence electrons	6	5	5	6	5	5
Lone pair electrons	4	0	4	6	0	2
$\frac{1}{2}$ shared electrons	2	4	2	1	4	3
Formal charge	0	+1	−1	−1	+1	0

Formal charges are low in both structures, but the second structure is preferred because it has the negative charge on the more electronegative atom (O) rather than on the less electronegative N, as in the first structure.

(b)

	Structure 1			Structure 2		
	O	Cl	Cl	Cl	O	Cl
Valence electrons	6	7	7	7	6	7
Lone pair electrons	6	4	6	6	4	6
$\frac{1}{2}$ shared electrons	1	2	1	1	2	1
Formal charge	−1	+1	0	0	0	0

In both cases, the total formal charge is zero, which it should be for a neutral molecule. Structure 2 is preferred because it has all zero formal charges.

PROBLEM-SOLVING PRACTICE 8.8

A third Lewis structure can be written for N_2O, which also obeys the octet rule. Write this other Lewis structure and determine the formal charges on its atoms.

CONCEPTUAL EXERCISE 8.8 Formal Charge

Determine the formal charge of each atom in hydrazine, H_2NNH_2.

8.9 Lewis Structures and Resonance

The troposphere is the region of the atmosphere closest to Earth's surface. The stratosphere lies above the troposphere (Chapter 10, Section 10.1.)

Ozone, O_3, an allotropic form of oxygen (***← p. 26***), is an unstable, pale blue, diamagnetic gas with a pungent odor. Depending on its location, ozone can be either beneficial or harmful. A low-concentration, but very important, layer of ozone in the stratosphere protects Earth and its inhabitants from intense ultraviolet solar radiation (Section 10.9), but ozone pollution in the lower atmosphere (troposphere) causes respiratory problems (Section 10.10).

As you have seen, the number of bonding electron pairs between two atoms is important in determining bond length and strength. The experimentally measured lengths for the two oxygen-oxygen bonds in ozone are the same, 127.8 pm, implying that both bonds contain the same number of bond pairs. However, using the guidelines for writing Lewis structures, you might come to a different conclusion. Two possible Lewis structures are

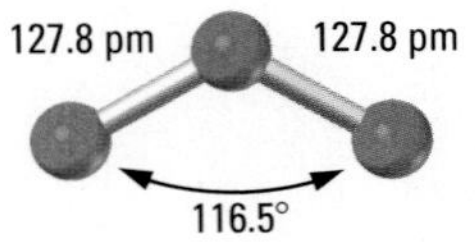
127.8 pm
127.8 pm
116.5°
ozone

:Ö=Ö—Ö̤: and :Ö̤—Ö=Ö:

Each structure shows a double bond on one side of the central O atom and a single bond on the other side. If either one were the actual structure of O_3, then one bond, O=O, should be shorter than the other, O—O, but this is not the case. That the oxygen-to-oxygen bonds in ozone are neither double bonds nor single bonds is supported by the fact that the experimentally determined 127.8 pm bond length is longer than that for O=O (112 pm), but shorter than that for O—O (132 pm). Therefore, no *single* Lewis structure can be written that is consistent with all of the experimental data. When this situation arises, the concept of *resonance* is invoked to reconcile the experimental observation with two or more Lewis structures for the same molecule. Each of the Lewis structures, called **resonance structures,** is thought of as contributing to the true structure that cannot be written. The actual structure of O_3 is neither of the Lewis structures above, but a composite called a **resonance hybrid.** It is conventional to connect the resonance structures with a double-headed arrow, ↔, to emphasize that the actual bonding is a composite of these structures.

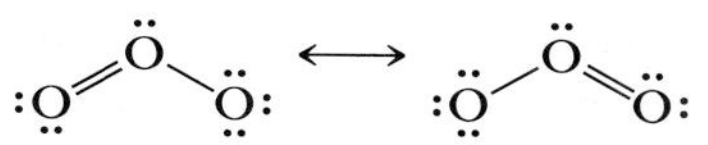

Resonance structures
of ozone

The double-headed arrow does not imply that one resonance structure converts to the other.

The resonance concept is useful whenever there is a choice about which of two or three atoms contribute lone pairs to achieve an octet of electrons about a central atom by multiple bond formation.

When applying the concept of resonance, keep several important things in mind:

- Lewis structures contributing to the resonance hybrid structure differ only in the assignment of electron pair positions; *the positions of the atoms don't change.*
- Contributing Lewis structures differ in the number of bond pairs between one or more specific pairs of atoms.
- The resonance hybrid structure represents a single composite structure, not different structures that are continually changing back and forth.

"Resonance" is an unfortunate term, because it implies that the molecule somehow "resonates," moving in some way to form different kinds of molecules, which is not true. There is only one kind of ozone molecule.

To further illustrate the use of resonance, consider what happens in writing the Lewis structure of the carbonate ion, CO_3^{2-}, which has 24 valence electrons (four from C, 18 from three O atoms, and two for the 2− charge). Writing the skeleton structure and putting in lone pairs so that each O has an octet uses all 24 electrons but leaves carbon without an octet:

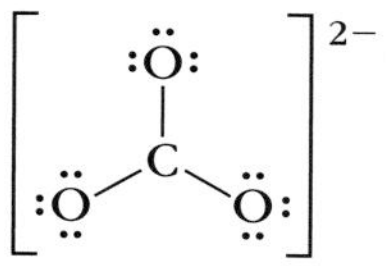

To give carbon an octet requires changing a single bond to a double bond, which can be done in three equivalent ways:

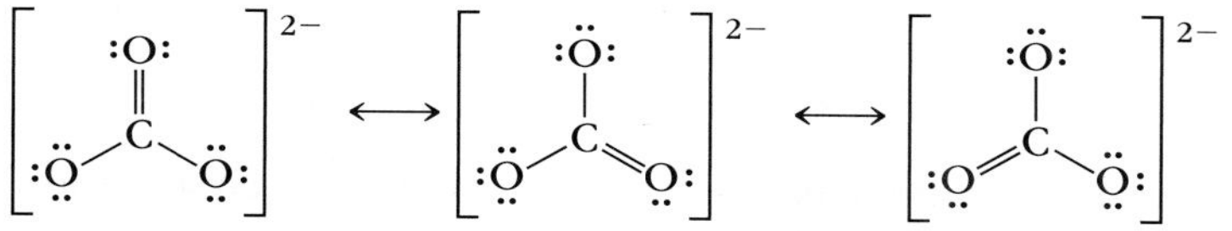

Writing the Lewis structures of oxygen-containing anions often requires using resonance.

The three resonance structures contribute to the resonance hybrid. As with ozone, experiments confirm that there are no single and double bonds in the carbonate ion. All three carbon-oxygen bond distances are 129 pm, intermediate between the C—O single bond (143 pm) and the C═O double bond (122 pm) distances.

Resonance and the Structure of Benzene

To nineteenth-century chemists, the C_6H_6 molecular formula of benzene implied that it was an unsaturated compound because it lacked the ratio of carbon to hydrogen found in saturated noncyclic hydrocarbons, C_nH_{2n+2}. A six-membered ring structure with alternating double bonds uses all the available valence electrons and gives each carbon atom an octet of valence electrons.

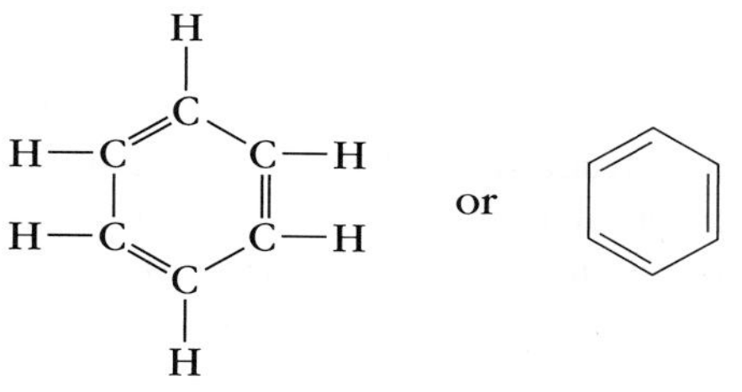

This structure implies that alternating C—C single bonds and C=C double bonds are present in benzene. But, carbon compounds that contain C=C double bonds react with bromine, Br_2, by adding bromine atoms directly to the double bond.

Addition reaction (occurs only with C=C double bonds):

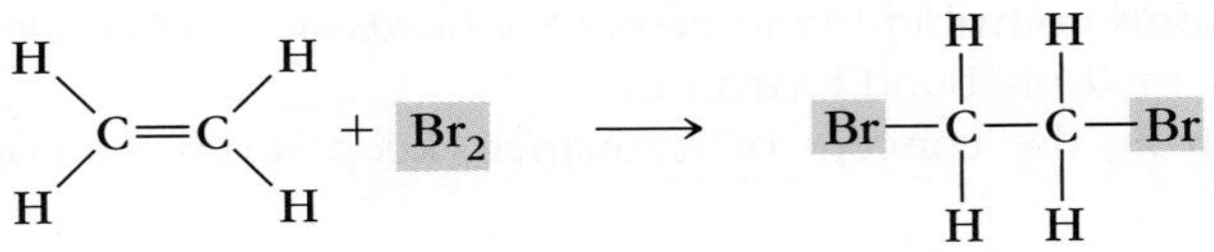

Benzene, however, does *not* react this way under the same conditions. Rather, benzene reacts with bromine by a *substitution* reaction in which a bromine atom replaces (substitutes for) a hydrogen atom of benzene to produce bromobenzene. The displaced hydrogen atom combines with the remaining bromine atom to form hydrogen bromide, HBr. The fact that benzene reacts by substitution rather than addition is strong experimental evidence that benzene does not contain C=C bonds.

Substitution reaction:

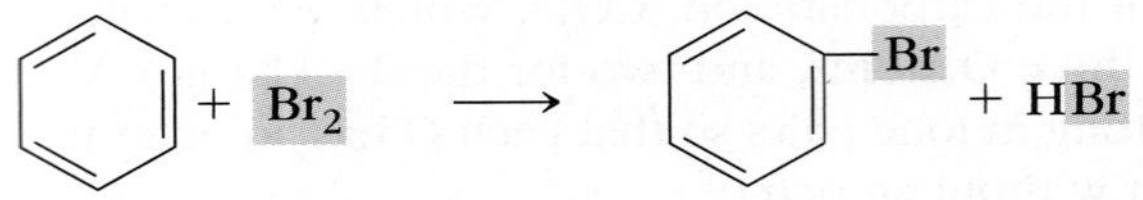

In 1872 Friedrich A. Kekulé proposed that benzene could be represented by a combination of two structures, now called resonance structures, indicated by the double-headed arrow.

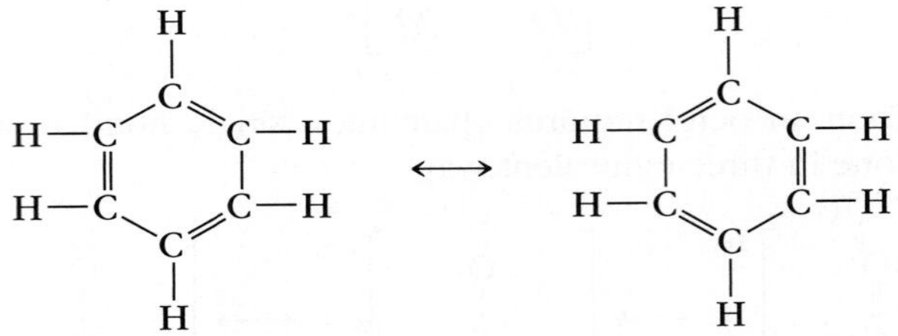

Bonding in benzene can be described better using the molecular orbital theory (see Section 8.12).

Neither of these alternating single- and double-bond resonance structures, however, accurately represents benzene. Experimental structural data for benzene indicate a planar, symmetric molecule in which all six carbon-carbon bonds are equivalent. Each carbon-carbon bond is 139 pm long, intermediate between the length of a C—C single bond (154 pm) and a C=C double bond (134 pm). Benzene is a resonance hybrid of these resonance structures—it is a molecule in which the six electrons of the suggested three double bonds are spread (delocalized) uniformly around the ring. The six *delocalized electrons* are shared equally by all six carbon atoms. When hydrogen and carbon atoms are not shown, the benzene ring is sometimes written as a hexagon with a circle in the middle. The circle represents the six delocalized electrons spread evenly over all of the carbon atoms. Each corner in the hexagon represents one carbon atom and one hydrogen atom, and each line represents a single C—C bond.

Sometimes a dotted circle rather than a solid circle is used to represent the delocalized electrons.

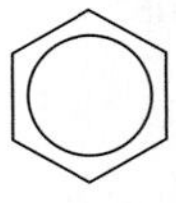 rather than 

Unlike the resonance structures of CO_3^{2-} and benzene, which are similar and contribute equally to the resonance hybrid, some resonance structures are significantly different and cannot make equal contributions to the hybrid. As demonstrated in Problem-Solving Example 8.9, formal charge is useful in determining which resonance structures are most plausible.

PROBLEM-SOLVING EXAMPLE 8.9 Writing Resonance Structures

The Lewis structure for cyanate ion was given previously as $[:\mathrm{N}\equiv\mathrm{C}-\ddot{\mathrm{O}}:]^-$ (p. 350). This is just one of three possible resonance structures for cyanate ion.

(a) Draw the Lewis structures for the two other resonance structures.

(b) Use formal charges to determine which resonance structure is preferred.

Answer

(a) Structure 2: $[:\ddot{\mathrm{N}}-\mathrm{C}\equiv\mathrm{O}:]^-$ Structure 3: $[:\ddot{\mathrm{N}}=\mathrm{C}=\ddot{\mathrm{O}}:]^-$

(b) The initial structure, $[:\mathrm{N}\equiv\mathrm{C}-\ddot{\mathrm{O}}:]^-$, is preferred.

Strategy and Explanation

(a) The cyanate ion has 16 valence electrons: five from N, four from C, six from O, and one to account for the 1− charge. Both structures 2 and 3 have 16 valence electrons and differ only in the placement of the multiple bonds, showing that they are correct resonance structures.

(b) The formal charges on the atoms in structures 2 and 3 are shown in the table below. The table for the initial structure is on p. 350.

	Structure 2			Structure 3		
	N	C	O	N	C	O
Valence electrons	5	4	6	5	4	6
Lone pair electrons	6	0	2	4	0	4
$\frac{1}{2}$ shared electrons	1	4	3	2	4	2
Formal charge	−2	0	+1	−1	0	0

The formal charges in structures 2 and 3 add up to −1, as they must for the 1− cyanate ion. Oxygen, the most electronegative atom, should bear the negative charge. Structure 2 would be least preferred because it has an atom with a high formal charge (−2 on N) and a positive charge on oxygen. The initial structure $[:\mathrm{N}\equiv\mathrm{C}-\ddot{\mathrm{O}}:]^-$ is the one in which oxygen has the negative charge (p. 350). Thus, it is the preferred one.

PROBLEM-SOLVING PRACTICE 8.9

The nitrogen-oxygen bond lengths in NO_2^- are both 124 pm. Compare this with the bond distances given in Table 8.1 for N—O and N═O bond lengths. Account for any difference.

CONCEPTUAL EXERCISE 8.9 Resonance Structures

Why is $[:\ddot{\mathrm{N}}-\mathrm{O}\equiv\mathrm{C}:]^-$ not a resonance structure for cyanate ion?

8.10 Exceptions to the Octet Rule

Many molecules and polyatomic ions have structures that are not consistent with the octet rule (p. 332). Three kinds of exceptions occur: (1) molecules or ions with central atoms having fewer than eight electrons; (2) molecules or ions with an odd

CHEMISTRY IN THE NEWS Self-Darkening Eyeglasses

Photochromic eyeglass lenses reversibly darken when exposed to UV light and fade back to clear when UV light is not present. Transitions Optical's proprietary photochromic technology developed and used to create Transitions® lenses involves a dye called a naphthopyran containing a bond that breaks when struck by UV light photons. The molecule then rearranges to a molecular structure that absorbs light at a longer wavelength, in the visible region of the spectrum, darkening the lens and blocking 100% of UVA/UVB rays. The lens lightens automatically and fades back to a fully clear state when away from a UV source, such as indoors.

The dye coats the surface of a plastic lens made of a proprietary polymer. Such a specialty polymer must be used because photochromic dyes do not work well with common plastic lens materials. The nature of the proprietary lens materials affects how fast the lenses darken and lighten.

A different version of photochromic lenses does not use organic dyes, but the reversible reduction of Ag^+ ions in a silver halide crystal to metallic silver embedded in a glass matrix. Although the glass photochromic lenses are heavier than plastic ones, glass lenses resist scratching better and have a longer lifetime of being able to switch reversibly from dark to light.

One common issue to note with both types of lenses is that neither works well for driving. Federal law requires that auto windshields contain an inner layer of a UV-absorbing substance to decrease glare. Once the driver is inside a vehicle, most of the UV radiation is blocked and prevented from reaching the photochromic lenses to activate or darken them. To avoid this difficulty, specialty lenses such as polarized lenses have been developed and are widely available. Another option is to keep a pair of quality sunglasses handy that are labeled as providing 100% UV protection.

Courtesy of Transitions Optical

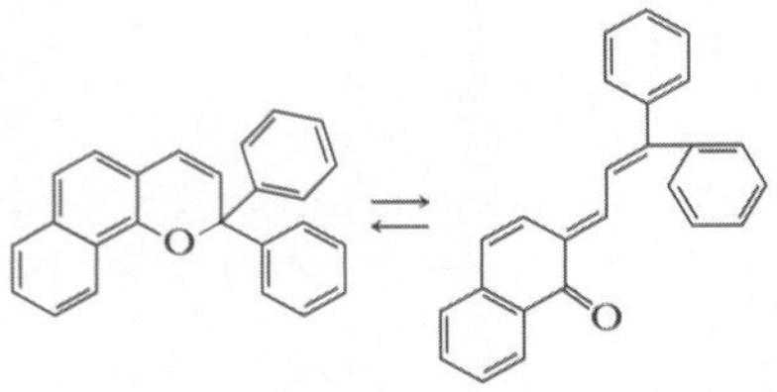

Absorbs UV light

Absorbs UV and visible light

Photochromic reaction. When a naphthopyran dye is exposed to UV light, a weak bond (red) breaks, and the molecule rearranges to a species that absorbs light at longer wavelengths.

Source: From *Chemical and Engineering News*, April 6, 2009, p. 54. Copyright © 2009 American Chemical Society. Reprinted by permission.

number of valence electrons; and (3) molecules or ions with central atoms having more than an octet of electrons.

Fewer Than Eight Valence Electrons

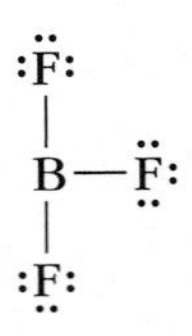

The type of bond in which both electrons are provided by the same atom is known as a coordinate covalent bond. It is discussed further in Section 16.9.

Boron trifluoride, BF_3, is a molecule with less than an octet of valence electrons around the central boron atom. Boron, a Group 3A element, has only three valence electrons; each fluorine contributes seven, for a total of 24 valence electrons. Although the Lewis structure has an octet around each fluorine atom, there are only six electrons around the B atom, an exception to the octet rule. Because it lacks an octet around boron, BF_3 is very reactive—for example, it readily combines with NH_3 to form a compound with the formula H_3NBF_3. The bonding between BF_3 and NH_3 can be explained by using the lone pair of electrons on N to form a covalent bond with B in BF_3. In this case, the nitrogen lone pair provides *both* of the shared electrons, resulting in an octet of electrons for both B and N.

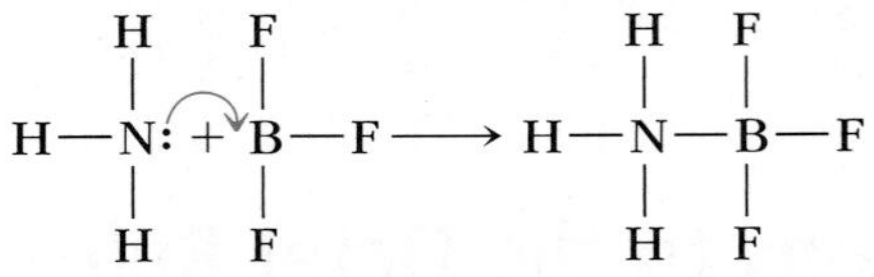

Odd Number of Valence Electrons

All the molecules we have discussed up to this point have contained only *pairs* of valence electrons. However, a few stable molecules have an odd number of valence electrons. For example, NO has 11 valence electrons, and NO_2 has 17 valence electrons. The most plausible Lewis structures for these molecules are

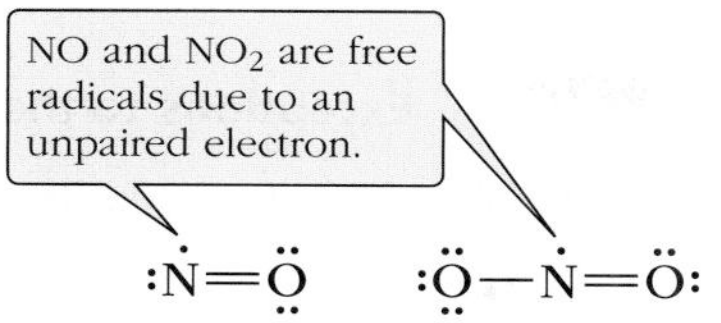

Atoms and molecules that have an unpaired electron are known as **free radicals.** How do unpaired electrons affect reactivity? Simple free radicals such as atoms of H· and Cl· are very reactive and readily combine with other atoms to give molecules such as H_2, Cl_2, and HCl. Therefore, we would expect free radical molecules to be more reactive than molecules that have all electrons paired, and they are. A free radical either combines with another free radical to form a more stable molecule in which the electrons are paired, or it reacts with other molecules to produce new free radicals. These kinds of reactions are central to the formation of addition polymers (Section 12.7) and air pollutants (Section 10.10). For example, when gaseous NO and NO_2 are released in vehicle exhaust, the colorless NO reacts with O_2 in the air to form brown NO_2. The NO_2 decomposes in the presence of sunlight to give NO and O, both of which are free radicals.

The free O atom reacts with O_2 in the air to give ozone, O_3, a tropospheric air pollutant that affects the respiratory system (Section 10.10). Free radicals also have a tendency to combine with themselves to form dimers, substances made from two smaller units. For example, when NO_2 gas is cooled it dimerizes to N_2O_4.

As expected, NO and NO_2 are paramagnetic ***(← p. 305)*** because of their odd numbers of electrons. Experimental evidence indicates that O_2 is also paramagnetic (Figure 8.8) with two unpaired electrons and a double bond. The predicted Lewis structure for O_2 shows a double bond, but in that case all the electrons would be paired. It is impossible to write a conventional Lewis structure of O_2 that is in agreement with the experimental results. The molecular orbital theory (Section 8.12) explains bonding in a way that accounts for the paramagnetism of O_2.

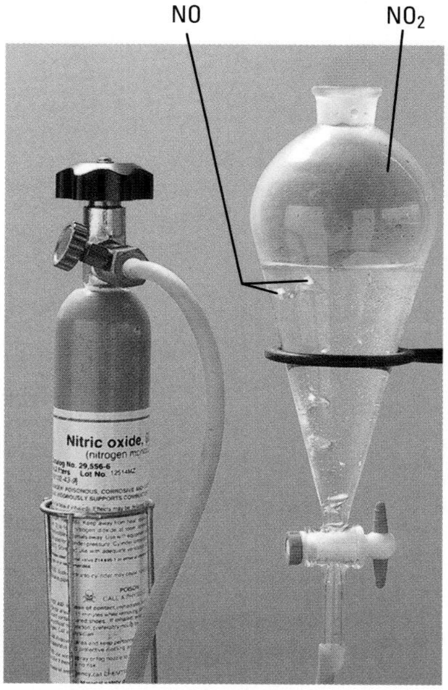

© Cengage Learning/Charles D. Winters

Nitrogen dioxide formation. As it emerges from water, colorless nitrogen monoxide (also called nitric oxide) reacts with oxygen (in air) to form reddish-brown NO_2.

NO_2 is a free radical; N_2O_4 is not because it has no unpaired electrons.

See Suidan, L., *et al. Journal of Chemical Education,* Vol. 72, 1995; p. 583 for a discussion of bonding in exceptions to the octet rule.

More Than Eight Valence Electrons

Exceptions to the octet rule are most common among molecules or ions with an "expanded octet"—that is, more than eight electrons in the valence shell around a central atom. For example, sulfur and phosphorus commonly form stable molecules and ions in which the Lewis structure has S or P surrounded by more than an octet of valence electrons. Expanded octets are almost never found with elements in the first or second period, where central atoms are too small to accommodate more than four terminal atoms. The octet rule reliably predicts formulas for all stable molecules in which the central atom is C, N, O, or F; when the central atom is from the third period or below, the octet rule predicts some, but not all, of the formulas for stable molecules.

Compounds with expanded octets occur mainly with elements in the third period and beyond. For example, the Period 3 elements P and S form the known compounds PF_5 and SF_4, but their Period 2 analogs, NF_5 and OF_4, do not exist. Phosphorus as a central atom can violate the octet rule to form PF_5 and PF_6^-.

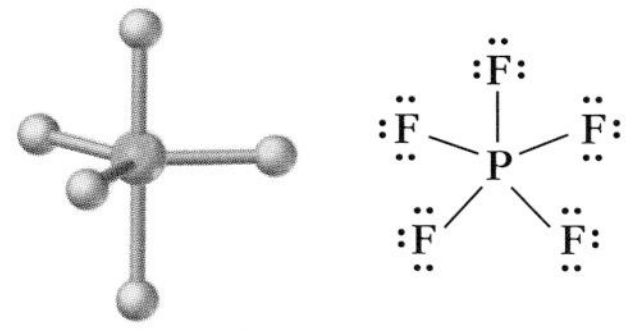

phosphorus pentafluoride

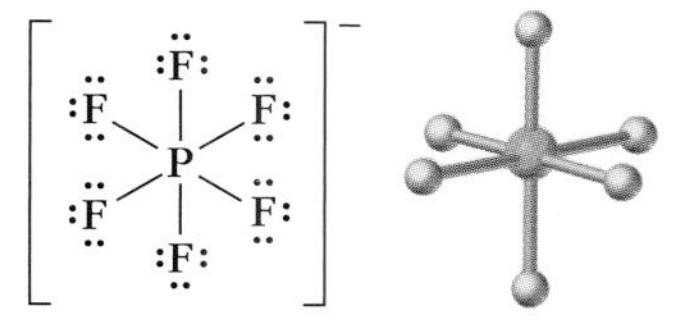

phosphorus hexafluoride ion

© Cengage Learning/Charles D. Winters

Figure 8.8 Paramagnetism of liquid oxygen. Liquid oxygen, the boiling liquid between the poles of the magnet, can be suspended between the poles of a magnet because O_2 is paramagnetic. Paramagnetic substances are attracted into a magnetic field.

Table 8.3 illustrates molecules and ions of central atoms beyond the second period that violate the octet rule.

PROBLEM-SOLVING EXAMPLE 8.10 Exceptions to the Octet Rule

Write the Lewis structure for (a) tellurium tetrabromide, $TeBr_4$; (b) triiodide ion, I_3^-; and (c) boric acid, $B(OH)_3$.

Answer

(a) 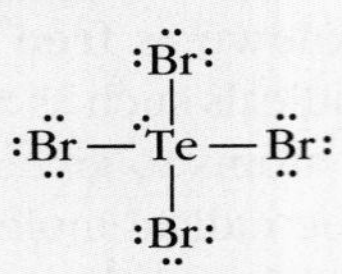

(b) 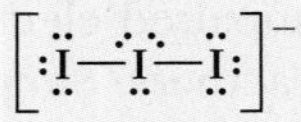

(c) 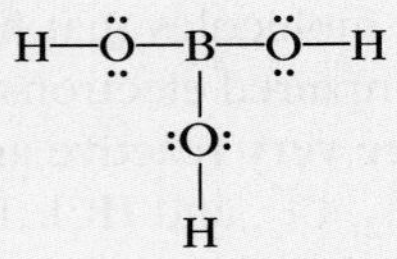

Strategy and Explanation

(a) $TeBr_4$ has 34 valence electrons, eight of which are distributed among four Te—Br bonds. Of the remaining 26 lone pair electrons, 24 complete octets for the Br atoms. The other two electrons form a lone pair on Te, which has a total of ten electrons (five pairs: four shared, one unshared) around it, acceptable for a Period 5 element.

(b) There are 22 valence electrons: seven from each iodine atom and one for the 1− charge on the ion. Forming two I—I single bonds with the central I atom uses two of the 11 electron pairs. Distributing six of the remaining nine electron pairs as lone pairs on terminal I atoms to satisfy the octet rule uses a total of eight electron pairs. The remaining three electron pairs are placed on the central iodine, which can accommodate more than eight electrons because it is from the fifth period.

(c) $B(OH)_3$ uses six of its 24 valence electrons to form three B—O bonds and an additional six electrons for three O—H bonds. The remaining 12 electrons complete two lone pairs on each oxygen atom, giving each oxygen atom an octet. This leaves boron with only three electron pairs, an exception to the octet rule.

Guideline 4 (p. 332) says to place any remaining electrons on the central atom once the terminal atoms all have an octet.

PROBLEM-SOLVING PRACTICE 8.10

Write the Lewis structure for:

(a) BeF_2 (b) ClO_2 (c) PCl_5

(d) BH_2^+ (e) IF_7

Indicate which are exceptions to the octet rule, and why.

Table 8.3 Examples of Lewis Structures with More Than Eight Electrons Around the Central Atom*

	Group 4A	Group 5A	Group 6A	Group 7A	Group 8A
Central atoms with five valence pairs		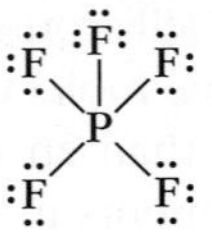	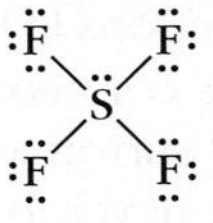	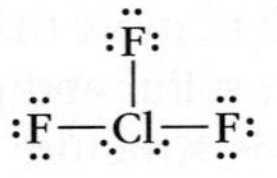	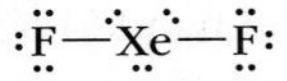
	—	PF_5	SF_4	ClF_3	XeF_2
Bonding pairs	—	5	4	3	2
Lone pairs	—	0	1	2	3
Central atoms with six valence pairs	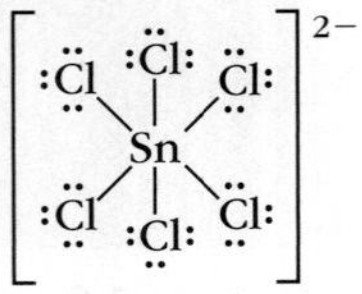	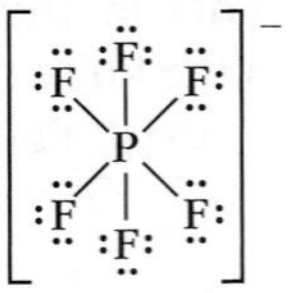	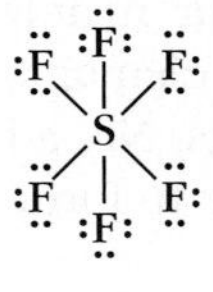	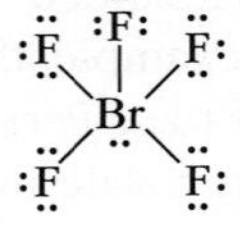	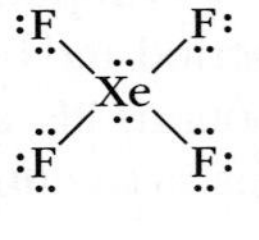
	$SnCl_6^{2-}$	PF_6^-	SF_6	BrF_5	XeF_4
Bonding pairs	6	6	6	5	4
Lone pairs	0	0	0	1	2

*In each case, the numbers of bond pairs and lone pairs about the central atom are given.

8.11 Aromatic Compounds

Benzene, C_6H_6, is an important industrial chemical that is used, along with its derivatives, to manufacture plastics, detergents, pesticides, drugs, and other organic chemicals. It is the simplest member of a very large family of compounds known as **aromatic compounds,** which are compounds containing one or more benzene or benzene-like rings. The word "aromatic" is derived from "aroma," which describes the rather strong and often pleasant odors of these compounds.

As noted earlier (p. 353), the benzene ring can be represented by a hexagon with a circle in the middle. Whenever you see a formula with one or more carbon rings, each with a central circle or alternating double and single bonds throughout, the compound is aromatic, like benzene. Benzaldehyde and toluene are examples of the many aromatic compounds with functional groups or alkyl groups bonded to the aromatic ring. Naphthalene is representative of a large group of aromatic compounds with more than one ring joined by common carbon-carbon bonds. Benzaldehyde is used in synthetic almond and cherry food flavoring, toluene and benzene boost the octane rating of gasoline, and naphthalene is a moth repellant in one kind of moth balls.

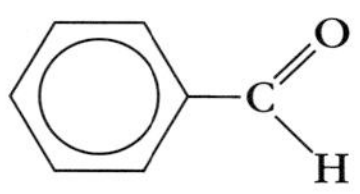

benzaldehyde

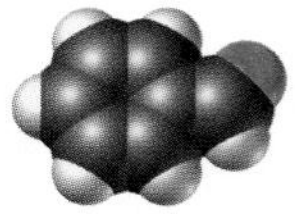

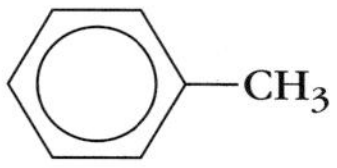

toluene

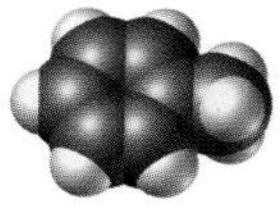

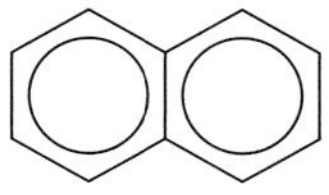
naphthalene

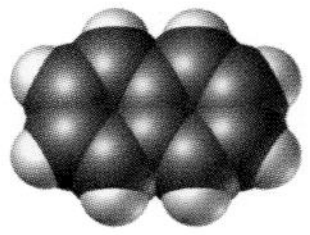

EXERCISE 8.10 Aromatic Compounds

Write the Lewis structures for the resonance structures of toluene.

Constitutional Isomers of Aromatic Compounds

Because benzene is a planar molecule, constitutional isomers are possible when two or more groups (or atoms) are substituted for hydrogen atoms on the benzene ring. If two groups are substituted for two hydrogen atoms on the benzene ring, three constitutional isomers are possible. When the two groups are methyl groups, the compound is called xylene. The prefixes *ortho-*, *meta-*, and *para-* are used to differentiate the three isomers of any disubstituted benzene. These constitutional isomers differ in melting point, boiling point, density, and other physical properties.

Ortho—a prefix indicating that two substituents are on *adjacent* carbon atoms on a benzene ring. *Meta*—two substituents attached to carbon atoms *separated by one carbon atom* on a benzene ring. *Para*—two substituents attached to carbon atoms *separated by two carbon atoms* on a benzene ring.

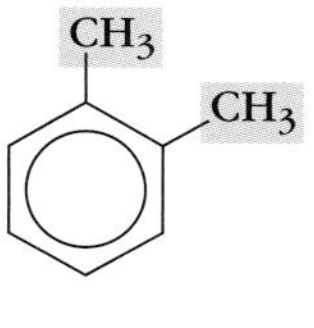

ortho-xylene

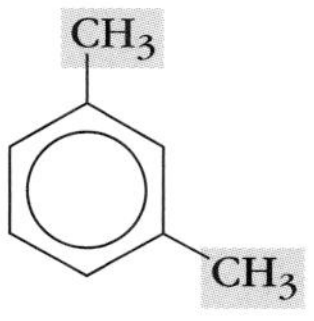

meta-xylene

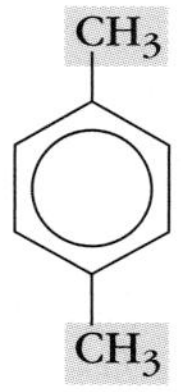

para-xylene

Physical Property	*ortho*-xylene	*meta*-xylene	*para*-xylene
Melting point, °C	−25.2	−47.8	13.2
Boiling point, °C	144.5	139.1	138.4
Density, g/mL	0.876	0.860	0.857

If more than two groups are attached to the benzene ring, numbers must be used to identify them and their positions, as for the three trichlorobenzenes:

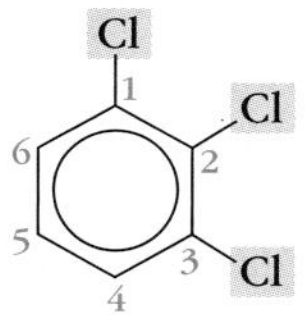

1,2,3-trichlorobenzene

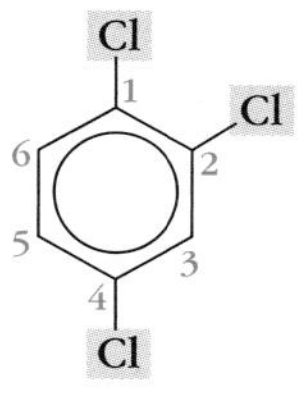

1,2,4-trichlorobenzene

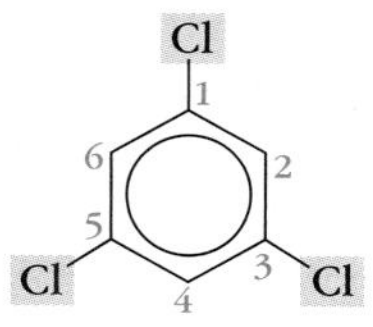

1,3,5-trichlorobenzene

Because there is no other way to attach three chlorine atoms to a benzene ring, only three trichlorobenzenes exist.

EXERCISE 8.11 Constitutional Isomers of Aromatic Compounds

Write the structural formula of 1,2,4-trimethylbenzene. Explain why 1,5,6-trimethylbenzene is not different from 1,2,3-trimethylbenzene

Throughout this chapter, you have seen simple as well as complex structural formulas. The structural formulas of many biochemically active molecules are large and seemingly quite complicated. But, by closely examining the structural formulas, you can recognize the individual parts of the molecule. For example, by looking at the structural formulas of vitamins A and E that follow, you should recognize that the molecules are almost entirely assembled from structural parts that you now understand—rings and chains of carbon atoms, including an aromatic ring, C=C double bonds, and an alcohol group. There is also a six-membered ring containing one oxygen atom in the place of one of the carbon atoms.

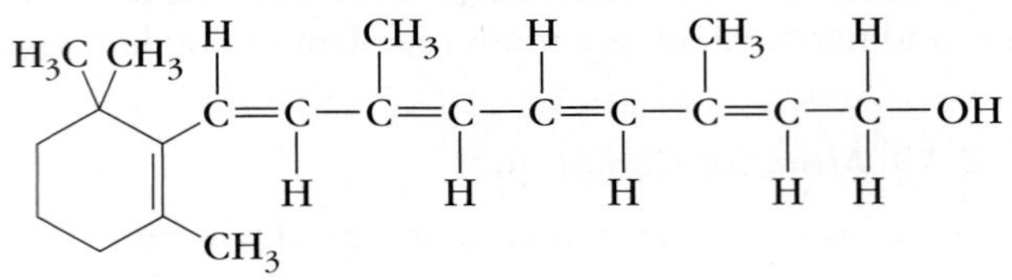

vitamin A

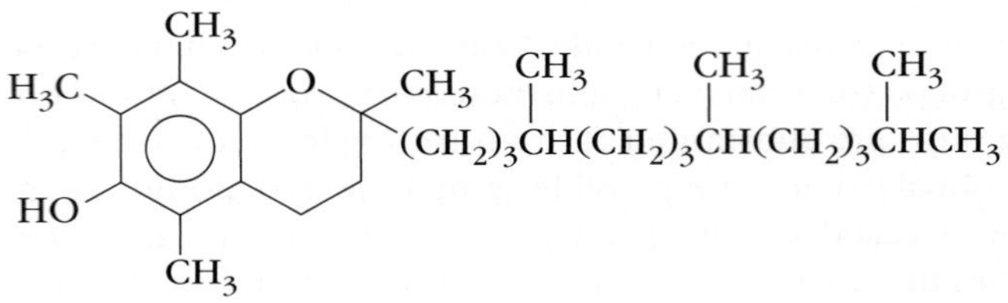

vitamin E

CONCEPTUAL EXERCISE 8.12 Knowing Your Vitamins

1. Use the structural formula for vitamin A to answer these questions.
   (a) How many carbon atoms and how many hydrogen atoms does it contain?
   (b) Locate the carbon atom that is bonded directly to four carbon atoms.
   (c) How many C=C double bonds are there in this molecule?
2. Consider the structural formula for vitamin E.
   (a) What is its molecular formula?
   (b) How many C=C double bonds does it have?
   (c) Which is likely to be the most polar bond in the molecule?

8.12 Molecular Orbital Theory

Earlier in this chapter, we used electron-pair bonds (valence bond theory) to explain bonding in molecules, accounting qualitatively for the stability of the covalent bond in terms of the overlap of atomic orbitals. Where more than one Lewis structure could be drawn for the same arrangement of atomic nuclei, as in SO_2, the concept of resonance (p. 352) was used to account for observed properties of molecules.

A major weakness of the valence bond theory presented earlier in this chapter is that the theory does not always correctly predict the magnetic properties of substances. An important example is O_2, which is paramagnetic. This means that O_2 molecules must have unpaired electrons. Diatomic oxygen has an even number of valence electrons (12), and the octet rule predicts that all of these electrons should be paired. According to valence bond theory, O_2 should be diamagnetic, but experimental measurements show that it is not.

Molecular Orbitals

This discrepancy between experiment and theory for O_2 (and other molecules) can be resolved by using an alternative model of covalent bonding, the **molecular orbital** (MO) approach. *Molecular orbital theory* treats bonding in terms of molecular orbitals that can extend over an entire molecule. Thus, the molecular orbitals are not confined to two atoms at a time. The MO approach involves three basic operations.

Step 1: Combine the valence atomic orbitals of all atoms in the molecule to give a new set of molecular orbitals (MOs) that are characteristic of the molecule as a whole. *The number of MOs formed is equal to the number of atomic orbitals combined.* For example, when two H atoms combine to form H_2, two *s atomic* orbitals, one from each atom, yield two *molecular* orbitals in H_2. For O_2 there are one *s* atomic orbital and three *p* atomic orbitals in the valence shell of each of the two oxygen atoms. This gives eight atomic orbitals, which combine to give eight MOs in O_2.

Step 2: Arrange the MOs in order of increasing energy, as will be illustrated later in this section for second-period elements (Figure 8.11, p. 363). The relative energies of MOs are usually deduced from experiments involving spectral and magnetic properties.

Step 3: Distribute the *valence electrons* of the molecule among the available MOs, filling the lowest-energy MO first and continuing to build up the electron configuration of the molecule in the same way that electron configurations of atoms are built up ***(← p. 296).*** As electrons are added,

(a) *Each MO can hold a maximum of two electrons.* In a filled molecular orbital the two electrons have opposed spins, in accordance with the Pauli exclusion principle ***(← p. 292).***

(b) *Electrons go into the lowest-energy MO available.* A higher-energy molecular orbital starts to fill only when each molecular orbital below it has its quota of two electrons.

(c) *Hund's rule is obeyed.* When two molecular orbitals of equal energy are available to two electrons, one electron goes into each, giving two half-filled molecular orbitals.

Molecular Orbitals for Diatomic Molecules

To illustrate molecular orbital theory, we apply it to the diatomic molecules of the elements in the first two periods of the periodic table.

Hydrogen and Helium (Combining Two 1*s* Atomic Orbitals) When two hydrogen atoms (or two helium atoms) come close together, two 1*s* atomic orbitals overlap and combine to give two MOs. One MO has an energy lower than that of the atomic orbitals from which it was formed; the other MO is of higher energy (Figure 8.9). A molecule that has two electrons in the lower-energy MO is lower in energy (more stable) than the isolated atoms. That lowering of energy corresponds to the bond energy. For that reason the lower-energy MO in Figure 8.9a is called a **bonding molecular orbital.** If electrons are placed in the higher-energy MO, the molecule's

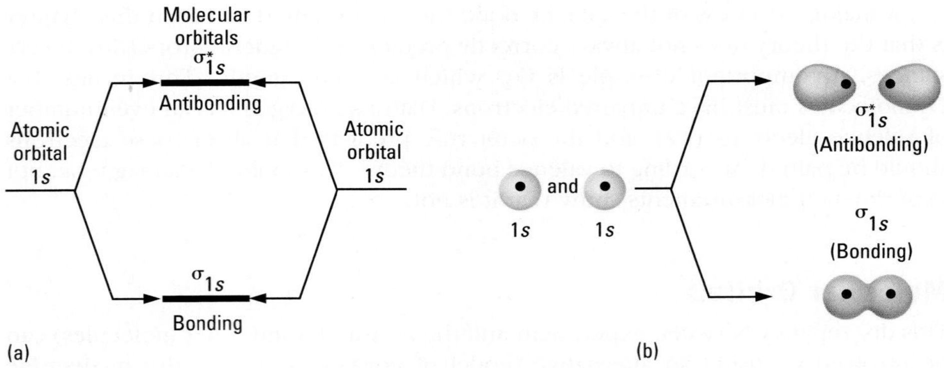

Figure 8.9 Molecular orbital formation. Two MOs are formed by combining two 1*s* atomic orbitals (a) Energy level diagram. (b) Orbital overlap diagram. (Black dots in (b) represent atomic nuclei.)

When two atomic orbitals of equal energy combine, the resulting MOs lie above and below the energy of the atomic orbitals.

energy is higher than the energy of the isolated atoms. This unstable situation is the opposite of bonding; the higher-energy MO is called an **antibonding molecular orbital.**

The electron density in these MOs is shown in Figure 8.9b. The bonding MO has higher electron density between the nuclei than the sum of the individual atomic orbitals. This attracts the nuclei together and forms the bond. In the antibonding orbital, the electron density between the nuclei is smaller than it would have been in the atoms alone. The positively charged nuclei have less electron "glue" to hold them together, they repel each other, and the molecule flies apart—the opposite of a bond.

The electron density in both MOs is symmetrical along a line connecting the two nuclei, which means that both are sigma, σ, orbitals. In MO notation, the 1*s* bonding orbital is designated as σ_{1s}. The antibonding orbital is given the symbol σ^*_{1s}. An asterisk designates an antibonding molecular orbital.

In the H_2 molecule, there are two 1*s* electrons. They fill the σ_{1s} orbital, giving a single bond. If an He_2 molecule could form, there would be four electrons—two from each atom. These would fill the bonding and antibonding orbitals. One bond and one antibond give a *bond order* (number of bonds) of zero in He_2. The bond order is calculated as

$$\text{Bond order} = \text{number of bonds} = \frac{n_B - n_A}{2}$$

where n_B is the number of electrons in bonding molecular orbitals, n_A is the number of electrons in antibonding molecular orbitals, and dividing by 2 accounts for the fact that two electrons are needed for a bond. In H_2, $n_B = 2$ and $n_A = 0$, so we have one bond. For He_2, $n_B = n_A = 2$, so the number of bonds is zero. The MO theory predicts that the He_2 molecule should not exist, and it does not.

Second-Period Elements (Combining 2*s* and 2*p* Atomic Orbitals) Three of the elements in the second period form familiar diatomic molecules: N_2, O_2, and F_2. Less common, but also known, are Li_2, B_2, and C_2, which have been observed as gases. The molecules Be_2 and Ne_2 are either highly unstable or nonexistent. To see what MO theory predicts about the stability of diatomic molecules from the second period, consider the valence atomic orbitals, 2*s* and 2*p*.

Combining two 2*s* atomic orbitals, one from each atom, gives two MOs. These are very similar to the ones shown in Figure 8.9. They are designated as σ_{2s} (sigma, bonding, 2*s*) and σ^*_{2s} (sigma, antibonding, 2*s*).

In an isolated atom, there are three 2*p* atomic orbitals, oriented at right angles to each other. We call these atomic orbitals p_x, p_y, and p_z (Figure 8.10a). Assume that two atoms approach along the *x*-axis. The two p_x atomic orbitals overlap head to head to form two orbitals that are symmetric around the line connecting the two nuclei (Figure 8.10b). That is, they form a sigma bonding orbital, σ_{2p}, and a sigma antibonding orbital, σ^*_{2p}.

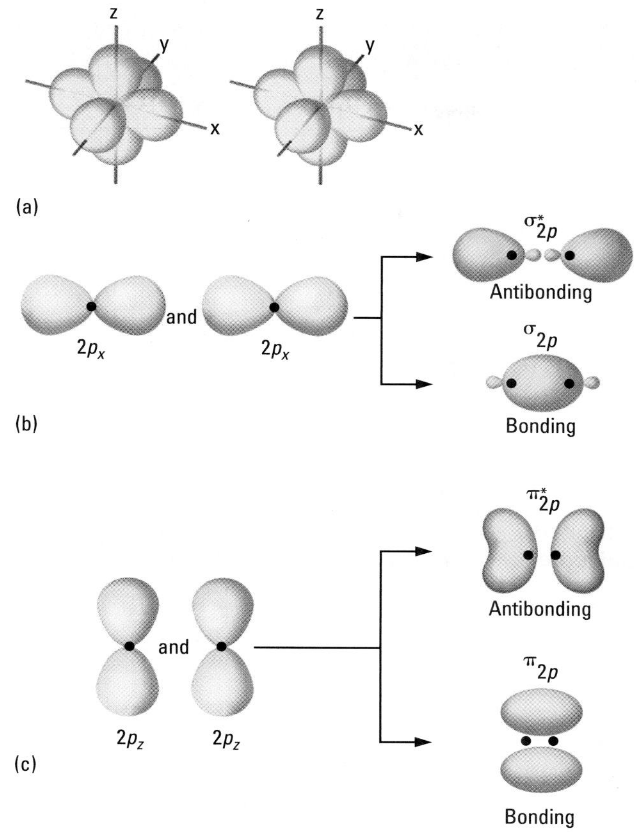

Figure 8.10 Forming molecular orbitals from *p* atomic orbitals. When *p* atomic orbitals from two different atoms (a) overlap, there are two quite different possibilities. If they overlap head to head (b), two σ MOs are produced. If, on the other hand, they overlap side to side (c), two π MOs result. In this example, both p_z and p_y atomic orbitals can overlap to form π MOs. Only p_z overlap is shown. (Black dots represent atomic nuclei.)

The situation is quite different when the p_z atomic orbitals overlap. Because they are oriented parallel to one another, they overlap side to side (Figure 8.10c). The two MOs formed in this case are pi, π, orbitals; one is a bonding MO, π_{2p}, and the other is an antibonding MO, π^*_{2p}. Similarly, the p_y atomic orbitals of the two atoms interact to form another pair of pi MOs, π_{2p} and π^*_{2p}, which are not shown in Figure 8.10.

If you look down the internuclear axis, a π molecular orbital looks like a *p* atomic orbital.

The relative energies of the MOs available for occupancy by the valence electrons of diatomic molecules formed from second-period atoms are shown in Figure 8.11. To obtain the MO structure of the diatomic molecules of the elements in the second period, we fill the available MOs in order of increasing energy. This order of energies applies to Li_2 through N_2, but for O_2 and F_2 the energy of the σ_{2p} orbital drops below the energy of the two π_{2p} orbitals. This does not affect the filling order, because in O_2 and F_2 the σ_{2p} and π_{2p} orbitals are filled. The results are shown in Table 8.4. Note that MO theory correctly predicts the number of unpaired electrons in each molecule and the bonding in the molecule.

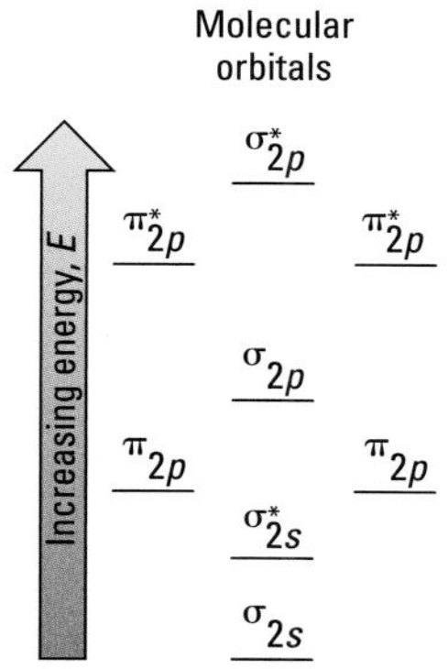

Figure 8.11 Relative order of filling molecular orbitals. The energies of MOs formed by combining 2*s* and 2*p* atomic orbitals increase from bottom to top of the diagram.

PROBLEM-SOLVING EXAMPLE 8.11 Filling Molecular Orbitals

Without consulting Table 8.4, apply the energy diagram in Figure 8.11 to C_2 and fill the molecular orbitals appropriately. Verify that the results correspond to the number of unpaired electrons and the bond order listed for C_2 in Table 8.4.

Answer See Table 8.4.

Table 8.4 Predicted and Observed Properties of Diatomic Molecules of Second-Period Elements

	Occupancy of Orbitals							
	σ_{2s}	σ^*_{2p}	π_{2p}	π_{2p}	σ_{2p}	π^*_{2p}	π^*_{2p}	σ^*_{2p}
Li_2	(↑↓)	( )	( )	( )	( )	( )	( )	( )
Be_2	(↑↓)	(↑↓)	( )	( )	( )	( )	( )	( )
B_2	(↑↓)	(↑↓)	(↑)	(↑)	( )	( )	( )	( )
C_2	(↑↓)	(↑↓)	(↑↓)	(↑↓)	( )	( )	( )	( )
N_2	(↑↓)	(↑↓)	(↑↓)	(↑↓)	(↑↓)	( )	( )	( )
O_2	(↑↓)	(↑↓)	(↑↓)	(↑↓)	(↑↓)	(↑)	(↑)	( )
F_2	(↑↓)	(↑↓)	(↑↓)	(↑↓)	(↑↓)	(↑↓)	(↑↓)	( )
Ne_2	(↑↓)	(↑↓)	(↑↓)	(↑↓)	(↑↓)	(↑↓)	(↑↓)	(↑↓)

	Predicted Properties		Observed Properties	
	Number of Unpaired e^-	Bond Order	Number of Unpaired e^-	Bond Energy (kJ/mol)
Li_2	0	1	0	105
Be_2	0	0	0	Unstable
B_2	2	1	2	289
C_2	0	2	0	598
N_2	0	3	0	946
O_2	2	2	2	498
F_2	0	1	0	158
Ne_2	0	0	0	Nonexistent

Strategy and Explanation First, determine the number of valence electrons and then use the MO diagram in Figure 8.11 to fill the MOs in proper order of increasing energy. There are eight valence electrons, four from each carbon atom. The correct MO diagram is given in Table 8.4. The correct bond order is 2, as calculated from $(6 - 2)/2 = 2$. There are no unpaired electrons.

PROBLEM-SOLVING PRACTICE 8.11

Repeat Problem-Solving Example 8.11, but use N_2 instead of C_2.

There is also a general correlation between the predicted bond order, $(n_B - n_A)/2$, and the experimental bond energy. The bond order of two for C_2 and O_2 implies a double bond, which is expected to be stronger than the single bonds in Li_2, B_2, and F_2 (bond order = 1). The bond order of three for N_2 (triple bond) implies a still stronger bond. A major triumph of MO theory is its ability to explain the properties of O_2. The bond order of two and the presence of two unpaired electrons in the two π^*_{2p} MOs explain how the molecule can have a double bond and at the same time be paramagnetic.

PROBLEM-SOLVING EXAMPLE 8.12 Electron Structure of Peroxide Ion

Using MO theory, predict the bond order and number of unpaired electrons in the peroxide ion, O_2^{2-}.

Answer Bond order = 1; there are no unpaired electrons.

Strategy and Explanation First, find the number of valence electrons. Then, construct a molecular orbital diagram, filling the available MOs (Figure 8.11) in order of increasing energy.

Recall that oxygen is in Group 6A of the periodic table, so an oxygen atom has six valence electrons. The peroxide ion has two oxygen atoms and two extra electrons (2− charge). The total number of valence electrons is 2(6) + 2 = 14. The molecular orbital diagram is

	σ_{2s}	σ^*_{2s}	π_{2p}	π_{2p}	σ_{2p}	π^*_{2p}	π^*_{2p}	σ^*_{2p}
O_2^{2-}	(↑↓)	(↑↓)	(↑↓)	(↑↓)	(↑↓)	(↑↓)	(↑↓)	( )

The bond order is (8 − 6)/2 = 1. There are no unpaired electrons. These conclusions are in agreement with the Lewis structure (valence bond theory) for the peroxide ion: $(:\ddot{\underset{..}{O}}-\ddot{\underset{..}{O}}:)^{2-}$

PROBLEM-SOLVING PRACTICE **8.12**

Use MO theory to predict the bond order and the number of unpaired electrons in the superoxide ion, O_2^-.

Molecular orbital theory can also be applied to predict the properties of diatomic molecules containing different kinds of atoms, such as NO. There are 11 valence electrons in this molecule (five from nitrogen and six from oxygen). These 11 valence electrons are placed into molecular orbitals of increasing energy in the molecular orbital diagram:

	σ_{2s}	σ^*_{2s}	π_{2p}	π_{2p}	σ_{2p}	π^*_{2p}	π^*_{2p}	σ^*_{2p}
NO	(↑↓)	(↑↓)	(↑↓)	(↑↓)	(↑↓)	(↑)	( )	( )

There are eight electrons in bonding orbitals (two in σ_{2s}, four in π_{2p}, two in σ_{2p}) and three electrons in antibonding orbitals (two in σ^*_{2s}, one in π^*_{2p}). Therefore there is one unpaired electron (in the π^*_{2p} molecular orbital). The bond order is

$$\text{Number of bonds} = \frac{8-3}{2} = \frac{5}{2} = 2.5$$

Polyatomic Molecules; Delocalized π Electrons

The bonding in molecules containing more than two atoms can also be described in terms of MOs. We will not attempt to do this; the energy level structure is considerably more complex than that for diatomic molecules. However, one point is worth mentioning: In polyatomic species, *the MOs can be spread over the entire molecule* rather than being localized between two atoms.

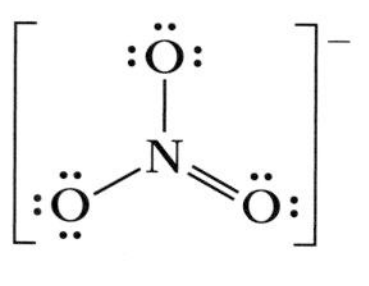

The nitrate ion, NO_3^-, is known from experiment to be triangular planar, with three N—O bonds whose lengths are equal. However, a single Lewis structure for nitrate ion does not predict three equal-length bonds. Valence bond theory invokes three contributing resonance structures to explain the fact that the three N—O bonds are identical. In each contributing structure there is a double bond from N to a different O atom. MO theory, on the other hand, considers that the pi electron pair (the second pair in the double bond) is in a pi molecular orbital that is delocalized; it is shared by all of the atoms in the ion. That is, this pair of electrons is shared equally among the three O atoms. According to MO theory, a similar interpretation applies with all of the resonance hybrids described in this chapter.

Ozone, O_3, is a molecule for which two resonance structures can be written (p. 353), each of which contains two sigma bonds and one pi bond. The sigma bonds are localized electron pairs between adjacent oxygen atoms. The pi electrons

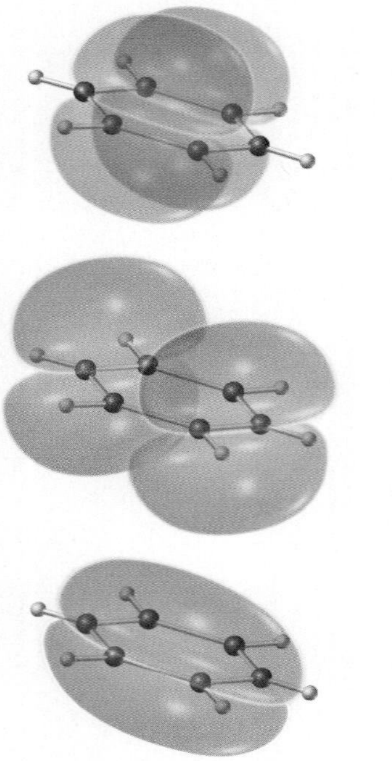

Figure 8.12 Electron delocalization in benzene. Three of the 15 valence electron pairs in the benzene molecule occupy three π MOs that are delocalized in regions above and below the plane of the benzene ring. These three electron pairs in π MOs add 0.5 to the bond order between each pair of carbon atoms, giving a total of $6 \times 0.5 = 3$ additional bonds.

can be considered to be delocalized over the entire three-oxygen chain of the molecule.

Benzene, C_6H_6, is another species in which delocalized pi molecular orbitals play an important role. There are 30 valence electrons—15 electron pairs—in this molecule; 12 of the electron pairs form the sigma-bond framework:

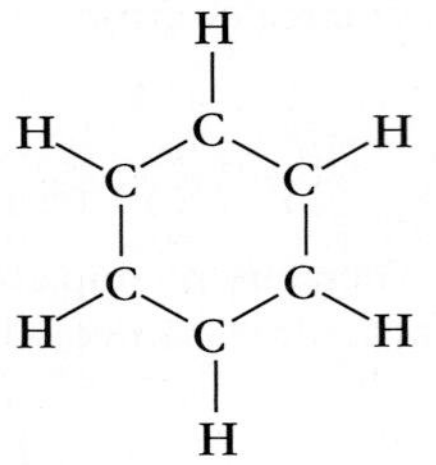

The remaining six electrons—three electron pairs—are located in three π orbitals, which according to MO theory extend over more than two atoms. Figure 8.12 shows the three π MOs. These delocalized electrons are spread over all six C atoms and are represented simply as

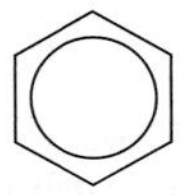

SUMMARY PROBLEM

Glycine, $C_2H_5NO_2$, is a naturally occurring amino acid with this Lewis structure,

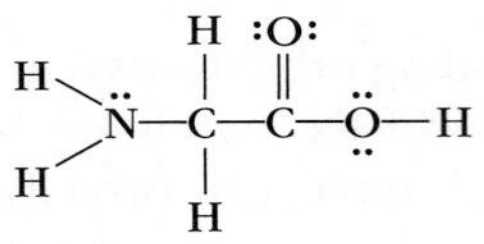

that contains the amine functional group, —NH_2, and a carboxylic acid functional group, —COOH.

(a) List the C—H, O—H, C—N, N—H, C—C, C—O, and C═O bonds in glycine in order of increasing bond length.

(b) List the C—H, O—H, C—N, C—C, C—O, and C═O bonds in glycine in order of increasing bond strength.

(c) Arrange these bonds in glycine in order of increasing bond polarity: C—N, C—O, C—H, N—H, O—H, and C—C in order of increasing bond polarity.

(d) The ionization of glycine in solution forms H^+(aq) and aqueous glycinate ions. Write the structural formula for glycinate ion.

(e) Glycinate ion is an example of a resonance hybrid. Write two Lewis structures that contribute to this resonance hybrid.

(f) Glycine also forms a zwitterion. In the zwitterion an H^+ is released by the carboxylic acid group and attaches to the nitrogen of the amine group. Write the Lewis structure of the zwitterion form of glycine.

(g) Using the Lewis structure drawn for part (f), calculate the formal charges on each of its atoms. What is the net charge of the zwitterion?

IN CLOSING

Having studied this chapter, you should be able to . . .

- Recognize the different types of covalent bonding (Sections 8.1–8.3 and 8.9).
- Use Lewis structures to represent covalent bonds in molecules and polyatomic ions (Sections 8.1–8.3, 8.8, 8.10). End-of-chapter questions: 14, 18
- Describe multiple bonds in alkenes and alkynes (Section 8.5). Question 26
- Recognize molecules that can have *cis-trans* isomerism (Section 8.5). Questions 28, 30
- Predict bond lengths from periodic trends in atomic radii (Section 8.6). Questions 34, 38, 82
- Relate bond energy to bond length (Section 8.6). Questions 82, 86
- Use bond enthalpies to calculate the enthalpy of a reaction (Section 8.6). Question 42
- Predict bond polarity from electronegativity trends (Section 8.7). Questions 46, 48, 78
- Use formal charges to compare Lewis structures (8.8). Questions 52, 56
- Use resonance structures to model multiple bonding in molecules and polyatomic ions (Section 8.9). Questions 60, 65
- Recognize three types of exceptions to the octet rule (Section 8.10). Question 67
- Describe bonding and constitutional isomerism in aromatic compounds (Section 8.11).
- Use molecular orbital theory to explain bonding in diatomic molecules and ions (Section 8.12). Question 75

OWL and go Chemistry

Sign in at **www.cengage.com/owl** to:

- View tutorials and simulations, develop problem-solving skills, and complete online homework assigned by your professor.
- For quick review and exam prep, download Go Chemistry mini lecture modules from OWL (or purchase them at **www.CengageBrain.com**).

KEY TERMS

alkenes *(Section 8.5)*

alkynes *(8.5)*

aromatic compounds *(8.11)*

antibonding molecular orbital *(8.12)*

bond enthalpy (bond energy) *(8.6)*

bond length *(8.6)*

bonding electrons *(8.2)*

bonding molecular orbital *(8.12)*

cis **isomer** *(8.5)*

cis-trans **isomerism** *(8.5)*

covalent bond *(8.1)*

double bond *(8.4)*

electronegativity *(8.7)*

formal charge *(8.8)*

free radicals *(8.10)*

functional group *(8.3)*

Lewis structure *(8.2)*

lone pair electrons *(8.2)*

molecular orbital *(8.12)*

multiple covalent bonds *(8.4)*

nonpolar covalent bond *(8.7)*

octet rule *(8.2)*

polar covalent bond *(8.7)*

resonance hybrid *(8.9)*

resonance structures *(8.9)*

saturated hydrocarbons *(8.3)*

single covalent bond *(8.2)*

trans **isomer** *(8.5)*

triple bond *(8.4)*

unsaturated hydrocarbons *(8.5)*

QUESTIONS FOR REVIEW AND THOUGHT

OWL Interactive versions of these problems are assignable in OWL.

Blue-numbered questions have short answers at the back of this book in Appendix M and fully worked solutions in the *Student Solutions Manual.*

Review Questions

These questions test vocabulary and simple concepts.

1. Explain the difference between an ionic bond and a covalent bond.
2. What kind of bonding (ionic or covalent) would you predict for the products resulting from the following combinations of elements?
   (a) Na + I_2 (b) C + S_8
   (c) Mg + Br_2 (d) P_4 + Cl_2
3. What characteristics must atoms A and X have if they are able to form a covalent bond A—X with each other? A polar covalent bond with each other?
4. Indicate how alkanes, alkenes, and alkynes differ by giving the structural formula of a compound in each class that contains three carbon atoms.
5. Refer to Table 8.3 and answer these questions:
   (a) Do any molecules with more than eight valence electrons have a second-period element as the central atom?
   (b) What is the maximum number of bond pairs and lone pairs that surround the central atom in any of these molecules?
6. While sulfur forms the compounds SF_4 and SF_6, no equivalent compounds of oxygen, OF_4 and OF_6, are known. Explain.
7. Which of these molecules have an odd number of valence electrons: NO_2, SCl_2, NH_3, NO_3?
8. Write resonance structures for NO_2^-. Predict a value for the N—O bond length based on bond lengths given in Table 8.1, and explain your answer.
9. Consider these structures for the formate ion, HCO_2^-. Designate which two are resonance structures and which is equivalent to one of the resonance structures.

(a)
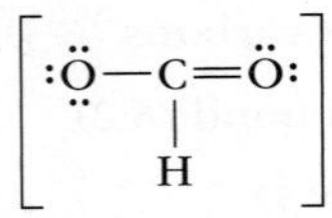
(b)
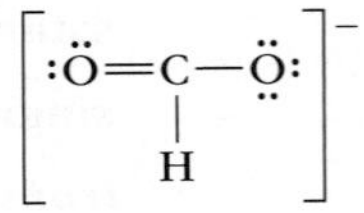
(c)
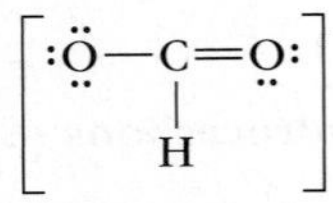

10. Consider a series of molecules in which the C atom is bonded to atoms of second-period elements: C—O, C—F, C—N, C—C, and C—B. Place these bonds in order of increasing bond length.
11. What are the trends in bond length and bond energy for a series of related bonds—for instance, single, double, and triple carbon-to-oxygen bonds?
12. Why is *cis-trans* isomerism not possible for alkynes?

Topical Questions

These questions are keyed to the major topics in the chapter. Usually a question that is answered at the back of this book is paired with a similar one that is not.

Lewis Structures (Sections 8.2, 8.4)

13. Write the Lewis structures of (a) dichlorine monoxide, Cl_2O; (b) hydrogen peroxide, H_2O_2; (c) borohydride ion, BH_4^-; (d) phosphonium ion, PH_4^+; and (e) PCl_5.
14. Write Lewis structures for these molecules or ions.
   (a) ClF (b) H_2Se
   (c) BF_4^- (d) PO_4^{3-}
15. Write Lewis structures for these molecules.
   (a) $CHClF_2$, one of several chlorofluorocarbons that has been used in refrigeration
   (b) Methyl alcohol, CH_3OH
   (c) Methyl amine, CH_3NH_2
16. Write Lewis structures for these molecules or ions.
   (a) CH_3Cl (b) SiO_4^{4-}
   (c) ClF_4^+ (d) C_2H_6
17. Write Lewis structures for these molecules.
   (a) Formic acid, HCOOH, in which atomic arrangement is

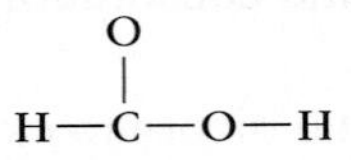

   (b) Acetonitrile, CH_3CN
   (c) Vinyl chloride, CH_2CHCl, the molecule from which PVC plastics are made
18. Write Lewis structures for these molecules.
   (a) Tetrafluoroethylene, C_2F_4, the molecule from which Teflon is made
   (b) Acrylonitrile, CH_2CHCN, the molecule from which Orlon is made
19. Write Lewis structures for:
   (a) N_2^+ (b) XeF_7^- (c) tetracyanoethane, C_6N_4
20. Write Lewis structures for:
   (a) Disulfur dichloride, S_2Cl_2
   (b) Nitrous acid, HNO_2, which has no oxygen-to-oxygen bonds
   (c) ClF_6^+
21. Which of these are correct Lewis structures and which are incorrect? Explain what is wrong with the incorrect ones.

(a)
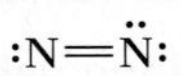
N_2

(b)
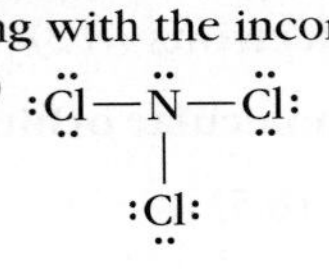
NCl_3

(c)
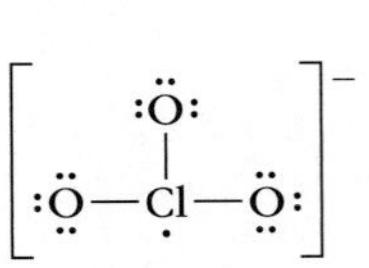
ClO_3^-

(d)
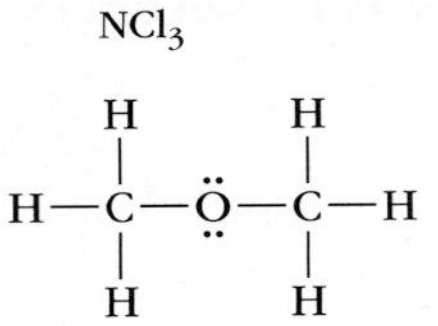
$(CH_3)_2O$

(e)
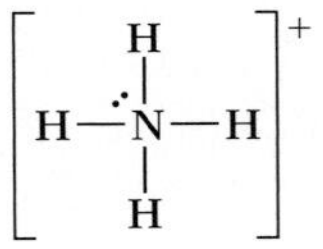
NH_4^+

22. Which of these are correct Lewis structures and which are incorrect? Explain what is wrong with the incorrect ones.

(a) 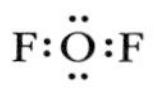

OF_2

(b) 

O_2

(c) 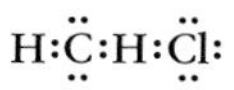

CH_3Cl

(d) 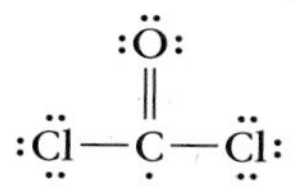

CCl_2O

(e) 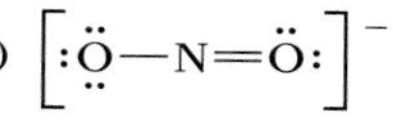

NO_2^-

Bonding in Hydrocarbons (Sections 8.3, 8.5)

23. Write the structural formulas for all the branched-chain compounds with the molecular formula C_4H_{10}.
24. Write the structural formulas for all the branched-chain compounds with the molecular formula C_6H_{14}.
25. Write structural formulas for two straight-chain alkenes with the formula C_5H_{10}. Are these the only two structures that meet these specifications?
26. From their molecular formulas, classify each of these straight-chain hydrocarbons as an alkane, an alkene, or an alkyne.
(a) C_5H_8 (b) $C_{24}H_{50}$ (c) C_7H_{14}
27. From their molecular formulas, classify each of these straight-chain hydrocarbons as an alkane, an alkene, or an alkyne.
(a) $C_{21}H_{44}$ (b) C_4H_6 (c) C_8H_{16}
28. Write the structural formulas for the *cis* and *trans* isomers of 2-pentene.
29. The structural formulas are given for these *cis* or *trans* alkenes.

(a) 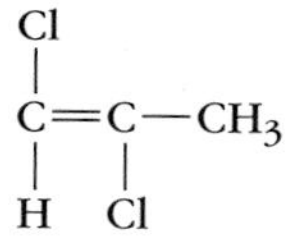 *trans*-1,2-dichloropropene

(b) 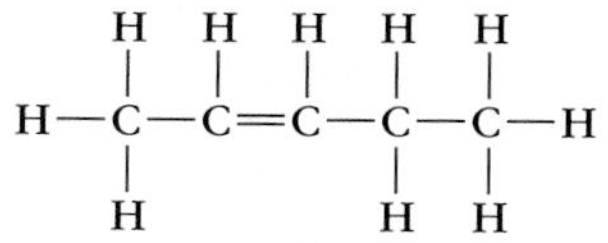 *cis*-2-pentene

(c) 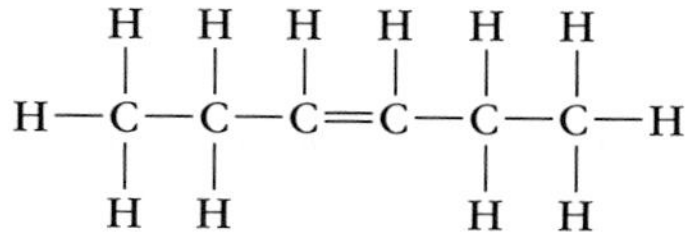 *cis*-3-hexene

(d) 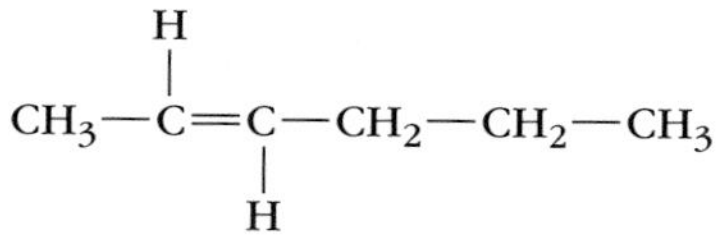 *trans*-2-hexene

Write the structural formula for
(a) *cis*-1,2-dichloropropene (b) *trans*-2-pentene
(c) *trans*-3-hexene (d) *cis*-2-hexene
30. In each case, tell whether *cis* and *trans* isomers exist. If they do, write structural formulas for the two isomers and label each *cis* or *trans*.
(a) Br_2CH_2
(b) $CH_3CH_2CH{=}CHCH_2CH_3$
(c) $CH_3CH{=}CHCH_3$
(d) $CH_2{=}CHCH_2CH_3$
31. Which of these molecules can have *cis* and *trans* isomers? For those that do, write the structural formulas of the two isomers and label each *cis* or *trans*. For those that cannot have these isomers, explain why.
(a) $CH_3CH_2BrC{=}CBrCH_3$
(b) $(CH_3)_2C{=}C(CH_3)_2$
(c) $CH_3CH_2IC{=}ClCH_2CH_3$
(d) $CH_3ClC{=}CHCH_3$
(e) $(CH_3)_2C{=}CHCH_3$
32. Oxalic acid has this structural formula.

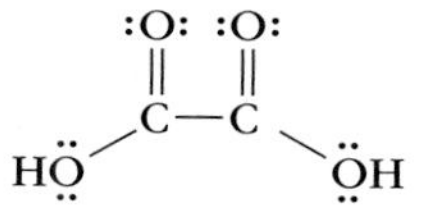

Is *cis-trans* isomerism possible for oxalic acid? Explain your answer.
33. 2-methyl propene has this Lewis structure.

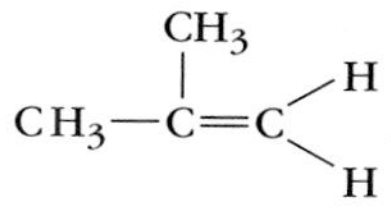

Is *cis-trans* isomerism possible for this molecule? Explain your answer.

Bond Properties (Sections 8.6, 8.7)

34. For each pair of bonds, predict which is the shorter.
(a) B—Cl or Ga—Cl (b) C—O or Sn—O
(c) P—S or P—O
(d) The C=C or the C=O bond in acrolein

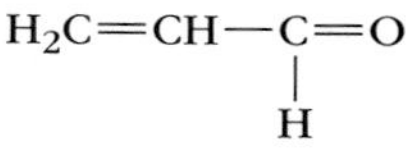

35. For each pair of bonds, predict which is the shorter.
(a) Si—N or P—O (b) Si—O or C—O
(c) C—F or C—Br
(d) The C=C or the C≡N bond in acrylonitrile, $H_2C{=}CH{-}C{\equiv}N$
36. Using only a periodic table (not a table of bond energies), decide which of these is likely to be the strongest bond.
(a) Si—F (b) P—S (c) Si—O
37. Compare the nitrogen-nitrogen bonds in hydrazine, N_2H_4, and in "laughing gas," N_2O. In which molecule is the nitrogen-nitrogen bond shorter? In which should the nitrogen-nitrogen bond be stronger?
38. Consider the carbon-oxygen bonds in formaldehyde, H_2CO, and in carbon monoxide, CO. In which molecule is the carbon-oxygen bond shorter?
39. Which bond will require more energy to break, the carbon-oxygen bond in formaldehyde, H_2CO, or the CO bond in carbon monoxide, CO?
40. Compare the carbon-oxygen bond lengths in the formate ion, HCO_2^-, and in the carbonate ion, CO_3^{2-}. In which ion is the bond longer? Explain briefly.
41. Compare the nitrogen-oxygen bond lengths in NO_2^+ and in NO_3^-. In which ion are the bonds longer? Explain briefly.

Bond Energies and Enthalpy Changes (Section 8.6)

42. Estimate $\Delta H°$ for forming 2 mol ammonia from molecular nitrogen and molecular hydrogen. Is this reaction exothermic or endothermic? (N_2 has a triple bond.)
43. Estimate $\Delta H°$ for the conversion of 1 mol carbon monoxide to carbon dioxide by combination with molecular oxygen. Is this reaction exothermic or endothermic? (CO has a triple bond.)
44. Which of the four molecules HF, HCl, HBr, and HI has the strongest chemical bond? Using bond energies, estimate $\Delta H°$ for the reaction of molecular hydrogen with each of the gaseous molecular halogens: fluorine, chlorine, bromine, and iodine. Which is the most exothermic reaction?

Electronegativity and Bond Polarity (Section 8.7)

45. For each pair of bonds, indicate the more polar bond and use $\delta+$ or $\delta-$ to show the partial charge on each atom.
    (a) C—O and C—N (b) B—O and P—S
    (c) P—H and P—N (d) B—H and B—Cl
46. Given the bonds C—N, C—H, C—Br, and S—O,
    (a) Which atom in each is the more electronegative?
    (b) Which of these bonds is the most polar?
47. For each pair of bonds, identify the more polar one and use $\delta+$ or $\delta-$ to indicate the partial charge on each atom.
    (a) B—Cl and B—O (b) O—F and O—Se
    (c) S—Cl and B—F (d) N—H and N—F
48. The molecule below is urea, a compound used in plastics and fertilizers.

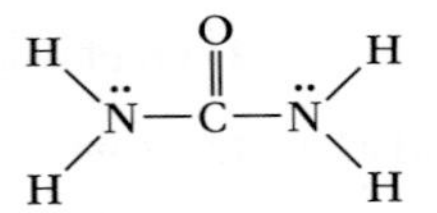

urea

    (a) Which bonds in the molecule are polar and which are nonpolar?
    (b) Which is the most polar bond in the molecule? Which atom is the partial negative end of this bond?
49. The molecule below is acrolein, the starting material for certain plastics.

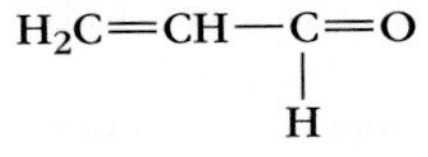

acrolein

    (a) Which bonds in the molecule are polar and which are nonpolar?
    (b) Which is the most polar bond in the molecule? Which atom is the partial negative end of this bond?

Formal Charge (Section 8.8)

50. What is the total formal charge of a molecule? Of an ion?
51. What is the relationship between formal charge and electronegativity?
52. Write correct Lewis structures and assign a formal charge to each atom.
    (a) SO_3
    (b) C_2N_2 (atoms bonded in the order NCCN)
    (c) NO_2^-
53. Write correct Lewis structures and assign a formal charge to each atom.
    (a) OCS
    (b) HNC (atoms bonded in that order)
    (c) CH_3^-
54. Write correct Lewis structures and assign a formal charge to each atom.
    (a) CH_3CHO (b) N_3^- (c) CH_3CN
55. Write correct Lewis structures and assign a formal charge to each atom.
    (a) KrF_4 (b) ClO_3^- (c) SO_2Cl_2
56. Write the correct Lewis structure and assign a formal charge to each atom in fulminate ion, $(CNO)^-$.
57. Peroxydisulfate ion, $S_2O_8^{2-}$, contains an —O—O— linkage between two sulfate ions, SO_4^{2-}. Write the correct Lewis structures and assign a formal charge to each atom in the peroxydisulfate and sulfate ions.
58. Two Lewis structures can be written for nitrosyl chloride, which contains one nitrogen, one oxygen, and one chlorine atom per molecule. Write the two Lewis structures and assign a formal charge to each atom.
59. Two Lewis structures can be written for nitrosyl fluoride, which contains one nitrogen, one oxygen, and one fluorine atom per molecule. Write the two Lewis structures and assign a formal charge to each atom.

Resonance (Section 8.9)

60. These have two or more resonance structures. Write all the resonance structures for each.
    (a) Nitric acid

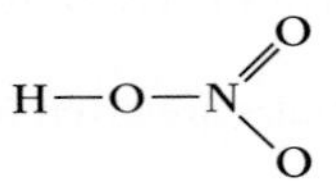

    (b) Nitrate ion, NO_3^-
61. These have two or more resonance structures. Write all the resonance structures for:
    (a) SO_3 (b) SCN^-
62. Several Lewis structures can be written for perbromate ion, BrO_4^-, the central Br with all single Br—O bonds, or with one, two, or three Br=O double bonds. Draw the Lewis structures of these possible resonance forms, and use formal charges to predict the most plausible one.
63. Use formal charges to predict which of the resonance forms is most plausible for
    (a) SO_3 and (b) HNO_3 (See Questions 60 and 61.)
64. Write the resonance structures for adenine, which is the nitrogen-containing organic base portion of ATP (adenosine triphosphate).

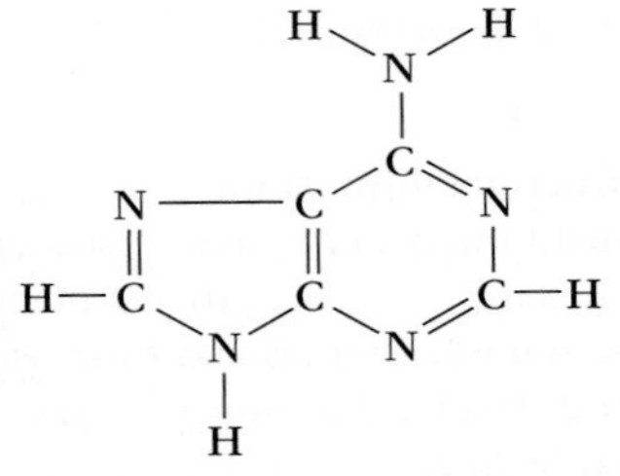

65. Several Lewis structures can be written for thiosulfate ion, $S_2O_3^{2-}$. Write the Lewis structures of these possible resonance forms.
66. Use formal charges to predict the most plausible Lewis structure for thiosulfate ion, $S_2O_3^{2-}$. (See Question 65.)

Exceptions to the Octet Rule (Section 8.10)

67. Write the Lewis structure for:
(a) BrF_5 (b) IF_5 (c) IBr_2^-

68. Write the Lewis structure for:
(a) BrF_3 (b) I_3^- (c) XeF_4

69. Which of these atoms can form molecules with five or six pairs of valence electrons surrounding the atom?
(a) C (b) P (c) O
(d) F (e) Cl (f) B
(g) Se (h) Sn

Aromatic Compounds (Section 8.11)

70. Carbon-to-carbon double bonds, C=C, react by addition. Cite experimental evidence that benzene does not have C=C bonds.

71. All carbon-to-carbon bond lengths are identical in benzene. Does this argue for or against the presence of C=C bonds in benzene? Explain.

72. Three dibromobenzenes are known. Write the Lewis structure and name for each compound.

73. The structural formula of anthracene is

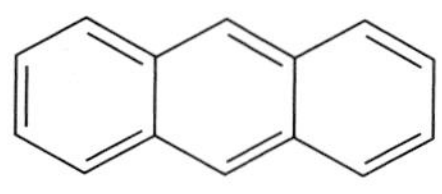

anthracene

What is its molecular formula?

74. Write the structural formula for 1,2-diiodobenzene (also known as *ortho*-diiodobenzene). Write the structural formulas for the *meta* and *para* isomers as well.

Molecular Orbital Theory (Section 8.12)

75. Use MO theory to predict the MO diagram, the number of bonds, and the number of unpaired electrons in the nitrosyl ion, NO^+.

76. Use MO theory to predict the number of electrons in each of the 2*s* and 2*p* molecular orbitals, the number of bonds, and the number of unpaired electrons in:
(a) CO (b) F_2^- (c) NO^-

77. Use MO theory to predict the number of bonds and the number of unpaired electrons in each of these ions:
(a) B_2^+ (b) Li_2^+ (c) O_2^+

General Questions

These questions are not explicitly keyed to chapter topics; many require integration of several concepts.

78. Using just a periodic table (not a table of electronegativities), decide which of these is likely to be the most polar bond. Explain your answer.
(a) C—F (b) S—F
(c) Si—F (d) O—F

79. The C—Br bond length in CBr_4 is 191 pm; the Br—Br distance in Br_2 is 228 pm. Estimate the radius of a C atom in CBr_4. Use this value to estimate the C—C distance in ethane, $H_3C—CH_3$. How does your calculated bond length agree with the measured value of 154 pm? Are radii of atoms exactly the same in every molecule?

80. Is it a good generalization that elements that are close together in the periodic table form covalent bonds, whereas elements that are far apart form ionic bonds? Why or why not?

81. Write Lewis structures for:
(a) CS_2 (b) NH_2OH
(c) S_2Cl_2 (contains S—S bond) (d) NS_2^+

82. The molecule pictured below is acrylonitrile, the building block of the synthetic fiber Orlon.

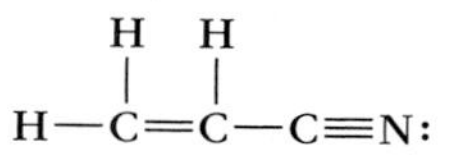

(a) Which is the shorter carbon-carbon bond?
(b) Which is the stronger carbon-carbon bond?
(c) Which is the most polar bond and what is the partial negative end of the bond?

83. In nitryl chloride, NO_2Cl, there is no oxygen-oxygen bond. Write a Lewis structure for the molecule. Write any resonance structures for this molecule.

84. Write Lewis structures for:
(a) SCl_2 (b) Cl_3^+
(c) $ClOClO_3$ (contains Cl—O—Cl bond) (d) $SOCl_2$

85. Arrange these bonds in order of increasing length (shortest first). List all the factors responsible for each placement: O—H, O—O, Cl—O, O=O, O=C.

86. List the bonds in Question 85 in order of increasing bond *strength*.

87. Chlorine trifluoride, ClF_3, is one of the most reactive compounds known. Write the Lewis structure for ClF_3.

88. Judging from the number of carbon and hydrogen atoms in their formulas, which of these formulas represent alkanes? Which are likely aromatic? Which fall into neither category? (It may help to write structural formulas.)
(a) C_8H_{10} (b) $C_{10}H_8$ (c) C_6H_{12}
(d) C_6H_{14} (e) C_8H_{18} (f) C_6H_{10}

Applying Concepts

These questions test conceptual learning.

89. A student drew this incorrect Lewis structure for ClO_3^-. What errors were made when determining the number of valence electrons?

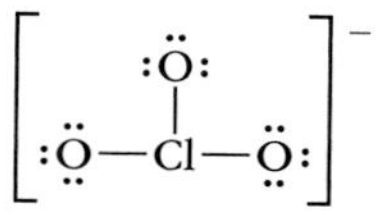

90. This Lewis structure for SF_5^+ is drawn incorrectly. What error was made when determining the number of valence electrons?

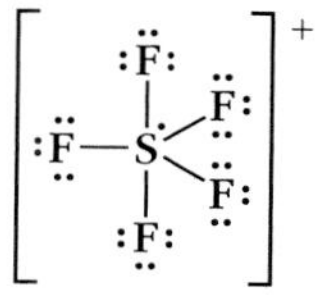

91. Tribromide, Br_3^-, and triiodide, I_3^-, ions exist, but trifluoride ion, F_3^-, is not known. Explain.

92. Explain why nonmetal atoms in Period 3 and beyond can accommodate greater than an octet of electrons and those in Period 2 cannot do so.

93. When asked to give an example of resonance structures, a student drew these Lewis structures. Why is this example incorrect?

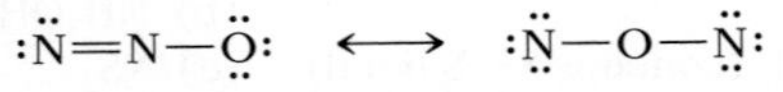

94. Why is this not an example of resonance structures?

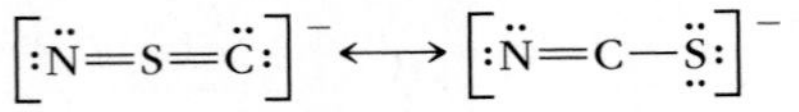

95. How many bonds would you expect the elements in Groups 3A through 7A to form if they obeyed the octet rule?

96. In another universe, elements try to achieve a nonet (nine valence electrons) instead of an octet when forming chemical bonds. As a result, covalent bonds form when a trio of electrons is shared between two atoms. Draw Lewis structures for the compounds that would form between (a) hydrogen and oxygen and (b) hydrogen and fluorine.

97. Elemental phosphorus has the formula P_4. Propose a Lewis structure for this molecule. [*Hints:* (1) Each phosphorus atom is bonded to three other phosphorus atoms. (2) Visualize the structure three-dimensionally, not flat on a page.]

98. The elements As, Br, Cl, S, and Se have electronegativity values of 2.6, 2.3, 2.1, 2.7, and 2.4, but not in that order. Using the periodic trend for electronegativity, assign the values to the elements. Which assignments are you certain about? Which are you not?

99. A substance is analyzed and found to contain 85.7% carbon and 14.3% hydrogen by weight. A gaseous sample of the substance is found to have a density of 1.87 g/L, and 1 mol of it occupies a volume of 22.4 L. What are two possible Lewis structures for molecules of the compound? (*Hint:* First determine the empirical formula and molar mass of the substance.)

100. Below is a grid with 12 numbered boxes, each of which contains an item that is used to answer the questions that follow. Items may be used more than once and there may be more than one correct item in response to a question.

1 Si—Br > P—Br > S—Br > Cl—Br	**2** Si—F	**3** S—Cl < Se—Cl < O—Cl
4 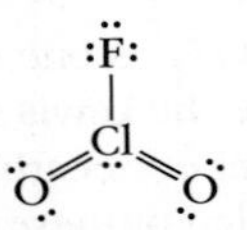	**5** Cl—Br > S—Br > P—Br > Si—Br	**6** 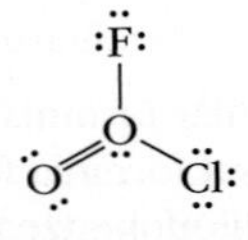
7 Si—H	**8** Si—Si	**9** 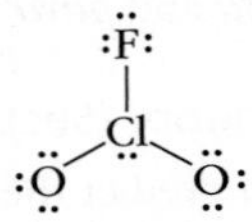
10 O—Cl > S—Cl > Se—Cl	**11** C—F > C=N > C—P	**12** C—P < C—F < C=N

Place the number(s) of the correct selection(s) on the appropriate line.

(a) Shortest Si-containing bond. ______________

(b) Correct sequence of bond lengths for Period 3 bromides. ______________

(c) Resonance hybrid structure(s) of chloryl fluoride, $FClO_2$. ______________

(d) Correct order of polarity of bonds in chlorides of Group 6A. ______________

(e) Correct sequence of bond enthalpies of carbon-containing bonds. ______________

(f) Strongest silicon-containing bond (of those given).

____

101. Below is a grid with 12 numbered boxes, each of which contains an item that is used to answer the questions that follow. Items may be used more than once and there may be more than one correct item in response to a question.

1	2	3
C≡C < N≡N < C≡O	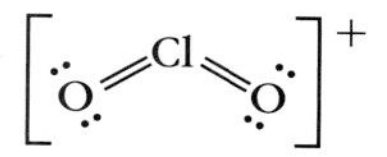	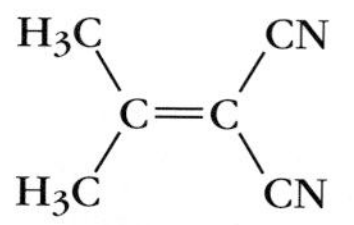
4	**5**	**6**
C—H < N—H < O—H < F—H	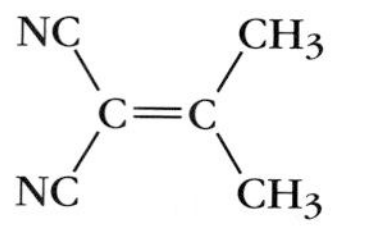	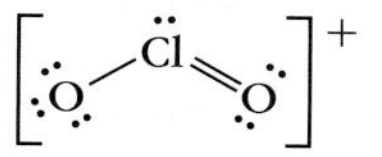
7	**8**	**9**
N≡N < C≡O < C≡C	C—H > N—H > O—H > F—H	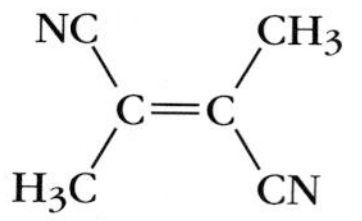
10	**11**	**12**
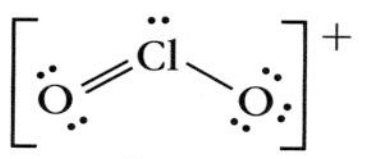	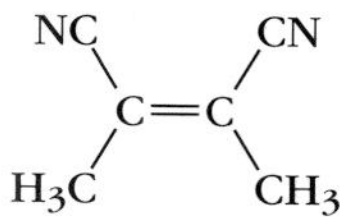	C≡C > N≡N > C≡O

Place the number(s) of the correct selection(s) on the appropriate line.

(a) Correct sequence of bond enthalpies for multiple bonds. ______________

(b) Correct sequence of bond lengths for multiple bonds. ______________

(c) Resonance hybrid structure(s) of $[ClO_2]^+$. ______________

(d) Correct order of bond polarity of hydrides of Period 2. ______________

(e) *cis* and *trans* forms (identify the *cis* then the *trans* form). ______________

102. Which of these molecules is least likely to exist: NF_5, PF_5, SbF_5, or IF_5? Explain why.

103. When we estimate $\Delta H°$ from bond enthalpies we assume that all bonds of the same type (single, double, triple) between the same two atoms have the same energy, regardless of the molecule in which they occur. The purpose of this problem is to show you that this is only an approximation. You will need these standard enthalpies of formation:

C(g)	$\Delta H_f^\circ = 716.7$ kJ/mol
CH(g)	$\Delta H_f^\circ = 596.3$ kJ/mol
CH_2(g)	$\Delta H_f^\circ = 392.5$ kJ/mol
CH_3(g)	$\Delta H_f^\circ = 146.0$ kJ/mol
H(g)	$\Delta H_f^\circ = 218.0$ kJ/mol

(a) What is the average C—H bond energy in methane, CH_4?

(b) Using bond enthalpies, estimate $\Delta H°$ for the reaction

$$CH_4(g) \longrightarrow C(g) + 2\,H_2(g)$$

(c) By heating CH_4 in a flame it is possible to produce the reactive gaseous species CH_3, CH_2, CH, and even carbon atoms, C. Experiments give these values of $\Delta H°$ for the reactions shown:

$$CH_3(g) \longrightarrow C(g) + H_2(g) + H(g) \qquad \Delta H° = 788.7 \text{ kJ}$$

$$CH_2(g) \longrightarrow C(g) + H_2(g) \qquad \Delta H° = 324.2 \text{ kJ}$$

$$CH(g) \longrightarrow C(g) + H(g) \qquad \Delta H° = 338.3 \text{ kJ}$$

For each of the reactions in part (c), draw a diagram similar to Figure 8.4. Then, calculate the average C—H bond energy in CH_3, CH_2, and CH. Comment on any trends you see.

104. Write the Lewis structure for nitrosyl fluoride, FNO. Using only a periodic table, identify:

(a) Which is the longer bond.

(b) Which is the stronger bond.

(c) Which is the more polar bond.

105. Write the Lewis structure for nitrosyl chloride, ClNO. Using only a periodic table, identify:

(a) Which is the longer bond.

(b) Which is the stronger bond.

(c) Which is the more polar bond.

106. Methylcyanoacrylate is the active ingredient in "super" glues. Its Lewis structure is

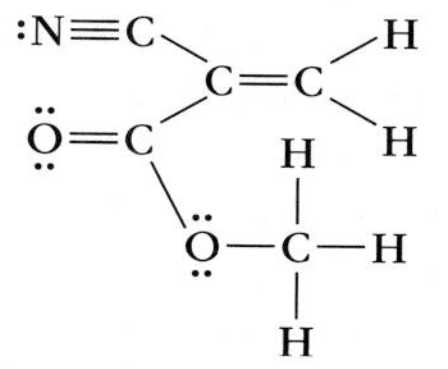

In this molecule, which is the:

(a) Weakest carbon-containing bond?

(b) Strongest carbon-containing bond?

(c) Most polar bond?

107. Acrylonitrile is used to make synthetic carpets and wigs. Its Lewis structure is

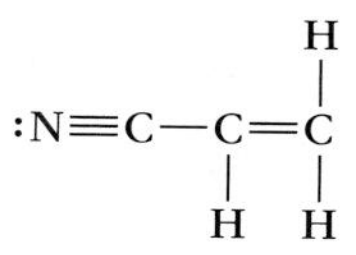

In this molecule, which is the:

(a) Strongest carbon-containing bond?

(b) Second strongest carbon-containing bond?

(c) Most polar bond?

More Challenging Questions

These questions require more thought and integrate several concepts.

108. Write the Lewis structure for:
 (a) tetraboron tetrachloride, B_4Cl_4
 (b) hyponitrous acid, $H_2N_2O_2$
109. Nitrosyl azide, N_4O, is a pale yellow solid first synthesized in 1993. Write the Lewis structure for nitrosyl azide.
110. Write the Lewis structures for:
 (a) $(Cl_2PN)_3$ (b) $(Cl_2PN)_4$
111. Write the Lewis structures for:
 (a) dithiocyanogen, $(NCS)_2$ (b) tetrahydrofuran, C_4H_8O
112. In sulfuric acid production, sulfur trioxide is combined with sulfuric acid to produce oleum, which contains $H_2S_2O_7$. Write the Lewis structure for an oleum molecule.
113. Tetrasulfur tetranitride, S_4N_4, can be converted to disulfur dinitride, S_2N_2. Write the Lewis structures for these molecules.
114. Nitrous oxide, N_2O, is a linear molecule that has the two nitrogen atoms adjacent to each other. Using concepts from this chapter, explain why the structure is not NON with the nitrogen atoms each bonded to the oxygen.
115. The azide ion, N_3^-, has three resonance hybrid structures.
 (a) Write the Lewis structure of each.
 (b) Use formal charges to determine which is the most favorable resonance structure.
116. Hydrazoic acid, HN_3, has three resonance hybrid structures.
 (a) Write the Lewis structure of each.
 (b) Use formal charges to determine which is the least favorable resonance structure.
117. Write the Lewis structures for:
 (a) hyponitrite ion, $(N_2O_2)^{2-}$ (b) $(SO_3)_3$
118. Write the Lewis structures for:
 (a) S_2F_{10} (b) $FXeN(SO_2F)_2$
119. Experimental evidence indicates the existence of HC_3N molecules in interstellar clouds. Write a plausible Lewis structure for this molecule.
120. Molecules of SiC_3 have been discovered in interstellar clouds. Write a plausible Lewis structure for this molecule.
121. For many people, tears flow when they cut raw onions. The tear-inducing agent is thiopropionaldehyde-S-oxide, C_3H_6OS. Write the Lewis structure for this molecule.
122. One of the structural isomers of C_3H_6OS is the compound that is the offending odorant released by striped skunks. Write Lewis structures for two isomers of this molecule.
123. Phosphorus and sulfur form a series of compounds, one of which is tetraphosphorus trisulfide. Write the Lewis structure for this molecule.
124. Piperine, the active ingredient in black pepper, has the structural formula given below.

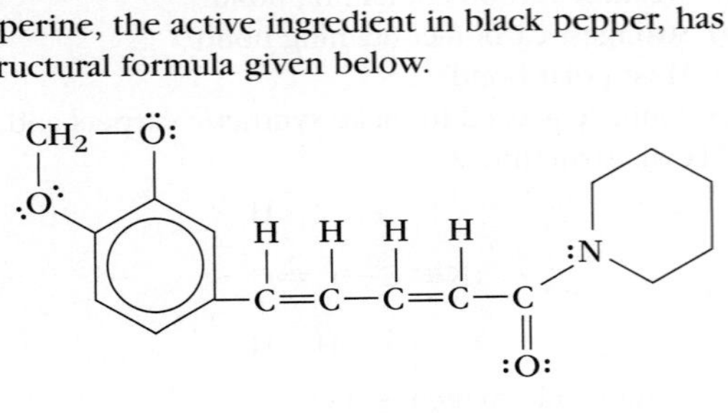

 (a) Write the molecular formula of piperine.
 (b) Identify the shortest carbon-to-carbon bond in piperine.
 (c) Identify the shortest carbon-to-oxygen bond in piperine.
 (d) Identify the stongest carbon-to-carbon bond in piperine.
 (e) Identify the most polar bond in piperine.
125. Sulfur and oxygen form a series of 2− anions including sulfite, SO_3^{2-}, and sulfate, SO_4^{2-}. In addition to these, there are three other more complex anions—dithionite, $S_2O_4^{2-}$, dithionate, $S_2O_6^{2-}$, and tetrathionate, $S_4O_6^{2-}$. Write the correct Lewis structure for dithionite, dithionate, and tetrathionate ions.
126. Tetrasulfur tetranitride reacts with disulfur dichloride to form S_4N_3Cl, a salt.

$$3\ S_4N_4 + 2\ S_2Cl_2 \rightarrow 4\ S_4N_3Cl$$

 Write a plausible Lewis structure for the two reactants and the cation of the salt.
127. Gaseous molecules in the ground vibrational state have a vibrational energy given by the equation $E = \frac{1}{2}h\nu$ where h is Planck's constant and ν is the frequency of the vibration. Consider a water vapor molecule, HOH, in which one of the lighter isotope hydrogen atoms, H, is replaced by the heavier isotope deuterium atom, D, to form HOD.
 (a) Will an HOD molecule have a higher or lower ground vibrational state energy than an HOH molecule? Explain your answer.
 (b) Which bond is stronger: O—H or O—D? Explain your answer.
128. Gamma-aminobutanoic acid, $C_4H_9O_2N$, is a neurotransmitter. The molecule contains a four-carbon chain.
 (a) Write the Lewis structure for this molecule.
 (b) Which bond is the longest?
 (c) Which bond is the most polar?
 (d) Which bond is the strongest?
129. Suppose in building up molecular orbitals, the π_{2p} were placed above the σ_{2p}. Prepare a diagram similar to Figure 8.11 based on these changes. For which species in Table 8.4 would this change in relative energies of the MOs affect the prediction of number of bonds and number of unpaired electrons?

Conceptual Challenge Problems

These rigorous, thought-provoking problems integrate conceptual learning with problem solving and are suitable for group work.

CP8.A (Section 8.2) What deficiency is acknowledged by chemists when they write the formula for silicon dioxide as $(SiO_2)_n$ but write CO_2 for carbon dioxide?

CP8.B (Section 8.7) Without referring to the periodic table, write an argument to predict how the electronegativities of elements change based on the composition of their atoms.

CP8.C (Section 8.7) How would you rebut the statement, "There are no ionic bonds, only polar covalent bonds"?

9 Molecular Structures

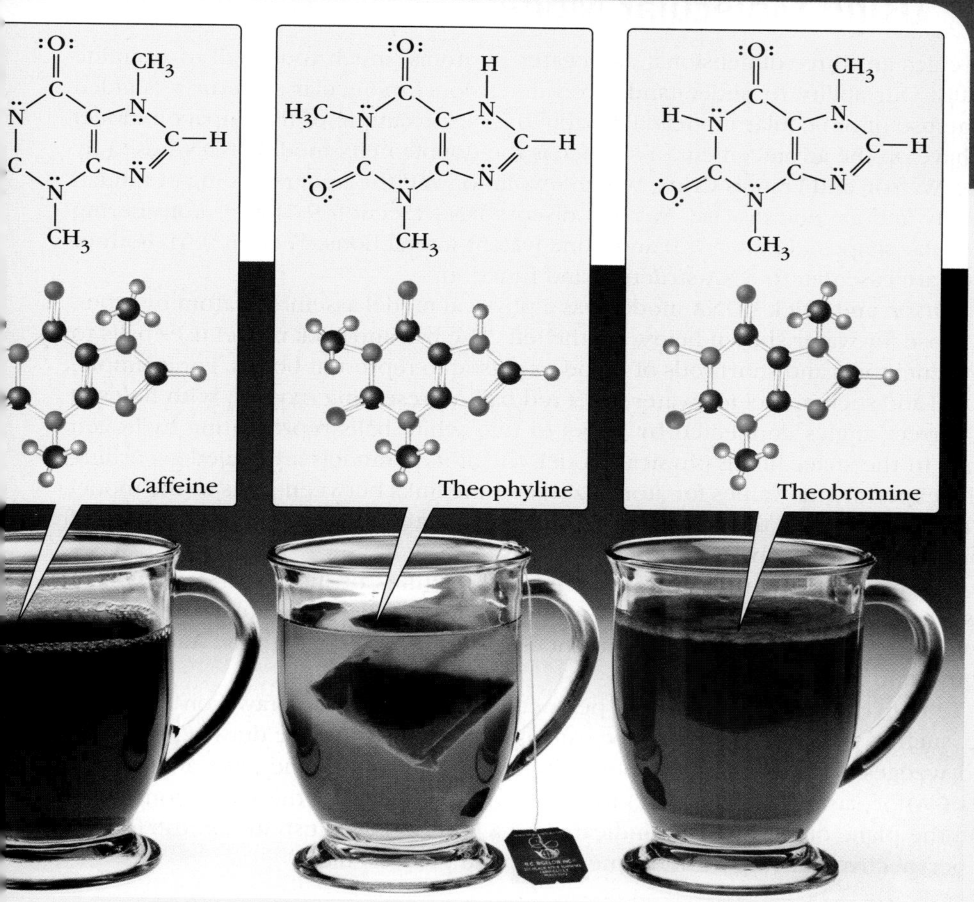

© Cengage Learning/Charles D. Winters

Molecular shapes are important. Coffee (left), tea (center), and hot chocolate (right) contain alkaloid molecules, compounds characterized by a double-ring structure of carbon and nitrogen atoms. Look carefully at the structural formulas and molecular models of the particular alkaloid associated with each beverage. Although they are structurally similar, caffeine (in coffee), theophyline (in tea), and theobromine (in chocolate) have molecular structures that differ sufficiently to help create the distinctive flavor and taste of the particular beverage. This chapter discusses molecular shapes and their role in the polarity of molecules and the forces that exist between molecules.

Although the composition, empirical formula, molecular formula, and Lewis structure of a substance provide important information, they are not sufficient to predict or explain the properties of most molecular compounds. Also important are the arrangement of the atoms and how they occupy three-dimensional space—the shape of a molecule. Because of differences in the sequence in which their atoms are joined, molecules can have the same numbers of the same kinds of atoms, yet have different properties. For example, ethanol (in alcoholic beverages) and dimethyl ether (a refrigerant) have the same molecular formula, C_2H_6O. But the C, H, and O atoms of the two compounds are arranged so differently that their melting points (m.p.) differ by 27 °C and their boiling points (b.p.) by 103 °C.

The ideas about molecular shape presented in this chapter are crucial to understanding the relationships between the structure and the properties of molecules, properties related to the behavior of molecules in living organisms, the design of molecules that are effective drugs, and many other aspects of modern chemistry.

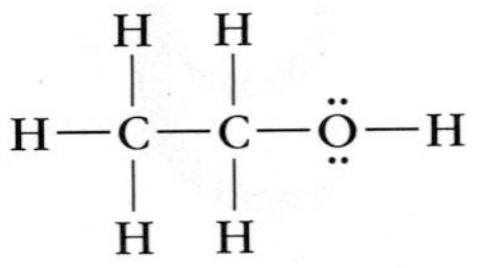

ethanol
m.p. = −114 °C;
b.p. = 78.3 °C

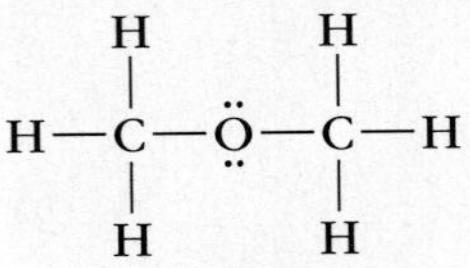

dimethyl ether
m.p. = −141.5 °C;
b.p. = −24.8 °C

OWL Sign in to OWL at **www.cengage.com/owl** to view tutorials and simulations, develop problem-solving skills, and complete online homework assigned by your professor.

go Chemistry Download mini lecture videos for key concept review and exam prep from OWL or purchase them from **www.CengageBrain.com.**

Companion Website

Visit this book's companion website at **www.cengage.com/chemistry/moore** to work interactive modules for the Estimation boxes and Active Figures in this text.

9.1 Using Molecular Models

Molecules are three-dimensional aggregates of atoms, much too small to examine visually. Our ability to understand three-dimensional molecular structures is aided by the use of molecular models. Probably the best example of the impact a model can have on the advancement of science is the double helix model of DNA used by James Watson and Francis Crick, which revolutionized the understanding of human heredity and genetic disease. We will discuss DNA (Section 9.7) after considering molecular shapes (Section 9.2) and noncovalent interactions (Section 9.6), both of which are essential to DNA structure and function.

Watson and Crick's DNA model was a physical model assembled atom by atom, like those for water shown below on the left. The ball-and-stick model uses balls to represent atoms and short rods of wood or plastic to represent bonds. For example, the ball-and-stick model for water has a red ball representing oxygen, with holes at the correct angles connected by sticks to two white balls representing hydrogen atoms. In the space-filling physical model, the atomic models are scaled according to the experimental values for atom sizes, and the links between parts of the model are not visible when the model is assembled. This gives a better representation of the actual space occupied by atoms.

This book uses many computer-generated pictures of molecular models, like those for water shown below. The computer programs that generate these pictures rely on the most accurate experimentally derived data on atomic radii, bond lengths, and bond angles.

To convey a three-dimensional perspective for a molecule drawn on a flat surface, such as the page of a book, we can also make a perspective drawing that uses solid wedges (▶) to represent bonds extending in front of the page and dashed lines (---) or dashed wedges (IIIII) to represent bonds behind the page. Bonds that lie in the plane of the page are indicated by a line (—), as illustrated in the following perspective drawing for methane, a tetrahedral molecule:

Ball-and-stick model kits are available in many campus bookstores. The models are easy to assemble and will help you to visualize the molecular geometries described in this chapter. Ball-and-stick models are relatively inexpensive compared with space-filling models.

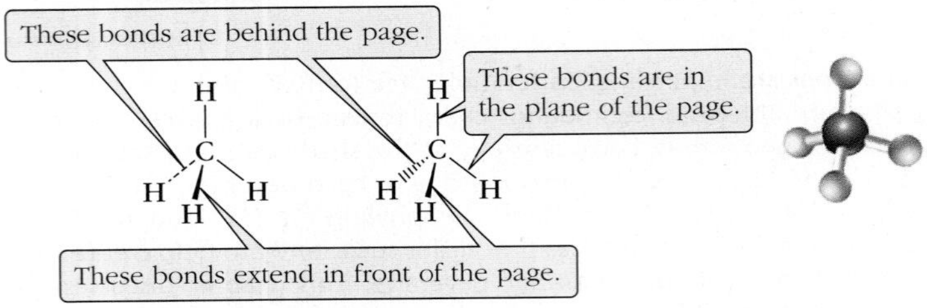

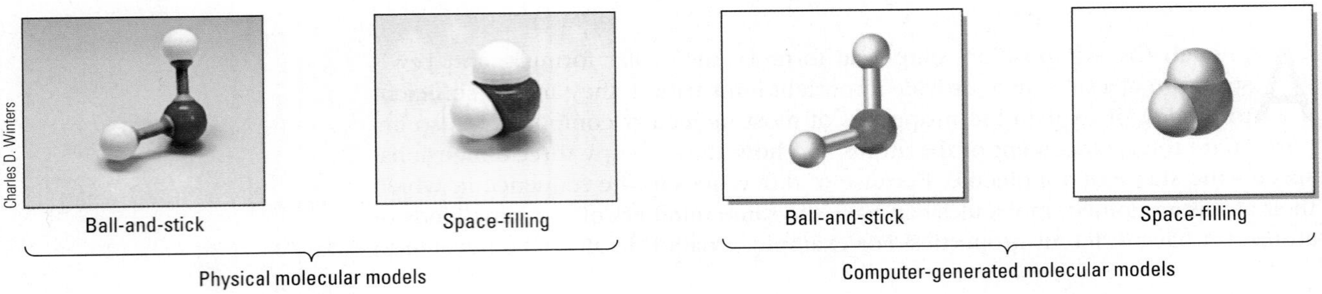

© Cengage Learning/Charles D. Winters

Advances in computer graphics have made it possible to draw scientifically accurate pictures of extremely complex molecules, as well as to study interactions between molecules. Figure 9.1 is a computer-generated model of hemoglobin, the remarkable and essential protein in the blood that takes up and releases oxygen. This vital protein is made up of two folded sets of pairs of long chains of atoms, indicated by the red ribbons. In a hemoglobin molecule, oxygen binds to Fe^{2+} ions (yellow) at four sites shown in green, red, and blue in Figure 9.1. Computer graphics can rotate such computer-generated models to provide different views of the molecule. This is illustrated in Figure 9.2 for ethane, a simple alkane molecule.

Figure 9.1 **Hemoglobin, a protein essential to human life.** An Fe^{2+} ion shown in yellow is at the center of each green heme cluster.

9.2 Predicting Molecular Shapes: VSEPR

Because molecular shape is so important to the reactivity and function of molecules, it is essential to have a simple, reliable method for predicting the shapes of molecules and polyatomic ions. The **valence-shell electron-pair repulsion (VSEPR)** model is such a method. The VSEPR model is based on the idea that repulsions occur among regions of electron density, such as the repulsions among pairs of bonding and lone pair electrons. Such repulsions control the angles between bonds from a central atom to other atoms surrounding it.

How do repulsions among electron pairs result in different shapes? Imagine that the volume of a balloon represents the repulsive force of an electron pair that prevents other electron pairs from occupying the same space. When two or more balloons are tied together at a central point, the cluster of balloons assumes the shapes shown in Figure 9.3. The central point represents the nucleus and the core electrons of a central atom; each balloon represents an electron pair. The arrangements of the balloons are those that minimize interactions among them, and among the electron pairs that the balloons represent.

To minimize their repulsions, electron pairs (regions of electron density) are oriented as far apart as possible. Because electrons are constrained by electrostatic attraction to be near the nucleus, the shapes resulting from minimizing the electron-pair repulsions are those predicted by VSEPR and illustrated in the balloon analogy (Figure 9.3).

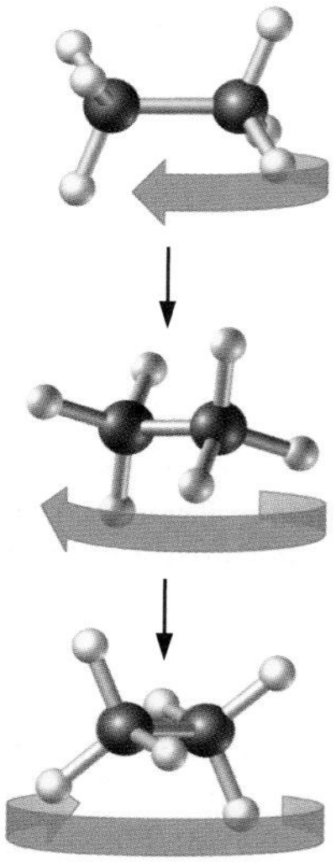

Figure 9.2 **Rotation of an ethane molecule, H_3CCH_3, around an axis perpendicular to the C—C bond axis.**

Central Atoms with Only Bonding Pairs

The simplest application of VSEPR is to molecules and polyatomic ions having a central atom that has only shared electron pairs to bond it to other atoms (terminal atoms) by only single bonds. In such cases, the central atom has no lone pairs. We can summarize this using an AXE-based shorthand notation: A (the central atom) is bonded to X_n (the terminal atoms); E_m represents the number of lone pair electrons on the central atom. When there are no lone pairs, the symbolism is E_0. For example, an AX_3E_0 designation denotes a central atom (A) with no lone pairs (E_0) and whose shared pairs are bonded to three terminal atoms (X_3), such as in BF_3.

Linear | Triangular planar | Tetrahedral | Triangular bipyramidal | Octahedral

Figure 9.3 **Balloon models of the geometries predicted by the VSEPR model.** The positions of the balloons are as far apart as possible. (See also Figure 9.4.)

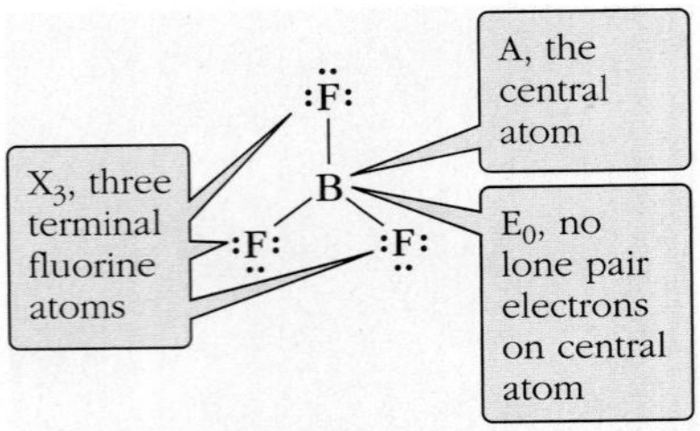

Using the AX_3E_0 designation for boron trifluoride.

The term "molecular shape" is sometimes used rather than "molecular geometry."

Figure 9.4 illustrates the geometries predicted by the VSEPR model for molecules of types AX_2E_0 to AX_6E_0.

Two categories of geometries are given in Figure 9.4. Consider these different types of geometries:

- **Electron-pair geometry**, which is determined by the number of electron pairs around the *central* atom in a molecule or a polyatomic ion.
- **Molecular geometry**, which is the arrangement of atoms in space around the central atom in a molecule or a polyatomic ion.

With regard to the two categories it is important to note that

- *For molecules or polyatomic ions whose central atoms have no lone pair electrons, the electron-pair geometry and the molecular geometry are the same.*
- *When lone electron pairs are present on the central atom, the electron-pair and molecular geometries are* not *the same, as will be discussed shortly.*

Bond angles, as shown in Figure 9.4, are the angles between the bonds of two atoms that are bonded to the same third atom. In a methane molecule, CH_4, for example, all the H—C—H bond angles are 109.5°.

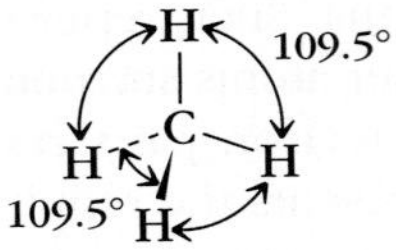

methane, CH_4, a tetrahedral molecule

The predicted bond angles given in the examples in Figure 9.4 are in agreement with experimental values obtained from structural studies.

Central atoms from Period 2 are too small to accommodate more than four atoms bonded to them.

The geometries illustrated in Figure 9.4 are the ones found experimentally for typical molecules and polyatomic ions in which there are no lone pairs on the central atom. The tetrahedral geometry (middle column in Figure 9.4) is by far the most common, because, with four bonding electron pairs, the central atom obeys the octet rule. The central atom has less than an octet in the linear and triangular planar geometries on the left side of the figure; in these structures there are two bonding pairs and three bonding pairs respectively. The central atom has more than an octet in the trigonal bipyramidal and octahedral geometries in the last two columns on the right side of the figure; these two geometries have five bonding pairs and six bonding pairs respectively. The trigonal bipyramidal and octahedral structures are expected only when the central atom is an element in Period 3 or higher (← *p. 356*).

Suppose you want to use VSEPR concepts to predict the shape of $SiCl_4$. First, draw the Lewis structure, with Si as the central atom. From the Lewis structure, you can see that there are four bonding pairs from the terminal Cl atoms to the central Si atom forming four single bonds to Si, and no lone pairs on Si. In this case, A is Si, X is Cl, and X_4 indicates four Cl atoms bonded to A; E_0 denotes no lone pairs on

Molecular model	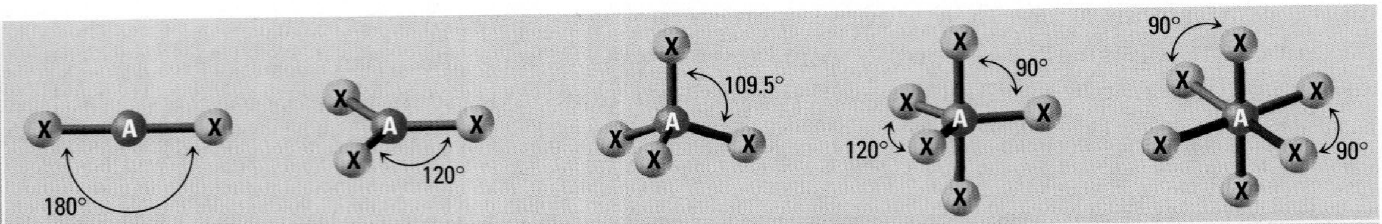				
Type	AX_2E_0	AX_3E_0	AX_4E_0	AX_5E_0	AX_6E_0
Electron-pair geometry	Linear	Triangular planar	Tetrahedral	Triangular bipyramidal	Octahedral
Molecular geometry	Linear	Triangular planar	Tetrahedral	Triangular bipyramidal	Octahedral
Example molecule	BeF_2, CO_2	BF_3, CO_3^{2-}	CH_4, $SiCl_4$	PCl_5, PF_5	SF_6

Figure 9.4 **Geometries predicted by the VSEPR model for molecules of types AX_2E_0 through AX_6E_0 that contain *only* single covalent bonds and no lone pairs on the central atom, A.**

A (Si), the central atom. You would predict a tetrahedral electron-pair geometry and a tetrahedral molecular geometry for the molecule (Figure 9.4). This is in agreement with structural studies of $SiCl_4$, which indicate a tetrahedral molecule with all Cl—Si—Cl bond angles 109.5°.

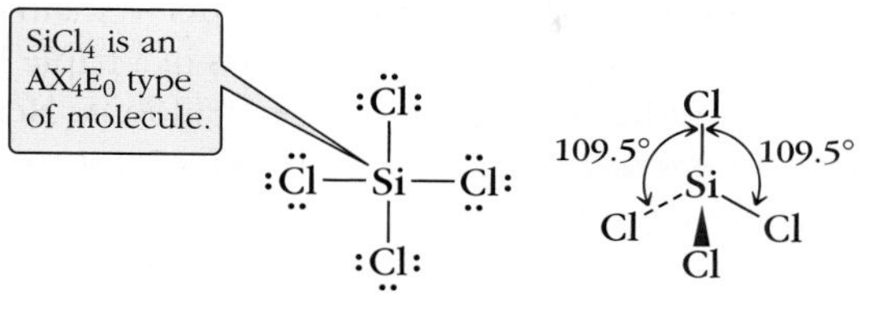

In summary, molecules and polyatomic ions whose central atoms have *no* lone pair electrons (E_0) have the *same* electron-pair *and* molecular geometries. These geometries and bond angles are

- **Linear:** AX_2E_0—bond angles 180°.
- **Triangular planar:** AX_3E_0—bond angles 120°.
- **Tetrahedral:** AX_4E_0—bond angles 109.5°.
- **Triangular bipyramidal:** AX_5E_0—bond angles 90° and 120°.
- **Octahedral:** AX_6E_0—bond angles 90°.

PROBLEM-SOLVING EXAMPLE **9.1** **Molecular Geometry**

Use the VSEPR model to predict the electron-pair geometry, molecular geometry, and bond angles of (a) BF_3 and (b) CBr_4.

Answer

(a) Triangular planar electron-pair and molecular geometries with 120° F—B—F bond angles

(b) Tetrahedral electron-pair and molecular geometries with 109.5° Br—C—Br bond angles

Strategy and Explanation The first step in using VSEPR is to write the correct Lewis structure and classify the molecule according to the number of bonding electron pairs and lone pairs on the central atom.

(a) The Lewis structure of BF_3 is

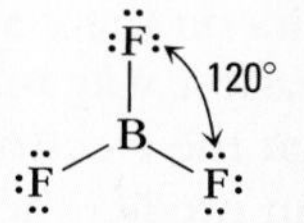

BF_3 is an AX_3E_0-type molecule (three bonding pairs, no lone pairs around the central B atom; see Figure 9.4). The molecule has three B—F single bonds around the central boron atom, arranged at 120° angles from each other. Therefore, BF_3 has both a triangular planar electron-pair geometry and molecular geometry with all the atoms lying in the same plane.

(b) Being an AX_4E_0-type molecule, CBr_4 will be a tetrahedral molecule with 109.5° Br—C—Br bond angles. The electron pair and molecular geometries are both tetrahedral.

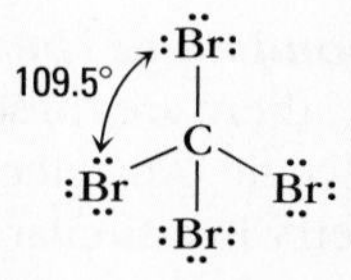

PROBLEM-SOLVING PRACTICE **9.1**

Identify the electron-pair geometry, the molecular geomety, and the bond angles for BeF_2.

The **PROBLEM-SOLVING STRATEGY** in this book is

- Analyze the problem
- Plan a solution
- Execute the plan
- Check that the result is reasonable

Appendix A.1 explains this in detail.

PROBLEM-SOLVING PRACTICE answers are provided at the back of this book in Appendix K.

Multiple Bonds and Molecular Geometry

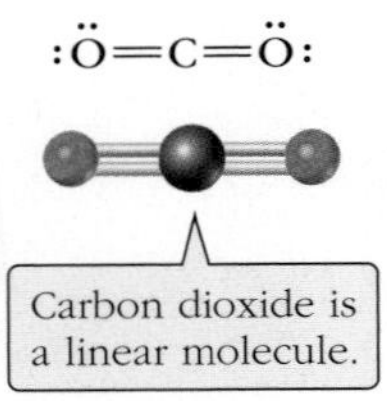
Carbon dioxide is
a linear molecule.

Although double bonds and triple bonds are shorter and stronger than single bonds, this has only a minor effect on predictions of bond angles. Why? Electron pairs involved in a multiple bond are all shared between the same two nuclei and therefore occupy the same region of space. Because they must remain in that region, two electron pairs in a double bond or three electron pairs in a triple bond are like a single, slightly fatter balloon, rather than two or three balloons. Hence, ***for the purpose of determining molecular geometry, the electron pairs in a multiple bond constitute a single region of electron density as in a single bond.*** For example, compare BeF_2 (Figure 9.4) with CO_2. BeF_2 is a linear molecule with the two Be—F single bonds 180° apart. In CO_2, each C═O double bond is a single region of electron density, just as each Be—F single bond is, so the structure of CO_2 is also linear.

When resonance structures are possible, the geometry can be predicted from the Lewis resonance structures or from the resonance hybrid structure. For example, the geometry of the CO_3^{2-} ion is predicted to be triangular planar because the carbon atom has three sets of equivalent bonds and no lone pairs. Each of the three bonds is intermediate between a C—O single bond and a C═O double bond. The geometry can be predicted from the Lewis resonance structure.

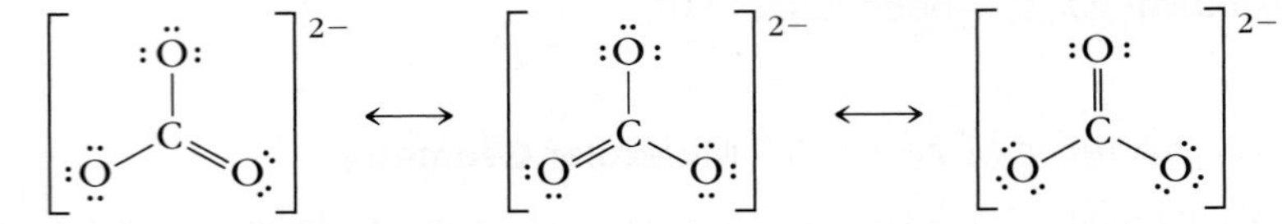

Answers to **EXERCISES** are provided at the back of this book in Appendix L.

EXERCISES that are labeled **CONCEPTUAL** are designed to test your understanding of one or more concepts; they usually involve qualitative rather than quantitative thinking.

CONCEPTUAL EXERCISE 9.1 Geometries

Based on the discussion so far, identify a characteristic that is common to all situations where electron-pair geometry and molecular geometry are the same for a molecule or a polyatomic ion.

Central Atoms with Bonding Pairs and Lone Pairs

How does the presence of lone pairs on the central atom affect the geometry of a molecule or polyatomic ion? The easiest way to visualize this situation is to return to the balloon model and notice that the electron pairs on the central atom do not all have to be bonding pairs. We can predict the electron-pair geometry, molecular geometry, and bond angles by applying the VSEPR model to the total number of valence electron pairs—that is, bonding pairs as well as lone pairs around the central atom. We will use NH_3, ammonia, to illustrate guidelines for doing so.

1. ***Draw the Lewis structure.*** The Lewis structure for NH_3 is

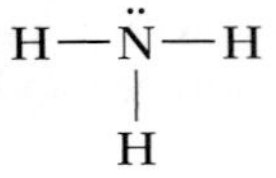

In ammonia, A is N, X is H, and X_3 indicates three H atoms bonded to A, the central atom; E_1 denotes one lone pair on A.

2. ***Determine the number of bonds and the number of lone pairs around each atom.*** In the case of NH_3, there are three bonding pairs and one lone pair; thus, NH_3 is an AX_3E_1-type molecule. Any electron pairs in a multiple bond contribute to the molecular geometry in a similar manner as those in a single bond. The central N atom in NH_3 has no multiple bonds, only three H—N single covalent bonds.
3. ***Pick the appropriate electron-pair geometry around each central atom, and then choose the molecular shape that matches the total number of bonds and lone pairs.*** The *electron-pair geometry* around a cen-

tral atom includes the spatial positions of all bonds and lone pairs. Because there are three bonds and one lone pair for a total of four pairs of electrons, we predict that the electron-pair geometry of NH_3 is tetrahedral. To represent this geometry, draw a tetrahedron with N as the central atom and the three bonds represented by lines, wedges, or dashed lines, since they are single covalent bonds. The lone pair is drawn as a balloon shape to indicate its spatial position in the tetrahedron.

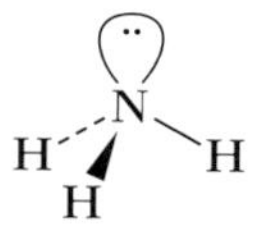

The positions of the lone pairs are *not* specified when describing the molecular geometry of molecules. The *molecular geometry* of ammonia is described as a triangular pyramid because the three hydrogen atoms form a triangular base with the nitrogen atom at the apex of the pyramid. (This can be seen by covering up the lone pair of electrons and looking at the molecular geometry—the location of the three H atoms and the N atoms.)

4. ***Predict the bond angles, remembering that lone pairs occupy more space around the central atom than do bonding pairs.*** Bonding pairs are concentrated in the bonding region between two atoms by the strong attractive forces of two positive nuclei and are, therefore, relatively compact, or "skinny." For a lone pair, there is only one nucleus attracting the electron pair. As a result, lone pairs are less compact. Using the balloon analogy, a lone pair is like a fatter balloon that takes up more room and squeezes the thinner balloons closer together. The relative strengths of electron-pair repulsions are

The success of the VSEPR model in predicting molecular shapes indicates that it is appropriate to account for the effects of lone pairs in this way.

lone pair–lone pair > lone pair–bonding pair > bonding pair–bonding pair

This listing predicts that lone pairs will force bonding pairs closer together and decrease the angles between the bonding pairs. Recognizing this, we can predict that *bond angles adjacent to lone pairs will be smaller than those predicted for perfect geometric shapes* (those with no lone pairs). The determination of electron-pair geometry and molecular geometry for NH_3 is summarized by this sequence:

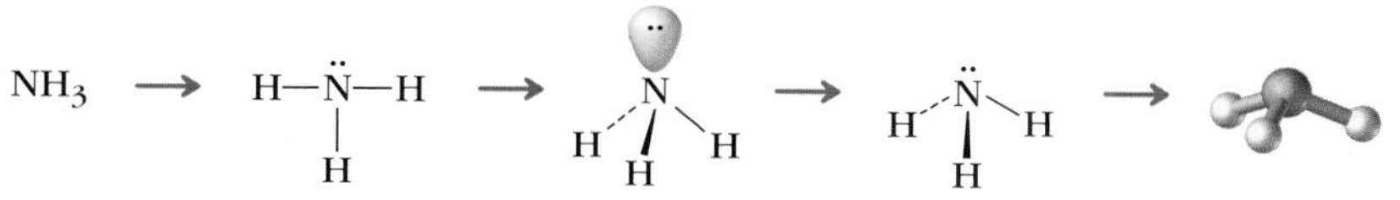

Molecular formula	Lewis structure	Electron-pair geometry (tetrahedral)	Molecular geometry (triangular pyramidal)

Because the electron-pair geometry is tetrahedral, we would expect the H—N—H bond angles to be 109.5°. However, the experimentally determined bond angles in NH_3 are 107.5°. This is attributed to the bulkier lone pair forcing the bonding pairs closer together, thereby reducing the bond angle from 109.5° to 107.5°. Figures 9.5 (p. 382) and 9.6 (p. 384) summarize the electron-pair geometry and molecular geometry for cases of one, two, or three lone pairs on the central atom.

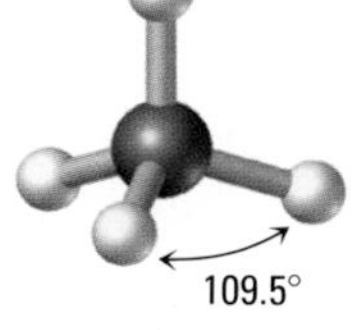

methane
CH_4

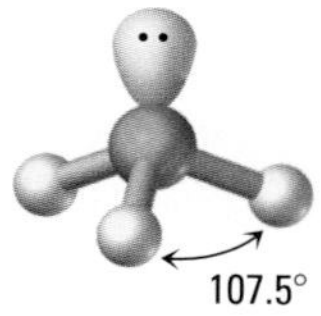

ammonia
NH_3

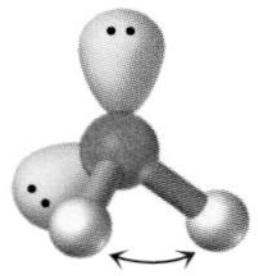

water
H_2O

The effect that increasing numbers of lone pairs on the central atom have on bond angles is shown in the margin for CH_4, NH_3, and H_2O. Note from the series of molecules that *as the number of lone pairs on the central atom increases, the bond angle decreases*. The three molecules all have tetrahedral electron-pair geometry because each has four electron pairs around the central atom. Each molecule,

Sometimes the term "bent" is used rather than "angular" to describe this molecular geometry.

however, has slightly different bond angles due to the lone pair effect. There are bonding pair-bonding pair repulsions in all three molecules. In addition, there are also bonding pair-lone pair repulsions in NH_3, as well as lone pair-lone pair repulsions in water. This results in tetrahedral molecular geometry for CH_4 (109.5°), triangular pyramidal molecular geometry for NH_3 (107.5°), and angular (bent) molecular geometry for H_2O (104.5°).

Methane, which has a tetrahedral shape, is the smallest member of the large family of saturated hydrocarbons called alkanes *(← p. 334)*. It is important to recognize that every carbon atom in an alkane has four bonds and hence a tetrahedral environment. For example, the carbon atoms in propane and in the much longer carbon chain of hexadecane do not lie in a straight line because of the tetrahedral geometry about each carbon atom.

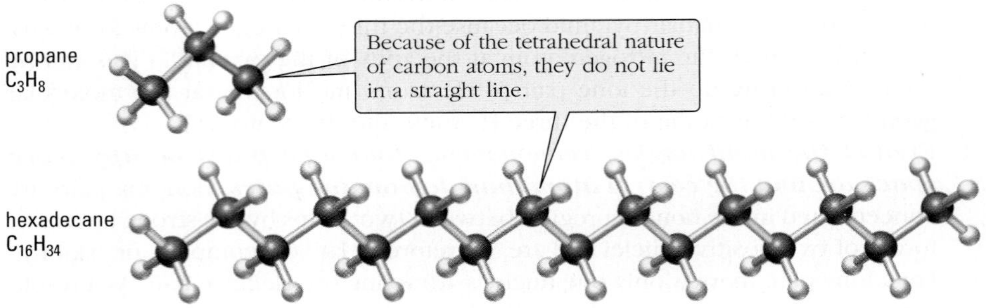
propane
C_3H_8
Because of the tetrahedral nature of carbon atoms, they do not lie in a straight line.
hexadecane
$C_{16}H_{34}$

The tetrahedron is arguably the most important shape in chemistry because of its predominance in the chemistry of carbon and the carbon-based chains that form the backbone of many biomolecules.

Figure 9.5 gives additional examples of electron-pair and molecular geometries for molecules or ions with three and four electron pairs around the central atom with at least one lone pair. (See Figure 9.4 for examples having no lone pairs.) To check your understanding of the VSEPR model, try to explain the electron-pair geometry, the molecular geometry, and bond angles given in Figure 9.5.

Molecular model		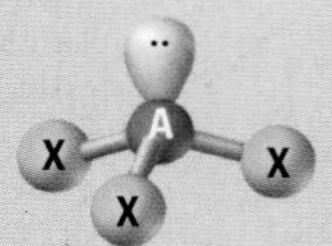	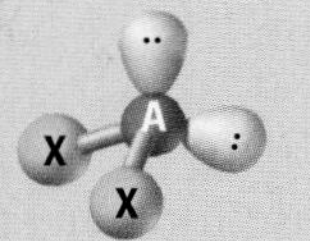
	***Three* electron pairs; *one* lone pair**	***Four* electron pairs; *one* lone pair**	***Four* electron pairs; *two* lone pairs**
Type	AX_2E_1	AX_3E_1	AX_2E_2
Electron-pair geometry	Triangular planar	Tetrahedral	Tetrahedral
Molecular geometry	Angular (bent)	Triangular pyramidal	Angular (bent)
Examples	$SnCl_2$, $GeCl_2$	NCl_3	H_2O, OF_2, OCl_2

Figure 9.5 Three and four electron pairs around a central atom. Each central atom has at least *one* lone pair. For AX_nE_0 species see Figure 9.4.

PROBLEM-SOLVING EXAMPLE **9.2** **Molecular Structure**

Use the VSEPR model to predict the electron-pair geometry and the molecular geometry of (a) ClF_2^+, (b) $SnCl_3^-$, (c) SeO_3, and (d) CS_2.

Answer

Electron-Pair Geometry	Molecular Geometry
(a) Tetrahedral	Angular
(b) Tetrahedral	Triangular pyramidal
(c) Triangular planar	Triangular planar
(d) Linear	Linear

Strategy and Explanation Writing the correct Lewis structures identifies the central atom in each ion or molecule. The Lewis structures are

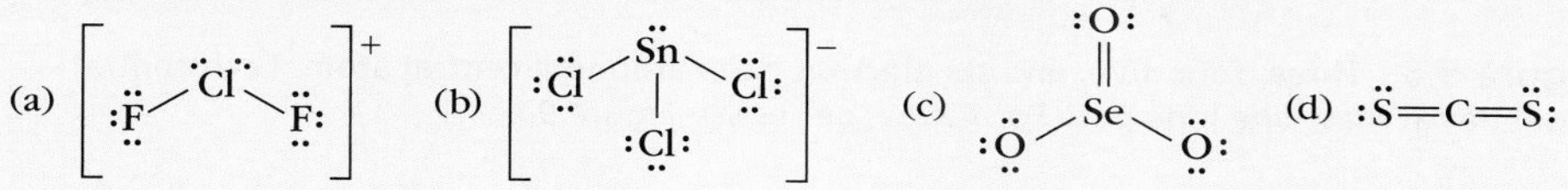

(a) The Lewis structure of ClF_2^+ reveals that the central Cl atom has four electron pairs: two bonding pairs to terminal fluorine atoms and two lone pairs. Consequently, the ion is an AX_2E_2 type and has a tetrahedral electron-pair geometry and an angular (bent) molecular geometry.

(b) The central tin atom is surrounded by three bonding pairs and one lone pair (AX_3E_1 type). These four electron pairs give a tetrahedral electron-pair geometry. The molecular geometry is triangular pyramidal. The lone pair pushes the Sn—Cl bonds closer together than the purely 109.5° tetrahedral angle expected for four bonding pair-bonding pair repulsions.

(c) The two bonding pairs in the Se═O double bond are counted as one region of electron density for determining the geometry of the molecule, giving three bonding regions around Se (AX_3E_0 type). There are three resonance structures (one shown here) that give three equivalent regions of electron density around the Se atom. Because there are no lone pairs around Se, the electron-pair and molecular geometries are the same, triangular planar.

(d) For determining molecular structure, each C═S double bond is treated as one region of electron density. There are no lone pairs on the central C atom (AX_2E_0 type). Therefore, the electron-pair geometry and molecular geometry are both linear.

PROBLEM-SOLVING PRACTICE **9.2**

Use Lewis structures and the VSEPR model to determine electron-pair and molecular geometries for (a) BrO_3^-, (b) SeF_2, and (c) NO_2^-.

Expanded Octets: Central Atoms with Five or Six Electron Pairs

Analysis of electron-pair and molecular geometries becomes more complicated if a central atom has five or six electron pairs, some of which are lone pairs. Let's look first at the entries in Figure 9.6 for the case of five electron pairs. The three angles in the triangular plane are all 120°. The angles between any of the pairs in this plane and an upper or lower pair are only 90°. Thus, the *triangular bipyramidal* structure has two sets of positions that are *not* equivalent. Because the positions in the triangular plane lie in the equator of an imaginary sphere around the central atom, they are called **equatorial positions.** The north and south poles are called **axial positions.** Each equatorial position is next to only two electron pairs at 90° angles, the axial ones, while an axial position is proximate to three electron pairs, the equatorial ones. This means that to minimize repulsions *any lone pairs will occupy*

Triangular bipyramidal structure. The axial and equatorial positions are *not* equivalent.

Molecular model	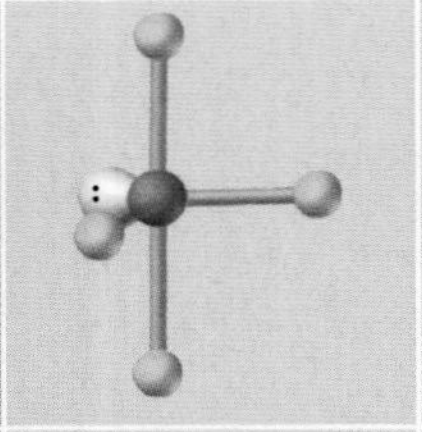	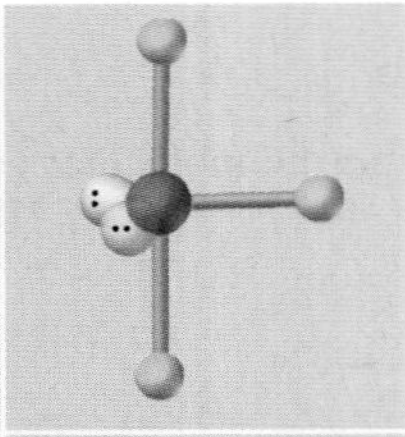	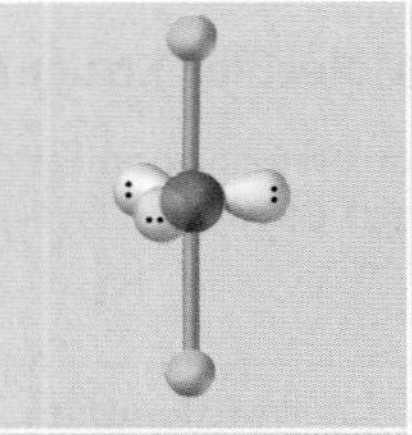	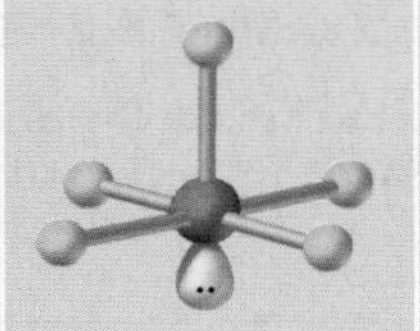	
	Five **electron pairs; *one* lone pair**	***Five*** **electron pairs; *two* lone pairs**	***Five*** **electron pairs; *three* lone pairs**	***Six*** **electron pairs; *one* lone pair**	***Six*** **electron pairs; *two* lone pairs**
Type	AX_4E_1	AX_4E_1	AX_2E_3	AX_5E_1	AX_4E_2
Electron-pair geometry	Triangular bipyramidal	Triangular bipyramidal	Triangular bipyramidal	Octahedral	Octahedral
Molecular geometry	Seesaw	T-shaped	Linear	Square pyramidal	Square planar
Examples	SF_4	ClF_3	XeF_2	BrF_5	$[ICl_4]^-$

Figure 9.6 Three, four, five, and six electron pairs around a central atom. Each central atom has at least *one* lone pair. For AX_nE_0 species see Figure 9.4.

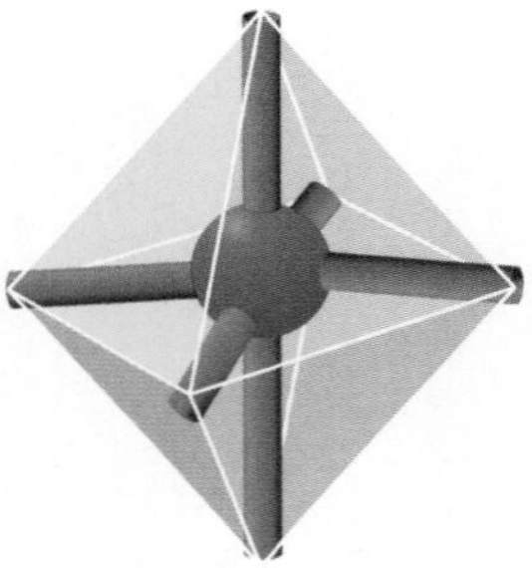

Octahedral structure. All six positions are equivalent and are at angles of 90° and 180° to each other. There is no differentiation into axial and equatorial positions. The name "octahedron" comes from the eight faces of the structure, not from the six bonds to the central atom.

In Figure 9.6, the lone pair in BrF_5 could have been located in any one of the six positions around Br. We arbitrarily chose the "down" position.

equatorial positions rather than axial positions because lone pairs are more extensive than bonding pairs. For example, consider the ClF_3 molecule, which has three bonding pairs and two lone pairs, as shown in Figure 9.6. The two lone pairs in ClF_3 are equatorial; two bonding pairs are axial, with a third occupying an equatorial position, so the molecular geometry is T-shaped (Figure 9.6). (Given our viewpoint of axial positions lying on a vertical line, the T of the molecule is actually on its side ⊢.)

An *octahedron* is the electron-pair geometry for six electron pairs around a central atom with each angle 90°.

- All six positions around the central atom in an octahedron are equivalent. Unlike the trigonal bipyramid, the octahedron has no distinct axial or equatorial positions.
- The six terminal atoms bonded to a central atom in an octahedron can occupy three axes at right (90°) angles to each other (the *x*-, *y*-, and *z*- axes; Figure 9.6).
- All six terminal atoms are *equidistant* from the central atom.

Thus, if a molecule with octahedral electron-pair geometry has one lone pair on the central atom, it makes no difference which apex the lone pair occupies. An example is BrF_5, whose molecular geometry is *square pyramidal* (Figure 9.6). There is a *square plane* containing the central atom and four of the atoms bonded to it; the other bonded atom is directly above the central atom and equidistant from the first four. If the central atom has two lone pairs, as in ICl_4^-, each of the lone pairs needs as much room as possible. This is best achieved by placing the lone pairs at a 180° angle to each other, above and below the square plane that contains the I atom and the four Cl atoms. Thus, the molecular geometry of ICl_4^- is *square planar* (Figure 9.6).

An example of the power of the VSEPR model is its use to correctly predict the shape of XeF_4. At one time the noble gases were not expected to form compounds because their atoms have a stable octet of valence electrons **(← p. 329)**. The synthesis of XeF_4 was a surprise to chemists, but you can use the VSEPR model to predict the correct geometry. The molecule has 36 valence electrons (eight from Xe and seven from each of four F atoms). There are eight electrons in four bonding pairs around Xe, and a total of 24 electrons in the lone pairs on the four F atoms. That leaves four electrons in two lone pairs on the Xe atom.

Because Xe is in Period 5, it can accommodate more than an octet of electrons (p. 378). The total of six electron pairs on Xe leads to a prediction of an octahedral *electron-pair geometry* (AX_4E_2 type). Where do you put the lone pairs? As explained earlier for the ICl_4^- ion, the lone pairs are placed at opposite corners of the octahedron to minimize repulsion by keeping them as far apart as possible. The result is a square planar *molecular geometry* for the XeF_4 molecule (cover the lone pairs in the drawing for ICl_4^- in Figure 9.6 to see this). This shape agrees with experimental structural results. A number of other xenon compounds have been prepared, and the VSEPR model has been useful in predicting their geometries as well.

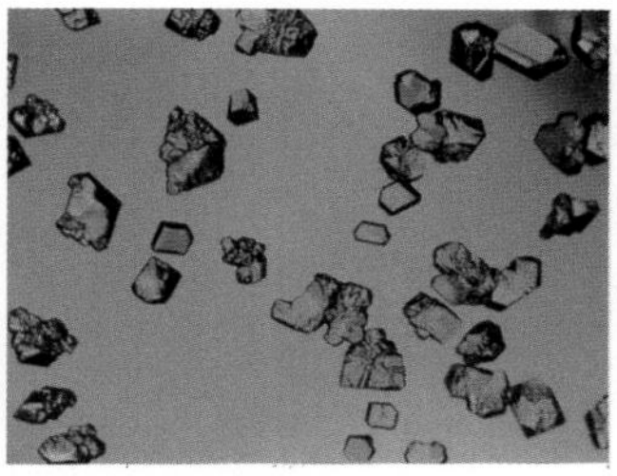

Argonne National Laboratory, managed and operated by UChicago Argonne, LLC, for the U.S. Dept of Energy under contract No. DE-AC02-06CH11357

Xenon tetrafluoride crystals.

CONCEPTUAL EXERCISE 9.2 Two Dissimilar Shapes

Triangular bipyramidal and square pyramidal molecular geometries each have a central atom with five terminal atoms around it, but the molecular shapes are different. Explain how and why these two shapes differ.

The VSEPR model also can be used to predict the geometry around atoms in molecules with more than one central atom. Consider, for example, lactic acid, a compound important in carbohydrate metabolism. The molecular structure of lactic acid is

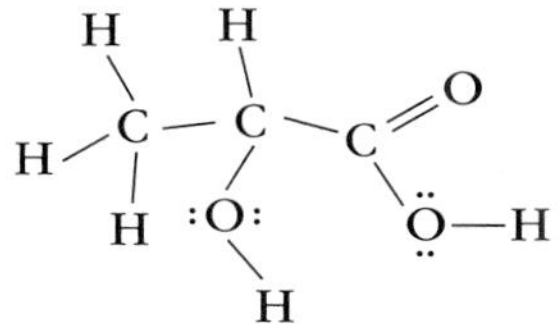

Around C in the $-CH_3$ group are four bonding pairs and no lone pairs, so its molecular geometry is tetrahedral (109.5° bond angles). The middle carbon also has four bonding pairs associated with it and so also has tetrahedral geometry (109.5° bond angles). The remaining carbon atom has four bonding pairs—two in single bonds and two in a C=O double bond—giving, for molecular geometry purposes, three bonding regions around it and a triangular planar geometry (120° bond angles). The single-bonded oxygens, in each case, have two bonding pairs and two lone pairs, creating an angular molecular geometry around each single-bonded oxygen atom (with a bond angle of less than 109.5°). The combined effect of these geometries is shown in the molecular models below.

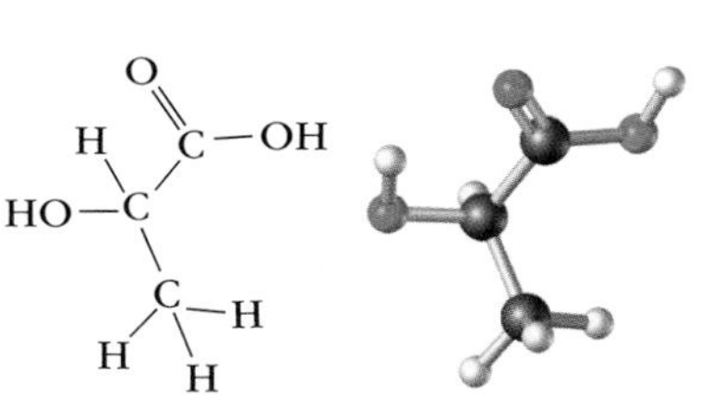

lactic acid

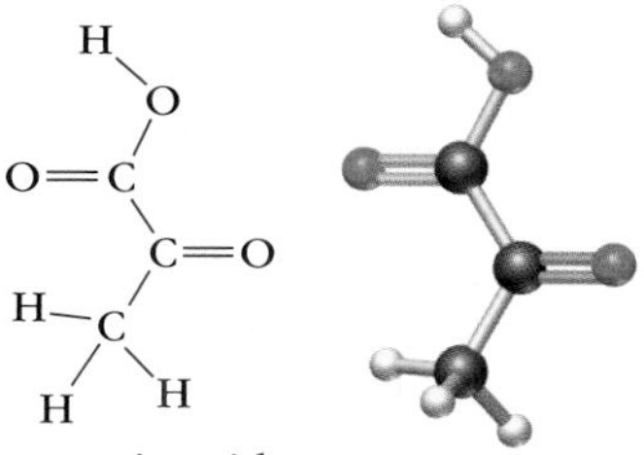

pyruvic acid

Lactic acid, a waste product of glucose metabolism, builds up rapidly in muscles during strenuous short-term exercise such as sprinting (100 to 800 m) and swimming (100 or 200 m). During such exercise, the accumulated lactic acid irritates the muscle tissue, causing muscle fatigue and soreness. But after the exercise stops, the lactic acid is converted to pyruvic acid, which is removed from the muscles, and the soreness and fatigue dissipate.

© Cengage Learning/Charles D. Winters

Lactic acid has built up in the cells of this lacrosse player.

PROBLEM-SOLVING EXAMPLE 9.3 VSEPR and Molecular Shape

What are the electron-pair and the molecular geometries of (a) SeF_6, (b) IF_3, (c) TeF_4, and (d) XeF_5^+?

Answer

(a) Octahedral electron-pair and molecular geometries

(b) Triangular bipyramidal electron-pair geometry and T-shaped molecular geometry

(c) Triangular bipyramidal electron-pair geometry and seesaw-shaped molecular geometry

(d) Octahedral electron-pair geometry and square pyramidal molecular geometry

Strategy and Explanation First, draw the correct Lewis structures, then use them as a guide to determine the electron-pair and molecular geometries. The Lewis structures are

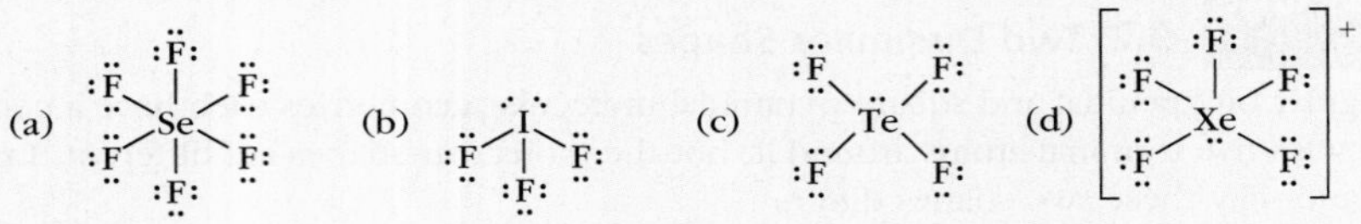

(a) The 48 valence electrons are used for six single bonds from the central Se to the terminal fluorine atoms and three lone pairs around each fluorine. Because SeF_6 is an AX_6E_0-type molecule, there are six bonding pairs around Se and no lone pairs. The electron-pair geometry and the molecular geometry are octahedral (Figure 9.4).

(b) IF_3 is an AX_3E_2-type molecule. There are five electron pairs around I, giving a triangular bipyramidal electron-pair geometry. As noted in Figure 9.6, the two lone pairs will be in equatorial positions. One bonding pair is equatorial, while the remaining two bonding pairs are axial. The molecule is T-shaped.

TOOLS OF CHEMISTRY Infrared Spectroscopy

How are molecular structures determined? Many methods rely on the interaction of electromagnetic radiation with matter. Probing matter with electromagnetic radiation is called **spectroscopy,** and each area of the electromagnetic spectrum ***(← p. 273).*** can be used as the basis for a particular spectroscopic method. Recall from Section 7.1 that electromagnetic radiation is emitted or absorbed in quantized packets of energy called photons and that the energy of the photon is represented by $E = h\nu$, where ν is the frequency of the light. Molecules may absorb several different electromagnetic radiation frequencies depending on the energy differences between their allowed energy levels. Each frequency absorbed must provide the exact packet of energy needed to lift a molecule from one energy level to another. Infrared (IR) spectroscopy uses the interaction of infrared radiation with matter to study molecular structure. It is particularly useful for learning about molecular structure because the energy of the internal motions of molecules is similar to the energy of photons whose frequency is in the infrared region.

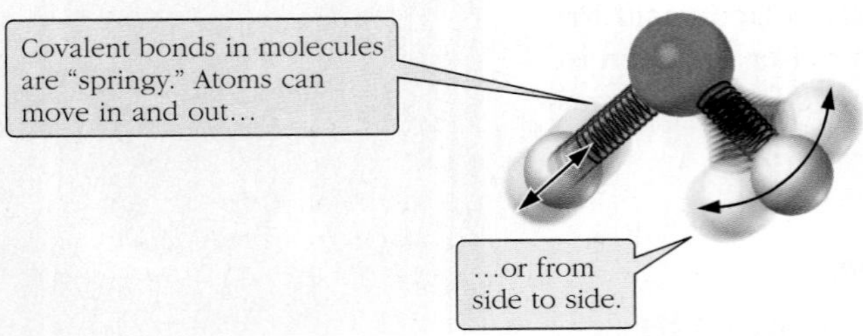
Covalent bonds in molecules are "springy." Atoms can move in and out...
...or from side to side.

Covalent bonds between atoms in a molecule can be considered to be like springs that can bend or stretch only in specified amounts. For example, bending or stretching of the bonds of the water molecule, as shown, occurs at specific frequencies corresponding to specific energy levels. The

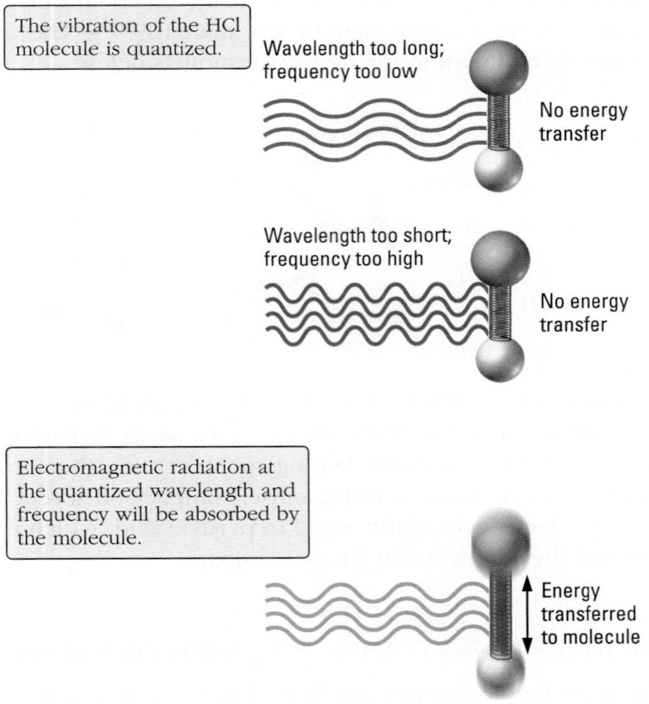
The vibration of the HCl molecule is quantized.
Wavelength too long; frequency too low
No energy transfer
Wavelength too short; frequency too high
No energy transfer
Electromagnetic radiation at the quantized wavelength and frequency will be absorbed by the molecule.
Energy transferred to molecule

(c) In TeF_4, an AX_4E_1-type molecule, the central Te atom is surrounded by five electron pairs, giving it a triangular bipyramidal electron-pair geometry. The lone pair is in an equatorial position, creating a seesaw-shaped molecule (Figure 9.6).

(d) In XeF_5^+, an AX_5E_1-type polyatomic ion, the central Xe atom has a lone pair of electrons plus five bonding pairs to terminal fluorine atoms. The ion has octahedral electron-pair geometry and a square pyramidal molecular geometry (Figure 9.6).

PROBLEM-SOLVING PRACTICE 9.3

What is the electron-pair geometry and the molecular geometry of (a) ClF_2^- and (b) XeO_3?

EXERCISE 9.3 Electron Pairs and Molecular Shapes

Classify each of these species according to its AX_nE_m type and give the molecular geometry of each.

(a) XeF_2 (b) NI_3 (c) I_3^-

Chiral Molecules

Are you right-handed or left-handed? Regardless of the preference, we learn at a very early age that a right-handed glove doesn't fit the left hand, and vice versa. Our

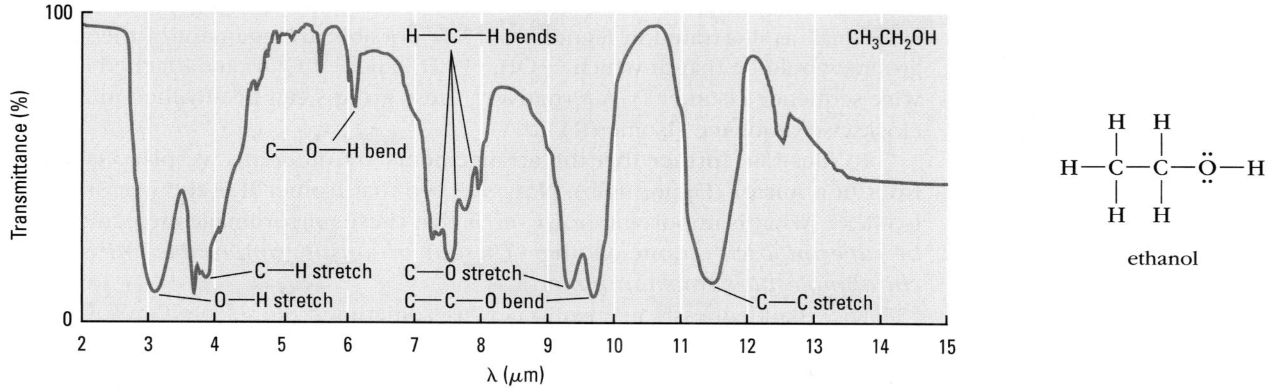

Ethanol and its infrared absorption spectrum.

strength of the covalent bonds determines what frequency of infrared light is necessary for changing from one stretching or bending energy level to another. The molecule must be excited from one of these allowed energy states to another by an exact quantity of energy.

As another example, a hydrogen chloride molecule vibrates at a specific energy. Photons that are too low or too high in energy do not cause vibration at the next higher energy level, and the radiation passes through the molecule without being absorbed. In contrast, photons of the proper energy are absorbed and increase the molecule's vibrational energy.

Because the covalent bonds in molecules differ in strength and number, the molecular motions and the number of vibrational energy levels vary; hence, the wavelengths of infrared radiation that are absorbed by different molecules differ. As a result, infrared spectroscopy can be used to learn about the structures of molecules and even to analyze an unknown material by matching its infrared spectrum with that of a known compound. In fact, the infrared frequencies absorbed by a molecule are so characteristic of the bending and stretching of various bonds that the infrared spectrum of a molecule is regarded as its *fingerprint.*

An example of the use of infrared spectroscopy in the identification of ethanol is shown above, where each of the absorption peaks is labeled with the type of bond action that causes it. To obtain such an IR absorption spectrum, infrared radiation is passed through a sample. The frequency of the IR radiation striking the sample is varied and the sample absorbs or transmits the IR radiation at specific frequencies. Peaks (drops in transmittance) occur when the sample absorbs IR radiation of a particular frequency. The overall result is an absorption spectrum, as shown for ethanol.

Figure 9.7 Nonsuperimposable mirror images. Right- and left-handed seashells are mirror images, as are the hands holding them.

In *A Midsummer Night's Dream*, Shakespeare wrote of the chirality of honeysuckle and woodbine plants. Queen Titania says to Bottom, ". . . Sleep thou, and I will wind thee in my arms . . . So doth the woodbine the sweet honeysuckle gently entwist. . . ." Woodbine spirals clockwise and honeysuckle twists in a counterclockwise direction.

In *Through the Looking Glass* (the companion volume to *Alice in Wonderland*), Alice speculates to her cat that ". . . perhaps looking glass milk is not good to drink."

hands are mirror images of one another and are not superimposable (Figure 9.7). *An object that cannot be superimposed on its mirror image is called* **chiral**. Objects that are superimposable on their mirror images are **achiral**. Stop and think about the extent to which chirality is a part of our everyday life. We've already discussed the chirality of hands (and feet). Helical seashells are chiral (most spiral to the right like a right-handed screw), and many creeping vines show a chirality when they wind around a tree or post.

What is not as well known is that a large number of the molecules in plants and animals are chiral, and usually only one form (left-handed or right-handed) of the chiral molecule is found in nature. For example, all but one of the 20 amino acids from which proteins are made are chiral, and the left-handed amino acids predominate in nature! Most natural sugars are right-handed, including glucose and sucrose and deoxyribose, the sugar found in DNA.

A chiral molecule and its *nonsuperimposable* mirror image are called **enantiomers;** they are two different molecules, just as your left and right hands are different. The simplest case of chirality is a tetrahedral carbon atom that is bonded to four *different* atoms or groups of atoms. Such a carbon atom is said to be a *chiral center,* and a molecule with just one chiral center is always a chiral molecule.

Lactic acid is one of the compounds found in nature in both enantiomeric forms under different circumstances. During the contraction of muscles, the body produces only one enantiomer of lactic acid. The other enantiomer is produced when milk sours. Lactic acid has its central C atom bonded to four different groups: —CH_3, —OH, —H, and —COOH (Figure 9.8a). As a result of the tetrahedral geometry around the central carbon atom, it is possible to have two different arrangements of the four groups. If a lactic acid molecule is placed so that the C—H bond is vertical, as illustrated in Figure 9.8a, one possible arrangement of the remaining groups would be that in which —OH, —CH_3, and —COOH are attached in a clockwise sequence (Isomer I). Alternatively, these groups can be attached in a counterclockwise sequence (Isomer II).

To illustrate further that the arrangements are different, we place Isomer I in front of a mirror (Figure 9.8b). Now you see that Isomer II is the mirror image of Isomer I. What is important, however, is that these mirror image molecules *cannot be superimposed* on one another. *These two nonsuperimposable, mirror image chiral molecules are enantiomers.*

The "handedness" of enantiomers is sometimes represented by D for right-handed (D stands for "dextro" from the Latin *dexter,* meaning "right") and L for left-handed (L stands for "levo" from the Latin *laevus,* meaning "left"). Nature's prefer-

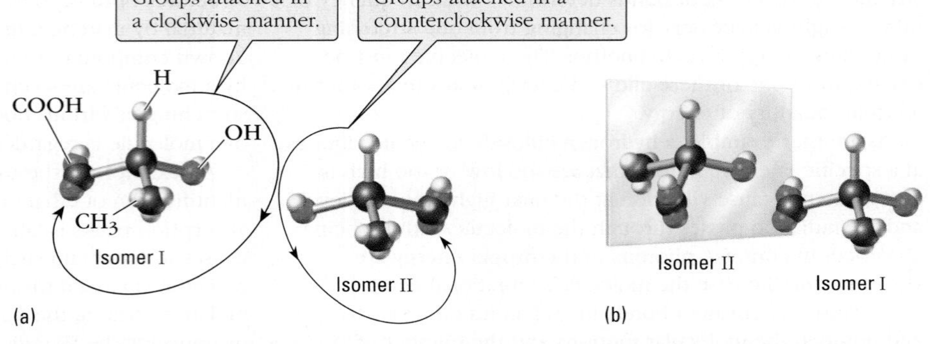

Figure 9.8 The enantiomers of lactic acid. (a) Isomer I: —OH, —CH_3, and —COOH, are attached in a clockwise manner. Isomer II: —OH, —CH_3, and —COOH are attached in a counterclockwise manner. (b) Isomer I is placed in front of a mirror, and its mirror image is Isomer II. You could superimpose C, H, and CH_3, but then OH and COOH would not superimpose.

ence for one enantiomer of amino acids (the L form) has provoked much discussion and speculation among scientists since Louis Pasteur's discovery of molecular chirality in 1848. However, there is still no widely accepted explanation of this preference.

Large organic molecules may have a number of chiral carbon atoms within the same molecule. At each such carbon atom (a chiral center), two arrangements of the molecule are possible. The maximum number of possible isomers increases exponentially with the number of different chiral carbon atoms; with n different chiral carbon atoms there are up to 2^n possible isomers. (The words "up to" are used because in some cases a mirror image may not be different from the mirrored structure.) The widely used artificial sweetener aspartame (NutraSweet) has two chiral centers and two pairs of enantiomers. One enantiomer has a sweet taste, while another enantiomer is bitter, indicating that the receptor sites on our taste buds must be chiral, because they respond differently to the "handedness" of aspartame enantiomers! In another example, D-glucose is sweet and nutritious, while L-glucose is tasteless and cannot be metabolized by the body.

PROBLEM-SOLVING EXAMPLE 9.4 Handedness in Asparagine

Asparagine is a naturally occurring amino acid first isolated as a bitter-tasting white powder from asparagus juice. Later, a sweet-tasting second form of asparagine was isolated from a sprouting vetch plant. D-Asparagine is bitter; the L isomer is sweet. The Lewis structure of asparagine is

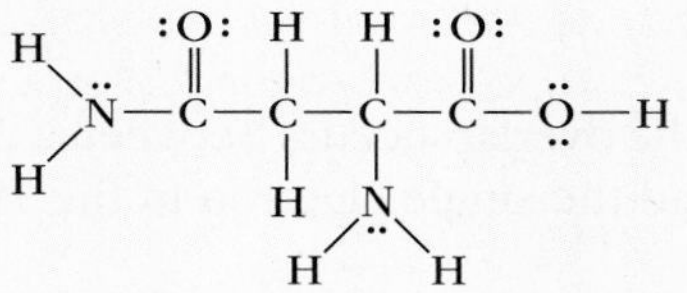

Identify which carbon atom in asparagine is the chiral center (chiral carbon).

Answer

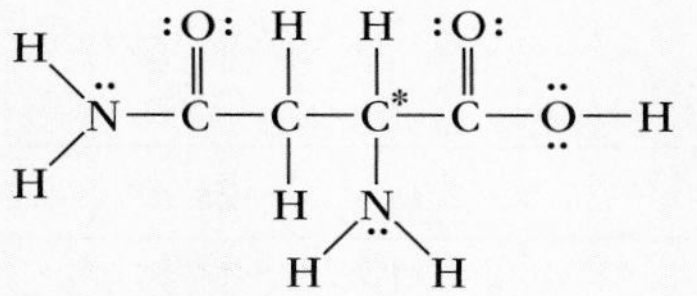

*Chiral center

Strategy and Explanation Carefully examine the Lewis structure to identify a carbon atom with four different atoms or groups attached to it; that carbon is the chiral center. The carbon next to the —COOH group is the chiral center. It has four different atoms or groups of atoms attached to it: (1) a hydrogen atom, (2) a —COOH group, (3) an $—NH_2$ group, and (4) the rest of the molecule, starting with the $—CH_2—$ group.

PROBLEM-SOLVING PRACTICE 9.4

Aspartame, a widely used sugar substitute, is a chiral molecule with the structural formula

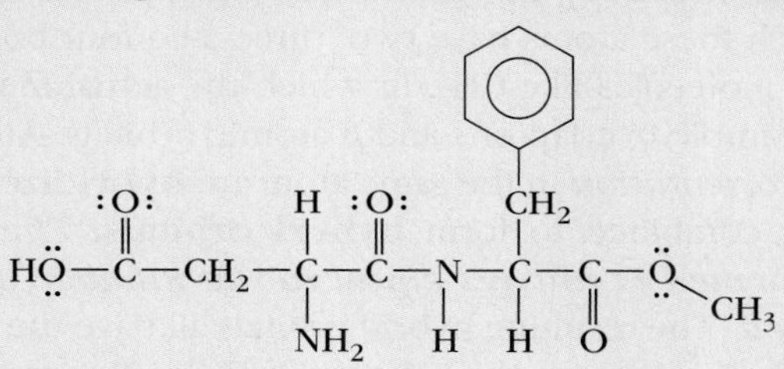

Identify the asymmetric (chiral) carbon atoms.

9.3 Atomic Orbitals Consistent with Molecular Shapes: Hybridization

go Chemistry Module 14: Hybrid Atomic Orbitals covers concepts in this section.

Although Lewis structures are helpful in assigning molecular geometries, they indicate nothing about the orbitals occupied by the bonding and lone pair electrons. A theoretical model of covalent bonding, referred to as the **valence bond model,** does so by describing a covalent bond as the result of an overlap of orbitals on each of two bonded atoms. For H_2, for example, the bond is a shared electron pair located in the overlapping *s* atomic orbitals.

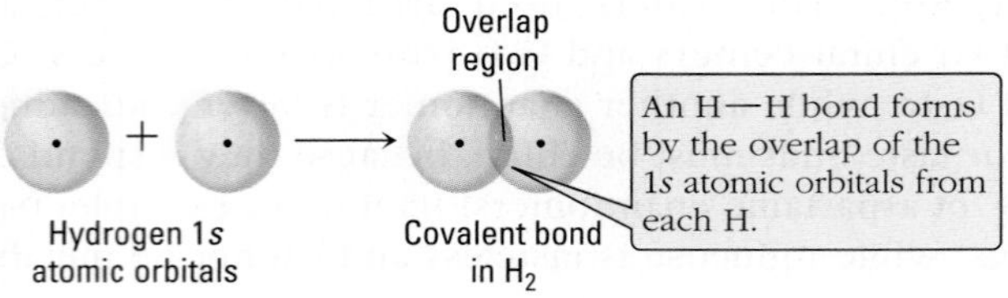

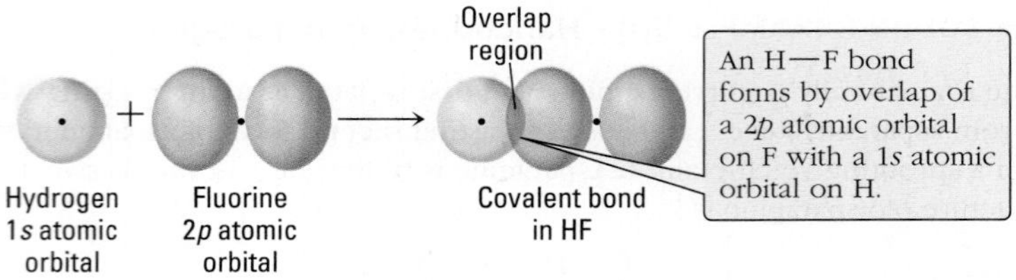

In hydrogen fluoride, HF, the overlap occurs between a 2*p* atomic orbital with a single electron on fluorine and the single electron in the 1*s* atomic orbital of a hydrogen atom.

The simple valence bond model of overlapping *s* and *p* atomic orbitals, however, must be modified to account for bonding in molecules with central atoms, such as Be, B, and C. Consider the ground-state electron configurations for Be, B, and C atoms:

			1s	**2s**		**2p**	
Be	$1s^22s^2$		↑↓	↑↓			
B	$1s^22s^22p^1$		↑↓	↑↓	↑		
C	$1s^22s^22p^2$	$1s^22s^22p_x^12p_y^1$	↑↓	↑↓	↑	↑	

The simple valence bond model would predict that Be, with an outer s^2 configuration like He, should form no covalent bonds; B, with one unpaired electron ($2p^1$) should form only one bond; and C ($2p^2$) should form only two bonds with its two unpaired 2*p* electrons. But BeF_2, BF_3, and CF_4 all exist, as well as other Be, B, and C compounds in which these atoms have two, three, and four bonds, respectively.

To account for molecules like CH_4, in which the actual molecular geometry is incompatible with simple overlap of *s* and *p* atomic orbitals. Atomic orbitals of the *proper energy and orientation* in the same atom are **hybridized,** meaning that the atomic orbitals are combined to form **hybrid orbitals.** ***The total number of hybrid orbitals formed is always equal to the number of atomic orbitals that are hybridized.*** The resulting hybrid orbitals all have the same shape and are oriented in the same directions as the balloons in Figure 9.3 (p. 377). Consequently, they are better able to overlap with bonding orbitals on other atoms. Hybrid atomic orbitals result in more bonds, stronger bonds, or both between atoms than do the unhybridized atomic orbitals from which they are formed. We turn now to a discussion of several types of hybrid orbitals.

Hybridization is described as a combination of the mathematical functions that represent *s* and *p* atomic orbitals to give new functions that predict shapes of the hybrid orbitals.

Another theory, the *molecular orbital theory,* is also used to describe chemical bonding, as well as the magnetic properties of molecules. Molecular orbital theory was discussed in Section 8.12 ***(← p. 360).***

sp Hybrid Orbitals

The simplest hybrid orbitals are ***sp* hybrid orbitals** formed by the combination of one *s* atomic orbital and one *p* atomic orbital; these *two* atomic orbitals combine to form *two sp* hybrid orbitals (Figure 9.9).

One *s* atomic orbital + one *p* atomic orbital → two *sp* hybrid orbitals

In BeF_2, for example, a 2*s* and a 2*p* atomic orbitals on Be are hybridized to form two *sp* hybrid orbitals that are 180° from each other. That is, the hybrid orbitals are oriented exactly the same way we predicted earlier for two regions of electron density.

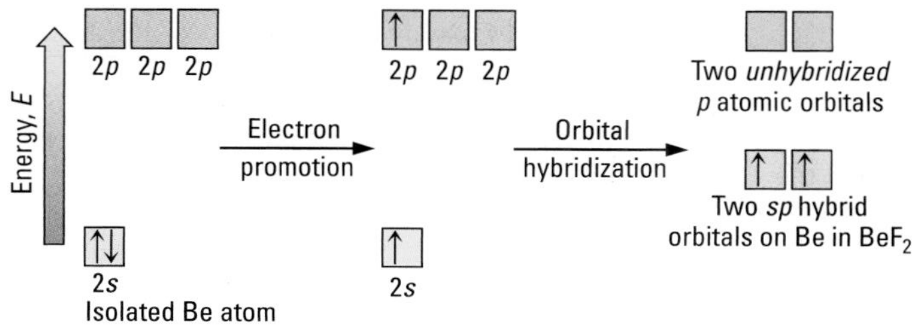

Each *sp* hybrid orbital has one electron that is shared with an electron from a fluorine atom to form two equivalent Be—F bonds. The remaining two 2*p* atomic orbitals in Be are *unhybridized* and are 90° to each other and to the two *sp* hybrid orbitals.

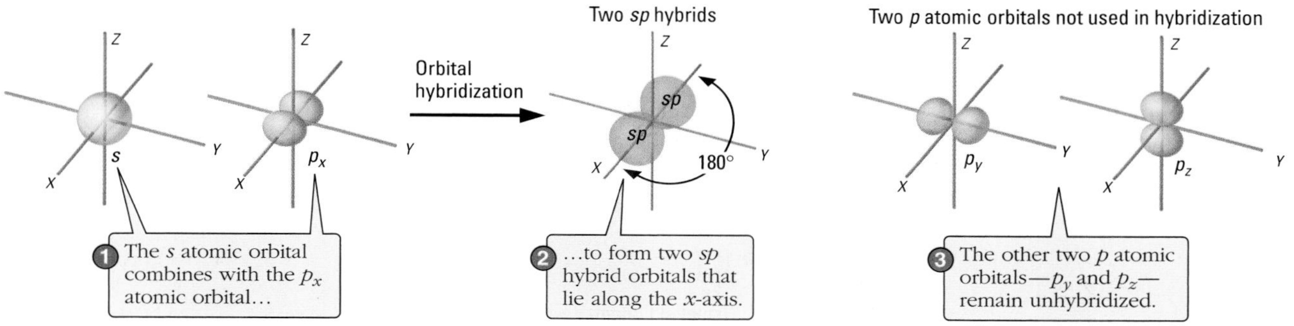

Active Figure 9.9 Formation of two *sp* hybrid orbitals. An *s* atomic orbital combines with a *p* atomic orbital, say, p_x, in the same atom to form two *sp* hybrid orbitals that lie along the *x*-axis. The angle between the two *sp* orbitals is 180°. The other two *p* atomic orbitals remain unhybridized. Visit this book's companion website at **www.cengage.com/chemistry/moore** to test your understanding of the concepts in this figure.

*sp*² Hybrid Orbitals

In BF_3, three atomic orbitals on the central B atom—a 2*s* and two 2*p* atomic orbitals—are hybridized to form *three* ***sp*² hybrid orbitals** (Figure 9.10). The superscript sum (1 + 2 = 3) indicates the number of atomic orbitals that have hybridized—three in this case: one *s* and two *p* atomic orbitals.

One *s* atomic orbital + two *p* atomic orbitals → three sp^2 hybrid orbitals

The three sp^2 hybridized orbitals on boron are 120° apart in a plane, each with an electron shared with a fluorine atom electron to form three equivalent B—F bonds. One of the boron 2*p* atomic orbitals remains unhybridized.

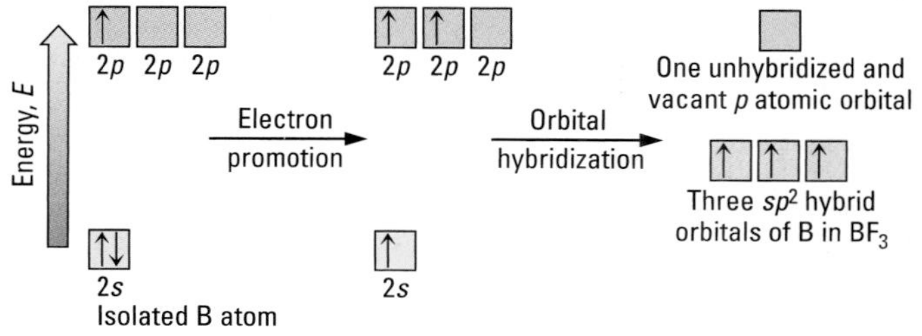

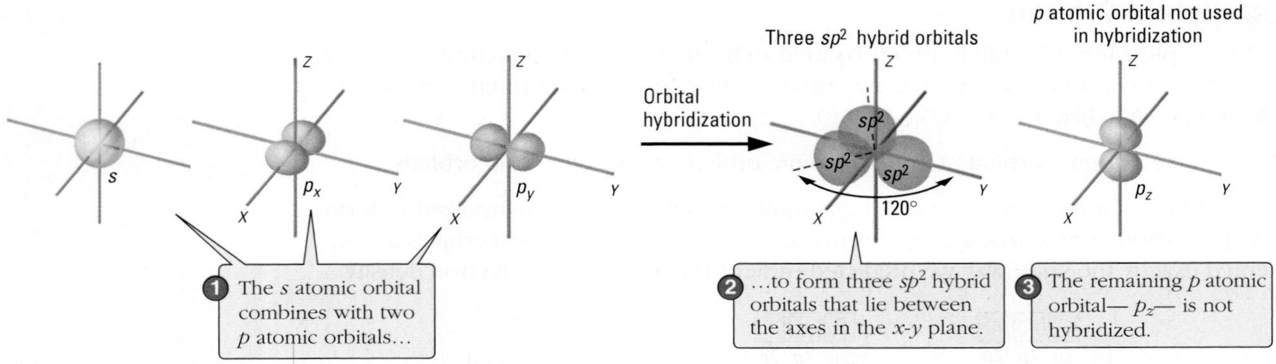
p atomic orbital not used in hybridization
Three sp^2 hybrid orbitals
Orbital hybridization
s
p_x
p_y
p_z
120°
1 The s atomic orbital combines with two p atomic orbitals...
2 ...to form three sp^2 hybrid orbitals that lie between the axes in the x-y plane.
3 The remaining p atomic orbital— p_z— is not hybridized.

Active Figure 9.10 Formation of three sp^2 hybrid orbitals. An *s* atomic orbital combines with two *p* atomic orbitals in the same atom say p_x and p_y, to form three sp^2 hybrid orbitals that lie in the *x-y* plane. The angle between the three sp^2 orbitals is 120°. The other *p* atomic orbital is unhybridized. Visit this book's companion website at **www.cengage.com/chemistry/moore** to test your understanding of the concepts in this figure.

sp^3 Hybrid Orbitals

If an *s* and the three *p* atomic orbitals on a central atom are hybridized, *four* hybrid orbitals, called ***sp*^{3} hybrid orbitals,** are formed (Figure 9.11).

One *s* atomic orbital + three *p* atomic orbitals → four sp^3 hybrid orbitals

The four sp^3 hybrid orbitals are equivalent and directed to the corners of a tetrahedron.

There are no *unhybridized* 2*s* and 2*p* atomic orbitals in the sp^3 case; all 2*s* and 2*p* atomic orbitals are hybridized.

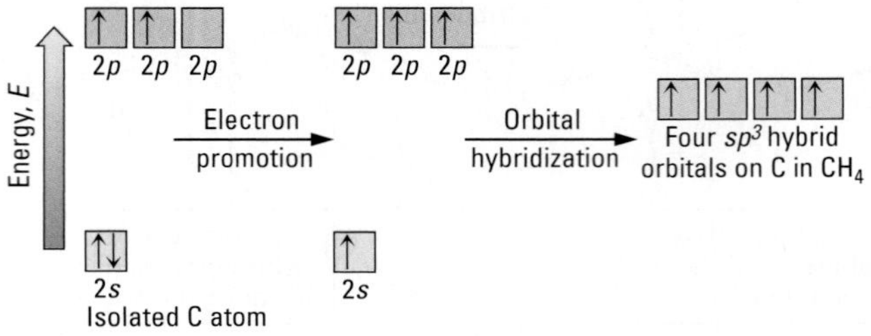
Energy, E
2p 2p 2p
2s
Isolated C atom
Electron promotion
2p 2p 2p
2s
Orbital hybridization
Four sp^3 hybrid orbitals on C in CH_4

The sp^3 hybridization is consistent with the fact that carbon forms four tetrahedral bonds in CH_4 and in all other single-bonded carbon compounds. Overlap of each half-filled sp^3 hybrid orbital on carbon with half-filled orbitals from four hydrogen atoms forms four equivalent C—H bonds. The central atoms of Periods 2 and 3 elements that obey the octet rule commonly have sp^3 hybridization.

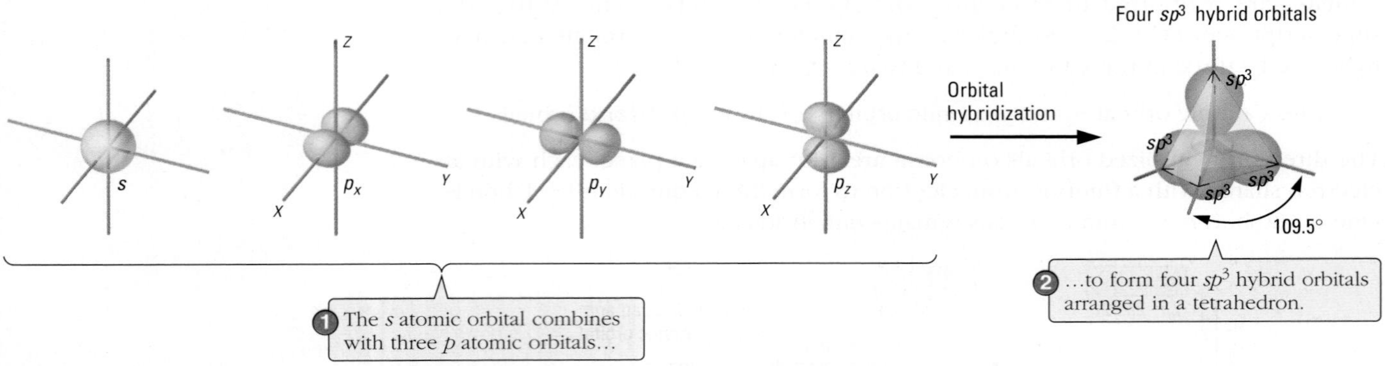
Four sp^3 hybrid orbitals
Orbital hybridization
s
p_x
p_y
p_z
109.5°
1 The s atomic orbital combines with three p atomic orbitals...
2 ...to form four sp^3 hybrid orbitals arranged in a tetrahedron.

Active Figure 9.11 Formation of four sp^3 hybrid orbitals. An *s* atomic orbital combines with three *p* atomic orbitals to form four hybrid sp^3 orbitals that are directed to the corners of a tetrahedron. The angle between the four sp^3 orbitals is 109.5°. The bonds of carbon atoms with four single bonds are described as formed by sp^3 hybrid orbitals. Visit this book's companion website at **www.cengage.com/chemistry/moore** to test your understanding of the concepts in this figure.

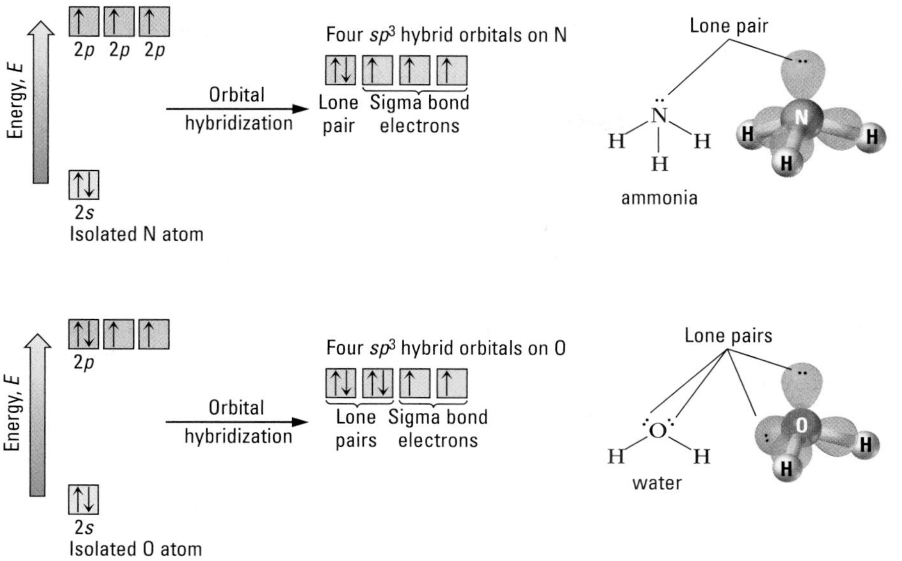

Only the five valence electrons of N are shown.

Only the six valence electrons of O are shown.

Figure 9.12 Hybridization of nitrogen in ammonia and oxygen in water. Nitrogen and oxygen are both sp^3 hybridized in these ammonia and water molecules. The lone pair electrons are in sp^3 hybrid orbitals.

Because the bond angles in NH_3 and H_2O are close to those in CF_4 and CH_4, this suggests that in general, lone pairs as well as bonding electron pairs can occupy hybrid orbitals. For example, according to the valence bond model, the four electron pairs surrounding the nitrogen atom in ammonia and the oxygen atom in water occupy sp^3 hybrid orbitals like those in CF_4 and CH_4. In NH_3, three shared pairs on nitrogen occupy three of these orbitals, with a lone pair filling the fourth. In H_2O, two of the four sp^3 hybrid orbitals on oxygen contain bonding pairs, and two contain lone pairs (Figure 9.12).

Geometry	Hybrid Orbitals
Linear	Two *sp*
Triangular planar	Three sp^2
Tetrahedral	Four sp^3

It is important for you to recognize that hybrid orbitals are consistent with the molecular geometries predicted using VSEPR theory. Table 9.1 summarizes information about hybrid orbitals formed from *s* and *p* atomic orbitals.

Bonds in which there is head-to-head orbital overlap so that the electron density of the bond lies along the bonding axis are called **sigma bonds** (σ bonds). The single bonds in diatomic molecules such as H_2 and Cl_2 and also those in BeF_2, BF_3, and CF_4 are all sigma bonds. The N—H and O—H bonds in ammonia and water, respectively, are also sigma bonds.

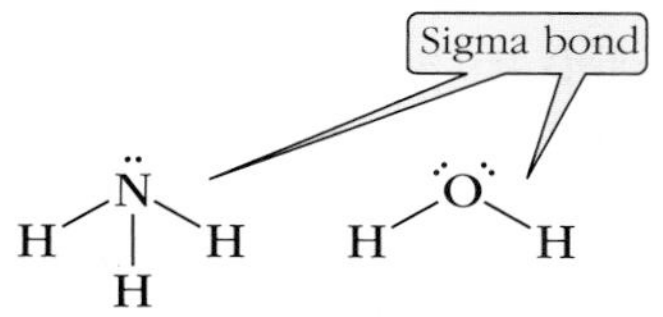

Table 9.1 Hybrid Orbitals and Their Geometries

	Linear	Triangular Planar	Tetrahedral
Atomic orbitals mixed	One *s* and one *p*	One *s* and two *p*	One *s* and three *p*
Hybrid orbitals formed	Two *sp*	Three sp^2	Four sp^3
Unhybridized orbitals remaining	Two *p*	One *p*	None

PROBLEM-SOLVING EXAMPLE 9.5 Hybridization of Atomic Orbitals

Describe the hybridization around the central atom and the bonding in
(a) NCl_3 (b) BCl_3
(c) CH_3OH (hybridization of C and O) (d) $AlCl_4^-$

Answer

(a) sp^3 with three N—Cl sigma bonds (b) sp^2 with three B—Cl sigma bonds (c) sp^3 on C as well as O with three C—H sigma bonds, one C—O sigma bond, and one O—H sigma bond (d) sp^3 with four sigma bonds between Al and Cl atoms

Strategy and Explanation Begin by writing the correct Lewis structure of each molecule or ion. Each Lewis structure provides a way to determine the number of bonding pairs and lone pairs around the central atom, a number that indicates how many hybrid orbitals are required.

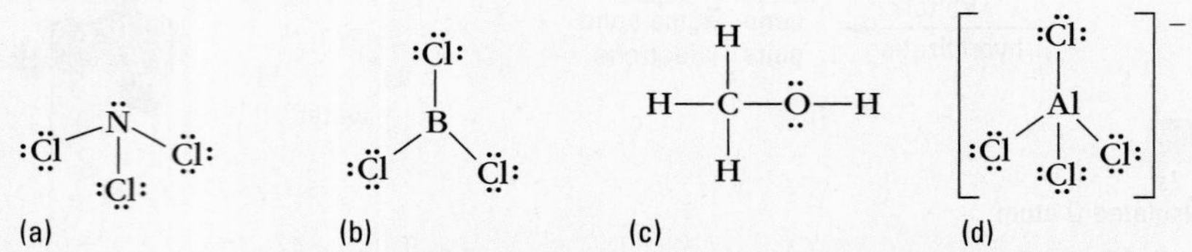

(a) There are four electron pairs around nitrogen in NCl_3, a compound analogous to NH_3. Three of the pairs are bonding pairs and the other is a lone pair. Each electron pair is in an sp^3 hybrid orbital. The three N—Cl bonds are sigma bonds formed by the head-to-head overlap of three half-filled sp^3 hybrid orbitals on nitrogen with those of chlorine. The lone pair on nitrogen is in a filled sp^3 hybrid orbital.
(b) BCl_3 is analogous to BF_3 and, therefore, boron has sp^2 orbital hybridization with boron forming three sigma bonds to chlorine.
(c) Carbon forms four sigma bonds in CH_3OH (three to hydrogen atoms and one to oxygen) and so has four bonding pairs, all in sp^3 hybrid orbitals. Oxygen has two sigma bonds formed by the head-to-head overlap of two half-filled sp^3 hybrid orbitals on oxygen with one on carbon and one on hydrogen. The two lone pairs on oxygen are in filled sp^3 hybrid orbitals.
(d) The four sigma bonds in $AlCl_4^-$ are due to four bonding pairs being in sp^3 hybrid orbitals on aluminum that overlap head-to-head with the chlorine atoms.

PROBLEM-SOLVING PRACTICE 9.5

Describe the hybridization around the central atom and the bonding in
(a) $BeCl_2$ (b) NH_4^+
(c) CH_3OCH_3 (hybridization of C and O) (d) BF_4^-

Expanded Octets and Hybridization

This discussion of hypervalent molecules nicely illustrates the continuing development of the theory of chemical bonding and molecular structure. How we explain bonding in hypervalent molecules has changed over time, and may change again.

Molecules with expanded octets (p. 383) are sometimes described as being *hypervalent* because in a Lewis diagram the central atom has more electron pairs than can be accommodated by an octet of electrons in the valence shell. This situation occurs with elements from the third period and beyond, such as in PF_5 and SF_6. The structures of hypervalent molecules can be predicted based on the number of electron regions around the central atom using VSEPR theory; however, using quantum mechanics to explain the structures is more complicated.

Until recently, the favored explanation was that *d* atomic orbitals, as well as *s* and *p* atomic orbitals, are involved in hybrid orbitals on the central atom in a hypervalent molecule. Two types of hybrid orbitals are dsp^3 (the hybridized combination of one *d* atomic orbital, one *s* atomic orbital, and three *p* atomic orbitals, which allows for 10 valence electrons) and d^2sp^3 (the hybridized combination of two *d* atomic orbitals, one *s* atomic orbital, and three *p* atomic orbitals, which allows for 12 electrons). However, a growing number of publications based on quantum mechanical calculations report that there is *little to no* hybridization of *d* atomic orbitals in hypervalent molecules or polyatomic ions where a *nonmetal* atom is the

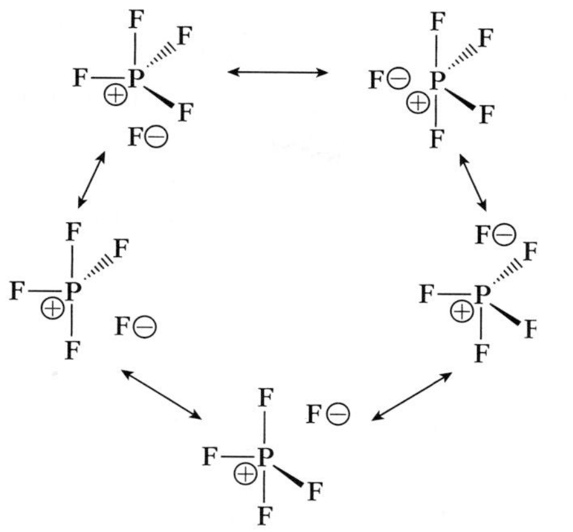

Figure 9.13 Resonance structures of PF_5 in which the octet rule is obeyed. (From T. A. Mitchell, et al., *Journal of Chemical Education,* Vol. 84, p. 629, 2007. Copyright © 2007, Division of Chemical Education, Inc. Reprinted by permission.)

central atom. Such molecules or ions include triangular bipyramidal or octahedral ones, such as PF_5, PF_6^-, and SF_6. Quantum chemistry calculations using both valence bond ***(← p. 329)*** and molecular orbital ***(← p. 360)*** theories indicate that for hypervalent nonmetal central atoms, the energy of their *d* atomic orbitals is too high for them to hybridize with lower-energy *s* and *p* atomic orbitals. Therefore, participation of *d* atomic orbitals is very small and designations such as dsp^3 and d^2sp^3 are not appropriate. The stability of species such as PF_5, PF_6^- , and SF_6 is more directly related to the size of the central atom and the high electronegativity of the terminal atoms than to *d* atomic orbital involvement.

See, for example:

Gillespie, R. J., and Popelier, P. L. A. *Chemical Bonding and Molecular Geometry*. London: Oxford University Press, 2001.

Mitchell, T. A., *et al. Journal of Chemical Education,* Vol. 84, 2007; p. 629.

Galbraith, J. M. *Journal of Chemical Education,* Vol. 84, 2007; p. 783.

These findings raise the point of whether the Lewis structure and bonding in a molecule such as PF_5 can be described in such a way that the octet rule is not exceeded, that is, without an expanded octet. An alternate approach that does so involves five equivalent resonance structures, each of which has four covalent bonds and one ionic bond as shown in Figure 9.13.

The highly electronegative terminal fluorine atoms draw sufficient electron density away from the much less electronegative phosphorus atom to leave a net of eight electrons around the phosphorus. These eight electrons are shared in four covalent bonds between phosphorus and four fluorine atoms, conforming to the octet rule. In this model, the P atom has a formal charge of 1+ ***(← p. 350)***; each resonance form also has a countervailing 1− fluoride ion to establish the ionic bond. (Note that the formal charges are not the minimum possible values; this is because of the very strong electron-attracting ability of the terminal fluorine atoms.) This method can be extended to SF_6 as well as to polyatomic ions such as SO_4^{2-} and ClO_4^-, where Lewis structures are frequently written that have more than eight electrons in the valence shell of the central nonmetal atom.

For example, the Lewis structure for SO_4^{2-} is often written

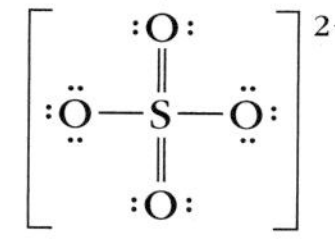

, which indicates 12 valence electrons around S.

It is important to note that there is ample experimental evidence that *d* atomic orbitals do participate in bonding when the central atom is a transition metal ion or atom. Such *d*-orbital involvement is described in detail in Chapter 22.

9.4 Hybridization in Molecules with Multiple Bonds

When using hybrid orbitals to account for molecular shapes, it must be recognized that in a multiple bond, the shared electron pairs do *not* all occupy hybrid orbitals. On a central atom with a multiple bond, the hybrid orbitals contain

- The electron pairs in single bonds to central atoms
- All lone pairs
- Only *one* of the shared electron pairs in a double or triple bond

The electrons in *unhybridized* atomic orbitals are used to form the second bond of a double bond and the second and third bonds in a triple bond. Such bonds, called **pi bonds (π bonds),** are formed by the *sideways* (edgewise) overlap of *parallel p* atomic orbitals, as shown in Figure 9.14. In contrast to sigma bonds, in which there is orbital overlap along the bond axis, pi bonds result when *parallel p* atomic orbitals (such as a p_y on one atom and a p_y on the other atom) overlap sideways above and below the bond axis.

It is also possible for *d* atomic orbitals to overlap above and below a bond axis, forming a pi bond.

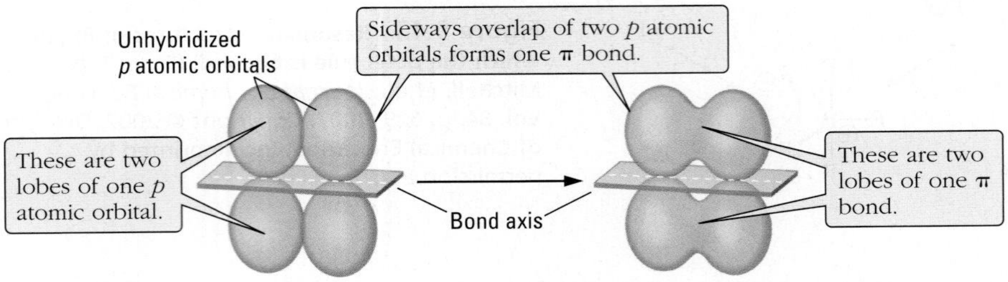
Unhybridized
p atomic orbitals
Sideways overlap of two p atomic orbitals forms one π bond.
These are two lobes of one p atomic orbital.
Bond axis
These are two lobes of one π bond.

Figure 9.14 Pi bonding. Pi bonding occurs by sideways overlap of two adjacent *p* atomic orbitals to make one pi (π) bond.

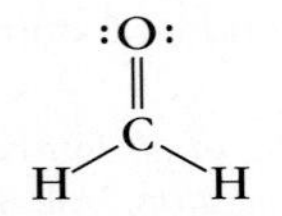
:O:
C
H
H

formaldehyde

The bonding in formaldehyde, H_2CO, exemplifies sigma and pi bonding. Formaldehyde has two C—H single bonds and one C═O double bond, as shown by the Lewis structure in the margin. The triangular planar electron-pair geometry suggests sp^2 hybridization of the carbon atom to supply three sp^2 hybrid orbitals for three sigma bonds.

As seen in Figure 9.15, the C and O atoms in formaldehyde are sp^2 hybridized.

- Two of the half-filled sp^2 hybrid orbitals on carbon each form a C—H sigma bond with a half-filled H 1*s* atomic orbital.
- The third half-filled sp^2 hybrid orbital on carbon forms a C—O sigma bond with a half-filled sp^2 hybrid orbital on oxygen by head-to-head overlap.
- Lone pairs occupy the two remaining sp^2 hybrid orbitals on oxygen.
- The half-filled *unhybridized* 2*p* atomic orbital on carbon overlaps *sideways* with the *unhybridized* 2*p* atomic orbital on oxygen to form a pi bond.
- The C═O double bond consists of one sigma plus one pi bond.

Bond energies are given in Table 8.2 (← *p. 345*).

Because pi bonds have less orbital overlap than sigma bonds, pi bonds are generally weaker than sigma bonds. Thus, a C═C double bond is stronger (and shorter) than a C—C single bond, but not twice as strong; correspondingly, a C≡C triple bond, although stronger (and shorter) than a C═C double bond, is not three times stronger than a C—C single bond.

Consider what would happen if one end of a pi bond were to rotate around the sigma bond axis relative to the other end. Because the *p* atomic orbitals need to be

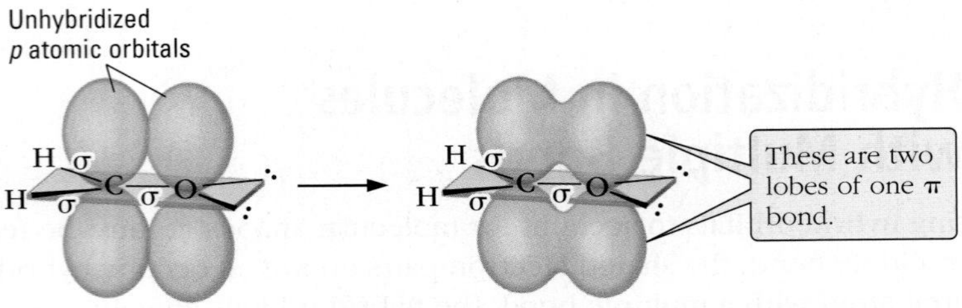
Unhybridized
p atomic orbitals
H σ
H
σ
C
σ
O
These are two lobes of one π bond.

Figure 9.15 Sigma and pi bonding in formaldehyde. In formaldehyde, the C and O atoms are sp^2 hybridized. The C—O sigma bond forms from the end-to-end overlap of two sp^2 hybrid orbitals, one from C and one from O. The sp^2 hybridization leaves a half-filled unhybridized *p* atomic orbital on each C and O atom. These unhybridized 2*p* atomic orbitals overlap edgewise to form the carbon-oxygen pi bond.

parallel to overlap, the pi bond would be broken and the molecule containing the pi bond would be less stable by an amount of energy equal to the strength of the pi bond. The resulting instability means that significant energy is required to rotate around a pi bond. For this reason, as we stated earlier, at room temperature there is no rotation around a double bond. A consequence of this barrier to rotation is *cis-trans isomerism* (← p. 340).

PROBLEM-SOLVING EXAMPLE 9.6 Hybrid Orbitals; Sigma and Pi Bonding

Use hybridized orbitals and sigma and pi bonding to describe bonding in
(a) ethane, C_2H_6 (b) ethylene, C_2H_4 (c) acetylene, C_2H_2

Answer

(a) On each carbon, sp^3 hybridized orbitals form sigma bonds to three H atoms and to the other C atom.

(b) On each carbon, sp^2 hybridized orbitals form sigma bonds to two H atoms and to the other C atom; the unhybridized $2p$ atomic orbitals form one pi bond between C atoms.

(c) The *sp* hybridized orbitals on each carbon atom form sigma bonds to one H atom and to the other C; the two unhybridized $2p$ atomic orbitals on each carbon atom form two pi bonds between C atoms.

Strategy and Explanation The Lewis structures can be used to derive the number of sigma and pi bonds, which indicate where and what type of hybridization must occur. The Lewis structures for the three compounds are

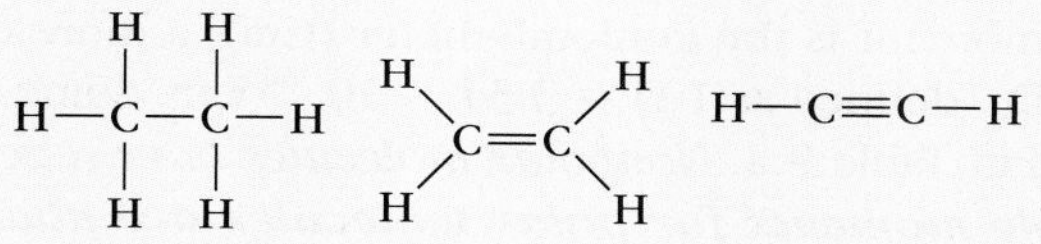

ethane ethylene acetylene

(a) Consistent with the tetrahedral bond angles for single-bonded carbon atoms, each carbon atom in ethane is sp^3 hybridized and all bonds are sigma bonds.

(b) The C=C double bond in ethylene, like the C=O double bond in formaldehyde, requires sp^2 hybridization of carbon. Two of the three sp^2 hybrid orbitals on each carbon form sigma bonds with hydrogens; the third sp^2 hybrid orbital overlaps head-to-head with an sp^2 hybrid orbital on the other carbon, creating a C—C sigma bond. The double bond between the carbon atoms is completed by the sideways overlap of parallel unhybridized *p* atomic orbitals from each carbon to form a pi bond. Thus, the C=C double bond (like all double bonds) consists of a sigma bond and a pi bond.

(c) In acetylene, *sp* hybridization of each carbon atom is indicated by the linear molecular geometry. One of the two *sp* hybrid orbitals on each carbon forms a sigma bond between C and H; the other *sp* hybrid orbital on the carbons joins the carbon atoms by a sigma bond. The two unhybridized *p* atomic orbitals on each carbon overlap sideways to form two pi bonds, completing the triple bond. Therefore, the triple bond consists of one sigma and two pi bonds in which the pi bonds are at right angles (90°) to each other (Figure 9.16).

PROBLEM-SOLVING PRACTICE 9.6

Using hybridization and sigma and pi bonding, explain the bonding in (a) HCN and (b) H_2CNH.

CONCEPTUAL EXERCISE 9.4 Pi Bonding

Explain why pi bonding is not possible for an sp^3 hybridized carbon atom.

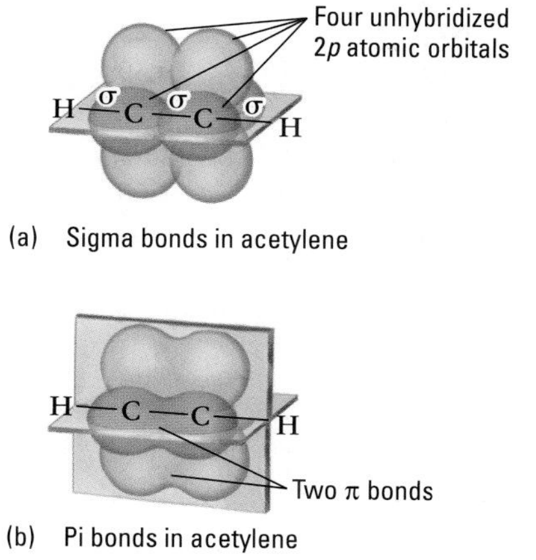

Figure 9.16 Sigma and pi bonding in acetylene. (a) The carbon atoms in acetylene are *sp* hybridized. (b) Thus, each contains two unhybridized *p* atomic orbitals, one from each carbon atom, which overlap sideways to form two pi bonds. There is also a sigma bond formed by the head-to-head overlap of two *sp* hybrid orbitals.

go Chemistry **Module 13: Molecular Polarity** covers concepts in this section.

Figure 9.17 Nonpolar (Cl_2) and polar (HCl) molecules. The Cl_2 molecule is nonpolar and so has no net dipole ($\mu = 0$). The planar HCl molecule has a net dipole ($\mu = 1.07$).

Table 9.2 Dipole Moments of Selected Molecules

Molecule	Dipole Moment, μ (D)
H_2	0
HF	1.78
HCl	1.07
HBr	0.79
HI	0.38
ClF	0.88
BrF	1.29
BrCl	0.52
H_2O	1.85
H_2S	0.95
CO_2	0
NH_3	1.47
NF_3	0.23
NCl_3	0.39
CH_4	0
CH_3Cl	1.92
CH_2Cl_2	1.60
$CHCl_3$	1.04
CCl_4	0

9.5 Molecular Polarity

Polar bonds ***(← p. 347)*** occur when two bonded atoms differ in their electronegativities ***(← p. 347).*** Such is the case in HCl where chlorine (2.7) is more electronegative than hydrogen (2.1) resulting in a polar H—Cl bond. As seen in Figure 9.17, the partial negative charge (δ−) is on the more electronegative atom (Cl); the partial positive charge (δ+) resides on the less electronegative atom (H). The bond polarity is indicated by an arrow (↦), with the plus sign indicating the partial positive charge (δ+) on the less electronegative atom, and the arrow pointing to the partial negative charge (δ−) end of the bond. A molecule that has polar bonds may or may not be polar. If electron density is concentrated at one end of the molecule, as in HCl, the result is a **polar molecule.** A polar molecule is a *permanent* dipole with a partial negative charge (δ−) where the electron density is concentrated, and a partial positive (δ+) charge at the opposite end, as illustrated in Figure 9.17. If the polar bonds are oriented so that their polarities cancel each other, such as in CO_2, a **nonpolar molecule** results.

Before examining the factors that determine whether a molecule is polar or nonpolar, let's look at the experimental measurement of the polarity of molecules. Polar molecules experience a force in an electric field that tends to align them with the field (Figure 9.18). In an electric field created by a pair of oppositely charged plates, the partial positive end of each polar molecule is attracted toward the negative plate, and the partial negative end is attracted toward the positive plate. The extent to which the molecules line up with the field depends on their **dipole moment** (μ), which is defined as the product of the magnitude of the partial charges (δ−) and (δ+) times the distance of separation between them. The derived unit of the dipole moment is the coulomb-meter (Cm); a more convenient derived unit is the debye (D), defined as $1\ D = 3.34 \times 10^{-30}$ Cm. Some typical experimental values are listed in Table 9.2. *Nonpolar molecules have a zero dipole moment ($\mu = 0$); **the dipole moments for polar molecules are always greater than zero and increase with greater molecular polarity.***

To predict whether a molecule is polar, we need to consider two factors:

- **Whether the molecule has polar bonds, and**
- **How those bonds are positioned relative to one another.**

The net dipole moment of a molecule is the sum of its bond dipoles. A diatomic molecule is, of course, linear. If its atoms differ in electronegativity, then the bond and the molecule will be polar, with the partial negative charge at the more electronegative atom, as in HCl (Figure 9.17). In polyatomic molecules, the individual

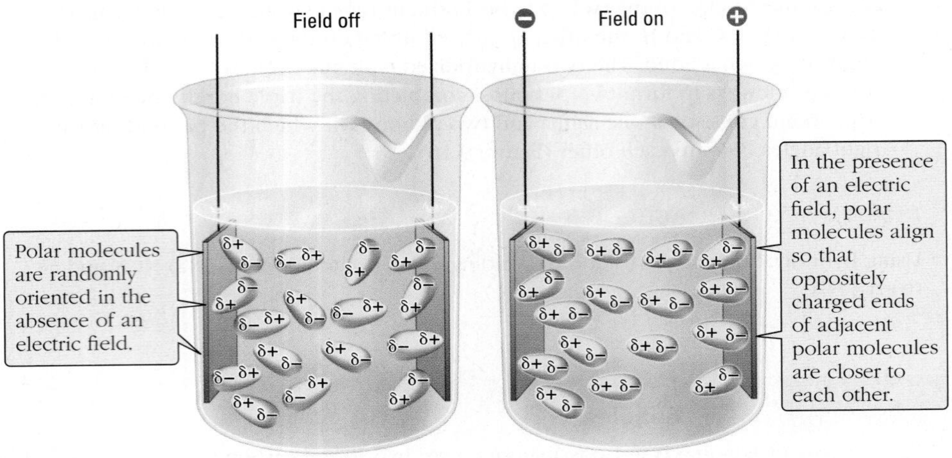

Figure 9.18 Polar molecules in an electric field.

bonds can be polar, but because of the molecular shape, there may be no net dipole ($\mu = 0$); if so, the *molecule* is nonpolar. In other nondiatomic molecules, the bond polarity and molecular shape combine to give a net dipole ($\mu > 0$) and a polar molecule. We can correlate the types of molecular geometry with dipole moment by applying a general rule to a molecule of the type AX_n where A is the central atom, X represents a terminal atom or group of atoms, and *n* is the number of terminal atoms or groups. Such a molecule will *not* be polar if it meets *both* of the following conditions:

- All the terminal atoms or groups are the same, *and*
- All the terminal atoms or groups are symmetrically arranged around the central atom, A, in AX_nE_0 molecules with the molecular geometries given in Figure 9.4.

These conditions mean that AX_nE_0 molecules with the molecular geometries given in Figure 9.4 will never be polar if all their terminal atoms or groups are the same. Any dipoles of the individual bonds will cancel due to the symmetry of the molecular structure.

Consider, for example, carbon dioxide, CO_2, a linear triatomic molecule. Each C=O bond is polar because O is more electronegative than C, so O is the partial negative end of the bond dipole. The dipole moment contribution from each bond (the bond dipole) is represented by the symbol ↦. We can use the arrows to help estimate whether a molecule is polar. In CO_2, the O atoms are at the same distance from the C atom, both have the same $\delta-$ charge, and they are symmetrically on opposite sides of C. Therefore, their bond dipoles cancel each other, resulting in a molecule with a zero molecular dipole moment (no *net* dipole). Even though each C=O bond is polar, CO_2 is a nonpolar molecule due to its linear shape.

The situation is different for water, which is an angular, triatomic molecule. Here, both O—H bonds are polar, with the H atoms having the same $\delta+$ charge (Figure 9.19). Note, however, that the two bond dipoles are not symmetrically arranged; they do not point directly toward or away from each other, as in CO_2, but augment each other to give a molecular dipole moment of 1.85 D (Table 9.2). Thus, water is a polar molecule.

Using a microwave oven to make popcorn or to heat dinner is an everyday application of the fact that water is a polar molecule. Most foods have a high water content. When water molecules in foods absorb microwave radiation, they rotate, turning so that the dipoles align with the crests and troughs of the oscillating microwave radiation. The radiation generator (the magnetron) in the oven creates microwave radiation that oscillates at 2.45 GHz (2.45 gigahertz, $2.45 \times 10^9\ s^{-1}$), very nearly the optimum frequency to rotate water molecules. So the leftover pizza warms up in a hurry for a late-night snack.

Now consider the differing dipole moments of CF_4 ($\mu = 0$ D) and CF_3H ($\mu = 1.60$ D). Both molecules have the same geometry, with their atoms tetrahedrally arranged around a central carbon atom (Figure 9.19). The terminal F atoms are all the same in CF_4 and thus have the same partial charges. However, the terminal atoms in CF_3H are not the same; F is more electronegative than C, but H is less

Hulton Archive/Getty Images

Peter Debye

1884–1966

Peter Debye was one of the leading figures in physical chemistry during the early- to mid-twentieth century. His many contributions included the first major X-ray diffraction studies with randomly oriented microcrystals, the seminal theoretical work in conjunction with Hückel on interactions of ionic solutes in solution, and the defining research on molecular polarities. After receiving the 1936 Nobel Prize in Chemistry, Debye left Germany in 1940 due to increasing Nazi influence, accepting an invitation to join the chemistry faculty at Cornell University where he remained until the end of his career.

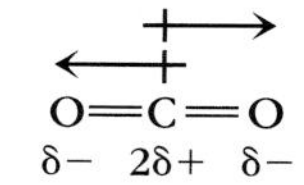

no *net* dipole; the bond dipoles cancel

Combining bond dipoles to obtain a molecular dipole involves addition of vectors (the bond dipoles) to obtain a resultant vector (the molecular dipole).

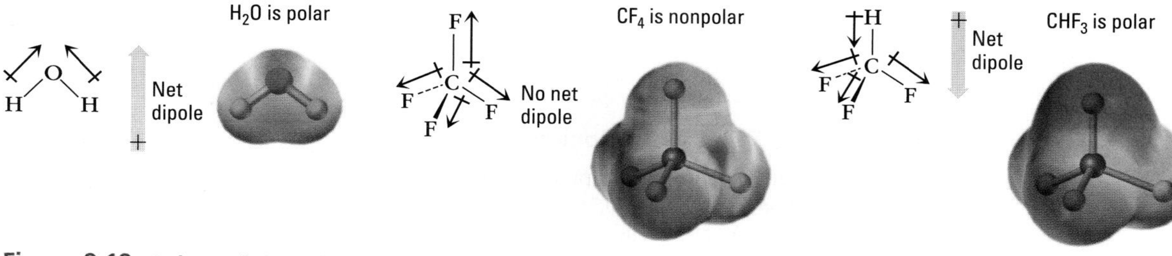

Figure 9.19 Polar and nonpolar molecules. Water ($\mu = 1.85$) and CHF_3 ($\mu = 1.65$) are polar molecules; CF_4 ($\mu = 0$) is a nonpolar molecule because its terminal atoms are all alike and are arranged symmetrically around the central carbon.

electronegative than C, so that the bond dipoles reinforce each other in CF_3H. Consequently, CF_4 is a nonpolar molecule and CF_3H is polar.

PROBLEM-SOLVING EXAMPLE 9.7 Molecular Polarity

Are boron trifluoride, BF_3, and dichloromethane, CH_2Cl_2, polar or nonpolar? If polar, indicate the direction of the net dipole.

Answer BF_3 is nonpolar; CH_2Cl_2 is polar, with the partial negative end at the Cl atoms.

Strategy and Explanation To decide whether a molecule is polar requires knowing the geometry of the molecule, the polarity of its bonds, and whether the bonds, if polar, are arranged symmetrically or unsymmetrically. As predicted for a molecule with three bonding pairs and no lone pairs around the central atom, BF_3 is triangular planar.

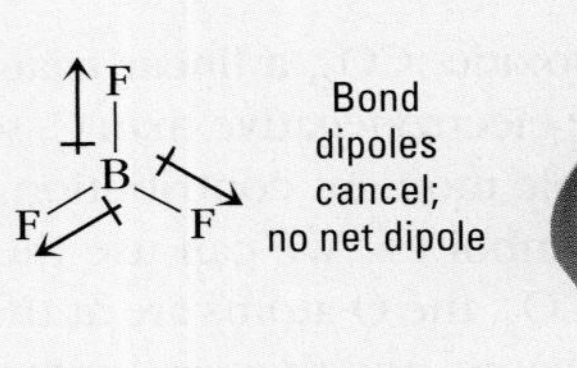

Because F is more electronegative than B, the B—F bonds are polar, with F being the partial negative end. The molecule is nonpolar, though, because the three terminal F atoms are identical, they have the same partial charge, and they are arranged symmetrically around the central B atom. The bond dipole of each B—F is canceled by the bond dipoles of the other two B—F bonds.

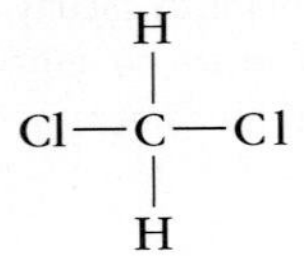

From the Lewis structure for CH_2Cl_2 in the margin, you might be tempted to say that CH_2Cl_2 is nonpolar because the hydrogens appear across from each other, as do the chlorines. In fact, CH_2Cl_2 is a tetrahedral molecule in which neither hydrogen nor chlorine is directly across from each other.

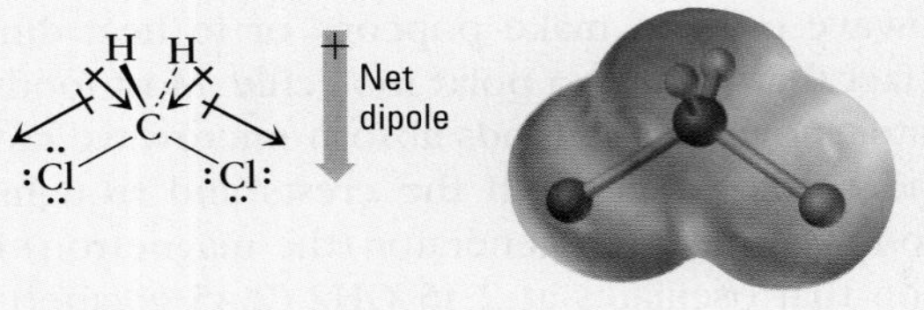

Because Cl is more electronegative than C, while H is less electronegative, negative charge is drawn away from H atoms and toward Cl atoms. As a result, CH_2Cl_2 has a net dipole (μ = 1.60 D), with the partial negative end at the Cl atoms and the partial positive end between the H atoms.

PROBLEM-SOLVING PRACTICE 9.7

For each of these molecules, decide whether the molecule is polar and, if so, which region is partially positive and which is partially negative.

(a) $BFCl_2$ (b) NH_2Cl (c) SCl_2

EXERCISE 9.5 Dipole Moments

Explain the differences in the dipole moments of

(a) HI (0.38 D) and HBr (0.79 D)

(b) CH_3Cl (1.92 D), CH_3Br (1.81 D), and CH_3I (1.62 D)

TOOLS OF CHEMISTRY Ultraviolet-Visible Spectroscopy

When a molecule absorbs radiation in the ultraviolet (UV) or visible region, the absorbed energy promotes an electron from a lower energy orbital in the ground state molecule to a higher energy orbital in the excited state molecule. The maximum UV-visible absorption occurs at a wavelength characteristic of the molecule's structure. This absorption can be determined from a UV-visible spectrum, a plot of the absorbance, which is the intensity of radiation absorbed by the molecule, versus the wavelength at which the absorbance occurred.

Ultraviolet-visible spectral data can be used to account for several structural features, including (1) differentiation of *cis* from *trans* isomers, because a *trans* isomer generally absorbs UV-visible radiation at a longer wavelength than the *cis* isomer, (2) confirmation of the presence (or absence) of carbonyl >C=O and aromatic groups in organic compounds, because they absorb UV radiation at characteristic wavelengths, and (3) specific transitions of electrons between *d* atomic orbitals in transition metal compounds.

Ultraviolet-visible spectroscopy can also be used to study the molecular structures of colored compounds, those that absorb radiation in the visible region of the spectrum. Generally, organic compounds containing only sigma bonds are colorless; that is, they do not absorb light in the visible region. On the other hand, brightly colored pigments, such as beta-carotene, have an extended sequence of alternating single and double bonds, called a *conjugated* system. (The conjugated double bonds are highlighted in the beta-carotene structure below.) When visible light is absorbed, electron transitions occur among the pi electrons in the conjugated double bonds. Carrots look yellow-orange because electron transitions in their beta-carotene absorb in the 400- to 500-nm region, which is blue-green. We then see the remaining unabsorbed (reflected) portion of the visible spectrum, which is yellow-orange.

beta-carotene

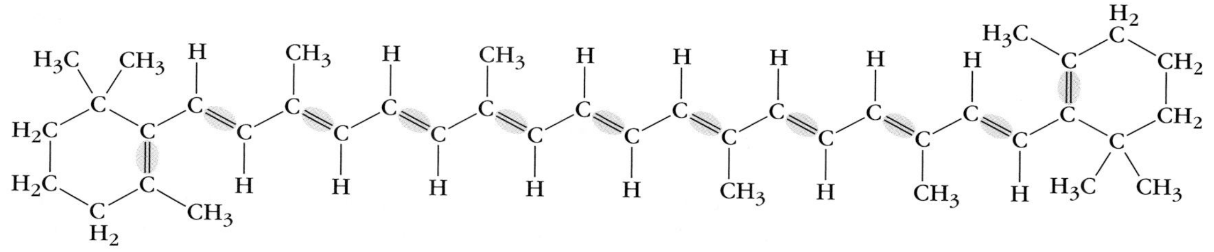

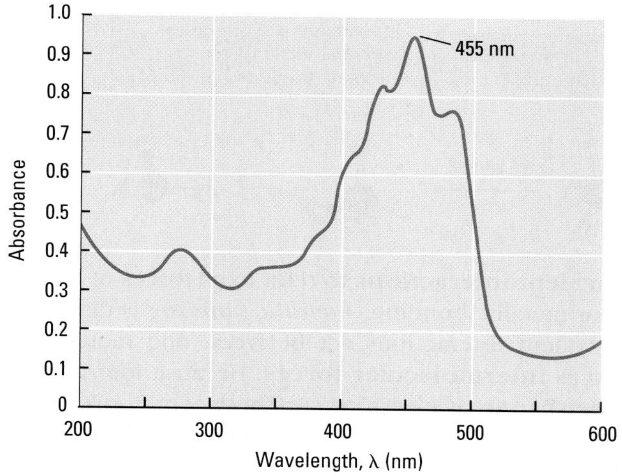

The UV-visible spectrum of beta-carotene. Beta-carotene absorbs in the blue-green region of the visible spectrum, indicated by a maximum absorbance at 455 nm. This peak is caused by transitions of pi electrons in the conjugated double bonds of the beta-carotene molecule.

© Cengage Learning/Charles D. Winters

Some vegetables and fruits that are rich in beta-carotene.

go Chemistry **Module 17: Intermolecular Forces** covers concepts in this section.

9.6 Noncovalent Interactions and Forces Between Molecules

Molecules are attracted to one another. If there were no attractions, then there would be no liquid or solid molecular compounds, only gaseous ones. The electrostatic attraction between molecules is always the result of an attraction between opposite charges, regardless of whether the charges are permanent or temporary. Polar molecules attract each other due to their permanent dipoles. Even nonpolar molecules, those without a permanent dipole, have some attraction for one another, as you will see in this section.

Atoms within the same molecule are held together by covalent chemical bonds with strengths ranging from 150 to 1000 kJ/mol. Covalent bonds are strong forces. For example, it takes 1656 kJ to break 4 mol C—H covalent bonds and separate the one C atom and four H atoms in all the molecules in 1 mol methane molecules:

$CH_4(g) \longrightarrow C(g) + 4\ H(g)$

$\Delta H^\circ_{reaction} = +1656$ kJ/mol $CH_4(g)$

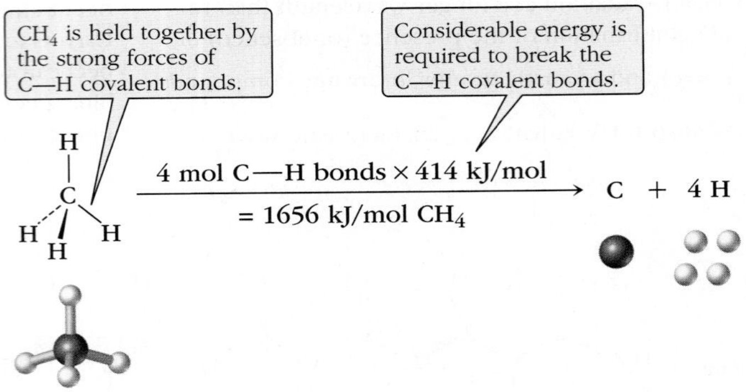

CH4 is held together by the strong forces of C—H covalent bonds.
Considerable energy is required to break the C—H covalent bonds.
4 mol C—H bonds × 414 kJ/mol
= 1656 kJ/mol CH4
C + 4 H

Weaker forces attract one molecule to another molecule. In contrast to the 1656 kJ/mol it takes to atomize methane, only 8.9 kJ is required to pull 1 mol of methane molecules that are close together in liquid methane away from each other to evaporate liquid methane to a gas.

$CH_4(\ell) \longrightarrow CH_4(g)$

$\Delta H^\circ_{vaporization} = +8.9$ kJ/mol $CH_4(\ell)$

It requires about 184 times more energy to decompose 1 mol methane gas into its atoms than to evaporate that amount of liquid methane.

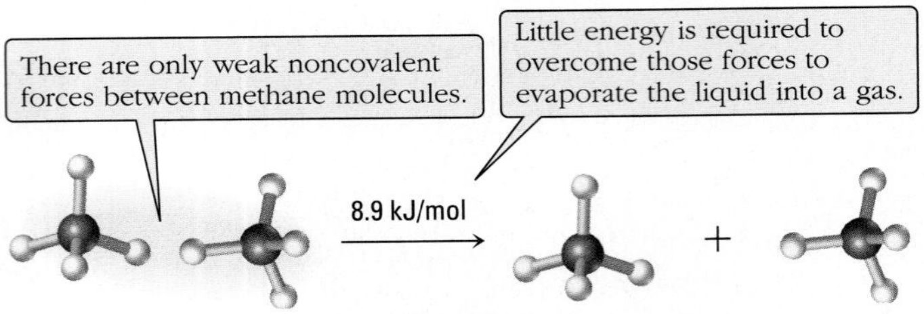

There are only weak noncovalent forces between methane molecules.
Little energy is required to overcome those forces to evaporate the liquid into a gas.
8.9 kJ/mol
+

We use the term **noncovalent interactions** to refer to all forces of attraction *other than* covalent, ionic, or metallic bonding (*metallic bonding* is discussed in Section 11.9). When noncovalent interactions act between one molecule and another, they are referred to as **intermolecular forces.** Because intermolecular forces do not result from the sharing of electron pairs between atoms, they are weaker than covalent bonds. The strengths of noncovalent interactions between molecules (*inter*molecular forces) account for melting points, boiling points, and other physical properties of molecular substances. In large molecules, noncovalent interactions can occur between different parts of the same large molecule (*intra*molecular forces). Such intermolecular and intramolecular noncovalent interactions maintain biologically important molecules in the exact shapes required to carry out their functions. Chymotrypsin, an enzyme with a molar mass of more than 22,000 g/mol, is folded into its biochemically active shape with the assistance of noncovalent interactions between strategically placed atoms. Figure 9.20 shows a ribbon model of chymotrypsin. Intramolecular noncovalent interactions help maintain the shapes of thousands of different protein molecules in our bodies.

*Intra*molecular means within a single molecule; *inter*molecular means *between* two or more separate molecules.

The next few sections explore three types of noncovalent interactions: *London forces, dipole-dipole attractions, and hydrogen bonding.*

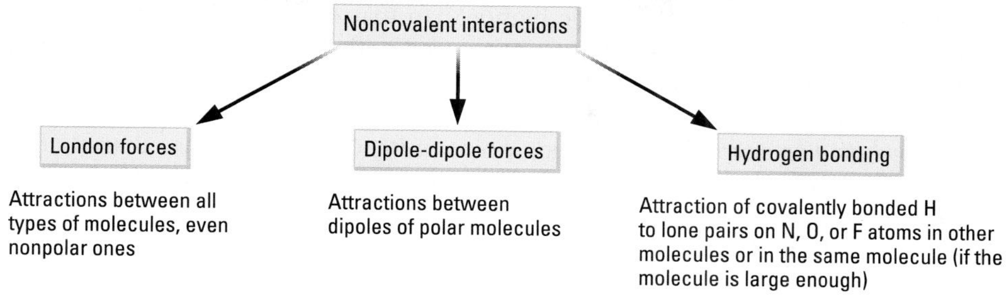
Noncovalent interactions
London forces
Dipole-dipole forces
Hydrogen bonding
Attractions between all types of molecules, even nonpolar ones
Attractions between dipoles of polar molecules
Attraction of covalently bonded H to lone pairs on N, O, or F atoms in other molecules or in the same molecule (if the molecule is large enough)

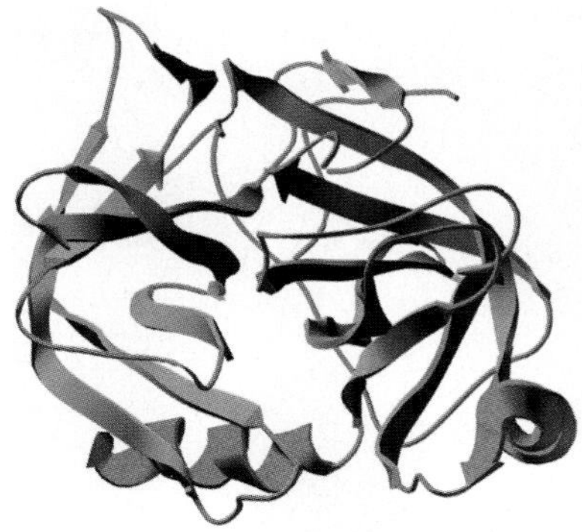

Figure 9.20 A ribbon model of a chymotrypsin molecule. Noncovalent interactions help fold chymotrypsin into its biochemically active shape. Red and green ribbons show portions where the protein chain is folded into helical or sheet structures, respectively; the narrow purple ribbons represent simpler, unfolded protein regions.

London (Dispersion) Forces

London forces, also known as *dispersion forces,* occur in *all* molecular substances. They result from the attraction between the positive and negative ends of *induced* (nonpermanent) dipoles in adjacent molecules. An **induced dipole** is caused in one molecule when the electrons of a neighboring molecule are momentarily unequally distributed, resulting in a *temporary* dipole in the first molecule. Figure 9.21 illustrates how one H_2 molecule with a momentary unevenness in its electron distribution can induce a dipole in a neighboring H_2 molecule. This kind of shift in electron distribution in a molecule is known as **polarization.** Due to polarization, even noble gas atoms, molecules of diatomic gases such as oxygen, nitrogen, and chlorine (all of which must be nonpolar), and nonpolar hydrocarbon molecules such as CH_4 and C_2H_6 have these fleeting dipoles. *Such London forces are the* only *noncovalent interactions among nonpolar molecules.*

London forces are named to recognize the work of Fritz London, who extensively studied the origins and nature of such forces.

London forces range in energy from approximately 0.05 to 40 kJ/mol. Their strength depends on how readily electrons in a molecule can be polarized, which depends on the number of electrons in a molecule and how tightly they are held by nuclear attraction. In general, electrons are more easily polarized when the molecule contains more electrons and the electrons are less tightly attracted to the nucleus; *London forces increase with increased number of electrons in a molecule.* Thus, large molecules with many electrons, such as Br_2 and I_2, are relatively polarizable. In contrast, smaller molecules (F_2, N_2, O_2) are less polarizable because they have fewer electrons.

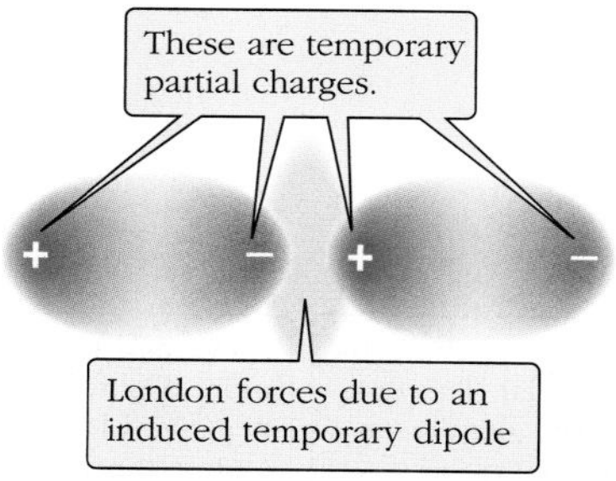
These are temporary partial charges.
+
−
+
−
London forces due to an induced temporary dipole

Figure 9.21 Origin of London forces. Such attractive forces originate when electrons are momentarily distributed unevenly in a molecule, such as the one on the right. The molecule has a temporary positive charge that is close to the molecule on the left. The positive charge attracts electrons in the left-hand molecule, temporarily creating an induced dipole (an unbalanced electron distribution) in that molecule. The two temporary dipoles attract each other.

When we look at the boiling points of several groups of nonpolar molecules (Table 9.3), the effect of the total number of electrons becomes readily apparent: *Boiling points increase as the total number of electrons increases.* (This effect also correlates with molar mass—the heavier an atom or molecule, the more electrons it has.) For a liquid to boil, its molecules must have enough energy to overcome

Table 9.3 Effect of Number of Electrons on Boiling Points of Nonpolar Molecular Substances

Noble Gases			Halogens			Hydrocarbons		
	No. e's	bp (°C)		No. e's	bp (°C)		No. e's	bp (°C)
He	2	−269	F_2	18	−188	CH_4	10	−161
Ne	10	−246	Cl_2	34	−34	C_2H_6	18	−88
Ar	18	−186	Br_2	70	59	C_3H_8	26	−42
Kr	36	−152	I_2	106	184	C_4H_{10}*	34	0

*Butane.

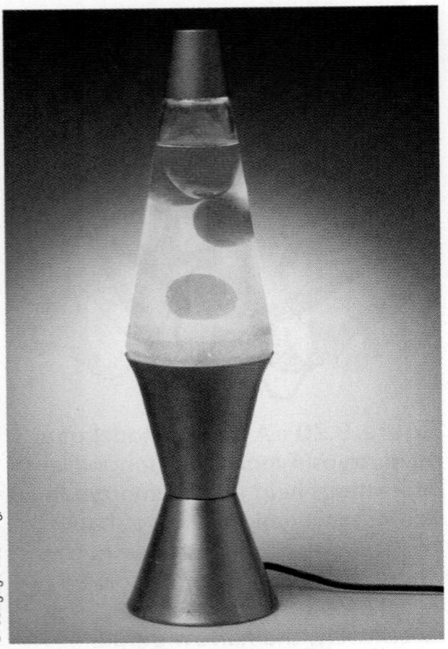
© Cengage Learning/Charles D. Winters

Lava lamps take advantage of differences in the intermolecular forces within the red liquid and within the yellow liquid in the lamp, and those intermolecular forces between the liquids.

noncovalent intermolecular attractive forces among the molecules. Thus, *the boiling point of a liquid depends on the nature and strength of intermolecular forces.* If more energy is required to overcome the intermolecular attractions between molecules of liquid A than the intermolecular attractions between molecules of liquid B, then the boiling point of A will be higher than that of B. As noted in Table 9.3, such is the case for Cl_2 (bp = −34 °C) compared with F_2 (bp = −188 °C) due to weaker London forces between diatomic fluorine molecules than between diatomic chlorine molecules. Conversely, *weaker intermolecular attractions result in lower boiling points.* For example, the boiling point of Br_2 (59 °C) is higher than that of Cl_2 (−34 °C), indicating stronger London forces among Br_2 molecules than among Cl_2 molecules.

Interestingly, molecular shape can also play a role in London forces. Two of the isomers of pentane—straight-chain pentane and 2,2-dimethylpropane (both with the molecular formula C_5H_{12})—differ in boiling point by 27 °C. The linear shape of the pentane molecule allows close contact with adjacent molecules over its entire length, resulting in stronger London forces, while the more compact 2,2-dimethylpropane molecule does not allow as much close contact.

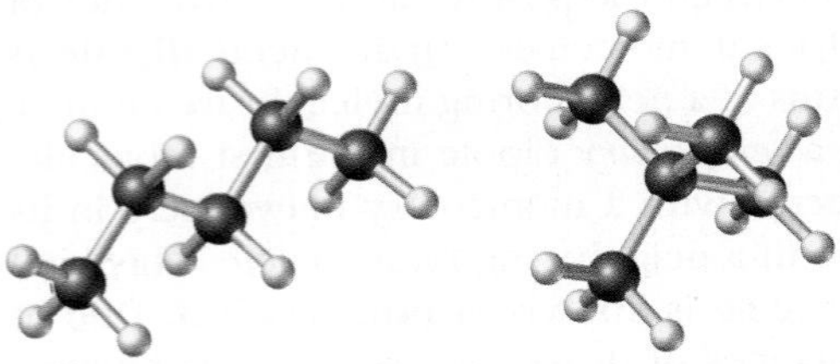

pentane, bp = 36.0 °C   2,2-dimethylpropane, bp = 9.5 °C

Structure and boiling point. The boiling points of pentane and 2,2-dimethylpropane differ because of differences in their molecular structures even though the total number of electrons in each molecule is the same.

Dipole-Dipole Attractions

Because polar molecules have *permanent* dipoles, a noncovalent interaction called a **dipole-dipole attraction** is created between two *polar* molecules or between two polar groups in the same large molecule. Molecules that are permanent dipoles attract each other when the partial positive region of one is close to the partial negative region of another (Figure 9.22).

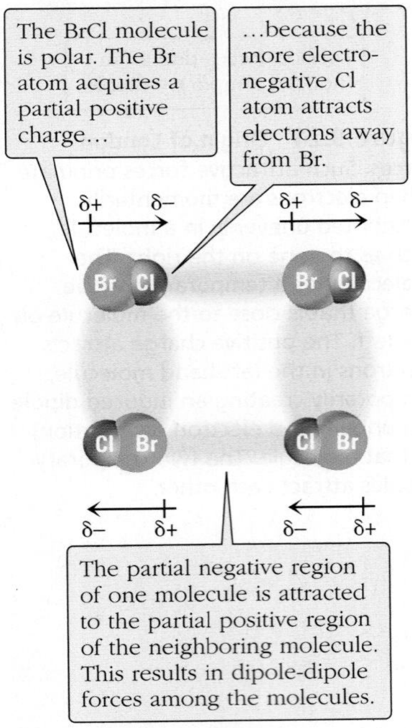

Figure 9.22 Dipole-dipole attractions between polar BrCl molecules.

The boiling points of several nonpolar and polar substances with comparable numbers of electrons, and therefore comparable London forces, are given in Table 9.4. In general, *the more polar its molecules, the higher the boiling point of a substance, provided the London forces are similar* (molecules with similar numbers of electrons). The lower boiling points of nonpolar substances compared with those of polar substances in Table 9.4 reflect this relationship. Dipole-dipole forces range from 5 to 25 kJ/mol, but London forces (0.05 to 40 kJ/mol) can be stronger. For example, the greater London forces in HI cause it to have a higher boiling point (−36 °C) than HCl (−85 °C), even though HCl is more polar. When London forces are similar, however, a more polar substance will have stronger intermolecular attractions than a less polar one. An example of this is the difference in boiling

Table 9.4 Numbers of Electrons and Boiling Points of Nonpolar and Polar Substances

Nonpolar Molecules			Polar Molecules		
	No. e's	bp (°C)		No. e's	bp (°C)
SiH_4	18	−112	PH_3	18	−88
GeH_4	36	−90	AsH_3	36	−62
Br_2	70	59	ICl	70	97

points of polar ICl (97 °C) with dipole-dipole attractions and nonpolar Br_2 (59 °C) with only London forces, even though the substances have approximately the same London forces due to their similar number of electrons (Table 9.4).

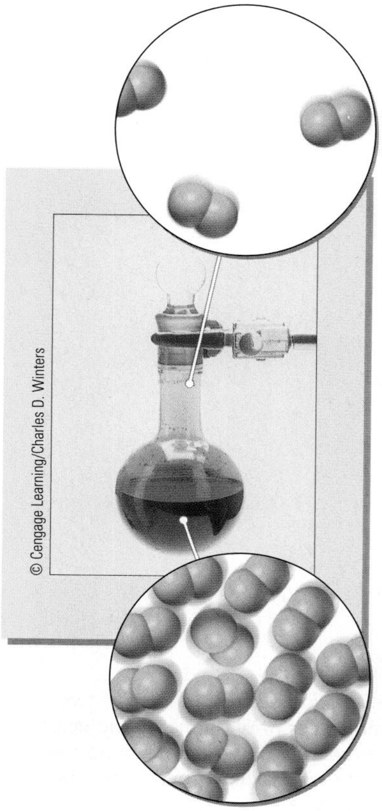
© Cengage Learning/Charles D. Winters

Bromine vapor and liquid bromine.

CONCEPTUAL EXERCISE 9.6 Dipole-Dipole Forces

Draw a sketch, like that in Figure 9.22, of four CO molecules to indicate dipole-dipole forces between the CO molecules.

PROBLEM-SOLVING EXAMPLE 9.8 Molecular Forces

Which forces must be overcome to
(a) Evaporate gasoline, a mixture of hydrocarbons?
(b) Convert solid carbon dioxide (Dry Ice) into a vapor?
(c) Decompose ammonium nitrate, NH_4NO_3, into N_2 and H_2O?
(d) Boil ClF?
(e) Convert P_4 to P_2?

Answer
(a) and (b) London forces
(c) Covalent bonds between N and H atoms, and O and N atoms; ionic attractions between NH_4^+ and NO_3^- ions.
(d) London forces and dipole-dipole forces
(e) Covalent bonds between phosphorus atoms

Strategy and Explanation Noncovalent intermolecular attractions between molecules must be broken when a molecular substance changes physical state. In such a change:

- Dipole-dipole forces are involved if the molecules are polar.
- Depending on the molecular structure, hydrogen bonding may possibly be involved if the molecules are polar.
- London forces are the *only* noncovalent intermolecular forces present between nonpolar molecules; such forces are present to some extent between all molecules, polar as well as nonpolar ones.

The decomposition of a molecular substance is drastically different from a change of physical state. Such decomposition requires breaking covalent bonds *within* the molecules.
(a) Gasoline is a mixture of hydrocarbons, which are nonpolar molecules. To evaporate, the hydrocarbon molecules must overcome London forces among them in the liquid to escape from each other and enter the gas phase.
(b) Solid carbon dioxide converts to a vapor when the thermal energy becomes great enough to overcome some of the noncovalent forces of attraction (London forces) among the nonpolar CO_2 molecules. Noncovalent forces still exist in the gaseous CO_2.
(c) Decomposing ammonium nitrate into N_2 and H_2O requires breaking the N—H and N—O covalent bonds in the NH_4^+ and NO_3^- polyatomic ions and rearranging the atoms to form nitrogen and water. These bond-breaking changes involve much greater energy (the bond enthalpy) than that needed to melt the compound, which involves overcoming ionic interactions.
(d) Dipole-dipole forces, like those between BrCl molecules in Figure 9.22, attract the polar ClF molecules to each other in the liquid. London forces are also present. These collective forces must be overcome so that ClF molecules can get away from their neighbors at the surface of the liquid and enter the gaseous state.
(e) Covalent bonds must be broken between phosphorus atoms in P_4.

PROBLEM-SOLVING PRACTICE 9.8

Explain the principal type of forces that must be overcome for
(a) Kr to melt.
(b) Propane to release C and H_2.

Hydrogen Bonds

A hydrogen bond is an especially significant type of noncovalent interaction due to a special kind of dipole-dipole force. A **hydrogen bond** is the attraction between a partially positive hydrogen atom covalently bonded to F, N, or O and a lone electron pair on a small, very electronegative atom (generally F, O, or N). The H atom and the lone pair may be in two different molecules or in different parts of the same large molecule. Electron density within molecules shifts toward F, O, or N because of their high electronegativity, giving these atoms *partial negative charges.* As a result, a hydrogen atom bonded to the nitrogen, oxygen, or fluorine atom acquires a *partial positive charge.* In a hydrogen bond, a partially positive hydrogen atom bonded covalently to one of the electronegative atoms (X) is attracted electrostatically to the negative charge of a lone pair on the other atom (Z). Hydrogen bonds are typically shown as dotted lines (· · ·) between the atoms:

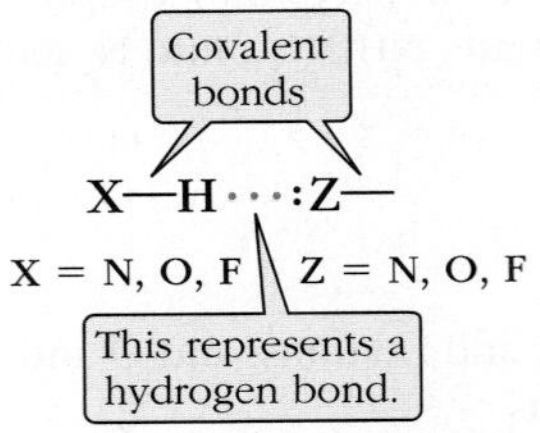
Covalent bonds
X—H···:Z—
X = N, O, F
Z = N, O, F
This represents a hydrogen bond.

Hydrogen bonds can form between molecules or within large molecules. These are known as *inter*molecular and *intra*molecular hydrogen bonds, respectively.

The hydrogen bond forms a "bridge" between a hydrogen atom and a highly electronegative atom. This type of bridge from hydrogen to a lone pair on nitrogen or on oxygen plays an essential role in determining the folding (three-dimensional structure) of large protein molecules.

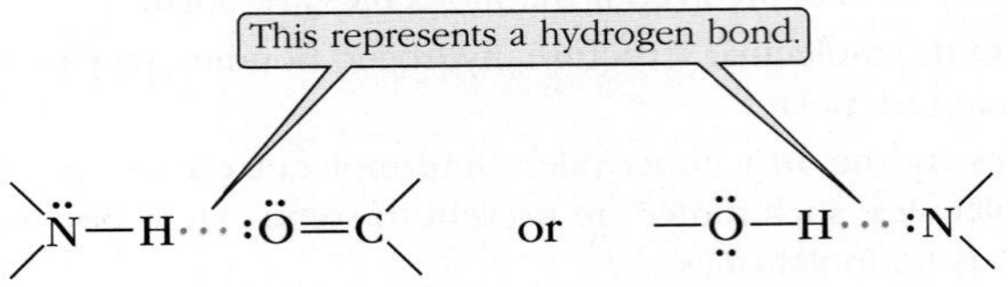
This represents a hydrogen bond.
N—H···:O=C
or
—O—H···:N

Type of Force	Magnitude of Force (kJ/mol)
Covalent bond	150–1000
London force	0.05–40
Dipole-dipole	5–25
Hydrogen bond	10–40

The greater the electronegativity of the atom connected to H, the greater the partial positive charge on H and hence the stronger the hydrogen bond.

The H atom is very small and its partial positive charge is concentrated in a very small volume, so it can come very close to the lone pair to form an especially strong dipole-dipole force through hydrogen bonding. Hydrogen-bond strengths range from 10 to 40 kJ/mol, which is less than for covalent bonds. However, a great many hydrogen bonds often occur in a sample of matter, and the overall effect can be very dramatic. An example of this effect can be seen in the melting and boiling points of ethanol. This chapter began by noting the very different melting and boiling points of ethanol and dimethyl ether, both of which have the same molecular formula, C_2H_6O, and thus the same number of electrons and roughly the same London forces. Both molecules are polar (dipole moment > 0), so there are dipole-dipole forces in each case. The differences in melting and boiling points arise because of hydrogen bonding in ethanol. The O—H bonds in ethanol make intermolecular hydrogen bonding possible, while this is not possible in dimethyl ether because it has no O—H bonds (p. 407).

The hydrogen halides also illustrate the significant effects of hydrogen bonding (Figure 9.23). The boiling point of hydrogen fluoride, HF, the lightest hydrogen halide, is much higher than expected. This is attributed to hydrogen bonding, which does not occur significantly in the other hydrogen halides.

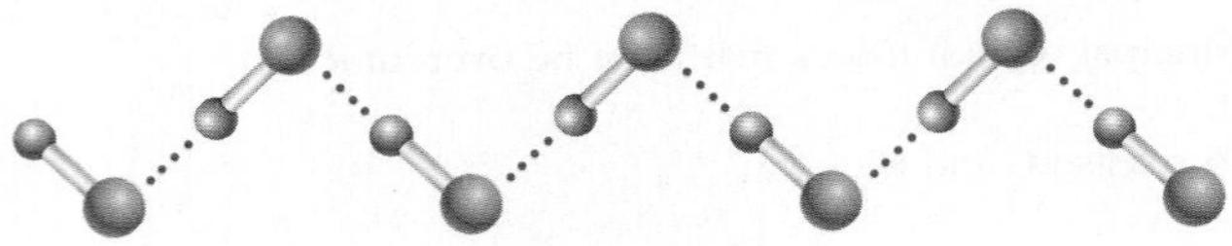

Hydrogen bonding in solid HF

Molecular model	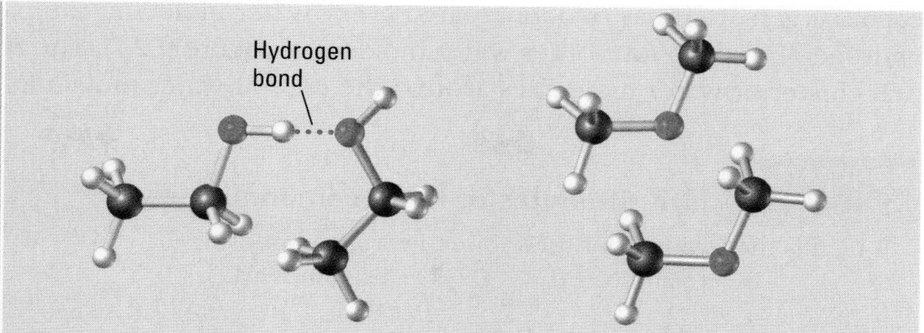 Hydrogen bond	
	ethanol	dimethyl ether
	CH_3CH_2OH	CH_3OCH_3
Dipole moment, D	1.69	1.30
Melting point, °C	−114.1	−141.5
Boiling point, °C	78.29	−24.8

Noncovalent interactions in ethanol and dimethyl ether. The molecules have the same number of electrons, so London forces are roughly the same. An ethanol molecule has an —O—H group, which means that there are both dipole forces and hydrogen bond forces attracting ethanol molecules to each other. A dimethyl ether molecule is polar, but without an —O—H group there is no hydrogen bonding, so the noncovalent intermolecular noncovalent forces are weaker than in ethanol.

Hydrogen bonding is especially strong among water molecules and is responsible for many of the unique properties of water (Section 11.4). Hydrogen compounds of oxygen's neighbors and family members in the periodic table are gases at room temperature: CH_4, NH_3, H_2S, H_2Se, H_2Te, PH_3, HF, and HCl. But H_2O is a liquid at room temperature, indicating a strong degree of intermolecular attraction. Figure 9.23 shows that the boiling point of H_2O is about 200 °C higher than would be predicted if hydrogen bonding were not present.

In liquid and solid water, where the molecules are close enough to interact, the hydrogen atom on one water molecule is attracted to the lone pair of electrons on the oxygen atom of an adjacent water molecule. Because each hydrogen atom can form a hydrogen bond to an oxygen atom in another water molecule, and because

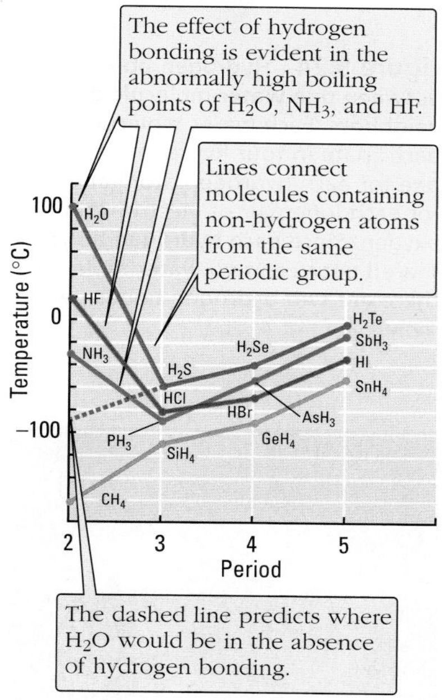
The effect of hydrogen bonding is evident in the abnormally high boiling points of H₂O, NH₃, and HF.
Lines connect molecules containing non-hydrogen atoms from the same periodic group.
Temperature (°C)
100
0
−100
H2O
HF
NH3
CH4
H2S
HCl
PH3
SiH4
H2Se
HBr
GeH4
AsH3
H2Te
SbH3
HI
SnH4
2
3
4
5
Period
The dashed line predicts where H₂O would be in the absence of hydrogen bonding.

Figure 9.23 Boiling points of some simple hydrogen-containing binary compounds.

CHEMISTRY IN THE NEWS Icy Pentagons

Typically, when water freezes, the water molecules in ice are arranged in puckered hexagons. Recently, researchers led by Angelos Michaelides of the London Center for Nanotechnology found that when water freezes on a copper surface, the water molecules are arranged in a pentagonal pattern. Using scanning tunneling microscopy, reflection absorption infrared spectroscopy, and density functional theory (DTF), the researchers found chains of water molecules on the copper surface. The chains were arranged into face-sharing pentagons. Calculations using DFT indicated that the pentagons are favored over the normal hexagons because the pentagons maximize the proportion of oxygen atoms that interact with copper, while retaining the extended hydrogen-bonding network among the water molecules.

Source: From "A One-dimensional Ice Structure Built from Pentagons" by J. Carrasco *et al.*, *Nature Materials,* Vol. 8 (March 2009), p. 427–431. "Surface water: Pentagonal ice in chains" by Peter J. Feibelman. *Nature Materials,* Vol. 8 (May 2009), p. 372–373. Published by the Nature Publishing Group **(www.nature.com).** Reprinted by permission.

Adapted from Nat. Mater.

Chains of water molecules on a copper surface.

each oxygen atom has two lone pairs, every water molecule can participate in four hydrogen bonds to four other water molecules (Figure 9.24). The result is a tetrahedral cluster of water molecules around the central water molecule.

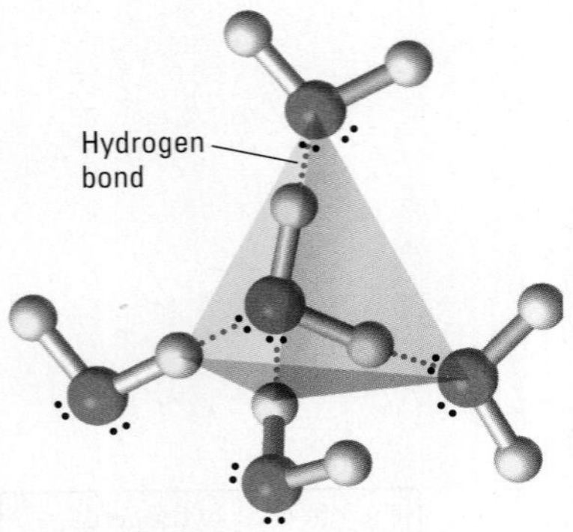

Figure 9.24 Hydrogen bonding between one water molecule and its neighbors. Each water molecule can participate in four hydrogen bonds—one for each hydrogen atom and one for each lone pair of electrons on oxygen. Since each hydrogen bond is shared between two water molecules, there are two hydrogen bonds per molecule ($\frac{1}{2} \times 4$).

© Scott Camazine/Photo Researchers, Inc.

Ice crystals. The crystal shapes reflect the hexagonal orientation of water molecules in ice due to their hydrogen bonds.

CONCEPTUAL EXERCISE 9.7 Strengths of Hydrogen Bonds

Which is the strongest of these three hydrogen bonds, F—H · · · F—H, O—H · · · N—C, and N—H · · · O—C ? Explain why.

PROBLEM-SOLVING EXAMPLE 9.9 Molecular Forces

What are the types of forces, in addition to London forces, that are overcome in these changes? Using structural formulas, make a sketch representing the major type of force in each case.

(a) The evaporation of liquid ethanol, CH_3CH_2OH
(b) The decomposition of ammonia, NH_3, into N_2 and H_2
(c) The sublimation of iodine, I_2, that is, its conversion from a solid directly into a vapor
(d) The boiling of liquid NH_3

Answer

(a) Hydrogen bonding and dipole forces:

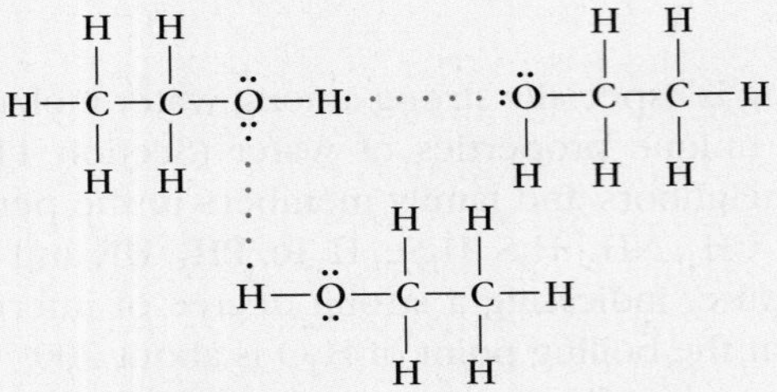

(b) Covalent bonds between N—H:

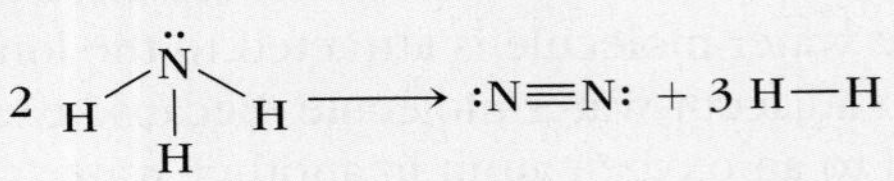

(c) London forces between I_2 molecules:

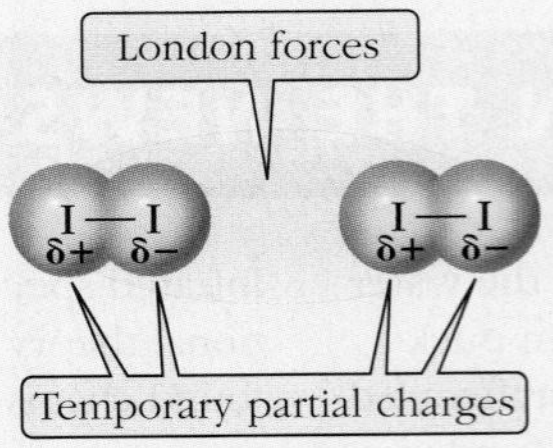

(d) Hydrogen bonding and dipole forces:

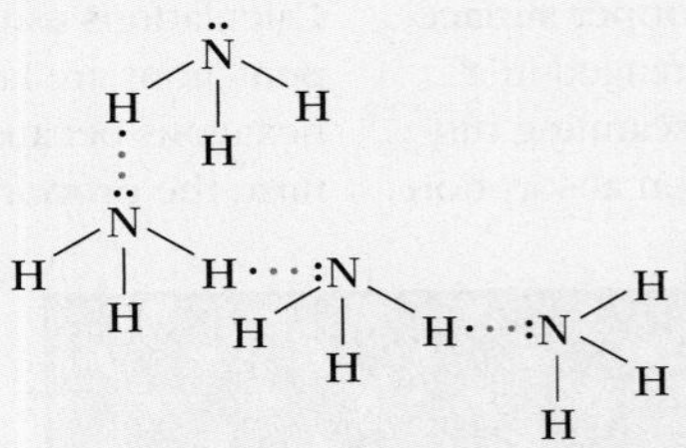

Strategy and Explanation In each case identify the force to be overcome. Where the change involves an intermolecular force, consider the particular kind of force.

(a) Ethanol molecules have a hydrogen atom covalently bonded to a highly electronegative oxygen atom with a lone pair of electrons. Therefore, ethanol molecules can hydrogen bond to each other. The bent —O—H group is polar also.

(b) This is not a case of overcoming an intermolecular force, but rather an *intra*molecular one, covalent bonds. The decomposition of ammonia involves breaking the N—H covalent bonds in the NH_3 molecule.
(c) Iodine is composed of nonpolar I_2 molecules, which are held to each other in the solid by London forces. Therefore, these forces must be overcome for the iodine molecules in the solid to escape from one another and become gaseous.
(d) Ammonia molecules are held together in the liquid by hydrogen bonds, which must be broken for the molecules at the surface of the liquid to escape into the vapor. (Figure 9.23 shows the much higher boiling point of NH_3 compared to PH_3.) Because NH_3 is polar, dipole-dipole forces must also be overcome.

PROBLEM-SOLVING PRACTICE 9.9

Decide what types of intermolecular forces are involved in the attraction between molecules of (a) N_2 and N_2, (b) CO_2 and H_2O, and (c) CH_3OH and NH_3.

250 ml
PYREX

Sublimation of iodine. When heated, solid iodine in the bottom of the beaker sublimes directly to violet-colored iodine vapor. When it contacts the cold surface of the ice-filled, round flask at the top of the beaker, the iodine vapor deposits as dark-colored iodine crystals.

Noncovalent Forces in Living Cells

Noncovalent forces are important in biological structures, including cell membranes whose basic structure, a **lipid bilayer,** is two aligned layers of phospholipids. Phospholipids are glycerol derivatives having two long, nonpolar fatty acid chains and a polar phosphate derivative at the head of the molecule (Figure 9.25).

The polar end of phospholipid molecules is **hydrophilic,** that is, "water-loving," because of its attraction to water molecules. The polar end is oriented toward the aqueous environments of the cell and its surroundings. The long, nonpolar chains of the fatty acids are **hydrophobic,** or "water-hating." They point away from the aqueous environments and are held together by London forces. The result is separation of the interior of a cell from the surrounding fluids, a separation that allows control over which ions or molecules enter a cell and which chemical reactions can occur there.

In our bodies, noncovalent interactions bring metabolic products, drugs, hormones, and neurotransmitters (nervous system messengers) into contact with cellular receptors. The receptors are often proteins embedded in the cell membrane's lipid bilayer. Dipole-dipole forces, hydrogen bonding, and London forces help to align the target molecule with its particular site on the receptor surface. Once a chemical messenger connects with its receptor, the messenger may enter the cell to initiate a chemical reaction, or a reaction may occur as a result of changes initiated by the receptor.

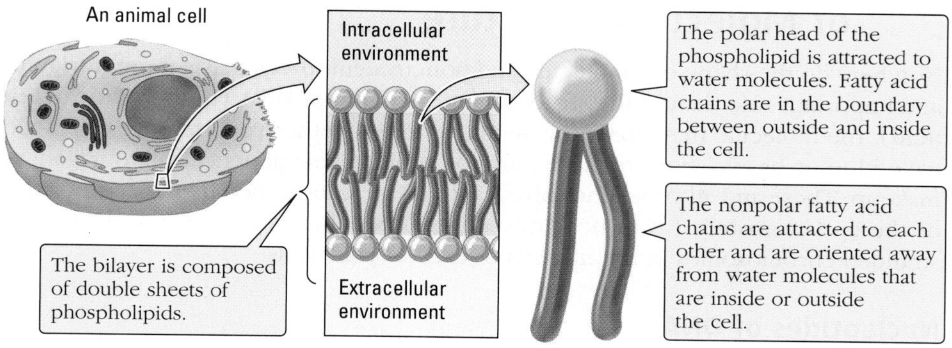
An animal cell
Intracellular environment
The polar head of the phospholipid is attracted to water molecules. Fatty acid chains are in the boundary between outside and inside the cell.
The nonpolar fatty acid chains are attracted to each other and are oriented away from water molecules that are inside or outside the cell.
The bilayer is composed of double sheets of phospholipids.
Extracellular environment

Figure 9.25 Lipid bilayer in cell membranes.

CHEMISTRY YOU CAN DO Molecular Structure and Biological Activity

Noncovalent forces (London forces, hydrogen bonds, and dipole-dipole forces) are important in biochemical reactions for bringing together reactants in proper orientations. This is particularly the case for enzymes—biomolecules that help to speed up the rates of biochemical reactions (Section 13.9). Noncovalent forces often bind the reactant to an enzyme's *active site,* the particular region of the enzyme where its reactivity occurs.

Lactase, which acts on lactose, is such an enzyme. Lactose, a disaccharide (Section 12.7), is also known as milk sugar because it is present in the milk of humans, cows, and other mammals. Lactose can make up to about 40% of an infant's diet during the first year. Lactase accelerates the decomposition of lactose to its component monosaccharides, glucose and galactose. A portion of the lactase molecule has a region whose shape matches only that of lactose. Therefore, the enzyme acts selectively on lactose and not on sucrose or maltose, two other dietary disaccharides whose shapes do not match that required by lactase.

Infants have a more active form of the enzyme than do adults. Consequently, nearly 20% of adults in the United States and nearly 70% worldwide suffer some degree of lactose intolerance, the inability to rapidly digest lactose in milk and dairy products. Lactaid, an over-the-counter product that contains lactase, can be used by lactose-intolerant individuals to break down lactose in milk.

In this activity, you will use Lactaid to study the conversion of lactose in milk to glucose and galactose. Do this experiment with several friends. You will need:

- One quart of 2% milk
- One package of Lactaid drops (not caplets), available at most pharmacies
- Two one-pint (16-oz) containers that can be closed
- Four to eight small glasses

Taste a drop of Lactaid and record its taste.

Make a sample of milk with Lactaid and a control of pure milk. For the samples pour half the milk into one container as a control and close the container. Pour the remainder of the milk into the second container, add 3 to 4 drops of Lactaid, close the container, and shake it to mix its contents. Mark the Lactaid container to identify it. Refrigerate all containers for about 48 hours.

Lactaid.

Take the containers from the refrigerator and pour samples of the Lactaid-treated and untreated milk into separate glasses. Each person should taste a sample of Lactaid-treated milk and a sample of pure milk and record any differences in the taste.

Think about these questions:

1. What difference in taste was there between the untreated and the Lactaid-treated samples?
2. Explain the difference in taste to the others who have done this exercise.

Source: Richman, R. "Detection of Catalysis by Taste." *Journal of Chemical Education,* Vol. 75, 1998; p. 315.

9.7 Biomolecules: DNA and the Importance of Molecular Structure

Nowhere do the shape of a molecule and noncovalent forces play a more intriguing and important role than in the structure and function of **DNA (deoxyribonucleic acid),** the molecule that stores the genetic code. Whether you are male or female, have blue or brown eyes, or have curly or straight hair depends on your genetic makeup. These and all your other physical traits are determined by the composition of the approximately 1.8 m of tightly coiled DNA that makes up the 23 double-stranded chromosomes in the nucleus of each of your cells.

Nucleotides of DNA

A DNA nucleotide is a combination of a phosphate group, deoxyribose, and a nitrogen base.

DNA is a **polymer,** a molecule composed of many small, repeating units bonded together. Each repeating unit in DNA, called a **nucleotide,** has the three connected parts shown in Figure 9.26—one sugar unit, one phosphate group (phosphoric acid unit), and a cyclic nitrogen compound known as a *nitrogen base.* In DNA the sugar is always deoxyribose and the base is one of four bases—*adenine (A), thymine (T), guanine (G),* or *cytosine (C).* The bases are often referred to by their single-letter abbreviations.

A DNA segment with nucleotides bonded together is shown in Figure 9.26. The phosphate units join nucleotides into a polynucleotide chain that has a backbone of alternating deoxyribose and phosphate units in a long strand with the various nitrogenous bases extending out from the sugar-phosphate backbone. The order of the nucleotides (and thus the particular sequence of bases) along the DNA strand carries the genetic code from one generation to the next.

DOE/Science Source/Photo Researchers, Inc.

Francis H. C. Crick *(right)* and James Watson. Working in the Cavendish Laboratory at Cambridge, England, Watson and Crick built a scale model of the double-helical structure, based on X-ray data. Knowing distances and angles between atoms, they compared the task to working on a three-dimensional jigsaw puzzle. Watson, Crick, and Maurice Wilkins received the 1962 Nobel Prize in Physiology or Medicine for their work relating to the structure of DNA.

The Double Helix: The Watson-Crick Model

In the 1950s, Erwin Chargaff made the important discovery that the adenine/thymine and guanine/cytosine ratios are constant in any given organism, from a human genius to a bacterium. Based on this analysis, it was apparent that in any given organism,

- The base composition is the same in all cells of an organism and is characteristic of that organism.
- The amounts of adenine and thymine are equal, as are the amounts of guanine and cytosine.
- The amount of the bases adenine plus guanine equals that of the bases cytosine plus thymine.

This information implied that the bases occurred in pairs: adenine with thymine and guanine with cytosine.

Using X-ray crystallography data gathered by Rosalind Franklin and Maurice Wilkins on the relative positions of atoms in DNA, plus the concept of A-T and G-C base pairs, the American biologist James D. Watson and the British physicist Francis H. C. Crick proposed a double helix structure for DNA in 1953.

The deoxyribose units in the backbone chain are joined through phosphate diester linkages. Phosphoric acid is H_3PO_4, $O{=}P(OH)_3$.

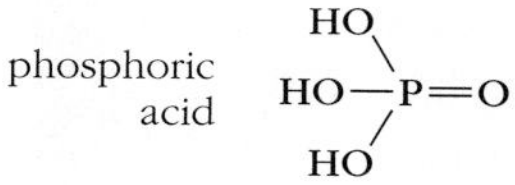

Replacing two H atoms with two CH_3 groups gives a phosphate diester, $O{=}P(OH)(OCH_3)_2$.

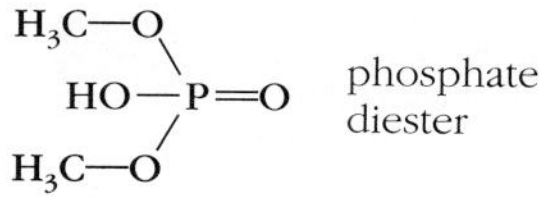

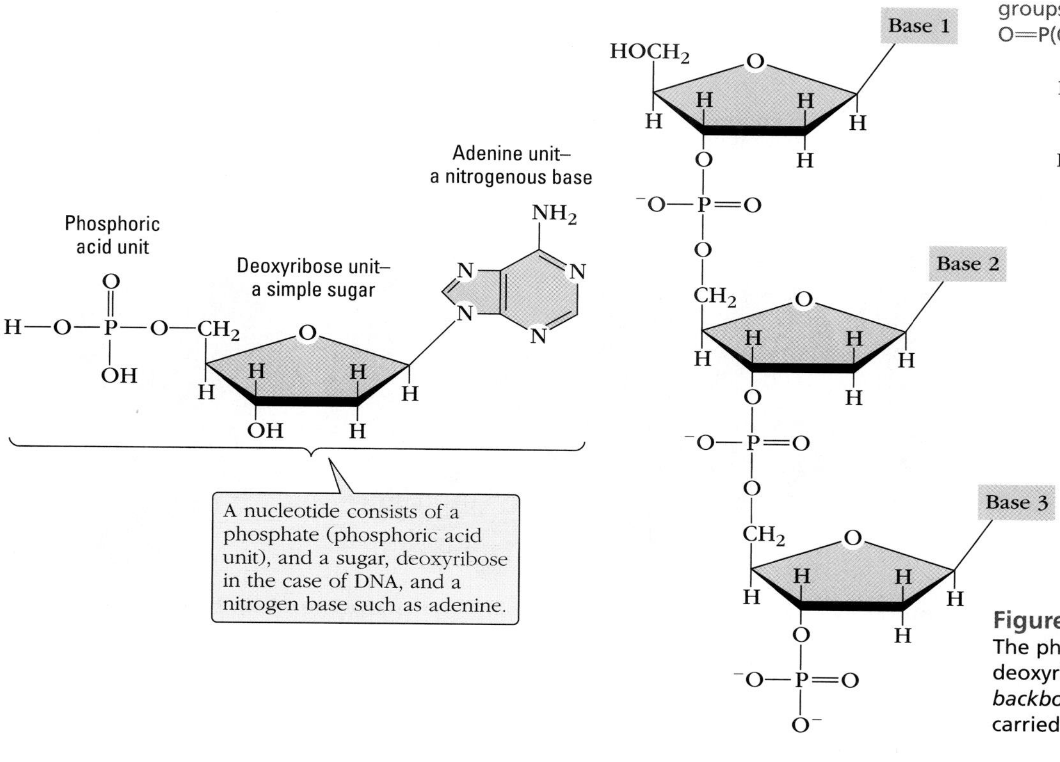

Figure 9.26 Nucleotides and DNA. The phosphate groups and the deoxyribose sugar groups form the *backbone* of DNA. The genetic code is carried by the nitrogenous bases.

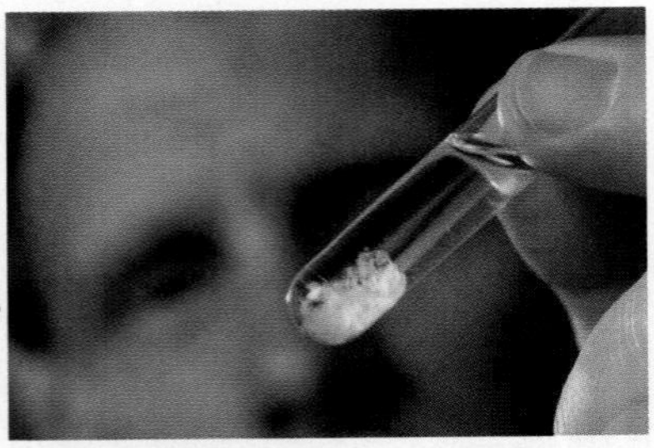

DNA. Solid DNA is precipitated from solution.

In the three-dimensional Watson-Crick structure, two polynucleotide DNA strands wind around each other to form a double helix. Remarkable insight into how hydrogen bonding could stabilize DNA ultimately led Watson and Crick to propose the double helix structure. Hydrogen bonds form between specific base pairs lying opposite each other in the two polynucleotide strands. Adenine is hydrogen-bonded to thymine and guanine is hydrogen-bonded to cytosine to form **complementary base pairs;** that is, the bases of one strand hydrogen-bond strongly to those of the other (Figure 9.27). The result is that the base pairs (A on one strand with T on the other strand, or C with G) are stacked one above the other in the interior of the double helix (*blue* in Figure 9.27).

Two hydrogen bonds occur between every adenine and thymine pair; three occur between each guanine and cytosine. These hydrogen bonds hold the double helix together. Note in Figure 9.28 the similar structures of thymine and cytosine, and of adenine and guanine. If adenine and guanine try to pair, there is insufficient space between the strands to accommodate the bulky pair; thymine and cytosine are too small to pair and to align properly.

CONCEPTUAL EXERCISE 9.8 Hydrogen Bonding and DNA

Only one hydrogen bond is possible between G and T or between C and A. Use the structural formulas for these compounds to indicate where such hydrogen bonding can occur in these base pairs (see Figure 9.28).

Rosalind Franklin

1920–1958

Rosalind Franklin received her degrees from Cambridge University. Her most famous work was the first X-ray photographs of DNA, done in collaboration with Maurice Wilkins at King's College in London, showing two forms of DNA, one of them a helix. By interpreting these X-ray data, James Watson and Francis Crick derived their now-famous double helix model of DNA for which they, along with Wilkins, received the 1962 Nobel Prize in Physiology or Medicine. Rosalind Franklin's untimely death of ovarian cancer at the age of 37 resulted in her not sharing that award. Nobel Prizes are not awarded posthumously.

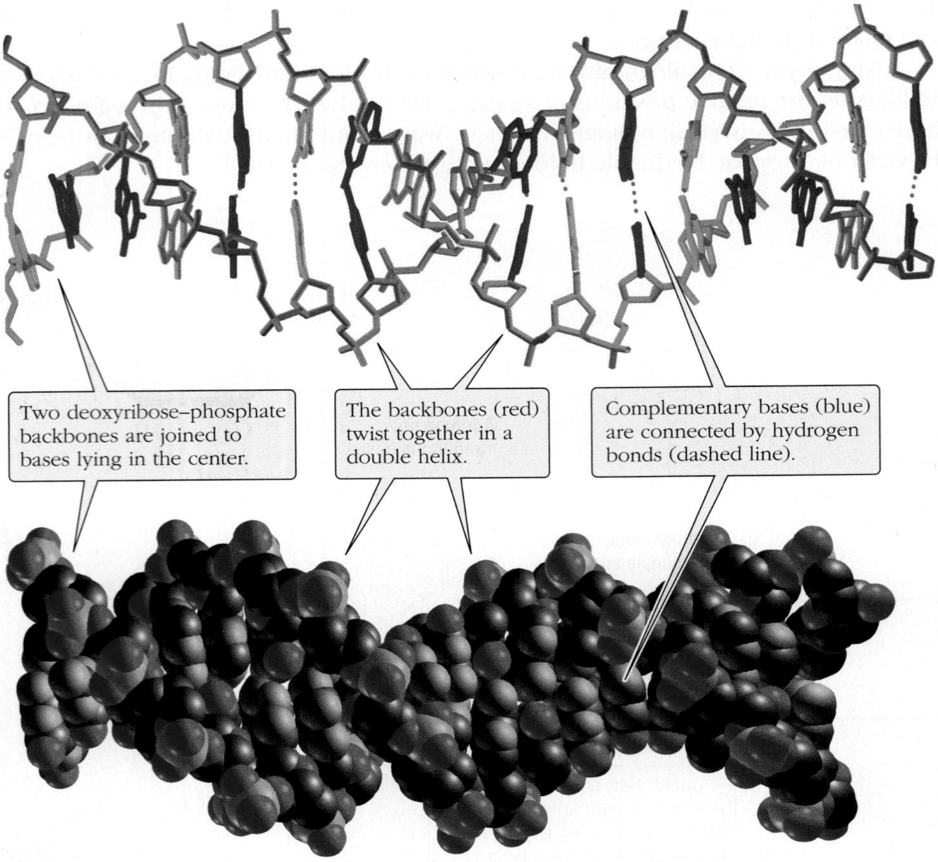
Two deoxyribose–phosphate backbones are joined to bases lying in the center.
The backbones (red) twist together in a double helix.
Complementary bases (blue) are connected by hydrogen bonds (dashed line).

Figure 9.27 Double-stranded DNA. The illustration shows the deoxyribose-phosphate backbone of each strand as phosphate groups (orange and red) linking five-membered rings of deoxyribose joined to bases. The complementary bases (*shown in blue*) are connected by hydrogen bonds: two between adenine and thymine, and three for guanine and cytosine.

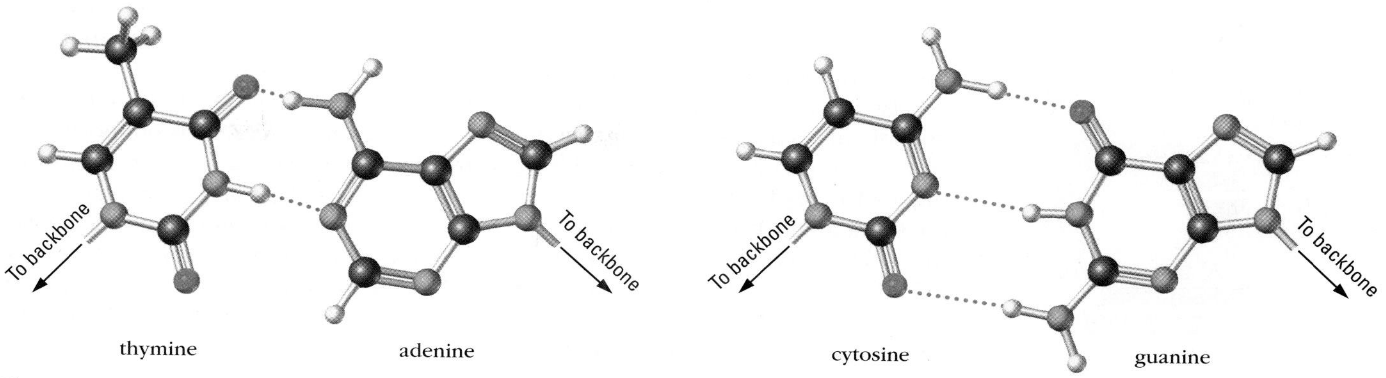

Figure 9.28 Hydrogen bonding (· · · · ·) between T-A and C-G in DNA.

The Genetic Code and DNA Replication

In the historic scientific journal article describing their revolutionary findings, Watson and Crick wrote, "It has not escaped our notice that the specific pairing we have postulated immediately suggests a possible copying mechanism for the genetic material." This bit of typical British understatement belies the enormous impact that deciphering DNA's structure had on establishing the molecular basis of heredity. It was now clear that the DNA sequence of base pairs in the nucleus of a cell represents a genetic code that controls the inherited characteristics of the next generation, as well as most of the life processes of an organism.

The Double Helix by James Watson chronicles this discovery.

In humans, the double-stranded DNA forms 46 chromosomes (23 chromosome pairs). The **gene** that accounts for a particular hereditary trait is usually located in the same position on the same chromosome. Each gene is distinguished by a unique

The total sequence of base pairs in a plant or animal cell is called its genome.

ESTIMATION Base Pairs and DNA

Human DNA contains about three billion base pairs, which are an average distance of 0.34 nm apart in the DNA molecule. Only about 2% of this DNA consists of unique genes. The number of genes is estimated to be 3×10^4.

(a) What is the average number of base pairs per gene?

(b) Approximately how long (m) is a DNA molecule?

(a) We begin by calculating the average number of base pairs/gene, starting with approximately 10^9 base pairs, of which there are 2 base pairs in genes per 100 base pairs. We get

$$\frac{10^9 \text{ base pairs}}{\text{DNA molecule}} \times \frac{2 \text{ base pairs in genes}}{100 \text{ base pairs}} = \frac{2 \times 10^7 \text{ base pairs in genes}}{\text{DNA molecule}}$$

Thus, there are about 2×10^7 base pairs in genes/DNA. Using this information and the fact that DNA contains about 3×10^4 genes, there are roughly 1×10^3 base pairs/gene.

$$\frac{2 \times 10^7 \text{ base pairs in genes}}{\text{DNA molecule}} \times \frac{1 \text{ DNA molecule}}{3 \times 10^4 \text{ genes}} \approx \frac{10^3 \text{ base pairs}}{\text{gene}}$$

(b) To approximate how many meters long a DNA molecule is, we use the information that the distance between base pairs in DNA is 0.34 nm and 1 m = 10^9 nm. Therefore,

$$\frac{10^9 \text{ base pairs}}{\text{DNA molecule}} \times \frac{0.34 \text{ nm}}{\text{base pair}} \times \frac{1 \text{ m}}{10^9 \text{ nm}} \approx \frac{0.34 \text{ m}}{\text{DNA molecule}}$$

This distance, approximately 0.34 m (about a foot), would be much too large to fit into a cell. Therefore, the DNA molecule must be tightly coiled, which it is.

Visit this book's companion website at www.cengage.com/chemistry/moore to work an interactive module based on this material.

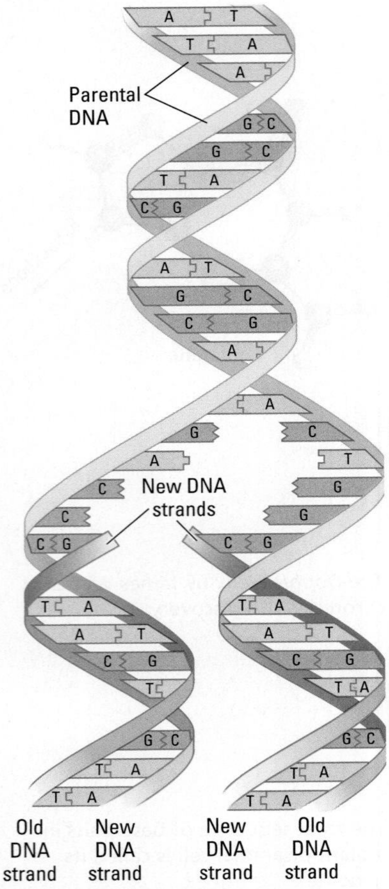
Parental DNA
New DNA strands
Old DNA strand
New DNA strand
New DNA strand
Old DNA strand

Figure 9.29 DNA replication. When the original DNA helix (*orange*) unwinds, each half is a template on which to assemble nucleotides from the cell environment to produce a complementary strand (*green*).

sequence of nitrogen bases that codes for the synthesis of a single protein within the body. The protein then goes on to play its role in the growth and functioning of the individual. (Proteins are discussed in Section 12.7.)

Each organism (except viruses) begins life as a single cell. The DNA in that cell consists of a single strand from each parent. During regular cell division, both DNA strands are accurately copied, with a remarkably low incidence of error. The copying, called **replication,** takes place as the DNA helix unzips and new nucleotides are sequentially brought into the proper places on the new strand. Thus, each original DNA strand serves as a template from which the complementary strand is produced (Figure 9.29). The process is termed *semiconservative* because in each of the two new cells, each chromosome consists of one DNA strand from the parent cell and one newly made strand.

CONCEPTUAL EXERCISE **9.9 Replication and Base Pairing in DNA**

How easily would the base pairs in DNA unpair during replication if they were linked by covalent bonds? How is it that the helix unzips so readily?

SUMMARY PROBLEM

Part I

Write the Lewis structures and give the electron-pair geometry, molecular geometry, and bond angles, and the hybridization of the central atom of these polyatomic ions and molecule.

(a) BrF_2^+ (b) $OCCl_2$ (c) CH_3^+ (d) SeCS (e) CH_3^-

Part II

Aspartame is a commonly used artificial sweetener (NutraSweet®) that was discovered accidentally by a chemist in the laboratory carelessly licking his fingers after synthesizing a new compound (aspartame). The sweetener's popularity stems from the fact that aspartame tastes over 100 times sweeter than sugar. Aspartame has this structural formula.

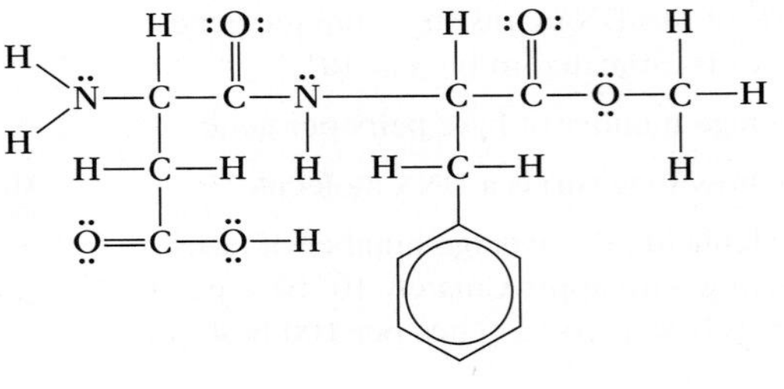

(a) Write the molecular formula of aspartame.

(b) Identify the hybridization of each nitrogen atom.

(c) Identify the hybridization of carbon in the CH_2 groups.

(d) Identify the hybridization of the carbon atom in the C=O group.

(e) Identify the hybridization of the oxygen atom to which a methyl (—CH_3) group is attached.

(f) Label each of the sigma and pi bonds involving carbon and those involving nitrogen.

Part III

The structural formula for the open-chain form of glucose is

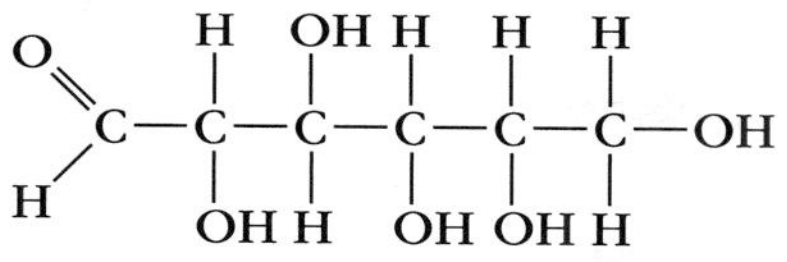

Glucose dissolves readily in water. Use molecular structure principles to explain why glucose is so water-soluble.

IN CLOSING

Having studied this chapter, you should be able to . . .

- Recognize the various ways that the shapes of molecules are represented by models and on the printed page (Section 9.1).
- Predict shapes of molecules and polyatomic ions by using the VSEPR model (Section 9.2). End-of-chapter questions: 17, 19, 21, 23
- Define and describe the nature of chiral molecules and enantiomers (Section 9.2). Questions 27, 29
- Determine the orbital hybridization of a central atom and the associated molecular geometry (Section 9.3). Questions 35, 37, 44
- Describe in terms of sigma and pi bonding the orbital hybridization of a central atom and the associated molecular geometry for molecules containing multiple bonds (Section 9.4). Question 46
- Describe covalent bonding between two atoms in terms of sigma or pi bonds, or both (Section 9.3). Question 48
- Use molecular structure and electronegativities to predict the polarities of molecules (Section 9.5). Questions 50, 52, 56
- Describe the different types of noncovalent interactions and use them to explain melting points and boiling points (Section 9.6). Questions 62, 66
- Identify the major components in the structure of DNA (Section 9.7). Question 73
- Describe the basis of infrared spectroscopy *(Tools of Chemistry)* and UV-visible spectroscopy *(Tools of Chemistry)* and explain how they are used to determine molecular structures. Question 70

OWL and Go Chemistry

Sign in at **www.cengage.com/owl** to:

- View tutorials and simulations, develop problem-solving skills, and complete online homework assigned by your professor.
- For quick review and exam prep, download Go Chemistry mini lecture modules from OWL (or purchase them at **www.CengageBrain.com**).

KEY TERMS

achiral *(Section 9.2)*
axial positions *(9.2)*
bond angle *(9.2)*
chiral *(9.2)*
complementary base pairs *(9.7)*
deoxyribonucleic acid (DNA) *(9.7)*
dipole-dipole attraction *(9.6)*
dipole moment *(9.5)*
electron-pair geometry *(9.2)*
enantiomers *(9.2)*
equatorial positions *(9.2)*
gene *(9.7)*
hybrid orbital *(9.3)*
hybridized *(9.3)*
hydrogen bond *(9.6)*
hydrophilic *(9.6)*
hydrophobic *(9.6)*
induced dipole *(9.6)*
intermolecular forces *(9.6)*
lipid bilayer *(9.6)*
London forces *(9.6)*
molecular geometry *(9.2)*
noncovalent interactions *(9.6)*
nonpolar molecule *(9.5)*
nucleotide *(9.7)*
pi bond, π bond *(9.4)*
polar molecule *(9.5)*
polarization *(9.6)*
polymer *(9.7)*
replication *(9.7)*

sigma bond, σ bond *(9.3)*

sp hybrid orbital *(9.3)*

$\mathbf{sp^2}$ **hybrid orbital** *(9.3)*

$\mathbf{sp^3}$ **hybrid orbital** *(9.3)*

spectroscopy *(Tools of Chemistry, pp. 386, 401)*

valence bond model *(9.3)*

valence-shell electron-pair repulsion (VSEPR) model *(9.2)*

QUESTIONS FOR REVIEW AND THOUGHT

OWL Interactive version of these problems are assignable in OWL.

Blue-numbered questions have short answers at the back of this book in Appendix M and fully worked solutions in the *Student Solutions Manual.*

Review Questions

These questions test vocabulary and simple concepts.

1. What is the VSEPR model? What is the physical basis of the model?
2. What is the difference between the electron-pair geometry and the molecular geometry of a molecule? Use the water molecule as an example in your discussion.
3. Designate the electron-pair geometry for each case from two to six electron pairs around a central atom.
4. What are the molecular geometries for each of these?

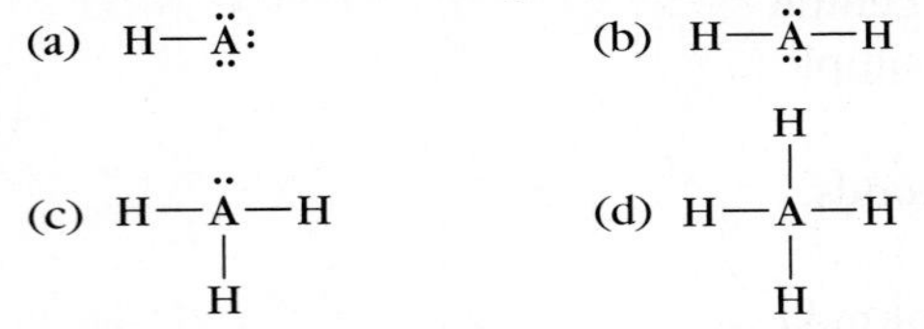

Give the H—A—H bond angle for each of the last three.

5. If you have three electron pairs around a central atom, how can you have a triangular planar molecule? An angular molecule? What bond angles are predicted in each case?
6. Use VSEPR to explain why ethylene is a planar molecule.
7. How can a molecule with polar bonds be nonpolar? Give an example.
8. Explain the significance of Erwin Chargaff's research to the understanding of the structure of DNA.
9. (a) Describe how nucleotides are joined into a polynucleotide chain in DNA.
   (b) Explain how the nitrogen bases are joined in the DNA helix.
10. Explain why the infrared spectrum of a molecule is referred to as its "fingerprint."
11. For infrared energy to be absorbed by a molecule, what frequency of motion must the molecule have?

Topical Questions

These questions are keyed to the major topics in the chapter. Usually a question that is answered at the back of this book is paired with a similar one that is not.

Molecular Shape (Section 9.2)

12. Use the various molecular modeling techniques (ball-and-stick, space-filling, two-dimensional pictures using wedges and dashed lines) to illustrate these simple molecules:
    (a) NH_3 (b) H_2O
    (c) CO_2
13. All of these molecules have central atoms with only bonding pairs of electrons. After drawing the Lewis structure, identify the molecular shape of each molecule.
    (a) BeH_2 (b) CH_2Cl_2
    (c) BH_3 (d) $SeCl_6$
    (e) PF_5
14. Draw the Lewis structure for each of these molecules or ion. Describe the electron-pair geometry and the molecular geometry.
    (a) NH_2Cl (b) OF_2
    (c) SCN^- (d) HOF
15. Use the VSEPR model to predict the electron-pair geometry and the molecular geometry of
    (a) PH_4^+ (b) OCl_2
    (c) SO_3 (d) H_2CO
16. In each of these molecules or ions, two oxygen atoms are attached to a central atom. Draw the Lewis structure for each, and then describe the electron-pair geometry and the molecular geometry. Comment on similarities and differences in the series.
    (a) CO_2 (b) NO_2^-
    (c) SO_2 (d) O_3
    (e) ClO_2^-
17. In each of these ions, three oxygen atoms are attached to a central atom. Draw the Lewis structure for each ion, and then describe the electron-pair geometry and the molecular geometry. Comment on similarities and differences in the series.
    (a) BO_3^{3-} (b) CO_3^{2-}
    (c) SO_3^{2-} (d) ClO_3^-
18. These are examples of molecules and ions that do not obey the octet rule. After drawing the Lewis structure, describe the electron-pair geometry and the molecular geometry for each.
    (a) ClF_2^- (b) ClF_3
    (c) ClF_4^- (d) ClF_5
19. These are examples of molecules and an ion that do not obey the octet rule. After drawing the Lewis structure, describe the electron-pair geometry and the molecular geometry for each.
    (a) SiF_6^{2-} (b) SF_4
    (c) PF_5 (d) XeF_4
20. Iodine forms three compounds with chlorine: ICl, ICl_3, and ICl_5. Draw the Lewis structures and determine the molecular shapes of these three molecules.

21. Give the approximate values for the indicated bond angles. The figures given are not the actual molecular shapes, but are used only to show the bonds being considered.
(a) O—S—O angle in SO_2 (b) F—B—F angle in BF_3

(c) 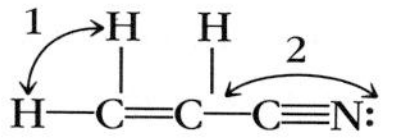

(d) 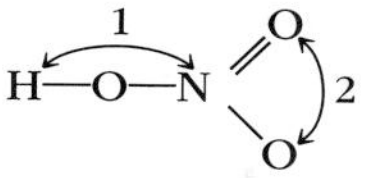

22. Give approximate values for the indicated bond angles. The figures given are not the actual molecular shapes, but are used only to show the bonds being considered.
(a) Cl—S—Cl angle in SCl_2 (b) N—N—O angle in N_2O

(c) 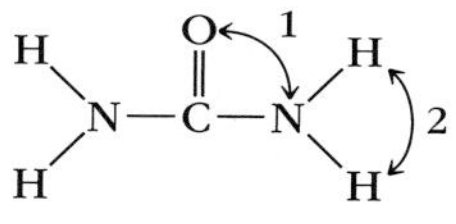

(d) 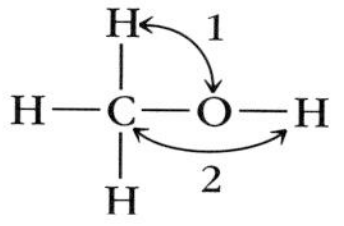

23. Give approximate values for the indicated bond angles.
(a) F—Se—F angles in SeF_4
(b) O—S—F angles in SOF_4 (The O atom is in an equatorial position.)
(c) F—Br—F angles in BrF_5

24. Give approximate values for the indicated bond angles.
(a) F—S—F angles in SF_6
(b) F—Xe—F angle in XeF_2
(c) F—Cl—F angle in ClF_2^-

25. Which would have the greater O—N—O bond angle, NO_2 or NO_2^+? Explain your answer.

26. Compare the F—Cl—F angles in ClF_2^+ and ClF_2^-. From Lewis structures, determine the approximate bond angle in each ion. Explain which ion has the greater angle and why.

Chirality in Organic Compounds (Section 9.2)

27. Circle the chiral centers, if any, in these molecules.

(a) 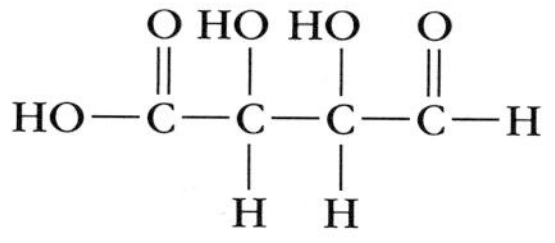

(b) 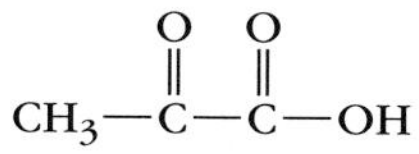

(c) 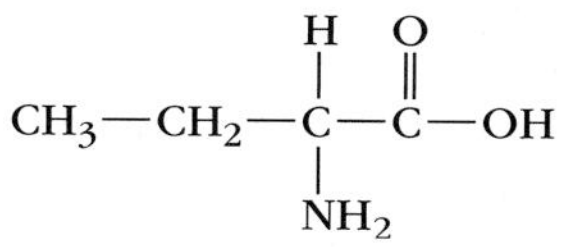

28. Circle the chiral centers, if any, in these molecules.

(a) 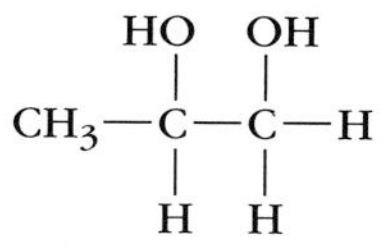

(b) 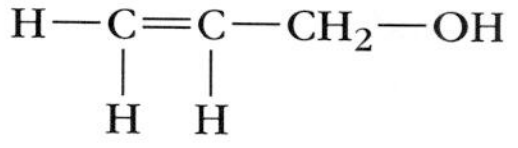

(c) 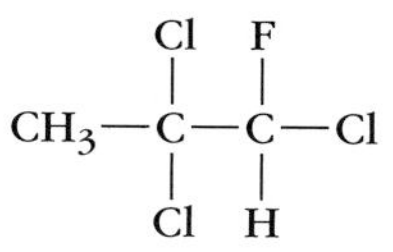

29. Circle the chiral centers, if any, in these compounds.

(a) 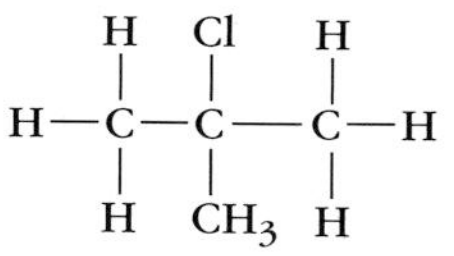

(b) 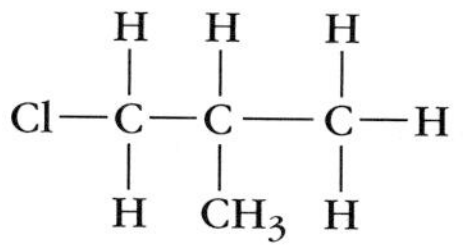

(c) 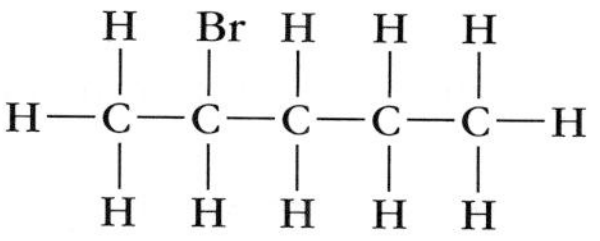

(d) 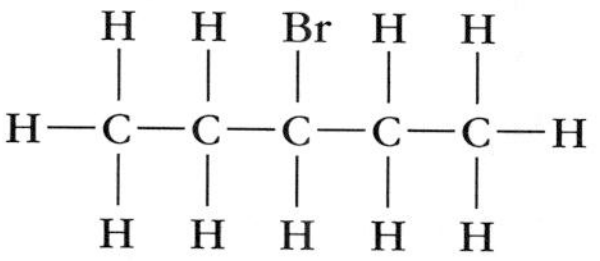

30. Circle the chiral centers, if any, in these compounds.

(a) 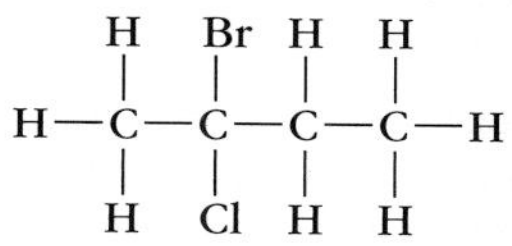

(b) 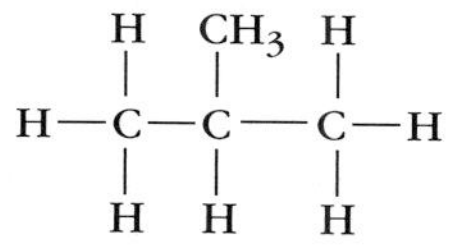

(c) 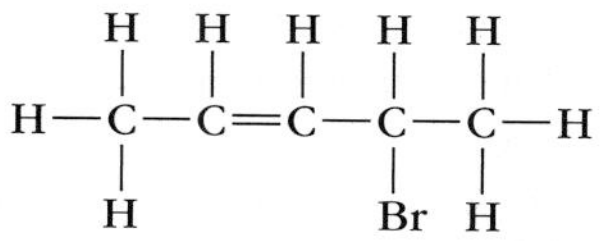

31. How can you tell from its structural formula whether a compound can exist as a pair of enantiomers?

32. Which of these is not superimposable on its mirror image?
(a) Nail (b) Screw
(c) Shoe (d) Sock
(e) Golf club (f) Football
(g) Your ear (h) Helix
(i) Baseball bat (j) Sweater

Hybridization (Section 9.3)

33. What designation is used for the hybrid orbitals formed by these combinations of atomic orbitals?
(a) One *s* and three *p*
(b) One *s* and one *p*
(c) One *s* and two *p*

34. How many hybrid orbitals are formed in each case in Question 33?

35. Describe the geometry and hybridization of carbon in chloroform, $CHCl_3$.

36. Describe the geometry and hybridization for each inner atom in ethylene glycol, $HOCH_2CH_2OH$, the main component in antifreeze.

37. Describe the hybridization around the central sulfur atom and the bonding in SCl_2 and OCS.

38. Determine the geometry and bond angles of the three phosphorus-chlorine species: PCl_4^+, PCl_5, and PCl_6^-.

39. Describe the bond angles generally associated with each of the hybrid orbitals described in Question 33.
40. The hybridization of the two carbon atoms differs in an acetic acid, CH_3COOH, molecule.
    (a) Designate the correct hybridization for each carbon atom in this molecule.
    (b) What is the approximate bond angle around each carbon?
41. The hybridization of the two nitrogen atoms differs in NH_4NO_3.
    (a) Designate the correct hybridization for each nitrogen atom.
    (b) What is the approximate bond angle around each nitrogen?
42. What are the hybridization and approximate bond angles for the N, C, and O atoms in alanine, an amino acid, whose Lewis structure is

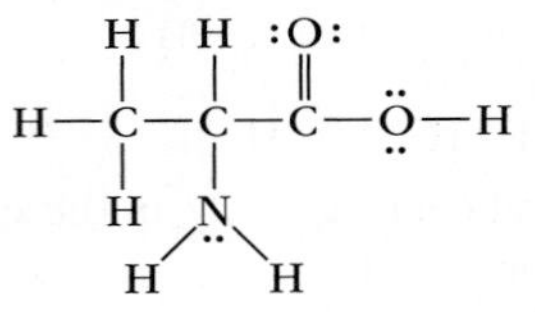

43. Identify the type of hybridization, geometry, and approximate bond angle for each carbon atom in an alkane.
44. (a) Identify the type of hybridization and approximate bond angle for each carbon atom in CH_3CH_2CCH.
    (b) Which is the shortest carbon-to-carbon bond length in this molecule?
    (c) Which is the strongest carbon-to-carbon bond in this molecule?
45. Write the Lewis structure and designate which are sigma and pi bonds in each of these molecules.
    (a) HCN (b) N_2H_2 (c) HN_3
46. Write the Lewis structure and designate which are sigma and pi bonds in each of these molecules.
    (a) OCS (b) NH_2OH
    (c) CH_2CHCHO (d) $CH_3CH(OH)COOH$
47. Methylcyanoacrylate is the active ingredient in "super glues." Its Lewis structure is

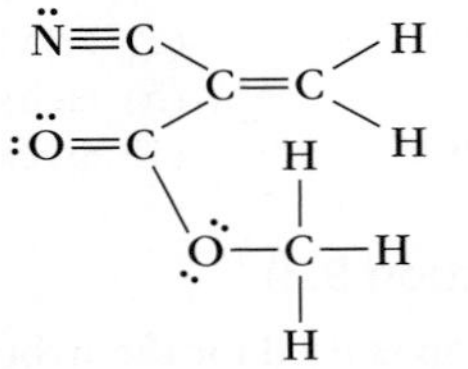

    (a) How many sigma bonds are in the molecule?
    (b) How many pi bonds are in the molecule?
    (c) What is the hybridization of the carbon atom bonded to nitrogen?
    (d) What is the hybridization of the carbon atom bonded to oxygen?
    (e) What is the hybridization of the double-bonded oxygen?
48. Acrylonitrile is polymerized to manufacture carpets and wigs. Its Lewis structure is

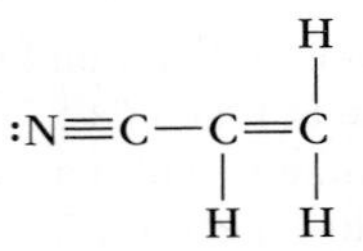

    (a) Locate the sigma bonds in the molecule. How many are there?
    (b) Locate the pi bonds in the molecule. How many are there?
    (c) What is the hybridization of the carbon atom that is bonded to nitrogen?
    (d) What is the hybridization of the nitrogen atom?
    (e) What is the hybridization of the hydrogen-bearing carbon atoms?

Molecular Polarity (Section 9.5)

49. Consider these molecules: CH_4, NCl_3, BF_3, CS_2.
    (a) In which compound are the bonds most polar?
    (b) Which compounds in the list are not polar?
50. Consider these molecules: H_2O, NH_3, CO_2, ClF, CCl_4.
    (a) In which compound are the bonds most polar?
    (b) Which compounds in the list are not polar?
    (c) Which atom in ClF has a partial negative charge?
51. Which of these molecules is (are) not polar? Which molecule has the most polar bonds?
    (a) CO (b) PCl_3
    (c) BCl_3 (d) GeH_4
    (e) CF_4
52. Which of these molecules is (are) polar? For each polar molecule, what is the direction of polarity; that is, which is the partial negative end and which is the partial positive end of the molecule?
    (a) CO_2 (b) HBF_2
    (c) CH_3Cl (d) SO_3
53. Which of these molecules has a net dipole moment? For each of these polar molecules, indicate the direction of the dipole in the molecule.
    (a) XeF_2 (b) H_2S
    (c) CH_2Cl_2 (d) HCN
54. Explain the differences in the dipole moments of
    (a) BrF (1.29 D) and BrCl (0.52 D)
    (b) H_2O (1.86 D) and H_2S (0.95 D)
55. Which of these molecules has a net dipole moment? For each of the polar molecules, indicate the direction of the dipole in the molecule.
    (a) Nitrosyl fluoride, FNO
    (b) Disulfur difluoride, S_2F_2
56. Which of these molecules is polar? For each of the polar molecules, indicate the direction of the dipole in the molecule.
    (a) Hydroxylamine, NH_2OH
    (b) Sulfur dichloride, SCl_2, an unstable, red liquid

Noncovalent Interactions (Section 9.6)

57. Explain in terms of noncovalent interactions why water and ethanol are miscible, but water and cyclohexane are not.
58. Construct a table covering all the types of noncovalent interactions and comment about the distance dependence for each. (In general, the weaker the force, the closer together the molecules must be to feel the attractive force of nearby molecules.) You should also include an example of a substance that exhibits each type of noncovalent interaction in the table.

59. Explain the trends seen in the following diagram for the boiling points of some main group hydrogen compounds.
(a) Group 4A: CH_4, SiH_4, GeH_4, SnH_4
(b) Group 5A: NH_3, PH_3, AsH_3, SbH_3
(c) Group 6A: H_2O, H_2S, H_2Se, H_2Te
(d) Group 7A: HF, HCl, HBr, HI

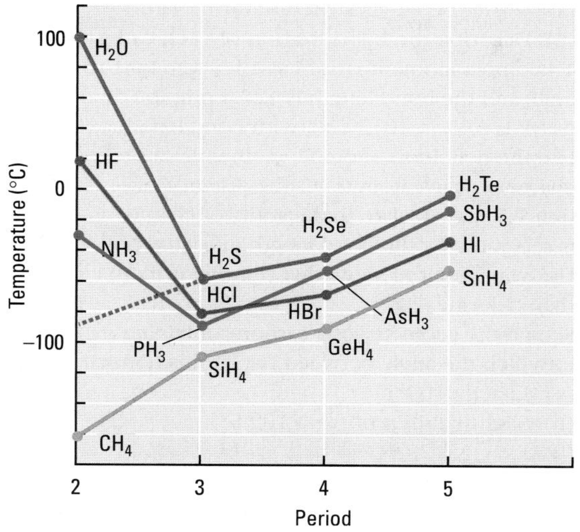

60. Explain why water "beads up" on a freshly waxed car, but not on a dirty, unwaxed car.
61. Explain why water will not remove tar from your shoe, but kerosene will.
62. Which of these will form hydrogen bonds?
(a) CH_2Br_2 (b) $CH_3OCH_2CH_3$
(c) H_2NCH_2COOH (d) H_2SO_3
(e) CH_3CH_2OH
63. Arrange the noble gases in order of increasing boiling point. Explain your reasoning.
64. The structural formula for vitamin C is

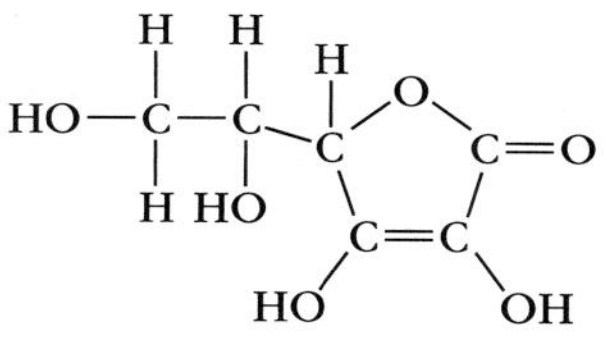

Give a molecular-level explanation why vitamin C is a water-soluble rather than a fat-soluble vitamin.
65. Which of these would you expect to be most soluble in cyclohexane, C_6H_{12}? The least soluble? Explain your reasoning.
(a) NaCl (b) CH_3CH_2OH (c) C_3H_8
66. What are the types of forces that must be overcome in these changes?
(a) The sublimation of solid $C_{10}H_8$
(b) The melting of propane, C_3H_8
(c) The decomposition of water into H_2 and O_2
(d) The evaporation of liquid PCl_3
(e) Unzipping the DNA double helix during replication
67. The structural formula of alpha-tocopherol, a form of vitamin E, is

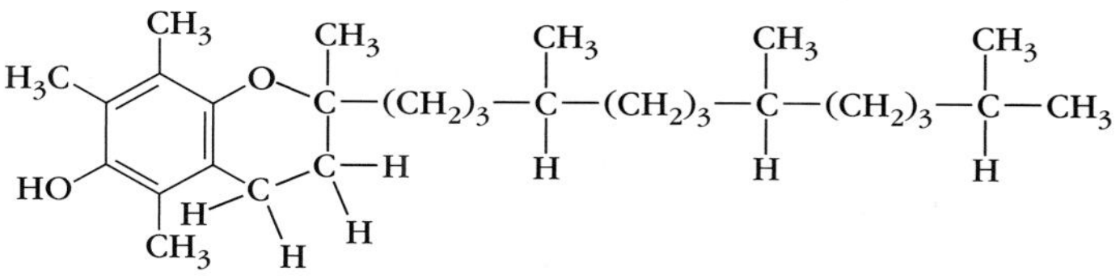

Give a molecular-level explanation why alpha-tocopherol dissolves in fat, but not in water.

Molecular Structure Determination by Spectroscopy

68. Ultraviolet and infrared spectroscopies probe different aspects of molecules.
(a) Ultraviolet radiation is energetic enough to cause transitions in the energies of which electrons in an atom?
(b) What does infrared spectroscopy tell us about a given molecule?
69. The infrared spectrum of ethanol is given on page 387.
(a) Which stretching requires the lowest amount of energy?
(b) The highest amount of energy?
70. Compare IR and UV spectroscopy in these ways:
(a) Energy of radiation
(b) Wavelength of radiation
(c) Frequency of radiation
71. The UV-visible spectrum of beta-carotene given on page 401 shows a maximum absorbance at 455 nm. Estimate the wavelengths of two other peaks in the spectrum and indicate which peak identifies the more energetic electron transition.

Biomolecules (Section 9.7)

72. Discuss the differences between the extent of hydrogen bonding for the pairs G-C and A-T in nucleic acids. If a strand of DNA has more G-C pairs than A-T pairs in the double helix, the melting point (unwinding point) increases. The melting point for strands with more A-T pairing will decrease in comparison. Explain.
73. One strand of DNA contains the base sequence T-C-G. Draw a structure of this section of DNA that shows the hydrogen bonding between the base pairs of this strand and its complementary strand.

General Questions

These questions are not explicitly keyed to chapter topics; many require integration of several concepts.

74. The formula for nitryl chloride is NO_2Cl. Draw the Lewis structure for the molecule, including all resonance structures. Describe the electron-pair and molecular geometries, and give values for all bond angles.

75. Methylcyanoacrylate is the active ingredient in "super glues." Its Lewis structure is

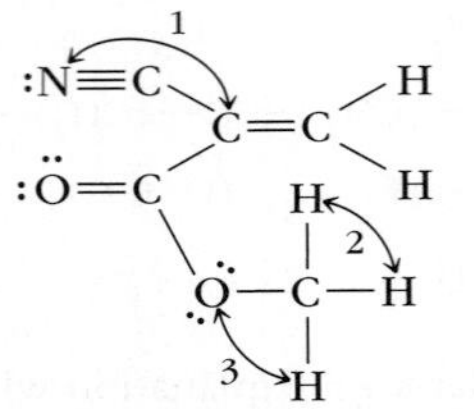

(a) Give values for the three bond angles indicated.
(b) Indicate the most polar bond in the molecule.
(c) Circle the shortest carbon-oxygen bond.

76. Acrylonitrile is polymerized to manufacture carpets and wigs. Its Lewis structure is

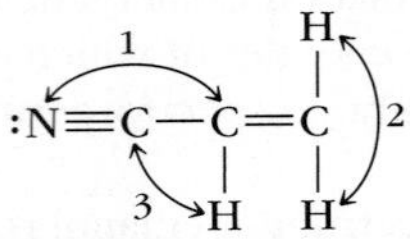

(a) Give approximate values for the indicated bond angles.
(b) Which is the most polar bond in the molecule?

77. In addition to CO and CO_2, there are other carbon oxides. One is tricarbon dioxide, C_3O_2, also called carbon suboxide, a foul-smelling gas.
(a) Write the Lewis structure of this compound.
(b) What is the value of the C-to-C-to-O bond angle in carbon suboxide?
(c) What is the value of the C-to-C-to-C bond angle in tricarbon dioxide?

78. In addition to CO, CO_2, and C_3O_2, there is another molecular oxide of carbon, pentacarbon dioxide, C_5O_2, a yellow solid.
(a) What is the value of the C-to-C-to-O bond angle in pentacarbon dioxide?
(b) What is the value of the C-to-C-to-C bond angle in this compound?

79. Cyanidin chloride is an anthocyanin found in strawberries, apples, and cranberries.

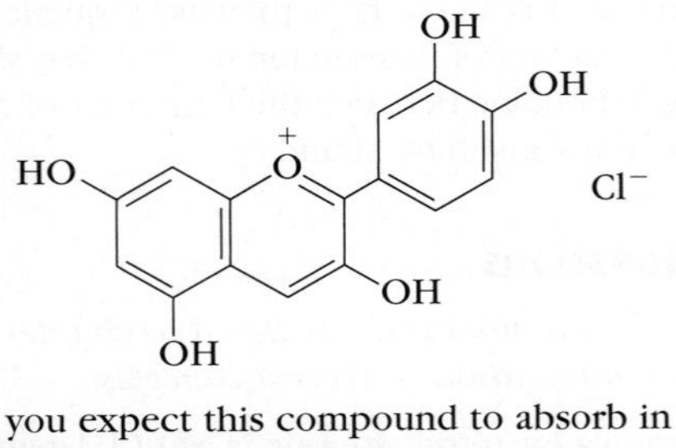

Would you expect this compound to absorb in the visible region of the spectrum? Explain your answer.

80. Each of these molecules has fluorine atoms attached to an atom from Group 1A or 3A to 6A. Draw the Lewis structure for each one and then describe the electron-pair geometry and the molecular geometry. Comment on similarities and differences in the series.
(a) BF_3 (b) CF_4 (c) NF_3
(d) OF_2 (e) HF

81. Carbon and oxygen form the squarate ion, a polyatomic ion with the formula $C_4O_4^{2-}$.
(a) Write the Lewis structure for this ion.
(b) Describe the hybridization of each carbon atom.

82. In addition to carbonate ion, carbon and oxygen form the croconate ion, a polyatomic ion with the formula $C_5O_5^{2-}$.
(a) Write the Lewis structure for this ion.
(b) Describe the hybridization of each carbon atom.

83. The dipole moment is 3.43×10^{-30} Cm for the HCl molecule, and the bond length is 127.4 pm; the dipole moment of HF is 6.37×10^{-30} Cm, with bond length of 91.68 pm. Use the definition of dipole moment as a product of partial charge on each atom times the distance of separation (Section 9.5) to calculate the quantity of charge in coulombs that is separated by the bond length in each dipolar molecule. Use your result to show that fluorine is more electronegative than chlorine.

84. Sketch the geometry of a carbon-containing molecule or ion in which the angle between two atoms bonded to carbon is
(a) Exactly 109.5°
(b) Slightly different from 109.5°
(c) Exactly 120°
(d) Exactly 180°

85. In the gas phase, positive and negative ions form ion pairs that are like molecules. An example is KF, which is found to have a dipole moment of 28.7×10^{-30} Cm and a distance of separation between the two ions of 217.2 pm. Use this information and the definition of dipole moment to calculate the partial charge on each atom. Compare your result with the expected charge, which is the charge on an electron, -1.62×10^{-19} C. Based on your result, is KF really completely ionic?

86. How can a diatomic molecule be nonpolar? Polar?

87. How can a molecule have polar bonds yet have a dipole moment of zero?

88. Below is a grid with nine lettered boxes, each of which contains an item that is used to answer the questions that follow. Items may be used more than once and there may be more than one correct item in response to a question.

A PH_4^+	**B** PF_3	**C** PF_5
D O=C=S	**E** O=C=O	**F** PF_3 or PF_5
G ClF or ClF_3	**H** NO_3^-	**I** H_2NOH

Place the letter(s) of the correct selection(s) on the appropriate line.
(a) Electron-pair geometry is the same as the molecular geometry ________
(b) Nonpolar molecule ________
(c) Linear molecular geometry ________
(d) Triangular planar molecular geometry ________
(e) Central atom is sp^3 hybridized ________
(f) Central atom is sp^2 hybridized ________
(g) Has dipole-dipole and hydrogen bonding intermolecular forces ________

89. Below is a grid with nine lettered boxes, each of which contains an item that is used to answer the questions that follow. Items may be used more than once and there may be more than one correct item in response to a question.

A HCN	**B** PO_4^{3-}	**C** PH_3 or PF_3
D SiH_4	**E** Cl_2O	**F** NH_2Cl
G HF or F_2	**H** CH_4	**I** OF_2

Place the letter(s) of the correct selection(s) on the appropriate line.

(a) Electron-pair geometry is the same as the molecular geometry _________
(b) Nonpolar molecule _________
(c) Linear molecular geometry _________
(d) Angular (bent) molecular geometry _________
(e) Central atom is sp^3 hybridized _________
(f) Central atom is sp hybridized _________
(g) Which one in each pair of compounds has the lower boiling point _________
(h) Which one in each pair of compounds has the higher vapor pressure _________
(i) Which one in each pair of compounds has the higher dipole moment _________
(j) Has dipole-dipole and hydrogen bonding intermolecular forces _________

Applying Concepts

These questions test conceptual learning.

90. The nitride ion, N_3^-, has the same number of electrons as (is isoelectronic with) carbon dioxide.
(a) Write correct Lewis structures for two resonance hybrids of nitride ion.
(b) Identify the hybridization of the central N atom in each case.

91. Azidotrifluoromethylmethane, CF_3N_3, is a colorless gas that is stable at room temperature.
(a) Write a plausible Lewis structure for this compound and estimate the N—N—C bond angle.
(b) Estimate the N—N—N bond angle.
(c) Identify the hybridization of the central nitrogen atom and the hybridization of the carbon atom.
(d) An intermediate compound with the molecular formula $CF_3N_3H_2$ forms during the synthesis of azidotrifluoromethylmethane. Draw the correct Lewis structures for two plausible resonance hybrids of this intermediate. Identify the hybridization of each nitrogen atom in each of the two structures.

92. Complete this table.

Molecule or Ion	Electron-Pair Geometry	Molecular Geometry	Bond Angle
ICl_2^+			
I_3^-			
ICl_3			
ICl_4^-			
IO_4^-			
IF_4^+			
IF_5			
IF_6^+			

93. Complete this table.

Molecule or Ion	Electron-Pair Geometry	Molecular Geometry	Bond Angle
SO_2			
SCl_2			
SO_3			
SO_3^{2-}			
SF_4			
SO_4^{2-}			
SF_5^+			
SF_6			

94. What are the types of forces, in addition to London forces, that are overcome in these changes? Using structural formulas, make a sketch representing the major type of force in each case.
(a) The evaporation of liquid methanol, CH_3OH
(b) The decomposition of hydrogen peroxide, H_2O_2, into water and oxygen
(c) The melting of urea, H_2NCONH_2
(d) The boiling of liquid HCl

95. Name a Group 1A to 8A element that could be the central atom (X) in these compounds.
(a) XH_3 with one lone pair of electrons
(b) XCl_3 (c) XF_5
(d) XCl_3 with two lone pairs of electrons

96. Name a Group 1A to 8A element that could be the central atom (X) in these compounds.
(a) XCl_2
(b) XH_2 with two lone pairs of electrons
(c) XF_4 with one lone pair of electrons
(d) XF_4

97. What is the maximum number of water molecules that could hydrogen-bond to an acetic acid molecule? Draw in the water molecules and use dotted lines to show the hydrogen bonds.

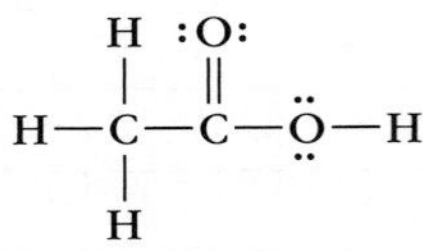

98. What is the maximum number of water molecules that could hydrogen-bond to an ethylamine molecule? Draw in the water molecules and use dotted lines to show the hydrogen bonds.

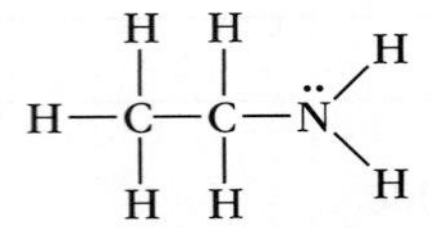

99. These are responses students wrote when asked to give an example of hydrogen bonding. Which is/are correct?

(a) H—H···H—H

(b) 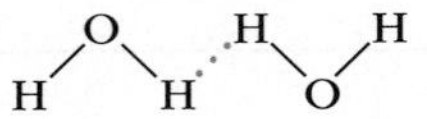

(c) H—F

(d) 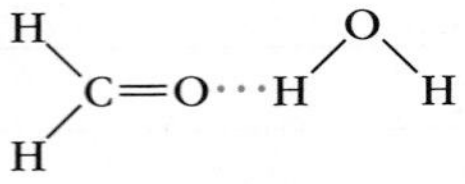

100. Which of these are examples of hydrogen bonding?

(a) 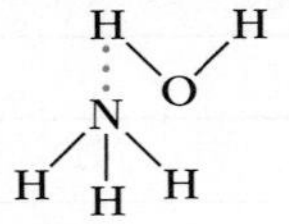

(b) H—H

(c) 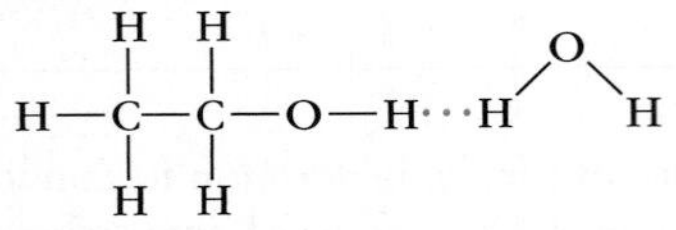

(d) 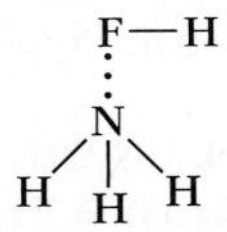

101. In another universe, elements try to achieve a nonet (nine valence electrons) instead of an octet when forming chemical bonds. As a result, covalent bonds form when a trio of electrons is shared between two atoms. Two compounds in this other universe are H_3O and H_2F. Draw their Lewis structures, then determine their electron-trio geometry and molecular geometry.

More Challenging Questions

These questions require more thought and integrate several concepts.

102. One of the three isomers of dichlorobenzene, $C_6H_4Cl_2$, has a dipole moment of zero. Draw the structural formula of the isomer and explain your choice.

103. There are three isomers with the formula $C_6H_6O_2$. Each isomer contains a benzene ring to which two —OH groups are attached.
(a) Write the Lewis structures for the three isomers.
(b) Taking their molecular structure and the likelihood of hydrogen bonding into account, list them in order of increasing melting point.

104. Halomethane, which had been used as an anesthetic, has the molecular formula $CHBrClCF_3$.
(a) Write the Lewis structure for halomethane.
(b) Is halomethane a polar molecule? Explain your answer.
(c) Can hydrogen bonding occur in halomethane? Explain.

105. Ketene, C_2H_2O, is a reactant for synthesizing cellulose acetate, which is used to make films, fibers, and fashionable clothing.
(a) Write the Lewis structure of ketene. Ketene does not contain an —OH bond.
(b) Identify the geometry around each carbon atom and all the bond angles in the molecule.
(c) Identify the hybridization of each carbon and oxygen atom.
(d) Is the molecule polar or nonpolar? Use appropriate data to support your answer.

106. Gamma hydroxybutyric acid, GHB, infamous as a "date rape" drug, is used illicitly because of its effects on the nervous system. The condensed molecular formula for GHB is $HO(CH_2)_3COOH$.
(a) Write the Lewis structure for GHB.
(b) Identify the hybridization of the carbon atom in the CH_2 groups and of the terminal carbon.
(c) Is hydrogen bonding possible in GHB? If so, write Lewis structures to illustrate the hydrogen bonding.
(d) Which carbon atoms are involved in sigma bonds? In pi bonds?
(e) Which oxygen atom is involved in sigma bonds? In pi bonds?

107. There are two compounds with the molecular formula HN_3. One is called hydrogen azide; the other is cyclotriazene.
(a) Write the Lewis structure for each compound.
(b) Designate the hybridization of each nitrogen in hydrogen azide.
(c) What is the hybridization of each nitrogen in cyclotriazene?
(d) How many sigma bonds are in hydrogen azide? In cyclotriazene?
(e) How many pi bonds are in hydrogen azide? In cyclotriazene?
(f) Give approximate values for the N-to-N-to-N bond angles in each molecule.

108. Nitrosyl azide, a yellow solid first synthesized in 1993, has the molecular formula N_4O.
(a) Write its Lewis structure.
(b) What is the hybridization on the terminal nitrogen?
(c) What is the hybridization on the "central" nitrogen?
(d) Which is the shortest nitrogen-nitrogen bond?
(e) Give the approximate bond angle between the three nitrogens, beginning with the nitrogen that is bonded to oxygen.
(f) Give the approximate bond angle between the last three nitrogens, those not involved in bonding to oxygen.
(g) How many sigma bonds are there? How many pi bonds?

109. Piperine, the active ingredient in black pepper, has this Lewis structure.

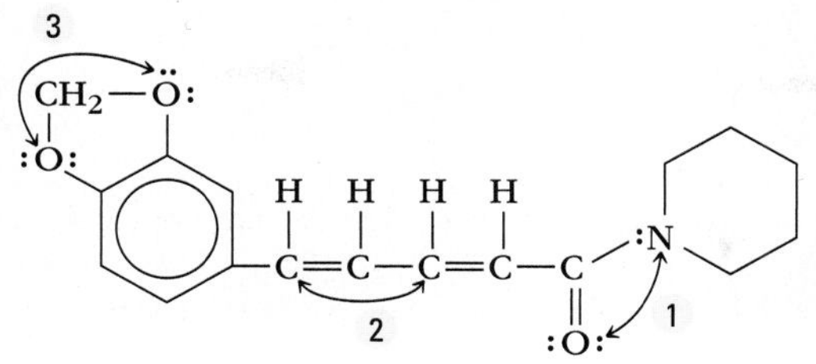

(a) Give the values for the indicated bond angles.
(b) What is the hybridization of the nitrogen?
(c) What is the hybridization of the oxygens?

110. Three compounds have the molecular formula N_2H_2.
(a) Write the correct Lewis structure for each compound.
(b) If there are geometric isomers, properly label each.

111. Two compounds have the molecular formula N_3H_3. One of the compounds, triazene, contains an N═N bond; the other compound, triaziridene, does not.
(a) Write the correct Lewis structures for each compound.
(b) Approximate the bond angle between the three nitrogen atoms in each compound.

112. This compound is commonly called acetylacetone. As shown, it exists in two forms, one called the *enol* form and the other called the *keto* form.

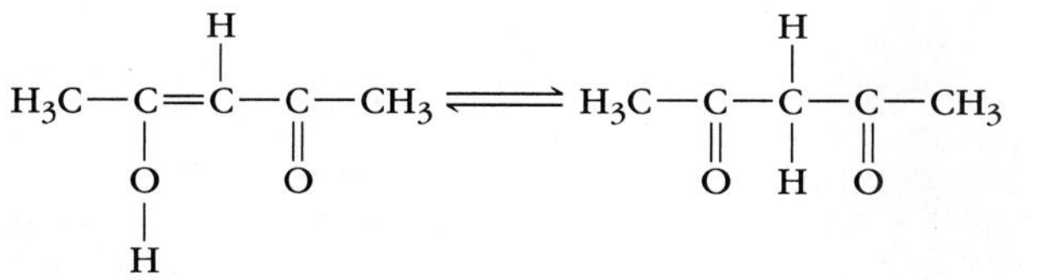

enol form　　　　*keto* form

While in the *enol* form, the molecule can lose H^+ from the —OH group to form an anion. One of the most interesting aspects of this anion (sometimes called the *acac* ion) is that one or more of them can react with a transition metal cation to give very stable, highly colored compounds.

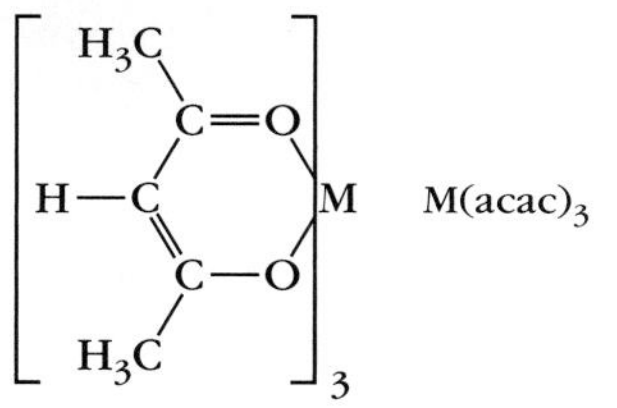

$M(acac)_3$

(a) Using bond enthalpies, calculate the enthalpy change for the *enol* to *keto* change. Is the reaction exothermic or endothermic?
(b) What are the electron-pair and "molecular" geometries around each C atom in the *keto* and *enol* forms? What changes (if any) occur when the *keto* form changes to the *enol* form?
(c) If you wanted to prepare 15.0 g deep red $Cr(acac)_3$ using this reaction,

$$CrCl_3 + 3\ H_3C{-}C(OH){=}CH{-}C(OH){-}CH_3 + 3\ NaOH \rightarrow Cr(acac)_3 + 3\ H_2O + 3\ NaCl$$

how many grams of each of the reactants would you need?

113. The compound S_4N_4 is a bright orange solid that is insoluble in water. Its eight-membered ring has several resonance hybrid structures, one of which is

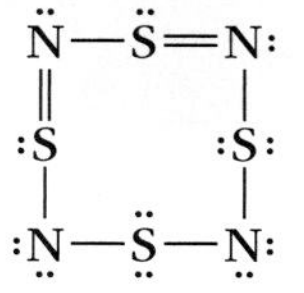

(a) Identify the hybridization of single-bonded N atoms.
(b) Identify the hybridization of double-bonded S atoms.
(c) Identify the hybridization of double-bonded N atoms.
(d) How many sigma bonds and how many pi bonds are in this molecule?
(e) Identify where the sigma and the pi bonds are in this molecule.
(f) Write the Lewis structure for at least one other resonance hybrid structure.

Conceptual Challenge Problems

These rigorous, thought-provoking problems integrate conceptual learning with problem solving and are suitable for group work.

CP9.A (Section 9.2) What advantages does the VSEPR model of chemical bonding have compared with the Lewis dot formulas predicted by the octet rule?

CP9.B (Section 9.2) The VSEPR model does not differentiate between single bonds and double bonds for predicting molecular shapes. What experimental evidence supports this?

CP9.C (Section 9.3) What evidence could you present to show that two carbon atoms joined by a single sigma bond are able to rotate about an axis that coincides with the bond, but two carbon atoms bonded by a double bond cannot rotate about an axis along the double bond?

10 Gases and the Atmosphere

The Breitling Orbiter 3 became the first balloon to fly around the world nonstop in March 1999. The 20-day trip covered 28,431 miles. During its flight, the balloon attained an altitude of 36,000 feet and was driven by the jet stream. The balloon combined an outer hot-air chamber with an inner helium-filled chamber. As the temperature and pressure of the air and the helium changed, the volumes occupied by the gases changed, and the balloon rose and fell in elevation in response to these changes. The black object hanging beneath the gondola is a set of solar panels that supplied the necessary power for telecommunication and navigation instruments during the flight. The Breitling Orbiter 3 gondola in this picture is now on display at the Smithsonian National Air and Space Museum in Washington, DC.

Early chemists studied gases and their chemistry extensively. They carried out reactions that generated gases, bubbled the gases through water into glass containers, and transferred the gases to animal bladders for storage. They mixed gases together to see whether the gases would react and, if so, how much of each gas would be consumed or formed. They discovered that gases have many properties in common, much more so than liquids or solids. All gases are transparent, although some are colored. (F_2 is light yellow, Cl_2 is greenish yellow, Br_2 and NO_2 are reddish brown, and I_2 vapor is violet.) Gases are mobile; that is, they expand to fill all space available, and they mix together physically in any proportions. The volume of a gas sample can be altered by changing its temperature or pressure, or both. Thus, gas densities (mass/volume) are quite variable, depending on the conditions. At first this might seem to make the study of gases complicated, but it really does not. The way that gas volume depends on temperature and pressure is essentially the same for all gases, so it is easy to formulate a mathematical expression—the ideal gas law—that accounts for the dependence.

Earth is surrounded by a mixture of gases we call the atmosphere. A vast number of chemical reactions take place in this gas mixture, many of them driven by energy from solar photons. Although many of these reactions are beneficial to Earth's inhabitants, some produce undesirable products. This chapter contains fundamental facts and concepts about gases as well as information about atmospheric reactions. The relationships among pressure, temperature, and amount of a gas and its volume are expressed in the ideal gas law. Other quantitative treatments of gases in chemical reactions are introduced, and adjustments to the ideal gas law to account for real gas behaviors are described. The chapter concludes with discussions of chemistry in the atmosphere—ozone in the stratosphere and atmospheric carbon dioxide.

OWL Sign in to OWL at **www.cengage.com/owl** to view tutorials and simulations, develop problem-solving skills, and complete online homework assigned by your professor.

go Chemistry Download mini lecture videos for key concept review and exam prep from OWL or purchase them from **www.CengageBrain.com.**

Companion Website

Visit this book's companion website at **www.cengage.com/chemistry/moore** to work interactive modules for the Estimation boxes and Active Figures in this text.

10.1 The Atmosphere

Earth is enveloped by a few vertical miles of atoms and molecules that compose the atmosphere. This gaseous medium is a perfect place to begin the study of gases. Everything on the surface of Earth experiences the atmosphere's pressure. The atmosphere's total mass is approximately 5.3×10^{15} metric tons, a huge figure, but still only about one millionth of Earth's total mass. Close to Earth's surface the atmosphere is mostly nitrogen and life-sustaining oxygen, but a fraction of a percent of other chemicals can make a difference in the quality of life, even its very existence. Extra water in the atmosphere can support a rain forest, a little less water produces a balanced rainfall, and practically no water results in a desert. Air pollutants, which are present in such small quantities that they are outnumbered by oxygen and nitrogen molecules by 100,000 to 1 or more, can lower our quality of life and generally prove harmful to people with respiratory disorders.

A metric ton is 1000 kg.

The 5.3×10^{15} metric tons of the atmosphere has about the same mass as a cube of water 100 km on a side.

The two major chemicals in our atmosphere are nitrogen, a rather unreactive gas, and oxygen, a highly reactive one. In dry air at sea level, nitrogen is the most abundant atmospheric gas, followed by oxygen, and then 13 other gases, each at less than 1% by volume (Table 10.1). For every 100 volume units of air, 21 units are oxygen. When it is pure, oxygen supports very rapid combustion (Figure 10.1). When it is diluted with nitrogen at moderate temperatures such as in the atmosphere, however, oxygen's oxidizing capability is tamed somewhat. The atmosphere is our only source for most of the noble gases—argon, neon, krypton, and xenon.

© Cengage Learning/Charles D. Winters

Figure 10.1 Combustion in pure oxygen. When a glowing wood splint is inserted into the mouth of a vessel containing pure oxygen, the high oxygen concentration causes the splint to burn much faster.

Due to the force of gravity, molecules making up the atmosphere are most concentrated near Earth's surface. In addition to the percentages by volume used in Table 10.1, *parts per million (ppm)* and *parts per billion (ppb)* by volume are used to describe the concentrations of components of the atmosphere. Since volume is proportional to the number of molecules, ppm and ppb units also give a ratio of molecules of one kind to another kind. For example, "10 ppm SO_2" means that for every 1 million molecules in air, 10 of them are SO_2 molecules. This may not sound like much until you consider that in just 1 cm^3 of air there are about 2.7×10^{13} million molecules. If this 1 cm^3 of air contains 10 ppm SO_2, then it contains 2.7×10^8 million SO_2 molecules.

To convert percent to ppm, multiply by 10,000. Divide by 10,000 to convert ppm to percent.

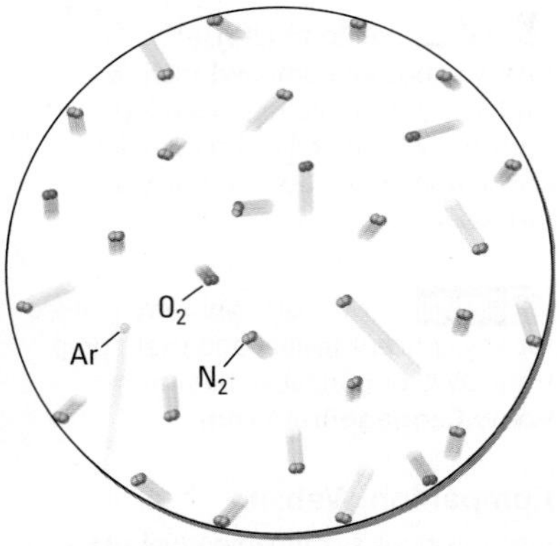

Nanoscale view of air. (Not to scale.) This instantaneous view of a sample of air at the nanoscale shows nitrogen, oxygen, and argon in rapid, random motion.

Table 10.1 The Composition of Dry Air at Sea Level*

Gas	Percentage by Volume	Gas	Percentage by Volume
Nitrogen	78.084	Krypton	0.0001
Oxygen	20.948	Carbon monoxide‡	0.00001
Argon	0.934	Xenon	0.000008
Carbon dioxide†	0.033	Ozone‡	0.000002
Neon	0.00182	Ammonia	0.000001
Hydrogen	0.0010	Nitrogen dioxide‡	0.0000001
Helium	0.00052	Sulfur dioxide‡	0.00000002
Methane†	0.0002		

*The data in this table are for dry air because the percentage of water in air varies. Therefore, water has been omitted even though it is an important component of the atmosphere.
†The greenhouse gases carbon dioxide and methane are discussed in Section 10.11 as they relate to fuels and the burning of fuels for energy production.
‡Trace gases of environmental importance.

Answers to **EXERCISES** are provided at the back of this book in Appendix L.

EXERCISE 10.1 Calculating the Mass of a Gas

Calculate the mass of SO_2 molecules found in 1 cm^3 of air that contains 10 ppm SO_2.

The troposphere was named by the British meteorologist Sir Napier Shaw from the Greek word *tropos,* meaning "turning."

The stratosphere was named by the French meteorologist Leon Phillipe Treisserenc de Bort, who believed this region consisted of orderly layers with no turbulence or mixing. *Stratum* is a Latin word meaning "layer."

The Troposphere and the Stratosphere

Earth's atmosphere can be roughly divided into layers, as shown in Figure 10.2. From the surface to an altitude of about 10 km, in the region named the **troposphere,** the temperature of the atmosphere decreases with increasing altitude. In this region the most violent mixing of air and the biggest variations in moisture content and temperature occur. Winds, clouds, storms, and precipitation are the result, the phenomena we know as weather. The troposphere is where we live. A commercial jet airplane flying at an altitude of about 10 km (approximately 33,000 ft) is still in the troposphere, although near its upper limits. The composition of the tropo-

ESTIMATION Thickness of Earth's Atmosphere

In the photograph of Earth on page 427, Earth's diameter is about 5 cm. About 99.9% of the atmosphere is contained within 50 km of Earth's surface. To draw a circle around the photo properly relating the diameter of Earth with the thickness of the atmosphere, how thick should the line be drawn?

The diameter of Earth is about 7900 miles, which is 12,700 km. Set up an equation that equates the ratio of the diameter of the photo to the diameter of Earth with the ratio of the sought line width (x) to the thickness of the atmosphere.

$$\frac{5\text{ cm}}{12{,}700\text{ km}} = \frac{x}{50\text{ km}}$$

$$x = 0.020\text{ cm} = 0.2\text{ mm}$$

If the circle representing Earth is 5 cm in diameter, then the line width that properly shows the relative thickness of the atmosphere would be only 0.2 mm, less than the width of the period at the end of this sentence. The proportions are about the same as the skin on an apple in relation to the apple. The atmosphere of Earth is really quite thin.

Visit this book's companion website at www.cengage.com/chemistry/moore to work an interactive module based on this material.

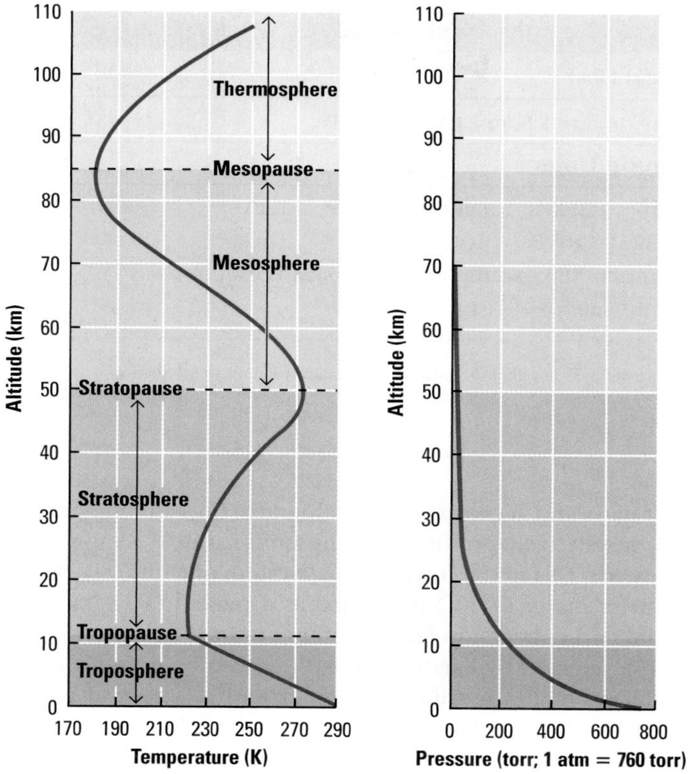

Figure 10.2 **Pressure and temperature changes in Earth's atmosphere.**

NASA

Earth.

sphere is roughly that of dry air near sea level (see Table 10.1), but the concentration of water vapor varies considerably, with an average of about 10 ppm.

Just above the troposphere, from about 12 to 50 km above Earth's surface, is the **stratosphere.** The pressures in the stratosphere are extremely low, and little mixing occurs between the stratosphere and the troposphere. The lower limit of the stratosphere varies from night to day over the globe, and in the polar regions it may be as low as 8 to 9 km above Earth's surface. About 75% of the mass of the atmosphere is in the troposphere, and 99.9% of the atmosphere's mass is below 50 km, in the troposphere and stratosphere.

10.2 Gas Pressure

Many molecular compounds are gases under ordinary conditions of temperature and pressure. The air we breathe consists mostly of diatomic nitrogen and oxygen molecules. Other elements that exist as gases at normal temperature and pressure include H_2, Cl_2, F_2, and the noble gases. Molecular compounds that are gases include: methane, CH_4; acetylene, C_2H_2; nitrous oxide, N_2O; ammonia, NH_3; carbon dioxide, CO_2; and sulfur dioxide, SO_2. At sufficiently high temperatures, many compounds that are normally liquids or solids become gases.

One of the most important properties of any gas is its pressure. The firmness of a balloon filled with air indicates that the gas inside exerts pressure. This pressure is caused by gas molecules striking the balloon's inner surface with each collision exerting a force on the surface. The force per unit area is the **pressure** of the gas. A gas exerts pressure on every surface it contacts, no matter what the direction of contact.

$$\text{Pressure} = \frac{\text{force}}{\text{area}}$$

A force can accelerate an object, and the force equals the mass of the object times its acceleration.

$$\text{Force} = \text{mass} \times \text{acceleration}$$

Table 10.2 Pressure Units

SI Unit: Pascal (Pa)
1 Pa = 1 kg m^{-1} s^{-2} = 1 N/m^2; 1 kPa = 10^3 Pa
Other Common Units
1 bar = 10^5 Pa = 100 kPa; 1 millibar (mbar) = 10^2 Pa
1 atm = 1.01325 × 10^5 Pa = 101.325 kPa
1 atm = 760 torr = 760 mm Hg (this conversion is exact)*
1 atm = 14.7 $lb/inch^2$ (psi) = 1.01325 bar

*Exact conversion factors do not limit the number of significant figures in calculations.

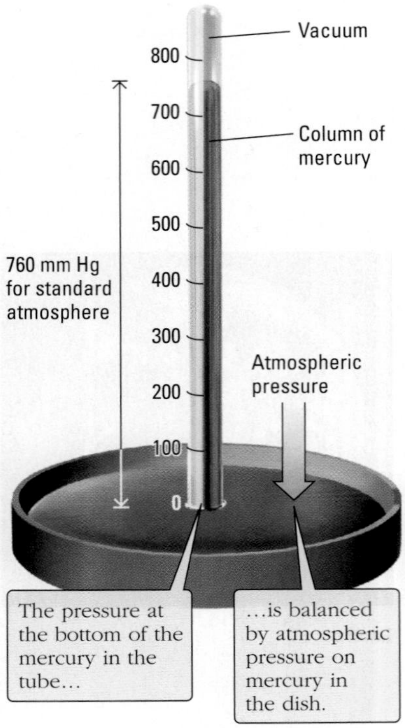

Figure 10.3 **A Torricellian barometer.**

The SI units for mass and acceleration are kilograms (kg) and meters per second per second (m/s^2), respectively, so force has the units kg m/s^2. A force of 1 kg m/s^2 is defined as a **newton (N)** in the International System of Units (SI). A pressure of one newton per square meter (N/m^2) is defined as a **pascal (Pa).** Table 10.2 provides units of pressure and conversion factors among units.

The pressure exerted by Earth's atmosphere can be measured with a **barometer,** which can be made by filling a tube closed at one end with a liquid and then inverting the tube in a dish containing the same liquid. Figure 10.3 shows a mercury barometer—a glass tube filled with mercury, inverted, and placed in a container of mercury. At sea level the height of the mercury column is 760 mm above the surface of the mercury in the dish. The pressure at the bottom of a 760-mm column of mercury is balanced by the pressure at the bottom of the column of air surrounding the dish—a column that extends to the top of the atmosphere. Pressure measured with a mercury barometer is usually reported in **millimeters of mercury (mm Hg),** a unit that is also called the **torr** after Evangelista Torricelli, who invented the mercury barometer in 1643. The **standard atmosphere (atm)** is defined as

1 standard atmosphere = 1 atm = 760 mm Hg (exactly) = 760 torr = 101.325 kPa

The pressure of the atmosphere at sea level is about 101,300 Pa (101.3 kPa). A related unit, the **bar,** equal to 100,000 Pa, is sometimes used for atmospheric pressure. The standard thermodynamic properties ***(← p. 246)*** are given for a gas pressure of 1 bar.

The **PROBLEM-SOLVING STRATEGY** in this book is

- Analyze the problem
- Plan a solution
- Execute the plan
- Check that the result is reasonable

Appendix A.1 explains this in detail.

PROBLEM-SOLVING EXAMPLE 10.1 Converting Pressure Units

Convert a pressure reading of 0.988 atm to (a) mm Hg, (b) kPa, (c) bar, (d) torr, and (e) psi.

Answer (a) 751. mm Hg (b) 100. kPa (c) 1.00 bar (d) 751 torr (e) 14.5 psi

Strategy and Explanation We use the unit equivalencies shown in Table 10.2 as conversion factors.

(a) $0.988 \text{ atm} \times \dfrac{760 \text{ mm Hg}}{1 \text{ atm}} = 751 \text{ mm Hg}$

(b) $0.988 \text{ atm} \times \dfrac{101.31 \text{ kPa}}{1 \text{ atm}} = 100. \text{ kPa}$

(c) $0.988 \text{ atm} \times \dfrac{1.013 \text{ bar}}{1 \text{ atm}} = 1.00 \text{ bar}$

(d) $0.988 \text{ atm} \times \dfrac{760 \text{ torr}}{1 \text{ atm}} = 751 \text{ torr}$

(e) $0.988 \text{ atm} \times \dfrac{14.7 \text{ psi}}{1 \text{ atm}} = 14.5 \text{ psi}$

PROBLEM-SOLVING PRACTICE 10.1

A TV weather person says the barometric pressure is "29.5 inches of mercury." What is this pressure in (a) atm, (b) mm Hg, (c) bar, and (d) kPa?

PROBLEM-SOLVING PRACTICE answers are provided at the back of this book in Appendix K.

Any liquid can be used in a barometer, but the height of the column depends on the density of the liquid. The atmosphere can support a column of mercury that is 760 mm high, or a column of water almost 34 ft high since water (1.0 g/cm^3) is much less dense than mercury (13.55 g/cm^3). A water barometer would be between 33 and 34 ft high, far too tall to be practical. Many water wells (especially in regions of the world without readily available electricity) bring up water from rock strata less than 33 ft below the surface. A simple hand pump can reduce the pressure at the top of the well, enabling the atmospheric pressure acting on the water at the bottom of the well to force the water upward. Before the invention of submersible electric pumps, well diggers knew from experience that if water was found deeper than 33 ft below the surface, a hole large enough for a bucket would have to be dug. The bucket would then have to be lowered and filled with water because atmospheric pressure alone could not push the water higher than 33 ft.

An old well pump. Old-fashioned water pumps like this one depend on atmospheric pressure to lift water from the well. They can only lift water from a depth of about 33 feet at sea level. At higher elevations this depth is lessened because the atmospheric pressure is lower.

10.3 Kinetic-Molecular Theory

All gases have a number of properties in common:

- ***Gases can be compressed.*** We often pump compressed air into an automobile or bicycle tire. Compressed air occupies less volume than the noncompressed air (Figure 10.4).
- ***Gases exert pressure on whatever surrounds them.*** A gas sample inside a balloon or a storage cylinder exerts pressure on its surroundings. If the container cannot sustain the pressure, some of the gas sample will escape, perhaps by bursting the container.
- ***Gases expand into whatever volume is available.*** The gaseous contents of a container will escape when it is opened.
- ***Gases mix completely with one another.*** Once gases are mixed, they do not separate spontaneously. The gases in the atmosphere are an example.

go Chemistry Module 16: Gas Laws and the Kinetic Molecular Theory covers concepts in this section.

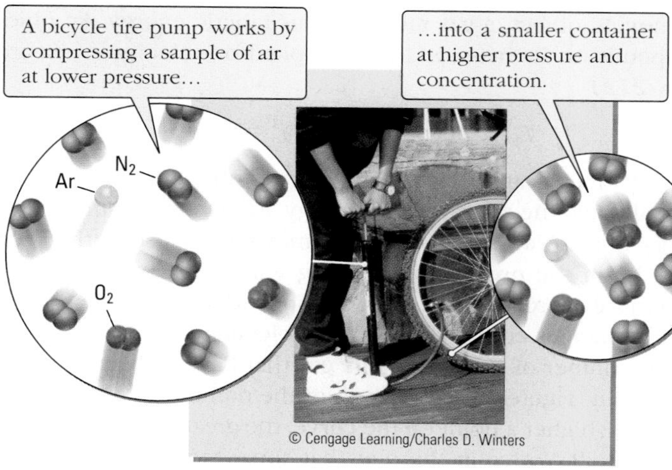

Figure 10.4 Compression of gases.

- ***Gases are described in terms of their temperature and pressure, the volume occupied, and the amount (numbers of molecules or moles) of gas present.*** For example, a hot gas in a fixed volume exerts a greater pressure than does the same sample of gas when it is cold.

In the kinetic-molecular theory the word "molecule" is taken to include atoms of the monatomic noble gases He, Ne, Ar, Kr, and Xe.

To explain why gases behave as they do, we first look at the nanoscale behavior of gas molecules. The fact that all gases behave in very similar ways can be interpreted by means of the *kinetic-molecular theory* (***← p. 19***), a theory that applies to the properties of liquids and solids as well as gases. According to the kinetic-molecular theory, a gas consists of tiny molecules in constant, rapid, random motion. The pressure a gas exerts on the walls of its container results from the continual bombardment of the walls by rapidly moving gas molecules (Figure 10.5).

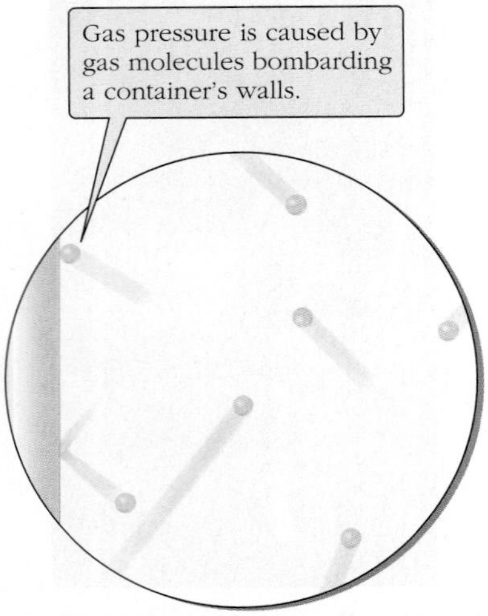
Gas pressure is caused by gas molecules bombarding a container's walls.

Figure 10.5 Kinetic-molecular theory and pressure.

Four fundamental postulates form the foundation of the kinetic-molecular theory, and a fifth is closely associated with it. Each is consistent with the results of experimental studies of gases.

1. ***A gas is composed of molecules whose size is much smaller than the distances between them.*** This concept accounts for the ease with which gases can be compressed and for the fact that gases at ordinary temperature and pressure mix completely with each other. These facts imply that there must be much unoccupied space in gases that provides substantial room for additional molecules in a sample of gas.
2. ***Gas molecules move randomly at various speeds and in every possible direction.*** This concept is consistent with the fact that gases quickly and completely fill any container in which they are placed.
3. ***Except when gas molecules collide, forces of attraction and repulsion between them are negligible.*** This concept is consistent with the fact that all gases behave in the same way, regardless of the types of noncovalent interactions among their molecules.
4. ***When collisions between molecules occur, the collisions are elastic.*** In an elastic collision, the speeds of colliding molecules may change, but the total kinetic energy of two colliding molecules is the same after a collision as before the collision. This concept is consistent with the fact that a gas sample at constant temperature never "runs down," with all molecules falling to the bottom of the container.
5. ***The average kinetic energy of gas molecules is proportional to the absolute temperature.*** Though not part of the kinetic-molecular theory, this useful concept is consistent with the fact that gas molecules escape through a tiny hole faster as the temperature increases, and with the fact that rates of chemical reactions are faster at higher temperatures.

Like any moving object, a gas molecule has kinetic energy. An object's kinetic energy, E_k, depends on its mass, m, and its speed or velocity, v, according to the equation (***← p. 212***)

$$E_k = \frac{1}{2}(\text{mass})(\text{speed})^2 = \frac{1}{2}mv^2$$

All the molecules in a gas are moving, but they do not all move at the same speed, so they do not all have the same kinetic energy. At any given time a few molecules are moving very quickly, a large number are moving at close to the average speed, and a few may be in the process of colliding with a surface, in which case their speed is momentarily zero. The speed of any individual molecule changes as it collides with and exchanges energy with other molecules.

The relative number of molecules of a gas that have a given speed can be measured experimentally. Figure 10.6 is a graph of the number of molecules plotted versus their speed. The higher a point on the curve, the greater the number of molecules moving at that speed. Notice in the plot that some molecules are moving quickly (have high kinetic energy) and some are moving slowly (have low kinetic energy). The maximum in the distribution curve is the most probable speed. For oxygen gas

CHEMISTRY IN THE NEWS Nitrogen in Tires

Usually, we fill our automobile tires with compressed air because it is cheap. But race cars, jet aircraft, and some heavy trucks use nitrogen. For the ordinary motorist, many tire shops now offer to fill car's tires with nitrogen for about $5.00 each. Is it worth it? What advantages might nitrogen have over air?

Many bogus claims are being made about such fillings. For example, some claim that nitrogen will leak out of your tires more slowly than air because the molecules are bigger, which is false.

Some good reasons for using nitrogen also exist. For example, nitrogen is relatively inert and will not support combustion in case of a fire around the tires. Pure nitrogen contains no oxygen, whereas air does, so corrosion of rims and the insides of tires is said to be decreased. The generation units used to provide the nitrogen produce very dry nitrogen, whereas the air going through a typical gas station compressor contains substantial amounts of moisture, which could aid in corrosion.

During 2006–2007, *Consumer Reports* performed a year-long study of the potential benefits of nitrogen in automobile tires and concluded that the nitrogen was of little benefit to consumers. Most professionals recommend using compressed air to fill automobile tires except in specialized applications.

Landing gear on a Boeing 747. Jet aircraft use nitrogen in the tires.

Sources: Chemical & Engineering News, October 25, 2004; p. 88; **http://blogs.consumerreports.org/cars/2007/10/tires-nitrogen-.html**

Moran, T. "Deflated Hype for Nitrogen." *New York Times*, March 3, 2008.

at 25 °C, for example, the maximum in the curve occurs at a speed of about 400 m/s (1000 mph), and most of the molecules' speeds are in the range from 200 m/s to 700 m/s. Notice that the curves are not symmetric. A consequence of this asymmetry is that the average speed (shown in Figure 10.6 for O_2) is a little faster than the most probable speed. Molecules with speeds higher than the most probable speed outnumber those with speeds lower than the most probable speed, moving the average speed toward a somewhat higher value.

Also notice in Figure 10.6 that as temperature increases, the most probable speed increases, and the number of molecules moving increases very quickly. The areas under the two curves representing the two oxygen samples at different temperatures are the same because the total number of oxygen molecules is the same in both samples.

Graphs of molecular speeds (or energies) versus numbers of molecules are called Boltzmann distribution curves. They are named after Ludwig Boltzmann (1844–1906), an Austrian physicist who helped develop the kinetic-molecular theory of gases.

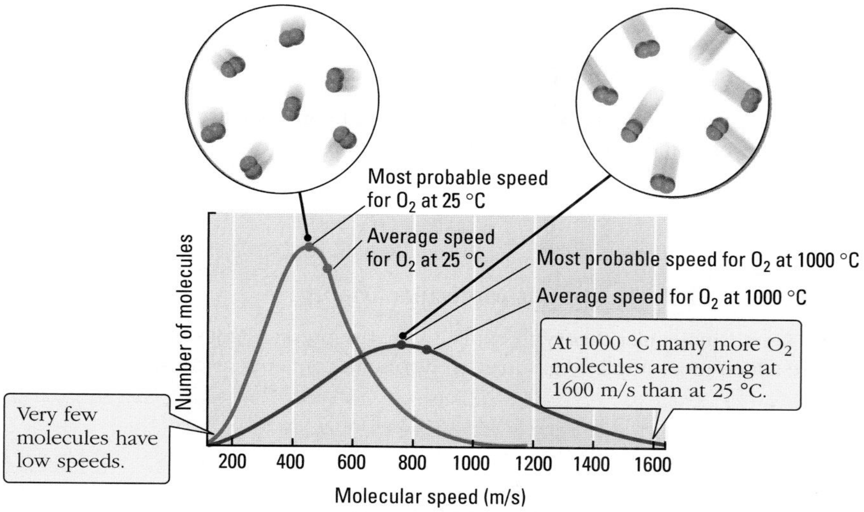
Most probable speed for O2 at 25 °C
Average speed for O2 at 25 °C
Most probable speed for O2 at 1000 °C
Average speed for O2 at 1000 °C
At 1000 °C many more O2 molecules are moving at 1600 m/s than at 25 °C.
Very few molecules have low speeds.
Number of molecules
200 400 600 800 1000 1200 1400 1600
Molecular speed (m/s)

Active Figure 10.6 Distribution of molecular speeds. A plot of the relative number of oxygen molecules with a given speed versus that speed. Visit this book's companion website at **www.cengage.com/chemistry/moore** to test your understanding of the concepts in this figure.

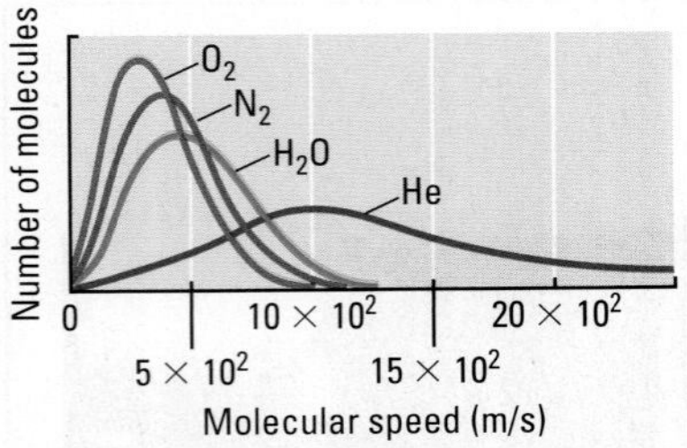

Figure 10.7 The effect of molar mass on the distribution of molecular speeds at a given temperature. On average, heavier molecules move slower than lighter ones.

Since $E_k = \frac{1}{2}mv^2$ and the average kinetic energy of the molecules of any gas is the same at a given temperature, the larger m is, the smaller the average v must be. That is, heavier molecules have slower average speed than lighter ones *at the same temperature*. Figure 10.7 illustrates this relationship. The peaks in the curves for the heavier molecules, O_2 and N_2, occur at a much lower speed than that for the lightest molecule, He. You can also see from the graph that average speeds range from a few hundred to a few thousand meters per second.

PROBLEM-SOLVING EXAMPLE 10.2 Kinetic-Molecular Theory

Consider a sample of oxygen gas, O_2, at 25 °C in a rigid container. Use the kinetic-molecular theory to answer these questions:

(a) How would the pressure of the gas change if the temperature were raised to 40 °C?
(b) If some of the O_2 molecules were removed from the sample (at constant T), how would the pressure change?

Answer (a) The pressure would increase. (b) The pressure would decrease.

Strategy and Explanation We use the fundamental concepts of kinetic-molecular theory to answer each question.

(a) Raising the temperature of the O_2 molecules causes them to move at higher average speed and higher average kinetic energy. At the higher temperature the molecules would hit the container walls harder and more often, which would yield a higher pressure.
(b) Molecules in the new sample of O_2, one with fewer gas molecules than originally present but at the same temperature, would still have the same average speed and average energy, so they would hit the walls with the same impact after the removal of part of the sample. However, the gas molecules would hit the walls less frequently, so the pressure would be lower after the removal of some of the gas.

PROBLEM-SOLVING PRACTICE 10.2

(a) How does kinetic-molecular theory explain the change in pressure when a sample of gas has its volume decreased while the temperature remains constant?
(b) How does kinetic-molecular theory explain the change in pressure when gas molecules are added to a sample of gas in a fixed-volume container at constant temperature?

EXERCISES that are labeled **CONCEPTUAL** are designed to test your understanding of one or more concepts; they usually involve qualitative rather than quantitative thinking.

CONCEPTUAL EXERCISE 10.2 Seeing Through Gases

Explain why gases are transparent to light. (*Hint:* You may need to refer to Chapter 7 in addition to using some of the concepts in this section.)

CONCEPTUAL EXERCISE 10.3 The Kinetic-Molecular Theory

Use the kinetic-molecular theory to explain why the pressure goes up when gas molecules are added to a sample of gas in a fixed-volume container at constant temperature.

EXERCISE 10.4 Molecular Kinetic Energies

Arrange these gaseous substances in order of increasing average kinetic energy of their molecules at 25 °C: Cl_2, H_2, NH_3, SF_6.

CONCEPTUAL EXERCISE 10.5 Molecular Kinetic Energies

Using Figure 10.7 as a source of information, first draw a plot of the number of molecules versus molecular speed for a sample of helium at 25 °C. Now assume that an equal number of molecules of argon, also at 25 °C, are added to the helium. What would the distribution curve for the mixture of gases look like (one distribution curve for all molecules in the mixture)?

CONCEPTUAL EXERCISE 10.6 Molecular Motion

Suppose you have two helium-filled balloons of about equal size. Put one of them in the freezer compartment of your refrigerator, and leave the other one out in your room. After a few hours you take the balloon from the freezer and compare it to the one left out in the room. Based on the kinetic-molecular theory, what differences would you expect to see (a) immediately after taking the balloon from the freezer and (b) after the cold balloon warms to room temperature?

The dependence of molecular speeds on molecular mass can be used to explain some interesting phenomena. One is **effusion,** the escape of gas molecules from a container through a tiny hole into a vacuum. The rate of effusion of a gas is inversely proportional to the square root of its molar mass: rate $\propto 1/\sqrt{M}$. (The symbol $\propto$ means "proportional to.") Therefore, He will effuse much faster than N_2, and a balloon filled with He will deflate faster than one filled with air. A related phenomenon is **diffusion,** the spread of gas molecules of one type through those of another type—for example, a pleasant odor spreading throughout a room. Diffusion is more complicated than effusion because the spreading gas molecules are colliding with the second type of gas molecules. Nevertheless, diffusion follows the same relationship between rate and molar mass. Diffusion was employed in the Manhattan project during World War II to purify uranium for the atomic bomb because the gaseous UF_6 molecules diffuse at slightly different rates due to the difference in masses of the two uranium isotopes, ^{235}U and ^{238}U.

Ballons just after filling

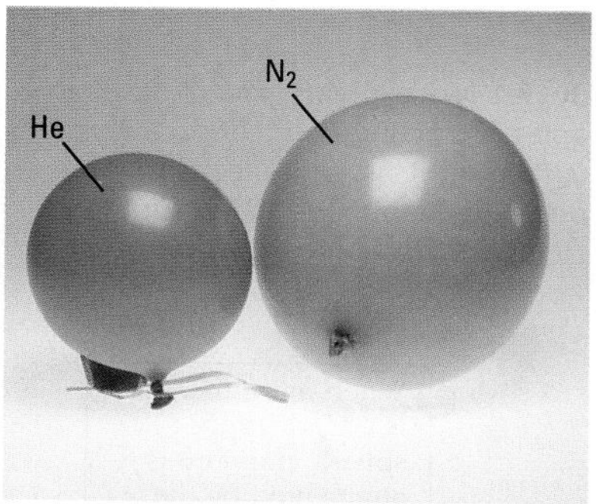

Several hours later at the same temperature

Helium, with its lower molar mass, effuses through tiny holes in a balloon surface faster than nitrogen, with its higher molar mass.

10.4 The Behavior of Ideal Gases

The kinetic-molecular theory explains gas behavior on the nanoscale. On the macroscale, gases have been studied for hundreds of years, and the properties that all gases display have been summarized into *gas laws* that are named for their discoverers. Using the variables pressure, volume, temperature, and amount (number of moles), we can write equations that explain how gases behave. A gas that behaves exactly as described by these equations is called an **ideal gas.** At room temperature and atmospheric pressure, most gases behave nearly ideally. However, at pressures much higher than 1 atm or at temperatures just above the boiling point, gases deviate substantially from ideal behavior (Section 10.8). Each of the gas laws can be explained by the kinetic-molecular theory.

go Chemistry **Module 16: Gas Laws and the Kinetic Molecular Theory** covers concepts in this section.

The Pressure-Volume Relationship: Boyle's Law

Boyle's law *states that the volume* (V) *of an ideal gas varies inversely with the applied pressure* (P) *when temperature* (T) *and amount* (n, *moles) are constant.*

$$V \propto \frac{1}{P} \qquad V = \text{constant} \times \frac{1}{P} \qquad PV = \text{constant} \quad (\text{unchanging } T \text{ and } n)$$

The value of the constant depends on the temperature and the amount of the gas. The inverse relationship between V and P is shown graphically in Figure 10.8a.

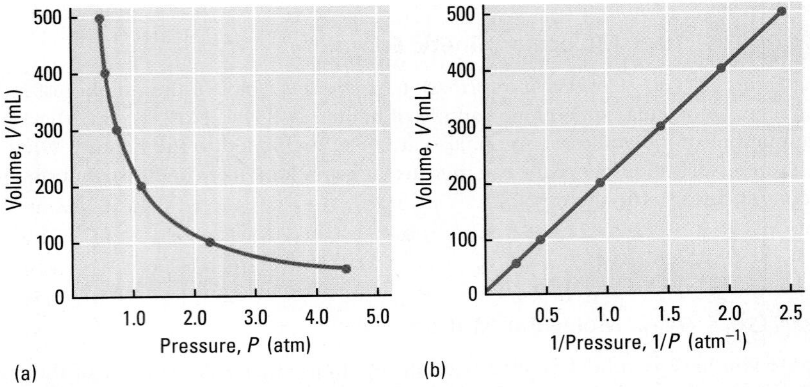

Volume, V (mL)
500
400
300
200
100
0
1.0 2.0 3.0 4.0 5.0
Pressure, P (atm)
(a)
Volume, V (mL)
500
400
300
200
100
0
0.5 1.0 1.5 2.0 2.5
1/Pressure, 1/P (atm⁻¹)
(b)

Figure 10.8 Graphical illustration of Boyle's law. (a) Volume (V) versus pressure (P). This curve shows the inverse proportionality between volume and pressure. As pressure increases, volume decreases. (b) V versus $1/P$. A linear plot results.

Boyle's law explains how expanding and contracting your chest cavity (changing its volume) leads to pressure changes which in turn lead to inhalation and exhalation of air.

Plotting V versus $1/P$ yields a linear relationship, as shown in Figure 10.8b. For a gas sample under two sets of pressure and volume conditions, Boyle's law can be written as $P_1V_1 = P_2V_2$.

In terms of the kinetic-molecular theory, a decrease in volume of a gas increases its pressure because there is less room for the gas molecules to move around before they collide with the walls of the container. Thus, there are more frequent collisions with the walls. These collisions produce pressure on the container walls, and more collisions mean a higher pressure. When you pump up a tire with a bicycle pump, the gas in the pump is squeezed into a smaller volume by application of pressure. This property is called *compressibility.* In contrast to gases, liquids and solids are only slightly compressible.

Robert Boyle studied the compressibility of gases in 1661 by pouring mercury into a J-shaped tube containing a sample of trapped gas. Each time he added more mercury, at constant temperature, the volume of the trapped gas decreased (Figure 10.9). The mercury additions increased the pressure on the gas and changed the gas volume in a predictable fashion.

The left side of Figure 10.9 shows an amount of trapped gas in the tube having a volume $V_1 = 30.0\ \text{cm}^3$ at a pressure of 760 mm Hg. The right side of Figure 10.9 shows that when mercury is added, the pressure increases to 760 mm Hg + 305 mm Hg = 1065 mm Hg, and the volume of the gas decreases according to Boyle's law.

$$V_2 = V_1 \times \frac{P_1}{P_2} = 30.0\ \text{cm}^3 \times \frac{760\ \text{mm Hg}}{1065\ \text{mm Hg}} = 21.4\ \text{cm}^3$$

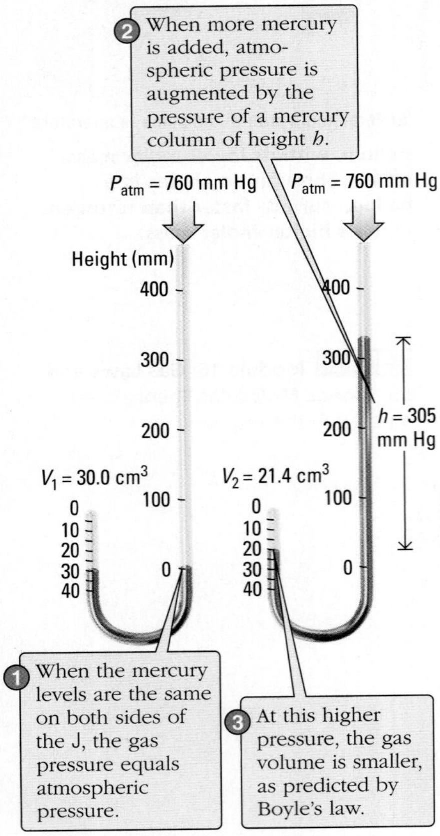

2 When more mercury is added, atmospheric pressure is augmented by the pressure of a mercury column of height h.
P_atm = 760 mm Hg
P_atm = 760 mm Hg
Height (mm)
400
300
200
100
0
V₁ = 30.0 cm³
0
10
20
30
40
V₂ = 21.4 cm³
h = 305 mm Hg
1 When the mercury levels are the same on both sides of the J, the gas pressure equals atmospheric pressure.
3 At this higher pressure, the gas volume is smaller, as predicted by Boyle's law.

Figure 10.9 Boyle's law. Boyle's experiment showing the compressibility of gases.

CONCEPTUAL EXERCISE 10.7 Visualizing Boyle's Law

Many cars have gas-filled shock absorbers to give the car and its occupants a smooth ride. If a four-passenger car is loaded with four NFL linemen, describe the gas inside the shock absorbers compared with that when the car has no passengers.

The Temperature-Volume Relationship: Charles's Law

***Charles's law** states that the volume* (V) *of an ideal gas varies directly with absolute temperature* (T) *when pressure* (P) *and amount* (n, *moles*) *are constant.*

$$V \propto T \qquad V = \text{constant} \times T \qquad \frac{V}{T} = \text{constant} \qquad (\text{unchanging } P \text{ and } n)$$

The value of the constant depends on pressure and the amount of the gas. If the volume, V_1, and temperature, T_1, of a sample of gas are known, then the volume, V_2, at some other temperature, T_2, at the same pressure is given by

$$\frac{V_1}{T_1} = \frac{V_2}{T_2} \qquad (P \text{ and } n \text{ constant})$$

In terms of the kinetic-molecular theory, higher temperature means faster molecular motion and a higher average kinetic energy. The more rapidly moving molecules therefore strike the walls of a container more often, and each collision exerts greater force. For the pressure to remain constant, the volume of the container must expand.

When using the gas law relationships, temperature must be expressed in terms of the **absolute temperature scale.** The zero on this scale is the lowest possible temperature. The SI unit of the absolute temperature scale is the kelvin, symbol K (with no degree sign). The relationship between the absolute scale and the Celsius scale is shown in Figure 10.10. The kelvin is the same size as a degree Celsius. Thus, when ΔT is calculated by subtracting one temperature from another, the result is the same on both scales, even though the numbers involved are different. The lowest possible temperature, 0 K, known as **absolute zero,** is −273.15 °C. Temperature on the Celsius scale T(°C) and on the absolute scale T(K) are related as follows:

$$T(\text{K}) = T(°\text{C}) + 273.15$$

Thus, 25.00 °C (a typical room temperature) is the same as 298.15 K.

In 1787 Jacques Charles discovered that the volume of a fixed quantity of a gas at constant pressure increases with increasing temperature. Figure 10.11 shows how the volume of 1.0 mol H_2 and the volume of 0.55 mol O_2 change with temperature (pressure and amount of gas remain constant). When the plots of volume versus temperature for different gases are extended toward lower temperatures, they all reach zero volume at the same temperature, −273.15 °C, which is 0 K.

Suppose you want to calculate the new volume when 450.0 mL of a gas is cooled from 60.0 °C to 20.0 °C, at constant pressure. First, convert the temperatures to kelvins by adding 273.15 to the Celsius values: 60.0 °C becomes 333.2 K and 20.0 °C becomes 293.2 K. Using Charles's law for the two sets of conditions at constant pressure gives

$$V_2 = \frac{V_1 T_2}{T_1} = \frac{450.0 \text{ mL} \times 293.2 \text{ K}}{333.2 \text{ K}} = 396.0 \text{ mL}$$

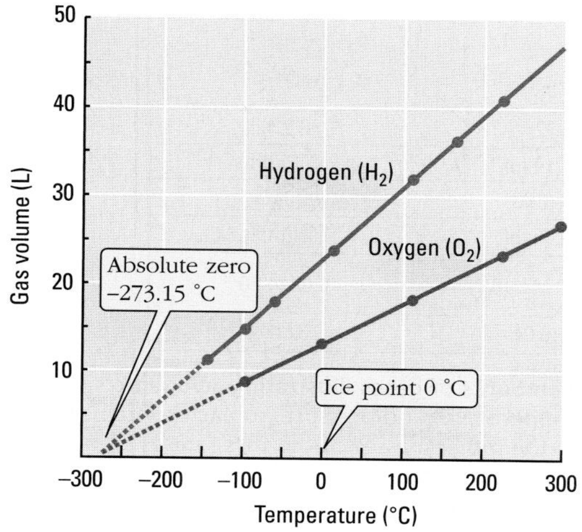

Figure 10.11 Charles's law. The volumes of two different samples of gases decrease with decreasing temperature (at constant pressure and constant molar amount). These graphs (as would those of all gases) intersect the temperature axis at about −273 °C.

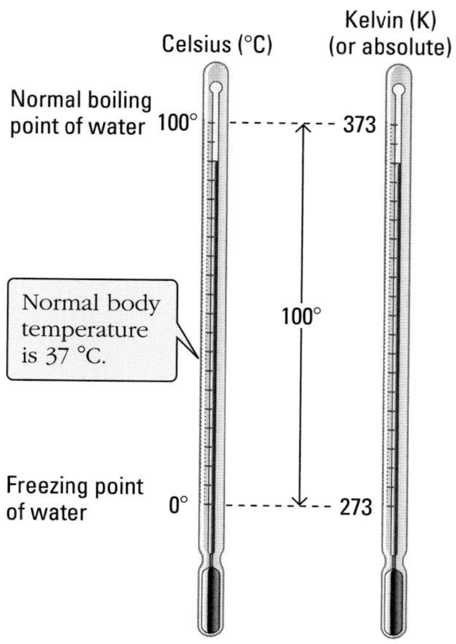

Figure 10.10 Temperature scales. Zero on the absolute scale is the lowest possible temperature (0 K or −273.15 °C).

The absolute temperature scale is also called the **Kelvin temperature scale.**

1746–1823

The French scientist Jacques Charles was most famous in his lifetime for his experiments in ballooning. In August 1783 Charles filled a tarred silk balloon with hydrogen. Inflating the bag to its final diameter (10 m) took several days and required nearly 500 lb acid and 1000 lb iron to generate the required volume of hydrogen gas. The balloon rose to 50 m, stayed aloft for almost 45 min, and traveled about 15 miles. When it landed in a village, the people there were so terrified that they tore the balloon to shreds.

CONCEPTUAL EXERCISE 10.8 Visualizing Charles's Law

Consider a collection of gas molecules at temperature T_1. The temperature increases to T_2. Use the ideas of the kinetic-molecular theory and explain why the volume would have to be larger if the pressure remains constant. What would have to be done to maintain a constant volume if the temperature increased?

The Amount-Volume Relationship: Avogadro's Law

***Avogadro's law** states that the volume* (V) *of an ideal gas varies directly with amount* (n) *when temperature* (T) *and pressure* (P) *are constant.*

$$V \propto n \qquad V = \text{constant} \times n \qquad \frac{V}{n} = \text{constant} \qquad (\text{unchanging } T \text{ and } P)$$

The value of the constant depends on the temperature and the pressure. Avogadro's law means, for example, that at constant temperature and pressure, if the number of moles of gas doubles, the volume doubles. It also means that at the same temperature and pressure, the volumes of two different amounts of gases are related as follows:

$$\frac{V_1}{n_1} = \frac{V_2}{n_2} \qquad (T \text{ and } P \text{ constant})$$

In terms of the kinetic-molecular theory, increasing the number of gas molecules at a constant temperature means that the added molecules have the same average molecular kinetic energy as the molecules to which they were added. As a result, the number of collisions with the container walls increases in proportion to the number of molecules. This would increase the pressure if the volume were held constant. To maintain constant pressure, the volume must increase.

Another way to state the relationship between the amount of gas and its volume is *Avogadro's hypothesis: Equal volumes of gases contain equal numbers of molecules at the same temperature and pressure.* Experiments show that at 0 °C and 1 atm pressure, 22.4 L of any gas will contain 1 mol of the gas (6.02×10^{23} gas molecules). This is illustrated in Figure 10.12.

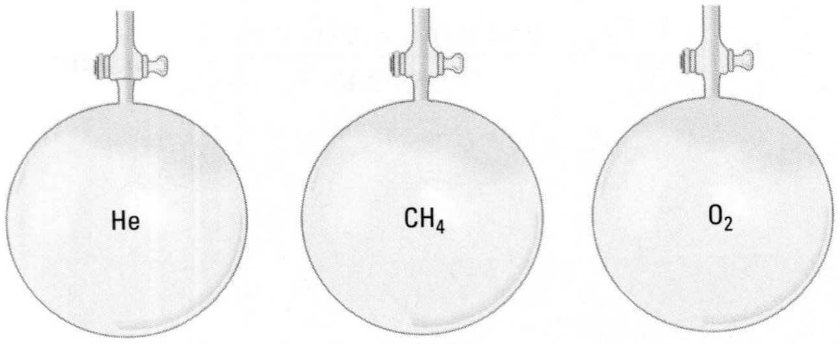

V	22.4 L	22.4 L	22.4 L
P	1 atm	1 atm	1 atm
T	0 °C	0 °C	0 °C
Amount of gas	1 mol	1 mol	1 mol
No. of gas molecules	6.02×10^{23}	6.02×10^{23}	6.02×10^{23}
Mass of gas	4.00 g	16.04 g	32.00 g

Figure 10.12 Comparison of volumes and amounts of gases illustrating Avogadro's hypothesis. Each gas has the same volume, temperature, and pressure, so each container holds the same number of molecules. Because the molar masses of the three gases are different, the masses of gas in each container are different.

The Law of Combining Volumes

In 1809 the French scientist Joseph Gay-Lussac (1778–1850) conducted experiments in which he measured the volumes of gases reacting with one another to form gaseous products. He found that, at constant temperature and pressure, the volumes of reacting gases were always in the *ratios of small whole numbers.* This is known as the **law of combining volumes.** For example, the reaction of 2 L H_2 with 1 L O_2 produces 2 L water vapor. Similarly, the reaction of 4 L H_2 with 2 L O_2 produces 4 L water vapor.

	$2\ H_2(g)$	+	$O_2(g)$	→	$2\ H_2O(g)$
Experiment 1	2 L		1 L		2 L
Experiment 2	4 L		2 L		4 L

In 1811 Amadeo Avogadro suggested that Gay-Lussac's observations actually showed that equal volumes of all gases under the same temperature and pressure conditions contain the same number of molecules. Viewed on a molecular scale, if a tiny volume contained only one molecule of a gas, then twice that volume would contain two molecules, and so on (Figure 10.13).

John Dalton (who devised the atomic theory) strongly opposed Avogadro's ideas and never accepted them. It took about 50 years—long after Avogadro and Dalton had died—for Avogadro's explanation of Gay-Lussac's experiments to be generally accepted.

PROBLEM-SOLVING EXAMPLE 10.3 Using Avogadro's Law and the Law of Combining Volumes

Nitrogen and hydrogen gases react to form ammonia gas in the reaction

$$N_2(g) + 3\ H_2(g) \longrightarrow 2\ NH_3(g)$$

If 500. mL N_2 at 1 atm and 25 °C were available for reaction, what volume of H_2, at the same temperature and pressure, would be required in the reaction?

Answer 1.50 L H_2

Strategy and Explanation We want to find the volume of H_2, and the information given is the volume of the N_2 at the same temperature and pressure. Therefore, we can

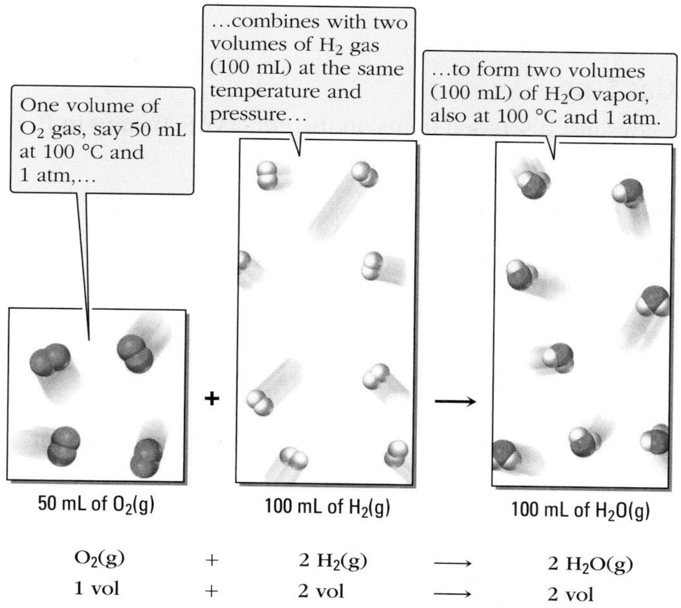
One volume of O_2 gas, say 50 mL at 100 °C and 1 atm,...
...combines with two volumes of H_2 gas (100 mL) at the same temperature and pressure...
...to form two volumes (100 mL) of H_2O vapor, also at 100 °C and 1 atm.
50 mL of $O_2(g)$
+
100 mL of $H_2(g)$
→
100 mL of $H_2O(g)$
$O_2(g)$ + $2\ H_2(g)$ ⟶ $2\ H_2O(g)$
1 vol + 2 vol ⟶ 2 vol

Figure 10.13 Law of combining volumes. When gases at the same temperature and pressure combine with one another, their volumes are in the ratio of small whole numbers.

solve the problem using the law of combining volumes and the reaction coefficients of 3 for H_2 and 1 for N_2 as the volume ratio.

$$\text{Volume of } H_2 = 500.\text{ mL } N_2 \times \frac{3 \text{ mL } H_2}{1 \text{ mL } N_2} = 1.50 \text{ L } H_2$$

PROBLEM-SOLVING PRACTICE **10.3**

Nitrogen monoxide, NO, combines with oxygen to form nitrogen dioxide.

$$2\,NO(g) + O_2(g) \longrightarrow 2\,NO_2(g)$$

If 1.0 L oxygen gas at 30.25 °C and 0.975 atm is used and there is an excess of NO, what volume of NO gas at the same temperature and pressure will be converted to NO_2?

CONCEPTUAL EXERCISE **10.9 Gas Burner Design**

The gas burner in a stove or furnace admits enough air so that methane gas can react completely with oxygen in the air according to the equation

$$CH_4(g) + 2\,O_2(g) \longrightarrow CO_2(g) + 2\,H_2O(g)$$

Air is one-fifth oxygen by volume. Both air and methane gas are supplied to the flame by passing them through separate small tubes. Compared with the tube for the methane gas, how much bigger does the cross section of the tube for the air need to be? Assume that both gases are at the same T and P.

EXERCISE **10.10 Filling Balloons**

You have 100 balloons of equal volume filled with a total of 26.8 g helium gas at 23.0 °C and 748 mm Hg. The total volume of these balloons is 168 L. You are given 150 more balloons of the same size and 41.8 g He gas. The temperature and pressure remain the same. Determine by calculation whether you will be able to fill all the balloons with the He you have available.

The Ideal Gas Law

The three gas laws just discussed focus on the effects of changes in P, T, or n on gas volume:

- ***Boyle's law and pressure*** $(V \propto 1/P)$; $V_1P_1 = V_2P_2$
- ***Charles's law and temperature*** $(V \propto T)$; $\dfrac{V_1}{T_1} = \dfrac{V_2}{T_2}$
- ***Avogadro's law and amount (mol)*** $(V \propto n)$; $\dfrac{V_1}{n_1} = \dfrac{V_2}{n_2}$

These three gas laws can be combined to give the ideal gas law, which summarizes the relationships among volume, temperature, pressure, and amount.

$$V \propto \frac{nT}{P} \quad \text{or} \quad PV \propto nT$$

To make this proportionality into an equation, a proportionality constant, R, named the **ideal gas constant,** is used. The equation becomes

$$V = R\frac{nT}{P}$$

and, on rearranging, gives the equation called the **ideal gas law.**

$$PV = nRT$$

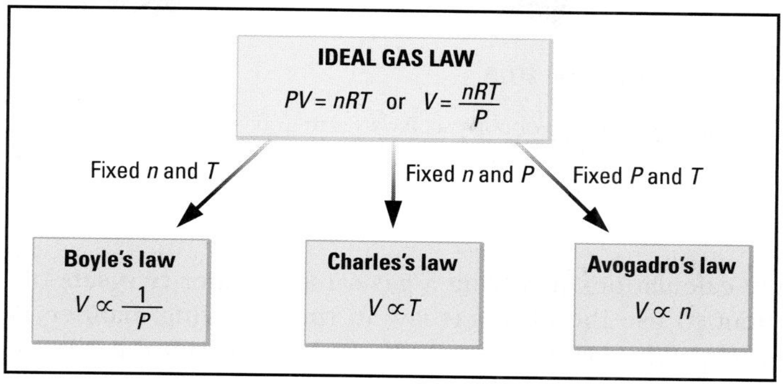

The ideal gas law correctly predicts the amount, pressure, volume, and temperature for samples of most gases at pressures of a few atmospheres or less and at temperatures well above their boiling points. The constant R can be calculated from the experimental fact that at 0 °C and 1 atm the volume of 1 mol gas is 22.414 L. This temperature and pressure are called **standard temperature and pressure (STP),** and the volume is called the **standard molar volume.** Solving the ideal gas law for R, and substituting, gives

$$R = \frac{PV}{nT} = \frac{(22.414\ \text{L})(1\ \text{atm})}{(1\ \text{mol})(273.15\ \text{K})} = 0.082057\ \text{L atm mol}^{-1}\ \text{K}^{-1}$$

which is usually rounded to 0.0821 L atm mol^{-1} K^{-1}. The ideal gas constant has different numerical values in different units, as shown in Table 10.3.

The ideal gas law can be used to calculate P, V, n, or T whenever three of the four variables are known, provided that the conditions of temperature and pressure are not extreme.

Table 10.3 Values of *R*, in Different Units

$R = 0.08206 \frac{\text{L atm}}{\text{mol K}}$
$R = 62.36 \frac{\text{mm Hg L}}{\text{mol K}}$
$R = 8.314 \frac{\text{kPa dm}^3}{\text{mol K}}$
$R = 8.314 \frac{\text{J}}{\text{mol K}}$

PROBLEM-SOLVING EXAMPLE 10.4 Using the Ideal Gas Law

Calculate the volume that 0.40 g methane, CH_4, occupies at 25 °C and 1.0 atm.

Answer 0.61 L or 6.1×10^2 mL CH_4

Strategy and Explanation The given information is the mass of methane and its temperature (T) and pressure (P). We can convert the mass of methane to the number of moles (n) and then use the ideal gas equation to find the volume (V). When we use $PV = nRT$, we must have all the variables in the same units as the gas constant, $R = 0.0821$ L atm mol^{-1} K^{-1}. That is, temperature must be in kelvins, pressure in atmospheres, volume in liters, and the amount of gas in moles.

We begin by converting the mass of methane to moles.

$$(0.40\ \text{g CH}_4) \times \frac{1\ \text{mol CH}_4}{16.04\ \text{g CH}_4} = 0.0249\ \text{mol CH}_4$$

Next, we convert the temperature to kelvins.

$$(25 + 273.15)\ \text{K} = 298\ \text{K}$$

Now we can solve for V in the ideal gas equation. The result will be in liters.

$$V = \frac{nRT}{P} = \frac{(0.0249\ \text{mol})(0.0821\ \text{L atm mol}^{-1}\ \text{K}^{-1})(298\ \text{K})}{1.0\ \text{atm}}$$

$$= 0.61\ \text{L or } 6.1 \times 10^2\ \text{mL CH}_4$$

Notice how the units cancel. Also, notice that the answer is given with two significant digits because the mass of methane and the pressure had only two significant digits.

☑ **Reasonable Answer Check** The temperature and pressure are close to STP. The mass of methane is about 2.5% of a mole, so the volume should also be about 2.5% of the volume of a mole of gas (22.4 L × 0.025 = 0.56 L). This approximate result is close to our more exact calculation.

PROBLEM-SOLVING PRACTICE 10.4

What volume will 2.64 mol N_2 occupy at 0.640 atm and 31 °C?

For many calculations involving a gas sample under two sets of conditions, it is convenient to use the ideal gas law in the following manner. For two sets of conditions (n_1, P_1, V_1, and T_1; n_2, P_2, V_2, and T_2), the ideal gas law can be written as

$$R = \frac{P_1V_1}{n_1T_1} \quad \text{and} \quad R = \frac{P_2V_2}{n_2T_2}$$

Since in both sets of conditions the quotient is equal to R, we can set the two quotients equal to each other.

$$\frac{P_1V_1}{n_1T_1} = \frac{P_2V_2}{n_2T_2}$$

When the amount of gas, n, is constant, so that $n_1 = n_2$, this equation simplifies to what is known as the **combined gas law:**

$$\frac{P_1V_1}{T_1} = \frac{P_2V_2}{T_2}$$

PROBLEM-SOLVING EXAMPLE 10.5 Pressure and Volume

Consider a gas sample with a volume of 1.00 L at a pressure of 0.500 atm. What volume will this gas sample occupy if the pressure is increased to 2.00 atm? Assume that the temperature remains constant and the gas sample obeys the ideal gas law.

Answer 0.250 L or 250. mL

Strategy and Explanation The temperature (T) and the amount of gas (n) are constant. The given conditions are $P_1 = 0.500$ atm, $P_2 = 2.00$ atm, and $V_1 = 1.00$ L. We must find V_2, which can be calculated by canceling T in the combined gas law and rearranging to solve for V_2.

$$\frac{P_1V_1}{\cancel{T_1}} = \frac{P_2V_2}{\cancel{T_2}}$$

$$V_2 = \frac{P_1V_1}{P_2} = \frac{(0.500\ \text{atm})(1.00\ \text{L})}{2.00\ \text{atm}} = 0.250\ \text{L} = 250.\ \text{mL}$$

☑ **Reasonable Answer Check** The inverse relationship between P and V tells us that V_2 will be smaller than V_1 because P_2 is larger than P_1. The answer is reasonable.

PROBLEM-SOLVING PRACTICE 10.5

At a pressure of exactly 1 atm and some temperature, a gas sample occupies 400. mL. What will be the volume of the gas at the same temperature if the pressure is decreased to 0.750 atm?

PROBLEM-SOLVING EXAMPLE 10.6 The Combined Gas Law

Helium-filled balloons are used to carry scientific instruments high into the atmosphere. Suppose that such a balloon is launched on a summer day when the temperature at ground level is 22.5 °C and the barometer reading is 754 mm Hg. If the balloon's volume is 1.00×10^6 L at launch, what will its volume be at a height of 37 km, where the pressure is 76.0 mm Hg and the temperature is 240. K? (Assume that no helium escapes.)

Answer 8.05×10^6 L

Strategy and Explanation Only T, P, and V change. Using subscript 1 to indicate initial conditions (at launch) and subscript 2 to indicate final conditions (high in the atmosphere) gives

Initial: $P_1 = 754$ mm Hg $T_1 = (22.5 + 273.15)$ K $= 295.6$ K $V_1 = 1.00 \times 10^6$ L

Final: $P_2 = 76.0$ mm Hg $T_2 = 240.$ K

Solving the combined gas law for V_2 gives

$$V_2 = \frac{P_1V_1T_2}{P_2T_1} = \frac{(754 \text{ mm Hg})(1.00 \times 10^6 \text{ L})(240. \text{ K})}{(76.0 \text{ mm Hg})(295.6 \text{ K})} = 8.05 \times 10^6 \text{ L}$$

Thus, the final volume is about eight times larger. The volume has increased because the pressure has dropped. For this reason, weather balloons are never fully inflated at launch. A great deal of room has to be left so that the helium can expand at high altitudes.

☑ **Reasonable Answer Check** The pressure is dropping by a factor of about 10, and T is changing only 20%, so the volume should increase by a factor of about 20% less than 10, that is, 8, and it does.

PROBLEM-SOLVING PRACTICE 10.6

A small sample of a gas is prepared in the laboratory and found to occupy 21 mL at a pressure of 710. mm Hg and a temperature of 22.3 °C. The next morning the temperature has changed to 26.5 °C, and the pressure is found to be 740. mm Hg. No gas has escaped from the container.

(a) What volume does the sample of gas now occupy?

(b) Assume that the pressure does not change. What volume would the gas occupy at the new temperature?

A weather balloon filled with helium. As it ascends into the atmosphere, does the volume increase or decrease?

It is important to remember that Boyle's, Charles's, and Avogadro's laws do not depend on the identity of the gas being studied. These laws reflect properties of all gases and therefore describe the behavior of any gaseous substance.

In summary, for problems involving gases, there are two useful equations:

- ***When three of the variables* P, V, n, *and* T *are given, and the value of the fourth variable is needed, use the ideal gas law.***

$$PV = nRT$$

- ***When one set of conditions is given for a single gas sample and one of the variables under a new set of conditions is needed, then use the combined gas law and cancel any of the three variables that do not change.***

$$\frac{P_1V_1}{T_1} = \frac{P_2V_2}{T_2} \quad (n \text{ constant})$$

CONCEPTUAL EXERCISE 10.11 Predicting Gas Behavior

Name three ways the volume occupied by a sample of gas can be decreased.

10.5 Quantities of Gases in Chemical Reactions

The law of combining volumes and the ideal gas law make it possible to use volumes as well as masses or molar amounts in calculations based on reaction stoichiometry (← *p. 128*). Consider this balanced equation for the combustion reaction of methane, CH_4, and oxygen to produce carbon dioxide and water vapor.

$$CH_4(g) + 2\,O_2(g) \longrightarrow CO_2(g) + 2\,H_2O(g)$$

Here are some questions that might be asked about this reaction. For each question, try to recognize what information is known and what is needed. Then decide what relationship connects the two.

1. *How many liters of CO_2 are produced from 0.5 L methane and sufficient O_2?* The information known is the volume of CH_4 (and that there is plenty of O_2). The information needed is the volume of CO_2. The coefficients of the balanced chemical equation represent volumes of gases, so the law of combining volumes leads to the answer.
   Answer: 0.5 L CO_2, because the balanced equation shows 1 mol CO_2 is produced for every mole of CH_4 consumed—a 1:1 ratio.

$$0.5\text{ L }CH_4 \times \frac{1\text{ L }CO_2}{1\text{ L }CH_4} = 0.5\text{ L }CO_2$$

2. *How many moles of O_2 are required to react completely with 10.0 g CH_4?* Here the known information is the mass of one reactant (methane), and the needed information is amount of oxygen required. The connection is provided by the stoichiometric factors in the balanced equation; no gas law equation is needed.
   Answer: 1.25 mol O_2, because 10.0 g CH_4 is 0.624 mol CH_4, and 2 mol O_2 is required for every 1 mol CH_4.

$$10.0\text{ g }CH_4 \times \frac{1\text{ mol }CH_4}{16.03\text{ g }CH_4} \times \frac{2\text{ mol }O_2}{1\text{ mol }CH_4} = 1.25\text{ mol }O_2$$

3. *What volume (in liters) of O_2 at STP is required to react with 10.0 g CH_4?* Because the needed information is the volume of a gaseous reactant, O_2, the law of combining volumes can lead to the answer, but first the number of moles of methane must be found. The standard molar volume of any gas at STP is 22.4 L.
   Answer: 27.9 L O_2, because 10.0 g CH_4 is 0.624 mol CH_4, which requires 1.25 mol O_2.

$$10.0\text{ g }CH_4 \times \frac{1\text{ mol }CH_4}{16.03\text{ g }CH_4} \times \frac{22.4\text{ L }CH_4}{1\text{ mol }CH_4} \times \frac{2\text{ L }O_2}{1\text{ L }CH_4} = 27.9\text{ L }O_2$$

4. *What volume (in liters) of O_2 at 747 mm Hg and 21.0 °C is required to react with 10.0 g CH_4?* Realizing that 10.0 g CH_4 is 0.624 mol CH_4 (Part 2) and that 1.25 mol O_2 will be required, you then use the ideal gas law to calculate the volume occupied by 1.25 mol O_2 at the temperature and pressure given.
   Answer: The units must be consistent with those of R, in this case L atm mol^{-1} K^{-1}. Convert the Celsius temperature to kelvins to get 294.2 K, and convert the pressure to atmospheres to get 0.983 atm. Use the ideal gas law to calculate the volume of O_2 required.

$$V_{O_2} = \frac{nRT}{P} = \frac{(1.25\text{ mol})(0.0821\text{ L atm mol}^{-1}\text{ K}^{-1})(294.2\text{ K})}{0.983\text{ atm}} = 30.7\text{ L }O_2$$

PROBLEM-SOLVING EXAMPLE 10.7 Gases and Stoichiometry

If 13.8 g Mg is reacted with excess HCl, how many liters of H_2 gas are produced at 28.0 °C and 675 torr?

$$Mg(s) + 2\,HCl(aq) \longrightarrow MgCl_2(aq) + H_2(g)$$

Answer 15.8 L H_2

Strategy and Explanation First, find the amount of H_2 generated; then use the ideal gas law to calculate the volume. Use the coefficients from the balanced equation and the molar mass of magnesium to obtain the amount of H_2.

$$13.8 \text{ g Mg} \times \frac{1 \text{ mol Mg}}{24.305 \text{ g Mg}} \times \frac{1 \text{ mol } H_2}{1 \text{ mol Mg}} = 0.568 \text{ mol } H_2$$

Next find the volume of H_2 from the ideal gas law, $PV = nRT$. Solve the ideal gas law for V and substitute the values for P, T, and n. The units of V, P, and T must be compatible with the units of R. This means that we must express the pressure in atmospheres and the temperature in kelvins (28.0 + 273.2 = 301.2 K).

$$P = (675 \text{ mm Hg}) \times \frac{1 \text{ atm}}{760 \text{ mm Hg}} = 0.888 \text{ atm}$$

$$V = \frac{nRT}{P} = \frac{(0.568 \text{ mol } H_2)(0.0821 \text{ L atm mol}^{-1} \text{ K}^{-1})(301.2 \text{ K})}{0.888 \text{ atm}} = 15.8 \text{ L } H_2$$

☑ **Reasonable Answer Check** We started with 13.8/24.3 or about 0.6 mol Mg, so about the same amount of H_2 should be produced. Six-tenths of 22.4 L (the volume of one mole of gas at STP) is about 13.4 L, which is comparable to our more exact answer.

© Cengage Learning/Charles D. Winters

Magnesium ribbon reacting with hydrochloric acid. The reaction produces H_2 gas that bubbles to the surface.

PROBLEM-SOLVING PRACTICE **10.7**

Ammonium nitrate, NH_4NO_3, is an explosive that undergoes this decomposition.

$$2 \text{ } NH_4NO_3(s) \longrightarrow 4 \text{ } H_2O(g) + O_2(g) + 2 \text{ } N_2(g)$$

If 10.0 g NH_4NO_3 explodes, how many liters of gas are generated at 25 °C and 1 atm?

PROBLEM-SOLVING EXAMPLE **10.8** **Stoichiometry with Gases—Air Bags**

Automobile air bags are filled with N_2 from the decomposition of sodium azide.

$$2 \text{ } NaN_3(s) \longrightarrow 2 \text{ Na}(s) + 3 \text{ } N_2(g)$$

How many grams of sodium azide are needed to generate 60.0 L N_2 at 1 atm and 26.0 °C?

Answer 106 g NaN_3

Strategy and Explanation Calculate the amount of N_2, use a mole ratio from the balanced equation to calculate the amount of NaN_3, and then calculate the mass of NaN_3 using the molar mass.

$$n_{N_2} = \frac{PV}{RT} = \frac{(1 \text{ atm})(60.0 \text{ L})}{(0.0821 \text{ L atm mol}^{-1} \text{ K}^{-1})(299.15 \text{ K})} = 2.44 \text{ mol } N_2$$

$$n_{NaN_3} = 2.44 \text{ mol } N_2 \times \frac{2 \text{ mol } NaN_3}{3 \text{ mol } N_2} = 1.63 \text{ mol } NaN_3$$

$$m_{NaN_3} = 1.63 \text{ mol } NaN_3 \times \frac{65.01 \text{ g } NaN_3}{1 \text{ mol } NaN_3} = 106 \text{ g } NaN_3$$

☑ **Reasonable Answer Check** Dividing the 106 g NaN_3 by its molar mass of 65 g/mol gives an answer of about 1.6 mol NaN_3. We get $\frac{3}{2}$ as many moles of N_2, or about 2.4 mol N_2, which is about 54 L N_2. This is close to the required volume of N_2.

© TRL Ltd./Photo Researchers, Inc.

Air bags deploying on crash test dummies. Air bags have saved numerous lives in automobile collisions.

Air bags use additional reactants because the reaction in Example 10.8 would produce elemental sodium, which is much too hazardous to remain in a consumer product. However, the actual reactions have nearly the same stoichiometric ratio of NaN_3 to N_2 as the simpler reaction given in this example.

PROBLEM-SOLVING PRACTICE **10.8**

Lithium hydroxide is used in spacecraft to absorb CO_2 exhaled by astronauts.

$$2 \text{ LiOH}(s) + CO_2(g) \longrightarrow Li_2CO_3(s) + H_2O(\ell)$$

What volume of CO_2 at 22 °C and 1 atm is absorbed per gram of LiOH?

10.6 Gas Density and Molar Mass

The ideal gas law can be used to relate gas density to molar mass. The definition of density is mass per unit volume (← p. 8). The densities of gases are extremely variable because the volume of a gas sample (but not its mass) varies with temperature and pressure. However, once T and P are specified, the density of a gas can be calculated from the ideal gas law. Additionally, because equal volumes of gas at the same T and P contain equal numbers of molecules, the densities of different gases are directly proportional to their molar masses. As a result, experimental gas densities can be used to determine molar masses.

To derive the relationship between gas density and molar mass, start with the ideal gas law, $PV = nRT$. For any substance, the number of moles (n) equals its mass (m) divided by its molar mass (M), so we will substitute m/M for n in the ideal gas equation:

$$PV = \frac{m}{M}RT$$

Density (d) is mass divided by volume (m/V), and we can rearrange the equation so that m/V is on one side:

$$d = \frac{m}{V} = \frac{PM}{RT}$$

Thus, the density of a gas is directly proportional to its molar mass, M.

Consider the densities of the three pure gases He, O_2, and SF_6. If we consider 1 mol of each of these gases at 25 °C and 0.750 atm, we can see from the table below that density (at the same conditions) increases with molar mass.

	He	O_2	SF_6
Molar mass, g/mol	4.003	31.999	146.06
Density (25 °C), g/L	0.123	0.981	4.48

Let's use the density equation to estimate the density of air at STP. The density of nitrogen, N_2, at STP is

$$d_{N_2} = \frac{PM}{RT} = \frac{(1\text{ atm})(28.01\text{ g/mol})}{(0.0821\text{ L atm mol}^{-1}\text{ K}^{-1})(273.15\text{ K})} = 1.249\text{ g/L}$$

The density of oxygen, O_2, at STP is

$$d_{O_2} = \frac{PM}{RT} = \frac{(1\text{ atm})(31.999\text{ g/mol})}{(0.0821\text{ L atm mol}^{-1}\text{ K}^{-1})(273.15\text{ K})} = 1.427\text{ g/L}$$

The density of air is estimated as the weighted average of the densities of nitrogen and oxygen

$$d_{air} = (0.80)(\text{density of N}_2) + (0.20)(\text{density of O}_2)$$
$$= (0.80)(1.249\text{ g/L}) + (0.020)(1.427\text{ g/L}) = 1.28\text{ g/L}$$

Air has a density of approximately 1.28 g/L at STP.

EXERCISE 10.12 Calculating Gas Densities

Calculate the densities of Cl_2 and of SO_2 at 25 °C and 0.750 atm. Then calculate the density of Cl_2 at 35 °C and 0.750 atm and the density of SO_2 at 25 °C and 2.60 atm.

ESTIMATION Helium Balloon Buoyancy

How large a helium balloon is needed to lift a weather-instruments package weighing 500 lb? The buoyancy of He is the difference in density between air and He at the given temperature and pressure.

First, we must estimate the density of helium. Let's assume a temperature of 21 °C and a pressure of 1 atm. Then 1 L He contains

$$n = \frac{(1\text{ atm})(1\text{ L})}{(0.0821\text{ L atm mol}^{-1}\text{ K}^{-1})[(273 + 21)\text{ K}]}$$

$$= 0.0414\text{ mol He}$$

To calculate helium's density, we need to know the mass of He in 1 L, so

$$0.0414\text{ mol He} \times 4.00\text{ g/mol} = 0.166\text{ g He}$$

The density of He at 21 °C and 1 atm is 0.166 g/L. At these same conditions of temperature and pressure, the density of air (which we estimated on p. 444) is 1.28 g/L. The mass difference between 1 L He and 1 L air is 1.28 g/L − 0.166 g/L = 1.11 g/L. This is the buoyant (lifting) force of He per liter.

We can now estimate the size of the balloon.

$$500\text{ lb} \times 454\text{ g/lb} = 2.27 \times 10^5\text{ g}$$

Since He can lift 1.11 g/L, the total weight can be lifted by

$$2.27 \times 10^5\text{ g} \times \frac{1\text{ L}}{1.11\text{ g}} = 2.04 \times 10^5\text{ L He}$$

Let's change this to cubic meters.

$$2.04 \times 10^5\text{ L He} \times 10^{-3}\text{ m}^3\text{/L} = 204\text{ m}^3\text{ He}$$

The volume and diameter of a spherical balloon are related by

$$V = \tfrac{4}{3}\pi r^3 = \tfrac{1}{6}\pi d^3$$

Rearranging this equation allows us to calculate the diameter of the balloon.

$$d = \left(\frac{6V}{\pi}\right)^{1/3} = \left(\frac{6 \times 204\text{ m}^3}{3.14}\right)^{1/3} = 390^{1/3}\text{ m} = 7.30\text{ m}$$

The diameter of a spherical He balloon that can lift 500 lb is approximately 7.3 m (approximately 24 ft).

Visit this book's companion website at www.cengage.com/chemistry/moore to work an interactive module based on this material.

CONCEPTUAL EXERCISE 10.13 Comparing Densities

Express the gas density of He (0.166 g/L) in grams per milliliter (g/mL) and compare that value with the density of metallic lithium (0.53 g/mL). The mass of a Li atom (7 amu) is less than twice that of a He atom (4 amu). What do these densities tell you about how closely Li atoms are packed compared with He atoms when each element is in its standard state? To which of the concepts of the kinetic-molecular theory does this comparison apply?

CONCEPTUAL EXERCISE 10.14 Densities of Gas Mixtures

Assume a mixture of equal volumes of the two gases nitrogen and oxygen. Would the density of this gas mixture be higher, lower, or the same as the density of air at the same temperature and pressure? Explain your answer.

PROBLEM-SOLVING EXAMPLE 10.9 Using the Ideal Gas Law to Calculate Molar Mass

Suppose you have a 1.02-g sample of a gas containing C, H, and F atoms, and you suspect that it is $C_2H_2F_4$. The gas sample has a pressure of 750. mm Hg in a 250.-mL vessel at 25.0 °C. From this information, calculate the molar mass of the compound to determine whether your molecular formula is correct.

Answer 101 g/mol. The molecular formula is correct.

CHEMISTRY YOU CAN DO Helium-Filled Balloon in Car

A helium-filled balloon floats in air because helium is less dense than air. We suggest a simple experiment with a helium-filled balloon that will surprise you. Performing the experiment couldn't be simpler. Just suspend a small, helium-filled balloon inside a closed car by taping its string to a seat. The experiment consists of observing what happens to the balloon when the car accelerates or decelerates. Have an observer watch the balloon while you drive. You may be surprised by what the observer sees. Use your knowledge of gas behavior to explain your observations.

Think about these questions:

1. What would happen to you and your companion observer if you had to slam on the brakes?
2. What did you expect to happen to the balloon when you applied the brakes?
3. Do your observations conform to your prior expectations or contradict them?
4. What role does the air trapped in the car play?

Strategy and Explanation Begin by organizing the data and changing units where necessary.

$$V = 0.250\text{ L} \quad P = 750.\text{ mm Hg} \times \frac{1\text{ atm}}{760\text{ mm Hg}} = 0.987\text{ atm} \quad T = 298.1\text{ K}$$

Use the ideal gas law equation to find n.

$$n = \frac{PV}{RT} = \frac{(0.987\text{ atm})(0.250\text{ L})}{(0.0821\text{ L atm mol}^{-1}\text{ K}^{-1})(298.1\text{ K})} = 0.0101\text{ mol gas}$$

The mass of the sample is 1.02 g, so the molar mass according to this gas-law calculation is

$$\frac{1.02\text{ g}}{0.0101\text{ mol}} = 101\text{ g/mol}$$

Now sum the atomic weights of the molecular formula, $C_2H_2F_4$. This gives a molar mass of 102.0 g/mol, which is in close agreement with that determined using the ideal gas law.

☑ **Reasonable Answer Check** 250. mL of gas is about 0.250 L/22.4 L = 0.01 mol gas. If 1.0 g of gas is 0.01 mol, then the molar mass is about 1.0 g/0.01 mol = 100 g/mol, which is close to the more accurate answer.

PROBLEM-SOLVING PRACTICE **10.9**

A flask contains 1.00 L of a pure gas at 0.850 atm and 20. °C. The mass of the gas is 1.13 g. What is the molar mass of the gas? What is its identity?

10.7 Gas Mixtures and Partial Pressures

Our atmosphere is a mixture of nitrogen, oxygen, argon, carbon dioxide, water vapor, and small amounts of several other gases (p. 426, Table 10.1). What we call atmospheric pressure is the sum of the pressures exerted by all these individual gases. The same is true of every gas mixture. Consider the mixture of nitrogen and oxygen illustrated on the right side in Figure 10.14. The pressure exerted by the mixture is equal to the sum of the pressures that the nitrogen alone and the oxygen alone would exert in the same volume at the same temperature and pressure. The pressure of one gas in a mixture of gases is called the **partial pressure** of that gas.

John Dalton was the first to observe that ***the total pressure exerted by a mixture of gases is the sum of the partial pressures of the individual gases in the mixture.*** This statement, known as **Dalton's law of partial pressures,** is a consequence of the fact that gas molecules behave independently of one another. As a demonstration of Dalton's law, consider the main components of our atmosphere, for which the total number of moles is approximately

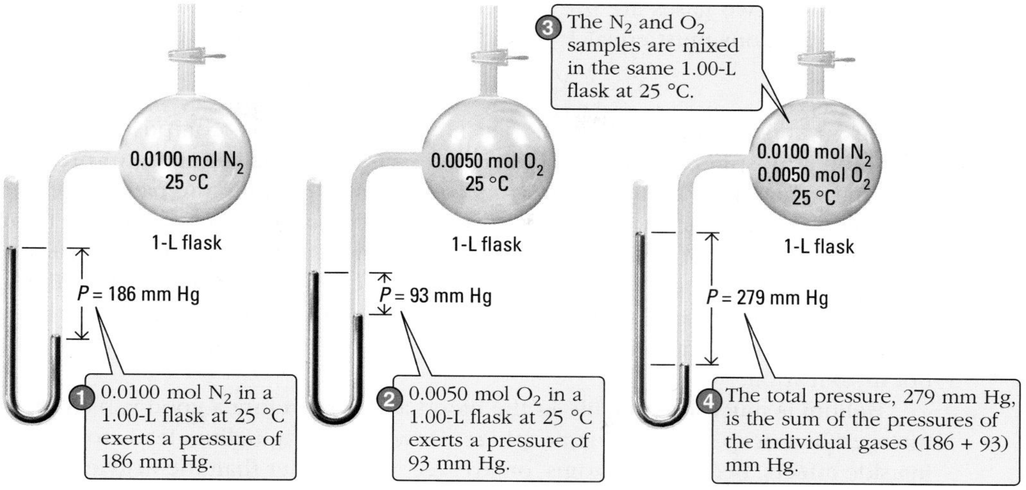

Figure 10.14 Dalton's law of partial pressures.

$$n_{\text{total}} = n_{N_2} + n_{O_2} + n_{Ar}$$

If we replace n in the ideal gas law with n_{total}, the summation of the individual numbers of moles of gases, the equation becomes

$$P_{\text{total}}V = n_{\text{total}}RT$$

$$P_{\text{total}} = \frac{n_{\text{total}}RT}{V} = \frac{(n_{N_2} + n_{O_2} + n_{Ar})RT}{V}$$

Expanding the right side of this equation and rearranging gives

$$P_{\text{total}} = \frac{n_{N_2}RT}{V} + \frac{n_{O_2}RT}{V} + \frac{n_{Ar}RT}{V} = P_{N_2} + P_{O_2} + P_{Ar}$$

The quantities P_{N_2}, P_{O_2}, and P_{Ar} are the partial pressures of the three major components of the atmosphere. Dalton's law means that the pressure exerted by the atmosphere is the sum of the pressures due to nitrogen, oxygen, argon, and the other much less abundant components.

We can write a ratio of the partial pressure of one of the components, A, of a gas mixture over the total pressure,

$$\frac{P_A}{P_{\text{total}}} = \frac{n_A(RT/V)}{n_{\text{total}}(RT/V)}$$

On canceling terms on the right-hand side of this equation, we get

$$\frac{P_A}{P_{\text{total}}} = \frac{n_A}{n_{\text{total}}}$$

The ratio of the pressures is the same as the ratio of the number of moles of gas A to the total number of moles. This ratio (n_A/n_{total}) is called the **mole fraction** of A and is given the symbol X_A. Hence, rearranging the equation gives

$$P_A = X_A P_{\text{total}}$$

In the gas mixture in Figure 10.14, the mole fractions of nitrogen and oxygen are

$$X_{N_2} = \frac{0.010 \text{ mol } N_2}{0.010 \text{ mol } N_2 + 0.0050 \text{ mol } O_2} = 0.67$$

$$X_{O_2} = \frac{0.0050 \text{ mol } O_2}{0.010 \text{ mol } N_2 + 0.0050 \text{ mol } O_2} = 0.33$$

Because these two gases are the only components of the mixture, the sum of the two mole fractions must equal 1:

$$X_{N_2} + X_{O_2} = 0.67 + 0.33 = 1.00$$

An interesting application of partial pressures is the composition of the breathing atmosphere in deep-sea–diving vessels. If normal air at 1 atm pressure, with an oxygen mole fraction of 0.21, is compressed to 2 atm, the partial pressure of oxygen becomes about 0.42 atm. Such high oxygen partial pressures are toxic, so a diluting gas must be added to lower the oxygen partial pressure to near-normal values. Nitrogen gas might seem the logical choice because it is the diluting gas in the atmosphere. The problem is that nitrogen is fairly soluble in the blood and at high concentrations causes *nitrogen narcosis,* a condition similar to alcohol intoxication.

Helium is much less soluble in the blood and is therefore a good substitute for nitrogen in a deep-sea-diving atmosphere. However, using helium leads to interesting side effects. Because He atoms, on average, move faster than the heavier N_2 molecules at the same temperature (Figure 10.7, p. 432), they strike a diver's skin more often than would the N_2 molecules and are therefore more efficient at transferring energy away. This effect causes divers to complain of feeling chilled while breathing a helium/oxygen mixture.

Scuba diver under water.

PROBLEM-SOLVING EXAMPLE 10.10 Calculating Partial Pressures

Mixtures of oxygen and helium are used in scuba diving tanks to provide the proper oxygen pressure for divers. Say that you want to fill a 10.0-L scuba tank with He and O_2 so that the partial pressure of the oxygen is 20% of the total pressure (the same as in air at 1.0 atm). You start with 56.0 L He at 25 °C and 1.00 atm. How many liters of O_2 at the same T and P should you add to get the desired mixture? What will the partial pressure of each gas be in the scuba tank?

Answer 14.0 L O_2; $P_{O_2} = 1.4$ atm; $P_{He} = 5.6$ atm

Strategy and Explanation To solve the problem of partial pressures, calculate the number of moles of each gas in the final mixture. Use $PV = nRT$ to calculate the number of moles of He from the initial conditions.

$$n_{He} = \frac{PV}{RT} = \frac{(1.00 \text{ atm})(56.0 \text{ L})}{(0.0821 \text{ L atm mol}^{-1} \text{ K}^{-1})(298 \text{ K})} = 2.29 \text{ mol He}$$

The partial pressure of He in the scuba tank is

$$P_{He} = \frac{nRT}{V} = \frac{(2.29 \text{ mol})(0.0821 \text{ L atm mol}^{-1} \text{ K}^{-1})(298 \text{ K})}{10.0 \text{ L}} = 5.60 \text{ atm}$$

If the mixture in the scuba tank is desired to be 20% O_2, then it will be 80% He. If 5.60 atm of He is 80% of the pressure in the scuba tank, then the total pressure is

$$P_{total} = \frac{5.60 \text{ atm}}{0.80} = 7.00 \text{ atm}$$

The partial pressure of O_2 is

$$P_{O_2} = P_{total} - P_{He} = 7.00 - 5.60 = 1.40 \text{ atm}$$

To find the volume of oxygen at the starting conditions that would produce this partial pressure in the scuba tank, we need to know how many moles of oxygen we have.

$$n = \frac{PV}{RT} = \frac{(1.40 \text{ atm})(10.0 \text{ L})}{(0.0821 \text{ L atm mol}^{-1} \text{ K}^{-1})(298 \text{ K})} = 0.572 \text{ mol } O_2$$

Now we calculate what volume of O_2 at the original conditions would contain this many moles of O_2.

$$V = \frac{nRT}{P} = \frac{(0.572 \text{ mol})(0.0821 \text{ L atm mol}^{-1} \text{ K}^{-1})(298 \text{ K})}{1.0 \text{ atm}} = 14.0 \text{ L } O_2$$

The total amount of gas is 2.29 mol He + 0.572 mol O_2 = 2.86 mol. The partial pressure of O_2 in the mixture can also be obtained from the mole ratio:

$$X_{O_2} = \frac{n_{O_2}}{n_{total}} = \frac{0.572 \text{ mol}}{2.86 \text{ mol}} = 0.20 \qquad P_{O_2} = 0.20 \times P_{total} = 0.20 \times 7.00 \text{ atm} = 1.40 \text{ atm}$$

☑ **Reasonable Answer Check** The amount of O_2 needed is 20% of the final mixture, or one-fourth of the amount of He, (56 L)/4 = 14 L, which agrees with our more exact calculation.

The partial pressure of oxygen in a scuba tank will be the same under water as it is in air at atmospheric pressure.

PROBLEM-SOLVING PRACTICE 10.10

A mixture of 7.0 g N_2 and 6.0 g H_2 is confined in a 5.0-L reaction vessel at 500. °C. Assume that no reaction occurs and calculate the total pressure. Then calculate the mole fraction and partial pressure of each gas.

CONCEPTUAL EXERCISE 10.15 Pondering Partial Pressures

What happens to the partial pressure of each gas in a mixture when the volume is decreased by (a) lowering the temperature or (b) increasing the total pressure?

EXERCISE 10.16 Partial Pressures

A sealed 355-mL flask contains 0.146 g neon gas, Ne, and an unknown amount of argon gas, Ar, at 35 °C and a total pressure of 626 mm Hg. How many grams of Ar are in the flask?

PROBLEM-SOLVING EXAMPLE 10.11 Partial Pressure

The reaction between nitrogen monoxide, NO, and oxygen gas to form NO_2 is environmentally important.

$$2\ NO(g) + O_2(g) \longrightarrow 2\ NO_2(g)$$

Nitrogen oxides are discussed in Section 10.10, p. 457.

Suppose that the two reactant gases are kept in separate containers, as shown below. Then suppose the valve is opened and the reaction proceeds to completion. Assume a constant temperature of 25 °C.

(a) What gases remain at the end of the reaction, and what amount of each of them is present?

(b) What are their partial pressures, and what is the total pressure in the system?

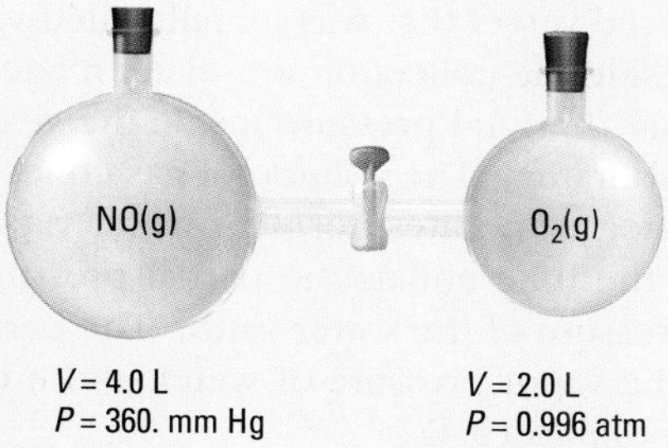
NO(g)
O2(g)
V = 4.0 L
P = 360. mm Hg
V = 2.0 L
P = 0.996 atm

Answer

(a) No NO remains; 0.0427 mol O_2 and 0.0774 mol NO_2 remain

(b) P_{NO_2} = 0.316 atm; P_{O_2} = 0.174 atm; P_{total} = 0.490 atm

Strategy and Explanation

(a) First, we must identify the limiting reactant by finding the number of moles of each gas and comparing these values with the coefficients of the balanced equation.

$$\frac{(360./760)\ \text{atm}\ (4.0\ \text{L})}{(0.0821\ \text{L atm mol}^{-1}\ \text{K}^{-1})(298\ \text{K})} = 0.0774\ \text{mol NO}$$

$$\frac{(0.996\ \text{atm})(2.0\ \text{L})}{(0.0821\ \text{L atm mol}^{-1}\ \text{K}^{-1})(298\ \text{K})} = 0.0814\ \text{mol } O_2$$

NO is the limiting reactant because 0.0744 mol is less than the amount required by the O_2 (2 × 0.0814 mol = 0.163 mol). The NO will be completely consumed by the reaction. The amount of oxygen consumed will be

$$0.0774\ \text{mol NO} \times \frac{1\ \text{mol } O_2}{2\ \text{mol NO}} = 0.0387\ \text{mol } O_2$$

Subtracting the amount used from the original amount gives 0.0814 − 0.0387 = 0.0427 mol O_2 left when the reaction is complete. The amount of NO_2 formed is

$$0.0774\ \text{mol NO} \times \frac{2\ \text{mol } NO_2}{2\ \text{mol NO}} = 0.0774\ \text{mol } NO_2 \text{ formed}$$

At the end of the reaction, we have 0.0427 mol O_2, 0.0774 mol NO_2, and no NO gas.

(b) The pressures of the two gases are

$$P_{NO_2} = \frac{(0.0774\ \text{mol})(0.0821\ \text{L atm mol}^{-1}\ \text{K}^{-1})(298\ \text{K})}{6.0\ \text{L}} = 0.316\ \text{atm } NO_2$$

$$P_{O_2} = \frac{(0.0427\ \text{mol})(0.0821\ \text{L atm mol}^{-1}\ \text{K}^{-1})(298\ \text{K})}{6.0\ \text{L}} = 0.174\ \text{atm } O_2$$

The total pressure of the system is the sum of the partial pressures.

$$P_{total} = 0.316\ \text{atm} + 0.174\ \text{atm} = 0.490\ \text{atm}$$

PROBLEM-SOLVING PRACTICE **10.11**

Consider the case with the same reactants in the same vessels as above, but with a NO pressure of 1.00 atm and O_2 pressure of 0.400 atm. What would the final pressure in the system be after the reaction is complete?

Collecting a Gas over Water

While studying chemical reactions that produce a gas as a product, it is often necessary to determine the number of moles of the product gas. One convenient way to do this, for gases that are insoluble in water, involves collecting the gas over water (Figure 10.15). The gas bubbles through the water and is collected by displacing the water in an inverted vessel that was initially filled with water. The levels of the water inside and outside the collection vessel are made equal at the end of the experiment to ensure that the total pressure inside the vessel is equal to the barometric pressure in the laboratory. The volume of gas collected is then determined. Because some of the water evaporates, forming water vapor, the total pressure of the mixture of gases in the tube equals the partial pressure of the gas being collected plus the partial pressure of the water vapor. The partial pressure of water in the gaseous mixture is the vapor pressure of water, and it depends on the temperature of the liquid water (Table 10.4).

To calculate the amount of gaseous product, we first subtract the vapor pressure of the water from the total gas pressure (the barometric pressure), to give the partial pressure of the gaseous product. Substituting this partial pressure and the known volume and temperature into the ideal gas law allows the calculation of the number of moles of the gaseous product.

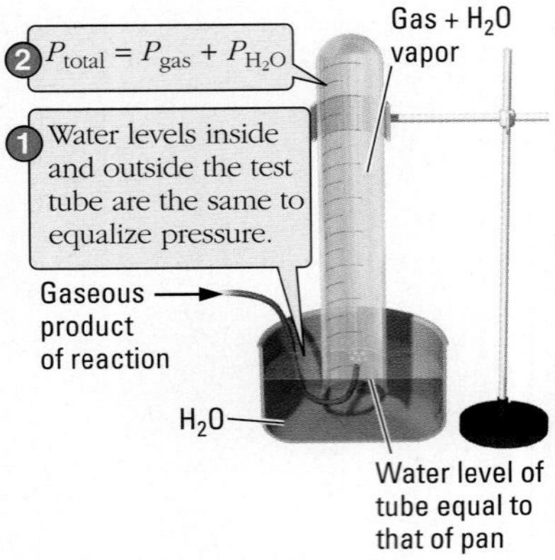

Gas + H_2O vapor
2 $P_{total} = P_{gas} + P_{H_2O}$
1 Water levels inside and outside the test tube are the same to equalize pressure.
Gaseous product of reaction
H_2O
Water level of tube equal to that of pan

Figure 10.15 Collecting a gas over water.

PROBLEM-SOLVING EXAMPLE 10.12 Collecting a Gas over Water

Before electric lamps were available, the reaction between calcium carbide, CaC_2, and water was used to produce acetylene, C_2H_2, to generate a bright flame in miners' lamps.

$$CaC_2(s) + 2\,H_2O(\ell) \longrightarrow C_2H_2(g) + Ca(OH)_2(aq)$$

At a barometric pressure of 745 mm Hg and a temperature of 24.0 °C, 625 mL of gas was collected over water. How many milligrams of acetylene were produced? The vapor pressure of water at 24.0 °C is 22.4 mm Hg.

Answer 635 mg acetylene

Strategy and Explanation We will need to calculate the number of moles and then milligrams of acetylene, so we must first find the partial pressure of the acetylene. The total pressure of the gas mixture is 745 mm Hg, and the vapor pressure of water at 24.0 °C is 22.4 mm Hg, so the partial pressure of acetylene is

$$P_{acetylene} = P_{total} - P_{water} = 745 \text{ mm Hg} - 22.4 \text{ mm Hg} = 723 \text{ mm Hg}$$

Now we calculate the molar amount and mass of acetylene

$$n_{C_2H_2} = \frac{PV}{RT} = \frac{(723/760 \text{ atm})(0.625 \text{ L})}{(0.0821 \text{ L atm mol}^{-1}\text{ K}^{-1})(273.15 + 24.0) \text{ K}}$$

$$= 0.0244 \text{ mol acetylene}$$

$$m_{C_2H_2} = 0.0244 \text{ mol} \times 26.04 \text{ g/mol} = 0.635 \text{ g} = 635 \text{ mg acetylene}$$

☑ **Reasonable Answer Check** The quantities entered into the ideal gas equation can be approximated as $n = \frac{(1)(0.6)}{(0.082)(300)} = 0.02$ mol, which is close to our more exact answer.

PROBLEM-SOLVING PRACTICE 10.12

Zinc metal reacts with HCl to produce hydrogen gas, H_2.

$$2\,HCl(aq) + Zn(s) \longrightarrow ZnCl_2(aq) + H_2(g)$$

The H_2 can be collected over water. If you collected 260. mL H_2 at 23 °C, and the total pressure was 740. mm Hg, how many milligrams of H_2 were collected?

H—C≡C—H

acetylene

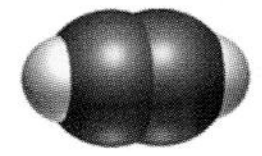

Table 10.4 Vapor Pressure of Water at Different Temperatures

T (°C)	P_{water} (mm Hg)
0	4.6
10	9.2
20	17.5
21	18.6
22	19.8
23	21.1
24	22.4
25	23.8
30	31.8
35	42.2
40	55.3
50	92.5
60	149.4
70	233.7
80	355.1
90	525.8
100	760.0

10.8 The Behavior of Real Gases

The ideal gas law provides accurate predictions for the pressure, volume, temperature, and amount of a gas well above its boiling point. At STP (0 °C and 1 atm), most gases deviate only slightly from ideal behavior. At much higher pressures or much lower temperatures, however, the ideal gas law does not work nearly as well. We can illustrate this departure from ideality by plotting PV/nRT for a gas as a function of pressure (Figure 10.16). For a gas that follows ideal gas behavior, $PV = nRT$, the ratio PV/nRT must equal 1. But Figure 10.16 shows that for real gases, the ratio PV/nRT deviates from 1, first dipping to values lower than 1 and then rising to values higher than 1 as pressure increases. Thus, the product of the measured pressure times the measured volume of the real gas is smaller than expected at medium pressures, and it is higher than expected at high pressures.

To see what causes these deviations, we must revisit two of the fundamental concepts of kinetic-molecular theory (p. 430). The theory assumes that the molecules of a gas occupy no appreciable volume themselves and that gas molecules have no appreciable attraction for one another.

At STP, the volume occupied by a single molecule is very small relative to its share of the total gas volume. Recall that there are 6.02×10^{23} molecules in a mole

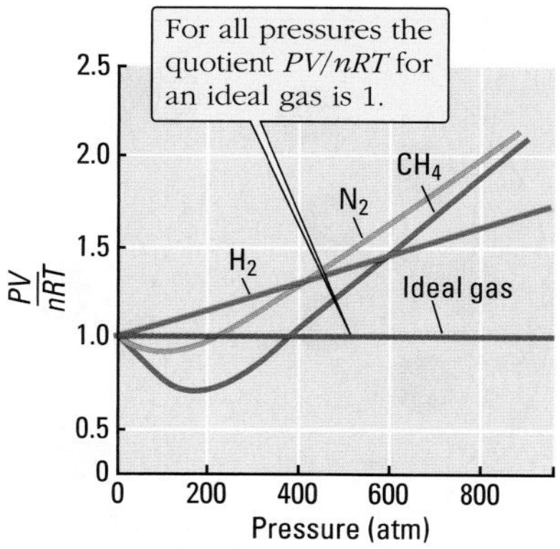

Figure 10.16 The nonideal behavior of real gases compared with an ideal gas.

The volume of a sphere is given by $\frac{4}{3}\pi r^3$. Solving for r gives

$$r = \sqrt[3]{\frac{3V}{4\pi}}$$

$$= \sqrt[3]{\frac{3(3.72 \times 10^{-26}\ \text{m}^3)}{4(3.14)}}$$

$$r = 2.07 \times 10^{-9}\ \text{m} = 2070\ \text{pm}$$

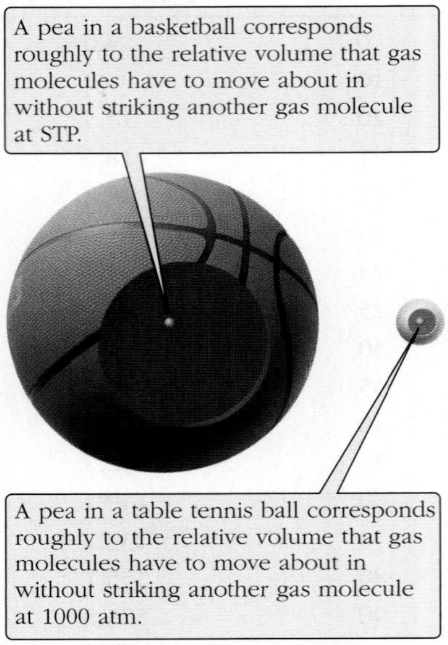
A pea in a basketball corresponds roughly to the relative volume that gas molecules have to move about in without striking another gas molecule at STP.
A pea in a table tennis ball corresponds roughly to the relative volume that gas molecules have to move about in without striking another gas molecule at 1000 atm.

A cut-away view of a pea inside a basketball and inside a table tennis ball.

and that 1 mol of a gas occupies about 22.4 L (22.4×10^{-3} m^3) at STP (p. 436). The volume, V, that each molecule has to move around in is given by

$$V = \frac{22.4 \times 10^{-3}\ \text{m}^3}{6.02 \times 10^{23}\ \text{molecules}} = 3.72 \times 10^{-26}\ \text{m}^3/\text{molecule}$$

If this volume is assumed to be a sphere, then the radius, r, of the sphere is about 2000 pm. The radius of the smallest gas molecule, the helium atom, is 31 pm ***(← p. 307),*** so a helium atom has a space to move around in that is similar to the room a pea has inside a basketball. Now suppose the pressure is increased significantly, to 1000 atm. The volume available to each molecule is now a sphere only about 200 pm in radius, which means the situation is now like that of a pea inside a sphere a bit larger than a table tennis ball. The volume occupied by the gas molecules themselves relative to the volume of the sphere is no longer negligible. This violates the first concept of the kinetic-molecular theory. The kinetic-molecular theory and the ideal gas law deal with the volume available for the molecules to move around in, not the volume of the molecules themselves. However, the measured volume of the gas must include both. Therefore, at very high pressures, the measured volume will be larger than predicted by the ideal gas law, and the value of PV/nRT will be greater than 1.

At medium pressures, the product PV for a real gas is smaller than the predicted value of nRT for an ideal gas (Figure 10.16). Consider a gas molecule that is about to hit the wall of the container, as shown in Figure 10.17. The kinetic-molecular theory assumes that the other molecules exert no forces on the molecule, but in fact such forces do exist ***(← p. 403).*** Their influence increases as higher pressures push gas molecules closer to each other. This means that when a molecule is about to hit the wall of the container, the other molecules tend to pull it *away* from the wall, which causes the molecule to hit the wall with less impact. The collision is softer than if there were no attraction among the molecules. Since all collisions with the walls are softer, the internal pressure is less than that predicted by the ideal gas law, and PV/nRT is less than 1. As the external pressure increases, the gas volume decreases, the molecules are squeezed closer together, and the attraction among the molecules grows stronger, which increases this deviation from ideal behavior. Eventually, the pressure gets so high that the loss of empty space becomes the dominating factor, and PV/nRT becomes larger than 1 (Figure 10.16).

The result of these considerations is that

- **The measured pressure for a real gas is less than the ideal pressure due to intermolecular attractive forces.**
- **The measured volume of a real gas is larger than the ideal volume due to the volume of the molecules themselves.**

The **van der Waals equation** predicts the behavior of n moles of a real gas quantitatively by taking these two effects into account. It assumes that $P_{\text{ideal}} V_{\text{ideal}} = nRT$, but expresses the left-hand side in terms of *measured P* and *V*.

$$\left(P_{\text{measured}} + \frac{n^2 a}{V^2_{\text{measured}}}\right)(V_{\text{measured}} - nb) = nRT$$

Correction for molecular attraction (adjusts measured P up)

Correction for volume of molecules (adjusts measured V down)

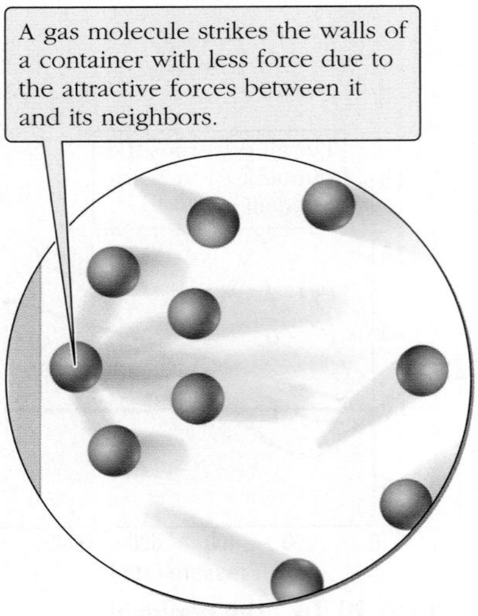
A gas molecule strikes the walls of a container with less force due to the attractive forces between it and its neighbors.

Figure 10.17 Nonideal gas behavior.

The measured volume is decreased by the factor nb, which accounts for the volume occupied by n moles of the gas molecules themselves. The van der Waals constant

b has units of L/mol, and it is larger for larger molecules. The measured pressure is increased by the factor n^2a/V^2, which accounts for the attractive forces between molecules. The van der Waals constant a has units of L^2 atm/mol^2. The a pressure correction term has this form because the attractive forces being accounted for depend on interactions between *pairs* of molecules; the a term is therefore proportional to the square of the number of molecules per unit volume, $(n/V)^2$. The value of a shows how strongly the gas molecules attract one another; the stronger the intermolecular interactions, the larger the value of a. The constants a and b are different for each gas (Table 10.5) and must be determined experimentally. In general, the values of a and b increase as the gas molecule's size increases and as the molecule's complexity increases.

Note that for an ideal gas, where the kinetic-molecular theory of gases holds, the attractive forces between molecules are negligible, so $a \approx 0$, and $P_{measured} = P_{ideal}$. The volume of the molecules themselves is negligible compared with the container volume, so $b \approx 0$ and $V_{measured} = V_{ideal}$. Therefore, for ordinary conditions of P and T, the van der Waals equation simplifies to the ideal gas law.

Table 10.5 Van der Waals Constants for Some Common Gases

Gas	$a\left(\frac{L^2\ atm}{mol^2}\right)$	$b\left(\frac{L}{mol}\right)$
He	0.034	0.0237
Ne	0.211	0.0171
Ar	1.35	0.0322
H_2	0.244	0.0266
N_2	1.39	0.0391
O_2	1.36	0.0318
Cl_2	6.49	0.0562
CO_2	3.59	0.0427
CH_4	2.25	0.0428
NH_3	4.17	0.0371
H_2O	5.46	0.0305

PROBLEM-SOLVING EXAMPLE 10.13 Using the van der Waals Equation

A 2.50-mol sample of CO_2 gas is in a sealed 2.00-L container at 0 °C.
(a) What is the pressure according to the ideal gas law?
(b) What is the pressure calculated with the van der Waals equation? (The van der Waals constants for CO_2 are in Table 10.5.)
(c) Calculate the percentage difference between pressures (a) and (b) as a fraction of the ideal gas pressure.

Answer (a) 28.0 atm (b) 24.0 atm (c) 14.3% difference

Strategy and Explanation

(a) Use the ideal gas law equation to calculate the pressure if the CO_2 behaved ideally.

$$P = \frac{nRT}{V} = \frac{(2.50\ \text{mol})(0.0821\ \text{L atm mol}^{-1}\ \text{K}^{-1})(273\ \text{K})}{2.00\ \text{L}} = 28.0\ \text{atm}$$

(b) Calculate the pressure using the van der Waals equation.

$$P = \left(\frac{nRT}{V - nb}\right) - \left(\frac{n^2a}{V^2}\right)$$

$$P = \left(\frac{(2.50\ \text{mol})(0.0821\ \text{L atm mol}^{-1}\ \text{K}^{-1})(273\ \text{K})}{(2.00\ \text{L}) - (2.50\ \text{mol})(0.0427\ \text{L/mol})}\right)$$

$$- \left(\frac{(2.50\ \text{mol})^2(3.59\ \text{L}^2\ \text{atm/mol}^2)}{(2.00\ \text{L})^2}\right) = 24.0\ \text{atm}$$

(c) Percentage difference in pressure $= \frac{28.0\ \text{atm} - 24.0\ \text{atm}}{28.0\ \text{atm}} \times 100\% = 14.3\%$

The first term in the van der Waals equation containing the constant b equals 29.60 atm, which differs from the ideal gas pressure by 1.6 atm. The second term in the van der Waals equation, containing the constant a, equals 5.61 atm. Thus, the term with a that depends on intermolecular forces accounts for most of the difference in pressure calculated by the van der Waals equation compared with the ideal gas law equation in this case.

PROBLEM-SOLVING PRACTICE 10.13

What is the percentage difference between the pressures predicted by the ideal gas law and the van der Waals equation if the volume in Problem-Solving Example 10.13 is increased by a factor of 10 to 20.0 L?

EXERCISE 10.17 Errors Caused by Deviations from Ideal Gas Behavior

In Problem-Solving Example 10.9 (p. 445), the molar mass calculated using the ideal gas law was slightly larger than that calculated using atomic weights. This means that the number of moles, n, must have been too small. Look at the Strategy and Explanation for Problem-Solving Example 10.9 and explain why the number of moles is less than it should be.

10.9 Ozone and Stratospheric Ozone Depletion

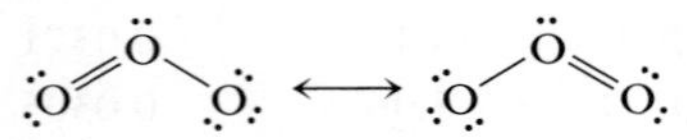

ozone

Ozone resonance structures.

Ozone, O_3, is an important substance in the stratosphere, and its presence there is essential for life on Earth. A hybrid of two resonance structures, ozone has a bent (angular) geometry ***(← p. 353)*** with a bond angle of 116.8°. The ΔH_f° of 142.3 kJ/mol shows that ozone is a higher-energy molecule than O_2 (which is assigned an enthalpy of formation of zero).

$$\tfrac{3}{2}\,O_2(g) \longrightarrow O_3(g) \qquad \Delta H_f^\circ = 142.3\ \text{kJ/mol}$$

Ozone is highly reactive and does not survive long after formation.

In the stratosphere, ozone is formed by photodissociation of O_2. **Photodissociation** is a process in which a molecule absorbs an ultraviolet photon and produces two free radicals as products. Ultraviolet radiation with λ <242 nm splits an O_2 molecule into two oxygen atoms, each a free radical. The oxygen atoms then combine with other O_2 molecules to form ozone.

Remember that the oxygen atom has the electron configuration [He] $2s^2 2p_x^2 2p_y^1 2p_z^1$, and it is a free radical with two unpaired electrons.

The bond energy of O_2 is 495 kJ/mol, so a calculation using $E = h\nu = hc/\lambda$ shows that the energy in a photon of $\lambda \leq 242$ nm is needed to break the O_2 bond.

Natural formation of O_3 $\quad O_2(g) \xrightarrow[\lambda\,<\,242\ \text{nm}]{\text{UV radiation}} \cdot O\cdot(g) + \cdot O\cdot(g)$

$$\cdot O\cdot(g) + O_2(g) \longrightarrow O_3(g)$$

The region of maximum ozone concentration in the upper atmosphere, known as the **stratospheric ozone layer,** is at an altitude of 25 to 30 km, where the ozone concentration can reach 10 ppm. Under natural conditions, the ozone concentration in the stratosphere is maintained at a constant level by a reaction that destroys ozone at the same rate at which it is produced.

Natural destruction of O_3 $\quad O_3 + \cdot O\cdot(g) \longrightarrow 2\,O_2(g)$

Because ozone molecules and oxygen atoms are at very low concentrations, this reaction is relatively slow. However, it consumes ozone fast enough to maintain a balance with the ozone-forming reaction so that under normal circumstances the concentration of ozone in the stratosphere remains relatively constant. Every day, approximately 3×10^8 tons of stratospheric ozone are formed and destroyed by these two competing natural processes.

Protection of Earth's surface

Stratospheric ozone is important because it absorbs ultraviolet radiation while undergoing a photodissociation reaction. This prevents damaging ultraviolet radiation from reaching Earth's surface. Wavelengths in the range of 200 to 310 nm are absorbed during the decomposition reaction. The oxygen atoms from the photodissociation of O_3 react with O_2 to regenerate O_3, so this reaction results in no net O_3 loss.

$$O_3(g) \xrightarrow[\lambda\ 200\ \text{to}\ 310\ \text{nm}]{\text{UV radiation}} \cdot O\cdot(g) + O_2(g)$$

Absorption of ultraviolet radiation in the stratosphere is *essential* for living things on Earth. Stratospheric ozone prevents 95% to 99% of the sun's ultraviolet radiation from reaching Earth's surface. Photons in this 200- to 310-nm range have enough energy to cause skin cancer in humans and damage to living plants. For every 1% decrease in the stratospheric ozone, an additional 2% of this damaging radiation reaches Earth's surface. Stratospheric ozone depletion therefore has the potential to drastically damage our environment.

This answers the question posed in Chapter 1 (← *p. 2*): "Why is stratospheric ozone depletion harmful? I thought too much ozone was harmful for your lungs."

Chlorofluorocarbons

Chlorofluorocarbons (CFCs) are small molecules with halogen atoms bonded to central carbons. CFCs play a major role in ozone depletion. Specific examples of CFCs include $CFCl_3$ (CFC-11) and CF_2Cl_2 (CFC-12). These compounds are nonflammable, relatively inert, volatile yet readily liquefied, and nontoxic—an extremely useful set of properties. CFCs have been used as coolants for refrigeration, in foam plastics manufacture, as aerosol propellants, and as industrial solvents. Unfortunately, the same properties that make them useful are also the reason for their destructive effect on stratospheric ozone. Once gaseous CFCs are released into the atmosphere, they persist for a very long time in the troposphere because of their chemical nonreactivity. On average, a CF_2Cl_2 molecule survives roughly 100 years in the troposphere. Eventually, atmospheric mixing causes CFCs released into the troposphere to rise to the stratosphere, where they are decomposed by high-energy solar radiation. The decomposition products of CFCs participate in reactions in the stratosphere that result in a lowering of the concentration of ozone there.

$CFCl_3$ (CFC-11)

CF_2Cl_2 (CFC-12)

Destruction of the stratospheric ozone layer by CFCs begins when a photon with sufficient energy breaks a carbon-chlorine bond in a CFC molecule. This produces a chlorine atom, as shown here using CFC-12 as an example.

$$CF_2Cl_2(g) \xrightarrow{h\nu} CF_2Cl\cdot(g) + Cl\cdot(g)$$

The chlorine atom, a free radical, then participates in what is called a *chain reaction mechanism*. It first combines with an ozone molecule, producing a chlorine monoxide, $ClO\cdot$, radical and an oxygen molecule.

Step 1: $Cl\cdot(g) + O_3(g) \longrightarrow ClO\cdot(g) + O_2(g)$

Thus, an ozone molecule has been destroyed. If this were the only reaction that particular CFC molecule caused, there would be little danger to the ozone layer. However, once Step 1 has occurred twice, the two $ClO\cdot$ radicals produced can react further.

Step 2: $ClO\cdot + ClO\cdot \longrightarrow ClOOCl$

Step 3: $ClOOCl \xrightarrow{h\nu} \cdot ClOO + \cdot Cl$

Step 4: $\cdot ClOO \xrightarrow{h\nu} \cdot Cl + O_2$

The net reaction obtained by adding two Step 1 reactions to Steps 2, 3, and 4 and canceling species that appear as both reactants and products is the conversion of two ozone molecules to three oxygen molecules.

$$\text{Net reaction:} \quad 2\,O_3(g) \longrightarrow 3\,O_2(g)$$

That is, this chain reaction increases the rate at which stratospheric ozone is destroyed, but it does not affect the rate at which ozone is formed. It is called a chain reaction because the reaction steps can repeat over and over. The two chlorine atoms that react when Step 1 occurs twice are regenerated in Steps 3 and 4, so there is no net change in concentration of chlorine atoms. It has been estimated that a single chlorine atom can destroy as many as 100,000 molecules of O_3 before it is inactivated or returned to the troposphere (probably as HCl). This sequence of ozone-destroying reactions upsets the balance in the stratosphere because ozone is being destroyed faster than it is being produced. As a consequence, the concentration of stratospheric ozone decreases.

F. Sherwood Rowland

1927–

Born in Ohio, Sherwood Rowland entered Ohio Wesleyan University at age 12. He received his Ph.D. from the University of Chicago in 1952. He held faculty appointments at several institutions and joined the newly formed University of California at Irvine in 1964 as Chair of the Chemistry Department. In 1974 he and Mario Molina recognized that CFCs could deplete the ozone layer in the atmosphere. Their work was received skeptically at first, but it was confirmed by others in later experiments, notably in 1985 by satellite data showing an "ozone hole" over Antarctica. Rowland, Molina, and Paul Crutzen shared the 1995 Nobel Prize in Chemistry ". . . for their work in atmospheric chemistry, particularly concerning the formation and decomposition of ozone."

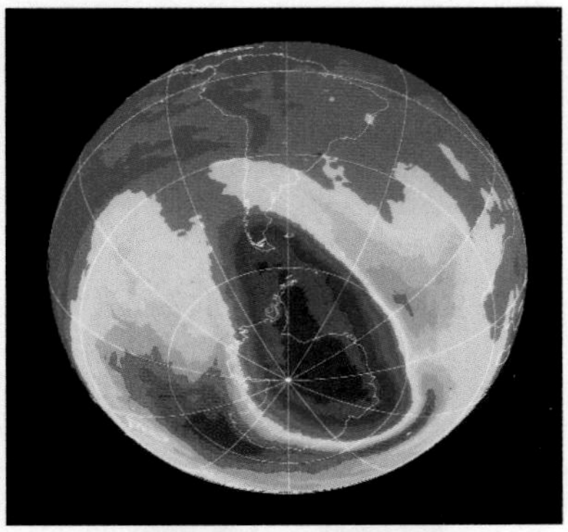

NASA

Figure 10.18 The Antarctic ozone hole. The darker colors indicate where stratospheric ozone is depleted.

The most prominent manifestation of stratospheric ozone depletion is the Antarctic **ozone hole.** In 1985 the British Antarctic Survey reported a startlingly large depletion of ozone in September and October, at the end of the Antarctic winter. (Seasons are reversed in the southern hemisphere.) This ozone hole has reappeared at this season each year (Figure 10.18). Subsequent measurements by several research teams showed that loss of ozone was correlated with high concentrations of ClOOCl, supporting the theory that a chain reaction involving chlorine atoms was responsible. However, the huge depletion of ozone in the Antarctic could not be explained solely by the reaction steps given above. In the dark Antarctic winter, a vortex of intensely cold air containing ice crystals, unique to the region, builds up. On the surfaces of these crystals additional reactions produce hydrogen chloride and chlorine nitrate, $ClONO_2$, which can react with each other to form chlorine molecules.

$$HCl(g) + ClONO_2(g) \longrightarrow Cl_2(g) + HNO_3(g)$$

When sunlight returns in the spring, the Cl_2 molecules are readily photodissociated into chlorine atoms, which can then become involved in the ozone destruction reactions.

$$Cl_2(g) \xrightarrow{h\nu} 2\,Cl\cdot(g)$$

The first direct confirmation of the relationship between stratospheric ozone depletion and increased ultraviolet intensity on Earth was reported in the September 10, 1999, issue of *Science* by scientists from the New Zealand National Institute of Water and Atmospheric Research. They found that ". . . over the past 10 years, peak levels of skin-frying and DNA-damaging ultraviolet (UV) rays have been increasing in New Zealand, just as the concentrations of protective stratospheric ozone have decreased." Figure 10.19 shows a plot of declining stratospheric ozone concentration and increasing UV radiation over a period of 20 years, as reported by the New Zealand scientists.

Carlye Calvin. Courtesy of Susan Solomon

1956–

A senior scientist at NOAA (National Oceanic and Atmospheric Administration), Susan Solomon was the first to propose a good explanation for the Antarctic ozone hole. Her chemist's intuition told her that the ice crystals that form during the Antarctic winter could provide a surface on which chemical reactions of CFC decomposition products could take place. In 1986, NSF chose Solomon (then 30 years old) to lead a team to Antarctica. Experiments performed during that visit showed that her cloud theory was correct and provided the first solid proof of a connection between CFCs and stratospheric ozone depletion. Solomon, a member of the National Academy of Sciences, won a National Medal of Science in March 2000, the 2002 Weizmann Women & Science Award, and the 2004 Blue Planet Prize for ". . . achievements . . . that have contributed to the resolution of global environmental problems." Solomon has said that her winters as a young girl in Chicago prepared her for her visits to Antarctica.

Dealing with the Problem

Stratospheric ozone depletion is a global problem, and it has led to global cooperation designed to combat the problem. In an effort to reduce the harm done by CFCs and other chemicals to the stratospheric ozone layer, the Montreal Protocol on Substances that Deplete the Ozone Layer (1987) was implemented. The Protocol has been amended several times and has now been signed by 194 countries. This proto-

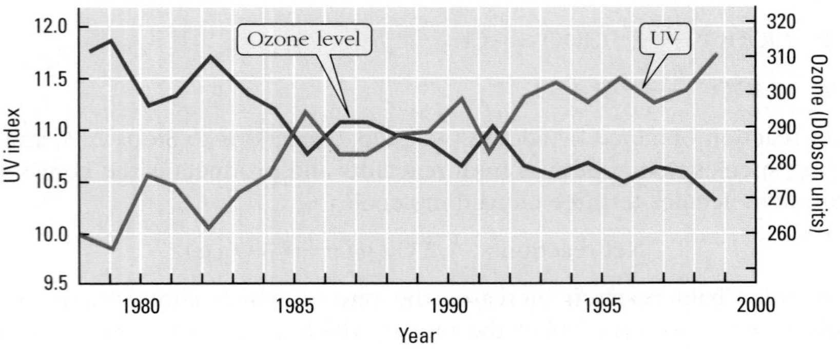

Figure 10.19 As the stratospheric ozone level decreases, the UV intensity increases. These measurements were made in Lauder, New Zealand. The red line indicates ultraviolet (UV) radiation at noon calculated from measured ozone levels, shown by the blue line. Dobson units: 1 DU = 2.69×10^{16} molecules/cm^2. One Dobson unit refers to a layer of ozone that would be 10 micrometers thick under standard temperature and pressure. The UV index is a standardized way to report UV radiation to the public. The higher the UV index, the more intense the UV radiation. (*Source:* Reprinted by permission from McKenzie, R., Connor, B., and Bodecker, G. "Increased Summertime UV Radiation in New Zealand in Response to Ozone Loss." *Science,* Vol. 285, September 10, 1999; p. 1709.)

col, and its amendments, form the basis of amended Title VI of the Clean Air Act in the United States. CFC production was banned as of 1996 in the developed countries. Other chemicals that deplete the stratospheric ozone layer were also banned. For example, halons, used as fire-suppression agents, were banned in the developed countries as of 1994, and methyl bromide, used as a pesticide, was banned as of 2001.

The efforts to halt production of stratospheric ozone-depleting chemicals are beginning to show positive results. The United Nations Environment Programme (UNEP) and the World Meteorological Organization (WMO) periodically issue reports, the latest of which is *UNEP/WMO Scientific Assessment of Ozone Depletion: 2006*. Levels of ozone-depleting compounds in the atmosphere are slowly declining from the peak values reached in the 1992–1994 period. However, the seasonal Antarctic ozone hole is expected to continue for decades and Antarctic stratospheric ozone levels are not expected to return to pre-1980 levels until 2060–2075. Overall, the report concludes, "The Montreal Protocol is working: There is clear evidence of a decrease in the atmospheric burden of ozone-depleting substances and some early signs of stratospheric ozone recovery."

The United Nations Environment Programme website reports activities relevant to stratospheric ozone control at **www.unep.org/ozone**.

The U.S. EPA website **(www.epa.gov/ozone)** and its many links provide current information about stratospheric ozone depletion.

10.10 Chemistry and Pollution in the Troposphere

An **air pollutant** is any substance that degrades air quality. Natural processes pollute the air on a massive scale with volcanic ash, sulfur oxides, mercury vapor, hydrogen chloride, and hydrogen sulfide from volcanoes and with reactive, odorous organic compounds from coniferous plants such as pine trees. Decaying vegetation, ruminant animals, and even termites add large quantities of methane gas to the atmosphere, and decaying animal carcasses and other protein materials add dinitrogen monoxide, N_2O. But automobiles, electric power plants, smelting and other metallurgic processes, and petroleum refining also add significant quantities of unwanted substances to the atmosphere. Atmospheric pollutants, whatever their origin, can have adverse effects, including burning eyes, coughing, decreased lung capacity, harm to vegetation, and the destruction of ancient monuments. Millions of tons of soot, dust, smoke particles, and chemicals are discharged directly into the atmosphere every year. Such pollutants, which enter the environment directly from their sources, are called **primary pollutants.**

Primary Pollutants

Particle Pollutants Pollutant particles range in size from particles that are big enough to see, to individual molecules, ions, or atoms. Many pollutants are incorporated into water droplets and form **aerosols,** which are colloids (Section 15.8) consisting of liquid droplets or finely divided solids dispersed in a gas. Fogs and smoke are common examples of aerosols. Larger solid particles in the atmosphere are called **particulates.** The solids in an aerosol or particulate may be metal oxides, soil particles, sea salt, fly ash from electrical generating plants and incinerators, elemental carbon, or even small metal particles. Aerosol particles can range in diameter from about 1 nm to about 10,000 nm (10 μm) and can contain as many as 10^{12} atoms, ions, or small molecules. Particles in the 2000-nm range are largely responsible for the deterioration of visibility often observed in highly populated urban centers such as Houston and Mexico City.

In 1997 the Environmental Protection Agency (EPA) revised its national ambient air quality standards (NAAQS) for particulate matter and set standards for the first time for particles with diameters less than 2.5 μm. The fine-particle rule set an annual limit on the mass per unit volume of particulates with diameters less than 2.5 μm in the air to 15 $\mu g/m^3$. A series of court cases in 1997–2002 ultimately led

A table of the current NAAQS for many pollutants, including particulates, can be found at **http://epa.gov/air/criteria.html**.

to the upholding of the U.S. EPA standards by the U.S. Supreme Court, and these standards are now in place.

To *adsorb* is to attract firmly to a surface. To *absorb* is to draw into the bulk of a solid or liquid.

Aerosols are small enough to remain suspended in the atmosphere for long periods. Such small particles are easily breathable and can cause lung disease. Because of their relatively large surface area, aerosol particles have great capacities to *adsorb* and concentrate chemicals on their surfaces. Liquid aerosols or particles covered with a thin coating of water may also *absorb* air pollutants, thereby concentrating them and providing a medium in which reactions may occur.

Oil-burning electrical generation plants produce quantities of SO_2 comparable to coal-burning facilities, since fuel oils can also contain up to 4% sulfur. The sulfur in oil is in the form of odorous organic compounds such as CH_3CH_2SH in which sulfur atoms are bound to carbon and hydrogen atoms (the —SH functional group). A mixture of such low-molar-mass compounds is used as an odorant in natural gas so that gas leaks can be detected by smell.

Sulfur Dioxide Sulfur dioxide, SO_2, the pollutant that is a major contributor to industrial smog and acid rain, is produced when sulfur or sulfur-containing compounds are burned in air. Most of the coal burned in the United States contains sulfur in the form of the mineral pyrite, FeS_2 and as part of some of the organic compounds making up the coal itself. The weight percent of sulfur in this coal ranges from 1% to 4%. The pyrite is oxidized as the coal is burned, producing sulfur dioxide.

$$4\ FeS_2(s) + 11\ O_2(g) \longrightarrow 2\ Fe_2O_3(s) + 8\ SO_2(g)$$

EXERCISE 10.18 Calculating SO_2 Emissions from Burning Coal

Large quantities of coal are burned in the United States to generate electricity. A 1000-MW coal-fired generating plant will burn 3.06×10^6 kg of coal per hour. For coal that contains 4% sulfur by mass, calculate the mass of SO_2 released (a) per hour and (b) per year.

Once in the atmosphere, SO_2 can be oxidized by reactions with O_2 or ozone, for example, to form SO_3, which has a strong affinity for water. The SO_3 will dissolve in aqueous aerosol droplets to form sulfuric acid, which contributes to acid rain (Section 17.3).

$$SO_3(g) + H_2O(\ell) \longrightarrow H_2SO_4(aq)$$

Sulfur dioxide can be physiologically harmful to both plants and animals, although most healthy adults can tolerate fairly high levels of SO_2 without apparent lasting ill effects. Exposure to SO_2 levels exceeding 0.2 ppm more than 3% of the time leads to respiratory symptoms in a community of people. Exposure to SO_2 levels exceeding 0.3 ppm for 8 hours can cause vegetation damage. People with chronic respiratory difficulties such as bronchitis or asthma tend to be much more sensitive to SO_2,

CHEMISTRY YOU CAN DO Particle Size and Visibility

A common feature of all aerosols is that they decrease visibility. This problem can be observed in a city or along a busy highway, for example. Fog is an extreme example of disruption of visibility by aerosol particles. Here is a way to simulate the effect of air pollutants on visibility. You will need a flashlight, a little milk, a transparent container (if possible, with flat, parallel sides) full of water, and something with which to stir the water.

Turn off the room lights and shine the beam of the flashlight through the container perpendicular to the flat sides. What do you observe? Can you see the beam? Now add a couple of drops of milk to the container and stir. Can you see the flashlight beam now? How does the beam's appearance change as the angle of viewing changes? What color is it? What color is the light that passes through the milky water? Keep adding milk dropwise, stirring and observing until the beam of the flashlight is no longer visible from the far side of the container.

Think about these questions:

1. Which wavelengths of light are scattered more by the particles? Which are scattered less?
2. Based on your observations of the milky water, devise an explanation of the fact that at midday on a sunny day the sun appears to be white or yellow, while at sunset it appears orange or red.

accounting for many of the deaths during episodes of industrial smog. To reduce the hazards, many facilities have installed equipment that removes SO_2 from emitted gases.

EXERCISE 10.19 Calculations Involving Air Pollutants

Suppose the air on a smoggy day contains 5 ppm SO_2. What volume percent is that?

Nitrogen Oxides Most of the nitrogen oxides, $\mathbf{NO_x}$, found in the atmosphere originate from the primary pollutant nitrogen monoxide, NO, formed whenever nitrogen and oxygen—always present in the atmosphere—are raised to high temperatures, as in an internal combustion engine. This nitrogen monoxide reacts rapidly with atmospheric oxygen to produce NO_2.

The common oxides of nitrogen found in air, NO and NO_2, are collectively called NO_x.

Vast quantities of nitrogen oxides are formed each year throughout the world, resulting in a global atmospheric concentration of NO_2 of a few parts per billion or less. In the United States, most oxides of nitrogen are produced during fossil fuel combustion, including that of gasoline in automobile engines, with significantly less coming from natural sources such as the action of lightning. Elsewhere in the world, large amounts are produced by the burning of trees and other biomass. Most NO_2, either from human activities or from natural causes, eventually washes out of the atmosphere in precipitation. In the troposphere, however, especially around urban centers, excessive NO_2 causes problems.

Normally nitrogen dioxide has a lifetime of about three days in the atmosphere.

In laboratory studies, nitrogen dioxide in concentrations of 25 to 250 ppm inhibits plant growth and causes defoliation. Breathing NO_2 at a concentration of 3 ppm for 1 hour causes bronchial constriction in humans, and short exposures at high levels (150 to 220 ppm) causes changes in the lungs that can be fatal.

EXERCISE 10.20 Air Pollutant Stoichiometry

If 400. metric tons N_2 is converted to NO and then to HNO_3, what mass (in metric tons) of HNO_3 is produced?

Ozone: A Secondary Pollutant in the Troposphere

In a complex series of reactions involving O_2, NO_2, and oxygen-containing organic compounds, ozone is produced in the troposphere. Ozone is a **secondary pollutant,** one that is formed by reaction of a primary pollutant. Its pungent odor can be detected at concentrations as low as 0.02 ppm. We often smell the ozone produced by photocopiers and sparking electric appliances, or after a thunderstorm when lightning-caused ozone washes out with the rainfall. In the troposphere (the air we breathe), ozone is harmful because it is a component of photochemical smog and because it can damage human health and decompose materials such as plastics and rubber.

The most significant chemical reaction producing ozone in the atmosphere is the combination of molecular oxygen and atomic oxygen. In the troposphere the major source of oxygen atoms is the photodissociation of NO_2.

$$\cdot NO_2(g) \xrightarrow{h\nu} \cdot NO(g) + \cdot O \cdot (g)$$

Ozone photodissociates at any altitude to give an oxygen atom and an oxygen molecule whenever it is struck by photons in the near-ultraviolet range (240 to 310 nm).

$$O_3(g) \xrightarrow{h\nu} O_2(g) + \cdot O \cdot (g)$$

In the troposphere, oxygen atoms react with water to produce hydroxyl radicals, ·OH.

$$\cdot O\cdot(g) + H_2O(g) \longrightarrow 2\ \cdot OH(g)$$

In the daytime, when they are produced in large numbers, hydroxyl radicals can react with nitrogen dioxide to produce nitric acid.

$$\cdot NO_2(g) + \cdot OH(g) \longrightarrow HNO_3(g)$$

This three-step process is the primary means of removing NO_2 from the atmosphere. Summing the three steps gives the net reaction

$$O_3(g) + 2\ \cdot NO_2(g) + H_2O(g) \longrightarrow 2\ HNO_3(g) + O_2(g)$$

© William Taufic/Corbis

Automobile emissions testing. Such testing is mandated in many cities that have failed to meet EPA's ozone standards. Cars that fail the emissions standards for hydrocarbons and carbon monoxide are required to be repaired. Often the local government will not reissue vehicle registration until satisfactory emissions levels are achieved.

Ozone is a very difficult pollutant to control because its formation depends on sunlight as well as NO_2 and hydrocarbons, which almost every automobile emits to some degree. In cities with high ozone concentrations, the cause is always related to emissions of nitrogen oxides from automobiles, buses, and trucks. Most major urban areas operate vehicle inspection centers for passenger automobiles in an effort to control emissions of nitrogen oxides as well as those of carbon monoxide and unburned hydrocarbons.

As difficult as they are to attain, established ozone standards may not be low enough for good health. Exposure to concentrations of ozone at or near 0.12 ppm has been shown to lower the volume of air a person breathes out per second; studies of children who were exposed to ozone concentrations slightly below the former EPA standard (0.12 ppm) showed a significant decrease in exhaled air volume. In 2008 the EPA lowered the ozone standard to 0.075 ppm for an eight-hour average exposure. No matter what the official standard, ozone concentrations in many urban areas represent health hazards to children at play, joggers, others doing outdoor labor or exercise, and older persons who may have diminished respiratory capabilities.

EXERCISE **10.21 Photodissociation Reactions**

Write the equations for the two photodissociation reactions that produce atomic oxygen. Then write the equation for a reaction in which ozone is formed.

Urban Air Pollution—Industrial Smog

The poisonous mixture of smoke (particulate matter), fog (an aerosol), air, and other chemicals was first called **smog** in 1911 by Dr. Harold de Voeux in his report on a London air pollution disaster that caused the deaths of 1150 people. The smog de Voeux identified is the *chemically reducing type* that is derived largely from the combustion of coal and oil and contains sulfur dioxide (a strong reducing agent) mixed with soot, fly ash, smoke, and partially oxidized organic compounds. This industrial smog is common in many cities in the world where heavy industry and power plants are found. Thanks to the U.S. Clean Air Act, industrial smog is becoming less common in the United States as more pollution controls are installed. It is, unfortunately, still a major problem in some cities of the world.

Urban Air Pollution—Photochemical Smog

The other major kind of smog is the *chemically oxidizing type,* which contains strong oxidizing agents such as ozone and oxides of nitrogen, NO_x. It is known as **photochemical smog** because light—in this instance sunlight—is important in initiating the reactions that cause it.

Photochemical smog is typical in cities where sunshine is abundant and internal combustion engines exhaust large quantities of pollutants to the atmosphere (Figure 10.20). This type of smog is practically free of sulfur dioxide but contains substantial amounts of nitrogen oxides, ozone, the products of hydrocarbon oxidation, organic peroxides, and hydrocarbons of varying complexity. The concentrations of these substances vary during the day, building up in the morning hours and dropping off at night (Figure 10.21).

Figure 10.20 Photochemical smog over Houston.

A city's atmosphere is an enormous mixing bowl of interrelated chemical reactions. Identifying the exact chemical reactions that produce photochemical smog was a challenge, but we now know that photochemical reactions are essential to the smog-making process and that aerosols serve to keep the primary pollutants together long enough to form secondary pollutants. Photons in the ultraviolet region of the spectrum are primarily responsible for the formation of photochemical smog.

The reaction scheme by which primary pollutants are converted into the secondary pollutants found in photochemical smog (Figure 10.22) is thought to begin with the photodissociation of nitrogen dioxide (Reaction 1). The very reactive atomic oxygen next reacts with molecular oxygen to form ozone, O_3 (Reaction 2), which is then consumed by reacting with nitrogen monoxide to form the original reactant—nitrogen dioxide (Reaction 3).

$$\cdot NO_2(g) \xrightarrow{h\nu} \cdot NO(g) + \cdot O \cdot (g) \qquad (1)$$

$$\cdot O \cdot (g) + O_2(g) \longrightarrow O_3(g) \qquad (2)$$

$$O_3(g) + \cdot NO(g) \longrightarrow \cdot NO_2(g) + O_2(g) \qquad (3)$$

The net effect of these three reactions is absorption of energy without any net chemical change. If this were all that happened there would be no problem. Unfortunately, some of the reactive ozone molecules and oxygen atoms go on to react with other species in the atmosphere, which leads to photochemical smog formation.

The air in an urban environment contains unburned hydrocarbons released from car exhausts, evaporation, and spillage at filling stations. Atomic oxygen produced in Reaction 1 reacts with the more reactive of these hydrocarbons—unsaturated compounds and aromatics—to form other free radicals. These radicals, in turn, react to form yet other radicals and secondary pollutants such as aldehydes (for example, formaldehyde, HCHO). In addition, reactions of ·OH (hydroxyl radical), ·NO, O_2, and other species can produce other compounds that contribute to

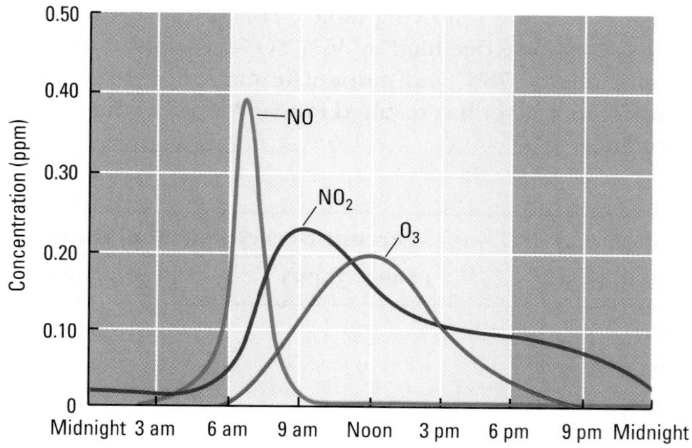

Figure 10.21 The average concentrations of the pollutants NO, NO_2, and O_3 on a smoggy day in Los Angeles, California. The NO concentration builds up during the morning rush hour. Later in the day the concentrations of NO_2 and O_3, which are secondary pollutants, build up.

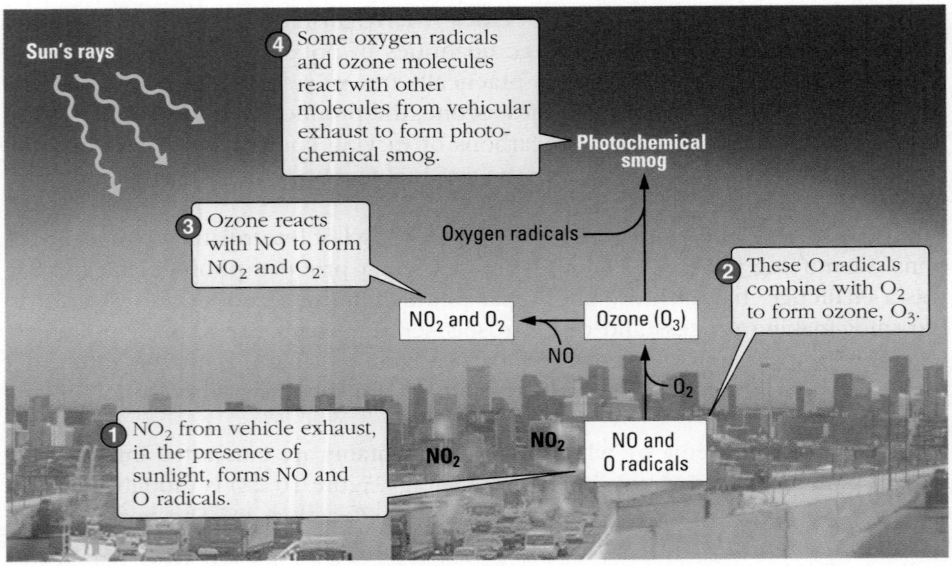

Figure 10.22 **Photochemical smog formation.**

the smog. Some of those compounds stabilize NO_2 so that it can be carried over great distances by prevailing winds. Through a variety of chemical reactions coupled with wind, urban pollution in the form of NO_2 can be carried to outlying areas, where it may do additional damage to vegetation, human tissue, and fabrics.

EXERCISE 10.22 Smog Ingredients

Write the formulas and give sources for three ingredients of photochemical smog.

Where Do We Stand on Air Quality Now?

Average air quality across the United States is improving. Under the Clean Air Act, the EPA establishes air quality standards for six principal air pollutants: NO_2, O_3, SO_2, PM (particulate matter), CO, and Pb. The EPA measures the levels of these pollutants at thousands of monitoring stations. Yearly summaries of current air pollution status are provided. Data show that these six primary pollutants are declining in the United States (see the following table). For example, over the period from 1990 to 2008, NO_2 emissions declined by 35%, SO_2 emissions by 59%, CO emissions by 68%, lead emissions by 78%, and particulate matter by 19%. The EPA has concluded that "the Clean Air Act has resulted in many improvements in the quality of air in the United States."

Principal Pollutants	Percent Decrease in Emissions (1980–2008)	(1990–2008)
CO	79	68
Pb	92	78
NO_2	46	35
SO_2	71	59
Particulates, < 2.5 μm	—	19

Source: **http://www.epa.gov/airtrends/sixpoll.html**.

10.11 Atmospheric Carbon Dioxide, the Greenhouse Effect, and Global Warming

The Greenhouse Effect

An important function of our atmosphere is to moderate Earth's surface temperature. The energy that heats Earth comes from the sun as electromagnetic radiation. Some of the electromagnetic radiation from the sun is reflected by the atmosphere back into space, and some is absorbed by the atmosphere. The remainder reaches the surface, warming its land and oceans. The warmed surfaces then reradiate this energy into the troposphere as infrared radiation ***(← p. 272).*** Carbon dioxide, water vapor, methane, and tropospheric ozone all absorb radiation in various portions of the infrared region. By absorbing this reradiated energy they warm the atmosphere, creating what is called the **greenhouse effect** (Figure 10.23). Thus, all four are "greenhouse gases." Such gases constitute an absorbing blanket that reduces the quantity of energy radiated back into space. Thanks to the greenhouse effect, Earth's average temperature is a comfortable 15 °C (59 °F). By comparison, the moon, with no moderating atmosphere, has a surface mean temperature that fluctuates from approximately 107 °C when exposed to sunlight and −153 °C when in the dark.

The greenhouse effect derives its name by analogy with a botanical greenhouse. However, the warming in a botanical greenhouse is much more dependent on the fact that glass reduces convection than that it blocks infrared radiation from leaving the greenhouse.

Earth has such a vast reservoir of water in the oceans that human activity has a negligible influence on the concentration of water vapor in the atmosphere. In addition, methane is produced by natural processes in such large quantities that human contributions are negligible. Tropospheric ozone is present in such small concentrations that its contribution to the greenhouse effect is small. Among the four greenhouse gases, most attention is focused on CO_2.

Radiation from objects on Earth to outer space answers the question, "Why does frost form on top of a parked car in winter, but not on the sides?" that was posed in Chapter 1 *(← p. 2).*

Carbon dioxide is a greenhouse gas because it absorbs infrared radiation, causing C═O bonds in the molecule to stretch and bend, much like springs attached to balls. Stretching and compressing the C═O bonds requires more energy (shorter

$1\ \mu m = 1 \times 10^{-6}\ m = 1 \times 10^{3}\ nm$

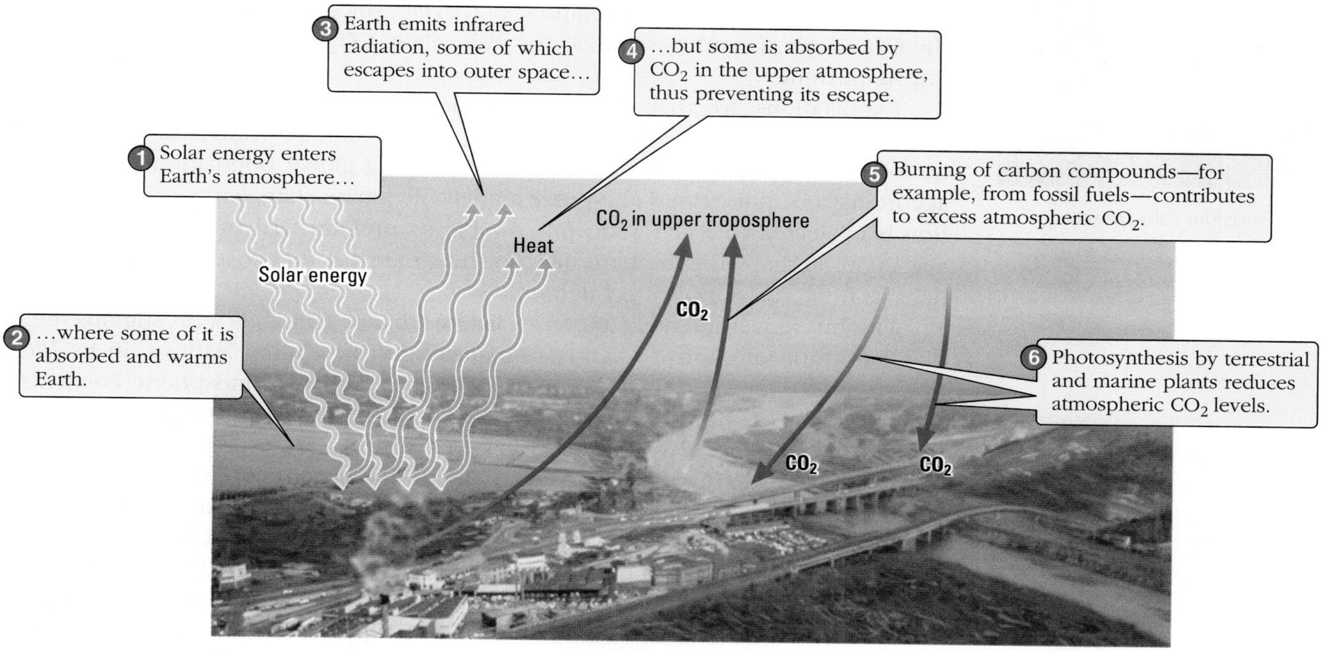

Figure 10.23 The greenhouse effect. Greenhouse gases form an effective barrier that prevents some heat from escaping Earth's surface. Without the greenhouse effect, Earth's average temperature would be much lower.

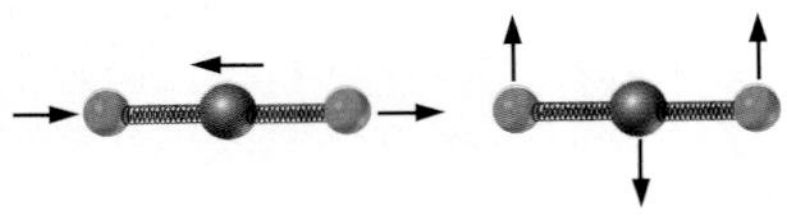

Recall from Section 7.1 that energy and wavelength are inversely related.

Worldwide, nearly one third of all atmospheric CO_2 is released as a by-product from fossil fuel-burning electric power plants.

wavelength) than does bending them. Thus, C=O stretching and compressing occurs when infrared radiation with a wavelength of 4.257 μm is absorbed, whereas the C=O bending vibrations occur at lower energy (longer wavelength) when the molecule absorbs 15.000-μm infrared radiation.

Each year, combustion of fossil fuels (coal, petroleum, natural gas) worldwide puts billions of metric tons of carbon into the atmosphere as CO_2. About 45% is removed from the atmosphere by natural processes—some by plants during photosynthesis and the rest by dissolving in rainwater and the oceans to form hydrogen carbonates and carbonates.

$$CO_2(g) + 2\,H_2O(\ell) \longrightarrow H_3O^+(aq) + HCO_3^-(aq)$$

$$HCO_3^-(aq) + H_2O(\ell) \longrightarrow H_3O^+(aq) + CO_3^{2-}(aq)$$

The other 55% of the carbon dioxide from fossil fuel combustion remains in the atmosphere, increasing the global CO_2 concentration.

EXERCISE **10.23 Sources of CO_2**

List as many natural sources of CO_2 as you can. List as many sources of CO_2 from human activities as you can.

Parts per million (ppm) is a convenient way to express low concentrations. One ppm means one part of something in one million parts. A CO_2 concentration of 360 ppm means that for every million molecules of air, 360 are CO_2 molecules.

Stephen Ferry/Liaison/Getty Images

Clearing a Brazilian rain forest.

Without human influences, the flow of carbon dioxide among the air, plants, animals, and the oceans would be roughly balanced. In 1750, during the preindustrial era, CO_2 concentration in the atmosphere was 277 parts per million (ppm). During the next 130 years, as the Industrial Revolution progressed, the concentration increased to 291 ppm, a 5% increase. Since 1900, however, the rate of increase of CO_2 has reflected the rapid increase in the use of fossil fuel combustion for industrial and domestic purposes, especially for motor vehicle transportation. Between 1959 and 2007, the atmospheric concentration of CO_2 increased from 316 to 384 ppm, an increase of 20% (Figure 10.24). Population pressure is also contributing heavily to increased CO_2 concentrations. In the Amazon region of Brazil, for example, forests are being cut and burned to create cropland. This activity places a double burden on the natural CO_2 cycle, since there are fewer trees to use the CO_2 in photosynthesis and, at the same time, CO_2 is added to the atmosphere during burning.

Expectations are that the atmospheric CO_2 concentration will continue to increase at the current rate of about 1.5 ppm per year, due principally to industrialization, agricultural production, and an expanding global population. As a result, according to sophisticated computer models of climate change, the CO_2 concentration is projected to reach 550 ppm, double its preindustrial value, between 2030 and 2050. Perhaps the increase may not be so great if fuel costs increase substantially as fossil fuel supplies get tighter.

The Intergovernmental Panel on Climate Change (IPCC) is a prestigious multinational group of climate scientists organized by the United Nations. In its *IPCC Fourth Assessment Report* of 2007, the IPCC noted that, "Atmospheric concentrations of CO_2 (379 ppm) . . . in 2006 exceed by far the natural range over the last 650,000 years." The IPCC noted earlier that nearly 75% of CO_2 emissions from human activities in the past 20 years are due to the burning of fossil fuels.

To see how easily everyday activities affect the quantity of CO_2 being released into the atmosphere, consider a round-trip flight from New York to Los Angeles. Each passenger pays for about 200 gal jet fuel, which weighs 1400 lb. When burned, each pound of jet fuel produces about 3.14 lb carbon dioxide. So 4400 lb, or 2 metric tons, of carbon dioxide are produced per passenger during that trip.

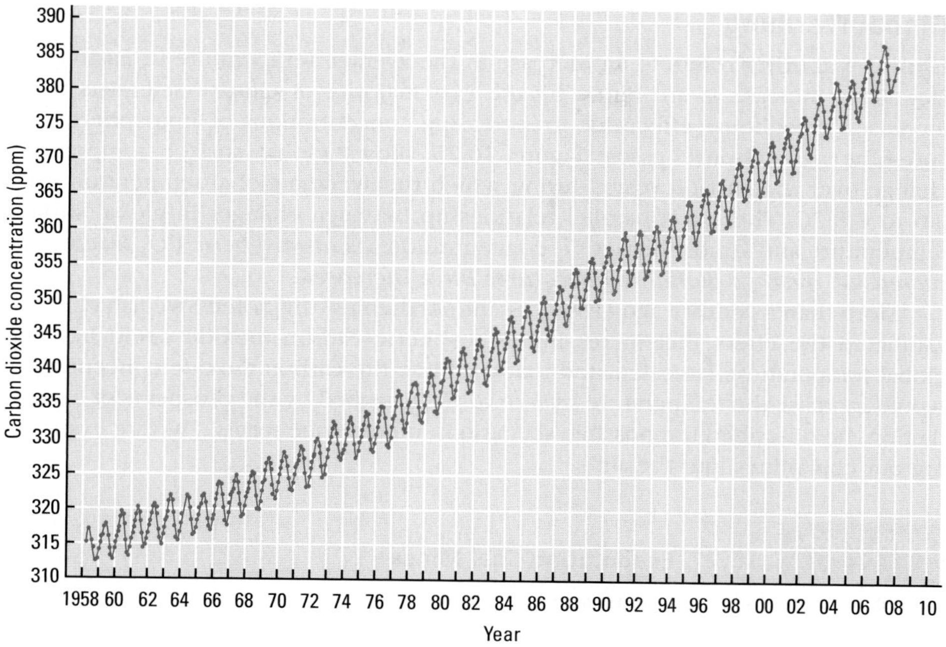

Figure 10.24 **Atmospheric carbon dioxide concentration measured at Mauna Loa, Hawaii.** These data have been carefully collected by a standardized methodology for more than 50 years, and they are the longest continuous record of atmospheric CO_2 concentrations available. The Mauna Loa record shows an increase in CO_2 concentration, from 316 ppm to 384 ppm, over the period 1959 to 2007. (*Source:* **http://www.esrl.noaa.gov/gmd/ccgg/trends/co2_data_mlo.html.**)

EXERCISE 10.24 CO_2 Changes

Using data from Figure 10.24, calculate the percent increase in atmospheric CO_2 from
(a) 1960 to 1975 (b) 1975 to 1990 (c) 1990 to 2008
Which period showed the greatest percent increase in CO_2?

CONCEPTUAL EXERCISE 10.25 Annual CO_2 Changes

During a year, atmospheric CO_2 concentration fluctuates, building up to a high value, then dropping to a low value before building up again (Figure 10.24). Explain what causes this fluctuation and when during the year the high and the low occur in the Northern Hemisphere.

CONCEPTUAL EXERCISE 10.26 Carbon Dioxide and Air Travel

Use an Internet search engine to find the annual airline passenger miles for a recent year and calculate how many tons of CO_2 were produced by those flights. Compare this quantity to the quantity of CO_2 produced for the same number of miles traveled in automobiles. (*Hint:* Assume a miles-per-gallon value and an average number of passengers per vehicle. Then use the same numbers for gasoline as were used for jet fuel in the last paragraph on page 464.)

Global Warming

Global warming refers to an increase in the average temperature of Earth's surface. This increase is attributed by the vast majority of climate scientists to an increased greenhouse effect due to increasing concentrations of atmospheric CO_2 and other greenhouse gases. Figure 10.25 shows global surface temperature change over the past 60 years, during which the average global temperature rose 0.6 to 0.8 °C. There is a parallel relationship between changes in global surface temperature and atmospheric CO_2 concentration: *As atmospheric CO_2 levels have increased, average global temperatures have increased.*

Arrhenius had not foreseen the advent of motor vehicles and their impact on increasing atmospheric carbon dioxide.

More than a century ago (1896), Svante Arrhenius, the noted Swedish chemist, estimated the effect that increasing atmospheric carbon dioxide could have on global temperature. He calculated that doubling the carbon dioxide concentration would increase the temperature by 5 to 6 °C. Disturbed by the prospects of his prediction, he described the possible scenario in picturesque terms: "We are evaporating our coal mines into the air."

CONCEPTUAL EXERCISE 10.27 Evaporating Coal Mines

Explain Arrhenius's quote in terms of global warming.

A 1.5 °C rise in global temperature is a significant change, requiring a very large input of energy.

The 2007 Nobel Peace Prize was awarded to former Vice President Al Gore and the IPCC "for their efforts to build up and disseminate greater knowledge about man-made climate change."

Many computer models predict that increasing atmospheric CO_2 to 600 ppm will increase average global temperature, and most climate experts who do global warming research agree. What is uncertain is the extent of the temperature increase, with estimates varying from 1.5 to 4.5 °C (2.7 to 8.1 °F). The IPCC estimates a 1.0 to 3.5 °C (2 to 6 °F) increase by the year 2100. Warming by as little as 1.5 °C would produce the warmest climate seen on Earth in the past 6000 years; an increase of 4.5 °C would produce world temperatures higher than any since the Mesozoic era, the time of the dinosaurs.

Many scientists are concerned that rising temperatures are causing ice caps in Greenland and Antarctica to melt, raising the sea level. In conjunction with this, atmospheric currents will change and produce significant changes in weather and agricultural productivity. Since the 1940s, average summertime temperatures in Antarctica have increased 2.5 °C to just above 0 °C.

Computer models used to predict future global temperature changes have become more sophisticated and accurate over the past decade. When factors in addition to greenhouse gases, such as the presence of aerosols and changes in the sun's irradiance are taken into account, the computer models predict temperature changes close to what has been observed since 1860 (Figure 10.26). Such work

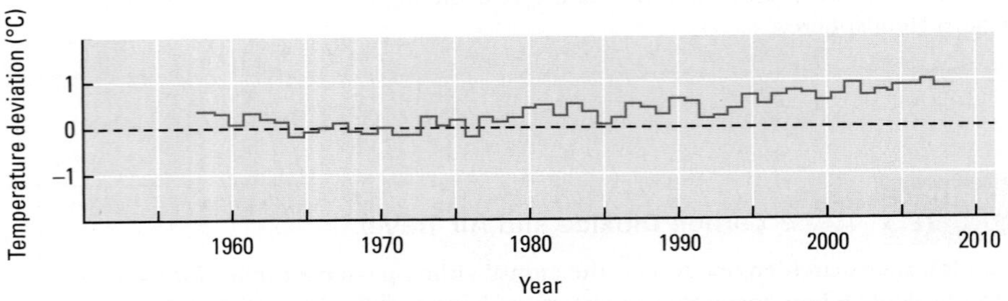

Figure 10.25 Deviations in global surface temperatures 1958–2008. The warmest year in the 51-year record was 2007, having an average surface temperature 0.97 °C above the long-term mean temperature. (*Source:* Angell, J.K. 2009. Global, hemispheric, and zonal temperature deviations derived from radiosonde records. In Trends Online: A Compendium of Data on Global Change. Carbon Dioxide Information Analysis Center, Oak Ridge National Laboratory, U.S. Department of Energy, Oak Ridge, Tennessee, U.S.A. **http://cdiac.ornl.gov/trends/temp/angell/angell.html.**)

adds credibility to the predictions of temperature increases of 1.5 to 4.5 °C due to global warming.

Global warming has become an acknowledged worldwide problem. Concerns about global warming prompted delegates from 159 countries to convene in 1997 in Kyoto, where the Kyoto Protocol was negotiated. Goals were set to reduce greenhouse gas emissions below 1990 levels. Enough countries ratified these protocols to make them official in February 2005. The only industrialized country that did not ratify them was the United States. The Kyoto Protocol is due to expire in 2012, and successor agreements to the Kyoto Protocol are now under consideration. The United Nations Climate Change Conference 2009 was held in Copenhagen in December 2009 with the aim of providing a framework for climate change mitigation beyond 2012.

The most obvious measure to reduce global warming is to control CO_2 emissions that result from human activities throughout the world. Given our dependence on fossil fuels, however, getting and keeping these emissions under control will be very difficult. One direct way to decrease the quantity of atmospheric CO_2 is to plant more trees and to replace those in deforested areas.

The projected extent of global warming remains a thorny issue. The uncertainty in the temperature changes attributable to global warming is related to three fundamental, but difficult, questions:

1. To what levels will CO_2 and other greenhouse gases rise during the next decades?
2. How responsive is Earth's climate to the warming created by the greenhouse effect?
3. To what extent is the current global warming being caused by human activities, or could it be part of a natural cycle of global temperature changes?

Credible climate scientists worldwide think that enhanced global warming has already occurred and that human activities are contributing to this warming. In 2006 an expert panel convened by the U.S. National Academy of Sciences, the Committee of Surface Temperature Reconstruction for the Last 2,000 Years, produced a report that stated, "It can be said with a high level of confidence that global mean surface temperature was higher during the last few decades of the 20th century than during any comparable period during the preceding four centuries."

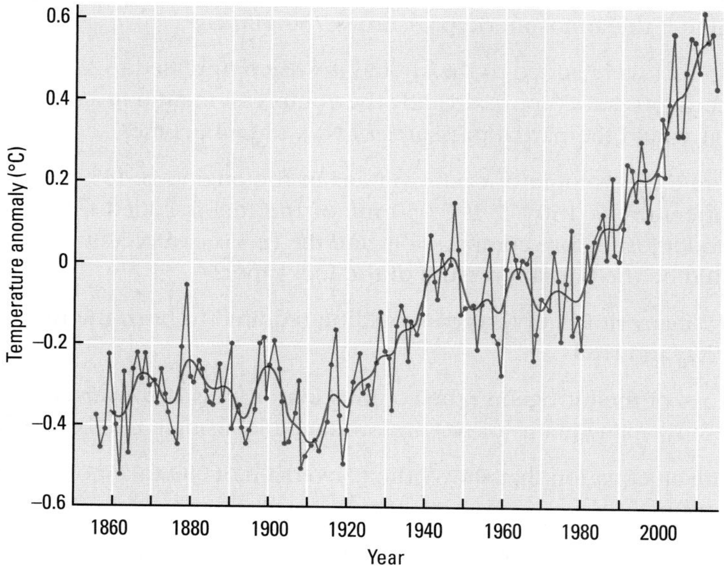

Figure 10.26 Global temperature change over the past 160 years. The blue points show the annual average temperatures, and the red line shows the five-year average temperatures. Temperature anomaly refers to the annual temperature minus the average temperature during the period 1961–1990. The global average temperature has risen approximately 0.6 to 0.8 °C over the past 160 years. (*Source*: **http://en.wikipedia.org/wiki/Global_warming.**)

CHEMISTRY IN THE NEWS Removing CO_2 from the Air

The buildup of CO_2 in Earth's atmosphere is a global problem that demands worldwide attention. Cutting CO_2 emissions will help in the long run, but methods to remove CO_2 directly from the air would be a big advance. For decades, small devices for removing CO_2 from air have been used in closed environments such as spacecraft and submarines, but these devices are too expensive to use for large-scale applications. Now several newer devices are being developed for capture of CO_2 directly from ambient air, where the CO_2 concentration is quite low, 0.04%. In these devices, air is swept over a *sorbent* that binds the CO_2. The CO_2 is later separated ("stripped") from the sorbent in a separate processing step.

One design, developed by David Keith's research team at the University of Calgary, employs the sorbent in the form of a fine mist, so it has a very large surface area and the reaction with CO_2 is more efficient. The CO_2 is separated from the sorbent in a separate processing step. While the technology is at an early stage, a spray tower using this mist approach has been demonstrated to be effective.

Another design, developed by Klaus Lackner's research team at Columbia University in conjunction with Global Research Technologies, uses a solid sorbent coated on thin sheets. The CO_2 is trapped as air is swept over the sheets. The CO_2 is separated from the sorbent-coated sheets in a second processing step. Several stripping approaches are under investigation.

At present, these air-capture approaches are experimental. It remains for future development to make them economical on a larger scale for widespread commercial use.

Sources: "Scrubbing the Skies," *The Economist*, March 7, 2009; p. 22.

Adam, D. "Could US Scientist's 'CO_2 Catcher' Help to Slow Warming?" *The Guardian*, May 31, 2008.

SUMMARY PROBLEM

The air that enters an automobile engine contains the oxygen that reacts with the hydrocarbon fuel vapors to provide the energy needed to move the vehicle. Prior to the combustion process the fuel-air mixture is compressed, and then it is ignited with a spark. Most of the fuel is completely burned to CO_2 and H_2O, and some of the nitrogen in air is converted to nitrogen monoxide, NO. For simplicity, assume the fuel has a formula of C_8H_{18} and a density of 0.760 g/mL.

(a) What are the partial pressures of N_2 and O_2 in the air before it goes into the engine if the atmospheric pressure is 734 mm Hg?

(b) If no fuel were added to the air and it was compressed to seven times atmospheric pressure (approximately the compression ratio in a modern engine), what would the partial pressures of N_2 and O_2 become?

(c) Assume the volume of each cylinder in the engine is 485 mL and that the temperature is 150 °C. If 0.050 mL of the fuel is added to the air in each cylinder just before compression, and the fuel is completely vaporized, what would be the partial pressure of the fuel vapor?

(d) How many moles of oxygen would be required to burn the fuel completely to CO_2 and H_2O?

(e) If 10.% of the nitrogen in the combustion process is converted to NO, calculate how many grams of NO are produced in a single combustion reaction.

(f) Write an equation that shows the conversion of NO to a more photoreactive compound.

(g) Calculate the mass of the more photoreactive compound that would be formed from the NO produced in part (e).

(h) What additional information would you need to calculate the NO emissions for an entire city for a year?

IN CLOSING

Having studied this chapter, you should be able to . . .

- Describe the components of the atmosphere (Section 10.1). End-of-chapter question 13
- Describe the properties of gases (Section 10.2). Questions 16, 114, 115
- State the fundamental concepts of the kinetic-molecular theory and use them to explain gas behavior (Section 10.3). Question 24
- Solve problems using the appropriate gas laws (Section 10.4). Questions 30, 32, 37, 41, 43, 45
- Calculate the quantities of gaseous reactants and products involved in chemical reactions (Section 10.5). Questions 53, 55, 59
- Apply the ideal gas law to finding gas densities and molar masses (Section 10.6). Questions 63, 66, 70
- Perform calculations using partial pressures of gases in mixtures (Section 10.7). Questions 71, 73
- Describe the differences between real and ideal gases (Section 10.8). Question 84
- Explain the main features of stratospheric ozone depletion and the role of CFCs in it (Section 10.9). Question 90
- Describe the main chemicals found in industrial pollution and urban pollution and the reactions producing industrial pollution and urban pollution (Section 10.10). Questions 15, 103
- Relate atmospheric CO_2 concentration to the greenhouse effect and to enhanced global warming (Section 10.11). Question 108

OWL and go Chemistry

Sign in at **www.cengage.com/owl** to:

- View tutorials and simulations, develop problem-solving skills, and complete online homework assigned by your professor.
- For quick review and exam prep, download Go Chemistry mini lecture modules from OWL (or purchase them at **www.CengageBrain.com**).

KEY TERMS

absolute temperature scale *(Section 10.4)*

absolute zero *(10.4)*

aerosols *(10.10)*

air pollutant *(10.10)*

Avogadro's law *(10.4)*

bar *(10.2)*

barometer *(10.2)*

Boyle's law *(10.4)*

Charles's law *(10.4)*

chlorofluorocarbons (CFCs) *(10.9)*

combined gas law *(10.4)*

Dalton's law of partial pressures *(10.7)*

diffusion *(10.3)*

effusion *(10.3)*

ideal gas *(10.4)*

ideal gas constant *(10.4)*

ideal gas law *(10.4)*

global warming *(10.11)*

greenhouse effect *(10.11)*

Kelvin temperature scale *(10.4)*

law of combining volumes *(10.4)*

millimeters of mercury (mm Hg) *(10.2)*

mole fraction *(10.7)*

newton (N) *(10.2)*

NO_x *(10.10)*

ozone hole *(10.9)*

partial pressure *(10.7)*

particulates *(10.10)*

pascal (Pa) *(10.2)*

photochemical smog *(10.10)*

photodissociation *(10.9)*

pressure *(10.2)*

primary pollutant *(10.10)*

secondary pollutant *(10.10)*

smog *(10.10)*

standard atmosphere (atm) *(10.2)*

standard molar volume *(10.4)*

standard temperature and pressure (STP) *(10.4)*

stratosphere *(10.1)*

stratospheric ozone layer *(10.9)*

torr *(10.2)*

troposphere *(10.1)*

van der Waals equation *(10.8)*

QUESTIONS FOR REVIEW AND THOUGHT

OWL Interactive versions of these problems are assignable in OWL.

Blue-numbered questions have short answers at the back of this book in Appendix M and fully worked solutions in the *Student Solutions Manual.*

Review Questions

These questions test vocabulary and simple concepts.

1. Name the three gas laws and explain how they interrelate *P*, *V*, *T*, and *n*. Explain the relationships in words and with equations.
2. What are the conditions represented by STP?
3. What is the volume occupied by 1 mol of an ideal gas at STP?
4. What is the definition of pressure?
5. State Avogadro's law. Explain why two volumes of hydrogen react with one volume of oxygen to form two volumes of steam.
6. State Dalton's law of partial pressures. If the air we breathe is 78% N_2 and 21% O_2 on a mole basis, what is the mole fraction of O_2? What is the partial pressure of O_2 if the total pressure is 720 mm Hg?
7. Explain Boyle's law on the basis of the kinetic-molecular theory.
8. Explain why gases at low temperature and high pressure do not obey the ideal gas equation as well as gases at high temperature and low pressure.
9. Gaseous water and carbon dioxide each absorb infrared radiation. Does either of them absorb ultraviolet radiation? Explain your answer.
10. Describe the difference between primary and secondary air pollutants and give two examples of each.

Topical Questions

These questions are keyed to the major topics in the chapter. Usually a question that is answered at the back of this book is paired with a similar one that is not.

The Atmosphere (Section 10.1)

11. Explain the major roles played by nitrogen in the atmosphere. Do the same for oxygen.
12. Beginning at Earth's surface and proceeding upward, name the first two layers or regions of the atmosphere. Describe, in general, the kinds of chemical reactions that occur in each layer.
13. Convert all of the "percentage by volume" figures in Table 10.1 into (a) parts per million and (b) parts per billion. Which atmospheric gases are present at concentrations of less than 1 ppb? Between 1 ppb and 1 ppm? Greater than 1 ppm?
14. The mass of Earth's atmosphere is 5.3×10^{15} metric tons. The atmospheric abundance of helium is 0.7 ppm when expressed as a fraction by *mass* instead of by *volume,* as in Table 10.1. How many metric tons of helium are there in the atmosphere? How many moles of helium is this?
15. Coal contains about 2.0% sulfur by mass, and when coal is burned the sulfur is all converted to SO_2. In one year, 3.1×10^9 metric tons of coal were burned worldwide. How many tons of SO_2 were added to the atmosphere? How many tons of SO_2 are currently in the atmosphere? (*Note:* The mass fraction of SO_2 in air is 0.4 ppb.)

Properties of Gases (Section 10.2)

16. Gas pressures can be expressed in units of mm Hg, atm, torr, and kPa. Convert these pressure values.
    (a) 720. mm Hg to atm (b) 1.25 atm to mm Hg
    (c) 542. mm Hg to torr (d) 740. mm Hg to kPa
    (e) 700. kPa to atm
17. Convert these pressure values.
    (a) 120. mm Hg to atm (b) 2.00 atm to mm Hg
    (c) 100. kPa to mm Hg (d) 200. kPa to atm
    (e) 36.0 kPa to atm (f) 600. kPa to mm Hg
18. Mercury has a density of 13.96 g/cm^3. A barometer is constructed using an oil with a density of 0.75 g/cm^3. If the atmospheric pressure is 1.0 atm, what will be the height in meters of the oil column in the barometer?
19. A vacuum pump is connected to the top of an upright tube whose lower end is immersed in a pool of mercury. At sea level, how high will the mercury rise in the tube when the pump is turned on?
20. Why can't a hand-driven pump on a water well pull underground water from depths more than 33 ft? Would it help to have a motor-driven vacuum pump?
21. A scuba diver taking photos of a coral reef 60 ft below the ocean surface breathes out a stream of bubbles. What is the total gas pressure of these bubbles at the moment they are released? What is the gas pressure in the bubbles when they reach the surface of the ocean?

Kinetic-Molecular Theory (Section 10.3)

22. List the five basic postulates of the kinetic-molecular theory. Which assumption is incorrect at very high pressures? Which one is incorrect at low temperatures? Which assumption is probably most nearly correct?
23. You are given two flasks of equal volume. Flask A contains H_2 at 0 °C and 1 atm pressure. Flask B contains CO_2 gas at 0 °C and 2 atm pressure. Compare these two samples with respect to each of these properties.
    (a) Average kinetic energy per molecule
    (b) Average molecular velocity
    (c) Number of molecules
24. Place these gases in order of increasing average molecular speed at 25 °C: Kr, CH_4, N_2, CH_2Cl_2.
25. Arrange these four gases in order of increasing average molecular speed at 25 °C: Cl_2, F_2, N_2, O_2.
26. If equal amounts of the four inert gases Ar, Ne, Kr, and Xe are released at the same time at the end of a long, evacuated tube, which gas will reach the end of the tube first?

27. The reaction of SO_2 with Cl_2 to give dichlorine oxide is

$$SO_2(g) + 2\ Cl_2(g) \longrightarrow SOCl_2(g) + Cl_2O(g)$$

Place all molecules in the equation in order of increasing average molecular speed.

28. From Table 10.1 list all the gases in the atmosphere in order of *decreasing* average molecular speed. Concern has been expressed about one of these gases escaping into outer space because a significant fraction of its molecules have velocities large enough to break free from Earth's gravitational field. Which gas is it? (Assume equal T for all gases.)

Gas Behavior and the Ideal Gas Law (Section 10.4)

29. How many moles of CO are present in 1.0 L air at STP that contains 950 ppm CO by volume?
30. If 50.75 g of a gas occupies 10.0 L at STP, what volume will 129.3 g of the gas occupy at STP?
31. A sample of gaseous CO exerts a pressure of 45.6 mm Hg in a 56.0-L flask at 22 °C. If the gas is released into a 2.70×10^4-L room, what is the partial pressure of the CO in the room at 22 °C?
32. A sample of a gas has a pressure of 100. mm Hg in a sealed 125-mL flask. If this gas sample is transferred to another flask with a volume of 200. mL, what will be the new pressure? Assume that the temperature remains constant.
33. A sample of a gas in a sealed 256-mL flask exerts a pressure of 75.0 mm Hg. What is the pressure of this gas if it is transferred to a 125-mL flask? (The temperature stays constant.)
34. A sample of gas has a pressure of 62 mm Hg in a 100-mL flask. This sample of gas is transferred to another flask, where its pressure is 29 mm Hg. What is the volume of the new flask? (The temperature does not change.)
35. Some butane, the fuel used in backyard grills, is placed in a sealed 3.50-L container at 25 °C; its pressure is 735 mm Hg. If you transfer the gas to a sealed 15.0-L container, also at 25 °C, what is the pressure of the gas in the larger container?
36. A sample of gas at 30. °C has a pressure of 2.0 atm in a sealed 1.0-L container. What pressure will it exert in a 4.0-L container? The temperature does not change.
37. Suppose you have a sample of CO_2 in a gas-tight syringe with a movable piston. The gas volume is 25.0 mL at a room temperature of 20. °C. What is the final volume of the gas if you hold the syringe in your hand to raise the gas temperature to 37 °C?
38. A balloon is inflated with helium to a volume of 4.5 L at 23 °C. If you take the balloon outside on a cold day (−10. °C), what will be the new volume of the balloon?
39. A sample of gas has a volume of 2.50 L at a pressure of 670. mm Hg and a temperature of 80. °C. If the pressure remains constant but the temperature is decreased, the gas occupies 1.25 L. What is this new temperature, in degrees Celsius?
40. A sample of 9.0 L CO_2 at 20 °C and 1 atm pressure is cooled so that it occupies a volume of 8.0 L at some new temperature. The pressure remains constant. What is the new temperature, in kelvins?
41. A bicycle tire is inflated to a pressure of 3.74 atm at 15 °C. If the tire is heated to 35 °C, what is the pressure in the tire? Assume the tire volume doesn't change.
42. An automobile tire is inflated to a pressure of 3.05 atm on a rather warm day when the temperature is 40. °C. The car is then driven to the mountains and parked overnight. The morning temperature is −5.0 °C. What will be the pressure of the gas in the tire? Assume the volume of the tire doesn't change.
43. A sample of gas occupies 754 mL at 22 °C and a pressure of 165 mm Hg. What is its volume if the temperature is raised to 42 °C and the pressure is raised to 265 mm Hg? (The number of moles does not change.)
44. A balloon is filled with helium to a volume of 1.05×10^3 L on the ground, where the pressure is 745 mm Hg and the temperature is 20. °C. When the balloon ascends to a height of 2 miles, where the pressure is only 600. mm Hg and the temperature is −33 °C, what is the volume of the helium in the balloon?
45. What is the pressure exerted by 1.55 g Xe gas at 20. °C in a sealed 560-mL flask?
46. A 1.00-g sample of water is allowed to vaporize completely inside a sealed 10.0-L container. What is the pressure of the water vapor at a temperature of 150. °C?
47. Which of these gas samples contains the largest number of molecules and which contains the smallest?
    (a) 1.0 L H_2 at STP
    (b) 1.0 L N_2 at STP
    (c) 1.0 L H_2 at 27 °C and 760. mm Hg
    (d) 1.0 L CO_2 at 0 °C and 800. mm Hg
48. Ozone molecules attack rubber and cause cracks to appear. If enough cracks occur in a rubber tire, for example, it will be weakened, and the tread will wear away much faster. As little as 0.020 ppm O_3 will cause cracks to appear in rubber in about 1 hour. Assume that a 1.0-cm^3 sample of air containing 0.020 ppm O_3 is brought in contact with a sample of rubber that is 1.0 cm^2 in area. Calculate the number of O_3 molecules that are available to collide with the rubber surface. The temperature of the air sample is 25 °C and the pressure is 0.95 atm.

Quantities of Gases in Chemical Reactions (Section 10.5)

49. When a commercial drain cleaner containing sodium hydroxide and small pieces of aluminum is poured, along with water, into a clogged drain, this reaction occurs:

$$2\ Al(s) + 2\ NaOH(aq) + 6\ H_2O(\ell) \rightarrow 2\ NaAl(OH)_4(aq) + 2\ H_2(g)$$

If 6.5 g Al and excess NaOH are reacted, what volume of H_2 gas measured at 742 mm Hg and 22.0 °C will be produced?

50. If 2.7 g Al metal is reacted with excess HCl, how many liters of H_2 gas are produced at 25 °C and 1.00 atm pressure?
51. The yeast in rising bread dough converts sugar (sucrose, $C_{12}H_{22}O_{11}$) into carbon dioxide and ethanol, C_2H_5OH, in a mole ratio of 1:1. That is, 1 mol sucrose produces 4 mol CO_2 and 4 mol C_2H_5OH. A popular recipe for two loaves of French bread requires 1 package of yeast and $\frac{1}{4}$ teaspoon (about 2.4 g) of sugar. What volume of CO_2 at STP is produced by the complete conversion of this quantity of sucrose into CO_2 by the yeast? Compare this volume with the typical volume of two loaves of bread.

52. Water can be made by combining gaseous O_2 and H_2. If you begin with 1.5 L $H_2(g)$ at 360. mm Hg and 23 °C, what volume in liters of $O_2(g)$ will you need for complete reaction if the O_2 gas is also measured at 360. mm Hg and 23 °C?

53. Gaseous silane, SiH_4, ignites spontaneously in air according to the equation

$$SiH_4(g) + 2\,O_2(g) \longrightarrow SiO_2(s) + 2\,H_2O(g)$$

If 5.2 L SiH_4 reacts with O_2, what volume in liters of O_2 is required for complete reaction? What volume of H_2O vapor is produced? Assume all gases are measured at the same temperature and pressure.

54. Hydrogen can be made in the "water gas reaction."

$$C(s) + H_2O(g) \longrightarrow H_2(g) + CO(g)$$

If you begin with excess C(s) and 250. L gaseous water at 120. °C and 2.00 atm, what mass in grams of H_2 can be prepared?

55. If the boron hydride, B_4H_{10}, is treated with pure oxygen, it burns to give B_2O_3 and H_2O.

$$2\,B_4H_{10}(s) + 11\,O_2(g) \longrightarrow 4\,B_2O_3(s) + 10\,H_2O(g)$$

If a 0.050-g sample of the boron hydride burns completely in O_2, what will be the pressure of the gaseous water in a 4.25-L flask at 30. °C?

56. If 1.0×10^3 g uranium metal is converted to gaseous UF_6, what pressure of UF_6 would be observed at 62 °C in a chamber that has a volume of 3.0×10^2 L?

57. Ten liters of F_2 gas at 1.00 atm and 100.0 °C reacts with 99.9 g $CaBr_2$ to form CaF_2 and bromine gas. What volume of Br_2 gas will form at this temperature and pressure?

58. Metal carbonates decompose to the metal oxide and CO_2 on heating according to this general equation.

$$M_x(CO_3)_y(s) \longrightarrow M_xO_y(s) + y\,CO_2(g)$$

You heat 0.158 g of a white, solid carbonate of a Group 2A metal and find that the evolved CO_2 has a pressure of 69.8 mm Hg in a 285-mL flask at 25 °C. What is the molar mass of the metal carbonate?

59. Nickel carbonyl, $Ni(CO)_4$, can be made by the room-temperature reaction of finely divided nickel metal with gaseous CO. This is the basis for purifying nickel on an industrial scale. If you have CO in a sealed 1.50-L flask at a pressure of 418 mm Hg at 25.0 °C, what is the maximum mass in grams of $Ni(CO)_4$ that can be made?

60. Assume that a car burns octane, C_8H_{18} ($d = 0.703$ g/cm^3).
    (a) Write the balanced equation for burning octane in air, forming CO_2 and H_2O.
    (b) If the car has a fuel efficiency of 32 miles per gallon of octane, what volume of CO_2 at 25 °C and 1.0 atm is generated when the car goes on a 10.-mile trip?

61. Follow the directions in the previous question, but use methanol, CH_3OH ($d = 0.791$ g/cm^3) as the fuel. Assume the fuel efficiency is 20. miles per gallon.

62. Zinc reacts with aqueous sulfuric acid to form hydrogen gas.

$$Zn(s) + H_2SO_4(aq) \longrightarrow ZnSO_4(aq) + H_2(g)$$

In an experiment, 201 mL H_2 is collected over water at 27 °C and a barometric pressure of 733 torr. The vapor pressure of water at 27 °C is 26.74 torr. How many grams of H_2 are produced?

Gas Density and Molar Mass (Section 10.6)

63. What is the molar mass of a gas that has a density of 5.75 g/L at STP?

64. A sample of gaseous SiH_4 weighing 0.425 g is placed in a 580-mL container. The resulting pressure is 0.61 atm. What is the temperature in °C?

65. To find the volume of a flask, the flask is evacuated so it contains no gas. Next, 4.4 g CO_2 is introduced into the flask. On warming to 27 °C, the gas exerts a pressure of 730. mm Hg. What is the volume of the flask in milliliters?

66. What mass of helium in grams is required to fill a 5.0-L balloon to a pressure of 1.1 atm at 25 °C?

67. Consider two 5.0-L containers, each filled with gas at 25 °C. One container is filled with helium and the other with N_2. The density of gas in the two containers is the same. What is the relationship between the pressures in the two containers?

68. A hydrocarbon with the general formula C_xH_y is 92.26% carbon. Experiment shows that 0.293 g hydrocarbon fills a 185-mL flask at 23 °C with a pressure of 374 mm Hg. What is the molecular formula for this compound?

69. Forty miles above Earth's surface the temperature is −23 °C, and the pressure is only 0.20 mm Hg. What is the density of air (molar mass = 29.0 g/mol) at this altitude?

70. A newly discovered gas has a density of 2.39 g/L at 23.0 °C and 715 mm Hg. What is the molar mass of the gas?

Partial Pressures of Gases (Section 10.7)

71. What is the total pressure exerted by a mixture of 1.50 g H_2 and 5.00 g N_2 in a sealed 5.00-L vessel at 25 °C?

72. At 298 K, a sealed 750-mL vessel contains equimolar amounts of O_2, H_2, and He at a total pressure of 3.85 atm. What is the partial pressure of the H_2 gas?

73. A sample of the atmosphere at a total pressure of 740. mm Hg is analyzed to give these partial pressures: $P(N_2) = 575$ mm Hg; $P(Ar) = 6.9$ mm Hg; $P(CO_2) = 0.2$ mm Hg; $P(H_2O) = 4.0$ mm Hg. No other gases except O_2 have appreciable partial pressures.
    (a) What is the partial pressure of O_2?
    (b) What is the mole fraction of each gas?
    (c) What is the composition of this sample in percentage by volume? Compare your results with those of Table 10.1.

74. Gaseous CO exerts a pressure of 45.6 mm Hg in a 56.0-L tank at 22.0 °C. If this gas is released into a room with a volume of 2.70×10^4 L, what is the partial pressure of CO (in mm Hg) in the room at 22 °C?

75. Three flasks are connected as shown in the figure on p. 473. The starting conditions, with the stopcocks closed, are shown. Assume T does not change.
    (a) What is the final pressure inside the system when all the stopcocks are open?
    (b) What is the partial pressure of each of the three gases? Assume that the connecting tube has negligible volume.

76. The density of air at 20.0 km above Earth's surface is 92 g/m^3. The pressure is 42 mm Hg and the temperature is −63 °C. Assuming the atmosphere contains only O_2 and N_2, calculate
    (a) The average molar mass of the air at 20.0 km.
    (b) The mole fraction of each gas.

O_2
$V = 3.00$ L
$P = 1.46$ atm
N_2
$V = 2.00$ L
$P = 0.908$ atm
Ar
$V = 5.00$ L
$P = 2.71$ atm

77. Benzene has acute health effects. For example, it causes mucous membrane irritation at a concentration of 100 ppm and fatal narcosis at 20,000 ppm (by volume). Calculate the partial pressures in atmospheres at STP corresponding to these concentrations.
78. On a humid, rainy summer day the partial pressure of water vapor in the atmosphere can be as high as 25 mm Hg. What is the mole fraction of water vapor in the atmosphere under these conditions? Compare your results with the composition of dry air given in Table 10.1.
79. The mean fraction *by mass* of water vapor and cloud water in Earth's atmosphere is about 0.0025. Assume that the atmosphere contains two components: "air," with a molar mass of 29.2 g/mol, and water vapor. What is the *mean* mole fraction of water vapor in Earth's atmosphere? What is the *mean* partial pressure of water vapor? Why is this so much smaller than the typical partial pressure of water vapor at Earth's surface on a rainy summer day?
80. Acetylene can be made by reacting calcium carbide with water.

$$CaC_2(s) + 2\ H_2O(\ell) \longrightarrow C_2H_2(g) + Ca(OH)_2(aq)$$

Assume that you place 2.65 g CaC_2 in excess water and collect the acetylene over water. The volume of the acetylene and water vapor is 795 mL at 25.0 °C and a barometric pressure of 735.2 mm Hg. Calculate the percent yield of acetylene. The vapor pressure of water at 25 °C is 23.8 mm Hg.
81. Potassium chlorate, $KClO_3$, can be decomposed by heating.

$$2\ KClO_3(s) \longrightarrow 2\ KCl(s) + 3\ O_2(g)$$

If 465 mL gas was collected over water at a total pressure of 750. mm Hg and a temperature of 23 °C, how many grams of O_2 were collected?

The Behavior of Real Gases (Section 10.8)

82. From the density of liquid water and its molar mass, calculate the volume that 1 mol liquid water occupies. If water were an ideal gas at STP, what volume would a mole of water vapor occupy? Can we achieve the STP conditions for water vapor? Why or why not?
83. At high temperatures and low pressures, gases behave ideally, but as the pressure is increased the product *PV* becomes greater than the product *nRT*. Give a molecular-level explanation of this fact.
84. At low temperatures and very low pressures, gases behave ideally, but as the pressure is increased the product *PV* becomes less than the product *nRT*. Give a molecular-level explanation of this fact.
85. The densities of liquid noble gases and their normal boiling points are given in this table.

Gas	Normal Boiling Point (K)	Liquid Density (g/cm^3)
He	4.2	0.125
Ne	27.1	1.20
Ar	87.3	1.40
Kr	120.	2.42
Xe	165	2.95

Calculate the volume occupied by 1 mol of each of these liquids. Comment on any trend that you see. What is the volume occupied by exactly 1 mol of each of these substances as an ideal gas at STP? On the basis of these calculations, which gas would you expect to show the largest deviations from ideality at room temperature? Why?
86. Use the van der Waals constants in Table 10.5 to predict whether N_2 or CO_2 will behave more like an ideal gas at high pressures.
87. Calculate the pressure of 7.0 mol CO_2 in a sealed 2.00-L vessel at 50 °C using
(a) The ideal gas equation.
(b) The van der Waals equation.

Ozone and Stratospheric Ozone Depletion (Section 10.9)

88. Write the products for these reactions that take place in the stratosphere.
(a) $CF_3Cl \xrightarrow{h\nu}$ (b) $\cdot Cl + \cdot O\cdot \longrightarrow$
(c) $ClO\cdot + \cdot O\cdot \longrightarrow$
89. The molecule CF_4 is not implicated as having ozone-depletion potential. Explain why.
90. The molecule CH_3F has much less ozone-depletion potential than the corresponding molecule CH_3Cl. Explain why.
91. Write the chemical formulas of all possible CFCs with one carbon atom.
92. Are CFCs toxic? Compare the toxicity of CFCs with that of compounds used for refrigeration before CFCs were invented. Look up the toxicity of these compounds on the Internet.
93. Can CFCs catalyze the destruction of ozone in the stratosphere at night? Explain.
94. Can ozone form in the stratosphere at night? Explain why or why not.

Chemistry and Pollution in the Troposphere (Section 10.10)

95. Define air pollution in terms of the kinds of pollutants, their sources, and the ways they are harmful.
96. Explain how particulates can contribute to air pollution.
97. What is adsorption? What is absorption?

98. Assume that limestone, $CaCO_3$, is used to remove 90.% of the sulfur from 4.0 metric tons of coal containing 2.0% S. The product is $CaSO_4$.

$$CaCO_3(s) + SO_3(g) \longrightarrow CaSO_4(s) + CO_2(g)$$

Calculate the mass of limestone required. Express your answer in metric tons.

99. Approximately 65 million metric tons of SO_2 enter the atmosphere every year from the burning of coal. If coal, on average, contains 2.0% S, how many metric tons of coal were burned to produce this much SO_2? A 1000-MW power plant burns about 700. metric tons of coal per hour. Calculate the number of hours the quantity of coal will burn in one of these power plants.

100. What mass of gasoline must be burned according to the reaction

$$C_8H_{18}(\ell) + 8.5\ O_2(g) \longrightarrow 8\ CO(g) + 9\ H_2O(g)$$

to raise the CO concentration to 1000. ppm in a garage that measures 7.00 m × 3.00 m × 3.00 m? (Assume STP conditions.)

Urban Air Pollution (Section 10.10)

101. Give an example of a photochemical reaction. Do all photons of light have sufficient energy to cause photochemical reactions? Explain with examples.

102. What two reactions account for the production of ·O· free radicals in the troposphere?

103. The reducing nature of industrial (London) smog is due to what oxide? The burning of which two fuels produces this oxide? Write an equation showing this oxide being further oxidized.

104. Photochemical smog contains which two oxidizing gases? What is the energy source for photochemical smog?

105. What atmospheric reaction produces nitrogen monoxide, NO? Explain how the formation of NO in a combustion chamber is similar to the formation of NH_3 in a reactor designed to manufacture ammonia.

106. Give an example of a situation where atmospheric ozone is beneficial and an example of a situation where it is harmful. Explain how ozone is beneficial and how it is harmful.

107. The air pollutant sulfur dioxide, SO_2, is known to increase mortality in people exposed to it for 24 hours at a concentration of 0.175 ppm.
(a) What is the partial pressure of SO_2 when its concentration is 0.175 ppm?
(b) What is the mole fraction of SO_2 at the same concentration?
(c) Assuming the air is at STP, what mass in micrograms of SO_2 is present in 1 m^3?

Greenhouse Gases and Global Warning (Section 10.11)

108. What is the difference between the greenhouse effect and global warming? How are they related?

109. Name four greenhouse gases, and explain why they are called that.

110. Carbon dioxide is known to be a major contributor to the greenhouse effect. List some of its sources in our atmosphere and some of the processes that remove it. Currently, which predominates—the production of CO_2 or its removal?

111. Name a favorable effect of the global increase of CO_2 in the atmosphere.

General Questions

These questions are not explicitly keyed to chapter topics; many require integration of several concepts.

112. HCl can be made by the direct reaction of H_2 and Cl_2 in the presence of light. Assume that 3.0 g H_2 and 140. g Cl_2 are mixed in a 10-L flask at 28 °C, and the flask is sealed.
Before the reaction:
(a) What are the partial pressures of the two reactants?
(b) What is the total pressure due to the gases in the flask?
After the reaction:
(c) What is the total pressure in the flask?
(d) What reactant remains in the flask? How many moles of it remain?
(e) What are the partial pressures of the gases in the flask?
(f) What will be the pressure inside the flask if the temperature is increased to 40. °C?

113. Worldwide, about 100. million metric tons of H_2S are produced annually from sources that include the oceans, bogs, swamps, and tidal flats. One of the major sources of SO_2 in the atmosphere is the oxidation of H_2S, produced by the decay of organic matter. The reaction in which H_2S molecules are oxidized to SO_2 involves O_3. Write an equation showing that one molecule of each reactant combines to form two product molecules, one of them being SO_2. Then calculate the annual production in tons of H_2SO_4, assuming all of this SO_2 is converted to sulfuric acid.

Applying Concepts

These questions test conceptual learning.

114. Consider a sample of N_2 gas under conditions in which it obeys the ideal gas law exactly. Which of these statements is/are true?
(a) A sample of Ne(g) under the same conditions must obey the ideal gas law exactly.
(b) The speed at which one particular N_2 molecule is moving changes from time to time.
(c) Some N_2 molecules are moving more slowly than some of the molecules in a sample of O_2(g) under the same conditions.
(d) Some N_2 molecules are moving more slowly than some of the molecules in a sample of Ne(g) under the same conditions.
(e) When two N_2 molecules collide, it is possible that both may be moving faster after the collision than they were before.

115. Which graph below would best represent the distribution of molecular speeds for the gases acetylene, C_2H_2, and N_2? Both gases are in the same sealed flask with a total pressure of 750. mm Hg. The partial pressure of N_2 is 500. mm Hg.

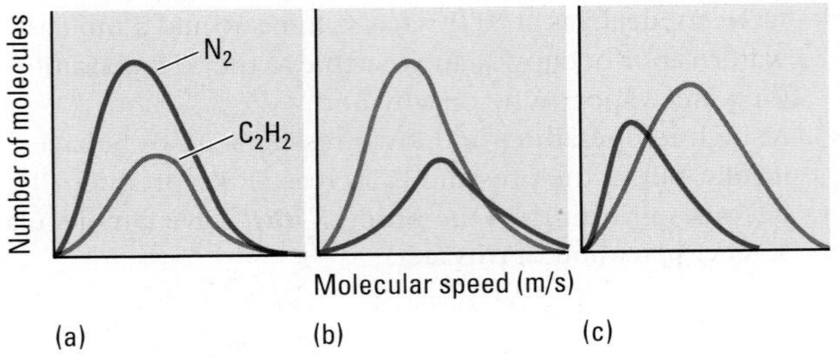

116. Draw a graph representing the distribution of molecular speeds for the gases ethane, C_2H_6, and F_2 when both are in the same sealed flask with a total pressure of 720. mm Hg and a partial pressure of 540. mm Hg for F_2.

117. In this chapter Boyle's, Charles's, and Avogadro's laws were presented as word statements and mathematical relationships. Express each of these laws graphically.

118. Consider these four samples of helium atoms, all at the same temperature. The larger boxes have twice the volume of the smaller boxes. Rank the gas samples with respect to: (a) pressure, (b) density, (c) average kinetic energy, (d) average molecular speed.

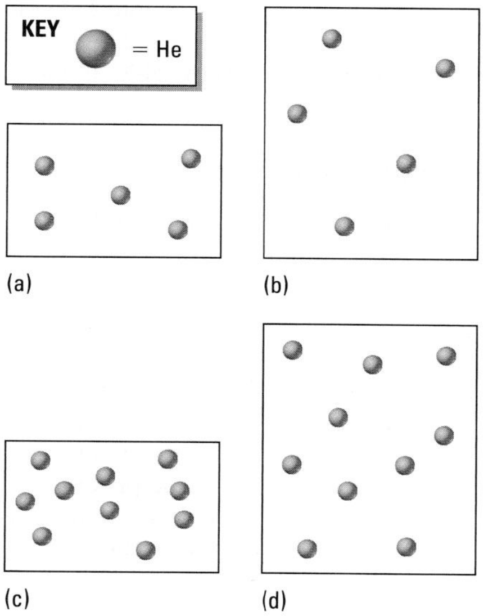

119. Consider these four gas samples, all at the same temperature. The larger boxes have twice the volume of the smaller boxes. Rank the gas samples with respect to: (a) pressure, (b) density, (c) average kinetic energy, (d) average molecular speed.

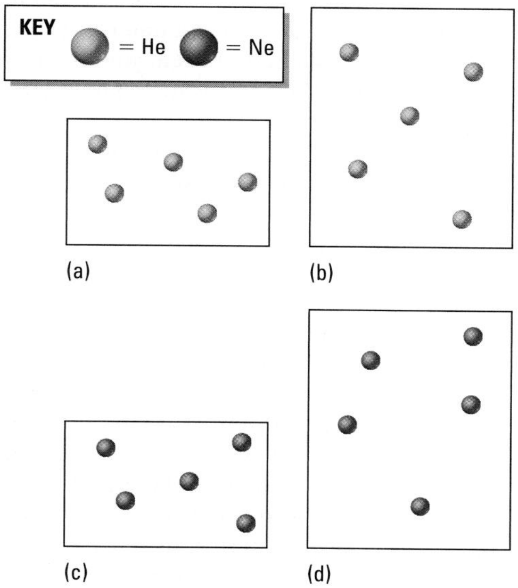

120. Consider these four gas samples, all at the same temperature. The larger boxes have twice the volume of the smaller boxes. Rank the gas samples with respect to: (a) pressure, (b) density, (c) average kinetic energy, (d) average molecular speed.

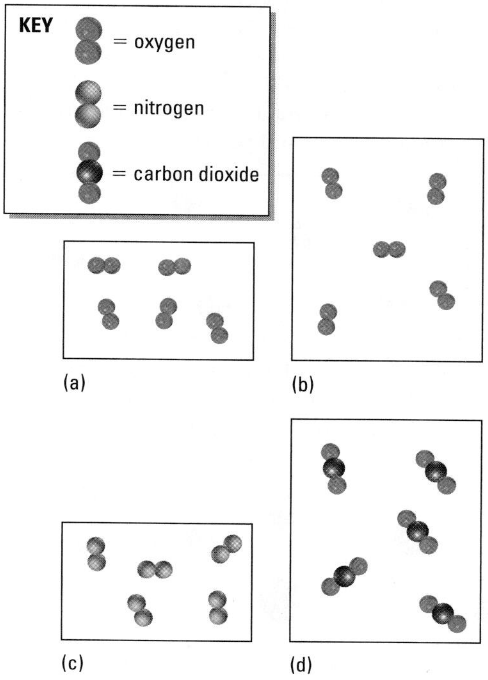

121. The drawing below represents a gas collected in a syringe (the needle end was sealed after collecting) at room temperature and pressure. Redraw the syringe and gas to show what it would look like under these conditions. Assume that the plunger can move freely, but no gas can escape.
(a) The temperature of the gas is decreased by one half.
(b) The pressure of the gas is decreased to one half of its initial value.
(c) The temperature of the gas is tripled and the pressure is doubled.

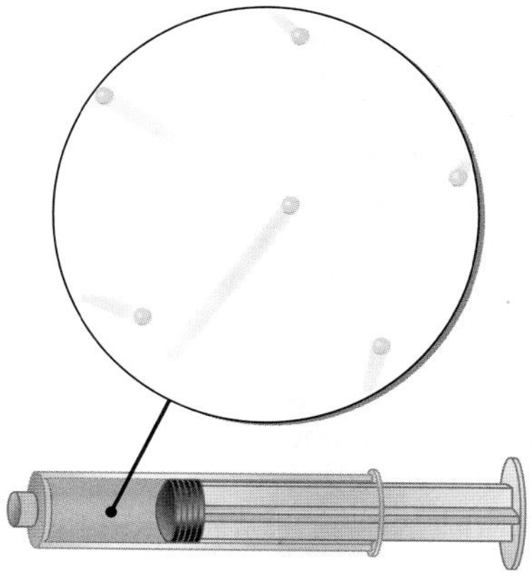

122. A gas phase reaction takes place in a syringe at a constant temperature and pressure. If the initial volume is 40. cm^3 and the final volume is 60. cm^3, which of these general reactions took place? Explain your reasoning.
(a) $A(g) + B(g) \rightarrow AB(g)$
(b) $2\ A(g) + B(g) \rightarrow A_2B(g)$
(c) $2\ AB_2(g) \rightarrow A_2(g) + 2\ B_2(g)$
(d) $2\ AB(g) \rightarrow A_2(g) + B_2(g)$
(e) $2\ A_2(g) + 4\ B(g) \rightarrow 4\ AB(g)$

123. The gas molecules in the box below undergo a reaction at constant temperature and pressure.

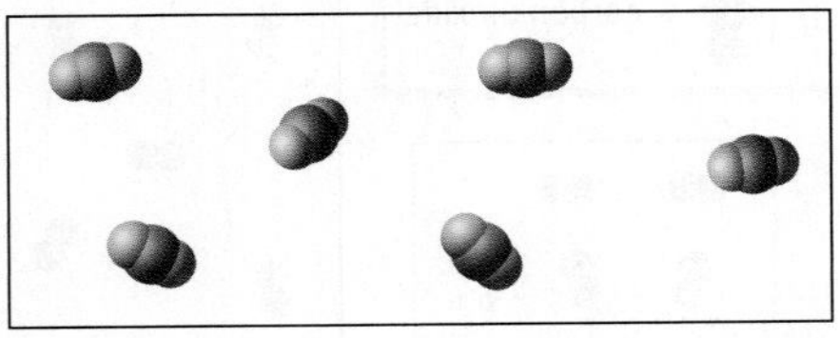

If the initial volume is 1.8 L and the final volume is 0.9 L, which of the boxes below could be the products of the reaction? Explain your reasoning.

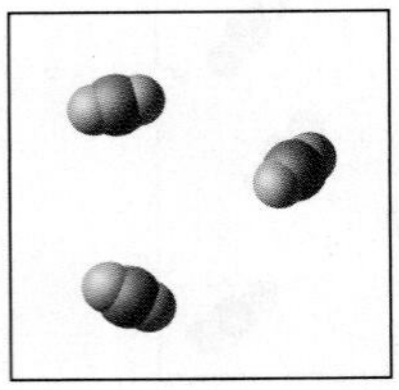

(a)

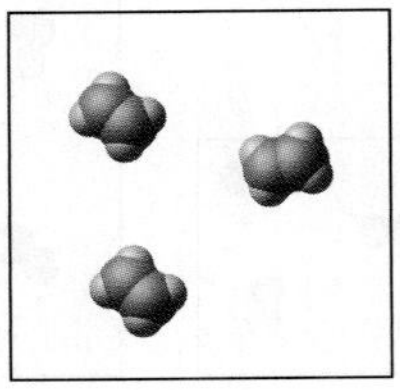

(b)

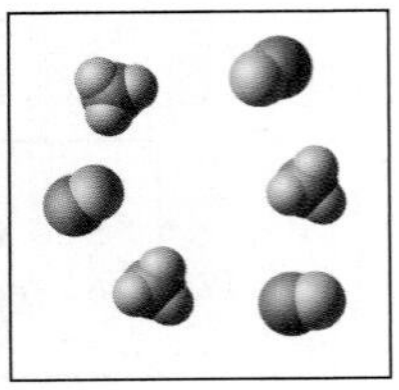

(c)

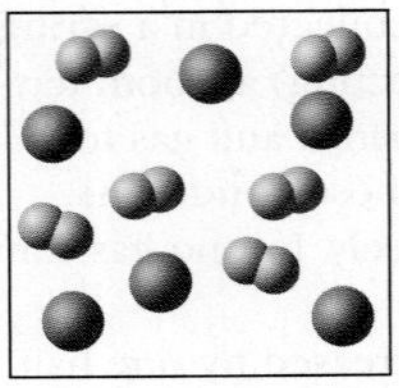

(d)

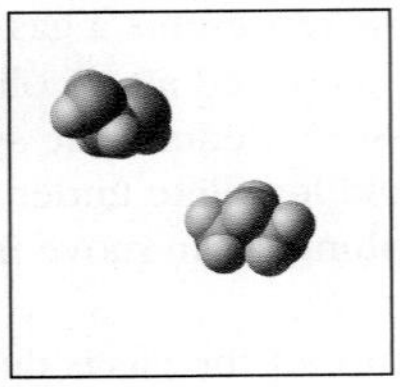

(e)

124. A substance is analyzed and found to contain 85.7% carbon and 14.3% hydrogen by mass. A gaseous sample of the substance is found to have a density of 1.87 g/L at STP.
(a) What is the molar mass of the compound?
(b) What are the empirical and molecular formulas of the compound?
(c) What are two possible Lewis structures for molecules of the compound?

125. A compound consists of 37.5% C, 3.15% H, and 59.3% F by mass. When 0.298 g of the compound is heated to 50. °C in an evacuated 125-mL flask, the pressure is observed to be 750. mm Hg. The compound has three isomers.
(a) What is the molar mass of the compound?
(b) What are the empirical and molecular formulas of the compound?
(c) Draw the Lewis structure for each isomer of the compound.

126. One very cold winter day you and a friend purchase a helium-filled balloon. As you leave the store and walk down the street, your friend notices the balloon is not as full as it was a moment ago in the store. He says the balloon is defective and he is taking it back. Do you agree with him? Explain why you do or do not agree.

127. For these forms of nitrogen most prevalent in the environment, write chemical formulas and give the oxidation number of the nitrogen.
(a) Ammonia
(b) Ammonium ion
(c) Dinitrogen monoxide
(d) Gaseous nitrogen
(e) Nitric acid
(f) Nitrous acid
(g) Nitrogen dioxide

128. Complete and balance these equations (all reactants and products are gases).
(a) $\cdot O \cdot + O_2 \longrightarrow$
(b) $O_3 \xrightarrow{h\nu}$
(c) $\cdot O \cdot + SO_2 \longrightarrow$
(d) $H_2O \xrightarrow{h\nu}$
(e) $\cdot NO_2 + \cdot NO_2 \longrightarrow$

129. Complete and balance these equations (all reactants and products are gases).
(a) $\cdot O \cdot + H_2O \longrightarrow$
(b) $\cdot NO_2 + \cdot OH \longrightarrow$
(c) $RH + \cdot OH \longrightarrow$

More Challenging Questions

These questions require more thought and integrate several concepts.

130. What is the total pressure in a sealed 8.7-L vessel that contains 3.44 g carbon dioxide, 1.88 g methane, and 4.28 g Ar, all at 37 °C?

131. An ideal gas was contained in a glass vessel of unknown volume with a pressure of 0.960 atm. Some of the gas was withdrawn from the vessel and used to fill a 25.0-mL glass bulb to a pressure of 1.00 atm. The pressure of the gas remaining in the vessel of unknown volume was 0.882 atm. All the measurements were done at the same temperature. What is the volume of the vessel?

132. You are holding two balloons, an orange balloon and a blue balloon, both at the same temperature and pressure. The orange balloon is filled with neon gas and the blue balloon is filled with argon gas. The orange balloon has twice the volume of the blue balloon. What is the mass ratio of Ne to Ar in the two balloons?

133. A sealed bulb filled with a gas is immersed in an ice bath, and its pressure is measured as 444 mm Hg. Then the same bulb is placed in a heated oven, and the pressure is measured as 648 mm Hg. What is the temperature of the oven in °C?

134. Nitroglycerine, the explosive ingredient in dynamite, decomposes violently when shocked.

$$4\ C_3H_5(NO_3)_3(\ell) \longrightarrow 12\ CO_2(g) + 10\ H_2O(\ell) + 6\ N_2(g) + O_2(g)$$

(a) How many total moles of gas are produced when 1.00 kg nitroglycerine explodes?
(b) What volume would the gases occupy at 1.0 atm and 25 °C?
(c) At P_{total} = 1.00 atm, what would be the partial pressure of each product gas?

135. Liquid oxygen, which has many industrial uses, is produced by cooling dry air to −183 °C. To produce 200. L liquid oxygen, how many liters of dry air at STP would be needed? (The density of liquid oxygen is 1.14 g/mL.)

136. A container of gas has a pressure of 550. torr. A chemical change then occurs that consumes half of the molecules present at the start and produces two new molecules for each three consumed. What is the new pressure in the container? T and V are unchanged.

137. A sealed balloon filled with helium contains 500. mol He at 1.00 atm and 25 °C.
(a) What is the volume of the balloon?
(b) What is the mass of air that it displaces? (Assume the molar mass of air to be 29.0 g/mol.)
(c) What is the mass of He in the balloon?
(d) What is the buoyancy of the balloon?

138. Imagine a tiny gas bubble with a volume of exactly one cubic millimeter being formed at the bottom of a lake at a pressure of 4.4 atm. When the bubble reaches the surface of the lake, where the pressure is 740. torr, what will its volume be? Assume that the temperature is constant.

139. Propane, $CH_3CH_2CH_3$, is a gaseous fuel that can be liquified when put under modest pressure.
(a) How many moles of propane can be stored in a container with a volume of 150. L at a pressure of 3.00 atm and a temperature of 25 °C?
(b) How many moles of liquid propane can be stored in a vessel of the same volume? (Liquid propane's density is approximately 0.60 g/mL.)

140. The effects of intermolecular interactions on gas properties depend on T and P. Are these effects more or less significant when these changes occur? Why?
(a) A sealed container of gas is compressed to a smaller volume at constant temperature.
(b) A container of gas has more gas added into the same volume at constant temperature.
(c) The gas in a container of variable volume is heated at constant pressure.

141. Formaldehyde, CH_2O, is a volatile organic compound that is sometimes released from insulation used in home construction, and it can be trapped and build up inside the home. When this happens, people exposed to the formaldehyde can suffer adverse health effects. The U.S. federal standard for the maximum allowable concentration of formaldehyde in air is 1.0 ppbv (parts per billion volume) defined as

$$\text{ppbv} = \frac{\text{volume of component X at STP}}{\text{total volume of air at STP}} \times 10^9$$

(a) What is the partial pressure of formaldehyde at the maximum allowable level of 1.0 ppbv?
(b) How many molecules of formaldehyde are present in each cubic centimeter of air when formaldehyde is present at 1.0 ppbv?
(c) How many total molecules of formaldehyde would be present in a room: 15.0 ft long × 10.0 ft wide × 8.00 ft high?

Conceptual Challenge Problems

These rigorous, thought-provoking problems integrate conceptual learning with problem solving and are suitable for group work.

CP10.A (Section 10.1) How would you quickly estimate whether the mass of Earth's atmosphere is 5.3×10^{15} metric tons?

CP10.B (Section 10.4) Under what conditions would you expect to observe that the pressure of a confined gas at constant temperature and volume is *not* constant?

CP10.C (Section 10.4) Suppose that the gas constant, R, were defined as 1.000 L atm mol^{-1} deg^{-1} where the "deg" referred to a newly defined Basic temperature scale. What would be the melting and boiling temperatures of water in degrees Basic (°B)?

11

Liquids, Solids, and Materials

Peter Jurs

An ice floe floats in a cold, northern lake with about 10% of the floe's mass above the surface of the surrounding water. As described in this chapter, the hydrogen bonding of water molecules in ice makes the solid slightly less dense than liquid water, an unusual property of water. This property of water, however, is crucial for life on Earth.

The properties of liquids and solids, like those of gases, can be understood on the molecular, nanoscale level. In common liquids and solids—water, salt crystals, metals—the molecules, ions, or atoms are close enough to have strong interactions with each other. The strength of these interactions accounts for the physical properties of liquids and solids. For example, the strong repulsions when the molecules, ions, or atoms in a liquid or solid are forced very close together account for the fact that solids and liquids can be compressed very little—in contrast to gases. Liquids are like gases in that their molecules, ions (in the case of molten ionic compounds), or atoms can move freely. Liquids are therefore able to flow and fill a container to a certain level. Solids, on the other hand, are composed of molecules, ions, or atoms in relatively fixed positions, which explains the fact that solids have definite shapes. When the molecules, ions, or atoms making up a solid have enough energy to overcome the attractive forces holding them in place, the solid melts or sublimes.

In molecular liquids and solids, molecules having sufficient energy can leave the surface of the liquid or solid and enter the gaseous state. In the reverse process, as temperature decreases, gases condense into liquids, and liquids solidify. All of these properties of solids and liquids have great practical importance.

Sublimation is the direct conversion of a solid to its vapor without becoming a liquid.

Some solids decompose when heated, so they do not get a chance to melt or sublime.

OWL Sign in to OWL at **www.cengage.com/owl** to view tutorials and simulations, develop problem-solving skills, and complete online homework assigned by your professor.

go Chemistry Download mini lecture videos for key concept review and exam prep from OWL or purchase them from **www.CengageBrain.com.**

Companion Website

Visit this book's companion website at **www.cengage.com/chemistry/moore** to work interactive modules for the Estimation boxes and Active Figures in this text.

11.1 The Liquid State

At low enough temperatures, gases condense to liquids. Condensation occurs when most molecules no longer have enough kinetic energy to overcome their intermolecular attractions ***(← p. 402)***. Most liquids, such as water, alcohol, and gasoline, are substances whose condensation temperatures are above room temperature. Some liquids, such as molten salts and liquid polymers, have no corresponding vapor state, so condensation of their vapors has no meaning.

In a liquid, the molecules are much closer together than are those in a gas. Nevertheless, the molecules remain mobile enough that the liquid flows. At the nanoscale, liquids have a regular structure only in very small regions, and most of the molecules continue to move about randomly. Because confined liquids are difficult to compress and their molecules move in all directions, they can transmit applied pressure equally in all directions. This property has a practical application in the hydraulic fluids that operate mechanical devices such as the hydraulic lifts that raise and lower autos in repair shops.

The resistance of a liquid to flow is called **viscosity.** Water flows smoothly and quickly (low viscosity), while motor oil and honey flow more slowly (higher viscosity; Figure 11.1). The viscosity of a liquid is related to its intermolecular forces, which determine how easily the molecules can move past each other, and also the degree to which the molecules can become entangled. Viscosity decreases as temperature increases because the molecules have more kinetic energy at higher temperature and the attractive intermolecular forces are more easily overcome. Examples of common materials whose viscosity decreases as their temperature rises include motor oil, cooking oil, and honey.

Unlike gases, liquids have *surface properties.* Molecules beneath the surface of the liquid are completely surrounded by other liquid molecules and experience intermolecular attractions in all directions. By contrast, molecules at the liquid surface have intermolecular interactions only with molecules below or beside them, but not above them (Figure 11.2). This unevenness of attractive forces at the liquid surface causes the surface to contract. The energy required to expand a liquid surface is called its **surface tension,** and it is higher for liquids that have stronger

Surface tension answers the question posed in Chapter 1 ***(← p. 2)***, "Why does rain fall as drops instead of cubes or cylinders?"

(a) Low compressibility. The hydraulic lifts are raised and lowered by pumping hydraulic fluid into and out of the vertical cylinders. Low compressibility of the liquid makes this application possible.

(b) Viscosity. Honey is viscous, so it builds up rather than spreading out as less viscous liquids would.

(c) Surface tension. A water bug is supported by the high surface tension of water. This answers the question posed in Chapter 1 ***(← p. 2)***, "Why can some insects walk on water?"

Figure 11.1 Properties of liquids.

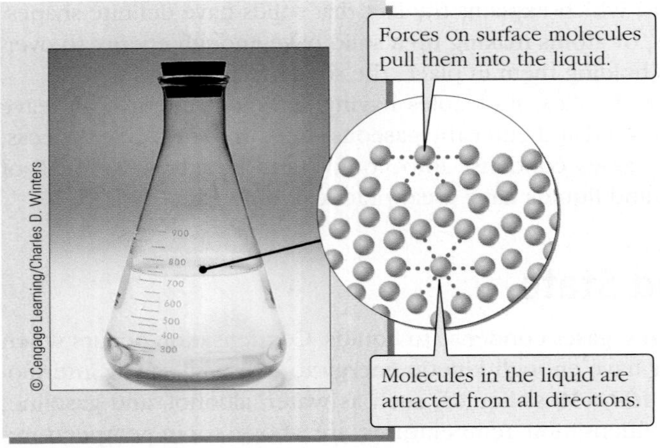
Forces on surface molecules pull them into the liquid.
Molecules in the liquid are attracted from all directions.
© Cengage Learning/Charles D. Winters

Figure 11.2 Surface tension. Surface tension is the energy required to increase the surface area of a liquid.

Table 11.1 Surface Tensions of Some Liquids

Name, Formula	Surface Tension (J/m^2 at 20 °C)
Octane, C_8H_{18}	2.16×10^{-2}
Ethanol, CH_3CH_2OH	2.23×10^{-2}
Chloroform, $CHCl_3$	2.68×10^{-2}
Benzene, C_6H_6	2.85×10^{-2}
Water, H_2O	7.29×10^{-2}
Mercury, Hg	46×10^{-2}

intermolecular attractions. For example, water's surface tension is high compared with those of most other liquids (Table 11.1) because extensive hydrogen bonding causes the water molecules to attract each other strongly (← p. 406). The surface tension of water, 7.29×10^{-2} J/m^2, is the quantity of energy required to increase the surface area of water by 1 m^2 at 20 °C. The very high surface tension of mercury (six times that of water) is due to the much stronger metallic bonding (Section 11.9) that holds Hg atoms together in the liquid.

Even though their densities are greater than that of water, water bugs can walk on the surface of water and small metal objects can float. Surface tension prevents the objects from breaking through the surface and sinking. Surface tension also accounts for the nearly spherical shape of rain droplets, as well as the rounded shape of water droplets that bead up on waxy surfaces such as leaves. A sphere has less surface area per unit volume than any other shape, so a spherical rain droplet has fewer surface H_2O molecules, minimizing the surface energy.

Answers to **EXERCISES** are provided at the back of this book in Appendix L.

EXERCISES that are labeled **CONCEPTUAL** are designed to test your understanding of one or more concepts; they usually involve qualitative rather than quantitative thinking.

CONCEPTUAL EXERCISE 11.1 Explaining Differences in Surface Tension

At 20 °C, chloroform, $CHCl_3$, has a surface tension of 2.68×10^{-2} J/m^2, while bromoform, $CHBr_3$, has a surface tension of 4.11×10^{-2} J/m^2. Explain this difference in terms of intermolecular forces.

CONCEPTUAL EXERCISE 11.2 Predicting Surface Tension

Predict which substance listed in Table 11.1 has a surface tension most similar to that of

(a) Glycerol, $C_3H_8O_3$, which has three —OH groups in each molecule

(b) Decane, $C_{10}H_{22}$

Explain your answer in each case.

The teardrop shape of raindrops you see in illustrations of rain bears little resemblance to actual raindrops, which are spherical. Air pressure begins to distort the sphere into an egg shape for 2- to 3-mm-radius raindrops. Larger raindrops assume other shapes, but they never look like teardrops.

The chemicals known as surfactants (soaps and detergents are examples) can dissolve in water and dramatically lower its surface tension. When this happens, water becomes "wetter" and does a better job of cleaning. You can see the effect of a surfactant by comparing how water alone beads on a surface such as the waxed hood of a car and how a soap solution fails to bead on the same surface. Surfactants are discussed in Section 15.10.

When a glass tube with a small diameter is put into water, the water rises in the tube due to **capillary action.** The glass walls of the tube are largely silicon dioxide, SiO_2, so the water molecules form hydrogen bonds to the oxygen atoms of the glass. This attractive force is stronger than the attractive forces among water molecules (also hydrogen bonding), so water creeps up the wall of the tube. Simultaneously, the surface tension of the water tries to keep the water's surface area small. The combination of the forces raises the water level in the tube. The liquid surface—concave upward for water—is called a ***meniscus*** (Figure 11.3a). The water rises in the tube until the force of the water-to-wall hydrogen bonds is balanced by the pull of gravity. Capillary action is crucial to plant life because it helps water, with its dissolved nutrients, to move upward through plant tissues against the force of gravity.

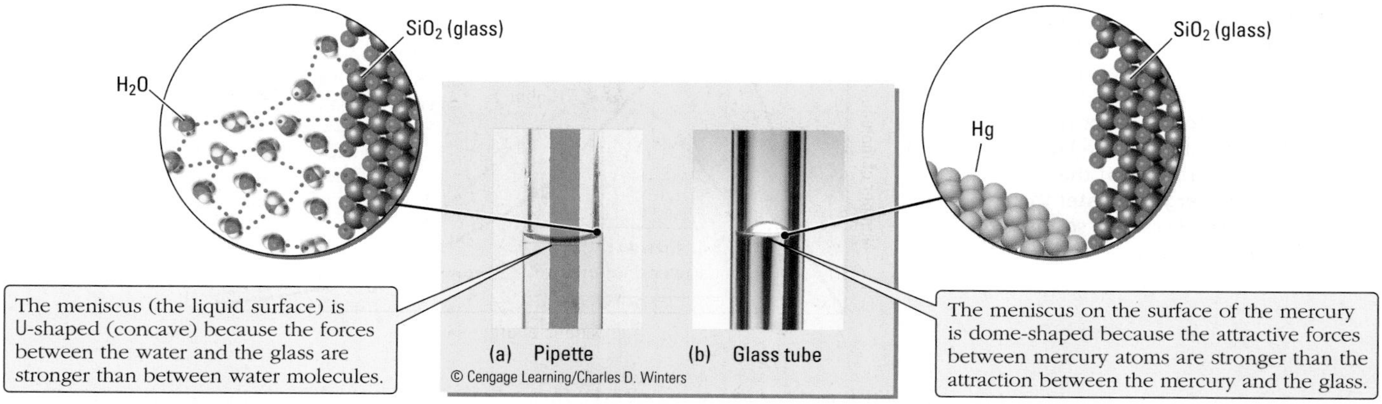

Figure 11.3 Noncovalent forces at surfaces. (a) The meniscus of water in a pipette is concave upward. (b) The meniscus of mercury in a glass tube is dome-shaped (convex upward).

Mercury behaves oppositely from water. Mercury in a small-diameter tube has a dome-shaped (convex upward) meniscus because the attractive forces between mercury atoms are stronger than the attraction between the mercury and the glass wall (Figure 11.3b).

11.2 Vapor Pressure

The tendency of a liquid to vaporize, its **volatility,** increases as temperature increases. Everyday experiences such as heating water or soup in a pan on a stove, or evaporation of rain from hot pavement, demonstrate this phenomenon. Conversely, the volatility of a liquid is lower at lower temperatures.

The change of volatility with temperature can be explained by considering what is happening at the nanoscale. The molecules in a liquid have varying kinetic energies and speeds (Figure 11.4). At any time, some fraction of the molecules have sufficient energy to escape from the liquid into the gas phase. When the temperature of the liquid is raised, a larger number of molecules exceeds the energy thresh-

CHEMISTRY IN THE NEWS Surface Tension and Bird Feeding

Surface tension allows insects to walk on water, droplets of water to form on the surface of your car, and a meniscus to form at the top of a water column in a tube. It also enables shorebirds with long, thin beaks to feed, according to a study by Manu Prakash and co-workers of the Massachusetts Institute of Technology.

The study focused on shorebirds named phalaropes. The birds feed by pecking, which captures small droplets of water (about 2 mm in diameter) that contain tiny invertebrates. Since the bird's beak is facing straight down, the researchers wondered how the droplets move upward toward the bird's mouth. The answer is that the birds open and close their beaks repeatedly, like tweezers, and this motion moves the droplets upward. When the beak is closing and the droplet is compressed, the inside edge of the droplet moves somewhat toward the bird's mouth, whereas the outside edge remains where it is. Then, when the beak is opening and the droplet is stretched, the inward edge remains where it is and the outward edge moves upward toward the bird's mouth. These incremental movements of the droplet, termed ratcheting, are aided by the surface tension of the water. The researchers studied movies of the birds and also generated a mathematical model that confirmed their observations. A video showing the ratcheting of the droplet is available at: **http://www.sciencemag.org/content/vol320/issue5878/images/data/931/DC1/1156023s2.mov**

From Prakash et al, SCIENCE 320:931 (2008) Photo Courtesy of Robert Lewis/John W. M. Bush. Reprinted with permission from AAAS.

Shorebird with inset showing close-up of beak with droplet of water.

Sources: Fountain, H., "Birds Rely on Surface Tension to Eat Dinner." *The New York Times*, May 20, 2008.

Prakash, M., Quere, D., and Bush, J. W. M. "Surface Tension Transport of Prey by Feeding Shorebirds: The Capillary Ratchet." *Science*, Vol. 230, May 16, 2008; p. 931.

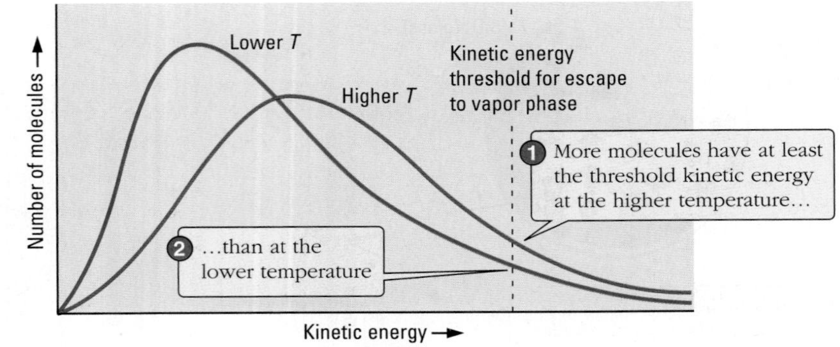

Figure 11.4 Kinetic energy and evaporation. At any given temperature, a fraction of the molecules at the surface of a liquid have sufficient energy to escape to the vapor phase. When the temperature of the liquid is raised, a larger fraction of the molecules have energies greater than the kinetic energy threshold, so the vapor pressure increases.

old, and evaporation proceeds more rapidly. Thus, hot water evaporates more quickly than cold water.

An evacuated container is specified so that only the pure liquid is present.

A liquid in an open container will eventually evaporate completely because air currents and diffusion take away most of the gas phase molecules before they can reenter the liquid phase. In a closed container, however, no molecules can escape. If a liquid is injected into an evacuated, closed container, the rate of vaporization (number of molecules entering the gas phase per unit time) will at first far exceed the rate of condensation. Over time the pressure of the gas above the liquid will increase as the number of gas-phase molecules increases. Eventually the system will attain a state of *dynamic equilibrium,* in which molecules are entering and leaving the liquid state at equal rates. At this point the pressure of the gas will no longer increase; this pressure is known as the **equilibrium vapor pressure** (or just the **vapor pressure**) of the liquid.

As shown in Figure 11.5 for three liquids, the vapor pressure of a liquid increases with increasing temperature. The differences in vapor pressures among the three liquids at the same temperature are due to differences in the strengths of the intermolecular interactions that attract their molecules together in the liquid phase. Consider the vapor pressures of ethyl alcohol and water at 60 °C (see Figure 11.5). At that temperature, ethyl alcohol has a higher vapor pressure, which means that it has weaker intermolecular interactions than those in water. Water's stronger intermolecular interactions (due to more hydrogen bonds) require a higher temperature to disrupt them and volatilize liquid water ***(← p. 406)***. At a particular temperature, the higher the vapor pressure of a liquid, the weaker its intermolecular forces, and the lower its boiling point.

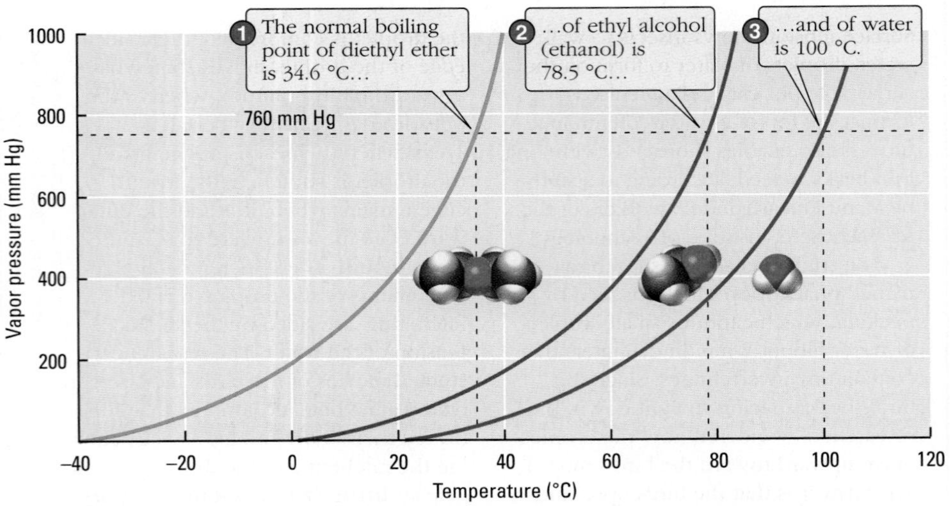

Figure 11.5 Vapor pressure curves for diethyl ether, $C_2H_5OC_2H_5$; ethyl alcohol, C_2H_5OH; and water. Each curve represents the conditions of T and P where the two phases (pure liquid and its vapor) are in equilibrium.

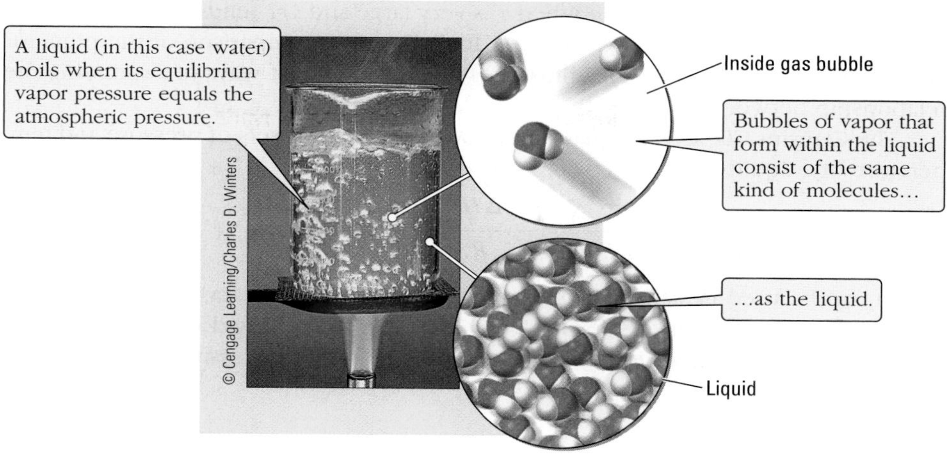

A liquid (in this case water) boils when its equilibrium vapor pressure equals the atmospheric pressure.
Inside gas bubble
Bubbles of vapor that form within the liquid consist of the same kind of molecules...
...as the liquid.
Liquid
© Cengage Learning/Charles D. Winters

Figure 11.6 Boiling liquid. A liquid is boiling when bubbles of vapor form throughout the liquid and rise to the surface. The gas bubbles contain the same kind of molecules that are in the liquid, but the molecules are much further apart.

If a liquid is placed in an open container and heated, a temperature eventually is reached at which the vapor pressure of the liquid is equal to the atmospheric pressure. Below this temperature, only molecules at the surface of the liquid can go into the gas phase. But when the vapor pressure equals the atmospheric pressure, the liquid begins vaporizing throughout. Bubbles of vapor form and immediately rise to the surface due to their lower density. The liquid is said to be **boiling** (Figure 11.6). The temperature at which the equilibrium vapor pressure equals the atmospheric pressure is the **boiling point** of the liquid. When the atmospheric pressure is 1 atm (760 mm Hg), the temperature is designated the **normal boiling point.** The temperatures where the vapor pressure curves cross the 760 mm Hg (1 atm) line are the normal boiling points of the three liquids (Figure 11.5).

To shorten cooking times, one can use a pressure cooker, a sealed pot (with a relief valve for safety) that allows water vapor to build up pressures slightly greater than the external atmospheric pressure. At the higher pressure, the boiling point of water is higher, and foods cook faster.

CONCEPTUAL EXERCISE 11.3 Estimating Boiling Points

Use Figure 11.5 to estimate the boiling points of: (a) ethyl alcohol at 400 mm Hg; (b) diethyl ether at 200 mm Hg; (c) water at 400 mm Hg.

CONCEPTUAL EXERCISE 11.4 Explaining Bubbles

One of your classmates believes that the bubbles in a boiling liquid are air bubbles. Explain to him what is wrong with that idea and what the bubbles actually are. Suggest an experiment to show that the bubbles are not air.

Clausius-Clapeyron Equation

As seen in Figure 11.5, plotting the vapor pressure of a substance versus its temperature results in a curved line. The relationship between vapor pressure and temperature is given by the **Clausius-Clapeyron equation:**

$$\ln P = \frac{-\Delta H_{vap}}{RT} + C = \left(\frac{-\Delta H_{vap}}{R}\right)\left(\frac{1}{T}\right) + C$$

which relates the natural logarithm of the vapor pressure P of the liquid to the absolute temperature T, the universal gas constant R (8.314 J mol^{-1} K^{-1}), the molar enthalpy of vaporization ΔH_{vap} (J/mol), and a constant C that is characteristic of the liquid. The form of the Clausius-Clapeyron equation shows that a graph of ln P versus $1/T$ is a straight line with its slope equal to $-\Delta H_{vap}/R$ and a y-axis intercept of C.

Note that R is in units of J mol^{-1} K^{-1} here.

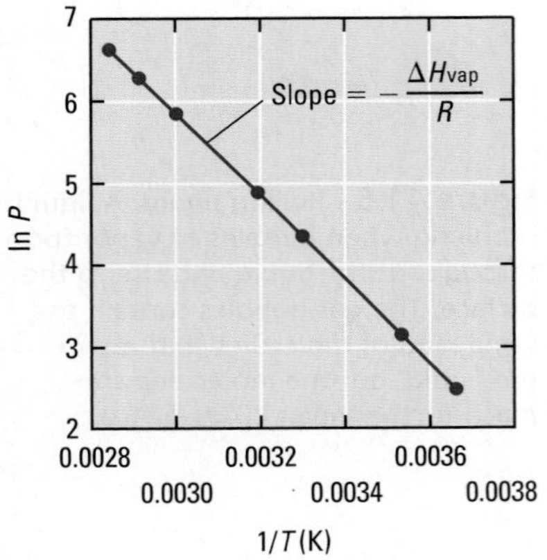

Figure 11.7 Clausius-Clapeyron equation plot for ethanol.

That is, ln P equals C in the limit when T is very large and $1/T$ tends toward zero. Figure 11.7 shows a plot of ln P versus $1/T$ for ethanol. A comparison of the plots for ethanol in Figures 11.5 and 11.7 illustrates the two different ways of expressing the relationship between T and P.

The Clausius-Clapeyron equation can be recast for two sets of pressure and temperature of a liquid to give

$$\ln\left(\frac{P_2}{P_1}\right) = \frac{-\Delta H_{vap}}{R}\left[\frac{1}{T_2} - \frac{1}{T_1}\right]$$

which can be used to calculate ΔH_{vap} from the two sets of data: T_1, T_2 and P_1, P_2. If ΔH_{vap} is known, then this form of the equation can be used to find the vapor pressure of a liquid at a new temperature given its vapor pressure at some other temperature.

The **PROBLEM-SOLVING STRATEGY** in this book is

- Analyze the problem
- Plan a solution
- Execute the plan
- Check that the result is reasonable

Appendix A.1 explains this in detail.

PROBLEM-SOLVING EXAMPLE 11.1 Clausius-Clapeyron Equation

Ethanol has a vapor pressure of 44.22 mm Hg at 20.0 °C and 350.8 mm Hg at 60.0 °C. Determine ΔH_{vap} for ethanol.

Answer 42.04 kJ/mol

Strategy and Explanation Use the Clausius-Clapeyron equation in the form that contains the four known quantities: T_1, T_2, P_1, and P_2. The temperatures must be converted to kelvins. Substitute the given values into the equation and solve for ΔH_{vap}.

$$\ln\left(\frac{350.8 \text{ mm Hg}}{44.22 \text{ mm Hg}}\right) = \frac{-\Delta H_{vap}}{8.314 \text{ J mol}^{-1}\text{ K}^{-1}}\left[\frac{1}{333.15 \text{ K}} - \frac{1}{293.15 \text{ K}}\right]$$

$$\ln(7.933) = \frac{-\Delta H_{vap}}{8.314 \text{ J mol}^{-1}\text{ K}^{-1}}(-4.096 \times 10^{-4}\text{ K}^{-1})$$

$$\Delta H_{vap} = \frac{\ln(7.933) \times 8.314 \text{ J mol}^{-1}\text{ K}^{-1}}{4.096 \times 10^{-4}\text{ K}^{-1}} = 42037 \text{ J/mol} = 42.04 \text{ kJ/mol}$$

PROBLEM-SOLVING PRACTICE answers are provided at the back of this book in Appendix K.

PROBLEM-SOLVING PRACTICE 11.1

At 30.0 °C the vapor pressure of carbon tetrachloride is 143.0 mm Hg. Determine its vapor pressure at 60.0 °C. The ΔH_{vap} of carbon tetrachloride is 32.1 kJ/mol.

Vapor Pressure of Water and Relative Humidity

Water has an appreciable vapor pressure in the temperature range 10 °C to 25 °C, normal outdoor temperatures. Therefore, water vapor is present in the atmosphere at all times. The equilibrium vapor pressure depends on the temperature as shown in Table 10.4 (← *p. 451*). The vapor pressure of water in the atmosphere can be expressed as the **relative humidity,** which is the ratio of the actual partial pressure of the water vapor in the atmosphere, P_{H_2O}, to the equilibrium partial pressure of water vapor at the relevant temperature, $P^{\circ}_{H_2O}$. The factor of 100 changes a fraction into a percentage.

$$\text{Relative humidity} = \frac{P_{H_2O}}{P^{\circ}_{H_2O}} \times 100\%$$

As an example, consider a typical warm day with an outdoor temperature of 25 °C. The equilibrium water vapor pressure at this temperature is 23.8 mm Hg (Table 10.5). If the actual partial pressure of water vapor were 17.8 mm Hg, then the relative humidity would be

$$\text{Relative humidity} = \frac{17.8 \text{ mm Hg}}{23.8 \text{ mm Hg}} \times 100\% = 75\%$$

CHEMISTRY IN THE NEWS Stopping Windshields from Fogging

When the humidity is very high, the windshield of your car may fog and reduce visibility. The same effect is seen when a bathroom mirror fogs when a hot shower is on. Fogging happens when warm, humid air contacts a colder surface. Water condenses on the cold surface and forms tiny droplets that scatter light, decreasing visibility. The droplets bead on the cold surface due to surface tension of the water. Cars overcome windshield fogging by directing a stream of warm air at it, evaporating the condensed water and improving visibility. If the glass surface were coated with a material that greatly reduced the surface tension so that the droplets couldn't form, then fogging would not occur.

Junhui He and co-workers at the Chinese Academy of Sciences in Beijing have developed a new anti-mist coating for glass. The new coating consists of nanoparticles that greatly reduce the surface tension of water, causing water to form a thin, transparent coating on the glass rather than droplets. After testing different materials and shapes and sizes of particles, the researchers found that silica-coated polymer spheres worked best. The researchers plan to develop a commercial process so that windshield fogging will no longer be a problem.

Source: "Not Misty for Me." *The Economist*, June 11, 2009.

If the temperature were lowered (which would lower the equilibrium vapor pressure of water) and the partial pressure of water vapor remained constant, the relative humidity would rise. Eventually, it would reach 100%. The temperature at which the actual partial pressure of water vapor equals the equilibrium vapor pressure is the **dew point.** At this temperature, water vapor will condense as water droplets to form fog or dew. When the dew point rises above about 20 °C we consider it to be somewhat muggy; when it rises above about 25 °C we consider the day to be extremely uncomfortable. In desert conditions the relative humidity can fall to less than 10%; this, too, can be uncomfortable for some people.

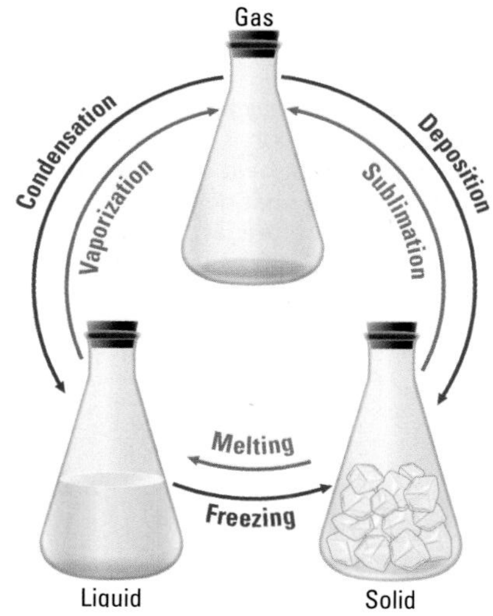

Figure 11.8 The three states of matter and their six interconversions. Red arrows indicate endothermic conversions—processes that require an input of energy. Blue arrows indicate exothermic conversions—processes that transfer energy to the surroundings.

11.3 Phase Changes: Solids, Liquids, and Gases

A substance in one state of matter—solid, liquid, or gas—can change to another. Water evaporates from the street after a rain shower. Ice cubes melt in your cold drink, and water vapor in the air condenses to water droplets on the outside wall of the glass. Each of the three states of matter can be converted to the other two states through six important processes (Figure 11.8).

Vaporization and Condensation

Like molecules in a gas, molecules in a liquid are in constant motion and have a range of kinetic energies, as shown in Figure 11.4. Some of the molecules in the liquid have enough kinetic energy to overcome the intermolecular attractive forces among the liquid molecules. If such a molecule is at the surface of the liquid and moving in the right direction, it will leave the liquid phase and enter the gas phase (Figure 11.9). This process is called **vaporization** or **evaporation.**

As the high-energy molecules leave the liquid, they take some energy with them. Therefore, vaporization is *endothermic;* the enthalpy change is called the enthalpy of vaporization ***(← p. 228).*** The rate of vaporization increases with temperature because an increasingly larger fraction of the molecules have enough energy to vaporize.

Enthalpy of vaporization is also referred to as heat of vaporization.

CONCEPTUAL EXERCISE 11.5 Evaporative Cooling

In some countries where electric refrigeration is not readily available, drinking water is chilled by placing it in porous clay water pots. Water slowly passes through the clay, and when it reaches the outer surface, it evaporates. Explain how this process cools the water inside.

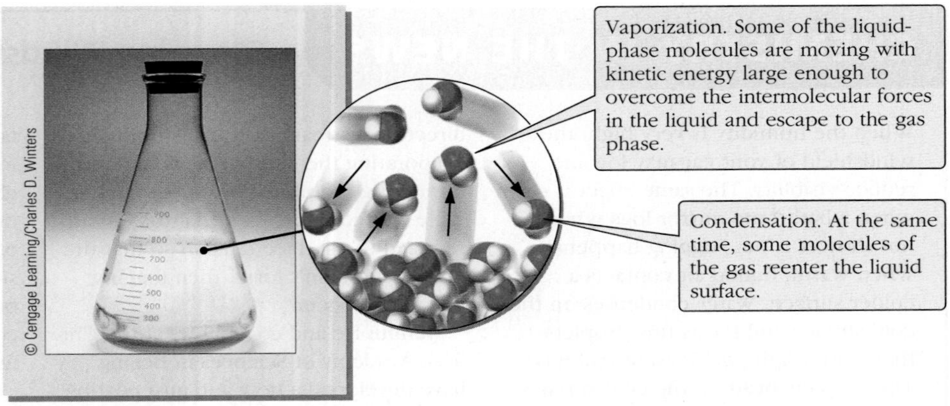

Figure 11.9 Molecules in the liquid and gas phases. All the molecules in both phases are moving, although the distances traveled in the liquid before collision with another molecule are much smaller.

When a molecule in the gas phase hits the liquid's surface, it can transfer some of its energy to the other liquid particles and remain in the liquid, a process called **condensation.** The overall effect of molecules reentering the liquid phase is a release of thermal energy, making condensation an exothermic process. The energy transferred out of the system upon condensation is equal to the energy transferred into the system upon vaporization.

The thermal energy absorbed during evaporation is sometimes called the latent heat of vaporization. It is somewhat dependent on the temperature: at 100 °C, ΔH°_{vap} $H_2O(\ell)$ = 40.7 kJ/mol, but at 25 °C the value is 44.0 kJ/mol.

$$\text{Liquid} \underset{\text{heat of condensation}}{\overset{\text{heat of vaporization}}{\rightleftharpoons}} \text{Gas} \qquad \Delta H_{\text{vaporization}} = -\Delta H_{\text{condensation}}$$

For example, the enthalpy change required to completely vaporize 1 mol liquid water once it has reached the boiling point of 100 °C at 1 bar (the *molar enthalpy of vaporization*) and the enthalpy change when 1 mol water vapor at 100 °C condenses to liquid water at 100 °C (the *molar enthalpy of condensation*) have the same values but opposite signs:

$$H_2O(\ell) \rightarrow H_2O(g) \qquad \Delta H^\circ = \Delta H_{vap} = +40.7 \text{ kJ/mol}$$

$$H_2O(g) \rightarrow H_2O(\ell) \qquad \Delta H^\circ = \Delta H_{cond} = -40.7 \text{ kJ/mol}$$

Table 11.2 illustrates the influence of noncovalent forces on enthalpies of vaporization and boiling points. In the series of nonpolar molecules and noble gases, the increasing noncovalent forces **(← p. 403)** with increasing molecular size and numbers of electrons are shown by the increasing ΔH°_{vap} and boiling point values. Comparison of HF and H_2O with CH_4, which have the same number of electrons, shows the effect of molecular polarity as well as hydrogen bonding.

PROBLEM-SOLVING EXAMPLE 11.2 Enthalpy of Vaporization

To vaporize 50.0 g carbon tetrachloride, $CCl_4(\ell)$, requires 9.69 kJ at its normal boiling point of 76.7 °C and 1 bar pressure. Calculate the molar enthalpy of vaporization of CCl_4 under these conditions.

Answer $\Delta H^\circ_{vap} = 29.8$ kJ/mol

Strategy and Explanation Use the molar mass of CCl_4 along with the given enthalpy change to calculate the enthalpy change per mole of CCl_4. The molar mass of CCl_4 is 153.8 g/mol, so

$$50.0 \text{ g } CCl_4 \times \frac{1 \text{ mol } CCl_4}{153.8 \text{ g } CCl_4} = 0.325 \text{ mol } CCl_4$$

Table 11.2 Molar Enthalpies of Vaporization and Boiling Points for Some Common Substances

Substance	Number of Electrons	ΔH°_{vap} (kJ/mol)*	Boiling Point (°C)†
Noble gases			
He	2	0.08	−269.0
Ne	10	1.8	−246.0
Ar	18	6.5	−185.9
Xe	54	12.6	−107.1
Nonpolar molecules			
H_2	2	0.90	−252.8
O_2	16	6.8	−183.0
F_2	18	6.54	−188.1
Cl_2	34	20.39	−34.6
Br_2	70	29.54	59.6
CH_4 (methane)	10	8.9	−161.5
$CH_3—CH_3$ (ethane)	18	15.7	−88.6
$CH_3—CH_2—CH_3$ (propane)	26	19.0	−42.1
$CH_3—CH_2—CH_2—CH_3$ (butane)	34	24.3	−0.5
Polar molecules			
HF	10	25.2	19.7
HCl	18	17.5	−84.8
HBr	36	19.3	−66.5
HI	54	21.2	−35.1
NH_3	10	25.1	−33.4
H_2O	10	40.7	100.0
SO_2	32	26.8	−10.0

*ΔH°_{vap} is given at the normal boiling point of the liquid.
†Boiling point is the temperature at which vapor pressure = 760 mm Hg

The enthalpy change per mole of $CCl_4(\ell)$ is given by

$$\Delta H^\circ_{vap} = \frac{9.69 \text{ kJ}}{0.325 \text{ mol}} = 29.8 \text{ kJ/mol}$$

☑ **Reasonable Answer Check** We have approximately one-third mol $CCl_4(\ell)$, so the enthalpy required to vaporize this amount of CCl_4 should be approximately one third of the molar enthalpy of vaporization, and it is.

PROBLEM-SOLVING PRACTICE 11.2

Using data from Table 11.2, calculate the thermal energy transfer required to vaporize 0.500 mol Br_2 at its normal boiling point.

PROBLEM-SOLVING EXAMPLE 11.3 Enthalpy of Vaporization

You empty the contents of a one-liter bottle of water into a pan at 100 °C and all the water evaporates. How much thermal energy was required to evaporate this water at 100 °C and 1 bar? The density of liquid water at 100 °C is 0.958 g/mL.

Answer 2.16×10^3 kJ

Strategy and Explanation Three properties of water are needed to solve this problem:
(a) $\Delta H^\circ_{vap} = 40.7$ kJ/mol for water at 100 °C (Table 11.2).
(b) The density of water at 100 °C is 0.958 g/mL.
(c) The molar mass of water is 18.02 g/mol.

First, determine the mass of 1.00 L water and then the molar amount of 1.00 L water.

$$1.00 \times 10^3 \text{ mL water} \times \frac{0.958 \text{ g}}{1 \text{ mL}} \times \frac{1 \text{ mol}}{18.02 \text{ g}} = 53.16 \text{ mol water}$$

Now calculate the heat required for vaporization.

$$53.16 \text{ mol water} \times \frac{40.7 \text{ kJ}}{1 \text{ mol}} = 2164 \text{ kJ} = 2.16 \times 10^3 \text{ kJ}$$

This quantity of thermal energy is equivalent to approximately one fourth the energy supplied by the daily food intake of an average person in the United States.

☑ **Reasonable Answer Check** One liter of water is about 1 kg water. The molar mass is about 20 g/mol, so we have about $\frac{1000 \text{ g}}{20 \text{ g/mol}} = 50$ mol water. Each mole of water requires about 40 kJ to evaporate, so 50 mol × 40 kJ/mol is about 2000 kJ, which is close to our more carefully calculated answer. The answer is reasonable.

PROBLEM-SOLVING PRACTICE **11.3**

How many grams of ethanol, CH_3CH_2OH, can be vaporized at its boiling point of 78.4 °C with 500. kJ thermal energy? The ΔH°_{vap} of ethanol is 38.6 kJ/mol at this temperature.

CONCEPTUAL EXERCISE **11.6 Understanding Boiling Points**

(a) Chlorine and bromine are both diatomic. Explain the difference in their boiling points (Table 11.2).
(b) Methane and ammonia have the same number of electrons. Explain the difference in their boiling points (Table 11.2).

Melting and Freezing

When a solid is heated, its temperature increases until the solid begins to melt. Unless the solid decomposes first, a temperature is reached at which the kinetic energies of the molecules or ions are sufficiently high that the intermolecular interactions in the solid are no longer strong enough to keep the particles in their fixed positions. The solid's nanoscale structure collapses, and the solid melts (Figure 11.10). This temperature is the *melting point* of the solid. Melting requires transfer of energy, the *enthalpy of fusion,* from the surroundings into the system, so it is always endothermic. The *molar enthalpy of fusion* **(← p. 227)** is the enthalpy change required to melt 1 mol of a pure solid. Solids with high enthalpies of fusion usually melt at high temperatures, and solids with low enthalpies of fusion usually melt at low temperatures. The reverse of melting—called *solidification, freezing,* or **crystallization**—is always an exothermic process. The *molar enthalpy of crystallization* has the same magnitude as the molar enthalpy of fusion, but the opposite sign.

Some solids do not have measurable melting points, and some liquids do not have measurable boiling points, because increasing temperature causes them to decompose before they melt or boil. Making peanut brittle provides an example: Melted sucrose chars before it boils, but produces great-tasting candy when melted correctly.

$$\text{Solid} \underset{\text{enthalpy of crystallization}}{\overset{\text{enthalpy of fusion}}{\rightleftharpoons}} \text{Liquid} \qquad \Delta H_{\text{fusion}} = -\Delta H_{\text{crystallization}}$$

Table 11.3 lists melting points and enthalpies of fusion for examples of three classes of compounds: (a) nonpolar molecular solids, (b) polar molecular solids,

Table 11.3 Melting Points and Enthalpies of Fusion of Some Solids

Solid	Melting Point (°C)	Enthalpy of Fusion (kJ/mol)	Intermolecular Noncovalent Forces
Molecular solids: Nonpolar molecules			
O_2	−248	0.445	These molecules have only London forces (which increase with the number of electrons).
F_2	−220	1.020	
Cl_2	−103	6.406	
Br_2	−7.2	10.794	
Molecular solids: Polar molecules			
HCl	−114	1.990	All of these molecules have London forces enhanced significantly by dipole-dipole forces. H_2O also has significant hydrogen bonding.
HBr	−87	2.406	
HI	−51	2.870	
H_2O	0	6.020	
H_2S	−86	2.395	
Ionic solids			
NaCl	800	30.21	All ionic solids have strong attractions between oppositely charged ions.
NaBr	747	25.69	
NaI	662	21.95	

some capable of hydrogen bonding, and (c) ionic solids. Solids composed of low-molecular-weight nonpolar molecules have the lowest melting temperatures, because their intermolecular attractions are weakest. These molecules are held together by London forces only ***(← p. 403)***, and they form solids with melting points so low that we seldom encounter them in the solid state at normal temperatures. Melting points and enthalpies of fusion of nonpolar molecular solids increase with increasing number of electrons (which corresponds to increasing molar mass) as the London forces become stronger. The ionic compounds in Table 11.3 have the highest melting points and enthalpies of fusion because of the very strong ionic bonding that holds the oppositely charged ions together in the solid. The polar molecular solids have intermediate melting points.

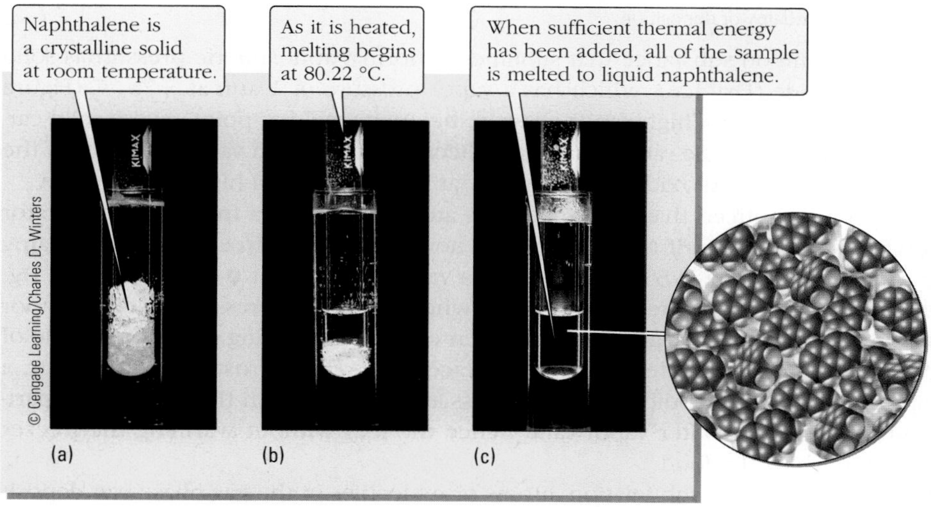
Naphthalene is a crystalline solid at room temperature.
As it is heated, melting begins at 80.22 °C.
When sufficient thermal energy has been added, all of the sample is melted to liquid naphthalene.
KIMAX
KIMAX
(a)
(b)
(c)
© Cengage Learning/Charles D. Winters

Figure 11.10 **The melting of naphthalene, $C_{10}H_8$, at 80.22 °C.**

PROBLEM-SOLVING EXAMPLE 11.4 Enthalpy of Fusion

The molar enthalpy of fusion of NaCl is 30.21 kJ/mol at its melting point. How much thermal energy will be transferred into the system when 10.00 g NaCl melts?

Answer 5.169 kJ

Strategy and Explanation Use the molar mass of NaCl and the given molar enthalpy of fusion, ΔH_{fusion}, to calculate the required enthalpy change. The molar mass of NaCl is 58.443 g/mol, so

$$10.00 \text{ g NaCl} \times \frac{1 \text{ mol NaCl}}{58.443 \text{ g NaCl}} = 0.1711 \text{ mol NaCl}$$

Now we can use the molar enthalpy of fusion to calculate the enthalpy change to melt this sample of NaCl.

$$0.1711 \text{ mol NaCl} \times \frac{30.21 \text{ kJ}}{1 \text{ mol NaCl}} = 5.169 \text{ kJ}$$

☑ **Reasonable Answer Check** We have approximately one-sixth mol NaCl, so the enthalpy change required to melt this amount of NaCl should be approximately one sixth of the molar enthalpy of fusion, and it is.

PROBLEM-SOLVING PRACTICE 11.4

Calculate the thermal energy transfer required to melt 0.500 mol NaI at its normal melting point.

EXERCISE 11.7 Heat Liberated upon Crystallization

Which would transfer more thermal energy to the surroundings, the crystallization of 2 mol liquid bromine or the crystallization of 1 mol liquid water?

© Cengage Learning/Charles D. Winters

Figure 11.11 Dry Ice. The cold vapors of Dry Ice, CO_2, cause moisture to condense into tiny droplets of liquid water that appear as wispy white clouds. The CO_2 vapors are denser than air at the same temperature, so they carry the cloud of water droplets downward to the table top. This answers the question posed in Chapter 1 (← p. 3), "Why is Dry Ice always surrounded by fog?"

"Freezer burn" refers to loss of water by sublimation from the surface of food stored in a freezer.

Sublimation and Deposition

Atoms or molecules can escape directly from the solid to the gas phase, a process known as **sublimation.** The enthalpy change is the *enthalpy of sublimation.* The reverse process, in which a gas is converted directly to a solid, is called **deposition.** The enthalpy change for this exothermic process (the enthalpy of deposition) has the same magnitude as the enthalpy of sublimation, but the opposite sign.

$$\text{Solid} \underset{\text{enthalpy of deposition}}{\overset{\text{enthalpy of sublimation}}{\rightleftharpoons}} \text{Gas} \qquad \Delta H_{sublimation} = -\Delta H_{deposition}$$

One common substance that sublimes at normal atmospheric pressure is solid carbon dioxide (Dry Ice), which has a vapor pressure of 1 atm at −78 °C (Figure 11.11). Having such a high vapor pressure below its melting point causes solid carbon dioxide to sublime rather than melt. Because of the high vapor pressure of the solid, liquid carbon dioxide can exist only at pressures much higher than 1 atm.

Have you noticed that snow outdoors and ice cubes in a frost-free refrigerator slowly disappear even if the temperature never gets above freezing? The enthalpy of sublimation of ice is 51 kJ/mol, and its vapor pressure at 0 °C is 4.60 mm Hg. Therefore, ice sublimes readily in dry air when the partial pressure of water vapor is below 4.60 mm Hg (Figure 11.12). Given enough air passing over it, a sample of ice will sublime completely, leaving no trace behind. In a frost-free refrigerator, a current of dry air periodically blows across any ice formed in the freezer compartment, taking away water vapor (and hence the ice) without warming the freezer enough to thaw the food.

In the reverse of sublimation, atoms or molecules in the gas phase can deposit (solidify directly) on the surface of a solid. Deposition is used to form thin coatings

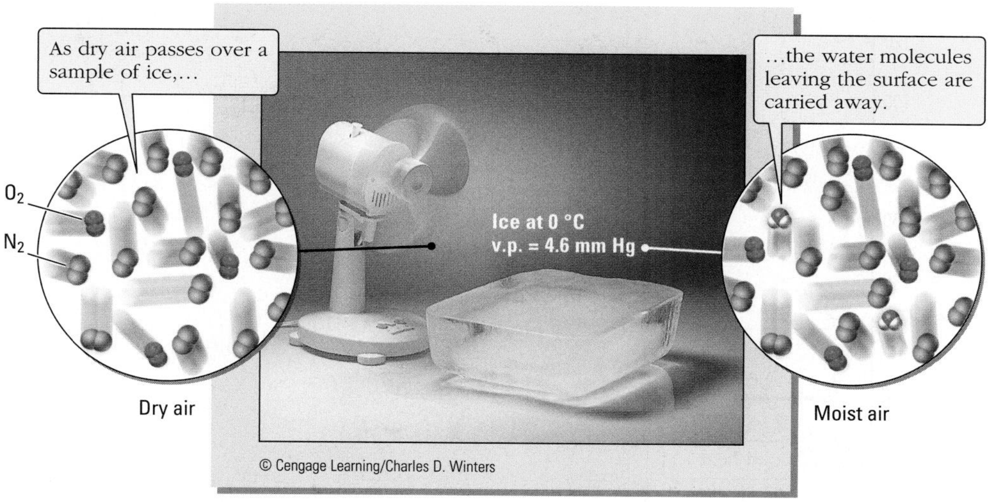

Figure 11.12 Ice subliming.

of metal atoms on surfaces. Audio CD discs and CD-ROM discs, for example, have shiny metallic surfaces of deposited aluminum or gold atoms. To make such discs, a metal filament is heated in a vacuum to a temperature at which metal atoms begin to sublime rapidly from the surface of the filament. The plastic compact disc is cooler than the filament, so the metal atoms in the gas phase quickly deposit on the cool surface. The purpose of the metal coating is to provide a reflective surface for the laser beam that reads the pits and lands (unpitted areas) containing the digital audio or data information ***(← p. 285)***.

CONCEPTUAL EXERCISE 11.8 Frost-Free Refrigeration

Sometimes, because of high humidity, a frost-free refrigerator doesn't work as efficiently as it should. Explain why.

CONCEPTUAL EXERCISE 11.9 Purification by Sublimation

Sublimation is an excellent means of purification for compounds that will readily sublime. Explain how purification by sublimation works at the nanoscale.

Heating Curve

Heating a solid or liquid increases its temperature as long as its phase does not change and the sample does not decompose. The size of this temperature increase is governed by the quantity of energy added (in joules), the mass of the substance (in grams), and its specific heat capacity ($J\ g^{-1}\ °C^{-1}$) ***(← p. 221)***. The temperature rises until a phase-transition temperature is reached—for example, 100 °C for liquid water. At this point, the temperature no longer rises. Instead, the added energy goes into changing the phase of the substance. Once all of the substance has changed phase, then the addition of still more energy causes the temperature to rise. A plot of the temperature of a substance versus the quantity of energy added is called a **heating curve** (Figure 11.13).

For example, consider how much thermal energy is required to heat 100. g water from −20 °C to 120 °C, as illustrated in Figure 11.13. These constants are used to construct the heating curve: specific heat capacity of liquid water = $4.184\ J\ g^{-1}\ °C^{-1}$;

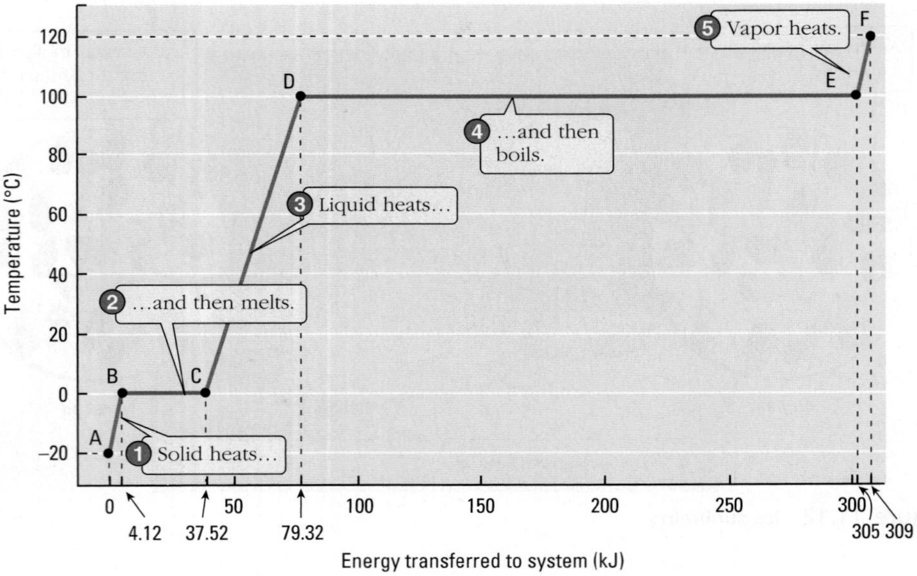

Notice that there is no temperature change during a phase change.

Figure 11.13 Heating curve for water.

specific heat capacity of solid water = 2.06 J g^{-1} $°C^{-1}$; specific heat capacity of water vapor = 1.84 J g^{-1} $°C^{-1}$; $\Delta H°_{fusion}$ = 6.020 kJ/mol; $\Delta H°_{vaporization}$ = 40.7 kJ/mol.

This heating curve for water has five distinct parts. Three parts relate to heating the water in its three states (red lines). Two parts relate to phase changes (horizontal blue lines).

1. *Heat the ice from −20 °C to 0 °C (A → B).*

$$H_2O(s) \text{ at } -20\ °C \rightarrow H_2O(s) \text{ at } 0\ °C$$

$$\Delta H° = (100.\ g)(2.06\ J\ g^{-1}\ °C^{-1})(20.0\ °C) = 4120\ J = 4.12\ kJ$$

2. *Melt the ice at 0 °C (B → C).*

$$H_2O(s) \text{ at } 0\ °C \rightarrow H_2O(\ell) \text{ at } 0\ °C$$

$$\Delta H° = 100.\ g \times \frac{1\ mol}{18.015\ g} \times \frac{6.020\ kJ}{1\ mol} = 33.4\ kJ$$

3. *Heat the water from 0 °C to 100 °C (C → D).*

$$H_2O(\ell) \text{ at } 0\ °C \rightarrow H_2O(\ell) \text{ at } 100\ °C$$

$$\Delta H° = (100.\ g)(4.184\ J\ g^{-1}\ °C^{-1})(100.\ °C) = 41{,}800\ J = 41.8\ kJ$$

4. *Boil the water at 100 °C (D → E).*

$$H_2O(\ell) \text{ at } 100\ °C \rightarrow H_2O(g) \text{ at } 100\ °C$$

$$\Delta H° = 100.\ g \times \frac{1\ mol}{18.015\ g} \times \frac{40.7\ kJ}{1\ mol} = 226\ kJ$$

5. *Heat the water vapor from 100 °C to 120 °C (E → F).*

$$H_2O(g) \text{ at } 100\ °C \rightarrow H_2O(g) \text{ at } 120\ °C$$

$$\Delta H° = (100.\ g)(1.84\ J\ g^{-1}\ °C^{-1})(20.0\ °C) = 3680\ J = 3.68\ kJ$$

The total energy required to complete the transformation is 309 kJ, the sum of the five steps. Notice that the largest portion of the energy, 226 kJ (73%), goes into vaporizing water at 100 °C to steam at 100 °C.

Phase Diagrams

The three phases (states of matter) and the six interconversions among them can be represented in a **phase diagram** (Figure 11.14). The phase diagram contains much information:

- Each of the three phases—gas, liquid, solid—is represented by an area in the diagram.
- The areas represent the range of temperature and pressure at which each phase exists.
- Each solid line in the phase diagram represents the conditions of temperature and pressure at which the two phases on either side of the line coexist in equilibrium.
- The line between liquid and solid regions is the melting point curve.
- The line between the gas and liquid regions is the vapor pressure curve.
- The **triple point** is the temperature-pressure value at which all three phases are in equilibrium.
- The gas/liquid curve ends at the **critical point**, where the liquid and gas phases become indistinguishable.

Every pure substance that exists in all three phases has a characteristic phase diagram.

The phase diagram for water is shown in Figure 11.15 (p. 494). The temperature scale has been exaggerated to better illustrate some of the features of water's phase diagram. The *vapor pressure curve,* the line *AD*, is the same one shown in Figure 11.5 (p. 482). The triple point for water occurs at $P = 4.58$ mm Hg and $T = 0.01$ °C.

You are most likely familiar with the common situation in which ice and liquid water are in equilibrium because of keeping food cool in an ice chest. If you fill an ice chest with ice, some of the ice melts to produce liquid. (Thermal energy is slowly transferred into the ice chest, no matter how well it is insulated.) The ice-liquid water mixture remains at approximately 0 °C until all of the ice has melted. This equilibrium between solid water and liquid water is represented in the phase diagram (Figure 11.15) by point *C*, at which $P = 760$ mm Hg and $T = 0$ °C.

On a very cold day, ice can have a temperature well below 0 °C. Look along the temperature axis of the phase diagram for water and notice that below 1 atm and 0 °C the only equilibrium possible is between ice and water vapor. As a result, if the

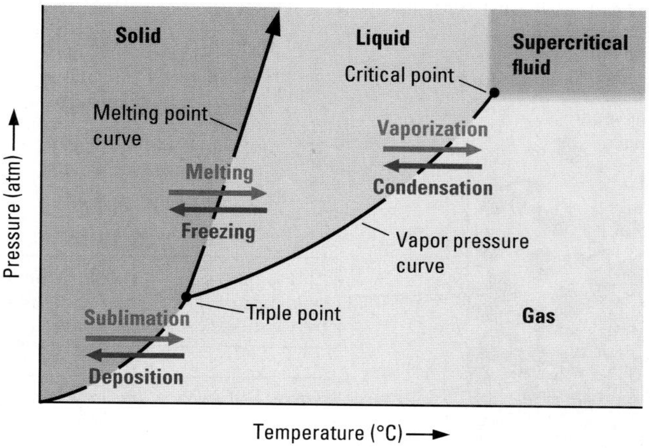

Figure 11.14 Generic phase diagram. The phase diagram shows the pressure-temperature regions in which the substance is solid, liquid, or gas (vapor). It also shows the melting point curve, the vapor pressure curve, and the six interconversions among the three phases.

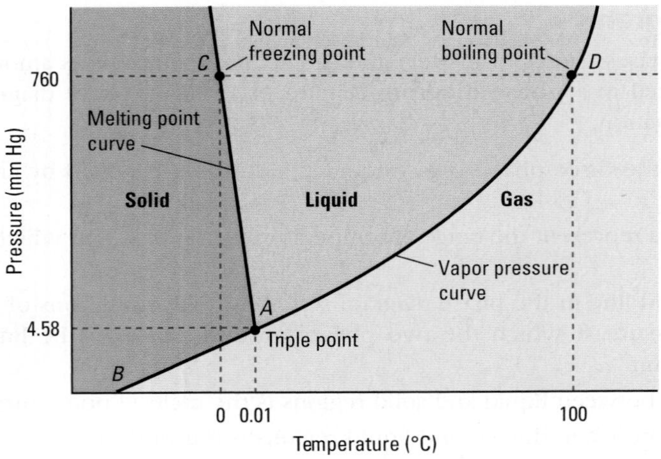

Figure 11.15 The phase diagram for water. The temperature and pressure scales are nonlinear for emphasis. Note that the melting point curve, *AC*, has a negative slope, a very unusual property of water. (Axes not to scale.)

partial pressure of water in the air is low enough, ice will sublime on a cold day. Sublimation, which is endothermic, takes place more readily when solar radiation provides energy to the ice. The sublimation of ice allows snow or ice to gradually disappear even though the temperature does not climb above freezing.

The phase diagram for water is unusual in that the melting point curve *AC* slopes in the opposite direction from that seen for almost every other substance. The right-to-left, or negative, slope of the solid/liquid water equilibrium line is a consequence of the lower density of ice compared to that of liquid water (which is discussed in the next section). Thus, when ice and water are in equilibrium, one way to melt the ice is to apply greater pressure. This is evident from Figure 11.15. If you start at the normal freezing point (0 °C, 1 atm) and increase the pressure, you will move into the area of the diagram that corresponds to liquid water.

It was long thought that one's body weight applied to a very small area of an ice-skating blade caused sufficient pressure to form liquid water, thus making the ice under the blade slippery. However, more recent investigations show that the water molecules on the surface of ice are actually moving much more than in the main part of the solid—nearly as much as they do in liquid water. This is due to "surface melting" (Figure 11.16), which arises because the surface water molecules have interactions with fewer neighboring water molecules than do those molecules deeper in the solid. The ice lattice becomes less and less ordered closer to the surface because the surface molecules have the fewest hydrogen bonds holding them in place. Ice is made much more slippery when a thin coating of liquid water is present on its surface. This thin coat of liquid water acts as a lubricant to help the skates move easily

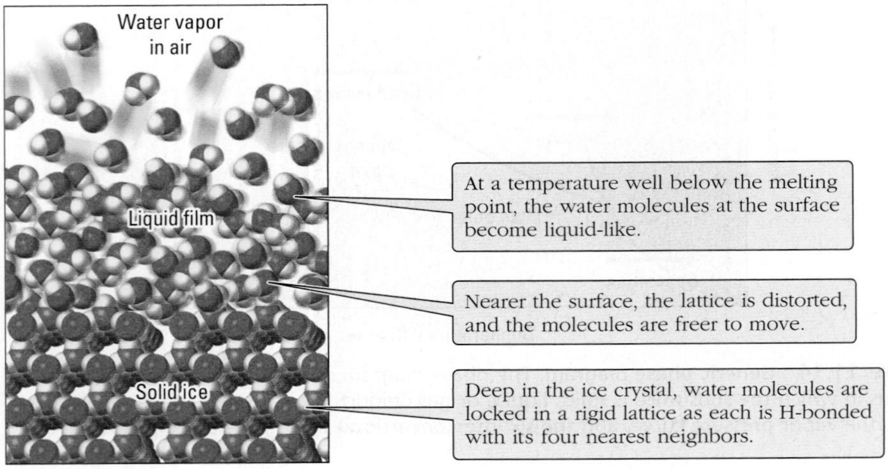

Figure 11.16 The surface of ice is slick due to a liquid-like film of water.

over the ice. Although increasing the pressure on ice does cause it to melt, the effect is too small to contribute significantly in the context of ice skating.

The phase diagram for CO_2 (Figure 11.17) is similar to that of water, except for one very important difference—the solid/liquid equilibrium line has a positive slope, as it does for most substances. Solid CO_2 is more dense than liquid CO_2. Notice also that the values of the pressure and temperature axes in the phase diagram for CO_2 are much different from those in the diagram for water. For pressures below about 5 atm, the only equilibrium that can exist is between solid and gaseous CO_2. This means solid CO_2 sublimes when heated at atmospheric pressure. Liquid CO_2 can be produced only at pressures above 5 atm. The temperature range for liquid CO_2 is narrower than that for water. Tanker trucks marked "liquid carbonic" carry liquid CO_2, a convenient source of CO_2 gas for making carbonated beverages.

PROBLEM-SOLVING EXAMPLE 11.5 Phase Diagrams

Use the phase diagram for CO_2 (Figure 11.17) to answer these questions.

(a) What is the temperature and pressure at the triple point?

(b) Starting at the triple point, what phase exists when the pressure is held constant and the temperature is increased to 65 °C?

(c) Starting at $T = -70$ °C and $P = 4$ atm, what phase change occurs when the pressure is held constant and the temperature is increased to -30 °C?

(d) Starting at $T = -30$ °C and $P = 5$ atm, what phase change occurs when the temperature is held constant and the pressure is increased to 10 atm?

Answer (a) $T = -57$ °C, $P = 5.2$ atm (b) Gas (c) Solid to gas (sublimation) (d) Gas to liquid (condensation)

Strategy and Explanation In each case, consult the location on the CO_2 phase diagram that corresponds to the given temperature and pressure conditions.

(a) The phase diagram shows that the triple point occurs at a temperature of -57 °C and a pressure of 5.2 atm.

(b) Increasing the temperature and holding the pressure constant means moving somewhat to the right from the triple point. This is the gas region of the phase diagram.

(c) At the starting temperature and pressure, the CO_2 is a solid. Increasing the temperature while holding the pressure constant causes the solid to sublime and change directly to the gas phase.

(d) At the starting temperature and pressure, the CO_2 is a gas. Increasing the pressure while holding the temperature constant causes the gas to condense to the liquid phase.

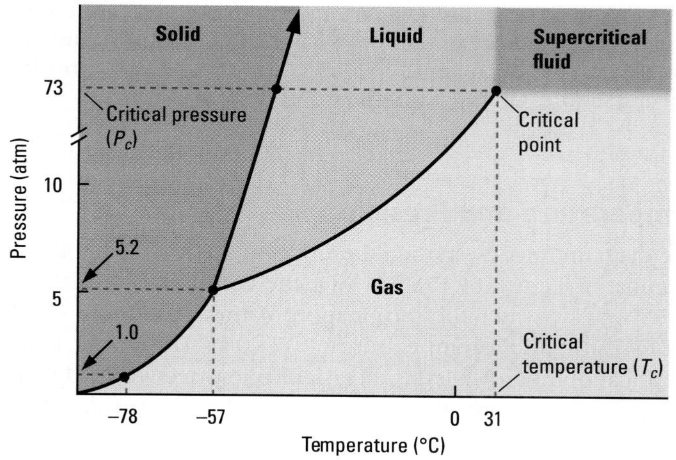

Figure 11.17 The phase diagram for carbon dioxide. The temperature and pressure scales are nonlinear for emphasis. Note that CO_2 cannot be a liquid at 1 atm or any pressure less than 5.2 atm.

CHEMISTRY YOU CAN DO Melting Ice with Pressure

Do this experiment in a sink. Obtain a piece of thin, strong, single-strand wire about 50 cm long, two weights of about 1 kg each, and a piece of ice about 25 cm by 3 cm by 3 cm. The ice can be either a cylinder or a bar and can be made by pouring some water into a mold made of several thicknesses of aluminum foil and then placing it in a freezer. A piece of plastic pipe, not tightly sealed, will also work fine as a mold. You can use metal weights or make weights by filling two 1-qt milk jugs with water. (One quart is about 1 L, which is 1 kg of water.)

Fasten one weight to each end of the wire. Support the bar of ice so that, without breaking the ice, you can hang the wire over it with one weight on each side. Observe the bar of ice and the wire every minute or so and record what happens. Suggest an explanation for what you have observed. The effect of pressure on the melting of ice and the enthalpy of fusion of ice are good ideas to consider when thinking about your explanation.

© Cengage Learning/Charles D. Winters

The apparatus used to melt ice with pressure.

Think about these questions:

1. Explain how the ice froze once again after the wire had passed through it.
2. What would be the effect of using a thinner piece of wire? Of using heavier weights?

PROBLEM-SOLVING PRACTICE 11.5

On the phase diagram for CO_2 (Figure 11.17), the point that corresponds to a temperature of −13 °C and a pressure of 7.5 atm is on the vapor pressure curve where liquid phase and gas phase are in equilibrium. What phase of CO_2 exists when the pressure remains the same and the temperature is increased by several degrees?

EXERCISE 11.10 Using a Phase Diagram

Consider the phase diagram for CO_2 (Figure 11.17). A curve connects the triple point and the critical point; what is the name of this curve? What phase transition occurs when you traverse this curve at its midpoint going upward, that is, increasing the pressure at constant temperature? Downward? What phase transition occurs when you traverse this curve at its midpoint going from left to right, that is, increasing the temperature at constant pressure? Right to left?

Critical Temperature and Pressure

On the phase diagram for CO_2, notice the upper point where the liquid/gas equilibrium line terminates (Figure 11.17). For CO_2 the conditions at this critical point are 73 atm and 31 °C. For water the termination of the solid/liquid curve occurs at a pressure of 217.7 atm and a temperature of 374.0 °C (Figure 11.15). These conditions at the critical point are called the **critical pressure** (P_c) and **critical temperature** (T_c). At any temperature above T_c the molecules of a substance have sufficient kinetic energy to overcome any attractive forces, and no amount of pressure can cause the substance to act like a liquid again. Above T_c and P_c the substance becomes a **supercritical fluid,** which has a density characteristic of a liquid but the flow properties of a gas, thereby enabling it to diffuse through many substances easily.

The term "fluid" describes a substance that will flow. It is used for both liquids and gases.

Supercritical CO_2 is present in a carbon-dioxide fire extinguisher under certain conditions. As the phase diagram shows, at any temperature below 31 °C, the pressurized CO_2 (usually at about 73 atm) is a liquid that can slosh around in the container. On a hot day, however, the CO_2 becomes a supercritical fluid and cannot be heard sloshing. In fact, under these conditions the only way to know whether CO_2 is in the extinguisher—without discharging it—is to weigh it and compare its mass with the mass of the empty container, which is usually indicated on a tag attached to the fire extinguisher.

Supercritical fluids are excellent solvents. Because supercritical CO_2 is nonpolar and is therefore a good solvent for nonpolar substances, it is used for dry cleaning. Supercritical CO_2 is also used to extract caffeine from coffee beans and for extracting toxic components from hazardous industrial wastes.

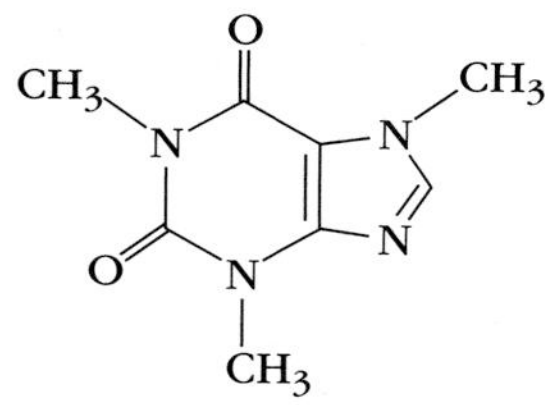

caffeine

© Cengage Learning/Charles D. Winters

Coffee beans and the molecular structure of caffeine.

> **EXERCISE 11.11 The Behavior of CO_2**
>
> If liquid CO_2 is slowly released to the atmosphere from a cylinder, what state will the CO_2 be in? If the liquid is suddenly released, as in the discharge of a CO_2 fire extinguisher, why is solid CO_2 seen? Can you explain this phenomenon on the basis of the phase diagram alone, or do you need to consider other factors?

11.4 Water: An Important Liquid with Unusual Properties

Earth is sometimes called the blue planet because the large quantities of water on its surface make it look blue from outer space. Three quarters of the globe is covered by oceans, and vast ice sheets cover the poles. Large quantities of water are present in soils and rocks on the surface. Water is essential to almost every form of life, has played a key role in human history, and is a significant factor in weather and climate. Water's importance is linked to its unique properties (Table 11.4). Ice floats on water; in contrast, when most substances freeze, the solid sinks in its liquid. More thermal energy must be transferred to melt ice, heat water, and vaporize water than to accomplish these changes for the same mass of almost any other substance. Water has the largest thermal conductivity and the highest surface tension of any molecular substance in the liquid state. In other words, this most common of substances in our daily lives has properties that are highly unusual. These properties are crucial for the welfare of our planet and our species.

"If there is magic on this planet, it is contained in water . . . Its substance reaches everywhere."
—Loren Eiseley

Most of water's unusual properties can be attributed to its molecules' unique capacity for hydrogen bonding. As seen from Figure 9.24 (← ***p. 408***), one water

Table 11.4 Unusual Properties of Water

Property	Comparison with Other Substances	Importance in Physical and Biological Environment
Specific heat capacity ($4.18\ J\ g^{-1}\ °C^{-1}$)	Highest of all liquids and solids except NH_3	Moderates temperature in the environment and in organisms; climate affected by movement of water (e.g., Gulf Stream)
Heat of fusion (333 J/g)	Highest of all molecular solids except NH_3	Freezing water releases large quantity of thermal energy; used to save crops from freezing by spraying them with liquid water
Heat of vaporization (2250 J/g)	Highest of all molecular substances	Condensation of water vapor in clouds releases large quantities of thermal energy, fueling storms
Surface tension ($7.3 \times 10^{-2}\ J/m^2$)	Highest of all molecular liquids	Contributes to capillary action in plants; causes formation of spherical droplets; supports insects on water surfaces
Thermal conductivity ($0.6\ J\ s^{-1}\ m^{-1}\ °C^{-1}$)	Highest of all molecular liquids	Provides for transfer of thermal energy within organisms; rapidly cools organisms immersed in cold water, causing hypothermia

The six-sided symmetry of snowflakes corresponds to the symmetry of these hexagonal rings.

molecule can participate in four hydrogen bonds to other water molecules. When liquid water freezes, a three-dimensional network of water molecules forms to accommodate the maximum hydrogen bonding. In the crystal lattice of ice, the oxygen atoms lie at the corners of puckered, six-sided rings. Considerable open space is left within the rings, forming empty channels that run through the entire crystal lattice, making the solid's density less than the liquid's.

When ice melts to form liquid water, approximately 15% of the hydrogen bonds are broken and the rigid ice lattice collapses. This makes the density of liquid water greater than that of ice at the melting point. The density of ice at 0 °C is 0.917 g/cm^3, and that of liquid water at 0 °C is 0.9998 g/cm^3. The density difference is not large, but it is enough so that ice floats on the surface of the liquid. This explains why about 90% of an iceberg is submerged and about 10% is above water.

As the liquid water is warmed further, more hydrogen bonds break, and more empty space is filled by water molecules. The density continues to increase until a temperature of 3.98 °C and a density of 1.000 g/cm^3 are reached. As the temperature rises beyond 3.98 °C, increased molecular motion causes the molecules to push each other aside more vigorously, the empty space between molecules increases, and the density decreases by about 0.0001 g/cm^3 for every 1 °C temperature rise above 3.98 °C (Figure 11.18).

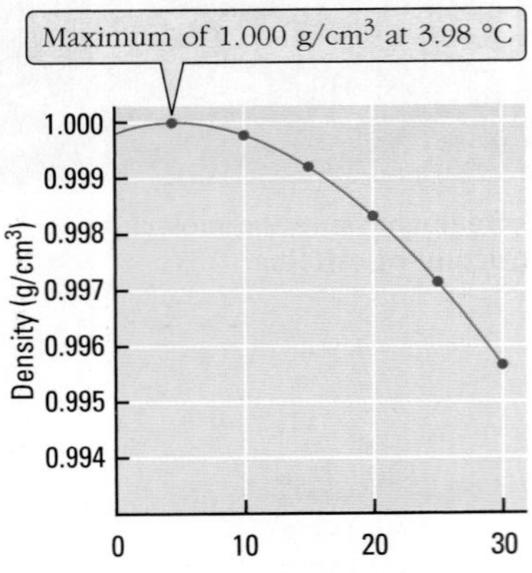
Maximum of 1.000 g/cm3 at 3.98 °C
1.000
0.999
0.998
0.997
0.996
0.995
0.994
Density (g/cm3)
0
10
20
30
Temperature (°C)

Figure 11.18 The density of liquid water between 0 °C and 30 °C. The density of ice at 0 °C and 1 bar is 0.917 g/cm^3, but the density of liquid water at those conditions is 0.9998 g/cm^3. Ice is about 10% less dense than water at 0 °C.

Because of this unusual variation of density with temperature, when water in a lake is cooled to 3.98 °C, the higher density causes the cold water to sink to the bottom. Water cooled below 3.98 °C is less dense and stays on the surface, where it can be cooled even further. Consequently, the water at the bottom of the lake remains at 3.98 °C, while that on the surface freezes. Ice on the surface insulates the remaining liquid water from the cold air, and, unless the lake is quite shallow, not all of the water freezes. This allows fish and other organisms to survive without being frozen in the winter. When water at 3.98 °C sinks to the bottom of a lake in the fall, it carries with it dissolved oxygen. Nutrients from the bottom are brought to the surface by the water it displaces. This is called "turnover" of the lake. The same thing happens in the spring, when the ice melts and water on the surface warms to 3.98 °C. Spring and fall turnovers are essential to maintain nutrient and oxygen levels required by fish and other lake-dwelling organisms.

When water is heated, the increased molecular motion breaks additional hydrogen bonds. The strength of these intermolecular forces requires that considerable energy be transferred to raise the temperature of 1 g water by 1 °C. That is, water's specific heat capacity is quite large. Many hydrogen bonds are broken when liquid water vaporizes, because the molecules are completely separated. This gives rise to water's very large enthalpy of vaporization, which is a factor in humans' ability to

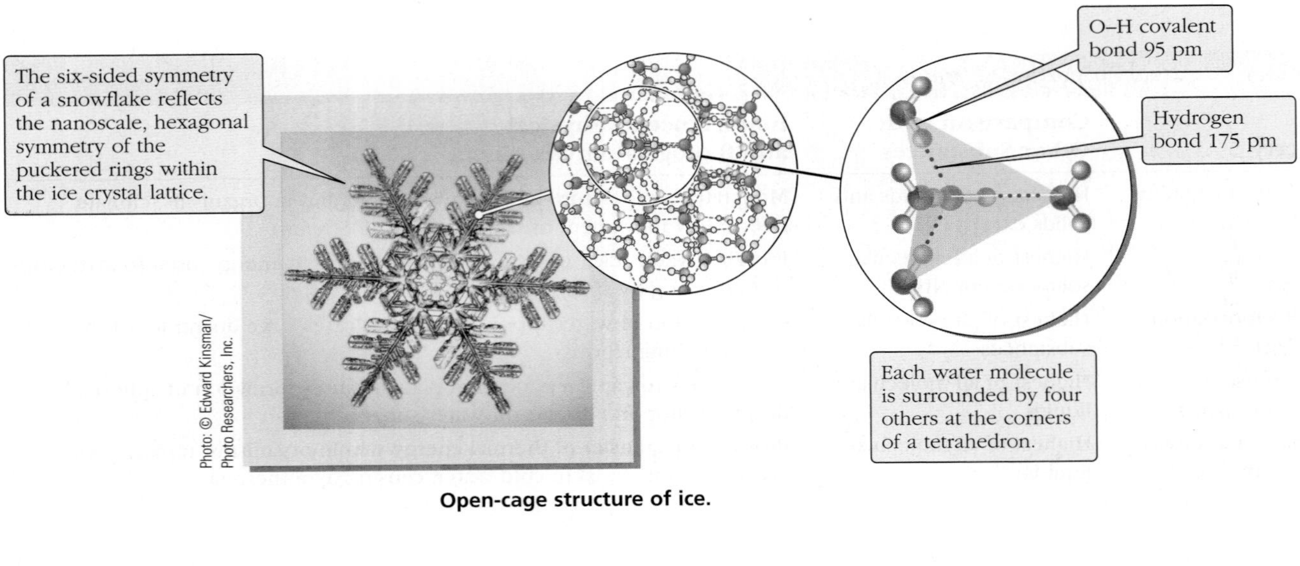
The six-sided symmetry of a snowflake reflects the nanoscale, hexagonal symmetry of the puckered rings within the ice crystal lattice.
O–H covalent bond 95 pm
Hydrogen bond 175 pm
Each water molecule is surrounded by four others at the corners of a tetrahedron.
Photo: © Edward Kinsman/ Photo Researchers, Inc.

Open-cage structure of ice.

regulate their body temperature by evaporation of sweat. As we have already described, hydrogen bonds are broken when ice melts, and this requires a large enthalpy of fusion. Its larger-than-normal enthalpy changes upon vaporization and freezing, together with its large specific heat capacity, allow water to moderate climate and influence weather by a much larger factor than other liquids could. In the vicinity of a large body of water, summer temperatures do not get as high and winter temperatures do not get as low (at least until the water freezes over) as they do far away from water ***(← p. 221)***. Seattle is farther north than Minneapolis, for example, but Seattle has a much more moderate climate because it borders the Pacific Ocean and Puget Sound.

11.5 Types of Solids

go Chemistry **Module 18: The Solid State** covers concepts in this section.

The relationship of nanoscale structure to macroscale properties is a central theme of this text. Nowhere is the influence of the nanoscale arrangement of atoms, molecules, or ions on properties more evident than in the study of solids. Chemists, in collaboration with physicists, engineers, and other scientists, explore such relationships as they work in the field of materials science. The nature of solid substances is determined by the type of forces holding their nanoscale particles together.

- *Ionic solids* are held together by electrostatic interactions between cations and anions. Example: table salt, NaCl ***(← p. 85)***
- *Metallic solids* are held together by attractions among positively charged metal atom cores and valence electrons (Section 11.9).
- *Molecular solids* are held together by intermolecular interactions such as London forces, dipole-dipole forces, and hydrogen bonding. Example: water ice ***(← p. 76)***
- *Network solids* are bonded together into infinite molecules by covalent bonds between atoms. Example: diamond (Section 11.7)
- *Amorphous solids* are held together by covalent bonds (as in network solids) but there is no long-range repeating pattern (Section 11.11).

Table 11.5 summarizes the characteristics and physical properties of the major types of solid substances. By classifying a substance as one of these types of solid, you will be able to form a reasonably good idea of what general physical properties to expect, even for a substance that you have never encountered before.

A solid is rigid, having its own shape rather than assuming the shape of its container as a liquid does. Solids have varying degrees of hardness that depend on the kinds of atoms in the solid and the types of forces that hold the atoms, molecules, or ions together. For example, talc (soapstone, Figure 11.19a), which is used as a lubricant and in talcum powder, is one of the softest solids known. At the atomic level, talc consists of layered sheets containing silicon, magnesium, and oxygen atoms. Attractive forces between these sheets are very weak, so one sheet of talc can slide along another and be removed easily from the rest. In contrast, diamond (Figure 11.19b) is one of the hardest solids known. In diamond, each carbon atom is covalently bonded to four neighbors in a tetrahedral arrangement. Each of those neighbor atoms is in turn strongly bonded to four carbon atoms, and so on throughout the solid (a network solid, Section 11.7). Because of the number and strength of the bonds holding each carbon atom to its neighbors, diamond is so hard that it can scratch or cut almost any other solid. Diamonds are used in cutting tools and abrasives, which are more important commercially than diamonds used as gemstones.

Figure 11.19 **(a) Talc.** Although they do not appear to be soft, these talc crystals are so soft that they can be crushed between one's fingers. **(b) Diamond.** One of the hardest substances known.

Although all solids consist of atoms, molecules, or ions in relatively immobile positions, some solids exhibit a greater degree of internal order than others.

- **Crystalline solids** have an ordered, long-range arrangement of the individual particles, which is reflected in the planar faces and sharp angles of the crystals. Examples: salt crystals, minerals, gemstones, and water ice

Table 11.5 Structures and Properties of Various Types of Solid Substances

Type	Examples	Structural Units
Ionic (← *p. 85*)	NaCl, K_2SO_4, $CaCl_2$, $(NH_4)_3PO_4$	Positive and negative ions (some polyatomic); no discrete molecules
Metallic (Section 11.9)	Iron, silver, copper, other metals and alloys	Metal atoms (positive metal ions surrounded by an electron sea)
Molecular (← *p. 76*)	H_2, O_2, I_2, H_2O, CO_2, CH_4, CH_3OH, CH_3COOH	Molecules with covalent bonds
Network (Section 11.7)	Graphite, diamond, quartz, feldspars, mica	Atoms held in an infinite one-, two-, or three-dimensional network
Amorphous (glassy) (Section 11.11)	Glass, polyethylene, nylon	Covalently bonded networks of atoms or collections of large molecules with short-range order only

- **Amorphous solids** exhibit very little long-range order, yet they are hard and have definite shapes. Examples: ordinary glass and organic polymers such as some forms of polyethylene and polystyrene

The remainder of this chapter is devoted to explaining the nanoscale structures that give rise to the properties of solids summarized in Table 11.5.

PROBLEM-SOLVING EXAMPLE 11.6 Types of Solids

What types of solids are these substances?

(a) Sucrose, $C_{12}H_{22}O_{11}$ (table sugar), has a melting point of about 185 °C. It has poor electrical conductance both as a solid and as a liquid.

(b) Solid Na_2SO_4 has a melting point of 884 °C and has low electrical conductivity that increases dramatically when the solid melts.

Answer (a) Molecular solid (b) Ionic solid

Strategy and Explanation In each case compare the stated properties of the substance to the known properties of ionic solids and molecular solids.

(a) Sucrose molecules contain covalently bonded atoms. Sucrose's properties correspond to those of a molecular solid since it has poor electrical conductance both as a solid and as a liquid.

(b) Sodium sulfate is composed of Na^+ and SO_4^{2-} ions. The given properties are consistent with those of ionic solids. Ionic solids have high melting points and are poor conductors as solids but good conductors as liquids.

PROBLEM-SOLVING PRACTICE 11.6

What types of solids are these substances?

(a) The hydrocarbon decane, $C_{10}H_{22}$, has a melting point of −31 °C and is a poor electrical conductor.

(b) Solid $MgCl_2$ has a melting point of 780 °C and conducts electricity only when melted.

CONCEPTUAL EXERCISE 11.12 Lead into Gold?

Imagine a sample of gold and a sample of lead, each with a smoothly polished surface. The surfaces of these two samples are placed in contact with one another and held in place, under pressure, for about one year. After that time the two surfaces are analyzed. The gold surface is tested for the presence of lead, and the lead surface is tested for the presence of gold. Predict what the outcome of these two tests will be and explain what has happened, if anything.

Table 11.5 continued

Forces Holding Units Together	Typical Properties
Ionic bonding; attractions among charges on positive and negative ions	Hard; brittle; high melting point; poor electrical conductor as solid, but good conductor as liquid; often water-soluble
Metallic bonding; electrostatic attraction among metal ions and electrons	Malleable; ductile; good electrical conductor in solid and liquid; good heat conductor; wide range of hardness and melting point
London forces, dipole-dipole forces, hydrogen bonds	Low to moderate melting point and boiling point; soft; poor electrical conductor in solid and liquid
Covalent bonds (directional electron-pair bonds)	Wide range of hardness and melting point (three-dimensional bonding > two-dimensional bonding > one-dimensional bonding); poor electrical conductor with some exceptions
Covalent bonds (directional electron-pair bonds)	Noncrystalline; wide temperature range for melting; poor electrical conductor, with some exceptions

11.6 Crystalline Solids

go Chemistry **Module 18: The Solid State** covers concepts in this section.

The beautiful regularity of ice crystals, crystalline salts, and gemstones suggests that they must have some *internal order.* Toward the end of the eighteenth century, scientists found that shapes of crystals can be used to identify minerals. The angles at which crystal faces meet are characteristic of a crystal's composition. The shape of each crystalline solid reflects the shape of its **crystal lattice**—an orderly, repeating arrangement of points in three-dimensional space in which each point has surroundings identical to those of all other points. The arrangement of atoms, molecules, or ions about each lattice point defines the position of each individual particle ***(← p. 95).*** Hence, at a given lattice point, each atom, molecule, or ion is surrounded by neighbors in exactly the same arrangement. In the structures of metals and salts, atoms or ions usually occupy lattice points, but in some structures there are no particles located at the lattice points. A crystal is built up from a three-dimensional repetition of the same pattern, which gives the crystal long-range order throughout.

Unit Cells

A convenient way to describe and classify the repeating pattern of atoms, molecules, or ions in a crystal is to define a small segment of a crystal lattice as a **unit cell**—a small part of the lattice that, when repeated along the directions defined by its edges, reproduces the entire crystal structure. To help understand the idea of the unit cell, look at the simple two-dimensional array of circles shown in Figure 11.20. Each circle is centered on a lattice point. The same size circle is repeated over and over, but a circle is not a proper unit cell because it gives no indication of its relationship to all the other circles. A better choice is to recognize that the lattice points at the centers of four adjacent circles lie at the corners of a square and to draw four lines connecting those lattice points. A square unit cell results. Figure 11.20 shows this simple unit cell superimposed on the circles. As you look at the unit cell drawn in darker blue, notice that each of four circles contributes one quarter of itself to the unit cell, so a net of one circle is located within the unit cell. When this two-dimensional unit cell is repeated by moving the square parallel to its edges (that is, when unit cells are placed to the left and right and above and below the first one), the two-dimensional lattice is generated. Notice that the corners of a unit cell (lattice points) are equivalent to each other and that collectively they define the crystal lattice.

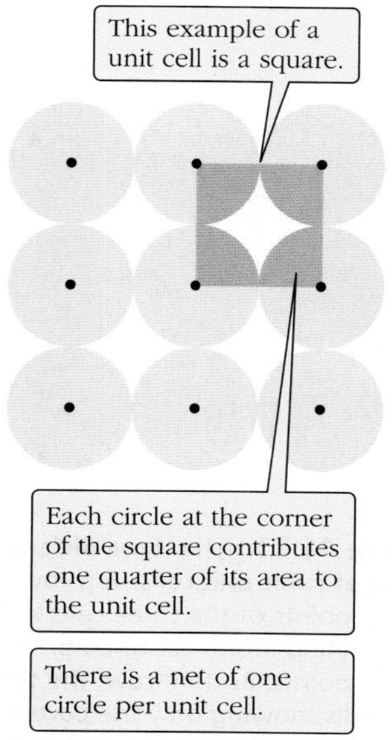
This example of a unit cell is a square.
Each circle at the corner of the square contributes one quarter of its area to the unit cell.
There is a net of one circle per unit cell.

Figure 11.20 Unit cell. For this two-dimensional arrangement of flat, circular objects, such as coins, a suitable lattice is the square array of points shown at the centers of the circles. The unit cell is a square; when repeated along the two directions defined by its perpendicular edges, the square generates the entire lattice.

The three-dimensional unit cells from which all known crystal lattices can be constructed fall into only seven categories. These seven types of unit cells differ because the edges (which connect lattice points) have different relative lengths and meet at different angles. We will discuss only one of these types—cubic unit cells—and only cubic unit cells composed of atoms or monatomic ions. Such unit cells are quite common in nature and are easier to visualize than the other types of unit cells. The principles illustrated, however, apply to all unit cells and all crystal structures, including those composed of polyatomic ions and large molecules. In these more complicated cases the lattice points are often not occupied by atoms, molecules, or ions.

Cubic unit cells (Figure 11.21), those with edges of equal length that meet at 90° angles, are of three types:

- *primitive cubic (pc)*
- *body-centered cubic (bcc)*
- *face-centered cubic (fcc)*

Examples of metals with cubic unit cells:

Primitive cubic: Po

Body-centered cubic: alkali metals, V, Cr, Mo, W, Fe

Face-centered cubic: Ni, Cu, Ag, Au, Al, Pb

Many metals and ionic compounds crystallize in cubic unit cells. In metals, all three types of cubic unit cells have identical atoms centered at lattice points at each corner of the cube. When the cubes pack into three-dimensional space, an atom at a corner is shared among eight cubes (Figure 11.21); thus only one eighth of each corner atom is actually within the unit cell. In the pc unit cell there are only lattice points at the corners of the cube. Since a cube has eight corners and since one eighth of the atom at each corner belongs to the unit cell, the net result is $8 \times \frac{1}{8} = 1$ *composite atom (or lattice point) per primitive cubic unit cell.*

In the body-centered cubic unit cell, an additional lattice-point atom is at the center of the cube—entirely within the unit cell. This atom, combined with the net of one atom from the corners, gives a total of *two composite atoms (or lattice points) per body-centered cubic unit cell.*

pc: 1 atom/unit cell
bcc: 2 atoms/unit cell
fcc: 4 atoms/unit cell

In the face-centered cubic unit cell six atoms (lattice points) lie in the centers of the six faces of the cube. One half of each of these atoms (lattice points) belongs to the unit cell (Figure 11.21). In this case there is a net of $6 \times \frac{1}{2} = 3$ composite atoms (lattice points) per unit cell, plus the net of one atom (lattice point) contributed by the corners, for a total of *four atoms per face-centered cubic unit cell.* The number of atoms per unit cell helps to determine the density of a solid.

EXERCISE 11.13 Counting Atoms in Unit Cells of Metals

Crystalline polonium has a primitive cubic unit cell, lithium has a body-centered cubic unit cell, and calcium has a face-centered cubic unit cell. How many Po atoms belong to one unit cell? How many Li atoms belong to one unit cell? How many Ca atoms belong to one unit cell? Draw each unit cell. Indicate on your drawing what fraction of each atom lies within the unit cell.

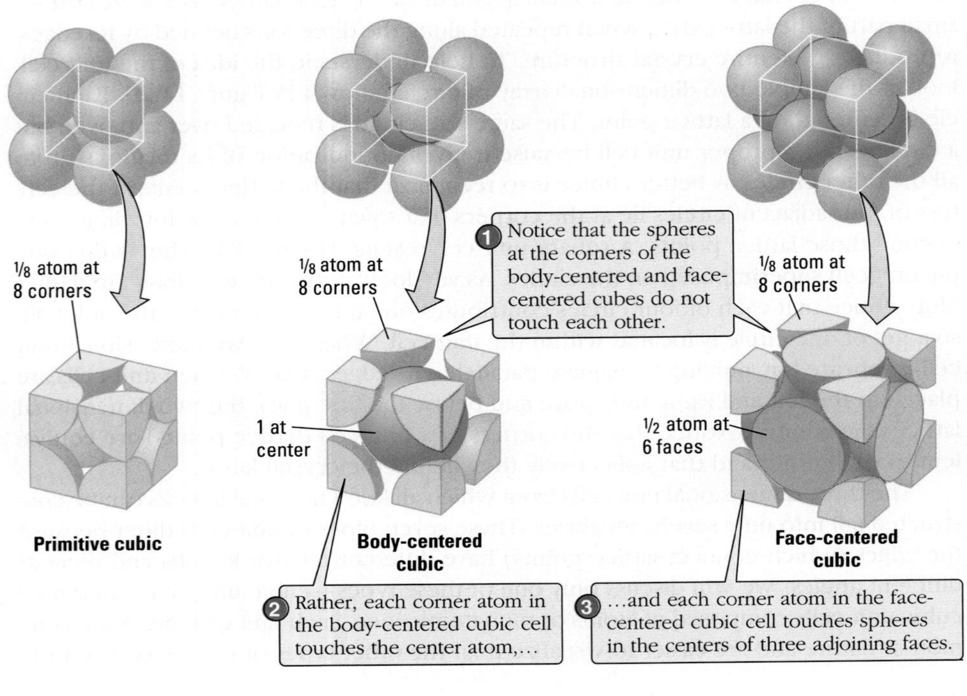

Figure 11.21 The three different types of cubic unit cells. Top row: lattice points of the three cells with space-filling atoms centered on the lattice points. Bottom row: the three unit cells showing only the parts of each atom that lie within the unit cell.

Unit Cells and Density

The information given so far about unit cells allows us to check whether a proposed unit cell is reasonable. Because a unit cell can be replicated to give the entire crystal lattice, the unit cell should have the same density as the crystal. As an example, we will consider platinum.

For each of the three cubic unit cells, there is a close relationship between the radius of the atom forming the unit cell and the unit cell size. For the primitive cubic unit cell, radii of two atoms form the edge of the unit cell. For the face-centered cubic unit cell, one complete atom diameter and radii of two other atoms form the face diagonal. For the body-centered cubic unit cell, one complete atom diameter and radii of two other atoms form the body diagonal of the unit cell. To summarize:

- **Primitive cubic:** $2 \times$ (atomic radius) = edge
- **Face-centered cubic:** $4 \times$ (atomic radius) = edge $\times \sqrt{2}$ = face diagonal
- **Body-centered cubic:** $4 \times$ (atomic radius) = edge $\times \sqrt{3}$ = body diagonal

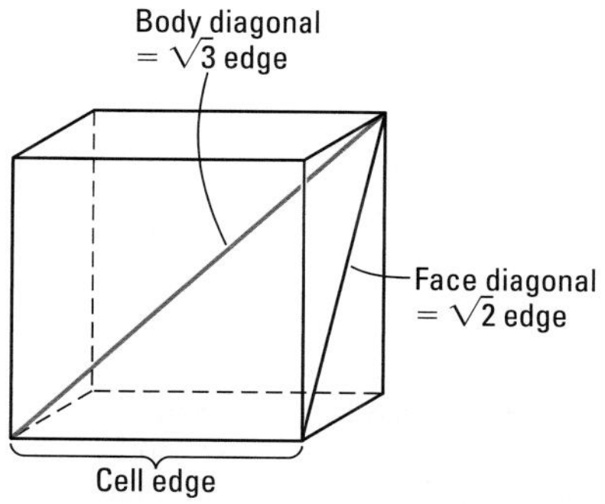

Dimensions of a cubic unit cell. The edge, face diagonal, and body diagonal of the unit cell are directly related to the atomic radius for the pc, fcc, and bcc unit cells, respectively.

PROBLEM-SOLVING EXAMPLE 11.7 Unit Cell Dimension, Type, and Density

Platinum crystals are fcc, with a density of 21.45 g/cm^3. Platinum has a molar mass of 195.08 g/mol, and platinum atoms have a 139-pm radius. What is the length of a unit cell edge? Is this consistent with Pt being fcc?

Answer 392 pm. Yes, it is consistent with fcc.

Strategy and Explanation Starting with the radius of a Pt atom, we calculate the length of the face diagonal of the unit cell, the length of an edge of the unit cell, and the unit cell's volume. From the mass of Pt atoms per unit cell and the unit cell's volume we calculate the density and compare this to the known density.

The mass of one Pt atom is

$$\frac{195.08 \text{ g Pt}}{1 \text{ mol Pt}} \times \frac{1 \text{ mol Pt}}{6.022 \times 10^{23} \text{ atoms Pt}} = 3.239 \times 10^{-22} \text{ g}$$

In an fcc unit cell there are four atoms per unit cell, so the mass of Pt atoms per unit cell is

$$4 \times 3.239 \times 10^{-22} \text{ g} = 1.296 \times 10^{-21} \text{ g Pt per unit cell}$$

Now we calculate the volume of the unit cell. To start we calculate the face diagonal length (see figure in margin).

$$\text{Face diagonal length} = 4 \times \text{Pt atom radius} = 4 \times 139 \text{ pm} = 556 \text{ pm}$$

For an fcc unit cell, the relationship between the edge length of the unit cell and its face diagonal length is (See figure in margin)

$$(\text{Diagonal length})^2 = \text{edge}^2 + \text{edge}^2 = 2(\text{edge})^2$$

so

$$\text{Edge} = \frac{\text{diagonal length}}{\sqrt{2}} = \frac{556 \text{ pm}}{1.414} = 393 \text{ pm}$$

The volume of the unit cell is

$$V = (\text{edge})^3 = (393 \text{ pm})^3 = 6.07 \times 10^{-23} \text{ cm}^3$$

$$(1 \text{ pm} = 10^{-12} \text{ m} = 10^{-10} \text{ cm})$$

The density is the mass of Pt atoms per unit cell divided by the unit cell's volume.

$$d = \frac{m}{V} = \frac{1.296 \times 10^{-21} \text{ g Pt}}{6.07 \times 10^{-23} \text{ cm}^3} = 21.4 \text{ g/cm}^3$$

This density value compares well with the given value, and the calculation shows that the data are consistent with an fcc unit cell for the Pt crystal.

☑ **Reasonable Answer Check** The density, atomic weight, unit cell edge length and diagonal length, and Pt atom radius are all in agreement.

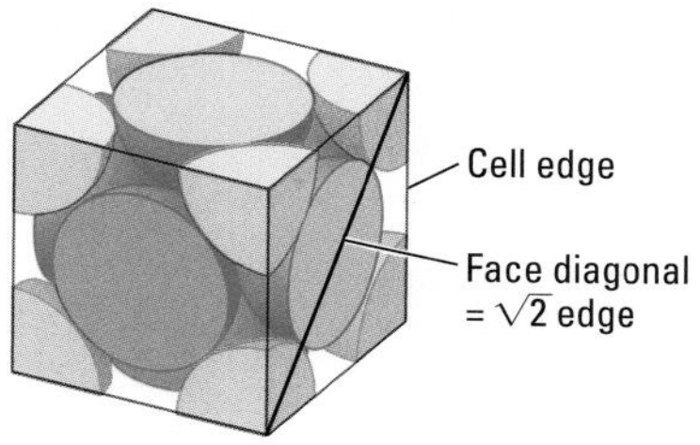

A face-centered cubic (fcc) unit crystal.

PROBLEM-SOLVING PRACTICE 11.7

Gold crystals have a bcc structure. The radius of a gold atom is 144 pm, and the atomic weight of gold is 196.97 amu. Calculate the density of gold.

Ionic Crystal Structures

Remember that negative ions are usually larger than positive ions. Therefore, building an ionic crystal is a lot like placing marbles (smaller positive ions) in the spaces among table tennis balls (larger negative ions).

It is not appropriate to describe the CsI structure as bcc. In a bcc lattice, the species at the center of the unit cell must be the same as the species at the corner of the cell. In the CsI unit cell described here, there are iodide ions at the corners and a cesium ion at the center. Another, equally valid, unit cell for CsI has cesium ions at the corners and an iodide ion at the center.

The crystal structures of many ionic compounds can be described as primitive cubic or face-centered cubic lattices of spherical negative ions, with smaller positive ions occupying spaces (called holes) among the larger negative ions. The number and locations of the occupied holes are the keys to understanding the relation between the lattice structure and the formula of an ionic compound. The simplest example is an ionic compound in which the hole at the center of a primitive cubic unit cell (Figure 11.21) is occupied. The ionic compound cesium iodide, CsI, has such a structure. In it, each cube of I^- ions has a Cs^+ ion at its center (Figure 11.22). The spaces occupied by the Cs^+ ions are called cubic holes, and each Cs^+ has eight nearest neighbor I^- ions. The CsI unit cell contains one Cs^+ ion and $\frac{1}{8} \times 8 = 1\ I^-$ ion for a formula of CsI.

The structure of sodium chloride, NaCl, is a very common ionic crystal lattice (Figures 11.23 and 11.24). If you look carefully at Figure 11.23, it is possible to determine the number of Na^+ and Cl^- ions in the NaCl unit cell. There is one eighth of a Cl^- ion at each corner of the unit cell (lattice point) and one half of a Cl^- ion at each lattice point in the middle of each face. The total number of Cl^- ions within the unit cell is

$\frac{1}{8}\ Cl^-$ per corner $\times$ 8 corners $= 1\ Cl^-$

$\frac{1}{2}\ Cl^-$ per face $\times$ 6 faces $= 3\ Cl^-$

Total of 4 Cl^- in a unit cell

Since the edge of a cube in a cubic lattice is surrounded by four cubes, one fourth of a spherical ion at the midpoint of an edge is within any one of the cubes.

There is one fourth of a Na^+ at the midpoint of each edge and a whole Na^+ in the center of the unit cell. For Na^+ ions, the total is

$\frac{1}{4}\ Na^+$ per edge $\times$ 12 edges $= 3\ Na^+$

1 Na^+ per center $\times$ 1 center $= 1\ Na^+$

Total of 4 Na^+ in a unit cell

Thus, the unit cell contains four Na^+ and four Cl^- ions, a 1:1 ratio. This result agrees with the formula of NaCl for sodium chloride.

As shown in Figure 11.24, the NaCl crystal lattice consists of an fcc lattice of the larger Cl^- ions, in which Na^+ ions occupy so-called *octahedral* holes—octahedral because each Na^+ ion is surrounded by six Cl^- ions at the corners of an octahedron. Likewise, each Cl^- ion is surrounded by six Na^+ ions. Figure 11.24 also shows a space-filling model of the NaCl lattice, in which each ion is drawn to scale based on its ionic radius.

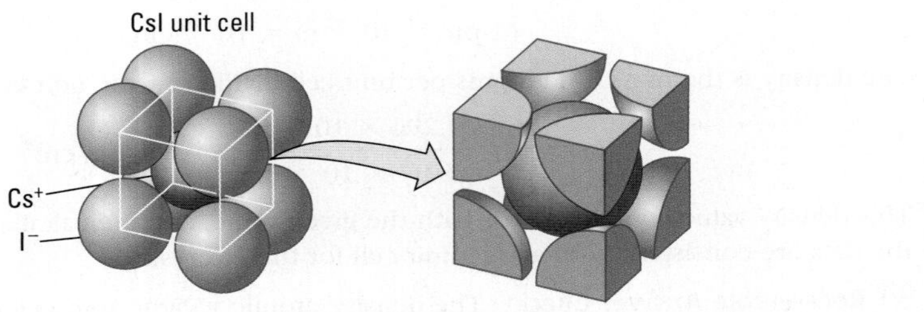

Figure 11.22 **Unit cell of the cesium iodide (CsI) crystal lattice.**

EXERCISE 11.14 Formulas and Unit Cells

Cesium chloride has a cubic unit cell, like the cesium iodide unit cell in Figure 11.22. Show that the formula for the salt must be CsCl.

PROBLEM-SOLVING EXAMPLE 11.8 Calculating the Volume and Density of a Unit Cell

The fcc unit cell of NaCl is shown in Figure 11.23. The ionic radii for Na^+ and Cl^- are 116 pm and 167 pm, respectively. Calculate the density of NaCl.

Answer 2.14 g/cm^3

Strategy and Explanation To calculate the density, determine the mass and volume of the unit cell. Calculate the mass from the number of formula units in a unit cell and the molar mass of NaCl. Then calculate the length of an edge of the unit cell and its volume.

First, calculate the edge length and volume of the unit cell. As seen in Figure 11.23, the Na^+ and Cl^- ions touch along the edge of the unit cell. Thus, the edge length is equal to two Cl^- radii plus two Na^+ radii.

$$\text{Edge} = 167 \text{ pm} + (2 \times 116 \text{ pm}) + 167 \text{ pm} = 566 \text{ pm}$$

The volume of the fcc unit cell is the cube of the edge length.

$$\text{Volume of unit cell} = (\text{edge})^3 = (566 \text{ pm})^3 = 1.81 \times 10^8 \text{ pm}^3 = 1.81 \times 10^{-22} \text{ cm}^3$$

Next calculate the mass of a unit cell. With four NaCl formula units per unit cell,

$$4 \text{ NaCl formula units} \times \frac{58.44 \text{ g}}{\text{mol NaCl}} \times \frac{1 \text{ mol}}{6.022 \times 10^{23} \text{ formula units}} = 3.88 \times 10^{-22} \text{ g}$$

This means the density of a NaCl unit cell is

$$\frac{3.88 \times 10^{-22} \text{ g}}{1.81 \times 10^{-22} \text{ cm}^3} = 2.14 \text{ g/cm}^3$$

☑ **Reasonable Answer Check** The experimental density is 2.164 g/cm^3, which is in reasonable agreement with the calculated value. Remember, however, that all experiments have uncertainties associated with them. The density of NaCl calculated from

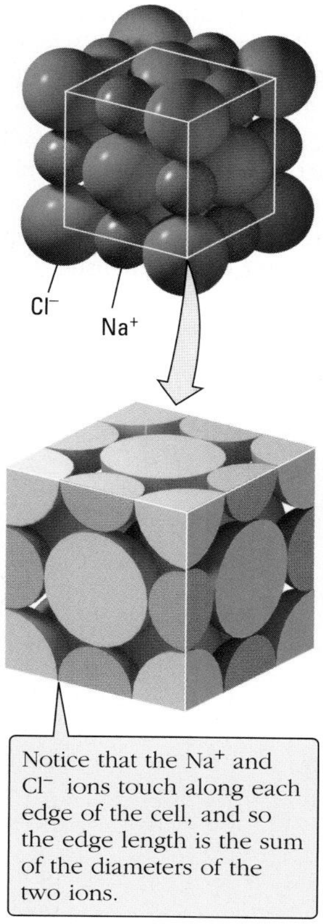

Cl⁻
Na⁺
Notice that the Na⁺ and Cl⁻ ions touch along each edge of the cell, and so the edge length is the sum of the diameters of the two ions.

Figure 11.23 An NaCl unit cell.

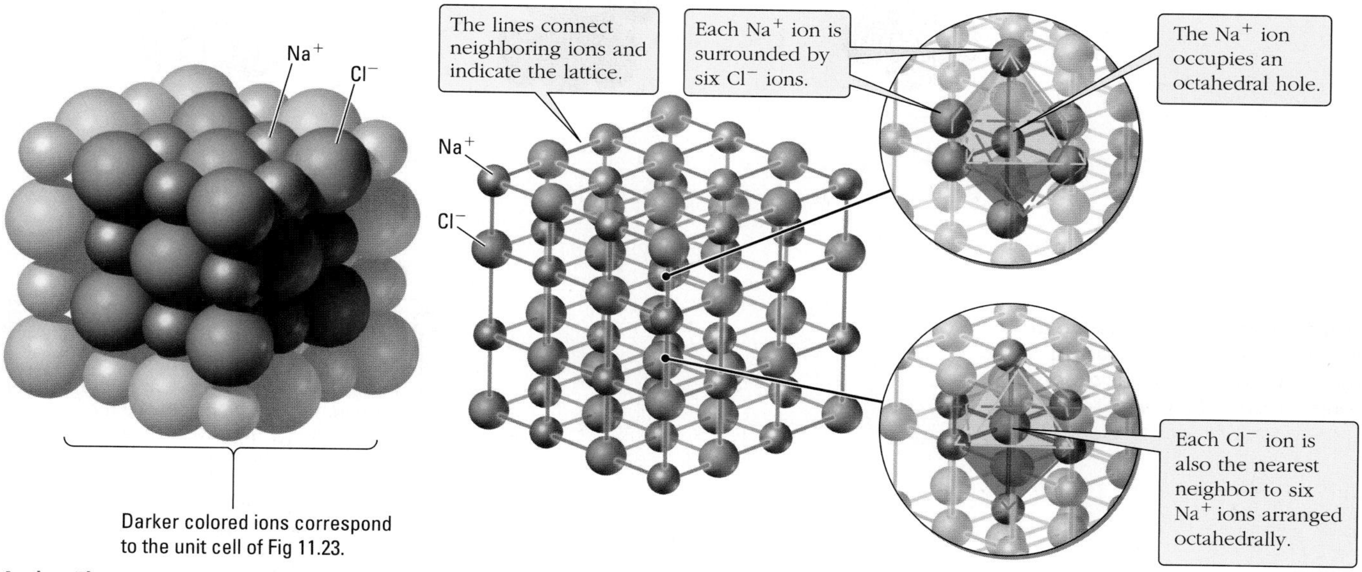

Na⁺
Cl⁻
Darker colored ions correspond to the unit cell of Fig 11.23.
The lines connect neighboring ions and indicate the lattice.
Na⁺
Cl⁻
Each Na⁺ ion is surrounded by six Cl⁻ ions.
The Na⁺ ion occupies an octahedral hole.
Each Cl⁻ ion is also the nearest neighbor to six Na⁺ ions arranged octahedrally.

Active Figure 11.24 The NaCl crystal lattice. Visit this book's companion website at **www.cengage.com/chemistry/moore** to test your understanding of the concepts in this figure.

The general relationships among the unit cell type, ion or atom size, and solid density for most metallic or ionic solids are as follows:

Mass of 1 formula unit (e.g., NaCl) × number of formula units per unit cell (4) = mass of unit cell

Mass of unit cell ÷ unit cell volume (= edge^3 for NaCl) = density

unit cell dimensions could easily have given a value closer to the experimental density if the tabulated radii for the Na^+ and Cl^- ions were slightly smaller. There is, of course, a slight uncertainty in the published radii of all ions.

PROBLEM-SOLVING PRACTICE 11.8

KCl has the same crystal structure as NaCl. Calculate the volume of the unit cell for KCl, given that the ionic radii are K^+ = 152 pm and Cl^- = 167 pm. Compute the density of KCl. Which has the larger unit cell, NaCl or KCl?

Closest Packing of Spheres

In crystalline solids, the atoms, molecules, or ions are often arranged as closely as possible so that their interactions are maximized, which results in a stable structure. This arrangement is most easily illustrated for metals, in which the individual particles are identical atoms that can be represented as spheres. The arrangement known as **closest packing** can be built up as a series of layers of atoms. In each layer of equal-sized spheres, as shown in Figure 11.25, part 1, each sphere is surrounded by

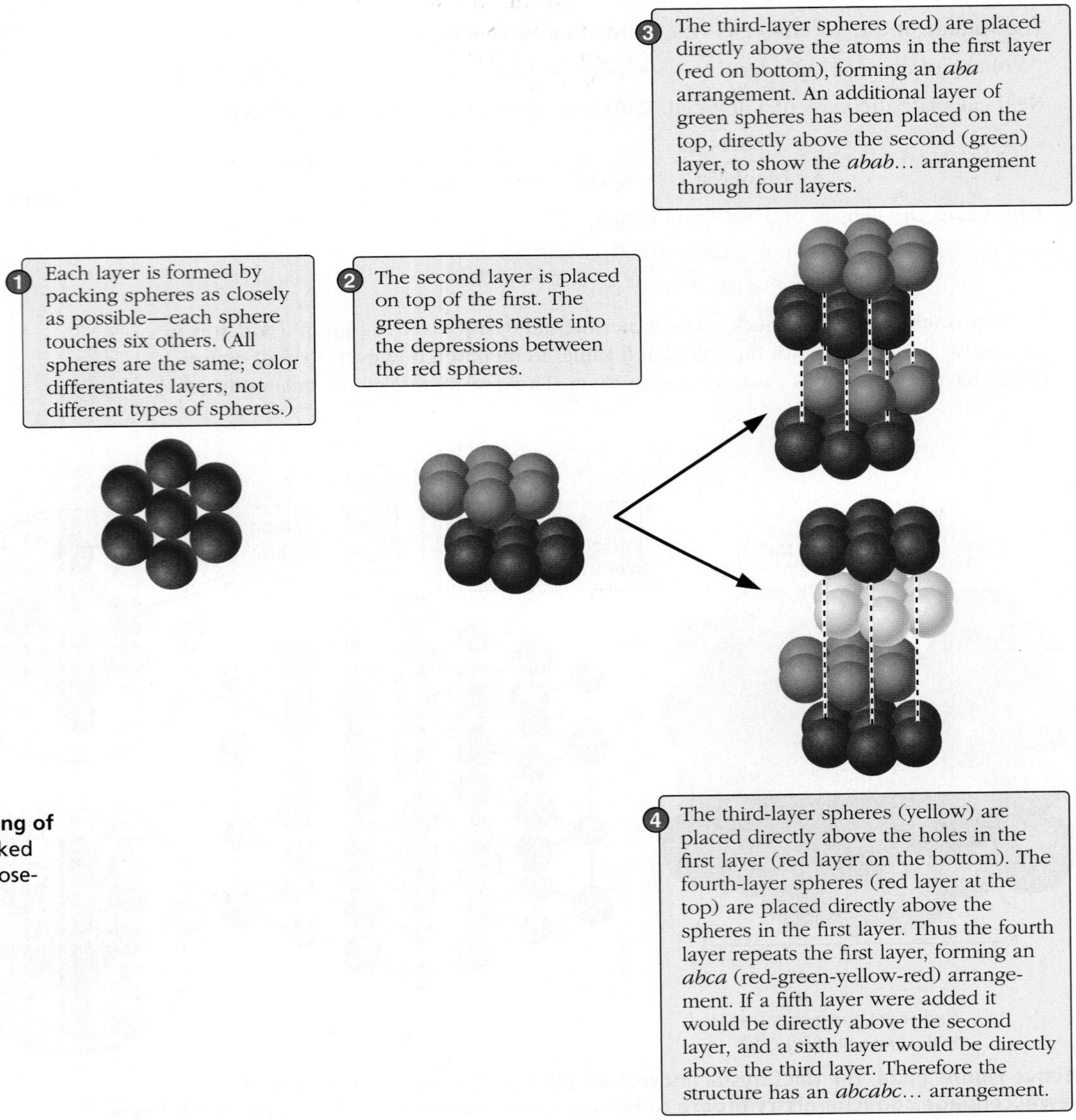

3 The third-layer spheres (red) are placed directly above the atoms in the first layer (red on bottom), forming an *aba* arrangement. An additional layer of green spheres has been placed on the top, directly above the second (green) layer, to show the *abab*... arrangement through four layers.
1 Each layer is formed by packing spheres as closely as possible—each sphere touches six others. (All spheres are the same; color differentiates layers, not different types of spheres.)
2 The second layer is placed on top of the first. The green spheres nestle into the depressions between the red spheres.
4 The third-layer spheres (yellow) are placed directly above the holes in the first layer (red layer on the bottom). The fourth-layer spheres (red layer at the top) are placed directly above the spheres in the first layer. Thus the fourth layer repeats the first layer, forming an *abca* (red-green-yellow-red) arrangement. If a fifth layer were added it would be directly above the second layer, and a sixth layer would be directly above the third layer. Therefore the structure has an *abcabc*... arrangement.

Figure 11.25 Closest packing of spheres. Hexagonal close-packed structure (part 3) and cubic close-packed structure (part 4).

six neighbors. A second layer can be put on top of the first layer so the spheres nestle into the depressions of the first layer (Figure 11.25, part 2). Then a third layer can be added. There are two ways to add the third layer, each of which yields a different structure, as shown in Figure 11.25, parts 3 and 4. If the third layer is directly above the first layer, an *abab*... arrangement results, called a **hexagonal close-packed** (hcp) structure. In this *abab*... structure, the centers of the spheres of the third layer are directly above the centers of the spheres of the first layer (Figure 11.25, part 3). If the third layer is *not* directly above the first layer, an *abcabc*... arrangement results, giving a **cubic close-packed** (ccp) structure (Figure 11.25, part 4). In this structure, the centers of the spheres of the third layer are directly above the *holes* between the spheres of the first layer, and the centers of the spheres of the fourth layer are directly above the *centers* of the spheres of the first layer. The unit cell for a cubic close-packed structure is face-centered cubic (fcc).

In *abab*... packing, the third layer repeats the first layer; in *abcabc*... packing, the fourth layer repeats the first layer.

The stacked fruits in markets are placed in closest packed arrangements for efficiency.

CHEMISTRY YOU CAN DO Closest Packing of Spheres

In a metal, the crystal lattice is occupied by identical atoms, which can be pictured as identical spheres. Obtain a bag of marbles, some expanded polystyrene foam spheres, some table-tennis balls, or some other set of reasonably small, identical spheres. Use them to construct models of each situation described here.

Consider how identical spheres can be arranged in two dimensions. Pack your spheres into a layer so that each sphere touches four other spheres; this should resemble the arrangement in part (a) of the figure. You will find this somewhat hard to do because the spheres are rather inefficiently packed, and they readily adopt a more efficient packing in which each sphere touches six other spheres, part (b).

To make a three-dimensional lattice requires stacking layers of spheres. Start with the layer of spheres in a square array and stack another identical layer directly on top of the first; then stack another on that one, and so on. This is really hard and will require vertical walls to hold the spheres as you stack the layers. This three-dimensional lattice will consist of *primitive cubic* unit cells; in such a lattice 52.4% of the available space is occupied by the spheres.

Next, make the more efficiently packed hexagonal layer, in which each sphere is surrounded by six others, as shown in part (b) of the figure. In this arrangement, the holes between spheres are smaller than they were in the square array; in fact, they are as small as possible. For this reason the layer you have made is called a close-packed layer. Now add a second layer, and let the spheres nestle into the holes in the first layer. Now add a third layer. If you look carefully, you will see that there are two ways to put down the third layer. One of these places spheres directly above *holes* in the first layer. This is called *cubic close packing,* and the unit cell is face-centered cubic (fcc) (Figure 11.24). The other arrangement, also found for many metals, places third-layer spheres directly above *spheres* in the first layer. This is *hexagonal close packing,* and its unit cell is *not* cubic. Cubic and hexagonal close packing are the most efficient ways possible for filling space with spheres; 74% of the available space is occupied. The atoms in most metals are arranged in cubic close packing, in hexagonal close packing, or in lattices composed of body-centered cubic unit cells. Efficient packing of the atoms or ions in the crystal lattice allows stronger bonding, which gives greater stability to the crystal.

Think about these questions:

1. In either the cubic close-packed or the hexagonal close-packed arrangement, each atom has six neighbors in the same plane. How many closest neighbors does each atom have, considering all three dimensions?
2. How many closest neighbors does each atom have in the primitive cubic structure?

One layer of marbles in (a) a square arrangement and (b) a hexagonal close-packed arrangement.

(a) (b)

A rigorous mathematical proof has been developed that showed the impossibility of packing spheres to get a packing fraction greater than 74%.

In each of these two close-packed structures (hcp and ccp), each sphere has 12 nearest-neighbor spheres that are equidistant from it; 74% of the total volume of the structure is occupied by spheres and 26% is empty. Compare these percentages with those for the body-centered cubic structure, which is 68% occupied, and for the primitive cubic structure, which is only 52% occupied.

11.7 Network Solids

A number of solids are composed of nonmetal atoms connected by a network of covalent bonds. Such **network solids** really consist of huge molecules in which all the atoms are connected to all the others via the network of covalent bonds. Separate small molecules do not exist in a network solid.

The most important network solids are the *silicates.* Many bonding patterns exist among the silicates, but extended arrays of covalently bonded silicon and oxygen atoms are common, such as quartz, SiO_2. Silicates will be discussed in depth in Chapter 21.

Graphite, Diamond, and Fullerenes

The fact that planar layers in graphite can slide past one another explains why part of the "lead" is transferred to the paper when we write with a pencil.

Graphite, diamond, and fullerenes are allotropes of carbon **(← p. 26).** Graphite's name comes from the Greek *graphein,* meaning "to write," because one of its earliest uses was for writing on parchment. Artists today still draw with charcoal, an impure form of graphite, and we write with pencil leads that contain graphite. Graphite is an example of a *planar network solid* (Figure 11.26). Each carbon atom is covalently bonded to three other carbon atoms. The planes consist of six-membered rings of carbon atoms (like those in benzene) **(← p. 359).** Each hexagon shares all six of its sides with other hexagons around it, forming a *two-dimensional network* resembling chicken wire. Some of the bonding electrons are able to move freely around this network, so graphite readily conducts electricity. Within a plane, there are strong covalent bonds between carbon atoms, but attractions between the planes are due to London forces and hence are weaker. Because of this, the planes can easily slip across one another, which makes graphite an excellent solid lubricant for uses such as in locks, where greases and oils are undesirable.

The structure of diamond is built with tetrahedral carbon atoms, each bonded to four others by single covalent bonds. This structure forms a *three-dimensional network* (Figure 11.27). Because of the tetrahedral arrangement of bonds around

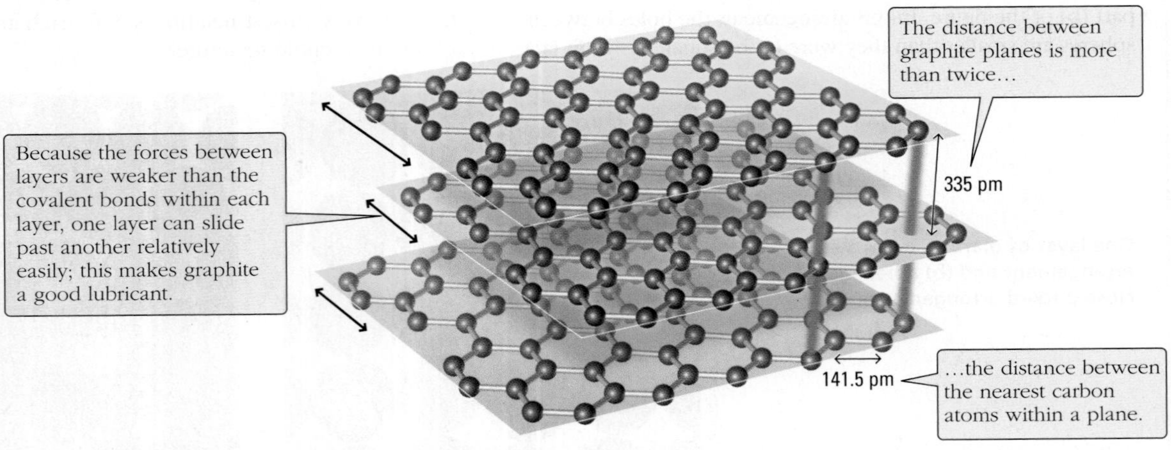
The distance between graphite planes is more than twice...
Because the forces between layers are weaker than the covalent bonds within each layer, one layer can slide past another relatively easily; this makes graphite a good lubricant.
335 pm
141.5 pm
...the distance between the nearest carbon atoms within a plane.

Figure 11.26 The structure of graphite. Three of the many layers of six-membered carbon rings are shown. These layers can slide past one another relatively easily, making graphite a good lubricant. In addition, some of the carbon valence electrons in the layers are delocalized, allowing graphite to be a good conductor of electricity parallel to the layers.

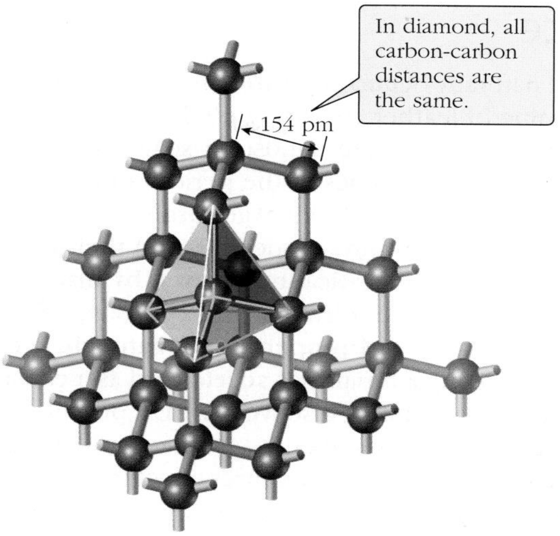
In diamond, all carbon-carbon distances are the same.
154 pm

Figure 11.27 The structure of diamond. Each carbon atom is covalently bonded to four other carbon atoms in an extended tetrahedral arrangement.

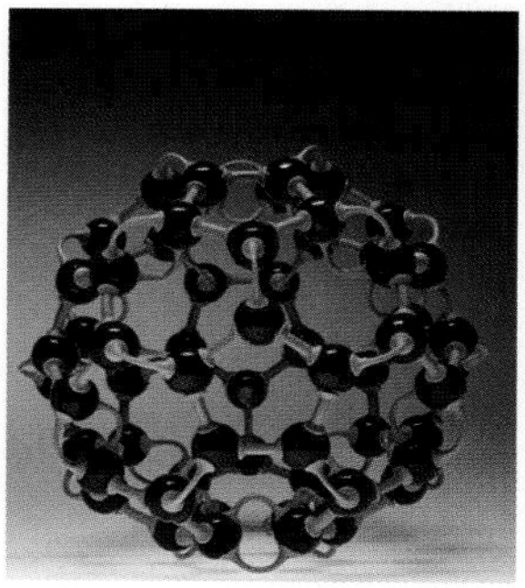

The C_{60} fullerene molecule, known as a buckyball.

each carbon atom, there are puckered six-membered rings in the diamond structure. Diamond (3.51 g/cm^3) is much denser than graphite (2.22 g/cm^3) because the graphite layers are more than twice as far apart as the atoms within a layer. Also, because its valence electrons are localized in covalent bonds between carbon atoms, diamond does not conduct electricity. Diamond is one of the hardest materials and also one of the best conductors of heat known. It is also transparent to visible, infrared, and ultraviolet radiation.

Synthetic diamonds were first produced in the 1950s.

Buckyballs, the prototype of the third allotrope of carbon, were discovered in the 1980s. Since then, a variety of carbon structures have been made, all classified as **fullerenes.** The prototype structure, C_{60}, is shown in the margin; it is a sphere composed of pentagons and hexagons of carbon atoms similar to a soccer ball ***(Figure 1.22c, ← p. 26).*** Larger cage-like structures have also been synthesized. In fullerenes, each carbon atom has three nearest-neighbor carbon atoms. Each carbon atom has sp^2 hybridization, and has one unhybridized p atomic orbital. Electrons can be delocalized over these p atomic orbitals, just as they can in graphite. Unlike graphite, each carbon atom and its three bonds are not quite planar, resulting in the curved fullerene structures. Another fullerene geometry is elongated tubes of carbon atoms, called **nanotubes** because of their small size. Nanotubes have been formed with a single wall of carbon atoms similar to a layer of graphite rolled into a tube, resembling rolled-up chicken wire. The nanotube is capped at each end with a truncated buckyball (Figure 11.28). Nanoscale materials such as nanotubes have physical and chemical properties different from bulk materials. Scientists are now exploring the possible useful properties of nanotubes and other similar materials.

Dorothy Crowfoot Hodgkin

1910–1994

Dorothy Crowfoot Hodgkin was born in Egypt, was schooled in England, and graduated from Oxford University in 1931. She was fascinated by the study of crystals and went to Cambridge University, where she eventually joined the faculty.

Hodgkin's first major achievement in crystallography was the determination of the structure of penicillin in 1945. In 1964 she received the Nobel Prize in Chemistry for determining the molecular structure of vitamin B-12. Much of her work required pioneering and painstaking experimental techniques that, as one biographer put it, "transformed crystallography from a black art into an indispensable scientific tool."

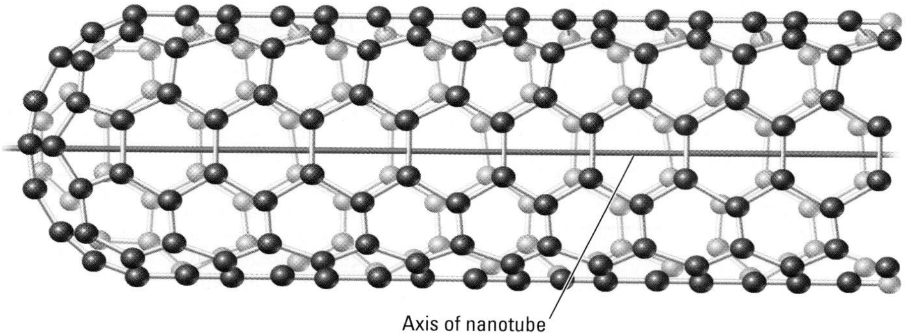
Axis of nanotube

Figure 11.28 A single-walled carbon nanotube composed of hexagons of carbon atoms.

11.8 Materials Science

Humans have long adapted or altered naturally occurring materials for use in clothing, structures, and devices. The tanning of leather and firing of clay pots predate recorded history. Operations such as these alter the nanoscale structure of the material to improve its resulting properties. The names of the periods of early civilization reflect the defining materials of the time—Stone Age, Bronze Age, Iron Age. This same close relationship between the nature of societies and their materials continues to hold in the modern world. The critical role played by materials is a cornerstone of life today. Only recently has the *scientific* understanding of the relationships among the processing, structure, and properties of materials been developed and exploited; the creation of new materials has accelerated as a consequence. Furthermore, new phenomena have been discovered that permit the invention of new materials. For example, superconducting ceramic oxides that conduct electricity without resistance at critical temperatures greater than the boiling point of liquid nitrogen may pave the way to astounding advances in information technology and computing in the coming decades. The study of materials on the nanoscale has led to an entire field of science and technology and the creation of nanostructures such as carbon nanotubes.

TOOLS OF CHEMISTRY X-Ray Crystallography

X-ray crystallography is used to determine the absolute atomic arrangement of the atoms in solid samples. One of the most powerful tools of solid-state chemistry, it is routinely applied to solids as varied as crystalline minerals, proteins, and semiconductors. When X-rays strike crystals, the X-rays are scattered in different directions and with different intensities due to interference effects. X-rays have wavelengths on the order of 1 nm, the same scale as atomic dimensions. When the interference is *constructive,* the waves combine to produce X-rays that can be observed indirectly because they expose a spot on photographic film or create an electrical signal in a detector. When the interference is *destructive,* the waves partly or entirely cancel each other, so little or no X-ray intensity is recorded.

Interference effects among the scattered X-rays from the different atoms cause the intensity of the scattered radiation to show maxima and minima in various directions, creating a diffraction pattern. The particular diffraction pattern obtained depends on the X-ray wavelength and the distances between the planes of atoms in the crystal. The Bragg equation relates the X-ray wavelength (λ), the spacing between atomic layers in the crystal (d), the angle of scattering (θ), and an integer (n, usually considered as 1):

$$n\lambda = 2d \sin \theta$$

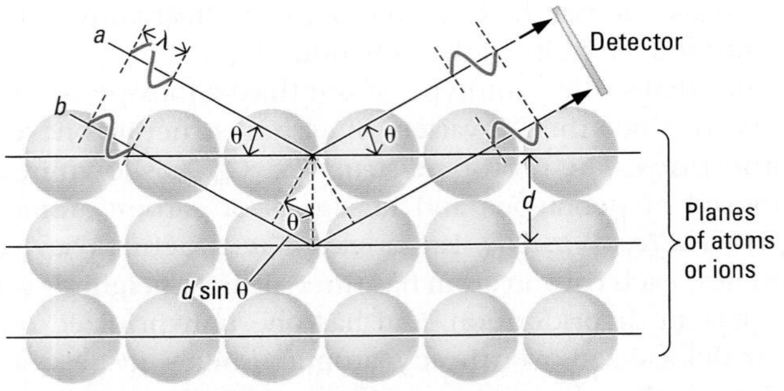

X-ray diffraction from a solid. The relationships among the variables in the Bragg equation are shown. For wave *b* to reach the detector in phase with wave *a*, *b* must travel farther by $2d \sin \theta$, and this distance must be an integer multiple of the X-ray wavelength.

Waves *a* and *b* reinforce each other and produce a spot or electrical signal when the Bragg equation is satisfied—that is, when the additional distance traveled by wave *b* to reach the detector is an integer multiple of the X-ray wavelength. Therefore, the X-ray diffraction pattern can be related back to the nanoscale arrangement of the atoms in the crystal. The intensity of the scattered X-ray beam depends on the electron density of the atoms in the crystal. Therefore, a hydrogen atom is

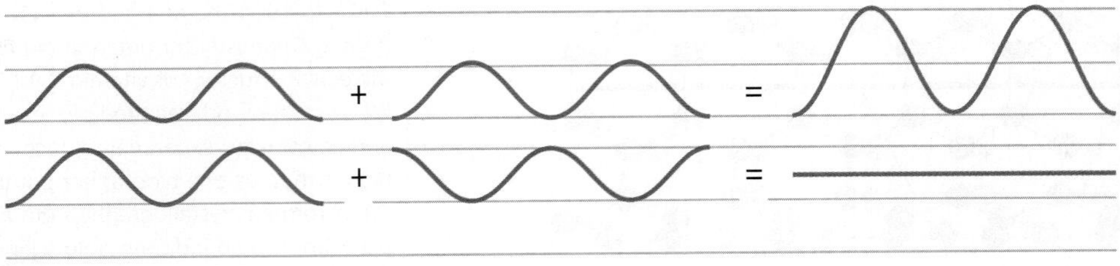

Constructive interference occurs when two in-phase waves combine to produce a wave of greater amplitude.

Destructive interference results from the combination of two waves of equal magnitude that are exactly out of phase. The result is zero amplitude.

Just what are *materials* anyway? By materials we mean matter used to construct the devices and macroscopic structures of our highly developed technology. Application areas include transportation, communication and information technology, infrastructure, consumer goods, health care, and biomaterials. Properties that can be important in materials cover the entire range of characteristics of matter: electronic, optical, magnetic, superconducting, strength, and flexibility.

The four major classes of materials are metals, polymers, ceramics, and composites. Materials in each of the four major categories tend to have specific properties that are similar:

- **Metals** Opaque with shiny surface (unless covered by an oxidized coating, as in rusted iron). Usually crystalline. Generally ductile—can be drawn into wires (copper). Strong and stiff (steel). Most are not magnetic, except Fe, Ni, and Co or alloys. Usually conduct heat and electricity well (Cu). Form alloys such as stainless steel, bronze, and brass.
- **Polymers** Both natural (wool, cotton, rubber) and synthetic (polyethylene, polystyrene). Most are flexible and can be bent, but some are brittle. Thermoplastics become more flexible when heated; thermoset polymers become hard and brittle after heating. Usually thermal and electrical insulators. Usually not magnetic.

Polymers are discussed in detail in Chapter 12.

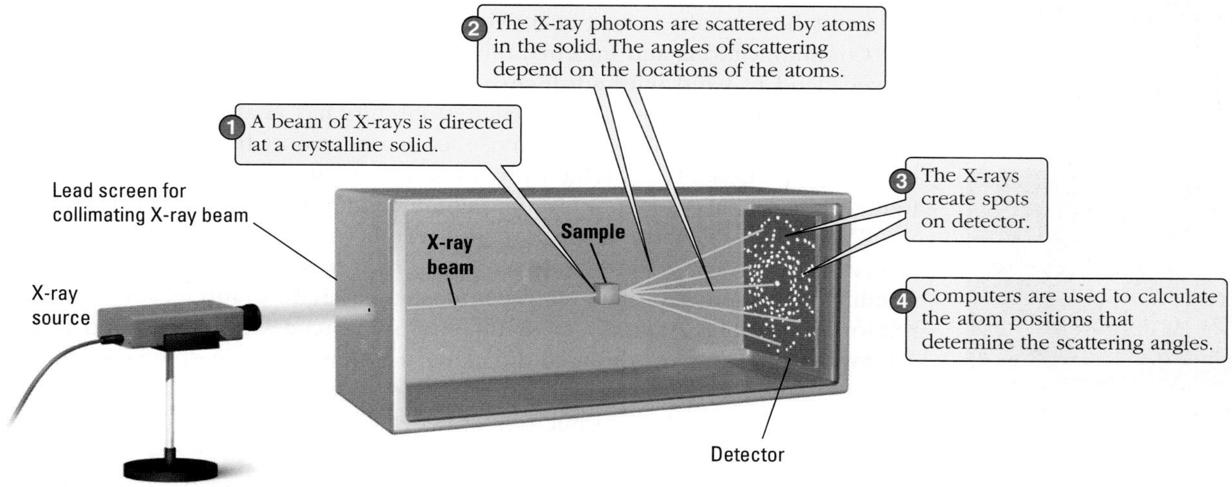
2 The X-ray photons are scattered by atoms in the solid. The angles of scattering depend on the locations of the atoms.
1 A beam of X-rays is directed at a crystalline solid.
Lead screen for collimating X-ray beam
X-ray beam
Sample
3 The X-rays create spots on detector.
X-ray source
4 Computers are used to calculate the atom positions that determine the scattering angles.
Detector

An X-ray diffraction experiment.

least effective in causing X-rays to scatter, while heavy atoms such as lead or mercury are quite effective.

Photographic film was used to record the diffraction patterns in early instruments. Today, crystallographers use computer-controlled instruments, highly sensitive solid-state detectors, and other advanced instrumentation to measure the angles of reflection and the intensities of X-ray beams diffracted by crystals. Sophisticated software allows the atomic arrangements in crystalline materials to be determined quickly enough that X-ray crystallography can be used routinely.

In one of the most famous uses of X-ray crystallography, James Watson and Francis Crick in 1953 interpreted X-ray data obtained by Maurice Wilkins and Rosalind Franklin to discover the double helix structure of DNA, the biological macromolecule that carries genetic information ***(← p. 410).*** Watson, Crick, and Wilkins were awarded the Nobel Prize in Physiology or Medicine in 1962 for this discovery (Franklin was deceased and therefore could not share the prize).

- **Ceramics** Nonmetallic materials, often based on clays. Porcelain, bricks, glass, and cement are examples. Usually brittle and cannot be bent. Poor thermal conductors and good thermal insulators; can withstand high temperatures. Usually poor electrical conductors and good electrical insulators. Some ceramics are magnetic. Usually crystalline. Many are transparent (glass). Semiconductors are ceramics with special electrical properties. Glasses, a special class of ceramics, are brittle yet often strong and are amorphous.
- **Composites** Materials combining components from the other three categories with properties that blend the best of both components depending on the application.

Materials science is the science of the relationships between the structure and the chemical and physical *properties* of materials. How strong is the material? How dense? How easily formed? How does it react to acid, stress, magnetic fields, high pressure, heat? Does it conduct electricity? *Structure* refers to the organization of the material on scales ranging from the nanoscale (electronic structure, arrangement of atoms, crystal arrangement) through the microscale and to the macroscale.

The major breakthrough of the second half of the twentieth century with respect to materials was that scientists and engineers can now often *design* a material for a specific application. That is, scientific understanding at the nanoscale permits synthesis and fabrication of new materials with precise properties desired for a particular application. Coupling this capability with inventiveness and technological know-how results in cost-effective manufacture of materials and devices.

Communications and information systems provide examples of advances that moved hand in hand with advances in materials. The sequential development of vacuum tubes, transistors, and integrated circuits has enabled our technology to advance. Lasers, which provide intense, coherent, monochromatic light, and optical fibers (Section 11.11), which provide a means of carrying optical signals over long distances, have revolutionized communications and many other fields.

The twentieth-century transportation revolution was also enabled by advances in materials. Aircraft require light yet strong materials, and the advances from aluminum and its alloys to titanium to composite materials (graphite- and boron fiber-reinforced polymers, aluminum or titanium reinforced with fibers, and graphite-reinforced epoxy) have advanced the modern aircraft industry. Similar materials are used in high-performance mountain bike frames, tennis racquets, skis, and other sporting equipment. At the same time, the development of materials that can withstand the extreme temperatures and stresses in rocket and jet engine turbine blades has made modern aircraft and rockets possible. Automobiles have an ever-growing number of parts made from aluminum alloys, engineered plastics, and ceramics.

Modern materials in action. Modern materials are employed in many ways to improve our lives—for example, in hip replacements, bicycle parts, and fleece garments.

11.9 Metals, Semiconductors, and Insulators

All metals are solids at room temperature except mercury (m.p. −38.8 °C), and they exhibit **common properties that we call metallic.**

- **High electrical conductivity** Metal wires carry electricity from power plants to homes and offices because electrons in metals are highly mobile.
- **High thermal conductivity** We learn early in life not to touch any part of a hot metal pot because it will transfer thermal energy rapidly and painfully.
- **Ductility and malleability** Most metals are easily drawn into wire (ductility) or hammered into thin sheets (malleability); some metals (gold, for example) are more easily formed into shapes than others.
- **Luster** Polished metal surfaces reflect light. Most metals have a silvery-white color because they reflect all wavelengths equally well.

- **Insolubility in water and other common solvents** No metal dissolves in water, but a few (mainly from Groups 1A and 2A) react with water to form hydrogen gas and solutions of metal hydroxides.

As you will see, these properties are explained by the kinds of bonding that hold atoms together in metals.

The enthalpies of fusion and melting points of metals vary greatly. Low melting points correlate with low enthalpies of fusion, which implies weaker attractive forces holding the metal atoms together. Mercury, which is a liquid at room temperature, has an enthalpy of fusion of only 2.3 kJ/mol. The alkali metals and gallium also have very low enthalpies of fusion and notably low melting points (Table 11.6). Compare these values with those in Table 11.3 for nonmetals (p. 489).

Figure 11.29 shows the relative enthalpies of fusion of the metals related to their positions in the periodic table. The transition metals, especially those in the third transition series, have very high melting points and extraordinarily high enthalpies of fusion. Tungsten (W) has the highest melting point (3410 °C) of all the metals, and among all the elements it is second only to carbon as graphite, which has a melting point of 3550 °C. Pure tungsten is used in incandescent light bulbs as the filament, the wire that glows white-hot. No other material has been found to be better since the invention of commercially successful, long-lived light bulbs in 1908 by Thomas Edison and his co-workers.

Table 11.6 Enthalpies of Fusion and Melting Points of Some Metals

Metal	ΔH°_{fusion} (kJ/mol)	Melting Point (°C)
Hg	2.3	−38.8
Ga	7.5	29.78*
Na	2.59	97.9
Li	3.0	180.5
Al	10.7	660.4
U	12.6	1132.1
Fe	13.8	1535.0
Ti	20.9	1660.1
Cr	16.9	1857.0
W	35.2	3410.1

*This means that gallium metal will melt in the palm of your hand from the warmth of your body (37 °C). It happens that gallium is a liquid over the largest range of temperature of any metal. Its boiling point is approximately 2250 °C.

PROBLEM-SOLVING EXAMPLE **11.9** **Calculating Enthalpies of Fusion**

Use data from Table 11.6 to calculate the thermal energy transfer required to melt 10.0 g chromium at its melting point.

Answer 3.25 kJ

Strategy and Explanation First, we determine how many moles of chromium are present. Then we calculate the energy required from the ΔH°_{fusion} value in Table 11.6.

$$10.0 \text{ g Cr} \times \frac{1 \text{ mol Cr}}{52.00 \text{ g Cr}} = 0.192 \text{ mol Cr}$$

$$0.192 \text{ mol Cr} \times \frac{16.9 \text{ kJ}}{1 \text{ mol Cr}} = 3.25 \text{ kJ}$$

☑ **Reasonable Answer Check** We have about 0.2 mol chromium, so the energy required should be about 0.2 of its molar enthalpy of fusion, about 17 kJ/mol, or about 3.4 kJ, which compares well with our calculated answer.

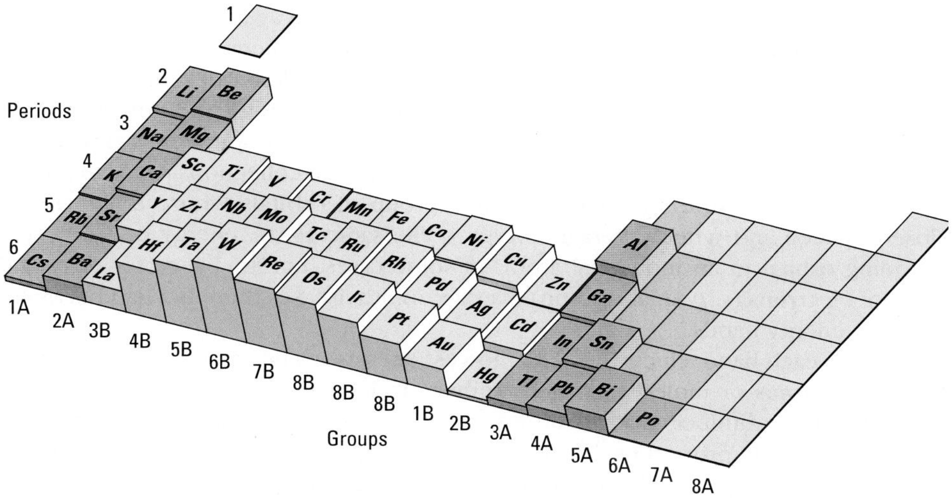

Figure 11.29 Relative enthalpies of fusion for the metals in the periodic table. See Table 11.6 for some numerical values of enthalpies of fusion.

PROBLEM-SOLVING PRACTICE 11.9

Use data from Table 11.6 to calculate the thermal energy transfer required to melt 1.45 g aluminum.

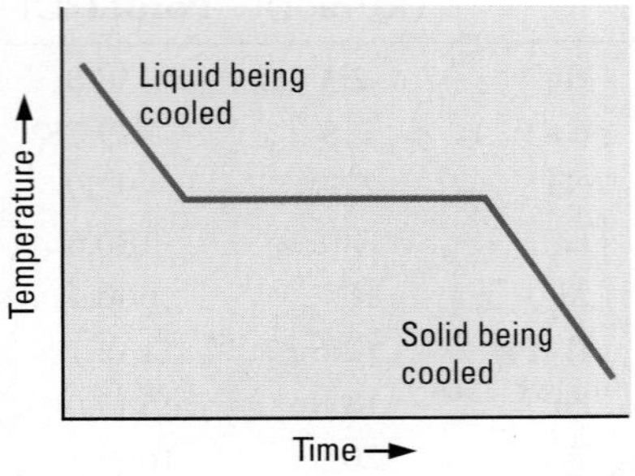

CONCEPTUAL EXERCISE 11.15 Cooling a Liquid Metal Until It Solidifies

When a liquid metal is cooled at a constant rate to the temperature at which it solidifies, and the solid is then cooled to an even lower temperature, the cooling curve—a plot of temperature against time—looks like the graph to the left. Account for the shape of this graph. Would all substances exhibit similar graphs?

CONCEPTUAL EXERCISE 11.16 Heats of Fusion and Electronic Configuration

Look in Appendix D and compare the electron configurations shown there with the heats of fusion for the metals shown in Table 11.6. Is there any correlation between these configurations and this property? Does strength of attraction among metal atoms correlate with number of valence electrons? Explain.

Electrons in Metals, Semiconductors, and Insulators

Metals behave as though metal cations exist in a "sea" of mobile electrons—the valence electrons of all the metal atoms. **Metallic bonding** is the nondirectional attraction between positive metal ions and the surrounding sea of negative charge (valence electrons). Each metal ion has a large number of near neighbors. The valence electrons are spread throughout the metal's crystal lattice, holding the positive metal ions together. When an electric field is applied to a metal, these valence electrons move toward the positive end of the field, and the metal conducts electricity.

The mobile valence electrons provide a uniform charge distribution in a metal lattice, so the positions of the positive ions can be changed without destroying the attractions among positive ions and electrons. Thus, most metals can be bent and drawn into wire. Conversely, when we try to deform an ionic solid, which consists of a lattice of positive and negative ions, the crystal usually shatters because the balance of positive ions surrounded by negative ions, and vice versa, is disrupted.

To visualize how bonding electrons behave in a metal, first consider the arrangement of electrons in an individual atom far enough away from any neighbor so that no bonding occurs. In such an atom the electrons occupy orbitals that have definite energy levels. In a large number of separated, identical atoms, all of the energy levels are identical. If the atoms are brought closer together, however, they begin to influence one another. The identical energy levels shift up or down and become bands of energy levels characteristic of the large collections of metal atoms (Figure 11.30). An **energy band** is a large group of orbitals whose energies are closely spaced and whose average energy is the same as the energy of the corresponding orbital in an individual atom. In some cases, energy bands for different types of electrons (*s*, *p*, *d*, and so on) overlap; in other cases there is a gap between different energy bands.

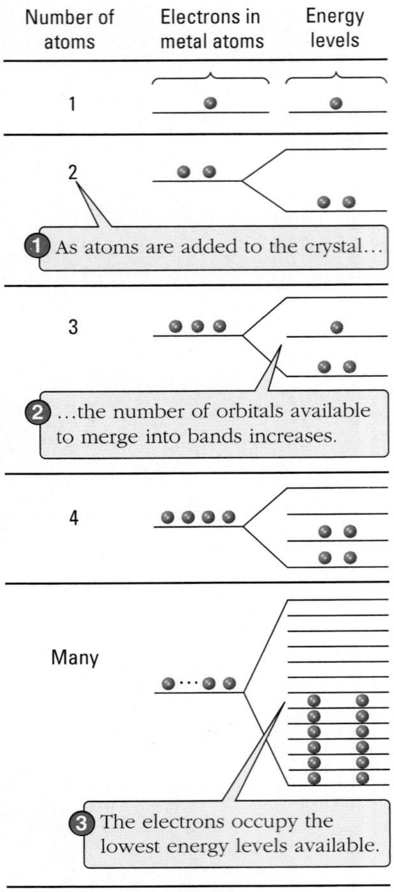

Figure 11.30 Formation of bands of electron orbitals in a metal crystal.

Within each band, electrons fill the lowest energy orbitals much as electrons fill orbitals in atoms or molecules. The number of electrons in a given energy band depends on the number of metal atoms in the crystal. In considering conductivity and other metallic properties, it is usually necessary to consider only valence electrons, as other electrons all occupy completely filled bands in which two electrons occupy every orbital. In these bands, no electron can move from one orbital to another, because there is no empty spot for it.

A band containing the valence electrons is called the **valence band**. If the valence band is partially filled it requires little added energy to excite a valence electron to a slightly higher energy orbital. Such a small increment of energy can be provided by applying an electric field, for example. The presence of low-energy, empty orbitals into which electrons can move allows the electrons to be mobile and to conduct an electric current (Figure 11.31).

Another band containing higher energy orbitals (the **conduction band**) exists at an average energy above the valence band. In a metal, the valence band and the conduction band overlap, so electrons can move from the valence band to the conduction band freely; this ability explains metals' electrical conductivity. Such a metal with overlapping valence and conduction bands is an electrical **conductor.**

When the conduction band is close in energy to the valence band, the electrons can absorb a wide range of wavelengths in the visible region of the spectrum. As the excited electrons fall back to their lower energy states, they emit their extra energy as visible light, producing the luster characteristic of metals.

The energy-band theory also explains why some solids are electrical **insulators,** and do not conduct electricity. In an insulator the valence band and conduction band do not overlap. Rather, there is a large band gap between them (Figure 11.31). Very few electrons have enough energy to move across the large gap from a filled lower energy band to an empty higher energy band, so no current flows through an insulator when an external electric field is applied.

In a **semiconductor** a very *narrow* energy gap separates the valence band and the conduction band (Figure 11.31). At quite low temperatures, electrons remain in the filled lower energy valence band, and semiconductors are not good conductors. At higher temperatures, or when an electric field is applied, some electrons have enough energy to jump across the band gap into the conduction band. This allows an electric current to flow. This property of semiconductors—to switch from insulator to conductor with the application of an external electric field—is the basis for the operation of transistors, the cornerstone of modern electronics.

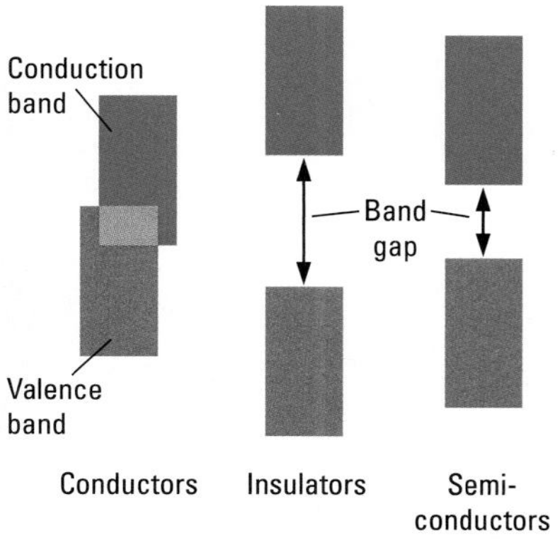

Figure 11.31 Differences in the energy bands of available orbitals in conductors, insulators, and semiconductors.

Fred Leonero/Shutterstock

Ceramic insulators are used on high-voltage electrical transmission lines.

Superconductors

One interesting property of metals is that their electrical conductivity *decreases* with increasing temperature. In Figure 11.32b a metal in a burner flame shows higher resistance (lower conductivity) than the same metal at room temperature (Figure 11.32a). The lower conductivity of metals at higher temperatures can be explained by considering valence electrons in the lattice of metal ions as waves. As an electron wave moves through the metal crystal under the influence of an electrical voltage, the wave encounters lattice positions where the metal ions are close enough together to scatter it. This scattering is analogous to the scattering of X-rays caused by atoms in crystals. The scattered electron wave moves off in another direction, only to be scattered again when it encounters some other occupied lattice position. All of this scattering *lowers* the conductivity of the metal. At higher temperatures, the metal ions vibrate more, and the distances between lattice positions change more from their average values. This effect causes more scattering of electron waves as they move through the crystal, because there are now more possibilities of unfavorable lattice spacings. Hence, electrical conductivity is lower.

Recall from Section 7.4 (← ***p. 285***) that electrons have wave-like properties as well as particle-like properties.

From this picture of electrical conductivity, it might be expected that the conductivity of a metal crystal at absolute zero (0 K or −273.15 °C) might be very large. In fact, the conductivity of a pure metal crystal approaches infinity as absolute zero is approached. In some metals, however, a more interesting phenomenon occurs. At a superconducting transition temperature (T_c) that is low but finite, the conductivity abruptly increases to infinity, which means that the resistance drops to zero (Figure 11.33). As a result, the metal becomes a **superconductor** of electricity. A superconductor offers no resistance whatsoever to electric current. Once a current has been started in a superconducting circuit, it continues to flow indefinitely. No clear theory explaining superconductivity has yet emerged, but it appears that the

Superconductivity was discovered in 1911 by the Dutch physicist Kamerlingh Onnes, who won the 1913 Nobel Prize in Physics for his work.

Information about superconductors is available at this website: **http://superconductors.org**.

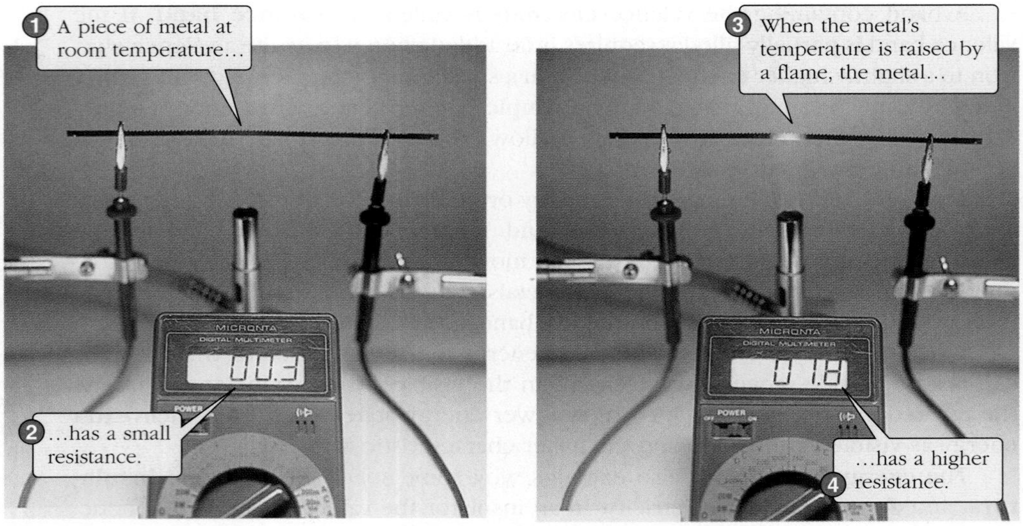

© Cengage Learning/Charles D. Winters

(a) Metal at room temperature (b) After metal is heated

Figure 11.32 Resistance, temperature, and electrical conductivity.

scattering of electron waves by vibrating atoms is replaced in a superconductor by some cooperative action that allows the electrons to move through the crystal unhindered.

If a material could be made superconducting at a high enough temperature, the material would find uses in the transmission of electrical energy, in high-efficiency motors, and in computers and other devices. Table 11.7 lists some metals that display superconductivity. While some useful devices can be fabricated from these metals, the very low temperatures at which they become superconductors make them impractical for most applications. Shortly after superconductivity of metals was discovered, alloys were prepared that had higher transition temperatures. Niobium alloys were the best, but still required cooling to below 23 K (−250 °C) to exhibit superconductivity. Maintenance of such a low temperature requires liquid helium, which is expensive.

Müller and Bednorz received the 1987 Nobel Prize in Physics for their discovery of the superconducting properties of $LaBa_2Cu_3O_x$.

This picture abruptly changed in January 1986, when Alex Müller and Georg Bednorz, IBM scientists in Switzerland, discovered that a lanthanum-barium-copper oxide became superconducting at 35 K (−238 °C). This type of mixed metal oxide is a ceramic with the same structure as the mineral perovskite, $CaTiO_3$, and, therefore, would be expected to have insulating properties. The announcement of the

Table 11.7 Superconducting Transition Temperatures of Some Metals*

Metal	Superconducting Transition Temperature (K)
Gallium	1.10
Aluminum	1.15
Tin	3.72
Mercury	4.15
Lanthanum	4.9
Lead	7.2

*Not all metals have superconducting properties. Those that can become superconductors at atmospheric pressure are Al, Ti, Zn, Ga, Zr, Mo, Tc, Ru, Cd, In, Sn, La, Hf, Ta, W, Re, Os, Ir, Hg, Tl, Pb, Th, Pa, and U.

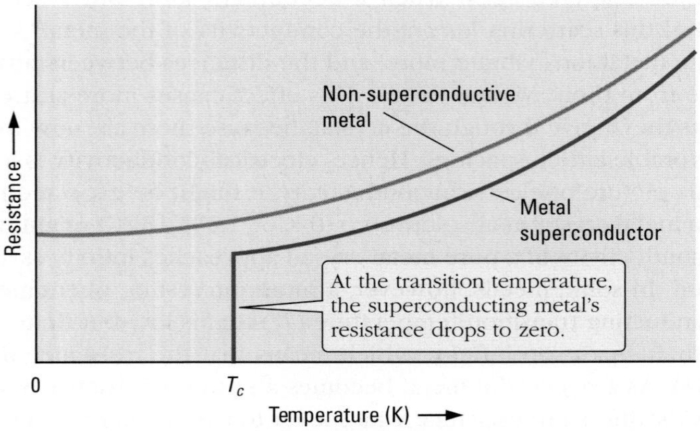

Figure 11.33 The resistance of a superconducting metal and a non-superconducting metal as a function of temperature.

superconducting properties of $LaBa_2Cu_3O_x$ (x means that varying amounts of oxygen are incorporated) excited the world of science and provoked a flurry of activity to prepare related compounds in the hope that even higher transition temperatures could be found. Within four months, a material, $YBa_2Cu_3O_7$, that became superconducting at 90 K (−183 °C) was announced. Its identification was a major breakthrough because 90 K exceeds the boiling point of liquid nitrogen (bp = 77 K or −196 °C), which is easily obtainable and cheap. The current record transition temperature of 138 K (−135 °C) at atmospheric pressure is for a thallium-doped mercuric-cuprate ceramic with the composition $Hg_{0.8}Tl_{0.2}Ba_2Ca_2Cu_3O_{8.23}$.

Why the great excitement over the potential of superconductivity? Superconducting materials would allow the building of more powerful electromagnets such as those used in nuclear particle accelerators and in magnetic resonance imaging (MRI) machines for medical diagnosis (← p. 552). Such electromagnets would permit higher energies to be maintained for longer periods of time (and at lower cost) in the case of the particle accelerators, and support better imaging of problem areas in a patient's body. Currently, the main barrier to their wider application is the cost of cooling the magnets used in these devices.

Many scientists say that the discovery of high-temperature superconducting materials is more important than the discovery of the transistor because of its potential effect on electrical and electronic technology. For example, the use of superconducting materials for transmission of electric power could save as much as 30% of the energy now lost because of the resistance of the wire. Superchips for computers could be as much as 1000 times faster than existing silicon chips. Since a superconductor repels magnetic materials (Figure 11.34), trains can be levitated above a track and move with no friction other than air resistance to slow them down. When a maglev (magnetic levitation) rail line in Japan was tested on April 4, 1999, the test train attained a speed of 343 mi/h (552 km/h). On December 2, 2003, a Japanese three-car test train attained a maximum speed of 581 km/h. On December 31, 2002, the world's first commercial maglev train service ran at 430 km/h over a 30-km link from Pudong airport to Pudong (adjacent to Shanghai).

IBM Research/Peter Arnold, Inc.

Figure 11.34 Levitation of a magnet above a superconductor. A magnet moving above a superconductor induces a current in the superconductor, which induces a magnetic field. This magnetic field and that of the magnet oppose one another; the repulsive magnetic field strength is balanced by the force of gravity.

11.10 Silicon and the Chip

Silicon is known as an "intrinsic" semiconductor because the element itself is a semiconductor. Silicon of about 98% purity can be obtained by heating silica (purified sand) and coke (an impure form of carbon) at 3000 °C in an electric arc furnace.

$$SiO_2(s) + 2\ C(s) \xrightarrow{\text{heat}} Si(s) + 2\ CO(g)$$

Silicon of this purity can be alloyed with aluminum and magnesium to increase the hardness and durability of the metals and is used for making silicone polymers. For use in electronic devices, however, a much higher degree of purification is needed. Impure silicon can be reacted with chlorine to make $SiCl_4(\ell)$, which can be purified by distillation. High-purity silicon can be prepared by reducing $SiCl_4$ with magnesium.

$$SiCl_4(\ell) + 2\ Mg(s) \longrightarrow Si(s) + 2\ MgCl_2(s)$$

Magnesium chloride, which is water-soluble, is washed from the silicon. The final purification of the silicon takes place by a melting process called **zone refining** (Figure 11.35), which produces silicon containing less than one part per billion of impurities such as boron, aluminum, and arsenic. Zone refining takes advantage of the fact that impurities are often more soluble in the liquid phase than in the solid phase. As a hot molten zone is moved through a sample being purified, the impurities move along in the liquefied portion of the sample. As the heated zone cools, the sample that resolidifies is purer than it was. Multiple passes of the hot molten zone are usually necessary to achieve the degree of purity necessary to fabricate semiconducting devices.

Like all semiconductors, high-purity silicon fails to conduct an electric current until a certain electrical voltage is applied, but at higher voltages it conducts mod-

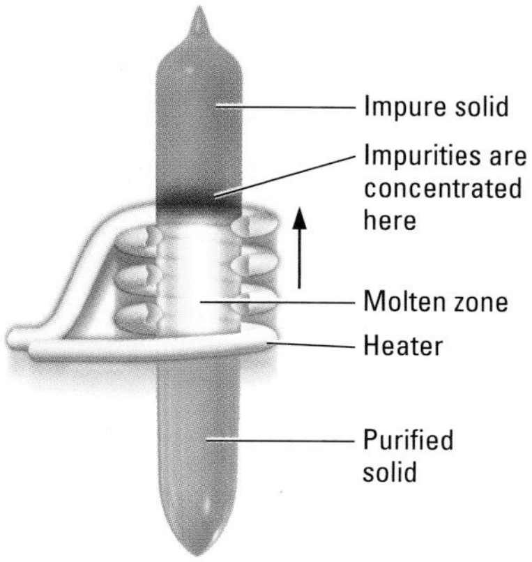

Impure solid
Impurities are concentrated here
Molten zone
Heater
Purified solid

Figure 11.35 Schematic of the zone refining process for silicon.

erately well. Silicon's semiconducting properties can be improved dramatically by a process known as doping. **Doping** is the addition of a tiny amount of some other element (a *dopant*) to the silicon.

For example, consider what happens when a few atoms of a Group 5A element such as arsenic are substituted for a tiny fraction of the silicon atoms in solid silicon. Arsenic has five valence electrons, whereas silicon has four. Thus, only four of the five valence electrons of As are used for bonding with four Si atoms, leaving one electron relatively free to move (Figure 11.36). This type of doped silicon is referred to as negative-type or *n-type silicon* because it has extra (negative) valence electrons. In ***n*-type semiconductors**, electrons are the charge carriers.

On the other hand, consider what happens when a small number of boron atoms (or atoms of some other Group 3A element) replace silicon atoms (Group 4A) in solid silicon. Boron has only three valence electrons. This leaves a deficiency of one electron around the B atom, creating what is called a *hole* for every B atom added. Hence silicon doped in this manner is referred to as positive-type or *p-type silicon* (Figure 11.36). In ***p*-type semiconductors**, holes are the charge carriers. It is important to remember that such holes in semiconductors are the absence of electrons. Thus, the flow of holes in one direction is in reality the flow of electrons in the opposite direction.

A positive hole in an energy band of a semiconductor is a place where there is one less electron than normal and hence extra positive charge.

When *p*-type and *n*-type semiconductors are brought together, a ***p-n* junction** results (Figure 11.37). Such a junction allows current to flow in one direction but not the other. When the two materials are joined, some of the excess electrons in the *n*-type material migrate across the junction and some of the holes in the *p*-type material migrate in the opposite direction. The result is the buildup of a negative charge on the *p*-type region and a positive charge on the *n*-type region. This charge buildup is called the *junction potential,* and it prevents the further migration of electrons or holes. However, if an external potential is applied to the *p-n* junction, one of two effects can result. If a negative charge is connected to the *p*-type side and a positive charge to the *n*-type side, then the electrons and holes migrate away from the junction, and no current flows through the junction (Figure 11.37, *bottom*). If a negative charge is connected to the *n*-type side and a positive charge to the *p*-type side, then both electrons and holes flow across the junction, and current flows through the device (Figure 11.37, *top*). Thus, the *p-n* junction acts to pass current one way but not the other.

These *p-n* junctions can be joined into larger composite structures to create transistors and integrated circuits. A single integrated circuit often contains thousands or even millions of transistors, as well as the circuits to carry the electrical signals. These devices, in the form of computer memories and central processing units (CPUs), permeate our society.

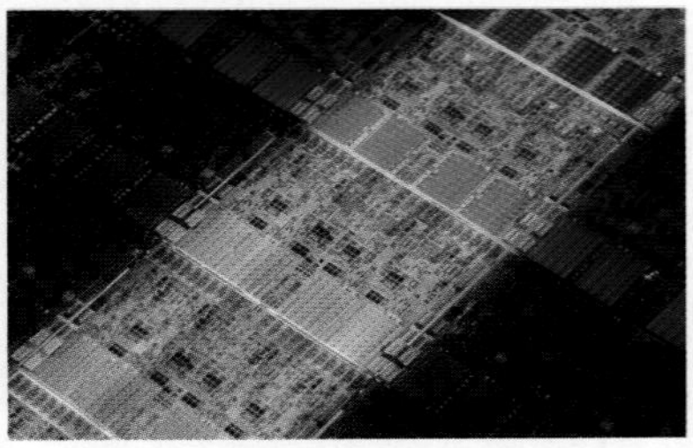

Courtesy of Intel

Intel microprocessor.

Doped silicon is also the basis of the **solar cell** (Figure 11.38). When a layer of *n*-type doped silicon is next to a *p*-type layer, there is a strong tendency for the extra electrons in the *n*-type layer to pair with the unpaired electrons in the holes of the *p*-type layer. If the two layers are connected by an external electrical circuit, light that strikes the silicon provides enough energy to cause an electrical current to

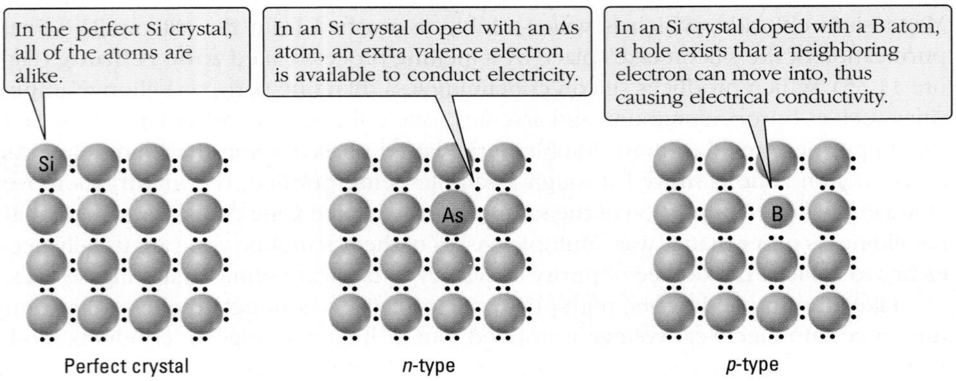
In the perfect Si crystal, all of the atoms are alike.
In an Si crystal doped with an As atom, an extra valence electron is available to conduct electricity.
In an Si crystal doped with a B atom, a hole exists that a neighboring electron can move into, thus causing electrical conductivity.
Si
As
B
Perfect crystal
n-type
p-type

Figure 11.36 Schematic drawing of semiconductor crystals derived from silicon.

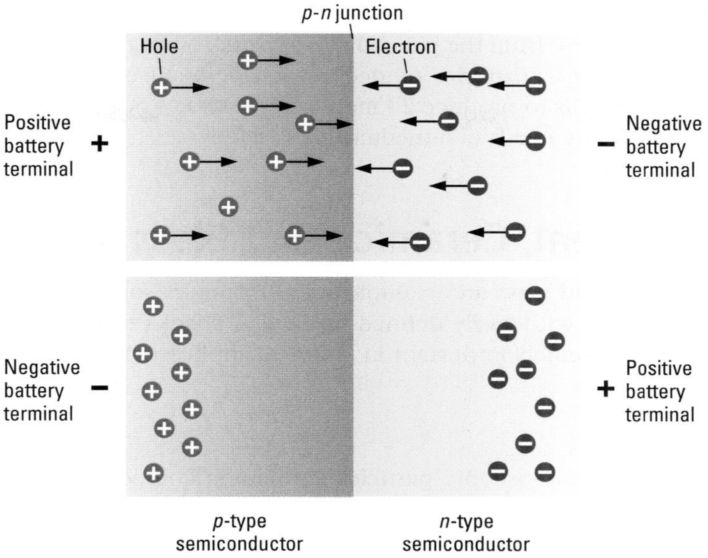
p-n junction
Hole
Electron
Positive battery terminal +
Negative battery terminal −
Negative battery terminal −
Positive battery terminal +
p-type semiconductor
n-type semiconductor

Figure 11.37 A *p-n* junction. Placing a *p*-type semiconductor next to an *n*-type semiconductor creates a *p-n* junction. The charge carriers in the *p*-type region are holes, and the charge carriers in the *n*-type region are electrons. *Top:* The negative battery terminal is connected to the *n*-type region, and the positive battery terminal is connected to the *p*-type region. The holes and electrons move as indicated by the arrows. Electric current flows through the junction and the overall device. *Bottom:* If the battery terminals are reversed, the holes and electrons are attracted to the edges of the regions, charge depletion occurs near the junction, and no current flows through the junction or the overall device.

National Renewable Energy Laboratory

© Cengage Learning/Charles D. Winters

Solar power can provide electricity in remote areas. These Brazilian houses and this water pump in upstate New York are powered by solar-generated electricity.

flow. When a photon is absorbed, it excites an electron to a higher-energy orbital, allowing the electron to leave the *n*-type silicon layer and flow through the external circuit to the *p*-type layer. As the *p*-type layer becomes more negative (because of added electrons), the extra electrons are repelled *internally* back into the *n*-type layer (which has become positive because of the loss of electrons via the circuit). This process can continue as long as the silicon layers are exposed to sunlight and the circuit remains closed.

Solar cells are on the threshold of a major technological breakthrough, perhaps comparable to the computer chip. Although experimental solar-powered automo-

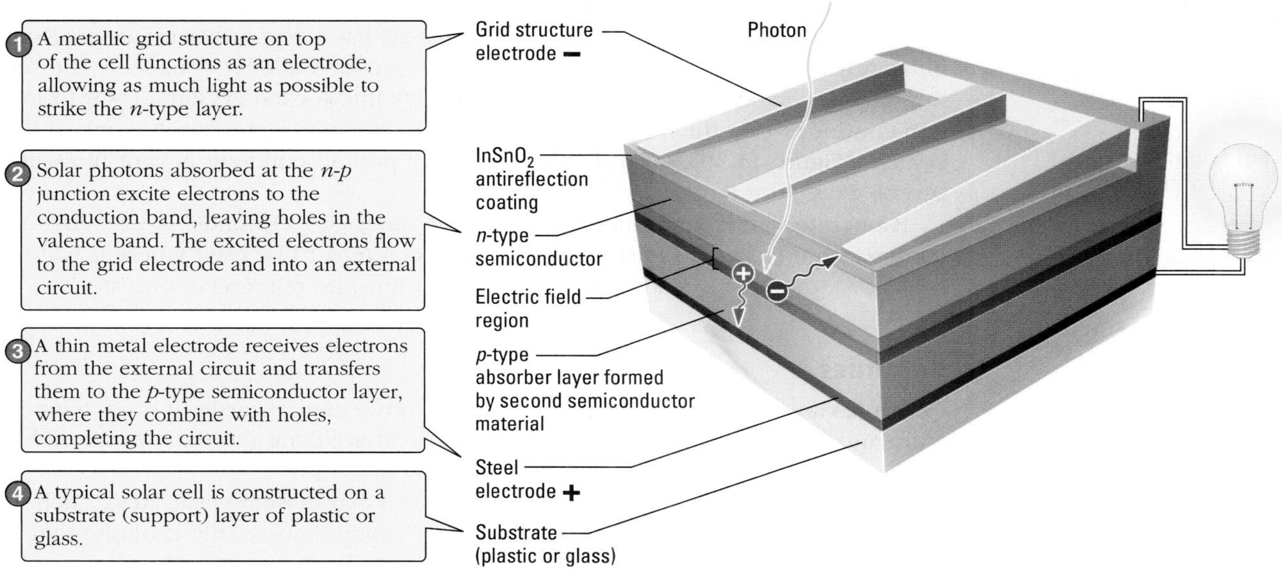
1 A metallic grid structure on top of the cell functions as an electrode, allowing as much light as possible to strike the n-type layer.
2 Solar photons absorbed at the n-p junction excite electrons to the conduction band, leaving holes in the valence band. The excited electrons flow to the grid electrode and into an external circuit.
3 A thin metal electrode receives electrons from the external circuit and transfers them to the p-type semiconductor layer, where they combine with holes, completing the circuit.
4 A typical solar cell is constructed on a substrate (support) layer of plastic or glass.
Grid structure electrode −
Photon
InSnO2 antireflection coating
n-type semiconductor
Electric field region
p-type absorber layer formed by second semiconductor material
Steel electrode +
Substrate (plastic or glass)

Figure 11.38 Typical photovoltaic cell using layers of doped silicon.

biles are now available and many novel applications of solar cells exist, the real breakthrough will come from the general use of banks of solar cells at utility power plants to produce huge quantities of electricity. One plant operating in California uses banks of solar cells to produce 20 megawatts (MW) of power—enough to supply the daily electricity needs of a medium-sized city.

11.11 Cement, Ceramics, and Glass

Cement, ceramics, and glass are examples of amorphous solids; that is, they lack crystalline structures with easily defined unit cells. Thanks to their useful properties, all three are extremely important in practical applications.

Cement

Courtesy of J.H. Findorff & Son

Figure 11.39 Pouring concrete. After the concrete has been formed and exposed surfaces smoothed, it is allowed to set, a process hastened by moisture, in which strong bonds are formed between many of the tiny particles originally present in the dry concrete.

An article on concrete is in *Journal of Chemical Education*, Vol. 83, 2006; pp. 1425–1427. An informative overview of cement chemistry is found in *Journal of Chemical Education*, Vol. 80, 2003; pp. 623–635.

An article describing new uses for concrete is found in *Science News*, January 1, 2005; p. 7.

Compressive strength refers to a solid's resistance to shattering under compression.

Tensile strength refers to a substance's resistance to stretching.

Cement consists of microscopic particles containing compounds in varying proportions of calcium, iron, aluminum and silicon with oxygen. In the presence of water, cement forms hydrated particles with large surface areas, which subsequently undergo recrystallization and reaction to bond to themselves as well as to the surfaces of bricks, stone, or other silicate materials.

Cement is made by roasting a powdered mixture of calcium carbonate (limestone or chalk), silica (sand), aluminosilicate mineral (kaolin, clay, or shale), and iron oxide at a temperature of up to 870 °C in a rotating kiln. As the materials pass through the kiln, they lose water and carbon dioxide and ultimately form a "clinker," in which the materials are partially fused. A small amount of calcium sulfate is added, and the cooled clinker is then ground to a very fine powder. A typical composition of cement is 60% to 67% CaO, 17% to 25% SiO_2, 3% to 8% Al_2O_3, up to 6% Fe_2O_3, and small amounts of magnesium oxide, magnesium sulfate, and oxides of potassium and sodium.

Cement is usually mixed with other substances. *Mortar* is a mixture of cement, sand, water, and lime. **Concrete** is a mixture of cement, sand, and aggregate (crushed stone or pebbles) in proportions that vary according to the application and the strength required.

In cement, the oxides are not isolated into ionic crystals, but an entire nanoscale structure forms that is a complex network of ions, each satisfying its charge requirements with ions of opposite charge. Many different reactions occur during the setting of cement. Various constituents react with water and with carbon dioxide in the air. The initial reaction of cement with water involves the hydrolysis of the calcium silicates, which forms a gel that sticks to itself and to the other particles (sand, crushed stone, or gravel). This gel has a very large surface area and ultimately is responsible for the great strength of concrete once it has set. The setting process involves formation of small, densely interlocked crystals. Their formation continues after the initial setting and increases the compressive strength of the cement (Figure 11.39). For this reason, freshly poured concrete is kept moist for several days.

Concrete, like many other materials containing Si—O bonds, is very hard to compress but lacks tensile strength. When concrete is to be used where it will be subject to tension, as in a bridge or building, it must be reinforced with steel.

Ceramics

Ceramics are generally fashioned from clay or other natural earths at room temperature and then permanently hardened by heat in a baking ("firing") process that binds the particles together.

Silicate ceramics include objects made from clays (aluminosilicates), such as pottery, bricks, and table china. China clay, or kaolin, is primarily kaolinite that is practically free of iron, which otherwise imparts a red color to the clay. As a result, china clay is white and particularly valuable in making fine pottery. Clays mixed with water form a moldable paste consisting of tiny silicate sheets that can easily slide past one another. When heated, the water is driven off, and new Si—O—Si bonds form so that the mass becomes permanently rigid.

Oxide ceramics are produced from powdered metal oxides such as alumina, Al_2O_3, and magnesia, MgO, by heating the solids under pressure, causing the particles to bind to one another and thereby form a rigid solid. Because it has a high electrical resistivity, alumina is used in spark plug insulators. High-density alumina has very high mechanical strength, so it is also used in armor plating and in high-speed cutting tools for machining metals. Magnesia is an insulator with a high melting point (2800 °C), so it is often used as insulation in electric heaters and electric stoves.

A third class of ceramics includes the *nonoxide ceramics* such as silicon nitride, Si_3N_4; silicon carbide, SiC; and boron nitride, BN. Heating the solids under pressure forms ceramics that are hard and strong, but brittle. Boron nitride has the same average number of electrons per atom as does elemental carbon and exists in the graphite structure or the diamond structure, making it comparable in hardness with diamond and more resistant to oxidation. For this reason, boron nitride cups and tubes are used to contain molten metals that are being evaporated. Silicon carbide (trade name Carborundum) is a widely used abrasive; its structure is the diamond structure with half of the C atoms replaced by Si atoms.

The one severely limiting problem in using ceramics is their brittleness. Ceramics deform very little before they fail catastrophically, with the failure resulting from a weak point in the bonding within the ceramic matrix. Such weak points are not consistent from sample to sample, so the failure is not readily predicted. Stress failure of ceramic composites occurs due to nanoscale irregularities resulting from impurities or disorder in the atomic arrangements, so much attention is now being given to using purer starting materials and more strictly controlling the processing steps. Adding fibers to ceramic composites makes them less susceptible to brittleness and sudden fracture.

© Fulvio Roiter/Corbis

Colored glass. Adding small amounts of metal oxides to colorless glass creates various colors.

Glasses

One of the more common **glasses** is soda-lime glass, which is clear and colorless if the purity of the ingredients is carefully controlled. Various substances give color to the glass (Table 11.8).

Table 11.8 Substances Used to Color Glass

Substance	Color
Calcium fluoride	Milky white
Cobalt(II) oxide	Blue
Copper(I) oxide	Red, green, blue
Finely divided gold	Red, purple, blue
Iron(II) compounds	Green
Iron(III) compounds	Yellow
Manganese(IV) oxide	Violet
Tin(IV) oxide	Opaque
Uranium compounds	Yellow, green

The simplest glass is amorphous silica, SiO_2 (known as *vitreous silica*); it is prepared by melting and quickly cooling either quartz or cristobalite. Such glass is built up of corner-sharing SiO_4 tetrahedra linked into a three-dimensional network that lacks symmetry or long-range order. Another common glass is *borosilicate glass* or Pyrex, which contains SiO_2, B_2O_3, and Al_2O_3. It is made from melting boric acid, H_3BO_3; soda ash, Na_2CO_3; silica sand, SiO_2; alumina, Al_2O_3; and borax, $Na_2B_4O_7 \cdot 10\ H_2O$, and is valuable because of its low expansion with temperature and high chemical durability.

If another oxide is added to SiO_2, the melting point of the SiO_2 mixture is lowered considerably (from 1800 °C for quartz to about 800 °C if about 25 mole percent of Na_2O is added). The resulting melt cools to form a glass that is somewhat water-soluble and definitely soluble in strongly basic solutions. It is also prone to convert back to a crystalline solid. If other metal oxides such as CaO, MgO, or Al_2O_3 are added, the mixture still melts at a fairly low temperature, but it becomes resistant to chemical attack. Common glass like that used for windows, bottles, and lamps contains these metal oxides in addition to SiO_2.

Mole percent is just another way of expressing concentration. If 0.75 mol SiO_2 and 0.25 mol Na_2O are mixed, 1.00 mol matter is present. The mole percent of SiO_2 is 75%, and the mole percent of Na_2O is 25%.

It is important that glass be *annealed* properly during the manufacturing process. Annealing means cooling the glass slowly as it passes from a viscous liquid state to a solid at room temperature. If a glass is cooled too quickly, bonding forces become uneven because small regions of crystallinity develop. Poorly annealed glass may crack or shatter when subjected to mechanical shocks or sudden temper-

CHEMISTRY IN THE NEWS Glassy Metals?

A recent study at the U.S. National Institute of Standards and Technology (NIST) has found that metals and ceramics have more in common with glass than was previously thought. Metals, such as structural steel, are in the form of polycrystals—tiny metal crystals that have come together as the molten steel solidified during cooling. Each tiny crystal, called a grain, has high internal order, but the space between grains is not well ordered. Grain boundaries are a focus of materials research since their behavior affects the properties of the steel. As an example, polycrystalline materials usually fracture along grain boundaries, so understanding grain boundaries is key to understanding the strength of materials.

New research by Jack Douglas and James Warren of NIST has found that the atoms in grain boundaries in polycrystals show behavior very similar to the atomic motion of atoms found in glass-forming liquids. The atoms at the grain boundaries form strings, and these atoms engage in collective atomic motion—just as the atoms in glass-forming liquids do. The researchers concluded that grain boundaries in polycrystalline materials are a type of glass.

Source: **http://www.world-science.net/othernews/090618_polycrystal**

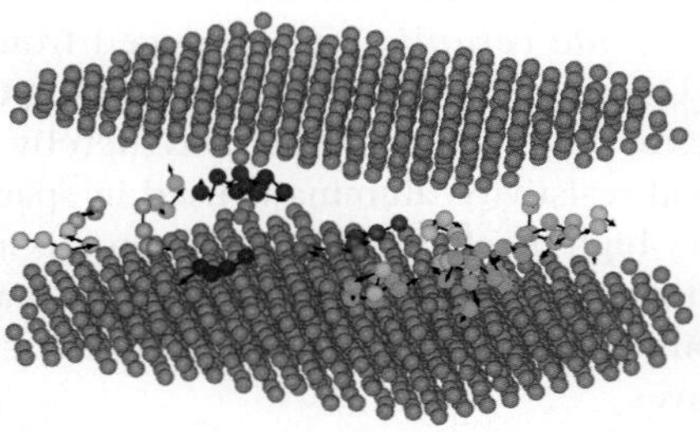

NIST

Two polycrystalline grains (red sheets) with a grain boundary containing disordered atoms (blue and green atoms).

ature changes. High-quality glass, such as that used in optics, must be annealed very carefully. For example, the 200-in mirror for the telescope at Mt. Palomar, California, was annealed from 500 °C to 300 °C over a period of nine months.

The refractive index of a material expresses the angle at which light is diffracted upon entering the material from a vacuum.

Optical Fibers

Our technological culture's dependence on instantaneous communication relies on the properties of many materials, but none more so than optical fibers. These fibers have rapidly replaced copper wire for the transmission of data such as telephone conversations as well as Internet-based digital data.

An **optical fiber** has three parts: core glass, which carries the light; cladding glass, which surrounds the core; and a polymer on the outside for physical protection and strength (Figure 11.40). The cladding glass has a slightly lower refractive index than the core glass, so there is total internal reflection of light traveling through the fiber as long as the light hits the surface of the core at a sufficiently small angle.

Figure 11.40 Total internal reflection in an optical fiber.

A good optical fiber must have two properties: (1) It must contain the lightthat is, it must produce total internal reflection and have no defects on the surface that would let light out of the fiber; and (2) it must pass light over long distances without significant loss of intensity. To pass light with no attenuation, the core glass must be purein fact, ultrapure. Before 1965, when light was put into the best available optical fiber 1000 m long, no light came out, because it was all absorbed. In the decades that followed, sophisticated manufacturing methods were developed that allowed the generation of core glass appropriate for optical fiber. Pure silica, SiO_2, is evaporated into a rod that can be heated and pulled (like taffy) to make a tiny optical fiber. The attenuation of the light signals was decreased by as much as 100 orders of magnitude by the materials scientists working on this problem. At present, optical fibers carrying signals over long distances need to have optical amplifiers only every 50 km or so to keep the signals from becoming too weak. Using such methodology, very long fiber optic cables are now in routine use.

Macs Peter/Shutterstock

A bundle of optical fibers. Optical fibers carry information in the form of light waves.

The two main advantages of optical fibers for transmission of voice and electronic information are the capability of the fibers to carry many more streams of data than copper wire can and the decreased interference when fibers rather than wires are used. In addition, optical fibers are made from cheap, abundant materials.

SUMMARY PROBLEM

Part I

Use the vapor pressure curves shown in Figure 11.5 to answer these questions.

(a) What is the vapor pressure of diethyl ether at 0 °C?

(b) Does diethyl ether have stronger or weaker intermolecular attractions than ethanol?

(c) At what temperature does diethyl ether have a vapor pressure of 600 mm Hg?

(d) If the normal boiling point of diethyl ether is 34.6 °C and it has a vapor pressure of 410 mm Hg at 20 °C, calculate ΔH°_{vap} for diethyl ether.

Part II

Consider the phase diagram for xenon shown below. Answer these questions.

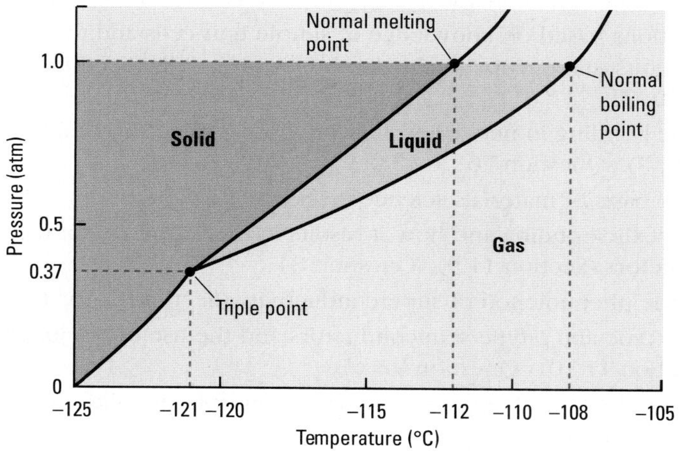

(a) In what phase is xenon found at room temperature and a pressure of 1.0 atm?

(b) If the pressure exerted on a sample of xenon is 0.75 atm and the temperature is −114 °C, in what phase does xenon exist?

(c) If the vapor pressure of a sample of liquid xenon is 375 mm Hg, what is the temperature of the liquid phase?

(d) What is the vapor pressure of solid xenon at −122 °C?

(e) Which is the denser phase, solid or liquid? Explain.

Part III

Consider the CsI unit cell shown in Figure 11.22.

(a) How many Cs^+ ions are there per unit cell?

(b) How many I^- ions are there per unit cell?

(c) How many I^- ions share each face of the unit cell?

IN CLOSING

OWL and go Chemistry

Sign in at **www.cengage.com/owl** to:

- View tutorials and simulations, develop problem-solving skills, and complete online homework assigned by your professor.
- For quick review and exam prep, download Go Chemistry mini lecture modules from OWL (or purchase them at **www.CengageBrain.com**).

Having studied this chapter, you should be able to . . .

- Explain the properties of surface tension, capillary action, vapor pressure, and boiling point, and describe how these properties are influenced by intermolecular forces (Sections 11.1, 11.2). End-of-chapter questions: 13, 25, 28, 44, 46
- Calculate the energy transfers associated with vaporization and fusion (Section 11.3). Questions 21, 32, 34, 40, 42
- Describe the phase changes that occur among solids, liquids, and gases (Section 11.3).
- Use phase diagrams to predict what happens when temperature and pressure are changed for a sample of matter (Section 11.3). Questions 49, 51, 106
- Understand critical temperature and critical pressure (Section 11.3).
- Describe and explain the unusual properties of water (Section 11.4). Questions 53, 55
- Differentiate among the major types of solids (Section 11.5). Questions 61, 63
- Do calculations based on knowledge of simple unit cells and the dimensions of atoms and ions that occupy positions in those unit cells (Section 11.6). Questions 67, 70, 74
- Explain the bonding in network solids and how it results in their properties (Section 11.7). Question 76
- Explain the basis of materials science (Section 11.8).
- Explain metallic bonding and how it results in the properties of metals and semiconductors (Section 11.9). Question 84
- Describe the phenomenon of superconductivity (Section 11.9). Question 88
- Describe *n*-type and *p*-type semiconductors and the use of *p-n* junctions as rectifiers (Section 11.10). Question 92
- Explain how the lack of regular structure in amorphous solids affects their properties (Section 11.11). Question 95

KEY TERMS

amorphous solids *(Section 11.5)*
boiling *(11.2)*
boiling point *(11.2)*
capillary action *(11.1)*
cement *(11.11)*
ceramics *(11.11)*
Clausius-Clapeyron equation *(11.2)*
closest packing *(11.6)*
concrete *(11.11)*
condensation *(11.3)*
conduction band *(11.9)*
conductor *(11.9)*
critical point *(11.3)*
critical pressure (P_c) *(11.3)*
critical temperature (T_c) *(11.3)*
crystal lattice *(11.6)*
crystalline solids *(11.5)*
crystallization *(11.3)*
cubic close packing *(11.6)*
cubic unit cell *(11.6)*
deposition *(11.3)*
dew point *(11.2)*
doping *(11.10)*
energy band *(11.9)*
equilibrium vapor pressure *(11.2)*
evaporation *(11.3)*
fullerenes *(11.7)*
glass *(11.11)*
heating curve *(11.3)*
hexagonal close packing *(11.6)*
insulator *(11.9)*
materials science *(11.8)*
meniscus *(11.1)*
metallic bonding *(11.9)*
***n*-type semiconductor** *(11.10)*
nanotube *(11.7)*
network solids *(11.7)*
normal boiling point *(11.2)*
optical fiber *(11.11)*
***p-n* junction** *(11.10)*
***p*-type semiconductor** *(11.10)*
phase diagram *(11.3)*
relative humidity *(11.2)*
semiconductor *(11.9)*
solar cell *(11.10)*

sublimation *(11.3)*

superconductor *(11.9)*

supercritical fluid *(11.3)*

surface tension *(11.1)*

triple point *(11.3)*

unit cell *(11.6)*

valence band *(11.9)*

vapor pressure *(11.2)*

vaporization *(11.3)*

viscosity *(11.1)*

volatility *(11.2)*

X-ray crystallography (*Tools of Chemistry,* p. 510)

zone refining *(11.10)*

QUESTIONS FOR REVIEW AND THOUGHT

OWL Interactive versions of these problems are assignable in OWL.

Blue-numbered questions have short answers at the back of this book in Appendix M and fully worked solutions in the *Student Solutions Manual.*

Review Questions

These questions test vocabulary and simple concepts.

1. Name three properties of solids that are different from those of liquids. Explain the differences for each.
2. List the postulates of the kinetic-molecular theory that apply to liquids.
3. What causes surface tension in liquids? Name a substance that has a very high surface tension. What kinds of intermolecular forces account for the high value?
4. Explain how the equilibrium vapor pressure of a liquid might be measured.
5. Define boiling point and normal boiling point.
6. Define the heat of crystallization of a substance. How is it related to the substance's heat of fusion?
7. Define sublimation.
8. Which processes are endothermic?
   (a) Condensation (b) Melting (c) Evaporation
   (d) Sublimation (e) Deposition (f) Freezing
9. Define the unit cell of a crystal.
10. Assuming the same substance could form crystals with its atoms or ions in either simple cubic packing or hexagonal closest packing, which form would have the higher density? Explain.
11. How does conductivity vary with temperature for (a) a metallic conductor, (b) a nonconductor, (c) a semiconductor, and (d) a superconductor? In your answer, begin at high temperatures and come down to low temperatures.

Topical Questions

These questions are keyed to the major topics in the chapter. Usually a question that is answered at the back of this book is paired with a similar one that is not.

The Liquid State (Section 11.1)

12. Predict which substance in Table 11.1 has a surface tension most similar to that of each liquid:
    (a) Ethylene glycol, $HOCH_2—CH_2OH$
    (b) Hexane, C_6H_{14}
    (c) Gallium metal at 40 °C
13. The surface tension of a liquid decreases with increasing temperature. Using the idea of intermolecular attractions, explain why this is so.
14. Explain on the molecular scale the processes of condensation and vaporization.
15. How would you convert a sample of liquid to vapor without changing the temperature?
16. Define the heat of vaporization of a liquid. How is it related to the heat of condensation of that liquid? Using the idea of intermolecular attractions, explain why the process of vaporization is endothermic.
17. After exercising on a hot summer day and working up a sweat, you often become cool when you stop. What is the molecular-level explanation of this phenomenon?
18. The Ne atom and the molecules HF, H_2O, NH_3, and CH_4 all have the same number of electrons. In a thought experiment, you can make HF from Ne by removing a single proton a short distance from the nucleus and having the electrons follow the new arrangement of nuclei so as to make a new chemical bond. You can do the same for each of the other molecules. (Of course, none of these thought experiments can actually be done because of the enormous energies required to remove protons from nuclei.) For all of these substances, make a plot of (a) the boiling point in kelvins versus the number of hydrogen atoms and (b) the molar heat of vaporization versus the number of hydrogen atoms. Explain any trend that you see in terms of intermolecular forces.
19. How much thermal energy transfer is required to vaporize 1.0 metric ton of ammonia? (1 metric ton = 10^3 kg.) The ΔH_{vap} for ammonia is 25.1 kJ/mol.
20. The chlorofluorocarbon CCl_3F has an enthalpy of vaporization of 24.8 kJ/mol. To vaporize 1.00 kg of the compound, how much thermal energy transfer is required?
21. The molar enthalpy of vaporization of methanol is 38.0 kJ/mol at 25 °C. How much thermal energy transfer is required to convert 250. mL of the alcohol from liquid to vapor? The density of CH_3OH is 0.787 g/mL at 25 °C.
22. Some camping stoves contain liquid butane, C_4H_{10}. They work only when the outside temperature is warm enough to allow the butane to have a reasonable vapor pressure (so they are not very good for camping in temperatures below about 0 °C). Assume the enthalpy of vaporization of butane is 24.3 kJ/mol. If the camp stove fuel tank contains 190. g liquid C_4H_{10}, how much thermal energy transfer is required to vaporize all of the butane?
23. Mercury is highly toxic. Although it is a liquid at room temperature, it has a high vapor pressure and a low enthalpy of vaporization (294 J/g). What quantity of ther-

mal energy transfer is required to vaporize 0.500 mL mercury at 357 °C, its normal boiling point? The density of Hg (ℓ) is 13.6 g/mL. Compare this thermal energy transfer with the amount needed to vaporize 0.500 mL water. See Table 11.2 for the molar enthalpy of vaporization of H_2O.

24. Rationalize the observation that 1-propanol, $CH_3CH_2CH_2OH$, has a boiling point of 97.2 °C, whereas a compound with the same empirical formula, ethyl methyl ether, $CH_3CH_2OCH_3$, boils at 7.4 °C.

25. Briefly explain the variations in the boiling points in this table. In your discussion be sure to mention the types of intermolecular forces involved.

Compound	Boiling Point (°C)
NH_3	−33.4
PH_3	−87.5
AsH_3	−62.4
SbH_3	−18.4

Vapor Pressure (Section 11.2)

26. Give a molecular-level explanation of why the vapor pressure of a liquid increases with temperature.

27. Methanol, CH_3OH, has a normal boiling point of 64.7 °C and a vapor pressure of 100 mm Hg at 21.2 °C. Formaldehyde, $H_2C{=}O$, has a normal boiling point of −19.5 °C and a vapor pressure of 100 mm Hg at −57.3 °C. Explain why these two compounds have different boiling points and require different temperatures to achieve the same vapor pressure.

28. The vapor pressure curves for four substances are shown in the plot. Which one of these four substances has the greatest intermolecular attractive forces at 25 °C? Explain your answer.

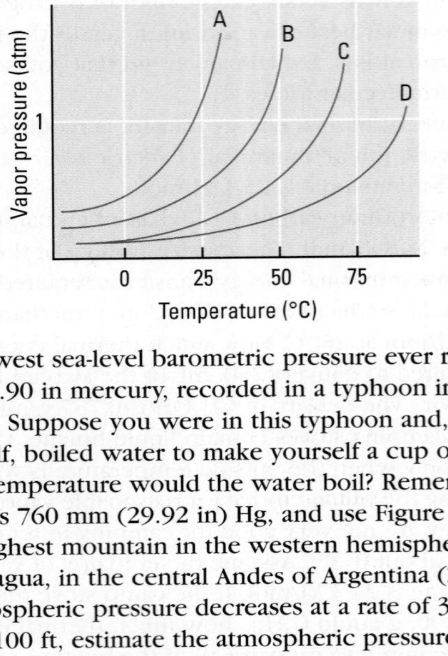

29. The lowest sea-level barometric pressure ever recorded was 25.90 in mercury, recorded in a typhoon in the South Pacific. Suppose you were in this typhoon and, to calm yourself, boiled water to make yourself a cup of tea. At what temperature would the water boil? Remember that 1 atm is 760 mm (29.92 in) Hg, and use Figure 11.5.

30. The highest mountain in the western hemisphere is Mt. Aconcagua, in the central Andes of Argentina (22,834 ft). If atmospheric pressure decreases at a rate of 3.5 millibar every 100 ft, estimate the atmospheric pressure at the top of Mt. Aconcagua, and then estimate from Figure 11.5 the temperature at which water would boil at the top of the mountain.

31. A liquid has a ΔH_{vap} of 38.7 kJ/mol and a boiling point of 110 °C at 1 atm pressure. Calculate the vapor pressure of the liquid at 97 °C.

32. A liquid has a ΔH_{vap} of 44.0 kJ/mol and a vapor pressure of 370 mm Hg at 90 °C. Calculate the vapor pressure of the liquid at 130 °C.

33. The vapor pressure of ethanol, C_2H_5OH, at 50.0 °C is 233 mm Hg, and its normal boiling point at 1 atm is 78.3 °C. Calculate the ΔH_{vap} of ethanol.

34. Calculate the ΔH_{vap} for a substance whose vapor pressure doubled when its temperature was raised from 70.0 °C to 80.0 °C.

Phase Changes: Solids, Liquids, and Gases (Section 11.3)

35. What does a low enthalpy of fusion for a solid tell you about the solid (its bonding or type)?

36. What does a high melting point and a high enthalpy of fusion tell you about a solid (its bonding or type)?

37. Which would you expect to have the higher enthalpy of fusion, N_2 or I_2? Explain your choice.

38. The enthalpy of fusion for H_2O is about 2.5 times larger than the enthalpy of fusion for H_2S. What does this say about the relative strengths of the forces between the molecules in these two solids? Explain.

39. Benzene is an organic liquid that freezes at 5.5 °C and forms beautiful, feather-like crystals. How much heat transfer occurs when 15.5 g benzene freezes at 5.5 °C? The enthalpy of fusion of benzene is 127 J/g. If the 15.5-g sample is remelted, again at 5.5 °C, what quantity of heat transfer is required to convert it to a liquid?

40. What is the total quantity of heat energy transfer required to change 0.50 mol ice at −5 °C to 0.50 mol steam at 100 °C?

41. How much thermal energy is needed to melt a 36.00-g ice cube that is initially at −10 °C and bring it to room temperature (20 °C)? The solid ice and liquid water have heat capacities of 2.06 J g^{-1} $°C^{-1}$ and 4.184 J g^{-1} $°C^{-1}$, respectively. The enthalpy of fusion for solid ice is 6.02 kJ/mol and the enthalpy of vaporization of liquid water is 40.7 kJ/mol.

42. The chlorofluorocarbon CCl_2F_2 was once used as a refrigerant. What mass of this substance must evaporate to freeze 2 mol water initially at 20 °C? The enthalpy of vaporization for CCl_2F_2 is 289 J/g. The enthalpy of fusion for solid ice is 6.02 kJ/mol and specific heat capacity for liquid water is 4.184 J g^{-1} $°C^{-1}$.

43. The ions of NaF and MgO all have the same number of electrons, and the internuclear distances are about the same (235 pm and 212 pm). Why, then, are the melting points of NaF and MgO so different (992 °C and 2642 °C, respectively)?

44. For the pair of compounds LiF and CsI, tell which compound is expected to have the higher melting point, and briefly explain why.

45. Which of these substances has the highest melting point? The lowest melting point? Explain your choices briefly.
(a) LiBr (b) CaO
(c) CO (d) CH_3OH

46. Which of these substances has the highest melting point? The lowest melting point? Explain your choices briefly.
(a) SiC (b) I_2
(c) Rb (d) $CH_3CH_2CH_2CH_3$
47. Why is solid CO_2 called Dry Ice?
48. During thunderstorms, very large hailstones can fall from the sky. To preserve some of these hailstones, you place them in the freezer compartment of your frost-free refrigerator. A friend, who is a chemistry student, tells you to put the hailstones in a tightly sealed plastic bag. Why?
49. In this phase diagram, make these identifications:

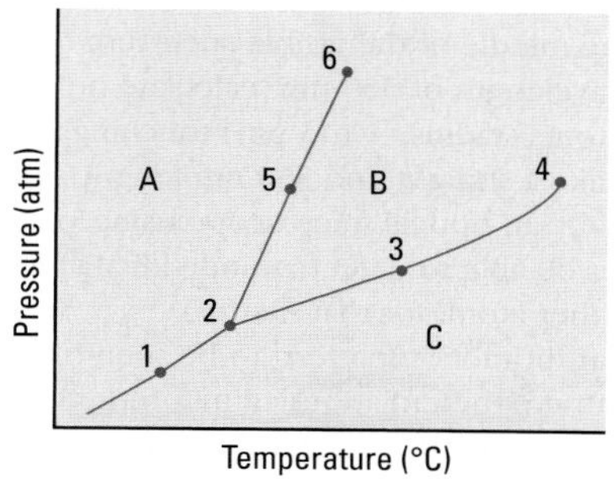

(a) What phase is present in region A? Region B? Region C?
(b) What phases are in equilibrium at point 1? Point 2? Point 3? Point 5?
50. From memory, sketch the phase diagram of water. Label all the regions as to the physical state of water. Draw either horizontal (constant pressure) or vertical (constant temperature) paths (i.e., lines with arrows indicating a direction) for these changes of state:
(a) Sublimation (b) Condensation to a liquid
(c) Melting (d) Vaporization
(e) Crystallization
51. Consult the phase diagram of CO_2 in Figure 11.17. What phase or phases are present under these conditions:
(a) $T = -70$ °C and $P = 1.0$ atm
(b) $T = -40$ °C and $P = 15.5$ atm
(c) $T = -80$ °C and $P = 4.7$ atm
52. At the critical point for carbon dioxide, the substance is very far from being an ideal gas. Prove this statement by calculating the density of an ideal gas in g/cm^3 at the conditions of the critical point and comparing it with the experimental value. Compute the experimental value from the fact that a mole of CO_2 at its critical point occupies 94 cm^3.

Water (Section 11.4)

53. The boiling point of water is relatively high for a compound of such low molar mass. Explain why this is so.
54. The molecular structure of water makes it a good solvent for many substances. (a) For what types of substances is water a good solvent? Explain why. (b) For what types of substances is water a poor solvent? Explain why.
55. For most substances, the density of the solid phase is larger than for the liquid phase, but for water the reverse is true. What is the molecular-scale reason for this property of water? Why is this property important?
56. Water can participate in hydrogen-bonding with other water molecules. In liquid water, how many hydrogen bonds does each water molecule engage in? What three-dimensional shape do these bonds assume?
57. The presence of large amounts of water (as a lake or ocean) has a moderating effect on the climate near the water. What molecular property of water accounts for this observation?

Types of Solids (Section 11.5)

58. Classify each of these solids as ionic, metallic, molecular, network, or amorphous.
(a) KF (b) I_2
(c) SiO_2 (d) BN
59. Classify each of these solids as ionic, metallic, molecular, network, or amorphous.
(a) Tetraphosphorus decaoxide (b) Brass
(c) Graphite (d) Ammonium phosphate
60. On the basis of the description given, classify each of these solids as molecular, metallic, ionic, network, or amorphous, and explain your reasoning.
(a) A brittle, yellow solid that melts at 113 °C; neither the solid nor the liquid conducts electricity
(b) A soft, silvery solid that melts at 40 °C; both the solid and the liquid conduct electricity
(c) A hard, colorless, crystalline solid that melts at 1713 °C; neither the solid nor the liquid conducts electricity
(d) A soft, slippery solid that melts at 63 °C; neither the solid nor the liquid conducts electricity
61. On the basis of the description given, classify each of these solids as molecular, metallic, ionic, network, or amorphous, and explain your reasoning.
(a) A soft, slippery solid that has no definite melting point but decomposes at temperatures above 250 °C; the solid does not conduct electricity.
(b) Violet crystals that melt at 114 °C and whose vapor irritates the nose; neither the solid nor the liquid conducts electricity.
(c) Hard, colorless crystals that melt at 2800 °C; the liquid conducts electricity, but the solid does not.
(d) A hard solid that melts at 3410 °C; both the solid and the liquid conduct electricity.
62. Describe how each of these materials would behave if it were deformed by a hammer strike. Explain why the materials behave as they do.
(a) A metal, such as gold
(b) A nonmetal, such as sulfur
(c) An ionic compound, such as NaCl
63. What type of solid exhibits each of these sets of properties?
(a) Melts below 100 °C and is insoluble in water
(b) Conducts electricity only when melted
(c) Insoluble in water and conducts electricity
(d) Noncrystalline and melts over a wide temperature range

Crystalline Solids (Section 11.6)

64. Each diagram below represents an array of like atoms that would extend indefinitely in two dimensions. Draw a two-dimensional unit cell for each array. How many atoms are in each unit cell?

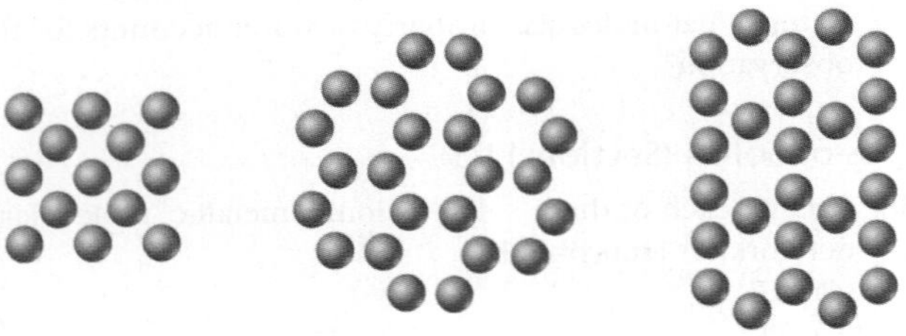

65. Name and draw the three cubic unit cells. Describe their similarities and differences.
66. Explain how the volume of a primitive cubic unit cell is related to the radius of the atoms in the cell.
67. Solid xenon forms crystals with a face-centered unit cell that has an edge of 620 pm. Calculate the atomic radius of xenon.
68. Gold (atomic radius = 144 pm) crystallizes in an fcc unit cell. What is the length of a side of the cell?
69. Using the NaCl structure shown in Figure 11.23, how many unit cells share each of the Na^+ ions in the front face of the unit cell? How many unit cells share each of the Cl^- ions in this face?
70. The ionic radii of Cs^+ and Cl^- are 181 and 167 pm, respectively. What is the length of the body diagonal in the CsCl unit cell? What is the length of the side of this unit cell? (CsCl has the same unit cell as CsI, shown in Figure 11.22.)
71. You know that thallium chloride, TlCl, crystallizes in either a primitive cubic or a face-centered cubic lattice of Cl^- ions with Tl^+ ions in the holes. If the density of the solid is 7.00 g/cm^3 and the edge of the unit cell is 3.85×10^{-8} cm, what is the unit cell geometry?
72. Could $CaCl_2$ possibly have the NaCl structure? Explain your answer briefly.
73. A primitive cubic unit cell is formed so that the spherical atoms or ions just touch one another along the edge. Prove mathematically that the percentage of empty space within the unit cell is 47.6%. (The volume of a sphere is $\frac{4}{3}\pi r^3$, where r is the radius of the sphere.)
74. Metallic lithium has a body-centered cubic structure, and its unit cell is 351 pm along an edge. Lithium iodide has the same crystal lattice structure as sodium chloride. The cubic unit cell is 600 pm along an edge.
    (a) Assume that the metal atoms in lithium touch along the body diagonal of the body-centered cubic unit cell, and estimate the radius of a lithium atom.
    (b) Assume that in lithium iodide the I^- ions touch along the face diagonal of the cubic unit cell and that the Li^+ and I^- ions touch along the edge of the cube; calculate the radius of an I^- ion and of an Li^+ ion.
    (c) Compare your results in parts (a) and (b) for the radius of a lithium atom and a lithium ion. Are your results reasonable? If not, how could you account for the unexpected result? Could any of the assumptions that were made be in error? Explain.

Network Solids (Section 11.7)

75. Explain why diamond is denser than graphite.
76. Determine, by looking up data in a reference such as the *Handbook of Chemistry and Physics,* whether the examples of network solids given in the text are soluble in water or other common solvents. Explain your answer in terms of the chemical bonding in network solids.
77. Explain why diamond is an electrical insulator and graphite is an electrical conductor.

Tools of Chemistry: X-Ray Crystallography

78. Taking the middle of the visible spectrum to be green light with a wavelength of 550 nm, calculate how many aluminum atoms (radius = 143 pm) touching its neighbors would make a straight line 550 nm long. Using this result, explain why an optical microscope using visible radiation will never be able to detect an individual aluminum atom (or any other atom, for that matter).
79. For a clear diffraction pattern to be seen from a regularly spaced lattice, the radiation falling on the lattice must have a wavelength less than the lattice spacing. From the unit-cell size of the NaCl crystal, estimate the maximum wavelength of the radiation that would be diffracted by this crystal. Calculate the frequency of the radiation and the energy associated with (a) one photon and (b) one mole of photons of the radiation. In what region of the spectrum is this radiation?
80. The first-order Bragg reflection (n = 1) from an aluminum crystal for X-rays with a wavelength of 154 pm is 19.3°. What is the spacing between the planes of aluminum atoms?
81. The second-order Bragg reflection (n = 2) from a copper crystal for X-rays with a wavelength of 166 pm is 27.35°. What is the spacing between the planes of copper atoms?
82. If the first-order Bragg reflection (n = 1) from a NaCl crystal with a spacing of 282 pm is seen at 23.0°, what is the wavelength of the X-ray radiation used?

Metals, Semiconductors, and Insulators (Section 11.9)

83. What is the principal difference between the orbitals that electrons occupy in individual, isolated atoms and the orbitals they occupy in solids?
84. In terms of band theory, what is the difference between a conductor and an insulator? Between a conductor and a semiconductor?
85. Name three properties of metals, and explain them by using a theory of metallic bonding.
86. Which substance has the greatest electrical conductivity? The smallest electrical conductivity? Explain your choice briefly.
    (a) Si (b) Ge (c) Ag (d) P_4
87. Which substance has the greatest electrical conductivity? The smallest electrical conductivity? Explain your choices briefly.
    (a) RbCl(ℓ) (b) NaBr(s) (c) Rb (d) Diamond
88. Define the term "superconductor." Give the chemical formulas of two kinds of superconductors and their associated transition temperatures.
89. What is the main technological or economic barrier to the widespread use of superconductors?

Silicon and the Chip (Section 11.10)

90. What are the two main chemical reactions involved in the production of electronic-grade silicon? Identify the elements being reduced and being oxidized.
91. Extremely high-purity silicon is required to manufacture semiconductors such as the memory chips found in calculators and computers. If a silicon wafer is 99.99999999% pure, approximately how many atoms of some other element are present per gram of high-purity silicon?
92. What is the process of doping, as applied to semiconductors? Why are Group 3A and Group 5A elements used to dope silicon?
93. Explain the difference between *n*-type semiconductors and *p*-type semiconductors.

Cement, Ceramics, and Glass (Section 11.11)

94. Define the term "amorphous."
95. What makes a glass different from a solid such as NaCl? Under what conditions could NaCl become glass-like?
96. A typical cement contains, by weight, 65% CaO, 20% SiO_2, 5% Al_2O_3, 6% Fe_2O_3, and 4% MgO. Determine the mass percent of each type of atom present. Then determine an empirical formula of the material from the percent composition, setting the subscript of the least abundant element to 1.00.
97. Give two examples of (a) oxide ceramics and (b) nonoxide ceramics.

General Questions

These questions are not explicitly keyed to chapter topics; many require integration of several concepts.

98. The chlorofluorocarbon CCl_2F_2 was once used in air conditioners as the heat transfer fluid. Its normal boiling point is −30 °C, and its enthalpy of vaporization is 165 J g^{-1}. The gas and the liquid have specific heat capacities of 0.61 J g^{-1} $°C^{-1}$ and 0.97 J g^{-1} $°C^{-1}$, respectively. How much thermal energy is evolved when 10.0 g CCl_2F_2 is cooled from 40 °C to −40 °C?
99. Liquid ammonia, $NH_3(\ell)$, was used as a refrigerant fluid before the discovery of the chlorofluorocarbons and is still widely used today. Its normal boiling point is −33.4 °C, and its enthalpy of vaporization is 23.5 kJ/mol. The gas and liquid have specific heat capacities of 2.2 J g^{-1} K^{-1} and 4.7 J g^{-1} K^{-1}, respectively. How much thermal energy transfer is required to raise the temperature of 10.0 kg liquid ammonia from −50.0 °C to −33.4 °C, and then to 0.0 °C?
100. Potassium chloride and rubidium chloride both have the sodium chloride structure. X-ray diffraction experiments indicate that their cubic unit cell dimensions are 629 pm and 658 pm, respectively.
    (i) One mol KCl and 1 mol RbCl are ground together to a very fine powder in a mortar and pestle, and the X-ray diffraction pattern of the pulverized solid is measured. Two patterns are observed, each corresponding to a cubic unit cell—one with an edge length of 629 pm and one with an edge length of 658 pm. Call this Sample 1.
    (ii) One mol KCl and 1 mol RbCl are heated until the entire mixture is molten and then cooled to room temperature. A single X-ray diffraction pattern indicates a cubic unit cell with an edge length of roughly 640 pm. Call this Sample 2.
    (a) Suppose that Samples 1 and 2 were analyzed for their chloride content. What fraction of each sample is chloride? Could the samples be distinguished by means of chemical analysis?
    (b) Interpret the two X-ray diffraction results in terms of the structures of the crystal lattices of Samples 1 and 2.
    (c) What chemical formula should you write for Sample 1? For Sample 2?
    (d) Suppose that you dissolved 1.00 g Sample 1 in 100 mL water in a beaker and did the same with 1.00 g Sample 2. Which sample would conduct electricity better, or would both be the same? What ions would be present in each solution at what concentrations?
101. Sulfur dioxide, SO_2, is found in polluted air.
    (a) What types of forces are responsible for binding SO_2 molecules to one another in the solid or liquid phase?
    (b) Using the information below, place the compounds listed in order of increasing intermolecular attractions.

Compound	Normal Boiling Point (°C)
SO_2	−10
NH_3	−33.4
CH_4	−161.5
H_2O	100

Applying Concepts

These questions test conceptual learning.

102. Refer to Figure 11.5 when answering these questions.
    (a) What is the equilibrium vapor pressure for ethyl alcohol at room temperature?
    (b) At what temperature (°C) does diethyl ether have an equilibrium vapor pressure of 400 mm Hg?
    (c) If water is boiling at a temperature of 95 °C, what is the atmospheric pressure?
    (d) At 200 mm Hg and 60 °C, which of the three substances are gases?
    (e) If you put a couple of drops of each substance on your hand, which would immediately evaporate, and which would remain as a liquid?
    (f) Which of the three substances has the greatest intermolecular attractions?
103. The normal boiling point of SO_2 is 263.1 K and that of NH_3 is 239.7 K. At −40 °C, would you predict that ammonia has a vapor pressure greater than, less than, or equal to that of sulfur dioxide? Explain.
104. Butane is a gas at room temperature; however, if you look closely at a butane lighter you see it contains liquid butane. How is this possible?
105. While camping with a friend in the Rocky Mountains, you decide to cook macaroni for dinner. Your friend says the macaroni will cook faster in the Rockies because the lower atmospheric pressure will cause the water to boil at a lower temperature. Do you agree with your friend? Explain your reasoning.

106. Examine the nanoscale diagrams and the phase diagram below. Match each particulate diagram (1 through 7) to its corresponding point (A through H) on the phase diagram.

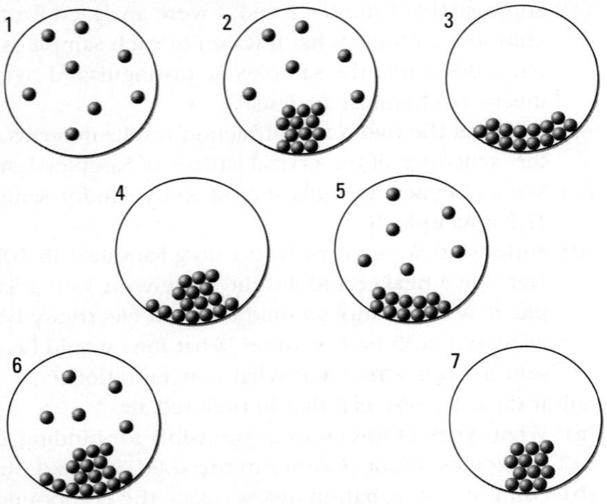

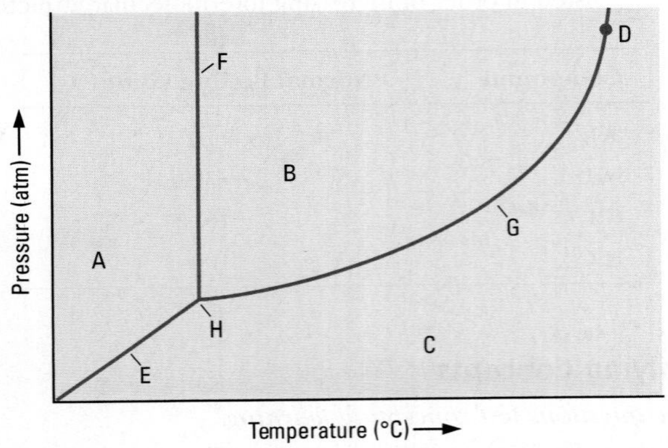

107. Consider the phase diagram below. Draw corresponding heating curves for T_1 to T_2 at pressures P_1 and P_2. Label each phase and phase change on your heating curves.

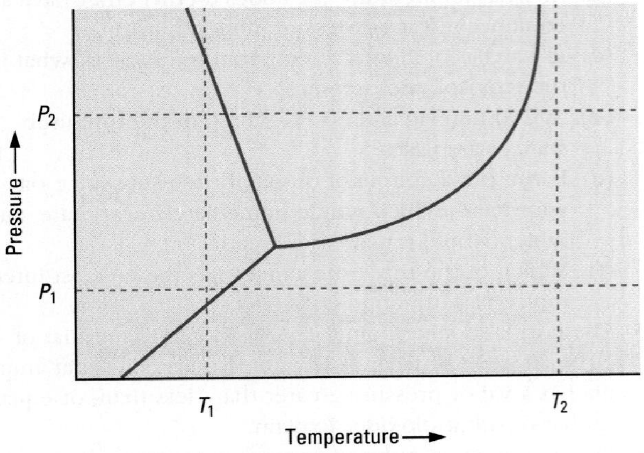

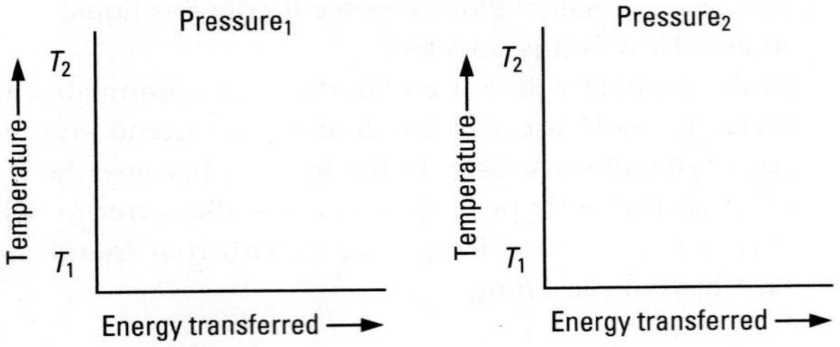

108. Consider three boxes of equal volume. One is filled with tennis balls, another with golf balls, and the third with marbles. If a closest-packing arrangement is used in each box, which one has the most occupied space? Which one has the least occupied space? (Disregard the difference in filling space at the walls, bottom, and top of the box.)

More Challenging Questions

These questions require more thought and integrate several concepts.

109. If you get boiling water at 100 °C on your skin, it burns. If you get 100 °C steam on your skin, it burns much more severely. Explain why this is so.

110. If water at room temperature is placed in a flask that is connected to a vacuum pump and the vacuum pump then lowers the pressure in the flask, we observe that the volume of the water decreases and the remaining water turns into ice. Explain what has happened.

111. We hear reports from weather forecasters of "relative humidity," the ratio of the partial pressure of water in the air to the equilibrium vapor pressure of water at the same temperature. (The vapor pressure of water at 32.2 °C is 36 mm Hg.)
   (a) On a sticky, humid day, the relative humidity may reach 90% with a temperature of 90. °F (32.2 °C). What is the partial pressure of water under these conditions? How many moles per liter are present in the air? How many water molecules per cm^3 are present in such air?
   (b) On a day in a desert, the relative humidity may be 5.% with the same temperature of 90. °F. How many water molecules per cm^3 are present in such air?

112. (a) In the diagram below, for a substance going from point F (initial state) to point G (final state), what changes in phase occur?
   (b) What does point A represent?
   (c) What does the curve from point A to point B represent?

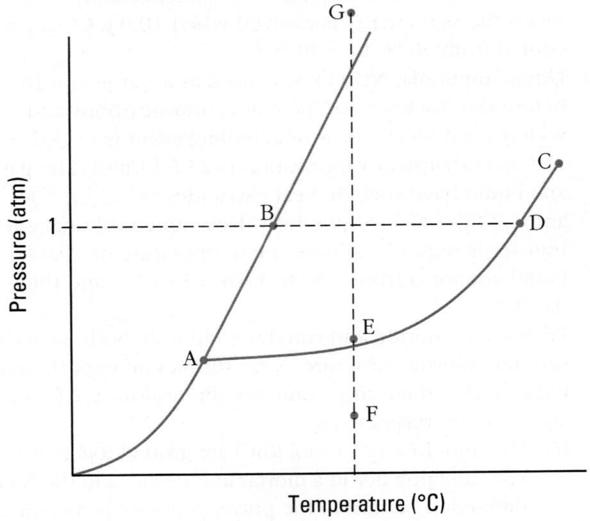

113. Suppose that liquid A has stronger intermolecular forces than liquid B at room temperature.
(a) Which substance will have the greater surface tension?
(b) Which substance will have the greater vapor pressure?
(c) Which substance will have the greater viscosity?

114. Use the vapor pressure curves shown in the figure below for methyl ethyl ether, $CH_3OCH_2CH_3$; carbon disulfide, CS_2; and benzene, C_6H_6, to answer these questions.
(a) What is the vapor pressure of methyl ethyl ether at 0 °C?
(b) Which of these three liquids has the strongest intermolecular attractions?
(c) At what temperature does benzene have a vapor pressure of 600 mm Hg?
(d) What are the normal boiling points of these three liquids?

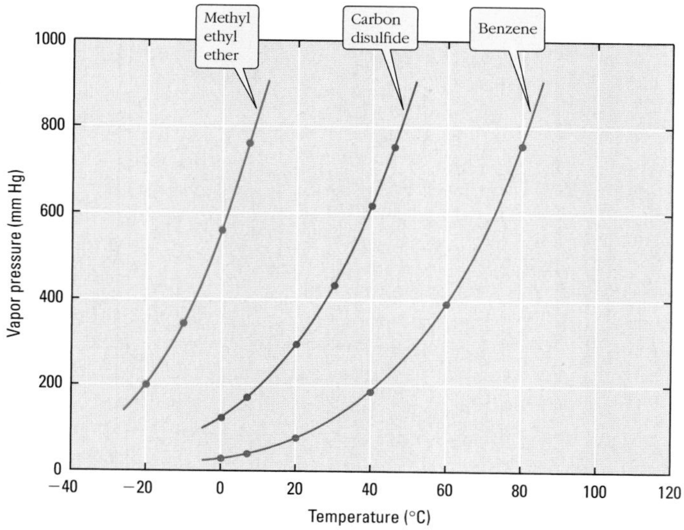

115. Will a closed container of water at 70 °C or an open container of water at the same temperature cool faster on a cold winter day? Explain why.

116. Calculate the boiling point of water at 24 mm Hg. (The ΔH_{vap} of water is 40.7 kJ/mol.)

117. What is the concentration in mol/L of water vapor in air at 25 °C at saturation?

118. Solid lithium has a body-centered cubic unit cell with the length of the edge of 351 pm at 20 °C. Calculate the density of lithium at this temperature.

119. Tungsten has a body-centered cubic unit cell and an atomic radius of 141 pm. What is the density of solid tungsten?

120. Copper is an important metal in the U.S. economy. Most of it is mined in the form of the mineral chalcopyrite, $CuFeS_2$.
(a) To obtain one metric ton (1000. kilograms) of copper metal, how many metric tons of chalcopyrite would you have to mine?
(b) If the sulfur in chalcopyrite is converted to SO_2, how many metric tons of the gas would you get from one metric ton of chalcopyrite?
(c) Copper crystallizes as a face-centered cubic lattice. Knowing that the density of copper is 8.95 g/cm^3, calculate the radius of the copper atom.

121. Organic compounds with structures based on benzene, C_6H_6, can be formed by substituting an atom or a group of atoms for one of the hydrogens. Such substituted benzenes have their own properties, different from benzene and from each other. Explain the order of experimental boiling points for these four compounds:
C_6H_6 (80 °C), C_6H_5Cl (131 °C),
C_6H_5Br (156 °C), C_6H_5OH (182 °C).

Conceptual Challenge Problems

These rigorous, thought-provoking problems integrate conceptual learning with problem solving and are suitable for group work.

CP11.A (Section 11.2) In Section 11.2 you read that the enthalpy of vaporization of water "is somewhat dependent on the temperature." At 100 °C this value is 40.7 kJ/mol, but at 25 °C it is 44.0 kJ/mol, a difference of 3.3 kJ/mol. List four enthalpy changes whose sum would equal this difference. Remember, the sum of the changes for a cyclic process must be zero because the system is returned to its initial state.

CP11.B (Section 11.3) For what reasons would you propose that two of the substances listed in Table 11.2 be considered better refrigerants for use in household refrigerators than the others listed there?

CP11.C (Section 11.3) A table of enthalpies of sublimation is not given in Section 11.3, but the enthalpy of sublimation of ice at 0 °C is given as 51 kJ/mol. How was this value obtained? Tables 11.2 and 11.3 list the enthalpies of vaporization and fusion, respectively, for several substances. Determine from data in these tables the ΔH_{sub} for ice. Using the same method, estimate the enthalpies of sublimation of HBr and HI at their melting points.

12

Fuels, Organic Chemicals, and Polymers

AP Photo/Greg Baker

The architecturally stunning aquatic center at the 2008 Beijing Summer Olympics is sheathed in a blue, synthetic polymer containing C—F bonds. Both synthetic and natural polymers are discussed in this chapter.

The combustion of *fossil fuels*—coal, natural gas, and petroleum—provides nearly 85% of all the energy used in the world. When these substances burn, the carbon they contain is released into the atmosphere as CO_2. Photosynthesis converts CO_2 back into other carbon-containing compounds. Many of the carbon compounds produced by photosynthesis are directly or indirectly very useful as energy sources for humans and animals. But such carbon compounds are not as convenient to use in power plants or auto-

OWL Sign in to OWL at **www.cengage.com/owl** to view tutorials and simulations, develop problem-solving skills, and complete online homework assigned by your professor.

go Chemistry Download mini lecture videos for key concept review and exam prep from OWL or purchase them from **www.CengageBrain.com.**

Companion Website

Visit this book's companion website at **www.cengage.com/chemistry/moore** to work interactive modules for the Estimation boxes and Active Figures in this text.

mobiles as are fossil fuels, which are burned in prodigious quantities daily. As fossil fuels become scarcer, and if stringent conservation measures are not taken, conventional petroleum reserves could last for no more than into the last half of the 21st century. If this occurs, fuel costs will inevitably rise due to the increasing scarcity of petroleum.

Fossil fuels are also, by far, the largest source of starting material for the syntheses of most of the organic chemicals used to make consumer products such as plastics, pharmaceuticals, synthetic rubber, synthetic fibers, and hundreds of other products we rely on. For this reason, the organic chemical industry is often referred to as the petrochemical industry. In this chapter we will discuss a few of the major classes of organic compounds and some of their reactions, especially those used to furnish energy and to make synthetic and natural polymers.

12.1 Petroleum

The great Russian chemist Dimitri Mendeleev recognized the importance of petroleum as a source from which to make valuable carbon compounds and not merely to be used as a fuel. On visiting the oil fields of Pennsylvania and Azerbaijan, he supposedly remarked that burning petroleum as a fuel "would be akin to firing up a kitchen stove with bank notes."

Petroleum is a complex mixture of alkanes, cycloalkanes, alkenes, and aromatic hydrocarbons formed from the remains of plants and animals from millions of years ago. More than 20,000 compounds, almost all of them hydrocarbons, are present in *crude oil,* the form of petroleum that is pumped from the ground. Crude oil's composition and color vary with the location in which it is found. Pennsylvania crude oils are primarily straight-chain hydrocarbons, whereas California crude oil contains a larger portion of aromatic hydrocarbons.

Report of Rock Oil or Petroleum from Venango County, PA, for the Pennsylvania Rock Oil Company, 1855. Benjamin J. Silliman, Professor of General and Applied Chemistry, Yale University. Pennsylvania Rock Oil Company was bought by Seneca Oil Company prior to the initial drilling under the supervision of Edwin Drake.

In 1859 the first commercial oil well was drilled at Titusville, Pennsylvania, by the Seneca Oil Company in response to favorable reports about the likelihood of establishing a commercial oil drilling operation there. Dr. Benjamin Silliman, a Yale University chemistry professor, was commissioned to analyze petroleum samples from natural surface seepages in the area. In an 1855 report to the Pennsylvania Rock Oil Company he stated, ". . . In conclusion, gentlemen, it appears to me that your company have in your possession a raw material from which by a simple and not expensive process they may manufacture very valuable products." This classic example of understatement helped to launch the petroleum revolution.

The early uses for petroleum components were mainly for lubrication and as kerosene burned in lamps. The development of automobiles with internal combustion engines created the need for liquid fuels that would burn efficiently in these engines. To meet our need for gasoline and other petroleum products, it is necessary to refine crude oil—that is, to separate the various useful components from the complex mixture and to modify their properties.

Petroleum Refining

Petroleum contains about as many compounds as there are genes in the human body.

Substances that have sufficiently different boiling points can be separated from liquid mixtures and solutions by distillation. For example, simple distillation can separate a mixture of the liquids cyclohexane (b.p. 80.7 °C) and toluene (b.p. 110.6 °C). The mixture is heated to slightly above 80.7 °C, at which point cyclohexane boils, but toluene does not. The cyclohexane vapor is condensed as a pure liquid and collected in a separate container (Figure 12.1).

The difference between simple distillation and fractional distillation is the degree of separation achieved.

Petroleum fractions are each mixtures of thousands of hydrocarbons with boiling points within a certain temperature range.

Petroleum contains thousands of different hydrocarbons and their separation as individual pure compounds is neither economically feasible nor necessary. Instead, petroleum refining uses *fractional distillation* to separate petroleum into petroleum fractions. A **petroleum fraction** is a mixture of hundreds of hydrocarbons that have boiling points within a specified range. Such separation is possible because the boiling point increases as the number of carbon atoms increases: Larger hydrocarbon molecules (larger number of electrons and greater polarizability) have greater noncovalent intermolecular forces and higher boiling points than smaller ones ***(← p. 404).***

In fractional distillation, the crude oil is first heated to about 400 °C to produce a hot vapor mixture that enters the bottom of the fractionating tower (Figure 12.2).

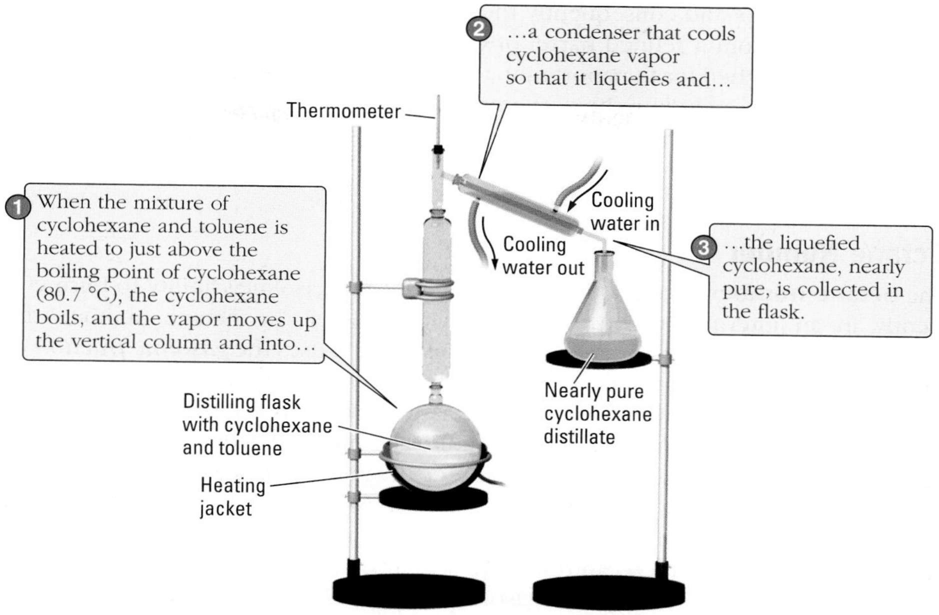

Figure 12.1 Simple laboratory distillation apparatus. Distillation separates cyclohexane from toluene due to differences in their boiling points.

A petroleum fractional distillation tower.

The temperature decreases as the vapor rises up the tower. Consequently, different gaseous components condense at various points in the tower. The more volatile, lower-boiling petroleum fractions remain in the vapor state longer than the less volatile, higher-boiling fractions. The hydrocarbons consisting of the smallest molecules do not condense and are drawn off the top of the tower as gases. The largest hydrocarbon molecules are collected at the bottom of the tower as liquids and dissolved solids.

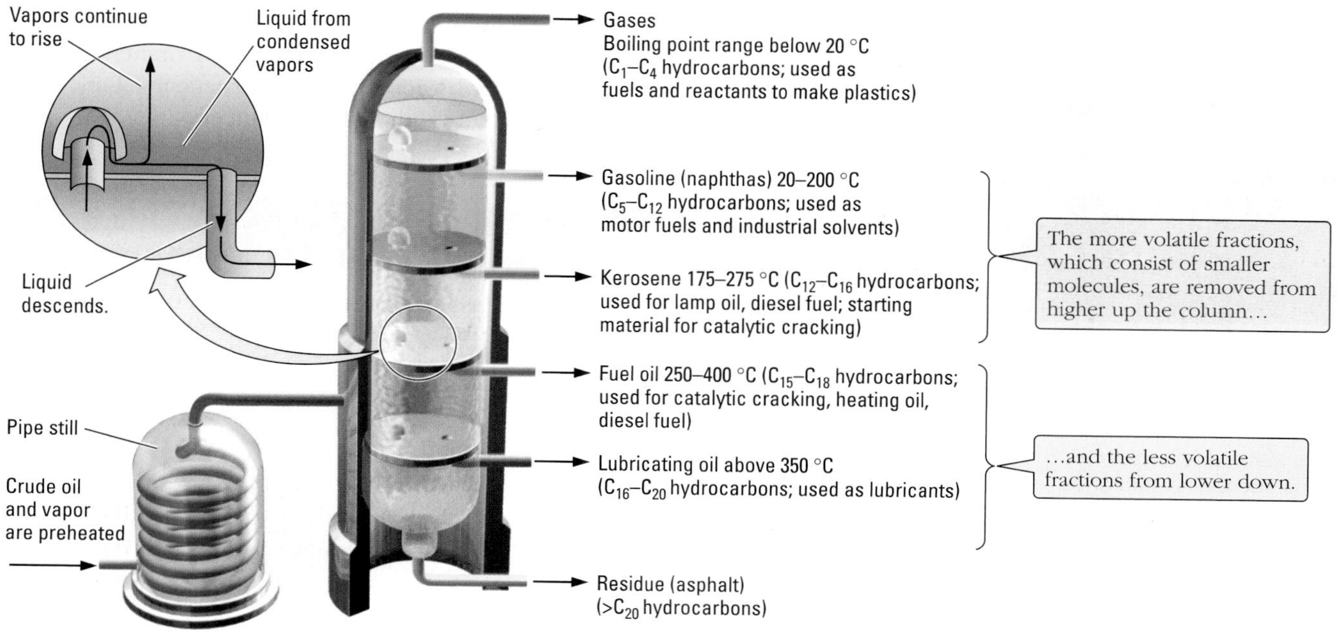

Figure 12.2 Petroleum fractional distillation. Crude oil is first heated to 400 °C in the pipe still. The vapors then enter the fractionation tower. As the vapors rise, different fractions can be removed at different heights in the tower based on boiling point ranges. The smallest (lightest) molecules, C_1–C_4, do not condense.

Today petroleum is seldom seen in barrels, but rather is shipped in pipelines and ocean-going tankers. Nevertheless, the barrel (42 U.S. gallons) remains as the common unit of volume measure for petroleum.

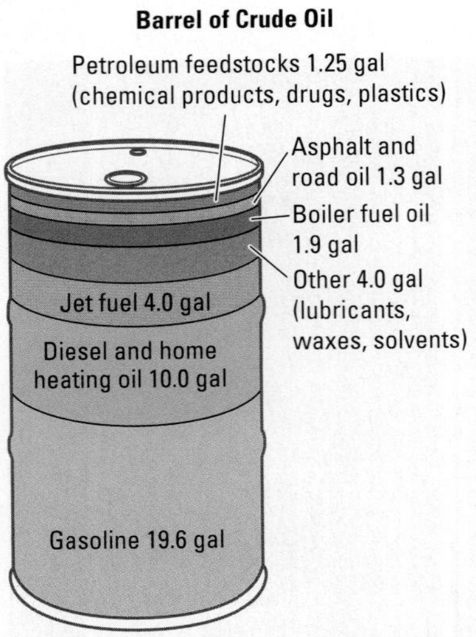
Barrel of Crude Oil
Petroleum feedstocks 1.25 gal (chemical products, drugs, plastics)
Asphalt and road oil 1.3 gal
Boiler fuel oil 1.9 gal
Other 4.0 gal (lubricants, waxes, solvents)
Jet fuel 4.0 gal
Diesel and home heating oil 10.0 gal
Gasoline 19.6 gal

End-uses for products from refining of one barrel of crude oil (42 U.S. gallons).

Diesel engines do not have spark plugs but use autoignition to ignite the diesel fuel, which consists of larger hydrocarbon molecules that have lower autoignition temperatures. Because autoignition is associated with rougher combustion, diesel engines are noisier and must be stronger and heavier than gasoline engines.

6
REGULAR
MID-GRADE
PREMIUM
87
89
93

Typical octane ratings of commercially available gasoline.

Since the method for determining octane numbers was established, fuels that are superior to 2,2,4-trimethylpentane have been developed and thus have octane numbers greater than 100. Note in Table 12.1 that octane burns even less smoothly than heptane and has a negative octane number.

The properties and consequently the uses of various fractions differ, as shown in Figure 12.2. From a refined barrel of crude oil (42 gallons), about 83% (35 gallons) is simply burned for transportation and heating. The remaining 17% is used for nonfuel purposes, including the very important petroleum components needed as reactants to make fabrics, plastics, pharmaceuticals, and other synthetic organic chemicals on which we depend.

Octane Number

The **octane number** of a gasoline is a measure of the gasoline's ability to burn efficiently in an internal combustion engine; the higher the octane number, the smoother the combustion. A typical automobile engine uses the *gasoline fraction* of refined petroleum—a mixture of C_5 to C_{12} hydrocarbons used primarily as motor fuels. This hydrocarbon mixture contains relatively small molecules and has a fairly high autoignition temperature, the temperature at which the hydrocarbon vapor will ignite and burn without a spark or other source of ignition. Because of this fairly high autoignition temperature, a gasoline engine requires a source of ignition—a spark plug—to ignite its fuel. Once ignited, the fuel burns smoothly. If autoignition occurs, combustion is rough, which can reduce power or damage the engine.

The octane-number rating of a gasoline is determined by comparing the burning characteristics when the gasoline burns in a one-cylinder test engine with those obtained for mixtures of heptane and 2,2,4-trimethylpentane, a multibranched chain isomer of straight-chain octane. Heptane does not burn smoothly and is assigned by general consensus an octane number of 0, whereas 2,2,4-trimethylpentane burns smoothly and is assigned an octane number of 100 (Table 12.1). A gasoline assigned an octane number of 87 has the same premature autoignition characteristics as a mixture of 13% heptane and 87% 2,2,4-trimethylpentane. This 87-octane rating corresponds to the octane number of regular unleaded gasoline currently available in the United States. Other, higher grades of gasoline available at gas stations have octane numbers of 89 (regular plus) and 92 (premium).

Straight-chain alkanes are less thermally stable and burn less smoothly than branched-chain alkanes. For example, the "straight-run" gasoline fraction obtained directly from the fractional distillation of petroleum has an octane number of 50 and is a poor motor fuel. It needs additional refining because it consists primarily of straight-chain hydrocarbons that autoignite too readily. The octane number of a gasoline can be increased either by increasing the percentage of branched-chain and aromatic hydrocarbon components or by adding octane enhancers. Table 12.1 lists octane numbers for some hydrocarbons and octane enhancers.

Table 12.1 Octane Numbers of Some Hydrocarbons and Gasoline Additives; Higher Octane Number Indicates Smoother Combustion

Name	Class of Compound	Octane Number
Octane	Alkane	−20
Heptane	Alkane	0
Hexane	Alkane	25
Pentane	Alkane	62
1-Pentene	Alkene	91
2,2,4-Trimethylpentane (isooctane)	Branched-chain alkane	100
Benzene	Aromatic hydrocarbon	106
Methanol	Alcohol	107
Ethanol	Alcohol	108
Tertiary-butyl alcohol	Alcohol	113
Toluene	Aromatic hydrocarbon	118

In addition to fractional distillation, petroleum refining also includes converting the components of various fractions into more economically important products through catalytic cracking and catalytic reforming. Among these products are compounds that can increase octane number.

Catalytic Cracking

During petroleum refining, the percentage of each fraction collected is adjusted to match the market demand. For example, there is greater demand for gasoline than for kerosene and diesel fuel. Demands also vary seasonally. In winter, the need for home heating oil is high; in summer, when more people take vacations, demand for gasoline is higher. In summer, refiners use chemical reactions in a process called "cracking" to convert some of the larger, kerosene-fraction molecules (C_{12} to C_{16}) into smaller molecules in the gasoline range (C_5 to C_{12}). **Catalytic cracking** uses a catalyst, high temperature, and high pressure to break long-chain hydrocarbons into shorter-chain hydrocarbons that include alkanes and alkenes, many in the gasoline range. A **catalyst** is a substance that increases the rate of a chemical reaction without being consumed as a reactant would be. (The role of catalysts in chemical reactions is further discussed in Section 13.8.)

Catalytic cracking breaks larger alkanes into smaller molecules.

$$\underset{\text{An alkane}}{C_{16}H_{34}} \xrightarrow{\text{catalyst, pressure, and heat}} \underset{\text{An alkene}}{C_8H_{16}} + \underset{\text{An alkane}}{C_8H_{18}}$$

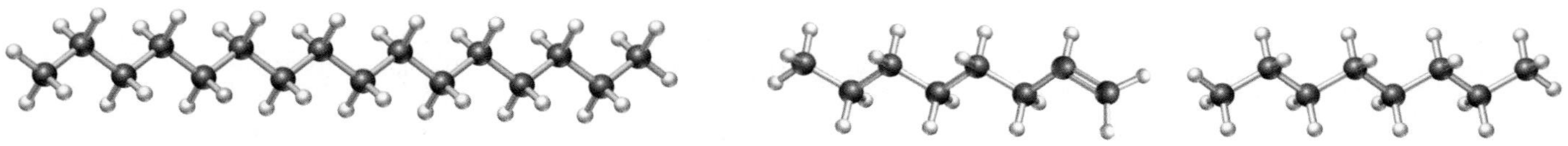

Since alkenes have higher octane numbers than alkanes, catalytic cracking also increases the octane number of the mixture. Catalytic cracking is beneficial in another way. Unlike alkanes, alkenes have C=C bonds, which makes them much more reactive than alkanes. Thus, as seen in Table 12.2, alkenes such as ethylene and propylene can be used as starting materials to make other organic compounds, many of which are the raw materials for making plastics (Section 12.7). Catalytic cracking of natural gas or petroleum produces acetylene, butylene, ethylene, and propylene that are converted chemically into chemical raw materials and many kinds of commercial products (Table 12.2).

Catalytic Reforming

The octane rating of gasoline can also be increased by using **catalytic reforming** to convert straight-chain hydrocarbons to branched-chain hydrocarbons and aromatics. In catalytic reforming, certain catalysts, such as finely divided platinum on a support

Table 12.2 Some Chemical Raw Materials and Commercial Products Made from Petroleum or Natural Gas

Unsaturated Hydrocarbon	Acetylene	Butylene	Ethylene	Propylene
Chemical raw materials	Neoprene	Butadiene	Ethyl alcohol, polyethylene, ethylene oxide	Polypropylene
Commercial products	Synthetic rubber, paints, adhesives, fibers, solvents	Synthetic rubber, car bumpers, computer cases	Plastics, pharmaceuticals, ethylene glycol, antifreeze, synthetic fibers	Plastics, pharmaceuticals, antifreeze, detergents, drycleaning fluids

of Al_2O_3, reform straight-chain hydrocarbons with low octane numbers into their branched-chain isomers, which have higher octane numbers.

Catalytic reforming changes the form of molecules (re-forms them) from straight-chain to branched-chain or aromatic forms.

$$\underset{\substack{\text{pentane}\\ \text{62 octane}\\ (C_5H_{12})}}{CH_3CH_2CH_2CH_2CH_3} \xrightarrow{\text{catalyst}} \underset{\substack{\text{2-methylbutane}\\ \text{94 octane}\\ (C_5H_{12})}}{CH_3\overset{\overset{\displaystyle CH_3}{|}}{C}HCH_2CH_3}$$

Catalytic reforming converts straight-chain hydrocarbons to commercially more valuable branched-chain and aromatic ones.

By using different catalysts and petroleum mixtures, catalytic reforming also produces aromatic hydrocarbons, which have high octane numbers. For example, catalytic reforming converts a high percentage of straight-run gasoline, kerosene, and light oil fractions into a mixture of aromatic hydrocarbons including benzene, toluene, and xylenes. This process can be represented by the equation for converting hexane into benzene.

The additional cost of higher-octane gasoline is due to the extra processing required to make higher-octane compounds.

$$\underset{\substack{\text{hexane}\\ \text{25 octane}\\ (C_6H_{14})}}{CH_3CH_2CH_2CH_2CH_2CH_3} \xrightarrow{\text{catalyst}} \underset{\substack{\text{benzene}\\ \text{106 octane}\\ (C_6H_6)}}{C_6H_6} + 4\ H_2$$

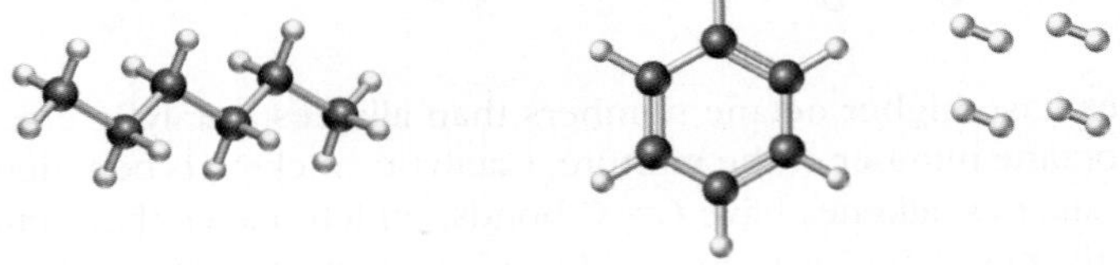

It is no exaggeration to state that our economy relies on the hydrocarbons present in petroleum or derived from it. In addition to being the source of our fuels for transportation, petroleum provides raw materials to make an impressive array of consumer products, as shown in Table 12.2. The hydrogen produced by catalytic reforming is also important for synthesizing ammonia, our main source of nitrogen fertilizer (see Section 14.8).

Answers to **EXERCISES** are provided at the back of this book in Appendix L.

EXERCISES that are labeled **CONCEPTUAL** are designed to test your understanding of one or more concepts; they usually involve qualitative rather than quantitative thinking.

CONCEPTUAL EXERCISE 12.1 Rearranging Hydrocarbons

Heptane, C_7H_{16}, can be catalytically reformed to make toluene, $C_6H_5CH_3$, another seven-carbon molecule.

(a) How many hydrogen molecules are produced for every toluene molecule derived from heptane?

(b) Write a balanced chemical equation for this reaction.

(c) Why would it be profitable to convert heptane into toluene?

Octane Enhancers

Autoignition is also known as "knocking," or "premature ignition."

The octane number of a given blend of gasoline is also increased by adding octane enhancers to prevent autoignition. In the United States, prior to 1975, the most widely used anti-knock agent was tetraethyllead, $(C_2H_5)_4Pb$.

The exhaust emissions of internal combustion engines contain carbon monoxide, nitrogen oxides, and unburned hydrocarbons, all of which contribute to air pollution. Because urban air pollution had worsened, Congress passed the Clean Air Act of 1970, which required that 1975-model-year cars emit no more than 10% of the carbon monoxide and hydrocarbons emitted by 1970 models. The solution to lowering these emissions was a platinum-based catalytic converter, which accelerates the conversion of carbon monoxide to carbon dioxide and the more complete burning of hydrocarbons. A major problem was that the catalyic converter required lead-free gasolines, since lead deactivates the platinum catalyst by coating its surface. As a result, automobiles manufactured since 1975 have been required to use lead-free gasoline to protect the catalytic converter. Because tetraethyllead can no longer be used in the United States and a few other countries, other octane enhancers are now added to gasoline. These include toluene, 2-methyl-2-propanol (also called *tertiary*-butyl alcohol), methanol, and ethanol.

A catalytic converter. Semiprecious metals, like platinum and rhodium, coat the honey combed ceramic surface of the catalytic converter. The metals catalyze the complete combustion of CO to carbon dioxide in the exhaust. Other metal catalysts in the converter speed up the rate at which nitrogen oxides in the exhaust are decomposed to nitrogen and oxygen.

As little as two tanks of leaded gasoline can destroy the activity of a catalytic converter.

Regrettably, many countries of the world still allow the use of leaded gasoline, putting millions of children at risk for lead poisoning from automobile exhaust fumes.

Oxygenated and Reformulated Gasolines

The 1990 amendments to the Clean Air Act of 1970 required cities with excessive carbon monoxide pollution to use oxygenated gasoline during the winter. *Oxygenated gasoline* is a blend of gasoline to which organic compounds that contain oxygen, such as methanol, ethanol, and *tertiary*-butyl alcohol, have been added. Tests conducted by the U.S. Environmental Protection Agency (EPA) indicate that in cold weather oxygenated gasoline burns more completely than nonoxygenated gasoline, thus potentially reducing carbon monoxide emissions in urban areas by up to 17%.

Reformulated gasoline (RFG) is oxygenated gasoline that contains a lower percentage of aromatic hydrocarbons and has a lower volatility than ordinary (nonoxygenated) gasoline. Reformulated gasoline is now used in 17 states and the District of Columbia. The 1990 regulations require RFG to contain a minimum of 2.7% oxygen by mass. Ethanol is the only gasoline oxygenate additive now used in RFG by U.S. refineries.

Oxygenated gasoline is produced by adding oxygen-containing organic compounds to refined gasoline. Reformulated gasoline, which contains such oxygenates, also requires changes in the refining process to alter its percentages of various hydrocarbons, particularly alkenes and aromatics.

All gasolines are highly volatile and have vapors that can be ignited, allowing you to start your car even in the coldest of weather. However, this volatility means that some hydrocarbons get into the atmosphere as a result of accidental spills and evaporation during normal filling operations at the gas station, especially so during warm weather. Recent federal regulations require that the vapor pressure of RFG meet a limit of 9.0 psi from June 1 to September 15, the summer ozone season. More stringent requirements (7.8 psi) may apply in areas not meeting federal guidelines for air quality. Because atmospheric hydrocarbons play an important role in a series of reactions that contribute to urban air pollution (← *p. 460*) including tropospheric ozone formation (especially in heavy-traffic metropolitan areas), reduction of hydrocarbon emissions improves air quality. The EPA estimates that RPG use has reduced smog-forming emissions by 105 million tons, the equivalent of such emissions from 16 million vehicles.

Psi (pounds per square inch) is a pressure unit.

EXERCISE 12.2 Percent Oxygen in Ethanol

Calculate the percent oxygen by mass in ethanol, CH_3CH_2OH. Is it less than 3%? How can ethanol help to meet the 2.7% oxygen requirement for RFG?

Engineering improvements such as emission control systems have succeeded in decreasing the emissions of ozone-forming nitrogen oxides and hydrocarbons from automobiles per mile traveled. Today's cars typically emit 70% less nitrogen oxides and 80% to 90% less hydrocarbons over their lifetimes compared with automobiles produced 30 to 40 years ago. Nevertheless, tropospheric ozone levels remain high

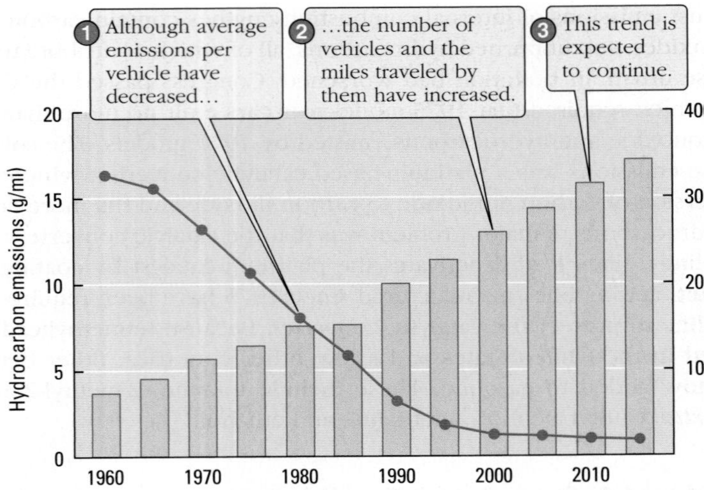

Figure 12.3 Average per-vehicle emissions (grams of hydrocarbons/mile) and mileage increases since 1960.

for two reasons: (1) Since 1970 the number of vehicles in the United States has doubled, as has the number of miles they travel (Figure 12.3), and (2) The EPA attributes a major portion (approximately 25%) of ozone-forming hydrocarbons to a relatively small percentage of "super-dirty" vehicles with faulty emission control systems.

The **PROBLEM-SOLVING STRATEGY** in this book is

- Analyze the problem
- Plan a solution
- Execute the plan
- Check that the result is reasonable

Appendix A.1 explains this in detail.

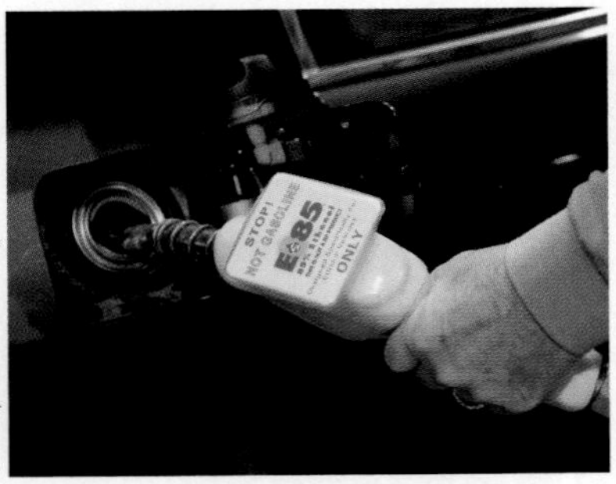

Courtesy of Peter McGaney

Ethanol 85 (E85). E85 is a mixture of 85% ethanol and 15% gasoline by volume.

PROBLEM-SOLVING EXAMPLE 12.1 The Enthalpy of Combustion of Ethanol Compared to That of Octane

Calculate the enthalpy of combustion of ethanol in kilojoules per mole (kJ/mol) and compare the value with that of octane. Then, using the densities of the liquids, calculate the thermal energy liberated on burning a liter of each liquid fuel. The densities of octane and ethanol are 0.703 g/mL and 0.789 g/mL, respectively. The enthalpy of formation of octane(g) is −208.0 kJ/mol.

Answer

Ethanol: $\Delta H_{comb} = -1277.37$ kJ/mol $= -2.19 \times 10^4$ kJ/L

Octane: $\Delta H_{comb} = -5116.4$ kJ/mol $= -3.15 \times 10^4$ kJ/L

Strategy and Explanation All internal combustion engines burn fuel vapors.

- The balanced equation for the complete combustion of ethanol vapor is

$$C_2H_5OH(g) + 3\,O_2(g) \longrightarrow 2\,CO_2(g) + 3\,H_2O(g)$$

Using the standard molar enthalpies of formation from Appendix J and Hess's law (← *p. 242*) for both the products and the reactants, we find that the enthalpy of combustion, ΔH_{comb}, for ethanol is

$$\Delta H_{comb} = \{2[\Delta H_f^\circ\ CO_2(g)] + 3[\Delta H_f^\circ\ H_2O(g)]\} - \{1[\Delta H_f^\circ\ C_2H_5OH(g)] + 3[\Delta H_f^\circ\ O_2(g)]\}$$

$$= \{2(-393.509\ \text{kJ/mol}) + 3(-241.818\ \text{kJ/mol})\} - \{(-235.10\ \text{kJ/mol}) + 0\}$$

$$= -1277.37\ \text{kJ/mol}$$

- Calculate the enthalpy of combustion of octane vapor using standard enthalpies of formation and Hess's law. The balanced combustion reaction for octane vapor is

$$C_8H_{18}(g) + \tfrac{25}{2}\,O_2(g) \longrightarrow 8\,CO_2(g) + 9\,H_2O(g)$$

$$\Delta H_{comb} = \{8[\Delta H_f^\circ\ CO_2(g)] + 9[\Delta H_f^\circ\ H_2O(g)]\} - \{1[\Delta H_f^\circ\ C_8H_{18}(g)] + \tfrac{25}{2}[\Delta H_f^\circ\ O_2(g)]\}$$

$$= \{8(-393.509\ \text{kJ/mol}) + 9(-241.818\ \text{kJ/mol})\} - \{(-208.0\ \text{kJ/mol}) + 0\}$$

$$= -5116.4\ \text{kJ/mol}$$

- Using molar masses and density values, compare the fuels by calculating the energy released, in terms of kJ/L, by the combustion of each fuel. The molar masses of ethanol and octane are 46.069 g/mol and 114.23 g/mol, respectively.

$$\text{Ethanol:}\quad -1277.37\ \text{kJ/mol} \times \frac{1\ \text{mol}}{46.069\ \text{g}} \times \frac{0.789\ \text{g}}{\text{mL}} \times \frac{1000\ \text{mL}}{\text{L}} = -2.19 \times 10^4\ \text{kJ/L}$$

$$\text{Octane:}\quad -5116.4\ \text{kJ/mol} \times \frac{1\ \text{mol}}{114.23\ \text{g}} \times \frac{0.703\ \text{g}}{\text{mL}} \times \frac{1000\ \text{mL}}{\text{L}} = -3.15 \times 10^4\ \text{kJ/L}$$

☑ **Reasonable Answer Check** There are about 17 mol ethanol in a liter of ethanol [(789 g ethanol/L ethanol)(1 mol ethanol/46 g ethanol) = 17 mol] and about 6 mol octane in a liter of octane [(703 g octane/L octane)(1 mol octane/114 g octane) = 6 mol]. But nearly four times as much energy is liberated by burning a mole of octane. Applying this factor on a liter-to-liter basis:

$$\frac{5000\ \text{kJ/mol octane}}{1200\ \text{kJ/mol ethanol}} \times \frac{6\ \text{mol octane/L}}{17\ \text{mol ethanol/L}} = \frac{1.5\ \text{kJ/L octane}}{1.0\ \text{kJ/L ethanol}} = 1.5$$

The 1.5 indicates that about 50% more energy is released by combustion of 1 L octane than from 1 L ethanol. This is close to the 44% difference in the energy per liter calculated for the two fuels, $(3.15 \times 10^4/2.19 \times 10^4) \times 100\% = 1.44$.

PROBLEM-SOLVING PRACTICE 12.1

Calculate the enthalpy of combustion of methanol and the thermal energy released per liter using data from Appendix J. The density of methanol is 0.791 g/mL.

PROBLEM-SOLVING PRACTICE answers are provided at the back of this book in Appendix K.

EXERCISE 12.3 Carbon Monoxide from Ethanol and from Toluene

(a) The combustion of C_2H_5OH can produce carbon monoxide and water. Use the balanced chemical equation for this reaction as the basis to calculate the mass in grams of CO produced per gram of ethanol burned.

(b) Repeat balancing the equation and the calculation this time using toluene, $C_6H_5CH_3$, instead of ethanol.

(c) What conclusions can you draw from your answers about using ethanol to reduce carbon monoxide emissions?

12.2 U.S. Energy Sources and Consumption

The fossil fuels coal, natural gas, and crude oil continue to be the primary U.S. energy sources (Figure 12.4). From the data illustrated in this figure it should be noted that the relative percentages of energy derived from among these fossil fuels have not changed significantly from 2004 to 2007. There was a very slight increase in overall energy derived from nonrenewable energy sources (hydroelectric, geothermal, wind, solar, biomass). We continue to rely on fossil fuels as our predominant energy sources.

The energy provided by the sources overall is consumed by various sectors as shown in Figure 12.5. Note from Figure 12.5 that electric power, transportation, and industry used almost 90% of the energy in 2004 and 2007. During this time, pattterns of energy consumption remained relatively unchanged. During 2008, unusually high fuel prices in the United States, averaging greater than \$4.00/gallon, decreased gasoline usage and energy consumption by transportation for several months. Subsequent lower fuel prices increased gasoline consumption to usual levels.

Natural Gas

Natural gas is a mixture of low-molar-mass hydrocarbons and other gases trapped with petroleum in Earth's crust. It can be recovered from oil wells or from gas wells to which the gases have migrated through the surrounding rock. The nat-

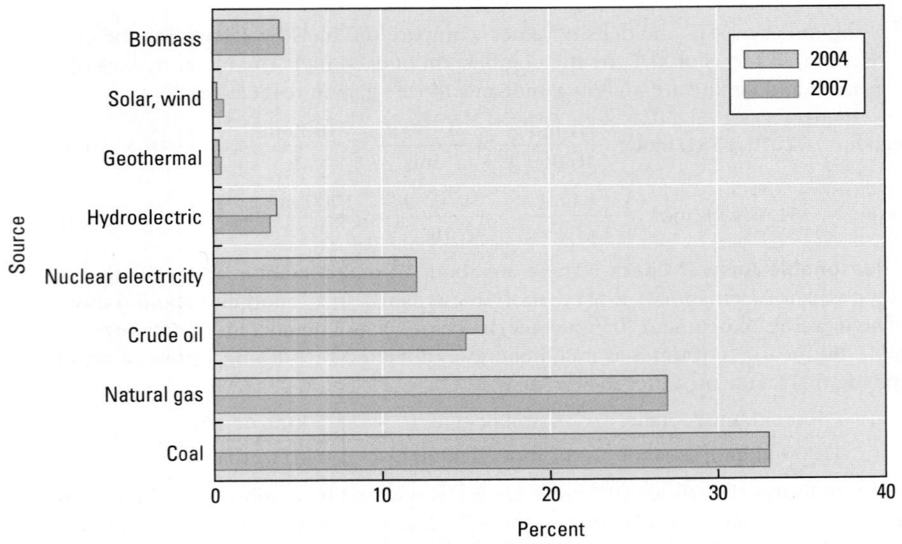

Figure 12.4 **U.S. energy sources 2004 and 2007.** Total energy production (quadrillion kJ, 1 quadrillion kJ = 1.0×10^{15} kJ): 2004 = 70.4 quadrillion kJ; 2007 = 71.5 quadrillion kJ. Fossil fuels are the primary energy sources. (*Source:* **HTTP://EIA.DOE.GOV** Energy Information Administration)

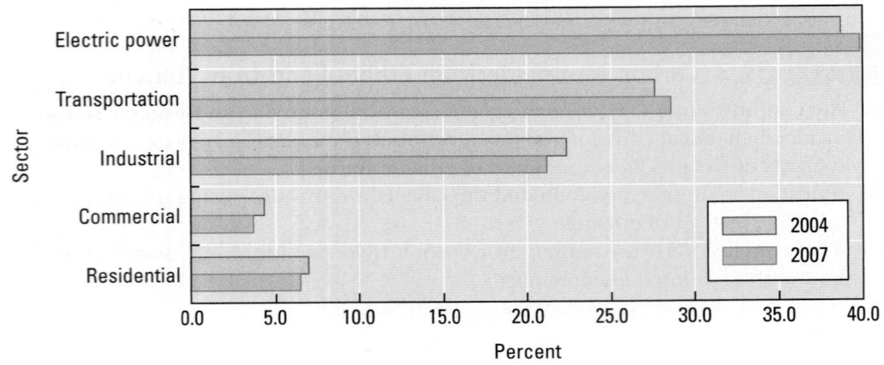

Figure 12.5 **U.S. energy consumption by sector 2004 and 2007.** Most of the energy (almost 90%) was consumed by electric power, transportation, and industry. (*Source:* **HTTP://EIA.DOE.GOV** Energy Information Administration)

ural gas found in North America is a mixture of C_1 to C_4 alkanes [methane, CH_4 (60% to 90%); ethane, C_2H_6 (5% to 9%); propane, C_3H_8 (3% to 18%); and butane, C_4H_{10} (1% to 2%)] with small and varying amounts of other gases, such as CO_2, N_2, H_2S, and the noble gases, mainly helium. In Europe and Japan, natural gas is essentially only methane. The widespread use of natural gas as a fuel did not occur until a vast network of underground pipelines was built to distribute it from natural gas wells to consumers. Before then, most of it was simply burned at the wellheads.

For more than four decades, the U.S. production of natural gas has exceeded that of U.S.-produced oil. Currently, 98% of natural gas consumed in the United States is produced domestically. Approximately 32% of this natural gas is used industrially, 23% for home heating and cooling, 14% in commercial applications, and 31% as a fuel to produce electricity (Figure 12.5). About half of the homes in the United States are heated by natural gas.

Worldwide, more than 7 million vehicles are powered by natural gas. In the United States, 14% of school buses and 10,000 public transit buses are fueled by natural gas, which has the equivalent of a 130-octane rating. Test results have shown that vehicles powered by natural gas emit significantly less carbon monoxide, hydrocarbons, and nitrogen oxides per mile compared with gasoline-powered cars and trucks. The main disadvantages of natural gas vehicles include the need for a pressurized gas tank and the lack of refueling stations that sell the gas in liquefied form.

© Peter McGahey

A natural gas–powered bus. Natural gas powers this bus, which can refuel at a natural gas pump.

CONCEPTUAL EXERCISE 12.4 Natural Gas as a CO Source

It has been asserted that burning natural gas produces 30% less carbon monoxide than the burning of gasoline. Assume that natural gas is only methane and that gasoline is only octane, C_8H_{18}. Based on the assumption that the same fraction of the carbon in each fuel is converted to CO, do calculations that either confirm or refute the assertion.

Coal

Known world coal reserves are far greater than petroleum reserves. About 90% of U.S. annual coal production is burned to produce electricity. The use of coal as the fuel for commercial power plants is on the rise. But, the burning of coal for home heating has declined because it is a relatively dirty fuel and bulky to handle. Burning coal is a major cause of air pollution because of coal's 3% to 4% sulfur content and because coal has a high percentage of carbon and produces more CO_2.

Coal consists of a complex and irregular array of partially hydrogenated six-membered carbon rings and other structures, some of which contain oxygen, nitrogen, and sulfur atoms. Like petroleum, coal supplies raw materials to the chemical industry. Most of the useful compounds obtained from coal are aromatic hydrocarbons. Heating coal at high temperatures in the absence of air, a process called *pyrolysis,* produces a mixture of coke (mostly carbon), coal tar, and coal gas. Fractional distillation of coal tar produces aromatic hydrocarbons such as benzene, toluene, and xylene. They serve as the starting materials for a large variety of important commercial products such as paints, solvents, pharmaceuticals, plastics, dyes, and synthetic detergents.

The mass percent of carbon in coal varies from about 40% to more than 80%, depending on the type of coal.

The usefulness of coal tar to make other materials is summed up in this bit of doggerel from a 1905 organic chemistry textbook:

You can make anything
From a salve to a star
If you only know how
From black coal tar*

*Cornish, *Organic Chemistry,* 1905.

ESTIMATION Burning Oil

A large oil refinery refines 400,000 barrels of crude oil daily. Assume that all the refined crude oil refined in a day is burned to produce electrical energy. Estimate how many toasters this amount of electrical energy would operate for a year assuming the conversion of thermal energy to electrical energy is 50% efficient (a typical value) and that each toaster uses 39 kilowatt hours (kWh) per year. 1.00 kWh = 3.60×10^3 kJ; the energy released by burning one barrel of crude oil is equivalent to 5.9×10^6 kJ.

The energy derived from burning the crude oil is approximately

$$(4 \times 10^5 \text{ barrels}) \times \left(\frac{6 \times 10^6 \text{ kJ}}{\text{barrel}}\right) = 2 \times 10^{12} \text{ kJ}$$

Assuming 50% efficiency in conversion from thermal to electrical energy and rounding values to one significant figure,

$$(2 \times 10^{12} \text{ kJ}) \times \frac{50 \text{ kJ converted}}{100 \text{ kJ released}} \times \left(\frac{1 \text{ kWh}}{4 \times 10^3 \text{ kJ}}\right) = 3 \times 10^8 \text{ kWh}$$

This amount of electrical energy could operate almost 10 million toasters for a year.

$$(3 \times 10^8 \text{ kWh}) \times \left(\frac{1 \text{ toaster}}{40 \text{ kWh}}\right) = 8 \times 10^6 \text{ toasters}$$

Visit this book's companion website at **www.cengage.com/chemistry/moore** to work an interactive module based on this material.

TOOLS OF CHEMISTRY Gas Chromatography

Gas chromatography (GC) is an important method to separate, identify, and quantify chemical compounds in complex mixtures. Widely used in chemistry, biochemistry, environmental analysis, and forensic science, GC is applicable to compounds that have appreciable volatility and are thermally stable at temperatures up to several hundred degrees Celsius. We focus here on capillary column gas chromatography.

An analysis begins by injecting the sample into an inlet at the head of the GC column, where the sample is vaporized. A flow of inert carrier gas, usually helium, sweeps the vaporized sample along the column where the components are separated. The carrier gas stream containing the separated components passes through a detector, which allows identification of individual compounds in the gas stream. The resulting signal produces a *chromatogram*—a plot of concentration of each component versus time.

For capillary GC, the carrier gas usually flows at a rate in the range of 0.5 to 15 mL/min. A measured volume of sample (0.1 to 10 μL) is injected using a microsyringe. Capillary GC columns are made from specially purified silica tubes, typically 10 to 100 m long, that are drawn to have an inside diameter of 0.1 to 0.5 mm. The column is flexible, coiled, and put inside an oven for temperature control. The inside of the column tube is coated with an immobilized liquid, the *stationary phase.* Many factors determine the coating used to separate a particular mixture, but ultimately the choice is dictated by the molecular structure of the stationary liquid phase material in relation to that of the components being separated.

The detector must respond to very low concentrations of separated compounds in the flowing carrier gas. Two widely used detectors are (a) flame-ionization detectors, which atomize and ionize the sample and measure the flow of current that results, and (b) thermal conductivity detectors, which measure the thermal conductivity of the gas stream. Mass spectrometers **(← *p. 56*)** are also used as detectors. The resulting chromatogram, a series of peaks, is a plot of the concentration of each component as a function of time. The position of each peak on the time (x) axis qualitatively identifies the components in the sample; the area under each peak provides a quantitative measure of the amount of each compound.

Compounds in a mixture are separated by GC due to their interactions with the stationary liquid phase on the column. Molecular structure dictates how quickly each component moves along the column to the detector. This depends on the magnitude of intermolecular interactions—polar interactions and London forces **(← *p. 403*)**—between components of the mixture and the stationary phase. Components having strong attractions for the stationary phase are retained longer on the column; those with weaker interactions reach the detector sooner. Thus, as the sample moves along the column, the components are progressively separated before they reach the detector.

If polar compounds are to be separated from a mixture, then a polar stationary phase works best, one that contains polar functional groups such as —CN, >C=O, and —OH. A nonpolar stationary phase contains hydrocarbon groups only. When the polarity of the stationary phase is matched well with that of the sample, the components exit from the column in order of their boiling points.

Capillary column GC can readily identify and measure components in gasoline. In the chromatogram below, 31 compounds ranging in size and complexity from butane and pentane to larger, ring compounds such as xylenes and toluene are separated from a gasoline sample.

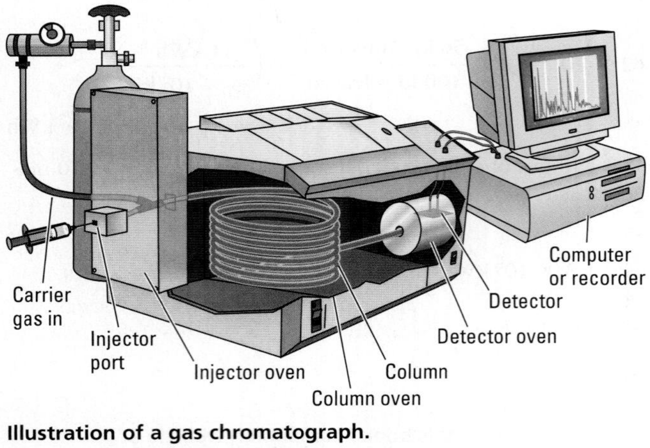

Illustration of a gas chromatograph.

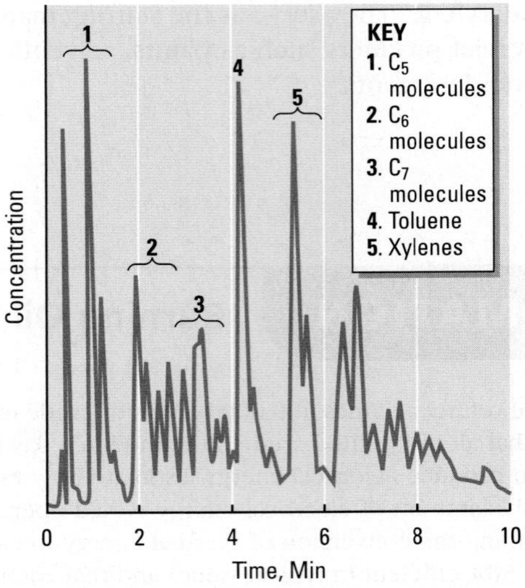

The chromatogram for a gasoline analysis. (Column: Petrocol A; Carrier: helium, 2.5 mL/min; Detector: flame ionization.) Peaks 1 are for components with five-carbon molecules, peaks 2 for six-carbon components, and peaks 3 for seven-carbon components. The last two peaks are for aromatic compounds—peak 4 indicates the presence of toluene, $C_6H_5CH_3$, and peak 5 the various xylenes, $C_6H_4(CH_3)_2$ **(← *p. 359*).** These more complex aromatic hydrocarbons are held on the column longer than the nonaromatic hydrocarbons.

12.3 Organic Chemicals

Organic chemicals were once obtained only from plants and animals. A few living organisms are still direct sources of useful hydrocarbons. Rubber trees produce latex that contains rubber, a familiar hydrocarbon, and other plants produce an oil that burns almost as well in a diesel engine as diesel fuel. As important as these examples are, the development of synthetic organic chemistry has led to cheaper methods of making copies of naturally occurring substances and to many substances that have no counterpart in nature.

Prior to 1828, it was widely believed that chemical compounds found in living matter could not be made without living matter—a "vital force" was thought to be necessary for their synthesis. In 1828 a young German chemist, Friedrich Wöhler, contradicted the vital force theory when he prepared the organic compound urea, a major product in urine, by heating an aqueous solution of ammonium cyanate, a compound obtained from mineral sources. Soon thereafter, other chemists began to prepare more and more organic chemicals without using a living system.

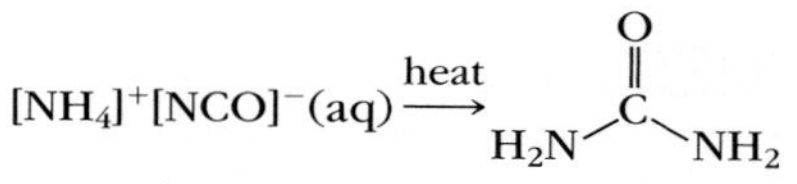

ammonium cyanate urea

Because about 85% of all known compounds are organic compounds, it is natural to ask: Why are there so many organic compounds, almost all of which contain carbon and hydrogen atoms (and commonly other kinds of atoms as well)? As discussed in earlier sections on bonding and isomerism (← *p. 334, p. 340*), two reasons are (1) the ability of as many as thousands of carbon atoms to be linked to each other in a single molecule by stable C—C bonds, and (2) the occurrence of struc-

CHEMISTRY IN THE NEWS

Small Molecules, Big Results: Molecular Possibilities for Drug Development

Pharmaceutical chemists worldwide design experiments to synthesize new compounds that could cure deadly diseases, without significant adverse side effects to patients. Synthesis and purification of the target compounds can be difficult. Once synthesized, the compounds must go through rigorous, sequential testing at various levels before being approved for patient use.

Usually, useful drugs are relatively small molecules with relatively low molar mass and not many different kinds of atoms. How many combinations of such atoms are possible that could lead to a miracle drug? In a recent study reported in the *Journal of the American Chemical Society,* Swiss chemists Jean-Louis Reymond and Lorenz Blum point out that the rules of chemical bonding allow simple elements such as carbon, hydrogen, oxygen, nitrogen, and chlorine to potentially form millions of different molecules. This so-called chemical universe or chemical space has an enormous potential for drug discovery, particularly for identifying so-called small molecules, those containing 10 to 50 atoms.

Using an extensive new database they developed, GDB-13, Reymond and Blum comprehensively analyzed the molecular structures of likely molecules that could populate "chemical space" and potentially lead to new drugs. The new searchable database (**http://www.gdb.unibe.ch**) consists of all molecules containing up to 13 atoms of carbon, nitrogen, oxygen, sulfur, and chlorine under rules that define chemical stability and synthetic feasibility. Through GDB-13 analyses, the researchers identified more than 970 million possible molecular structures suitable for study in the design and production of potential new drugs to fight disease. The vast majority of the suggested compounds have never been synthesized in the lab.

Source: Journal of the American Chemical Society, Vol. 131, 2009; pp. 8732–8733.

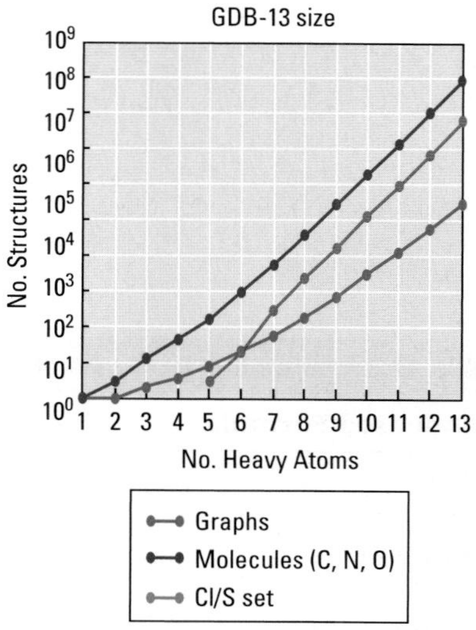

GDB-13 enumerates small organic molecules containing up to 13 atoms of C, N, O, S, and Cl following simple chemical stability and synthetic feasibility rules.

A table of functional groups and further information on how organic compounds are named is given in Appendix E.

The alcohol, aldehyde, ketone, carboxylic acid, ester, amine, and amide functional groups are discussed in Sections 12.4–12.7.

go Chemistry **Module 15: Naming Organic Compounds** covers concepts in this section.

tural isomers. A third reason—the variety of functional groups that bond to carbon atoms—is further illustrated in this chapter. A **functional group** is a distinctive grouping of atoms that, as part of an organic molecule, imparts specific properties to the molecule. In addition to these three reasons, many molecules also display optical isomerism ***(← p. 388)***, a very subtle type of molecular isomerism related to "handedness."

All organic molecules can be viewed as derived from hydrocarbons, and many of them are prepared by chemists starting with hydrocarbons. Among some of the most useful of these compounds are alcohols, carboxylic acids, esters, amines, and the natural and synthetic polymers that can be made from them.

12.4 Alcohols and Their Oxidation Products

Alcohols are a major class of organic compounds. All alcohols, both natural and synthetic, contain the characteristic —OH functional group bonded directly to a carbon atom. Some alcohols contain more than one —OH group. Examples of commercially important alcohols and their uses are listed in Table 12.3.

Methanol

Methanol, CH_3OH, the simplest of all alcohols, has just one carbon atom. Because methanol is so useful in making other products, more than 8 billion pounds of it are produced annually in the United States by the reaction of carbon monoxide with hydrogen in the presence of a catalyst at 300 °C.

$$CO(g) + 2\ H_2(g) \xrightarrow{\text{catalyst, 300 °C}} CH_3OH(g)$$

Methanol is sometimes called *wood alcohol* because the old method of producing it was by heating a hardwood such as beech, hickory, maple, or birch in the absence of air. Methanol is highly toxic. Drinking as little as 30 mL can cause death, and smaller quantities (10 to 15 mL) can cause blindness.

Table 12.3 Some Important Alcohols

Condensed Formula	b.p. (°C)	Systematic Name	Common Name	Use
CH_3OH	65.0	Methanol	Methyl alcohol	Fuel, gasoline additive, making formaldehyde
CH_3CH_2OH	78.5	Ethanol	Ethyl alcohol	Beverages, gasoline additive, solvent
$CH_3CH_2CH_2OH$	97.4	1-propanol	Propyl alcohol	Industrial solvent
$CH_3CH(OH)CH_3$ (OH on the middle C)	82.4	2-propanol	Isopropyl alcohol	Rubbing alcohol
$CH_2(OH)CH_2(OH)$ (OH on each C)	198	1,2-ethanediol	Ethylene glycol	Antifreeze
$CH_2(OH)—CH(OH)—CH_2(OH)$ (OH on each C)	290	1,2,3-propanetriol	Glycerol (glycerin)	Moisturizer in foods

About 50% of methanol is used in the production of formaldehyde, HCHO, which is used to make plastics, embalming fluid, germicides, and fungicides; 30% is used in the production of other chemicals; and the remaining 20% is used for jet fuels, antifreeze solvent mixtures, and gasoline additives.

Since methanol is relatively cheap, its potential as a fuel and as a starting material for the synthesis of other chemicals is receiving more attention. The technology for methanol-powered vehicles has existed for many years, particularly for racing cars that burn methanol because of its high octane rating (107). Methanol has both advantages and disadvantages as a gasoline replacement. As a motor fuel, methanol burns more cleanly than gasoline, but methanol has only about half the energy density of the same volume of gasoline. Therefore, twice as much methanol must be burned to give the same distance per tankful as gasoline. This disadvantage is partially compensated for by the fact that methanol costs about half as much to produce as gasoline.

Some racing cars burn methanol.

PROBLEM-SOLVING EXAMPLE 12.2 Methanol and MTBE

Methanol reacts with 2-methylpropene to yield methyl *tert*-butyl ether (MTBE), which was used as an octane enhancer and in reformulated gasolines.

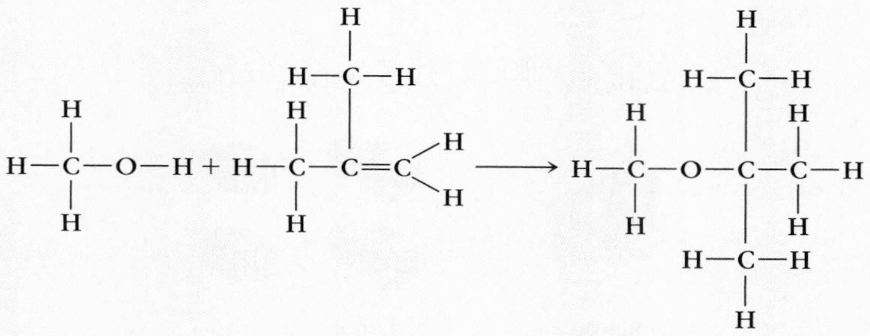

methanol 2-methylpropene methyl *tert*-butyl ether (MTBE)

Given the bond enthalpy data in the margin, estimate the enthalpy change, ΔH, for the synthesis of MTBE. Is the reaction endothermic or exothermic?

Bond	Bond Enthalpy (kJ/mol)
C—H	416
C—O	336
O—H	467
C—C	356
C=C	598

Answer −43.0 kJ; the reaction is exothermic.

Strategy and Explanation We can think of the reaction as being one in which bonds in the reactants are broken (endothermic process) and then bonds are formed in the products (exothermic). The balanced chemical equation for the reaction is given using Lewis structures, so we can directly count the number of moles of each kind of bond to be broken in the reactants and those to be formed in the product. We then multiply these amounts by their respective bond enthalpies, D, to estimate the enthalpy change for the reaction ***(Section 8.6, ← p. 345).***

$$\begin{aligned}\Delta H^\circ &= \sum[(\text{moles of bonds}) \times D(\text{bonds broken})] \\ &\quad - \sum[(\text{moles of bonds}) \times D(\text{bonds formed})] \\ &= \{[(3\text{ mol C—H})(416\text{ kJ/mol}) + (1\text{ mol C—O})(336\text{ kJ/mol}) \\ &\quad + (1\text{ mol O—H})(467\text{ kJ/mol})] + [(8\text{ mol C—H})(416\text{ kJ/mol}) \\ &\quad + (2\text{ mol C—C})(356\text{ kJ/mol}) + (1\text{ mol C=C})(598\text{ kJ/mol})]\} \\ &\quad - \{[(12\text{ C—H})(416\text{ kJ/mol}) + (2\text{ mol C—O})(336\text{ kJ/mol}) \\ &\quad + (3\text{ mol C—C})(356\text{ kJ/mol})]\} \\ &= (6689\text{ kJ}) - (6732) = -43.0\text{ kJ}\end{aligned}$$

The negative sign indicates that the reaction is exothermic.

☑ **Reasonable Answer Check** The net change in the reaction is the conversion of 1 mol C=C bonds and 1 mol O—H bonds into 1 mol C—H bonds, 1 mol C—C bonds, and 1 mol C—O bonds. The other bonds are unchanged. The enthalpy change for the conversions is

$$\Delta H^\circ = [(1 \text{ mol C}{=}\text{C})(598 \text{ kJ/mol}) + (1 \text{ mol O—H})(467 \text{ kJ/mol})]$$
$$- [(1 \text{ mol C—H})(416 \text{ kJ/mol}) + (1 \text{ mol C—C})\ (356 \text{ kJ/mol})$$
$$+ (1 \text{ mol (C—O)}(336 \text{ kJ/mol})]$$

which equals −43.0 kJ, the correct answer.

PROBLEM-SOLVING PRACTICE 12.2

Using Table 8.2 (*← p. 345*), of Average Bond Enthalpies, estimate the enthalpy change for the industrial synthesis of methanol by the catalyzed reaction of carbon monoxide with hydrogen.

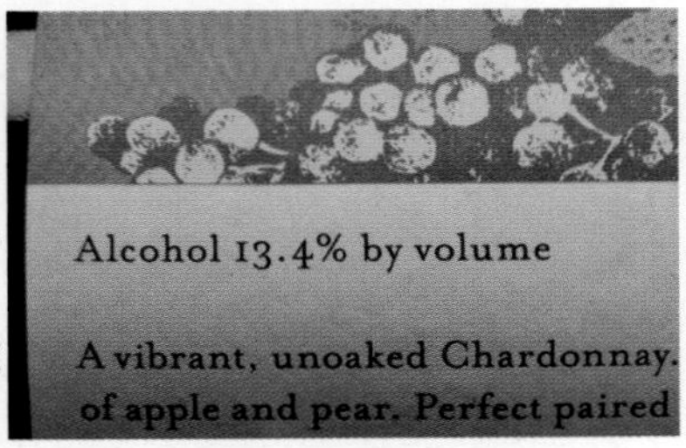
Alcohol 13.4% by volume
A vibrant, unoaked Chardonnay.
of apple and pear. Perfect paired

Alcohol concentration in wine.

Table 12.4 Blood Alcohol Levels and Their Effects

% by Volume	Effect
0.05–0.15	Lack of coordination
0.08	Commonly defined point for "driving while intoxicated"
0.15–0.20	Obvious intoxication
0.30–0.40	Unconsciousness
0.50	Possible death

Ninety-five percent ethanol is the maximum ethanol concentration that can be obtained by distillation of alcohol/water mixtures because ethanol and water form a mixture that boils without changing composition.

Ethanol as a vehicle fuel is also available as Ethanol85, a mixture of 85% ethanol and 15% gasoline by volume. There are now more than six million FlexFuel vehicles in the United States using gasohol and Ethanol85 in conjunction with regular gasoline.

Ethanol

Ethanol, also called *ethyl alcohol* or *grain alcohol,* is the "alcohol" of alcoholic beverages. For millennia it has been prepared for this purpose by fermentation of carbohydrates (starch, sugars) from a wide variety of plant sources. For example, glucose is converted into ethanol and carbon dioxide by the action of yeast in the absence of oxygen.

$$C_6H_{12}O_6 \xrightarrow{\text{Yeast}} 2\ C_2H_5OH + 2\ CO_2$$

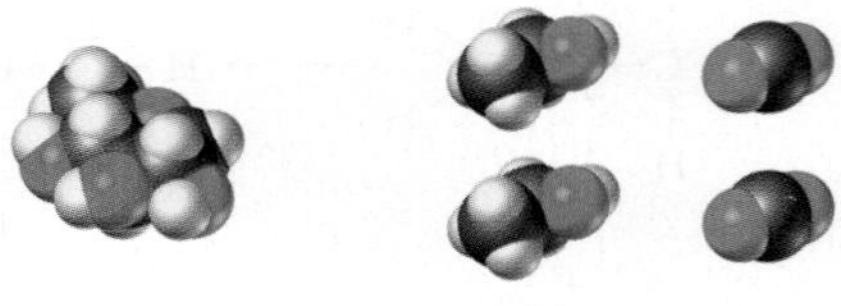

glucose ethanol

Fermentation eventually stops when the alcohol concentration reaches a level sufficient to inhibit the yeast cells. The "proof" of an alcoholic beverage is twice the volume percent of ethanol; 80-proof vodka, for example, contains 40% ethanol by volume. Ethanol is not as toxic as methanol, but one pint of pure ethanol, rapidly ingested, will kill most people. Ethanol is a depressant; the effects of various blood levels of alcohol are shown in Table 12.4. Rapid consumption of two 1-oz "shots" of 90-proof whiskey, two 12-oz beers, or two 4-oz glasses of wine can cause one's blood alcohol level to reach 0.05%.

In the United States, the federal tax on alcoholic beverages is about $13 per gallon, which is about 20 cents per ounce of ethanol in the beverage. Since the cost of producing ethanol is only about $1 per gallon, ethanol intended for industrial use must be *denatured* to avoid the beverage tax and to prevent people from drinking the alcohol. Ethanol is denatured by adding to it small amounts of a toxic substance, such as methanol or gasoline, that cannot be removed easily by chemical or physical means.

Ethanol, like methanol, is receiving increased attention as an alternative fuel. At present, most of it is used in a blend of 90% gasoline and 10% ethanol by mass (known as *gasohol* when introduced in the 1970s). Ethanol is also used as an oxygenated fuel to add to gasoline. The fermentation of corn, an abundant (but not limitless) source of carbohydrate, is used to produce ethanol for fuel use.

Hydrogen Bonding in Alcohols

The physical properties of liquid alcohols are a direct result of the effects of hydrogen bonding among alcohol molecules. Alcohols can be considered as compounds in which alkyl groups such as CH_3— and CH_3CH_2— are substituted for one of the hydrogens of water. The change from the hydrogen in water to the alkyl group in an alcohol reduces the influence of hydrogen bonding in the progression from water, HOH, to methanol, CH_3OH, to ethanol, CH_3CH_2OH, and higher-molar-mass alcohols. The boiling points of methanol (32 g/mol; 65 °C) and ethanol (46 g/mol; 78.5 °C) are lower than that of water (18 g/mol; 100 °C) because methanol and ethanol have only one —OH hydrogen atom available for hydrogen bonding; water has two. The higher boiling point of ethylene glycol (198 °C) can be attributed to the presence of two —OH groups per molecule. Glycerol, with three —OH groups, has an even higher boiling point (290 °C). Hydrogen bonding among ethanol molecules explains why it is a liquid, while propane, which has a similar number of electrons (and molar mass) but no hydrogen bonding, is a gas at the same temperature. The alcohols listed in Table 12.3 are very water-soluble because of hydrogen bonding between water molecules and the —OH alcohol group.

See Table 12.3 (p. 546) for the structural formulas of these compounds.

CONCEPTUAL EXERCISE 12.5 Water Solubility of Alcohols

What would you expect concerning the water solubility of an alcohol containing ten carbon atoms and one —OH group? Recall that hydrocarbons such as octane are not water-soluble.

CONCEPTUAL EXERCISE 12.6 One Alcohol in Another

Methanol dissolves in glycerol. Use Lewis structures to illustrate the hydrogen bonding between methanol and glycerol molecules.

Oxidation of Alcohols

Alcohols are classified according to the number of carbon atoms *directly* bonded to the —C—OH carbon. If no carbon atom or one carbon atom is bonded directly, the compound is a *primary* alcohol. If two are bonded directly, it is a *secondary* alcohol; with three it is a *tertiary* alcohol, as illustrated below. The use of R, R′, and R″ to represent alkyl groups indicates that the alkyl groups can be different.

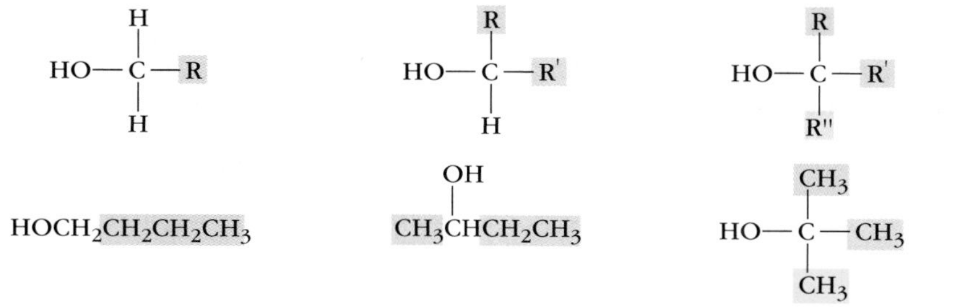

1-butanol, a primary alcohol | **2-butanol, a secondary alcohol** | **2-methyl-2-propanol, a tertiary alcohol**

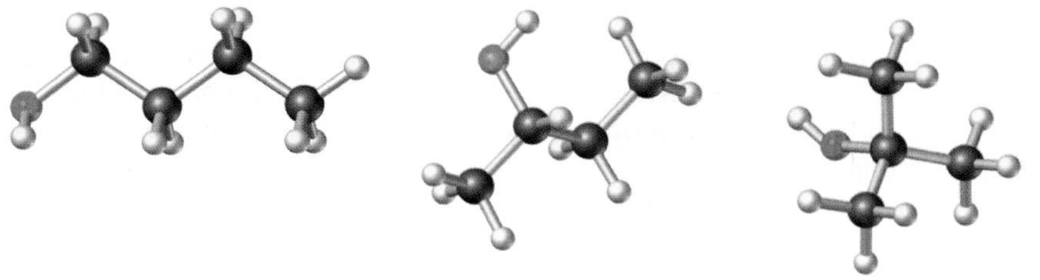

In naming alcohols, the longest chain of carbon atoms is numbered so that the carbon with the —OH attached has the lowest possible number. A table of functional groups and further information on how organic compounds are named is given in Appendix E.

Controlled oxidation of primary alcohols initially produces compounds called aldehydes; aldehydes can be further oxidized to carboxylic acids.

$$\text{Primary alcohol} \xrightarrow{\text{oxidation}} \text{aldehyde} \xrightarrow{\text{oxidation}} \text{carboxylic acid}$$

Controlled oxidation of organic compounds usually involves either the loss of two hydrogen atoms or the gain of one oxygen atom. For example, the stepwise oxidation of ethanol with aqueous potassium permanganate as the oxidizing agent first produces acetaldehyde, which contains a —CHO functional group and belongs to the class of compounds called **aldehydes.**

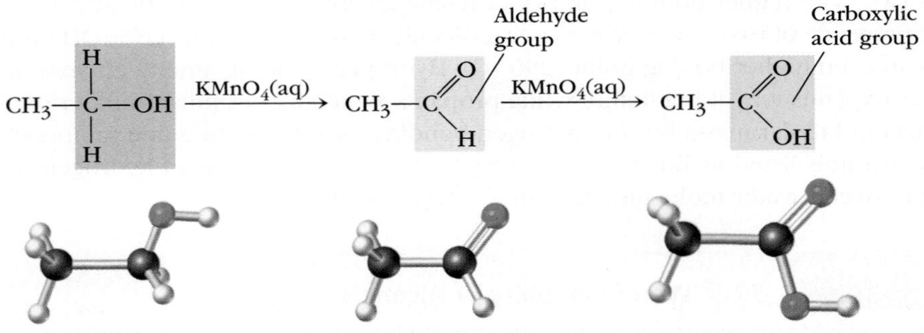

ethanol (a primary alcohol) acetaldehyde acetic acid

Vinegar is a 5% solution of acetic acid in water.

The acetaldehyde is then oxidized to acetic acid, a member of compounds called **carboxylic acids,** which contains the —COOH functional group (also represented as —CO_2H). When ethanol is ingested, enzymes in the liver produce the same products. Acetaldehyde, the intermediate product, contributes to the toxic effects of alcoholism. In the presence of oxygen in air, ethanol in wine is oxidized naturally to acetic acid, converting the wine from a beverage into something more suitable for a salad dressing.

CONCEPTUAL EXERCISE 12.7 Looking at the Oxidation of Primary Alcohols

Look carefully at the formulas for ethanol and acetaldehyde. Explain the change in structure when ethanol is oxidized to acetaldehyde. Do the same for the oxidation of acetaldehyde to acetic acid. How do your explanations differ for the different products?

Compound Class	Functional Group	Example
Aldehyde	—C(=O)H	Formaldehyde H—C(=O)H
Carboxylic acid	—C(=O)—OH	Acetic acid CH_3—C(=O)—OH
Ketone	—C(=O)—	Acetone CH_3—C(=O)—CH_3

Oxidation of organic compounds is usually the removal of two hydrogen atoms or the addition of one oxygen atom; reduction is usually the addition of two hydrogens or the removal of one oxygen.

Alcohols, aldehydes, carboxylic acids, and ketones are important biologically. *Oxidation of secondary alcohols by the loss of two hydrogens produces* **ketones,** which contain the C=O functional group. In the laboratory an acidic solution of potassium dichromate, $K_2Cr_2O_7$, is a common oxidizing agent for this reaction.

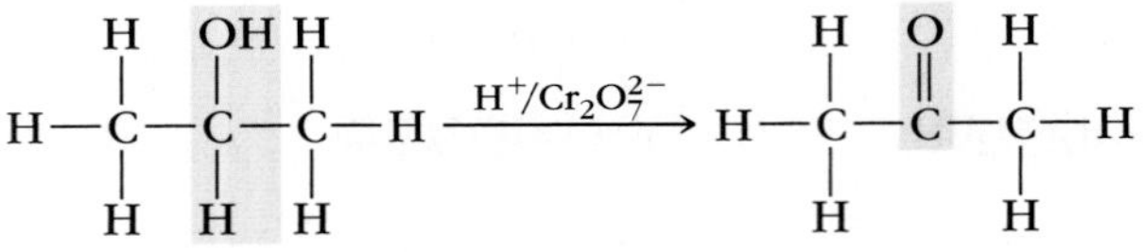

Secondary alcohol ⟶ ketone

Oxidation of a secondary alcohol to a ketone is important during strenuous exercise, when the lungs and circulatory system are unable to deliver sufficient oxygen to the muscles. Under these conditions, lactic acid builds up in muscle tissue, causing soreness and exhaustion. After the exercising is finished, enzymes help to oxidize the secondary alcohol group in lactic acid to a ketone group, thereby converting the lactic acid to pyruvic acid. Pyruvic acid is an important biological compound that is further metabolized to provide energy.

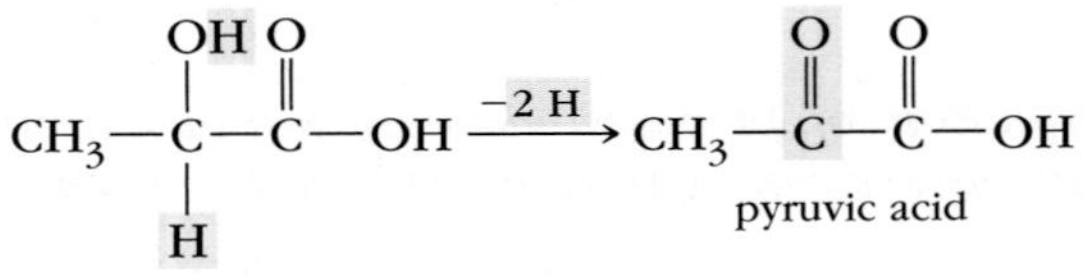

pyruvic acid

lactic acid

Tertiary alcohols, which have no hydrogen atoms directly bonded to the carbon bearing the —OH group, are not oxidized easily.

PROBLEM-SOLVING EXAMPLE 12.3 Oxidation of Alcohols

Write the condensed structural formulas of the alcohols that can be oxidized to make these compounds:

(a) $CH_3CH_2CH_2\overset{\overset{\displaystyle O}{\|}}{C}—H$ (b) $CH_3CH_2\overset{\overset{\displaystyle O}{\|}}{C}—CH_3$ (c) $CH_3CH_2CH_2CH_2\overset{\overset{\displaystyle O}{\|}}{C}—OH$

Answer

(a) $CH_3CH_2CH_2CH_2OH$ (b) $CH_3CH_2\underset{}{\overset{\overset{\displaystyle OH}{|}}{C}}HCH_3$ (c) $CH_3CH_2CH_2CH_2CH_2OH$

Strategy and Explanation An aldehyde or carboxylic acid group is formed by the oxidation of a primary alcohol in which the —OH group is at the end of the molecule. A ketone group is made by the oxidation of a secondary alcohol, a compound in which the —OH group is not on a terminal (end) carbon atom.

(a) The oxidation of the four-carbon primary alcohol, 1-butanol, will produce this aldehyde.
(b) To produce this ketone, choose the secondary alcohol, 2-butanol, which has two carbons to the left and one carbon to the right of the C—OH group.
(c) The oxidation of the five-carbon primary alcohol, 1-pentanol, will produce this carboxylic acid.

PROBLEM-SOLVING PRACTICE 12.3

Draw the structural formulas of the expected oxidation products of these compounds.

(a) $CH_3CH_2CH_2OH$ (b) $CH_3\underset{\underset{\displaystyle OH}{|}}{C}HCH_2CH_3$

CONCEPTUAL EXERCISE 12.8 Aldehydes and Combustion Products

Write the equation for the formation of an aldehyde by the oxidation of methanol. Critics of the use of methanol as a fuel or fuel oxygenate have cited the formation of this aldehyde as a major health threat. Use the Internet to find some of the toxic properties of the aldehyde.

CONCEPTUAL EXERCISE 12.9 Working Backward

When oxidized, a certain alcohol yields an aldehyde containing three carbon atoms. Further oxidation forms a three-carbon acid with the acid functional group located on the number-one carbon of the chain. What are the structural formula and the name of the alcohol?

DePauw University Archives

Percy Lavon Julian

1899–1975

After receiving a doctorate in chemistry in Vienna in 1931, Percy Julian spent 18 years as a research director in the chemical industry, and then directed his own research institute. He was granted more than 100 patents and was the first to synthesize hydrocortisone (a steroid) and to isolate from soybean oil the compounds used to make the first synthetic sex hormone (progesterone). These amazing accomplishments came from a person who had to leave his home in Montgomery, Alabama, after eighth grade for further studies—no more public education was available there for an African American man. Julian enrolled as a "subfreshman" at DePauw University in Indiana. On his first day, a white student welcomed him with a handshake. Julian later related his reaction: "In the shake of a hand my life was changed. I soon learned to smile and act like I believed they all liked me, whether they wanted to or not."

Biologically Important Compounds Containing Alcohol Groups

Many natural organic compounds are cyclic. They can be considered to be derivatives of hydrocarbons that consist of aromatic rings (← p. 359), cycloalkane rings (← p. 335), or cycloalkene rings. In many of these compounds, two or more rings are fused, that is, the rings share carbon atoms. Steroids, all of which have the four-ring structure shown on the next page, are an example of such fused-ring structures.

Cycloalkenes are nonaromatic ring compounds that contain at least one C=C double bond.

TOOLS OF CHEMISTRY Nuclear Magnetic Resonance and Its Applications

Like the electron (*← p. 292*), the nuclei of certain isotopes have quantized spin. For example, a ^1H nucleus (the hydrogen nucleus, a proton) can spin in either of two directions. In the absence of a magnetic field, the two spin states have the same energy. When a strong external magnetic field is applied, those ^1H nuclei with spins aligned with the external magnetic field are slightly lower in energy than those that are aligned against the magnetic field. The value of the energy difference, ΔE, is small enough that a photon of radio wave radiation (***Figure 7.1,*** *← p. 273*) has sufficient energy to change the direction of nuclear spin.

Consider the aligned hydrogen nuclei *absorbing* a particular radio frequency ($h\nu$) that changes their spin to the less stable direction. When the nuclei return to the more stable spin direction, the same radio frequency is *emitted* and can be measured with a radio receiver. This phenomenon, known as **nuclear magnetic resonance (NMR),** provides an extremely valuable tool for studying molecular structure.

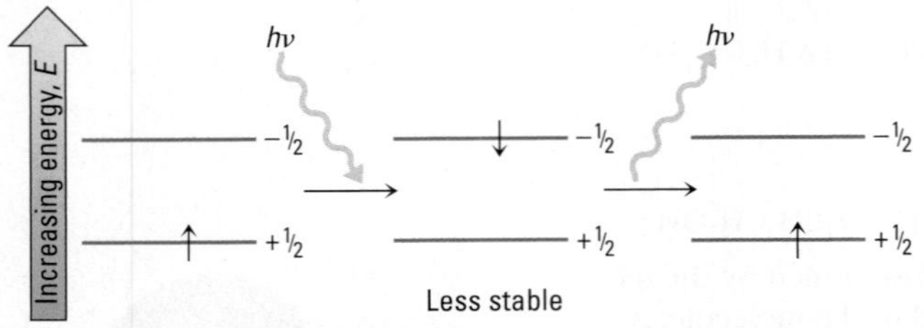

Absorption of radio frequency energy changes the direction and energy of spin of a proton. Emission of the energy returns the proton to its original spin.

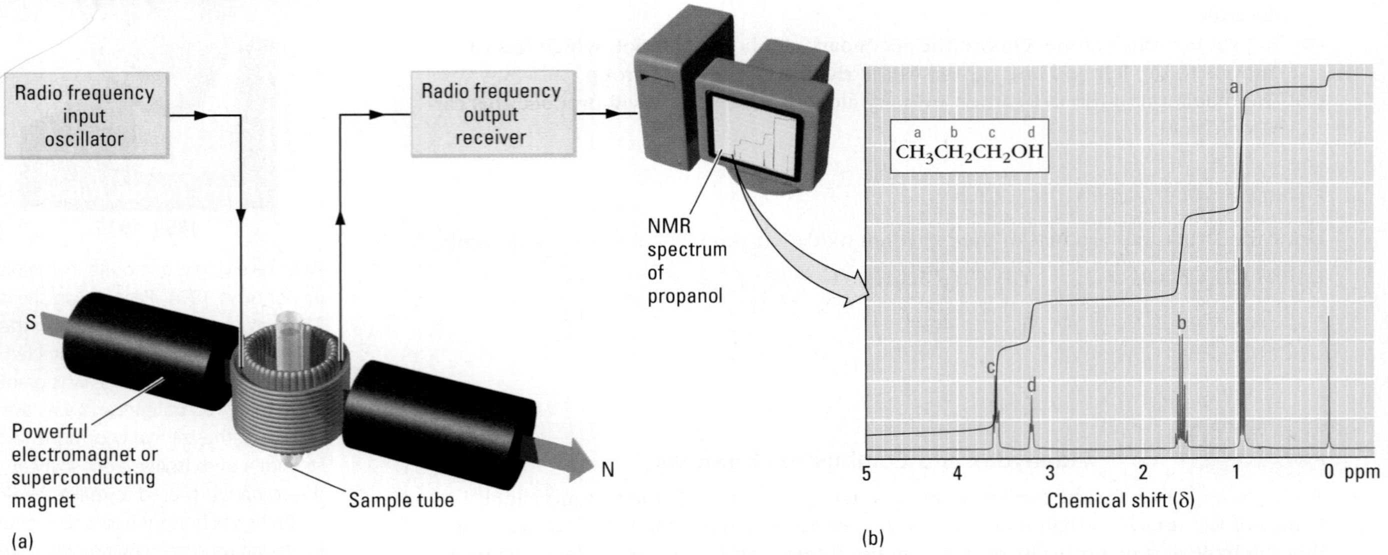

A modern NMR spectrometer. (a) Schematic diagram. (b) The spectrum pictured here is of propanol, showing the four different types of protons (hydrogen nuclei)—a, b, c, d—present in different bonding environments in the molecule. The chemical shift (symbolized by δ and given in parts per million) is the position of the absorption relative to a defined zero point; chemical shifts depend on the strength of the magnetic field of the instrument.

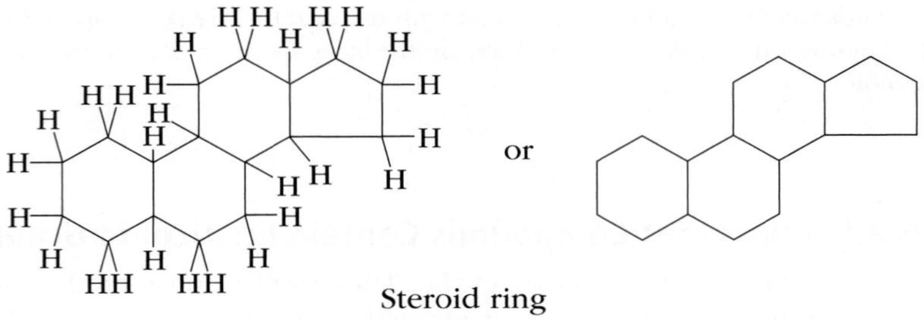

Steroid ring

Many steroids—including cholesterol, the female sex hormones estradiol and estrone, and the male sex hormone testosterone—contain alcohol functional groups. Cholesterol is the most abundant steroid in the human body, where it serves as a major component of cell membranes and as the starting point for the production of the steroid hormones. The elongated, rigid structure of cholesterol helps to

NMR is used extensively by chemists, especially organic chemists, because the radio frequency absorbed and then emitted depends on the chemical environment of the atoms with nuclear spin states in the sample. The study of hydrogen atoms by NMR (known as *proton* nuclear magnetic resonance, ^{1}H NMR) is an indispensable structural and analytical tool. Plots of the intensity of energy absorption versus the magnetic field strength are called *NMR spectra* from which chemists deduce the kinds of atoms bonded to hydrogen as well as the number of hydrogen atoms present in a molecule.

Because hydrogen is the most common element in the body and H atoms give a strong NMR signal, ^{1}H NMR can be used in medical imaging. NMR imaging in medicine is based on the time it takes for the protons in the unstable high-energy nuclear spin position to "relax" or return to the low-energy nuclear spin position. *Relaxation times* for protons in muscle, blood, and bone differ due to differences in the chemical environments. Based on these differences, a computer can produce a magnetic resonance image showing different tissues as different colors.

NMR medical imaging is now referred to as magnetic resonance imaging (MRI). The name "nuclear" was dropped because it frightened some people due to its association with weapons and radioactivity. In MRI, the patient is placed in the opening of a large magnet whose magnetic field aligns the magnetic spins of the protons, as well as those of other magnetic nuclei. A radio frequency transmitter coil is placed in position near the region of the body to be examined. The radio frequency energy absorbed by the spinning hydrogen nuclei causes the aligned nuclei to flip to the less stable, high-energy spin direction. When the nuclei flip back to the more stable spin state, photons of radio frequency electromagnetic radiation are emitted and detected by a radio receiver coil, which then sends the signal to a computer for construction of the MRI image. Because the relaxation time is related to the type of tissue scanned, MRI can readily image organs, detect small tumors and blockage in blood vessels, and assess damage to bones and vertebrae.

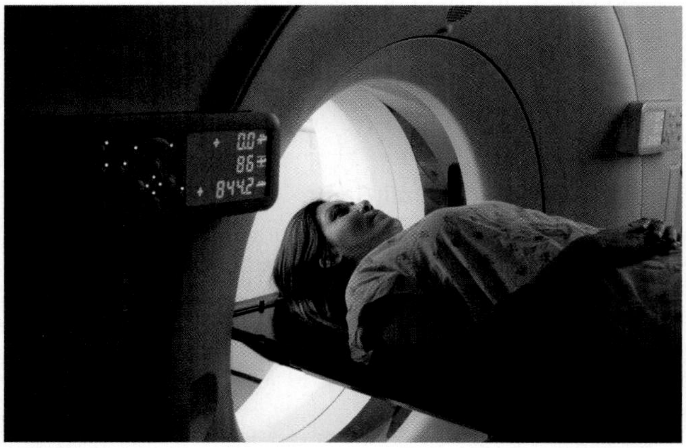
Mark Kostich/iStockphoto

The MRI method. A patient undergoing an MRI scan.

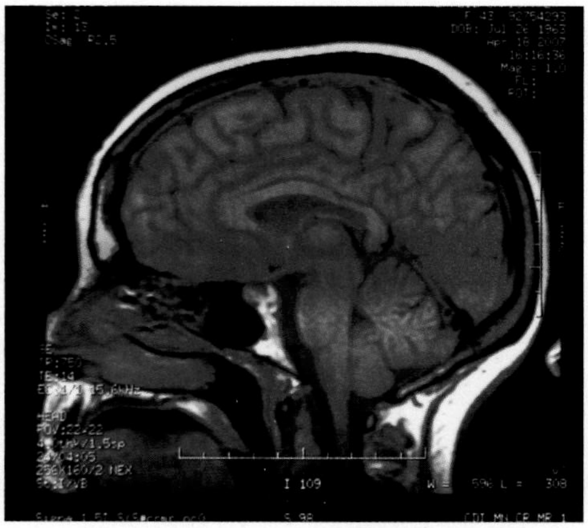
Mark Herreid/Shutterstock

An MRI image. An MRI scan of a normal human brain.

make cell membranes sturdier. However, elevated levels of cholesterol in the blood are associated with *atherosclerosis,* a thickening of arterial walls that can lead to medical problems such as strokes and heart attacks.

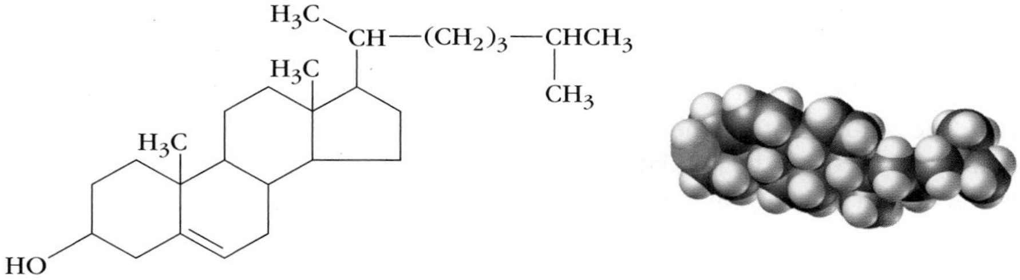

cholesterol

The presence and position of the C=C bond and the presence of an aromatic ring distinguish different steroids, as seen from the structural formulas of cholesterol, estradiol, and testosterone. Estradiol and estrone are female sex hormones responsible for the development of secondary sexual characteristics during puberty; testosterone is the male counterpart. From their structural formulas you can see that steroids are relatively large molecules made up predominantly of carbon and hydrogen. Thus, steroids are not soluble in water.

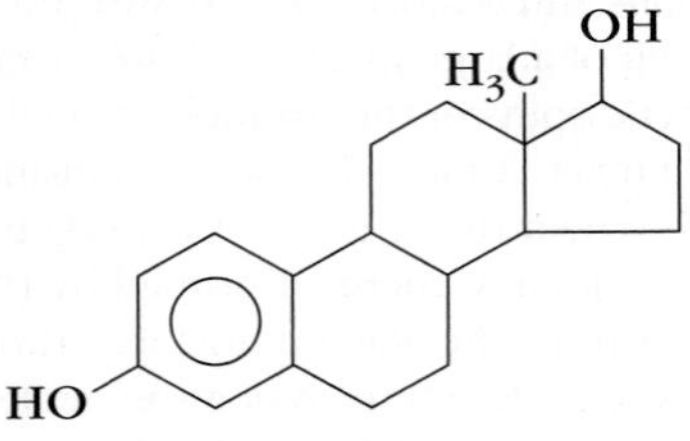

estradiol
(female hormone, an estrogen)

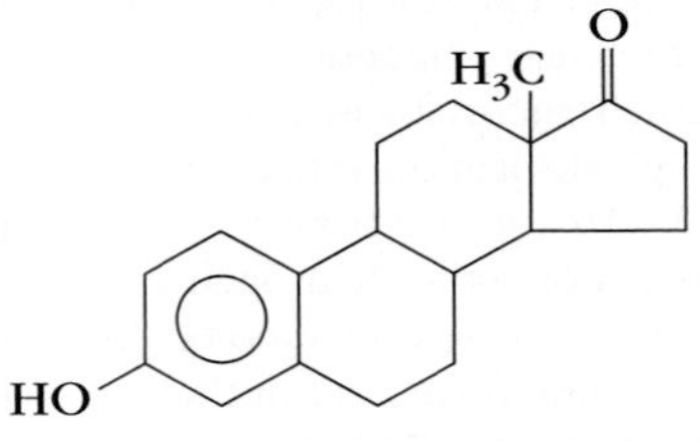

estrone
(female hormone, an estrogen)

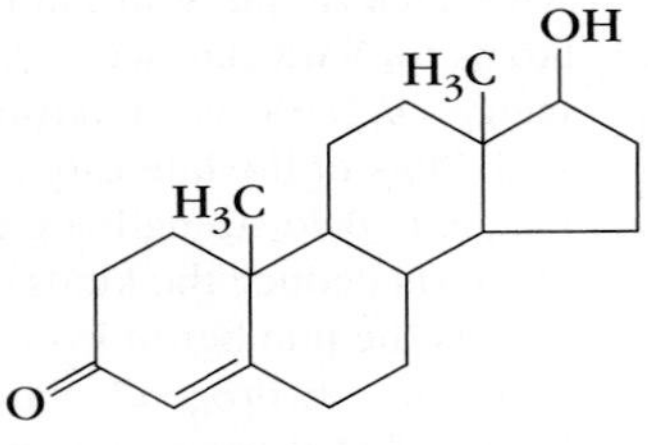

testosterone
(male hormone)

CONCEPTUAL EXERCISE 12.10 A Closer Look

Look carefully at the structures for estradiol and estrone.

(a) What does the *-diol* suffix indicate in estradiol?

(b) What kind of alcohol group is on the five-membered ring in estradiol?

(c) What process converts the alcohol group of the five-membered ring in estradiol to the functional group present in that same ring in estrone?

(d) Write a description of the differences in the molecular structures between the male and female sex hormones, testosterone and estradiol, respectively.

EXERCISE 12.11 Take Me Out to the Ball Game

Home run slugger Manny Ramirez, cyclist Floyd Landis, and other athletes have been accused of using synthetic anabolic steroids to enhance their athletic performance. Androstenedione is such a steroid; its molecular structure is

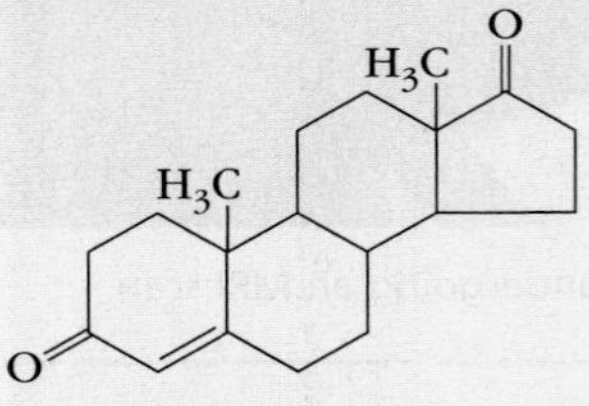

androstenedione

Compare it with the molecular structure of testosterone. In the body, what simple chemical process converts testosterone to androstenedione?

go Chemistry Module 15: Naming Organic Compounds covers concepts in this section.

12.5 Carboxylic Acids and Esters

Carboxylic Acids

Carboxylic acids contain the $-\overset{\overset{\displaystyle O}{\|}}{C}-OH$ ($-COOH$ or $-CO_2H$) functional group and are prepared by the oxidation of aldehydes or the complete oxidation of primary alcohols (p. 549). All carboxylic acids react with bases to form salts; for example, sodium lactate is formed by the neutralization reaction ***(← p. 172)*** of lactic acid and sodium hydroxide.

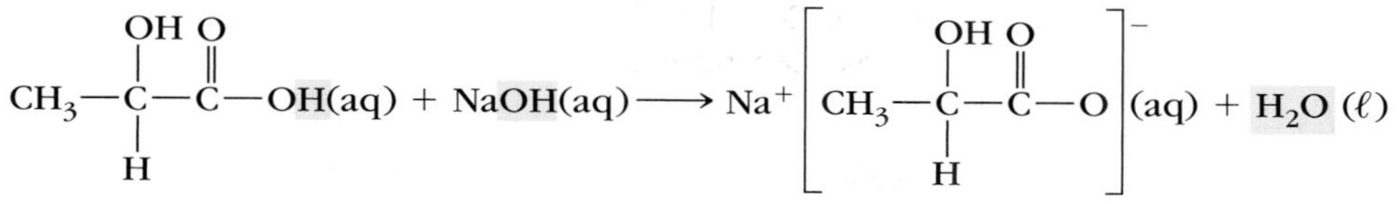

lactic acid sodium lactate

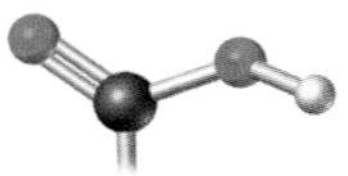

carboxylic acid group

The carboxylic acid group is present in all organic acids. Acids of all kinds are an important class of compounds, as discussed in Chapter 5.

Carboxylic acid molecules are polar and form hydrogen bonds with each other. This hydrogen bonding results in relatively high boiling points for the acids—even higher than those of alcohols of comparable molecular size. For example, formic acid (46 g/mol) has a boiling point of 101 °C, while ethanol (46 g/mol) has a boiling point of only 78.5 °C. Both formic acid and ethanol form hydrogen bonds, but those in formic acid are stronger.

EXERCISE 12.12 Hydrogen Bonding in Formic Acid

Use the structural formula of formic acid given in Table 12.5 to illustrate hydrogen bonding in formic acid.

A large number of carboxylic acids are found in nature and have been known for many years. As a result, some of the familiar carboxylic acids are almost always referred to by their common names (Table 12.5).

Carboxylic acids in foods such as citrus fruits, strawberries, and vinegar cause them to have a sour taste.

Three commercially important carboxylic acids—adipic acid, terephthalic acid, and phthalic acid—have two acid groups per molecule and are known as *dicarboxylic acids.* These three acids are used as starting materials to make vast quantities of synthetic polymers (Section 12.6).

Table 12.5 Some Naturally Occurring Carboxylic Acids

Structure	Systematic Name	Common Name	b.p. (°C)	Natural Source
	Methanoic acid	Formic acid	101	Ants
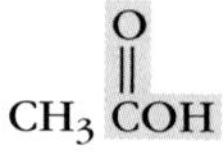	Ethanoic acid	Acetic acid	118	Fermented fruit
	Propanoic acid	Propionic acid	141	Dairy products
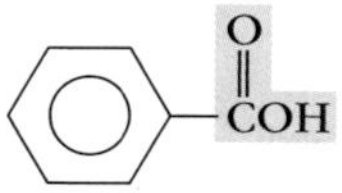	Benzoic acid	Benzoic acid	250	Berries
			m.p. (°C)	
$HOOC—CH_2—C(OH)(COOH)—CH_2—COOH$		Citric acid	153	Citrus fruits
$HOOC—CH_2—CH(OH)—COOH$		Malic acid	131	Apples
$HOOC—CH(OH)—CH(OH)—COOH$		Tartaric acid	168–170	Grape juice, wine

In *Les Miserables,* Victor Hugo comments about naturally occurring acids when he writes, "Comrades, we will overthrow the government, as sure as there are fifteen acids intermediate between margaric acid and formic acid." Formic acid, the simplest organic acid, has just one carbon atom. Margaric acid, from which margarine is derived, has seventeen carbon atoms.

Goodshoot/Corbis

Carboxylic acids in foods. Citrus fruits contain citric acid as well as other naturally occurring acids.

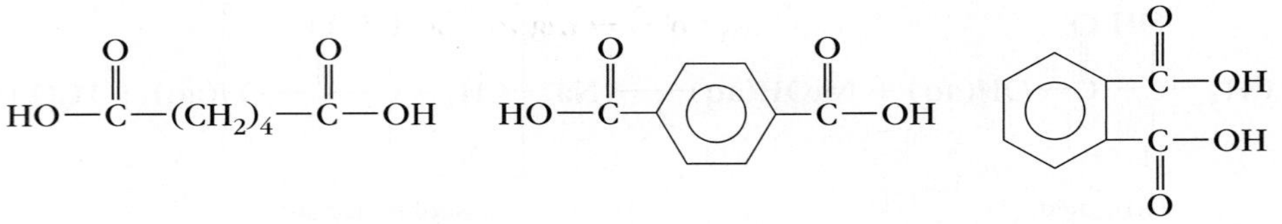

adipic acid   terephthalic acid   phthalic acid

Esters

Carboxylic acids react with alcohols in the presence of strong acids (such as sulfuric acid) to produce **esters,** which contain the

$$-\overset{\overset{\displaystyle O}{\|}}{C}-O-R$$

functional group. In an ester, the —OH of the carboxylic acid is replaced by the —OR group from the alcohol. The general equation for ester formation (esterification) is

Esterification is catalyzed by the presence of an acid (H^+ donor.)

$$R-OH + H-O-\overset{\overset{\displaystyle O}{\|}}{C}-R' \xrightarrow{H^+} R-O-\overset{\overset{\displaystyle O}{\|}}{C}-R' + H_2O$$

alcohol + organic acid   ester + water

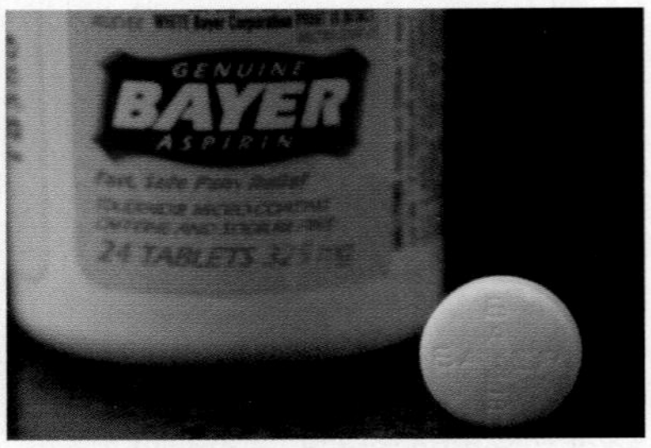

Aspirin (acetylsalicylic acid).

Esterification is an example of a condensation reaction. In a **condensation reaction,** two molecules combine to form a larger molecule (in esterification it is the ester) while simultaneously producing a small molecule (water in this case). For example, just over a century ago, Felix Hoffmann, a chemist at the Bayer Chemical Company in Germany, synthesized aspirin, the world's most common pain reliever, by esterifying an alcohol group on salicylic acid with acetic acid in strong acid solution to form acetylsalicylic acid, which is aspirin. The ester group makes aspirin less irritating to the stomach lining than salicylic acid. The carboxylic acid group originally in salicylic acid remains intact in aspirin.

The systematic name for asprin (acetylsalicylic acid) is 2-(acetyloxy)-benzoic acid.

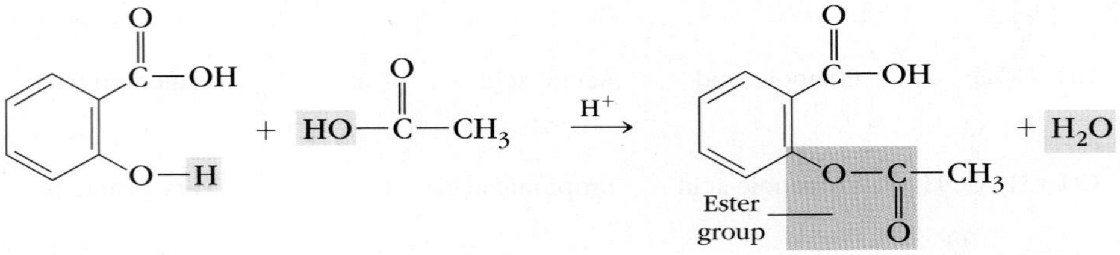

salicylic acid   acetic acid   aspirin (acetylsalicyclic acid)   water

Triglycerides: Biologically Important Esters

One benefit of understanding organic molecules is in applying this understanding to the molecules involved in life—organic molecules that we eat, that provide the structure of our bodies, and that allow our bodies to function. Fats and oils are such molecules. *Fats* are solids at room temperature; *oils* are liquids at this temperature. Edible fats and oils are all **triglycerides** because they share the common structural feature of a three-carbon backbone from glycerol to which three long-chain fatty acids are bonded. Fatty acids usually have an even number of carbon atoms, ranging from 4 to 20. Triglycerides are formed by the esterification reaction of three moles of fatty acids with one mole of glycerol.

3 Fatty acid + glycerol $\longrightarrow$ triglyceride + 3 H_2O

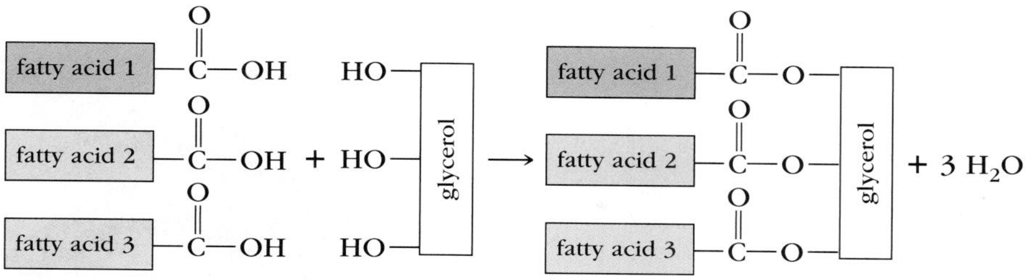

To illustrate this process, consider the reaction of a glycerol molecule with three molecules of stearic acid to produce tristearin, a very common animal fat. Stearic acid is representative of fatty acids, most of which are molecules with long chains of carbon atoms with their attached hydrogens and, at the end of the carbon chain, the —COOH group characteristic of all organic acids. In tristearin formation, the carboxylic acid group in each of three molecules of stearic acid reacts with an —OH group (hydroxyl group) of a glycerol molecule. A water molecule is formed and released as each stearic acid molecule is joined to the glycerol molecule:

Glycerol + 3 stearic acid $\longrightarrow$ tristearin + 3 water

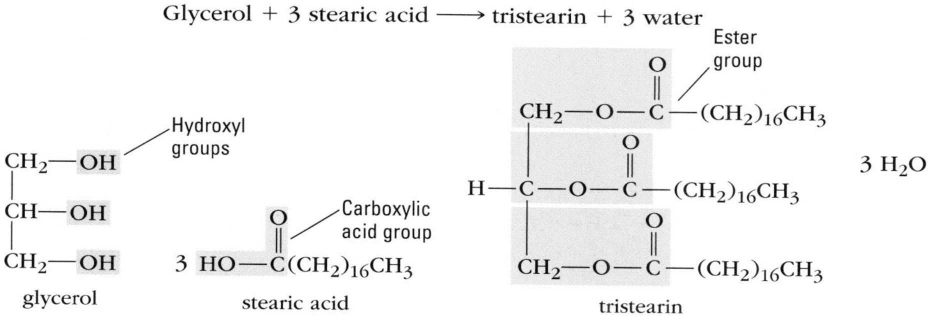

© Cengage Learning/Charles D. Winters

Cooking oils are liquids at room temperature. Liquid oils and solid fats differ in the bonds among the carbon atoms within their molecular structures.

Animal and vegetable fats and oils vary considerably because their fatty acids differ in the length of the hydrocarbon chain and the number of double bonds in the chain. **Saturated fats,** such as stearic acid, contain only C—C single bonds in their hydrocarbon chains along with C—H bonds; **unsaturated fats** contain one or more C=C double bonds in their hydrocarbon chains. Oleic acid, with one C=C bond per molecule, is classified as a *monounsaturated* fatty acid, whereas fatty acids such as linoleic acid, with two or more C=C double bonds per molecule, are termed *polyunsaturated.*

The three fatty acids in a triglyceride can be the same (stearic acid in tristearin) or can be different such as stearic acid, oleic acid, and linoleic acid.

The discussion in this section answers the question posed in Chapter 1 (← *p. 2*), "What is the difference between a saturated fat, an unsaturated fat, and a polyunsaturated fat?"

Linoleic and linolenic acids are essential fatty acids; we must have them in our diet. The body cannot manufacture them.

Fatty acid 1

$$HOOC{-}CH_2CH_2CH_2CH_2CH_2CH_2CH_2CH_2CH_2CH_2CH_2CH_2CH_2CH_2CH_2CH_2CH_3$$

stearic acid, a saturated fatty acid

Fatty acid 2

$$HOOC{-}(CH_2)_7{-}CH{=}CH(CH_2)_7{-}CH_3$$

oleic acid, a monounsaturated fatty acid

Fatty acid 3

$$HOOC{-}(CH_2)_7{-}CH{=}CH{-}CH_2{-}CH{=}CH{-}(CH_2)_4{-}CH_3$$

linoleic acid, a polyunsaturated fatty acid

The resulting triglyceride is

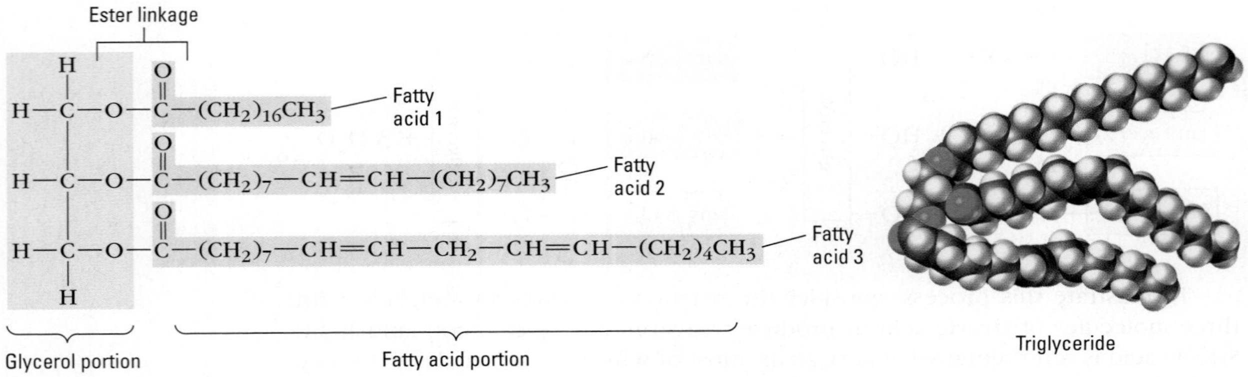
Ester linkage
Fatty acid 1
Fatty acid 2
Fatty acid 3
Glycerol portion
Fatty acid portion
Triglyceride

PROBLEM-SOLVING EXAMPLE 12.4 Putting Together a Triglyceride

Using structural formulas, write the equation for the formation of a triglyceride by the reaction of 1 mol glycerol with 1 mol stearic acid and 2 mol oleic acid.

Answer

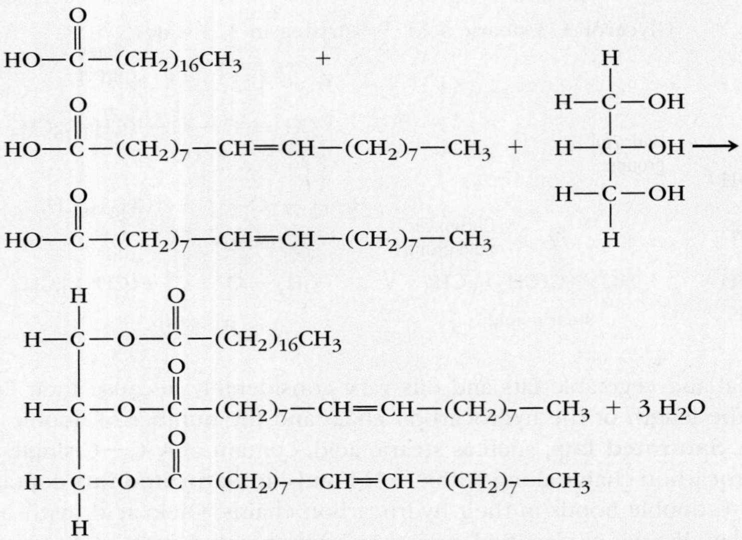

Strategy and Explanation Use the structural formulas given previously for the acids and glycerol and combine them through an esterification (condensation) reaction. A water molecule is eliminated for each ester bond formed. Several different triglycerides are possible in this case depending on the position of the component fatty acids in the glycerol carbon chain. In the example shown in the answer, stearic acid is attached to the "first" (top) carbon in glycerol, with oleic acid attached to carbons 2 and 3. See Answer above.

PROBLEM-SOLVING PRACTICE 12.4

Use structural formulas to write the equation for the formation of a triglyceride different from the triglyceride given in Problem-Solving Example 12.4 by the reaction of 1 mol glycerol with 1 mol stearic acid and 2 mol oleic acid.

Diets consisting of moderate amounts of fats and oils containing mono- and polyunsaturated fatty acids are considered better for good health than diets containing only saturated fats. In spite of this fact, there is a demand for solid or semisolid fats because of their texture and spreadability. The figure that follows indicates the relative proportions of saturated, monounsaturated, and polyunsaturated fats in some familiar cooking oils and solids.

The American Heart Association recommends that no more than 30% of dietary calories come from fat. Fats account for almost 40% of the calories in the average American's diet.

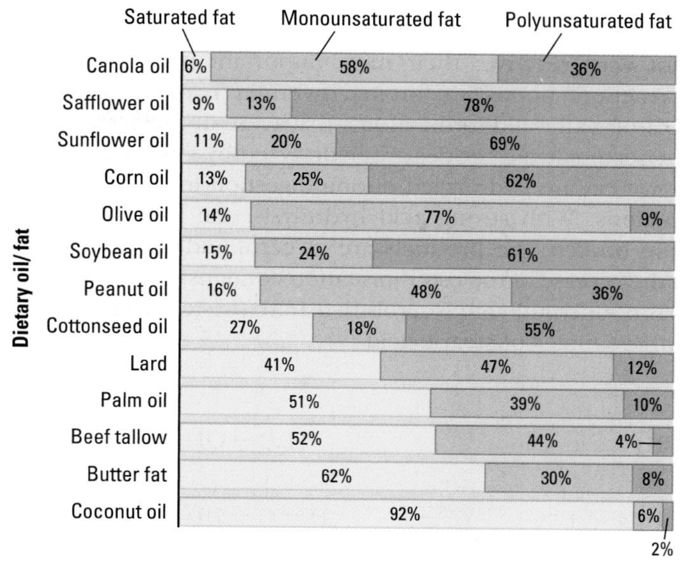

Saturated fats: No C═C

Monounsaturated fats: One C═C bond per molecule

Polyunsaturated fats: Several C═C bonds per molecule

As seen from the diagram above, most dietary vegetable oils are high in unsaturated fatty acids, while most dietary animal fats contain mainly saturated fatty acids. In general, the greater the percentage of unsaturated fatty acids in a triglyceride, the lower its melting point. Thus, highly unsaturated fats are liquids at room temperature (oils); animal fats high in saturated fats are solids at that temperature. The melting points of fats reflect the shape of the molecules. The C—C single bonds in the hydrocarbon portion of saturated fats are all the same, and they allow the molecule to adopt a rather linear shape, which fits nicely into a solid packing arrangement. The C═C double bonds in unsaturated fats create a different shape. Most natural unsaturated fats are in the *cis* configuration at the C═C double bond ***(Section 8.5, ← p. 339)***, which puts a "kink" into the molecule, preventing unsaturated fat molecules from packing regularly into a solid. *Trans* fatty acid molecules are more linear, similar to those of saturated fats. This has raised health concerns about the widespread dietary use of foods containing *trans* fatty acids, and *trans* fats are banned from being used in certain processed foods.

The space-filling model on p. 558 shows the "kink" put into the triglyceride molecule due to a *cis* configuration at the C═C double bond of fatty acid 2 (oleic acid).

Vegetable oils can be converted to semisolids by *hydrogenation*—reaction of the oil with H_2 in the presence of a metal catalyst such as palladium, which accelerates the reaction. The hydrogen reacts by addition to some of the C═C double bonds, converting them to C—C single bonds, as shown to the right. The degree of saturation in the partially hydrogenated product is limited by carefully controlling reaction conditions so that the product has the proper softness and spreadability. Margarine, shortening (for example, Crisco), and many snack foods contain partially hydrogenated fats, noted on the label as "partially hydrogenated soybean oil" (or cottonseed or other vegetable oils).

Hydrogenation converts

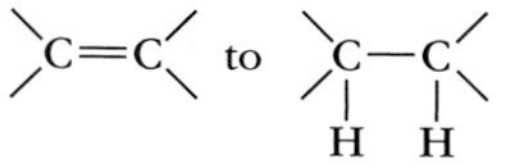

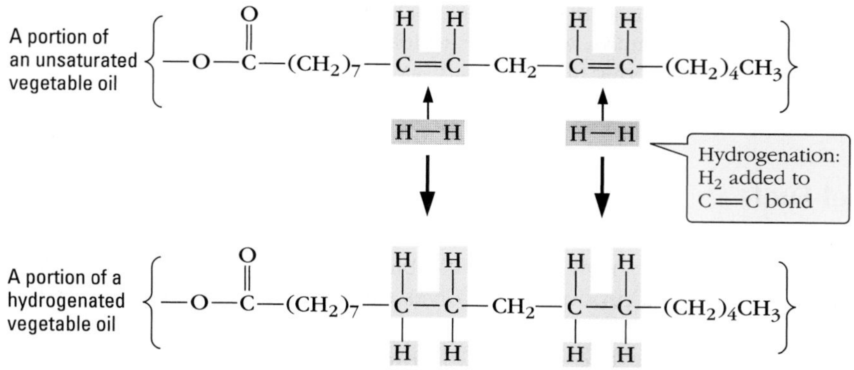

Diagram of a hydrogenation reaction.

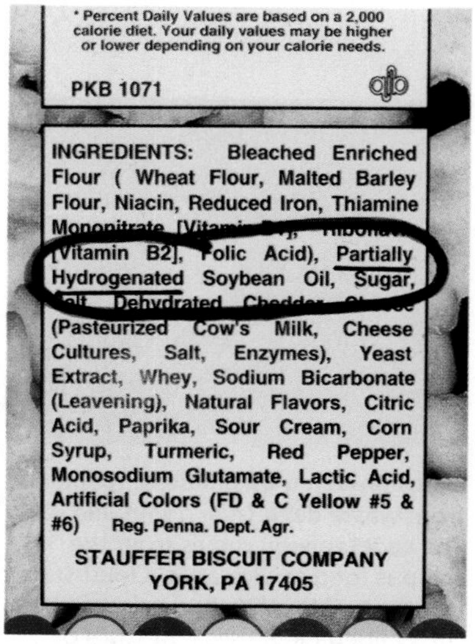

© Cengage Learning/George Semple

Partially hydrogenated fats. Most snack foods contain partially hydrogenated fats.

Hydrolysis of Esters

Hydrolysis is the reverse of the condensation reaction by which an ester is formed.

Esters are not very reactive; their most important reaction is hydrolysis. In a **hydrolysis** reaction, bonds are broken by their reaction with water, and the H— and —OH of water add to the atoms that were in the bond broken by hydrolysis. Like other esters, triglycerides are hydrolyzed by strong acid or aqueous base. Triglycerides we eat are hydrolyzed during digestion in the same way by enzyme-catalyzed reactions. With strong acid hydrolysis, the three ester linkages in the triglyceride are broken; the products are glycerol and three moles of fatty acids, which is just the reverse of the ester formation condensation reaction. For example, acid hydrolysis of one mole of glycerol tristearate (tristearin) produces one mole of glycerol and three moles of stearic acid.

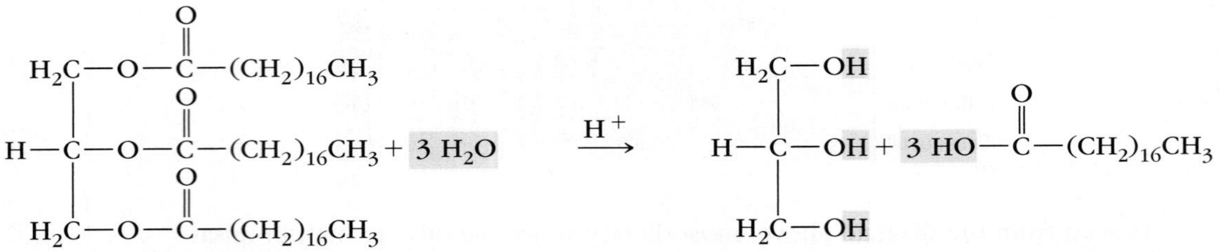

glycerol tristearate (tristearin) glycerol stearic acid

Hydrolysis of a triglyceride with an aqueous base such as NaOH forms glycerol; the fatty acids released then react with the base converting the fatty acids to their salts. In aqueous base, the hydrolysis process is called **saponification,** and the salts formed are called *soaps*. In biological systems, enzymes assist with saponification reactions that digest fats and oils. The saponification of one mole of glycerol tristearate forms one mole of glycerol and three moles of sodium stearate, a soap. Note that the saponification reaction requires three moles of base per mole of triglyceride.

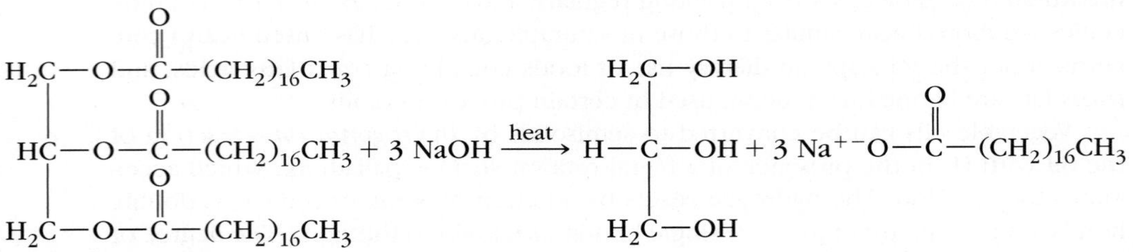

glycerol tristearate, a fat NaOH glycerol sodium stearate, a soap

Courtesy of Don Neu

A bus powered by biodiesel fuel made from waste deep fryer vegetable oil. The vegetable oil comes from the campus food service at St. Cloud State University (MN). The bus, termed the Husky Fried Ride, uses a mixture of 80% recycled vegetable oil biodiesel and 20% conventional diesel fuel.

EXERCISE 12.13 Triglyceride Structures

(a) Draw the structural formula for the triglyceride formed when glycerol reacts with linoleic acid. Circle the ester linkages in this triester molecule.

(b) Using structural formulas, write the equation for the hydrolysis in aqueous NaOH of the triglyceride formed in part (a).

Biodiesel Fuel

Biodiesel is a fuel made from biological sources, principally vegetable oils. Waste vegetable oils, such as those from fast-food establishments and campus dining halls, can be used. Biodiesel fuel production involves the base-catalyzed *transesterification reaction* of triglycerides in vegetable oils with low molar mass alcohols, typically methanol or ethanol. The reaction produces glycerol (glycerine) and the biodiesel fuel, which is a mixture of esters of the fatty acids initially part of the triglyceride. Using ethanol as the alcohol, this is the reaction:

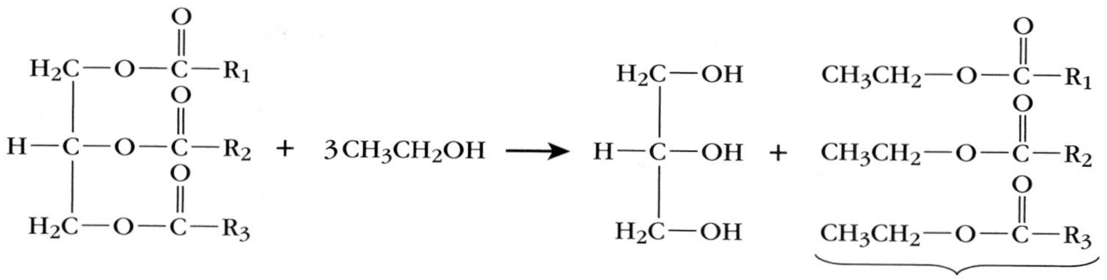

Triglyceride (in the vegetable oil) + 3 Ethanol ⟶ Glycerol + Ethyl esters of fatty acids (biodiesel fuel)

The waste oil is filtered to remove solid wastes and water is removed from the filtrate by heating. A measured sample of the cleaned oil is titrated with a standardized solution of a strong base to evaluate the concentration of fatty acids in the oil, those not part of the triglycerides. Sufficient ethanol is used to provide the required stoichiometric ratio of three moles of ethanol per mole of triglyceride in the oil. A slight excess of NaOH is added as a catalyst. The NaOH/ethanol mixture is added to the oil and heated for several hours to complete the reaction. Once the reaction is complete, the more dense glycerol settles to the bottom and can be drawn off. The remaining biodiesel fuel is washed to remove the NaOH catalyst and any soaps that might have been formed during the reaction. After further drying, the biodiesel fuel is ready to use.

12.6 Synthetic Organic Polymers

Simple organic molecules are important, not only because of their own properties, but also because they are starting materials to make an important class of compounds called polymers. **Polymers** (*poly,* "many"; *mer,* "part") are large molecules composed of smaller repeating units, usually arranged in a chain-like structure. For example, polyethylene consists of long chains of $—CH_2—CH_2—$ groups made from repeating ethylene units, $CH_2{=}CH_2$. Polymers occur in nature and are synthesized by chemists as well. Some useful synthetic polymers have resulted from copying natural polymers. Synthetic rubber, used in almost every automobile tire, is a copy of the molecule found in natural rubber. Other useful synthetic polymers, such as polystyrene, nylon, Teflon, and Dacron, have no natural analogs. It is virtually impossible for us to get through a day without using a dozen or more synthetic organic polymers such as those used to make our clothes, package our foods, and build our computers, phones, and cars. You, a friend, or a family member may be alive because of a medical application of synthetic polymers. Synthetic polymers are so important that approximately 80% of the organic chemical industry is devoted to their production.

Nature makes many different polymers, including cellulose and starch in plants and proteins in both plants and animals.

The discussion in this section answers the question posed in Chapter 1 (← *p. 2*), "How are plastics made and why are there so many different kinds?"

There are two broad categories of the synthetic polymers commonly known as "plastics." One type, called **thermoplastics,** softens and flows when heated; when it is cooled, it hardens again. Common plastic materials that undergo such reversible changes when heated and cooled are polyethylene (milk jugs), polystyrene (inexpensive sunglasses and toys), and polycarbonates (CD discs). The other general type is **thermosetting plastics.** When first heated, a thermosetting plastic flows like a thermoplastic. When heated further, however, it forms a rigid structure that will not remelt. Bowling balls, football helmets, and some kitchen countertops are examples of thermosetting plastics.

Common household items made from plastic.

Both synthetic and natural polymers are made by chemically joining many small molecules, called **monomers,** into giant polymer molecules, known as **macro-**

A macromolecule is a molecule with a very high molar mass.

molecules, that have molar masses ranging from thousands to millions. In nature, polymerization reactions usually are controlled by enzymes, and in animal cells biopolymer synthesis takes place rapidly at body temperature. Making synthetic polymers often requires high temperatures and pressures and lengthy reaction times. Synthetic and natural polymers are formed by one of two main reaction types:

- **Addition polymers are made by adding monomer units directly together;** no other products are formed.
- **Condensation polymers are produced when monomer units combine to form the polymer and also release a small molecule,** usually water. One small molecule is formed and released for each new linkage formed in the polymer.

Addition Polymers

In addition polymerization, the *monomer units are added directly to each other,* hooking together like boxcars on a train to form the polymer. The monomers for making addition polymers usually contain one or more C=C double bonds.

Polyethylene The simplest monomer for addition polymerization is ethylene, $CH_2{=}CH_2$, which polymerizes to form *polyethylene*. When heated to 100 to 250 °C at a pressure of 1000 to 3000 atm in the presence of a catalyst, ethylene forms polyethylene chains with molar masses of up to many thousands. Polymerization involves three steps: initiation, propagation, and chain termination.

An organic peroxide, RO—OR, produces two free radicals, RO·, each with an unpaired electron.

- **Initiation.** The first step, *initiation* of the polymerization, involves chemicals such as organic peroxides (R—O—O—R) that are unstable and easily break apart into free radicals, ·OR, each with an unpaired electron (← *p. 355*). The free radicals react readily with molecules containing carbon-carbon double bonds to produce new free radicals.

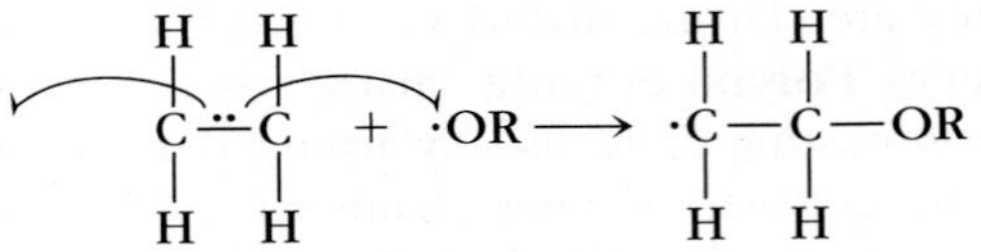

- **Propagation.** The polyethylene chain begins to grow as the unpaired electron radical formed in the initiation step bonds to a double-bond electron in another ethylene molecule.

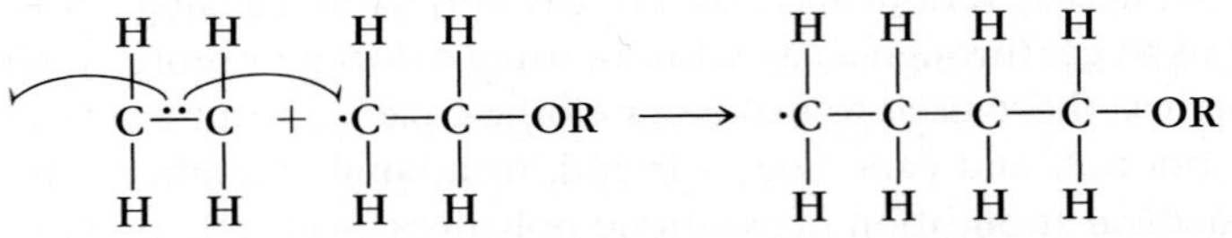

This forms another radical containing an unpaired electron that can bond with yet another ethylene molecule. The process continues, with the chain growing to form a huge polymer molecule.

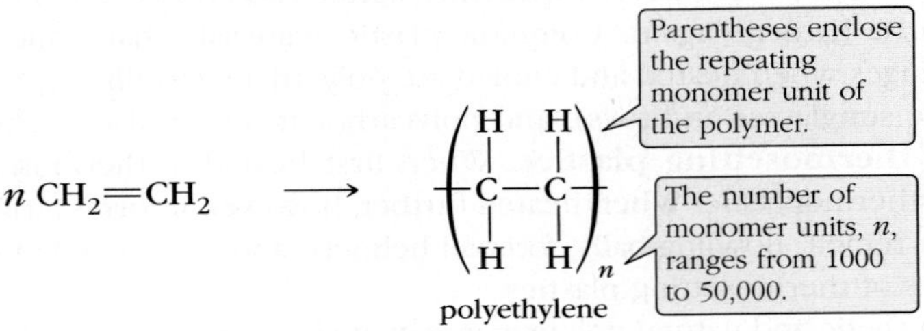
n CH₂=CH₂
polyethylene
Parentheses enclose the repeating monomer unit of the polymer.
The number of monomer units, n, ranges from 1000 to 50,000.

The C=C double bonds in the ethylene monomer have been changed to C—C single bonds in the polyethylene chain.

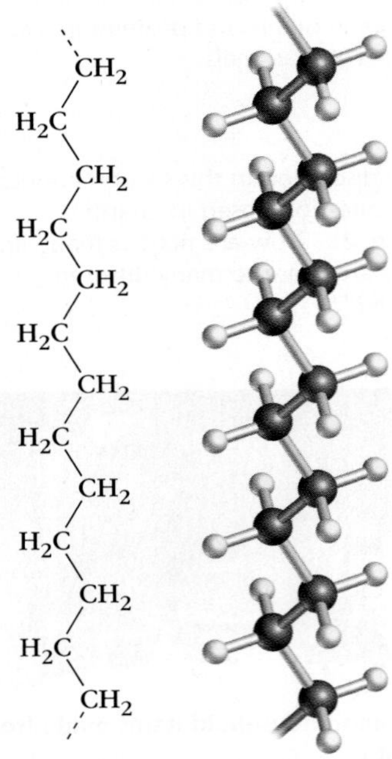

A portion of a polyethylene molecule.

- **Chain Termination.** Eventually, production of the polymer chain stops. This occurs when the supply of monomer is consumed or when no more free radicals are produced, such as when two chains containing free radicals combine. The value of n is called the **degree of polymerization,** the number of repeating units in the polymer chain.

$$\text{RO}\!\leftarrow\!\text{CH}_2\text{CH}_2\!\rightarrow_n\!\text{CH}_2\cdot \; + \; \cdot\text{CH}_2\!\leftarrow\!\text{CH}_2\text{CH}_2\!\rightarrow_{n*}\!\text{OR}' \longrightarrow$$

$$\text{RO}\!\leftarrow\!\text{CH}_2\text{CH}_2\!\rightarrow_n\!\text{CH}_2\text{CH}_2\!\leftarrow\!\text{CH}_2\text{CH}_2\!\rightarrow_{n*}\!\text{OR}'$$

n and n^* represent different numbers of CH_2CH_2 monomer units.

CONCEPTUAL EXERCISE 12.14 What Is at the Ends of the Polymer Chains?

What do you think is attached at the ends of the polymer chains when all of the ethylene monomer molecules have been polymerized to form polyethylene?

Changing pressures and catalytic conditions produces polyethylenes of different molecular structures and hence different physical properties. For example, chromium oxide as a catalyst yields almost exclusively the linear polyethylene shown in the margin—a polymer with no branches on the carbon chain. Because the bonds around each carbon in the saturated polyethylene chain are in a tetrahedral arrangement, the zigzag structure (shown on page 562) more closely represents the shape of the chain than does a linear drawing. When ethylene is heated to 230 °C at a pressure of 200 atm without the chromium oxide catalyst, free radicals attack the chain at random positions, causing irregular branching.

Polyethylene is the world's most widely used polymer (Figure 12.6). Long linear chains of polyethylene can pack closely to give a material with high density (0.97 g/mL) and high molar mass, referred to as high-density polyethylene (HDPE). This material is hard, tough, and semirigid; it is used in plastic milk jugs. Branched chains of polyethylene cannot pack as closely as the linear chains in HDPE, so the resulting material has a lower density (0.92 g/mL) and is called low-density polyethylene (LDPE). This material is soft and flexible because of its weaker intermolecular forces. It is used to make sandwich bags, for example. If the linear chains of

Linear polyethylene (high density)

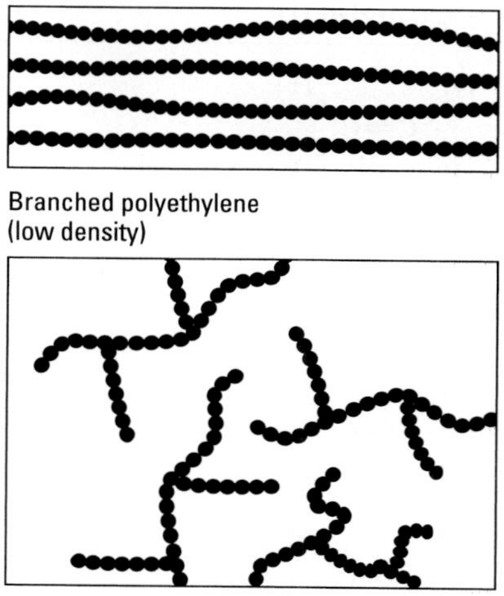

Branched polyethylene (low density)

Linear and branched polyethylene.

Figure 12.6 Polyethylene. (a) Production of polyethylene film. The film is manufactured by forcing the polymer through a thin space and inflating it with air. (b) The wide range of properties of different structural types of polyethylene leads to a wide variety of applications.

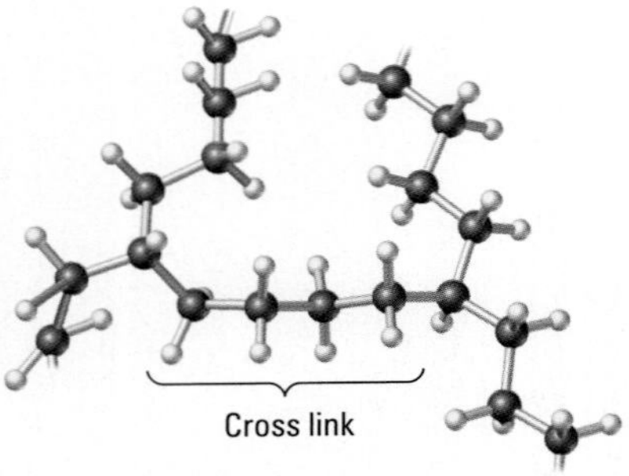

Figure 12.7 Model of cross-linked polyethylene.

X is an aromatic ring in styrene, the monomer for making polystyrene.

polyethylene are treated in a way that causes short chains of —CH_2— groups to connect adjacent chains, the result is cross-linked polyethylene (CLPE), a very tough material, used for synthetic ice rinks and soft-drink bottle caps (Figure 12.7).

Other Addition Polymers Many different kinds of addition polymers are made from monomers in which one or more of the hydrogen atoms in ethylene have been replaced with a halogen atom or an organic group. Table 12.6 gives information on some of these monomers and their addition polymers. In each case, the replacement atom or group is represented by *X* in this general equation.

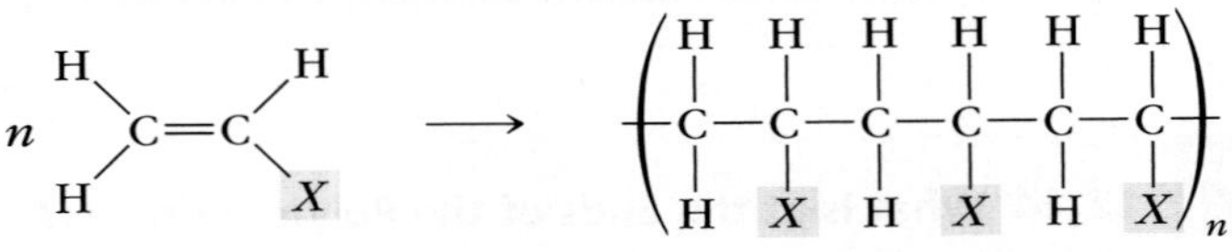

The growing chain

Table 12.6 Ethylene Derivatives That Undergo Addition Polymerization

Formula (*X* atom or group is highlighted)	Monomer Common Name	Polymer Name (trade names)	Uses
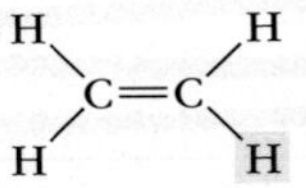	Ethylene	Polyethylene (Polythene)	Squeeze bottles, bags, films, toys and molded objects, electrical insulation
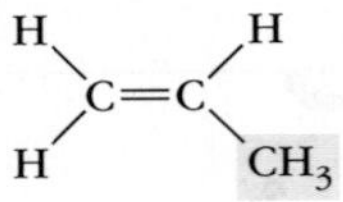	Propylene	Polypropylene (Vectra, Herculon)	Bottles, films, indoor-outdoor carpets
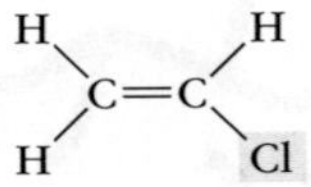	Vinyl chloride	Poly(vinyl chloride) (PVC)	Floor tile, raincoats, pipe
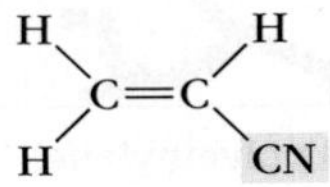	Acrylonitrile	Polyacrylonitrile (Orlon, Acrilan)	Rugs, fabrics
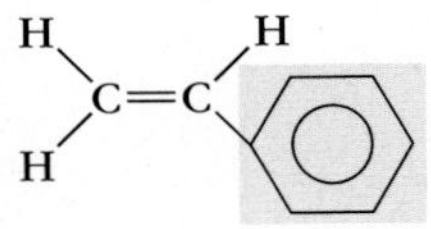	Styrene	Polystyrene (Styrene, Styrofoam, Styron)	Food and drink coolers, building material insulation
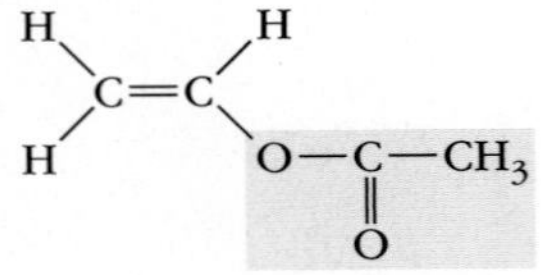	Vinyl acetate	Poly(vinyl acetate) (PVA)	Latex paint, adhesives, textile coatings
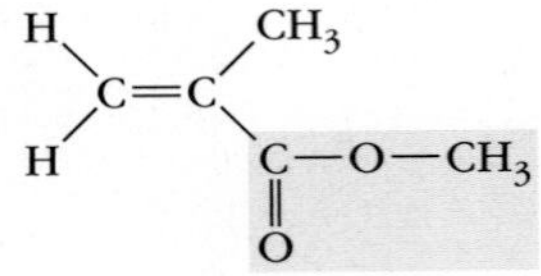	Methyl methacrylate	Poly(methyl methacrylate) (Plexiglas, Lucite)	High-quality transparent objects, latex paints, contact lenses
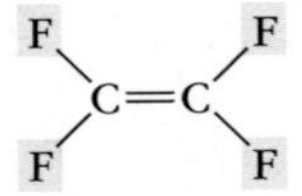	Tetrafluoroethylene	Polytetrafluoroethylene (Teflon)	Gaskets, insulation, bearings, cooking pan coatings

X can be an atom, such as Cl in vinyl chloride, or a group of atoms, such as $—CH_3$ in propylene or —CN in acrylonitrile. For example, vinyl chloride units combine to form poly(vinyl chloride). Each monomer is incorporated directly into the growing polymer chain.

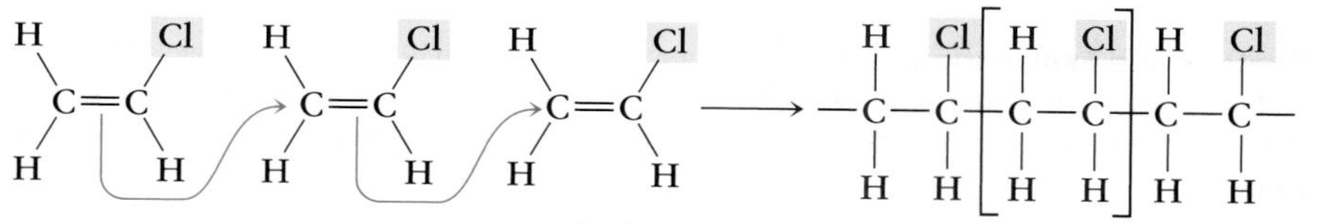

Figure 12.8 Polystyrene. Expanded polystyrene (Styrofoam®) coffee cup *(left)* is soft. Clear polystyrene cup *(right)* is brittle.

In polystyrene, n is typically about 5700. Polystyrene is a clear, hard, colorless, solid thermoplastic that can be molded easily at 250 °C. Nearly seven billion pounds of polystyrene are used annually in the United States alone to make food containers, toys, electrical parts, and many other items. The variation in properties shown by polystyrene products is typical of synthetic polymers. For example, a clear polystyrene drinking glass that is brittle and breaks into sharp pieces somewhat like glass is quite different from an expanded polystyrene coffee cup that is soft and pliable (Figure 12.8).

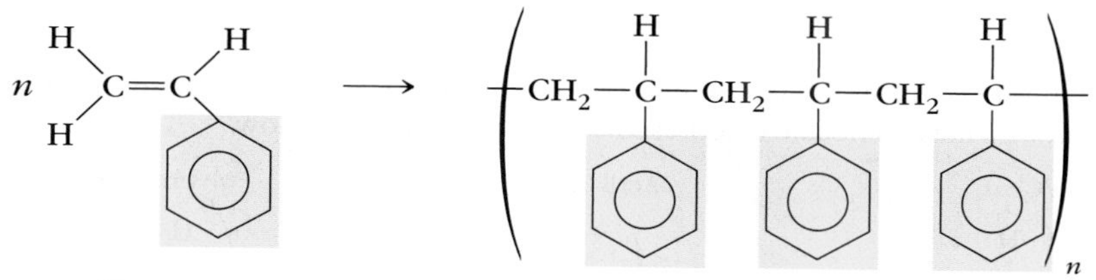

A major use of polystyrene is in the production of Styrofoam by "expansion molding." In this process, polystyrene beads are placed in a mold and heated with steam or hot air. The tiny beads contain 4%–7% by weight of a low-boiling liquid such as pentane. The steam causes the low-boiling liquid to vaporize and expand the beads. As the foamed particles expand, they are molded into the shape of the mold cavity. Styrofoam is used for meat trays, coffee cups, and many kinds of packing material.

The numerous variations in chain length, branching, and crosslinking make it possible to produce a variety of properties for each type of addition polymer. Chemists and chemical engineers can fine-tune the properties of the polymer to match the desired properties by appropriate selection of monomer and reaction conditions, thus accounting for the widespread and growing use of synthetic polymers.

PROBLEM-SOLVING EXAMPLE 12.5 Adding Up the Monomers

Kel-F is an addition polymer made from the monomer $FClC{=}CF_2$.

(a) Write the Lewis structure of this monomer.

(b) Write the Lewis structure of a portion of the Kel-F polymer containing three monomer units.

Answer

(a) 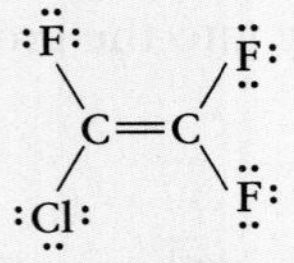

(b) 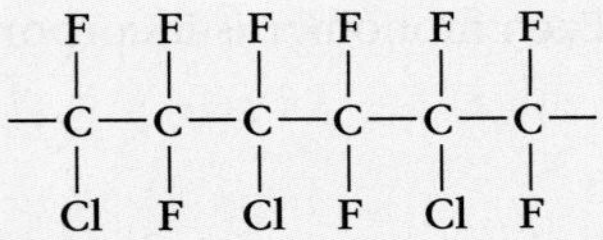

Strategy and Explanation This monomer, which is derived from ethylene, forms an addition polymer. Three monomer units combine to form the beginning of a long polymer chain with each monomer unit coupled directly to the others. No other products form.

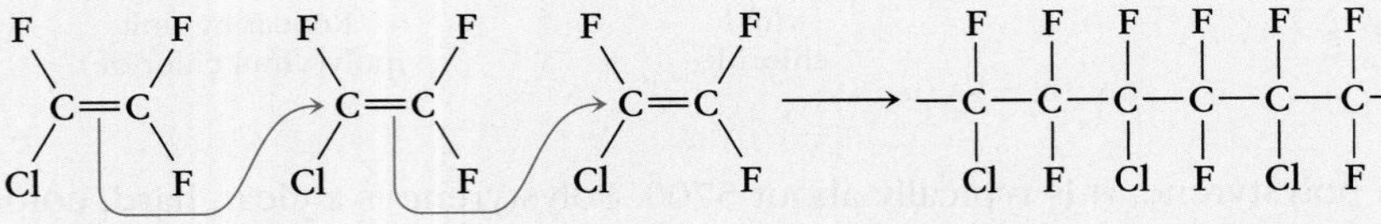

PROBLEM-SOLVING PRACTICE 12.5

Draw the structural formula of a polymer chain formed by the combination of three vinyl acetate monomers. The structural formula of vinyl acetate is in Table 12.6.

PROBLEM-SOLVING EXAMPLE 12.6 Identify the Monomer

Use Table 12.6 to identify the monomer used to make each of the addition polymers, a portion of whose molecule is represented by the structures below.

PVC

(a) 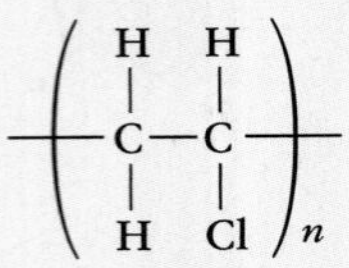

Acrilan

(b) 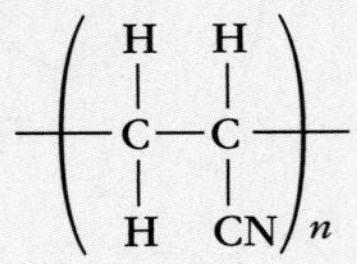

Polypropylene

(c) 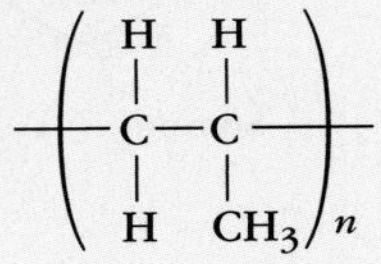

Answer

(a) Vinyl chloride, $CH_2{=}CHCl$ (b) Acrylonitrile, $CH_2{=}CHCN$
(c) Propylene, $CH_2{=}CHCH_3$

Strategy and Explanation Each addition polymer has a repeating unit that is derived from its monomer. Each of the monomers has a C=C double bond, but they differ in the nonhydrogen atom or group attached to one carbon atom.

(a) $CH_2{=}CHCl$ (b) $CH_2{=}CHCN$
(c) $CH_2{=}CHCH_3$

PROBLEM-SOLVING PRACTICE 12.6

Draw the structural formula of the repeating unit for each of these addition polymers.
(a) Polypropylene (b) Poly(vinyl acetate) (c) Poly(methyl methacrylate)

Saran Wrap is an example of a copolymer of vinyl chloride with 1,1-dichloroethylene.

The suffix "diene" in butadiene indicates that there are two C=C bonds in the molecule.

Copolymers Many commercially important addition polymers are **copolymers,** polymers obtained by polymerizing a mixture of two or more different monomers. A copolymer of styrene with butadiene is the most important synthetic rubber produced in the United States. More than 1.5 million tons of styrene-butadiene rubber (SBR) are produced each year in the United States for making tires. A 3:1 mole ratio of butadiene to styrene is used to make SBR.

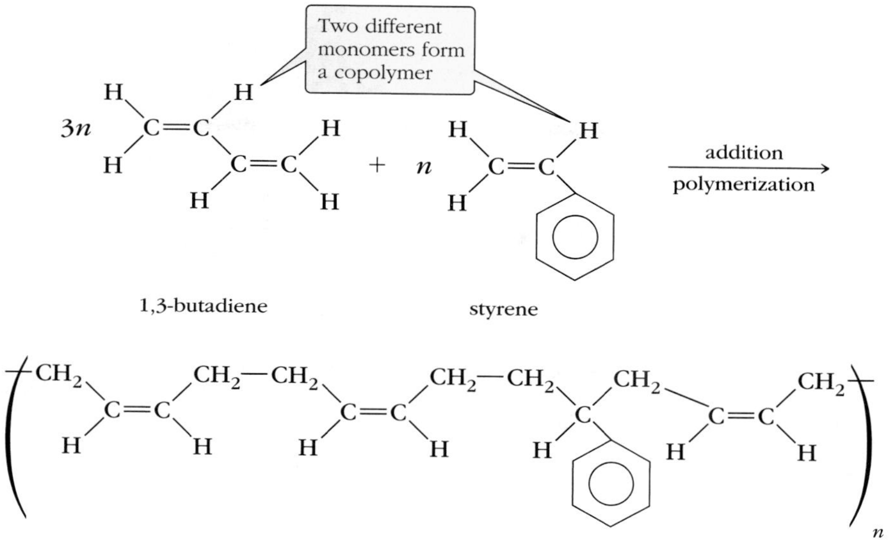
Two different monomers form a copolymer
3n
+ n
addition
polymerization
1,3-butadiene
styrene
n

styrene-butadiene rubber (SBR)

Each 1,3-butadiene monomer molecule has two double bonds (four pi electrons). Only two of the double-bond electrons are involved in bonds between monomer units. Thus, one double bond remains in the polymer backbone for each 1,3-butadiene monomer.

Another important copolymer is made by polymerizing mixtures of acrylonitrile, butadiene, and styrene (ABS) to produce a sturdy material used in car bumpers and computer cases.

PROBLEM-SOLVING EXAMPLE 12.7 Macromolecular Masses

Polytetrafluoroethylene (Teflon) is an addition polymer made of chains of $CF_2{=}CF_2$ monomer units. This polymer has a molar mass of 1.0×10^6 g/mol. Calculate the number of such units in the polymer—that is, the degree of polymerization.

Answer 10,000 monomer units

Strategy and Explanation Each $CF_2{=}CF_2$ monomer unit has a molar mass of 100.0 g/mol. Therefore, the number of monomer units can be calculated using the molar mass of the polymer.

$$\frac{1 \times 10^6\ \text{g}}{1\ \text{mol polymer}} \times \frac{1\ \text{monomer unit}}{100.0\ \text{g}} = \frac{10{,}000\ \text{monomer units}}{1\ \text{mol polymer}}$$

PROBLEM-SOLVING PRACTICE 12.7

Calculate the degree of polymerization of poly(vinyl chloride) that has a molar mass of 150,000 g/mol.

© GRANDADAM/Photo Researchers, Inc.

Natural latex coming from a rubber tree. Natural rubber is an addition polymer whose monomer is 2-methyl-1,3,-butadiene (isoprene).

Natural and Synthetic Rubbers Natural rubber is a hydrocarbon whose monomer unit has the empirical formula C_5H_8. When rubber is decomposed in the absence of oxygen, the monomer 2-methyl-1,3-butadiene (isoprene) is obtained. Natural rubber occurs as *latex* (an emulsion of rubber particles in water) that oozes from rubber trees when the bark is cut. Precipitation of the rubber particles yields a gummy mass that is not only elastic and water-repellent but also very sticky, especially when warm. In 1839, after five years' work on natural rubber, Charles Goodyear (1800–1860) discovered that heating gum rubber with sulfur produces a material that is no longer sticky but is still elastic, water-repellent, and resilient.

Vulcanized rubber contains short chains of sulfur atoms that join and align the natural rubber polymer chains and reduce their unsaturation. The sulfur crosslinks allow the rubber to spring back to its original shape and size after it is stretched and the stress is removed. Substances that behave this way are called *elastomers.*

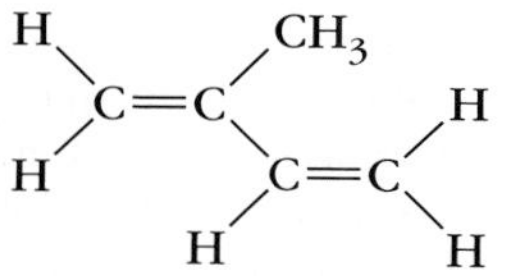

2-methyl-1,3-butadiene (isoprene)

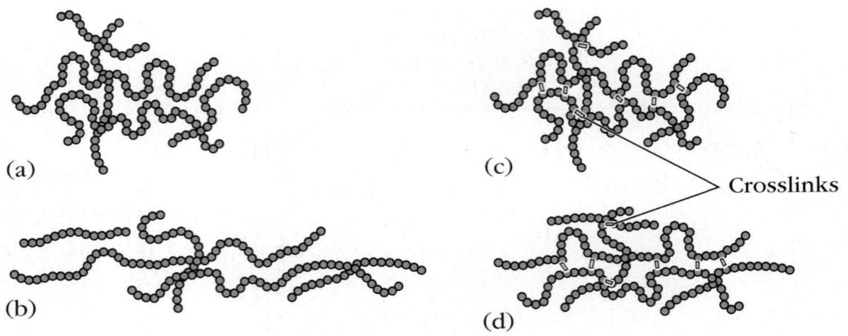

Vulcanization. The four parts illustrate polymer chains: (a) and (b) are in unvulcanized rubber; (c) and (d) are in vulcanized rubber. (b) and (d) illustrate the different effects of applying a horizontal stretching force to the rubber. Because the crosslinks (yellow) in the vulcanized rubber hold some of the polymer chains together, it is stronger and harder to stretch (d) than the unvulcanized rubber (b). The diagram is schematic. In the real polymer there would be many more chains and each chain would be much longer. There would not be as many crosslinks. (*Source:* From *Journal of Chemical Education,* 85(10) p. 1324, 2008. Copyright © 2008, Division of Chemical Education, Inc. Reprinted by permission.)

The behavior of natural rubber is due to the specific molecular geometry within the polymer chain. Natural rubber (poly-*cis*-isoprene) has the $—CH_2CH_2—$ groups on the same side of the double bond in a *cis* arrangement **(← *p. 339*)**,

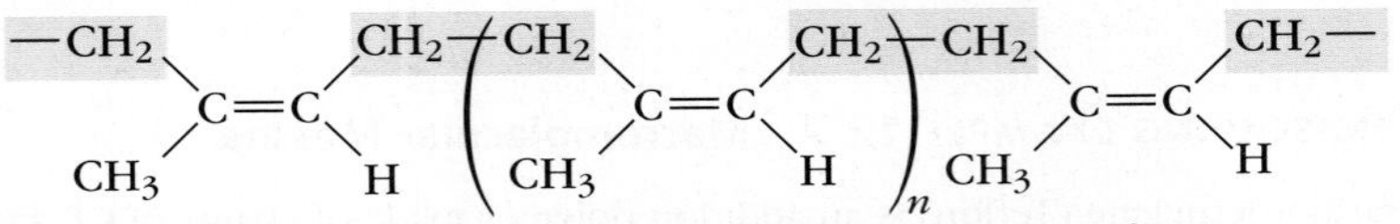

poly-*cis*-isoprene (the $—CH_2—CH_2—$ groups are *cis*)

Polyisoprene with the $—CH_2CH_2—$ groups on opposite sides of the double bond (the *trans* arrangement) also occurs in nature.

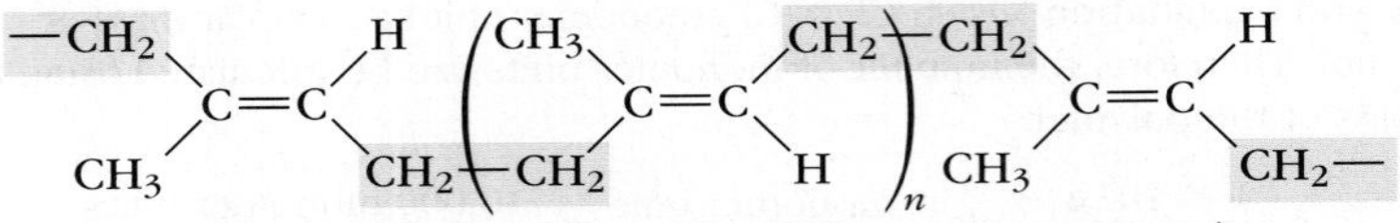

poly-*trans*-isoprene (the $—CH_2—CH_2—$ groups are *trans*)

In 1955 chemists at the Goodyear and Firestone companies almost simultaneously discovered how to prepare synthetic poly-*cis*-isoprene, a material structurally identical to natural rubber. Today, synthetic poly-*cis*-isoprene can be manufactured cheaply and is used when natural rubber is in short supply.

CHEMISTRY YOU CAN DO — Making "Gluep"

White school glue, such as Elmer's glue, contains poly(vinyl acetate) and other ingredients. A "gluep" similar to Silly Putty can be made by mixing $\frac{1}{2}$ cup of glue with $\frac{1}{2}$ cup of water and then adding $\frac{1}{2}$ cup of liquid starch and stirring the mixture. After stirring it, work the mixture in your hands until it has a putty consistency. Roll it into a ball and let it sit on a flat surface. Shape a piece into a ball and drop it on a hard surface.

The "gluep" can be stored in a sealed plastic bag for several weeks. Although "gluep" does not readily stick to clothes, it leaves a water mark on wooden furniture, so be careful where you set it. Mold will form on the "gluep" after a few weeks, but adding a few drops of Lysol to it will retard mold formation.

Think about these questions:

1. What did you observe about "gluep" when it was left sitting on a flat surface?
2. Did the ball of "gluep" bounce when it was dropped?

Condensation Polymers

Polyesters The reactions of alcohols with carboxylic acids to produce esters (p. 556) are examples of condensation reactions. Condensation polymers are formed by condensation polymerization reactions between monomers. Unlike addition polymerization reactions, condensation polymerization reactions do not depend on the presence of a double bond in the reacting molecules. Rather, condensation polymerization reactions generally require *two different functional groups present in the same monomer or in two different monomers.* Each of the two monomers has a functional group at each end of the monomer. As the functional groups at each end of one monomer react with the groups of the other monomer, long-chain condensation polymers are produced.

For example, molecules with two carboxylic acid groups (—COOH), such as terephthalic acid, and other molecules with two alcohol groups, such as ethylene glycol, can react with each other at both ends to form ester linkages (shaded part of the structure). Water is the other product of this condensation polymerization reaction; it is produced during the formation of each ester linkage in the polymer.

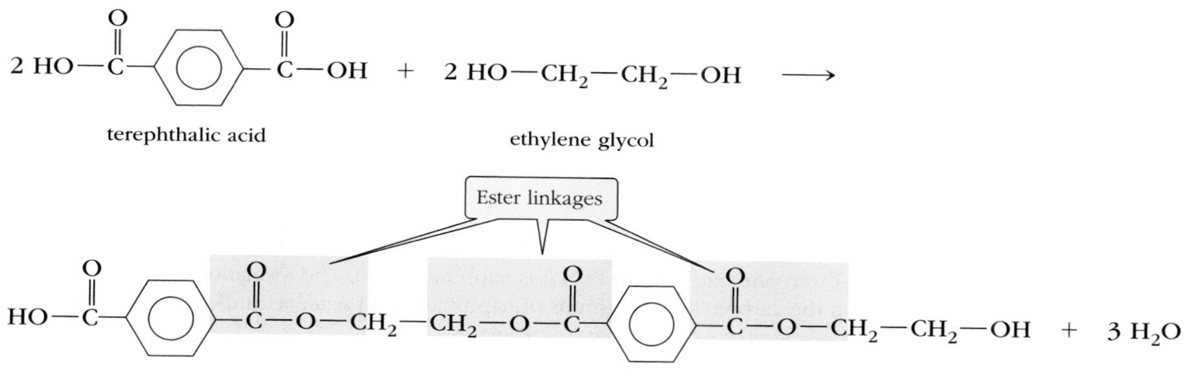

The growing terephthalic acid-ethylene glycol polymer

Because this product has a —COOH group on one end and an —OH group on the other end, the —COOH group can react with an —OH group of another ethylene glycol molecule. Similarly, the remaining alcohol group on the ester can react with another terephthalic acid molecule. This process continues, forming long chains of poly(ethylene terephthalate), a **polyester** commonly known as PET.

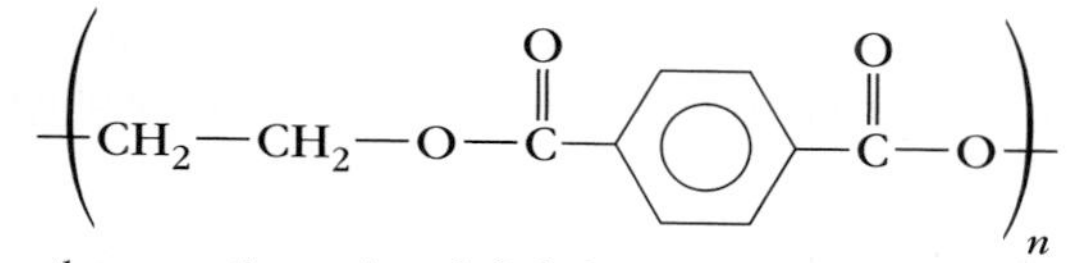

the repeating unit poly(ethylene terephthalate), PET

Each year more than one billion pounds of PET are produced in the United States for making beverage bottles, apparel, tire cord, food packaging, coatings for microwave and conventional ovens, and home furnishings. A variety of trade names is associated with these applications. PET textile fibers are marketed under such names as Dacron and Terylene. Films of the same polyester, when magnetically coated, are used to make TV tapes. This film, Mylar, has unusual strength and can be rolled into sheets one thirtieth the thickness of a human hair. The inert, nontoxic, noninflammatory, and non-blood-clotting characteristics of PET polymers make Dacron tubing an excellent substitute for human blood vessels in heart bypass operations. Dacron sheets are also used as a skin substitute for burn victims.

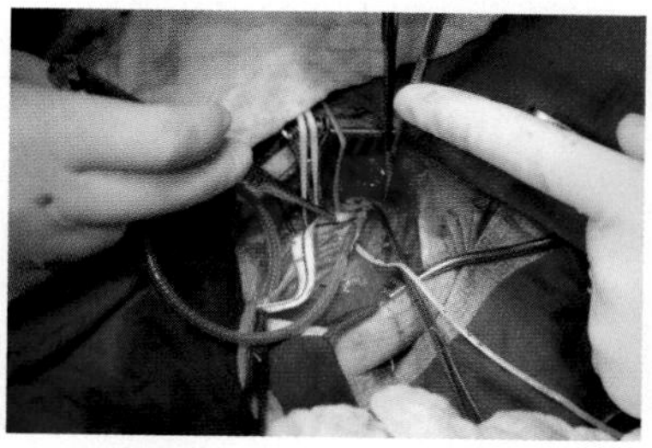

© Barry Slaven/The Medical File/ Peter Arnold, Inc.

Medical uses of Dacron. A Dacron patch is used to close an atrial septal defect in a heart patient.

PROBLEM-SOLVING EXAMPLE 12.8 Condensation Polymerization

Poly(ethylene naphthalate) (PEN) is used for bar code labels. This condensation polymer is made by the reaction between naphthalic acid and ethylene glycol,

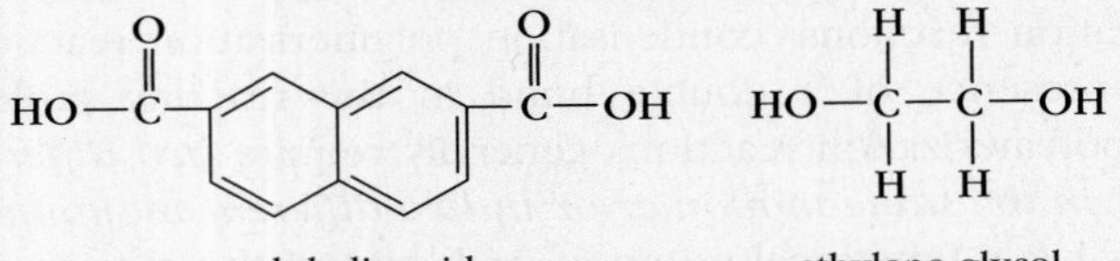

(a) Write the structural formula of the molecule formed after two ethylene glycol molecules have polymerized with two naphthalic acid molecules.
(b) Write the structural formula of the repeating unit of the polymer.

Answer
(a)

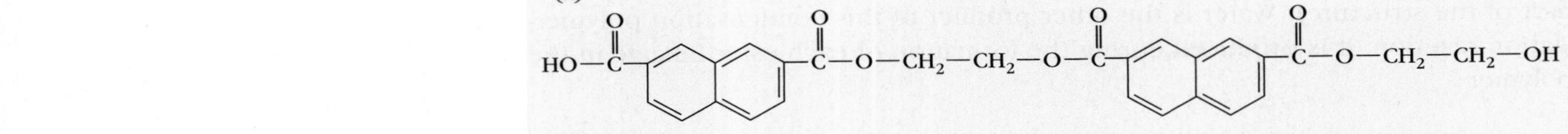

(b)

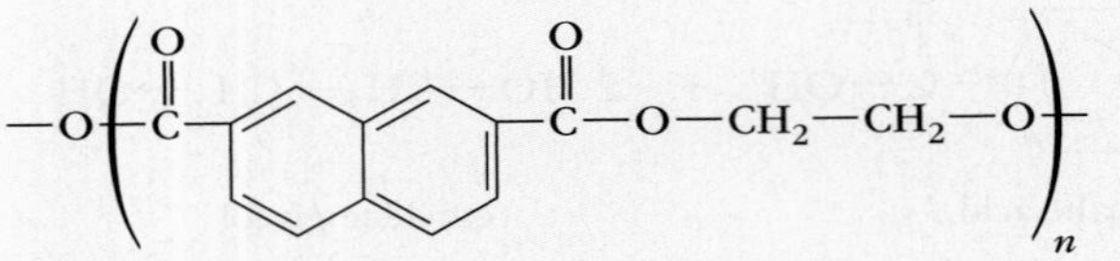

Strategy and Explanation
(a) The formation of PEN is similar to that of PET; both are copolymers and polyesters. Ethylene glycol is one of the monomers in both of them. The other monomer in PET is terephthalic acid; in PEN it is naphthalic acid. The formation of PEN occurs when the carboxylic acid groups of naphthalic acid react with the alcohol groups of ethylene glycol to form ester linkages.

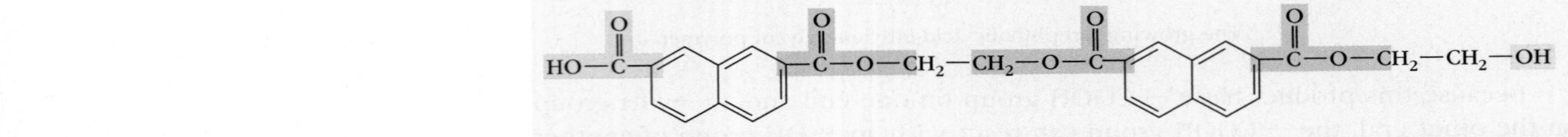

Notice that there are unreacted carboxylic acid and alcohol groups at each end of the molecule. These react with other ethylene glycol and naphthalic acid groups, respectively. The growing polymer chain continues reacting at each end until it becomes very long.
(b) The repeating unit is the smallest unit that contains both monomers. In this case, the repeating unit is formed by combining one naphthalic acid molecule with one ethylene glycol molecule.

PROBLEM-SOLVING PRACTICE 12.8

A condensation polymer can be made from the single monomer glycolic acid. A portion of the polymer is given below.

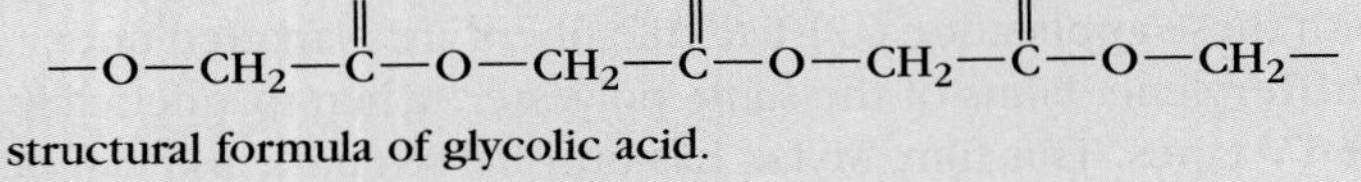

Write the structural formula of glycolic acid.

Amines, Amides, and Polyamides **Amines** are organic compounds that can be thought of as derived from the structure of ammonia, NH_3. Amines are classified as primary, secondary, or tertiary amines according to how many of the H atoms in NH_3 are replaced by alkyl groups. There can be one, two, or three alkyl groups covalently bonded to the nitrogen atom, for example as in methylamine, dimethylamine, and trimethylamine:

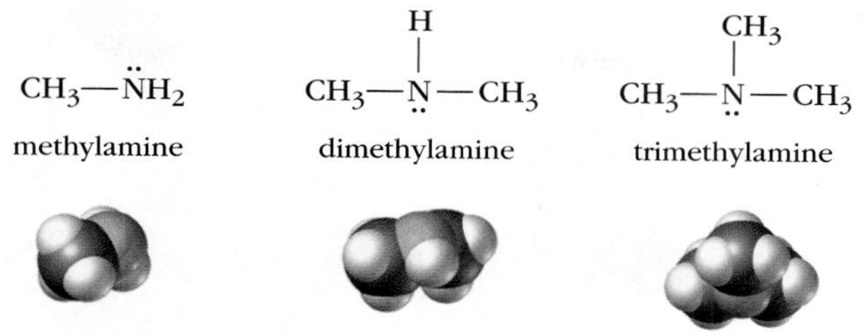

RNH_2 is primary; R_2NH is secondary; and R_3N is tertiary.

An important *condensation* reaction occurs when a primary or secondary amine reacts with a carboxylic acid at high temperature to split out a water molecule and form an **amide:**

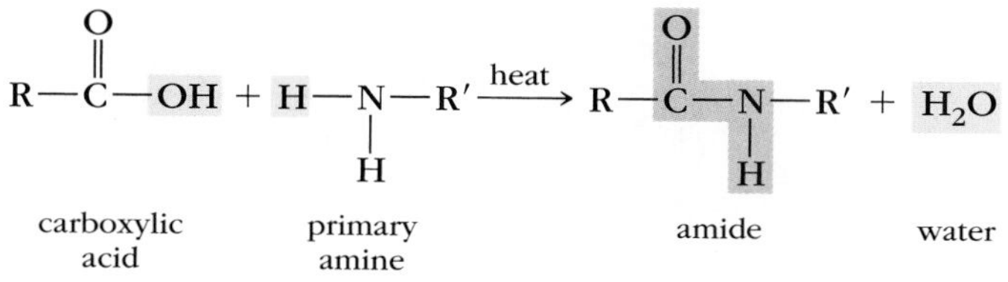

Amides contain the —C(=O)—N(H)— functional group.

Dr. Wallace Carothers of the DuPont Company discovered a useful and important type of condensation polymerization reaction that occurs when *diamines* (compounds containing two $—NH_2$ groups) react with *dicarboxylic acids* (compounds containing two —COOH groups) to form polymers called **polyamides** or nylons. In February 1935 his research yielded a product known as nylon-66 prepared from adipic acid (a dicarboxylic acid) and hexamethylenediamine (a diamine) (Figure 12.9).

The name of nylon-66 is based on the number of carbon atoms in the diamine and diacid, respectively, that are used to make the polymer. Since hexamethylenediamine and adipic acid each have six carbon atoms, the product is called nylon-66.

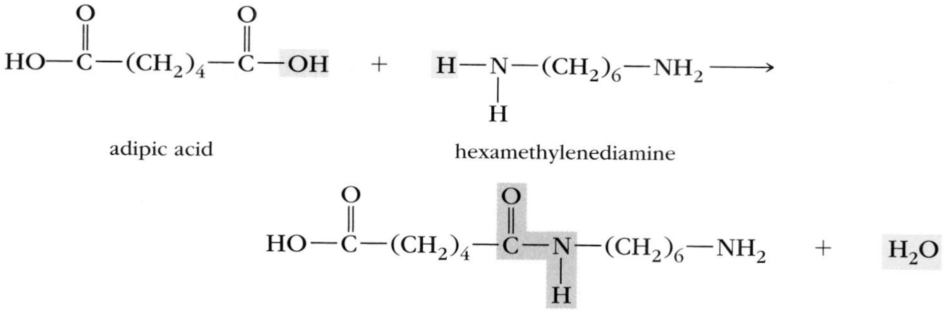

Additional adipic acid and hexamethylenediamine molecules react to extend the polymer chain, which consists of alternating adipic acid and hexamethylenediamine units. The overall equation is

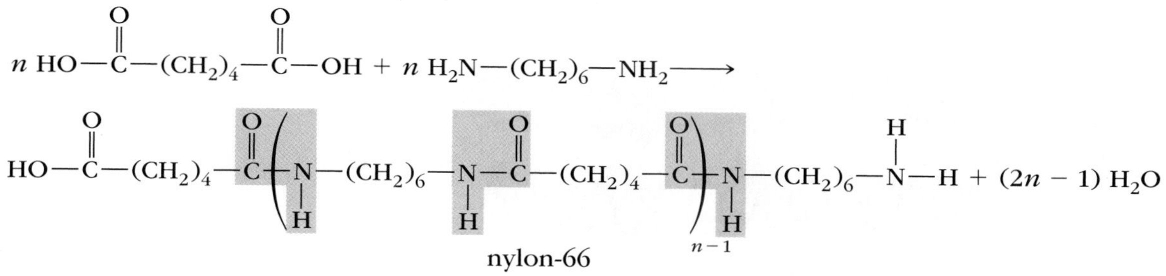

The functional group —C(=O)—N(H)— between the monomers is the **amide linkage;** hence, the polymers are known as polyamides.

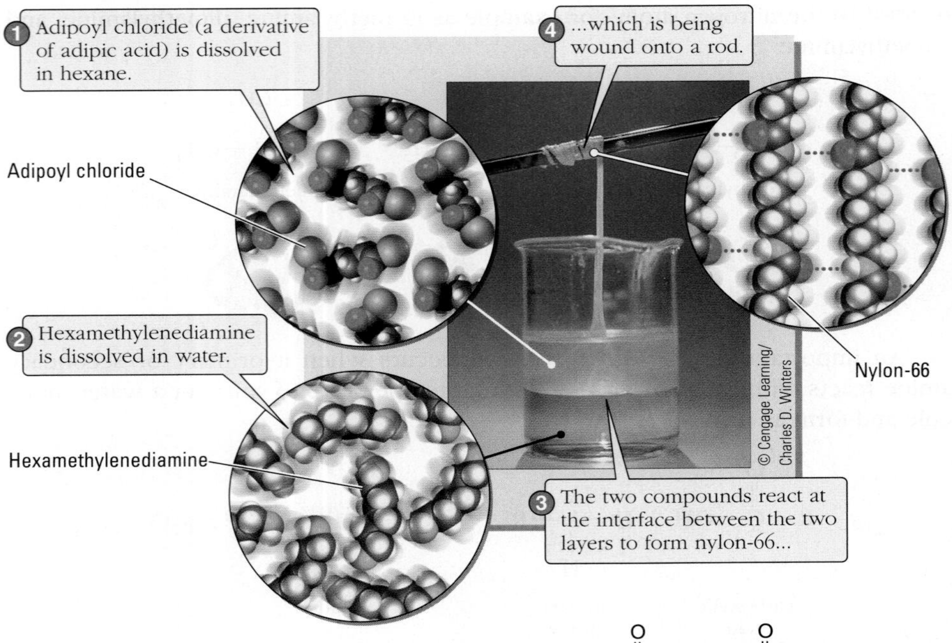

Active Figure 12.9 Making nylon-66. Adipoyl chloride is $Cl{-}\overset{O}{\overset{\|}{C}}{-}(CH_2)_4{-}\overset{O}{\overset{\|}{C}}{-}Cl$ is used here in place of adipic acid. Adipoyl chloride also contains six carbon atoms, like adipic acid. (For clarity, water molecules and hexane molecules are not shown.) Visit this book's companion website at **www.cengage.com/chemistry/moore** to test your understanding of the concepts in this figure.

Figure 12.10 illustrates another facet of the structure of nylon—intermolecular hydrogen bonding—that explains why nylon makes such good fibers. To have good tensile strength, the chains of atoms in a polymer should be able to attract one another, but not so strongly that the polymer molecules cannot initially be extended

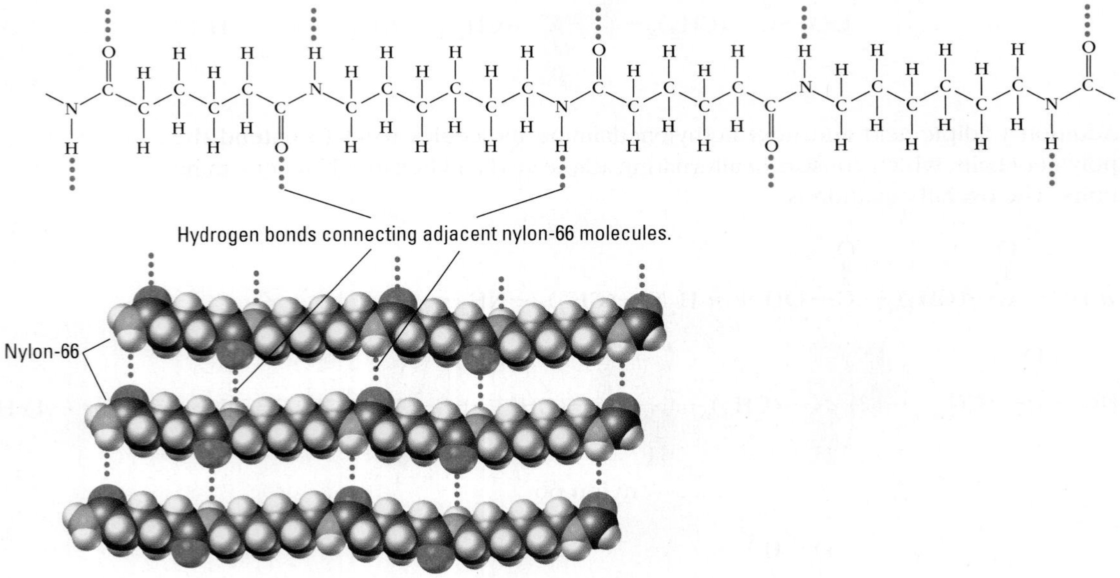

Figure 12.10 Hydrogen bonding in nylon-66.

to form the fibers. Linking the chains together with covalent bonds would be too strong. Hydrogen bonds, with a strength about one tenth that of an ordinary covalent bond, join the chains in the desired manner.

© Reuters/Corbis

A Kevlar helmet. A bullet creased, but did not penetrate, the helmet of this U.S. soldier.

PROBLEM-SOLVING EXAMPLE 12.9 Kevlar, a Condensation Polyamide

Kevlar, used to make bulletproof vests and helmets, sturdier canoes, and baseball batting gloves, is made by the condensation polymerization of the monomers *p*-phenylenediamine and terephthalic acid.

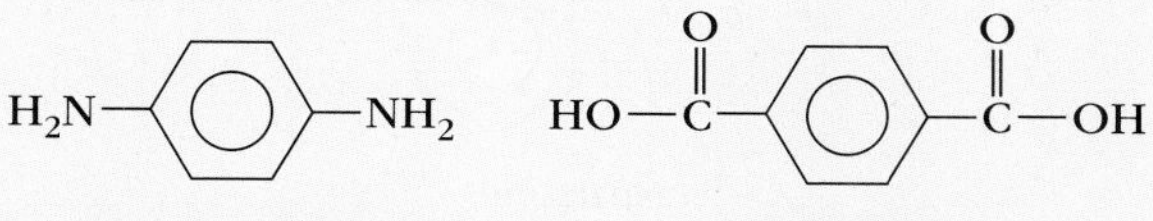

p-phenylenediamine terephthalic acid

(a) Using structural formulas, write the chemical equation for the formation of a segment of a Kevlar molecule containing three amide linkages.
(b) Identify the repeating unit for Kevlar.

Answer

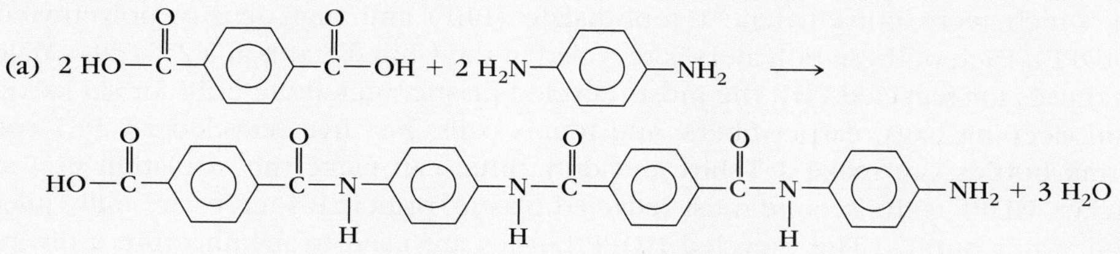

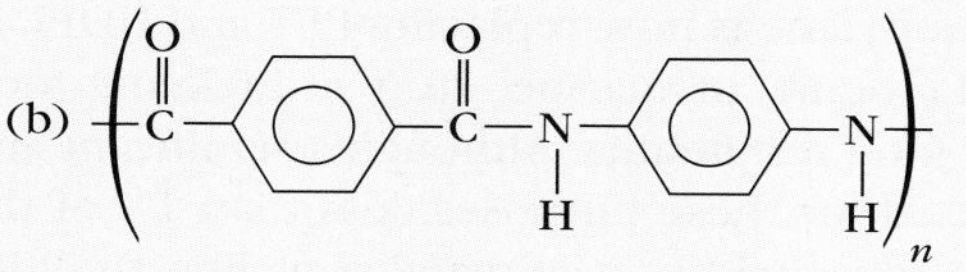

Strategy and Explanation
(a) The amide linkage joining Kevlar units is formed by the condensation reaction between the amine functional group of *p*-phenylenediamine and the carboxylic acid group of terephthalic acid. The reaction continues between additional amine and acid groups to form three amide linkages as shown in Answer (a).
(b) The repeating unit in Kevlar contains an amide linkage and the remaining portions of the reacting acid and amine molecules (Answer [b]).

PROBLEM-SOLVING PRACTICE 12.9

How many moles of water are formed from the condensation reaction between 6 mol *p*-phenylenediamine and 10 mol terephthalic acid?

Courtesy DuPont Company

Stephanie Louise Kwolek
1923–

In 1946, Stephanie Kwolek received a Bachelor of Science degree from Carnegie Tech (now Carnegie Mellon University). Although she wanted to study medicine, she couldn't afford it and decided to take a temporary job at DuPont. She liked her work so well that she stayed for 40 years, retiring in 1986. Kwolek is best known for the development of Kevlar fiber, which is five times stronger than steel on an equal weight basis. She has received many awards, including the Perkin Medal in 1997, considered one of the most prestigious awards a chemist can receive in the United States.

CONCEPTUAL EXERCISE 12.15 Nylon from 3-Aminopropanoic Acid

Polyamides can also be formed from a single monomer that contains both an amine and a carboxylic acid group. For example, the compound 3-aminopropanoic acid can polymerize to form a nylon. Write the general formula for this polymer. Write a formula for the other product that is formed.

$$H_2N-CH_2-CH_2-\overset{\overset{\displaystyle O}{\|}}{C}-OH$$

3-aminopropanoic acid

EXERCISE 12.16 Functional Groups

We have discussed several functional groups. Shown below is the structural formula of aspartame, an artificial sweetener. Identify each of the numbered functional groups.

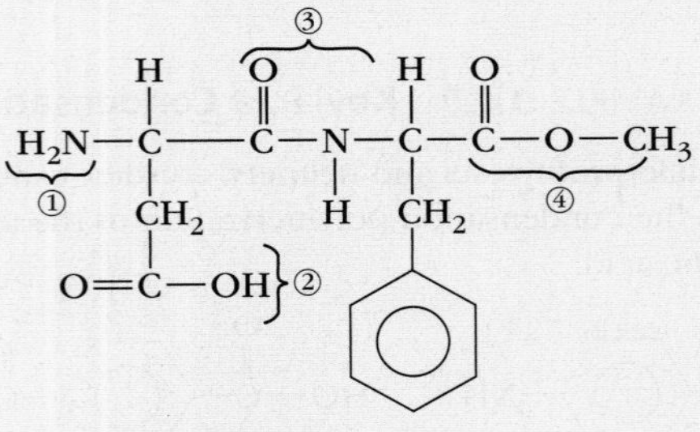

aspartame

Recycling Plastics

The manufacturing of plastic bottles for bottled water uses an estimated 50 million barrels of oil worldwide annually. U.S. bottled water consumption in 2007 required the equivalent of over 40 million barrels of oil for the production and transportation of the bottles.

These jackets are made from recycled PET soda bottles.

More than 2.4 million plastic bottles were recycled in the United States in 2007, 99% of which were polyethylene terephthalate (PET) and high-density polyethylene (HDPE). Each of these polymers is recycled in the United States at a 25% rate. Major end uses for recycled PET, the most recycled plastic, include fiberfill for ski jackets and sleeping bags, carpet fibers, and tennis balls. Just five recycled 2-L PET soft-drink bottles can make a T-shirt and only three can make the insulation in a ski jacket. HDPE is the second most recycled plastic; principal sources are milk, juice, and water bottles. The recycled HDPE bottles are used to manufacture a diverse range of products (Figure 12.11). Polypropylene is now replacing PET and HDPE in the manufacturing of commercial food (yogurt, margarine, etc.) and takeout food containers, as well as drink cups in fast-food restaurants. Although a significant and growing amount of polypropylene is used for these purposes, less than 1% of the polypropylene containers and cups are recycled because many recycling facilities only recycle plastic bottles.

Codes are stamped on plastic containers to help consumers identify and sort their recyclable plastics (Table 12.7). There has been a dramatic increase in the recycling of plastics in recent years. Still further increases require a sufficient demand for

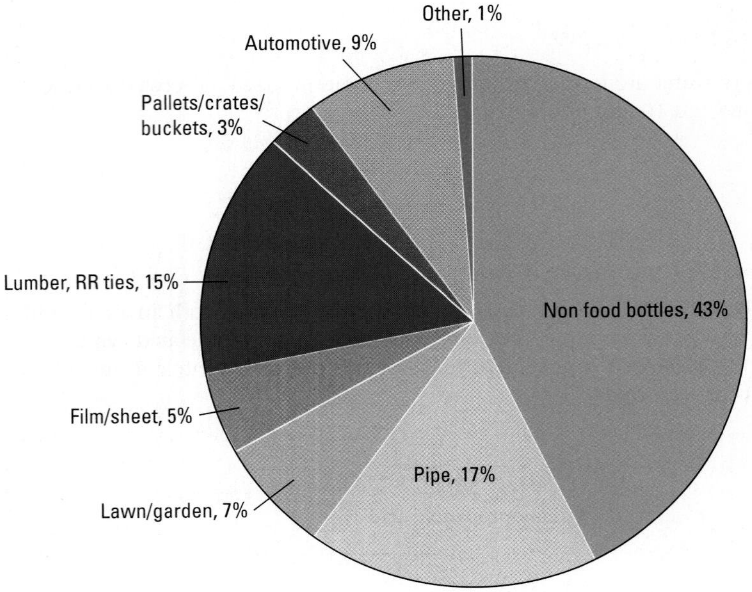

Figure 12.11 Domestic recycled HDPE bottle end use, 2008. Recycled HDPE bottles are used to make a wide variety of products. (*Source:* American Chemistry Council 2008 United States National Post-Consumer Plastics Bottle Recycling Report, Figure 5, p. 9. Reprinted by permission.)

Table 12.7 Plastic Container Codes

Code	Material	Code	Material
1 PETE	Polyethylene terephthalate (PET)*; beverage bottles, clothing	5 PP	Polypropylene; bottles, containers, auto battery cases
2 HDPE	High-density polyethylene; milk and water jugs, gasoline tanks, Tyvek wrap	6 PS	Polystyrene; thermal insulation, drinking glasses
3 V	Poly(vinyl chloride) (PVC)*; garden hoses, plumbing pipe	7 OTHER	All other resins and layered multimaterial; cell phone and computer casings
4 LDPE	Low-density polyethylene; plastic bags, toys, electrical insulation		

*Bottle codes are different from standard industrial identification to avoid confusion with registered trademarks.

John Sohiden/Visuals Unlimited, Inc.

A Tyvek-wrapped house under construction.

products made partially or completely from recycled materials. The use of some recycled, postconsumer plastics has been mandated by law in some places. Since 1995 in California, for example, all HDPE packaging must contain 25% recycled material. A real challenge lies ahead in finding economically viable methods to recycle the plastics from the growing, massive numbers of obsolete personal computers, and cellular phones. In the United States 31,000 *tons* of laptop computers and 12,000 *tons* of cell phones were discarded in 2005. Only 26% of the laptops and 19% of the cell phones were recycled. The rest were incinerated or put into landfills.

12.7 Biopolymers: Polysaccharides and Proteins

Biopolymers—naturally occurring polymers—are an integral part of living things. Many advances in creating and understanding synthetic polymers came from studying biopolymers.

Cellulose and starch, made in plants by condensation polymerization reactions, resemble a synthetic polymer in that the monomer molecules—glucose—are all alike. On the other hand, proteins, which are made by both plants and animals, are very different from synthetic polymers because proteins include many different monomers. Also, the occurrences of the different monomers along the protein polymer chain are anything but regular. As a result, proteins are extremely complex copolymers formed by condensation polymerization.

Monosaccharides to Polysaccharides

Nature makes an abundance of compounds with the general formula $C_x(H_2O)_y$, in which x and y are whole numbers, such as in glucose, $C_6H_{12}O_6$, where $x = 6$, $y = 6$. These compounds are variously known as sugars, carbohydrates, and mono-, di-, or polysaccharides (from the Latin *saccharum,* "sugar"—because they taste sweet). As noted in Figure 12.12, the simplest carbohydrates, such as glucose, are **monosaccharides,** also called simple sugars, because they contain only a single saccharide molecule. **Disaccharides** consist of two monosaccharide units joined together, such as glucose and fructose linked to form sucrose (table sugar) or two glucose units to form maltose. **Polysaccharides** are polymers containing many monosaccharide units, up to several thousand.

The monomer units of disaccharides and polysaccharides are joined by a C—O—C arrangement called a **glycosidic linkage** between carbons 1 and 4 of ad-

1,6-glycosidic linkages are also possible.

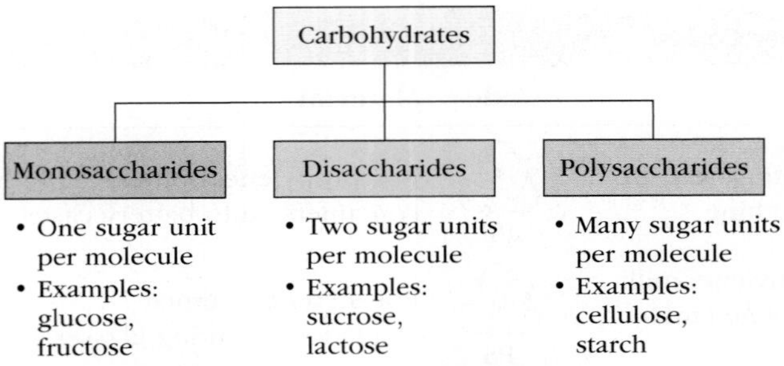

Figure 12.12 **The classes of carbohydrates.**

jacent monosaccharide units. Polysaccharides contain many monosaccharide monomers joined together via glycosidic linkages into a very large polymer. The glycosidic linkages in disaccharides and polysaccharides are formed by a condensation reaction like the condensation reactions by which synthetic polymers are formed. A water molecule is released during the formation of each glycosidic bond from the reaction between the —OH groups of the monosaccharide monomers.

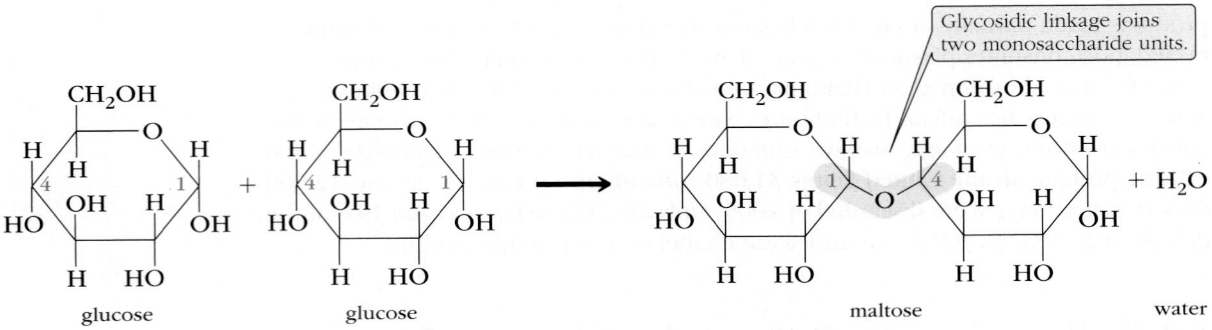

Notice that maltose has two fewer hydrogen atoms and one less oxygen atom than the combined formulas of two glucose molecules. This difference arises because a water molecule is eliminated as a product of the condensation reaction.

Starches and cellulose are the most abundant natural polysaccharides. D-glucose is the monomer in each of these polymers, which can contain as many as 5000 glucose units.

CONCEPTUAL EXERCISE **12.17 Sucrose Solubility**

On a chemical basis, explain why table sugar (sucrose) is soluble in water.

Polysaccharides: Starches and Glycogen

Plant starch is stored in protein-covered granules until glucose is needed for synthesis of new molecules or for energy production. If these granules are ruptured by high temperature, they yield two starches—*amylose* and *amylopectin.* Natural starches contain about 75% amylopectin and 25% amylose, both of which are polymers of glucose units joined by glycosidic linkages. Structurally, amylose is a condensation polymer with an average of about 200 glucose monomers per molecule arranged in a straight chain, like pearls on a necklace. Amylopectin is a branched-chain polymer analogous to the branched-chain synthetic polymers discussed earlier. A typical amylopectin molecule has about 1000 glucose monomers arranged into branched chains. The branched chains of glucose units give amylopectin properties different from those of the unbranched amylose. The main difference is their water solubilities. A family of enzymes called amylases helps to break down starches sequentially into a mixture of small branched-chain polysaccharides called dextrins

Amylose turns blue-black when tested with iodine solution, whereas amylopectin turns red.

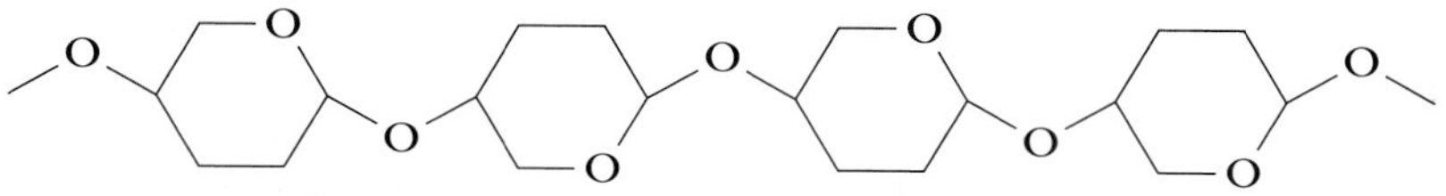

Figure 12.13 The structure of cellulose. About 280 glucose units are bonded together to form a chain structure in cellulose. The glycosidic linkages between glucose units alternate in their orientation in cellulose, unlike those in amylose.

and ultimately into glucose. Dextrins are used as food additives and in mucilage, paste, and finishes for paper and fabrics.

In animals, *glycogen* serves the same storage function as starch does in plants. Glycogen, the reserve carbohydrate in the body, is stored in the liver and muscle tissues and provides glucose for "instant" energy until the process of fat metabolism can take over and serve as the energy source (← p. 254). The chains of D-glucose units in glycogen are more highly branched than those in amylopectin.

Animals store energy as fats rather than carbohydrates because fats have more energy per gram.

Cellulose, a Polysaccharide

Cellulose is the most abundant organic compound on Earth, found as the woody part of trees and the supporting material in plants and leaves. Cotton is the purest natural form of cellulose. Like amylose, cellulose is composed of D-glucose monomer units. The difference between the structures of cellulose and amylose lies in the orientation of the glycosidic linkages between the glucose monomer units. In cellulose, the —OH groups at carbons 1 and 4 are in the *trans* position, so the glycosidic linkages between glucose units alternate in direction; thus, every other glucose unit is turned over (Figure 12.13). In amylose, the —OH groups at carbons 1 and 4 are in the *cis* position, so all the glycosidic linkages are in the same direction. This subtle structural difference allows humans to digest starch, but not cellulose; we lack the enzyme necessary to break the *trans* glycosidic linkages in cellulose. However, termites, a few species of cockroaches, and ruminant mammals such as cows, sheep, goats, and camels do have the proper internal chemistry for this purpose. Because cellulose is so abundant, it would be advantageous if humans could use it, as well as starch, for food.

Paper and cotton are essentially cellulose.

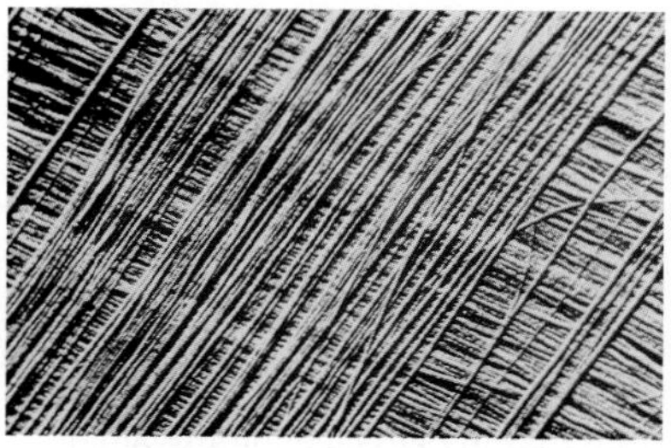

Cabisco/Visuals Unlimited, Inc.

Parallel strands of cellulose in a plant fiber.

CONCEPTUAL EXERCISE 12.18 Digestion of Cellulose

Explain why humans cannot digest cellulose. Consult a reference on the Internet and explain why ruminant animals can digest cellulose.

CONCEPTUAL EXERCISE 12.19 What If Humans Could Digest Cellulose?

Think of some of the implications if humans could digest cellulose. What would be some desirable consequences? What would be some undesirable ones?

Amino Acids to Proteins

Amino acids, the monomer units in proteins, each contain a carboxylic acid group and an amine group. The 20 naturally occurring amino acids from which proteins are made have the general formula

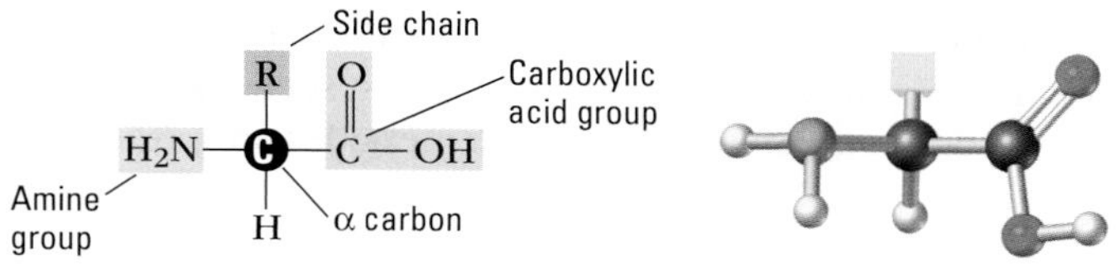

They are described as α-amino acids because the amine, —NH_2, group is attached to the **alpha carbon,** the first carbon next to the carboxylic acid, —COOH, group. Each amino acid has a different R group, called a side chain (Table 12.8, p. 579). Glycine, the simplest amino acid, has just hydrogen as its R group. Note from the table that some amino acid R groups contain only carbon and hydrogen, while others contain carboxylic acid, amine, or other functional groups. The amino acids are grouped in Table 12.8 according to whether the R group is nonpolar, polar, acidic, or basic. Each amino acid has a three-letter abbreviation for its name.

CONCEPTUAL EXERCISE 12.20 Hydrogen Bonding Between Amino Acids in Proteins

Pick two amino acids from Table 12.8 whose R groups can hydrogen-bond with one another if they were close together in a protein chain or in two adjacent protein chains. Then pick two whose R groups would not hydrogen-bond under similar circumstances.

Peptide linkages are also referred to as peptide bonds.

Like nylon, proteins are polyamides. The amide bond in a protein is formed by the *condensation* polymerization reaction between the amine group of one amino acid and the carboxylic acid group of another. In proteins, the amide linkage is called a **peptide linkage.** Relatively small amino acid polymers (up to about 50 amino acids) are known as **polypeptides.** Proteins are polypeptides containing hundreds to thousands of amino acids bonded together.

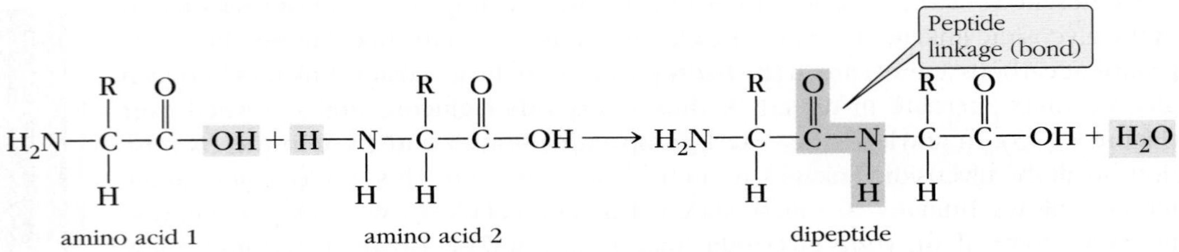

Note that in glycylalanine the amine group of glycine is unreacted as is the carboxylic acid group of alanine. This differs from alanylglycine, which is formed by the condensation reaction of the amine group of glycine (its carboxylic acid group is unreacted) and the carboxylic acid group of alanine (its amine group is unreacted).

Any two amino acids can react to form two different dipeptides, depending on which amine and acid groups react from each amino acid. For example, glycine and alanine can react and join in either of the following two ways:

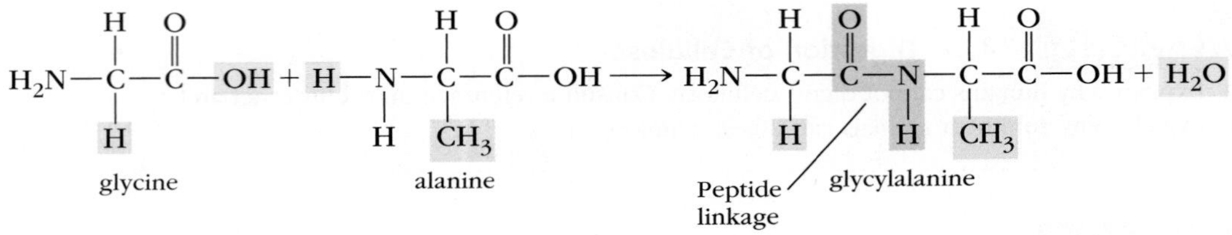

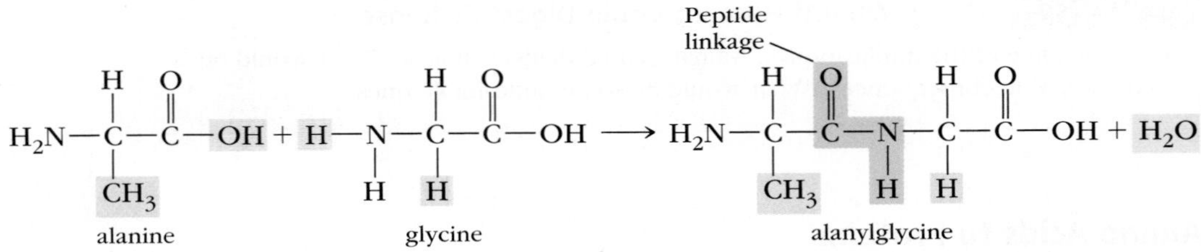

Either of these dipeptides can react with other amino acids at both ends. The extensive chains of amino acid units in proteins are built up by such condensation polymerization reactions.

As the number of amino acids in the chain increases and the polypeptide chain lengthens, the number of variations quickly increases to a degree of complexity not generally found in synthetic polymers. As we have just seen, two different dipeptides can form from two different amino acids. Six *tri*peptides are possible if three differ-

Table 12.8 Common L-Amino Acids Found in Proteins†

Amino Acid	Abbreviation	Structure
Nonpolar R groups		
Glycine	Gly	$H-CH(NH_2)-COOH$
Alanine	Ala	$CH_3-CH(NH_2)-COOH$
*Valine	Val	$CH_3-CH(CH_3)-CH(NH_2)-COOH$
*Leucine	Leu	$CH_3-CH(CH_3)-CH_2-CH(NH_2)-COOH$
*Isoleucine	Ile	$CH_3-CH_2-CH(CH_3)-CH(NH_2)-COOH$
Proline	Pro	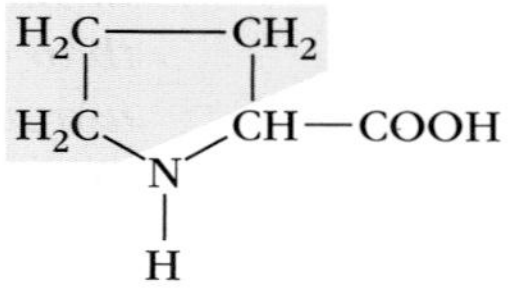
*Phenylalanine	Phe	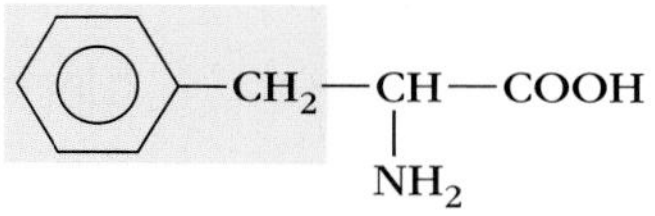
*Methionine	Met	$CH_3-S-CH_2CH_2-CH(NH_2)-COOH$
*Tryptophan	Trp	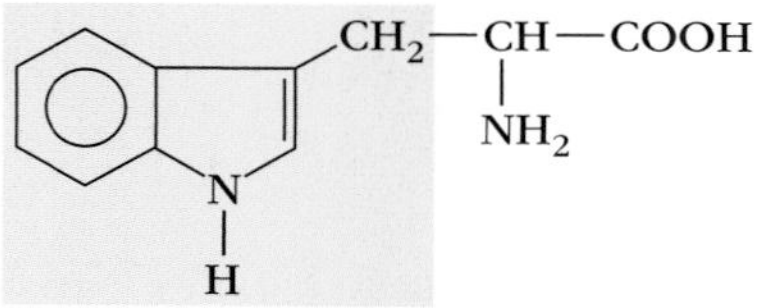
Polar but neutral R groups		
Serine	Ser	$HO-CH_2-CH(NH_2)-COOH$
*Threonine	Thr	$CH_3-CH(OH)-CH(NH_2)-COOH$
Cysteine	Cys	$HS-CH_2-CH(NH_2)-COOH$
Asparagine	Asn	$H_2N-C(=O)-CH_2-CH(NH_2)-COOH$
Glutamine	Gln	$H_2N-C(=O)-CH_2CH_2-CH(NH_2)-COOH$
Tyrosine	Tyr	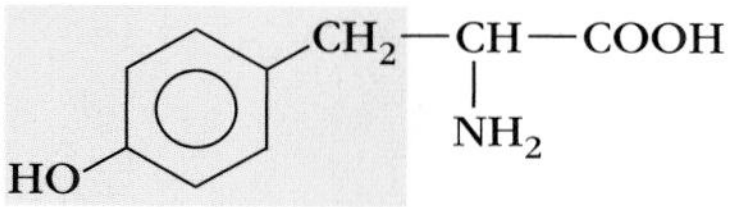
Acidic R groups		
Glutamic acid	Glu	$HO-C(=O)-CH_2CH_2-CH(NH_2)-COOH$
Aspartic acid	Asp	$HO-C(=O)-CH_2-CH(NH_2)-COOH$
Basic R groups		
*Lysine	Lys	$H_2N-CH_2CH_2CH_2CH_2-CH(NH_2)-COOH$
‡Arginine	Arg	$H_2N-C(=NH)-NH-CH_2CH_2CH_2-CH(NH_2)-COOH$
Histidine	His	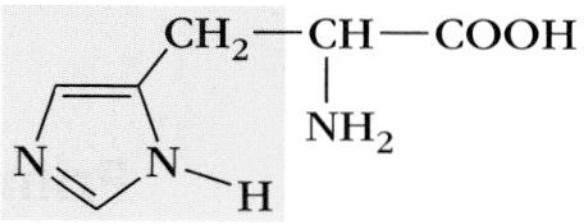

*Essential amino acids that must be part of the human diet. The other amino acids can be synthesized by the body.
†The R group in each amino acid is highlighted.
‡Growing children also require arginine in their diet.

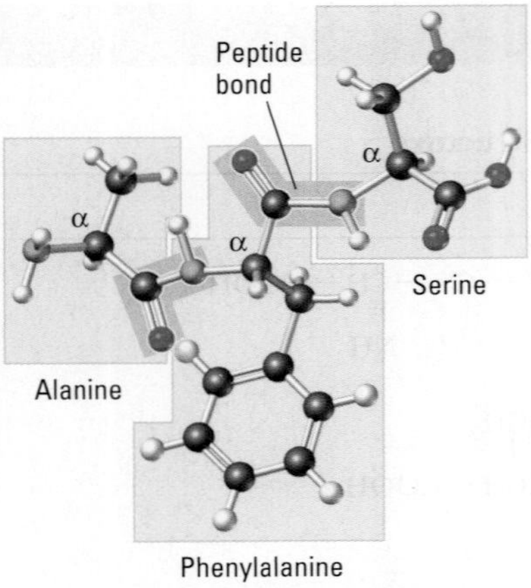

The Ala-Phe-Ser tripeptide

ent amino acids (for example, phenylalanine, Phe; alanine, Ala; and serine, Ser) are linked in combinations that contain all three amino acids, if each is used only once:

Phe-Ala-Ser	Ser-Ala-Phe	Ala-Ser-Phe
Phe-Ser-Ala	Ser-Phe-Ala	Ala-Phe-Ser

If n different amino acids are present, the number of arrangements is $n!$ (n factorial). For four different amino acids, the number of different tetrapeptides is 4!, or $4 \times 3 \times 2 \times 1 = 24$. If all 20 different naturally occurring amino acids were bonded in one polypeptide, the sequences would make $20! = 2.43 \times 10^{18}$ (2.43 quintillion) unique 20-monomer molecules. *Because proteins can also include more than one molecule of a given amino acid, the number of possible combinations is astronomical.* It is truly remarkable that of the many different proteins that could be made from a set of amino acids, a living cell makes only the relatively small number of proteins it needs.

PROBLEM-SOLVING EXAMPLE 12.10 Peptides

Using information from Table 12.8, draw the structural formula of the tripeptide represented by Ala-Ser-Gly. Explain why it is a different compound from the tripeptide with the amino acids joined in the order Gly-Ala-Ser.

Answer

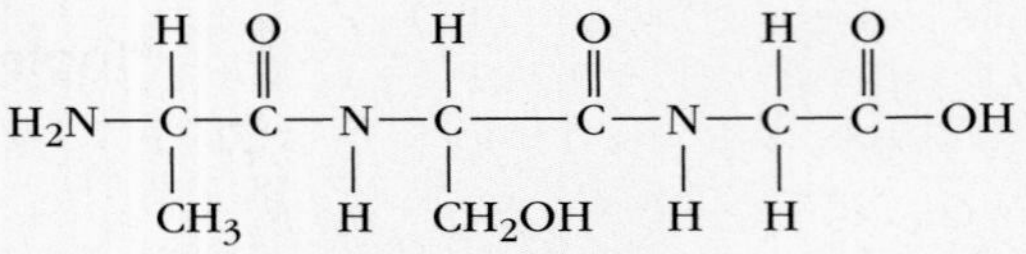
Ala-Ser-Gly

The structure Gly-Ala-Ser differs because the free $—NH_2$ group is on the glycine part of the molecule and the free —COOH group is on the serine part of the molecule.

Strategy and Explanation The amino acid sequence in the abbreviated name shows that alanine should be written at the left with a free $H_2N—$ group, glycine should be written at the right with a free —COOH group, and both should be connected to serine by peptide bonds. Writing the structure of Gly-Ala-Ser shows how the two tripeptides are different.

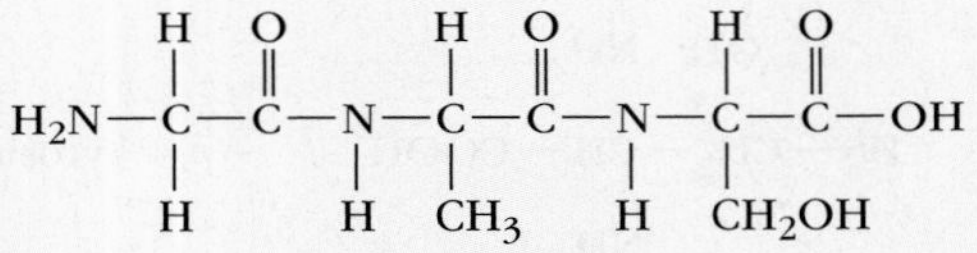
Gly-Ala-Ser

PROBLEM-SOLVING PRACTICE 12.10

Draw the structural formula of the tetrapeptide Cys-Phe-Ser-Ala.

EXERCISE 12.21 Peptide Sequences

Draw the structural formula of the tetrapeptide Ala-Ser-Phe-Cys.

Primary, Secondary, and Tertiary Structure of a Protein

The three-dimensional shape of a protein, and consequently its function, is determined by the sequence of amino acids along the chain, known as the **primary structure** of the protein. All proteins, no matter how large, have a peptide backbone of amino acid units covalently bonded to each other through peptide linkages.

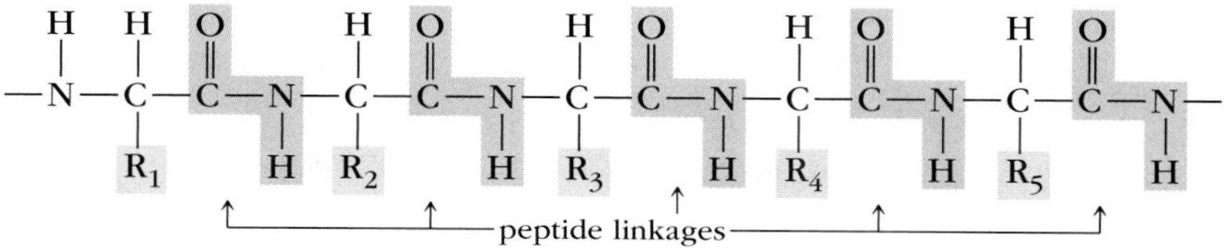
peptide linkages

The distinctions that determine the shape and function of a protein lie in the sequence of amino acids and the order of the R groups along the backbone. Even one amino acid out of place can create dramatic changes in the shape of a protein, which can lead to serious medical conditions. For example, sickle cell anemia results from an alteration in the primary structure of hemoglobin, the molecule that carries oxygen in red blood cells. Glutamic acid, the seventh amino acid in a 146-amino acid chain in hemoglobin, is replaced by valine. Replacement of the carboxylic acid side chain of glutamic acid with the nonpolar side chain of valine (Table 12.8) disturbs the intramolecular attractions in hemoglobin and changes its shape. As a result, hemoglobin molecules gather into fibrous chains that distort the red blood cells into a sickle shape.

The change in structure of hemoglobin that causes sickle-cell anemia answers the question, "How can a disease be caused or cured by a tiny change in a molecule?" that was posed in Chapter 1 ***(← p. 2).***

Protein chains are not long and floppy, as we might imagine. Instead, regular patterns of **secondary structure** are created by hydrogen bonding. There are two major types of secondary protein structure: (1) the alpha (α)-helix, which occurs within a protein molecule, and (2) the beta (β)-pleated-sheet, which occurs between adjacent protein chains. In the alpha-helix, hydrogen bonding occurs between the N—H hydrogen of an amine group in a peptide bond with a lone electron pair on a C=O oxygen of a peptide bond four amino acid units farther down the peptide backbone (Figure 12.14). In effect, the hydrogen in the hydrogen bond "looks over its shoulder," causing the protein to curl into a helix, much like the coiling of a spring. In the alpha-helix, the R groups on the amino acid units are on the outside of the helix. In con-

Wool is primarily the protein keratin, which has an alpha-helix secondary structure. Fibroin, the principal protein in silk, is largely beta-pleated-sheets.

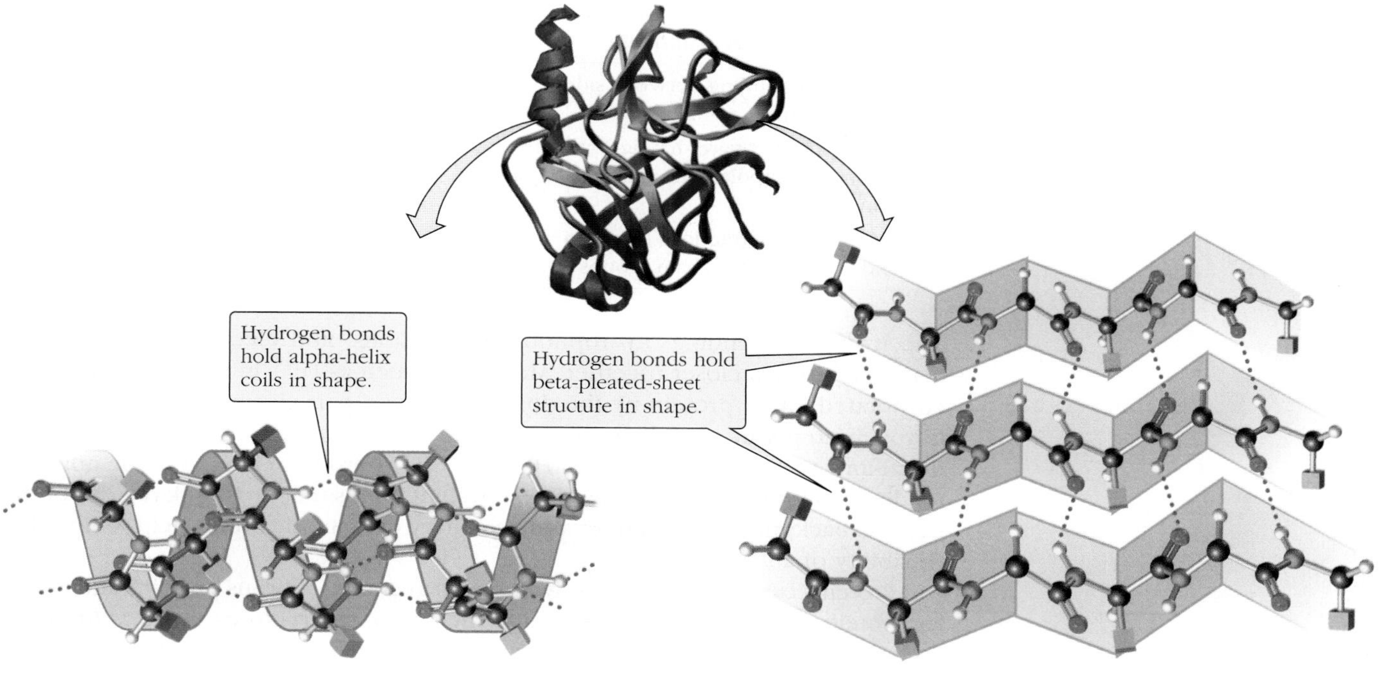
Hydrogen bonds hold alpha-helix coils in shape.
Hydrogen bonds hold beta-pleated-sheet structure in shape.
Alpha-helix
Beta-pleated-sheet

Figure 12.14 Alpha-helix and beta-pleated-sheet structures of proteins. The enzyme molecule has regions of both alpha-helix and beta-pleated-sheet protein structures. R groups are represented by green blocks.

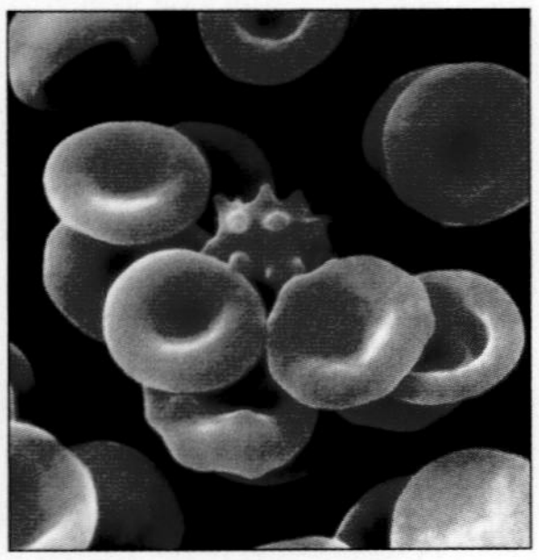

(a) Normal cells

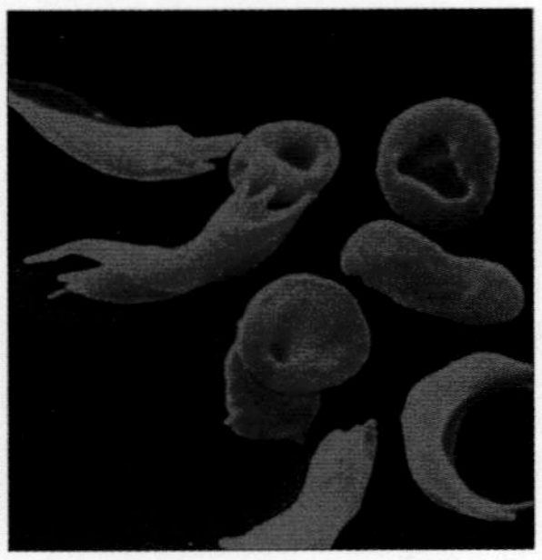

(b) Sickled cells

Red blood cells.

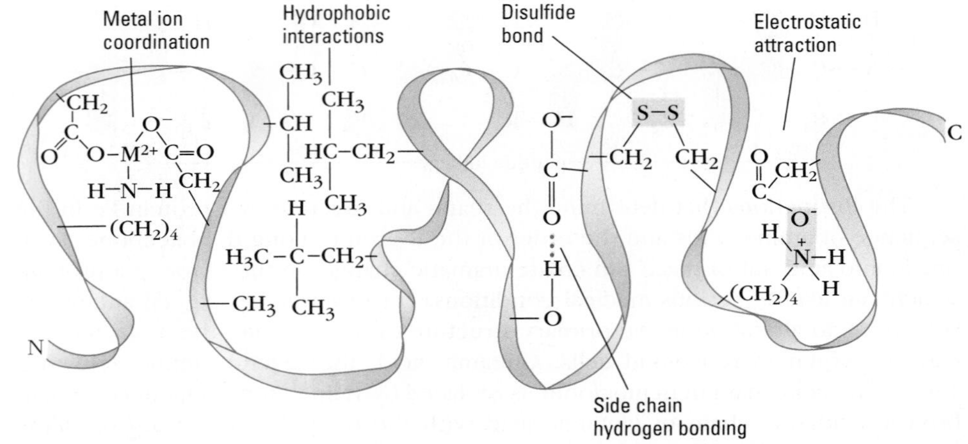

Figure 12.15 Noncovalent forces and bonds that stabilize protein tertiary structure. The tertiary structure of a protein is stabilized by several different forces.

trast, hydrogen bonding in the beta-pleated sheet occurs between the peptide backbones of neighboring protein chains, creating a zigzag pattern resembling a pleated sheet with the R groups above and below the sheet (Figure 12.14).

In many proteins, the backbone is folded into a globular structure as shown at the top of Fig. 12.14. Within the globular structure there may be regions of α-helix and β-pleated-sheet. The overall three-dimensional arrangement that accounts for all the twists and turns and folding of a protein is called its **tertiary structure.** The twists and turns of the tertiary structure result in a protein molecule of maximum stability. Tertiary structure is determined by interactions among the side chains strategically placed along the backbone of the chain. These interactions include noncovalent forces of attraction and covalent bonding, as illustrated in Figure 12.15.

Noncovalent Attractions	**Covalent Bonds**
Hydrogen bonding between side-chain groups	Disulfide bonds, —S—S—
Hydrophobic (water-hating) interactions between nonpolar side-chain hydrocarbon groups	Coordinate covalent bonding between metal ions and electron pairs of side-chain N and O atoms
Electrostatic attractions between $—NH_3^+$ and $—COO^-$ side-chain groups	

The hydrogen bonding, hydrophobic interactions, and electrostatic attractions bring the groups closer together. Metal ions, such as Fe^{2+} in hemoglobin, are incorporated into a protein by the donation of lone pair electrons from oxygen or nitrogen atoms in side-chain groups to form coordinate covalent bonds to the metal ions **(← *p. 356*).** By loss of H, the —SH groups in nearby cysteine units are oxidized to form disulfide, —S—S—, covalent bonds that cross-link regions of the peptide backbone. The number and proximity of disulfide bonds help to limit the flexibility of a protein.

Proteins can be divided into two broad categories—fibrous proteins and globular proteins—that reflect differences in their tertiary structures. *Fibrous proteins,* such as hair, muscle fibers, and fingernails have little tertiary structure and are rodlike, with the coils or sheets of protein aligned into parallel bundles, making for tough, water-insoluble materials. In contrast, *globular proteins* are highly folded, with hydrophilic (water-loving) side chains on the outside, making these proteins water-soluble. Hemoglobin and chymotrypsin (Figure 12.16) are globular proteins, as are most enzymes.

Because of the complexity and the variety of properties provided by the different R groups and associated molecules or ions, proteins are able to perform widely diverse functions in the body. Consider some of them.

Class of Protein	Example	Description of Function
Hormones	Insulin, growth hormone	Regulate vital processes according to needs
Muscle tissue	Myosin	Does mechanical work
Transport proteins	Albumin	Carry fatty acids and other hydrophobic molecules through the bloodstream
Clotting proteins	Fibrin	Form blood clots
Enzymes	Chymotrypsin	Catalyze biochemical reactions
Messengers	Endorphins	Transmit nerve impulses
Immune system proteins	Immunoglobulins	Fight off disease

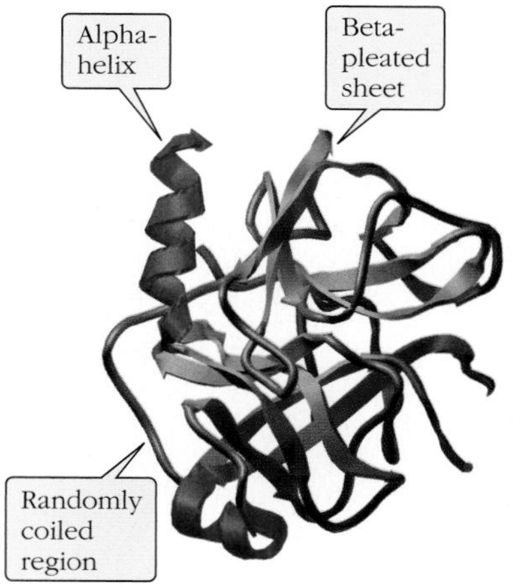

Figure 12.16 Protein folding. The protein chymotrypsin is shown in a ribbon structure that illustrates only the polypeptide backbone. Alpha-helix regions are shown in blue, beta-pleated sheet regions are shown in green, and randomly coiled regions are shown in copper.

SUMMARY PROBLEM

Part I

This chapter has described many kinds of carbon-containing compounds from a variety of sources. Several classes of carbon compounds have molecules with a carbon-to-oxygen bond, which gives the compound certain chemical properties. Use the information in the chapter to complete the sequence of chemical changes for the compound

$$CH_3(CH_2)_8CH_2OH$$

(a) Suppose this compound reacts with aqueous $KMnO_4$. Write the structural formulas of the initial and final oxidation products of this compound.

(b) Use structural formulas to write the equation for the reaction of the initial reactant in Question (a) with the final oxidation product in Question (a).

(c) Into what class of compounds does the product of the reaction in Question (b) fall?

(d) The final product in Question (b) reacts with aqueous sodium hydroxide. Write the equation for this reaction. What types of compounds are formed by this reaction? What would the products be if the reaction took place under acidic conditions instead of with aqueous base?

(e) The final oxidation product in Question (a) reacts with dimethylamine, $(CH_3)_2NH$. Write the structural formula of the product of this reaction. What class of compound is formed?

(f) Can the two reactants in Question (b) form a polymer? Explain.

Part II

The citric acid cycle is a series of steps involving the degradation and reformation of oxaloacetic acid. Two steps in the cycle involve a reaction studied in this chapter. One step converts isocitric acid into oxalosuccinic acid.

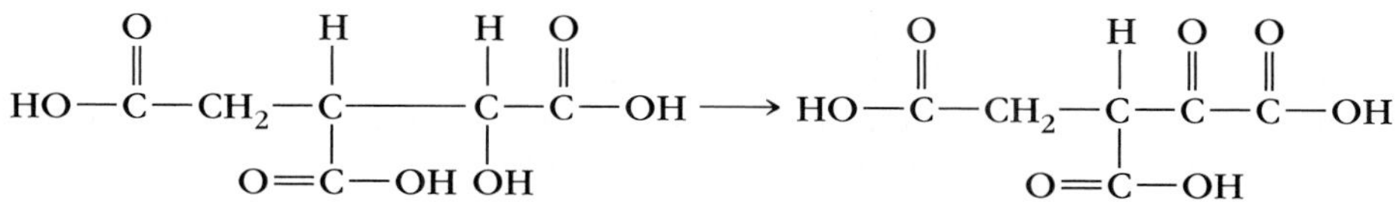

isocitric acid oxalosuccinic acid

In the last step of the cycle, L-malic acid is converted into oxaloacetic acid.

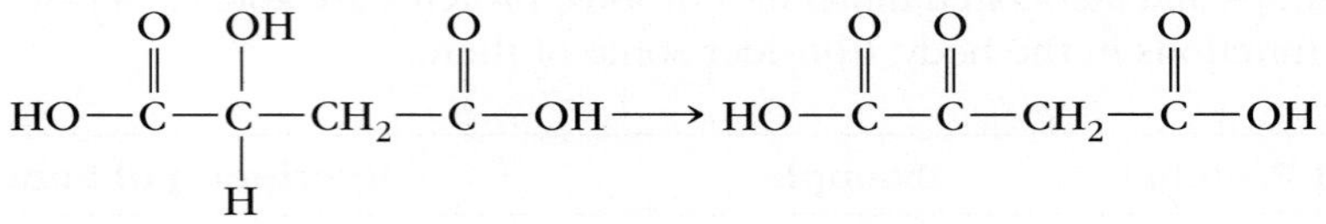

L-malic acid   oxaloacetic acid

(a) Identify the general type of reaction in each step.

(b) What do the two steps have in common?

(c) Identify any chiral carbon atoms ***(← p. 388)*** in the structures above in Part II.

Part III

Select any four amino acids from Table 12.8. Write the structural formulas of four different tetrapeptides that could be formed from these four amino acids.

IN CLOSING

Having studied this chapter, you should be able to . . .

- Describe petroleum refining and methods used to improve the gasoline fraction (Section 12.1).
- Identify the major U.S. energy sources (Section 12.2).
- Identify organic chemicals derived from coal (Section 12.2).
- Identify major organic chemicals of industrial and economic importance (Section 12.3).
- Name and draw the structures of three functional groups produced by the oxidation of alcohols (Section 12.4). Questions 31, 33
- Name and give examples of the uses of some important alcohols (Section 12.4).
- Explain how nuclear magnetic resonance works and how it is used in chemical analysis and medical diagnosis *(Tools of Chemistry)*.
- List some properties of carboxylic acids, and write equations for the formation of esters from carboxylic acids and alcohols (Section 12.5). Questions 41, 43
- Use Lewis structures to illustrate triglyceride formation (12.5).
- Differentiate among saturated, monounsaturated, and polyunsaturated fats (Section 12.5).
- Explain the formation of polymers by addition or condensation polymerization; give examples of synthetic polymers formed by each type of reaction (Section 12.6). Questions 47, 55, 65, 94
- Draw the structures of the repeating units in some common types of synthetic polymers and identify the monomers that form them (Section 12.6). Question 52
- Identify or write the structures of the functional groups in alcohols, aldehydes, ketones, carboxylic acids, esters, and amines (Sections 12.4–12.6). Question 71
- Identify the types of plastics being recycled most successfully (Section 12.6). Question 68
- Identify polysaccharides, their sources, the different ways they are linked, and the different uses resulting from these linkages (Section 12.7).
- Illustrate the basics of protein structures and how peptide linkages hold amino acids together in proteins (Section 12.7). Question 72
- Differentiate among the primary, secondary, and tertiary structures of proteins (Section 12.7).

OWL and **go Chemistry**

Sign in at **www.cengage.com/owl** to:

- View tutorials and simulations, develop problem-solving skills, and complete online homework assigned by your professor.
- For quick review and exam prep, download Go Chemistry mini lecture modules from OWL (or purchase them at **www.CengageBrain.com**).

KEY TERMS

addition polymer *(Section 12.6)*
aldehyde *(12.4)*
alpha carbon *(12.7)*
amide *(12.6)*
amide linkage *(12.6)*
amine *(12.6)*
amino acid *(12.7)*
carboxylic acid *(12.4)*
catalyst *(12.1)*
catalytic cracking *(12.1)*
catalytic reforming *(12.1)*
condensation polymer *(12.6)*
condensation reaction *(12.5)*
copolymer *(12.6)*
degree of polymerization, *n* *(12.6)*
disaccharide *(12.7)*
ester *(12.5)*
functional group *(12.3)*
glycosidic linkage *(12.7)*
hydrolysis *(12.5)*
ketone *(12.4)*
macromolecule *(12.6)*
monomer *(12.6)*
monosaccharide *(12.7)*
nuclear magnetic resonance (NMR) *(Tools of Chemistry, p. 552)*
octane number *(12.1)*
peptide linkage *(12.7)*
petroleum fractions *(12.1)*
polyamides *(12.6)*
polyester *(12.6)*
polymer *(12.6)*
polypeptide *(12.7)*
polysaccharide *(12.7)*
primary structure (proteins) *(12.7)*
saponification *(12.5)*
saturated fats *(12.5)*
secondary structure (proteins) *(12.7)*
tertiary structure (proteins) *(12.7)*
thermoplastics *(12.6)*
thermosetting plastics *(12.6)*
triglycerides *(12.5)*
unsaturated fats *(12.5)*

QUESTIONS FOR REVIEW AND THOUGHT

OWL Interactive versions of these problems are assignable in OWL.

Blue-numbered questions have short answers at the back of this book in Appendix M and fully worked solutions in the *Student Solutions Manual.*

Review Questions

These questions test vocabulary and simple concepts.

1. Why is the organic chemical industry referred to as the petrochemical industry?
2. What products are produced by the petrochemical industry?
3. What is the difference between catalytic cracking and catalytic reforming?
4. Explain how the octane number of a gasoline is determined.
5. Why is coal receiving increased attention as a source of organic compounds?
6. What is the difference between oxygenated gasoline and reformulated gasoline? Why are they being produced?
7. Explain why world use of natural gas is predicted to double between 1990 and 2010.
8. Table 12.1 lists several compounds with octane numbers above 100 and one compound with an octane number below zero. Explain why such values are possible.
9. Explain why methanol has a lower boiling point (65.0 °C) than water (100 °C).
10. Explain why ethylene glycol has a higher boiling point than ethanol.
11. Outline the steps necessary to obtain 89-octane gasoline, starting with a barrel of crude oil.
12. What is the major difference between crude oil and coal as a source of hydrocarbons?
13. Write the structural formula of a representative compound for each of these classes of organic compounds: alcohols, aldehydes, ketones, carboxylic acids, esters, and amines.
14. Explain why esters have lower boiling points than carboxylic acids with the same number of electrons.
15. What structural feature must a molecule have to undergo addition polymerization?
16. What feature do all condensation polymerization reactions have in common?
17. Give examples of (a) a synthetic addition polymer, (b) a synthetic condensation polymer, and (c) a natural addition polymer.
18. How does *cis-trans* isomerism affect the properties of rubber?
19. Discuss which two plastics are currently being recycled the most successfully, and give examples of some products being made from these recycled plastics.
20. What is the difference between the formation of an addition polymer and a condensation polymer?

Topical Questions

These questions are keyed to the major topics in the chapter. Usually a question that is answered at the back of this book is paired with a similar one that is not.

Petroleum and Fuels (Sections 12.1, 12.2)

21. What are petroleum fractions? What process is used to produce them?

22. (a) What is the boiling point range for the petroleum fraction containing the hydrocarbons that will provide fuel for your car?
    (b) Would you expect the octane rating from the "straight-run" gasoline obtained by fractional distillation of petroleum to be greater than 87 octane? Explain your answer.
    (c) Would you use this fraction to fuel your car? Why or why not?
23. (a) Draw the Lewis structure for the hydrocarbon that is assigned an octane rating of 0.
    (b) Draw the Lewis structure for the hydrocarbon that is assigned an octane rating of 100.
    (c) What is the boiling point for each of these hydrocarbons?
24. Explain what is meant by this statement: "All gasolines are highly volatile."
25. What would be the advantage of removing the higher-octane components such as aromatics and alkenes from oxygenated gasolines?
26. Compare and contrast catalytic cracking and catalytic reforming. (a) How are they similar? (b) How do they differ?

Alcohols (Section 12.4)

27. Give an example of (a) a primary alcohol, (b) a secondary alcohol, and (c) a tertiary alcohol. Draw Lewis structures for each example.
28. Classify each of these alcohols as primary, secondary, or tertiary.

(a) $CH_3CH_2CH_2CH_2OH$

(b) 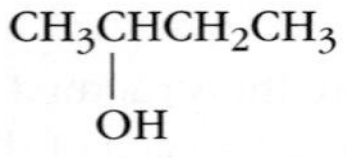

(c) 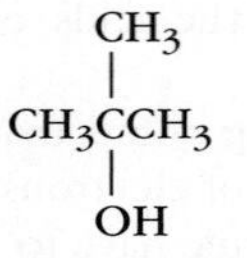

(d) 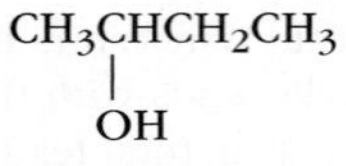

(e) 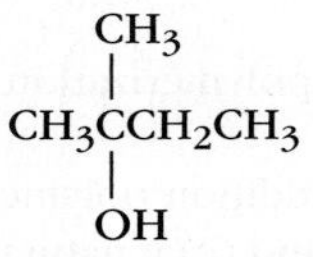

29. Classify each of these alcohols as primary, secondary, or tertiary.

(a) 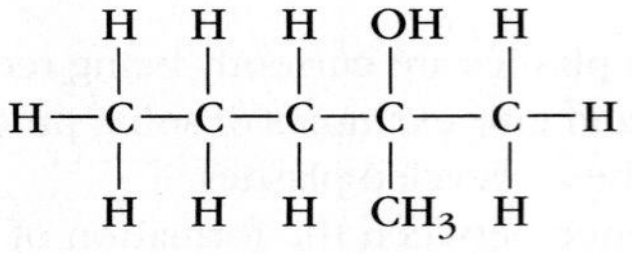

(b) 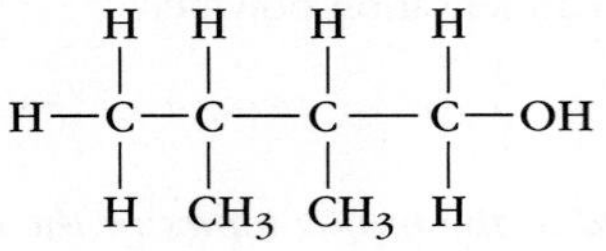

(c) 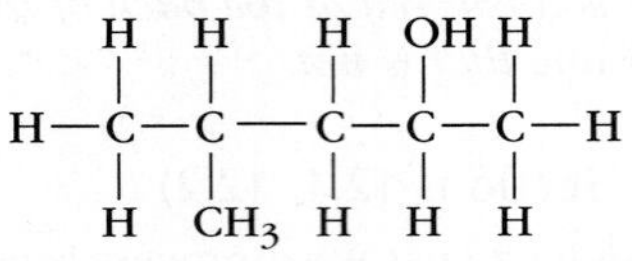

(d) 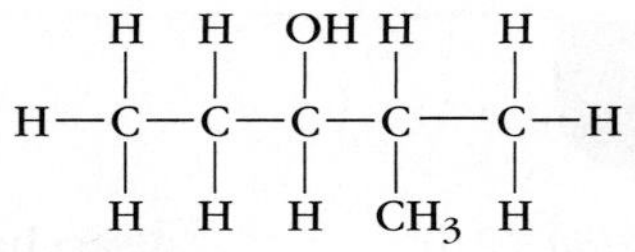

(e) 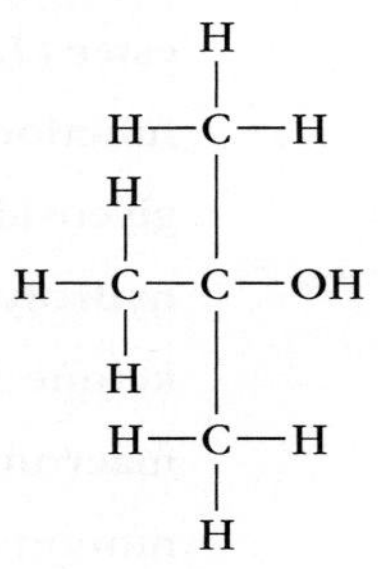

(f) 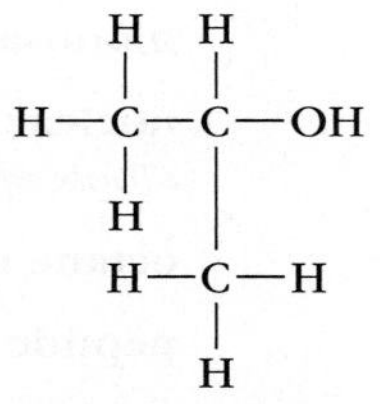

30. Explain what *oxidation* of organic compounds usually involves. What is meant by *reduction* of organic compounds?
31. Draw the structures of the first two oxidation products of each of these alcohols.
    (a) CH_3CH_2OH (b) $CH_3CH_2CH_2CH_2OH$
32. Draw the structures of the oxidation products of each of these alcohols.

(a) 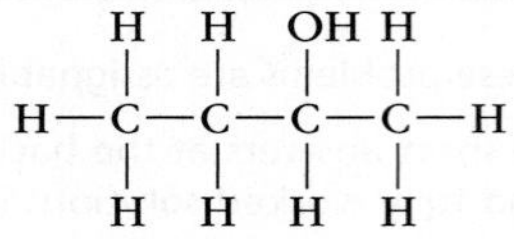

(b) 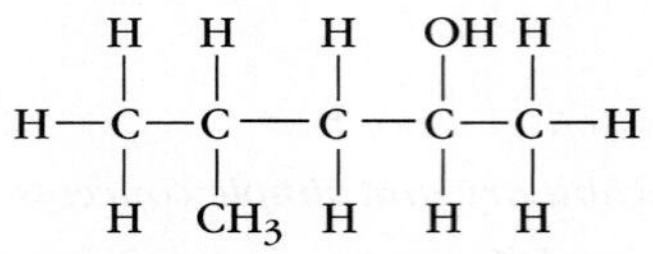

33. Write the condensed structural formula of the alcohols that can be oxidized to make these compounds.

(a) 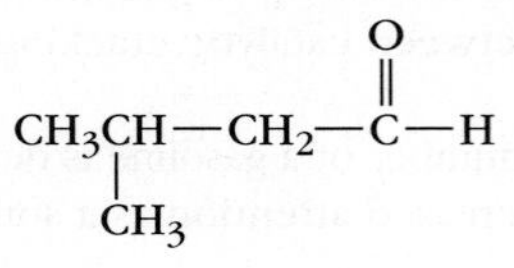

(b) 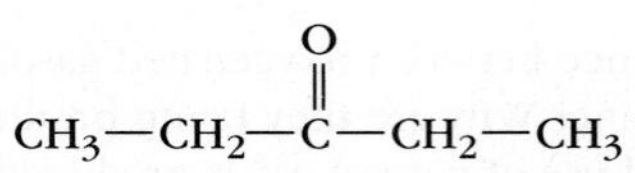

(c) 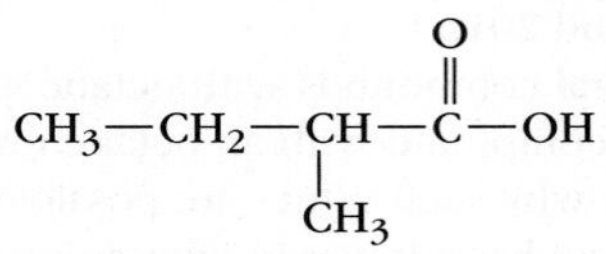

34. What is the percentage of ethanol in 90-proof vodka?
35. Explain how the common name *grain alcohol* for ethanol came about.
36. What is denatured alcohol? Why is it made?
37. Many biological molecules, including steroids and carbohydrates, contain many —OH groups. What need might biological systems have for this particular functional group?

Tools of Chemistry: NMR and MRI

38. What kind of electromagnetic radiation is used in nuclear magnetic resonance (NMR)?
39. In magnetic resonance imaging (MRI), the intensity of the emitted signal is related to the ________ of hydrogen nuclei and the relaxation time is related to the type of tissue being examined.

Carboxylic Acids and Esters (Section 12.5)

40. Explain why the boiling points for carboxylic acids are higher than those for alcohols with comparable molar masses.
41. Write the structural formula of the esters that can be formed from these reactions.
    (a) $CH_3COOH + CH_3CH_2OH$
    (b) $CH_3CH_2COOH + CH_3CH_2CH_2OH$
    (c) $CH_3CH_2COOH + CH_3OH$
42. Write the structural formula for the esters that can be produced by these reactions.
    (a) Formic acid + methanol
    (b) Butyric acid + ethanol
    (c) Acetic acid + 1-butanol
    (d) Propanoic acid + 2-propanol
43. Write the condensed formula of the alcohol and acid that will react to form each of these esters.
    (a) $CH_3CH_2C(=O)—OCH_3$
    (b) $HC(=O)—OCH_2CH_3$
    (c) $CH_3C(=O)—OCH_2CH_3$
44. Explain why carboxylic acids are more soluble in water than are esters with the same molar mass.

Synthetic Organic Polymers (Section 12.6)

45. What are some examples of thermoplastics? What are the properties of thermoplastics when heated and cooled?
46. What are some examples of thermosetting plastics? What are the properties of thermosetting plastics when heated and cooled?
47. Draw the structure of the repeating unit in a polymer in which the monomer is
    (a) 1-butene (b) 1,1-dichloroethylene
    (c) Vinyl acetate
48. What is the principal structural difference between low-density and high-density polyethylene? Is polyethylene an addition or a condensation polymer?
49. Methyl methacrylate has the structural formula shown in Table 12.6 (p. 564). When polymerized, it is very transparent, and it is sold in the United States under the trade names Lucite and Plexiglas.
    (a) Use structural formulas to write the chemical equation for the formation of a segment of poly(methyl methacrylate) having four monomer units.
    (b) Identify the repeating unit in the polymer.
50. Use structural formulas and data from Table 12.6 (p. 564) to write the chemical equation for the formation of a polymer of five monomer units from these monomers.
    (a) Acrylonitrile
    (b) Propylene
51. Use structural formulas and data from Table 12.6 (p. 564) to write the chemical equation for the formation of a polymer of five monomer units from these monomers.
    (a) Vinylidene chloride, $H_2C{=}CCl_2$
    (b) Tetrafluoroethylene
52. What monomers are used to prepare these polymers?
    (a) $—CH_2CH_2CH_2CH_2CH_2CH_2CH_2CH_2CH_2—$
    (b) 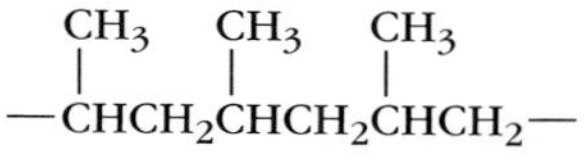
    (c) 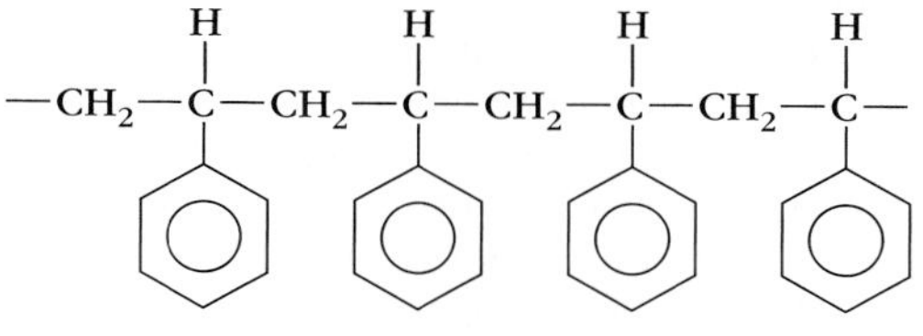
    (d) 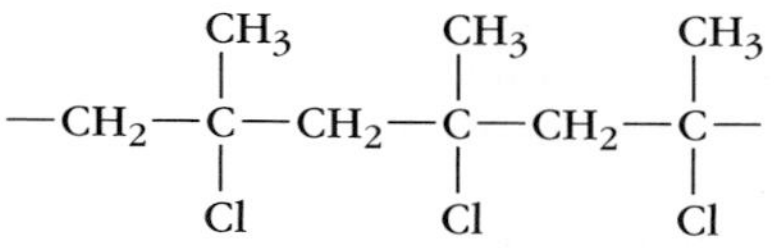
    (e) 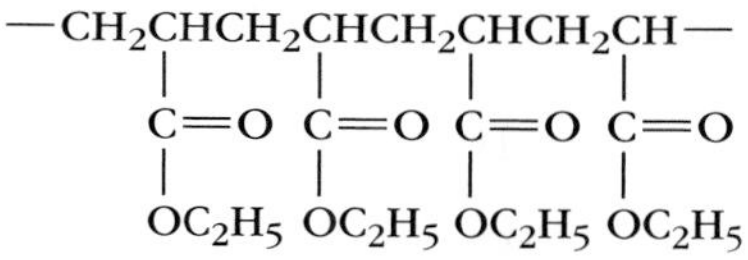
53. What is the monomer in *natural rubber?* Which isomer is present in natural rubber, *cis* or *trans?*
54. Write the structural formula of four units of the polymer made from the reaction of $H_2N—(CH_2)_4—NH_2$ and HOOCCOOH. Indicate the repeating unit of the polymer.
55. 4-Hydroxybenzoic acid can form a condensation polymer. Use structural formulas to write the chemical equation for the formation of four units of the polymer. Indicate the repeating unit of the polymer.

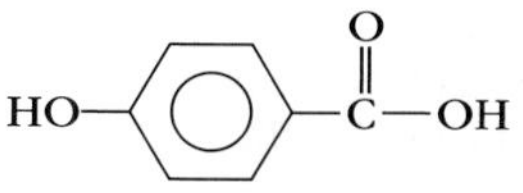

4-hydroxybenzoic acid

56. What are the two monomers used to make SBR?
57. The formation of polyesters involves which two functional groups?
58. Name one important polyester polymer and its uses.
59. Polyamides are made by condensing which functional groups? Name the most common example of this class of synthetic polymers.
60. How are amide linkages and peptide linkages similar? How are they different?
61. State one major difference between proteins and unstructured polyamides.

62. Draw structures of monomers that could form each of these condensation polymers.

(a) 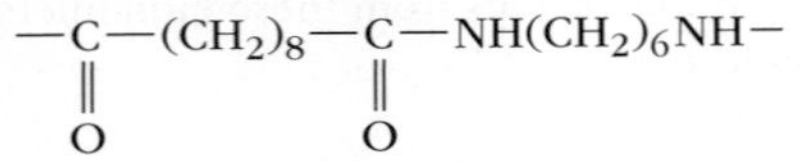

(b) 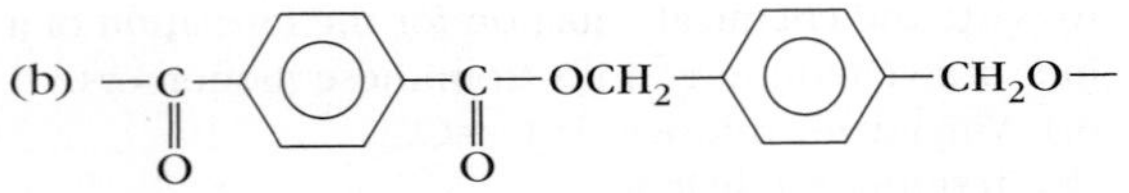

63. Orlon has this polymeric chain structure:

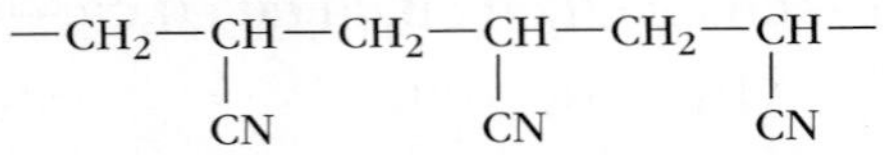

What is the monomer from which this structure can be made?

64. How many ethylene units are in a polyethylene molecule that has a molecular weight of approximately 42,000?

65. A sample of high-molecular-weight (HMW) polyethylene has a molecular weight of approximately 450,000. Determine its degree of polymerization.

66. Ultrahigh-molecular-weight polyethylene has high wear resistance and is used in conveyor belts. A sample of the polymer has approximately 150,000 monomer units per chain. Calculate the approximate molecular weight of this polymer.

67. Write the structural formulas for the repeating units of these compounds.
(a) Natural rubber
(b) Neoprene 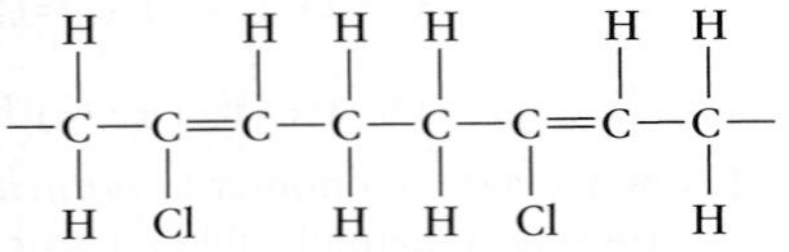
(c) Polybutadiene

68. What are some major end uses for recycled PET and HDPE polymers?

Polysaccharides and Proteins (Section 12.7)

69. Which biological molecules have monomer units that are all alike, as in synthetic polymers?

70. Which biological molecules have monomer units that are not all alike, as in synthetic copolymers?

71. Identify and name all the functional groups in this tripeptide.

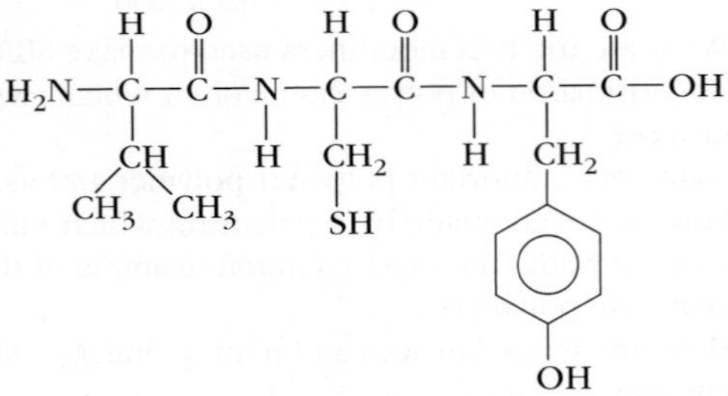

72. Draw the structural formula of alanylglycylphenylalanine.

73. Draw the structural formula of leucylmethionylalanylserine.

74. Explain the difference between (a) monosaccharides and disaccharides, and (b) disaccharides and polysaccharides.

75. What is the chief function of glycogen in animal tissue?

76. What polysaccharides yield only D-glucose upon complete hydrolysis?

77. (a) How do amylose and amylopectin differ?
(b) How are they similar?
(c) Are amylose and glycogen similar?

78. (a) Explain why humans can use glycogen but not cellulose for energy.
(b) Why can cows digest cellulose?

Tools of Chemistry: Gas Chromatography

79. If you were to analyze an oxygenated gasoline using GC, would you use a less polar or more polar stationary phase than the one used for unoxygenated gasoline?

80. Would polar compounds appear earlier or later on a chromatogram if a nonpolar stationary phase were used? Explain your answer.

General Questions

These questions are not explicitly keyed to chapter topics; many require integration of several concepts.

81. Compounds A and B both have the molecular formula C_2H_6O. The boiling points of compounds A and B are 78.5 °C and −23.7 °C, respectively. Use the table of functional groups in Appendix E and write the structural formulas and names of the two compounds.

82. Explain why ethanol, CH_3CH_2OH, is soluble in water in all proportions, but decanol, $CH_3(CH_2)_9OH$, is almost insoluble in water.

83. Nitrile rubber (Buna N) is a copolymer of two parts 1,3-butadiene to one part acrylonitrile. Draw the repeating unit of this polymer.

84. How are rubber molecules modified by vulcanization?

85. Write the condensed structural formula for 3-ethyl-5-methyl-3-hexanol. Is this a primary, secondary, or tertiary alcohol?

86. Using structural formulas, write a reaction for the hydrolysis of a triglyceride that contains fatty acid chains, each consisting of 16 total carbon atoms.

87. Is the plastic wrap used in covering food a thermoplastic or thermosetting plastic? Explain.

88. Assume that a car burns pure octane,

$$C_8H_{18}\ (d = 0.703\ g/cm^3)$$

(a) Write the balanced equation for burning octane in air, forming CO_2 and H_2O.
(b) If the car has a fuel efficiency of 32 miles per gallon of octane, what volume of CO_2 at 25 °C and 1.0 atm is generated when the car goes on a 10.0-mile trip? (The volume of 1 mol $CO_2(g)$ at 25 °C and 1 atm is 24.5 L.)

89. Perform the same calculations as in Question 88, but use methanol, CH_3OH (d = 0.791 g/cm³), as the fuel. Assume the fuel efficiency is 20.0 miles per gallon.

90. Show structurally why glycogen forms granules when stored in the liver, but cellulose is found in cell walls as sheets.

91. Polytetrafluoroethylene (Teflon) is made by first treating HF with chloroform, then cracking the resultant difluorochloromethane.

$$CHCl_3 + 2\ HF \longrightarrow CHClF_2 + 2\ HCl$$
$$2\ CHClF_2 + \text{heat} \longrightarrow F_2C{=}CF_2 + 2\ HCl$$
$$F_2C{=}CF_2 \xrightarrow{\text{peroxide catalyst}} \text{Teflon}$$

If you wish to make 1.0 kg Teflon, what mass of chloroform and HF must you use to make the starting material, $CHClF_2$? (Although it is not realistic, assume that each reaction step proceeds to a 100% yield.)

Applying Concepts

These questions test conceptual learning.

92. Hydrogen bonds can form between propanoic acid molecules and between 1-butanol molecules. Draw all the propanoic acid molecules that can hydrogen-bond to the one shown below. Draw all the 1-butanol molecules that can hydrogen-bond to the one shown below. Use dotted lines to represent the hydrogen bonds.

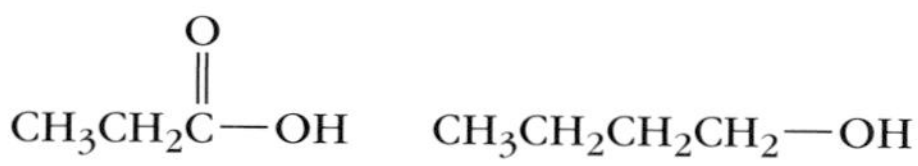

propanoic acid 1-butanol

Based on your drawings, which should have the higher boiling point, propanoic acid or 1-butanol? Explain your reasoning.

93. Both propanoic acid and ethyl methanoate form hydrogen bonds with water. Draw all the water molecules that can hydrogen-bond to these molecules. Use dotted lines to represent the hydrogen bonds.

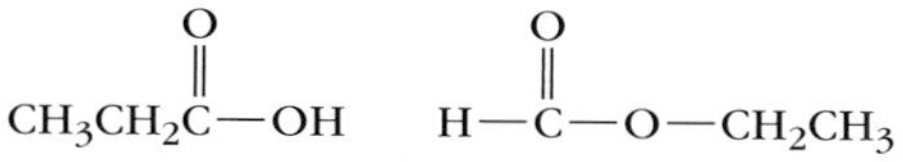

propanoic acid ethyl methanoate

Based on your drawings, which should be more soluble in water: propanoic acid or ethyl methanoate? Explain your reasoning.

94. What monomer formed this polymer?

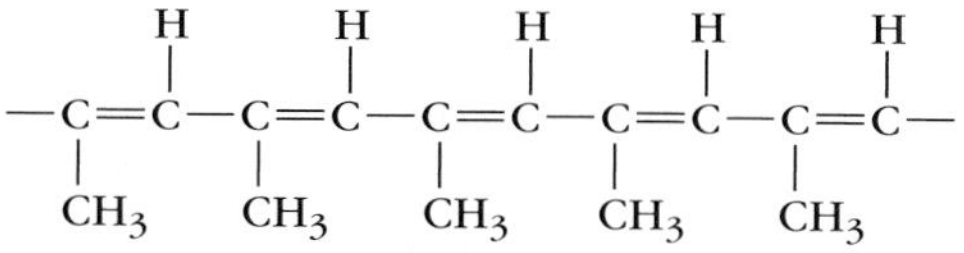

95. The illustrations at the top of the next column represent two different samples of polyethylene, each with the same number of monomer units. Based on the concept of density and not structure, which one is high-density polyethylene and which is low-density polyethylene? Write a brief explanation.

(a)

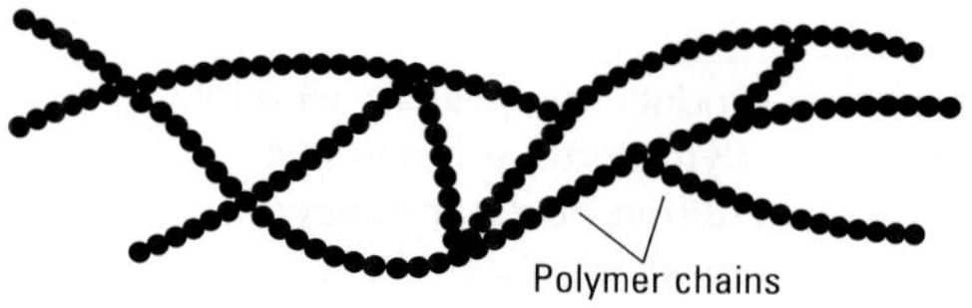

(b)

96. The backbone of a DNA molecule is a polymer of alternating sugar (deoxyribose) and phosphoric acid units held together by a phosphate ester bond. Draw a segment of the polymer consisting of at least two sugar and two phosphate units. Circle the phosphate ester bonds.

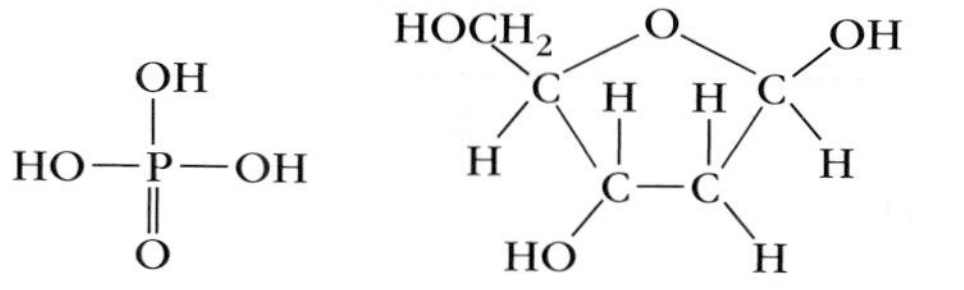

phosphoric acid deoxyribose

97. Draw the structure of a molecule that could undergo a condensation reaction with itself to form a polyester. Draw a segment of the polymer consisting of at least five monomer units.

98. It has been asserted that the photosynthesis of the trees in a forest the size of Australia would be needed to compensate for the additional CO_2 put into the atmosphere each year from burning fossil fuels. Identify the data that would be required to check whether this assertion is valid.

99. Towels made from polyester fibers are less water absorbent than towels made from cotton. Explain this difference on a molecular basis.

100. Below is a grid with nine lettered boxes, each of which contains an item that is used to answer the questions that follow. Items may be used more than once and there may be more that one correct item in response to each question (see p. 590).

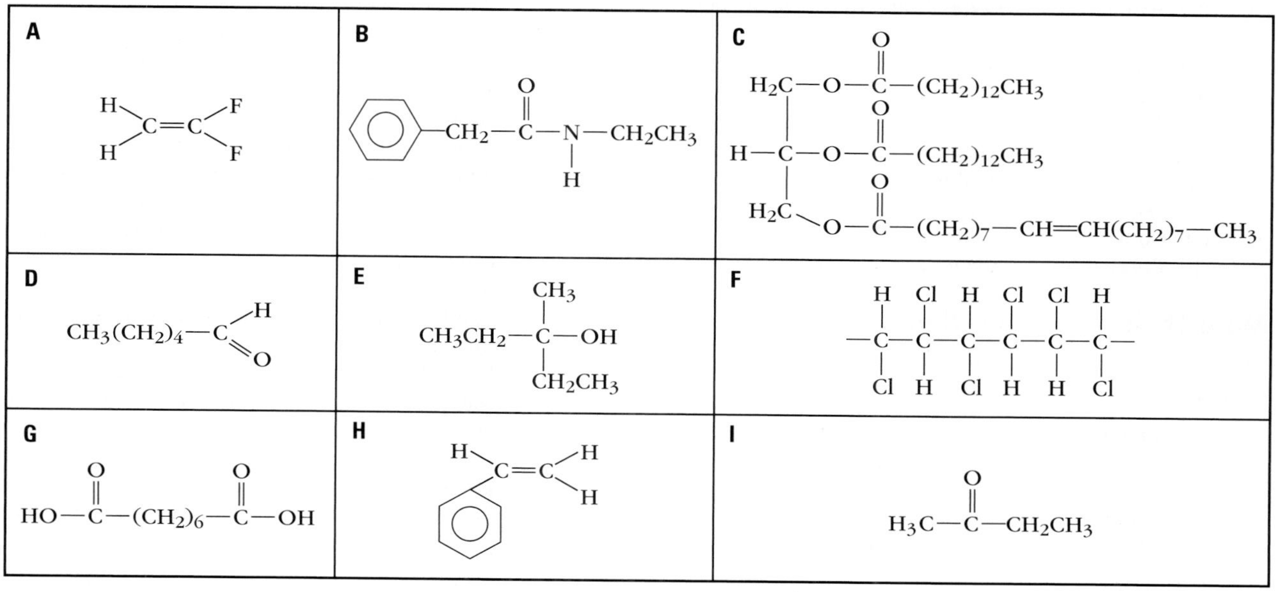

Place the letter(s) of the correct selection(s) (from the grid on p. 589) on the appropriate line.
(a) A triglyceride ________
(b) Final oxidation product of a primary alcohol ________
(c) Condensation polymerization monomer ________
(d) Segment of an addition polymer molecule ________
(e) Initial oxidation product of a primary alcohol ________
(f) Contains an amide bond ________
(g) Addition polymerization monomer ________
(h) Contains an ester linkage ________
(i) Final oxidation product of a secondary alcohol ________
(j) A tertiary alcohol ________

101. Below is a grid with nine lettered boxes, each of which contains an item that is used to answer the questions that follow. Items may be used more than once and there may be more that one correct item in response to a question.

A $CH_3CH_2CH_2OH$	**B** $H_2N—(CH_2)_8—NH_2$	**C** 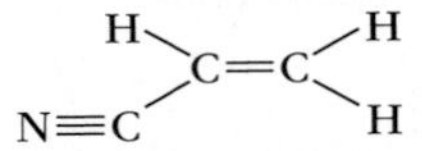
D 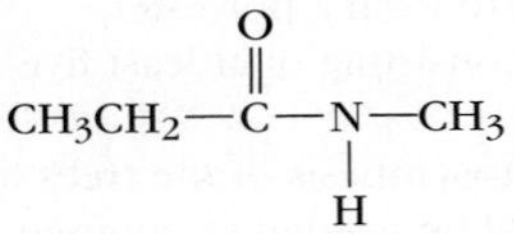	**E** 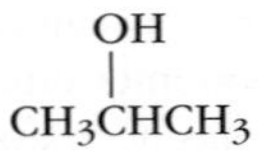	**F** 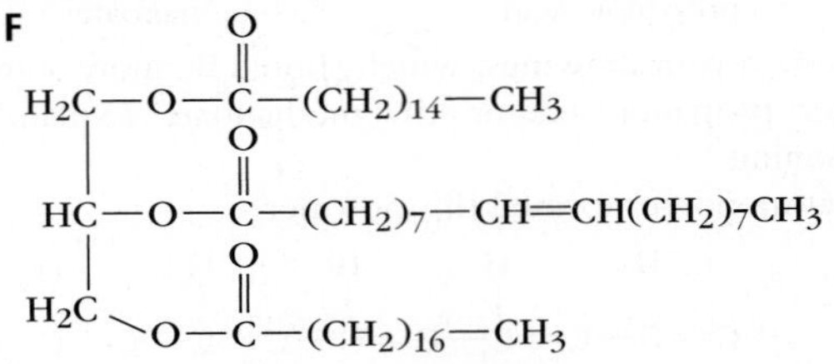
G 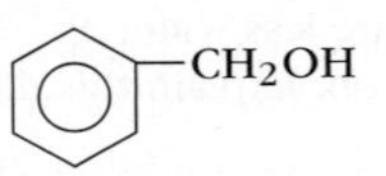	**H** 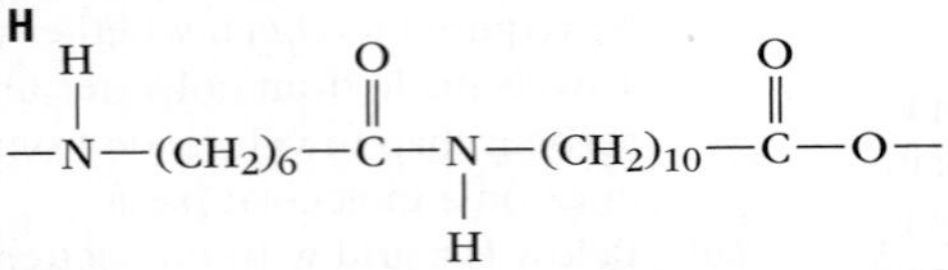	**I** 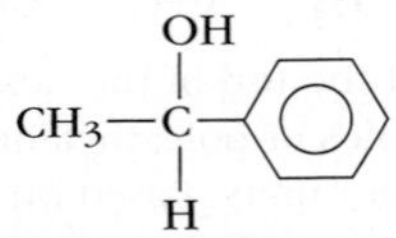

Place the letter(s) of the correct selection(s) on the appropriate line.
(a) Can be oxidized to an aldehyde ________
(b) Segment of a condensation polymer molecule ________
(c) Can be oxidized to a carboxylic acid ________
(d) A primary alcohol ________
(e) A triglyceride ________
(f) Addition polymerization monomer ________
(g) Contains an amide linkage ________
(h) Oxidized to form a ketone ________
(i) Contains an ester linkage ________
(j) Condensation polymerization monomer ________

More Challenging Questions

These questions require more thought and integrate several concepts.

102. The total mass of carbon in living systems on Earth is estimated to be 7.5×10^{17} g; the total mass of carbon in all forms on Earth is estimated as 7.5×10^{22} g. Calculate the concentration of organic carbon as a fraction of total carbon. Express your answer as percent and in parts per million (ppm).

103. In the laboratory, a student was given three bottles labeled X, Y, and Z. One of the bottles contained acetic acid, the second acetaldehyde, and the third ethanol. The student ran a series of experiments to determine the contents of each bottle. Substance X reacted with substance Y to form an ester under certain conditions and substance Y formed an acidic solution when dissolved in water. (a) Identify which compound is in which bottle, and write the Lewis structural formula for each compound. (b) Using structural formulas, write a chemical equation for the ester formation reaction and for the ionization of the acid. (c) The student confirmed the identifications by treating each compound with a strong oxidizing agent and found that compound X required roughly twice the amount of oxidizing agent as compound Z. Explain why this was the case.

104. In his 1989 essay "The End of Nature," William McKibben states that ". . . a clean burning automobile engine will emit 5.5 lb of carbon in the form of carbon dioxide for every gallon of gasoline it consumes." Assume that the gasoline is C_8H_{18} and its density is 0.703 g/mL.
(a) Write a balanced equation for the complete combustion of gasoline.

(b) Do calculations that either corroborate or refute McKibben's assertion. Is his assertion refuted or corroborated?

105. A 1.685-g sample of a hydrocarbon is burned completely to form 5.287 g carbon dioxide and 2.164 g water.
(a) Determine the empirical formula of the hydrocarbon.
(b) Identify whether it is an alkane or an alkene.
(c) Write a plausible Lewis structure for it.

106. Suppose 2.511 g of a hydrocarbon is burned completely to form 7.720 g carbon dioxide and 3.612 g water.
(a) Determine the empirical formula of the hydrocarbon.
(b) Identify whether it is an alkane or an alkene.
(c) Write a plausible Lewis structure for it.

107. Consider these data:

Hydrocarbon	Formula	Enthalpy of Combustion (kJ/mol)
Butane	C_4H_{10}	2853.9
Pentane	C_5H_{12}	3505.8
Hexane	C_6H_{14}	4159.5
Heptane	C_7H_{16}	4812.8
Octane	C_8H_{18}	5465.7

Graph these data with enthalpy of combustion (kJ/mol C) on the *y*-axis and number of carbon atoms on the *x*-axis. Use the graph to estimate the enthalpy of combustion of propane, C_3H_8, nonane, C_9H_{20}, and hexadecane, $C_{16}H_{34}$.

108. Two different compounds each have the formula $C_5H_{10}O_2$. Each contains a C=O group. Write the Lewis structure of each compound and identify the functional group in each one.

109. Using structural formulas, show how vinyl chloride can polymerize to form three types of PVC polymers.

110. Explain why an amino acid has a higher boiling point than an amine with the same number of electrons.

111. Silicones are Si-containing polymers used in waterproof caulking. They do not contain carbon-to-carbon bonds.
(a) Silly Putty is a polymer made from the monomer dimethylsilanol, $(CH_3)_2Si(OH)_2$. Use structural formulas to write the structure of four units of this polymer.
(b) Identify the type of polymerization that occurs.

112. Polyacetylene is an electrically conducting polymer of acetylene, HC≡CH. Now used in batteries, the polymer was discovered accidentally in 1970 when far too much catalyst was used in an experimental attempt to polymerize acetylene into synthetic rubber. Write a plausible Lewis structure for a portion of a polyacetylene molecule.

113. Glycolic acid, $HOCH_2COOH$, and lactic acid form a copolymer used as absorbable stitches in surgery. Use structural formulas to write a chemical equation showing the formation of four units of the copolymer.

114. A particular condensation copolymer forms a silk-like synthetic fabric. A portion of the polymer's structure is shown below.

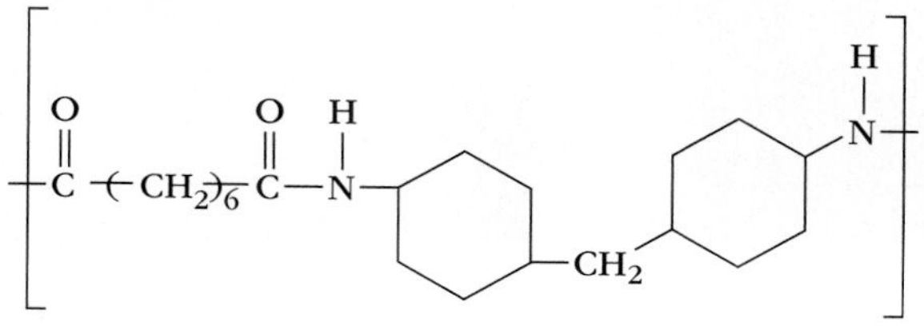

Write the structural formulas of the two monomers used to synthesize the polymer.

Conceptual Challenge Problems

These rigorous, thought-provoking problems integrate conceptual learning with problem solving and are suitable for group work.

CP12.A (Section 12.1) How are the boiling points of hydrocarbons during the distillation of petroleum related to their molecular size?

CP12.B (Section 12.4) Even though millions of organic compounds exist and each compound may have 10, 100, or even thousands of atoms bonded together to make one molecule, the reactions of organic compounds can be studied and even predicted for compounds yet to be discovered. What characteristic of organic compounds allows their reactions to be studied and predicted?

CP12.C (Section 12.5) What is the advantage of animals storing chemical potential energy in their bodies as triesters of glycerol and long-chain fatty acids, known as fats, instead of as carbohydrates?

13

Chemical Kinetics: Rates of Reactions

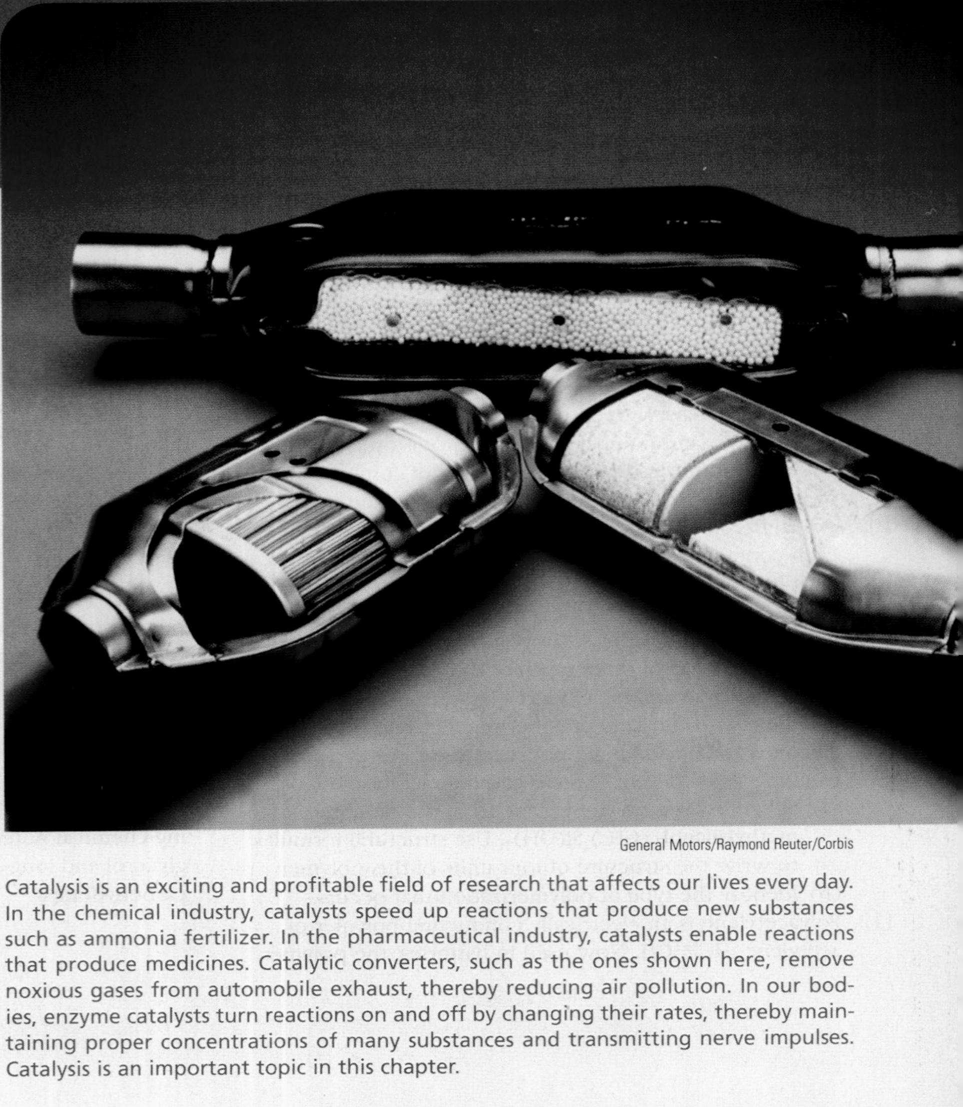

General Motors/Raymond Reuter/Corbis

Catalysis is an exciting and profitable field of research that affects our lives every day. In the chemical industry, catalysts speed up reactions that produce new substances such as ammonia fertilizer. In the pharmaceutical industry, catalysts enable reactions that produce medicines. Catalytic converters, such as the ones shown here, remove noxious gases from automobile exhaust, thereby reducing air pollution. In our bodies, enzyme catalysts turn reactions on and off by changing their rates, thereby maintaining proper concentrations of many substances and transmitting nerve impulses. Catalysis is an important topic in this chapter.

Turn on the valve of a Bunsen burner in your laboratory, bring up a lighted match, and a rapid combustion reaction begins with a whoosh:

$$CH_4(g) + 2\,O_2(g) \longrightarrow CO_2(g) + 2\,H_2O(g) \qquad \Delta H^\circ = -802.34\text{ kJ}$$

What would happen if you didn't put a lighted match in the methane-air stream? Nothing obvious. At room temperature the reaction of methane with oxygen is so slow that the two potential reactants can be mixed in a closed flask and stored unreacted for centuries. These facts about combustion of methane lie within the realm of **chemical kinetics**—*the study of* *the speeds of reactions and the nanoscale pathways or rearrangements by which atoms and molecules are transformed from reactants to products.*

Chemical kinetics is extremely important, because knowing about kinetics enables us to control many kinds of reactions in addition to combustion. In pharmaceutical chemistry an important problem is devising drugs that remain in their active form long enough to get to the site in the body where they are intended to act. Consequently, it is important to know whether a drug will react with other substances in the body and how long it will take to do so. In environmental chemistry, there was more than a decade of controversy over whether stratospheric ozone is being depleted by chlorofluorocarbons. Much of this debate hinged on verifying the sequence and rates of reactions by which stratospheric ozone is produced and consumed. Their careful studies of such reactions led to a Nobel Prize in Chemistry for Sherwood Rowland, Mario Molina, and Paul Crutzen ***(← p. 455)***.

This chapter focuses on the factors that affect the speeds of reactions, the nanoscale basis for understanding those factors, and their importance in modern society, from industrial plants to cars to the cells of our bodies.

OWL Sign in to OWL at **www.cengage.com/owl** to view tutorials and simulations, develop problem-solving skills, and complete online homework assigned by your professor.

go Chemistry Download mini lecture videos for key concept review and exam prep from OWL or purchase them from **www.CengageBrain.com.**

Companion Website
Visit this book's companion website at **www.cengage.com/chemistry/moore** to work interactive modules for the Estimation boxes and Active Figures in this text.

13.1 Reaction Rate

For a chemical reaction to occur, reactant molecules must come together so that their atoms can be exchanged or rearranged. Atoms and molecules are more mobile in the gas phase or in solution than in the solid phase, so reactions are often carried out in a mixture of gases or among solutes in a solution. For a **homogeneous reaction,** one in which reactants and products are all in the same phase (gas or solution, for example), four factors affect the speed of a reaction:

- The *properties* of reactants and products—in particular, their molecular structure and bonding
- The *concentrations* of the reactants and sometimes the products
- The *temperature* at which the reaction occurs
- The *presence* of a catalyst ***(← p. 537)*** and, if one is present, its concentration

A catalyst (described in Section 13.8) speeds up a reaction but undergoes no net chemical change itself. The catalytic converter in an automobile speeds up reactions that remove pollutants from the exhaust gases.

Many important reactions, including the ones in catalytic converters that remove air pollutants from automobile exhaust, are **heterogeneous reactions.** They take place at a surface—at an interface between two different phases (solid and gas, for example). The speed of a heterogeneous reaction depends on the four factors listed above as well as on the area and nature of the surface at which the reaction occurs. For example, very finely divided metal powder can burn very rapidly, whereas a pile of powder with much less surface exposed to oxygen in the air is difficult to ignite (Figure 13.1). The much more rapid reaction when greater surface is exposed has been responsible for explosions in grain elevators, sugar mills, and coal mines where finely divided, combustible solids in the air are exposed to a spark or flame.

© Cengage Learning/Charles D. Winters

Sugar mill explosion. Finely powdered sugar ignited and exploded, causing the destruction of this sugar mill.

The speed of any process is expressed as its **rate,** which is *the change in some measurable quantity per unit of time*. A car's rate of travel, for example, is found by measuring the change in its position, Δx, during a given time interval, Δt. Suppose you are driving on an interstate highway. If you pass mile marker 43 at 2:00 PM and mile marker 173 at 4:00 PM, $\Delta x = (173 - 43)$ mi $= 130$ mi and $\Delta t = 2.00$ h. You are traveling at an average rate of $\Delta x/\Delta t = 65$ mi/h. For a chemical process, the **reaction rate** is defined as *the change in concentration of a reactant or product per unit time*. (Time can be measured in seconds, hours, days, or whatever unit is most convenient for the speed of the reaction.)

Change in concentration is used (rather than change in total amount of reactant) because using change in concentration makes the rate independent of the volume of the reaction mixture.

As an example of measurements made in chemical kinetics, consider Figure 13.2, which shows the violet-colored dye crystal violet reacting with aqueous sodium hydroxide to form a colorless product. The dye's violet color disappears over time, and the intensity of color can be used to determine the concentration of

Recall that the Greek letter Δ (delta) means that a change in some quantity has been measured ***(← p. 219)***. As usual, Δ means to subtract the initial value of the quantity from the final value.

© Cengage Learning/Charles D. Winters

Figure 13.1 Combustion of iron powder. (a) The very finely divided metal powder burns slowly in a pile where only a relatively small surface area is exposed to air. (b) Pouring the same powder through the air greatly increases the exposed surface. When the powder enters a flame, combustion is rapid—even explosive.

the dye. Crystal violet consists of polyatomic positive ions that we abbreviate as Cv^+, and chloride ions, Cl^-. Its formula is therefore represented by CvCl. The Cv^+ ions cause the beautiful violet color. They combine with hydroxide ions, OH^-, to form a colorless, uncharged product abbreviated as CvOH. The reaction can be represented by the net ionic equation

$$Cv^+(aq) + OH^-(aq) \longrightarrow CvOH(aq) \qquad [13.1]$$

The rate at which this reaction occurs can be calculated by dividing the change in concentration of crystal violet cation, $\Delta[Cv^+]$, by the elapsed time, Δt. (Concentration is indicated by square brackets.) For example, if the concentration of crystal violet is measured at some time t_1 to give $[Cv^+]_1$, and the measurement is repeated at a subsequent time t_2 to give $[Cv^+]_2$, then the rate of reaction is

$$\text{Rate of change of concentration of } Cv^+ = \frac{\text{change in concentration of } Cv^+}{\text{elapsed time}}$$

$$= \frac{\Delta[Cv^+]}{\Delta t} = \frac{[Cv^+]_2 - [Cv^+]_1}{t_2 - t_1}$$

The experimentally measured concentration of crystal violet as a function of time is shown in Table 13.1. Because the concentration of crystal violet decreases

Table 13.1 Concentration-Time Data for Reaction of Crystal Violet with 0.10-M NaOH(aq) at 23 °C

Time, t (s)	Concentration of Crystal Violet Cation, $[Cv^+]$ (mol/L)	Average Rate (mol L^{-1} s^{-1})
0.0	5.000×10^{-5}	
		13.2×10^{-7}
10.0	3.680×10^{-5}	
		9.70×10^{-7}
20.0	2.710×10^{-5}	
		7.20×10^{-7}
30.0	1.990×10^{-5}	
		5.30×10^{-7}
40.0	1.460×10^{-5}	
		3.82×10^{-7}
50.0	1.078×10^{-5}	
		2.85×10^{-7}
60.0	0.793×10^{-5}	
		1.82×10^{-7}
80.0	0.429×10^{-5}	
		0.985×10^{-7}
100.0	0.232×10^{-5}	

© Cengage Learning/Charles D. Winters

Figure 13.2 Disappearance of a dye. Violet-colored crystal violet dye in aqueous solution is poured into aqueous sodium hydroxide. They react, converting the dye into a colorless product. The intensity of the solution's color decreases and eventually the color disappears. The rate of the reaction can be determined by repeated, simultaneous measurements of the intensity of color and the time. From the intensity of color the concentration of dye can be calculated, so concentration can be determined as a function of time.

as time increases, $[Cv^+]_2$ is smaller than $[Cv^+]_1$. Therefore $\Delta[Cv^+]/\Delta t$ is negative. *By convention, reaction rate is defined as a positive quantity.* Therefore, for the crystal violet reaction the rate is defined as

$$\text{Reaction rate} = -\frac{\Delta[Cv^+]}{\Delta t}$$

The negative sign converts a negative $\Delta[Cv^+]/\Delta t$ value to a positive one.

Table 13.1 also shows calculated values of reaction rates for each time interval. Because calculating the rate involves dividing a concentration difference by a time difference, the units of reaction rate are units of concentration divided by units of time, in this case mol/L divided by s, mol L^{-1} s^{-1}.

PROBLEM-SOLVING EXAMPLE 13.1 Calculating Average Rates

Using the data in the first two columns of Table 13.1, calculate $\Delta[Cv^+]$, Δt, and the average rate of reaction for each time interval given. Use the numbers given in the third column to check your results. (A good way to do so is to use a computer spreadsheet program.)

Answer See the third column in Table 13.1.

Strategy and Explanation The time interval from 80.0 s to 100.0 s provides an example of the calculation.

$$\Delta[Cv^+] = 0.232 \times 10^{-5}\text{ mol/L} - 0.429 \times 10^{-5}\text{ mol/L} = -0.197 \times 10^{-5}\text{ mol/L}$$

$$\Delta t = 100.0\text{ s} - 80.0\text{ s} = 20.0\text{ s}$$

$$\text{Rate} = -\frac{\Delta[Cv^+]}{\Delta t} = -\frac{(-0.197 \times 10^{-5}\text{ mol/L})}{20.0\text{ s}} = 0.985 \times 10^{-7}\text{ mol L}^{-1}\text{ s}^{-1}$$

Do the other calculations in a similar way.

☑ **Reasonable Answer Check** The rates of reaction should be positive numbers and should have units of concentration divided by time (such as mol L^{-1} s^{-1}). Both of these conditions are met.

The **PROBLEM-SOLVING STRATEGY** in this book is

- Analyze the problem
- Plan a solution
- Execute the plan
- Check that the result is reasonable

Appendix A.1 explains this in detail.

PROBLEM-SOLVING PRACTICE 13.1

For the reaction of crystal violet with NaOH(aq), the measured rate of reaction is 1.27×10^{-6} mol L^{-1} s^{-1} when the concentration of crystal violet cation is 4.13×10^{-5} mol/L.

(a) Estimate how long it will take for the concentration of crystal violet to drop from 4.30×10^{-5} mol/L to 3.96×10^{-5} mol/L.

(b) Could you use the same method to make an accurate estimate of how long it would take for the concentration of crystal violet to drop from 4.30×10^{-5} mol/L to 0.43×10^{-5} mol/L? Explain why or why not.

PROBLEM-SOLVING PRACTICE answers are provided at the back of this book in Appendix K.

EXERCISE 13.1 Rates of Reaction

(a) From data in Table 13.1, calculate the rate of reaction for each time interval: (i) from 40.0 s to 60.0 s; (ii) from 20.0 s to 80.0 s; (iii) from 0.0 to 100.0 s.

(b) Use all of the data in the first two columns of Table 13.1 to draw a graph with time on the horizontal (x) axis and concentration on the vertical (y) axis. Draw a smooth curve through the data. On the graph, draw lines that correspond to $\Delta[Cv^+]/\Delta t$ for each interval.

(c) Why is the rate not the same for each time interval in part (a), even though the average time for each interval is 50.0 s? (That is, for interval i, the average time is (40.0 s + 60.0 s)/2 = 50.0 s.) Write an explanation of the reason for a friend who is taking this course, and ask your friend to evaluate what you have written.

Answers to **EXERCISES** are provided at the back of this book in Appendix L.

Reaction Rates and Stoichiometry

From the stoichiometry of the crystal violet reaction (Reaction 13.1), it is clear that for every mole of crystal violet that reacts, a mole of CvOH product is formed, because the coefficients of $Cv^+(aq)$, $OH^-(aq)$, and CvOH(aq) are the same. Thus the rate of appearance of CvOH(aq) equals the rate of disappearance of $Cv^+(aq)$. In many reactions, however, the coefficients are not all the same. For example, in the reaction

$$2\ NO_2(g) \longrightarrow 2\ NO(g) + O_2(g) \qquad [13.2]$$

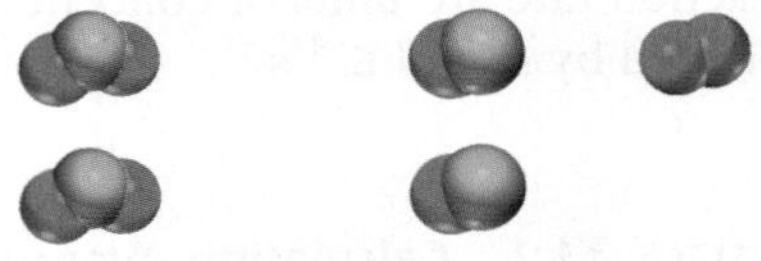

for every mole of O_2 formed, two moles of NO_2 react. Thus the rate of disappearance of NO_2 is twice as great as the rate of appearance of O_2. We would like to define the rate of reaction in a way that does not depend on which substance's concentration change is measured. If we multiply $-\frac{\Delta[NO_2]}{\Delta t}$ by $\frac{1}{2}$ in the definition of the rate, then the rate is the same whether expressed in terms of $[O_2]$ or $[NO_2]$. Notice that $\frac{1}{2}$ is the reciprocal of the stoichiometric coefficient of NO_2 in Equation 13.2. For the general reaction equation

$$a\,A + b\,B \longrightarrow c\,C + d\,D$$

where A, B, C, and D represent formulas of substances and a, b, c, and d are coefficients, the reaction rate can be defined uniformly in terms of each substance as

$$\text{Rate} = -\frac{1}{a}\frac{\Delta[A]}{\Delta t} = -\frac{1}{b}\frac{\Delta[B]}{\Delta t} = \frac{1}{c}\frac{\Delta[C]}{\Delta t} = \frac{1}{d}\frac{\Delta[D]}{\Delta t} \qquad [13.3]$$

That is, *the rate of change in concentration of any of the reactants or products is multiplied by the reciprocal of the stoichiometric coefficient to find the rate of reaction*. Because the concentrations of reactants decrease with time, their rates of change are given negative signs.

Because reaction rates are related to Δconcentration/Δt as shown in Equation 13.3, it is important to know the exact chemical equation for which a rate is reported. If the coefficients in the equation are changed, for example, by doubling all of them, then the definition of reaction rate also changes.

PROBLEM-SOLVING EXAMPLE 13.2 Rates and Stoichiometry

Equation 13.2 shows the decomposition of $NO_2(g)$ to form NO(g) and $O_2(g)$.

(a) Define the rate of reaction in terms of the rate of change in concentration of each reactant and product.

(b) If the rate of appearance of $O_2(g)$ is 0.023 mol L^{-1} s^{-1}, what is the rate of appearance of NO(g)?

Answer

(a) $\text{Rate} = \frac{1}{2}\frac{\Delta[NO]}{\Delta t} = \frac{\Delta[O_2]}{\Delta t} = -\frac{1}{2}\frac{\Delta[NO_2]}{\Delta t}$

(b) 0.046 mol L^{-1} s^{-1}

Strategy and Explanation

(a) Define the rate of reaction as change in concentration divided by time elapsed (Δt). Multiply the rate for each substance by the reciprocal of the stoichiometric coefficient, and place a negative sign in front of the rate for each reactant.

(b) From the coefficients, 2 mol NO forms for every 1 mol O_2. Therefore, the rate of formation of NO is twice the rate of formation of O_2, which equals the rate of reaction. Algebraically,

$$\frac{1}{2}\frac{\Delta[NO]}{\Delta t} = \frac{1}{1}\frac{\Delta[O_2]}{\Delta t}$$

and

$$\frac{\Delta[NO]}{\Delta t} = 2 \times \frac{\Delta[O_2]}{\Delta t} = 2 \times 0.023\ \text{mol L}^{-1}\,\text{s}^{-1} = 0.046\ \text{mol L}^{-1}\,\text{s}^{-1}$$

PROBLEM-SOLVING PRACTICE 13.2

For the reaction

$$4\ NO_2(g) + O_2(g) \longrightarrow 2\ N_2O_5(g)$$

(a) Express the rate of formation of N_2O_5 in terms of the rate of disappearance of O_2.
(b) If the rate of disappearance of O_2 is 0.0037 mol L^{-1} s^{-1}, what is the rate of disappearance of NO_2?

Average Rate and Instantaneous Rate

A reaction rate calculated from a change in concentration divided by a change in time is called the **average reaction rate** over the time interval from which it was calculated. For example, the average reaction rate at 23 °C for the crystal violet reaction over the interval from 0.0 to 10.0 s is 13.2×10^{-7} mol L^{-1} s^{-1}. The data in Table 13.1 indicate that as the concentration of Cv^+ decreases, the average rate also decreases. Most reactions are like this: The average rate becomes smaller as the concentration of one or more reactants decreases. Your results in Exercise 13.1 should have been different for each range of time over which you calculated. Because the average reaction rate changes over time, the rate you calculate depends on when, and for what range of time, you calculate. If you want to know the rate that corresponds to a particular concentration of Cv^+ (and therefore to a particular time after the reaction began), the average rate is not appropriate, because it depends on the size of the time interval.

The **instantaneous reaction rate** is *the rate at a particular time after a reaction has begun*. To obtain it, the rate must be calculated over a very small interval around the time or concentration for which the rate is desired. For example, to calculate the rate at which Cv^+ is disappearing when its concentration is 4.29×10^{-6} mol/L, you would need to calculate $\Delta[Cv^+]/\Delta t$ at exactly 80 s from the start of the reaction. A good way to do so is shown in Figure 13.3. ***The instantaneous rate is the slope of a line tangent to the concentration-time curve at the point corresponding to the specified concentration and time.*** For a particular concentration of the same reactant at the same temperature and the same concentrations of other species, the instantaneous rate has a specific value. As you saw in Exercise 13.1c, the value of the average rate depends on the size of the Δt used to calculate the average rate.

If you are familiar with calculus, then you may recognize that in the limit of very small time intervals, $\Delta[A]/\Delta t$, where A represents a substance, becomes the same as the derivative of concentration with respect to time. That is,

$$\lim_{\Delta t \to 0} \frac{\Delta[A]}{\Delta t} = \frac{d[A]}{dt}$$

This also means that the rate of reaction at any time can be found from the *slope* (at that time) of the tangent to a curve of concentration versus time, such as the curve in Figure 13.3. Appendix A.8 discusses how to determine the slope and intercept of a graph. The slope of a tangent to the graph at a given point is referred to as the derivative of the graph at that point and can be obtained using scientific graphing programs.

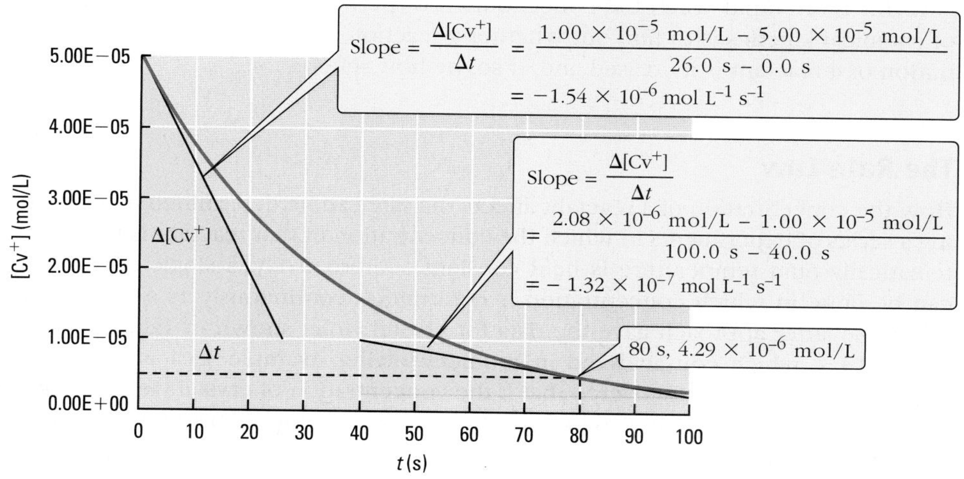
Slope = Δ[Cv+]/Δt = (1.00 × 10^-5 mol/L − 5.00 × 10^-5 mol/L)/(26.0 s − 0.0 s)
= −1.54 × 10^-6 mol L^-1 s^-1
Slope = Δ[Cv+]/Δt
= (2.08 × 10^-6 mol/L − 1.00 × 10^-5 mol/L)/(100.0 s − 40.0 s)
= − 1.32 × 10^-7 mol L^-1 s^-1
80 s, 4.29 × 10^-6 mol/L
Δ[Cv+]
Δt
[Cv+] (mol/L)
5.00E−05
4.00E−05
3.00E−05
2.00E−05
1.00E−05
0.00E+00
0
10
20
30
40
50
60
70
80
90
100
t (s)

Figure 13.3 Instantaneous reaction rates. The experimentally measured concentration of Cv^+ is plotted as a function of time during the reaction in which Cv^+ reacts with OH^- in aqueous solution. The slopes at time 0 s and at 80 s are indicated on the graph. From these slopes the instantaneous rates 1.54×10^{-6} mol L^{-1} s^{-1} and 1.32×10^{-7} mol L^{-1} s^{-1} can be obtained.

EXERCISES that are labeled **CONCEPTUAL** are designed to test your understanding of one or more concepts; they usually involve qualitative rather than quantitative thinking.

CONCEPTUAL EXERCISE 13.2 Instantaneous Rates

Instantaneous rates for the reaction of hydroxide ion with Cv^+ can be determined from the slope of the curve in Figure 13.3 at various concentrations. They are

(1) At 4.0×10^{-5} mol/L, rate = 12.3×10^{-7} mol L^{-1} s^{-1}
(2) At 3.0×10^{-5} mol/L, rate = 9.25×10^{-7} mol L^{-1} s^{-1}
(3) At 2.0×10^{-5} mol/L, rate = 6.16×10^{-7} mol L^{-1} s^{-1}
(4) At 1.5×10^{-5} mol/L, rate = 4.60×10^{-7} mol L^{-1} s^{-1}
(5) At 1.0×10^{-5} mol/L, rate = 3.09×10^{-7} mol L^{-1} s^{-1}

(a) What is the relationship between the rates in (1) and (3)? Between (2) and (4)? Between (3) and (5)?
(b) What is the relationship between the concentrations in each of these cases?
(c) Is the rate of the reaction proportional to the concentration of Cv^+?

CONCEPTUAL EXERCISE 13.3 Graphing Concentrations versus Time

Consider the decomposition of N_2O_5(g),

$$2\ N_2O_5(g) \longrightarrow 4\ NO_2(g) + O_2(g)$$

Assume that the initial concentration of N_2O_5(g) is 0.0200 mol/L and that none of the products are present. Make a graph that shows concentrations of N_2O_5(g), NO_2(g), and O_2(g) as a function of time, all on the same set of axes and roughly to scale.

Figure 13.4 Reaction of aqueous potassium permanganate with aqueous hydrogen peroxide. The rate of the reaction depends on the permanganate concentration. In both cases the temperature is the same, so the difference in rate must be due to concentration of permanganate ions.

13.2 Effect of Concentration on Reaction Rate

The rates of most reactions change when reactant concentrations change, just as we found for the crystal violet reaction. Figure 13.4 shows another example. The oxidation of hydrogen peroxide by permanganate ion in acidic aqueous solution

$$\underset{\text{purple}}{2\ MnO_4^-(aq)} + 5\ H_2O_2(aq) + 6\ H_3O^+(aq) \longrightarrow 2\ Mn^{2+}(aq) + 5\ O_2(g) + 14\ H_2O(\ell)$$

is visibly more rapid when the concentration of permanganate is higher. One goal of chemical kinetics is to find out whether a reaction speeds up when the concentration of a reactant is increased and, if so, by how much.

The Rate Law

How the concentration of a reactant affects the rate can be determined by performing a series of experiments in which the concentration of that reactant is varied systematically (and temperature is held constant). Alternatively, a single experiment can be done in which concentration is determined continuously as a function of time. The latter approach gave the data for crystal violet shown in Table 13.1 and Figure 13.3, which you analyzed in Problem-Solving Example 13.1 and Exercise 13.2. You should have discovered that if the concentration of crystal violet cation is halved, the reaction rate is also halved. If the concentration of crystal violet cation is doubled, then the reaction rate is doubled. This leads to the expression

$$\text{Rate} \propto [Cv^+]$$

It says that the rate is directly proportional to (symbol ∝) the concentration of one of the reactants, crystal violet cation.

This proportionality can be changed to a mathematical equation by including a proportionality constant, k. *A mathematical equation that summarizes the relationship between reactant concentration and reaction rate* is called a **rate law** (or *rate equation*). For the crystal violet reaction the rate law is

$$\text{Rate} = k \times [\text{Cv}^+]$$

The proportionality constant, k, is called the **rate constant.** The rate constant is independent of concentration, but it has different values at different temperatures, usually becoming larger as the temperature increases. The rate constant applies only to the specific reaction being studied and it applies at a specific temperature. Thus the chemical equation and the temperature for the reaction should be given along with the rate constant. In this case we write

$$\text{Cv}^+(\text{aq}) + \text{OH}^-(\text{aq}) \longrightarrow \text{CvOH}(\text{aq}) \qquad k = 3.07 \times 10^{-2}\ \text{s}^{-1}\ (\text{at } 25\ °\text{C})$$

Determining Rate Laws from Initial Rates

The relation between rate and concentration (the rate law) must be determined experimentally. One way to do so was illustrated in Exercise 13.3, but it is difficult to determine rates from tangents to a curve such as that in Figure 13.3. Another way is to measure initial rates. The **initial rate** of a reaction is *the instantaneous rate determined at the very beginning of the reaction*. A good approximation to the initial rate is to calculate $-\Delta[\text{reactant}]/\Delta t$ after no more than 2% of the limiting reactant has been consumed.

An advantage of measuring initial rates is that the concentrations of products are low early in the process. As a reaction proceeds, more and more products are formed. In some cases products can alter the rate; comparing initial rates with rates when products are present can reveal such a complication.

Many reactions can be started by mixing two different solutions or two different gas samples. Usually the concentrations of the reactants are known before they are mixed, so the initial rate corresponds to a known set of reactant concentrations. Several experiments can then be done in which initial concentrations are varied, and the change in the reaction rate can be correlated with changes in the concentration of each reactant. As an example, consider the reaction of a base with methyl acetate, CH_3COOCH_3, an ester (***← p. 556***). This reaction produces acetate ion and methanol.

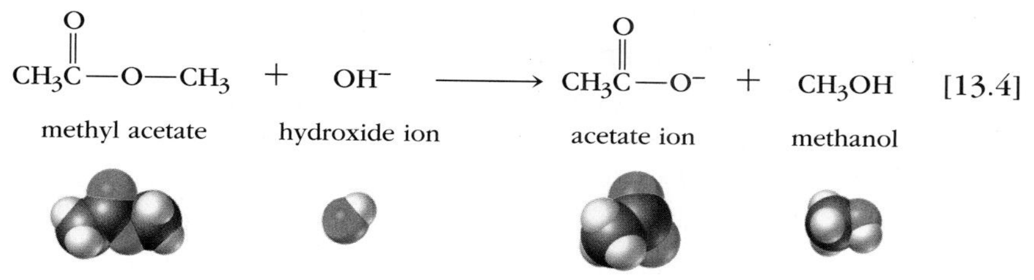
O
CH₃C—O—CH₃ + OH⁻ ⟶ CH₃C—O⁻ + CH₃OH [13.4]
methyl acetate
hydroxide ion
acetate ion
methanol

To control for the effect of temperature on rate, several experiments were done at the same temperature:

Experiment	Initial Concentration (mol/L) [CH_3COOCH_3]	[OH^-]	Initial Rate (mol L^{-1} s^{-1})
1	0.040	0.040	0.00022
	no change	× 2	× 2
2	0.040	0.080	0.00045
	× 2	no change	× 2
3	0.080	0.080	0.00090

Notice that in Experiments 1 and 2 the initial concentration of methyl acetate is the same. In Experiments 2 and 3 the initial concentration of hydroxide is the same.

Rate laws *must* be determined from experimental data such as these.

To determine the rate law, compare two experiments in which only a single initial concentration changed. In Experiments 1 and 2, the [CH_3COOCH_3] remained constant and the [OH^-] doubled. The rate also doubled, which means that the rate is *directly proportional* to the [OH^-]. In Experiments 2 and 3, the [OH^-] remained the same, the [CH_3COOCH_3] doubled, and the rate doubled, indicating that the rate is also proportional to the $[CH_3COOCH_3]$. Therefore, the experimental data show that the rate is proportional to the *product* of the two concentrations, and the rate law is

$$\text{Rate} = k\,[CH_3COOCH_3][OH^-]$$

This equation also tells us that doubling both initial concentrations at the same time would cause the rate to go up by a factor of 4, which it does from Experiment 1 to Experiment 3.

Another way to approach this problem involves proportions. As before, choose two experiments in which one concentration did not change. Then calculate the ratio of the other concentrations and the ratio of rates. For the methyl acetate reaction, using Experiments 1 and 2 where the $[CH_3COOCH_3]$ was constant,

$$\frac{[OH^-]_2}{[OH^-]_1} = \frac{0.080\text{ M}}{0.040\text{ M}} = 2.0 \quad \text{and} \quad \frac{\text{rate}_2}{\text{rate}_1} = \frac{0.00045\text{ mol L}^{-1}\text{ s}^{-1}}{0.00022\text{ mol L}^{-1}\text{ s}^{-1}} = 2.0$$

it is clear that both the concentrations and the rates change in the same proportion. This same method could be applied to analyze results of Experiments 2 and 3, where the initial $[OH^-]$ was constant and the initial $[CH_3COOCH_3]$ changed.

Once the rate law is known, a value for k, the rate constant, can be found by substituting rate and initial concentration data for any one experiment into the rate law. For example, a value of k for the methyl acetate-hydroxide ion reaction could be obtained from data for the first experiment,

rate | $[CH_3COOCH_3]$ | $[OH^-]$

$$0.00022\text{ mol L}^{-1}\text{ s}^{-1} = k\,(0.040\text{ mol/L})(0.040\text{ mol/L})$$

$$k = \frac{0.00022\text{ mol L}^{-1}\text{ s}^{-1}}{(0.040\text{ mol/L})(0.040\text{ mol/L})} = 0.14\text{ L mol}^{-1}\text{ s}^{-1}$$

Note that the units of the rate constant depend on the number of concentration terms multiplied. In this case the reaction has two concentration terms and the units of k are $\text{L mol}^{-1}\text{ s}^{-1}$.

A more precise value for k can be obtained by using all available experimental data—that is, by calculating a k for each experiment and then averaging the k values to obtain an overall result.

EXERCISE 13.4 Rates and Concentrations

Use the rate law for the reaction of methyl acetate with OH^- to predict the effect on the rate of reaction if the concentration of methyl acetate is doubled and the concentration of hydroxide ions is halved.

PROBLEM-SOLVING EXAMPLE 13.3 Rate Law from Initial Rates

Initial rates $\left(-\frac{\Delta[Cl_2]}{\Delta t}\right)$ for the reaction of nitrogen monoxide and chlorine

$$2\,NO(g) + Cl_2(g) \longrightarrow 2\,NOCl(g)$$

were measured at 27 °C starting with various concentrations of NO and Cl_2. These data were collected.

Experiment	Initial Concentrations (mol/L) [NO]	$[Cl_2]$	Initial Rate ($\text{mol L}^{-1}\text{ s}^{-1}$)
1	0.020	0.010	8.27×10^{-5}
2	0.020	0.020	1.65×10^{-4}
3	0.020	0.040	3.31×10^{-4}
4	0.040	0.020	6.60×10^{-4}
5	0.010	0.020	4.10×10^{-5}

(a) What is the rate law? (b) What is the value of the rate constant k?

For the reaction as written in Problem-Solving Example 13.3,

$$\text{Rate} = -\frac{1}{2}\frac{\Delta[NO]}{\Delta t} = -\frac{\Delta[Cl_2]}{\Delta t} = +\frac{1}{2}\frac{\Delta[NOCl]}{\Delta t}$$

Answer

(a) Rate = $k[Cl_2][NO]^2$ (b) $k = 2.1 \times 10^1 \text{ mol}^{-2} \text{ L}^2 \text{ s}^{-1}$

Strategy and Explanation (a) Analyze data from experiments in which one concentration remains the same. In Experiments 1, 2, and 3, the concentration of NO is constant, while the Cl_2 concentration increases from 0.010 to 0.020 to 0.040 mol/L. Each time $[Cl_2]$ is doubled, the initial rate also doubles. For example, when $[Cl_2]$ is doubled from 0.020 to 0.040 mol/L in Experiments 2 and 3, the initial rate doubles from 1.65×10^{-4} to 3.31×10^{-4} mol L^{-1} s^{-1}. The initial rate is directly proportional to $[Cl_2]$.

In Experiments 2, 4, and 5, $[Cl_2]$ is constant, while [NO] varies. In Experiments 2 and 4, [NO] is doubled, but the initial rate increases by a factor of 4, or 2^2.

$$\frac{\text{Experiment 4 rate}}{\text{Experiment 2 rate}} = \frac{6.60 \times 10^{-4} \text{ mol L}^{-1} \text{ s}^{-1}}{1.65 \times 10^{-4} \text{ mol L}^{-1} \text{ s}^{-1}} = \frac{4}{1} = \frac{2^2}{1}$$

This same result is found in Experiments 2 and 5. Thus the initial rate is proportional to the *square* of [NO]. Therefore, the rate law is

$$\text{Rate} = k[Cl_2][NO]^2$$

(b) Once the rate law is known, the rate constant k can be calculated. For Experiment 1, for example,

$$8.27 \times 10^{-5} \text{ mol L}^{-1} \text{ s}^{-1} = k(0.010 \text{ mol/L})(0.020 \text{ mol/L})^2$$

$$k = \frac{8.27 \times 10^{-5} \text{ mol L}^{-1} \text{ s}^{-1}}{(0.010 \text{ mol/L})(0.020 \text{ mol/L})^2} = 2.1 \times 10^1 \text{ mol}^{-2} \text{ L}^2 \text{ s}^{-1}$$

For Experiments 2, 3, 4, and 5, the rate constants are 2.1×10^1, 2.1×10^1, 2.1×10^1, and 2.0×10^1 mol^{-2} L^2 s^{-1}, respectively. The average of these values is the rate constant determined from this series of experiments, 2.1×10^1 mol^{-2} L^2 s^{-1}. It can be used to calculate the rate for any set of NO and Cl_2 concentrations at 27 °C.

☑ **Reasonable Answer Check** The five calculated k values are nearly equal. If the rate law were incorrect, or if an error were made in one or more calculations, some k values would be quite different from the others.

When comparing one experiment with another, as is done in the Strategy and Explanation of Problem-Solving Example 13.3, it is usually convenient to put the larger rate in the numerator. Otherwise fractions are obtained instead of whole numbers.

For this reaction the rate is proportional to one concentration and to the square of another. The rate constant equals the rate (units of mol L^{-1} s^{-1}) divided by three concentration terms multiplied together (units of mol^3 L^{-3}). Thus, the rate constant has units of

$$\frac{\text{mol L}^{-1} \text{ s}^{-1}}{\text{mol}^3 \text{ L}^{-3}} = \text{mol}^{-2} \text{ L}^2 \text{ s}^{-1}.$$

PROBLEM-SOLVING PRACTICE 13.3

At 23 °C, these data were collected for the crystal violet reaction

$$Cv^+(aq) + OH^-(aq) \longrightarrow CvOH(aq)$$

Experiment	Initial Concentrations (mol/L) $[Cv^+]$	$[OH^-]$	Initial Rate (mol L^{-1} s^{-1})
1	4.3×10^{-5}	0.10	1.3×10^{-6}
2	2.2×10^{-5}	0.10	6.7×10^{-7}
3	1.1×10^{-5}	0.10	3.3×10^{-7}

(a) Is it possible to determine the complete rate law from the data given? Why or why not?

(b) Assume that the rate does not depend on the concentration of hydroxide ion. What is the rate law?

(c) Calculate the rate constant, again assuming that the rate does not depend on the concentration of hydroxide ion.

(d) Calculate the initial rate of reaction when the concentration of crystal violet cation is 0.00045 M and the concentration of hydroxide ion is 0.10 M. Report your results in mol L^{-1} s^{-1}.

(e) Calculate the rate when the concentration of Cv^+ is half the initial value of 0.00045 M.

EXERCISE 13.5 Determining the Rate Law Using Logarithms

For the reaction in Problem-Solving Example 13.3, assume that the rate law is of the form Rate = $k[NO]^x[Cl_2]^y$. Show mathematically that by taking logarithms of both sides of the rate law and comparing Experiments 2 and 4, where the concentration of chlorine is the same, x, is given by

$$x = \log\left(\frac{\text{Rate}_4}{\text{Rate}_2}\right) \Big/ \log\left(\frac{[NO]_4}{[NO]_2}\right)$$

go Chemistry **Module 20: Chemical Kinetics** covers concepts in this section.

13.3 Rate Law and Order of Reaction

For many (but not all) homogeneous reactions, the rate law has the general form

The rate constant k always has the same value at a given temperature

m and n are the orders of the reaction with respect to substances A and B.

$$\text{Rate} = k[A]^m[B]^n \ldots$$

where concentrations of substances, [A], [B], . . . are raised to powers, m, n, ... The substances A, B, . . . might be reactants, products, or catalysts. The exponents m, n, . . . are usually positive whole numbers but might be negative numbers or fractions. These exponents define the **order of the reaction** with respect to each reactant. If n is 1, for example, the reaction is *first*-order with respect to B; if m is 2, then the reaction is *second*-order with respect to A. The sum of m and n (plus the exponents on any other concentration terms in the rate equation) gives the **overall reaction order.** (The reaction in Problem-Solving Example 13.3 is first-order in Cl_2, second-order in NO, and third-order overall.) A very important point to remember is that ***the rate law and reaction orders must be determined experimentally; they cannot be predicted from stoichiometric coefficients in the balanced overall chemical equation.***

None of the reactions in Problem-Solving Example 13.4 has a rate law that can be derived correctly from the stoichiometric equation. For example, H_2 has a coefficient of 2 in Reaction (a), but the rate law involves $[H_2]$ to the first power.

PROBLEM-SOLVING EXAMPLE 13.4 Reaction Order and Rate Law

For each reaction and experimentally determined rate law listed below, determine the order with respect to each reactant and the overall order.

(a) $2\ NO(g) + 2\ H_2(g) \longrightarrow N_2(g) + 2\ H_2O(g)$ Rate = $k[NO]^2[H_2]$

(b) $14\ H_3O^+(aq) + 2\ HCrO_4^-(aq) + 6\ I^-(aq) \longrightarrow 2\ Cr^{3+}(aq) + 3\ I_2(aq) + 22\ H_2O(\ell)$
Rate = $k[HCrO_4^-][I^-]^2[H_3O^+]^2$

(c) *cis*-2-butene(g) $\longrightarrow$ *trans*-2-butene(g) (catalytic concentration of I_2 present)
Rate = k[*cis*-2-butene]$[I_2]^{1/2}$

Answer

(a) First-order in H_2, second-order in NO, third-order overall
(b) First-order in $HCrO_4^-$, second-order in I^-, second-order in H_3O^+, fifth-order overall
(c) First-order in *cis*-2-butene, 0.5-order in I_2, 1.5-order overall

Strategy and Explanation Use the exponents in the rate law—not the stoichiometric coefficients—to determine the order.

(a) The rate law has a single term that is raised to the first power and another term that is squared, so the reaction is first-order in H_2, second-order in NO, and third-order overall.
(b) The rate law contains three terms. Since the $HCrO_4^-$ term is raised to the first power, the reaction is first-order in $HCrO_4^-$. The other two terms are squared, so the reaction is second-order in I^- and second-order in H_3O^+. The exponents sum to five, so the reaction is fifth-order overall.

(c) In this case the rate of reaction depends on the concentration of the reactant and also on the square root ($\frac{1}{2}$ power) of the concentration of a catalyst, I_2. The reaction is therefore first-order in *cis*-2-butene, 0.5-order in I_2, and 1.5-order overall.

PROBLEM-SOLVING PRACTICE 13.4

In Problem-Solving Example 13.3 the rate law for reaction of NO with Cl_2 was found to be

$$2\,NO(g) + Cl_2(g) \longrightarrow 2\,NOCl(g) \qquad \text{Rate} = k[NO]^2[Cl_2]$$

(a) What is the order of the reaction with respect to NO? With respect to Cl_2?
(b) Suppose that you triple the concentration of NO and simultaneously decrease the concentration of Cl_2 by a factor of 8. Will the reaction be faster or slower under the new conditions? How much faster or slower? (Assume that the temperature is the same in both sets of conditions.)

The Integrated Rate Law

Another approach to experimental determination of the rate law and rate constant for a reaction uses calculus to derive what is called the integrated rate law. As an example of the integrated rate law method, consider a hypothetical reaction in which a single substance A reacts to form products.

We have already seen that the rate law allows us to calculate the rate of reaction from the concentration of the reactants (and perhaps other substances). The integrated rate law allows us to calculate the concentration of a reactant (or perhaps another substance) as a function of time.

$$A \longrightarrow \text{products}$$

First-Order Reaction If the rate law is first-order, then

$$\text{Rate} = -\frac{\Delta[A]}{\Delta t} = k[A]$$

This expression can be transformed, using calculus, to the integrated first-order rate law,

$$\ln[A]_t = -kt + \ln[A]_0 \qquad [13.5]$$

where $[A]_t$ represents the concentration of A at time t, $[A]_0$ represents the initial concentration of A (when $t = 0$), and ln represents the natural logarithm function. (Logarithms are discussed in Appendix A.6.)

Equation 13.5 has the same form as the general equation for a straight line, $y = mx + b$, in which m is the slope and b is the y-intercept.

y-axis variable | slope | *x*-axis variable | *y*-intercept

$$\ln[A]_t = -kt + \ln[A]_0 \qquad [13.5]$$

$$y \quad = \quad mx + \quad b$$

You do not have to know calculus to use the results that constitute the integrated rate law method. If you do know calculus, however, you will be able to derive the results for yourself.

Using calculus,

$$-\frac{d[A]}{dt} = k[A] \quad \text{and} \quad \frac{d[A]}{[A]} = -k\,dt$$

$$\int_{[A]_0}^{[A]_t} \frac{d[A]}{[A]} = -k\int_0^t dt$$

$$\ln[A]_t - \ln[A]_0 = -k(t_t - t_0)$$

If the reaction starts at time t_0, then $t_t - t_0 = t$, the elapsed time, and $\ln[A]_t - \ln[A]_0 = -kt$ or $\ln[A]_t = -kt + \ln[A]_0$.

If the reaction is actually first-order, then a graph of ln[A] on the vertical (y) axis versus t on the horizontal (x) axis should be a straight line. A linear graph, such as the one in Figure 13.5a, is evidence that the reaction is first-order.

Second-Order Reaction For the same reaction

$$A \longrightarrow \text{products}$$

suppose that the rate depends on the square of the concentration of the reactant; that is, suppose the rate law is second-order.

$$\text{Rate} = k[A]^2$$

The integrated rate law derived using calculus is

$$\frac{1}{[A]_t} = kt + \frac{1}{[A]_0}$$

This equation is also of the form $y = mx + b$. If a reaction is second-order, a graph of $1/[A]_t$ versus t will be linear with slope $= k$ and y-intercept $= 1/[A]_0$. Such a

Using calculus,

$$-\frac{d[A]}{dt} = k[A]^2 \quad \text{and} \quad \frac{d[A]}{[A]^2} = -k\,dt$$

$$\int_{[A]_0}^{[A]_t} \frac{d[A]}{[A]} = -k\int_0^t dt$$

$$-\frac{1}{[A]_t} - \left(-\frac{1}{[A]_0}\right) = -k(t_t - t_0)$$

If the reaction starts at time t_0, then $t_t - t_0 = t$, the elapsed time, and

$$\frac{1}{[A]_t} = kt + \frac{1}{[A]_0}$$

First-order reaction

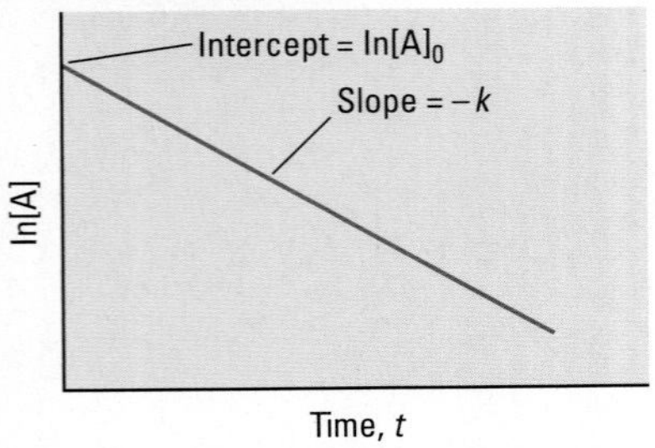

(a) If a reaction is first-order in reactant A, plotting $\ln[A]_t$ versus t gives a straight line.

Second-order reaction

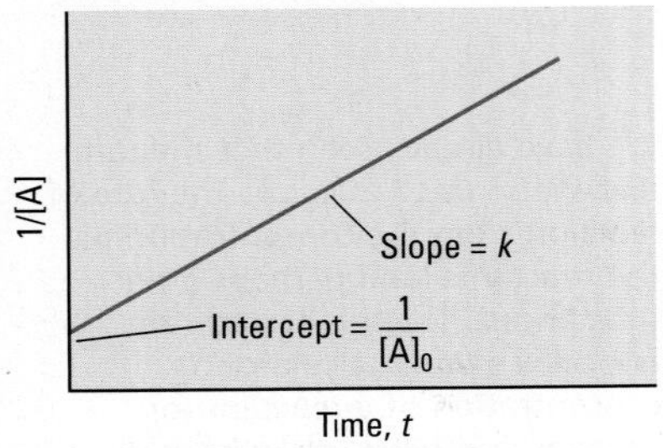

(b) If a reaction is second-order in reactant A, plotting $1/[A]_t$ versus t gives a straight line.

Zeroth-order reaction

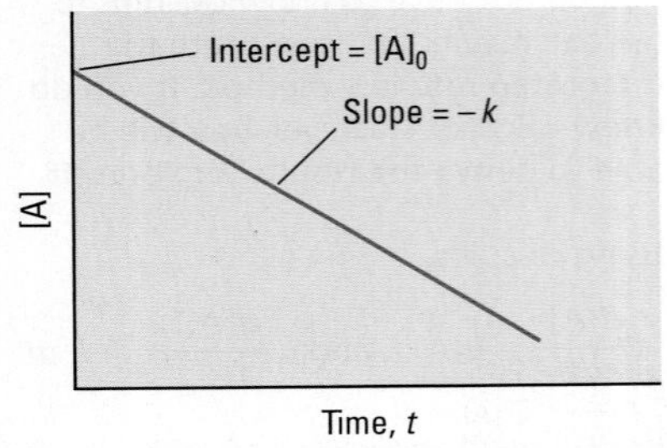

(c) If a reaction is zeroth-order in reactant A, plotting $[A]_t$ versus t gives a straight line.

Figure 13.5 First-order, second-order, and zeroth-order plots.

straight-line graph is evidence that a reaction is second-order. An example graph is shown in Figure 13.5b.

Zeroth-Order Reaction There are a few reactions for which the rate does not depend on the concentration of a reactant at all. These are called zeroth-order reactions, because the rate law can be written as a rate constant k times a concentration to the zeroth power. (Recall that anything raised to the zeroth power equals 1.)

$$\text{Rate} = -\frac{\Delta[A]}{\Delta t} = k[A]^0 = k$$

This rate law says that for a zeroth-order reaction the rate is always the same no matter what the concentration of reactant. The rate is equal to the rate constant k. For a zeroth-order reaction you can derive the integrated form without calculus. Simply use algebra to rearrange the rate law just given.

$$\Delta[A] = -k\Delta t$$

$$[A]_t - [A]_0 = -k(t_t - t_0) = -kt$$

$$[A]_t = -kt + [A]_0$$

Again, the equation is of the form $y = mx + b$. If a reaction is zeroth-order, graphing A_t versus t gives a straight line with slope $= -k$ and y-intercept $= [A]_0$, as shown in Figure 13.5c.

To summarize these three situations, a rate law that involves powers of the reactant concentration can be written as

$$\text{Rate} = -\frac{\Delta[A]}{\Delta t} = k[A]^m$$

where m is the order of the reaction as defined earlier. The integrated rate law depends on the value of m, and the results for m equal to 0; 1; and 2 are given in Table 13.2.

To determine the order of reaction, then, we collect concentration-time data and make the three plots listed in Table 13.2 and shown in Figure 13.5. Only one (or perhaps none) of the plots will be a straight line. If one is straight, then it indicates the order, and the rate constant can be calculated from its slope. The units of the rate constant are those of the slope of the line; they are given in the last column of Table 13.2. The units of the rate constant depend on the order of the reaction.

Table 13.2 Integrated Rate Laws

Order	Rate Equals	Integrated Rate Law*	Straight-line Plot	Slope of Plot	Units of k†
0	$k[A]^0 = k$	$[A]_t = -kt + [A]_0$	$[A]_t$ vs t	$-k$	conc. time^{-1}
1	$k[A]$	$\ln[A]_t = -kt + \ln[A]_0$	$\ln[A]_t$ vs t	$-k$	time^{-1}
2	$k[A]^2$	$\frac{1}{[A]_t} = kt + \frac{1}{[A]_0}$	$\frac{1}{[A]_t}$ vs t	k	conc.$^{-1}$ time^{-1}

*In the table, $[A]_0$ indicates the initial concentration of substance A, that is, the concentration of A at $t = 0$, the time when the reaction was started.

†If you know the units of k you can use the units to figure out whether the reaction is zeroth-order, first-order, or second-order because the units are different in each case.

PROBLEM-SOLVING EXAMPLE **13.5** **Reaction Order and Rate Constant from Integrated Rate Law**

The data on the next page were obtained for decomposition of cyclopentene at 825 K.

$$C_5H_8(g) \longrightarrow C_5H_6(g) + H_2(g)$$

Cyclopentene has this structure

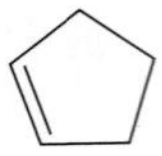

Time (s)	$[C_5H_8]$ (mol/L)	Time (s)	$[C_5H_8]$ (mol/L)
0	0.0200	300.	0.0084
20.	0.0189	400.	0.0063
50.	0.0173	500.	0.0047
100.	0.0149	700.	0.0027
200.	0.0112	1000.	0.0011

Obtain the order of the reaction and the rate constant.

Answer The reaction is first-order in C_5H_8; $k = 2.88 \times 10^{-3}\ s^{-1}$.

Strategy and Explanation Use the integrated rate law method, making graphs to test for zeroth-, first-, and second-order. Such plots are shown in Figure 13.6. The zeroth-order and second-order plots are curved, while the first-order plot is a straight line. Thus the reaction must be *first-order*.

From the first-order plot, calculate the slope using the points marked on the graph as red, open circles. The points are (910, −6.56) and (74, −4.15).

$$\text{Slope} = \frac{\{-6.56 - (-4.15)\}}{\{910 - 74\}\ s} = -2.88 \times 10^{-3}\ s^{-1}$$

The slope of $-2.88 \times 10^{-3}\ s^{-1}$ is the negative of the rate constant (see Table 13.2), which means that $k = 2.88 \times 10^{-3}\ s^{-1}$.

☑ **Reasonable Answer Check** The units are s^{-1}, which corresponds to the reciprocal time units indicated in Table 13.2 for a first-order rate constant.

It is important to use two points on the straight line through the experimental data (open red circles in Figure 13.6b), not two of the data points themselves, when you calculate the slope. Making a graph is similar to averaging, because the straight line and its slope are based on all ten points in the data table, not just two of them.

PROBLEM-SOLVING PRACTICE **13.5**

Use the concentration-time data in Table 13.1 for the reaction of Cv^+ with OH^- (p. 594) to deduce the order of the reaction with respect to Cv^+. Determine the rate constant. (Assume that $[OH^-]$ is constant during the reaction.)

Calculating Concentration or Time from Rate Law

Once the rate law has been determined experimentally, it provides a way to calculate the concentration of a reactant or product at any time after the reaction has begun. All that is needed is the integrated rate law (from Table 13.2), the value of the rate constant, and the initial concentration of reactant or product. These are related by the equations given in Table 13.2.

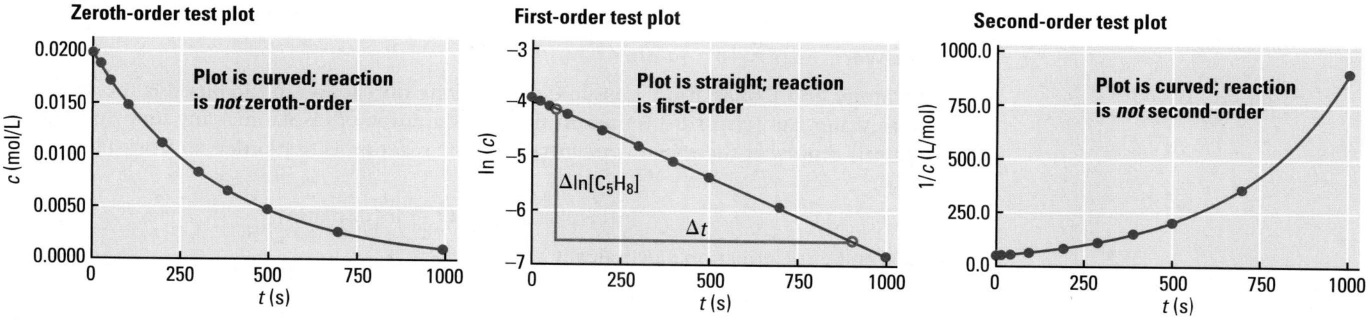

Figure 13.6 Integrated rate law plots for cyclopentene decomposition reaction at 825 K. Only the first-order plot is linear. The rate constant is calculated from two widely separated points on the line (red circles), not from two experimental points (which would ignore all of the other experimental points).

CHEMISTRY YOU CAN DO Simulating First-Order and Zeroth-Order Reactions

Obtain 100 pennies, several dripless birthday candles, a rigid piece of cardboard or plastic at least six inches square, matches, a watch that displays seconds, and a ruler. You will do two separate experiments.

In the first experiment, shake the pennies in a container and toss them onto a flat surface. Count how many pennies land heads-up. Record the number of tosses (n_t = 1 so far) and the number of heads (n_h). Pick up the heads-up pennies and set them aside. Repeat the toss with the remaining pennies, again counting the number of heads, setting them aside, and recording the results. Continue this process until the number of pennies remaining is three or less. Graph the number of heads as a function of the number of tosses. Also graph the natural logarithm of the number of heads as a function of number of tosses.

In the second experiment, melt the bottom of a birthday candle a little and stick it to the rigid piece of cardboard so that it stands vertically. Fasten the ruler vertically and close enough to the candle so that you can measure the length of the candle (the wax part, not counting the wick). Record the length of the candle. Make certain that no combustible material (including the ruler) is close enough to the candle that a flame might ignite it. Light the candle and record the time to the nearest second. Read the length of the candle every 30 s and record your measurement. After five minutes, extinguish the candle. Plot the length of the candle wax as a function of time. Plot the natural logarithm of the length of the candle wax versus time.

© Cengage Learning/Charles D. Winters

Think about these questions:

1. In the experiment with the pennies, which variable corresponds with time in a real kinetics experiment? Which variable corresponds with concentration?
2. What is the order of the penny-tossing process?
3. For each data point in the penny-tossing process, calculate $\Delta n_h/n_h$, where Δn_h is the difference between the number of heads for the current toss and the previous one. (Use 100 for the number of heads in the zeroth toss.) What do you observe about the values of $\Delta n_h/n_h$? Why do you think these values are the way they are?
4. What is the order of the candle-burning process?
5. What can you observe about the change in the length of the candle for each 30-s interval? Is this consistent with your answer to Question 4? If so, why?

Sources: Based on Sanger, M. J., Wiley, R. A. Jr., Richter, E. W., and Phelps, A. J. *Journal of Chemical Education*, Vol. 79, 2002; pp. 989–991; and Sanger, M. J. *Journal of Chemical Education*, Vol. 80, 2003; pp. 304A–304B.

Cisplatin has the formula $Pt(NH_3)_2Cl_2$. Because one of the two chloride ions is replaced by a water molecule in the reaction, the cisplatin reactant has been written as cisplatinCl in the chemical equation to show where the chloride ion comes from.

PROBLEM-SOLVING EXAMPLE 13.6 Calculating Concentrations

The first-order rate constant is $1.87 \times 10^{-3}\ \text{min}^{-1}$ at 37 °C (body temperature) for reaction of cisplatin, a cancer chemotherapy agent, with water. The reaction is

$$\text{cisplatinCl} + H_2O \longrightarrow \text{cisplatinOH}_2^+ + Cl^-$$

Suppose that the concentration of cisplatin in the bloodstream of a cancer patient is 4.73×10^{-4} mol/L. What will the concentration be exactly 24 hours later?

Answer 3.20×10^{-5} mol/L

Strategy and Explanation Assume that the rate of reaction in the blood is the same as in water. Let [cisplatin] represent the concentration of cisplatin at any time and [cisplatin]$_0$ represent the initial concentration. The reaction is first-order, so from Table 13.2,

$$\ln[\text{cisplatin}] = -kt + \ln[\text{cisplatin}]_0$$

Rearrange the equation algebraically to

$$\ln[\text{cisplatin}] - \ln[\text{cisplatin}]_0 = -kt$$

$$\ln\left\{\frac{[\text{cisplatin}]}{[\text{cisplatin}]_0}\right\} = -kt$$

$$\frac{[\text{cisplatin}]}{[\text{cisplatin}]_0} = \text{anti}\ \ln(-kt) = e^{-kt}$$

$$[\text{cisplatin}] = [\text{cisplatin}]_0\, e^{-kt}$$

$$[\text{cisplatin}] = (4.73 \times 10^{-4}\ \text{mol/L})\ e^{-(1.87\times 10^{-3}\ \text{min}^{-1}\times 24\ \text{h})}$$

$$[\text{cisplatin}] = (4.73 \times 10^{-4}\ \text{mol/L})\ e^{-(1.87\times 10^{-3}\ \text{min}^{-1}\times 24\ \text{h}\times 60\ \text{min h}^{-1})}$$

$$[\text{cisplatin}] = (4.73 \times 10^{-4}\ \text{mol/L})(6.77 \times 10^{-2})$$

$$[\text{cisplatin}] = 3.20 \times 10^{-5}\ \text{mol/L}$$

☑ **Reasonable Answer Check** If the drug is to be effective, it almost certainly needs to be in the body for some time—several minutes to hours or more. The concentration has dropped to a little less than 10% of its initial value in 24 hours, which is reasonable.

Mathematical operations involving logarithms and exponentials (antilogarithms) are discussed in Appendix A.6.

Because the solution to this example involves a ratio of concentrations in which the units cancel, the same approach can be taken in problems that involve the number of moles, the mass, or the number of atoms or molecules at two different times.

PROBLEM-SOLVING PRACTICE 13.6

The first-order rate constant for decomposition of an insecticide in the environment is $3.43 \times 10^{-2}\ \text{d}^{-1}$. How long does it take for the concentration of insecticide to drop to $\frac{1}{10}$ of its initial value?

Half-Life

The **half-life** of a reaction, $t_{1/2}$, is *the time required for the concentration of a reactant A to fall to one half of its initial value.* That is, $[\text{A}]_{t_{1/2}} = \frac{1}{2}[\text{A}]_0$. For a *first-order reaction* the half-life has the same value, no matter what the initial concentration is. For other reaction orders this is *not* true.

The half-life is related to the first-order rate constant. To see how, use algebra to rearrange Equation 13.5 (which was associated with a reaction A ⟶ products):

$$\ln[\text{A}]_t = -kt + \ln[\text{A}]_0 \qquad [13.5]$$

$$\ln[\text{A}]_t - \ln[\text{A}]_0 = -kt$$

$$\ln[\text{A}]_{t_{1/2}} - \ln[\text{A}]_0 = -kt_{1/2} \qquad [13.6]$$

Remember that Equation 13.5 was derived for a reaction of the form A ⟶ products. Had the equation been 2 A ⟶ products, that is, had the coefficient of A been 2 in the chemical equation, the result would be slightly different. (See Question 158 at the end of this chapter.)

Because $[\text{A}]_{t_{1/2}} = \frac{1}{2}[\text{A}]_0$, Equation 13.5 can be rewritten as

$$\ln([\text{A}]_0/2) - \ln[\text{A}]_0 = -kt_{1/2}$$

and so

$$-kt_{1/2} = \ln[\text{A}]_0 - \ln(2) - \ln[\text{A}]_0 = -\ln 2$$

$$t_{1/2} = \frac{-\ln 2}{-k} = \frac{0.693}{k} \qquad [13.7]$$

This means that measuring the half-life of a first-order reaction determines the rate constant, and vice versa. Radioactive decay (Section 20.4) is a first-order process, and half-life is typically used to report the rate of decay of radioactive nuclei.

PROBLEM-SOLVING EXAMPLE 13.7 Half-Life and Rate Constant

In Problem-Solving Example 13.5, the rate constant for decomposition of cyclopentene at 825 K was found to be $2.88 \times 10^{-3}\ \text{s}^{-1}$. The reaction gave a linear first-order plot. Determine the half-life in seconds for this reaction.

Answer 241 s

Strategy and Explanation The reaction is first-order and is of the form A ⟶ products, so Equation 13.7 can be used.

$$t_{1/2} = \frac{0.693}{k} = \frac{0.693}{2.88 \times 10^{-3}\ \text{s}^{-1}} = 2.41 \times 10^2\ \text{s} = 241\ \text{s}$$

☑ **Reasonable Answer Check** The rate constant is about $3 \times 10^{-3}\ s^{-1}$. If the concentration of cyclohexene were $1.0\ mol\ L^{-1}$, then the reaction rate would be $0.003\ s^{-1} \times 1.0\ mol\ L^{-1} = 0.003\ mol\ L^{-1}\ s^{-1}$. This means that $0.003\ mol\ L^{-1}$ would react every second. For the concentration to drop to half of $1.0\ mol\ L^{-1}$, the change in concentration would be $0.500\ mol\ L^{-1}$. Therefore, it would take at least $\frac{0.500\ mol\ L^{-1}}{0.003\ mol\ L^{-1}\ s^{-1}} = 167\ s$ for the concentration to drop to half its initial value, and the half-life should be at least 167 s. Because the rate decreases as concentration decreases, it will take longer than 167 s. A half-life of 241 s is reasonable.

PROBLEM-SOLVING PRACTICE 13.7

From Figure 13.3 determine the time required for the concentration of Cv^+ to fall to one half the initial value. Verify that the same period is required for the concentration to fall from one half to one fourth the initial value. From this half-life, calculate the rate constant.

13.4 A Nanoscale View: Elementary Reactions

Macroscale experimental observations reveal that reactant concentrations, temperature, and catalysts can affect reaction rates. But how can we interpret such observations in terms of nanoscale models? We will use the ***kinetic-molecular theory of matter,*** which was first introduced in Section 1.8 (← *p. 17*) and developed further in Section 10.3 (← *p. 429*), together with the ideas about molecular structure developed in Chapters 8 and 9. These concepts provide a good basis for understanding how atoms and molecules move and chemical bonds are made or broken during the very short time it takes for reactant molecules to be converted into product molecules.

According to kinetic-molecular theory, molecules are in constant motion. In a gas or liquid they bump into one another; in a solid they vibrate about specific locations. Molecules also rotate, flex, or vibrate around or along the bonds that hold the atoms together. These motions produce the transformations of molecules that occur during chemical reactions.

There are only two important types of molecular transformations:

- **Unimolecular reactions** are those in which **the structure of a *single* particle (molecule or ion) rearranges to produce a different particle or particles.** A unimolecular reaction might involve breaking a bond (or bonds) and forming two new molecules, or it might involve rearrangement of one isomeric structure into another. A unimolecular reaction has a molecularity of one.
- **Bimolecular reactions** are those in which ***two*** **particles (atoms, molecules, or ions) collide and rearrange bonds to form products.** In a bimolecular reaction new bonds may be formed between the reactant particles and existing bonds may be broken. Sometimes the two particles combine to form a new, larger particle. Sometimes two or more new particles are formed from the two reactant particles. A bimolecular reaction has a molecularity of two.

All chemical reactions can be understood in terms of simple reactions such as those just described. Very complicated reactions can be built up from combinations of unimolecular and bimolecular reactions, just as complicated compounds can be built from chemical elements. For example, hundreds of such reactions are needed to understand how smog is produced in a city such as Los Angeles or to understand

ESTIMATION Pesticide Decay

There are usually several different ways that a pesticide can decompose in an ecosystem. For this reason, it is difficult to define an accurate rate of decomposition and even more difficult to define a rate law. Often it is assumed that decomposition is first-order, and an approximate half-life is reported.

Organochlorine pesticides such as DDT, lindane, and dieldrin may have half-lives as long as 10 years in the environment. The maximum contaminant level (MCL) for lindane is 0.2 ppb (parts per billion). Suppose that an ecosystem has been contaminated with lindane at a concentration of 200 ppb. How long would you have to wait before it would be safe to enter the ecosystem without protection from the pesticide?

The level of contamination is 200 ppb/0.2 ppb = 1000 times the MCL. Presumably it would be safe to wait until the level had dropped to 1/1000 of its initial value. You can estimate how long this would be by using powers of 2, because the number of half-lives required is n, where $(1/2)^n = 1/1000$. That is, as soon as 2^n exceeds 1000, you have waited long enough. Computer scientists, who deal with binary arithmetic, can easily tell you that $2^{10} = 1024$, so $n = 10$. You can verify this using your calculator's y^x-key, or you could simply raise 2 to a power until a value greater than 1000 was calculated. If you did not have a calculator, you could multiply $2 \times 2 \times 2 \ldots$ in your head until you had enough factors to multiply out to a number bigger than 1000. Ten half-lives means 10×10 years, or 100 years, so after 100 years the ecosystem would be free of significant lindane contamination.

Visit this book's companion website at www.cengage.com/chemistry/moore to work an interactive module based on this material.

why chlorofluorocarbons can deplete stratospheric ozone (← *p. 455*). Like the chemical elements, the simplest nanoscale reactions are building blocks, so they are referred to as **elementary reactions.** ***The equation for an elementary reaction shows exactly which molecules, atoms, or ions take part in the elementary reaction.*** The next two sections describe the two important types of elementary reactions in more detail.

EXERCISE 13.6 Unimolecular and Bimolecular Reactions

For each of the nanoscale molecular diagrams below, write a balanced equation using chemical formulas. Which of the reactions are unimolecular reactions? Which are bimolecular?

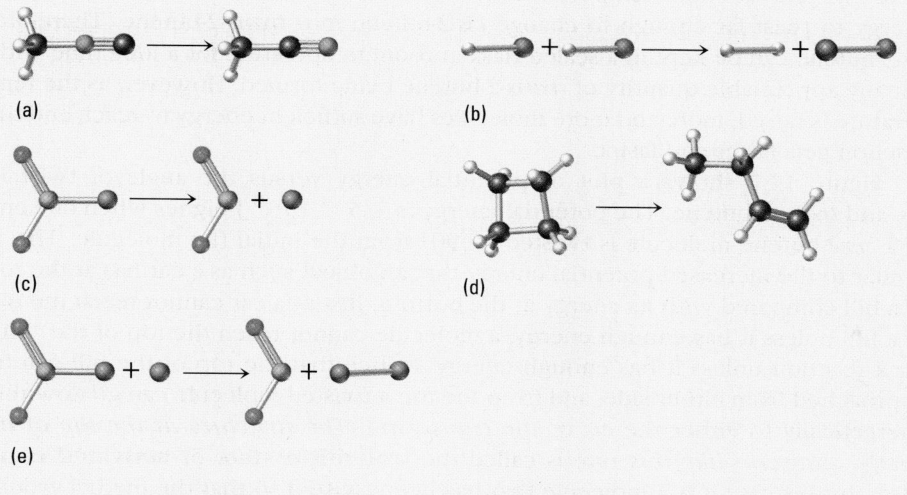
(a)
(b)
(c)
(d)
(e)

The color code for atoms is

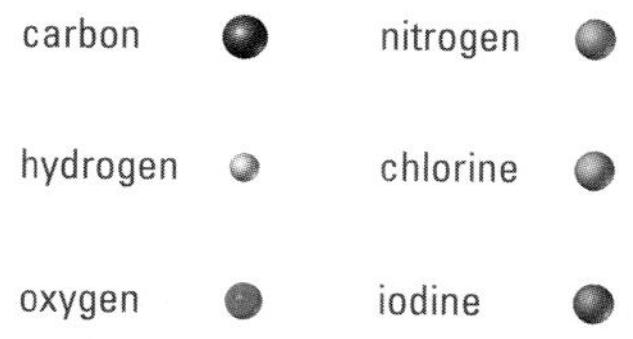
carbon
nitrogen
hydrogen
chlorine
oxygen
iodine

Unimolecular Reactions

An example of a unimolecular reaction is the conversion of *cis*-2-butene to *trans*-2-butene.

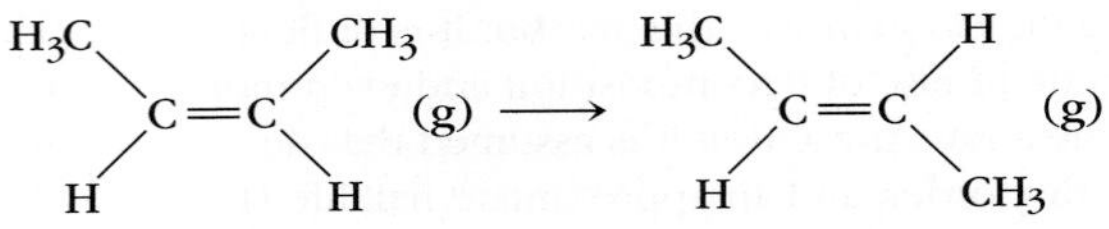

cis-2-butene *trans*-2-butene

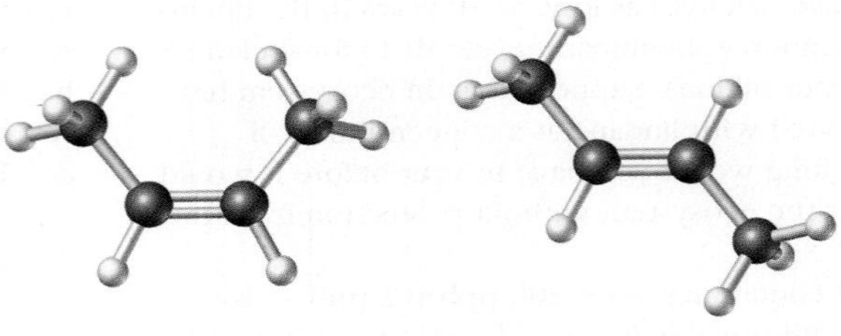

Cis-2-butene and *trans*-2-butene are *cis-trans* stereoisomers (**← *p. 340***), and all four carbon atoms are in the same plane. The difference between the two molecules is the orientation of the methyl groups, which are on the same side of the double bond in the plane of the flat molecule for the *cis* structure and on opposite sides for the *trans* structure. If we could grab one end of the molecule and twist it 180° around the axis of the double bond, we would get the other molecule. Thus, it is a reasonable hypothesis that the molecular pathway by which *cis*-2-butene changes to *trans*-2-butene involves twisting the molecule around the double bond (twisting one methyl group out of the plane of the molecule). The angle of twist around the double-bond axis measures the progress of the reaction on the nanoscale. The greater the angle, the less the molecule is like *cis*-2-butene and the more it is like *trans*-2-butene, until an angle of 180° is reached and it has become *trans*-2-butene. Figure 13.7 shows several structures with different angles of rotation, illustrating this process.

To convert one isomer into the other, such a twist requires that the reactant molecule have sufficient energy. Chemical bonds are like springs. They can be stretched, twisted, and bent, but these changes raise the potential energy. Consequently, some kinetic energy must be converted to potential energy when one end of the *cis*-2-butene molecule twists relative to the other, just as it would if a spring were twisted. At room temperature most of the molecules do not have enough energy to twist far enough to change *cis*-2-butene into *trans*-2-butene. Therefore, *cis*-2-butene can be kept in a sealed flask at room temperature for a long time without any appreciable quantity of *trans*-2-butene being formed. However, as the temperature is raised, more and more molecules have sufficient energy to react, and the reaction gets faster and faster.

Figure 13.7 shows a plot of potential energy versus the angle of twist in *cis*- and *trans*-2-butene. The potential energy is 435×10^{-21} J higher when one end of a *cis*-2-butene molecule is twisted by 90° from the initial flat molecule. This is similar to the increased potential energy that an object such as a car has at the top of a hill compared with its energy at the bottom. Just as a car cannot reach the top of a hill unless it has enough energy, a molecule cannot reach the top of the "hill" for a reaction unless it has enough energy. Notice that the top of the hill can be approached from either side, and from the top a twisted molecule can go downhill energetically to either the *cis* or the *trans* form. *The structure at the top of an energy diagram like this one* is called the **transition state** or **activated complex.** In this case it is a molecule that has been twisted so that the methyl groups are at a 90° angle.

Since molecules are very small, the energy required to twist one *cis*-2-butene molecule is very small. However, if we wanted to twist a mole of molecules all at once, it would take a lot of energy. The energy required to reach the top of the "hill" is often reported per mole of molecules—that is, as (435×10^{-21} J/molecule) $\times$ (1 kJ/10^3 J) $\times$ (6.022×10^{23} molecules/mol) = 262 kJ/mol.

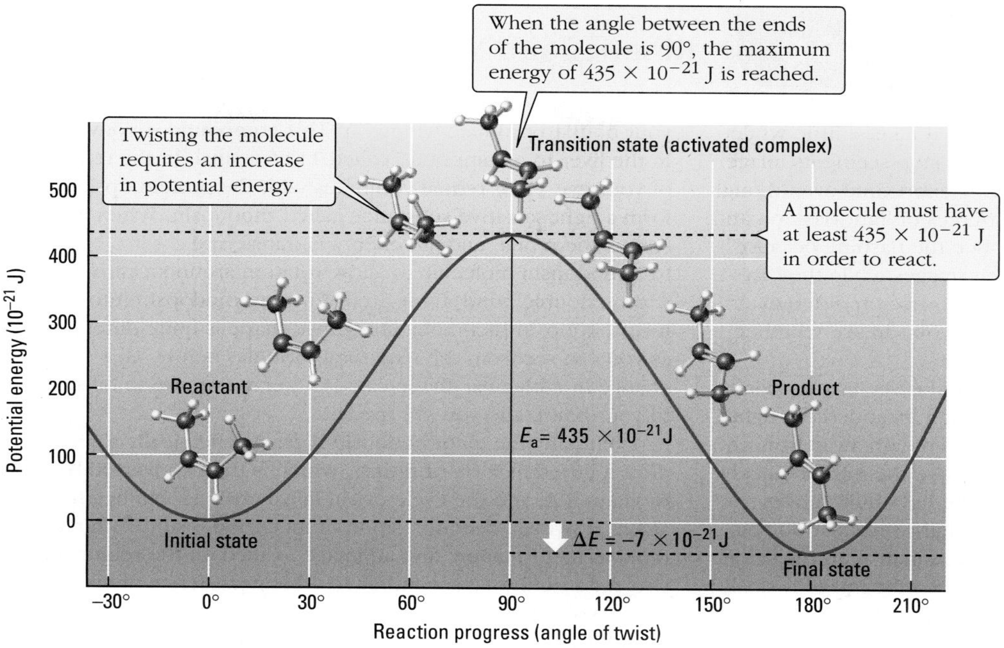
When the angle between the ends of the molecule is 90°, the maximum energy of 435 × 10⁻²¹ J is reached.
Transition state (activated complex)
Twisting the molecule requires an increase in potential energy.
A molecule must have at least 435 × 10⁻²¹ J in order to react.
Reactant
Product
E_a = 435 × 10⁻²¹ J
ΔE = −7 × 10⁻²¹ J
Initial state
Final state
Potential energy (10⁻²¹ J)
500
400
300
200
100
0
−30°
0°
30°
60°
90°
120°
150°
180°
210°
Reaction progress (angle of twist)

Active Figure 13.7 Energy diagram for the conversion of *cis*-2-butene to *trans*-2-butene. Twisting one end of the molecule with respect to the other requires an increase in potential energy because the double bond between the two central C atoms is like a spring and resists twisting. Twisting in either direction requires the same increase in potential energy, so the energy for +30° equals that for −30°. A molecule of *cis*-2-butene must have at least 435×10^{-21} J before it can twist past 90° to 180°, which converts it to *trans*-2-butene. The progress of the reaction (the change in the structure of this single molecule) is measured by the angle of twist. Visit this book's companion website at **www.cengage.com/chemistry/moore** to test your understanding of the concepts in this figure.

Figure 13.7 is similar to Figure 6.15 (← *p. 237*), which showed the energy change when bonds were broken and formed as H_2 and Cl_2 changed into HCl.

CONCEPTUAL EXERCISE 13.7 Transition State

Methyl isonitrile reacts to form acetonitrile in a single-step elementary reaction.

$$CH_3NC(g) \longrightarrow CH_3CN(g)$$

During the reaction the nitrogen atom and one of the carbon atoms exchange places, but the rest of the molecule is unchanged. Suggest a structure for the transition state for this reaction. Draw this structure as a Lewis structure and as a ball-and-stick molecular model.

Almost every chemical reaction has an energy barrier that must be surmounted as reactant molecules change into product molecules. The heights of such barriers vary greatly—from almost zero to hundreds of kilojoules per mole. *At a given temperature, the higher the energy barrier, the slower the reaction.* The *minimum energy required to surmount the barrier* is called the **activation energy, E_a,** for the reaction. For the *cis*-2-butene ⟶ *trans*-2-butene reaction the activation energy is 435×10^{-21} J/molecule, or 262 kJ/mol (see Figure 13.7).

Another interesting relationship shown in Figure 13.7 connects kinetics and thermodynamics. The energy of the product, one molecule of *trans*-2-butene, is 7×10^{-21} J *lower* than that of the reactant, one molecule of *cis*-2-butene. This means that the *cis* ⟶ *trans* reaction is *exothermic* by 7×10^{-21} J/molecule, which translates to 4 kJ/mol. Also, *cis*-2-butene is higher in energy by 7×10^{-21} J/molecule, so the reverse reaction requires that 4 kJ/mol be absorbed from the surroundings; it is *endothermic*. The height of the energy hill that must

The generalization that higher activation energy results in slower reaction applies best if the reactions are similar. For example, it applies to a group of reactions that all involve twisting around a double bond. It also applies to reactions that involve collisions of one molecule with each of a group of similar molecules. It would be less applicable if we were comparing one reaction that involved collision of two molecules with another reaction that involved twisting around a bond.

CHEMISTRY YOU CAN DO Kinetics and Vision

Vision is not an instantaneous process. It takes a little while after a bright light goes out before you stop seeing its image. The flash of a camera blinds you for a short time even though it is on for only an instant. Some sources of light flash on and off very rapidly, but you do not notice the flashing because your eyes continue to perceive their images while they are off. However, if you can focus such a source on different parts of your retina at different times, you can see whether it is flashing. Here's how.

Find a small mirror that you can hold easily in your hand. Use the mirror to reflect the image of an incandescent light bulb onto your eye. You should be far enough away from the light so that its image is small. Now move the mirror quickly back and forth so that the image of the light bulb moves quickly across your eyeball. Does the light smear or do you see individual dots? Try the same experiment with the screen of a TV set. (Get really far away from it so the image is small.) Do you see separate images or just a smear of light? If you see individual images, it means that the light is flashing. Each time it flashes on, the moving mirror has caused it to hit a different part of your retina, and you see a separate image.

Repeat this experiment with as many different light sources as you can, and classify them as flashing or continuous. Try street lights of various kinds, car headlights, neon signs, fluorescent lights, and anything else you can think of, but *don't* try a very bright light, such as a laser pointer, that might damage your eyes. Record your observations.

Rotation around a double bond, as in the interconversion of *cis*- and *trans*-2-butene, occurs in the reactions that allow you to see. A yellow-orange compound called β-carotene, the natural coloring agent in carrots (← p. 401), breaks down in your body to produce vitamin A. This compound is converted in the liver to a compound called 11-*cis*-retinal. In the retina of your eye 11-*cis*-retinal combines with the protein opsin to form a light-sensitive substance called rhodopsin. When light strikes the retina, enough energy is transferred to a rhodopsin molecule to allow rotation around a carbon-carbon double bond, C═C, transforming rhodopsin into metarhodopsin II, a molecule whose shape is quite different, as you can see from the structural formulas below. This change in molecular shape causes a nerve impulse to be sent to your brain, and you see the light.

Eventually the metarhodopsin II reacts chemically to produce a different form of retinal, which is then converted back to vitamin A, and the cycle of chemical changes can begin again. However, decomposition of metarhodopsin II is not as rapid as its formation, and an image formed on the retina persists for a tenth of a second or so. This persistence of vision allows you to perceive videos as continuously moving images, even though they actually consist of separate pictures, each painted on a screen for a thirtieth of a second.

Think about these questions:

1. For each light source that you observed, what would happen if you focused a camera on the light source, opened the shutter, and moved the camera quickly while the shutter was open?
2. What would happen if you moved the camera more slowly or very quickly?

Note: This *Chemistry You Can Do* relates to the question, "What are the molecules in my eyes doing when I watch a movie?" that was posed in Chapter 1 (← p. 2).

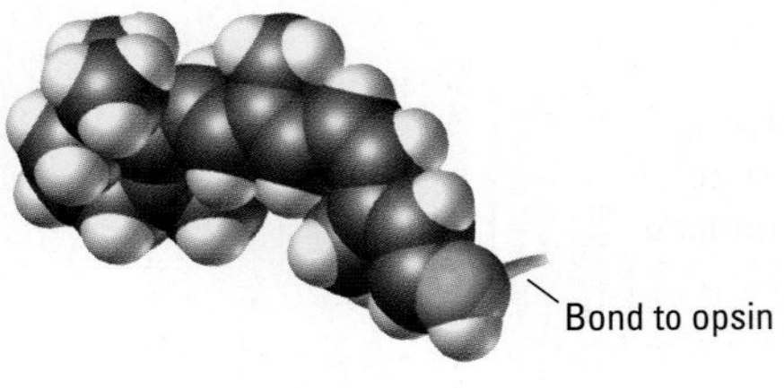

rhodopsin

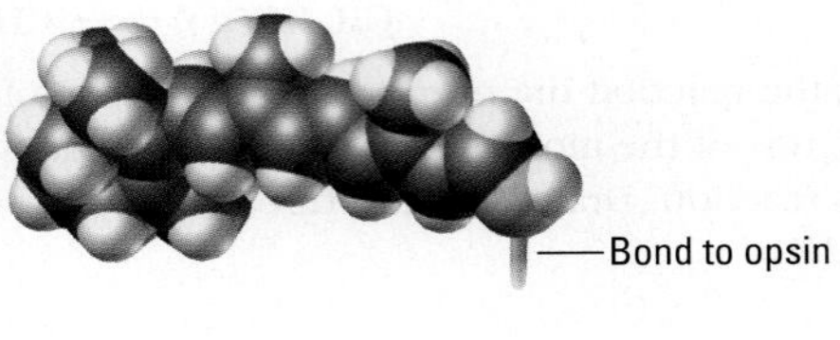

metarhodopsin II

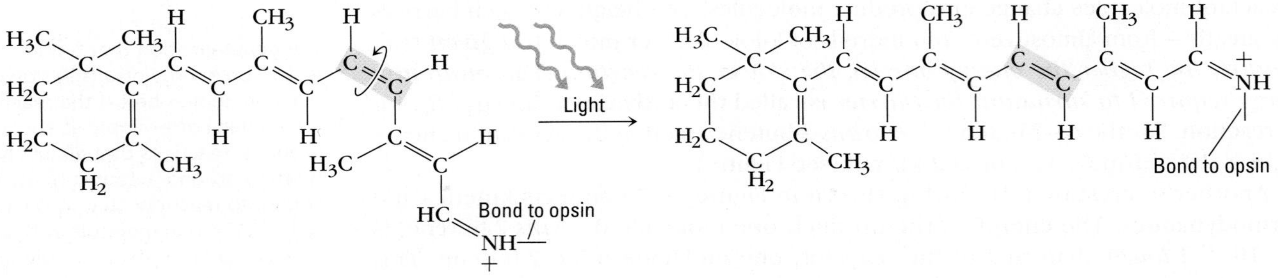

be climbed when the reverse reaction occurs is $(435 + 7) \times 10^{-21}$ J/molecule or 442×10^{-21} J/molecule (266 kJ/mol). Thus, the activation energy for the forward reaction is 4 kJ/mol less than that for the reverse reaction. For almost all reactions the activation energy for a forward reaction will differ from the activation energy of the reverse reaction, and the difference is $\Delta E°$ for the reaction.

The actual relation is $\Delta E° = E_a(\text{forward}) - E_a(\text{reverse})$. Since $\Delta E°$ differs from $\Delta H°$ only when there is a change in volume of the reaction system (under constant pressure), the difference in activation energies is often equated with the enthalpy change *(← p. 227).*

Bimolecular Reactions

An example of a bimolecular process is the reaction of iodide ion, I^-, with methyl bromide, CH_3Br, in aqueous solution.

$$I^-(aq) + CH_3Br(aq) \longrightarrow ICH_3(aq) + Br^-(aq)$$

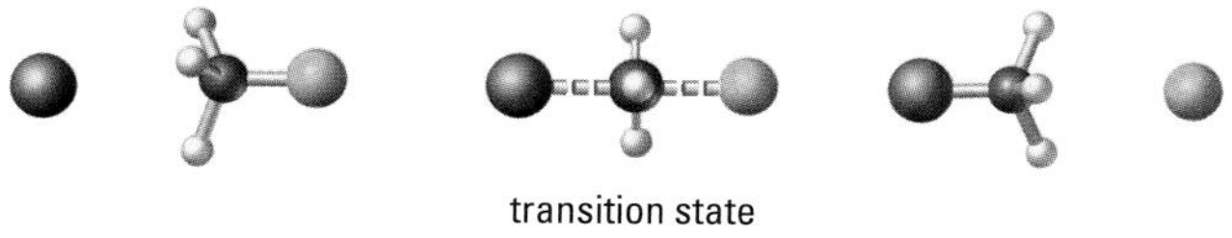

Here the equation for the elementary reaction shows that an iodide ion must collide with a methyl bromide molecule for the reaction to occur. The carbon-bromine bond does not break until after the iodine-carbon bond has begun to form. This makes sense, because just breaking a carbon-bromine bond would require a large increase in potential energy. Partially forming a carbon-iodine bond while the other bond is breaking lowers the potential energy. This makes the activation energy hill lower than it would otherwise be. Figure 13.8 shows the energy-versus-reaction progress diagram for this reaction.

The methyl bromide molecule has a tetrahedral shape that is distorted because the Br atom is much larger than the H atoms. Numerous experiments suggest that the reaction occurs most rapidly in solution when the I^- ion approaches the methyl bromide from the side of the tetrahedron opposite the bromine atom. That is, approach to only one of the four sides of CH_3Br can be effective, which limits reaction to only one fourth of all the collisions at most. This factor of one fourth is called a **steric factor** because it depends on the three-dimensional shapes of the reacting molecules. For molecules much more complicated than methyl iodide, such geometry restraints mean that only a very small fraction of the total collisions can lead to reaction. No wonder some chemical reactions are slow.

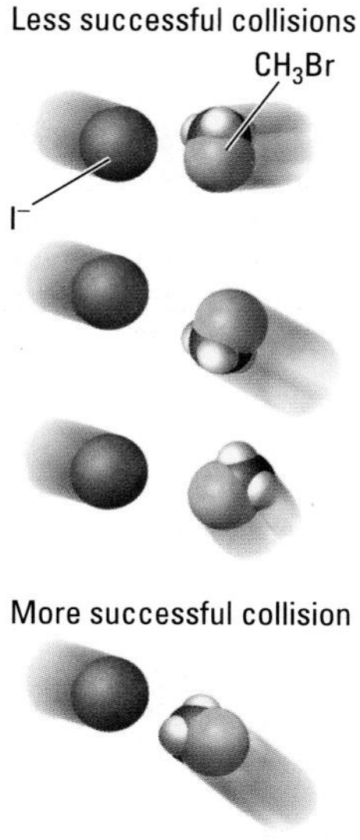

Less successful collisions. In the first three collisions shown, the iodide ion does not approach the methyl bromide molecule from the side opposite the bromine atom. None of these collisions is as likely to result in a reaction as is the collision shown at the bottom.

The word *steric* comes from the same root as the prefix *stereo-*, which means "three-dimensional."

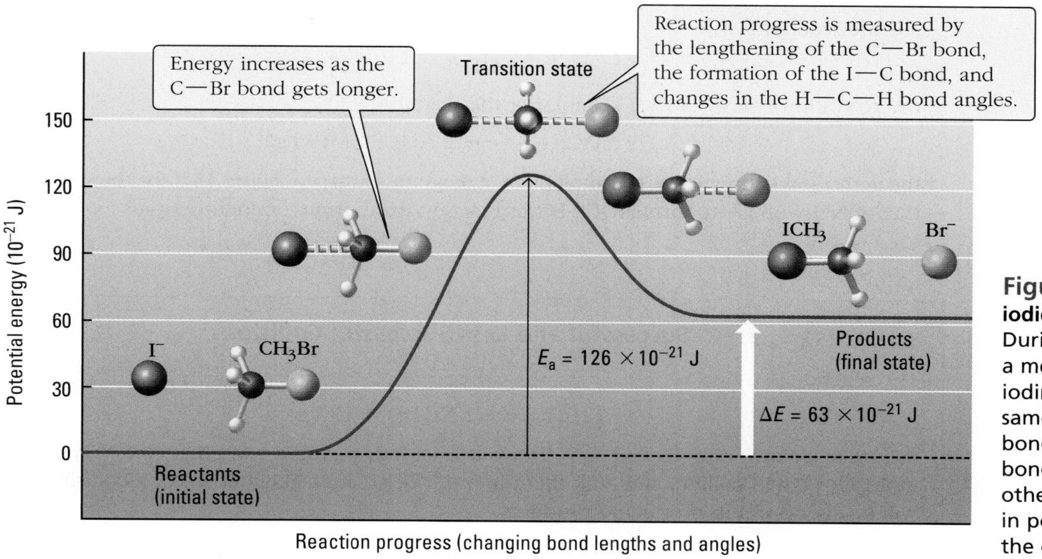

Figure 13.8 Energy diagram for iodide-methyl bromide reaction. During collision of an iodide ion with a methyl bromide molecule, a new iodine-carbon bond forms at the same time that the carbon-bromine bond is breaking. Forming the new bond partially compensates for what otherwise would be a large increase in potential energy due to breaking the carbon-bromine bond.

PROBLEM-SOLVING EXAMPLE 13.8 Reaction Energy Diagrams

A reaction by which ozone is destroyed in the stratosphere (← p. 454) is

$$O_3(g) + O(g) \longrightarrow 2\,O_2(g)$$

(O represents atomic oxygen, which is formed in the stratosphere when photons of ultraviolet light from the sun split oxygen molecules in two.) The activation energy for ozone destruction is 19 kJ/mol of O_3 consumed. Use standard enthalpies of formation from Appendix J to calculate the enthalpy change for this reaction. Then construct an energy diagram for the reaction. Draw vertical arrows to indicate the sizes of $\Delta H°$, E_a(forward), and E_a(reverse) for the reaction.

Remember that because there are equal numbers of moles of gas phase reactants and products, there is no volume change at constant temperature, and $\Delta E° = \Delta H°$ for this reaction (← p. 227).

Answer $\Delta H° = \Delta E° = -392$ kJ/mol of O_3 consumed, E_a(forward) = 19 kJ/mol O_3 consumed, E_a(reverse) = 411 kJ/mol O_3 formed; the energy diagram is shown below.

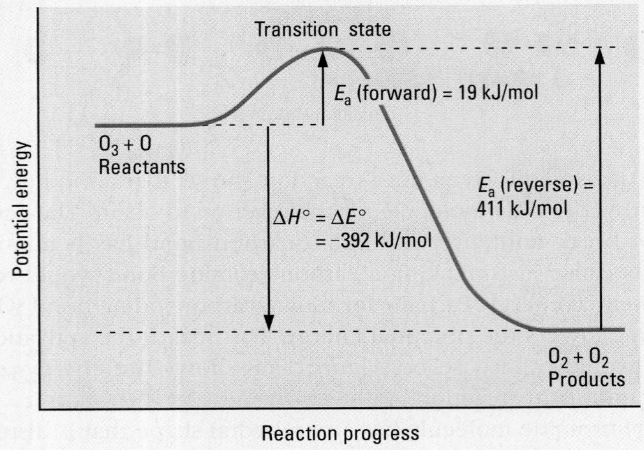

Strategy and Explanation Standard enthalpies of formation are 249.2 kJ/mol for ozone, 0 kJ/mol for diatomic oxygen, and 142.7 kJ/mol for atomic oxygen. Using these enthalpy values, we get $\Delta H° = 0 - (249.2 + 142.7)$ kJ/mol = −391.9 kJ/mol of O_3 consumed. The negative sign of $\Delta H°$ indicates that the reaction is exothermic, so the products must be lower in energy than the reactants by 391.9 kJ/mol. Since E_a(forward) = 19 kJ/mol, the transition state must be this much higher in energy than the reactants. Thus, the diagram can be drawn based on the E_a arrow and the $\Delta H°$ arrow. Then the third arrow (from products to the transition state) can be drawn. It indicates that

$$E_a(\text{reverse}) = -\Delta H° + E_a(\text{forward}) = (391.9 + 19)\ \text{kJ/mol} = 411\ \text{kJ/mol}$$

PROBLEM-SOLVING PRACTICE 13.8

For the reaction

$$Cl_2(g) + 2\,NO(g) \longrightarrow 2\,NOCl(g)$$

the activation energy is 18.9 kJ/mol. For the reaction

$$2\,NOCl(g) \longrightarrow 2\,NO(g) + Cl_2(g)$$

the activation energy is 98.1 kJ/mol. Draw a diagram similar to Figure 13.7 for the reaction of NO with Cl_2 to form NOCl. Is this reaction exothermic or endothermic? Explain.

CONCEPTUAL EXERCISE 13.8 Successful and Less Successful Collisions

The reaction

$$2\,NOCl(g) \longrightarrow 2\,NO(g) + Cl_2(g)$$

occurs in a single bimolecular step. Draw at least four possible ways that two NOCl molecules could collide, and rank them in order of greatest likelihood that a collision will be successful in producing products.

CHEMISTRY IN THE NEWS Bimolecular Collisions Can Be Complicated

In 2008 it was reported that bimolecular collisions are more complicated than chemists had thought. For the reaction of iodide ion with methyl bromide shown in the figure on p. 613, reaction is most likely when the iodide ion approaches the side of the methyl bromide molecule opposite the bromine atom. However, other types of collisions can also be successful. Scientists at the University of Freiburg, Germany, and at Texas Tech University report that for a closely related reaction,

$$Cl^-(g) + CH_3I(g) \longrightarrow ClCH_3(g) + I^-(g)$$

with collision energies well above the activation energy, a successful reaction can occur when the chloride ion hits the side, not the back, of the CH_3I molecule.

The experiments were done in a reactor at a very low gas pressure. The experiments involved forming a narrow beam of CH_3I molecules and shooting a beam of chloride ions across the path of the CH_3I beam. When the beams cross, individual chloride ions hit individual CH_3I molecules and react. Because the system is at a very low pressure there are very few molecules present and it is unlikely that the product molecules will collide with anything else until they reach a detector. After the reaction occurs, the product iodide ions can be detected at different angles from the direction of the original methyl iodide beam. (This is similar to, but more sophisticated than, the Rutherford-Geiger-Marsden experiment that revealed the nuclear atom *(← p. 43)*. The detector can report both the angle by which the collision deflects the iodide ion and how fast the iodide ion is moving (its kinetic energy). The chloride ions in the beam had energies within a narrow range, so measuring the speeds of the iodide ions and applying energy conservation (total energy of all particles in a collision has to remain constant) allowed the researchers to figure out how much energy was left in the $ClCH_3$ molecules after the reaction.

In about one of every ten successful collisions the product iodide ions had a lot less energy than they should have. This implied that the $ClCH_3$ molecules from those collisions had a lot more energy than expected. To account for this extra energy, the researchers proposed that the chloride ions hit the side of the CH_3I molecules, spun the CH_3I completely around (through 360°), and then replaced the iodine atom. This was dubbed the "roundabout" reaction, and it is shown schematically in the figure. After the collision the product $ClCH_3$ molecule is spinning as well as moving through space (like a top skittering across the floor). This means extra energy for each spinning $ClCH_3$ molecule and consequently less energy in the iodide ion.

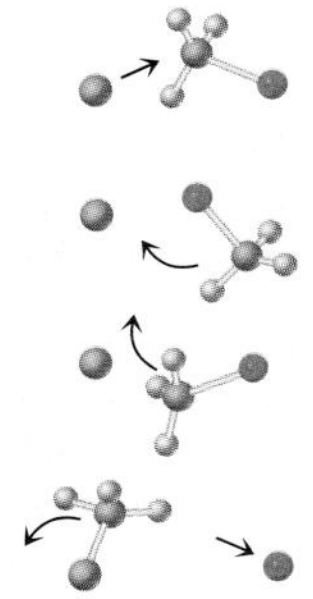

That a collision with the side of the CH_3I would result in reaction was amazing to scientists who study chemical kinetics, because for more than half a century it was thought that the only pathway for this kind of bimolecular reaction was the one shown in Figure 13.8. This new research also reveals the power of the tools now available for studying the changes in molecular structure during a chemical reaction. Those tools are likely to reveal many other secrets of nature in the near future. Some of those new revelations will certainly result in new ways to control chemical reactions for the benefit of humanity.

Sources: Based on information from *Science*, January 11, 2008; pp. 183–186; *Chemical & Engineering News*, January 14, 2008; p. 9; videos showing both the standard collision and the new roundabout collision are available at **http://pubs.acs.org/cen/news/86/i02/8602notw1.html**.

13.5 Temperature and Reaction Rate: The Arrhenius Equation

The most common way to speed up a reaction is to increase the temperature. A mixture of methane and air can be ignited by a lighted match, which raises the temperature of the mixture of reactants. This increases the reaction rate, the thermal energy evolved maintains the high temperature, and the reaction continues at a rapid rate. Reactions that speed up when the temperature is raised must slow down when the temperature is lowered. Foods are stored in refrigerators or freezers because the reactions in cells of microorganisms that produce spoilage occur more slowly at the lower temperature.

As a rough rule of thumb, the reaction rate increases by a factor of 2 to 4 for each 10-K rise in temperature.

Reaction rates increase with temperature because at a higher temperature a greater fraction of reactant molecules has enough energy to surmount the activation energy barrier. Consider again the conversion of *cis*- to *trans*-2-butene (Figure 13.7, p. 611). You learned in Section 10.3 *(← p. 429)* that gas phase molecules are constantly in motion and have a wide distribution of speeds and energies. At room

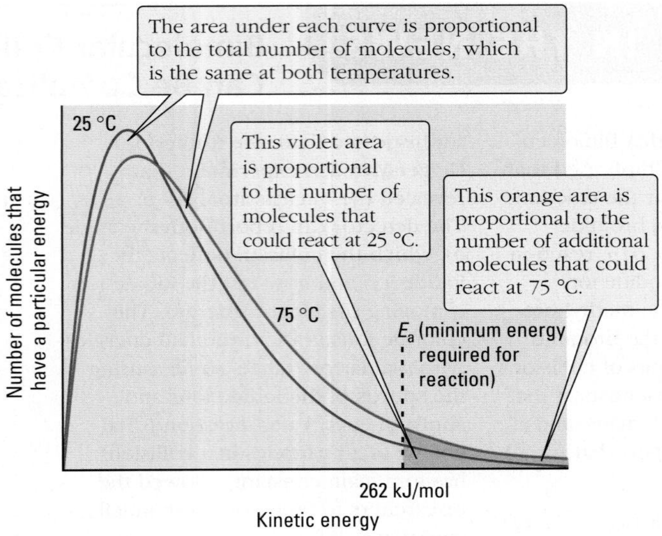

Figure 13.9 Energy distribution curves. The vertical axis gives the number of molecules that have the energy shown on the horizontal axis. Assume that a molecule reacts if it has more energy than the activation energy. The number of reactive molecules is given by the area under each curve to the right of the activation energy (262 kJ/mol). The two curves show that at 75 °C many more molecules have energies of 262 kJ/mol or higher than at 25 °C. (This graph is not to scale. Actually, 262 kJ/mol would be much farther to the right on the graph.)

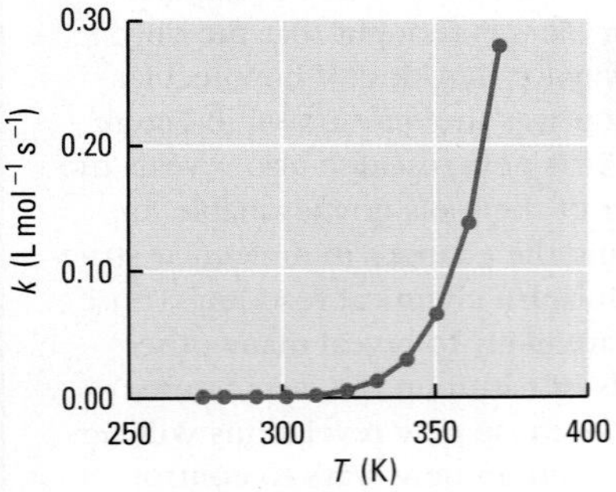

Figure 13.10 Effect of temperature on rate constant. The rate constant for the reaction of iodide ion with methyl bromide in aqueous solution is plotted as a function of temperature. The rate constant increases very rapidly with temperature, and the shape of the curve is characteristic of exponential increase.

Table 13.3 Temperature Dependence of Rate Constant for Iodide Plus Methyl Bromide Reaction

T(K)	k(L mol^{-1} s^{-1})	T(K)	k(L mol^{-1} s^{-1})	T(K)	k(L mol^{-1} s^{-1})
273	4.18×10^{-5}	310.	2.31×10^{-3}	350.	6.80×10^{-2}
280.	9.68×10^{-5}	320.	5.82×10^{-3}	360.	1.41×10^{-1}
290.	2.00×10^{-4}	330.	1.39×10^{-2}	370.	2.81×10^{-1}
300.	8.60×10^{-4}	340.	3.14×10^{-2}		

temperature relatively few *cis*-2-butene molecules have sufficient energy to surmount the energy barrier. However, as the temperature goes up, the number of molecules that have enough energy goes up rapidly, so the reaction rate increases rapidly.

The number of *cis*-2-butene molecules that have a given energy is shown by the curves in Figure 13.9. One curve is for 25 °C; the other is for 75 °C. The higher a point is on either curve, the greater is the number of molecules that have the energy corresponding to that point. The areas under the two curves are the same and represent the same total number of molecules. With a 50 °C rise in temperature, the number of molecules whose energy exceeds the activation energy is much higher, and so the reaction rate is also higher.

A reaction is faster at a higher temperature because its rate constant is larger. That is, ***a rate constant is constant only for a given reaction at a given temperature.*** For example, for the reaction of iodide ion with methyl bromide, the data shown in Table 13.3 are found for the rate constant at different temperatures.

Exponential curves also represent the growth of populations (such as human population) over time and are important in many other scientific fields. Logarithms and exponentials are discussed in Appendix A.6.

When the data from Table 13.3 are graphed (Figure 13.10), it is obvious that the rate constant increases very rapidly as temperature increases. A graph that is shaped like this is called an exponential curve. It can be represented by the equation

$$k = Ae^{-E_a/RT} \qquad \text{[Arrhenius equation]}$$

R is the constant found in the ideal gas law, *PV* = *nRT* (***← p. 438***). Here it is expressed in units of J mol^{-1} K^{-1} instead of L atm mol^{-1} K^{-1}, so the numerical value is different.

where *A* is the **frequency factor,** *e* is the base of the natural logarithm system (2.718 . . .), E_a is the activation energy, *R* is the gas law constant and has the value 8.314 J mol^{-1} K^{-1}, and *T* is the absolute (Kelvin) temperature (***← p. 435***). This equation is called the **Arrhenius equation** after its discoverer, Svante Arrhenius, a Swedish chemist.

The Arrhenius equation can be interpreted as follows. The frequency factor, A, depends on how often molecules collide when all concentrations are 1 mol/L and on whether the molecules are properly oriented when they collide. For example, in the case of the iodide-methyl bromide reaction, A includes the steric factor of $\frac{1}{4}$ that resulted because only one of the four sides of the CH_3Br molecule was appropriate for iodide to approach. The other term in the equation, $e^{-E_a/RT}$, gives the fraction of all the reactant molecules that have sufficient energy to surmount the activation energy barrier.

1946–

For his studies of the transition states of chemical reactions using femtosecond spectroscopy, Ahmed H. Zewail of the California Institute of Technology received the 1999 Nobel Prize in Chemistry. Zewail, who holds joint Egyptian and U.S. citizenship, pioneered the use of extremely short laser pulses—on the order of femtoseconds (10^{-15} s)—to study chemical kinetics. His technique has been called the world's fastest camera, and his research has enhanced understanding of many reactions, among them those involving rhodopsin and vision.

Determining Activation Energy

The activation energy and frequency factor can be obtained from experimental measurements of rate constants as a function of temperature (such as those in Table 13.3). When a large number of experimental data pairs are given, a graph is usually a good way of obtaining information from the data. This is easier to do if the graph is linear. The activation energy equation can be modified by taking natural logarithms of both sides so that its graph is linear.

$$k = Ae^{-E_a/RT}$$

$$\ln(k) = \ln(A) + \ln(e^{-E_a/RT}) = \ln(A) + (-E_a/RT)$$

Rearranging this equation gives the equation of a straight line.

$$\ln(k) = -\left(\frac{E_a}{R} \times \frac{1}{T}\right) + \ln(A)$$

$$y = (m \times x) + b$$

That is, if we graph $\ln(k)$ on the vertical (y) axis and $1/T$ on the horizontal (x) axis, the result should be a straight line whose slope is $-E_a/R$ and whose y-intercept is $\ln(A)$. For the data in Table 13.3, such a graph is shown in Figure 13.11. It is linear, its slope is -9.29×10^3 K, and the y-intercept is 23.85. Since the slope $= -E_a/R$, the activation energy can be calculated as

$$E_a = -(\text{slope}) \times R = -(-9.29 \times 10^3\ \text{K})\left(\frac{8.314\ \text{J}}{\text{mol K}}\right)\left(\frac{1\ \text{kJ}}{1000\ \text{J}}\right) = 77.2\ \text{kJ/mol}$$

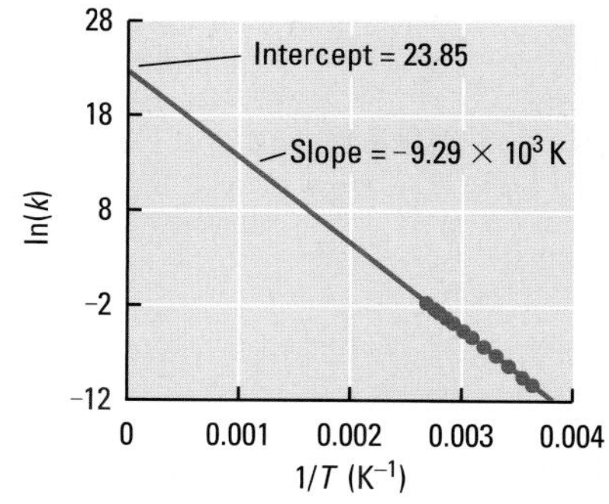

Figure 13.11 Determining activation energy graphically. A graph of ln(k) versus 1/T gives a straight line for the iodide-methyl bromide reaction. The activation energy can be obtained from the slope of the line, and the frequency factor A from its y-intercept. Because the line through the experimental data has to be extrapolated a long way to reach the y-intercept, determining A accurately requires measuring k over a wide range of temperatures.

The vertical axis plots $\ln(k)$, and k has units of L $\text{mol}^{-1}\ \text{s}^{-1}$. At the y-intercept, $\ln(A)$ equals $\ln(k)$, and therefore A, the frequency factor, must have the same units as k. Since $\ln(A) = 23.85$, the frequency factor, A, is

$$e^{23.85}\ \text{L mol}^{-1}\ \text{s}^{-1} = 2.28 \times 10^{10}\ \text{L mol}^{-1}\ \text{s}^{-1}$$

The Arrhenius equation can be used to calculate the rate constant at any temperature. For example, the rate constant for the reaction of iodide ion with methyl bromide can be calculated at 50. °C by substituting the temperature (in kelvins), the frequency factor, the activation energy, and the constant R into the equation.

$$k = Ae^{-E_a/RT} = (2.28 \times 10^{10}\ \text{L mol}^{-1}\ \text{s}^{-1})e^{(-77{,}200\ \text{J/mol})/(8.314\ \text{J K}^{-1}\ \text{mol}^{-1})(273.15 + 50.)\ \text{K}}$$

$$= (2.28 \times 10^{10}\ \text{L mol}^{-1}\ \text{s}^{-1})e^{-28.7} = 7.56 \times 10^{-3}\ \text{L mol}^{-1}\ \text{s}^{-1}$$

As a means of calculating rate constants, the Arrhenius equation works best within the range of temperatures over which the activation energy and frequency factor were determined. (For the reaction of iodide with methyl bromide, that range was 273 to 370 K.)

Be sure to use k_1 as the rate constant for T_1 and k_2 as the rate constant for T_2.

PROBLEM-SOLVING EXAMPLE 13.9 Temperature Dependence of Rate Constant

The experimental rate constant for the reaction of iodide ion with methyl bromide is 7.70×10^{-3} L mol^{-1} s^{-1} at 50.00 °C and 4.25×10^{-5} L mol^{-1} s^{-1} at 0.00 °C. Calculate the frequency factor and activation energy.

Answer $A = 1.72 \times 10^{10}$ L mol^{-1} s^{-1}; $E_a = 7.64 \times 10^4$ J mol^{-1}

Strategy and Explanation This problem could be solved by making a graph of ln(k) versus $1/T$, but with only two data pairs, a graph is not the best way. Instead, write two equations, one for each data pair.

$$k_1 = Ae^{-E_a/RT_1} \qquad k_2 = Ae^{-E_a/RT_2}$$

Now divide the first equation by the second to eliminate A.

$$\frac{k_1}{k_2} = \frac{Ae^{-E_a/RT_1}}{Ae^{-E_a/RT_2}} = e^{-E_a/RT_1} \times e^{+E_a/RT_2}$$

$$\frac{k_1}{k_2} = e^{\frac{E_a}{R}\left(\frac{1}{T_2} - \frac{1}{T_1}\right)}$$

Next take the natural logarithm of both sides.

$$\ln\left(\frac{k_1}{k_2}\right) = \frac{E_a}{R}\left(\frac{1}{T_2} - \frac{1}{T_1}\right)$$

This equation can be solved for E_a.

$$E_a = \frac{R\ln\left(\frac{k_1}{k_2}\right)}{\frac{1}{T_2} - \frac{1}{T_1}} = \frac{(8.314 \text{ J mol}^{-1} \text{ K}^{-1})\ln\left(\frac{7.70 \times 10^{-3}}{4.25 \times 10^{-5}}\right)}{\frac{1}{273.15 \text{ K}} - \frac{1}{323.15 \text{ K}}}$$

$$= \frac{43.23 \text{ J mol}^{-1} \text{ K}^{-1}}{5.664 \times 10^{-4} \text{ K}^{-1}} = 7.64 \times 10^4 \text{ J mol}^{-1}$$

Finally, using the calculated value of E_a, solve one of the rate constant expressions for A.

$$k_1 = Ae^{-E_a/RT_1}$$

$$A = \frac{k_1}{e^{-E_a/RT_1}} = k_1 e^{E_a/RT_1}$$

$$= (7.70 \times 10^{-3} \text{ L mol}^{-1} \text{ s}^{-1}) e^{\left(\frac{7.64 \times 10^4 \text{ J mol}^{-1}}{(8.314 \text{ J mol}^{-1} \text{ K}^{-1})(323.15 \text{ K})}\right)}$$

$$= (7.70 \times 10^{-3} \text{ L mol}^{-1} \text{ s}^{-1})(2.238 \times 10^{12}) = 1.72 \times 10^{10} \text{ L mol}^{-1} \text{ s}^{-1}$$

PROBLEM-SOLVING PRACTICE 13.9

Calculate the rate constant for the reaction of iodide ion with methyl bromide at a temperature of 75 °C.

Recall that the collision frequency contribution to the frequency factor is for 1-M concentrations. Equation 13.8 summarizes the effects of both temperature and concentration on rate of a reaction. The temperature effect depends primarily on the large increase in the number of sufficiently energetic collisions as the temperature increases, which shows up as larger values of k at higher temperatures. The effect of concentration is clearly indicated by the concentration terms in the rate law. If the rate law is known for a reaction, and if both the A and E_a values are known, then the rate can be calculated over a wide range of conditions.

The Arrhenius equation enables us to calculate the rate constant as a function of temperature, and the rate law shows how the rate depends on concentration. To obtain a single equation that summarizes the effects of both temperature and concentration, substitute k from the Arrhenius equation into the rate law for the iodide-methyl bromide reaction, giving

$$\text{Rate} = k \times [\text{I}^-] \times [\text{CH}_3\text{Br}]$$

$$\text{Rate} = A \times e^{-E_a/RT} \times [\text{I}^-] \times [\text{CH}_3\text{Br}] \qquad [13.8]$$

Collision frequency × steric factor | Fraction of sufficiently energetic molecules | Concentrations of colliding molecules

EXERCISE 13.9 Activation Energy and Experimental Data

The frequency factor A is 6.31×10^8 L mol^{-1} s^{-1} and the activation energy is 10. kJ/mol for the gas phase reaction

$$NO(g) + O_3(g) \longrightarrow NO_2(g) + O_2(g)$$

which is important in the chemistry of stratospheric ozone depletion.

(a) Calculate the rate constant for this reaction at 370. K.

(b) Assuming that this is an elementary reaction, calculate the rate of the reaction at 370. K if [NO] = 0.0010 M and $[O_3]$ = 0.00050 M.

13.6 Rate Laws for Elementary Reactions

An elementary reaction is a one-step process whose equation describes which nanoscale particles break apart, rearrange their positions, or collide to make a reaction occur. Therefore it is possible to figure out what the rate law and reaction order are for an elementary reaction, without doing an experiment. By contrast, when an equation represents a reaction that we do not understand at the nanoscale, rate laws and reaction orders must be determined experimentally (p. 602). The macroscale rate law can then be used to help develop a hypothesis about how a particular reaction takes place at the nanoscale.

Rate Law for a Unimolecular Reaction

In Section 13.4 we used the isomerization of *cis*-2-butene as an example of a reaction in which a single reactant molecule was converted to a product molecule or molecules—a unimolecular reaction (p. 610).

$$cis\text{-2-butene} \longrightarrow trans\text{-2-butene}$$

Suppose a flask contains 0.0050 mol/L of *cis*-2-butene vapor at room temperature. The molecules have a wide range of energies, but only a few of them have enough energy at this temperature to get over the activation energy barrier. Thus, during a given period only a few molecules twist sufficiently to become *trans*-2-butene. Now suppose that we double the concentration of *cis*-2-butene in the flask to 0.0100 mol/L, while keeping the temperature the same. The fraction of molecules with enough energy to cross over the barrier remains the same. However, as there are now twice as many molecules, twice as many must be crossing the barrier in any given time. Therefore, the rate of the *cis* $\longrightarrow$ *trans* reaction is twice as great. That is, the reaction rate is proportional to the concentration of *cis*-2-butene, and the rate law must be

Suppose that the fraction of molecules that have enough energy to react is 0.1%, or 0.001. If there are 10,000 molecules in a given volume, then $0.001 \times 10{,}000$ gives only 10 that have enough energy to react. If there are twice as many molecules in the same volume—that is, 20,000 molecules—then $0.001 \times 20{,}000$ gives 20 with enough energy to react, and the number reacting per unit volume (the rate) is twice as great.

$$\text{Rate} = k[cis\text{-2-butene}]$$

In the general case of any unimolecular elementary reaction,

$$\text{A} \longrightarrow \text{products} \qquad \text{the rate law is} \qquad \text{Rate} = k[\text{A}]$$

For any unimolecular reaction the nanoscale mechanism predicts that a first-order rate law will be observed in a macroscale laboratory experiment.

Rate Law for a Bimolecular Reaction

A good example of a reaction in which two molecules collide [a bimolecular reaction, (p. 613)] is the gas phase reaction of nitrogen monoxide and ozone that is involved in stratospheric ozone depletion and was mentioned in Exercise 13.9.

$$NO(g) + O_3(g) \longrightarrow NO_2(g) + O_2(g) \qquad [13.9]$$

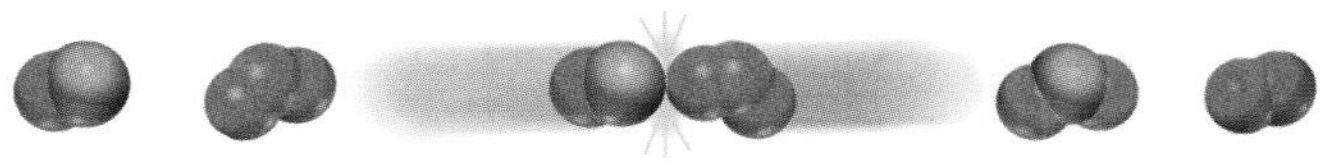

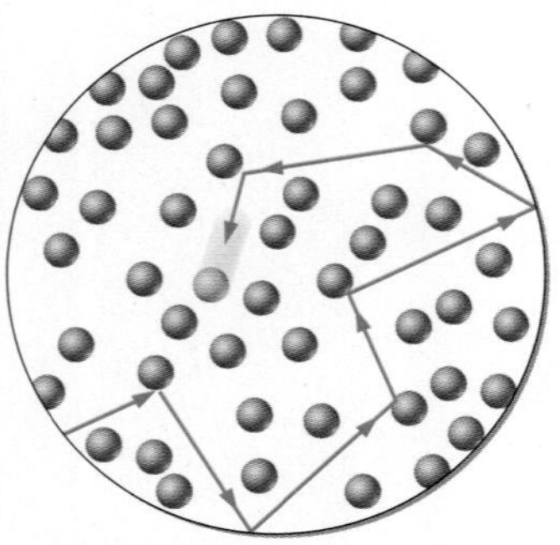

(a) A single green molecule moves among 50 purple molecules and collides with five of them per second.

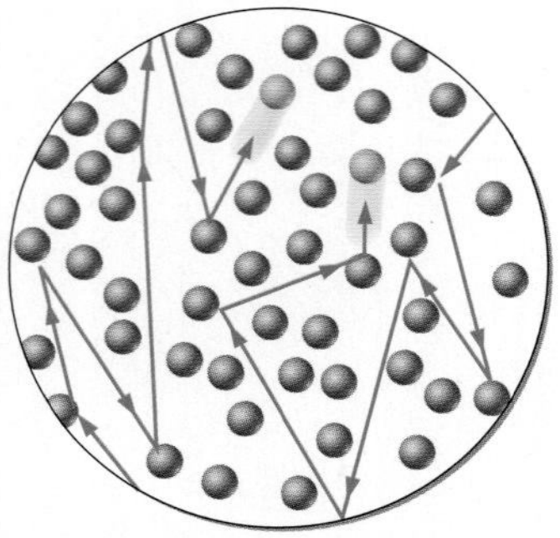

(b) Two green molecules now move among 50 purple molecules, and there are 10 green-purple collisions per second.

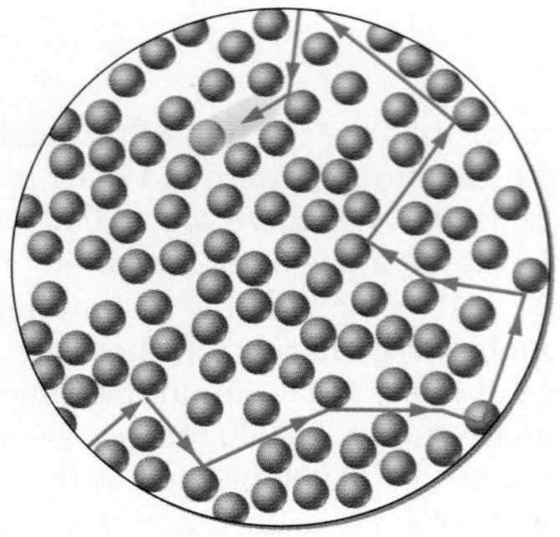

(c) If the number of purple molecules is doubled to 100, the frequency of green-purple collisions is also doubled, to 10 per second.

Figure 13.12 Effect of concentration on frequency of bimolecular collisions. The number of collisions is proportional to *both* the concentration of green molecules *and* the concentration of purple molecules. (In a real sample both the green and the purple molecules would be moving, but in these diagrams the motion of the purple molecules is not shown.)

Here the equation shows that the elementary reaction involves the collision of one NO molecule and one O_3 molecule. Since the molecules must collide to exchange atoms, the rate depends on the number of collisions per unit time.

Figure 13.12a represents one NO molecule (the green ball) and many O_3 molecules (the purple balls) in a tiny region within a flask where Reaction 13.9 is taking place. In a given time, the NO molecule collides with five O_3 molecules. If the concentration of NO molecules is doubled to two NO molecules in the same portion of the flask (Figure 13.12b), *each* NO molecule collides with five different O_3 molecules. Doubling the concentration of NO has doubled the number of collisions. This also doubles the rate, because the rate is proportional to the number of collisions (Equation 13.8 [p. 618]). The number of collisions also doubles when the O_3 concentration is doubled (Figure 13.12c). Thus, the rate law for this reaction must be

$$\text{Rate} = k[\text{NO}][\text{O}_3]$$

This description of the NO + O_3 reaction applies in general to bimolecular elementary reactions, even if the two molecules that must collide are of the same kind. That is, for the elementary reaction

$$\text{A} + \text{B} \longrightarrow \text{products} \qquad \text{the rate law is} \qquad \text{Rate} = k[\text{A}][\text{B}]$$

and for the elementary reaction

$$\text{A} + \text{A} \longrightarrow \text{products} \qquad \text{the rate law is} \qquad \text{Rate} = k[\text{A}]^2$$

For the NO + O_3 reaction the experimentally determined rate law is the same as the one we just derived by assuming the reaction occurs in one step. The chemical equation for the reaction is

$$\text{NO(g)} + \text{O}_3\text{(g)} \longrightarrow \text{NO}_2\text{(g)} + \text{O}_2\text{(g)}$$

and the experimental rate law is

$$\text{Rate} = k[\text{NO}][\text{O}_3]$$

This experimental observation suggests, but does not prove, that the reaction does take place in a single step. Other experimental evidence also suggests that this reaction is bimolecular.

In contrast, for the decomposition of hydrogen peroxide the chemical equation is

$$2\,\text{H}_2\text{O}_2\text{(aq)} \longrightarrow 2\,\text{H}_2\text{O}(\ell) + \text{O}_2\text{(g)}$$

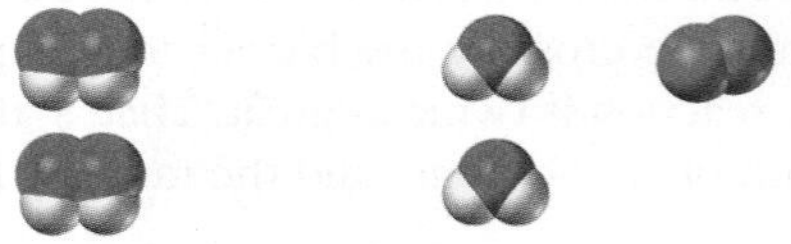

and the experimental rate law is Rate = $k[H_2O_2]$. This rate law proves that this reaction *cannot* occur in a single step that involves collision of two H_2O_2 molecules. A single-step bimolecular reaction would have a second-order rate law, but the observed rate law is first-order. This means that more than a single elementary step is needed when hydrogen peroxide decomposes.

CONCEPTUAL EXERCISE 13.10 Rate Law and Elementary Reactions

For the reaction

$$\text{NO}_2\text{(g)} + \text{CO(g)} \longrightarrow \text{NO(g)} + \text{CO}_2\text{(g)}$$

the experimentally determined rate law is

$$\text{Rate} = k[\text{NO}_2]^2$$

Does this reaction occur in a single step? Explain why or why not.

13.7 Reaction Mechanisms

Most chemical reactions do not take place in a single step. Instead, they involve a sequence of unimolecular or bimolecular elementary reactions. For each elementary reaction in the sequence we can write an equation. A set of such equations is called a **reaction mechanism** (or just a *mechanism*). For example, iodide ion can be oxidized by hydrogen peroxide in acidic solution to form iodine and water according to this overall equation:

$$2\,I^-(aq) + H_2O_2(aq) + 2\,H_3O^+(aq) \longrightarrow I_2(aq) + 4\,H_2O(\ell) \qquad [13.10]$$

When the acid concentration is between 10^{-3} M and 10^{-5} M, experiments show that the rate law is

$$\text{Rate} = k[I^-][H_2O_2]$$

The reaction is first-order in the concentrations of I^- and H_2O_2, and second-order overall.

Looking at Equation 13.10 for the oxidation of iodide ion by hydrogen peroxide, you might think that two iodide ions, one hydrogen peroxide molecule, and two hydronium ions would all have to come together at once. However, the rate law corresponds with a bimolecular collision of I^- and H_2O_2. It is highly unlikely that at the same time five ions or molecules would all be at the same place, be properly oriented, and have enough energy to react. Instead, chemists who have studied this reaction propose that initially one H_2O_2 (HOOH) molecule and one I^- ion come together:

Step 1:

$$HOOH + I^- \xrightarrow{\text{slow}} OH^- + HOI$$

This first step forms hypoiodous acid, HOI, and hydroxide ion, both known substances. The HOI then reacts with another I^- to form the product I_2:

Step 2:

$$HOI + I^- \xrightarrow{\text{fast}} OH^- + I_2$$

In each of Steps 1 and 2, a hydroxide ion was produced. Since the solution is acidic, these OH^- ions react immediately with H_3O^+ ions to form water.

Step 3:

$$2\,OH^- + 2\,H_3O^+ \xrightarrow{\text{fast}} 4\,H_2O$$

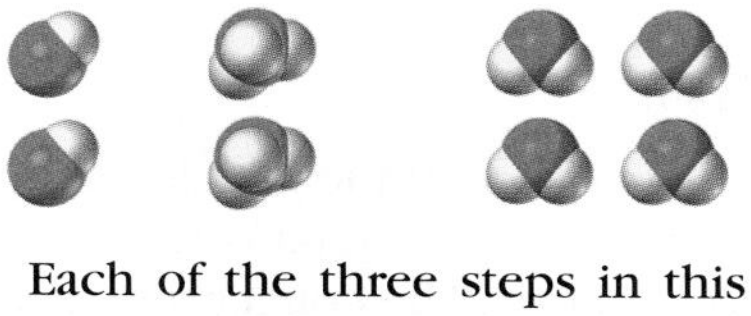

Each of the three steps in this mechanism is an elementary reaction. Each has its own activation energy, E_a, and its own rate constant, k. When the three steps are summed (by putting all the reactants on the left, putting all the products on the right, and eliminating formulas that appear as both reactants and products), the overall stoichiometric equation (Equation 13.10) is obtained. ***Any valid mechanism must consist of a series of unimolecular or bimolecular elementary reaction steps that sum to the overall reaction.***

$$\textit{Step 1:} \quad HOOH + I^- \xrightarrow{\text{slow}} \cancel{HOI} + \cancel{OH^-}$$

$$\textit{Step 2:} \quad \cancel{HOI} + I^- \xrightarrow{\text{fast}} I_2 + \cancel{OH^-}$$

$$\textit{Step 3:} \quad \cancel{2\,OH^-} + 2\,H_3O^+ \xrightarrow{\text{fast}} 4\,H_2O$$

$$\textit{Overall:} \quad 2\,I^- + HOOH + 2\,H_3O^+ \longrightarrow I_2 + 4\,H_2O \qquad [13.10]$$

Step 1 of the mechanism is slow, while Steps 2 and 3 are fast. Step 1 is called the **rate-limiting step;** because it is the slowest in the sequence, it limits the rate at which I_2 and H_2O can be produced. Steps 2 and 3 are rapid and therefore not rate-limiting. As soon as some HOI and OH^- are produced by Step 1, they are transformed into I_2 and H_2O by Steps 2 and 3. ***The rate of the overall reaction is limited by, and equal to, the rate of the slowest step in the mechanism.***

An analogy to the rate-limiting or rate-determining step is that no matter how quickly you shop in the supermarket, it seems that the time it takes to get out of the store depends on the rate at which you move through the checkout line.

Step 1 is a bimolecular elementary reaction. Therefore its rate must be first-order in HOOH and first-order in I^-. The mechanism predicts that the rate law should be

$$\text{Reaction rate} = k[\text{HOOH}][I^-]$$

which agrees with the experimentally observed rate law. ***A valid mechanism must correctly predict the experimentally observed rate law.***

If significant concentrations of an intermediate build up while a reaction is occurring, then reactants may disappear faster than products are formed (because buildup of the intermediate stores some of the used reactant before it is converted to final product). In such a case, the definition of reaction rate given in Equation 13.3 will not be correct. One way to detect formation of an intermediate is to notice that products are not formed as fast as reactants are used up.

The species HOI and OH^- are produced in Step 1 and used up in Step 2 or 3. In a mechanism, atoms, molecules, or ions that are produced in one step and consumed in a later step (or later steps) are called **reaction intermediates** (or just **intermediates**). Very small concentrations of HOI and OH^- are produced while the reaction is going on. Once the HOOH, the I^-, or both are used up, the intermediates HOI and OH^- are consumed by Steps 2 and 3 and disappear. HOI and OH^- are crucial to the reaction mechanism, but neither of them appears in the overall stoichiometric equation. If an experimenter is proficient enough to demonstrate that a particular intermediate was present, this provides additional evidence that a mechanism involving that intermediate is the correct one.

In summary, **a valid reaction mechanism should**

- **Consist of a series of unimolecular or bimolecular elementary reactions;**
- **Consist of reaction steps that sum to the overall reaction equation; and**
- **Correctly predict the experimentally observed rate law.**

EXERCISE 13.11 Rate Law for an Elementary Reaction

What is the rate law for Step 2 of the mechanism for the reaction of hydrogen peroxide and iodide ion?

Mechanisms with a Fast Initial Step

An experimental rate law should include only concentrations that can be measured experimentally. Usually this means concentrations of reactants (and perhaps products), but not the concentrations of intermediates. (The concentrations of intermediates usually are very small and therefore hard to measure.) The first step in the mechanism in the previous section was rate-limiting and involved two of the reactants, so it was easy to relate the overall rate to the concentrations of reactants. But what happens if the rate-limiting step is the second or a subsequent step?

Consider the reaction

$$2\,NO(g) + Br_2(g) \longrightarrow 2\,NOBr(g) \qquad \text{Rate (experimental)} = k[\text{NO}]^2[Br_2]$$

for which the currently accepted mechanism is

Step 1: (fast) $NO(g) + Br_2(g) \rightleftharpoons NOBr_2(g)$

Step 2: (slow) $NOBr_2(g) + NO(g) \longrightarrow 2\ NOBr(g)$

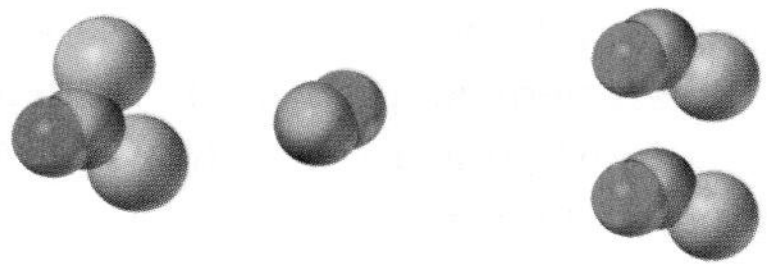

Because the second step is slow, $NOBr_2$ can often break apart and reform $NO + Br_2$ before it reacts to form products. We say that Step 1 is reversible, which is indicated by the double arrow ($\rightleftharpoons$). Once $NOBr_2$ forms, it can react in either of two ways: back to the reactants, $NO + Br_2$, or forward (in Step 2) to form the product, NOBr.

The overall rate of the reaction is the rate of the slow (rate-limiting) step (Step 2):

$$\text{Rate} = k[NOBr_2][NO]$$

However, this is not a valid rate law to compare with experimental results, because an experimental rate law should include only concentrations that can be measured experimentally. (Usually, the concentration of an intermediate, such as $NOBr_2$, cannot be measured.) Therefore, we need a way to relate the concentration of $NOBr_2$ to the concentrations of the reactants, NO and Br_2. This can be done by recognizing that in the mechanism there are three reactions and three rate constants:

Step +1 (forward), rate constant k_1

Step −1 (backward), rate constant k_{-1}

Step 2, rate constant k_2

The concentration of $NOBr_2$ depends on all three reactions. It is increased by Step +1 and decreased by Step −1 and Step 2. Initially, the concentration of $NOBr_2$ builds up because Step +1 produces $NOBr_2$. However, when $[NOBr_2]$ gets big enough, the rates of Step −1 and Step 2 get bigger, and eventually the $[NOBr_2]$ reaches what is called a steady state—it neither increases nor decreases until the reaction is nearly over.

In the steady state, the rate of the reaction by which $NOBr_2$ is formed must equal the sum of the two rates by which it is destroyed; that is,

$$\text{Rate of Step } +1 = \text{rate of Step } -1 + \text{rate of Step } 2$$

$$k_1[NO][Br_2] = k_{-1}[NOBr_2] + k_2[NOBr_2][NO] \qquad [13.11]$$

Equation 13.11 can usually be simplified because Step 2 is much slower than either Step 1 or Step −1. This means that $k_2[NOBr_2][NO] \ll k_{-1}[NOBr_2]$ and $k_2[NOBr_2][NO]$ can be neglected in the calculation.

$$k_1[NO][Br_2] \cong k_{-1}[NOBr_2] \qquad [13.12]$$

Equation 13.12 can be rearranged algebraically to show that $[NOBr_2]$ is proportional to the concentration of each reactant.

$$[NOBr_2] = \frac{k_1}{k_{-1}}[NO][Br_2]$$

Suppose that $k_{-1}[NOBr_2] = 0.20\ \text{mol L}^{-1}\ \text{s}^{-1}$ and $k_2[NOBr_2][NO] = 0.00020\ \text{mol L}^{-1}\ \text{s}^{-1}$, which is much smaller. Then $k_{-1}[NOBr_2] - k_2[NOBr_2][NO]$ $= (0.20 - 0.00020)\ \text{mol L}^{-1}\ \text{s}^{-1}$ $= 0.20\ \text{mol L}^{-1}\ \text{s}^{-1}$. That is, the rate for Step 2 is negligible and can be ignored in the calculation.

This allows the rate to be expressed in terms of the concentrations of reactants.

$$\text{Rate} = k_2[\text{NOBr}_2][\text{NO}] = k_2\left(\frac{k_1}{k_{-1}}[\text{NO}][\text{Br}_2]\right)[\text{NO}]$$

$$= \frac{k_1 k_2}{k_{-1}}[\text{NO}]^2[\text{Br}_2] = k'[\text{NO}]^2[\text{Br}_2] \quad [13.13]$$

Equation 13.13 shows that the rate constant k' is actually a quotient of rate constants for three elementary reactions, but the rate is proportional to the concentration of Br_2 and to the square of the concentration of NO. That is, the mechanism predicts that the reaction is second-order in NO and first-order in Br_2, which agrees with the experimental rate law. For mechanisms in which the rate-limiting step is the second or a subsequent step, a mathematical relationship such as Equation 13.13 can usually be found that relates an overall rate constant, k', to the concentrations of the reactants and the rate constants for the steps up to and including the rate-limiting step.

Kinetics and Mechanism

Studying the kinetics of a chemical reaction involves collecting data on the concentrations of reactants as a function of time. From such data the rate law for the reaction and a rate constant can usually be obtained. The reaction can also be studied at several different temperatures to determine its activation energy. This allows us to predict how fast the macroscale reaction will be under a variety of experimental conditions, but it does not provide definitive information about the nanoscale mechanism by which the reaction takes place. A reaction mechanism is an educated guess—a hypothesis—about the way the reaction occurs. If the mechanism predicts correctly the overall stoichiometry of the reaction *and* the experimentally determined rate law, then it is a reasonable hypothesis. However, it is impossible to prove for certain that a mechanism is correct. Sometimes several mechanisms can agree with the same set of experiments. This is what makes kinetic studies one of the most interesting and rewarding areas of chemistry, but it also can provoke disputes among scientists who favor different possible mechanisms for the same reaction.

PROBLEM-SOLVING EXAMPLE 13.10 Rate Law and Reaction Mechanism

The gas phase reaction between nitrogen monoxide and oxygen,

$$2\,\text{NO(g)} + \text{O}_2\text{(g)} \longrightarrow 2\,\text{NO}_2\text{(g)}$$

is found experimentally to obey the rate law

$$\text{Rate} = k[\text{NO}]^2[\text{O}_2]$$

Decide which of these mechanisms is (are) compatible with this rate law.

(a) $NO + NO \rightleftharpoons N_2O_2$ fast
$N_2O_2 + O_2 \longrightarrow 2\,NO_2$ slow

(b) $NO + NO \longrightarrow NO_2 + N$ slow
$N + O_2 \longrightarrow NO_2$ fast

(c) $NO + O \rightleftharpoons NO_2$ fast
$NO_2 + NO \longrightarrow N_2O_3$ fast
$N_2O_3 + O \longrightarrow 2\,NO_2$ slow

(d) $NO + O_2 \rightleftharpoons NO_3$ fast
$NO_3 + NO \longrightarrow 2\,NO_2$ slow

(e) $NO + O_2 \longrightarrow NO_2 + O$ slow
$NO + O \longrightarrow NO_2$ fast

(f) $2\,NO + O_2 \longrightarrow 2\,NO_2$

Answer Mechanisms (a) and (d) are compatible with the rate law and stoichiometry.

Strategy and Explanation Examine each mechanism to see whether it (1) consists only of unimolecular and bimolecular steps, (2) agrees with the overall stoichiometry, and (3) predicts the experimental rate law. Eliminate those that do not. The remaining mechanism(s) may be correct.

Mechanism (f) involves collision of three molecules: two NO and one O_2. It can be eliminated because it does not consist of unimolecular or bimolecular steps. All of the other mechanisms consist of bimolecular steps.

Mechanism (c) does not have O_2 as a reactant in the overall stoichiometry, so it can be eliminated. All other mechanisms predict the observed overall stoichiometry.

In mechanism (a) the first step is fast and reversible. Applying the idea that the rates are approximately equal for the forward and reverse reactions in that first step gives

$$k_1[\mathrm{NO}]^2 = k_{-1}[\mathrm{N_2O_2}] \quad \text{and} \quad [\mathrm{N_2O_2}] = \frac{k_1}{k_{-1}}[\mathrm{NO}]^2$$

Since the overall rate equals the rate of the rate-limiting step,

$$\text{Rate} = k_2[\mathrm{N_2O_2}][\mathrm{O_2}] = k_2\frac{k_1}{k_{-1}}[\mathrm{NO}]^2[\mathrm{O_2}] = k'[\mathrm{NO}]^2[\mathrm{O_2}]$$

Consequently, mechanism (a) could be the actual mechanism.

The slow first step in mechanism (b) implies an overall rate $= k[\mathrm{NO}]^2$, which eliminates it from consideration.

Continuing this kind of reasoning, mechanism (d) is seen to be a possibility, but mechanism (e) predicts rate $= k[\mathrm{NO}][\mathrm{O_2}]$ and therefore can be eliminated. Because there are still two possible mechanisms, (a) and (d), additional experiments need to be done to try to distinguish between them.

PROBLEM-SOLVING PRACTICE 13.10

The Raschig reaction produces the industrially important reducing agent hydrazine, N_2H_4, from ammonia, NH_3, and hypochlorite ion, OCl^-, in basic aqueous solution. A proposed mechanism is

Step 1: $\mathrm{NH_3(aq) + OCl^-(aq) \xrightarrow{slow} NH_2Cl(aq) + OH^-(aq)}$

Step 2: $\mathrm{NH_2Cl(aq) + NH_3(aq) \xrightarrow{fast} N_2H_5^+(aq) + Cl^-(aq)}$

Step 3: $\mathrm{N_2H_5^+(aq) + OH^-(aq) \xrightarrow{fast} N_2H_4(aq) + H_2O(\ell)}$

(a) What is the overall stoichiometric equation?
(b) Which step is rate-limiting?
(c) What reaction intermediates are involved?
(d) What rate law is predicted by this mechanism?

EXERCISE 13.12 Rate Law and Mechanism

Consider the reaction mechanism

$$\mathrm{ICl(g) + H_2(g) \rightleftharpoons HI(g) + HCl(g)} \quad \text{fast}$$

$$\mathrm{HI(g) + ICl(g) \longrightarrow HCl(g) + I_2(g)} \quad \text{slow}$$

(a) What is the overall reaction equation?
(b) Derive the rate law predicted by this mechanism.
(c) Does the rate law depend on the concentration of one of the products of the reaction?
(d) Would the rate constant determined from the initial rate of this reaction equal the rate constant determined at a time when 80% of the reactants had been consumed? Explain why or why not.

13.8 Catalysts and Reaction Rate

Raising the temperature increases a reaction rate because it increases the fraction of molecules that are energetic enough to surmount the activation energy barrier. Increasing reactant concentrations can also increase the rate because it increases the number of molecules per unit volume. A third way to increase reaction rates is to add a catalyst.

For example, an aqueous solution of hydrogen peroxide can decompose to water and oxygen.

$$2\,H_2O_2(aq) \longrightarrow O_2(g) + 2\,H_2O(\ell)$$

At room temperature the rate of the decomposition reaction is exceedingly slow. If the hydrogen peroxide solution is stored in a cool, dark place in a clean plastic container, it is stable for months. However, in the presence of a manganese salt, an iodide-containing salt, or a biological catalyst *(an enzyme;* see p. 629), the decomposition reaction can occur quite rapidly (Figure 13.13a).

Ammonium nitrate is used as fertilizer and is stable at room temperature. At higher temperatures and in the presence of chloride ion as a catalyst, however, ammonium nitrate can explode with tremendous force (Figure 13.13b). Approximately 600 people were killed in Texas City, Texas, in 1947 when workers tried to use salt water (which contains ~0.5-M Cl^-) to extinguish a fire in the hold of the ship *Grandcamp* and the ammonium nitrate cargo exploded (Figure 13.13c).

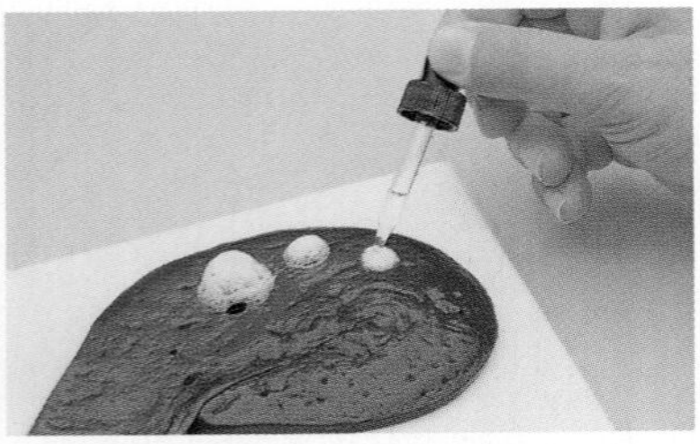

(a) A 30% aqueous solution of H_2O_2 is dropped onto a piece of liver. The liquid foams as H_2O_2 rapidly decomposes to O_2 and H_2O. Liver contains an enzyme that catalyzes the decomposition of H_2O_2.

(b) Laboratory-scale explosion of a sample of ammonium nitrate, NH_4NO_3, catalyzed by chloride ion.

(c) Scene following explosion of the ship *Grandcamp* in Texas City, Texas, April 16, 1947.

Figure 13.13 Catalysis in action.

How does a catalyst or an enzyme help a reaction to go faster? It does so by participating in the reaction mechanism. That is, *the mechanism for a catalyzed reaction is different from the mechanism of the same reaction without the catalyst.* The rate-limiting step in the catalyzed mechanism has a lower activation energy and therefore is faster than the slow step for the uncatalyzed reaction. To see how this works, let us again consider conversion of *cis*- to *trans*-2-butene in the gas phase.

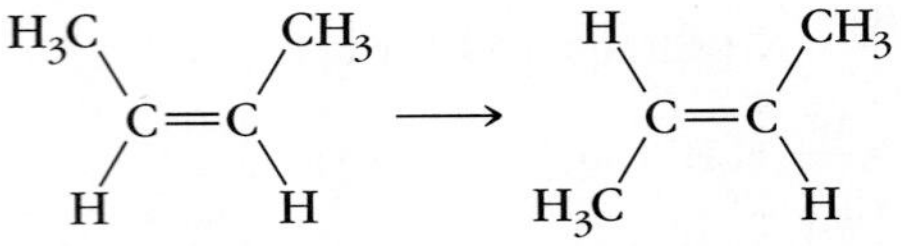

cis-2-butene → *trans*-2-butene

Rate = k [*cis*-2-butene]

If a trace of gaseous molecular iodine, I_2, is added to a sample of *cis*-2-butene, the iodine accelerates the change to *trans*-2-butene. The iodine is neither consumed nor produced in the overall reaction, so it does not appear in the overall balanced equation. However, because the reaction rate depends on the concentration of I_2, there is a term involving concentration of I_2 in the rate law for the catalyzed reaction.

$$\text{Rate} = k[\textit{cis}\text{-2-butene}][I_2]^{1/2}$$

The exponent of $\frac{1}{2}$ for the concentration of I_2 in the rate law indicates the square root of the concentration. A square root dependence usually means that only half a molecule—in this case a single iodine atom—is involved in the mechanism.

The rate of the conversion of *cis*- to *trans*-2-butene changes because the presence of I_2 somehow changes the reaction mechanism. The best hypothesis is that iodine molecules first dissociate to form iodine atoms.

***Step 1:* I_2 dissociation**

$$\tfrac{1}{2}[\,I_2(g) \longrightarrow 2\,I(g)\,]$$

(This equation is multiplied by $\frac{1}{2}$ because only one of the two I atoms from the I_2 molecule is needed in subsequent steps of the mechanism.) An iodine atom then attaches to the *cis*-2-butene molecule, breaking half of the double bond between the two central carbon atoms and allowing the ends of the molecule to twist freely relative to each other.

Step 2: **Attachment of I atom to *cis*-2-butene**

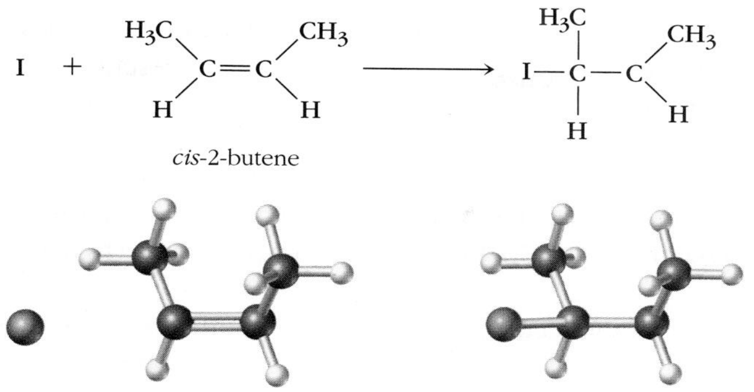

Step 3: **Rotation around the C—C bond**

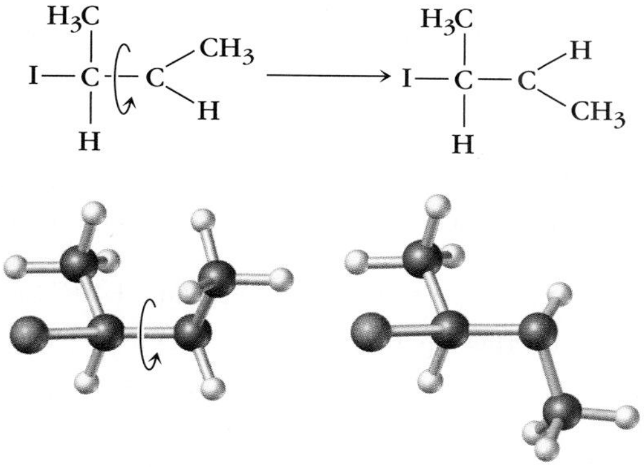

Step 4: **Loss of an I atom and re-formation of the carbon-carbon double bond**

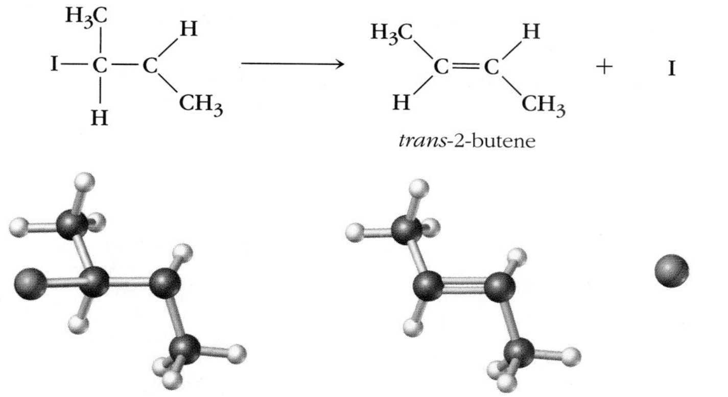

After the new double bond forms to give *trans*-2-butene and the iodine atom falls away, two iodine atoms come together to regenerate molecular iodine.

Step 5: **I_2 regeneration**

$$\frac{1}{2}[\,2\,I(g) \longrightarrow I_2(g)\,]$$

There are five important points concerning this mechanism.

- The I_2 dissociates to atoms and then reforms. To an "outside" observer the concentration of I_2 is unchanged; I_2 is not involved in the balanced stoichiometric

Because $k = Ae^{-E_a/RT}$,

$$\frac{k_2}{k_1} = \frac{Ae^{-E_{a_2}/RT}}{Ae^{-E_{a_1}/RT}} = e^{-(E_{a_2}-E_{a_1})/RT}$$

$$= e^{\left(\frac{(262{,}000-115{,}000)\text{ J mol}^{-1}}{8.314\text{ J K}^{-1}\text{ mol}^{-1}\times 500.\text{ K}}\right)}$$

$$= e^{3.536\times 10^{1}} = 2.28\times 10^{15}$$

equation even though it has appeared in the mechanism. *This is generally true of catalysts.*

- Figure 13.14 shows that the activation energy barrier is significantly lower for the catalyzed reaction (because the mechanism is different). Consequently the reaction rate is much faster. Dropping the activation energy from 262 kJ/mol for the uncatalyzed reaction to 115 kJ/mol for the catalyzed process makes the catalyzed reaction 10^{15} times faster at a temperature of 500. K.
- The catalyzed mechanism has five reaction steps, and its energy-versus-reaction-progress diagram (Figure 13.14) has five energy barriers (five humps appear in the curve).
- The catalyst I_2 and the reactant *cis*-2-butene are both in the gas phase during the reaction. When a catalyst is present in the same phase as the reacting substance or substances, it is called a **homogeneous catalyst.**
- Although the mechanism is different, the initial and final energies for the catalyzed reaction are the same as for the uncatalyzed reaction. This means that ΔE and ΔH are the same for the catalyzed as for the uncatalyzed reaction.

EXERCISE 13.13 Catalysis

The oxidation of thallium(I) ion by cerium(IV) ion in aqueous solution has the equation

$$2\,Ce^{4+}(aq) + Tl^{+}(aq) \longrightarrow 2\,Ce^{3+}(aq) + Tl^{3+}(aq)$$

The accepted mechanism for this reaction is

Step 1: $Ce^{4+}(aq) + Mn^{2+}(aq) \longrightarrow Ce^{3+}(aq) + Mn^{3+}(aq)$

Step 2: $Ce^{4+}(aq) + Mn^{3+}(aq) \longrightarrow Ce^{3+}(aq) + Mn^{4+}(aq)$

Step 3: $Mn^{4+}(aq) + Tl^{+}(aq) \longrightarrow Mn^{2+}(aq) + Tl^{3+}(aq)$

(a) Verify that this mechanism predicts the overall reaction.
(b) Identify all intermediates in this mechanism.
(c) Identify the catalyst in this mechanism.
(d) Suppose that the first step in this mechanism is rate-limiting. What would the rate law be?
(e) Suppose that the second step in this mechanism is rate-limiting. What would the rate law be?

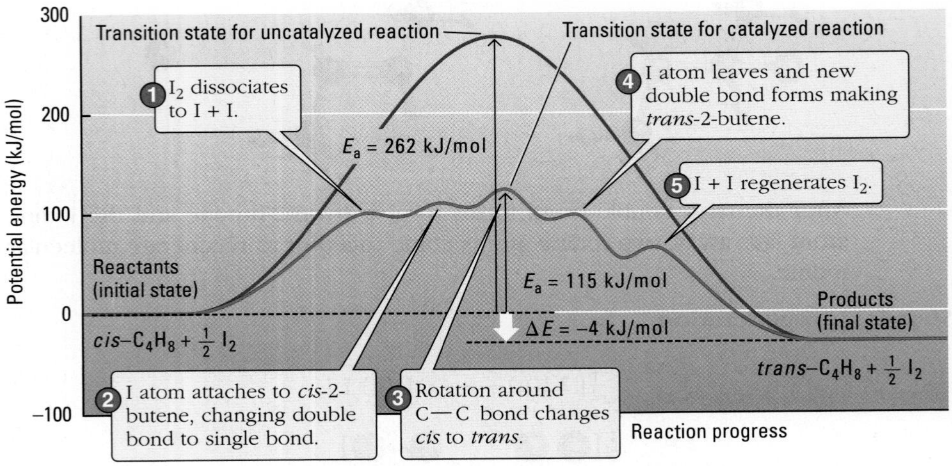

Figure 13.14 Energy diagrams for catalyzed and uncatalyzed reactions. A catalyst accelerates a reaction by altering the mechanism so that the activation energy is reduced. With a smaller barrier to overcome, more reactant molecules have enough energy to cross the barrier, and reaction occurs more readily. (The steps involved are described in the text.) Notice that the shape of the barrier has changed because the mechanism has changed. This changes the activation energy, but not ΔE for the reaction.

13.9 Enzymes: Biological Catalysts

Your body is a chemical factory of cells that can manufacture a broad range of compounds that are needed so that you can move, breathe, digest food, see, hear, smell, and even think. But did you ever consider how the reactions that make those compounds are controlled? And how they can all occur reasonably quickly at the relatively low body temperature of 37 °C? Oxidation of glucose powers all the systems of your body, but you would not want it to take place at the temperature it does when cellulose ***(a polymer of glucose, ← p. 577)*** in wood burns in a fireplace. The chemical reactions of your body are catalyzed by enzymes. An **enzyme** is *a highly efficient catalyst for one or more chemical reactions in a living system.* The presence or absence of appropriate enzymes turns these reactions on or off by speeding them up or slowing them down. This allows your body to maintain nearly constant temperature and nearly constant concentrations of a variety of molecules and ions, an absolute necessity if you are to continue functioning.

Enzymes are usually proteins, but other biological macromolecules, such as RNA, can also increase the rates of reactions in living systems. Most enzymes are globular proteins, polymers of amino acids ***(← p. 577)*** in which one or more long chains of amino acids fold into a nearly spherical shape. The shape of a globular protein is determined by noncovalent interactions ***(← p. 582)*** among the amino acid components (hydrogen bonds, attractions of opposite ionic charges, dipole-dipole and ion-dipole forces), a few weak covalent bonds, and the fact that nonpolar (hydrophobic) amino acid side groups congregate in the middle of the molecule, avoiding the surrounding aqueous solution.

Enzymes are among the most effective catalysts known. They can increase reaction rates by factors of 10^9 to 10^{19}. For example, essentially every collision of the enzyme carbonic anhydrase with a carbonic acid molecule results in decomposition, and the enzyme can decompose about 36 million H_2CO_3 molecules every minute.

$$H_2CO_3(aq) \xrightarrow{\text{carbonic anhydrase}} CO_2(g) + H_2O(\ell)$$

Most enzymes are highly specific catalysts. Some act on only one or two of the hundreds of different substances found in living cells. For example, carbonic anhydrase catalyzes only the decomposition of carbonic acid. Other enzymes can speed up several reactions, but usually these reactions are all of the same type.

Some enzymes can act as catalysts entirely on their own. Others require one or more inorganic or organic molecules or ions called **cofactors** to be present before their catalytic activity becomes fully available. For example, many enzymes require nicotinamide adenine dinucleotide ion, NAD^+ (niacinamide ion). Molecules or ions that are cofactors are often derived from small quantities of minerals and vitamins in our diets ***(← p. 108).*** If the cofactor needed for an enzyme to catalyze a reaction is not available because of dietary deficiency, that reaction cannot occur when it is needed, and a bodily function will be impaired.

Enzyme Activity and Specificity

A **substrate** is a molecule whose reaction is catalyzed by an enzyme. In some cases there may be more than one substrate, as when an enzyme catalyzes transfer of a group of atoms from one molecule to another. Enzyme catalysis is extremely effective and specific because the structure of the enzyme is finely tuned to minimize the activation energy barrier. Usually one part of the enzyme molecule, called the **active site,** interacts with the substrate via the same kinds of noncovalent attractions that hold the enzyme in its globular structure. The nanoscale structure of an enzyme's active site is specifically suited to attract and bind a substrate molecule and to help the substrate react.

Oxidation of cellulose in wood.

The 1989 Nobel Prize in Chemistry was given to Thomas Cech and Sidney Altman for the discovery that RNA molecules as well as proteins can be biological catalysts. In fact, the ribosome, the large protein/RNA complex that is responsible for all polymerization of amino acids into intact proteins, uses RNA to accomplish its reactions.

The weak covalent bonds in some globular proteins are disulfide bonds. They occur between sulfur atoms in side chains of the amino acid cysteine. A cysteine side chain at one point in the protein can become bonded to a cysteine side chain much farther along the protein backbone ***(← p. 580).***

CHEMISTRY YOU CAN DO Enzymes: Biological Catalysts

Raw potatoes contain an enzyme called *catalase,* which converts hydrogen peroxide to water and oxygen. You can demonstrate this by performing the following experiment.

Purchase a small bottle of hydrogen peroxide at a pharmacy or find one in your medicine chest. The peroxide is usually sold as a 3% solution in water. Pour about 50 mL of the peroxide solution into a clear glass or plastic cup. Add a small slice of a fresh potato to the cup. (Since potato is less dense than water, the potato will float.)

Almost immediately you will see bubbles of oxygen gas on the potato slice. (To make the bubbles more obvious you can add some dishwashing soap.)

Think about these questions:

1. Does the rate of evolution of oxygen change with time? If so, how does it change?
2. If you cool the hydrogen peroxide solution in a refrigerator and then do the experiment, is there a perceptible change in the initial rate of O_2 evolution?
3. Is there a difference between the time at which O_2 evolution begins for warm and for cold hydrogen peroxide?
4. What happens to the rate of oxygen evolution if you heat the slice of potato on a stove or in an oven before adding it to the peroxide solution?

When a substrate binds to an enzyme, the structures of both molecules can change. Each structure adjusts to fit closely with the other, and the structures become complementary. The change in shape of either the enzyme, the substrate, or both molecules when they bind is called **induced fit.** Enzymes catalyze reactions of only certain molecules because the structures of most molecules are not close enough to the structure of the active site for an induced fit to occur. The induced fit of a substrate to an enzyme also can lower the activation energy for a reaction. For example, it may distort the substrate and stretch a bond that will be broken in the desired reaction. A schematic example of how this can work is shown in Figure 13.15. To see how it works in a specific case, consider the enzyme lysozyme, whose structure is shown in Figure 13.16 as a space-filling model with substrate in the active site. Lysozyme catalyzes hydrolysis reactions of polysaccharides ***(← p. 576)*** found in bacterial cell walls. The reaction involved is

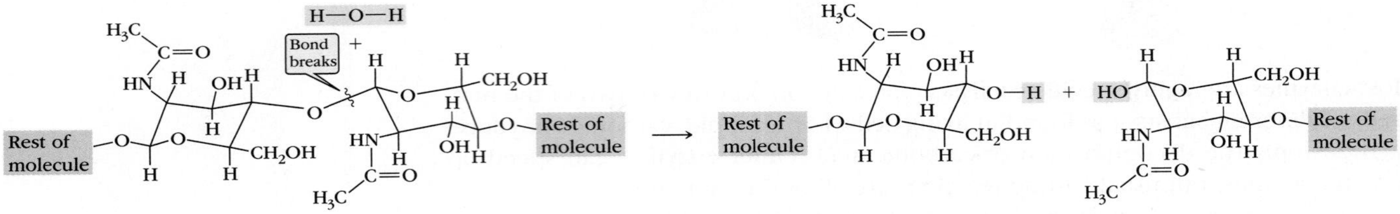

In Section 12.6 ***(← p. 560)*** hydrolysis was described as a reaction in which a water molecule and some other molecule react, with both molecules splitting in two. In the lysozyme-catalyzed reaction, the H from the water ends up with one part of the substrate molecule, and the OH ends up with the other part.

A hydrolysis reaction is the opposite of a condensation reaction. Most biopolymers are formed by condensation. Breaking them into their building block molecules requires hydrolysis, and many important enzymes catalyze hydrolysis reactions.

The section of polysaccharide shown in Figure 13.16 fits nicely into the cleft along the surface of the lysozyme, but many other long-chain molecules, such as

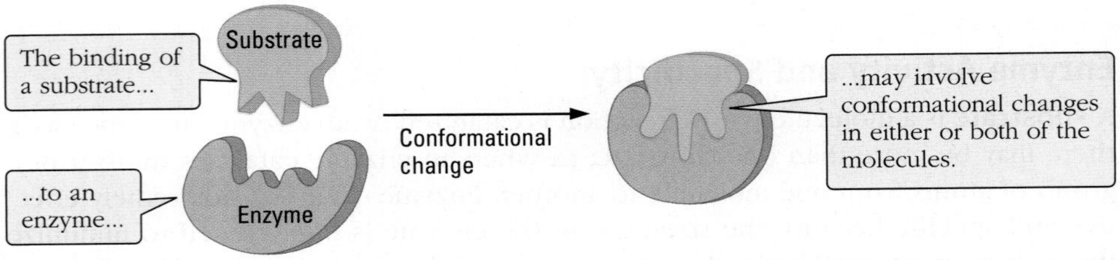

Figure 13.15 Induced fit of substrate to enzyme. Binding of a substrate to an enzyme may involve changing the shape of either or both molecules, thereby inducing them to fit together. In some cases a substrate molecule may be stretched or strained, helping bonds to break and reaction to occur.

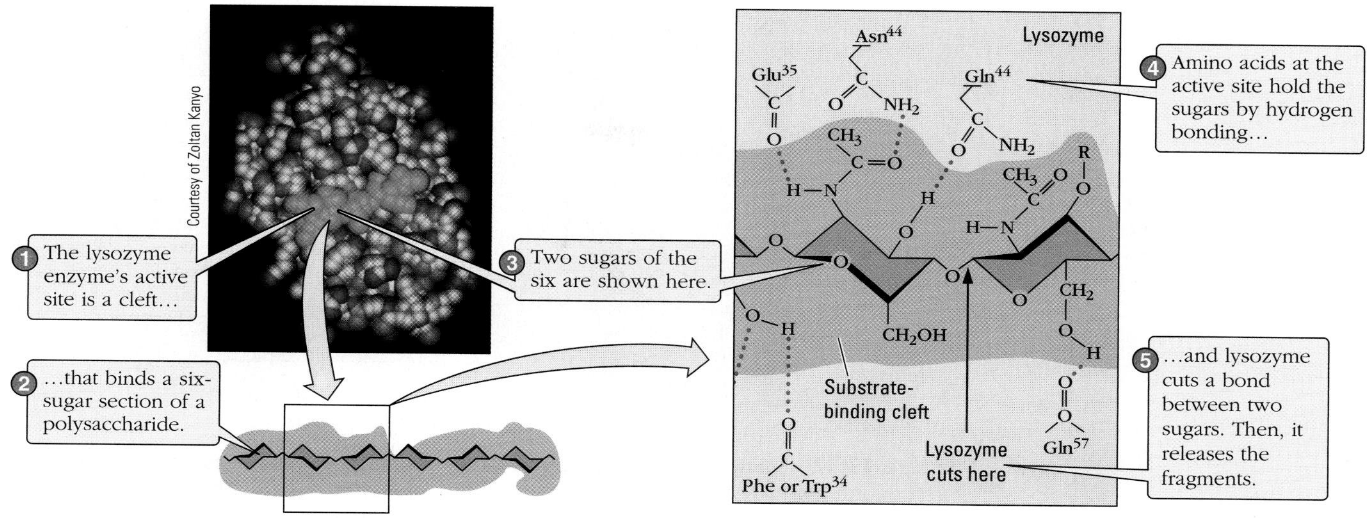

Figure 13.16 Lysozyme with substrate in the active site. The structure of lysozyme is shown at the left, as a space-filling model. The active site is occupied by a portion of a polysaccharide molecule. The diagram at the right shows noncovalent interactions (*red dotted lines*) that hold the substrate to the enzyme.

polypeptides, might fit there as well. Shape is important, but so are noncovalent attractions and their positioning so that the substrate can make the most effective use of them. The enlarged portion of Figure 13.16 shows many hydrogen bonds between enzyme and substrate. It should be clear that the specificity of the enzyme depends not only on the shape of the active site, but also on the positions of hydrogen-bonding groups and groups that participate in other noncovalent interactions so that they can adjust to complementary sites on the substrate.

To summarize, enzymes are extremely effective as catalysts for several reasons:

- Enzymes bring substrates into close proximity and hold them there while a reaction occurs.
- Enzymes hold substrates in the shape that is most effective for reaction.
- Enzymes can act as acids and bases during reaction, donating or accepting hydrogen ions from the substrate quickly and easily.
- The potential energy of a bond distorted by the induced fit of the substrate to the enzyme is already partway up the activation energy hill that must be surmounted for reaction to occur.
- Enzymes sometimes contain metal ions that are needed to help catalyze oxidation-reduction and hydrolysis reactions.

Enzyme Kinetics

An enzyme changes the mechanism of a reaction, as does any catalyst. The first step in the mechanism for any enzyme-catalyzed reaction is binding of the substrate and the enzyme, which is referred to as formation of an **enzyme-substrate complex.** Representing enzyme by E, substrate by S, and products by P, we can write a single-step uncatalyzed mechanism and a two-step enzyme-catalyzed mechanism as follows.

Uncatalyzed mechanism: $S \longrightarrow P$

Enzyme-catalyzed mechanism:

Step 1 (fast): $S + E \rightleftharpoons ES$ (formation of enzyme-substrate complex)

Step 2 (slow): $ES \longrightarrow P + E$ (formation of products and regeneration of enzyme)

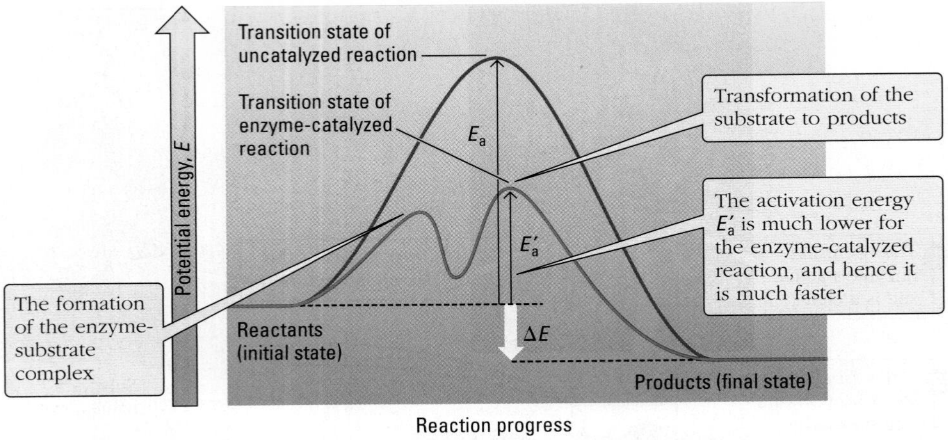

Figure 13.17 Energy diagram for enzyme-catalyzed reaction. The red curve is the energy profile for a typical reaction in a living system with no enzyme present. The green energy profile is drawn to the same scale for the same reaction with enzyme catalysis.

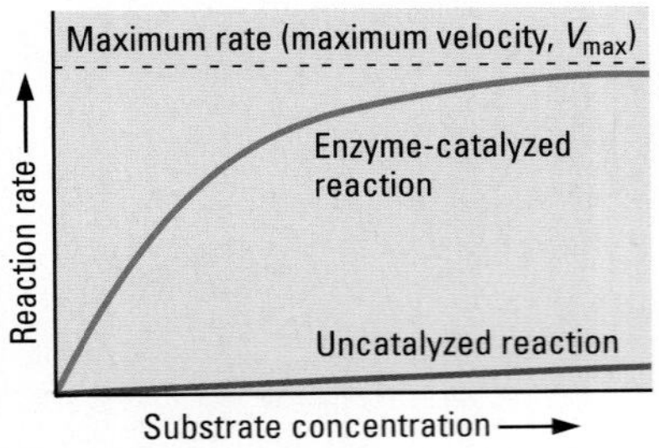

Figure 13.18 Maximum velocity for an enzyme-catalyzed reaction. Because there is only a limited quantity of enzyme available, increasing substrate concentration beyond the point at which the enzyme becomes the limiting reactant does not increase the rate further. There is a maximum rate (maximum velocity) for any enzyme-catalyzed reaction.

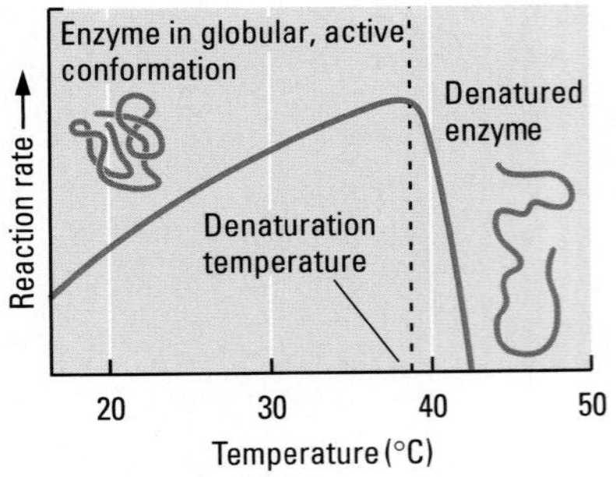

Figure 13.19 Enzyme activity destroyed by high temperature. At a temperature somewhat above normal body temperature, there is sufficient molecular motion to overcome the noncovalent interactions that maintain protein structure. This disrupts the structure of an enzyme, thereby destroying its catalytic activity. The process by which the enzyme structure becomes disrupted is called denaturation.

That the enzyme is a catalyst is evident from the fact that it is a reactant in the first step and is regenerated in the second. This mechanism applies to nearly all enzyme-catalyzed reactions. Because the second step is slow, the enzyme-substrate complex can often separate and reform S + E before it reacts to form products. This possibility is indicated by the double arrow in the first step. This mechanism for enzyme catalysis is similar to the mechanisms of reactions with a rapid, reversible first step that were discussed in Section 13.7 (p. 622), and it can be analyzed mathematically in the same way as those mechanisms were.

Because of the noncovalent interactions between enzyme and substrate, the activation energy is significantly lower for the enzyme-catalyzed reaction than it would be for the uncatalyzed process. This situation is shown in Figure 13.17. Even at temperatures only a little above room temperature, significant numbers of molecules have enough energy to surmount this lower barrier. Thus, enzyme-catalyzed reactions can occur reasonably quickly at body temperature.

Special Features of Enzyme Catalysis

Enzyme-catalyzed reactions obey the same principles of chemical kinetics that we discussed earlier in this chapter. Nevertheless, both the enzyme itself and the mechanism of enzyme catalysis have some special features that you should be aware of. First, because of the form of the mechanism, either the enzyme or the substrate may be the limiting reactant in the first step. If the substrate is limiting, increasing the concentration of substrate produces more enzyme-substrate complex and makes the reaction go faster. This is the expected behavior: Increasing the concentration of a reactant should increase the rate proportionately. If the enzyme becomes the limiting reactant, however, it can become completely converted to enzyme-substrate complex, leaving no enzyme available for additional substrate. If this happens, a further increase in the concentration of substrate will not increase the rate of reaction. As a result there is a *maximum rate* (those who study enzyme kinetics call this the maximum velocity) for an enzyme-catalyzed reaction. The behavior of rate with increasing substrate concentration is shown in Figure 13.18.

Enzyme-catalyzed reactions also behave unusually with respect to temperature. The rate does increase with increasing temperature, but if the temperature becomes high enough, there is a sudden decrease in rate, as shown in Figure 13.19. This happens because there is increased molecular and atomic motion as the temperature increases, and that motion can disrupt the structures of enzymes and other proteins. This change in protein structure is called **denaturation.** It occurs, for example, when an egg is boiled or fried. When a globular protein is denatured it loses its coiled globular structure, and its solubility and other properties change. Once an enzyme's structure has changed, the active site is no longer available, enzyme catalysis is seriously impaired, and the reaction rate falls to its uncatalyzed value. Notice that this happens only a little above 37 °C, which is body temperature for humans. Enzymes have evolved to produce maximum rates at body temperature, and slightly higher temperatures cause most of them to denature.

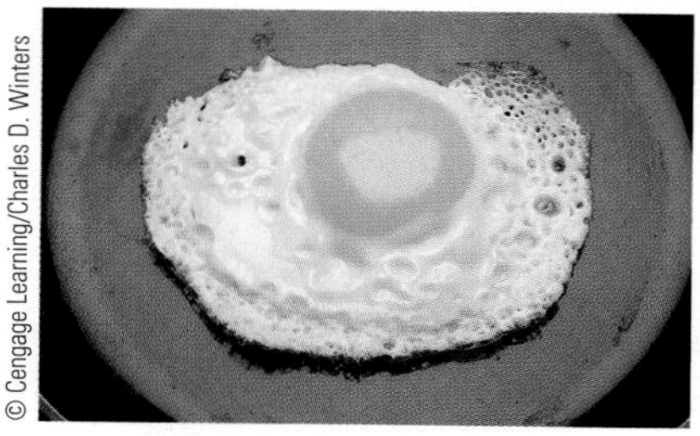

Denaturation. Frying an egg causes protein in the white to denature and precipitate from the colorless, nearly clear original solution to form a white flexible solid. Denaturation of enzymes answers the question, "What happens to an egg when I cook it?" that was posed in Chapter 1 (← *p. 2*).

Inhibition of Enzymes

There is another way that the activity of an enzyme can be destroyed. Some molecules or ions can fit an enzyme's active site, but remain there unreacted. Such a molecule is called an **inhibitor**. An inhibitor bound to an enzyme decreases its effective concentration, thereby decreasing the rate of the reaction that the enzyme catalyzes. If sufficient inhibitor becomes bound to an enzyme, the enzyme provides little catalytic effect because the concentration of available active sites becomes very small. An example of enzyme inhibition is the action of sulfa drugs on bacteria. Bacteria use *para*-aminobenzoic acid and an enzyme called dihydropteroate synthetase to synthesize folic acid, which is essential to their metabolism. Sulfa drugs bind to this enzyme, inhibit synthesis of folic acid, and destroy bacterial populations.

PROBLEM-SOLVING EXAMPLE 13.11 Enzyme Inhibition

The label of a container of methanol (methyl alcohol) invariably indicates that its contents are poisonous and should not be taken internally. Methanol, CH_3OH, which is not very toxic, is metabolized by the enzyme methanol oxidase to formaldehyde, $H_2C{=}O$, which is very toxic. Methanol poisoning is sometimes treated by giving the patient ethanol, CH_3CH_2OH, which inhibits the enzyme. Identify similarities and differences in the structures of methanol and ethanol that could account for ethanol's acting as an inhibitor.

methanol

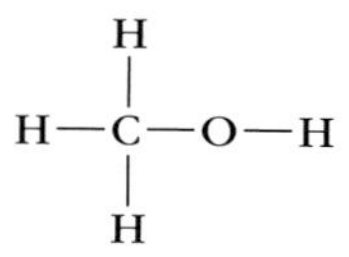

ethanol

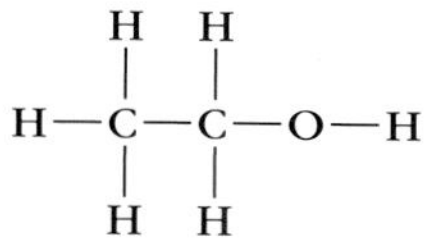

Answer Both molecules are alcohols and can hydrogen-bond. Methanol has three hydrogens on the carbon next to the —OH group. Ethanol has only two hydrogens on the carbon adjacent to the —OH, and it has one more carbon and two more hydrogens.

Strategy and Explanation Because its shape is similar to that of methanol and because it can form hydrogen bonds of similar strength, ethanol binds to the active sites of some of the methanol oxidase catalyst molecules. Because of the extra carbon atom or the difference in number of hydrogens adjacent to the —OH group, the catalyst is unable to oxidize the ethanol, and ethanol remains bound. The smaller concentration of catalyst molecules decreases the rate of conversion of methanol to formaldehyde in the body, and the harmful effect is smaller.

PROBLEM-SOLVING PRACTICE 13.11

Bacteria need to use *p*-aminobenzoic acid to help synthesize folic acid in order to survive. Sulfa drugs interfere with this process. The structures of *p*-aminobenzoic acid and folic acid are

p-aminobenzoic acid

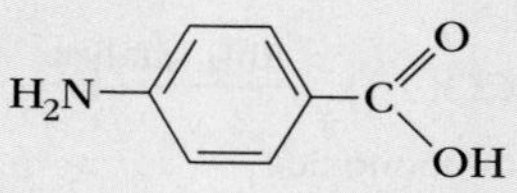

folic acid

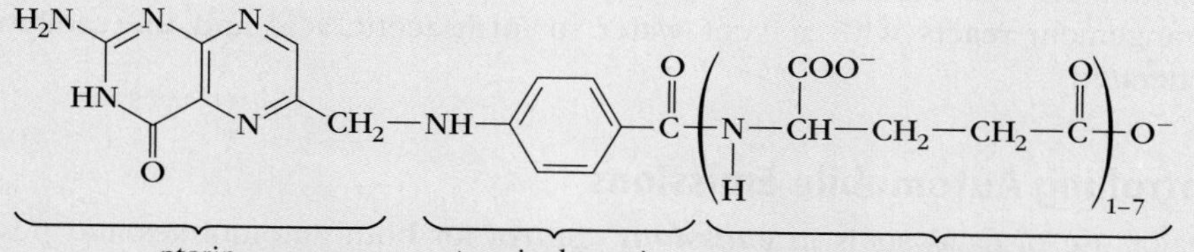

pterin (2-amino-4-oxopteridine) — *p*-aminobenzoic acid — glutamates

Which of these structures is most likely a sulfa drug? Explain your choice.

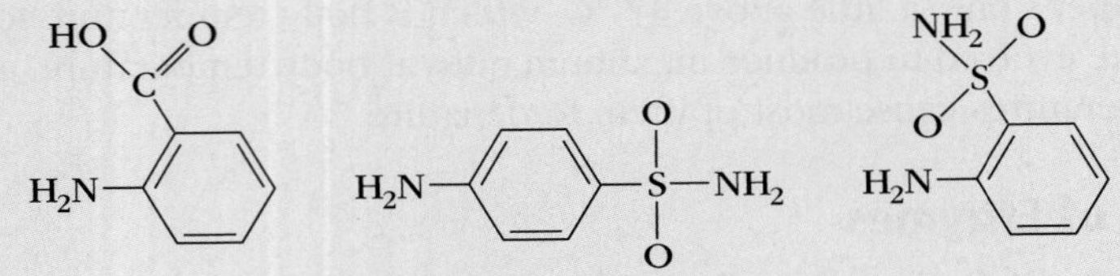

13.10 Catalysis in Industry

An expert in the field of industrial chemistry has said that every year more than one trillion dollars' worth of goods is manufactured with the aid of manmade catalysts. Without them, fertilizers, pharmaceuticals, fuels, synthetic fibers, solvents, and detergents would be in short supply. Indeed, 90% of all manufactured items use catalysts at some stage of production. The major areas of catalyst use are in petroleum refining, industrial production of chemicals, and environmental controls. In this section we provide a few examples of the many important industrial reactions that depend on catalysis.

Many industrial reactions use **heterogeneous catalysts.** Such catalysts are present in a different phase from that of the reactants being catalyzed. Usually the catalyst is a solid and the reactants are in the gaseous or liquid phase. Heterogeneous catalysts are used in industry because they are more easily separated from the products and leftover reactants than are homogeneous catalysts. Catalysts for chemical processing are generally metal-based and often contain precious metals such as platinum and palladium. In the United States more than $600 million worth of such catalysts are used annually by the chemical processing industry, almost half of them in the preparation of polymers.

Manufacture of Acetic Acid

The importance of acetic acid, CH_3COOH, in the organic chemicals industry is comparable to that of sulfuric acid in the inorganic chemicals industry; annual production of acetic acid in the United States exceeds 5 billion pounds. Acetic acid is used widely in industry to make plastics and synthetic fibers, as a fungicide, and as the starting material for preparing many dietary supplements. One way of synthesizing the acid is an excellent example of homogeneous catalysis: Aqueous rhodium(III) iodide is used to speed up the combination of carbon monoxide and methyl alcohol, both inexpensive chemicals, to form acetic acid.

$$\underset{\text{methyl alcohol}}{CH_3OH(aq)} + \underset{\text{carbon monoxide}}{CO(aq)} \xrightarrow{RhI_3\text{ catalyst}} \underset{\text{acetic acid}}{CH_3\overset{\overset{\displaystyle O}{\|}}{C}{-}OH(aq)}$$

The role of the rhodium(III) iodide catalyst in this reaction is to bring the reactants together and allow them to rearrange to form the products. Carbon monoxide and the methyl group from the alcohol become attached to the rhodium atom, which helps transfer the methyl group to the CO. The intermediate formed by this rearrangement reacts with solvent water to form acetic acid and the catalyst is regenerated.

Controlling Automobile Emissions

A major use of catalysts is in *emissions control* for both automobiles and power plants. This market uses very large quantities of platinum group metals: platinum, palladium, rhodium, and iridium. In 2008, approximately 1.2×10^5 kg platinum and

1.3×10^5 kg palladium were sold worldwide for automotive uses. Demand was also high for rhodium for this same purpose (about 2.6×10^4 kg). All three metals are also used in chemical processing as catalysts, and the petroleum industry uses platinum and rhodium to catalyze refining processes.

The purpose of the catalysts in the exhaust system of an automobile is to ensure that the combustion of carbon monoxide and hydrocarbons is complete (Figure 13.20).

$$2\ CO(g) + O_2(g) \xrightarrow{\text{Pt-NiO catalyst}} 2\ CO_2(g)$$

$$2\ C_8H_{18}(g) + 25\ O_2(g) \xrightarrow{\text{Pt-NiO catalyst}} 16\ CO_2(g) + 18\ H_2O(g)$$

2,2,4-trimethylpentane, a component of gasoline

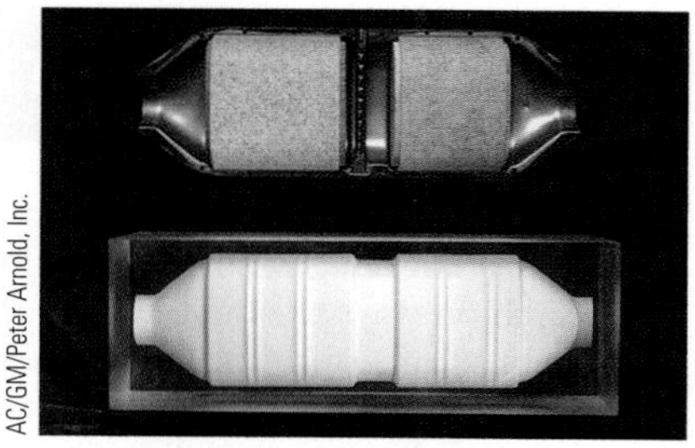
AC/GM/Peter Arnold, Inc.

Figure 13.20 Automobile catalytic converter. Catalytic converters, such as the one in the lower part of this photo, are standard equipment on the exhaust systems of all new automobiles. This one contains two catalysts (the two slightly different colored, lighter areas in the upper part of the photo). One catalyst converts nitrogen monoxide to nitrogen and oxygen. The other catalyst converts carbon monoxide and hydrocarbons to carbon dioxide and water.

and to convert nitrogen oxides to molecules that are less harmful to the environment. At the high temperature of combustion, some N_2 from air reacts with O_2 to give NO, a serious air pollutant. Thermodynamic data show that nitrogen monoxide is unstable and should revert to N_2 and O_2. But remember that thermodynamics says nothing about rate. Unfortunately, the rate of reversion of NO to N_2 and O_2 is slow. Fortunately, catalysts have been developed that greatly speed this reaction.

$$2\ NO(g) \xrightarrow{\text{catalyst}} N_2(g) + O_2(g)$$

The role of the heterogeneous catalyst in the preceding reactions is probably to weaken the bonds of the reactants and to assist in product formation. For example, Figure 13.21 shows how NO molecules can dissociate into N and O atoms on the surface of a platinum metal catalyst.

Converting Methane to Liquid Fuel

Your home may be heated with natural gas, which consists largely of methane, CH_4. Although widely used, it is also widely wasted because much of it is found in geographical areas far removed from where fuels are consumed, and because transporting the flammable gas is expensive and dangerous. One solution to making methane

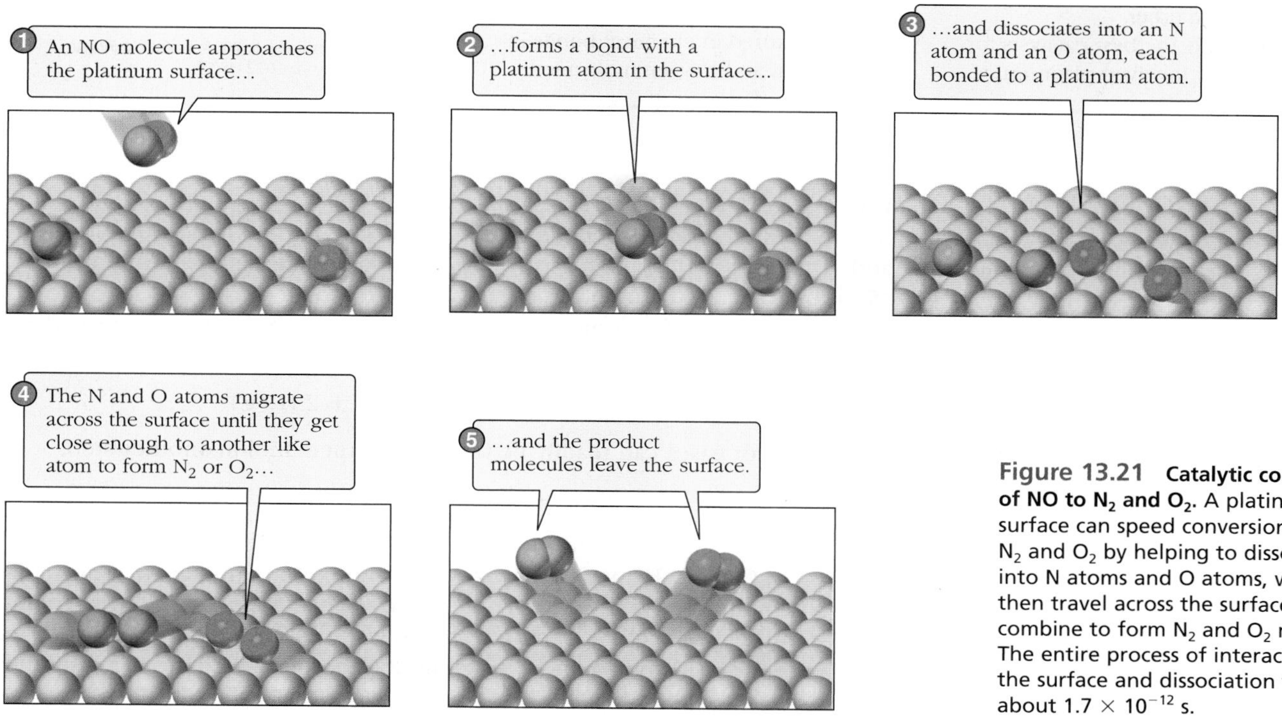
1 An NO molecule approaches the platinum surface...
2 ...forms a bond with a platinum atom in the surface...
3 ...and dissociates into an N atom and an O atom, each bonded to a platinum atom.
4 The N and O atoms migrate across the surface until they get close enough to another like atom to form N2 or O2...
5 ...and the product molecules leave the surface.

Figure 13.21 Catalytic conversion of NO to N_2 and O_2. A platinum surface can speed conversion of NO to N_2 and O_2 by helping to dissociate NO into N atoms and O atoms, which then travel across the surface and combine to form N_2 and O_2 molecules. The entire process of interaction with the surface and dissociation takes about 1.7×10^{-12} s.

CHEMISTRY IN THE NEWS Catalysis and Hydrogen Fuel

Storing solar energy in chemical bonds has many advantages, so there is a lot of research on ways to obtain fuels from sunlight. Plants store solar energy using photosynthesis to make carbohydrates and oxygen from carbon dioxide and water. Could we do something similar without using the CO_2? That is, could we make hydrogen and oxygen directly from water? This would be a great advance because hydrogen burns without producing the greenhouse gas, carbon dioxide. This idea has led to much research on catalysts that can help sunlight to split water into hydrogen and oxygen. A great deal of solar energy falls on Earth's surface, but it not very concentrated. To use solar energy will require collectors with a very large surface area covered with catalytic material. This means that the catalyst must be relatively inexpensive.

It has been known for 200 years that water can be split into hydrogen and oxygen by an electric current (electrolysis, ← *p. 125*). In electrolysis, hydrogen is produced in one part of the apparatus and oxygen in another; keeping them separate prevents them from recombining explosively. A similar separation is needed for catalytic water splitting. The most difficult part is oxidizing water to form oxygen—making the hydrogen we want is easier, but you cannot produce one without the other.

Until recently the best catalysts for oxidizing water involved platinum, which is relatively rare and very expensive, but that may be a thing of the past. Daniel Nocera and Matthew Kanan at the Massachusetts Institute of Technology have found a very promising catalyst that is much less expensive. They start with a surface of indium tin oxide, which is an electrical conductor, and place it into a neutral aqueous solution of Co^{2+} ions and potassium dihydrogen phosphate. Then they pass an electric current through the solution. This causes the Co^{2+} ions to be oxidized to Co^{3+} ions, which precipitate $CoPO_4(s)$ on to the indium tin oxide. The cobalt phosphate forms a stable, hard surface. It appears that the Co^{3+} can be oxidized further to cobalt in the +4 oxidation state, but this is not yet known for sure.

When sunlight hits the surface, the cobalt ions are reduced from +3 or +4 oxidation states to +2 by gaining electrons from hydrogen atoms in the water molecules. This breaks the H—O bonds, leaving O atoms on the cobalt phosphate surface and forming hydrogen ions in the solution. Pairs of O atoms bond to form O_2 molecules, and bubbles of oxygen form. The hydrogen ions (actually hydrogen ions combined with water molecules, H_3O^+) in the solution migrate to a different part of the solar collector where they combine with electrons to form H_2 molecules. As a consequence, energy from the solar photons is stored in chemical bonds in the hydrogen and oxygen molecules.

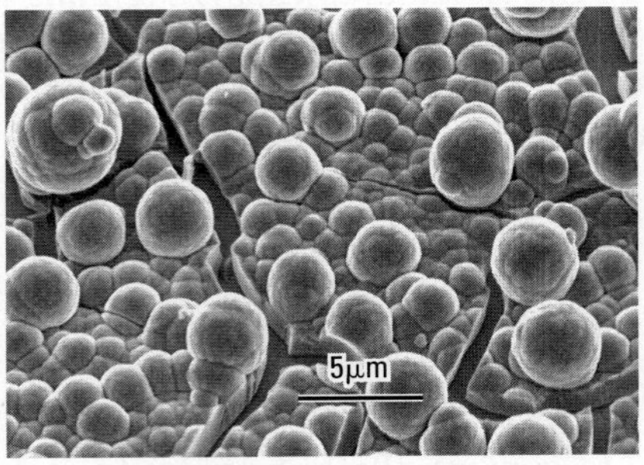

Daniel G. Nocera

Scanning electron microscope image of catalytic surface. The indium tin oxide surface can be seen through several cracks in the surface deposit of cobalt(III) phosphate.

Storage of energy is an important point, because the sun is available only in daylight and when there are no clouds. If we could not store the solar energy there would have to be another energy source for when the sun is not shining, which is more than half the time.

The new catalyst is not perfect. Much more research and development will be needed before it can help to split water molecules on a commercial scale. But it is an important breakthrough in capturing solar energy.

Source: *Science*, August 1, 2008, p. 620; *Science*, August 22, 2008; pp. 1072–1075.

useful is to convert it, where it is found, to a more readily transportable substance such as liquid methanol, CH_3OH. The methanol can then be used directly as a fuel, added to gasoline [as is currently done in some areas of the United States (← *p. 546*)], or used to make other chemicals.

It has been known for some time that methane can be converted to carbon monoxide and hydrogen,

$$CH_4(g) + \frac{1}{2}\,O_2(g) \longrightarrow CO(g) + 2\,H_2(g)$$

and this mixture of gases can readily be turned into methanol in another step.

$$CO(g) + 2\,H_2(g) \longrightarrow CH_3OH(\ell)$$

Chemical engineers at the University of Minnesota discovered that methane can be converted to CO and H_2 under very mild conditions of temperature. They simply found the right catalyst. Figure 13.22 shows what happens when a room-temperature mixture of methane and oxygen flows through a heated, sponge-like ceramic disk coated with platinum or rhodium. Rather than oxidizing the methane all the way to water and carbon dioxide, the process produces a hot mixture of CO and H_2, which can be converted in good yield to methanol. It is also possible to produce other partially oxidized hydrocarbons by a similar catalytic process.

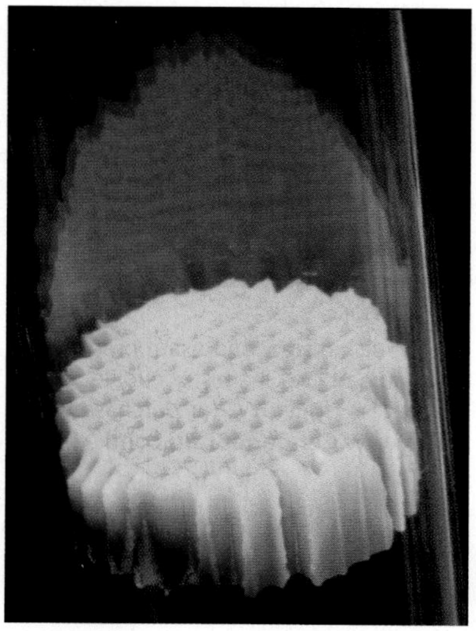

Schmidt/University of Minnesota

Figure 13.22 Methane flowing through a catalyst. The square grid of the catalyst glows yellow because of energy transferred to its surface by catalyzed combustion of the methane.

CONCEPTUAL EXERCISE 13.14 Catalysis

Which of these statements is (are) true? If any are false, change the wording to make them true.

(a) The concentration of a homogeneous catalyst may appear in the rate law.
(b) A catalyst is always consumed in the overall reaction.
(c) A catalyst must always be in the same phase as the reactants.

SUMMARY PROBLEM

An excellent way to make highly pure nickel metal for use in specialized steel alloys is to decompose $Ni(CO)_4$ by heating it in a vacuum to slightly above room temperature.

$$Ni(CO)_4(g) \longrightarrow Ni(s) + 4\ CO(g)$$

The reaction is proposed to occur in four steps, the first of which is

$$Ni(CO)_4(g) \longrightarrow Ni(CO)_3(g) + CO(g)$$

Kinetic studies of this first-order decomposition reaction have been carried out between 47.3 °C and 66.0 °C to give the results in the table in the margin.*

Temperature (°C)	Rate Constant (s^{-1})
47.3	0.233
50.9	0.354
55.0	0.606
60.0	1.022
66.0	1.873

(a) What is the activation energy for this reaction?

(b) $Ni(CO)_4$ is formed by the reaction of nickel metal with carbon monoxide. If you have 2.05 g CO and you combine it with 0.125 g nickel metal, what is the maximum quantity of $Ni(CO)_4$ (in grams) that can be formed?

The replacement of CO by another molecule in $Ni(CO)_4$ (in the nonaqueous solvents toluene and hexane) was also studied to understand the general principles that govern the chemistry of such compounds.*

$$Ni(CO)_4 + P(CH_3)_3 \longrightarrow Ni(CO)_3P(CH_3)_3 + CO$$

A detailed study of the kinetics of the reaction led to the mechanism

$$Ni(CO)_4 \longrightarrow Ni(CO)_3 + CO \qquad \text{slow}$$

$$Ni(CO)_3 + P(CH_3)_3 \longrightarrow Ni(CO)_3P(CH_3)_3 \qquad \text{fast}$$

(c) Which step in the mechanism is unimolecular? Which is bimolecular?

(d) Add the steps of the mechanism to show that the result is the balanced equation for the observed reaction.

(e) Is there an intermediate in this reaction? If so, what is it?

(f) It was found that doubling the concentration of $Ni(CO)_4$ increased the reaction rate by a factor of 2. Doubling the concentration of $P(CH_3)_3$ had no effect on the reaction rate. Based on this information, write the rate equation for the reaction.

(g) Does the experimental rate equation support the proposed mechanism? Why or why not?

*See Day, J. P., Basolo, F., and Pearson, R. G. *Journal of the American Chemical Society*, Vol. 90, 1968; p. 6933.

OWL and go Chemistry

Sign in at **www.cengage.com/owl** to:

- View tutorials and simulations, develop problem-solving skills, and complete online homework assigned by your professor.
- For quick review and exam prep, download Go Chemistry mini lecture modules from OWL (or purchase them at **www.CengageBrain.com**).

IN CLOSING

Having studied this chapter, you should be able to . . .

- Define reaction rate and calculate average rates (Section 13.1). End-of-chapter questions: 11, 13, 21
- Describe the effect that reactant concentrations have on reaction rate, and determine rate laws and rate constants from initial rates (Section 13.2). Questions 23, 25, 29, 32
- Determine reaction orders from a rate law, and use the integrated rate law method to obtain orders and rate constants (Section 13.3). Questions 34, 40
- Calculate concentration at a given time, time to reach a certain concentration, and half-life for a reaction (Section 13.3). Questions 38, 48, 50, 51, 53, 55
- Define and give examples of unimolecular and bimolecular elementary reactions (Section 13.4). Question 56
- Show by using an energy profile what happens as two reactant molecules interact to form product molecules (Section 13.4). Question 77
- Define activation energy and frequency factor, and use them to calculate rate constants and rates under different conditions of temperature and concentration (Section 13.5). Questions 62, 64, 65, 69, 73
- Derive rate laws for unimolecular and bimolecular elementary reactions (Section 13.6). Questions 75, 83
- Define reaction mechanism and identify rate-limiting steps and intermediates (Section 13.7). Questions 85, 87, 89
- Given several reaction mechanisms, decide which is (are) in agreement with experimentally determined stoichiometry and rate law (Section 13.7). Questions 89, 93, 95
- Explain how a catalyst can speed up a reaction; draw energy profiles for catalyzed and uncatalyzed reaction mechanisms (Section 13.8). Question 99
- Define the terms *enzyme, substrate,* and *inhibitor,* and identify similarities and differences between enzyme-catalyzed reactions and uncatalyzed reactions (Section 13.9). Questions 101, 104, 108
- Describe several important industrial processes that depend on catalysts (Section 13.10).

KEY TERMS

activated complex *(Section 13.4)*
activation energy (E_a) *(13.4)*
active site *(13.9)*
Arrhenius equation *(13.5)*
average reaction rate *(13.1)*
bimolecular reaction *(13.4)*
chemical kinetics *(Introduction)*
cofactor *(13.9)*
denaturation *(13.9)*
elementary reaction *(13.4)*
enzyme *(13.9)*
enzyme-substrate complex *(13.9)*
frequency factor *(13.5)*
half-life *(13.3)*
heterogeneous catalyst *(13.10)*
heterogeneous reaction *(13.1)*
homogeneous catalyst *(13.8)*
homogeneous reaction *(13.1)*
induced fit *(13.9)*
inhibitor *(13.9)*
initial rate *(13.2)*
instantaneous reaction rate *(13.1)*
intermediate *(13.7)*
order of reaction *(13.3)*
overall reaction order *(13.3)*
rate *(13.1)*
rate constant *(13.2)*
rate law *(13.2)*
rate-limiting step *(13.7)*
reaction intermediate *(13.7)*
reaction mechanism *(13.7)*
reaction rate *(13.1)*
steric factor *(13.4)*
substrate *(13.9)*
transition state *(13.4)*
unimolecular reaction *(13.4)*

QUESTIONS FOR REVIEW AND THOUGHT

OWL Interactive versions of these problems are assignable in OWL.

Blue-numbered questions have short answers at the back of this book in Appendix M and fully worked solutions in the *Student Solutions Manual.*

Review Questions

These questions test vocabulary and simple concepts.

1. Which of these is appropriate for determining the rate law for a chemical reaction?
   (a) Theoretical calculations based on balanced equations
   (b) Measuring the rate of the reaction as a function of the concentrations of the reacting species
   (c) Measuring the rate of the reaction as a function of temperature
2. Name at least three factors that affect the rate of a chemical reaction.
3. Using the rate law, rate $= k[A]^2[B]$, define the order of the reaction with respect to A and B and the overall reaction order.
4. Define the terms "enzyme," "substrate," and "inhibitor," and give an example of each kind of molecule.
5. Explain the difference between a homogeneous and a heterogeneous catalyst. Give an example of each.
6. Define the terms "unimolecular elementary reaction" and "bimolecular elementary reaction," and give an example of each.
7. Define the terms "activation energy" and "frequency factor." Write an equation that relates activation energy and frequency factor to reaction rate.
8. Why is reaction kinetics important to our understanding of depletion of stratospheric ozone ***(← p. 454)***?

Topical Questions

These questions are keyed to the major topics in the chapter. Usually a question that is answered at the back of this book is paired with a similar one that is not.

Reaction Rate (Section 13.1)

9. Consider the dissolving of sugar as a simple process in which kinetics is important. Suppose that you dissolve an equal mass of each kind of sugar listed. Which dissolves the fastest? Which dissolves the slowest? Explain why in terms of rates of heterogeneous reactions. (If you are not sure which is fastest or slowest, try them all.)
   (a) Rock candy sugar (large sugar crystals)
   (b) Sugar cubes
   (c) Granular sugar
   (d) Powdered sugar
10. A cube of aluminum 1.0 cm on each edge is placed into 9-M NaOH(aq), and the rate at which H_2 gas is given off is measured.
    (a) By what factor will this reaction rate change if the aluminum cube is cut exactly in half and the two halves are placed in the solution? Assume that the reaction rate is proportional to the surface area, and that all of the surface of the aluminum is in contact with the NaOH(aq).
    (b) If you had to speed up this reaction as much as you could without raising the temperature, what would you do to the aluminum?
11. Experimental data in the table are for the hypothetical reaction

$$A \longrightarrow 2\,B$$

Time (s)	[A] (mol/L)
0.00	1.000
10.0	0.833
20.0	0.714
30.0	0.625
40.0	0.555

    (a) Make a graph of concentration as a function of time, draw a smooth curve through the points, and calculate the rate of change of [A] for each 10-s interval from 0.00 to 40.0 s. Why might the rate of change decrease from one time interval to the next?
    (b) How is the rate of change of [B] related to the rate of change of [A] in the same time interval?
    (c) Calculate the rate of change of [B] for the time interval from 10.0 to 20.0 s.
12. A compound called phenyl acetate reacts with water according to the equation

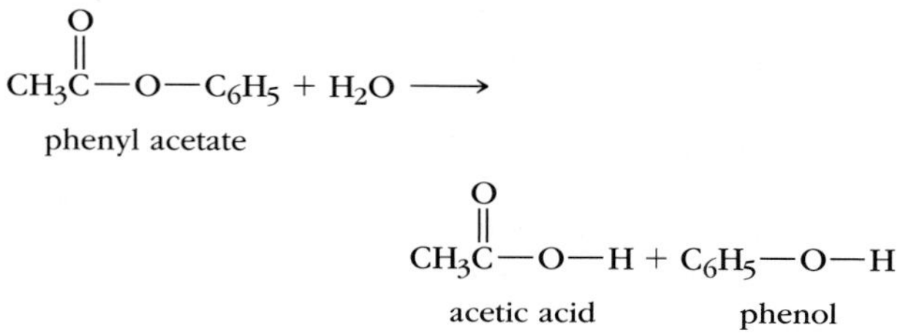
CH3C—O—C6H5 + H2O
phenyl acetate
CH3C—O—H + C6H5—O—H
acetic acid
phenol

These data were collected at 5 °C.

Time (min)	[Phenyl acetate] (mol/L)
0.00	0.55
0.25	0.42
0.50	0.31
0.75	0.23
1.00	0.17
1.25	0.12
1.50	0.082

    (a) Make a graph of concentration as a function of time, describe the shape of the curve, and compare it with Figure 13.3 on p. 597.
    (b) Calculate the rate of change of the concentration of phenylacetate during the period from 0.20 to 0.40 min, and then during the period from 1.2 to 1.4 min.

Compare the values and explain why one is smaller than the other.

(c) What is the rate of change of the phenol concentration during the period from 1.00 to 1.25 min?

13. Experimental data are listed here for the reaction

$$A \longrightarrow 2\,B$$

Time (s)	[B] (mol/L)
0.00	0.000
10.0	0.326
20.0	0.572
30.0	0.750
40.0	0.890

(a) Prepare a graph from these data, connect the points with a smooth line, and calculate the rate of change of [B] for each 10-s interval from 0.0 to 40.0 s. Does the rate of change decrease from one time interval to the next? Suggest a reason for this result.

(b) How is the rate of change of [A] related to the rate of change of [B] in each time interval? Calculate the rate of change of [A] for the time interval from 10.0 to 20.0 s.

(c) What is the instantaneous rate, $\Delta[B]/\Delta t$, when $[B] = 0.750$ mol/L?

14. Cyclobutane can decompose to form ethylene:

$$C_4H_8(g) \longrightarrow 2\,C_2H_4(g)$$

The cyclobutane concentration can be measured as a function of time by mass spectrometry (a graph follows).

(a) Write an expression for the rate of reaction in terms of a changing concentration.

(b) Calculate the average rate of reaction between 10. and 30. s.

(c) Calculate the instantaneous rate of reaction after 20. s.

(d) Calculate the initial rate of reaction.

(e) Calculate the instantaneous rate of formation of ethylene 40. s after the start of the reaction.

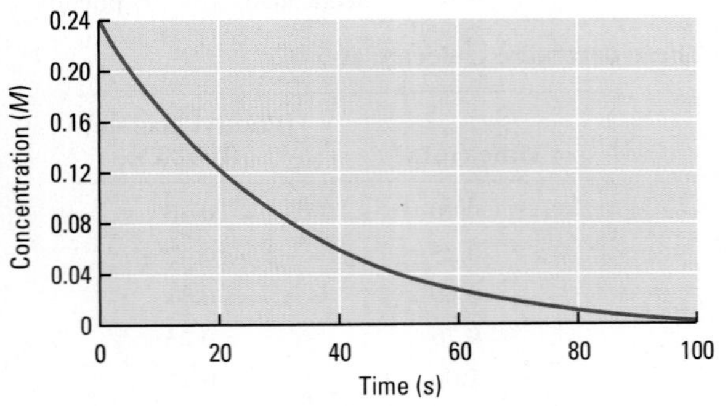

15. Using data given in the table for the reaction

$$N_2O_5 \longrightarrow 2\,NO_2 + \tfrac{1}{2}\,O_2$$

calculate the average rate of reaction during each of these intervals:

(a) 0.00 to 0.50 h (b) 0.50 to 1.0 h

(c) 1.0 to 2.0 h (d) 2.0 to 3.0 h

(e) 3.0 to 4.0 h (f) 4.0 to 5.0 h

Time (h)	$[N_2O_5]$ (mol/L)	Time (h)	$[N_2O_5]$ (mol/L)
0.00	0.849	3.00	0.352
0.50	0.733	4.00	0.262
1.00	0.633	5.00	0.196
2.00	0.472		

16. Using data from Question 15, calculate the average rate over the interval 0 to 5.0 h. Compare your result with the average rates over the intervals 1.0 to 4.0 h and 2.0 to 3.0 h, all of which have the same midpoint (2.5 h from the start).

17. Using all your calculated rates from Question 15,

(a) Show that the reaction obeys the rate law

$$\text{Rate} = -\frac{\Delta[N_2O_5]}{\Delta t} = k[N_2O_5]$$

(b) Evaluate the rate constant k as an average of the values obtained for the six intervals.

18. Using the rate law and the rate constant you calculated in Question 17, calculate the reaction rate exactly 2.5 h from the start. Do your results from this and Question 16 agree with the statement in the text that the smaller the time interval, the more accurate the average rate?

19. For the reaction

$$2\,NO_2(g) \longrightarrow 2\,NO(g) + O_2(g)$$

make qualitatively correct plots of the concentrations of $NO_2(g)$, NO(g), and $O_2(g)$ versus time. Draw all three graphs on the same axes; assume that you start with $NO_2(g)$ at a concentration of 1.0 mol/L. Explain how you would determine, from these plots,

(a) The initial rate of the reaction

(b) The final rate (i.e., the rate as time approaches infinity)

20. For the reaction

$$O_3(g) + O(g) \longrightarrow 2\,O_2(g)$$

make qualitatively correct plots of the concentrations of $O_3(g)$, O(g), and $O_2(g)$ versus time. Draw all three graphs on the same axes, assume that you start with $O_3(g)$ and O(g), each at a concentration of 1.0 μmol/L. Explain how you would determine, from these plots,

(a) The initial rate of the reaction

(b) The final rate (i.e., the rate as time approaches infinity)

21. Express the rate of the reaction

$$2\,N_2H_4(\ell) + N_2O_4(\ell) \longrightarrow 3\,N_2(g) + 4\,H_2O(g)$$

in terms of

(a) $\Delta[N_2O_4]$ (b) $\Delta[N_2]$

22. Ammonia is produced by the reaction between nitrogen and hydrogen gases.

(a) Write a balanced equation using smallest whole-number coefficients for the reaction.

(b) Write an expression for the rate of reaction in terms of $\Delta[NH_3]$.

(c) The concentration of ammonia increases from 0.257 M to 0.815 M in 15.0 min. Calculate the average rate of reaction over this time interval.

Effect of Concentration on Reaction Rates (Section 13.2)

23. If a reaction has the experimental rate law, rate = $k[A]^2$, explain what happens to the rate when
(a) The concentration of A is tripled
(b) The concentration of A is halved

24. A reaction has the experimental rate law, rate = $k[A]^2[B]$. If the concentration of A is doubled and the concentration of B is halved, what happens to the reaction rate?

25. The reaction of $CO(g) + NO_2(g)$ is second-order in NO_2 and zeroth-order in CO at temperatures less than 500 K.
(a) Write the rate law for the reaction.
(b) How will the reaction rate change if the NO_2 concentration is halved?
(c) How will the reaction rate change if the concentration of CO is doubled?

26. Nitrosyl bromide, NOBr, is formed from NO and Br_2.

$$2\ NO(g) + Br_2(g) \longrightarrow 2\ NOBr(g)$$

Experiment shows that the reaction is first-order in Br_2 and second-order in NO.
(a) Write the rate law for the reaction.
(b) If the concentration of Br_2 is tripled, how will the reaction rate change?
(c) What happens to the reaction rate when the concentration of NO is doubled?

27. For the reaction of $Pt(NH_3)_2Cl_2$ with water,

$$Pt(NH_3)_2Cl_2 + H_2O \longrightarrow Pt(NH_3)_2(H_2O)Cl^+ + Cl^-$$

the rate law is rate = $k[Pt(NH_3)_2Cl_2]$ with $k = 0.090\ h^{-1}$.
(a) Calculate the initial rate of reaction when the concentration of $Pt(NH_3)_2Cl_2$ is
(i) 0.010 M (ii) 0.020 M (iii) 0.040 M
(b) How does the rate of disappearance of $Pt(NH_3)_2Cl_2$ change with its initial concentration?
(c) How is this related to the rate law?
(d) How does the initial concentration of $Pt(NH_3)_2Cl_2$ affect the rate of appearance of Cl^- in the solution?

28. Methyl acetate, CH_3COOCH_3, reacts with base to break one of the C—O bonds.

$$CH_3\overset{O}{\overset{\|}{C}}—O—CH_3(aq) + OH^-(aq) \longrightarrow CH_3\overset{O}{\overset{\|}{C}}—O^-(aq) + HO—CH_3(aq)$$

The rate law is rate = $k[CH_3COOCH_3][OH^-]$ where $k = 0.14\ L\ mol^{-1}\ s^{-1}$ at 25 °C.
(a) Calculate the initial rate at which the methyl acetate disappears when both reactants, CH_3COOCH_3 and OH^-, have a concentration of 0.025 M.
(b) How rapidly (i.e., at what rate) does the methyl alcohol, CH_3OH, initially appear in the solution?

29. Rate data were obtained at 25 °C for this reaction. What is the rate law for this reaction?

$$A + 2\ B \longrightarrow C + 2\ D$$

Expt.	Initial [A] (mol/L)	Initial [B] (mol/L)	Initial Rate of Formation of C (M min^{-1})
1	0.10	0.10	3.0×10^{-4}
2	0.30	0.30	9.0×10^{-4}
3	0.10	0.30	3.0×10^{-4}
4	0.20	0.40	6.0×10^{-4}

30. Measurements of the initial rate of reaction between two compounds, triphenylmethyl hexachloroantimonate, **I**, and bis-(9-ethyl-3-carbazolyl)methane, **II**, in 1,2-dichloroethane at 40 °C yielded these data:

Initial Concentration $\times 10^5$ (mol/L)		Initial Rate $\times 10^9$ (mol L^{-1} s^{-1})
[I]	[II]	
1.65	10.6	1.50
14.9	10.6	17.7
14.9	7.10	11.2
14.9	3.52	6.30
14.9	1.76	3.10
4.97	10.6	4.52
2.48	10.6	2.70

(a) Derive the rate law for this reaction.
(b) Calculate the rate constant k and express it in appropriate units.

31. A study of the hypothetical reaction $2\ A + B \longrightarrow C + D$ gave these results:

Experiment	Initial Concentration (mol/L) [A]	[B]	Initial Rate (mol L^{-1} s^{-1})
1	0.10	0.050	6.0×10^{-3}
2	0.20	0.050	1.2×10^{-2}
3	0.30	0.050	1.8×10^{-2}
4	0.20	0.150	1.1×10^{-1}

(a) What is the rate law for this reaction?
(b) Calculate the rate constant k and express it in appropriate units.

32. For the reaction

$$2\ NO(g) + 2\ H_2(g) \longrightarrow N_2(g) + 2\ H_2O(g)$$

these data were obtained at 1100 K:

[NO] (mol/L)	$[H_2]$ (mol/L)	Initial Rate (mol L^{-1} s^{-1})
5.00×10^{-3}	2.50×10^{-3}	3.0×10^{-3}
15.0×10^{-3}	2.50×10^{-3}	9.0×10^{-3}
15.0×10^{-3}	10.0×10^{-3}	3.6×10^{-2}

(a) What is the order with respect to NO? With respect to H_2?
(b) What is the overall order?
(c) Write the rate law.
(d) Calculate the rate constant.
(e) Calculate the initial rate of this reaction at 1100 K when $[NO] = [H_2] = 8.0 \times 10^{-3}$ mol L^{-1}.

33. The hypothetical reaction

$$2\,A + 2\,B \longrightarrow C + 3\,D$$

was studied by measuring the initial rate of appearance of C. These data were obtained:

[A] (mol/L)	[B] (mol/L)	Initial Rate (mol L^{-1} s^{-1})
6.0×10^{-3}	1.0×10^{-3}	0.012
6.0×10^{-3}	2.0×10^{-3}	0.024
2.0×10^{-3}	1.5×10^{-3}	0.0020
4.0×10^{-3}	1.5×10^{-3}	0.0080

(a) What is the order of the reaction with respect to substance A?
(b) What is the order with respect to B?
(c) What is the overall order?
(d) What is the rate law?
(e) Calculate the rate constant.
(f) If at a given instant A is disappearing at a rate of 0.034 mol L^{-1} s^{-1}, what is the rate of appearance of C? What is the rate of appearance of D?

Rate Law and Order of Reaction (Section 13.3)

34. For each of these rate laws, state the reaction order with respect to the hypothetical substances A and B, and give the overall order.
(a) Rate $= k[A][B]^3$ (b) Rate $= k[A][B]$
(c) Rate $= k[A]$ (d) Rate $= k[A]^3[B]$

35. For each of the rate laws below, what is the order of the reaction with respect to the hypothetical substances X, Y, and Z? What is the overall order?
(a) Rate $= k[X][Y][Z]$ (b) Rate $= k[X]^2[Y]^{1/2}[Z]$
(c) Rate $= k[X]^{1.5}[Y]^{-1}$ (d) Rate $= k[X]/[Y]^2$

36. A reaction $A + B \longrightarrow$ products is found to be second-order in B. Which rate equation cannot be correct?
(a) Rate $= k[A][B]$
(b) Rate $= k[A][B]^2$
(c) Rate $= k[B]^2$

37. The reaction

$$2\,NO(g) + 2\,H_2(g) \longrightarrow N_2(g) + 2\,H_2O(g)$$

is found to be first-order in $H_2(g)$. Which rate equation cannot be correct?
(a) Rate $= k[NO]^2[H_2]$ (b) Rate $= k[H_2]$
(c) Rate $= k[NO]^2[H_2]^2$

38. The decomposition of ammonia on tungsten at 1100 °C is zeroth-order with a rate constant of 2.5×10^{-4} mol L^{-1} min^{-1}.
(a) Write the rate expression.
(b) Calculate the rate when $[NH_3] = 0.075$ M.
(c) At what concentration of ammonia is the rate equal to the rate constant?

39. For the reaction of phenyl acetate with water the concentration as a function of time was given in Question 12. Assume that the concentration of water does not change during the reaction. Analyze the data from Question 12 to determine
(a) The rate law
(b) The order of the reaction with respect to phenyl acetate
(c) The rate constant
(d) The rate of reaction when the concentration of phenyl acetate is 0.10 mol/L (assuming that the concentration of water is the same as in the experiments in the table in Question 12)

40. Determine the rate constant and order from the concentration-time dependence for the reaction $A \longrightarrow P$.

Time (s)	[A] (M)
0	0.0350
10.	0.0223
20.	0.0142
50.	0.0037
70.	0.0015

41. When phenacyl bromide and pyridine are both dissolved in methanol, they react to form phenacylpyridinium bromide.

$$C_6H_5{-}\overset{\overset{\large O}{\|}}{C}{-}CH_2Br + C_5H_5N \longrightarrow C_6H_5{-}\overset{\overset{\large O}{\|}}{C}{-}CH_2NC_5H_5^+ + Br^-$$

When equal concentrations of reactants were mixed in methanol at 35 °C, these data were obtained:

Time (min)	[Reactant] (mol/L)	Time (min)	[Reactant] (mol/L)
0	0.0385	500.	0.0208
100.	0.0330	600.	0.0191
200.	0.0288	700.	0.0176
300.	0.0255	800.	0.0163
400.	0.0220	1000.	0.0143

(a) What is the rate law for this reaction?
(b) What is the overall order?
(c) What is the rate constant?
(d) What is the rate of this reaction when the concentration of each reactant is 0.030 mol/L?

42. The transfer of an oxygen atom from NO_2 to CO has been studied at 540 K:

$$CO(g) + NO_2(g) \longrightarrow CO_2(g) + NO(g)$$

These data were collected:

Initial Rate ($mol\ L^{-1}\ h^{-1}$)	Initial Concentration (mol/L)	
	[CO]	$[NO_2]$
5.1×10^{-4}	0.35×10^{-4}	3.4×10^{-8}
5.1×10^{-4}	0.70×10^{-4}	1.7×10^{-8}
5.1×10^{-4}	0.18×10^{-4}	6.8×10^{-8}
1.0×10^{-3}	0.35×10^{-4}	6.8×10^{-8}
1.5×10^{-3}	0.35×10^{-4}	10.2×10^{-8}

Use the data in the table to

(a) Write the rate law.

(b) Determine the reaction order with respect to each reactant.

(c) Calculate the rate constant, and express it in appropriate units.

43. The compound *p*-methoxybenzonitrile *N*-oxide, which has the formula $CH_3OC_6H_4CNO$, reacts with itself to form a dimer—a molecule that consists of two *p*-methoxybenzonitrile *N*-oxide units connected to each other $(CH_3OC_6H_4CNO)_2$. The reaction can be represented as

$$A + A \longrightarrow B \quad \text{or} \quad 2\,A \longrightarrow B$$

where A represents *p*-methoxybenzonitrile *N*-oxide and B represents the dimer $(CH_3OC_6H_4CNO)_2$. For the reaction in carbon tetrachloride at 40 °C with an initial concentration of 0.011 M, these data were obtained:

Time (min)	Percent Reaction	Time (min)	Percent Reaction
0	0	942	60.9
60.	9.1	1080.	64.7
120.	16.7	1212	66.6
215	26.5	1358	68.5
325	32.7	1518	70.3
565	47.3		

(a) Determine the rate law for the reaction.

(b) Determine the rate constant.

(c) Determine the order of the reaction with respect to A.

44. The bromination of acetone is catalyzed by acid.

$$CH_3COCH_3(aq) + Br_2(aq) + H_2O(\ell) \xrightarrow{\text{acid catalyst}} CH_3COCH_2Br(aq) + H_3O^+(aq) + Br^-(aq)$$

The rate of disappearance of bromine was measured for several different initial concentrations of acetone, bromine, and hydronium ion.

Initial Concentration (mol/L)			Initial Rate of Change of $[Br_2]$ ($mol\ L^{-1}\ s^{-1}$)
$[CH_3COCH_3]$	$[Br_2]$	$[H_3O^+]$	
0.30	0.050	0.050	5.7×10^{-5}
0.30	0.10	0.050	5.7×10^{-5}
0.30	0.050	0.10	12.0×10^{-5}
0.40	0.050	0.20	31.0×10^{-5}
0.40	0.050	0.050	7.6×10^{-5}

(a) Deduce the rate law for the reaction and give the order with respect to each reactant.

(b) What is the numerical value of *k*, the rate constant?

(c) If $[H_3O^+]$ is maintained at 0.050 M, whereas both $[CH_3COCH_3]$ and $[Br_2]$ are 0.10 M, what is the rate of the reaction?

45. One of the major eye irritants in smog is formaldehyde, CH_2O, formed by reaction of ozone with ethylene.

$$C_2H_4(g) + O_3(g) \longrightarrow 2\,CH_2O(g) + \tfrac{1}{2}\,O_2(g)$$

These data were collected:

Initial Concentration (mol/L)		Initial Rate of Formation of CH_2O ($mol\ L^{-1}\ s^{-1}$)
$[O_3]$	$[C_2H_4]$	
0.50×10^{-7}	1.0×10^{-8}	1.0×10^{-12}
1.5×10^{-7}	1.0×10^{-8}	3.0×10^{-12}
1.0×10^{-7}	2.0×10^{-8}	4.0×10^{-12}

(a) Determine the rate law for the reaction using the data in the table.

(b) Determine the reaction order with respect to O_3. Determine the order with respect to C_2H_4.

(c) Calculate the rate constant, *k*.

(d) Calculate the rate of reaction when $[C_2H_4]$ and $[O_3]$ are both 2.0×10^{-7} M.

46. This question requires working with the equations of Table 13.2, p. 604. Using an initial concentration $[A]_0$ of 1.0 mol/L and a rate constant *k* with a numerical value of 1.0 in appropriate units, make plots of [A] versus time over the time interval 0 to 5 s for each type of integrated rate law. Compare your results with Figure 13.5.

47. Studies of radioactive decay of nuclei show that the *decay rate* of a radioactive sample is proportional to the *amount* of the radioactive species present. Once half the radioactivity has disappeared, the radioactive decay *rate* is only half of its original value. Is radioactive decay a zeroth-order, first-order, or second-order process?

48. Radioactive gold-198 is used in the diagnosis of liver problems. ^{198}Au decays in a first-order process, emitting a β particle (electron). The half-life of this isotope is 2.7 days. If you begin with a 5.6-mg sample of the isotope, how much of this sample remains after 1.0 day?

49. The compound $Xe(CF_3)_2$ decomposes in a first-order reaction to elemental Xe with a half-life of 30. min. If you place 7.50 mg of $Xe(CF_3)_2$ in a flask, how long must you wait until only 0.25 mg of $Xe(CF_3)_2$ remains?

50. The first-order rate constant for the radioactive decay of radium-223 is 0.0606 day^{-1}. Calculate the half-life of radium-223.

51. If the initial concentration of the reactant in a first-order reaction A ⟶ products is 0.64 mol/L and the half-life is 30. s,

(a) Calculate the concentration of the reactant 60. s after initiation of the reaction.

(b) How long would it take for the concentration of the reactant to drop to one-eighth its initial value?

(c) How long would it take for the concentration of the reactant to drop to 0.040 $mol\ L^{-1}$?

52. If the initial concentration of the reactant in a first-order reaction A $\longrightarrow$ products is 0.50 mol/L and the half-life is 400. s,
    (a) Calculate the concentration of the reactant 1600 s after initiation of the reaction.
    (b) How long would it take for the concentration of the reactant to drop to one-sixteenth its initial value?
    (c) How long would it take for the concentration of the reactant to drop to 0.062 mol L^{-1}?
53. The compound SO_2Cl_2 decomposes in a first-order reaction

$$SO_2Cl_2(g) \longrightarrow SO_2(g) + Cl_2(g)$$

that has a half-life of 1.47×10^4 s at 600. K. If you begin with 1.6×10^{-3} mol of pure SO_2Cl_2 in a 2.0-L flask, at what time will the amount of SO_2Cl_2 be 1.2×10^{-4} mol?
54. The hypothetical compound A decomposes in a first-order reaction that has a half-life of 2.3×10^2 s at 450. °C. If the initial concentration of A is 4.32×10^{-2} M, how long will it take for the concentration of A to drop to 3.75×10^{-3} M?
55. The first-order rate constant for the decomposition of a certain hormone in water at 25 °C is 3.42×10^{-4} day^{-1}.
    (a) If a 0.0200-M solution of the hormone is stored at 25 °C for two months, what will its concentration be at the end of that period?
    (b) How long will it take for the concentration of the solution to drop from 0.0200 M to 0.00350 M?
    (c) What is the half-life of the hormone?

A Nanoscale View: Elementary Reactions (Section 13.4)

56. Which of these reactions are unimolecular and elementary, which are bimolecular and elementary, and which are not elementary?
    (a) $CH_4(g) + 2\,O_2(g) \longrightarrow CO_2(g) + 2\,H_2O(g)$
    (b) $O_3(g) + O(g) \longrightarrow 2\,O_2(g)$
    (c) $Mg(s) + 2\,H_2O(\ell) \longrightarrow H_2(g) + Mg(OH)_2(s)$
    (d) $O_3(g) \longrightarrow O_2(g) + O(g)$
57. Which of these reactions are unimolecular and elementary, which are bimolecular and elementary, and which are not elementary?
    (a) $HCl(g) + H_2O(g) \longrightarrow H_3O^+(g) + Cl^-(g)$
    (b) $I^-(g) + CH_3Cl(g) \longrightarrow ICH_3(g) + Cl^-(g)$
    (c) $C_2H_6(g) \longrightarrow C_2H_4(g) + H_2(g)$
    (d) $N_2(g) + 3\,H_2(g) \longrightarrow 2\,NH_3(g)$
    (e) $O_2(g) + O(g) \longrightarrow O_3(g)$
58. Assume that each gas phase reaction occurs via a single bimolecular step. For which reaction would you expect the steric factor to be more important? Why?

$$Cl + O_3 \longrightarrow ClO + O_2 \quad \text{or} \quad NO + O_3 \longrightarrow NO_2 + O_2$$

59. Assume that each gas phase reaction occurs via a single bimolecular step. For which reaction would you expect the steric factor to be more important? Why?

$$H_2C{=}CH_2 + H_2 \longrightarrow H_3C{-}CH_3 \quad \text{or}$$

$$(CH_3)_2C{=}CH_2 + HBr \longrightarrow (CH_3)_2CBr{-}CH_3$$

Temperature and Reaction Rate (Section 13.5)

60. From Problem-Solving Example 13.8 (p. 614), where the energy profile of the ozone plus atomic oxygen reaction was derived, obtain the activation energy. Then determine the ratio of the reaction rate for this reaction at 50. °C to the reaction rate at room temperature (25 °C). Assume that the initial concentrations are the same at both temperatures.
61. Suppose a reaction rate constant has been measured at two different temperatures, T_1 and T_2, and its value is k_1 and k_2, respectively.
    (a) Write the Arrhenius equation at each temperature.
    (b) By combining these two equations, derive an expression for the ratio of the two rate constants, k_1/k_2. Use this formula to answer the next two questions.
62. Suppose a chemical reaction rate constant has an activation energy of 76 kJ/mol, as in the example in Figure 13.11. By what factor is the rate of the reaction at 50. °C increased over its rate at 25 °C?
63. A chemical reaction has an activation energy of 30. kJ/mol. If you had to slow down this reaction a thousandfold by cooling it from room temperature (25 °C), what would the temperature be?
64. What is the activation energy for a reaction if its rate constant is found to triple when the temperature is raised from 600. K to 610. K?
65. These data were obtained for the rate constant for reaction of an unknown compound with water:

T (°C)	k (s^{-1})	T (°C)	k (s^{-1})
25.0	7.95×10^{-8}	56.2	1.04×10^{-5}
30.0	2.37×10^{-7}	78.2	1.45×10^{-4}

    (a) Calculate the activation energy and frequency factor for this reaction.
    (b) Estimate the rate constant of the reaction at a temperature of 100.0 °C.
66. *p*-Methylphenyl acetate reacts with imidazole to produce *p*-methylphenol and acetyl imidazole. The rate constants for this second-order reaction at a series of temperatures are given in the table.

T (°C)	k (L mol^{-1} s^{-1})
10	2.34×10^{-2}
18	3.25×10^{-2}
25	4.5×10^{-2}
35	5.83×10^{-2}
42	7.5×10^{-2}
60	1.52×10^{-1}

    (a) Calculate the activation energy and frequency factor for this reaction.
    (b) Estimate the rate constant for this reaction at a temperature of 100.0 °C.
67. For the reaction of iodine atoms with hydrogen molecules in the gas phase, these rate constants were obtained experimentally.

$$2\,I(g) + H_2(g) \longrightarrow 2\,HI(g)$$

T (K)	$10^{-5}\, k$ ($L^2\ mol^{-2}\ s^{-1}$)
417.9	1.12
480.7	2.60
520.1	3.96
633.2	9.38
666.8	11.50
710.3	16.10
737.9	18.54

(a) Calculate the activation energy and frequency factor for this reaction.
(b) Estimate the rate constant of the reaction at 400.0 K.

68. Make an Arrhenius plot and calculate the activation energy for the gas phase reaction

$$2\ NOCl(g) \longrightarrow 2\ NO(g) + Cl_2(g)$$

T (K)	Rate Constant ($L\ mol^{-1}\ s^{-1}$)
400.	6.95×10^{-4}
450.	1.98×10^{-2}
500.	2.92×10^{-1}
550.	2.60
600.	16.3

69. The activation energy E_a is 139.7 kJ mol^{-1} for the gas phase reaction

$$HI + CH_3I \longrightarrow CH_4 + I_2$$

Calculate the fraction of the molecules whose collisions would be energetic enough to react at
(a) 100. °C (b) 200. °C
(c) 500. °C (d) 1000. °C

70. The activation energy E_a is 10. kJ mol^{-1} for the gas phase reaction

$$NO + O_3 \longrightarrow NO_2 + O_2$$

Calculate the fraction of the molecules whose collisions would be energetic enough to react at
(a) 400. °C (b) 600. °C
(c) 800. °C (d) 1000. °C

71. For the gas phase reaction

$$CH_3CH_2I(g) \longrightarrow CH_2CH_2(g) + HI(g)$$

the activation energy E_a is 221 kJ/mol and the frequency factor A is $1.2 \times 10^{14}\ s^{-1}$. If the concentration of CH_3CH_2I is 0.012 mol/L, what is the rate of reaction at
(a) 400. °C? (b) 800. °C?

72. For the gas phase reaction

$$cis\text{-}CHClCHCl(g) \longrightarrow trans\text{-}CHClCHCl(g)$$

the activation energy E_a is 234 kJ/mol and the frequency factor A is $6.3 \times 10^{12}\ s^{-1}$. If the concentration of *cis*-CHClCHCl is 0.0043 mol/L, what is the rate of reaction at
(a) 400. °C? (b) 800. °C?

73. For the reaction

$$N_2O_5(g) \longrightarrow 2\ NO_2(g) + + \tfrac{1}{2}\ O_2(g)$$

the rate constant k at 25 °C is $3.46 \times 10^{-5}\ s^{-1}$ and at 55 °C it is $1.5 \times 10^{-3}\ s^{-1}$. Calculate the activation energy, E_a.

74. For a hypothetical reaction the rate constant is $4.63 \times 10^{-3}\ s^{-1}$ at 25 °C and $2.37 \times 10^{-2}\ s^{-1}$ at 43 °C. Calculate the activation energy.

Rate Laws for Elementary Reactions (Section 13.6)

75. What is the rate law for each of these elementary reactions?
(a) $Cl(g) + ICl(g) \longrightarrow I(g) + Cl_2(g)$
(b) $O(g) + O_3(g) \longrightarrow 2\ O_2(g)$
(c) $2\ NO_2(g) \longrightarrow N_2O_4(g)$

76. Use the diagram to answer these questions.

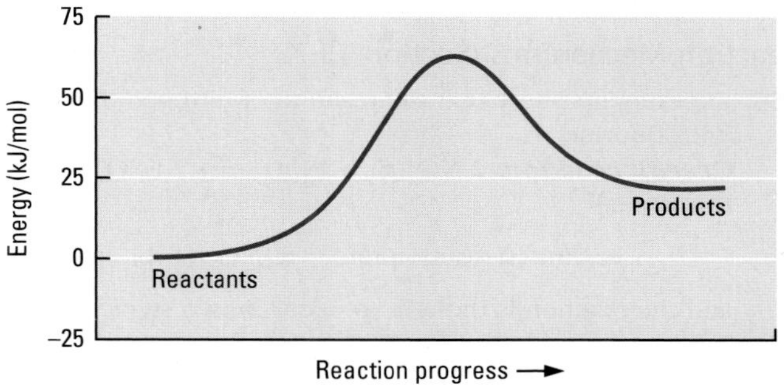

(a) Is the reaction exothermic or endothermic?
(b) Determine the approximate value of ΔE for the forward reaction.
(c) What is the activation energy in each direction?
(d) A catalyst is found that lowers the activation energy of the reaction by about 10 kJ/mol. How will this catalyst affect the rate of the reverse reaction?

77. Draw an energy versus reaction progress diagram (similar to the one in Question 76) for each of the reactions whose activation energy and enthalpy change are given below.
(a) $\Delta H° = -145$ kJ mol^{-1}; $E_a = 75$ kJ mol^{-1}
(b) $\Delta H° = -70$ kJ mol^{-1}; $E_a = 65$ kJ mol^{-1}
(c) $\Delta H° = 70$ kJ mol^{-1}; $E_a = 85$ kJ mol^{-1}

78. Draw an energy versus reaction progress diagram (similar to the one in Question 76) for each of the reactions whose activation energy and enthalpy change are given below.
(a) $\Delta H° = 105$ kJ mol^{-1}; $E_a = 175$ kJ mol^{-1}
(b) $\Delta H° = -43$ kJ mol^{-1}; $E_a = 95$ kJ mol^{-1}
(c) $\Delta H° = 15$ kJ mol^{-1}; $E_a = 55$ kJ mol^{-1}

79. Which of the reactions in Question 77 would be expected to (a) Occur fastest? (b) Occur slowest? (Assume equal temperatures, equal concentrations, equal frequency factors, and the same rate law for all reactions.)

80. Which of the reactions in Question 78 would be expected to (a) Occur fastest? (b) Occur slowest? (Assume equal temperatures, equal concentrations, equal frequency factors, and the same rate law for all reactions.)

81. For which of the reactions in Question 77 would the *reverse* reaction (a) Be fastest? (b) Be slowest? (Assume equal temperatures, equal concentrations, equal frequency factors, and the same rate law for all reactions.)

82. For which of the reactions in Question 78 would the *reverse* reaction (a) Be fastest? (b) Be slowest? (Assume equal temperatures, equal concentrations, equal frequency factors, and the same rate law for all reactions.)

83. Assuming that each reaction is elementary, predict the rate law.
(a) $NO(g) + NO_3(g) \longrightarrow 2\ NO_2(g)$
(b) $O(g) + O_3(g) \longrightarrow 2\ O_2(g)$
(c) $(CH_3)_3CBr(aq) \longrightarrow (CH_3)_3C^+(aq) + Br^-(aq)$
(d) $2\ HI(g) \longrightarrow H_2(g) + I_2(g)$

84. Assuming that each reaction is elementary, predict the rate law.
(a) $Cl(g) + ICl(g) \longrightarrow I(g) + Cl_2(g)$
(b) $Cl(g) + H_2(g) \longrightarrow HCl(g) + H(g)$
(c) $2\ NO_2(g) \longrightarrow N_2O_4(g)$
(d) Cyclopropane(g) $\longrightarrow$ propene(g)

Reaction Mechanisms (Section 13.7)

85. Experiments show that the reaction of nitrogen dioxide with fluorine
Overall reaction: $2\ NO_2(g) + F_2(g) \longrightarrow 2\ FNO_2(g)$ has the rate law

$$\text{Initial reaction rate} = k[NO_2][F_2]$$

and the reaction is thought to occur in two steps.

Step 1: $NO_2(g) + F_2(g) \longrightarrow FNO_2(g) + F(g)$

Step 2: $NO_2(g) + F(g) \longrightarrow FNO_2(g)$

(a) Show that the sum of this sequence of reactions gives the balanced equation for the overall reaction.
(b) Which step is rate-determining?

86. Nitrogen monoxide is reduced by hydrogen to give water and nitrogen

$$2\ H_2(g) + 2\ NO(g) \longrightarrow N_2(g) + 2\ H_2O(g)$$

and one possible mechanism for this reaction is a sequence of three elementary steps.

Step 1: $2\ NO(g) \rightleftharpoons N_2O_2(g)$ fast

Step 2: $N_2O_2(g) + H_2(g) \longrightarrow N_2O(g) + H_2O(g)$ slow

Step 3: $N_2O(g) + H_2(g) \longrightarrow N_2(g) + H_2O(g)$ fast

(a) Show that the sum of these steps gives the net reaction.
(b) What is the rate law for this reaction?

87. The reaction of $NO_2(g)$ and $CO(g)$ is thought to occur in two steps:

Step 1: $NO_2(g) + NO_2(g) \longrightarrow NO(g) + NO_3(g)$ slow

Step 2: $NO_3(g) + CO(g) \longrightarrow NO_2(g) + CO_2(g)$ fast

(a) Show that the elementary steps add up to give the overall, stoichiometric equation.
(b) What is the molecularity of each step?
(c) For this mechanism to be consistent with kinetic data, what must be the experimental rate equation?
(d) Identify any intermediates in this reaction.

88. Write the overall reaction and the rate expressions that correspond to these reaction mechanisms. Be sure to eliminate intermediates from the answers:

(a) $A + B \rightleftharpoons C + D$ fast
$C + E \longrightarrow F$ slow
(b) $A \rightleftharpoons B + C$ fast
$C + D \rightleftharpoons E$ fast
$E \longrightarrow F$ slow

89. For the reaction

$$2\ NO(g) + Cl_2(g) \longrightarrow 2\ NOCl(g)$$

the currently accepted mechanism is

$NO + Cl_2 \rightleftharpoons NOCl_2$ fast

$NOCl_2 + NO \longrightarrow 2\ NOCl$ slow

(a) Determine the rate law for this mechanism. (Be sure to express it in terms of concentrations of reactants or products of the overall reaction, not in terms of intermediates.)
(b) Suggest another mechanism that agrees with the same rate law.
(c) Suggest another mechanism that does not agree with the same rate law.

90. For the reaction

$$2\ N_2O_5(g) \longrightarrow 4\ NO_2(g) + O_2(g)$$

the currently accepted mechanism is

$N_2O_5 \rightleftharpoons NO_2 + NO_3$ fast

$NO_2 + NO_3 \longrightarrow NO_2 + O_2 + NO$ slow

$NO + NO_3 \longrightarrow 2\ NO_2$ fast

Determine the rate law for this reaction.

91. For the reaction mechanism

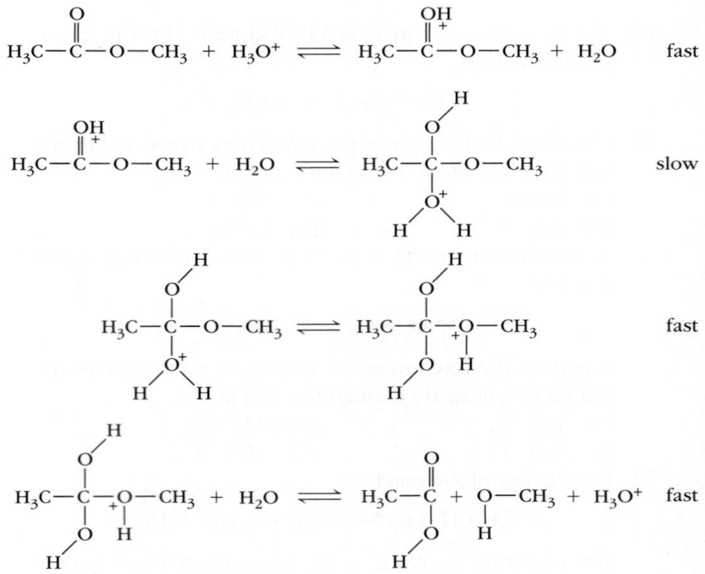

(a) Write the chemical equation for the overall reaction.
(b) Write the rate law for the reaction.
(c) Is there a catalyst involved in this reaction? If so, what is it?
(d) Identify all intermediates in the reaction.

92. For the reaction mechanism

$A + B \rightleftharpoons C$ fast

$C + A \longrightarrow 2\ D$ slow

(a) Write the chemical equation for the overall reaction.
(b) Write the rate law for the reaction.
(c) Is there a catalyst involved in this reaction? If so, what is it?
(d) Identify all intermediates in the reaction.

93. For the reaction

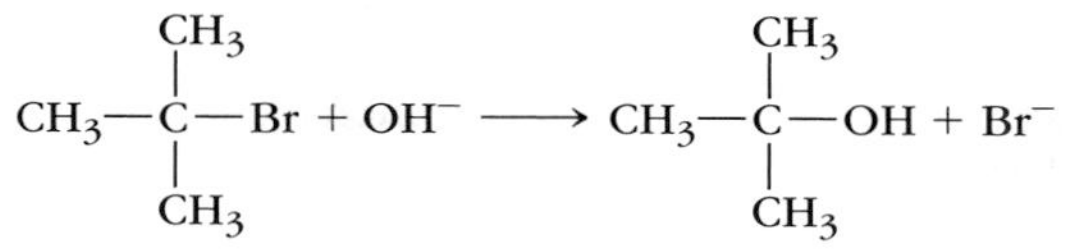

the rate law is

$$\text{Rate} = k[(CH_3)_3CBr]$$

Identify each mechanism that is compatible with the rate law.

(a) $(CH_3)_3CBr \longrightarrow (CH_3)_3C^+ + Br^-$ slow
$(CH_3)_3C^+ + OH^- \longrightarrow (CH_3)_3COH$ fast

(b) $(CH_3)_3CBr + OH^- \longrightarrow (CH_3)_3COH + Br^-$

(c) $(CH_3)_3CBr + OH^- \longrightarrow (CH_3)_2(CH_2)CBr^- + H_2O$ fast
$(CH_3)_2(CH_2)CBr^- \longrightarrow (CH_3)_2(CH_2)C + Br^-$ slow
$(CH_3)_2(CH_2)C + H_2O \longrightarrow (CH_3)_3COH$ fast

94. Which of these mechanisms (more than one may be chosen) is compatible with the rate law?

$$\text{Rate} = k[Cl_2]^{3/2}[CO]$$

(a) $\frac{1}{2}\,Cl_2 \rightleftharpoons Cl$ fast
$Cl + Cl_2 \longrightarrow Cl_3$ fast
$Cl_3 + CO \longrightarrow COCl_2 + Cl$ slow
$Cl \rightleftharpoons \frac{1}{2}\,Cl_2$ fast

(b) $Cl_2 + CO \longrightarrow CCl_2 + O$ slow
$O + Cl_2 \longrightarrow Cl_2O$ fast
$Cl_2O + CCl_2 \longrightarrow COCl_2 + Cl_2$ fast

(c) $\frac{1}{2}\,Cl_2 \rightleftharpoons Cl$ fast
$Cl + CO \rightleftharpoons COCl$ fast
$COCl + Cl_2 \longrightarrow COCl_2 + Cl$ slow
$Cl \rightleftharpoons \frac{1}{2}\,Cl_2$ fast

(d) $Cl_2 + CO \rightleftharpoons COCl + Cl$ fast
$COCl + Cl_2 \longrightarrow COCl_2 + Cl$ slow
$Cl + Cl \longrightarrow Cl_2$ fast

95. The mechanism for the reaction of CH_3OH and HBr is believed to involve two steps. The overall reaction is exothermic.

Step 1: $CH_3OH + H^+ \rightleftharpoons CH_3OH_2^+$ fast, endothermic

Step 2: $CH_3OH_2^+ + Br^- \longrightarrow CH_3Br + H_2O$ slow

(a) Write an equation for the overall reaction.
(b) Draw a reaction energy diagram for this reaction.
(c) Show that the rate law for this reaction is

$$\text{Rate} = k[CH_3OH][H^+][Br^-]$$

96. For the reaction between NO and Br_2 this mechanism has been proposed.

$2\,NO \rightleftharpoons N_2O_2$ fast
$N_2O_2 + Br_2 \longrightarrow 2\,NOBr$ slow

Is this mechanism consistent with the observation that the reaction is second-order in NO and first-order in Br_2?

Catalysts and Reaction Rate (Section 13.8)

97. Which of these statements is (are) true? If a statement is false, reword it so that it becomes true.
(a) The concentration of a homogeneous catalyst may appear in the rate law.
(b) A catalyst is always consumed in the reaction.
(c) A catalyst must always be in the same phase as the reactants.
(d) A catalyst can change the course of a reaction and allow different products to be produced.

98. Hydrogenation reactions—processes in which H_2 is added to a molecule—are usually catalyzed. An excellent catalyst is a very finely divided metal suspended in the reaction solvent. Tell why finely divided rhodium, for example, is a much more efficient catalyst than a small block of the metal that has the same mass.

99. Which of these reactions appear to involve a catalyst? In those cases where a catalyst is present, tell whether it is homogeneous or heterogeneous.
(a) $CH_3COOCH_3(aq) + H_2O(\ell) \longrightarrow CH_3COOH(aq) + CH_3OH(aq)$
Rate $= k[CH_3COOCH_3][H_3O^+]$
(b) $H_2(g) + I_2(g) \longrightarrow 2\,HI(g)$
Rate $= k[H_2][I_2]$
(c) $2\,H_2(g) + O_2(g) \longrightarrow 2\,H_2O(g)$
Rate $= k[H_2][O_2]$ (area of Pt surface)
(d) $H_2(g) + CO(g) \longrightarrow H_2CO(g)$
Rate $= k[H_2]^{1/2}[CO]$

100. In acid solution, methyl formate forms methyl alcohol and formic acid.

$$\underset{\text{methyl formate}}{HCO_2CH_3(aq)} + H_2O(\ell) \longrightarrow \underset{\text{formic acid}}{HCOOH(aq)} + \underset{\text{methyl alcohol}}{CH_3OH(aq)}$$

The rate law is as follows: rate $= k[HCO_2CH_3][H_3O^+]$. Why does H_3O^+ appear in the rate law but not in the overall equation for the reaction?

Enzymes: Biological Catalysts (Section 13.9)

101. A catalyst lowers the activation energy of a reaction from 215 kJ to 206 kJ. By what factor would you expect the reaction-rate constant to increase at 25 °C? Assume that the frequency factors (A) are the same for both reactions.

102. Write a one- or two-sentence definition in your own words for each term:

active site	monomer
cofactor	polypeptide
enzyme	polysaccharide
inhibition	protein
lysozyme	substrate

103. Write a one- or two-sentence definition in your own words for each term:

dimer	hydrolysis
denaturation	induced fit
enzyme-substrate complex	maximum velocity
globular protein	

104. When enzymes are present at very low concentration, their effect on reaction rate can be described by first-order kinetics. By what factor does the rate of an enzyme-catalyzed reaction change when the enzyme concentration is changed from 1.5×10^{-7} to 4.5×10^{-6} M?

105. When substrates are present at relatively high concentration and are catalyzed by enzymes, the effect on reaction rate of changing substrate concentration can be described by zeroth-order kinetics. By what factor does the rate of an enzyme-catalyzed reaction change when the substrate concentration is changed from 1.5×10^{-3} to 4.5×10^{-2} M?

106. The reaction

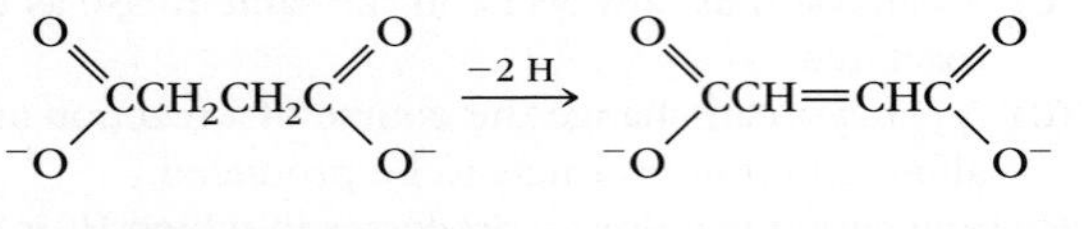

is catalyzed by the enzyme succinate dehydrogenase. When malonate ions or oxalate ions are added to the reaction mixture, the rate decreases significantly. Try to account for this observation in terms of the description of enzyme catalysis given in the text. The structures of malonate and oxalate ions are

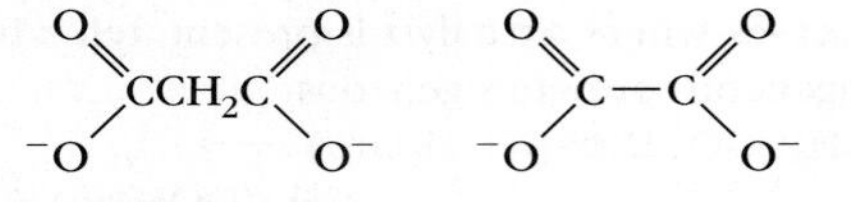

malonate ion oxalate ion

107. Some enzymes can be inhibited by heavy metal ions such as Hg^{2+} and Pb^{2+}. These metal ions have a large affinity for sulfur-containing groups and can react with molecules such as CH_3CH_2SH to form compounds such as $Pb(CH_3CH_2S)_2$. Based on the structures of the amino acids given in Table 12.8 (← *p. 579*), suggest at least one amino acid that is likely to be present in enzymes that are inhibited by heavy metals.

108. Many biochemical reactions are catalyzed by acids. A typical mechanism consistent with the experimental results (in which HA is the acid and X is the reactant) is

Step 1: reversible $HA + H_2O \rightleftharpoons H_3O^+ + A^-$ fast

Step 2: reversible $X + H_3O^+ \rightleftharpoons XH^+ + H_2O$ fast

Step 3: $XH^+ \longrightarrow$ products slow

What rate law is derived from this mechanism? What is the order of the reaction with respect to HA? How would doubling the concentration of HA affect the reaction?

Catalysis in Industry (Section 13.10)

109. In the first paragraph of Section 13.10 of the text, an expert in the field of industrial chemistry is quoted. Explain the expert's statement, in view of your understanding of the nature of catalysts. Why are catalysts so important?

110. Why are homogeneous catalysts harder to separate from products and leftover reactants than are heterogeneous reactants?

111. In an automobile catalytic converter the catalysis is accomplished on a surface consisting of platinum and other precious metals. The metals are deposited as a thin layer on a honeycomb-like ceramic support (see the photo).

Johnson Matthey Platinum Today

(a) Why is the ceramic support arranged in the honeycomb geometry?
(b) Why are the metals deposited on the ceramic surface instead of being used as strips or rods?

112. Find all examples of reactions described in this chapter that are catalyzed by metals. Are these metals main group metals or transition metals? What type of chemical reactions are they: acid-base or oxidation-reduction? What conclusions can be drawn about metal-catalyzed chemical reactions from these examples?

General Questions

These questions are not explicitly keyed to chapter topics; many require integration of several concepts.

113. For the hypothetical reaction $A + B \longrightarrow C + D$, the activation energy is 32 kJ/mol. For the reverse reaction $(C + D \longrightarrow A + B)$, the activation energy is 58 kJ/mol. Is the reaction $A + B \longrightarrow C + D$ exothermic or endothermic?

114. Draw a reaction energy diagram for an exothermic process. Mark the positions of reactants, products, and activated complex. Indicate the activation energies of the forward and reverse processes and explain how ΔE for the reaction can be calculated from the diagram.

115. Draw a reaction energy diagram for an endothermic process. Mark the positions of reactants, products, and activated complex. Indicate the activation energies of the forward and reverse processes and explain how ΔE for the reaction can be calculated from the diagram.

116. Under certain conditions, biphenyl, $C_{12}H_{10}$, can be produced by the decomposition of cyclohexane, C_6H_{12}:

$$2\ C_6H_{12} \longrightarrow C_{12}H_{10} + 7\ H_2$$

This table represents part of the data obtained in the kinetics experiment.

Time (s)	$[C_6H_{12}]$	$[C_{12}H_{10}]$	$[H_2]$
0.0	0.200	0.000	0.000
1.00	0.159	0.021	—
2.00	0.132	—	—
3.00	—	0.044	—

(a) Fill in the missing concentrations.
(b) Calculate the rate of reaction at 1.5 s.

117. The statements below relate to the reaction for the formation of HI:

$$H_2(g) + I_2(g) \longrightarrow 2\ HI(g) \qquad \text{Rate} = k[H_2][I_2]$$

Determine which statements are true. If a statement is false, indicate why it is incorrect.
(a) The reaction must occur in a single step.
(b) This is a second-order reaction overall.
(c) Raising the temperature will cause the value of *k* to decrease.
(d) Raising the temperature lowers the activation energy for this reaction.
(e) If the concentrations of both reactants are doubled, the rate will double.
(f) Adding a catalyst in the reaction will cause the initial rate to increase.

118. Indicate whether each of these statements is true or false. Change the wording of each false statement to make it true.
(a) It is possible to change the rate constant for a reaction by changing the temperature.
(b) The reaction rate remains constant as a first-order reaction proceeds at a constant temperature.
(c) The rate constant for a reaction is independent of reactant concentrations.
(d) As a second-order reaction proceeds at a constant temperature, the rate constant changes.

119. Consider the class of substances known as catalysts.
(a) Define "catalyst."
(b) What effect does a catalyst have on the energy barrier for a reaction?
(c) What special characteristics do enzymes have that distinguish them from other catalysts?

120. Nitrogen monoxide can be reduced with hydrogen.

$$2\ H_2(g) + 2\ NO(g) \longrightarrow 2\ H_2O(g) + N_2(g)$$

Experiment shows that when the concentration of H_2 is halved, the reaction rate is halved. Furthermore, raising the concentration of NO by a factor of 3 raises the rate by a factor of 9. Write the rate equation for this reaction.

121. One reaction that may occur in air polluted with nitrogen monoxide is

$$2\ NO(g) + O_2(g) \longrightarrow 2\ NO_2(g)$$

Using the data in the table, answer the questions that follow. Assume that all experiments are done at the same temperature.

Experiment	Initial Concentration (mol/L)		Initial Rate of Formation of $[NO_2]$ ($mol\ L^{-1}\ s^{-1}$)
	[NO]	$[O_2]$	
1	0.0010	0.0010	7.0×10^{-6}
2	0.0010	0.0020	1.4×10^{-5}
3	0.0010	0.0030	2.1×10^{-5}
4	0.0020	0.0030	8.4×10^{-5}
5	0.0030	0.0030	1.9×10^{-4}

(a) Determine the order of reaction with respect to each reactant.
(b) Write the rate law for the reaction.
(c) Calculate the rate of formation of NO_2 when $[NO] = [O_2] = 0.0050$ mol/L.

122. For the reaction of NO and O_2 at 660 K,

$$2\ NO(g) + O_2(g) \longrightarrow 2\ NO_2(g)$$

Concentration (mol/L)		Rate of Disappearance of NO ($mol\ L^{-1}\ s^{-1}$)
[NO]	$[O_2]$	
0.010	0.010	2.5×10^{-5}
0.020	0.010	1.0×10^{-4}
0.010	0.020	5.0×10^{-5}

(a) Determine the order of the reaction for each reactant.
(b) Write the rate equation for the reaction.
(c) Calculate the rate constant.
(d) Calculate the rate when $[NO] = 0.025$ mol/L and $[O_2] = 0.050$ mol/L.
(e) If O_2 disappears at a rate of $1.0 \times 10^{-4}\ mol\ L^{-1}\ s^{-1}$, what is the rate at which NO disappears? What is the rate at which NO_2 is forming?

123. Nitryl fluoride is an explosive compound that can be made by oxidizing nitrogen dioxide with fluorine:

$$2\ NO_2(g) + F_2(g) \longrightarrow 2\ NO_2F(g)$$

Several kinetics experiments, all done at the same temperature and involving formation of nitryl fluoride are summarized in this table:

Experiment	Initial Concentration (mol/L)			Initial Rate ($mol\ L^{-1}\ s^{-1}$)
	$[NO_2]$	$[F_2]$	$[NO_2F]$	
1	0.0010	0.0050	0.0020	2.0×10^{-4}
2	0.0020	0.0050	0.0020	4.0×10^{-4}
3	0.0020	0.0020	0.0020	1.6×10^{-4}
4	0.0020	0.0020	0.0010	1.6×10^{-4}

(a) Write the rate law for the reaction.
(b) What is the order of the reaction with respect to each reactant and each product?
(c) Calculate the rate constant *k* and express it in appropriate units.

124. The deep blue compound $CrO(O_2)_2$ can be made from the chromate ion by using hydrogen peroxide in an acidic solution.

$$HCrO_4^-(aq) + 2\ H_2O_2(aq) + H_3O^+(aq) \longrightarrow CrO(O_2)_2(aq) + 4\ H_2O(\ell)$$

The kinetics of this reaction have been studied, and the rate equation is found to be

$$\text{Rate of disappearance of } HCrO_4^- = k[HCrO_4^-]\ [H_2O_2][H_3O^+]$$

One of the mechanisms suggested for the reaction is

$$HCrO_4^- + H_3O^+ \rightleftharpoons H_2CrO_4 + H_2O$$

$$H_2CrO_4 + H_2O_2 \longrightarrow H_2CrO(O_2)_2 + H_2O$$

$$H_2CrO(O_2)_2 + H_2O_2 \longrightarrow CrO(O_2)_2 + 2\ H_2O$$

(a) Give the order of the reaction with respect to each reactant.
(b) Show that the steps of the mechanism agree with the overall equation for the reaction.
(c) Which step in the mechanism is rate-limiting?

125. How does a chemical reaction mechanism differ from other types of mechanisms—for example, the gear-changing mechanism of a bicycle, or the mechanism of an elevator? How is it similar to these mechanisms?

126. Two mechanisms are proposed for the reaction

$$2\ NO(g) + O_2(g) \longrightarrow 2\ NO_2(g)$$

Mechanism 1: $NO + O_2 \rightleftharpoons NO_3$ fast
$NO_3 + NO \longrightarrow 2\ NO_2$ slow

Mechanism 2: $NO + NO \rightleftharpoons N_2O_2$ fast
$N_2O_2 + O_2 \longrightarrow 2\ NO_2$ slow

Show that each mechanism is consistent with the observed rate law: rate $= k[NO]^2 \times [O_2]$.

127. The reaction between nitrogen dioxide and ozone,

$$2\ NO_2 + O_3 \longrightarrow N_2O_5 + O_2$$

has been studied at 231 K. The experimental rate equation is rate $= k[NO_2][O_3]$.
(a) What is the order of the reaction?
(b) Is either of these proposed mechanisms consistent with the rate law? Show how you arrived at your answer.

Mechanism 1: $NO_2 + NO_2 \rightleftharpoons N_2O_4$ fast
$N_2O_4 + O_3 \longrightarrow N_2O_5 + O_2$ slow

Mechanism 2: $NO_2 + O_3 \longrightarrow NO_3 + O_2$ slow
$NO_3 + NO_2 \longrightarrow N_2O_5$ fast

128. Why is chemical kinetics important in understanding the environmental problem posed by chlorofluorocarbons in the ozone layer (***← p. 455***)?

129. A reaction between molecules A and B ($A + B \longrightarrow$ products) is found to be first-order in A. Which rate equation cannot be correct?
(a) Rate $= k[A][B]$ (b) Rate $= k[A][B]^2$
(c) Rate $= k[B]^2$

Applying Concepts

These questions test conceptual learning.

130. The graph shows the change in concentration as a function of time for the reaction

$$2\ H_2O_2(g) \longrightarrow 2\ H_2O(g) + O_2(g)$$

What do each of the curves A, B, and C represent?

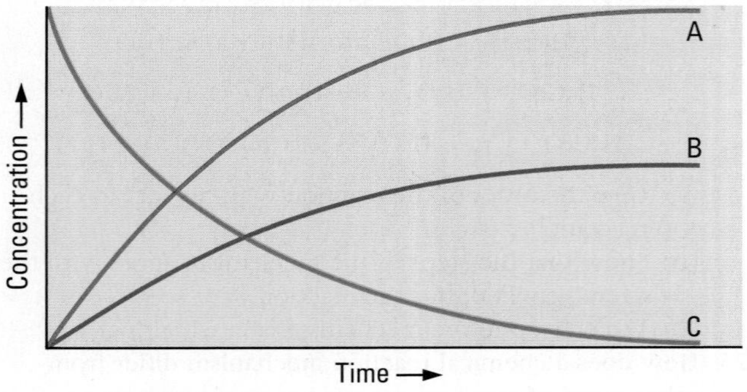
A
B
C
Concentration →
Time →

131. Draw a graph similar to the one in Question 130 for the reaction

$$2\ N_2O_5(g) \longrightarrow 4\ NO_2(g) + O_2(g)$$

132. The illustration below is a "snapshot" of the reactants at time = 0 for the reaction

$$H_2(g) + I_2(g) \longrightarrow 2\ HI(g)$$

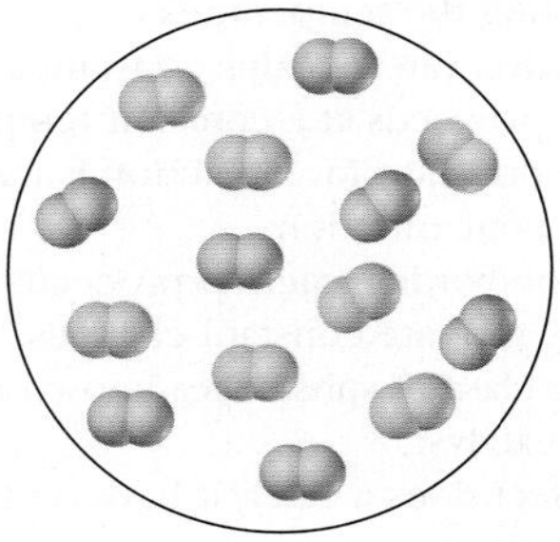

Suppose the reaction is carried out at two different temperatures and that another snapshot is taken after a constant time has elapsed.

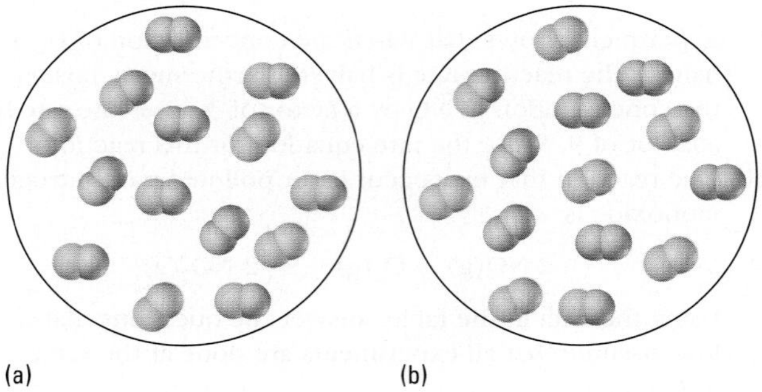

(a) (b)

Which of these two snapshots corresponds to the lower temperature reaction condition?

133. Consider Question 132 again, only this time a catalyst is used instead of a lower temperature. Which of the two snapshots corresponds to the presence of a catalyst?

134. Initial rates for the reaction $A + B + C \longrightarrow D + E$ were measured with various concentrations of A, B, and C as represented in the pictures below. Based on these data, what is the rate law?

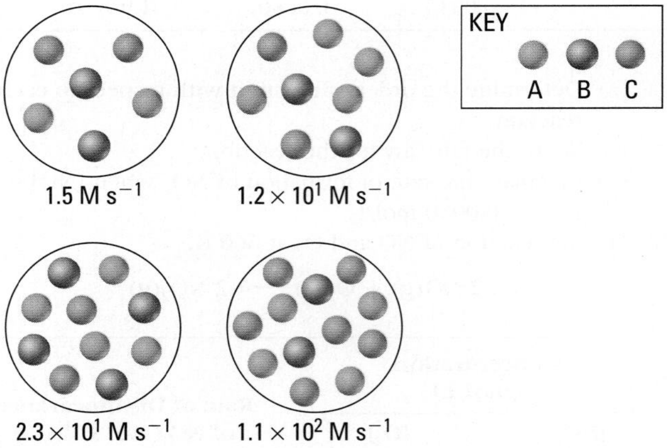
KEY
A B C
1.5 M s^{-1}
1.2 × 10^{1} M s^{-1}
2.3 × 10^{1} M s^{-1}
1.1 × 10^{2} M s^{-1}

135. Using a molecular model kit, build models of *cis*-2-butene, *trans*-2-butene, and the transition state, or activated complex. How much force do you need to apply to the models to change the reactant into the product by passing through the activated complex? The answer will, of course, depend on the kind of model kit that you use.

136. Examine the reaction coordinate diagram given here.

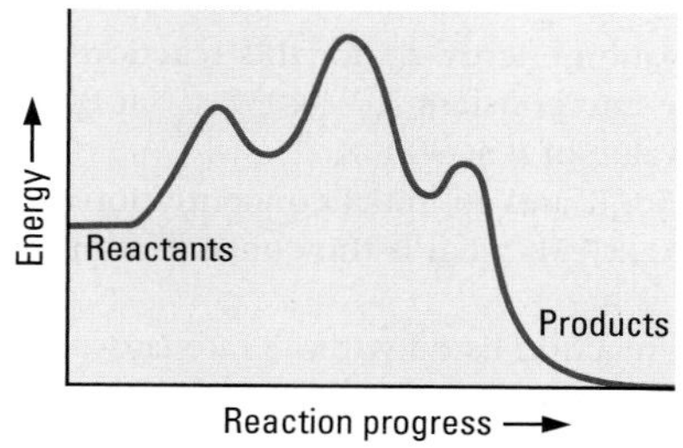

(a) How many steps are in the mechanism for the reaction described by this diagram?
(b) Is the reaction overall exothermic or endothermic?

137. The rate of decay of a radioactive solid is independent of the temperature of that solid—at least for temperatures easily obtained in the laboratory. What does this observation imply about the activation energy for this process?

138. Platinum metal is used as a catalyst in the decomposition of NO(g) into N_2(g) and O_2(g). A graph of the rate of the reaction as a function of NO concentration is shown below. Explain why the rate stops increasing and levels out.

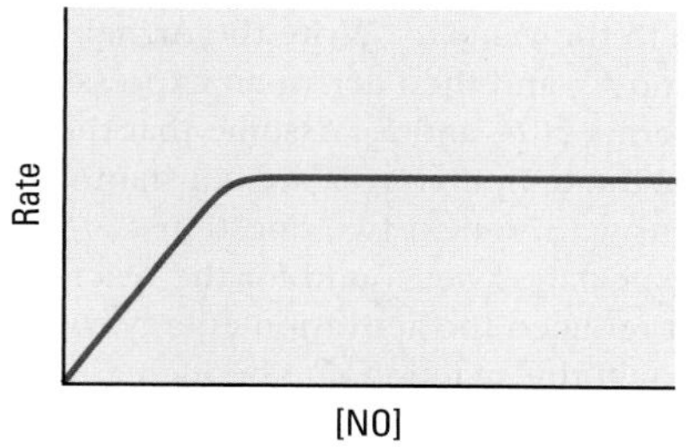

139. During the 1970s, at the time when catalytic converters began to be installed on automobiles, tetraethyllead, which had been added to gasoline to improve the octane rating, was gradually phased out. In the combustion chamber of the engine the four ethyl groups in tetraethyllead are converted to carbon dioxide and water, leaving either elemental lead or lead(II) oxide. In a modern automobile, burning a single tank of leaded gasoline can ruin the catalytic converter, destroying nearly all of the catalytic activity. Based on what you learned about the mechanism of reactions in a catalytic converter, suggest a reason that lead in gasoline might destroy the catalyst's ability to speed up reactions.

140. Calculate the half-life of a first-order reaction if 30.5 s after the reaction starts the concentration of the reactant is 0.0451 M and 45.0 s after the reaction starts it is 0.0321 M. How many seconds after the start of the reaction does it take for the reactant concentration to decrease to 0.0100 M?

More Challenging Questions

These questions require more thought and integrate several concepts.

141. Let x represent the number of half-lives that have elapsed during the course of a first-order reaction. That is, the elapsed time t is x times the half-life $t_{1/2}$: $t = xt_{1/2}$.
(a) Show that at time t the concentration $[A]_t$ of reactant is related to the initial concentration $[A]_0$ by

$$[A]_t = [A]_0(\tfrac{1}{2})^x$$

(b) Use your result in part (a) to show that

$$\log\left(\frac{[A]_t}{[A]_0}\right) = \frac{t}{t_{1/2}}(-\log 2) = \frac{t}{t_{1/2}}(-0.301)$$

(c) Use your result in part (b) to show that for any first-order reaction a plot of log $[A]_t$ versus t will be a straight line and that half-life can be obtained from the slope of the line.

142. Measurements of the initial rate of hydrolysis of benzenesulfonyl chloride in aqueous solution at 15 °C in the presence of fluoride ion yielded the results in the table for a fixed concentration of benzenesulfonyl chloride of 2×10^{-4} M. The reaction rate is known to be proportional to the concentration of benzenesulfonyl chloride.

$[F^-] \times 10^2$ (mol/L)	Initial Rate $\times 10^7$ (mol L^{-1} s^{-1})
0	2.4
0.5	5.4
1.0	7.9
2.0	13.9
3.0	20.2
4.0	25.2
5.0	32.0

Note that some reaction must be occurring in the absence of any fluoride ion, because at zero concentration of fluoride the rate is not zero. This residual rate should be subtracted from each observed rate to give the rate of the reaction being studied.
(a) Derive the complete rate law for the reaction.
(b) Calculate the rate constant k and express it in appropriate units.

143. Given these data and your result from Question 141, determine the half-life and the initial concentration of the reactant for the reaction

$$trans\text{-CHClCHCl(g)} \longrightarrow cis\text{-CHClCHCl(g)}$$

[*trans*-CHClCHCl] (mol/L)	Time (s)
9.23×10^{-4}	30
8.51×10^{-4}	60
7.86×10^{-4}	90
7.25×10^{-4}	120
6.19×10^{-4}	180
3.82×10^{-4}	360

144. When formic acid is heated, it decomposes to hydrogen and carbon dioxide in a first-order decay.

$$HCOOH(g) \longrightarrow CO_2(g) + H_2(g)$$

The rate of reaction is monitored by measuring the total pressure in the reaction container.

Time (s)	*P* (torr)
0	220
50	324
100	379
150	408
200	423
250	431
300	435

Calculate the rate constant and half-life, in seconds, for the reaction. At the start of the reaction (time = 0), only formic acid is present. (*Hint*: Find the partial pressure of formic acid [use Dalton's law of partial pressure (← *p. 446*) and the reaction stoichiometry to find P_{HCOOH} at each time]).

145. For a reaction involving the decomposition of Y, these data are obtained:

Rate (mol L^{-1} min^{-1})	0.288	0.245	0.202	0.158
[Y]	0.200	0.170	0.140	0.110

(a) Determine the order of the reaction.
(b) Write the rate expression for the decomposition of Y.
(c) Calculate *k* for the experiment above.

146. The decomposition of N_2O_5 is first-order with a rate constant of 2.5×10^{-4} s^{-1}.
(a) What is the half-life for decomposition of N_2O_5?
(b) How long does it take for the concentration of N_2O_5 to drop to $\frac{1}{32}$ of its original value?
(c) How long does it take for the concentration of N_2O_5 to drop from 3.4×10^{-3} mol/L to 2.3×10^{-5} mol/L?

147. The rate constant for decomposition of azomethane at 425 °C is 0.68 s^{-1}.

$$CH_3N{=}NCH_3(g) \longrightarrow N_2(g) + C_2H_6(g)$$

(a) Based on the units of the rate constant, is the reaction zeroth-, first-, or second-order?
(b) If 2.0 g azomethane is placed in a 2.0-L flask and heated to 425 °C, what mass of azomethane remains after 5.0 s?
(c) How long does it take for the mass of azomethane to drop to 0.24 g?
(d) What mass of nitrogen would be found in the flask after 0.50 s of reaction?

148. Cyclopropane isomerizes to propene when heated. Rate constants for the reaction

$$\text{cyclopropane} \longrightarrow \text{propene}$$

are 1.10×10^{-4} s^{-1} at 470. °C and 1.02×10^{-3} s^{-1} at 510. °C.
(a) Calculate the activation energy, E_a, for this reaction.
(b) At 500. °C and [cyclopropane] = 0.10 M, how long does it take for the concentration to drop to 0.023 M?

149. When heated, cyclobutane, C_4H_8, decomposes to ethylene, C_2H_4.

$$C_4H_8(g) \longrightarrow 2\ C_2H_4(g)$$

The activation energy, E_a, for this reaction is 262 kJ/mol.
(a) If the rate constant $k = 0.032$ s^{-1} at 800. K, what is the value of *k* at 900. K?
(b) At 850. K and an initial concentration of cyclobutane of 0.0427 M, what is the concentration of cyclobutane after 2 h?

150. For each reaction listed with its rate law, propose a reasonable mechanism.
(a) $CH_3COOCH_3(aq) + H_2O(\ell) \longrightarrow CH_3COOH(aq) + CH_3OH(aq)$
Rate = $k[CH_3COOCH_3][H_3O^+]$
(b) $H_2(g) + I_2(g) \longrightarrow 2\ HI(g)$
Rate = $k[H_2][I_2]$
(c) $2\ H_2(g) + O_2(g) \longrightarrow 2\ H_2O(g)$
Rate = $k[H_2][I_2]$ (area of Pt surface)
(d) $H_2(g) + CO(g) \longrightarrow H_2CO(g)$
Rate = $k[H_2]^{1/2}[CO]$

151. Suppose a rate constant for an uncatalyzed reaction is described by the Arrhenius equation with an activation energy E_a. The introduction of a catalyst lowers the activation energy to the value E_a' and thus increases the rate constant to the value k'. Write the Arrhenius equation for both k and k', and then derive an expression for the ratio k'/k in terms of E_a' and E_a. Assume that the frequency factor *A* and the temperature *T* are constant. Use this equation to answer the next two questions.

152. Suppose a catalyst was found for the reaction in Exercise 13.9 that reduced the activation energy to zero. By what factor would this reaction rate be increased at 370. K?

153. In the discussion of Figure 13.14, this statement is made: "Dropping the activation energy from 262 kJ/mol for the uncatalyzed reaction to 115 kJ/mole for the catalyzed process makes the catalyzed reaction 10^{15} times faster." Prove this statement.

154. When a molecule of *cis*-2-butene has rotated 90° around the axis through the two central carbon atoms, the π bond between these atoms is essentially broken. Use bond energies to estimate the activation energy for rotation around the π bond in ethylene, C_2H_4. Compare your result with the activation energy value in Figure 13.7.

155. Consider the reaction mechanism for iodine-catalyzed isomerization of *cis*-2-butene presented in Section 13.8. Use bond enthalpies (Table 8.2) to estimate the energy change for each step in the mechanism. (In Step 3 assume that rotation around the single bond requires about 5 kJ/mol.) Do your estimates agree with the energy values for the intermediates in Figure 13.14 (p. 628)?

156. The iodine-catalyzed isomerization of *cis*-2-butene can be speeded up still further if a bright light with wavelength less than about 800 nm shines on the reaction mixture. Consider the mechanism presented in Section 13.8 and

suggest a reason that the reaction is even faster in the presence of light than in the dark. Also explain why wavelengths longer than 800 nm are not effective in speeding up the reaction.

157. In a time-resolved picosecond spectroscopy experiment, Sheps, Crowther, Carrier, and Crim (*J. Phys. Chem. A*, Vol. 110, 2006; pp. 3087–3092) generated chlorine atoms in the presence of pentane. The pentane was dissolved in dichloromethane, CH_2Cl_2. The chlorine atoms are free radicals and are very reactive. After a nanosecond the chlorine atoms have reacted with pentane molecules, removing a hydrogen atom to form HCl and leaving behind a pentane radical with a single unpaired electron. The equation is

$$Cl(dcm) + C_5H_{12}(dcm) \longrightarrow HCl(dcm) + C_5H_{11}\cdot(dcm)$$

where (dcm) indicates that a substance is dissolved in dichloromethane. Measurements of the concentration of chlorine atoms were made as a function of time at three different concentrations of pentane in the dichloromethane. These results are shown in the table.

	Concentration of Chlorine Atoms (M) for Different C_5H_{12} Concentrations		
Time (ps)	**0.15-M C_5H_{12}**	**0.30-M C_5H_{12}**	**0.60-M C_5H_{12}**
100.0	4.42×10^{-5}	3.11×10^{-5}	1.77×10^{-5}
140.0	3.823×10^{-5}	2.39×10^{-5}	1.15×10^{-5}
180.0	3.38×10^{-5}	1.94×10^{-5}	8.48×10^{-6}
220.0	3.03×10^{-5}	1.49×10^{-5}	6.04×10^{-6}
260.0	2.68×10^{-5}	1.19×10^{-5}	4.12×10^{-6}
300.0	2.42×10^{-5}	9.45×10^{-6}	3.14×10^{-6}
340.0	2.08×10^{-5}	7.75×10^{-6}	2.38×10^{-6}
380.0	1.91×10^{-5}	6.35×10^{-6}	1.75×10^{-6}
420.0	1.71×10^{-5}	4.58×10^{-6}	1.61×10^{-6}
460.0	1.53×10^{-5}	3.77×10^{-6}	9.98×10^{-7}

(a) What is the order of the reaction with respect to chlorine atoms?
(b) Does the reaction rate depend on the concentration of pentane in the dichloromethane? If so, what is the order of the reaction with respect to pentane?
(c) Why does the concentration of pentane not affect the analysis of data that you performed in part a?
(d) Write the rate law for the reaction and calculate the rate of reaction for a concentration of chlorine atoms equal to 1.0 μM and a pentane concentration of 0.23 M.
(e) Sheps, Crowther, Carrier, and Crim found that the rate of formation of HCl matched the rate of disappearance of Cl. From this they concluded that there were no intermediates and side reactions were not important. Explain the basis for this conclusion.

158. If you know some calculus, derive the integrated fi[illegible] order rate law for the reaction

$$2\ A \longrightarrow \text{products}$$

by following these steps.
(i) Define the reaction rate in terms of [A].
(ii) Write the rate law in terms of [A], *k*, and *t*.
(iii) Separate variables in the rate law.
(iv) Integrate the rate law.
(v) Write the integrated equation in the form $y = mx + b$.
(vi) Derive the half-life as was done on p. 607.

159. If you know some calculus, derive the integrated second-order rate law for the reaction

$$2\ A \longrightarrow \text{products}$$

Follow the same steps listed in Question 158.

Conceptual Challenge Problems

These rigorous, thought-provoking problems integrate conceptual learning with problem solving and are suitable for group work.

CP13.A (Section 13.5) A rule of thumb is that for a typical reaction, if concentrations are unchanged, a 10-K rise in temperature increases the reaction rate by two to four times. Use an average increase of three times to answer the questions below.

(a) What is the approximate activation energy of a "typical" chemical reaction at 298 K?
(b) If a catalyst increases a chemical reaction's rate by providing a mechanism that has a lower activation energy, then what change do you expect a 10-K increase in temperature to make in the rate of a reaction whose uncatalyzed activation energy of 75 kJ/mol has been lowered to one-half this value (at 298 K) by addition of a catalyst?

CP13.B (Section 13.7) A sentence in an introductory chemistry textbook reads, "Dioxygen reacts with itself to form trioxygen, ozone, according to the equation, $3\ O_2 \longrightarrow 2\ O_3$." As a student of chemistry, what would you write to criticize this sentence?

CP13.C (Section 13.7) A classmate consults you about a problem concerning the reaction of nitrogen monoxide and dioxygen in the gas phase. She has been told that the reaction is second-order in nitrogen monoxide and first-order in dioxygen; hence, the rate law may be written as rate $= k[NO]^2[O_2]$. She has been asked to propose a mechanism for this reaction. She proposes that the mechanism is this single equation:

$$NO + NO + O_2 \longrightarrow NO_2 + NO_2$$

She asks your opinion about whether this is correct. What should you tell her to explain why the answer is correct or incorrect?

CP13.D (Section 13.8) Polypropylene ***(Table 12.6, ← p. 564)*** is an important type of plastic, and over 3 million tons of it is produced every year in the United States. Properties of polypropylene can be changed by the way the polymer is made.

between 130 °C and 160 °C can be
priate catalyst to polymerize propy-
wo important types of polypropylene
the methyl groups are all on the same
chain, and syndiotactic, in which the
rnate between one side and the other of the
cha

Su hat you are part of a team designing a new catalyst to poly erize propylene. Your catalyst will have a zirconium atom at the center of a structure consisting of carbon and hydrogen atoms. The zirconium atom will hold the growing polypropylene chain by bonding to one end of it. The metal atom will also attract a propylene molecule and bond to it before transferring the growing polypropylene chain to the other end of the new propylene molecule. The process is shown here.

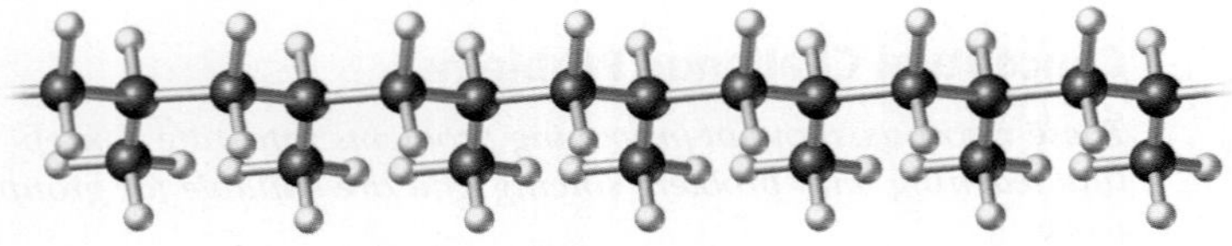

isotactic polypropylene

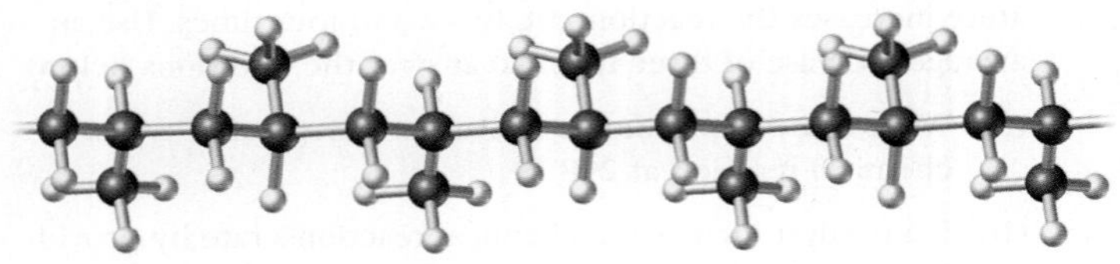

syndiotactic polypropylene

What would be a reasonable shape for the rest of the catalyst molecule surrounding the metal atom so that isotactic polypropylene would be produced? It may help to build molecular models to see how each new propylene molecule needs to be added to the growing polymer chain to get all the methyl groups on the same side.

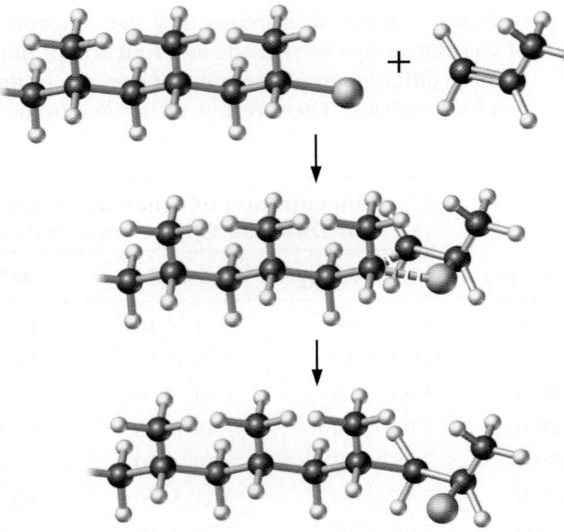

14

Chemical Equilibrium

© Ken Lucas/Visuals Unlimited, Inc.

This pinecone fish, *Monocentris gloriamaris,* has a bioluminescent area on its lower jaw. The bioluminescent area glows orange in the daytime and blue-green at night. It apparently helps to attract zooplankton upon which the pinecone fish feeds. In this daytime photograph, the orange area is clearly visible below and between the eye and the mouth. The bioluminescent area consists of a colony of rod-shaped bacteria, *Vibrio fischeri,* that are common in marine environments around the world. Colonies of *Vibrio fischeri* become luminescent only when they reach a minimum population and the bacteria in the colony sense that a sufficiently large population has been reached. This biological sensing depends on a chemical equilibrium involving acylhomoserine lactone, a substance that signals to one bacterium that others are nearby (see p. 680).

Chemical reactions that involve only pure solids or pure liquids are simpler than those that occur in the gas phase or in a solution. Either no reaction occurs between the solid and liquid reactants, or a reaction occurs in which at least one reactant is completely converted into products. [If one or more of the reactants are present in excess, those reactants

ew
velop
complete
ed by your

go Chem... nload mini lecture videos for key co... pt review and exam prep from OWL or purchase them from **www.CengageBrain.com.**

Companion Website
Visit this book's companion website at **www.cengage.com/chemistry/moore** to work interactive modules for the Estimation boxes and Active Figures in this text.

will be left over, but the limiting reactant **(← *p. 137*)** will be completely used up.] This happens because the concentration of a pure solid or a pure liquid does not change during a reaction, provided the temperature remains constant. If the initial concentrations of reactants are large enough to cause the reaction to occur, those same concentrations will be present throughout the reaction, and the reaction will not stop until the limiting reactant is used up.

When a reaction occurs in the gas phase or in a solution, concentrations of reactants decrease as the reaction takes place **(← *p. 597*).** Eventually the concentrations decrease to the point at which conversion of reactants to products is no longer favored. Then the concentrations of reactants and products stop changing, but none of the concentrations has become zero. At least a tiny bit (and often a lot) of each reactant and each product is present in the reaction mixture. Because the concentrations have stopped changing, it is often relatively easy to measure them, thus providing quantitative information about *how much* product can be obtained from the reaction. It is also possible to predict how changes in temperature, pressure, and concentrations will affect the quantity of product produced. This kind of information, combined with what you learned in Chapter 13 about factors that affect the rates of chemical reactions, enables us to predict how chemical reactions maintain living systems in balance (homeostasis) and which reactions will be useful for manufacturing a broad range of substances that enhance our quality of life.

The United States manufactures about 20 billion pounds of NH_3 per year and imports an additional 12 billion pounds.

As an example of the importance of such information, consider ammonia. The United States uses approximately 30 billion pounds of liquefied ammonia per year, mostly as fertilizer to provide nitrogen needed to support growth of a broad range of crops. Therefore, ammonia is a very important factor in providing people with food. Ammonia is synthesized directly from nitrogen and hydrogen by the *Haber-Bosch process* (Section 14.8, p. 689).

$$N_2(g) + 3\,H_2(g) \rightleftharpoons 2\,NH_3(g) \qquad \Delta H° = -92.2\text{ kJ}$$

The chemists and chemical engineers who operate ammonia manufacturing plants do their best to obtain the maximum quantity of ammonia with the minimum input of reactants and the minimum consumption of energy resources. The German chemist Fritz Haber won the 1918 Nobel Prize in Chemistry for research that showed how to determine the best conditions for carrying out this reaction. The German engineer Carl Bosch received the Nobel Prize in Chemistry in 1931 (together with Friedrich Bergius) for his pioneering chemical engineering work that enabled large-scale ammonia synthesis to be successful.

In this chapter you will learn the same principles that Haber used. With these principles you will be able to make both qualitative and quantitative predictions about how much product will be formed under a given set of reaction conditions.

14.1 Characteristics of Chemical Equilibrium

When the concentrations of reactants stop decreasing and the concentrations of products stop increasing, we say that a chemical reaction has reached equilibrium. In a **chemical equilibrium,** *there are finite concentrations of reactants and products, and these concentrations remain constant (unless reaction conditions change).* An equilibrium reaction always results in smaller amounts of products than the theoretical yield predicts **(← *p. 142*),** and sometimes hardly any products are produced. The concentrations of reactants and products at equilibrium provide a quantitative way of determining how successful a reaction has been. *When products predominate over reactants,* the reaction is **product-favored.** *When the equilibrium mixture consists mostly of reactants* with very little product, the reaction is **reactant-favored.**

EXERCISE 14.1 Concentrations of Pure Solids and Liquids

The introduction to this chapter states that at a given temperature the concentration of a pure solid or liquid does not depend on the quantity of substance present. Verify this assertion by calculating the concentration (in mol/L) of these solids and liquids at 20 °C. Obtain densities from Table 1.1 (← p. 9).

(a) Aluminum (b) Benzene (c) Water (d) Gold

Answers to **EXERCISES** are provided at the back of this book in Appendix L.

Equilibrium Is Dynamic

When a chemical reaction reaches equilibrium and concentrations of reactants and products remain constant, it appears that the reaction has stopped. This is true only of the net, macroscopic reaction, however. On the nanoscale both forward and reverse reactions continue, but *the rate of the forward reaction exactly equals the rate of the reverse reaction*. To emphasize that *chemical equilibrium involves a balance between opposite reactions*, it is often referred to as a **dynamic equilibrium,** and an equilibrium reaction is usually written with a double arrow ($\rightleftharpoons$) between reactants and products.

The *equi* in the word *equilibrium* means "equal." It refers to equal rates of forward and reverse reactions, not to equal quantities or concentrations of the substances involved. The *librium* part of the word comes from *libra*, meaning "balance." Chemical equilibrium is an equal balance between two reaction rates.

A good example of chemical equilibrium is provided by weak acids (← p. 169), which ionize only partially in water.

A set of double arrows, $\rightleftharpoons$, in an equation indicates a dynamic equilibrium in which forward and reverse reactions are occurring at equal rates; it also indicates that the reaction should be thought of in terms of the concepts of chemical equilibrium.

Double arrows indicate dynamic equilibrium.

$$\underset{\text{acetic acid}}{CH_3COOH(aq)} + \underset{\text{water}}{H_2O(\ell)} \rightleftharpoons \underset{\text{acetate ion}}{CH_3COO^-(aq)} + \underset{\text{hydronium ion}}{H_3O^+(aq)}$$

After equilibrium has been reached at room temperature, more than 90% of the acetic acid remains in molecular form, CH_3COOH, and the equilibrium concentrations of acetate ions and hydronium ions are each less than one tenth the concentration of acetic acid molecules. Nevertheless, both the forward and reverse reactions continue after equilibrium has been reached. Evidence to support this idea can be obtained by adding a tiny quantity of sodium acetate in which radioactive carbon-14 has been substituted into the CH_3COO^- ion to give $^{14}CH_3COO^-$. Almost immediately the radioactivity can be found in acetic acid molecules as well. This would not happen if the reaction had come to a halt, but it does happen because the reverse reaction,

$$^{14}CH_3COO^-(aq) + H_3O^+(aq) \longrightarrow {}^{14}CH_3COOH(aq) + H_2O(\ell)$$

is still taking place. To a macroscopic observer nothing seems to be happening because the reverse reaction and the forward reaction are occurring at equal rates and there is no *net* change in the concentrations of reactants or products.

Equilibrium Is Independent of Direction of Approach

Another important characteristic of chemical equilibrium is that, *for a specific reaction at a specific temperature, the equilibrium state will be the same, no matter what the direction of approach to equilibrium*. As an example, consider again the synthesis of ammonia from N_2 and H_2.

$$N_2(g) + 3\ H_2(g) \rightleftharpoons 2\ NH_3(g)$$

Suppose that you introduce 1.0 mol $N_2(g)$ and 3.0 mol $H_2(g)$ into an empty (evacuated), 1.00-L container at 472 °C and seal the container. Some (but not all) of the N_2 reacts with the H_2 to form NH_3. After equilibrium is established, you would find that the concentration of H_2 has fallen from its initial value of 3.0 mol/L to an equilibrium value of 0.89 mol/L. You would also find equilibrium concentrations of 0.30 mol/L for N_2 and 1.4 mol/L for NH_3. This is shown schematically in Figure 14.1.

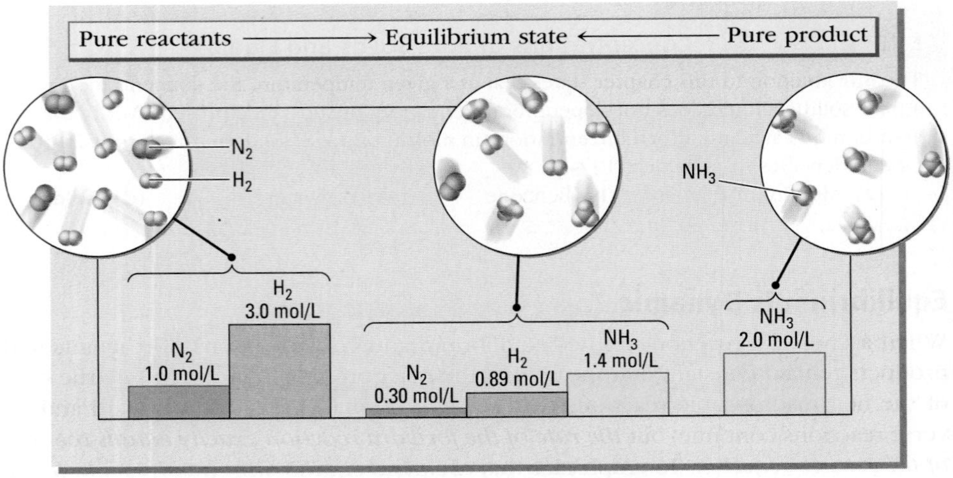

Figure 14.1 Reactants, products, and equilibrium. In the ammonia synthesis reaction, $N_2 + 3\ H_2 \rightleftharpoons 2\ NH_3$, as in any equilibrium, it is possible to start with reactants or products and achieve the same equilibrium state. Here 1.0 mol N_2 reacts with 3.0 mol H_2 in a 1.0-L container to give the same equilibrium concentrations as when 2.0 mol NH_3 is introduced into the same container at the same temperature (472 °C).

Now consider a second experiment at the same temperature in which you introduce 2.0 mol $NH_3(g)$ into an empty 1.00-L container at 472 °C and seal the container. The 2.0 mol NH_3 consists of 2.0 mol N atoms and 6.0 mol H atoms—the same number of N atoms and H atoms contained in the 1.0 mol N_2 and 3.0 mol H_2 used in the first experiment. Because only NH_3 is present initially, the reverse reaction occurs, producing some N_2 and H_2. Measuring the concentrations at equilibrium reveals that the concentration of NH_3 dropped from the initial 2.0 mol/L to 1.4 mol/L, and the concentrations of N_2 and H_2 built up to 0.30 mol/L and 0.89 mol/L (see Figure 14.1). These equilibrium concentrations are the same as those achieved in the first experiment. Thus, *whether you start with reactants or products, the same equilibrium state is achieved*—as long as the number of atoms of each type, the volume of the container, and the temperature are the same.

Catalysts Do Not Affect Equilibrium Concentrations

Another important characteristic of chemical equilibrium is that *if a catalyst is present, the same equilibrium state will be achieved, but more quickly*. A catalyst speeds up the forward reaction, but it also speeds up the reverse reaction equally. The overall effect is to produce exactly the same concentrations at equilibrium, whether or not a catalyst is in the reaction mixture. A catalyst can be used to speed up production of products in an industrial process, but it will not result in greater equilibrium concentrations of products, nor will it reduce the concentration of product present when the system reaches equilibrium.

EXERCISES that are labeled **CONCEPTUAL** are designed to test your understanding of one or more concepts; they usually involve qualitative rather than quantitative thinking.

CONCEPTUAL EXERCISE 14.2 Recognizing an Equilibrium State

A mixture of hydrogen gas and oxygen gas is maintained at 25 °C for 1 year. On the first day of each month the mixture is sampled and the concentrations of hydrogen and oxygen measured. In every case they are found to be 0.50 mol/L H_2 and 0.50 mol/L O_2; that is, the concentrations do not change over a long period. Is this mixture at equilibrium? If you think not, how could you do an experiment to prove it?

14.2 The Equilibrium Constant

Consider again the isomerization reaction of *cis*-2-butene to *trans*-2-butene, whose rate we discussed previously (***← p. 610***).

cis-2-butene *trans*-2-butene

At 500 K the reaction reaches an equilibrium in which the concentration of *trans*-2-butene is 1.65 times the concentration of *cis*-2-butene. This is a single-step, elementary process (***← p. 610***). Both the forward and reverse reactions in this system involve only a single molecule and, therefore, are unimolecular and first-order (***← p. 619***). Thus, the rate equations for forward and reverse reactions can be derived from the reaction equations.

$$\text{Rate}_{\text{forward}} = k_{\text{forward}}[\textit{cis}\text{-2-butene}] \qquad \text{Rate}_{\text{reverse}} = k_{\text{reverse}}[\textit{trans}\text{-2-butene}]$$

Suppose that we start with 0.100 mol *cis*-2-butene in a 5-L closed flask at 500 K. The *cis*-2-butene begins to react at a rate given by the forward rate equation. Initially no *trans*-2-butene is present, so the initial rate of the reverse reaction is zero. As the forward reaction proceeds, *cis*-2-butene is converted to *trans*-2-butene. The concentration of *cis*-2-butene decreases, so the forward rate decreases. As soon as some *trans*-2-butene has formed, the reverse reaction begins. As the concentration of *trans*-2-butene builds up, the reverse rate gets faster. The forward rate slows down (and the reverse rate speeds up) until the two rates are equal. At this time equilibrium has been achieved, and in the macroscopic system no further change in concentrations will be observed (Figure 14.2).

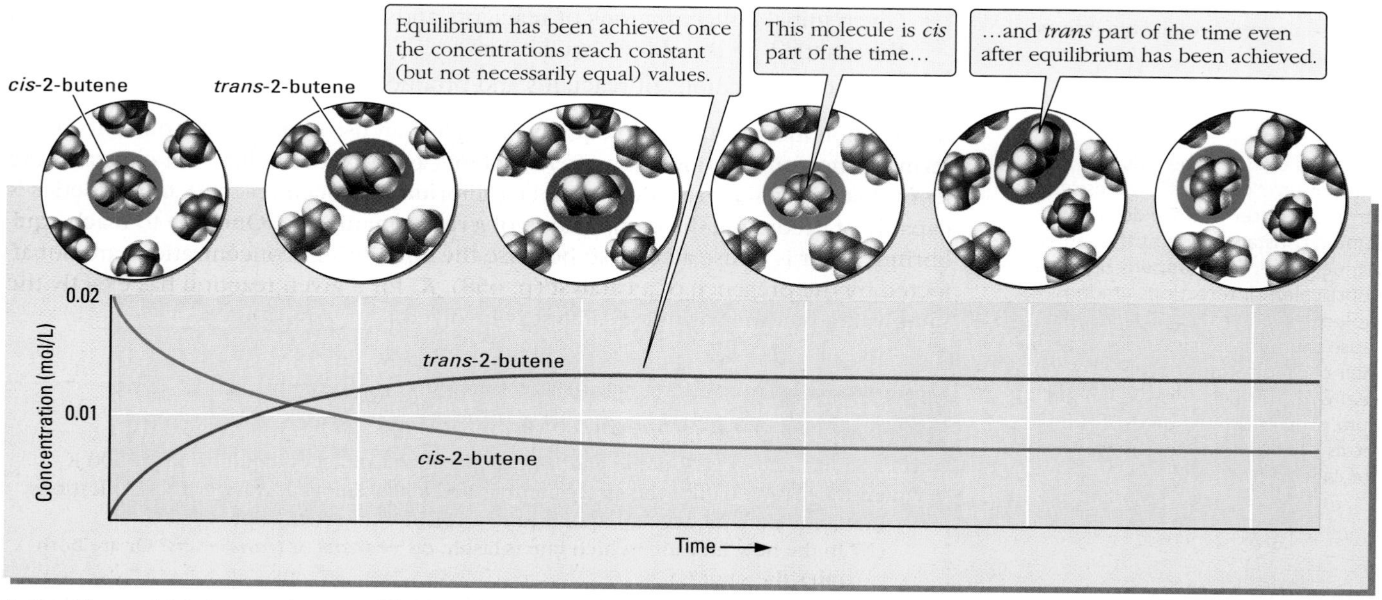

Active Figure 14.2 Approach to equilibrium. The graph shows the concentrations of *cis*-2-butene and *trans*-2-butene as the *cis* compound reacts to form the *trans* compound at 500 K. The nanoscale diagrams above the graph show "snapshots" of the composition of a tiny portion of the reaction mixture. The same molecule has been circled in each diagram. Visit this book's companion website at **www.cengage.com/chemistry/moore** to test your understanding of the concepts in this figure.

On the nanoscale, when equilibrium has been achieved, both reactions are still occurring, but the forward and reverse rates are equal. Therefore we can equate the two rates to give

$$\text{Rate}_{\text{forward}} = \text{Rate}_{\text{reverse}}$$

and, by substituting from the two previous rate equations,

$$k_{\text{forward}}[cis\text{-2-butene}] = k_{\text{reverse}}[trans\text{-2-butene}]$$

In this and subsequent equations in this chapter, square brackets designate *equilibrium* concentrations. The equation can be rearranged so that both rate constants are on one side and both equilibrium concentrations are on the other side.

$$\frac{k_{\text{forward}}}{k_{\text{reverse}}} = \frac{[trans\text{-2-butene}]}{[cis\text{-2-butene}]}$$

This expression shows that the ratio of equilibrium concentrations is equal to a ratio of rate constants. Because a ratio of two constants is also a constant, the ratio of equilibrium concentrations must also be constant. We call this ratio K_c, where the capital letter K is used to distinguish it from the rate constants, k_{forward} and k_{reverse}, and the subscript c indicates that it is a ratio of equilibrium *concentrations*. At 500 K the experimental value of K_c is 1.65 for the *cis* $\rightleftharpoons$ *trans* butene reaction.

$$K_c = \frac{k_{\text{forward}}}{k_{\text{reverse}}} = \frac{[trans\text{-2-butene}]}{[cis\text{-2-butene}]} = 1.65 \qquad \text{(at 500 K)}$$

Because the values of the rate constants vary with temperature (**← p. 615**)*, the value of* K_c *also varies with temperature.* For the butene isomerization reaction, K_c is 1.47 at 600 K and 1.36 at 700 K.

An **equilibrium constant,** K_c, *is a quotient of equilibrium concentrations of product and reactant substances that has a constant value for a given reaction at a given temperature.* Equilibrium constants can be used to answer three important questions about a reaction:

- When equilibrium has been achieved, do products predominate over reactants?
- Given initial concentrations of reactants and products, in which direction will the reaction go to achieve equilibrium?
- What concentrations of reactants and products are present at equilibrium?

If a reaction moves quickly to equilibrium, you can use equilibrium constants to determine the composition soon after reactants are mixed. Equilibrium constants are less valuable for slow reactions. Until equilibrium has been reached, only kinetics is capable of predicting the composition of a reaction mixture. One way to reach equilibrium faster is to use a catalyst; because the equilibrium concentrations are not affected by the presence of a catalyst (p. 658), K_c for a given reaction has exactly the same value whether or not a catalyst is involved.

It is important to distinguish equilibrium concentrations, which do not change over time, from the changing concentrations of reactants and products before equilibrium is achieved. Later in this chapter we will use the notation "conc. X" to indicate the concentration of X in a system that is not at equilibrium. This distinction between concentrations is important because *only* equilibrium concentrations can be calculated from equilibrium constants.

In Chapter 13 we noted that the rate law for an overall chemical reaction cannot be derived from the balanced equation, but must be determined experimentally by kinetic studies. This is not true of the equilibrium constant expression. When a reaction takes place by a mechanism that consists of a sequence of steps, the equilibrium constant expression can be obtained by multiplying together the rate constants for the forward reactions in all steps and then dividing by the rate constants for the reverse reactions in all steps. This process gives the same equilibrium constant expression that can be obtained from the coefficients of the balanced overall equilibrium equation.

It turns out that when concentrations are large enough, the value of the equilibrium "constant" expressed in terms of concentrations does not remain constant, even at the same temperature. This happens because of noncovalent interactions among molecules, and especially ions, that cause them to behave differently as their concentrations become larger. To deal with this behavior, true equilibrium constants must be expressed in terms of corrected concentrations that are called *activities.*

CONCEPTUAL EXERCISE

14.3 Properties of Equilibrium

After a mixture of *cis*-2-butene and *trans*-2-butene has reached equilibrium at 600 K, where $K_c = 1.47$, half of the *cis*-2-butene is suddenly removed. Answer these questions:

(a) Is the new mixture at equilibrium? Explain why or why not.

(b) In the new mixture, which rate is faster, *cis* → *trans* or *trans* → *cis*? Or are both rates the same?

(c) In an equilibrium mixture, which concentration is larger, *cis*-2-butene or *trans*-2-butene?

(d) If the concentration of *cis*-2-butene at equilibrium is 0.10 mol/L, what will be the concentration of *trans*-2-butene?

Writing Equilibrium Constant Expressions

The **equilibrium constant expression** for any reaction is *a mathematical expression that relates the equilibrium concentrations of reactants and products.* It has concentrations of products in the numerator and concentrations of reactants in the denominator. Each concentration is raised to the power of its stoichiometric coefficient in the balanced equation. *The only concentrations that appear in an equilibrium constant expression are those of gases and of solutes in dilute solutions,* because these are the only concentrations that can change as a reaction occurs. Concentrations of pure solids, pure liquids, and solvents in dilute solutions ***do not*** appear in equilibrium constant expressions.

To illustrate how this works, consider the general equilibrium reaction

$$a\,\mathrm{A} + b\,\mathrm{B} \rightleftharpoons c\,\mathrm{C} + d\,\mathrm{D} \qquad [14.1]$$

By convention we write the equilibrium constant expression for this reaction as

$$K_c = \frac{[\mathrm{C}]^c[\mathrm{D}]^d}{[\mathrm{A}]^a[\mathrm{B}]^b} \qquad [14.2]$$

Equilibrium constant

c is for concentration

Product concentrations raised to powers of coefficients

Reactant concentrations raised to powers of coefficients

Let's apply these ideas to the combination of nitrogen and oxygen gases to form nitrogen monoxide.

$$N_2(g) + O_2(g) \rightleftharpoons 2\,NO(g)$$

The reaction of nitrogen with oxygen occurs in automobile engines and other high-temperature combustion processes where air is present.

Because all substances in the reaction are gaseous, all concentrations appear in the equilibrium constant expression. The concentration of the product NO(g) is in the numerator and is squared because of its coefficient 2 in the balanced chemical equation. The concentrations of the reactants $N_2(g)$ and $O_2(g)$ are in the denominator. Each is raised to the first power because each has a coefficient of 1 in the balanced chemical equation.

$$\text{Equilibrium constant} = K_c = \frac{[NO]^2}{[N_2][O_2]}$$

Equilibria Involving Pure Liquids and Solids

As another example, consider the combustion of solid yellow sulfur, which consists of S_8 molecules. The combustion reaction produces sulfur dioxide gas.

$$\tfrac{1}{8}\,S_8(s) + O_2(g) \rightleftharpoons SO_2(g)$$

This reaction occurs whenever a material that contains sulfur burns in air, and it is responsible for a good deal of sulfur dioxide air pollution. It is also the first reaction in a sequence by which sulfur is converted to sulfuric acid, the number-one industrial chemical in the world.

Because sulfur is a solid, the number of molecules per unit volume is fixed by the density of sulfur at any given temperature (see Exercise 14.1, p. 657). Therefore, the sulfur concentration is not changed either by reaction or by addition or removal of solid sulfur. It is an experimental fact that, as long as some solid sulfur is present, the equilibrium concentrations of O_2 and SO_2 are not affected by changes in the amount of sulfur. Therefore, the equilibrium constant expression for this reaction is properly written as

$$K_c = \frac{[SO_2(g)]}{[O_2(g)]} \qquad \text{At 25 °C, } K_c = 4.2 \times 10^{52}$$

Concentration of S_8(s) is not included in K_c.

It is very important to remember that concentrations of pure solids, pure liquids, and solvents for dilute solutions do not appear in the equilibrium constant expression. Any substance designated as (s) or (ℓ) in the equilibrium equation does not appear in the equilibrium constant expression.

Equilibria in Dilute Solutions

Consider an aqueous solution of the weak base ammonia, which contains a small concentration of ammonium ions and hydroxide ions because ammonia reacts with water.

$$NH_3(aq) + H_2O(\ell) \rightleftharpoons NH_4^+(aq) + OH^-(aq)$$

If the concentration of ammonia molecules (and consequently of ammonium ions and hydroxide ions) is small, the number of water molecules per unit volume remains essentially the same as in pure water. Because the molar concentration of water is effectively constant for reactions involving dilute solutions, the concentration of water should not be included in the equilibrium constant expression. Thus, for this reaction we write

$$K_c = \frac{[NH_4^+][OH^-]}{[NH_3]}$$

and the concentration of water is not included in the denominator. At 25 °C, $K_c = 1.8 \times 10^{-5}$ for reaction of ammonia with water.

Notice that no units were included in the equilibrium constant for the ammonia ionization reaction. There are two concentrations in the numerator and only one in the denominator, which ought to give units of mol/L. However, if we always express concentrations in mol/L, the units of the equilibrium constant can be figured out from the equilibrium constant expression. Therefore it is customary to omit the units from the equilibrium constant value, even if the equilibrium constant should have units. We follow that custom in this book.

The **PROBLEM-SOLVING STRATEGY** in this book is

- Analyze the problem
- Plan a solution
- Execute the plan
- Check that the result is reasonable

Appendix A.1 explains this in detail.

PROBLEM-SOLVING EXAMPLE 14.1 Writing Equilibrium Constant Expressions

Write an equilibrium constant expression for each chemical equation.

(a) $4\ NO_2(g) + O_2(g) + 2\ H_2O(g) \rightleftharpoons 4\ HNO_3(g)$
(b) $BaSO_4(s) \rightleftharpoons Ba^{2+}(aq) + SO_4^{2-}(aq)$
(c) $6\ NH_3(aq) + Ni^{2+}(aq) \rightleftharpoons Ni(NH_3)_6^{2+}(aq)$
(d) $2\ BaO_2(s) \rightleftharpoons 2\ BaO(s) + O_2(g)$
(e) $NH_3(g) \rightleftharpoons NH_3(\ell)$

Answer

(a) $K_c = \frac{[HNO_3]^4}{[NO_2]^4[O_2][H_2O]^2}$ (b) $K_c = [Ba^{2+}][SO_4^{2-}]$

(c) $K_c = \frac{[Ni(NH_3)_6^{2+}]}{[NH_3]^6[Ni^{2+}]}$ (d) $K_c = [O_2]$

(e) $K_c = \frac{1}{[NH_3]}$

Strategy and Explanation Equilibrium concentrations of products go in the numerator of the fraction, and equilibrium concentrations of reactants go in the denominator. Each concentration is raised to the power of the stoichiometric coefficient of the species. In part (b), one species, $BaSO_4(s)$, is a pure solid and does not appear in the expression. In part (d), two species, $BaO_2(s)$ and $BaO(s)$, are solids and do not appear. In part (e), one species, $NH_3(\ell)$, is a pure liquid and does not appear.

PROBLEM-SOLVING PRACTICE answers are provided at the back of this book in Appendix K.

PROBLEM-SOLVING PRACTICE 14.1

Write an equilibrium constant expression for each equation.

(a) $CaCO_3(s) \rightleftharpoons CaO(s) + CO_2(g)$
(b) $HCl(g) + LiH(s) \rightleftharpoons H_2(g) + LiCl(s)$
(c) $CH_4(g) + H_2O(g) \rightleftharpoons CO(g) + 3\ H_2(g)$
(d) $CN^-(aq) + H_2O(\ell) \rightleftharpoons HCN(aq) + OH^-(aq)$

Equilibrium Constant Expressions for Related Reactions

Consider the equilibrium involving nitrogen, hydrogen, and ammonia.

$$N_2(g) + 3\,H_2(g) \rightleftharpoons 2\,NH_3(g) \qquad K_{c_1} = 3.5 \times 10^8 \quad \text{(at 25 °C)}$$

Two other cases are related to this chemical equation and its K_c value:

- We could also write the equation so that 1 mol NH_3 is produced.

$$\tfrac{1}{2}N_2(g) + \tfrac{3}{2}H_2(g) \rightleftharpoons NH_3(g) \qquad K_{c_2} = ?$$

Coefficients half as big

Is the value of the equilibrium constant, K_{c_2}, for the second equation the same as the value of the equilibrium constant, K_{c_1}, for the first equation? To see the relation between K_{c_1} and K_{c_2}, write the equilibrium constant expression for each balanced equation.

Concentrations raised to powers half as big

$$K_{c_1} = \frac{[NH_3]^2}{[N_2][H_2]^3} = 3.5 \times 10^8 \quad \text{and} \quad K_{c_2} = \frac{[NH_3]}{[N_2]^{1/2}[H_2]^{3/2}} = ?$$

This makes it clear that K_{c_1} is the square of K_{c_2}; that is, $K_{c_1} = (K_{c_2})^2$. Therefore, the answer to our question about the value of K_{c_2} is

$$K_{c_2} = (K_{c_1})^{1/2} = (3.5 \times 10^8)^{1/2} = 1.9 \times 10^4$$

Whenever the stoichiometric coefficients of a balanced equation are multiplied by some factor, the equilibrium constant for the new equation (K_{c_2} in this case) is the old equilibrium constant (K_{c_1}) raised to the power of the multiplication factor.

- What is the value of K_{c_3}, the equilibrium constant for the decomposition of ammonia to the elements, which is the reverse of the first equation?

$$2\,NH_3(g) \rightleftharpoons N_2(g) + 3\,H_2(g) \qquad K_{c_3} = ? = \frac{[N_2][H_2]^3}{[NH_3]^2}$$

Concentration of NH_3 is in denominator

K_{c_3} is the reciprocal of K_{c_1}; that is, $K_{c_3} = 1/K_{c_1} = 1/(3.5 \times 10^8) = 2.9 \times 10^{-9}$. ***The equilibrium constant for a reaction and that for its reverse are the reciprocals of one another.*** If a reaction has a very large equilibrium constant, the reverse reaction will have a very small one. That is, if a reaction is strongly product-favored, then its reverse is strongly reactant-favored. In the case of the production of ammonia from its elements at room temperature, the forward reaction has a large equilibrium constant, 3.5×10^8. As expected, the reverse reaction, decomposition of ammonia to its elements, has a small equilibrium constant, 2.9×10^{-9}.

EXERCISE 14.4 Manipulating Equilibrium Constants

The balanced equation for conversion of oxygen to ozone has a very small value of K_c.

$$3\,O_2(g) \rightleftharpoons 2\,O_3(g) \qquad K_c = 6.25 \times 10^{-58}$$

(a) Calculate the value of K_c if the equation is written as

$$\tfrac{3}{2}O_2(g) \rightleftharpoons O_3(g)$$

(b) Calculate the value of K_c for the conversion of ozone to oxygen.

$$2\,O_3(g) \rightleftharpoons 3\,O_2(g)$$

Equilibrium Constant for a Reaction That Combines Two or More Other Reactions

If two chemical equations can be combined to give a third, the equilibrium constant for the combined reaction can be obtained from the equilibrium constants for the two original reactions. For example, air pollution is produced when nitrogen monoxide forms from nitrogen and oxygen and then combines with additional oxygen to form nitrogen dioxide **(← p. 459)**.

$$(1) \quad N_2(g) + O_2(g) \rightleftharpoons 2\,NO(g) \qquad K_{c_1} = \frac{[NO]^2}{[N_2][O_2]}$$

$$(2) \quad 2\,NO(g) + O_2(g) \rightleftharpoons 2\,NO_2(g) \qquad K_{c_2} = \frac{[NO_2]^2}{[NO]^2[O_2]}$$

The sum of these two equations is

Sum of Equations 1 and 2

$$(3) \quad N_2(g) + 2\,O_2(g) \rightleftharpoons 2\,NO_2(g)$$

Product of equilibrium constants K_{c_1} and K_{c_2}

$$K_{c_3} = \frac{[NO_2]^2}{[N_2][O_2]^2} = \frac{\cancel{[NO]^2}}{[N_2][O_2]} \times \frac{[NO_2]^2}{\cancel{[NO]^2}[O_2]} = K_{c_1} \times K_{c_2}$$

Because K_{c_1} and K_{c_2} were known experimentally and could be used to calculate K_{c_3}, there is no need to measure K_{c_3} experimentally.

If two chemical equations can be summed to give a third, the equilibrium constant for the overall equation equals the product of the two equilibrium constants for the equations that were summed. This is a powerful tool for obtaining equilibrium constants without having to measure them experimentally for each individual reaction.

PROBLEM-SOLVING EXAMPLE 14.2 Manipulating Equilibrium Constants

Given these equilibrium reactions and constants,

$$(1)\ \tfrac{1}{4}\,S_8(s) + 2\,O_2(g) \rightleftharpoons 2\,SO_2(g) \qquad K_{c_1} = 1.86 \times 10^{105}$$

$$(2)\ \tfrac{1}{8}\,S_8(s) + \tfrac{3}{2}\,O_2(g) \rightleftharpoons SO_3(g) \qquad K_{c_2} = 1.77 \times 10^{53}$$

calculate the equilibrium constant, K_{c_3}, for this reaction, which is important in the formation of acid rain air pollution.

$$2\,SO_2(g) + O_2(g) \rightleftharpoons 2\,SO_3(g)$$

Answer $K_{c_3} = 16.8$

Strategy and Explanation Compare the given equations with the target equation to see how each given equation needs to be manipulated.

- The target equation has SO_2 on the left, so Equation 1 needs to be reversed and we need to take the reciprocal of the equilibrium constant.
- The target equation has SO_3 on the right. Equation 2 need not be reversed, but each coefficient must be multiplied by 2, which means squaring K_{c_2}.
- Once this has been done, the two new equations sum to the target equation, and we can multiply their equilibrium constants.

$$(1)'\ 2\,SO_2(g) \rightleftharpoons \tfrac{1}{4}\,S_8(s) + 2\,O_2(g) \qquad K'_{c_1} = \frac{1}{1.86 \times 10^{105}} = 5.38 \times 10^{-106}$$

$$(2)'\ \tfrac{1}{4}\,S_8(s) + 3\,O_2(g) \rightleftharpoons 2\,SO_3(g) \qquad K'_{c_2} = (1.77 \times 10^{53})^2 = 3.13 \times 10^{106}$$

$$2\,SO_2(g) + O_2(g) \rightleftharpoons 2\,SO_3(g) \qquad K_{c_3} = K'_{c_1} \times K'_{c_2} = 16.8$$

☑ **Reasonable Answer Check** For Equation (1) the equilibrium constant was quite large, so for the reverse reaction it should be quite small, which it is. Also check the order of magnitude of the answer by checking the powers of 10. In the square, the power of 10 should be doubled, which it is. The sum of the powers of 10 for the two equilibrium constants that were multiplied should be close to the power of 10 in the answer, which it is.

PROBLEM-SOLVING PRACTICE 14.2

When carbon dioxide dissolves in water it reacts to produce carbonic acid, $H_2CO_3(aq)$, which can ionize in two steps.

$$H_2CO_3(aq) + H_2O(aq) \rightleftharpoons HCO_3^-(aq) + H_3O^+(aq) \qquad K_{c_1} = 4.2 \times 10^{-7}$$

$$HCO_3^-(aq) + H_2O(aq) \rightleftharpoons CO_3^{2-}(aq) + H_3O^+(aq) \qquad K_{c_2} = 4.8 \times 10^{-11}$$

Calculate the equilibrium constant for the reaction

$$H_2CO_3(aq) + 2\,H_2O(aq) \rightleftharpoons CO_3^{2-}(aq) + 2\,H_3O^+(aq)$$

Equilibrium Constants in Terms of Pressure

In a constant-volume system, when the concentration of a gas changes, the partial pressure *(← p. 446)* of the gas also changes. This follows from the ideal gas equation

$$PV = nRT$$

Solving for the partial pressure P_A of a gaseous substance, A,

[A] = n_A/V, the number of moles of A per unit volume

$$P_A = \frac{n_A}{V}RT = [A]RT \qquad [14.3]$$

Equation 14.3 allows us to express the equilibrium constant for the general reaction in Equation 14.1 (p. 661) in a form similar to Equation 14.2, but in terms of partial pressures as

$$K_P = \frac{P_C^c \times P_D^d}{P_A^a \times P_B^b} \qquad [14.4]$$

Product pressures raised to powers of coefficients

Reactant pressures raised to powers of coefficients

The subscript on K_P indicates that the equilibrium constant has been expressed in terms of partial pressures. For some gas phase equilibria $K_c = K_P$; for many others it does not. Therefore it is useful to be able to relate one type of equilibrium constant to the other. This can be done by combining Equations 14.2, 14.3, and 14.4 to give

$\Delta n = c + d - a - b$ is the number of moles of gaseous products minus the number of moles of gaseous reactants

$$K_P = \frac{P_C^c \times P_D^d}{P_A^a \times P_B^b} = \frac{\{[C]RT\}^c\{[D]RT\}^d}{\{[A]RT\}^a\{[B]RT\}^b} = \frac{[C]^c[D]^d}{[A]^a[B]^b}(RT)^{c+d-a-b} = K_c(RT)^{\Delta n} \qquad [14.5]$$

Because K_c is related to K_P for the same gas phase reaction, either can be used to calculate the composition of a gas-phase equilibrium mixture. Most examples in this chapter involve K_c, but the same rules apply to solving problems with K_P.

Note that $K_c = K_p$ *only* when $\Delta n = 0$; that is, *only* when the number of gas molecules in the reactants is the same as in the products.

As an example of this relation, consider the equilibrium

$$2\,NOCl(g) \rightleftharpoons 2\,NO(g) + Cl_2(g) \qquad K_c = 4.02 \times 10^{-2}\ \text{mol/L at 298 K}$$

For this reaction $\Delta n = 2 + 1 - 2 = 3 - 2 = 1$, because there are three moles of gas phase product molecules and only two moles of gas phase reactants. Therefore, for this reaction

$$K_P = K_c \times (RT)^1$$

$$= (4.02 \times 10^{-2}\ \text{mol/L}) \times (0.0821\ \text{L atm K}^{-1}\ \text{mol}^{-1}) \times (298\ \text{K}) = 0.984\ \text{atm}$$

EXERCISE 14.5 Relating K_c and K_P

For each of these reactions, calculate K_P from K_c.

(a) $N_2(g) + 3\ H_2(g) \rightleftharpoons 2\ NH_3(g)$ $K_c = 3.5 \times 10^8$ (at 25 °C)
(b) $2\ H_2(g) + O_2(g) \rightleftharpoons 2\ H_2O(g)$ $K_c = 3.2 \times 10^{81}$ (at 25 °C)
(c) $N_2(g) + O_2(g) \rightleftharpoons 2\ NO(g)$ $K_c = 1.7 \times 10^{-3}$ (at 2300 K)
(d) $2\ NO_2(g) \rightleftharpoons N_2O_4(g)$ $K_c = 1.7 \times 10^2$ (at 25 °C)

go Chemistry **Module 21: Chemical Equilibrium** covers concepts in this section.

14.3 Determining Equilibrium Constants

To determine the value of an equilibrium constant it is necessary to know all of the equilibrium concentrations that appear in the equilibrium constant expression. This is most commonly done by allowing a system to reach equilibrium and then measuring the **equilibrium concentration** of one or more of the reactants or products. Algebra and stoichiometry are then used to obtain the numerical value of K_c.

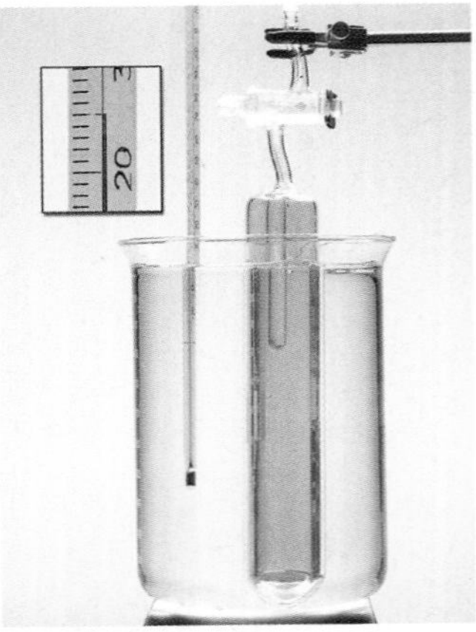

(a) room temperature, less NO_2

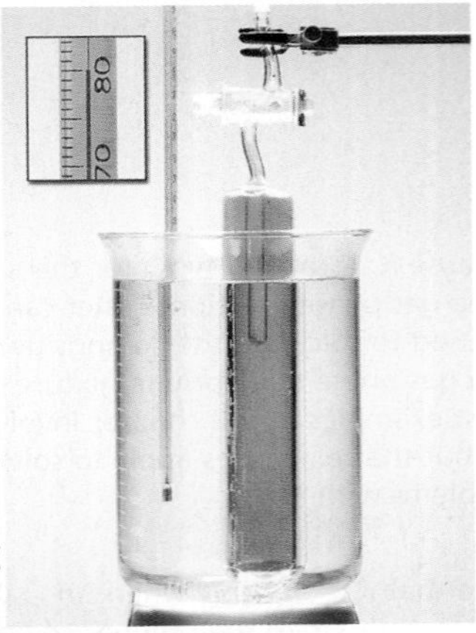

(b) 80 °C, more NO_2.

© Cengage Learning/Charles D. Winters

Figure 14.3 Formation of NO_2 from N_2O_4. Sealed glass tubes containing an equilibrium mixture of dinitrogen tetraoxide (colorless) and nitrogen dioxide (red-brown) are shown immersed in water (a) at room temperature and (b) at 80 °C. Notice the much darker color at the higher temperature, indicating that some dinitrogen tetraoxide has reacted to form nitrogen dioxide.

Reaction Tables, Stoichiometry, and Equilibrium Concentrations

A systematic approach to calculations involving equilibrium constants involves making a table that shows initial conditions, changes that take place when a reaction occurs, and final (equilibrium) conditions. As an example, consider the colorless gas dinitrogen tetraoxide, $N_2O_4(g)$. When heated it dissociates to form red-brown $NO_2(g)$ according to the equation

$$N_2O_4(g) \rightleftharpoons 2\ NO_2(g)$$

Suppose that 2.00 mol $N_2O_4(g)$ is placed into an empty 5.00-L flask, the flask is sealed, and the N_2O_4 is heated to 407 K. Almost immediately a dark red-brown color appears, indicating that much of the colorless gas has been transformed into NO_2 (Figure 14.3). By measuring the intensity of color, it can be determined that the concentration of NO_2 at equilibrium is 0.525 mol/L. To use this information to calculate the equilibrium constant, follow these steps:

1. ***Write the balanced equation for the equilibrium reaction. From it derive the equilibrium constant expression.*** The balanced equation and equilibrium constant expression are

$$N_2O_4(g) \rightleftharpoons 2\ NO_2(g) \qquad K_c = \frac{[NO_2]^2}{[N_2O_4]}$$

2. ***Set up a table containing initial concentration, change in concentration, and equilibrium concentration for each substance included in the equilibrium constant expression. Enter all known information into this reaction table.*** In this case the number of moles of $N_2O_4(g)$ and the volume of the flask were given, so we first calculate the initial concentration of $N_2O_4(g)$ as (conc. N_2O_4) $= \frac{2.00\ \text{mol}}{5.00\ \text{L}} = 0.400$ mol/L. Because the flask initially contained no NO_2, the initial concentration of NO_2 is zero. After the reaction took place and equilibrium was reached, the equilibrium concentration of NO_2 was measured as 0.525 mol/L. The reaction table looks like this:

	$N_2O_4(g)$	$\rightleftharpoons$	$2\ NO_2(g)$
*I*nitial concentration (mol/L)	0.400		0
*C*hange as reaction occurs (mol/L)	______		______
*E*quilibrium concentration (mol/L)	______		0.525

3. ***Use x to represent the change in concentration of one substance. Use the stoichiometric coefficients in the balanced equilibrium equation to calculate the other changes in terms of x.*** When the reaction proceeds from left to right, the concentrations of reactants decrease. Therefore the change in concentration of a reactant is negative. The concentrations of products increase, so the change in concentration of a product is positive. Usually it is best to begin with the reactant or product column that contains the most information. In this case that is the NO_2 column, where both initial and equilibrium concentrations are known. Therefore, we let x represent the unknown change in concentration of NO_2.

	$N_2O_4(g)$	$\rightleftharpoons$	$2\ NO_2(g)$
*I*nitial concentration (mol/L)	0.400		0
*C*hange as reaction occurs (mol/L)	______		x
*E*quilibrium concentration (mol/L)	______		0.525

Tables like this one are often called ICE tables from the initial letters of the labels on the rows: *I*nitial, *C*hange, and *E*quilibrium.

Next, we use the mole ratio from the balanced equation to find the change in concentration of N_2O_4 in terms of x.

$$\Delta(\text{conc. } N_2O_4) = \frac{x \text{ mol } NO_2 \text{ formed}}{\text{L}} \times \frac{1 \text{ mol } N_2O_4 \text{ reacted}}{2 \text{ mol } NO_2 \text{ formed}}$$

$$= \tfrac{1}{2}x \text{ mol } N_2O_4 \text{ reacted/L}$$

For every 1 mol N_2O_4 that decomposes, 2 mol NO_2 forms. Therefore the mole ratio of N_2O_4:NO_2 is $\frac{1}{2}$:1.

The sign of the change in concentration of N_2O_4 is *negative*, because the concentration of N_2O_4 *decreases*. The table becomes

	$N_2O_4(g)$	$\rightleftharpoons$	$2\ NO_2(g)$
*I*nitial concentration (mol/L)	0.400		0
*C*hange as reaction occurs (mol/L)	$-\frac{1}{2}x$		x
*E*quilibrium concentration (mol/L)	______		0.525

4. ***From initial concentrations and the changes in concentrations, calculate the equilibrium concentrations in terms of x and enter them in the table.*** The concentration of N_2O_4 at equilibrium, $[N_2O_4]$, is the sum of the initial 0.400 mol/L of N_2O_4 and the change due to reaction $-\frac{1}{2}x$ mol/L; that is, $[N_2O_4] = (0.400 - \frac{1}{2}x)$ mol/L. Similarly, the equilibrium concentration of NO_2 (which is already known to be 0.525 mol/L) is $0 + x$, and the table becomes

	$N_2O_4(g)$	$\rightleftharpoons$	$2\ NO_2(g)$
*I*nitial concentration (mol/L)	0.400		0
*C*hange as reaction occurs (mol/L)	$-\frac{1}{2}x$		x
*E*quilibrium concentration (mol/L)	$0.400 - \frac{1}{2}x$		$0.525 = 0 + x$

5. ***Use the simplest possible equation to solve for x. Then use x to calculate the unknown you were asked to find.*** (Usually the unknown is K_c or a concentration.) In this case the simplest equation to solve for x is the last entry in the table, $0.525 = 0 + x$, and it is easy to see that $x = 0.525$. Calculate $[N_2O_4] = (0.400 - \frac{1}{2}x)$ mol/L $= (0.400 - \frac{1}{2} \times 0.525)$ mol/L $= 0.138$ mol/L. The problem stated that $[NO_2] = 0.525$ mol/L, so K_c is given by

$$K_c = \frac{[NO_2]^2}{[N_2O_4]} = \frac{(0.525 \text{ mol/L})^2}{(0.138 \text{ mol/L})} = 2.00 \qquad (\text{at } 407 \text{ K})$$

6. ***Check your answer to make certain it is reasonable.*** In this case the equilibrium concentration of product is larger than the concentration of reactant. Since the products are in the numerator of the equilibrium constant expression, we expect a value greater than 1, which is what we calculated.

PROBLEM-SOLVING EXAMPLE 14.3 Determining an Equilibrium Constant Value

Consider the gas phase reaction

$$H_2(g) + I_2(g) \rightleftharpoons 2\,HI(g)$$

Suppose that a sealed flask containing H_2 and I_2 has been heated to 425 °C and the initial concentrations of H_2 and I_2 were each 0.0175 mol/L. With time, the concentrations of H_2 and I_2 decline and the concentration of HI increases. At equilibrium [HI] = 0.0276 mol/L. Use this experimental information to calculate the equilibrium constant.

Note that products form at the expense of reactants.

Answer $K_c = 56$

Strategy and Explanation Use the information given and follow the six steps. In the table below, each entry has been color-coded to match the numbered steps.

1. ***Write the balanced equation and equilibrium constant expression.***

$$H_2(g) + I_2(g) \rightleftharpoons 2\,HI(g) \qquad K_c = \frac{[HI]^2}{[H_2][I_2]}$$

2. ***Construct a reaction (ICE) table (see below) and enter known information.***

3. ***Represent changes in concentration in terms of* x.**
The best choice is to enter x in the third column, because both initial and equilibrium concentrations of HI are known, and this provides a simple way to calculate x. Next, derive the rest of the concentration changes in terms of x. If the concentration of HI increases by a given quantity, the mole ratios say that the concentrations of H_2 and I_2 must decrease only half as much:

$$\frac{x\ \text{mol}\ \cancel{\text{HI produced}}}{\text{L}} \times \frac{1\ \text{mol}\ H_2\ \text{consumed}}{2\ \cancel{\text{mol HI produced}}} = \frac{1}{2}x\ \text{mol/L}\ H_2\ \text{consumed}$$

Because the coefficients of H_2 and I_2 are equal, each of their concentrations decreases by $\frac{1}{2}x$ mol/L. The entries in the table for change in concentration of H_2 and I_2 are negative, because their concentrations decrease.

4. ***Calculate equilibrium concentrations and enter them in the table.***

	$H_2(g)$	+	$I_2(g)$	$\rightleftharpoons$	$2\,HI(g)$
Initial concentration (mol/L)	0.0175		0.0175		0
Change as reaction occurs (mol/L)	$-\frac{1}{2}x$		$-\frac{1}{2}x$		x
Equilibrium concentration (mol/L)	$0.0175 - \frac{1}{2}x$		$0.0175 - \frac{1}{2}x$		$0.0276 = 0 + x$

Color-coded entries in the table in this example correspond with color-coded steps in the explanation section above the table.

5. ***Solve the simplest equation for* x.** The last row and column in the table contains $0.0276 = 0 + x$, which gives $x = 0.0276$. Substitute this value into the other two equations in the last row of the table to get the equilibrium concentrations, and substitute them into the equilibrium constant expression:

$$K_c = \frac{[HI]^2}{[H_2][I_2]} = \frac{(0.0276)^2}{(0.0175 - [\frac{1}{2} \times 0.0276])(0.0175 - [\frac{1}{2} \times 0.0276])}$$

$$= \frac{(0.0276)^2}{(0.0037)(0.0037)} = 56 \qquad \text{(at 424 °C)}$$

☑ **Reasonable Answer Check** The equilibrium constant is larger than 1, so there should be more products than reactants when equilibrium is reached. The equilibrium concentration of the product (0.0276 mol/L) is larger than those of the reactants (0.0037 mol/L each).

PROBLEM-SOLVING PRACTICE 14.3

Measuring the conductivity of an aqueous solution in which 0.0200 mol CH_3COOH has been dissolved in 1.00 L of solution shows that 2.96% of the acetic acid molecules have ionized to CH_3COO^- ions and H_3O^+ ions. Calculate the equilibrium constant for ionization of acetic acid and compare your result with the value given in Table 14.1.

Saying that 2.96% of the acetic acid molecules have ionized means that at equilibrium the concentration of acetate ions is 2.96/100 = 0.0296 times the initial concentration of acetic acid molecules.

Experimentally determined equilibrium constants for a few reactions are given in Table 14.1. These reactions occur to widely differing extents, as shown by the wide range of K_c values.

14.4 The Meaning of the Equilibrium Constant

The numerical value of the equilibrium constant tells how far a reaction has proceeded by the time equilibrium has been achieved. In addition, it can be used to calculate how much product will be present at equilibrium. There are three important cases to consider.

Case 1. $K_c \gg 1$: *Reaction is strongly product-favored; equilibrium concentrations of products are much greater than equilibrium concentrations of reactants.*

The symbol $\gg$ means "much greater than."

A large value of K_c *means that reactants have been converted almost entirely to products when equilibrium has been achieved.* That is, the products are strongly favored over the reactants. An example is the reaction of NO(g) with $O_3(g)$, which is one way that ozone is destroyed in the stratosphere (← p. 454).

$$NO(g) + O_3(g) \rightleftharpoons NO_2(g) + O_2(g) \qquad K_c = \frac{[NO_2][O_2]}{[NO][O_3]} = 6 \times 10^{34} \quad (\text{at } 25\ °C)$$

Table 14.1 Selected Equilibrium Constants at 25 °C

Reaction	K_c	K_P
Gas phase reactions		
$\frac{1}{8}S_8(s) + O_2(g) \rightleftharpoons SO_2(g)$	4.2×10^{52}	4.2×10^{52}
$2\ H_2(g) + O_2(g) \rightleftharpoons 2\ H_2O(g)$	3.2×10^{81}	1.3×10^{80}
$N_2(g) + 3\ H_2(g) \rightleftharpoons 2\ NH_3(g)$	3.5×10^{8}	5.8×10^{5}
$N_2(g) + O_2(g) \rightleftharpoons 2\ NO(g)$	4.5×10^{-31}	4.5×10^{-31}
$N_2(g) + O_2(g) \rightleftharpoons 2\ NO(g)$ at 2300 K	1.7×10^{-3}	
$H_2(g) + I_2(g) \rightleftharpoons 2\ HI(g)$	2.5×10^{1}	2.5×10^{1}
$2\ NO_2(g) \rightleftharpoons N_2O_4(g)$	1.7×10^{2}	7.0
$CH_4(g) + H_2O(g) \rightleftharpoons CO(g) + 3\ H_2(g)$	2.0×10^{-28}	1.25×10^{-25}
cis-2-butene(g) $\rightleftharpoons$ *trans*-2-butene(g)	3.2	3.2
Weak acids and bases		
Formic acid: $HCOOH(aq) + H_2O(\ell) \rightleftharpoons H_3O^+(aq) + HCOO^-(aq)$	3.0×10^{-4}	—
Acetic acid: $CH_3COOH(aq) + H_2O(\ell) \rightleftharpoons H_3O^+(aq) + CH_3COO^-(aq)$	1.8×10^{-5}	—
Carbonic acid: $H_2CO_3(aq) + H_2O(\ell) \rightleftharpoons H_3O^+(aq) + HCO_3^-(aq)$	4.3×10^{-7}	—
Ammonia (weak base): $NH_3(aq) + H_2O(\ell) \rightleftharpoons NH_4^+(aq) + OH^-(aq)$	1.8×10^{-5}	—
Very slightly soluble solids		
$CaCO_3(s) \rightleftharpoons Ca^{2+}(aq) + CO_3^{2-}(aq)$	2.8×10^{-9}	—
$AgCl(s) \rightleftharpoons Ag^+(aq) + Cl^-(aq)$	1.8×10^{-10}	—
$AgI(s) \rightleftharpoons Ag^+(aq) + I^-(aq)$	8.3×10^{-17}	—

The reaction goes essentially to completion when equilibrium has been reached. This could take a long time if the reaction is slow.

The very large value of K_c tells us that if 1 mol NO and 1 mol O_3 are mixed in a sealed flask at 25 °C and allowed to come to equilibrium, $[NO_2][O_2] \gg [NO][O_3]$. Virtually none of the reactants will remain, and essentially only NO_2 and O_2 will be found in the flask. For practical purposes, this reaction goes to completion, and it would not be necessary to use the equilibrium constant to calculate the quantities of products that would be obtained. The simpler methods developed in Chapter 4 **(← p. 131)** would work just fine in this case.

The symbol $\ll$ means "much less than."

Case 2. $K_c \ll 1$: *Reaction is strongly reactant-favored; equilibrium concentrations of reactants are greater than equilibrium concentrations of products.*

Conversely, *an extremely small* K_c *means that when equilibrium has been achieved, very little of the reactants have been transformed into products*. The reactants are favored over the products at equilibrium.

$$3\,O_2(g) \rightleftharpoons 2\,O_3(g) \qquad K_c = \frac{[O_3]^2}{[O_2]^3} = 6.25 \times 10^{-58} \quad \text{(at 25 °C)}$$

This means that $[O_3]^2 \ll [O_2]^3$ and if O_2 is placed in a flask at 25 °C and the flask is sealed, *very* little O_3 will be found when equilibrium is achieved. The concentration of O_2 would remain essentially unchanged. In the terminology of Chapter 5, we would write "N.R." and say that no reaction occurs.

The symbol $\cong$ means "approximately equal to."

Case 3. $K_c \cong 1$: *Equilibrium mixture contains significant concentrations of reactants and products; calculations are needed to determine equilibrium concentrations.*

If K_c *is neither extremely large nor extremely small, the equilibrium constant must be used to calculate how far a reaction proceeds toward products*. In contrast with the reactions in Case 1 and Case 2, dissociation of dinitrogen tetraoxide has neither a very large nor a very small equilibrium constant. At 391 K the value is 1.00, which means that significant concentrations of both N_2O_4 and NO_2 are present at equilibrium.

Recall that the equation for dissociation of dinitrogen tetraoxide is

$$N_2O_4(g) \rightleftharpoons 2\,NO_2(g)$$

$$K_c = 1.00 = \frac{[NO_2]^2}{[N_2O_4]}, \text{ so } [NO_2]^2 = [N_2O_4] \quad \text{(at 391 K)} \qquad [14.6]$$

What range of equilibrium constants represents this middle ground depends on how small a concentration is significant and on the form of the equilibrium constant. If the concentrations of N_2O_4 and NO_2 at equilibrium are both 1.0 mol/L, then the ratio of $[NO_2]^2/[N_2O_4]$ does equal the K_c value of 1.00 at 391 K. But what if the concentrations were much smaller? Would they still be equal? You can verify this by using Equation 14.6. If the numeric value of the equilibrium concentration of NO_2 is 0.01, then the concentration of N_2O_4 must be $(0.01)^2$, which equals 0.0001. Thus, even though $K_c = 1.00$, the concentration of one substance can be much larger than the concentration of the other. This happens because there is a squared term in the numerator of the equilibrium constant expression and a term to a different power (the first power) in the denominator. Whenever the total of the exponents in the numerator differs from the total in the denominator, it becomes very difficult to say whether the concentrations of the products will exceed those of the reactants without doing a calculation.

By contrast, if the total of the exponents is the same, then if $K_c > 1$, products predominate over reactants, and if $K_c < 1$, reactants predominate over products. Examples in which this is true include

$$cis\text{-2-butene(g)} \rightleftharpoons trans\text{-2-butene(g)} \qquad K_c = \frac{[trans]}{[cis]} = 3.2 \quad (\text{at } 25\ ^\circ\text{C})$$

and

$$2\ \text{HI(g)} \rightleftharpoons \text{H}_2\text{(g)} + \text{I}_2\text{(g)} \qquad K_c = \frac{[\text{H}_2][\text{I}_2]}{[\text{HI}]^2} = 0.040 \quad (\text{at } 25\ ^\circ\text{C})$$

Figure 14.4 diagrams reactant and product concentrations as a function of K_c for the isomerization of *cis*-2-butene in the gas phase.

You might wonder whether reactant-favored systems in which small quantities of products form are important. Many are. Examples include the acids and bases listed in Table 14.1. For acetic acid, the acidic ingredient in vinegar, the reaction is

$$\text{CH}_3\text{COOH(aq)} + \text{H}_2\text{O}(\ell) \rightleftharpoons \text{H}_3\text{O}^+\text{(aq)} + \text{CH}_3\text{COO}^-\text{(aq)}$$

$$K_c = \frac{[\text{H}_3\text{O}^+][\text{CH}_3\text{COO}^-]}{[\text{CH}_3\text{COOH}]} = 1.8 \times 10^{-5} \quad (\text{at } 25\ ^\circ\text{C})$$

The value of K_c for acetic acid is small, and at equilibrium the concentrations of the products (acetate ions and hydronium ions) are small relative to the concentration of the reactant (acetic acid molecules). This confirms that acetic acid is a weak acid. Nevertheless, vinegar tastes sour because a small percentage of the acetic acid molecules react with water to produce H_3O^+(aq).

If the form of the equilibrium constant is the same for two or more different reactions, then the degree to which each of those reactions is product-favored can be compared quantitatively in a very straightforward way. For example, the equilibrium reactions are similar and the equilibrium constant expressions all have the same form for formic acid, acetic acid, and carbonic acid:

$$K_c = \frac{[\text{H}_3\text{O}^+][\text{anion}^-]}{[\text{acid}]}$$

where anion$^-$ is $HCOO^-$, CH_3COO^-, or HCO_3^-, and acid is HCOOH, CH_3COOH, or H_2CO_3. From data in Table 14.1, we can say that formic acid is stronger (has a larger K_c value) than acetic acid and carbonic acid is the weakest of the three acids.

If a reaction has a large tendency to occur in one direction, then the reverse reaction has little tendency to occur. This means that the equilibrium constant for the reverse of a strongly product-favored reaction will be extremely small. Table 14.1 shows that combustion of hydrogen to form water vapor has an enormous equilibrium constant (3.2×10^{81}). This reaction is strongly product-favored. We say that it goes to completion.

The reverse reaction, decomposition of water to its elements,

$$2\ \text{H}_2\text{O(g)} \rightleftharpoons 2\ \text{H}_2\text{(g)} + \text{O}_2\text{(g)} \qquad K_c = \frac{[\text{H}_2]^2[\text{O}_2]}{[\text{H}_2\text{O}]^2} = 3.1 \times 10^{-82} \quad (\text{at } 25\ ^\circ\text{C})$$

is strongly reactant-favored, as indicated by the *very* small value of K_c.

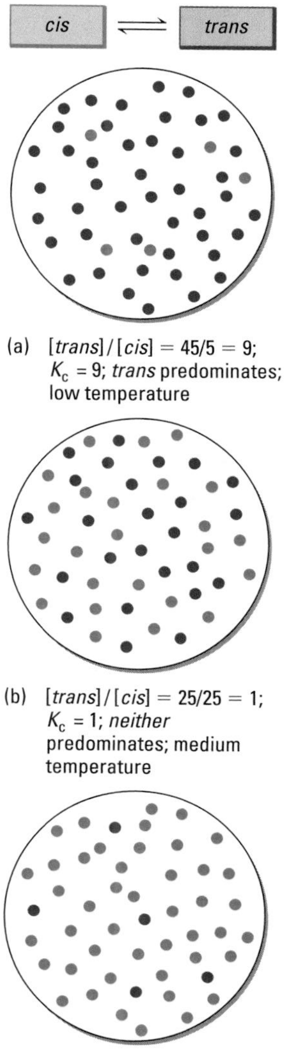

(a) [*trans*]/[*cis*] = 45/5 = 9; K_c = 9; *trans* predominates; low temperature

(b) [*trans*]/[*cis*] = 25/25 = 1; K_c = 1; *neither* predominates; medium temperature

(c) [*trans*]/[*cis*] = 5/45 = 1/9; K_c = 1/9; *cis* predominates; high temperature

Figure 14.4 Equilibrium constants and concentrations of reactants and products. For the system *cis*-2-butene(g) ⇌ *trans*-2-butene(g) the equilibrium constant decreases as temperature increases. The smaller the equilibrium constant is, the smaller the fraction of *trans*-2-butene (product) molecules becomes.

PROBLEM-SOLVING EXAMPLE 14.4 Using Equilibrium Constants

Use equilibrium constants (Table 14.1) to predict which of the reactions below will be product-favored at 25 °C. Place all of the reactions in order from most reactant-favored to least reactant-favored.

(a) $NH_3(aq) + H_2O(\ell) \rightleftharpoons NH_4^+(aq) + OH^-(aq)$
(b) $HCOOH(aq) + H_2O(\ell) \rightleftharpoons H_3O^+ + HCOO^-(aq)$
(c) $N_2O_4(g) \rightleftharpoons 2\ NO_2(g)$

Answer All reactions are reactant-favored. The order from most reactant-favored to least reactant-favored is (a), (b), (c).

Strategy and Explanation Check whether the reactions all have equilibrium constant expressions of the same form. If they do, then the smaller the equilibrium constant is, the less product-favored (more reactant-favored) the reaction is. The equilibrium constant expressions are all of the form

$$K_c = \frac{[\text{product 1}][\text{product 2}]}{[\text{reactant}]}$$

because $H_2O(\ell)$ does not appear in the expressions for (a) and (b). The equilibrium constants for reactions (a) and (b) are 1.8×10^{-5} and 1.8×10^{-4}, respectively. The equilibrium constant for reaction (c) is not given in Table 14.1, but K_c for the reverse reaction is given as 1.7×10^{2}. Because the reaction is reversed it is necessary to take the reciprocal (p. 663) which gives an equilibrium constant for reaction (c) of 5.8×10^{-3}. Therefore the most reactant-favored reaction (smallest K_c) is (a), the next smallest K_c is for reaction (b), and the largest K_c is for reaction (c), which is the least reactant-favored.

PROBLEM-SOLVING PRACTICE 14.4

Suppose that solid AgCl and AgI are placed in 1.0 L water in separate beakers.
(a) In which beaker would the silver ion concentration, $[Ag^+]$, be larger?
(b) Does the volume of water in which each compound dissolves affect the equilibrium concentration?

CONCEPTUAL EXERCISE 14.6 Manipulating Equilibrium Constants

The equilibrium constant is 1.8×10^{-5} for reaction of ammonia with water.

$$NH_3(aq) + H_2O(\ell) \rightleftharpoons NH_4^+(aq) + OH^-(aq)$$

(a) Is the equilibrium constant large or small for the reverse reaction, the reaction of ammonium ion with hydroxide ion to give ammonia and water?
(b) What is the value of K_c for the reaction of ammonium ions with hydroxide ions?
(c) What does the value of this equilibrium constant tell you about the extent to which a reaction can occur between ammonium ions and hydroxide ions?
(d) Predict what would happen if you added a 1.0-M solution of ammonium chloride to a 1.0-M solution of sodium hydroxide. What observations might allow you to test your prediction in the laboratory?

go Chemistry **Module 22: Equilibrium** covers concepts in this section.

14.5 Using Equilibrium Constants

Because equilibrium constants have numeric values, they can be used to predict quantitatively in which direction a reaction will proceed and how far it will go. Such quantitative predictions involve the **reaction quotient,** Q, which *has the same mathematical form as the equilibrium constant expression but is a ratio of actual concentrations in the mixture, instead of equilibrium concentrations.* For the general reaction

In the expression for Q, we have used (conc. N_2O_4) to represent the *actual* concentration of N_2O_4 at a given time. We use $[N_2O_4]$ to represent the *equilibrium* concentration of N_2O_4. When the reaction is at equilibrium (conc. N_2O_4) = $[N_2O_4]$.

$$a\,A + b\,B \rightleftharpoons c\,C + d\,D \qquad K_c = \frac{[C]^c[D]^d}{[A]^a[B]^b} \quad \text{and} \quad Q = \frac{(\text{conc. C})^c(\text{conc. D})^d}{(\text{conc. A})^a(\text{conc. B})^b}$$

K_c is a ratio of equilibrium concentrations, indicated by [A], [B], [C], and [D], whereas…

…Q is a ratio of actual concentrations, indicated by (conc. A), (conc. B), (conc. C), and (conc. D), with the reaction not necessarily at equilibrium.

Predicting the Direction of a Reaction

Suppose that you have a mixture of 50. mmol $NO_2(g)$ and 100. mmol $N_2O_4(g)$ at 25 °C in a sealed container with a volume of 10. L. Is the system at equilibrium? If not, in which direction will it react to achieve equilibrium? The reaction is

Recall that the metric prefix "m" means $\frac{1}{1000} = 10^{-3}$. Therefore, 1 mmol = 1×10^{-3} mol.

$$2\ NO_2(g) \rightleftharpoons N_2O_4(g) \qquad K_c = \frac{[N_2O_4]}{[NO_2]^2} = 1.7 \times 10^2 \quad (\text{at } 25\ °C)$$

$$Q = \frac{(\text{conc. } N_2O_4)}{(\text{conc. } NO_2)^2} = \frac{(100. \times 10^{-3}\ \text{mol}/10.\ \text{L})}{(50. \times 10^{-3}\ \text{mol}/10.\ \text{L})^2} = \frac{1.0 \times 10^{-2}}{(5.0 \times 10^{-3})^2} = 4.0 \times 10^2$$

- *If Q is equal to K_c, then the reaction is at equilibrium.* The concentrations will not change.
- *If Q is less than K_c, then the concentrations of products are not as large as they would be at equilibrium.* The reaction will proceed from left to right to increase the product concentrations until they reach their equilibrium values.
- *If Q is greater than K_c, then the product concentrations are higher than they would be at equilibrium.* The reaction will proceed from right to left, increasing reactant concentrations, until equilibrium is achieved.

These relationships are shown schematically in Figure 14.5. In the case of the NO_2/N_2O_4 mixture described above, Q is greater than K_c, so, to establish equilibrium, some N_2O_4 will react to form NO_2. As the reverse reaction takes place, Q becomes smaller and eventually becomes equal to K_c.

PROBLEM-SOLVING EXAMPLE 14.5 Predicting Direction of Reaction

Consider this equilibrium, which is used industrially to generate hydrogen gas.

$$CH_4(g) + H_2O(g) \rightleftharpoons CO(g) + 3\ H_2(g) \qquad K_c = 3.01 \times 10^{-10} \text{ at } 600\ K$$

If 1.0 mol CH_4, 1.0 mol H_2O, 2.0 mol H_2, and 0.50 mol CO are mixed in a sealed 10.0-L container at 25 °C, will the concentration of H_2O be greater or less than 0.10 mol/L when equilibrium is reached?

Answer Greater than 0.10 mol/L

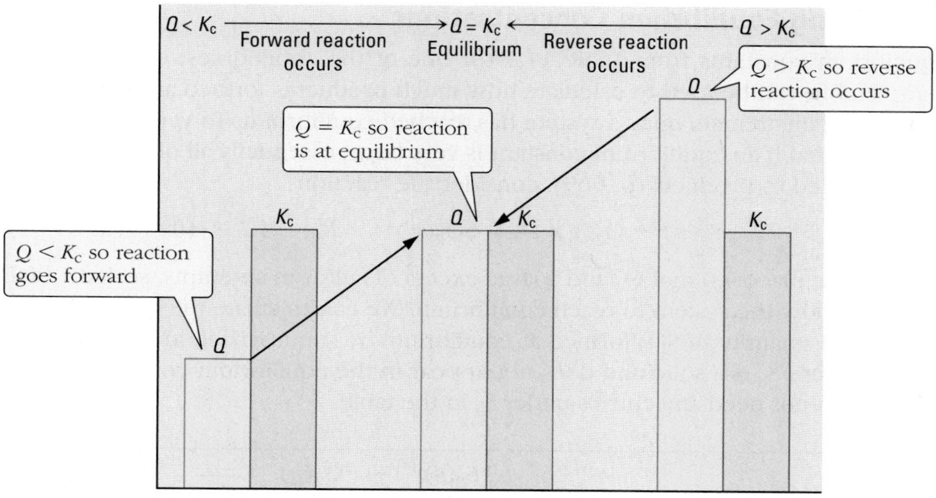

Figure 14.5 Predicting direction of a reaction. The relative sizes of the reaction quotient, Q, and the equilibrium constant, K_c, determine in which direction a mixture of substances will react to achieve equilibrium. The reaction will proceed until $Q = K_c$.

Strategy and Explanation Calculate the initial concentration of each gas and thus evaluate Q. Then compare Q with K_c.

$$(\text{conc. CH}_4) = \frac{1.0\text{ mol}}{10.0\text{ L}} = 0.10\text{ mol/L} \qquad (\text{conc. H}_2) = \frac{2.0\text{ mol}}{10.0\text{ L}} = 0.20\text{ mol/L}$$

$$(\text{conc. H}_2\text{O}) = \frac{1.0\text{ mol}}{10.0\text{ L}} = 0.10\text{ mol/L} \qquad (\text{conc. CO}) = \frac{0.50\text{ mol}}{10.0\text{ L}} = 0.050\text{ mol/L}$$

Because the units are defined by the equilibrium constant expression, leave them out.

$$Q = \frac{(\text{conc. CO})(\text{conc. H}_2)^3}{(\text{conc. CH}_4)(\text{conc. H}_2\text{O})} = \frac{(0.050)(0.20)^3}{(0.10)(0.10)} = 0.040$$

which is much larger than 3.01×10^{-10}, the value of K_c. Because $Q > K_c$, the reverse reaction—reaction of CO with H_2 to form CH_4 and H_2O—will occur until the equilibrium concentrations are reached. The initial concentration of H_2O was 0.10 mol/L; when CO reacts with H_2, the H_2O concentration increases.

☑ **Reasonable Answer Check** Because all concentrations are fractions, multiplying concentrations will give smaller fractions. The concentrations are all approximately 10^{-1}, but the numerator has concentration to the fourth power and the denominator to the second power overall. This means that Q should be much less than 1, and it is.

PROBLEM-SOLVING PRACTICE **14.5**

For the equilibrium

$$2\,SO_2(g) + O_2(g) \rightleftharpoons 2\,SO_3(g) \qquad K_c = 245 \quad (\text{at } 1000\text{ K})$$

the equilibrium concentrations are $[SO_2] = 0.102$, $[O_2] = 0.0132$, and $[SO_3] = 0.184$. Suppose that the concentration of SO_2 is suddenly doubled. Calculate Q and use it to show that the forward reaction would take place to reach a new equilibrium.

CONCEPTUAL EXERCISE **14.7 Reaction Quotient and Pressure Equilibrium Constant**

Is it possible to apply the idea of the reaction quotient to gas phase reactions where the equilibrium constant is given in terms of pressure (as in Equation 14.4, p. 665)? Define Q for such a reaction and give an appropriate set of rules by which you can predict in which direction a gas phase reaction will go to achieve equilibrium.

go Chemistry **Module 21: Chemical Equilibrium** covers concepts in this section.

Calculating Equilibrium Concentrations

Equilibrium constants from Table 14.1 (or one of the appendixes, or a reference compilation) can be used to calculate how much product is formed and how much of the reactants remain once a system has reached equilibrium. To verify our earlier statement that if an equilibrium constant is very large, essentially all of the reactants are converted to products (p. 669), consider the reaction

$$\tfrac{1}{8}\,S_8(s) + O_2(g) \rightleftharpoons SO_2(g) \qquad K_c = 4.2 \times 10^{52} \quad (\text{at } 25\text{ °C})$$

Suppose we place 4.0 mol O_2 and a large excess of sulfur in an empty, sealed 1.00-L flask and allow the system to reach equilibrium. We can calculate the quantity of O_2 left and the quantity of SO_2 formed at equilibrium by summarizing information in a table. Because S_8 is a solid and does not appear in the equilibrium constant expression, we do not need any entries under S_8 in the table.

	$\tfrac{1}{8}S_8(s)$	+	$O_2(g)$	$\rightleftharpoons$	$SO_2(g)$
*I*nitial concentration (mol/L)			4.0		0.0
*C*hange as reaction occurs (mol/L)			$-x$		$+x$
*E*quilibrium concentration (mol/L)			$4.0 - x$		$0.0 + x$

We know the concentrations of reactant and product before the reaction, but we do not know how many moles per liter of O_2 are consumed during the reaction, and so we designate this as x mol/L. (There is a minus sign in the table because O_2 is consumed.) Since the mole ratio is (1 mol SO_2)/(1 mol O_2), we know that x mol/L SO_2 is formed when x mol/L O_2 is consumed. To calculate the concentration of O_2 we take what was present initially, 4.0 mol/L, minus what was consumed in the reaction, x mol/L. The equilibrium concentration of SO_2 must be the initial concentration, 0 mol/L, plus what was formed by the reaction, x mol/L. Putting these values into the equilibrium constant expression, we have

$$K_c = \frac{[SO_2]}{[O_2]} = \frac{x}{4.0 - x} = 4.2 \times 10^{52}$$

Solving algebraically for x (and following the usual rules for significant figures), we find

$$x = (4.2 \times 10^{52})(4.0 - x)$$
$$x = (16.8 \times 10^{52}) - (4.2 \times 10^{52})x$$
$$x + (4.2 \times 10^{52})x = 16.8 \times 10^{52}$$

Notice that $x + (4.2 \times 10^{52})x = (1 + 4.2 \times 10^{52})x$, which to a very good approximation is equal to $(4.2 \times 10^{52})x$. Thus,

Because 4.2×10^{52} is so much larger than 1, adding 1 to it makes no appreciable change in the very large number.

$$x = \frac{16.8 \times 10^{52}}{4.2 \times 10^{52}} = 4.0$$

The equilibrium concentration of SO_2 is x = 4.0 mol/L and that of O_2 is $(4.0 - x)$ mol/L, or 0 mol/L. That is, within the precision of our calculation, all the O_2 has been converted to SO_2. As stated earlier, a very large K_c value (4.2×10^{52} in this case) implies that essentially all of the reactants have been converted to products. The reaction is strongly product-favored and goes to completion, so the calculation could have been done using the methods in Section 4.4 **(← *p. 131*)**.

The fact that samples of sulfur at 25 °C can be exposed to oxygen in the air for long periods without being converted to sulfur dioxide shows the importance of chemical kinetics. This reaction is very slow at room temperature, so only a faint odor of sulfur dioxide is noticeable in the vicinity of solid sulfur.

PROBLEM-SOLVING EXAMPLE 14.6 Calculating Equilibrium Concentrations

One way to generate hydrogen gas is to react carbon monoxide with steam.

$$CO(g) + H_2O(g) \rightleftharpoons CO_2(g) + H_2(g) \qquad K_c = 10. \text{ at } 420\ °C$$

Suppose you place 2.5×10^{-3} mol/L CO and 2.5×10^{-3} mol/L H_2O into a container, seal it, and heat the mixture to 420 °C. What is the concentration of each of the four substances when equilibrium is reached?

Answer $[CO] = [H_2O] = 6.0 \times 10^{-4}$ mol/L; $[CO_2] = [H_2] = 1.9 \times 10^{-3}$ mol/L

Strategy and Explanation Follow the procedure on p. 666. The chemical equation written in the statement of the problem shows that all of the mole ratios are 1:1. This tells us that equal numbers of moles of CO and H_2O are consumed as the reaction proceeds to equilibrium. Since both substances are in the same flask, equal numbers of moles *per liter* (equal concentrations) of reactants must also be consumed, and we designate each of these as $-x$. Mole ratios also tell us that if the concentration of CO decreases by x mol/L, then the concentration of H_2 (and the concentration of CO_2) must increase by the same quantity, x mol/L. Because their initial concentrations were the same and their mole ratio is 1:1, the equilibrium concentrations of H_2O and CO must be the same. Each is equal to the initial concentration plus the change (the change for reactants is $-x$, so this means subtracting x) as the reactants are consumed. The reaction table is

	CO	**+**	**H_2O**	**⇌**	**CO_2**	**+**	**H_2**
*I*nitial concentration (mol/L)	2.5×10^{-3}		2.5×10^{-3}		0.000		0.000
*C*hange as reaction occurs (mol/L)	$-x$		$-x$		$+x$		$+x$
*E*quilibrium concentration (mol/L)	$0.0025 - x$		$0.0025 - x$		x		x

Substituting these values into the expression for K_c, we have

$$K_c = 10.0 = \frac{[H_2][CO_2]}{[H_2O][CO]} = \frac{(x)(x)}{(0.0025 - x)(0.0025 - x)} = \frac{x^2}{(0.0025 - x)^2}$$

Solving this equation for x is not as difficult as it might seem at first glance. Because the right-hand side is a perfect square, we can take the square root of both sides, giving

$$\sqrt{K_c} = \sqrt{10.0} = 3.16 = \frac{x}{(0.0025 - x)}$$

and then solve for x.

$$x = (3.16)(0.0025 - x) = 0.0079 - 3.16x$$
$$4.16x = 0.0079; \quad x = 1.9 \times 10^{-3}$$

The concentrations of the products are both equal to x mol/L—that is, to 1.9×10^{-3} mol/L—while the concentrations of the reactants that remain are both $(0.0025 - x)$ mol/L = (0.0025 − 0.0019) mol/L = 6.0×10^{-4} mol/L. This result demonstrates quantitatively that when $K_c > 1$ a reaction is product-favored because at equilibrium the product concentrations are larger than the reactant concentrations.

☑ **Reasonable Answer Check** To check the calculation, substitute these concentrations into the equilibrium constant expression and compare the calculated K_c with the value given in the statement of the problem.

$$K_c = \frac{[H_2][CO_2]}{[H_2O][CO]} = \frac{(1.9 \times 10^{-3})^2}{(6.0 \times 10^{-4})^2} = 10.0$$

which equals the value given.

PROBLEM-SOLVING PRACTICE **14.6**

The equilibrium constant for dissolving the insoluble substance gold(I) iodide, AuI(s), in aqueous solution is 1.6×10^{-23} at 25 °C. Write the equilibrium constant expression and calculate the concentration of Au^+(aq) and I^-(aq) ions in a solution in which 0.345 g AuI(s) is in equilibrium with the aqueous ions.

PROBLEM-SOLVING EXAMPLE **14.7** **Calculating Equilibrium Concentrations**

When colorless hydrogen iodide gas is heated to 745 K, a beautiful purple color appears. This shows that some iodine gas has been formed, which implies that some HI has been decomposed to its elements.

$$2\,HI(g) \rightleftharpoons H_2(g) + I_2(g) \qquad K_c = 0.0200 \quad (\text{at } 745\text{ K})$$

Suppose that a mixture of 1.00 mol HI(g) and 1.00 mol H_2(g) is sealed into a 10.0-L flask and heated to 745 K. What will be the concentrations of all three substances when equilibrium has been achieved?

Answer [HI] = 0.096 M, $[I_2]$ = 0.0018 M, $[H_2]$ = 0.102 M

Strategy and Explanation Follow the usual steps. The balanced equation was given, and the equilibrium constant expression follows the table (next page). Because amounts are given instead of concentrations, divide each number of moles by the volume of the flask to get (conc. HI) = (conc. H_2) = 1.00 mol/10.0 L = 0.100 mol/L. Because no I_2 is present at the beginning, $Q = 0$. This is much less than K_c, even though K_c is small. Therefore the reaction must go from left to right, so it makes sense to let x mol/L be the concentration of I_2 when equilibrium is reached. Since the coefficients of H_2 and I_2 are the same, if x mol/L of I_2 is produced, then x mol/L of H_2 must also be produced. Therefore the change in concentration of both H_2 and I_2 is $+x$ mol/L. Because the coefficient of HI is twice the coefficient of I_2, twice as many

Iodine vapor has a beautiful purple color. (Conditions shown here differ from those in Problem-Solving Example 14.7.)

moles of HI must react as moles of I_2 formed; the change in concentration of HI is $-2x$ mol/L. The reaction table is then

	$2\ HI(g)$	$\rightleftharpoons$	$H_2(g)$	+	$I_2(g)$
*I*nitial concentration (mol/L)	0.100		0.100		0.000
*C*hange as reaction occurs (mol/L)	$-2x$		$+x$		$+x$
*E*quilibrium concentration (mol/L)	$(0.100 - 2x)$		$(0.100 + x)$		x

In solving this problem we might have chosen $-x$ to be the change in concentration of HI, in which case the changes in concentrations of H_2 and I_2 would each have been $+\frac{1}{2}x$.

Now write the equilibrium constant expression in terms of the equilibrium concentrations calculated in the third row of the table.

$$K_c = \frac{[H_2][I_2]}{[HI]^2} = \frac{(0.100 + x)x}{(0.100 - 2x)^2} = 0.0200$$

The ratio of terms involving x is not a perfect square, so you cannot take a square root as was done in Problem-Solving Example 14.6. To solve the equation directly, multiply out the numerator and denominator to obtain

$$\frac{0.100x + x^2}{0.0100 - 0.400x + 4x^2} = 0.0200$$

Multiply both sides by the denominator and then multiply out the terms. This gives

$$0.100x + x^2 = 0.0200 \times (0.0100 - 0.400x + 4x^2)$$

$$0.100x + x^2 = 0.000200 - 0.00800x + 0.0800x^2$$

Collecting terms in x^2 and x, we have

$$0.9200x^2 + 0.10800x - 0.000200 = 0$$

This is a quadratic equation of the form $ax^2 + bx + c = 0$, where $a = 0.9200$, $b = 0.10800$, and $c = -0.000200$. The equation can be solved using the quadratic formula (Appendix A.7).

$$x = \frac{-b \pm \sqrt{b^2 - 4ac}}{2a} = \frac{-0.10800 \pm \sqrt{0.10800^2 - 4 \times 0.9200 \times (-0.000200)}}{2 \times 0.9200}$$

$$x = \frac{-0.10800 + 0.11135}{1.840} = 1.83 \times 10^{-3}$$

The other root, $x = -0.119$, can be eliminated because it would result in a negative concentration of I_2 at equilibrium, which is clearly impossible. From the equilibrium concentration row of the table,

$$[HI] = (0.100 - 2x)\text{mol/L} = 0.100 - (2 \times 1.83 \times 10^{-3}) = 0.0963\ \text{mol/L}$$

$$[H_2] = (0.100 + x)\text{mol/L} = 0.100 + (1.83 \times 10^{-3}) = 0.102\ \text{mol/L}$$

$$[I_2] = x\ \text{mol/L} = 0.00183\ \text{mol/L}$$

☑ **Reasonable Answer Check** Solving this type of problem involves a lot of algebra, and it would be easy to make a mistake. It is very important to check the result by substituting the equilibrium concentrations into the equilibrium constant expression and verifying that the correct value of K_c (0.0200) results.

$$K_c = \frac{[H_2][I_2]}{[HI]^2} = \frac{(0.102)(0.00183)}{(0.0963)^2} = 0.0201$$

which is acceptable agreement because it differs from K_c by 1 in the last significant figure.

PROBLEM-SOLVING PRACTICE 14.7

Obtain the equilibrium constant for dissociation of dinitrogen tetraoxide to form nitrogen dioxide from Table 14.1. If 1.00 mol N_2O_4 and 0.500 mol NO_2 are initially placed in a container with a volume of 4.00 L and the container is sealed, calculate the concentrations of $N_2O_4(g)$ and $NO_2(g)$ present when equilibrium is achieved at 25 °C.

14.6 Shifting a Chemical Equilibrium: Le Chatelier's Principle

Le Chatelier is pronounced "luh SHOT lee ay."

Suppose you are an environmental engineer, biologist, or geologist, and you have just measured the concentration of hydronium ion, H_3O^+, in a lake. You know that the H_3O^+ ions are involved in many different equilibrium reactions in the lake. How can you predict the influence of changing conditions? For example, what happens if there is a large increase in acid rainfall that has a hydronium ion concentration different from that of the lake? Or what happens if lime (calcium oxide), a strong base, is added to the lake? These questions and many others like them can be answered qualitatively by applying a useful guideline known as **Le Chatelier's principle:** ***If a system is at equilibrium and the conditions are changed so that it is no longer at equilibrium, the system will react to reach a new equilibrium in a way that partially counteracts the change.*** To adjust to a change, a system reacts in either the forward direction (producing more products) or the reverse direction (producing more reactants) until a new equilibrium state is achieved. Le Chatelier's principle applies to changes in the concentrations of reactants or products that appear in the equilibrium constant expression, the pressure or volume of a gas phase equilibrium, and the temperature. Changing conditions, thereby changing the equilibrium concentrations of reactants and products, is called **shifting an equilibrium.** If the reaction occurs in the *forward direction,* we say that the equilibrium reaction has *shifted to the right.* If the system reacts in the *reverse direction,* the reaction has *shifted to the left.*

Henri Le Chatelier (1850–1936) was a French chemist who, as a result of his studies of the chemistry of cement, developed his ideas about how altering conditions affects an equilibrium system.

Changing Concentrations of Reactants or Product

If the concentration of a reactant or a product that appears in the equilibrium constant expression is changed, a system can no longer be at equilibrium because Q must have a different value from K_c. The equilibrium will shift to use up a substance that was added, or to replenish a substance that was removed. The shift will occur in whichever direction causes the value of Q to approach the value of K_c.

- If the concentration of a *reactant* is *increased,* the system will react in the *forward* direction.
- If the concentration of a *reactant* is *decreased,* the system will react in the *reverse* direction.
- If the concentration of a *product* is *increased,* the system will react in the *reverse* direction.
- If the concentration of a *product* is *decreased,* the system will react in the *forward* direction.

To see why this happens, consider the simple equilibrium discussed in Section 14.2.

$$cis\text{-2-butene(g)} \rightleftharpoons trans\text{-2-butene(g)} \qquad K_c = \frac{[trans]}{[cis]} = 1.5 \quad \text{(at 600 K)}$$

As long as the temperature remains the same, the value of the equilibrium constant also remains the same. Adding or removing a reactant or a product does not change the equilibrium constant value. If the concentration of the substance added or removed appears in the equilibrium constant expression, however, the value of Q changes. Because Q no longer equals K_c, the system is no longer at equilibrium and must react to achieve a new equilibrium.

Suppose that 2 mmol *cis*-2-butene is placed into a 1.0-L container at 1000 K and the container is sealed. The forward reaction will be faster than the reverse reaction until the concentration of *trans*-2-butene builds up to 1.5 times the concentration of *cis*-2-butene. Then equilibrium is achieved. Now suppose that half of the *cis*-2-butene is instantaneously removed from the container (see Exercise 14.3, p. 660). Because the concentration of *cis*-2-butene suddenly drops to half its former value, the forward reaction rate will also drop to half its former value. The reverse rate will not be affected, because the concentration of *trans*-2-butene has not changed. This means that the reverse reaction is twice as fast as the forward reaction, and *cis*-2-butene molecules are being formed twice as fast as they are reacting away. Therefore the concentration of *cis*-2-butene will increase and the concentration of *trans*-2-butene

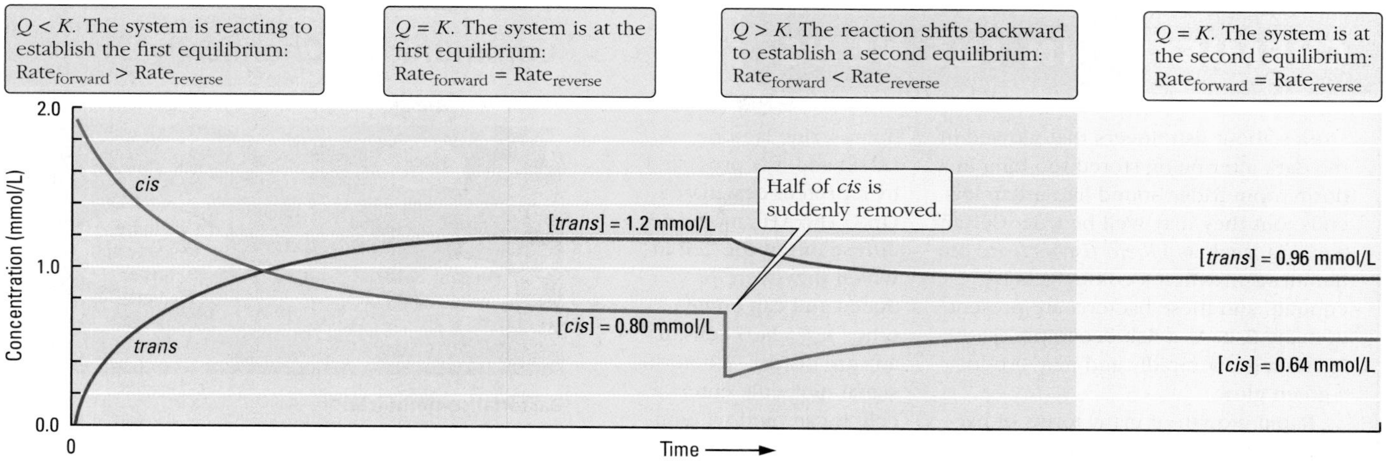

Figure 14.6 Approach to new equilibrium after a change in conditions. The reaction is *cis*-2-butene $\rightleftharpoons$ *trans*-2-butene. Note that K = [*trans*]/[*cis*] and has the same value, 1.5, at all times at this temperature.

will decrease until the forward and reverse rates are again equal. The graph in Figure 14.6 illustrates this situation.

The effect of changing concentration has many important consequences. For example, when the concentration of a substance needed by your body falls slightly, several enzyme-catalyzed chemical equilibria shift so as to increase the concentration of the essential substance. In industrial processes, reaction products are often continuously removed. This shifts one or more equilibria to produce more products and thereby maximize the yield of the reaction.

In nature, slight changes in conditions are responsible for effects such as the formation of limestone stalactites and stalagmites in caves and the crust of limestone that slowly develops in a tea kettle if you boil hard water in it. Both of these examples involve calcium carbonate. Limestone, a form of calcium carbonate, $CaCO_3$, is present in underground deposits, a leftover of the ancient oceans from which it precipitated long ago. Limestone dissolves when it reacts with an aqueous solution of CO_2.

Hard water contains dissolved CO_2 and metal ions such as Ca^{2+} and Mg^{2+}.

In some cases organisms, such as coral, induce precipitation of $CaCO_3$.

$$CaCO_3(s) + CO_2(aq) + H_2O(\ell) \rightleftharpoons Ca^{2+}(aq) + 2\,HCO_3^-(aq) \quad [14.7]$$

Figure 14.7 Reaction of $CaCl_2$(aq) with $NaHCO_3$(aq). (This is the equilibrium of Equation 14.7 being shifted in the reverse direction.)

CHEMISTRY IN THE NEWS Bacteria Communicate Chemically

Stories about fish fingers that glowed in the dark after being stored too long in a dorm-room fridge sound like urban legends, but they may well be true. Certain bacteria, such as *Vibrio fischeri*, are bioluminescent when a colony gets big enough, and these bacteria are present in many fish. As a fish decomposes, the bacteria grow rapidly and may produce a green glow.

Rapid growth of many forms of bacteria involves communication among individual bacterial cells. Only after a colony reaches a certain size and the bacteria in the colony realize it has reached that size is there really rapid growth. In this group state, pathogenic bacteria initiate the majority of infections in plants and animals. How do bacteria sense when they have reached an adequate group size? And could bacterial infections be treated by preventing bacteria from realizing they actually are in a group? Answers to these questions could be very beneficial to humankind, and scientists are working to find them.

Gram-negative bacteria communicate by secreting so-called signal molecules that can diffuse out of one bacterial cell and into another. In the case of *V. fischeri*, the signal molecule is an acyl-homoserine lactone (AHL) and it is produced by a protein designated LuxI. The AHL molecules diffuse out of the cell in which they were produced and can enter other *V. fischeri* cells, if any are nearby. When the signal molecule enters a cell, it can bind with another protein, LuxR, in an equilibrium process:

$$\text{LuxR} + \text{AHL} \rightleftharpoons \text{LuxR—AHL}$$

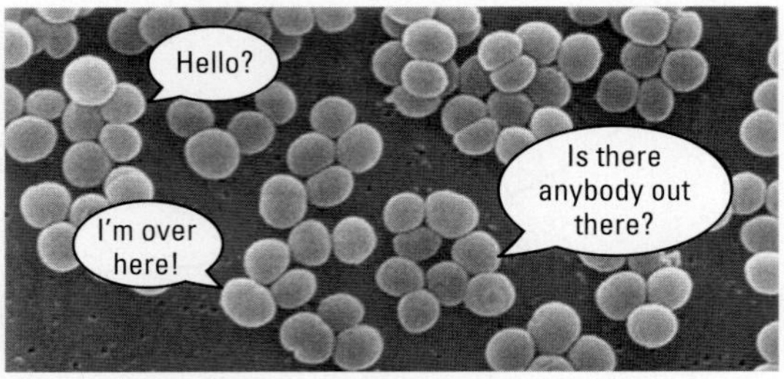

Cartoon by Greg Dahlmann; Photo micrograph: Courtesy of CDC

Bacterial communication.

If only a few *V. fischeri* cells are close together, most of the AHL diffuses away, and its concentration remains small. But once a bacterial colony reaches a certain size, most of the cells are surrounded by other cells, all of which are secreting AHL. The increased concentration of AHL shifts the equilibrium to the right. This forms sizable concentrations of the LuxR-AHL combination, which binds to DNA and switches on genes that encode the enzyme luciferase. In the presence of O_2, luciferase produces luminescence like that of a firefly. The LuxR-AHL combination also switches on genes that code for formation of LuxI, which produces more AHL. Therefore, AHL is said to *autocatalyze* its own synthesis, and this explains why the colony of bacteria suddenly grows very rapidly (and produces the green glow) after an initial period of slow growth.

This process is referred to as *quorum sensing,* because it requires a significant number of bacteria (a quorum) to initiate the autocatalysis of formation of AHL. Many bacteria display quorum-sensing behavior, and often their rapid growth has much worse consequences than a green-glowing fish stick. Quorum-sensing behavior is thought to account for more than 50% of all crop diseases and as much as 80% of all human bacterial infections. Therefore scientists are looking for a way to prevent quorum-

If groundwater that is saturated with CO_2 encounters a bed of limestone below the surface of Earth, the forward reaction can occur until equilibrium is reached, and the water subsequently contains significant concentrations of aqueous Ca^{2+} and HCO_3^- ions in addition to dissolved CO_2. These ions, often accompanied by Mg^{2+}(aq), constitute hard water (Section 15.10).

As with all equilibria, a reverse reaction is occurring in addition to the forward reaction. This reverse reaction can be demonstrated by mixing aqueous solutions of $CaCl_2$ and $NaHCO_3$ (salts containing the Ca^{2+} and HCO_3^- ions) in an open beaker (Figure 14.7). You will eventually see bubbles of CO_2 gas and a precipitate of solid $CaCO_3$. Because the beaker is open to the air, any gaseous CO_2 that escapes the solution is swept away. Its removal reduces the concentration of CO_2(aq) and causes the equilibrium to shift to the left to produce more CO_2. Eventually all of the dissolved Ca^{2+} and HCO_3^- ions are converted to gaseous CO_2, solid $CaCO_3$, and water.

Suppose that water containing dissolved CO_2, Ca^{2+}, and HCO_3^- contacts the air in a cave (or hard water contacts air in your tea kettle). Carbon dioxide bubbles out of the solution, the concentration of CO_2 decreases on the reactant side of Equation 14.7, and the equilibrium shifts toward the reactants. The reverse reaction forms CO_2(aq), compensating partially for the reduced concentration of CO_2(aq). Some of the calcium ions and hydrogen carbonate ions combine, and

Photo by Jason Varney Photography, Courtesy of Helen Blackwell

Helen Blackwell.

sensing equilibria from shifting right and forming the enzyme–signal molecule combination. Substances that could do this are called quorum-sensing antagonists. They have great potential to alleviate problems of bacterial resistance to antibiotics and could bring great benefit to humankind.

One way to find quorum-sensing antagonists is being explored by the research group of Professor Helen Blackwell of the University of Wisconsin-Madison. Blackwell and her co-workers make many different compounds whose molecular structures are closely related and test those compounds for their ability to inhibit quorum-sensing behavior in bacteria. To help make the compounds Blackwell uses a laboratory version of a microwave oven such as those found in kitchens to heat food. Microwave irradiation heats reactant chemicals quickly and results in greater yields of higher-purity products. In a single day Blackwell's research group can prepare as many as 100 to 500 related compounds. These can subsequently be tested for activity, and the rate at which new biologically active compounds can be discovered is greatly accelerated. Recently Blackwell and her students have identified quorum-sensing antagonists that are among the most potent discovered to date.

Knowledge of quorum sensing also allows chemists to enhance the growth of bacteria, and in some cases that may be a good thing. For example, changing the concentrations of autocatalyst substances might be a way of controlling the differentiation of stem cells into different type of tissues. Or, a system can be imagined in which bacterial populations could be maintained at a desired level. A quorum-sensing signal molecule could turn on a suicide gene if the colony got too large, but when the population decreased the signal concentration would become smaller, thus turning off the gene and not completely wiping out the colony. Exploring the reasons for that green glow in the refrigerator might have incredibly beneficial consequences.

Courtesy of Dr. Patrick Hickey, LUX Biotechnology, Edinburgh, UK

Glowing *Vibrio fischeri*.

Sources: Based on information from *New York Times*, January 17, 2006; *The Health Show*, April 21, 2005, **http://www.healthshow.org/archive/week_2005_04_17.shtml;** Blackwell laboratory, **http://www.chem.wisc.edu/blackwell/index.htm;** *Wisconsin State Journal*, November 4, 2007; *Science*, October 5, 2007; and *Scientific American*, May 2009, p. 19.

some $CaCO_3(s)$ precipitates (Figure 14.8) as a beautiful formation in the cave (or as scale in your kettle).

© Carlyn Iverson/Photo Researchers, Inc.

Figure 14.8 Stalactites in a limestone cave. Stalactites hanging from the ceiling and stalagmites growing from the floor both consist of limestone, $CaCO_3$. The process that produces these lovely formations is an excellent example of chemical equilibrium.

CONCEPTUAL EXERCISE 14.8 Effect of Adding a Substance

Solid phosphorus pentachloride decomposes when heated to form gaseous phosphorus trichloride and gaseous chlorine. Write the equation for the equilibrium that is set up when solid phosphorus pentachloride is introduced into a container, the container is evacuated and sealed, and the solid is heated. Once the system has reached equilibrium at a given temperature, what will be the effect on the equilibrium of

(a) Adding chlorine to the container?
(b) Adding phosphorus trichloride to the container?
(c) Adding a small quantity of solid phosphorus pentachloride to the container?

EXERCISE 14.9 *K* Is a Constant

Use concentrations read from the graph in Figure 14.6 to show that K_c has the same value before and after half of the *cis*-2-butene is removed from the first equilibrium mixture.

Changing Volume or Pressure in Gaseous Equilibria

One way to change the pressure of a gaseous equilibrium mixture is to keep the volume constant and to add or remove one or more of the substances whose concentrations appear in the equilibrium constant expression. The effect of adding or removing a substance has just been discussed. We consider here other ways of changing pressure or volume.

The pressures of all substances in a gaseous equilibrium can be changed by changing the volume of the container. Consider the effect of tripling the pressure on the equilibrium

$$N_2O_4(g) \rightleftharpoons 2\ NO_2(g) \qquad K_c = \frac{[NO_2]^2}{[N_2O_4]}$$

by reducing the volume of the container to one third of its original value (at constant temperature). This situation is shown in Figure 14.9. Decreasing the volume increases the pressures of N_2O_4 and NO_2 to three times their equilibrium values. Decreasing the volume also increases the concentrations of N_2O_4 and NO_2 to three times their equilibrium values. Because $[NO_2]$ is squared in the equilibrium constant expression but $[N_2O_4]$ is not, tripling both concentrations increases the numerator of Q by $3^2 = 9$ but increases the denominator by only 3.

$$Q = \frac{(\text{conc. } NO_2)^2}{(\text{conc. } N_2O_4)} = \frac{(3 \times [NO_2])^2}{(3 \times [N_2O_4])} = \frac{9}{3} \times \frac{[NO_2]^2}{[N_2O_4]} = 3 \times K_c$$

Because Q is larger than K_c under the new conditions, the reaction should produce more reactant; that is, the equilibrium should shift to the left.

The same prediction is made using Le Chatelier's principle: The reaction should shift to partially compensate for the increase in pressure. That means decreasing the pressure, which can happen if the total number of gas phase molecules decreases. In the case of the N_2O_4/NO_2 equilibrium, the reverse reaction should occur, because one N_2O_4 molecule is produced for every two NO_2 molecules that react. A shift to the left reduces the number of gas phase molecules and hence the pressure.

Remember that the pressure of an ideal gas is proportional to the number of moles of gas, and therefore to the number of molecules of gas (← p. 438).

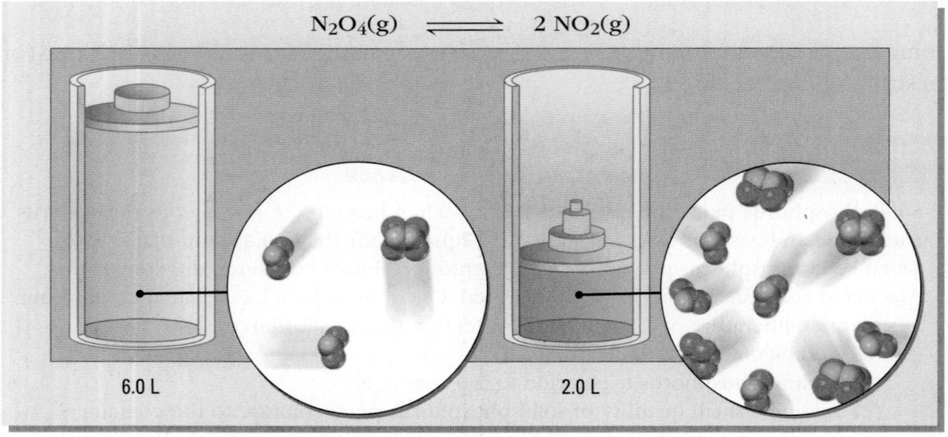

Figure 14.9 Shifting an equilibrium by changing pressure and volume. If the volume of an equilibrium mixture of NO_2 and N_2O_4 is decreased from 6.0 to 2.0 L, the equilibrium shifts toward the smaller number of molecules in the gas phase (the N_2O_4 side) to partially compensate for the increased pressure. When the new equilibrium is achieved, the concentrations of both NO_2 and N_2O_4 have increased, but the NO_2 concentration is less than three times as great, while the N_2O_4 concentration is more than three times as great. The total pressure goes from 1.00 to 2.62 atm (rather than to 3.00 atm, which was the new pressure before the equilibrium shifted).

However, consider the situation with respect to another equilibrium we have already mentioned in Problem-Solving Example 14.3 (p. 668). At 425 °C all substances are in the gas phase.

$$2\ HI(g) \rightleftharpoons H_2(g) + I_2(g)$$

Suppose that the pressure of this system were tripled by reducing its volume to one third of the original volume. What would happen to the equilibrium? In this case, all the concentrations triple, but because equal numbers of moles of gaseous substances appear on both sides of the equation, Q still has the same numeric value as K_c. That is, the system is still at equilibrium, and no shift occurs. Thus, *changing pressure by changing the volume shifts an equilibrium only if the sum of the coefficients for gas-phase reactants is different from the sum of the coefficients for gas-phase products.*

Finally, consider what happens if the pressure of the N_2O_4/NO_2 equilibrium system is increased by adding an inert gas such as argon while retaining exactly the same volume. The total pressure of the system would increase, but since neither the amounts of N_2O_4 and NO_2 nor the volume would change, the concentrations of N_2O_4 and NO_2 would remain the same. Q still equals K_c, the system is still at equilibrium, and no shift occurs (or needs to). Thus, changing the pressure of an equilibrium system must change the concentration of at least one substance in the equilibrium constant expression if a shift in the equilibrium is to occur.

An inert gas in this context is one that does not react with the other gases already present.

CONCEPTUAL EXERCISE 14.10 Changing Volume Does Not Always Shift an Equilibrium

In Problem-Solving Example 14.7 you found that for the reaction

$$2\ HI(g) \rightleftharpoons H_2(g) + I_2(g) \qquad K_c = 0.0200 \quad (\text{at } 745\ K)$$

the equilibrium concentrations were [HI] = 0.0963 mol/L, $[H_2]$ = 0.102 mol/L, and $[I_2]$ = 0.00183 mol/L. Use algebra to show that if each of these concentrations is tripled by reducing the volume of this equilibrium system to one third of its initial value, the system is still at equilibrium, and therefore the pressure change causes no shift in the equilibrium.

EXERCISE 14.11 Effect of Changing Volume

Verify the statement in the text that, for the N_2O_4/NO_2 equilibrium system, decreasing the volume to one third of its original value increases the equilibrium concentration of N_2O_4 by more than a factor of 3 while it increases the equilibrium concentration of NO_2 by less than a factor of 3. Start with the equilibrium conditions you calculated in Problem-Solving Practice 14.7. Then decrease the volume of the system from 4.00 L to 1.33 L, and calculate the new concentrations of N_2O_4 and NO_2 assuming the shift in the equilibrium has not yet taken place. Set up the usual table of initial concentrations, change, and equilibrium concentrations, and calculate the concentrations at the new equilibrium.

Changing Volume by Adding Solvent

Just as the concentrations (and pressures) of gas-phase reactants and products can be changed by changing the volume of a closed system, the concentrations of reactants and products in a solution can be changed by either adding solvent or removing solvent by evaporating it. For a reaction in which there is a change in the total number of solute particles, such a volume change will cause a shift in the equilibrium. For example, consider the reaction

$$HSO_4^-(aq) + H_2O(\ell) \rightleftharpoons H_3O^+(aq) + SO_4^{2-}(aq)$$

$$K_c = \frac{[H_3O^+][SO_4^{2-}]}{[HSO_4^-]} = 1.2 \times 10^{-2}$$

Suppose that the volume of the solution is doubled by adding water. This reduces the concentrations of all solutes (HSO_4^-, H_3O^+, and SO_4^{2-}) to one half their previous values. For example, suppose that $[H_3O^+] = [SO_4^{2-}] = 3.4 \times 10^{-2}$ M and $[HSO_4^-] =$ 0.10 M before the water was added. After the volume is doubled, (conc H_3O^+) = (conc SO_4^{2-}) = 1.7×10^{-2} M and (conc HSO_4^-) = 0.050 M. This gives

$$Q = \frac{(1.7 \times 10^{-2})(1.7 \times 10^{-2})}{5.0 \times 10^{-2}} = 5.8 \times 10^{-3}$$

Thus $Q < K_c$ immediately after the solution volume is doubled. Some reactant HSO_4^- must be converted to product H_3O^+ and SO_4^{2-} to reach a new equilibrium, and the reaction shifts to the right. $Q < K_c$ because there are two concentrations in the numerator of K_c and only one in the denominator; that is, the halved concentrations are squared in the numerator, but only raised to the first power in the denominator. As in the case of the gas-phase equilibria discussed in the preceding section, if there had been equal numbers of solute particles on each side of the chemical equation, then after the solution volume doubled Q would still equal K_c, the system would still be at equilibrium, and no shift would occur. An example of a reaction in solution for which there would be no shift upon volume change is

$$Cu^{2+}(aq) + Zn(s) \rightleftharpoons Cu(s) + Zn^{2+}(aq)$$

For this reaction, in which there is one solute particle on each side of the equation, diluting the solution has no effect on Q, and therefore there is no shift in the equilibrium.

Changing Temperature

When temperature changes, the values of most equilibrium constants also change, and systems will react to achieve new equilibria consistent with new values of K_c. You can make a qualitative prediction about the effect of temperature on an equilibrium if you know whether the reaction is exothermic or endothermic. As an example, consider the endothermic, gas phase reaction of N_2 with O_2 to give nitrogen monoxide, NO.

For this and other endothermic reactions, the equilibrium constant increases as the temperature increases.

$$N_2(g) + O_2(g) \rightleftharpoons 2\ NO(g) \qquad \Delta H^\circ = 180.5 \text{ kJ}$$

$$K_c = \frac{[NO]^2}{[N_2][O_2]} \qquad \begin{array}{ll} K_c = 4.5 \times 10^{-31} & \text{at 298 K} \\ K_c = 6.7 \times 10^{-10} & \text{at 900 K} \\ K_c = 1.7 \times 10^{-3} & \text{at 2300 K} \end{array}$$

In this case the equilibrium constant increases very significantly as the temperature increases. At 298 K the equilibrium constant is so small that essentially no reaction occurs. Suppose that a room-temperature equilibrium mixture were suddenly heated to 2300 K. What would happen? The equilibrium should shift to partially compensate for the temperature increase. This happens if the reaction shifts in the endothermic direction, since that change would involve a transfer of energy into the reaction system, cooling the surroundings. For the $N_2 + O_2$ reaction the forward process is endothermic ($\Delta H^\circ > 0$), so at the higher temperature N_2 and O_2 should react to produce more NO. At the new equilibrium, the concentration of NO should be higher and the concentrations of N_2 and O_2 should be lower. Since this makes the numerator in the K_c expression bigger and the denominator smaller, K_c should be larger at the higher temperature. This outcome corresponds with the experimental result.

The effect on K_c of raising the temperature of the $N_2 + O_2$ reaction leads us to a general conclusion:

- ***For an endothermic reaction, an increase in temperature always means an increase in K_c; the reaction will become more product-favored at higher temperatures.***

When equilibrium is achieved at the higher temperature, the concentration of products is greater and that of the reactants is smaller. Likewise, and as illustrated by Problem-Solving Example 14.8, the opposite is true for an exothermic reaction:

- ***For an exothermic reaction, an increase in temperature always means a decrease in K_c; the reaction will become less product-favored at higher temperatures.***

The effect of temperature on the reaction of N_2 with O_2 has important consequences. This reaction produces NO in Earth's atmosphere when lightning suddenly raises the temperature of the air along its path. Because the reverse reaction is slow at room temperature and because after the lightning bolt is over, the air rapidly cools to normal temperatures, much of the NO that is produced does not react to re-form N_2 and O_2 as it would at equilibrium. This situation provides one natural mechanism by which nitrogen in the air can be converted into a form that can be used by plants. (Converting nitrogen into a useful form is called nitrogen fixation; see Section 21.3.) Humans have tried to use the same kind of process to produce NO and from it HNO_3 for use in manufacturing fertilizer. About a century ago scientists and many in the general public were worried that Earth's farmland could not grow enough food to support a growing population. Consequently, strenuous efforts were made to adjust the conditions of the $N_2 + O_2$ reaction so that significant yields of NO could be obtained. At the end of the nineteenth century, a chemical plant at Niagara Falls, New York (where there was plentiful electric power), operated an electric arc process for fixing nitrogen for several years. This electric arc plant was important because it was the first attempt to deal with the limitations on plant growth caused by lack of sufficient nitrogen in soils that had been heavily farmed.

Another consequence of the shift toward products in the $N_2 + O_2$ reaction at high temperatures is that automobile engines emit small concentrations of NO. The NO is rapidly oxidized to brown NO_2 in the air above cities, and the NO_2 in turn produces many further reactions that create air pollution problems. Air pollution involving NO was discussed in Section 10.10 (← ***p. 457***). One of the functions of catalytic converters in automobiles is to speed up reduction of these nitrogen oxides back to elemental nitrogen (← ***p. 635***).

PROBLEM-SOLVING EXAMPLE 14.8 Le Chatelier's Principle

Consider an equilibrium mixture of nitrogen, hydrogen, and ammonia in which the reaction is

$$N_2(g) + 3\,H_2(g) \rightleftharpoons 2\,NH_3(g) \qquad \Delta H^\circ = -92.2 \text{ kJ} \quad \text{at } 25\ ^\circ\text{C}$$

For each of the changes listed below, tell whether the value of K_c increases or decreases, and tell whether more NH_3 or less NH_3 is present at the new equilibrium established after the change.

(a) More H_2 is added (at a constant temperature of 25 °C and constant volume).
(b) The temperature is increased.
(c) The volume of the container is doubled (at constant temperature).

Answer

(a) K_c stays the same; more NH_3 is present. (b) K_c decreases; less NH_3 is present.
(c) K_c stays the same; less NH_3 is present.

Strategy and Explanation

(a) Since the temperature does not change, the value of K_c does not change. Adding a reactant to the equilibrium mixture will shift the equilibrium toward the product, producing more NH_3. This can be seen in another way by considering the reaction quotient.

$$Q = \frac{(\text{conc. } NH_3)^2}{(\text{conc. } N_2)(\text{conc. } H_2)^3}$$

When more H_2 is added, the denominator becomes larger. This makes Q smaller than K_c, which predicts that the reaction should produce more product; that is, some of the added H_2 reacts with N_2 to make more NH_3. Notice also that the concentration of N_2 decreases, because some reacts with the H_2, but the concentration of H_2 in the new equilibrium is still greater than it was before.

(b) The reaction is exothermic ($\Delta H^\circ < 0$). Increasing the temperature shifts the equilibrium in the endothermic direction—that is, to the left (toward the reactants). This shift leads to a decrease in the NH_3 concentration, an increase in the concentrations of H_2 and N_2, and a decrease in the value of K_c.

(c) Since the temperature is constant, the value of K_c must be constant. Doubling the volume should cause the reaction to shift toward a greater number of moles of gaseous substance—that is, toward the left. Doubling the volume would normally halve the pressure, but the shift of the equilibrium partially compensates for this effect, and the final equilibrium will be at a pressure somewhat more than half the pressure of the initial equilibrium.

PROBLEM-SOLVING PRACTICE 14.8

Consider the equilibrium between N_2O_4 and NO_2 in a closed system.

$$N_2O_4(g) \rightleftharpoons 2\ NO_2(g)$$

Draw Lewis structures for the molecules involved in this equilibrium. Based on the bonding in the molecules, predict whether the reaction is exothermic or endothermic; hence, predict whether the concentration of N_2O_4 is larger in an equilibrium system at 25 °C or at 80 °C. Verify your prediction by looking at Figure 14.3.

ESTIMATION Generating Gaseous Fuel

The reaction of coke (mainly carbon) with steam is called the water-gas reaction. It was used for many years to generate gaseous fuel from coal. The thermochemical expression is

$$C(s) + H_2O(g) \rightleftharpoons CO(g) + H_2(g) \qquad \Delta H° = 131.293\ \text{kJ}$$

The equilibrium constant K_P has the value 9.5×10^{-17} at 298 K, 1.9×10^{-7} at 500 K, 2.60×10^{-2} at 800 K, 18.8 at 1200 K, and 500. at 1600 K. Suppose that you have an equilibrium mixture in which the partial pressure of steam is 1.00 atm. Estimate the temperature at which the partial pressures of CO and H_2 would also equal 1 atm. (The reaction would need to be carried out at temperatures roughly this high to produce appreciable quantities of products.)

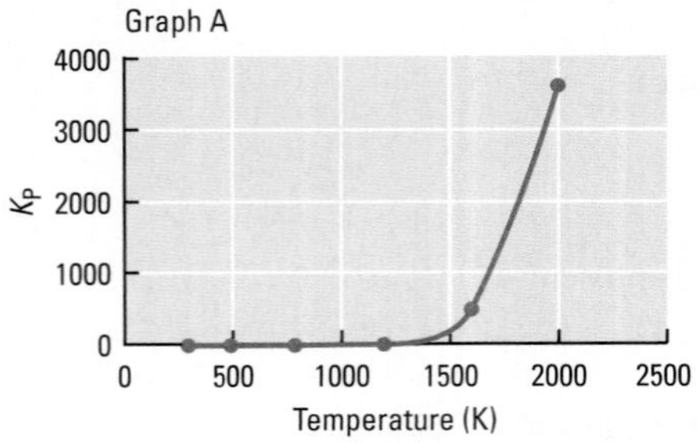

A good way to make this estimation is to use a spreadsheet program to plot the data and see which temperature corresponds to $K_P = 1$. This has been done in Graph A. It is clear that the range of values for the equilibrium constant is so wide that an accurate temperature cannot be read corresponding to $K_P = 1$. An estimate would be easier to make if the graph were a straight line. Because the values of K_P rise very rapidly, it is possible that taking the logarithm of each K_P value would generate a linear graph. This has been done in Graph B, where the natural logarithm of the equilibrium constant, $\ln(K_P)$, has been plotted on the vertical axis. Since $\ln(K_P) = 0$ when $K_P = 1$, we are looking for the temperature at which the graph crosses zero on the vertical axis. Although the graph is not linear, the appropriate temperature can be read from the graph and is slightly less than 1000 K.

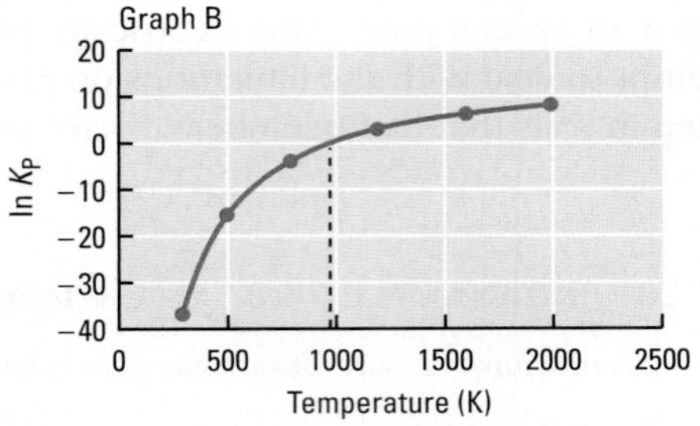

Experimenting with different functions of K_P on the vertical axis and different functions of T on the horizontal axis reveals that a linear graph is obtained when $\ln(K_P)$ is plotted against $1/T$, as in Graph C. (Note that the spreadsheet graph displays too many significant figures.) From this graph the equation of the straight line is found to be

$$\ln(K_P) = (-1.58 \times 10^4\ \text{K})\left(\frac{1}{T}\right) + 16.1$$

Substituting $K_P = 1$, that is, $\ln(K_P) = 0$, we find that

$$T = \frac{1.58 \times 10^4\ \text{K}}{16.1} = 980\ \text{K}$$

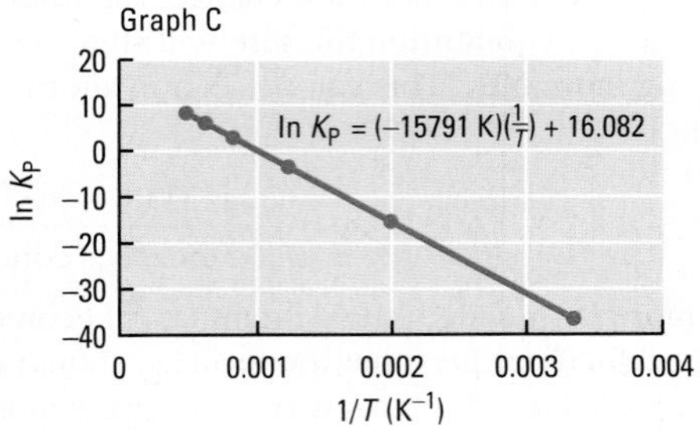

Visit this book's companion website at www.cengage.com/chemistry/moore to work an interactive module based on this material.

CONCEPTUAL EXERCISE 14.12 Summarizing Le Chatelier's Principle

Construct a table to summarize your understanding of Le Chatelier's principle. Consider these changes in conditions:

(a) Addition of a reactant
(b) Removal of a reactant
(c) Addition of a product
(d) Removal of a product
(e) Increasing pressure by decreasing volume
(f) Decreasing pressure by increasing volume
(g) Increasing temperature
(h) Decreasing temperature

For each of these changes in conditions, indicate (1) how the reaction system changes to achieve a new equilibrium, (2) in which direction the equilibrium reaction shifts, and (3) whether the value of K_c changes and, if so, in which direction. For some of these changes there are qualifications. For example, increasing pressure by decreasing volume does not always shift an equilibrium. List as many of these qualifications as you can.

14.7 Equilibrium at the Nanoscale

In Section 14.2 (p. 659), we used the isomerization of *cis*-2-butene to show that both forward and reverse reactions occur simultaneously in an equilibrium system.

$$cis\text{-2-butene(g)} \rightleftharpoons trans\text{-2-butene(g)} \qquad K_c = \frac{[trans]}{[cis]} = 2.0 \qquad \text{(at 415 K)}$$

Because the equilibrium constant is 2.0, there are twice as many *trans* molecules as *cis* molecules in an equilibrium mixture at 415 K. In other words, two thirds of the molecules have the *trans* structure and one third have the *cis* structure. Because the molecules are continually reacting in both the forward and the reverse directions, another way to think about this situation is that each molecule is *trans* two thirds of the time and *cis* one third of the time.

Based on its molecular structure, you might think that a 2-butene molecule ought to be just as likely to be *cis* as *trans,* so why is the *trans* isomer favored at 415 K? In Figure 13.7 **(← p. 611)** we noted that one molecule of the *trans* isomer is 7×10^{-21} J lower in energy than one molecule of the *cis* isomer, and this difference is all important. For a mole of molecules, the difference in energy is $(-7 \times 10^{-21}\ \text{J})(6.022 \times 10^{23}\ \text{mol}^{-1}) = -4 \times 10^3\ \text{J/mol} = -4\ \text{kJ/mol}$, giving the thermochemical expression

$$cis\text{-2-butene(g)} \rightleftharpoons trans\text{-2-butene(g)} \qquad \Delta H° = -4\ \text{kJ}$$

Consider the rate constants for the forward and reverse reactions in the isomerization of 2-butene. Based on Figure 13.7, E_a (forward) = 262 kJ/mol and E_a (reverse) = 266 kJ/mol. This means that the rate constant for the reverse reaction will be smaller than the rate constant for the forward reaction. At equilibrium the forward and reverse rates are equal, so

Larger *k* value and smaller concentration → k_{forward}

Smaller *k* value and larger concentration → $[trans]$

$$k_{\text{forward}} \times [cis] = k_{\text{reverse}} \times [trans]$$

If k_{reverse} is smaller than k_{forward}, then the concentration of *trans* must be larger than the concentration of *cis,* or the rates would not be equal. Because *cis*-2-butene is 4 kJ/mol higher in energy than *trans*-2-butene, it occurs only half as often as *trans*-2-butene at 415 K. We can generalize that ***in an equilibrium system, molecules that are higher in energy occur less often.*** We shall refer to this tendency as the energy effect on the position of equilibrium.

A second energy factor also affects how large an equilibrium constant is. It depends on how spread out in space the total energy of products is compared with the total energy of reactants. This factor can be illustrated by the dissociation of dinitrogen tetraoxide.

$$N_2O_4(g) \rightleftharpoons 2\ NO_2(g) \qquad \Delta H° = 57.2\ kJ$$

In a constant-pressure system (such as a reaction at atmospheric pressure), the two moles of product molecules occupy twice the volume of the one mole of reactant molecules. This means that the energy of the NO_2 molecules is spread over twice the volume that would have been occupied by the N_2O_4 molecules. A thermodynamic quantity called **entropy** provides a quantitative measure of how much energy is spread out when something happens. Entropy will be defined more completely in Section 18.3. For now it suffices to say that ***if there are more product molecules than reactant molecules, entropy favors the products in an equilibrium system.*** Although it depends on spreading out of energy, we shall refer to this tendency as the entropy effect on the position of equilibrium.

Despite the fact that the entropy factor favors NO_2 in the dissociation of N_2O_4, when data from Table 14.1 are used to calculate K_c at 25 °C (298 K) the result is $1/(1.7 \times 10^2) = 5.9 \times 10^{-3}$. This is because the dissociation of N_2O_4 is endothermic ($\Delta H° = 57.2$ kJ). The enthalpy of 2 mol NO_2 is 57.2 kJ higher than the enthalpy of 1 mol N_2O_4. From our first generalization that molecules with more energy occur less often in an equilibrium system, we expect that 2 NO_2 is less likely than N_2O_4 because of the energy effect. Therefore the equilibrium constant should be smaller than the entropy argument predicts. *Both the energy effect and the entropy effect must be taken into account to predict an equilibrium constant value.*

$N_2O_4(g) \rightleftharpoons 2\ NO_2(g)$

T (K)	K_c
298	5.9×10^{-3}
350	1.3×10^{-1}
400	1.5×10^{0}
500	4.6×10^{1}
600	4.6×10^{2}

These ideas can also help us to understand the temperature dependence of the equilibrium constant. K_c values for the dissociation of N_2O_4 are given in the table in the margin. As the temperature rises, K_c becomes much larger. At high temperatures the molecules have lots of energy, and when the volume doubles, much more energy is dispersed. The energy difference between reactants and products (ΔH) is smaller relative to the average energy per molecule. ***The higher the temperature is, the less important the energy effect (ΔH) becomes, and the more the entropy effect (dispersal of energy) determines the position of equilibrium.***

We have shown that if the number of gas phase molecules increases when a reaction occurs, then the products will be favored by entropy. This is not the only way for the products of a reaction to have greater entropy than the reactants, but it is one of the most important. In Chapter 18 we will show how entropy can be measured and tabulated. That discussion will enable us to use enthalpy changes and entropy changes for a reaction to calculate its equilibrium constant at a variety of temperatures.

PROBLEM-SOLVING EXAMPLE 14.9 Energy and Entropy Effects on Equilibria

For the equilibrium

$$CH_4(g) + H_2O(g) \rightleftharpoons CO(g) + 3\ H_2(g)$$

(a) Estimate whether the entropy increases, decreases, or remains the same when products form.
(b) Does the entropy effect favor reactants or products?
(c) Use data from Appendix J to calculate $\Delta H°$.
(d) Does the energy effect favor reactants or products?
(e) Is the reaction likely to be product-favored at high temperatures? Why or why not?

Answer

(a) Entropy increases
(b) Entropy favors products
(c) $\Delta H° = 206.10$ kJ
(d) Reactants
(e) Yes. Entropy favors products and the energy effect is less important at high temperatures.

Strategy and Explanation Consider how many gas-phase reactants and products there are and apply the ideas about entropy in the previous paragraphs.

(a) Because there are 4 mol gas-phase products and only 2 mol gas phase reactants, the products will have higher entropy.

(b) The products have higher entropy, and the products are favored.

(c) $\Delta H^\circ = \sum\{(\text{moles product})\Delta H_f^\circ(\text{product})\} - \sum\{(\text{moles reactant})\Delta H_f^\circ(\text{reactant})\}$

$= \{(1\ \text{mol})(-110.525\ \text{kJ/mol})\}$

$- \{(1\ \text{mol})(-74.81\ \text{kJ/mol}) + (1\ \text{mol})(-241.818\ \text{kJ/mol})\}$

$= -110.525\ \text{kJ} - (-316.628\ \text{kJ})$

$= 206.10\ \text{kJ}$

(d) Because the reactants are lower in energy, they are favored by the energy effect.

(e) The reaction is product-favored at high temperatures because the energy effect favoring reactants becomes less important, and there is a relatively large entropy effect (four product molecules for every two reactant molecules in the gas phase).

☑ **Reasonable Answer Check** In part (c) it is reasonable that the reaction is endothermic, because six bonds are broken and only four are formed when reactant molecules are changed into product molecules (*← p. 238*).

PROBLEM-SOLVING PRACTICE 14.9

For the ammonia synthesis reaction

$$N_2(g) + 3\ H_2(g) \rightleftharpoons 2\ NH_3(g) \qquad \Delta H^\circ = -92.2\ \text{kJ}$$

(a) Does the entropy effect favor products?

(b) Does the energy effect favor products?

(c) Will there be a greater concentration of $NH_3(g)$ at high temperature or at low temperature? Explain.

14.8 Controlling Chemical Reactions: The Haber-Bosch Process

The principles that allow us to control a reaction are based on our understanding of both equilibrium systems and the rates of chemical reactions. Some generalizations about equilibrium systems are

- ***A product-favored reaction has an equilibrium constant larger than 1.***
- ***If a reaction is exothermic, this energy factor favors the products.***
- ***If there is an increase in entropy when a reaction occurs, this entropy factor favors the products.***
- ***Product-favored reactions at low temperatures are usually exothermic.***
- ***Product-favored reactions at high temperatures are usually ones in which the entropy increases (energy becomes more dispersed).***

Using these general rules about equilibria, we can often predict whether a reaction is capable of yielding products. But it is also important that those products be produced rapidly. Recall these useful generalizations about reaction rates from Chapter 13.

- ***Reactions in the gas phase or in solution, where molecules of one reactant are completely mixed with molecules of another, occur more rapidly than do reactions between pure liquids or solids that do not dissolve in one another (← p. 593).***
- ***Reactions occur more rapidly at high temperatures than at low temperatures (← p. 615).***
- ***Reactions are faster when the reactant concentrations are high than when they are low (← p. 598).***

For gases, higher concentration corresponds to higher partial pressure.

Because finely divided solids react more rapidly, coal is ground to a powder before it is burned to generate electricity.

The current interest in biotechnology is largely driven by the fact that naturally occurring enzymes are among the most effective catalysts known *(← p. 629)*.

- ***Reactions between a solid and a gas, or between a solid and something dissolved in solution, are usually much faster when the solid particles are as small as possible (← p. 593).***
- ***Reactions are faster in the presence of a catalyst. Often the right catalyst makes the difference between success and failure in industrial chemistry (← p. 625).***

One of the best examples of the application of the principles of chemical reactivity is the chemical reaction used for the synthesis of ammonia from its elements. Even though Earth is bathed in an atmosphere that is about 80% N_2 gas, nitrogen cannot be used by most plants until it has been fixed—that is, converted into biologically useful forms. Although nitrogen fixation is done naturally by organisms such as cyanobacteria and some field crops such as alfalfa and soybeans, most plants cannot fix N_2. They must instead obtain nitrogen from cyanobacteria, some other organism, or fertilizer. Proper fertilization is especially important for recently developed varieties of wheat, corn, and rice that have resulted in much improved food production.

It is estimated that 40% to 60% of the nitrogen in the average human body has come from ammonia produced by the Haber-Bosch process and that increased agricultural productivity resulting from this process supports about 40% of the world's population.

Direct combination of nitrogen and oxygen was used at the beginning of the twentieth century to provide fertilizer, but this process was not very efficient (p. 685). A much better way of manufacturing ammonia was devised by Fritz Haber and Carl Bosch, who chose the *direct synthesis of ammonia from its elements* as the basis for an industrial process.

$$N_2(g) + 3\,H_2(g) \rightleftharpoons 2\,NH_3(g)$$

CONCEPTUAL EXERCISE 14.13 Ammonia Synthesis

For the ammonia synthesis reaction, predict

(a) Whether the reaction is exothermic or endothermic.
(b) Whether the reaction product is favored by entropy.
(c) Whether the reaction produces more products at low or high temperatures.
(d) What would happen if you tried to increase the rate of the reaction by increasing the temperature.

Science Museum/Science & Society Picture Library

Fritz Haber

1868–1934

The industrial chemical process by which ammonia is manufactured was developed by Fritz Haber, a chemist, and Carl Bosch, an engineer. Haber's studies in the early 1900s revealed that direct ammonia synthesis should be possible. In 1914 the engineering problems were solved by Bosch. Haber's contract with the manufacturer of ammonia called for him to receive 1 pfennig (one hundredth of a German mark—similar to a penny) per kilogram of ammonia, and he soon became not only famous but also rich. In 1918 he was awarded a Nobel Prize in Chemistry for the ammonia synthesis, but the choice was criticized because of his role in developing the use of poison gases for Germany during World War I.

At first glance this reaction might seem to be a poor choice. Hydrogen is available naturally only in combined form—for example, in water or hydrocarbons—meaning that hydrogen must be extracted from these compounds at considerable expense in energy resources and money. As you discovered in Problem-Solving Practice 14.9, the ammonia synthesis reaction becomes less product-favored at higher temperatures. But higher temperatures are needed for ammonia to be produced fast enough for the process to be efficient and economical. Nonetheless, the **Haber-Bosch process** (shown schematically in Figure 14.10) has been so well developed that ammonia is inexpensive (less than $600 per ton). For this reason it is widely used as a fertilizer and is often among the top five chemicals produced in the United States. Annual U.S. production of NH_3 by the Haber-Bosch process is approximately 20 billion pounds, and about 12 billion pounds is imported each year.

Both the thermodynamics and the kinetics of the direct synthesis of ammonia have been carefully studied and fine-tuned by industry so that the maximum yield of product is obtained in a reasonable time and at a reasonable cost of both money and energy resources.

- The reaction is exothermic, and there is a decrease in entropy when it takes place. Therefore this reaction is predicted to be product-favored at low temperatures, but reactant-favored at high temperatures. (You should have made this prediction in Problem-Solving Practice 14.9, and it is in accord with Le Chatelier's principle for an exothermic process.)

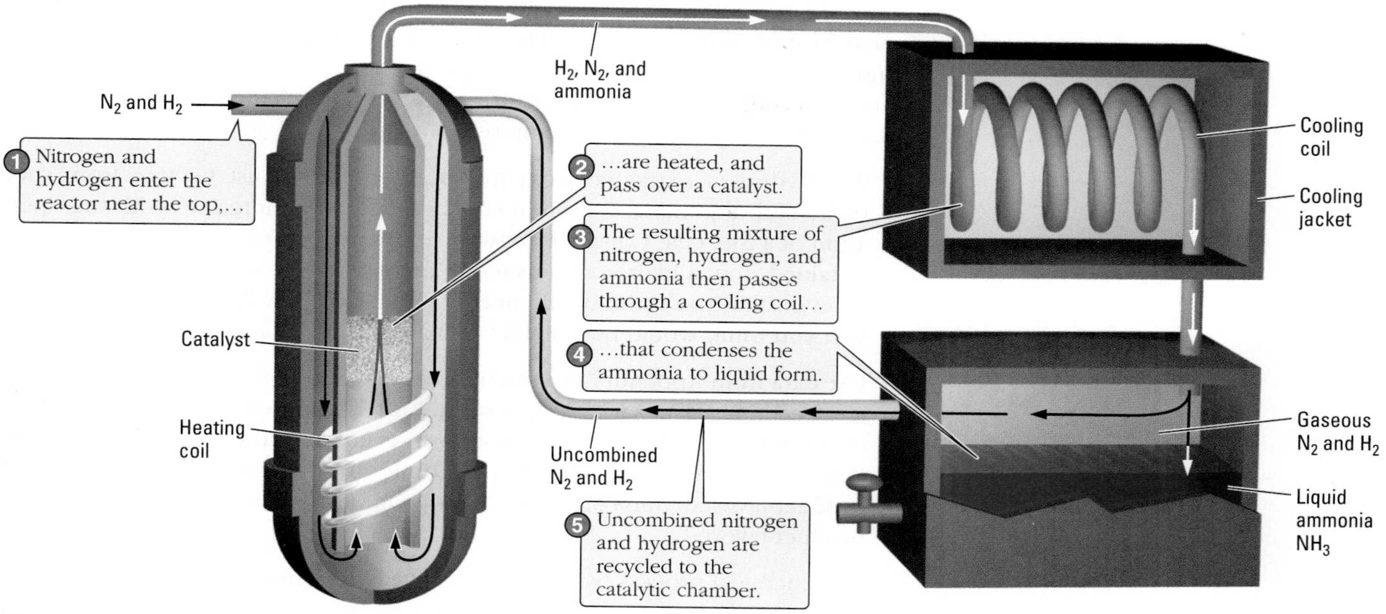
H$_2$, N$_2$, and ammonia
N$_2$ and H$_2$
1 Nitrogen and hydrogen enter the reactor near the top,...
2 ...are heated, and pass over a catalyst.
3 The resulting mixture of nitrogen, hydrogen, and ammonia then passes through a cooling coil...
4 ...that condenses the ammonia to liquid form.
5 Uncombined nitrogen and hydrogen are recycled to the catalytic chamber.
Catalyst
Heating coil
Uncombined N$_2$ and H$_2$
Cooling coil
Cooling jacket
Gaseous N$_2$ and H$_2$
Liquid ammonia NH$_3$

Figure 14.10 Haber-Bosch process for synthesis of ammonia (schematic).

- To increase the equilibrium concentration of NH_3, the reaction is carried out at high pressure (200 atm). This does not change the value of K_c, or K_P, but an increase in pressure can be compensated for by converting N_2 and H_2 to NH_3; 2 mol NH_3(g) exerts less pressure than a total of 4 mol gaseous reactants [N_2(g) + 3 H_2(g)] in the same-sized container.
- Ammonia is continually liquefied and removed from the reaction vessel, which reduces the concentration of the product of the reaction and shifts the equilibrium toward the right.
- The reaction is quite slow at room temperature, so the temperature must be raised to increase the rate. Although the rate increases with increasing temperature, the equilibrium constant declines. Thus, the faster the reaction, the smaller the yield.
- The temperature cannot be raised too much in an attempt to increase the rate, but a rate increase can be achieved with a catalyst. An effective catalyst for the Haber-Bosch process is Fe_3O_4 mixed with KOH, SiO_2, and Al_2O_3. Since the catalyst is not effective below about 400 °C, the optimum temperature, considering all the factors controlling the reaction, is about 450 °C.

Making predictions about chemical reactivity is part of the challenge, the adventure, and the art of chemistry. Many chemists enjoy the challenge of making useful new materials, which usually means choosing to make them by reactions that we believe will be product-favored and reasonably rapid. Such predictions are based on the ideas outlined in this chapter and Chapter 13.

SUMMARY PROBLEM

One approach to achieving cleaner-burning fuels and more-efficient automobiles is to extract hydrogen from gasoline and other liquid fossil fuels. The extracted hydrogen could be combined with oxygen in fuel cells (Section 19.9) like those currently used in spacecraft to generate electricity. The electricity could be used in electric motors to power automobiles and to provide air conditioning and other amenities expected by auto buyers. Because electric motors are far more efficient than current automobile engines, such a car might get 80 miles per gallon of fuel. By answering these questions, you can explore how the ideas of chemical equilibrium and chemical kinetics can be applied to motive power for cars.

1. The hydrogen extracted from hydrocarbon fuels must be free from soot (solid carbon) and carbon monoxide, which would interfere with the operation of a fuel cell. Consider possible reactions by which hydrogen could be obtained from a hydrocarbon such as octane, C_8H_{18}. Write an equation for a reaction that you think would not be suitable and write an equation for one that you think would be suitable. Explain your choice in each case.

2. Use data from Appendix J to calculate the change in enthalpy for each of the two reactions you wrote in Question 1. Predict whether entropy increases or decreases when each reaction occurs. Is either of them ruled out because it is not product-favored? If not, continue. If either of the reactions is not product-favored, think about whether conditions could be altered to make it product-favored.

3. The chemical process by which hydrogen is obtained for use in synthesizing ammonia (Haber-Bosch process) involves treating methane (from natural gas) with steam. The first step in this process is

$$CH_4(g) + H_2O(g) \rightleftharpoons CO(g) + 3\ H_2(g)$$

(a) Write the equilibrium constant expression for this reaction.

(b) What is the relation between K_c and K_P for this reaction?

(c) Calculate the enthalpy change for this reaction.

(d) Based on the equation, predict the sign of the entropy change for this reaction.

(e) Is the reaction product-favored at high temperatures but not at lower temperatures? Or the other way around? Explain.

4. To remove carbon monoxide from the hydrogen destined for the Haber-Bosch process, this reaction is used:

$$CO(g) + H_2O(g) \rightleftharpoons CO_2(g) + H_2(g)$$

(a) Use bond enthalpies to estimate the enthalpy change for this reaction.

(b) At 2300 K, K_P for this reaction is 2.643×10^{-3}. Calculate K_c.

(c) Suppose that 0.100 mol CO and 0.100 mol H_2O were introduced into an empty 10.0-L flask at 450 °C and the flask was sealed. Determine the concentration of $H_2(g)$ in the flask once equilibrium has been achieved.

(d) What is the concentration of CO(g) remaining in the flask in part (c)? Is it low enough that we can say that the hydrogen is free of carbon monoxide?

5. If you were in charge of designing a system for generating hydrogen gas for use in the Haber-Bosch process, how might you obtain pure hydrogen? Assume that the process is based on the reactions given in Questions 3 and 4. Suggest a chemical reagent that could be used to react with $CO_2(g)$ and thereby remove it from the hydrogen generated as a product in each of the two reactions. Would this same reagent work if you needed to remove $SO_2(g)$?

6. To get the highest-purity $H_2(g)$ and the maximum yield from the hydrogen-generating process, what reactant concentration(s) would you increase or decrease? How would you adjust product concentrations?

7. In the hypothetical fuel cell system for an electric automobile, hydrocarbon fuel is vaporized and partially oxidized in a limited quantity of air. In a second step the products of the first reaction are treated with steam over copper oxide and zinc oxide catalysts. In a final purification step, more air is introduced and a platinum catalyst helps convert carbon monoxide to carbon dioxide. Write a balanced chemical equation for each of these three steps. Assume that the hydrocarbon fuel is octane.

8. What are the advantages of generating hydrogen gas from hydrocarbon fuel in an automobile rather than storing hydrogen in a fuel tank? What are the disadvantages of storing hydrogen in a car? What advantages does combining hydrogen with oxygen in a fuel cell have, as opposed to burning a hydrocarbon fuel in an internal combustion engine?

IN CLOSING

Having studied this chapter, you should be able to . . .

- Recognize a system at equilibrium and describe the properties of equilibrium systems (Section 14.1). End-of-chapter question: 10
- Describe the dynamic nature of equilibrium and the changes in concentrations of reactants and products that occur as a system approaches equilibrium (Sections 14.1, 14.2). Questions 8, 10, 12
- Write equilibrium constant expressions, given balanced chemical equations (Section 14.2). Questions 14, 16, 17, 19, 21, 23, 25
- Obtain equilibrium constant expressions for related reactions from the expression for one or more known reactions (Section 14.2). Questions 27, 29, 31, 35
- Calculate K_P from K_c or K_c from K_P for the same equilibrium (Section 14.2). Questions 33, 37
- Calculate a value of K_c for an equilibrium system, given information about initial concentrations and equilibrium concentrations (Section 14.3). Questions 39, 41, 43, 45, 47, 48, 50, 52, 54
- Make qualitative predictions about the extent of reaction based on equilibrium constant values—that is, predict whether a reaction is product-favored or reactant-favored based on the size of the equilibrium constant (Section 14.4). Questions 55, 57, 58
- Calculate concentrations of reactants and products in an equilibrium system if K_c and initial concentrations are known (Section 14.5). Questions 60, 62, 64, 66, 67, 69, 71
- Use the reaction quotient Q to predict in which direction a reaction will go to reach equilibrium (Section 14.5). Questions 72, 73, 76, 78
- Use Le Chatelier's principle to show how changes in concentrations, pressure or volume, and temperature shift chemical equilibria (Section 14.6). Questions 80, 82, 84, 85, 87, 89, 91, 93, 95, 97
- Use the change in enthalpy and the change in entropy qualitatively to predict whether products are favored over reactants (Section 14.7). Questions 99, 101
- List the factors affecting chemical reactivity, and apply them to predicting optimal conditions for producing products (Section 14.8). Questions 103, 105

OWL and go Chemistry

Sign in at **www.cengage.com/owl** to:

- View tutorials and simulations, develop problem-solving skills, and complete online homework assigned by your professor.
- For quick review and exam prep, download Go Chemistry mini lecture modules from OWL (or purchase them at **www.CengageBrain.com**).

KEY TERMS

chemical equilibrium *(Section 14.1)*

dynamic equilibrium *(14.1)*

entropy *(14.7)*

equilibrium concentration *(14.3)*

equilibrium constant *(14.2)*

equilibrium constant expression *(14.2)*

Haber-Bosch process *(14.8)*

Le Chatelier's principle *(14.6)*

product-favored *(14.1)*

reactant-favored *(14.1)*

reaction quotient *(14.5)*

shifting an equilibrium *(14.6)*

QUESTIONS FOR REVIEW AND THOUGHT

OWL Interactive versions of these problems are assignable in OWL.

Blue-numbered questions have short answers at the back of this book in Appendix M and fully worked solutions in the *Student Solutions Manual.*

Review Questions

These questions test vocabulary and simple concepts.

1. Define the terms *chemical equilibrium* and *dynamic equilibrium.*
2. If an equilibrium is product-favored, is its equilibrium constant large or small with respect to 1? Explain.
3. List three characteristics that you would need to verify in order to determine that a chemical system is at equilibrium.
4. Decomposition of ammonium dichromate is shown in the series of photos below. In a closed container this process reaches an equilibrium state. Write a balanced chemical equation for the equilibrium reaction. How is the equilibrium affected if
   (a) More ammonium dichromate is added to the equilibrium system?
   (b) More water vapor is added?
   (c) More chromium(III) oxide is added?

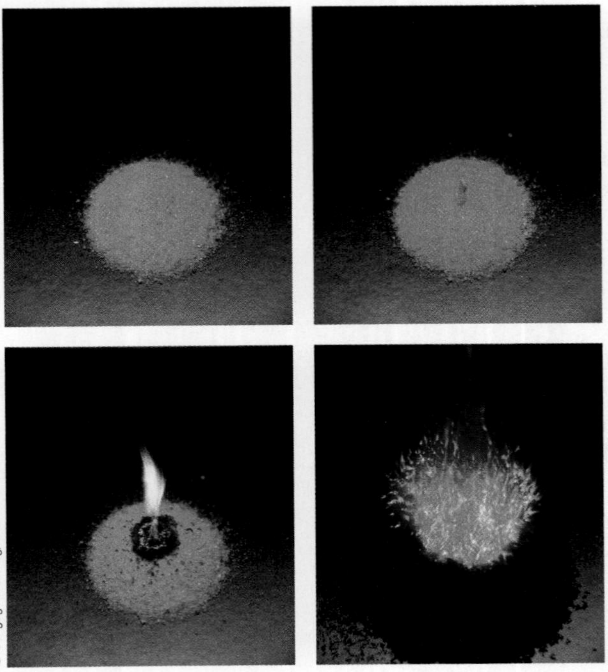

© Cengage Learning/Charles D. Winters

Decomposition of $(NH_4)_2Cr_2O_7$. Orange, solid $(NH_4)_2Cr_2O_7$ (upper left) can be ignited with a wick (upper right), which initiates decomposition (lower left) forming Cr_2O_3 (the dark green solid, lower right), N_2 gas, and water vapor. Energy is transferred to the surroundings by the process.

5. For the equilibrium reaction in Question 4, write the expression for the equilibrium constant.
   (a) How would this equilibrium constant change if the total pressure on the system were doubled?
   (b) How would the equilibrium constant change if the temperature were increased?
6. Indicate whether each statement below is true or false. If a statement is false, rewrite it to produce a closely related statement that is true.
   (a) For a given reaction, the magnitude of the equilibrium constant is independent of temperature.
   (b) If there is an increase in entropy and a decrease in enthalpy when reactants in their standard states are converted to products in their standard states, the equilibrium constant for the reaction will be negative.
   (c) The equilibrium constant for the reverse of a reaction is the reciprocal of the equilibrium constant for the reaction itself.
   (d) For the reaction

$$H_2O_2(\ell) \rightleftharpoons H_2O(\ell) + \tfrac{1}{2}\,O_2(g)$$

   the equilibrium constant is one-half the magnitude of the equilibrium constant for the reaction

$$2\,H_2O_2(\ell) \rightleftharpoons 2\,H_2O(\ell) + O_2(g)$$

7. If the reaction quotient is larger than the equilibrium constant, what will happen to the reaction as it approaches equilibrium? What will happen if $Q < K$?

Topical Questions

These questions are keyed to the major topics in the chapter. Usually a question that is answered at the back of this book is paired with a similar one that is not.

Characteristics of Chemical Equilibrium (Section 14.1)

8. Think of an experiment you could do to demonstrate that the equilibrium

$$2\,NO_2(g) \rightleftharpoons N_2O_4(g)$$

is a dynamic process in which the forward and reverse reactions continue to occur after equilibrium has been achieved. Describe how such an experiment might be carried out.

9. Discuss this statement: "No true chemical equilibrium can exist unless reactant molecules are constantly changing into product molecules, and vice versa."

10. Suppose you place a large piece of ice into a well-insulated thermos with some water in it, and it comes to equilibrium with part of the ice melted.
(a) What is the temperature of the equilibrium system?
(b) Is this a static or a dynamic equilibrium? Explain.

11. The atmosphere consists of about 80% N_2 and 20% O_2, yet there are many oxides of nitrogen that are stable and can be isolated in the laboratory.
(a) Is the atmosphere at chemical equilibrium with respect to forming NO?
(b) If not, why doesn't NO form? If so, how is it that NO can be made and kept in the laboratory for long periods?

The Equilibrium Constant (Section 14.2)

12. Consider the gas phase reaction of $N_2 + O_2$ to give 2 NO and the reverse reaction of 2 NO to give $N_2 + O_2$, discussed in Section 14.2. An equilibrium mixture of NO, N_2, and O_2 at 5000. K that contains equal concentrations of N_2 and O_2 has a concentration of NO about half as great. Make qualitatively correct plots of the concentrations of reactants and products versus time for these two processes, showing the initial state and the final dynamic equilibrium state. Assume a temperature of 5000. K. Don't do any calculations—just sketch how you think the plots will look.

13. After 0.1 mol of pure *cis*-2-butene is allowed to come to equilibrium with *trans*-2-butene in a closed 5.0-L flask at 25 °C, another 0.1 mol *cis*-2-butene is suddenly added to the flask.
(a) Is the new mixture at equilibrium? Explain why or why not.
(b) In the new mixture, immediately after addition of the *cis*-2-butene, which rate is faster: *cis* → *trans* or *trans* → *cis*? Or are both rates the same?
(c) After the second 0.1 mol *cis*-2-butene has been added and the system is at equilibrium, if the concentration of *trans*-2-butene is 0.01 mol/L, what is the concentration of *cis*-2-butene?

14. Write equilibrium constant expressions for these reactions. For gases, use either pressures or concentrations.
(a) $3\,O_2(g) \rightleftharpoons 2\,O_3(g)$
(b) $Fe(s) + 5\,CO(g) \rightleftharpoons Fe(CO)_5(g)$
(c) $(NH_4)_2CO_3(s) \rightleftharpoons 2\,NH_3(g) + CO_2(g) + H_2O(g)$
(d) $Ag_2SO_4(s) \rightleftharpoons 2\,Ag^+(aq) + SO_4^{2-}(aq)$

15. Write equilibrium constant expressions for these reactions:
(a) $CH_4(g) + H_2O(\ell) \rightleftharpoons CO(g) + 3\,H_2(g)$
(b) $4\,NH_3(g) + 5\,O_2(g) \rightleftharpoons 4\,NO(g) + 6\,H_2O(g)$
(c) $BaCO_3(s) \rightleftharpoons BaO(s) + CO_2(g)$
(d) $NH_3(g) + HCl(g) \rightleftharpoons NH_4Cl(s)$

16. Write the expression for K_c for each of these reactions:
(a) $CO_2(g) + H_2(g) \rightleftharpoons CO(g) + H_2O(g)$
(b) $SrCO_3(s) \rightleftharpoons SrO(s) + CO_2(g)$
(c) $2\,CHCl_3(g) + 3\,H_2(g) \rightleftharpoons 2\,CH_4(g) + 3\,Cl_2(g)$
(d) $H_2(g) + I_2(g) \rightleftharpoons 2\,HI(g)$
(e) $2\,NOCl(g) \rightleftharpoons 2\,NO(g) + Cl_2(g)$

17. Write the equilibrium constant expression for each reaction.
(a) $2\,H_2O_2(g) \rightleftharpoons 2\,H_2O(g) + O_2(g)$
(b) $PCl_3(g) + Cl_2(g) \rightleftharpoons PCl_5(g)$
(c) $SiO_2(s) + 3\,C(s) \rightleftharpoons SiC(s) + 2\,CO(g)$
(d) $H_2(g) + \frac{1}{8}\,S_8(s) \rightleftharpoons H_2S(g)$

18. Write the equilibrium constant expression for each reaction.
(a) $3\,O_2 \rightleftharpoons 2\,O_3(g)$
(b) $SiH_4(g) + 2\,O_2(g) \rightleftharpoons SiO_2(s) + 2\,H_2O(g)$
(c) $MgO(s) + SO_2(g) + \frac{1}{2}\,O_2(g) \rightleftharpoons MgSO_4(s)$
(d) $2\,PbS(s) + 3\,O_2(g) \rightleftharpoons 2\,PbO(s) + 2\,SO_2(g)$

19. Write equilibrium constant expressions for these reactions:
(a) $Ni(s) + 4\,CO(g) \rightleftharpoons Ni(CO)_4(g)$
(b) $HF(aq) + H_2O(\ell) \rightleftharpoons H_3O^+(aq) + F^-(aq)$
(c) $Cl_2(g) + 2\,Br^-(aq) \rightleftharpoons Br_2(\ell) + 2\,Cl^-(aq)$

20. Write the equilibrium constant expression for each reaction.
(a) $TlCl_3(s) \rightleftharpoons TlCl(s) + Cl_2(g)$
(b) $CuCl_4^{2-}(aq) \rightleftharpoons Cu^{2+}(aq) + 4\,Cl^-(aq)$
(c) $CO(g) + H_2O(g) \rightleftharpoons CO_2(g) + H_2(g)$
(d) $4\,H_3O^+(aq) + 2\,Cl^-(aq) + MnO_2(s) \rightleftharpoons Mn^{2+}(aq) + 6\,H_2O(\ell) + Cl_2(aq)$

21. Write the expression for K_c for each reaction.
(a) $PCl_5(s) \rightleftharpoons PCl_3(g) + Cl_2(g)$
(b) $Co(H_2O)_6^{2+}(aq) + 4\,Cl^-(aq) \rightleftharpoons CoCl_4^{2-}(aq) + 6\,H_2O(\ell)$
(c) $CH_3COOH(aq) + H_2O(\ell) \rightleftharpoons CH_3COO^-(aq) + H_3O^+(aq)$
(d) $2\,F_2(g) + H_2O(g) \rightleftharpoons OF_2(g) + 2\,HF(g)$

22. Write the equilibrium constant expression for each reaction.
(a) The oxidation of ammonia with ClF_3 in a rocket motor
$$NH_3(g) + ClF_3(g) \rightleftharpoons 3\,HF(g) + \tfrac{1}{2}\,N_2(g) + \tfrac{1}{2}\,Cl_2(g)$$
(b) The simultaneous oxidation and reduction of a chlorite ion
$$3\,ClO_2^-(aq) \rightleftharpoons 2\,ClO_3^-(aq) + Cl^-(aq)$$
(c) $IO_3^-(aq) + 6\,OH^-(aq) + Cl_2(aq) \rightleftharpoons IO_6^{5-}(aq) + 2\,Cl^-(aq) + 3\,H_2O(\ell)$

23. Write the equilibrium constant expression for each of these heterogeneous systems.
(a) $CaSO_4\cdot 5\,H_2O(s) \rightleftharpoons CaSO_4\cdot 3\,H_2O(s) + 2\,H_2O(g)$
(b) $SiF_4(g) + 2\,H_2O(g) \rightleftharpoons SiO_2(s) + 4\,HF(g)$
(c) $LaCl_3(s) + H_2O(g) \rightleftharpoons LaClO(s) + 2\,HCl(g)$

24. Write the equilibrium constant expression for each of these heterogeneous systems.
(a) $N_2O_4(g) + O_3(g) \rightleftharpoons N_2O_5(s) + O_2(g)$
(b) $C(s) + 2\,N_2O(g) \rightleftharpoons CO_2(g) + 2\,N_2(g)$
(c) $H_2O(\ell) \rightleftharpoons H_2O(g)$

25. Write a chemical equation for an equilibrium system that would lead to each expression (a–c) for *K*.
(a) $K = \dfrac{(P_{H_2S})^2\,(P_{O_2})^3}{(P_{SO_2})^2\,(P_{H_2O})^2}$ (b) $K = \dfrac{(P_{F_2})^{1/2}(P_{I_2})^{1/2}}{(P_{IF})}$
(c) $K = \dfrac{[Cl^-]^2}{(P_{Cl_2})[Br^-]^2}$

26. Consider this reaction at 122 °C:
$$2\,SO_3(g) \rightleftharpoons 2\,SO_2(g) + O_2(g)$$

(a) Write an equilibrium constant expression for the reaction and call the constant K_1.
(b) Write an equilibrium constant expression for the decomposition of 1 mol SO_3 to SO_2 and O_2 and call the constant K_2.
(c) Relate K_1 and K_2.

27. Consider these two equilibria involving $SO_2(g)$ and their corresponding equilibrium constants.

$$SO_2(g) + \tfrac{1}{2}O_2(g) \rightleftharpoons SO_3(g) \qquad K_{c_1}$$

$$2\,SO_3(g) \rightleftharpoons 2\,SO_2(g) + O_2(g) \qquad K_{c_2}$$

Which of these expressions correctly relates K_{c_1} to K_{c_2}?
(a) $K_{c_2} = K_{c_1}^2$ (b) $K_{c_2}^2 = K_{c_1}$
(c) $K_{c_2} = 1/K_{c_1}$ (d) $K_{c_2} = K_{c_1}$
(e) $K_{c_2} = 1/K_{c_1}^2$

28. The reaction of hydrazine, N_2H_4, with chlorine trifluoride, ClF_3, has been used in experimental rocket motors.

$$N_2H_4(g) + \tfrac{4}{3}ClF_3(g) \rightleftharpoons 4\,HF(g) + N_2(g) + \tfrac{2}{3}Cl_2(g)$$

How is the equilibrium constant, K_P, for this reaction related to K_P' for the reaction written this way?

$$3\,N_2H_4(g) + 4\,ClF_3(g) \rightleftharpoons 12\,HF(g) + 3\,N_2(g) + 2\,Cl_2(g)$$

(a) $K_P = K_P'$ (b) $K_P = 1/K_P'$
(c) $K_P^3 = K_P'$ (d) $K_P = (K_P')^3$
(e) $3K_P = K_P'$

29. At 627 °C, $K_c = 0.76$ for the reaction

$$2\,SO_2(g) + O_2(g) \rightleftharpoons 2\,SO_3(g)$$

Calculate K_c at 627 °C for
(a) synthesis of 1 mol sulfur trioxide gas.
(b) decomposition of 2 mol SO_3

30. Hydrogen can react with elemental sulfur to give the smelly, toxic gas H_2S according to the reaction

$$H_2(g) + \tfrac{1}{8}S_8(s) \rightleftharpoons H_2S(g)$$

If the equilibrium constant K_c for this reaction is 7.6×10^5 at 25 °C, determine the value of the equilibrium constant for the reaction written as

$$8\,H_2(g) + S_8(s) \rightleftharpoons 8\,H_2S(g)$$

31. At standard temperature and pressure, the reaction indicated by the following equation has an equilibrium constant, K_c, equal to 0.021.

$$2\,HI(g) \rightleftharpoons H_2(g) + I_2(g)$$

Calculate the equilibrium constant, K_c, for the reverse equation.

32. At 450 °C, the equilibrium constant K_c for the Haber-Bosch synthesis of ammonia is 0.16 for the reaction written as

$$3\,H_2(g) + N_2(g) \rightleftharpoons 2\,NH_3(g)$$

Calculate the value of K_c for the same reaction written as

$$\tfrac{3}{2}H_2(g) + \tfrac{1}{2}N_2(g) \rightleftharpoons NH_3(g)$$

33. For each reaction in Question 17, write the equilibrium constant expression for K_P.

34. For each reaction in Question 18, write the equilibrium constant expression for K_P.

35. Given these data at a certain temperature,

$$2\,N_2(g) + O_2(g) \rightleftharpoons 2\,N_2O(g) \qquad K = 1.2 \times 10^{-35}$$

$$N_2O_4(g) \rightleftharpoons 2\,NO_2(g) \qquad K = 4.6 \times 10^{-3}$$

$$\tfrac{1}{2}N_2(g) + O_2(g) \rightleftharpoons NO_2(g) \qquad K = 4.1 \times 10^{-9}$$

calculate K for the reaction between 1 mol dinitrogen oxide gas and oxygen gas to give dinitrogen tetraoxide gas.

36. Given these data at a certain temperature,

$$2\,H_2(g) + O_2(g) \rightleftharpoons 2\,H_2O(g) \qquad K_c = 3.2 \times 10^{81}$$

$$N_2(g) + 3\,H_2(g) \rightleftharpoons 2\,NH_3(g) \qquad K_c = 3.5 \times 10^{8}$$

calculate K_c for the reaction of ammonia with oxygen to give $N_2(g)$ and $H_2O(g)$.

37. The vapor pressure of water at 80. °C is 0.467 atm. Find the value of K_c for the process

$$H_2O(\ell) \rightleftharpoons H_2O(g)$$

at this temperature.

38. The value of K_c for the reaction

$$N_2(g) + 3\,H_2(g) \rightleftharpoons 2\,NH_3(g)$$

is 2.00 at 400. °C. Find the value of K_P for this reaction at this temperature using atmospheres as units.

Determining Equilibrium Constants (Section 14.3)

39. A mixture of CO and Cl_2 is placed in a reaction flask that is then sealed. Initially [CO] = 0.0102 mol/L and $[Cl_2]$ = 0.00609 mol/L. When the reaction

$$CO(g) + Cl_2(g) \rightleftharpoons COCl_2(g)$$

reaches equilibrium at 600 K, $[Cl_2]$ = 0.00301 mol/L.
(a) Calculate the concentrations of CO and $COCl_2$ at equilibrium.
(b) Calculate K_c.

40. Consider the equilibrium

$$2\,A(g) \rightleftharpoons B(g)$$

At equilibrium, [A] = 0.056 M and [B] = 0.21 M. Calculate the equilibrium constant for the reaction as written.

41. At elevated temperatures, BrF_5 establishes this equilibrium.

$$2\,BrF_5(g) \rightleftharpoons Br_2(g) + 5\,F_2(g)$$

The equilibrium concentrations of the gases at 1500 K are 0.0064 mol/L for BrF_5, 0.0018 mol/L for Br_2, and 0.0090 mol/L for F_2. Calculate K_c.

42. This reaction was examined at 250 °C.

$$PCl_5(g) \rightleftharpoons PCl_3(g) + Cl_2(g)$$

At equilibrium, $[PCl_5] = 4.2 \times 10^{-5}$ M, $[PCl_3] = 1.3 \times 10^{-2}$ M, and $[Cl_2] = 3.9 \times 10^{-3}$ M. Calculate the equilibrium constant K_c for the reaction.

43. At high temperature, hydrogen and carbon dioxide react to give water and carbon monoxide.

$$H_2(g) + CO_2(g) \rightleftharpoons H_2O(g) + CO(g)$$

Laboratory measurements at 986 °C show that there is 0.11 mol each of CO and water vapor and 0.087 mol each of H_2 and CO_2 at equilibrium in a sealed 1.0-L container. Calculate the equilibrium constant K_P for the reaction at 986 °C.

44. Carbon dioxide reacts with carbon to give carbon monoxide according to the equation

$$C(s) + CO_2(g) \rightleftharpoons 2\ CO(g)$$

At 700. °C, a 2.0-L sealed flask is found to contain at equilibrium 0.10 mol CO, 0.20 mol CO_2, and 0.40 mol C. Calculate the equilibrium constant K_P for this reaction at the specified temperature.

45. Isomer A is in equilibrium with isomer B, as in the reaction

$$A(g) \rightleftharpoons B(g)$$

Three experiments are done, each at the same temperature, and equilibrium concentrations are measured. For each experiment, calculate the equilibrium constant, K_c.
(a) [A] = 0.74 mol/L, [B] = 0.74 mol/L
(b) [A] = 2.0 mol/L, [B] = 2.0 mol/L
(c) [A] = 0.01 mol/L, [B] = 0.01 mol/L

46. Carbon tetrachloride can be produced by this reaction:

$$CS_2(g) + 3\ Cl_2(g) \rightleftharpoons S_2Cl_2(g) + CCl_4(g)$$

Suppose 1.2 mol CS_2 and 3.6 mol Cl_2 are placed in a 1.00-L flask and the flask is sealed. After equilibrium has been achieved, the mixture contains 0.90 mol CCl_4. Calculate K_c.

47. Consider the decomposition of ammonium hydrogen sulfide:

$$NH_4HS(s) \rightleftharpoons NH_3(g) + H_2S(g)$$

In a sealed flask at 25 °C are 10.0 g NH_4HS, ammonia with a partial pressure of 0.692 atm, and H_2S with a partial pressure of 0.0532 atm. When equilibrium is established, it is found that the partial pressure of ammonia has increased by 12.4%. Calculate K_P for the decomposition of NH_4HS at 25 °C.

48. Assume you place 0.010 mol $N_2O_4(g)$ in a sealed 2.0-L flask at 50. °C. After the system reaches equilibrium, $[N_2O_4]$ = 0.00090 M. Calculate the value of K_c for this reaction.

$$N_2O_4(g) \rightleftharpoons 2\ NO_2(g)$$

49. Nitrosyl chloride, NOCl, decomposes to NO and Cl_2 at high temperatures.

$$2\ NOCl(g) \rightleftharpoons 2\ NO(g) + Cl_2(g)$$

Suppose you place 2.00 mol NOCl in a 1.00-L flask, seal it, and raise the temperature to 462 °C. When equilibrium has been established, 0.66 mol NO is present. Calculate the equilibrium constant K_c for the decomposition reaction from these data.

50. An equilibrium mixture contains 3.00 mol CO, 2.00 mol Cl_2, and 9.00 mol $COCl_2$ in a sealed 50.-L reaction flask at 800. K. Calculate the value of the equilibrium constant K_c for the reaction

$$CO(g) + Cl_2(g) \rightleftharpoons COCl_2(g)$$

at this temperature.

51. At 667 K, HI is found to be 11.4% dissociated into its elements.

$$2\ HI(g) \rightleftharpoons H_2(g) + I_2(g)$$

If 1.00 mol HI is placed in a 1.00-L container and the container is sealed and heated to 667 K, calculate (a) the equilibrium concentration of all three substances and (b) the value of K_c for this equilibrium at this temperature.

52. Suppose 0.086 mol Br_2 is placed in a 1.26-L flask. The flask is sealed and heated to 1756 K, a temperature at which the Br_2 dissociates to atoms

$$Br_2(g) \rightleftharpoons 2\ Br(g)$$

If Br_2 is 3.7% dissociated at this temperature, calculate K_c.

53. H_2 gas and I_2 vapor are mixed in a flask. The flask is sealed and heated to 700 °C. The initial concentration of each gas is 0.0088 mol/L, and 78.6% of the I_2 has reacted when equilibrium is achieved according to the equation

$$H_2(g) + I_2(g) \rightleftharpoons 2\ HI(g)$$

Calculate K_c for this reaction.

54. Chemists carried out a study of the high temperature reaction of sulfur dioxide with oxygen in which a sealed reactor initially contained 0.0076-M SO_2, 0.0036-M O_2, and no SO_3. After equilibrium was achieved, the SO_2 concentration decreased to 0.0032 M. Calculate K_c at this temperature for

$$2\ SO_2(g) + O_2(g) \rightleftharpoons 2\ SO_3(g)$$

The Meaning of the Equilibrium Constant (Section 14.4)

55. Using the data of Table 14.1, predict which of these reactions will be product-favored at 25 °C. Then place all the reactions in order from most reactant-favored to most product-favored.
(a) $2\ NH_3(g) \rightleftharpoons N_2(g) + 3\ H_2(g)$
(b) $NH_4^+(aq) + OH^-(aq) \rightleftharpoons NH_3(aq) + H_2O(\ell)$
(c) $2\ NO(g) \rightleftharpoons N_2(g) + O_2(g)$

56. Using the data of Table 14.1, predict which of these reactions will be product-favored at 25 °C. Then place all the reactions in order from most reactant-favored to most product-favored.
(a) $2\ NO_2(g) \rightleftharpoons N_2O_4(g)$
(b) $H_2CO_3(aq) + H_2O(\ell) \rightleftharpoons HCO_3^-(aq) + H_3O^+(aq)$
(c) $AgI(s) \rightleftharpoons Ag^+(aq) + I^-(aq)$

57. On the basis of the equilibrium constant values, choose the reactions in which the *reactants* are favored.
(a) $2\ H_2O(\ell) \rightleftharpoons H_3O^+(aq) + OH^-(aq)$ $K = 1.0 \times 10^{-14}$
(b) $[AlF_6]^{3-}(aq) \rightleftharpoons Al^{3+}(aq) + 6\ F^-(aq)$ $K = 2 \times 10^{-24}$
(c) $Ca_3(PO_4)_2(s) \rightleftharpoons 3\ Ca^{2+}(aq) + 2\ PO_4^{3-}(aq)$ $K = 1 \times 10^{-25}$
(d) $2\ Fe^{3+}(aq) + 3\ S^{2-}(aq) \rightleftharpoons Fe_2S_3(s)$ $K = 1 \times 10^{88}$

58. The equilibrium constants for dissolving silver sulfate and silver sulfide in water are 1.7×10^{-5} and 6×10^{-30}, respectively.
(a) Write the balanced dissociation reaction equation and the associated equilibrium constant expression for each process.
(b) Which compound is more soluble?
(c) Which compound is less soluble?

59. The equilibrium constants for dissolving calcium carbonate, silver nitrate, and silver chloride in water are 3.8×10^{-9}, 2.0×10^2, and 1.8×10^{-10}, respectively.
(a) Write the balanced dissociation reaction equation and the associated equilibrium constant expression for each process.
(b) Which compound is most soluble?
(c) Which compound is least soluble?

Using Equilibrium Constants (Section 14.5)

60. The hydrocarbon C_4H_{10} can exist in two gaseous forms: butane and 2-methylpropane. The value of K_c for conversion of butane to 2-methylpropane is 2.5 at 25 °C.

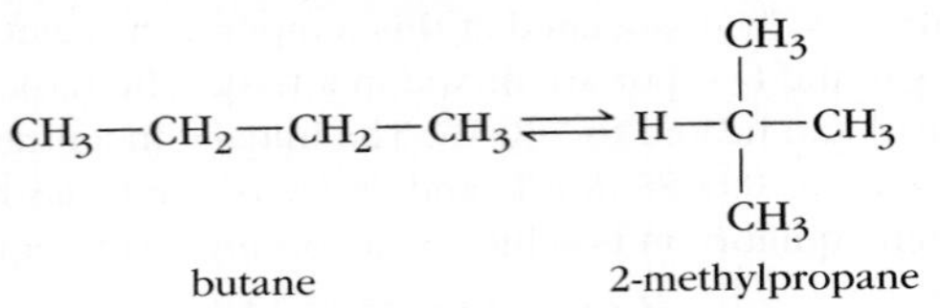

(a) Suppose that the initial concentrations of butane and 2-methylpropane are each 0.100 mol/L. Make up a table of initial concentrations, change in concentrations, and equilibrium concentrations for this reaction.
(b) Write the equilibrium constant expression in terms of x, the change in the concentration of butane, and then solve for x.
(c) If you place 0.017 mol butane in a 0.50-L sealed flask at 25 °C, what will be the equilibrium concentrations of the two isomers?

61. A mixture of butane and 2-methylpropane at 25 °C has [butane] = 0.025 mol/L and [2-methylpropane] = 0.035 mol/L. Is this mixture at equilibrium? If the *equilibrium* concentration of butane is 0.025 mol/L, what must [2-methylpropane] be at equilibrium? (See the reaction and K_c value in Question 60.)

62. The hydrocarbon cyclohexane, C_6H_{12}, can isomerize, changing into methylcyclopentane, a compound with the same formula but with a different molecular structure.

$$\underset{\text{cyclohexane}}{C_6H_{12}(g)} \rightleftharpoons \underset{\text{methylcyclopentane}}{C_5H_9CH_3(g)}$$

The equilibrium constant K_c has been estimated to be 0.12 at 25 °C. If you had originally placed 3.79 g cyclohexane in an empty 2.80-L flask and sealed the flask, how much cyclohexane (in grams) would be present when equilibrium is established?

63. At room temperature, the equilibrium constant K_c for the reaction

$$2\ NO(g) \rightleftharpoons N_2(g) + O_2(g)$$

is 1.4×10^{30}.
(a) Is this reaction product-favored or reactant-favored?
(b) In the atmosphere at room temperature the concentration of N_2 is 0.33 mol/L, and the concentration of O_2 is about 25% of that value. Calculate the equilibrium concentration of NO in the atmosphere produced by the reaction of N_2 and O_2.
(c) How does this affect your answer to Question 11?

64. Hydrogen gas and iodine gas react via the equation

$$H_2(g) + I_2(g) \rightleftharpoons 2\ HI(g) \qquad K_c = 76 \quad (\text{at } 600.\ K)$$

If 0.050 mol HI is placed in an empty, sealed 1.0-L flask at 600. K, what are the equilibrium concentrations of HI, I_2, and H_2?

65. Consider the equilibrium

$$N_2(g) + O_2(g) \rightleftharpoons 2\ NO(g)$$

At 2300 K the equilibrium constant $K_c = 1.7 \times 10^{-3}$. If 0.15 mol NO(g) is placed into an empty, sealed 10.0-L flask and heated to 2300 K, what are the equilibrium concentrations of all three substances?

66. The equilibrium constant, K_c, for the reaction

$$Br_2(g) + F_2(g) \rightleftharpoons 2\ BrF(g)$$

is 55.3. What are the equilibrium concentrations of all these gases if the initial concentrations of bromine and fluorine were both 0.220 mol/L? (Assume constant-volume conditions.)

67. The equilibrium constant K_c for the reaction

$$H_2(g) + I_2(g) \rightleftharpoons 2\ HI(g)$$

has the value 50.0 at 745 K.
(a) When 1.00 mol I_2 and 3.00 mol H_2 are allowed to come to equilibrium at 745 K in a sealed flask of volume 10.00 L, what amount (in moles) of HI will be produced?
(b) What amount of HI is produced in a 5.00-L flask?
(c) What total amount of HI is present at equilibrium if an additional 3.00 mol H_2 is added to the 10.00-L flask?

68. The equilibrium constant K_c for the *cis-trans* isomerization of gaseous 2-butene has the value 1.50 at 580. K.

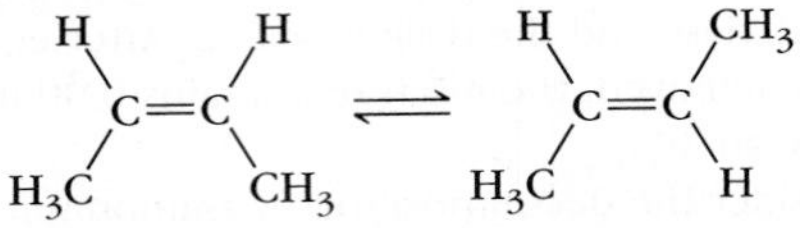

(a) Is the reaction product-favored at 580. K?
(b) Calculate the amount (in moles) of *trans* isomer produced when 1 mol *cis*-2-butene is heated to 580. K in the presence of a catalyst in a sealed, 1.00-L flask and reaches equilibrium.
(c) What would the answer be if the flask had a volume of 10.0 L?

69. The equilibrium constant K_c for the reaction

$$CO(g) + H_2O(g) \rightleftharpoons CO_2(g) + H_2(g)$$

has the value 2.64×10^{-3} at 2300. K. If a mixture of 1.00 mol CO and 1.00 mol H_2O is allowed to come to equilibrium in a sealed, 1.00-L flask at 2300. K,
(a) Calculate the final concentrations of all four species: CO, H_2O, CO_2, and H_2.
(b) What would be the equilibrium concentrations if an additional 1.00 mol each of CO and H_2O were added to the flask?

70. At 503 K the equilibrium constant K_c for the dissociation of N_2O_4,

$$N_2O_4(g) \rightleftharpoons 2\ NO_2(g)$$

has the value 40.0.
(a) Calculate the fraction of N_2O_4 left undissociated when 1.00 mol of this gas is heated to 503 K in a 10.0-L sealed container.
(b) If the volume is now reduced to 2.0 L, what will be the new fraction of N_2O_4 that is undissociated?

71. Carbonyl bromide decomposes to carbon monoxide and bromine.

$$COBr_2(g) \rightleftharpoons CO(g) + Br_2(g)$$

K_c is 0.190 at 73 °C. If you place 0.500 mol $COBr_2$ in a 2.00-L flask, seal the flask, and heat it to 73 °C,
(a) What are the equilibrium concentrations of $COBr_2$, CO, and Br_2?
(b) What percentage of the original $COBr_2$ decomposed at this temperature?

72. Consider the equilibrium at 25 °C

$$2\ SO_3(g) \rightleftharpoons 2\ SO_2(g) + O_2(g) \qquad K_c = 3.58 \times 10^{-3}$$

Suppose that 0.15 mol SO_3(g), 0.015 mol SO_2(g), and 0.0075 mol O_2(g) are placed into a 10.0-L flask at 25 °C and the flask is sealed.
(a) Is the system at equilibrium?
(b) If the system is not at equilibrium, in which direction must the reaction proceed to reach equilibrium? Explain your answer.

73. The reaction

$$2\ NO_2(g) \rightleftharpoons N_2O_4(g)$$

has an equilibrium constant, K_c, of 170 at 25 °C. If 2.0×10^{-3} mol NO_2 is present in a sealed 10.-L flask along with 1.5×10^{-3} mol N_2O_4, is the system at equilibrium? If it is not at equilibrium, does the concentration of NO_2 increase or decrease as the system proceeds to equilibrium?

74. At 2300 K the equilibrium constant for the formation of NO(g) is 1.7×10^{-3}.

$$N_2(g) + O_2(g) \rightleftharpoons 2\ NO(g)$$

(a) Analysis of the contents of a sealed flask at 2300 K shows that the concentrations of N_2 and O_2 are both 0.25 M and that of NO is 0.0042 M. Is the system at equilibrium?
(b) If the system is not at equilibrium, in which direction does the reaction proceed?
(c) When the system is at equilibrium, what are the equilibrium concentrations?

75. Consider the equilibrium

$$CH_4(g) + H_2O(g) \rightleftharpoons CO(g) + 3\ H_2(g) \qquad K_c = 9.4 \times 10^{-1}$$

Suppose that 0.25 mol CH_4(g), 0.15 mol H_2O(g), 0.25 mol CO(g), and 0.15 mol H_2(g) are placed into a 10.0-L flask at 25 °C and the flask is sealed.
(a) Is the system at equilibrium?
(b) If the system is not at equilibrium, in which direction must the reaction proceed to reach equilibrium? Explain your answer.

76. Consider the equilibrium

$$N_2(g) + O_2(g) \rightleftharpoons 2\ NO(g)$$

At 2300 K the equilibrium constant $K_c = 1.7 \times 10^{-3}$. Suppose that 0.015 mol NO(g), 0.25 mol N_2(g), and 0.25 mol O_2(g) are placed into a 10.0-L flask, sealed, and heated to 2300 K.
(a) Is the system at equilibrium?
(b) If not, in which direction must the reaction proceed to reach equilibrium?
(c) Calculate the equilibrium concentrations of all three substances.

77. Consider the equilibrium

$$H_2(g) + I_2(g) \rightleftharpoons 2\ HI(g)$$

At 745 K the equilibrium constant $K_c = 50.0$. Suppose that 0.75 mol HI(g), 0.025 mol H_2(g), and 0.025 mol I_2(g) are placed into a sealed 20.0-L flask and heated to 745 K.
(a) Is the system at equilibrium?
(b) If not, in which direction must the reaction proceed to reach equilibrium?
(c) Calculate the equilibrium concentrations of all three substances.

78. $K_c = 19.9$ for the reaction

$$Cl_2(g) + F_2(g) \rightleftharpoons 2\ ClF(g)$$

Consider a reaction mixture originally containing $[Cl_2] = 0.50$ mol/L, $[F_2] = 0.20$ mol/L, and $[ClF] = 7.3$ mol/L in a sealed container.
(a) Is the system at equilibrium?
(b) In which direction must the reaction go to achieve equilibrium?

79. The value of K_c is 3.7×10^{-23} at 25 °C for

$$C(graphite) + CO_2(g) \rightleftharpoons 2CO(g)$$

Describe what will happen if 3.5 mol CO and 3.5 mol CO_2 are mixed in a 1.5-L sealed graphite container with a suitable catalyst so that the reaction rate is rapid at this temperature.

Shifting a Chemical Equilibrium: Le Chatelier's Principle (Section 14.6)

80. K_p for this reaction is 0.16 at 25 °C:

$$2\ NOBr(g) \rightleftharpoons 2\ NO(g) + Br_2(g) \qquad \Delta H° = 16.3\ kJ$$

Predict the effect of each change on the position of the equilibrium; that is, state which way the equilibrium will shift (left, right, or no change) when each change is made. Assume constant-volume conditions for parts (a), (b), and (c).
(a) Add more Br_2(g) (b) Remove some NOBr(g)
(c) Decrease the temperature
(d) Increase the container volume

81. The decomposition of NH_4HS is endothermic.

$$NH_4HS(s) \rightleftharpoons NH_3(g) + H_2S(g)$$

(a) Using Le Chatelier's principle, explain how increasing the temperature would affect the equilibrium.
(b) If more NH_4HS is added to a sealed flask in which this equilibrium exists, how is the equilibrium affected?
(c) What if some additional NH_3 is placed in a sealed flask containing an equilibrium mixture?
(d) What will happen to the partial pressure of NH_3 if some H_2S is removed from the flask?

82. Consider this equilibrium, established in a sealed 2.0-L flask at 25 °C:

$$N_2O_4(g) \rightleftharpoons 2\ NO_2(g) \qquad \Delta H° = +57.2\ kJ$$

What will happen to the concentration of N_2O_4 if the temperature is increased? Explain your choice.
(a) It will increase. (b) It will decrease.
(c) It will not change.
(d) It is not possible to tell from the information provided.

83. Solid barium sulfate is in equilibrium with barium ions and sulfate ions in solution.

$$BaSO_4(s) \rightleftharpoons Ba^{2+}(aq) + SO_4^{2-}(aq)$$

What will happen to the barium ion concentration if more solid $BaSO_4$ is added to the flask? Explain your choice.
(a) It will increase. (b) It will decrease.
(c) It will not change.
(d) It is not possible to tell from the information provided.

84. Predict the direction in which each equilibrium will shift if the pressure on the system is decreased by expansion.
(a) $Ni(s) + 4\ CO(g) \rightleftharpoons Ni(CO)_4(g)$
(b) $ClF_5(g) \rightleftharpoons ClF_3(g) + F_2(g)$
(c) $HBr(g) \rightleftharpoons \frac{1}{2}\ H_2(g) + \frac{1}{2}\ Br_2(g)$

85. Assume that the reaction

$$ClF_5(g) \rightleftharpoons ClF_3(g) + F_2(g)$$

is at equilibrium and the equilibrium conditions are changed as described. Indicate whether the forward or the reverse reaction rate is faster immediately after the change and explain your choice.
(a) Some $F_2(g)$ is added without changing the total volume.
(b) Some $ClF_3(g)$ is removed without changing the total volume.
(c) The total volume of the system is doubled.

86. Assume that the reaction

$$2\ HBr(g) \rightleftharpoons H_2(g) + Br_2(g)$$

is at equilibrium and the equilibrium conditions are changed as described. Indicate whether the forward or the reverse reaction rate is faster immediately after the change and explain your choice.
(a) Some HBr(g) is added without changing the total volume.
(b) Some $Br_2(g)$ is removed without changing the total volume.
(c) The total volume of the system is halved.

87. Hydrogen, bromine, and HBr in the gas phase are in equilibrium in a container of fixed volume.

$$H_2(g) + Br_2(g) \rightleftharpoons 2\ HBr(g) \qquad \Delta H° = -103.7\ kJ$$

How will each of these changes affect the indicated quantities? Write "increase," "decrease," or "no change."

Change	$[Br_2]$	[HBr]	K_c	K_P
Some H_2 is added to the container.	____	____	____	____
The temperature of the gases in the container is increased.	____	____	____	____
The pressure of HBr is increased.	____	____	____	____

88. Nitrogen, oxygen, and nitrogen monoxide are in equilibrium in a container of fixed volume.

$$N_2(g) + O_2(g) \rightleftharpoons 2\ NO(g) \qquad \Delta H° = 180.5\ kJ$$

How will each of these changes affect the indicated quantities? Write "increase," "decrease," or "no change."

Change	$[N_2]$	[NO]	K_c	K_P
Some NO is added to the container.	____	____	____	____
The temperature of the gases in the container is decreased.	____	____	____	____
The pressure of N_2 is decreased.	____	____	____	____

89. The equilibrium constant K_c for this reaction is 0.16 at 25 °C, and the standard enthalpy change is 16.1 kJ.

$$2\ NOBr(g) \rightleftharpoons 2\ NO(g) + Br_2(\ell)$$

Predict the effect of each of these changes on the position of the equilibrium; that is, state which way the equilibrium will shift (left, right, or no change) when each of these changes is made for a constant-volume system.
(a) Adding more Br_2
(b) Removing some NOBr
(c) Lowering the temperature

90. The formation of hydrogen sulfide from the elements is exothermic.

$$H_2(g) + \tfrac{1}{8}\ S_8(s) \rightleftharpoons H_2S(g) \qquad \Delta H° = -20.6\ kJ$$

Predict the effect of each of these changes on the position of the equilibrium; that is, state which way the equilibrium will shift (left, right, or no change) when each change is made in a constant-volume system.
(a) Adding more sulfur
(b) Adding more H_2
(c) Raising the temperature

91. The oxidation of NO to NO_2,

$$2\ NO(g) + O_2(g) \rightleftharpoons 2\ NO_2(g)$$

is exothermic. Predict the effect of each of these changes on the position of the equilibrium; that is, state which way the equilibrium will shift (left, right, or no change) when each change is made in a constant-volume system.
(a) Adding more O_2
(b) Adding more NO_2
(c) Lowering the temperature

92. Phosphorus pentachloride is in equilibrium with phosphorus trichloride and chlorine in a sealed flask.

$$PCl_5(s) \rightleftharpoons PCl_3(g) + Cl_2(g)$$

What will happen to the concentration of Cl_2 if additional $PCl_5(s)$ is added to the flask?
(a) It will increase. (b) It will decrease.
(c) It will not change.
(d) It is impossible to tell from the information provided.

93. Heating a metal carbonate leads to decomposition.

$$BaCO_3(s) \rightleftharpoons BaO(s) + CO_2(g)$$

Predict the effect on the equilibrium of each change listed below. Answer by choosing (i) no change, (ii) shifts left, or (iii) shifts right. (In parts (a), (b), and (c), assume that the volume is constant.)
(a) Add $BaCO_3$ (b) Add CO_2
(c) Add BaO (d) Raise the temperature
(e) Increase the volume of the reaction flask

94. Consider the equilibrium

$$PbCl_2(s) \rightleftharpoons Pb^{2+}(aq) + 2\ Cl^-(aq)$$

(a) Will the equilibrium concentration of aqueous lead(II) ion increase, decrease, or remain the same if some solid NaCl is added to a sealed flask?
(b) Make a graph like the one in Figure 14.6 to illustrate what happens to each of the concentrations after the NaCl is added.

95. Consider the system

$$4\ NH_3(g) + 3\ O_2(g) \rightleftharpoons 2\ N_2(g) + 6\ H_2O(\ell) \qquad \Delta H° = -1530.4\ kJ$$

(a) How will the amount of ammonia at equilibrium be affected by
(1) Removing $O_2(g)$ without changing the total volume?
(2) Adding $N_2(g)$ without changing the total volume?

(3) Adding water without changing the total volume?
(4) Expanding the container?
(5) Increasing the temperature?

(b) Which of the above changes (1–5) will increase the value of K? Which will decrease it?

96. Consider the transformation of butane into 2-methylpropane (see Question 60). The system is originally at equilibrium at 25 °C in a 1.0-L sealed flask with [butane] = 0.010 M and [2-methylpropane] = 0.025 M. Suppose that 0.0050 mol of 2-methylpropane is suddenly added to the flask, and the system shifts to a new equilibrium.
(a) What is the new equilibrium concentration of each gas?
(b) Make a graph like the one in Figure 14.6 to show how the concentrations of the isomers change when the 2-methylpropane is added.

97. Phosphorus pentachloride decomposes at high temperatures.

$$PCl_5(g) \rightleftharpoons PCl_3(g) + Cl_2(g)$$

An equilibrium mixture at some temperature consists of 3.120 g PCl_5, 3.845 g PCl_3, and 1.787 g Cl_2 in a sealed 1.00-L flask.
(a) If you add 1.418 g Cl_2 without changing the volume, how will the equilibrium be affected?
(b) What will the concentrations of PCl_5, PCl_3, and Cl_2 be when equilibrium is reestablished?

98. Predict whether the equilibrium for the photosynthesis reaction described by the equation

$$6\ CO_2(g) + 6\ H_2O(\ell) \rightleftharpoons C_6H_{12}O_6(s) + 6\ O_2(g) \quad \Delta H° = 2801.69\ kJ$$

would (i) shift to the right, (ii) shift to the left, or (iii) remain unchanged for each of these changes
(a) decrease $[CO_2]$ at constant volume
(b) increase the partial pressure of O_2 at constant volume
(c) remove one half of the $C_6H_{12}O_6$
(d) decrease the total pressure by increasing the volume
(e) increase the temperature
(f) introduce a catalyst into a constant-volume system

Equilibrium at the Nanoscale (Section 14.7)

99. For each of these reactions at 25 °C, indicate whether the entropy effect, the energy effect, both, or neither favors the reaction.
(a) $N_2(g) + 3\ F_2(g) \rightleftharpoons 2\ NF_3(g)$ $\Delta H° = -249$ kJ
(b) $N_2F_4(g) \rightleftharpoons 2\ NF_2(g)$ $\Delta H° = 93.3$ kJ
(c) $N_2(g) + 3\ Cl_2(g) \rightleftharpoons 2\ NCl_3(g)$ $\Delta H° = 460$ kJ

100. For each of these processes at 25 °C, indicate whether the entropy effect, the energy effect, both, or neither favors the process.
(a) $C_3H_8(g) + 5\ O_2(g) \rightleftharpoons 3\ CO_2(g) + 4\ H_2O(g)$ $\Delta H° = -2045$ kJ
(b) $Br_2(g) \rightleftharpoons Br_2(\ell)$ $\Delta H° = -31$ kJ
(c) $2\ Ag(s) + 3\ N_2(g) \rightleftharpoons 2\ AgN_3(s)$ $\Delta H° = 621$ kJ

101. For each of these chemical reactions, predict whether the equilibrium constant at 25 °C is greater than 1 or less than 1, or state that insufficient information is available. Also indicate whether each reaction is product-favored or reactant-favored.
(a) $2\ NO(g) + O_2(g) \rightleftharpoons 2\ NO_2(g)$ $\Delta H° = -115$ kJ
(b) $2\ O_3(g) \rightleftharpoons 3\ O_2(g)$ $\Delta H° = -285$ kJ
(c) $N_2(g) + 3\ Cl_2(g) \rightleftharpoons 2\ NCl_3(g)$ $\Delta H° = 460$ kJ

102. For each of these chemical reactions, predict whether the equilibrium constant at 25 °C is greater than 1 or less than 1, or state that insufficient information is available. Also indicate whether each reaction is product-favored or reactant-favored.
(a) $2\ NaCl(s) \rightleftharpoons 2\ Na(s) + Cl_2(g)$ $\Delta H° = 823$ kJ
(b) $2\ CO(g) + O_2(g) \rightleftharpoons 2\ CO_2(g)$ $\Delta H° = -566$ kJ
(c) $3\ CO_2(g) + 4\ H_2O(g) \rightleftharpoons C_3H_8(g) + 5\ O_2(g)$ $\Delta H° = 2045$ kJ

Controlling Chemical Reactions: The Haber-Bosch Process (Section 14.8)

103. Considering both the enthalpy effect and the entropy effect for the Haber-Bosch process, explain why choosing the temperature at which to run this reaction is very important.

104. Explain in your own words why it was so important to find a highly effective catalyst for the ammonia synthesis reaction before the Haber-Bosch process could become successful.

105. Although ammonia is made in enormous quantities by the Haber-Bosch process, sulfuric acid is made in even greater quantities by the *contact process.* A simplified version of this process can be represented by these three reactions.

$$S(s) + O_2(g) \rightleftharpoons SO_2(g)$$
$$2\ SO_2(g) + O_2(g) \rightleftharpoons 2\ SO_3(g)$$
$$SO_3(g) + H_2O(\ell) \rightleftharpoons H_2SO_4(\ell)$$

(a) Use data from Appendix J to calculate $\Delta H°$ for each reaction.
(b) Which reactions are exothermic? Which are endothermic?
(c) In which of the reactions does entropy increase? In which does it decrease? In which does it stay about the same?
(d) For which reaction(s) do low temperatures favor formation of products?

106. Lime, CaO(s), can be produced by heating limestone, $CaCO_3(s)$, to decompose it.
(a) Write a balanced equation for the reaction.
(b) Predict the sign of the enthalpy change for the reaction.
(c) From the data in Appendix J, calculate $\Delta H°$ for this reaction at 25 °C to verify or contradict your prediction in part (b).
(d) Predict the entropy effect for this reaction.
(e) Is the reaction favored by entropy, energy, both, or neither?
(f) Explain in terms of Le Chatelier's principle why limestone must be heated to make lime.

General Questions

These questions are not explicitly keyed to chapter topics; many require integration of several concepts.

107. Write equilibrium constant expressions, in terms of reactant and product concentrations, for each of these reactions.

$H_2O(\ell) \rightleftharpoons H^+(aq) + OH^-(aq)$ $K_c = 1.0 \times 10^{-14}$

$CH_3COOH(aq) \rightleftharpoons CH_3COO^-(aq) + H^+(aq)$ $K_c = 1.8 \times 10^{-5}$

$N_2(g) + 3\ H_2(g) \rightleftharpoons 2\ NH_3(g)$ $K_c = 3.5 \times 10^8$

Assume that all gases and solutes have initial concentrations of 1.0 mol/L. Then let the *first* reactant in each reaction change its concentration by $-x$.

(a) Using the reaction table (ICE table) approach, write equilibrium constant expressions in terms of the unknown variable x for each reaction.
(b) Which of these expressions yield quadratic equations?
(c) How would you go about solving the others for x?

108. Write equilibrium constant expressions, in terms of reactant and product concentrations, for each of these reactions.

$2\ O_3(g) \rightleftharpoons 3\ O_2(g) \qquad K_c = 7 \times 10^{56}$

$2\ NO_2(g) \rightleftharpoons N_2O_4(g) \qquad K_c = 1.7 \times 10^2$

$HCOO^-(aq) + H^+(aq) \rightleftharpoons HCOOH(aq) \qquad K_c = 5.6 \times 10^3$

$Ag^+(aq) + I^-(aq) \rightleftharpoons AgI(s) \qquad K_c = 6.7 \times 10^{15}$

Assume that all gases and solutes have initial concentrations of 1.0 mol/L. Then let the *first* reactant in each reaction change its concentration by $-x$.

(a) Using the reaction table (ICE table) approach, write equilibrium constant expressions in terms of the unknown variable x for each reaction.
(b) Which of these expressions yield quadratic equations?
(c) How would you go about solving the others for x?

109. The equilibrium constant K_c is 1.6×10^5 at 1297 K and 3.5×10^4 at 1495 K for the reaction

$$H_2(g) + Br_2(g) \rightleftharpoons 2\ HBr(g)$$

(a) Is $\Delta H°$ for this reaction positive or negative?
(b) Find K_c at 1297 K for the reaction

$$\tfrac{1}{2}\ H_2(g) + \tfrac{1}{2}\ Br_2(g) \rightleftharpoons HBr(g)$$

(c) Pure HBr is placed into an evacuated container of constant volume. The container is sealed and heated to 1297 K. What percentage of the HBr is decomposed to H_2 and Br_2 at equilibrium?

110. Many common nonmetallic elements exist as diatomic molecules at room temperature. When these elements are heated to 1500. K, the molecules break apart into atoms. A general equation for this type of reaction is

$$E_2(g) \rightleftharpoons 2\ E(g)$$

where E stands for an atom of each element. Equilibrium constants for dissociation of these molecules at 1500. K are

Species	K_c	Species	K_c
Br_2	8.9×10^{-2}	H_2	3.1×10^{-10}
Cl_2	3.4×10^{-3}	N_2	1×10^{-27}
F_2	7.4	O_2	1.6×10^{-11}

(a) Assume that 1.00 mol of each diatomic molecule is placed in a separate 1.0-L container, sealed, and heated to 1500. K. Calculate the equilibrium concentration of the atomic form of each element at 1500. K.
(b) From these results, predict which of the diatomic elements has the lowest bond dissociation energy, and compare your results with thermochemical calculations and with Lewis structures.

111. The total pressure for a mixture of N_2O_4 and NO_2 is 1.5 atm. If $K_P = 7.0$ (at 25 °C), calculate the partial pressure of each gas in the mixture.

$$2\ NO_2(g) \rightleftharpoons N_2O_4(g)$$

112. The equilibrium constant, K_c, for the reaction

$$N_2O_4(g) \rightleftharpoons 2\ NO_2(g)$$

at 25 °C is 5.8×10^{-3}. Suppose 15.6 g N_2O_4 is placed in a 5.00-L flask at 25 °C and the flask is sealed. Calculate:
(a) The amount of NO_2 (mol) present at equilibrium.
(b) The percentage of the original N_2O_4 that is dissociated.

113. The chemistry of compounds composed of a transition metal and carbon monoxide has been an interesting area of research for more than 70 years. $Ni(CO)_4$ is formed by the reaction of nickel metal with carbon monoxide.

(a) If you have 2.05 g CO, and you combine it with 0.125 g nickel metal, how many grams of $Ni(CO)_4$ can be formed?
(b) An excellent way to make pure nickel metal is to decompose $Ni(CO)_4$ in a vacuum at a temperature slightly higher than room temperature. If the molar enthalpy of formation of $Ni(CO)_4$ gas is −602.9 kJ/mol, what is the enthalpy change for this decomposition reaction?

$$Ni(CO)_4(g) \longrightarrow Ni(s) + 4\ CO(g)$$

In an experiment at 100. °C it is determined that with 0.010 mol $Ni(CO)_4(g)$ initially present in a sealed 1.0-L flask, only 0.000010 mol remains after decomposition.

(c) What is the equilibrium concentration of CO in the flask?
(d) What is the value of the equilibrium constant K_c for this reaction at 100. °C?
(e) Predict whether there is an increase or a decrease in entropy when this reaction occurs.
(f) Calculate the equilibrium constant K_P for this reaction at 100. °C.

114. At 500 °C, K_P for the formation of ammonia from nitrogen and hydrogen gases is 1.5×10^{-5}.

$$N_2(g) + 3\ H_2(g) \rightleftharpoons 2\ NH_3(g)$$

Calculate the equilibrium partial pressure of hydrogen if the equilibrium partial pressures of ammonia and nitrogen are 0.015 atm and 1.2 atm, respectively.

115. A small sample of *cis*-dichloroethene in which one carbon atom is the radioactive isotope ^{14}C is added to an equilibrium mixture of the *cis* and *trans* isomers at a certain temperature. Eventually, 40% of the radioactive molecules are found to be in the *trans* configuration at any given time.

(a) What is the value of K_c for the *cis* $\rightleftharpoons$ *trans* equilibrium?
(b) What would have happened if a small sample of radioactive *trans* isomer had been added instead of the *cis* isomer?

116. Solid ammonium iodide decomposes to ammonia and hydrogen iodide gases at sufficiently high temperatures.

$$NH_4I(s) \rightleftharpoons NH_3(g) + HI(g)$$

The equilibrium constant, K_P, for the decomposition at 673 K is 0.215. A 15.0-g sample of ammonium iodide are sealed in a 5.00-L flask and heated to 673 K.

(a) What is the total pressure in the flask at equilibrium?
(b) How much ammonium iodide decomposes?

117. In a 0.0020-M solution of acetic acid, any acetate species spends 9% of its time as an acetate ion, CH_3COO^-, and the remaining 91% of its time as acetic acid, CH_3COOH. What is the value of K_c for the reaction?

$$CH_3COOH + H_2O \rightleftharpoons CH_3COO^- + H_3O^+$$

118. These amounts of HI, H_2, and I_2 are introduced into a 10.00-L flask. The flask is sealed and heated to 745 K.

	n_{HI} (mol)	n_{H_2} (mol)	n_{I_2} (mol)
Case a	1.0	0.10	0.10
Case b	10.	1.0	1.0
Case c	10.	10.	1.0
Case d	5.62	0.381	1.75

The equilibrium constant for the reaction

$$2\ HI(g) \rightleftharpoons H_2(g) + I_2(g)$$

has the value 0.0200 at 745 K. In which cases will the concentration of HI increase as equilibrium is attained, and in which cases will the concentration of HI decrease?

119. These amounts of CO(g), H_2O(g), CO_2(g), and H_2(g) are introduced into a 10.00-L flask. The flask is sealed and heated to a very high temperature.

	n_{CO} (mol)	n_{H_2O} (mol)	n_{CO_2} (mol)	n_{H_2O} (mol)
Case a	1.0	0.10	0.10	0.10
Case b	10.	1.0	1.0	1.0
Case c	10.	10.	1.0	1.0
Case d	5.62	0.381	1.75	1.75

The equilibrium constant for the reaction

$$CO(g) + H_2O(g) \rightleftharpoons CO_2(g) + H_2(g)$$

has the value $K_c = 4.00$ at this temperature. For which cases will the concentration of CO increase as equilibrium is attained, and in which cases will the concentration of CO decrease?

120. Carbonylbromide, $COBr_2$, can be formed by combining carbon monoxide and bromine gas.

$$CO(g) + Br_2(g) \rightleftharpoons COBr_2(g)$$

When equilibrium is established at 346 K, the partial pressures (in atm) of $COBr_2$, CO, and Br_2 are 0.12, 1.00, and 0.65, respectively.

(a) What is K_P at 346 K?

(b) Enough bromine condenses to decrease its partial pressure to 0.50 atm. What are the equilibrium partial pressures of all gases after equilibrium is re-established?

121. Mustard gas, used in chemical warfare in World War I, has been found to be an effective agent in the chemotherapy of Hodgkin's disease. It can be produced according to this reaction:

$$SCl_2(g) + 2\ C_2H_4(g) \rightleftharpoons S(CH_2CH_2Cl)_2(g)$$

An evacuated 5.00-L flask at 20.0 °C is filled with 0.258 mol SCl_2 and 0.592 mol C_2H_4 and sealed. After equilibrium is established, 0.0349 mol mustard gas is present.

(a) Calculate the partial pressure of each gas at equilibrium.

(b) Calculate K_c at 20.0 °C.

Applying Concepts

These questions test conceptual learning.

122. Two molecules of A react to form one molecule of B, as in the reaction

$$2\ A(g) \rightleftharpoons B(g)$$

Three experiments are done at different temperatures and equilibrium concentrations are measured. For each experiment, calculate the equilibrium constant, K_c.

(a) [A] = 0.74 mol/L, [B] = 0.74 mol/L

(b) [A] = 2.0 mol/L, [B] = 2.0 mol/L

(c) [A] = 0.01 mol/L, [B] = 0.01 mol/L

What can you conclude about this statement: "If the concentrations of reactants and products are equal, then the equilibrium constant is always 1.0."

123. Suppose that you have heated a mixture of *cis*- and *trans*-2-pentene to 600. K, and after 1 h you find that the composition is 40% *cis.* After 4 h the composition is found to be 42% *cis,* and after 8 h it is 42% *cis.* Next, you heat the mixture to 800. K and find that the composition changes to 45% *cis.* When the mixture is cooled to 600. K and allowed to stand for 8 h, the composition is found to be 42% *cis.* Is this system at equilibrium at 600. K? Or would more experiments be needed before you could conclude that it was at equilibrium? If so, what experiments would you do?

124. In Section 14.2 the equilibrium constant for the reaction

$$\tfrac{1}{8}\ S_8(s) + O_2(g) \rightleftharpoons SO_2(g)$$

is given as 4.2×10^{52}. If this reaction is so product-favored, why can large piles of yellow sulfur exist in our environment (as they do in Louisiana and Texas)?

125. When pressure greater than atmospheric pressure is applied to ice at 0.00 °C, the ice melts. Explain how this is an example of Le Chatelier's principle.

126. For the reaction

$$cis\text{-2-butene} \rightleftharpoons trans\text{-2-butene}$$

K_c is 1.65 at 500. K, 1.47 at 600. K, and 1.36 at 700. K. Predict whether the conversion from the *cis* to the *trans* isomer of 2-butene is exothermic or endothermic.

Use this table to answer Questions 127 and 128.

Equilibrium Constants K_c for Some *Cis-Trans* Interconversions

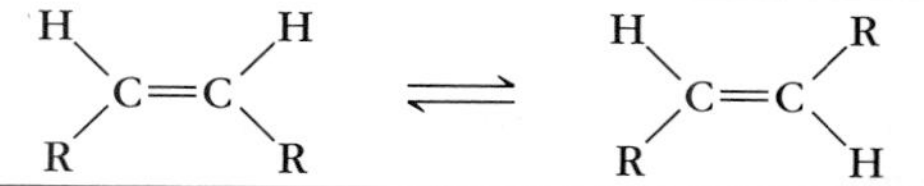

Temperature (K)	R is F	R is Cl	R is CH_3
500.	0.420	0.608	1.65
600.	0.491	0.678	1.47
700.	0.549	0.732	1.36

127. Based on the data in the table and these reaction equations,

(i) *cis*-(R is F) $\rightleftharpoons$ *trans*-(R is F)

(ii) *trans*-(R is Cl) $\rightleftharpoons$ *cis*-(R is Cl)

(iii) *cis*-(R is CH_3) $\rightleftharpoons$ *trans*-(R is CH_3)

(a) Which reaction is most exothermic?

(b) Which reaction is most endothermic?

128. Based on the data in the table and these reaction equations,
(i) *cis*-(R is F) $\rightleftharpoons$ *trans*-(R is F)
(ii) *trans*-(R is Cl) $\rightleftharpoons$ *cis*-(R is Cl)
(iii) *cis*-(R is CH_3) $\rightleftharpoons$ *trans*-(R is CH_3)
(a) Which reaction is most product-favored at 500. K?
(b) Which reaction is most reactant-favored at 700. K?

129. Figure 14.3 shows the equilibrium mixture of N_2O_4 and NO_2 at two different temperatures. Imagine that you can shrink yourself down to the size of the molecules in the two glass tubes and observe their behavior for a short period of time. Write a brief description of what you observe in each of the flasks.

130. Imagine yourself to be the size of ions and molecules inside a beaker containing this equilibrium mixture with a K_c greater than 1.

$$\underset{\text{pink}}{Co(H_2O)_6^{2+}(aq)} + 4\,Cl^-(aq) \rightleftharpoons \underset{\text{blue}}{CoCl_4^{2-}(aq)} + 6\,H_2O(\ell)$$

Write a brief description of what you observe around you before and after additional water is added to the mixture.

131. Which of the diagrams represent equilibrium mixtures for the reaction

$$A_2(g) + B_2(g) \rightleftharpoons 2\,AB(g)$$

at temperatures where $10^2 > K_c > 0.1$?

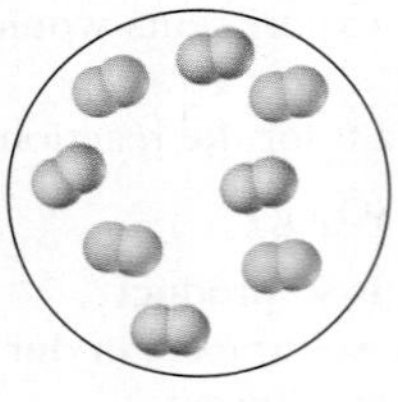
(a)
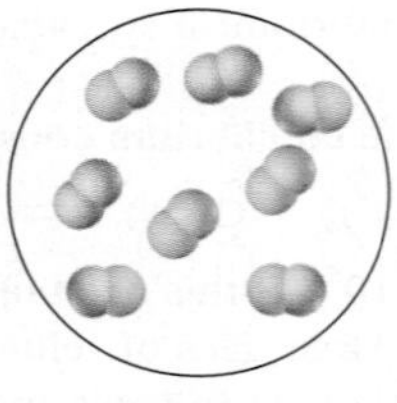
(b)
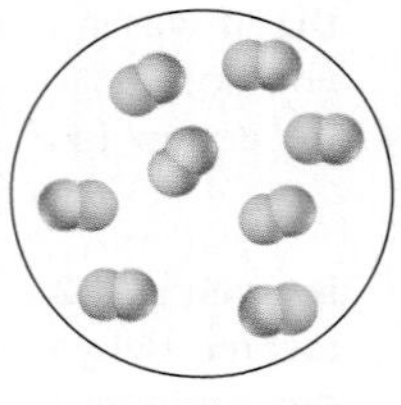
(c)
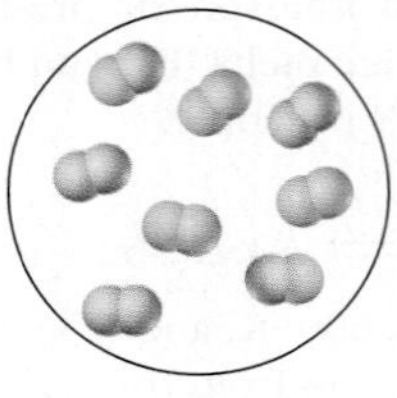
(d)
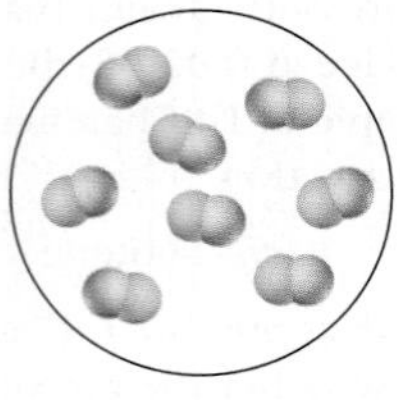
(e)

132. Which diagram in Question 131 best represents an equilibrium mixture with an equilibrium constant of
(a) 0.44? (b) 4.0? (c) 36?

133. Draw a nanoscale (particulate) level diagram for an equilibrium mixture of

$$CO(g) + H_2O(g) \rightleftharpoons CO_2(g) + H_2(g) \qquad K_c = 4.00$$

134. The diagram below represents an equilibrium mixture for the reaction

$$N_2(g) + O_2(g) \rightleftharpoons 2\,NO(g)$$

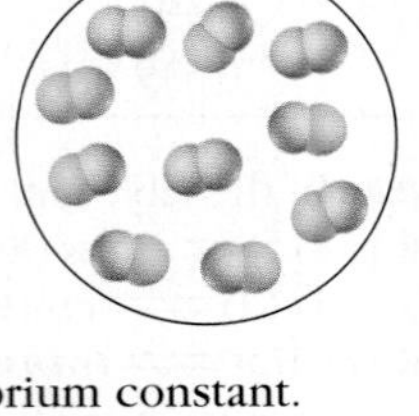

Estimate the equilibrium constant.

135. The equilibrium constant, K_c, is 1.05 at 350 K for the endothermic reaction

$$2\,CH_2Cl_2(g) \rightleftharpoons CH_4(g) + CCl_4(g)$$

Which diagram below represents an equilibrium mixture at 350 K?

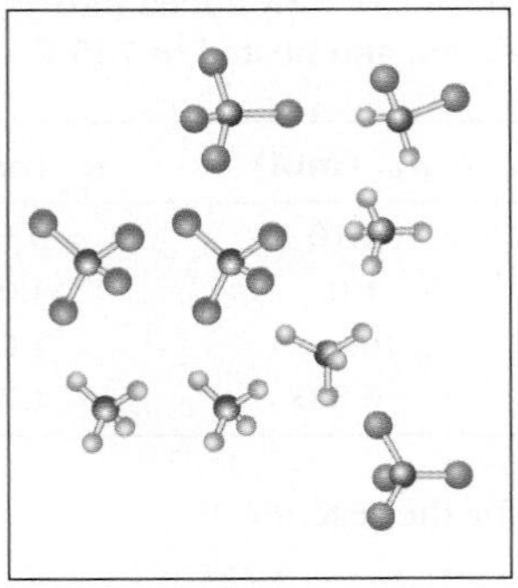
(a)
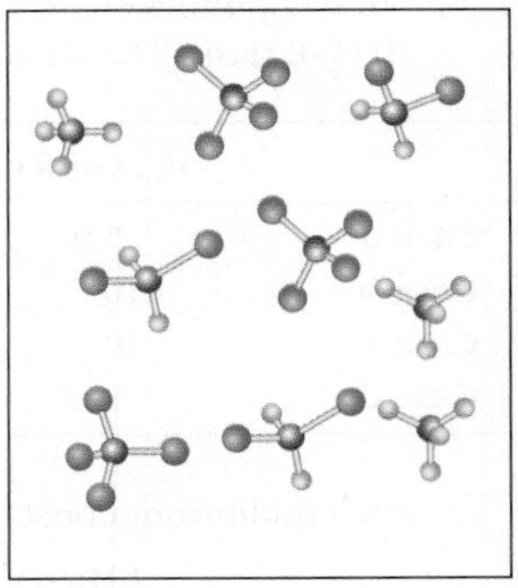
(b)
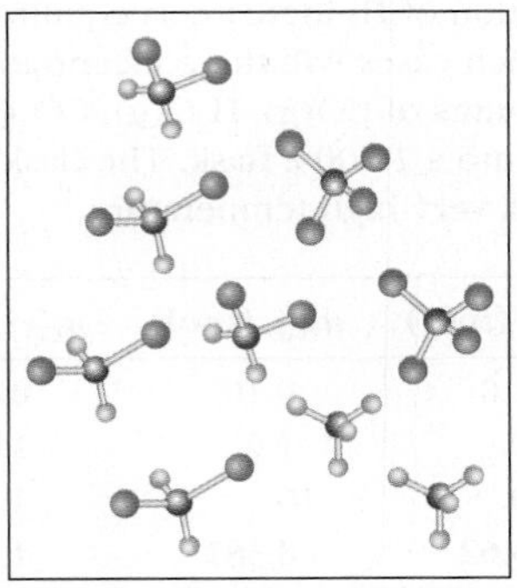
(c)

136. For the reaction in Question 135, which diagram in Question 135 represents an equilibrium mixture at
(a) a temperature below 350 K?
(b) a temperature above 350 K?

137. A solid sample of benzoic acid, a carboxylic acid, is in equilibrium with an aqueous solution of benzoic acid. A tiny quantity of D_2O, water containing the isotope ^2H, deuterium, is added to the solution. The solution is allowed to stand at constant temperature for several hours, after which some of the solid benzoic acid is removed and analyzed. The benzoic acid is found to contain a tiny quantity of deuterium, D, and the formula of the deuterium-containing molecules is C_6H_5COOD. Explain how this can happen.

138. In a second experiment with the benzoic acid in Question 137, a tiny quantity of water that contains the isotope ^{18}O is added to the solution of benzoic acid in water. When some of the solid benzoic acid is analyzed, no ^{18}O is found in the benzoic acid. Compare this situation with the experiment involving deuterium, and explain how the results of the two experiments can differ as they do.

139. Samples of N_2O_4 can be prepared in which both nitrogen atoms are the heavier isotope ^{15}N. Designating this isotope as N*, we can write the formula of the molecules in such a sample as $O_2N^*{-}N^*O_2$ and the formula of typical N_2O_4 as $O_2N{-}NO_2$. When a tiny quantity of $O_2N^*{-}N^*O_2$ is introduced into an equilibrium mixture of N_2O_4 and NO_2,

the ^{15}N immediately becomes distributed among both N_2O_4 and NO_2 molecules, and in the N_2O_4 it is invariably in the form $O_2N^*—NO_2$. Explain how this observation supports the idea that equilibrium is dynamic.

140. For the equilibrium

$$\underset{\text{pink}}{Co(H_2O)_6^{2+}(aq)} + 4\ Cl^-(aq) \rightleftharpoons \underset{\text{blue}}{CoCl_4^{2-}(aq)} + 6\ H_2O(\ell)$$

K_c is somewhat greater than 1. If water is added to a blue solution of $CoCl_4^{2-}(aq)$, the color changes from blue to pink.

(a) Does water appear in the equilibrium constant expression for this reaction?

(b) How can adding water shift the equilibrium to the left?

(c) Is this shift in the equilibrium in accord with Le Chatelier's principle? Why or why not?

More Challenging Questions

These questions require more thought and integrate several concepts.

141. A sealed 15.0-L flask at 300. K contains 64.4 g of a mixture of NO_2 and N_2O_4 in equilibrium. What is the total pressure in the flask? (For $2\ NO_2(g) \rightleftharpoons N_2O_4(g)$ $K_P = 6.67$.)

142. A sample of nitrosyl bromide is heated to 100. °C in a sealed 10.00-L container to decompose it partially according to the equation

$$2\ NOBr(g) \rightleftharpoons 2\ NO(g) + Br_2(g)$$

The container is found to contain 6.44 g NOBr, 3.15 g NO, and 8.38 g Br_2 at equilibrium.

(a) Find the value of K_c at 100. °C.

(b) Find the total pressure exerted by the mixture of gases.

(c) Calculate K_P for this reaction at 100. °C.

143. When heated, lanthanum oxalate decomposes to lanthanum oxide, CO, and CO_2.

$$La_2(C_2O_4)_3(s) \rightleftharpoons La_2O_3(s) + 3\ CO(g) + 3\ CO_2(g)$$

(a) If, at equilibrium, the total pressure in a sealed 10.0-L flask is 0.200 atm, what is the value of K_p?

(b) Suppose 0.100 mol $La_2(C_2O_4)_3$ was originally placed in the 10.0-L flask. What quantity of $La_2(C_2O_4)_3$ remains unreacted at equilibrium at 373 K?

144. Exactly 5.0 mol ammonia was placed in a 2.0-L flask that was then sealed and heated to 473 K. When equilibrium was established, 0.20 mol nitrogen had been formed according to the decomposition reaction

$$2\ NH_3(g) \rightleftharpoons N_2(g) + 3\ H_2(g)$$

(a) Calculate the value of the equilibrium constant K_c for this reaction at 473 K.

(b) Calculate the total pressure exerted by the mixture of gases inside the 2.0-L flask at this temperature.

(c) Calculate K_P for this reaction at 473 K.

145. The equilibrium constant K_c has a value of 3.30 at 760. K for the decomposition of phosphorus pentachloride,

$$PCl_5(g) \rightleftharpoons PCl_3(g) + Cl_2(g)$$

(a) Calculate the equilibrium concentrations of all three species arising from the decomposition of 0.75 mol PCl_5 in a sealed 5.00-L vessel.

(b) Calculate the equilibrium concentrations of all three species resulting from an initial mixture of 0.75 mol PCl_5 and 0.75 mol PCl_3 in a sealed 5.00-L vessel.

146. A 1.00-mol sample of CO_2 is heated to 1000. K with excess solid graphite in a sealed, 40.0-L container. At 1000. K, K_c is 2.11×10^{-2} for the reaction

$$C(graphite) + CO_2(g) \rightleftharpoons 2\ CO(g)$$

(a) What is the composition of the equilibrium mixture at 1000. K?

(b) If the volume of the flask is changed and a new equilibrium established in which the amount of CO_2 is equal to the amount of CO, what is the new volume of the flask? (Assume the temperature remains 1000. K.)

147. The reaction of hydrogen gas and iodine gas to give hydrogen iodide gas, when written with the smallest whole-number coefficients, has an equilibrium constant, K_c, of 56 at 435 °C.

(a) What is the value of K_P?

(b) Suppose 0.45 mol H_2 and 0.45 mol I_2 are sealed in a 10.0-L flask at 435 °C. Calculate the total pressure of the mixture before and after equilibrium is achieved.

(c) Calculate the partial pressure of each gas at equilibrium.

148. Limestone decomposes at high temperatures.

$$CaCO_3(s) \rightleftharpoons CaO(s) + CO_2(g)$$

At 1000 °C, $K_P = 3.87$. Pure $CaCO_3$ is placed into an empty 5.00-L flask. The flask is sealed and heated to 1000. °C. What mass of $CaCO_3$ must decompose to achieve the equilibrium pressure of CO_2?

149. Use the fact that the equilibrium constant K_c equals the ratio of the forward rate constant divided by the reverse rate constant, together with the Arrhenius equation $k = Ae^{-E_a/RT}$, to show that a catalyst does not affect the value of an equilibrium constant even though the catalyst increases the rates of forward and reverse reactions. Assume that the frequency factors *A* for forward and reverse reactions do not change, and that the catalyst lowers the activation barrier for the catalyzed reaction.

150. Consider the reaction mechanism given for formation of NOBr in Section 13.7.

Step 1: $NO(g) + Br_2(g) \rightleftharpoons NOBr_2(g)$ fast

Step 2: $NOBr_2(g) + NO(g) \longrightarrow 2\ NOBr(g)$ slow

(a) What is the overall stoichiometric equation?

(b) Assume that when the reaction reaches equilibrium, each step in the mechanism also reaches equilibrium. Write the equilibrium constant for each step in terms of the forward and reverse rate constants for that step.

(c) Use your result in part (b) to derive the equilibrium constant for the overall stoichiometric equation in terms of the rate constants for the forward and reverse reactions of each step in the mechanism.

(d) Does your result confirm the statement made on p. 660 in the margin that the equilibrium constant can be obtained by taking the product of the rate constants for all forward steps and dividing by the product of the rate constants for all reverse steps?

151. A reaction important in smog formation is

$$O_3(g) + NO(g) \rightleftharpoons O_2(g) + NO_2(g) \qquad K_c = 6.0 \times 10^{34}$$

(a) If the initial concentrations are $[O_3] = 1.0 \times 10^{-6}$ M, $[NO] = 1.0 \times 10^{-5}$ M, $[NO_2] = 2.5 \times 10^{-4}$ M, and

$[O_2] = 8.2 \times 10^{-3}$ M, is the system at equilibrium? If not, in which direction does the reaction proceed?

(b) If the temperature is increased, as on a very warm day, will the concentrations of the products increase or decrease?

152. When a mixture of hydrogen and bromine is maintained at normal atmospheric pressure and heated above 200. °C in a closed container, the hydrogen and bromine react to form hydrogen bromide and a gas-phase equilibrium is established.
   (a) Write a balanced chemical equation for the equilibrium reaction.
   (b) Use bond enthalpies from Table 8.2 (← **p. 345**) to estimate the enthalpy change for the reaction.
   (c) Based on your answers to parts (a) and (b), which is more important in determining the position of this equilibrium, the entropy effect or the energy effect?
   (d) In which direction will the equilibrium shift as the temperature increases above 200. °C? Explain.
   (e) Suppose that the pressure were increased to triple its initial value. In which direction would the equilibrium shift?
   (f) Why is the equilibrium not established at room temperature?

153. Neither $PbCl_2$ nor PbF_2 is appreciably soluble in water. If solid $PbCl_2$ and solid PbF_2 are placed in equal volumes of water in separate beakers, in which beaker is the concentration of Pb^{2+}(aq) greater? Equilibrium constants for these solids dissolving in water are

$$PbCl_2(s) \rightleftharpoons Pb^{2+}(aq) + 2\ Cl^-(aq) \qquad K_c = 1.7 \times 10^{-5}$$
$$PbF_2(s) \rightleftharpoons Pb^{2+}(aq) + 2\ F^-(aq) \qquad K_c = 3.7 \times 10^{-8}$$

154. A sample of pure SO_3 weighing 0.8312 g was placed in a 1.00-L flask, sealed, and heated to 1100. K to decompose it partially.

$$2\ SO_3(g) \rightleftharpoons 2\ SO_2(g) + O_2(g)$$

If a total pressure of 1.295 atm was developed, calculate the value of K_c for this reaction at this temperature.

155. At 25 °C the vapor pressure of water is 0.03126 atm.
   (a) Calculate K_P and K_c for

$$H_2O(\ell) \rightleftharpoons H_2O(g)$$

   (b) What is the value of K_P for this same reaction at 100. °C?
   (c) Suggest a general rule for calculating K_P for any liquid in equilibrium with its vapor at its normal boiling point.

156. A student studies the equilibrium

$$I_2(g) \rightleftharpoons 2\ I(g)$$

at a high temperature. She finds that the total pressure at equilibrium is 40.% greater than it was originally, when only I_2 was present at a pressure of 1.00 atm in the same sealed container. Calculate K_p for this reaction at that temperature.

157. The equilibrium constant K_c for the reaction

$$N_2(g) + 3\ H_2(g) \rightleftharpoons 2\ NH_3(g)$$

has the value 5.97×10^{-2} at 500. °C. If 1.00 mol N_2 gas and 1.00 mol H_2 gas are heated to 500. °C in a 10.00-L sealed flask together with a catalyst, calculate the percentage of N_2 converted to NH_3. (*Hint:* Assume that only a very small fraction of the reactants is converted to products. Obtain an approximate answer and use it to obtain a more accurate result.)

158. What percentage of N_2 would be converted to NH_3 in Question 157 if the volume of the flask were 5.00 L?

Conceptual Challenge Problems

These rigorous, thought-provoking problems integrate conceptual learning with problem solving and are suitable for group work.

Conceptual Challenge Problems CP14.A, CP14.B, CP14.C, CP14.D, and CP14.E are related to the information in this paragraph. Aqueous iron(III) ions, Fe^{3+}(aq), are nearly colorless. If their concentration is 0.001 M or lower, a person cannot detect their color. Thiocyanate ions, SCN^-(aq), are colorless also, but monothiocyanatoiron(III) ions, $Fe(SCN)^{2+}$(aq), can be detected at very low concentrations because of their color. These ions are light amber in very dilute solutions, but as their concentration increases, the color intensifies and appears blood-red in more concentrated solutions. Suppose you prepared a stock solution by mixing equal volumes of 1.0×10^{-3}-M solutions of both iron(III) nitrate and potassium thiocyanate solutions. The equilibrium reaction is

$$\underset{\text{colorless}}{Fe^{3+}(aq)} + \underset{\text{colorlesss}}{SCN^-(aq)} \rightleftharpoons \underset{\text{amber}}{Fe(SCN)^{2+}(aq)}$$

CP14.A (Section 14.1) Describe how you would use 5-mL samples of the stock solution and additional solutions of 0.010-M Fe^{3+}(aq) and 0.010-M SCN^-(aq) to show experimentally that the reaction between Fe^{3+}(aq) and SCN^-(aq) does not go to completion but instead reaches an equilibrium state in which appreciable quantities of reactants and product are present. (Refer to the first paragraph for further information.)

CP14.B (Section 14.1) Suppose that you added 1 drop of 0.010-M Fe^{3+}(aq) to a 5-mL sample of the stock solution, followed by 10 drops of 0.010-M SCN^-(aq). You treated a second 5-mL sample of the stock solution by first adding 10 drops of 0.010-M SCN^-(aq), followed by 1 drop of Fe^{3+}(aq). How would the color intensity of these two solutions compare after the same quantities of the same solutions were added in reverse order? (Refer to the first paragraph for further information.)

CP14.C (Section 14.6) Predict what will happen if you add a small crystal of sodium acetate to a 5-mL sample of the stock solution (described in the first paragraph) so that some acetatoiron(III) ion, a coordination complex, is formed.

CP14.D (Section 14.6) Predict what will happen if you begin to add a 0.010-M solution of Fe^{3+}(aq) drop by drop to a 5-mL sample of the stock solution (described in the first paragraph) until the total volume becomes 10 mL. (A 0.010-M solution of Fe^{3+} ions is pale yellow.) Predict what will happen if you do the same experiment but add 0.010-M SCN^-(aq) to the stock solution. Predict what will happen if you mix 0.010-M Fe^{3+}(aq) with 0.010-M SCN^-(aq). Would the results of these three experiments be similar or different? Explain.

CP14.E (Section 14.6) Predict what will happen if you put a 5-mL sample of the stock solution (described in the first paragraph) in a hot water bath. Predict what will happen if the solution is placed in an ice bath.

CP14.F (Section 14.4) Consider the equilibrium reaction between dioxygen and trioxygen (ozone). What is the minimum volume of air (21% dioxygen by volume) at 1.00 atm and 25 °C that you would predict to have at least one molecule of trioxygen, if the only source of trioxygen were its formation from dioxygen and if the atmospheric system were at equilibrium?

$$3\ O_2(g) \rightleftharpoons 2\ O_3(g) \qquad K_c = 6.3 \times 10^{-58} \quad (\text{at } 25\ °C)$$

(The volume of 1 mol air at 1 atm and 25 °C is 24.45 L.)

15

The Chemistry of Solutes and Solutions

Borodaev/Shutterstock

Oceans are vast aqueous solutions containing Na^+, Mg^{2+}, and Cl^- ions plus other solutes that give ocean water its characteristic salinity, vapor pressure, and freezing point. This chapter considers the macroscale to nanoscale links that govern how solutes dissolve in solvents, how to express solute concentration in a solution, and how concentration affects the freezing point, boiling point, and osmotic pressure of a solution.

Every day we all encounter many solutions, such as soft drinks, juices, coffee, and gasoline. A *solution* is a homogeneous mixture of two or more substances **(← *p. 13*)**. The component present in greatest quantity is usually called the *solvent;* the other components are *solutes* **(← *p. 190*)**. In sweetened iced tea, for example, water is the solvent; sugar and soluble extracts of tea are the solutes.

Although solids dissolved in liquids or mixtures of liquids are the most common types of solutions, other kinds are possible as well, encompassing the three physical states of matter (Table 15.1). Chemistry often focuses on liquid solutions, and in particular on those in which water is the solvent

OWL Sign in to OWL at **www.cengage.com/owl** to view tutorials and simulations, develop problem-solving skills, and complete online homework assigned by your professor.

go Chemistry Download mini lecture videos for key concept review and exam prep from OWL or purchase them from **www.CengageBrain.com.**

Companion Website

Visit this book's companion website at **www.cengage.com/chemistry/moore** to work interactive modules for the Estimation boxes and Active Figures in this text.

Table 15.1 Types of Solutions

Type of Solution	Example
Gas in gas	Air—a mixture principally of N_2 and O_2 but containing other gases as well
Gas in liquid	Carbonated beverages (CO_2 in water)
Gas in solid	Hydrogen in palladium metal
Liquid in liquid	Motor oil—a mixture of liquid hydrocarbons; ethanol in water
Solid in liquid	The oceans (dissolved Na^+, Cl^-, and other ions)
Solid in solid	Bronze (copper and tin); pewter (tin, antimony, and lead)

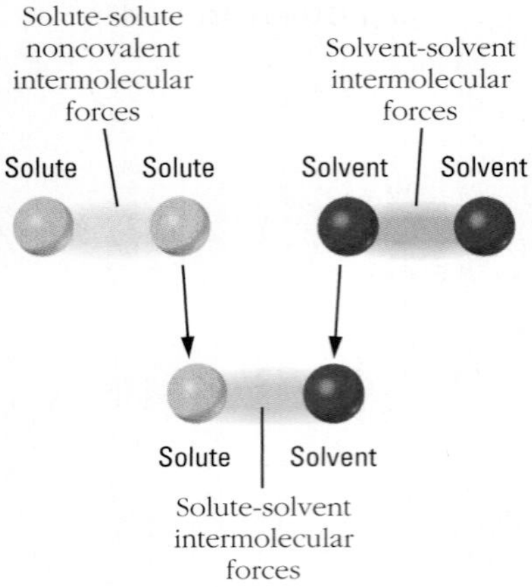

(aqueous solutions). Much of this chapter is devoted to aqueous solutions because water is the most important solvent on our planet.

In this chapter we explore in some detail the macroscale to nanoscale connections between solutes and solvents to answer questions such as:

- Why does a particular solvent readily dissolve one kind of solute, but not another? Water, for example, dissolves NaCl, but does not dissolve gasoline.
- In what ways can the concentration of a dissolved solute be expressed?
- Is thermal energy released or absorbed when a solute dissolves?
- How do changes in temperature or pressure affect the solubility of a solute in a given solvent? Why, for example, does a cold, carbonated beverage become "flat" when it is opened and warmed to room temperature?

To answer these questions, we will apply what you learned from Chapter 9 about the noncovalent forces that act between molecules—London forces, dipole-dipole forces, and hydrogen bonding (← *p. 406*)—to the interactions among solute and solvent molecules. We will also utilize the thermodynamic and equilibrium principles presented in Chapters 6 and 14. These principles are important to understand the effect that solutes have on the vapor pressures, melting points, and boiling points of solvents. In addition, the nature of unwanted solutes in natural and polluted water is discussed.

15.1 Solubility and Intermolecular Forces

In every solution, the interplay between solute and solvent particles determines whether a solute will dissolve in a particular solvent and how much can dissolve.

Solute-Solvent Interactions

An old adage correctly states that "oil and water don't mix." Chemists use a similar saying about solubility: "Like dissolves like," where "like" refers to solutes and solvents whose molecules attract each other by similar types of noncovalent intermolecular forces. Such forces exist between solute particles, as well as between solvent particles. Dissolving a solute in a solvent is favored when the solute-solvent intermolecular forces are stronger than the solute-solute or the solvent-solvent intermolecular forces.

Consider, for example, dissolving the hydrocarbon octane, C_8H_{18}, in carbon tetrachloride, CCl_4 (Figure 15.1a). Both are nonpolar liquids that dissolve in each other because of the similar London forces between molecules in each compound. Liquids such as these that dissolve in each other in any proportion are said to be **miscible.** In contrast, gasoline, a mixture of nonpolar hydrocarbons, is not miscible with water, a polar substance (Figure 15.1b). The nonpolar hydrocarbons cannot hydrogen bond to water molecules, but rather stay attracted to each other through London forces (solute-solute attractions); the water molecules remain hydrogen-bonded to each other (solvent-solvent attractions). Liquids such as gasoline and water, with noncovalent attractions so different between their molecules that they do not dissolve in each other, are described as **immiscible.**

The differing solubilities of various alcohols in water further illustrate the "like dissolves like" principle and the role of noncovalent intermolecular forces. Simple, low-molar-mass alcohols dissolve in water due to hydrogen bonding between water molecules and the —OH group of the alcohol. The hydrogen bonding forces in the solution are the same forces as those in pure water or pure ethanol. The hydrogen bonding between ethanol and water is illustrated in Figure 15.2.

Alcohol molecules contain a polar region, the —OH group, and a nonpolar portion, the hydrocarbon part. The polar region is *hydrophilic* ("water loving"); any polar part of a molecule will be hydrophilic because of its attraction to polar water

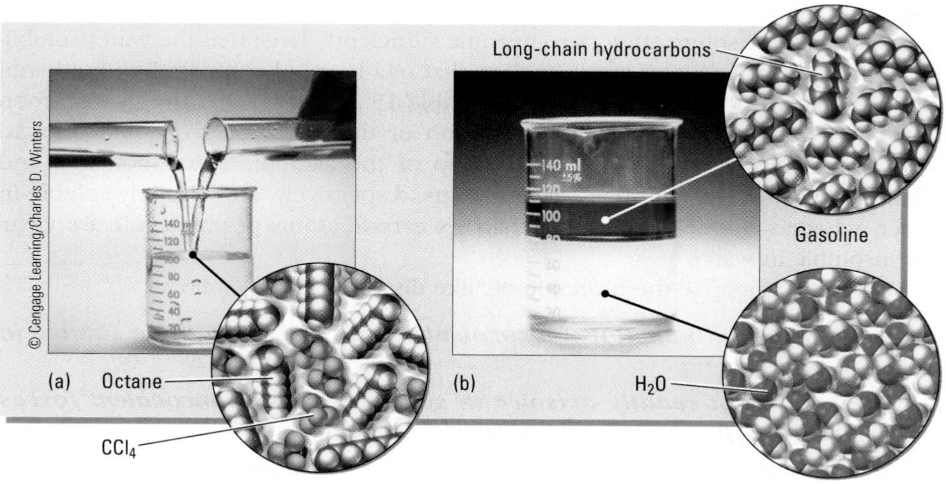

Active Figure 15.1 (a) Miscible and (b) immiscible liquids. When carbon tetrachloride and octane, both colorless, clear liquids, are mixed (a), each dissolves completely in the other. They are miscible and there is no sign of an interface or boundary between them. (b) When gasoline and water are put together, the mixture remains as two distinct layers because the two liquids are immiscible; they do not dissolve in each other. Visit this book's companion website at **www.cengage.com/chemistry/moore** to test your understanding of the concepts in this figure.

There is a dye in the gasoline to make its color different from that of water.

molecules. The nonpolar hydrocarbon region is *hydrophobic* ("water hating") ***(← p. 409).*** In a low-molar-mass alcohol such as methanol or ethanol (Figure 15.2), hydrogen bonding of the —OH group with water molecules is stronger than the London forces between the nonpolar hydrocarbon-like parts of the alcohol molecules. As the hydrophobic hydrocarbon chain of the alcohol increases in length, the molecular structure of the alcohol becomes less and less like that of water and more and more like that of a hydrocarbon. The London forces between the nonpolar hydrocarbon portion of the alcohol molecules become stronger, so a point is

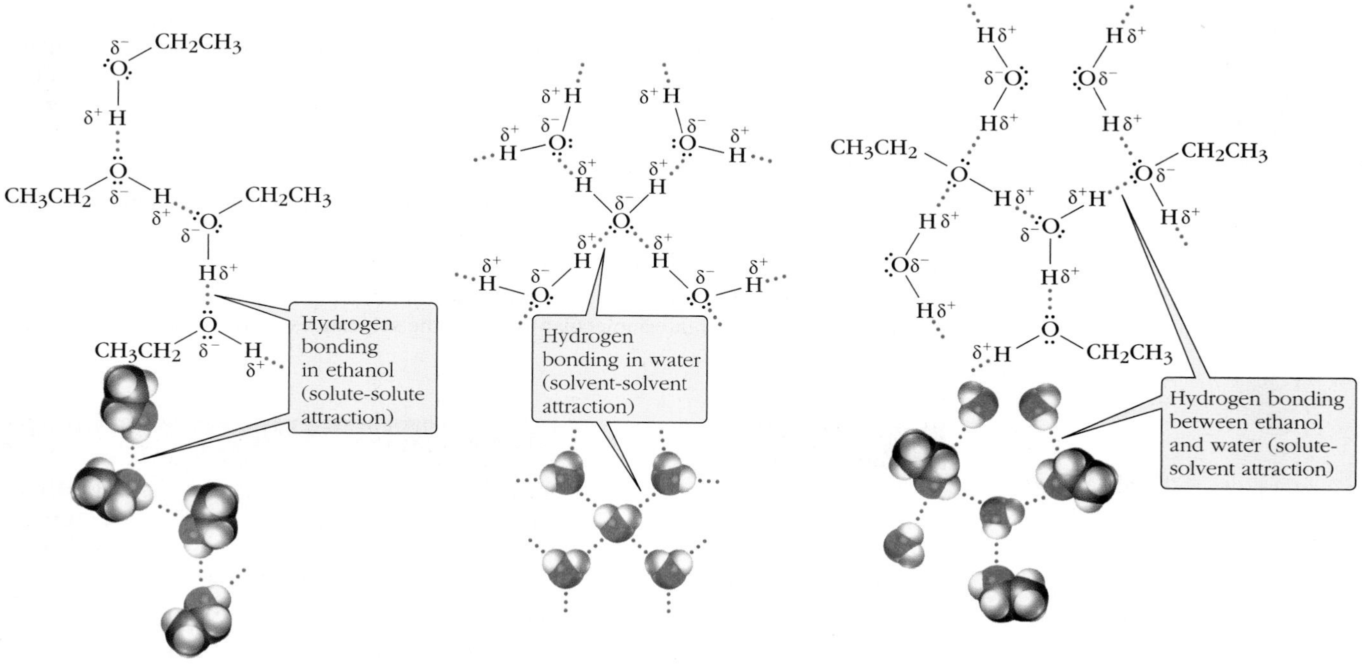

(a) among ethanol molecules (solute-solute attraction)

(b) among water molecules (solvent-solvent attraction)

(c) among ethanol and water molecules (solute-solvent attraction)

Figure 15.2 Hydrogen bonding. Hydrogen bonding is illustrated in (a) pure ethanol, (b) pure water, and (c) an ethanol-water solution.

Natalie B. Fobes/National Geographic/Getty Images

Alaskan oil spill. This oil spill in Prince William Sound, Alaska, is a large-scale example of the nonsolubility of crude oil, a nonpolar material, in water, a polar substance. The oil floats on top of the water, but does not dissolve in it.

Answers to **EXERCISES** are provided at the back of this book in Appendix L.

EXERCISES that are labeled **CONCEPTUAL** are designed to test your understanding of one or more concepts; they usually involve qualitative rather than quantitative thinking.

The **PROBLEM-SOLVING STRATEGY** in this book is

- Analyze the problem
- Plan a solution
- Execute the plan
- Check that the result is reasonable

Appendix A.1 explains this in detail.

reached at which the London forces between the hydrocarbon portion of alcohol molecules (solute-solute attraction) become sufficiently large that the water solubility of the alcohol becomes small, such as that of 1-hexanol compared with ethanol (Table 15.2). The condensed formulas in Table 15.2 indicate that the hydrocarbon portion becomes a relatively larger portion of the molecule from methanol to 1-hexanol compared with the —OH group of the alcohol. Thus, methanol and ethanol, with just one and two carbon atoms, respectively, are infinitely soluble in water, whereas alcohols with more than six carbon atoms per molecule are virtually insoluble in water.

We can summarize the principle of "like dissolves like" as follows:

- ***Substances with similar noncovalent forces are likely to be soluble in each other.***
- ***Solutes do not readily dissolve in solvents whose noncovalent forces are quite different from their own.***
- ***Stronger solute-solvent attractions favor solubility. Stronger solute-solute or solvent-solvent attractions reduce solubility.***

CONCEPTUAL EXERCISE 15.1 Predicting Water Solubility

How could the data in Table 15.2 be used to predict the solubility in water of 1-octanol or 1-decanol?

CONCEPTUAL EXERCISE 15.2 Predicting Solubility

You have a sample of 1-octanol and a sample of methanol. Which is more water-soluble? Which is more soluble in gasoline? Explain your choices in terms of "like dissolves like."

PROBLEM-SOLVING EXAMPLE 15.1 Predicting Solubilities

A beaker initially contains three layers (Figure 15.3): In part (a) the colorless top layer is heptane, C_7H_{16} (density = 0.684 g/mL); the green middle layer is an aqueous solution of $NiCl_2$ (density = 1.10 g/mL); the colorless bottom layer is carbon tetrachloride, CCl_4 (density = 1.59 g/mL). After the liquids are mixed, two layers remain—a lower, colorless layer and an upper green-colored layer. Using the principle of "like dissolves like," explain what resulted when the liquids were mixed.

Answer The nonpolar liquids carbon tetrachloride and heptane dissolved in each other and the solution's density is greater than that of $NiCl_2(aq)$; neither the nickel chloride nor the water dissolved in the nonpolar liquids.

Strategy and Explanation To apply the "like dissolves like" principle requires identifying the polarity and intermolecular forces of the substances in the beaker. Hexane is a

Table 15.2 Solubilities of Some Alcohols in Water

Name	Formula	Solubility (g/100 g H_2O at 20 °C)	Name	Formula	Solubility (g/100 g H_2O at 20 °C)
Methanol	CH_3OH	Miscible	1-Butanol	$CH_3(CH_2)_3OH$	7.9
Ethanol	CH_3CH_2OH	Miscible	1-Pentanol	$CH_3(CH_2)_4OH$	2.7
1-Propanol	$CH_3(CH_2)_2OH$	Miscible	1-Hexanol	$CH_3(CH_2)_5OH$	0.6

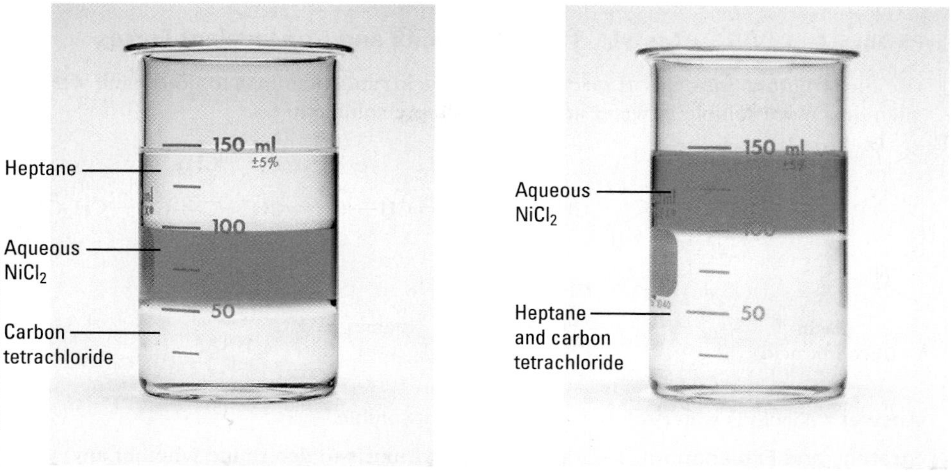

(a) Three-component mixture (b) After mixing

Figure 15.3 **Miscibility of liquids.** (a) This mixture was prepared by carefully layering three liquids of differing densities: a colorless bottom layer of carbon tetrachloride; a middle layer of green aqueous nickel(II) chloride; and a colorless upper layer of heptane. (b) The mixture after the components were thoroughly mixed.

hydrocarbon and thus nonpolar with London forces between hexane molecules. Likewise, carbon tetrachloride is a nonpolar molecule (no *net* dipole even though each C—Cl bond is polar) due to the symmetry of its tetrahedral shape ***(← p. 377).*** Nickel(II) chloride is a green, ionic solid that dissolves in water to form a green solution due to the attraction of polar water molecules for the Ni^{2+} cations and Cl^- anions in the compound. Thus, there is little tendency for the polar aqueous solution to mix with either of the nonpolar liquids above and below the aqueous layer. Once the mixture is stirred to mix the components, the nonpolar liquids dissolve in each other to form a bottom layer denser than the upper aqueous layer, $NiCl_2$ (aq).

PROBLEM-SOLVING PRACTICE answers are provided at the back of this book in Appendix K.

PROBLEM-SOLVING PRACTICE 15.1

Using the principle of "like dissolves like," predict whether
(a) Ethylene glycol, $HOCH_2CH_2OH$, dissolves in gasoline
(b) Molecular iodine dissolves in carbon tetrachloride
(c) Motor oil dissolves in carbon tetrachloride

Explain your predictions.

Solids such as quartz, SiO_2, that are held together by an extensive network of covalent bonds ***(← p. 508)*** are generally insoluble in polar or nonpolar solvents. In quartz, the silicon and oxygen atoms form SiO_4 tetrahedra linked through covalent bonds to shared oxygen atoms. These strong covalent bonds are not broken by weaker attractions to solvent molecules. Thus, quartz (and sand derived from it) is insoluble in water or any other solvent at room temperature. Sandy beaches do not dissolve in ocean water.

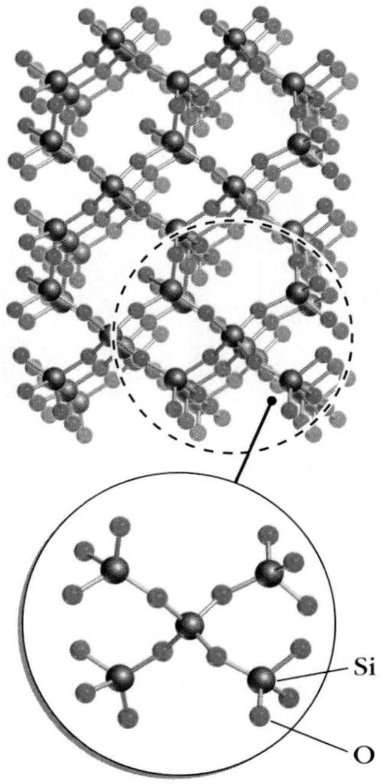

A portion of a quartz, SiO_2, structure. The structure is based on SiO_4 tetrahedra linked through shared oxygen atoms.

Solubility of a substance in water or in the triglycerides—nonpolar fats or oily substances ***(← p. 556)***—plays an important role in our body chemistry. Vitamins, for example, are either water-soluble or fat-soluble. A major significance of this difference is the danger of overdosing on fat-soluble vitamins because they are stored in fatty tissues and may accumulate to harmful levels. By contrast, overdosing on water-soluble vitamins is difficult and uncommon because these vitamins are not stored in the body and any excess of them is excreted in urine.

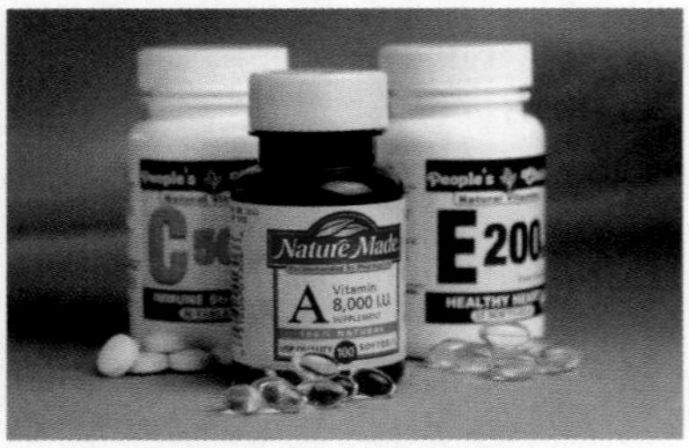

Some vitamins are fat-soluble; some are water-soluble.

PROBLEM-SOLVING EXAMPLE 15.2 Solubility and Noncovalent Forces

Use the structural formulas of niacin (nicotinic acid) and vitamin A to determine which vitamin is more soluble in water and which is more soluble in fat.

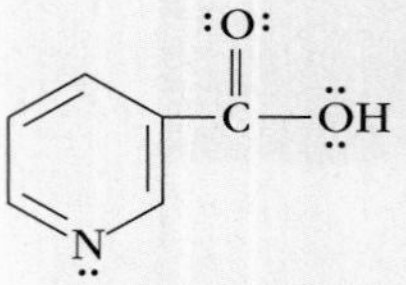

niacin
(nicotinic acid)

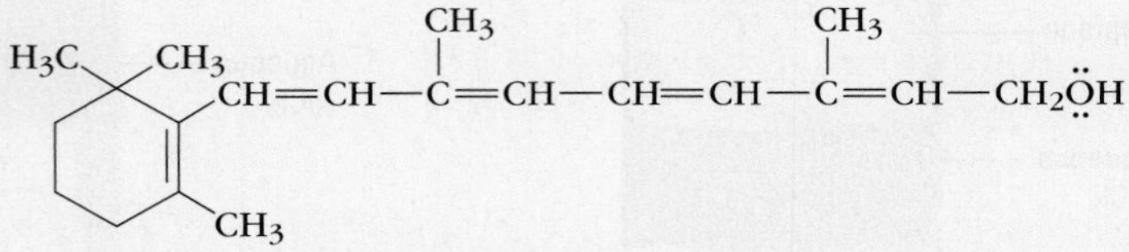

vitamin A

Answer Niacin is water-soluble; vitamin A is fat-soluble.

Strategy and Explanation Use the structural formulas to determine whether any water-soluble or fat-soluble portions exist in the molecules.

Niacin is water-soluble because its —COOH group and nitrogen atom can hydrogen bond with water molecules.

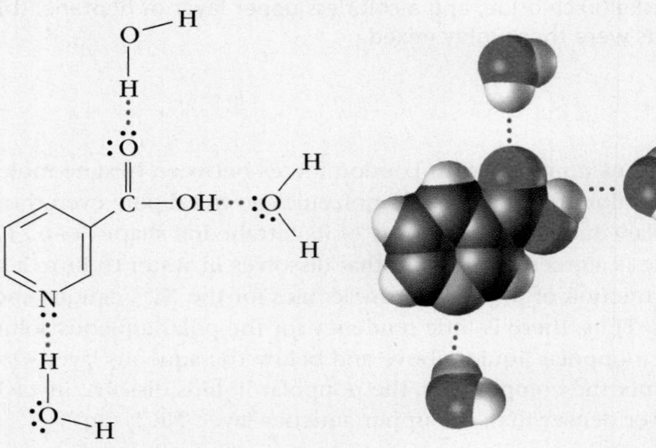

niacin
(nicotinic acid)

Niacin dissolves because the solute-solvent interaction between niacin and water is stronger than hydrogen bonding between water molecules or the dipole-dipole attractions and hydrogen bonding between niacin molecules.

Vitamin A has an —OH group that can hydrogen bond with water. However, as is the case with other long-chain alcohols, the hydrocarbon portion of the molecule is large enough to make the molecule hydrophobic. London forces between the hydrocarbon portions are more important than the hydrogen bonding, and, thus, vitamin A is insoluble in water. On the other hand, the extended hydrocarbon portion of the molecule is soluble in the long-chain hydrocarbon portion of fats.

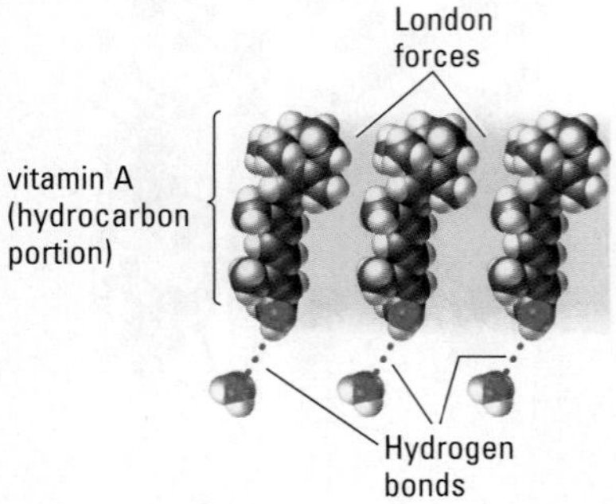

PROBLEM-SOLVING PRACTICE 15.2

Explain why vitamin C, which has the structure shown in the margin, is water-soluble.

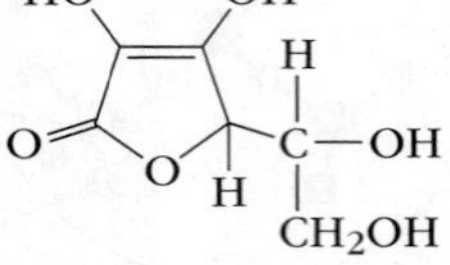

Vitamin C

15.2 Enthalpy, Entropy, and Dissolving Solutes

A solution forms when atoms, molecules, or ions of one kind mix with atoms, molecules, or ions of a different kind. For the solution to form, intermolecular forces attracting solvent molecules to each other, and such forces attracting solute molecules or ions to each other must be overcome (*steps a and b* in Figure 15.4).

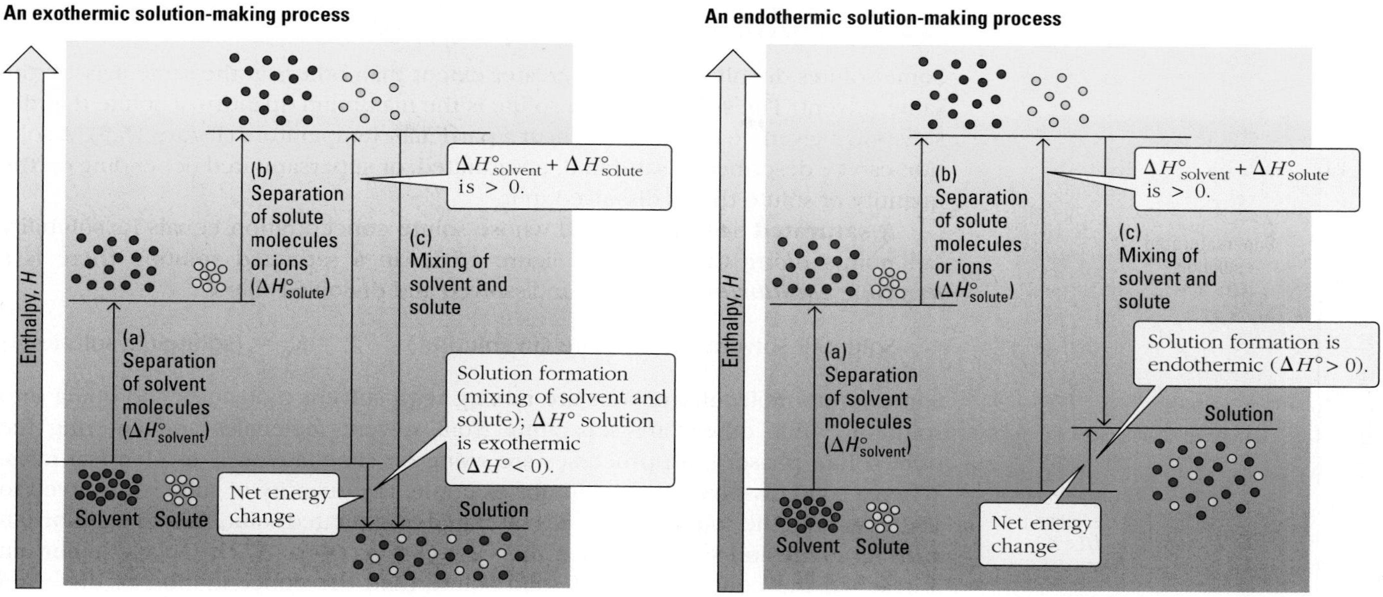

Figure 15.4 **The solution-making process.**

- **Step (a)** The enthalpy of the collection of solvent molecules increases due to their separation. This is an *endothermic* process with a positive ΔH ($\Delta H > 0$).
- **Step (b)** The separation of solute ions or molecules is also an *endothermic* process ($\Delta H > 0$).
- **Step (c)** When solute and solvent particles mix, they attract each other and a decrease in enthalpy occurs; ΔH is negative ($\Delta H < 0$). Energy is released in this *exothermic* step.

If the enthalpy of the final solution is lower than that of the initial solute and solvent, as shown in Figure 15.4 (*left*), a net release of thermal energy to the surroundings occurs and the solution-making process is *exothermic.* When the enthalpy of the final solution is greater than the enthalpies of the initial solute and solvent, the process is *endothermic* and the surroundings transfer thermal energy to the system (Figure 15.4, *right*). The net energy change, either exothermic or endothermic, is called the **enthalpy of solution,** ΔH_{soln}.

The enthalpy of solution is also known as the heat of solution.

$$\text{Enthalpy of solution} = \Delta H_{soln} = \Delta H_{step\ a} + \Delta H_{step\ b} + \Delta H_{step\ c}$$

After the solute and solvent mix, the solvent-solute forces (indicated by the magnitude of $\Delta H_{step\ c}$) may not be strong enough to overcome the solute-solute and solvent-solvent attractions ($\Delta H_{step\ a} + \Delta H_{step\ b}$). Under these conditions, dissolving is not favored by the enthalpy effect ***(← p. 688)***.

When solutes and solvents mix to form solutions, solute and solvent molecules usually occupy a larger volume. This spreads their energy over a larger volume and results in an increase in entropy ***(← p. 688)***. As described in Section 14.7, *entropy* is a measure of the energy dispersal of a system; an increase in entropy indicates a process where energy is spread out more, and dispersal of energy corresponds with dispersal of matter. *Processes in which the entropy of the system increases tend to be product-favored.* Therefore, in many cases of solution making, a large entropy increase occurs as solvent and solute molecules mix. When the enthalpy of solution is rather small, this increase in entropy results in solution formation, such as when octane and carbon tetrachloride are mixed. In other cases, even though the enthalpy of solution is positive and significantly large ($\Delta H_{soln} > 0$), the entropy increase is large enough to cause the solution to form. This is true for some ionic solutes, such as ammonium nitrate, NH_4NO_3 (Section 15.3). Because molecules in gases are always much more dispersed than liquids, there is a significant decrease in entropy as gas solute molecules are brought closer together between solvent molecules.

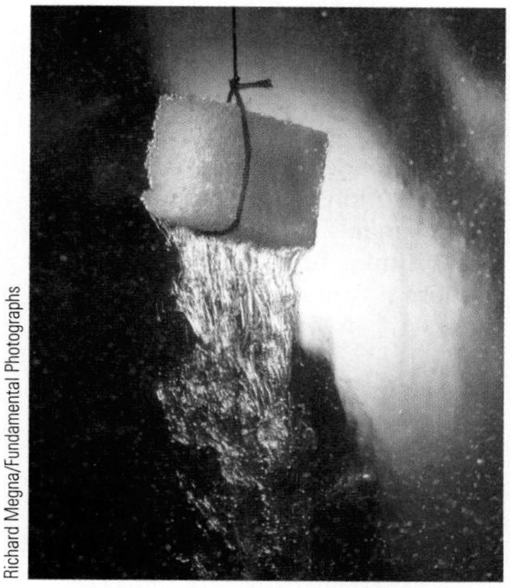

A solute (sugar) dissolving in a solvent (water). As the solute dissolves, concentration gradients are observed as wavy lines.

15.3 Solubility and Equilibrium

Some solutes dissolve to a much greater extent than others in the same mass of the same solvent. The **solubility** of a solute is the maximum quantity of solute that dissolves in a given quantity of solvent at a particular temperature (Figure 15.5). A solution can be described as saturated, unsaturated, or supersaturated depending on the quantity of solute that is dissolved in it.

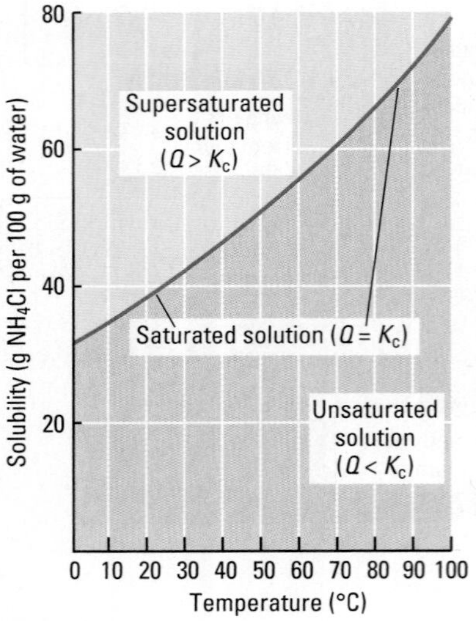

Figure 15.5 Types of solutions. The solubility of solid ammonium chloride and three types of ammonium chloride solutions—saturated (the curve), unsaturated (below the curve), and supersaturated (above the curve).

A **saturated solution** is one whose solute concentration equals its solubility (all points *along* the curve in Figure 15.5). In a saturated solution there is a *dynamic equilibrium* between undissolved and dissolved solute.

$$\text{Solute} + \text{solvent} \rightleftharpoons \text{solute (in solution)} \qquad K_c = [\text{solute (in solution)}]$$

Remember that the concentration of a pure substance (solute or solvent) does not appear in the K_c expression.

Some solute molecules or ions are mixing with solvent molecules and going into solution, while others are separating from solvent molecules and entering the pure solute phase. Both processes are going on simultaneously at identical rates. When a solid dissolves in a liquid, for example, the quantity of solid is observed to decrease until the solution becomes saturated. A saturated solution is in equilibrium with its solute and solvent, and in this case $Q = K_c$ **(← p. 672)**. Once equilibrium is reached the concentration of solid solute (and the concentration of dissolved solute) stops changing.

An **unsaturated solution** is one in which the *solute concentration is less than its solubility*; that is, *an unsaturated solution can accommodate additional solute at a given temperature* (region *under* the curve in Figure 15.5). Thus, for an unsaturated solution, $Q < K_c$, and more solute can dissolve. If solute continues to be added to an unsaturated solution at a given temperature, a point is reached where the solution becomes saturated, at which point $Q = K_c$.

Unsaturated solution: $Q < K_c$
saturated solution: $Q = K_c$
supersaturated solution: $Q > K_c$

For some solutes, there is a third case. It is possible to prepare solutions that contain *more* than the equilibrium concentration of solute at a given temperature; such a solution is **supersaturated** (the region *above* the curve in Figure 15.5). For example, a supersaturated solution of ammonium chloride can be made by first making a saturated solution at 90 °C. This is done by adding 72 g solid NH_4Cl to 100 g water, heating the mixture to 90 °C, and stirring. At that point the solution is saturated, as seen in Figure 15.5. To make the supersaturated solution, the temperature of the saturated solution is lowered very slowly to 25 °C. If this is done carefully, none of the NH_4Cl will crystallize out of solution, and the resulting solution will be supersaturated, holding more dissolved NH_4Cl than a saturated solution at that temperature; 72 g of the solute is dissolved at 25 °C, whereas the solution should hold a maximum of 40 g at that temperature (Figure 15.5). In equilibrium terms, $Q > K_c$, and excess solute should crystallize out of solution, but it is very slow to do so. Some supersaturated solutions, such as honey or fudge, can remain so for days or months. For the dissolved solute to form a crystal, several solute ions or molecules must move into an arrangement much like the appropriate crystal lattice positions, and it can take a long time for such an alignment to occur by chance. However, precipitation of a solid from a supersaturated solution occurs rapidly if a tiny crystal of the solute is added to the solution (Figure 15.6). The lattice of the added crystal provides a template onto which more ions or molecules can be added.

Fudge is made using a supersaturated sugar solution. In smooth fudge, the sugar remains uncrystallized; poor-quality fudge has a gritty texture because it contains crystallized excess sugar.

Sometimes other actions, such as stirring a supersaturated solution or scratching the inner walls of its container, will cause solute to precipitate rapidly.

EXERCISE 15.3 Crystallizing Out of Solution

Using Figure 15.5, how many grams of excess NH_4Cl would crystallize out of the supersaturated NH_4Cl solution described previously at 25 °C?

(a) (b) (c) (d) (e)

Figure 15.6 Crystallization from a supersaturated solution. Sodium acetate, $Na(CH_3COO)$, easily forms supersaturated solutions in water. The supersaturated solution was prepared by dissolving sufficient sodium acetate in water at a much higher temperature and cooling the solution very slowly. After a tiny seed crystal of sodium acetate (a) is added, some of the excess dissolved sodium acetate immediately begins to crystallize (b). Very soon (c through e), numerous sodium acetate crystals can be seen. If the solution remains uncovered for a long time, all of the water will evaporate, leaving behind the solid sodium acetate.

EXERCISE 15.4 Solubility

Refer to Figure 15.5 to determine whether each of these NH_4Cl solutions in 100 g H_2O is unsaturated, saturated, or supersaturated:

(a) 30 g NH_4Cl at 70 °C (b) 60 g NH_4Cl at 60 °C (c) 50 g NH_4Cl at 50 °C

Dissolving Ionic Solids in Liquids

Sodium chloride is an ionic compound whose crystal lattice consists of Na^+ and Cl^- ions in a cubic array ***(← p. 505).*** Strong electrostatic attractions between oppositely charged ions hold the ions tightly in the lattice. The enthalpy change when 1 mol Na^+ and Cl^- ions is completely separated from a crystal lattice is referred to as the lattice energy of the ionic compound ***(← p. 315).*** The large lattice energy of NaCl (788 kJ/mol) accounts for sodium chloride's high melting point (800 °C). It is possible for solvent molecules to attract Na^+ ions away from Cl^- ions in a crystal, but they have to be the right kind of solvent molecules. Trying to dissolve NaCl (or any other ionic compound) with carbon tetrachloride or hexane (both nonpolar solvents) is a futile exercise because nonpolar molecules have very little attraction for ions. On the other hand, water dissolves NaCl.

Water is a good solvent for an ionic compound because water molecules are small and highly polar ***(← p. 399).*** As shown in Figure 15.7 on p. 716, the partially negative oxygen atoms of water molecules are attracted to cations and help pull them away from the crystal lattice, while the partially positive hydrogen atoms of other water molecules are attracted to the anions in the lattice and help pull them away from the lattice. This process, in which water molecules surround cations and anions, is called **hydration** (Figure 15.8). Energy known as the *enthalpy of hydration* is released when these new attractions form between ions and water molecules as they mix and get close enough to one another. Energy is always required to separate the ions to overcome their attraction, and energy is always released when ions become hydrated because noncovalent forces attract the molecules to the ions.

Whether dissolving a particular ionic compound is exothermic (ΔH_{soln} is negative) or endothermic (ΔH_{soln} is positive) depends on the relative sizes of the lattice energy ***(← p. 316)*** of the ionic compound and the hydration enthalpies of its cations and anions. The relationship between the enthalpy of solution, the lattice energy of the ionic compound, and the enthalpies of hydration of the ions is

$$\Delta H_{soln} = -\text{lattice energy} + \Delta H_{hydration}\,(\text{cations}) + \Delta H_{hydration}\,(\text{anions})$$

Figure 15.9 shows how lattice energy and the enthalpies of hydration combine to give the enthalpy of solution.

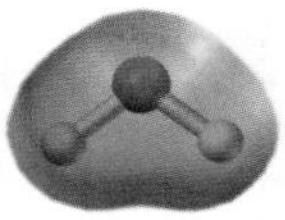

Electron density of a water molecule. The red area has high electron density and partial negative charge; the blue area has low electron density and partial positive charge.

Enthalpies of Hydration of Selected Ions (kJ/mol)

Cations		Anions	
H^+	−1130	F^-	−483
Li^+	−558	Cl^-	−340
Na^+	−444	Br^-	−309
Mg^{2+}	−2003		
Ca^{2+}	−1557		
Al^{3+}	−2537		

Energy is always released when particles that attract each other get closer together, so the lattice energy is always negative. Energy is always required to separate particles that attract each other.

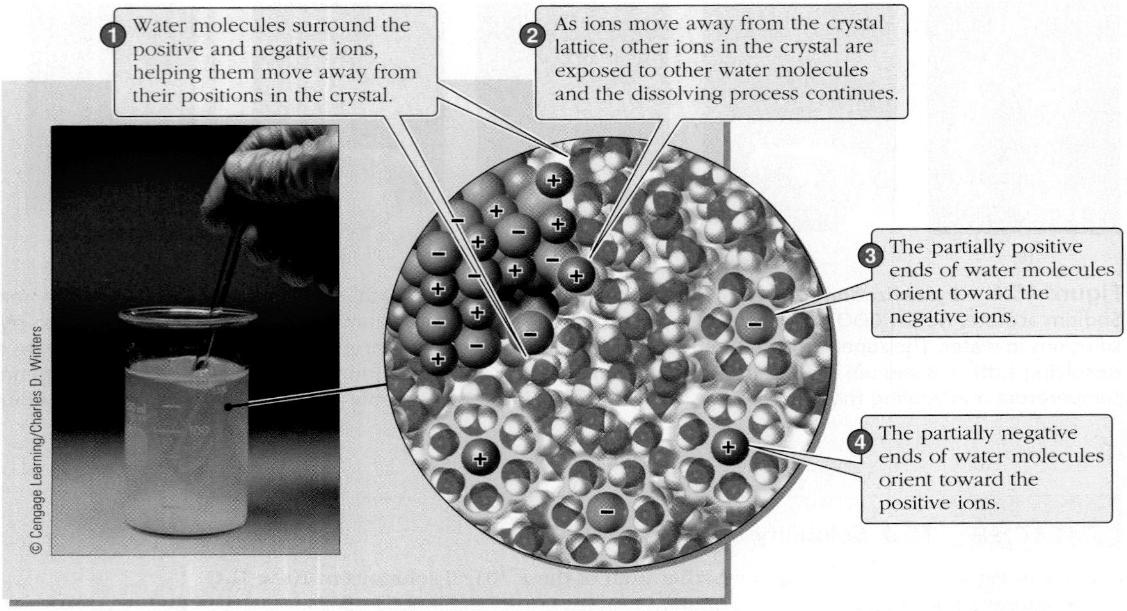

Figure 15.7 Water dissolving an ionic solid. Water molecules surround the positive (gray) and negative (green) ions, helping them move away from their positions in the crystal.

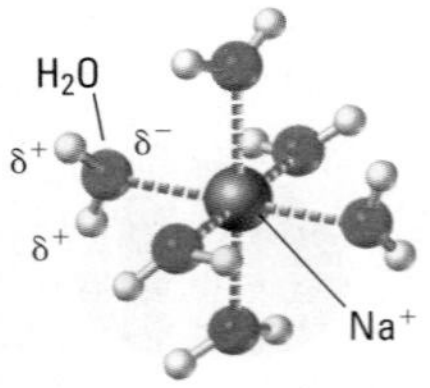

Figure 15.8 Hydration of a sodium cation. The arrangement of water molecules around this Na^+ cation is highly ordered. The dashed lines, - - - -, represent ion-dipole attraction between the central metal cation and the polar water molecules.

Practical applications of endothermic and exothermic dissolution include cold packs containing NH_4NO_3 used to treat athletic injuries and hot packs containing $CaCl_2$ used to warm hands and foods.

The solubility rules for ionic compounds in water *(← p. 163)* remind us that not all ionic compounds are highly water-soluble even though there are strong attractions between water molecules and ions. For some ionic compounds, the lattice energy is so large that water molecules cannot effectively pull ions away from the lattice. As a result, such compounds have large positive enthalpies of solution and usually are only slightly soluble.

Entropy and the Dissolving of Ionic Compounds in Water

The spreading of energy introduced when a crystal lattice breaks down, and the spreading of energy introduced by the mixing of ions with solvent molecules, both

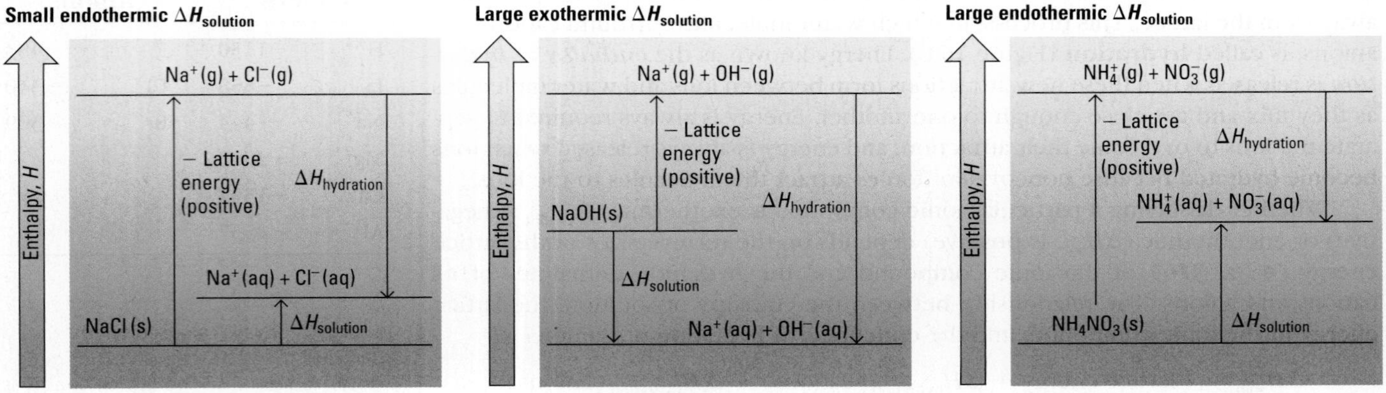

(a) For NaCl, the lattice energy that must be overcome is larger than the energy released on hydration of the ions. This causes the ΔH_{soln} of NaCl to be positive (endothermic).

(b) The lattice energy for NaOH is much smaller than the energy released when the ions become hydrated. Consequently, the ΔH_{soln} of NaOH is negative (highly exothermic).

(c) For NH_4NO_3, the heat of hydration of the ions is much smaller than the lattice energy, so the resulting ΔH_{soln} has a large positive value (highly endothermic).

Figure 15.9 Enthalpies of solution of equal amounts of three different ionic compounds dissolving in equal volumes of water.

favor the dissolving process. This entropy increase is counteracted by the ordering of solvent molecules around the ions, a negative entropy change (Figure 15.8). For 1+ and 1− charged ions, the overall entropy change is positive, and dissolving is favored. For some salts that contain 2+ or 3+ ions, the charges on the ions are so large and the ions so small that water molecules are aligned in a highly organized manner around the ions. When a large number of water molecules are locked into place by this strong hydration, the entropy of solution may be negative, which does not favor solubility. Calcium oxide (only 0.131 g CaO dissolves in 100 mL water at 10 °C) and aluminum oxide Al_2O_3 (insoluble) exemplify this effect.

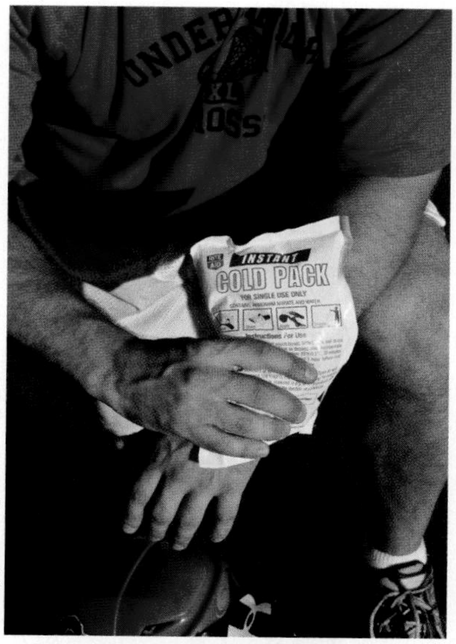

© Cengage Learning/Charles D. Winters

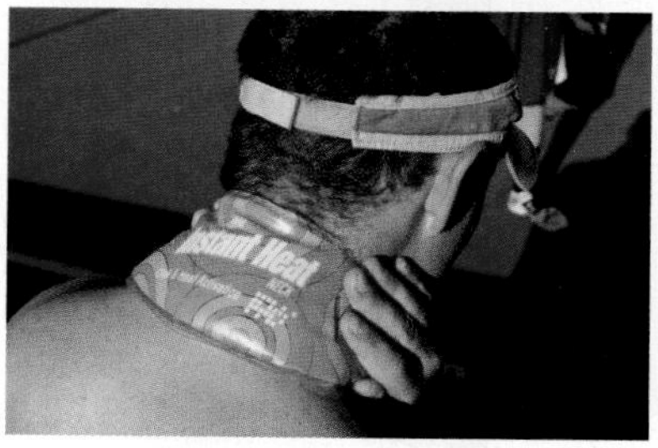

© Cengage Learning/Charles D. Winters

Commercial cold and hot packs. (*Top*) When the desired cooling is needed, ammonium nitrate in the inner container of this cold pack is brought into contact with water in the outer container by breaking the seal of the inner container. (*Bottom*) This hot pack warms its surroundings when the inner pouch containing either $CaCl_2$ or $MgSO_4$ is punctured and the compound dissolves in the water in the outer container.

15.4 Temperature and Solubility

Solubility of Gases

To understand how temperature affects gas solubility, we apply Le Chatelier's principle (***← p. 678***) to the equilibrium between a pure solute gas, the solvent, and a saturated solution of the gas.

$$\text{Gas} + \text{solvent} \rightleftharpoons \text{saturated solution} \qquad \text{Usually, } \Delta H_{\text{soln}} < 0 \text{ (exothermic)}$$

When a gas dissolves to form a saturated liquid solution, the process is almost always *exothermic*. Molecules that were relatively far apart in the gas phase are brought much closer in the solution to other molecules that attract them, lowering the potential energy of the system and releasing some energy to the surroundings.

If the temperature of a solution of a gas in a liquid increases, the equilibrium shifts in the direction that partially counteracts the temperature rise. That is, the equilibrium shifts to the left in the preceding equation. Thus, *a dissolved gas becomes less soluble with increasing temperature*. Conversely, cooling a solution of a gas that is at equilibrium with undissolved gas will cause the equilibrium to shift to the right in the direction that liberates heat, so more gas dissolves. This is illustrated in Figure 15.10 with data for the solubility of oxygen gas in water.

Cooler water in contact with the atmosphere contains more dissolved oxygen at equilibrium than water at a higher temperature. For this reason fish seek out cooler (usually deeper) waters in the summer. Fish have an easier time obtaining oxygen when its concentration in the water is higher. The decrease in gas solubility as temperature increases makes *thermal pollution* a problem for aquatic life in rivers and streams. Natural heating of water by sunlight and by warmer air can usually be accommodated. But excess heat from extended periods of very hot weather or from sources such as industrial facilities and electrical power plants can warm

Although generally useful, Le Chatelier's principle does not always correctly predict how the solubility of ionic solutes changes with temperature.

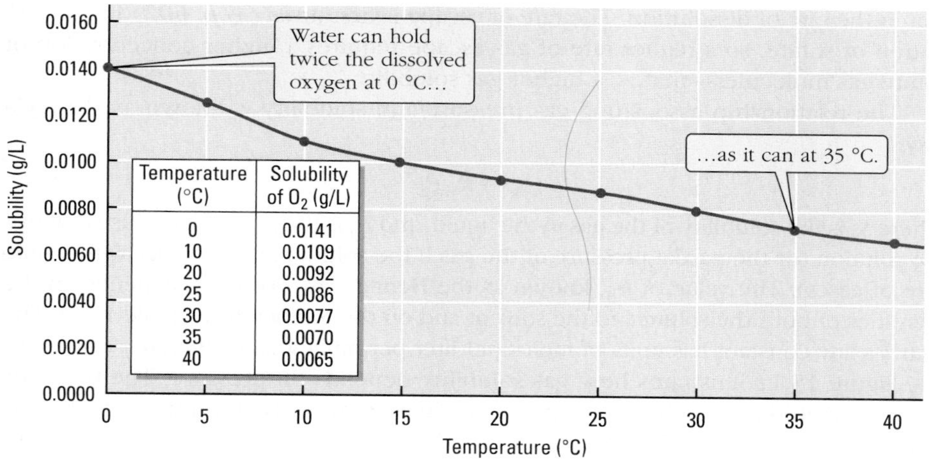

Temperature (°C)	Solubility of O_2 (g/L)
0	0.0141
10	0.0109
20	0.0092
25	0.0086
30	0.0077
35	0.0070
40	0.0065

Figure 15.10 Solubility of oxygen gas in water at various temperatures. The solubility of oxygen, like that of other gases, decreases with increasing temperature.

Mike Goldwater/Alamy

A fish kill caused by a lack of dissolved oxygen.

the water sufficiently that the concentration of dissolved oxygen is reduced to the point where some species of fish die.

CONCEPTUAL EXERCISE 15.5 Carbonated Beverages

Explain on a molecular basis why a carbonated beverage goes "flat" once it is opened and warms to room temperature.

CONCEPTUAL EXERCISE 15.6 Warming Water

Explain why water that has been used to cool a reactor in a nuclear power plant, and thus is at a relatively high temperature, must be cooled before it is put back into the lake or river from which it came.

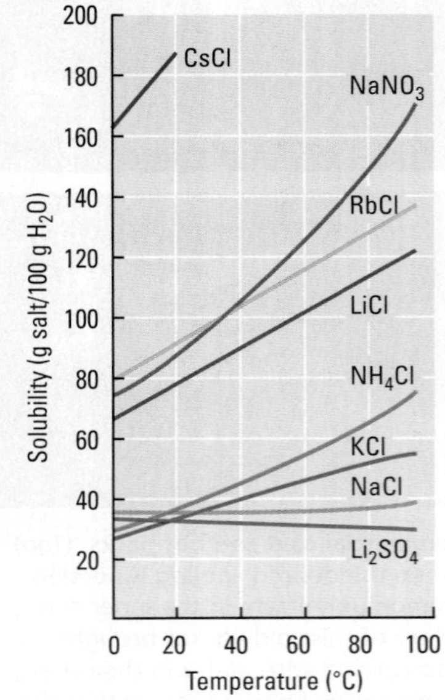

Figure 15.11 Solubility of ionic compounds versus temperature. The solubility of most ionic compounds in water increases with temperature.

Solubility of Solids

Common experience tells us that more sugar (sucrose) can dissolve in hot coffee or hot tea than in cooler coffee or tea. This is an example of the fact that the aqueous solubility of most solid solutes, including ionic compounds, increases with increasing temperature (Figure 15.11). Although predictions of solubility based on enthalpies of solution usually work, there are notable exceptions, such as Li_2SO_4.

CONCEPTUAL EXERCISE 15.7 Temperature and Solubility

If a substance has a positive enthalpy of solution, which would likely cause more of it to dissolve, hot solvent or cold solvent? Explain.

15.5 Pressure and Dissolving Gases in Liquids: Henry's Law

Pressure does not measurably affect the solubilities of solids or liquids in liquid solvents, but *the solubility of any gas in a liquid increases as the partial pressure of the gas increases*. A dynamic equilibrium is established when a gas is in contact with a liquid—the rate at which gas molecules enter the liquid phase equals the rate at which gas molecules escape from the liquid. If the pressure is increased, gas molecules strike the surface of the liquid more often, increasing the rate of dissolution of the gas. A new equilibrium is established when the rate of escape increases to match the rate of dissolution. The rate of escape is first order (← *p. 602*) in concentration of solute, so a higher rate of gas escape requires a higher concentration of solute gas molecules—that is, a higher gas solubility.

The relationship associating gas pressure and solubility is known as **Henry's law:**

$$S_g = k_H P_g$$

where S_g is the solubility of the gas in the liquid, and P_g is the pressure of the gas above the solution (or the partial pressure of the gas if the solution is in contact with a mixture of gases). The value of k_H, known as the Henry's law constant, depends on the identities of both the solute and the solvent and on the temperature (Table 15.3). The Henry's law constant has units of moles per liter per millimeter of mercury (mm Hg).

Figure 15.12 illustrates how gas solubility depends on pressure. The behavior of a carbonated drink when the cap is opened is an everyday illustration of the solubility of gases in liquids under pressure. The drink fizzes when opened because the partial pressure of CO_2 over the solution drops, the solubility of the gas decreases, and dissolved gas escapes from the solution.

Henry's law. The greater the partial pressure of CO_2 over the soft drink in the bottle, the greater the concentration of dissolved CO_2. When the bottle is opened, the partial pressure of CO_2 drops and CO_2 bubbles out of the solution.

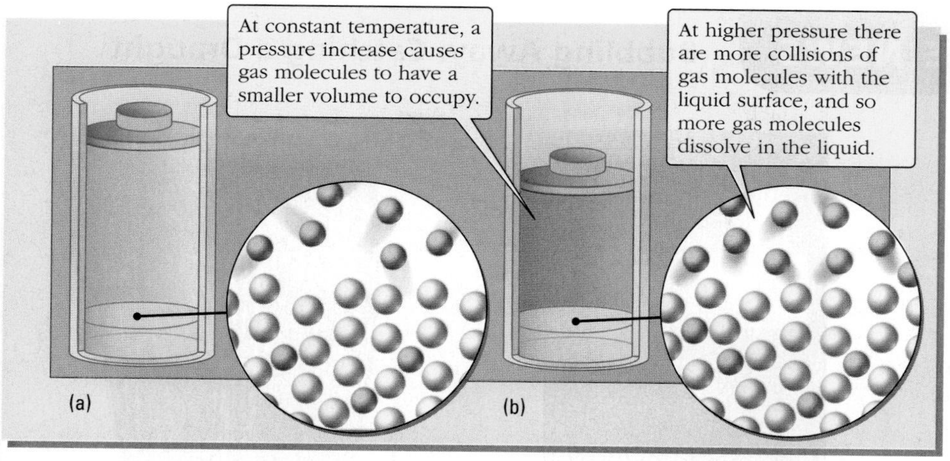

Figure 15.12 Henry's law. (a) A gas sample in a liquid under pressure in a closed container. (b) The pressure is increased at constant temperature, causing more gas to dissolve.

The partial pressure of a gas in a mixture of gases is the pressure that a pure sample of the gas would exert if it occupied the same volume as the mixture. Partial pressure is proportional to the mole fraction of the gas (← p. 447).

PROBLEM-SOLVING EXAMPLE 15.3 Using Henry's Law

The "bends" is a condition that can occur in scuba divers due to the solubility of N_2 gas in the blood. For a saturated solution, calculate
(a) the number of grams of dissolved N_2 in 1 L blood at 25 °C.
(b) the volume (mL) of $N_2(g)$ that can be dissolved in 1 L blood at 25 °C.

The Henry's law constant for N_2 in water at 25 °C is 8.42×10^{-7} mol L^{-1} mm Hg^{-1}. Assume that the solubility of N_2 in blood is approximately that of N_2 in water. The partial pressure of N_2 in the atmosphere is 0.78 atm.

Answer (a) 1.40×10^{-2} g N_2 (b) 16 mL of N_2

Strategy and Explanation Use Henry's law to calculate the solubility of N_2 gas in blood in mol/L. Then use the molar mass of N_2 to convert mol/L to g/L. Then use the ideal gas law to calculate the volume of N_2 gas from the number of moles dissolved in 1 L.

(a) We assume that the dissolved N_2 is in equilibrium with nitrogen in the air. Converting atm to mm Hg we have

$$0.78 \text{ atm} \times \frac{760 \text{ mm Hg}}{1 \text{ atm}} = 593 \text{ mm Hg}$$

Use Henry's law to calculate the solubility of N_2:

$$S_{N_2} = k_H P_{N_2} = (8.42 \times 10^{-7} \text{ mol L}^{-1} \text{ mm Hg}^{-1})(593 \text{ mm Hg}) = 4.99 \times 10^{-4} \text{ mol/L}$$

$$\text{mass of } N_2 = (4.99 \times 10^{-4} \text{ mol/L})\left(\frac{28.02 \text{ g } N_2}{1 \text{ mol } N_2}\right) = 1.40 \times 10^{-2} \text{ g } N_2\text{/L}$$

(b) Apply the ideal gas law to calculate the volume (mL) of the 4.99×10^{-4} mol N_2 gas dissolved in 1 L blood. Remember that the temperature must be in kelvins.

$$V = \frac{nRT}{P} = \frac{(4.99 \times 10^{-4} \text{ mol})(0.0821 \text{ L atm mol}^{-1} \text{ K}^{-1})(298 \text{ K})}{0.78 \text{ atm}}$$

$$= 0.0157 \text{ L} = 16 \text{ mL}$$

This is a significant volume of nitrogen gas dissolved in one liter of blood. When such a large volume of gaseous nitrogen is released from the blood as the scuba diver rises to the surface of the water, it can cause serious pain.

PROBLEM-SOLVING PRACTICE 15.3

The Henry's law constant for oxygen in water at 25 °C is 1.66×10^{-6} mol L^{-1} mm Hg^{-1}. Suppose that a trout stream at 25 °C is in equilibrium with air at normal atmospheric pressure. Calculate the concentration of O_2 in this stream. Express the result in milligrams per liter (mg/L). The mole percent of oxygen in air is 21%.

Table 15.3 Henry's Law Constants, 25 °C; Water Is the Solvent

Gas	k_H mol L^{-1} mm Hg^{-1}
N_2	8.42×10^{-7}
O_2	1.66×10^{-6}
CO_2	4.4×10^{-5}
He	4.6×10^{-7}
Air	8.6×10^{-4}

CHEMISTRY IN THE NEWS Bubbling Away: Catching a Draught

In the movie *Gigi,* Maurice Chevalier sings of "the night they invented champagne," that fizzy, celebratory beverage. As bubbles rise in a glass of champagne, their fizz and pop capture our attention. Researchers in France and Australia have investigated more closely the bubbles in champagne and in beer (the body of the liquid, not the foamy "head" on beer). In both drinks the bubbles are carbon dioxide, produced as a by-product of the natural process that produces alcohol via fermentation of sugars in grapes (wine, champagne) or grains (beer).

$$\underset{\text{glucose}}{C_6H_{12}O_6} \longrightarrow 2\ \underset{\text{ethanol}}{CH_3CH_2OH} + 2\ CO_2$$

The properties of the bubbles are affected by the concentration of dissolved gas, the alcohol content of the drink, and the concentration of large molecules, including proteins, present in the drink. Champagne has three times the CO_2 concentration and twice the alcohol content of beer. Beer has as much as 30 times more large molecules than champagne.

Bubbles in beverages generally start to grow from what are called *nucleation sites*—places where the gas first comes out of solution—such as on dust particles or on tiny imperfections on the inside walls of a glass. When a bubble gets sufficiently buoyant, it breaks away from the nucleation site. As CO_2 diffuses into it, a bubble gets larger, its buoyancy increases, and it moves more quickly to the surface. Gerard Liger-Belair and his colleagues at the University of Reims in France have used high-speed cameras to measure the growth and ascent rate of champagne bubbles, whose radii grow at the rate of 120 to 240 μm per second as they ascend from the sides of a champagne flute.

© Cengage Learning/George Semple
(a) Glass of Guinness draught.

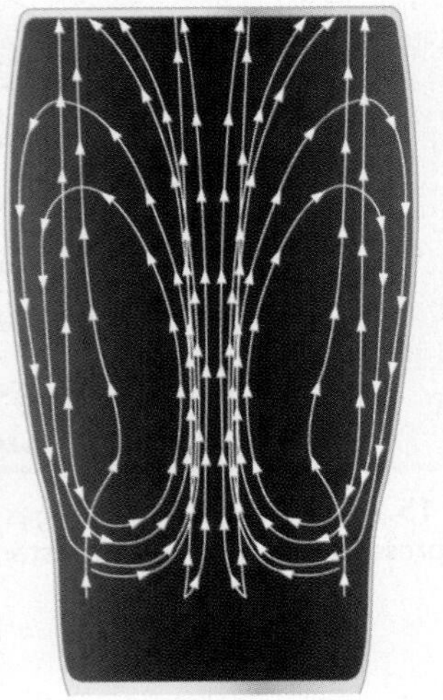
© Cengage Learning/George Semple
(b) Computer simulation of bubbles traveling in Guinness draught.

© Cengage Learning/Charles D. Winters
Rising bubbles in champagne.

Whereas bubbles in champagne always rise, those in the Irish beer Guinness draught rise and fall within the body of the beer, seemingly defying the laws of nature. For years, pub regulars argued whether the bubbles in Guinness draught actually were falling as well as rising. Was it perhaps an optical illusion or an alcohol-induced error of the observer? The higher protein concentration in beer causes more "drag" on the bubbles, slowing their rate of ascent much more than in champagne. At the higher protein concentration, the bubbles in beer act as rigid spheres because they collect a protein "skin" as they rise. The lower protein content in champagne does not create bubbles that are as rigid. Champagne bubbles would have to travel more than 10 cm (about 4 in) to pick up enough protein to become as rigid as beer bubbles. This is longer than a typical champagne flute. That explains differences for rising bubbles. But do the bubbles in the draught actually descend? Using computer modeling software, Australian researchers Clive Fletcher and colleagues demonstrated that, unlike those in champagne, bubbles in Guinness draught ascend *and* descend. The critical issue is the rigidity, size, and bouyancy of the bubbles. Beer bubbles with a radius of about 30 μm sink near the sides of the glass while 500-μm ones only rise. The barrel-like shape of the typical Guinness glass also seems to be a factor.

Richard Zare, a Stanford University chemistry professor who also has published articles about beer bubbles, points out that "Once you begin to learn about the nature of beer bubbles, you will never again look at a glass of beer in quite the same way."

Source: Weiss, P. "The Physics of Fizz." *Science News*, Vol. 157, May 6, 2000; p. 300.

http://news-service. Stanford.edu/news/2004/March 17/Beer-video-317.html

Fountain, H. "Flavor and Aroma Rise in Champagne Bubbles." *New York Times,* 29 September 2009, p. D3.

15.6 Solution Concentration: Keeping Track of Units

The terms *unsaturated, saturated*, and *supersaturated* describe a solution with respect to the quantity of solute in a given quantity of solvent at a certain temperature. Sometimes the terms *concentrated* or *dilute* are also used to describe a solution. Although useful, these are all qualitative descriptors of solution concentration, and it is often important to specify the concentration of a solute in a solution more precisely. Several concentration units are used to do so, including mass fraction, weight percent, molarity, and molality. Often it is useful to be able to quantitatively describe quantities of solute across a wide range of concentrations, from rather large to very small. For example, many different units are used to express the concentrations of dilute solutions containing unwanted—even potentially harmful—solute species in drinking water, such as those containing lead, mercury, selenium, and nitrate, and organic compounds. These units are also useful clinically in discussing solute concentrations in blood or urine.

Mass Fraction and Weight Percent

The **mass fraction** of a solute is the fraction of the total mass of the solution that a solute contributes—that is, the mass of a single solute divided by the total mass of all the solutes *and* the solvent. Mass fraction is commonly expressed as a percentage and called mass percent or **weight percent,** which is the mass fraction multiplied by 100%. Weight percent is the same as the number of grams of solute per 100 g of solution. For example, the mass fraction of sucrose in a solution consisting of 25.0 g sucrose, 10.0 g fructose, and 300. g water is

The symbol % means "per hundred," that is, divided by 100. Thus, 100% is $\frac{100}{100} = 1$. Weight percent could also be called parts per hundred.

$$\text{Mass fraction of sucrose} = \frac{\text{mass of sucrose}}{\text{mass of sucrose} + \text{mass of fructose} + \text{mass of water}}$$

$$= \frac{25.0\text{ g}}{25.0\text{ g} + 10.0\text{ g} + 300.\text{ g}} = 0.0746$$

The weight percent of sucrose in the solution is $(0.0746) \times 100\% = 7.46\%$.

Weight percent can be thought of as parts per hundred.

PROBLEM-SOLVING EXAMPLE 15.4 Mass Fraction and Weight Percent

Sterile saline solutions containing NaCl in water are often used in medicine. Calculate the weight percent of NaCl in a solution made by dissolving 4.6 g NaCl in 500. g pure water.

Answer 0.91%

Strategy and Explanation We apply the definitions of mass fraction and weight percent to the solute and solution stated in the problem.

$$\text{Mass fraction of NaCl} = \frac{4.6\text{ g NaCl}}{4.6\text{ g NaCl} + 500.\text{ g } H_2O} = 0.0091$$

$$\text{Weight percent of NaCl} = \text{mass fraction of NaCl} \times 100\%$$

$$= 0.0091 \times 100\% = 0.91\%$$

☑ **Reasonable Answer Check** The mass fraction is about 5 g NaCl in about 500 g solution, or about 0.01, which agrees with the more accurate result.

PROBLEM-SOLVING PRACTICE 15.4

Calculate the weight percent of glucose in a solution containing 21.5 g glucose, $C_6H_{12}O_6$, in 750. g pure water.

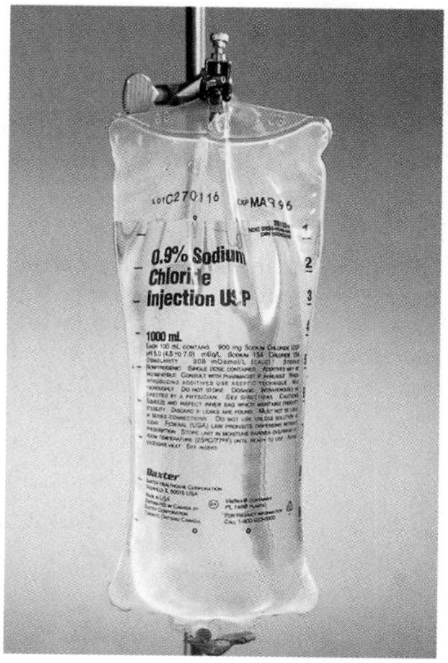

© Cengage Learning/Charles D. Winters

Sterile saline solution. Saline solutions like this one are routinely given to patients who have lost body fluids.

The density of liquid water is essentially 1 g/mL, 10^3 g/L, or 1 kg/L.

EXERCISE 15.8 Mass Fraction and Weight Percent

Ringer's solution is used in physiology experiments. One liter of the solution contains 6.5 g NaCl, 0.20 g $NaHCO_3$, 0.10 g $CaCl_2$, and 0.10 g KCl. Assume that one liter of water has been used to prepare the solution and that the density of water is 1.00 g/mL.
(a) Calculate the mass fraction and weight percent of $NaHCO_3$ in the solution.
(b) Which solute has the lowest weight percent in the solution?

Parts per Million, Billion, and Trillion

1 ppm is equivalent to one penny in \$10,000; 1 ppb is one penny in \$10,000,000.

1 μg (microgram) = 10^{-6} g; 1 ng (nanogram) = 10^{-9} g.

Solutes in very dilute solutions have very low mass fractions. Consequently, the mass fraction in such solutions is often expressed in **parts per million** (abbreviated ppm). One part per million is equivalent to one gram of solute per one million grams of solution, or proportionally, one milligram of solute per one thousand grams of solution (1 mg/kg). Thus, a commercial bottled water with a calcium ion concentration of 66 ppm contains 66 mg Ca^{2+} per 1000 g solution, essentially one liter. For even smaller mass fractions, scientists often use **parts per billion** (1 ppb = one gram of solute per one billion grams of solution or one microgram, μg, of solute per thousand grams of solution) and **parts per trillion** (1 ppt = one gram of solute per one trillion grams of solution or one nanogram, ng, of solute in one thousand grams of solution). As the names imply, a mass fraction converts to parts per billion by multiplying by 10^9 ppb and to parts per trillion by multiplying by 10^{12} ppt.

PROBLEM-SOLVING EXAMPLE 15.5 ppm, ppb, and Mass Fraction

Many water wells in the country of Bangladesh are contaminated with arsenic, a toxic material in sufficient dose, with levels well above the World Health Organization's (WHO) guideline maximum value of 0.010 mg As/L. The WHO estimates that 28–35 million Bangladesh citizens have been exposed to high arsenic levels by drinking water from wells in which the arsenic concentration in the water is at least 0.050 mg/L.
(a) Calculate the mass of arsenic per liter of solution (g/L) in a 0.050 mg/L solution. Assume that the solution is almost entirely water and the density of the solution is 1.0 g/mL.
(b) Calculate the concentration of arsenic in ppb.

Answer (a) 5.0×10^{-5} g arsenic/L solution (b) 50. ppb

Strategy and Explanation

(a) Assuming that the solution is almost entirely water, 1 L of the solution has a mass of 1.0×10^3 g. The mass of arsenic in this 1-L water sample is calculated from its mass fraction, expressed as the ratio of grams of arsenic per 10^6 g of solution.

$$\left(\frac{0.050 \text{ g arsenic}}{1 \times 10^6 \text{ g solution}}\right)\left(\frac{1.0 \times 10^3 \text{ g solution}}{\text{L solution}}\right) = 5.0 \times 10^{-5} \text{ g arsenic/L solution}$$

To express the mass fraction as parts per billion, multiply it by 10^9 ppb. This makes a very small number bigger and easier to handle.

(b) This answer can be converted to ppb by first converting it to micrograms of arsenic per liter.

$$\left(\frac{5.0 \times 10^{-5} \text{ g arsenic}}{1 \text{ L}}\right)\left(\frac{1 \text{ μg}}{1 \times 10^{-6} \text{ g}}\right) = 50. \text{ μg arsenic/L}$$

Thus, a mass fraction of 5.0×10^{-2} g arsenic per 10^6 g of solution (0.050 ppm) corresponds to a concentration of 50. μg arsenic/L, which is the equivalent of 50. ppb.

☑ **Reasonable Answer Check** Used correctly, the conversion factors change 0.050 mg/L directly to 5.0×10^{-5} g arsenic/L and 50 ppb arsenic.

PROBLEM-SOLVING PRACTICE 15.5

Drinking water often contains small concentrations of selenium (Se). If a sample of water contains 30 ppb Se, how many micrograms of Se are present in 100. mL of this water?

CONCEPTUAL EXERCISE 15.9 Lead in Drinking Water

One drinking-water sample has a lead concentration of 20. ppb; another has a concentration of 0.0030 ppm.

(a) Which sample has the higher lead concentration?

(b) The current U.S. Environmental Protection Agency acceptable limit for lead in drinking water is 0.015 mg/L. Compare the lead concentration of each water sample with the acceptable limit.

At the height of the Roman Empire, worldwide lead production was about 80,000 tons per year. Today it is about 3 million tons annually. Lead was first used for water pipes in ancient Rome. The Latin name for lead, *plumbum,* gave us the word "plumber."

EXERCISE 15.10 Bottled Water as a Magnesium Source

A 500-mL bottle of Evian bottled water contains 12 mg magnesium. The recommended daily allowance of magnesium for adult women is 280 mg/day. How many 1-L bottles of Evian would a woman have to drink to obtain her total daily allowance of magnesium solely in this way?

EXERCISE 15.11 Striking It Rich in the Oceans?

The concentration of gold in seawater is about 1×10^{-3} ppm. Earth's oceans contain 3.5×10^{20} gal seawater. Approximately how many pounds of gold are in the oceans? 1 gal = 3.785 L; 1 lb = 454 g.

Molarity

As defined in Section 5.6 ***(← p. 190),*** the *molarity* of a solution is

$$\text{Molarity} = \frac{\text{number of moles of solute}}{\text{number of liters of solution}}$$

Multiplying the volume of a solution (L) by its molarity yields the number of moles of solute in that volume of solution. For example, the number of moles of KNO_3 in 250. mL of 0.0200-M KNO_3 is

$$0.250\text{ L} \times \left(\frac{0.0200\text{ mol KNO}_3}{1\text{ L}}\right) = 5.00 \times 10^{-3}\text{ mol KNO}_3$$

$$\left(\frac{\text{moles solute}}{\text{L solution}}\right) \times (\text{solution volume, L}) = \text{number moles solute}$$

from which the number of grams can be determined using the molar mass of potassium nitrate.

$$5.00 \times 10^{-3}\text{ mol KNO}_3 \times \frac{101.1\text{ g KNO}_3}{1\text{ mol KNO}_3} = 0.506\text{ g KNO}_3$$

Thus, to make 250. mL of 0.0200-M KNO_3 solution, you would weigh 0.506 g KNO_3, put it into a 250-mL volumetric flask, and add to it sufficient water to bring the volume of the solution to 250. mL ***(← p. 191).*** Note that as long as the *ratio* of moles of solute to liters of solution remains the same, the molarity of the solute does not change. For example,

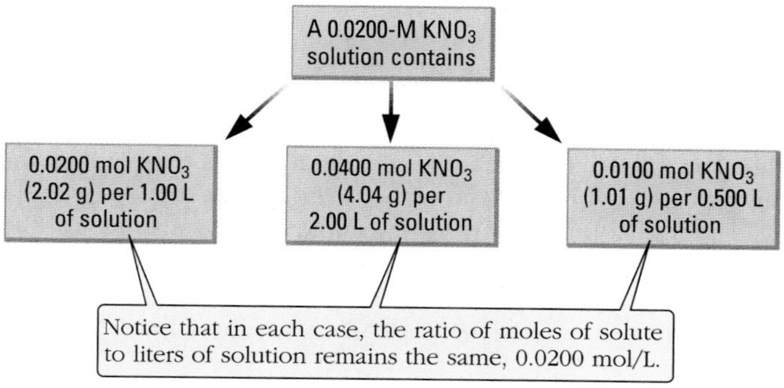

PROBLEM-SOLVING EXAMPLE 15.6 Molarity

(a) Calculate the mass in grams of $NiCl_2$ needed to prepare 500. mL of 0.125-M $NiCl_2$.
(b) Calculate the volume (mL) of this solution required to prepare 250. mL of 0.0300-M $NiCl_2$.

Answer (a) 8.10 g $NiCl_2$ (b) 60.0 mL

Strategy and Explanation To solve this problem, apply the definition of molarity in part (a) and use the dilution relationship ***(← p. 192)*** in part (b).

(a) • To find the number of grams of $NiCl_2$, the number of moles of $NiCl_2$ must first be calculated. This is done by multiplying the volume (L) times the molarity.

$$0.500 \text{ L} \times \frac{0.125 \text{ mol NiCl}_2}{1 \text{ L}} = 0.0625 \text{ mol NiCl}_2$$

• The number of grams of $NiCl_2$ is determined from the number of moles by using the molar mass of $NiCl_2$, 129.6 g/mol.

$$0.0625 \text{ mol NiCl}_2 \times \frac{129.6 \text{ g NiCl}_2}{1 \text{ mol NiCl}_2} = 8.10 \text{ g NiCl}_2$$

(b) • A more concentrated solution is being diluted to a less concentrated one, so the number of moles of $NiCl_2$ remains the same; only the volume of solution changes.

• To calculate the required volume of the more concentrated (undiluted) solution, use the relation

$$\text{Molarity}_{\text{conc}} \times V_{\text{conc}} = \text{Molarity}_{\text{dil}} \times V_{\text{dil}}$$

$\text{Molarity}_{\text{conc}}$ and V_{conc} represent the molarity and volume of the initial (undiluted) solution and $\text{Molarity}_{\text{dil}}$ and V_{dil} are the molarity and volume of the final (diluted) solution ***(← p. 192)***. In this case,

$$\text{Molarity}_{\text{conc}} = 0.125 \text{ M} \qquad V_{\text{conc}} = \text{volume to be determined}$$

$$\text{Molarity}_{\text{dil}} = 0.0300 \text{ M} \qquad V_{\text{dil}} = 250 \text{ mL.} = 0.250 \text{ L}$$

• Solve for V_{conc}

$$V_{\text{conc}} = \frac{\text{Molarity}_{\text{dil}} \times V_{\text{dil}}}{\text{Molarity}_{\text{conc}}} = \frac{(0.0300 \text{ M})(0.250 \text{ L})}{0.125 \text{ M}} = 0.0600 \text{ L} = 60.0 \text{ mL}$$

☑ **Reasonable Answer Check** The molar mass of $NiCl_2$ is approximately 130 g/mol, so 1 L of a 0.125-M $NiCl_2$ solution would contain about 0.125 mol or 16 g $NiCl_2$. One half of a liter (500 mL) of that solution would contain one half as much $NiCl_2$, or about 8 g, which is close to the calculated value of 8.10 g. The diluted solution is 0.0300 M, which is only about one fourth the undiluted concentration. Therefore, in the dilution the volume must increase by a factor of about 4, from 60.0 mL to 250. mL.

PROBLEM-SOLVING PRACTICE 15.6

(a) Calculate the mass in grams of NaBr needed to prepare 250. mL of 0.0750-M NaBr.
(b) Calculate the volume in milliliters of the solution in part (a) required to prepare 500. mL of 0.00150-M NaBr.

PROBLEM-SOLVING EXAMPLE 15.7 Molarity and ppm

Currently, about 60% of the U.S. population drinks fluoridated water from municipal water supplies. The U.S. Environmental Protection Agency's current limit for fluoride, F^-, in drinking water is 4.0 mg/L. Assume that a fluoridated drinking water sample is principally water and that the density of the sample is 1.0 g/mL. Calculate
(a) the concentration of fluoride in the sample in ppm.
(b) the molarity of fluoride in the sample.

Answer (a) 4.0 ppm (b) 2.1×10^{-4} mol F^-/L

Strategy and Explanation Apply the concentration expressions discussed previously to calculate the values requested.

(a) To calculate the fluoride concentration in ppm, determine the mass (grams) of fluoride in 10^6 g solution.

$$\frac{4.0 \text{ mg F}^-}{1 \text{ L solution}} \times \frac{1 \text{ g F}^-}{10^3 \text{ mg F}^-} \times \frac{1 \text{ L solution}}{10^3 \text{ g solution}} = \frac{4.0 \text{ g F}^-}{10^6 \text{ g solution}} = 4.0 \text{ ppm}$$

(b) The molarity of fluoride equals moles F^-/L solution, so the milligrams of fluoride must be converted to moles. The volume of the solution is already expressed in liters.

$$\frac{4.0 \text{ mg F}^-}{1 \text{ L solution}} \times \frac{1 \text{ g F}^-}{10^3 \text{ mg F}^-} \times \frac{1 \text{ mol F}^-}{18.998 \text{ g F}^-} = \frac{2.1 \times 10^4 \text{ mol F}^-}{\text{L solution}} = 2.1 \times 10^{-4} \text{ mol/L}$$

☑ **Reasonable Answer Check** A liter of solution, which is about 1000 g of solution, contains about 0.004 g F^-. This is equivalent to 4 g F^- in 10^6 g solution or 4 ppm fluoride, close to the calculated answer. A 4-g sample of F^- would be approximately 0.2 mol F^-, so 4 mg F^- would be approximately 0.0002 mol F^-, nearly equal to the calculated value. The answers are reasonable.

PROBLEM-SOLVING PRACTICE 15.7

The concentration of magnesium ion in seawater is 0.0556 M making magnesium the second most abundant cation in oceans. Express the Mg^{2+} concentration as (a) mass fraction and (b) ppm. Assume that the density of seawater is 1.03 g/mL.

PROBLEM-SOLVING EXAMPLE 15.8 Weight Percent and Molarity

Hydrochloric acid is sold as a concentrated aqueous solution of HCl with a density of 1.18 g/mL. The concentrated acid contains 38% HCl by mass. Calculate the molarity of hydrochloric acid in this solution.

Answer 12.4 M

Strategy and Explanation To calculate the molarity, the number of moles of HCl and the volume of the solution in liters must be determined.

- The 38.0% hydrochloric acid solution means that 100. g solution contains 38.0 g HCl. Therefore,

$$\text{Amount (moles) of HCl} = 38.0 \text{ g HCl} \times \frac{1 \text{ mol HCl}}{36.5 \text{ g HCl}} = 1.04 \text{ mol HCl}$$

- The molarity of the solution can be calculated using the number of moles of HCl and the density of the solution to determine its volume.

$$\text{Molarity} = \frac{1.04 \text{ mol}}{100. \text{ g solution}} \times \frac{1.18 \text{ g solution}}{1 \text{ mL solution}} \times \frac{1000 \text{ mL solution}}{1 \text{ L solution}}$$

$$= 12.4 \frac{\text{mol}}{\text{L}} = 12.4 \text{ M}$$

☑ **Reasonable Answer Check** The molar mass of HCl, 36.5 g/mol, is close to the number of grams of HCl in 100 g solution. The density of the solution is about 1, so there is about 1 mol HCl in 100 mL of this solution. Consequently, there is approximately 10 mol HCl in 1 L solution—approximately a 10-M HCl solution. The calculated answer is a bit higher because we rounded the actual density (1.18 g/mL) to 1.

PROBLEM-SOLVING PRACTICE 15.8

The density of a commercial 30.0% hydrogen peroxide, H_2O_2, solution is 1.11 g/mL at 25 °C. Calculate the molarity of hydrogen peroxide in this solution.

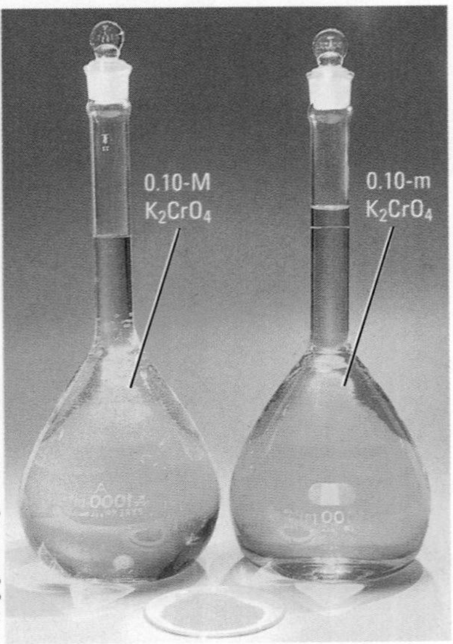

Molarity and molality. The photo shows a 0.10-*molar* solution (0.10 mol/L of potassium chromate (*flask at left*) and a 0.10-*molal* solution (0.10 mol/L) of potassium chromate (*flask at right*). Each solution contains 0.10 mol (19.4 g) of K_2CrO_4 (yellow solid), shown in the dish at the front. The 0.10-molar (0.10 mol/L) solution on the left was made by placing the solid in the flask and adding enough water to make exactly 1 L solution. The 0.10-molal (0.10 mol/kg) solution on the right was made by placing the solid in the flask and adding 1000 g (1 kg) water. Adding 1 kg water produces a solution that has a volume greater than 1 L.

Molality

Solute concentration can also be expressed in terms of the moles of solute in relation to the mass of the solvent. **Molality** (abbreviated m), is defined as the amount (moles) of solute per kilogram of *solvent* (*not kilograms of solution*).

$$\text{Molality of solute A} = m_A = \frac{\text{moles of solute A}}{\text{kilograms of solvent}}$$

For example, we can calculate the molality of a solution prepared by dissolving 0.413 g methanol, CH_3OH, in 1.50×10^3 g water by first determining the number of moles of methanol solute and the number of kilograms of solvent, water.

$$0.413 \text{ g methanol} \times \frac{1 \text{ mol methanol}}{32.042 \text{ g methanol}} = 0.0129 \text{ mol methanol}$$

The mass of water is 1.50×10^3 g, which is 1.50 kg. Substituting these values into the molality equation gives

$$\text{Molality} = m = \frac{0.0129 \text{ mol methanol}}{1.50 \text{ kg water}} = 8.60 \times 10^{-3}$$

Molality will be used in Section 15.7 when we deal with the effect of solute concentration on decreasing the freezing point and raising the boiling point of a solution.

PROBLEM-SOLVING EXAMPLE 15.9 Molarity and Molality

Automobile lead storage batteries contain an aqueous solution of sulfuric acid. The solution has a density of 1.230 g/mL at 25 °C and contains 368 g H_2SO_4 per liter. Calculate

(a) the molarity of sulfuric acid in the solution.
(b) the molality of sulfuric acid in the solution.

Answer (a) Molarity = 3.75 mol H_2SO_4/L (b) Molality = 4.35 mol H_2SO_4/kg solvent

Strategy and Explanation Use the definitions of molarity and molality.

(a) Determine the number of moles of solute in a liter of solution and calculate the molarity.

$$368 \text{ g } H_2SO_4 \times \frac{1 \text{ mol } H_2SO_4}{98.1 \text{ g } H_2SO_4} = 3.75 \text{ mol } H_2SO_4$$

This is the amount of solute in 1 L solution; the molarity is 3.75 mol/L, 3.75 M.

(b) To calculate the molality we first need to determine the mass (kg) of solvent (water) and then divide that mass into the moles of solute. The density of the sulfuric acid solution is 1.230 g/mL, which is equivalent to 1.230×10^3 g solution/L; 1 L of the sulfuric acid solution contains 1.230×10^3 g of the water and sulfuric acid mixture, 368 g of which is sulfuric acid. Therefore, the mass of water can be determined.

$$1230 \text{ g solution} - 368 \text{ g } H_2SO_4 = 862 \text{ g water} = 0.862 \text{ kg water}$$

We now calculate the molality of sulfuric acid in the solution.

$$\text{Molality of } H_2SO_4 = \frac{\text{moles of } H_2SO_4}{\text{kg of solvent (water)}} = \frac{3.75 \text{ mol } H_2SO_4}{0.862 \text{ kg solvent}}$$

$$= 4.35 \text{ mol } H_2SO_4\text{/kg solvent}$$

☑ **Reasonable Answer Check** The molar mass of sulfuric acid is about 100 g/mol so 368 g of the acid would be about 3.7 mol. There is about 4 mol of the acid in about a kilogram of water, so the molality is about 4, close to the more accurate answer of 4.35 mol/kg. Because the density of the solution is not 1.00 g/ml, the mass of solvent and the volume of the solution are not equal, so the molarity and the molality differ.

Molality and molarity are *not* the same, although the difference becomes negligibly small for dilute solutions, those less than 0.01 mol/L.

PROBLEM-SOLVING PRACTICE 15.9

Calculate the molarity and the molality of NaCl in a 20% aqueous NaCl solution whose density is 1.148 g/mL at 25 °C.

Table 15.4 summarizes the ways concentration units have been described so far.

Table 15.4 Comparison of Concentration Expressions

Concentration Expression	Concentration Expressed as
Mass fraction	None
Weight percent	Percent
Parts per million	ppm
Parts per billion	ppb
Parts per trillion	ppt
Molarity	$\frac{\text{Moles solute}}{\text{liter solution}}$
Molality	$\frac{\text{Moles solute}}{\text{kg solvent}}$

EXERCISE 15.12 Concentration Units

A 12-oz (355-mL) Diet Coke contains 46.3 mg caffeine (molar mass = 194.2 g/mol). Assume that the Diet Coke, essentially water, has a density of 1.01 g/mL. For such a Diet Coke, calculate: (a) its caffeine concentration in ppm; (b) its molarity of caffeine; and (c) the molality of caffeine.

CONCEPTUAL EXERCISE 15.13 Molality and Molarity

(a) What information is required to calculate the molality of a solution?
(b) What information is needed to calculate the molarity of a solution if the solution's composition is given in weight percent?

15.7 Vapor Pressures, Boiling Points, Freezing Points, and Osmotic Pressures of Solutions

go Chemistry **Module 19: Colligative Properties** covers concepts in this section.

Up to this point, we have discussed solutions in terms of the nature of the solute and the nature of the solvent. Some properties of solutions do not depend on the nature of the solute, but rather depend only on the *number* of dissolved solute particles—ions or molecules—per known quantity of solvent or solution.

In liquid solutions, solute molecules or ions disrupt solvent-solvent noncovalent attractions, causing changes in solvent properties that depend on these attractions ***(← p. 481)***. As a result, when a solute is added to a solvent, the freezing point of the resulting solution is lower and its boiling point is higher than those of the pure solvent. For a particular solvent, how much the properties of the solution differ from those of the pure solvent depends *only* on the concentration of the solute particles. **Colligative properties** of solutions are those that *depend only on the concentration of solute particles* (ions or molecules) in the solution, regardless of what kinds of particles are present. We will consider four colligative properties: vapor pressure lowering, boiling point elevation, freezing point depression, and osmotic pressure. These are important in the world around us.

Vapor Pressure Lowering

In a closed container a dynamic equilibrium exists between a pure liquid and its vapor—the rate at which molecules escape the liquid phase equals the rate at which vapor phase molecules return to the liquid. This equilibrium gives rise to a vapor pressure that depends on the temperature ***(← p. 481)***. But the vapor pressure of a pure liquid differs from that of the liquid when a solute has been dissolved in it. Compare a small portion of the liquid/vapor boundary for pure water with that for seawater (mainly an aqueous sodium chloride solution), as illustrated at the molecular scale in Figure 15.13. For an aqueous solution such as seawater, in which sodium ions and chloride ions (and many other kinds of ions and molecules) are dissolved, the vapor pressure of water in the solution is lower than for a sample of pure water.

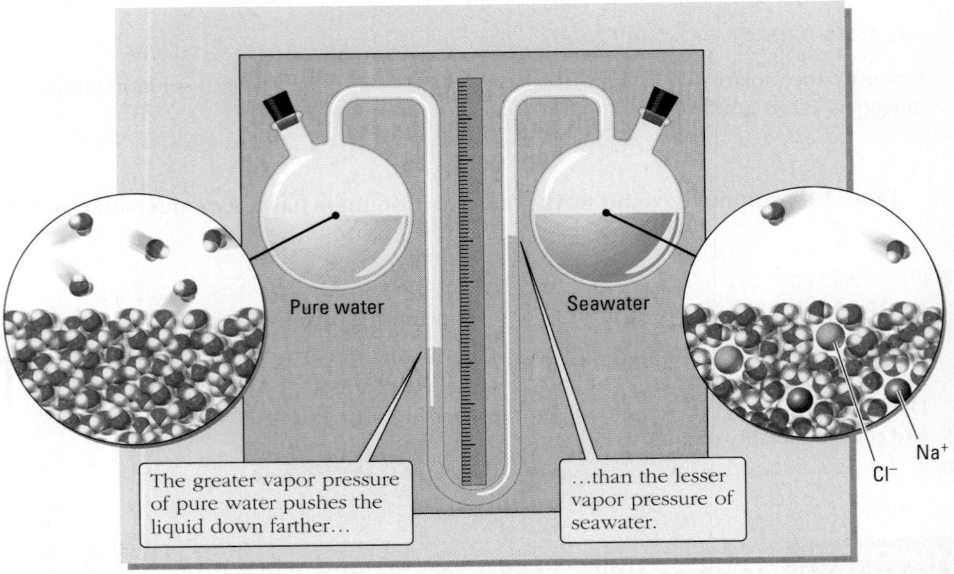

Figure 15.13 The vapor pressure of pure water and that of seawater. Seawater is an aqueous solution of NaCl and many other soluble salts. The vapor pressure over an aqueous solution is not as great as that over pure water at the same temperature.

Raoult's law is most accurate with dilute solutions. Deviations from Raoult's law occur when solute-solvent intermolecular forces are either much weaker or much stronger than the solvent-solvent and solute-solute intermolecular forces.

The vapor pressure of any pure solvent will be lowered by the addition of a nonvolatile solute to the solvent. How much a dissolved solute lowers the vapor pressure of a solvent depends on the solute concentration and is expressed by **Raoult's law:**

$$P_1 = X_1 P_1^0$$

where P_1 is the vapor pressure of the solvent over the *solution,* P_1^0 is the vapor pressure of the *pure* solvent at the same temperature, and X_1 is the mole fraction of *solvent* in the solution. For example, suppose you want to know the vapor pressure over a sucrose solution at 25 °C in which the mole fraction of water is 0.986. The vapor pressure of pure water at 25 °C is 23.76 mm Hg. From these data, the vapor pressure of water over the solution can be calculated using Raoult's law.

$$P_{\text{water}} = (X_{\text{water}})(P^0_{\text{water}}) = (0.986)(23.76 \text{ mm Hg}) = 23.42 \text{ mm Hg}$$

Raoult's law can also be applied to solutions in which the solvent and the solute are both volatile so that an appreciable amount of each can be in the vapor above the solution. We will not consider such cases.

The vapor pressure has been lowered by (23.76 − 23.42) mm Hg = 0.34 mm Hg. Therefore, the vapor pressure of the solution is only 98.6% $\left(\frac{23.42}{23.76} \times 100\%\right)$ that of pure water at 25 °C.

PROBLEM-SOLVING EXAMPLE 15.10 Raoult's Law

Ethylene glycol, $HOCH_2CH_2OH$, is used as an antifreeze. Calculate the vapor pressure of water above a solution of 100.0 mL ethylene glycol and 100.0 mL water at 90 °C. Densities: ethylene glycol, 1.15 g/mL; water, 1.00 g/mL. The vapor pressure of pure water at 90 °C is 525.8 mm Hg.

Answer 394 mm Hg

Strategy and Explanation To determine the vapor pressure of water over the solution using Raoult's law, we must first calculate the mole fraction of water in the solution. For the mole fraction, the milliliters of ethylene glycol and water must be converted to grams and then to moles.

ethylene glycol

$HOCH_2CH_2OH$

$$100.0 \text{ mL eth. gly.} \times \frac{1.15 \text{ g eth. gly.}}{1.00 \text{ mL eth. gly.}} \times \frac{1 \text{ mol eth. gly.}}{62.0 \text{ g eth. gly.}} = 1.86 \text{ mol eth. gly.}$$

$$100.0 \text{ mL water} \times \frac{1.00 \text{ g water}}{1.00 \text{ mL water}} \times \frac{1 \text{ mol water}}{18.0 \text{ g water}} = 5.56 \text{ mol water}$$

$$X_{\text{water}} = \frac{5.56}{5.56 + 1.86} = \frac{5.56}{7.42} = 0.749$$

Applying Raoult's law:

$$P_{\text{water}} = (X_{\text{water}})(P^0_{\text{water}}) = (0.749)(525.8 \text{ mm Hg}) = 394 \text{ mm Hg}$$

☑ **Reasonable Answer Check** Because there are nearly three times the number of moles of water as there are moles of ethylene glycol, the mole fraction of water should be greater than the mole fraction of ethylene glycol, and it is (mole fraction of ethylene glycol = 1 − 0.749 = 0.251). Therefore, the vapor pressure of water over the solution should be about 75% that of pure water (394/525.8 × 100%), which it is. The answer is reasonable.

Mole fraction of A

$$X_A = \frac{\text{moles of A}}{\text{total number of moles}} = \frac{\text{moles A}}{\text{moles A} + \text{moles B} + \cdots}$$

PROBLEM-SOLVING PRACTICE 15.10

The vapor pressure of an aqueous solution of urea, CH_4N_2O, is 291.2 mm Hg at a measured temperature. The vapor pressure of pure water at that temperature is 355.1 mm Hg. Calculate the mole fraction of each component.

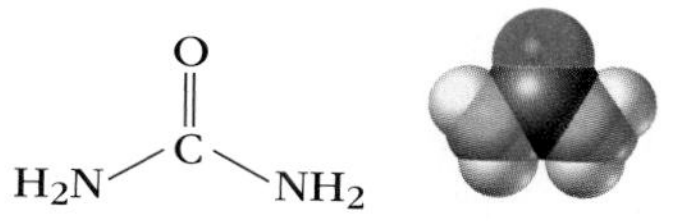

urea

EXERCISE 15.14 Vapor Pressure of a Mixture

Calculate the vapor pressure of water over a solution containing 50.0 g sucrose, $C_{12}H_{22}O_{11}$, and 100.0 g water at 45 °C. The vapor pressure of pure water at this temperature is 71.88 mm Hg.

Entropy plays a role in vapor pressure lowering. Consider the entropy change for the vaporization of pure water with that for the vaporization of a corresponding quantity of water from a sodium chloride solution (Figure 15.14). Because NaCl has a very low vapor pressure, very few sodium or chloride ions escape from the solution and the vapor in equilibrium with the salt water consists almost entirely of water molecules. Thus, the entropy of a given amount of the vapor is approximately the same in both cases (pure water and salt water). The entropy within the salt solution, however, is greater than that in pure water.

Now consider what happens when water vaporizes from pure water and from the salt solution. As with any change from a liquid to a gas, there is a significant increase in entropy of the water vaporizing from either source (pure water or salt solution). But, the entropy of vaporization is not the same in each case. As illustrated in Figure 15.14, the entropy of vaporization is larger for the water vapor from pure water than from the salt solution because the entropy of pure water was already smaller than that of the salt solution. The main point is that a larger entropy increase corresponds to a more product-favored process. The result is that the vaporization from pure water creates a higher pressure of water vapor (that is, more water molecules per unit volume) in equilibrium with pure water than does the vaporization of water from a salt solution.

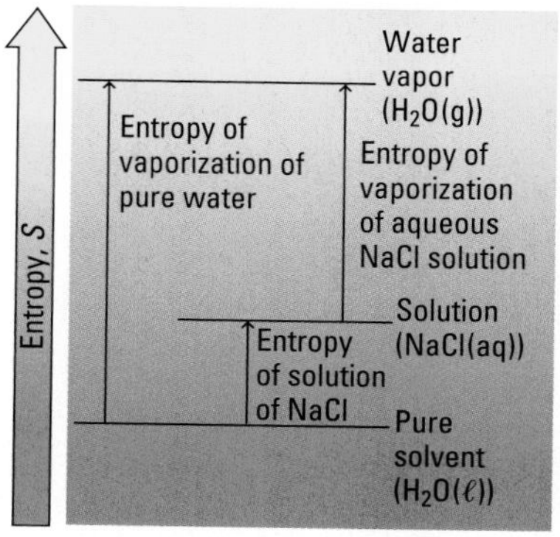

Figure 15.14 Vapor pressure lowering and entropies of solution and vaporization. The entropy of vaporization of pure water is greater than the entropy of solution.

Boiling Point Elevation

The normal boiling point of pure water is the temperature at which the vapor pressure of water is 760 mm Hg. As a result of vapor pressure lowering, the vapor pressure of an *aqueous* solution of a nonvolatile solute at 100 °C is less than 760 mm Hg (1 atm), that of pure water. For the solution to boil, it must be heated

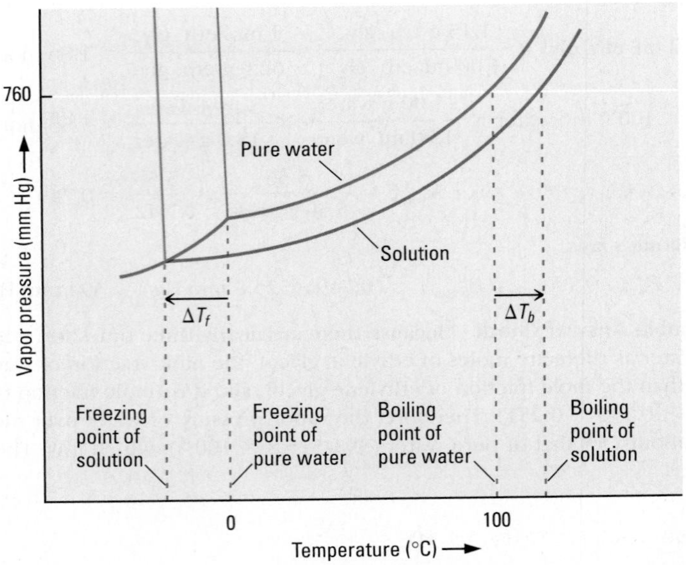

Figure 15.15 Boiling point elevation (ΔT_b) and freezing point lowering (ΔT_f) for aqueous solutions. Comparing the liquid-vapor equilibrium for pure water with that for a solution (where the vapor pressure is lower) shows that addition of solute to a pure solvent raises its boiling point and lowers its freezing point. Note that this diagram is a phase diagram like the ones described in Section 11.3 (← *p. 493*).

above 100 °C. The **boiling point elevation,** ΔT_b, is the difference between the normal boiling point of pure water and the higher boiling point of an aqueous solution of a nonvolatile nonelectrolyte solute (Figure 15.15).

The increase in boiling point is proportional to the concentration of the solute, expressed as *molality,* and can be calculated from this relationship.

$$\Delta T_b = T_b\ (\text{solution}) - T_b\ (\text{solvent}) = K_b m_{\text{solute}}$$

where ΔT_b is the boiling point of solution minus the boiling point of pure solvent. The value of K_b, the *molal boiling point elevation constant* of the *solvent,* depends only on the solvent. For example, K_b for water is 0.52 °C kg mol^{-1}; that of benzene is 2.53 °C kg mol^{-1}.

The units of the molal boiling point elevation constant are abbreviated °C kg mol^{-1}.

Molality, rather than molarity, is used in boiling point elevation determinations because the molality of a solution does not change with temperature changes, but the molarity of a solution does. Molality is based on the masses of solute *and* solvent, which are unaffected by temperature changes. The volume of a solution, which is used in molarity, expands or contracts when the solution is heated or cooled, respectively.

EXERCISE 15.15 Calculating the Boiling Point of a Solution

The boiling point elevation constant for benzene is 2.53 °C kg mol^{-1}. The boiling point of pure benzene is 80.10 °C. If a solute's concentration in benzene is 0.10 mol/kg, calculate the boiling point of the solution.

Freezing Point Lowering

A pure liquid begins to freeze when the temperature is lowered to the substance's freezing point and the first few molecules cluster into a crystal lattice forming a tiny quantity of solid. As long as both solid and liquid phases are present and the temperature is at the freezing point, there is a dynamic equilibrium in which the rate of crystallization equals the rate of melting. When a *solution* freezes, a few molecules of solvent cluster together to form pure solid *solvent* (Figure 15.16), and a dynamic equilibrium is set up between solution and pure solid solvent.

The molecules or ions in the liquid in contact with the frozen solvent in a freezing solution are not all solvent molecules. This causes a slower rate at which particles move from solution to solid than the rate in the pure liquid solvent. To achieve dynamic equilibrium, a correspondingly slower rate of escape of molecules from the solid crystal lattice must occur. According to the kinetic-molecular theory, this

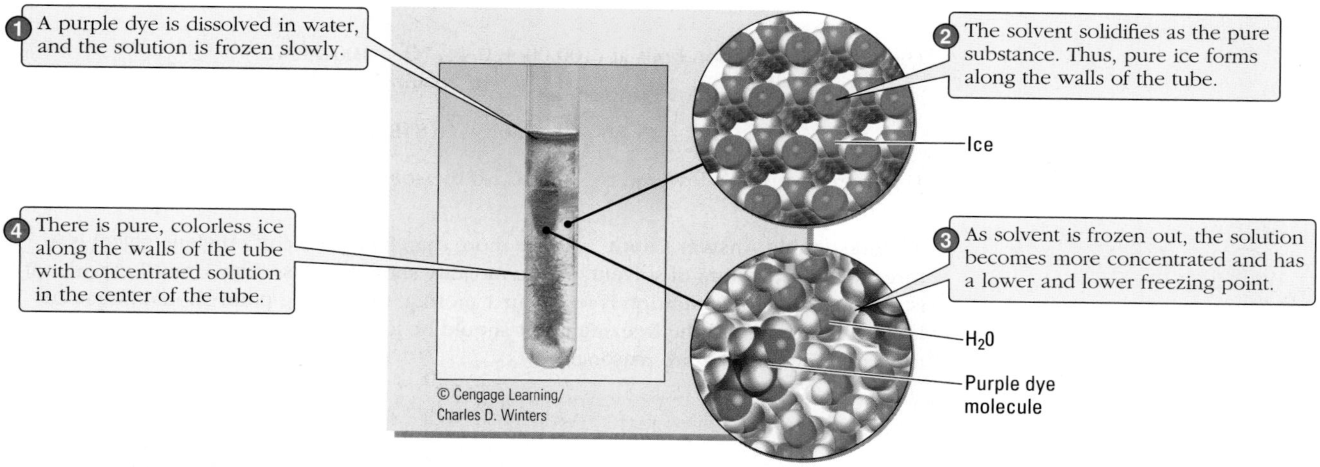
1 A purple dye is dissolved in water, and the solution is frozen slowly.
2 The solvent solidifies as the pure substance. Thus, pure ice forms along the walls of the tube.
Ice
4 There is pure, colorless ice along the walls of the tube with concentrated solution in the center of the tube.
3 As solvent is frozen out, the solution becomes more concentrated and has a lower and lower freezing point.
H20
Purple dye molecule
© Cengage Learning/ Charles D. Winters

Figure 15.16 **Solvent freezing.**

slower rate occurs at a lower temperature, so the freezing point of the solution is lower than that of the pure liquid solvent (Figure 15.15).

The **freezing point lowering**, ΔT_f, is proportional to the concentration of the solute (molality) in the same way as the boiling point elevation.

$$\Delta T_f = K_f m_{solute}$$

As with K_b, the proportionality constant K_f depends only on the solvent, not the type of solute. For water, the freezing point constant is 1.86 °C kg mol^{-1}; by comparison, that of benzene is 5.10 °C kg mol^{-1} and that of cyclohexane is 20.2 °C kg mol^{-1}.

As in the boiling point elevation equation, molality is also used in the freeing point depression equation.

Using ethylene glycol, $HOCH_2CH_2OH$, a relatively nonvolatile alcohol, in automobile cooling systems is a practical application of boiling point elevation and freezing point lowering. Ethylene glycol raises the boiling temperature of the coolant mixture of ethylene glycol and water to a level that prevents engine overheating in hot weather. Ethylene glycol also lowers the freezing point of the coolant, thereby keeping the solution from freezing in the winter.

Another practical application of freezing point lowering is adding salt, NaCl, to ice when making homemade ice cream. Lowering the freezing temperature of the ice-salt water mixture freezes the ice cream more quickly.

PROBLEM-SOLVING EXAMPLE 15.11 Boiling Point Elevation and Freezing Point Lowering

Calculate the boiling and freezing points of an aqueous solution containing 39.5 g ethylene glycol, $HOCH_2CH_2OH$, dissolved in 750. mL water. Assume the density of water to be 1.00 g/mL; its K_b = 0.52 °C kg mol^{-1} and its K_f = 1.86 °C kg mol^{-1}.

Answer Boiling point = 100.44 °C; freezing point = −1.58 °C

Strategy and Explanation To use the equations for freezing point and boiling point changes, the molality of the solution must be determined. The molar mass of ethylene glycol is 62.07 g/mol, and the number of moles of ethylene glycol is

$$39.5\ \text{g} \times \frac{1\ \text{mol}}{62.07\ \text{g}} = 0.636\ \text{mol}$$

The mass of solvent is

$$750.\ \text{mL} \times \frac{1.00\ \text{g}}{\text{mL}} \times \frac{1\ \text{kg}}{10^3\ \text{g}} = 0.750\ \text{kg}$$

The molality of the solution is

$$\frac{0.636\ \text{mol}}{0.750\ \text{kg}} = 0.848\ \text{mol/kg}$$

The boiling point elevation is

$$\Delta T_b = (0.52\ °\text{C kg mol}^{-1})(0.848\ \text{mol/kg}) = 0.44\ °\text{C}$$

Therefore, the solution boils at (100.00 + 0.44) °C = 100.44 °C.

The freezing point lowering is

$$\Delta T_f = (1.86\ °C\ kg\ mol^{-1})(0.848\ mol/kg) = 1.58\ °C$$

The freezing point is lowered by 1.58 °C, so the solution freezes at (0.00 − 1.58) °C = −1.58 °C.

☑ **Reasonable Answer Check** A little more than half a mole of ethylene glycol is in three-fourths kilogram of solvent, so the molality should be less than 1 mol/kg, which it is. Because the concentration is less than 1 mol/kg, the boiling point should be raised less than 0.52 °C and the freezing point should be lowered less than 1.86 °C, which they are. The answers are reasonable.

PROBLEM-SOLVING PRACTICE 15.11

A water tank contains 6.50 kg water. Will the addition of 1.20 kg ethylene glycol be sufficient to prevent the solution from freezing if the temperature drops to −25 °C? Show calculations that support your answer.

Ethylene glycol is toxic and should not be allowed to get into drinking water supplies.

EXERCISE 15.16 Protection Against Freezing

Suppose that you are closing a cabin in the north woods for the winter and you do not want the water in the toilet tank to freeze. You know that the temperature might get as low as −30. °C, and you want to protect about 4.0 L water in the toilet tank from freezing. Calculate the volume of ethylene glycol (density = 1.113 g/mL; molar mass = 62.1 g/mol) you should add to the 4.0 L water.

PROBLEM-SOLVING EXAMPLE 15.12 Molar Mass from Freezing Point Lowering

A researcher prepares a new compound and uses freezing point depression measurements to determine the molar mass of the compound. The researcher dissolves 1.50 g of the new compound in 75.0 g pure cyclohexane, which has a freezing point of 6.50 °C and a freezing point depression constant of 20.2 °C kg mol^{-1}. The freezing point of the solution is measured as 2.70 °C. Use these data to calculate the molar mass of the new compound.

Answer 106 g/mol

Strategy and Explanation To calculate the molar mass, we need to calculate the change in freezing point and the molality of the solution.

- The change in the freezing point, ΔT_f, is 6.50 °C − 2.70 °C = 3.80 °C.
- Rearranging the freezing point lowering equation, solve for the molality of the solution.

$$\text{Molality} = \frac{\Delta T_f}{K_f} = \frac{3.80\ °C}{20.2\ °C\ kg\ mol^{-1}} = 0.188\ mol/kg$$

- From the molality we can calculate the molar mass of the new compound from the relation

$$\text{Molality, } m = \frac{\text{moles solute}}{\text{kg solvent}} = \frac{\text{grams solute/molar mass}}{\text{kg solvent}}$$

- Solve this equation for the molar mass, g/mol, of the unknown compound.

$$\text{Molar mass} = \frac{\text{grams solute}}{(\text{molality})(\text{kg solvent})} = \frac{1.50\ g}{(0.188)(0.0750)\ mol} = 106\ g/mol$$

☑ **Reasonable Answer Check** About 1.5 g compound in 0.0750 kg solvent, equivalent to 20 g compound per kilogram of solvent, forms a 0.188 mol/kg solution.

Thus, a solution containing 1 mol compound per kilogram would contain

$20\ \text{g/kg solvent} \times \dfrac{1\ \text{kg solvent}}{0.188\ \text{mol}}$, which equals 106 g/mol.

PROBLEM-SOLVING PRACTICE 15.12

A student determines the freezing point to be 5.15 °C for a solution made from 0.180 g of a nonelectrolyte in 50.0 g benzene. Calculate the molar mass of the solute. The freezing point constant of benzene is 5.10 °C kg mol^{-1}. The freezing pont of benzene is 5.50 °C.

Edgar Fahs Smith Memorial Collection/ University of Pennsylvania

Jacobus Henricus van't Hoff

1852–1911

Van't Hoff was one of the founders of physical chemistry. Physical chemistry is the branch of chemistry that applies the laws of physics to the understanding of chemical phenomena. Van't Hoff conducted seminal experimental studies in chemical kinetics, chemical equilibrium, osmotic pressure, and chemical affinity. While still a graduate student, he proposed an explanation of optical isomerism based on the tetrahedral nature of the carbon atom (← *p. 388*). Van't Hoff received the first Nobel Prize in Chemistry (1901) for his fundamental discoveries in physical chemistry, including his work on the colligative properties of solutions.

Colligative Properties of Electrolytes

Experimentally, the vapor pressures of 1-M aqueous solutions of sucrose, NaCl, and $CaCl_2$ are all less than that of water at the same temperature. This is to be expected because solutes lower the vapor pressure of the pure solvent. However,

$$\text{vp pure water} > \text{vp 1-M sucrose} > \text{vp 1-M NaCl} > \text{vp 1-M CaCl}_2$$

Because colligative properties of dilute solutions are proportional to the concentration of solute *particles,* this vapor pressure order is not surprising. Due to their dissociation, electrolytes such as NaCl and $CaCl_2$ contribute more particles per mole than do nonelectrolytes such as sucrose or ethanol, which do not dissociate. Whereas 1 mol sucrose contributes 1 mol particles (sucrose molecules) to solution, 1 mol NaCl contributes 2 mol particles (1 mol Na^+ and 1 mol Cl^-), and 1 mol $CaCl_2$ produces 3 mol particles (1 mol Ca^{2+} and 2 mol Cl^-). Therefore, electrolytes have a greater effect on boiling and freezing points than nonelectrolytes have.

For solutions of electrolytes, the boiling point elevation equation can be written as

$$\Delta T_b = K_b\, m_{solute}\, i_{solute}$$

and the freezing point lowering equation becomes

$$\Delta T_f = K_f\, m_{solute}\, i_{solute}$$

The i_{solute} factor gives the *number of particles per formula unit of solute*. It is called the van't Hoff factor, named after Jacobus Henricus van't Hoff. The *value of* i *is 1 for nonelectrolytes* because these molecular solutes, such as ethanol, sucrose, benzene, and carbon tetrachloride, do *not* dissociate in solution. *For soluble ionic solutes* (strong electrolytes), i *equals the number of ions per formula unit* of the ionic compound. In extremely dilute solutions $i_{solute} = 2$ for NaCl and $i_{solute} = 3$ for calcium chloride. For more concentrated solutions, the actual i_{solute} value must be determined experimentally. The theoretical i value assumes that the ions act independently in solution, which is achieved *only* in extremely dilute solutions where the interaction between cations and anions is minimal. In more concentrated solutions, cations and anions interact and $i_{expt} < i_{theor}$. For example, in aqueous $MgSO_4$ solutions, $i_{theor} = 2$, whereas in 0.50 M $MgSO_4$, $i_{expt} = 1.07$; in 0.005 M $MgSO_4$, $i_{expt} = 1.72$.

Because $i = 1$ for nonelectrolytes, $\Delta T_b = K_b\, m_{solute}$ and $\Delta T_f = K_f\, m_{solute}$.

Another practical application of freezing point lowering can be seen in areas where winter weather produces lots of frozen precipitation. To remove snow and particularly ice, roads and walkways are often salted. Although sodium chloride is usually used, calcium chloride is particularly good for this purpose because it has three ions per formula unit and dissolves exothermically. Not only is the freezing point of water lowered, but the heat of solution helps melt the ice.

A buildup of dissolved NaCl or $CaCl_2$ along roads is environmentally hazardous because the excessive Cl^- concentration is harmful to roadside plants. This answers the question, "Why does salt help to clear ice and snow from roads?" that was posed in Chapter 1 (← *p. 2*).

CONCEPTUAL EXERCISE 15.17 Freezing Point Lowering

The freezing point of a 2.0 mol/kg $CaCl_2$ solution is measured as −4.78 °C. Calculate the i_{solute} factor and use it to approximate the degree of dissociation of $CaCl_2$ in this solution.

Osmotic Pressures of Solutions

A *membrane* is a thin layer of material that allows molecules or ions to pass through it. A **semipermeable membrane** allows only certain kinds of molecules or ions to pass through while excluding others (Figure 15.17). Examples of semipermeable membranes are animal bladders, cell membranes in plants and animals, and cellophane, a polymer derived from cellulose. When two solutions containing the same solvent are separated by a membrane permeable only to solvent molecules, osmosis will occur. **Osmosis** is *the movement of a solvent through a semipermeable membrane from a region of higher solvent concentration (lower solute concentration) to a region of lower solvent concentration (higher solute concentration).* The **osmotic pressure** of a solution is *the pressure that must be applied to the solution to stop osmosis from a sample of pure solvent.*

Consider the osmosis example shown in Figure 15.18. A 5% aqueous sugar solution in a bag is attached to a glass tube. The bag is made of a semipermeable membrane that allows water, but not sugar molecules, to pass through it. When the bag is submerged in pure water, the pure water flows into the bag by osmosis and raises the liquid level in the tube. When the bag is first submerged, more collisions of solvent molecules per unit area of the membrane occur on the pure solvent (water) side than there are on the solution side (where there are fewer solvent molecules per unit volume). Hence, pure water moves from the beaker where water is in greater concentration through the membrane into the solution in the bag, where the water concentration is lower. As this continues, the number of water molecules in the solution increases, the number of collisions of water molecules within the solution increases, and water rises in the tube as pressure builds up in the bag. A dynamic equilibrium is achieved when the pressure in the bag equals the osmotic pressure, at which point the rate of water molecules passing through the membrane is the same in both directions. The height of the water column then remains unchanged, and is a measure of the osmotic pressure.

Osmotic pressure—like vapor pressure lowering, boiling point elevation, and freezing point lowering—results from the unequal rates at which solvent molecules pass through an interface or boundary. In the case of evaporation and boiling, it is the solution/vapor interface; for freezing, it is the solution/solid interface. The semipermeable membrane is the interface for osmosis.

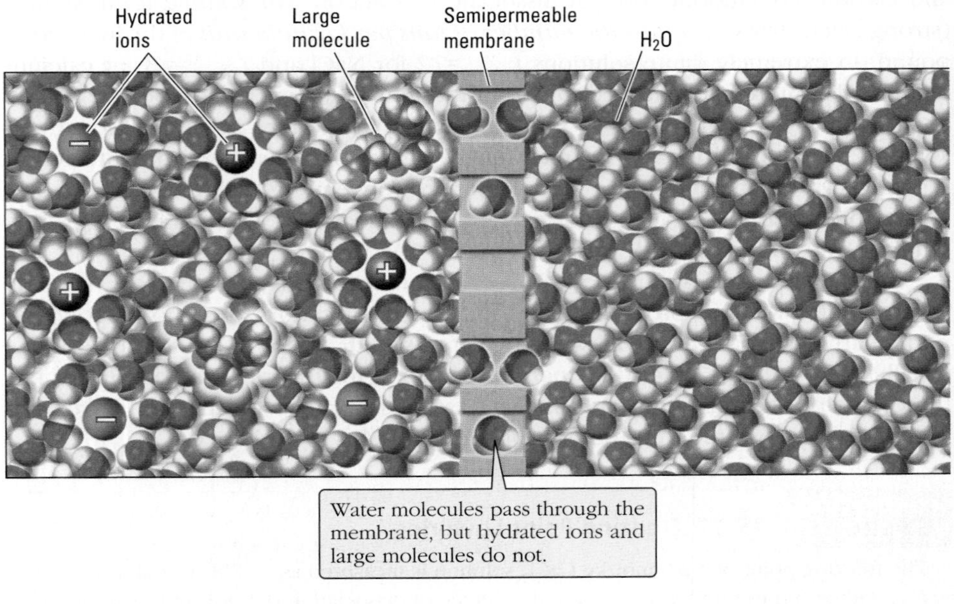
Hydrated ions
Large molecule
Semipermeable membrane
H2O
Water molecules pass through the membrane, but hydrated ions and large molecules do not.

Figure 15.17 Osmotic flow of a solvent through a semipermeable membrane to a solution. The semipermeable membrane is shown acting as a sieve. Many membranes operate in different ways, but the ultimate effect is the same.

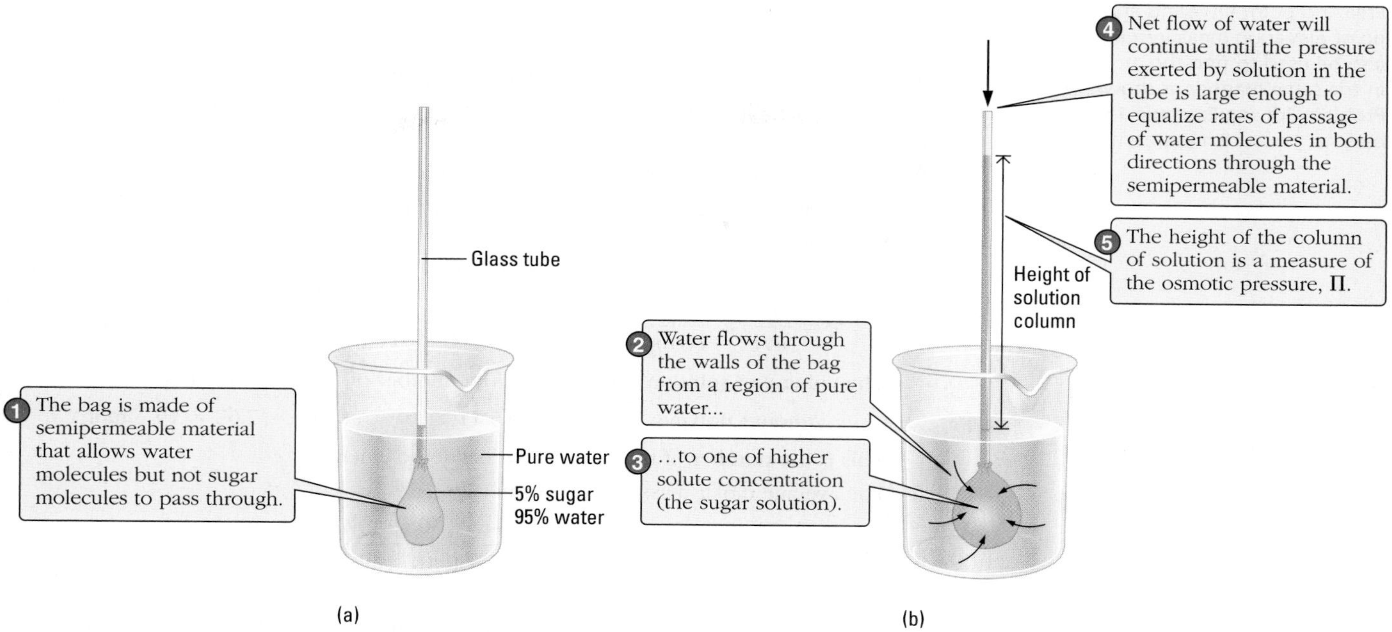

Figure 15.18 **Demonstration of osmotic pressure.**

All colligative properties can be understood in terms of differences in entropy between a pure solvent and a solution. This is perhaps most easily seen in the case of osmosis. When solvent and solute molecules mix, entropy usually increases. If pure solvent is added to a solution, a higher entropy state will be achieved as solvent and solute molecules diffuse among one another to form a more dilute solution. Unless strong noncovalent intermolecular forces exist between the solute and solvent, there will be a negligible enthalpy change, so the increase in entropy makes mixing of solvent and solution a product-favored process. A semipermeable membrane prevents solute molecules from passing into pure solvent, so the only way mixing can occur (and entropy can increase) is for solvent to flow into the solution, and it does.

The more concentrated the solution, the more product-favored the mixing is and the greater is the pressure required to prevent it. Osmotic pressure (Π) is proportional to the *molarity* of the solution, c,

$$\Pi = cRTi$$

where R is the gas constant, T is the absolute temperature (in kelvins), and i is the number of particles per formula unit of solute.

Even though the solution concentration is small, osmotic pressure can be quite large. For example, the osmotic pressure of a 0.020-M solution of a molecular solute ($i = 1$) at 25 °C is

$$\Pi = cRTi = \left(\frac{0.020 \text{ mol}}{\text{L}}\right)(0.0821 \text{ L atm mol}^{-1} \text{ K}^{-1})(298 \text{ K})(1) = 0.49 \text{ atm}$$

This pressure would support a water column more than 15 ft high. One way to determine osmotic pressure is to measure the height of a column of solution in a tube, as shown in Figure 15.18. Heights of a few centimeters can be measured accurately, so quite small concentrations can be determined by osmotic pressure experiments. If the mass of solute dissolved in a measured volume of solution is known, it is possible to calculate the molar mass of the solute by using the definition of molar concentration, $c = n/V$ = amount (mol)/volume (L).

The osmotic pressure equation is similar to the ideal gas law equation, $PV = nRT$, which can be rearranged to $P = (n/V)RT = cRT$, where n/V is the molar concentration of the gas.

Note that molarity, not molality, is used to express solute concentration in this equation.

Freezing point lowering and boiling point elevation measurements can also be used to find the molar mass in the same manner as shown in Problem-Solving Example 15.13 for osmotic pressure measurements.

PROBLEM-SOLVING EXAMPLE 15.13 Molar Mass from Osmotic Pressure

A solution containing 2.50 g of a nonelectrolyte polymer dissolved in 150. mL solution has an osmotic pressure of 1.25×10^{-2} atm at 25 °C. Calculate the molar mass of the polymer.

Answer 3.26×10^4 g/mol

Strategy and Explanation We calculate the molarity of the solution using the osmotic pressure equation.

- Because the polymer is a nonelectrolyte, $i = 1$.

$$c = \frac{\Pi}{RTi} = \frac{1.25 \times 10^{-2}\ \text{atm}}{(0.0821\ \text{L atm mol}^{-1}\ \text{K}^{-1})\ (298\ \text{K})(1)} = 5.11 \times 10^{-4}\ \text{mol/L}$$

The volume of solution is 0.150 L, so the number of moles of polymer is

$$\text{Amount polymer} = (0.150\ \text{L})(5.11 \times 10^{-4}\ \text{mol/L}) = 7.67 \times 10^{-5}\ \text{mol}$$

This is the number of moles in 2.50 g of the polymer.

- The average molar mass of the polymer is

$$\text{Average molar mass of polymer} = \frac{2.50\ \text{g polymer}}{7.67 \times 10^{-5}\ \text{mol polymer}} = 3.26 \times 10^4\ \text{g/mol}$$

☑ **Reasonable Answer Check** Because the molarity is low, the molar mass of the polymer must be relatively large to create an osmotic pressure of 1.25×10^{-2} atm. In Section 12.6, polymers were described as long chains of linked monomer units, so a molar mass of over 30,000 g/mol is not unreasonable for the polymer.

PROBLEM-SOLVING PRACTICE 15.13

The osmotic pressure of a solution of 5.0 g of horse hemoglobin (a protein) in 1.0 L water is 1.8×10^{-3} atm at 25 °C. Calculate the molar mass of the hemoglobin.

Blood and other fluids inside living cells contain many different solutes, and the osmotic pressures of these solutions play an important role in the distribution and balance of solutes within the body. Dehydrated patients are often given water and nutrients intravenously. However, pure water cannot simply be dripped into a patient's veins (Fig. 15.19). The water would flow into the red blood cells by osmosis, causing

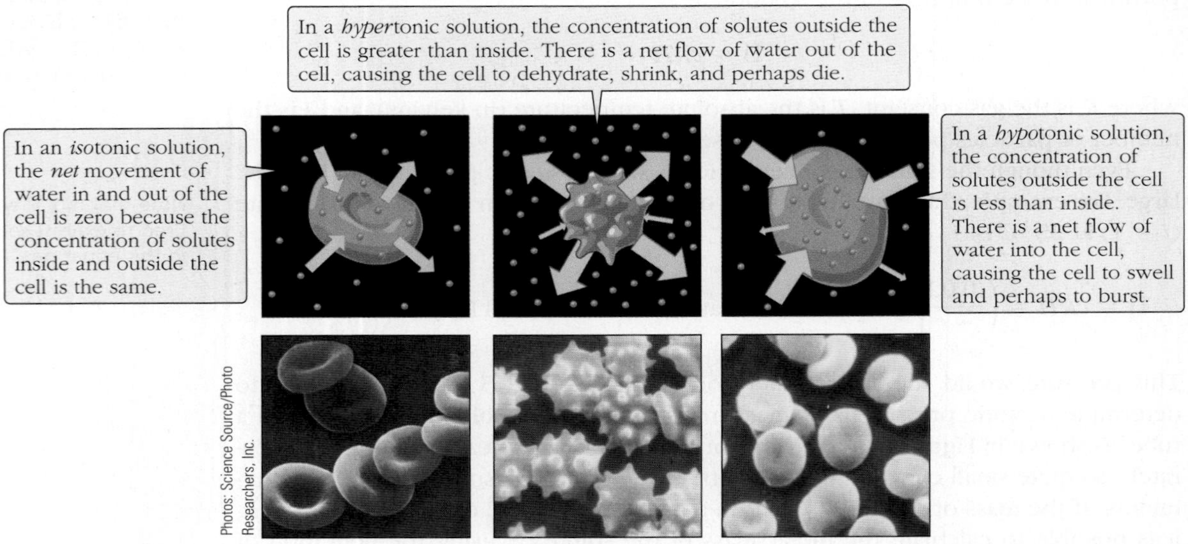
In a *hyper*tonic solution, the concentration of solutes outside the cell is greater than inside. There is a net flow of water out of the cell, causing the cell to dehydrate, shrink, and perhaps die.
In an *iso*tonic solution, the *net* movement of water in and out of the cell is zero because the concentration of solutes inside and outside the cell is the same.
In a *hypo*tonic solution, the concentration of solutes outside the cell is less than inside. There is a net flow of water into the cell, causing the cell to swell and perhaps to burst.
Photos: Science Source/Photo Researchers, Inc.

(a) Isotonic solution (b) Hypertonic solution (c) Hypotonic solution

Figure 15.19 Osmosis and the living cell.

them to burst (Figure 15.19c). A solution that causes this condition is called a **hypotonic** solution. To prevent cells from bursting, an **isotonic** (or iso-osmotic) intravenous solution must be used. Such a solution has the same total concentration of solutes and therefore the same osmotic pressure as the patient's blood (Figure 15.19a). A solution of 0.9% sodium chloride is isotonic with fluids inside cells in the body.

If an intravenous solution more concentrated than the solution inside a red blood cell were added to blood, the cell would lose water to the solution and shrivel up. A solution that causes this condition is a **hypertonic** solution (Figure 15.19b). Cell-shriveling by osmosis happens when vegetables or meats are cured in *brine,* a concentrated solution of NaCl. If you put a fresh cucumber into brine, water will flow out of its cells and into the brine, leaving behind a shriveled vegetable (Figure 15.20). With proper spices added to the brine, a cucumber becomes a tasty pickle.

© Cengage Learning/Charles D. Winters

Figure 15.20 Osmosis in vegetables. A cucumber soaked in a concentrated salt solution (*right*) has lost much water and shrivels into a pickle. A cucumber soaked in pure water (*left*) is affected very little.

Reverse Osmosis

Reverse osmosis occurs when pressure greater than the osmotic pressure is applied and solvent is forced to flow through a semipermeable membrane from a concentrated solution to a dilute solution. In effect, the semipermeable membrane serves as a filter with very tiny pores through which only solvent can pass. Reverse osmosis can be used to remove small molecules or ions to obtain highly purified water. Seawater contains a high concentration of dissolved salts; its osmotic pressure is 24.8 atm. If a pressure greater than 24.8 atm is applied to a chamber containing seawater, water molecules can be forced to flow from seawater through a semipermeable membrane to a region containing purer water (Figure 15.21). Pressures up to 100 atm are used to provide reasonable rates of seawater purification. Seawater, which contains upward of 35,000 ppm (3.5%) of dissolved salts (Table 15.5), can be purified by reverse osmosis to between 400 and 500 ppm of solutes, which is well within the World Health Organization's limits for drinking water. Large reverse osmosis plants in places like the Persian Gulf countries and Florida can purify more than 100 million gallons of water per day. Nearly 50% of Saudi Arabia's fresh water is provided by the world's largest desalination plant at Jubai. Reverse osmosis purifies nearly 15 million gallons of brackish underground water daily for the city of Cape Coral, Florida. (Brackish water is a mixture of salt water and fresh water.) That facility is one of more than 100 in the state of Florida. Reverse osmosis is also used to further purify fresh water when ultrapure water is essential, as in the production of pharmaceuticals or in laboratory research.

Table 15.5 Ions Present in Seawater at 100 ppm or More

Ion	Mass Fraction g/kg	Mass Fraction ppm
Cl^-	19.35	19,350
Na^+	10.76	10,760
SO_4^{2-}	2.710	2710
Mg^{2+}	1.290	1290
Ca^{2+}	0.410	410
K^+	0.400	400
HCO_3^-, CO_3^{2-}	0.106	106
Total	35.026	35,026

Small reverse osmosis units are used to make ultrapure water for some "spotless" car washes.

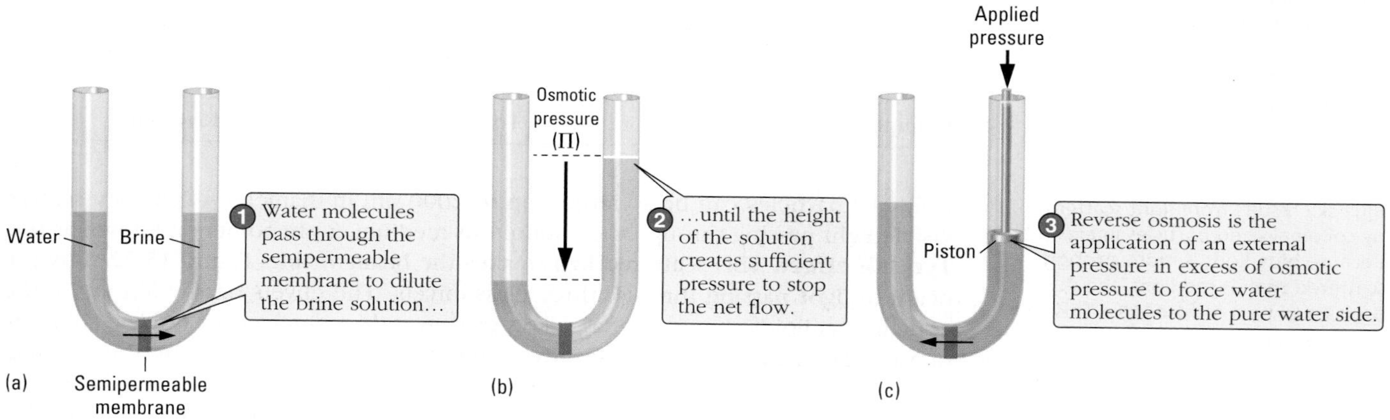

Figure 15.21 Normal and reverse osmosis. Normal osmosis is represented in (a) and (b). Reverse osmosis is represented in (c).

CHEMISTRY IN THE NEWS Thirsty Southern California to Test Desalination

The increasing need for additional fresh water has recently led the San Diego regional water authority to approve a vast $320 million desalination plant to be built in Carlsbad, north of San Diego. The plant, to be built by Poseidon Resources and completed by 2011, will be the first large-scale desalination operation on the West Coast and the largest in the hemisphere. About 100 million gallons of water from the Pacific Ocean will be desalinated daily by reverse osmosis to produce 50 million gallons of fresh water, one of the biggest capacities outside the Middle East. The process will provide fresh water to nine municipal water authorities in northern San Diego County, filling 10% of the county's drinking water needs.

Peter Gleick, cofounder and president of the Pacific Institute, said about the plant, "If they build it well and it operates well and the price is right, we will see more. I think that there is going to be some hesitancy to really expand desalination until this plant is up and running. There's going to be hesitancy on the part of everyone—regulators, water agencies, and municipalities." Environmentalists battled the project with lawsuits citing concerns about how many fish would be killed by the pumping process and potential changes to the aquatic ecosystem when waste brine (sea salt) is returned to the ocean. Poseidon Resources plans to create 55 acres of new coastal wetlands to mitigate fish losses.

Source: Barringer, F. *New York Times*, May 15, 2009, **http://www.nytimes.com/2009/05/15/science/earth/15water.html**

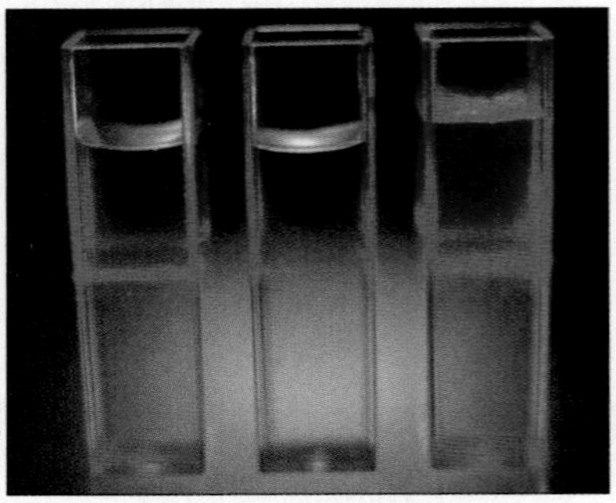

© Cengage Learning/Charles D. Winters

Figure 15.22 Light scattering by a colloidal dispersion. Entering from the side of the photo, a narrow laser beam of light half way up from the bottom of each cuvet passes through a colloidal mixture (*left*), then through a salt solution, and finally through another colloidal mixture. The Tyndall effect is seen in the colloidal mixtures, but not in the salt solution. This illustrates the light-scattering ability of colloid-sized

15.8 Colloids

Around 1860 Thomas Graham found that starch, gelatin, glue, and egg albumin diffuse only very slowly in water and do not diffuse through a thin membrane, but sugar or salt do. He coined the word "colloid" to describe a class of substances distinctly different from sugar and salt and similar materials. **Colloids** are now understood to be mixtures in which relatively large particles, the **dispersed phase,** are distributed uniformly throughout a solvent-like medium, the **continuous phase,** or dispersing medium. Like true solutions, colloids are found in the gas, liquid, and solid states. Although both true solutions and colloids appear homogeneous to the naked eye, at the nanoscale level colloids are not homogeneous.

In colloids, the dispersed-phase particles might be as large as 10 to nearly 1000 times the size of a single small molecule. Colloidal particles are larger than those found in true solutions and smaller than the particles in suspensions.

	Smaller Particles ⟶		Larger Particles
	True Solution	**Colloidal Dispersion**	**Suspension**
Particles	Ions and molecules	Colloids	Large-sized particles
Particle size	0.2–2.0 nm	2–2000 nm	>2000 nm
Properties	• Don't settle out on standing • Not filterable	• Don't settle out on standing • Not separable using ordinary filters	• Settle out on standing • Filterable
Example	Seawater	Fog	River silt

Blue sky results from light scattering by colloidal-sized particles in air. Because blue light is more strongly scattered than is red light, the sky appears to be a diffuse blue. This answers the question posed in Chapter 1 (← p. 2), "Why is the sky blue?"

Colloidal particles can be so large—up to 2000 nm in diameter—that they scatter visible light passing through the continuous medium, a phenomenon known as the **Tyndall effect.** This scattered light makes the beam visible. Figure 15.22 shows a beam of light passing through three glass cuvets. The cuvets on the left and right contain a colloidal mixture of a gelatin in water; the center cuvet holds a solution of NaCl. The light scattering is clearly seen in the colloidal suspensions; the Na^+ and Cl^- ions in the NaCl solution are too small to scatter the light, and the Tyndall effect is not seen. Colloidal particles of dust and smoke in the air of a room can easily be

CHEMISTRY YOU CAN DO Curdled Colloids

Regular (whole) milk contains about 4% fat; skim milk contains considerably less. In addition, milk contains proteins. Both the fat and the proteins are in the form of colloids.

Add about 2 tablespoons of vinegar or lemon juice to about 100 mL of whole milk (do the same with skim milk and 1% or 2% butterfat milk, if you have it), stir, and watch what happens. Let each sample stand overnight at room temperature. Observe and record the results.

An additional experiment you can try is adding salt to similar samples of milk and recording your observations.

When you finish, discard this milk down the drain. You should *not* drink it because it has been unrefrigerated and might contain harmful bacteria.

Think about these questions:

1. Write an explanation for what you observed.
2. Does the salt have the same effect on milk as the acid has?

observed in a beam of sunlight because they scatter the light; you have probably seen such a well-defined sunbeam many times. A common colloid, *fog,* consists of water droplets (the dispersed phase) in air (the continuous phase, and itself a solution). The lack of visibility in fog is due to the Tyndall effect.

© Galen Rowell/Corbis

The Tyndall effect. Shafts of light are visible through the mist in this forest because fog particles are large enough to scatter light.

Types of Colloids

Colloids are classified according to the state of the dispersed phase (solid, liquid, or gas) and the state of the continuous phase. Table 15.6 lists several types of colloids and some examples of each. Liquid-liquid colloids form only in the presence of an emulsifier—a third substance that coats and stabilizes the particles of the dispersed phase. Such colloidal dispersions are called **emulsions.** In mayonnaise, for example, egg yolk contains a protein that stabilizes the tiny drops of oil that are dispersed in the aqueous continuous phase. As you can see from examples listed in Table 15.6, colloids are very common in everyday life.

Colloids with water as the continuous phase are either hydrophilic or hydrophobic. In a hydrophilic colloid there is *a strong attraction between the dispersed phase and the continuous (aqueous) phase*. Hydrophilic colloids form when the molecules of the dispersed phase have multiple sites that interact with water through hydrogen bonding and dipole-dipole attractions. Proteins in aqueous media are hydrophilic colloids.

Table 15.6 Types of Colloids

Continuous Phase	Dispersed Phase	Type	Examples
Gas	Liquid	Aerosol	Fog, clouds, aerosol sprays
Gas	Solid	Aerosol	Smoke, airborne viruses, automobile exhaust
Liquid	Gas	Foam	Shaving cream, whipped cream
Liquid	Liquid	Emulsion	Mayonnaise, milk, face cream
Liquid	Solid	Sol	Gold in water, milk of magnesia, mud
Solid	Gas	Foam	Foam rubber, sponge, pumice
Solid	Liquid	Gel	Jelly, cheese, butter
Solid	Solid	Solid sol	Milk glass, many alloys such as steel, some colored gemstones

NASA/Peter Arnold, Inc.

Figure 15.23 Aerial view of silt formation in a river delta from colloidal soil particles. Silt forms at a river delta (lower left, lighter blue color) as colloidal particles in the river water meet the salt water as the river enters an ocean or a salt-water bay. The higher salt concentration causes the colloidal particles to coagulate.

In a hydrophobic colloid there is a *lack of attraction between the dispersed phase and the continuous phase*. Although you might assume that such colloids would tend to separate quickly, hydrophobic colloids can be quite stable once they are formed. A colloidal dispersion (sol) of gold nanoparticles prepared in 1857 is still preserved in the British Museum.

A stable hydrophobic colloid coagulates when ions come into contact with the dispersed phase. Soil particles carried in rivers are hydrophobic. When river water containing large amounts of colloidal soil particles meets seawater with a high ion concentration, the colloidal particles coagulate to form silt. The deltas of the Mississippi and Nile Rivers have been and continue to be formed in this way (Figure 15.23).

15.9 Surfactants

Molecules that have a hydrophobic part and a hydrophilic part are called **surfactants** (surface-active agents) because they tend to act at the surface of a substance that is in contact with the solution that contains the surfactant. Soap, the classic surfactant, dates back to the Sumerians in 2500 BC. Chemically, soaps are salts of fatty acids and have always been made by the reaction of a fat with an alkali, a process known as *saponification* (← p. 560).

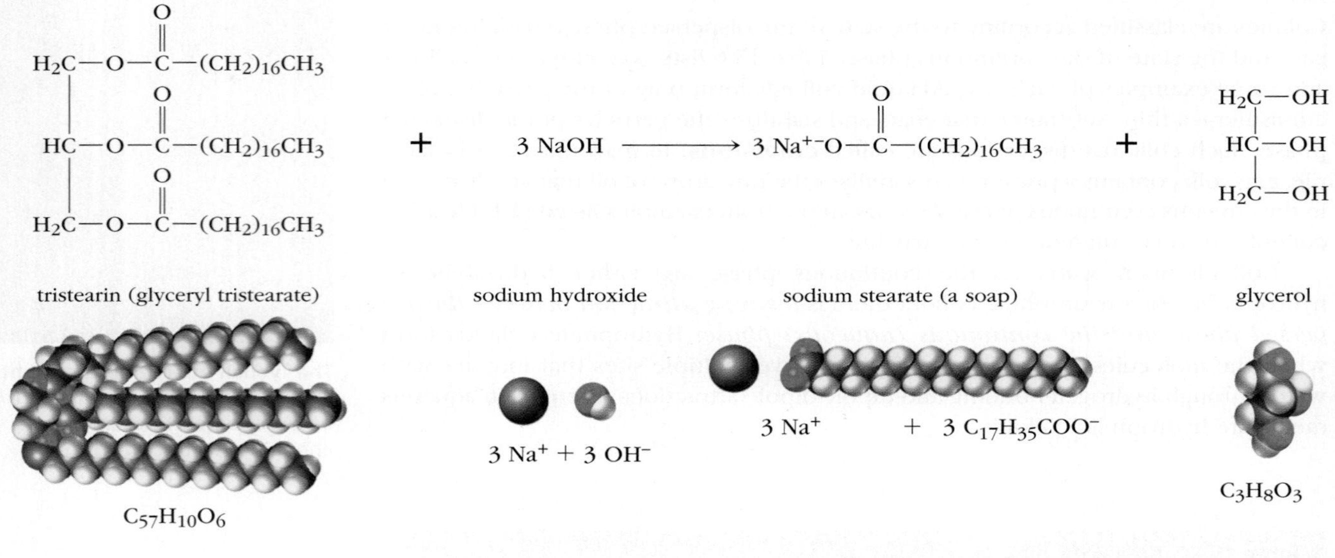

$H_2C-O-C(=O)-(CH_2)_{16}CH_3$
$HC-O-C(=O)-(CH_2)_{16}CH_3$
$H_2C-O-C(=O)-(CH_2)_{16}CH_3$
+
3 NaOH
$3\ Na^{+-}O-C(=O)-(CH_2)_{16}CH_3$
+
H_2C-OH
$HC-OH$
H_2C-OH
tristearin (glyceryl tristearate)
sodium hydroxide
sodium stearate (a soap)
glycerol
$3\ Na^+ + 3\ OH^-$
$3\ Na^+$
$+\ 3\ C_{17}H_{35}COO^-$
$C_{57}H_{110}O_6$
$C_3H_8O_3$

Sodium stearate is a typical soap. The long-chain hydrocarbon part of the molecule is hydrophobic, while the polar carboxylate group is hydrophilic.

Hand soaps are pure soaps to which dyes and perfumes are added.

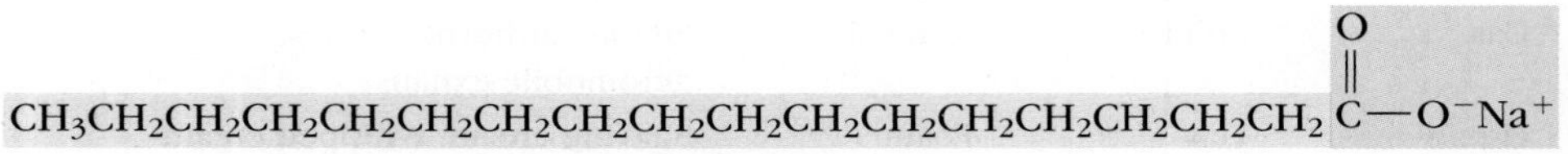

$CH_3CH_2CH_2CH_2CH_2CH_2CH_2CH_2CH_2CH_2CH_2CH_2CH_2CH_2CH_2CH_2C(=O)-O^-Na^+$
Hydrophobic end
Hydrophilic end

sodium stearate

Water, oil, and a surfactant together form an emulsion. The surfactant acts as the emulsifying agent, such as in the case of bile salts emulsifying dietary fats.

Bile salts are important surfactants in the body that help to break up ingested fats. The hydrophobic portion of bile salts is incorporated into the surface of the ingested fat molecules, where it acts as an emulsifying agent, keeping fat molecules dispersed from each other. The hydrophilic portion of the bile salt keeps the emulsified fats in solution.

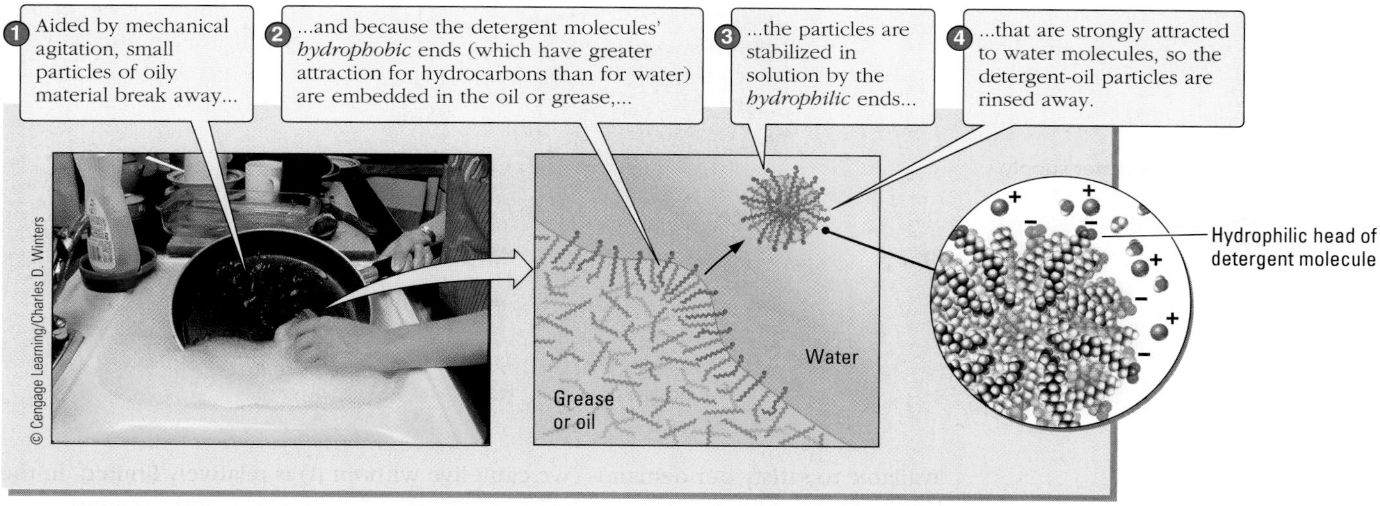

Figure 15.24 **The cleansing action of soaps and detergents depends on micelle formation.**

Synthetic surfactants called **detergents** are made from refined petroleum or coal products. Detergents have a long hydrocarbon chain that is hydrophobic and a polar end that is hydrophilic, somewhat like those of soaps. One common detergent is sodium lauryl sulfate, which is used in many shampoos.

$$CH_3CH_2CH_2CH_2CH_2CH_2CH_2CH_2CH_2CH_2CH_2CH_2OSO_3^- Na^+$$

sodium lauryl sulfate

CONCEPTUAL EXERCISE **15.18 Estimating Osmotic Pressure**

Which solution would have the higher osmotic pressure, a 0.02-M sucrose solution or a 0.02-M typical soap solution? Explain your answer.

In water solutions, surfactants tend to aggregate to form hollow, colloid-sized spherical particles called *micelles* (Figure 15.24) that can transport various materials within them. The hydrophobic ends of the surfactant molecules point inward to the center of the micelle, while their hydrophilic heads point outward so that they interact with water molecules. Ordinary soap, a surfactant, forms micelles in water. Soaps cleanse because hydrophobic oil on clothing or skin joins to the hydrophobic centers of the soap micelles and is rinsed away (Figure 15.24). When ordinary soaps are used with hard water (p. 744), the soaps react with Ca^{2+}, Mg^{2+}, and Fe^{2+} ions in the hard water to produce insoluble salts that form an unsightly scum around the bathtub or on clothes. Detergents do not form these insoluble salts, even with hard water.

This discussion of soaps and detergents answers the question posed in Chapter 1 (***← p. 2):*** "How does soap help to get clothes clean?"

15.10 Water: Natural, Clean, and Otherwise

More than 97% of water, the most abundant substance on Earth's surface, is found in the oceans, which cover about three quarters of Earth's surface (Figure 15.25, next page). Glaciers, ice caps, and snow pack account for 1.72% of water. *Surface water*—lakes, rivers, streams, and reservoirs—makes up only a very small portion (0.008%). *Groundwater*, water that is held in large underground natural *aquifers*, makes up 0.77% of water.

"... Water, water everywhere, nor any drop to drink." *The Rime of the Ancient Mariner* by Samuel Taylor Coleridge alludes to the fact that the dissolved salt in seawater renders it useless for drinking.

Note from Figure 15.25 that fresh water makes up less than 3% of Earth's surface water, with most of it tied up in glaciers. Indeed, the amount of fresh water

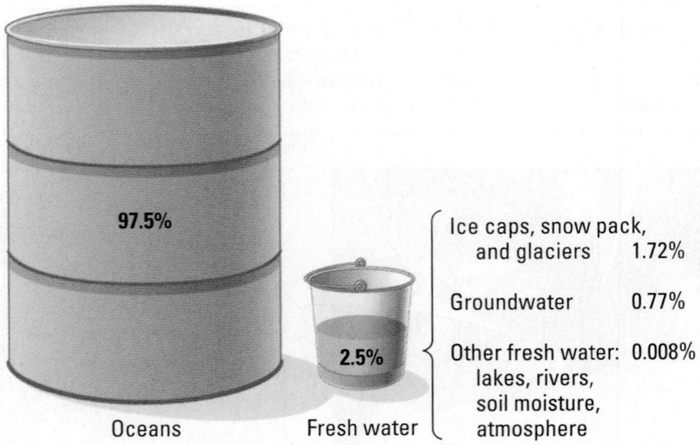
97.5%
2.5%
Oceans
Fresh water
Ice caps, snow pack, and glaciers 1.72%
Groundwater 0.77%
Other fresh water: 0.008%
lakes, rivers, soil moisture, atmosphere

Figure 15.25 Earth's water supply. Only 2.5% is fresh water, and less than 1% is available as groundwater or surface water.

available to satisfy our demands (we can't live without it) is relatively limited. In the United States, a bit more than half (57%) of fresh water is used in industry and about a third (34%) in agriculture to irrigate crops; only 9% is available for domestic and municipal purposes. Even when a sufficient quantity of water is available, the quality of the water needs to be assured before it is safe to drink (potable). Water must first be treated to make it potable.

Municipal Drinking Water Purification

Fresh water, even from natural sources, contains dissolved materials that must be removed entirely or partially to make water fit for domestic use, or for agricultural and industrial purposes. The Safe Water Drinking Act of 1974 established required standards of purity and safety for U.S. public water supplies. The U.S. EPA sets limits for contaminants that may be present in drinking water and requires continual monitoring of municipal water supplies.

Municipal water purification takes place in a series of steps (Figure 15.26). After a coarse filter (a screen) removes large objects such as tires, tree limbs, bottles, and so on, the water goes into a settling tank, where small clay and dirt particles settle out. To speed up the sedimentation, $Al_2(SO_4)_3$ (alum) and CaO (lime) are

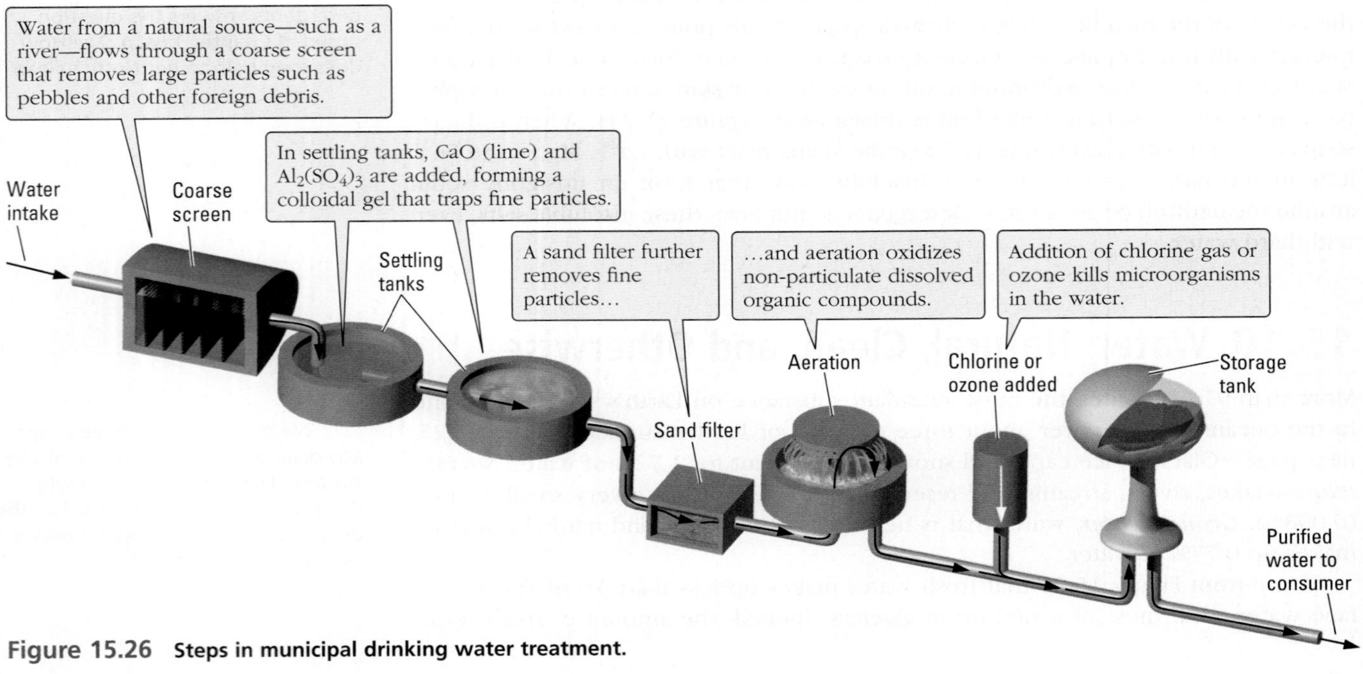
Water from a natural source—such as a river—flows through a coarse screen that removes large particles such as pebbles and other foreign debris.
In settling tanks, CaO (lime) and Al2(SO4)3 are added, forming a colloidal gel that traps fine particles.
A sand filter further removes fine particles...
...and aeration oxidizes non-particulate dissolved organic compounds.
Addition of chlorine gas or ozone kills microorganisms in the water.
Water intake
Coarse screen
Settling tanks
Sand filter
Aeration
Chlorine or ozone added
Storage tank
Purified water to consumer

Figure 15.26 Steps in municipal drinking water treatment.

added. These compounds react to form a sticky gelatinous precipitate of aluminum hydroxide.

Calcium hydroxide (slaked lime) forms when lime, CaO, reacts with water.

$$Al_2(SO_4)_3(aq) + 3\ Ca(OH)_2(aq) \longrightarrow 2\ Al(OH)_3(s) + 3\ CaSO_4(aq)$$

The $Al(OH)_3$ gel collects suspended clay and dirt particles as it sinks slowly in the settling tank. Particles not settled out are removed by passing the water from the settling tank through a sand filter.

In the next step, the water is aerated—sprayed into the air to oxidize dissolved organic substances. To this point, nothing has been done to remove potentially harmful bacteria. These bacteria are killed in the final step by chemically treating the water. *Chlorination* is the most common bactericidal method used in the United States. Chlorination is done by adding chlorine gas, sodium hypochlorite, NaOCl, or calcium hypochlorite, $Ca(OCl)_2$. In all three cases, the antibacterial agent generated is HOCl, hypochlorous acid. In the case of Cl_2, the HOCl forms by the reaction of chlorine with water.

$$Cl_2(g) + H_2O(\ell) \longrightarrow HOCl(aq) + H^+(aq) + Cl^-(aq)$$

The extent of chlorination is adjusted so that between 0.075 and 0.600 ppm of HOCl remains in solution as the water leaves the treatment plant, which is sufficient to ensure that bacterial contamination does not occur before the water reaches the user.

Chlorination was first used for drinking water supplies in the early 1900s, with a resulting drop in the number of deaths in the United States caused by typhoid and other water-borne diseases from 35/100,000 population in 1900 to 3/100,000 population in 1930. Chlorination is the principal means of preventing water-borne diseases spread by bacteria, including cholera, typhoid, paratyphoid, and dysentery.

In spite of chlorination, most city water supplies are not bacteria-free, but only rarely do these surviving bacteria cause disease. In the United States the most common water-borne bacterial disease is giardiasis, a gastrointestinal disorder. Most often this disease is caused by bacteria in surface water that has leaked into drinking water supplies, but on occasion it can be traced to city water systems.

Chlorination is not without its own small risk because of by-products it forms. Even the best-designed purification systems allow some organic compounds to pass through; these then become chlorinated. In particular, humic acids, breakdown products of plant materials always present in surface water, react with residual HOCl. The reaction forms a class of compounds known as trihalomethanes (THMs), such as chloroform, $CHCl_3$. Most drinking water meets the current maximum contaminant level standard of 80 ppb THMs. Chloroform is the trihalomethane of chief concern because it is suspected of causing liver cancer. Information is still being evaluated about the seriousness of this potential threat, especially given the fact that the THM level in drinking water is normally much less than 80 ppb.

The 80 ppb standard for THMs was established in 1998 by the U.S. Environmental Protection Agency. The previous standard had been 100 ppb. The national average for THMs in drinking water is 51 ppb.

Gaseous ozone, O_3, is used in many European cities to disinfect municipal water supplies. Ozone is an even more effective bactericide than chlorine, so less of it is needed to purify the water. The disadvantages of ozone are that it must be generated on site, and that ozone does not remain in the water as long as chlorine, allowing the possibility of recontamination.

Disinfection of water by using ultraviolet radiation is becoming more popular. It is fast, is economical for small installations such as rural homes, and leaves no residual by-products. Like ozonation, UV disinfection does not protect against bacterial contamination after water leaves the treatment site unless the appropriate amount of chlorine is added.

EXERCISE 15.19 Drinking Water

Selenium poisoning can occur in individuals who ingest more than 400 μg selenium per day. Calculate the mass of selenium ingested daily by an individual who drinks 3.0 qt water containing the maximum contaminant level of selenium, 0.050 ppm.

EXERCISE 15.20 Lead in Drinking Water

If the Pb^{2+} concentration in tap water is 0.0250 ppm, calculate how many liters of this water contain 100.0 μg Pb^{2+}.

Hard Water: Natural Impurities

Degrees of hardness:
Soft water: < 65 mg of metal ion/gal
Slightly hard: 65–228 mg/gal
Moderately hard: 228–455 mg/gal
Hard: 455–682 mg/gal
Very hard: > 682 mg/gal

A relatively high concentration of Ca^{2+}, Mg^{2+}, Fe^{2+}, or Mn^{2+} ions imparts "hardness" to water. Water hardness is objectionable because it (1) causes precipitates (called scale) to form in boilers and hot-water pipes, (2) causes soaps to form insoluble curds, and (3) may impart a disagreeable taste to the water.

Water hardness is produced when surface water containing carbon dioxide trickles through limestone or dolomite, releasing calcium or magnesium ions as their soluble hydrogen carbonates.

$$\underset{\text{limestone}}{CaCO_3(s)} + CO_2(g) + H_2O(\ell) \longrightarrow Ca^{2+}(aq) + 2\,HCO_3^-(aq)$$

$$\underset{\text{dolomite}}{CaCO_3 \cdot MgCO_3(s)} + 2\,CO_2(g) + 2\,H_2O(\ell) \longrightarrow Ca^{2+}(aq) + Mg^{2+}(aq) + 4\,HCO_3^-(aq)$$

Such hard water can be softened by removing these ions by two principal methods: (1) the lime-soda process or (2) ion exchange.

In the lime-soda process, slaked lime, $Ca(OH)_2$, and soda, Na_2CO_3, are added to the water. Several reactions take place, which can be summarized as:

In hard water **Added**

$$HCO_3^-(aq) + OH^-(aq) \longrightarrow CO_3^{2-}(aq) + H_2O(\ell)$$

$$Ca^{2+}(aq) + Na_2CO_3(aq) \longrightarrow CaCO_3(s) + 2\,Na^+(aq)$$

$$Mg^{2+}(aq) + 2\,OH^-(aq) \longrightarrow Mg(OH)_2(s)$$

The same equation applies if Mg^{2+} is substituted for Ca^{2+}.

The lime-soda process works because calcium carbonate, $CaCO_3$, is much less soluble than calcium hydrogen carbonate, $Ca(HCO_3)_2$, and because magnesium hydroxide, $Mg(OH)_2$, is much less soluble than magnesium hydrogen carbonate, $Mg(HCO_3)_2$. The overall result of the lime-soda process is to precipitate almost all the calcium and magnesium ions and to leave sodium ions (which do not make water hard) as replacements.

For persons on low-sodium diets, lime-soda–treated water may provide too high a daily dose of Na^+.

Iron present as Fe^{2+} and manganese present as Mn^{2+} can be removed from water by aeration to produce higher oxidation states. If the water is neutral or slightly alkaline (either naturally or from the addition of lime), insoluble compounds $Fe(OH)_3$ and $MnO_2(H_2O)_x$ form and precipitate from solution.

Ion exchange is another way to remove ions that cause water hardness. In a cation-exchange resin containing either H^+ or Na^+ ions, these ions replace the ions that cause hard water. Home water treatment ion-exchange units usually replace hardness ions with Na^+ ions. The exchange resin is a polymer containing numerous negatively charged $—SO_3^-$ functional groups that have Na^+ ions attached. When smaller, more positive ions like Mg^{2+} and Ca^{2+} in hard water flow over the resin, they displace Na^+ ions from the resin. The process happens many times, resulting in water that contains Na^+ ions in place of the Ca^{2+} and Mg^{2+} ions, which are bound to the resin. Two $—SO_3^-$ groups are required for every 2+ ion removed from solution; two Na^+ ions are released for every 2+ ion removed.

$$\underset{\text{ion-exchange resin}}{2\,(\text{Polymer—}SO_3^-)Na^+} + \underset{\text{from hard water}}{Ca^{2+}} \rightleftharpoons (\text{Polymer—}SO_3^-)_2Ca^{2+} + 2\,Na^+(aq)$$

Because Na^+ ions do not harden water, the resulting water is called "soft" water.

When all its sodium ions have been replaced, the ion-exchange resin becomes saturated with hard water ions. The resin must be regenerated by treating it with a highly concentrated NaCl solution. This reverses the process given in the previous equation and the hard water ions released from the resin are rinsed down the drain.

A home water softener. Hard water is softened by substituting sodium ions for calcium and magnesium ions.

SUMMARY PROBLEM

You are asked to prepare three mixtures at 25 °C.

Mixture I: 25.0 g CCl_4 and 100. mL water

Mixture II: 15.0 g $CaCl_2$ in 125 mL water

Mixture III: 21 g ethylene glycol, $HOCH_2CH_2OH$, in 150. mL water

Answer these questions about these mixtures. If one of the solutes fails to dissolve in water, some of the questions will not be applicable.

(a) Calculate the mass fraction of the mixture.

(b) Calculate the weight percent of the mixture.

(c) Is a solution formed? (If a solution is formed, answer the remaining questions. Assume a density of the solution of 1.0 g/mL.)

(d) Name the dissolved species in solution and draw a diagram representing how the solvent (water) molecules interact with these species.

(e) Express the concentration of the solution in ppm.

(f) Express the concentration of the solution in molality.

(g) Calculate the vapor pressure of water in equilibrium with the solution.

(h) Calculate the boiling point of the solution.

(i) Calculate the freezing point of the solution.

(j) Calculate the osmotic pressure of the solution.

IN CLOSING

Having studied this chapter, you should be able to . . .

- Describe how liquids, solids, and gases dissolve in a solvent (Section 15.1). End-of-chapter question: 1
- Predict solubility based on properties of solute and solvent (Section 15.1). Questions 25, 27, 31
- Interpret the dissolving of solutes in terms of enthalpy and entropy changes (Section 15.2). Question 30
- Differentiate among unsaturated, saturated, and supersaturated solutions (Section 15.3). Questions 5, 110
- Describe how ionic compounds dissolve in water (Section 15.3).
- Predict how temperature affects the solubility of ionic compounds (Section 15.4).
- Predict the effects of temperature (Section 15.4) and pressure on the solubility of gases in liquids (Section 15.5). Questions 8, 35
- Describe the compositions of solutions in terms of weight percent, mass fraction, parts per million, parts per billion, parts per trillion, molarity, and molality (Section 15.6). Questions 38, 40, 48, 60
- Interpret vapor pressure lowering in terms of Raoult's law (Section 15.7). Question 71

OWL and Go Chemistry

Sign in at **www.cengage.com/owl** to:

- View tutorials and simulations, develop problem-solving skills, and complete online homework assigned by your professor.
- For quick review and exam prep, download Go Chemistry mini lecture modules from OWL (or purchase them at **www.CengageBrain.com**).

- Use molality to calculate the colligative properties: freezing point lowering and boiling point elevation (Section 15.7). Questions 65, 75, 79, 121
- Differentiate the colligative properties of nonelectrolytes and electrolytes (Section 15.7). Questions 67, 102
- Explain the phenomena of osmosis and reverse osmosis, and calculate osmotic pressure (Section 15.7). Question 81
- Describe the various kinds of colloids and their properties (Section 15.8).
- Explain how surfactants work (Section 15.9).
- Discuss Earth's water supply and the sources of fresh water (Section 15.10).
- Discuss how municipal drinking water is purified (Section 15.10). Question 90
- Describe what causes hard water and explain how water can be softened (Section 15.10). Question 88

KEY TERMS

boiling point elevation *(Section 15.7)*
colligative properties *(15.7)*
colloid *(15.8)*
continuous phase *(15.8)*
detergents *(15.9)*
dispersed phase *(15.8)*
emulsion *(15.8)*
enthalpy of solution *(15.2)*
freezing point lowering *(15.7)*
Henry's law *(15.5)*
hydration *(15.3)*
hypertonic *(15.7)*
hypotonic *(15.7)*
immiscible *(15.1)*
isotonic *(15.7)*
mass fraction *(15.6)*
miscible *(15.1)*
molality *(15.6)*
osmosis *(15.7)*
osmotic pressure *(15.7)*
parts per billion *(15.6)*
parts per million *(15.6)*
parts per trillion *(15.6)*
Raoult's law *(15.7)*
reverse osmosis *(15.7)*
saturated solution *(15.3)*
semipermeable membrane *(15.7)*
solubility *(15.3)*
supersaturated solution *(15.3)*
surfactants *(15.9)*
Tyndall effect *(15.8)*
unsaturated solution *(15.3)*
weight percent *(15.6)*

QUESTIONS FOR REVIEW AND THOUGHT

OWL Interactive versions of these problems are assignable in OWL.

Blue-numbered questions have short answers at the back of this book in Appendix M and fully worked solutions in the *Student Solutions Manual.*

Review Questions

These questions test vocabulary and simple concepts.

1. Which of these general types of substances would you expect to dissolve readily in water?
   (a) Alcohols (b) Hydrocarbons
   (c) Metals (d) Nonpolar molecules
   (e) Polar molecules (f) Salts
2. Explain on a molecular basis why the components of blended motor oils remain dissolved and do not separate.
3. Explain why gasoline and motor oil are miscible, as in the fuel mixtures used in two-cycle lawn mower engines.
4. Describe the differences among solutions that are unsaturated, saturated, and supersaturated in terms of amount of solute.
5. Describe the differences among unsaturated, saturated, and supersaturated solutions in terms of Q and K_c.
6. State Henry's law. Name three factors that govern the solubility of a gas in a liquid.
7. In general, how does the water solubility of most ionic compounds change as the temperature is increased?
8. How does the solubility of gases in liquids change with increased temperature? Explain why.
9. Which is the highest solute concentration: 50 ppm, 500 ppb, or 0.05% by weight?
10. Estimate your concentration on campus in parts per million and parts per thousand.
11. Define molality. How does it differ from molarity?
12. Explain the difference between the mass fraction and the mole fraction of solute in a solution.
13. Explain why the vapor pressure of a solvent is lowered by the presence of a nonvolatile solute.
14. Why is a higher temperature required for boiling a solution containing a nonvolatile solute than for boiling the pure solvent?

15. Which would have the lowest freezing point?
    (a) A 1.0 mol/kg NaCl solution in water
    (b) A 1.0 mol/kg $CaCl_2$ solution in water
    (c) A 1.0 mol/kg methanol solution in water
    Explain your choice.
16. Write the osmotic pressure equation and explain all terms.
17. Explain the difference between (a) a hypotonic and an isotonic solution; (b) an isotonic and a hypertonic solution.
18. Explain how reverse osmosis works.
19. How do colloids differ from suspensions?
20. Explain why the Tyndall effect is not observed with solutions.
21. How can the presence of a strong electrolyte cause a hydrophobic colloid to coagulate?
22. Sketch an illustration of a soap molecule. Based on its structure, why is it considered a surfactant?
23. Surfactant molecules have what common structural features?

Topical Questions

These questions are keyed to the major topics in the chapter. Usually a question that is answered at the back of this book is paired with a similar one that is not.

How Substances Dissolve (Sections 15.1–15.5)

24. Explain why some liquids are miscible in each other while other liquids are immiscible. Using only three liquids, give an example of a miscible pair and an immiscible pair.
25. Why would the same solid readily dissolve in one liquid and be almost insoluble in another liquid? Give an example of such behavior.
26. The beakers (a), (b), and (c) are representations of tiny sections (not to scale) of mixtures made from pure glycerol and pure cyclohexane. Select which beaker gives proper representation of the result when the two pure substances are mixed.

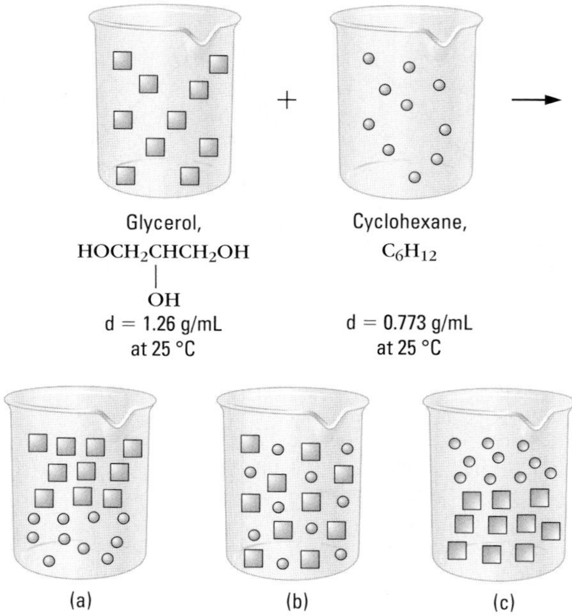

27. The beakers (a), (b), and (c) in the next column are representations of tiny sections (not to scale) of mixtures made from pure benzene and pure water. Select which beaker gives proper representation of the result when the two pure substances are mixed.

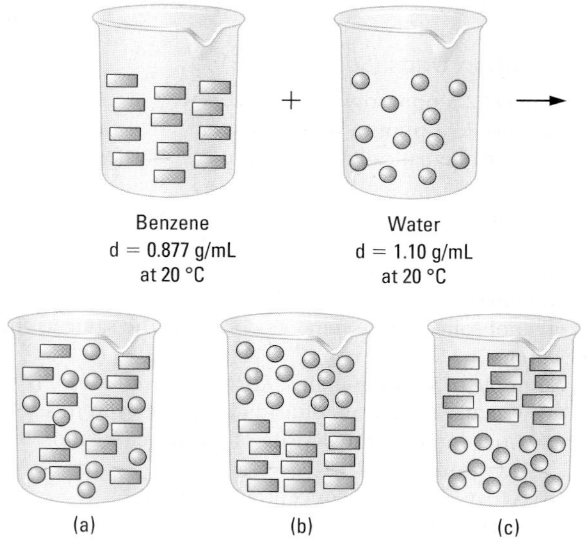

28. Knowing that the solubility of oxalic acid at 25 °C is 1 g per 7 g water, how would you prepare 1 L of a saturated oxalic acid solution?
29. A saturated solution of NH_4Cl was prepared by adding solid NH_4Cl to water until no more solid NH_4Cl would dissolve. The resulting mixture felt very cold and had a layer of undissolved NH_4Cl on the bottom. When the mixture reached room temperature, no solid NH_4Cl was present. Explain what happened. Was the solution still saturated?
30. The lattice energy of $CaCl_2$ is −2258 kJ/mol, and its enthalpy of hydration is +2175 kJ/mol. Is the process of dissolving $CaCl_2$ in water endothermic or exothermic?
31. Simple acids such as formic acid, HCOOH, and acetic acid, CH_3COOH, are very soluble in water; however, fatty acids such as stearic acid, $CH_3(CH_2)_{16}COOH$, and palmitic acid, $CH_3(CH_2)_{14}COOH$, are water-insoluble. Based on what you know about the solubility of alcohols, explain the solubility of these organic acids.
32. If a solution of a certain salt in water is saturated at some temperature and a few crystals of the salt are added to the solution, what do you expect will happen? What happens if the same quantity of the same salt crystals is added to an unsaturated solution of the salt? What would you expect to happen if the temperature of this second salt solution is slowly lowered?
33. Describe what happens when an ionic solid dissolves in water. Sketch an illustration that includes at least three positive ions, three negative ions, and a dozen or so water molecules in the vicinity of the ions.
34. The partial pressure of O_2 in your lungs varies from 25 mm Hg to 40 mm Hg. Calculate the concentration of O_2 (in grams per liter) that can dissolve in water at 37 °C when the O_2 partial pressure is 40. mm Hg. The Henry's law constant for O_2 at 37 °C is 1.5×10^{-6} mol L^{-1} mm Hg^{-1}.
35. The Henry's law constant for nitrogen in blood serum is approximately 8×10^{-7} mol L^{-1} mm Hg^{-1}. Calculate the N_2 concentration in a diver's blood at a depth where the total pressure is 2.5 atm. The air the diver is breathing is 78% N_2 by volume.

36. The Henry's law constant for N_2 in water at 25 °C is 8.4×10^{-7} mol L^{-1} mm Hg^{-1}. Calculate the solubility of N_2 in mol/L if its partial pressure is 1520 mm Hg. Calculate the solubility when the N_2 partial pressure is 20. mm Hg.

Concentration Units (Section 15.6)

37. Convert 2.5 ppm to weight percent.
38. Convert 73.2 ppm to weight percent.
39. Show mathematically how 1 ppm is equivalent to 1 mg/1 kg.
40. Show mathematically how 1 ppb is equivalent to 1 μg/1 kg.
41. Calculate the mass (in grams) of sucrose in 1.0 kg of a 0.25% sucrose solution.
42. How many grams of ethanol are in 750. mL of a 12% ethanol solution? (Assume its density is the same as that of water.)
43. A sample of water contains 0.010 ppm lead ions, Pb^{2+}.
    (a) Calculate the mass of lead ions per liter in this solution. (Assume the density of the water solution is 1.0 g/mL.)
    (b) What is the lead concentration in ppb?
44. A sample of lead-based paint contains 60.5 ppm lead. The density of the paint is 8.0 lb/gal. Calculate the mass of lead (in grams) that would be present in 50. gal of this paint.
45. A paint contains 200. ppm lead. Calculate what mass of lead (in grams) will be in 1.0 cm^2 of this paint (density = 8.0 lb/gal) when 1 gal is uniformly applied to a 500. ft^2 wall.
46. Calculate the mass in grams of solute needed to prepare each of these solutions.
    (a) 250. mL of 0.50-M NaCl
    (b) 0.50 L of 0.15-M sucrose, $C_{12}H_{22}O_{11}$
    (c) 200. mL of 0.20-M $NaHCO_3$
47. Calculate the mass in grams of solute required to prepare each of these solutions.
    (a) 750. mL of 4.00-M NH_4Cl
    (b) 1.50 L of 0.750-M KCl
    (c) 150. mL of 0.350-M Na_2SO_4
48. Calculate the molarity of the solute in a solution containing
    (a) 14.2 g KCl in 250. mL solution
    (b) 5.08 g K_2CrO_4 in 150. mL solution
    (c) 0.799 g $KMnO_4$ in 400. mL solution
    (d) 15.0 g $C_6H_{12}O_6$ in 500. mL solution
49. Calculate the molarity of the solute in a solution containing
    (a) 6.18 g $MgNH_4PO_4$ in 250. mL solution
    (b) 16.8 g $NaCH_3COO$ in 300. mL solution
    (c) 2.50 g CaC_2O_4 in 750. mL solution
    (d) 2.20 g $(NH_4)_2SO_4$ in 400. mL solution
50. Concentrated nitric acid is a 70.0% solution of nitric acid, HNO_3, in water. The density of the solution is 1.41 g/mL at 25 °C. Calculate the molarity of nitric acid in this solution.
51. Concentrated sulfuric acid has a density of 1.84 g/cm^3 and is 18 M. Calculate the weight percent of H_2SO_4 in the solution.
52. Consider a 13.0% solution of sulfuric acid, H_2SO_4, whose density is 1.090 g/mL.
    (a) Calculate the molarity of this solution.
    (b) To what volume should 100. mL of this solution be diluted to prepare a 1.10-M solution?
53. A 0.6-mL teardrop contains 4 mg NaCl. Calculate the molarity of NaCl in the teardrop.
54. You need a 0.050 mol/kg aqueous solution of methanol, CH_3OH. Calculate the mass of methanol you would need to dissolve in 500. g water to make this solution.
55. You want to prepare a 1.0 mol/kg solution of ethylene glycol, $C_2H_4(OH)_2$, in water. Calculate the mass of ethylene glycol you would need to mix with 950. g water.
56. A 23.2% by weight aqueous solution of sucrose has a density of 1.127 g/mL. Calculate the molarity of sucrose in this solution.
57. An aqueous beverage has a lead concentration of 25 ppb. Express this lead concentration as molarity.
58. Calculate the number of grams of KI required to prepare 100. mL of 0.0200-M KI. How many milliliters of this solution are required to produce 250. mL of 0.00100-M KI?
59. Calculate the molality of a solution made by dissolving 6.58 g NaCl in 250. mL water.
60. A 12-oz (355-mL) Pepsi contains 38.9 mg caffeine (molar mass = 194.2 g/mol). Assume that the Pepsi, mainly water, has a density of 1.01 g/mL. For such a Pepsi, calculate: (a) its caffeine concentration in ppm; (b) its molarity of caffeine; and (c) the molality of caffeine.
61. A concentration of 0.05 to 0.07 mg F^-/kg body weight is the optimum range for maximum protection against tooth decay with a minimum of risk. Assume that in one day you eat these items, whose fluoride content is given in ppm: one 12-oz cup brewed coffee (0.91); one 3-oz hot dog (0.48); one 12-oz Diet Coke (0.60); one 12-oz glass tap water (0.71); one 4-oz chocolate ice cream cone (0.23). Use calculations to determine whether the cumulative fluoride concentration in what you ate that day falls within the optimum range. 1.0 kg = 2.2 lb.

Colligative Properties (Section 15.7)

62. Which of these graphs correctly represents Raoult's Law? Explain your choice.

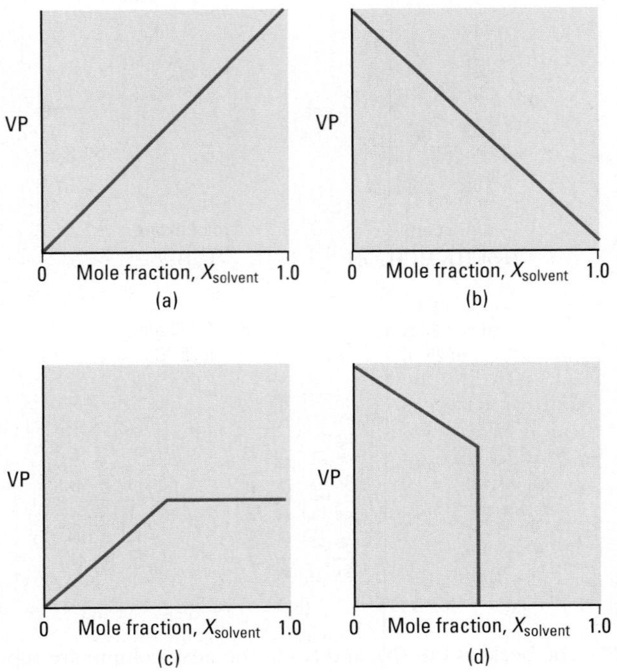

63. The two curves in the graph below represent pure benzene and a solution in which benzene is the solvent. Which curve is for pure benzene and which for the solution? Explain your answer.

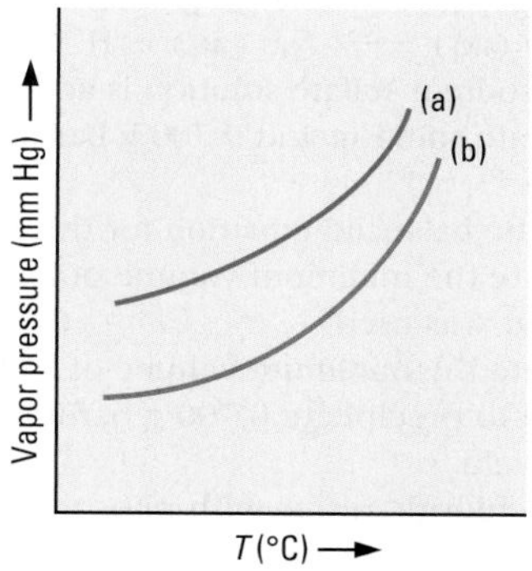

64. Calculate the boiling point of a solution containing 0.200 mol of a nonvolatile nonelectrolyte solute in 100. g benzene. The normal boiling point of benzene is 80.10 °C, and K_b = 2.53 °C kg mol^{-1}.
65. Calculate the boiling point of a solution composed of 15.0 g urea, $(NH_2)_2CO$, in 0.500 kg water.
66. Place these aqueous solutions in order of increasing boiling point.
    (a) 0.10 mol KCl/kg (b) 0.10 mol glucose/kg
    (c) 0.080 mol $MgCl_2$/kg
67. List these aqueous solutions in order of decreasing freezing point.
    (a) 0.10 mol methanol/kg (b) 0.10 mol KCl/kg
    (c) 0.080 mol $BaCl_2$/kg (d) 0.040 mol Na_2SO_4/kg
    (Assume that all of the salts dissociate completely into their ions in solution.)
68. Calculate the boiling point and the freezing point of these solutions at 760 mm Hg.
    (a) 20.0 g citric acid, $C_6H_8O_7$, in 100.0 g water
    (b) 3.00 g CH_3I in 20.0 g benzene (K_b benzene = 2.53 °C kg mol^{-1}; K_f benzene = 5.10 °C kg mol^{-1})
69. Calculate the freezing and boiling points (at 760 mm Hg) of a solution of 4.00 g urea, $CO(NH_2)_2$, dissolved in 75.0 g water.
70. Calculate the mass in grams of urea, $CO(NH_2)_2$, that must be added to 150. g water to give a solution whose vapor pressure is 2.5 mm Hg less than that of pure water at 40 °C (vp H_2O at 40 °C = 55.34 mm Hg).
71. At 60 °C the vapor pressure of pure water is 149.44 mm Hg and that above an aqueous sucrose, $C_{12}H_{22}O_{12}$, solution is 119.55 mm Hg. Calculate the mole fraction of water and the mass in grams of sucrose in the solution if the mass of water is 150. g.
72. At 760 mm Hg, a solution of 5.52 g glycerol in 40.0 g water has a boiling point of 100.777 °C. Calculate the molar mass of glycerol.
73. The boiling point of benzene is increased by 0.65 °C when 5.0 g of an unknown organic compound (a nonelectrolyte) is dissolved in 100. g benzene. Calculate the approximate molar mass of the organic compound. (K_b benzene = 2.53 °C kg mol^{-1}.)
74. The freezing point of *p*-dichlorobenzene is 53.1 °C, and its K_f is 7.10 °C kg mol^{-1}. A solution of 1.52 g of the drug sulfanilamide in 10.0 g *p*-dichlorobenzene freezes at 46.7 °C. Calculate the molar mass of sulfanilamide.
75. You add 0.255 g of an orange crystalline compound with an empirical formula of $C_{10}H_8Fe$ to 11.12 g benzene. The boiling point of the solution is 80.26 °C. The normal boiling point of benzene is 80.10 °C and its K_b = 2.53 °C kg mol^{-1}. Determine the molar mass and molecular formula of the compound.
76. If you use only water and pure ethylene glycol, $HOCH_2CH_2OH$, in your car's cooling system, what mass (in grams) of the glycol must you add to each quart of water to give freezing protection down to −31.0 °C? (1.00 L = 1.057 qt.)
77. Anthracene, a hydrocarbon obtained from coal, has an empirical formula of C_7H_5. To find its molecular formula you dissolve 0.500 g anthracene in 30.0 g benzene. The boiling point of the solution is 80.34 °C. The normal boiling point of benzene is 80.10 °C, and K_b = 2.53 °C kg mol^{-1}. Determine the molar mass and molecular formula of anthracene.
78. A 1.00 mol/kg aqueous sulfuric acid solution, H_2SO_4, freezes at −4.04 °C. Calculate *i*, the van't Hoff factor, for sulfuric acid in this solution.
79. Some ethylene glycol, $HOCH_2CH_2OH$, was added to your car's cooling system along with 5.0 kg water.
    (a) If the freezing point of the solution is −15.0 °C, what mass (in grams) of the glycol must have been added?
    (b) Calculate the boiling point of the coolant mixture.
80. Calculate the concentration of nonelectrolyte solute particles in human blood if the osmotic pressure is 7.53 atm at 37 °C, the temperature of the body.
81. The blood of cold-blooded animals and fish is isotonic with seawater. If seawater freezes at −2.3 °C, calculate the osmotic pressure of the blood of these animals at 20.0 °C. (Assume the density is that of pure water.)
82. The molar mass of a polymer was determined by measuring the osmotic pressure, 7.6 mm Hg, of a solution containing 5.0 g of the polymer dissolved in 1.0 L benzene. Calculate the molar mass of the polymer. Assume a temperature of 298.15 K.
83. The osmotic pressure at 25 °C is 1.79 atm for a solution prepared by dissolving 2.50 g sucrose, empirical formula $C_{12}H_{22}O_{11}$, in enough water to give a solution volume of 100 mL. Use the osmotic pressure equation to show that the empirical formula for sucrose is the same as its molecular formula.

Water: Purification and Solutions (Section 15.10)

84. Some dietitians recommend drinking six 8-oz glasses of water each day. If your drinking water contains the maximum contamination level for arsenic, 0.050 ppm, how much arsenic would you consume in a week following this recommendation?
85. When lead is present in drinking water, in what form does it exist?
86. The maximum contamination level (MCL) for chlordane is 0.002 ppm. A sample of well water contained 5 ppb chlordane. Is the sample within the MCL for chlordane?
87. In a home, 200. gal hard water containing 500. mg Ca^{2+}/gal passed through the Na^+-based ion-exchange water softener. If the ion-exchange resin operates at 100% efficiency, what mass of Na^+ ions is displaced from the resin?

88. How do the lime-soda and ion-exchange processes differ in treating hard water?
89. Explain how hard water produces "ring around the bathtub."
90. During municipal drinking water treatment, water is sprayed into the air. Why is this done?
91. Discuss the risks and benefits of using ozone to treat municipal drinking water.
92. Why is it necessary to bubble air through an aquarium?

General Questions

These questions are not explicitly keyed to chapter topics; many require integration of several concepts.

93. What is the difference between solubility and miscibility?
94. If 5 g solvent, 0.2 g solute A, and 0.3 g solute B are mixed to form a solution, what is the weight percent concentration of A?
95. A chemistry classmate tells you that a supersaturated solution is also saturated. Is the student correct? What would you tell the student about her/his statement?
96. In *The Rime of the Ancient Mariner* the poet Samuel Taylor Coleridge wrote, ". . . Water, water, everywhere/And all the boards did shrink. . . ." Explain this effect in terms of osmosis.
97. A 10.0-M aqueous solution of NaOH has a density of 1.33 g/cm^3 at 20 °C. Calculate the weight percent of NaOH in the solution.
98. Concentrated aqueous ammonia is 14.8 M and has a density of 0.90 g/cm^3. Calculate the weight percent of NH_3 in the solution.
99. (a) Calculate the molality of a solution made by dissolving 115.0 g ethylene glycol, $HOCH_2CH_2OH$, in 500. mL water. The density of water at this temperature is 0.978 g/mL.
(b) Calculate the molarity of the solution.
100. Dimethylglyoxime (DMG) reacts with nickel(II) ion in aqueous solution to form a bright red compound. However, DMG is insoluble in water. To get it into aqueous solution where it can encounter Ni^{2+} ions, it must first be dissolved in a suitable solvent, such as ethanol. Suppose you dissolve 45.0 g DMG ($C_4H_8N_2O_2$) in 500. mL ethanol (C_2H_5OH; density = 0.7893 g/mL). Determine the molality and weight percent of DMG in this solution.
101. Arrange these aqueous solutions in order of increasing boiling point. (Assume theoretical values for *i*.)
(a) 0.20 mol ethylene glycol/kg (b) 0.12 mol K_2SO_4/kg
(c) 0.10 mol $BaCl_2$/kg (d) 0.12 mol KBr/kg
102. Arrange these aqueous solutions in order of decreasing freezing point. (Assume theoretical values for *i*.)
(a) 0.20 mol ethylene glycol/kg (b) 0.12 mol Na_2SO_4/kg
(c) 0.10 mol NaBr/kg (d) 0.12 mol KI/kg
103. The solubility of NaCl in water at 100 °C is 39.1 g/100. g water. Calculate the boiling point of a saturated solution NaCl.
104. The organic salt $[(C_4H_9)_4N][ClO_4]$ consists of the ions $(C_4H_9)_4N^+$ and ClO_4^-. The salt dissolves in chloroform. What mass (in grams) of the salt must have been dissolved if the boiling point of a solution of the salt in 25.0 g chloroform is 63.20 °C? The normal boiling point of chloroform is 61.70 °C and K_b = 3.63 °C kg mol^{-1}. Assume that the salt dissociates completely into its ions in solution.
105. A solution, prepared by dissolving 9.41 g $NaHSO_3$ in 1.00 kg water, freezes at −0.33 °C. From these data, decide which of these equations is the correct expression for the dissociation of the salt.
(a) $NaHSO_3(aq) \longrightarrow Na^+(aq) + HSO_3^-(aq)$
(b) $NaHSO_3(aq) \longrightarrow Na^+(aq) + H^+(aq) + SO_3^{2-}(aq)$
106. A 0.250-M sodium sulfate solution is added to a 0.200-M barium nitrate solution and 0.700 g barium sulfate precipitates.
(a) Write the balanced equation for this reaction.
(b) Calculate the minimum volume of barium nitrate solution that was used.
(c) Calculate the minimum volume of sodium sulfate needed to precipitate 0.700 g barium sulfate. Assume 100% yield.
107. Aluminum chloride reacts with phosphoric acid to give aluminum phosphate, $AlPO_4$, which is used industrially as the basis of adhesives, binders, and cements.
(a) Write a balanced equation for the reaction of aluminum chloride and phosphoric acid.
(b) If you begin with 152. g aluminum chloride and 3.00 L of 0.750-M phosphoric acid, calculate the mass of $AlPO_4$ that can be isolated.
(c) Use the solubility rules in Table 5.1 (← ***p. 163***) to predict the solubility of $AlPO_4$.

Applying Concepts

These questions test conceptual learning.

108. Using these symbols,

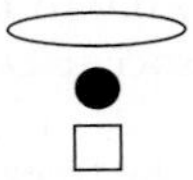
Sugar
Water
Carbon tetrachloride

draw nanoscale diagrams for the contents of a beaker containing
(a) Water and sugar
(b) Carbon tetrachloride and sugar
109. Using these symbols,

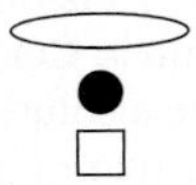
Ethanol
Water
Carbon tetrachloride

draw nanoscale diagrams for the contents of a beaker containing
(a) Water and ethanol
(b) Water and carbon tetrachloride
110. Refer to Figure 15.11 (p. 718) to determine whether these situations would result in an unsaturated, saturated, or supersaturated solution.
(a) 40. g NH_4Cl is added to 100. g H_2O at 80 °C.
(b) 100. g LiCl is dissolved in 100. g H_2O at 30 °C.
(c) 120. g $NaNO_3$ is added to 100. g H_2O at 40 °C.
(d) 50. g Li_2SO_4 is dissolved in 200. g H_2O at 50 °C.
111. Refer to Figure 15.11 (p. 718) to determine whether these situations would result in an unsaturated, saturated, or supersaturated solution.
(a) 120. g RbCl is added to 100. g H_2O at 50 °C.
(b) 30. g KCl is dissolved in 100. g H_2O at 70 °C.
(c) 20. g NaCl is dissolved in 50. g H_2O at 60 °C.
(d) 150. g CsCl is added to 100. g H_2O at 10 °C.

112. Complete this table.

Compound	Mass of Compound	Mass of Water	Mass Fraction of Solute	Weight Percent of Solute	ppm of Solute
Lye (NaOH)	______	125 g	0.375	______	______
Glycerol	33 g	200. g	______	______	______
Acetylene	0.0015 g	______	______	0.0009%	______

113. Complete this table.

Compound	Mass of Compound	Mass of Water	Mass Fraction	Weight Percent	ppm of Solute
Table salt	52 g	175 g	______	______	______
Glucose	15 g	______	______	______	7×10^4
Methane	______	100. g	______	0.0025%	______

114. What happens on the molecular level when a liquid freezes? What effect does a nonvolatile solute have on this process? Comment on the purity of water obtained by melting an iceberg.

115. If KI is added to a saturated solution of SrI_2, will the amount of solid SrI_2 present decrease, increase, or remain unchanged? What about the concentration of Sr^{2+} ion in solution?

116. In your own words, explain why
(a) Seawater has a lower freezing point than fresh water.
(b) Salt is added to the ice in an ice cream maker to freeze the ice cream faster.

117. Criticize these statements.
(a) A saturated solution is always a concentrated one.
(b) A 0.10 mol/kg sucrose solution and a 0.10 mol/kg KCl solution have the same osmotic pressure.

118. The illustration shows two beakers and their contents in a sealed container at some starting time, t_0. Both beakers contain the same solvent—pure solvent in one case, and solvent as part of a solution with a nonvolatile solute in the other case.

Pure solvent
Solvent + solute

Sketch the beakers and their contents in the sealed container to indicate any changes that have occurred after some time has passed.

More Challenging Questions

These questions require more thought and integrate several concepts.

119. In chemical research, newly synthesized compounds are often sent to commercial laboratories for analysis that determines the weight percent of C and H by burning the compound and collecting the evolved CO_2 and H_2O (← *p. 146*). The molar mass is determined by measuring the osmotic pressure of a solution of the compound. Calculate the empirical and molecular formulas of a compound, C_xH_yCr, given this information:
(a) The compound contains 73.94% C and 8.27% H; the remainder is chromium.
(b) At 25 °C, the osmotic pressure of 5.00 mg of the unknown dissolved in 100. mL of chloroform solution is 3.17 mm Hg.

120. An osmotic pressure of 5.15 atm is developed by a solution containing 4.80 g dioxane (a nonelectrolyte) dissolved in 250. mL water at 15.0 °C. The empirical formula of dioxane is C_2H_4O. Use the osmotic pressure data to show that the empirical formula and the molecular formula of dioxane are not the same.

121. The "proof" of an alcohol-containing beverage is twice the volume percentage of ethanol, C_2H_5OH (density = 0.789 g/mL), in the beverage. A bottle of 100-proof vodka is left outside where the temperature is −15 °C.
(a) Will the vodka freeze?
(b) Calculate the boiling point of the vodka.

122. A martini is a 5-oz (142-g) cocktail containing 30% by mass of alcohol. When the martini is consumed, about 15% of it passes directly into the bloodstream (blood volume = 7.0 L in an adult). Consider an adult who drinks two martinis with lunch. Estimate the blood alcohol concentration in this person after the two martinis have been consumed. An adult with a blood alcohol concentration of 3.0×10^{-3} g/mL or more is considered intoxicated. Is the person intoxicated?

123. Osmosis is responsible for sap rising in trees. Calculate the approximate height to which it could rise if the sap were 0.13 M in sugar and the dissolved solids in the water outside the tree were at a concentration of 0.020 M. *Hint:* A column of liquid exerts a pressure directly proportional to its density.

124. A 0.100 mol/kg NaCl solution has a van't Hoff factor of 1.87. Calculate the freezing point, the boiling point, and the osmotic pressure of this solution, whose density is 1.01 g/mL.

125. The osmotic pressure of human blood at 37 °C is 7.7 atm. Calculate what the molarity of a glucose solution should be if it is to be safely administered intravenously.

126. A 0.109 mol/kg aqueous solution of formic acid, HCOOH, freezes at −0.210 °C. Calculate the percent dissociation of formic acid.

127. Consider these data for a solution of water and ethanol, C_2H_5OH.

Mass Percent Ethanol	Boiling Point (°C)
25.0	84.5
50.0	81.2
67.0	80.0
85.0	78.4

(a) Plot these data and use the graph to determine the boiling point of a solution that is 59.0% ethanol by mass.
(b) Calculate the molarity and the molality of the 67.0% solution.

128. A 0.050 mol/kg aqueous solution of iodic acid, HIO_3, freezes at −3.12 °C; a 0.20 mol/kg solution of the acid freezes at −2.71 °C; a 1.0 mol/kg solution freezes at −1.72 °C.
    (a) Plot these data and use the graph to determine the boiling point of a solution that is 0.5 mol iodic acid/kg.
    (b) Calculate the percent dissociation of iodic acid in each solution.

129. Consider these data for aqueous solutions of ammonium chloride, NH_4Cl.

Molality (mol/kg)	Freezing Point (°C)
0.0050	−3.617
0.020	−3.544
0.20	−3.392
1.0	−3.33

    (a) Plot these data and from the graph determine the freezing point of a 0.50 mol/kg ammonium chloride solution.
    (b) Calculate the percent dissociation of ammonium chloride in each solution.

130. Maple syrup sap is 3% sugar (sucrose) and 97% water by mass. Maple syrup is produced by heating the sap to evaporate a certain amount of the water.
    (a) Describe what happens to the composition and boiling point of the solution as evaporation takes place.
    (b) A rule of thumb among maple syrup producers is that the finished syrup should boil about 4 °C higher than the original sap being boiled. Explain the chemistry behind this guideline.
    (c) If the finished product boils 4 °C higher than the original sap, calculate the concentration of sugar in the final product. Assume that sugar is the only solute and the operation is done at 1 atm pressure.

131. A 0.63% by weight aqueous tin(II) fluoride, SnF_2, solution is used as an oral rinse in dentistry to decrease tooth decay.
    (a) Calculate the SnF_2 concentration in ppm and ppb.
    (b) Calculate the molarity of SnF_2 in the solution.
    (c) Tin(II) fluoride is produced commercially by a series of steps that begin with reduction of cassiterite, SnO_2, the principal tin ore, with carbon.

$$SnO_2(s) + 2\,C(s) \longrightarrow Sn(s) + 2\,CO(g)$$

    Once purified, the tin is reacted with hydrogen fluoride vapor to produce SnF_2. Consider that one metric ton of cassiterite ore is reduced with sufficient carbon to tin metal in 80% yield. The tin is purified and reacted with sufficient hydrogen fluoride to produce SnF_2 in 94% yield. Calculate how many 250.-mL bottles of 0.63% SnF_2 solution could be prepared from the one metric ton of cassiterite by these steps.

Conceptual Challenge Problems

These rigorous, thought-provoking problems integrate conceptual learning with problem solving and are suitable for group work.

CP15.A (Section 15.6) Concentrations expressed in units of parts per million and parts per billion often have no meaning for people until they relate these small and large numbers to their own experiences.
(a) What time in seconds is 1 ppm of a year?
(b) What time in seconds is 1 ppb of a 70-year lifetime?

CP15.B (Section 15.4) Bodies of water with an abundance of nutrients that support a blooming growth of plants are said to be eutrophoric. In general, fish do not thrive for long in eutrophoric waters because little oxygen is available for them. Suppose someone asked you why this was true, given the fact that growing plants produce oxygen as a product of photosynthesis. How would you respond to this person's inquiry?

CP15.C (Section 15.7) Suppose that you want to produce the lowest temperature possible by using ice, sodium chloride, and water to chill homemade ice cream made in a 1.5-L metal cylinder surrounded by a coolant held in a wooden bucket. You have all the ice, salt, and water you want. How would you plan to do this?

Acids and Bases

16

© Cengage Learning/Charles D. Winters

Monster Energy drink has a pH of 2.87 as indicated by the pH meter. Measurement of the acidity or alkalinity of substances, expressed as pH, is important in commerce, medicine, and studies of the environment. The measurement and uses of pH are discussed in this chapter.

It is difficult to overstate the importance of acids and bases. Aqueous solutions abound in our environment and in all living organisms and are almost always acidic or basic to some degree. Photosynthesis and respiration, the two most important biological processes on Earth, depend on acid-base reactions. Carbon dioxide, CO_2, is the most important acid-producing compound in nature. Rainwater is usually slightly acidic because of dissolved CO_2, and acid rain results from further acidification of rainwater by acids formed by the gaseous pollutants SO_2 and NO_2. The oceans are slightly basic, as are many ground and surface waters. Natural waters can

OWL Sign in to OWL at **www.cengage.com/owl** to view tutorials and simulations, develop problem-solving skills, and complete online homework assigned by your professor.

go Chemistry Download mini lecture videos for key concept review and exam prep from OWL or purchase them from **www.CengageBrain.com.**

Companion Website
Visit this book's companion website at **www.cengage.com/chemistry/moore** to work interactive modules for the Estimation boxes and Active Figures in this text.

also be acidic; the more acidic the water, the more easily metals such as lead can be dissolved from water pipes or soldered joints in older buildings.

Because of their importance in much of chemistry and biochemistry the properties of acids and bases have been studied extensively. In 1677 Antoine Lavoisier proposed that oxygen made an acid acidic. He even derived the name *oxygen* from Greek words meaning "acid former." But in 1808 it was discovered that the gaseous compound HCl, which dissolves in water to give hydrochloric acid, contains only hydrogen and chlorine. It later became clear that hydrogen, not oxygen, is common to acids in aqueous solution. It was also shown that aqueous solutions of both acids and bases conduct an electrical current because acids and bases are electrolytes—they release ions into solution (← *p. 97*). In 1887, the Swedish chemist Svante Arrhenius proposed that acids *ionize in aqueous solution to produce hydrogen ions (protons) and anions;* bases *ionize to produce hydroxide ions and cations.* We now rely on a more general acid-base concept: ***hydronium ions, H_3O^+, are responsible for the properties of acidic aqueous solutions, and hydroxide ions, OH^-, are responsible for the properties of basic aqueous solutions.***

Because of their small size and high charge density, hydrogen ions (protons) are always associated with water molecules in aqueous solution and are usually represented as H_3O^+, the hydronium ion.

$$H^+ + H_2O \longrightarrow H_3O^+$$

16.1 The Brønsted-Lowry Concept of Acids and Bases

A major problem with the Arrhenius acid-base concept is that certain substances, such as ammonia, NH_3, produce basic solutions and react with acids, yet contain no hydroxide ions. In 1923, J. N. Brønsted in Denmark and T. M. Lowry in England independently proposed a new way of defining acids and bases in aqueous solutions:

- **Brønsted-Lowry acids** are hydrogen ion donors.
- **Brønsted-Lowry bases** are hydrogen ion acceptors.

According to the Brønsted-Lowry concept, an acid can donate a hydrogen ion, H^+ (a proton), to another substance, while a base can accept an H^+ ion from another substance. In **acid-base reactions,** acids *donate* H^+ ions, and bases *accept* them. To accept an H^+ and serve as a Brønsted-Lowry base, a molecule or ion must have an *unshared (lone) pair of electrons.* For example, in aqueous solutions, H^+ ions from acids such as nitric acid, HNO_3, react with water molecules, which accept protons to form hydronium ions, H_3O^+.

Color is used functionally throughout this book. In this and subsequent chapters, a pinkish color screen is used to designate an acid; a bluish color screen indicates a base.

Acid: H^+ donor | Base: H^+ acceptor

$$HNO_3(aq) + H_2O(\ell) \longrightarrow H_3O^+(aq) + NO_3^-(aq)$$

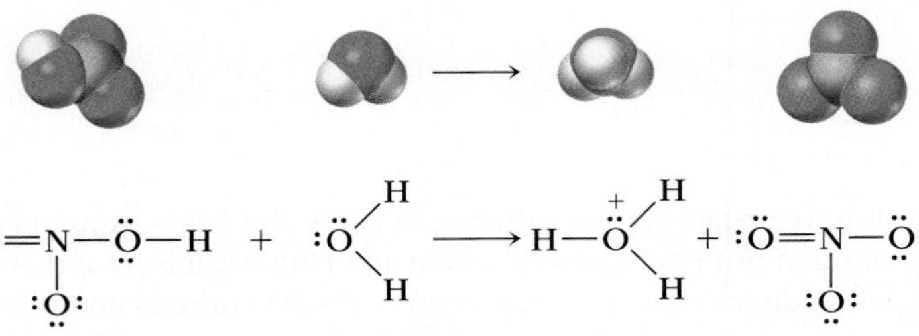

In reacting with an acid, water acts as a Brønsted-Lowry base by using an unshared electron pair to accept the H^+. Because nitric acid is a strong acid, it is 100% ionized in aqueous solution, and the equation above is strongly product-favored.

An ionic substance that dissolves in water is a strong electrolyte (← *p. 97*). A strong acid ionizes completely and is a strong electrolyte.

In contrast, a weak acid does *not* ionize completely and therefore is a weak electrolyte and at equilibrium appreciable quantities of un-ionized acid molecules

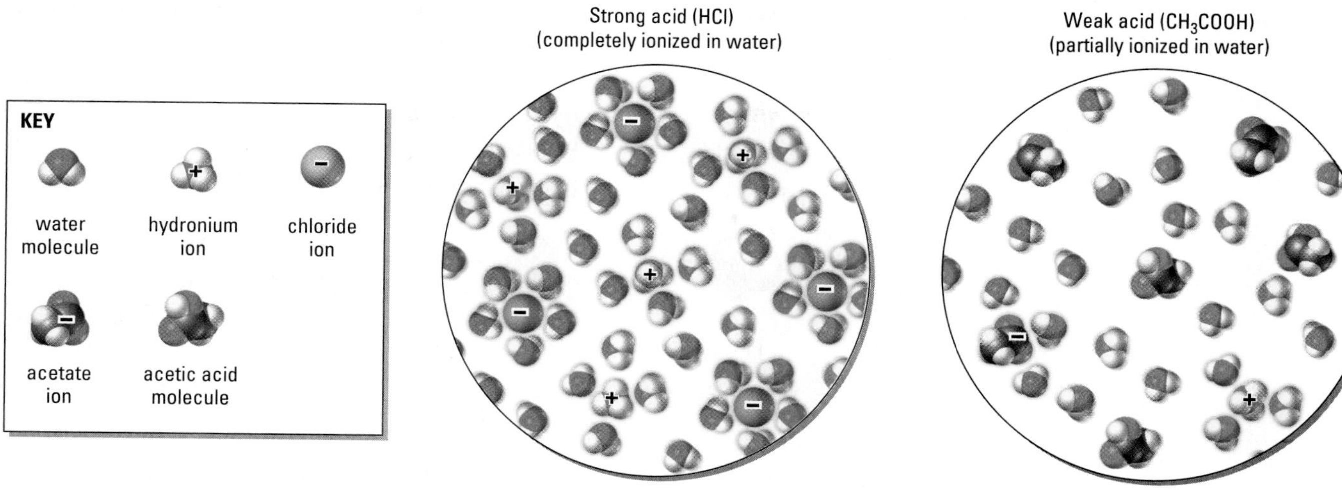
Strong acid (HCl)
(completely ionized in water)
Weak acid (CH_3COOH)
(partially ionized in water)
KEY
water molecule
hydronium ion
chloride ion
acetate ion
acetic acid molecule

Ionization of acids in water. A strong acid such as hydrochloric acid, HCl, is completely ionized in water; a weak acid such as acetic acid, CH_3COOH, is only partially ionized in water. To make other molecules and ions more visible, not all of the water molecules are shown.

are present. For example, hydrofluoric acid, HF, is a weak acid, and its ionization in water is written as

$$\underset{\text{hydrofluoric acid, a weak acid}}{HF(aq)} + H_2O(\ell) \rightleftharpoons H_3O^+(aq) + F^-(aq)$$

The double arrow indicates an equilibrium between the reactants and the products *(← p. 657).* Because ionization of HF is much less than 100%, this means that, at equilibrium, most HF molecules are un-ionized and there are relatively few hydronium and fluoride ions in solution. Thus, the ionization of *weak* acids is a reactant-favored process.

CHEMISTRY IN THE NEWS — HCl Dissociation at the Smallest Scale

Using laser-based infrared spectroscopy and quantum mechanics calculations, researchers at Ruhr University in Germany have determined that at least four water molecules are required to spontaneously dissociate one HCl molecule. Under conditions near 1 K in nonreactive liquid helium droplets, one HCl molecule reacts with four water molecules to form an $H_3O^+(H_2O)_3Cl^-$ ion pair. The ion pair results from a fourth water molecule adding to a cyclic $HCl(H_2O)_3$ cluster, which is formed by stepwise addition of water molecules to HCl. The helium droplets absorb the HCl and water molecules where they form the ion pairs.

Applying *ab initio* quantum mechanics calculations, the researchers calculated the structure of the $H_3O^+(H_2O)_3Cl^-$ ion pair. The calculated structure shows hydrogen bonding between the lone pair electrons on oxygen atoms (red in the figure) of three water molecules to the hydrogen atoms of H_3O^+. Each polar water molecule orients one of its hydrogen atoms toward the negatively charged chloride ion (green).

This research has broad implications for fundamental studies of factors affecting acid strength and also ionization of HCl on ice surfaces, such as in stratospheric ozone depletion conditions *(← p. 454).*

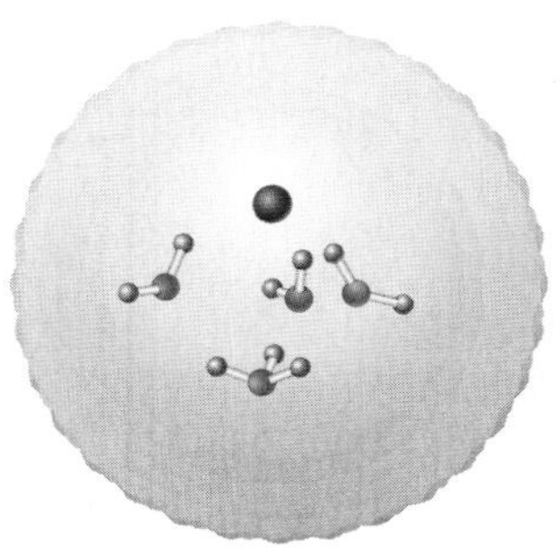

Calculated Structure of $H_3O^+(H_2O)_3Cl^-$. Hydrogen bonding occurs between the water molecules and the hydronium ion. There are ion-dipole attractions between the water molecules and the chloride ion. (From *Science*, Vol. 324, June 19, 2009, p. 1523. Reprinted by permission.)

Sources: Gutberlet, A., *et al.*, "Aggregation-Induced Dissociation of $HCl(H_2O)_4$ Below 1 K: The Smallest Droplet of Acid." *Science*, Vol. 324, June 19, 2009; pp. 1545–1548.

Zwier, T. S. "Squeezing the Water Out of HCl(aq)." *Science*, Vol. 324, June 19, 2009; pp. 1522–1523.

Ammonia, NH_3, is a compound consisting of electrically neutral molecules; the ammonium ion, NH_4^+, is a polyatomic ion.

Ammonia, NH_3, is a base because it accepts an H^+ from water (an acid) to form an ammonium ion, NH_4^+. Water, having donated an H^+, is converted into a hydroxide ion, OH^-.

Base: H^+ acceptor — Acid: H^+ donor

$$NH_3(g) + H_2O(\ell) \rightleftharpoons NH_4^+(aq) + OH^-(aq)$$

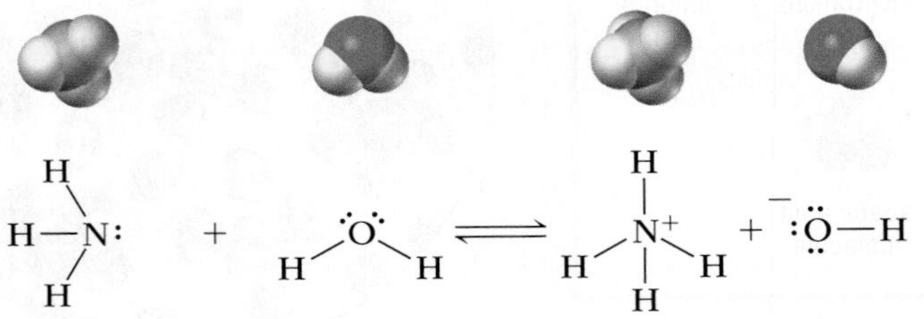

Ammonia establishes an equilibrium with water, ammonium ions, and hydroxide ions and is therefore a weak base (**←** ***p. 171***). At equilibrium, the solution contains far more ammonia molecules than ammonium and hydroxide ions. Therefore, ammonium hydroxide is not an appropriate name for an aqueous solution of ammonia.

Answers to **EXERCISES** are provided at the back of this book in Appendix L.

EXERCISES that are labeled **CONCEPTUAL** are designed to test your understanding of one or more concepts; they usually involve qualitative rather than quantitative thinking.

CONCEPTUAL EXERCISE 16.1 Brønsted-Lowry Acids and Bases

Identify each molecule or ion as a Brønsted-Lowry acid or base.

(a) HBr (b) Br^- (c) HNO_2 (d) CH_3NH_2

CONCEPTUAL EXERCISE 16.2 Using Le Chatelier's Principle

Use Le Chatelier's principle (**←** ***p. 678***) to explain why a larger percentage of NH_3 will react with water in a very dilute solution than in a more concentrated solution.

Water's Role as Acid or Base

In aqueous solution, all Brønsted-Lowry acids and bases react with water molecules. As we have seen, a water molecule *accepts* an H^+ from an acid such as nitric acid, while a water molecule *donates* an H^+ to a base such as an ammonia molecule. According to the Brønsted-Lowry definitions, water serves as a base (an H^+ acceptor) when an acid is present and as an acid (an H^+ donor) when a base is present. Therefore, water displays both acid and base properties—*it can donate or accept H^+ ions,* depending on its chemical environment. The general reactions of water with acids, HA, and molecular bases, B, are

Water acting as a base, an H^+ acceptor:

$$\underset{\text{acid}}{HA} + \underset{\text{base}}{H_2O} \rightleftharpoons H_3O^+ + A^-$$

Water acting as an acid, an H^+ donor:

If the base is an anion, A^-, it accepts H^+ from water to form AH.

$$\underset{\text{base}}{B} + \underset{\text{acid}}{H_2O} \rightleftharpoons BH^+ + OH^-$$

A substance, like water, that can donate or accept H^+ is said to be **amphiprotic.**

EXERCISE 16.3 Acids and Bases

Complete these equations. (*Hint:* CH_3NH_2 and $(CH_3)_2NH$ are amines, so they are bases.)

(a) $HCN + H_2O \longrightarrow$ (b) $HBr + H_2O \longrightarrow$

(c) $CH_3NH_2 + H_2O \longrightarrow$ (d) $(CH_3)_2NH + H_2O \longrightarrow$

Conjugate Acid-Base Pairs

Whenever an acid donates H^+ to a base, a new acid and a new base are formed. A pair of molecules or ions *related to each other by the loss or gain of a single* H^+ is called a **conjugate acid-base pair.** Every Brønsted-Lowry acid has its conjugate base, and every Brønsted-Lowry base has its conjugate acid.

We illustrate this using the reaction between acetic acid, CH_3COOH, and water (Figure 16.1). Acetic acid is an H^+ ion donor (an acid), and water is an H^+ ion acceptor (a base). The products of the reaction are a new acid, H_3O^+, and a new base, CH_3COO^-. In the reverse reaction, H_3O^+ acts as an acid (H^+ donor), and acetate ion as a base (H^+ acceptor). The structures of CH_3COOH and CH_3COO^- differ from one another by only a single H^+, just as the structures of H_2O and H_3O^+ do.

Removing an H^+ ion from the acid forms the conjugate base making the charge of the remaining portion of the acid one unit more negative. For example, hydrofluoric acid, HF, has the F^- (fluoride) ion as its conjugate base; HF and F^- are a conjugate acid-base *pair.*

ACID
Donates H^+ ⟶

$$HF(aq) + H_2O(\ell) \rightleftharpoons H_3O^+(aq) + F^-(aq)$$

ACID base acid BASE

BASE
⟵ Accepts H^+

In the forward reaction, HF is a Brønsted-Lowry acid. It donates an H^+ to water, which, by accepting the H^+, acts as a Brønsted-Lowry base. In the reverse reaction, fluoride ion is the Brønsted-Lowry base, accepting an H^+ from H_3O^+, the acid. As noted in the equation, there are two conjugate acid-base pairs: (1) HF and F^- and (2) H_2O and H_3O^+. *One member of a conjugate acid-base pair is always a reactant and the other is always a product; they are never both products or both reactants.*

The conjugate acid formula can be derived from the formula of its conjugate base by adding an H^+ ion to the conjugate base and making the charge of the resulting conjugate acid one unit more positive than the conjugate base. Therefore, the conjugate acid of Cl^- ion is HCl; the conjugate acid of NH_3 is NH_4^+.

BASE
Accepts H^+ ⟶

$$NH_3(aq) + H_2O(\ell) \rightleftharpoons NH_4^+(aq) + OH^-(aq)$$

BASE acid ACID base

ACID
⟵ Donates H^+

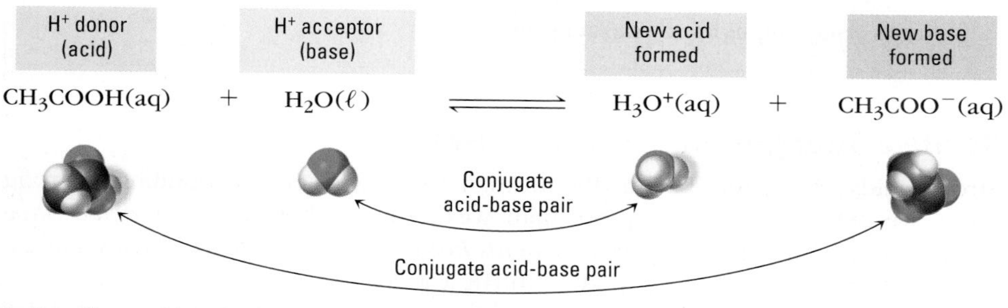
H⁺ donor (acid)
H⁺ acceptor (base)
New acid formed
New base formed
CH₃COOH(aq) + H₂O(ℓ) ⇌ H₃O⁺(aq) + CH₃COO⁻(aq)
Conjugate acid-base pair
Conjugate acid-base pair

Active Figure 16.1 Conjugate acid-base pairs. Transfer of an H^+ converts a conjugate acid into its conjugate base. Visit this book's companion website at **www.cengage.com/chemistry/moore** to test your understanding of the concepts in this figure.

The **PROBLEM-SOLVING STRATEGY** in this book is

- Analyze the problem
- Plan a solution
- Execute the plan
- Check that the result is reasonable

Appendix A.1 explains this in detail.

PROBLEM-SOLVING EXAMPLE 16.1 Conjugate Acid-Base Pairs

Complete this table by identifying the correct conjugate acid or conjugate base.

Acid	Its Conjugate Base	Base	Its Conjugate Acid
HCOOH		CN^-	_____
H_2S	_____	_____	HSO_4^-
PH_4^+	_____	_____	H_2SO_3
_____	ClO^-	S^{2-}	_____

Answer

Acid	Its Conjugate Base	Base	Its Conjugate Acid
HCOOH	$HCOO^-$	CN^-	HCN
H_2S	HS^-	SO_4^{2-}	HSO_4^-
PH_4^+	PH_3	HSO_3^-	H_2SO_3
HClO	ClO^-	S^{2-}	HS^-

Strategy and Explanation In each of these cases, apply the relationship

ACID
Donates H^+ ⟶

Conjugate acid ⇌ Conjugate base

BASE
⟵ Accepts H^+

The conjugate acid can be identified by adding H^+ to the conjugate base; the conjugate base forms by loss of H^+ from the conjugate acid. For example, because CN^- has no H^+ to donate, it must be a base—the conjugate base of HCN, its conjugate acid. Likewise, HSO_3^- is the conjugate base of its conjugate acid, H_2SO_3. The other conjugate acid-base pairs can be worked out similarly.

PROBLEM-SOLVING PRACTICE answers are provided at the back of this book in Appendix K.

PROBLEM-SOLVING PRACTICE 16.1

Complete the table.

Acid	Its Conjugate Base	Base	Its Conjugate Acid
$H_2PO_4^-$	_____	_____	HPO_4^{2-}
_____	H^-	NH_2^-	_____
HSO_3^-	_____	ClO_4^-	_____
HF	_____	_____	HBr

EXERCISE 16.4 HSO_4^- as a Base

(a) Write the equation for HSO_4^- ion acting as a base in water.
(b) Identify the conjugate acid-base pairs.
(c) Is HSO_4^- amphiprotic? Explain your answer.

Relative Strengths of Acids and Bases

Strong acids are better H^+ ion donors than weak acids. Correspondingly, strong bases are better H^+ ion acceptors than weak bases. Thus, *stronger acids have weaker conjugate bases and weaker acids have stronger conjugate bases.* For example, compare HCl, a strong acid, with HF, a weak acid.

$$HCl(aq) + H_2O(\ell) \longrightarrow H_3O^+(aq) + Cl^-(aq)$$

$$HF(aq) + H_2O(\ell) \rightleftharpoons H_3O^+(aq) + F^-(aq)$$

At concentrations <1.0 M the ionization of HCl is virtually 100%; essentially all of the HCl molecules react with water to form H_3O^+ and Cl^-. The Cl^- ion exhibits virtually no tendency to accept H^+ from H_3O^+; Cl^- is an extremely weak base. On the other hand, the reverse of the ionization of HF is significant; F^- ion readily accepts H^+ from H_3O^+, and hydrofluoric acid is a weak acid that is mainly un-ionized (remains largely in its molecular form, HF).

By measuring the extent to which various acids donate H^+ ions to water, chemists have developed an extensive tabulation of the relative strengths of acids and their conjugate bases. An abbreviated list is given in Figure 16.2, where the strongest acids are at the top left and the weakest bases at the top right. The weakest acids are at the bottom left, with the strongest bases at the bottom right. From Figure 16.2 we can draw two important generalizations regarding conjugate acid-base pairs:

- ***As acid strength decreases, conjugate base strength increases; the weaker the acid, the stronger its conjugate base.***
- ***As base strength decreases, conjugate acid strength increases; the weaker the base, the stronger its conjugate acid.***

Knowing the relative acid and base strengths of the reactants, we can predict the direction of an acid-base reaction. ***The stronger acid and the stronger base will always react to form a weaker conjugate base and a weaker conjugate acid.***

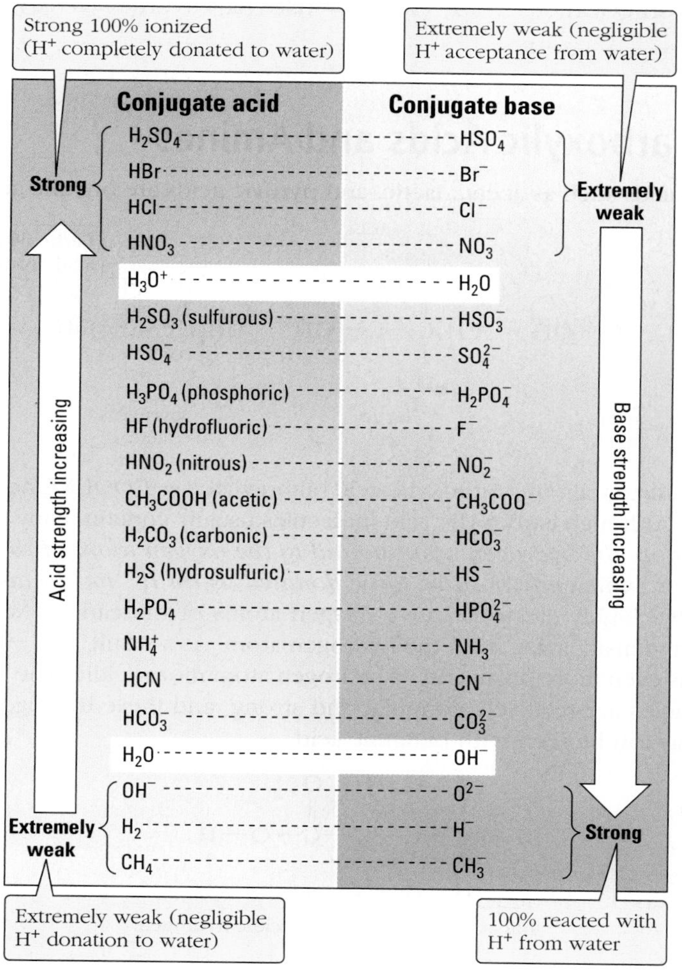

Figure 16.2 **Relative strengths of conjugate acids and bases in water.**

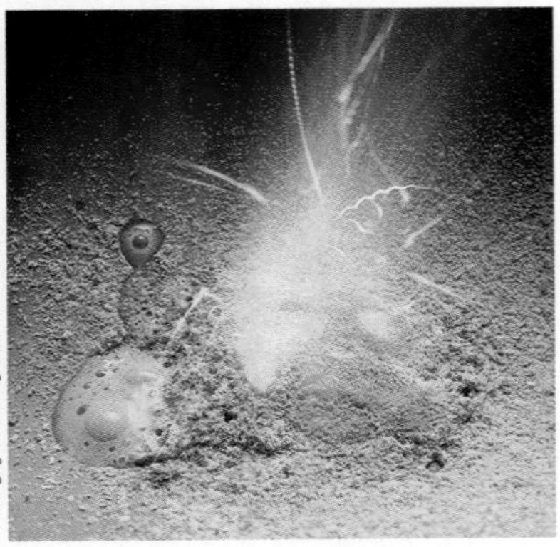

The strongly basic properties of hydride ion. The reaction of calcium hydride with water is highly exothermic due to the strongly basic hydride ion.

Strong Brønsted-Lowry bases such as hydride ion, H^-; sulfide ion, S^{2-}; oxide ion, O^{2-}; amide ion, NH_2^-; and hydroxide ion, OH^-, readily accept H^+ ions, while weaker Brønsted-Lowry bases do so less readily. For example, the reaction of calcium hydride, CaH_2, with water is highly exothermic because of the extremely strong basic properties of the hydride ion, H^-, which avidly accepts H^+ from water to produce H_2, the extremely weak conjugate acid of hydride ion. In this reaction, hydride ion is a stronger base than OH^-, and water is a stronger acid than H_2, so the forward reaction is favored.

$$H^-(aq) + H_2O(\ell) \longrightarrow H_2(g) + OH^-(aq)$$

A great many weak bases are anions, such as CN^- (cyanide), F^- (fluoride), and CH_3COO^- (acetate).

We can apply information from Figure 16.2 to consider whether the forward or the reverse reaction is favored in an equilibrium. For example, consider the equilibrium

$$HSO_4^-(aq) + CO_3^{2-}(aq) \rightleftharpoons HCO_3^-(aq) + SO_4^{2-}(aq)$$

From Figure 16.2 we note that HSO_4^- is a stronger acid than HCO_3^-, and CO_3^{2-} is a stronger base than SO_4^{2-}. Since acid-base reactions favor going from the stronger to the weaker member of each conjugate acid-base pair, the forward reaction is favored and H^+ will be transferred from HSO_4^- to CO_3^{2-} to form hydrogen carbonate, HCO_3^-, and sulfate, SO_4^{2-}, ions.

CONCEPTUAL EXERCISE 16.5 Conjugate Acid-Base Strength

Use Figure 16.2 to predict whether the forward or reverse reaction is favored for the equilibrium

$$CH_3COOH(aq) + SO_4^{2-}(aq) \rightleftharpoons HSO_4^-(aq) + CH_3COO^-(aq)$$

16.2 Carboxylic Acids and Amines

Many weak acids such as acetic, lactic, and pyruvic acids are organic acids.

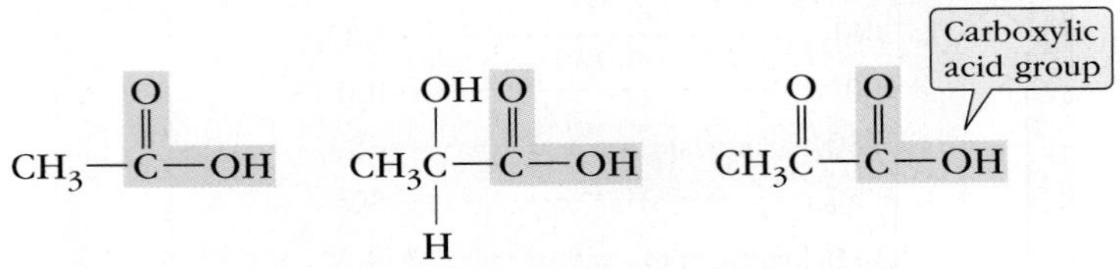
Carboxylic acid group

acetic acid — lactic acid — pyruvic acid

Note that although the carboxylic acid group is abbreviated —COOH, the oxygen atoms are not bonded to each other, but to carbon.

These acids all contain the carboxylic acid (abbreviated —COOH) functional group **(← *p. 554*)**. Although carboxylic acid molecules usually contain many other hydrogen atoms, *only the hydrogen atom bound to the oxygen atom of the carboxylic acid group is sufficiently positive to be donated as an H^+ ion in aqueous solution.* The two highly electronegative oxygen atoms of the carboxylic acid group pull electron density away from the hydrogen atom. As a result, the —O—H bond of the acid is even more polar, and its hydrogen atom more acidic. The C—H bonds in organic acids are relatively nonpolar and strong, and these hydrogen atoms are *not* acidic, as can be seen with butanoic acid.

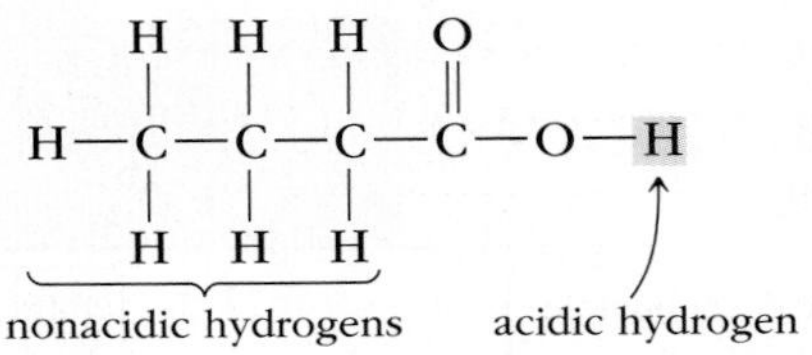
nonacidic hydrogens
acidic hydrogen

butanoic acid

Anions formed by loss of an H^+ from a —COOH group, such as acetate ion, CH_3COO^-, from acetic acid, CH_3COOH, are stabilized by resonance *(← p. 352)*.

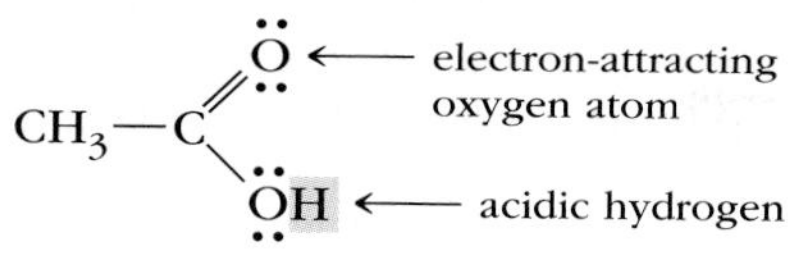
electron-attracting
oxygen atom
acidic hydrogen
acetic acid

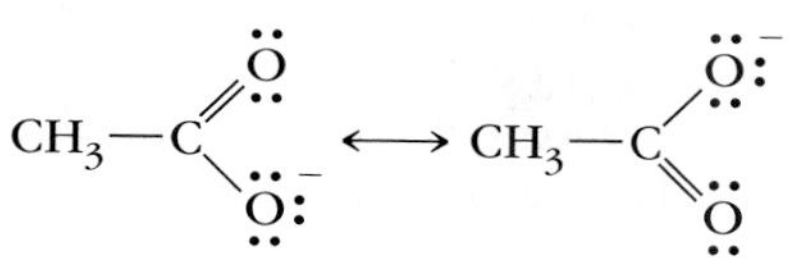
acetate ion

Amines, such as methylamine, CH_3NH_2, are compounds that, like ammonia, have a nitrogen atom with three of its valence electrons in covalent bonds and an unshared electron pair on the nitrogen atom. The lone pair of electrons can accept an H^+, and so, like ammonia, amines react as weak bases with water, accepting an H^+ from water.

$CH_3—\ddot{N}H_2$

methylamine

H^+-accepting lone pair

$$R—\ddot{N}H_2(aq) + H_2O(\ell) \rightleftharpoons R—NH_3^+(aq) + OH^-(aq)$$

Amines can have one, two, or three groups covalently bonded to the nitrogen atom *(← p. 570)*, for example, as in methylamine, dimethylamine, and trimethylamine. Many natural biochemicals and drugs are amines, such as epinephrine (adrenaline, a natural hormone) and Novocain (a local anesthetic).

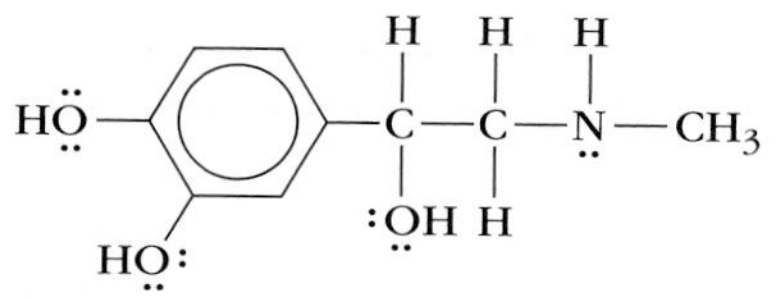
epinephrine

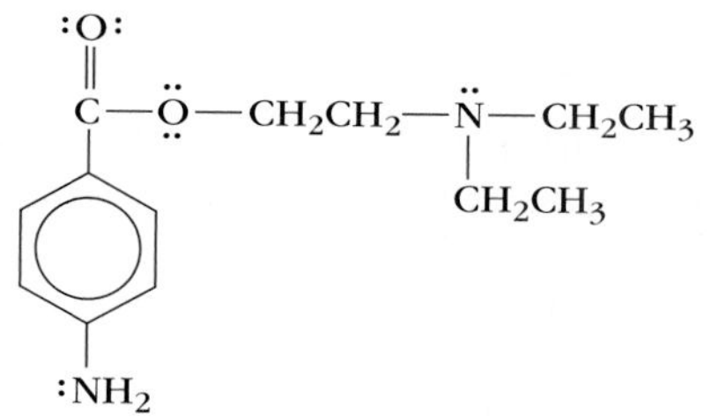
Novocain

CONCEPTUAL EXERCISE 16.6 Piperidine, an Analog of Ammonia

Write the equation for the reaction of piperidine, a component of pepper, with (a) water and with (b) hydrochloric acid.

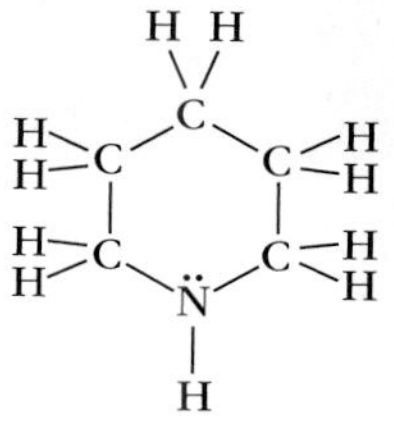
piperidine

EXERCISE 16.7 Classifying Compounds

Identify each of these molecules as a carboxylic acid, an amine, or neither.

(a) 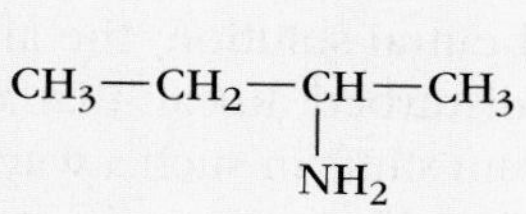

(b) 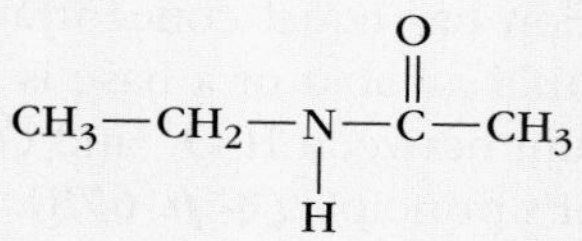

(c) 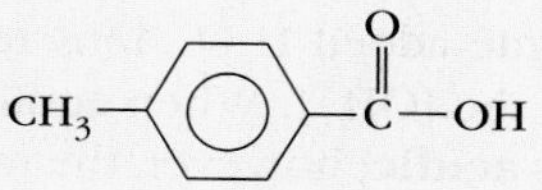

(d) 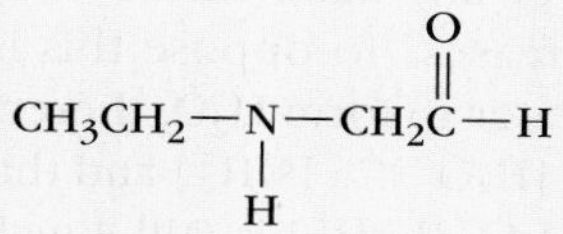

(e) 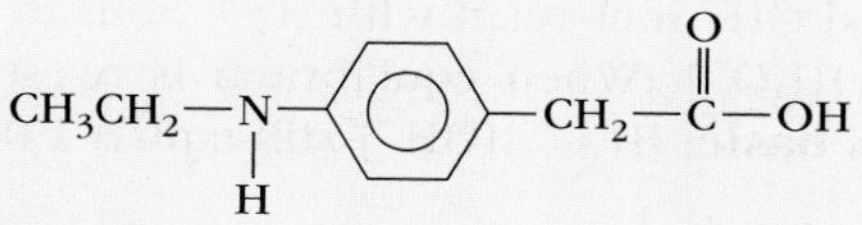

16.3 The Autoionization of Water

Even highly purified water conducts a very small electrical current, which indicates that pure water contains a very small concentration of ions. These ions are formed when water molecules react to produce hydronium ions and hydroxide ions in a process called **autoionization.**

$$H_2O(\ell) + H_2O(\ell) \rightleftharpoons H_3O^+(aq) + OH^-(aq)$$

BASE acid ACID base

In this reaction, one water molecule serves as an H^+ acceptor (base) while the other is an H^+ donor (acid). The equilibrium between the water molecules and the hydronium and hydroxide ions is very reactant-favored. Therefore, the concentrations of these ions in pure water are *very* low. Nevertheless, autoionization of water is very important in understanding how acids and bases function in aqueous solutions. As in the case of any equilibrium reaction, an equilibrium constant expression can be written for the autoionization of water.

Like all equilibrium constant expressions, the one for K_w includes concentrations of solutes but not the concentration of the solvent, which in this case is water.

$$2\,H_2O(\ell) \rightleftharpoons H_3O^+(aq) + OH^-(aq) \qquad K_w = [H_3O^+][OH^-]$$

This equilibrium constant K_w is known as the **ionization constant for water.** From electrical conductivity measurements of pure water, we know that $[H_3O^+] = [OH^-] = 1.0 \times 10^{-7}$ M at 25 °C. Hence

$$K_w = [H_3O^+][OH^-] = (1.0 \times 10^{-7})(1.0 \times 10^{-7}) = 1.0 \times 10^{-14} \qquad \text{(at 25 °C)}$$

In aqueous solutions, the H_3O^+ and OH^- concentrations are inversely related; as one increases, the other must decrease. Their product always equals 1.0×10^{-14} at 25 °C.

The equation $K_w = [H_3O^+][OH^-]$ applies to pure water and all aqueous solutions. Like other equilibrium constants, the value of K_w is temperature-dependent (Table 16.1).

According to the K_w expression, the product of the hydronium ion concentration times the hydroxide ion concentration will always remain the same at a given temperature. If the hydronium ion concentration increases (because an acid was added to the water, for example), then the hydroxide ion concentration must decrease, and vice versa. The equation also tells us that if we know one concentration, we can calculate the other.

Table 16.1 Temperature Dependence of K_w for Water

T (°C)	K_w
10	0.29×10^{-14}
15	0.45×10^{-14}
20	0.68×10^{-14}
25	1.01×10^{-14}
30	1.47×10^{-14}
50	5.48×10^{-14}

The relative concentrations of H_3O^+ and OH^- also indicate the acidic, neutral, or basic nature of the aqueous solution. **For all aqueous solutions** there are three possibilities.

Neutral solution: $[H_3O^+] = [OH^-]$ both equal to 1.0×10^{-7} M at 25 °C

Acidic solution: $[H_3O^+] > 1.0 \times 10^{-7}$ M $[OH^-] < 1.0 \times 10^{-7}$ M at 25 °C

Basic solution: $[H_3O^+] < 1.0 \times 10^{-7}$ M $[OH^-] > 1.0 \times 10^{-7}$ M at 25 °C

When a solution has equal concentrations of $[H_3O^+]$ and $[OH^-]$, it is said to be **neutral.** If either an acid or a base is added to a neutral solution, the autoionization equilibrium between H_3O^+ and OH^- will be disturbed. Recall that according to Le Chatelier's principle ***(← p. 678),*** an equilibrium shifts in such a way as to offset the effect of any disturbance. When an acid is added, the concentration of H_3O^+ ions increases. To oppose this increase, some added H_3O^+ ions react with OH^- ions in water to form H_2O, thereby reducing the $[OH^-]$. When equilibrium is re-established, $[H_3O^+] > [OH^-]$ and the solution is **acidic;** however, the mathematical product $[H_3O^+][OH^-]$ is still equal to 1.0×10^{-14} at 25 °C. Similarly, if a base is added to water, some of the added OH^- ions react with H_3O^+ ions in water to form H_2O, thereby decreasing the $[H_3O^+]$. When equilibrium is re-established, $[H_3O^+] < [OH^-]$ and the solution is **basic;** $[H_3O^+][OH^-]$ still equals 1.0×10^{-14} at 25 °C.

The term *alkaline* is also used to describe basic solutions.

PROBLEM-SOLVING EXAMPLE 16.2 Hydronium and Hydroxide Concentrations

Calculate:

(a) The hydroxide ion concentration at 25 °C in 0.10-M HCl, a strong acid.

(b) The hydronium ion concentration at 25 °C in 0.010-M KOH, a strong base.

Based on your calculations, explain why the 0.10-M HCl solution is acidic and the 0.010-M KOH solution is basic.

Answer (a) $[OH^-] = 1.0 \times 10^{-13}$ M (b) $[H_3O^+] = 1.0 \times 10^{-12}$ M

The 0.10-M HCl solution is acidic because its hydronium concentration is greater than its hydroxide concentration: $[H_3O^+] > [OH^-]$. The 0.010-M KOH solution is basic because $[OH^-] > [H_3O^+]$.

Strategy and Explanation In each case, the relationship $[H_3O^+][OH^-] = 1.0 \times 10^{-14}$ applies.

(a) Being a strong acid, hydrochloric acid is 100% ionized and so the $[H_3O^+]$ is 1.0×10^{-1} M. The $[OH^-]$ can readily be calculated.

$$[H_3O^+][OH^-] = 1.0 \times 10^{-14} = (1.0 \times 10^{-1}\ \text{M})\ [OH^-]$$

$$[OH^-] = \frac{1.0 \times 10^{-14}}{1.0 \times 10^{-1}} = 1.0 \times 10^{-13}\ \text{M}$$

(b) The strong base KOH is completely ionized and the $[OH^-]$ is 1.0×10^{-2} M. The hydronium concentration can be calculated directly:

$$[H_3O]^+ = \frac{1.0 \times 10^{-14}}{[OH^-]} = \frac{1.0 \times 10^{-14}}{1.0 \times 10^{-2}} = 1.0 \times 10^{-12}\ \text{M}$$

PROBLEM-SOLVING PRACTICE 16.2

Which is more acidic, a solution whose H_3O^+ concentration is 5.0×10^{-4} M or one that has an OH^- concentration of 3.0×10^{-8} M?

PROBLEM-SOLVING EXAMPLE 16.3 $[H_3O^+]$ and $[OH^-]$ Concentrations

Calculate the hydroxide ion concentration at 25 °C in 6.0-M HNO_3 and the hydronium ion concentration in 6.0-M NaOH.

Answer $[OH^-]$ of 6.0-M HNO_3 = 1.7×10^{-15} M; $[H_3O^+]$ of 6.0-M NaOH = 1.7×10^{-15} M

Strategy and Explanation Assume that nitric acid and sodium hydroxide, both strong electrolytes, are 100% ionized. Therefore, a 6.0-M nitric acid solution has $[H_3O^+] =$ 6.0 M and its $[OH^-]$ can be calculated.

$$[H_3O^+][OH^-] = (6.0)[OH^-] = 1.0 \times 10^{-14}$$

$$[OH^-] = \frac{1.0 \times 10^{-14}}{6.0} = 1.7 \times 10^{-15}\ \text{M}$$

Because NaOH is a strong base, 6.0-M sodium hydroxide has $[OH^-] = 6.0$ M.

$$[H_3O^+][OH^-] = [H_3O^+](6.0) = 1.0 \times 10^{-14}$$

$$[H_3O^+] = \frac{1.0 \times 10^{-14}}{6.0} = 1.7 \times 10^{-15}\ \text{M}$$

☑ **Reasonable Answer Check** Note that at the high hydronium ion concentration of 6.0-M nitric acid, the hydroxide ion concentration is very, very low, which is to be expected of a highly acidic solution. In contrast, the hydronium ion concentration is exceedingly low in 6.0-M NaOH, a highly basic solution.

PROBLEM-SOLVING PRACTICE 16.3

Which is more basic, a solution whose H_3O^+ concentration is 2.0×10^{-5} M or one that has an OH^- concentration of 5.0×10^{-9} M?

go Chemistry **Modules 9a and b: pH (Parts 1 and 2)** cover concepts in this section.

The definition pH = $-\log[H_3O^+]$ is accurate only for very low concentrations of hydronium ions. A more accurate definition is pH = $-\log a_{H_3O^+}$, where $a_{H_3O^+}$ represents the *activity* of hydronium ions. Activity represents an effective concentration that has been corrected for noncovalent interactions among ions and molecules in solutions. For examples where the definition involving concentration fails, see McCarty, C. G.; Vitz, E. *Journal of Chemical Education*, Vol. 83, 2006; pp. 752–757. A complete definition of pH is quite complicated (see http://www.iupac.org/goldbook/P04524.pdf and Galster, H. *pH Measurement: Fundamentals, Methods, Applications, Instrumentation;* VCH: New York, 1991).

The p in pH is derived from French "puissance" meaning "power." Thus, pH is the "power of hydrogen."

See Appendix A.6 for more about using logarithms.

Notice that, as in the case of equilibrium constants, the concentration units of mol/L are not included when the logarithm is taken. It is not possible to take the logarithm of a unit.

16.4 The pH Scale

The $[H_3O^+]$ and $[OH^-]$ in an aqueous solution vary widely depending on the acid or base present and its concentration. In general, the $[H_3O^+]$ in aqueous solutions can range from about 10 mol/L down to about 10^{-15} mol/L. The $[OH^-]$ can also vary over the same range in aqueous solution.

Because these concentrations can be so small, they have very large negative exponents. It is more convenient to express these concentrations using logarithms. We define the **pH** of a solution as ***the negative of the base 10 logarithm (log) of the hydronium ion concentration (mol/L).***

$$\mathbf{pH = -\log[H_3O^+]}$$

The *negative* logarithm of the small concentration values is used since it gives a positive pH value. Thus, the pH of pure water at 25 °C is given by

$$\text{pH} = -\log[1.0 \times 10^{-7}] = -[\log(1.0) + \log(10^{-7})]$$
$$= -[0 + (-7.00)] = 7.00$$

In terms of pH, for aqueous solutions at 25 °C we can write

Neutral solution	pH = 7.00
Acidic solution	pH < 7.00
Basic (alkaline) solution	pH > 7.00

Figure 16.3 shows the pH values along with the corresponding H_3O^+ and OH^- concentrations of some common solutions. Notice that, for example, $-\log(1 \times 10^{-x}) = x$.

$$\text{Lemon juice:}\quad [H_3O^+] = 1 \times 10^{-2}\ \text{M};\quad \text{pH} = -\log(1 \times 10^{-2}) = 2$$

$$\text{Black coffee:}\quad [H_3O^+] = 1 \times 10^{-5}\ \text{M};\quad \text{pH} = -\log(1 \times 10^{-5}) = 5$$

Keep in mind that a change of *one* pH unit represents a *ten-fold* change in H_3O^+ concentration, a change of two pH units represents a 100-fold change, and so on. Thus, according to Figure 16.3, the $[H_3O^+]$ in lemon juice (pH = 2) is 1000 times *greater* than that in black coffee (pH = 5).

For solutions in which $[H_3O^+]$ or $[OH^-]$ has a value other than an exact power of 10 ($1, 1 \times 10^{-1}, 1 \times 10^{-2}, \ldots$) a calculator is convenient for finding the pH. For example, the pH is 2.35 for a solution that contains 0.0045 mol of the strong acid HNO_3 per liter of solution.

$$\text{pH} = -\log(4.50 \times 10^{-3}) = 2.35$$

PROBLEM-SOLVING EXAMPLE 16.4 Calculating pH from $[H_3O^+]$

Calculate the pH of an aqueous HNO_3 solution that has a volume of 250. mL and contains 0.4649 g HNO_3.

Answer 1.530 (*Note:* This pH has been calculated to three significant figures, those to the right of the decimal point. In actual measurements, pH values are seldom measurable to this degree of accuracy.)

Strategy and Explanation Nitric acid is a strong acid, so every mole of HNO_3 that dissolves produces a mole of H_3O^+ and a mole of NO_3^-. First, determine the number of moles of HNO_3 and then calculate the H_3O^+ concentration.

$$0.465 \text{ g } HNO_3 \times \frac{1 \text{ mol } HNO_3}{63.012 \text{ g } HNO_3} = 0.007380 \text{ mol } HNO_3$$

$$[H_3O^+] = \frac{0.007380 \text{ mol } HNO_3}{0.250 \text{ L}} = 0.0295 \text{ M}$$

Then, express this concentration as pH.

$$\text{pH} = -\log(2.95 \times 10^{-2}) = 1.530$$

☑ **Reasonable Answer Check** If the $[H_3O^+]$ were 0.10 M, the pH would be 1.00; the pH would be 2.00 for an H_3O^+ concentration of 0.010 M. Therefore, a solution with an H_3O^+ concentration of 0.0295 M, which is between these two values, should have a pH between 1.00 and 2.00, which it does.

The digits to the left of the decimal point in a pH represent a power of 10. *Only the digits to the right of the decimal point are significant.* In Problem-Solving Example 16.4, where pH $= -\log(2.95 \times 10^{-2})$ $= -\log(2.95) + (-\log 10^{-2})$ $= -0.470 + 2.000 = 1.530$, there are three significant figures in 1.530, the result, because there are three significant figures in 2.95. See Appendix A.3 for a discussion of significant figures.

PROBLEM-SOLVING PRACTICE 16.4

Calculate the pH of a 0.040-M NaOH solution.

The calculation done in Problem-Solving Example 16.4 can be reversed; the hydronium ion concentration of a solution can be calculated from its pH value as shown in Problem-Solving Example 16.5.

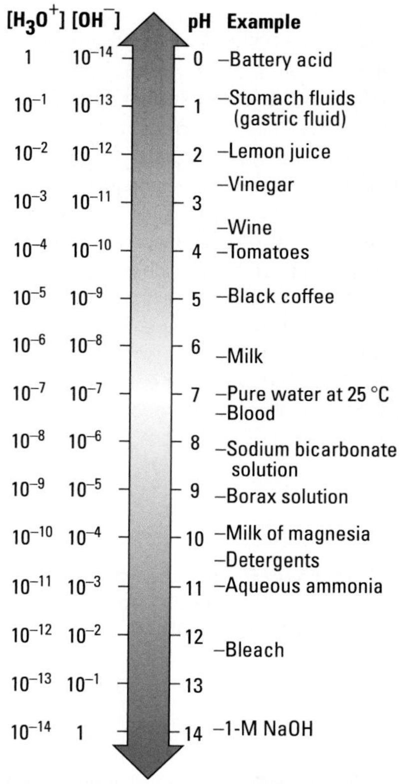

Figure 16.3 The pH of aqueous solutions of some common substances.

PROBLEM-SOLVING EXAMPLE 16.5 Calculating $[H_3O^+]$ from pH

A hospital patient's blood sample has a pH of 7.40.
(a) Calculate the sample's H_3O^+ concentration.
(b) What is its OH^- concentration?
(c) Is the sample acidic, neutral, or basic?

Answer (a) 4.0×10^{-8} M (b) 2.5×10^{-7} M (c) Basic

Strategy and Explanation Use the pH value to find the hydronium ion concentration. From that concentration, calculate the OH^- concentration from the $[H_3O^+][OH^-] = 1.0 \times 10^{-14}$ relationship.

(a) Substituting into the definition of pH,

$$-\log[H_3O^+] = 7.40, \quad \text{so} \quad \log[H_3O^+] = -7.40$$

By the rules of logarithms, $10^{\log(x)} = x$, so we can write $10^{\log[H_3O^+]} = 10^{-\text{pH}} = [H_3O^+]$. Finding $[H_3O^+]$ therefore requires finding the antilogarithm of −7.40 (Appendix A.6).

$$[H_3O^+] = 10^{-7.40} = 4.0 \times 10^{-8} \text{ M}$$

(b) Rearrange the $[H_3O^+][OH^-] = 1.0 \times 10^{-14}$ equation to solve for $[OH^-]$.

$$[OH^-] = \frac{1.0 \times 10^{-14}}{[H_3O^+]} = \frac{1.0 \times 10^{-14}}{4.0 \times 10^{-8}} = 2.5 \times 10^{-7} \text{ M}$$

(c) Because the pH is greater than 7.0, the sample is basic.

☑ **Reasonable Answer Check** Because the pH of 7.40 is slightly above 7.00, the hydroxide ion concentration should be a bit higher than 1.0×10^{-7} M, that of a neutral solution, which it is. Therefore, the sample is slightly basic.

PROBLEM-SOLVING PRACTICE 16.5

In a hospital laboratory the pH of a bile sample is measured as 7.90.
(a) Calculate the H_3O^+ concentration.
(b) Is the sample acidic or basic?

Arnold Beckman

1900–2004

With his invention of the first electronic pH meter in 1934, Arnold Beckman revolutionized pH measurement. Beckman, a professor at the California Institute of Technology at the time, developed the instrument in response to a request from the California Fruit Growers' Association for a quicker, more accurate way to measure the acidity of lemon juice. He went on to found the highly successful Beckman Instrument Company, a firm that invented the first widely used infrared and ultraviolet spectrophotometers and other laboratory instruments. Arnold Beckman and his wife Mabel have donated millions of dollars to advance chemical research and education nationwide.

CONCEPTUAL EXERCISE 16.8 pH of Solutions of Different Acids

Would the pH of a 0.1-M solution of the strong acid HNO_3 be the same as the pH of a 0.1-M solution of the strong acid HCl? Explain.

EXERCISE 16.9 Acidity and pH

A sample of carbonated beverage has a pH of 2.10. A sample of lime juice has a hydronium ion concentration of 1.3×10^{-2} mol/L. Determine which sample is more acidic.

The OH^- concentration can also be expressed in exponential terms as pOH.

$$\text{pOH} = -\log[\text{OH}^-]$$

The $[OH^-]$ of pure water at 25 °C is 1.0×10^{-7} M, and therefore its pOH is

$$\text{pOH} = -\log(1 \times 10^{-7}) = -(-7.00) = 7.00$$

Because the values of $[H_3O^+]$ and $[OH^-]$ are related by the K_w expression, for all aqueous solutions at 25 °C, we can write

$$K_w = [\text{H}_3\text{O}^+][\text{OH}^-] = 1.0 \times 10^{-14}$$

This equation can be rewritten by taking $-\log$ of each side

$$-\log K_w = -\log[\text{H}_3\text{O}^+] + (-\log[\text{OH}^-]) = -\log(1.0 \times 10^{-14})$$

or

$$\mathbf{p}K_w = \mathbf{pH} + \mathbf{pOH} = \mathbf{14.00}$$

The relation between pH and pOH can be used to find one value when the other is known. A 0.0010-M solution of the strong base NaOH, for example, has an OH^- concentration of 0.0010 M and a pOH given by

$$\text{pOH} = -\log(1.0 \times 10^{-3}) = 3.00$$

and therefore, at 25 °C,

$$\text{pH} = 14.00 - \text{pOH} = 14.00 - 3.00 = 11.00$$

If you know the pH, then the pOH is just 14.00 − pH; if you know the pOH, the pH is 14.00 − pOH.

EXERCISE 16.10 pOH and pH

Which solution is more basic, one that has a pH of 5.5 or one with a pOH of 8.5? Calculate the H_3O^+ concentration in each solution.

Measuring pH

The pH of a solution is readily measured using a pH meter (Figure 16.4). This device consists of a pair of electrodes (often in a single probe) that detect the H_3O^+ concentration of the test solution, convert it into an electrical signal, and display it directly as the pH value. The meter is initially calibrated using standard solutions of known pH. The pH of body fluids, soil, environmental and industrial samples, and other substances can be measured easily and accurately with a pH meter.

Acid-base indicators are a much older and less precise (but more convenient) method to determine the pH of a sample. Such indicators are substances that change color within a narrow pH range, generally 1 to 2 pH units, by the loss or gain of an H^+ ion that changes the indicator's molecular structure so that it absorbs light in different regions of the visible spectrum. The indicator is one color at a lower pH (its "acid" form), and a different color at a higher pH (its "base" form).

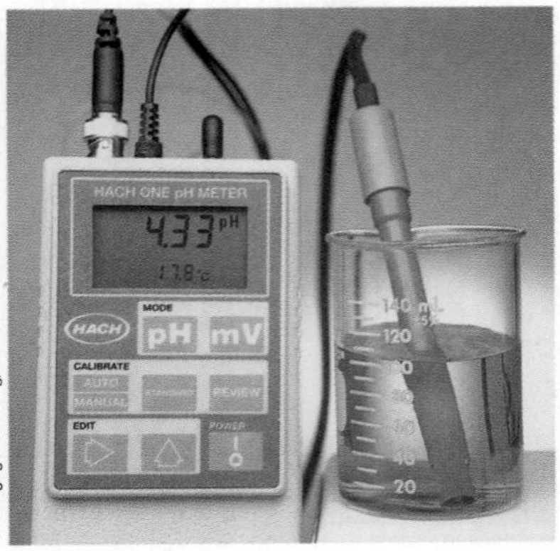

Figure 16.4 A pH meter. A pH meter can quickly and accurately determine the pH of a sample. How a pH meter works is described in Section 19.7.

Consider the indicator bromthymol blue (Figure 16.5). At or below pH 6 it is yellow (its acid form); at pH 8 and above it is blue (its base form). Between pH 6 and 7, the indicator changes from pure yellow to a yellow-green color. At pH 7, it is a mixture of 50% yellow and 50% blue, so it appears green. As the pH changes from 7 to 8, the color becomes pure blue. Thus, the pH of a sample that turns bromthymol blue to green has a pH of about 7. If the indicator color is blue, the pH of the sample is at least 8, and it could be much higher.

Strips of paper impregnated with acid-base indicators are also used to test the pH of many substances. The color of the paper after it has been dampened by the solution to be tested is compared with a set of colors at known pHs.

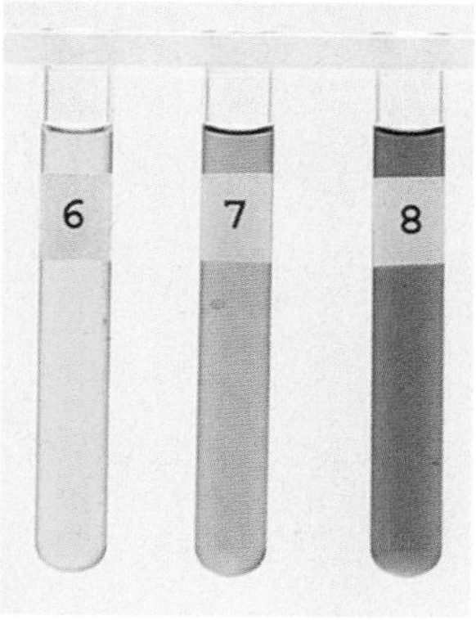

Figure 16.5 Bromthymol blue indicator. At or below a pH of 6 the indicator is yellow. At pH 7 it is pale green, and at pH 8 and above, the color is blue.

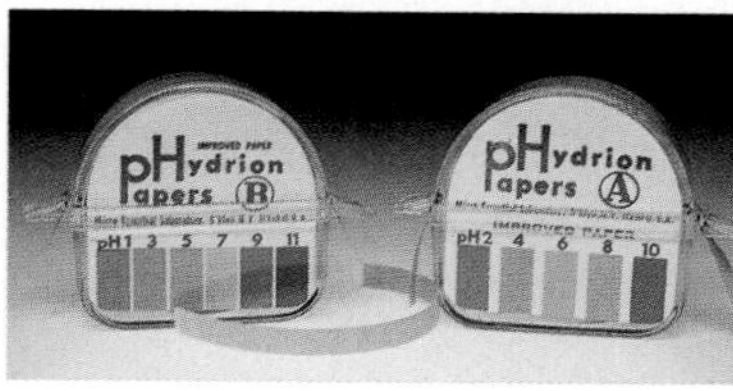

Indicator paper strips. Strips of paper impregnated with indicator are used to find an approximate pH.

16.5 Ionization Constants of Acids and Bases

You learned earlier that the greater the value of the equilibrium constant for a reaction, the more product-favored that reaction is. In an acid-base reaction, the stronger the reactant acid and the reactant base, the more product-favored the reaction (← *p. 656*). Consequently, the magnitude of equilibrium constants can give us an idea about the relative strengths of weak acids and bases. For example, *the larger the equilibrium constant for an acid's ionization, the stronger the acid.*

Acid Ionization Constants

An ionization equation for the transfer to water of H^+ from any acid represented by the general formula HA is

$$\underset{\text{conjugate acid}}{HA(aq)} + H_2O(\ell) \rightleftharpoons H_3O^+(aq) + \underset{\text{conjugate base}}{A^-(aq)}$$

The corresponding **acid ionization constant expression** is

$$K_a = \frac{[H_3O^+][\text{conjugate base}]}{[\text{conjugate acid}]} = \frac{[H_3O^+][A^-]}{[HA]}$$

Remember that the solvent (water, in this case) does not appear in the equilibrium constant expression.

The equilibrium constant K_a is called the **acid ionization constant.** In the acid ionization constant expression:

- The *equilibrium* concentrations of conjugate base and hydronium ion appear in the numerator.
- The *equilibrium* concentration of *un-ionized* conjugate acid appears in the denominator.
- As with other equilibrium constant expressions, pure solids and liquids, such as water, are not included.

Acid ionization constants are also called acid dissociation constants.

As more acid ionizes, the [HA] denominator term in the acid ionization constant expression gets smaller as the numerator terms increase. Consequently, the ratio gets larger. *For strong acids* such as hydrochloric acid, the equilibrium is so product-favored that *the acid ionization constant value is much larger than 1.* In contrast with strong acids, weak acids such as acetic acid ionize to a much smaller extent, establishing equilibria in which significant concentrations of un-ionized weak acid molecules are still present in the solution. *All weak acids have* K_a *values less than 1* because the ionization of a weak acid is reactant-favored; *weak acids are weak electrolytes.*

The larger its acid ionization constant, the stronger the acid.

The common strong acids are

Nitric, HNO_3	Hydrochloric, HCl
Sulfuric, H_2SO_4	Hydrobromic, HBr
Perchloric, $HClO_4$	Hydroiodic, HI

PROBLEM-SOLVING EXAMPLE 16.6 Acid Ionization Constant Expressions

Write the ionization equation and the ionization constant expression for these weak acids.

(a) HF (b) HBrO (c) $H_2PO_4^-$

Answer

(a) $HF(aq) + H_2O(\ell) \rightleftharpoons H_3O^+(aq) + F^-(aq) \qquad K_a = \dfrac{[H_3O^+][F^-]}{[HF]}$

(b) $HBrO(aq) + H_2O(\ell) \rightleftharpoons H_3O^+(aq) + BrO^-(aq) \qquad K_a = \dfrac{[H_3O^+][BrO^-]}{[HBrO]}$

(c) $H_2PO_4^-(aq) + H_2O(\ell) \rightleftharpoons H_3O^+(aq) + HPO_4^{2-}(aq) \qquad K_a = \dfrac{[H_3O^+][HPO_4^{2-}]}{[H_2PO_4^-]}$

Strategy and Explanation In each case, the ionization equation represents the transfer of an H^+ ion from an acid to water, creating a hydronium ion and the conjugate base of the acid. In the K_a expression, the product concentrations at equilibrium are divided by the reactant concentrations at equilibrium; $[H_2O]$ is not included. For example, in the case of HF:

$$HF(aq) + H_2O(\ell) \rightleftharpoons H_3O^+(aq) + F^-(aq) \qquad K_a = \frac{[H_3O^+][F^-]}{[HF]}$$

PROBLEM-SOLVING PRACTICE 16.6

Write the ionization equation and ionization constant expression for each of these weak acids:

(a) Hydrazoic acid, HN_3 (b) Formic acid, HCOOH (c) Chlorous acid, $HClO_2$

That weak acids are only slightly ionized can be shown by measuring the pH of their aqueous solutions. The pH of a 0.10-M acetic acid solution is 2.88, which corresponds to a concentration of H_3O^+ of only 1.3×10^{-3} M. Compare this value with the 0.10-M concentration of H_3O^+ ions in a 0.10-M solution of HCl, a strong acid that has pH = 1.00. In a 0.10-M acetic acid solution, only 1.3% of the initial concentration of acetic acid is ionized:

Stronger acid than CH_3COOH [points to H_3O^+] Stronger base than H_2O [points to CH_3COO^-]

$$CH_3COOH(aq) + H_2O(\ell) \rightleftharpoons H_3O^+(aq) + CH_3COO^-(aq)$$

$$\%\ \text{ionization} = \frac{[H_3O^+]\ \text{at equilibrium}}{\text{initial acid conc.}} \times 100\% = \frac{1.3 \times 10^{-3}}{1.0 \times 10^{-1}} \times 100\% = 1.3\%$$

Therefore, almost 99% of the acetic acid remains in the un-ionized molecular form, CH_3COOH. This is why weak acids (and bases) are weak electrolytes.

In an acetic acid solution or an aqueous solution of any weak acid, two different bases compete for H^+ ions that can be donated from two different acids. In the equation above, the two bases are water and acetate ion; the two acids are acetic acid and hydronium ion. Since acetic acid is a weak acid, its K_a is much less than 1 and the equilibrium favors the reactants. The acetate ion must be a stronger H^+ acceptor than the water molecule. Another way of looking at the same reaction is that the hydronium ion must be a stronger H^+ donor than the acetic acid molecule. Both of these statements are true. Recall from Section 16.1 that acid-base reactions favor going from the stronger to the weaker member of each conjugate acid-base pair. Thus, the acetic acid equilibrium is reactant-favored; a significant concentration of un-ionized acetic acid molecules is present in solution at equilibrium.

Base Ionization Constants

A general equation analogous to that for the donation of H^+ to water by acids can be written for the *acceptance* of an H^+ *from* water by a molecular base, B, to form its conjugate acid, BH^+.

$$\underset{\text{conjugate base}}{B(aq)} + H_2O(\ell) \rightleftharpoons \underset{\text{conjugate acid}}{BH^+(aq)} + OH^-(aq)$$

If the base B were NH_3, then BH^+ would be NH_4^+. The corresponding equilibrium constant expression is

$$K_b = \frac{[\text{conjugate acid}][OH^-]}{[\text{conjugate base}]} = \frac{[BH^+][OH^-]}{[B]}$$

The equilibrium constant K_b is called the **base ionization constant,** a term that can be misleading. Notice from the chemical equation above that the base does not ionize. Rather, K_b and its equilibrium constant expression refer to *the reaction in which a base forms its conjugate acid by removing an H^+ ion from water.*

When the base is an anion, A^- (such as the anion of a weak acid), the general equation is

$$\underset{\text{conjugate base}}{A^-(aq)} + H_2O(\ell) \rightleftharpoons \underset{\text{conjugate acid}}{HA(aq)} + OH^-(aq)$$

If the base A^- were CH_3COO^-, then HA would be CH_3COOH. For acetate ion:

$$CH_3COO^-(aq) + H_2O(\ell) \rightleftharpoons CH_3COOH(aq) + OH^-(aq)$$

The corresponding **base ionization constant expression** is

$$K_b = \frac{[\text{conjugate acid}][OH^-]}{[\text{conjugate base}]} = \frac{[HA][OH^-]}{[A^-]}$$

The magnitude of the K_b value indicates the extent to which the base removes H^+ ions from water to produce OH^- ions. The larger the base ionization constant, K_b, is

- The stronger the base is,
- The more product-favored is the H^+ transfer reaction from water to the base,
- The greater the OH^- concentration produced by the transfer reaction.

For a strong base, the base ionization constant is greater than 1. For a weak base, the ionization constant is less than 1, sometimes considerably less than 1, because at equilibrium there is a significant concentration of unreacted weak conjugate base and a much smaller concentration of its conjugate acid and OH^- ions in solution.

PROBLEM-SOLVING EXAMPLE 16.7 Base Ionization

For each of these weak bases, write the equation for the reaction of the base with water and the companion K_b expression.

(a) C_5H_5N (b) NH_2OH (c) F^-

Answer

(a) $C_5H_5N(aq) + H_2O(\ell) \rightleftharpoons C_5H_5NH^+(aq) + OH^-(aq) \quad K_b = \frac{[C_5H_5NH^+][OH^-]}{[C_5H_5N]}$

(b) $NH_2OH(aq) + H_2O(\ell) \rightleftharpoons NH_3OH^+(aq) + OH^-(aq) \quad K_b = \frac{[NH_3OH^+][OH^-]}{[NH_2OH]}$

(c) $F^-(aq) + H_2O(\ell) \rightleftharpoons HF(aq) + OH^-(aq) \quad K_b = \frac{[HF][OH^-]}{[F^-]}$

Strategy and Explanation The general reaction is the same for each of the bases: The base removes an H^+ from water to form the corresponding conjugate acid. In parts (a) and (b), the base is a neutral molecule to which an H^+ is added, forming a positively charged conjugate acid. In part (c), the H^+ adds to a negatively charged ion, F^-, resulting in a conjugate acid, HF, with no net charge.

PROBLEM-SOLVING PRACTICE 16.7

Write the ionization equation and the K_b expression for these weak bases.
(a) CH_3NH_2 (b) Phosphine, PH_3 (c) NO_2^-

Values of Acid and Base Ionization Constants

Table 16.2 summarizes the ionization constants for a number of acids and their conjugate bases. The ionization constants for strong acids (those above H_3O^+ in Table 16.2) and strong bases (those below OH^- in Table 16.2) are too large to be measured easily. Fortunately, because their ionization reactions are virtually complete, these K_a

Table 16.2 Ionization Constants for Some Acids and Their Conjugate Bases at 25 °C*

Acid Name	Acid	$K_a = \frac{[H_3O^+][\text{conj base}]}{[\text{conj acid}]}$	Base Name	Base	$K_b = \frac{[\text{conj acid}][OH^-]}{[\text{conj base}]}$
Perchloric acid	$HClO_4$	Large	Perchlorate ion	ClO_4^-	Very small
Sulfuric acid	H_2SO_4	Large	Hydrogen sulfate ion	HSO_4^-	Very small
Hydrochloric acid	HCl	Large	Chloride ion	Cl^-	Very small
Nitric acid	HNO_3	≈20	Nitrate ion	NO_3^-	$\approx 5 \times 10^{-16}$
Hydronium ion	H_3O^+	1.0	Water	H_2O	1.0×10^{-14}
Sulfurous acid	H_2SO_3	1.7×10^{-2}	Hydrogen sulfite ion	HSO_3^-	5.9×10^{-13}
Hydrogen sulfate ion	HSO_4^-	1.1×10^{-2}	Sulfate ion	SO_4^{2-}	9.1×10^{-13}
Phosphoric acid	H_3PO_4	7.2×10^{-3}	Dihydrogen phosphate ion	$H_2PO_4^-$	1.4×10^{-12}
Hydrofluoric acid	HF	6.8×10^{-4}	Fluoride ion	F^-	1.5×10^{-11}
Nitrous acid	HNO_2	4.5×10^{-4}	Nitrite ion	NO_2^-	2.2×10^{-11}
Formic acid	HCOOH	3.0×10^{-4}	Formate ion	$HCOO^-$	3.3×10^{-11}
Benzoic acid	C_6H_5COOH	1.2×10^{-4}	Benzoate ion	$C_6H_5COO^-$	8.3×10^{-11}
Acetic acid	CH_3COOH	1.8×10^{-5}	Acetate ion	CH_3COO^-	5.6×10^{-10}
Propanoic acid	CH_3CH_2COOH	1.4×10^{-5}	Propanoate ion	$CH_3CH_2COO^-$	7.1×10^{-10}
Carbonic acid	H_2CO_3	4.3×10^{-7}	Hydrogen carbonate ion	HCO_3^-	2.3×10^{-8}
Dihydrogen phosphate ion	$H_2PO_4^-$	6.3×10^{-8}	Hydrogen phosphate ion	HPO_4^{2-}	1.6×10^{-7}
Hydrogen sulfide	H_2S	1×10^{-7}	Hydrogen sulfide ion	HS^-	1×10^{-7}
Hydrogen sulfite ion	HSO_3^-	6.3×10^{-8}	Sulfite ion	SO_3^{2-}	1.6×10^{-7}
Hypochlorous acid	HClO	6.8×10^{-8}	Hypochlorite ion	ClO^-	1.5×10^{-7}
Ammonium ion	NH_4^+	5.6×10^{-10}	Ammonia	NH_3	1.8×10^{-5}
Boric acid	$B(OH)_3(H_2O)$	5.8×10^{-10}	Tetrahydroxoborate ion	$B(OH)_4^-$	1.7×10^{-5}
Hydrocyanic acid	HCN	3.3×10^{-10}	Cyanide ion	CN^-	3.0×10^{-5}
Hydrogen carbonate ion	HCO_3^-	4.7×10^{-11}	Carbonate ion	CO_3^{2-}	2.1×10^{-4}
Hydrogen phosphate ion	HPO_4^{2-}	4.6×10^{-13}	Phosphate ion	PO_4^{3-}	2.2×10^{-2}
Water	H_2O	1.0×10^{-14}	Hydroxide ion	OH^-	1.0
Hydrogen sulfide ion	HS^-	1×10^{-19}	Sulfide ion	S^{2-}	1×10^5
Ethanol	C_2H_5OH	Very small	Ethoxide ion	$C_2H_5O^-$	Large
Ammonia	NH_3	Very small	Amide ion	NH_2^-	Large
Hydrogen	H_2	Very small	Hydride ion	H^-	Large
Methane	CH_4	Very small	Methide ion	CH_3^-	Large

Decreasing Acid Strength

Increasing Base Strength

*Taken from Högfeldt, E., Perrin, D. D. Stability Constants of Metal-Ion Complexes, 1st ed. Oxford; New Pergamon, 1979–1982. International Union of Pure and Applied Chemistry, Commission on Equilibrium.

and K_b values are hardly ever needed. For weak acids, K_a values show relative strengths quantitatively; for weak bases, K_b values do the same.

Consider acetic acid and boric acid. Boric acid is below acetic acid in Table 16.2, so boric acid must be a weaker acid than acetic acid; the K_a values tell us how much weaker. The K_a for boric acid is 5.8×10^{-10}; that for acetic acid is 1.8×10^{-5}, which shows that boric acid is somewhat more than 10^4 times weaker than acetic acid. In fact, boric acid is such a weak acid that a dilute solution of it can be used safely as an eyewash. Don't try that with acetic acid!

- The smaller the K_a value, the weaker the acid;
- The smaller the K_b value, the weaker the base.

CONCEPTUAL EXERCISE 16.11 Acid Strengths

The K_a of lactic acid is 1.5×10^{-4}; that of pyruvic acid is 3.2×10^{-3}.
(a) Which of these acids is the stronger acid?
(b) Which acid's ionization reaction is more reactant-favored?

K_a Values for Polyprotic Acids

So far we have concentrated on **monoprotic acids** such as hydrogen fluoride, HF, hydrogen chloride, HCl, and nitric acid, HNO_3—*acids that can donate a single H^+ per molecule.*

sulfuric acid

Some acids, called **polyprotic acids,** can donate more than one H^+ per molecule. These include sulfuric acid, H_2SO_4; carbonic acid, H_2CO_3; and phosphoric acid, H_3PO_4. Oxalic acid, $H_2C_2O_4$ or HOOC—COOH, and other organic acids with two or more carboxylic acid, —COOH, groups are also polyprotic acids (Table 16.3).

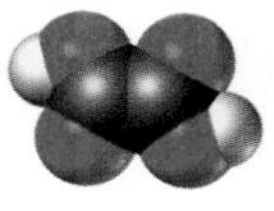
oxalic acid

In aqueous solution, a polyprotic acid donates its H^+ ions to water molecules in a stepwise manner. In the first step for sulfuric acid, hydrogen sulfate ion, HSO_4^-, is formed. Sulfuric acid is a strong acid, so this first ionization is essentially complete.

phosphoric acid

$$\text{First Ionization: } \underset{\text{ACID}}{H_2SO_4(aq)} + \underset{\text{base}}{H_2O(\ell)} \longrightarrow \underset{\text{acid}}{H_3O^+(aq)} + \underset{\text{BASE}}{HSO_4^-(aq)}$$

Hydrogen sulfate ion is the conjugate base of sulfuric acid.

In the next ionization step, hydrogen sulfate ion acting as an acid donates an H^+ ion to another water molecule. In this case, hydrogen sulfate ion is a weak acid ($K_a < 1$) and, as with other weak acids, an equilibrium is established. Sulfate ion, SO_4^{2-}, is the conjugate base of HSO_4^-, its conjugate acid.

Many chemical reactions occur in steps that can be represented by individual equations. Sometimes only the overall equation is written.

$$\text{Second Ionization: } \underset{\text{ACID}}{HSO_4^-(aq)} + \underset{\text{base}}{H_2O(\ell)} \rightleftharpoons \underset{\text{acid}}{H_3O^+(aq)} + \underset{\text{BASE}}{SO_4^{2-}(aq)}$$

When all acidic protons have been donated by a polyprotic acid, the result is a polyprotic base, such as SO_4^{2-}. A polyprotic base can accept more than one H^+ per molecule of base.

Table 16.3 Polyprotic Acids

Acid Form	Conjugate Base Form
H_2S, hydrosulfuric acid	HS^-, hydrogen sulfide or bisulfide ion
H_3PO_4, phosphoric acid	$H_2PO_4^-$, dihydrogen phosphate ion
$H_2PO_4^-$, dihydrogen phosphate ion	HPO_4^{2-}, monohydrogen phosphate ion
H_2CO_3, carbonic acid	HCO_3^-, hydrogen carbonate or bicarbonate ion
$H_2C_2O_4$, oxalic acid	$HC_2O_4^-$, hydrogen oxalate ion
$C_3H_5(COOH)_3$, citric acid	$C_3H_5(COOH)_2COO^-$, hydrogen citrate ion

CONCEPTUAL EXERCISE 16.12 Explaining Acid Strengths

Look at the charge on the hydrogen sulfate ion. What does the charge have to do with the fact that this ion is a weaker acid than H_2SO_4?

The weak acid H_3PO_4 can donate three H^+ ions per molecule through these three ionization reactions.

First ionization

$$H_3PO_4(aq) + H_2O(\ell) \rightleftharpoons H_3O^+(aq) + H_2PO_4^-(aq) \qquad K_a = 7.2 \times 10^{-3}$$

Second ionization

$$H_2PO_4^-(aq) + H_2O(\ell) \rightleftharpoons H_3O^+(aq) + HPO_4^{2-}(aq) \qquad K_a = 6.3 \times 10^{-8}$$

Third ionization

$$HPO_4^{2-}(aq) + H_2O(\ell) \rightleftharpoons H_3O^+(aq) + PO_4^{3-}(aq) \qquad K_a = 4.6 \times 10^{-13}$$

Both $H_2PO_4^-$ ion and HPO_4^{2-} ion are amphiprotic because they can gain or lose a proton.

The successive K_a values for the ionization of a polyprotic acid *decrease* by a factor of 10^4 to 10^5, indicating that each ionization step occurs to a *lesser* extent than the one before it. The $H_2PO_4^-$ ion ($K_a = 6.3 \times 10^{-8}$) is a much weaker acid than phosphoric acid ($K_a = 7.2 \times 10^{-3}$), and the HPO_4^{2-} ion ($K_a = 4.6 \times 10^{-13}$) is an even weaker acid than $H_2PO_4^-$. The K_a values indicate that it is more difficult to remove H^+ from a negatively charged $H_2PO_4^-$ ion than from a neutral H_3PO_4 molecule and even more difficult to remove H^+ from a doubly negative HPO_4^{2-} ion.

EXERCISE 16.13 Polyprotic Acids

Write equations for the stepwise ionization in aqueous solution of (a) oxalic acid and (b) citric acid. (Formulas for these acids are given in Table 16.3.)

16.6 Molecular Structure and Acid Strength

If all acids donate H^+ ions, why are some acids strong while others are weak? Why is there such a broad range of K_a values? To answer these questions we turn to the relationship of an acid's strength to its molecular structure. In doing so, we will consider a wide range of acids, from simple binary ones like HBr to structurally more complex ones containing oxygen, carbon, and other elements.

Factors Affecting Acid Strength

All acids have their acidic hydrogen bonded to some other atom, call it A, which can be bonded to other atoms as well. The H—A bond must be broken for the acid to transfer its hydrogen as an H^+ to water, and that will occur only if the H—A bond is polar. An H—A bond with very little polarity, such as the H—C bond in methane, CH_4, makes the hydrogens nonacidic, and methane is not a significant H^+ donor to water.

$\delta^+ \ \delta^-$
$\longleftrightarrow$
H—A

$\delta^+ \ \delta^-$
$\longleftrightarrow$
H—Br

The simplest case of an acid is a *binary acid,* such as HBr, one that contains just hydrogen and one other element. In this case, A is bromine. The H—Br bond is polar and HBr is a strong acid ($K_a \approx 10^8$). For acids with A atoms from the same group in the periodic table, for example HF, HCl, HBr, and HI, the H—A bond energies determine the relative acid strengths for binary acids. *As H—A bond energies decrease down a group, the H—A bond weakens, and binary acid strengths increase.*

For a series of binary acids in which the A atoms are in the same period, H—A bond energies do not vary much, and therefore the H—A bond polarity is the principal factor affecting acid strengths for binary acids. As the electronegativity of A increases across a period, the H—A bond polarity also gets larger due to greater electronegativity differences. Correspondingly, the hydrogen becomes more acidic, as seen for Period 3 nonmetals:

- The H—Si bond is relatively nonpolar and SiH_4 is nonacidic.
- The H—P bond is only slightly polar and the unshared electron pair on phosphorus makes PH_3 accept H^+ ions rather than donate them.
- The H—S bond is slightly polar and H_2S is a weak acid.
- The H—Cl bond is very polar and HCl is a strong acid.

Strengths of Oxoacids

Acids in which the acidic hydrogen is bonded directly to oxygen in an H—O— bond are called **oxoacids.** Three of the strong acids—nitric, HNO_3; perchloric, $HClO_4$; and sulfuric, H_2SO_4—are oxoacids.

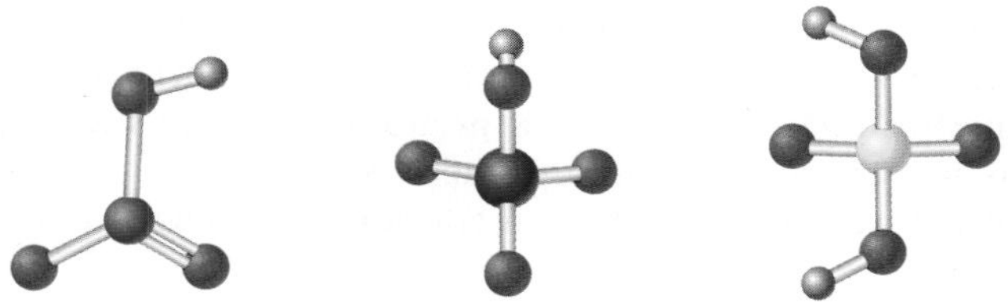

nitric acid   perchloric acid   sulfuric acid

Like other oxoacids, they have at least one hydrogen atom bonded to an oxygen and have the general formula

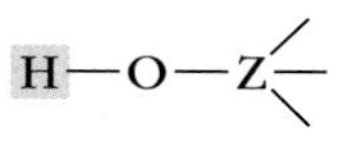

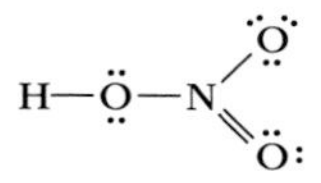

In nitric acid, HNO_3, Z is nitrogen.

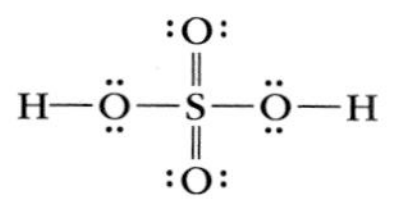

In sulfuric acid, H_2SO_4, Z is sulfur.

The nature of Z and other atoms that may be attached to it are important in determining the strength of the H—O bond and thus the strength of an oxoacid. In general, *acid strength decreases with the decreasing electronegativity of Z.* This is reflected in the differences among the K_a values of HOCl, HOBr, and HOI, as the electronegativity of the halogen decreases from chlorine (2.7) to bromine (2.6) to iodine (2.1).

Acid:	HOCl	HOBr	HOI
	hypochlorous acid	hypobromous acid	hypoiodous acid

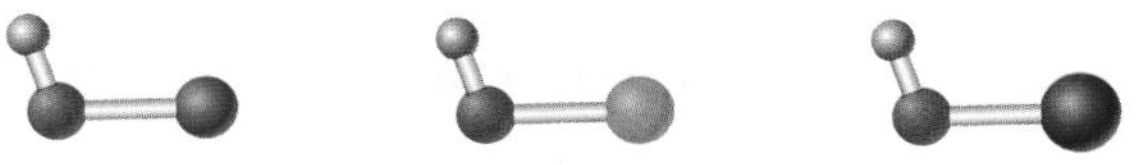

K_a:	6.8×10^{-8}	2.5×10^{-9}	2.3×10^{-11}

The number of oxygen atoms attached to Z also significantly affects the strength of the H—O bond and oxoacid strength: ***The acid strength increases as the number of oxygen atoms attached to Z increases.*** The terminal oxygen atoms (those not in an H—O bond) are sufficiently electronegative, along with Z, to withdraw electron density from the H—O bond. This weakens the bond, promoting the transfer of an H^+ ion to water. The more terminal oxygen atoms present, the greater the electron density shift and the greater the acid strength. A particularly striking example of this trend is seen with the oxoacids of chlorine from the weakest, hypochlorous acid, HOCl, to the strongest, perchloric acid, $HClO_4$.

The chemical formula for perchloric acid can be considered as $HOClO_3$ in relation to its structural formula.

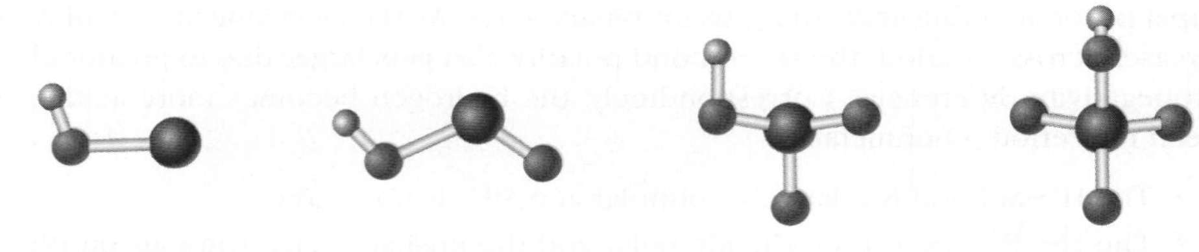

$HOCl$	$HClO_2$	$HClO_3$	$HClO_4$
hypochlorous acid	chlorous acid	chloric acid	perchloric acid
K_a: 6.8×10^{-8}	1.1×10^{-2}	$\approx 10^3$	$\approx 10^8$

To be a strong acid, an inorganic oxoacid must have *at least* two more oxygen atoms than acidic hydrogen atoms in the molecule. Thus, sulfuric acid is a strong acid. In contrast, the inorganic oxoacid phosphoric acid, H_3PO_4, has only four oxygen atoms for three hydrogen atoms and is a weak acid.

Strengths of Carboxylic Acids

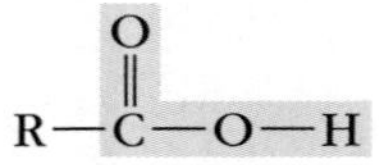

Carboxylic acid group

All carboxylic acids contain the carboxylic acid group, —COOH, although all such acids do not have the same acid strength. The differences in carboxylic acid strength are due to differences in the R group attached to the —COOH group. When R is simply a hydrocarbon group, there is little effect on acid strength; K_a values are similar, as seen by comparing acetic acid, CH_3COOH, and hexanoic acid, $CH_3(CH_2)_4COOH$.

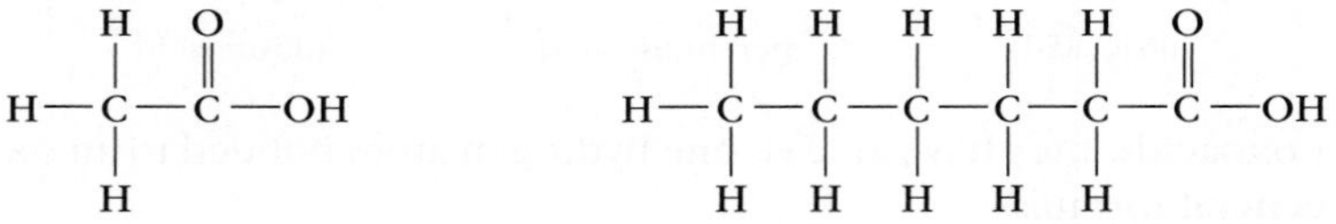

acetic acid $K_a = 1.8 \times 10^{-5}$ | hexanoic acid $K_a = 1.4 \times 10^{-5}$

Acid strength, however, is affected by the addition of a highly electronegative atom to the R group. The electronegative atom causes electron density to shift away from the O—H bond, weakening it and thus increasing the acid's strength. For example, replacing a hydrogen in the CH_3 group of acetic acid with chlorine, a more electronegative atom, forms chloroacetic acid, $ClCH_2COOH$. The K_a for this acid is 1.4×10^{-3}, nearly 100 times larger than that of acetic acid. Replacing all three hydrogens of the CH_3 group in acetic acid with chlorines converts acetic acid to trichloroacetic acid, Cl_3CCOOH, which is about 10,000 times stronger than acetic acid and 100 times stronger than chloroacetic acid.

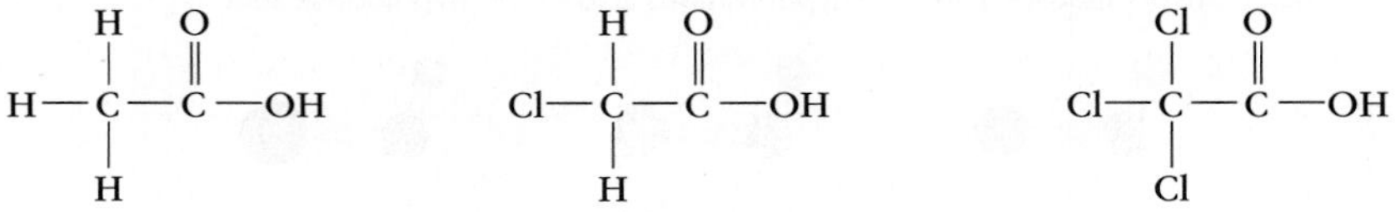

acetic acid $K_a = 1.8 \times 10^{-5}$ | chloroacetic acid $K_a = 1.4 \times 10^{-3}$ | trichloroacetic acid $K_a = 3 \times 10^{-1}$

CONCEPTUAL EXERCISE 16.14 Molecular Structure and Acid Strength

Which has the larger K_a,

(a) Fluorobenzoic acid, C_6H_4FCOOH, or benzoic acid, C_6H_5COOH?

(b) Chloroacetic acid or bromoacetic acid, $BrCH_2COOH$?

Explain your answer.

CONCEPTUAL EXERCISE **16.15 Molecular Structure and Acid Strength**

Consider acetic acid, CH_3COOH, and oxalic acid, HOOC—COOH. Which is the stronger acid? (Consider only the first ionization.) Explain your answer.

Amino Acids and Zwitterions

Amino acids are the monomers from which proteins are assembled *(← p. 577)*. The general structural formula for amino acids is

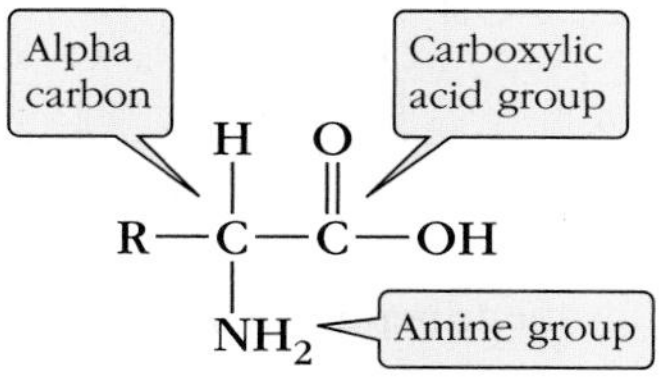

The amino acids are called alpha (α) amino acids because the R group is on the alpha carbon—the one next to the —COOH group.

Unlike acetic acid, CH_3COOH, and ethylamine, $CH_3CH_2NH_2$, which are liquids at 25 °C, amino acids are crystalline solids that generally melt above 200 °C. What makes amino acids more like salts than simple organic compounds? The general structural formula given for amino acids does not predict such properties. Consider, however, the functional groups in an amino acid. All amino acids contain at least one acidic carboxylic acid group *and* at least one basic amine group. An intramolecular Brønsted-Lowry acid-base reaction occurs in amino acids in which H^+ is transferred from the carboxylic acid group to the amine group. This resulting dipolar structure, called a **zwitterion,** is a salt-like substance. In the case of alanine the change is

The hydrogen ion is transferred from the carboxylic acid group to the lone pair on the nitrogen atom of the amine group.

The term "zwitterion" is derived from the German word *zwitter,* meaning "a hybrid."

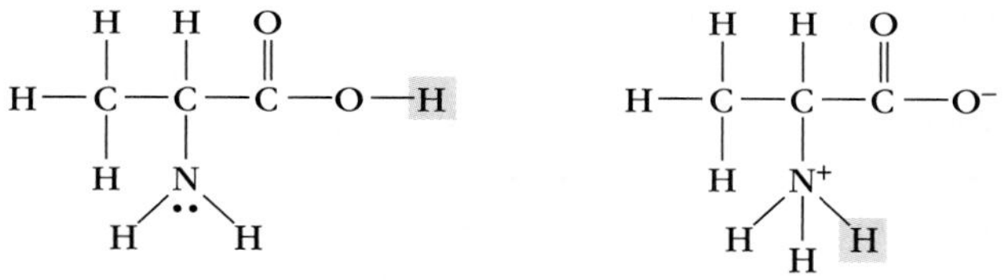

alanine, molecular form alanine, zwitterion form

The alanine zwitterion, for example, has no net charge, although regions of positive ($—NH_3^+$) and negative charge ($—O^-$) exist in the molecule.

Alanine and other amino acids in solution can undergo additional acid-base reactions depending on the pH of the solution. Under acidic conditions, the H^+ concentration is large enough to add an H^+ ion to the negative carboxylate group, $—COO^-$, resulting in the formation of a —COOH group. Under these conditions, the amino acid has a net positive charge. In basic solution, the concentration of OH^- ions is sufficient to remove the H^+ attached to the nitrogen of the zwitterion, resulting in a net negative charge on alanine.

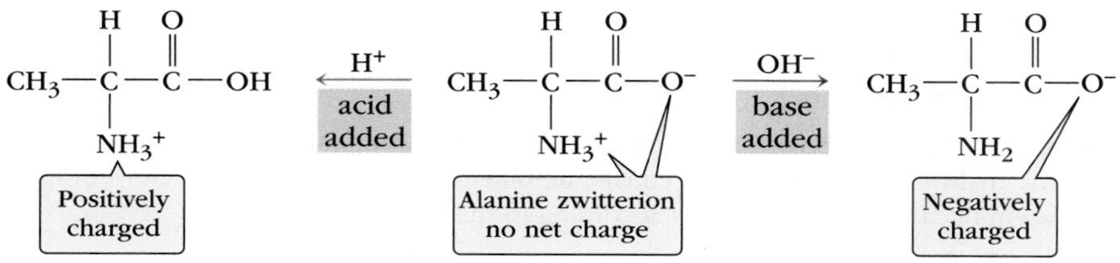

The relative concentrations of the three alanine forms, or the analogous forms of any amino acid, depend on the pH of the body fluids or other aqueous solutions in which they are present.

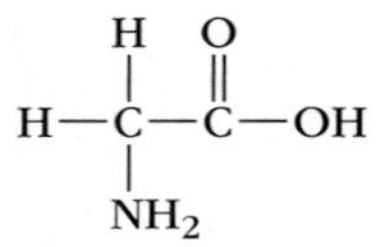

Glycine

CONCEPTUAL EXERCISE 16.16 Glycine and Its Zwitterion

Glycine is the simplest amino acid.

(a) Write the structural formula for its zwitterion form.

(b) Write the structural formula for glycine in solution at a pH of 2 and at a pH of 10.

CONCEPTUAL EXERCISE 16.17 H^+ Transfers

This tripeptide, made from three amino acids, is shown in its neutral form.

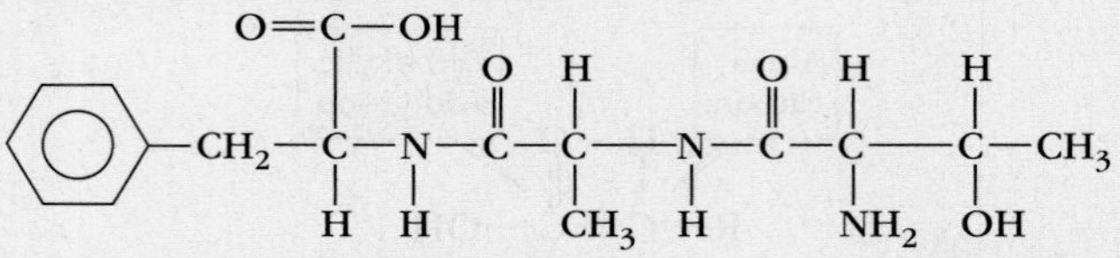

Write the structural formula for the tripeptide with all the acidic or basic groups in their charged forms.

16.7 Problem Solving Using K_a and K_b

Calculations with K_a or K_b follow the same patterns as those of other equilibrium calculations illustrated earlier (← p. 666). Similar important relationships apply to these calculations.

- *Starting with only reactants, equilibrium can be achieved only if some amount of the reactants is converted to products; that is,* ***products are formed at the expense of reactants.***
- *The chemical equilibrium equation for the ionization of the acid or the reaction of the base with water is the basis for the acid ionization or base ionization constant expression, respectively.*
- *The concentrations in the acid ionization or base ionization expression, expressed as molarity (mol/L), are those* ***at equilibrium.***
- *The magnitude of the K_a or K_b value indicates how far the forward reaction occurs at equilibrium (K_a: H^+ donation* ***to*** *water by an acid; K_b: H^+ gain by a base* ***from*** *water).*

There are several experimental methods for determining acid or base ionization constants. The simplest is based on measuring the pH of an acid solution of known concentration. If both the acid concentration and the pH are known, the K_a for the acid can be calculated, as illustrated in Problem-Solving Example 16.8.

PROBLEM-SOLVING EXAMPLE 16.8 K_a from pH

Calculate the K_a of butanoic acid, $CH_3CH_2CH_2COOH$, a weak organic acid, if a 0.025-M butanoic acid solution has a pH of 3.21 at 25 °C.

Answer $K_a = 1.6 \times 10^{-5}$

Strategy and Explanation We can determine the hydronium concentration from the pH. The hydronium concentration equals the butanoate ion, $CH_3CH_2CH_2COO^-$, concentration because the acid ionizes according to the balanced equation

$$CH_3CH_2CH_2COOH(aq) + H_2O(\ell) \rightleftharpoons H_3O^+(aq) + CH_3CH_2CH_2COO^-(aq)$$

Using the definition of pH, the equilibrium concentration of H_3O^+ is calculated to be 0.00062 M; $[H_3O^+] = 10^{-pH} = 10^{-3.21} = 0.00062$ M.

Butanoic acid dissociates to give equal concentrations of hydronium ions and butanoate ions, in this case 0.00062 M. At equilibrium the concentration of un-ionized butanoic acid equals the original concentration minus the amount that dissociated. The equilibrium concentrations of the species are represented by using a reaction (ICE) table.

	$CH_3CH_2CH_2COOH$	H_3O^+	$CH_3CH_2COO^-$
*I*nitial concentration (mol/L)	0.025	1.0×10^{-7} (from water)*	0
*C*hange as reaction occurs (mol/L)	−0.00062	+0.00062	+0.00062
*E*quilibrium concentration (mol/L)	0.025 − 0.00062	0.00062	0.00062

*This concentration can be ignored because it is so small.

From the measured pH we have calculated H_3O^+ to be 6.2×10^{-4} mol/L, which is also the $CH_3CH_2CH_2COO^-$ concentration at equilibrium because the ions are formed in equal amounts as butanoic acid ionizes. The concentration of the un-ionized acid at equilibrium is 0.025 − 0.00062 = 0.0244 M. Using these values, we can now calculate K_a for butanoic acid.

$$K_a = \frac{[H_3O^+][CH_3CH_2CH_2COO^-]}{[CH_3CH_2CH_2COOH]}$$

$$= \frac{[0.00062][0.00062]}{0.0244} = 1.6 \times 10^{-5}$$

☑ **Reasonable Answer Check** This K_a is small, indicating that butanoic acid is a weak acid, as reflected by the fact that a 0.025-M butanoic acid solution has an $[H_3O^+]$ of just 0.00062 M. Thus, the answer makes sense; butanoic acid is only slightly ionized. Butanoic acid is similar to acetic acid in strength, as expected from its similar structure, and a 0.025-M solution has a pH identical to that of 0.025-M acetic acid (pH = 3.21).

PROBLEM-SOLVING PRACTICE 16.8

Lactic acid is a monoprotic acid that occurs naturally in sour milk and also forms by metabolism in the human body. A 0.10-M aqueous solution of lactic acid, $CH_3CH(OH)COOH$, has a pH of 2.43. Calculate the value of K_a for lactic acid. Is lactic acid stronger or weaker than propanoic acid?

Acid-base ionization constants such as those in Table 16.2 can be used to calculate the pH of a solution of a weak acid or a weak base from its concentration.

PROBLEM-SOLVING EXAMPLE 16.9 pH from K_a

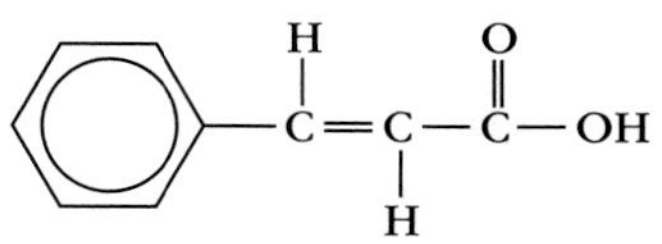

trans-cinnamic acid, $C_9H_8O_2$

(a) Calculate the pH of a 0.020-M solution of *trans*-cinnamic acid, which has $K_a = 3.6 \times 10^{-5}$ at 25 °C. For simplicity, we will symbolize *trans*-cinnamic acid as HtCA.

(b) Calculate the percent of the acid that has ionized in this solution.

Answer (a) pH = 3.07 (b) 4.3% ionized

Strategy and Explanation (a) We first must relate the information to the ionization of HtCA to release hydronium ions and *trans*-cinnamate ions, (tCA^-), into solution. Start by writing the equilibrium equation and equilibrium constant expression for *trans*-cinnamic acid.

$$HtCA(aq) + H_2O(\ell) \rightleftharpoons H_3O^+(aq) + tCA^-(aq) \qquad K_a = \frac{[H_3O^+][tCA^-]}{[HtCA]}$$

© Cengage Learning/Charles D. Winters

Cinnamon.

Next, define equilibrium concentrations and organize the known information in the usual ICE table. In this case, let x equal the H_3O^+ concentration at equilibrium. At equilibrium, the tCA^- ion concentration is also equal to x because the reaction produces H_3O^+ and tCA^- in equal amounts.

	HtCA	H_3O^+	tCA^-
Initial concentration (mol/L)	0.020	1.0×10^{-7} (from water)*	0
Change as reaction occurs (mol/L)	$-x$	$+x$	$+x$
Equilibrium concentration (mol/L)	$0.020 - x$	x	x

*This concentration can be ignored because it is so small.

The equilibrium constant expression can be applied using the values from the ICE table where all equilibrium concentrations are defined in terms of the single unknown, x.

$$K_a = 3.6 \times 10^{-5} = \frac{[H_3O^+][tCA^-]}{[HtCA]} = \frac{[x][x]}{0.020 - x}$$

Because K_a is very small, the reaction is reactant-favored and therefore, not very much product will form. Consequently, at equilibrium the concentrations of H_3O^+ and tCA^- will be very small. Therefore, x must be quite small compared with 0.020. When x is subtracted from 0.020, the result will still be almost exactly 0.020, and so we can approximate $0.020 - x$ as 0.020 to get

$$\frac{x^2}{0.020} = 3.6 \times 10^{-5}$$

Solving for x gives

$$x = \sqrt{(3.6 \times 10^{-5})(0.020)} = \sqrt{7.2 \times 10^{-7}} = 8.5 \times 10^{-4} = [H_3O^+]$$

$$pH = -\log[H_3O^+] = -\log(8.5 \times 10^{-4}) = 3.07$$

The solution is acidic (pH $<$ 7.0).

(b) The ionization of the acid is the major source of H_3O^+ ions (the concentration from water is insignificant). Therefore, the percent ionization is calculated by comparing the H_3O^+ concentration at equilibrium with the initial concentration of the acid.

$$\%\ \text{ionization} = \frac{[H_3O^+]}{(HtCA)_{\text{initial}}} \times 100\% = \frac{8.5 \times 10^{-4}}{0.020} \times 100\% = 4.3\%$$

☑ **Reasonable Answer Check** Its K_a value of 3.6×10^{-5} indicates that *trans*-cinnamic acid is a weak acid, similar to acetic acid ($K_a = 1.8 \times 10^{-5}$) in strength and should be only slightly ionized. That 0.020-M *trans*-cinnamic acid is only 4.3% ionized and has a $[H_3O^+]$ of 8.5×10^{-4} M and a pH of 3.07 is reasonable. To check if the approximation was valid, substitute the equilibrium values into the equilibrium constant expression. The calculated result should equal the K_a.

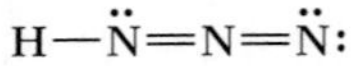

hydrazoic acid

PROBLEM-SOLVING PRACTICE 16.9

(a) Calculate the pH of a 0.015-M solution of hydrazoic acid, HN_3 ($K_a = 1.9 \times 10^{-5}$), at 25 °C.

(b) Calculate the percent of the acid that has ionized in this solution.

EXERCISE 16.18 The pH of Aqueous Nitrous Acid

Nitrous acid, HNO_2, has a $K_a = 4.5 \times 10^{-4}$ at 25 °C. Calculate the pH of a 0.495-M aqueous nitrous acid solution.

An analogous calculation can be done to find the pH of a solution of a weak base, such as piperidine, a component of pepper.

© Cengage Learning/Charles D. Winters

Black pepper contains piperidine.

PROBLEM-SOLVING EXAMPLE 16.10 pH of a Weak Base from K_b

Piperidine, $C_5H_{11}N$, a nitrogen-containing base analogous to ammonia, has a $K_b = 1.3 \times 10^{-3}$. Calculate the OH^- concentration and the pH of a 0.025-M solution of piperidine.

Answer $[OH^-] = 5.1 \times 10^{-3}$ M; pH = 11.71

Strategy and Explanation Use the balanced equilibrium chemical equation to provide the K_b expression. Using the value of K_b, calculate $[OH^-]$ and then pOH. The pH can then be derived using the relation pH + pOH = 14.

Piperidine reacts with water to transfer H^+ ions from water to piperidine to form hydroxide ions and piperidinium ions, $C_5H_{11}NH^+$, according to the equation

$$C_5H_{11}N(aq) + H_2O(\ell) \rightleftharpoons C_5H_{11}NH^+(aq) + OH^-(aq)$$

$$K_b = \frac{[C_5H_{11}NH^+][OH^-]}{[C_5H_{11}N]} = 1.3 \times 10^{-3}$$

Piperidinium ion, $C_5H_{11}NH^+$, is analogous to ammonium ion, NH_4^+.

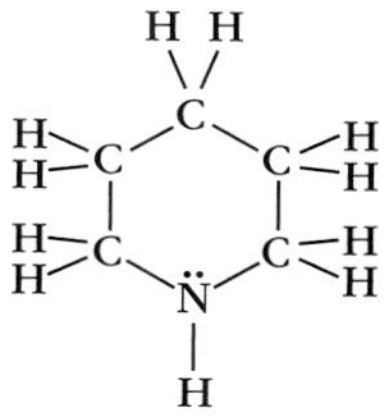

piperidine, $C_5H_{11}N$

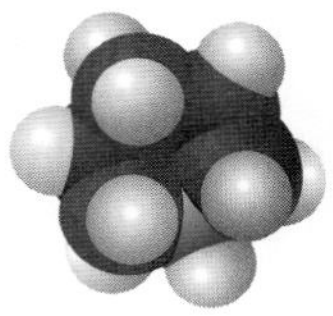

An ICE table can be set up like the one in Problem-Solving Example 16.9, letting x be the concentration of OH^- and of piperidinium ions at equilibrium because the forward reaction produces them in equal amounts. The equilibrium concentration of *unreacted* piperidine will be its initial concentration, 0.025 mol/L, minus x, the amount per liter that has reacted with water.

	$C_5H_{11}N$	$C_5H_{11}NH^+$	OH^-
*I*nitial concentration (mol/L)	0.025	0	1.0×10^{-7}*
*C*hange as reaction occurs (mol/L)	$-x$	$+x$	$+x$
*E*quilibrium concentration (mol/L)	$(0.025 - x)$	x	x

*The low concentration can be ignored, as it was in the K_a calculations.

Substitution into the base ionization constant expression gives

$$K_b = \frac{[C_5H_{11}NH^+][OH^-]}{[C_5H_{11}N]} = \frac{x^2}{0.025 - x} = 1.3 \times 10^{-3}$$

Generally in K_a and K_b calculations, if $\frac{x}{\text{initial concentration}} \times 100\% > 5\%$, the x term *cannot be dropped* from the denominator in the equilibrium constant expression and the quadratic equation is used. As seen below, in this case x is not negligible compared with 0.025 because piperidine reacts with water sufficiently and so x, the OH^- concentration, must be found by using the quadratic formula.

- Multiply the terms to give Equation A.

$$x^2 = (0.025 - x)(1.3 \times 10^{-3}) = 3.3 \times 10^{-5} - (1.3 \times 10^{-3}x) \quad \text{(A)}$$

- Rearrange Equation A into the quadratic form $ax^2 + bx + c = 0$ to give Equation B.

$$x^2 + (1.3 \times 10^{-3}x) - (3.3 \times 10^{-5}) = 0 \quad \text{(B)}$$

Appendix A.6 reviews the use of the quadratic formula.

- Solve Equation B for x using the quadratic formula.

$$x = \frac{-(1.3 \times 10^{-3}) \pm \sqrt{(1.3 \times 10^{-3})^2 - (4 \times 1)(-3.3 \times 10^{-5})}}{2(1)}$$

$$= \frac{-(1.3 \times 10^{-3}) \pm \sqrt{1.34 \times 10^{-4}}}{2}$$

$$= \frac{-(1.3 \times 10^{-3}) \pm (1.16 \times 10^{-2})}{2}$$

$$= \frac{1.03 \times 10^{-2}}{2} = 5.1 \times 10^{-3}$$

Therefore x, the OH^- concentration, equals 5.1×10^{-3} molar. (The negative root in the solution of the quadratic equation is disregarded because concentration cannot be negative; you can't have less than 0 mol/L of a substance.)

Note that $\frac{5.1 \times 10^{-3}}{0.025} \times 100\% = 20\%$, which is far greater than 5%, so the quadratic equation was necessary in this case; the simplifying approximation of

$$\frac{x^2}{0.025 - x} \approx \frac{x^2}{0.025}$$

would not have given the correct answer.

The pOH can be calculated from the OH^- concentration.

$$\text{pOH} = -\log(5.1 \times 10^{-3}) = 2.29$$

$$\text{pH} = 14.00 - \text{pOH} = 14.00 - 2.29 = 11.71$$

Therefore, piperidine reacts sufficiently with water to generate a fairly basic solution.

☑ **Reasonable Answer Check** Piperidine is a weak base, so the pH should be less than that of a 0.025-M solution of a strong base like NaOH, which would be 12.40.

$$[OH^-] = 0.025 \text{ M; pOH} = -\log(0.025) = 1.60\text{; pH} = 12.40$$

A pH of 11.71 for 0.025-M piperidine is reasonable.

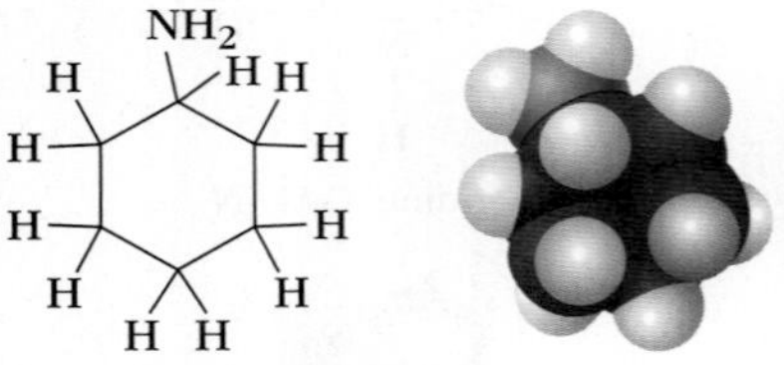

cyclohexylamine

PROBLEM-SOLVING PRACTICE 16.10

Calculate the OH^- concentration and the pH of a 0.015-M solution of cyclohexylamine, $C_6H_{11}NH_2$. $K_b = 4.6 \times 10^{-4}$.

Relationship between K_a and K_b Values

The right-hand side of Table 16.2 (p. 770) gives K_b values for the conjugate base of each acid. Try an experiment with these data: Multiply a few of the K_a values by K_b values for their conjugate bases. What do you find? Within a very small error you ought to find that $K_a \times K_b = 1.0 \times 10^{-14}$. This value is the same as K_w, the autoionization constant for water. To see why, multiply the equilibrium constant expressions for K_a and K_b.

$$K_a \times K_b = \left(\frac{[H_3O^+][A^-]}{[HA]}\right)\left(\frac{[HA][OH^-]}{[A^-]}\right)$$

Canceling like terms in the numerator and denominator of this expression gives

$$K_a \times K_b = \left(\frac{[H_3O^+]\cancel{[A^-]}}{\cancel{[HA]}}\right)\left(\frac{\cancel{[HA]}[OH^-]}{\cancel{[A^-]}}\right) = [H_3O^+][OH^-] = K_w$$

$K_a \times K_b = K_w$ is an inverse relationship; as K_a increases (stronger acid), K_b decreases (weaker conjugate base). This is noted in Table 16.2 (p. 770).

This relation shows that if you know K_a for an acid, you can find K_b for its conjugate base by using K_w. Furthermore, the larger the K_a, the smaller the K_b, and vice versa (because when multiplied they always have to give the same product, K_w). For example, K_a for HCN is 4.0×10^{-10}. The value of K_b for the conjugate base, CN^-, is

$$K_b\,(\text{for } CN^-) = \frac{K_w}{K_a\,(\text{for HCN})} = \frac{1.0 \times 10^{-14}}{3.3 \times 10^{-10}} = 3.0 \times 10^{-5}$$

HCN has a relatively small K_a and lies fairly far down in Table 16.2, which means it is a relatively weak weak acid. However, CN^- is a fairly strong weak base; its K_b of 3.0×10^{-5} is nearly the same as the K_b for ammonia (1.7×10^{-5}), making CN^- a slightly stronger base than ammonia. *In general, if* $K_a > K_b$, *the acid is stronger than its conjugate base*. Alternatively, *if* $K_b > K_a$, *the conjugate base is stronger than its conjugate acid*. For example, hypochlorite ion, OCl^- ($K_b = 1.8 \times 10^{-7}$) is a stronger base than hypochlorous acid, HOCl, is an acid ($K_a = 6.8 \times 10^{-8}$).

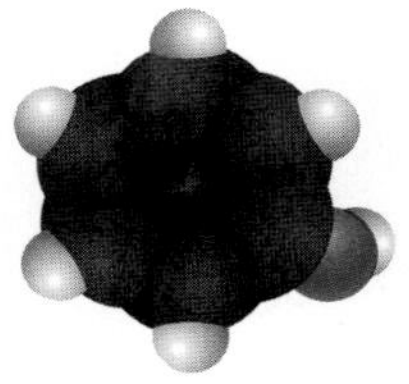

phenol, C_6H_5OH

EXERCISE 16.19 K_b from K_a

Phenol, or carbolic acid, C_6H_5OH, is a weak acid, $K_a = 1.3 \times 10^{-10}$. Calculate K_b for the phenolate ion, $C_6H_5O^-$. Which base in Table 16.2 (p. 770) is closest in strength to the phenolate ion? Explain how you made your choice.

16.8 Acid-Base Reactions of Salts

An exchange reaction between an acid and a base produces a salt plus water ***(← p. 172).*** The salt's positive ion comes from the base and its negative ion comes from the acid. In the case of a metal hydroxide as a base, the salt-forming general reaction is

$$\underset{\text{acid}}{HX(aq)} + \underset{\text{base}}{MOH(aq)} \longrightarrow \underset{\text{salt}}{MX(aq)} + HOH(\ell)$$

Now that you know more about the Brønsted-Lowry acid-base concept and the strengths of acids and bases, it is useful to consider acid-base reactions and salt formation in more detail.

Salts of Strong Bases and Strong Acids

Strong acids react with strong bases to form *neutral* salts. Consider the reaction of the strong acid HCl with the strong base NaOH to form the salt NaCl. If the amounts of HCl and NaOH are in the correct stoichiometric ratio (1 mol HCl per 1 mol NaOH), this reaction occurs with the complete neutralization of the acidic properties of HCl and the basic properties of NaOH. The reaction can be described first by an overall equation, then by a complete ionic equation, and finally by a net ionic equation ***(← p. 166).*** Each of these equations contains useful information.

$$HCl(aq) + NaOH(aq) \longrightarrow NaCl(aq) + H_2O(\ell) \quad \text{Overall equation}$$

$$H_3O^+(aq) + Cl^-(aq) + Na^+(aq) + OH^-(aq) \longrightarrow Na^+(aq) + Cl^-(aq) + 2\ H_2O(\ell) \quad \text{Complete ionic equation}$$

$$\underset{\text{ACID}}{H_3O^+(aq)} + \underset{\text{base}}{OH^-(aq)} \longrightarrow \underset{\text{BASE}}{H_2O(\ell)} + \underset{\text{acid}}{H_2O(\ell)} \quad \text{Net ionic equation}$$

- The overall equation shows the substances that were dissolved or that could be recovered at the end of the reaction.
- The complete ionic equation indicates all of the ions present before and after the reaction.
- The net ionic equation emphasizes that a Brønsted-Lowry acid, H_3O^+, is reacting with a Brønsted-Lowry base, OH^-; the spectator ions, Na^+ and Cl^-, are omitted from the net ionic equation.

This reaction goes to completion because H_3O^+ is a strong acid, OH^- is a strong base, and water is a very weak acid and a very weak base.

Table 16.4 Some Salts Formed by Neutralization of Strong Acids with Strong Bases

Strong Acid	Strong Base NaOH	KOH	$Ba(OH)_2$
	Salts	Salts	Salts
HCl	NaCl	KCl	$BaCl_2$
HNO_3	$NaNO_3$	KNO_3	$Ba(NO_3)_2$
H_2SO_4	Na_2SO_4	K_2SO_4	$BaSO_4$
$HClO_4$	$NaClO_4$	$KClO_4$	$Ba(ClO_4)_2$

The resulting solution contains only sodium ions and chloride ions, with a few more water molecules than before. The properties of the solution are the same as if it had been prepared by simply dissolving some NaCl(s) in water. The solution has a neutral pH because it contains no significant concentrations of acids or bases. The Cl^- ion is the conjugate base of a strong acid and hence is such a weak base that it does not react with water. The Na^+ ion also does not react as either an acid or a base with water. Examples of some other salts of this type are given in Table 16.4. These salts all form *neutral* solutions.

Salts of Strong Bases and Weak Acids

Strong bases react with weak acids to form *basic* salts. Suppose, for example, that 0.010 mol NaOH is added to 0.010 mol of the weak acid acetic acid in 1 L of solution. The three equations are

Overall equation $$CH_3COOH(aq) + NaOH(aq) \longrightarrow NaCH_3COO(aq) + H_2O(\ell)$$

Complete ionic equation $$CH_3COOH(aq) + Na^+(aq) + OH^-(aq) \longrightarrow Na^+(aq) + CH_3COO^-(aq) + H_2O(\ell)$$

Net ionic equation $$\underset{\text{weak acid}}{CH_3COOH(aq)} + \underset{\text{strong base}}{OH^-(aq)} \longrightarrow \underset{\text{base}}{CH_3COO^-(aq)} + \underset{\text{acid}}{H_2O(\ell)}$$

In this case, acetate ion, a weaker base than OH^-, has been formed by the reaction. Therefore, the solution is slightly basic (pH > 7), even though exactly the stoichiometric amount of acetic acid was added to the sodium hydroxide. The reaction that makes the solution basic is the reaction of water with acetate ion, a weak Brønsted-Lowry base.

$$CH_3COO^-(aq) + H_2O(\ell) \rightleftharpoons CH_3COOH(aq) + OH^-(aq)$$

This is a **hydrolysis** reaction, one in which a water molecule is split—in this case, into an H^+ ion and an OH^- ion. The H^+ ion is donated to the acetate ion to form acetic acid. The extent of hydrolysis is determined by the value of K_b for acetate ion.

All of the weak bases in Table 16.2 (p. 770), except for the very weak bases (NO_3^-, Cl^-, HSO_4^-, and ClO_4^-) above water in the next to last column, undergo hydrolysis reactions in aqueous solution. *The larger the* K_b *value of a base, the more basic the solutions it produces.* The pH of a solution of a salt of a strong base and a weak acid can be calculated from K_b, as shown in Problem-Solving Example 16.11.

The term "hydrolysis" is derived from *hydro,* meaning "water," and *lysis,* meaning "to break apart." The hydrolysis reaction of a molecular compound results in the addition of H— and —OH to the molecules produced by breaking a covalent bond.

As K_b of the base increases, the pH of the solution increases.

PROBLEM-SOLVING EXAMPLE 16.11 pH of a Salt Solution

Sodium benzoate, $NaC_7H_5O_2$, is used as a preservative in foods. Calculate the pH of a 0.025-M solution of $NaC_7H_5O_2$ ($K_b = 8.3 \times 10^{-11}$).

Answer 8.14

Strategy and Explanation To calculate the pH we must first calculate the hydroxide ion concentration in the solution. Sodium benzoate is a basic salt that could be synthesized from a strong base, NaOH, and a weak acid, benzoic acid, $HC_7H_5O_2$. The Na^+ ion does not react with water, but benzoate ion, $C_7H_5O_2^-$, the conjugate base of benzoic acid reacts with water to produce a basic solution.

$$C_7H_5O_2^-(aq) + H_2O(\ell) \rightleftharpoons HC_7H_5O_2(aq) + OH^-(aq)$$

$$K_b = 8.3 \times 10^{-11} = \frac{[\text{conjugate acid}][OH^-]}{[\text{conjugate base}]} = \frac{[HC_7H_5O_2][OH^-]}{[C_7H_5O_2^-]}$$

The concentrations of benzoate ion, benzoic acid, and hydroxide ion initially and at equilibrium are summarized in the ICE table. We let x be equal to the equilibrium concentration of OH^- as well as that of $C_7H_5O_2^-$, because they are formed in equal amounts.

	$C_7H_5O_2^-$	$HC_7H_5O_2$	OH^-
*I*nitial concentration (mol/L)	0.025	0	1.0×10^{-7} (from water)
*C*hange as reaction occurs (mol/L)	$-x$	$+x$	$+x$
*E*quilibrium concentration (mol/L)	$0.025 - x$	x	x

Benzoate ion has a very small K_b (8.3×10^{-11}) and thus is a very weak base, so it is reasonable to assume that x is negligibly small compared with 0.025, and $0.025 - x \approx 0.025$.

$$K_b = 8.3 \times 10^{-11} \approx \frac{x^2}{0.025}$$

Solving for x gives $x = 1.4 \times 10^{-6}$, which is equal to the hydroxide and benzoate ion concentrations. Because $0.025 - 1.4 \times 10^{-6} = 0.025$ (using the significant figures rules), our assumption that x is negligible compared with 0.025 is justified. Therefore, at equilibrium

$$OH^- = [HC_7H_5O_2] = 1.4 \times 10^{-6}\ \text{mol/L};\ [C_7H_5O_2^-] = 0.025\ \text{mol/L}$$

Finally, the pH of the solution can be calculated.

$$K_w = [H_3O^+][OH^-] = [H_3O^+](1.4 \times 10^{-6}) = 1.0 \times 10^{-14}$$

$$[H_3O^+] = \frac{1.0 \times 10^{-14}}{1.4 \times 10^{-6}} = 7.1 \times 10^{-9}$$

$$\text{pH} = -\log(7.1 \times 10^{-9}) = 8.14$$

As expected, the solution is basic.

☑ **Reasonable Answer Check** The reaction of benzoate ion with water produces hydroxide ions in addition to those from the dissociation of water. The excess hydroxide ions cause the solution to become basic, as indicated by the pH greater than 7. This is expected because the salt is formed from a strong base and a weak acid.

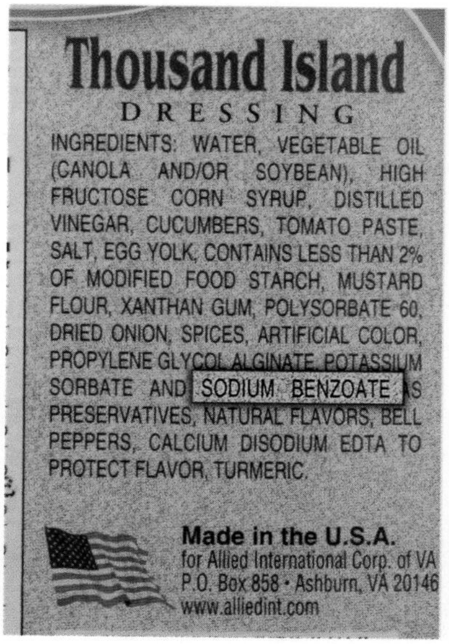

© Cengage Learning/Charles D. Winters

PROBLEM-SOLVING PRACTICE 16.11

Sodium carbonate is an environmentally benign paint stripper. It is water-soluble, and carbonate ion is a strong enough base to loosen paint so it can be scraped off. Calculate the pH of a 1.0-M solution of Na_2CO_3.

CONCEPTUAL EXERCISE 16.20 pH of Soap Solutions

Ordinary soaps are often sodium salts of fatty acids, which are weak organic acids (← p. 560). Would you expect the pH of a soap solution to be greater than or less than 7? Explain your answer.

Salts of Weak Bases and Strong Acids

When a weak base reacts with a strong acid, the resulting salt solution is *acidic*. The conjugate acid of the weak base determines the pH of the solution. For example, suppose equal volumes of 0.10-M NH_3 and 0.10-M HCl are mixed. The reaction, shown in overall, complete ionic, and net ionic equations sequentially is

$$NH_3(aq) + HCl(aq) \longrightarrow NH_4Cl(aq)$$

$$NH_3(aq) + H_3O^+(aq) + Cl^-(aq) \longrightarrow NH_4^+(aq) + Cl^-(aq) + H_2O(\ell)$$

$$\underset{\text{weak base}}{NH_3(aq)} + \underset{\text{strong acid}}{H_3O^+(aq)} \longrightarrow \underset{\text{acid}}{NH_4^+} + \underset{\text{base}}{H_2O(\ell)}$$

As soon as it is formed, the weak acid NH_4^+ reacts with water and establishes an equilibrium. The resulting solution is slightly acidic, because the reaction produces hydronium ions.

$$NH_4^+(aq) + H_2O(\ell) \rightleftharpoons NH_3(aq) + H_3O^+(aq)$$

- Strong base + strong acid yields a neutral salt (solution pH = 7.0)
- Strong base + weak acid yields a basic salt (solution pH > 7.0)
- Strong acid + weak base yields an acidic salt (solution pH < 7.0)

PROBLEM-SOLVING EXAMPLE 16.12 pH of Ammonium Nitrate Solution

Ammonium nitrate, NH_4NO_3, is a salt used in fertilizers and in making matches. The salt is made by the reaction of ammonia, NH_3, and nitric acid, HNO_3. Calculate the pH of a 0.15-M solution of ammonium nitrate; $K_a(NH_4^+) = 5.6 \times 10^{-10}$.

Answer 5.04

Strategy and Explanation You should first recognize that ammonium nitrate is the salt of a weak base and a strong acid so the solution will be acidic, with a pH less than 7.0. Because ammonium ion is a weak acid, ammonium ions react with water to form ammonia and hydronium ions, making the solution acidic.

$$NH_4^+(aq) + H_2O(\ell) \rightleftharpoons NH_3(aq) + H_3O^+(aq)$$

The ICE table summarizes the concentrations of ammonium ion, ammonia, and hydronium ions initially and at equilibrium. We let x equal the equilibrium concentration of H_3O^+ as well as that of NH_3 because they are formed in equal amounts.

	NH_4^+	NH_3	H_3O^+
*I*nitial concentration (mol/L)	0.15	0	1.0×10^{-7} (from water)
*C*hange as reaction occurs (mol/L)	$-x$	$+x$	$+x$
*E*quilibrium concentration (mol/L)	$0.15 - x$	x	x

As in prior problems, we substitute the equilibrium values into the K_a expression.

$$K_a = \frac{[NH_3][H_3O^+]}{[NH_4^+]} = \frac{(x)(x)}{(0.15 - x)} = 5.6 \times 10^{-10}$$

Because ammonium ion is such a weak acid, as indicated by its very low K_a, we can simplify the equation by approximating that $0.15 - x \approx 0.15$

$$K_a = \frac{(x)(x)}{(0.15)} \approx 5.6 \times 10^{-10}$$

x also equals the NH_3 concentration at equilibrium.

and solve for x, the H_3O^+ concentration.

$$x^2 = (0.15)(5.6 \times 10^{-10})$$

$$x = 9.2 \times 10^{-6} = [H_3O^+]$$

The pH is calculated from the hydronium ion concentration.

$$pH = -\log[H_3O^+] = -\log(9.2 \times 10^{-6}) = 5.04$$

☑ **Reasonable Answer Check** As predicted, the pH of the solution is less than 7.0, which it should be because of the release of hydronium ions by the forward reaction.

PROBLEM-SOLVING PRACTICE 16.12

Calculate the pH of a 0.10-M aqueous solution of ammonium chloride, NH_4Cl.

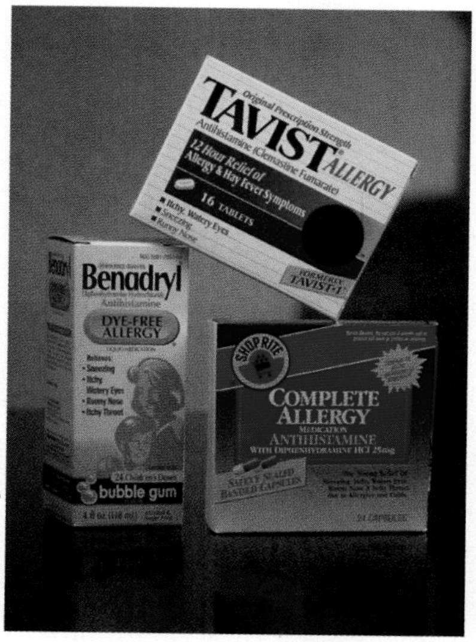

Various pharmaceutical amines.

Many drugs are high-molecular-weight amines that are weak bases (p. 761). Such amines are not soluble in water, which limits the ways they can be administered and means that they are not soluble in body fluids such as blood plasma and cerebrospinal fluid. By reaction with hydrochloric acid, the amines are converted to water-soluble hydrochloride salts that can be administered by injection or dissolved in liquid oral medications. The resulting hydrochloride salts have the general formula BH^+Cl^-, where B represents the basic amine. This formula is like that of ammonium chloride, $NH_4^+Cl^-$. Two examples are

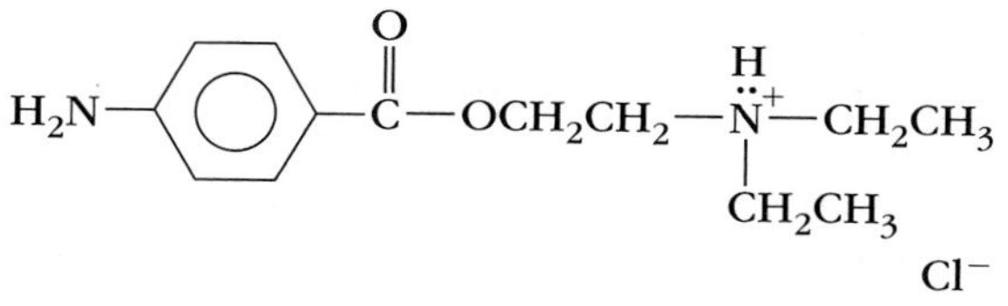

novocain hydrochloride, a local anesthetic

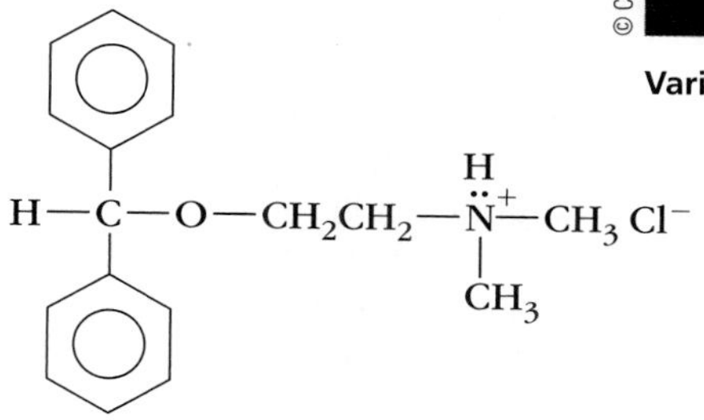

diphenhydramine hydrochloride (Benadryl), an antihistamine

The amine hydrochloride salt of a drug is much more water-soluble than the amine form of the drug. For example, only 0.5 g Novocain dissolves in 100 g water, whereas 100 g Novocain hydrochloride dissolves in 100 g water.

EXERCISE 16.21 Forming a Drug Hydrochloride

To stymie illicit methamphetamine synthesis, phenylephrine is substituted for pseudoephedrine, an active ingredient in over-the-counter decongestants. Use structural formulas to write the equation for the formation of phenylephrine hydrochloride from phenylephrine.

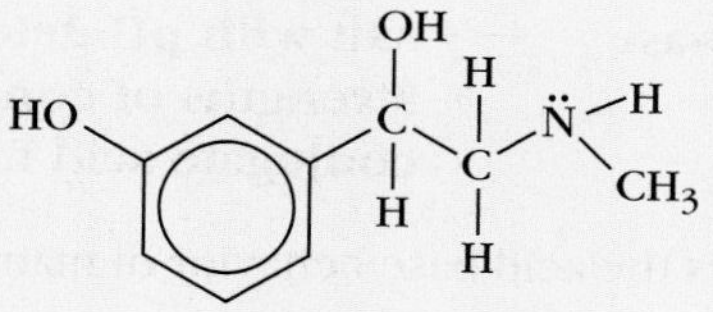

phenylephrine

Salts of Weak Bases and Weak Acids

What is the pH of a solution of a salt such as NH_4F containing an acidic cation and a basic anion? The salt could be formed by the reaction of a weak acid and a weak base. There are two possible reactions that can determine the pH of the NH_4F solution: formation of H_3O^+ by H^+ transfer from the cation;

$$NH_4^+(aq) + H_2O(\ell) \rightleftharpoons H_3O^+(aq) + NH_3(aq) \qquad K_a(NH_4^+) = 5.6 \times 10^{-10}$$

and formation of OH^- by hydrolysis of the anion.

$$F^-(aq) + H_2O(\ell) \rightleftharpoons HF(aq) + OH^-(aq) \qquad K_b(F^-) = 1.4 \times 10^{-11}$$

Table 16.5 Acid-Base Properties of Typical Ions in Aqueous Solution

	Neutral		Basic			Acidic
Anions	Cl^-	NO_3^-	CH_3COO^-	CN^-	SO_4^{2-}	HSO_4^-
	Br^-	ClO_4^-	$HCOO^-$	PO_4^{3-}	HPO_4^{2-}	$H_2PO_4^-$
	I^-		CO_3^{2-}	HCO_3^-	SO_3^{2-}	HSO_3^-
			S^{2-}	HS^-	ClO^-	
			F^-	NO_2^-		
Cations	Li^+	Mg^{2+}		*None*		Al^{3+}
	Na^+	Ca^{2+}				NH_4^+
	K^+	Ba^{2+}				Transition metal ions

The acidity of aqueous Al^{3+} and transition metal ions is discussed on pp. 787–788.

Since $K_a(NH_4^+) > K_b(F^-)$, the reaction of ammonium ions with water to produce hydronium ions is the more favorable reaction. Therefore, the resulting solution is slightly acidic.

In general, the K_a of the weak acid and the K_b of the weak base need to be considered to determine whether the aqueous solution of a salt of a weak acid and weak base will be acidic or basic.

CONCEPTUAL EXERCISE 16.22 Hydrolysis of a Salt of a Weak Acid and a Weak Base

Name a salt of a weak acid and a weak base where $K_a = K_b$. What should the pH of a solution of this salt be?

These generalizations can be made about acid-base neutralization reactions in aqueous solution and the pH of the resulting salt solutions at 25 °C.

- **Strong acid + strong base $\longrightarrow$ salt with pH = 7 (neutral)**
- **Strong acid + weak base $\longrightarrow$ salt with pH < 7 (acidic)**
- **Weak acid + strong base $\longrightarrow$ salt with pH > 7 (basic)**
- **Weak acid + weak base $\longrightarrow$ salt with pH determined by relative strengths of conjugate base and conjugate acid formed**

Table 16.5 summarizes the acid-base behavior of many different ions in aqueous solution.

16.9 Lewis Acids and Bases

In 1923, when Brønsted and Lowry independently proposed their acid-base definition, Gilbert N. Lewis also was developing a new definition of acids and bases. By the early 1930s Lewis had proposed definitions of acids and bases that are more general than those of Brønsted and Lowry because they are based on sharing of electron pairs rather than on H^+ ion transfers.

- A **Lewis acid** is *a substance that can accept a pair of electrons to form a new bond.*
- A **Lewis base** is *a substance that can donate a pair of electrons to form a new bond.*

Those definitions mean that in the Lewis sense, an acid-base reaction occurs when a molecule (or ion) that has a lone pair of electrons that can be donated (a Lewis base) reacts with a molecule (or ion) that can accept an electron pair (a Lewis acid). When both electrons in an electron-pair bond were originally associated with one of the bonded atoms (the Lewis base), the bond is called a **coordinate covalent bond** *(← p. 356)*.

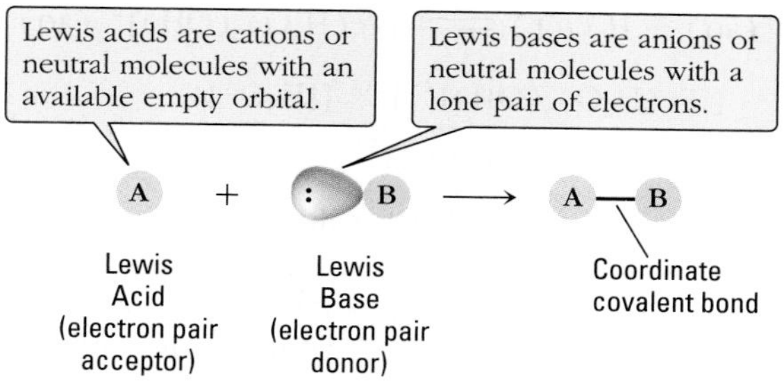

A simple example of a Lewis acid-base reaction is the formation of a hydronium ion from H^+ and water. The H^+ ion has no electrons, while the water molecule has two lone pairs of electrons on the oxygen atom. One of the lone pairs can be shared between H^+ and oxygen, thus forming an O—H bond.

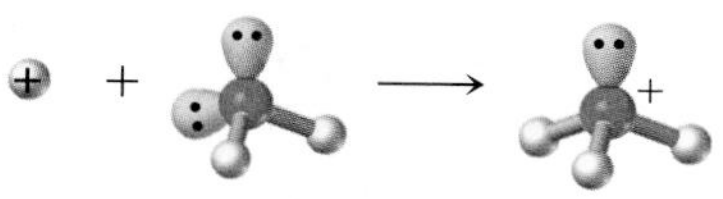

$$H^+ + H_2O \longrightarrow H_3O^+$$

hydronium ion

A Brønsted-Lowry base (H^+ ion acceptor) must also be a Lewis base by donating an electron pair to bond with the H^+.

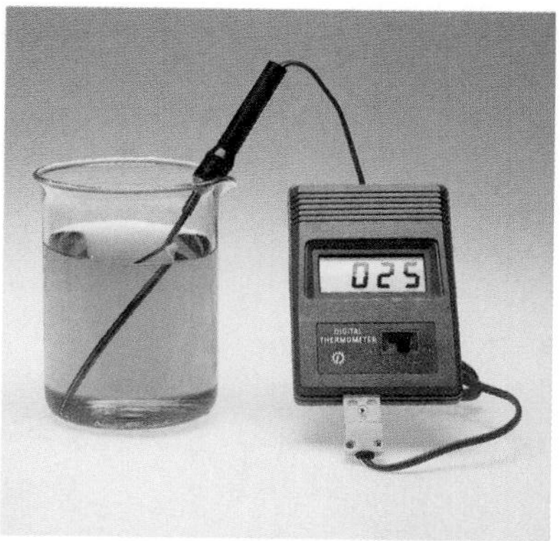

Acidic pH of an aqueous copper(II) sulfate solution. The blue solution of this copper salt is acidic (pH = 2.5) due to hydrolysis of the $[Cu(H_2O)_6]^{2+}$ ion.

Positive Metal Ions as Lewis Acids

All metal cations are potential Lewis acids. Not only do the positively charged metal cations attract electrons, but all such cations have at least one empty orbital. This empty atomic orbital can accommodate an electron pair donated by a base, thereby forming a coordinate covalent bond. Consequently, metal ions readily form coordination complexes (Section 22.6) and also are hydrated in aqueous solution. When a metal ion such as Fe^{3+} becomes hydrated, one of the lone pairs on the oxygen atom in each of several water molecules forms a coordinate covalent bond to the metal ion; *the metal ion acts as a Lewis acid, and water acts as a Lewis base*. The combination of a metal ion and a Lewis base forms a **complex ion,** for example $[Fe(H_2O)_6]^{3+}$ (Figure 16.6). A hydrated metal ion often has six water molecules bonded to it, $[M(H_2O)_6]^{n+}$, where M represents the metal ion whose charge is $n+$.

Hydrated metal ions, especially those of transition metal ions, are weak acids (Table 16.6). As shown by the structure in Figure 16.6, there are M—O—H bonds in the hydrated metal ion. Transition metal ions and Al^{3+} ions have large enough

Complex ions are discussed in detail in Section 22.6.

Table 16.6 Ionization Constants for Hydrated Metal Complex Ions and Their Conjugate Bases at 25 °C

Acid	K_a	Base	K_b
$[Fe(H_2O)_6]^{3+}$	6.3×10^{-3}	$[Fe(H_2O)_5(OH)]^{2+}$	1.6×10^{-12}
$[Al(H_2O)_6]^{3+}$	7.9×10^{-6}	$[Al(H_2O)_5(OH)]^{2+}$	1.3×10^{-9}
$[Cu(H_2O)_6]^{2+}$	1.6×10^{-7}	$[Cu(H_2O)_5(OH)]^{1+}$	6.3×10^{-8}
$[Pb(H_2O)_6]^{2+}$	1.5×10^{-8}	$[Pb(H_2O)_5(OH)]^{1+}$	6.7×10^{-7}
$[Co(H_2O)_6]^{2+}$	1.3×10^{-9}	$[Co(H_2O)_5(OH)]^{1+}$	7.7×10^{-6}
$[Fe(H_2O)_6]^{2+}$	3.2×10^{-10}	$[Fe(H_2O)_5(OH)]^{1+}$	3.1×10^{-5}
$[Ni(H_2O)_6]^{2+}$	2.5×10^{-11}	$[Ni(H_2O)_5(OH)]^{1+}$	4.0×10^{-4}

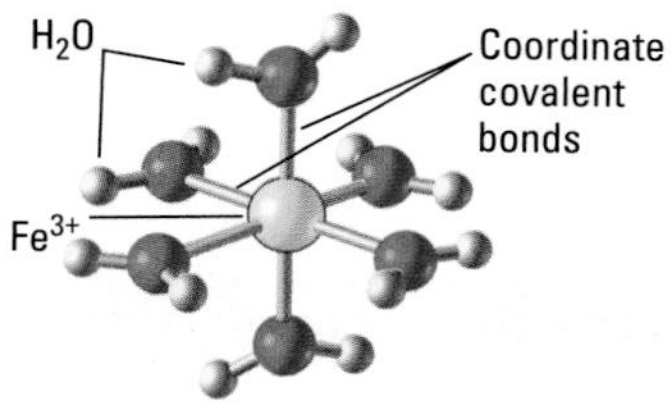

Figure 16.6 Water molecules bonded to an Fe^{3+} ion in $[Fe(H_2O)_6]^{3+}$. Square brackets in the formula indicate that ions or molecules within the brackets are bonded directly to the metal ion.

charges and small enough sizes to attract the shared pair of the M—O bond to the metal ion. This attraction weakens the O—H bond making the hydrogen of the M—O—H bond more acidic than it would be in the O—H bond of a water molecule that is not bonded to a metal ion. Thus, the $[M(H_2O)_6]^{n+}$ complex ion can donate H^+ and the solution becomes acidic. The ionization reaction and acid dissociation constant expression for a hydrated Fe^{3+} ion can be written as

Look carefully at the formulas of the complex ions in the chemical equation. There are six water molecules in the reactant ion, but only five in the product ion. The other water molecule has lost an H^+ to become a hydroxide ion that is part of the product complex ion. Thus, the net charge on the product complex ion is one less than that on the reactant complex ion.

$$[Fe(H_2O)_6]^{3+}(aq) + H_2O(\ell) \rightleftharpoons [Fe(H_2O)_5(OH)]^{2+}(aq) + H_3O^+(aq)$$

$$K_a = \frac{\{[Fe(H_2O)_5(OH)]^{2+}\} \times \{[H_3O^+]\}}{\{[Fe(H_2O)_6]^{3+}\}} = 6.3 \times 10^{-3}$$

This K_a value shows that an aqueous $FeCl_3$ solution will have about the same pH as a solution of phosphoric acid ($K_a = 7.2 \times 10^{-3}$) of equal concentration. The acidic properties of aqueous metal ions are important in the environment. For example, Al^{3+} ions in soils react with water to produce an acidic environment that can be detrimental to tree growth.

Metal ions (Lewis acids) also form many complex ions with the Lewis base ammonia, $:NH_3$. For example, silver ion readily forms a water-soluble, colorless complex ion in liquid ammonia or in aqueous ammonia.

$$Ag^+(aq) + 2\ :NH_3(aq) \longrightarrow [H_3N:Ag:NH_3]^+(aq)$$

Indeed, this complex is so stable that the very water-insoluble compound AgCl can be dissolved in aqueous ammonia.

$$AgCl(s) + 2\ :NH_3(aq) \longrightarrow [H_3N:Ag:NH_3]^+(aq) + Cl^-(aq)$$

The hydroxide ion, OH^-, is an excellent Lewis base and so it binds readily to metal cations to give metal hydroxides. An important feature of the chemistry of many metal hydroxides is that they are **amphoteric,** meaning that they can react as both a base and an acid (see Table 16.7). Amphoteric aluminum hydroxide, for example, behaves as a Lewis acid when it dissolves in a basic solution by forming a complex ion containing one additional OH^- ion, a Lewis base.

$$Al(OH)_3(s) + OH^-(aq) \rightleftharpoons [Al(OH)_4]^-(aq)$$

This reaction is shown in Figure 16.7 (b). Aluminum hydroxide behaves as a Brønsted-Lowry base when it reacts with a Brønsted-Lowry acid, as seen in Figure 16.7 (c).

$$Al(OH)_3(s) + 3\ H_3O^+(aq) \rightleftharpoons Al^{3+}(aq) + 6\ H_2O(\ell)$$

Neutral Molecules as Lewis Acids

Lewis's ideas about acids and bases account nicely for the fact that oxides of nonmetals behave as acids. Two important examples are carbon dioxide and sulfur dioxide, whose Lewis structures are

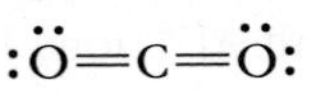

carbon dioxide

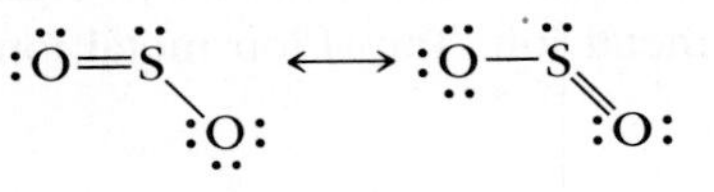

sulfur dioxide

Table 16.7 Some Common Amphoteric Metal Hydroxides

Hydroxide	Reaction as a Brønsted-Lowry Base	Reaction as a Lewis Acid
$Al(OH)_3$	$Al(OH)_3(s) + 3\ H_3O^+(aq) \longrightarrow Al^{3+}(aq) + 6\ H_2O(\ell)$	$Al(OH)_3(s) + OH^-(aq) \longrightarrow [Al(OH)_4]^-(aq)$
$Zn(OH)_2$	$Zn(OH)_2(s) + 2\ H_3O^+(aq) \longrightarrow Zn^{2+}(aq) + 4\ H_2O(\ell)$	$Zn(OH)_2(s) + 2\ OH^-(aq) \longrightarrow [Zn(OH)_4]^{2-}(aq)$
$Sn(OH)_4$	$Sn(OH)_4(s) + 4\ H_3O^+(aq) \longrightarrow Sn^{4+}(aq) + 8\ H_2O(\ell)$	$Sn(OH)_4(s) + 2\ OH^-(aq) \longrightarrow [Sn(OH)_6]^{2-}(aq)$
$Cr(OH)_3$	$Cr(OH)_3(s) + 3\ H_3O^+(aq) \longrightarrow Cr^{3+}(aq) + 6\ H_2O(\ell)$	$Cr(OH)_3(s) + OH^-(aq) \longrightarrow [Cr(OH)_4]^-(aq)$

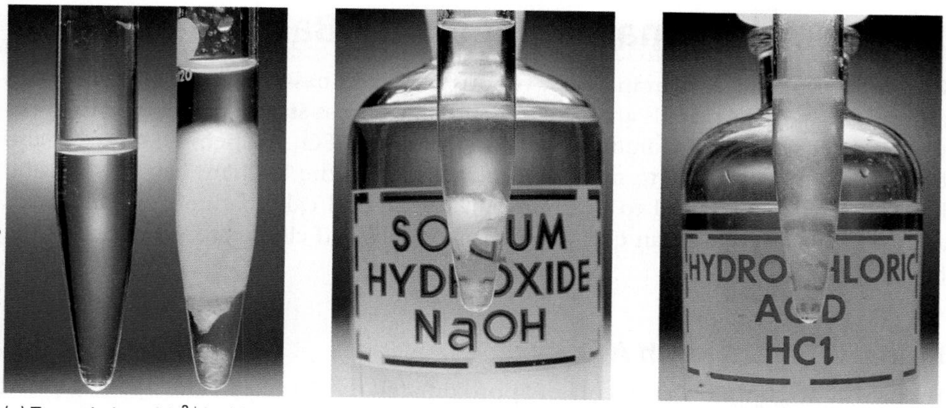

Photos: © Cengage Learning/Charles D. Winters

(a) To a solution of Al^{3+}(aq) (left test tube) is added aqueous ammonia, NH_3(aq); a white precipitate of $Al(OH)_3$(s) forms (right test tube).

(b) Adding a strong base, NaOH, to the $Al(OH)_3$ dissolves the precipitate. Here the aluminum hydroxide acts as a Lewis acid toward the Lewis base OH^- and forms a soluble salt of the complex ion $[Al(OH)_4]^-$.

(c) If we begin again with freshly precipitated $Al(OH)_3$, it dissolves as strong acid, HCl, is added. In this case $Al(OH)_3$ acts as a Brønsted-Lowry base and forms a soluble aluminum salt and water.

Figure 16.7 The amphoteric nature of $Al(OH)_3$.

In each case, there is a double bond; an "extra" pair of electrons is being shared between an oxygen atom and the central atom. Because oxygen is highly electronegative, electrons in these bonds are attracted away from the central atom, which becomes slightly positively charged. This makes the central atom a likely site to attract a pair of electrons. A Lewis base such as OH^- can bond to the carbon atom in CO_2 to give hydrogen carbonate ion, HCO_3^-. This bonding displaces one double-bond pair of electrons onto an oxygen atom.

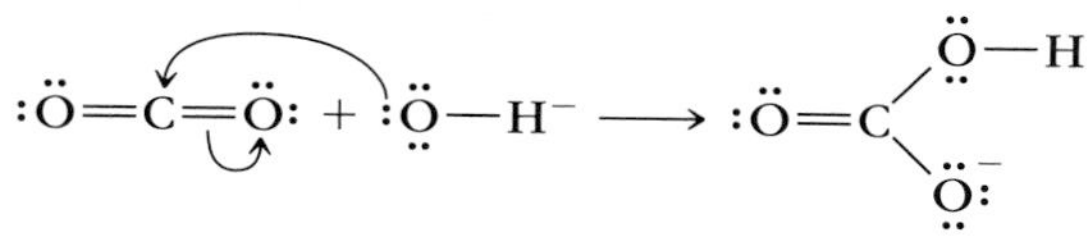

hydrogen carbonate ion

Carbon dioxide from the air can react to form sodium carbonate around the mouth of a bottle of sodium hydroxide. Sulfur dioxide can react similarly with hydroxide ion.

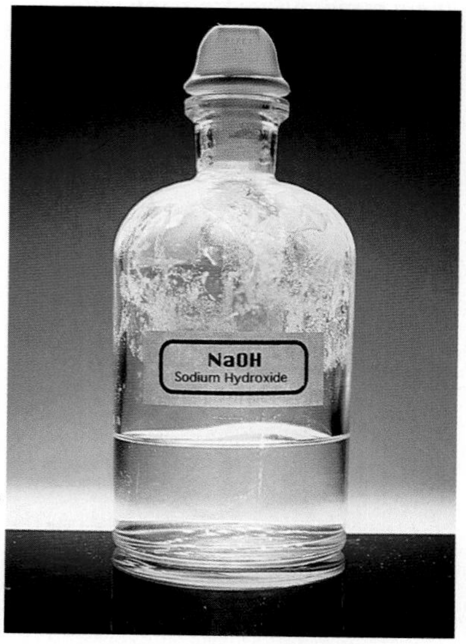

© Cengage Learning/Charles D. Winters

Carbon dioxide in air reacts with spilled base such as aqueous NaOH to form Na_2CO_3, the white solid coating the upper part of the bottle.

CONCEPTUAL EXERCISE 16.23 Lewis Acids and Bases

Predict whether each of these is a Lewis acid or a Lewis base. (*Hint:* Drawing a Lewis structure for a molecule or ion is often helpful in making such a prediction.)

(a) PH_3 (b) BCl_3 (c) H_2S (d) NO_2 (e) Ni^{2+} (f) CO

EXERCISE 16.24 Salt of Weak Acid and a Weak Base

Use data given in the chapter to determine whether an aqueous solution of nickel(II) acetate has a pH < 7.0, $= 7.0$, or > 7.0. Explain your answer.

EXERCISE 16.25 The pH of a Solution of $Ni(NO_3)_2$

When anhydrous nickel(II) nitrate dissolves in water, the Ni^{2+} ions become hydrated, forming $[Ni(H_2O)_6]^{2+}$ ions. Calculate the pH of a solution that is 0.15 M in nickel nitrate. K_a for $[Ni(H_2O)_6]^{2+}$ is 2.5×10^{-11}.

The pH of some household substances. The flasks contain club soda (left), vinegar (middle), and a household cleaner (right). The color of an acid-base indicator in each flask shows the acidity or alkalinity of the liquid—vinegar (red) is more acidic than club soda (orange); the cleaner (green) is much more basic than either of the other two liquids.

16.10 Additional Applied Acid-Base Chemistry

In addition to their uses in industry, various acids and bases find many applications around the home. Antacids are used to neutralize excess stomach acidity. Gardeners use acid salts such as sodium hydrogen sulfate, $NaHSO_4$, to help acidify soil and bases such as lime, CaO, to make soil more basic. In the kitchen, baking soda and baking powders are used to make biscuit dough and cake batter rise. Mild acids and bases are used to clean everything from dishes and clothes to vehicles and the family dog.

Neutralizing Stomach Acidity

Human stomach fluids have a pH of approximately 1. This very acidic pH is caused by HCl secreted by thousands of cells in the stomach lining that specialize in transporting $H_3O^+(aq)$ and $Cl^-(aq)$ from the blood. The main purpose of this acid is to suppress the growth of bacteria and to aid in the digestion of certain foods. The hydrochloric acid does not harm the stomach because its inner lining is replaced at the rate of about half a million cells per minute. However, when too much food is eaten and the stomach is stretched, or when the stomach is irritated by very spicy food, some of its acidic contents can flow back into the esophagus (gastroesophageal reflux); the burning sensation that results is called *heartburn.*

Chloride ions secreted by the stomach lining come mostly from the salty foods we eat and the salt we add to our foods.

An antacid is a base that is used to neutralize excess stomach acid. The recommended dose is the amount of the base required to neutralize *some,* but not all, of the stomach acid. Several antacids and their acid-base reactions are shown in Table 16.8. People who need to restrict the quantity of sodium (Na^+) in their diets should avoid sodium-containing antacids such as sodium hydrogen carbonate.

CONCEPTUAL EXERCISE 16.26 Strong Antacids?

Explain why strong bases such as NaOH or KOH are never used as antacids.

PROBLEM-SOLVING EXAMPLE 16.13 Neutralizing Stomach Acid

Calculate how many moles and what mass of HCl could be neutralized by 0.750 g of the antacid $CaCO_3$.

Answer 1.50×10^{-2} mol HCl; 0.546 g HCl

Table 16.8 The Acid-Base Chemistry of Some Antacids

Compound	Reaction in Stomach	Examples of Commercial Products
Milk of magnesia: $Mg(OH)_2$ in water	$Mg(OH)_2(s) + 2\ H_3O^+(aq) \longrightarrow Mg^{2+}(aq) + 4\ H_2O(\ell)$	Phillips' Milk of Magnesia
Calcium carbonate: $CaCO_3$	$CaCO_3(s) + 2\ H_3O^+(aq) \longrightarrow Ca^{2+}(aq) + 3\ H_2O(\ell) + CO_2(g)$	Tums, Di-Gel
Sodium hydrogen carbonate: $NaHCO_3$	$NaHCO_3(s) + H_3O^+(aq) \longrightarrow Na^+(aq) + 2\ H_2O(\ell) + CO_2(g)$	Baking soda, Alka-Seltzer
Aluminum hydroxide: $Al(OH)_3$	$Al(OH)_3(s) + 3\ H_3O^+(aq) \longrightarrow Al^{3+}(aq) + 6\ H_2O(\ell)$	Amphojel
Dihydroxyaluminum sodium carbonate: $NaAl(OH)_2CO_3$	$NaAl(OH)_2CO_3(s) + 4\ H_3O^+(aq) \longrightarrow Na^+(aq) + Al^{3+}(aq) + 7\ H_2O(\ell) + CO_2(g)$	Rolaids

ESTIMATION Using an Antacid

Estimate how many Rolaids tablets (Table 16.8) it would take to neutralize the acidity in one glass (250 mL) of a regular cola drink. Assume the pH of the cola is 3.0. One Rolaids tablet contains 334 mg $NaAl(OH)_2CO_3$.

With a pH of 3.0, the cola has 1×10^{-3} mol acid per liter of cola, so 0.250 L cola has one fourth that much acid, or about 3×10^{-4} mol acid. To neutralize this amount of acid requires 3×10^{-4} mol base (1 mol base for every 1 mol acid). There are two bases in Rolaids—hydroxide ions and carbonate ions. Each mole of hydroxide neutralizes 1 mol acid, and each mole of carbonate neutralizes 2 mol acid.

$$H^+(aq) + OH^-(aq) \longrightarrow H_2O(\ell)$$

$$2\,H^+(aq) + CO_3^{2-}(aq) \longrightarrow H_2O(\ell) + CO_2(g)$$

Because each mole of $NaAl(OH)_2CO_3$ contains 2 mol OH^- ions and 1 mol CO_3^{2-} ions, 1 mol $NaAl(OH)_2CO_3$ neutralizes 4 mol acid. The molar mass of $NaAl(OH)_2CO_3$ is 144 g/mol, so one Rolaids tablet contains about 0.002 mol of the antacid.

$$\frac{0.334 \text{ g antacid}}{1 \text{ antacid tablet}} \times \frac{1 \text{ mol antacid}}{144 \text{ g antacid}} = 0.00232 \text{ mol antacid/tablet}$$

This tablet can neutralize four times that many moles of acid, or about 0.01 mol acid. To neutralize the 3×10^{-4} mol acid in the cola requires about 0.03 tablet.

$$3 \times 10^{-4} \text{ mol antacid} \times \frac{1 \text{ tablet}}{0.01 \text{ mol acid}} \approx 0.03 \text{ tablet}$$

It would take only a small portion of a tablet to do the job.

Visit this book's companion website at www.cengage.com/chemistry/moore to work an interactive module based on this material.

Strategy and Explanation The balanced equation for this reaction is given in Table 16.8. From the equation we see that 1 mol antacid reacts with 2 mol HCl, molar mass = 36.46 g/mol.

$$0.750 \text{ g } CaCO_3 \times \frac{1 \text{ mol } CaCO_3}{100.1 \text{ g } CaCO_3} = 7.49 \times 10^{-3} \text{ mol } CaCO_3$$

$$7.49 \times 10^{-3} \text{ mol } CaCO_3 \times 2 \text{ mol HCl}/1 \text{ mol } CaCO_3 = 1.50 \times 10^{-2} \text{ mol HCl}$$

Using the stoichiometric mole ratio,

$$7.49 \times 10^{-3} \text{ mol } CaCO_3 \times \frac{2 \text{ mol HCl}}{1 \text{ mol } CaCO_3} \times \frac{36.46 \text{ g HCl}}{1 \text{ mol HCl}} = 0.546 \text{ g HCl}$$

☑ **Reasonable Answer Check** Two moles of HCl are required to react with each mole of $CaCO_3$, so to neutralize 0.00749 mol $CaCO_3$ requires 0.0150 mol HCl, which is just over 0.5 g (0.546 g) HCl.

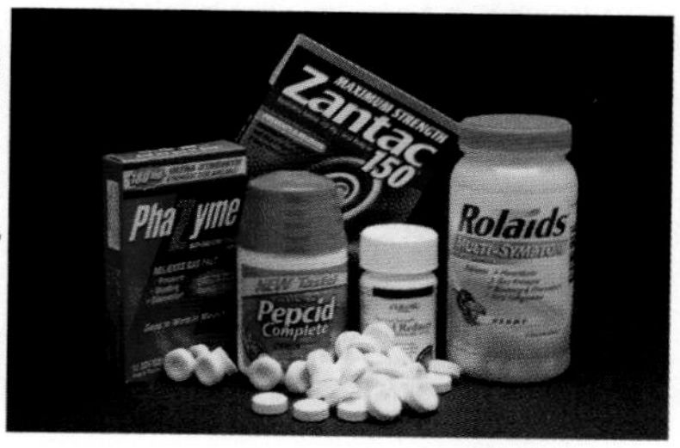

© Cengage Learning/Charles D. Winters

Commercial antacid products. These antacids neutralize excess stomach acid and relieve heartburn.

PROBLEM-SOLVING PRACTICE **16.13**

Using the reactions in Table 16.8, determine which antacid, on a per gram basis, can neutralize the most stomach acid (assume 0.10-M HCl).

Acid-Base Chemistry in the Kitchen

Cooking and baking use chemical reactions, often involving acids and bases. Vinegar is an approximately 4%–5% aqueous acetic acid solution present in almost all salad dressings. Lemon juice, handy for flavoring tea and cooked fish and for making salad dressings, contains citric acid, as do all citrus fruits.

One of the most useful substances in the kitchen is carbon dioxide gas, produced by chemical reactions as it is needed. Pockets of the gas are generated in bread dough and cake batter. The expanding gas makes the resulting biscuits, breads, and cakes rise, producing lighter and more palatable baked goods.

Various methods are used to generate CO_2. One is the addition of yeast, which causes bread dough to rise by catalyzing the fermentation of carbohydrates to produce ethyl alcohol and carbon dioxide.

$$C_6H_{12}O_6 \xrightarrow{\text{yeast}} 2\ CH_3CH_2OH + 2\ CO_2$$

Many commercial breads and homemade dinner rolls use this method to make their doughs rise.

Because CO_2 production by fermentation is slow, it is sometimes necessary to use another method, the reaction of a bicarbonate salt such as sodium hydrogen carbonate, $NaHCO_3$ (also known as *baking soda*), with acid. But which acid should be used? A weak acid is needed; if a strong acid were used, complete neutralization of the acid would be required to make the food safe to eat. Although vinegar could be used, it would impart an undesirable taste. Long ago it was discovered that lactic acid, $CH_3CH(OH)COOH$, present in milk and formed in larger quantities when milk sours to form buttermilk, is a good source of acid for reacting with hydrogen carbonate.

© Cengage Learning/Charles D. Winters

Action of baking powder. Baking powder contains the weak acid dihydrogen phosphate ion and the weak base hydrogen carbonate ion. When water is added, the acid and base react to produce carbon dioxide gas, which is seen bubbling out of the aqueous solution in the picture.

$$\underset{\text{lactic acid}}{CH_3\underset{\displaystyle OH}{\underset{|}{C}}HCOOH(aq)} + HCO_3^-(aq) \longrightarrow \underset{\text{lactate ion}}{CH_3\underset{\displaystyle OH}{\underset{|}{C}}HCOO^-(aq)} + H_2CO_3(aq)$$

$$\underset{\text{carbonic acid}}{H_2CO_3(aq)} \longrightarrow CO_2(g) + H_2O(\ell)$$

When buttermilk is not available, or when a different taste is desired, dihydrogen phosphate ion, $H_2PO_4^-$, is a convenient acid to react with the hydrogen carbonate ion. Baking powders are a mixture of sodium or potassium dihydrogen phosphate and sodium hydrogen carbonate. When dry, the two salts in baking powder do not react with one another. But, when mixed with water in the dough, they react to produce CO_2, a reaction that occurs even more quickly in a heated oven.

$$H_2PO_4^-(aq) \rightleftharpoons H^+(aq) + HPO_4^{2-}(aq)$$

$$H^+(aq) + HCO_3^-(aq) \longrightarrow H_2O(\ell) + CO_2(g)$$

$$\text{Net reaction: } H_2PO_4^-(aq) + HCO_3^-(aq) \longrightarrow H_2O(\ell) + CO_2(g) + HPO_4^{2-}(aq)$$

Buttermilk Blueberry Muffins

- $2\frac{1}{2}$ C flour
- $1\frac{1}{2}$ tsp baking powder
- $\frac{1}{2}$ tsp baking soda
- $\frac{1}{2}$ C sugar
- $\frac{1}{4}$ tsp salt
- 2 eggs, beaten
- 1 C buttermilk
- 3 oz butter
- $1\frac{1}{2}$ C blueberries (added for flavor)

EXERCISE **16.27 Acids and Muffins**

Consider the list of ingredients for buttermilk blueberry muffins in the margin. There are two sources of acid and two sources of hydrogen carbonate, the reaction of which produces the CO_2 to make these muffins rise. What are the sources?

Household Cleaners

Most cleaning compounds such as dishwashing detergents, scouring powders, laundry detergents, and oven cleaners are basic (sometimes highly basic). Synthetic detergents are derived from organic molecules designed to have even better cleaning action than soaps, but less reaction with the doubly positive ions (Mg^{2+} and Ca^{2+}) found in hard water.

The molecular structure of a synthetic detergent molecule, like that of a soap, consists of a long oil-soluble (hydrophobic) group, and a water-soluble (hydrophilic) group ***(← p. 741)***.

A Typical Synthetic Detergent Molecule

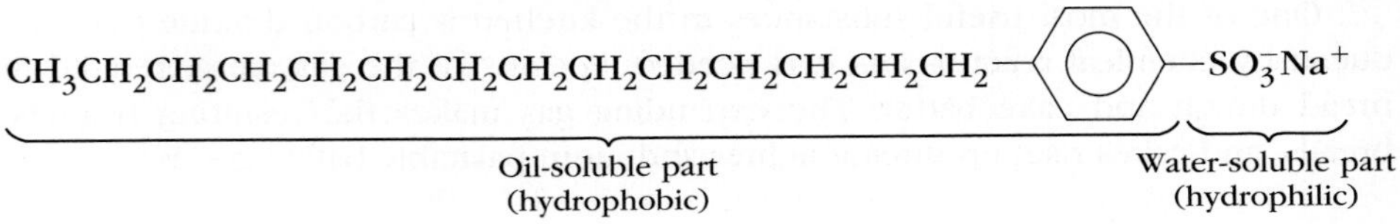
CH3CH2CH2CH2CH2CH2CH2CH2CH2CH2CH2CH2CH2CH2
SO3−Na+
Oil-soluble part
(hydrophobic)
Water-soluble part
(hydrophilic)

Typical hydrophilic groups in detergents include negatively charged sulfate, $—OSO_3^-$; sulfonate, $—SO_3^-$; and phosphate, $—OPO_3^{3-}$, groups. Compounds with these groups are called *anionic surfactants.* There are also *cationic surfactants,* almost all of which are quaternary ammonium halides in which a central nitrogen atom has four organic groups attached to it, one of which is a long hydrocarbon chain and another that frequently includes an —OH group. Some detergents are *nonionic;* they have an uncharged hydrophilic polar group attached to a large organic group of low polarity, for example,

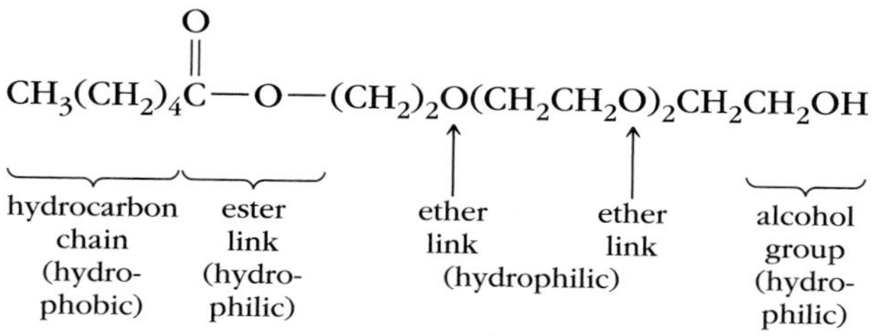
O
CH3(CH2)4C—O—(CH2)2O(CH2CH2O)2CH2CH2OH
hydrocarbon chain (hydro- phobic)
ester link (hydro- philic)
ether link
ether link
(hydrophilic)
alcohol group (hydro- philic)

The carbon chain in this molecule is oil-soluble and hydrophobic; the rest of the molecule is hydrophilic, the combination of properties needed for a molecule to be a detergent. Nonionic detergents have several advantages over ionic detergents. With no ionic groups, nonionic detergents cannot form salts with calcium, magnesium, and iron ions; consequently they are totally unaffected by hard water. For the same reason, nonionic detergents do not react with acids and may be used even in relatively strong acid solutions, which makes them useful in toilet bowl cleaners.

Many decades ago homemade lye soap was used to clean clothes as well as people's skins. This type of soap was made using either pure lye (NaOH, also called caustic soda) or sodium carbonate (Na_2CO_3, also called soda ash) and potassium carbonate, K_2CO_3 (also called potash) from wood ashes. Most of the soap made this way contained considerable amounts of unreacted base. This base was considered desirable because it helped raise the pH and break up the heavy soil particles common on fabrics in those days. In addition, the fabrics then were much more durable than fabrics today.

The closest thing to lye soap we see today is the typical dishwashing detergent. The first three ingredients in Table 16.9 contain anions that react with water to produce OH^-. Together, these three salts produce solutions inside the dishwasher that have a pH near 12.5, high enough to quickly break away animal and vegetable oils from surfaces during the agitation cycle. The detergent helps to dissolve these oily particles and carry them away in the rinse water.

Table 16.9 Formulation of a Dishwashing Detergent

Ingredient	% by Weight
Sodium carbonate, Na_2CO_3	37.5
Sodium tripolyphosphate, $Na_5P_3O_{10}$	30
Sodium metasilicate, Na_2SiO_3	30
Low-foam surfactant	0.5
Sodium dichloroisocyanurate (Cl_2 source)	1.5
Other ingredients such as colorants	0.5

EXERCISE 16.28 pH of a Basic Cleaning Solution

Calculate the pH of a solution that is 5.2 M in sodium carbonate.

Corrosive Household Cleaners

Really tough cleaning jobs around the home require chemically aggressive cleaners. In those circumstances, either very acidic or highly basic cleaners are used to get rid of dirt, grease, or stains. Hydrochloric acid along with phosphoric acid and oxalic acid are used in toilet bowl cleaners to help get rid of stains. This combination of acids makes the pH of such cleaners very low, around 2. They should be handled cautiously because of their high acidity and reactivity. For example, labels on bottles of bleach warn against mixing bleach with other cleaners such as acidic toilet bowl cleaners. Bleach commonly contains sodium hypochlorite, NaOCl, and generally has a pH above 8; OCl^- and Cl^- are present in the bleach solution. In the presence of acidic toilet bowl cleaners, however, OCl^- and Cl^- are converted to toxic Cl_2 gas, which can erupt from the mixture. Bleach also should not be mixed with any cleaning agents containing ammonia. The chlorine in the bleach reacts

Muriatic acid (hydrochloric acid) is used for cleaning bricks and concrete in new home construction or when remodeling is done. The strong acid should be handled with extreme caution.

This answers the question posed in Chapter 1 (← *p. 2*), "Why is there a warning on a container of household bleach that says not to mix the bleach with other cleaners, such as toilet-bowl cleaner?"

with ammonia to produce fumes of chloramines such as NH_2Cl and $NHCl_2$, which can cause respiratory distress.

Drain and oven cleaners are at the other end of the pH spectrum, having very basic pHs of 12 or higher. Deposits of hair, grease, and fats build up inside pipes and eventually clog the drain. When this occurs, the drain has to be dismantled and cleaned (a messy job), or a drain cleaner is added to dissolve the clog.

Drain cleaners generally contain a strong base such as NaOH that reacts with fats and grease to form a soluble soap ***(← pp. 560, 740).***

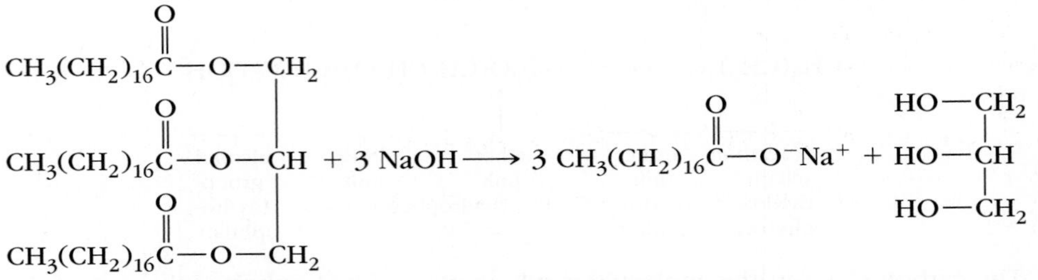

tristearin (glyceryl tristearate) sodium stearate (a soluble soap) glycerol

This reaction converts the grease or fat into water-soluble products (a sodium salt and glycerol) that are washed down the now-opened drain. The reaction is exothermic, and the released heat helps soften the grease or fats, which helps their removal. The strong base also decomposes and rinses away any hair trapped in the clog.

Grocery and home-improvement stores have an abundance of sodium hydroxide-based drain cleaners. Some are almost pure solid NaOH (pellets or flakes). Liquid drain cleaners are often 50% or more NaOH by weight in water. These solutions, being more dense than water, sink to the bottom of the drain trap to start working quickly. Although easy to use, the liquid cleaners become diluted when running water is put into the drain, reducing their efficiency. A particularly aggressive form of drain cleaner contains small bits of aluminum metal along with solid NaOH. In water, the aluminum reacts with the sodium hydroxide to produce hydrogen gas, whose bubbles help to unseat the clogged material. The net ionic equation for the reaction is

$$2\ Al(s) + 2\ OH^-(aq) + 6\ H_2O(\ell) \longrightarrow 3\ H_2(g) + 2\ [Al(OH)_4]^-(aq)$$

The hydrogen gas is flammable, so no flames or sparks should be present when this type of cleaner is used. If a pipe is weak and the clog very strong, sufficient gas pressure can build up to rupture the pipe (one way—but not a very desirable one—to unclog the drain).

Spray-on oven cleaners contain an NaOH solution mixed with a detergent and a propellant to apply the mixture to the soiled places. This mixture is sufficiently viscous to adhere to the oven surfaces long enough for the strongly basic solution to react with the baked-on food. If baked at a high temperature, the food has probably carbonized. If so, the cleaner will not be effective, and only scraping will remove the deposits.

You should always be cautious with household acids and bases. They are usually just as concentrated and harmful as industrial chemicals. All strongly acidic and basic solutions, both in the lab and in the home, can be hazardous. Acids, interestingly, are somewhat less dangerous than solutions of bases because the H_3O^+ ion tends to *denature* ***(← p. 632)*** proteins in skin. The denatured proteins harden, forming a protective layer that prevents further attack by the acid, unless it is hot or highly concentrated. Basic solutions, by contrast, tend to dissolve proteins slowly and so produce little, if any, pain, causing considerable harm before any problem is noticed. If you should get acids or bases on your skin, wash with water for at least 15 min. If acid or base splashes into your eyes, have someone call a physician while you begin gently washing the affected area with lots of water. The international warning placard that is required for shipments of acids and bases in quantities of 1001 lb or more illustrates schematically the personal dangers of these substances.

A corrosive chemical warning placard. This type of placard is required by the U.S. Department of Transportation on loads of 1001 lb or more of acids or bases transported by highway or rail. Such a placard makes fairly clear the reactions with human skin and metals.

CHEMISTRY YOU CAN DO Aspirin and Digestion

Aspirin is a potent drug capable of relieving pain, fever, and inflammation. Recent studies indicate that it may also decrease blood clotting and heart disease. It is made from salicylic acid, which is found naturally in a variety of plants. The effect of pure salicylic acid on your stomach, however, makes it quite unpleasant as a pain remedy. Commercial aspirin is a derivative of salicylic acid, called acetylsalicylic acid, that has all the benefits of salicylic acid with less discomfort.

Aspirin is still somewhat acidic, however, and can sometimes cause discomfort or worse in people susceptible to stomach irritation. As a result, several different forms of aspirin are marketed today. The first and most common is plain aspirin. For people with stomach problems, there is also buffered aspirin which includes a buffer in the aspirin tablet to lessen the effect of aspirin's acidity (Section 17.1). A more recent development is enteric-coated aspirin, which is plain aspirin in a tablet with a coating that prevents the aspirin from dissolving in stomach acid, but does allow it to dissolve in the small intestine fluids, which are alkaline.

For this experiment, obtain at least three tablets of each kind of aspirin. Examples of regular aspirin are Bayer and Anacin. Buffered aspirin can be found as Bufferin or Bayer Plus. Enteric aspirin is most commonly available as Bayer Enteric or Ecotrin. Fill three transparent cups or glasses with water. Drop one intact tablet of plain aspirin into the first glass, one tablet of buffered aspirin into the second glass, and one tablet of enteric aspirin into the third glass. Note the changes in the tablets at 1-min intervals until no further changes occur. Now repeat this experiment using vinegar instead of water to dissolve the tablets. Observe each of the tablets in the vinegar at 30-s intervals until no further change is seen. For the final experiment, fill the glasses with water and add 2 tsp of baking soda and stir until dissolved. Once you have made your baking soda solution, add one of each type of aspirin tablet and observe what happens.

Think about these questions:

1. How did each of the tablets react to each of the different solutions? The acidity of the vinegar should have mimicked the acidity of your stomach. The basic nature of the baking soda should have mimicked the environment of your intestines, which follow your stomach in the digestive system.
2. How do you suppose each type of aspirin works according to its ability to dissolve in your experiments?
3. Does this lead you to think about the kind of aspirin you take?

SUMMARY PROBLEM

Lactic acid, $CH_3CH(OH)COOH$, is a weak monoprotic acid with a melting point of 53 °C. It exists as two enantiomers (← *p. 388*) that have slightly different K_a values. The L form has a K_a of 1.6×10^{-4} and the D form has a K_a of 1.5×10^{-4}. The D form is found in molasses, beer, wines, and souring milk. The L form is produced in muscle cells during anaerobic metabolism in which glucose molecules are broken down into lactic acid and molecules of adenosine triphosphate (ATP) are formed. When lactic acid builds up too rapidly in muscle tissue, severe pain results.

(a) Which form of lactic acid (D or L) is the stronger acid? Explain your answer.

(b) What would be the measured pK_a of a 50:50 mixture of the two forms of lactic acid? $pK_a = -\log K_a$

(c) A solution of D-lactic acid is prepared. Use HL as a general formula for lactic acid, and write the equation for the ionization of lactic acid in water.

(d) If 0.1-M solutions of these two acids (D and L) were prepared, calculate what the pH of each solution would be.

(e) Before any lactic acid dissolves in the water, what reaction determines the pH?

(f) Calculate the pH of a solution made by dissolving 4.46 g D-lactic acid in 500. mL of water.

(g) Calculate the number of milliliters of a 1.15-M NaOH solution required to completely neutralize 4.46 g of pure lactic acid.

(h) Calculate the pH of the solution made by the neutralization if the lactic acid were the D form. The L form. A 50:50 mixture of the two forms.

IN CLOSING

OWL and go Chemistry

Sign in at **www.cengage.com/owl** to:

- View tutorials and simulations, develop problem-solving skills, and complete online homework assigned by your professor.
- For quick review and exam prep, download Go Chemistry mini lecture modules from OWL (or purchase them at **www.CengageBrain.com**).

Having studied this chapter, you should be able to . . .

- Describe water's role in aqueous acid-base chemistry (Section 16.1). End-of-chapter questions: 12, 16
- Identify the conjugate base of an acid, the conjugate acid of a base, and the relationship between conjugate acid and base strengths (Section 16.1). Questions 26, 28
- Recognize how amines act as bases and how carboxylic acids ionize in aqueous solution (Section 16.2). Questions 16, 24
- Use the autoionization of water and show how this equilibrium takes place in aqueous solutions of acids and bases (Section 16.3).
- Classify an aqueous solution as acidic, neutral, or basic based on its concentration of H_3O^+ or OH^- and its pH or pOH (Section 16.4). Questions 33, 43
- Calculate pH (or pOH) given $[H_3O^+]$, or $[OH^-]$ given pH (or pOH) (Section 16.4). Questions 35, 37
- Estimate acid and base strengths from K_a and K_b values (Section 16.5). Question 51
- Write the ionization steps of polyprotic acids (Section 16.5). Question 31
- Describe the relationships between acid strength and molecular structure (16.6). Questions 20, 128
- Explain the nature of zwitterions (Section 16.6).
- Calculate pH from K_a or K_b values and solution concentration (Section 16.7). Questions 56, 60, 64
- Describe the hydrolysis of salts in aqueous solution (Section 16.8). Questions 76, 78
- Recognize Lewis acids and bases and describe how they react (Section 16.9). Questions 82, 86
- Describe the acidic behavior of hydrated metal ions (Section 16.9).
- Apply acid-base principles to the chemistry of antacids, kitchen chemistry, and household cleaners (Section 16.10). Question 92

KEY TERMS

acid-base reaction *(Section 16.1)*

acid ionization constant, K_a *(16.5)*

acid ionization constant expression *(16.5)*

acidic solution *(16.3)*

amines *(16.2)*

amphiprotic *(16.1)*

amphoteric *(16.9)*

autoionization *(16.3)*

base ionization constant, K_b *(16.5)*

base ionization constant expression *(16.5)*

basic solution *(16.3)*

Brønsted-Lowry acid *(16.1)*

Brønsted-Lowry base *(16.1)*

complex ion *(16.9)*

conjugate acid-base pair *(16.1)*

coordinate covalent bond *(16.9)*

hydrolysis *(16.8)*

ionization constant for water, K_w *(16.3)*

Lewis acid *(16.9)*

Lewis base *(16.9)*

monoprotic acids *(16.5)*

neutral solution *(16.3)*

oxoacids *(16.6)*

pH *(16.4)*

polyprotic acids *(16.5)*

zwitterion *(16.6)*

QUESTIONS FOR REVIEW AND THOUGHT

OWL Interactive versions of these problems are assignable in OWL.

Blue-numbered questions have short answers at the back of this book in Appendix M and fully worked solutions in the *Student Solutions Manual*.

Review Questions

These questions test vocabulary and simple concepts.

1. Define a Brønsted-Lowry acid and a Brønsted-Lowry base.
2. Explain in your own words what 100% ionization means.
3. Write the chemical equation for the autoionization of water. Write the equilibrium constant expression for this reaction. What is the value of the equilibrium constant at 25 °C? What is this constant called?
4. When OH^- is the base in a conjugate acid-base pair, the acid is ________; when OH^- is the acid, the base is ________.
5. Write balanced chemical equations that show phosphoric acid, H_3PO_4, ionizing stepwise as a polyprotic acid.
6. Write ionization equations for a weak acid and its conjugate base. Show that adding these two equations gives the autoionization equation for water.
7. Designate the acid and the base on the left side of these equations, and designate the conjugate partner of each on the right side.
   (a) $HNO_3(aq) + H_2O(\ell) \longrightarrow H_3O^+(aq) + NO_3^-(aq)$
   (b) $NH_4^+(aq) + CN^-(aq) \longrightarrow NH_3(aq) + HCN(aq)$
8. Dissolving ammonium bromide in water gives an acidic solution. Write a balanced equation showing how that can occur.
9. Solution A has a pH of 8 and solution B a pH of 10. Which has the greater hydronium ion concentration? How many times greater is its concentration?
10. Contrast the main ideas of the Brønsted-Lowry and Lewis acid-base concepts. Name and write the formula for a substance that behaves as a Lewis acid but not as a Brønsted-Lowry acid.

Topical Questions

These questions are keyed to major topics in the chapter. Usually a question that is answered at the back of the book is paired with a similar one that is not.

The Brønsted-Lowry Concept of Acids and Bases (Section 16.1)

11. Write a chemical equation to describe the proton transfer that occurs when each of these acids is added to water.
    (a) HBr (b) CF_3COOH
    (c) HSO_4^- (d) HNO_2
12. Write a chemical equation to describe the proton transfer that occurs when each of these acids is added to water.
    (a) HCO_3^- (b) HCl
    (c) CH_3COOH (d) HCN
13. Write a chemical equation to describe the proton transfer that occurs when each of these acids is added to water.
    (a) HClO (b) CH_3CH_2COOH
    (c) $HSeO_3^-$ (d) HO_2^-
14. Write a chemical equation to describe the proton transfer that occurs when each of these acids is added to water.
    (a) HIO (b) $CH_3(CH_2)_4COOH$
    (c) HOOCCOOH (d) $CH_3NH_3^+$
15. Write a chemical equation to describe the proton transfer that occurs when each of these bases is added to water.
    (a) H^- (b) HCO_3^- (c) NO_2^-
16. Write a chemical equation to describe the proton transfer that occurs when each of these bases is added to water.
    (a) HSO_4^- (b) CH_3NH_2
    (c) I^- (d) $H_2PO_4^-$
17. Write a chemical equation to describe the proton transfer that occurs when each of these bases is added to water.
    (a) PO_4^{3-} (b) SO_3^{2-} (c) HPO_4^{2-}
18. Write a chemical equation to describe the proton transfer that occurs when each of these bases is added to water.
    (a) AsO_4^{3-} (b) S^{2-} (c) N_3^-
19. Based on formulas alone, classify each of the following oxoacids as strong or weak.
    (a) H_3PO_4 (b) H_2SO_4
    (c) HClO (d) $HClO_4$
    (e) HNO_3 (f) H_2CO_3
    (g) HNO_2
20. Based on formulas alone, which is the stronger acid?
    (a) H_2CO_3 or H_2SO_4 (b) HNO_3 or HNO_2
    (c) $HClO_4$ or H_2SO_4 (d) H_3PO_4 or $HClO_3$
    (e) H_2SO_3 or H_2SO_4
21. Write the formula and name for the conjugate partner for each acid or base.
    (a) CN^- (b) SO_4^{2-}
    (c) HS^- (d) S^{2-}
    (e) HSO_3^- (f) HCOOH (formic acid)
22. Write the formula and name for the conjugate partner for each acid or base.
    (a) HI (b) NO_3^-
    (c) CO_3^{2-} (d) H_2CO_3
    (e) HSO_4^- (f) SO_3^{2-}
23. Which are conjugate acid-base pairs?
    (a) H_2O and H_3O^+ (b) H_3O^+ and OH^-
    (c) NH_2^- and NH_4^+ (d) NH_3 and NH_4^+
    (e) O^{2-} and H_2O
24. Which are conjugate acid-base pairs?
    (a) NH_2^- and NH_4^+ (b) NH_3 and NH_2^-
    (c) H_3O^+ and H_2O (d) OH^- and O^{2-}
    (e) H_3O^+ and OH^-
25. Identify the acid and the base that are reactants in each equation; identify the conjugate base and conjugate acid on the product side of each equation.
    (a) $HI(aq) + H_2O(\ell) \longrightarrow H_3O^+(aq) + I^-(aq)$
    (b) $OH^-(aq) + NH_4^+(aq) \longrightarrow H_2O(\ell) + NH_3(aq)$
    (c) $NH_3(aq) + H_2CO_3(aq) \longrightarrow NH_4^+(aq) + HCO_3^-(aq)$
26. Identify the acid and the base that are reactants in each equation; identify the conjugate base and conjugate acid on the product side of the equation.
    (a) $HS^-(aq) + H_2O(\ell) \longrightarrow H_2S(aq) + OH^-(aq)$
    (b) $S^{2-}(aq) + NH_4^+(aq) \longrightarrow NH_3(g) + HS^-(aq)$
    (c) $HCO_3^-(aq) + HSO_4^-(aq) \longrightarrow H_2CO_3(aq) + SO_4^{2-}(aq)$
    (d) $NH_3(aq) + NH_2^-(aq) \longrightarrow NH_2^-(aq) + NH_3(aq)$

27. Identify the acid and the base that are reactants in each equation; identify the conjugate base and conjugate acid on the product side of the equation.
(a) $H_2PO_4^-(aq) + HCO_3^-(aq) \longrightarrow H_2CO_3(aq) + HPO_4^{2-}(aq)$
(b) $NH_3(aq) + NH_2^-(aq) \longrightarrow NH_2^-(aq) + NH_3(aq)$
(c) $HSO_4^-(aq) + CO_3^{2-}(aq) \longrightarrow SO_4^{2-}(aq) + HCO_3^-(aq)$
28. Identify the acid and the base that are reactants in each equation; identify the conjugate base and conjugate acid on the product side of the equation.
(a) $CN^-(aq) + CH_3COOH(aq) \longrightarrow CH_3COO^-(aq) + HCN(aq)$
(b) $O^{2-}(aq) + H_2O(\ell) \longrightarrow 2\ OH^-(aq)$
(c) $HCO_2^-(aq) + H_2O(\ell) \longrightarrow HCOOH(aq) + OH^-(aq)$
29. Identify the conjugate acid or base of these acids or bases.
(a) $HClO$ (b) O^{2-} (c) $HCOOH$
(d) OH^- (e) IO_3^- (f) PH_3
30. Write stepwise chemical equations for protonation or deprotonation of each of these polyprotic acids and bases in water.
(a) H_2SO_3 (b) S^{2-}
(c) $NH_3CH_2COOH^+$ (glycinium ion, a diprotic acid)
31. Write stepwise chemical equations for protonation or deprotonation of each of these polyprotic acids and bases in water.
(a) CO_3^{2-} (b) H_3AsO_4
(c) $NH_2CH_2COO^-$ (glycinate ion, a diprotic base)

pH Calculations (Section 16.3)

32. The pH of a popular soft drink is 3.30. Calculate its hydronium ion concentration. Is the drink acidic or basic?
33. Milk of magnesia, $Mg(OH)_2$, has a pH of 10.5. Calculate the hydronium ion concentration of the solution. Is this solution acidic or basic?
34. A sample of coffee has a pH of 4.3. Calculate the hydronium ion concentration in this coffee.
35. A solution of lactic acid has a pH of 2.44. Calculate its hydronium ion concentration.
36. Calculate the pH of a 0.0013-M solution of HNO_3. Calculate the pOH of this solution.
37. Calculate the pH of a solution that is 0.025-M in NaOH. Calculate the pOH of this solution.
38. A solution of benzyl amine, $C_7H_7NH_2$, has a hydroxide ion concentration of 2.4×10^{-3} M. Calculate the pH of the solution. Calculate its pOH.
39. The hydronium ion concentration of a cyanoacetic acid solution is 0.032 M. Calculate its pOH.
40. The pH of a $Ba(OH)_2$ solution is 10.66 at 25 °C. Calculate the hydroxide ion concentration of this solution. If the solution volume is 250. mL, calculate how many grams of $Ba(OH)_2$ must have been used to make this solution.
41. A 1000.-mL solution of hydrochloric acid has a pH of 1.3. Calculate the mass in grams of HCl dissolved in the solution.
42. Make these interconversions. In each case tell whether the solution is acidic or basic.

	pH	$[H_3O^+]$ (M)	$[OH^-]$ (M)
(a)	1.00	______	______
(b)	10.5	______	______
(c)	______	1.8×10^{-4}	______
(d)	______	______	2.3×10^{-5}

43. Make these interconversions. In each case tell whether the solution is acidic or basic.

	pH	$[H_3O^+]$ (M)	$[OH^-]$ (M)
(a)	______	6.1×10^{-7}	______
(b)	______	______	2.2×10^{-9}
(c)	4.67	______	______
(d)	______	2.5×10^{-2}	______
(e)	9.12	______	______

44. Figure 16.3 shows the pH of some common solutions. How many times more acidic or basic are these compared with a neutral solution?
(a) Milk (b) Seawater
(c) Blood (d) Battery acid
45. Figure 16.3 shows the pH of some common solutions. How many times more acidic or basic are these compared with a neutral solution?
(a) Black coffee (b) Household ammonia
(c) Baking soda (d) Vinegar
46. The value of K_w varies with temperature. Use Table 16.1 to calculate the pH of a neutral solution at:
(a) 30 °C. (b) 50 °C.
47. The measured pH of a sample of seawater is 8.30.
(a) Calculate the H_3O^+ concentration.
(b) Is the sample acidic or basic?

Acid-Base Strengths (Section 16.6)

48. Write ionization equations and ionization constant expressions for these acids and bases.
(a) CH_3COOH (b) HCN
(c) SO_3^{2-} (d) PO_4^{3-}
(e) NH_4^+ (f) H_2SO_4
49. Write ionization equations and ionization constant expressions for these acids and bases.
(a) F^- (b) NH_3
(c) H_2CO_3 (d) H_3PO_4
(e) CH_3COO^- (f) S^{2-}
50. Which solution is more acidic?
(a) 0.10-M H_2CO_3 or 0.10-M NH_4Cl
(b) 0.10-M HF or 0.10-M $KHSO_4$
(c) 0.1-M $NaHCO_3$ or 0.1-M Na_2HPO_4
(d) 0.1-M H_2S or 0.1-M HCN
51. Which solution is more basic?
(a) 0.10-M NH_3 or 0.10-M NaF
(b) 0.10-M K_2S or 0.10-M K_3PO_4
(c) 0.10-M $NaNO_3$ or 0.10-M $NaCH_3COO$
(d) 0.10-M NH_3 or 0.10-M KCN
52. Without doing any calculations, assign each of these 0.10-M aqueous solutions to one of these pH ranges: pH 2; pH between 2 and 6; pH between 6 and 8; pH between 8 and 12; pH 12.
(a) HNO_2 (b) NH_4Cl (c) NaF
(d) $Mg(CH_3COO)_2$ (e) BaO (f) $KHSO_4$
(g) $NaHCO_3$ (h) $BaCl_2$
53. Write the chemical equation and the K_b expression for these bases.
(a) CN^- (b) $C_6H_5NH_2$ (c) HS^-

Using K_a and K_b (Section 16.7)

54. Calculate the pH of each solution in Question 52 to verify your prediction.
55. A 0.015-M solution of cyanic acid has a pH of 2.67. Calculate the ionization constant, K_a, of the acid.
56. Calculate the K_a of butyric acid if a 0.025-M butyric acid solution has a pH of 3.21.
57. Calculate the equilibrium concentrations of H_3O^+, acetate ion, and acetic acid in a 0.20-M aqueous solution of acetic acid, CH_3COOH.
58. The ionization constant of a very weak acid, HA, is 4.0×10^{-9}. Calculate the equilibrium concentrations of H_3O^+, A^-, and HA in a 0.040-M solution of the acid.
59. (a) Calculate the pH of a 0.050-M solution of benzoic acid, C_6H_5COOH; $K_a = 1.2 \times 10^{-4}$ at 25 °C.
    (b) Calculate the percent of the acid that has ionized in this solution.
60. The pH of a 0.10-M solution of propanoic acid, CH_3CH_2COOH, a weak organic acid, is measured at equilibrium and found to be 2.93 at 25 °C. Calculate the K_a of propanoic acid.
61. The weak base methylamine, CH_3NH_2, has $K_b = 5.0 \times 10^{-4}$. It reacts with water according to the equation

$$CH_3NH_2(aq) + H_2O(\ell) \rightleftharpoons CH_3NH_3^+(aq) + OH^-(aq)$$

Calculate the pH of a 0.23-M methylamine solution.

62. Calculate the pH of a 0.12-M aqueous solution of the base aniline, $C_6H_5NH_2$; $K_b = 3.9 \times 10^{-10}$.
63. Calculate the $[OH^-]$ and the pH of a 0.024-M methylamine solution; $K_b = 5.0 \times 10^{-4}$.
64. Piperidine, $C_5H_{11}N$, is an oily liquid with a peppery smell. The pH of a 0.0250-M piperidine solution is 11.70. Calculate the K_b of piperidine.
65. Pyridine, C_5H_5N, is a weak base; $K_a = 6.3 \times 10^{-6}$. A 0.2-M solution of pyridine has a pH = 8.5.
    (a) Write a balanced chemical equation to represent why the solution has this pH.
    (b) Calculate the concentration of unreacted pyridine in this solution.
66. Amantadine, $C_{10}H_{15}NH_2$, is used in the treatment of Parkinson's disease. For amantadine: $K_a = 7.9 \times 10^{-11}$.
    (a) Write the chemical equation for the reaction of amantadine with water.
    (b) Calculate the pH of a 0.0010-M aqueous solution of amantadine at 25 °C.
67. After doing K_a calculations, you may wish you had an aspirin. Aspirin is a weak acid with $K_a = 3.27 \times 10^{-4}$ for the reaction

$$HC_9H_7O_4(aq) + H_2O(\ell) \rightleftharpoons H_3O^+(aq) + C_9H_7O_4^-(aq)$$

Two aspirin tablets, each containing 0.325 g aspirin (along with a nonreactive "binder" to hold the tablet together), are dissolved in 200.0 mL of water. Calculate the pH of this solution.

68. Lactic acid, $C_3H_6O_3$, occurs in sour milk as a result of the metabolism of certain bacteria. Calculate the pH of a solution of 56. mg lactic acid in 250. mL water. K_a for lactic acid is 1.4×10^{-4}.
69. Boric acid is a weak acid often used as an eyewash. K_a for boric acid is 5.8×10^{-10}. Calculate the pH of a 0.10-M solution of boric acid.
70. Calculate the pH of a 0.0050-M solution of dimethylamine, $(CH_3)_2NH$, whose K_b is 5.8×10^{-4}.

Acid-Base Reactions of Salts (Section 16.8)

71. Complete each of these reactions by filling in the blanks. Predict whether each reaction is product-favored or reactant-favored, and explain your reasoning.
    (a) ______(aq) + Br^-(aq) $\rightleftharpoons$ NH_3(aq) + HBr(aq)
    (b) CH_3COOH(aq) + CN^-(aq) $\rightleftharpoons$ ______(aq) + HCN(aq)
    (c) ______(aq) + $H_2O(\ell)$ $\rightleftharpoons$ NH_3(aq) + OH^-(aq)
72. Complete each of these reactions by filling in the blanks. Predict whether each reaction is product-favored or reactant-favored, and explain your reasoning.
    (a) ______(aq) + HSO_4^-(aq) $\rightleftharpoons$ HCN(aq) + SO_4^{2-}(aq)
    (b) H_2S(aq) + $H_2O(\ell)$ $\rightleftharpoons$ H_3O^+(aq) + ______(aq)
    (c) H^-(aq) + $H_2O(\ell)$ $\rightleftharpoons$ OH^-(aq) + ______(g)
73. Predict which of these acid-base reactions are product-favored and which are reactant-favored. In each case write a balanced equation for any reaction that might occur, even if the reaction is reactant-favored. Consult Table 16.2 if necessary.
    (a) $H_2O(\ell)$ + HNO_3(aq) (b) H_3PO_4(aq) + $H_2O(\ell)$
    (c) CN^-(aq) + HCl(aq) (d) NH_4^+(aq) + F^-(aq)
74. Predict which of these acid-base reactions are product-favored and which are reactant-favored. In each case write a balanced equation for any reaction that might occur, even if the reaction is reactant-favored. Consult Table 16.2 if necessary.
    (a) NH_4^+(aq) + HPO_4^{2-}(aq)
    (b) CH_3COOH(aq) + OH^-(aq)
    (c) HSO_4^-(aq) + $H_2PO_4^-$(aq)
    (d) CH_3COOH(aq) + F^-(aq)
75. For each salt, predict whether an aqueous solution will have a pH less than, equal to, or greater than 7. Explain your answer.
    (a) $NaHSO_4$ (b) NH_4Br (c) $KClO_4$
76. For each salt, predict whether an aqueous solution will have a pH less than, equal to, or greater than 7. Explain your answer.
    (a) $AlCl_3$ (b) Na_2S (c) $NaNO_3$
77. For each salt, predict whether an aqueous solution will have a pH less than, equal to, or greater than 7. Explain your answer.
    (a) NaH_2PO_4 (b) NH_4NO_3 (c) $SrCl_2$
78. For each salt, predict whether an aqueous solution will have a pH less than, equal to, or greater than 7. Explain your answer.
    (a) Na_2HPO_4 (b) $(NH_4)_2S$ (c) KCH_3COO
79. Explain why $BaCO_3$ is soluble in aqueous HCl, but $BaSO_4$, which is used to make the intestines visible in X-ray photographs, remains sufficiently insoluble in the HCl in a human stomach so that poisonous barium ions do not get into the bloodstream.
80. For which of these substances would the solubility be greater at pH = 2 than at pH = 7?
    (a) $Cu(OH)_2$ (b) $CuSO_4$ (c) $CuCO_3$
    (d) CuS (e) $Cu_3(PO_4)_2$

Lewis Acids and Bases (Section 16.9)

81. Which of these is a Lewis acid? A Lewis base?
(a) NH_3 (b) $BeCl_2$ (c) BCl_3
82. Which of these is a Lewis acid? A Lewis base?
(a) O^{2-} (b) CO_2 (c) H^-
83. Which of these is a Lewis acid? A Lewis base?
(a) Al^{3+} (b) H_2O (c) SCN^-
84. Which of these is a Lewis acid? A Lewis base?
(a) Cr^{3+} (b) SO_3 (c) CH_3NH_2
85. Identify the Lewis acid and the Lewis base in each reaction.
(a) $H_2O(\ell) + SO_2(aq) \longrightarrow H_2SO_3(aq)$
(b) $H_3BO_3(aq) + OH^-(aq) \longrightarrow B(OH)_4^-(aq)$
(c) $Cu^{2+}(aq) + 4\ NH_3(aq) \longrightarrow [Cu(NH_3)_4]^{2+}(aq)$
(d) $2\ Cl^-(aq) + SnCl_2(aq) \longrightarrow SnCl_4^{2-}(aq)$
86. Identify the Lewis acid and the Lewis base in each reaction.
(a) $I_2(s) + I^-(aq) \longrightarrow I_3^-(aq)$
(b) $SO_2(g) + BF_3(g) \longrightarrow O_2SBF_3(s)$
(c) $Au^+(aq) + 2\ CN^-(aq) \longrightarrow [Au(CN)_2]^-(aq)$
(d) $CO_2(g) + H_2O(\ell) \longrightarrow H_2CO_3(aq)$
87. Trimethylamine, $(CH_3)_3N$:, reacts readily with diborane, B_2H_6. The diborane dissociates to two BH_3 fragments, each of which can react with trimethylamine to form a complex, $(CH_3)_3N{:}BH_3$. Write an equation for this reaction and interpret it in terms of Lewis acid-base theory.
88. Draw a Lewis structure for ICl_3. Predict the shape of this molecule. Does it function as a Lewis acid or base when it reacts with chloride ion to form ICl_4^-? What is the structure of this ion?
89. Write the formula for the conjugate base of $[Zn(H_2O)_3(OH)]^{1+}$.
90. Write the formula for the conjugate acid of $[Zn(OH)_4]^{2-}$.

Additional Applied Acid-Base Chemistry (Section 16.10)

91. Double-acting baking powder contains two salts, sodium hydrogen carbonate and potassium dihydrogen phosphate, whose anions react in water to form CO_2 gas. Write a balanced chemical equation for the reaction. Which anion is the acid and which is the base?
92. Common soap is made by reacting sodium carbonate with stearic acid, $CH_3(CH_2)_{16}COOH$. Write a balanced equation for the reaction.
93. If 1 g of each antacid in Table 16.8 reacted with equal volumes of stomach acid, calculate which would neutralize the most stomach acid.
94. Why is it not a good idea to substitute dishwashing detergent for automobile-washing detergent?
95. Why do cleaning products containing sodium hydroxide feel slippery when you get them on your skin?
96. Some cooked fish have a "fishy" odor due to amines. Explain why putting lemon juice on the fish reduces this odor.

General Questions

These questions are not explicitely keyed to chapter topics; many require integration of several concepts.

97. Classify each of these as a strong acid, weak acid, strong base, weak base, amphiprotic substance, or neither acid nor base.
(a) HCl (b) NH_4^+ (c) H_2O
(d) CH_3COO^- (e) CH_4 (f) CO_3^{2-}
98. Classify each of these as a strong acid, weak acid, strong base, weak base, amphiprotic substance, or neither acid nor base.
(a) CH_3COOH (b) Na_2O (c) H_2SO_4
(d) NH_3 (e) $Ba(OH)_2$ (f) $H_2PO_4^-$
99. Several acids and their respective equilibrium constants are:

$$HF(aq) + H_2O(\ell) \rightleftharpoons H_3O^+(aq) + F^-(aq) \qquad K_a = 6.8 \times 10^{-4}$$

$$HS^-(aq) + H_2O(\ell) \rightleftharpoons H_3O^+(aq) + S^{2-}(aq) \qquad K_a = 1 \times 10^{-19}$$

$$CH_3COOH(aq) + H_2O(\ell) \rightleftharpoons H_3O^+(aq) + CH_3COO^-(aq) \qquad K_a = 1.8 \times 10^{-5}$$

(a) Which is the strongest acid? Which is the weakest acid?
(b) Which acid has the weakest conjugate base?
(c) Which acid has the strongest conjugate base?
100. State whether equal molar amounts of these would have a pH equal to 7, less than 7, or greater than 7.
(a) A weak base and a strong acid react.
(b) A strong base and a strong acid react.
(c) A strong base and a weak acid react.
101. Sulfurous acid, H_2SO_3, is a weak diprotic acid ($K_{a_1} = 1.7 \times 10^{-2}$, $K_{a_2} = 6.3 \times 10^{-8}$). Calculate the pH of a 0.45-M solution of H_2SO_3. (Assume that only the first ionization is important in determining pH.)
102. Ascorbic acid (vitamin C, $C_6H_8O_6$) is a diprotic acid ($K_{a_1} = 7.9 \times 10^{-5}$, $K_{a_2} = 1.6 \times 10^{-12}$). Calculate the pH of a solution that contains 5.0 mg acid per mL water. (Assume that only the first ionization is important in determining pH.)
103. Does the pH of the solution increase, decrease, or stay the same when you
(a) Add solid ammonium chloride to 100. mL of 0.10-M NH_3?
(b) Add solid sodium acetate to 50.0 mL of 0.015-M acetic acid?
(c) Add solid NaCl to 25.0 mL of 0.10-M NaOH?
104. Does the pH of the solution increase, decrease, or stay the same when you
(a) Add solid sodium oxalate, $Na_2C_2O_4$, to 50.0 mL of 0.015-M oxalic acid?
(b) Add solid ammonium chloride to 100. mL of 0.016-M HCl?
(c) Add 20.0 g NaCl to 1.0 L of 0.012-M sodium acetate, $NaCH_3COO$?
105. (a) Solid ammonium bromide, NH_4Br, is added to water. Will the resulting solution be acidic, basic, or neutral? Use a chemical equation to explain your answer.
(b) An aqueous ammonium bromide solution has a pH of 5.50. Calculate the ammonium bromide concentration of this solution.
106. Sodium hypochlorite, NaOCl, is used as a source of chlorine in some laundry bleaches, swimming pool disinfectants, and water treatment plants. Calculate the pH of a 0.010-M solution of NaOCl ($K_b = 1.5 \times 10^{-7}$).

Applying Concepts

These questions test conceptual learning.

107. When a 0.1-M aqueous ammonia solution is tested with a conductivity apparatus (← *p. 97*), the bulb glows dimly. When a 0.1-M hydrochloric acid solution is tested, the bulb glows brightly. As water is added to each of the solutions, would you expect the bulb to glow brighter, stop glowing, or stay the same? Explain your reasoning.

108. Using Table 16.1, calculate what the pH of pure water is at 10 °C, 25 °C, and 50 °C. Classify the water at each temperature as either acidic, neutral, or basic.

109. If you evaporate the water in a sodium hydroxide solution, you will end up with solid sodium hydroxide. However, if you evaporate the water in an ammonium hydroxide solution, you will not end up with solid ammonium hydroxide. Explain why. What will remain after the water is evaporated?

110. For each aqueous solution, predict what ions and molecules will be present. Without doing any calculations, list the ions and molecules in order of decreasing concentration.
(a) HCl (b) $NaClO_4$ (c) HNO_2
(d) NaClO (e) NH_4Cl (f) NaOH

111. The diagrams below are nanoscale representations of different acids.

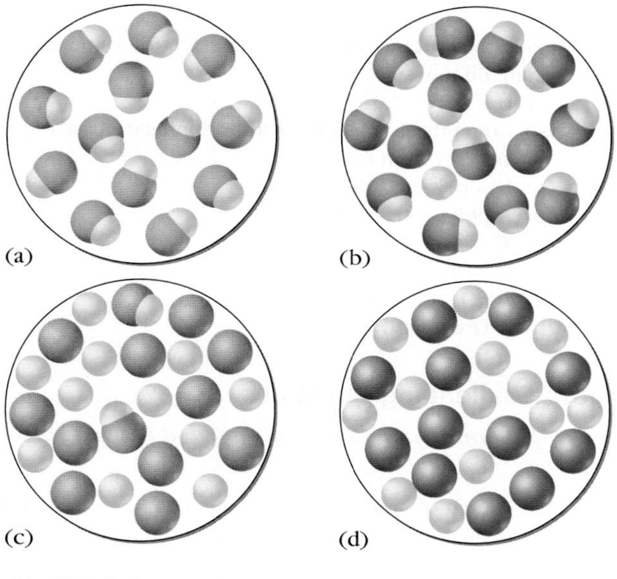

(a) Which diagram best represents hydrochloric acid? (The yellow spheres are H^+ ions and the other spheres are Cl^- ions.)
(b) Which diagram best represents acetic acid? (The yellow spheres are H^+ ions and the other spheres are CH_3COO^- ions.)

112. When asked to identify the conjugate acid-base pairs in the reaction

$$HCO_3^-(aq) + HSO_4^-(aq) \rightleftharpoons H_2CO_3(aq) + SO_4^{2-}(aq)$$

a student incorrectly wrote: "HCO_3^- is a base and HSO_4^- is its conjugate acid. H_2CO_3 is an acid and SO_4^- is its conjugate base." Write a brief explanation to the student telling why the answer is incorrect.

113. A 0.1-M aqueous solution has a pH of 5.5. Write a chemical equation that represents why the solution is acidic.

114. Explain how the Arrhenius acid-base theory and the Brønsted-Lowry theory of acids and bases are explained by the Lewis acid-base theory.

115. Calculate the K_b for the conjugate base of gallic acid, found in tea; $K_a = 3.9 \times 10^{-5}$. Identify a base from Table 16.2 with a K_b close to that of the conjugate base of gallic acid.

116. Trichloroacetic acid, CCl_3COOH, has an acid dissociation constant of 3.0×10^{-1} at 25 °C.
(a) Calculate the pH of a 0.0100-M trichloroacetic acid solution.
(b) How many times greater is the H_3O^+ concentration in this solution than in a 0.0100-M acetic acid?

117. Explain why a 0.1-M NH_4NO_3 aqueous solution has a pH of 5.1.

118. Is the dissociation of water

$$2\ H_2O(\ell) \rightleftharpoons H_3O^+(aq) + OH^-(aq)$$

exothermic or endothermic? Refer to Table 16.1. Explain your answer.

119. The illustrations below are nanoscale representations of aqueous solutions of three monoprotic acids—HX, HY, and HZ. Water molecules are excluded for clarity.

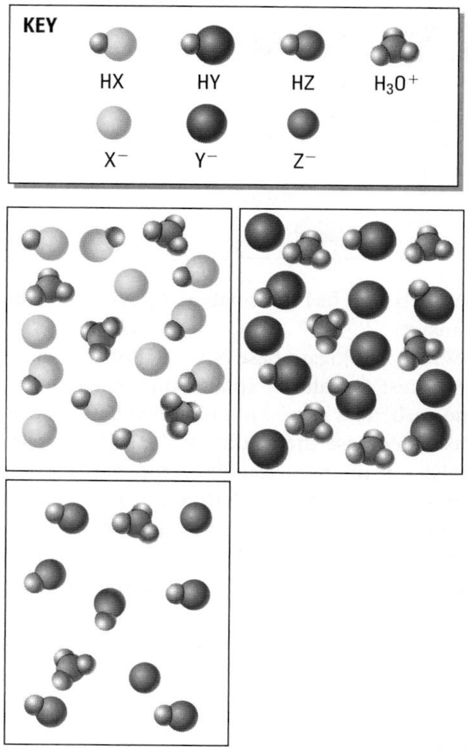

(a) Which acid has the largest K_a value? Explain your answer.
(b) Which conjugate base—X^-, Y^-, or Z^-—has the largest K_b value? Explain your answer.

120. The illustrations below are nanoscale representations of aqueous solutions of three monoprotic acids—HA, HB, and HC. Water molecules are excluded for clarity.

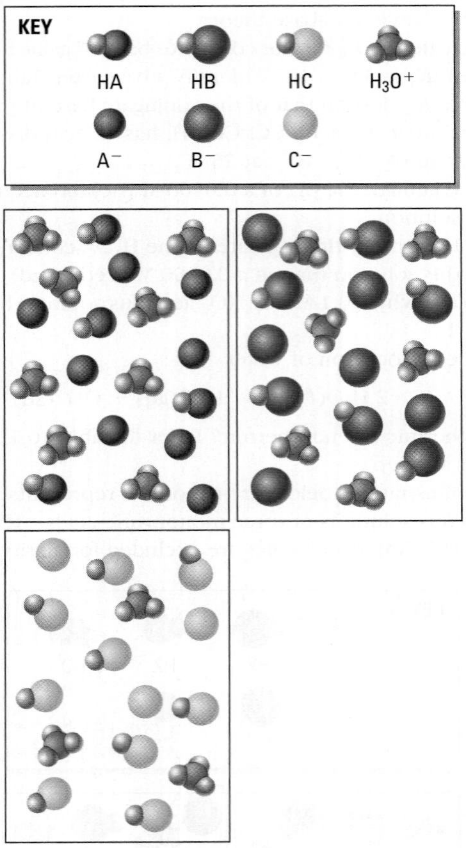

(a) Which acid has the smallest K_a value? Explain your answer.
(b) Which conjugate base—A^-, B^-, or C^-—has the smallest K_b value? Explain your answer.

121. Home gardeners spread aluminum sulfate powder around plants to increase the acidity of the soil. Describe the chemistry at the nanoscale level for such a soil treatment.

122. Below is a grid with nine lettered boxes, each of which contains an item that is used to answer the questions that follow. Items may be used more than once and there may be more than one correct item in response to a question.

A K_2CO_3	**B** Cu^{2+}	**C** $[OH^-] = 3.0 \times 10^{-5}$ M
D H_2SO_3	**E** $H_2PO_4^-$	**F** 0.10-M LiOH
G $[H_3O^+] = 2.00 \times 10^{-8}$ M	**H** $NaNO_2$	**I** 0.010-M $HClO_4$

Place the letter(s) of the correct selection(s) on the appropriate line. Assume $T = 298$ K.
(a) Has pH > 7.0 ________
(b) $[H_3O^+] > 1.0 \times 10^{-7}$ M ________
(c) A basic salt ________
(d) $K_a > K_b$ ________
(e) A Lewis acid ________
(f) Has pH = 2.00 ________
(g) A polyprotic acid ________

123. Below is a grid with nine lettered boxes, each of which contains an item that is used to answer the questions that follow. Items may be used more than once and there may be more than one correct item in response to a question.

A NH_4Br	**B** Zn^{2+}	**C** 0.0010-M KOH
D 0.010-M HBr	**E** H_2S	**F** H_2O
G K_2HPO_4	**H** Na_3PO_4	**I** KF

Place the letter(s) of the correct selection(s) on the appropriate line. Assume $T = 298$ K.
(a) Has pH < 7.0 ________
(b) Contains a conjugate acid or conjugate base of another item in the grid. ________
(c) Has pH = 11.00 ________
(d) A polyprotic acid ________
(e) $K_a > K_b$ ________
(f) A basic salt ________
(g) $[H_3O^+] < 1.0 \times 10^{-7}$ M ________
(h) A Lewis acid ________

More Challenging Questions

These questions require more thought and integrate several concepts.

124. If 1 g each of vinegar, lemon juice, and lactic acid react with equal masses of baking soda, which will produce the most CO_2 gas?

125. A person claimed that his stomach ruptured when he took a teaspoonful of baking soda in a glass of water to relieve heartburn after a full meal ($\frac{1}{2}$ tsp = 2.5 g $NaHCO_3$). Assume that the pH of stomach acid is 1 and that the stomach has a volume of 1 L when expanded fully. Body temperature is 37 °C. Calculate the volume of carbon dioxide gas generated by the reaction of baking soda with stomach acids. Might his stomach have ruptured from this volume of CO_2?

126. A chilled carbonated beverage is opened and warmed to room temperature. Will the pH change and, if so, how will it change?

127. For an experiment a student needs a pH = 9.0 solution. He plans to make the solution by diluting 6.0-M HCl until the H_3O^+ concentration equals 1.0×10^{-9} M.
(a) Will this plan work? Explain your answer.
(b) Will diluting 1.00-M NaOH work? Explain your answer.

128. Explain why $BrNH_2$ is a weaker base than ammonia, NH_3. Which will have the smaller K_b value? Will $ClNH_2$ be a stronger or weaker base than $BrNH_2$? Explain your answer.

129. The conjugate acid of aniline has $pK_a = 4.59$; the conjugate acid of 3-bromoaniline has $pK_a = 3.58$. Which is the stronger base? Explain your answer.

130. It is determined that 0.1-M solutions of the sodium salts NaM, NaQ, and NaZ have pH values of 7.0, 8.0, and 9.0, respectively. Arrange the acids HM, HQ, and HZ in order of decreasing strength. Where possible, find the K_a values of these acids.

131. Use the data in Appendix J to calculate the enthalpy change for the reaction of:
(a) 100. mL of 0.100-M NaOH with 100. mL of 1.00-M HCl.
(b) 25.0 mL of 1.00-M H_2SO_4 with 75.0 mL of 1.0-M NaOH.

132. Hydrogen peroxide, HOOH, is a powerful oxidizing agent that, in diluted form, is used as an antiseptic. For HOOH: $K_a = 2.1 \times 10^{-12}$; for OOH^-: $K_b = 4.8 \times 10^{-3}$.
(a) Is hydrogen peroxide a strong or weak acid?
(b) Is OOH^- a strong or weak base?
(c) What is the relationship between HOOH and OOH^-?
(d) Calculate the pH of 0.100-M aqueous hydrogen peroxide.

133. *Para*-bromoaniline is used in dye production. Two different pK_b values have been reported for this compound: 10.25 and 9.98. Using each value, calculate the pH for a 0.010-M solution of *para*-bromoaniline. What is the difference in pH between the two pK_b values?

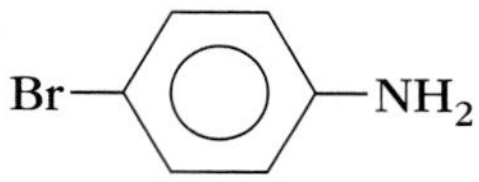

134. At 25 °C, a 0.10% aqueous solution of adipic acid, $C_5H_9O_2COOH$, has a pH of 3.2. A saturated solution of the acid, which contains 1.44 g acid per 100. mL of solution, has a pH = 2.7. Calculate the percent dissociation of adipic acid in each solution.

135. Niacin, a B vitamin, can act as an acid and as a base.
(a) Write chemical equations showing the action of niacin as an acid in water. Symbolize niacin as HNc.
(b) Write chemical equations showing the action of niacin as a base in water.
(c) A 0.020-M aqueous solution of niacin has a pH = 3.26. Calculate the K_a of niacin.
(d) Calculate the K_b of niacin.
(e) Write the chemical equation that accompanies the K_b expression for niacin.

136. The loss of CO_2 from lysine forms cadaverine, an aptly named compound with an exceptionally revolting smell.

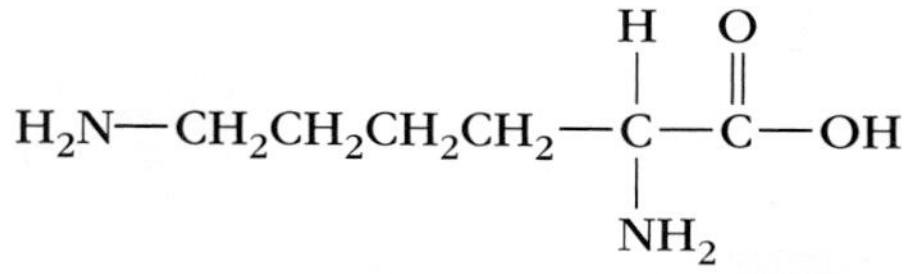

lysine

(a) Write the structural formula for cadaverine.
(b) Write the structural formula for cadaverine in its fully protonated form.
(c) Cadaverine has two pK_a values: 10.25 and 9.13; $pK_a = -\log(K_a)$. Which one is for cadaverine in its fully protonated form?

137. At normal body temperature, 37 °C, $K_w = 2.5 \times 10^{-14}$.
(a) Calculate the pH of a neutral solution at this temperature.
(b) The pH of blood at this temperature ranges from 7.35 to 7.45. Assuming a pH of 7.40, calculate the H_3O^+ concentration of the blood at this temperature.
(c) By how much does the H_3O^+ concentration at this temperature differ from that at 25 °C?

138. The structural formula for lysine is given in Question 136. Explain why lysine has three pK_a values: 2.18, 8.95, and 10.53. $pK_a = -\log(K_a)$.

Conceptual Challenge Problems

These rigorous, thought-provoking problems integrate conceptual learning with problem solving and are suitable for group work.

CP16.A (Section 16.4) Is it possible for an aqueous solution to have a pH of 0 or even less than 0? Explain your answer mathematically as well as practically based on what you know about acid solubilities.

CP16.B (Section 16.4) What is the pH of water at 200 °C? Liquid water this hot must be under a pressure greater than 1.0 atm and might be found in a pressurized water reactor located in a nuclear power plant.

CP16.C (Section 16.5) Develop a set of rules by which you could predict the pH for solutions of strong or weak acids and strong or weak bases without using a calculator. Your predictions need to be accurate to ±1 pH units. Assume that you know the concentration of the acid or base and that for the weak acids and bases you can look up the pK_a ($-\log K_a$) or K_a values. What rules would work to predict pH?

17

Additional Aqueous Equilibria

Nick Stone/iStockphoto

A portion of a geode is shown above. Geodes have a rocky outer shell, usually limestone, that encloses an internal cavity containing crystals of quartz, SiO_2, or other minerals. Once the outer shell hardens, silicate and carbonate minerals dissolved in groundwater migrate through the shell and into the cavity where they crystallize. In the geode here, colorless quartz crystals at the perimeter surround purple crystals of amethyst, a variety of quartz. Amethyst's coloration is due to Fe^{3+} and Al^{3+} ions in ppm concentration within the SiO_2 framework. Factors affecting the solubility and precipitation of solutes, such as minerals, are described in this chapter.

In environments as different as the interior of red blood cells, coral reefs, ocean waters, and clouds high above Earth's surface, important interactions occur among solutes in aqueous solution. This chapter extends the discussion of aqueous solutions begun in Chapters 15 and 16 (conjugate acid-base behavior, acid-base neutralization, the link between solubility and precipitation) to quantitative aspects related to

- Buffers, which are combinations of a weak acid and its conjugate base, or a weak base with its conjugate acid.
- Acid-base titrations, which involve acid-base neutralization reactions.
- Equilibria associated with solutions of slightly soluble salts.

17.1 Buffer Solutions

Adding a small amount of acid or base to pure water radically changes the pH. Consider what happens if 0.010 mol HCl is added to 1 L pure water. The pH changes from 7.00 to 2.00 because $[H_3O^+]$ changes from 1.0×10^{-7} M to 1.0×10^{-2} M. This pH change represents a *100,000-fold increase* in H_3O^+ concentration. Similarly, if 0.010 mol NaOH is added to 1 L pure water, the pH goes from 7.00 to 12.00, a *100,000-fold decrease* in $[H_3O^+]$ and a *100,000-fold increase* in $[OH^-]$. Most aquatic organisms could not survive such dramatic pH changes; the organisms can survive only within a narrow pH range. For example, if acid rain lowers the pH of a lake or stream sufficiently, fish such as trout may die.

Unlike pure water and aqueous solutions of NaOH or HCl, some aqueous solutions maintain a relatively constant pH when limited amounts of base or acid are added to them. Such solutions contain a **buffer**—*a chemical system that resists change in pH*—and are referred to as **buffer solutions.** For example, a solution that contains 0.50 mol acetic acid, CH_3COOH, and 0.50 mol sodium acetate, $NaCH_3COO$, in 1.0 L of solution is a buffer with a pH of 4.74. When 0.010 mol of a strong acid is added to it, the pH changes from 4.74 to 4.72, only 0.02 pH unit; adding 0.010 mol of strong base to this buffer changes the pH from 4.74 to 4.76 (Figure 17.1). These are only slight pH changes, clearly much less than those that occur when the same number of moles of HCl or NaOH are added to water, as described above. How do the sodium acetate-acetic acid buffer and other buffers offset such additions of acid or base without significant change in pH?

OWL Sign in to OWL at **www.cengage.com/owl** to view tutorials and simulations, develop problem-solving skills, and complete online homework assigned by your professor.

go Chemistry Download mini lecture videos for key concept review and exam prep from OWL or purchase them from **www.CengageBrain.com.**

Companion Website

Visit this book's companion website at **www.cengage.com/chemistry/moore** to work interactive modules for the Estimation boxes and Active Figures in this text.

go Chemistry **Module 23: Understanding Acid-Base Buffers** covers concepts in this section.

We will discuss shortly how to calculate the pH of a buffer.

Buffer Action

A buffer is a chemical system that maintains a relatively constant pH because the buffer contains *a weak acid that can react with added base,* and the buffer also contains *a weak base that can react with added acid.* In addition, the acid and base components of a buffer solution must not react with each other. A conjugate acid-base pair, such as acetic acid and acetate ion (from sodium acetate), satisfies this requirement. In a conjugate pair, if the acid and base react with each other they just produce conjugate base and conjugate acid—no observable

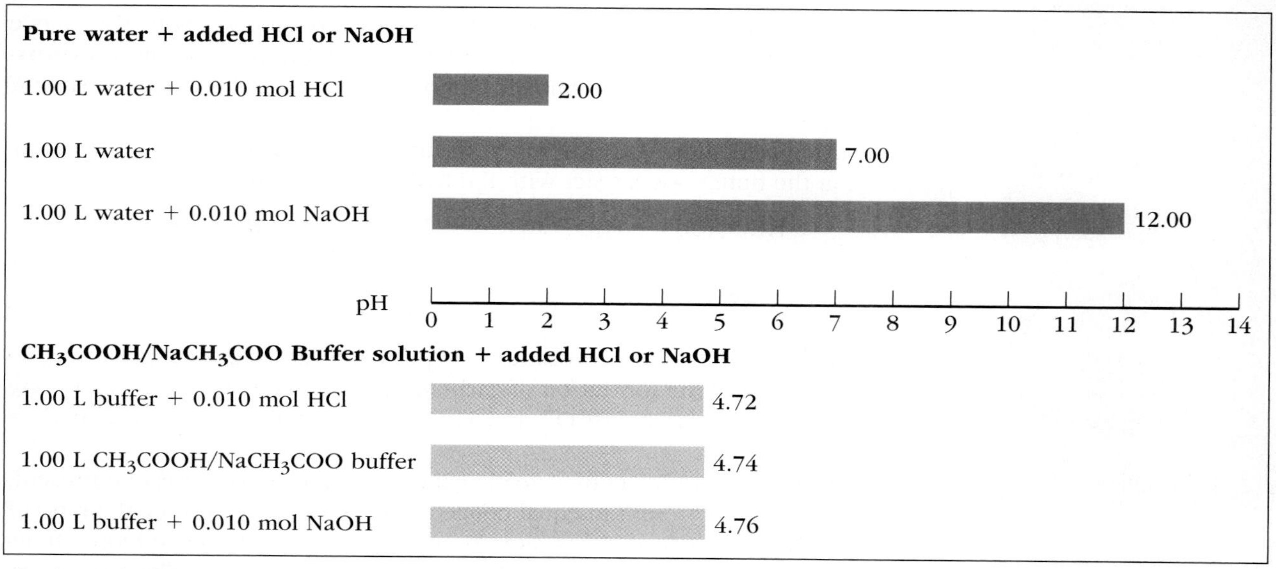

Figure 17.1 pH changes. The pH of water changes dramatically when 0.010 mol of acid (HCl) or base (NaOH) is added to it. By contrast, the pH of a sodium acetate-acetic acid buffer containing 0.50 mol acetic acid and 0.50 mol sodium acetate remains relatively constant when 0.010 mol HCl or NaOH is added to it. Only a slight change in pH occurs.

change occurs. For example, acetic acid reacts with acetate ion to form acetate ion and acetic acid.

$$CH_3COOH(aq) + CH_3COO^-(aq) \rightleftharpoons CH_3COO^-(aq) + CH_3COOH(aq)$$

CONJ ACID   conj base   CONJ BASE   CONJ BASE

A buffer usually consists of approximately equal quantities of a weak acid and its conjugate base, or a weak base and its conjugate acid.

The blood of a mammal is an aqueous solution that maintains a constant pH, that is, a buffer. Consider human blood, whose normal pH is 7.40 ± 0.05. If the pH decreases below 7.35, a condition known as *acidosis* occurs; increasing the pH above 7.45 causes *alkalosis.* Both conditions can be life-threatening. Acidosis, for example, causes a decrease in oxygen transport by hemoglobin and depresses the central nervous system, leading in extreme cases to coma and death by creating weak and irregular cardiac contractions—symptoms of heart failure. To prevent such problems, your body must keep the pH of your blood nearly constant.

The term acidosis is used medically even though the pH is not less than 7.0.

Carbon dioxide provides the most important blood buffer (but not the only one). In solution, CO_2 reacts with water to form H_2CO_3, which ionizes to produce H_3O^+ and HCO_3^- ions. The equilibria are

$$CO_2(aq) + H_2O(\ell) \rightleftharpoons H_2CO_3(aq)$$

$$H_2CO_3(aq) + H_2O(\ell) \rightleftharpoons H_3O^+(aq) + HCO_3^-(aq)$$

H_2CO_3: $K_a = 4.3 \times 10^{-7}$
HCO_3^-: $K_b = 2.3 \times 10^{-8}$

Since H_2CO_3 is a weak acid and HCO_3^- is its conjugate weak base, together they constitute a buffer. The normal concentrations of H_2CO_3 and HCO_3^- in blood are 0.0025 M and 0.025 M, respectively—a 1:10 ratio. As long as the ratio of H_2CO_3 to HCO_3^- concentrations remains about 1 to 10, the pH of the blood remains near 7.4. (We will calculate this pH in Problem-Solving Example 17.1.)

If a strong base such as NaOH is added to this buffer, carbonic acid—the conjugate acid—in the buffer will react with the added OH^-.

$$H_2CO_3(aq) + OH^-(aq) \longrightarrow HCO_3^-(aq) + H_2O(\ell)$$

$$K = \frac{1}{K_b(HCO_3^-)} = \frac{1}{2.3 \times 10^{-8}} = 4.3 \times 10^{7}$$

Here the equilibrium constant is $1/K_b$ of hydrogen carbonate ion, because the reaction is the reverse of the reaction of hydrogen carbonate ion with H_2O. Since OH^- is the strongest base that can exist in water solution, the reaction of hydroxide ions with carbonic acid is essentially complete, as indicated by the very large K value of 4.3×10^7.

If a strong acid such as HCl is added to this buffer, the HCO_3^- ion—the conjugate base in the buffer—will react with the hydronium ions from HCl.

$$HCO_3^-(aq) + H_3O^+(aq) \longrightarrow H_2CO_3(aq) + H_2O(\ell)$$

$$K = \frac{1}{K_a(H_2CO_3)} = \frac{1}{4.3 \times 10^{-7}} = 2.3 \times 10^{6}$$

In this case, the equilibrium constant is $1/K_a$ of carbonic acid, because the reaction is the reverse of the ionization of carbonic acid. Since H_3O^+ is such a strong acid, the reaction between HCO_3^- and H_3O^+ is essentially complete and $K = 2.3 \times 10^6$.

To form a buffer, both a conjugate acid and its conjugate base must be present, but they need not be present in equal concentrations. In many buffers, the ratio of conjugate base : conjugate acid concentrations is about 1 : 1. In human blood, however, the $[HCO_3^-]/[H_2CO_3]$ ratio is about 10 to 1, with good reason. There are more acidic by-products of metabolism in the blood that must be neutralized than there are basic ones.

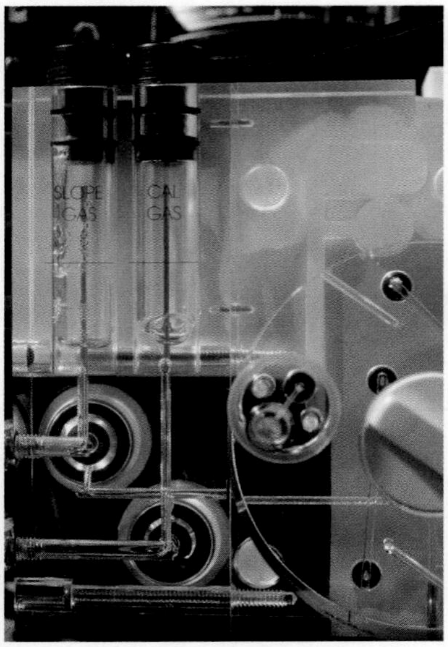

Hank Morgan/Photo Researchers, Inc.

A blood gas analyzer. This instrument measures CO_2 level in blood, blood pH, and oxygen level. These values are related and must be within a narrow range for good health.

CONCEPTUAL EXERCISE 17.1 Possible Buffers?

Could a solution of equimolar amounts of HCl and NaCl be a buffer? What about a solution of equimolar amounts of KOH and KCl? Explain each of your answers.

Answers to **EXERCISES** are provided at the back of this book in Appendix L.

EXERCISES that are labeled **CONCEPTUAL** are designed to test your understanding of one or more concepts; they usually involve qualitative rather than quantitative thinking.

The pH of Buffer Solutions

The pH of a buffer solution can be calculated in two ways:

- By using $[H_3O]^+$ in the K_a expression if the K_a value and the concentrations of conjugate acid and conjugate base are known (Problem-Solving Example 17.1)
- By using the Henderson-Hasselbalch equation (Problem-Solving Example 17.2)

PROBLEM-SOLVING EXAMPLE 17.1 The pH of a Buffer from K_a

To mimic a blood buffer, a scientist prepared 1.00 L buffer containing 0.0025 mol carbonic acid and 0.025 mol hydrogen carbonate ion. Calculate the pH of the buffer. The K_a of carbonic acid is 4.3×10^{-7}.

Answer 7.37

Strategy and Explanation Carbonic acid (conjugate acid) and hydrogen carbonate ion (conjugate base) are a conjugate acid-conjugate base pair. Use the ionization constant equation for carbonic acid and its K_a value and expression to find $[H_3O^+]$ and then pH.

$$H_2CO_3(aq) + H_2O(\ell) \rightleftharpoons H_3O^+(aq) + HCO_3^-(aq)$$

$$K_a = 4.3 \times 10^{-7} = \frac{[H_3O^+][\text{conj base}]}{[\text{conj acid}]} = \frac{[H_3O^+][HCO_3^-]}{[H_2CO_3]}$$

We can set up an "ICE" table such as we have done for other equilibrium calculations.

	H_2CO_3	+	H_2O	$\rightleftharpoons$	H_3O^+	+	HCO_3^-
*I*nitial concentration (mol/L)	0.0025				≈0		0.025
*C*hange as reaction occurs (mol/L)	$-x$				$+x$		$+x$
*E*quilibrium concentration (mol/L)	$0.0025 - x$				$+x$		$0.025 + x$

The symbol ≈ means "approximately equal to."

Given the small value of K_a, we will assume that $x \ll 0.0025$ and, thus, $0.0025 - x \approx 0.0025$ and $0.025 + x \approx 0.025$. These values can be substituted into the K_a expression to calculate $[H_3O^+]$ and then pH.

$$K_a = 4.3 \times 10^{-7} = \frac{x(0.025 + x)}{(0.0025 - x)} \approx \frac{x(0.025)}{(0.0025)}$$

$$x = [H_3O^+] = \frac{(4.3 \times 10^{-7})(0.0025)}{0.025} = 4.3 \times 10^{-8}$$

$$pH = -\log(4.3 \times 10^{-8}) = 7.37$$

Note that $x = H_3O^+ = 4.3 \times 10^{-8}$ M, which is $\ll 0.0025$. Therefore, the assumption that $x \ll 0.0025$ was valid.

☑ **Reasonable Answer Check** From the K_a expression, we see that when $[HCO_3^-] = [H_2CO_3]$, the $[H_3O^+] = K_a$. But, in this example, the concentration of HCO_3^-, the conjugate base, is ten times that of H_2CO_3, the conjugate acid. Therefore, the $[H_3O^+]$ should be ten times less than the K_a, which it is.

The **PROBLEM-SOLVING STRATEGY** in this book is

- Analyze the problem
- Plan a solution
- Execute the plan
- Check that the result is reasonable

Appendix A.1 explains this in detail.

PROBLEM-SOLVING PRACTICE answers are provided at the back of this book in Appendix K.

PROBLEM-SOLVING PRACTICE 17.1

Calculate the pH of blood containing 0.0020-M carbonic acid and 0.025-M hydrogen carbonate ion.

The Henderson-Hasselbalch Equation

The *Henderson-Hasselbalch equation* can be used conveniently to calculate the pH of a buffer containing known concentrations of conjugate base and conjugate acid. The equation can also be applied to determine the ratio of conjugate base to conjugate acid concentrations needed to achieve a buffer of a given pH. This equation is derived by writing the acid ionization constant expression for a weak acid, HA, and solving for $[H_3O^+]$; A^- is the conjugate base of HA.

$$HA(aq) + H_2O(\ell) \rightleftharpoons H_3O^+(aq) + A^-(aq)$$

$$K_a = \frac{[H_3O^+][\text{conj base}]}{[\text{conj acid}]} = \frac{[H_3O^+][A^-]}{[HA]}$$

$$[H_3O^+] = \frac{K_a[\text{conj acid}]}{[\text{conj base}]} = \frac{K_a[HA]}{[A^-]}$$

This result is more useful when expressed in terms of pH. To do so

- Take the base 10 logarithm of each side of this equation to give

$$\log[H_3O^+] = \log K_a + \log\frac{[\text{conj acid}]}{[\text{conj base}]} = \log K_a + \log\frac{[HA]}{[A^-]}$$

- Multiply both sides of the equation by -1 and use the relation $-\log(x) = \log(1/x)$ to get

$$-\log[H_3O^+] = -\log K_a + \log\frac{[\text{conj base}]}{[\text{conj acid}]} = -\log K_a + \log\frac{[A^-]}{[HA]}$$

- Use the definition of pH and define $-\log K_a$ as pK_a (analogous to the definition of pH); the equation becomes the Henderson-Hasselbalch equation.

Because $-\log(x) = \log\left(\frac{1}{x}\right)$, $-\log\frac{[\text{conj acid}]}{[\text{conj base}]} = \log\frac{[\text{conj base}]}{[\text{conj acid}]}$ and $-\log\frac{[HA]}{[A^-]} = \log\frac{[A^-]}{[HA]}$.

$$pH = pK_a + \log\frac{[\text{conj base}]}{[\text{conj acid}]} = pK_a + \log\frac{[A^-]}{[HA]}$$ **Henderson-Hasselbalch equation.**

The Henderson-Hasselbalch equation can be used to determine the blood pH calculated in Problem-Solving Example 17.1, giving the same results.

$$pH = pK_a + \log\frac{[\text{conj base}]}{[\text{conj acid}]} = -\log K_a + \log\frac{[HCO_3^-]}{[H_2CO_3]}$$

$$= -\log(4.3 \times 10^{-7}) + \log\left(\frac{0.0250}{0.00250}\right) = -\log(4.3 \times 10^{-7}) + \log(10)$$

$$= 6.37 + \log(10) = 6.37 + 1.00 = 7.37$$

Note that since the concentration of conjugate base is ten times that of the conjugate acid, the pH should be greater than the pK_a by 1 pH unit, and it is.

It is important to note that application of the Henderson-Hasselbalch equation has its limits. Two important rules for when it can be applied to buffer solutions are

1. The value of the $\frac{[\text{conj base}]}{[\text{conj acid}]}$ ratio *must* be between 0.1 and 10.
2. The [conj base] and [conj acid] must *each* exceed the K_a by a factor of 100 or more.

EXERCISE 17.2 Blood pH

Calculate the pH of blood containing 0.0025-M HPO_4^{2-} (aq) and 0.0015-M $H_2PO_4^-$(aq). K_a of $H_2PO_4^- = 6.3 \times 10^{-8}$. (Assume that this is the only blood buffer.)

PROBLEM-SOLVING EXAMPLE 17.2 A Buffer Solution and Its pH

Calculate the pH of a buffer containing 0.050 mol/L pyruvic acid, $CH_3COCOOH$, and 0.060 mol/L sodium pyruvate, $Na^+CH_3COCOO^-$. K_a of pyruvic acid $= 3.2 \times 10^{-3}$.

Answer 2.57

Strategy and Explanation We solve for the pH in two ways. First, we use the K_a expression to find the H_3O^+ concentration from which the pH can be calculated. Alternatively, we use the Henderson-Hasselbalch equation to determine pH.

Using the K_a expression, we first write the chemical equation for the equilibrium of pyruvic acid with pyruvate ions, and its corresponding K_a expression.

$$CH_3COCOOH(aq) + H_2O(\ell) \rightleftharpoons H_3O^+(aq) + CH_3COCOO^-(aq)$$

$$K_a = 3.2 \times 10^{-3} = \frac{[H_3O^+][CH_3COCOO^-]}{[CH_3COCOOH]} = \frac{[H_3O^+](0.060)}{(0.050)}$$

To calculate $[H_3O^+]$ and then pH, rearrange the equation to give

$$[H_3O^+] = \frac{3.2 \times 10^{-3}(0.050)}{0.060} = 2.67 \times 10^{-3}$$

$$pH = -\log(2.67 \times 10^{-3}) = 2.57$$

Applying the Henderson-Hasselbalch equation:

$$pH = pK_a + \log\frac{[\text{conj base}]}{[\text{conj acid}]} = -\log(3.2 \times 10^{-3}) + \log\frac{[CH_3COCOO^-]}{[CH_3COCOOH]}$$

$$pH = 2.49 + \log\left(\frac{0.060}{0.050}\right) = 2.49 + \log(1.2) = 2.49 + 0.079 = 2.57$$

☑ **Reasonable Answer Check** The concentration of conjugate base is a bit greater than the concentration of conjugate acid. Therefore the pH should be slightly higher than the pK_a, which it is. The answer is reasonable.

PROBLEM-SOLVING PRACTICE 17.2

Calculate the ratio of $[HPO_4^{2-}]$ to $[H_2PO_4^-]$ in blood at a normal pH of 7.40. Assume that this is the only buffer system present. For $H_2PO_4^-$, $K_a = 6.3 \times 10^{-8}$.

From the Henderson-Hasselbalch equation note that when the concentrations of conjugate base and conjugate acid are equal,

$$\frac{[\text{conj base}]}{[\text{conj acid}]} = \frac{[A^-]}{[HA]} = 1 \qquad \log\frac{[\text{conj base}]}{[\text{conj acid}]} = \log\frac{[A^-]}{[HA]} = \log(1) = 0$$

so pH = pK_a. Thus, ***a buffer's pH equals the pK$_a$ of its weak acid when the concentrations of the acid and its conjugate base in the buffer are equal.***

In Problem-Solving Example 17.2, the [conjugate base]/[conjugate acid] ratio was > 1.0, so pH > pK_a.

Selecting an Appropriate Conjugate Acid-Base Buffer Pair

A buffer for maintaining a desired pH can be chosen readily by examining pK_a values, which are often tabulated along with K_a values. *Choose a conjugate acid-base pair whose conjugate acid has a pK$_a$ near the desired pH.* Table 17.1 lists pK_a

Table 17.1 Buffer Systems That Are Useful at Various pH Values

Desired pH	Weak Conjugate Acid	Weak Conjugate Base	K_a of Weak Conjugate Acid	pK_a
4	Lactic acid, $CH_3CHOHCOOH$	Lactate ion, $CH_3CHOHCOO^-$	1.4×10^{-4}	3.85
5	Acetic acid, CH_3COOH	Acetate ion, CH_3COO^-	1.8×10^{-5}	4.74
6	Carbonic acid, H_2CO_3	Hydrogen carbonate ion, HCO_3^-	4.3×10^{-7}	6.37
7	Dihydrogen phosphate ion, $H_2PO_4^-$	Monohydrogen phosphate ion, HPO_4^{2-}	6.3×10^{-8}	7.20
8	Hypochlorous acid, HClO	Hypochlorite ion, ClO^-	6.8×10^{-8}	7.17
9	Ammonium ion, NH_4^+	Ammonia, NH_3	5.6×10^{-10}	9.25
10	Hydrogen carbonate ion, HCO_3^-	Carbonate ion, CO_3^{2-}	4.7×10^{-11}	10.33

- As K_a decreases, the pK_a increases.
- The weaker the acid, the larger its pK_a.

values of several common acids that could be used with their conjugate bases to prepare buffers over the pH range from 4 to 10.

To have comparable quantities of both acid and conjugate base in a buffer solution, the ratio of conjugate base to conjugate acid cannot get much smaller than 1 : 10 or much bigger than 10 : 1. ***The pH range of a buffer is limited to about one pH unit above or below the pK_a of the conjugate acid.*** In the carbonic acid/hydrogen carbonate case, for example, that would be a pH from 5.37 to 7.37 because $pK_a = -\log K_a = -\log 4.3 \times 10^{-7} = 6.37$. Other conjugate acid-base pairs, such as those in Table 17.1, can be used to prepare buffers with much different pH ranges, as determined by the pK_a value of the acid in the buffer (Table 17.1).

PROBLEM-SOLVING EXAMPLE 17.3 Selecting an Acid-Base Pair for a Buffer Solution of Known pH

You are doing an experiment that requires a buffer solution with a pH of 4.00. Available to you are solutions used to make buffers with these acid-base pairs: CH_3COOH/CH_3COO^-; H_2CO_3/HCO_3^-; and $CH_3CHOHCOOH/CH_3CHOHCOO^-$. Which acid-base pair should you use to make a pH 4.00 buffer solution, and what mole ratio of the compounds should you use? Use K_a and pK_a values from Table 17.1.

Answer The $CH_3CHOHCOOH/CH_3CHOHCOO^-$ conjugate acid-base pair with a mole ratio of 0.10 mol/L lactic acid to 0.14 mol/L lactate would work.

Strategy and Explanation Because you want to have a buffer with a pH of 4.00, you need an acid-base pair whose conjugate acid has a pK_a near the desired pH of 4.00. You can use Table 17.1 to evaluate and select the conjugate acid-base pair whose pK_a is closest to pH = 4.00. Once that acid-base pair has been selected, you can use the Henderson-Hasselbalch equation to calculate the necessary conjugate acid and conjugate base concentrations to give a buffer with pH 4.00.

Lactic acid, $CH_3CHOHCOOH$, with $pK_a = 3.85$, is closest to the target pH of 4.00; the pK_a values of the other acids are not close enough to 4.00.

Substituting this pK_a and the lactic acid–lactate conjugate acid-base pair concentration terms into the Henderson-Hasselbalch equation gives

$$pH = 4.00 = 3.85 + \log \frac{[CH_3CHOHCOO^-]}{[CH_3CHOHCOOH]}$$

$$\log \frac{[CH_3CHOHCOO^-]}{[CH_3CHOHCOOH]} = 4.00 - 3.85 = 0.15; \frac{[CH_3CHOHCOO^-]}{[CH_3CHOHCOOH]} = 10^{0.15} = 1.41$$

This means that the required concentration of lactate ion, $CH_3CHOHCOO^-$, will be roughly 1.41 times that of lactic acid, $CH_3CHOHCOOH$. The buffer could be made using lactic acid and a soluble lactate salt, such as sodium lactate, $Na^+CH_3CHOHCOO^-$. For example, if the concentration of lactic acid is 0.10 mol/L, the required concentration of sodium lactate is 1.41(0.10 mol/L) = 0.14 mol/L.

☑ **Reasonable Answer Check** The target pH is 4.00. We can see from the Henderson-Hasselbalch equation that with equal concentrations of conjugate acid and base, the pH would equal the pK_a, 3.85. Therefore, to reach the target pH of 4.00, the concentration of lactate ion, the conjugate base, must be greater than that of lactic acid, the conjugate acid. Adding more conjugate base than conjugate acid raises the pH of the buffer pair from 3.85 to 4.00.

PROBLEM-SOLVING PRACTICE 17.3

Use the data in Table 17.1 to select a conjugate acid-base pair you could use to make buffer solutions having each of these hydrogen ion concentrations.

(a) 3.2×10^{-4} M (b) 5.0×10^{-5} M

(c) 7.0×10^{-8} M (d) 6.0×10^{-11} M

EXERCISE 17.3 Making a Buffer Solution

Use data from Table 17.1 to calculate the molar ratio of sodium acetate and acetic acid needed to make a buffer of pH 4.68.

EXERCISE 17.4 Buffers and pH

Use data from Table 17.1 to calculate the pH of these buffers.

(a) H_2CO_3 (0.10 M)/HCO_3^- (0.25 M) (b) $H_2PO_4^-$ (0.10 M)/HPO_4^{2-} (0.25 M)

Addition of Acid or Base to a Buffer

The pH of a buffer changes when acid or base is added to the buffer due to the changes in the amounts of conjugate acid and conjugate base in the buffer. **When acid (H_3O^+) is added to a buffer, the acid reacts with the conjugate base of the buffer to form its conjugate acid:**

Conjugate base in buffer + H_3O^+ added ⟶ conjugate acid + water

When base (OH^-) is added to a buffer, the base reacts with the conjugate acid of the buffer, which is converted into its conjugate base:

Conjugate acid in buffer + OH^- added ⟶ conjugate base + water

Figure 17.2 summarizes these relationships. For a buffer made from acetic acid and sodium acetate, the changes are

$$CH_3COO^-(aq) + H_3O^+(aq) \longrightarrow CH_3COOH(aq) + H_2O(\ell)$$

Conjugate base of the buffer	added	Conjugate acid of the buffer

$$CH_3COOH(aq) + OH^-(aq) \longrightarrow CH_3COO^-(aq) + H_2O(\ell)$$

Conjugate acid of the buffer	added	Conjugate base of the buffer

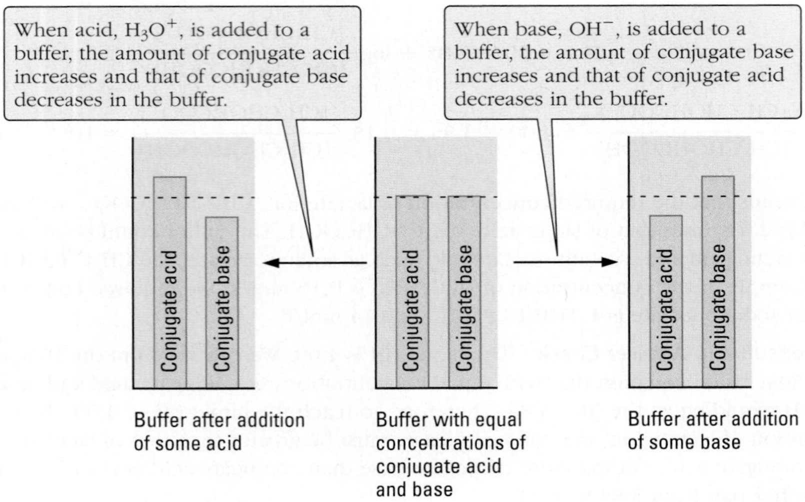

Figure 17.2 The effects of adding acid or base to a buffer. When H_3O^+ or OH^- ions are added to a buffer, the amounts of conjugate acid and conjugate base in the buffer change.

The pH Change on Addition of Acid or Base to a Buffer

The extent of the pH change depends on the amount of acid or base added and on the amounts of conjugate acid or conjugate base remaining in the buffer to offset additional amounts of acid or base to be added. As shown in Problem-Solving Example 17.4, the change in pH of a buffer when acid or base is added to it can be calculated in two ways: (1) by using the K_a expression and the K_a value, or (2) by using the Henderson-Hasselbalch equation and pK_a.

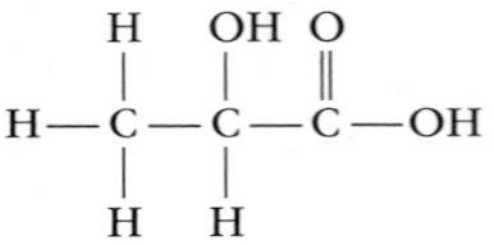

lactic acid

PROBLEM-SOLVING EXAMPLE 17.4 pH Changes in a Buffer

Sodium lactate, $Na^+CH_3CHOHCOO^-$, and lactic acid, $CH_3CHOHCOOH$, are used to prepare three buffer solutions. Each of the three buffer solutions differs in its concentration of conjugate base (lactate ion) and conjugate acid (lactic acid). The concentrations are

Buffer	[Lactate$^-$], mol/L	[Lactic acid], mol/L
I	0.15	0.15
II	0.20	0.15
III	0.15	0.25

From Table 17.1, K_a lactic acid = 1.4×10^{-4}; pK_a = 3.85. For 1.00 L of each buffer,

(a) Calculate the initial pH of *each* buffer solution.

(b) Calculate the pH of *each* buffer after 0.050 mol HCl has been added separately to each buffer (neglect volume changes).

(c) Calculate the pH of *each* buffer after 0.10 mol NaOH has been added separately to each buffer (neglect volume changes).

Answer (a) pH: Buffer I = 3.85; II = 3.97; III = 3.63; (b) pH: Buffer I = 3.55; II = 3.85; III = 3.15; (c) pH: Buffer I = 4.45; II = 4.37; III = 3.89

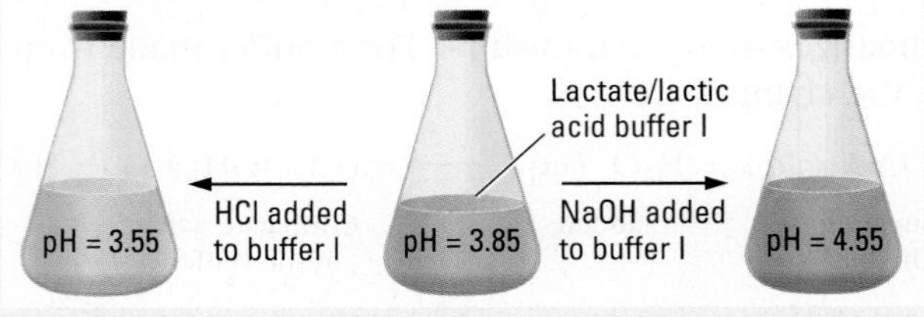

Comparison of pH of lactate/lactic acid buffer after addition of HCl or NaOH.

Strategy and Explanation These calculations can be done in one of two ways:

- Use the K_a value and expression to calculate H_3O^+ concentration and pH.
- Apply the Henderson-Hasselbalch equation.

The same chemical equation, K_a expression, and K_a value apply for *each* buffer:

$$CH_3CHOHCOOH(aq) + H_2O(\ell) \rightleftharpoons H_3O^+(aq) + CH_3CHOHCOO^-(aq)$$

$$K_a = 1.4 \times 10^{-4} = \frac{[H_3O^+][CH_3CHOHCOO^-]}{[CH_3CHOHCOOH]}$$

Solving for $[H_3O^+]$ gives

$$[H_3O^+] = (1.4 \times 10^{-4})\frac{[CH_3CHOHCOOH]}{[CH_3CHOHCOO^-]}$$

The pH of *each* buffer can also be calculated using the Henderson-Hasselbalch equation.

$$pH = pK_a + \log\frac{[CH_3CHOHCOO^-]}{[CH_3CHOHCOOH]}$$

An example calculation is provided for Buffer I using both methods.

(a) For Buffer I, $[CH_3CHOHCOO^-] = 0.15$ mol/L and $[CH_3CHOHCOOH] = 0.15$ mol/L; therefore, using the K_a expression, we have

$$[H_3O^+] = 1.4 \times 10^{-4}\ M \times \frac{0.15\ M}{0.15\ M} = 1.4 \times 10^{-4}\ M$$

$$pH = -\log(1.4 \times 10^{-4}) = 3.85$$

Using the Henderson-Hasselbalch equation gives

$$pH = pK_a + \log\frac{0.15\ M}{0.15\ M} = 3.85 + \log(1.00) = 3.85$$

The initial pH values of Buffers II and III can be calculated similarly using either method.

(b) In the case of each buffer, the 0.050 mol HCl added reacts with 0.050 mol lactate ion to form 0.050 mol lactic acid and water.

$$CH_3CHOHCOO^-(aq) + H_3O^+(aq) \rightleftharpoons CH_3CHOHCOOH(aq) + H_2O(\ell)$$

0.050 mol from buffer	0.050 mol added from HCl	0.050 mol formed by reaction

This changes the original lactate/lactic acid ratio in each buffer, so the pH changes because of the added HCl. For *Buffer I*:

	Moles Lactate	Moles Lactic Acid
Before reaction	0.15	0.15
After reaction	0.15 − 0.050 = 0.10	0.15 + 0.050 = 0.20

$$[H_3O^+] = 1.4. \times 10^{-4}\,\frac{[CH_3CHOHCOOH]}{[CH_3CHOHCOO^-]}$$

$$= \frac{[H_3O^+][0.10]}{[0.20]} = [H_3O^+](0.50)$$

$$[H_3O^+] = \frac{1.4 \times 10^{-4}}{0.50} = 2.8 \times 10^{-4}$$

$$pH = -\log(2.8 \times 10^{-4}) = 3.55$$

The Henderson-Hasselbalch equation may be used to calculate the pH directly.

$$pH = pK_a + \log\frac{[CH_3CHOHCOO^-]}{[CH_3CHOHCOOH]} = 3.85 + \log\frac{(0.10)}{(0.20)} = 3.85 + (-0.30) = 3.55$$

The pH drops from 3.85 to 3.55. Note that 0.10 mol lactate remains to react with additional acid that might be added.

(c) To offset the addition of 0.10 mol NaOH requires 0.10 mol lactic acid from *each* buffer.

$$CH_3CHOHCOOH(aq) + OH^-(aq) \rightleftharpoons CH_3CHOHCOO^-(aq) + H_2O(aq)$$

0.10 mol from buffer	0.10 mol added from NaOH	0.10 mol formed in buffer by the reaction

For Buffer I

	Moles Lactic Acid	Moles Lactate
Before reaction	0.15	0.15
After reaction	0.15 − 0.10 = 0.05	0.15 + 0.10 = 0.25

Addition of NaOH changes the initial lactate/lactic acid ratio in Buffer I from $\frac{0.15}{0.15} = 1.0$ to $\frac{0.25}{0.050} = 5.0$ and a change in pH occurs.

We calculate the new pH

$$K_a = 1.4 \times 10^{-4} = \frac{[H_3O^+][CH_3CHOHCOO^-]}{[CH_3CHOHCOOH]}$$

$$= \frac{[H_3O^+][0.25]}{[0.05]} = [H_3O^+]\,(5.0)$$

$$[H_3O^+] = \frac{1.4 \times 10^{-4}}{5.0} = 2.8 \times 10^{-5};\ pH = -\log(2.8 \times 10^{-5}) = 4.55$$

Using the Henderson-Hasselbach equation, the pH after NaOH addition to Buffer I is

$$pH = pK_a + \log\frac{[CH_3CHOHCOO^-]}{[CH_3CHOHCOOH]} = 3.85 + \log\frac{(0.25)}{(0.05)} = 3.85 + 0.70 = 4.55$$

☑ **Reasonable Answer Check** (a) The pH of the initial buffer, 3.85, is reasonable because the concentrations of lactate (conjugate base) and lactic acid (conjugate acid) are the same, therefore $pH = pK_a$. The initial pH of Buffer II should be a bit higher than 3.85 because the lactate/lactic acid ratio is > 1. For Buffer III, the opposite case is present; the initial lactate/lactic acid ratio is < 1 and the pH should be lower than 3.85. (b) When acid is added to the buffer, we expect a decrease in pH, and pH does decrease in each of the three cases. (c) More moles of NaOH were added than moles of HCl, so the pH changes should be greater than for the addition of HCl, and they are. As expected upon addition of base to a buffer, the pH rose, but by less than 1 pH unit in each case. The answers are reasonable.

PROBLEM-SOLVING PRACTICE 17.4

For the three lactate/lactic acid buffer solutions just given, calculate the pH when each quantity of acid or base is added separately to each buffer (neglect volume changes).
(a) 0.075 mol HCl (b) 0.065 mol NaOH

CONCEPTUAL EXERCISE 17.5 Blood Buffer Reaction

If an abnormally high CO_2 concentration is present in blood, which phosphorus-containing ion, $H_2PO_4^-$ or HPO_4^{2-}, can counteract the presence of excess CO_2? Explain your answer.

Buffer Capacity

The *amounts* of conjugate acid and conjugate base in the buffer solution determine the **buffer capacity**—the quantity of added acid or base that the buffer can accommodate without undergoing significant pH change (more than 1 pH unit).

When the conj base:conj acid ratio changes to 1:10, the pH *decreases* by 1 unit. When the ratio changes to 10:1, the pH *increases* by 1 unit.

- When nearly all of the conjugate acid in a buffer has reacted with added base, adding just a little more base can increase the pH significantly because there is not enough conjugate acid left in the buffer to neutralize the added base.
- If enough acid is added to a buffer to react with all of the buffer's conjugate base and excess unreacted acid remains, the pH will decrease significantly.
- In each of the cases above, the buffer capacity has been exceeded.

For example, 1 L of a buffer solution that is 0.25 M in CH_3COOH and 0.25 M in CH_3COO^- contains 0.25 mol CH_3COOH and 0.25 mol CH_3COO^-. This buffer can accommodate the addition of up to 0.25 mol H_3O^+ or OH^-, at which point it has used up its buffer capacity. Thus, the initial buffer *cannot* accommodate the addition of 0.30 mol of strong acid *or* 0.30 mol of strong base without undergoing a major change in pH. Such additions would use up all of the buffer's conjugate base or all of its conjugate acid, respectively, and exceed the buffer's capacity. The pH would drop or rise accordingly.

PROBLEM-SOLVING EXAMPLE 17.5 Buffer Capacity

A buffer is prepared using 0.25 mol $H_2PO_4^-$ and 0.15 mol HPO_4^{2-} in 500. mL of solution.
(a) Determine whether the buffer capacity will be exceeded if 6.2 g KOH is added. Calculate the pH of the solution after the addition.
(b) Determine whether the buffer capacity will be exceeded if 23.0 mL of 6.0-M HCl is added to the original buffer. Calculate the pH of the resulting solution.

Answer (a) No, the buffer capacity will not be exceeded; pH = 6.81. (b) Yes, the buffer capacity is exceeded; pH = 4.95.

Strategy and Explanation The initial pH of the buffer can be calculated by substituting the appropriate concentrations into the Henderson-Hasselbalch equation. From Table 17.1, the pK_a of $H_2PO_4^-$ is 7.20.

$$\text{pH} = 7.20 + \log\frac{[HPO_4^{2-}]}{[H_2PO_4^-]} = 7.20 + \log\frac{(0.15/0.500)}{(0.25/0.500)}$$

$$= 7.20 + \log(0.60) = 7.20 - 0.22 = 6.98$$

(a) The added OH^- ions are neutralized by $H_2PO_4^-$ ions, the conjugate acid in the buffer. The 6.2 g KOH contributes 0.11 mol OH^- to be neutralized.

$$6.2\text{ g KOH}\left(\frac{1\text{ mol KOH}}{56.1\text{ g KOH}}\right)\left(\frac{1\text{ mol OH}^-}{1\text{ mol KOH}}\right) = 0.11\text{ mol OH}^-$$

The 0.11 mol OH^- added to the buffer is neutralized by reacting with 0.11 mol $H_2PO_4^-$, to form 0.11 mol HPO_4^{2-}.

$$H_2PO_4^-(aq) + OH^- \longrightarrow HPO_4^{2-}(aq) + H_2O(\ell)$$

0.11 mol from buffer	0.11 mol from KOH	0.11 mol formed

The reaction changes the amounts of HPO_4^{2-} and $H_2PO_4^-$ remaining in the buffer:

	Before Reaction	After Reaction
Moles HPO_4^{2-}	0.15	0.15 + 0.11 = 0.26
Moles $H_2PO_4^-$	0.25	0.25 − 0.11 = 0.14

Because some $H_2PO_4^-$ still remains (0.14 mol), the buffer's capacity was not exceeded by adding 6.2 g KOH. However, after the KOH addition, the pH changes because the addition of base changed the $[HPO_4^{2-}]/[H_2PO_4^-]$ ratio:

$$\text{pH} = 7.20 + \log\frac{[HPO_4^{2-}]}{[H_2PO_4^-]} = 7.20 + \log\frac{(0.26/0.500)}{(0.14/0.500)}$$

$$= 7.20 + \log(1.9) = 7.20 + 0.27 = 7.47$$

(b) The added H_3O^+ ions react with HPO_4^{2-} ions, the conjugate base of the buffer. Adding 0.0200 L of 6.0-M HCl provides:

$$(0.0230\text{ L})\left(\frac{6.0\text{ mol HCl}}{\text{L}}\right)\left(\frac{1\text{ mol H}_3\text{O}^+}{1\text{ mol HCl}}\right) = 1.4 \times 10^{-1}\text{ mol H}_3\text{O}^+$$

$$HPO_4^{2-}(aq) + H_3O^+(aq) \longrightarrow H_2PO_4^-(aq) + H_2O(\ell)$$

0.14 mol from buffer	0.14 mol from HCl	0.14 mol formed

	Before Reaction	After Reaction
Moles HPO_4^{2-}	0.15	0.15 − 0.14 = 0.01
Moles $H_2PO_4^-$	0.25	0.25 + 0.14 = 0.39

Thus, the composition of the buffer changes. The number of moles of HPO_4^{2-} *decreases* to 0.01 mol; the moles of $H_2PO_4^-$ *increase* from 0.25 to 0.39. The volume of the resulting solution is 0.523 L due to the addition of 0.0230 L HCl. The pH after addition of HCl is

$$\text{pH} = 7.20 + \log\frac{[HPO_4^{2-}]}{[H_2PO_4^-]} = 7.20 + \log\frac{(0.01/0.523)}{(0.39/0.523)}$$

$$= 7.20 + \log(0.026) = 7.20 - 1.59 = 5.61$$

The pH changed by 1.37 units, from 6.98 to 5.61, showing that the buffer capacity was exceeded by the addition of that much HCl. The final conjugate base: conjugate acid ratio was 0.026, smaller than 1:10.

The pH calculated using the K_a method is the same as determined here using the Henderson-Hasselbalch equation, even though the conjugate base: conjugate acid ratio is <0.1. This result is because $\left(\frac{0.01}{0.523} - x\right)\Big/\left(\frac{0.39}{0.523} - x\right)$ is essentially $\frac{(0.01/0.523)}{(0.39/0.523)}$ since $x << \left(\frac{0.01}{0.523}\right)$.

☑ **Reasonable Answer Check**

(a) Addition of base to a buffer should increase the pH to some extent depending on the amount of base added. The amount of KOH (0.11 mol) added was less than the amount of conjugate acid (0.25 mol) in the initial buffer. The conjugate base: conjugate acid ratio changed from 0.60 to 1.9, a change that did not exceed ten. Therefore, the buffer's capacity was not exceeded. The pH change should be less than 1 pH unit, which it is.

(b) The amount of acid added (0.14 mol) lowered the amount of conjugate base remaining in the buffer to 0.01 mol, while the amount of conjugate acid increased to 0.39 mol, thereby making the conjugate acid : conjugate base ratio greater than ten. Consequently, the pH rose by more than 1 pH unit and the buffer capacity was exceeded.

PROBLEM-SOLVING PRACTICE 17.5

Calculate the minimum mass (grams) of KOH that would have to be added to the initial buffer in Problem-Solving Example 17.5 to exceed its buffer capacity.

17.2 Acid-Base Titrations

In Section 5.8 ***(← p. 198)*** acid-base titrations were described as a method by which the concentration of an acid or a base could be determined. An acid-base titration is carried out by slowly adding a measurable amount of an aqueous solution of a base or acid whose concentration is known to a known volume of an aqueous acid or base whose concentration is to be determined. For example, a standardized base could be added from a buret to a known volume of acid whose concentration is to be determined, as in Figure 17.3. The solution in the buret is known as the **titrant**, which is a *standard solution* ***(← p. 198)***, one whose concentration is known accurately. The titrant is added into the flask from the buret, a device that allows the required volume of standardized solution (titrant) to be measured accurately. The *equivalence point* ***(← p. 198)*** is reached when the stoichiometric amount of titrant has been added, the amount that exactly neutralizes the acid or base being titrated.

It is also possible to determine the concentration of the titrant, provided that the concentration of the solution to which it is added is known.

© Cengage Learning/Charles D. Winters

(a) A buret calibrated in 0.1-mL divisions contains NaOH(aq) of known concentration; the flask contains an acid sample of unknown concentration and a few drops of phenolphthalein indicator.

(b) The NaOH(aq) is added slowly from the buret to the flask; where the base has reacted with the acid, the phenolphthalein indicator turns pink. Stirring the solution removes the pink color.

(c) When all of the acid has reacted, stirring no longer causes the color to disappear. A very slightly pink solution indicates that exactly enough base has been added to react with all of the acid. This is called the equivalence point.

Figure 17.3 Laboratory setup for titrating an acid sample (in the flask) with NaOH(aq) (in the buret).

Detection of the Equivalence Point

For an acid-base titration to be done successfully, a method is needed to detect the equivalence point. This can be done by using a pH meter (Section 19.7), a device that electronically monitors the pH of the solution as the titration proceeds. Alternatively, the color change of an acid-base indicator can be used to detect when sufficient titrant has been added (Figure 17.3). The **end point** of a titration occurs when the indicator changes color. An indicator must be used that gives an end point very close to the equivalence point. The color changes of three acid-base indicators are shown in Figure 17.4.

Acid-base indicators, described in Section 16.4 (← ***p. 766***) are typically weak organic acids (HIn) that differ in color from their conjugate bases (In^-).

$$\underset{\text{Color 1}}{HIn(aq)} + H_2O(\ell) \rightleftharpoons H_3O^+(aq) + \underset{\text{Color 2}}{In^-(aq)}$$

Removal of an H^+ from the indicator molecule changes its structure so that the indicator molecule absorbs light in a different region of the visible spectrum, thus changing the color of the indicator. For methyl red the reaction is

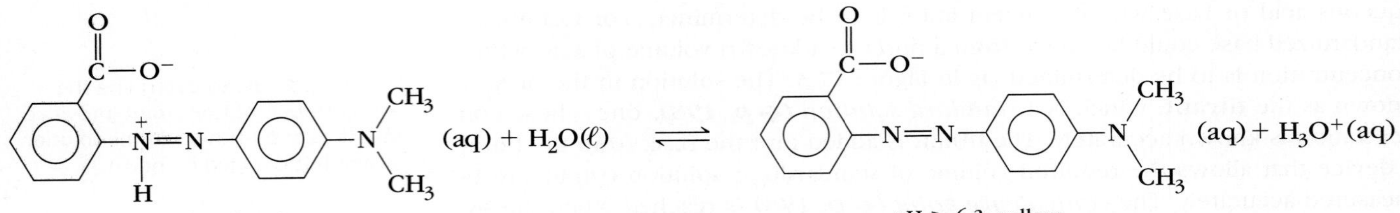

pH ≤ 4; red   pH ≥ 6.3; yellow

Methyl red changes color from red to yellow as the pH changes from 4 to 6.3 (Figure 17.4a); bromthymol blue, for example, changes from yellow to blue as the pH changes from 6 to 8 (Figure 17.4b).

The color observed for the indicator during an acid-base titration depends on the ratio $\frac{[HIn]}{[In^-]}$, for which three cases apply:

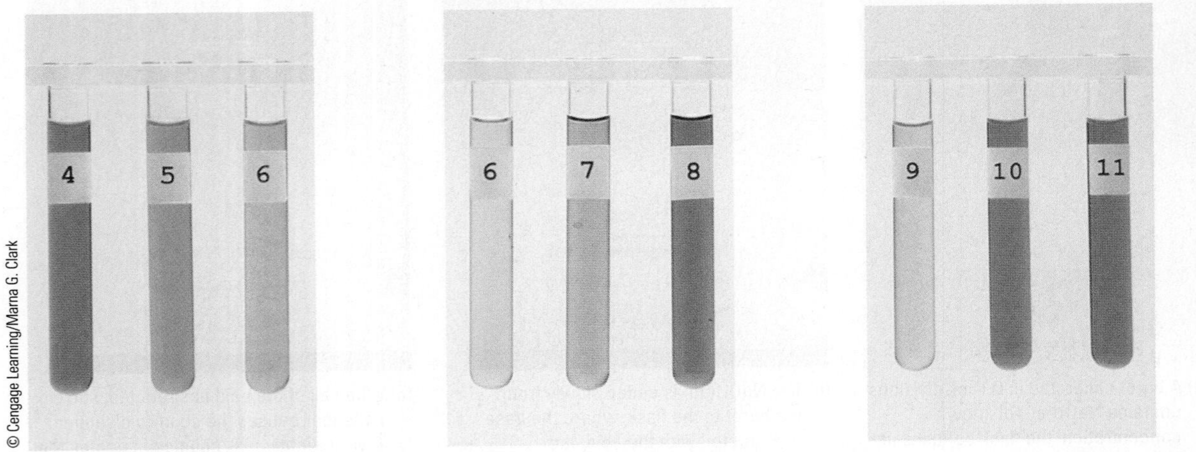

© Cengage Learning/Marna G. Clark

(a) Methyl red is red at pH 4 or lower, orange at pH 5, and yellow at pH 6.3 and higher.

(b) Bromthymol blue changes from yellow to blue as the pH changes from 6 to 8.

(c) Phenolphthalein is colorless below a pH of 8.3 and changes to red between 8.3 and 11.

Figure 17.4 Acid-base indicators. Acid-base indicators are compounds that change color in a particular pH range.

- When $\frac{[HIn]}{[In^-]} \geq 10$, the indicator solution is the acid color (HIn).
- When $\frac{[HIn]}{[In^-]} \leq 0.1$, the indicator solution is the conjugate base color (In^-).
- When $\frac{[HIn]}{[In^-]} \approx 1$, the indicator solution color is intermediate between the acid and the conjugate base colors.

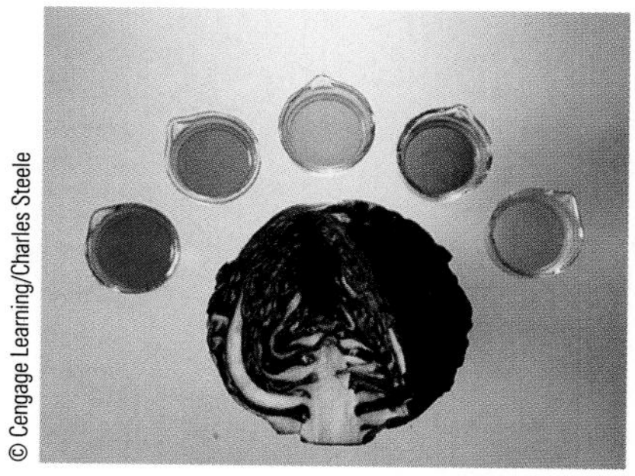

Red cabbage juice is a naturally occurring acid-base indicator. From left to right are solutions of pH 1, 4, 7, 10, and 13, respectively.

We will consider in some detail three types of acid-base titrations:

1. A strong acid, HCl, titrated by a strong base, NaOH.
2. A weak acid, CH_3COOH, titrated by a strong base, NaOH.
3. A weak base, NH_3, titrated by a strong acid, HCl.

For each titration we will examine its **titration curve,** a graph of pH as a function of the volume of titrant added. In each case, we will be interested in the pH particularly at four stages of the titration:

- Prior to the addition of titrant
- After addition of titrant, but prior to the equivalence point
- At the equivalence point
- After the equivalence point

Titration of a Strong Acid with a Strong Base

The titration curve for the titration of 50.0 mL of 0.100-M HCl, a strong acid, with 0.100-M NaOH, a strong base, as the titrant, is given in Figure 17.5. *The titration of a strong acid with a strong base produces a neutral salt and pH = 7.0 at the equivalence point.*

In general, prior to the equivalence point in a strong acid-strong base titration, the $[H_3O^+]$ can be calculated for any volume of base added by the relation

Equation 17.1 assumes that the added volumes equal the exact volume of the mixed solutions.

$$[H_3O^+] = \frac{\text{original moles acid} - \text{total moles base added}}{\text{volume acid (L)} + \text{volume base added (L)}} \qquad [17.1]$$

Problem-Solving Example 17.6 illustrates calculations for the four points marked on the curve.

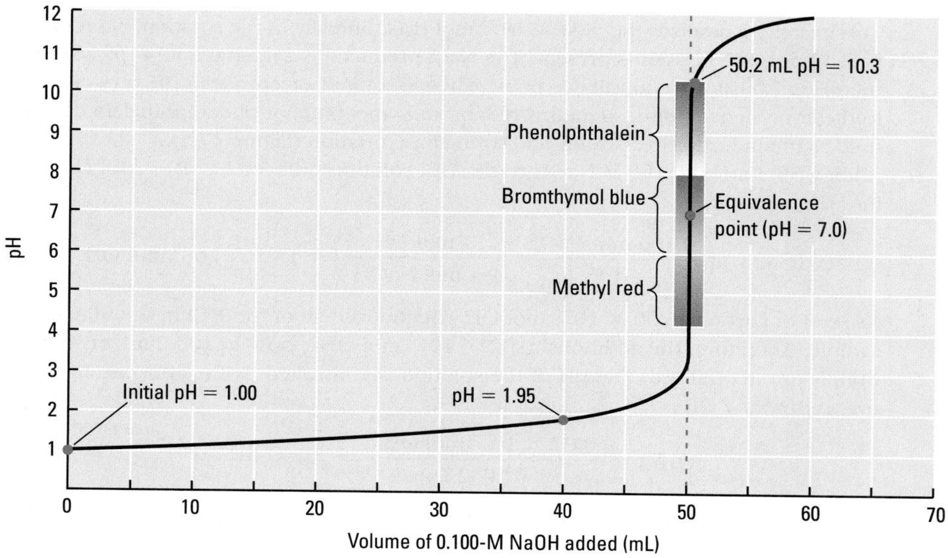

Figure 17.5 Curve for titration of 0.100-M HCl with 0.100-M NaOH. This strong acid reacts with this strong base to form a solution with a pH of 7.0 at the equivalence point. Any one of the three indicators can be used for a strong acid-strong base titration.

PROBLEM-SOLVING EXAMPLE 17.6 Titration of HCl with NaOH

A 0.100-M NaOH solution is used to titrate 50.0 mL of 0.100-M HCl. Calculate the pH of the solution at these four points:

(a) Before any titrant is added
(b) After 40.0 mL of titrant has been added
(c) After 50.0 mL of NaOH has been added. What indicator—methyl red, bromthymol blue, or phenolphthalein—can be used to detect the equivalence point? See Figure 17.4.
(d) After 50.2 mL NaOH has been added

Answer

(a) 1.00 (b) 1.95
(c) 7.00; methyl red, bromthymol blue, or phenolphthalein
(d) 10.30

Strategy and Explanation Initially, only hydrochloric acid is present so the pH can be calculated from the concentration of HCl, a strong acid (part a). As the titration proceeds, NaOH is added and the H_3O^+ concentration decreases. In a strong acid-strong base titration, the $[H_3O^+]$ can be calculated prior to the equivalence point for any volume of base added by using Equation 17.1. We will apply that equation in part (b).

We will round the pH values to two significant figures. See Appendix A.6 for treatment of significant figures when using logarithms.

(a) Because HCl is a strong acid, the initial H_3O^+ concentration is 0.100 M and the pH is $-\log(0.100) = 1.00$.

(b) The initial 50.0-mL solution of HCl contains

$$(0.0500\ \text{L})(0.100\ \text{mol/L}) = 5.00 \times 10^{-3}\ \text{mol}\ H_3O^+$$

As NaOH is added, the number of moles of H_3O^+ decreases due to the reaction of added OH^- ions with H_3O^+ ions in the acid.

$$\underset{\text{from HCl}}{H_3O^+(aq)} + \underset{\text{from NaOH}}{OH^-(aq)} \longrightarrow H_2O(\ell)$$

After 40.0 mL of 0.100-M NaOH has been added to the original 50.0 mL of 0.100-M HCl, the $[H_3O^+]$ can be calculated from Equation 17.1:

$$[H_3O^+] = \frac{(5.00 \times 10^{-3}\ \text{mol}) - (4.00 \times 10^{-3}\ \text{mol})}{0.0500\ \text{L} + 0.0400\ \text{L}} = 1.11 \times 10^{-2}\ \text{M}; \qquad \text{pH} = 1.95$$

(c) At the equivalence point, exactly 50.0 mL of 0.100-M NaOH has been added:

$$(0.0500\ \text{L})\left(\frac{0.100\ \text{mol NaOH}}{1\ \text{L}}\right)\left(\frac{1\ \text{mol OH}^-}{1\ \text{mol NaOH}}\right) = 5.00 \times 10^{-3}\ \text{mol OH}^-.$$

This exactly neutralizes the 5.00×10^{-3} mol H_3O^+ initially in the solution. No residual HCl or excess NaOH is present. The NaCl produced is a neutral salt (← ***p. 781***) so the pH is 7.00 at the equivalence point. There is a rapid, steep rise in pH very close to the equivalence point, so methyl red, bromthymol blue, or phenolphthalein can be used as the indicator in a strong acid-strong base titration (Figure 17.5).

(d) Adding 50.2 mL of 0.100-M NaOH to the HCl solution puts 5.02×10^{-3} mol OH^- into the solution:

$$(0.0502\ \text{L})\left(\frac{0.100\ \text{mol NaOH}}{1\ \text{L}}\right)\left(\frac{1\ \text{mol OH}^-}{1\ \text{mol NaOH}}\right) = 5.02 \times 10^{-3}\ \text{mol OH}^-.$$

As seen in part (c), 5.00×10^{-3} mol OH^- neutralized all of the HCl in the initial sample. Therefore, the additional 0.02×10^{-3} mol OH^-, now in 100.2 mL of solution, is not neutralized. The pH of the solution is controlled by the concentration of *un-neutralized* OH^-.

$$[OH^-] = \frac{0.02 \times 10^{-3}\ \text{mol OH}^-}{0.1002\ \text{L}} = 2.0 \times 10^{-4}\ \text{M}$$

$$\text{pOH} = -\log(2.0 \times 10^{-4}) = 3.70; \qquad \text{pH} = 14.00 - \text{pOH} = 14.00 - 3.70 = 10.30$$

A volume of 0.2 mL is approximately 4 drops.

Notice that the addition of just 0.2 mL of excess NaOH to the solution dramatically raises the pH (from 7.0 at the equivalence point to 10.3), as seen from the titration curve (Figure 17.5).

PROBLEM-SOLVING PRACTICE 17.6

For the titration of HCl(aq) with NaOH(aq) described on pp. 819–820, calculate the pH when these volumes of NaOH have been added:
(a) 10.0 mL (b) 25.00 mL (c) 45.0 mL (d) 50.5 mL

EXERCISE 17.6 Titration Curve

Draw the titration curve for the titration of 50.0 mL of 0.100-M NaOH using 0.100-M HCl as the titrant.

Titration of a Weak Acid with a Strong Base

As noted in Section 16.8 *(← p. 782)* the reaction of a weak acid, such as acetic acid, with a strong base, like NaOH, produces a salt—sodium acetate in this case—that has a basic anion. Like other basic anions, the acetate ions react with water to produce hydroxide ions due to the hydrolysis reaction:

$$CH_3COO^-(aq) + H_2O(\ell) \rightleftharpoons CH_3COOH(aq) + OH^-(aq)$$

As a result, *when a weak acid is titrated with a strong base, the pH of the solution at the equivalence point will be greater than 7.0 due to hydrolysis of the basic anion formed by the titration reaction.* The titration curve in Figure 17.6 for the titration of 50.0 mL of 0.100-M acetic acid with 0.100-M NaOH represents this type of titration. Problem-Solving Example 17.7 illustrates the calculations associated with the titration curve.

Notice in Figure 17.6 that the initial pH of 0.100-M acetic acid, 2.87, is higher than that of 0.100-M HCl, 1.00, in Figure 17.5. This is to be expected because acetic acid is a much weaker acid than HCl. Acetic acid is only slightly ionized ($K_a = 1.8 \times 10^{-5}$), and the pH of 0.100-M acetic acid is 2.78, greater than that of 0.100-M HCl, a strong acid. Also notice from Figure 17.6 that the rapidly rising portion of the titration curve near the equivalence point is shorter than it is for the titration of HCl with NaOH (Figure 17.5), limiting which indicators can be used. The equivalence point pH in Figure 17.6 is 8.72, so methyl red is an unsuitable indi-

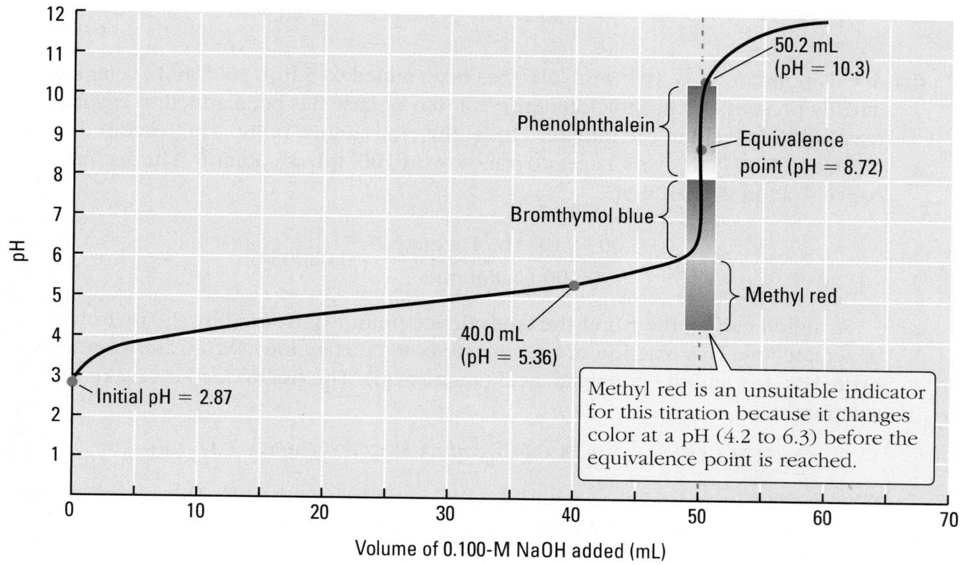

Figure 17.6 Curve for titration of 50.0 mL of 0.100-M acetic acid with 0.100-M NaOH.

cator because its color changes (pH 4 to 6.3) before the equivalence point is reached. Bromthymol blue or phenolphthalein can be used.

PROBLEM-SOLVING EXAMPLE 17.7 Titration of CH_3COOH with NaOH

A 0.100-M NaOH solution is used to titrate 50.0 mL of 0.100-M acetic acid ($K_a = 1.8 \times 10^{-5}$). Calculate the pH of the solution at these three points:
(a) After 40.0 mL of titrant has been added
(b) After 50.0 mL of NaOH has been added
(c) After 50.2 mL of NaOH has been added

Answer (a) 5.34 (b) 8.72 (c) 10.30

Strategy and Explanation As NaOH is added, it reacts with acetic acid to produce sodium acetate. The pH will be controlled by the moles of acetic acid remaining, the concentration of the sodium acetate at the equivalence point, and the OH^- concentration beyond it. The initial acetic acid sample contains (0.0500 L)(0.100 mol/L) = 5.00×10^{-3} mol acetic acid.

(a) Adding 40.0 mL of 0.100-M NaOH puts 4.00×10^{-3} mol OH^- ions into the solution, neutralizing 4.00×10^{-3} mol acetic acid. The reaction forms 4.00×10^{-3} mol acetate ions; 1.00×10^{-3} mol acetic acid remains un-neutralized.

$$CH_3COOH(aq) + OH^-(aq) \longrightarrow CH_3COO^-(aq) + H_2O(\ell)$$

$CH_3COOH(aq)$	$OH^-(aq)$	$CH_3COO^-(aq)$
5.00×10^{-3} mol in acid soln	4.00×10^{-3} mol added	4.00×10^{-3} mol formed

The total volume of the solution is now 90.0 mL and the concentrations are

$$\text{Concentration of acetic acid} = \frac{1.00 \times 10^{-3}\ \text{mol}}{0.0900\ \text{L}} = 0.0111\ \text{M}$$

$$\text{Concentration of acetate ion} = \frac{4.00 \times 10^{-3}\ \text{mol}}{0.0900\ \text{L}} = 0.0444\ \text{M}$$

The pH can be calculated using the K_a value and expression for acetic acid to solve for $[H_3O^+]$ and then pH.

$$K_a = \frac{[H_3O^+][CH_3COO^-]}{[CH_3COOH]} = 1.8 \times 10^{-5}$$

$$= \frac{[H_3O^+][0.00400\ \text{mol}/0.0900\ \text{L}]}{[0.00100\ \text{mol}/0.0900\ \text{L}]}$$

$$= [H_3O^+](4.00)$$

$$[H_3O^+] = \frac{1.8 \times 10^{-5}}{4.00} = 4.50 \times 10^{-6}; \qquad pH = -\log(4.50 \times 10^{-6}) = 5.34$$

The pH can also be calculated using the Henderson-Hasselbalch equation.

$$pH = 4.74 + \log\frac{(0.0444)}{(0.0111)}$$
$$= 4.74 + \log(4.00)$$
$$= 4.74 + 0.602 = 5.34$$

(b) At this point, 5.00×10^{-3} mol OH^- has been added to 5.0×10^{-3} mol acetic acid initially present, so the stoichiometric amount of base has been added to exactly react with the acid in the sample. This is the equivalence point. The reaction has produced 5.00×10^{-3} mol acetate ion, now in 100. mL of solution. The acetate concentration is 0.0500 M.

$$\frac{5.00 \times 10^{-3}\ \text{mol acetate}}{0.100\ \text{L solution}} = 0.0500\ \text{M}$$

As noted earlier, the pH at the equivalence point is governed by the hydrolysis of acetate ion. This reaction of a conjugate base (acetate ion) with water produces hydroxide ion and the conjugate acid (acetic acid). The hydroxide ions make the solution basic.

$$CH_3COO^-(aq) + H_2O(\ell) \rightleftharpoons CH_3COOH(aq) + OH^-(aq)$$

We use the K_b expression to calculate $[OH^-]$ and from it the pH. The K_b for acetate ion can be calculated from K_w and the K_a for acetic acid (← p. 780).

$$K_b = \frac{K_w}{K_a} = \frac{1.0 \times 10^{-14}}{1.8 \times 10^{-5}} = 5.6 \times 10^{-10}$$

Substituting into the K_b expression we let $x = [OH^-] = [CH_3COOH]$.

$$K_b = 5.6 \times 10^{-10} = \frac{[CH_3COOH][OH^-]}{[CH_3COO^-]} = \frac{x^2}{0.0500 - x}$$

Because K_b is small, we approximate $0.0500 - x$ to be 0.0500. Solving for x,

$$K_b = 5.6 \times 10^{-10} \approx \frac{x^2}{0.0500}; \qquad x \approx 5.3 \times 10^{-6} = [OH^-]$$

$$pOH = -\log(5.3 \times 10^{-6}) = 5.28$$

$$pH = 14.00 - 5.28 = 8.72.$$

This pH is in marked contrast to the equivalence point of 7.00 for the NaOH-HCl titration, where neutral NaCl was the titration product.

(c) The pH beyond the equivalence point is controlled by the OH^- concentration from excess NaOH, which is greater than the OH^- contributed by the hydrolysis of acetate ion. Therefore, the calculation for the pH beyond the equivalence point is like that for the titration with excess NaOH. An excess 0.2 mL NaOH has been added:

$$(0.2 \times 10^{-3}\ \text{L})\left(\frac{0.100\ \text{mol OH}^-}{\text{L}}\right) = 0.02 \times 10^{-3}\ \text{mol OH}^-$$

$$\text{Final OH}^-\ \text{concentration} = \frac{0.02 \times 10^{-3}\ \text{mol of OH}^-}{0.1002\ \text{L}} = 2.0 \times 10^{-4}\ \text{M}$$

$$pOH = -\log(2.0 \times 10^{-4}) = 3.70; \qquad pH = 14.00 - pOH = 14.00 - 3.70 = 10.30$$

PROBLEM-SOLVING PRACTICE 17.7

Calculate the pH when these volumes of 0.100-M NaOH have been added when titrating 50.0 mL of 0.100-M acetic acid:

(a) 10.0 mL (b) 25.00 mL (c) 45.0 mL (d) 51.0 mL

As seen from Figures 17.5 and 17.6 and their associated Problem-Solving Examples, the titration curve for a strong base with a strong acid differs from that of a strong base with a weak acid of equal concentration. Comparing Figures 17.5 and 17.6 we see in particular that

- Before the titration, the initial pH of the solution is higher for the weak acid.
- Very near the equivalence point, the length of the rapid rise of the curve is shorter for the weak acid.
- The pH at the equivalence point is higher for the weak acid.

A shorter rise in pH means that greater care is needed to select an appropriate indicator.

These features are shown in Figure 17.7. Note especially that near the equivalence point, the weaker the acid, the higher the pH and the shorter the rise in pH.

EXERCISE 17.7 Titration of Acetic Acid with NaOH

Use the K_a expression and value for acetic acid to calculate the pH after 30.0 mL of 0.100-M NaOH has been added to 50.0 mL of 0.100-M acetic acid.

CONCEPTUAL EXERCISE 17.8 Shape of the Titration Curve

Explain why the curve for the titration of acetic acid with NaOH in Figure 17.6 has a relatively flat region between ~10.0 and ~40.0 mL of NaOH added.

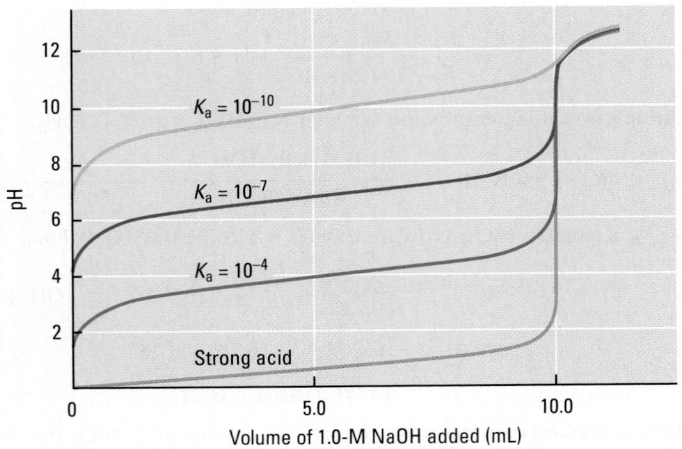

Figure 17.7 The effect of acid strength on the shape of the titration curve. Each curve is for titration of 10.0 mL of a 1.0-M acid with 1.0-M NaOH.

The pH of 0.100-M NaOH is

$$-\log\left(\frac{10^{-14}}{10^{-1}}\right) = 13$$

Titration of a Weak Base with a Strong Acid

The titration curve for the titration of the weak base NH_3 with the strong acid HCl has the shape shown in Figure 17.8. The reaction produces ammonium chloride, NH_4Cl. Notice that *because NH_3 is a weak base, the starting pH is greater than 7.0, but less than it would be for 0.100-M NaOH (13.00).* Also notice that *the pH at the equivalence point, 5.28, is less than 7.0 because ammonium chloride is an acidic salt.*

$$NH_4^+(aq) + H_2O(\ell) \rightleftharpoons NH_3(aq) + H_3O^+(aq)$$

Beyond the equivalence point the pH continues to drop as excess acid is added. Although methyl red is a suitable indicator for this titration, phenolphthalein or bromthymol blue are *not* suitable because their color changes occur before the equivalence point.

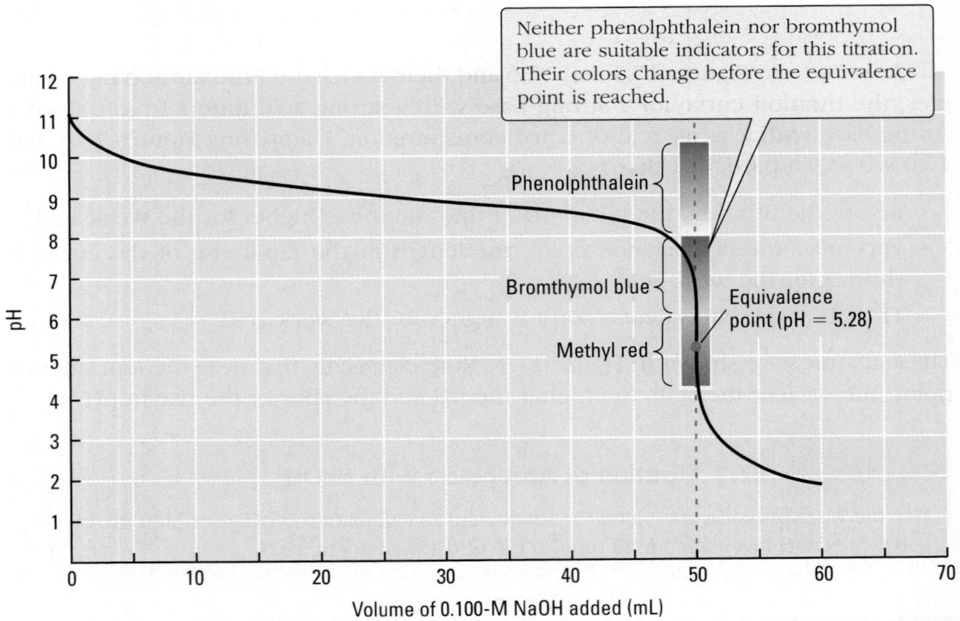

Figure 17.8 The titration of a weak base, NH_3, by a strong acid, HCl. This titration curve is for the titration of 50.0 mL of 0.100-M NH_3 with 0.100-M HCl. It is essentially the inverse of the curve for the titration of a weak acid by a strong base (Figure 17.6). Methyl red is a suitable indicator because it changes color near the equivalence point.

Titration of a Polyprotic Acid with Base

Polyprotic acids—those with more than one ionizable hydrogen—react stepwise when titrated with bases, one step for each ionizable hydrogen. If the K_a values of the ionizable forms of the acid are sufficiently different, the titration curve has an equivalence point for each of the ionizable hydrogens removed from the acid molecule by titration. For example, maleic acid, HOOC—CH=CH—COOH, is a diprotic acid with two ionizable hydrogens, one from each —COOH group. Its titration with NaOH occurs in two steps:

$$HOOC{-}CH{=}CH{-}COOH(aq) + OH^-(aq) \rightleftharpoons HOOC{-}CH{=}CH{-}COO^-(aq) + H_2O(\ell)$$

$$HOOC{-}CH{=}CH{-}COO^-(aq) + OH^-(aq) \rightleftharpoons {}^-OOC{-}CH{=}CH{-}COO^-(aq) + H_2O(\ell)$$

As shown in Figure 17.9, the two equivalence points occur at pH = 4.1 and pH = 9.4.

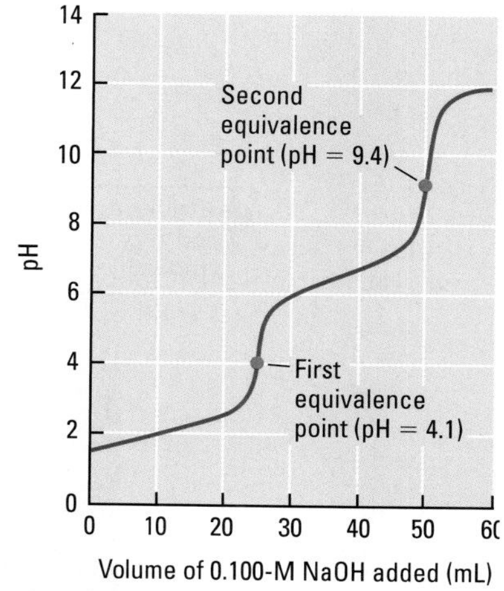

Figure 17.9 Titration of a polyprotic acid. The curve is for the titration of 25.00 mL of 0.100-M maleic acid with 0.100-M NaOH.

17.3 Acid Rain

The term *acid rain* was first used in 1872 by Robert Angus Smith, an English chemist and climatologist. In his book *Air and Rain,* Smith used the term to describe the acidic precipitation that fell on Manchester, England, at the start of the Industrial Revolution. **Acid rain** is now defined as any precipitation with pH < 5.6, which is the pH of natural, unpolluted rain. Although neutral water has a pH of 7, water in rain and in some lakes and rivers becomes acidified *naturally* from dissolved carbon dioxide, a normal component of the atmosphere. The carbon dioxide reacts reversibly with water to form a solution of carbonic acid, a weak acid, which ionizes into hydronium and hydrogen carbonate ions.

Although the term "acid rain" is commonly used, the more accurate term is "acid deposition," which takes into account acidic snow, sleet, rain, and fog.

$$2\,H_2O(\ell) + CO_2(g) \rightleftharpoons H_2CO_3(aq) + H_2O(\ell) \rightleftharpoons H_3O^+(aq) + HCO_3^-(aq)$$

The pH of water in equilibrium with CO_2 from the air is about 5.6, so *natural, unpolluted rainwater is slightly acidic.*

Nitrogen dioxide, NO_2, from industrial as well as natural sources reacts with water in the atmosphere to produce acids *(← p. 459);* NO_2 produces nitric acid, HNO_3, and nitrous acid, HNO_2.

$$2\,NO_2(g) + H_2O(\ell) \longrightarrow HNO_3(aq) + HNO_2(aq)$$

Atmospheric sulfur dioxide, SO_2, produced from burning sulfur-containing fossil fuels, reacts with water to produce sulfurous acid, H_2SO_3, and, if oxygen is present, sulfuric acid, H_2SO_4.

$$SO_2(g) + H_2O(\ell) \longrightarrow H_2SO_3(aq)$$

$$2\,SO_2(g) + O_2(g) \longrightarrow 2\,SO_3(g)$$

$$SO_3(g) + H_2O(\ell) \longrightarrow H_2SO_4(aq)$$

Approximately 160 million metric tons of SO_2 and 85 million metric tons of nitrogen oxides are emitted annually into the atmosphere from human activities, largely the burning of fossil fuels. Each year about 43 million metric tons of SO_2 and 59 million metric tons of nitrogen oxides are put into the atmosphere by natural sources.

The resulting acidic water droplets precipitate as rain or snow with a pH less than 5.6. Ice core samples taken in Greenland and dating back to 1900 contain sulfate, SO_4^{2-}, and nitrate, NO_3^-, ions indicating that acid rain has been commonplace, at least from 1900 onward.

Acid rain is a problem today due to the large amounts of these acidic oxides being put into the atmosphere by human activities every year (Figure 17.10). When such precipitation falls on areas without naturally occurring bases, such as limestone and other carbonate minerals, that offset the acidity, serious environmental damage can occur. The average annual pH of precipitation falling on much of the northeastern United States and northeastern Europe is between 4 and 4.5 (Figure 17.11). Rain in those areas has had pH values as low as 1.5. To further complicate

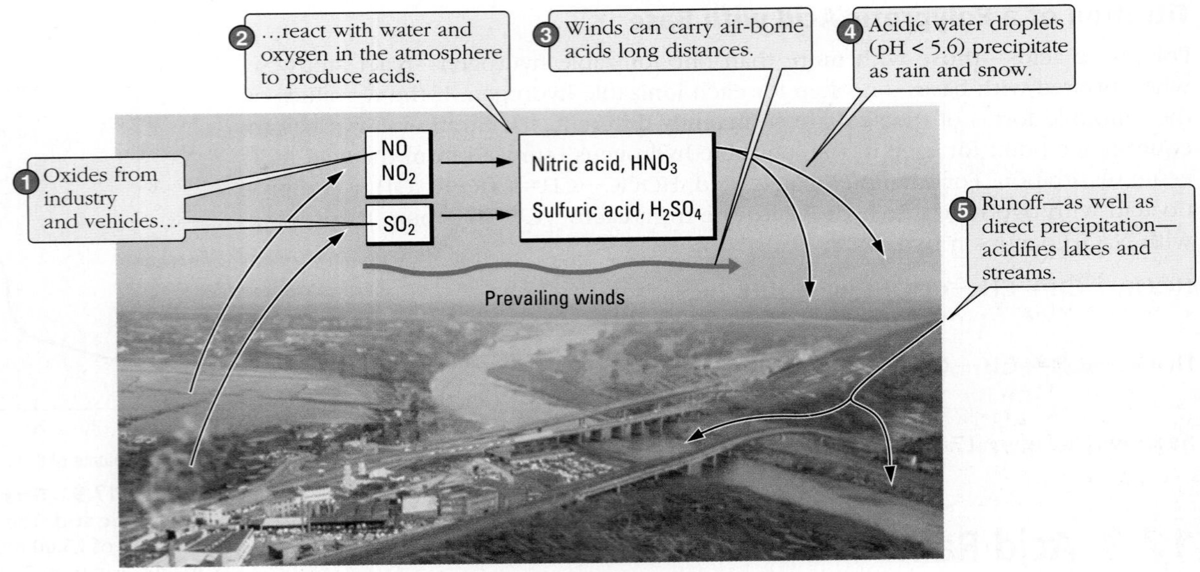

Figure 17.10 **How acid deposition occurs.**

Dan Kunkle Lehigh Gap Nature Center

Area damaged by acid rain and SO_2 emissions from nearby zinc smelter, 1898–1980.

Dan Kunkle Lehigh Gap Nature Center

The same area recovered by remediation (2007) using warm-season grasses. The grasses build and retain soil by growing in the damaged area without taking up toxic metal ions.

matters, acid rain is an international problem because precipitation carried by winds does not observe international borders. Canadian residents are offended by the fact that much of the acid precipitation falling on Canadian cities and forests results from acidic oxides produced in the United States.

Acid rain can be caused by SO_2 produced by smelting sulfide-containing ores, such as ZnS, to form ZnO and SO_2. In addition to SO_2, the smelting releases toxic metal ions such as Zn^{2+}, Cd^{2+}, and Pb^{2+} into the atmosphere. Eventually, compounds of these ions precipitate into the surrounding soil. The adjacent photos show acid rain damage to the area around the Zinc Company smelting facility near Palmerton, PA between 1898 and 1980 when smelting stopped, and the significant remediation that has occurred recently on that same site.

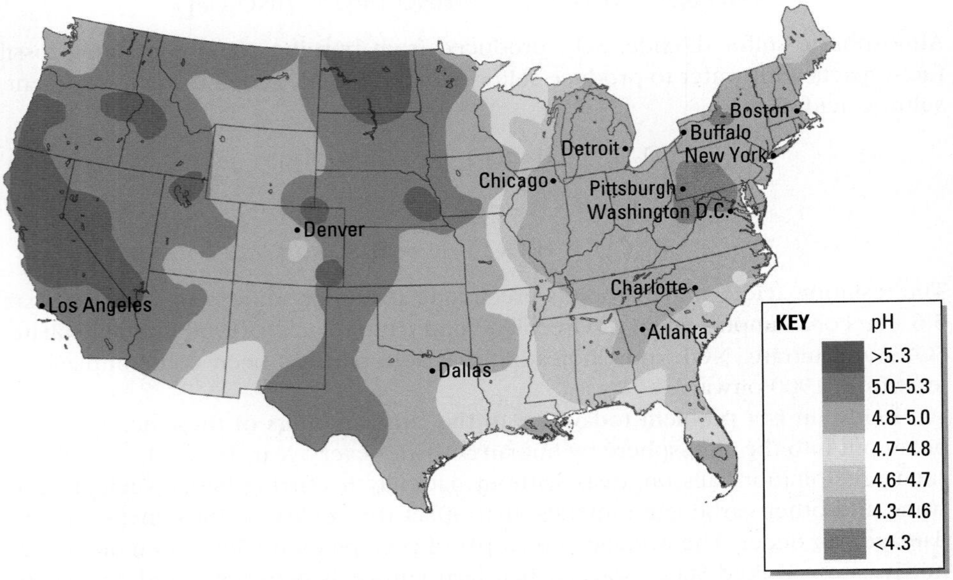

Figure 17.11 **pH of precipitation within the contiguous United States.**

17.4 Solubility Equilibria and the Solubility Product Constant, K_{sp}

Many ionic compounds, such as NH_4Cl and $NaNO_3$, are very soluble in water. Many other ionic compounds, however, are only modestly or slightly water-soluble; they produce *saturated* solutions of 0.001 M or less, far less in some cases ***(← p. 162).*** Consider the case of a saturated aqueous solution of silver bromide, AgBr. When sufficient AgBr is added to water, some of it dissolves to form a saturated solution, and some undissolved AgBr is present. The dissolved AgBr forms aqueous Ag^+ and Br^- ions in solution that are in equilibrium with the undissolved solid AgBr.

$$AgBr(s) \rightleftharpoons Ag^+(aq) + Br^-(aq)$$

This balanced chemical equation represents a solubility equilibrium. As with other equilibria ***(← p. 659),*** we can derive an equilibrium constant from the chemical equation. In this case, the equilibrium constant is called the **solubility product constant, K_{sp}**. The magnitude of K_{sp} indicates the extent to which the solid solute dissolves to give ions in solution. To evaluate the equilibrium constant, we first must write a **solubility product constant expression.** In general, the balanced chemical equation for dissolving a slightly soluble salt with the general formula A_xB_y is

$$A_xB_y(s) \rightleftharpoons x\,A^{n+}(aq) + y\,B^{m-}(aq)$$

This results in the general K_{sp} expression

$$K_{sp} = [A^{n+}]^x[B^{m-}]^y$$

For example, the solubility product expression for AgBr is

$$K_{sp} = [Ag^+][Br^-].$$

Notice that

- The chemical equation related to the solubility product constant expression is written for a solid solute compound as a reactant and its aqueous ions as the products.
- The concentration of the pure solid solute reactant is omitted from the K_{sp} expression because the concentration of pure solid remains unchanged as the reaction occurs.
- The K_{sp} value equals the product of the *equilibrium* molar concentrations of the cation and the anion, each raised to the power given by the coefficient in the balanced chemical equation representing the solubility equilibrium.

The solubility product constant is commonly called just the solubility product.

A solubility product expression has the same general form that other equilibrium constant expressions have, except there is no denominator in the K_{sp} expression, because the reactant is always a pure solid.

PROBLEM-SOLVING EXAMPLE 17.8 Writing K_{sp} Expressions

Write the K_{sp} expression for each of these slightly soluble salts:
(a) $Fe(OH)_2$ (b) MgC_2O_4 (c) Ag_3PO_4

Answer

(a) $K_{sp} = [Fe^{2+}][OH^-]^2$ (b) $K_{sp} = [Mg^{2+}][C_2O_4^{2-}]$ (c) $K_{sp} = [Ag^+]^3[PO_4^{3-}]$

Strategy and Explanation To write the correct K_{sp} expression we first must write the balanced equation for the dissociation of the solute, and then substitute the equilibrium concentrations of the products appropriately into the K_{sp} expression.

(a) The equilibrium reaction for the solubility of $Fe(OH)_2$ in water is

$$Fe(OH)_2(s) \rightleftharpoons Fe^{2+}(aq) + 2\,OH^-(aq)$$

In this example, the cation:anion relationship is 2:1; for every Fe^{2+} ion there are two OH^- ions produced, so the hydroxide ion concentration is raised to the power of 2 in the K_{sp} expression: $K_{sp} = [Fe^{2+}][OH^-]^2$.

(b) The equilibrium reaction for the solubility of MgC_2O_4 in water is

$$MgC_2O_4(s) \rightleftharpoons Mg^{2+}(aq) + C_2O_4^{2-}(aq)$$

Since the equilibrium chemical equation shows the cations and anions in a one-to-one relationship, the equilibrium constant expression is $K_{sp} = [Mg^{2+}][C_2O_4^{2-}]$.

(c) The equilibrium reaction for the solubility of Ag_3PO_4 in water is

$$Ag_3PO_4(s) \rightleftharpoons 3\ Ag^+(aq) + PO_4^{3-}(aq)$$

Since three Ag^+ ions are produced for every PO_4^{3-} ion, the K_{sp} expression is written $K_{sp} = [Ag^+]^3[PO_4^{3-}]$

PROBLEM-SOLVING PRACTICE 17.8

Write the K_{sp} expression for each of these slightly soluble salts:
(a) CuBr (b) HgI_2 (c) $SrSO_4$

Solubility and K_{sp}

The solubility of a sparingly soluble solute and its solubility product constant, K_{sp}, are not the same thing, but they are related. The solubility is the amount of solute per unit volume of solution (mol/L) that dissolves to form a *saturated* solution. On the other hand, the solubility product constant is the equilibrium constant for the chemical equilibrium that exists between a solid ionic solute and its ions in a saturated solution. In general, solutes with very low solubility have very small K_{sp} values.

If the equilibrium concentrations of the ions are known, they can be used to calculate the K_{sp} value for the solute. For example, in a saturated AgCl solution at 10 °C, the molar concentrations of Ag^+ and Cl^- each are experimentally determined to be 6.3×10^{-6} M. This means that K_{sp} at 10 °C is

$$K_{sp} = [Ag^+][Cl^-] = [6.3 \times 10^{-6}][6.3 \times 10^{-6}] = 4.0 \times 10^{-11}$$

The K_{sp} values for selected ionic compounds at 25 °C are listed in Table 17.2. A more extensive listing is in Appendix H.

Table 17.2 K_{sp} Values for Some Slightly Soluble Salts

Compound	K_{sp} at 25 °C
AgBr	5.0×10^{-13}
AgCl	1.8×10^{-10}
AgI	8.3×10^{-17}
Ag_2SO_4	1.4×10^{-5}
AuI	1.6×10^{-23}
AuI_3	1×10^{-46}
$BaCrO_4$	1.2×10^{-10}
$BaSO_4$	1.1×10^{-10}
$Fe(OH)_2$	8.0×10^{-16}
Hg_2Br_2*	5.6×10^{-23}
Hg_2Cl_2*	1.3×10^{-18}
Hg_2I_2*	4.5×10^{-29}
Hg_2SO_4*	7.4×10^{-7}
$Mn(OH)_2$	1.9×10^{-13}
$PbBr_2$	4.0×10^{-5}
$PbCl_2$	1.6×10^{-5}
PbI_2	7.1×10^{-9}
$PbSO_4$	1.6×10^{-8}
$SrSO_4$	3.2×10^{-7}

*These compounds contain the diatomic Hg_2^{2+} ion.

PROBLEM-SOLVING EXAMPLE 17.9 Solubility and K_{sp}

The K_{sp} of $BaSO_4$ is 1.1×10^{-10} at 25 °C. Calculate the solubility of $BaSO_4$, expressing the result in molarity, moles per liter.

Answer 1.0×10^{-5} M

Strategy and Explanation The solubility product constant expression for barium sulfate is derived from the chemical equation

$$BaSO_4(s) \rightleftharpoons Ba^{2+}(aq) + SO_4^{2-}(aq) \qquad K_{sp} = [Ba^{2+}][SO_4^{2-}] = 1.1 \times 10^{-10}$$

The ionization of solid $BaSO_4$ forms Ba^{2+} ions and SO_4^{2-} ions in equal amounts. Therefore, if we let S equal the solubility of $BaSO_4$, then at equilibrium the concentration of Ba^{2+} and SO_4^{2-} ions will each be S.

$$1.1 \times 10^{-10} = (S)(S) = S^2$$

$$S = \sqrt{1.1 \times 10^{-10}} = 1.0 \times 10^{-5}\ M$$

Consequently, the aqueous solubility of $BaSO_4$ at 25 °C is 1.0×10^{-5} M.

PROBLEM-SOLVING PRACTICE 17.9

The K_{sp} of AgBr at 100 °C is 5×10^{-10}. Calculate the solubility of AgBr at that temperature in moles per liter.

A note of caution is in order. It might seem straightforward to calculate the solubility of an ionic compound from its K_{sp}, the calculation in Problem-Solving Example 17.9, or to do the reverse, that is, calculate the K_{sp} from the solubility. Doing so, however, will often lead to incorrect results. This approach overlooks several complicating factors that we list here using $PbCl_2$ as an example:

- Ionic solids such as $PbCl_2$ dissociate stepwise, so that $PbCl^+$ ions as as well as Pb^{2+} and Cl^- ions are present in a $PbCl_2$ solution.
- Ion pairs, such as $PbCl^+Cl^-$ can exist, reducing the concentrations of unassociated Pb^{2+} and Cl^- ions.
- The solubilities of some solutes, such as metal hydroxides, depend on the acidity or alkalinity of the solution.
- Solutes containing anions such as CO_3^{2-} and PO_4^{3-} that react with water are more soluble than predicted by their K_{sp} values.

$$PbCl_2(s) \rightleftharpoons PbCl^+(aq) + Cl^-(aq)$$
$$PbCl^+(aq) \rightleftharpoons Pb^{2+}(aq) + Cl^-(aq)$$
$$PbCl_2(s) \rightleftharpoons Pb^{2+}(aq) + 2\,Cl^-(aq)$$

Solubilities calculated from K_{sp} values, and K_{sp} values calculated from solubilities, best agree with the experimentally measured solubilities of compounds with 1+ and 1− charged ions and ions that do not react with water.

PROBLEM-SOLVING EXAMPLE 17.10 Solubility and K_{sp}

A saturated aqueous solution of lead(II) iodate, $Pb(IO_3)_2$, contains 3.1×10^{-5}-M Pb^{2+} at 25 °C. Assuming that the ions do not react with water, calculate the K_{sp} of lead(II) iodate at that temperature.

Answer $K_{sp} = 1.2 \times 10^{-13}$

Strategy and Explanation Write the chemical equation for the dissociation of the solute, derive the K_{sp} expression correctly from it, and substitute concentrations of Pb^{2+} and IO_3^- ions into the expression to obtain K_{sp}. In the saturated solution, the chemical equilibrium is

$$Pb(IO_3)_2(s) \rightleftharpoons Pb^{2+}(aq) + 2\,IO_3^-(aq)$$

and the solubility product expression is $K_{sp} = [Pb^{2+}][IO_3^-]^2$.

We next determine the *molar* concentrations of Pb^{2+} and IO_3^-. From the balanced chemical equation we see that when dissociation of the solid lead(II) iodate occurs, one mole of lead ions is produced for two moles of iodate ions. Since the Pb^{2+} concentration is given as 3.1×10^{-5} M, the iodate ion concentration is 6.2×10^{-5} M, twice that of lead ion. The K_{sp} value can be calculated directly from these equilibrium concentrations and the solubility product equilibrium constant expression.

You can review molarity calculations in Section 5.7 (← *p. 196*).

$$K_{sp} = [Pb^{2+}][IO_3^-]^2 = (3.1 \times 10^{-5})(6.2 \times 10^{-5})^2 = 1.2 \times 10^{-13}$$

☑ **Reasonable Answer Check** Given the fact that the iodate ion concentration is twice that of the lead ion concentration, substituting these values correctly into the proper K_{sp} expression gives the calculated answer, which is reasonable.

PROBLEM-SOLVING PRACTICE 17.10

A saturated solution of silver oxalate, $Ag_2C_2O_4$, contains 6.9×10^{-5}-M $C_2O_4^{2-}$ at 25 °C. Calculate the K_{sp} of silver oxalate at that temperature, assuming that the ions do not react with water.

CONCEPTUAL EXERCISE 17.9 Solubility and Le Chatelier's Principle

At 25 °C, 0.014 g calcium carbonate dissolves in 100 mL water. Two equilibria are present in this solution.

(a) $CaCO_3(s) \rightleftharpoons Ca^{2+}(aq) + CO_3^{2-}(aq)$

(b) $CO_3^{2-}(aq) + H_2O(\ell) \rightleftharpoons HCO_3^-(aq) + OH^-(aq)$

Suppose Reaction (b) occurs to an appreciable extent. Use Le Chatelier's principle (← *p. 678*) to predict how the extent of Reaction (b) will affect the solubility of $CaCO_3$.

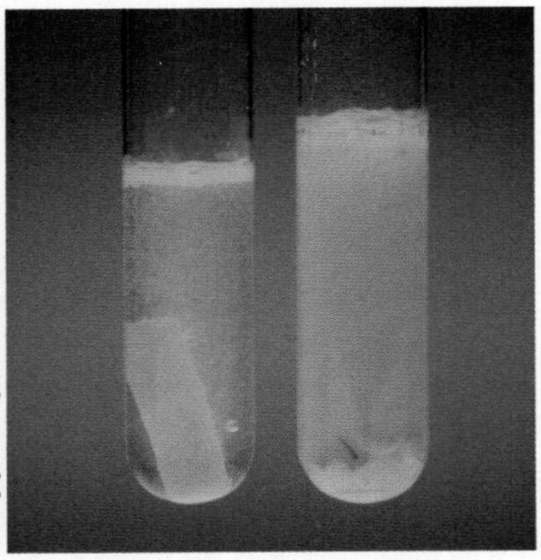

Reaction of calcium carbonate with acid. The left tube contains a piece of chalk (calcium carbonate) that is dissolving by reacting with hydrochloric acid. In the right tube, bubbles of CO_2 gas being formed by the reaction fill the test tube.

17.5 Factors Affecting Solubility

The aqueous solubility of ionic compounds is affected by several factors, some of which have already been mentioned—temperature ***(← p. 718)***, the formation of ion pairs, and competing equilibria. In this section we will consider four other factors affecting the aqueous solubility of ionic compounds:

- The effect of acids and pH
- The presence of common ions
- The formation of complex ions
- Amphoterism

pH and Dissolving Slightly Soluble Salts Using Acids

As noted earlier in the discussion of solubility rules, many salts are only slightly soluble in water ***(← p. 163)***. An acid can dissolve an insoluble salt that contains a moderately basic ion. As an example, consider calcium carbonate, $CaCO_3$, which is found in minerals such as limestone and marble. $CaCO_3$ is not very soluble in pure water.

$$\text{(a)}\ CaCO_3(s) \rightleftharpoons Ca^{2+}(aq) + CO_3^{2-}(aq) \qquad K_{sp} = 8.7 \times 10^{-9}$$

Since the solubility of calcium carbonate is so low, the equilibrium concentrations of Ca^{2+} and CO_3^{2-} must also be small. However, if acid is added to the solution, calcium carbonate dissolves and CO_2 is released from the solution. Adding acid adds hydronium ions, which react with carbonate and hydrogen carbonate ions.

$$\text{(b)}\ CO_3^{2-}(aq) + H_3O^+(aq) \rightleftharpoons HCO_3^-(aq) + H_2O(\ell)$$

$$\text{(c)}\ HCO_3^-(aq) + H_3O^+(aq) \rightleftharpoons H_2CO_3(aq) + H_2O(\ell)$$

$$\text{(d)}\ H_2CO_3(aq) \rightleftharpoons CO_2(g) + H_2O(\ell) \qquad K \approx 10^5$$

Reaction (b) is the reaction of a fairly strong weak base, CO_3^{2-}, with a strong acid, so nearly all of the carbonate is converted to hydrogen carbonate ion. Reaction (c) produces a product, carbonic acid, that is unstable and breaks down to CO_2 gas and water (Reaction (d)). Reaction (d) is a very product-favored reaction, as indicated by the large K value, $\approx 10^5$.

Reactions (a) through (d) are linked through carbonate, hydrogen carbonate, and carbonic acid:

- As CO_2 gas escapes from the solution, the H_2CO_3 concentration decreases (Reaction (d)).
- This shifts Reaction (c) to the right, decreasing the concentration of HCO_3^-; decreasing the HCO_3^- concentration shifts Reaction (b) to the right, decreasing the concentration of CO_3^{2-}.
- To oppose this decrease in carbonate concentration, Reaction (a) shifts to the right, and more solid $CaCO_3$ dissolves.

Damage caused by acid rain. Over the years, acid rain has eroded this marble statue of George Washington. Effects of acid rain on materials can be seen at **http://pubs.usgs.gov/gip/acidrain/5.html** and **http://www.elmhurst.edu/~chm/vchembook/196buildings.html.**

Because the acidity of the solution determines the positions of equilibria (b) and (c), small changes in pH can cause limestone and marble to dissolve or precipitate ***(← p. 679)***. Acid rain can dissolve a marble statue; it can also dissolve underground limestone deposits, creating massive cave formations. Impressive stalactite and stalagmite formations in caves result from such changes. In addition, limestone has precipitated as layers of sedimentary rock on the ocean floor where a slight increase in the pH of seawater has occurred.

In general, ***insoluble salts containing anions that are Brønsted-Lowry bases dissolve in acidic solutions (those of low pH).*** This rule covers carbonates, sulfides (which produce $H_2S(g)$), phosphates, and other anions listed as bases in Table 16.5 ***(← p. 786)***. The principal exceptions to this rule are a few sulfides,

CHEMISTRY IN THE NEWS

Ocean Acidification, a Global pH Change Concern

The world's oceans absorb enormous amounts of atmospheric carbon dioxide, approximately 22 million metric tons *daily*, which is one-third of the total amount of daily CO_2 emissions from the burning of fossil fuels. Since the industrial revolution began almost 250 years ago, atmospheric CO_2 has increased by about 100 ppm. Over the past 200 years, the oceans have absorbed approximately 550 billion metric tons of CO_2,

Several basic anions in the oceans, principally CO_3^{2-}, neutralize the carbonic acid generated by the added CO_2, but cannot keep pace with the increased rate of CO_2 addition. Marine scientists are becoming more concerned by the drop in pH of the oceans' surface water, a phenomenon called *ocean acidification* that has occurred over the past several decades. The drop in pH from 8.2 to 8.1 represents a 25% increase in H^+ concentration (from 6.3×10^{-9} M to 7.9×10^{-9} M). If atmospheric CO_2 emissions from anthropogenic sources continue to increase at current rates, marine scientists predict that the oceans' pH will become 7.8 ($[H^+] = 1.6 \times 10^{-8}$ M) within 50 to 100 years, a substantial change in a relatively short time. Such increases in ocean acidification could have significant consequences in several ways.

Most shelled and spiny sea organisms (corals, clams, sea urchins) use Ca^{2+} and CO_3^{2-} ions to create their calciferous shells or spines. As the pH drops, the equilibria involving CO_2, HCO_3^-, and H_2CO_3 are disrupted; more carbonate ions are converted to hydrogen carbonate ions, thereby decreasing the amounts of carbonate for calcification. Thus, shells and spines are thinner or misformed. In addition, the increased H_3O^+ concentration dissolves the calcium carbonate in shells and spines by converting it into the more soluble calcium hydrogen carbonate.

Many reaction rates are pH dependent, meaning that ocean acidification could affect chemical processes such as enzyme activity and the rates at which metal ions are absorbed. Shifts in carbonate and hydroxide concentrations due to a decrease in pH can influence the chemical forms of various vital trace metals and whether they remain in solution. The hydrated "free" forms of the metal ions have a greater biological activity than the "bound" forms. In bound forms, the metal ions form coordination complexes by bonding to Lewis bases other than water, such as carbonate and hydroxide ions, as well as organic compounds. Free metal ion concentration would increase as carbonate and hydroxide concentrations decrease.

Ocean acidification could lead to an unusual unintended consequence; acidified oceans absorb less sound at lower frequencies. Recent experiments determined that such a phenomenon could affect mammals that are dependent on sonar to navigate and locate objects. By absorb less sound, the oceans would be noisier, which could disrupt communications between such mammals.

2
3
4
5
6
2
3
4
5
6

Photos: Justin B. Ries, Department of Marine Sciences, University of North Carolina at Chapel Hill

Changes in sea urchin spines due to increased acidification. A sea urchin (left) grown in seawater acidified at the current ocean CO_2 level, 400 ppm, produces large spines. The sea urchin in the right figure was grown in seawater acidified to 2,850 ppm CO_2 and exhibits truncated spines due to dissolution of $CaCO_3$ in the spines. This extremely high CO_2 level is used for such laboratory tests.

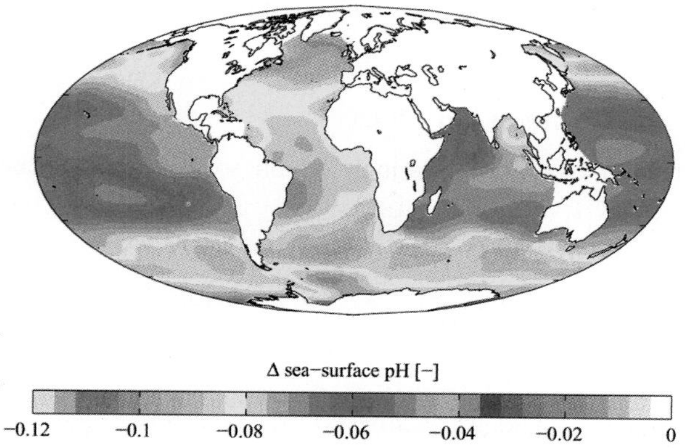
Δ sea−surface pH [−]
−0.12
−0.1
−0.08
−0.06
−0.04
−0.02
0

Plumbago/Wikipedia

Changes in sea surface pH. *Note:* A decrease of 0.01 pH unit represents a 2.4% increase in H^+ concentration.

Sources: Petkewich, R. "Off-Balance Ocean." *Chemical & Engineering News,* February 23, 2009. **http://www.cen-online.org.**

Science Daily. "Ocean Acidification: Another Undesired Side Effect of Global Warming." **http://www.sciencedaily.com/releases/2008/05/080521105251.htm.**

Pacific Science Association. "PSA Task Force on Ocean Acidification." **http://pacificscience.org/tfoceanacidification.html.**

such as HgS, CuS, and CdS, that have extremely low solubilities and therefore do not dissolve even when the pH is extremely low.

In contrast, an insoluble salt such as AgCl, which contains the conjugate base of a strong acid, is not soluble in strongly acidic solution because Cl^- is a very weak base and so does not react with H_3O^+.

Solubility and the Common Ion Effect

It is often desirable to remove a particular ion from solution by forming a precipitate of one of its insoluble compounds. For example, barium ions readily absorb X-rays and so are quite effective in making the intestinal tract visible when X-ray photographs are taken. But barium ions are poisonous and must not be allowed to dissolve in body fluids. Barium sulfate, an insoluble compound, can be ingested and used as an X-ray absorber, but both physician and patient want to be certain that no harmful amounts of barium ions will be in solution.

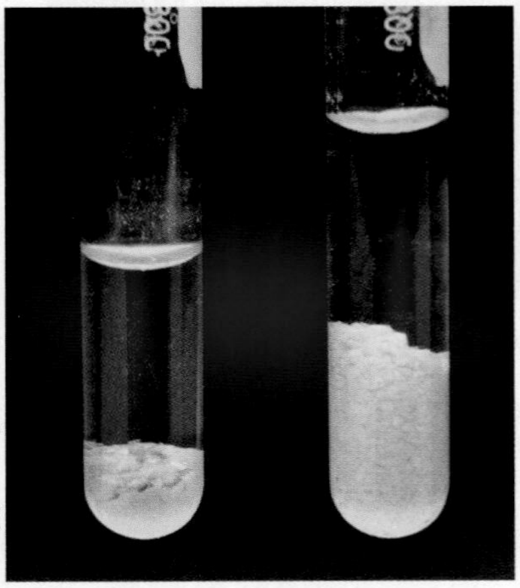

The common ion effect. The tube on the left contains a saturated solution of silver acetate, $AgCH_3COO$. When 1-M $AgNO_3$ is added to the tube, the equilibrium

$$AgCH_3COO(s) \rightleftharpoons Ag^+(aq) + CH_3COO^-(aq)$$

shifts to the left due to the addition of Ag^+ ions, causing more solid silver acetate to precipitate, evidenced by the tube on the right.

The solubility of $BaSO_4$ in water at 25 °C is 1.0×10^{-5} mol/L (p. 828), which means that the concentration of Ba^{2+} ions and of sulfate ions is 1.0×10^{-5} M.

$$BaSO_4(s) \rightleftharpoons Ba^{2+}(aq) + SO_4^{2-}(aq)$$

The concentration of aqueous Ba^{2+} ions can be reduced by adding a soluble sulfate salt, such as Na_2SO_4. The solubility of $BaSO_4$ decreases because of the increased concentration of SO_4^{2-} ions due to the sulfate ions from Na_2SO_4. The sulfate ion is called a "common ion" because it is common to both substances dissolved in the solution—barium sulfate and sodium sulfate. The common ion shifts the preceding equilibrium to the left by what is called the **common ion effect:** *The presence of a second solute that provides a common ion lowers the solubility of an ionic compound.*

$$Na_2SO_4(s) \longrightarrow 2\,Na^+(aq) + SO_4^{2-}(aq)$$ sulfate from $Na_2SO_4(s)$

$$BaSO_4(s) \rightleftharpoons Ba^{2+}(aq) + SO_4^{2-}(aq)$$ sulfate from $BaSO_4(s)$

← Solubility of $BaSO_4$ decreases; additional solid $BaSO_4$ forms

The common ion effect can be interpreted by using Le Chatelier's principle (← *p. 678*). Adding sodium sulfate to the saturated barium sulfate solution causes a stress (higher concentration of sulfate ions) on the equilibrium. To reduce the concentration of SO_4^{2-}, the equilibrium shifts to the left, reducing the concentration of Ba^{2+} ions as well, to form solid $BaSO_4$. The outcome is that less $BaSO_4$ is dissolved, lowering its solubility in the presence of sulfate, the common ion.

CONCEPTUAL EXERCISE **17.10 Common Ion Effect**

Consider 0.0010-M solutions of these sparingly soluble solutes in equilibrum with their ions.

$$BaSO_4(s) \rightleftharpoons Ba^{2+}(aq) + SO_4^{2-}(aq)$$

$$AgI(s) \rightleftharpoons Ag^+(aq) + I^-(aq)$$

$$PbI_2(s) \rightleftharpoons Pb^{2+}(aq) + 2\,I^-(aq)$$

(a) Using Le Chatelier's principle, explain what would likely happen if a saturated solution of sodium iodide were added to the last two solutions.

(b) What would likely occur if a saturated solution of potassium sulfate were added to the $BaSO_4$ solution?

PROBLEM-SOLVING EXAMPLE 17.11 Common Ion Effect

The solubility of AgCl in pure water is 1.3×10^{-5} mol/L at 25 °C. If you put some solid AgCl into a solution that is 0.55 M in NaCl, calculate what mass of AgCl will dissolve per liter of this solution. The K_{sp} of AgCl is 1.8×10^{-10} at 25 °C.

Answer About 4.7×10^{-8} g AgCl/L

Strategy and Explanation We must take into account that the solution has two sources of chloride ions—AgCl and NaCl. We predict that because of the common ion effect, the solubility of AgCl will be smaller when NaCl is present than it would be in pure water. Because AgCl is the only source of Ag^+ ions in the solution, the solubility of AgCl, S, equals the silver ion concentration $[Ag^+]$ at equilibrium.

$$AgCl(s) \rightleftharpoons Ag^+(aq) + Cl^-(aq)$$

If the AgCl(s) were dissolved in *pure* water, $[Ag^+]$ and $[Cl^-]$ would be equal because there is no other source of chloride ion. Thus, S would be equal to $[Ag^+] = [Cl^-]$. In water containing AgCl and NaCl, both are sources of the common ion Cl^-. Thus, the value of S is equal only to $[Ag^+]$ and not to $[Cl^-]$ because Cl^- comes from two sources.

The ICE table shows the concentrations of Ag^+ and Cl^- when equilibrium is attained in the presence of extra Cl^-.

	$AgCl(s) \rightleftharpoons$	$Ag^+(aq)$	+	$Cl^-(aq)$
*I*nitial concentration (mol/L)		0		0.55
*C*hange as reaction occurs (mol/L)		$+S$		$+S$
*E*quilibrium concentration (mol/L)		S		$S + 0.55$

The total chloride ion concentration at equilibrium is what came from AgCl (equals S) ***plus*** what was already there (0.55 M) from the NaCl. Because NaCl is a soluble salt, far more Cl^- comes from NaCl than from AgCl.

Using the equilibrium concentrations from the table gives

$$K_{sp} = 1.8 \times 10^{-10} = [Ag^+][Cl^-] = (S)(S + 0.55)$$

The easiest approach to solve such an equation is to approximate that S is *very* small compared to 0.55; that is, the answer will be approximately the same if we assume that $(S + 0.55) \approx 0.55$. Such an assumption is reasonable because we know that the solubility of AgCl equals only 1.3×10^{-5} mol/L *without* the Cl^- added from NaCl. When NaCl is added, it will further decrease the solubility of AgCl due to the presence of the common ion Cl^-. Therefore,

$$(S)(S + 0.55) \approx (S)(0.55) = K_{sp}$$

or

$$K_{sp} = (S)(0.55) = 1.8 \times 10^{-10}$$

Solving for S, we get

$$S = \frac{1.8 \times 10^{-10}}{0.55} = 3.3 \times 10^{-10}\ \text{M} = [Ag^+]$$

Therefore, the $[Ag^+]$, which is the same as S, is approximately 3.3×10^{-10} mol/L.

Use the molar mass for AgCl, 143.4 g/mol, to calculate the solubility of AgCl in 0.55-M NaCl.

$$\left(\frac{3.3 \times 10^{-10}\ \text{mol AgCl}}{1\ \text{L}}\right)\left(\frac{143.4\ \text{g AgCl}}{1\ \text{mol AgCl}}\right) = 4.7 \times 10^{-8}\ \text{g AgCl/L}$$

As predicted by Le Chatelier's principle, the solubility of AgCl (3.3×10^{-10} M) in the presence of Cl^- added from another source is clearly less than in pure water (1.3×10^{-5} M).

☑ **Reasonable Answer Check** To check the approximation we made, we substitute the approximate value of S into the exact expression $K_{sp} = (S)(S + 0.55)$. If the product $(S)(S + 0.55)$ is the same as the given value of K_{sp}, the approximation is valid.

$$K_{sp} = (S)(S + 0.55) = (3.3 \times 10^{-10})(3.3 \times 10^{-10} + 0.55) = 1.8 \times 10^{-10}$$

A more accurate solution to this problem can be obtained by solving for S using the quadratic equation described in Appendix A.7 and used in Problem-Solving Example 16.10 (← p. 779). When the quadratic equation is used, its answer, to two significant figures, is the same as that using the approximation.

PROBLEM-SOLVING PRACTICE **17.11**

Calculate the solubility of $PbCl_2$ at 25 °C in a solution that is 0.50 M in NaCl.

PROBLEM-SOLVING EXAMPLE 17.12 pH and Common Ion Effect

Manganese(II) hydroxide, $Mn(OH)_2$, is sparingly soluble in water: K_{sp} $Mn(OH)_2 = 1.9 \times 10^{-13}$ at 25 °C. Calculate the solubility of manganese(II) hydroxide at that temperature in (a) pure water and (b) at a pH of 11.00.

Answer (a) 3.6×10^{-5} M (b) 1.9×10^{-7} M

Strategy and Explanation In part (a) we apply solubility equilibrium concepts to the equilibrium between a solute and its ions in solution in the case where no common ion is present. In part (b) where the initial solution has a pH of 11.00, there will be a significant hydroxide ion concentration before any solid $Mn(OH)_2$ has been added to the solution. After the addition, there are two possible sources of hydroxide ion, the common ion: (1) the initial solution; and (2) that resulting from the dissolution of solid manganese(II) hydroxide.

(a) The chemical equilibrium is

$$Mn(OH)_2(s) \rightleftharpoons Mn^{2+}(aq) + 2\,OH^-(aq)$$

and the equilibrium constant expression is $K_{sp} = [Mn^{2+}][OH^-]^2 = 1.9 \times 10^{-13}$. Let S equal the solubility of $Mn(OH)_2$. From the solubility equation we see that the concentration of Mn^{2+} is S, that is, S moles per liter of Mn^{2+} for each mole per liter of $Mn(OH)_2$ that dissolves. Hydroxide ion concentration is $2S$ (Mn^{2+} and OH^- are in a 1:2 mole ratio). We summarize the equilibrium concentrations in an ICE table.

	$Mn(OH)_2(s)$ $\rightleftharpoons$	$Mn^{2+}(aq)$	+	$2\,OH^-(aq)$
*I*nitial concentration (mol/L)		0		≈0*
*C*hange as reaction occurs (mol/L)		$+S$		$+2S$
*E*quilibrium concentration (mol/L)		$+S$		$+2S$

*We assume that the contribution of OH^- from the ionization of water is negligible.

We substitute into the K_{sp} expression

$$K_{sp} = [Mn^{2+}][OH^-]^2 = (S)(2S)^2 = 1.9 \times 10^{-13}$$

and solve for S, the solubility of $Mn(OH)_2$.

$$(S)(2S)^2 = 4S^3 = 1.9 \times 10^{-13}$$

$$S = [Mn^{2+}] = \sqrt[3]{1.9 \times 10^{-13}/4} = 3.6 \times 10^{-5}\ \text{M}$$

Therefore, the solubility of $Mn(OH)_2$ equals 3.6×10^{-5} M, the Mn^{2+} concentration; the OH^- concentration is $2S$, which is 7.2×10^{-5} M.

(b) Before any $Mn(OH)_2$ is added to the solution, the pH is 11.00. Thus, initially the $[H_3O^+] = 1.0 \times 10^{-11}$ M and the $[OH^-] = \dfrac{1.0 \times 10^{-14}}{1.0 \times 10^{-11}} = 1.0 \times 10^{-3}$ M. When $Mn(OH)_2(s)$ is added, let s equal the concentration of $Mn(OH)_2$ that dissolves (the solubility) forming s mol/L of Mn^{2+} and $2s$ mol/L of OH^-. In the solution there are two possible sources of hydroxide ions: (1) the initial solution in which the hydroxide concentration is 1.0×10^{-3} M and (2) the hydroxide ions, $2s$, arising from the dissociation of dissolved $Mn(OH)_2$. Correspondingly, the OH^- concentration at equilibrium is $1.0 \times 10^{-3} + 2s$.

	$Mn(OH)_2(s) \rightleftharpoons$	$Mn^{2+}(aq)$	+	$2\ OH^-(aq)$
*I*nitial concentration (mol/L)		0		1.0×10^{-3}
*C*hange as addition of $Mn(OH)_2$ (mol/L)		$+s$		$+2s$
*E*quilibrium concentration (mol/L)		$+s$		$1.0 \times 10^{-3} + 2s$

To calculate the solubility of $Mn(OH)_2$ we substitute these into the K_{sp} expression.

$$K_{sp} = [Mn^{2+}][OH^-]^2 = (s)(1.0 \times 10^{-3} + 2s)^2 = 1.9 \times 10^{-13}$$

The solubility of $Mn(OH)_2$ in pure water is very slight, as calculated in part (a). In addition, the dissolution of solid $Mn(OH)_2$ will be suppressed by the presence of hydroxide ions (the common ion effect) in the solution at pH 11.00. Consequently, we will assume that $2s \ll 1.0 \times 10^{-3}$, which simplifies the equation to

$$(s)(1.0 \times 10^{-3})^2 \approx 1.9 \times 10^{-13}$$

$$s \approx \frac{1.9 \times 10^{-13}}{1.0 \times 10^{-6}} = 1.9 \times 10^{-7}\ \text{M}$$

Thus, the presence of hydroxide as the common ion decreased the solubility of $Mn(OH)_2$ from that in pure water, 3.6×10^{-5} M, to 1.9×10^{-7} M in a starting solution of pH 11.00.

☑ **Reasonable Answer Check** In part (a) the solubility is described by the expression $(S)(2S)^2 = 4S^3 = 1.9 \times 10^{-13}$. The answer calculated using this expression is reasonable considering how very small the K_{sp} is. We can check the approximation we made in part (b) by substituting the approximate value of s into the actual expression $K_{sp} = (s)(1.0 \times 10^{-3} + 2s)^2$. Doing so we find that the product $(s)(1.0 \times 10^{-3} + 2s)^2$ equals the given K_{sp} value, so the approximation is legitimate.

PROBLEM-SOLVING PRACTICE **17.12**

Calculate the solubility of $PbCl_2$ in (a) pure water and (b) 0.20-M NaCl. K_{sp} of $PbCl_2 = 1.6 \times 10^{-5}$.

Complex Ion Formation

As pointed out in Section 16.9 (← *p. 786*), all metal cations are potential Lewis acids because they can accept an electron pair donated by a Lewis base to form a complex ion. The reaction of Cu^{2+} ions with NH_3 is typical.

$$\underset{\text{Lewis acid}}{Cu^{2+}(aq)} + \underset{\text{Lewis base}}{4\ NH_3(aq)} \rightleftharpoons \underset{\text{complex ion}}{[Cu(NH_3)_4]^{2+}(aq)}$$

Many metal salts that are insoluble in water are brought into solution by complex ion formation with Lewis bases such as $S_2O_3^{2-}$, NH_3, OH^-, and CN^-. For example, the solubility of AgBr is very low in water, 1.35×10^{-4} g/L, equivalent to

© Cengage Learning/Charles D. Winters

The addition of aqueous ammonia to aqueous Cu^{2+} ions initially forms $Cu(OH)_2$, the light blue solid. Additional ammonia converts $Cu(OH)_2$ to the intense deep-blue/purple $[Cu(NH_3)_4]^{2+}$ complex ion.

7.19×10^{-7} M. However, AgBr dissolves readily in a sodium thiosulfate, $Na_2S_2O_3$, solution due to the formation of the $[Ag(S_2O_3)_2]^{3-}$ complex ion (Figure 17.12).

$$AgBr(s) + 2\,S_2O_3^{2-}(aq) \rightleftharpoons [Ag(S_2O_3)_2]^{3-}(aq) + Br^-(aq)$$

The dissolving of AgBr in this way can be considered as the sum of two reactions—the solubility equilibrium of aqueous AgBr and the formation of the complex ion.

$$AgBr(s) \rightleftharpoons Ag^+(aq) + Br^-(aq)$$

$$Ag^+(aq) + 2\,S_2O_3^{2-}(aq) \rightleftharpoons [Ag(S_2O_3)_2]^{3-}(aq)$$

$$\textit{Net reaction}: AgBr(s) + 2\,S_2O_3^{2-}(aq) \rightleftharpoons [Ag(S_2O_3)_2]^{3-}(aq) + Br^-(aq)$$

The extent to which complex ion formation occurs can be evaluated from the magnitude of the equilibrium constant for the formation of the complex ion, K_f, called the **formation constant.** For example, the formation constant for $[Ag(S_2O_3)_2]^{3-}$ is 2.0×10^{13}.

$$Ag^+(aq) + 2\,S_2O_3^{2-}(aq) \rightleftharpoons [Ag(S_2O_3)_2]^{3-}(aq)$$

$$K_f = \frac{[[Ag(S_2O_3)_2]^{3-}]}{[Ag^+][S_2O_3^{2-}]^2} = 2.0 \times 10^{13}$$

Formation constants for some metal complex ions are given in Table 17.3. The formation and structure of complex ions, which are very important in biochemistry and metallurgy, are considered in more detail in Chapter 22.

PROBLEM-SOLVING EXAMPLE 17.13 Solubility and Complex Ion Formation

The K_{sp} of AgCl is 1.8×10^{-10}. The K_f of $[Ag(CN)_2]^-$ is 1.4×10^{20}. Use these data to show that AgCl will dissolve in aqueous NaCN.

Answer The net equilibrium constant is 2.5×10^{10}, which indicates that dissolving AgCl by complex ion formation is highly favored and AgCl will dissolve in NaCN.

Strategy and Explanation To answer this question requires using the solubility product constant for AgCl and the formation constant for $[Ag(CN)_2]^-$ to determine the overall equilibrium constant for the reaction of CN^- with AgCl. The magnitude of an equilibrium constant indicates whether a reaction is product-favored; a $K >> 1$ indicates a very product-favored reaction ***(← p. 669).***

© Cengage Learning/Charles D. Winters

Active Figure 17.12 Sodium thiosulfate dissolves silver bromide. (a) Silver bromide (white solid) is insoluble in water. (b) When aqueous sodium thiosulfate is added, the AgBr dissolves. Water molecules and Na^+ ions have been omitted from the illustration for simplicity's sake. Visit this book's companion website at www.cengage.com/chemistry/moore to test your understanding of the concepts in this figure.

The net reaction for dissolving AgCl by $[Ag(CN)_2]^-$ complex ion formation is the sum of the K_{sp} and K_f equations.

$$AgCl(s) \rightleftharpoons Ag^+(aq) + Cl^-(aq) \qquad K_{sp} = 1.8 \times 10^{-10}$$

$$Ag^+(aq) + 2\,CN^-(aq) \rightleftharpoons [Ag(CN)_2]^-(aq) \qquad K_f = 1.4 \times 10^{20}$$

Net reaction: $AgCl(s) + 2\,CN^-(aq) \rightleftharpoons [Ag(CN)_2]^-(aq) + Cl^-(aq)$

Therefore, the equilibrium constant for the net reaction is the product of K_{sp} and K_f : $K_{net} = K_{sp} \times K_f = (1.8 \times 10^{-10})(1.4 \times 10^{20}) = 2.5 \times 10^{10}$. Because K_{net} is much greater than 1, the net reaction is product-favored, and AgCl is much more soluble in a NaCN solution than in water.

PROBLEM-SOLVING PRACTICE 17.13

The K_{sp} of AgCl is 1.8×10^{-10}. The K_f of $[Ag(S_2O_3)_2]^{3-}$ is 2.0×10^{13}. Use these data to show that dissolving AgCl by $[Ag(S_2O_3)_2]^{3-}$ complex ion formation is a product-favored process.

Amphoterism

The majority of metal hydroxides are insoluble in water, but many dissolve in highly acidic or basic solutions. This is because these hydroxides are ***amphoteric;*** that is, they can react with both H_3O^+ ions and OH^- ions **(← *p. 788*)**. Aluminum hydroxide, $Al(OH)_3$, is an example of an amphoteric hydroxide (Figure 17.13). When it reacts with acid, aluminum hydroxide dissolves by acting as a base, donating OH^- ions to react with hydronium ions from the acid to form water.

$$Al(OH)_3(s) + 3\,H_3O^+(aq) \longrightarrow Al^{3+}(aq) + 6\,H_2O(\ell)$$

Table 17.3 Formation Constants for Some Complex Ions in Aqueous Solution*

Formation Equilibrium	K_f
$Ag^+ + 2\,CN^- \rightleftharpoons [Ag(CN)_2]^-$	1.4×10^{20}
$Ag^+ + 2\,S_2O_3^{2-} \rightleftharpoons [Ag(S_2O_3)_2]^{3-}$	2.0×10^{13}
$Ag^+ + 2\,NH_3 \rightleftharpoons [Ag(NH_3)_2]^+$	1×10^{7}
$Al^{3+} + 4\,OH^- \rightleftharpoons [Al(OH)_4]^-$	4.0×10^{36}
$Au^+ + 2\,CN^- \rightleftharpoons [Au(CN)_2]^-$	4.0×10^{36}
$Cd^{2+} + 4\,CN^- \rightleftharpoons [Cd(CN)_4]^{2-}$	1.6×10^{19}
$Cd^{2+} + 4\,NH_3 \rightleftharpoons [Cd(NH_3)_4]^{2+}$	9.4×10^{6}
$Co^{2+} + 6\,NH_3 \rightleftharpoons [Co(NH_3)_6]^{2+}$	1.3×10^{4}
$Cu^{2+} + 4\,NH_3 \rightleftharpoons [Cu(NH_3)_4]^{2+}$	3.3×10^{12}
$Fe^{2+} + 6\,CN^- \rightleftharpoons [Fe(CN)_6]^{4-}$	3.2×10^{32}
$Hg^{2+} + 4\,Cl^- \rightleftharpoons [HgCl_4]^{2-}$	1.3×10^{15}
$Ni^{2+} + 4\,CN^- \rightleftharpoons [Ni(CN)_4]^{2-}$	1.7×10^{30}
$Ni^{2+} + 6\,NH_3 \rightleftharpoons [Ni(NH_3)_6]^{2+}$	5.5×10^{8}
$Zn^{2+} + 4\,OH^- \rightleftharpoons [Zn(OH)_4]^{2-}$	4.5×10^{17}
$Zn^{2+} + 4\,NH_3 \rightleftharpoons [Zn(NH_3)_4]^{2+}$	2.9×10^{9}

*Taken from Högfeldt, E., Perrin, D. D., *Stability Constants of Metal-Ion Complexes,* 1st ed., Oxford; New York: Pergamon, 1979–1982. International Union of Pure and Applied Chemistry Commission on Equilibrium Data. Values for $[AgBr_2]^-$, $[CO(NH_3)_6]^{2+}$, and $[Zn(NH_3)_4]^{2+}$, from Patnaik, Pradyot, *Dean's Analytical Chemistry Handbook,* 2nd ed., New York, McGraw-Hill, 2004, Table 4.2.

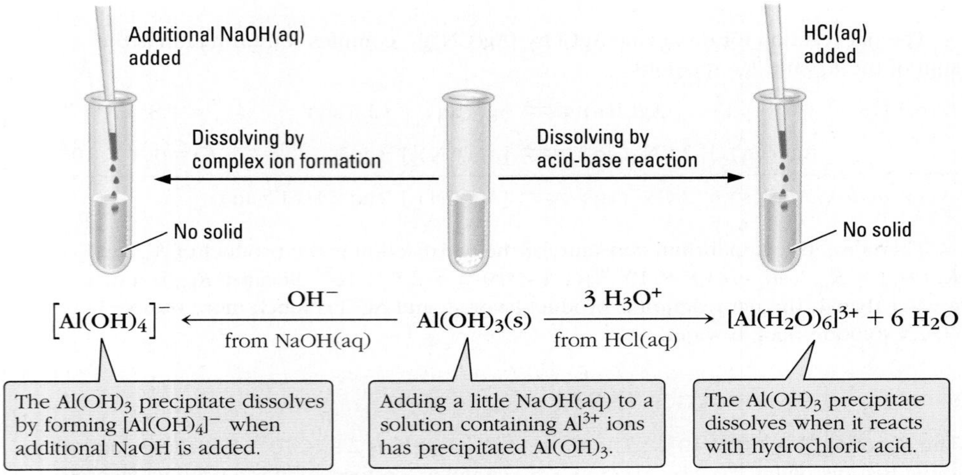
Additional NaOH(aq) added
HCl(aq) added
Dissolving by complex ion formation
Dissolving by acid-base reaction
No solid
No solid
$[Al(OH)_4]^- \xleftarrow[\text{from NaOH(aq)}]{OH^-} Al(OH)_3(s) \xrightarrow[\text{from HCl(aq)}]{3\ H_3O^+} [Al(H_2O)_6]^{3+} + 6\ H_2O$
The $Al(OH)_3$ precipitate dissolves by forming $[Al(OH)_4]^-$ when additional NaOH is added.
Adding a little NaOH(aq) to a solution containing Al^{3+} ions has precipitated $Al(OH)_3$.
The $Al(OH)_3$ precipitate dissolves when it reacts with hydrochloric acid.

Figure 17.13 **The amphoteric nature of $Al(OH)_3$.**

© Cengage Learning/Charles D. Winters

Precipitation of lead iodide. Adding a drop of aqueous potassium iodide solution to an aqueous lead(II) nitrate solution precipitates yellow lead(II) iodide, PbI_2. Potassium nitrate, a soluble salt, remains dissolved in the solution.

In highly basic solutions, $Al(OH)_3$ is dissolved through complex ion formation. In this case, Al^{3+} ions act as a Lewis acid by accepting electron pairs from OH^- ions, a Lewis base, to form $[Al(OH)_4]^-$.

$$Al(OH)_3(s) + OH^-(aq) \longrightarrow [Al(OH)_4]^-(aq)$$

17.6 Precipitation: Will It Occur?

Earlier, when writing net ionic equations, we used the solubility rules to predict whether a precipitate will form when ions in two solutions are mixed. Those rules apply to situations where the ions involved are at concentrations of 0.1 M or greater. If the ion concentrations are considerably less than 0.1 M, precipitation may or may not occur. The result depends on the concentrations of the ions in the resulting solution and the K_{sp} value for any precipitate that might form.

For example, AgBr might precipitate when a water-soluble silver salt, such as $AgNO_3$, is added to an aqueous solution of a bromide salt, such as KBr. The net ionic equation for the reaction is

$$Ag^+(aq) + Br^-(aq) \rightleftharpoons AgBr(s)$$

This is the reverse of the equation for K_{sp} of AgBr:

$$AgBr(s) \rightleftharpoons Ag^+(aq) + Br^-(aq)$$

To determine whether a precipitate will form, we compare the magnitude of the ion product, Q, with that of the solubility product constant, K_{sp}. The Q expression has the same form as that for K_{sp}. For Q, however, the *initial* concentrations are used, *not those at equilibrium as in* K_{sp}. For AgBr the two expressions are

Q is the reaction quotient introduced in Chapter 14 (← *p. 672*).

The bracket notation, [ ], represents molarity at *equilibrium*.

$$Q = (\text{conc. } Ag^+)(\text{conc. } Br^-) \qquad K_{sp} = [Ag^+][Br^-]$$

When the value of Q is compared with that of the K_{sp}, three cases are possible (Figure 17.14).

1. $Q < K_{sp}$ ***The solution is unsaturated and no precipitate forms.*** In this case, the solution contains ions at a concentration lower than required for equilibrium with the solid. An equilibrium is not established between a solid solute and its ions because no solid solute is present; more solute can be added to the solution before precipitation occurs. If solid were present, more solid would dissolve.

2. $Q > K_{sp}$ ***The solution contains a higher concentration of ions than it can hold at equilibrium; that is, the solution is supersaturated.*** To reach equilibrium, a precipitate forms, decreasing the concentration of ions until the ion product equals the K_{sp}.
3. $Q = K_{sp}$ ***The solution is saturated with ions and is at equilibrium and at the point of precipitation.***

Consider the case of two solutions, each made by combining $Pb(NO_3)_2$ and Na_2SO_4 solutions. In one case (solution 1) when the solutions are mixed the initial concentrations of Pb^{2+} and SO_4^{2-} are each 1.0×10^{-4} M. In the other case (solution 2) these concentrations are each 2.0×10^{-4}, twice that of the first solution. In each of the two solutions, the reaction products are $NaNO_3$ and $PbSO_4$. The solubility rules indicate that $NaNO_3$ is soluble and remains in solution as Na^+ and NO_3^- ions, whereas $PbSO_4$ is insoluble. Will a precipitate of $PbSO_4$ form in either or both solutions? We can determine this by using Q and K_{sp} for $PbSO_4$; K_{sp} of $PbSO_4 = 1.6 \times 10^{-8}$.

$$Q \text{ of solution 1} = (\text{conc. } Pb^{2+})(\text{conc. } SO_4^{2-})$$
$$= (1.0 \times 10^{-4}\text{ M})(1.0 \times 10^{-4}\text{ M}) = 1.0 \times 10^{-8} < K_{sp} = 1.6 \times 10^{-8}$$

$$Q \text{ of solution 2} = (\text{conc. } Pb^{2+})(\text{conc. } SO_4^{2-})$$
$$= (2.0 \times 10^{-4}\text{ M})(2.0 \times 10^{-4}\text{ M}) = 4.0 \times 10^{-8} > K_{sp} = 1.6 \times 10^{-8}$$

Since Q of solution 1 is less than the K_{sp}, no precipitate will form. In contrast, Q of solution 2 exceeds the K_{sp}, and precipitation will occur.

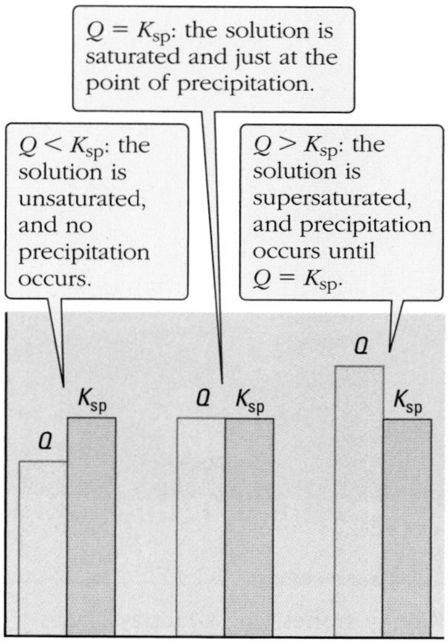

Figure 17.14 Predicting precipitation. When $Q < K_{sp}$, the solution is unsaturated, and no precipitation occurs. $Q = K_{sp}$: the solution is saturated and just at the point of precipitation. $Q > K_{sp}$: the solution is supersaturated, and precipitation occurs until $Q = K_{sp}$.

PROBLEM-SOLVING EXAMPLE 17.14 Q, K_{sp}, and Precipitation

A chemistry student mixes 20.0 mL of 4.5×10^{-3}-M $AgNO_3$ with 10.0 mL of 7.5×10^{-2}-M $NaBrO_3$. The final volume is 30.0 mL. Will a precipitate of $AgBrO_3$ form? The K_{sp} of $AgBrO_3 = 6.7 \times 10^{-5}$.

Answer Yes, $AgBrO_3$ precipitates.

Strategy and Explanation For a precipitate to form, Q must be greater than K_{sp}. The chemical equilibrium is

$$AgBrO_3(s) \rightleftharpoons Ag^+(aq) + BrO_3^-(aq)$$

The ion product expression is $Q = (\text{conc. } Ag^+)(\text{conc. } BrO_3^-)$; the K_{sp} expression is $K_{sp} = [Ag^+][BrO_3^-]$.

To determine whether a precipitate forms, substitute the original concentrations into the Q expression and compare the value with K_{sp}. In calculating the original concentrations, the total volume of 0.0300 L must be taken into account. The number of moles of Ag^+ in 0.0200 L of 4.5×10^{-3}-M $AgNO_3$ is

$$(0.0200\text{ L})\left(\frac{4.5 \times 10^{-3}\text{ mol } Ag^+}{1\text{ L}}\right) = 9.0 \times 10^{-5}\text{ mol } Ag^+$$

The Ag^+ concentration in the 0.0300-L mixture is

$$\frac{9.0 \times 10^{-5}\text{ mol } Ag^+}{0.0300\text{ L}} = 3.0 \times 10^{-3}\text{ M } Ag^+$$

Likewise, for the amount of BrO_3^- and its concentration in the 0.0300-L mixture:

$$(0.0100\text{ L})\left(\frac{7.5 \times 10^{-2}\text{ mol } BrO_3^-}{1\text{ L}}\right) = 7.5 \times 10^{-4}\text{ mol } BrO_3^-$$

$$\frac{7.5 \times 10^{-4}\text{ mol } BrO_3^-}{0.0300\text{ L}} = 2.5 \times 10^{-2}\text{ M } BrO_3^-$$

$$Q = (3.0 \times 10^{-3}\text{ M})(2.5 \times 10^{-2}\text{ M}) = 7.5 \times 10^{-5}$$

which is greater than the $K_{sp} = 6.7 \times 10^{-5}$. Precipitation occurs until the ion product equals K_{sp}.

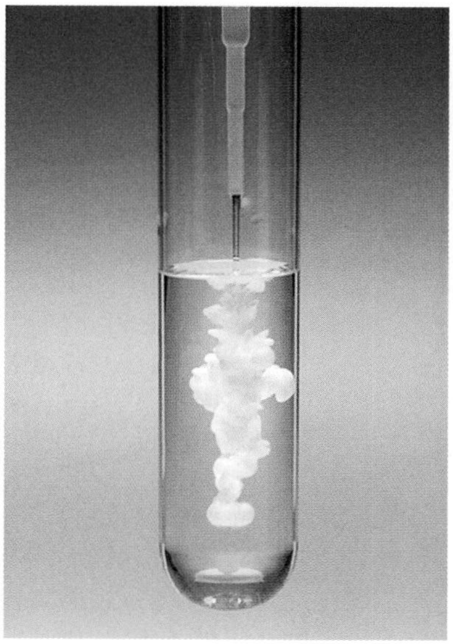
© Cengage Learning/Charles D. Winters

Precipitation of lead sulfate. The addition of sufficiently high concentrations of Pb^{2+} and SO_4^{2-} ions causes $Q > K_{sp}$ and white lead sulfate, $PbSO_4$, precipitates.

PROBLEM-SOLVING PRACTICE 17.14

(a) Will AgCl precipitate from a solution containing 1.0×10^{-5}-M Ag^+ and 1.0×10^{-5}-M Cl^-? K_{sp} AgCl $= 1.8 \times 10^{-10}$.

(b) An AgCl precipitate forms from a solution that is 1.0×10^{-5}-M Ag^+. What must be the minimum Cl^- concentration in this solution for precipitation to occur?

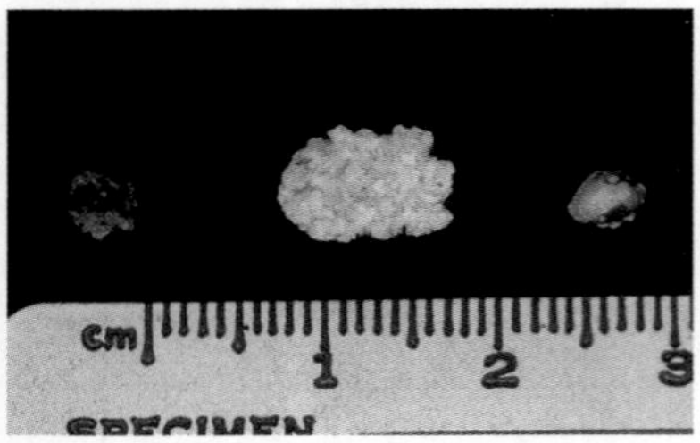

Kidney stones. These kidney stones were surgically removed from a patient.

Leaving chocolate out of a diet seems far more punishment than forgoing spinach.

Kidney Stones—Common Ion Effect and Le Chatelier's Principle

Many ions circulate in our bloodstream; some combinations of ions can precipitate to form kidney stones. Such stones can become large enough to be extremely painful and even life-threatening, requiring treatment by drugs, lasers, or surgery. Kidney stones usually consist of insoluble calcium and magnesium compounds such as calcium oxalate, CaC_2O_4; calcium phosphate, $Ca_3(PO_4)_2$; magnesium ammonium phosphate, $MgNH_4PO_4$; or a mixture of these. For calcium oxalate kidney stones, the equilibrium $CaC_2O_4(s) \rightleftharpoons Ca^{2+}(aq) + C_2O_4^{2-}(aq)$ applies. High intake of foods rich in calcium or oxalate can cause a rise in the urinary concentration of either ion (or both) sufficient to make $Q > K_{sp}$ so that the equilibrium shifts to the left. The result is precipitation of calcium oxalate as a kidney stone. Thus, foods rich in Ca^{2+}, such as milk, ice cream, or cheese, or those high in $C_2O_4^{2-}$, such as chocolate, spinach, celery, or black tea, can trigger the onset of a kidney stone through the common ion effect. Such foods are restricted in the diets of individuals prone to developing kidney stones. A high-sugar diet may also create kidney stones because excessive sugar promotes excretion of Ca^{2+} and Mg^{2+}, which increases the concentrations of these ions passing through the kidneys. This can cause kidney stone formation, such as through calcium phosphate precipitation:

$$3\,Ca^{2+}(aq) + 2\,PO_4^{3-}(aq) \longrightarrow Ca_3(PO_4)_2(s)$$

Selective Precipitation of Ions

If their solubilities are sufficiently different, ionic compounds can be precipitated selectively from solution. The more soluble compound remains in solution as the less soluble one starts to precipitate. For example, silver chloride, AgCl, and silver chromate, Ag_2CrO_4, are each only slightly soluble in water. The solubilities of these two solutes differ enough, however, that when they are both in the same solution, one can be precipitated, leaving the other behind in solution.

PROBLEM-SOLVING EXAMPLE 17.15 Selective Precipitation

Consider a solution containing 0.020-M Cl^- and 0.010-M CrO_4^{2-} ions to which Ag^+ ions are added slowly. Which precipitate forms first: AgCl or Ag_2CrO_4? K_{sp} AgCl $= 1.8 \times 10^{-10}$; K_{sp} $Ag_2CrO_4 = 1.1 \times 10^{-12}$.

Answer AgCl precipitates first.

Strategy and Explanation To answer this question, first find the minimum Ag^+ concentration required to precipitate each compound, which is the molar concentration product of its ions that just barely exceeds the K_{sp}. To precipitate AgCl,

$$K_{sp} \text{ of AgCl} = [Ag^+][Cl^-] = 1.8 \times 10^{-10}$$

$$[Ag^+] = \frac{1.8 \times 10^{-10}}{[Cl^-]} = \frac{1.8 \times 10^{-10}}{2.0 \times 10^{-2}} = 9.0 \times 10^{-9}\ \text{M}$$

An Ag^+ concentration of slightly greater than 9.0×10^{-9} M will precipitate some AgCl from the solution. To precipitate Ag_2CrO_4,

$$K_{sp} \text{ of } Ag_2CrO_4 = [Ag^+]^2[CrO_4^-] = 9 \times 10^{-12}$$

$$[Ag^+]^2 = \frac{1.1 \times 10^{-12}}{[CrO_4^{2-}]} = \frac{1.1 \times 10^{-12}}{1.0 \times 10^{-2}} = 1.1 \times 10^{-10} \text{ M}; \qquad [Ag^+] = 3 \times 10^{-5} \text{ M}$$

Silver chromate will precipitate when the Ag^+ concentration slightly exceeds 3×10^{-5} M. Because a *much* smaller concentration of Ag^+, 9.0×10^{-9} M, is required to precipitate AgCl, AgCl will precipitate before Ag_2CrO_4.

☑ **Reasonable Answer Check** The Ag^+ concentration required to precipitate Ag_2CrO_4 is approximately 10,000 times greater than that for AgCl (3×10^{-5} M versus 9×10^{-9} M). Therefore, the answer is reasonable; AgCl will precipitate first. In fact, the difference is so great that essentially all of the AgCl will precipitate before Ag_2CrO_4 precipitation begins.

PROBLEM-SOLVING PRACTICE 17.15

Hydrochloric acid is slowly added to a solution that is 0.10 M in Pb^{2+} and 0.010 M in Ag^+. Which precipitate is formed first, AgCl or $PbCl_2$?

SUMMARY PROBLEMS

1. (a) Describe how to prepare a pH 3.70 buffer using formic acid, HCOOH, and sodium formate, NaHCOO.
   (b) Calculate the pH of this buffer after the addition of 0.0050 mol HCl.
   (c) Calculate the mass in grams of NaOH that could be added to the initial buffer before its buffer capacity is just exceeded.
2. The K_a of nitrous acid, HNO_2, is 4.5×10^{-4}. In a titration, 50.0 mL of 1.00-M HNO_2 is titrated with 0.750-M NaOH.
   (a) Calculate the pH of the solution:
      (i) Before the titration begins
      (ii) When sufficient NaOH has been added to neutralize half of the nitrous acid originally present
      (iii) At the equivalence point
      (iv) When 0.05 mL NaOH less than that required to reach the equivalence point has been added
      (v) When 0.05 mL NaOH more than that required to reach the equivalence point has been added
   (b) Can bromthymol blue be used as the indicator for this titration? Explain your answer.
   (c) Will methyl red be a satisfactory indicator here?
   (d) Use data from part (a) to plot a graph of pH (y-axis) versus volume of titrant.
3. A 0.500-L solution contains 0.025 mol Ag^+.
   (a) Calculate the minimum mass of NaCl that must be added to precipitate AgCl from the solution.
   (b) If excess Cl^- is added to the solution, the AgCl precipitate dissolves due to the formation of $[Ag(Cl)_2]^-$; K_f of $[Ag(Cl)_2]^- = 1.8 \times 10^5$. Calculate the minimum amount of Cl^- that must be added to dissolve the precipitate.

In NaHCOO, $HCOO^-$ is the formate ion.

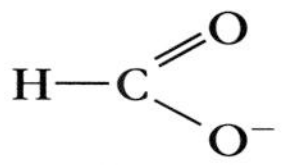

IN CLOSING

Having studied this chapter, you should be able to . . .

- Explain how buffers maintain pH, how to calculate their pH, how they are prepared, and the importance of buffer capacity (Section 17.1). End-of-Chapter Questions: 16, 22, 28

OWL and go Chemistry

Sign in at **www.cengage.com/owl** to:

- View tutorials and simulations, develop problem-solving skills, and complete online homework assigned by your professor.
- For quick review and exam prep, download Go Chemistry mini lecture modules from OWL (or purchase them at **www.CengageBrain.com**).

- Use the Henderson-Hasselbalch equation or the K_a expression to calculate the pH of a buffer and the pH change after acid or base has been added to the buffer (Section 17.1). Questions 26, 30
- Interpret acid-base titration curves and calculate the pH of the solution at various stages of the titration (Section 17.2). Questions 33, 36, 44, 94
- Explain how acid rain is formed and its effects on the environment (Section 17.3). Questions 48, 50
- Relate a K_{sp} expression to its chemical equation (Section 17.4). Question 54
- Use the solubility of a slightly soluble solute to calculate its solubility product (Section 17.4). Question 56
- Describe the factors affecting the aqueous solubility of ionic compounds and calculate their effect on solubility (Section 17.5). Questions 68, 70
- Apply Le Chatelier's principle to the common ion effect (Section 17.5). Question 88
- Use the solubility product to calculate the solubility of a sparingly soluble solute in pure water and in the presence of a common ion (Section 17.5). Question 63
- Describe the effect of complex ion formation on the solubility of a sparingly soluble ionic compound (Section 17.5).
- Relate Q, the ion product, to K_{sp} to determine whether precipitation will occur (Section 17.6).
- Predict which of two sparingly soluble ionic solutes will precipitate first (Section 17.6). Question 114

KEY TERMS

acid rain *(Section 17.3)*

buffer *(17.1)*

buffer capacity *(17.1)*

buffer solution *(17.1)*

common ion effect *(17.5)*

end point *(17.2)*

formation constant, K_f *(17.5)*

Henderson-Hasselbalch equation *(17.1)*

solubility product constant, K_{sp} *(17.4)*

solubility product constant expression *(17.4)*

titrant *(17.2)*

titration curve *(17.2)*

QUESTIONS FOR REVIEW AND THOUGHT

OWL Interactive versions of these problems are assignable in OWL.

Blue-numbered questions have short answers at the back of this book in Appendix M and fully worked solutions in the *Student Solutions Manual.*

Review Questions

These questions test vocabulary and simple concepts.

1. What is meant by the term "buffer capacity"?
2. Which would form a buffer?
   (a) HCl and CH_3COOH (b) NaH_2PO_4 and Na_2HPO_4
   (c) H_2CO_3 and $NaHCO_3$
3. Which would form a buffer?
   (a) NaOH and NaCl (b) NaOH and NH_3
   (c) Na_3PO_4 and Na_2HPO_4
4. Briefly describe how a buffer solution can control the pH of a solution when strong acid is added and when strong base is added. Use NH_3/NH_4Cl as an example of a buffer and HCl and NaOH as the strong acid and strong base.
5. What is the difference between the end point and the equivalence point in an acid-base titration?
6. What is meant by an indicator range for an acid-base indicator?
7. What are the characteristics of a good acid-base indicator?
8. A strong acid is titrated with a strong base, such as KOH. Describe the changes in the composition of the solution as the titration proceeds: prior to the equivalence point, at the equivalence point, and beyond the equivalence point.
9. Repeat the description for Question 8, but use a weak acid rather than a strong one.
10. Use Le Chatelier's principle to explain why $PbCl_2$ is less soluble in 0.010-M $Pb(NO_3)_2$ than in pure water.
11. Describe what a complex ion is and give an example.
12. What is amphoterism?

13. Distinguish between the ion product (*Q*) expression and the solubility product constant expression of a sparingly soluble solute.
14. Describe how the solubility of a sparingly soluble metal hydroxide can be changed.

Topical Questions

These questions are keyed to the major topics in the chapter. Usually a question that is answered at the back of this book is paired with a similar one that is not.

Buffer Solutions (Section 17.1)

15. Many natural processes can be studied in the laboratory but only in an environment of controlled pH. Which of these combinations would be the best choice to buffer the pH at approximately 7?
(a) H_3PO_4/NaH_2PO_4 (b) NaH_2PO_4/Na_2HPO_4
(c) Na_2HPO_4/Na_3PO_4
16. Which of these combinations would be the best to buffer the pH at approximately 9?
(a) $CH_3COOH/NaCH_3COO$ (b) HCl/NaCl
(c) NH_3/NH_4Cl
17. Without doing calculations, determine the pH of a buffer made from equimolar amounts of these acid-base pairs.
(a) Nitrous acid and sodium nitrite
(b) Ammonia and ammonium chloride
(c) Formic acid and potassium formate
18. Without doing calculations, determine the pH of a buffer made from equimolar amounts of these acid-base pairs.
(a) Phosphoric acid and sodium dihydrogen phosphate
(b) Sodium monohydrogen phosphate and sodium dihydrogen phosphate
(c) Sodium phosphate and sodium monohydrogen phosphate
19. Select from Table 17.1 a conjugate acid-base pair that would be suitable for preparing a buffer solution whose concentration of hydronium ions is
(a) 4.5×10^{-3} M (b) 5.2×10^{-8} M
(c) 8.3×10^{-6} M (d) 9.7×10^{-11} M
Explain your choices.
20. Select from Table 17.1 a conjugate acid-base pair that would be suitable for preparing a buffer solution with pH equal to
(a) 3.45 (b) 5.48
(c) 8.32 (d) 10.15
Explain your choices.
21. Calculate the mass of sodium acetate, $NaCH_3COO$, you should add to 500. mL of a 0.150-M solution of acetic acid, CH_3COOH, to buffer a solution at a pH of 4.57.
22. Calculate the mass in grams of ammonium chloride, NH_4Cl, that would have to be added to 500. mL of 0.10-M NH_3 solution to have a pH of 9.00.
23. A buffer solution can be made from benzoic acid, C_6H_5COOH, and sodium benzoate, NaC_6H_5COO. Calculate the mass in grams of the acid you would have to mix with 14.4 g of the sodium salt to have a liter of solution with a pH of 3.88.
24. A buffer solution is prepared from 5.15 g NH_4NO_3 and 0.10 L of 0.15-M NH_3; calculate the pH of the solution.
25. You dissolve 0.425 g NaOH in 2.00 L of a solution that originally had $[H_2PO_4^-] = [HPO_4^{2-}] = 0.132$ M. Calculate the resulting pH.
26. A buffer solution is prepared by adding 0.125 mol ammonium chloride to 500. mL of 0.500-M aqueous ammonia. Calculate the pH of the buffer. If 0.0100 mol HCl gas is bubbled into 500. mL buffer, calculate the new pH of the solution.
27. If added to 1 L of 0.20-M acetic acid, CH_3COOH, which of these would form a buffer?
(a) 0.10 mol $NaCH_3COO$ (b) 0.10 mol NaOH
(c) 0.10 mol HCl (d) 0.30 mol NaOH
Explain your answers.
28. If added to 1 L of 0.20-M NaOH, which of these would form a buffer?
(a) 0.10 mol acetic acid (b) 0.30 mol acetic acid
(c) 0.20 mol HCl (d) 0.10 mol $NaCH_3COO$
Explain your answers.
29. Calculate the pH change when 10.0 mL of 0.10-M NaOH is added to 90.0 mL pure water, and compare the pH change with that when the same amount of NaOH solution is added to 90.0 mL of a buffer consisting of 1.0-M NH_3 and 1.0-M NH_4Cl. Assume that the volumes are additive. K_b of $NH_3 = 1.7 \times 10^{-5}$.
30. Calculate the pH change when 1.0 mL of 1.0-M NaOH is added to 0.100 L of a solution of
(a) 0.10-M acetic acid and 0.10-M sodium acetate.
(b) 0.010-M acetic acid and 0.010-M sodium acetate.
(c) 0.0010-M acetic acid and 0.0010-M sodium acetate.
31. Calculate the pH change when 1.0 mL of 1.0-M HCl is added to 0.100 L of a solution of
(a) 0.10-M acetic acid and 0.10-M sodium acetate.
(b) 0.010-M acetic acid and 0.010-M sodium acetate.
(c) 0.0010-M acetic acid and 0.0010-M sodium acetate.
32. A buffer consists of 0.20-M propanoic acid ($K_a = 1.4 \times 10^{-5}$) and 0.30-M sodium propanoate.
(a) Calculate the pH of this buffer.
(b) Calculate the pH after the addition of 1.0 mL of 0.10-M HCl to 0.010 L of the buffer.
(c) Calculate the pH after the addition of 3.0 mL of 1.0-M HCl to 0.010 L of the buffer.

Acid-Base Titrations and Titration Curves (Section 17.2)

33. The titration curves for two acids with the same base are

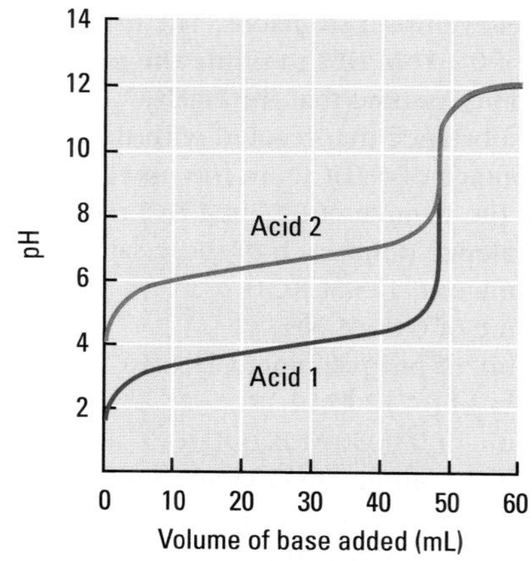

(a) Which is the curve for the weaker acid? Explain your choice.

(b) Give the approximate pH at the equivalence point for the titration of each acid.
(c) Explain why the pH at the equivalence point differs for each acid.
(d) Explain why the starting pH values of the two acids differ.
(e) Which indicator could be used for the titration of Acid 1 and for the titration of Acid 2? Explain your choices.

34. Explain why it is that the weaker the acid being titrated, the more alkaline the pH is at the equivalence point.
35. Sketch the titration curve for the titration of 20.0 mL of a 0.100-M solution of a strong acid by a 0.100-M weak base; that is, the base is the titrant. In particular, note the pH of the solution:
(a) Prior to the titration
(b) When half the required volume of titrant has been added
(c) At the equivalence point
(d) 10.0 mL beyond the equivalence point.
36. Consider all acid-base indicators discussed in this chapter. Which of these indicators would be suitable for the titration of
(a) NaOH with $HClO_4$ (b) Acetic acid with KOH
(c) NH_3 solution with HBr (d) KOH with HNO_3
Explain your choices.
37. Which of the acid-base indicators discussed in this chapter would be suitable for the titration of
(a) HNO_3 with KOH (b) KOH with acetic acid
(c) HCl with NH_3 (d) KOH with HNO_2
Explain your answers.
38. It required 22.6 mL of 0.0140-M $Ba(OH)_2$ solution to titrate a 25.0-mL sample of HCl to the equivalence point. Calculate the molarity of the HCl solution.
39. It took 12.4 mL of 0.205-M H_2SO_4 solution to titrate 20.0 mL of a sodium hydroxide solution to the equivalence point. Calculate the molarity of the original NaOH solution.
40. Vitamin C is a monoprotic acid. To analyze a vitamin C capsule weighing 0.505 g by titration took 24.4 mL of 0.110-M NaOH. Calculate the percentage of vitamin C, $C_6H_8O_6$, in the capsule. Assume that vitamin C is the only substance in the capsule that reacts with the titrant.
41. An acid-base titration was used to find the percentage of $NaHCO_3$ in 0.310 g of a powdered commercial product used to relieve upset stomachs. The titration required 14.3 mL of 0.101-M HCl to titrate the powder to the equivalence point. Assume that the $NaHCO_3$ in the powder is the only substance that reacted with the titrant. Calculate the percentage of $NaHCO_3$ in the powder.
42. Calculate the volume of 0.150-M HCl required to titrate to the equivalence point each of these samples.
(a) 25.0 mL of 0.175-M KOH
(b) 15.0 mL of 6.00-M NH_3
(c) 15.0 mL of propylamine, $CH_3CH_2CH_2NH_2$, which has a density of 0.712 g/mL
(d) 40.0 mL of 0.0050-M $Ba(OH)_2$
43. Calculate the volume of 0.225-M NaOH required to titrate to the equivalence point each of these samples.
(a) 20.0 mL of 0.315-M HBr
(b) 30.0 mL of 0.250-M $HClO_4$
(c) 6.00 g of concentrated acetic acid, CH_3COOH, which is 99.7% pure
44. A 30.00-mL solution of 0.100-M benzoic acid, a monoprotic acid, is titrated with 0.100-M NaOH. The K_a of benzoic acid is 1.2×10^{-4}. Determine the pH after each of these volumes of titrant has been added:
(a) 10.00 mL (b) 30.00 mL (c) 40.00 mL
45. The titration of 50.00 mL of 0.150-M NaOH with 0.150-M HCl (the titrant) is carried out in a chemistry laboratory. Calculate the pH of the solution after these volumes of the titrant have been added:
(a) 0.00 mL (b) 25.00 mL (c) 49.9 mL
(d) 50.00 mL (e) 50.1 mL (f) 75.00 mL
Use the results of your calculations to plot a titration curve for this titration. On the curve indicate the position of the equivalence point.
46. The titration of 50.00 mL of 0.150-M HCl with 0.150-M NaOH (the titrant) is carried out in a chemistry laboratory. Calculate the pH of the solution after these volumes of the titrant have been added:
(a) 0.00 mL (b) 25.00 mL (c) 49.9 mL
(d) 50.00 mL (e) 50.1 mL (f) 75.00 mL
Use the results of your calculations to plot a titration curve for this titration. On the curve indicate the position of the equivalence point.

Acid Rain (Section 17.3)

47. Explain why rain with a pH of 6.7 is not classified as acid rain.
48. Identify two oxides that are key producers of acid rain. Write chemical equations that illustrate how these oxides form acid rain.
49. Acid rain has been measured with a pH of 1.5. Calculate the hydrogen ion concentration of this rain.
50. Write a chemical equation that shows how limestone neutralizes acid rain.

Solubility Product (Section 17.4)

51. Write the K_{sp} expressions for these slightly soluble salts.
(a) $BaCrO_4$ (b) $Mn(OH)_2$ (c) Ag_2SO_4
52. The solubility of $PbBr_2$ is 3.74×10^{-3} g per 100.0 mL at 25 °C. Calculate the K_{sp} of $PbBr_2$, assuming that the solute dissociates completely into Pb^{2+} and Br^- ions and that these ions do not react with water.
53. Write a balanced chemical equation for the equilibrium occurring when each of these solutes is added to water, then write the K_{sp} expression for each solute.
(a) Lead(II) carbonate (b) Nickel(II) hydroxide
(c) Strontium phosphate (d) Mercury(I) sulfate
54. Write a balanced chemical equation for the equilibrium occurring when each of these solutes is added to water, then write the K_{sp} expression.
(a) Iron(II) carbonate (b) Silver sulfate
(c) Calcium phosphate (d) Mn(II) hydroxide
55. A saturated solution of silver arsenate, Ag_3AsO_4, contains 8.5×10^{-6} g Ag_3AsO_4 per mL. Calculate the K_{sp} of silver arsenate. Assume that there are no other reactions but the K_{sp} reaction.
56. At 20. °C, 2.03 g $CaSO_4$ dissolves per liter of water. From these data calculate the K_{sp} of calcium sulfate at 20. °C. Assume that there are no other reactions but the K_{sp} reaction.

57. The water solubility of strontium fluoride, SrF_2, is 0.011 g/100. mL. Calculate its solubility product constant. Assume that there are no reactions other than the K_{sp} reaction.
58. The solubility of silver chromate, Ag_2CrO_4, in water is 2.7×10^{-3} g/100. mL. Calculate the K_{sp} of silver chromate. Assume that there are no reactions other than the K_{sp} reaction.
59. Calculate the K_{sp} of HgI_2 given that its solubility in water is 2.0×10^{-11} M. Assume that there are no reactions other than the K_{sp} reaction.
60. The solubility of $PbCl_2$ in water is 1.62×10^{-2} M. Calculate the K_{sp} of $PbCl_2$. Assume that there are no reactions other than the K_{sp} reaction.
61. In a saturated CaF_2 solution at 25°C, the calcium concentration is analyzed to be 9.1 mg/L. Use this value to calculate the K_{sp} of CaF_2 assuming that the solute dissociates completely into Ca^{2+} and F^- ions, and that neither ion reacts with water.

Common Ion Effect (Section 17.5)

62. Calculate the molarity of Zn^{2+} ion in a saturated solution of $ZnCO_3$ that contains 0.25-M Na_2CO_3.
63. Calculate the solubility of $ZnCO_3$ in
    (a) Water (b) 0.050-M $Zn(NO_3)_2$
    (c) 0.050-M K_2CO_3
    K_{sp} of $ZnCO_3 = 1.4 \times 10^{-11}$
64. Calculate the solubility (mol/L) of $SrSO_4$ ($K_{sp} = 3.2 \times 10^{-7}$) in 0.010-M Na_2SO_4.
65. Iron(II) hydroxide, $Fe(OH)_2$, has a solubility in water of 6.0×10^{-1} mg/L at a given temperature.
    (a) From this solubility, calculate the K_{sp} of iron(II) hydroxide. Explain why the calculated K_{sp} differs from the experimental value of 8.0×10^{-16}.
    (b) Calculate the hydroxide concentration needed to precipitate Fe^{2+} ions such that no more than 1.0 μg Fe^{2+} per liter remains in the solution.
66. The solubility of $Mg(OH)_2$ in water is approximately 9 mg/L at a given temperature.
    (a) Calculate the K_{sp} of magnesium hydroxide.
    (b) Calculate the hydroxide concentration needed to precipitate Mg^{2+} ions such that no more than 5.0 μg Mg^{2+} per liter remains in the solution.
67. Calculate the Cl^- concentration (in mol/L) in a solution that is 0.05 M in $AgNO_3$ and contains some undissolved AgCl.

Factors Affecting the Solubility of Sparingly Soluble Solutes (Section 17.5)

68. Calculate the maximum concentration of Zn^{2+} in a solution of pH 10.00.
69. Determine the maximum concentration of Mn^{2+} in two solutions with pH:
    (a) 7.81 (b) 11.15
70. Hydrochloric acid is added to dissolve 5.00 g $Mg(OH)_2$ in a liter of water. Calculate the pH required to just dissolve all of the $Mg(OH)_2$.
71. When a few drops of 1×10^{-5}-M $AgNO_3$ are added to 0.01-M NaCl, a white precipitate immediately forms. When a few drops of 1×10^{-5}-M $AgNO_3$ are added to 5-M NaCl, no precipitate forms. Explain these observations.

Complex Ion Formation (Section 17.5)

72. Write the chemical equation for the formation of each complex ion, and write its formation constant expression.
    (a) $[Ag(CN)_2]^-$ (b) $[Cd(NH_3)_4]^{2+}$
73. Write the chemical equation for the formation of each complex ion and write its formation constant expression.
    (a) $[Co(Cl)_6]^{3-}$ (b) $[Zn(OH)_4]^{2-}$
74. Calculate the amount (moles) of $Na_2S_2O_3$ that must be added to dissolve 0.020 mol AgBr in 1.0 L water.
75. Gaseous ammonia is added to a 0.063-M solution of $AgNO_3$ until the aqueous ammonia concentration rises to 0.18 M. Calculate the concentrations of $[Ag(NH_3)_2]^+$ and Ag^+ in the solution.
76. Write chemical equations to illustrate the amphoteric behavior of
    (a) $Zn(OH)_2$ (b) $Sb(OH)_3$
77. Write chemical equations to illustrate the amphoteric behavior of
    (a) $Cr(OH)_3$ (b) $Sn(OH)_2$

General Questions

These questions are not explicitly keyed to chapter topics; many require integration of several concepts.

78. A buffer solution was prepared by adding 4.95 g sodium acetate to 250. mL of 0.150-M acetic acid.
    (a) What ions and molecules are present in the solution? List them in order of decreasing concentration.
    (b) Calculate the pH of the buffer solution.
    (c) Calculate the pH of 100. mL of the buffer solution if you add 80. mg NaOH. (Assume negligible change in volume.)
    (d) Write a net ionic equation for the reaction that occurs to change the pH.
79. Calculate the mass in grams of NH_4Cl that must be added to 400. mL of a 0.93-M solution of NH_3 to prepare a pH = 9.00 buffer.
80. Calculate the relative concentrations of *o*-ethylbenzoic acid ($pK_a = 3.79$) and potassium *o*-ethylbenzoate that are needed to prepare a pH = 4.0 buffer.
81. Calculate the relative concentrations of the amine aniline ($pK_b = 9.41$) and anilinium chloride that are required to prepare a buffer of pH 5.00.
82. A solution contains 7.50 g KNO_2 per liter. Calculate the amount (moles) HNO_2 required to prepare 1.00 L of a buffer of pH 4.00. (Assume there is no volume change.)
83. Which of these buffers involving a weak acid HA has the greater resistance to change in pH? Explain your answer.
    (i) [HA] = 0.100 M = [A^-]
    (ii) [HA] = 0.300 M = [A^-]
84. (a) Calculate the pH of a 0.15-M acetic acid solution.
    (b) If you add 83 g sodium acetate to 1.50 L of the 0.15-M acetic acid solution, calculate the new pH of the solution.
85. (a) Calculate the pH of a 0.050-M solution of HF.
    (b) Calculate the pH of the solution after addition of 1.58 g NaF to 250. mL of the 0.050-M solution.

86. When 40.00 mL of a weak monoprotic acid solution is titrated with 0.100-M NaOH, the equivalence point is reached when 35.00 mL base has been added. After 20.00 mL NaOH solution has been added, the titration mixture has a pH of 5.75. Calculate the ionization constant of the acid.

87. Each of the solutions in the table has the same volume and the same concentration, 0.1 M.

Acid	pH	Acid	pH
HCl	1.1	Acetic	2.9
Formic	2.3	HCN	5.1

Which solution requires the greatest volume of 0.1-M NaOH to titrate to the equivalence point? Explain your answer.

88. What is the effect on the equilibrium if more solid AgCl is added to a solution saturated with Ag^+ and Cl^- ions?

89. At 20. °C, 2.03 g $CaSO_4$ dissolves per liter of water. From these data calculate the K_{sp} of calcium sulfate at 20. °C.

90. The solubility of silver chromate, Ag_2CrO_4, in water is 2.7×10^{-3} g per 100. mL. Estimate the K_{sp} of silver chromate at that temperature.

Applying Concepts

These questions test conceptual learning.

91. The average normal concentration of Ca^{2+} in urine is 5.33 g/L.
    (a) Calculate the concentration of oxalate needed to precipitate calcium oxalate to initiate formation of a kidney stone. K_{sp} of calcium oxalate $= 2.3 \times 10^{-9}$.
    (b) Calculate the minimum phosphate concentration that would precipitate a calcium phosphate kidney stone. K_{sp} of calcium phosphate $= 2.0 \times 10^{-29}$.

92. Explain why even though an aqueous acetic acid solution contains acetic acid and acetate ions, it cannot be a buffer.

93. Vinegar must contain at least 4% acetic acid (0.67 M). A 5.00-mL sample of commercial vinegar required 33.5 mL of 0.100-M NaOH to reach the equivalence point. Do calculations to determine whether the vinegar meets the legal limit of 4% acetic acid.

94. An unknown acid is titrated with base, and the pH is 3.64 at the point where exactly half of the acid in the original sample has been neutralized. Calculate the value of the ionization constant of the acid.

95. When asked to prepare a carbonate buffer with pH = 10, a lab technician wrote this equation to determine the ratio of weak acid to conjugate base needed:

$$10 = 10.32 + \log \frac{[HCO_3^-]}{[H_2CO_3]}$$

What is wrong with this setup? If the technician prepared a solution containing equimolar concentrations of HCO_3^- and CO_3^{2-}, calculate the pH of the resulting buffer.

96. When you hold your breath, carbon dioxide gas is trapped in your body. Does this increase or decrease your blood pH? Does it lead to acidosis or alkalosis? Explain your answers.

97. Calcium fluoride, CaF_2, is used to fluoridate a municipal water supply. The water is extremely hard with a Ca^{2+} concentration of 0.070 M. Calculate the fluoride concentration in this solution. Calcium fluoride has $K_{sp} = 5.3 \times 10^{-9}$.

98. Apatite, $Ca_5(PO_4)_3OH$, is the mineral in teeth.

$$Ca_5(PO_4)_3OH(s) \rightleftharpoons 5\,Ca^{2+}(aq) + 3\,PO_4^{3-}(aq) + OH^-(aq)$$

    (a) On a chemical basis explain why drinking milk strengthens young children's teeth.
    (b) Sour milk contains lactic acid. Not removing sour milk from the teeth of young children can lead to tooth decay. Use chemical principles to explain why.

99. Calculate the maximum concentration of Mg^{2+} (molarity) that can exist in a solution of pH 12.00.

100. Choose the words that make this statement true: During a televised medical drama, a person went into cardiac arrest and stopped breathing. A doctor quickly injected sodium hydrogen carbonate solution into the heart. This would indicate that cardiac arrest leads to (acidosis or alkalosis) and that the sodium hydrogen carbonate helps to (increase or decrease) the pH. Explain your choices clearly.

101. Below is a grid with six lettered boxes, each of which contains an item that may be used to answer the questions below. Items may be used more than once and there may be more that one correct item in response to a question.

A pH >7.0	B pH < 5.6	C $[AlF_6]^-$
D Large increase in pH	E NaOH and excess $H_2PO_4^-$	F $Q < K_{sp}$

Place the letter(s) of the correct selection(s) on the appropriate line.
    (a) Add sufficient base to a buffer ________
    (b) Acid precipitation ________
    (c) Species formed by a Lewis acid-base reaction ________
    (d) Condition required for no precipitation ________
    (e) A buffer system ________
    (f) Equivalence point of a weak acid/strong base titration ________
    (g) Just prior to the equivalence point in the titration of lactic acid with KOH ________

102. Below is a grid with six lettered boxes, each of which contains an item that may be used to answer the questions that follow. Items may be used more than once and there may be more that one correct item in response to a question.

A $[Cr(OH)_4]^-$	B $Q > K_{sp}$	C NaOH
D Decreasing pH	E $pH = pK_a$	F Sufficient KCl added to CuCl(aq)

Place the letter(s) of the correct selection(s) on the appropriate line.
    (a) Could dissolve $Zn(OH)_2$ ________
    (b) Could be used to prepare a buffer from HPO_4^{2-} ________

(c) Halfway to the equivalence point in the titration of a weak, monoprotic acid with strong base __________
(d) SO_2-related atmospheric phenomenon __________
(e) Species formed by a Lewis acid-base reaction __________
(f) General condition required for precipitation to occur __________
(g) [conj. base] = [conj. acid] in a buffer __________

103. The diagrams below are nanoscale-level representations of the titration of the aqueous weak acid HX with aqueous NaOH, the titrant. Water molecules and Na^+ ions are omitted for clarity.

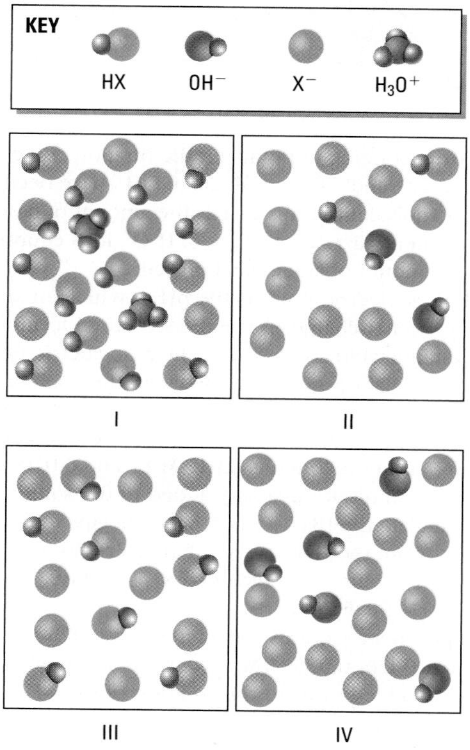

Which diagram corresponds to the situation:
(a) After a very small volume of titrant has been added to the initial HX solution.
(b) When enough titrant has been added to take the solution just past the equivalence point.
(c) Halfway to the equivalence point.
(d) At the equivalence point.

104. The diagrams at the top of the next column are nanoscale representations of the titration of the aqueous strong acid HA with aqueous NaOH, the titrant. Water molecules and Na^+ ions are omitted for clarity.
Which diagram corresponds to the situation:
(a) After a very small volume of titrant has been added to the initial HA solution.
(b) Halfway to the equivalence point.
(c) When enough titrant has been added to take the solution just past the equivalence point.
(d) At the equivalence point.

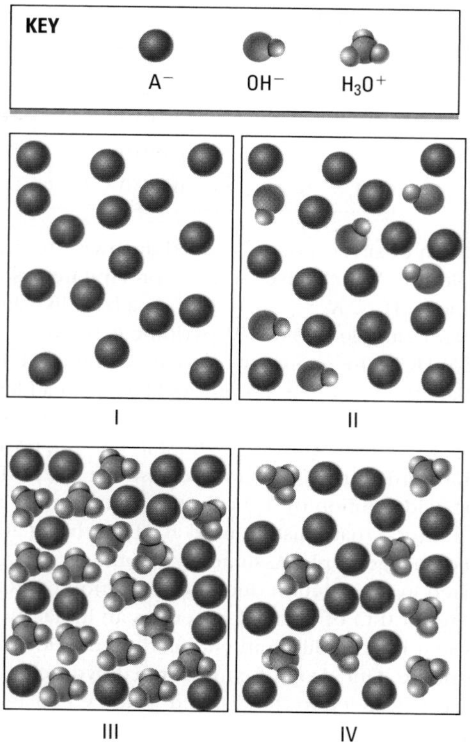

More Challenging Questions

These questions require more thought and integrate several concepts.

105. An experiment found that 0.0050 mol $Ca(OH)_2$ dissolved to form 0.100 L of a saturated aqueous solution.
(a) Calculate the pH of the solution.
(b) Calculate the K_{sp} of $Ca(OH)_2$. Explain why the calculated K_{sp} differs from the value of 5.5×10^{-6} given in Appendix H.

106. You are given four different aqueous solutions and told that they each contain NaOH, Na_2CO_3, $NaHCO_3$, or a mixture of these solutes. You do some experiments and gather these data about the samples.

Sample A: Phenolphthalein is colorless in the solution.
Sample B: The sample was titrated with HCl until the pink color of phenolphthalein disappeared, then methyl orange was added. The solution became pink. Methyl orange changes color from pH 3.01 (red) to pH 4.4 (orange).
Sample C: Equal volumes of the sample were titrated with standardized acid. Using phenolphthalein as an indicator required 15.26 mL of standardized acid to change the phenolphthalein color. The other sample required 17.90 mL for a color change using methyl orange as the indicator.
Sample D: Two equal volumes of the sample were titrated with standardized HCl. Using phenolphthalein as the indicator, it took 15.00 mL of acid to reach the equivalence point; using methyl orange as the indicator required 30.00 mL HCl to achieve neutralization.

Identify the solute in each of the solutions.

107. An aqueous solution contains these ions:

Ion	Concentration (M)
Br^-	0.0010
Cl^-	0.0010
CrO_4^{2-}	0.0010

A solution of silver nitrate is added dropwise to the aqueous solution. Which silver salt will precipitate first? Last?

108. Pyridine, C_5H_5N, $K_b = 1.6 \times 10^{-9}$, is a base like ammonia. A 25.0-mL sample of 0.085-M pyridine is titrated with 0.102-M HCl. The equivalence point occurs at 21.1 mL. Calculate the pH when 5.5 mL acid has been added.

109. You have 1.00 L of 0.10-M formic acid, HCOOH, whose $K_a = 3.0 \times 10^{-4}$. You want to bubble into the formic acid solution sufficient HCl gas to decrease the pH of the formic acid solution by 1.0 pH unit. Calculate the volume of HCl (liters) that must be used at STP to bring about the desired change in pH. Assume no volume change has occurred in the solution due to the addition of HCl gas.

110. One liter (1.0 L) of distilled water has an initial pH of 5.6 because it is in equilibrium with carbon dioxide in the atmosphere. One drop of concentrated hydrochloric acid, 12-M HCl, is added to the distilled water. Calculate the pH of the resulting solution. 20 drops = 1.0 mL.

111. A 0.0010-M aqueous solution of anisic acid, $C_8H_8O_3$, is prepared and 100. mL of it is left in an uncovered beaker. After some time, 50. mL water has evaporated from the solution in the beaker (assume no solute evaporated). The K_a of anisic acid is 3.38×10^{-5}.
(a) Calculate the pH of the initial solution and the pH after evaporation has occurred.
(b) Calculate the degree of dissociation of anisic acid in each of the solutions.

112. You want to prepare a pH 4.50 buffer using sodium acetate and glacial acetic acid. You have on hand 300. mL of 0.100-M sodium acetate. Calculate the mass in grams of glacial acetic acid (d = 1.05 g/mL) you should use to prepare the buffer.

113. The titration curves for the amino acids glutamic acid and lysine are given in the figure.
(a) Using the structural formulas given in Table 12.8, write the structural formulas for the forms of each amino acid corresponding to points on the titration curve when 0, 1.0, 2.0, and 3.0 equivalents of OH^- have been added. An equivalent in this case is the amount of base added that neutralizes an amount of H_3O^+ ions equivalent to the amount of amino acid present.
(b) The isoelectric point is the pH at which an amino acid has no net charge. Write the structural formulas of glutamic acid and lysine at their isoelectric points.
(c) What relationship exists between the conjugate acid and conjugate base forms of these amino acids at each of the pK points noted on the titration curve?

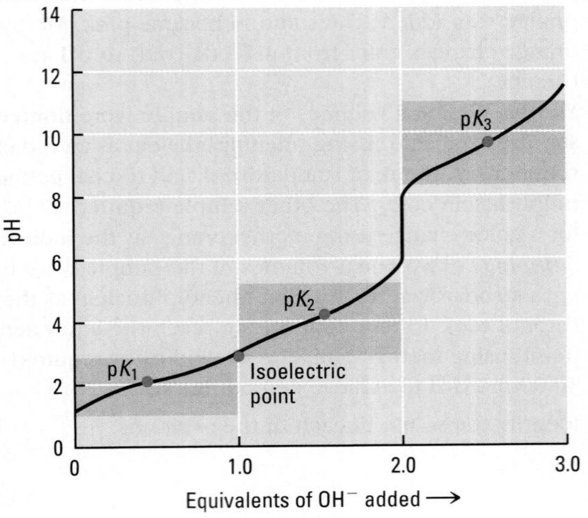

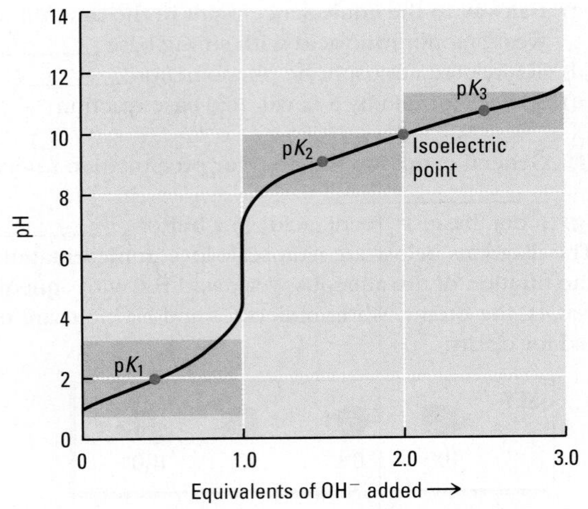

114. A 1.00-L solution contains 0.010-M F^- and 0.010-M SO_4^{2-}. Solid barium nitrate is slowly added to the solution.
(a) Calculate the $[Ba^{2+}]$ when $BaSO_4$ begins to precipitate.
(b) Calculate the $[Ba^{2+}]$ when BaF_2 starts to precipitate.
Assume no volume change occurs. K_{sp} values: $BaSO_4 = 1.1 \times 10^{-10}$; $BaF_2 = 1.0 \times 10^{-6}$.

115. An experiment requires the addition of 0.075 mol gaseous NH_3 to 1.0 L of 0.025-M $Mg(NO_3)_2$. Ammonium chloride, NH_4Cl, is added prior to the addition of the NH_3 to prevent precipitation of $Mg(OH)_2$. Calculate the minimum mass in grams of ammonium chloride that must be added. K_{sp} of $Mg(OH)_2 = 1.8 \times 10^{-11}$.

Conceptual Challenge Problems

These rigorous, thought-provoking problems integrate conceptual learning with problem solving and are suitable for group work.

CP17.A (Section 17.2) Suppose you were asked on a laboratory test to outline a procedure to prepare a buffered solution of pH 8.0 using hydrocyanic acid, HCN. You realize that a pH of 8.0 is basic, and you find that the K_a of hydrocyanic acid is 4.0×10^{-10}. What is your response?

CP17.B (Sections 17.4 and 17.5) Barium sulfate is swallowed to enhance X-ray studies of the gastrointestinal tract.
(a) Calculate the solubility of Ba^{2+} in mol/L in pure water. K_{sp} of $BaSO_4 = 1.1 \times 10^{-10}$.
(b) In the stomach, the HCl concentration is 0.10 M. Calculate the solubility of Ba^{2+} in mol/L in this solution. K_a of $HSO_4^- = 1.1 \times 10^{-2}$.
(c) The two calculations in parts (a) and (b) were done using data at 25 °C. Repeat part (b) at 37 °C, body temperature. At that temperature, K_{sp} of $BaSO_4 = 1.5 \times 10^{-10}$; K_a of $HSO_4^- = 7.1 \times 10^{-3}$.

Thermodynamics: Directionality of Chemical Reactions

18

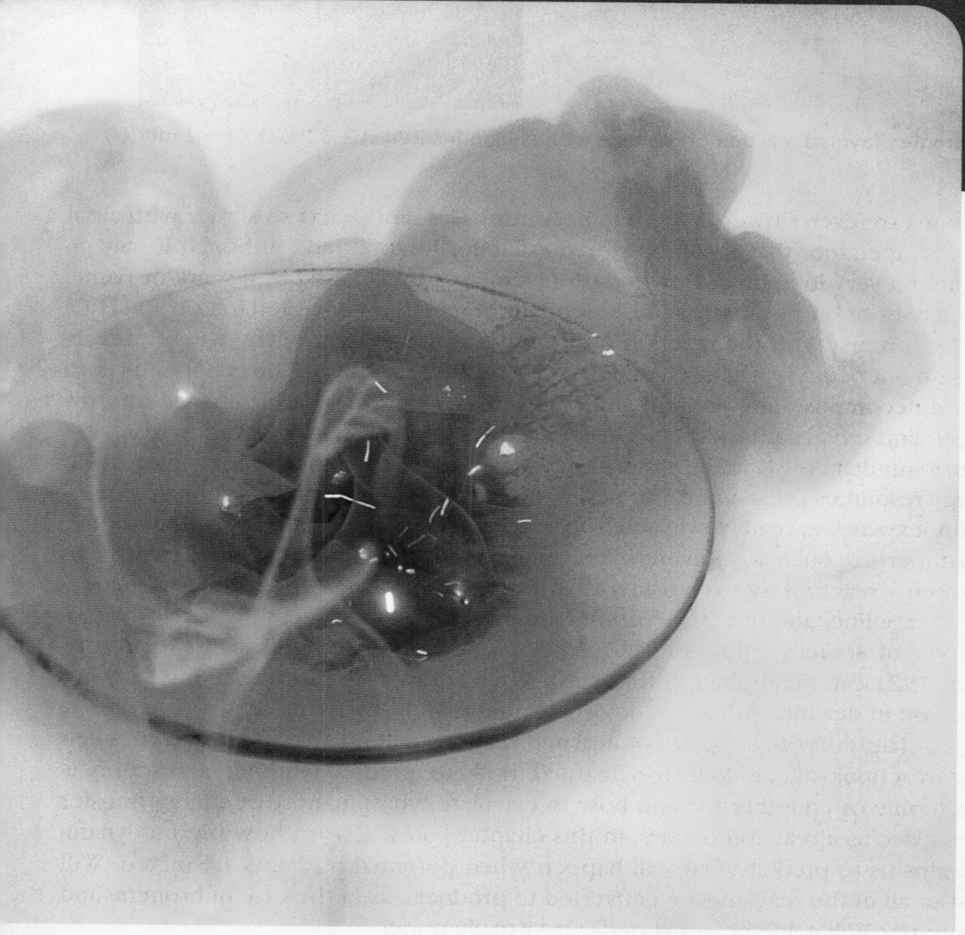

© Cengage Learning/Charles D. Winters

When liquid bromine is poured onto small pieces of aluminum, nothing appears to happen at first. Eventually the mixture becomes hot and a violent exothermic reaction occurs. Sparks fly, and you see flames. The reaction begins slowly as soon as the bromine contacts the aluminum, and its rate increases as the temperature rises. The reaction continues until either the bromine or the aluminum has been completely reacted. Chemical thermodynamics provides methods for predicting whether a reaction will be product-favored, like this one.

Many chemical reactions behave as the reaction of bromine and aluminum does (see photos on the next page). They begin when the reactants come into contact and continue until at least one reactant, (the limiting reactant, ← *p. 137*), is completely converted to products. Other reactions, such as the rusting of iron at room temperature, happen much more slowly, but reactants are still converted completely to products. After many years and enough flaking of hydrated iron(III) oxide (rust) from its surface, a piece of iron exposed to air will rust completely. Still other

Photos: © Cengage Learning/Charles D. Winters

The product-favored reaction of bromine with aluminum, $2\ Al(s) + 3\ Br_2(\ell) \longrightarrow 2\ AlBr_3(s)$.

Because air can slowly oxidize hydrocarbons to alkenes, which can polymerize and form insoluble gums, gasoline is usually drained from lawn mowers or cars that are going to be stored for many months or an antioxidant (such as N-nitrosodiethylamine) is added.

reactions are even slower at room temperature. Gasoline reacts so slowly with air at room temperature that it can be stored safely for long periods, although it may go bad after a very long time. However, if its temperature is raised by a spark or flame, gasoline vapor burns rapidly and is essentially completely converted to CO_2 and H_2O.

By contrast with the reaction of bromine and aluminum, many chemical reactions do not occur by themselves. For example, table salt, NaCl, does not of its own accord decompose into sodium and chlorine. Neither does water change into hydrogen and oxygen all by itself. These reactions take place only if another process occurs simultaneously and transfers energy to them. (A significant portion of world energy resources is used to cause desirable reactions to occur—reactions that transform inexpensive, readily available substances into new substances with more useful properties, such as polymers and medicines.) It is important to differentiate between a reaction that is so slow that it *appears* not to occur, such as air oxidation of gasoline, and one that *cannot* take place of its own accord, such as decomposition of sodium chloride. The principles of chemical kinetics in Chapter 13 (← *p. 592*) can be applied to find ways to speed up a slow reaction, but they are of no use in dealing with a reaction that cannot occur by itself.

In Chapter 6 (← *p. 232*) you learned that thermal energy is transferred when most reactions occur. You also learned how to predict whether a reaction is exothermic or endothermic and how to calculate what quantity of energy transfer takes place as a reaction occurs. In this chapter you will learn how thermodynamics helps us to predict what will happen when potential reactants are mixed. Will most or all of the reactants be converted to products, as in the case of bromine and aluminum? Will some be converted? Or virtually none?

18.1 Reactant-Favored and Product-Favored Processes

The term "product-favored" designates reactions that many scientists refer to as "spontaneous"; many people use the two terms interchangeably. We prefer "product-favored" because some reactions do begin spontaneously, but produce only tiny quantities of products when equilibrium is reached. "Product-favored" describes clearly a situation in which products predominate over reactants. Also, the nonscientific usage of "spontaneous" implies a rapid change; if the rate of a product-favored reaction is very slow, the reaction does not appear spontaneous at all.

In Chapter 14 (← *p. 656*) we introduced the idea that a chemical process can be described as reactant-favored or product-favored. *When products predominate over reactants, we designate the reaction as a product-favored process.* Examples include the reaction of bromine with aluminum, rusting of iron, and combustion of gasoline. If a process is product-favored, most or all of the reactants will eventually be converted to products without continuous outside intervention, although "eventually" may mean a very, very long time.

Other reactions have virtually no tendency to occur by themselves. Examples include the reactions for which we wrote "N.R." for "no reaction" in Chapter 5 (← *p. 161*). For example, nitrogen and oxygen have coexisted in Earth's atmosphere for at least a billion years without significant concentrations of nitrogen oxides such as N_2O, NO, or NO_2 building up. Similarly, deposits of salt, NaCl(s), have existed on Earth for millions of years without forming the elements Na(s) and Cl_2(g). *When equilibrium has been reached, if reactants predominate over products, we categorize a chemical reaction as a reactant-favored process.*

A reactant-favored process is always *exactly the opposite* of a product-favored process. For example, the equation

$$2\,Na(s) + Cl_2(g) \longrightarrow 2\,NaCl(s)$$

describes a product-favored reaction, because sodium metal and chlorine gas react readily to produce sodium chloride. However, if we had written the same equation in the reverse direction

$$2\,NaCl(s) \longrightarrow 2\,Na(s) + Cl_2(g)$$

the system would be designated as reactant-favored. This equation represents decomposition of sodium chloride to form sodium and chlorine, a reaction that does not occur of its own accord. The designations "product-favored" and "reactant-favored" indicate the direction in which a chemical reaction will take place—either forward or backward based on a given equation.

Unless there is some continuous outside intervention, a reactant-favored process does not produce large quantities of products. What do we mean by continuous outside intervention? Usually it is some flow of energy. For example, if enough energy is provided to a sample of air to keep it at a very high temperature, small but significant quantities of NO can be formed from the N_2 and O_2. Such high temperatures are found in lightning bolts and in combustion chambers of electric power generating plants and automobile engines. A power plant or a large number of automobiles can produce enough NO and other nitrogen oxides, albeit at low concentrations, to cause significant air pollution problems. Sodium chloride can be decomposed to its elements by continuously heating it to keep it molten and passing a direct electric current through it to separate the ions, carry out oxidation and reduction, and form the elements.

$$2\,NaCl(\ell) \xrightarrow[\text{current}]{\text{direct electric}} 2\,Na(\ell) + Cl_2(g)$$

In each case, a reactant-favored process can be forced to produce products if sufficient electrical energy is continuously supplied. This is in contrast to the situation for a product-favored process such as combustion of gasoline, which requires only a brief spark to initiate the reaction. Once started, gasoline combustion continues of its own accord without an additional continuous supply of energy from outside.

Although Earth's atmosphere is 78% N_2 and 21% O_2, the concentration of N_2O, the most abundant oxide of nitrogen in the atmosphere, is more than two million times smaller than the concentration of N_2.

The air in the immediate vicinity of a lightning bolt can be heated enough to cause a small fraction of the nitrogen and oxygen to combine to form NO, but this reaction takes place only while the lightning is present. A similar reaction can occur in the engine of an automobile, but again only a small fraction of the air is converted to nitrogen oxides, and only while the temperature remains high (>1000 °C).

The industrial process by which Na and Cl_2 are produced from NaCl is discussed in Sections 19.11 and 21.4.

EXERCISE 18.1 Reactant-Favored and Product-Favored Processes

Write a chemical equation for each process described below, and classify each as reactant-favored or product-favored.

(a) A puddle of water evaporates on a summer day.
(b) Silicon dioxide (sand) decomposes to the elements silicon and oxygen.
(c) Paper, which is mainly cellulose $(C_6H_{10}O_5)_n$, burns at a temperature of 451 °F.
(d) A pinch of sugar dissolves in water at room temperature.

Answers to **EXERCISES** are provided at the back of this book in Appendix L.

18.2 Chemical Reactions and Dispersal of Energy

The fundamental rule that governs whether a process is product-favored is that ***energy will spread out (disperse) unless it is hindered from doing so.*** A simple example is the one-way transfer of energy from a hotter sample to a colder sample ***(← p. 216)***. As a hot frying pan on a stove cools to room temperature, thermal energy that was concentrated in the pan spreads out over the atoms, molecules, and ions in the stove, the pan, and the surrounding air. When thermal equilibrium is reached, the room has become slightly warmer—energy has been dispersed and the process is product-favored.

A chemical reaction system is usually defined as the collection of atoms that make up the reactants. These same atoms also make up the products, but there they are bonded in a different way. Everything else is designated as the surroundings (← p. 218).

Most exothermic reactions are product-favored at room temperature for a similar reason. When an exothermic reaction takes place, energy is transferred from the system to the surroundings. (See, for example, the reaction of bromine with aluminum in the photo at the beginning of this chapter.) Chemical potential energy that has been stored in bonds between relatively few atoms, ions, and molecules (the reactants) spreads over many more atoms, ions, and molecules as the surroundings (as well as the products) are heated. Therefore it is usually true that after an exothermic reaction, energy is more dispersed than it was before.

Probability and Dispersal of Energy

Dispersal of energy occurs because the probability is much higher that energy will be spread over many particles than that it will be concentrated in a few. To better understand energy dispersal and probability, consider the hypothetical case of a very small sample of matter consisting of two atoms, A and B. Suppose that this sample contains two units of energy, each designated by *. The energy can be distributed over the two atoms in three ways: Atom A could have both units of energy, atom A and atom B could each have one unit, or atom B could have both units. Designate these three situations as

A** A*B* B**

Now suppose that atoms A and B come into contact with two other atoms, C and D, that have no energy. There are ten possibilities for distributing the two units of energy over four atoms.

A** A*B* A*C* A*D* B** B*C* B*D* C** C*D* D**

The low probability that a lot of energy will be associated with only a few particles makes a substance with a lot of chemical potential energy valuable. Humans call substances such as coal, oil, and natural gas "energy resources" and sometimes fight wars over them because of their concentrated energy.

Only three of these cases (A**, A*B*, and B**) have all the energy in atoms A and B, which was the initial situation. When all four atoms are in contact, there are seven chances out of ten that some energy will have transferred from A and B to C and D. Thus, there is a probability of $7/10 = 0.70$ that the energy will become spread out over more than just the two atoms A and B.

EXERCISES that are labeled **CONCEPTUAL** are designed to test your understanding of one or more concepts; they usually involve qualitative rather than quantitative thinking.

CONCEPTUAL EXERCISE 18.2 Probability of Energy Dispersal

Suppose that you have three units of energy to distribute over two atoms, A and B. Designate each possible arrangement. Now suppose that atoms A and B come into contact with three more atoms, C, D, and E. From the possible arrangements of energy over the five atoms, calculate the probability that all the energy will remain confined to atoms A and B.

For the same amount of substance or number of particles, higher thermal energy corresponds to higher temperature (← p. 218). Therefore, a substance at a higher temperature has greater energy per particle on average. Dispersal of energy over a larger number of particles corresponds to transfer of energy from a substance at a higher temperature to another substance at a lower temperature.

The probability that energy will become dispersed becomes overwhelming when large numbers of atoms or molecules are involved. For example, suppose that atoms A and B had been brought into contact with a mole of other atoms. There would still be only three arrangements in which all the energy was associated with atoms A and B, but there would be many, many more arrangements (more than 10^{47}) in which all the energy was associated with other atoms. In such a case it is essentially certain that energy will be transferred. ***If energy can be dispersed over a very much larger number of particles, it will be.***

Dispersal of Energy Accompanies Dispersal of Matter

Energy becomes more dispersed when a system consisting of atoms or molecules expands to occupy a larger volume. This kind of energy dispersal is associated with a characteristic property of gases: A gas expands until it fills a container. Recall that molecules in the gas phase are essentially independent of one another and that only weak forces attract the molecules to one another **(← *p. 429*)**. Suppose a sample of bromine gas at room temperature is confined within one flask that is connected through a tube with a barrier to a second flask of equal size from which all gas molecules have been removed (Figure 18.1). What happens if the barrier is removed? The confined bromine expands to fill the vacuum in the second flask.

The conclusions drawn about dispersal of bromine molecules on this page apply in general to particles of a gas, whether they are atoms or molecules.

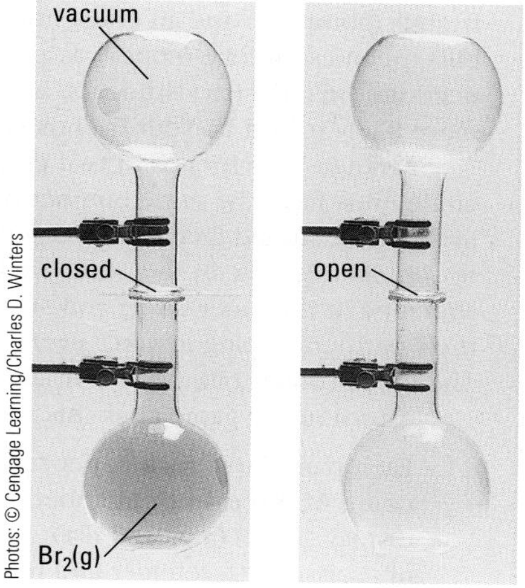

(a) Before (b) After

Figure 18.1 Expansion of a gas. A gas will expand to fill any container. (a) Bromine vapor is confined in the lower flask. There is a vacuum in the upper flask. (b) When the barrier between the flasks is removed, the bromine molecules rapidly rush into flask B, and eventually bromine is evenly distributed throughout both flasks.

That dispersal of atoms and molecules also involves dispersal of energy seems obvious, because energy accompanies the particles as they disperse. However, dispersal of energy refers to spreading of energy over a greater number of different *energy levels* **(*quantum levels*, ← *p. 280*)** of the atoms and molecules. When the volume of a gas-phase system increases, the energy levels associated with the motion of each atom or molecule get closer together; that is, there are more levels within the same range of energies. Therefore, at a given temperature, more different energy levels are accessible in the larger volume than in the smaller volume.

When matter spreads out, the number of ways of arranging the energy associated with that matter increases. The energy is more dispersed, and therefore the spreading out of matter is product-favored. This basic idea applies to mixing of different gases and dissolving of one liquid in another, as well as to expansion of a gas. We have already mentioned it in connection with dissolving of solids **(← *p. 716*)**, and it is also useful in understanding colligative properties **(← *p. 727*)**.

Expansion of a gas is product-favored, so the opposite process—compression of a gas—should be reactant-favored. If we wanted to reverse the expansion of a gas by concentrating all the particles into a smaller volume, a continuous outside influence such as a pump would be required—the pump could do work on the gas to force it into a less probable arrangement in which energy was less dispersed. The work done by the pump would be stored in the gas and could later be used for some other purpose.

To summarize, any physical or chemical process in which energy is concentrated in a few energy levels in the initial state and energy is dispersed over many energy levels in the final state is product-favored. Two important situations where this is true are

- an exothermic reaction, which disperses potential energy of chemical bonds to thermal energy of a much larger number of atoms or molecules, and
- a process where matter spreads out, which disperses energy as well as matter.

Atoms or molecules of a material that is a solid at room temperature (like the glass flask) do not disperse, because there are strong attractive forces between them. Their tendency to disperse becomes more obvious if the temperature is raised so that they can vaporize.

If *both* of these situations apply to a reaction, then it will *definitely* be *product-favored*, because the final distribution of energy will be more probable. On the other hand, a process that spreads out *neither energy nor matter* will be *reactant-favored*—the initial substances will remain no matter how long we wait. If one of these situations applies but not the other, then quantitative information is needed to decide which effect is greater. The remainder of this chapter develops that quantitative information.

18.3 Measuring Dispersal of Energy: Entropy

The nanoscale dispersal of energy in a sample of matter is measured by a thermodynamic quantity called *entropy*, symbolized by *S* **(← *p. 687*)**. Entropy changes can be measured with a calorimeter **(← *p. 238*)**, the same instrument used to measure the enthalpy change when a reaction occurs. For a process that takes place at constant temperature and pressure, the entropy change can be calculated by dividing the thermal energy transferred, q_{rev}, by the absolute temperature, T,

$$\Delta S = S_{final} - S_{initial} = \frac{q_{rev}}{T} \qquad [18.1]$$

The symbol Δ was defined in Section 6.2 **(← *p. 218*)**. The quantity of energy transferred by heating, *q*, was defined in Section 6.3 **(← *p. 221*)**.

CHEMISTRY YOU CAN DO Energy Distributions

To envision how energy can be distributed over nanoscale particles, try this game. You will need an even number of friends (about 20) and an equal number of tokens (dollar bills, pennies, or fake money) to play the game. The friends need not be chemistry students, but chemistry students are more likely to still be your friends when you finish!

Arrange your friends in two concentric circles. Each circle must have the same number of people. People in the inner circle should face outward and stand almost shoulder to shoulder. People in the outer circle should pair up with someone in the inner circle and stand directly in front of their partners, facing inward. Everyone should start with one token (dollar bill, penny, or fake money).

Courtesy of Robert M. Hanson/St. Olaf College. © *Journal of Chemical Education*. Vol. 83, 2006; p. 581.

A turn in the game goes this way.

1. Each pair plays "rock-paper-scissors" once. On the count of three, both members of a pair simultaneously display "rock" (a closed fist), "paper" (an open palm), or "scissors" (forefinger and middle finger in a V-shape). Rock breaks scissors (rock wins over scissors), scissors cuts paper, and paper wraps rock. If both people display the same thing, there is a tie.
2. If there is a winner and the loser has a token, one token is transferred from loser to winner. (Never exchange more than a single token.) If there is a tie, or if the loser has no token, there is no transfer.
3. When you say "next," each person in the inner circle moves one place to the left, to a new partner in the outer circle, and a new turn begins.

After every fifth turn, ask all players who have zero tokens to raise their hands. Record the number of raised hands. Also ask about one token, two tokens, and so on up to the highest number of tokens anyone has. Record how many players have each number of tokens. Continue playing the game and recording the results until you have recorded at least five sets of data (25 turns).

Think about the meaning of the data you have collected and answer these questions:

1. How does this game simulate what happens when lots of molecules exchange energy with each other? What role does probability play in the distribution of tokens? Of energy?
2. What do you think is the probability that a particular individual would end the game with more than a single token?
3. Why isn't the most probable distribution of tokens the distribution where everyone has a single token?
4. What do you think would happen if the game were repeated, but a single individual had all of the tokens at the beginning? Try it and find out whether your prediction was correct.
5. How are the two versions of the game (the original and the one in Question 4) related to the rule given in Chapter 14 that the same equilibrium state is reached (at a given temperature) whether you start with reactants or products?
6. Suppose you were the person who had all the tokens at the beginning. What do you think your chances are of coming away from the game with half as many tokens as you started with?

Source: Hanson, R. M. and Michalek, B. *Journal of Chemical Education*, Vol. 83, 2006; pp. 581–588.

The absolute temperature scale is also called the *Kelvin temperature scale* or *the thermodynamic temperature scale*. It was defined in Section 10.4 ***(← p. 435)***. The Kelvin scale should be used in all thermodynamic calculations involving temperature.

The subscript "rev" has been added to q to indicate that the equation applies only to processes that can be reversed by a very small change in conditions. An example of such a process is melting of ice at 0 °C and normal atmospheric pressure. If the temperature is just a tiny bit below 0 °C, so that energy is transferred from the water to its surroundings, the water will freeze. If the temperature is a tiny bit above 0 °C, the ice will melt. *Any process for which a very small change in conditions can reverse its direction* is called a **reversible process.**

Another case where Equation 18.1 can be used is when energy is transferred to or from a thermal reservoir at constant temperature and pressure. A thermal reservoir is any large sample of matter, such as everything in your dorm room or the surroundings of a chemical reaction. For example, when a cup of hot coffee cools to room temperature, the quantity of energy transferred is small enough that the temperature of the thermal reservoir (your room) hardly changes. This small quantity of energy could be transferred into or out of your room reversibly. Therefore, to calculate the entropy change of your room you could measure room temperature (in kelvins), calculate how much energy transferred to the room as the coffee cooled, and divide energy transfer by temperature. You could not do the same calculation to determine the entropy change of the coffee, because its temperature changed during the process.

PROBLEM-SOLVING EXAMPLE 18.1 Calculating Entropy Change

The melting point of glacial acetic acid (pure acetic acid) is 16.6 °C at 1 bar. Calculate $\Delta S°$ for the process

$$CH_3COOH(s) \longrightarrow CH_3COOH(\ell)$$

given that the molar enthalpy of fusion of acetic acid is 11.53 kJ/mol. Express your answer in units of joules per kelvin.

Answer $\Delta S° = 39.79$ J/K

Strategy and Explanation Melting acetic acid at its melting point of 16.6 °C is reversible, so use the equation $\Delta S = q_{rev}/T$. The pressure is 1 bar, the standard-state pressure ***(← p. 245),*** so $\Delta S° = \Delta S$. Remember that for a constant-pressure process, thermal energy transfer, q_{rev}, is the same as the enthalpy change, $\Delta H°$ ***(← p. 227).*** Therefore, the entropy change can be calculated from the enthalpy of fusion and the temperature.

$$\Delta S° \text{ (melting acetic acid)} = \frac{q_{rev}}{T} = \frac{\Delta H° \text{ (melting acetic acid)}}{T} = \frac{\Delta H_{fusion}}{T}$$

$$= \frac{11.53 \text{ kJ}}{(16.6 + 273.15) \text{ K}} = 3.979 \times 10^{-2} \text{ kJ/K} = 39.79 \text{ J/K}$$

☑ **Reasonable Answer Check** The result is positive, meaning that entropy increased when solid acetic acid was converted to liquid. Since larger entropy corresponds to greater spreading out of energy, and since molecules move greater distances and in more different ways in a liquid than in a solid ***(← p. 19),*** this is reasonable.

The **PROBLEM-SOLVING STRATEGY** in this book is

- Analyze the problem
- Plan a solution
- Execute the plan
- Check that the result is reasonable

Appendix A.1 explains this in detail.

Entropy changes are usually reported in units of joules per kelvin (J/K), whereas enthalpy changes are usually given in kilojoules (kJ). This means that you need to be careful about the units to avoid being wrong by a factor of 1000.

PROBLEM-SOLVING PRACTICE 18.1

A chemical reaction transfers 30.8 kJ to a thermal reservoir that has a temperature of 45.3 °C before and after the energy transfer. Calculate the entropy change for the thermal reservoir.

PROBLEM-SOLVING PRACTICE answers are provided at the back of this book in Appendix K.

CONCEPTUAL EXERCISE 18.3 The Importance of Absolute Temperature

Consider what would happen if the Celsius temperature scale were used when calculating entropy change by means of $\Delta S = q_{rev}/T$. Suppose, for example, that energy were transferred reversibly to $H_2O(s)$ at a temperature 10° below its melting point and we wanted to calculate the entropy change. Would the value calculated from the Celsius temperature agree with the fact that transfer of thermal energy to a sample always increases its entropy?

Absolute Entropy Values

In Chapter 6 we mentioned that there is no way to measure the total energy content of a sample of matter. Therefore, to summarize a large number of calorimetric measurements of enthalpy changes, we tabulated standard molar enthalpies of formation in Table 6.2 ***(← p. 246).*** The standard enthalpy of formation is the difference between the enthalpy of a substance in its standard state and the enthalpies of

On the next page we discuss absolute zero (0 K). Though it is impossible to cool anything all the way to absolute zero, it is possible to get very close. Temperatures of a few nanokelvins can be achieved in a Bose-Einstein condensate—the coldest thing known to science. (See **http://nobelprize.org/nobel_prizes/physics/laureates/2001/index.html**)

the elements that make up that substance, all in their standard states. The elements have enthalpy values, but we do not know what they are. Therefore, the standard enthalpies of formation of elements are by definition set to zero.

The idea that a perfect crystal of any substance at 0 K has minimum entropy is called the **third law of thermodynamics.** Even if absolute zero cannot be achieved, there are ways of estimating how much energy has been dispersed in a substance near 0 K. Thus, accurate entropy values can be obtained for many substances.

For entropy the situation is simpler, because it is possible to define conditions for which it is logical to assume the entropy of a substance has its lowest possible value—namely, zero. Measuring ΔS for a change from those zero-entropy conditions tells us the absolute entropy value for the substance under new conditions. This follows from the definition $\Delta S = S_{final} - S_{initial}$, because if $S_{initial} = 0$, then $\Delta S = S_{final}$, the absolute entropy of the substance. Because decreasing temperature corresponds to decreasing molecular motion, the minimum possible temperature can reasonably be expected to correspond to minimum motion and thus minimum dispersal of energy. Thus it is logical to assume that a perfect crystal of a substance at 0 K has an entropy value of zero. [In a perfect crystal every nanoscale particle is in exactly the right position in the crystal lattice ***(← p. 501)***, and there are no empty spaces or discontinuities.] To calculate the entropy change, start as close as possible to absolute zero, successively introduce small quantities of energy, and calculate ΔS from the equation $\Delta S = q_{rev}/T$ for each increment of energy at each temperature. Then sum these entropy changes to give the total (or absolute) entropy of a substance at any desired temperature.

The process of introducing small quantities of energy, calculating an entropy increase, and then summing these small entropy increases is actually done by measuring the heat capacity of a substance as a function of temperature and then using integral calculus to calculate the integral of the function q_{rev}/T between the limits of 0 K and the desired temperature.

The results of such measurements for several substances at 298.15 K are given in Table 18.1. These are *standard molar entropy values*, and so they apply to 1 mol of each substance at the standard pressure of 1 bar and the specified temperature of 25 °C. The units are joules per kelvin per mole (J K^{-1} mol^{-1}). Because there is a real zero on the entropy scale, the values in Table 18.1 are not measured relative to elements in their most stable form under standard-state conditions. Therefore, absolute entropies can be determined for elements as well as compounds. ***The standard molar entropy of a substance at temperature* T *is a sum of the quantities of energy that must be dispersed in that substance at successive temperatures up to* T*, that is, it is* ΔS *from 0 K to* T.**

Oesper Collection in the History of Chemistry, University of Cincinnati

Ludwig Boltzmann

1844–1906

An Austrian physicist, Ludwig Boltzmann, gave us the useful interpretation of entropy as probability. Engraved on his tombstone in Vienna is the equation

$$S = k \log W$$

It relates entropy, S, and thermodynamic probability, W—the number of different nanoscale arrangements of energy that correspond to a given macroscale system. The proportionality constant, k, is called Boltzmann's constant, and log stands for the natural logarithm (in modern symbolism, ln).

Table 18.1 Some Standard Molar Entropy Values at 298.15 K*

Compound or Element	Entropy, $S°$ (J K^{-1} mol^{-1})	Compound or Element	Entropy, $S°$ (J K^{-1} mol^{-1})
C(graphite)	5.740	O_2(g)	205.138
C(g)	158.096	NH_3(g)	192.45
CH_4(g)	186.264	HCl(g)	186.908
CH_3CH_3(g)	229.60	H_2O(g)	188.825
$CH_3CH_2CH_3$(g)	269.9	$H_2O(\ell)$	69.91
$CH_3OH(\ell)$	126.8	Ca(s)	41.42
CO(g)	197.674	NaF(s)	51.5
CO_2(g)	213.74	MgO(s)	26.94
F_2(g)	202.78	NaCl(s)	72.13
Cl_2(g)	223.066	KOH(s)	78.9
$Br_2(\ell)$	152.231	$MgCO_3$(s)	65.7
I_2(s)	116.135	NH_4NO_3(s)	151.08
Ar(g)	154.7	NaCl(aq)	115.5
H_2(g)	130.684	NH_4NO_3(aq)	259.8
N_2(g)	191.61	KOH(aq)	91.6

*Data from Wagman, D. D., Evans, W. H., Parker, V. B., Schumm, R. H., Halow, I., Bailey, S. M., Churney, K. L., and Nuttall, R. "The NBS Tables of Chemical Thermodynamic Properties." *Journal of Physical and Chemical Reference Data*, Vol. 11, Suppl. 2, 1982.

Qualitative Guidelines for Entropy

Some useful guidelines can be drawn from the data given in Table 18.1.

- ***Entropies of gases are usually much larger than those of liquids, which in turn are usually larger than those of solids.*** In a solid the particles can vibrate only around their lattice positions. When a solid melts, its particles can move around more freely, and molar entropy increases. When a liquid vaporizes, the position restrictions due to forces between the particles nearly disappear, and another large entropy increase occurs (Figure 18.2). For example, the entropies (in J K^{-1} mol^{-1}) of the halogens I_2(s), $Br_2(\ell)$, and Cl_2(g) are 116.1, 152.2, and 223.1, respectively. Similarly, the entropies of C(s, graphite) and C(g) are 5.7 and 158.1.
- ***Entropies of more complex molecules are larger than those of simpler molecules,*** especially in a series of closely related compounds. In a more complicated molecule there are more ways for the atoms to move about in three-dimensional space, and hence there is greater entropy. For example, the entropies (in J K^{-1} mol^{-1}) of the gases methane, CH_4; ethane, CH_3CH_3; and propane, $CH_3CH_2CH_3$, are 186.26, 229.6, and 269.9, respectively. For atoms or molecules of similar molar mass, compare the gases Ar, CO_2, and $CH_3CH_2CH_3$, which have entropies of 154.7, 213.74, and 269.9, respectively (Figure 18.3).
- ***Entropies of ionic solids that have similar formulas are larger when the attractions among the ions are weaker.*** The weaker such forces are, the easier it is for ions to vibrate about their lattice positions and the greater the

For a more detailed look at estimating entropy changes, see Craig, N. C. *Journal of Chemical Education*, Vol. 80, 2003; pp. 1432–1436.

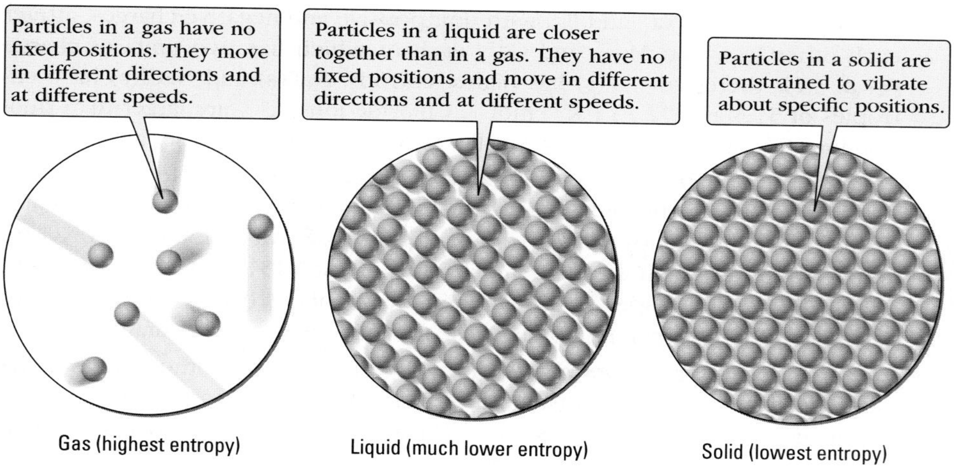
Particles in a gas have no fixed positions. They move in different directions and at different speeds.
Particles in a liquid are closer together than in a gas. They have no fixed positions and move in different directions and at different speeds.
Particles in a solid are constrained to vibrate about specific positions.
Gas (highest entropy)
Liquid (much lower entropy)
Solid (lowest entropy)

Figure 18.2 Entropies of solid, liquid, and gas phases. The entropy of each phase is related to the freedom of motion of the particles.

Name	Propane	Carbon dioxide	Argon
Molecular model	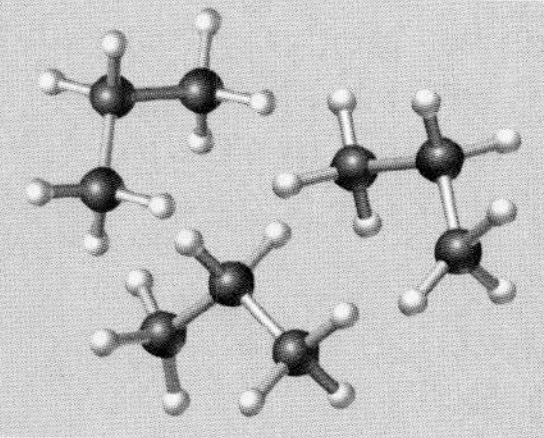	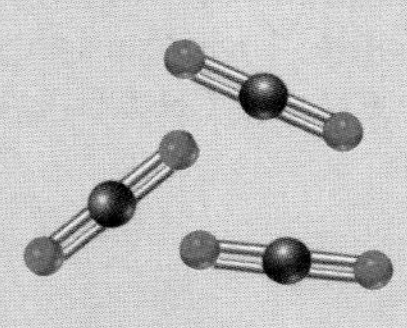	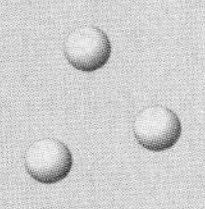
Entropy (J K^{-1} mol^{-1})	269.9	213.74	154.7
Molar mass (g/mol)	44.1	44.0	39.9

Figure 18.3 Entropy increases with molecular complexity. Three groups of particles of similar molar mass are shown. In propane, $CH_3CH_2CH_3$, many different conformations (different ways to arrange the atoms relative to one another within a molecule) and many different bond stretching and bending vibrations are possible (← p. 386) so the entropy is highest. In CO_2 there is a single conformation and many fewer vibrations. The individual Ar atoms can move about, but there are no conformations or vibrations possible so the entropy is lowest.

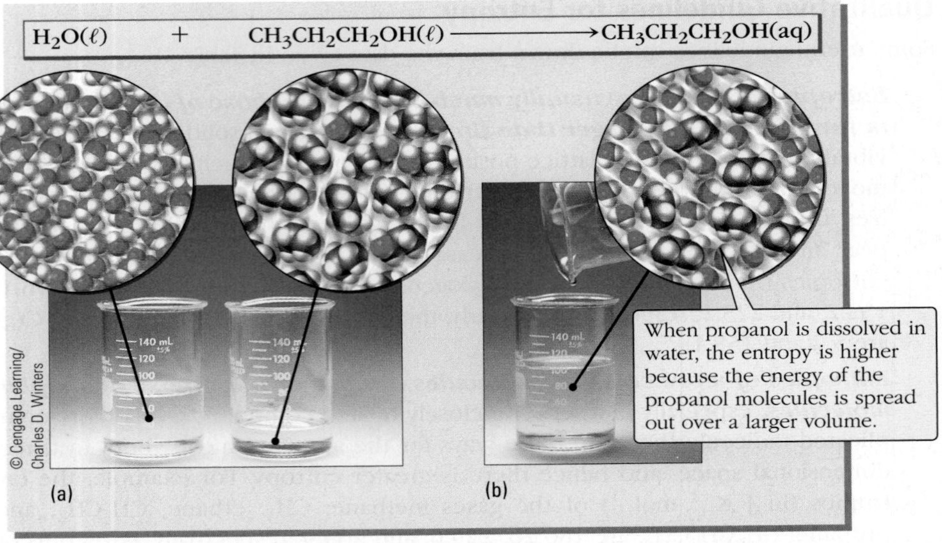
H2O(ℓ) + CH3CH2CH2OH(ℓ) → CH3CH2CH2OH(aq)
When propanol is dissolved in water, the entropy is higher because the energy of the propanol molecules is spread out over a larger volume.
© Cengage Learning/ Charles D. Winters
(a)
(b)

Active Figure 18.4 Entropy and dissolving. There is usually an increase in entropy when a solid or liquid dissolves in a liquid solvent, because, as solute particles become dispersed among solvent particles, the energy of motion of the particles is spread out over a larger volume. Visit this book's companion website at **www.cengage.com/chemistry/moore** to test your understanding of the concepts in this figure.

entropy is. The entropy of NaF(s) is 51.5 J K^{-1} mol^{-1}, and that of MgO(s) is 26.94 J K^{-1} mol^{-1}; Na^+ and F^-, with unit positive and negative charges, attract each other less than Mg^{2+} and O^{2-}, each of which has two units of charge ***(← p. 94)***; therefore NaF(s) has higher entropy. NaF(s) and NaCl(s) have entropies of 51.5 and 72.13 J K^{-1} mol^{-1}. Chloride ions, Cl^-, are larger than fluoride ions, F^-, and attractions are smaller when the ions are farther apart.

When ionic solids that consist of very small ions (such as Li^+) or of ions that carry two or more units of charge (such as Mg^{2+} or Al^{3+}) dissolve in water, there is often a decrease in entropy. This happens because very small ions and highly charged ions strongly attract water dipoles and are strongly hydrated ***(← p. 715)***. The water molecules surrounding such ions are held in a relatively rigid structure by ion-dipole forces and are no longer as free to move and rotate as they were in pure water. Although the ions become dispersed among the water molecules, the reduced dispersion of energy among the water molecules results in an overall decrease in entropy.

- ***Entropy usually increases when a pure liquid or solid dissolves in a solvent.*** Energy usually becomes more dispersed when different kinds of molecules mix together and occupy a larger volume (Figure 18.4). An example is NH_4NO_3(s) and NH_4NO_3(aq) with standard molar entropies of 151.08 J K^{-1} mol^{-1} and 259.8 J K^{-1} mol^{-1}, respectively. Some ionic compounds dissolving in water are exceptions to this generalization because the ions are strongly hydrated.
- ***Entropy decreases when a gas dissolves in a liquid.*** Although gas molecules are dispersed among solvent molecules in solution, the very large entropy of the gas phase is lost when the widely separated gas particles become crowded together with solvent particles in the liquid solution (Figure 18.5).

PROBLEM-SOLVING EXAMPLE 18.2 Relative Entropy Values

For each pair of substances below, predict which has greater entropy and give a reason for your choice. (Assume 1-mol samples at 25 °C and 1 bar.)

(a) $H_2C{=}CH_2$(g) or $CH_3CH_2CH_2CH_3$(g) (b) CO_2(aq) or CO_2(g)
(c) LiF(s) or RbBr(s) (d) N_2(ℓ) or N_2(s)

Answer

(a) $CH_3CH_2CH_2CH_3$(g) (b) CO_2(g)
(c) RbBr(s) (d) N_2(ℓ)

Strategy and Explanation Use the rules given in the preceding paragraphs.

(a) Larger, more complex molecules have greater entropy than similar smaller ones, so the entropy of $CH_3CH_2CH_2CH_3$(g) is greater.

(b) The molecules of a gas are free to move, rotate, and vibrate, so when a gas dissolves in a liquid, the entropy decreases; therefore $CO_2(g)$ has greater entropy.
(c) These are ionic solids, both with 1+ and 1− ions. The attractive forces are greater the closer the ions are to each other, and the ions are smaller in LiF ***(see table of ionic radii, ← p. 310),*** so RbBr has greater entropy.
(d) Entropy increases from solid to liquid to gas for the same substance, so $N_2(\ell)$ has greater entropy.

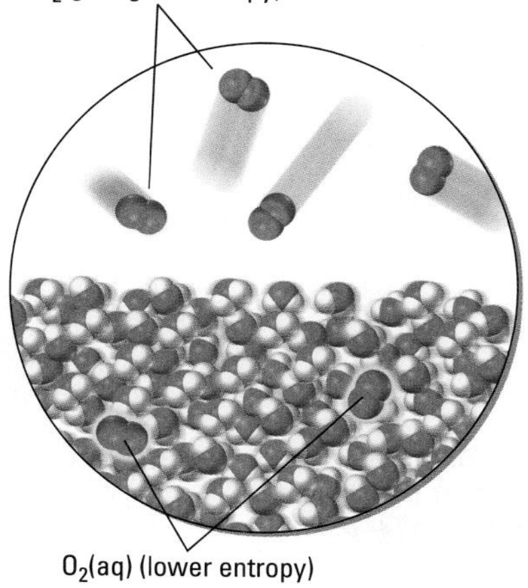
O2(g) (higher entropy)
O2(aq) (lower entropy)

Figure 18.5 Entropy of solution of a gas. The very large entropy of the gas exceeds that of the solution. Even though particles are dispersed among each other in the liquid solution, the gas particles have much more space to move in and have much higher entropy.

PROBLEM-SOLVING PRACTICE **18.2**

In each case, predict which of the two substances has greater entropy, assuming 1-mol samples at 25 °C and 1 bar. Check your prediction by looking up each substance's absolute entropy in Table 18.1.
(a) C(g) or C(s, graphite) (b) Ca(s) or Ar(g) (c) KOH(s) or KOH(aq)

Predicting Entropy Changes

The general guidelines about entropy can be used to predict whether an increase or decrease in entropy will occur when reactants are converted to products. For both of the processes

$$H_2O(s) \longrightarrow H_2O(\ell) \quad \text{and} \quad H_2O(\ell) \longrightarrow H_2O(g)$$

an entropy increase is expected. Water molecules in the solid phase are more restricted and their energy is more localized than in the liquid, and water molecules in the liquid cannot move, rotate, or vibrate as freely as they can in the gas. This is confirmed by the data in Table 18.1, where $S°\ (H_2O(g)) > S°\ (H_2O(\ell))$.

For the decomposition of iron(III) oxide to its elements,

$$2\ Fe_2O_3(s) \longrightarrow 4\ Fe(s) + 3\ O_2(g)$$

an increase in entropy is also predicted, because 3 mol gaseous oxygen is present in the products and the reactant is a solid. This is confirmed by the experimental $\Delta S°$, 551.7 J/K. Because gases have much higher entropy than solids or liquids, gaseous substances are most important in determining entropy changes.

Predicting entropy changes for chemical processes is usually easier than predicting enthalpy changes. For gas phase reactions, the enthalpy-change guideline is that having more bonds or stronger bonds (or both) in the products gives a negative $\Delta H°$ ***(← p. 238);*** however, a table of bond enthalpies is usually needed to tell which bonds are stronger.

An example where a decrease in entropy can be predicted is

$$2\ CO(g) + O_2(g) \longrightarrow 2\ CO_2(g)$$

Here there is 3 mol gaseous substance (2 mol CO and 1 mol O_2) at the beginning but only 2 mol gaseous substance at the end of the reaction. Two moles of gas almost always contains less entropy than three moles of gas, so $\Delta S°$ is negative (experimentally, $\Delta S° = -173.0$ J/K). Another example in which entropy decreases is the process

$$Ag^+(aq) + Cl^-(aq) \longrightarrow AgCl(s)$$

Here the reactant ions are free to move about among water molecules in aqueous solution, but those same ions are held in a crystal lattice in the solid, a situation with greater constraint and thus less spreading out of energy.

Predicting an entropy decrease when ions precipitate from aqueous solution does not always work, especially when the ions have more than single positive or negative charges, such as Mg^{2+} or SO_4^{2-}. The higher the charge on an ion the more tightly water-molecule dipoles are attracted to it. When water molecules are tightly attracted around an ion their motion is restricted and entropy decreases.

CONCEPTUAL EXERCISE **18.4 Predicting Entropy Changes**

For each process, tell whether entropy increases or decreases, and explain how you arrived at your prediction.
(a) $2\ CO_2(g) \longrightarrow 2\ CO(g) + O_2(g)$ (b) $NaCl(s) \longrightarrow NaCl(aq)$
(c) $MgCO_3(s) \xrightarrow{\text{heat}} MgO(s) + CO_2(g)$

18.4 Calculating Entropy Changes

The standard molar entropy values given in Table 18.1 can be used to calculate entropy changes for physical and chemical processes. Assume that each reactant and each product is at the standard pressure of 1 bar and at the temperature given (298.15 K). The number of moles of each substance is specified by its stoichiometric coefficient in the equation for the process. Multiply the entropy of each product substance by the number of moles of that product and add the entropies of all products. Calculate the total entropy of the reactants in the same way and subtract it from the total entropy of the products. This is summarized in the equation

Notice that the equation for calculating $\Delta S°$ has the same form as that for calculating $\Delta H°$ for a reaction (← p. 247).

$$\Delta S° = \Sigma\{(\text{moles product}) \times S°(\text{product})\}$$
$$- \Sigma\{(\text{moles reactant}) \times S°(\text{reactant})\}$$

Note that this calculation gives the entropy change for the chemical reaction *system* only. It tells whether the energy of the atoms that make up the system is more dispersed or less dispersed after the reaction. It does *not* account for any entropy change in the surroundings.

PROBLEM-SOLVING EXAMPLE 18.3 Calculating an Entropy Change from Tabulated Values

The reaction

$$CO(g) + 2\,H_2(g) \longrightarrow CH_3OH(\ell)$$

is being evaluated as a possible way to manufacture liquid methanol, $CH_3OH(\ell)$, for use in motor fuel. Calculate $\Delta S°$ for the reaction.

Answer $\Delta S° = -332.3$ J/K

Strategy and Explanation Use information in Table 18.1, subtracting the entropies of the reactants from the entropy of the product. Because all substances, including elements, have nonzero absolute entropy values, elements as well as compounds must be included.

$$\Delta S° = \Sigma\{(\text{moles of product}) \times S°(\text{product})\} - \Sigma\{(\text{moles of reactant}) \times S°(\text{reactant})\}$$
$$= \{(1\text{ mol}) \times S°[CH_3OH(\ell)]\} - \{(1\text{ mol}) \times S°[CO(g)] + (2\text{ mol}) \times S°[H_2(g)]\}$$
$$= \{(1\text{ mol}) \times (126.8\text{ J K}^{-1}\text{ mol}^{-1})\}$$
$$- \{(1\text{ mol}) \times (197.7\text{ J K}^{-1}\text{ mol}^{-1}) + (2\text{ mol}) \times (130.7\text{ J K}^{-1}\text{ mol}^{-1})\}$$
$$= -332.3\text{ J/K}$$

☑ **Reasonable Answer Check** There is a decrease in entropy, which is reasonable because 3 mol gaseous reactants is converted to 1 mol liquid phase product. The product molecule is more complicated, but the fact that it is a liquid makes its entropy much smaller than that of 1 mol gaseous carbon monoxide and 2 mol gaseous hydrogen.

PROBLEM-SOLVING PRACTICE 18.3

Use absolute entropies from Table 18.1 to calculate the entropy change for each of these processes, thereby verifying the predictions made in Conceptual Exercise 18.4.

(a) $2\,CO_2(g) \longrightarrow 2\,CO(g) + O_2(g)$ (b) $NaCl(s) \longrightarrow NaCl(aq)$

(c) $MgCO_3(s) \xrightarrow{\text{heat}} MgO(s) + CO_2(g)$

18.5 Entropy and the Second Law of Thermodynamics

A great deal of experience with many chemical reactions and other processes in which energy is transformed and transferred is consistent with the conclusion that *whenever a product-favored chemical or physical process occurs, energy becomes more dispersed or disordered.* This is summarized in the **second law of thermodynamics,** which states that the ***total entropy of the universe (a system plus its surroundings) is continually increasing.*** Evaluating whether the total entropy increases during a proposed chemical reaction allows us to predict whether reactants will form appreciable quantities of products.

Predicting whether a reaction is product-favored can be done in three steps:

1. Calculate how much entropy changes as a result of transfer of energy between system and surroundings ($\Delta S_{surroundings}$).
2. Calculate how much entropy changes as a result of dispersal of energy within the system (ΔS_{system}).
3. Add these two results to get $\Delta S_{universe} = \Delta S_{system} + \Delta S_{surroundings}$.

Steps 1 and 2 can be carried out by assuming that reactants under standard conditions of 1 bar and a specified temperature are converted to products under the same standard conditions.

Let us apply these steps to the reaction between carbon monoxide and hydrogen

$$CO(g) + 2\,H_2(g) \longrightarrow CH_3OH(\ell)$$

for which we calculated the entropy change in Problem-Solving Example 18.3. If the reaction is product-favored, it would be a good way to produce methanol for use as automotive fuel. The reactants can be obtained from plentiful resources: coal and water. We base our prediction upon having as reactants 1 mol CO(g) and 2 mol H_2(g) and as product 1 mol liquid methanol, with all substances at 1 bar and 298.15 K (25 °C). Then data from Table 6.2 ***(← p. 246)*** and Table 18.1 (or Appendix J) apply. If the entropy of the universe is predicted to be higher after the product has been produced, then the reaction is product-favored under these conditions and might be useful. If not, perhaps some other conditions could be used, or perhaps we should consider some other reaction altogether.

We will consider the effect of changing temperature later in this chapter. The effects of changing other conditions can often be predicted qualitatively by using Le Chatelier's principle ***(← p. 678).***

Step 1: ***Calculate the reaction's dispersal of energy to, or concentration of energy from, the surroundings by calculating ΔH° and assuming that this quantity of thermal energy is transferred reversibly.*** The entropy change for the surroundings can be calculated as

$$\Delta S^\circ_{surroundings} = \frac{q_{rev}}{T} = \frac{\Delta H_{surroundings}}{T} = \frac{-\Delta H^\circ_{system}}{T} = \frac{-\Delta H^\circ}{T}$$

The minus sign in this equation comes from the fact that the direction of energy transfer for the surroundings is opposite from the direction of energy transfer for the system. For an exothermic reaction (negative ΔH°) there will be an increase in entropy of the surroundings, a fact that we have already mentioned. For the proposed methanol-producing reaction, ΔH° (calculated from data in Appendix J) is −128.1 kJ, so the entropy change is

$$\Delta S^\circ_{surroundings} = \frac{-(-128.1\text{ kJ})}{298\text{ K}} \times \frac{1000\text{ J}}{\text{kJ}} = 430.\text{ J/K}$$

Step 2: ***Calculate the entropy change for dispersal of energy within the system*** (that is, for the atoms involved in the methanol-producing reaction). This entropy change can be evaluated from the absolute entropies of the products and reactants as described in the previous section and has already been calculated in Problem-Solving Example 18.3 to be

$$\Delta S^\circ_{system} = \Delta S^\circ = -332.3\text{ J/K}$$

Step 3: ***Calculate the total entropy change for the system and the surroundings, $\Delta S^\circ_{universe}$.*** Because the universe includes both system and surroundings, $\Delta S^\circ_{universe}$ is the sum of the entropy change for the system and the entropy change for the surroundings. (We assume that nothing else but our reaction happens, so there are no other entropy changes.) This total entropy change is

$$\Delta S^\circ_{\text{universe}} = \Delta S^\circ_{\text{surroundings}} + \Delta S^\circ_{\text{system}} = \frac{-\Delta H^\circ_{\text{system}}}{T} + \Delta S^\circ_{\text{system}} \quad [18.2]$$

$$= (430. - 332.3)\ \text{J/K} = 98\ \text{J/K}$$

Combination of carbon monoxide and hydrogen to form methanol increases the entropy of the universe. Because the process is product-favored, it could be useful for manufacturing methanol.

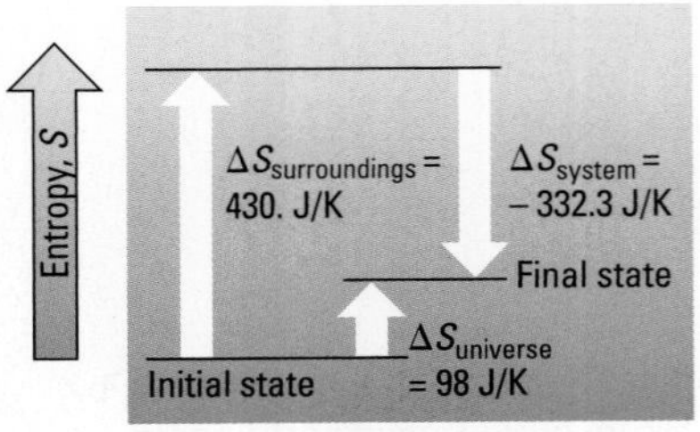

Entropy diagram for combination of carbon monoxide and hydrogen to form methanol.

CONCEPTUAL EXERCISE 18.5 Effect of Temperature on Entropy Change

The reaction of carbon monoxide with hydrogen to form methanol is quite slow at room temperature. As a general rule, reactions go faster at higher temperatures. Suppose that you tried to speed up this reaction by increasing the temperature.

(a) Assuming that ΔH° does not change very much as the temperature changes, what effect would increasing the temperature have on $\Delta S^\circ_{\text{surroundings}}$?

(b) Assuming that ΔS° for a reaction system does not change much as the temperature changes, what effect would increasing the temperature have on $\Delta S^\circ_{\text{universe}}$?

PROBLEM-SOLVING EXAMPLE 18.4 Calculating $\Delta S^\circ_{\text{universe}}$

When gasoline burns, one reaction is combustion of octane with oxygen from air.

$$2\ C_8H_{18}(g) + 25\ O_2(g) \longrightarrow 16\ CO_2(g) + 18\ H_2O(\ell)$$

Calculate $\Delta S^\circ_{\text{universe}}$ for this reaction, thereby confirming that the reaction is product-favored. (Assume that reactants and products are at 298.15 K and 1 bar; use data from Appendix J.)

Answer $\Delta S^\circ_{\text{universe}} = 35{,}591$ J/K

Strategy and Explanation A balanced chemical equation for the reaction is given. Calculate ΔH° and ΔS° by subtracting the sum of reactant values from the sum of product values.

$$\begin{aligned}\Delta H^\circ &= \Sigma\{(\text{moles of product}) \times \Delta H^\circ_f\ (\text{product})\}\\ &\quad - \Sigma\{(\text{moles of reactant}) \times \Delta H^\circ_f\ (\text{reactant})\}\\ &= \{16 \times (-393.509\ \text{kJ}) + 18 \times (-285.830\ \text{kJ})\} - \{2 \times (-208.447\ \text{kJ}) + 25 \times 0\ \text{kJ}\}\\ &= -11{,}024.2\ \text{kJ}\end{aligned}$$

$$\begin{aligned}\Delta S^\circ &= \Sigma\{(\text{moles of product}) \times S^\circ\ (\text{product})\} - \Sigma\{(\text{moles of reactant}) \times S^\circ\ (\text{reactant})\}\\ &= \{16 \times (213.74\ \text{J/K}) + 18 \times (69.91\ \text{J/K}\} - \{2 \times (466.835\ \text{J/K})\\ &\quad + 25 \times (205.138\ \text{J/K})\}\\ &= -1383.9\ \text{J/K}\end{aligned}$$

Now use Equation 18.2 to calculate $\Delta S^\circ_{\text{universe}}$.

$$\begin{aligned}\Delta S^\circ_{\text{universe}} &= \Delta S_{\text{surroundings}} + \Delta S_{\text{system}} = \frac{-\Delta H^\circ}{T} + \Delta S^\circ\\ &= \frac{-(-11{,}024.2\ \text{kJ})}{298.15\ \text{K}} + (-1383.9\ \text{J/K})\\ &= 36.975\ \text{kJ/K} - 1383.9\ \text{J/K}\\ &= 36{,}975\ \text{J/K} - 1383.9\ \text{J/K} = 35{,}591\ \text{J/K}\end{aligned}$$

Notice that for $-\Delta H^\circ/T$ the units were kilojoules per kelvin (kJ/K), while for ΔS° the units were joules per kelvin (J/K). Therefore it was necessary to convert kilojoules to joules by multiplying the first term by 1000 J/kJ.

Because $\Delta S^\circ_{\text{universe}}$ is positive, the reaction is product-favored.

☑ **Reasonable Answer Check** A positive result is reasonable, because you know that gasoline burns in air, which means that the reaction is product-favored. Even though the entropy change for the system is unfavorable, the reaction is highly product-favored because it is strongly exothermic.

PROBLEM-SOLVING PRACTICE 18.4

Use data from Appendix J to determine whether the synthesis of ammonia from nitrogen and hydrogen is product-favored at 298.15 K and 1 bar.

CONCEPTUAL EXERCISE 18.6 Variation of $\Delta H°$ and $\Delta S°$ with Temperature

Suppose that the combustion of gaseous octane is carried out at 150 °C.

(a) How would this affect the chemical equation for the reaction?

(b) What effect would the change in the chemical equation have on $\Delta H°$ and $\Delta S°$ for the reaction?

(c) When is it definitely *not* safe to assume that $\Delta H°$ and $\Delta S°$ for a reaction will be almost the same over a broad range of temperatures?

Predictions of the sort we have just made by calculating $\Delta S°_{universe}$ can also be made qualitatively, without calculating, if we know whether a reaction is exothermic and if we can predict whether energy is dispersed within the system when the reaction takes place.

- ***A reaction is certain to be product-favored if it is exothermic* and *the entropy of the product atoms, molecules, and ions is greater than the entropy of the reactants ($\Delta S_{system} > 0$).***
- ***A reaction is certain to be reactant-favored if it is endothermic* and *there is a decrease in entropy of the system ($\Delta S_{system} < 0$).***

There are two other possible cases, as indicated in Table 18.2, but they are more difficult to predict without quantitative information.

As examples, consider the reactions of carbonates with acids (← *p. 176*). These reactions are product-favored because they are exothermic and produce gases. Reaction of limestone with hydrochloric acid is typical.

$$CaCO_3(s) + 2\,HCl(aq) \longrightarrow CaCl_2(aq) + H_2O(\ell) + CO_2(g) \quad \text{(exothermic)}$$

Similarly, combustion reactions of hydrocarbons such as butane, $CH_3CH_2CH_2CH_3$, are product-favored because they are exothermic and produce a larger number of gas phase product molecules than there were gas phase reactant molecules.

$$2\,CH_3CH_2CH_2CH_3(g) + 13\,O_2(g) \longrightarrow 8\,CO_2(g) + 10\,H_2O(g) \quad \text{(exothermic)}$$

But what about a reaction such as the production of ethylene, $CH_2{=}CH_2$, from ethane, CH_3CH_3? Although entropy is predicted to increase (one gas phase molecule forms two), the reaction is very endothermic. (That the reaction is endothermic might be predicted on the basis of a decrease from seven bonds in the reactant molecule to six bonds—five single, one double—in the products.)

$$CH_3CH_3(g) \longrightarrow H_2(g) + CH_2{=}CH_2(g) \qquad \Delta H° = +137\text{ kJ}; \quad \Delta S° = +121\text{ J/K}$$

Enthalpy change predicts that this process is reactant-favored, while entropy change predicts the opposite. Which is more important? It depends on the temperature.

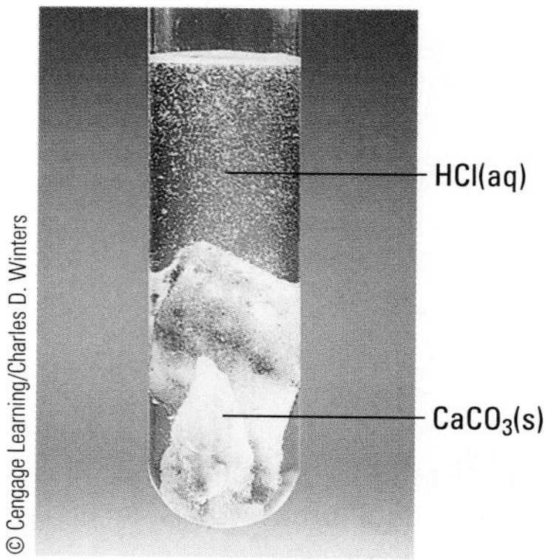
HCl(aq)
CaCO3(s)
© Cengage Learning/Charles D. Winters

A product-favored reaction. Carbonates react rapidly with acid to produce carbon dioxide and water.

Table 18.2 Predicting Whether a Reaction Is Product-Favored (at constant *T* and *P*)

Sign of ΔH_{system}	Sign of ΔS_{system}	Product-Favored?
Negative (exothermic)	Positive	Yes
Negative (exothermic)	Negative	Yes at low *T*; no at high *T*
Positive (endothermic)	Positive	No at low *T*; yes at high *T*
Positive (endothermic)	Negative	No

At 25 °C:

$$\Delta S^\circ_{surroundings} = \frac{-137 \text{ kJ}}{298 \text{ K}}$$
$$= -460 \text{ J/K}$$
$$\Delta S^\circ_{system} = 121 \text{ J/K}$$
$$\Delta S^\circ_{universe} = (-460 + 121) \text{ J/K}$$
$$= -339 \text{ J/K}$$

At 1000 °C:

$$\Delta S^\circ_{surroundings} = \frac{-137 \text{ kJ}}{(1000 + 273) \text{ K}}$$
$$= -108 \text{ J/K}$$
$$\Delta S^\circ_{system} = 121 \text{ J/K}$$
$$\Delta S^\circ_{universe} = 13 \text{ J/K}$$

Calculating $\Delta S^\circ_{surroundings}$ ($= -\Delta H^\circ/T$) requires dividing the enthalpy change by the temperature. Because ΔH° stays pretty much the same at different temperatures, the higher the temperature, the smaller the absolute value of $\Delta S^\circ_{surroundings}$ ($= -\Delta H^\circ/T$). At room temperature $\Delta S^\circ_{surroundings}$ is usually bigger in absolute value than ΔS°_{system}, so exothermic reactions are expected to be product-favored and endothermic reactions (like this one) are expected to be reactant-favored. The ethylene-producing reaction is reactant-favored at 25 °C, because $\Delta S^\circ_{universe} = -339$ J/K. To make a successful industrial process, chemical engineers have designed plants that carry out this reaction at about 1000 °C. At this higher temperature, $\Delta S^\circ_{surroundings}$ is smaller in magnitude than ΔS°_{system}. Thus $\Delta S^\circ_{universe} = 13$ J/K, and products are predicted to predominate over reactants.

In Chapter 14 (← **p. 687**) we stated that if a reaction is exothermic or involves an increase in entropy of the system, this favors the products. The entropy effect becomes more important the higher the temperature is.

EXERCISE **18.7 Predicting the Direction of a Reaction**

Using data from Appendix J, complete the table and then classify each reaction into one of the four types in Table 18.2. Predict whether each reaction is product-favored or reactant-favored at room temperature.

Reaction	ΔH°, 298 K (kJ)	ΔS°, 298 K (J/K)
(a) $C_2H_4(g) + 3\ O_2(g) \longrightarrow 2\ H_2O(\ell) + 2\ CO_2(g)$	______	______
(b) $2\ Fe_2O_3(s) + 3\ C(\text{graphite}) \longrightarrow 4\ Fe(s) + 3\ CO_2(g)$	______	______
(c) $C(\text{graphite}) + O_2(g) \longrightarrow CO_2(g)$	______	______
(d) $2\ Ag(s) + 3\ N_2(g) \longrightarrow 2\ AgN_3(s)$	______	______

EXERCISE **18.8 Product- or Reactant-Favored?**

(a) Is the combination reaction of hydrogen gas and chlorine gas to give hydrogen chloride gas (at 1 bar) predicted to be product-favored or reactant-favored at 298 K?
(b) Calculate the value for $\Delta S^\circ_{universe}$ for the reaction in part (a).

go Chemistry **Module 24: Gibbs Free Energy and Equilibrium** covers concepts in this section.

18.6 Gibbs Free Energy

Calculations of the sort done in the previous section would be simpler if we did not have to separately evaluate the entropy change of the surroundings from T and a table of ΔH°_f values and the entropy change of the system from a table of S° values. To simplify such calculations, a new thermodynamic function was defined by J. Willard Gibbs (1838–1903). It is now called **Gibbs free energy**, and it is given the symbol G. Gibbs defined his free energy so that $\Delta G_{system} = -T\Delta S_{universe}$. Because of the minus sign, if the entropy of the universe increases, the Gibbs free energy of the system must decrease. That is, ***a decrease in Gibbs free energy of a system is characteristic of a process that is product-favored at constant temperature and pressure.***

In the previous section we showed that the total entropy change accompanying a chemical reaction carried out at constant temperature and pressure is

$$\Delta S_{universe} = \Delta S_{surroundings} + \Delta S_{system} = \frac{-\Delta H_{system}}{T} + \Delta S_{system}$$

Combining this algebraically with Gibbs's definition of free energy, we have

$$\Delta G_{system} = -T\Delta S_{universe} = -T\left[\frac{-\Delta H_{system}}{T} + \Delta S_{system}\right] = \Delta H_{system} - T\Delta S_{system}$$

or, under standard-state conditions.

$$\Delta G° = \Delta H° - T\Delta S° \qquad [18.3]$$

This equation summarizes the ideas about chemical equilibrium that were developed in Section 14.7 (← p. 687). A *negative* value of $\Delta G°$ indicates that a reaction is *product-favored*, and the equation says that two conditions will make $\Delta G°$ more negative: (1) if the reaction is exothermic, $\Delta H°$ will be negative, thereby favoring the products; and (2) if the products have greater entropy than the reactants, then $\Delta S°$ will be positive and the $-T\Delta S°$ term will be negative, which favors the products. Because $\Delta S°$ is multiplied by T, the entropy of the system is more important at higher temperatures.

For $\Delta H°$, $\Delta S°$, and $\Delta G°$ you can assume that the values apply to the system—that is, $\Delta G° = \Delta G°_{system}$—unless a subscript is attached to indicate that a value is for the surroundings.

It is important to realize that $\Delta G°$ varies significantly as the temperature changes (because of the $-T\Delta S°$ term). Therefore, values of $\Delta G°$ calculated from Equation 18.4 apply only to the temperature specified in the table of $\Delta G_f°$ values. Appendix J specifies a temperature of 25 °C.

CONCEPTUAL EXERCISE 18.9 Predicting Whether a Process Is Product-Favored

Make a table similar to Table 18.2, but add a new column for the sign of $\Delta G°$. Based on the value of $\Delta G°$, predict whether the reaction is product-favored. If there is insufficient information, indicate whether the products would be favored more at high temperatures or at low temperatures. Check your results against Table 18.2.

The Gibbs free energy change provides a way of predicting whether a reaction will be product-favored that depends only on the system—that is, the chemical substances undergoing reaction. Therefore, we can tabulate values of the standard Gibbs free energy of formation, $\Delta G_f°$, for a variety of substances, and from them calculate

$$\Delta G° = \Sigma\{(\text{moles of product}) \times \Delta G_f°(\text{product})\} - \Sigma\{(\text{moles of reactant}) \times \Delta G_f°(\text{reactant})\} \qquad [18.4]$$

for a great many reactions. The calculation is similar to using $\Delta H_f°$ values from Table 6.2 or Appendix J to calculate $\Delta H°$ for a reaction (← p. 247). As was the case for $\Delta H_f°$ values, there are no $\Delta G_f°$ values for elements in their standard states, because forming an element from itself constitutes no change at all. Appendix J contains a table that includes $\Delta G_f°$ values for many compounds.

Beinecke Rare Book and Manuscript Library, Yale University

Josiah Willard Gibbs

1838–1903

Gibbs was the son of a Yale professor of sacred literature. Like many of his forebears Gibbs studied at Yale, where he received his Ph.D. in 1863. His interest in both mathematics and engineering were reflected in the title of his thesis, "On the Form of the Teeth of Wheels in Spur Gearing." He spent three years in Europe, studying in Paris, Berlin, and Heidelberg, before becoming the first professor of mathematical physics at Yale. His work was not well known until long after he published it, because his published articles were compactly written and very abstract. They were also profound and influential, providing the basis for chemical thermodynamics. (See **http://www.aip.org/history/gap/Gibbs/Gibbs.html** for a more detailed biography.)

PROBLEM-SOLVING EXAMPLE 18.5 Using Standard Gibbs Free Energies of Formation

Calculate the standard Gibbs free energy change for the combustion of octane using values of $\Delta G_f°$ from Appendix J. Assume that the initial and final states have the same temperature and pressure so that Equation 18.4 applies.

Answer $\Delta G° = -10{,}611$ kJ

Strategy and Explanation Write a balanced equation for the combustion reaction and look up $\Delta G_f°$ values in Appendix J.

	$2\,C_8H_{18}(g)$	$+\ 25\,O_2(g)$	$\longrightarrow 18\,H_2O(\ell)$	$+\ 16\,CO_2(g)$
$\Delta G_f°$(kJ/mol):	16.72	0	−237.1	−394.4

(Notice that elements in their standard states have $\Delta G_f° = 0$, just as they have $\Delta H_f° = 0$.)

Now calculate

$$\begin{aligned}\Delta G° &= \Sigma\{(\text{moles of product}) \times \Delta G_f°\ (\text{product})\} \\ &\quad - \Sigma\{(\text{moles of reactant}) \times \Delta G_f°\ (\text{reactant})\} \\ &= \{(18\ \text{mol} \times \Delta G_f°\ [H_2O(g)]\} + \{16\ \text{mol} \times \Delta G_f°[CO_2(g)]\} \\ &\quad - \{2\ \text{mol} \times \Delta G_f°[C_8H_{18}(g)]\} - \{25\ \text{mol} \times \Delta G_f°[O_2(g)]\} \\ &= \{18\ \text{mol} \times (-237.1\ \text{kJ/mol})\} + \{16\ \text{mol} \times (-394.4\ \text{kJ/mol})\} \\ &\quad - \{2\ \text{mol} \times (16.72\ \text{kJ/mol})\} - \{25\ \text{mol} \times (0\ \text{kJ/mol})\} \\ &= -10{,}611\ \text{kJ}\end{aligned}$$

Othmer Library of Chemical History, Chemical Heritage Foundation. Photo by James R. Voelkel

Scientific Communication. This is the first page of Josiah Willard Gibbs's paper describing results of his theoretical research on chemical equilibrium, entropy, and thermodynamics. The paper was 320 pages long and filled the entire third volume of the *Transactions of the Connecticut Academy of Arts and Sciences.* Because this journal was not read by most scientists, the publication went unnoticed for a long time, even though it was crucial for advancing our understanding of thermodynamics. This underscores how important it is to communicate scientific results widely.

The Gibbs free energy change, $\Delta G°$, is a large negative number, indicating that the reaction is product-favored under standard-state conditions.

☑ **Reasonable Answer Check** A large, negative Gibbs free energy change is reasonable for a combustion reaction, which, once initiated, occurs of its own accord.

PROBLEM-SOLVING PRACTICE 18.5

In the text we concluded that the reaction to produce methanol from CO and H_2 is product-favored.

$$CO(g) + 2\,H_2(g) \longrightarrow CH_3OH(\ell)$$

(a) Verify this result by calculating $\Delta G°$ from $\Delta H°$ and $\Delta S°$ for the system. Use values of $\Delta H_f°$ and $S°$ from Appendix J.

(b) Compare your result in part (a) with the calculated value of $\Delta G°$ obtained from $\Delta G_f°$ values from Appendix J.

(c) Is the sign of $\Delta G°$ positive or negative? Is the reaction product-favored? At all temperatures?

The Effect of Temperature on Reaction Direction

Many reactions are product-favored at some temperatures and reactant-favored at other temperatures. Thus, it might be possible to make such a reaction produce products by increasing or decreasing the temperature. There is a simple, approximate way to estimate the temperature at which a reactant-favored process becomes product-favored.

In Exercises 18.5 and 18.6 (p. 862 and p. 863) we developed the idea that $\Delta H°$ and $\Delta S°$ have nearly constant values over a broad range of temperatures, provided that each of the substances involved in a chemical reaction remains in the same state of matter (solid, liquid, or gas). Because $\Delta H°$ and $\Delta S°$ are nearly constant, the T on the right-hand side of the equation $\Delta G° = \Delta H° - T\Delta S°$ implies that $\Delta G°$ must vary with temperature. It also implies that if we know $\Delta H°$ and $\Delta S°$ at one temperature, we can estimate $\Delta G°$ over a range of temperatures. For example, suppose we want to know whether the reaction

$$2\,HgO(s) \longrightarrow 2\,Hg(\ell) + O_2(g)$$

will produce products at a temperature of 350. °C and a pressure of 1 bar. To find out, calculate $\Delta H°$ and $\Delta S°$ at 298 K and then estimate $\Delta G°$, assuming that $\Delta H°$ and $\Delta S°$ have the same values at 350. °C (623 K) that they do at 298 K. The boiling point of mercury is 356 °C, and mercury(II) oxide does not melt until well above 500. °C, so the substances are all in the same states at 350. °C that they were at 25 °C.

The effect of temperature on reaction direction can be seen in Equation 18.2 (p. 862). Because $\Delta S_{surroundings} = -\Delta H_{system}/T$ and ΔH is nearly constant over a broad range of temperatures, the larger T is the smaller (and therefore less important) the entropy change of the surroundings becomes relative to the entropy change of the system. At high temperatures ΔS_{system} governs the directionality of the reaction.

The notation "$\Delta S°$(298.15 K)" designates the entropy change at a temperature of 298.15 K.

$$\begin{aligned}
\Delta S°(298.15\text{ K}) &= \{2\text{ mol} \times S°[Hg(\ell)] + (1\text{ mol}) \times S°[O_2(g)]\} \\
&\quad - \{2\text{ mol} \times S°[HgO(s)]\} \\
&= \{2 \times 76.02\text{ J/K} + 205.138\text{ J/K}\} - \{2 \times 70.29\text{ J/K}\} = 216.60\text{ J/K} \\
\Delta H°(298.15\text{ K}) &= \{2\text{ mol} \times \Delta H_f°[Hg(\ell)] + 1\text{ mol} \times \Delta H_f°[O_2(g)]\} \\
&\quad - \{2\text{ mol} \times \Delta H_f°[HgO(s)]\} \\
&= \{0\text{ kJ} + 0\text{ kJ}\} - \{2 \times (-90.83)\text{ kJ}\} = 181.66\text{ kJ} \\
\Delta G°(623\text{ K}) &= \Delta H°(298.15\text{ K}) - T \times \Delta S°(298.15\text{ K}) \\
&= 181.66\text{ kJ} - 623\text{ K} \times 216.60\text{ J/K} \\
&= 181.66\text{ kJ} - 134{,}942\text{ J} = 181.66\text{ kJ} - 134.9\text{ kJ} = 47\text{ kJ}
\end{aligned}$$

Note that 134,942 J has been converted to 134.9 kJ in the last step.

Because $\Delta G°$ is positive, the reaction is not product-favored at 623 K. But at 623 K, $\Delta G°$ has a smaller positive value than at 298 K, where it has the value $\Delta G° = -(2\text{ mol}) \times \Delta G_f°(HgO[s]) = +117.1$ kJ. That is, the reaction is more product-favored at 623 K than it is at room temperature. Because the calculated values of $\Delta H°$ and $\Delta S°$ are both positive, the reaction is expected to be product-favored at high temperatures but not at low temperatures. Heating to an even higher temperature than 623 K does decompose mercury(II) oxide (see Figure 18.6).

Joseph Priestley's discovery of oxygen in 1774 involved heating mercury(II) oxide to decompose it.

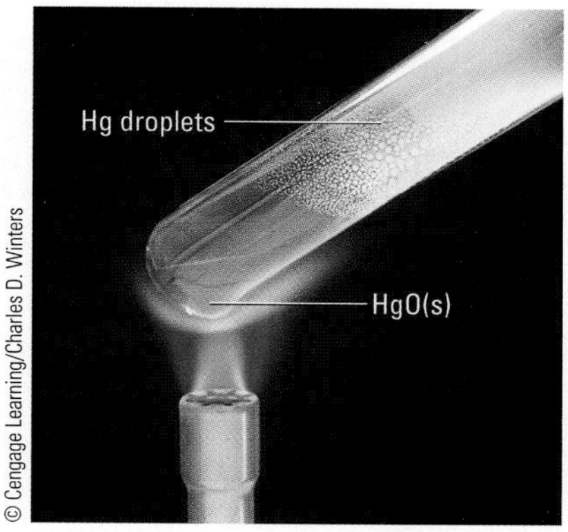

Figure 18.6 Decomposition of HgO(s). When heated, red mercury(II) oxide decomposes to liquid mercury metal and oxygen gas. Shiny droplets of mercury can be seen where they have condensed in the cooler part of the test tube.

CONCEPTUAL EXERCISE 18.10 High-Temperature Decomposition

Suppose that a sample of HgO(s) is heated above the boiling point of mercury (356 °C).

(a) Could you use the same method to estimate the Gibbs free energy change that was used in the preceding paragraph? Why or why not? (*Hint:* Write the equation for the process that would occur if the temperature were 400 °C.)

(b) At 400 °C, would you expect the reaction to be more or less product-favored than it was at 350 °C? Give two reasons for your choice.

For reactions such as decomposition of mercury(II) oxide that are reactant-favored at low temperatures and product-favored at high temperatures, it is possible to calculate the minimum temperature to which the system must be heated to make the reaction product-favored. Below that temperature $\Delta G°$ is positive, and above that temperature $\Delta G°$ is negative. Therefore, $\Delta G°$ must equal zero at the desired temperature. Because $\Delta G° = \Delta H° - T\Delta S°$ (Equation 18.3), we can set $\Delta H° - T\Delta S° = 0$. Solving for T gives

$$T(\text{at which } \Delta G° \text{changes sign}) = \frac{\Delta H°}{\Delta S°} \quad [18.5]$$

As an example, the contributions of $\Delta H°$ and $-T\Delta S°$ to $\Delta G°$ for the decomposition of silver(I) oxide are shown graphically in Figure 18.7. Equation 18.5 also applies to reactions that are product-favored at low temperatures and reactant-favored at high temperatures. For such reactions Equation 18.5 gives the temperature *below* which the reaction is product-favored. Heating the system above this temperature will probably result in insufficient quantities of products being produced.

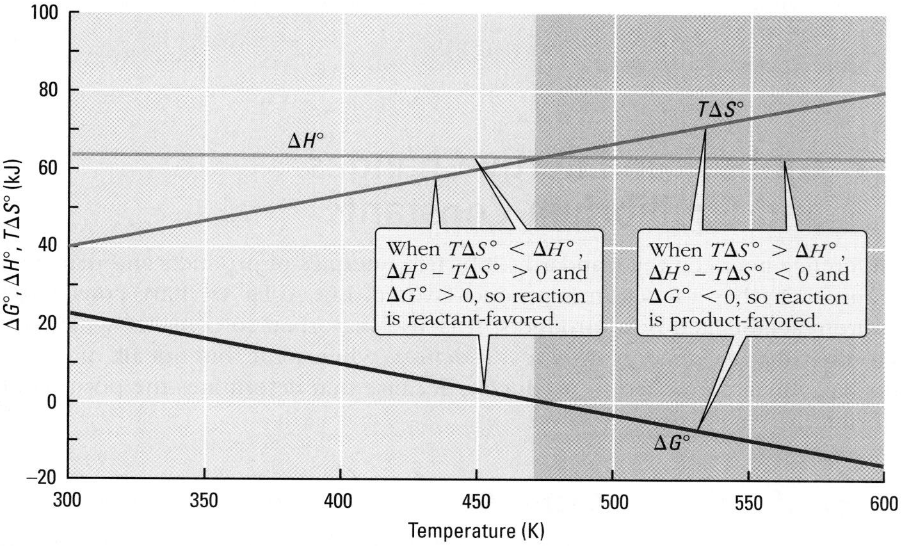

Figure 18.7 Effect of temperature on reaction spontaneity. The two terms that contribute to the Gibbs free energy change, $\Delta H°$ and $T\Delta S°$, are plotted as a function of temperature for the reaction of silver(I) oxide, $Ag_2O(s)$, to form silver metal and oxygen gas.

PROBLEM-SOLVING EXAMPLE 18.6 Effect of Temperature on Gibbs Free Energy Change

Calculate the temperature to which silver(I) oxide, $Ag_2O(s)$, must be heated to decompose it to oxygen gas and solid silver metal.

Answer 468 K (195 °C)

Strategy and Explanation Begin by writing the equation for the reaction, and use values from Appendix J to calculate $\Delta H°$ and $\Delta S°$ at 25 °C. Then use Equation 18.5 to calculate the desired temperature.

$$2\,Ag_2O(s) \longrightarrow 4\,Ag(s) + O_2(g)$$

$$\Delta H° = \{(4\text{ mol}) \times (0\text{ kJ/mol}) + (1\text{ mol}) \times (0\text{ kJ/mol})\} - \{(2\text{ mol}) \times (-31.05\text{ kJ/mol})\} = 62.10\text{ kJ}$$

$$\Delta S° = \{(4\text{ mol}) \times (42.55\text{ J K}^{-1}\text{ mol}^{-1}) + (1\text{ mol}) \times (205.138\text{ J K}^{-1}\text{ mol}^{-1})\} - \{(2\text{ mol}) \times (121.3\text{ J K}^{-1}\text{ mol}^{-1})\} = 132.7\text{ J/K}$$

Because both $\Delta H°$ and $\Delta S°$ are positive, the reaction will become more product-favored as temperature increases.

$$T = \frac{62.1\text{ kJ}}{132.7\text{ J/K}} \times \frac{1000\text{ J}}{1\text{ kJ}} = \frac{62.1\text{ kJ}}{0.1327\text{ kJ/K}} = 467.9\text{ K} = 468\text{ K}$$

Note that $\Delta H°$ has units of kJ, whereas $\Delta S°$ has units of J/K; this necessitates a unit conversion so that the energy units can cancel.

☑ **Reasonable Answer Check** The reaction produces a gas and a solid from a solid, so the entropy change should be positive. Silver is one of the few metals that can be found as the element in nature, so its oxide probably does not require an extremely high temperature to decompose; 468 K (195 °C) is a reasonable answer. Check that each reactant and each product is in the same physical state at 195 °C as it was at room temperature. The *CRC Handbook* gives the melting point of silver as 962 °C and reports that silver(I) oxide decomposes before it melts, so the approximation of nearly constant $\Delta S°$ and $\Delta H°$ is valid.

PROBLEM-SOLVING PRACTICE 18.6

For the reaction

$$2\,CO(g) + O_2(g) \longrightarrow 2\,CO_2(g)$$

(a) Predict the temperature at which the reaction changes from being product-favored to being reactant-favored.

(b) If you wanted this reaction to produce $CO_2(g)$, what temperature conditions would you choose?

18.7 Gibbs Free Energy Changes and Equilibrium Constants

go Chemistry **Module 24: Gibbs Free Energy and Equilibrium** covers concepts in this section.

The difference between the standard Gibbs free energies of products and reactants determines whether a reaction is product-favored, but so far we have considered only pure reactants and pure products. It is also important to consider what happens to the Gibbs free energy *during a reaction* (when some, but not all, of the reactants have been converted to products), because that determines the position of equilibrium.

Variation of Gibbs Free Energy During a Reaction

One way to remove carbon dioxide from air is to pass the air over solid sodium hydroxide.

$$NaOH(s) + CO_2(g) \longrightarrow NaHCO_3(s) \qquad \Delta G° \text{ (at 25 °C)} = -77.2\text{ kJ} \qquad [18.6]$$

Suppose that this reaction is carried out so that the pressure remains at 1 bar and the temperature remains at 25 °C throughout the change from NaOH and CO_2 to $NaHCO_3$. Then all three substances will remain in their standard states throughout the reaction. The concentration of CO_2, which is proportional to its pressure, remains constant, and the concentrations of the two solids also remain constant, as you showed in Exercise 14.1 (← *p. 657*).

When the reaction is halfway over, half of the reactants have been converted to products, so 0.5 mol NaOH(s) and 0.5 mol CO_2(g) have been converted to 0.5 mol $NaHCO_3$(s). The equation for what has happened so far is

$$0.5\ \mathrm{NaOH(s)} + 0.5\ \mathrm{CO_2(g)} \longrightarrow 0.5\ \mathrm{NaHCO_3(s)}$$

$$\Delta G°(\text{at } 25\ °\mathrm{C}) = (0.5)(-77.2\ \mathrm{kJ}) = -38.6\ \mathrm{kJ}$$

Recall that a thermochemical expression refers to the exact number of moles of reactants and products indicated by the coefficients.

At this point we say that the extent of reaction is 0.5. The **extent of reaction,** which we will represent by x, *is the fraction of the reactants that has been converted to products.* For the general equation

$$a\,\mathrm{A} + b\,\mathrm{B} \longrightarrow c\,\mathrm{C} + d\,\mathrm{D}$$

if z mol reactant A has been consumed, then the extent of reaction is z/a. If y mol B has been consumed, the extent of reaction is y/b. If w mol D has formed, the extent of reaction is w/d. For a reaction such as the combination of NaOH with CO_2, in which all substances remain in their standard states throughout the chemical change, an extent of reaction of x corresponds to a Gibbs free energy change of x times the Gibbs free energy change for the complete reaction. This is shown graphically in Figure 18.8.

If we start with *a* mol A, then the maximum quantity of A that can be consumed is *a* mol A. Therefore $z \le a$ and $x = z/a$ is always a fraction. Its maximum value is 1.

CONCEPTUAL EXERCISE 18.11 Gibbs Free Energy and Extent of Reaction

For Reaction 18.6,

(a) Calculate the Gibbs free energy change when the extent of reaction is 0.10, 0.40, and 0.80.

(b) Verify the statement in the text that an extent of reaction of x corresponds to a Gibbs free energy change of x times $\Delta G°$ for the complete reaction.

(c) Show that the slope of the line in Figure 18.8 is $\Delta G°$.

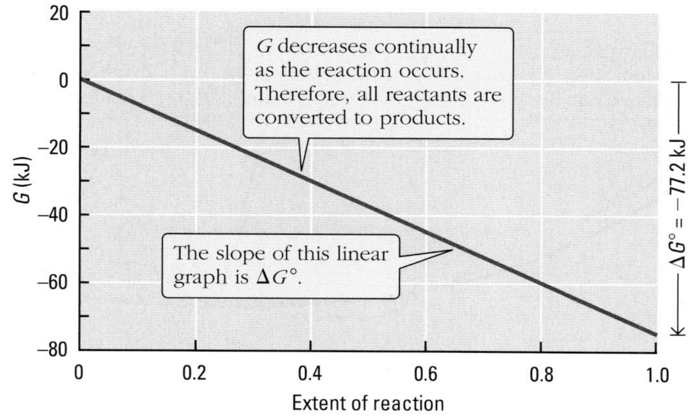

Figure 18.8 Gibbs free energy as a function of extent of reaction. For the reaction of NaOH(s) with CO_2(g), the Gibbs free energy decreases linearly as the reaction proceeds. A similar linear graph is obtained whenever all reactants and products are in their standard states throughout a reaction. The slope of such a graph is equal to $\Delta G°$. The Gibbs free energy of the reactants has been arbitrarily set to zero so that it is easier to see how big the differences in Gibbs free energy are.

Reactions That Reach Equilibrium

When a reaction occurs in which two or more substances form a solution, the situation is different. The reaction will reach an equilibrium state in which both reactants and products are present. Consider an equilibrium reaction we discussed in Chapter 14 **(← *p. 659*)**, the isomerization of *cis*-2-butene at constant pressure.

$$cis\text{-2-butene(g)} \rightleftharpoons trans\text{-2-butene(g)} \qquad \Delta G^\circ \text{ (at 500 K)} = -2.08 \text{ kJ}$$

If we start with 1.0 mol *cis*-2-butene at 500 K and 1 bar, then as soon as some *trans*-2-butene forms it will mix with (dissolve in) the *cis*-2-butene. When one gas dissolves in the other, each becomes less concentrated. Because the total pressure remains at 1 bar in the constant-pressure process, the partial pressure of each gas must be less than 1 bar, which means that neither gas is in its standard state, and the method you used in Exercise 18.11 to calculate ΔG as a function of extent of reaction no longer applies. However, we can tell something about how G varies as the reaction proceeds. Mixing two ideal gases involves no change in enthalpy, but it does involve an increase in entropy because each gas expands to fill the entire container, thereby causing increased dispersal of energy. (This is in addition to any difference in entropy between products and reactants.) Since $\Delta S_{\text{dilution}}$ is positive and $\Delta H_{\text{dilution}}$ is zero, then $\Delta G_{\text{dilution}}$ must be negative. Therefore, in Figure 18.9, there is *not* a straight line from reactants to products. Because there is a negative component ($\Delta G_{\text{dilution}}$) in addition to x times ΔG°, there is a curved line below where the straight line would have been.

Recall that one of the assumptions on which the ideal gas law is based is that there are no forces of attraction among the molecules. Therefore there is no enthalpy change when the gases mix.

The graph in Figure 18.9 is steeper at both ends than a straight line from $G_{\text{reactants}}$ to G_{products} and passes through a minimum at $x = 0.62$. This value of x means that the fraction of *trans* molecules is 0.62 and the fraction of *cis* molecules is $1 - 0.62 = 0.38$. Since both *cis*- and *trans*-2-butene occupy the same container, their concentrations and partial pressures are proportional to the number of molecules. Therefore, we can substitute these fractions into the equilibrium constant expression.

If there is the same number of gas phase molecules in the reactants as in the products, $K_c = K_P$.

$$K_P = K_c = \frac{P_{trans}}{P_{cis}} = \frac{[trans]}{[cis]} = \frac{0.62}{0.38} = 1.6$$

When $x = 0.62$, the system is at equilibrium and $Q = K$. When x differs from 0.62, $Q \neq K$ and the system must react to reach equilibrium.

This is the same value that was reported in Chapter 14 **(← *p. 660*)**. That is, the equilibrium concentrations correspond to the minimum in the graph of G versus extent of reaction. This correspondence makes sense, because any set of concentrations

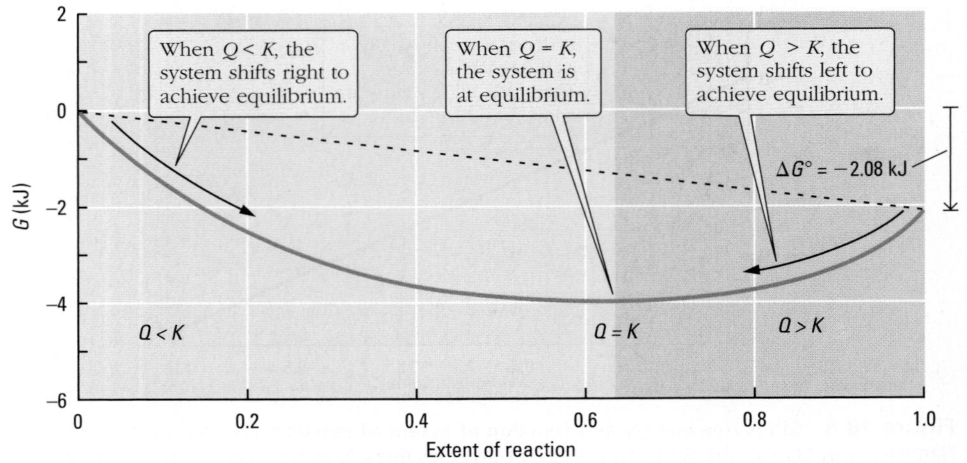

Figure 18.9 Gibbs free energy versus extent of reaction when there is mixing. For the isomerization of *cis*-2-butene, dilution of *cis*-2-butene with *trans*-2-butene increases entropy and decreases Gibbs free energy. This causes the graph of Gibbs free energy versus extent of reaction to curve below a straight line from reactants to products. As in Figure 18.8, the total Gibbs free energy of the reactants has been set arbitrarily to zero.

that differs from $x = 0.62$ has a higher value of G and therefore could change to the equilibrium concentrations with a decrease in Gibbs free energy (which corresponds to an increase in entropy of the universe).

At the far left of Figure 18.9 the graph drops faster than a straight line from $G_{\text{reactants}}$ to G_{products}. The slope at any point on the curve differs from the slope of the straight line ($\Delta G°$) by a factor that depends on the reaction quotient, Q (← *p. 673*).

$$\text{Slope} = \Delta G° + RT \ln Q \qquad [18.7]$$

At equilibrium the slope is zero, and $Q = K°$, where the superscript indicates the **standard equilibrium constant.** The expression for $K°$ is similar to the equilibrium constant as defined in Chapter 14 (← *p. 661*) except that each concentration is divided by the standard-state concentration of 1 mol/L and each pressure is divided by the standard-state pressure of 1 bar. That is, even if the equilibrium constant has units (of concentration or pressure), $K°$ is unitless. Substituting into Equation 18.7 gives

$$0 = \Delta G° + RT \ln K°$$

which rearranges to

$$\Delta G° = -RT \ln K° \qquad [18.8]$$

If the reaction occurs in solution, $K°$ has the same form (and the same value) as the concentration equilibrium constant. For gases, $K°$ relates pressures, not concentrations, and has the same value as K_P.

Regardless of the choice of standard state, Equation 18.8 indicates that the Gibbs free energy change for a reaction is the negative of a constant times the temperature times the natural logarithm of the equilibrium constant. If $K°$ is larger than 1, then $\ln K°$ is positive and $\Delta G°$ will be negative because of the minus sign. Both of these conditions, a negative $\Delta G°$ and $K° > 1$, indicate that the reaction is *product-favored* under standard-state conditions. Conversely, if $K° < 1$, then $\ln K°$ is negative and $\Delta G°$ must be positive, indicating a *reactant-favored* system.

$\Delta G°$ is the difference in Gibbs free energy between products in their standard states and reactants in their standard states. For example, for ionization of formic acid in water,

$$\underset{1\ \text{mol/L}}{\text{HCOOH(aq)}} \rightleftharpoons \underset{1\ \text{mol/L}}{\text{HCOO}^-\text{(aq)}} + \underset{1\ \text{mol/L}}{\text{H}^+\text{(aq)}} \qquad \Delta G° = 21.3\ \text{kJ}$$

the Gibbs free energy change of 21.3 kJ is for converting 1 mol HCOOH(aq) at a concentration of 1 mol/L into 1 mol $HCOO^-$(aq) and 1 mol H^+(aq), each at a concentration of 1 mol/L. Since $\Delta G°$ is positive, we predict that the process will be reactant-favored. This agrees with the fact that formic acid is a weak acid, is only slightly ionized, and therefore has an equilibrium constant much smaller than 1.

Relation between $\Delta G°$ and $K°$ at 25 °C

$\Delta G°$ (kJ/mol)	$K°$
200	9×10^{-36}
100	3×10^{-18}
10	2×10^{-2}
1	7×10^{-1}
0	1
−1	1.5
−10	6×10^{1}
−100	3×10^{17}
−200	1×10^{35}

$K°$	$\Delta G°$ ($-RT \ln K°$)	Product-Favored?
<1	Positive	No
>1	Negative	Yes
=1	0	Neither

For formic acid in water, the equilibrium constant is K_a, a ratio of concentrations (← *p. 767*).

PROBLEM-SOLVING EXAMPLE 18.7 Gibbs Free Energy and Equilibrium Constant

In the preceding paragraph you learned that $\Delta G° = 21.3$ kJ at 25 °C for the reaction

$$\text{HCOOH(aq)} \rightleftharpoons \text{HCOO}^-\text{(aq)} + \text{H}^+\text{(aq)}$$

Use this information to calculate the equilibrium constant, K_a, for ionization of formic acid in aqueous solution at 25 °C.

Answer $K_a = K_c = K° = 1.8 \times 10^{-4}$

Strategy and Explanation The relation between Gibbs free energy and $K°$ was given in Equation 18.8 as

$$\Delta G° = -RT \ln K°$$

Make certain to check units in calculations like this one. Standard Gibbs free energy changes involve kilojoules, and the gas constant R involves joules, so a unit conversion is needed.

To obtain $K°$ from $\Delta G°$, we first divide both sides of the equation by $-RT$:

$$-\Delta G°/RT = \ln K°$$

Next, we make use of the properties of logarithms (which are discussed in Appendix A.6). Since ln represents a logarithm to the base e, we can remove the logarithm function by using each side of the equation as an exponent of e.

$$e^{-\Delta G°/RT} = e^{\ln K°} = K°$$

Now we can substitute the known values into the equation.

$$K° = e^{-\Delta G°/RT} = e^{-(21.3\ \text{kJ/mol})(1000\ \text{J/kJ})/(8.314\ \text{J K}^{-1}\ \text{mol}^{-1})(298\ \text{K})} = 1.8 \times 10^{-4}$$

Thus, the positive value of $\Delta G°$ results in a value of $K°$ less than 1 and, indeed, indicates a reactant-favored system. Because this reaction occurs in aqueous solution, the standard states of reactants and products involve concentrations. Therefore $K° = K_c = K_a$.

☑ **Reasonable Answer Check** Formic acid is a weak acid and is not expected to have a very large equilibrium constant. The value calculated appears to be reasonable.

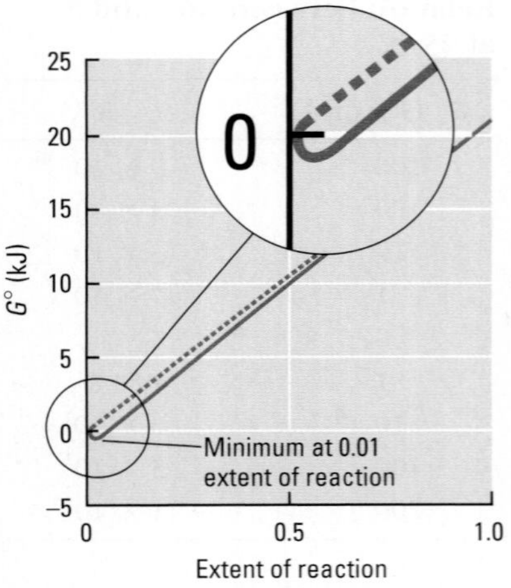

Graph of *G* as a function of extent of reaction for the reaction in Problem-Solving Example 18.7. $\Delta G° = 21.3$ kJ, a positive value. The minimum in the solid line is very close to zero extent of reaction; that is, the system is reactant-favored. The extent of reaction at equilibrium is 0.01. (The dotted line is a straight line from reactants to products. As shown in the enlarged circle, the solid curve has a minimum at a very small extent of reaction.)

PROBLEM-SOLVING PRACTICE 18.7

For each of these reactions, evaluate $K°$ at 298 K from the standard free energy change. If necessary, obtain data from Appendix J to calculate $\Delta G°$. Check your results against the K_c and K_P values in Table 14.1 (← p. 669). For which of these reactions is $K_c = K°$?

(a) $CaCO_3(s) \rightleftharpoons Ca^{2+}(aq) + CO_3^{2-}(aq)$
(b) $H_2CO_3(aq) \rightleftharpoons HCO_3^-(aq) + H^+(aq)$
(c) $2\ NO_2(g) \rightleftharpoons N_2O_4(g)$

Remember also that $\Delta G°$ can be calculated from the equation

$$\Delta G° = \Delta H° - T\Delta S°$$

If we know or can estimate changes in enthalpy and entropy for a reaction, then we can calculate or estimate the standard Gibbs free energy change and hence the equilibrium constant. And because $\Delta H°$ and $\Delta S°$ have nearly constant values over a wide range of temperatures, we can estimate equilibrium constants at a variety of temperatures, not just at 25 °C. Problem-Solving Example 18.8 shows how to do this.

Assuming constant values of $\Delta H°$ and $\Delta S°$ over a broad range of temperatures does not work well for reactions involving ions in aqueous solution, such as (a) and (b) in Problem-Solving Practice 18.7, because the extent of hydration of the ions varies with temperature.

This reaction is of great importance because it can take place to a significant extent in high-temperature combustion processes. If the temperature is high enough, nitrogen and oxygen react to form the air pollutant NO.

PROBLEM-SOLVING EXAMPLE 18.8 Estimating $K°$ at Different Temperatures

Use data from Appendix J to obtain values of $\Delta H°$ and $\Delta S°$ for the reaction

$$N_2(g) + O_2(g) \rightleftharpoons 2\ NO(g)$$

From these data estimate the value of $\Delta G°$ and hence the value of $K°$ at (a) 298 K, (b) 1000. K, and (c) 2300. K.

Answer

(a) $\Delta G° = 173.1$ kJ; $K° = 4.5 \times 10^{-31}$
(b) $\Delta G° = 155.7$ kJ; $K° = 7.3 \times 10^{-9}$
(c) $\Delta G° = 123.5$ kJ; $K° = 1.57 \times 10^{-3}$

Strategy and Explanation At each temperature, use the Gibbs equation to calculate $\Delta G° = \Delta H° - T\Delta S°$. Then calculate $K°$ as was done in Problem-Solving Example 18.7. Part (c) is done below to illustrate the calculations:

$$\Delta G° = \Delta H° - T\Delta S° = 180{,}500\ \text{J} - (2300.\ \text{K})(24.772\ \text{J/K})$$

$$= 1.235 \times 10^5\ \text{J} = 123.5\ \text{kJ}$$

$$K° = e^{-\Delta G°/RT} = e^{-(1.235 \times 10^5\ \text{J mol}^{-1})/(8.314\ \text{J mol}^{-1}\ \text{K}^{-1})(2300.\ \text{K})} = 1.57 \times 10^{-3}$$

☑ **Reasonable Answer Check** The reaction is endothermic, so Le Chatelier's principle predicts that the equilibrium will shift toward products as the temperature increases. This is reflected by the increasing values of $K°$ as the temperature rises.

PROBLEM-SOLVING PRACTICE 18.8

For the ammonia synthesis reaction,

$$N_2(g) + 3\,H_2(g) \rightleftharpoons 2\,NH_3(g)$$

estimate the equilibrium constant at (a) 298 K, (b) 450. K, and (c) 800. K.

Gibbs Free Energy Changes under Nonstandard-State Conditions

In previous sections we calculated $\Delta G°$ for reactions in which reactants in their standard states were converted to products in their standard states. However, substances usually are not at a pressure of 1 bar or at a concentration of 1 mol/L. How do we calculate ΔG if the reactants and products are not at standard concentration or standard pressure? A simple adjustment can be made to $\Delta G°$ to account for the difference between actual pressures or concentrations and standard-state pressures or concentrations. The equation from which ΔG (for *nonstandard-state* conditions) can be calculated from $\Delta G°$ (for *standard-state* conditions) is

$$\Delta G = \Delta G° + RT \ln Q \qquad [18.9]$$

For the standard state (concentration of 1 mol/L or pressure of 1 bar), $Q = 1$. Substituting this into Equation 18.9, we get $\ln Q = \ln(1) = 0$ and $\Delta G = \Delta G°$, the correct value for the standard state.

According to Equation 18.9, the bigger Q becomes, the more positive the correction factor $RT \ln Q$ becomes, and the more positive ΔG becomes. This makes sense, because the larger Q is, the more the concentrations (or pressures) of products exceed those of reactants. According to Le Chatelier's principle, increasing the concentrations of products (or decreasing the concentrations of reactants) causes the reaction to shift in the reverse direction. A shift toward more reactants is expected, because the more positive ΔG is, the more reactant-favored a process is.

To determine the change in Gibbs free energy when reactants at nonstandard-state concentrations or pressures are converted to products at nonstandard-state concentrations or pressures, we first write an appropriate chemical equation, then calculate $\Delta G°$ and Q, and finally use Equation 18.9 to correct the value of $\Delta G°$ for the nonstandard-state conditions, giving ΔG.

PROBLEM-SOLVING EXAMPLE 18.9 Gibbs Free Energy Change for Nonstandard-State Conditions

For the ammonia synthesis reaction at 25 °C, calculate the change in Gibbs free energy if 1 mol $N_2(g)$ at 0.230 bar and 3 mol $H_2(g)$ at 0.420 bar are converted to 2 mol $NH_3(g)$ at 1.45 bar.

Answer $\Delta G = -21.0$ kJ/mol

Strategy and Explanation First write a balanced equation, then calculate $\Delta G°$ and Q. Finally, use Equation 18.9 to calculate ΔG.

$$N_2(g) + 3\,H_2(g) \rightleftharpoons 2\,NH_3(g)$$

$$\Delta G° = \{2(\Delta G_f°(NH_3(g)))\} - \{(\Delta G_f°(N_2(g))) + 3(\Delta G_f°(H_2(g)))\}$$

$$= \{2(-16.45)\text{ kJ/mol}\} - \{0 + 0\} = -32.90\text{ kJ/mol}$$

$$Q = \frac{P^2_{NH_3}}{P_{N_2}P^3_{H_2}} = \frac{(1.45)^2}{(0.230)(0.420)^3} = 123.4$$

$$\Delta G = \Delta G° + RT \ln Q = -32.90\text{ kJ/mol} + (8.314\text{ J mol}^{-1}\text{ K}^{-1})(298.15\text{ K}) \times \ln(123.4)$$

$$= -32.90\text{ kJ/mol} + 11936\text{ J/mol}$$

$$= -32.90\text{ kJ/mol} + 11.936\text{ kJ/mol} = -21.0\text{ kJ/mol}$$

☑ **Reasonable Answer Check** The nonstandard-state conditions involve a concentration of ammonia (the product) well above standard pressure and concentrations of nitrogen and hydrogen (the reactants) well below standard pressure. According to Le Chatelier's principle, if an equilibrium is disturbed by increasing concentrations of products or decreasing concentrations of reactants (both of which apply here), then the equilibrium will shift toward the left—that is, in a reactant-favored direction. The value of ΔG (-20.96 kJ/mol) is negative, but it is less negative than the value of $\Delta G°$ (-32.90 kJ/mol). A less negative ΔG corresponds to a less product-favored (that is, more reactant-favored) process, which corresponds with the prediction from Le Chatelier's principle.

PROBLEM-SOLVING PRACTICE **18.9**

Calculate ΔG at 25 °C for a reaction in which $Ca^{2+}(aq)$ combines with $CO_3^{2-}(aq)$ to form a precipitate of $CaCO_3(s)$; the concentrations of $Ca^{2+}(aq)$ and $CO_3^{2-}(aq)$ are 0.023 M and 0.13 M, respectively.

18.8 Gibbs Free Energy, Maximum Work, and Energy Resources

The work represented by ΔG does not include work done by expansion of a system, pushing back the atmosphere.

An important interpretation of the Gibbs free energy is that **ΔG *represents the maximum useful work that can be done by a product-favored system on its surroundings under conditions of constant temperature and pressure.* ΔG *also represents the minimum work that must be done to cause a reactant-favored process to occur.*** Consider the product-favored reaction of hydrogen with oxygen to form liquid water under standard conditions.

$$2\,H_2(g) + O_2(g) \rightleftharpoons 2\,H_2O(\ell) \qquad \Delta G° = -474.258 \text{ kJ}$$

This thermochemical expression tells us that for every 2 mol $H_2O(\ell)$ produced, as much as 474.258 kJ of work could be done. The negative sign of $\Delta G°$ tells us that the work is done on the surroundings. (Because the system has less Gibbs free energy after the reaction than before it, the surroundings will have more energy.) Even if the reactants and the products are not at standard pressure or concentration, ΔG still equals $-w_{max}$, the maximum work the system can do on its surroundings.

It is important to remember that w_{max} is the maximum work the system can do *on* the *surroundings*. Therefore the sign of w_{max} is opposite to that of ΔG.

$$\Delta G = -w_{max} \text{ (work done on the surroundings)} \qquad [18.10]$$

Now consider the decomposition of water to form hydrogen and oxygen, which is the reverse of the previous reaction.

Because transformations of energy from one form to another are not 100% efficient, we seldom observe anything close to the maximum quantity of useful work given by the value of $\Delta G°$.

$$2\,H_2O(\ell) \rightleftharpoons 2\,H_2(g) + O_2(g) \qquad \Delta G° = 474.258 \text{ kJ}$$

The positive value of $\Delta G°$ indicates that this process is reactant-favored. Because the Gibbs free energy of the products is greater than the Gibbs free energy of the reactant, at least 474.258 kJ must be supplied for every 2 mol $H_2O(\ell)$ that decomposes. This 474.258 kJ is the minimum work that must be done to change 2 mol liquid water into hydrogen gas and oxygen gas. One way to supply this work is to use a direct electric current to carry out electrolysis of the water. In general, a continuous supply of energy is required for a reactant-favored process, such as decomposition of liquid water, to continue.

PROBLEM-SOLVING EXAMPLE **18.10** **Gibbs Free Energy Change and Maximum Work**

Use data from Appendix J to predict whether each reaction is product-favored or reactant-favored at 25 °C and 1 bar. For each product-favored reaction, calculate the maximum useful work the reaction could do. For each reactant-favored process, calculate the minimum work needed to cause it to occur.

(a) $2\,Al_2O_3(s) \longrightarrow 4\,Al(s) + 3\,O_2(g)$ (b) $Cl_2(g) + Mg(s) \longrightarrow MgCl_2(s)$

Answer

(a) Reactant-favored; at least 3164.6 kJ must be supplied

(b) Product-favored; can do up to 591.79 kJ of useful work

Strategy and Explanation Use data from Appendix J to calculate $\Delta G°$ for each reaction. If $\Delta G°$ is negative, the process is product-favored, and the value of $\Delta G°$ gives the maximum work that can be done. If $\Delta G°$ is positive, the process is reactant-favored, and the value tells the minimum work that has to be done to force the reaction to occur.

(a) $\Delta G° = 0 + 0 - 2(-1582.3)$ kJ $= 3164.6$ kJ; at least 3164.6 kJ is required.

(b) $\Delta G° = -591.79$ kJ $- 0 - 0 = -591.79$ kJ; up to 591.79 kJ useful work can be done.

☑ **Reasonable Answer Check** Reaction (a) is decomposition of an oxide to a metal and oxygen. Because metals are good reducing agents and oxygen is a strong oxidizing agent, the reverse of this reaction is likely to be product-favored, which would make Reaction (a) reactant-favored. This result agrees with the calculation. Reaction (b) is combination of an alkaline-earth element with a halogen, which should form a stable ionic compound. Therefore Reaction (b) should be product-favored, which agrees with the calculation. In both cases the value of $\Delta G°$ is large, which also would be expected based on the arguments just given.

PROBLEM-SOLVING PRACTICE **18.10**

Predict whether each reaction is reactant-favored or product-favored at 298 K and 1 bar, and calculate the minimum work that would have to be done to force it to occur, or the maximum work that could be done by the reaction.

(a) $2\ CO_2(g) \longrightarrow 2\ CO(g) + O_2(g)$ (b) $4\ Fe(s) + 3\ O_2(g) \longrightarrow 2\ Fe_2O_3(s)$

© Cengage Learning/George Semple

Charging a dead battery. A dead battery can be charged by using electricity from a power plant to cause a reactant-favored process to occur. After the battery has been charged, the reverse of that reactant-favored process (a product-favored process) can generate electricity to start the car.

Coupling Reactant-Favored Processes with Product-Favored Processes

A dead car battery will not charge itself. The process that takes place when a battery is charged is reactant-favored. But a battery can be charged if it is connected to a charger that is, in turn, powered by electricity generated in a power plant that burns coal. Coal, which is mainly carbon, burns in air according to the equation

$$C(s) + O_2(g) \longrightarrow CO_2(g) \qquad \Delta G° = -394.4 \text{ kJ}$$

If enough coal is burned, the large negative Gibbs free energy change for its combustion more than offsets the positive Gibbs free energy change of the battery-charging process. An overall decrease in Gibbs free energy occurs, even though the battery-charging part we are interested in has an increase. Once a battery has been charged, the chemicals in the battery can undergo a product-favored reaction that can supply electricity to start a car's engine. Some of the Gibbs free energy lost when the coal was burned has been stored in the car's battery for use later.

Charging a battery is an example of coupling a product-favored reaction with a reactant-favored process to cause the latter to take place. Both processes occur at the same time and in a way that allows the Gibbs free energy released by the product-favored reaction to be used by the reactant-favored reaction. Other examples include obtaining aluminum or iron from their ores; synthesizing large, complicated molecules from simple reactants to make medicines, plastics, and other useful materials; and keeping a house cool on a day when the outside temperature is above 100 °F. All involve decreasing entropy in the region of our interest, but all can be made to occur provided that there is a larger increase in entropy at a power plant or somewhere else.

The Gibbs free energy change indicates a chemical reaction's capacity to drive a different reactant-favored system to produce products. The word "free" in the name indicates not "zero cost," but rather "available." ***Gibbs free energy is available to do useful tasks that would not happen on their own.*** Another way of saying

Andy Olsen/iStockphoto

Coal-fired electric power plant.

this is that Gibbs free energy is a measure of the *quality* of the energy contained in a chemical reaction system. If it contains a lot of Gibbs free energy, a chemical system can do a lot of work for us; the energy is of high quality—potentially useful to humankind. When the system's reactants are transformed into products, that available free energy can do useful work, but *only* if the reaction is coupled to some other, reactant-favored process we want to carry out. If systems are *not* coupled, then the free energy released by a reaction will be wasted as thermal energy. All living things depend on coupling of chemical reactions to provide for their energy needs.

Thermite reaction. Exercise 18.12 deals with the reaction of aluminum with iron(III) oxide. The reaction is very product-favored and releases a large quantity of Gibbs free energy.

CONCEPTUAL EXERCISE 18.12 Coupling Reactions

One way to produce iron metal is to reduce iron(III) oxide with aluminum. This is called the thermite reaction. You can think of the reaction as occurring in two steps. The first is the loss of oxygen from iron(III) oxide,

(i) $Fe_2O_3(s) \longrightarrow 2\,Fe(s) + \frac{3}{2}\,O_2(g)$

The second is the combination of aluminum with the oxygen,

(ii) $2\,Al(s) + \frac{3}{2}\,O_2(g) \longrightarrow Al_2O_3(s)$

(a) Calculate the enthalpy, entropy, and Gibbs free energy changes for each step. Decide whether each step is product- or reactant-favored. Comment on the signs of $\Delta H°$, $\Delta S°$, and $\Delta G°$ for each step.

(b) What is the overall net reaction that occurs when aluminum is combined with iron(III) oxide? What are the enthalpy, entropy, and Gibbs free energy changes for the overall reaction? Is it product- or reactant-favored? Comment on the signs of $\Delta H°$, $\Delta S°$, and $\Delta G°$ for the overall reaction.

(c) Discuss briefly how coupling Reaction (i) with Reaction (ii) affects our ability to obtain iron metal from iron(III) oxide by reacting it with aluminum.

(d) Suggest a reaction other than oxidation of aluminum that might be used to reduce iron(III) oxide to iron. Test your selection by calculating the Gibbs free energy change for the coupled system.

18.9 Gibbs Free Energy and Biological Systems

Have you ever thought about how unlikely it is that a human being can exist? Your body contains about 100 trillion (10^{14}) cells, all working together to make you what you are. Each of those cells contains trillions of molecules, and many of the molecules contain tens of thousands of atoms. Those molecules and cells are arranged in structures such as organs, bones, and skin that provide for all the functions of your body and that determine its overall shape and size. When necessary, you can synthesize molecules on very short notice. For example, it does not take long to generate the surge of adrenalin your body makes when you are scared. Your body is a very highly organized system, which means that its entropy must be very low. This in turn means that, thermodynamically speaking, you are very, very improbable.

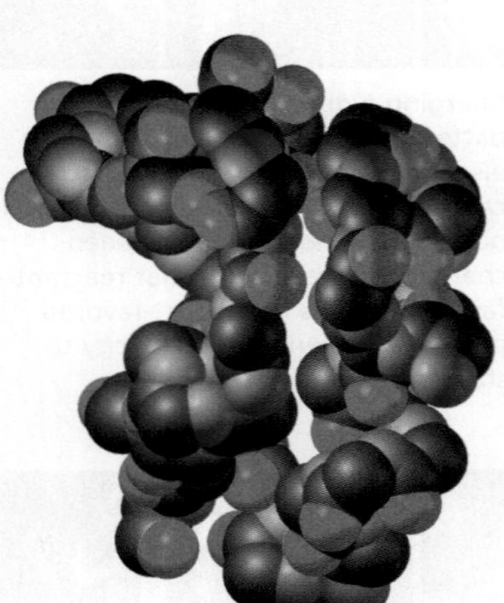
insulin

Insulin, a protein. Like all protein molecules, the insulin molecule is highly ordered. It contains 51 amino acids (← *p. 577*) connected in exactly the correct order and folded into exactly the molecular shape needed for its function in the metabolism of glucose. (Hydrogen atoms are not shown.)

Human Metabolism and Gibbs Free Energy

How can it be, then, that you exist? How can all the molecules from which you are made be synthesized and organized into the organs and other tissues of your body? The answer lies in the coupling of reactions described in the preceding section. Since you are a very-low-entropy system, you must be very high in Gibbs free energy. Your body obtains that Gibbs free energy by oxidizing the food you eat with oxygen you inhale. In the processes of metabolism, foods are oxidized, and the Gibbs free energy released by their oxidation causes other reactions to occur that store Gibbs free energy in specific molecules within your body. Later those molecules can release the Gibbs free energy to cause muscles to contract, nerve signals to be sent, important molecules to be synthesized, and other processes to occur. **Metabolism** refers to all of the chemical changes that occur as food nutrients are processed by an organism to release Gibbs free energy and form the complex chemical constituents of living cells. **Nutrients** are the chemical raw materials needed for survival of an organism.

As an example of metabolism, consider the single nutrient glucose, also known as dextrose or blood sugar ***(← p. 575).*** Glucose can be oxidized to carbon dioxide and water according to the equation

$$C_6H_{12}O_6(aq) + 6\ O_2(g) \longrightarrow 6\ CO_2(g) + 6\ H_2O(\ell) \qquad \Delta G°' = -2870\ \text{kJ}$$

Thus, a large quantity of Gibbs free energy can be released when glucose is oxidized. This reaction, which is strongly product-favored, is an example of an **exergonic** reaction—*one that releases Gibbs free energy.* The same quantity of Gibbs free energy is available whether glucose is burned in air or reacts in your body. However, burning glucose would release all of the Gibbs free energy as thermal energy. This process would not be appropriate in your body because it would raise the temperature rapidly, which in turn would kill cells. Instead, your body makes use of a large number of reactions that allow the Gibbs free energy to be released in small steps and stored in small quantities that can be used later.

By far the most important way in which Gibbs free energy is stored in your body is through formation of adenosine triphosphate (ATP) from adenosine diphosphate (ADP). The structures of ADP and ATP are shown in Figure 18.10; note that they are closely related.

$$ADP^{3-}(aq) + H_2PO_4^-(aq) \longrightarrow ATP^{4-}(aq) + H_2O(\ell) \qquad \Delta G°' = 30.5\ \text{kJ}$$

This is an example of an **endergonic** reaction—*one that consumes Gibbs free energy and is therefore reactant-favored.* In a typical bacterial cell, this reaction takes place 38 times for each molecule of glucose that is oxidized. In human cells, it takes place 32 times for each molecule of glucose that is oxidized. That is, in a human cell the Gibbs free energy released by the exergonic glucose oxidation is used to force the endergonic, reactant-favored process of forming ATP from ADP to occur 32 times. The overall process is

$$C_6H_{12}O_6(aq) + 6\ O_2(g) + 32\ ADP^{3-}(aq) + 32\ H_2PO_4^-(aq) \longrightarrow$$
$$6\ CO_2(g) + 32\ ATP^{4-}(aq) + 38\ H_2O(\ell) \qquad \Delta G°' = -1894\ \text{kJ}$$

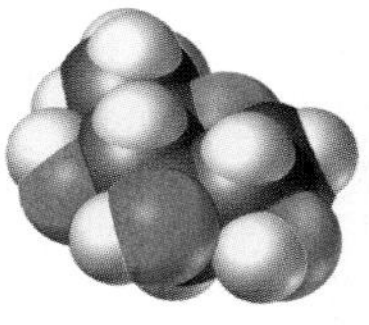

glucose

The prime symbol (′) on $\Delta G°'$ indicates that the value of the Gibbs free energy change is for pH = 7, the same concentration of $H_3O^+(aq)$ ions as in a typical cell. When aqueous solutions are involved, $\Delta G°$ values in tables such as Appendix J refer to H_3O^+ concentrations of 1 mol/L, but such a high concentration of acid would destroy most cells. Consequently, biochemists have calculated a set of $\Delta G°'$ values that apply to solutions at pH = 7. These values are usually reported for a temperature of 37 °C, human body temperature.

The words "exergonic" and "endergonic" have nearly the same prefixes as "exothermic" and "endothermic." In both cases *ex* means "out" and *end* means "into." *Thermic* indicates that thermal energy is released or taken up. *Ergonic* indicates that Gibbs free energy is released or used up.

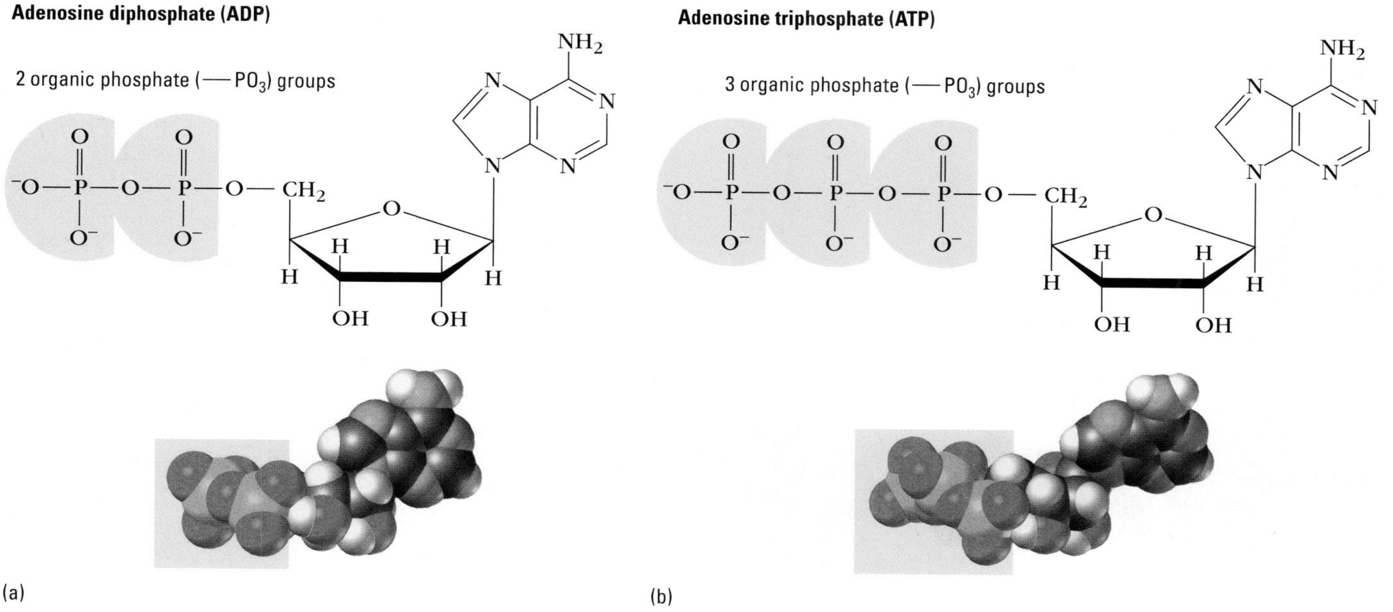

Figure 18.10 Biochemical storage of Gibbs free energy. Structures of (a) adenosine diphosphate (ADP) and (b) adenosine triphosphate (ATP). Notice that ATP has one more organic phosphate (—PO_3) group at the left end of the molecule, but otherwise the structures are identical. Notice also that ADP has three negatively charged oxygen atoms and ATP has four.

Because ATP is high in Gibbs free energy, it is said to be a high-energy molecule (or ion). Sometimes the bonds in ATP are called high-energy bonds, but this description is a misnomer. Actually, the bonds have low bond energies and can break easily to form ADP and release Gibbs free energy.

Since it is exergonic, this reaction must be product-favored, and therefore appreciable quantities of products can be obtained.

EXERCISE 18.13 Coupled Metabolic Reactions

Add the Gibbs free energy change for oxidation of glucose to the appropriate Gibbs free energy change for 32 conversions of ADP to ATP, thereby verifying that the Gibbs free energy change given above for the overall reaction is correct. What happens to the 1894 kJ of Gibbs free energy released by the overall reaction?

The citric acid cycle is also known as the Krebs cycle or the tricarboxylic acid (TCA) cycle.

The metabolic process by which the Gibbs free energy contained in nutrients is stored in ATP is far more complicated than the overall equation given above makes it seem. Metabolism can be divided into three stages that were first clearly identified by Hans Krebs. The first stage, digestion, breaks apart large molecules, such as carbohydrates (polysaccharides), fats, or proteins, into smaller molecules, such as glucose, glycerol and fatty acids, or amino acids. These smaller molecules are more easily transferred into the blood by the digestive system. In the second stage, the smaller molecules are changed into a few simple units that play a central role in metabolism. The most important of these is the acetyl group in acetyl coenzyme A (acetyl CoA). The structure of acetyl CoA is shown in Figure 18.11. The third stage consists of oxidation of the acetyl group from acetyl CoA to form carbon dioxide and water. This takes place in an eight-step cycle of reactions called the citric acid cycle, which also transforms ADP into ATP, a process called oxidative phosphorylation. The overall three-stage process is diagrammed in Figure 18.12.

Conversion of ATP to ADP is the source of the energy that causes muscles to contract. This answers the question, "Where does the energy come from to make my muscles work?" that was posed in Chapter 1 (← p. 2).

Because conversion of ADP to ATP is endergonic, ATP contains stored Gibbs free energy. In your body, ATP generated from glucose or other nutrients is a convenient and readily available Gibbs free energy resource, just as electricity generated from coal or natural gas is a convenient and readily available Gibbs free energy resource in modern society. ATP can release Gibbs free energy in packets of 30.5 kJ for each ATP converted to ADP. This size is convenient for driving many biochemical processes in your body. For example, as part of the metabolism of glucose, it is necessary to attach a phosphate group to the glucose molecule.

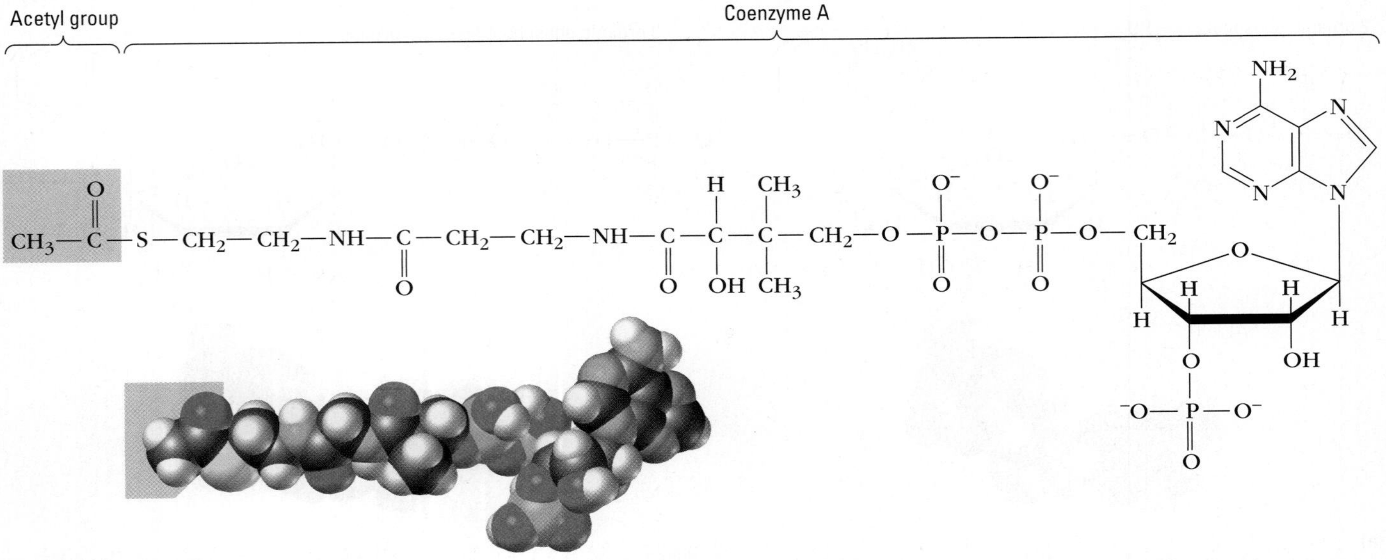

Figure 18.11 Structure of acetyl coenzyme A. Structural formula and space-filling model of acetyl coenzyme A. The acetyl group at the far left end of the molecule can be formed from glucose that came originally from starch.

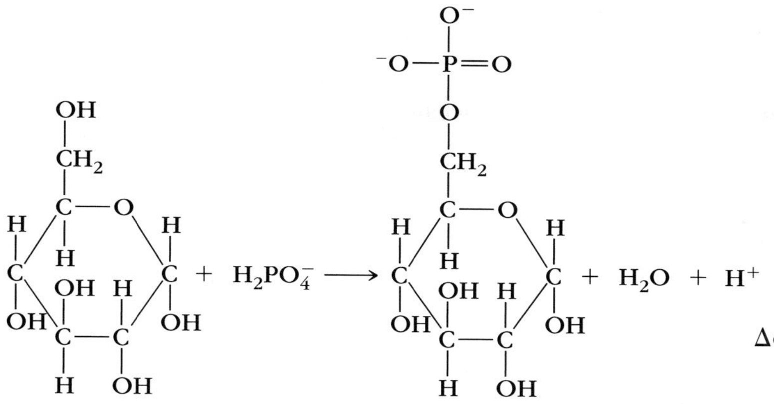

glucose   glucose 6-phosphate^{2-}

$\Delta G' = 13.8$ kJ

The "6" in glucose 6-phosphate indicates that the phosphate group has been added to the oxygen atom attached to carbon number 6 of glucose, the only carbon atom not in the ring structure.

This reaction is endergonic by 13.8 kJ and, therefore, is not product-favored. It will not occur unless forced to do so.

The endergonic reaction can be caused to occur by coupling it to the transformation of ATP to ADP.

$$ATP(aq)^{4-} + H_2O(\ell) \longrightarrow ADP^{3-}(aq) + H_2PO_4^-(aq) \qquad \Delta G^{\circ\prime} = -30.5 \text{ kJ}$$

$H_2PO_4^-$ produced by this reaction can react with glucose to form glucose 6-phosphate, coupling the two reactions directly. Also, water produced in the first reaction is used up in the second one. The overall process is

$$\text{Glucose} + ATP^{4-} \longrightarrow \text{glucose 6-phosphate}^{2-} + ADP^{3-} + H_3O^+$$
$$\Delta G^{\circ\prime} = (-30.5 + 13.8) \text{ kJ} = -16.7 \text{ kJ}$$

The negative value of $\Delta G^{\circ\prime}$ indicates that the overall process is exergonic and product-favored. Thus the ATP $\longrightarrow$ ADP transformation can cause glucose to undergo a condensation reaction with dihydrogen phosphate. The 16.7 kJ of Gibbs free energy released appears as thermal energy transferred from the system to its surroundings.

In biochemistry the convention is to write these equations in a shorthand notation that indicates that they are coupled. The process just described is represented as:

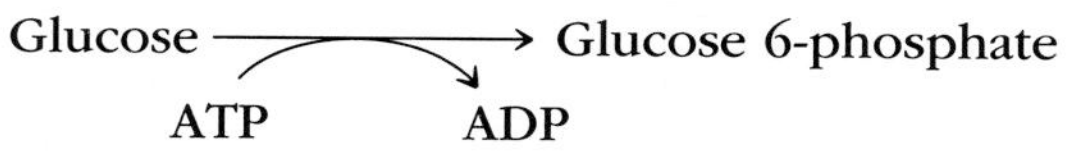

The curved line indicates that the transformation of ATP to ADP occurs simultaneously with the glucose reaction and that the two are coupled.

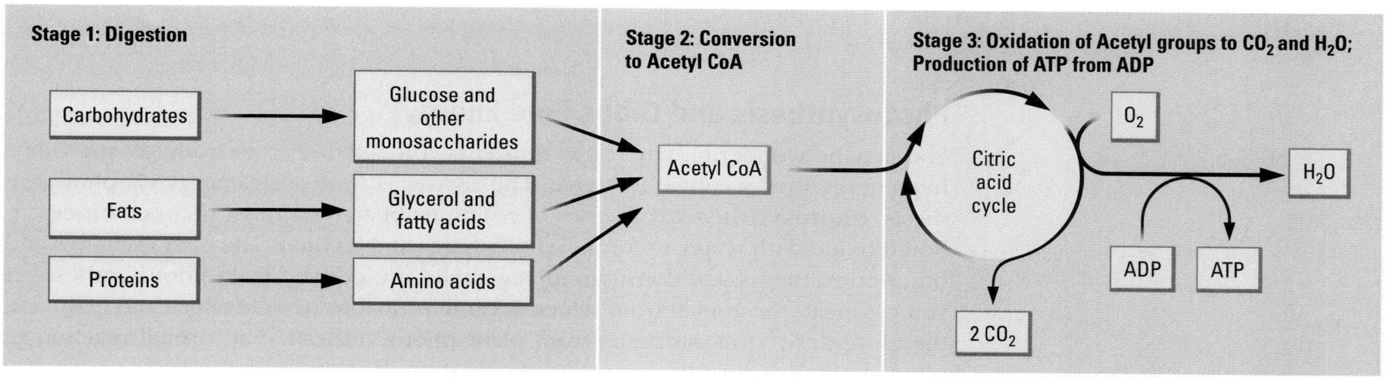

Figure 18.12 Gibbs free energy and nutrients. Extraction of Gibbs free energy from nutrients is a three-stage process. In stage 1 (digestion), large molecules are broken down into smaller ones. In stage 2, smaller molecules are converted to acetyl groups attached to coenzyme A. In stage 3 (the citric acid cycle), the acetyl groups are oxidized to carbon dioxide and water.

PROBLEM-SOLVING EXAMPLE 18.11 Biochemical Standard State

Many biochemical processes involve reactions that take place at a temperature of 37 °C and a pH of 7 in body fluids. Under these conditions the Gibbs free energy change is specified as $\Delta G°'$, where the prime specifies that all substances are at their standard-state concentrations except for H_3O^+, which is at a biological concentration of 10^{-7} mol/L (pH = 7). Calculate $\Delta G°$ (1 mol/L H_3O^+) for this reaction.

$$\text{Glucose} + ATP^{4-} \longrightarrow \text{glucose 6-phosphate}^{2-} + ADP^{3-} + H_3O^+ \qquad \Delta G°' = -16.7 \text{ kJ/mol}$$

Answer $\Delta G° = 24.8$ kJ/mol

Strategy and Explanation The $\Delta G°'$ value differs from $\Delta G°$ because one of the concentrations (that of H_3O^+) has a nonstandard value of 10^{-7} mol/L. That is, $\Delta G°'$ is ΔG for conditions such that every concentration is 1 mol/L except for the concentration of H_3O^+. Therefore, set $\Delta G = \Delta G°'$, calculate Q, and use Equation 18.9 (p. 873) to calculate $\Delta G°$.

$$\Delta G = \Delta G°' = -16.7 \text{ kJ/mol} = \Delta G° + RT \ln Q$$

$$\Delta G° = -16.7 \text{ kJ/mol} - RT \ln Q$$

$$Q = \frac{(\text{conc. glucose 6-phosphate}^{2-})(\text{conc. } ADP^{3-})(\text{conc. } H_3O^+)}{(\text{conc. glucose})(\text{conc. } ATP^{4-})}$$

$$= \frac{(1)(1)(1 \times 10^{-7})}{(1)(1)} = 1 \times 10^{-7}$$

$$\Delta G° = -16.7 \text{ kJ/mol} - RT \ln(1 \times 10^{-7})$$

$$= -16.7 \text{ kJ/mol} - (8.314 \text{ J mol}^{-1} \text{ K}^{-1})\{(273 + 37) \text{ K}\}(-16.12)$$

$$= -16.7 \text{ kJ/mol} + 41541 \text{ J/mol} = -16.7 \text{ kJ/mol} + 41.5 \text{ kJ/mol} = 24.8 \text{ kJ/mol}$$

☑ **Reasonable Answer Check** Using Le Chatelier's principle, we predict less shift toward products for a system in which the concentration of H_3O^+, a product, is 1 mol/L than there would be for a system in which the concentration of H_3O^+ is 1×10^{-7} mol/L. Less shift toward products means a more positive ΔG, and the value of $\Delta G°$ is indeed more positive than $\Delta G°'$.

PROBLEM-SOLVING PRACTICE 18.11

Will $\Delta G°$ be larger than, smaller than, or the same size as $\Delta G°'$ for this reaction?

$$C_6H_{12}O_6(aq) + 6\,O_2(g) \longrightarrow 6\,CO_2(g) + 6\,H_2O(\ell) \qquad \Delta G°' = -2870 \text{ kJ}$$

Explain why you chose the response you did.

Photosynthesis and Gibbs Free Energy

You may be wondering where the nutrients you take into your body get the Gibbs free energy they so obviously have. The answer is from solar energy via photosynthesis. **Photosynthesis** is a series of reactions in a green plant that combines carbon dioxide with water to form carbohydrates and oxygen. The carbohydrates and other constituents you consume in vegetables are derived from photosynthesis. If you eat meat, the animal from which it came probably ate vegetables and grain and therefore derived its nutrients from plant photosynthesis. The overall reaction in photosynthesis is just the opposite of oxidation of glucose.

$$6\,CO_2(g) + 6\,H_2O(\ell) \longrightarrow C_6H_{12}O_6(aq) + 6\,O_2(g) \qquad \Delta G°' = 2870 \text{ kJ}$$

Photosynthesis is endergonic and can occur only because of an influx of energy in the form of sunlight. That is, the energy in the sunlight causes this reactant-favored process to form appreciable quantities of products, and the sunlight's energy is stored as Gibbs free energy in the glucose and oxygen that are formed. This process is diagrammed in Figure 18.13.

Organisms that can carry out photosynthesis are called *phototrophs* (literally, "light-feeders") because they can use sunlight to supply needed energy. Phototrophs include all green plants, all algae, and some groups of bacteria. The phototrophs capture light by means of photosynthetic pigment systems and store the light energy in chemical bonds in molecules such as glucose. Nearly all other organisms belong to the class of *chemotrophs* (literally, "chemical-feeders"), which must depend on the chemical bonds created by the phototrophs for their energy. All animals, fungi, and most bacteria are chemotrophs. A world composed only of chemotrophs would not last long because without the phototrophs, food supplies would disappear almost immediately. Without sunlight and its ability to drive a reactant-favored system to form products (carbohydrate and oxygen), organisms such as humans and indeed almost the entire biosphere of planet Earth could not exist.

Some organisms, the chemautotrophs, which are found near deep-ocean volcanic vents, do not depend on phototrophs for their energy supply.

Both phototrophs and chemotrophs make use of the Gibbs free energy stored during photosynthesis by using oxidation of glucose to drive a large number of conversions of ADP to ATP and then using the ATP to couple to desired endergonic reactions and force them to occur. Thus, ATP is the minute-to-minute energy currency of living cells. The Gibbs free energy released in these reactions contributes to synthesis of molecules needed by the cell, causes some desirable process such as muscle contraction, or is dissipated as thermal energy. If more Gibbs free energy is

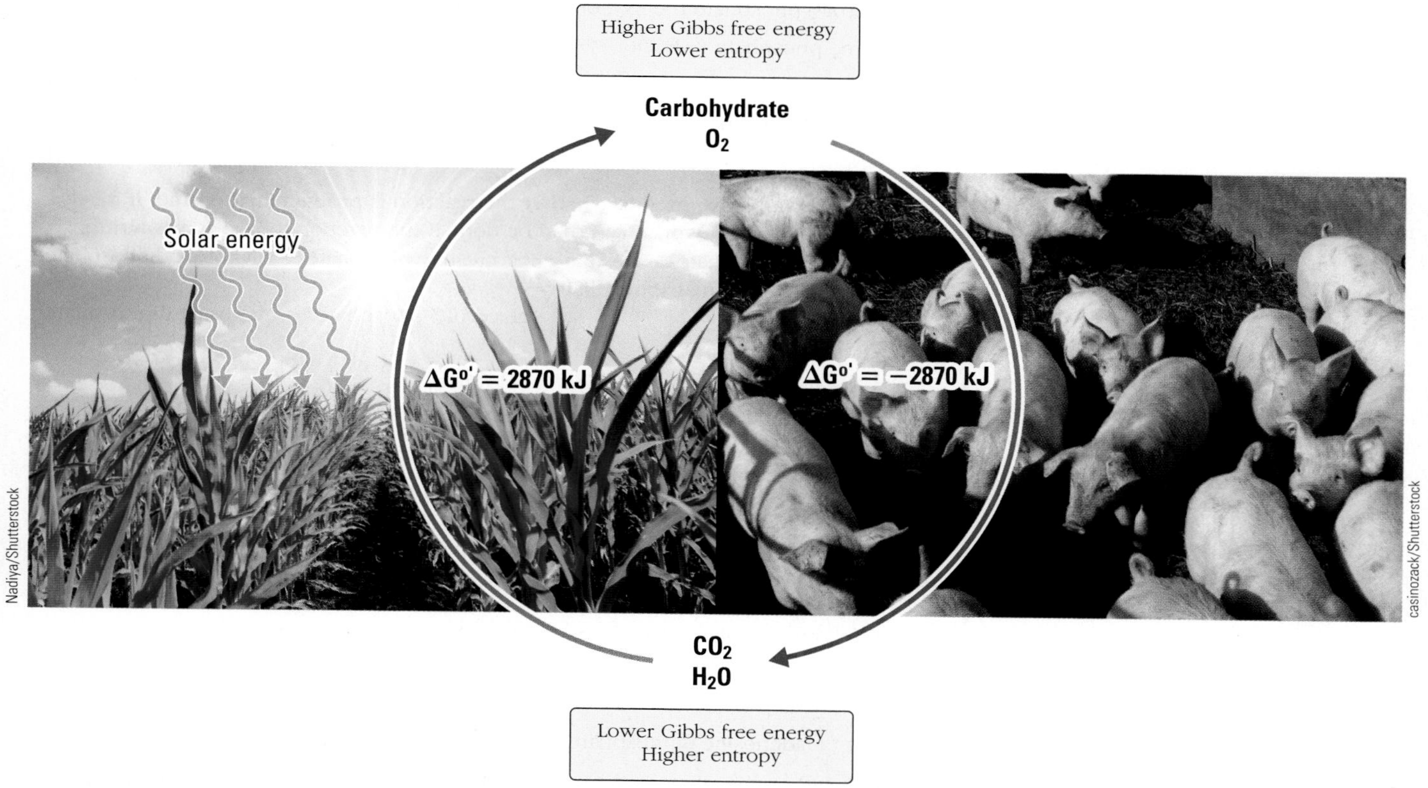

Figure 18.13 Solar energy storage by photosynthesis. The Gibbs free energy in the foods animals eat is derived from solar energy via photosynthesis. Organisms that can photosynthesize combine carbon dioxide with water to form carbohydrates and oxygen, which have a much higher Gibbs free energy. That Gibbs free energy is released when the carbohydrates are oxidized in metabolic processes.

taken in than the organism needs, then the excess Gibbs free energy can be stored long-term through the synthesis of fats, which have approximately twice as much Gibbs free energy as an equal mass of carbohydrate.

It is significant that when ATP reacts and causes other reactions to occur, the product ADP is very similar to the reactant ATP. ADP can easily be recycled to form ATP. A reasonable estimate of the quantity of ATP converted to ADP during one day in the life of an average human is 117 mol. Since the molar mass of the sodium salt of ATP is 551 g/mol, we can calculate that

$$117 \text{ mol} \times \frac{551 \text{ g}}{1 \text{ mol}} = 64{,}500 \text{ g ATP}$$

is converted to ADP every day. This is 64.5 kg, which is close to the 70-kg body weight of an average person. Obviously ATP is not a long-term storage molecule for Gibbs free energy. Instead it is recycled from ADP as needed and used almost immediately for some necessary process. The typical 70-kg human body contains only 50 g of ATP and ADP. If we actually had to take in 64.5 kg of ATP per day to provide Gibbs free energy, it would be a very expensive habit. The price of ATP from a laboratory supplier is currently about $50 per gram, which would put the cost of supplying each of us with our daily energy currency at more than two million dollars!

PROBLEM-SOLVING EXAMPLE 18.12 Coupling of Biological Reactions

ATP undergoes hydrolysis with release of Gibbs free energy according to the equation

(i) adenosine triphosphate + $H_2O(\ell) \longrightarrow$
adenosine diphosphate + dihydrogen phosphate $\quad \Delta G^{\circ\prime}\text{ (i)} = -30.5 \text{ kJ}$

Other organophosphates undergo similar hydrolysis reactions. For creatine phosphate and glycerol 3-phosphate the hydrolysis reactions are

(ii) creatine phosphate + $H_2O(\ell) \longrightarrow$ creatine + dihydrogen phosphate
$\Delta G^{\circ\prime}\text{ (ii)} = -43.1 \text{ kJ}$

(iii) glycerol 3-phosphate + $H_2O(\ell) \longrightarrow$ glycerol + dihydrogen phosphate
$\Delta G^{\circ\prime}\text{ (iii)} = -9.7 \text{ kJ}$

For each reaction below, predict whether the reaction is product-favored and, if it is, calculate the maximum work that could be done if the reaction took place as written.

(a) adenosine triphosphate + creatine $\longrightarrow$ creatine phosphate + adenosine diphosphate
(b) glycerol + adenosine triphosphate $\longrightarrow$
glycerol 3-phosphate + adenosine diphosphate

Answer

(a) Not product-favored
(b) Product-favored; $\Delta G^{\circ\prime} = -20.8$ kJ, so up to 20.8 kJ of work could be done on the surroundings

Strategy and Explanation Use the same procedure as for Hess's law calculations (← p. 242) to write overall reactions that couple two of the reactions for which $\Delta G^{\circ\prime}$ values are known. Then sum the $\Delta G^{\circ\prime}$ values to obtain $\Delta G^{\circ\prime}$ for the desired reaction. If $\Delta G^{\circ\prime}$ for the desired reaction is negative, the process is product-favored and the magnitude of $\Delta G^{\circ\prime}$ gives the maximum work. Use part (a) as an example. (The calculation for part (b) follows a similar procedure.) Because ATP is a reactant in the desired equation, use Reaction (i) as written.

(i) adenosine triphosphate + $H_2O(\ell) \longrightarrow$
adenosine diphosphate + dihydrogen phosphate $\quad \Delta G^{\circ\prime}\text{(i)} = -30.5 \text{ kJ}$

Because creatine phosphate is a product in the desired reaction, reverse Reaction (ii) and change the sign of $\Delta G^{\circ\prime}$.

reverse of (ii) creatine + dihydrogen phosphate $\longrightarrow$
creatine phosphate + $H_2O(\ell) \quad \Delta G^{\circ\prime} = -\Delta G^{\circ\prime}\text{ (ii)} = +43.1 \text{ kJ}$

The overall reaction is

$$\text{adenosine triphosphate} + \text{creatine} \longrightarrow \text{adenosine diphosphate} + \text{creatine phosphate}$$

$$\Delta G^{\circ\prime} = \Delta G^{\circ\prime}\,(\text{i}) - \Delta G^{\circ\prime}\,(\text{ii}) = -30.5\text{ kJ} + 43.1\text{ kJ} = 12.6\text{ kJ}$$

Therefore, the process (a) is reactant-favored, and at least 12.6 kJ would have to be supplied to force it to occur.

PROBLEM-SOLVING PRACTICE 18.12

ATP, creatine phosphate, and glycerol 3-phosphate could be thought of as phosphate donors (just as Brønsted-Lowry acids can be thought of as proton donors).
(a) Which of the three substances is the strongest phosphate donor?
(b) Which is the weakest?
(c) Explain your choices.

EXERCISE 18.14 Recycling of ATP

From the figures given previously for the daily quantity of ATP converted to ADP by an average human and the quantity of ATP and ADP actually present in the body, calculate the number of times each ADP molecule must be recycled to ATP each day.

18.10 Conservation of Gibbs Free Energy

When a ton of coal is burned its energy has not been used up. The law of conservation of energy ***(← p. 215)*** summarizes many experiments whose results verify that energy cannot be destroyed. When coal is burned in a power plant, its chemical energy is changed to an equal quantity of energy in other forms. These are electrical energy, which can be very useful, and thermal energy in the gases going up the smokestack and in the immediate surroundings of the plant, which is much less useful. However, an *energy resource* has been used up: the coal's ability to store energy and release it to do work. When coal burns in air, some of the Gibbs free energy that was in the coal and the oxygen that combined with it has been used up. This fact is indicated by the negative value of ΔG° for the combustion of coal. The same is true of any other product-favored reaction.

What we commonly refer to as **energy conservation** *is actually conservation of useful energy: Gibbs free energy.* Energy conservation does not mean conserving energy—nature takes care of conserving energy automatically. But nature does not automatically conserve Gibbs free energy. Substances with high Gibbs free energies are energy resources, and it is their *useful* energy that we must take pains to conserve. Once a product-favored reaction with a negative ΔG° has taken place, it cannot be reversed, thereby restoring the Gibbs free energy of its reactants, without coupling the reverse reaction with some other product-favored reaction. That is, once we have used an energy resource, it cannot be restored, except by using some other energy resource. Analysis of chemical systems in terms of Gibbs free energy can lead to important insights into how energy resources can be conserved effectively.

By comparing Gibbs free energy changes calculated using the equations in this chapter with the actual loss of Gibbs free energy in industrial processes, environmentalists and industrialists can suggest ways to minimize loss of Gibbs free energy. For example, there is a very large quantity of Gibbs free energy stored in aluminum metal and oxygen gas compared with aluminum oxide, Al_2O_3, from an ore. This can be seen from the thermochemical expression

$$2\,Al_2O_3(s) \longrightarrow 4\,Al(s) + 3\,O_2(g) \qquad \Delta G^\circ = 3164.6\text{ kJ}$$

which shows that the Gibbs free energy of 4 mol Al(s) and 3 mol $O_2(g)$ is 3164.6 kJ higher than the Gibbs free energy of 2 mol $Al_2O_3(s)$. If 4 mol Al(s) is oxidized to aluminum oxide, 3164.6 kJ of Gibbs free energy is lost—energy that was expended to

© Cengage Learning/Charles D. Winters

Diamond: a material resource. Energy resources are like other natural resources in that they contain high-quality, concentrated energy. An analogy is a material resource such as a diamond, which is pure carbon with the atoms bonded so that each is surrounded tetrahedrally by four others. A diamond is valuable because it consists of a single crystal with atoms arranged in a specific way. If you ground a diamond into dust and spread the carbon it was made of over the area of a city block, the carbon would be nearly worthless, because it would require tremendous expense to collect the carbon and convert it back to diamond. Similarly, an energy resource is valuable not for the energy it contains, but because that energy is concentrated and available to do useful work.

CHEMISTRY IN THE NEWS Ethanol Fuel and Energy

With gasoline prices increasing, and with the serious problems anticipated from global warming *(← p. 463)*, there is increasing interest in switching to fuels such as ethanol or biodiesel. Such fuels are said to require less fossil fuel input and to produce smaller quantities of greenhouse gases. They might also serve as sources for the many chemicals currently manufactured from petroleum *(← p. 537)*. But are the claims valid? Analyzing the net energy (Gibbs free energy) available from biofuels and the quantities of greenhouse gases emitted in processing the fuels allows us to find out.

To find the net energy, it is necessary to factor in all of the energy inputs, not just the obvious ones, that are required to manufacture the fuel and deliver the fuel to its intended use. For example, suppose that E85 ethanol *(← p. 540)* is made from corn. At first glance the principal energy input is sunlight, which causes the corn to grow. Also, some energy is required for the fermentation process that converts the cornstarch to ethanol and some for transporting the fuel to a gas station. Because corn plants take in carbon dioxide as they grow, the carbon dioxide produced when the ethanol burns will be recycled into the next crop of corn and will not contribute to global warming. So biofuels are great, right? Not necessarily!

This analysis left out the fact that the farmland on which the corn grew was tilled and fertilized. So tractors had to burn gasoline or diesel fuel to do the tilling and distribute the fertilizer, ammonia fertilizer had to be made from hydrogen that came from natural gas *(← p. 541)*, and other fertilizers such as phosphates and lime had to be manufactured. All of these processes consume Gibbs free energy. If the land was irrigated, water had to be pumped and coal had to be burned to produce the electricity to drive the pumps. These and many other processes are needed to create biofuels. All such processes consume Gibbs free energy (they have to, for anything to happen) and many of them create greenhouse gases. When corn is fermented to ethanol, by-products are produced (not all of the corn changes to ethanol), but this may be good. The by-products can be used as animal feed, for example, replacing animal feed that would otherwise have consumed Gibbs free energy during its production.

As you can imagine, there has been much scientific study of this issue. Some studies have indicated that burning corn-derived ethanol in a car actually consumes more Gibbs free energy than burning gasoline. That is, the biofuel was worse for the environment than the fossil fuel. Other studies found that ethanol as currently produced is better than gasoline, but no study found that it was environmentally benign. In January 2006, *Science* magazine published a study that examined and compared six previous studies and attempted to produce a definitive result. A summary of the results of this study is shown in the figure.

The figure shows relative energy inputs for gasoline and three different ways of generating ethanol: Ethanol Today—from corn as is done today; CO_2-intensive—from Nebraska corn shipped to a lignite-fueled processing plant in North Dakota; and cellulosic—from cellulose in plants such as switchgrass, not just from starch in corn kernels. The horizontal green bars represent the quantity of Gibbs free energy consumed in each process relative to one unit of Gibbs free energy delivered as fuel. Numbers at the right side of the figure (such as 94 for gasoline) indicate kg CO_2-equivalent per MJ (1 MJ = 10^6 J) of fuel and therefore measure greenhouse-gas (GHG) emissions.

It is clear from the left side of the figure that current methods for generating ethanol have a large potential for displacing petroleum (0.05 for ethanol

The molecules of a bathtub full of water at room temperature are moving in a variety of ways and therefore have a lot of energy, but that energy cannot easily be used to boil water or to generate electricity. The energy in the water is not useful in the same sense that energy stored in a fuel and oxygen is useful.

manufacture the aluminum from its ore is wasted if the aluminum is oxidized. It is not surprising, then, that major programs for recycling aluminum operate throughout the United States. A similar statement can be made about almost every metal: Once reduced from their ores, metals are storehouses of Gibbs free energy that should be maintained in their reduced forms to avoid repeating the expenditure of Gibbs free energy needed to separate them from chemical combination with oxygen.

Energy Conservation and Coupled Reactions

The previous section mentioned that in a typical human cell, oxidation of 1 mol glucose to carbon dioxide and water can cause 32 conversions of adenosine diphosphate to adenosine triphosphate (p. 877). In Exercise 18.13, you calculated the overall change in Gibbs free energy when 1 mol glucose is metabolized and 32 mol ADP is transformed into 32 mol ATP.

$$C_6H_{12}O_6(aq) + 6\,O_2(g) \longrightarrow 6\,CO_2(g) + 6\,H_2O(\ell) \qquad \Delta G^{\circ\prime} = -2870\text{ kJ}$$

$$32\,ADP^{3-}(aq) + 32\,H_2PO_4^-(aq) \longrightarrow 32\,ATP^{4-}(aq) + 32\,H_2O(\ell) \qquad \Delta G^{\circ\prime} = 976.0\text{ kJ}$$

$$C_6H_{12}O_6(aq) + 6\,O_2(g) + 32\,ADP^{3-}(aq) + 32\,H_2PO_4^-(aq) \longrightarrow 6\,CO_2(g) + 32\,ATP^{4-}(aq) + 38\,H_2O(\ell) \qquad \Delta G^{\circ\prime} = -1894\text{ kJ}$$

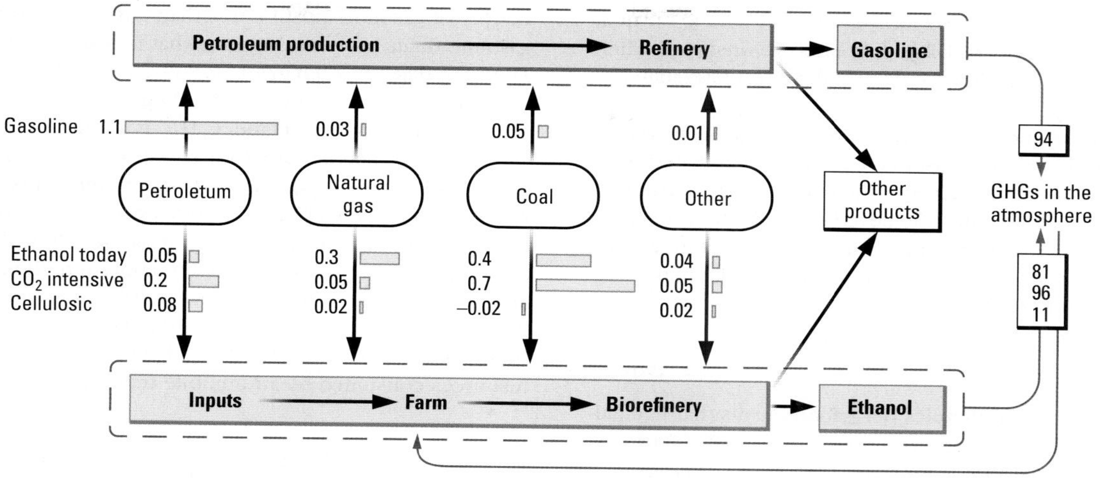

versus 1.1 for gasoline), but that producing the ethanol consumes large quantities of natural gas (0.3) and coal (0.4). To produce 1 MJ from ethanol fuel requires 0.05 + 0.3 + 0.4 + 0.04 = 0.79 MJ of energy input, so only 0.21 (about 20%) of the ethanol energy has actually displaced use of other fuels. That is, producing ethanol as we do today will reduce the need for petroleum, but increase the need for natural gas and coal. The GHG emissions are 81 versus 94 for gasoline, a reduction, but only by 14%—not the complete avoidance of GHG emissions that an incomplete analysis would have suggested.

Perhaps the most significant result of this study is the tremendous reduction in fossil fuel use and in GHG emissions that would accrue if an economically feasible process were developed by which any cellulosic material (wood, brush, grass) could be converted to ethanol. In that case the energy input is only 0.08 + 0.02 − 0.02 + 0.02 = 0.10 MJ per MJ of fuel produced. (The negative value for coal results because some electricity would be produced and this would reduce burning of coal.)

Even more important, a thermodynamic analysis of the entire system required to generate fuel provides important results that apply to societal decisions. Science is crucial to good public policy.

Sources: New York Times, November 15, 2005; p. D3; *New York Times*, February 28, 2006; p. C1; *New York Times*, June 29, 2009; p. B3; *Scientific American*, July 2009; pp. 52–59; from Farrell, et al., *Science*, Vol. 311, 2006, pp. 506–508. Reprinted with permission AAAS.

Some of the Gibbs free energy released when the glucose is oxidized does useful work by causing synthesis of ATP, the energy-storage medium of living cells. However, nearly two thirds of the Gibbs free energy does no useful work and ends up as thermal energy. That is, about two thirds of the original Gibbs free energy is lost, and one third is stored in ATP. The energy-storage process has an efficiency of about 33%, because only about 33% of the Gibbs free energy change actually does useful work.

The more efficient a biochemical process (or an industrial process) is, the less Gibbs free energy is lost and the more energy is used productively. Efficiency is defined as the useful work done per 100 units of energy input. Current energy-conversion systems have a broad range of efficiencies. For example, the generators in a large electrical generating plant convert about 99% of the mechanical energy input to electric energy output; based on the energy of the coal combustion reaction, however, the overall efficiency of such a plant is only about 40%. An incandescent electric light bulb converts only about 5% of the electrical energy it receives into light energy, but a fluorescent light converts four times as much—about 20%. The higher the energy efficiency of the devices we use, the less Gibbs free energy is lost.

Like ATP in your body, many compounds can store Gibbs free energy. An example is ethylene. About 50 million pounds of this gas is produced in the United States

ESTIMATION Gibbs Free Energy and Automobile Travel

Given that $\Delta G^\circ = -5295.74$ kJ per mole of octane burned, estimate the quantity of Gibbs free energy consumed when a typical car makes a 1000-mile round trip on interstate highways.

Assume that the typical car averages 20 miles per gallon and that combustion of gasoline can be approximated by combustion of octane. Because the trip is a round trip, the car ends up exactly where it started out, which means that it has done no useful thermodynamic work. Therefore all of the Gibbs free energy released by combustion of the fuel is lost. The combustion reaction is

$$C_8H_{18}(\ell) + \tfrac{25}{2}\,O_2(g) \longrightarrow 8\,CO_2(g) + 9\,H_2O(\ell) \qquad \Delta G^\circ = -5295.74\ \text{kJ}$$

Fuel economy of 20 miles per gallon means that five gallons of fuel will be used in 100 miles and 50 gallons in 1000 miles. One gallon is four quarts and a quart is about a liter, so the volume of octane is about $4 \times 50 = 200$ L. The density of gasoline must be less than that of water, because gasoline floats on water. Assume that it is about 80% as big. Then the density is 0.8 g/mL or 800 g/L. The 200 L of fuel weighs about $200 \times 800 = 160{,}000$ g.

The molar mass of octane, C_8H_{18}, is about $8 \times 12 + 18 = 114$ g/mol. To make the arithmetic easier, round this value to 100 g/mol. Then 160,000 g of octane corresponds to 1600 mol of octane. The Gibbs free energy released by combustion is about 5000 kJ for every mole of octane burned. Since $1600 \times 5000 = 8{,}000{,}000$ kJ, about 8 million kJ of useful energy is consumed for every 1000 miles a car is driven. Most of us drive ten times that far every year, and there are a lot of cars in the United States, so the energy resources consumed by automobile travel are huge.

Visit this book's companion website at www.cengage.com/chemistry/moore to work an interactive module based on this material.

press photo BASF

Figure 18.14 Ethylene production plant. A chemical manufacturing plant in Nanjing, China, that produces ethylene. From the ethylene the plant also manufactures other products, such as polyethylene, that are used in many consumer items *(← p. 562).*

every year from the dehydrogenation of ethane in chemical plants like the one shown in Figure 18.14.

$$C_2H_6(g) \longrightarrow H_2(g) + C_2H_4(g) \qquad \Delta G^\circ = 100.97\ \text{kJ}$$

When a mole of hydrogen and a mole of ethylene are produced from a mole of ethane, at least 100.97 kJ of Gibbs free energy must be supplied from an external source. This Gibbs free energy is then stored in the hydrogen and ethylene. Ethylene production is the largest single consumer of Gibbs free energy in the chemical industry, so there has been great interest in improving the process to save energy and money. Since 1960 the Gibbs free energy requirement per pound of ethylene produced has declined by 60%. Even so, the energy used to make 1 mol ethylene from 1 mol ethane (about 400 kJ) is four times the minimum required (100.97 kJ) because manufacture of ethylene is only 25% efficient. This is largely due to inefficiencies in energy transfer from external sources to the reaction system.

It is important to recognize that completely eliminating consumption of Gibbs free energy is impossible. Whenever anything happens, whether a chemical reaction or a physical process, the final state must have less Gibbs free energy than was available initially. This is the same as saying that the entropy of the universe must have increased during the change. This statement is true of any system in which the initial substances are changed into something new—any product-favored system. Thus, losses of Gibbs free energy are inevitable. The aim of energy conservation is to minimize—not eliminate—them. This can be done by maximizing the efficiency of coupling exergonic reactions to endergonic processes we want to cause to occur. The ideas of thermodynamics help us figure out how to accomplish that goal and are the most powerful tool we have for conserving energy while maintaining a high standard of living.

18.11 Thermodynamic and Kinetic Stability

Chemists often say that substances are "stable," but what exactly does that statement mean? Usually it means that the substance in question does not decompose or react with other substances that normally come in contact with it. Most chemists, for example, would say that the aluminum can that holds the soda you drink is stable. It will be around for quite a long time. The fact that aluminum cans do not decompose rapidly is one of the reasons you are encouraged to recycle them instead of throwing them away. Some aluminum cans have emerged almost unchanged from landfills after 40 or 50 years.

Strictly speaking, there are two kinds of stability. We discussed one of them earlier in this chapter. A substance is *thermodynamically stable* if it does not undergo product-favored reactions. Such reactions disperse energy and increase the entropy of their surroundings. Although we just said it was stable, the aluminum in a soda can is actually *thermodynamically unstable* compared with its oxide, because its reaction with oxygen in air has a negative Gibbs free energy change.

$$4\,Al(s) + 3\,O_2(g) \longrightarrow 2\,Al_2O_3(s) \qquad \Delta G^\circ = -3164.6\text{ kJ}$$

The aluminum exhibits a different kind of stability—it is ***kinetically stable.*** Although it has the potential to undergo a product-favored oxidation reaction, this reaction proceeds so slowly that the can remains essentially unchanged for a long time. This happens because a thin coating of aluminum oxide forms on the surface of the aluminum and prevents oxygen from reaching the rest of the aluminum atoms below the can's surface. If we grind the aluminum into a fine powder and throw it into a flame, the powder will burn and the evolved heat will lead to an entropy increase in the little piece of the universe around the burning metal. The reason an aluminum can is stable (does not oxidize away) is that the oxidation is slow (kinetics), not that formation of the oxide would not occur of its own accord (thermodynamics).

Iron displays kinetic stability. When exposed to the oxygen in air, iron rusts; iron is not thermodyamically stable under these conditions. The many iron objects we use—tools, cars, bridges, pipes—all will eventually rust completely, but the reaction rate is slow enough that we can use them for many years.

Another substance that is *thermodynamically unstable* but ***kinetically stable*** is diamond. If you look up the data in Appendix J, you will find that the conversion of diamond to graphite has a negative Gibbs free energy change; therefore it is product-favored. But diamonds don't change into graphite. Engagement rings contain diamonds precisely because the diamond (like the love it represents) is expected to last for a long time. It does so because there is a very high activation energy barrier ***(← p. 611)*** for the change from the diamond structure to the graphite structure. When a chemist says something is stable, it usually means that it is kinetically stable—only an activation energy barrier prevents it from reacting fast enough for us to see a change.

PROBLEM-SOLVING EXAMPLE 18.13 Thermodynamic and Kinetic Stability

Whenever air is heated to a very high temperature, the reaction between nitrogen and oxygen to form nitrogen monoxide occurs. It is an important source of nitrogen-containing air pollutants that can be formed in the cylinders of an automobile engine.

(a) Write a balanced equation with minimum whole-number coefficients for the equilibrium reaction of N_2 with O_2 to form NO.

(b) Is this reaction product-favored at room temperature? That is, is NO thermodynamically stable compared to N_2 and O_2?

(c) Estimate the temperature at which the standard equilibrium constant for this reaction equals 1.

(d) If NO is formed at high temperature in an automobile engine, why does it not all change back to N_2 and O_2 when the mixture of gases enters the exhaust system and its temperature falls?

(e) How might the concentration of NO in automobile exhaust be reduced?

Answer

(a) $N_2(g) + O_2(g) \rightleftharpoons 2\,NO(g)$ (b) No (c) 7301 K

(d) The reverse reaction is too slow. (e) Use a suitable catalyst.

Strategy and Explanation

(a) See answer.

(b) Calculate ΔG° at 25 °C using data from Appendix J.
$\Delta G^\circ = 2\{\Delta G_f^\circ[NO(g)]\} = 2(86.55\text{ kJ}) = 173.10\text{ kJ}$. Since $\Delta G^\circ > 0$, the reaction is not product-favored.

(c) If $K° = 1$, then $\Delta G° = -RT \ln K° = -RT \ln(1) = 0$.
Because $\Delta G° = \Delta H° - T\Delta S° = 0$, $\Delta H° = T\Delta S°$ and $T = \Delta H°/\Delta S°$.
Using data from Appendix J gives $\Delta H° = 2\{\Delta H_f° [NO(g)]\} = 2(90.25 \text{ kJ}) = 180.50 \text{ kJ}$, and

$$\Delta S° = 2\{S°[NO(g)]\} - \{S°[N_2(g)] + S°[O_2(g)]\}$$
$$= 2(210.76 \text{ J/K}) - 191.66 \text{ J/K} - 205.138 \text{ J/K} = 24.722 \text{ J/K}$$

Therefore $T = 180.50 \text{ kJ}/24.722 \text{ J/K} = 180{,}500 \text{ J}/24.722 \text{ J/K} = 7301 \text{ K}$.

(d) When the mixture of gases, which contains some NO as well as N_2 and O_2, leaves the cylinder of the automobile engine and enters the exhaust system, it cools very rapidly to a temperature below 500 K. The reverse reaction should occur, according to thermodynamics, but it does not. The activation energy for the decomposition of NO is quite high, because NO contains a double bond, and it is very difficult to separate the two atoms (which must be done to form N_2 and O_2). Therefore, the reaction rate is greatly affected by temperature. At low temperatures the reverse reaction is very slow, so significant concentrations of NO exist in automobile exhaust, even though N_2 and O_2 are thermodynamically stable compared to NO.

(e) With a suitable catalyst, decomposition of NO to its elements can take place at appreciable rates even at relatively low temperatures. Catalytic converters are installed in the exhaust systems of cars partly to reduce the concentration of NO. Because N_2 and O_2 are more stable thermodynamically, it is reasonable to use a catalyst to speed up their formation.

egd/Shutterstock

Automobile emissions. NO, CO, and unburned hydrocarbons emitted from automobile exhausts are major causes of air pollution.

☑ **Reasonable Answer Check** It is reasonable that $\Delta G°$ for the reaction of N_2 with O_2 is positive, because N_2 and O_2 are the principal components of the atmosphere where their partial pressures are close to standard pressure, and they do not react with each other. It is reasonable that $\Delta S°$ for the reaction is small and positive. The total number of gas phase molecules does not change, but the product molecules have two different atoms, and both reactant molecules have two atoms that are the same, making the product molecules slightly more probable. It is reasonable that the reaction is endothermic, because the reactant molecules have a triple bond and a double bond, and the product molecules have two double bonds. The bonds broken are therefore expected to be stronger than the bonds formed.

PROBLEM-SOLVING PRACTICE 18.13

All of these substances are stable with respect to decomposition to their elements at 25 °C. Which are kinetically stable and which are thermodynamically stable?

(a) MgO(s) (b) $N_2H_4(\ell)$ (c) C_2H_6(g) (d) N_2O(g)

Another example of kinetic versus thermodynamic stability is gasoline and other fuels. We want a fuel to be thermodynamically unstable compared with CO_2 and H_2O, but kinetically stable so that it will not burn unless we want it to burn.

Finally, think about whether you yourself are stable (thermodynamically or kinetically). From a thermodynamic standpoint, most of the substances you are made of are unstable with respect to oxidation to carbon dioxide, water, and other substances. That is, based on Gibbs free energy changes, most of the substances that you are made of should undergo product-favored reactions that would completely destroy them. Your proteins, fats, carbohydrates, and even DNA should spontaneously change into much smaller, simpler molecules. Fortunately for you, the reactions by which this change would happen are very slow at room temperature and body temperature. Only when enzymes catalyze those reactions do they occur with reasonable speed. It is the combination of thermodynamic instability and kinetic stability that allows those enzymes to control the reactions in your body or in any living organism. Were it not for the kinetic stability of a wide variety of substances, everything would be quickly converted to a small number of very thermodynamically stable substances. Life and the environment as we know them would then be impossible.

The roles of chemical thermodynamics and chemical kinetics in determining chemical reactivity can be summarized this way:

- ***Thermodynamics tells whether a reaction can produce predominantly products under standard conditions and, if it does, how much useful work can be accomplished by coupling the reaction to another process.***
- ***Thermodynamics tells how to calculate the standard equilibrium constant and allows quantitative prediction of how much product is formed for reactions that involve dilution of substances in the gas phase or in solution.***
- ***Thermodynamics can also be used to predict what will happen under nonstandard conditions.***
- ***Kinetics tells how fast a given reaction goes and indicates how we can control the rate of reaction.***

Together, thermodynamics and kinetics provide the intellectual foundation on which modern chemical industries are based and the principles upon which fundamental understanding of physiology and medicine depends.

SUMMARY PROBLEM

In a blast furnace for making iron from iron ore, large quantities of coke (which is mainly carbon) are dumped into the top of the furnace along with iron ore (which can be assumed to be Fe_2O_3) and limestone (which is used to help remove impurities from the iron). The overall process is

$$2\ Fe_2O_3(s) + 3\ C(s) \longrightarrow 4\ Fe(s) + 3\ CO_2(g)$$

This reaction can be thought of as a combination of several individual steps.

$$2\ Fe_2O_3(s) \longrightarrow 4\ FeO(s) + O_2(g)$$
$$2\ FeO(s) \longrightarrow 2\ Fe(s) + O_2(g)$$
$$2\ C(s) + O_2(g) \longrightarrow 2\ CO(g)$$
$$2\ CO(g) + O_2(g) \longrightarrow 2\ CO_2(g)$$

(a) Calculate the enthalpy change for each step, assuming a temperature of 25 °C. Which steps are exothermic and which are endothermic?

(b) Based on the equations, predict which of the individual steps would involve an increase and which a decrease in the entropy of the system.

(c) Based on your results in parts (a) and (b), what can you say about whether each step is reactant-favored or product-favored at room temperature? At a much higher temperature ($>$1000 K)?

(d) Calculate the entropy change and the Gibbs free energy change for each reaction step, assuming a temperature of 25 °C. (Obtain data from Appendix J.)

(e) Keeping in mind the equation $\Delta G° = \Delta H° - T\Delta S°$ and the fact that the enthalpy change and entropy change for a reaction do not vary much with temperature, what would be the slope of a graph of $\Delta G°$ versus T for each of the reactions? For which of the reactions does $\Delta G°$ become more negative as the temperature increases? For which does it become more positive? Does this agree with what you predicted in part (c)?

(f) For which of these reactions might the assumption of nearly constant $\Delta H°$ and $\Delta S°$ not be valid as the temperature increases from 25 °C? For each reaction you choose, explain why the assumption might not be correct.

(g) Use your results from previous parts of this problem to estimate the Gibbs free energy change for each of these reactions at a temperature of 1500 K.

(h) Which of the two iron oxides is more easily reduced at 1500 K? Which of the reactions involving carbon compounds is more product-favored at 1500 K? What chemical reactions are taking place in the hottest part of the blast furnace?

(i) In portions of the furnace where the temperature is about 800 K, would the same reactions be occurring as in the higher-temperature part of the furnace? Why or why not?

(j) Show that the individual steps can be combined to give the overall reaction. From the enthalpy, entropy, and Gibbs free energy changes already calculated, calculate these changes for the overall reaction.

(k) In a typical blast furnace every kilogram of iron produced requires 2.5 kg iron ore, 1 kg coke, and nearly 6 kg air (to provide oxygen for oxidation of the coke to heat the furnace and to combine with carbon in the coke, forming CO(g)). How much Gibbs free energy would be destroyed if the coke were simply burned to form carbon dioxide? Given the quantity of iron produced in a typical furnace, how much Gibbs free energy is stored by coupling the oxidation of coke to the reduction of iron oxides? What percentage of the Gibbs free energy available from combustion of coke is wasted per kilogram of iron produced?

IN CLOSING

OWL and go Chemistry

Sign in at **www.cengage.com/owl** to:

- View tutorials and simulations, develop problem-solving skills, and complete online homework assigned by your professor.
- For quick review and exam prep, download Go Chemistry mini lecture modules from OWL (or purchase them at **www.CengageBrain.com**).

Having studied this chapter, you should be able to . . .

- Understand and be able to use the terms “product-favored” and “reactant-favored” (Section 18.1). End-of-chapter question: 15
- Explain why there is a higher probability that energy will be dispersed than that it will be concentrated in a small number of nanoscale particles (Section 18.2). Questions 17, 19
- Calculate the entropy change for a process occurring at constant temperature (Section 18.3). Questions 35, 37
- Use qualitative rules to predict the sign of the entropy change for a process (Section 18.3). Questions 21, 23, 25, 27, 29, 31, 33
- Calculate the entropy change for a chemical reaction, given standard molar entropy values for elements and compounds (Section 18.4). Questions 39, 41, 43
- Use entropy and enthalpy changes to predict whether a reaction is product-favored (Section 18.5). Questions 45, 49, 51, 53
- Describe the connection between enthalpy and entropy changes for a reaction and the Gibbs free energy change; use this relation to estimate quantitatively how temperature affects whether a reaction is product-favored (Section 18.6). Questions 47, 55, 59, 61, 69, 71, 73
- Calculate the Gibbs free energy change for a reaction from values given in a table of standard molar Gibbs free energies of formation (Section 18.6). Questions 57, 63, 65, 67, 75
- Relate the Gibbs free energy change and the standard equilibrium constant for the same reaction and be able to calculate one from the other (Section 18.7). Questions 77, 79, 81, 83, 85
- Describe how a reactant-favored system can be coupled to a product-favored system so that a desired reaction can be carried out (Section 18.8). Questions 87, 89, 91, 93

- Explain how biological systems make use of coupled reactions to maintain the high degree of order found in all living organisms; give examples of coupled reactions that are important in biochemistry (Section 18.9). Questions 95, 97
- Explain the relationship between Gibbs free energy and energy conservation (Sections 18.8 and 18.10). Question 99
- Distinguish between thermodynamic stability and kinetic stability and describe the effect of each on whether a reaction is useful in producing products (Section 18.11). Questions 101, 103

KEY TERMS

endergonic *(Section 18.9)*

energy conservation *(18.10)*

exergonic *(18.9)*

extent of reaction *(18.7)*

Gibbs free energy *(18.6)*

metabolism *(18.9)*

nutrients *(18.9)*

photosynthesis *(18.9)*

reversible process *(18.3)*

second law of thermodynamics *(18.5)*

standard equilibrium constant *(18.7)*

third law of thermodynamics *(18.3)*

QUESTIONS FOR REVIEW AND THOUGHT

OWL Interactive versions of these problems are assignable in OWL.

Blue-numbered questions have short answers at the back of this book in Appendix M and fully worked solutions in the *Student Solutions Manual.*

Review Questions

These questions test vocabulary and simple concepts.

1. Define the terms "product-favored system" and "reactant-favored system." Give one example of each.
2. What are the two ways that a final chemical state of a system can be more probable than its initial state?
3. Define the term "entropy," and give an example of a sample of matter that has zero entropy. What are the units of entropy? How do they differ from the units of enthalpy?
4. State five useful qualitative rules for predicting entropy changes when chemical or physical changes occur.
5. State the second law of thermodynamics.
6. In terms of values of $\Delta H°$ and $\Delta S°$, under what conditions can you be sure that a reaction is product-favored? When can you be sure that it is not product-favored?
7. Define the Gibbs free energy change of a chemical reaction in terms of its enthalpy and entropy changes. Why is the Gibbs free energy change especially useful in predicting whether a reaction is product-favored?
8. Why are materials whose reactions release large quantities of Gibbs free energy useful to society? Give two examples of such materials.
9. How are materials whose reactions release large quantities of Gibbs free energy important to you? Give two examples of such materials.
10. Define the terms "endergonic" and "exergonic."
11. What is the citric acid cycle, and why is it important to organisms?
12. Define these important biochemistry terms: metabolism, nutrients, ATP, ADP, oxidative phosphorylation, coupled reactions, phototrophs, chemotrophs, photosynthesis.
13. Describe two ways to cause reactant-favored reactions to form products.
14. Describe the process by which sunlight is employed to convert high-entropy, low-Gibbs-free-energy substances into low-entropy, high-Gibbs-free-energy substances.

Topical Questions

These questions are keyed to major topics in the chapter. Usually a question that is answered at the back of this book is paired with a similar one that is not.

Reactant-Favored and Product-Favored Processes (Section 18.1)

15. For each process, write a chemical equation and classify the process as reactant-favored or product-favored.
    (a) Water decomposes to its elements, hydrogen and oxygen.
    (b) Gasoline spilled on the ground evaporates (use octane, C_8H_{18}, to represent gasoline).
    (c) Sugar dissolves in water at room temperature.
16. For each process, write a chemical equation and classify the process as reactant-favored or product-favored.
    (a) Carbon dioxide gas decomposes to its elements, carbon and oxygen.
    (b) The steel (mostly iron) body of an automobile rusts.
    (c) Gasoline reacts with oxygen to form carbon dioxide and water (use octane, C_8H_{18}, to represent gasoline).

Chemical Reactions and Dispersal of Energy (Section 18.2)

17. Suppose you flip a coin.
    (a) What is the probability that the coin will come up heads?
    (b) What is the probability that it will come up tails?
    (c) If you flip the coin 100 times, what is the most likely number of heads and tails you will see?
18. Suppose you make a tetrahedron and put numbers 1, 2, 3, and 4 on each of the four sides. You toss the tetrahedron in the air and observe it after it comes to rest.
    (a) What is the probability that the tetrahedron will come to rest with the numbers 2, 3, and 4 visible?
    (b) What is the probability that the tetrahedron will come to rest with the numbers 1, 2, and 3 visible?
    (c) If you toss the tetrahedron 100 times, what is the most likely number of times you will see a 1 after it comes to rest?
19. Consider two equal-sized flasks connected as shown in the figure.

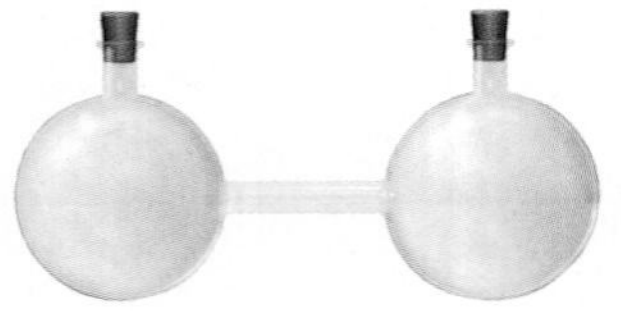

    (a) Suppose you put one molecule inside. What is the probability that the molecule will be in flask A? What is the probability that it will be in flask B?
    (b) If you put 100 molecules into the two-flask system, what is the most likely arrangement of molecules? Which arrangement has the highest entropy?
20. Suppose you have four identical molecules labeled 1, 2, 3, and 4. Draw 16 simple two-flask diagrams as in the figure for Question 19, and draw all possible arrangements of the four molecules in the two flasks. How many of these arrangements have two molecules in each flask? How many have no molecules in one flask? From these results, what is the most probable arrangement of molecules? Which arrangement has the highest entropy?

Measuring Dispersal of Energy: Entropy (Section 18.3)

21. For each process, tell whether the entropy change of the system is positive or negative.
    (a) Water vapor (the system) deposits as ice crystals on a cold windowpane.
    (b) A can of carbonated beverage loses its fizz. (Consider the beverage but not the can as the system. What happens to the entropy of the dissolved gas?)
    (c) A glassblower heats glass (the system) to its softening temperature.
22. For each process, tell whether the entropy change of the system is positive or negative.
    (a) Water boils.
    (b) A teaspoon of sugar dissolves in a cup of coffee. (The system consists of both sugar and coffee.)
    (c) Calcium carbonate precipitates out of water in a cave to form stalactites and stalagmites. (Consider only the calcium carbonate to be the system.)
23. For each situation described in Question 15, tell whether the entropy of the system increases or decreases.
24. For each situation described in Question 16, tell whether the entropy of the system increases or decreases.
25. For each pair of items, tell which has the higher entropy, and explain why.
    (a) Item 1, a sample of solid CO_2 at -78 °C, or item 2, CO_2 vapor at 0 °C
    (b) Item 1, solid sugar, or item 2, the same sugar dissolved in a cup of tea
    (c) Item 1, a 100-mL sample of pure water and a 100-mL sample of pure alcohol, or item 2, the same samples of water and alcohol after they had been poured together and stirred
26. For each pair of items, tell which has the higher entropy, and explain why.
    (a) Item 1, a sample of pure silicon (to be used in a computer chip), or item 2, a piece of silicon having the same mass but containing a trace of some other element, such as B or P
    (b) Item 1, an ice cube at 0 °C, or item 2, the same mass of liquid water at 0 °C
    (c) Item 1, a sample of pure I_2 solid at room temperature, or item 2, the same mass of iodine vapor at room temperature
27. Comparing the formulas or states for each pair of substances, tell which you would expect to have the higher entropy per mole at the same temperature, and explain why.
    (a) $NaCl(s)$ or $CaO(s)$
    (b) $Cl_2(g)$ or $P_4(g)$
    (c) $NH_4NO_3(s)$ or $NH_4NO_3(aq)$
28. Comparing the formulas or states for each pair of substances, tell which you would expect to have the higher entropy per mole at the same temperature, and explain why.
    (a) $CH_3NH_2(g)$ or $(CH_3)_2NH(g)$
    (b) $Au(s)$ or $Hg(\ell)$
    (c) $Kr(g)$ or $C_6H_{14}(g)$
29. From each pair of substances listed below, select the one having the larger standard molar entropy at 25 °C. Give reasons for your choice.
    (a) $Ga(s)$ or $Ga(\ell)$
    (b) $AsH_3(g)$ or $Kr(g)$
    (c) $NaF(s)$ or $MgO(s)$
30. From each pair of substances listed below, select the one having the larger standard molar entropy at 25 °C. Give reasons for your choice.
    (a) $H_2O(g)$ or $H_2S(g)$
    (b) $CH_3OH(\ell)$ or $C_2H_5OH(\ell)$
    (c) Butane or cyclobutane
31. Without doing a calculation, predict whether the entropy change will be positive or negative when each reaction occurs in the direction it is written.
    (a) $C_2H_4(g) + H_2(g) \longrightarrow C_2H_6(g)$
    (b) $CH_3OH(\ell) + \frac{3}{2}O_2(g) \longrightarrow CO_2(g) + 2\ H_2O(g)$
    (c) $N_2(g) + 3\ H_2(g) \longrightarrow 2\ NH_3(g)$
    (d) $CaCO_3(s) \longrightarrow CaO(s) + CO_2(g)$
32. Without doing a calculation, predict whether the entropy change will be positive or negative when each reaction occurs in the direction it is written.
    (a) $CH_3OH(\ell) \longrightarrow CO(g) + 2\ H_2(g)$
    (b) $Br_2(\ell) + H_2(g) \longrightarrow 2\ HBr(g)$
    (c) $C_3H_8(g) \longrightarrow C_2H_4(g) + CH_4(g)$
    (d) $Ag^+(aq) + I^-(aq) \longrightarrow AgI(s)$

33. Without consulting a table of standard molar entropies, predict whether ΔS°_{system} will be positive or negative for each of these reactions.
(a) $2\ CO(g) + O_2(g) \longrightarrow 2\ CO_2(g)$
(b) $2\ H_2(g) + O_2(g) \longrightarrow 2\ H_2O(\ell)$
(c) $2\ O_3(g) \longrightarrow 3\ O_2(g)$

34. Without consulting a table of standard molar entropies, predict whether ΔS°_{system} will be positive or negative for each of these reactions.
(a) $2\ NH_3(g) \longrightarrow N_2(g) + 3\ H_2(g)$
(b) $2\ Na(s) + Cl_2(g) \longrightarrow 2\ NaCl(s)$
(c) $H_2(g) + I_2(s) \longrightarrow 2\ HI(g)$

Calculating Entropy Changes (Section 18.4)

35. Calculate the entropy change, ΔS°, for the vaporization of ethanol, C_2H_5OH, at the boiling point of 78.3 °C. The heat of vaporization of the alcohol is 39.3 kJ/mol.

$$C_2H_5OH(\ell) \longrightarrow C_2H_5OH(g) \qquad \Delta S^\circ = ?$$

36. Diethyl ether, $(C_2H_5)_2O$, was once used as an anesthetic. Calculate the entropy change, ΔS°, for the vaporization of ether if its heat of vaporization is 26.0 kJ/mol at the boiling point of 35.0 °C.

37. Calculate ΔS° for each substance when the quantity of thermal energy indicated is transferred reversibly to the system at the temperature specified. Assume that you have enough of each substance so that its temperature remains constant as the thermal energy is transferred.
(a) $H_2(g)$, 0.775 kJ, 295 K
(b) KCl(s), 500. kJ, 500. K
(c) $N_2(g)$, 2.45 kJ, 1000. K

38. Calculate ΔS° for each of these substances when the quantity of thermal energy indicated is transferred reversibly to the system at the temperature specified. Assume that you have enough of each substance so that its temperature remains constant as the thermal energy is transferred.
(a) NaCl(s), 5.00 kJ, 500. K
(b) $N_2O(g)$, 0.30 kJ, 300. K

39. The standard molar entropy of methanol vapor, $CH_3OH(g)$, is 239.8 J K^{-1} mol^{-1}.
(a) Calculate the entropy change for the vaporization of 1 mol methanol (use data from Table 18.1 or Appendix J).
(b) Calculate the enthalpy of vaporization of methanol, assuming that ΔS° doesn't depend on temperature and taking the boiling point of methanol to be 64.6 °C.

40. The standard molar entropy of iodine vapor, $I_2(g)$, is 260.7 J K^{-1} mol^{-1} and the standard molar enthalpy of formation is 62.4 kJ/mol.
(a) Calculate the entropy change for vaporization of 1 mol of solid iodine (use data from Table 18.1 or Appendix J).
(b) Calculate the enthalpy change for sublimation of iodine.
(c) Assuming that ΔS° does not change with temperature, estimate the temperature at which iodine would sublime (change directly from solid to gas).

41. Check your predictions in Question 31 by calculating the entropy change for each reaction. Standard molar entropies not in Table 18.1 can be found in Appendix J.

42. Check your predictions in Question 32 by calculating the entropy change for each reaction. Standard molar entropies not in Table 18.1 can be found in Appendix J.

43. Check your predictions in Question 33 by calculating the entropy change for each reaction. Standard molar entropies not in Table 18.1 can be found in Appendix J.

44. Check your predictions in Question 34 by calculating the entropy change for each reaction. Standard molar entropies not in Table 18.1 can be found in Appendix J.

Entropy and the Second Law of Thermodynamics (Section 18.5)

45. Calculate ΔS°_{system} at 25 °C for the reaction

$$C_2H_4(g) + H_2O(g) \longrightarrow C_2H_5OH(\ell)$$

Can you tell from the result of this calculation whether this reaction is product-favored? If you cannot tell, what additional information do you need? Obtain that information and determine whether the reaction is product-favored.

46. Calculate ΔS°_{system} at 25 °C for the reaction

$$C_6H_6(\ell) + 4\ H_2(g) \longrightarrow C_6H_{14}(\ell)$$

Can you tell from the result of this calculation whether this reaction is product-favored? If you cannot tell, what additional information do you need? Obtain that information and determine whether the reaction is product-favored.

47. Is this reaction predicted to favor the products at low temperatures, at high temperatures, or both? Explain your answer briefly.

$$Mg(s) + \tfrac{1}{2}\ O_2(g) \longrightarrow MgO(s) \qquad \Delta H^\circ = -601.70\ kJ$$

48. Is this reaction predicted to favor the products at low temperatures, at high temperatures, or both? Explain your answer briefly.

$$MgCO_3(s) \longrightarrow MgO(s) + CO_2(g) \qquad \Delta H^\circ = 116.48\ kJ$$

49. Explain briefly why the exothermic combustion of propane is product-favored.

$$C_3H_8(g) + 5\ O_2(g) \longrightarrow 3\ CO_2(g) + 4\ H_2O(g)$$

50. Explain briefly why the exothermic reaction of a metal carbonate with an acid is product-favored.

$$CuCO_3(s) + H_2SO_4(aq) \longrightarrow CuSO_4(aq) + CO_2(g) + H_2O(\ell)$$

51. Sodium reacts violently with water according to the equation

$$Na(s) + H_2O(\ell) \longrightarrow NaOH(aq) + \tfrac{1}{2}\ H_2(g)$$

(a) Predict the signs of ΔH° and ΔS° for the reaction.
(b) Verify your predictions with calculations.

52. Once ignited, magnesium reacts vigorously with oxygen in air according to the equation

$$2\ Mg(s) + O_2(g) \longrightarrow 2\ MgO(s)$$

(a) Predict the signs of ΔH° and ΔS° for the reaction.
(b) Verify your predictions with calculations.

53. Hydrogen burns in air with considerable heat transfer to the surroundings. Consider the decomposition of water to gaseous hydrogen and oxygen. Without doing any calculations, and basing your prediction on the enthalpy change and the entropy change, is this reaction product-favored at 25 °C? Explain your answer briefly.

54. Hydrogen gas combines with chlorine gas in an exothermic reaction to form HCl(g). Consider the decomposition of gaseous hydrogen chloride to hydrogen and chlorine. Without doing any calculations, and basing your prediction on the enthalpy change and the entropy change, is this reaction product-favored at 25 °C? Explain your answer briefly.

55. For each reaction, calculate $\Delta H°$ and $\Delta S°$ and predict whether the reaction is always product-favored, product-favored only at low temperatures, product-favored only at high temperatures, or never product-favored.
(a) $Fe_2O_3(s) + 2\ Al(s) \longrightarrow 2\ Fe(s) + Al_2O_3(s)$
(b) $N_2(g) + 2\ O_2(g) \longrightarrow 2\ NO_2(g)$

56. For each reaction, calculate $\Delta H°$ and $\Delta S°$ and predict whether the reaction is always product-favored, product-favored only at low temperatures, product-favored only at high temperatures, or never product-favored.
(a) $C_6H_{12}O_6(s) + 6\ O_2(g) \longrightarrow 6\ CO_2(g) + 6\ H_2O(\ell)$
(b) $MgO(s) + C(graphite) \longrightarrow Mg(s) + CO(g)$

Gibbs Free Energy (Section 18.6)

57. Determine whether the combustion of ethane, C_2H_6, is product-favored at 25 °C.

$$C_2H_6(g) + \tfrac{7}{2}\ O_2(g) \longrightarrow 2\ CO_2(g) + 3\ H_2O(\ell)$$

(a) Calculate $\Delta S_{universe}$. Required values of $\Delta H_f°$ and $S°$ are in Appendix J.
(b) Verify your result by calculating the value of $\Delta G°$ for the reaction.
(c) Do your calculated answers in parts (a) and (b) agree with your preconceived idea of this reaction?

58. The reaction of magnesium with water can be used as a means for heating food.

$$Mg(s) + 2\ H_2O(\ell) \longrightarrow Mg(OH)_2(s) + H_2(g)$$

Determine whether this reaction is product-favored at 25 °C.
(a) Calculate $\Delta S_{universe}$. See Appendix J for the needed data.
(b) Verify your result by calculating $\Delta G°$ for the reaction.

59. Add a column for the sign of the Gibbs free energy to Table 18.2 (p. 863). For the first and last lines in the table, tell whether ΔG is positive or negative.

60. Based on your table from Question 59, when ΔH_{system} and ΔS_{system} are both negative, is ΔG is positive or negative or does the sign depend on temperature? If the sign depends on temperature, does the reaction become product-favored at high or low temperatures?

61. Use a mathematical equation to show how the statement leads to the conclusion cited: If a reaction is exothermic (negative ΔH) and if the entropy of the system increases (positive ΔS), then ΔG must be negative, and the reaction will be product-favored.

62. Use a mathematical equation to show how the statement leads to the conclusion cited: If ΔH and ΔS have the same sign, then the magnitude of T determines whether ΔG will be negative and whether the reaction will be product-favored.

63. Predict whether the reaction below is product-favored or reactant-favored by calculating $\Delta G°$ from the entropy and enthalpy changes for the reaction at 25 °C.

$$H_2(g) + CO_2(g) \longrightarrow H_2O(g) + CO(g)$$
$$\Delta H° = 41.17\ kJ \qquad \Delta S° = 42.08\ J/K$$

64. Predict whether this reaction is product-favored at 25 °C by calculating the change in standard Gibbs free energy from the entropy and enthalpy changes.

$$H_2(g) + I_2(g) \rightleftharpoons 2\ HI(g)$$
$$\Delta H° = 52.96\ kJ \qquad \Delta S° = 166.4\ J/K$$

65. If this reaction were product-favored, it would be a good way to make pure silicon, crucial in the semiconductor industry, from sand (SiO_2).

$$SiO_2(s) + C(s) \longrightarrow Si(s) + CO_2(g)$$

Calculate $\Delta G°$ from data in Appendix J and decide whether the reaction can be used to produce silicon at 25 °C.

66. From data in Appendix J, calculate $\Delta G°$ for the reactions of sand with hydrogen fluoride and hydrogen chloride. Explain why hydrogen fluoride attacks glass, whereas hydrogen chloride does not.

$$SiO_2(s) + 4\ HF(g) \longrightarrow SiF_4(g) + 2\ H_2O(g)$$
$$SiO_2(s) + 4\ HCl(g) \longrightarrow SiCl_4(g) + 2\ H_2O(g)$$

67. Use data from Appendix J to calculate $\Delta G°$ for each reaction at 25 °C. Which are product-favored?
(a) $C_2H_2(g) + H_2(g) \longrightarrow C_2H_4(g)$
(b) $2\ SO_3(g) \longrightarrow 2\ SO_2(g) + O_2(g)$
(c) $4\ NH_3(g) + 5\ O_2(g) \longrightarrow 4\ NO(g) + 6\ H_2O(g)$

68. Evaluate $\Delta H°$ for each reaction in Question 67. Use your results to calculate standard molar entropies at 25.00 °C for
(a) $C_2H_2(g)$ (b) $SO_3(g)$
(c) NO(g)

69. If a system falls within the second or third category in Table 18.2 (p. 863), then there must be a temperature at which it shifts from being reactant-favored to being product-favored. For each reaction, obtain data from Appendix J and calculate what that temperature is.
(a) $CO(g) + 2\ H_2(g) \rightleftharpoons CH_3OH(\ell)$
(b) $2\ Fe_2O_3(s) + 3\ C(graphite) \rightleftharpoons 4\ Fe(s) + 3\ CO_2(g)$

70. If a system falls within the second or third category in Table 18.2 (p. 863) then there must be a temperature at which it shifts from being reactant-favored to being product-favored. For each reaction, obtain data from Appendix J and calculate what that temperature is.
(a) $2\ H_2O(g) \rightleftharpoons 2\ H_2(g) + O_2(g)$
(b) $N_2(g) + 3\ H_2(g) \rightleftharpoons 2\ NH_3(g)$

71. Estimate $\Delta G°$ at 2000. K for each reaction in Question 69.

72. Estimate $\Delta G°$ at 2000. K for each reaction in Question 70.

73. Many metal carbonates can be decomposed to the metal oxide and carbon dioxide by heating.

$$CaCO_3(s) \longrightarrow CaO(s) + CO_2(g)$$

(a) Calculate the enthalpy, entropy, and Gibbs free energy changes for this reaction at 25.00 °C.
(b) Is it product-favored or reactant-favored?
(c) Based on the signs of $\Delta H°$ and $\Delta S°$, predict whether the reaction is product-favored at all temperatures.
(d) Predict the lowest temperature at which appreciable quantities of products can be obtained.

74. Some metal oxides, such as lead(II) oxide, can be decomposed to the metal and oxygen simply by heating.

$$PbO(s) \longrightarrow Pb(s) + \tfrac{1}{2}\ O_2(g)$$

(a) Is the decomposition of lead(II) oxide product-favored at 25 °C? Explain.
(b) If not, can it become so if the temperature is raised?
(c) As the temperature increases, at what temperature does the reaction first become product-favored?

75. Use the thermochemical expression

$$CaC_2(s) + 2\,H_2O(\ell) \longrightarrow C_2H_2(g) + Ca(OH)_2(aq) \qquad \Delta G° = -119.282\text{ kJ}$$

and data from Appendix J to calculate $\Delta G_f°$ for $Ca(OH)_2(aq)$ at 25 °C.

76. Use the thermochemical expression

$$PCl_3(g) + Cl_2(g) \longrightarrow PCl_5(g) \qquad \Delta G° = -37.2\text{ kJ}$$

and data from Appendix J to calculate $\Delta G_f°$ for $PCl_5(g)$.

Gibbs Free Energy Changes and Equilibrium Constants (Section 18.7)

77. Use data from Appendix J to obtain the equilibrium constant K_P for each reaction at 298.15 K.
(a) $2\,HCl(g) \rightleftharpoons H_2(g) + Cl_2(g)$
(b) $N_2(g) + O_2(g) \rightleftharpoons 2\,NO(g)$

78. Use data from Appendix J to obtain the equilibrium constant K_P for each of these reactions at 298 K.
(a) $CH_4(g) + 2\,O_2(g) \rightleftharpoons CO_2(g) + 2\,H_2O(g)$
(b) $2\,NO_2(g) \rightleftharpoons N_2O_4(g)$

79. Ethylene reacts with hydrogen to produce ethane.

$$H_2C{=}CH_2(g) + H_2(g) \rightleftharpoons H_3C{-}CH_3(g)$$

(a) Using the data in Appendix J, calculate $\Delta G°$ for the reaction at 25 °C. Is the reaction predicted to be product-favored under standard conditions?
(b) Calculate K_P from $\Delta G°$. Comment on the connection between the sign of $\Delta G°$ and the magnitude of K_P.

80. Use the data in Appendix J to calculate $\Delta G°$ and K_P at 25 °C for the reaction

$$2\,HBr(g) + Cl_2(g) \rightleftharpoons 2\,HCl(g) + Br_2(\ell)$$

Comment on the connection between the sign of $\Delta G°$ and the magnitude of K_P.

81. For each chemical reaction, calculate the standard equilibrium constant at 298 K and at 1000. K from the thermodynamic data in Appendix J. Indicate whether each reaction is product-favored or reactant-favored at each temperature.
(a) The conversion of nitric oxide to nitrogen dioxide in the atmosphere

$$2\,NO(g) + O_2(g) \rightleftharpoons 2\,NO_2(g)$$

(b) The reaction of an alkali metal with a halogen to produce an alkali metal halide salt

$$2\,Na(s) + Cl_2(g) \rightleftharpoons 2\,NaCl(s)$$

82. For each chemical reaction, calculate the standard equilibrium constant at 298 K and at 1000. K from the thermodynamic data in Appendix J. Indicate whether each reaction is product-favored or reactant-favored at each temperature.
(a) The oxidation of carbon monoxide to carbon dioxide

$$2\,CO(g) + O_2(g) \rightleftharpoons 2\,CO_2(g)$$

(b) The first step in the production of electronic-grade silicon from sand

$$SiO_2(s) + 2\,C(s) \rightleftharpoons Si(s) + 2\,CO(g)$$

83. For each reaction, estimate $K°$ at the temperature indicated.
(a) $2\,H_2(g) + O_2(g) \rightleftharpoons 2\,H_2O(g)$ at 800. K
(b) $2\,SO_2(g) + O_2(g) \rightleftharpoons 2\,SO_3(g)$ at 500. K
(c) $2\,HF(g) \rightleftharpoons H_2(g) + F_2(g)$ at 2000. K

84. For each reaction, estimate $K°$ at the temperature indicated.
(a) $H_2(g) + I_2(g) \rightleftharpoons 2\,HI(g)$ at 500. K
(b) $N_2(g) + 3\,H_2(g) \rightleftharpoons 2\,NH_3(g)$ at 400. K
(c) $CO(g) + 3\,H_2(g) \rightleftharpoons CH_4(g) + H_2O(g)$ at 800. K

85. For each reaction, an equilibrium constant at 298 K is given. Calculate $\Delta G°$ for each reaction.
(a) $Br_2(\ell) + H_2(g) \rightleftharpoons 2\,HBr(g)$ $\qquad K_P = 4.4 \times 10^{18}$
(b) $H_2O(\ell) \rightleftharpoons H_2O(g)$ $\qquad K_P = 3.17 \times 10^{-2}$
(c) $N_2(g) + 3\,H_2(g) \rightleftharpoons 2\,NH_3(g)$ $\qquad K_c = 3.5 \times 10^{8}$

86. For each reaction, an equilibrium constant at 298 K is given. Calculate $\Delta G°$ for each reaction.
(a) $\frac{1}{8}\,S_8(s) + O_2(g) \rightleftharpoons SO_2(g)$ $\qquad K_P = 4.2 \times 10^{52}$
(b) $2\,H_2(g) + O_2(g) \rightleftharpoons 2\,H_2O(g)$ $\qquad K_c = 3.3 \times 10^{81}$
(c) $CH_4(g) + H_2O(g) \rightleftharpoons CO(g) + 3\,H_2(g)$ $\qquad K_c = 9.4 \times 10^{-1}$

Gibbs Free Energy, Maximum Work, and Energy Resources (Section 18.8)

87. Which of these reactions are capable of being harnessed to do useful work at 298 K and 1 bar? Which require that work be done to make them occur?
(a) $2\,C_6H_6(\ell) + 15\,O_2(g) \longrightarrow 12\,CO_2(g) + 6\,H_2O(g)$
(b) $2\,NF_3(g) \longrightarrow N_2(g) + 3\,F_2(g)$
(c) $TiO_2(s) \longrightarrow Ti(s) + O_2(g)$

88. Which of these reactions are capable of being harnessed to do useful work at 298 K and 1 bar? Which require that work be done to make them occur?
(a) $Al_2O_3(s) \longrightarrow 2\,Al(s) + \frac{3}{2}\,O_2(g)$
(b) $2\,CO(g) + O_2(g) \longrightarrow 2\,CO_2(g)$
(c) $C_2H_6(g) \longrightarrow C_2H_4(g) + H_2(g)$

89. For each of the reactions in Question 87 that requires work to be done, calculate the minimum mass of graphite that would have to be oxidized to $CO_2(g)$ to provide the necessary work.

90. For each of the reactions in Question 88 that requires work to be done, calculate the minimum mass of hydrogen gas that would have to be burned to form water vapor to provide the necessary work.

91. Titanium is obtained from its ore, $TiO_2(s)$, by heating the ore in the presence of chlorine gas and coke (carbon) to produce gaseous titanium(IV) chloride and carbon monoxide.
(a) Write a balanced equation for this process.
(b) Calculate $\Delta H°$, $\Delta S°$, and $\Delta G°$ for the reaction.
(c) Is this reaction product-favored or reactant-favored at 25 °C?
(d) Does the reaction become more product-favored or more reactant-favored as the temperature increases?

92. To obtain a metal from its ore, the decomposition of the metal oxide to form the metal and oxygen is often coupled with oxidation of coke (carbon) to carbon monoxide. For each metal oxide listed, write a balanced equation for the decomposition of the oxide and for the overall reaction when the decomposition is coupled to oxidation of coke to carbon monoxide. Calculate the overall value of $\Delta G°$ for each coupled reaction at 25 °C. Which of the

metals could be obtained from these ores at 25 °C by this method?

(a) $CuO(s)$ (b) $Ag_2O(s)$
(c) $HgO(s)$ (d) $MgO(s)$
(e) $PbO(s)$

93. From which of the metal oxides in Question 92 could the metal be obtained by coupling reduction of the oxide with oxidation of coke to carbon monoxide at 800 °C?

94. From which of the metal oxides in Question 92 could the metal be obtained by coupling reduction of the oxide with oxidation of coke to carbon monoxide at 1500 °C?

Gibbs Free Energy in Biological Systems (Section 18.9)

95. The molecular structure of one form of glucose, $C_6H_{12}O_6$, looks like this:

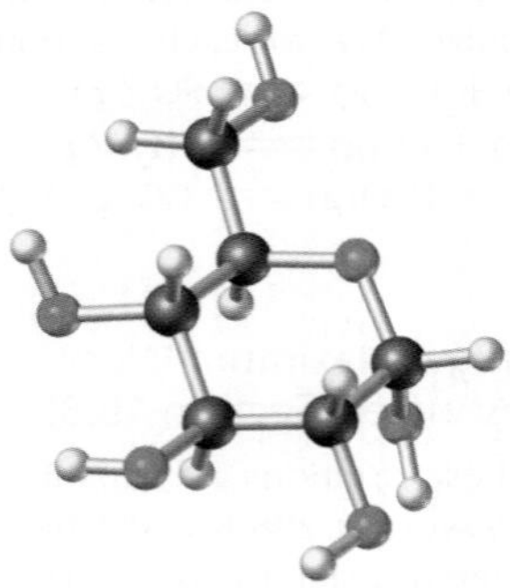

glucose

Glucose can be oxidized to carbon dioxide and water according to the equation

$$C_6H_{12}O_6(s) + 6\,O_2(g) \longrightarrow 6\,CO_2(g) + 6\,H_2O(\ell)$$

(a) Using the method described in Section 8.6 (← p. 342) for estimating enthalpy changes from bond energies, estimate $\Delta H°$ for the oxidation of this form of glucose. Make a list of all bonds broken and all bonds formed in this process.

(b) Compare your result with the experimental value of −2816 kJ for combustion of a mole of glucose. Why might there be a difference between this value and the one you calculated in part (a)?

96. Another step in the metabolism of glucose, which occurs after the formation of glucose 6-phosphate, is the conversion of fructose 6-phosphate to fructose 1,6-bisphosphate ("bis" means *two*):

$$\text{fructose 6-phosphate(aq)} + H_2PO_4^-(aq) \longrightarrow \text{fructose 1,6-bisphosphate (aq)} + H_3O^+(aq)$$

(a) This reaction has a Gibbs free energy change of +16.7 kJ per mole of fructose 6-phosphate. Is it endergonic or exergonic?

(b) Write the equation for the formation of 1 mol ADP from ATP, for which $\Delta G° = -30.5$ kJ.

(c) Couple these two reactions to get an exergonic process; write its overall chemical equation, and calculate the Gibbs free energy change.

97. In muscle cells under the condition of vigorous exercise, glucose is converted to lactic acid ("lactate"), $CH_3CHOHCOOH$, by the chemical reaction

$$C_6H_{12}O_6 \longrightarrow 2\,CH_3CHOHCOOH \qquad \Delta G°' = -197 \text{ kJ}$$

(a) If all of the Gibbs free energy from this reaction were used to convert ADP to ATP, how many moles of ATP could be produced per mole of glucose?

(b) The actual reaction involves the production of 3 mol ATP per mole of glucose. What is the $\Delta G°$ for this reaction?

(c) Is the overall reaction in part (b) reactant-favored or product-favored?

98. The biological oxidation of ethanol, C_2H_5OH, is also a source of Gibbs free energy.

(a) Does the oxidation of 1 g ethanol give more or less energy than the oxidation of 1 g glucose? (*Hint:* Write the balanced equation for the production of carbon dioxide and water from ethanol and oxygen, and use Appendix J.)

(b) Comment on potential problems of replacing glucose with ethanol in your diet.

Conservation of Gibbs Free Energy (Section 18.10)

99. What are the resources human society uses to supply Gibbs free energy? (*Hint:* Consider information you learned in Chapter 6.)

100. For one day, keep a log of all the activities you undertake that consume Gibbs free energy. Distinguish between Gibbs free energy provided by nutrient metabolism and that provided by other energy resources.

Thermodynamic and Kinetic Stability (Section 18.11)

101. Billions of pounds of acetic acid are made each year, much of it by the reaction of methanol with carbon monoxide.

$$CH_3OH(\ell) + CO(g) \longrightarrow CH_3COOH(\ell)$$

(a) By calculating the standard Gibbs free energy change, $\Delta G°$, for this reaction, show that it is product-favored.

(b) Determine the standard Gibbs free energy change, $\Delta G°$, for the reaction of acetic acid with oxygen to form gaseous carbon dioxide and liquid water.

(c) Based on this result, is acetic acid thermodynamically stable compared with $CO_2(g)$ and $H_2O(\ell)$?

(d) Is acetic acid kinetically stable compared with $CO_2(g)$ and $H_2O(\ell)$?

102. Determine the standard Gibbs free energy change, $\Delta G°$, for the reactions of liquid methanol, of CO(g), and of ethyne, $C_2H_2(g)$, with oxygen gas to form gaseous carbon dioxide and (if hydrogen is present) liquid water. Use your calculations to decide which of these substances are kinetically stable and which are thermodynamically stable: $CH_3OH(\ell)$, $CO(g)$, $C_2H_2(g)$, $CO_2(g)$, $H_2O(\ell)$.

103. There are millions of organic compounds known, and new ones are being discovered or made at a rate of more than 100,000 compounds per year. Organic compounds burn readily in air at high temperatures to form carbon dioxide and water. Several classes of organic compounds are listed, with a simple example of each. Write a balanced chemical equation for the combustion in O_2 of each of these compounds, and then use the data in Appendix J to show that each reaction is product-favored at room temperature.

Class of Organics	Simple Example
Aliphatic hydrocarbons	Methane, CH_4
Aromatic hydrocarbons	Benzene, C_6H_6
Alcohols	Methanol, CH_3OH

From these results, it is reasonable to hypothesize that *all* organic compounds are thermodynamically unstable in an oxygen atmosphere (i.e., their room-temperature reaction with O_2(g) to form CO_2(g) and $H_2O(\ell)$ is product-favored). If this hypothesis is true, how can organic compounds exist on Earth?

104. Actually, the carbon in CO_2(g) is thermodynamically unstable with respect to the carbon in calcium carbonate (limestone). Verify this by determining the standard Gibbs free energy change for the reaction of lime, CaO(s), with CO_2(g) to make $CaCO_3$(s).

General Questions

These questions are not explicitly keyed to chapter topics; many require integration of several concepts.

105. This problem will help you understand the dependence of the U.S. economy on energy. Referring to the figure, calculate the energy (in joules) used by the agriculture, mining, and construction industries

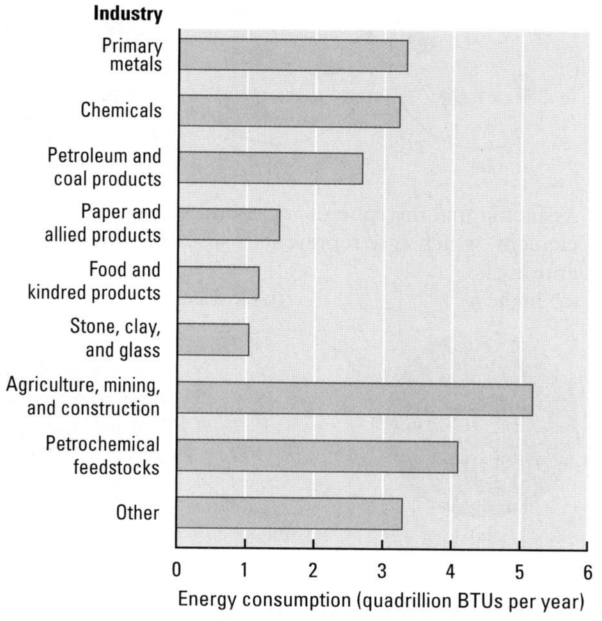

Energy consumption.

(a) in one year.
(b) in one day.
(c) in one second.
(d) Remembering that 1 watt is the expenditure of 1 joule every second, calculate the average power needs of these industries in watts.
(e) Assuming a U.S. population of 307 million people, calculate the power needed by the agriculture, mining, and construction industries *per person in the United States.*

106. Suppose you signed a contract to provide to the agriculture, mining, and construction industries the energy they use each year (see Question 105) by eating glucose and giving them the resulting energy from its oxidation in your body.
(a) How much glucose would you have to eat each day to meet your contract? Assume that it is someone else's job to figure out how to get the energy stored in your ATP to the industries!
(b) An Olympic sprinter uses energy at the rate of 700 to 900 watts in a sprint. Compare this figure with the one you calculated in part (a), and draw conclusions about the feasibility of fulfilling your contract.

Reaction	Chemical Equation	K_c	$\Delta H°$ (kJ)
1	$CH_3OH(g) + H_2(g) \rightleftharpoons CH_4(g) + H_2O(g)$	3.6×10^{20}	−115.4
2	$Mg(OH)_2(s) \rightleftharpoons MgO(s) + H_2O(g)$	1.24×10^{-5}	81.1
3	$2\ CH_4(g) \rightleftharpoons C_2H_6(g) + H_2(g)$	9.5×10^{-13}	64.9
4	$2\ H_2(g) + CO(g) \rightleftharpoons CH_3OH(g)$	3.76	−90.7
5	$H_2(g) + Br_2(g) \rightleftharpoons 2\ HBr(g)$	1.9×10^{24}	−103.7

107. The table above provides data at 25 °C for five reactions. For which (if any) of the reactions 1 through 5 is
(a) K_P greater than K_c?
(b) the reaction product-favored?
(c) there only a single concentration in the K_c expression?
(d) there an increase in the concentrations of products when the temperature increases?
(e) there a change in the sign of $\Delta G°$ if water is liquid instead of gas?

108. The table above provides data at 25 °C for five reactions. For which (if any) of the reactions 1 through 5 is
(a) K_P less than K_c?
(b) there a decrease in the concentrations of products when the pressure increases?
(c) the value of $\Delta S°$ positive?
(d) the sign of $\Delta G°$ dependent on temperature?

109. Consider the gas phase decomposition of sulfur trioxide to sulfur dioxide and oxygen.
(a) Calculate $\Delta G°$ for the reaction at 25 °C.
(b) Is the reaction product-favored under standard conditions at 25 °C?
(c) If the reaction is not product-favored at 25 °C, is there a temperature at which it will become so?
(d) Estimate K_P for the reaction at 1500. °C.
(e) Estimate K_c for the reaction at 1500. °C.

110. The Haber process for the synthesis of ammonia involves the reaction

$$N_2(g) + 3\ H_2(g) \rightleftharpoons 2\ NH_3(g)$$

Using data from Appendix J, estimate the amount (in moles) of NH_3(g) that would be produced from 1 mol N_2(g) and 3 mol H_2(g) once equilibrium is reached at 450 °C and a total pressure of 1000. atm.

111. Mercury is a poison, and its vapor is readily absorbed through the lungs. Therefore it is important that the partial pressure of mercury be kept as low as possible in any area where people could be exposed to it (such as a dentist's office). The relevant equilibrium reaction is

$$Hg(\ell) \rightleftharpoons Hg(g)$$

For Hg(g), $\Delta H_f^\circ = 61.4$ kJ/mol, $S^\circ = 175.0\ J\ K^{-1}\ mol^{-1}$, and $\Delta G_f^\circ = 31.8$ kJ/mol. Use data from Appendix J and these values to evaluate the vapor pressure of mercury at different temperatures. (Remember that concentrations of pure liquids and solids do not appear in the equilibrium constant expression, and for gases K° involves pressures in bars.)
(a) Calculate ΔG° for vaporization of mercury at 25 °C.
(b) Write the equilibrium constant expression for vaporization of mercury.
(c) Calculate K° for this reaction at 25 °C.
(d) What is the vapor pressure of mercury at 25 °C?
(e) Estimate the temperature at which the vapor pressure of mercury reaches 10 mm Hg.

Applying Concepts

These questions test conceptual learning.

112. A friend says that the boiling point of water is twice that of cyclopentane, which boils at 50 °C. Write a brief statement about the validity of this observation.

113. Using the second law of thermodynamics, explain why it is very difficult to unscramble an egg. Who was Humpty-Dumpty? Why did his moment of glory illustrate the second law of thermodynamics?

114. Appendix J lists standard molar entropies S°, not standard entropies of formation ΔS_f°. Why is this possible for entropy but not for internal energy, enthalpy, or Gibbs free energy?

115. When calculating ΔS° from S° values, it is necessary to look up all substances, including elements in their standard state, such as $O_2(g)$, $H_2(g)$, and $N_2(g)$. When calculating ΔH° from ΔH_f° values, however, elements in their standard state can be ignored. Why is the situation different for S° values?

116. In the *Chemistry You Can Do* experiment in Chapter 6 **(← p. 235)** you considered the heat generated when iron rusts to form iron oxide. Look at the enthalpies of formation of other metal oxides in Table 6.3 or Appendix J and comment on your observations. Are oxidations of metals generally endothermic or exothermic? Are they usually reactant-favored or product-favored?

117. Explain why the entropy of the system increases when solid NaCl dissolves in water.

118. Explain how the entropy of the universe increases when an aluminum metal can is made from aluminum ore. The first step is to extract the ore, which is primarily a form of Al_2O_3, from the ground. After it is purified by freeing it from oxides of silicon and iron, aluminum oxide is changed to the metal by an input of electrical energy.

$$2\ Al_2O_3(s) \xrightarrow{\text{electrical energy}} 4\ Al(s) + 3\ O_2(g)$$

119. Suppose that at a certain temperature T a chemical reaction is found to have a standard equilibrium constant K° of 1.0. Indicate whether each statement is true or false and explain why.
(a) The enthalpy change for the reaction, ΔH°, is zero.
(b) The entropy change for the reaction, ΔS°, is zero.
(c) The Gibbs free energy change for the reaction, ΔG°, is zero.
(d) ΔH° and ΔS° have the same sign.
(e) $\Delta H^\circ/T = \Delta S^\circ$ at the temperature T.

120. When you eat a candy bar, how does your body store the Gibbs free energy that is released during oxidation of the sugars (glucose and other carbohydrates) in the candy bar? What was the original source of the Gibbs free energy needed to synthesize the sugars before they went into the candy bar?

121. Explain how biological systems make use of coupled reactions to maintain the high degree of order found in all living organisms.

122. How can kinetically stable substances exist at all, if they are not thermodynamically stable?

123. Criticize this statement: Provided it occurs at an appreciable rate, any chemical reaction for which $\Delta G < 0$ will proceed until all reactants have been converted to products.

124. Reword the statement in Question 123 so that it is always true.

125. Assuming that the particles shown in the figures below represent molecules that have the same molar mass, in which case is the entropy
(a) highest (b) lowest

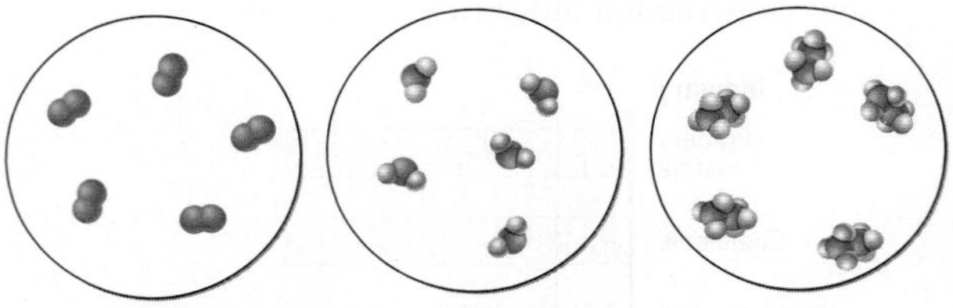

(a) (b) (c)

126. Assuming that the spheres represent atoms of the same element, which case represents a situation where the entropy is
(a) highest (b) lowest

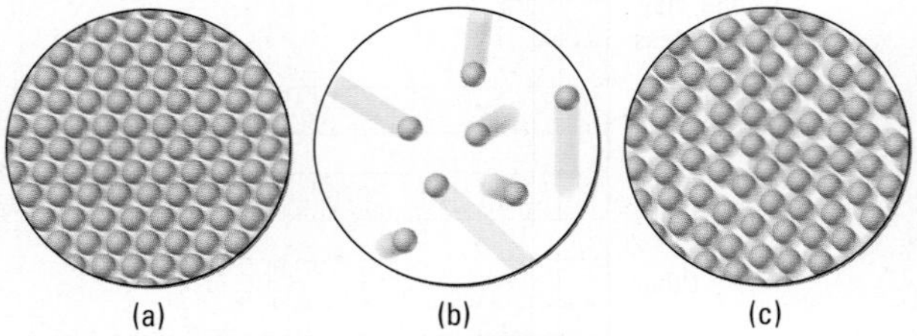

(a) (b) (c)

More Challenging Questions

These questions require more thought and integrate several concepts.

127. Calculate the entropy change for formation of exactly 1 mol of each of these gaseous hydrocarbons under standard conditions from carbon (graphite) and hydrogen. What trend do you see in these values? Does ΔS° increase or decrease on adding H atoms?
(a) acetylene, $C_2H_2(g)$ (b) ethylene, $C_2H_4(g)$
(c) ethane, $C_2H_6(g)$

128. Calcium hydroxide, $Ca(OH)_2(s)$, can be dehydrated to form lime, CaO, by heating. Without doing any calculations, and basing your prediction on the enthalpy change and the entropy change, is this reaction product-favored at 25 °C? Explain your answer briefly.

129. Octane is the product of adding hydrogen to 1-octene.

$$\underset{\text{1-octene}}{C_8H_{16}(g)} + H_2(g) \longrightarrow \underset{\text{octane}}{C_8H_{18}(g)}$$

The enthalpies of formation are

$$\Delta H_f^\circ\ C_8H_{16}(g) = -82.93\ \text{kJ/mol}$$
$$\Delta H_f^\circ\ C_8H_{18}(g) = -208.45\ \text{kJ/mol}$$

Predict whether this reaction is product-favored or reactant-favored at 25 °C and explain your reasoning.

130. This is a group project: Estimate or look up, to the nearest order of magnitude,
(a) the number of kilograms of CH_3OH made each year
(b) the number of kilograms of CO in the entire atmosphere
(c) the number of kilograms of CH_3COOH made each year
(d) the number of kilograms of H_2O on Earth
(e) the number of kilograms of CO_2 in the atmosphere
What do these facts tell you about the difference between kinetic stability and thermodynamic stability?

131. From data in Appendix J, estimate
(a) the boiling point of bromine.
(b) the boiling point of tin(IV) chloride.

132. From data in Appendix J, estimate
(a) the boiling point of titanium(IV) chloride.
(b) the boiling point of carbon disulfide, CS_2, which is a liquid at 25 °C and 1 bar.

133. Nitric oxide and chlorine combine at 25 °C to produce nitrosyl chloride, NOCl.

$$2\ NO(g) + Cl_2(g) \longrightarrow 2\ NOCl(g)$$

(a) Calculate the equilibrium constant K_P for the reaction.
(b) Is the reaction product-favored or reactant-favored?
(c) Calculate the equilibrium constant K_c for the reaction.

134. Hydrogen for use in the Haber-Bosch process for ammonia synthesis is generated from natural gas by the reaction

$$CH_4(g) + H_2O(g) \rightleftharpoons CO(g) + 3\ H_2(g)$$

(a) Calculate ΔG° for this reaction at 25 °C.
(b) Calculate K_P for the reaction at 25 °C.
(c) Is the reaction product-favored under standard conditions? If not, at what temperature will it become so?
(d) Estimate K_c for the reaction at 1000. K.

135. It would be very useful if we could use the inexpensive carbon in coal to make more complex organic molecules such as gaseous or liquid fuels. The formation of methane from coal and water is reactant-favored and thus cannot occur unless there is some energy transfer from outside. This problem examines the feasibility of other reactions using coal and water.
(a) Write three balanced equations for the reactions of coal (carbon) and steam to make ethane gas, $C_2H_6(g)$, propane gas, $C_3H_8(g)$, and liquid methanol, $CH_3OH(\ell)$, with carbon dioxide as a by-product.
(b) Using the data in Appendix J, calculate ΔH°, ΔS°, and ΔG° for each reaction, and then comment on whether any of them would be a feasible way to make the stated products.

136. You are exploring the marketing possibilities of a scheme by which every family in the United States produces enough water for its own needs by the combustion of hydrogen and oxygen. Would the release of Gibbs free energy from the combination of hydrogen and oxygen be sufficient to supply the family's energy needs? Do not try to collect the actual data you would use, but define the problem well enough so that someone else could collect the necessary data and do the calculations that would be needed.

137. Quite often a graph of ln K° versus $1/T$ is a straight line. Use Equation 18.8 (p. 871) to show how ΔH° and ΔS° can be determined from such a graph. Does the fact that such a graph is straight tell you anything about the dependence of ΔH° and ΔS° on temperature?

138. Assuming that ΔH° and ΔS° do not vary with temperature, use Equation 18.8 (p. 871) to derive a formula relating K_1° at temperature T_1 to K_2° at temperature T_2.

139. Without consulting tables of ΔH_f°, S°, or ΔG_f° values, predict which of these reactions is
(i) always product-favored.
(ii) product-favored at low temperatures, but not product-favored at high temperatures.
(iii) not product-favored at low temperatures, but product-favored at high temperatures.
(iv) never product-favored.

(a) $2\ NO_2(g) \longrightarrow N_2O_4(g)$
(b) $C_5H_{12}(g) + 8\ O_2(g) \longrightarrow 5\ CO_2(g) + 6\ H_2O(g)$
(c) $P_4(g) + 10\ F_2(g) \longrightarrow 4\ PF_5(g)$

[*Hint:* Use the qualitative rules regarding bond enthalpies in Section 6.7 (**← p. 238**) to predict the sign of ΔH°.]

140. Using the reactions

$$2\ H_2(g) + O_2(g) \longrightarrow 2\ H_2O(\ell)$$
$$2\ H_2(g) + O_2(g) \longrightarrow 2\ H_2O(g)$$

as an example, explain why it may be incorrect to assume for reactions involving solids or liquids that ΔS° and ΔH° do not change appreciably with increasing temperature.

141. For the reaction

$$CH_3OH(\ell) + \tfrac{3}{2}O_2(g) \longrightarrow 2\ H_2O(\ell) + CO_2(g)$$

the value of ΔG° is -702.35 kJ at 25 °C. Other data at 25 °C are

	ΔH_f° (kJ/mol)	S° (J mol^{-1} K^{-1})
$CH_3OH(\ell)$	−238.66	126.8
$H_2O(\ell)$	−285.83	69.91
$CO_2(g)$	−393.509	213.74

Calculate the standard molar entropy, S°, for $O_2(g)$.

142. The standard equilibrium constant is 2.1×10^9 for this reaction at 25 °C

$$Zn^{2+}(aq) + 4\ NH_3(aq) \rightleftharpoons Zn(NH_3)_4^{2+}(aq)$$

(a) Calculate ΔG° at this temperature.
(b) If standard-state concentrations of the reactants and products are combined, in which direction will the reaction proceed?
(c) Calculate ΔG when $[Zn(NH_3)_4^{2+}] = 0.010$ M, $[Zn^{2+}] = 0.0010$ M and $[NH_3] = 3.5 \times 10^{-4}$ M.

Conceptual Challenge Problems

These rigorous, thought-provoking problems integrate conceptual learning with problem solving and are suitable for group work.

CP18.A (Section 18.2) Suppose that you are invited to play a game as either the "player" or the "house." A pair of dice is used to determine the winner. Each die is a cube having a different number, one through six, showing on each face. The player rolls two dice and sums the numbers showing on the top side of each die to determine the number rolled. Obviously, the number rolled has a minimum value of 2 (both dice showing a 1) and a maximum of 12 (both dice showing a 6). The player begins the game with his or her initial roll of the dice. If the player rolls a 7 or an 11, he or she wins on the first roll and the house loses. If the player does not roll a 7 or an 11 on the initial roll, then whatever number was rolled is called the point, and the player must roll again. For the player to win, he or she must roll the point again before either a 7 or an 11 is rolled. Should the player roll a 7 or an 11 before rolling the point a second time, the house wins. Which would you choose to be, player or house? Explain clearly in terms of the probabilities of rolling the dice why you chose the role you did.

CP18.B (Section 18.2) Suppose a button is placed in the middle of an American football field and a penny is flipped to decide which direction to move the button, up or down the field. Each time the penny comes up heads, the button is moved 10 cm toward your opponent's goal line; and each time it comes up tails, the button is moved 10 cm toward your goal line. Your friend concludes that after many flips of the penny the button is likely to remain within 10 cm of the middle of the field, because numerous flips of the penny will produce heads just as often as tails. You doubt this because you know that perfume molecules and particles diffuse away from their original source, even though, like the button, they are just as likely to be hit from one direction as from any other by the moving molecules around them. How would you explain the error of your friend's conclusion about the movement of the button?

CP18.C (Section 18.3) When thermal energy is transferred to a substance at its standard melting point or boiling point, the substance melts or vaporizes, but its temperature does not change while it is doing so. It is clear then that temperature cannot be a measure of "how much energy is in a sample of matter" or the "intensity of energy in a sample of matter." In "Qualitative Guidelines for Entropy" (p. 857) we noted that atoms and molecules are not stationary, but rather are in constant motion. When heated, their motion increases. If this is true, what can you infer that temperature measures about a sample of matter?

CP18.D (Section 18.10) Suppose that you are a member of an environmental group and have been assigned to evaluate various ways of delivering milk to consumers with respect to Gibbs free energy conservation. Think of all the ways that milk could be delivered, the kinds of containers that could be used, and the ways they could be transported. Consider whether the containers could be reused (refilled) or recycled. Define the problem in terms of the kinds of information you would need to collect, how you would analyze the information, and which criteria you would use to decide which systems are more efficient in use of Gibbs free energy. Do not try to collect the actual data you would use, but define the problem well enough so that someone could collect the necessary data based on your statement of the problem.

CP18.E (Section 18.11) Consider planet Earth as a thermodynamic system. Is Earth thermodynamically or kinetically stable? Discuss your choice, providing as many arguments as you can to support it.

19 Electrochemistry and Its Applications

Mario Tama/Getty Images

Personal electronic devices, such as this iPhone, depend on advanced types of batteries to supply their electrical power; lithium-ion batteries power iPhones. Dependable, long-lasting, light-weight batteries are important for many modern conveniences such as cellular telephones, calculators, flashlights, laptop computers, MP3 players, cordless tools, and hybrid cars. In this chapter we discuss electrochemical reactions and their extraordinary range of applications.

Many of our modern devices are powered by batteries—portable storage devices for electrochemical energy that is released by product-favored redox reactions. Making these reactions do useful work is the goal of significant chemical research. What chemistry goes on inside a battery? How does it produce electricity? Is this chemistry similar to or different from the other kinds of reactions you have studied?

Oxidation-reduction (redox) reactions are an important class of chemical reactions ***(← p. 177)***. Redox reactions involve the transfer of electrons from one atom, molecule, or ion to another. **Electrochemistry** is the study

OWL Sign in to OWL at **www.cengage.com/owl** to view tutorials and simulations, develop problem-solving skills, and complete online homework assigned by your professor.

go Chemistry Download mini lecture videos for key concept review and exam prep from OWL or purchase them from **www.CengageBrain.com.**

Companion Website
Visit this book's companion website at **www.cengage.com/chemistry/moore** to work interactive modules for the Estimation boxes and Active Figures in this text.

of the relationship between electron flow and redox reactions, that is, the relationship between electricity and chemical changes. Applications of electrochemistry are numerous and important. In electrochemical cells (commonly called batteries), electrons from a product-favored redox reaction are released and transferred as an electrical current through an external circuit. The current depends on the amount of reactants converted to products, and how fast they are converted. We rely on batteries to power many useful devices, including CD players, cellular telephones, calculators, flashlights, laptop computers, heart pacemakers, golf carts, and hybrid cars.

The voltage produced by an electrochemical reaction depends on the strengths of the oxidizing agents and reducing agents used as reactants. A knowledge of the strengths of oxidizing and reducing agents helps in the design of better batteries. Product-favored electrochemical reactions are not always beneficial, however. Corrosion of iron, for example, is a product-favored redox reaction. Damage to materials as a result of corrosion is very costly, so preventing corrosion is an important goal.

By contrast, electroplating and electrolysis are applications of reactant-favored redox reactions. In an electrolysis cell, an external energy source creates an electrical current that forces a reactant-favored process to occur. Electrolysis is important in the manufacture of many products, such as aluminum metal and the chlorine used to disinfect water supplies.

19.1 Redox Reactions

Redox reactions form a large class of chemical reactions in which the reactants can be atoms, ions, or molecules. How do you know when a reaction involves oxidation-reduction?

- ***By identifying the presence of strong oxidizing or reducing agents as reactants (Table 5.3, ← p. 181).***
- ***By recognizing a change in oxidation number (← p. 183).*** This means you have to determine the oxidation number of each element as it appears in a reactant or a product.
- ***By recognizing the presence of an uncombined element as a reactant or product.*** Producing a free element or incorporating one into a compound almost always results in a change in oxidation number.

An uncombined element is always assigned an oxidation number of 0.

You may want to review the definitions of oxidation and reduction in Section 5.3 (***← p. 177***) and the rules for assigning oxidation numbers in Section 5.4.

To briefly review the definitions of oxidation and reduction, consider the displacement reaction between magnesium, a relatively reactive metal ***(Table 5.5, ← p. 187)***, and hydrochloric acid. The oxidation numbers of the species are shown above their symbols.

H$^+$ is reduced

$$\overset{0}{Mg(s)} + 2\ \overset{+1-1}{HCl(aq)} \longrightarrow \overset{+2-1}{MgCl_2(aq)} + \overset{0}{H_2(g)}$$

Mg is oxidized

Oxidation: loss of electron(s) and increase in oxidation number

Reduction: gain of electron(s) and decrease in oxidation number

The presence of the uncombined elements Mg and H_2 indicates a redox reaction, as do the changes in oxidation number. Mg(s) is oxidized, indicated by an *increase* in its oxidation number (from 0 to +2). Hydrogen ions from HCl are reduced to H_2, as shown by a *decrease* in oxidation number (from +1 to 0). Magnesium metal is the reducing agent because it causes the H^+ ions in HCl to be reduced (gain electrons). In the process of giving electrons to HCl, metallic Mg is oxidized to Mg^{2+} ions.

Hydrochloric acid is the oxidizing agent because it causes magnesium metal to be oxidized (lose electrons). In this and all redox reactions

- ***The oxidizing agent is reduced.***
- ***The reducing agent is oxidized.***

Oxidation and reduction *always* occur together, with one reactant acting as the oxidizing agent and another acting as the reducing agent. Oxidizing and reducing agents are *always* reactants, never products.

Redox reactions such as the one between Mg and HCl involve complete gain and loss of electrons by the reacting species—the type of reaction that is utilized in electrochemistry. A flow of electrons through the reaction system to make a complete electrical circuit is necessary for the reaction to proceed.

PROBLEM-SOLVING EXAMPLE 19.1 Identifying Oxidizing and Reducing Agents in Redox Reactions

In the thermite reaction, iron(III) oxide and aluminum metal react to give iron metal and aluminum oxide:

$$Fe_2O_3(s) + 2\,Al(s) \longrightarrow 2\,Fe(s) + Al_2O_3(s)$$

Is this a redox reaction? Identify the oxidation numbers of all the atoms that change oxidation number. What is reduced? What is oxidized? What is the oxidizing agent? What is the reducing agent?

Answer Yes, this is a redox reaction. Oxidation number changes: Fe, +3 to 0; Al, 0 to +3. The Fe^{3+} in iron(III) oxide is reduced to metallic iron. The aluminum metal is oxidized to Al^{3+} ions. The oxidizing agent is Fe^{3+} in iron(III) oxide and the reducing agent is aluminum metal.

Strategy and Explanation

First, determine the oxidation number of each element on the reactant side of the equation. Oxygen in compounds is normally −2 ***(← p. 181)***. Since the sum of the oxidation numbers of all the atoms in a formula must equal the charge on the formula, each iron in Fe_2O_3 is +3 and each oxygen is −2. The oxidation number for metallic Al is 0, as it is for all uncombined elements.

On the product side, the Al in Al_2O_3 has an oxidation number of +3. Iron is now in its uncombined form, so its oxidation number is now 0.

Thus, the oxidation number changes are

$$\overset{+3}{Fe_2}O_3(s) + 2\,\overset{0}{Al}(s) \longrightarrow 2\,\overset{0}{Fe}(s) + \overset{+3}{Al_2}O_3(s)$$

The oxidation number of iron decreased, while the oxidation number of aluminum increased. Thus, Fe^{3+} in Fe_2O_3 was reduced (+3 to 0) and aluminum was oxidized (0 to +3). Consequently, Fe^{3+} in Fe_2O_3 is the oxidizing agent and Al metal is the reducing agent. Oxygen is neither reduced nor oxidized in this reaction; its oxidation number remains −2.

PROBLEM-SOLVING PRACTICE 19.1

Give the oxidation number for each atom and identify the oxidizing and reducing agents in these balanced chemical equations.

(a) $2\,Fe(s) + 3\,Cl_2(g) \longrightarrow 2\,FeCl_3(s)$
(b) $2\,H_2(g) + O_2(g) \longrightarrow 2\,H_2O(\ell)$
(c) $Cu(s) + 2\,NO_3^-(aq) + 4\,H_3O^+(aq) \longrightarrow Cu^{2+}(aq) + 2\,NO_2(g) + 6\,H_2O(\ell)$
(d) $C(s) + O_2(g) \longrightarrow CO_2(g)$
(e) $6\,Fe^{2+}(aq) + Cr_2O_7^{2-}(aq) + 14\,H_3O^+(aq) \longrightarrow 6\,Fe^{3+}(aq) + 2\,Cr^{3+}(aq) + 21\,H_2O(\ell)$

The **PROBLEM-SOLVING STRATEGY** in this book is

- Analyze the problem
- Plan a solution
- Execute the plan
- Check that the result is reasonable

Appendix A.1 explains this in detail.

The thermite reaction in progress.

PROBLEM-SOLVING PRACTICE answers are provided at the back of this book in Appendix K.

19.2 Using Half-Reactions to Understand Redox Reactions

Consider the redox reaction between zinc metal and copper(II) ions shown in Figure 19.1. The net ionic equation is

reduction

$$Zn(s) + Cu^{2+}(aq) \longrightarrow Zn^{2+}(aq) + Cu(s)$$

oxidation

Zinc metal is oxidized to Zn^{2+} ions, and Cu^{2+} ions are reduced to copper metal. As Cu^{2+} ions are converted to Cu(s) in this half-reaction, the blue color of the solution becomes less intense and metallic copper forms on the zinc surface.

To see more clearly how electrons are transferred, this overall reaction can be thought of as the result of two simultaneous **half-reactions:** one half-reaction for the oxidation of Zn and one half-reaction for the reduction of Cu^{2+} ions. The oxidation half-reaction

$$Zn(s) \longrightarrow Zn^{2+}(aq) + 2\,e^-$$

Note that the sum of the charges on the left side of the reaction equals the sum of the charges on the right side, even for half-reactions. Otherwise, conservation of matter would be violated—electrons would be created or destroyed.

shows that each zinc atom loses two electrons when it is oxidized to a Zn^{2+} ion. These two electrons are accepted by a Cu^{2+} ion in the reduction half-reaction,

$$Cu^{2+}(aq) + 2\,e^- \longrightarrow Cu(s)$$

The net reaction is the sum of the oxidation and reduction half-reactions.

$$\begin{array}{rcll} Zn(s) & \longrightarrow & Zn^{2+}(aq) + 2\,e^- & \text{(oxidation half-reaction)} \\ Cu^{2+}(aq) + 2\,e^- & \longrightarrow & Cu(s) & \text{(reduction half-reaction)} \\ \hline Zn(s) + Cu^{2+}(aq) & \longrightarrow & Zn^{2+}(aq) + Cu(s) & \text{(net reaction)} \end{array}$$

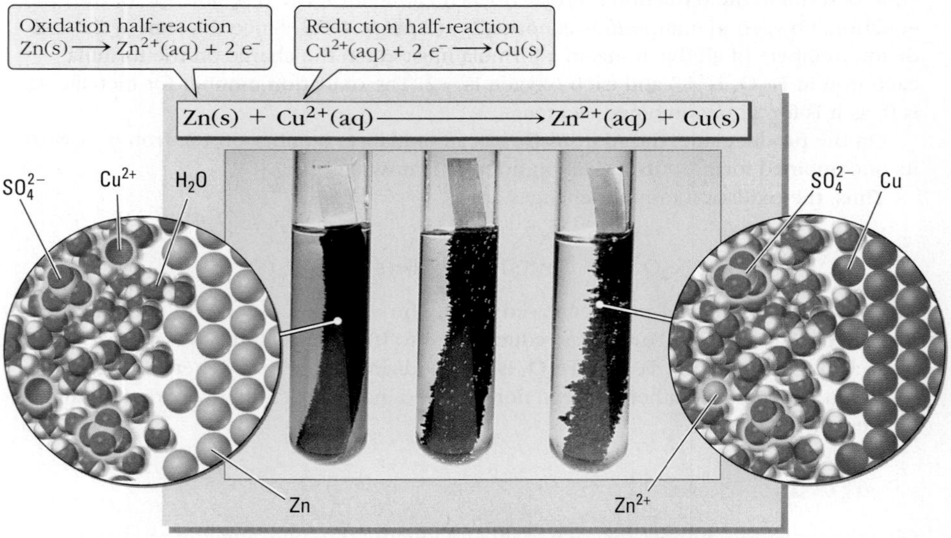
Oxidation half-reaction
Zn(s) ⟶ Zn²⁺(aq) + 2 e⁻
Reduction half-reaction
Cu²⁺(aq) + 2 e⁻ ⟶ Cu(s)
Zn(s) + Cu²⁺(aq) ⟶ Zn²⁺(aq) + Cu(s)
SO₄²⁻
Cu²⁺
H₂O
Zn
Zn²⁺
SO₄²⁻
Cu

Photos: © Cengage Learning/Charles D. Winters

Active Figure 19.1 An oxidation-reduction reaction. A strip of zinc has been placed in a solution of copper(II) sulfate (*left*). As time passes, the zinc metal reacts with the Cu^{2+} ions to produce copper metal (the brown-colored deposit on the zinc strip) and Zn^{2+} ions in solution.

$$Zn(s) + Cu^{2+}(aq) \longrightarrow Zn^{2+}(aq) + Cu(s)$$

As copper metal accumulates on the zinc strip, the blue color due to the aqueous Cu^{2+} ions gradually fades (*middle* and *right*) as Cu^{2+} ions are reduced to metallic copper. The Zn^{2+} ions in aqueous solution are colorless. Visit this book's companion website at **www.cengage.com/chemistry/moore** to test your understanding of the concepts in this figure.

Notice that no electrons appear in the equation for the net reaction because the number of electrons produced by the oxidation half-reaction must equal the number of electrons gained by the reduction half-reaction. This must always be true in a net reaction. Otherwise, electrons would be created or destroyed, violating the laws of conservation of mass and conservation of electrical charge.

Consider another example, as shown in Figure 19.2. A metallic copper screen is immersed in a solution of silver nitrate. As the reaction proceeds, the solution gradually turns blue, and fine, silvery, hair-like crystals form on the copper screen. Knowing that Cu^{2+} ions in aqueous solution appear blue, we can conclude that the copper metal is being oxidized to Cu^{2+}. Reduction must also be taking place, so it is reasonable to conclude that the hair-like crystals of silver result from the reduction of Ag^+ ions to metallic silver. The two half-reactions are

$$Cu(s) \longrightarrow Cu^{2+}(aq) + 2\,e^- \qquad \text{(oxidation half-reaction)}$$

$$Ag^+(aq) + e^- \longrightarrow Ag(s) \qquad \text{(reduction half-reaction)}$$

In this case, two electrons are produced in the oxidation half-reaction, but only one is needed for the reduction half-reaction. *One* atom of copper provides two electrons to reduce *two* Ag^+ ions, so the reduction half-reaction must occur twice every time the oxidation half-reaction occurs once. To indicate this relationship, we multiply the reduction half-reaction by 2.

$$2\,Ag^+(aq) + 2\,e^- \longrightarrow 2\,Ag(s) \qquad \text{(reduction half-reaction)} \times 2$$

Adding this reduction half-reaction to the oxidation half-reaction gives the net equation

$$\begin{array}{rcl} 2\,Ag^+(aq) + 2\,e^- & \longrightarrow & 2\,Ag(s) \\ Cu(s) & \longrightarrow & Cu^{2+}(aq) + 2\,e^- \\ \hline Cu(s) + 2\,Ag^+(aq) & \longrightarrow & Cu^{2+}(aq) + 2\,Ag(s) \end{array}$$

The method shown here is a general one. *A net equation can always be generated by writing oxidation and reduction half-reactions, using coefficients to adjust the half-reaction equations so the number of electrons released by the oxidation equals the number gained by the reduction, and then adding the two half-reactions to give the equation for the net reaction.*

© Cengage Learning/Charles D. Winters

Figure 19.2 **Copper metal screen in a solution of $AgNO_3$.** The blue color intensifies as more copper is oxidized to aqueous Cu^{2+} ion. Ag^+ ions are reduced to silver metal.

PROBLEM-SOLVING EXAMPLE 19.2 Determining Half-Reactions from Net Redox Reactions

Aluminum metal undergoes a redox reaction with $Zn^{2+}(aq)$ to produce $Al^{3+}(aq)$ and zinc metal.

(unbalanced equation) $Al(s) + Zn^{2+}(aq) \longrightarrow Al^{3+}(aq) + Zn(s)$

Write the oxidation half-reaction, the reduction half-reaction equations, and combine them to give the balanced equation for the net reaction.

Answer

Oxidation half-reaction: $Al(s) \longrightarrow Al^{3+}(aq) + 3\,e^-$
Reduction half-reaction: $Zn^{2+}(aq) + 2\,e^- \longrightarrow Zn(s)$
Net reaction: $2\,Al(s) + 3\,Zn^{2+}(aq) \longrightarrow 2\,Al^{3+}(aq) + 3\,Zn(s)$

Strategy and Explanation To identify the half-reactions, we first identify the species whose oxidation number increases (it is oxidized) and the species whose oxidation number decreases (it is reduced). The oxidation number of aluminum increases from 0 to +3, so aluminum is oxidized.

$$Al(s) \longrightarrow Al^{3+}(aq) + 3\,e^-$$

This half-reaction must have three electrons on the right side to balance the 3+ charge of the aluminum ion and yield equal charges on the two sides of the equation.

In this book, oxidation numbers are written as +1, +2, etc., whereas charges on ions are written as 1+, 2+, etc.

The oxidation number of zinc decreases from +2 to 0, so zinc is reduced.

$$Zn^{2+}(aq) + 2\,e^- \longrightarrow Zn(s)$$

This half-reaction must have two electrons on the left side to balance the 2+ charge of the zinc ion and yield equal charges on the two sides of the equation.

These two half-reactions include different numbers of electrons. The two half-reactions are multiplied by 2 and 3, respectively, so that six electrons appear in each half-reaction.

$$2\,[Al(s) \longrightarrow Al^{3+}(aq) + 3\,e^-] \text{ gives } 2\,Al(s) \longrightarrow 2\,Al^{3+}(aq) + 6\,e^-$$
$$3\,[Zn^{2+}(aq) + 2\,e^- \longrightarrow Zn(s)] \text{ gives } 3\,Zn^{2+}(aq) + 6\,e^- \longrightarrow 3\,Zn(s)$$
$$\text{Net reaction: } 2\,Al(s) + 3\,Zn^{2+}(aq) \longrightarrow 2\,Al^{3+}(aq) + 3\,Zn(s)$$

The net reaction is the sum of these two half-reactions, each multiplied by the proper whole number to make the number of electrons produced by the oxidation half-reaction equal to the number of electrons gained by the reduction half-reaction.

PROBLEM-SOLVING PRACTICE 19.2

Write oxidation and reduction half-reactions for these unbalanced net redox equations. For each, show that their sum is the net reaction.

(a) $Cd(s) + Cu^{2+}(aq) \longrightarrow Cd^{2+}(aq) + Cu(s)$

(b) $Mg(s) + Cr^{3+}(aq) \longrightarrow Mg^{2+}(aq) + Cr(s)$

(c) $Fe^{3+}(aq) + Cu^{+}(aq) \longrightarrow Fe^{2+}(aq) + Cu^{2+}(aq)$

go Chemistry **Module 25: Oxidation-Reduction Reactions** covers concepts in this section.

Balancing Redox Equations Using Half-Reactions

Two of the equations in Problem-Solving Practice 19.2 are unbalanced. While these particular redox equations could be easily balanced by inspection, this is not always the case. Equations for redox reactions often involve water, hydronium ions, or hydroxide ions as reactants or products. It is difficult to tell by observing the unbalanced equation how many H_2O, H_3O^+, and OH^- are involved, or whether they will be reactants, or products, or even present at all. Fortunately, there are systematic ways to figure this out.

Oxalic acid, HOOC—COOH, is the simplest organic acid containing two carboxylic acid groups.

Balancing Redox Equations in Acidic Solutions Consider the reaction of permanganate ion with oxalic acid in an acidic solution. The products are manganese(II) ions and carbon dioxide, so the *unbalanced* equation is

$$\text{(unbalanced equation)} \quad \underset{\text{oxalic acid}}{H_2C_2O_4(aq)} + \underset{\text{permanganate ion}}{MnO_4^-(aq)} \longrightarrow Mn^{2+}(aq) + CO_2(g)$$

If you try to balance this equation by trial and error, you will almost certainly have a hard time balancing hydrogen and oxygen. You have probably already noticed that no hydrogen-containing species appears on the product side of the unbalanced equation. Because the reaction takes place in an aqueous acidic solution, water and hydronium ions can be involved. Generating the balanced equation for a reaction like this one is best done by following a series of steps. In each step you must use what you know about the oxidation and reduction half-reactions, as well as conservation of matter and conservation of electrical charge. Problem-Solving Example 19.3 illustrates the steps that produce a balanced equation for a redox reaction occurring in acidic solution.

PROBLEM-SOLVING EXAMPLE 19.3 Balancing Redox Equations for Reactions in Acidic Solutions

Balance the equation for the oxidation of oxalic acid in an acidic permanganate solution. The products of this reaction are CO_2 and Mn^{2+} ions.

Answer $5\ H_2C_2O_4(aq) + 6\ H_3O^+(aq) + 2\ MnO_4^-(aq) \longrightarrow 10\ CO_2(g) + 2\ Mn^{2+}(aq) + 14\ H_2O(\ell)$

Strategy and Explanation It is best to follow a systematic approach listed in the steps below to balance the equation for this reaction (and all similar redox reactions).

Step 1: *Recognize whether the reaction is an oxidation-reduction process. If it is, then determine what is reduced and what is oxidized.* This is a redox reaction because the oxidation number of Mn changes from +7 in MnO_4^- to +2 in Mn^{2+}, so the Mn in MnO_4^- is reduced. The oxidation number of each C changes from +3 in $H_2C_2O_4$ to +4 in CO_2, so the C in $H_2C_2O_4$ is oxidized. The oxidation numbers of H (+1) and O (−2) are unchanged.

Step 2: *Break the overall unbalanced equation into half-reactions.*

$$H_2C_2O_4(aq) \longrightarrow CO_2(g) \qquad \text{(oxidation half-reaction)}$$

$$MnO_4^-(aq) \longrightarrow Mn^{2+}(aq) \qquad \text{(reduction half-reaction)}$$

Step 3: *Balance the atoms in each half-reaction.* First balance all atoms except for O and H, then balance O by adding H_2O and balance H by adding H^+. (Hydroxide ion, OH^-, cannot be used here because the reaction occurs in an acidic solution and the OH^- concentration is very low.)

In acidic solution, balance O by adding H_2O, and balance H by adding H^+.

Oxalic acid half-reaction: First, balance the carbon atoms in the half-reaction.

$$H_2C_2O_4(aq) \longrightarrow 2\ CO_2(g)$$

This step balances the O atoms as well (no H_2O needed here), so only H atoms remain to be balanced. Because the product side is deficient by two H, we put $2\ H^+$ there.

$$H_2C_2O_4(aq) \longrightarrow 2\ CO_2(g) + 2\ H^+(aq) \qquad \text{(oxalic acid half-reaction)}$$

Strictly speaking, we ought to use H_3O^+ instead of H^+, but this would result in adding water molecules to each side of the equation, which is rather cumbersome. It is simpler to add H^+ now and add the water molecules at the end.

Permanganate half-reaction: The Mn atoms are already balanced, but the oxygen atoms are not balanced until H_2O is added. Adding $4\ H_2O$ on the product side takes care of the needed oxygen atoms.

$$MnO_4^-(aq) \longrightarrow Mn^{2+}(aq) + 4\ H_2O(\ell)$$

Now there are 8 H atoms on the right and none on the left. To balance hydrogen atoms, $8\ H^+$ are placed on the left side of the half-reaction.

$$8\ H^+(aq) + MnO_4^-(aq) \longrightarrow Mn^{2+}(aq) + 4\ H_2O(\ell) \qquad \text{(permanganate half-reaction)}$$

Step 4: *Balance the half-reactions for charge using electrons (e^-).* The oxalic acid half-reaction has a net charge of 0 on the left side and 2+ on the right. The reactants have lost two electrons. To show this fact, $2\ e^-$ must appear on the right side.

$$H_2C_2O_4(aq) \longrightarrow 2\ CO_2(g) + 2\ H^+(aq) + 2\ e^-$$

This confirms that $H_2C_2O_4$ is the reducing agent (it loses electrons and is oxidized). The loss of two electrons is also in keeping with the increase in the oxidation number of each of two C atoms by 1, from +3 to +4. The $2\ e^-$ also balance the charge on the product side of the equation.

The MnO_4^- half-reaction has a charge of 7+ on the left and 2+ on the right. Therefore, to achieve a net 2+ charge on each side, $5\ e^-$ must appear on the left. The gain of electrons shows that MnO_4^- is the oxidizing agent; it is reduced.

$$5\ e^- + 8\ H^+(aq) + MnO_4^-(aq) \longrightarrow Mn^{2+}(aq) + 4\ H_2O(\ell)$$

Step 5: *Multiply the half-reactions by appropriate factors so that the oxidation half-reaction produces as many electrons as the reduction half-reaction accepts.* In this case, one half-reaction involves two electrons, and the other half-reaction involves five electrons. It takes ten electrons to balance the overall reaction. The oxalic acid half-reaction must be multiplied by 5, and the permanganate half-reaction by 2.

$$5\ [H_2C_2O_4(aq) \longrightarrow 2\ CO_2(g) + 2\ H^+(aq) + 2\ e^-]$$

$$2\ [5\ e^- + 8\ H^+(aq) + MnO_4^-(aq) \longrightarrow Mn^{2+}(aq) + 4\ H_2O(\ell)]$$

© Cengage Learning/Charles D. Winters

(a) At start

(b) Upon completion

Reaction of oxalic acid with permanganate in acidic solution. The purple color of the permanganate diminishes as the reaction proceeds.

***Step 6:** Add the half-reactions to give the overall reaction and cancel equal amounts of reactants and products that appear on both sides of the arrow.*

$$5\,H_2C_2O_4(aq) \longrightarrow 10\,CO_2(g) + 10\,H^+(aq) + 10\,e^-$$
$$10\,e^- + 16\,H^+(aq) + 2\,MnO_4^-(aq) \longrightarrow 2\,Mn^{2+}(aq) + 8\,H_2O(\ell)$$
$$5\,H_2C_2O_4(aq) + 16\,H^+(aq) + 2\,MnO_4^-(aq) \longrightarrow 10\,CO_2(g) + 10\,H^+(aq) + 2\,Mn^{2+}(aq) + 8\,H_2O(\ell)$$

Since 16 H^+ appear on the left and 10 H^+ appear on the right, 10 H^+ are canceled, leaving 6 H^+ on the left.

$$5\,H_2C_2O_4(aq) + 6\,H^+(aq) + 2\,MnO_4^-(aq) \longrightarrow 10\,CO_2(g) + 2\,Mn^{2+}(aq) + 8\,H_2O(\ell)$$

***Step 7:** Check the balanced equation to make sure both atoms and charge are balanced.*

Atom balance: Each side of the equation has 2 Mn, 28 O, 10 C, and 16 H atoms.

Charge balance: Each side has a net charge of 4+.
On the left side, (6 × 1+) + (2 × 1−) = 4+.
On the right side, 2(2+) = 4+.

***Step 8:** Add enough water molecules to both sides of the equation to convert all H^+ to H_3O^+.* In this case, six water molecules are needed ($6\,H_2O + 6\,H^+ \longrightarrow 6\,H_3O^+$). Six water molecules are added to each side of the equation, which increases the total to 14 on the product side.

$$5\,H_2C_2O_4(aq) + 6\,H_3O^+(aq) + 2\,MnO_4^-(aq) \longrightarrow 10\,CO_2(g) + 2\,Mn^{2+}(aq) + 14\,H_2O(\ell)$$

***Step 9:** Check the final result to make sure both atoms and charges are balanced.* The equation is balanced. In the final, balanced equation, the net charges on each side of the reaction are the same, and the numbers of atoms of each kind on each side of the reaction are equal.

PROBLEM-SOLVING PRACTICE 19.3

Balance this equation for the reaction of Zn with $Cr_2O_7^{2-}$ in acidic aqueous solution.

$$Zn(s) + Cr_2O_7^{2-}(aq) \longrightarrow Cr^{3+}(aq) + Zn^{2+}(aq)$$

CONCEPTUAL EXERCISE 19.1 Electrons Lost Equal Electrons Gained

Explain why the number of electrons lost must always equal the number gained in a redox reaction.

Answers to **EXERCISES** are provided at the back of this book in Appendix L.

EXERCISES that are labeled **CONCEPTUAL** are designed to test your understanding of one or more concepts; they usually involve qualitative rather than quantitative thinking.

Balancing Redox Equations in Basic Solutions For redox reactions that occur in basic solutions, the final electrochemical reaction must be completed with water and OH^- ions rather than water and H_3O^+ ions that we used for acidic solutions. During the balancing process, the half-reactions can be treated as if they occurred in acidic solution. Then, at the end of the series of steps, the H^+ ions can be neutralized by adding an equal number of OH^- ions to both sides of the electrochemical equation and, if necessary, canceling water molecules that appear as both reactants and products. Problem-Solving Example 19.4 illustrates how to balance a redox reaction in basic solution.

PROBLEM-SOLVING EXAMPLE 19.4 Balancing Redox Equations for Reactions in Basic Solutions

In a nickel-cadmium (nicad) battery, cadmium metal and nickel(III) oxide, Ni_2O_3 react in alkaline solution to form $Cd(OH)_2$ and $Ni(OH)_2$. Write the balanced equation for this reaction.

Answer $Cd(s) + Ni_2O_3(s) + 3\ H_2O(\ell) \longrightarrow Cd(OH)_2(s) + 2\ Ni(OH)_2(s)$

Strategy and Explanation Follow the systematic approach listed in the steps below to balance the equation for this reaction.

Step 1: *Recognize whether the reaction is an oxidation-reduction process. Then determine what is reduced and what is oxidized.* This is a redox reaction because the oxidation number of Cd changes from 0 in Cd metal to +2 in $Cd(OH)_2$, so Cd metal is oxidized. The oxidation number of each Ni changes from +3 in Ni_2O_3 to +2 in $Ni(OH)_2$, so the Ni is reduced.

Step 2: *Break the overall unbalanced equation into half-reactions.*

$$Cd(s) \longrightarrow Cd(OH)_2(s) \qquad \text{(oxidation half-reaction)}$$
$$Ni_2O_3(s) \longrightarrow Ni(OH)_2(s) \qquad \text{(reduction half-reaction)}$$

Step 3: *Balance the atoms in each half-reaction.* First, balance all atoms except the O and H atoms. Balance each half-reaction as if it were in an *acidic* solution; that is, balance O by adding H_2O and balance H by adding H^+.

In the Cd half-reaction, the Cd atoms are balanced. Adding two water molecules on the left balances the two O atoms on the right, but this leaves four H atoms on the left and only two on the right. Adding two H^+ ions on the right balances H atoms in this half-reaction.

$$2\ H_2O(\ell) + Cd(s) \longrightarrow Cd(OH)_2(s) + 2\ H^+(aq)$$

For the Ni_2O_3 half-reaction, a coefficient of 2 is needed for $Ni(OH)_2$ because there are two Ni atoms on the left. To balance the four O atoms and four H atoms now on the right with the three O atoms on the left requires one water molecule and two H^+ ions on the left.

$$2\ H^+(aq) + H_2O(\ell) + Ni_2O_3(s) \longrightarrow 2\ Ni(OH)_2(s)$$

Step 4: *Balance the half-reactions for charge using electrons.* The Cd half-reaction has two positive charges on the right, which requires 2 e^- as a product.

$$2\ H_2O(\ell) + Cd(s) \longrightarrow Cd(OH)_2(s) + 2\ H^+(aq) + 2\ e^- \qquad \text{(balanced)}$$

The Ni_2O_3 half-reaction requires 2 e^- as a reactant.

$$2\ H^+(aq) + H_2O(\ell) + Ni_2O_3(s) + 2\ e^- \longrightarrow 2\ Ni(OH)_2(s) \qquad \text{(balanced)}$$

Step 5: *Multiply the half-reactions by appropriate factors so that the reducing agent produces as many electrons as the oxidizing agent accepts.* The Cd half-reaction produces two electrons, and the Ni_2O_3 half-reaction accepts two, so the number of electrons lost in the oxidation equals the number gained in the reduction.

Step 6: *Because H^+ does not exist at any appreciable concentration in a basic solution, remove H^+ by adding an appropriate amount of OH^- to both sides of the equation. H^+ and OH^- react to form H_2O.* In the Cd half-reaction, add two OH^- ions to each side to get

$$2\ OH^-(aq) + 2\ H_2O(\ell) + Cd(s) \longrightarrow Cd(OH)_2(s) + 2\ H_2O(\ell) + 2\ e^-$$

(On the product side, two OH^- ions plus two H^+ ions form two H_2O molecules.) For the Ni_2O_3 half-reaction, add two OH^- ions to each side to get

$$3\ H_2O(\ell) + Ni_2O_3(s) + 2\ e^- \longrightarrow 2\ Ni(OH)_2(s) + 2\ OH^-(aq)$$

Step 7: *Add the half-reactions to give the overall reaction, and cancel reactants and products that appear in equal amounts on both sides of the reaction arrow.*

$$\cancel{2\ OH^-(aq)} + \cancel{2\ H_2O(\ell)} + Cd(s) \longrightarrow Cd(OH)_2(s) + \cancel{2\ H_2O(\ell)} + 2\ e^-$$
$$3\ H_2O(\ell) + Ni_2O_3(s) + 2\ e^- \longrightarrow 2\ Ni(OH)_2(s) + \cancel{2\ OH^-(aq)}$$
$$\overline{Cd(s) + Ni_2O_3(s) + 3\ H_2O(\ell) \longrightarrow Cd(OH)_2(s) + 2\ Ni(OH)_2(s)}$$

Step 8: *Check the final result to make sure both atoms and charge are balanced.* The equation is balanced. In the final equation, there are no net charges on either side of the reaction arrow, and the numbers of atoms of each kind on each side of the reaction arrow are equal.

PROBLEM-SOLVING PRACTICE 19.4

In a basic solution, aluminum metal forms $[Al(OH)_4]^-$ ion as the metal reduces NO_3^- ion to NH_3. Write the balanced equation for this reaction using the steps outlined in Problem-Solving Example 19.4.

19.3 Electrochemical Cells

Strictly speaking, a battery consists of several voltaic cells connected together, but the term "battery" has taken on the same meaning as "voltaic cell."

Electrochemical cells are sometimes referred to as *galvanic cells* in recognition of the work of Luigi Galvani, who discovered that a frog's leg placed in salt water would twitch when the leg was touched simultaneously by two dissimilar metals.

To identify the anode and cathode, remember that Oxidation takes place at the Anode (both words begin with vowels), and Reduction takes place at the Cathode (both words begin with consonants).

It is easy to see by the color changes in the two redox reactions shown in Figures 19.1 and 19.2 that these reactions favor the formation of products—as soon as the reactants are mixed, changes take place. All product-favored reactions release Gibbs free energy (← *p. 864*), energy that can do useful work. An electrochemical cell is a way to capture that useful work as electrical work.

In a redox reaction, electrons are *transferred* from one kind of atom, molecule, or ion to another. In an **electrochemical cell,** an oxidizing agent and a reducing agent are arranged in such a way that they can react only if electrons flow through an outside conductor. Such electrochemical cells are also known as **voltaic cells** or **batteries.** Figure 19.3 diagrams how a voltaic cell can be made from the Zn/Cu^{2+} reaction shown previously in Figure 19.1. The two half-reactions occur in separate beakers, each of which is called a **half-cell.** When Zn atoms are oxidized, the electrons that are given up by zinc metal pass through the wire and a light bulb to the copper metal. There the electrons are available to reduce Cu^{2+} ions from the solution. The metallic zinc and copper strips are called electrodes. An **electrode** conducts electrical current (electrons) into or out of something, in this case, a solution. An electrode is most often a metal plate or wire, but it can also be a piece of graphite or some other electrical conductor. An electrochemical cell contains two electrodes:

- **Anode:** the electrode where oxidation occurs
- **Cathode:** the electrode where reduction occurs

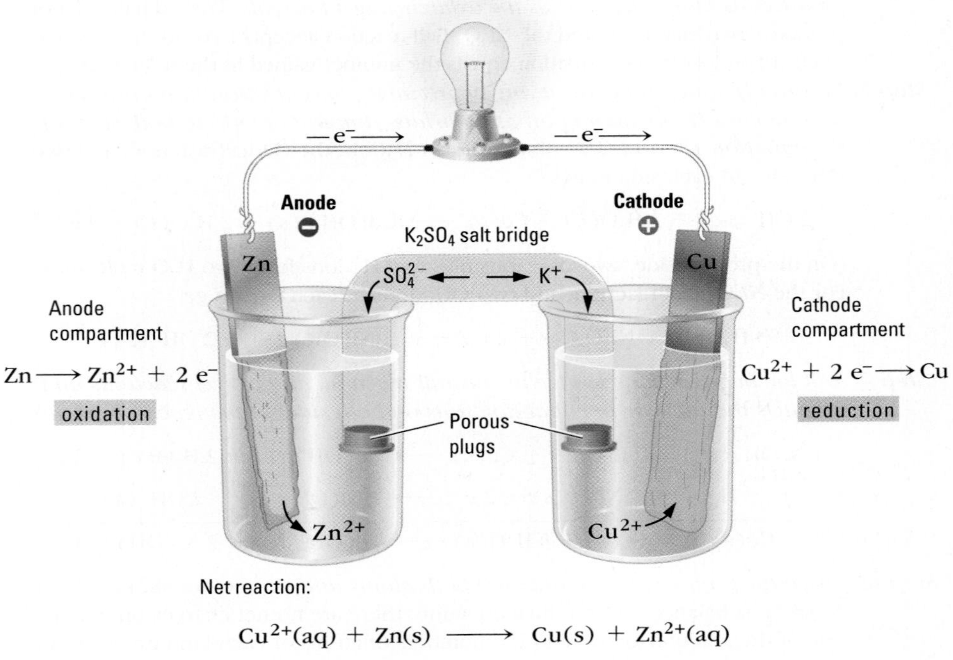

e⁻
e⁻
Anode
Cathode
K₂SO₄ salt bridge
SO₄²⁻
K⁺
Zn
Cu
Anode compartment
Cathode compartment
Zn → Zn²⁺ + 2 e⁻
oxidation
Cu²⁺ + 2 e⁻ → Cu
reduction
Porous plugs
Zn²⁺
Cu²⁺
Net reaction:
Cu²⁺(aq) + Zn(s) → Cu(s) + Zn²⁺(aq)

Figure 19.3 A simple electrochemical cell. The cell consists of a zinc electrode in a solution containing Zn^{2+} ions (*left*), a copper electrode in a solution containing Cu^{2+} ions (*right*), and a salt bridge that allows ions to flow into and out of the two solutions. When the two metal electrodes are connected by a conducting circuit, electrons flow through the wires and light bulb from the zinc electrode, where zinc is oxidized, to the copper electrode, where copper ions from the solution are reduced.

The voltaic cell is named after the Italian scientist Alessandro Volta. In about 1800 Volta constructed the first electrochemical cell, a stack of alternating zinc and silver disks separated by pieces of paper soaked in salt water (an electrolyte). Later, Volta showed that any two different metals and an electrolyte could be used to make a battery. Figure 19.4 shows a cell constructed by sticking a strip of zinc and a strip of copper metal into a grapefruit. In this case, the acidic solution of the grapefruit juice is the electrolyte.

In the electrochemical cell diagrammed in Figure 19.3, electrons are released at the anode by the oxidation half-reaction

$$Zn(s) \longrightarrow Zn^{2+}(aq) + 2\,e^- \qquad \text{(anode reaction)}$$

On a flashlight battery, the anode is marked "−" because an oxidation reaction produces electrons that make the anode negative. Conversely, the cathode is marked "+" because a reduction reaction consumes electrons, leaving the metal electrode positive.

They then flow from the anode through the filament in the bulb, causing it to glow, and eventually travel to the cathode, where they react with copper(II) ions in the reduction half-reaction

$$Cu^{2+}(aq) + 2\,e^- \longrightarrow Cu(s) \qquad \text{(cathode reaction)}$$

If nothing else but electron flow took place, the concentration of Zn^{2+} ions in the anode compartment would increase as zinc metal is oxidized, building up positive charge in the solution. The concentration of Cu^{2+} ions in the cathode compartment would decrease as Cu^{2+} ions are reduced to metallic copper. This makes that solution less positive due to the decrease of positive charge as Cu^{2+} ions are reduced. Excess negative charges from the now unbalanced SO_4^{2-} ions would be present. Because of this negative charge imbalance, the flow of electrons in the wires would very quickly stop. For the cell to work, there must be a way for the positive charge buildup in the anode compartment to be balanced by addition of negative ions or removal of positive ions, and vice versa for the cathode compartment.

The charge buildup can be avoided by using a salt bridge to connect the two compartments. A **salt bridge** is a solution of a salt (K_2SO_4 in Figure 19.3) arranged so that the bulk of that solution cannot flow into the cell solutions, but the ions (K^+ and SO_4^{2-}) can pass freely. As electrons flow through the wire from the zinc electrode to the copper electrode, negative ions (SO_4^{2-}) move from the salt bridge into the anode compartment solution, and positive ions (K^+) move in the opposite direction into the cathode compartment solution. In general, *anions from the salt bridge flow into the anode cell, and cations from the salt bridge flow into the cathode cell*. This flow of ions completes the electrical circuit, allowing current to flow. Without the salt bridge, the flow of electrons will stop.

© Cengage Learning/Charles D. Winters

Figure 19.4 A grapefruit battery. A voltaic cell can be made by inserting zinc and copper electrodes into a grapefruit. A potential of 0.95 V is obtained. (The water and citric acid of the fruit allow for ion conduction between electrodes.) This cell is more complicated than the one in Figure 19.3. To learn how the grapefruit battery works, see Goodisman, J. *Journal of Chemical Education,* Vol. 78, 2001; p. 516.

All voltaic cells and batteries have these characteristics:

- The oxidation-reduction reaction that occurs must be product-favored.
- There must be an external circuit through which electrons flow.
- There must be a salt bridge, porous barrier, or some other means of allowing ions to flow into and out of the electrode compartments to offset charge buildup.

In an electrochemical cell (Figure 19.5)

In commercial batteries, the salt bridge is often a porous polymer membrane.

- **Oxidation occurs at the anode.**
- **Reduction occurs at the cathode.**
- **Electrons move from the negative electrode (anode) to the positive electrode (cathode) through an external circuit.**
- **The electrical circuit is completed in the solution by movement of ions; anions move from the salt bridge compartment into the anode compartment, and cations move from the salt bridge compartment into the cathode compartment.**
- **The half-cells can be separated by either a salt bridge or a porous barrier.**

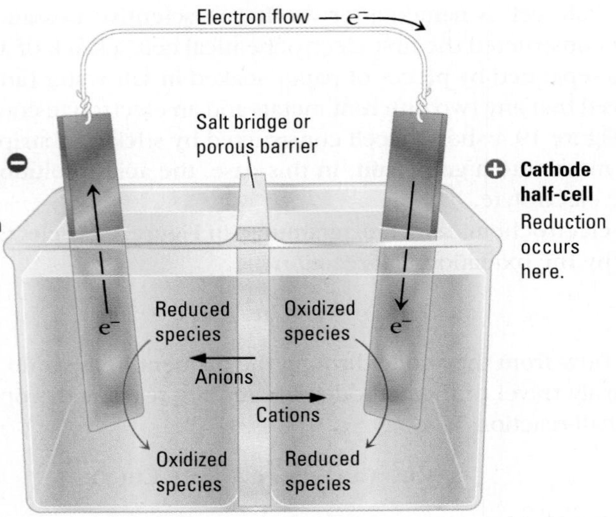
Electron flow — e⁻
Salt bridge or porous barrier
Anode half-cell
Oxidation occurs here.
Cathode half-cell
Reduction occurs here.
e⁻
Reduced species
Oxidized species
Anions
Cations
Oxidized species
Reduced species

Figure 19.5 Summary of the terminology used in voltaic cells.

PROBLEM-SOLVING EXAMPLE **19.5** **Electrochemical Cells**

A simple voltaic cell is assembled with Fe(s) and $Fe(NO_3)_2$(aq) in one compartment and Cu(s) and $Cu(NO_3)_2$(aq) in the other compartment. An external wire connects the two electrodes, and a salt bridge containing aqueous $NaNO_3$ solution connects the two solutions. The overall reaction is

$$Fe(s) + Cu^{2+}(aq) \longrightarrow Cu(s) + Fe^{2+}(aq)$$

For simplicity, this voltaic cell is shown without an electrical device in the external part of the circuit.

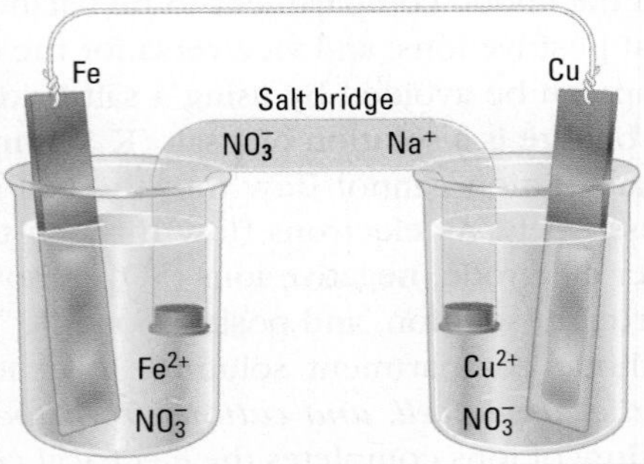
Fe
Cu
Salt bridge
NO_3^-
Na^+
Fe^{2+}
NO_3^-
Cu^{2+}
NO_3^-

What is the reaction at the anode? What is the reaction at the cathode? What is the direction of electron flow in the external wire? What is the direction of ion flow in the salt bridge? Draw a cell diagram, indicating the anode, the cathode, and the directions of electron flow and ion flow.

Answer

Anode reaction: $Fe(s) \longrightarrow Fe^{2+}(aq) + 2\,e^-$

Cathode reaction: $Cu^{2+}(aq) + 2\,e^- \longrightarrow Cu(s)$

The electrons flow through the wire from the anode to the cathode. Anions move from right to left, into and out of the salt bridge. Cations move from left to right, into and out of the salt bridge. The completed cell diagram is shown below.

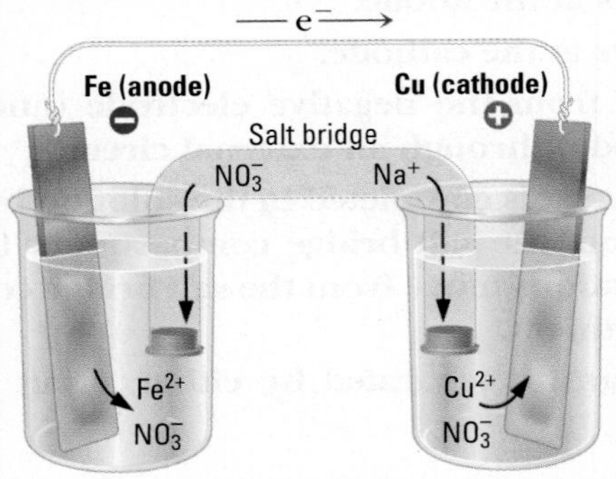
— e⁻⟶
Fe (anode)
Cu (cathode)
Salt bridge
NO_3^-
Na^+
Fe^{2+}
NO_3^-
Cu^{2+}
NO_3^-

Strategy and Explanation Use the overall equation given to decide (1) what is oxidized, (2) what is the product of the oxidation, (3) what is reduced, and (4) what is the product of the reduction. The anode is the electrode where oxidation occurs and the cathode is the electrode where reduction occurs. Since oxidation is loss of electrons, the anode will build up a negative charge, and electrons will flow through the external circuit to the cathode, where they will combine with positive ions to form uncharged atoms. A cell diagram can then be drawn.

The equation for the net reaction shows that iron is being oxidized from Fe(s) to Fe^{2+}(aq) and that Cu^{2+}(aq) is being reduced to Cu(s). Therefore, the iron electrode in the cell compartment containing Fe^{2+}(aq) is the anode, and the copper electrode in the cell compartment containing Cu^{2+}(aq) is the cathode. The half reactions are

$$\mathrm{Fe(s)} \longrightarrow \mathrm{Fe^{2+}(aq)} + 2\,e^- \qquad \text{(oxidation—anode)}$$

$$\mathrm{Cu^{2+}(aq)} + 2\,e^- \longrightarrow \mathrm{Cu(s)} \qquad \text{(reduction—cathode)}$$

In the external circuit electrons flow from the anode (oxidation is loss of electrons) to the cathode (reduction is gain of electrons). The electrons produced by the formation of Fe^{2+}(aq) from Fe flow through the external circuit to the cathode (Cu) where the electrons combine with Cu^{2+} ions to form Cu atoms. Because positive ions (Fe^{2+}) are being formed in the anode half cell, some of the excess positive ions (Fe^{2+}) enter the salt bridge and some negative ions (NO_3^-) leave the salt bridge; this keeps the total charge from positive ions equal to the total charge from negative ions in the anode compartment. Because positive ions (Cu^{2+}) are being depleted in the cathode half cell, negative ions begin to outnumber positive ions in the cathode compartment. This build-up of negative charge is counteracted by movement of excess negative ions (NO_3^-) into the salt bridge and movement of positive ions (Na^+) out of the salt bridge; this keeps the total charge from positive ions equal to the total charge from negative ions in the cathode compartment.

The cell diagram is partly given in the statement of the problem. Based on the discussion above, the left compartment contains the anode and the right contains the cathode. Electrons flow from anode to cathode in the external circuit. Negative ions move from right to left into and out of the salt bridge; positive ions move from left to right into and out of the salt bridge. (Note that the negative ions move clockwise through the circuit, as do the electrons; positive ions move the opposite direction, counterclockwise.)

To maintain electrical neutrality, ions move into or out of the salt bridge solution. Moving two NO_3^- ions out of the salt bridge solution is electrically equivalent to moving one Fe^{2+} ion into the salt bridge solution.

$$\text{Anode} \xrightarrow{e^-\text{ flow}} \text{Cathode}$$

Anode ⊖	Cathode ⊕
oxidation	reduction
(loss of e^-s)	(gain of e^-s)
$Fe \rightarrow Fe^{2+}$	$Cu^{2+} \rightarrow Cu$

PROBLEM-SOLVING PRACTICE 19.5

A voltaic cell is assembled to use this net reaction.

$$\mathrm{Ni(s)} + 2\,\mathrm{Ag^+(aq)} \longrightarrow \mathrm{Ni^{2+}(aq)} + 2\,\mathrm{Ag(s)}$$

(a) Write half-reactions for this cell, and indicate which is the oxidation reaction and which is the reduction reaction.

(b) Name the electrodes at which these reactions take place.

(c) What is the direction of flow of electrons in an external wire and electrical device connected between the electrodes?

(d) If a salt bridge connecting the two electrode compartments contains KNO_3, what is the direction of movement of the K^+ ions and the NO_3^- ions?

There is a shorthand notation for representing an electrochemical cell. For the cell shown in Figure 19.3 with the redox reaction

$$\mathrm{Zn(s)} + \mathrm{Cu^{2+}(aq)} \longrightarrow \mathrm{Zn^{2+}(aq)} + \mathrm{Cu(s)}$$

the representation is

anode electrode | salt bridge | cathode electrode

$$\underbrace{Zn(s) \mid Zn^{2+}(aq)}_{\text{anode compartment}} \parallel \underbrace{Cu^{2+}(aq) \mid Cu(s)}_{\text{cathode compartment}}$$

The anode half-cell is represented on the left, and the cathode half-cell is represented on the right. The electrodes are written on the extreme left (anode, Zn) and extreme right (cathode, Cu) of the notation. The single vertical lines denote boundaries between phases, and the double vertical lines denote the salt bridge and the separation between half-cells. Within each half-cell the reactants are written first, followed by the products. The electron flow is from left to right (anode to cathode).

CONCEPTUAL EXERCISE **19.2 Battery Design**

Devise an internal on-off switch for a battery that would not be a part of the flow of electrons.

19.4 Electrochemical Cells and Voltage

Because electrons flow from the anode to the cathode of an electrochemical cell, they can be thought of as being "driven" or "pushed" by an **electromotive force (emf).** The emf is produced by the *difference in electrical potential energy* between the two electrodes. Just as water flows downhill in response to a difference in gravitational potential energy, so an electron moves from an electrode of higher electrical potential energy to an electrode of lower potential energy. The moving water can do work, as can moving electrons; for example, the electrons could run a motor.

This is similar to comparing the quantity of work a few drops of water can do when falling 100 m with that possible when a few tons of water fall the same distance.

The quantity of electrical work done is proportional to the number of electrons that go from higher to lower potential energy as well as to the size of the potential energy difference.

$$\text{Electrical work} = \text{charge} \times \text{potential energy difference}$$

or

$$\text{Electrical work} = \text{number of electrons} \times \text{potential energy difference}$$

$$\frac{1.6022 \times 10^{-19}\ \text{C}}{1\ e^-} \times 6.24 \times 10^{18}\ e^- = 1\ \text{C}$$

Electrical charge is measured in coulombs. The charge on a single electron is very small (1.6022×10^{-19} C), so it takes 6.24×10^{18} electrons to produce 1 coulomb of charge. A **coulomb, C,** is the quantity of charge that passes a fixed point in an electrical circuit when a current of 1 ampere flows for 1 second. The **ampere, A,** is the unit of electrical current.

$$1\ \text{coulomb} = 1\ \text{ampere} \times 1\ \text{second}$$

When a single electron moves through a potential of 1 V, the work done is one electron-volt, abbreviated eV.

Electrical potential energy difference is measured in volts. The **volt, V,** is defined such that one joule of work is performed when one coulomb of charge moves through a potential difference of one volt:

$$1\ \text{volt} = \frac{1\ \text{joule}}{1\ \text{coulomb}} \quad \text{or} \quad 1\ \text{joule} = 1\ \text{volt} \times 1\ \text{coulomb}$$

The electromotive force of an electrochemical cell, commonly called its **cell voltage,** shows how much work a cell can do for each coulomb of charge that the chemical reaction transfers.

The voltage of an electrochemical cell depends on the temperature and substances that make up the cell, including their pressures if they are gases or their concentrations if they are solutes in solution. The quantity of charge (coulombs) depends on how much of each substance reacts. Look at the 1.5-V batteries shown in Figure 19.6. They have the same voltage because they have electrodes with the same potential energy difference between them. Yet, a larger battery is capable of far more work than a smaller one, because the larger one contains larger quantities of oxidizing and reducing agents. In this section and the next, we consider how cell voltage depends on the materials from which a cell is made. In Section 19.5, we will return to the question of how much electrical work a cell can do.

Figure 19.6 Dry cell batteries. The larger batteries are capable of more work since they contain more oxidizing and reducing agents.

CONCEPTUAL EXERCISE 19.3 Electrical Charges

Which has the larger charge, 1.0 C or Avogadro's number (1.0 mol) of electrons?

Cell Voltage

A cell's voltage is readily measured by inserting a high-resistance voltmeter into the circuit. Because the voltage varies with concentrations, **standard conditions** are defined for voltage measurements. These are the same as those used for ΔH° (← p. 230): *All reactants and products must be present as pure solids, pure liquids, gases at 1 bar pressure, or solutes at 1 M concentration.* Voltages measured under these conditions are **standard voltages,** symbolized by E°. Unless specified otherwise, all values of E° are for 25 °C (298 K). By definition, *cell voltages for product-favored electrochemical reactions are positive.* For example, the standard cell voltage for the product-favored Zn/Cu^{2+} cell discussed earlier is + 1.10 V at 25 °C.

It is important that the voltmeter have a very high electrical resistance so that very little chemical reaction takes place.

Since every redox reaction can be thought of as the sum of two half-reactions, it is convenient to assign a voltage to every possible half-reaction. Then the cell voltage for any reaction can be obtained by using the standard voltages of the half-reactions that occur at the cathode and the anode.

$$\boxed{E^\circ_{cell} = E^\circ_{cathode} - E^\circ_{anode}}$$

The two quantities on the right side of the equation, $E^\circ_{cathode}$ and E°_{anode}, are the standard reduction potential values for the half-reactions *written as reduction half-reactions.*

- **If E°_{cell} is positive, the reaction is product-favored.**
- **If E°_{cell} is negative, the reaction is reactant-favored.**

However, because only *differences* in potential energy can be measured, it is not possible to measure the voltage for a single half-reaction. Instead, one half-reaction is chosen as the standard, and then all others are compared to it. The half-reaction chosen as the standard is the one that occurs at the **standard hydrogen electrode,** in which hydrogen gas at a pressure of 1 bar is bubbled over a platinum electrode immersed in 1-M aqueous acid at 25 °C (Figure 19.7).

$$2\,H_3O^+(aq, 1\text{ M}) + 2\,e^- \longrightarrow H_2(g, 1\text{ bar}) + 2\,H_2O(\ell)$$

A voltage of exactly 0 V is *assigned* to this half-cell. In a cell that combines another half-reaction with the standard hydrogen electrode, the overall cell voltage is determined by the difference between the potentials of the two electrodes. Because the potential of the hydrogen electrode is assigned a value of 0, the overall cell voltage equals the voltage of the other electrode.

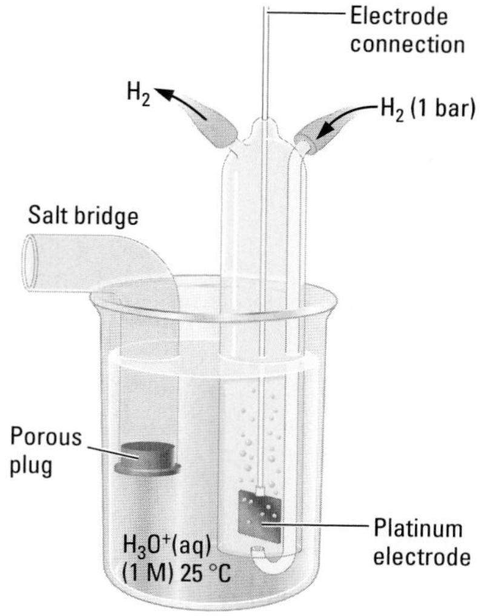

Figure 19.7 The standard hydrogen electrode. Hydrogen gas at 1 bar pressure bubbles over an inert platinum electrode immersed in a solution containing exactly 1-M H_3O^+ ions at 25 °C. The potential for this electrode is defined as exactly 0 V.

When the standard hydrogen electrode is paired with a half-cell that contains a better reducing agent than H_2, $H_3O^+(aq)$ is reduced to H_2.

$$H_3O^+ \text{ reduced: } 2\ H_3O^+(aq) + 2\ e^- \longrightarrow H_2(g,\ 1\ bar) + 2\ H_2O(\ell) \qquad E° = 0.0\ V$$

When the standard hydrogen electrode is paired with a half-cell that contains a better oxidizing agent than H_3O^+, then H_2 is oxidized to H_3O^+.

$$H_2 \text{ oxidized: } H_2(g,\ 1\ bar) + 2\ H_2O(\ell) \longrightarrow 2\ H_3O^+(aq,\ 1\ M) + 2\ e^- \qquad E° = 0.0\ V$$

The reaction that occurs at the standard hydrogen electrode can occur in either direction; H_3O^+ is reduced or H_2 is oxidized. In either case, the standard hydrogen electrode, by definition, has a potential of 0 V.

Figure 19.8 diagrams a cell in which one compartment contains a standard hydrogen electrode and the other contains a zinc electrode immersed in a 1 M solution of Zn^{2+}. The voltmeter connected between the two electrodes measures the difference in electrical potential energy. For this cell the voltage is +0.76 V. As the cell generates current, the zinc electrode decreases in mass as zinc is oxidized to $Zn^{2+}(aq)$. Therefore, the Zn electrode must be the anode; that is, oxidation takes place at this electrode. The hydrogen electrode must be the cathode, where reduction is taking place. The cell reaction is the sum of the half-cell reactions.

$$Zn(s) \longrightarrow Zn^{2+}(aq,\ 1\ M) + 2\ e^- \qquad E°_{anode} = ?$$

$$2\ H_3O^+(aq,\ 1\ M) + 2\ e^- \longrightarrow H_2(g,\ 1\ bar) + 2\ H_2O(\ell) \qquad E°_{cathode} = 0\ V$$

$$Zn(s) + 2\ H_3O^+(aq,\ 1\ M) \longrightarrow Zn^{2+}(aq,\ 1\ M) + H_2(g,\ 1\ bar) + 2\ H_2O(\ell)$$

$$E°_{cell} = +0.76\ V\ \text{(cell reaction)}$$

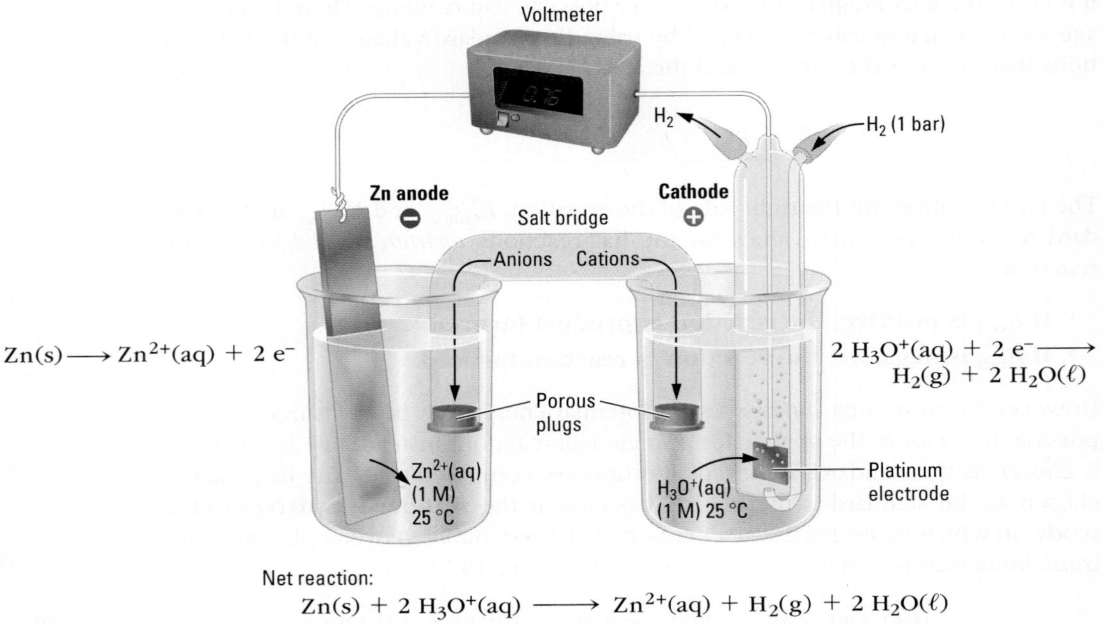

Voltmeter
0.76
H2
H2 (1 bar)
Zn anode
Cathode
Salt bridge
Anions
Cations
Zn(s) ⟶ Zn2+(aq) + 2 e−
2 H3O+(aq) + 2 e− ⟶ H2(g) + 2 H2O(ℓ)
Porous plugs
Zn2+(aq) (1 M) 25 °C
H3O+(aq) (1 M) 25 °C
Platinum electrode
Net reaction:
Zn(s) + 2 H3O+(aq) ⟶ Zn2+(aq) + H2(g) + 2 H2O(ℓ)

Figure 19.8 An electrochemical cell using a $Zn(s)|Zn^{2+}(aq)$ half-cell and a standard hydrogen electrode. In this cell, the zinc electrode is the anode and the standard hydrogen electrode is the cathode. The measured potential difference between the standard hydrogen electrode and the zinc electrode is +0.76 V. Zinc is the reducing agent and is oxidized to Zn^{2+}; H_3O^+ is the oxidizing agent and is reduced to H_2. In the standard hydrogen electrode, reaction occurs only where the three phases—gas, solution, and solid electrode—are in contact. The platinum electrode does not undergo any chemical change.

The voltmeter indicates that the electrical potential difference between the two electrodes is +0.76 V. The electrical potential difference is $E_{cathode} - E_{anode}$, so

$$E^\circ_{cell} = E^\circ_{cathode} - E^\circ_{anode}$$

and

$$0.76\ \text{V} = 0\ \text{V} - E^\circ_{anode}$$

$$E^\circ_{anode} = E^\circ_{Zn(s)|Zn^{2+}(aq,\ 1\ M)} = -0.76\ \text{V}$$

Thus, by using the standard hydrogen electrode, it is possible to assign a standard reduction potential E° value of −0.76 V to the $Zn(s)|Zn^{2+}(aq, 1\ M)$ electrode.

CONCEPTUAL EXERCISE 19.4 What Is Going On Inside the Electrochemical Cell?

Devise an experiment that would show that Zn is being oxidized in the electrochemical cell shown in Figure 19.8.

The half-cell standard reduction potentials of many different half-reactions can be measured by comparing them with the standard hydrogen electrode. For example, in a cell consisting of the $Cu^{2+}|Cu$ half-cell connected to a standard hydrogen electrode, the mass of the copper electrode increases and the voltmeter indicates +0.34 V (Figure 19.9). This means that the reactions are

The convention of assigning voltages to half-reactions is similar to the convention of tabulating standard enthalpies of formation; in both cases a relatively small table of data can provide information about a large number of different reactions.

$$H_2(g,\ 1\ \text{bar}) + 2\ H_2O(\ell) \longrightarrow 2\ H_3O^+(aq,\ 1\ M) + 2\ e^- \qquad E^\circ_{anode} = 0\ \text{V}$$

$$Cu^{2+}(aq,\ 1\ M) + 2\ e^- \longrightarrow Cu(s) \qquad E^\circ_{cathode} = ?$$

$$H_2(g,\ 1\ \text{bar}) + Cu^{2+}(aq,\ 1\ M) + 2\ H_2O(\ell) \longrightarrow 2\ H_3O^+(aq,\ 1\ M) + Cu(s) \qquad E^\circ_{cell} = +0.34\ \text{V}$$

$$E^\circ_{cell} = E^\circ_{cathode} - E^\circ_{anode} = E^\circ_{Cu(s)/Cu^{2+}(aq,\ 1\ M)} - 0\ \text{V} = +0.34\ \text{V}$$

The half-cell reduction potential for the $Cu^{2+}(aq, 1\ M) + 2\ e^- \longrightarrow Cu(s)$ reduction half-reaction must be +0.34 V. Note that in this cell the standard hydrogen electrode is the anode.

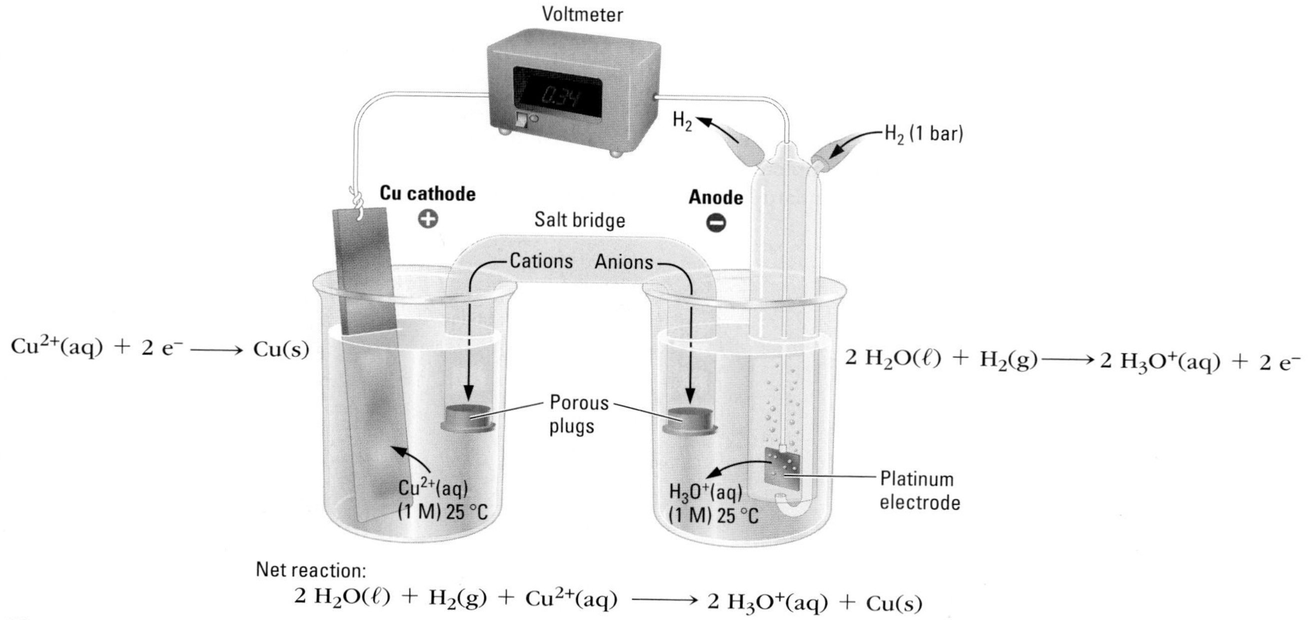

Voltmeter
0.34
H_2
H_2 (1 bar)
Cu cathode
Anode
Salt bridge
Cations
Anions
$Cu^{2+}(aq) + 2\ e^- \longrightarrow Cu(s)$
$2\ H_2O(\ell) + H_2(g) \longrightarrow 2\ H_3O^+(aq) + 2\ e^-$
Porous plugs
Cu^{2+}(aq) (1 M) 25 °C
H_3O^+(aq) (1 M) 25 °C
Platinum electrode
Net reaction:
$2\ H_2O(\ell) + H_2(g) + Cu^{2+}(aq) \longrightarrow 2\ H_3O^+(aq) + Cu(s)$

Figure 19.9 An electrochemical cell using the $Cu^{2+}|Cu$ half-cell and the standard hydrogen electrode. A voltage of +0.34 V is produced. In this cell, Cu^{2+} ions are reduced at the cathode to Cu metal, and H_2 is oxidized at the standard hydrogen electrode. The reaction at the standard hydrogen electrode is the opposite of that shown in Figure 19.8.

We now return to the first electrochemical cell we considered, in which Zn reduces Cu^{2+} ions to Cu. Using the standard reduction potentials for the half-reactions, we write

$$\begin{array}{lll} Zn(s) \longrightarrow Zn^{2+}(aq, 1\ M) + 2\ e^- & E^\circ_{anode} = -0.76\ V \\ Cu^{2+}(aq, 1\ M) + 2\ e^- \longrightarrow Cu(s) & E^\circ_{cathode} = +0.34\ V \\ \hline Zn(s) + Cu^{2+}(aq, 1\ M) \longrightarrow Zn^{2+}(aq, 1\ M) + Cu(s) & E^\circ_{cell} = ? \end{array}$$

When we combine the standard reduction potentials of the two half-reactions, we have the measured potential for the cell reaction.

$$E^\circ_{cell} = E^\circ_{cathode} - E^\circ_{anode} = (+0.34\ V) - (-0.76\ V) = +1.10\ V$$

The experimentally measured potential for this cell is 1.10 V, confirming that half-cell standard reduction potentials measured with the standard hydrogen electrode can be subtracted to obtain overall cell potentials.

PROBLEM-SOLVING EXAMPLE 19.6 Determining a Half-Cell Potential

The voltaic cell shown in the drawing below generates a potential difference under standard conditions of $E^\circ_{cell} = 0.36$ V at 25 °C. The net cell reaction is

$$Zn(s) + Cd^{2+}(aq, 1\ M) \longrightarrow Zn^{2+}(aq, 1\ M) + Cd(s)$$

The standard half-cell potential for $Zn(s)|Zn^{2+}(aq, 1\ M)$ is -0.76 V.

(a) Determine which electrode is the anode and which is the cathode.
(b) Show the direction of electron flow through the circuit outside the cell, and complete the cell diagram.
(c) Calculate the standard half-cell potential for $Cd^{2+}(aq) + 2\ e^- \longrightarrow Cd(s)$.

For simplicity, this voltaic cell is shown without an electrical device in the external circuit. To generate its standard voltage, a high resistance would have to be in the circuit.

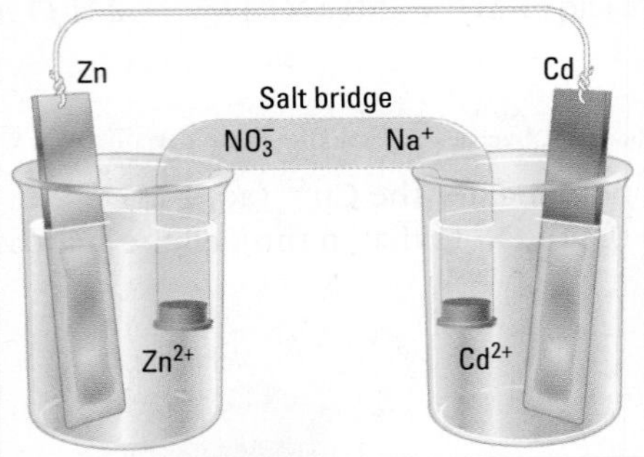

Answer

(a) Zinc is the anode, and cadmium is the cathode.
(b) The completed cell diagram is shown below.
(c) The standard half-cell potential is -0.40 V.

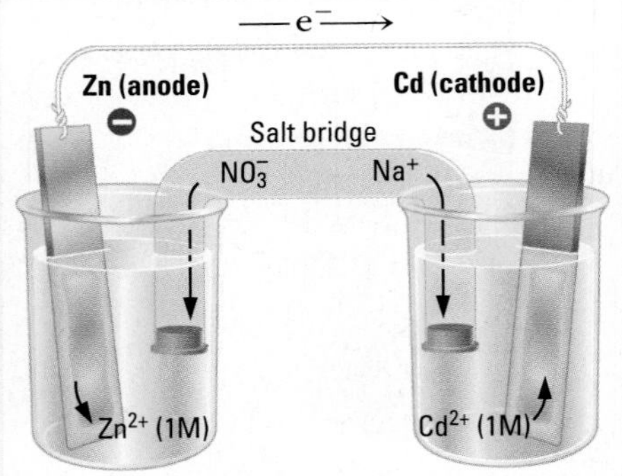

Strategy and Explanation The electrode where oxidation occurs is the anode. Because Zn(s) is oxidized to Zn^{2+}(aq), the Zn electrode is the anode. Since Cd^{2+} ions are reduced at the Cd electrode, the Cd electrode is the cathode.

The net cell potential and the standard reduction potential for the $Zn(s)|Zn^{2+}(aq, 1 M)$ half-cell are known, so the value of for $Cd^{2+}(aq, 1 M) + 2 e^- \longrightarrow Cd(s)$ can be calculated.

$$Zn(s) \longrightarrow Zn^{2+}(aq, 1 M) + 2 e^- \qquad E^\circ_{anode} = -0.76 \text{ V (anode)}$$
$$Cd^{2+}(aq, 1 M) + 2 e^- \longrightarrow Cd(s) \qquad E^\circ_{cathode} = ? \text{ V (cathode)}$$
$$Zn(s) + Cd^{2+}(aq, 1 M) \longrightarrow Zn^{2+}(aq, 1 M) + Cd(s) \qquad E^\circ_{cell} = +0.36 \text{ V}$$

Use $E^\circ_{cell} = E^\circ_{cathode} - E^\circ_{anode}$, and solve for $E^\circ_{cathode}$.

$$E^\circ_{cathode} = E^\circ_{cell} + E^\circ_{anode} = 0.36 \text{ V} + (-0.76 \text{ V}) = -0.40 \text{ V}$$

At 25 °C, the value of E° for the $Cd^{2+}(aq) + 2 e^- \longrightarrow Cd(s)$ half-reaction is −0.40 V.

PROBLEM-SOLVING PRACTICE 19.6

Given this reaction and its standard potential

$$Fe(s) + Cu^{2+}(aq, 1 M) \longrightarrow Fe^{2+}(aq, 1 M) + Cu(s) \qquad E^\circ_{cell} = +0.78 \text{ V}$$

calculate E° for the half-cell $Fe(s) \longrightarrow Fe^{2+}(aq) + 2 e^-$.

19.5 Using Standard Reduction Potentials

The results of a great many measurements of cell potentials such as the ones just described are summarized as **standard reduction potentials** in Table 19.1. A much longer and more complete list of standard reduction potentials appears in Appendix I. The values reported in the tables are called *standard reduction potentials* because they are the potentials, reported as voltages, that are measured for a cell in which a half-reaction *occurs as a reduction* when paired with the standard hydrogen electrode. *If a half-reaction occurs as an oxidation when paired with the standard hydrogen electrode, that half-reaction voltage has a negative sign.* For example, we saw the oxidation half-reaction

$$Zn(s) \longrightarrow Zn^{2+}(aq) + 2 e^-$$

in Figure 19.8. This reaction appears in Table 19.1 as the *reduction* half-reaction

$$Zn^{2+}(aq) + 2 e^- \longrightarrow Zn(s) \qquad E^\circ = -0.76 \text{ V}$$

Here are some important points to notice about Table 19.1:

1. ***Each half-reaction is written as a reduction.*** Thus, the species on the left-hand side of each half-reaction is in a higher oxidation state, and the species on the right-hand side is in a lower oxidation state. *Oxidizing agents are on the left side; reducing agents are on the right side.*
2. ***Each half-reaction can occur in either direction.*** A given substance can react at the anode or the cathode, depending on the conditions. For example, we have already seen cases in which H_2 is oxidized to H_3O^+ and others in which H_3O^+ is reduced to H_2 by different reactants.
3. ***The more positive the value of the standard reduction potential,* E°, *the more easily the substance on the left-hand side of a half-reaction can be reduced.*** When a substance is easy to reduce, it is a strong oxidizing agent. Thus, $F_2(g)$ is the best oxidizing agent in the table, and Li^+ is the poorest oxidizing agent in the table. Other strong oxidizing agents are at the top left of the table:

$$H_2O_2(aq), PbO_2(s), Au^{3+}(aq), Cl_2(g), O_2(g)$$

4. ***The less positive the value of the standard reduction potential,* E°, *the less likely the reaction will occur as a reduction, and the more likely an oxidation (the reverse reaction) will occur.*** The farther down we go in the table, the better the reducing (electron donating) ability of the atom, ion, or molecule on the right. Thus, Li(s) is the strongest reducing agent in the table, and F^- is the weakest reducing agent in the table. Other strong reducing agents are alkali and alkaline-earth metals and hydrogen at the lower right of the table.

F_2 is always reduced and Li(s) is always oxidized.

Recall that an oxidizing agent is reduced when it oxidizes something else.

Table 19.1 Standard Reduction Potentials in Aqueous Solution at 25 °C*

Reduction Half-Reaction		$E°$ (V)
Oxidizing Agents	**Reducing Agents**	
$F_2(g) + 2\ e^-$	$\longrightarrow 2\ F^-(aq)$	+2.87
$H_2O_2(aq) + 2\ H_3O^+(aq) + 2\ e^-$	$\longrightarrow 4\ H_2O(\ell)$	+1.763
$PbO_2(s) + SO_4^{2-}(aq) + 4\ H_3O^+(aq) + 2\ e^-$	$\longrightarrow PbSO_4(s) + 6\ H_2O(\ell)$	+1.690
$Au^{3+}(aq) + 3\ e^-$	$\longrightarrow Au(s)$	+1.52
$MnO_4^-(aq) + 8\ H_3O^+(aq) + 5\ e^-$	$\longrightarrow Mn^{2+}(aq) + 12\ H_2O(\ell)$	+1.51
$Cr_2O_7^{2-}(aq) + 14\ H_3O^+(aq) + 6\ e^-$	$\longrightarrow 2\ Cr^{3+}(aq) + 21\ H_2O(\ell)$	+1.36
$Cl_2(g) + 2\ e^-$	$\longrightarrow 2\ Cl^-(aq)$	+1.358
$O_2(g) + 4\ H_3O^+(aq) + 4\ e^-$	$\longrightarrow 6\ H_2O(\ell)$	+1.229
$Br_2(\ell) + 2\ e^-$	$\longrightarrow 2\ Br^-(aq)$	+1.066
$NO_3^-(aq) + 4\ H_3O^+(aq) + 3\ e^-$	$\longrightarrow NO(g) + 6\ H_2O(\ell)$	+0.96
$OCl^-(aq) + H_2O(\ell) + 2\ e^-$	$\longrightarrow Cl^-(aq) + 2\ OH^-(aq)$	+0.89
$Hg^{2+}(aq) + 2\ e^-$	$\longrightarrow Hg(\ell)$	+0.8535
$Ag^+(aq) + e^-$	$\longrightarrow Ag(s)$	+0.7991
$Hg_2^{2+}(aq) + 2\ e^-$	$\longrightarrow 2\ Hg(\ell)$	+0.7960
$Fe^{3+}(aq) + e^-$	$\longrightarrow Fe^{2+}(aq)$	+0.771
$I_2(s) + 2\ e^-$	$\longrightarrow 2\ I^-(aq)$	+0.535
$O_2(g) + 2\ H_2O(\ell) + 4\ e^-$	$\longrightarrow 4\ OH^-(aq)$	+0.401
$Cu^{2+}(aq) + 2\ e^-$	$\longrightarrow Cu(s)$	+0.340
$Sn^{4+}(aq) + 2\ e^-$	$\longrightarrow Sn^{2+}(aq)$	+0.15
$2\ H_3O^+(aq) + 2\ e^-$	$\longrightarrow H_2(g) + 2\ H_2O(\ell)$	0
$Sn^{2+}(aq) + 2\ e^-$	$\longrightarrow Sn(s)$	−0.1375
$Ni^{2+}(aq) + 2\ e^-$	$\longrightarrow Ni(s)$	−0.25
$PbSO_4(s) + 2\ e^-$	$\longrightarrow Pb(s) + SO_4^{2-}(aq)$	−0.3505
$Cd^{2+}(aq) + 2\ e^-$	$\longrightarrow Cd(s)$	−0.403
$Fe^{2+}(aq) + 2\ e^-$	$\longrightarrow Fe(s)$	−0.44
$Zn^{2+}(aq) + 2\ e^-$	$\longrightarrow Zn(s)$	−0.763
$2\ H_2O(\ell) + 2\ e^-$	$\longrightarrow H_2(g) + 2\ OH^-(aq)$	−0.8277
$Al^{3+}(aq) + 3\ e^-$	$\longrightarrow Al(s)$	−1.676
$Mg^{2+}(aq) + 2\ e^-$	$\longrightarrow Mg(s)$	−2.356
$Na^+(aq) + e^-$	$\longrightarrow Na(s)$	−2.714
$K^+(aq) + e^-$	$\longrightarrow K(s)$	−2.925
$Li^+(aq) + e^-$	$\longrightarrow Li(s)$	−3.045

Data from Bard, A. J., Parsons, R., and Jordan, J. *Standard Potentials in Aqueous Solution.* New York: Marcel Dekker: 1985. International Union of Pure and Applied Chemistry Commission on Electrochemistry and Electroanalytical Chemistry.

*In volts (V) versus the standard hydrogen electrode.

5. ***Under standard conditions, any species on the left of a half-reaction will oxidize any species that is below it on the right side of the table.*** For example, we can apply this rule to predict that $Fe^{3+}(aq)$ will oxidize Al(s), $Br_2(\ell)$ will oxidize Mg(s), and $Na^+(aq)$ will oxidize Li(s). The net reaction is found by adding the half-reactions, and the cell voltage can be calculated from the $E°_{cell} = E°_{cathode} - E°_{anode}$ equation, as illustrated below.

$$Br_2(\ell) + 2\ e^- \longrightarrow 2\ Br^-(aq) \qquad E°_{cathode} = +1.07\ V$$
$$Mg(s) \longrightarrow Mg^{2+}(aq) + 2\ e^- \qquad E°_{anode} = -2.36\ V$$
$$Br_2(\ell) + Mg(s) \longrightarrow Mg^{2+}(aq) + 2\ Br^-(aq) \qquad E°_{cell} = +3.45\ V$$
$$E°_{cell} = E°_{cathode} - E°_{anode} = 1.07\ V - (-2.36\ V) = +3.43\ V$$

A positive cell potential denotes a product-favored reaction.

CHEMISTRY YOU CAN DO Remove Tarnish the Easy Way

Silverware tarnishes when exposed to air because the silver reacts with hydrogen sulfide gas in the air to form a thin coating of black silver sulfide, Ag_2S. You can use chemistry to remove the tarnish from silverware and other silver utensils with a solution of baking soda and some aluminum foil. The chemical cleaning of silver is an electrochemical process in which electrons move from aluminum atoms to silver ions in the tarnish, reducing silver ions to silver atoms and oxidizing aluminum atoms to aluminum ions. Look up the position of Ag^+ ion with respect to aluminum metal in Table 19.1. The baking soda (sodium hydrogen carbonate) provides a conductive ionic solution for the flow of electrons and helps to remove the aluminum oxide coating from the surface of the aluminum foil.

Start by putting 1 to 2 L water in a large pan. Add 7 to 8 tbsp baking soda. Heat the solution, but do not boil it. Place some aluminum foil in the bottom of the pan, and put the tarnished silverware on the aluminum foil. Make sure the silverware is completely covered with the solution. Heat the solution almost to boiling. After a few minutes, remove the silverware and rinse it in running water.

This method of cleaning silverware is better than using polish, because polish *removes* the silver sulfide, which removes the silver it contains. The chemical process described here *restores* the silver from the tarnish to the surface. If you have aluminum pie pans or aluminum cooking pans, you can use them as both the container and the aluminum source. You may notice that devices for removing silver tarnish are sometimes advertised on television. These devices are actually little more than a piece of aluminum metal and some salt. Would you be willing to pay very much (plus shipping and handling) for such a device after you have done this simple experiment?

Think about these questions:

1. In the reaction between silver and H_2S, what is being oxidized? What is being reduced?
2. What metal other than aluminum could be used for this reaction?

6. ***Electrode potentials depend on the nature and concentration of reactants and products, but not on the quantity of each that reacts.*** Changing the stoichiometric coefficients for a half-reaction does *not* change the value of $E°$. For example, the reduction of Fe^{3+} has an $E°$ of +0.771 V whether the reaction is written as

$$Fe^{3+}(aq, 1\ M) + e^- \longrightarrow Fe^{2+}(aq, 1\ M) \qquad E° = +0.771\ V$$

or

$$2\ Fe^{3+}(aq, 1\ M) + 2\ e^- \longrightarrow 2\ Fe^{2+}(aq, 1\ M) \qquad E° = +0.771\ V$$

This fact about half-cell potentials seems unusual at first, but consider that a half-cell voltage is energy per unit charge (1 volt = 1 joule/1 coulomb). When a half-reaction is multiplied by some number, both the energy and the charge are multiplied by that number. Thus, the ratio of the energy to the charge (voltage) does not change.

In this respect, $E°$ values differ from $\Delta H°$ and $\Delta G°$ values, which do depend on the coefficients in thermochemical equations.

Using the preceding guidelines and the table of standard reduction potentials, we will make some *predictions* about whether reactions will occur and then check our results by calculating $E°_{cell}$.

PROBLEM-SOLVING EXAMPLE 19.7 Predicting Whether Redox Reactions Occur

(a) Will zinc metal react with a 1-M $Ag^+(aq)$ solution? If so, calculate $E°$ for the reaction.
(b) Will a 1-M $Fe^{2+}(aq)$ solution react with metallic tin? If so, calculate $E°$ for the reaction.

Answer
(a) Yes. $E°_{cell} = +1.56$ V
(b) No. $E°_{cell} = -0.30$ V

Strategy and Explanation Answer the questions by referring to Table 19.1 and comparing the positions of the reactants there. Use $E°_{cell} = E°_{cathode} - E°_{anode}$ to calculate $E°_{cell}$.

(a) $Ag^+(aq)$ is above $Zn^{2+}(aq)$ in Table 19.1, so it is a better oxidizing agent; we predict that Ag^+ can oxidize metallic zinc, causing zinc atoms to form $Zn^{2+}(aq)$ ions. To be certain, combine the half-cell reactions to give the balanced equation. Subtract the half-cell potentials.

$$Zn(s) \longrightarrow Zn^{2+}(aq, 1\ M) + 2\ e^- \qquad E^\circ_{anode} = -0.763 V$$
$$2\ [Ag^+(aq, 1\ M) + e^- \longrightarrow Ag(s)] \qquad E^\circ_{cathode} = +0.80\ V$$
$$Zn(s) + 2\ Ag^+(aq, 1\ M) \longrightarrow Zn^{2+}(aq, 1\ M) + 2\ Ag(s) \qquad E^\circ_{cell} = ?$$

Note that the half-reaction voltages are *not* multiplied by the balancing coefficients.

$$E^\circ_{cell} = E^\circ_{cathode} - E^\circ_{anode} = +0.80\ V - (-0.763\ V) = +1.56\ V$$

The positive value for E°_{cell} shows that this is a product-favored reaction.

(b) We evaluate the reaction between $Fe^{2+}(aq)$ and metallic Sn the same way. $Fe^{2+}(aq)$ is on the left in Table 19.1 below Sn(s), which is on the right. Therefore, $Fe^{2+}(aq)$ is not a strong enough oxidizing agent to oxidize Sn(s), and we predict that this reaction will not occur. The combined half-reactions are

$$Sn(s) \longrightarrow Sn^{2+}(aq) + 2e^- \qquad E^\circ_{anode} = -0.14\ V$$
$$Fe^{2+}(aq) + 2\ e^- \longrightarrow Fe(s) \qquad E^\circ_{cathode} = -0.44\ V$$
$$Sn(s) + Fe^{2+}(aq) \longrightarrow Sn^{2+}(aq) + Fe(s) \qquad E^\circ_{cell} = ?$$
$$E^\circ_{cell} = E^\circ_{cathode} - E^\circ_{anode} = -0.44\ V - (-0.14\ V) = -0.30\ V$$

The negative value for E°_{cell} shows that this process is reactant-favored, and it will not form appreciable quantities of products under standard conditions. In fact, iron metal will reduce Sn^{2+}, the reverse of the net reaction shown above.

PROBLEM-SOLVING PRACTICE 19.7

Look at Table 19.1 and determine which two half-reactions would produce the largest value of E°_{cell}. Write the two half-reactions and the overall cell reaction, and calculate the E°_{cell} for the reaction.

CONCEPTUAL EXERCISE 19.5 Using E° Values

Transporting chemicals is of great practical and economic importance. Suppose that you have a large volume of aqueous mercury(II) chloride solution, $HgCl_2$, that needs to be transported. A driver brings a tanker truck made of aluminum to the loading dock. Will it be okay to load the truck with your solution? Explain your answer fully.

Stainless steel is an alloy of iron.

Standard reduction potentials can be used to explain an annoying experience many of us have had—a pain in a tooth when a filling is touched with a stainless steel fork or a piece of aluminum foil. A common material for dental fillings is a dental amalgam—tin and silver dissolved in mercury to form solid solutions having compositions approximating Ag_2Hg_3, Ag_3Sn, and Sn_xHg (where x ranges from 7 to 9). All of these compounds can undergo electrochemical reactions; for example,

$$3\ Hg_2^{2+}(aq) + 4\ Ag(s) + 6\ e^- \longrightarrow 2\ Ag_2Hg_3(s) \qquad E^\circ = +0.85\ V$$
$$Sn^{2+}(aq) + 3\ Ag(s) + 2\ e^- \longrightarrow Ag_3Sn(s) \qquad E^\circ = -0.05\ V$$

The E° values in Table 19.1 indicate that both iron and aluminum have much more negative reduction potentials and, therefore, are much better reducing agents than any of the amalgam fillings. Consequently, if a piece of iron or aluminum comes in contact with a dental filling, the saliva and gum tissue will act as a salt bridge, resulting in an electrochemical cell. The iron or aluminum donates electrons, producing a tiny electrical current that activates a nerve and results in pain.

Before we leave this discussion of Table 19.1, consider again the activity series of metals shown in Table 5.5 (← p. 187), which contains many of the same elements as Table 19.1. Looking closely, however, you will notice that the most active metal, lithium, at the top of Table 5.5 is at the very bottom right of Table 19.1. That is because Table 19.1 is arranged by ***reduction potential,*** and lithium ion has the lowest tendency to be reduced. Table 5.5, on the other hand, lists the metals in order of activity, that is, their tendency to be oxidized. Since oxidation is the opposite of reduction, it is reasonable that lithium metal is in opposite positions in the two tables.

CONCEPTUAL EXERCISE 19.6 Predicting Redox Reactions Using $E°$ Values

Consider these reduction half-reactions:

Half-Reaction	$E°$ (V)
$Cl_2(g) + 2\,e^- \longrightarrow 2\,Cl^-(aq)$	+1.36
$I_2(s) + 2\,e^- \longrightarrow 2\,I^-(aq)$	+0.535
$Pb^{2+}(aq) + 2\,e^- \longrightarrow Pb(s)$	−0.125
$V^{2+}(aq) + 2\,e^- \longrightarrow V(s)$	−1.13

(a) Which is the weakest oxidizing agent?
(b) Which is the strongest oxidizing agent?
(c) Which is the strongest reducing agent?
(d) Which is the weakest reducing agent?
(e) Will Pb(s) reduce V^{2+}(aq) to V(s)?
(f) Will I_2(s) oxidize Cl^-(aq) to Cl_2(g)?
(g) Name the molecules or ions in the above reactions that can be reduced by Pb(s).

CONCEPTUAL EXERCISE 19.7 Predicting $E°$ Values

The elements lead and antimony are above and below hydrogen in Table 5.5, but they are not listed in Table 19.1. Indicate where they would appear in Table 19.1 and, based on values from the table, estimate the reduction potentials for their positive ions being reduced to the metal atom.

19.6 $E°$ and Gibbs Free Energy

The sign of $E°_{cell}$ indicates whether a redox reaction is product-favored (positive $E°$) or reactant-favored (negative $E°$). In Chapter 18 you learned another way to decide whether a reaction is product-favored: The change in standard Gibbs free energy, $\Delta G°$, must be negative ***(← p. 864).*** Since both $E°_{cell}$ and $\Delta G°$ tell something about whether a reaction will occur, it should be no surprise that a relationship exists between them.

The "free" in Gibbs free energy indicates that it is energy available to do work. The energy available for electrical work from an electrochemical cell can be calculated by multiplying the quantity of electrical charge transferred times the cell voltage, $E°$. The quantity of charge is given by the number of moles of electrons transferred in the overall reaction, n, multiplied by the number of coulombs per mole of electrons.

$$\text{Quantity of charge} = \text{moles of electrons} \times \text{coulombs per mole of electrons}$$

The charge on 1 mol of electrons can be calculated from the charge on one electron and Avogadro's number.

Science Photo Library/Photo Researchers, Inc.

Michael Faraday

1791–1867

As an apprentice to a London bookbinder, Michael Faraday became fascinated by science when he was a boy. At age 22 he was appointed as a laboratory assistant at the Royal Institution and became its director within 12 years. A skilled experimenter in chemistry and physics, he made many discoveries, the most important of which was electromagnetic induction, the basis of modern electromagnetic technology. Faraday built the first electric motor, generator, and transformer. A popular speaker and educator, he also performed chemical and electrochemical experiments and first synthesized benzene, C_6H_6.

$$\text{Charge on 1 mol of e}^- = \left(\frac{1.60218 \times 10^{-19}\ \text{C}}{\text{e}^-}\right)\left(\frac{6.02214 \times 10^{23}\ \text{e}^-}{1\ \text{mol e}^-}\right)$$
$$= 9.6485 \times 10^4\ \text{C/mol e}^-$$

The quantity 9.6485×10^4 C/mol of electrons (commonly rounded to 96,500 C/mol of electrons) is known as the **Faraday constant, *F*,** in honor of Michael Faraday, who first explored the quantitative aspects of electrochemistry.

The electrical work that can be done by a cell is equal to the Faraday constant (F) multiplied by the number of moles of electrons transferred (n) and by the cell voltage (E°_{cell}).

$$\text{Electrical work} = nFE^\circ_{\text{cell}}$$

Unlike the cell voltage, the electrical work a cell can do *does* depend on the quantity of reactants in the cell reaction. More reactants mean more moles of electrons transferred and hence more work. Equating the electrical work of a cell at standard conditions with ΔG°, we get

$$\Delta G^\circ = -nFE^\circ_{\text{cell}}$$

The negative sign on the right side of the equation accounts for the fact that ΔG° *is always negative for a product-favored process, but* E°_{cell} *is always positive for a product-favored process.* Thus, these values must have opposite signs.

Using this equation we can calculate ΔG° for the $Zn|Zn^{2+}||Cu^{2+}|Cu$ cell. This value represents the maximum work that the cell can do. For the reaction

$$Cu^{2+}(aq) + Zn(s) \longrightarrow Cu(s) + Zn^{2+}(aq) \qquad E^\circ_{\text{cell}} = +1.10\ \text{V}$$

2 mol electrons are transferred per mole of copper ions reduced; $n = 2$. The Gibbs free energy change when this quantity of reactants is converted is

$$\Delta G^\circ = -\left(\frac{2\ \text{mol e}^-}{\text{transferred}}\right)\left(\frac{9.65 \times 10^4\ \text{C}}{\text{mol e}^-}\right)\left(\frac{1\ \text{J}}{1\ \text{V} \times 1\ \text{C}}\right)\left(\frac{1\ \text{kJ}}{10^3\ \text{J}}\right)(+1.10\ \text{V})$$
$$= -212\ \text{kJ}$$

The large positive E°_{cell} and the significantly negative Gibbs free energy change values indicate a very product-favored reaction.

PROBLEM-SOLVING EXAMPLE 19.8 Determining E°_{cell} and ΔG°

Consider the redox reaction

$$Zn^{2+}(aq) + H_2(g) + 2\,H_2O(\ell) \longrightarrow Zn(s) + 2\,H_3O^+(aq)$$

Use the standard reduction potentials in Table 19.1 to calculate E°_{cell} and ΔG° and to determine whether the reaction as written is product-favored.

Answer $E^\circ_{\text{cell}} = -0.763$ V; $\Delta G^\circ = 147$ kJ. The reaction as written is not product-favored.

Strategy and Explanation The first step is to write the half-reactions for the oxidation and reduction that occur and to use their standard reduction potentials from Table 19.1.

Reduction: $Zn^{2+}(aq) + 2\,e^- \longrightarrow Zn(s) \qquad E^\circ_{\text{cathode}} = -0.763\ \text{V}$

Oxidation: $H_2(g) + 2\,H_2O(\ell) \longrightarrow 2\,H_3O^+(aq) + 2\,e^- \qquad E^\circ_{\text{anode}} = 0\ \text{V}$

We obtain the E°_{cell} for the reaction from the potentials for the two half-reactions.

$$E^\circ_{\text{cell}} = E^\circ_{\text{cathode}} - E^\circ_{\text{anode}} = -0.763\ \text{V} - 0\ \text{V} = -0.763\ \text{V}$$

Since the value of E°_{cell} is negative, the reaction is not product-favored in the direction written; Zn^{2+} will not oxidize H_2O. The reverse reaction is product-favored; that is, zinc metal reacts with acid.

From the calculated $E°_{cell}$ we can calculate $\Delta G°$.

$$\Delta G° = -nFE°_{cell}$$
$$= -(2 \text{ mol e}^-) \times \left(\frac{9.65 \times 10^4 \text{ C}}{1 \text{ mol e}^-}\right)\left(\frac{1 \text{ J}}{1 \text{ V} \times 1 \text{ C}}\right)(-0.763 \text{ V})$$
$$= 1.47 \times 10^5 \text{ J} = 147 \text{ kJ}$$

The positive value for $\Delta G°$ also indicates that the reaction as written is not product-favored.

☑ **Reasonable Answer Check** The given oxidation is the half-reaction at the standard hydrogen electrode with a potential of zero, so the overall reaction is governed by the Zn reduction potential. The negative cell potential and the positive $\Delta G°$ are consistent with the reaction not being product-favored as written.

PROBLEM-SOLVING PRACTICE 19.8

Using standard reduction potentials, determine whether this reaction is product-favored as written.

$$Hg^{2+}(aq) + 2\,I^-(aq) \longrightarrow Hg(\ell) + I_2(s)$$

$\Delta G°$, $E°_{cell}$, and $K°$

We have seen that the standard Gibbs free energy change is directly proportional to the $E°_{cell}$ for an electrochemical cell at standard conditions

$$\Delta G° = -nFE°_{cell}$$

Recall from Chapter 18 (*← p. 871*) that the standard Gibbs free energy change is directly proportional to the logarithm of the equilibrium constant of a reaction.

$$\Delta G° = -RT \ln K°$$

Putting these two equations together yields

$$-nFE°_{cell} = -RT \ln K°$$

which, when solved for $E°_{cell}$, yields

$$E°_{cell} = \frac{RT}{nF} \ln K°$$

Thus by measuring $E°_{cell}$, the values of $\Delta G°$ and $K°$ can be calculated. The relationships linking $\Delta G°$, $E°_{cell}$, and $K°$ are summarized in Figure 19.10.

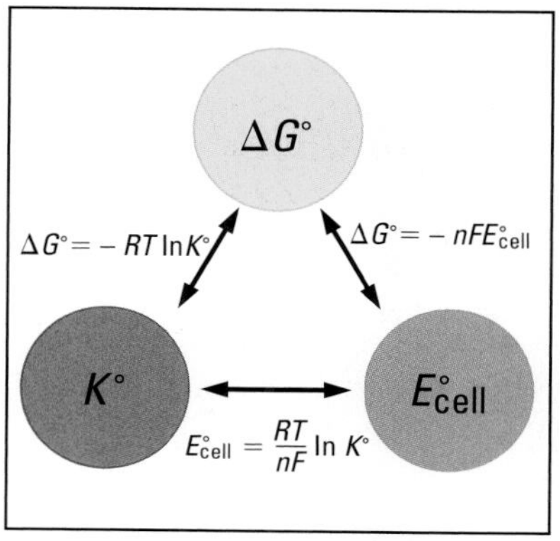

Figure 19.10 The relationships linking $\Delta G°$, $E°_{cell}$, and $K°$. Given any one of the values, the other two can be calculated.

The $E°_{cell}$ expression can be simplified by substituting numerical values for R (8.314 J mol^{-1} K^{-1}) and F (96,485 J V^{-1} mol^{-1}) and assuming a temperature of 298 K. For n electrons transferred we get

$F = 96{,}485$ C mol^{-1}
$= 96{,}485$ J V^{-1} mol^{-1}
because 1 J = 1 V × 1 C

$$E°_{cell} = \frac{RT}{nF} \ln K° = \frac{(8.314 \text{ J mol}^{-1} \text{ K}^{-1})(298 \text{ K})}{n\,(96{,}485 \text{ J V}^{-1}\text{mol}^{-1})} = \frac{0.0257 \text{ V}}{n} \ln K°$$

Changing from natural logarithms to base-10-logarithms (multiplying by 2.303) yields

$$E°_{cell} = \frac{0.0592 \text{ V}}{n} \log K° \quad \text{and} \quad \log K° = \frac{nE°_{cell}}{0.0592 \text{ V}} \quad \text{(at 298 K)}$$

These equations hold for standard states of all reactants and products at 25 °C.

PROBLEM-SOLVING EXAMPLE 19.9 Equilibrium Constant for a Redox Reaction

Calculate the equilibrium constant K_c for the reaction at 25 °C.

$$Fe(s) + Cd^{2+}(aq) \rightleftharpoons Fe^{2+}(aq) + Cd(s)$$

using the standard reduction potentials listed in Table 19.1.

Answer $K_c = K° = 22$

Strategy and Explanation First, calculate $E°_{cell}$. To do so, separate the reaction into its two half-reactions.

$$Fe(s) \longrightarrow Fe^{2+}(aq) + 2\,e^- \qquad E°_{anode} = -0.44\ V$$

$$Cd^{2+}(aq) + 2\,e^- \longrightarrow Cd(s) \qquad E°_{cathode} = -0.40\ V$$

$$Fe(s) + Cd^{2+}(aq) \longrightarrow Fe^{2+}(aq) + Cd(s) \qquad E°_{cell} = +0.04\ V$$

Two moles of electrons are transferred.

$$\log K° = \frac{nE°_{cell}}{0.0592\ V} = \frac{(2)(0.04\ V)}{0.0592\ V} = 1.35 \quad \text{and} \quad K° = 10^{1.35} = 22$$

The $K°$ value is larger than 1, which shows that the reaction is product-favored as written. Since the reaction occurs in aqueous solution, $K_c = K° = 22$.

☑ **Reasonable Answer Check** The value for $E°_{cell}$ is positive, which indicates a product-favored reaction, as does $K° > 1$.

PROBLEM-SOLVING PRACTICE 19.9

Using the standard reduction potentials listed in Table 19.1, calculate the equilibrium constant for the reaction

$$I_2(s) + Sn^{2+}(aq) \longrightarrow 2\,I^-(aq) + Sn^{4+}(aq)$$

19.7 Effect of Concentration on Cell Potential

All of the voltaic cells discussed so far are based on product-favored chemical reactions. As the reactions proceed in such an electrochemical cell, reactants are converted to products, so the concentrations of the species change continuously. As the reactant concentrations decrease, the cell voltage drops. The voltage eventually becomes zero when the reactants and products are at equilibrium.

The voltage of a voltaic cell is related to the concentrations of the reactants and products of its chemical reaction. We start with the relationship between Gibbs free energy change and concentration ***(← p. 873):***

$$\Delta G = \Delta G° + RT \ln Q$$

where Q is the reaction quotient ***(← p. 673).*** Q has the same form as the equilibrium constant, but it refers to a reaction mixture at a given instant in time, that is, a reaction mixture not necessarily at equilibrium.

Since $\Delta G = -nFE_{cell}$ and $\Delta G° = -nFE°_{cell}$,

$$-nFE_{cell} = -nFE°_{cell} + RT \ln Q$$

Solving for E_{cell} gives the relationship we seek:

$$E_{cell} = E°_{cell} - \frac{RT}{nF} \ln Q$$

$E°_{cell}$ is the voltage at standard-state concentrations and pressures; E_{cell} is the voltage at nonstandard conditions.

This equation is called the **Nernst equation.** We can change to base-10 logarithms (replace ln Q with 2.303 log Q) to get

$$E_{cell} = E°_{cell} - \frac{2.303\ RT}{nF} \log Q$$

In the Nernst equation, n is the number of moles of electrons transferred in the *balanced equation* of the process.

We can simplify this expression further by substituting numerical values for R (8.314 J mol^{-1} K^{-1}) and F (96,485 J V^{-1} mol^{-1}) and by assuming 25 °C (298 K) to get

$$E_{cell} = E^\circ_{cell} - \frac{(2.303)(8.314\ \text{J mol}^{-1}\ \text{K}^{-1})(298\ \text{K})}{n(96{,}485\ \text{J V}^{-1}\ \text{mol}^{-1})}\log Q$$

$$E_{cell} = E^\circ_{cell} - \frac{0.0592\ \text{V}}{n}\log Q \qquad (T = 298\ \text{K})$$

If the concentrations in Q all equal 1 (which is the standard state), then $Q = 1$ and $\log(1) = 0$, so the Nernst equation reduces to $E_{cell} = E^\circ_{cell}$.

The Nernst equation is used to calculate the voltage produced by an electrochemical cell under *nonstandard* conditions. It can also be used to calculate the concentration of a reactant or product in an electrochemical reaction from the measured value of the voltage produced.

PROBLEM-SOLVING EXAMPLE 19.10 Using the Nernst Equation

Consider this electrochemical reaction:

$$Zn(s) + Ni^{2+}(aq) \longrightarrow Zn^{2+}(aq) + Ni(s)$$

The standard cell potential $E^\circ_{cell} = 0.51$ V. Calculate the cell potential if the Ni^{2+} concentration is 5.0 M and the Zn^{2+} concentration is 0.050 M.

Answer $E_{cell} = 0.57$ V

Strategy and Explanation Use the Nernst equation to solve the problem. Two moles of electrons are transferred from 1 mol Zn to 1 mol Ni^{2+}, giving $n = 2$. At 298 K,

$$E_{cell} = 0.51\ \text{V} - \frac{0.0592\ \text{V}}{2}\log\frac{(\text{conc. }Zn^{2+})}{(\text{conc. }Ni^{2+})}$$

$$E_{cell} = 0.51\ \text{V} - \frac{0.0592\ \text{V}}{2}\log\left(\frac{0.050}{5.0}\right) = 0.51\ \text{V} - \frac{0.0592\ \text{V}}{2}\log(10^{-2})$$

$$= 0.51\ \text{V} - \frac{0.0592\ \text{V}}{2}(-2.00) = 0.57\ \text{V}$$

☑ **Reasonable Answer Check** The concentration of Ni^{2+} is 5.0 M, which is larger than the standard-state value of 1.0 M, and the concentration of Zn^{2+} is 0.050 M, much smaller than the standard-state value. Each of these departures from standard-state conditions tends to make the voltage slightly larger than the standard cell potential ($E^\circ_{cell} = 0.51$ V), and it is.

PROBLEM-SOLVING PRACTICE 19.10

What would the cell potential in the chemical system above become if (conc. Zn^{2+}) = 3.0 M and (conc. Ni^{2+}) = 0.010 M?

Concentration Cells

The voltaic cells discussed to this point have different reactions at the anode and the cathode. However, since the voltage of a cell depends on the concentrations of the reactants, we can construct a voltaic cell that uses the same species in both the anode and the cathode compartments but at different concentrations. A **concentration cell** is a voltaic cell in which the voltage is generated because of a difference in concentrations. As the cell operates, the concentration increases in the dilute cell and decreases in the concentrated cell.

Consider a concentration cell constructed with two identical $Cu|Cu^{2+}$ half-reactions occurring in separated compartments, as shown in Figure 19.11. If the same half-reactions occurred in the two compartments *at standard conditions* of 1-M concentrations, the potentials at the two electrodes would be the same ($E^\circ = +0.337$ V), so the cell potential would be zero. However, in a concentration cell the half-reactions are the same, but the concentrations in the two cells are different.

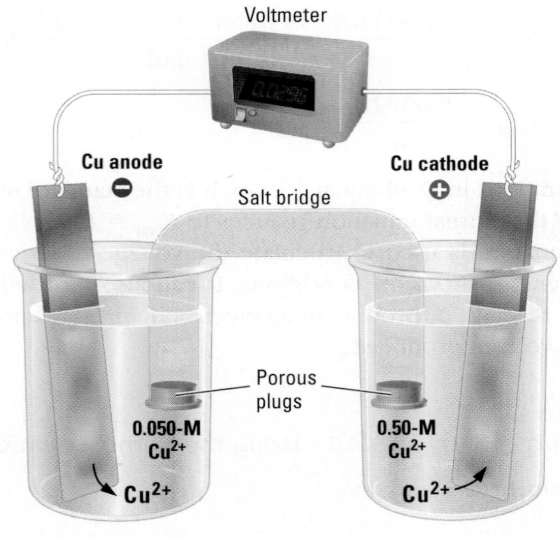

Figure 19.11 Concentration cell based on Cu|Cu^{2+} half-reactions. The cell has a positive net cell voltage and operates because the concentrations of Cu^{2+} ion are different in the two half-reaction compartments.

Let's take as an example Cu^{2+} concentrations of 0.050 M in the anode half-cell and 0.50 M in the cathode half-cell at 25 °C. The two half-reactions are

$$\begin{array}{rcll} Cu(s) & \longrightarrow & Cu^{2+}(aq, 0.050\ M) + 2\ e^- & \text{(anode, oxidation)} \\ Cu^{2+}(aq, 0.50\ M) + 2\ e^- & \longrightarrow & Cu(s) & \text{(cathode, reduction)} \\ \hline Cu^{2+}(aq, 0.50\ M) & \longrightarrow & Cu^{2+}(aq, 0.050\ M) & \text{(net reaction)} \end{array}$$

The cell potential is expressed by the Nernst equation:

$$\begin{aligned} E_{cell} &= E^\circ_{cell} - \frac{0.0592\ V}{2} \log \frac{(\text{conc. } Cu^{2+})_{dilute}}{(\text{conc. } Cu^{2+})_{concentrated}} \\ &= 0 - 0.0296 \times \log \frac{0.050\ M}{0.50\ M} \\ &= 0 - 0.0296 \times \log(10^{-1}) = 0 - 0.0296 \times (-1.00) = 0.0296\ V \end{aligned}$$

The cell potential is entirely determined by the ratio of concentrations of Cu^{2+} ions in the two half-reaction cells. The value of 0.0296 V is the potential of the cell when the concentrations of Cu^{2+} are at the initial conditions of 0.050 M and 0.50 M. As the cell operates, the concentration of Cu^{2+} in each half-cell changes; it increases in the dilute cell and decreases in the concentrated cell. Eventually, the two concentrations become equal, and the cell potential becomes zero.

Measurement of pH

This section uses H^+ for simplicity, but keep in mind that H_3O^+ is the species actually measured.

The H^+ concentration in a solution can be measured using the principles of a concentration cell. Consider a concentration cell based on the $H_2|H^+$ half-reaction. The cathode compartment contains a standard hydrogen electrode with known H^+ concentration (1.0 M), and the anode compartment contains the same type of electrode in contact with a solution of unknown H^+ concentration. The half-reactions and the overall reaction are

$$\begin{array}{rcll} H_2(g, 1\ bar) & \longrightarrow & 2\ H^+(aq, unknown) + 2\ e^- & \text{(anode, oxidation)} \\ 2\ H^+(aq, 1\ M) + 2\ e^- & \longrightarrow & H_2(g, 1\ bar) & \text{(cathode, reduction)} \\ \hline 2\ H^+(aq, 1\ M) & \longrightarrow & 2\ H^+(aq, unknown) & E_{cell} = ? \end{array}$$

The *standard* potential of the cell would be zero: $E^\circ_{cell} = 0$. However, the two half-cells have different hydrogen ion concentrations, so E_{cell} is not zero.

To analyze the cell further we assume a temperature of 25 °C and use the Nernst equation with $n = 2$.

$$E_{cell} = E^\circ_{cell} - \frac{0.0592\ V}{2} \times \log \frac{(\text{conc. } H^+)^2_{unknown}}{(\text{conc. } H^+)^2_{standard}}$$

$[H^+]_{standard} = 1$ M and $E^\circ_{cell} = 0$, so

$$E_{cell} = -\frac{0.0592 \text{ V}}{2} \times \log(\text{conc. } H^+)^2_{unknown}$$

Since $\log x^2 = 2 \log x$,

$$E_{cell} = \frac{-0.0592 \text{ V}}{2} \times 2 \log(\text{conc. } H^+)_{unknown}$$

$$= -0.0592 \text{ V} \times \log(\text{conc. } H^+)_{unknown}$$

Because $-\log [H^+] = \text{pH}$, the final expression is

$$E_{cell} = 0.0592 \text{ V} \times [-\log(\text{conc. } H^+)_{unknown}] = 0.0592 \text{ V} \times \text{pH}_{unknown}$$

$$\text{pH}_{unknown} = \frac{E_{cell}}{0.0592 \text{ V}}$$

Therefore, by measuring E_{cell}, the $\text{pH}_{unknown}$ can be determined.

PROBLEM-SOLVING EXAMPLE 19.11 Measuring pH with a Concentration Cell

Consider a concentration cell consisting of two hydrogen electrodes. One of the cells is a standard hydrogen electrode, and the other cell is in contact with an aqueous solution having an unknown pH. If the unknown concentration of H^+ is less than 1.0 M (which is generally true), then reduction occurs at the standard hydrogen electrode (cathode), and oxidation occurs at the nonstandard hydrogen electrode (anode). The measured potential of the cell is 0.426 V. Calculate the pH of the unknown solution at 25 °C.

Answer pH = 7.20

Strategy and Explanation We write the two half-reactions to start.

$$2\,H^+ (1\text{ M}) + 2\,e^- \longrightarrow H_2(g, 1\text{ bar}) \qquad \text{(reduction)}$$

$$H_2(g, 1\text{ bar}) \longrightarrow 2\,H^+ (\text{unknown M}) + 2\,e^- \qquad \text{(oxidation)}$$

$$2\,H^+(1\text{ M}) \longrightarrow 2\,H^+ (\text{unknown M}) \qquad E_{cell} = 0.426 \text{ V}$$

The Nernst equation at 25 °C is

$$E_{cell} = -\frac{0.0592 \text{ V}}{2} \log \frac{(\text{unknown conc. } H^+)^2}{(H^+\ 1\text{ M})^2}$$

$$= -\frac{0.0592 \text{ V}}{2} \times \log(\text{unknown conc. } H^+)^2 = 0.426 \text{ V}$$

$$= -\frac{0.0592 \text{ V}}{2} \times 2 \log(\text{unknown conc. } H^+) = 0.426 \text{ V}$$

$$E_{cell} = -0.0592 \text{ V} \times \log(\text{unknown conc. } H^+) = 0.426 \text{ V}$$

By definition $\text{pH} = -\log [H^+]$, so

$$0.0592 \text{ V} \times \text{pH} = 0.426 \text{ V} \qquad \text{and} \qquad \text{pH} = \frac{0.426 \text{ V}}{0.0592 \text{ V}} = 7.20$$

☑ **Reasonable Answer Check** The general relationship between the variables is $\text{pH} = E_{cell}/0.0592$ V, so using approximate values gives $0.43/0.06 \approx 7.2$, which is close to our more exact answer.

PROBLEM-SOLVING PRACTICE 19.11

If the same type of concentration cell were used with a solution of pH of 3.66, what E_{cell} would be measured?

The pH Meter

A concentration cell utilizing two hydrogen electrodes is not the best practical choice for routine pH measurement because it is bulky and difficult to maintain. Commercial pH meters are based on electrochemical principles similar to those described previously, but with more rugged and economical half-cells. A **pH meter** has two electrodes (Figure 19.12). One is a glass electrode using an Ag | AgCl half-cell dipped in an HCl solution of known concentration. At the tip of this indicator electrode is a very thin glass membrane that is sensitive to H^+ ion concentration differences. The other electrode is a reference electrode known as a *saturated calomel electrode.* It consists of a Pt wire dipped in a paste of Hg_2Cl_2 (calomel), liquid Hg, and saturated KCl solution. The glass electrode measures the H^+ ion concentration of the solution relative to its internal hydrogen ion concentration ***(← p. 764).*** The difference in voltage between the two electrodes is then converted electronically to give the pH of the solution.

In commercial instruments, the indicator and reference electrodes are generally combined in a *combination electrode.*

The pH of aqueous solutions is an extremely important indicator of their chemistry. Medical applications of pH measurements abound, as do environmental applications such as measuring the pH of acid rain ***(← p. 825).***

19.8 Neuron Cells

The central nervous system relies on specialized cells called **neurons** that communicate with each other through chemical signals generated by changes in the intracellular and extracellular ion concentrations (Figure 19.13). The communication between neurons also involves electrical signals generated by millisecond-long alterations in voltage due to changes in ion concentration. These electrical signals depend on the fact that the neuron's cell membrane separates different concentrations of ions inside and outside the cell as shown in Figure 19.14. While the cell is at rest—that is, while no signals are being sent—these different ion concentrations are maintained by a number of processes that move ions across the cell membrane. The main ions of importance are Na^+, K^+, Cl^-, and Ca^{2+}. The intracellular (inside the cell) concentrations of these four ions differ markedly from the extracellular (outside the cell) concentrations in mammalian cells.

Recall ***(← p. 409)*** that a cell membrane is composed of lipid molecules that separate the aqueous environment inside the cell from the aqueous environment outside the cell.

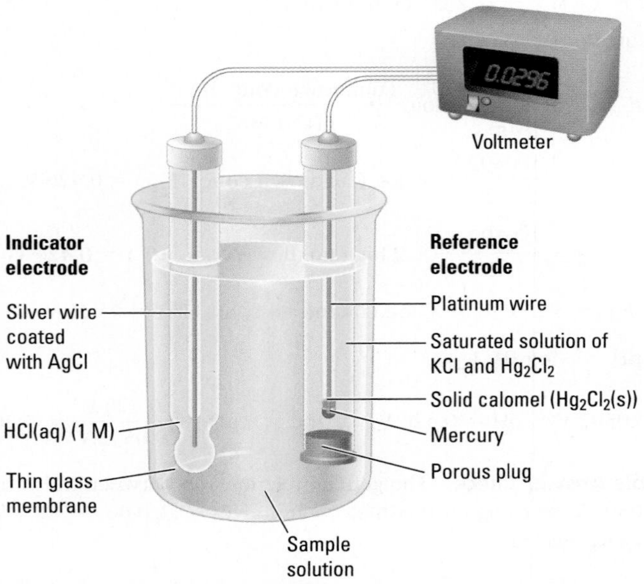
0.0296
Voltmeter
Indicator electrode
Silver wire coated with AgCl
HCl(aq) (1 M)
Thin glass membrane
Reference electrode
Platinum wire
Saturated solution of KCl and Hg2Cl2
Solid calomel (Hg2Cl2(s))
Mercury
Porous plug
Sample solution

Figure 19.12 The electrodes of a commercial pH meter. The glass electrode is a Ag|AgCl half-cell in a standard HCl solution that is enclosed by a glass membrane. It is sensitive to the external H_3O^+ in the solution relative to the H_3O^+ in the internal standard HCl solution. The saturated calomel electrode is the reference electrode.

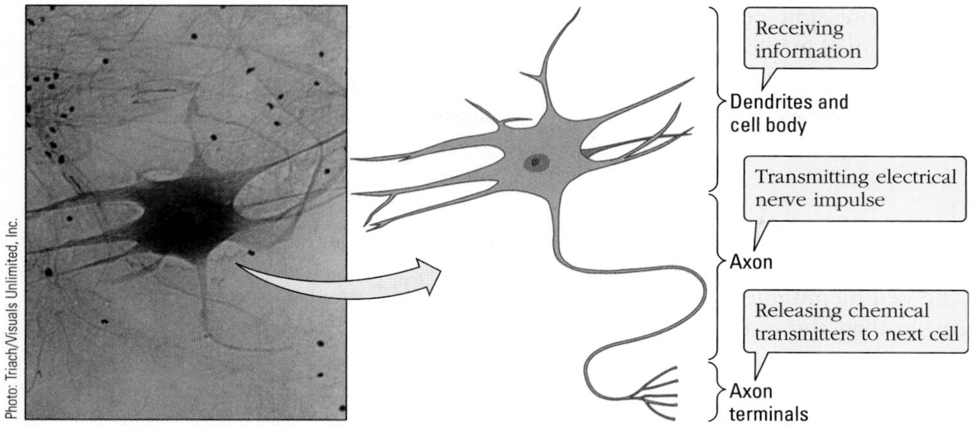

Figure 19.13 **A mammalian neuron cell.**

Ion	Intracellular Concentration (mM)	Extracellular Concentration (mM)	Potentials (mV)
Na^+	18	150	+56
K^+	135	3	−102
Cl^-	7	120	−76
Ca^{2+}	0.0001	1.2	+125

1 mM = 10^{-3} M
1 mV = 10^{-3} V

These differences in concentration create potentials across the neuron cell membrane. An associated equilibrium potential for each type of ion is given by this equation, which is based on the Nernst equation.

$$E_{\text{ion}} = \frac{2.303\ RT}{nF} \log \frac{(\text{conc. ion})_{\text{outside}}}{(\text{conc. ion})_{\text{inside}}}$$

Because this is a concentration cell, $E° = 0$, so $E°$ has been omitted from the equation.

Substituting numerical values for the constants, R and F, assuming $n = 1$ and a body temperature of 37 °C, and converting to millivolts, we can simplify this expression to

$$E_{\text{ion}} = (61.5\ \text{mV}) \log \frac{(\text{conc. ion})_{\text{outside}}}{(\text{conc. ion})_{\text{inside}}}$$

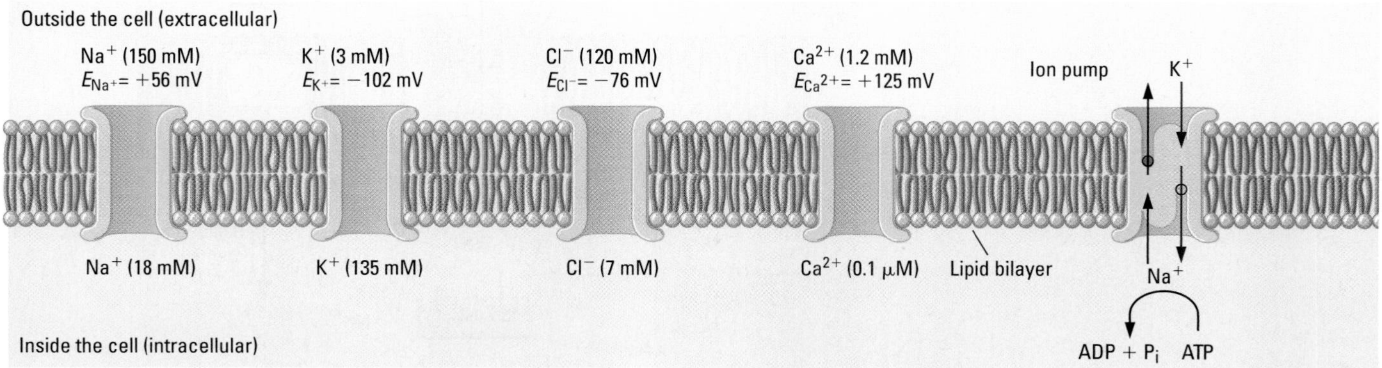

Figure 19.14 **Ion concentrations inside and outside a mammalian neuron cell.** Ion channels for Na^+, K^+, Cl^-, and Ca^{2+} are shown, as is the Na^+-K^+ ion pump. The sodium-potassium ion pump moves Na^+ and K^+ ions against their concentration gradients, by transferring Na^+ ions out of the cell and bringing K^+ ions into the cell. Concentrations are given in millimoles per liter, except for intracellular Ca^{2+}, which is given in micromoles per liter.

This expression computes the potential outside the cell membrane relative to the potential inside the cell membrane for each individual ion. For the specific case of K^+, where (conc. $K^+)_{outside}$ = 3 mM and (conc. $K^+)_{inside}$ = 135 mM,

$$E_{K^+} = (61.5 \text{ mV}) \log\left(\frac{(\text{conc. } K^+)_{outside}}{(\text{conc. } K^+)_{inside}}\right)$$

$$= (61.5 \text{ mV}) \log\left(\frac{3}{135}\right) = (61.5 \text{ mV})(-1.65) = -102 \text{ mV}$$

The cell membrane has a potential that is 102 mV (0.102 V) more negative on the inside than the outside due to the much higher K^+ concentration inside the cell (Figure 19.15).

Just as for K^+, each important type of ion shown in Figure 19.14 has a potential that depends on the concentrations of that type of ion inside and outside the cell membrane. The values for Na^+, K^+, Cl^-, and Ca^{2+} shown in the figure can be used with the Nernst equation to calculate the contribution that each ionic species makes to the final resting potential of the neuron. The *resting membrane potential* for a cell—that is, the potential when no nerve impulse is being transmitted—depends on each of the individual ion potentials. The equilibrium potentials for each of the ions involved are averaged in proportion to the relative permeability of the cell membrane for each ion. The resting membrane potential is different for different types of neuron cells and is in the range of −60 to −75 mV, with the inside of the cell being more negative than the outside.

How do the concentrations of ions inside and outside the cell membrane become different? Ions move across the cell membrane by several mechanisms, but all the ions undergo continual movement. In general, ions tend to move down concentration gradients, that is, from a region of higher to lower concentration. In addition, active *ion pumps* move Na^+ and K^+ *against* their concentration gradients, that is, from regions of lower to higher concentration. Thus, the ion pumps move Na^+ from inside to outside the cell membrane and move K^+ from outside to inside the cell membrane, a process called *active transport*. These active ion pumps require energy to perform this task, energy that comes from the hydrolysis of ATP ***(← p. 879).*** When a neuron is at rest, the passive movement of Na^+ and K^+ ions is exactly counterbalanced by the active movement of Na^+ and K^+ ions via the ion pumps. When a neuron is at rest, no *net* flow of ions occurs, so the ions are all at

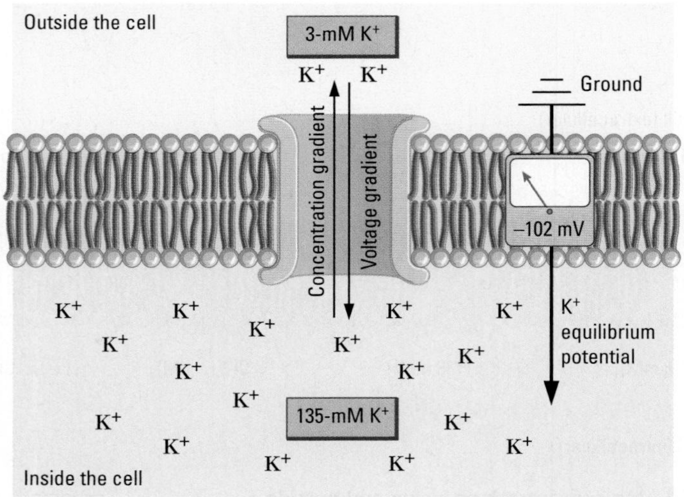

Figure 19.15 The equilibrium potential for K^+ across a neuron cell membrane. The K^+ concentration is higher inside the cell than outside it. The Nernst equation is used to calculate the −102 mV equilibrium potential for K^+.

equilibrium. The concentrations of the ions remain constant, although passive and active transport are both functioning at all times.

When a neuron is stimulated, a voltage change occurs from the resting value in the -60 to -75 mV range toward more positive values, and a large flow of Na^+ ions moves into the cell. This influx causes the membrane potential to become more positive. The positive change in the membrane potential generates an *action potential,* a brief electrical stimulus that results in chemical signaling between the neurons. For a cell to return to its resting potential, the cell membrane potential must return to a more negative value. The membrane potential is changed by a sustained flow of K^+ out of the cell, moving the potential back toward its resting value. The sequence leading to the generation of the action potential occurs very quickly in about 1 ms (1×10^{-3} s). Generation of action potentials by this mechanism forms the basis for the transmission of signals along nerve cells in the central nervous system. During this process, the bulk concentrations of K^+ and Na^+ change little, either inside or outside the cell membrane. The number of ions moving across the cell membrane is much less than the total number present. The action potential is an *electrochemical* event related to the change in ion concentrations inside and outside the cell, not to bulk concentrations of the ions.

EXERCISE 19.8 Neuron Equilibrium Potential

Calculate the membrane potential, E_{Na^+}, across a neuron cell membrane if Na^+ were the only ion to be considered.

19.9 Common Batteries

Used "dead" batteries.

Voltaic cells include the convenient, portable sources of energy that we call *batteries.* Some batteries, such as the common flashlight battery, consist of a single cell while others, such as automobile batteries, contain multiple cells. Batteries can be classified as primary or secondary depending on whether the reactions at the anode and cathode can be easily reversed. In a **primary battery** the electrochemical reactions cannot easily be reversed, so when the reactants are used up the battery is "dead" and must be discarded. In contrast, a **secondary battery** (sometimes called a storage battery or a rechargeable battery) uses an electrochemical reaction that can be reversed, so this type of battery can be recharged.

Primary Batteries

For many years the "dry cell," invented by Georges Leclanché in 1866, was the major source of energy for flashlights and toys. The container of the dry cell is made of zinc, which acts as the anode. The zinc is separated from the other chemicals by a liner of porous paper that functions as the salt bridge (Figure 19.16). In the center of the dry cell is a graphite cathode, which is unreactive, inserted into a moist mixture of ammonium chloride, NH_4Cl; zinc chloride, $ZnCl_2$; and manganese(IV) oxide, MnO_2. As electrons flow from the cell, the zinc is oxidized,

$$Zn(s) \longrightarrow Zn^{2+}(aq) + 2\,e^- \qquad \text{(anode, oxidation)}$$

and the ammonium ions are reduced,

$$2\,NH_4^+(aq) + 2\,e^- \longrightarrow 2\,NH_3(g) + H_2(g) \qquad \text{(cathode, reduction)}$$

The ammonia reacts with zinc ions to form a zinc-ammonia complex ion (Section 17.5 (← *p. 835*)), which prevents a buildup of gaseous ammonia.

$$Zn^{2+}(aq) + 2\,NH_3(g) \longrightarrow [Zn(NH_3)_2]^{2+}(aq)$$

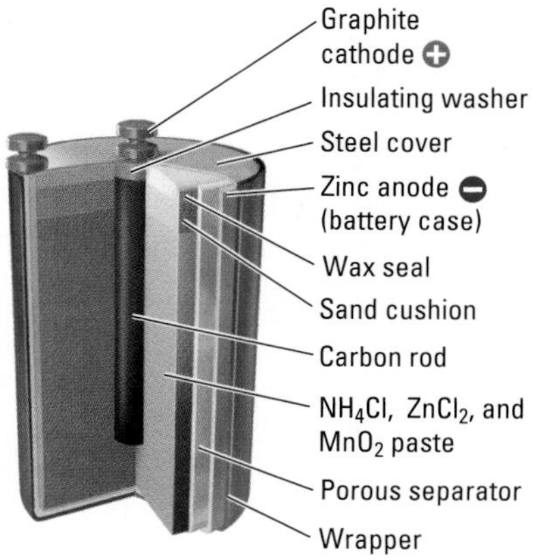

Figure 19.16 Leclanché dry cell. The cell contains a zinc anode (the battery container), a graphite cathode, and an electrolyte consisting of a moist paste made of NH_4Cl, $ZnCl_2$, and MnO_2.

The hydrogen produced at the cathode is oxidized by MnO_2 in the cell, which prevents hydrogen accumulation.

$$H_2(g) + 2\,MnO_2(s) \longrightarrow Mn_2O_3(s) + H_2O(\ell)$$

The overall cell reaction, which produces 1.5 V, is

$$2\,MnO_2(s) + 2\,NH_4^+(aq) + Zn(s) \longrightarrow Mn_2O_3(s) + H_2O^+(\ell) + [Zn(NH_3)_2]^{2+}(aq)$$

A major disadvantage of the Leclanché cell is the occurrence of a slow reaction even when current is not being drawn. As a consequence, stored cells run down and tend to have a short shelf life.

Some of the problems of the dry cell are overcome by the newer, more expensive alkaline battery. An *alkaline battery,* which produces 1.54 V, also uses the oxidation of zinc as the anode reaction, but under alkaline ($pH > 7$) conditions.

$$Zn(s) + 2\,OH^-(aq) \longrightarrow ZnO(aq) + H_2O(\ell) + 2\,e^- \qquad \text{(anode, oxidation)}$$

The electrons that pass through the external circuit are consumed by reduction of manganese(IV) oxide at the cathode.

$$MnO_2(s) + H_2O(\ell) + e^- \longrightarrow MnO(OH)(s) + OH^-(aq) \qquad \text{(cathode, reduction)}$$

Mercury batteries are sealed to prevent leakage of mercury and should never be heated. Heating increases the pressure of mercury vapor within the battery, ultimately causing the battery to explode.

In the *mercury battery* (Figure 19.17), the oxidation of zinc is again the anode reaction. The cathode reaction is the reduction of mercury(II) oxide.

$$HgO(s) + H_2O(\ell) + 2\,e^- \longrightarrow Hg(\ell) + 2\,OH^-(aq)$$

The voltage of this battery is about 1.35 V. Mercury batteries are used in calculators, watches, hearing aids, cameras, and other devices in which small size is an advantage. However, mercury and its compounds are poisonous, so proper disposal of mercury batteries is necessary.

Secondary Batteries

Secondary batteries are rechargeable because, as they discharge, the oxidation products remain at the anode and the reduction products remain at the cathode. As a result, if the direction of electron flow is reversed, the anode and cathode reactions are reversed and the reactants are regenerated. Under favorable conditions, secondary batteries may be discharged and recharged hundreds or even thousands of times. Examples of secondary batteries include automobile batteries, nicad (nickel-cadium) batteries, and lithium-ion batteries.

The lead-acid battery was first described to the French Academy of Sciences in 1860 by Gaston Planté.

Lead-Acid Storage Batteries The familiar automobile battery, the lead-acid storage battery, is a secondary battery consisting of six cells, each containing porous metallic lead electrodes and lead(IV) oxide electrodes immersed in aqueous sulfuric acid (Figure 19.18). When this battery produces an electric current, metallic lead is oxidized to lead(II) sulfate at the anode, and lead(IV) oxide is reduced to lead(II) sulfate at the cathode.

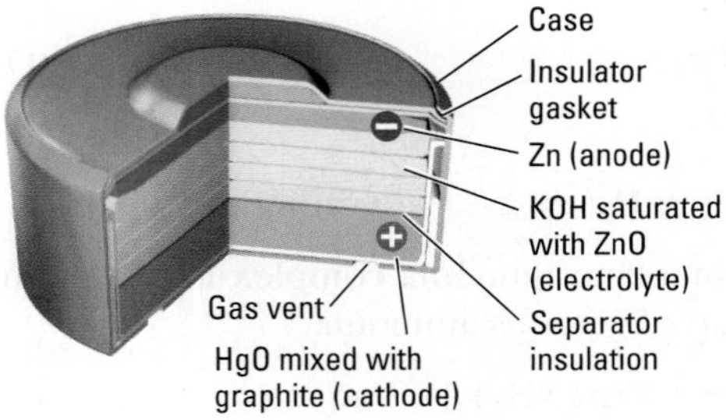
Case
Insulator gasket
Zn (anode)
KOH saturated with ZnO (electrolyte)
Separator insulation
Gas vent
HgO mixed with graphite (cathode)

Figure 19.17 Mercury battery. The reducing agent is zinc and the oxidizing agent is mercury(II) oxide, HgO.

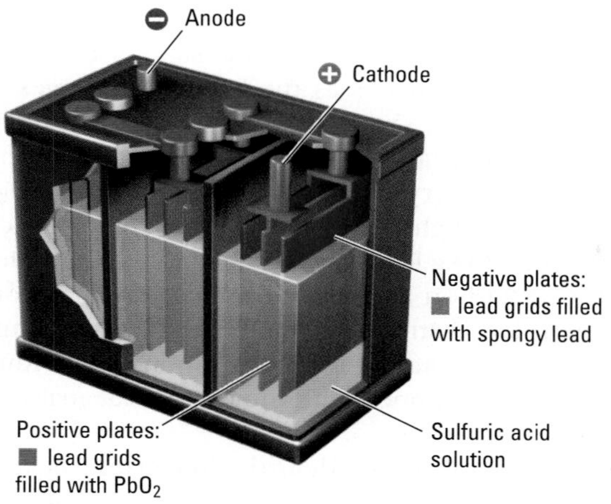
Anode
Cathode
Negative plates: lead grids filled with spongy lead
Positive plates: lead grids filled with PbO2
Sulfuric acid solution

Figure 19.18 Lead-acid storage battery. The anodes are lead grids filled with spongy lead. The cathodes are lead grids filled with lead(IV) oxide, PbO_2. Each cell produces a potential of about 2 V. Six cells connected in series produce the desired overall battery voltage.

$$Pb(s) + HSO_4^-(aq) + H_2O(\ell) \longrightarrow PbSO_4(s) + H_3O^+(aq) + 2\,e^- \qquad E^\circ_{anode} = -0.356\text{ V}$$

$$PbO_2(s) + 3\,H_3O^+(aq) + HSO_4^-(aq) + 2\,e^- \longrightarrow PbSO_4(s) + 5\,H_2O(\ell) \qquad E^\circ_{cathode} = 1.685\text{ V}$$

$$Pb(s) + PbO_2(s) + 2\,H_3O^+(aq) + 2\,HSO_4^-(aq) \longrightarrow 2\,PbSO_4(s) + 4\,H_2O(\ell) \qquad E^\circ_{cell} = +2.041\text{ V}$$

The six cells connected in series in a typical automobile battery produce a total voltage of 12 V.

To understand why the lead storage battery is rechargeable, consider that the lead sulfate formed at both electrodes is an insoluble compound that mostly *stays on the electrode surface.* As a result, it remains available for the reverse reaction. To recharge a secondary battery, a source of direct electrical current is supplied so that electrons are forced to flow in the direction opposite from when the battery was discharging. This causes the overall battery reaction to be reversed and regenerates the reactants that originally produced the battery's voltage and current. For the lead-acid storage battery, the overall redox reaction is

The process of charging a secondary battery involves electrolysis, which is discussed in Section 19.11.

Discharging battery produces electricity

$$Pb(s) + PbO_2(s) + 2\,HSO_4^-(aq) + 2\,H_3O^+(aq) \rightleftharpoons 2\,PbSO_4(s) + 4\,H_2O(\ell)$$

Charging battery requires electricity from an external source

Normal charging of an automobile lead-acid storage battery occurs during driving. In addition to reversing the overall battery reaction, charging reduces a little hydronium ion at the cathode and oxidizes a little water at the anode.

Reduction of hydronium ion: $4\,H_3O^+(aq) + 4\,e^- \longrightarrow 2\,H_2(g) + 4\,H_2O(\ell)$

Oxidation of water: $6\,H_2O(\ell) \longrightarrow O_2(g) + 4\,H_3O^+(aq) + 4\,e^-$

These reactions produce a hydrogen-oxygen mixture inside the battery, which, if accidentally ignited, can explode. Therefore, no sparks or open flames should be brought near a lead-acid storage battery, even the sealed kind.

When a car with a dead battery is jump-started, the last jumper cable connection should be to the car's frame—well away from the battery—to avoid igniting any H_2 in the battery with a spark.

As the battery operates, sulfuric acid is consumed in both the anode and the cathode reactions, thereby decreasing the concentration of the sulfuric acid electrolyte.

Before the introduction of modern sealed automotive batteries, the density of this battery acid was routinely measured to indicate the state of charge of the battery. The density of the battery acid decreases as the battery discharges. Consequently, the lower the density, the lower the charge of the battery. In modern sealed batteries, it is difficult to gain access to the acid to measure the electrolyte density.

The lead-acid storage battery is relatively inexpensive, reliable, and simple, and it has an adequate lifetime. High weight is its major fault. A typical automobile battery contains about 15 to 20 kg lead, which is required to provide the large number of electrons needed to start an automobile engine, especially on a cold morning. Another problem with lead batteries is that lead mining, manufacturing, and disposal of used batteries can contaminate air and groundwater. Auto batteries should be recycled by companies equipped with the proper safeguards.

The number of electrons that a battery can move from the anode to the cathode is proportional to the amount of reactants involved.

Nickel-Cadmium and Nickel-Metal Hydride Batteries Nickel-cadmium (nicad) secondary batteries are lightweight, can be quite small, and produce a constant voltage until completely discharged, which makes them useful in cordless appliances, video camcorders, portable radios, and other applications.

Nicad batteries can be recharged because the reaction products are insoluble hydroxides that remain at the electrode surfaces. The anode reaction during the discharge cycle is the oxidation of cadmium, and the cathode reaction is the reduction of nickel oxyhydroxide, NiO(OH).

$$\mathrm{Cd(s)} + 2\,\mathrm{OH^-(aq)} \longrightarrow \mathrm{Cd(OH)_2(s)} + 2\,\mathrm{e^-} \qquad E^\circ_{\mathrm{anode}} = -0.809\ \mathrm{V}$$

$$2[\mathrm{NiO(OH)(s)} + \mathrm{H_2O}(\ell) + e^- \longrightarrow \mathrm{Ni(OH)_2(s)} + \mathrm{OH^-(aq)}] \qquad E^\circ_{\mathrm{cathode}} = 0.490\ \mathrm{V}$$

$$\mathrm{Cd(s)} + 2\,\mathrm{NiO(OH)(s)} + 2\,\mathrm{H_2O}(\ell) \longrightarrow \mathrm{Cd(OH)_2(s)} + 2\,\mathrm{Ni(OH)_2(s)} \qquad E^\circ_{\mathrm{cell}} = 1.299\ \mathrm{V}$$

Like mercury batteries, nicad batteries should be disposed properly because of the toxicity of cadmium and its compounds.

Another nickel battery that uses the same cathode reaction but a different anode reaction is the *nickel-metal hydride* (NiMH) *battery,* which eliminates the use of cadmium. These batteries, shown being charged in Figure 19.19, are used in portable power tools, cordless shavers, and photoflash units. The anode is a metal (M) alloy, often nickel or a rare earth, in a basic electrolyte (KOH). The anode reaction oxidizes hydrogen absorbed in the metal alloy, and water is produced.

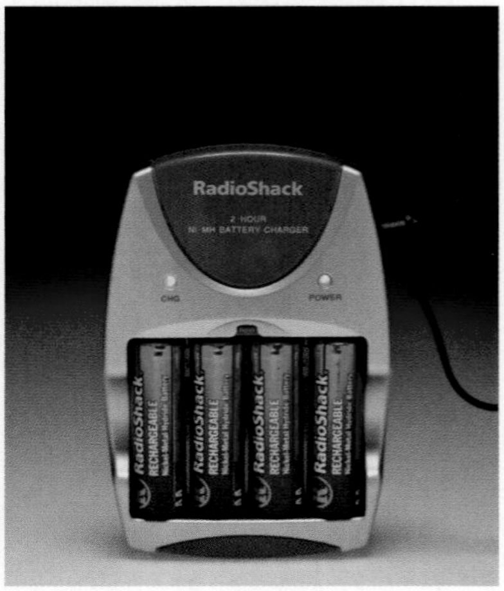

Figure 19.19 Nickel-metal hydride (NiMH) batteries in a battery charger.

$$\mathrm{MH(s)} + \mathrm{OH^-(aq)} \longrightarrow \mathrm{M(s)} + \mathrm{H_2O}(\ell) + \mathrm{e^-} \qquad \text{(anode reaction)}$$

The overall reaction is

$$\mathrm{MH(s)} + \mathrm{NiO(OH)(s)} \longrightarrow \mathrm{M(s)} + \mathrm{Ni(OH)_2(s)} \qquad E_{\mathrm{cell}} = 1.4\ \mathrm{V}$$

EXERCISE **19.9 Recharging a Nicad Battery**

Write the electrode reactions that take place when a nicad battery is recharged; identify the anode and cathode reactions.

Figure 19.20 Lithium-ion battery. This type of battery finds many uses in which a high energy density and low mass are desired.

Lithium-Ion Batteries Lithium-ion batteries (Figure 19.20) benefit from the low density and high reducing strength of lithium metal (Table 19.1). The anode in such a battery is made of lithium metal that has been mixed with a conducting carbon polymer. The polymer has tiny spaces in its structure that can hold the lithium atoms as well as the lithium ions formed by the oxidation reaction.

$$\mathrm{Li(s)\ (in\ polymer)} \longrightarrow \mathrm{Li^+(in\ polymer)} + \mathrm{e^-} \qquad \text{(anode reaction)}$$

CHEMISTRY IN THE NEWS Plug-in Hybrid Cars

Hybrid cars—cars that use a gasoline engine and an electric motor for power—have been on the road for years, and several models are currently for sale in the United States. Plug-in hybrid electric vehicles (PHEVs) are the newest trend. These cars can recharge their batteries from the external electrical grid. Because of this recharging capability, they can run for a large fraction of the time, maybe even all the time, on electricity and never, or almost never, need to engage their gasoline engines.

Major car companies are pushing ahead plans to introduce plug-in hybrids soon. General Motors plans to introduce the Chevy Volt, a PHEV sedan, in late 2010. Toyota has been a leader in hybrid car sales with the Prius since 2000 and plans a PHEV soon as well.

Hybrid cars use NiMH battery-powered electric motors for power, but when needed, the gasoline engine turns on and provides additional power. When hybrid cars brake or go downhill, the electric motor becomes a generator and recharges the batteries.

PHEVs require significantly more electrical power from their batteries than other hybrids, so car companies have turned to lithium-ion batteries which have a high power/mass ratio, even though they are expensive. The need for batteries that can store a lot of electrical energy, are not too massive, and are cheap enough to be economical has led to a new emphasis on battery research.

The Chevy Volt, a PHEV to be mass produced by 2010.

The cathode also contains lithium ions, but in the lattice of a metal oxide such as CoO_2. This oxide lattice, like the carbon-polymer electrode, has holes in it that accommodate Li^+ ions. The reduction reaction is

$$Li^+(\text{in } CoO_2) + e^- + CoO_2 \longrightarrow LiCoO_2 \qquad \text{(cathode reaction)}$$

The overall reaction in the lithium-ion battery is therefore

$$Li(s) + CoO_2(s) \longrightarrow LiCoO_2(s) \qquad E_{cell} = 3.4\ V$$

Lithium-ion batteries have a large voltage (3.4 V per cell) and very high energy output for their mass. They can be recharged many hundreds of times. Because of these desirable characteristics, lithium-ion batteries are used in cellular telephones, laptop computers, and digital cameras.

CONCEPTUAL EXERCISE 19.10 Emergency Batteries

You are stranded on an island and need to communicate your location to receive help. You have a battery-powered radio transmitter, but the lead batteries are discharged. There is a swimming pool nearby and you find a tank of chlorine gas and some plastic tubing that can withstand being exposed to chlorine. There is an iron chain-link fence around the pool, and you have some tools. Devise a battery that might be used to power the radio using these items.

19.10 Fuel Cells

A **fuel cell** is an electrochemical cell that converts the chemical energy of fuels directly into electricity. It functions somewhat like a battery, but in contrast to a battery, its reactants are continually supplied from an external reservoir. The best-known fuel cell is the alkaline fuel cell, which is used in the Space Shuttle. Alkaline fuel cells have been used since the 1960s by NASA to power electrical systems on spacecraft, but they are very expensive and unlikely to be commercialized. They require very pure hydrogen and oxygen to operate and are susceptible to contamination. Therefore, alternative fuel cells have been developed.

Wilson Greatbatch

1919–

In the 1960s, Wilson Greatbatch invented a lithium battery to power implantable cardiac pacemakers, which revolutionized them. One of the most prolific inventors of the twentieth century, with more than 150 patents, Greatbatch was educated at Cornell University and the University of Buffalo. As a young engineer, he worked on medical-technical instruments, which led to his work on cardiac pacemakers. He was awarded a patent for the world's first implantable pacemaker in the early 1960s. Subsequent pacemaker design improvements have been based on his pioneering development. Millions of people worldwide have benefited directly from Greatbatch's invention.

Proton-Exchange Membrane Fuel Cell

The hydrogen-oxygen fuel cell produces electrical power from the oxidation of hydrogen in an electrochemical cell. The two half-reactions are

$$H_2 \longrightarrow 2\,H^+ + 2\,e^- \qquad \text{(anode, oxidation)}$$

$$\tfrac{1}{2}O_2 + 2\,H^+ + 2\,e^- \longrightarrow H_2O \qquad \text{(cathode, reduction)}$$

Hydrogen gas is pumped to the anode, and oxygen gas or air is pumped to the cathode (Figure 19.21). The graphite electrodes are surrounded by catalysts containing platinum. The two electrodes are separated by a semipermeable membrane—a proton-exchange membrane (PEM), which is a thin plastic sheet—that allows the passage of H^+ ions but not electrons. Instead, the electrons must pass through an external circuit, where they can be used to perform work. The electrochemical reactions are aided by platinum catalysts on both sides of the PEM that are in contact with the anode and the cathode. The H^+ ions produced at the anode pass through the PEM to the cathode. At the cathode, hydroxide ions produced by the reduction of O_2 react with the protons to produce water. The overall reaction of the fuel cell

$$H_2(g) + \tfrac{1}{2}O_2(g) \longrightarrow H_2O(\ell)$$

produces approximately 0.7 V. To get a higher voltage, fuel cells are stacked and connected in series.

PEM fuel cells can operate at a temperature of 80 °C. Engineering and materials advances have produced PEM fuel cells with power densities such that a device the size of a suitcase can power a car.

The pure hydrogen needed by the PEM fuel cell poses a problem for their more general use. Pure hydrogen is flammable and is difficult to store and distribute. Other hydrogen sources are more attractive. A device called a *reformer* can turn a hydrocarbon (such as natural gas) or alcohol (such as methanol) fuel into hydrogen, which can then be fed to the fuel cell. However, a fuel cell with a reformer has lower overall efficiency and produces products in addition to water.

Hydrogen does not occur in nature as H_2. Therefore, H_2 fuel has to be manufactured. Currently, most H_2 is produced as a by-product of petroleum refining (*← p. 538*) or by treating methane with steam (*← p. 688*). Significant quantities of fossil-fuel energy are required to manufacture H_2.

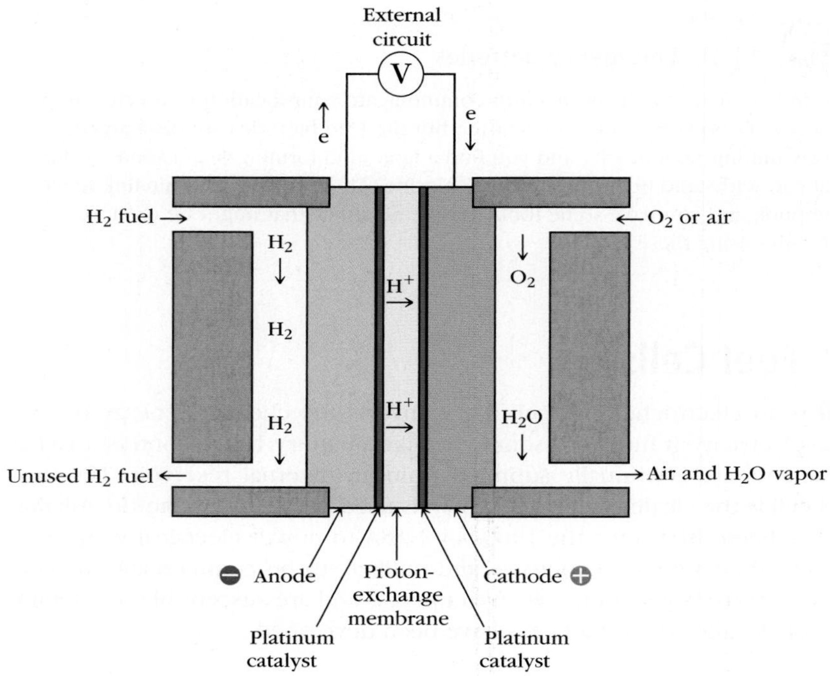

Figure 19.21 A proton-exchange membrane $H_2|O_2$ fuel cell. H_2 is oxidized in the anode chamber. O_2 is reduced in the cathode chamber.

19.11 Electrolysis—Causing Reactant-Favored Redox Reactions to Occur

Reactant-favored redox systems (positive $\Delta G°$) can be made to produce products if electrons are forced into the electrochemical system from an external source of electrical current, such as a battery. This process, called **electrolysis,** provides a way to carry out reactant-favored electrochemical reactions that will not take place by themselves. Electrolytic processes are even more important in our economy than the product-favored redox reactions that power batteries. Such electrolytic processes are used in the production and purification of many metals, including copper and aluminum (Sections 22.3 and 21.4), and in electroplating processes that produce a thin coating of metal on many different kinds of items.

Lysis means "splitting," so electrolysis means "splitting with electricity." Electrolysis reactions are chemical reactions caused by the flow of electricity.

Voltaic and *electrolytic* cells share some characteristics, but they also differ in important ways:

Properties of all electrochemical cells:

- Two electrodes are in contact with a conducting medium and an external circuit.
- Oxidation occurs at the anode.
- Reduction occurs at the cathode.

Properties of voltaic cells:

- The anode is the source of electrons and has a negative charge.
- The cathode is the sink for electrons and has a positive charge.
- Electrons flow through the external circuit from the anode to the cathode.
- The two half-reactions occur in separate cells, which are joined by a salt bridge.

Properties of electrolytic cells:

- The external circuit *must* contain a source of direct electrical current, for example, a battery. The battery forces electrons at a sufficiently high voltage into the anode (which becomes negative) and removes electrons from the cathode (which becomes positive).
- The two half-reactions often can be carried out in the same cell, in which case no separation is necessary and no salt bridge is necessary.
- The electrodes are often inert, their function being to provide a path for electrons to enter and leave the cell.

An example of electrolysis is the decomposition of molten sodium chloride. In this process, a pair of electrodes dips into molten NaCl (Figure 19.22). In the molten liquid, Na^+ and Cl^- ions are free to move. The Na^+ ions are attracted to the negative electrode, and the Cl^- ions are attracted to the positive electrode. Reduction of Na^+ ions to metallic Na atoms occurs at the cathode (negative electrode). Oxidation of Cl^- ions to $Cl_2(g)$ occurs at the anode (positive electrode).

$$2\,Na^+(\text{in melt}) + 2\,e^- \longrightarrow 2\,Na(\ell) \qquad \text{(cathode, reduction)}$$

$$2\,Cl^-(\text{in melt}) \longrightarrow Cl_2(g) + 2\,e^- \qquad \text{(anode, oxidation)}$$

$$2\,Na^+(\text{in melt}) + 2\,Cl^-(\text{in melt}) \longrightarrow 2\,Na(\ell) + Cl_2(g) \qquad \text{(net cell reaction)}$$

The E°_{cell} for this reaction as written is negative—the reaction is not product-favored. An external source of direct current at a potential higher than the negative potential of the reaction is required for the forward reaction to proceed.

The electrolysis of molten salts is an energy-intensive process because energy is needed to melt the salt as well as to cause the anode and cathode reactions to take place.

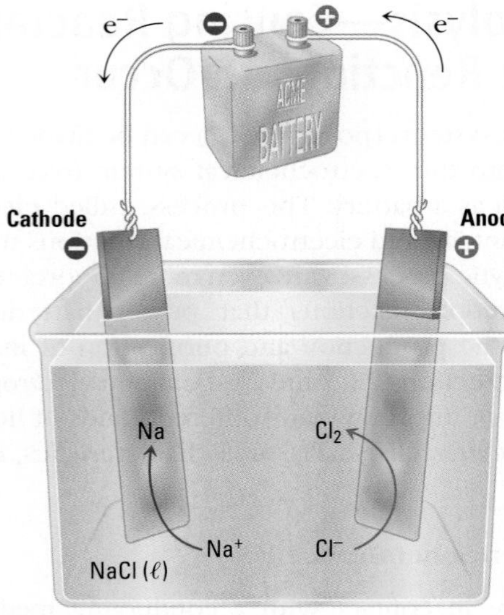

Figure 19.22 Electrolysis of molten sodium chloride. A direct current supplied by the battery drives the reaction in the product-favored direction. At the anode, Cl^- ions are oxidized to $Cl_2(g)$. At the cathode, Na^+ ions are reduced to $Na(\ell)$ atoms. The electrodes are made of inert materials.

What happens if a direct current is passed through an *aqueous solution* of a salt, such as potassium iodide, KI, rather than through the molten salt? To predict the outcome of the electrolysis we must first decide what in the solution can be oxidized and reduced. For KI(aq), the solution contains K^+ ions, I^- ions, and H_2O molecules. Potassium is already in its highest common oxidation state in K^+, so it cannot be oxidized. However, both the I^- ion and the H_2O could be oxidized. The possible anode half-reaction oxidations and their reduction potentials are

$$2\,I^-(aq) \longrightarrow I_2(s) + 2\,e^- \qquad E^\circ_{anode} = 0.535\ V$$

$$6\,H_2O(\ell) \longrightarrow O_2(g) + 4\,H_3O^+(aq) + 4\,e^- \qquad E^\circ_{anode} = 1.229\ V$$

Whenever two or more electrochemical reactions are possible at the same electrode, use Table 19.1 (p. 920) to decide which reaction is more likely to occur under standard-state conditions. Considering the two possible anode reactions, item 4 in the discussion of Table 19.1 indicates that the less positive the reduction potential, the more likely a half-reaction is to occur as an oxidation. That is, the farther toward the bottom of Table 19.1 a half-reaction is, the more likely it is to occur as an oxidation. Oxidation of I^- to I_2 has the less positive E°, so this is the more likely anode reaction.

Since I^- is the lowest common oxidation state of iodine, there are only two species that can be reduced at the cathode: K^+ ions and water molecules. The possible cathode half-reaction reductions and their reduction potentials are

$$K^+(aq) + e^- \longrightarrow K(s) \qquad E^\circ_{cathode} = -2.925\ V$$

$$2\,H_2O(\ell) + 2\,e^- \longrightarrow H_2(g) + 2\,OH^-(aq) \qquad E^\circ_{cathode} = -0.8277\ V$$

Point 3 in the discussion of Table 19.1 states that the more positive the reduction potential, the more easily a substance on the left-hand side of a half-reaction can be reduced. Since E° for H_2O is more positive (less negative) than E° for K^+, H_2O is more likely to be reduced. Therefore, in aqueous KI solution, the overall reaction and cell potential are

$$E^\circ_{cell} = E^\circ_{cathode} - E^\circ_{anode} = -0.8277\ V - (0.535\ V) = -1.363\ V$$

$$2\,I^-(aq) + 2\,H_2O(\ell) \longrightarrow I_2(s) + H_2(g) + 2\,OH^-(aq)$$

$$E^\circ_{cell} = -1.363\ V$$

The negative value of E°_{cell} indicates that the reaction is not product-favored and that an external energy source is needed for the reaction to occur.

An experiment in which electrical current is passed through aqueous KI (Figure 19.23) shows that this prediction is correct. At the anode (on the right in Figure 19.23a), the I^- ions are oxidized to I_2, which produces a yellow-brown color in the solution. At the cathode, water is reduced to gaseous hydrogen and aqueous hydroxide ions. The formation of excess OH^- ions is shown by the pink color of the phenolphthalein indicator added to the solution.

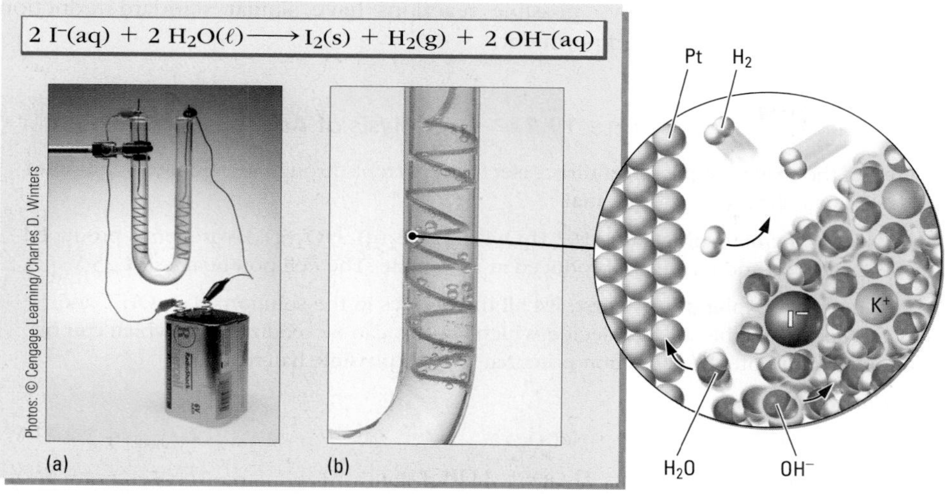

Figure 19.23 The electrolysis of aqueous potassium iodide. (a) Aqueous KI is found in both compartments of the cell, and both electrodes are platinum. At the anode (*left*), the I^- ion is oxidized to iodine, which gives the solution a yellow-brown color.

$$2\ I^-(aq) \longrightarrow I_2(aq) + 2\ e^-$$

At the cathode (*right*), water is reduced, and the presence of OH^- ion is indicated by the pink color of the acid-base indicator, phenolphthalein.

$$2\ H_2O(\ell) + 2\ e^- \longrightarrow H_2(g) + 2\ OH^-(aq)$$

(b) A close-up of the cathode shows bubbles of H_2 and evidence of generation of OH^-.

When electrolysis is carried out by passing a direct electrical current through an aqueous solution, the electrode reactions most likely to take place are those that require the least voltage, that is, the half-reactions that combine to give the least negative overall cell voltage. This means that in aqueous solution these conditions apply.

1. ***A metal ion or other species can be reduced if it has a reduction potential more positive than −0.8 V, the potential for reduction of water.*** Table 19.1 shows that many metal ions are in this category. If a species has a reduction potential more negative than −0.8 V, then water will preferentially be reduced to $H_2(g)$ and OH^- ions. Metal ions in this latter category include Na^+, K^+, Mg^{2+}, and Al^{3+}. Consequently, producing these metals from their ions requires electrolysis of a molten salt with no water present.
2. ***A species can be oxidized in aqueous solution if it has a reduction potential less positive than 1.2 V, the potential for reduction of $O_2(g)$ to water.*** If the reduction potential is less positive than for reduction of $O_2(g)$, then oxidation of the species on the right-hand side of a half-reaction is more likely than oxidation of water. Most of the half-equations in Table 19.1 are in this category. If a species has a reduction potential more positive than 1.2 V (that is, if its half-reaction is above the water-oxygen half-reaction in Table 19.1), water will be oxidized preferentially. For example, $F^-(aq)$ cannot be oxidized electrolytically to $F_2(g)$ because water will be oxidized to $O_2(g)$ instead.

The voltage that must be applied to an electrolysis cell is always somewhat greater than the voltage calculated from standard reduction potentials. An *overvoltage* is required, which is an additional voltage needed to overcome limitations in the electron transfer rate at the interface between electrode and solution. Another factor affecting the necessary voltage is that concentrations and pressures are usually not at standard-state conditions. Redox reactions that involve the formation of O_2 or H_2 are especially prone to have large overvoltages. Since overvoltages cannot be predicted accurately, the only way to determine with certainty which half-reaction will occur in an electrolysis cell when two possible reactions have similar standard reduction potentials is to perform the experiment.

PROBLEM-SOLVING EXAMPLE 19.12 Electrolysis of Aqueous NaOH

Predict the results of passing a direct electrical current through an aqueous solution of NaOH. Calculate the cell potential.

Answer The net cell reaction is $2\ H_2O(\ell) \rightarrow 2\ H_2(g) + O_2(g)$. Hydrogen is produced at the cathode and oxygen is produced at the anode. The cell potential is −1.23 V.

Strategy and Explanation First, list all the species in the solution: Na^+, OH^-, and H_2O. Next, use Table 19.1 to decide which species can be oxidized and which can be reduced, and note the reduction potential of each possible half-reaction.

Reductions:

$$Na^+(aq) + e^- \longrightarrow Na(s) \qquad E^\circ_{cathode} = -2.71\ V$$

$$2\ H_2O(\ell) + 2\ e^- \longrightarrow H_2(g) + 2\ OH^-(aq) \qquad E^\circ_{cathode} = -0.83\ V$$

Oxidations:

$$4\ OH^-(aq) \longrightarrow O_2(g) + 2\ H_2O(\ell) + 4\ e^- \qquad E^\circ_{anode} = +0.40\ V$$

$$6\ H_2O(\ell) \longrightarrow O_2(g) + 4\ H_3O^+(aq) + 4\ e^- \qquad E^\circ_{anode} = +1.229\ V$$

Water will be reduced to H_2 at the cathode because the potential for this half-reaction is more positive. At the anode, OH^- will be oxidized because the potential is smaller than that for water. The net cell reaction is

$$2\ H_2O(\ell) \longrightarrow 2\ H_2(g) + O_2(g)$$

and the cell potential under standard conditions is

$$E^\circ_{cell} = E^\circ_{cathode} - E^\circ_{anode} = (-0.83\ V) - (+0.40\ V) = -1.23\ V$$

PROBLEM-SOLVING PRACTICE 19.12

Predict the results of passing a direct electrical current through (a) molten NaBr, (b) aqueous NaBr, and (c) aqueous $SnCl_2$.

The electrolysis of aqueous NaCl (brine) is an extremely important industrial reaction in the chlor-alkali process. This process is the major commercial source of chlorine gas and sodium hydroxide; it is described in detail in Section 21.4.

CONCEPTUAL EXERCISE 19.11 Making F_2 Electrolytically

In 1886, Moissan was the first to prepare F_2 by the electrolysis of F^- ions. He electrolyzed KF dissolved in pure HF. No water was present, so only F^- ions were available at the anode. What was produced at the cathode? Write the half-equations for the oxidation and the reduction reactions, and then write the net cell reaction.

19.12 Counting Electrons

When a direct electric current is passed through an aqueous solution of the soluble salt $AgNO_3$, metallic silver is produced at the cathode. One mole of electrons is required for every mole of Ag^+ reduced.

$$Ag^+(aq) + e^- \longrightarrow Ag(s)$$

If a copper(II) salt in aqueous solution were reduced, 2 mol electrons would be required to produce 1 mol metallic copper from 1 mol copper(II) ions.

$$Cu^{2+}(aq) + 2\ e^- \longrightarrow Cu(s)$$

Each of these balanced half-reactions is like any other balanced chemical equation. That is, each illustrates the fact that both matter and charge are conserved in chemical reactions. Thus, if you could measure the number of moles of electrons flowing through an electrolysis cell, you would know the number of moles of silver or copper produced. Conversely, if you knew the amount of silver or copper produced, you could calculate the number of moles of electrons that had passed through the circuit.

Large electric currents, like those needed to run a hair dryer or refrigerator, are measured in amperes (C/s). Smaller currents, in the milliampere (mA) range, are more commonly used in laboratory electrolysis experiments.

The number of moles of electrons transferred during a redox reaction is usually determined experimentally by measuring the current flowing in the external electrical circuit during a given time. The product of the current (measured in amperes, A) and the time interval (in seconds, s) equals the electric charge (in coulombs, C) that flowed through the circuit.

$$\text{Charge} = \text{current} \times \text{time}$$

$$1 \text{ coulomb} = 1 \text{ ampere} \times 1 \text{ second}$$

The Faraday constant (96,500 C/mol of electrons) can then be used to find the number of moles of electrons from a known number of coulombs. This information is of practical significance in chemical analysis and synthesis.

Figure 19.24 shows the relationship between quantity of charge used and the quantities of substances that are oxidized or reduced during electrolysis.

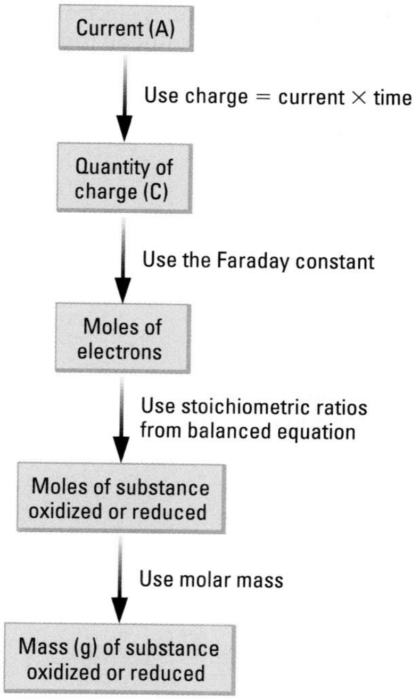

Figure 19.24 Calculation steps for electrolysis. These steps relate the quantity of electrical charge used in electrolysis to the amounts of substances oxidized or reduced.

PROBLEM-SOLVING EXAMPLE 19.13 Using the Faraday Constant

How many grams of copper will be deposited at the cathode of an electrolysis cell if an electric current of 15 mA is applied for 1.0 h through an aqueous solution containing excess Cu^{2+} ions? Assume the process is 100% efficient.

Answer 0.018 g C

Strategy and Explanation Use the strategy presented in Figure 19.24. First, write and balance the relevant half-reaction that occurs at the cathode.

$$Cu^{2+}(aq) + 2\,e^- \rightarrow Cu(s)$$

Then, calculate the quantity of charge transferred.

$$\text{Charge} = 15 \times 10^{-3}\,\text{A} \times 3600\,\text{s} = 15 \times 10^{-3}\,\text{C/s} \times 3600\,\text{s} = 54\,\text{C}$$

Finally, determine the mass of copper deposited.

$$(54\,\text{C})\left(\frac{1\text{ mol e}^-}{9.65 \times 10^4\,\text{C}}\right)\left(\frac{1\text{ mol Cu}}{2\text{ mol e}^-}\right)\left(\frac{63.5\text{ g Cu}}{1\text{ mol Cu}}\right) = 0.018\text{ g Cu}$$

PROBLEM-SOLVING PRACTICE 19.13

In the commercial production of sodium metal by electrolysis, the cell operates at 7.0 V and a current of 25×10^3 A. Calculate what mass of metallic sodium can be produced in 1 h.

CONCEPTUAL EXERCISE 19.12 How Many Moles of Electrons?

Which would require more moles of electrons?
(a) Making 1 mol Al from Al^{3+} (b) Making 2 mol Na from Na^+
(c) Making 2 mol Cu from Cu^{2+}

Electrolytic Production of Hydrogen

Hydrogen can be produced by the electrolysis of water to which a drop or two of sulfuric acid has been added to make the solution electrically conductive. The overall electrochemical reaction is

$$2\,H_2O(\ell) \longrightarrow 2\,H_2(g) + O_2(g) \qquad E^\circ_{cell} = -1.24\text{ V}$$

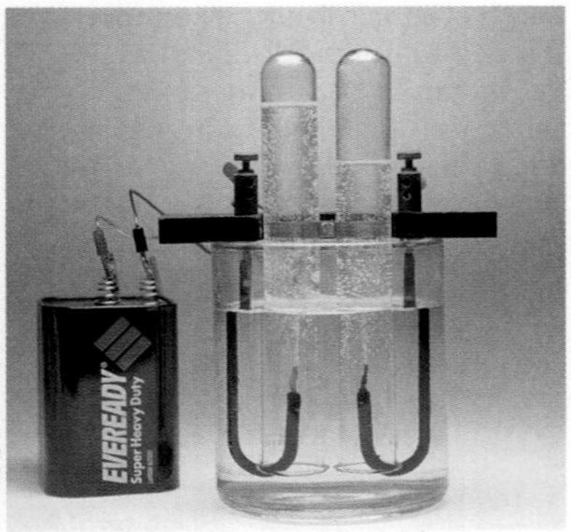

Electrolysis of water. A very dilute solution of sulfuric acid is electrolyzed to produce H_2 at the cathode (*right*) and O_2 at the anode (*left*).

Oxygen is produced at the anode and hydrogen at the cathode. The minimum voltage required for this reaction is 1.24 V (J/C), but in practice overvoltage requires a higher voltage of about 2 V.

Let's consider how much electrical energy would be required to produce 1.00 kg of gaseous H_2 (about 11,200 L at STP) and at what cost. First calculate the required charge in coulombs by using the Faraday constant; then use the fact that 1 joule = 1 volt × 1 coulomb.

The reduction half-reaction shows that 2 mol electrons produces 1 mol (2.02 g) $H_2(g)$.

$$2\ H_3O^+(aq) + 2\ e^- \longrightarrow H_2(g) + 2\ H_2O(\ell)$$

The amount (number of moles) of electrons required to produce 1.00 kg H_2 is calculated as follows:

$$1.00\ \text{kg H}_2 \times \left(\frac{1 \times 10^3\ \text{g}}{\text{kg}}\right)\left(\frac{1\ \text{mol H}_2}{2.016\ \text{g H}_2}\right)\left(\frac{2\ \text{mol e}^-}{1\ \text{mol H}_2}\right) = 9.92 \times 10^2\ \text{mol e}^-$$

Now we calculate the charge using the Faraday constant.

$$(9.92 \times 10^2\ \text{mol e}^-) \times \left(\frac{9.65 \times 10^4\ \text{C}}{1\ \text{mol e}^-}\right) = 9.57 \times 10^7\ \text{C}$$

The energy (in joules) is calculated from the charge and the cell voltage.

$$\text{Energy} = \text{charge} \times \text{voltage} = (9.57 \times 10^7\ \text{C})(1.24\ \text{J/C}) = 1.19 \times 10^8\ \text{J}$$

The kilowatt-hour (kWh) is a unit of energy: 1 kWh = 3.60×10^6 J.

We convert joules to kilowatt-hours (kWh), which is the unit we see when we pay the electric bill.

$$1.19 \times 10^8\ \text{J} \times \frac{1\ \text{kWh}}{3.60 \times 10^6\ \text{J}} = 33.1\ \text{kWh}$$

At a rate of 10 cents per kilowatt-hour, the production of 1.00 kg hydrogen would cost $3.31.

Hydrogen holds great promise as a fuel in our economy because it is a gas and can be transported through pipelines, it burns without producing pollutants, and it could be used in fuel cells to generate electricity on demand. Both water and sulfuric acid are in plentiful supply. The major problem with producing hydrogen in quantities large enough to meet the nation's energy demands is finding a cheap enough source of electricity to decompose water to its elements. Another challenging problem is the development of economical, reliable, and safe methods for hydrogen transfer and storage.

EXERCISE 19.13 Calculations Based on Electrolysis

In the production of aluminum metal, Al^{3+} is reduced to Al metal using currents of about 50,000 A and a low voltage of about 4.0 V. Calculate how much energy (in kilowatt-hours) is required to produce 2000. metric tons of aluminum metal.

CONCEPTUAL EXERCISE 19.14 How Many Joules?

Think of a battery you just purchased at the store as an energy source that can deliver some number of joules. Name the two pieces of information you need to calculate how many joules this battery can deliver. Which one is obviously available as you read the label on the battery? Devise a means of determining the other information needed.

Electroplating

If a metal or other electrical conductor serves as the cathode in an electrolysis cell, the conductor can be plated with another metal to decorate it or protect it against corrosion. To plate an object with copper, for example, we have only to make the object's surface conducting and use the object as the cathode in an electrolysis cell containing a solution of a soluble copper salt as a source of Cu^{2+} ions. The object will become coated with metallic copper, and the coating will thicken as the electrolysis continues and electrons reduce more Cu^{2+} ions to Cu atoms. If the plated object is a metal, it will conduct electricity by itself. If the object is a nonmetal, its surface can be lightly dusted with graphite powder to make it conducting.

Precious metals such as gold are often plated onto cheaper metals such as copper to make jewelry. If the current and duration of the plating reaction are known, it is possible to calculate the mass of gold that will be reduced onto the cathode surface. For example, suppose the object to be plated is immersed in an aqueous solution of $AuCl_3$ and is made a cathode by connecting it to the negative pole of a battery. The circuit is completed by immersing an inert electrode connected to the positive battery pole in the solution, and gold is reduced at the cathode for 60. min at a current of 0.25 A. The reduction half-reaction is

Oscar is gold-plated. The Oscar award and most gold jewelry are made by plating a thin coating of gold onto a much less expensive base metal.

$$Au^{3+}(aq) + 3\,e^- \longrightarrow Au(s)$$

The mass of gold that is reduced is calculated by

$$(0.25\ C/s)(60.\ min)\left(\frac{60\ s}{1\ min}\right)\left(\frac{1\ mol\ e^-}{9.65 \times 10^4\ C}\right)\left(\frac{1\ mol\ Au}{3\ mol\ e^-}\right)\left(\frac{197.\ g\ Au}{1\ mol\ Au}\right) = 0.61\ g\ Au$$

Assuming that gold is selling for \$900 per ounce, that's about \$19.25 worth of gold.

EXERCISE 19.15 Electroplating Silver

Calculate the mass of silver that could be plated from solution with a current of 0.50 A for 20. min. The cathode reaction is $Ag^+(aq) + e^- \rightarrow Ag(s)$.

ESTIMATION The Cost of Aluminum in a Beverage Can

How much does it cost to generate the mass of aluminum (14 g) in one beverage can? The aluminum in these cans is produced by reducing Al^{3+} to Al(s). The reaction is run commercially at 50,000 A and 4 V (4 J/C); 1 kWh of electricity costs about 10 cents, and 1 kWh = 3.60×10^6 J.

The charge needed to generate 14 g Al is

$$(14\ g\ Al)\left(\frac{1\ mol\ Al}{26.98\ g\ Al}\right)\left(\frac{3\ mol\ e^-}{1\ mol\ Al}\right)\left(\frac{96{,}500\ C}{1\ mol\ e^-}\right) = 1.5 \times 10^5\ C$$

The quantity of energy used is

$$(1.5 \times 10^5\ C)(4\ J/C)\left(\frac{1\ kWh}{3.60 \times 10^6\ J}\right) = 0.17\ kWh$$

The cost of 0.17 kWh at 10 cents per kWh is 1.7 cents. Thus, the mass of Al in one beverage can could be generated by electrolysis for less than 2 cents.

It is estimated that recycling of aluminum for cans requires overall less than 1% of the cost of extracting aluminum from its ore, which is an energy-intensive and costly process (Section 21.4). At 1.7 cents per can for the aluminum, a 5 cent per can deposit to encourage recycling is not a bad deal.

(For the cost of electricity, see **http://www.eia.doe.gov/neic/brochure/electricity/electricity.html**)

Visit this book's companion website at www.cengage.com/chemistry/moore to work an interactive module based on this material.

Figure 19.25 Rusting. The formation of rust destroys the structural integrity of objects made of iron and steel. Given time, this chain will completely rust.

Corrosion is so commonplace that about 25% of the annual steel production in the United States is used to replace material lost to corrosion.

19.13 Corrosion—Product-Favored Redox Reactions

Corrosion is the oxidation of a metal that is exposed to the environment. Visible corrosion on the steel supports of a bridge, for example, indicates possible structural failure. Corrosion reactions are invariably product-favored, which means that $E°$ for the reaction is positive and $\Delta G°$ is negative. Corrosion of iron, for example, takes place quite readily and is difficult to prevent. It produces the red-brown substance we call rust, which is hydrated iron(III) oxide ($Fe_2O_3 \cdot x\,H_2O$, where x varies from 2 to 4). The rust that forms when iron corrodes does not adhere to the surface of the metal, so it can easily flake off and expose fresh metal surface to corrosion (Figure 19.25). The corrosion of aluminum, a metal that is even more reactive than iron, is also very product-favored. However, the aluminum oxide that forms as a result of corrosion adheres tightly as a thin coating on the surface of the metal, creating a protective coating that prevents further corrosion.

For corrosion of a metal (M) to occur, the metal must have an anodic area where the oxidation can occur. The general reaction is

Anode reaction: $$M(s) \longrightarrow M^{n+} + n\,e^-$$

There must also be a cathodic area where electrons are consumed. Frequently, the cathode reactions are reductions of oxygen or water.

Cathode reactions:
$$O_2(g) + 2\,H_2O(\ell) + 4\,e^- \longrightarrow 4\,OH^-(aq)$$
$$2\,H_2O(\ell) + 2\,e^- \longrightarrow 2\,OH^-(aq) + H_2(g)$$

Anodic areas may occur at cracks in the oxide coating that protects the surfaces of many metals (such as aluminum) or around impurities. Cathodic areas may occur at the metal oxide coating, at less reactive metallic impurity sites, or around other metal compounds trapped at the surface, such as sulfides or carbides.

The other requirements for corrosion are an electrical connection between the anode and the cathode and an electrolyte in contact with both anode and cathode. Both requirements are easily fulfilled—the metal itself is the conductor, and ions dissolved in moisture from the environment provide the electrolyte.

In the corrosion of iron, the anodic reaction is the oxidation of metallic iron (Figure 19.26). If both water and O_2 gas are present, the cathode reaction is the reduction of oxygen, giving the net reaction

$$2\,[Fe(s) \longrightarrow Fe^{2+}(aq) + 2\,e^-] \qquad \text{(anode reaction)}$$
$$O_2(g) + 2\,H_2O(\ell) + 4\,e^- \longrightarrow 4\,OH^-(aq) \qquad \text{(cathode reaction)}$$
$$2\,Fe(s) + O_2(g) + 2\,H_2O(\ell) \longrightarrow \underset{\text{iron(II) hydroxide}}{2\,Fe(OH)_2(s)} \qquad \text{(net reaction)}$$

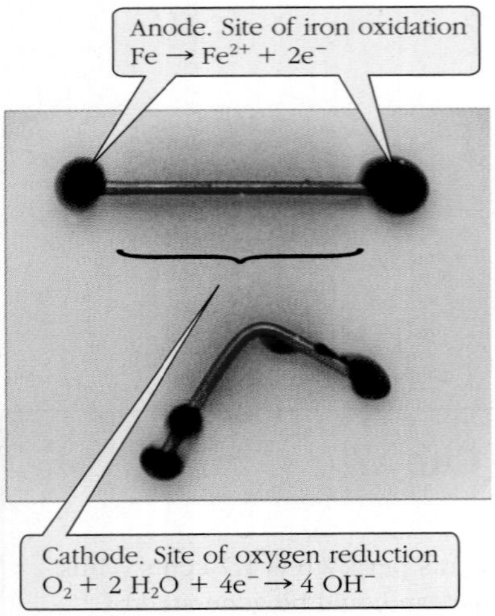
Anode. Site of iron oxidation
Fe → Fe²⁺ + 2e⁻
Cathode. Site of oxygen reduction
O₂ + 2 H₂O + 4e⁻ → 4 OH⁻

Figure 19.26 Corroding iron nails. Two nails were placed in an agar gel, which also contained the indicator phenolphthalein and $[Fe(CN)_6]^{3-}$. The nails began to corrode and produced Fe^{2+} ions at the tip and where the nail is bent. (These points of stress corrode more quickly.) These points are the anode, as indicated by the formation of the blue-colored compound of Fe^{2+} called Prussian blue, $Fe_3[Fe(CN)_6]_2$. The remainder of the nail is the cathode, where oxygen is reduced in water to give OH^-. The presence of OH^- ions causes the phenolphthalein to turn pink.

In the presence of an ample supply of oxygen and water, as in open air or flowing water, the iron(II) hydroxide is oxidized to the hydrated red-brown iron(III) oxide seen in Figure 19.25.

$$4\,Fe(OH)_2(s) + O_2(g) \longrightarrow \underset{\text{red-brown}}{2\,Fe_2O_3 \cdot 2\,H_2O(s)} + H_2O(\ell)$$

This hydrated iron oxide is the familiar rust you see on iron and steel objects and the substance that colors the water red in some mountain streams and home water pipes. Rust is easily removed from the metal surface by mechanical shaking, rubbing, or even the action of rain or freeze-thaw cycles, thus exposing more iron at the surface and allowing the objects to eventually deteriorate completely.

Other substances in air and water can hasten corrosion. Dissolved metal salts, such as the chlorides of sodium and calcium from sea air or from salt spread on roadways in the winter, function as salt bridges between anodic and cathodic regions, thus speeding up corrosion reactions.

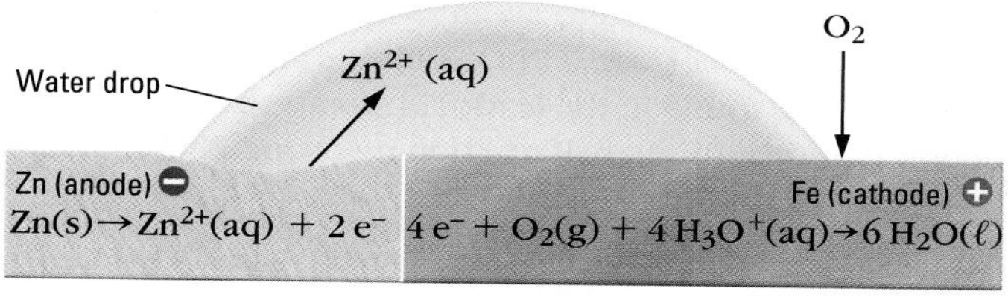

Figure 19.27 **Galvanized iron provides cathodic protection of an iron-containing object.** The iron is coated with a film of zinc (left side of figure), a metal more easily oxidized than iron. The zinc acts as the anode and forces iron to become the cathode, thereby preventing the corrosion of the iron.

CONCEPTUAL EXERCISE 19.16 Do All Metals Corrode?

Do all metals corrode as readily as iron and aluminum? Name three metals that you would expect to corrode about as readily as iron and aluminum, and name three metals that do not corrode readily. Name a use for each of the three noncorroding metals. Explain why metals fall into these two groups.

Corrosion Protection

How can metal corrosion be prevented? The general approaches are to (1) inhibit the anodic process, (2) inhibit the cathodic process, or (3) do both. The most common method is **anodic inhibition,** which directly limits or prevents the oxidation half-reaction by painting the metal surface, coating it with grease or oil, or allowing a thin film of metal oxide to form. This prevents oxygen and water from reaching the metal surface.

Cathodic protection is accomplished by forcing the metal to become the cathode instead of the anode. Usually, this goal is achieved by attaching another, more readily oxidized metal to the metal being protected. The best example involves galvanized iron, iron that has been coated with a thin film of zinc (Figure 19.27). The standard reduction potential for zinc is considerably more negative than that for iron (Zn is lower in Table 19.1 than Fe), so zinc is more easily oxidized. Therefore, the zinc metal film is oxidized before any of the iron and the zinc coating forms a *sacrificial anode.* In addition, when the zinc is corroded, $Zn(OH)_2$ forms an insoluble film on the surface (K_{sp} of $Zn(OH)_2 = 1.2 \times 10^{-17}$) that further slows corrosion.

Galvanized objects. A thin coating of zinc helps prevent the oxidation of iron.

CONCEPTUAL EXERCISE 19.17 Corrosion Rates

Rank these environments in terms of their relative rates of corrosion of iron. Place the fastest first. Explain your answers.

(a) Moist clay
(b) Sand by the seashore
(c) The surface of the moon
(d) Desert sand in Arizona

SUMMARY PROBLEM

If it were not for the density of lead, the lead-acid storage battery would find far greater application. Many electric vehicles use lead-acid storage batteries as their source of power, but automotive engineers continue to look longingly at other batteries because of their higher energy-to-mass ratios. When you review Table 19.1 and consider the chemical properties of all of the other oxidizing and reducing agents shown there, you might be tempted to create a hybrid battery that would combine some of the desirable features of, say, a PbO_2 cathode and some other kind of anode rather than the lead anode found in the lead-acid storage battery. In that way, at least the high reduction potential of the half-reaction involving PbO_2 might still be used.

(a) Calculate the $E°$ value of a cell made using the PbO_2 reduction reaction and magnesium metal as the reducing agent. Write the two half-reactions and the net cell reaction. Would this cell be a good choice for a secondary battery? Explain your answer.

(b) Calculate the $E°$ value of a cell made using the PbO_2 reduction reaction and nickel metal as the reducing agent. Write the two half-reactions and the net cell reaction. Would this cell have a voltage greater than or less than that of a single cell of a lead-acid storage battery? Would this cell be a good choice for a secondary battery? Explain. What could you do to the chemistry in the anode compartment to make it a secondary battery?

(c) If your Ni/PbO_2 hybrid battery were a success and it was manufactured for use in electric automobiles, how many amperes could it produce, assuming that 500.0 g Ni reacted in exactly 30 min? How much PbO_2 would be reduced during this same period of time?

(d) Of course, batteries must be recharged. How much time would be required to recharge your Ni/PbO_2 battery to its original state (the 500.0 g Ni being converted back to its original form) if a current of 25.5 A is passed through the battery?

(e) Just as you are getting ready to cash in on the success of your new battery, someone announces that it has some serious environmental problems. Explain what these problems could be.

IN CLOSING

Having studied this chapter, you should be able to . . .

- Identify the oxidizing and reducing agents in a redox reaction (Section 19.1). End-of-chapter question: 6
- Write equations for oxidation and reduction half-reactions, and use them to balance the net equation (Section 19.2). Questions 10, 14
- Identify and describe the functions of the parts of an electrochemical cell; describe the direction of electron flow outside the cell and the direction of the ion flow inside the cell (Section 19.3). Question 22
- Define standard reduction potentials; use them to predict whether a reaction will be product-favored as written (Sections 19.4 and 19.5). Questions 28, 32, 34
- Calculate $\Delta G°$ from the value of $E°$ for a redox reaction (Section 19.6). Questions 38, 42, 46
- Explain how product-favored electrochemical reactions can be used to do useful work, and list the requirements for using such reactions in rechargeable batteries (Section 19.6).
- Explain how the Nernst equation relates concentrations of redox reactants to E_{cell} (Section 19.7). Questions 48, 50
- Use the Nernst equation to calculate the potentials of cells that are not at standard conditions (Section 19.7). Question 52
- Explain the source of the equilibrium potential across a neuron-cell membrane (Section 19.8).

OWL and go Chemistry

Sign in at **www.cengage.com/owl** to:

- View tutorials and simulations, develop problem-solving skills, and complete online homework assigned by your professor.
- For quick review and exam prep, download Go Chemistry mini lecture modules from OWL (or purchase them at **www.CengageBrain.com**).

- Describe the chemistry of the dry cell, the mercury battery, and the lead-acid storage battery (Section 19.9).
- Describe how a fuel cell works, and indicate how it differs from a battery (Section 19.10). Question 58
- Use standard reduction potentials to predict the products of electrolysis of an aqueous salt solution (Section 19.11). Question 63
- Calculate the quantity of product formed at an electrode during an electrolysis reaction, given the current passing through the cell and the time during which the current flows (Section 19.12). Questions 65, 73, 77
- Explain how electroplating works (Section 19.12).
- Describe how corrosion occurs and how it can be prevented (Section 19.13).

KEY TERMS

ampere (A) *(Section 19.4)*
anode *(19.3)*
anodic inhibition *(19.13)*
battery *(19.3)*
cathode *(19.3)*
cathodic protection *(19.13)*
cell voltage *(19.4)*
concentration cell *(19.7)*
corrosion *(19.13)*
coulomb (C) *(19.4)*
electrochemical cell *(19.3)*
electrochemistry *(Introduction)*
electrode *(19.3)*
electrolysis *(19.11)*
electromotive force (emf) *(19.4)*
Faraday constant (*F*) *(19.6)*
fuel cell *(19.10)*
half-cell *(19.3)*
half-reaction *(19.2)*
Nernst equation *(19.7)*
neurons *(19.8)*
pH meter *(19.7)*
primary battery *(19.9)*
salt bridge *(19.3)*
secondary battery *(19.9)*
standard conditions *(19.4)*
standard hydrogen electrode *(19.4)*
standard reduction potentials *(19.5)*
standard voltages ($E°$) *(19.4)*
volt (V) *(19.4)*
voltaic cell *(19.3)*

QUESTIONS FOR REVIEW AND THOUGHT

OWL Interactive versions of these problems are assignable in OWL.

Blue-numbered questions have short answers at the back of this book in Appendix M and fully worked solutions in the *Student Solutions Manual.*

Review Questions

These questions test vocabulary and simple concepts.

1. Describe the principal parts of an electrochemical cell by drawing a hypothetical cell, indicating the cathode, the anode, the direction of electron flow outside the cell, and the direction of ion flow within the cell.
2. Explain how product-favored electrochemical reactions can be used to do useful work.
3. Explain how reactant-favored electrochemical reactions can be induced to make products.
4. Explain how electroplating works.
5. Tell whether each of these statements is true or false. If false, rewrite it to make it a correct statement.
   (a) Oxidation always occurs at the anode of an electrochemical cell.
   (b) The anode of a voltaic cell is the site of reduction and is negative.
   (c) Standard conditions for electrochemical cells are a concentration of 1.0 M for dissolved species and a pressure of 1 bar for gases.
   (d) The potential of a voltaic cell does not change with temperature.
   (e) All product-favored oxidation-reduction reactions have a standard cell voltage $E°_{cell}$, with a negative sign.

Topical Questions

These questions are keyed to the major topics in the chapter. Usually a question that is answered at the back of this book is paired with a similar one that is not.

Redox Reactions (Section 19.1)

6. In each of these reactions assign oxidation numbers to all species, and tell which substance is oxidized and which is

reduced. Tell which is the oxidizing agent and which is the reducing agent.

(a) $2\ Al(s) + 3\ Cl_2(g) \longrightarrow 2\ AlCl_3(s)$

(b) $8\ H_3O^+(aq) + MnO_4^-(aq) + 5\ Fe^{2+}(aq) \longrightarrow 5\ Fe^{3+}(aq) + Mn^{2+}(aq) + 12\ H_2O(\ell)$

(c) $FeS(s) + 3\ NO_3^-(aq) + 4\ H_3O^+(aq) \longrightarrow 3\ NO(g) + SO_4^{2-}(aq) + Fe^{3+}(aq) + 6\ H_2O(\ell)$

7. In each of these reactions assign oxidation numbers to all species, and tell which substance is oxidized and which is reduced. Tell which is the oxidizing agent and which is the reducing agent.

(a) $Fe(s) + Br_2(\ell) \longrightarrow FeBr_2(s)$

(b) $8\ HI(aq) + H_2SO_4(aq) \longrightarrow H_2S(aq) + 4\ I_2(s) + 4\ H_2O(\ell)$

(c) $H_2O_2(aq) + 2\ Fe^{2+}(aq) + 2\ H_3O^+(aq) \longrightarrow 2\ Fe^{3+}(aq) + 4\ H_2O(\ell)$

8. Choose four elements: a metal that is a representative element, a transition metal, a nonmetal, and a metalloid. Using the index to this text, find a chemical reaction in which each element occurs as a reactant. Assign oxidation numbers to all elements on the reactant and product sides, and identify the oxidizing agent and the reducing agent.

9. Answer Question 8 again, but this time find a chemical reaction in which each element is produced.

Using Half-Reactions to Understand Redox Reactions (Section 19.2)

10. Write half-reactions for these changes:

(a) Oxidation of zinc to Zn^{2+} ions

(b) Reduction of H_3O^+ ions to hydrogen gas

(c) Reduction of Sn^{4+} ions to Sn^{2+} ions

(d) Reduction of chlorine to Cl^- ions

(e) Oxidation of sulfur dioxide to sulfate ions in acidic solution

11. Write half-reactions for these changes:

(a) Reduction of MnO_4^- ion to Mn^{2+} ion in acid solution

(b) Reduction of $Cr_2O_7^{2-}$ ion to Cr^{3+} ion in acid solution

(c) Oxidation of hydrogen gas to H_3O^+ ions

(d) Reduction of hydrogen peroxide to water in acidic solution

(e) Oxidation of nitric oxide to nitrogen monoxide in acidic solution

12. For each reaction in Question 6, write balanced half-reactions.

13. For each reaction in Question 7, write balanced half-reactions.

14. Balance this redox reaction in a basic solution:

$$Zn(s) + NO_3^-(aq) \longrightarrow [Zn(OH)_4]^{2-}(aq) + NH_3(aq)$$

15. Balance this redox reaction in a basic solution:

$$NO_3^-(aq) + Al(s) \longrightarrow NH_3(aq) + [Al(OH)_4]^-(aq)$$

16. Balance these redox reactions, and identify the oxidizing agent and the reducing agent.

(a) $CO(g) + O_3(g) \longrightarrow CO_2(g)$

(b) $H_2(g) + Cl_2(g) \longrightarrow HCl(g)$

(c) $H_2O_2(aq) + Ti^{2+}(aq) \longrightarrow H_2O(\ell) + Ti^{4+}(aq)$ in acidic solution

(d) $Cl^-(aq) + MnO_4^-(aq) \longrightarrow Cl_2(g) + MnO_2(s)$ in acidic solution

(e) $FeS_2(s) + O_2(g) \longrightarrow Fe_2O_3(s) + SO_2(g)$

(f) $O_3(g) + NO(g) \longrightarrow O_2(g) + NO_2(g)$

(g) $Zn(Hg)$ (amalgam) $+ HgO(s) \longrightarrow ZnO(s) + Hg(\ell)$ in basic solution

17. Balance these redox reactions, and identify the oxidizing agent and the reducing agent.

(a) $FeO(s) + O_3(g) \longrightarrow Fe_2O_3(s)$

(b) $P_4(s) + Br_2(\ell) \longrightarrow PBr_5(\ell)$

(c) $H_2O_2(aq) + Co^{2+}(aq) \longrightarrow H_2O(\ell) + Co^{3+}(aq)$ in acidic solution

(d) $Cl^-(aq) + Cr_2O_7^{2-}(aq) \longrightarrow Cl_2(g) + Cr^{3+}(aq)$ in acidic solution

(e) $CuFeS_2(s) + O_2(g) \longrightarrow Cu_2S(s) + FeO(s) + SO_2(g)$

(f) $H_2CO(g) + O_2(g) \longrightarrow CO_2(g) + H_2O(\ell)$

(g) $C_3H_8(g) + O_2(g) \longrightarrow CO_2(g) + H_2O(\ell)$ in acidic solution

Electrochemical Cells (Section 19.3)

18. For the reaction $Cu^{2+}(aq) + Zn(s) \longrightarrow Cu(s) + Zn^{2+}(aq)$, why can't you generate electric current by placing a piece of copper metal and a piece of zinc metal in a solution containing $CuCl_2(aq)$ and $ZnCl_2(aq)$?

19. Explain the function of a salt bridge in an electrochemical cell.

20. Are standard half-cell reactions always written as oxidation reactions or reduction reactions? Explain.

21. Tell whether this statement is true or false. If false, rewrite it to make it a correct statement: The value of an electrode potential changes when the half-reaction is multiplied by a factor. That is, $E°$ for $Li^+ + e^- \longrightarrow Li$ is different from that for $2\ Li^+ + 2\ e^- \longrightarrow 2\ Li$.

22. A voltaic cell is assembled with $Pb(s)$ and $Pb(NO_3)_2(aq)$ in one compartment and $Zn(s)$ and $ZnCl_2(aq)$ in the other. An external wire connects the two electrodes, and a salt bridge containing $KNO_3(aq)$ connects the two solutions.

(a) In the product-favored reaction, zinc metal is oxidized to Zn^{2+}. Write a balanced net ionic equation for this reaction.

(b) Which half-reaction occurs at each electrode? Which is the anode and which is the cathode?

(c) Draw a diagram of the cell, indicating the direction of electron flow outside the cell and of ion flow within the cell.

23. A voltaic cell is assembled with $Sn(s)$ and $Sn(NO_3)_2(aq)$ in one compartment and $Ag(s)$ and $AgNO_3(aq)$ in the other. An external wire connects the two electrodes, and a salt bridge containing $KNO_3(aq)$ connects the two solutions.

(a) In the product-favored reaction, Ag^+ is reduced to silver metal. Write a balanced net ionic equation for this reaction.

(b) Which half-reaction occurs at each electrode? Which is the anode and which is the cathode?

(c) Draw a diagram of the cell, indicating the direction of electron flow outside the cell and of ion flow within the cell.

Electrochemical Cells and Voltage (Section 19.4)

24. You light a 25-W light bulb with the current from a 12-V lead-acid storage battery. After 1.0 h of operation, how much energy has the light bulb utilized? How many

coulombs have passed through the bulb? Assume 100% efficiency. (A watt is the transfer of 1 J of energy in 1 s.)

25. Draw a diagram of a standard hydrogen electrode and describe how it works.

26. Copper can reduce silver ion to metallic silver, a reaction that could, in principle, be used in a battery.

$$Cu(s) + 2\,Ag^+(aq) \longrightarrow Cu^{2+}(aq) + 2\,Ag(s)$$

(a) Write equations for the half-reactions involved.
(b) Which half-reaction is an oxidation and which is a reduction? Which half-reaction occurs in the anode compartment and which takes place in the cathode compartment?

27. Chlorine gas can oxidize zinc metal in a reaction that has been suggested as the basis of a battery. Write the half-reactions involved. Label which is the oxidation half-reaction and which is the reduction half-reaction.

Using Standard Reduction Potentials (Section 19.5)

28. What is the strongest oxidizing agent in Table 19.1 or Appendix I? What is the strongest reducing agent? What is the weakest oxidizing agent? What is the weakest reducing agent?

29. Using the standard reduction potentials in Table 19.1, place these elements in order of increasing ability to function as reducing agents:
(a) Cl_2 (b) Fe
(c) Ag (d) Na
(e) H_2

30. Using the standard reduction potentials in Table 19.1, place these elements in order of increasing ability to function as oxidizing agents:
(a) O_2 (b) H_2O_2
(c) $PbSO_4$ (d) H_2O

31. One of the most energetic redox reactions is that between F_2 gas and lithium metal.
(a) Write the half-reactions involved. Label the oxidation half-reaction and the reduction half-reaction.
(b) According to data from Table 19.1, what is E°_{cell} for this reaction?

32. Calculate the value of E°_{cell} for each of these reactions. Decide whether each is product-favored.
(a) $I_2(s) + Mg(s) \longrightarrow Mg^{2+}(aq) + 2\,I^-(aq)$
(b) $Ag(s) + Fe^{3+}(aq) \longrightarrow Ag^+(aq) + Fe^{2+}(aq)$
(c) $Sn^{2+}(aq) + 2\,Ag^+(aq) \longrightarrow Sn^{4+}(aq) + 2\,Ag(s)$
(d) $2\,Zn(s) + O_2(g) + 2\,H_2O(\ell) \longrightarrow 2\,Zn^{2+}(aq) + 4\,OH^-(aq)$

33. Consider these half-reactions:

Half-Reaction	$E°$ (V)
$Au^{3+}(aq) + 3\,e^- \longrightarrow Au(s)$	1.52
$Pt^{2+}(aq) + 2\,e^- \longrightarrow Pt(s)$	1.118
$Co^{2+}(aq) + 2\,e^- \longrightarrow Co(s)$	−0.277
$Mn^{2+}(aq) + 2\,e^- \longrightarrow Mn(s)$	−1.18

(a) Which is the weakest oxidizing agent?
(b) Which is the strongest oxidizing agent?
(c) Which is the strongest reducing agent?
(d) Which is the weakest reducing agent?
(e) Will Co(s) reduce $Pt^{2+}(aq)$ to Pt(s)?
(f) Will Pt(s) reduce $Co^{2+}(aq)$ to Co(s)?
(g) Which ions can be reduced by Co(s)?

34. Consider these half-reactions:

Half-Reaction	$E°$ (V)
$Ce^{4+}(aq) + e^- \longrightarrow Ce^{3+}(aq)$	1.72
$Ag^+(aq) + e^- \longrightarrow Ag(s)$	0.80
$Hg_2^{2+}(aq) + 2\,e^- \longrightarrow 2\,Hg(\ell)$	0.80
$Sn^{2+}(aq) + 2\,e^- \longrightarrow Sn(s)$	−0.14
$Ni^{2+}(aq) + 2\,e^- \longrightarrow Ni(s)$	−0.25
$Al^{3+}(aq) + 3\,e^- \longrightarrow Al(s)$	−1.68

(a) Which is the weakest oxidizing agent?
(b) Which is the strongest oxidizing agent?
(c) Which is the strongest reducing agent?
(d) Which is the weakest reducing agent?
(e) Will Sn(s) reduce $Ag^+(aq)$ to Ag(s)?
(f) Will $Hg(\ell)$ reduce $Sn^{2+}(aq)$ to Sn(s)?
(g) Name the ions that can be reduced by Sn(s).
(h) Which metals can be oxidized by $Ag^+(aq)$?

35. In principle, a battery could be made from aluminum metal and chlorine gas.
(a) Write a balanced equation for the reaction that would occur in a battery using $Al^{3+}(aq)/Al(s)$ and $Cl_2(g)/Cl^-(aq)$ half-reactions.
(b) Tell which half-reaction occurs at the anode and which at the cathode. What are the polarities of these electrodes?
(c) Calculate the standard potential, E°_{cell}, for the battery.

$E°$ and Gibbs Free Energy (Section 19.6)

36. Choose the correct answers: In a product-favored chemical reaction, the standard cell potential, E°_{cell}, is (greater/less) than zero, and the Gibbs free energy change, $\Delta G°$, is (greater/less) than zero.

37. For each of the reactions in Question 32, compute the Gibbs free energy change, $\Delta G°$.

38. Hydrazine, N_2H_4, can be used as the reducing agent in a fuel cell.

$$N_2H_4(aq) + O_2(aq) \longrightarrow N_2(g) + 2\,H_2O(\ell)$$

(a) If $\Delta G°$ for the reaction is −598 kJ, calculate the value of $E°$ expected for the reaction.
(b) Suppose the equation is written with all coefficients doubled. Determine $\Delta G°$ and $E°$ for this new reaction.

39. The standard cell potential for the oxidation of Mg by Br_2 is 3.45 V.

$$Br_2(\ell) + Mg(s) \longrightarrow Mg^{2+}(aq) + 2\,Br^-(aq)$$

(a) Calculate $\Delta G°$ for this reaction.
(b) Suppose the equation is written with all coefficients doubled. Determine $\Delta G°$ and $E°$ for this new reaction.

40. The standard cell potential, $E°$, for the reaction of Zn(s) and $Cl_2(g)$ is 2.12 V. Write the chemical equation for the reaction of 1 mol zinc. Calculate the standard Gibbs free energy change, $\Delta G°$, for this reaction.

41. Calculate the equilibrium constant K_c and $\Delta G°$ for the reaction between Cd(s) and Cu^{2+}(aq).
42. Calculate the equilibrium constant K_c and $\Delta G°$ for the reaction between I_2(s) and Br^-(aq).
43. Calculate the equilibrium constant K_c and $\Delta G°$ for the reaction between Ag(s) and Zn^{2+}(aq).
44. Calculate the equilibrium constant K_c and $\Delta G°$ for the reaction between Cl_2(g) and Br^-(aq).
45. Consider a voltaic cell with the following reaction. As the cell reaction proceeds, what happens to the values of E_{cell}, ΔG, and K_c. Explain your answers.

$$Cu^{2+}(aq, 1\ M) + Zn(s) \longrightarrow Cu(s) + Zn^{2+}(aq, 1\ M) \qquad E°_{cell} = 1.10\ V$$

46. Estimate the equilibrium constant K_c for this reaction.

$$Ni(s) + Co^{2+}(aq) \rightleftharpoons Ni^{2+}(aq) + Co(s) \qquad E°_{cell} = +0.046\ V$$

Effect of Concentration on Cell Potential (Section 19.7)

47. Consider the voltaic cell

$$Zn(s) + Cd^{2+}(aq) \longrightarrow Zn^{2+}(aq) + Cd(s)$$

operating at 298 K.
(a) Calculate the $E°_{cell}$ for this cell.
(b) If E_{cell} = 0.390 and (conc. Cd^{2+}) = 2.00 M, what is (conc. Zn^{2+})?
(c) If (conc. Cd^{2+}) = 0.068 M and (conc. Zn^{2+}) = 1.00 M, what is E_{cell}?

48. Consider the voltaic cell

$$2\ Ag^+(aq) + Cd(s) \longrightarrow 2\ Ag(s) + Cd^{2+}(aq)$$

operating at 298 K.
(a) Calculate the $E°_{cell}$ for this cell.
(b) If (conc. Cd^{2+}) = 2.0 M and (conc. Ag^+) = 0.25 M, what is E_{cell}?
(c) If E_{cell} = 1.25 V and (conc. Cd^{2+}) = 0.100 M, what is (conc. Ag^+)?

49. Consider the reaction

$$2\ H_2O(aq) + H_2(g) + Sn^{4+}(aq) \longrightarrow 2\ H_3O^+(aq) + Sn^{2+}(aq)$$

operating at 298 K.
(a) Calculate the $E°_{cell}$ for this cell.
(b) Calculate the E_{cell} for P_{H_2} = 1.0 bar, (conc. Sn^{2+}) = 6.0×10^{-4} M, (conc. Sn^{4+}) = 5.0×10^{-4} M, and pH = 3.60.

50. Calculate the cell potential of a concentration cell that contains two hydrogen electrodes if the cathode contacts a solution with pH = 7.8 and the anode contacts a solution with (conc. H_3O^+) = 0.05 M.
51. Calculate the potential of a cell with one electrode made from zinc metal immersed in a solution where (conc. Zn^{2+}) = 0.010 M and the other electrode is a standard hydrogen electrode.
52. For a voltaic cell with the reaction

$$Pb(s) + Sn^{2+}(aq) \longrightarrow Pb^{2+}(aq) + Sn(s)$$

at what ratio of concentrations of lead and tin ions will E_{cell} = 0?

Common Batteries (Section 19.9)

53. What are the advantages and disadvantages of lead-acid storage batteries?
54. Nicad batteries are rechargeable and are commonly used in cordless appliances. Although such batteries actually function under basic conditions, imagine an electrochemical cell using this setup.

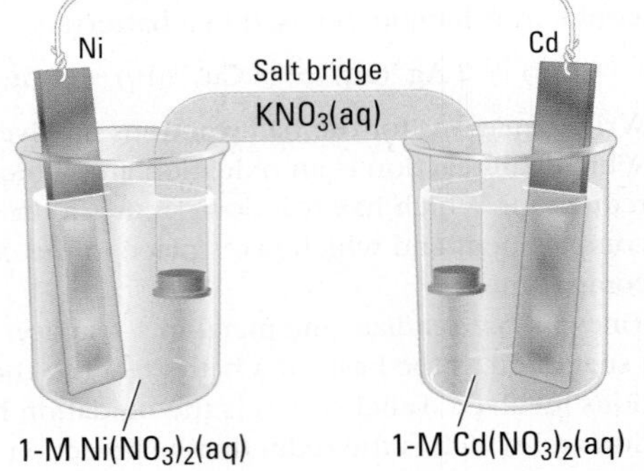
Ni
Cd
Salt bridge
KNO3(aq)
1-M Ni(NO3)2(aq)
1-M Cd(NO3)2(aq)

(a) Write a balanced net ionic equation depicting the reaction occurring in the cell.
(b) What is oxidized? What is reduced? What is the reducing agent and what is the oxidizing agent?
(c) Which is the anode and which is the cathode?
(d) Calculate $E°$ for the cell.
(e) What is the direction of electron flow in the external wire?
(f) If the salt bridge contains KNO_3, toward which compartment will the NO_3^- ions migrate?

55. Consider the nicad cell in Question 54.
(a) If the concentration of Cd^{2+} is reduced to 0.010 M, and (conc. Ni^{2+}) = 1.0 M, will the cell emf be smaller or larger than when the concentration of Cd^{2+}(aq) was 1.0 M? Explain your answer in terms of Le Chatelier's principle.
(b) Begin with 1.0 L of each of the solutions, both initially 1.0 M in dissolved species. Each electrode weighs 50.0 g at the start. If 0.050 A is drawn from the battery, calculate how long the battery will operate.

Fuel Cells (Section 19.10)

56. How does a fuel cell differ from a battery?
57. Describe the principal parts of an H_2/O_2 fuel cell. What is the reaction at the cathode? At the anode? What is the product of the fuel cell reaction?
58. Hydrazine, N_2H_4, has been proposed as the fuel in a fuel cell in which oxygen is the oxidizing agent. The reactions are

$$N_2H_4(aq) + 4\ OH^-(aq) \longrightarrow N_2(g) + 4\ H_2O(\ell) + 4\ e^-$$
$$O_2(g) + 2\ H_2O(\ell) + 4\ e^- \longrightarrow 4\ OH^-(aq)$$

(a) Which reaction occurs at the anode and which at the cathode?
(b) What is the net cell reaction?
(c) If the cell is to produce 0.50 A of current for 50.0 h, what mass in grams of hydrazine must be present?
(d) What mass in grams of O_2 must be available to react with the mass of N_2H_4 determined in part (c)?

Electrolysis: Reactant-Favored Redox Reactions (Section 19.11)

59. Consider the electrolysis of water in the presence of very dilute H_2SO_4. What species is produced at the anode? At

the cathode? What are the relative amounts of the species produced at the two electrodes?

60. Write chemical equations for the electrolysis of molten salts of three different alkali halides to produce the corresponding halogens and alkali metals.

61. From Table 19.1 write down all of the aqueous metal ions that can be reduced by electrolysis to the corresponding metal.

62. From Table 19.1 write down all of the aqueous species that can be oxidized by electrolysis, and determine the products.

63. What are the products of the electrolysis of a 1-M aqueous solution of NaBr? What species are present in the solution? What is formed at the cathode? What is formed at the anode?

64. For each of these solutions, tell what reactions take place at the anode and at the cathode during electrolysis.
(a) $NiBr_2(aq)$ (b) $NaI(aq)$
(c) $CdCl_2(aq)$ (d) $CuI_2(aq)$
(e) $MgF_2(aq)$ (f) $HNO_3(aq)$

Counting Electrons (Section 19.12)

65. A current of 0.015 A is passed through a solution of $AgNO_3$ for 155 min. What mass of silver is deposited at the cathode?

66. Current is passed through a solution containing $Ag^+(aq)$. How much silver was in the solution if all the silver was removed as Ag metal by electrolysis for 14.5 min at a current of 1.0 mA?

67. A current of 2.50 A is passed through a solution of $Cu(NO_3)_2$ for 2.00 h. What mass of copper is deposited at the cathode?

68. A current of 0.0125 A is passed through a solution of $CuCl_2$ for 2.00 h. What mass of copper is deposited at the cathode and what volume of Cl_2 gas (in mL at STP) is produced at the anode?

69. The major reduction half-reaction occurring in the cell in which molten Al_2O_3 and molten aluminum salts are electrolyzed is $Al^{3+}(aq) + 3\ e^- \rightarrow Al(s)$. If the cell operates at 5.0 V and 1.0×10^5 A, what mass (in grams) of aluminum metal can be produced in 8.0 h?

70. The vanadium(II) ion can be produced by electrolysis of a vanadium(III) salt in solution. How long must you carry out an electrolysis if you wish to convert completely 0.125 L of 0.0150-M $V^{3+}(aq)$ to $V^{2+}(aq)$ using a current of 0.268 A?

71. The reactions occurring in a lead-acid storage battery are given in Section 19.9. A typical battery might be rated at 50. ampere-hours (A-h). This means that it has the capacity to deliver 50. A for 1.0 h or 1.0 A for 50. h. If it does deliver 1.0 A for 50. h, what mass of lead would be consumed?

72. An effective battery can be built using the reaction between Al metal and O_2 from the air. If the Al anode of this battery consists of a 3-oz piece of aluminum (84 g), for how many hours can the battery produce 1.0 A of electricity?

73. A dry cell is used to supply a current of 250. mA for 20 min. What mass of Zn is consumed?

74. If the same current as in Question 73 were supplied by a mercury battery, what mass of Hg would be produced at the cathode?

75. Assume that the anode reaction for the lithium battery is

$$Li(s) \longrightarrow Li^+(aq) + e^-$$

and the anode reaction for the lead-acid storage battery is

$$Pb(s) + HSO_4^-(aq) + H_2O(\ell) \longrightarrow PbSO_4(s) + 2\ e^- + H_3O^+(aq)$$

Compare the masses of metals consumed when each of these batteries supplies a current of 1.0 A for 10. min.

76. A hydrogen-oxygen fuel cell operates on the simple reaction

$$2\ H_2(g) + O_2(g) \longrightarrow 2\ H_2O(\ell)$$

If the cell is designed to produce 1.5 A of current, how long can it operate if there is an excess of oxygen and only sufficient hydrogen to fill a 1.0-L tank at 200. bar pressure at 25 °C?

77. How long would it take to electroplate a metal surface with 0.500 g nickel metal from a solution of Ni^{2+} with a current of 4.00 A?

78. How much current is required to electroplate a metal surface with 0.400 g chromium metal from a solution of Cr^{3+} in 1.00 h?

Corrosion: Product-Favored Redox Reactions (Section 19.13)

79. Explain how rust is formed from iron materials by corrosion.

80. Why does iron corrode faster in salt water than in fresh water?

81. What common metal does not corrode readily under normal conditions?

82. Why does coating a steel object with chromium stop corrosion of the iron?

83. Explain how galvanizing iron stops corrosion of the underlying iron.

General Questions

These questions are not explicitly keyed to chapter topics; many require integration of several concepts.

84. A 12-V automobile battery consists of six cells of the type described in Section 19.9. The cells are connected in series so that the same current flows through all of them. Calculate the theoretical minimum electrical potential difference needed to recharge an automobile battery. (Assume standard-state concentrations.) How does this compare with the maximum voltage that could be delivered by the battery? Assuming that the lead plates in an automobile battery each weigh 2.50 kg and that sufficient PbO_2 is available, what is the maximum possible work that could be obtained from the battery?

85. Three electrolytic cells are connected in series, so that the same current flows through all of them for 20. min. In cell A, 0.0234 g Ag plates out from a solution of $AgNO_3(aq)$; cell B contains $Cu(NO_3)_2(aq)$; cell C contains $Al(NO_3)_3(aq)$. What mass of Cu will plate out in cell B? What mass of Al will plate out in cell C?

86. Fluorinated organic compounds are important commercially; they are used as herbicides, flame retardants, and fire-extinguishing agents, among other things. A reaction such as

$$CH_3SO_2F + 3\ HF \longrightarrow CF_3SO_2F + 3\ H_2$$

is actually carried out electrochemically in liquid HF as the solvent.
(a) Draw the structural formula for CH_3SO_2F. (S is the "central" atom with the O atoms, F atom, and CH_3 group bonded to it.) What is the geometry around the S atom? What are the O—S—O and O—S—F bond angles?
(b) If you electrolyze 150. g CH_3SO_2F, how many grams of HF are required and how many grams of each product can be isolated?
(c) Is H_2 produced at the anode or the cathode of the electrolysis cell?
(d) A typical electrolysis cell operates at 8.0 V and 250 A. How many kilowatt-hours of energy does one such cell consume in 24 h?

Applying Concepts

These questions test conceptual learning.

87. Four metals A, B, C, and D exhibit these properties:
(a) Only A and C react with 1.0-M HCl to give H_2 gas.
(b) When C is added to solutions of ions of the other metals, metallic A, B, and D are formed.
(c) Metal D reduces B^{n+} ions to give metallic B and D^{n+} ions.
On the basis of this information, arrange the four metals in order of increasing ability to act as reducing agents.

88. The table below lists the cell potentials for the ten possible electrochemical cells assembled from the elements A, B, C, D, and E and their respective ions in solutions. Using the data in the table, establish a standard reduction potential table similar to Table 19.1. Assign a reduction potential of 0.00 V to the element that falls in the middle of the series.

	A(s) in A^{n+}(aq)	B(s) in B^{n+}(aq)
E(s) in E^{n+}(aq)	+0.21 V	+0.68 V
D(s) in D^{n+}(aq)	+0.35 V	+1.24 V
C(s) in C^{n+}(aq)	+0.58 V	+0.31 V
B(s) in B^{n+}(aq)	+0.89 V	—

	C(s) in C^{n+}(aq)	D(s) in D^{n+}(aq)
E(s) in E^{n+}(aq)	+0.37 V	+0.56 V
D(s) in D^{n+}(aq)	+0.93 V	—
C(s) in C^{n+}(aq)	—	—
B(s) in B^{n+}(aq)	—	—

89. When this electrochemical cell runs for several hours, the green solution gets lighter and the yellow solution gets darker.

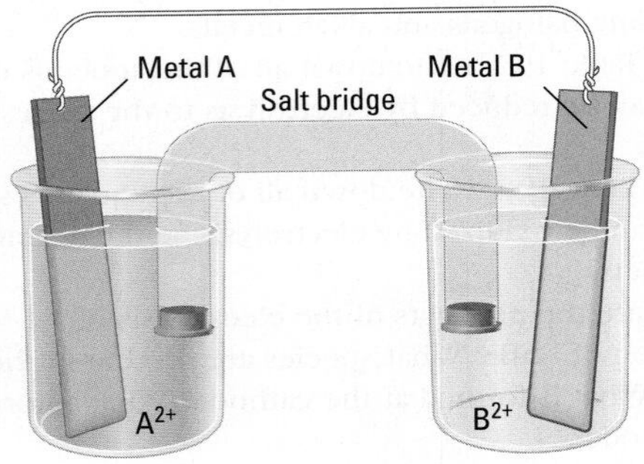

(a) What is oxidized, and what is reduced?
(b) What is the oxidizing agent, and what is the reducing agent?
(c) What is the anode, and what is the cathode?
(d) Write equations for the half-reactions.
(e) Which metal gains mass?
(f) What is the direction of the electron transfer through the external wire?
(g) If the salt bridge contains KNO_3(aq), into which solution will the K^+ ions migrate?

90. An electrolytic cell is set up with Cd(s) in $Cd(NO_3)_2$(aq) and Zn(s) in $Zn(NO_3)_2$(aq). Initially both electrodes weigh 5.00 g. After running the cell for several hours the electrode in the left compartment weighs 4.75 g.
(a) Which electrode is in the left compartment?
(b) Does the mass of the electrode in the right compartment increase, decrease, or stay the same? If the mass changes, what is the new mass?
(c) Does the volume of the electrode in the right compartment increase, decrease, or stay the same? If the volume changes, what is the new volume? (The density of Cd is 8.65 g/cm^3.)

91. Using data from Appendix I, show why
(a) Co^{3+} is not stable in aqueous solution.
(b) Fe^{2+} is not stable in air.

92. When H_2O_2 is mixed with Fe^{2+}, which redox reaction will occur—the oxidation of Fe^{2+} to Fe^{3+} or the reduction of Fe^{2+} to Fe? What are the E°_{cell} values for the electrochemical cells corresponding to the two reactions?

93. Calculate the potential of a cell consisting of two hydrogen electrodes, one immersed in a solution with (conc. H^+) = 1.0×10^{-8} M and the other with (conc. H^+) = 0.025 M.

94. The permanganate ion MnO_4^- can be reduced to the manganese(II) ion Mn^{2+} in aqueous acidic solution, and the reduction potential for this half-cell reaction is 1.52 V. If this half-cell is combined with a $Zn^{2+}|Zn$ half-cell to form a galvanic cell at standard conditions,
(a) Write the chemical equation for the half-reaction occurring at the anode.
(b) Write the chemical equation for the half-reaction occurring at the cathode.
(c) Write the overall balanced equation for the reaction.
(d) Calculate the cell voltage.

More Challenging Questions

These questions require more thought and integrate several concepts.

95. Fluorine, F_2, is made by the electrolysis of anhydrous HF.

$$2\ HF(\ell) \longrightarrow H_2(g) + F_2(g)$$

Typical electrolysis cells operate at 4000 to 6000 A and 8 to 12 V. A large-scale plant can produce about 9.0 metric tons of F_2 gas per day.
(a) What mass in grams of HF is consumed?
(b) Using the conversion factor of 3.60×10^6 J/kWh, how much energy in kilowatt-hours is consumed by a cell operating at 6.0×10^3 A at 12 V for 24 h?

96. What reaction would take place if a 1.0-M solution of $Cr_2O_7^{2-}$ was added to a 1.0-M solution of HBr?

97. If Cl_2 and Br_2 are added to an aqueous solution that contains Cl^- and Br^-, what product-favored reaction will occur?

98. This reaction occurs in a cell with H_2(g) pressure of 1.0 atm and (conc. Cl^-) = 1.0 M at 25 °C; the measured E_{cell} = 0.34 V. What is the pH of the solution?

$$2\ H_2O(\ell) + 2\ H_2(g) + 2\ AgCl(s) \longrightarrow 2\ H_3O^+(aq) + 2\ Cl^-(aq) + 2\ Ag(s)$$

99. An electric current of 2.00 A was passed through a platinum salt solution for 3.00 hours, and 10.9 g of metallic platinum was formed at the cathode. What is the charge on the platinum ions in the solution?

100. E_{cell} = 0.010 V for a galvanic cell with this reaction at 25 °C.

$$Sn(s) + Pb^{2+}(aq) \longrightarrow Sn^{2+}(aq) + Pb(s)$$

(a) What is the equilibrium constant K_c for the reaction?
(b) If a solution with (conc. Pb^{2+}) = 1.1 M had excess tin metal added to it, what would the equilibrium concentration of Sn^{2+} and Pb^{2+} be?

101. You wish to electroplate a copper surface having an area of 1200 mm^2 with a 1.0-μm-thick coating of silver from a solution of $Ag(CN)_2^-$ ions. If you use a current of 150.0 mA, what electrolysis time should you use? The density of metallic silver is 10.5 g/cm^3.

102. Will 1.0-M nitric acid, HNO_3, oxidize metallic gold to form a 1-M Au^{3+} solution? Explain why or why not.

103. A student wanted to measure the copper(II) concentration in an aqueous solution. For the cathode half-cell she used a silver electrode with a 1.00-M solution of $AgNO_3$. For the anode half-cell she used a copper electrode dipped into the aqueous sample. If the cell gave E_{cell} = 0.62 V at 25 °C, what was the copper(II) concentration of the solution?

104. In a mercury battery, the anode reaction is

$$Zn(s) + 2\ OH^-(aq) \longrightarrow ZnO(aq) + H_2O(\ell) + 2\ e^-$$

and the cathode reaction is

$$HgO(s) + H_2O(\ell) + 2\ e^- \longrightarrow Hg(\ell) + 2\ OH^-(aq)$$

The cell potential is 1.35 V. How many hours can such a battery provide power at a rate of 4.0×10^{-4} watt (1 watt = 1 J s^{-1}) if 1.25 g HgO is available?

Conceptual Challenge Problems

These rigorous, thought-provoking problems integrate conceptual learning with problem solving and are suitable for group work.

CP19.A (Section 19.6) Most automobiles run on internal combustion engines, in which the energy used to run the vehicle is obtained from the combustion of gasoline. The main component of gasoline is octane, C_8H_{18}. An automobile manufacturer has recently announced a chemical method for generating hydrogen gas from gasoline and proposes to develop a car in which an H_2/O_2 fuel cell powers an electric propulsion motor, thus eliminating the internal combustion engine with its problems (for example, the generation of unwanted by-products that pollute the air). The hydrogen for the fuel cell would be directly generated from gasoline on board the vehicle. There are two steps in this hydrogen generation process:
(i) Partial oxidation of octane by oxygen to carbon monoxide and hydrogen
(ii) Combination of carbon monoxide with additional gaseous water to form carbon dioxide and more hydrogen (the water-gas shift reaction)

(a) Write the chemical equation for the complete combustion of 1 mol octane.
(b) Write balanced chemical equations for the two-step hydrogen generation process. How many moles of H_2 are produced per mole of octane? (Remember that water is a reactant in the two-step process.)
(c) By combining these equations, show that the net *overall* reaction is the same as in the combustion of octane.
(d) Assuming that the entire Gibbs free energy change of the H_2/O_2 fuel cell reaction is available for use by the electric propulsion motor, calculate the energy produced by a fuel cell when it consumes all of the hydrogen produced from 1 mol of octane. Compare this energy with the Gibbs free energy change for the combustion of 1 mol of octane. (*Note:* The Gibbs free energy of formation, ΔG_f°, for $C_8H_{18}(\ell)$ is 6.14 kJ/mol.)

CP19.B (Section 19.4) People obtain energy by oxidizing food. Glucose, a typical foodstuff, is a carbohydrate. When metabolized, glucose is oxidized to water and carbon dioxide.

$$C_6H_{12}O_6(aq) + 6\ O_2(g) \longrightarrow 6\ CO_2(g) + 6\ H_2O(\ell)$$

The heat of combustion of glucose is 2.80×10^3 kJ/mol, which means that as glucose is oxidized, its electrons lose 2.80×10^3 kJ/mol as they give up potential energy in a complicated series of chemical steps.
(a) Assume that a person requires 2400 food Calories per day and that this energy is obtained from the oxidation of glucose. How much O_2 must a person breathe each day to react with this much glucose?
(b) Each mole of O_2 requires 4 mol electrons, regardless of whether the O atoms become part of CO_2 or H_2O. What would be the average electric current (C/s) in a human body using the above amount of energy described in part (a) per day?
(c) Use the answer from part (b) and calculate the electrical potential this current flows through in a day to produce the 2400 food Calories. (1 Calorie = 4.18 kJ)

CP19.C (Section 19.11) A piece of chromium metal is attached to a battery and dipped into 50 mL of 0.3-M KOH solution in a 250-mL beaker. A stainless steel electrode is connected to the other electrode of the battery and immersed in the same solution. A steady current of 0.50 A is maintained for exactly 2 hours. Several samples of a gas formed at the stainless steel electrode during the electrolysis are captured, and all are found to ignite in air. After the electrolysis, the chromium electrode is weighed and found to have decreased in weight by 0.321 g. The mass of the stainless steel electrode does not change.

After electrolysis, the KOH solution is neutralized with nitric acid to a pH of slightly less than 7, then is heated and reacted with 0.151-M lead(II) nitrate solution. As the lead(II) nitrate solution is added, a yellow precipitate quickly forms from the hot solution. The formation of precipitate stops after 40.4 mL of the lead(II) nitrate solution has been added. The yellow solid is then filtered, dried, and weighed. Its mass is 1.97 g.

(a) How much electrical charge passes through the cell?
(b) How many moles of Cr react?
(c) What is the oxidation state of the Cr after reacting?
(d) Assuming that the yellow compound that precipitates from the solution during the titration contains both Pb and Cr, what do you conclude to be the ratio of the numbers of atoms of Pb and Cr?
(e) If the yellow compound contains an element other than Pb and Cr, what is it and how much of it is in the compound? What is the formula for the yellow compound?

Nuclear Chemistry

20

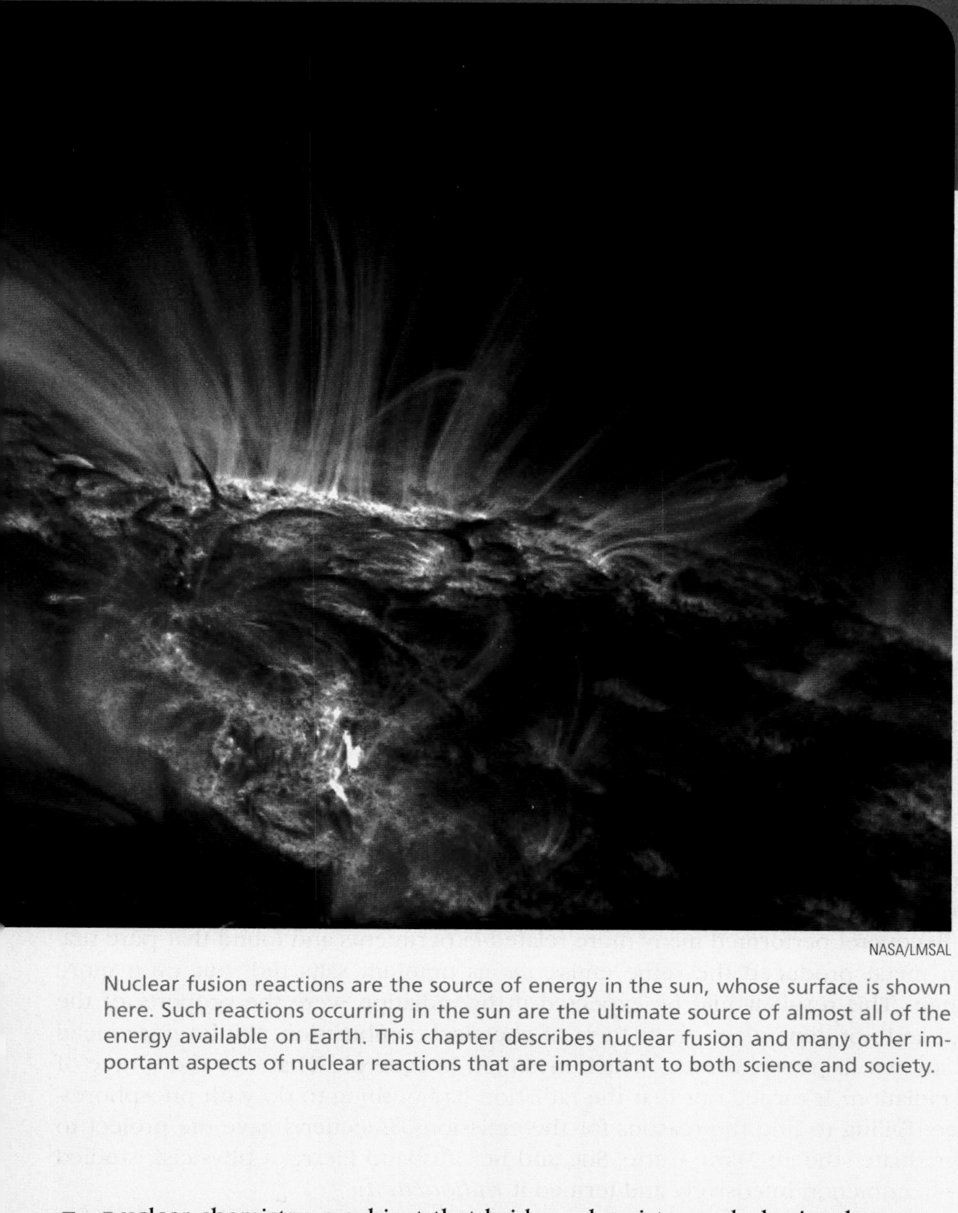

NASA/LMSAL

Nuclear fusion reactions are the source of energy in the sun, whose surface is shown here. Such reactions occurring in the sun are the ultimate source of almost all of the energy available on Earth. This chapter describes nuclear fusion and many other important aspects of nuclear reactions that are important to both science and society.

Nuclear chemistry, a subject that bridges chemistry and physics, has a significant impact on our society. No matter what your reason for taking a college course in chemistry—to prepare for a career in science or engineering or simply to gain knowledge as a concerned citizen—you should know about nuclear chemistry. Radioactive isotopes are widely used in medicine. PET (positron emission tomography) scans depend on radioactivity. Your room may be protected by a smoke detector that uses a radioactive element as part of its sensor, and research in all fields of science employs radioactive elements and their compounds. The national security of

OWL Sign in to OWL at **www.cengage.com/owl** to view tutorials and simulations, develop problem-solving skills, and complete online homework assigned by your professor.

go Chemistry Download mini lecture videos for key concept review and exam prep from OWL or purchase them from **www.CengageBrain.com.**

Companion Website

Visit this book's companion website at **www.cengage.com/chemistry/moore** to work interactive modules for the Estimation boxes and Active Figures in this text.

Einstein's original letter can be seen at **http://www.fdrlibrary.marist.edu/psf/box5/a64a01.html**.

The Alsos Digital Library for Nuclear Issues has a Web site at **http://alsos.wlu.edu** with a section on the Manhattan Project.

The Bulletin of the Atomic Scientists can be found on-line at **http://www.thebulletin.org/**.

the United States since World War II has depended on nuclear weapons, and more than 30 nations use nuclear reactors to produce electricity. This chapter considers several aspects of nuclear chemistry: changes in atomic nuclei and their effects, fission and fusion of nuclei and the energy that can be derived from such changes, units used to measure radioactivity, and beneficial applications of radioactive isotopes.

On August 2, 1939, with the world hovering on the brink of World War II, Albert Einstein sent a letter to President Franklin D. Roosevelt. In this letter, which profoundly changed the course of history, Einstein called attention to work being done on the physics of the atomic nucleus. He noted that he and others believed this work suggested the possibility that "uranium may be turned into a new and important source of energy . . . and [that it was] conceivable . . . that extremely powerful bombs of a new type may thus be constructed. . . ." Einstein's letter was the impetus for the Manhattan Project, which led to the detonation of the first atomic bomb at 5:30 AM on July 16, 1945, in the New Mexico desert.

Since World War II, more powerful nuclear weapons have been developed and stockpiled by a number of nations. With the end of the Cold War, fears of a nuclear disaster receded, and nuclear disarmament treaties were signed between the United States and the former Soviet Union for removing the plutonium-239 and other nuclear fuel from existing nuclear warheads. Unfortunately, those fears have been replaced by the concern that other nations have developed or acquired nuclear weapons. For many years, the respected journal *The Bulletin of the Atomic Scientists* has used the symbol of a clock with its hands near the fateful midnight hour (representing nuclear annihilation) to illustrate the danger faced by the world from atomic weapons. Even with the end of the Cold War, the hands have moved back only a little from midnight.

20.1 The Nature of Radioactivity

Many minerals, called phosphors, glow for some time after being exposed to sunlight or ultraviolet light. In 1896 French physicist Antoine Henri Becquerel was studying this phenomenon, called *phosphorescence,* when he made an important and totally unexpected observation that led him to the accidental discovery of radioactivity. While waiting for a sunny day, Becquerel stored a photographic plate wrapped in black paper along with a uranium salt (a material known to phosphoresce) in a dark drawer. To his amazement, the image of the uranium salt appeared on the plate that had been in the drawer, unexposed to sunlight. Becquerel realized that radiation from the uranium salt had penetrated the black paper and exposed the photographic plate even though the uranium salt had not been stimulated by light.

Becquerel performed many more related experiments and found that pure uranium metal produced the same emissions as uranium salts did, but even more strongly. This result would be expected if the radiation were the property of the metal and not dependent on its form of chemical combination. But no pure metal was known to phosphoresce, which mystified Becquerel. What was the source of this radiation? It turned out that the radiation had nothing to do with phosphorescence. Failing to find the reasons for the emissions, Becquerel gave the project to his graduate student Marie Curie. She and her husband Pierre, a physicist, studied the phenomenon intensively and termed it *radioactivity.*

For more on experiments done by Becquerel and the Curies, see Walton, H. F. *Journal of Chemical Education,* Vol. 69, 1992; p. 10.

One of Marie Curie's first findings was to confirm Becquerel's observation that uranium metal itself was radioactive and that the degree to which a uranium-containing sample was radioactive depended on the percentage of uranium present. When she tested pitchblende, a common ore containing uranium and other metals (such as lead, bismuth, and copper), she was surprised to find that it was even more radioactive than pure uranium. Only one explanation was possible: pitchblende contained an element (or elements) more radioactive than uranium. Eventually, the Curies discovered the element they named *polonium* after Marie's homeland of Poland. They also discovered radium, another highly radioactive element.

Table 20.1 Characteristics of α, β, and γ Emissions

Name	Symbol	Charge	Mass (g/particle)	Penetrating Power*
Alpha	${}^{4}_{2}He^{2+}$, ${}^{4}_{2}\alpha$, ${}^{4}_{2}He$	2+	6.65×10^{-24}	0.03 mm
Beta	${}^{0}_{-1}e$, ${}^{0}_{-1}\beta$	1−	9.11×10^{-28}	2 mm
Gamma	${}^{0}_{0}\gamma$, γ	0	0	10 cm

*Distance at which half the radiation has been stopped by water.

In the atomic representation ${}^{4}_{2}He$, the subscript denotes the atomic number (number of protons) of the atom, and the superscript denotes the mass number (number of protons plus number of neutrons) of the atom.

In England at about the same time, Sir J. J. Thomson and his graduate student Ernest Rutherford were studying the radiation from uranium and thorium *(← p. 41)*. Rutherford found that "There are present at least two distinct types of radiation—one that is readily absorbed, which will be termed for convenience alpha (α) radiation, and the other of a more penetrative character, which will be termed beta (β) radiation." **Alpha radiation,** he discovered, was composed of positively charged particles that, when passed through an electric field, were attracted to the negative side of the field *(← p. 41)*. Indeed, his later studies showed these **alpha (α) particles** to be helium nuclei, ${}^{4}_{2}He^{2+}$, which were ejected at high speeds from a radioactive element (Table 20.1). Alpha particles have limited penetrating power and can be absorbed by skin, clothing, or several sheets of ordinary paper.

In the same experiment with electric fields, Rutherford found that **beta radiation** must be composed of negatively charged particles, since the beam of beta radiation was attracted to the electrically positive plate of an electric field. Later work by Becquerel showed that these particles have an electric charge and mass equal to those of an electron. Thus, **beta (β) particles** are electrons ejected at high speeds from radioactive nuclei. They are more penetrating than alpha particles (Table 20.1), and a $\frac{1}{8}$-inch-thick piece of aluminum is necessary to stop them. Beta particles can penetrate 1 to 2 cm into living bone or tissue.

Encouraged by the Curies, Becquerel returned to the study of radiation. He found that the radiation from uranium was affected by magnetic fields and consisted of two kinds of particles, which we now know to be alpha and beta particles.

A third type of radiation was later discovered by P. Villard, a Frenchman, who named it **gamma (γ) radiation**, using the third letter in the Greek alphabet in keeping with Rutherford's scheme. Unlike alpha and beta radiation, which are particles, gamma rays are a form of electromagnetic radiation and are not affected by an electric field. Gamma radiation is the most penetrating, and it can pass completely through the human body. Thick layers of lead or concrete are required to stop a beam of gamma rays.

20.2 Nuclear Reactions

Equations for Nuclear Reactions

Ernest Rutherford found that radium not only emits alpha particles but also produces the radioactive gas radon. Such observations led Rutherford and Frederick Soddy, in 1902, to propose the revolutionary theory that *radioactivity is the result of a natural change of a radioactive isotope of one element into an isotope of a different element.* In such changes, called **nuclear reactions** or *transmutations,* an unstable nucleus (the *parent nucleus*) spontaneously emits radiation and is converted (decays) into a more stable nucleus of a different element (the *daughter product*). Thus, a nuclear reaction results in a change in atomic number (number of protons) and, in some cases, a change in mass number as well. For example, the reaction of radium studied by Rutherford can be written as

$$ {}^{226}_{88}Ra \longrightarrow {}^{4}_{2}He + {}^{222}_{86}Rn $$

In a chemical change, the atoms in molecules and ions are rearranged, but atoms are neither created nor destroyed; the number of atoms remains the same.

When a radioactive atom decays, the emission of a charged particle leaves behind a charged atom. Thus, when radium-226 decays, it gives a helium-4 cation ($^{4}_{2}He^{2+}$) and a radon-222 anion ($^{222}_{86}Rn^{2-}$). By convention, the ion charges are not shown in balanced nuclear equations.

Recall that the atomic number is the number of protons in an atom's nucleus (that is, the total positive charge on the nucleus), and the mass number is the sum of protons and neutrons in a nucleus.

Similarly, in nuclear reactions the total number of nuclear particles, or **nucleons** (protons and neutrons), remains the same. The essence of nuclear reactions, however, is that one nucleon can change into a different nucleon along with the release of energy. A proton can change to a neutron or a neutron can change to a proton, but the total number of nucleons remains the same.

- **The sum of the mass numbers of reacting nuclei must equal the sum of the mass numbers of the nuclei produced.**
- **The sum of the atomic numbers of the products must equal the sum of the atomic number of the reactants.**

These principles can be verified for the preceding nuclear equation.

	$^{226}_{88}Ra$	$\longrightarrow$	$^{4}_{2}He$	+	$^{222}_{86}Rn$
	radium-226		alpha particle		radon-222
Mass number:	226	$\longrightarrow$	4	+	222
Atomic number:	88	$\longrightarrow$	2	+	86

Alpha and Beta Particle Emission

One way a radioactive isotope can decay is to eject an alpha particle from the parent nucleus. This is illustrated by the radium-226 reaction above and by the conversion of uranium-234 to thorium-230 by alpha emission.

	$^{234}_{92}U$	$\longrightarrow$	$^{4}_{2}He$	+	$^{230}_{90}Th$
	uranium-234 (parent nucleus)		alpha particle		thorium-230 (daughter product)
Mass number:	234	$\longrightarrow$	4	+	230
Atomic number:	92	$\longrightarrow$	2	+	90

In alpha emission, *the atomic number of the parent nucleus decreases by two units and its mass number decreases by four units for each alpha particle emitted.*

Emission of a beta particle is another way for a radioactive isotope to decay. For example, loss of a beta particle by uranium-239 (parent nucleus) to form neptunium-239 (daughter product) is represented by

Note that in beta decay the mass number is uncharged.

	$^{239}_{92}U$	$\longrightarrow$	$^{0}_{-1}e$	+	$^{239}_{93}Np$
	uranium-239		beta particle		neptunium-239
Mass number:	239	$\longrightarrow$	0	+	239
Atomic number:	92	$\longrightarrow$	−1	+	93

How does a nucleus, composed only of protons and neutrons, increase its number of protons by ejecting an electron during beta emission? It is generally accepted that a series of reactions is involved, but the net process is

$$\underset{\text{neutron}}{^{1}_{0}n} \longrightarrow \underset{\text{electron}}{^{0}_{-1}e} + \underset{\text{proton}}{^{1}_{1}p}$$

If a neutron changes to a proton, conservation of charge requires that a negative particle (a beta particle) be created.

where we use the symbol p for a proton and n for a neutron. In this overall process, a neutron is converted to a proton and a beta particle is released. Therefore, *the ejection of a beta particle always means that a different element is formed because a neutron has been converted into a proton. The new element (daughter product) has an atomic number one unit greater than that of the decaying (parent) nucleus.* The mass number does not change, however, because no proton or neutron has been emitted.

In many cases, the emission of an alpha or beta particle results in the formation of a product nucleus that is also unstable and therefore radioactive. The new radioactive product may undergo a number of successive transformations until a stable, nonradioactive nucleus is finally produced. Such a series of reactions is called a

A nucleus formed as a result of alpha or beta emission is often in an excited state and therefore emits a gamma ray.

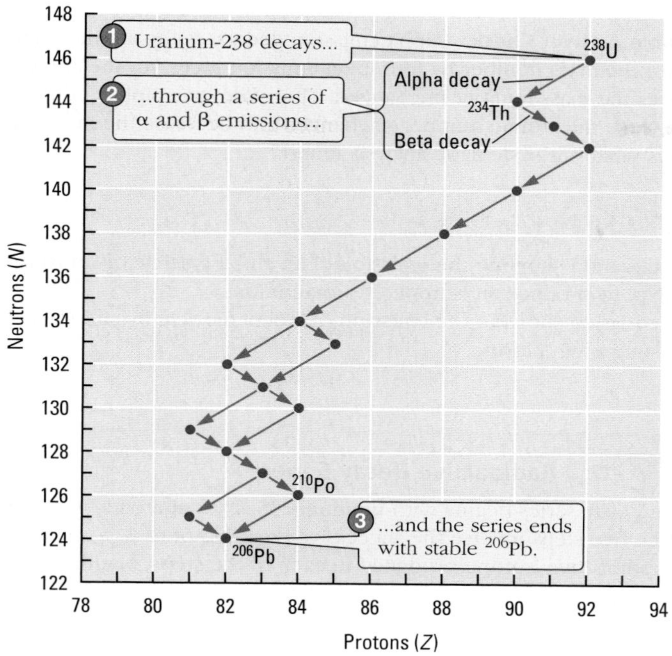

Figure 20.1 **The ^{238}U decay series.**

radioactive series. One such series begins with uranium-238 and ends with lead-206, as illustrated in Figure 20.1. The first step in the series is

$$^{238}_{92}U \longrightarrow {}^{4}_{2}He + {}^{234}_{90}Th$$

The final step, the conversion of polonium-210 to lead-206, is

$$^{210}_{84}Po \longrightarrow {}^{4}_{2}He + {}^{206}_{82}Pb$$

The **PROBLEM-SOLVING STRATEGY** in this book is

- Analyze the problem
- Plan a solution
- Execute the plan
- Check that the result is reasonable

Appendix A.1 explains this in detail.

PROBLEM-SOLVING EXAMPLE 20.1 Radioactive Series

An intermediate species in the uranium-238 decay series shown in Figure 20.1 is polonium-218. It emits an alpha particle, followed by emission of a beta particle, followed by the emission of a beta particle. Write the nuclear equations for these three reactions.

Answer $^{218}_{84}Po \longrightarrow {}^{4}_{2}He + {}^{214}_{82}Pb$ $\quad$ $^{214}_{82}Pb \longrightarrow {}^{0}_{-1}e + {}^{214}_{83}Bi$

$^{214}_{83}Bi \longrightarrow {}^{0}_{-1}e + {}^{214}_{84}Po$

Strategy and Explanation The starting point of these linked reactions is polonium-218. When it emits an alpha particle, the atomic number decreases by two and the mass number decreases by four to produce lead-214.

$$\underset{\text{polonium-218}}{^{218}_{84}Po} \longrightarrow {}^{4}_{2}He + \underset{\text{lead-214}}{^{214}_{82}Pb}$$

The second of these linked reactions begins with lead-214, which emits a beta particle. The atomic number of the daughter product increases by one and the mass number remains unchanged to produce bismuth-214.

$$\underset{\text{lead-214}}{^{214}_{82}Pb} \longrightarrow {}^{0}_{-1}e + \underset{\text{bismuth-214}}{^{214}_{83}Bi}$$

In the third reaction, bismuth-214 emits a beta particle, so the atomic number of the daughter product increases by one and the mass number remains unchanged to produce polonium-214.

$$\underset{\text{bismuth-214}}{^{214}_{83}Bi} \longrightarrow {}^{0}_{-1}e + \underset{\text{polonium-214}}{^{214}_{84}Po}$$

☑ **Reasonable Answer Check** Alpha emission decreases the atomic number by two and decreases the mass number by four. Beta emission increases the atomic number by one and leaves the mass number unchanged. Therefore, one alpha and two beta emissions would leave the atomic number unchanged and decrease the mass number by four, which is what our systematic analysis found.

PROBLEM-SOLVING PRACTICE answers are provided at the back of this book in Appendix K.

PROBLEM-SOLVING PRACTICE 20.1

(a) Write an equation showing the emission of an alpha particle by an isotope of neptunium, $^{237}_{93}Np$, to produce an isotope of protactinium.

(b) Write an equation showing the emission of a beta particle by sulfur-35, $^{35}_{16}S$, to produce an isotope of chlorine.

Answers to **EXERCISES** are provided at the back of this book in Appendix L.

EXERCISE 20.1 Radioactive Decay Series

The actinium decay series begins with uranium-235, $^{235}_{92}U$, and ends with lead-207, $^{207}_{82}Pb$. The first five steps involve the successive emission of α, β, α, α, and β particles. Identify the radioactive isotope produced in each of the steps, beginning with uranium-235.

Other Types of Radioactive Decay

The positron was discovered by Carl Anderson in 1932. It is sometimes called an "antielectron," one of a group of particles that have become known as "antimatter." Contact between an electron and a positron leads to mutual annihilation of both particles with production of two high-energy photons (gamma rays). This process is the basis of positron emission tomography (PET) scanning to detect tumors (Section 20.9, p. 985).

In addition to emission of alpha, beta, or gamma radiation, other nuclear decay processes are known. Some nuclei decay by emission of a **positron,** $^{0}_{+1}e$ or β^{+}, which is effectively a positively charged electron. For example, positron emission by polonium-207 leads to the formation of bismuth-207.

	$^{207}_{84}Po$	$\longrightarrow$	$^{0}_{+1}e$	+	$^{207}_{83}Bi$
	polonium-207		positron		bismuth-207
Mass number:	207	$\longrightarrow$	0	+	207
Atomic number:	84	$\longrightarrow$	+1	+	83

Notice that this process is the opposite of beta decay, because positron decay leads to a *decrease* in atomic number. Like beta decay, positron decay does not change the mass number because no proton or neutron is ejected.

Another nuclear process is **electron capture,** in which the atomic number is reduced by one but the mass number remains unchanged. In this process an inner-shell electron (for example, a 1*s* electron) is captured by the nucleus.

	$^{7}_{4}Be$	+	$^{0}_{-1}e$	$\longrightarrow$	$^{7}_{3}Li$
	beryllium-7		electron		lithium-7
Mass number:	7	+	0	$\longrightarrow$	7
Atomic number:	4	+	−1	$\longrightarrow$	3

In the old nomenclature of atomic physics, the innermost shell ($n = 1$ principal quantum number) was called the K-shell, so the electron capture mechanism is sometimes called *K-capture.*

In summary, radioactive nuclei can decay in four ways, shown in the chart in the margin.

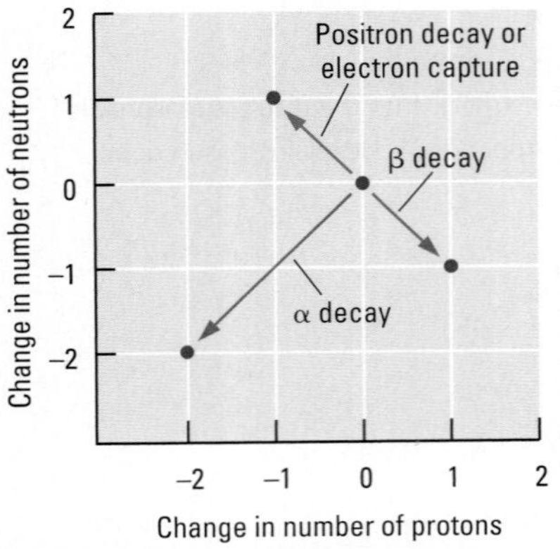

Effects of four radioactive decay processes. The chart shows the changes in the number of protons and neutrons during alpha decay, beta decay, positron emission, and electron capture.

PROBLEM-SOLVING EXAMPLE 20.2 Nuclear Equations

Complete these nuclear equations by filling in the missing symbol, mass number, and atomic number of the product species.

(a) $^{18}_{9}F \longrightarrow ^{18}_{8}O +$ ____ (b) $^{26}_{13}Al + ^{0}_{-1}e \longrightarrow$ ____

(c) $^{208}_{79}Au \longrightarrow ^{208}_{80}Hg +$ ____ (d) $^{218}_{84}Po \longrightarrow ^{4}_{2}He +$ ____

Answer

(a) ${}^{0}_{+1}e$ (b) ${}^{26}_{12}Mg$ (c) ${}^{0}_{-1}e$ (d) ${}^{214}_{82}Pb$

Strategy and Explanation In each case we deduce the missing species by comparing the atomic numbers and mass numbers before and after the reaction.

(a) The missing particle has a mass number of zero and a charge of +1, so it must be a positron, ${}^{0}_{+1}e$. When the positron is included in the equation, the atomic mass is 18 on each side, and the atomic numbers sum to 9 on each side.

(b) The missing nucleus must have a mass number of 26 + 0 = 26 and an atomic number of 13 − 1 = 12, so it is ${}^{26}_{12}Mg$.

(c) The missing particle has a mass number of zero and a charge of −1, so it must be a beta particle, ${}^{0}_{-1}e$.

(d) The missing nucleus has a mass number of 218 − 4 = 214 and an atomic number of 84 − 2 = 82, so it is ${}^{214}_{82}Pb$.

PROBLEM-SOLVING PRACTICE **20.2**

Complete these nuclear equations by filling in the missing symbol, mass number, and atomic number of the product species.

(a) ${}^{11}_{6}C \longrightarrow {}^{11}_{5}B + ?$ (b) ${}^{35}_{16}S \longrightarrow {}^{35}_{17}Cl + ?$

(c) ${}^{30}_{15}P \longrightarrow {}^{0}_{+1}e + ?$ (d) ${}^{22}_{11}Na \longrightarrow {}^{0}_{-1}e + ?$

EXERCISE **20.2 Nuclear Reactions**

Aluminum-26 can undergo either positron emission or electron capture. Write the balanced nuclear equation for each case.

20.3 Stability of Atomic Nuclei

The naturally occurring isotopes of elements from hydrogen to bismuth are shown in Figure 20.2, where the radioactive isotopes are represented by orange circles and the stable (nonradioactive) isotopes are represented by purple and green circles. It is surprising that so few stable isotopes exist. Why are there not hundreds more? To investigate this question, we will systematically examine the elements, starting with hydrogen.

In its simplest and most abundant form, hydrogen has only one nuclear particle, a single proton. In addition, the element has two other naturally occurring isotopes: nonradioactive deuterium, with one proton and one neutron, ${}^{2}_{1}H$ = D, and radioactive tritium, with one proton and two neutrons, ${}^{3}_{1}H$ = T. Helium, the next element, has two protons and two neutrons in its most stable isotope. At the end of the actinide series is element 103, lawrencium, one isotope of which has 154 neutrons and a mass number of 257. From hydrogen to lawrencium, except for ${}^{1}_{1}H$ and ${}^{3}_{2}He$, *the mass numbers of stable isotopes are always at least twice as large as the atomic number.* In other words, except for ${}^{1}_{1}H$ and ${}^{3}_{2}He$, every isotope of every element has a nucleus containing *at least* one neutron for every proton. Apparently the tremendous *repulsive* forces between the positively charged protons in the nucleus are moderated by the presence of neutrons, which have no electrical charge. Figure 20.2 illustrates a number of principles:

- **For light elements up to Ca (Z = 20), the stable isotopes usually have equal numbers of protons and neutrons, or perhaps one more neutron than protons.** Examples include ${}^{7}_{3}Li$, ${}^{12}_{6}C$, ${}^{16}_{8}O$, and ${}^{32}_{16}S$.
- **Beyond calcium the neutron-to-proton ratio becomes increasingly greater than 1.** The band of stable isotopes deviates more and more from the line $N = Z$ (number of neutrons equals number of protons). It is evident that

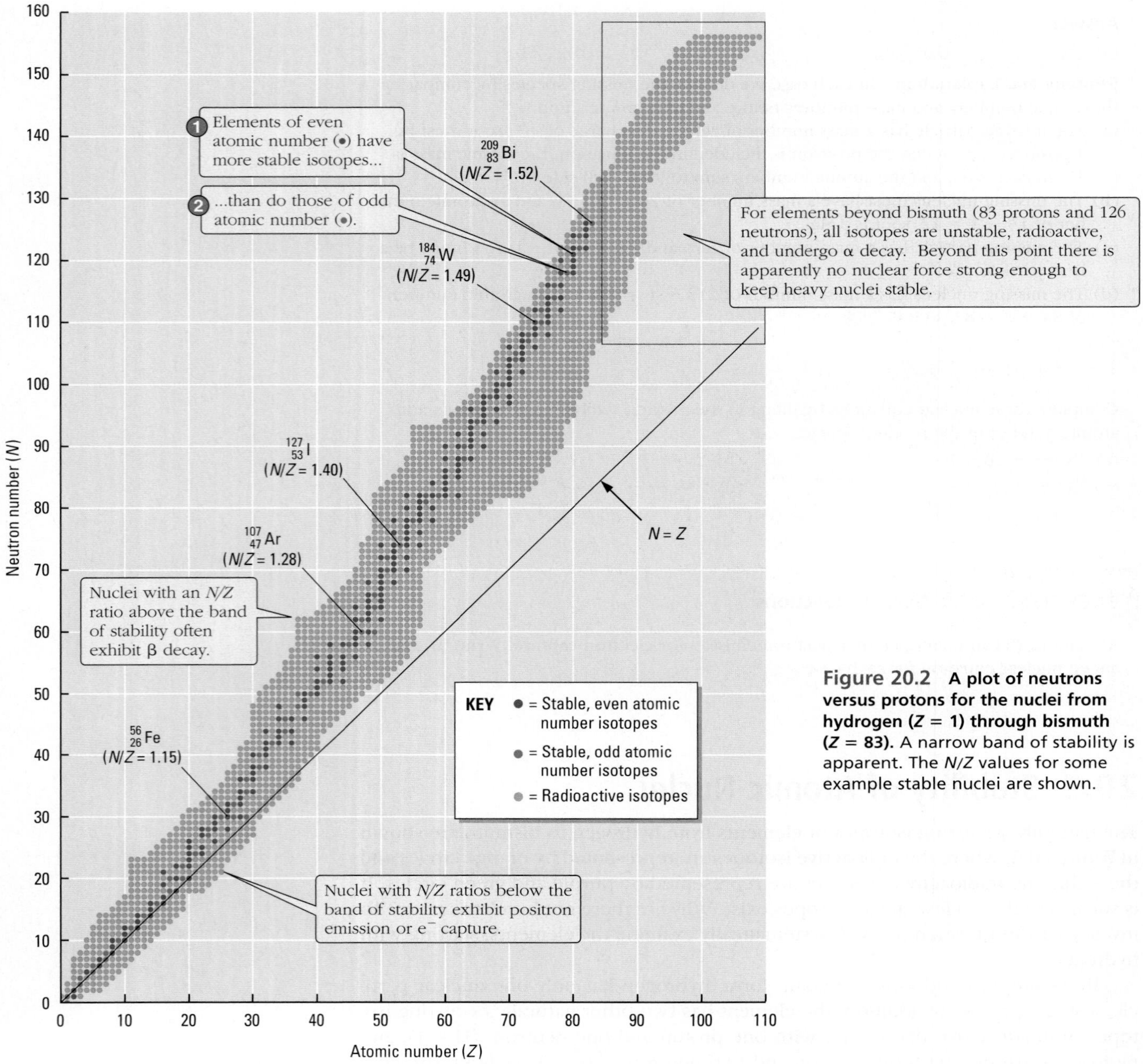

Elements of even atomic number (●) have more stable isotopes...
...than do those of odd atomic number (●).
$^{209}_{83}$Bi (N/Z = 1.52)
$^{184}_{74}$W (N/Z = 1.49)
$^{127}_{53}$I (N/Z = 1.40)
$^{107}_{47}$Ar (N/Z = 1.28)
$^{56}_{26}$Fe (N/Z = 1.15)
For elements beyond bismuth (83 protons and 126 neutrons), all isotopes are unstable, radioactive, and undergo α decay. Beyond this point there is apparently no nuclear force strong enough to keep heavy nuclei stable.
Nuclei with an N/Z ratio above the band of stability often exhibit β decay.
Nuclei with N/Z ratios below the band of stability exhibit positron emission or e⁻ capture.
N = Z
KEY
● = Stable, even atomic number isotopes
● = Stable, odd atomic number isotopes
● = Radioactive isotopes
Neutron number (N)
Atomic number (Z)

Figure 20.2 A plot of neutrons versus protons for the nuclei from hydrogen ($Z = 1$) through bismuth ($Z = 83$). A narrow band of stability is apparent. The N/Z values for some example stable nuclei are shown.

more neutrons are needed for nuclear stability in the heavier elements. For example, whereas one stable isotope of Fe has 26 protons and 30 neutrons ($N/Z = 1.15$), one stable isotope of platinum has 78 protons and 117 neutrons ($N/Z = 1.50$).

- **For elements beyond bismuth-209 (83 protons and 126 neutrons), all nuclei are unstable and radioactive.** Furthermore, the rate of disintegration becomes greater as the nuclear mass increases. For example, half of a $^{238}_{92}U$ sample decays in 4.5 billion years, whereas half of a $^{256}_{103}Lr$ sample decays in only 28 seconds.
- **Elements with an even atomic number have a greater number of stable isotopes than do those with an odd atomic number.**

- **For elements with an odd atomic number, the most stable isotope has an even number of neutrons.**
- **Of the nearly 270 stable isotopes shown in Figure 20.2, approximately 160 have an even number of neutrons *and* an even number of protons. Only about 110 have an odd number of either protons or neutrons. Only four isotopes ($^{2}_{1}H$, $^{6}_{3}Li$, $^{10}_{5}B$, and $^{14}_{7}N$) have odd numbers of *both* protons and neutrons.**

The Band of Stability and Type of Radioactive Decay

The narrow band of stable isotopes in Figure 20.2 (the purple and green circles) is sometimes called the *peninsula of stability* in a "sea of instability." Any nucleus (the orange circles) not on this peninsula will decay in such a way that the nucleus can come ashore on the peninsula. The chart can help us predict what type of decay will be observed.

The nuclei of all elements beyond Bi ($Z = 83$) are unstable—that is, radioactive—and most decay by *alpha particle emission.* For example, americium, the radioactive element used in smoke alarms, decays in this manner.

$$^{243}_{95}Am \longrightarrow {}^{4}_{2}He + {}^{239}_{93}Np$$

Beta emission occurs in isotopes that have too many neutrons to be stable—that is, isotopes above the peninsula of stability in Figure 20.2. When beta decay converts a neutron to a proton and an electron is ejected, the atomic number increases by one, and the mass number remains constant.

$$^{60}_{27}Co \longrightarrow {}^{0}_{-1}e + {}^{60}_{28}Ni$$

Conversely, lighter nuclei—below the peninsula of stability—that have too few neutrons attain stability by *positron emission* or by *electron capture,* because these processes convert a proton to a neutron in one step.

$$^{13}_{7}N \longrightarrow {}^{0}_{+1}e + {}^{13}_{6}C$$

$$^{41}_{20}Ca + {}^{0}_{-1}e \longrightarrow {}^{41}_{19}K$$

Decay by these two routes is observed for elements with atomic numbers ranging from 4 to greater than 100; as Z increases, electron capture becomes more likely than positron emission.

PROBLEM-SOLVING EXAMPLE 20.3 Nuclear Stability

For each of these unstable isotopes, write a nuclear equation for its probable mode of decay.

(a) Silicon-32, $^{32}_{14}Si$ (b) Titanium-43, $^{43}_{22}Ti$
(c) Plutonium-239, $^{239}_{94}Pu$ (d) Manganese-56, $^{56}_{25}Mn$

Answer

(a) $^{32}_{14}Si \longrightarrow {}^{0}_{-1}e + {}^{32}_{15}P$
(b) $^{43}_{22}Ti \longrightarrow {}^{0}_{+1}e + {}^{43}_{21}Sc$ or $^{43}_{22}Ti + {}^{0}_{-1}e \longrightarrow {}^{43}_{21}Sc$
(c) $^{239}_{94}Pu \longrightarrow {}^{4}_{2}He + {}^{235}_{92}U$
(d) $^{56}_{25}Mn \longrightarrow {}^{0}_{-1}e + {}^{56}_{26}Fe$

Strategy and Explanation Note the ratio of protons to neutrons. If there are excess neutrons, beta emission is probable. If there are excess protons, either electron capture or positron emission is probable. If the atomic number is greater than 83, then alpha emission is probable.

(a) Silicon-32 has excess neutrons, so beta decay is expected.
(b) Titanium-43 has excess protons, so either positron emission or electron capture is probable.
(c) Plutonium-239 has an atomic number greater than 83, so alpha decay is probable.
(d) Manganese-56 has excess neutrons, so beta decay is expected.

PROBLEM-SOLVING PRACTICE 20.3

For each of these unstable isotopes, write a nuclear equation for its probable mode of decay.

(a) $^{42}_{19}K$ (b) $^{234}_{92}U$ (c) $^{20}_{9}F$

Binding Energy

As demonstrated by Ernest Rutherford's alpha particle scattering experiment (← *p. 44*), the nucleus of the atom is extremely small. Yet the nucleus can contain up to 83 protons before becoming unstable, suggesting that there must be a very strong short-range binding force that can overcome the electrostatic repulsive force of a number of protons packed into such a tiny volume. A measure of the force holding the nucleus together is the nuclear **binding energy.** This energy (E_b) is defined as the negative of the energy change (ΔE) that would occur if a nucleus were formed directly from its component protons and neutrons. For example, if a mole of protons and a mole of neutrons directly formed a mole of deuterium nuclei, the energy change would be more than 200 million kJ, the equivalent of exploding 73 tons of TNT.

The nuclear binding energy is similar to the bond energy for a chemical bond (← *p. 344*). The binding energy is the energy that must be supplied to separate all of the particles (protons and neutrons) that make up the atomic nucleus; the bond energy is the energy per mole that must be supplied to separate two bonded atoms linked by the chemical bond. In both cases the energy change is positive, because work must be done to separate the particles, but nuclear binding energy is much greater per mole than chemical bond energy.

$$^{1}_{1}H + ^{1}_{0}n \longrightarrow ^{2}_{1}H \qquad \Delta E = -2.15 \times 10^{8}\ kJ$$

$$\text{Binding energy} = E_b = -\Delta E = 2.15 \times 10^{8}\ kJ$$

This nuclear synthesis reaction is highly exothermic (so E_b is very positive), an indication of the strong attractive forces holding the nucleus together. The deuterium nucleus is more stable than an isolated proton and an isolated neutron.

The enormous energy released during the formation of atomic nuclei occurs because the mass of a nucleus is always slightly less than the sum of the masses of its constituent protons and neutrons.

$$\underset{1.007825\ g/mol}{^{1}_{1}H} \quad + \quad \underset{1.008665\ g/mol}{^{1}_{0}n} \quad \longrightarrow \quad \underset{2.01410\ g/mol}{^{2}_{1}H}$$

Change in mass = Δm = mass of product − sum of masses of reactants

$$\Delta m = 2.01410\ g/mol - 1.008665\ g/mol - 1.007825\ g/mol$$

$$\Delta m = 2.01410\ g/mol - 2.016490\ g/mol$$

$$\Delta m = -0.00239\ g/mol = -2.39 \times 10^{-6}\ kg/mol$$

The missing mass, Δm, is released as energy, which we describe as the binding energy.

The relationship between mass and energy is contained in Albert Einstein's 1905 theory of special relativity, which holds that mass and energy are simply different manifestations of the same quantity. Einstein stated that the energy of a body is equivalent to its mass times the square of the speed of light, $E = mc^2$. To calculate the energy change in a process in which the mass has changed, the equation becomes

$$\Delta E = (\Delta m)c^2$$

For the formation of 1 mol deuterium nuclei from 1 mol protons and 1 mol neutrons, we have

$$\Delta E = (-2.39 \times 10^{-6}\ kg)(3.00 \times 10^{8}\ m/s)^2$$

$$= -2.15 \times 10^{11}\ J = -2.15 \times 10^{8}\ kJ$$

$1\ J = 1\ kg\ m^2/s^2$

This is the ΔE value given above for the change in energy when a mole of protons and a mole of neutrons form a mole of deuterium nuclei.

Consider another example, the formation of a helium-4 nucleus from two protons and two neutrons.

$$2\,{}^{1}_{1}\text{H} + 2\,{}^{1}_{0}\text{n} \longrightarrow {}^{4}_{2}\text{He} \qquad E_b = +2.73 \times 10^9 \text{ kJ/mol } {}^{4}_{2}\text{He nuclei}$$

This binding energy, E_b, is very large, even larger than that for deuterium. To compare nuclear stabilities more appropriately, nuclear scientists calculate the **binding energy per nucleon.** Each ${}^{4}_{2}$He nucleus contains four nucleons—two protons and two neutrons. Therefore, 1 mol ${}^{4}_{2}$He atoms contains 4 mol nucleons.

$$E_b \text{ per mol nucleons} = \frac{2.73 \times 10^9 \text{ kJ}}{\text{mol } {}^{4}_{2}\text{He nuclei}} \times \frac{1 \text{ mol } {}^{4}_{2}\text{He nuclei}}{4 \text{ mol nucleons}}$$
$$= 6.83 \times 10^8 \text{ kJ/mol nucleons}$$

The greater the binding energy per nucleon, the greater the stability of the nucleus. The binding energies per nucleon are known for a great number of nuclei and are plotted as a function of mass number in Figure 20.3. It is very interesting and important that the point of maximum stability occurs in the vicinity of iron-56, ${}^{56}_{26}$Fe. This means that *all nuclei are thermodynamically unstable with respect to iron-56.* That is, very heavy nuclei can split, or *fission,* to form smaller, more stable nuclei with atomic numbers nearer to that of iron, while simultaneously releasing enormous quantities of energy (Section 20.6, p. 975). In contrast, two very light nuclei can come together and undergo *nuclear fusion* exothermically to form heavier nuclei (Section 20.7, p. 980). Because of its high nuclear stability, *iron is the most abundant of the heavier elements in the universe.*

EXERCISE 20.3 Binding Energy

Calculate the binding energy, in kJ/mol, for the formation of lithium-6.

$$3\,{}^{1}_{1}\text{H} + 3\,{}^{1}_{0}\text{n} \longrightarrow {}^{6}_{3}\text{Li}$$

The necessary masses are ${}^{1}_{1}$H = 1.00783 g/mol, ${}^{1}_{0}$n = 1.00867 g/mol, and ${}^{6}_{3}$Li = 6.015125 g/mol. Is the binding energy greater than or less than that for helium-4? Compare the binding energy per nucleon of lithium-6 and helium-4. Which nucleus is more stable?

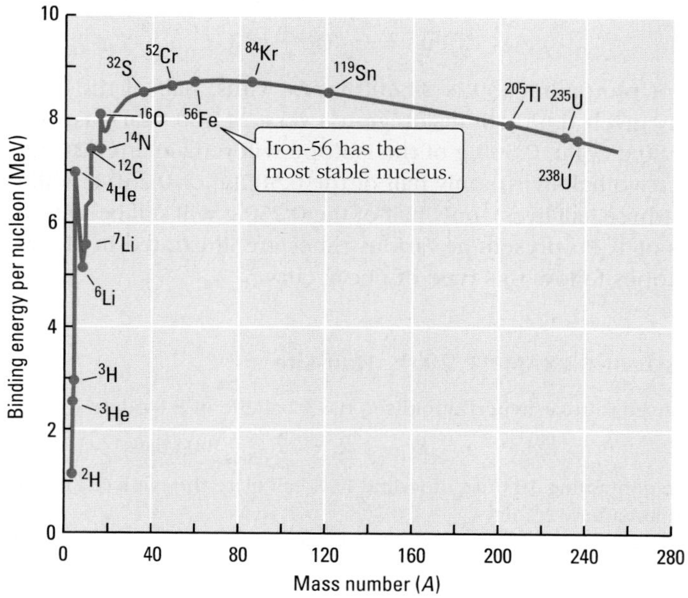

Figure 20.3 Binding energy per nucleon. The values plotted were derived by calculating the binding energy per nucleon in million electron volts (MeV) for the most abundant isotope of each element from hydrogen to uranium (1 MeV = 1.602×10^{-13} J). The nuclei at the top of the curve are most stable.

EXERCISES that are labeled **CONCEPTUAL** are designed to test your understanding of one or more concepts; they usually involve qualitative rather than quantitative thinking.

CONCEPTUAL EXERCISE 20.4 Binding Energy

By interpreting the shape of the curve in Figure 20.3, determine which is more exothermic per gram—fission or fusion. Explain your answer.

Table 20.2 Half-Lives of Some Common Radioactive Isotopes

Decay Process	Half-Life
$^{238}_{92}U \longrightarrow ^{234}_{90}Th + ^{4}_{2}He$	4.46×10^9 yr
$^{3}_{1}H \longrightarrow ^{3}_{2}He + ^{0}_{-1}e$	12.3 yr
$^{14}_{6}C \longrightarrow ^{14}_{7}N + ^{0}_{-1}e$	5730 yr
$^{131}_{53}I \longrightarrow ^{131}_{54}Xe + ^{0}_{-1}e$	8.04 d
$^{123}_{53}I + ^{0}_{-1}e \longrightarrow ^{123}_{52}Te$	13.2 h
$^{57}_{24}Cr \longrightarrow ^{57}_{25}Mn + ^{0}_{-1}e$	21 s
$^{28}_{15}P \longrightarrow ^{28}_{14}Si + ^{0}_{+1}e$	0.270 s
$^{90}_{38}Sr \longrightarrow ^{90}_{39}Y + ^{0}_{-1}e$	28.8 yr
$^{60}_{27}Co \longrightarrow ^{60}_{28}Ni + ^{0}_{-1}e$	5.26 yr

The relationship $t_{1/2} = \frac{0.693}{k}$ was introduced in the context of kinetics of first-order reactions *(← p. 607)*.

20.4 Rates of Disintegration Reactions

Cobalt-60 is radioactive and is used as a source of β particles and γ rays to treat malignancies in the human body. One-half of a sample of cobalt-60 will change via beta decay into nickel-60 in 5.26 years (Table 20.2). On the other hand, iodine-131, which is used to treat thyroid gland malfunction, decays much more rapidly; half of the radioactive iodine-131 decays in 8.04 days. These two radioactive isotopes are clearly different in their rates of decay.

Half-Life

The relative instability of a radioactive isotope is expressed as its **half-life,** the time required for one half of a given quantity of the isotope to undergo radioactive decay. In terms of reaction kinetics ***(← p. 602)***, radioactive decay is a first-order reaction. Therefore, the rate of decay is given by the first-order rate law equation

$$\ln[A]_t = -kt + \ln[A]_0$$

where $[A]_0$ is the initial concentration of isotope A, $[A]_t$ is the concentration of A after time t has passed, and k is the first-order rate constant. Because radioactive decay is first-order, the half-life ($t_{1/2}$) of an isotope is the same no matter what the initial concentration. It is given by

$$t_{1/2} = \frac{\ln 2}{k} = \frac{0.693}{k}$$

As illustrated by Table 20.2, isotopes have widely varying half-lives. Some take years, even millennia, for half of the sample to decay (^{238}U, ^{14}C), whereas others decay to half the original number of atoms in seconds or less (^{57}Cr, ^{28}P). The unit of half-life is whatever time unit is most appropriate—anything from years to seconds.

As an example of the concept of half-life, consider plutonium-239, an alpha-emitting isotope formed in nuclear reactors, which decays by alpha emission.

$$^{239}_{94}Pu \longrightarrow ^{4}_{2}He + ^{235}_{92}U$$

The half-life of plutonium-239 is 24,400 years. Thus, half of the quantity of $^{239}_{94}Pu$ present at any given time will disintegrate every 24,400 years. For example, if we begin with 1.00 g $^{239}_{94}Pu$, 0.500 g of the isotope will remain after 24,400 years. After 48,800 years (two half-lives), only half of the 0.500 g, or 0.250 g, will remain. After 73,200 years (three half-lives), only half of the 0.250 g will still be present, or 0.125 g. The amounts of $^{239}_{94}Pu$ present at various times are illustrated in Figure 20.4. All radioactive isotopes follow this type of decay curve.

PROBLEM-SOLVING EXAMPLE 20.4 Half-Life

Iodine-131, used to treat hyperthyroidism, has a half-life of 8.04 days.

$$^{131}_{53}I \longrightarrow ^{131}_{54}Xe + ^{0}_{-1}e \qquad t_{1/2} = 8.04 \text{ days}$$

For a sample containing 10.0 μg of iodine-131, calculate the mass (μg) of the isotope that will remain after 32.2 days.

Answer 0.625 μg

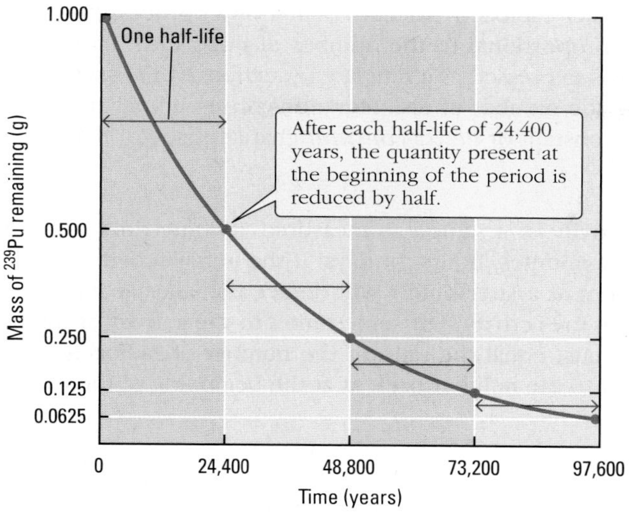

Figure 20.4 **The decay of 1.00 g plutonium-239.**

Strategy and Explanation First, we find the number of half-lives in the given 32.2-day time period. Since the half-life is 8.04 days, the number of half-lives is

$$32.2 \text{ days} \times \frac{1 \text{ half-life}}{8.04 \text{ days}} = 4.00 \text{ half-lives}$$

This means that the initial quantity of 10.0 μg is reduced by half four times.

$$10.0\ \mu g \times 1/2 \times 1/2 \times 1/2 \times 1/2 = 10.0\ \mu g \times 1/16 = 0.625\ \mu g$$

After 32.2 days, only one sixteenth of the original ^{131}I remains.

☑ **Reasonable Answer Check** After the passage of four half-lives, the remaining ^{131}I should be a small fraction of the starting amount, and it is.

PROBLEM-SOLVING PRACTICE **20.4**

Strontium-90 is a radioisotope ($t_{1/2}$ = 29 years) produced in atomic bomb explosions. Its long half-life and tendency to concentrate in bone marrow by replacing calcium make it particularly dangerous to people and animals.

(a) Strontium-90 decays through beta emission. Write a balanced equation showing the other product of decay.

(b) A ^{90}Sr sample emits 2000 β particles per minute. How many half-lives and how many years are necessary to reduce the emission to 125 β particles per minute?

EXERCISE **20.5 Half-Lives**

The radioactivity of formerly highly radioactive isotopes is essentially negligible after ten half-lives. What percentage of the original radioisotope remains after this amount of time (ten half-lives)?

Rate of Radioactive Decay

To determine the half-life of a radioactive element, its *rate of decay,* that is, the number of atoms that disintegrate in a given time—per second, per hour, per year—must be measured.

As noted earlier, radioactive decay is a first-order process (← *p. 602*), with a rate that is directly proportional to the number of radioactive atoms present (N). This proportionality is expressed as a rate law (Equation 20.1) in which A is the **activity** of the sample—the number of disintegrations observed per unit time—and k is the first-order rate constant or *decay constant* characteristic of that radioisotope.

$$A = kN \qquad [20.1]$$

Suppose the activity of a sample is measured at some time t_0 and then measured again after a few minutes, hours, or days. If the initial activity is A_0 at t_0, then a second measurement at a later time t will detect a smaller activity A. Using Equation 20.1, the ratio of the activity A at some time t to the activity at the beginning of the experiment A_0 must equal the ratio of the number of radioactive atoms N that are present at time t to the number present at the beginning of the experiment, N_0.

$$\frac{A}{A_0} = \frac{kN}{kN_0} \quad \text{or} \quad \frac{A}{A_0} = \frac{N}{N_0}$$

Thus, either A/A_0 or N/N_0 expresses the fraction of radioactive atoms remaining in a sample after some time has elapsed.

The activity of a sample can be measured with a device such as a Geiger counter (Figure 20.5). It detects radioactive emissions as they ionize a gas to form free electrons and cations that can be attracted to a pair of electrodes. In the Geiger counter, a metal tube is filled with low-pressure argon gas. The inside of the tube acts as the cathode. A thin wire running through the center of the tube acts as the anode. When radioactive emissions enter the tube through the thin window at the end, they collide with argon atoms; these collisions produce free electrons and argon cations. As the free electrons accelerate toward the anode, they collide with other argon atoms to generate more free electrons. The free electrons all go to the anode, and they constitute a pulse of current. This current pulse is counted, and the rate of pulses per unit time is the output of the Geiger counter.

The *curie* was named for Pierre Curie by his wife, Marie; the *becquerel* honors Henri Becquerel.

The **curie, Ci,** is commonly used as a unit of activity. One curie represents a decay rate of 3.7×10^{10} disintegrations per second (s^{-1}), which is the decay rate of 1 g radium. One millicurie, mCi, $= 10^{-3}$ Ci $= 3.7 \times 10^{7}\ s^{-1}$. Another unit of radio-

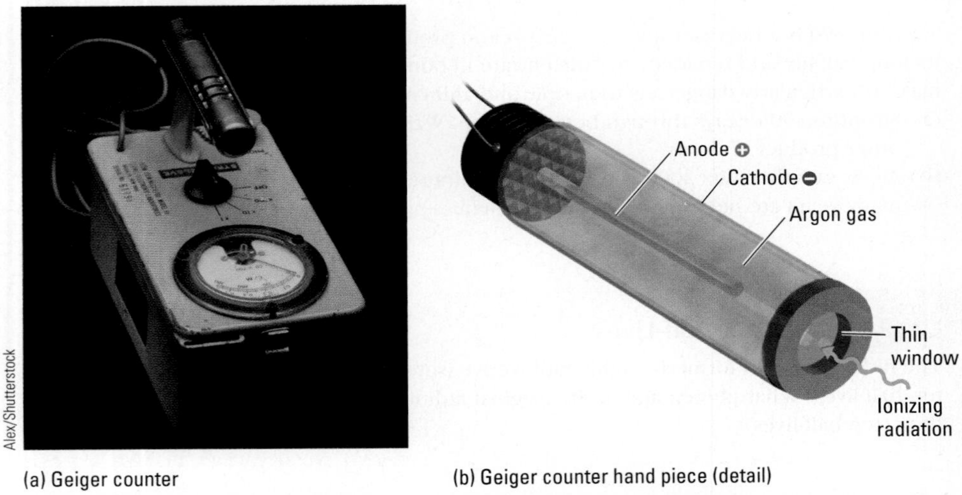

(a) Geiger counter (b) Geiger counter hand piece (detail)

Active Figure 20.5 A Geiger counter. Ionizing radiation passes through the thin window and momentarily ionizes the argon gas in the metal Geiger tube. The ions complete a circuit between the anode and the cathode, creating a momentary electrical pulse, which is amplified. The pulses are counted, and the pulse frequency indicates the radiation intensity. Go to **www.cengage.com/chemistry/moore** to test your understanding of the concepts in this figure.

activity is the **becquerel, Bq;** 1 becquerel is equal to one nuclear disintegration per second (1 Bq = 1 s^{-1}).

The change in activity of a radioactive sample over a period of time, or the fraction of radioactive atoms still present in a sample after some time has elapsed, can be calculated using the integrated rate equation for a first-order reaction

$$\ln A = -kt + \ln A_0$$

which can be rearranged to

$$\ln \frac{A}{A_0} = -kt \qquad [20.2]$$

where A/A_0 is the ratio of activity at time t to activity at time zero. Equation 20.2 can also be stated in terms of the fraction of radioactive atoms present in the sample at time t.

$$\ln \frac{N}{N_0} = -kt \qquad [20.3]$$

In words, Equation 20.3 says

$$\text{Natural logarithm}\left(\frac{\text{number of radioactive atoms at time } t}{\text{number of radioactive atoms at start of experiment}}\right)$$
$$= \text{natural logarithm(fraction of radioactive atoms remaining at time } t)$$
$$= -(\text{decay constant})(\text{time})$$

Notice the negative sign in Equation 20.3. The ratio N/N_0 is less than 1 because N is always less than N_0. This means that the logarithm of N/N_0 is negative, and the other side of the equation has a compensating negative sign because k and t are always positive.

As we have seen, the half-life of an isotope is inversely proportional to the first-order rate constant k:

$$t_{1/2} = \frac{0.693}{k}$$

Thus, the half-life can be found by calculating k from Equation 20.3 using N and N_0 from laboratory measurements over the time period t, as illustrated in Problem-Solving Example 20.5.

The unit for curie and becquerel is s^{-1} because each is a number (of disintegrations) per second.

Equation 20.2 can be derived from Equation 20.1 using calculus *(← p. 603)*.

As radioactive atoms decay, N becomes a smaller and smaller fraction of N_0.

PROBLEM-SOLVING EXAMPLE 20.5 Half-Life

A sample of ^{24}Na initially undergoes 3.50 × 10^4 disintegrations per second (s^{-1}). After 24.0 h, its disintegration rate has fallen to 1.16 × 10^4 s^{-1}. Calculate the half-life of ^{24}Na.

Answer 15.0 h

Strategy and Explanation We use Equation 20.2 relating activity (disintegration rate) at time zero and time t with the decay constant k. The experiment provided us with A, A_0, and the time.

$$\ln\left(\frac{1.16 \times 10^4 \text{ s}^{-1}}{3.50 \times 10^4 \text{ s}^{-1}}\right) = \ln(0.331) = -k(24.0 \text{ h})$$

$$k = -\frac{\ln(0.331)}{24.0 \text{ h}} = -\left(\frac{-1.104}{24.0 \text{ h}}\right) = 0.0460 \text{ h}^{-1}$$

From k we determine $t_{1/2}$.

$$t_{1/2} = \frac{0.693}{k} = \frac{0.693}{0.0460 \text{ h}^{-1}} = 15.0 \text{ h}$$

☑ **Reasonable Answer Check** The activity (disintegration rate) fell to between one half and one quarter of its initial value in 24 h, so the half-life must be less than 24 h, and this agrees with our more accurate calculation.

PROBLEM-SOLVING PRACTICE **20.5**

The decay of iridium-192, a radioisotope used in cancer radiation therapy, has a rate constant of $9.3 \times 10^{-3}\ d^{-1}$.

(a) Calculate the half-life of ^{192}Ir.

(b) Calculate the fraction of an ^{192}Ir sample that remains after 100 days.

PROBLEM-SOLVING EXAMPLE **20.6** **Time and Radioactivity**

A 1.00-mg sample of ^{131}I ($t_{1/2}$ = 8.04 days) has an initial disintegration rate of $4.7 \times 10^{12}\ s^{-1}$. How long will it take for the disintegration rate of the sample to fall to $2.9 \times 10^{11}\ s^{-1}$?

Answer 776 h

Strategy and Explanation We use the half-life of ^{131}I to find the decay constant, k, and convert the known half-life from days to hours.

$$t_{1/2} = 8.04 \text{ days} \times \frac{24 \text{ h}}{1 \text{ day}} = 193 \text{ h}$$

Then we calculate the decay constant, k.

$$k = \frac{0.693}{t_{1/2}} = \frac{0.693}{193 \text{ h}} = 3.59 \times 10^{-3}\ h^{-1}$$

We use this k value in the equation relating disintegration rate to time. The initial disintegration rate is $A_0 = 4.7 \times 10^{12}\ s^{-1}$, and the disintegration rate after the elapsed time is $A = 2.9 \times 10^{11}\ s^{-1}$. We can use Equation 20.2 to calculate the elapsed time t. Both disintegration rates are given in disintegrations per second, so we can use them as provided. If we converted them both to disintegrations per hour, we would get the same numerical result for the ratio.

$$\ln\left(\frac{2.9 \times 10^{11}\ s^{-1}}{4.7 \times 10^{12}\ s^{-1}}\right) = -kt = -(3.59 \times 10^{-3}\ h^{-1})t$$

$$t = \frac{-2.785}{3.59 \times 10^{-3}\ h^{-1}} = 776 \text{ h}$$

☑ **Reasonable Answer Check** The disintegration rate has fallen by approximately a factor of sixteen $((4.7 \times 10^{12})/(2.9 \times 10^{11}) = 16.2)$, so the elapsed time must be approximately four half-lives of ^{131}I, and it is.

PROBLEM-SOLVING PRACTICE **20.6**

In 1921 the women of America honored Marie Curie by giving her a gift of 1.00 g pure radium, which is now in Paris at the Curie Institute of France. The principal isotope, ^{226}Ra, has a half-life of 1.60×10^3 years. How many grams of radium-226 remain?

Bettmann/CORBIS

Willard Libby and his apparatus for carbon-14 dating.

Carbon-14 Dating

In 1946 Willard Libby developed a technique for measuring the age of archaeological objects using radioactive carbon-14. Carbon is an important building block of all living systems, and all organisms contain the three isotopes of carbon: ^{12}C, ^{13}C, and ^{14}C. The first two isotopes are stable (nonradioactive) and have been present for billions of years. Carbon-14, however, is radioactive and decays to nitrogen-14 by beta emission.

$$^{14}_{6}C \longrightarrow {}^{0}_{-1}e + {}^{14}_{7}N$$

The half-life of ^{14}C is known by experiment to be 5.73×10^3 years. The number of carbon-14 atoms (N) in a carbon-containing sample can be measured from the activity of the sample (A). If the number of carbon-14 atoms originally in the sample (N_0) can be determined, or if the initial activity (A_0) can be determined, the age of the sample can be found from Equation 20.2 or 20.3.

Willard Libby won the 1960 Nobel Prize in Chemistry for his discovery of radiocarbon dating.

This method of age determination clearly depends on knowing how much ^{14}C was originally in the sample. The answer to this question comes from work by physicist Serge Korff, who discovered in 1929 that ^{14}C is continually generated in the upper atmosphere. High-energy cosmic rays collide with gas molecules in the upper atmosphere and cause them to eject neutrons. These free neutrons collide with nitrogen atoms to produce carbon-14.

$$^{14}_{7}N + ^{1}_{0}n \longrightarrow ^{14}_{6}C + ^{1}_{1}H$$

Throughout the *entire* atmosphere, only about 7.5 kg ^{14}C is produced per year. However, this relatively small quantity of radioactive carbon is incorporated into CO_2 and other carbon compounds and then distributed worldwide as part of the carbon cycle. The continual formation of ^{14}C, transfer of the isotope within the oceans, atmosphere, and biosphere, and decay of living matter keep the supply of ^{14}C constant.

Plants absorb carbon dioxide from the atmosphere and convert it into food via photosynthesis ***(← p. 880)***. In this way, the ^{14}C becomes incorporated into living tissue, where radioactive ^{14}C atoms and nonradioactive ^{12}C atoms in CO_2 chemically react in the same way. The beta decay activity of carbon-14 in *living* plants and animals and in the air is constant at 15.3 disintegrations per minute per gram ($min^{-1}\ g^{-1}$) of carbon-14. When a plant or animal dies, however, carbon-14 disintegration continues *without the ^{14}C being replaced.* Consequently, the ^{14}C activity of the dead plant material decreases with the passage of time. The smaller the activity of carbon-14 in the plant, the longer the period of time between the death of the plant and the present. Assuming that ^{14}C activity in living plants was about the same hundreds or thousands of years ago as it is now, measurement of the ^{14}C beta activity of an artifact can be used to date an article containing carbon. The slight fluctuations of the ^{14}C activity in living plants for the past several thousand years have been measured by studying growth rings of long-lived trees, and the carbon-14 dates of objects can be corrected accordingly.

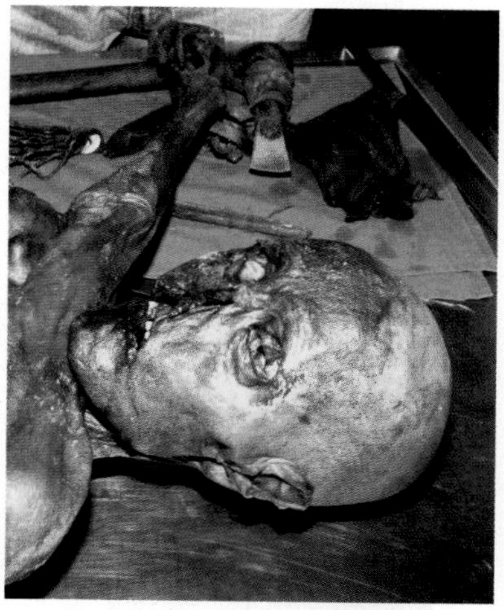
Reuters/CORBIS

The Ice Man. This human mummy was found in 1991 in glacial ice high in the Alps. Carbon-14 dating determined that he lived about 5300 years ago. The mummy is exhibited at the South Tyrol Archaeological Museum in Bolzano, Italy.

The time scale accessible to carbon-14 dating is determined by the value of the half-life of ^{14}C: 5.73×10^3 yr. Therefore, this method for dating objects can be extended back approximately 50,000 years. This span of time is almost nine half-lives, during which the number of disintegrations per minute per gram of carbon would fall by a factor of about $(\frac{1}{2})^9 = 1.95 \times 10^{-3}$ from about 15.3 $min^{-1}\ g^{-1}$ to about 0.030 $min^{-1}\ g^{-1}$, which is a disintegration rate so low that it is difficult to measure accurately.

PROBLEM-SOLVING EXAMPLE 20.7 Carbon-14 Dating

Charcoal fragments found in a prehistoric cave in Lascaux, France, had a measured disintegration rate of 2.40 $min^{-1}\ g^{-1}$ carbon. Calculate the approximate age of the charcoal.

Answer 1.53×10^4 years old

Strategy and Explanation We use Equation 20.2 to solve the problem

$$\ln\left(\frac{A}{A_0}\right) = -kt$$

where A is proportional to the known activity of the charcoal (2.40 $min^{-1}\ g^{-1}$) and A_0 is proportional to the activity of the carbon-14 in living material (15.3 $min^{-1}\ g^{-1}$). We first calculate k, the rate constant, using the half-life of carbon-14, 5.73×10^3 yr.

$$k = \frac{0.693}{t_{1/2}} = \frac{0.693}{5.73 \times 10^3\ \text{yr}} = 1.21 \times 10^{-4}\ \text{yr}^{-1}$$

© images-of-france/Alamy

Prehistoric cave paintings from Lascaux, France.

Now we calculate the time, *t*.

$$\ln\left(\frac{2.40\ \text{min}^{-1}\ \text{g}^{-1}}{15.3\ \text{min}^{-1}\ \text{g}^{-1}}\right) = -kt$$

$$\ln(0.157) = -(1.21 \times 10^{-4}\ \text{yr}^{-1})t$$

$$t = \frac{1.85}{1.21 \times 10^{-4}\ \text{yr}^{-1}} = 1.53 \times 10^{4}\ \text{yr}$$

Thus, the charcoal is approximately 15,300 years old.

☑ **Reasonable Answer Check** The disintegration rate has fallen a factor of six from the rate for living material, so more than two, but less than three, half-lives have elapsed. This agrees with our calculated result.

PROBLEM-SOLVING PRACTICE 20.7

Tritium, ^{3}H ($t_{1/2}$ = 12.3 yr), is produced in the atmosphere and incorporated in living plants in much the same way as ^{14}C. Estimate the age of a sealed sample of Scotch whiskey that has a tritium content 0.60 times that of the water in the area where the whiskey was produced.

EXERCISE 20.6 Radiochemical Dating

The radioactive decay of uranium-238 to lead-206 provides a method of radiochemically dating ancient rocks by using the ratio of ^{206}Pb atoms to ^{238}U atoms in a sample. Using this method, a moon rock was found to have a $^{206}Pb/^{238}U$ ratio of 100/109, that is, 100 ^{206}Pb atoms for every 109 ^{238}U atoms. No other lead isotopes were present in the rock, indicating that all of the ^{206}Pb was produced by ^{238}U decay. Estimate the age of the moon rock. The half-life of ^{238}U is 4.46×10^{9} years.

CONCEPTUAL EXERCISE 20.7 Radiochemical Dating

Ethanol, C_2H_5OH, is produced by the fermentation of grains or by the reaction of water with ethylene, which is made from petroleum. The alcohol content of wines can be increased fraudulently beyond the usual 12% available from fermentation by adding ethanol produced from ethylene. How can carbon dating techniques be used to differentiate the ethanol sources in these wines?

Courtesy of Lawrence Berkeley Laboratory

Glenn Seaborg

1912–1999

A pioneer in developing radioisotopes for medical use (Section 20.9, p. 985), Glenn Seaborg was the first to produce iodine-131, used subsequently to treat his mother's abnormal thyroid condition. As a result of Seaborg's further research, it became possible to predict accurately the properties of many of the as-yet-undiscovered transuranium elements. In a remarkable 21-year span (1940–1961), Seaborg and his colleagues synthesized ten new transuranium elements (plutonium to lawrencium). He received the Nobel Prize in 1951 for his creation of new elements. In the 1990s Seaborg was honored by having element 106 named for him.

20.5 Artificial Transmutations

In the course of his experiments, Rutherford found in 1919 that alpha particles ionize atomic hydrogen, knocking off an electron from each atom. Using atomic nitrogen instead, he found that bombardment with alpha particles *produced protons.* He correctly concluded that the alpha particles had knocked a proton out of the nitrogen nucleus and that a nucleus of another element had been produced. In other words, nitrogen had undergone a *transmutation* to oxygen.

$$^{4}_{2}\text{He} + ^{14}_{7}\text{N} \longrightarrow ^{17}_{8}\text{O} + ^{1}_{1}\text{H}$$

Rutherford had proposed that protons and neutrons are the fundamental building blocks of nuclei. Although his search for the neutron was not successful, it was later found by James Chadwick in 1932 as a product of the alpha particle bombardment of beryllium.

$$^{9}_{4}\text{Be} + ^{4}_{2}\text{He} \longrightarrow ^{12}_{6}\text{C} + ^{1}_{0}\text{n}$$

Changing one element into another by alpha particle bombardment has its limitations. Before a positively charged bombarding particle (such as the alpha particle) can be captured by a positively charged nucleus, the bombarding particle must have sufficient kinetic energy to overcome the repulsive forces developed as the particle approaches the nucleus. But the neutron is electrically neutral, so Enrico Fermi (in 1934) reasoned that a nucleus would not oppose its entry. By this approach, nearly every element has since been transmuted, and a number of *transuranium elements* (elements beyond uranium) have been prepared. For example, plutonium-239 forms americium-241 by neutron bombardment.

$$^{239}_{94}\text{Pu} + ^{1}_{0}\text{n} \longrightarrow ^{240}_{94}\text{Pu}$$

$$^{240}_{94}\text{Pu} + ^{1}_{0}\text{n} \longrightarrow ^{241}_{94}\text{Pu}$$

$$^{241}_{94}\text{Pu} \longrightarrow ^{241}_{95}\text{Am} + ^{0}_{-1}\text{e}$$

Of the elements currently known, all those with atomic numbers of 92 or less exist in nature except Tc and Pm. The transuranium elements, those with atomic numbers greater than 92, are all synthetic. Up to element 101, mendelevium, all the elements can be made by bombarding the nucleus of a lighter element with small particles such as $^{4}_{2}\text{He}$ or $^{1}_{0}\text{n}$. Beyond element 101, special techniques using heavier particles are required and are still being developed. For example, roentgenium, element 111, first synthesized in 1994, was made by bombarding bismuth-209 with nickel-64 nuclei.

$$^{64}_{28}\text{Ni} + ^{209}_{83}\text{Bi} \longrightarrow ^{272}_{111}\text{Rg} + ^{1}_{0}\text{n}$$

Roentgenium, Rg, element 111, honors Wilhelm Röntgen for his discovery of X-radiation.

The International Union of Pure and Applied Chemistry (IUPAC) has named element 110 darmstadtium (Ds) and element 111 roentgenium (Rg). Synthesis of element 112 has been confirmed by the IUPAC, and there have been reported syntheses of elements 113, 114, 115, 116, and 118, but these elements have not yet been named by the IUPAC.

EXERCISE 20.8 Nuclear Transmutations

Balance these equations for nuclear reactions, indicating the symbol, the mass number, and the atomic number of the remaining product.

(a) $^{13}_{6}\text{C} + ^{1}_{0}\text{n} \longrightarrow ^{4}_{2}\text{He} + ?$ (b) $^{14}_{7}\text{N} + ^{4}_{2}\text{He} \longrightarrow ^{1}_{0}\text{n} + ?$

(c) $^{253}_{99}\text{Es} + ^{4}_{2}\text{He} \longrightarrow ^{1}_{0}\text{n} + ?$

CONCEPTUAL EXERCISE 20.9 Element Synthesis

In 1996 a team of scientists reported the production of element 112. Bombardment of lead-208 produced a nucleus that emitted a neutron to yield element 112 with mass number = 277. Write a balanced nuclear equation for this process, and identify the bombarding nucleus.

20.6 Nuclear Fission

In 1938 the nuclear chemists Otto Hahn and Fritz Strassman were confounded when they isolated barium from a sample of uranium that had been bombarded with neutrons. Further work by Lise Meitner, Otto Frisch, Niels Bohr, and Leo Szilard confirmed that the bombarded uranium-235 nucleus had formed barium by the capture of a neutron and had undergone **nuclear fission,** a nuclear reaction in which the bombarded nucleus splits into two lighter nuclei (Figure 20.6).

Element 109 (Mt) is named in honor of Lise Meitner.

Figure 20.6 A neutron causes a ${}^{235}_{92}$U nucleus to undergo fission (Equation 20.4).

1926–

After achieving recognition for her investigation of the chemical properties of the heaviest elements at Los Alamos National Laboratory, the University of California, Berkeley, and the Lawrence Berkeley Laboratory, Darleane C. Hoffman was awarded the National Medal of Science in 1997 and the American Chemical Society's highest honor, the Priestley Medal, in 2000. Hoffman first became interested in nuclear chemistry as a student at Iowa State University and searched for new heavy elements and isotopes in nuclear test debris while working at Los Alamos National Laboratory. Later she moved to Berkeley to continue her research and became involved in the search for new, superheavy elements.

$$ {}^{235}_{92}\mathrm{U} + {}^{1}_{0}\mathrm{n} \longrightarrow {}^{236}_{92}\mathrm{U} \longrightarrow {}^{141}_{56}\mathrm{Ba} + {}^{92}_{36}\mathrm{Kr} + 3\,{}^{1}_{0}\mathrm{n} \qquad \Delta E = -2 \times 10^{10}\ \mathrm{kJ} \qquad [20.4] $$

This nuclear equation shows that bombardment with a single neutron produces three neutrons among the products. The fact that the fission reaction produces more neutrons than are required to begin the process is important. Each of the product neutrons is capable of inducing another fission reaction, so three neutrons would induce three fissions, which would release nine neutrons to induce nine more fissions, from which 27 neutrons are obtained, and so on. Since the neutron-induced fission of uranium-235 is extremely rapid, this sequence of reactions can lead to an explosive chain reaction, as illustrated in Figure 20.7.

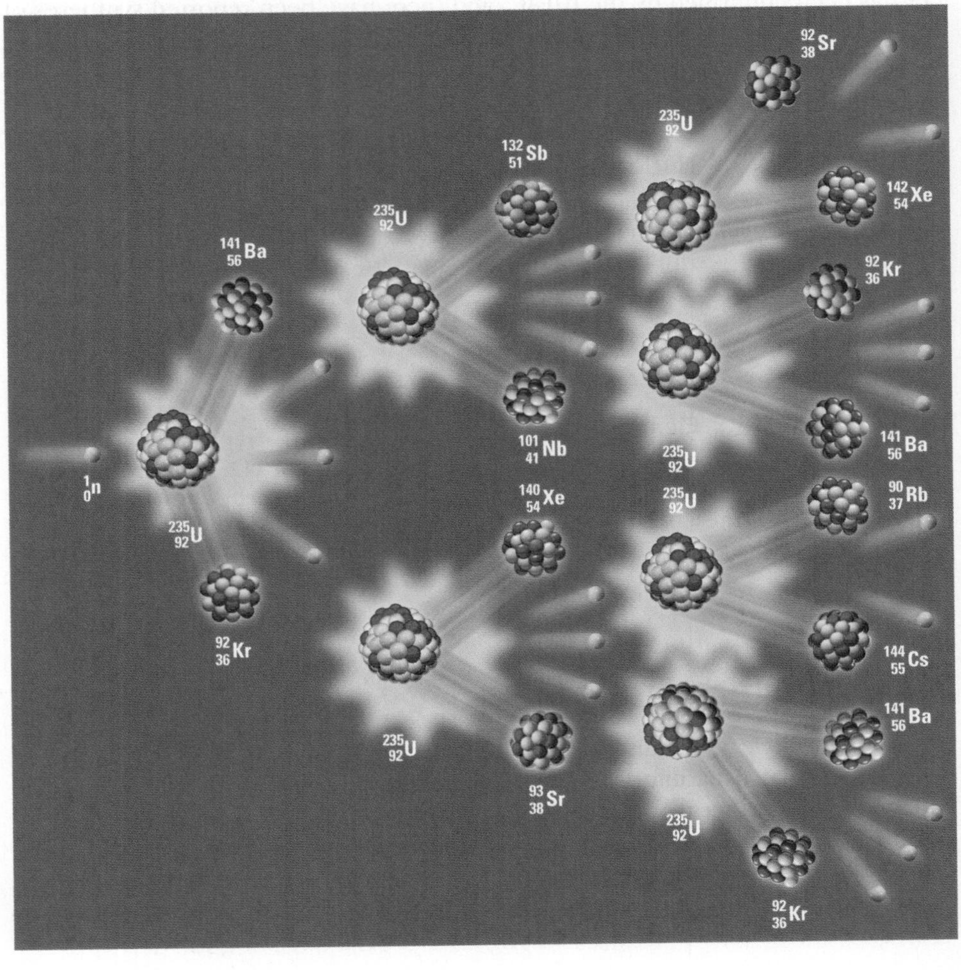

Figure 20.7 A self-propagating nuclear chain reaction initiated by capture of a neutron. Each fission event produces two lighter nuclei plus two or three neutrons.

Nuclear fission of uranium-235 produces a variety of products. Thirty-four elements have been detected among the fission products, including those shown in Figure 20.7. If the quantity of uranium-235 is small, then most of the neutrons escape without hitting a nucleus, and so few neutrons are captured by ^{235}U nuclei that the chain reaction cannot be sustained. The minimum mass of fissionable material required for a self-sustaining chain reaction is termed the **critical mass.** In an atomic bomb, two small masses of uranium-235, neither capable of sustaining a chain reaction, are brought together rapidly to form one mass capable of supporting a chain reaction. A nuclear fission explosion results.

Nuclear Reactors

Nuclear fission reactions are the thermal energy source in nuclear power plants (Figure 20.8). The nuclear reactor fuel is fissionable material, for example, uranium-235. In a **nuclear reactor,** the rate of fission is controlled by inserting rods of cadmium or other neutron absorbers into the reactor. The rods absorb the neutrons that would otherwise propagate fission reactions. The rate of the fission reaction can be increased or decreased by withdrawing or inserting the control rods, respectively. The materials that control the numbers of neutrons (by absorbing them) or control their energy (by absorbing some of their energy) are known as *moderators.* Thus, thermal energy from nuclear fission can be used safely in a nuclear power plant.

The uranium fuel in the reactor core is in the form of uranium dioxide, UO_2, pellets, which are about the size of the eraser on a pencil. The pellets are placed end-to-end in metal alloy tubes, which are then assembled into stainless steel–clad bundles. As pointed out earlier, once a fission reaction is started, it can be sustained as a chain reaction. However, a source of neutrons is needed to initiate the chain reaction. One means of generating these initial neutrons is a nuclear reaction source, such as beryllium-9, and a heavy, alpha-emitting element, such as plutonium or americium. The heavy element emits alpha particles; when they strike a beryllium-9 nucleus, the two nuclei combine to form a carbon-12 nucleus and a neutron is emitted.

$$^{238}_{94}\text{Pu} \longrightarrow {}^{4}_{2}\text{He} + {}^{234}_{92}\text{U}$$

$$^{4}_{2}\text{He} + {}^{9}_{4}\text{Be} \longrightarrow {}^{12}_{6}\text{C} + {}^{1}_{0}\text{n}$$

These neutrons then initiate the nuclear fission of uranium-235 in the reactor core.

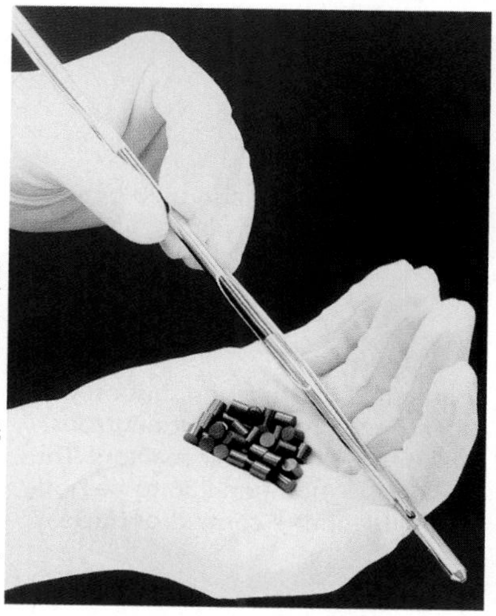

US Department of Energy/Photo Researchers, Inc.

Uranium oxide pellets used in nuclear fuel rods.

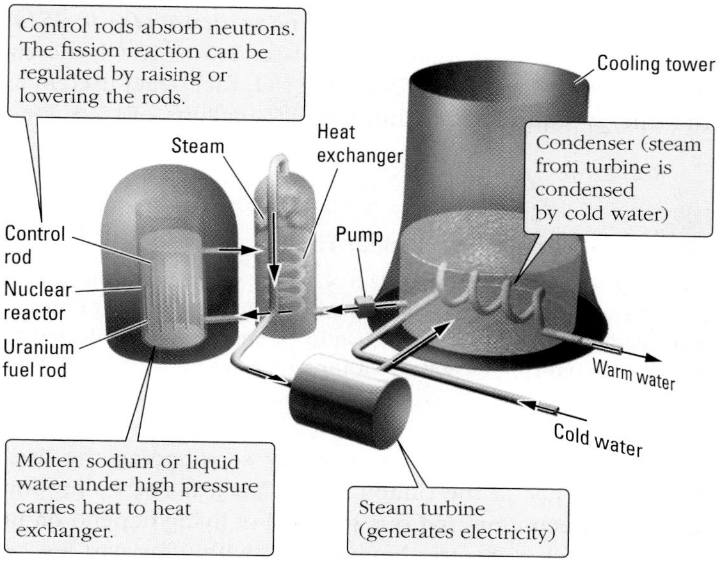

Figure 20.8 **Schematic diagram of a nuclear power plant.**

Hans F. Meier/iStockphoto

A nuclear power plant with two prominent cooling towers and one nuclear reactor building (the domed building).

Conventional (non-nuclear) power plants burn fossil fuel to generate the heat to produce steam to drive the turbine.

Cooling towers are also used by fossil fuel–burning power plants.

Uranium-238 can fission, but only when bombarded by fast neutrons, unlike those in nuclear reactors. Thus, we consider uranium-238 to be non-fissionable in the context of nuclear reactors.

Nuclear fuel rods that have become depleted of U-235 are known as *spent* fuel.

A sample is considered to be of weapons-grade quality only if its ^{235}U content is greater than 90%. Even in reactors using weapons-grade quality, the ^{235}U is still too dispersed to produce uncontrolled fission.

The tremendous heat generated by the nuclear fission reaction in the nuclear reaction vessel heats the primary coolant, a substance with a very high heat capacity, usually water. The primary coolant is at a high pressure of more than 150 atm, so it does not boil, even though the temperature is higher than its normal boiling point **(← *p. 483*)**. The hot primary coolant is pumped in a closed loop from the nuclear reaction vessel to the heat exchanger, where heat transfer to water that runs the steam generators lowers the temperature of the primary coolant, which then returns to the reactor to be reheated. This closed loop separates the nuclear reactor from the rest of the power plant.

In the heat exchanger, the water is vaporized to steam to turn the steam generators. This water is referred to as the secondary coolant. The steam strikes the large turbine blades, causing the turbine to spin. The turbine shaft is connected to the generator, which is surrounded by a magnetic field. The rapid spinning of the turbine shaft in a magnetic field produces electricity.

After striking the turbine blades, the steam must be condensed so that the heating/cooling cycle can be repeated to create additional electricity. To do so, cooling water is pumped from a neighboring river or lake to the secondary coolant loop. Enormous amounts of outside cooling water are needed to condense the vast quantity of steam produced by such power plants. For example, the nuclear power reactor at the Entergy Arkansas Unit 1 uses 750,000 gal of cooling water per minute. Having absorbed heat from the secondary coolant, the cooling water must then be cooled itself before being returned to its source. Such cooling is done in many nuclear power plants by passing the water through large concave evaporative cooling towers, which are often mistaken for the nuclear reactors themselves.

Not all nuclei can be made to fission on colliding with a neutron, but ^{235}U and ^{239}Pu are both fissionable isotopes. Natural uranium contains an average of only 0.72% of the fissionable ^{235}U isotope. More than 99% of the natural element is non-fissionable uranium-238. Since the percentage of natural ^{235}U is too small to sustain a chain reaction, uranium for nuclear power fuel must be enriched to about 3% uranium-235. To accomplish this goal, some of the ^{238}U isotope in a sample is effectively discarded, thereby raising the concentration of ^{235}U. If sufficient fissionable uranium-235 is present, it can capture enough neutrons to sustain the fission chain reaction. Approximately one third of the fuel rods in a nuclear reactor are replaced annually because fission by-products absorb neutrons, reducing the efficiency of the fission reactions.

Because the mass of uranium-235 in the fuel rods of a nuclear power plant is lower than the critical mass needed for an atomic bomb, the reactor core *cannot* undergo an uncontrolled explosive chain reaction.

Nuclear fission fuels have extremely large energy density **(← *p. 251*)**. For example, fission of 1.0 kg (2.2 lb) uranium-235 releases 9.0×10^{13} J, the equivalent of exploding 33,000 tons (33 kilotons) TNT. Each UO_2 fuel pellet used in a nuclear reactor has the energy equivalent to burning 136 gallons oil, 2.5 tons wood, or 1 ton coal.

EXERCISE 20.10 Energy of Nuclear Fission

Burning 1.0 kg high-grade coal produces 2.8×10^4 kJ, whereas fission of 1.0 mol uranium-235 generates 2.1×10^{10} kJ. How many metric tons of coal (1 metric ton = 10^3 kg) are needed to produce the same quantity of energy as that released by the fission of 1.0 kg uranium-235? (Assume that the processes have equal efficiency.)

There is, of course, substantial controversy surrounding the use of nuclear power plants, and not just in the United States. Nuclear energy proponents argue that the health of our economy and our standard of living depend on inexpensive, reliable, and safe sources of electrical energy. Just within the past few years the demand for electric power has at times exceeded the supply in the United States, and many believe nuclear power plants should be built to meet that demand. Nuclear power plants can be the source of "clean" energy, in that they do not pollute the atmosphere with ash, smoke, or oxides of sulfur, nitrogen, or carbon as coal-fired plants do. In addition, nuclear plants help to ensure that our supplies of fossil fuels will not be depleted as quickly in the near future, and they reduce our dependency on buying such fuels from other countries. Currently, 104 nuclear plants operate in the United States. Approximately 440 nuclear power plants worldwide in 31 countries produce about 17% of the world's electricity. The 104 nuclear plants in the United States supply about 19% of our nation's electric energy (Figure 20.9). Around the world, 16 countries use nuclear power to generate at least 25% of their electricity. France, for example, uses 59 nuclear plants to generate more than three out of every four kilowatts of electricity (77%) in that country. In contrast, 64% of the electricity in Lithuania is produced by just one nuclear plant, indicating the differences in demand for electricity between the two countries.

No new nuclear power plants have been built in the United States between 1979—the year of the accident at the Three Mile Island nuclear power plant near Harrisburg, Pennsylvania—and 2009. Currently, applications have been filed with the U.S. Nuclear Regulatory Commission requesting permission to build 34 additional nuclear power plants in this country. One problem associated with nuclear power plants is the highly radioactive fission products in the spent fuel. In the United States, tens of thousands of tons of spent fuel waste are being stored, and the amount is growing steadily. Although some of the products are put to various uses (Section 20.9, p. 985), many are unsuitable as a nuclear fuel or for other purposes. Because these products are often highly radioactive and some have long half-lives (plutonium-239, $t_{1/2} = 24{,}400$ yr), proper disposal of this high-level nuclear waste poses an enormous problem. One approach is to encase high-level radioactive wastes in a glassy material having a volume of about 2 m^3 per nuclear reactor per year. In 1996 the Department of Energy Savannah River site near Augusta, Georgia, began encapsulating radioactive waste in glass, a process called vitrification. A mixture of glass particles and radioactive waste is heated to 1200 °C. The molten mixture is poured into stainless steel canisters, cooled, and stored. Eventually, such high-level nuclear wastes may be stored underground in geological formations, such as salt deposits, that are known to be stable for hundreds of millions of years. A plan to store high-level nuclear waste in a site at Yucca Mountain, Nevada, the designated national repository for such waste, such as that from spent nuclear reactor cores, has now been shelved. No other potential repository has been designated.

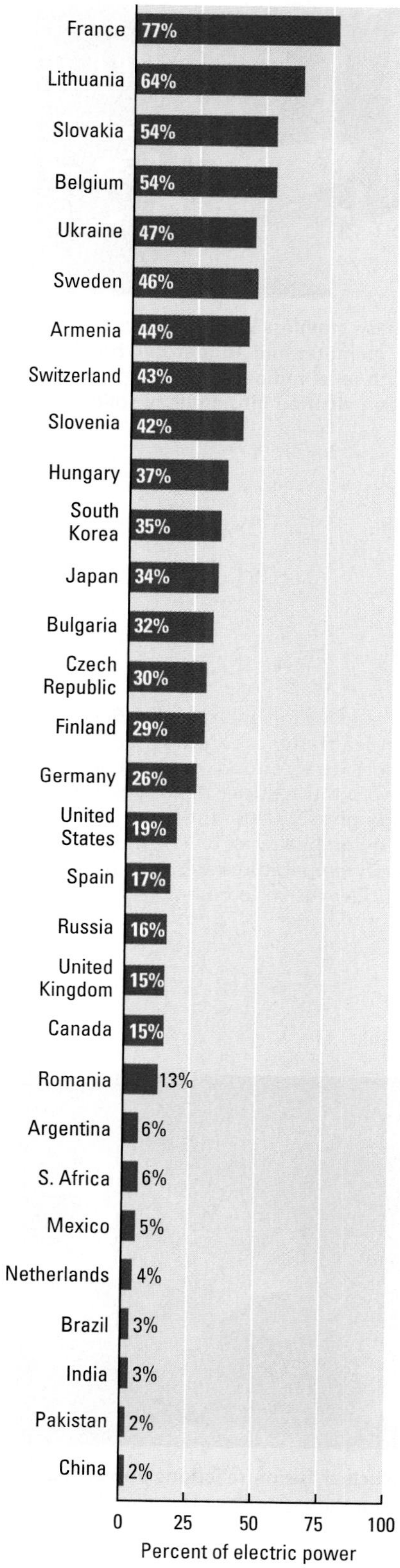

Figure 20.9 The approximate fraction of electricity generated by nuclear power in various countries. About 19% of the electricity in the United States is produced by nuclear power.

EXERCISE 20.11 Radioactive Decay of Fission Products

Unlike the 1979 incident at Three Mile Island, the 1986 accident at the Chernobyl nuclear plant in the former Soviet Union released significant quantities of radioisotopes into the atmosphere. One of those radioisotopes was strontium-90 ($t_{1/2} = 29.1$ yr). What fraction of strontium-90 released at that time remains?

EXERCISE 20.12 Nuclear Waste

Cesium-137 ($t_{1/2} = 30.2$ yr) is produced by ^{235}U fission. If ^{137}Cs is part of nuclear waste stored deep underground, how long will it take for the initial ^{137}Cs activity when it was first buried to drop (a) by 60%? and (b) by 90%?

In recent years, a new type of nuclear reactor design, termed a pebble bed reactor, has emerged. These are high-temperature, gas-cooled reactors. The fuel elements, "pebbles" the size of a tennis ball, contain approximately 4% fissionable uranium encased in silicon carbide and pyrolytic graphite, which act as the neutron moderators. The core of such a reactor contains hundreds of thousands of these

These stainless steel canisters (2 feet in diameter and 10 feet tall) hold high-level radioactive waste that has been vitrified into a glassy solid.

pebbles. They are cooled by flowing inert gas (helium, nitrogen, carbon dioxide), which is heated and then used to drive turbines to generate power. The gas temperature is in the range of 700 °C to 900 °C, which provides up to 50% thermal efficiency for power production. Pebbles are withdrawn from the pebble bed, tested to see whether they are spent, and replaced with new ones as needed. In this design, the fuel, its containment, and the moderator are all together in the pebble. This design has been advanced as being inherently safe. Even if the gas flow stops, the reactor core cools rather than increasing its temperature. One criticism of the design is that it produces more radioactive waste than traditional designs. A current pebble bed reactor research project is China's HTR-10 reactor at the Institute of Nuclear and New Energy Technology (INET) at Tsinghua University near Beijing. This research reactor has been running since 2000. China's energy plans for the next decade include a major move into the use of nuclear power.

20.7 Nuclear Fusion

Tremendous amounts of energy are generated when very light nuclei combine to form heavier nuclei in a reaction called **nuclear fusion.** One of the best examples is the fusion of hydrogen nuclei (protons) to give helium nuclei.

$$4\,{}^{1}_{1}\text{H} \longrightarrow {}^{4}_{2}\text{He} + 2\,{}^{\;\,0}_{+1}\text{e} \qquad \Delta E = -2.5 \times 10^{9}\text{ kJ}$$

The activation energy (← p. 611) for nuclear fusion is very large because there is very strong repulsion between two positive nuclei when they are brought very close together (Coulomb's law ← p. 94). Thus, very high temperatures are needed for nuclear fusion to occur.

The helium nucleus produced by this reaction is more stable (has higher binding energy per nucleon) than the reactant hydrogen nuclei, as shown in Figure 20.3 (p. 967). This fusion reaction is the source of the energy from our sun and other stars, and it is the beginning of the synthesis of the elements in the universe (Section 21.1). Temperatures of 10^6 to 10^7 K, found in the core and radioactive zone of the sun, are required to bring the positively charged ${}^{1}_{1}\text{H}$ nuclei together with enough kinetic energy to overcome nuclear repulsions and to react.

Deuterium can also be fused to give helium-3,

$${}^{2}_{1}\text{H} + {}^{2}_{1}\text{H} \longrightarrow {}^{3}_{2}\text{He} + {}^{1}_{0}\text{n} \qquad \Delta E = -3.2 \times 10^{8}\text{ kJ}$$

which can undergo further fusion with a proton to give helium-4.

$${}^{1}_{1}\text{H} + {}^{3}_{2}\text{He} \longrightarrow {}^{4}_{2}\text{He} + {}^{\;\,0}_{+1}\text{e} \qquad \Delta E = -1.9 \times 10^{9}\text{ kJ}$$

Nuclear fusion reactions power the sun, whose surface is shown here.

Each of these reactions releases an enormous quantity of energy, so it has been the dream of many nuclear physicists to harness them to provide energy for the people of the world.

Hydrogen bombs are based on fusion. At the very high temperatures that allow fusion reactions to occur rapidly, atoms do not exist as such. Instead, there is a **plasma,** which consists of unbound nuclei and electrons. To achieve the high temperatures required for the nuclear fusion reaction of the hydrogen bomb, a nuclear fission bomb (atomic bomb) is first set off. In one type of hydrogen bomb, lithium-6 deuteride (LiD, a solid salt) is placed around a ${}^{235}\text{U}$ or ${}^{239}\text{Pu}$ fission bomb, and the nuclear fission reaction is set off to initiate the process. Lithium-6 nuclei absorb neutrons produced by the fission and split into tritium and helium.

$${}^{1}_{0}\text{n} + {}^{6}_{3}\text{Li} \longrightarrow {}^{3}_{1}\text{H} + {}^{4}_{2}\text{He}$$

The temperature reached by the fission of uranium or plutonium is high enough to bring about the fusion of tritium and deuterium, accompanied by the release of 1.7×10^9 kJ per mole of ${}^{3}\text{H}$. A 20-megaton hydrogen bomb usually contains about 300 lb LiD, as well as a considerable mass of plutonium and uranium.

Development of nuclear fusion as a commercial energy source is appealing because hydrogen isotopes are available (from water), and fusion products are usually nonradioactive or have short half-lives, which eliminates the problems associated with the disposal of high-level radioactive fission reactor waste products. However, controlling a nuclear fusion reaction for peaceful commercial uses has proven to be extraordinarily difficult. Three critical requirements must be met to achieve controlled fusion. First, the temperature must be high enough for fusion to occur sufficiently rapidly. The fusion of deuterium and tritium, for example, requires a temperature of at least 100 million K. Second, the plasma must be confined long enough to generate a net output of energy. Third, the energy must be recovered in some usable form.

Magnetic "bottles" (enclosures bounded by magnetic fields) have confined the plasma so that controlled fusion has been achieved. But the energy generated by the fusion has been less than that required to produce the magnetic bottle and control the fusion reaction. Using more energy to produce less energy is not a commercially appealing investment. Thus, commercial fusion reactors are not likely in the near future without a dramatic breakthrough in fusion technology.

Containment is one of the biggest problems in developing controlled nuclear fusion.

EXERCISE 20.13 Nuclear Fusion

Complete the equations for these nuclear fusion reactions.

(a) ${}^{7}_{3}Li + {}^{1}_{1}H \longrightarrow {}^{1}_{0}n + \_\_\_\_$ (b) ${}^{2}_{1}H + \_\_\_\_ \longrightarrow {}^{4}_{2}He + {}^{1}_{1}H$

20.8 Nuclear Radiation: Effects and Units

The use of nuclear energy and radiation is a double-edged sword that carries both risks and benefits. It can be used to harm (nuclear armaments) or to cure (radioisotopes in medicine).

Alpha, beta, and gamma radiation disrupt normal cell processes in living organisms by interacting with key biomolecules, breaking their covalent bonds, and producing energetic free radicals and ions that can lead to further disruptive reactions. The potential for serious radiation damage to humans is well known. The biological effects of the atomic bombs exploded at Hiroshima and Nagasaki, Japan, at the close of World War II in 1945 have been well documented. However, controlled exposure to nuclear radiation can be beneficial in destroying malignant tissue, as in radiation therapy for treating some cancers.

All technologies carry risks as well as benefits. In the 1800s railroads were new and the poet William Wordsworth wrote of their risks and benefits in terms of "Weighing the mischief with the promised gain. . . ."

Radiation Units

The SI unit of radioactivity is the becquerel; 1 Bq = 1 s^{-1}. Another common unit of radioactivity is the curie; 1 Ci = 3.70×10^{10} s^{-1}. However, to measure the effects of radiation on tissue, units are needed for radiation dose that take into account the energy absorbed by tissue when radiation passes into it.

To quantify radiation and its effects, particularly on humans, several units have been developed. The SI unit of absorbed radiation dose is the **gray, Gy,** which is equal to the absorption of 1 J per kilogram of material; 1 Gy = 1 J/kg. The **rad** (*r*adiation *a*bsorbed *d*ose) also measures the quantity of radiation energy *absorbed;* 1 rad represents a dose of 1.00×10^{-2} J absorbed per kilogram of material. Thus, 1 Gy = 100 rad. Another unit is the **roentgen, R,** which corresponds to the deposition of 93.3×10^{-7} J per gram of tissue.

The roentgen is named to honor Wilhelm Röntgen, the German physicist who discovered X-rays.

The biological effects of radiation per rad or gray differ with the kind of radiation, which can be quantified more generally using the **rem** (*r*oentgen *e*quivalent in *m*an).

$$\text{Effective dose in rems} = \text{quality factor} \times \text{dose in rads}$$

The quality factor depends on the type of radiation and other factors. It is arbitrarily set as 1 for beta and gamma radiation. It is between 10 and 20 for alpha particles, depending on total dose, dose rate, and type of tissue. Since one rem is a large quantity of radiation, the millirem (mrem) is commonly used (1 mrem = 10^{-3} rem). The SI unit of effective dose is the **sievert, Sv,** which is defined similarly to the rem, except that the absorbed dose is in grays, not rads. Consequently, 1 Sv = 100 rem.

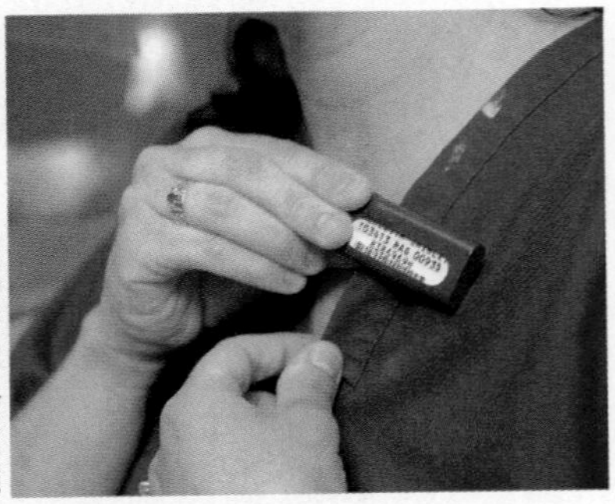

Cliff Moore/Photo Researchers

Individuals who work where there is potential danger from exposure to excessive nuclear radiation wear film badges to monitor their radiation dose.

Background Radiation

Humans are constantly exposed to natural and artificial **background radiation,** estimated to be collectively about 360 mrem per year (Figure 20.10), well below 500 mrem, the U.S. federal government's background radiation standard for the general public. Note that *most* background radiation, about 300 mrem per year (82%), comes from *natural* background radiation sources: cosmic radiation and radioactive elements and minerals found naturally in Earth, air, and materials around and within us. The remaining 18% comes from artificial sources.

Cosmic radiation, emitted by the sun and other stars, continually bombards Earth and accounts for about 8% of natural background radiation. The remainder comes from radioactive isotopes such as ^{40}K. Potassium is present to the extent of about 0.3 g per kilogram of soil and is essential to all living organisms. We all acquire some radioactive potassium from the foods we eat. For example, a hamburger contains 960 mg ^{40}K, giving off 29 disintegrations per second (s^{-1}); a hot dog contains 200 mg ^{40}K and gives off 6 s^{-1}; a serving of french fries has 650 mg ^{40}K and gives off 20 s^{-1}. Other radioactive elements found in some abundance on Earth include thorium-232 and uranium-238. Approximately 8% of the natural background radiation arises from Th-232 and U-238 in rocks and soil. Thorium, for example, is found to the extent of 12 g per 1000 kg soil. Most natural background radiation comes from radon, a by-product of radium decay, as discussed in the next subsection.

Burning fossil fuels (coal and oil) releases into the atmosphere considerable quantities of naturally occurring radioactive isotopes originally in the fossil fuel. Thus, fossil fuel plants add significantly to the background radiation. Far more thorium and uranium are released annually into the atmosphere from fossil fuel-burning plants than from nuclear power plants. The emission from nuclear power plants is essentially zero.

On average, roughly 15% of our annual exposure comes from medical procedures such as diagnostic X-rays and the use of radioactive compounds to trace the body's functions. Consumer products account for 3% of our total annual exposure. Contrary to popular belief, less than 1% comes from sources such as the radioactive fission products from testing nuclear explosives in the atmosphere, nuclear power plants and their wastes, nuclear weapons manufacture, and nuclear fuel processing.

There are no observable physiologic effects from a single dose of radiation less than 25 rem (25×10^3 mrem).

A report published by the U.S. National Research Council concludes that there is no safe level of radiation for humans, although the risks of low-dose radiation are small. The researchers studied doses of radiation of 0.1 Sv (10,000 mrem) or less, which is much greater than the 360 mrem per year natural background level discussed above. Rules for nuclear workers and others who are systematically exposed to elevated radiation levels may be affected by this type of research finding.

Radon

As shown in Figure 20.10, radon accounts for 55% of natural background radiation. Radon is a chemically inert gas in the same periodic table group as helium, neon, argon, and krypton. Radon-222 is produced in the decay series of uranium-238 (p. 961). Other isotopes of radon are products of other decay series. Although

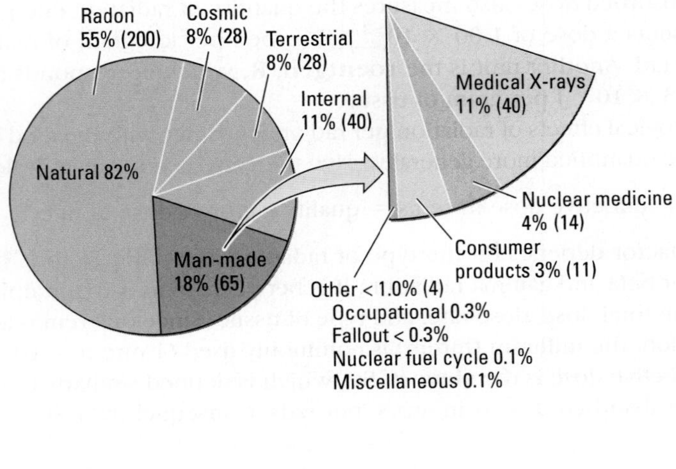
Radon
55% (200)
Cosmic
8% (28)
Terrestrial
8% (28)
Internal
11% (40)
Natural 82%
Man-made
18% (65)
Medical X-rays
11% (40)
Nuclear medicine
4% (14)
Consumer
products 3% (11)
Other < 1.0% (4)
Occupational 0.3%
Fallout < 0.3%
Nuclear fuel cycle 0.1%
Miscellaneous 0.1%

Figure 20.10 Sources of average background radiation exposure in the United States. The sources are expressed as percentages of the total, as well as in millirems per year, the values in parentheses. As seen from the figure, background radiation from natural sources far exceeds that from artificial sources.

ESTIMATION Counting Millirems: Your Radiation Exposure

In 1987 the Committee on Biological Effects of Ionizing Radiation of the U.S. National Academy of Sciences issued a report that contained a survey for an individual to evaluate his or her exposure to ionizing radiation. The table below is adapted from this report and updated. By adding up your exposure, you can compare your annual dose to the U.S. annual average of 360 mrem.

Source: Based on the BEIR Report III. National Academy of Sciences, Committee on Biological Effects of Ionizing Radiation. *The Effects on Populations of Exposure to Low Levels of Ionizing Radiation.* Washington, DC: National Academy of Sciences, 1987.

Visit this book's companion website at **www.cengage.com/chemistry/moore** to work an interactive module based on this material.

Common Sources of Radiation		Your Annual Dose (mrem)
Where you live	**Location:** Cosmic radiation at sea level	27
	For your elevation (in feet), add this number of mrem	______
	Elevation 1000 – *mrem* 2; 2000 – 5; 3000 – 9; 4000 – 15; 5000 – 21; 6000 – 29; 7000 – 40; 8000 – 53; 9000 – 70	
	Ground: U.S. average	26
	Radon: U.S. average	200
	House construction: For stone, concrete, or masonry building, add 7; for wood, add 30	______
What you eat, drink, and breathe	**Radioisotopes** in the body from	
	Food, air, water: U.S. average	40
	Weapons test fallout	4
How you live	**X-ray and radiopharmaceutical diagnosis**	
	Number of chest X-rays ______ × 10	______
	Number of lower gastrointestinal tract X-rays ______ × 500	______
	Number of radiopharmaceutical examinations ______ × 300	______
	(Average dose to total U.S. population = 53 mrem)	______
	Jet plane travel: For each 2500 miles add 1 mrem	______
	TV viewing: Number of hours per day ______ × 0.15	______
How close you live to a nuclear plant	**At site boundary:** Average number of hours per day ______ × 0.2	______
	One mile away: Average number of hours per day ______ × 0.02	______
	Five miles away: Average number of hours per day ______ × 0.002	______
	Over five miles away: None	______
	Note: Maximum allowable dose determined by "as low as reasonably achievable" (ALARA) criteria established by the U.S. Nuclear Regulatory Commission. Experience shows that your actual dose is substantially less than these limits.	
	Your total annual dose in mrem:	______

Elevation table:

Elevation	*mrem*	*Elevation*	*mrem*	*Elevation*	*mrem*
1000	2	4000	15	7000	40
2000	5	5000	21	8000	53
3000	9	6000	29	9000	70

Compare your annual dose with the U.S. annual average of 360 mrem.

chemically inert, radon is problematic because it is a radioactive gas that can be inhaled.

It should be kept in mind that radon occurs naturally in our environment. Because it comes from natural uranium deposits, the quantity of radon depends on the nature of the rocks and soil in a given locality. Furthermore, since the gas is chemically inert, it is not trapped by chemical processes in the soil or water. Thus, radon gas is free to seep up from the ground and into underground mines

An online radiation calculator is available at **http://www.epa.gov/radiation/understand/calculate.html**.

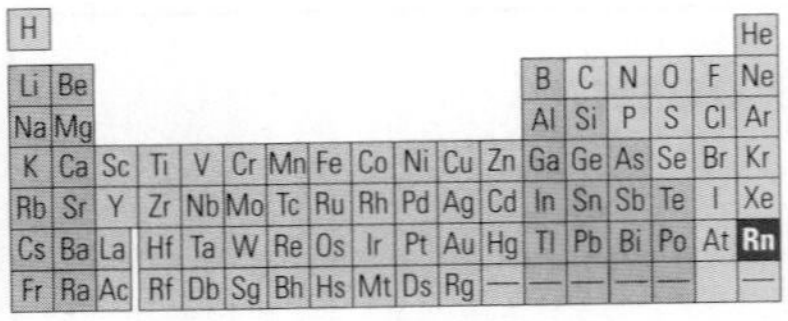

or into homes through pores in concrete block walls, cracks in basement floors or walls, and around pipes. Radon-222 decays to give polonium-218, a radioactive, heavy metal element that is not a gas but, unlike radon-222, is not chemically inert.

If radon is inhaled, decay to polonium will occur deep in the lungs, where ^{218}Po will be generated.

$$^{222}_{86}Rn \longrightarrow ^{4}_{2}He + ^{218}_{84}Po \qquad t_{1/2} = 3.82 \text{ days}$$

$$^{218}_{84}Po \longrightarrow ^{4}_{2}He + ^{214}_{82}Pb \qquad t_{1/2} = 3.10 \text{ minutes}$$

Polonium-218 can lodge in lung tissues, where it undergoes alpha decay to give lead-214, itself a radioactive isotope. The range of an alpha particle is quite small, perhaps 0.7 mm (about the thickness of a sheet of paper). However, this is approximately the thickness of the epithelial cells of the lungs, so the alpha radiation can damage these tissues and induce lung cancer.

This answers the question posed in Chapter 1 (← *p. 3*), "I've heard that most homes in the United States contain small quantities of a radioactive gas. How can I find out whether my home is safe?"

Most homes in the United States are believed to have some level of radon gas. There is currently a great deal of controversy over the level of radon that is considered safe. Estimates indicate that only about 6% of U.S. homes have radon levels above 4 picocuries per liter (pCi/L) of air, the action level standard set by the U.S. Environmental Protection Agency. Some believe that 1.5 pCi/L is more likely the average level and that only about 2% of the homes will contain more than 8 pCi/L. To test for the presence of radon, you can purchase testing kits of various kinds. If your home shows higher levels of radon gas than 4 pCi/L, you should probably have it tested further and perhaps take corrective actions such as sealing cracks around the foundation and in the basement. But, keep in mind the relative risks involved. A 1.5 pCi/L level of radon leads to a lung cancer risk about the same as the risk of your dying in an accident in your home.

1 picocurie, pCi = 10^{-12} Ci.

© Cengage Learning/Charles D. Winters

A commercially available kit to test for radon gas in a residence.

EXERCISE **20.14 Radon Levels**

Calculate how long it will take for the activity of a radon-222 sample ($t_{1/2}$ = 3.82 days) initially at 8 pCi to drop to
(a) 4 pCi, the EPA action level.
(b) 1.5 pCi, approximately the U.S. average.

CHEMISTRY IN THE NEWS Another Reason Not to Smoke

Cigarette smoke is radioactive; smokers' lungs are exposed to alpha radiation because cigarette smoke contains polonium-210, $^{210}_{84}Po$. The radiation exposure from polonium-210 for a pack-a-day smoker is equivalent to at least 300 chest X-rays per year. Some researchers think the number might be as high as 2000 chest X-rays annually.

Why do cigarettes contain polonium? It is part of the tobacco used to make cigarettes. Soils release some radon, and a decay product is ^{210}Po, which can settle as dust on tobacco leaves. Another radioactivity source is the calcium phosphate fertilizer used on tobacco crops. Calcium phosphate ore accumulates uranium, which by radioactive decay releases radon gas. In addition, decay of uranium-238 (Figure 20.1, p. 961) produces ^{210}Po. The amount of ^{210}Po on tobacco leaves depends a great deal on soil conditions and fertilizer use. Whatever its source, ^{210}Po ends up on tobacco leaves, in cigarettes, and in cigarette smoke. When tobacco burns, ^{210}Po is carried by the inhaled smoke deep into the smoker's lungs where it deposits and then decays by alpha emission,

$$^{210}_{84}Po \longrightarrow ^{206}_{82}Pb + ^{4}_{2}He$$

This process releases a million times more energy than that needed to break a chemical bond. If this broken bond is in a DNA molecule (← ***p. 376***), it can cause genetic damage to the smoker's lungs. Furthermore, polonium-210 compounds are soluble and can enter the bloodstream of the smoker from the lungs. The circulating radioactive polonium can then cause genetic damage elsewhere in the body.

Sources: Proctor, R. N. "Puffing on Polonium." *New York Times*, December 1, 2006.

Amato, I. "How Polonium Poisons." *Chemical & Engineering News*, December 4, 2006; p. 15.

20.9 Applications of Radioactivity

Food Irradiation, Cold Pasteurization

In some parts of the world, spoilage of stored food may claim up to 50% of the food crop. In developed nations, this figure is lowered considerably by refrigeration, canning, and preservatives. Still, there are problems with food spoilage, and food preservation costs are a substantial fraction of the final cost of food. Food and grains can be preserved by gamma irradiation, also known as cold pasteurization. Contrary to some popular opinion, such irradiation does *not* make foods radioactive, any more than a dental X-ray makes you radioactive.

In addition to preserving foods, gamma irradiation is used to sterilize bandages, contact lens solutions, and many cosmetics.

Food irradiation with gamma rays from ^{60}Co or ^{137}Cs sources is allowed in 40 countries and is endorsed by the World Health Organization and the American Medical Association. Astronauts' food has been preserved by gamma irradiation. The United States and several other countries require that foods preserved by irradiation be labeled with the international symbol for irradiated food, the radura.

The radura, the international symbol for irradiated food.

Bacteria, molds, and yeasts are killed or their growth is retarded by irradiation with gamma rays. As a result, the shelf life of irradiated foods during refrigeration is prolonged in much the same way that heat pasteurization protects milk. In recent years, outbreaks of foodborne illnesses caused by new types of harmful bacteria or inappropriate food handling have heightened interest in the benefits of irradiation as a safety measure, especially for use with meat.

The U.S. Food and Drug Administration (FDA) has approved irradiation of meat, poultry, and a variety of fresh fruits and vegetables and spices. Except for spices, however, irradiated foods are not yet widely available.

The FDA permits irradiation up to 300 kilorads for the pasteurization of meat and poultry. Radiation levels in the 1- to 5-Mrad range (1 Mrad = 10^6 rads) sterilize, killing every living organism. Foods irradiated at these levels will keep indefinitely when sealed in plastic or aluminum foil packages. However, FDA approval is

ESTIMATION Radioactivity of Common Foods

It may be surprising to you, but common foods, such as baked potatoes and bananas, are radioactive. But how radioactive are they?

Potassium ions are essential as a major mineral ***(Section 3.11, ← p. 107).*** They play an important role in conducting electric impulses that lead to muscle contraction and nerve transmission ***(← p. 930).*** Many fruits and vegetables are good sources of dietary potassium.

Potassium has three naturally occurring isotopes, ^{39}K, ^{40}K, and ^{41}K, of which ^{40}K is radioactive ($t_{1/2} = 1.26 \times 10^9$ yr) emitting beta and gamma radiation. The natural abundance of ^{40}K is 0.0117%. A typical banana or baked potato contains approximately 500 mg K^+, and therefore about (0.5 g)(0.000117) = 6×10^{-5} g $^{40}K^+$.

We calculate the rate constant of ^{40}K from its half-life, $k = \dfrac{0.693}{t_{1/2}} = 5.5 \times 10^{-10}\ \text{yr}^{-1}$. Convert the units from years to seconds:

$$k = 5.5 \times 10^{-10}\ \text{yr}^{-1} \times \frac{1\ \text{yr}}{3.15 \times 10^7\ \text{s}} = 1.7 \times 10^{-17}\text{s}^{-1}.$$

We will use the relationship $A = k\,N$ to calculate the disintegration rate, so we need to convert the mass of $^{40}K^+$ to number of $^{40}K^+$ ions:

$$N = \frac{6 \times 10^{-5}\ \text{g}}{40\ \text{g/mol}} \times 6.022 \times 10^{23}\ \text{ions/mol} = 9 \times 10^{17}\ \ ^{40}K^+\ \text{ions}$$

Now we can calculate the disintegration rate of $^{40}K^+$ in a baked potato or banana:

$$A = kN = -\ (1.7 \times 10^{-17}\text{s}^{-1})(9 \times 10^{17}\ ^{40}K^+\ \text{ions}) = 15\ \text{s}^{-1} = 15\ \text{Bq}$$

The $^{40}K^+$ in a typical baked potato or banana undergoes an average of 15 disintegrations per second. This is a small amount of radioactivity. The ^{243}Am in a typical household smoke detector has a disintegration rate (via alpha particle emission) about 2000 times larger than that of a baked potato or banana.

Sources: Ball, D. W. *Journal of Chemical Education,* Vol. 81, 2004; p. 1440. See also, **http://www.whfoods.com**.

Visit this book's companion website at **www.cengage.com/chemistry/moore** to work an interactive module based on this material.

Courtesy of Food Technology Service, Inc.

Strawberries preserved by gamma irradiation.

unlikely for irradiation sterilization of foods in the near future because of potential problems caused by as-yet-undiscovered, but possible, "unique radiolytic products." For example, irradiation sterilization might produce a substance that is capable of causing genetic damage. To prove or disprove the presence of these substances, animal feeding studies using foods sterilized by irradiation are now being conducted in the United States.

Radioactive Tracers

The chemical behavior of a radioisotope is essentially identical to that of the nonradioactive isotopes of the same element. Compounds containing radioactive atoms are formed and undergo chemical reactions in exactly the same way as compounds containing the same nonradioactive atoms. Therefore, chemists can use radioactive isotopes as **tracers,** radioisotopes used to track the pathway of a radioactive element in a chemical process. This works for both nonbiological and biological chemical reactions. To use a tracer, a chemist prepares a reactant compound in which one of the elements consists of both radioactive and stable (nonradioactive) isotopes, and uses the compound in a reaction (or feeds it to an organism). After the reaction, the chemist measures the radioactivity of the products (or determines which parts of the organism contain the radioisotope) using a Geiger counter or other radiation detector. Several radioisotopes commonly used as tracers are listed in Table 20.3.

For example, plants take up phosphorus-containing compounds from the soil through their roots. The use of the radioactive phosphorus isotope ^{32}P, a beta emitter, provides a way to detect the uptake of phosphorus by a plant, as well as to measure the speed of uptake under various conditions. Plant biologists can grow hybrid strains of plants that quickly absorb phosphorus, an essential nutrient. They can test the new plants by measuring their uptake of the radioactive ^{32}P tracer. This type of research can lead to faster-maturing crops, better yields per acre, and more food or fiber at less expense.

Melvin Calvin used ^{14}C to monitor the uptake and release of $^{14}CO_2$ to determine the basic biochemical pathways of photosynthesis. This groundbreaking work earned him the 1961 Nobel Prize in Chemistry.

Medical Imaging

Radioactive isotopes are used in **nuclear medicine** in two different areas: diagnosis and therapy. In the diagnosis of internal disorders such as tumors, physicians need information on the locations of abnormal tissue. This identification is done by imaging, a technique in which the radioisotope, either alone or combined with some other substance, accumulates at the site of the disorder. There, the radioisotope decays, emitting its characteristic radiation, which is then detected. Modern medical diagnostic instruments not only determine where the radioisotope is located in the patient's body, but also construct an image of where the radioisotope is concentrated in the body.

Four of the most common diagnostic radioisotopes are given in Table 20.4. Most are created in particle accelerators, devices in which heavy, charged nuclear particles bombard other target atoms. Each of these radioisotopes produces gamma

Table 20.3 Radioisotopes Used as Tracers

Isotope	Half-Life	Use
^{14}C	5730 years	CO_2 for photosynthesis research
^{3}H	12.33 years	Tag hydrocarbons
^{35}S	87.2 days	Tag pesticides, measure air flow
^{32}P	14.3 days	Measure phosphorus uptake by plants

Table 20.4 Diagnostic Radioisotopes

Radioisotope	Name	Half-Life (hours)	Site for Diagnosis
^{99m}Tc*	Technetium-99m	6.0	As $^{99m}TcO_4^-$ to the thyroid
^{201}Tl	Thallium-201	72.9	To the heart
^{123}I	Iodine-123	13.2	To the thyroid
^{67}Ga	Gallium-67	78.2	To various tumors and abscesses

*The technetium-99m isotope is the radioisotope most commonly used for diagnostic purposes. The *m* stands for "metastable."

radiation, which, in low doses, is less harmful to tissue than internal ionizing radiations such as beta or alpha particles because gamma rays pass through the tissue without being absorbed. When combined with special carrier compounds, these radioisotopes can be made to accumulate in specific areas of the body. For example, the pyrophosphate ion, $P_4O_7^{4-}$, can bond to the technetium-99m radioisotope. Together they accumulate in the skeletal structure where abnormal bone metabolism is occurring (Figure 20.11). The technetium-99m radioisotope is metastable, as denoted by the letter *m*; this term means that the nucleus loses energy by disintegrating to a more stable version of the same isotope,

$$^{99m}Tc \longrightarrow {}^{99}Tc + \gamma$$

and emitting a gamma ray. Images produced by detecting the gamma rays often pinpoint bone tumors.

The metastable ^{99m}Tc is in an excited nuclear energy state. This is analogous to a H atom in which an electron is in an excited atomic energy state (← p. 282).

EXERCISE 20.15 Rate of Radioactive Decay

Gallium citrate containing radioactive gallium-67 is used medically as a tumor-seeking agent. It has a half-life of 78.2 hours. How much time is needed for a gallium citrate sample to reach 10% of its original activity?

EXERCISE 20.16 Half-Life

Chromium-51 is a radioisotope ($t_{1/2}$ = 27.7 days) used to evaluate the lifetime of red blood cells; the radioisotope iron-59 ($t_{1/2}$ = 44.5 days) is used to assess bone marrow function. A hospital laboratory has 80 mg iron-59 and 100 mg chromium-51. After 90 days, which radioisotope is present in greater mass?

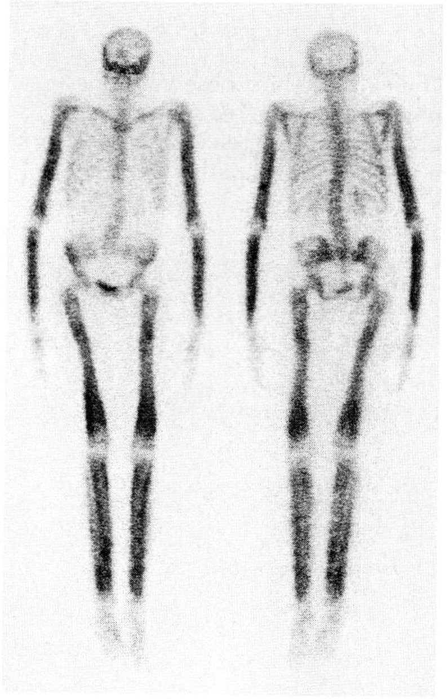

Susan Law Cain/iStockphoto

Figure 20.11 A whole-body scan. Phosphate with technetium-99m was injected into the blood and then absorbed by the bones and kidneys. This picture was taken three hours after injection.

Paradoxically, high-energy radiation, which can kill healthy cells, is used therapeutically to kill malignant, cancerous cells—those exhibiting rapid, uncontrolled growth. Because they divide more rapidly than normal cells, malignant cells are more susceptible to radiation damage. Thus, malignant cells are more likely to be killed than normal cells. For external radiation therapy, a narrow beam of high-energy gamma radiation from a cobalt-60 or cesium-137 source is focused on the cancerous cells. Internal radiation therapy uses gamma-emitting salts of radioisotopes such as ^{192}Ir ($t_{1/2}$ = 73.8 days). The radioactive salts are encapsulated in platinum or gold "seeds" or needles and surgically implanted into the body. Because the thyroid gland uses iodine, thyroid cancer can be treated internally by oral administration of a sodium iodide solution containing a relatively high concentration of radioactive iodine-131.

Positron emission tomography (PET) is a form of nuclear imaging that uses positron emitters, such as carbon-11, fluorine-18, nitrogen-13, or oxygen-15. These

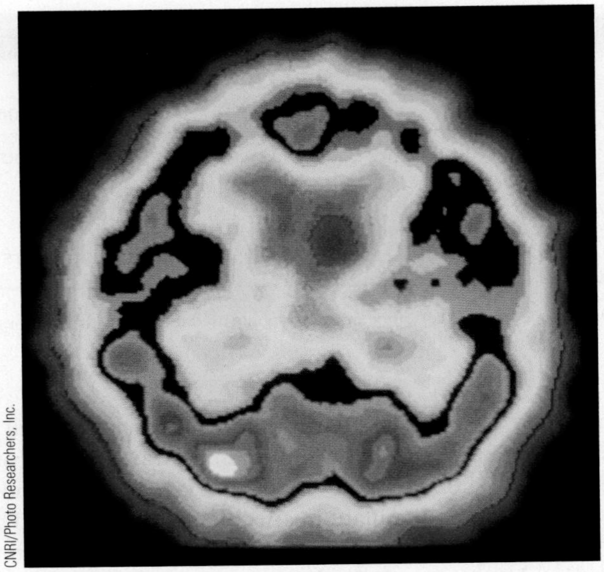

Figure 20.12 PET (positron emission tomography) scan of an axial section through a normal human brain. PET scans are obtained by injecting a tracer labeled with a short-lived radioisotope such as nitrogen-13 into the bloodstream. The isotope concentrates in brain tissue and emits positrons. The positrons react with electrons to create gamma rays that are recorded by a circular detector when the scan is performed. Here, radioactive methionine (an amino acid) has been used to show the level of activity of protein synthesis in the brain.

The neutrino, first observed experimentally in 1956, is a subatomic particle with zero electrical charge and a mass smaller than that of an electron.

radioisotopes are neutron-deficient, have short half-lives, and therefore must be prepared in a cyclotron immediately before use. When they decay, a proton is converted into a neutron, a positron, and a neutrino; the neutrino is generally not shown in the equation.

$$^{1}_{1}\text{p} \longrightarrow {}^{1}_{0}\text{n} + {}^{0}_{+1}\text{e}$$

Since matter is essentially transparent to neutrinos, they escape undetected. However, the positron, ${}^{0}_{+1}\text{e}$, travels on average less than a few millimeters before it encounters an electron, ${}^{0}_{-1}\text{e}$, and undergoes antimatter-matter annihilation.

$${}^{0}_{+1}\text{e} + {}^{0}_{-1}\text{e} \longrightarrow 2\,\gamma$$

The annihilation event produces two gamma rays that move in opposite directions and are detected by detectors located 180° apart in the PET scanner. Several million annihilation gamma rays can be detected within a circular field around the subject over approximately 10 min. Computerized signal-averaging techniques applied to these data generate an image of the region of tissue containing the radioisotope (Figure 20.12).

SUMMARY PROBLEM

(a) One of the species in the uranium-238 decay series (Figure 20.1) is radon-222, an alpha emitter. Write the nuclear equation for alpha emission by ^{222}Rn.

(b) Uranium-238 can be converted to plutonium-239 through a series of nuclear reactions involving absorption of a neutron followed by two beta emissions. Write these three nuclear reactions connecting ^{238}U and ^{239}Pu.

(c) Uranium-235 is the main fissionable nucleus used in nuclear reactors. When it fissions, it can produce ^{132}Sb and ^{101}Nb as products. Write the nuclear reaction for this fission reaction.

(d) Hydrogen bombs that use nuclear fusion reactions were developed following World War II. One reaction used in a hydrogen bomb was

$$^{2}_{1}H + ^{3}_{1}H \longrightarrow ^{4}_{2}He + ^{1}_{0}n$$

Calculate the energy released, in kilojoules per gram of reactants, for this fusion reaction. The necessary nuclear masses are $^{2}_{1}H$ = 2.01355 g/mol; $^{3}_{1}H$ = 3.01550 g/mol; $^{4}_{2}He$ = 4.00150 g/mol; $^{1}_{0}n$ = 1.00867 g/mol.

(e) The goal of recent nuclear arms treaties has been to dismantle the stockpiles of nuclear weapons built up by the United States and the former Soviet Union since World War II, including those containing plutonium-239 ($t_{1/2} = 2.44 \times 10^4$ years). How long will it take for the activity of plutonium-239 in a nuclear warhead to decrease (i) to 75% of its initial activity? (ii) To 10% of its initial activity?

(f) Deep underground burial has been proposed for long-term storage of the ^{239}Pu waste removed from nuclear weapons. Based on the answers to Summary Problem (e), comment on factors that need to be considered for the storage and burial of such nuclear waste.

(g) Iodine-131 emits both beta and gamma rays and is used in treating thyroid cancer. Its half-life is 8.04 d. (i) Write the nuclear reaction for beta decay of ^{131}I. (ii) If a ^{131}I sample had an activity of 5.0×10^{11} Bq, what would its activity be after 48 h?

IN CLOSING

Having studied this chapter, you should be able to . . .

- Characterize the three major types of radiation observed in radioactive decay: alpha (α), beta (β), and gamma (γ) (Section 20.1). End-of-chapter question: 11
- Write a balanced equation for a nuclear reaction or transmutation (Section 20.2). Questions 13, 17
- Decide whether a particular radioactive isotope will decay by α, β, positron emission, or by electron capture (Sections 20.2, 20.3). Question 20
- Calculate the binding energy for a particular isotope and understand what this energy means in terms of nuclear stability (Section 20.3). Question 22
- Use Equation 20.3, $\ln\frac{N}{N_0} = -kt$, which relates (through the decay constant k) the time period over which a sample is observed (t) to the number of radioactive atoms present at the beginning (N_0) and end (N) of the time period (Section 20.4). Questions 27, 35
- Calculate the half-life of a radioactive isotope ($t_{1/2}$) from the activity of a sample at two times, or use the half-life to find the time required for an isotope to decay to a particular activity (Section 20.4). Questions 31, 77
- Describe nuclear chain reactions, nuclear fission, and nuclear fusion (Sections 20.6, 20.7).
- Describe the basic functioning of a nuclear power reactor (Section 20.6).
- Describe some sources of background radiation and the units used to measure radiation (Section 20.8).
- Give examples of some uses of radioisotopes (Section 20.9).

OWL and go Chemistry

Sign in at **www.cengage.com/owl** to:

- View tutorials and simulations, develop problem-solving skills, and complete online homework assigned by your professor.
- For quick review and exam prep, download Go Chemistry mini lecture modules from OWL (or purchase them at **www.CengageBrain.com**).

KEY TERMS

activity *(Section 20.4)*
alpha (α) particles *(20.1)*
alpha radiation *(20.1)*
background radiation *(20.8)*
becquerel, Bq *(20.4)*
beta (β) particles *(20.1)*
beta radiation *(20.1)*
binding energy *(20.3)*
binding energy per nucleon *(20.3)*
critical mass *(20.6)*
curie, Ci *(20.4)*
electron capture *(20.2)*
gamma (γ) radiation *(20.1)*
gray, Gy *(20.8)*
half-life, $t_{1/2}$ *(20.4)*
nuclear fission *(20.6)*
nuclear fusion *(20.7)*
nuclear medicine *(20.9)*
nuclear reactions *(20.2)*
nuclear reactor *(20.6)*
nucleons *(20.2)*
plasma *(20.7)*
positron *(20.2)*
rad *(20.8)*
radioactive series *(20.2)*
rem *(20.8)*
roentgen, R *(20.8)*
sievert, Sv *(20.8)*
tracers *(20.9)*

QUESTIONS FOR REVIEW AND THOUGHT

OWL Interactive versions of these problems are assignable in OWL.

Blue-numbered questions have short answers at the back of this book in Appendix M and fully worked solutions in the *Student Solutions Manual.*

Review Questions

These questions test vocabulary and simple concepts.

1. Complete the tables.

	Symbol	Mass	Charge
α particle	______	______	______
β particle	______	______	______
γ radiation	______	______	______

	Ionizing Power	Penetrating Power
α particle	______	______
β particle	______	______
γ radiation	______	______

2. Compare nuclear and chemical reactions in terms of changes in reactants, type of products formed, and conservation of matter and energy.
3. What is meant by the "band of stability"?
4. What is the binding energy of a nucleus?
5. If the mass number of an isotope is much greater than twice the atomic number, what type of radioactive decay might you expect?
6. If the number of neutrons in an isotope is much less than the number of protons, what type of radioactive decay might you expect?
7. Define critical mass and chain reaction.
8. What is the difference between nuclear fission and nuclear fusion? Illustrate your answer with an example of each.
9. Use the World Wide Web to locate the nuclear reactor power plant nearest to your campus residence. Do you consider it to pose a threat to your health and safety? If so, why? If not, why not?
10. Name at least two uses of radioactive isotopes (outside of their use in power reactors and weapons).

Topical Questions

These questions are keyed to the major topics in the chapter. Usually a question that is answered at the back of this book is paired with a similar one that is not.

Nuclear Reactions (Section 20.2)

11. By what processes do these transformations occur?
(a) Thorium-230 to radium-226
(b) Cesium-137 to barium-137
(c) Potassium-38 to argon-38
(d) Zirconium-97 to niobium-97
12. By what processes do these transformations occur?
(a) Uranium-238 to thorium-234
(b) Iodine-131 to xenon-131
(c) Nitrogen-13 to carbon-13
(d) Bismuth-214 to polonium-214
13. Fill in the mass number, atomic number, and symbol for the missing particle in each nuclear equation.
(a) ${}^{242}_{94}\text{Pu} \longrightarrow {}^{4}_{2}\text{He} + \_\_\_\_$
(b) $\_\_\_\_ \longrightarrow {}^{32}_{16}\text{S} + {}^{0}_{-1}\text{e}$
(c) ${}^{252}_{98}\text{Cf} + \_\_\_\_ \longrightarrow 3\,{}^{1}_{0}\text{n} + {}^{259}_{103}\text{Lr}$
(d) ${}^{55}_{26}\text{Fe} + \_\_\_\_ \longrightarrow {}^{55}_{25}\text{Mn}$
(e) ${}^{15}_{8}\text{O} \longrightarrow \_\_\_\_ + {}^{0}_{+1}\text{e}$
14. Fill in the mass number, atomic number, and symbol for the missing particle in each nuclear equation.
(a) $\_\_\_\_ \longrightarrow {}^{22}_{10}\text{Ne} + {}^{0}_{+1}\text{e}$
(b) ${}^{122}_{53}\text{I} \longrightarrow {}^{122}_{54}\text{Xe} + \_\_\_\_$
(c) ${}^{210}_{84}\text{Po} \longrightarrow \_\_\_\_ + {}^{4}_{2}\text{He}$

(d) $^{195}_{79}Au$ + ____ $\longrightarrow$ $^{195}_{78}Pt$

(e) $^{241}_{94}Pu + ^{16}_{8}O \longrightarrow 5\ ^{1}_{0}n$ + ____

15. Write a balanced nuclear equation for each word statement.
    (a) Magnesium-28 undergoes β emission.
    (b) When uranium-238 is bombarded with carbon-12, four neutrons are emitted and a new element forms.
    (c) Hydrogen-2 and helium-3 react to form helium-4 and another particle.
    (d) Argon-38 forms by positron emission.
    (e) Platinum-175 forms osmium-171 by spontaneous radioactive decay.
16. Write a balanced nuclear equation for each word statement.
    (a) Einsteinium-253 combines with an alpha particle to form a neutron and a new element.
    (b) Nitrogen-13 undergoes positron emission.
    (c) Iridium-178 captures an electron to form a stable nucleus.
    (d) A proton and boron-11 fuse, forming three identical particles.
    (e) Nobelium-252 and six neutrons form when carbon-12 collides with a transuranium isotope.
17. One radioactive series that begins with uranium-235 and ends with lead-207 undergoes this sequence of emission reactions: α, β, α, β, α, α, α, α, β, β, α. Identify the radioisotope produced in each of the *first five steps.*
18. One radioactive series that begins with uranium-235 and ends with lead-207 undergoes this sequence of emission reactions: α, β, α, β, α, α, α, α, β, β, α. Identify the radioisotope produced in each of the *last six steps.*
19. Radon-222 is unstable, and its presence in homes may constitute a health hazard. It decays by this sequence of emissions: α, α, β, β, α, β, β, α. Write the sequence of nuclear reactions leading to the final product nucleus, which is stable.

Nuclear Stability (Section 20.3)

20. Write a nuclear equation for the type of decay each of these unstable isotopes is most likely to undergo.
    (a) Neon-19 (b) Thorium-230
    (c) Bromine-82 (d) Polonium-212
21. Write a nuclear equation for the type of decay each of these unstable isotopes is most likely to undergo.
    (a) Silver-114 (b) Sodium-21
    (c) Radium-226 (d) Iron-59
22. Boron has two stable isotopes, ^{10}B (abundance = 19.78%) and ^{11}B (abundance = 80.22%). Calculate the binding energies per nucleon of these two nuclei and compare their stabilities.

$$5\ ^{1}_{1}H + 5\ ^{1}_{0}n \longrightarrow ^{10}_{5}B$$
$$5\ ^{1}_{1}H + 6\ ^{1}_{0}n \longrightarrow ^{11}_{5}B$$

The required masses (in g/mol) are $^{1}_{1}H$ = 1.00783; $^{1}_{0}n$ = 1.00867; $^{10}_{5}B$ = 10.01294; and $^{11}_{5}B$ = 11.00931.

23. Calculate the binding energy in kJ per mole of P for the formation of $^{30}_{15}P$ and $^{31}_{15}P$.

$$15\ ^{1}_{1}H + 15\ ^{1}_{0}n \longrightarrow ^{30}_{15}P$$
$$15\ ^{1}_{1}H + 16\ ^{1}_{0}n \longrightarrow ^{31}_{15}P$$

Which is the more stable isotope? The required masses (in g/mol) are $^{1}_{1}H$ = 1.00783; $^{1}_{0}n$ = 1.00867; $^{30}_{15}P$ = 29.97832; and $^{31}_{15}P$ = 30.97376.

24. The most abundant isotope of uranium is U-238, which has an isotopic mass of 238.0508 g/mol. Calculate its nuclear binding energy in kJ/mol and binding energy per nucleon.
25. Calculate the nuclear binding energy in kJ/mol and binding energy per nucleon of chlorine-35, which has an isotopic mass of 34.9689 g/mol.
26. Calculate the nuclear binding energy and binding energy per nucleon in kJ/mol of iodine-127, which has an isotopic mass of 126.9004 g/mol.

Rates of Disintegration Reactions (Section 20.4)

27. Sodium-24 is a diagnostic radioisotope used to measure blood circulation time. How much of a 20-mg sample of sodium-24 remains after 1 day and 6 hours if sodium-24 has $t_{1/2}$ = 15 hours?
28. Iron-59 in the form of iron(II) citrate is used in iron metabolism studies. Its half-life is 45.6 days. If you start with 0.56 mg iron-59, how much would remain after 1 year?
29. Iodine-131 is used in the form of sodium iodide to treat cancer of the thyroid.
    (a) The isotope decays by ejecting a β particle. Write a balanced equation to show this process.
    (b) The isotope has a half-life of 8.04 days. If you begin with 25.0 mg of radioactive $Na^{131}I$, what mass remains after 32.2 days?
30. Phosphorus-32 is used in the form of $Na_2H^{32}PO_4$ in the treatment of chronic myeloid leukemia, among other things.
    (a) The isotope decays by emitting a β particle. Write a balanced equation to show this process.
    (b) The half-life of ^{32}P is 14.3 days. If you begin with 9.6 mg radioactive $Na_2H^{32}PO_4$, what mass remains after 28.6 days?
31. Calculate the half-life of a radioisotope if it decays to 12.5% of its radioactivity in 12 years.
32. After 2 hours, tantalum-172 has $\frac{1}{16}$ of its initial radioactivity. How long is its half-life?
33. Radioisotopes of iodine are widely used in medicine. For example, iodine-131 ($t_{1/2}$ = 8.04 days) is used to treat thyroid cancer. If you ingest a sample of $Na^{131}I$, how much time is required for the isotope to decrease to 5.0% of its original activity?
34. The noble gas radon has been the focus of much attention because it may be found in homes. Radon-222 emits α particles and has a half-life of 3.82 days.
    (a) Write a balanced equation to show this process.
    (b) How long does it take for a sample of radon to decrease to 10.0% of its original activity?
35. A sample of wood from a Thracian chariot found in an excavation in Bulgaria has a ^{14}C activity of 11.2 disintegrations per minute per gram. Estimate the age of the chariot and the year it was made. ($t_{1/2}$ for ^{14}C is 5.73×10^3 years, and the activity of ^{14}C in living material is 15.3 disintegrations per minute per gram.)
36. A piece of charred bone found in the ruins of a Native American village has a $^{14}C/^{12}C$ ratio of 0.72 times that found in living organisms. Calculate the age of the bone fragment. (See Question 35 for required data on carbon-14.)

37. How long will it take for a sample of plutonium-239 with a half-life of 2.4×10^4 years to decay to 1% of its original activity?
38. A 1.00-g sample of wood from an archaeological site gave 4100 disintegrations of ^{14}C in a 10-hour measurement. In the same time, a 1.00-g modern sample gave 9200 disintegrations. Calculate the age of the wood.

Artificial Transmutations (Section 20.5)

39. There are two isotopes of americium, both with half-lives sufficiently long to allow the handling of large quantities. Americium-241, an alpha emitter, has a half-life of 248 years; it is used in gauging the thickness of materials and in smoke detectors. This isotope is formed from ^{239}Pu by absorption of two neutrons followed by emission of a β particle. Write a balanced equation for this process.
40. Americium-240 is made by bombarding plutonium-239 atoms with α particles. In addition to ^{240}Am, the products are a proton and two neutrons. Write a balanced equation for this process.
41. To synthesize the heavier transuranium elements, one must bombard a lighter nucleus with a relatively large particle. If you know that the products of a bombardment reaction are californium-246 and four neutrons, with what particle would you bombard uranium-238 atoms?
42. The officially named element with the highest atomic number is $^{272}_{111}Rg$. To try to make heavier elements, attempts have been made to force calcium-40 and curium-248 to merge. What would be the atomic number of the element formed?

Nuclear Fission and Fusion (Sections 20.6 & 20.7)

43. Name the fundamental parts of a nuclear fission reactor and describe their functions.
44. Explain why it is easier for a nucleus to capture a neutron than to force a nucleus to capture a proton.
45. What is the missing product in each of these fission equations?
(a) $^{235}_{92}U + ^{1}_{0}n \longrightarrow$ ____ $+ ^{93}_{38}Sr + 3\,^{1}_{0}n$
(b) $^{235}_{92}U + ^{1}_{0}n \longrightarrow$ ____ $+ ^{132}_{51}Sb + 3\,^{1}_{0}n$
(c) $^{235}_{92}U + ^{1}_{0}n \longrightarrow$ ____ $+ ^{141}_{56}Ba + 3\,^{1}_{0}n$
46. Explain why no commercial fusion reactors are in operation today.
47. The average energy output of a barrel of oil is 5.9×10^6 kJ/barrel. Fission of 1 mol ^{235}U releases 2.1×10^{10} kJ of energy. Calculate the number of barrels of oil needed to produce the same energy as 1.0 lb ^{235}U.
48. A concern in the nuclear power industry is that, if nuclear power becomes more widely used, there may be serious shortages in worldwide supplies of fissionable uranium. One solution is to build breeder reactors that manufacture more fuel than they consume. One such cycle works as follows:
(i) A ^{238}U nucleus collides with a neutron to produce ^{239}U.
(ii) ^{239}U decays by β emission ($t_{1/2} = 24$ minutes) to give an isotope of neptunium.
(iii) The neptunium isotope decays by β emission to give a plutonium isotope.
(iv) The plutonium isotope is fissionable. On its collision with a neutron, fission occurs and gives energy, at least two neutrons, and other nuclei as products.

Write an equation for each of these steps, and explain how this process can be used to breed more fuel than the reactor originally contained and still produce energy.

Effects of Nuclear Radiation (Section 20.8)

49. Two common units of radiation used in newspaper and news magazine articles are the rad and rem. What does each measure? Which would you use in an article describing the damage an atomic bomb would inflict on a human population? What relationship does the gray have with these units?
50. Which electrical power plant—fossil fuel or nuclear—exposes a community to more nuclear radiation? Explain why.
51. Explain how our own bodies are sources of nuclear radiation.
52. What is the source of radiation exposure during jet plane travel?

Uses of Radioisotopes (Section 20.9)

53. Why are foods irradiated with gamma rays instead of alpha or beta particles?
54. X-rays and PET scans are two medical imaging techniques. How are they similar and how are they different?
55. To measure the volume of the blood system of an animal, this experiment was done. A 1.0-mL sample of an aqueous solution containing tritium with an activity of $2.0 \times 10^6\ s^{-1}$ was injected into the bloodstream. After time was allowed for complete circulatory mixing, a 1.0-mL blood sample was withdrawn and found to have an activity of $1.5 \times 10^4\ s^{-1}$. Calculate the volume of the circulatory system. (The half-life of tritium is 12.3 years, so this experiment assumes that only a negligible quantity of tritium has decayed during the experiment.)
56. Radioactive isotopes are often used as "tracers" to follow an atom through a chemical reaction. Acetic acid reacts with methanol, CH_3OH, by eliminating a molecule of H_2O to form methyl acetate, CH_3COOCH_3. Explain how you would use the radioactive isotope ^{18}O to show whether the oxygen atom in the water product comes from the —OH of the acid or the —OH of the alcohol.

$$\underset{\text{acetic acid}}{CH_3COOH} + \underset{\text{methanol}}{CH_3OH} \longrightarrow \underset{\text{methyl acetate}}{CH_3COOCH_3} + H_2O$$

General Questions

These questions are not explicitly keyed to chapter topics; many require integration of several concepts.

57. Complete these nuclear equations.
(a) $^{214}Bi \longrightarrow$ ____ $+ ^{214}Po$
(b) $4\,^{1}_{1}H \longrightarrow$ ____ + 2 positrons
(c) ^{249}Es + neutron $\longrightarrow$ 2 neutrons + ____ $+ ^{161}Gd$
(d) $^{220}Rn \longrightarrow$ ____ + alpha particle
(e) ^{68}Ge + electron $\longrightarrow$ ____
58. Complete these nuclear equations.
(a) ____ + neutron $\longrightarrow$ 2 neutrons $+ ^{137}Tc + ^{97}Zr$
(b) $^{45}Ti \longrightarrow$ ____ + positron
(c) ____ $\longrightarrow$ beta particle $+ ^{59}Co$

(d) ^{24}Mg + neutron $\longrightarrow$ _____ + proton

(e) ^{131}Cs + _____ $\longrightarrow$ ^{131}Xe

59. Radioactive nitrogen-13 has a half-life of 10. minutes. After an hour, how much of this isotope remains in a sample that originally contained 96 mg?

60. The half-life of molybdenum-99 is 67.0 hours. How much of a 1.000-mg sample of ^{99}Mo is left after 335 hours? How many half-lives did it undergo?

61. The oldest known fossil cells were found in South Africa. The fossil has been dated by the reaction

$$^{87}Rb \longrightarrow {}^{87}Sr + {}^{0}_{-1}e \qquad t_{1/2} = 4.9 \times 10^{10} \text{ years}$$

If the ratio of the present quantity of ^{87}Rb to the original quantity is 0.951, calculate the age of the fossil cells.

62. Cobalt-60 is a therapeutic radioisotope used in treating certain cancers. A sample of cobalt-60 initially disintegrates at a rate of $4.3 \times 10^6 \text{ s}^{-1}$ and after 21.2 years the rate has dropped to $2.6 \times 10^5 \text{ s}^{-1}$. Calculate the half-life of cobalt-60.

63. Balance these equations used for the synthesis of transuranium elements.

(a) $^{238}_{92}U + {}^{14}_{7}N \longrightarrow$ _____ $+ 5\,{}^{1}_{0}n$

(b) $^{238}_{92}U$ + _____ $\longrightarrow {}^{249}_{100}Fm + 5\,{}^{1}_{0}n$

(c) $^{253}_{99}Es$ + _____ $\longrightarrow {}^{256}_{101}Md + {}^{1}_{0}n$

(d) $^{246}_{96}Cm$ + _____ $\longrightarrow {}^{254}_{102}No + 4\,{}^{1}_{0}n$

(e) $^{252}_{98}Cf$ + _____ $\longrightarrow {}^{257}_{103}Lr + 5\,{}^{1}_{0}n$

64. On December 2, 1942, the first man-made self-sustaining nuclear fission chain reactor was operated by Enrico Fermi and others under the University of Chicago stadium. In June 1972, natural fission reactors, which operated billions of years ago, were discovered in Oklo, Gabon. At present, natural uranium contains 0.72% ^{235}U. How many years ago did natural uranium contain 3.0% ^{235}U, sufficient to sustain a natural reactor? ($t_{1/2}$ for $^{235}U = 7.04 \times 10^8$ years.)

Applying Concepts

These questions test conceptual learning.

65. If a radioisotope is used for diagnosis (e.g., detecting cancer), it should decay by gamma radiation. However, if its use is therapeutic (e.g., treating cancer), it should decay by alpha or beta radiation. Explain why in terms of ionizing and penetrating power.

66. During the Three Mile Island incident, people in central Pennsylvania were concerned that strontium-90 (a beta emitter) released from the reactor could become a health threat (it did not). Where would this isotope collect in the body? If so, what types of problems could it cause?

67. Classify the isotopes ^{17}Ne, ^{20}Ne, and ^{23}Ne as stable or unstable. What type of decay would you expect the unstable isotope(s) to have?

68. This demonstration was carried out to illustrate the concept of a nuclear chain reaction. Explain the connections between the demonstration and the reaction.

Eighty mousetraps are arranged side by side in eight rows of ten traps each. Each trap is set with two rubber stoppers for bait. A small plastic mouse is tossed into the middle of the traps, setting off one trap, which in turn sets off two traps and so on until all the traps are sprung.

69. Most students have no trouble understanding that 1.5 g of a 24-g sample of a radioisotope would remain after 8 h if it had $t_{1/2}$ = 2 h. What they don't always understand is where the other 22.5 g went. How would you explain this disappearance to another student?

70. Nuclear chemistry is a topic that raises many debatable issues. Briefly discuss your views on the following.

(a) Twice a year the general public is allowed to visit the Trinity Site in Alamogordo, New Mexico, where the first atomic bomb was tested. If you had the opportunity to do so, would you visit the site? Explain your answer.

(b) Now that the Cold War has ended, should the United States stockpile nuclear weapons? Explain your answer.

(c) The practice of sterilizing food by irradiation is controversial in the United States. What are some of the possible advantages and disadvantages of this practice? Explain your answer.

More Challenging Questions

These questions require more thought and integrate several concepts.

71. All radioactive decays are first-order. Why is this so?

72. An average 70.0-kg adult contains about 170 g potassium. Potassium-40, with a relative abundance of 0.0118%, undergoes beta decay with a half-life of 1.28×10^9 years. Calculate the total activity due to beta decay of ^{40}K for this average person.

73. Earth receives 7.0×10^{14} kJ/s of energy from the sun. Calculate the mass of solar material lost per hour to supply this amount of energy.

74. A sample of the alpha emitter ^{222}Ra had an initial activity, A_0, of 7.00×10^4 Bq. After 10.0 days its activity, A, had fallen to 1.15×10^4 Bq. Calculate the decay constant and half-life of radon-222.

75. When a wine was analyzed for its tritium, ^3H, content, it was found to contain 1.45% of the tritium originally present when the wine was produced. How old is the wine? ($t_{1/2}$ of ^3H = 12.3 years.)

76. A chemist is setting up an experiment using ^{47}Ca, which has a half-life of 4.5 days. He needs 10.0 μg of the calcium. How many μg of $^{47}CaCO_3$ must he order if the delivery time is 50 hours?

77. To determine the age of the charcoal found at an archaeological site, this sequence of experiments was done: The charcoal sample was burned in excess oxygen and the CO_2 obtained was captured by bubbling it through lime water, $Ca(OH)_2$, to form a precipitate of $CaCO_3$. This precipitate was filtered, dried, and weighed. A sample of 1.14 g $CaCO_3$ produced 2.17×10^{-2} Bq from carbon-14. Modern carbon produces 15.3 disintegrations per minute per gram. Calculate the age of the charcoal. (The half-life of carbon-14 is 5730 years.)

Conceptual Challenge Problems

These rigorous, thought-provoking problems integrate conceptual learning with problem solving and are suitable for group work.

CP20.A (Section 20.4) The half-life for the alpha decay of uranium-238 to thorium-234 is 4.5×10^9 years, which happens to be the estimated age of Earth.

(a) How many atoms were decaying per second in a 1.0-g sample of uranium-238 that existed 1.0×10^6 years ago?

(b) Explain how you would find the number of atoms now decaying per second in this sample.

CP20.B (Section 20.4) If Earth is 4.5×10^9 years old and the quantity of radioactivity in a sample becomes smaller with time, how is it possible for there to be any radioactive elements on Earth that have half-lives less than a few million years?

CP20.C (Section 20.4) Using experiments based on a sample of living wood, a nuclear chemist estimates that the uncertainty of her measurements of the carbon-14 radioactivity in the sample is 1.0%. The half-life of carbon-14 is 5730 years.

(a) How long must a sample of wood be separated from a living tree before the chemist's radioactivity measurements on the sample provide credible evidence for the time when the tree died?

(b) Suppose that the chemist's uncertainty in the radioactivity of carbon-14 continues to be 1.0% of the radioactivity of living wood. How long must a sample of wood be dead before the chemist's measurements support the claim that the time since the wood was separated from the tree is not changing?

CP20.D (Section 20.8) You have read that alpha radiation is the least penetrating type of radiation, followed by beta radiation. Gamma radiation penetrates matter well, and thick samples of matter are required to contain gamma radiation. Knowing these facts, what can you correctly deduce about the harmful effects of these three types of radiation on living tissue?

CP-20.E (Section 20.8) Death will probably occur within weeks for a 150-lb person who receives 500,000 mrem of radiation over a short time, an exposure that is 1000 times the U.S. government's standard for 1 year (500 mrem/yr). A student realizes that 500,000 mrem is 500 rem, and that 500 rem has the effect of depositing 317 J of energy on the body of the 150-lb person. The student is puzzled. How can the deposition of only 317 J of energy from nuclear radiation—much less energy than that deposited by cooling a cup of coffee 1 °C within a person's body—have such a disastrous effect on the person?

21

The Chemistry of the Main Group Elements

Wim van Egmond/Visuals Unlimited, Inc.

Diatoms are microscopic unicellular algae whose cell wall is constructed of hydrated SiO_2. This photomicrograph, entitled "The Glass Forest," shows green, triangular diatoms, *Licmorpha ehrenbergii*, 30 μm across, clinging to a marine invertebrate. Silicon and oxygen are the two most abundant elements in Earth's crust.

Elements have been discussed throughout this book. Their names, symbols, physical properties, and chemical reactivity have been noted, demonstrating their enormous diversity. Some elements, such as sodium and fluorine, react violently with many other substances; other elements, the noble gases, are so quiescent that they enter into very few or no chemical combinations. In spite of this diversity, elements in each group of the periodic table have predictable chemical similarities based on their number of valence electrons.

OWL Sign in to OWL at **www.cengage.com/owl** to view tutorials and simulations, develop problem-solving skills, and complete online homework assigned by your professor.

go Chemistry Download mini lecture videos for key concept review and exam prep from OWL or purchase them from **www.CengageBrain.com.**

Companion Website

Visit this book's companion website at **www.cengage.com/chemistry/moore** to work interactive modules for the Estimation boxes and Active Figures in this text.

. . . "I now mean by elements certain primitive and simple bodies, which not being made of any other bodies or of one another, are the ingredients of which all those perfectly mixed bodies are immediately compounded and into which they are ultimately resolved.". . .

Robert Boyle, *The Sceptical Chymist,* 1661

Expansion is a cooling process; contraction is a heating process.

The term "nuclear burning" arises from the fact that nuclear fusion reactions are highly exothermic and that we think of the sun and other stars as burning in the sky.

A fundamental question that has not yet been discussed concerns the origin of the elements: How did they form? How, for example, did magnesium atoms acquire a different number of protons from atoms of calcium or helium? This chapter will answer those questions. It will also describe how ten selected main group elements—those in Groups 1A through 8A—are extracted from natural sources, the chemical principles associated with the extraction processes, and the properties of those three metallic and seven nonmetallic elements. The chapter concludes with an overview of the main group elements, those in Groups 1A through 8A, describing their properties and uses, and relating these to the periodic table. The transition elements, all metals, are discussed in Chapter 22.

21.1 Formation of the Elements

Cosmologists—scientists who study the formation of the universe—use spectral evidence as well as knowledge of nuclear reactions to develop theories about the origin of the universe. Cosmologists theorize that about 15 billion years ago all the matter in the universe was contained in a pinpoint-sized region that exploded at inconceivably high temperatures, estimated to be about 10^{32} K, in what is called the "Big Bang." This explosion produced a universe expanding so quickly that in one second it cooled to 10^9 K, a temperature at which the fundamental subatomic particles formed—neutrons, protons, and electrons. Within two hours, the temperature had dropped to 10^7 K, temperatures suitable for the formation of light nuclei—^{2}H, ^{3}He, and ^{4}He.

Nuclear Burning

The elements from hydrogen to iron are formed inside stars by **nuclear burning,** a sequence of nuclear *fusion* reactions ***(← p. 980)*** not to be confused with chemical combustion. The fusion of protons (hydrogen-1 nuclei) to form helium-4 nuclei is called **hydrogen burning.**

$$4\,{}^1_1\text{H} \longrightarrow {}^4_2\text{He} + 2\,{}^{\ 0}_{+1}\text{e} + 2\,\gamma$$

After billions of years of hydrogen burning, the star contracts and the core becomes sufficiently dense and hot for **helium burning,** the fusion of helium-4 nuclei, to occur.

$${}^4_2\text{He} + {}^4_2\text{He} \longrightarrow {}^8_4\text{Be}$$

Beryllium-8 is unstable, but by fusing with another helium-4 nucleus, beryllium-8 is converted to stable carbon-12.

$${}^8_4\text{Be} + {}^4_2\text{He} \longrightarrow {}^{12}_{6}\text{C}$$

The low natural abundance of beryllium is evidence of the instability of beryllium-8 ***(← p. 965).***

When helium burning stops, the star contracts further with the result that the temperature becomes high enough for heavier nuclei to form by fusion. Starting with carbon-12, three successive fusions with helium-4 nuclei form oxygen-16, then neon-20, and then magnesium-24. The process continues up to the formation of calcium-40. Carbon and oxygen burning also occurs:

$${}^{12}_{6}\text{C} + {}^{12}_{6}\text{C} \longrightarrow {}^{23}_{11}\text{Na} + {}^1_1\text{H}$$

$${}^{12}_{6}\text{C} + {}^{16}_{8}\text{O} \longrightarrow {}^{28}_{14}\text{Si}$$

Starting with silicon-28, fusion reactions build up heavier nuclei all the way to iron-56 and nickel-58, which are very stable nuclei with the highest binding energies per nucleon ***(← p. 967).*** Elements heavier than iron cannot be formed by such nuclear fusion reactions.

EXERCISE 21.1 Fusing Nuclei

Write balanced nuclear equations representing the formation of oxygen-16, neon-20, and magnesium-24 starting from carbon-12 and helium-4 nuclei.

Answers to **EXERCISES** are provided at the back of this book in Appendix L.

Formation of Heavier Elements

Figure 21.1 indicates the amount of time a star spends in the various stages of elemental synthesis in relation to the central temperature of the star. Following helium burning, elements heavier than iron form by neutron capture in massive stars in which the core collapses rapidly. Stable nuclei such as those of iron decompose into neutrons and protons, and the protons are converted into additional neutrons by capturing electrons. The result is the formation of a neutron star whose outer layers explode away as a supernova.

Heavier elements form during supernova generation by one of two processes. In the *s* (slow) *process,* the slow capture of neutrons takes place over thousands of years. Because this capture shifts the neutron/proton ratio, eventually the nuclei become beta emitters. As noted in Section 20.3 ***(← p. 963),*** beta emission causes an increase in the atomic number, so a new element is formed from the parent nucleus. Such is the case for $^{98}_{42}Mo$, for example, which is converted to technetium-99 by this two-step process.

$$^{98}_{42}Mo + ^{1}_{0}n \longrightarrow ^{99}_{42}Mo$$

$$^{99}_{42}Mo \longrightarrow ^{99}_{43}Tc + ^{0}_{-1}e$$

Isotopes with mass numbers up to 209 can form by the *s process.*

Some elements, including the radioactive actinides, are produced by the very rapid *r process,* which occurs during the explosive stage of a star. Because of the speed at which a series of neutrons can be captured sequentially in the *r* process, new elements can be produced from nuclei with very short half-lives, too short to react by the *s* process. In the *r* process, a nucleus may capture many neutrons in an extremely short time (0.01 to 10 s) in a series of reactions that produce a nucleus much heavier than the original one. Suppose, for example, that $^{130}_{48}Cd$ is produced in the *r* process. This isotope is highly unstable because it has far too many neutrons; cadmium-116 is the most stable known isotope of cadmium. The cadmium-130 nucleus can undergo a rapid series of beta decays, increasing in atomic number until it reaches tellurium-130, the most abundant isotope of tellurium.

Nuclear fusion reactions have very high activation energy barriers ***(← p. 611).*** because two positively charged nuclei repel each other strongly. Thus, high temperatures and high kinetic energies of colliding particles are necessary to overcome the high activation energy barrier. The greater the nuclear charge, the higher the energy barrier and the higher the temperature required to overcome the barrier.

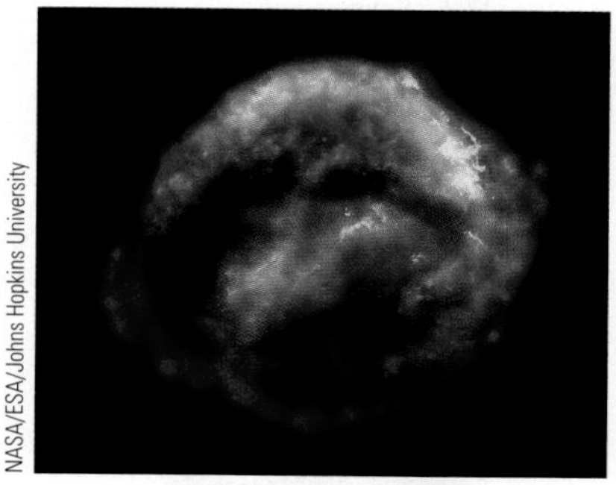

Kepler's supernova remnant. A bubble-shaped cloud of gas and dust envelops Kepler's supernova remnant. This fast-moving material from an exploded star (supernova) is partly composed of newly synthesized chemical elements.

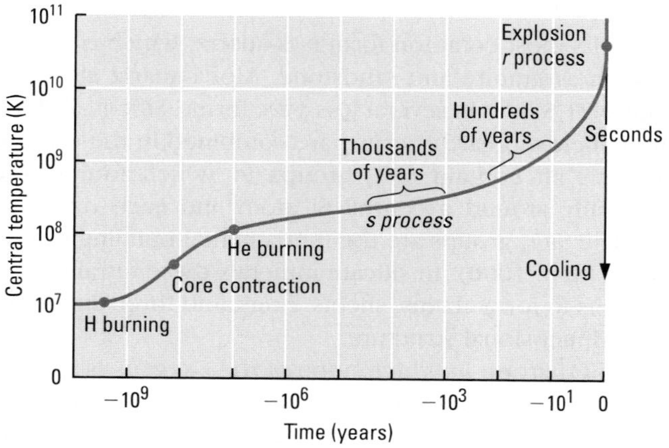

Figure 21.1 The timescale for various stages of elemental syntheses in stars.

EXERCISES that are labeled **CONCEPTUAL** are designed to test your understanding of one or more concepts; they usually involve qualitative rather than quantitative thinking.

EXERCISE 21.2 Cadmium-130 Decay

Cadmium-130 decays to tellurium-130 through four consecutive beta decays. Write balanced equations for these reactions, starting with cadmium-130 and ending with tellurium-130.

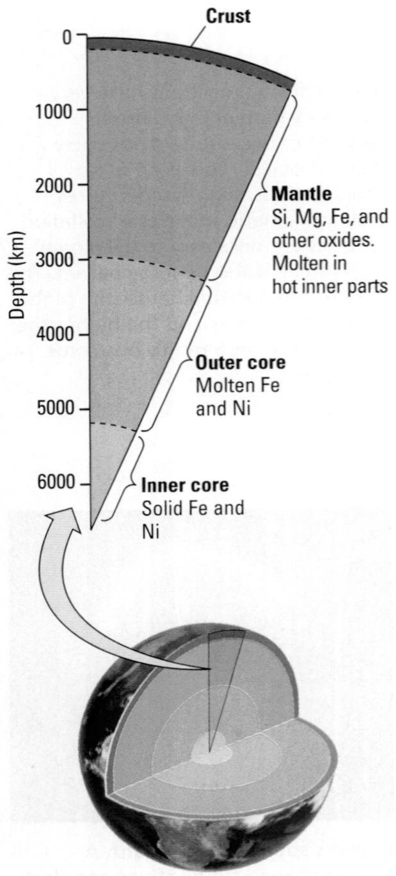

Figure 21.2 A cross section of Earth. Note the thinness of the crust compared with that of the mantle and the core. Most of the iron on Earth is in the core and the mantle, not the crust.

21.2 Terrestrial Elements

In this and Sections 21.3 to 21.6, we describe how nitrogen, oxygen, sulfur, sodium, chlorine, magnesium, aluminum, phosphorus, bromine, and iodine are obtained from their natural sources; we also consider the properties and uses of these elements.

We obtain large quantities of nitrogen and oxygen from the atmosphere. Ocean water is treated to extract commercial quantities of magnesium, bromine, and sodium chloride. Earth's crust is an indispensable source of most of the other elements. Figure 21.2 illustrates the relation between Earth's crust, mantle, and core. The crust, which extends from the surface to a depth of about 35 km, is but a tiny fraction of the entire depth of Earth. If you think of Earth as an apple, the crust is akin to the thickness of the skin of the apple.

The average composition of Earth's crust is given in Figure 21.3. All the elements shown in the pie chart are in compounds; because of their reactivity, these elements do not exist in their "free" (elemental) form in Earth's crust. Note the preponderance of oxygen and silicon; they are the major components of silicate minerals, clays, and sand. Aluminum is the most abundant metal ion, followed by ions of iron and several alkali and alkaline-earth metals—calcium, sodium, potassium, and magnesium.

Most elements in the crust of Earth are in chemically combined forms as complex ionic solids known as minerals. A **mineral** is commonly defined as a naturally occurring inorganic compound with a characteristic composition and crystal structure. The major chemical form in which each element occurs on Earth's surface as a source of the element is shown in Figure 21.4. In particular, note the predominance of oxygenated compounds, either binary ones such as MgO and TiO_2, or more complex ones such as carbonates, phosphates, and silicates. The preponderance of such compounds is testimony to the abundance of oxygen and silicon in Earth's surface. The lanthanides and naturally occurring actinides also form oxide minerals. Many transition metals and heavier elements of Groups 3A (13) to 6A (16) are found as sulfides, such as ZnS and Sb_2S_3.

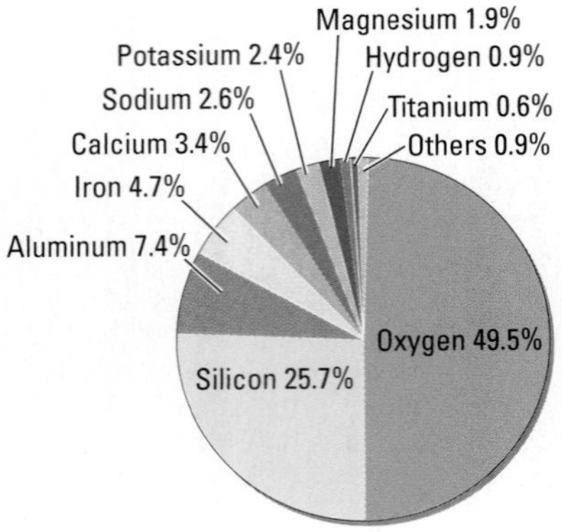

Figure 21.3 Elemental composition of Earth's crust by mass. Note that 76% of Earth's crust is composed of silicon and oxygen.

Silica and Silicates

Silica is pure SiO_2. Its most common form is α-quartz, which is a major component of many rocks such as granite and sandstone. Alpha-quartz also occurs as a pure rock crystal (Figure 21.5) and in several less pure forms. Silicon and oxygen, the two most abundant elements in Earth's crust, are combined in the crust as silicate minerals. Such minerals all contain SiO_4 groups in which four oxygen atoms are arranged tetrahedrally around a central Si atom and each oxygen is bonded to another Si atom. The SiO_4 groups are the fundamental building block for all silicate minerals (Figure 21.6; p. 1000). In silicate minerals these tetrahedra typically share one or more oxygens to form chains, sheets, rings, and three-dimensional networks; quartz has a three-dimensional structure.

An individual SiO_4 group in which none of the oxygens is shared with another silicon has a charge of 4−. An example is found in the mineral olivine, Mg_2SiO_4, which contains two Mg^{2+} ions for each SiO_4^{4-} ion to balance the charge. The simple silicate mineral willemite, Zn_2SiO_4, and the gemstone garnet contain discrete SiO_4^{4-} units that do not share oxygens.

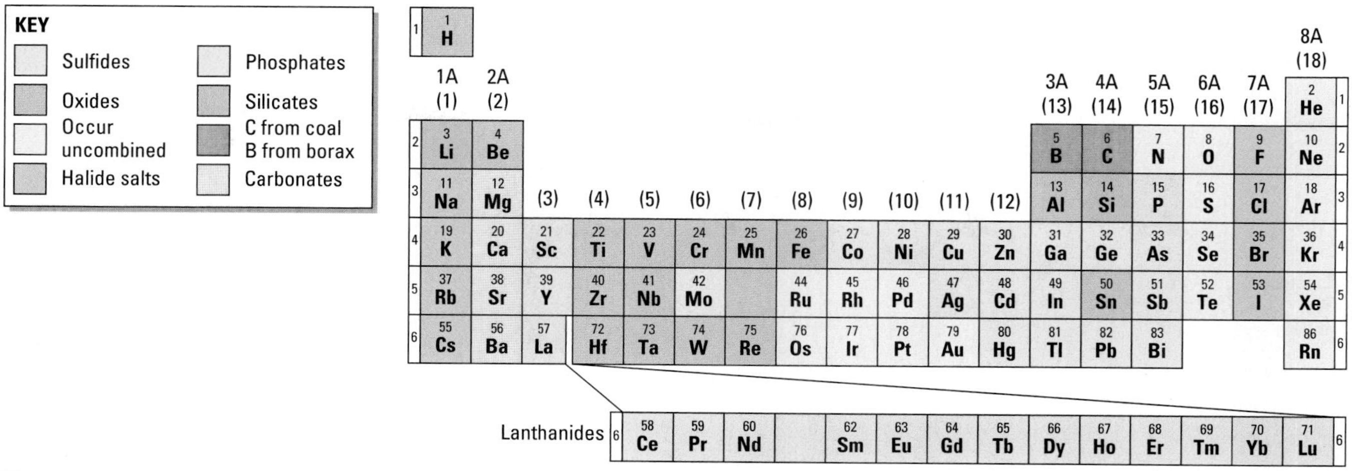

Figure 21.4 **Main types of minerals in Earth's crust used as sources of elements.**

Condensed silicates contain SiO_4 tetrahedra in which oxygen atoms are shared. *Pyroxenes* contain extended chains of linked SiO_4 tetrahedra, each sharing two oxygen atoms (Figure 21.6). The repeating unit in the pyroxene polymer is SiO_3, so pyroxene appears to contain the metasilicate ion, SiO_3^{2-}; a typical formula is Na_2SiO_3. Pyroxenes are abundant in the ocean's floor and Earth's mantle. If two pyroxene chains are laid side by side, they can link by sharing oxygen atoms in adjoining chains to form a type of silicate called an *amphibole,* with the typical repeating unit $Si_4O_{11}^{6-}$ (Figure 21.6).

An excellent example of an amphibole is crocidolite, one form of *asbestos.* What is called "asbestos" is not a single substance; rather, this term applies broadly to a family of naturally occurring hydrated silicates that crystallize as fibers. Asbestos minerals are generally subdivided into two forms: serpentine and amphibole fibers. Approximately 4 million tons of the serpentine form of asbestos, chrysotile, are mined each year, chiefly in Canada and the former Soviet Union. Chrysotile is the only form widely used commercially in the United States. Another form, the amphibole crocidolite, is mined in small quantities, mainly in South Africa. The two asbestos minerals differ greatly in composition, color, shape, solubility, and persistence in human tissue. This last property is important in determining their toxicity. Crocidolite is blue, is relatively insoluble, and persists in tissue. Its long, thin, straight fibers can penetrate narrow lung passages. In contrast, chrysotile is white, and it tends to be soluble and disappear in tissue. Chrysotile fibers are curly, so they ball up like yarn and are more easily rejected by the body.

Figure 21.5 **A pure quartz crystal (pure SiO_2).** The formula is derived from the fact that each oxygen atom of a SiO_4 tetrahedron is shared by two silicon atoms. Correspondingly, only half of the oxygens "belong" to a given Si, making the formula SiO_2, not SiO_4. Quartz crystals are used as oscillators in watches, radios, VCRs, and computers.

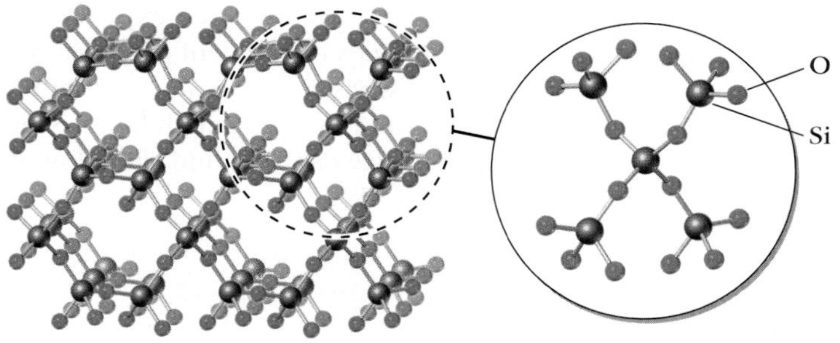

Three-dimensional structure of quartz. The right-hand circle shows part of the top layer.

Class	Independent tetrahedra	Single chains; pyroxenes	Double chains; amphiboles	Sheet silicates; mica
Unit composition	SiO_4^{4-}	$(SiO_3^{2-})_n$	$(Si_4O_{11}^{6-})_n$	$(Si_2O_5^{2-})_n$
Arrangement of SiO_4 tetrahedra				

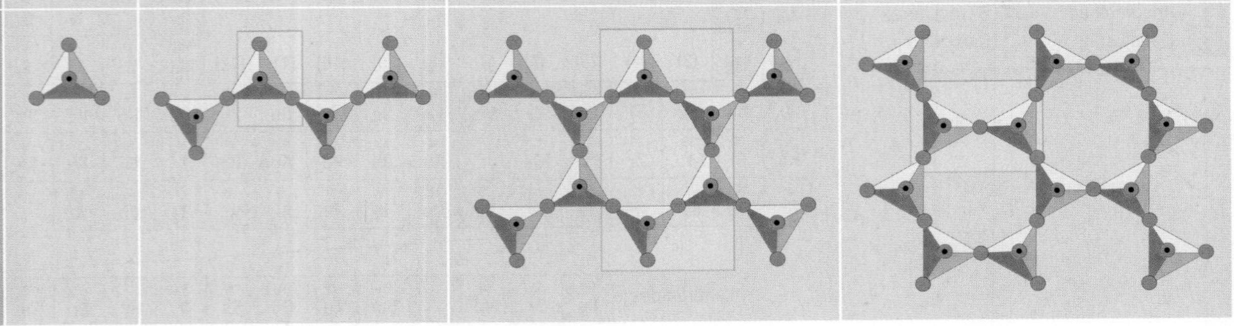

Figure 21.6 Silicate structures. These structures are all based on the tetrahedral SiO_4 unit. The repeating unit of each structure is shown with a tan background.

© Cengage Learning/Charles D. Winters

Garnet is a gemstone containing SiO_4^{4-} tetrahedra.

© Cengage Learning/Charles D. Winters

Sheets of mica (Figure 21.6) can be peeled away from each other using a razor blade because the sheets are weakly bonded to each other.

© AP Images

Figure 21.7 A mudslide caused by shifting clay. During heavy rains, clays become saturated with water, causing the aluminosilicate layers to shift, sliding over each other.

Long-term occupational exposure to certain asbestos minerals can lead to lung cancer. Although some disagreement persists in the medical and scientific communities, evidence strongly suggests that amphiboles such as crocidolite are much more potent cancer-causing agents than the serpentines such as chrysotile. Most asbestos in public buildings is the chrysotile type, so initiatives to remove asbestos insulation may be misguided overreaction in many cases. Nevertheless, most asbestos-containing materials have been removed from the U.S. market, and strict standards now exist for their handling and use.

When silicate chains continue to link in two dimensions, extended sheets of SiO_4 tetrahedral units result (Figure 21.6), characterized by the repeating unit $Si_2O_5^{2-}$. All of the atoms within each sheet are strongly covalent bonded, but each sheet is only weakly bonded to those above and below it. Various clay minerals and mica have this sheet-like silicate structure. Mica, for example, is used to prepare "metallic"-looking paint on new automobiles.

Clays are essential components of soils that come from the weathering of igneous rocks. Since early in human history, clays have been used for pottery, bricks, tiles, and writing materials. Clays are actually *aluminosilicates.* In an aluminosilicate, some of the tetrahedral groups are AlO_4 instead of SiO_4, and some O atoms are shared between an Al atom and a Si atom. An example is feldspar, $KAlSi_3O_8$, a component of many rocks. When Al^{3+} ions are replaced by other 3+ metal ions, the clay may become colored. For example, a red clay contains Fe^{3+} ions in place of some Al^{3+} ions.

Artists who work with clay first wet it and then mold the clay into a shape. Water molecules strongly interact with the oxygen atoms as well as the metal ions near the surface of clay particles, and the silicate layers slide over one another, making the clay pliable. After the clay has been formed into the desired shape, the object is heated to remove the water. Bonds form between exposed oxygen atoms and aluminum or silicon atoms on the surfaces of adjacent particles, which causes the clay to harden. Too much water in the wet clay can make it unstable. This instability occurs not only in clay for pottery, but also on a much larger scale in nature. During very heavy rains, entire hillsides of clay can shift and slide downhill, causing massive destruction of property (Figure 21.7).

CONCEPTUAL EXERCISE 21.3 Linking Tetrahedra

Explain how the silicate unit in pyroxenes has the general formula SiO_3^{2-}, not SiO_4^{4-}.

Table 21.1 Methods for Extraction of Elements from Their Ores

Extraction Method	Examples of Elements Extracted by This Method
Carbon reduction of oxide	Si, Fe, Sn
Oxidation with Cl_2	Br, I
Reaction of sulfide with O_2	Cu, Hg
Conversion of sulfide to oxide, then reduction with C	Zn, Pb
Halide reduction with sodium or other highly reactive metal	K, Ti, Cr, Cs, U
Halide or oxide reduction with H_2	B, Ni, Mo, W
Electrolysis of solution or molten salt	H, Li, F, Na, Ca, Al, Cl

Methods for Obtaining Pure Elements

Relatively few elements are available in nature directly in their uncombined form: the noble gases, mercury, gold, silver, copper, and sulfur. Nitrogen and oxygen occur as diatomic molecules. All other elements needed for practical applications must be extracted from their compounds. The types of extraction methods used are listed in Table 21.1. In minerals, metal cations must be reduced to their elemental form. Therefore, chemical and electrochemical oxidation-reduction reactions are needed in the production of metals.

In the subsequent sections of this chapter we describe how some elements are extracted from their naturally occurring forms by

- physical methods (Section 21.3: nitrogen, oxygen, and sulfur),
- electrochemical redox reactions (Section 21.4: sodium, chlorine, magnesium, and aluminum), and
- chemical redox reactions (Section 21.5: phosphorus, bromine, and iodine).

An **ore** is a mineral that contains a sufficiently high concentration of an element to make its extraction profitable. Because not all elements are used to the same extent, and the quantity used can vary with market demands, a metal is extracted from its ore in response to such demands. Current known reserves of some common elements such as aluminum and iron are sufficient to last hundreds of years at the current rate of use, whereas the known reserves of other widely used elements, such as copper, tin, and lead, are rather slim (Table 21.2). Notice from Table 21.2

Table 21.2 Known Reserves of Selected Elements

Element	Reserves (10^9 kg)	Lifetime (yr)	Locations of Major Reserves
Al	20,000	220	Australia, Brazil, Guinea
Fe	66,000	120	Australia, Canada, CIS*
Mn	800	100	CIS,* Gabon, S. Africa
Cr	400	100	CIS,* S. Africa, Zimbabwe
Cu	300	36	Chile, CIS,* USA, Zaïre
Zn	150	21	Australia, Canada, USA
Pb	71	20	Australia, Canada, CIS,* USA
Ni	47	55	Canada, CIS,* Cuba, New Caledonia
Sn	5	28	Brazil, China, Indonesia, Malaysia
U	2.8	58	Australia, CIS,* S. Africa, USA

*No individual breakdown is available for nations constituting the Commonwealth of Independent States (formerly the USSR).

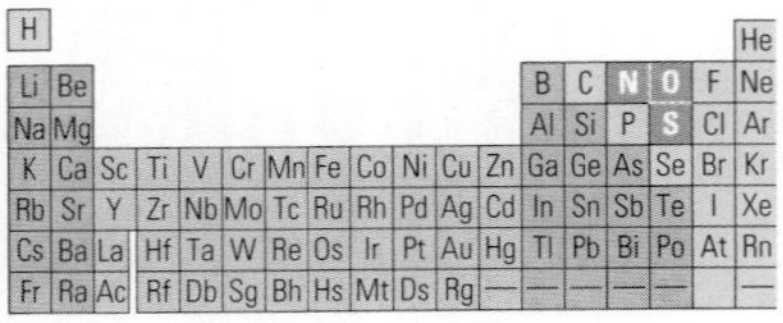

that the United States does not have major reserves of several critical metals, for example, the chromium and manganese needed for making steel and other alloys. Thus, we must import and stockpile such metals.

21.3 Some Main Group Elements Extracted by Physical Methods: Nitrogen, Oxygen, and Sulfur

Some elements occur in nature in their elemental form, that is, not combined with any other element. These metals are aloof—gold, silver, mercury, and copper; so is sulfur, a nonmetal. Clearly, nitrogen and oxygen, the principal components of the atmosphere, are such elements, as are the noble gases in the atmosphere. Large quantities of nitrogen, oxygen, and, to a lesser extent, argon are extracted from the atmosphere by the liquefaction of air.

Table 21.3 Composition of Clean, Dry Air at Sea Level

Component	Percent by Volume
N_2	78.09
O_2	20.948
Ar	0.93
CO_2	0.03*
He, Ne, Kr, Xe	0.002
CH_4	0.00015*
H_2	0.00005
All others combined	<0.00004

*Variable.

Elements from the Atmosphere

The composition of the atmosphere given in Table 21.3 shows that nitrogen is by far the most abundant component, its concentration being nearly four times that of oxygen, the next most abundant component. The gases of the atmosphere can be separated from one another by liquefying and fractionally distilling air in a process similar to the separation of petroleum fractions (***← p. 535***), except at much lower temperatures and higher pressures.

At low temperatures and high pressures, gases no longer behave ideally and the attractive forces between molecules cause the gases in air to condense to liquids. Because of their different boiling points, the liquid components can then be separated from one another by distillation. Before pure oxygen and nitrogen can be obtained from air, water vapor and carbon dioxide are removed (Figure 21.8). The dry air is compressed by more than 100 times normal atmospheric pressure, cooled to room temperature, and allowed to expand into a chamber. This expansion produces a cooling effect (the *Joule-Thomson effect*) because energy is required to overcome intermolecular forces as the molecules move farther apart. The expanding gas absorbs kinetic energy from the motion of its own molecules, which cools the gas. If this expansion is repeated and controlled properly, the expanding air cools to the point of liquefaction.

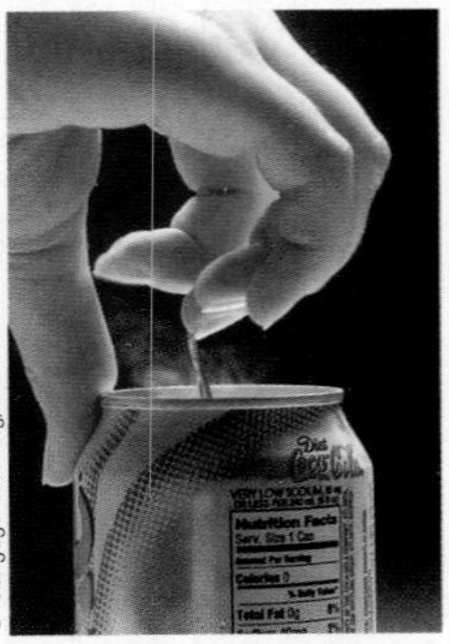

© Cengage Learning/Charles D. Winters

The Joule-Thomson effect. When the tab is opened on a can of carbonated beverage, the gases in the liquid are expelled rapidly enough to cool the water vapor to a liquid in the vicinity of the mouth of the can. The water vapor cools to form a tiny, visible "cloud."

The temperature of the liquid air is usually well below the normal boiling points of nitrogen (−195.8 °C), oxygen (−183 °C), and argon (−189 °C). The very cold liquid air is again allowed to vaporize partially. Since N_2 is more volatile and has a lower boiling point than O_2 or Ar, the N_2 evaporates first and the remaining liquid becomes more concentrated in O_2 and Ar. This process, known as the *Linde process*, produces high-purity nitrogen (>99.5%) and oxygen (99.5%). Further processing produces pure Ar.

EXERCISE 21.4 Liquefied Gases

A cryogenic flask contains 5.0 L of liquid oxygen, which has a density of 1.4 g/mL. Calculate what volume this oxygen will occupy at STP if it is allowed to boil.

Ottmar Bierwagen/Spectrum Stock

Pure sulfur. Large piles of recently mined sulfur are in a harbor awaiting shipment.

Sulfur

Sulfur, the element known biblically as brimstone, is a bright yellow solid. Very pure sulfur has been obtained from large deposits in salt domes along the coast of the Gulf of Mexico in the United States and Mexico and in underground

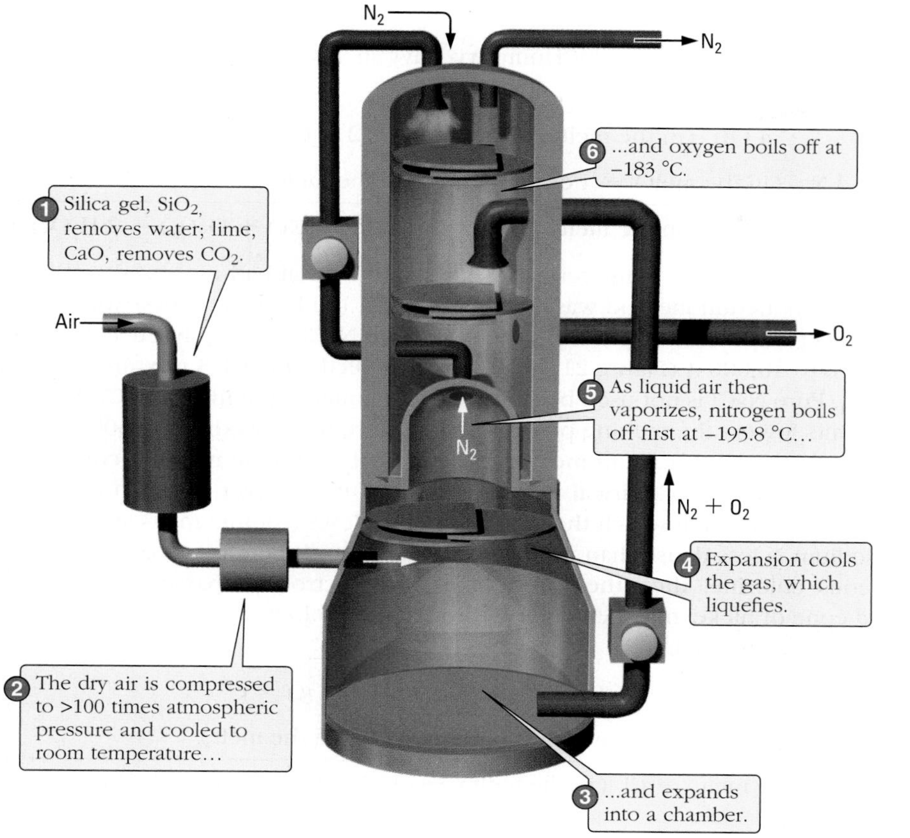

Figure 21.8 Fractional distillation of air. Air can be liquefied by using low temperatures and high pressure. The components of the liquefied air are then separated by taking advantage of their distinctly different boiling points.

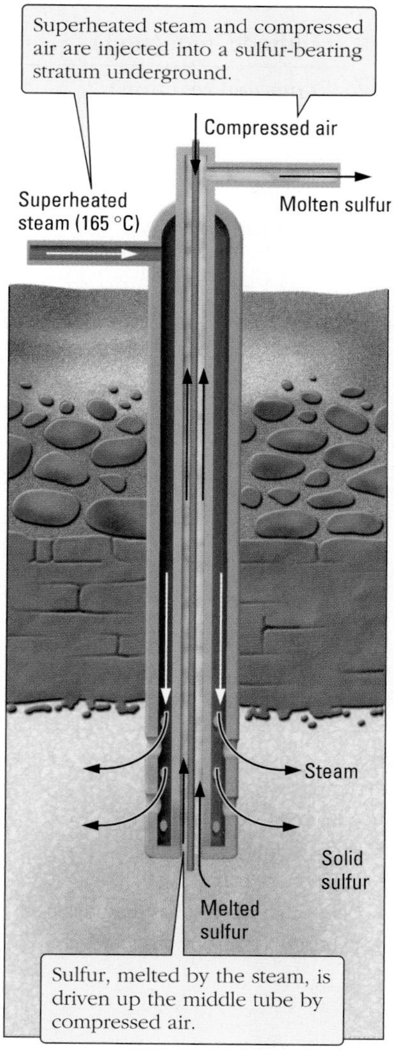

Figure 21.9 The Frasch process for mining sulfur. The molten sulfur froths up out of the middle pipe. Most sulfur is used to manufacture sulfuric acid.

deposits in Poland. Such sulfur deposits are believed to have been formed by bacterial reduction of sulfur in the naturally occurring mineral gypsum, $CaSO_4 \cdot 5\ H_2O$. Millions of tons of sulfur have been recovered from such deposits by the **Frasch process,** developed in the 1890s by Herman Frasch, a petroleum engineer (Figure 21.9). Most sulfur is now produced by extracting it from petroleum and natural gas. Removing the sulfur avoids the formation of sulfur dioxide, an atmospheric pollutant produced when petroleum burns. High-sulfur natural gas from Alberta, Canada, an especially large source of recovered sulfur, has now displaced the Frasch process as the chief source of sulfur.

21.4 Some Main Group Elements Extracted by Electrolysis: Sodium, Chlorine, Magnesium, and Aluminum

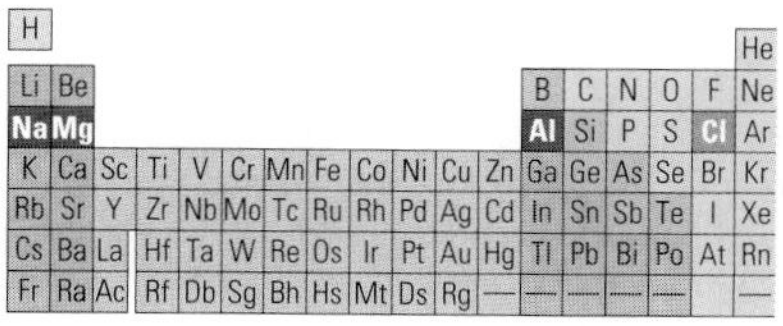

Chapter 19 described how electrolysis is used to force reactant-favored chemical reactions to occur *(← p. 939)*. Electrolysis is applied commercially on a vast scale to extract the elements sodium, magnesium, aluminum, and chlorine from their natural sources. These reactive elements exist naturally only in ionic form. Consequently, the metals must be obtained by reduction of their cations from compounds, and chlorine must be oxidized from Cl^- to Cl_2.

Sodium

Sodium metal was discovered by Humphry Davy in 1807; he electrolyzed molten NaOH. The half-reactions are

(anode, oxidation)	$4\ OH^-(\text{in the melt}) \longrightarrow O_2(g) + 2\ H_2O(g) + 4\ e^-$
(cathode, reduction)	$4\ Na^+(\text{in the melt}) + 4\ e^- \longrightarrow 4\ Na(\text{in the melt})$
(net cell reaction)	$4\ Na^+(\text{in the melt}) + 4\ OH^-(\text{in the melt}) \longrightarrow 4\ Na(\text{in the melt}) + O_2(g) + 2\ H_2O(g)$

By the early 1900s, commercial uses for sodium metal had increased so that a large-scale production method was needed. In 1921, the Downs process was developed to meet this demand. In a Downs cell, molten NaCl is electrolyzed at 7 to 8 V and 25,000 to 40,000 A (Figure 21.10). The cell is filled with a 1:3 mixture of NaCl and $CaCl_2$. Pure NaCl is not used because of its high melting point (800 °C). Mixing the two salts lowers the melting point of the mixture to approximately 600 °C.

In the Downs cell, sodium metal is produced at a cathode made of copper or iron that surrounds a cylindrical graphite anode. Directly over the cathode is an inverted trough through which the molten sodium flows (sodium melts at 97.8 °C); liquid sodium is less dense than the molten mixture and therefore floats on top of it. Gaseous chlorine, the other product of the electrolysis, passes through an inverted cone of nickel metal extending through the molten salt mixture and is collected, cooled, and liquefied.

(anode, oxidation)	$Cl^-(\text{in the melt}) \longrightarrow \frac{1}{2}\ Cl_2(g) + e^-$
(cathode, reduction)	$Na^+(\text{in the melt}) + e^- \longrightarrow Na(\text{in the melt})$
(net cell reaction)	$Na^+(\text{in the melt}) + Cl^-(\text{in the melt}) \longrightarrow Na(\text{in the melt}) + \frac{1}{2}\ Cl_2(g)$

The **PROBLEM-SOLVING STRATEGY** in this book is

- Analyze the problem
- Plan a solution
- Execute the plan
- Check that the result is reasonable

Appendix A.1 explains this in detail.

PROBLEM-SOLVING EXAMPLE 21.1 Titanium Production

Assume that the annual production of sodium metal in the United States is 76,000 tons. If half of this amount were used to produce titanium from $TiCl_4$, calculate how many tons of titanium could be produced.

$$TiCl_4(\ell) + 4\ Na(\ell) \longrightarrow Ti(\ell) + 4\ NaCl(\ell)$$

Answer 2.0×10^4 tons

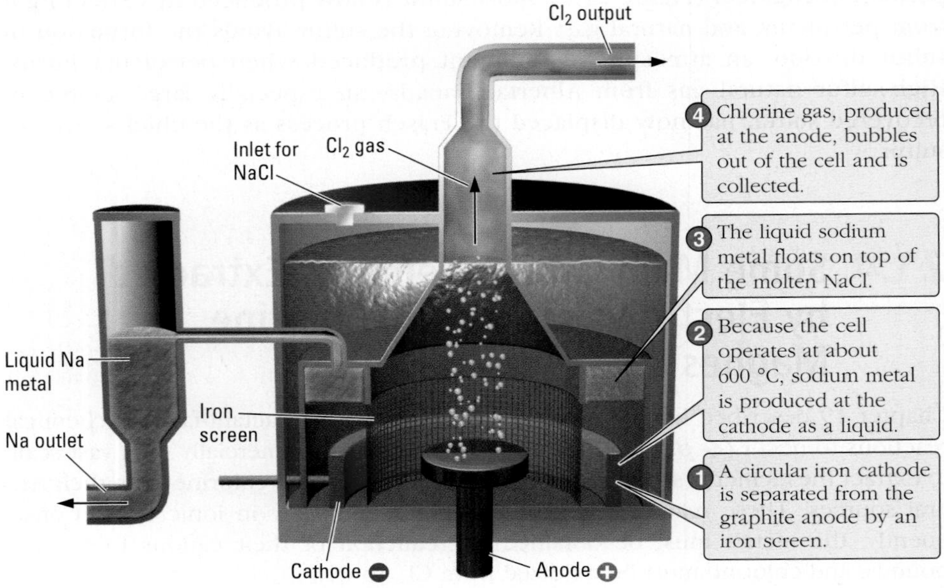

Cl_2 output
Inlet for NaCl
Cl_2 gas
4 Chlorine gas, produced at the anode, bubbles out of the cell and is collected.
3 The liquid sodium metal floats on top of the molten NaCl.
2 Because the cell operates at about 600 °C, sodium metal is produced at the cathode as a liquid.
1 A circular iron cathode is separated from the graphite anode by an iron screen.
Liquid Na metal
Na outlet
Iron screen
Cathode
Anode

Figure 21.10 The Downs cell for the electrolysis of molten NaCl.

Strategy and Explanation Half of the sodium produced would be 38,000 tons. From the balanced equation, we see that 4 mol sodium are needed to produce 1 mol titanium. Using this mole ratio we can calculate the number of moles of titanium, from which we can then find the mass of titanium.

$$3.8 \times 10^4 \text{ ton Na}\left(\frac{2000 \text{ lb Na}}{1 \text{ ton Na}}\right)\left(\frac{454 \text{ g Na}}{1 \text{ lb Na}}\right)\left(\frac{1 \text{ mol Na}}{23.0 \text{ g Na}}\right) = 1.5 \times 10^9 \text{ mol Na}$$

$$1.5 \times 10^9 \text{ mol Na}\left(\frac{1 \text{ mol Ti}}{4 \text{ mol Na}}\right) = 3.8 \times 10^8 \text{ mol Ti}$$

and

$$3.8 \times 10^8 \text{ mol Ti}\left(\frac{47.9 \text{ g Ti}}{1 \text{ mol Ti}}\right)\left(\frac{1 \text{ lb Ti}}{454 \text{ g Ti}}\right)\left(\frac{1 \text{ ton Ti}}{2000 \text{ lb Ti}}\right) = 2.0 \times 10^4 \text{ ton Ti}$$

☑ **Reasonable Answer Check** It takes about 100 g sodium to produce about 50 g titanium, or approximately 100 tons of sodium per 50 tons of titanium, or about half the mass of titanium per mass of sodium. Therefore, 38,000 tons of sodium would produce about 19,000 tons of Ti, which is close to the calculated answer.

PROBLEM-SOLVING PRACTICE 21.1

Calculate how many tons of sodium chloride are produced under the same conditions as in Problem-Solving Example 21.1.

PROBLEM-SOLVING PRACTICE answers are provided at the back of this book in Appendix K.

Manufacturing facilities in the United States have the capacity to produce about 76,000 tons of sodium metal per year. Much of the manufacturing is located near Niagara Falls, New York, because of the relatively low-cost electricity available from hydroelectric plants.

Nada Pecnik/Visuals Unlimited

Hydroelectric power. Water that came from the top of Niagara Falls flows out of a hydroelectric plant near the base of the falls. The falling water spins turbines to generate electricity.

EXERCISE 21.5 The Downs Cell

Calculate how many tons of sodium can be produced in one day by a Downs cell operating at 2.0×10^4 A. Determine how many tons of Cl_2 are produced in this same time.

Chlorine and Sodium Hydroxide

Most chlorine is produced by the electrolysis of aqueous sodium chloride in the **chlor-alkali process;** the alkali produced by this process is sodium hydroxide. Approximately 25 billion pounds of sodium hydroxide are produced annually in the United States, along with a similar quantity of chlorine by this process. These large amounts testify to the usefulness of these two products. The oxidizing and bleaching ability of chlorine is utilized in many industrial and everyday applications, and this element is a raw material in the manufacture of chlorine-containing chemicals. Sodium hydroxide is the base of choice in many industrial chemistry applications because it is inexpensive. It is also used widely to produce soaps, detergents, and other compounds.

The chlor-alkali process electrolyzes brine (saturated aqueous NaCl), as illustrated in Figure 21.11. Chloride ions are oxidized at the anode, and water is reduced at the cathode. The anode and cathode compartments are separated by a special polymeric membrane that allows only cations to pass through it. The brine solution is added to the anode compartment, and sodium ions pass through the membrane into the cathode compartment. The half-reactions are

The reduction potential of Na^+ is more negative than that of water, so water, not sodium ions, is reduced.

$$2\,Cl^-(aq) \longrightarrow Cl_2(g) + 2\,e^- \qquad \text{(anode, oxidation)}$$

$$2\,H_2O(\ell) + 2\,e^- \longrightarrow 2\,OH^-(aq) + H_2(g) \qquad \text{(cathode, reduction)}$$

$$2\,Cl^-(aq) + 2\,H_2O(\ell) \longrightarrow Cl_2(g) + 2\,OH^-(aq) + H_2(g) \qquad \text{(net cell reaction)}$$

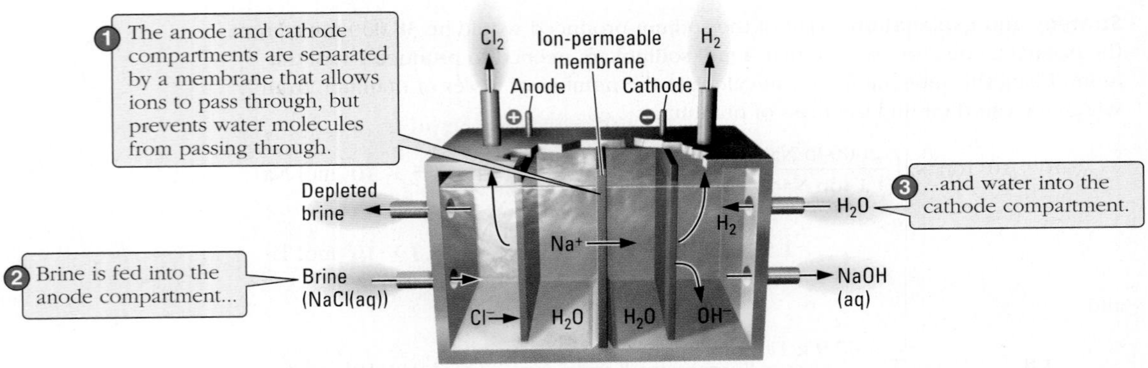

Figure 21.11 **A membrane cell used in the chlor-alkali process.**

The anode is specially treated titanium, and the cathode is stainless steel or nickel. The membrane is not permeable to water and acts as a salt bridge. Thus, as chloride ions are oxidized in the anode compartment, sodium ions must migrate from there to the cathode compartment to maintain charge balance. The resulting NaOH solution in the cathode compartment is 21% to 30% NaOH by weight.

The membrane cell was developed to replace the mercury cell that had been used previously in the chlor-alkali process. A major problem with mercury cells is the environmental damage caused by loss of mercury into the environment during normal operation of the cells. In the past, when mercury cells were cleaned, mercury was routinely allowed to run into neighboring bodies of water.

EXERCISE 21.6 NaOH Production

A chlor-alkali membrane cell operates at 2.0×10^4 A for 100. hours. Calculate how many tons of NaOH are produced in this time.

Magnesium from Seawater

With a concentration of 1.35 mg Mg^{2+} per liter, the oceans provide a nearly limitless supply of magnesium, containing approximately 6 thousand tons of it per cubic mile. As with other reactive metals, the conversion of the metal ion to the metal is not a product-favored reaction, so electrolysis is required to extract magnesium metal.

The *Dow process* is used to reduce Mg^{2+} ions in seawater into magnesium metal (Figure 21.12). It begins with the precipitation of Mg^{2+} as its insoluble hydroxide ($K_{sp} = 1.8 \times 10^{-11}$). Hydroxide ions come from an inexpensive base, $Ca(OH)_2$, produced by roasting calcium carbonate from seashells to form calcium oxide, which then reacts with water to produce calcium hydroxide. The calcium hydroxide is added to seawater to precipitate magnesium hydroxide.

$Mg(OH)_2$ ($K_{sp} = 1.8 \times 10^{-11}$) is much less soluble than $Ca(OH)_2$ ($K_{sp} = 6 \times 10^{-6}$), thus allowing the separation of solid magnesium hydroxide from calcium hydroxide in solution.

$$\underset{\text{seashells}}{CaCO_3(s)} \xrightarrow{\text{heat}} \underset{\text{lime}}{CaO(s)} + CO_2(g)$$

$$CaO(s) + H_2O(\ell) \longrightarrow Ca(OH)_2(aq)$$

$$Mg^{2+}(aq) + Ca(OH)_2(aq) \longrightarrow Mg(OH)_2(s) + Ca^{2+}(aq)$$

The magnesium hydroxide is filtered and neutralized by hydrochloric acid, another inexpensive chemical, to produce magnesium chloride.

$$Mg(OH)_2(s) + 2\ HCl(aq) \longrightarrow MgCl_2(aq) + 2\ H_2O(\ell)$$

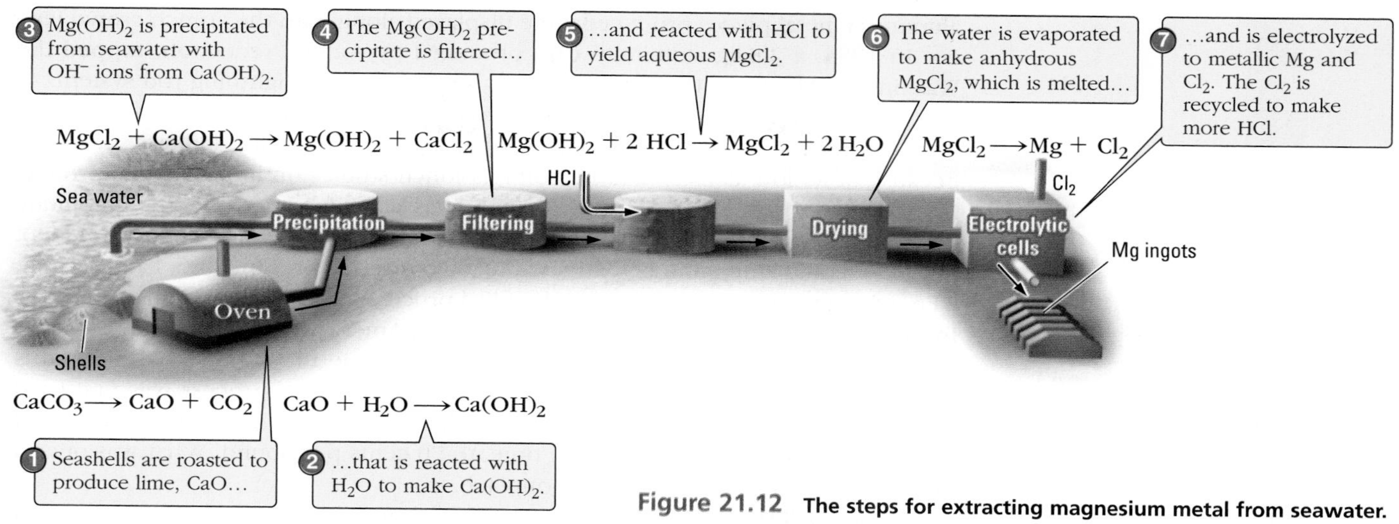
3 $Mg(OH)_2$ is precipitated from seawater with OH^- ions from $Ca(OH)_2$.
4 The $Mg(OH)_2$ precipitate is filtered...
5 ...and reacted with HCl to yield aqueous $MgCl_2$.
6 The water is evaporated to make anhydrous $MgCl_2$, which is melted...
7 ...and is electrolyzed to metallic Mg and Cl_2. The Cl_2 is recycled to make more HCl.
$MgCl_2 + Ca(OH)_2 \rightarrow Mg(OH)_2 + CaCl_2$
$Mg(OH)_2 + 2\ HCl \rightarrow MgCl_2 + 2\ H_2O$
$MgCl_2 \rightarrow Mg + Cl_2$
Sea water
Precipitation
Filtering
HCl
Drying
Cl_2
Electrolytic cells
Mg ingots
Oven
Shells
$CaCO_3 \rightarrow CaO + CO_2$
$CaO + H_2O \rightarrow Ca(OH)_2$
1 Seashells are roasted to produce lime, CaO...
2 ...that is reacted with H_2O to make $Ca(OH)_2$.

Figure 21.12 **The steps for extracting magnesium metal from seawater.**

The dried, anydrous magnesium chloride is melted and electrolyzed in a steel pot, which serves as the cathode (Figure 21.13). The electrode reactions are

$$2\ Cl^-(\text{in the melt}) \longrightarrow Cl_2(g) + 2\ e^- \qquad \text{(anode, oxidation)}$$

$$Mg^{2+}(\text{in the melt}) + 2\ e^- \longrightarrow Mg(\text{in the melt}) \qquad \text{(cathode, reduction)}$$

$$Mg^{2+}(\text{in the melt}) + 2\ Cl^-(\text{in the melt}) \longrightarrow Mg(\text{in the melt}) + Cl_2(g) \qquad \text{(net cell reaction)}$$

Molten magnesium is less dense than molten $MgCl_2$ and the molten metal floats at the surface, where the metal can be removed. Chlorine produced at the anode is converted to HCl by mixing Cl_2 with methane from natural gas and burning the mixture.

$$4\ Cl_2(g) + 2\ CH_4(g) + O_2(g) \longrightarrow 2\ CO(g) + 8\ HCl(g)$$

The HCl is recycled to neutralize $Mg(OH)_2$, which forms $MgCl_2$.

Aluminum Production

Aluminum is the most abundant metal in Earth's surface (7.4%), where it is present as Al^{3+} ions, from which the metal must be obtained by reduction. Aluminum was first isolated in metallic form in 1825 by an expensive and potentially dangerous method—using metallic sodium or potassium to reduce Al^{3+} ions in aluminum chloride, $AlCl_3$. Thus, metallic aluminum was very expensive and considered to be a precious metal, like gold or platinum. An early use of aluminum was in jewelry, including the Danish crown. In the 1855 Exposition in Paris, some of the first

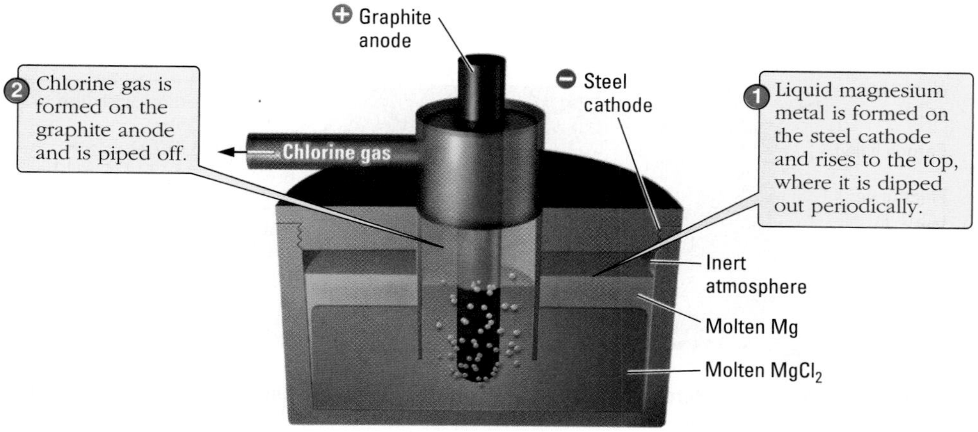
+ Graphite anode
2 Chlorine gas is formed on the graphite anode and is piped off.
Chlorine gas
− Steel cathode
1 Liquid magnesium metal is formed on the steel cathode and rises to the top, where it is dipped out periodically.
Inert atmosphere
Molten Mg
Molten $MgCl_2$

Figure 21.13 **Electrolysis of molten magnesium chloride.**

aluminum metal pieces produced were displayed along with the French crown jewels. In 1884, a 2.8-kg aluminum cap, produced by sodium reduction, topped the Washington Monument as ornamentation and the tip of a lightning rod system. At that time, the aluminum cap cost considerably more than the same mass of silver.

Napoleon II saw the advantages of using aluminum for military purposes because of its low density, and he commissioned studies on improving its production. Near the town of Les Baux, France, was a ready source of the aluminum-containing ore bauxite (Al_2O_3 combined with oxides of Si, Fe, and other elements); but, how could aluminum be extracted readily from the ore? In 1886, Paul Héroult, a Frenchman, conceived an electrochemical process to do so that is still used today. In a curious coincidence, Charles Martin Hall, an American, independently came up with the identical process two months earlier. Hence, the commercial method is known as the *Hall-Héroult process.* Just five years after the process was first used to produce aluminum commercially, the price of the metal plummeted from $12 per pound, a substantial sum at that time, to 70 cents per pound. What was once a precious metal soon became commonplace.

In the Hall-Héroult process, metallic aluminum is obtained by electrolysis of Al_2O_3 dissolved in molten cryolite, Na_3AlF_6. The cryolite allows the electrolysis to be carried out at a lower temperature (1000 °C) than would be required for molten Al_2O_3 (m.p. 2030 °C). The aluminum oxide-cryolite mixture is electrolyzed in a cell using carbon anodes and a carbon cell lining that serves as the cathode (Figure 21.14). The half-reactions for extracting aluminum are

(anode, oxidation) $3\ C(s) + 3\ O_2(g) \longrightarrow 3\ CO_2(g) + 12\ e^-$

(cathode, reduction) $4\ Al^{3+}(\text{in the melt}) + 12\ e^- \longrightarrow 4\ Al(\text{in the melt})$

(net cell reaction) $4\ Al^{3+}(\text{in the melt}) + 3\ C(s) + 3\ O_2(g) \longrightarrow 4\ Al(\text{in the melt}) + 3\ CO_2(g)$

As the cell operates, molten aluminum deposits on the cathode and sinks to the bottom of the cell, from which it is removed periodically. Such cells operate at a very low voltage of 4.0 to 5.5 V, but at a very high current of 50,000 to 150,000 A.

Aluminum production uses extremely large quantities of electricity, so aluminum production plants are located near hydroelectric power sources, such as those in the Pacific Northwest, because electricity from hydroelectric plants is generally less expensive than that from fossil fuel power plants. Production of each kilogram of aluminum requires about 13 to 16 kWh (4.7×10^4 to 5.8×10^4 kJ) of electric energy, excluding that required to heat the molten mixture. Because of the high energy cost to extract aluminum metal from its ore, there is widespread interest in recycling aluminum beverage containers and other aluminum objects. It takes far less energy to process recycled aluminum than to produce the metal from bauxite. Putting this into perspective, you could run your television set for three hours on the energy saved by recycling just one aluminum can!

Bettmann/Corbis

Charles Martin Hall

1863–1914

While a student at Oberlin College (OH), Charles Martin Hall became intrigued with trying to separate aluminum from its ores cheaply. When just 22 years old, using batteries and a blacksmith's forge, Hall succeeded in reducing Al_2O_3 dissolved in cryolite to metallic aluminum. To take advantage of his discovery, he formed the Aluminum Corporation of America (ALCOA), an enterprise that made Hall a multimillionaire.

Graphite anodes ⊕
Solid electrolyte crust
Carbon lining
Electrolyte (Al_2O_3 in Na_3AlF_6 (ℓ))
Molten aluminum
Carbon-coated steel cathode ⊖
At the anodes, oxidation of carbon occurs: $3\ C(s) + 3\ O_2(g) \longrightarrow 3\ CO_2(g) + 12\ e^-$
At the cathode, Al^{3+} is reduced: $4\ Al^{3+} + 12\ e^- \longrightarrow 4\ Al$
Tap to remove molten Al

Figure 21.14 A Hall-Héroult process electrolytic cell. Molten aluminum is drawn off from the bottom of the cell into molds.

PROBLEM-SOLVING EXAMPLE 21.2 Aluminum Production

If electricity costs \$0.100 per kilowatt-hour (kWh), calculate how much the electricity costs to produce 1.00 ton aluminum in a Hall-Héroult cell operating at 5.00 V. 1 kWh = 3.60×10^6 J; 1 V = 1 J/C.

Answer \$1350

Strategy and Explanation First, calculate the moles of aluminum produced.

$$1.00 \text{ ton Al}\left(\frac{2000 \text{ lb Al}}{1 \text{ ton Al}}\right)\left(\frac{454 \text{ g Al}}{1 \text{ lb Al}}\right)\left(\frac{1 \text{ mol Al}}{26.98 \text{ g Al}}\right) = 3.37 \times 10^4 \text{ mol Al}$$

Because the reduction reaction is $Al^{3+} + 3\,e^- \rightarrow Al$, 1 mol aluminum requires 3 mol electrons. Therefore,

$$\text{Total charge} = 3.37 \times 10^4 \text{ mol Al} \times \frac{3 \text{ mol e}^-}{1 \text{ mol Al}} \times \frac{9.65 \times 10^4 \text{ C}}{1 \text{ mol e}^-} = 9.74 \times 10^9 \text{ C}$$

The number of kilowatt-hours is

$$9.74 \times 10^9 \text{ C} \times \frac{5.00 \text{ J}}{1 \text{ C}} \times \frac{1 \text{ kWh}}{3.60 \times 10^6 \text{ J}} = 1.35 \times 10^4 \text{ kWh}$$

$$\text{Cost} = 1.35 \times 10^4 \text{ kWh} \times \frac{\$0.100}{1 \text{ kWh}} = \$1.35 \times 10^3, \text{ or } \$1350.$$

PROBLEM-SOLVING PRACTICE 21.2

Calculate how long it would take a Hall-Héroult cell operating at 1.00×10^5 A to produce 1.00 ton aluminum metal.

William Haynes Portrait Collection, Chemical Heritage Foundation

Paul Louis-Toussaint Héroult

1863–1914

While at a boarding school near Paris, Paul Héroult became interested in extracting aluminum from its ores after reading a book on electrolytic reduction of Al^{3+} ions. The 22-year-old Héroult convinced his mother to give him 50,000 francs, a large sum of money at the time, to buy a 400-A, 30-V battery to use in electrolytic reduction of aluminum ores. Like Charles Martin Hall, Héroult used molten cryolite to dissolve the ores and produced metallic aluminum. He filed his French patent on April 23, 1886 and his U.S. patent May 22, 1886, just months before Hall filed his patent on July 6, 1886. Hall received precedence because he demonstrated the electrolytic extraction of aluminum metal on February 9, 1886, about two months before Héroult's patent filing. The two men had an amicable relationship. Born in the same year, they also shared the same year of death; Héroult died just eight days after Hall.

21.5 Some Main Group Elements Extracted by Chemical Oxidation-Reduction: Phosphorus, Bromine, and Iodine

Phosphorus, bromine, and iodine are all produced by chemical redox reactions.

Phosphorus

Elemental phosphorus is extracted from phosphate-bearing rock by heating the rock with sand (SiO_2) and coke in an electric furnace (Figure 21.15). At 1400 to 1500 °C, a redox reaction produces gaseous phosphorus that evaporates from the mixture, leaving behind insoluble calcium silicate.

$$2\,Ca_3(PO_4)_2(\ell) + 10\,C(s) + 6\,SiO_2(\ell) \longrightarrow P_4(g) + 10\,CO(g) + 6\,CaSiO_3(\ell)$$

The mixture of phosphorus vapor and carbon monoxide gas is passed through water, where the phosphorus condenses and the CO bubbles out.

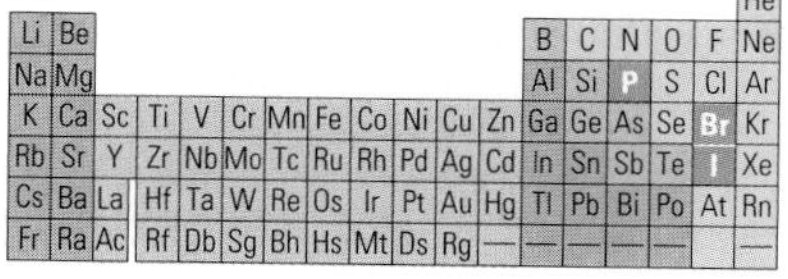

CONCEPTUAL EXERCISE 21.7 Phosphorus Extraction

The extraction of phosphorus from phosphate rock involves oxidation and reduction. Identify which element is oxidized and which is reduced.

About 90% of the elemental phosphorus produced is oxidized subsequently in air to P_4O_{10}, which reacts with water to produce phosphoric acid, H_3PO_4.

$$P_4(s) + 5\,O_2(g) \longrightarrow P_4O_{10}(s)$$

$$P_4O_{10}(s) + 6\,H_2O(\ell) \longrightarrow 4\,H_3PO_4(aq)$$

Some phosphoric acid is used in soft drinks, baking powder, and detergents.

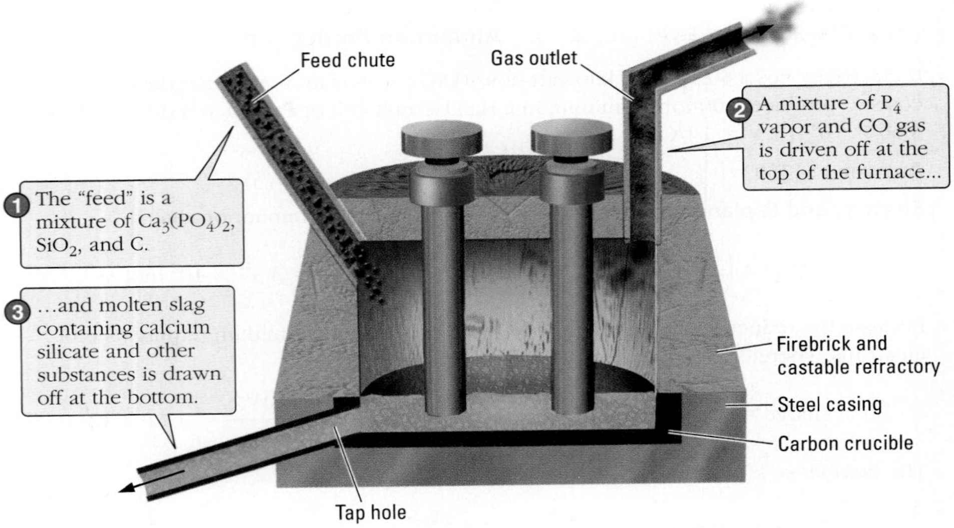
Feed chute
Gas outlet
2 A mixture of P_4 vapor and CO gas is driven off at the top of the furnace...
1 The "feed" is a mixture of $Ca_3(PO_4)_2$, SiO_2, and C.
3 ...and molten slag containing calcium silicate and other substances is drawn off at the bottom.
Firebrick and castable refractory
Steel casing
Carbon crucible
Tap hole

Figure 21.15 The production of phosphorus in an electric furnace.

The principal use of phosphate rock is to make fertilizers directly rather than to produce elemental phosphorus (Figure 21.16). Phosphate rock is reacted with sulfuric acid and converted into a soluble fertilizer. The mixture of hydrated calcium dihydrogen phosphate and hydrated calcium sulfate is called "superphosphate."

$$\underset{\text{phosphate rock}}{Ca_3(PO_4)_2(s)} + 2\,H_2SO_4(aq) + 5\,H_2O(\ell) \longrightarrow \underset{\text{superphosphate}}{Ca(H_2PO_4)_2 \cdot H_2O(s) + 2\,CaSO_4 \cdot 2\,H_2O(s)}$$

Fertilizer is also made from phosphoric acid by neutralizing the acid with ammonia to form ammonium hydrogen phosphate, $(NH_4)_2HPO_4$.

EXERCISE 21.8 Phosphorus in Phosphate Rock

Calculate the mass percent of phosphorus present in another form of phosphate rock, hydroxyapatite, $Ca_5(PO_4)_3OH$.

Bromine and iodine. The left flask contains a trace of liquid bromine (dark droplets) and bromine vapor; the right flask contains solid iodine and iodine vapor.

Bromine and Iodine

Bromine and iodine are halogens with similar, but different properties. Like the other halogens, they are too reactive to be found uncombined in nature. Consequently, Br_2 and I_2 must be produced by the oxidation of their anions.

Bromine and iodine are produced from seawater or *brines* (underground natural salt water deposits) by treating the solution with chlorine gas, which oxidizes Br^- to Br_2 and I^- to I_2. This is a case of a more reactive halogen, chlorine, displacing a less reactive one, bromide or iodide, from solution.

$$Cl_2(g) + 2\,Br^-(aq) \longrightarrow Br_2(\ell) + 2\,Cl^-(aq) \qquad E^\circ = +0.292\text{ V}$$

$$Cl_2(g) + 2\,I^-(aq) \longrightarrow I_2(aq) + 2\,Cl^-(aq) \qquad E^\circ = +0.823\text{ V}$$

The large positive voltages indicate both as very product-favored reactions.

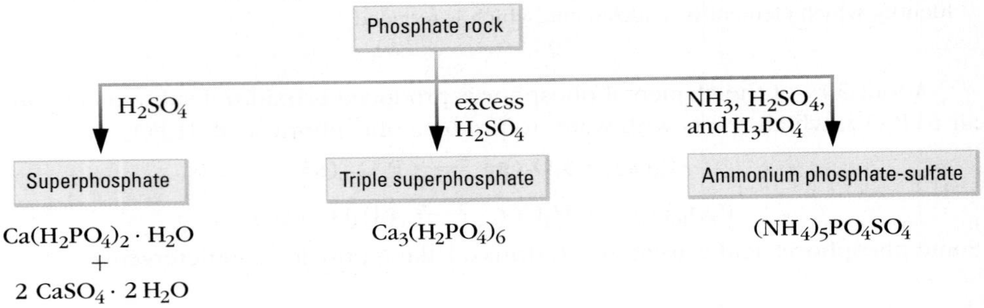
Phosphate rock
H_2SO_4
excess H_2SO_4
NH_3, H_2SO_4, and H_3PO_4
Superphosphate
Triple superphosphate
Ammonium phosphate-sulfate
$Ca(H_2PO_4)_2 \cdot H_2O$ + $2\,CaSO_4 \cdot 2\,H_2O$
$Ca_3(H_2PO_4)_6$
$(NH_4)_5PO_4SO_4$

Figure 21.16 Fertilizers produced from phosphate rock.

Another source of iodine is iodate ions, IO_3^-, in Chilean ore deposits. Iodate is converted to I_2 in a two-step process using hydrogen sulfite ions.

Step 1: $2\ IO_3^-(aq) + 6\ HSO_3^-(aq) \longrightarrow 2\ I^-(aq) + 6\ HSO_4^-(aq)$

Step 2: $5\ I^-(aq) + IO_3^-(aq) + 3\ H_3O^+ + 3\ HSO_4^-(aq) \longrightarrow$
$3\ I_2(aq) + 3\ SO_4^{2-}(aq) + 3\ H_2O(\ell)$

PROBLEM-SOLVING EXAMPLE 21.3 Oxidation-Reduction Reactions

Identify the oxidizing and reducing agents in Step 1 of the extraction of iodine from IO_3^--bearing ores.

Answer Oxidizing agent: IO_3^-; reducing agent: HSO_3^-

Strategy and Explanation Recall from Chapter 19 that reduction involves a decrease in oxidation number due to a gain of electrons. The reduction requires a reducing agent, which in this case is hydrogen sulfite ion, HSO_3^-, which donates the electrons. The +4 oxidation number (state) of sulfur in HSO_3^- is increased to +6 in HSO_4^-. This is oxidation, an increase in oxidation number, indicating a loss of electrons. Thus, the reducing agent, HSO_3^-, is oxidized (+4 sulfur to +6 sulfur); simultaneously, iodine in the oxidizing agent, IO_3^-, is reduced (from +5 to −1).

PROBLEM-SOLVING PRACTICE 21.3

Identify the oxidizing and reducing agents in Step 2 of the extraction of I_2 from Chilean ores.

CONCEPTUAL EXERCISE 21.9 Bromine Conversion

Use the terms *oxidation, reduction, oxidizing agent,* and *reducing agent* to explain the extraction of bromine from brines.

EXERCISE 21.10 Iodine Reaction

Calculate $E°$ for the reaction of I_2(s) with Br^-(aq). What does the value of $E°$ indicate about using I_2(s) to oxidize Br^-(aq) to $Br_2(\ell)$?

Post Street Archives, Michigan

Herbert H. Dow

1866–1930

Herbert H. Dow was the first to produce bromine by the electrolysis of brine (1891). In the 1920s, the demand for bromine rose sharply in order to make ethylene bromide, which was starting to be used in the higher octane gasoline required by high-performance automobile engines. Dow realized that ethylene bromide demand would be so great that brine sources could not supply enough bromine. He told the head of General Motors that to meet the demand, ". . . we'll have to go to sea and extract bromine from ocean water."* Herbert Dow died four years before achieving this goal, which was accomplished by his son Willard.

*Brandt, E. N. *Chemical Heritage*, Vol. 18, Number 3, Fall 2000; p. 39.

(a) Chlorine gas is bubbled into a colorless NaBr or NaI solution. (b) I^- is oxidized by Cl_2 to give I_2. (c) Br^- is oxidized by Cl_2 to give Br_2.

Displacement of Br^- and I^- by Cl_2. Carbon tetrachloride, a dense liquid, is added and extracts the Br_2 and I_2 from the upper aqueous layer into the bottom CCl_4 layer, concentrating the Br_2 *(orange)* and I_2 *(violet)*.

Chunks of metallic potassium in oil. A layer of oil is used as protective covering to prevent the potassium from reacting with oxygen in the air.

21.6 A Periodic Perspective: The Main Group Elements

In this section we discuss some of the properties and uses of the main group elements, Groups 1A–8A (1, 2, 13–18), from the perspective of their positions in the periodic table. In addition to providing a general overview of each group, one or more elements in each group will be highlighted for more detailed description.

Group 1A(1): The Alkali Metals

The alkali metals make up the leftmost group of the periodic table. Their densities, melting points, boiling points, atomic radii, and ionic radii are given below. The densities are low because the alkali metals have relatively large atomic radii compared to their molar masses. Weak metallic bonding (← *p. 514*) is responsible for their softness and relatively low melting points.

The chemical behavior of the alkali metals is dominated by loss of the ns^1 outer electron leading solely to the formation of M^+ cations. Therefore, most alkali metal compounds are ionic, except for organometallic compounds containing an alkali metal-to-carbon bond. The Group 1A elements react with air, water, and most nonmetals. The reactions of the heavier alkali metals are particularly vigorous, even explosive. Reactions of alkali metals with oxygen, sulfur, the halogens, and water are summarized in Table 21.4. The uses of some alkali metal compounds are given in Table 21.5.

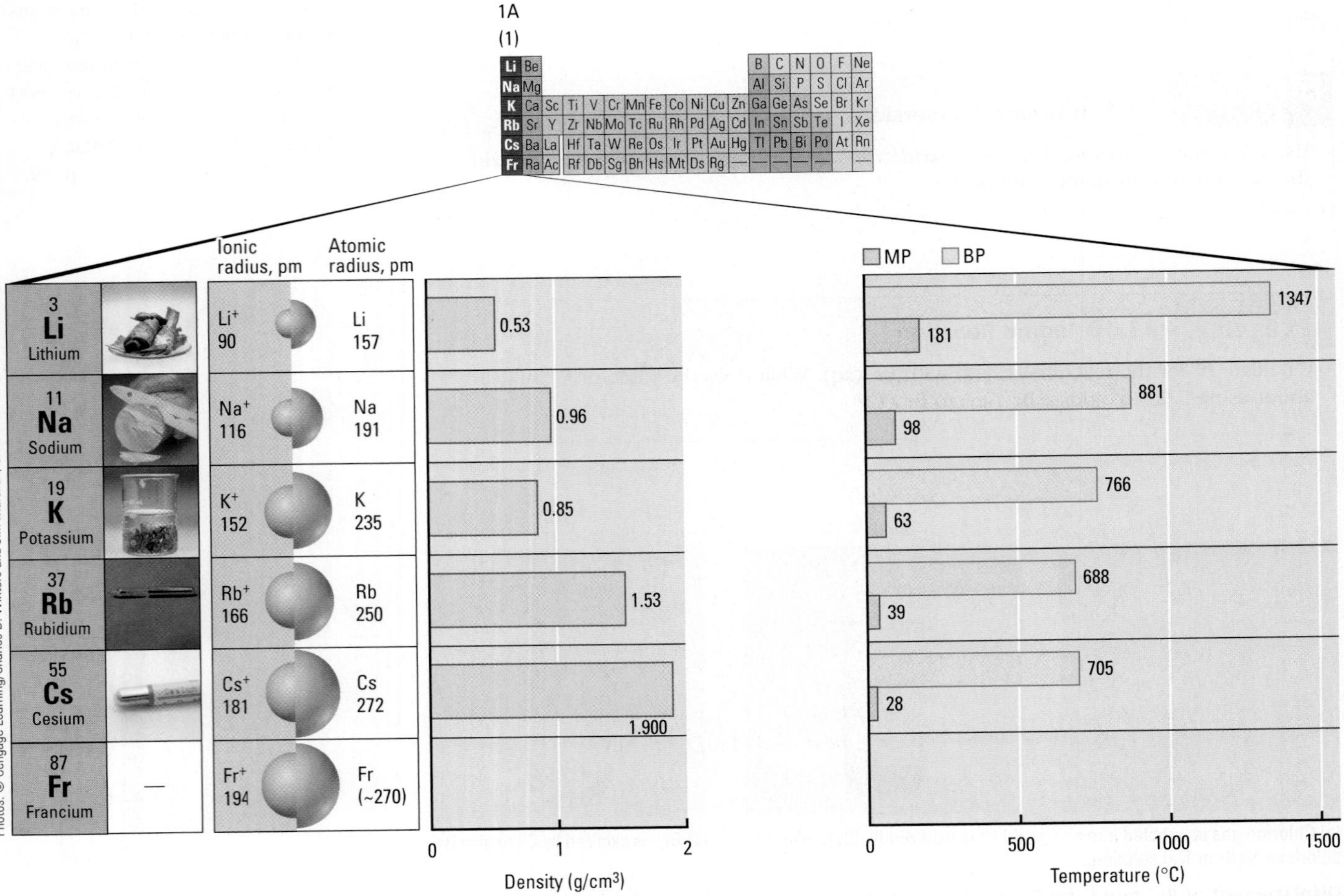

Group 1A(1) elements, the alkali metals: Li, Na, K, Rb, Cs, Fr.

Table 21.4 Reactions of Alkali Metals

Group 1A Metal (M)	Combining Substance	Reaction
Li	Oxygen	$4\ Li(s) + O_2(g) \longrightarrow 2\ Li_2O(s)$
Na	Oxygen	$2\ Na(s) + O_2(g) \longrightarrow Na_2O_2(s)$
K, Rb, Cs	Oxygen	$M(s) + O_2(g) \longrightarrow MO_2(s)$
All	Halogens	$2\ M(s) + X_2 \longrightarrow 2\ MX(s)$; X = F, Cl, Br, I
All	Sulfur	$2\ M(s) + S(s) \longrightarrow M_2S(s)$
Li	Nitrogen	$6\ Li(s) + N_2(g) \longrightarrow 2\ Li_3N(s)$
All	Water	$2\ M(s) + 2\ H_2O(\ell) \longrightarrow 2\ M^+(aq) + 2\ OH^-(aq) + H_2(g)$

The product of the reaction of an alkali metal with oxygen is dependent on the alkali metal, as seen from Table 21.4. Lithium is the only Group 1A metal that reacts directly with oxygen to form in good yield the normal oxide, M_2O, containing M^+ metal ions and O^{2-} oxide ions. In contrast, sodium reacts directly with oxygen to form predominantly sodium peroxide, Na_2O_2, which contains Na^+ and O_2^{2-}, peroxide, ions. The remaining alkali metals produce the metal superoxide, MO_2, in which M^+ and superoxide O_2^- ions are present. The peroxide and superoxide ions contain two covalently bonded oxygen atoms, with superoxide ions having one fewer electron than peroxide ions. Potassium superoxide is a convenient source of oxygen used in emergency breathing apparatus, such as for firefighters and miners in rescue circumstances where the concentration of oxygen is low. Water vapor in the breath reacts with superoxide ions to produce oxygen and potassium hydroxide; the latter removes exhaled carbon dioxide.

The molecular orbital theory (← p. 360) can be used to describe bonding in O_2^{2-} and O_2^- ions.

$$4\ KO_2(s) + 2\ H_2O(\ell) \longrightarrow 4\ KOH(s) + 3\ O_2(g)$$

$$2\ KOH(s) + CO_2(g) \longrightarrow K_2CO_3(s) + H_2O(\ell)$$

EXERCISE 21.11 Lewis Structures

Write the Lewis structures for oxide, peroxide, and superoxide ions. Write the molecular orbital diagrams for peroxide ion and superoxide ion.

Table 21.5 Uses of Alkali Metals and Some of Their Compounds

Element or Compound	Uses
Lithium	Lithium-ion batteries for computers, cell phones
Lithium carbonate, Li_2CO_3	Treatment of bipolar disorder
Sodium	Nuclear reactor coolant, manufacture of Ti
Sodium chloride, NaCl	Production of sodium metal, chlorine, NaOH
Sodium hydroxide, NaOH	Soaps and detergents, pulp and paper industry, bleach preparation, widely used industrial base
Sodium carbonate, Na_2CO_3	Glass manufacturing, water softening, detergents, reduction of SO_2 stack gas emission
Sodium hydrogen carbonate, $NaHCO_3$	Baking powder, baking soda, fire extinguishers, pharmaceuticals
Potassium nitrate, KNO_3	Gunpowder, fireworks; strong oxidizing agent
Potassium superoxide, KO_2	Oxygen source in emergency breathing apparatus
Rubidium and cesium	Photoelectric cells

Figure 21.17 Reaction of potassium with water. When water is dripped on to potassium metal, a violent reaction occurs.

Halogens react directly with alkali metals to produce stable binary halide salts whose lattice energies are substantial *(← p. 315)*. Examples include NaCl, KBr, and CsI. Likewise, the Group 1A metals all react directly with sulfur to form ionic sulfides with the general formula M_2S. Lithium is the only Group 1A metal that reacts directly with nitrogen gas, forming an ionic nitride, Li_3N.

Water reacts vigorously with the alkali metals, especially the heavier ones (Figure 21.17), to produce hydrogen gas plus a solution of the metal hydroxide. The reaction of sodium with water is exemplary:

$$Na(s) + 2\,H_2O(\ell) \longrightarrow 2\,Na^+(aq) + 2\,OH^-(aq) + H_2(g)$$

This reaction is highly exothermic: $\Delta H° = -367.6$ kJ/mol.

Because sodium metal is a strong reducing agent, it is used to obtain other metals from their metal halides. In particular, titanium, an element essential in aircraft production, can be prepared from its chloride by reduction with sodium.

$$TiCl_4(s) + 4\,Na(s) \longrightarrow Ti(s) + 4\,NaCl(s)$$

A major use for sodium metal was in the production of tetraethyllead, $Pb(C_2H_5)_4$, once used as an octane enhancer in leaded gasoline. Although leaded gasoline is still used in some countries, tetraethyllead is banned as a gasoline additive in the United States. Consequently, sodium production has declined.

Liquid sodium has high thermal conductivity and an anomalously high heat capacity. Metallic sodium has a low melting point and can be liquefied easily. These properties make liquid sodium an excellent heat-exchange liquid in some types of nuclear reactors *(← p. 977)*.

Group 2A(2): The Alkaline-Earth Metals

Like their Group 1A neighbors, the alkaline-earth metals (Group 2A) are silvery white, ductile, and malleable metals that are a bit harder than the alkali metals. The densities, melting points, boiling points, atomic radii, and ionic radii of the alkaline-earth metals are given on page 1015. The Group 2A elements, like the adjacent Group 1A elements, show regular changes in properties down the group. Chemical reactivity increases down the group.

Using pulsed-laser vaporization of beryllium metal, researchers recently created and characterized Be_2 in its ground state for the first time.

The toxicity of Be compounds is considered to be due to beryllium displacing Mg^{2+} ions from Mg-based enzymes.

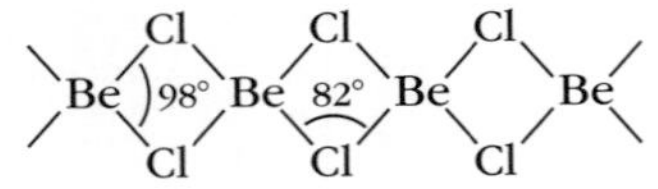

The polymeric structure of solid beryllium chloride.

The alkaline-earth metals are characterized chemically by the loss of the ns^2 outer electrons to yield 2+ ions. Beryllium is the exception, forming no predominantly ionic compounds due to the high charge density of Be^{2+} ions, which are found only as hydrated ions such as $[Be(H_2O)_4]^{2+}$. Anhydrous beryllium compounds are covalent and many are polymeric solids, such as $BeCl_2$, which has a Cl-bridging chain structure. Beryllium and its compounds are poisonous.

Some reactions of the alkaline-earth metals are summarized in Table 21.6. Like the alkali metals, the lighter alkaline-earth metals, (Mg and Ca), react with oxygen to form oxides, while the heavier metals, (Sr and Ba), form peroxides. Beryllium

Table 21.6 Some Reactions of Alkaline-Earth Elements

Group 2A Metal (M)	Combining Substance	Reaction
Be, Mg, Ca	Oxygen	$2\,M(s) + O_2(g) \longrightarrow 2\,MO(s)$
Sr, Ba	Oxygen	$M(s) + O_2(g) \longrightarrow MO_2(s)$
All	Halogens	$M(s) + X_2 \longrightarrow MX_2(s)$; X = F, Cl, Br, I
All (high temp.)	Nitrogen	$3\,M(s) + N_2(g) \longrightarrow M_3N_2(s)$
Ca, Sr, Ba	Water	$M(s) + 2\,H_2O(\ell) \longrightarrow M(OH)_2(aq) + H_2(g)$
Mg, Ca, Sr, Ba	Hydrogen	$M(s) + H_2(g) \longrightarrow MH_2(s)$
Mg, Ca, Sr, Ba	Carbon	$M(s) + 2\,C(s) \longrightarrow MC_2(s)$

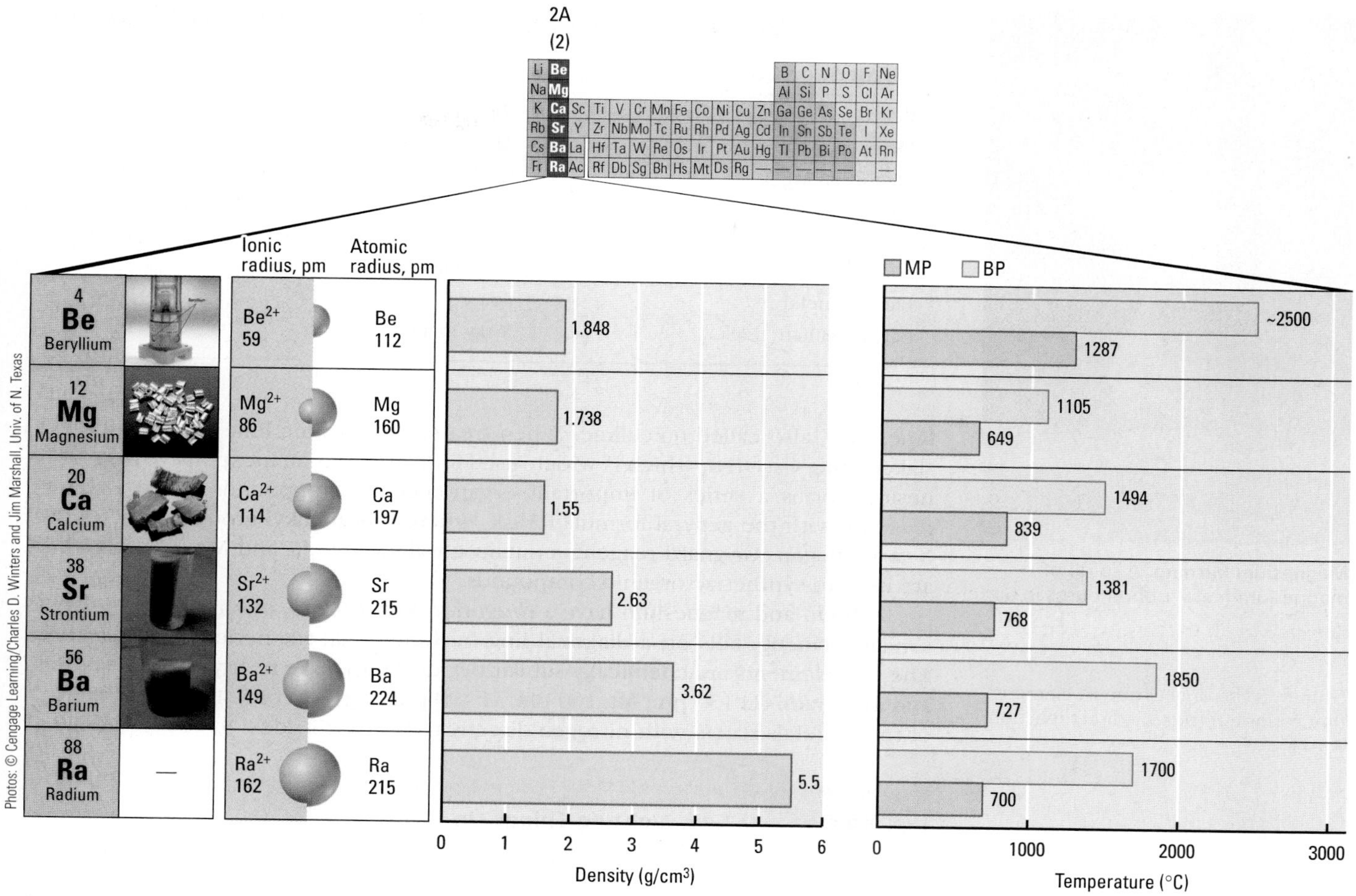

Group 2A(2) elements, the alkaline earths.

oxide forms directly only at temperatures higher than 600 °C. The oxides, except BeO, react with water to form the corresponding hydroxides. The vigorousness of the reaction of the metals with water increases down the group: beryllium does not react; magnesium does so only with steam above 100 °C; calcium and strontium react slowly with liquid water at room temperature, but barium reacts rapidly. All of the Group 2A elements react directly with nitrogen to form nitrides, which react with water to form aqueous hydroxides and release ammonia. Direct halogenation of the metals forms ionic halide salts of the general formula MX_2. The alkaline-earth metals, except beryllium, react directly with hydrogen to form hydrides, MH_2, and with carbon to form carbides, MC_2.

The carbide ion, C_2^{2-}, has the Lewis structure $:C{\equiv}C:^{2-}$.

EXERCISE 21.12 Group 2A Compounds

Write the formulas for these compounds.
(a) Calcium oxide (b) Barium peroxide (c) Strontium nitride (d) Calcium carbide

Magnesium metal has limited use in fireworks and flares because the metal burns with a brilliant white light. Its most important use is in making alloys, principally with aluminum. Magnesium is the least dense structural material; lightweight, strong magnesium alloys are used to make aircraft wheels, truck bodies, and ladders, among other things (Table 21.7).

Calcium and magnesium compounds are used extensively, as noted in Table 21.7. Limestone, $CaCO_3$, is abundant and, when heated, decomposes to produce

Table 21.7 Uses of Alkaline-Earth Elements and Some of Their Compounds

Element or Compound	Uses
Beryllium metal	X-ray tube windows
Magnesium metal	Alloys for building materials
Magnesia, MgO	Firebrick manufacturing, thermal insulator
Lime, CaO	Steel and paper manufacturing, water treatment
Calcium carbonate, $CaCO_3$	Limestone and marble for building materials, toothpaste abrasive, antacid
Barium metal	Spark plugs (alloy)
Barium sulfate, $BaSO_4$	X-ray imaging

Magnesium burning. A spiral of magnesium metal ribbon burns in air.

Lithium is the only Group IA element that forms a nitride by direct reaction with N_2.

lime, CaO, also called quicklime. When treated with water, lime is converted into slaked lime, $Ca(OH)_2$, which is widely used industrially as an inexpensive base. Magnesium forms a series of important organometallic compounds called Grignard reagents with the general formula RMgX, where R is an alkyl group ***(← p. 84)*** and X is a halogen. Grignard reagents contain covalent Mg—C and Mg—X bonds and are used to synthesize organic compounds.

Lithium and magnesium have a *diagonal relationship*, one in which a pair of similar elements falls on a diagonal line between adjacent periodic table groups. The two elements are chemically similar because they have nearly the same atomic and ionic radii (Li 157 pm, Mg 160 pm; Li^+ 90 pm, Mg^{2+} 86 pm). For example, each of these metals reacts with nitrogen directly to form a nitride.

EXERCISE **21.13 Lighting Things Up**

When magnesium metal burns in air, magnesium nitride and magnesium oxide are produced.

(a) Write the formula for magnesium nitride.

(b) Write a balanced chemical equation for the formation of magnesium nitride from the elements.

Group 3A(13): Boron, Aluminum, Gallium, Indium, Thallium

All elements of Group 3A have an ns^2np^1 outer electron configuration leading to +3 as the stable oxidation state except thallium, for which the +1 oxidation state is the more stable. *Where multiple oxidation states are possible for elements in a group, the lower oxidation state is usually the more favored in the heavier elements of the group.*

Some of the physical and atomic properties of the elements in Group 3A are given on p. 1017. The elements of this group exhibit a wider range of properties than those in Group 1A or 2A. Boron, the first member of the group, is a metalloid, an anomaly in a group where all other elements are silvery white metals. Aluminum is more representative of the group and for that reason is discussed in more detail here. The extraction of aluminum metal from bauxite ore was described in Section 21.4. Aluminum is the most abundant metal ion, Al^{3+}, in Earth's crust, exceeded in elemental abundance only by oxygen and silicon. As with Li and Mg, a diagonal relationship exists between Be and Al and between B and Si due to similarities in electronegativity and effective nuclear charge ***(← p. 309)*** within each pair of elements. Table 21.8 summarizes some reactions of the Group 3A elements.

Boron and hydrogen form an extensive series of covalently bonded hydrides called *boranes* with the general formulas B_nH_{n+4} or B_nH_{n+6} such as B_2H_6 and B_5H_{11},

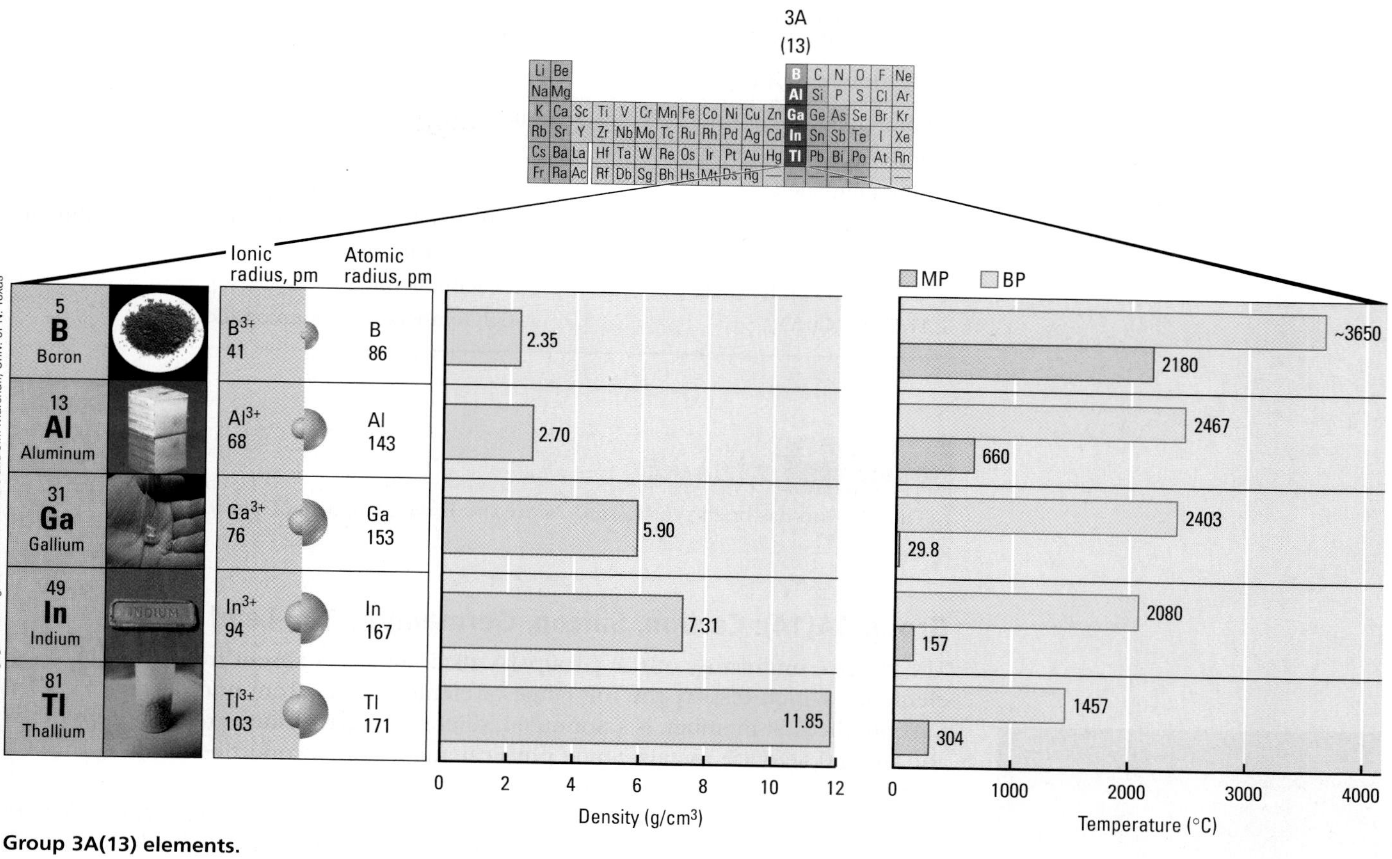
3A
(13)
Li Be B C N O F Ne
Na Mg Al Si P S Cl Ar
K Ca Sc Ti V Cr Mn Fe Co Ni Cu Zn Ga Ge As Se Br Kr
Rb Sr Y Zr Nb Mo Tc Ru Rh Pd Ag Cd In Sn Sb Te I Xe
Cs Ba La Hf Ta W Re Os Ir Pt Au Hg Tl Pb Bi Po At Rn
Fr Ra Ac Rf Db Sg Bh Hs Mt Ds Rg
Ionic radius, pm
Atomic radius, pm
5 B Boron
13 Al Aluminum
31 Ga Gallium
49 In Indium
81 Tl Thallium
INDIUM
B3+ 41
B 86
Al3+ 68
Al 143
Ga3+ 76
Ga 153
In3+ 94
In 167
Tl3+ 103
Tl 171
2.35
2.70
5.90
7.31
11.85
0 2 4 6 8 10 12
Density (g/cm3)
MP BP
~3650
2180
2467
660
2403
29.8
2080
157
1457
304
0 1000 2000 3000 4000
Temperature (°C)

Photos: © Cengage Learning/Charles D. Winters and Jim Marshall, Univ. of N. Texas

Group 3A(13) elements.

respectively. These compounds contain boron atoms bridged by hydrogen atoms in what is described as a three-center-two-electron bond.

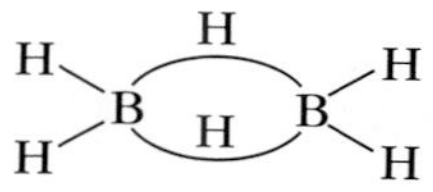
H
H
H
B
H
B
H
H

The B—H—B bridge bonding in diborane, B_2H_6.

Aluminum is an economically important, useful metal because of its low density (2.70 g/cm^3) and high strength when alloyed. It can be fashioned into wire, food wrapping sheets, stepladders, aircraft and automotive parts, and many other useful items. Aluminum metal resists corrosion because a transparent, chemically inactive film of aluminum oxide clings avidly to the metal's surface and protects the metal beneath it from further oxidation.

$$4\ \mathrm{Al(s)} + 3\ \mathrm{O_2(g)} \longrightarrow 2\ \mathrm{Al_2O_3(s)}$$

Aluminum oxide occurs as the mineral corundum, which is used widely as an abrasive in sandpaper and toothpaste (Table 21.9). A number of precious gems are primarily Al_2O_3 with small amounts of other metal ions strategically substituted for aluminum ions. Red rubies contain Cr^{3+} ions, blue sapphires contain Fe^{2+} and Fe^{3+} ions, and green emeralds contain Cr^{3+} and V^{3+} ions.

© Cengage Learning/Charles D. Winters

Gemstones: sapphire, ruby, and emerald.

Table 21.8 Some Reactions of Group 3A Elements

Group 3A Element (M)	Combining Substance	Reaction
Al, Ga, In	Oxygen	$4\ \mathrm{M(s)} + 3\ \mathrm{O_2(g)} \longrightarrow 2\ \mathrm{M_2O_3(s)}$
B, Al, Ga, In	Halogens	$2\ \mathrm{M(s)} + 3\ \mathrm{X_2} \longrightarrow 2\ \mathrm{MX_3}$; X = F, Cl, Br, I
Ga, In (high temperature)	Water	$2\ \mathrm{M(s)} + 6\ \mathrm{H_2O(g)} \longrightarrow 2\ \mathrm{M^{3+}(aq)} + 6\ \mathrm{OH^-(aq)} + 3\ \mathrm{H_2(g)}$

Table 21.9 Some Group 3A Compounds and Their Uses

Element or Compound	Uses
Boron oxide	Borosilicate glass
Boric acid, H_3BO_3	Eyewash, astringent
Aluminum metal	Foil wrap, alloys, structural material
Aluminum oxide	Pigments, fireworks, refractory bricks, toothpaste
Aluminum sulfate	Water purification
Gallium arsenide, GaAs	Semiconductor
$Tl_2Ba_2Ca_2Cu_3O_{10}$	High-temperature superconductor

CONCEPTUAL EXERCISE 21.14 Al_4^{4-}

The Al_4^{4-} ion has been synthesized. Write the Lewis structure of this ion.

Group 4A(14): Carbon, Silicon, Germanium, Tin, Lead

The relative uniformity of the Group 1A alkali metals is absent from the Group 4A elements, which display the full range of element types from nonmetals to metals. Carbon, the first member, is a nonmetal; silicon and germanium are both metalloids; and tin and lead are metals. Some physical and atomic properties of the Group 4A elements are shown on page 1019.

Chlorides of tin and lead. $SnCl_2$ (left, front) and $PbCl_2$ (right, front) are crystalline solids; $SnCl_4$ (left, back) and $PbCl_4$ (right, back) are volatile liquids.

The Group 4A elements all have an ns^2np^2 outer electron configuration. Promotion of the ns^2 electrons into empty np orbitals allows for hybridization and, through electron sharing, the formation of four bonds as in compounds such as CH_4, $SiBr_4$, $SnCl_4$, and $Pb(C_2H_5)_4$. The bonding in Group 4A compounds shifts from predominantly covalent in earlier members of the group to more ionic with tin and lead. The lower oxidation state (+2) is more important for tin and lead than the +4 state. Sn(II) and Pb(II) compounds are white, crystalline solids, whereas the Sn(IV) and Pb(IV) analogs are volatile liquids consisting of covalently bonded molecules. The np^2 electrons are used to form the lower oxidation state compounds. In such cases, the ns^2 electrons are not involved in bonding, a phenomenon sometimes referred to as the *"inert pair effect."* Sn(II) compounds are reducing agents, being converted to Sn(IV) by oxidizing agents. In contrast, Sn(IV) compounds are oxidizing agents that are reduced to Sn(II) by reducing agents.

Carbon compounds, and to a lesser extent silicon compounds, exhibit **catenation,** in which bonds between atoms of the same element form chains or rings. Hydrocarbons containing carbon-to-carbon bonds exemplify this phenomenon (← *p. 80*). The chemistry of carbon and its compounds has been described in Chapters 3 and 12. Silicon chemistry was discussed in Section 21.2.

Tin and lead have been known since ancient times; lead lined the aqueducts to ancient Rome. In contrast, germanium was not discovered until 1886 after Mendeleev predicted its expected properties in 1871, calling the proposed element *ekasilicon.* Like diamond and silicon, germanium has a covalent network structure and was used in the first transistors due to its semiconductor properties. Because ultrapure silicon is cheaper and more rugged, it has replaced germanium for this application.

Pewter and bronze articles.

Tin is used to make pewter, an alloy of 85% tin and the remainder a combination of copper, zinc and antimony, or lead. Bronze, an alloy of tin (20%) and copper (80%), revolutionized tool making and weaponry because bronze can be fabricated into a sharp, hard edge. Its use for this purpose ushered in the Bronze Age. Tin was available because cassiterite, SnO_2, its ore, can be reduced easily using a charcoal fire.

$$SnO_2(s) + 2\,C(s) \longrightarrow Sn(\ell) + 2\,CO(g)$$

The molten tin was recovered readily as it flowed from the fire.

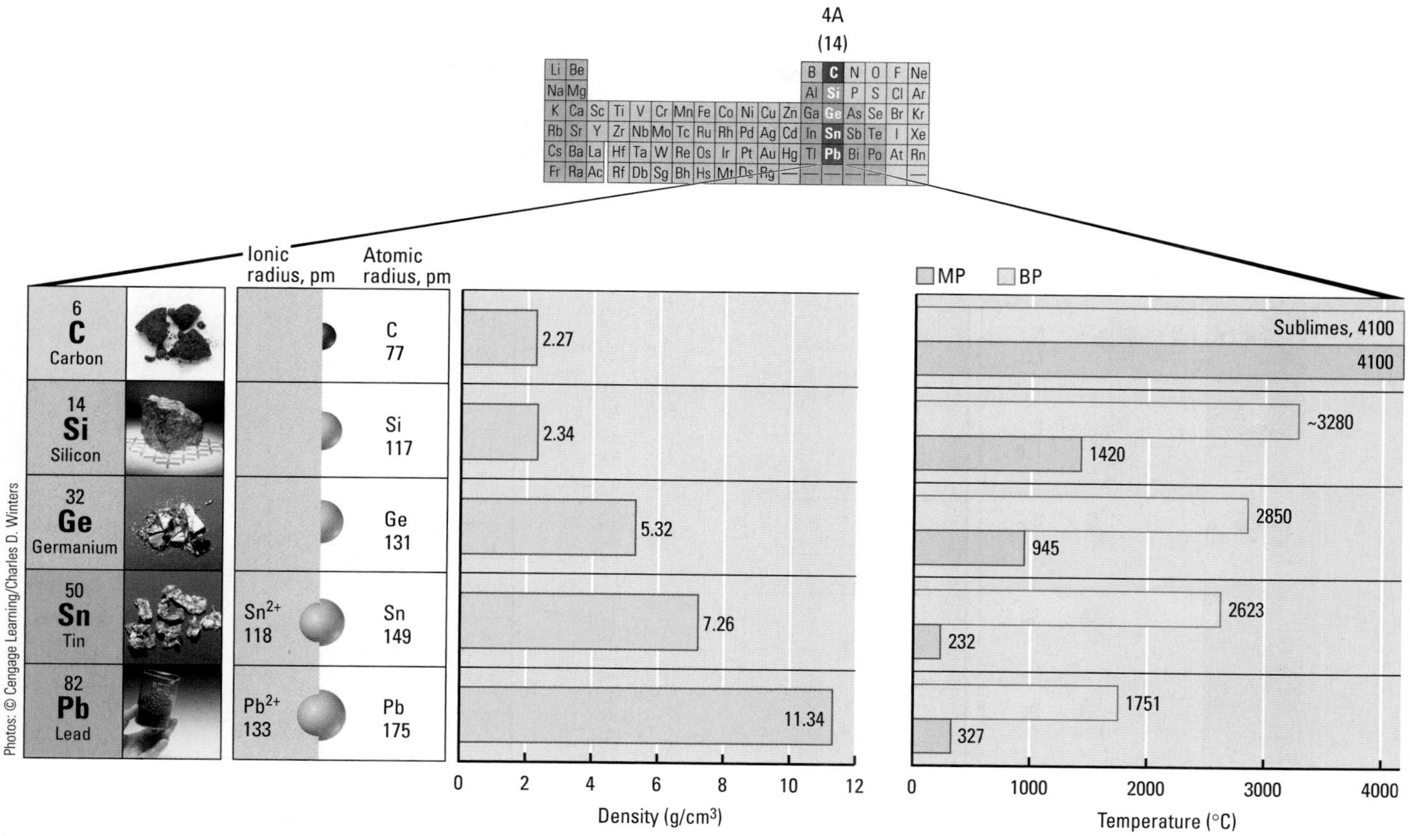
4A
(14)
Ionic radius, pm
Atomic radius, pm
6 C Carbon
14 Si Silicon
32 Ge Germanium
50 Sn Tin
82 Pb Lead
C 77
Si 117
Ge 131
Sn²⁺ 118
Sn 149
Pb²⁺ 133
Pb 175
2.27
2.34
5.32
7.26
11.34
0 2 4 6 8 10 12
Density (g/cm³)
MP
BP
Sublimes, 4100
4100
~3280
1420
2850
945
2623
232
1751
327
0 1000 2000 3000 4000
Temperature (°C)
Photos: © Cengage Learning/Charles D. Winters

Group 4A(14) elements.

Lead was used to produce, tetraethyllead, $(C_2H_5)_4Pb$, as a gasoline additive to enhance octane rating ***(← p. 536)***, but this use has been phased out in the United States due to the release of toxic lead compounds into the atmosphere and the fact that lead destroys the catalytic effect of automobile catalytic converters. Lead reacts with oxygen and carbon dioxide to form an oxide or carbonate coating on the surface that protects the metal from further reaction. Lead reacts with and dissolves slowly in water. Because lead compounds are toxic, they must not be allowed in water used for human consumption.

Group 5A(15): Nitrogen, Phosphorus, Arsenic, Antimony, Bismuth

Like the elements in Group 4A, the lightest to heaviest members of Group 5A range from typical nonmetals (N and P) to metalloids (As and Sb) and then to a metal (Bi). Some physical properties of the Group 5A elements are given on page 1020.

All of these elements have an ns^2np^3 outer electron configuration. Except for nitrogen, all can form pentavalent compounds such as PCl_5 and $BiCl_5$. Nitrogen forms only trivalent compounds such as NH_3 and NCl_3, because the nitrogen atom is too small to accommodate five bonding pairs of electrons around it.

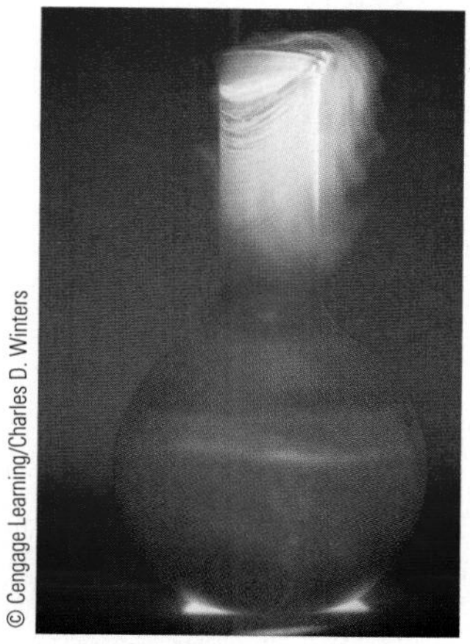
© Cengage Learning/Charles D. Winters

Liquid nitrogen.

Uses of Nitrogen and Its Compounds The temperature of liquid nitrogen, −196 °C, is very low, making it a **cryogen.** It is used in cryosurgery, for example, to cool an area of skin prior to removal of a wart or other unwanted or pathogenic tissue. Because of its low boiling point and inertness at low temperatures, liquid nitrogen has found wide use in frozen-food preparation and preservation during transit. Since nitrogen is so chemically unreactive at room temperature, it is used as an inert atmosphere for applications such as welding.

Cryogen, from the Greek word *kryos* meaning "icy cold."

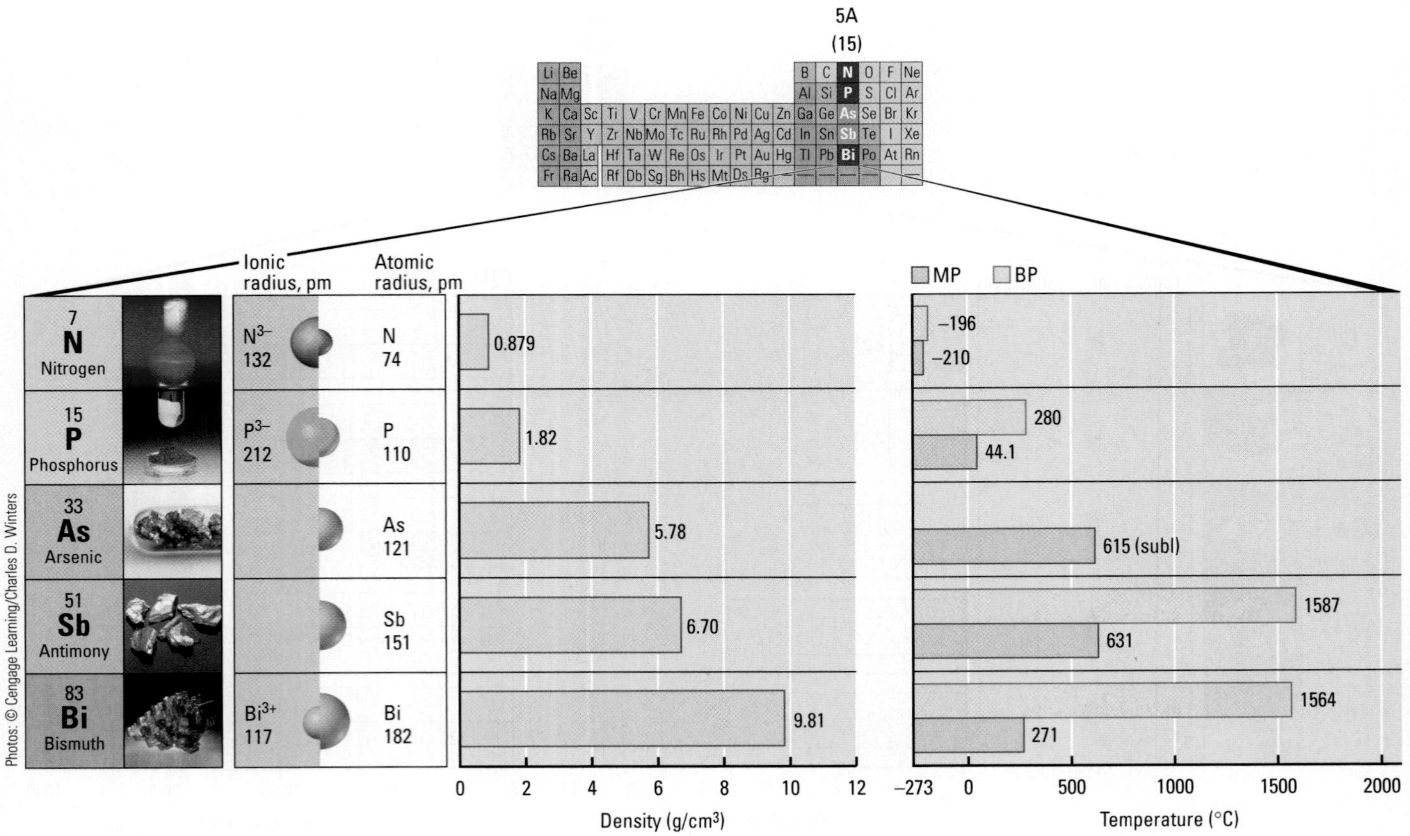

Group 5A(15) elements.

Nitrogen, phosphorus, and potassium are primary nutrients for plants. Although bathed in an atmosphere containing abundant nitrogen, most plants are unable to use the air directly as a supply of this vital element due to the energy required to break the N≡N triple bond, one of the strongest known. **Nitrogen fixation** is the process of changing atmospheric nitrogen into water-soluble compounds that can be absorbed through the plant roots, and assimilated by the plant. Nitrogen fixation is part of the **nitrogen cycle** (Figure 21.18), a natural cycle of chemical pathways involving nitrogen. In nitrogen fixation, nitrogen-fixing bacteria convert N_2 into NH_3. Ammonia is converted by other bacteria into nitrate, NO_3^-, which is used by plants. When an organism dies, bacteria reverse the process by converting nitrate to N_2 and organic nitrogen compounds to ammonia.

Most plants thrive on soils rich in nitrates, but many plants that grow in swamps, where there is a lack of oxidized materials, can use reduced forms of nitrogen such as the ammonium ion. The nitrate ion is the most highly oxidized form of combined nitrogen, and the ammonium ion is the most reduced form of nitrogen.

CONCEPTUAL EXERCISE 21.15 Chemically Combined Nitrogen

Show that the nitrate ion is the most highly oxidized form of combined nitrogen and that the ammonium ion is the most reduced form of combined nitrogen.

Nitrogen is fixed by natural processes on a massive scale in two ways. In the first method, nitrogen is oxidized under highly energetic conditions in the discharge of lightning or, to a lesser extent, in a fire. The initial atmospheric reaction

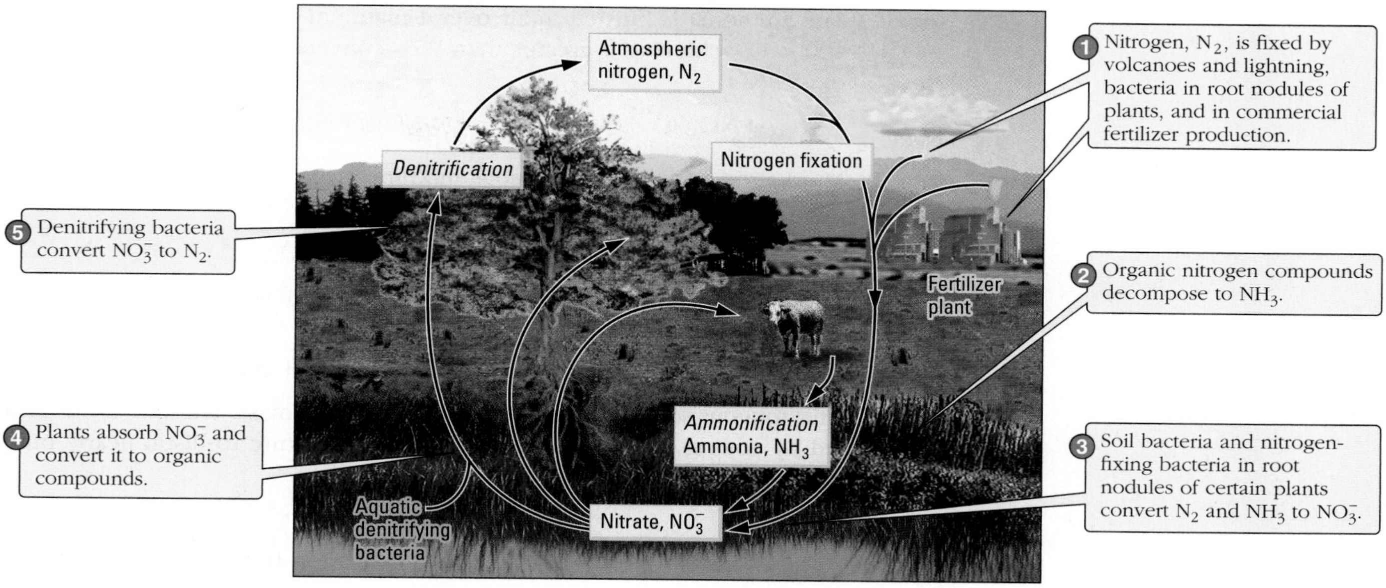

Figure 21.18 Natural nitrogen chemical pathways: the nitrogen cycle. Nitrogen-fixing bacteria convert N_2 into NH_3, which is converted by other bacteria into nitrate, NO_3^-, which is used by plants. When an organism dies, bacteria reverse the process, converting nitrate to N_2 and nitrogen compounds to ammonia.

is that of nitrogen with oxygen to form nitrogen monoxide, NO, a colorless, reactive gas.

$$N_2(g) + O_2(g) \rightleftharpoons 2\,NO(g) \qquad K_c = 1.7 \times 10^{-3} \text{ (at 2300 K)}$$

Once formed, nitrogen monoxide is easily oxidized in air to nitrogen dioxide, NO_2, which dissolves in water to form nitrous acid, HNO_2 and nitric acid, HNO_3.

$$H_2O(\ell) + 2\,NO_2(g) \longrightarrow HNO_2(aq) + HNO_3(aq)$$

These acids are readily soluble in rain, clouds, or ground moisture, and thus increase nitrogen concentration in soil. They also contribute to the formation of acid rain **(← *p. 825*)**.

In the second natural method of nitrogen fixation, bacteria that live on the roots of plants called *legumes,* such as clover, beans, and peas, convert atmospheric nitrogen into ammonia. This complex series of reactions depends on enzyme catalysis. Under ideal conditions, legume fixation can add more than 100 lb of nitrogen per acre of soil in one growing season.

The main industrial use of nitrogen at present is in the Haber-Bosch process **(← *p. 689*)**, which synthetically fixes nitrogen by combining it with hydrogen to form ammonia.

$$N_2(g) + 3\,H_2(g) \rightleftharpoons 2\,NH_3(g)$$

Millions of tons of ammonia are produced annually by this method. Pure gaseous ammonia is condensed and the liquid anhydrous ammonia is applied directly to fields as a fertilizer. Ammonia is also reacted with nitric acid to produce ammonium nitrate, the major solid fertilizer in the world.

About 15% of the ammonia made by the Haber-Bosch process is converted to nitric acid through a process developed by a German chemist, Wilhelm Ostwald. This three-step process is carried out at pressures of 1 to 10 atm.

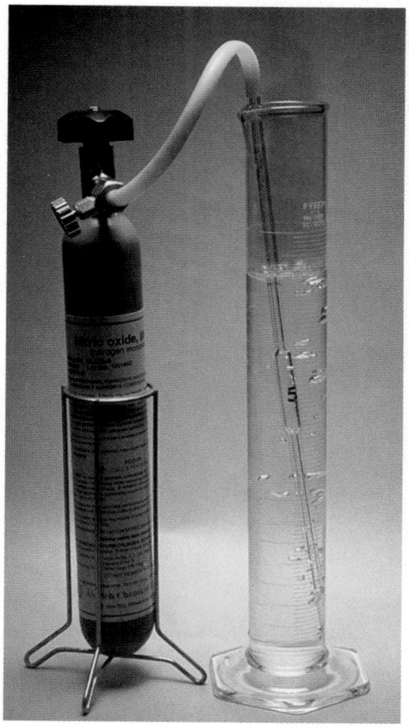

© Cengage Learning/Charles D. Winters

Conversion of colorless NO to red-brown NO_2. Colorless NO is bubbled through water. When NO emerges from the water into the air above the water, oxygen in the air reacts with NO to form reddish-brown nitrogen dioxide, NO_2.

U.S. production (2008):
NH_3, 9.6×10^6 tons;
NH_4NO_3, 7.3×10^6 tons

Step 1: The ammonia is burned in air over a platinum-rhodium catalyst at about 1000 °C, achieving a greater than 95% conversion of ammonia to nitric oxide, NO.

$$4\,NH_3(g) + 5\,O_2(g) \xrightarrow[\text{catalyst}]{1000\ °C} 4\,NO(g) + 6\,H_2O(g)$$

Step 2: More air is added to the gaseous mixture, which lowers the temperature and causes oxidation of NO.

$$2\,NO(g) + O_2(g) \rightleftharpoons 2\,NO_2(g)$$

Step 3: The nitrogen dioxide produced in the second step is passed through water to produce nitric acid.

$$3\,NO_2(g) + H_2O(\ell) \longrightarrow 2\,HNO_3(aq) + NO(g)$$

The resulting aqueous solution is about 60% nitric acid by mass. The anhydrous acid is produced by adding sulfuric acid and boiling the mixture to distill nearly pure nitric acid from it.

PROBLEM-SOLVING EXAMPLE 21.4 Another Nitrogen-Hydrogen Compound

Compounds of nitrogen and hydrogen other than ammonia exist. For example, *trans*-tetrazene, N_4H_4, has four nitrogen atoms in a chain with the terminal nitrogens each bonded to two hydrogen atoms. Write the Lewis structure of *trans*-tetrazene.

Answer

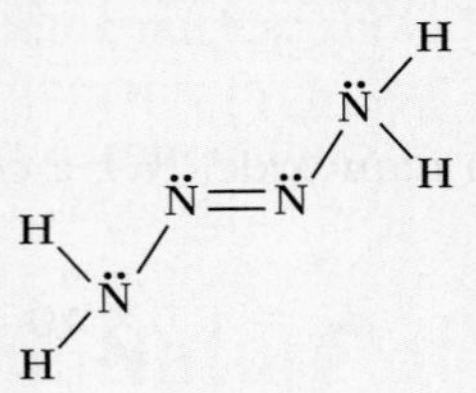

Strategy and Explanation The prefix *trans* indicates that the $-NH_2$ groups must be across the plane of a double bond from each other, an N=N double bond between the interior nitrogen atoms. The correct Lewis structure is shown above.

PROBLEM-SOLVING PRACTICE 21.4

Dinitrogen pentaoxide is the acid anhydride of nitric acid.

(a) Write the Lewis structure of dinitrogen pentaoxide.

(b) Write the balanced chemical equation for the formation of nitric acid from the reaction of dinitrogen pentaoxide with water.

PROBLEM-SOLVING EXAMPLE 21.5 Producing Nitric Acid

Consider the second step in the Ostwald process.

$$2\,NO(g) + O_2(g) \rightleftharpoons 2\,NO_2(g) \qquad \Delta H° = -113.0\ kJ$$

What would happen to the yield of NO_2 at equilibrium if

(a) The pressure were increased by decreasing the volume?

(b) The temperature were increased? (c) A catalyst were used?

Answer

(a) The yield should increase. (b) The yield should decrease. (c) No effect on yield.

Strategy and Explanation Apply Le Chatelier's principle (← ***p. 678***) in each case.

(a) The yield would increase because the pressure change favors the formation of fewer moles of gas: 3 mol gaseous reactants, 2 mol gaseous products.

(b) An increase in temperature favors the endothermic reverse reaction, decreasing the yield.

(c) A catalyst will speed up both the forward and reverse reactions equally, so it has no effect on yield.

PROBLEM-SOLVING PRACTICE **21.5**

Use Le Chatelier's principle to explain why in the manufacturing of nitric acid
(a) Lowering the temperature after Step 1 favors NO_2 formation.
(b) Coupling Step 2 with Step 3 of the Ostwald process favors the formation of nitric acid.

Richard Olivier/Corbis

Expanding air bags.

A Lifesaving Use of N_2: Automobile Air Bags Air bags save the lives of thousands of motorists annually in the United States. This is done through the application of a simple decomposition reaction of sodium azide, an ionic compound containing sodium ions and azide ions, N_3^-, which decomposes rapidly to liberate N_2 gas. An uninflated air bag has a small cylinder containing a carefully formulated mixture of the solids sodium azide, NaN_3, potassium nitrate, KNO_3, and silicon dioxide, SiO_2. When a car decelerates rapidly, as in a collision, a sensor sends an electrical signal to the mixture, rapidly igniting and decomposing the sodium azide and releasing nitrogen gas, which inflates the air bag.

$$2\ NaN_3(s) \longrightarrow 2\ Na(s) + 3\ N_2(g)$$

In addition to being used in automobile air bags, azides are used as explosives.

Residual sodium metal must be removed because it could react vigorously with water. Removal in this case is achieved by reacting the residual sodium with potassium nitrate in a reaction that produces additional nitrogen gas to inflate the air bag.

$$10\ Na(s) + 2\ KNO_3(s) \longrightarrow K_2O(s) + 5\ Na_2O(s) + N_2(g)$$

The heat released by these reactions melts the solid products and the silicon dioxide (sand), fusing them into an unreactive glass.

$$K_2O(s) + Na_2O(s) + SiO_2(s) \xrightarrow{\text{heat}} \text{glass}$$

EXERCISE **21.16 Expanding Air Bags**

Calculate the volume of N_2 released in an air bag at STP when 150. g sodium azide decompose.

The Discovery of Phosphorus, 1775, William Pether Engraving after Joseph Wright painting. Fisher Collections. Chemical Heritage Foundation. Photo by Will Brown.

The discovery of phosphorus in 1669 by Herman Brand by extraction from urine. In a very limited oxygen supply, white phosphorus glows with a greenish light, the source of the term "phosphorescence."

Phosphorus and Its Compounds Phosphorus has two main allotropes, *white* phosphorus and *red* phosphorus, which have very different properties. White phosphorus is highly reactive, igniting spontaneously in air at room temperature. For this reason, white phosphorus is stored under water. The waxy, nonpolar, solid white phosphorus is soft and easily cut, reflecting the fact that it consists of P_4 tetrahedra held together by weak noncovalent intermolecular forces ***(← p. 402)***. Because it is nonpolar, white phosphorus is not soluble in water, but dissolves readily in nonpolar liquids such as carbon disulfide, CS_2, or hexane, C_6H_{14}. Red phosphorus, unlike white phosphorus, does not ignite in air at room temperature and is much less toxic. A third, less common allotrope called black phosphorus is produced by heating white phosphorus at high pressure.

Although white phosphorus is toxic, phosphorus is an essential dietary mineral because of the many ways it is used by the body. Phosphorus is part of phosphate groups that link alternately with deoxyribose units to form the backbone of the DNA double helix ***(← p. 410)***. Phosphate anhydride linkages, which have this structure,

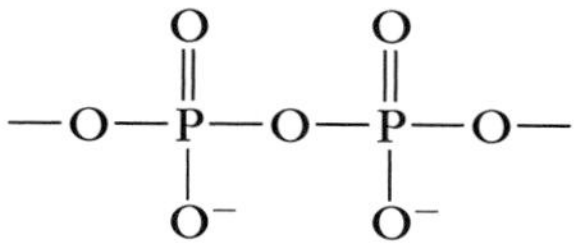

are responsible for how cellular energy is stored in ATP ***(← p. 877)***.

Michael Davidson/Photo Researchers, Inc.

ATP in cells (electron microscope photo).

CHEMISTRY IN THE NEWS Air-Stable White Phosphorus

White phosphorus, P_4, is typically stored under water to prevent the white phosphorus from reacting spontaneously with air at room temperature. A team of researchers from England and Finland recently developed an organic cage to fit around the P_4 molecules, which renders them unreactive to oxygen. The "host" cage, consisting of six organic iron-tethered organic groups, is just the right volume to accommodate the "guest" P_4 molecule. The initial research on the host compound used cyclohexane as the guest molecule. Jonathan Nitschke of Cambridge University, one the group co-leaders said, "We looked around to see what else we might be able to put in there, and we noticed that the volume of P_4 is about the same as that of cyclohexane. We gave it a try, and it turned out to be a very good guest." Rutgers University chemistry professor Ralf Warmuth said, ". . . the air insensitivity of encapsulated P_4 is remarkable."

This host-guest arrangement is stable because the reaction of P_4 with O_2 would create an intermediate product too large to fit into the host molecule's cavity. The caged P_4 doesn't decompose, even after contact with the atmosphere for more than four months. The caged P_4 can be removed from the host by adding a competing guest, such as benzene, which displaces the P_4, making it available for reaction. Potential applications for this host-guest system include using it to control the rate of release of P_4 into reaction systems, and regulating the reactivity of P_4 to produce useful phosphorus compounds.

Sources: Mal, P. "White Phosphorus Is Air Stable Within a Self-Assembled Tetrahedral Capsule." *Science*, Vol. 324, June 26, 2009; pp. 1697–1700.
Borman, S. "Controlling Flammability." *Chemical & Engineering News*, June 29, 2009; p 7. **www.cen-online.org**

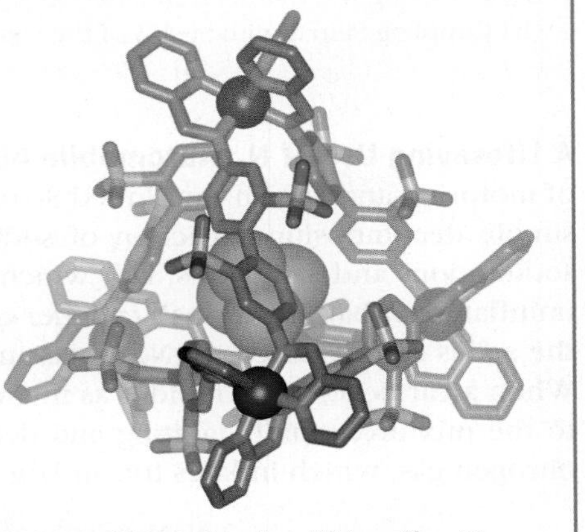

A "guest" P_4 molecule within a "host" cage of six organic groups. Fe, violet; N, blue; C, gray; O, red; S, yellow; P, orange. The cage has just the right volume for the tetrahedral P_4 to fit into it.

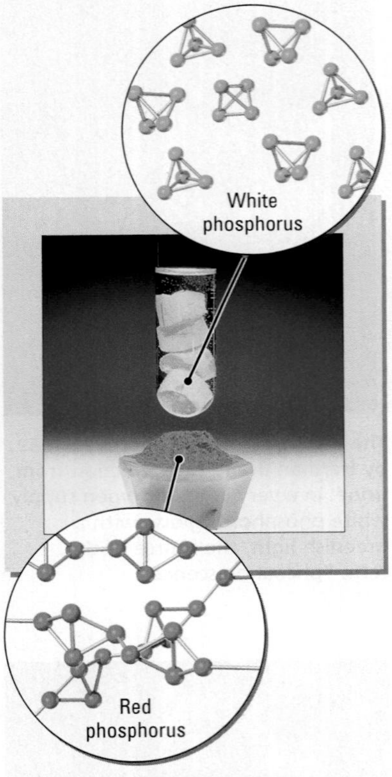

Photos: © Cengage Learning/Charles D. Winters

Phosphorus allotropes. Because white phosphorus *(top)* reacts with air at room temperature, it must be stored under water. Red phosphorus *(bottom)* does not react with air at room temperature.

Tooth enamel and bone contain the mineral hydroxyapatite, $Ca_5(PO_4)_3OH$. Water fluoridation reduces tooth decay because fluoride ions from the fluoridated water substitute for OH^- ions in tooth enamel to form fluoroapatite, $Ca_5(PO_4)_3F$, which is more resistant to decay than hydroxyapatite.

PROBLEM-SOLVING EXAMPLE 21.6 Phosphorus Pentachloride

Although the empirical formula for phosphorus pentachloride is PCl_5, in the solid state the compound actually consists of PCl_4^+ and PCl_6^- ions. Write the Lewis structures of these ions and predict their shapes.

Answer

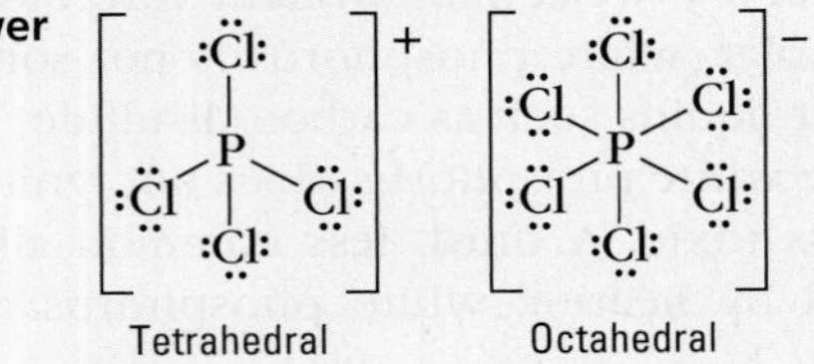

Strategy and Explanation Count the number of valence electrons and allow for the charge of the ion in each case. The PCl_4^+ ion has 32 valence electrons; PCl_6^- has 48 valence electrons. As a central atom, phosphorus can exceed an octet of electrons. The Lewis structures are shown above.

PROBLEM-SOLVING PRACTICE 21.6

Pure phosphoric acid, H_3PO_4, occurs only as a solid. When it melts, phosphoric acid units gradually lose water. Write the Lewis structure for the other product of this dehydration.

Arsenic, Antimony, and Bismuth Arsenic and antimony each have two allotropes, a metallic form and an amorphous form. Amorphous arsenic is yellow and, like white phosphorus, is soluble in carbon disulfide, where it exists as tetrahedral As_4 molecules. Antimony and arsenic are used to harden lead alloys such as those used in automobile batteries and in bullets. Arsenic compounds are poisonous, a fact often put to use in mystery novels. A 0.1 g or greater dose of As_2O_3 is fatal to humans. All of the trihalides of arsenic and antimony are known, such as AsF_3, a colorless liquid, and SbI_3, a red crystalline solid. By contrast, only four pentahalides are known: AsF_5, $AsCl_5$, SbF_5, and $SbCl_5$. That the other pentahalides have not been found is likely due to the strong oxidizing nature of the +5 oxidation state.

Bismuth is a yellowish metal that has limited uses. Chief among them are low-melting alloys for automatic sprinkler systems and electrical fuses. One physical property of bismuth is worth noting: it is one of the few substances that expands on freezing. Thus, bismuth was used in low-melting alloys to make printing type that expands on solidification when cast in a mold. The expanded metal gives a very sharp edge to the type, which produces a clear ink image on the printed page. Printing type made of these alloys was first used in the Middle Ages shortly after the initial Gutenberg printing press was developed (1440). Bismuth is commonly found in the lower +3 oxidation state, whereas the lighter elements (P, As, and Sb) are often found in the higher +5 oxidation state. This is another example of the inert pair effect, as noted previously for the heavier Group 3A and 4A elements.

Group 6A(16): Oxygen, Sulfur, Selenium, Tellurium, Polonium

Group 6A is the oxygen family of elements; S, Se, and Te are also known as the *chalcogens*. Some physical properties of the Group 6A elements are given below. Like the two preceding groups, Group 6A is diverse, consisting of true nonmetals (O, S, and Se), a metalloid (Te), and a metal (Po, a rare, radioactive element).

The term *chalcogen* is derived from the Greek word *khalos* for copper because copper ores contain compounds of these elements.

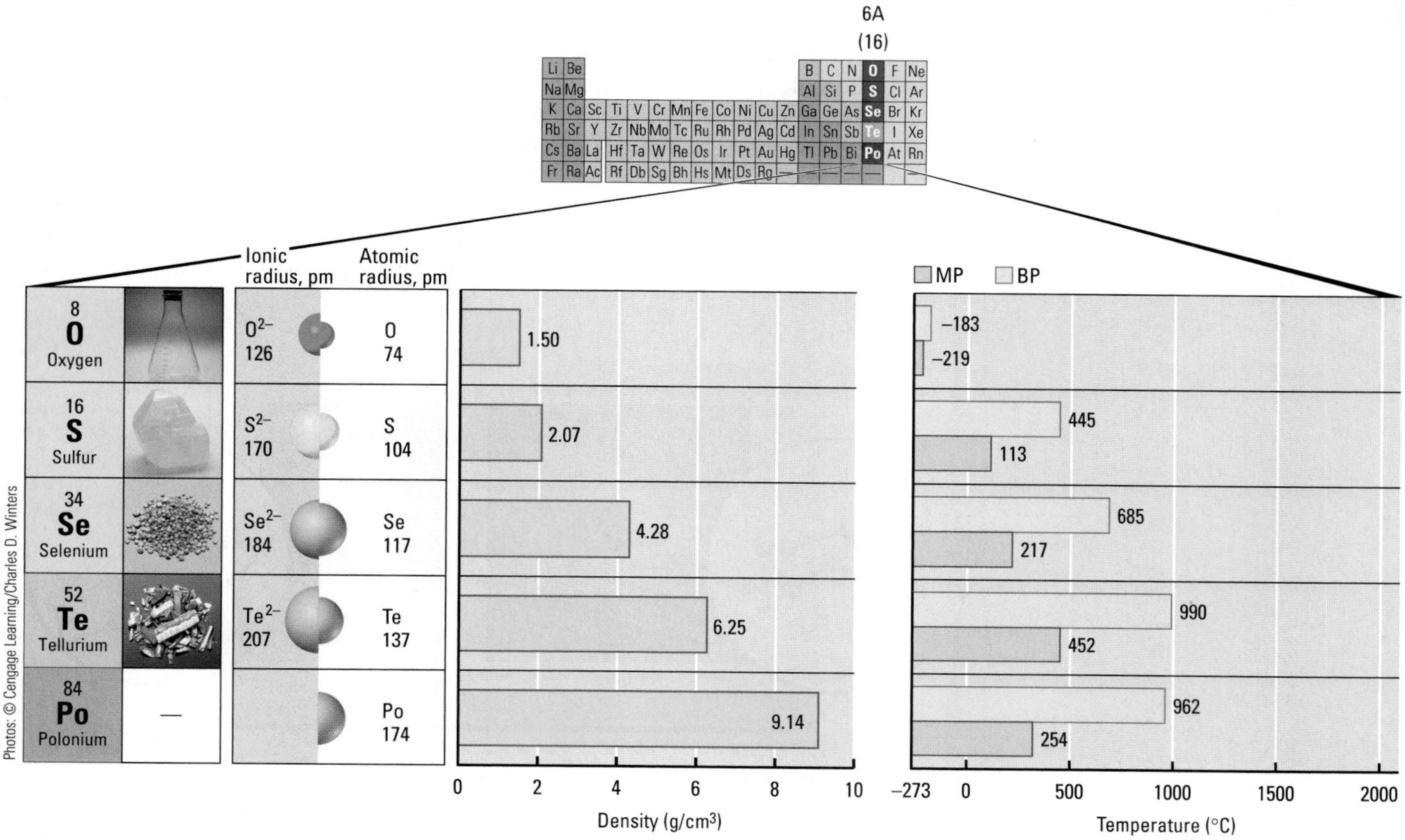

Group 6A(16) elements.

Properties and Uses of Oxygen Most of the oxygen produced by fractional distillation of liquid air is used as an oxidizing agent and in steel making (Section 22.2), although some is used in rocket propulsion and in controlled oxidation reactions of other types. Liquid oxygen (LOX) can be shipped and stored at its boiling temperature of −183 °C under atmospheric pressure. Cryogens, such as liquid nitrogen and liquid oxygen, present special hazards since contact with them produces instantaneous frostbite of flesh and brittleness in structural materials such as plastics, rubber gaskets, and some metals, causing materials to fracture easily at these low temperatures. The high oxygen concentration present in liquid oxygen can, in spite of the low temperature, accelerate oxidation reactions to the point of explosion. For this reason, contact between liquid oxygen and substances that will ignite and burn in air must be prevented.

Special cryogenic containers holding liquid oxygen incorporate huge vacuum-walled bottles much like those used to carry hot soup or hot coffee. These special containers can be seen outside hospitals and industrial complexes, on highways and railroads, and even aboard ocean-going vessels. In hospitals as well as in homes, supplemental oxygen is used to help patients who have difficulty breathing.

Most atmospheric oxygen comes from photosynthesis, in which green plants use energy from sunlight to convert water and carbon dioxide into glucose and oxygen. The concentration of oxygen in the atmosphere has upper and lower limits that are essential for our safety. If the concentration exceeds 25%, the rates of oxidation reactions would increase significantly, potentially endangering us by the increased rates of oxygen-requiring metabolic processes. With too little atmospheric oxygen, less than 17%, we would suffocate.

A cryogenic container of liquid oxygen.

Properties and Uses of Sulfur Sulfur exists in two common allotropic forms—rhombic (mp 115 °C) and monoclinic (mp 119 °C), both consisting of S_8 rings in the solid. When sulfur is heated above 150 °C, the S_8 rings break open, forming chains that become entangled, thereby increasing the viscosity of the molten sulfur. Upon continued heating, the color of sulfur changes from yellow to dark red because of unpaired electrons at the ends of the chains. If heated to 210 °C and poured into cold water, the sulfur forms an uncrystallized polymer called "plastic sulfur," which reverts back to the common crystalline forms at 25 °C (Figure 21.19).

Sulfur is a critical element in the body, necessary for the formation of methionine, an essential amino acid *(← p. 577)*. Sulfur atoms form —S—S— disulfide linkages among chains of amino acids; the disulfide linkages help to create the essential molecular shapes of proteins and enzymes *(← p. 581)*. Sulfur is also used to cross-link polymer chains in the vulcanization of rubber *(← p. 568)*. The sulfur helps to

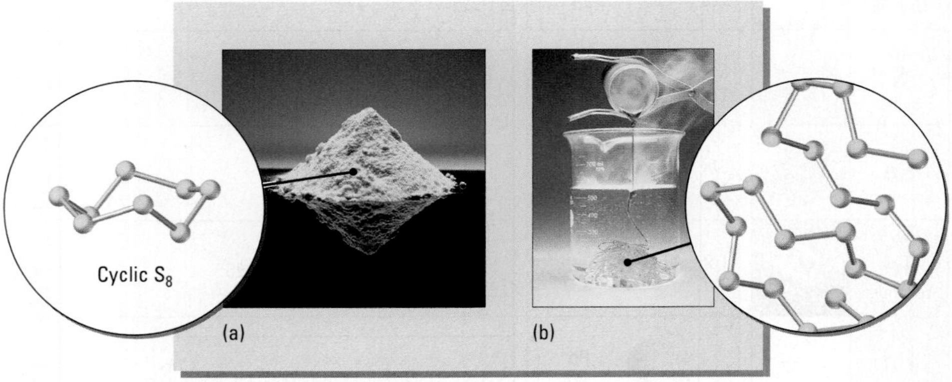

Active Figure 21.19 Sulfur allotropes. (a) At room temperature, sulfur is a bright yellow solid. At the nanoscale it consists of rings of eight sulfur atoms. (b) When melted, the rings break open to form long chains. Visit this book's companion website at **www.cengage.com/chemistry/moore** to test your understanding of the concepts in this figure.

align the polymer chains, which makes the rubber more elastic and prevents it from becoming sticky in warm weather.

Sulfuric Acid Production Most sulfur is used to produce sulfuric acid, the workhorse industrial chemical used in steel production, in automobile batteries, in the petroleum industry, and in the manufacture of fertilizers, plastics, drugs, dyes, and many other products. Since sulfuric acid costs less to make than any other acid, it is the first to be considered when an acid is needed in an industrial process.

Sulfur is converted to sulfuric acid in four steps, collectively called the *contact process.* In the first step, sulfur is burned in air to give mostly sulfur dioxide.

$$S_8(s) + 8\,O_2(g) \longrightarrow 8\,SO_2(g)$$

The SO_2 is then converted to SO_3 over a heated catalyst, such as platinum metal or vanadium(V) oxide.

$$2\,SO_2(g) + O_2(g) \xrightarrow{\text{catalyst}} 2\,SO_3(g)$$

The next step converts the sulfur trioxide to sulfuric acid by the addition of water. The best way to do this is to pass the SO_3 into H_2SO_4 to form pyrosulfuric acid, $H_2S_2O_7$, and then to dilute the $H_2S_2O_7$ with water. The net reaction is 1 mol H_2SO_4 for every 1 mol SO_3.

$$SO_3(g) + H_2SO_4(\ell) \longrightarrow H_2S_2O_7(\ell)$$

$$H_2S_2O_7(\ell) + H_2O(\ell) \longrightarrow 2\,H_2SO_4(aq)$$

$$\textit{Net reaction}: SO_3(g) + H_2O(\ell) \longrightarrow H_2SO_4(aq)$$

Sulfur dioxide for the contact process can also be obtained as a by-product from copper or lead smelting. Unless this sulfur dioxide is recovered, it pollutes the atmosphere ***(← p. 458).***

PROBLEM-SOLVING EXAMPLE 21.7 Sulfur and Sulfuric Acid

In a recent year, 1.3×10^{10} kg sulfur were produced in the United States. If all of this had been converted to sulfuric acid, calculate how many kilograms of sulfuric acid it would have produced.

Answer 4.0×10^{10} kg H_2SO_4

Strategy and Explanation The equations for the formation of sulfuric acid provide the mole-to-mole relationships for the conversion of S_8 to H_2SO_4. The net reaction is the formation of one mole of H_2SO_4 per mole of SO_3. Each mole of SO_3 requires 1 mol SO_2, and each mole of S_8 forms 8 mol SO_2.

$$1.3 \times 10^{10}\,\text{kg S}_8\left(\frac{10^3\,\text{g S}_8}{1\,\text{kg S}_8}\right)\left(\frac{1\,\text{mol S}_8}{256.5\,\text{g S}_8}\right)\left(\frac{8\,\text{mol SO}_2}{1\,\text{mol S}_8}\right)\left(\frac{1\,\text{mol SO}_3}{1\,\text{mol SO}_2}\right)$$
$$= 4.05 \times 10^{11}\,\text{mol SO}_3$$

This amount can be converted to kilograms of sulfuric acid.

$$4.05 \times 10^{11}\,\text{mol SO}_3\left(\frac{1\,\text{mol H}_2\text{SO}_4}{1\,\text{mol SO}_3}\right)\left(\frac{98.08\,\text{g H}_2\text{SO}_4}{1\,\text{mol H}_2\text{SO}_4}\right)\left(\frac{1\,\text{kg H}_2\text{SO}_4}{10^3\,\text{g H}_2\text{SO}_4}\right)$$
$$= 4.0 \times 10^{10}\,\text{kg H}_2\text{SO}_4$$

☑ **Reasonable Answer Check** Sulfuric acid is approximately 33% sulfur by mass ([32 g S/98 g H_2SO_4] × 100%). Therefore, 33 kg sulfur forms about 100 kg sulfuric acid, and 1×10^{10} kg S forms about 3×10^{10} kg sulfuric acid, which is close to the calculated value of 4.0×10^{10} kg H_2SO_4.

PROBLEM-SOLVING PRACTICE 21.7

Calculate the number of kilograms of SO_3 produced from 1.3×10^{10} kg S_8.

Group 7A(17): The Halogens

The halogens exhibit a trend in physical states not found in any other group: The lightest halogen, fluorine, is a gas and the heaviest, iodine, is a solid at room temperature. Some physical properties of the Group 7A elements are given below. At room temperature, fluorine and chlorine, the first two elements, are diatomic pale yellow and yellow-green gases, respectively; bromine, the next element, is a reddish-brown liquid; and then iodine is a violet-black solid. These physical states reflect the increasing strengths of noncovalent intermolecular forces with increasing numbers of electrons in the diatomic molecules **(← *p. 403*)**. Astatine is an intensely radioactive element with isotopes of very short half-lives, accounting for the scarcity of astatine in nature.

All halogens have an ns^2np^5 outer electron configuration that leads to a gain of one electron to form a 1− ion or a sharing of one electron in a pair to complete an octet. Fluorine is extremely reactive, evidence of its exceptionally high electronegativity, the highest of any element, and of the very weak F—F bond in F_2 (bond enthalpy 158 kJ/mol). The halogens are oxidizing agents in many reactions including the oxidation of a heavier halide ion by a lighter halogen. For example, chlorine is a strong oxidizing agent that oxidizes bromide ions to diatomic bromine.

$$Cl_2(g) + 2\,Br^-(aq) \longrightarrow 2\,Cl^-(aq) + Br_2(aq)$$

This reaction is used to extract bromine from seawater, which contains 30 ppm bromide ions (Section 21.5). Some uses of the halogens and their compounds are given in Table 21.10.

Chlorine, a toxic gas with an irritating odor, is the most important halogen used in industry. Chlorine is used to purify water **(← *p. 743*)**, to bleach paper and textiles, to manufacture herbicides, insecticides, and other chlorinated organic

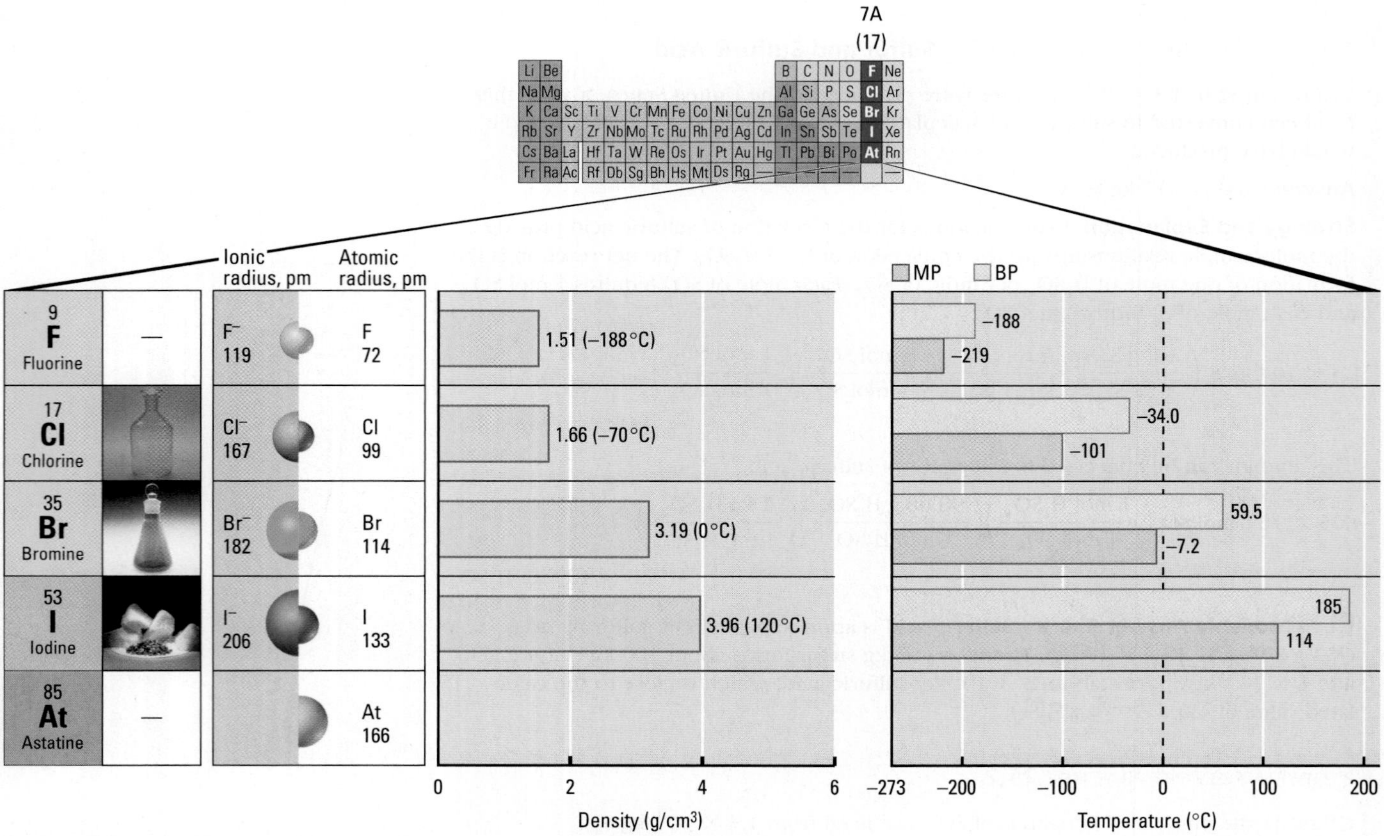

Group 7A(17) elements, the halogens.

Table 21.10 Uses of Some Halogen Compounds

Halogen Compound	Uses
Hydrogen fluoride	Frosting light bulbs and television tubes, production of uranium hexafluoride
Uranium hexafluoride, UF_6	Separation of fissionable U-235 from U-238 during processing of uranium ores
Hydrochloric acid, HCl	Magnesium manufacturing, manufacture of vinyl chloride and chlorinated solvents; human stomach acid (0.1 M)
Ammonium perchlorate, NH_4ClO_4	Solid propellant for Space Shuttle
Methyl bromide, CH_3Br	Pesticide and fire extinguisher
Potassium iodide, KI	Salt additive to prevent goiter, a thyroid condition

compounds, to produce polyvinyl chloride, and to extract titanium metal from its ores.

Bromine is the only nonmetal that is a liquid at room temperature. It is used to prepare methyl bromide, CH_3Br, an efficient fire extinguisher and pesticide.

Iodine is the only common halogen that is a solid at room temperature. It is a gray-black metallic-looking solid that sublimes to a violet-colored vapor. Iodine was discovered by burning dried seaweed, which contains a relatively high concentration of iodide ions. Iodine is an essential dietary mineral for humans because the iodide ions are necessary for the production of thyroxine, a growth-controlling hormone produced by the thyroid gland. Insufficient dietary iodine causes enlargement of the thyroid gland, a condition known as *goiter.* Potassium iodide (0.01%) is added to table salt (iodized salt) to prevent goiter.

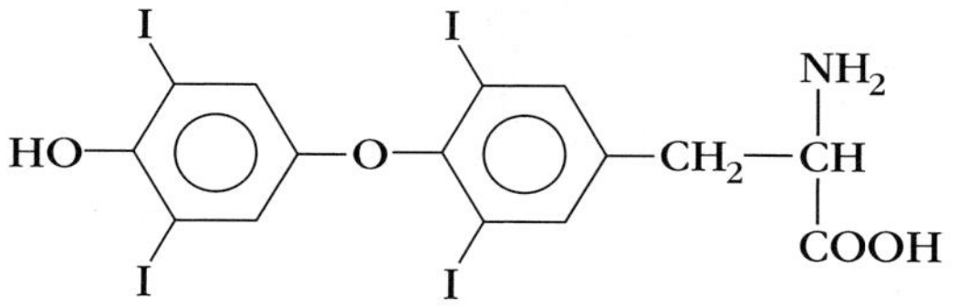
I
I
HO
O
NH2
CH2
CH
COOH
I
I

thyroxine

Ice
Condensed
I2(s)
I2(g)
I2(s)

© Cengage Learning/Charles D. Winters

Sublimation of iodine. Solid iodine is a gray, metallic-looking solid. When heated, it converts directly by sublimation into gaseous iodine, a purple vapor.

Because it is so rare in nature, astatine or its compounds are not available in easily manipulated amounts. Elegant tracer experiments have established the existence of the astatide ion, At^-, in keeping with Group 7A behavior.

PROBLEM-SOLVING EXAMPLE 21.8 Reactive Fluorine

In a Teflon vessel, water reacts with hypofluorous acid, HOF, to produce hydrogen fluoride, hydrogen peroxide, and oxygen.

(a) Write a balanced chemical equation for this reaction.

(b) Why is a glass reaction vessel not used?

Answer

(a) $4\ HOF(aq) + 2\ H_2O(\ell) \longrightarrow 4\ HF(aq) + H_2O_2(aq) + O_2(g)$

(b) The HF produced would etch the glass vessel and break it.

Strategy and Explanation

(a) Balance the equation by writing the correct formulas and coefficients for the reactants and products. Hypofluorous acid is analogous to hypochlorous acid, HOCl.

$$4\ HOF(aq) + 2\ H_2O(\ell) \longrightarrow 4\ HF(aq) + H_2O_2(aq) + O_2(g)$$

(b) Teflon is used because it is nonreactive. Glass would react with the HF produced by the reaction.

PROBLEM-SOLVING PRACTICE 21.8

Classify each of these equations as being either a redox reaction or a non-redox reaction. If a redox reaction occurs, also identify the oxidizing agent and the reducing agent.

(a) $2\ NaF(s) + H_2SO_4(aq) \longrightarrow 2\ HF(g) + Na_2SO_4(aq)$

(b) $S_8(s) + 24\ F_2(g) \longrightarrow 8\ SF_6(g)$

Group 8A(18): The Noble Gases

Some physical properties of the Group 8A elements are given below. Once called the "rare gases," these elements are not rare. Though relatively low in terrestrial abundance, helium is the second most abundant element in the universe. Argon makes up nearly 1% of Earth's atmosphere.

Evidence of helium was first obtained in 1868 when a new yellow line was observed spectroscopically from Earth during an eclipse of the sun. Proof of the existence of terrestrial helium first came from observation of the same line in the spectrum of gases released during the eruption of Mount Vesuvius in 1881. Subsequently, the remaining noble gases except radon were isolated from liquid air and characterized by J. Rayleigh, W. Ramsay, and M. Travers between 1895 and 1898. Radon was first isolated and identified in 1902 by Rutherford and Soddy. These findings led to the addition of a new group, 8A, to the periodic table.

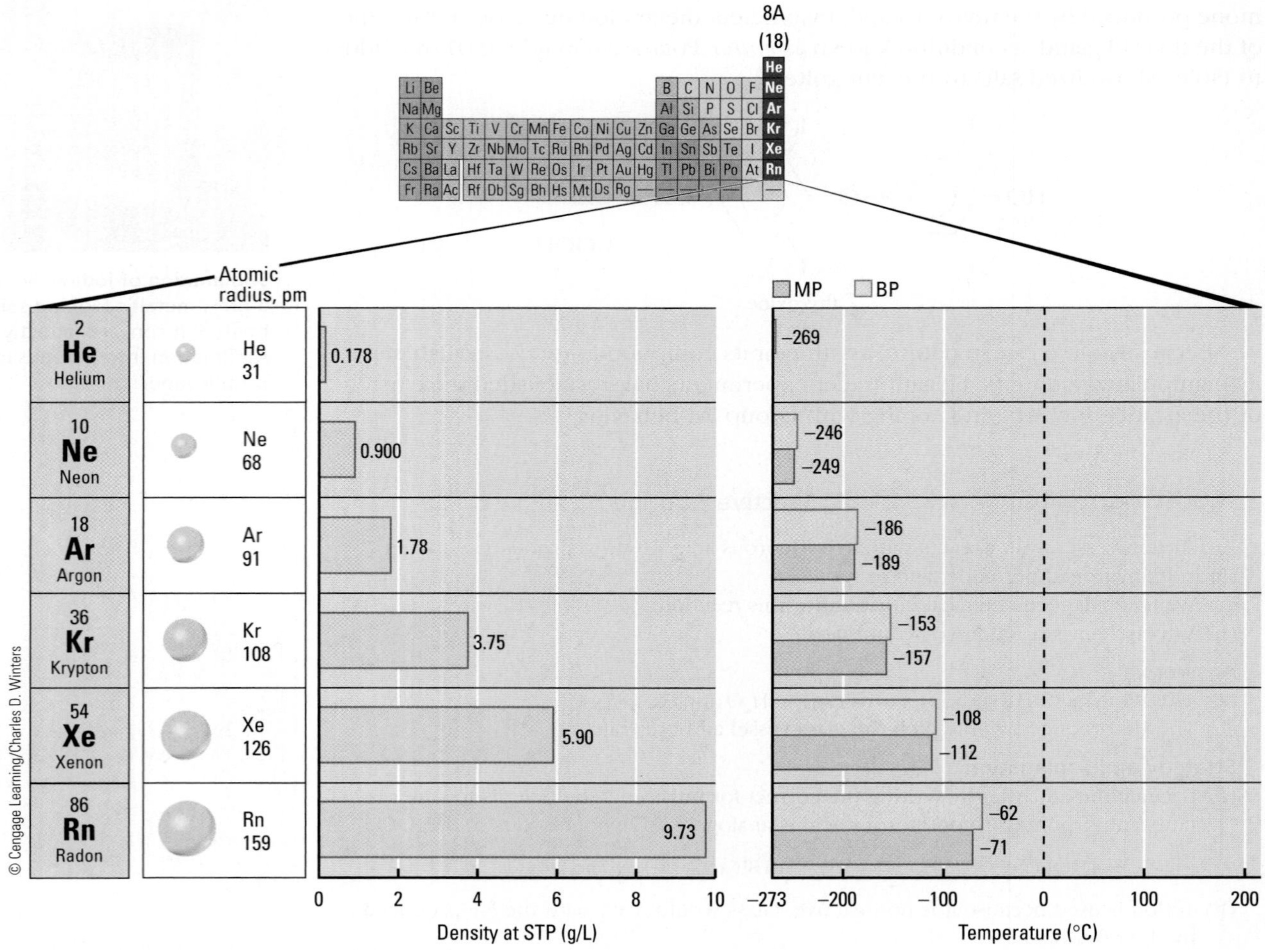

Group 8A(18) elements, the noble gases.

The reluctance of Group 8A elements to form compounds is to be expected given their complete octets of ns^2np^6 outer electrons (only ns^2 for helium). In fact, it was a long-standing accepted chemical truism that these elements, then called the "inert elements," formed no compounds. That view was overthrown dramatically in 1962 with the synthesis of a series of xenon compounds. Since then, compounds of xenon containing oxygen and fluorine have been synthesized, as well as fluorides of krypton and radon. Spectroscopic evidence exists for the argon compound HArF.

Helium is used as an inert atmosphere in welding, as a nitrogen substitute in gas mixtures for underwater diving, as a liquid cryogen, and to fill meteorological balloons. Argon is used as a protective atmosphere for arc-welding aluminum, in airport runway lights, and as the gas in incandescent light bulbs, where it decreases the evaporation of the tungsten metal filament.

SUMMARY PROBLEM

(a) In 2008, the United States used about 7.1×10^9 barrels of crude oil, enough for 24 barrels per person (1 barrel = 42 gallons; density of crude oil = 0.83 g/mL). Assume that the crude oil is 3% sulfur by mass and that all of the sulfur was removed from the crude oil before it was used. Calculate how many liters of SO_2 at 25 °C and 1 atm from just your share of crude oil in 2008 would have been prevented from entering the atmosphere.

(b) Consider the conversion of $SO_2(g)$ to $SO_3(g)$.

$$SO_2(g) + \tfrac{1}{2}O_2(g) \longrightarrow SO_3(g)$$

(i) Use data from Appendix J to calculate $\Delta H°$ for this reaction.

(ii) The reaction reaches equilibrium. Identify the effect on the concentration of sulfur trioxide produced if these conditions are applied to the equilibrium.

1. The pressure is increased.
2. The temperature is decreased.
3. A catalyst is used.
4. Sulfur dioxide is added.
5. Sulfur trioxide is removed as it forms.

(iii) At 1000. K, there initially were 0.250 mol SO_2, 0.125 mol O_2, and no SO_3 in a 10.0-L reaction chamber. At equilibrium there are 0.136 mol SO_3 and 0.00570 mol O_2. Calculate the equilibrium constant.

(c) Using data from Appendix J, calculate $\Delta G°$ for the conversion of sulfur dioxide gas to sulfur trioxide gas. Then, calculate the value for K_P at 800. °C and at 1000. °C.

IN CLOSING

Having studied this chapter, you should be able to . . .

- Give a general explanation of how elements form in stars (Section 21.1).
- Know the principal elements in Earth's crust (Section 21.2).
- Describe the general structure of silicates (Section 21.2).

OWL and go Chemistry

Sign in at **www.cengage.com/owl** to:

- View tutorials and simulations, develop problem-solving skills, and complete online homework assigned by your professor.
- For quick review and exam prep, download Go Chemistry mini lecture modules from OWL (or purchase them at **www.CengageBrain.com**).

- Identify the general methods by which elements are extracted from Earth's crust (Section 21.2). End-of-chapter questions: 20, 22, 28
- Identify the major components of the atmosphere (Section 21.3).
- Apply chemical principles to the processes for extracting and purifying elements, and the reactions they undergo (Sections 21.3-21.5). Questions 24, 34, 51
- Describe the Frasch process for obtaining sulfur (Section 21.3).
- Describe how electrolysis is used to obtain sodium, chlorine, magnesium, and aluminum (Section 21.4). Questions 20, 22
- Explain how chemical redox reactions are used to extract bromine, iodine, and phosphorus from compounds (Section 21.5).
- Explain how elements are obtained by the liquefaction of air (Section 21.6).
- Explain how sulfuric acid is produced (Section 21.6).
- Interpret the periodic trends among the main group elements (Section 21.6).

KEY TERMS

catenation *(Section 21.6)*

chlor-alkali process *(21.4)*

cryogen *(21.6)*

Frasch process *(21.3)*

helium burning *(21.1)*

hydrogen burning *(21.1)*

mineral *(21.2)*

nitrogen cycle *(21.6)*

nitrogen fixation *(21.6)*

nuclear burning *(21.1)*

ore *(21.2)*

QUESTIONS FOR REVIEW AND THOUGHT

OWL Interactive versions of these problems are assignable in OWL.

Blue-numbered questions have short answers at the back of this book in Appendix M and fully worked solutions in the *Student Solutions Manual.*

Review Questions

These questions test vocabulary and simple concepts.

1. What is meant by hydrogen burning and helium burning in relation to the formation of elements?
2. Identify the most abundant nonmetallic element in Earth's crust. Identify the most abundant metallic element in Earth's crust.
3. Describe the difference between an ore and a mineral.
4. Give a simple explanation for the abundance of clay minerals in Earth's crust.
5. Differentiate among pyroxenes, amphiboles, and silica.
6. Explain how the silicate unit in amphiboles has the general formula $Si_4O_{11}^{6-}$.
7. Explain how the silicate unit in mica and other sheet silicates has the general formula $Si_2O_5^{2-}$.
8. Identify two major differences between white phosphorus and red phosphorus.
9. Identify
   (a) Two elements obtained from the atmosphere.
   (b) Two elements obtained from the sea.
   (c) Two elements obtained from Earth's crust.
10. Write balanced equations for the recovery of magnesium from seawater. Begin with the precipitation of magnesium hydroxide by addition of calcium hydroxide to seawater.
11. Why are nitrogen and oxygen important industrial chemicals?
12. Describe how nature fixes nitrogen. Why is nitrogen fixation necessary?
13. Briefly explain why different products are obtained from the electrolysis of molten NaCl and the electrolysis of aqueous NaCl.
14. Identify two uses of phosphate rock.
15. Describe the structural changes that occur in sulfur as it is heated from room temperature to 210 °C.
16. Identify the substance or substances produced by each of these commercial processes.
    (a) Hall-Héroult (b) Contact
    (c) Ostwald (d) Dow
17. Identify the substance or substances produced by each of these commercial processes.
    (a) Frasch (b) Chlor-alkali
    (c) Downs cell
18. Why is phosphate rock not applied directly as a phosphorus fertilizer?

Topical Questions

These questions are keyed to the major topics in the chapter. Usually a question that is answered at the back of this book is paired with a similar one that is not.

Electrolytic Methods (Section 21.4)

19. (a) Write the balanced chemical equation for the electrolysis of aqueous NaCl.
    (b) In 2002, 8.98×10^9 kg NaOH and 1.14×10^{10} kg chlorine were produced in the United States. Does the ratio of these masses agree with the ratio of masses from the balanced chemical equation? If not, what does that suggest about the ways that NaOH and Cl_2 are produced?
20. To produce magnesium metal, 1000. kg molten $MgCl_2$ are electrolyzed.
    (a) At which electrode is magnesium produced?
    (b) What is produced at the other electrode?
    (c) Calculate how many moles of electrons are used in the process.
    (d) An industrial process uses 8.4 kWh per pound of Mg. Calculate how much energy is required per mole of magnesium.
21. A Downs cell operates at 7.0 V and 4.0×10^4 A.
    (a) Calculate how much Na(s) and Cl_2(g) can be produced in 24 hours by such a cell.
    (b) Assuming 100% efficiency, calculate the energy consumption (kWh) of this cell.
22. Calculate how much energy (kWh) is required to prepare a ton of sodium in a typical Downs cell operating at 25,000 A and 7.0 V.
23. Calculate the mass in grams of aluminum that can be produced when 6.0×10^4 A is passed through a series of 100 Hall-Héroult electrolytic cells operating at 85% efficiency for 24 hours.
24. Calculate what mass in grams of aluminum can be produced from the electrolysis of molten $AlCl_3$ in an electrolytic cell operating at 100. A for 2.00 hr.

General Questions

These questions are not explicitly keyed to chapter topics; many require integration of several concepts.

25. Complete this table.

Formula	Name	Oxidation State of Nitrogen
________	Nitrogen	________
NH_3	________	________
________	Nitrous acid	________
________	Nitrogen dioxide	________
NH_4^+	________	________
________	Ammonium nitrate	________

26. Complete this table.

Formula	Name	Oxidation State of Phosphorus
________	Phosphorus	________
$(NH_4)_2HPO_4$	________	________
________	Phosphoric acid	________
________	Tetraphosphorus decaoxide	________
$Ca_3(PO_4)_2$	________	________
________	Calcium dihydrogen phosphate	________

27. Molten NaCl is electrolyzed in a Downs cell operating at 1.00×10^4 A for 24 hr.
    (a) Calculate the mass in grams of sodium produced.
    (b) Calculate the volume of Cl_2 in liters that is collected from the outlet tube at 20. °C and 15 atm.
28. Bauxite, the principal source of aluminum oxide, contains 55% Al_2O_3. Calculate how much bauxite is required to produce the 5.0×10^6 tons of Al produced annually by electrolysis.
29. Write a plausible Lewis structure for azide ion, N_3^-.
30. Write a plausible Lewis structure for P_4O_{10}.
31. Write the chemical equation for the
    (a) Combustion of white phosphorus.
    (b) Reaction of the combustion product with water.
32. Calculate the temperature at which the conversion of white phosphorus to red phosphorus occurs. $\Delta H° = -17.6$ kJ/mol; $\Delta S° = -18.3$ J(mol K) at 25 °C.
33. There are two common oxides of sulfur. Name these oxides, and write chemical equations for the reaction of each with water. Identify the products.
34. What raw materials are used to produce sulfuric acid? Write chemical equations to represent the steps in the contact process to produce sulfuric acid.
35. Write Lewis structures for all the resonance forms of sulfuric acid.
36. Write Lewis structures for all the resonance forms of nitric acid.
37. Iodine trichloride, ICl_3, is an interhalogen compound.
    (a) Write the Lewis structure of ICl_3.
    (b) Does the central atom have more than an octet of valence electrons?
    (c) Using VSEPR theory, predict the molecular shape of ICl_3.
38. At some temperature, a gaseous mixture in a 1.00-L vessel originally contained 1.00 mol SO_2 and 5.00 mol O_2. When equilibrium was reached, 77.8% of the SO_2 had been converted to SO_3. Calculate the equilibrium constant (K_c) for this reaction at this temperature.
39. White phosphorus is soluble in carbon disulfide. At 10. °C, 800. g P_4 dissolves in 100. g CS_2. The density of carbon disulfide at 10. °C is 1.26 g/mL.
    (a) Calculate the molarity of phosphorus in the solution.
    (b) Calculate the molality of phosphorus in the solution.
40. Drug stores sell 3% aqueous hydrogen peroxide that is used as an antiseptic. Hydrogen peroxide, H_2O_2, decomposes to water and oxygen. Calculate the volume of

oxygen produced if 250. mL of 3% hydrogen peroxide decomposes fully at 750. mm Hg and 22 °C.

41. Calculate the volume of concentrated (98%) sulfuric acid that is needed to produce two tons of phosphoric acid from the reaction of sulfuric acid with sufficient phosphate-bearing rock. Density of conc. sulfuric acid = 1.84 g/mL.

$$Ca_3(PO_4)_2(s) + 3\,H_2SO_4(\ell) \longrightarrow 2\,H_3PO_4(\ell) + 3\,CaSO_4(s)$$

Applying Concepts

These questions test conceptual learning.

42. The K_{sp} of $Ca(OH)_2$ is 5.5×10^{-6}; that of $Mg(OH)_2$ is 1.8×10^{-11}. Calculate the equilibrium constant for the reaction

$$Ca(OH)_2(s) + Mg^{2+}(aq) \longrightarrow Ca^{2+}(aq) + Mg(OH)_2(s)$$

Use it to explain why this reaction can be used commercially to extract magnesium from seawater.

43. Commercial concentrated nitric acid contains 69.5 mass percent HNO_3 and has a density of 1.42 g/mL.
(a) Calculate the molarity of this solution.
(b) Calculate what volume of the concentrated acid must be used to prepare 10.0 L of 6.00-M HNO_3.

44. The compound nitrosyl azide, N_4O, is a covalent compound with an NNNNO atomic arrangement. Write a plausible Lewis structure for this compound.

45. Hydrazoic acid, HN_3, is very explosive in its pure state but can be studied in aqueous solution. The acid is prepared by the reaction of hydrazine with nitrous acid.

$$N_2H_4(\ell) + HNO_2(aq) \longrightarrow HN_3(aq) + 2\,H_2O(\ell)$$

(a) Determine the oxidation states of nitrogen in the compounds in this reaction.
(b) What is the oxidizing agent in this reaction?
(c) The K_a of hydrazoic acid is 1.0×10^{-5} at 25 °C. Calculate the pH of a 0.010-M solution of HN_3.

46. (a) Write two plausible resonance structures for hydrazoic acid, HN_3.
(b) Use bond energy data (Table 8.2) to calculate the ΔH_f° for each resonance form; $\Delta H_f^\circ = 218.0$ kJ/mol for H(g) and 472.7 kJ/mol for N(g). (See Appendix J.)

47. Given the reaction

$$Cl_2(g) + H_2O(\ell) \rightleftharpoons H^+(aq) + Cl^-(aq) + HOCl(aq)$$

(a) Identify the oxidizing agent and the reducing agent.
(b) Write the equilibrium constant expression for the reaction.
(c) Calculate the concentration of HOCl in equilibrium with $Cl_2(g)$ at 1.0 atm. $K_c = 2.7 \times 10^{-5}$.

48. Dinitrogen trioxide, N_2O_3, is a blue liquid formed by the reaction of NO_2 and NO.
(a) Write a balanced chemical equation for the formation of N_2O_3.
(b) Write the Lewis structure of N_2O_3 and any plausible resonance forms.
(c) Predict the O—N—O and the N—N—O bond angles.

49. (a) Write the resonance forms of SO_3.
(b) Predict the molecular shape of SO_3 and the O—S—O bond angle.

50. Iodine can be produced by the oxidation of iodide ion with permanganate ion.

$$MnO_4^- + 2\,I^-(aq) + 8\,H^+(aq) \longrightarrow I_2(s) + Mn^{2+}(aq) + 4\,H_2O(\ell)$$

Excess HI is added to 0.200 g MnO_4^-. Assuming 100% yield, calculate how many grams of iodine are produced.

51. In a Downs cell, molten NaCl is electrolyzed to sodium metal and chlorine gas.

$$2\,NaCl(\ell) \longrightarrow 2\,Na(\ell) + Cl_2(g)$$

ΔH° and ΔS° for the reaction are +820 kJ and +180 J/K, respectively.
(a) Calculate ΔG° at 600. °C, the electrolysis temperature.
(b) Calculate the voltage required for the electrolysis.

52. A 425-gal tank of water contains 175 g NaI. Calculate the number of liters of chlorine gas at 758 mm Hg and 25 °C required to convert all the iodide to iodine.

53. The phase diagram for sulfur is given below. The solid forms of sulfur are rhombic and monoclinic.

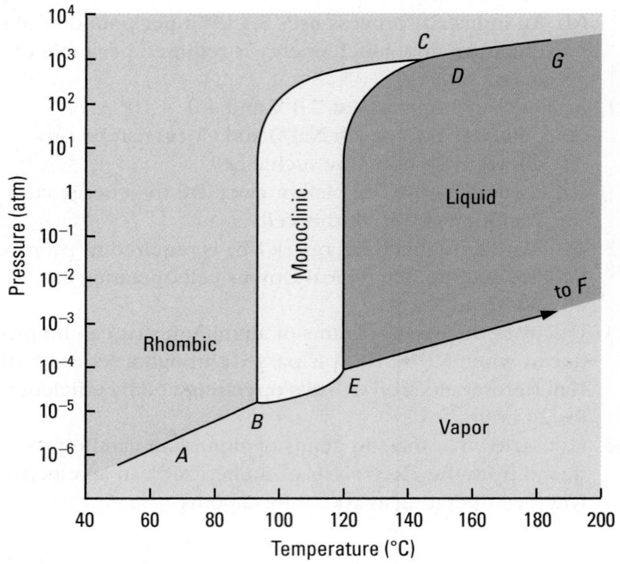

Phase diagram of sulfur.

(a) How many triple points does sulfur have? Indicate the approximate temperature and pressure at each.
(b) Which physical states are present at equilibrium under these conditions?
(i) 10^{-2} atm and 80 °C
(ii) 10^{-1} atm and 140 °C
(iii) 10^{-3} atm and 110 °C
(iv) 10^{-4} atm and 160 °C

54. Refer to the phase diagram in Question 53. Explain how you would sublime monoclinic sulfur.

55. In the laboratory, small amounts of bromine can be produced by the reaction of hydrobromic acid with MnO_2. The unbalanced equation for the reaction is

$$MnO_2(s) + H_3O^+(aq) + Br^-(aq) \longrightarrow Mn^{2+}(aq) + Br_2(\ell) + H_2O(\ell)$$

(a) Balance the equation.
(b) If the reaction is in 100% yield, calculate how many moles of bromide ions must react to produce 6.50 g bromine.

(c) Calculate how many grams of MnO_2 are required in part (b).

56. Lapis lazuli is an aluminum silicate whose brilliant blue color is due to the presence of S_3^- ions. Write a plausible Lewis structure for this ion.

Lapis lazuli jewelry.

57. Bromine is prepared by bubbling 0.240 mol gaseous chlorine into 0.500 L of a solution that is 0.500 M in bromide ion.
    (a) Determine the limiting reactant.
    (b) Calculate the theoretical number of moles of bromine that could be produced.
    (c) If 0.124 mol bromine is produced, calculate the percent yield.
58. Magnesium can be extracted from dolomite, a mineral that contains 13.2% Mg, 21.7% Ca, 13.0% C, and 52.1% O. Determine the simplest formula for this ionic compound.
59. A natural brine found in Arkansas has a bromide ion concentration of 5.00×10^{-3} M. If 210. g Cl_2 were added to 1.00×10^3 L of the brine,
    (a) Determine the limiting reactant.
    (b) Calculate the theoretical yield of Br_2 ($d = 3.12$ g/mL).
60. Chlorine gas was first prepared by Karl Scheele in 1774 by the reaction of sodium chloride, manganese(IV) oxide, and sulfuric acid. In addition to chlorine, the reaction produces water, sodium sulfate, and manganese(II) sulfate. Write the balanced equation for this reaction.
61. Mercury(II) azide, $Hg(N_3)_2$, is an unstable compound used as a detonator in blasting caps. Calculate the volume (L) of nitrogen produced at 1 atm and 25 °C when 2.50 g mercury azide decomposes to liquid mercury and nitrogen.

More Challenging Questions

These questions require more thought and integrate several concepts.

62. Use the Clausius-Clapeyron equation to calculate the heat of sublimation (J/mol) of monoclinic sulfur. See Question 53.
63. At 20. °C the vapor pressure of white phosphorus is 0.0254 mm Hg; at 40. °C it is 0.133 mm Hg. Use the Clausius-Clapeyron equation to estimate the heat of sublimation (J/mol) of white phosphorus.
64. The density of sulfur vapor at 700. °C and 1.00 atm is 0.8012 g/L. What is the molecular formula of sulfur in the vapor?
65. Assume that the radius of Earth is 6400 km, the crust is 50. km thick, the density of the crust is 3.5 g/cm^3, and 25.7% of the crust is silicon by mass. Calculate the total mass of silicon in the crust of Earth.
66. A 5.00-g sample of white phosphorus is burned in excess oxygen and the product is dissolved in sufficient water to form 250. mL of solution.
    (a) Write the balanced chemical equation for the burning of phosphorus in excess oxygen.
    (b) Calculate the pH of the resulting solution.
    (c) An excess of aqueous calcium nitrate is added to the solution causing a white precipitate to form. Write a balanced chemical equation for this reaction and calculate the mass of precipitate formed.
    (d) An excess of zinc is added to the remaining solution. The reaction generates a colorless gas. Identify the gas and calculate its volume at STP.
67. Bromine, Br_2, reacts vigorously with hydrogen, H_2, to form hydrogen bromide. At STP, 100. mL hydrogen gas reacts with a stoichiometric amount of bromine. The resulting hydrogen bromide is dissolved in sufficient water to form 250. mL of solution. Calculate the pH of the solution.
68. A solid-fuel rocket booster is used to lift a rocket from its launching pad. The solid fuel contains a mixture of ammonium perchlorate, NH_4ClO_4, and powdered aluminum metal that reacts in the presence of an Fe_2O_3 catalyst.

$$3\,NH_4ClO_4(s) + 3\,Al(s) \longrightarrow Al_2O_3(s) + AlCl_3(s) + 6\,H_2O(g) + 3\,NO(g)$$

    (a) Which element is oxidized? Which is reduced?
    (b) Use the data in Appendix J and calculate the enthalpy change for the reaction. ΔH_f° $NH_4ClO_4(s) = -295$ kJ/mol.
69. At 1 atm and approximately 1800 °C, 50% of P_4 is dissociated into 2 P_2. If equilibrium is established under these conditions, calculate the equilibrium constant.
70. The reaction for the production of white phosphorus is exothermic; $\Delta H^\circ = -3060$ kJ/mol P_4.

$$2\,Ca_3(PO_4)_2(s) + 6\,SiO_2(s) + 10\,C(s) \longrightarrow 6\,CaSiO_3(s) + 10\,CO(g) + P_4(g)$$

    Calculate the ΔH_f° of $CaSiO_3(s)$; ΔH_f° $(Ca_3(PO_4)_2(s)) = -4138$ kJ. The enthalpy of sublimation of phosphorus = 13.06 kJ/mol.
71. A typical electric furnace using 500. V and a certain amperage produces four tons of phosphorus per hour using the reaction

$$2\,Ca_3(PO_4)_2(s) + 6\,SiO_2(s) + 10\,C(s) \longrightarrow 6\,CaSiO_3(s) + 10\,CO(g) + P_4(g)$$

    Calculate the amperage needed to produce this much phosphorus.
72. The Gibbs free energy of formation, ΔG_f°, of HI is +1.70 kJ/mol at 25 °C. Calculate the equilibrium constant for the reaction $HI(g) \rightleftharpoons \frac{1}{2}H_2(g) + \frac{1}{2}I_2(g)$.
73. Predict whether the ionization of the alkaline-earth elements is easier or harder than that of the alkali metals. Explain your answer.

74. Use a Born-Haber cycle to calculate the lattice energy of MgF_2 using these thermodynamic data.

$$\Delta H_{\text{sublimation}}\ Mg(s) = +146\ kJ/mol;$$
$$B.E.\ F_2(g) = +158\ kJ/mol;$$
$$I.E._1\ Mg(g) = +738\ kJ/mol;$$
$$I.E._2\ Mg^+(g) = +1450\ kJ/mol;$$
$$E.A.\ F(g) = -328\ kJ/mol;$$
$$\Delta H_f^\circ\ MgF_2(s) = -1124\ kJ/mol.$$

Compare this lattice energy with that of SrF_2, −2496 kJ/mol. Explain the difference in the values in structural terms.

75. Aluminum metal reacts rapidly and completely with hydrochloric acid, but only incompletely with nitric acid. Explain.

76. Elemental analysis of a borane indicates this composition: 84.2% B and 15.7% H. The compound has a molar mass of 76.7 g/mol. Determine the molecular formula of the borane.

77. The reaction of iodine with excess liquid chlorine produces the interhalogen compound I_2Cl_6.
   (a) Write the Lewis structure of this molecule.
   (b) Use VSEPR theory to predict the structure of the molecule.

78. Hydroxyapatite is the important compound in tooth enamel. It dissociates according to the equation

$$Ca_5(PO_4)_3OH(s) \rightleftharpoons 5\ Ca^{2+}(aq) + 3\ PO_4^{3+}(aq) + OH^-(aq)$$

Children drink milk to obtain calcium, but the fermentation of the milk produces lactic acid, which remains on the teeth.
   (a) Use the equation above to explain how drinking milk helps babies to produce "strong" teeth.
   (b) Explain why the lactic acid inhibits formation of strong teeth.

Chemistry of Selected Transition Elements and Coordination Compounds

22

© David Toase/Photodisc Green/Getty Images

Absorption and transmittance of light in the visible region of the spectrum by *d*-to-*d* electron transitions in transition metal ions cause the vivid colors of stained glass windows, such as this one. Transition metal oxides are added to colorless molten glass, which then solidifies to produce colored glass. The oxides and their colors are: Cu_2O, red; Cr_2O_3, green; Co_2O_3, blue; and MnO_2, violet. Transition metals, their complex ions, and coordination compounds are discussed in this chapter.

It is hard to overstate how important metals have been to the development of civilizations. Transitions from the Stone Age (400,000 to 7000 BC) to the Bronze Age (4000 to 3500 BC) to the Iron Age (1800s and beyond) and to the Computer Age (late twentieth century) have been marked by humans' ability to extract metals from ores and to process the metals into tools and objects useful in industry, warfare, and dwellings. In this chapter we consider the transition metals, the *d*-block of elements of the periodic table. Some of these elements and their compounds are of major economic importance. The precious metals gold, silver, and platinum are transition elements used in coinage and jewelry. Others, such as iron and its alloy, steel, are valuable for their structural uses.

We first consider the transition metals in overview and then look more closely at a few of them, including iron, the most economically important. The chapter closes by describing coordination compounds. In such compounds ions or molecules are bonded to transition metal ions or atoms. These compounds run the gamut from being responsible for the vivid colors of famous oil paintings to their role in significant biomolecules such as hemoglobin, vitamin B_{12}, and critical enzymes.

22.1 Properties of the Transition (*d*-Block) Elements

The four series of *d*-block elements called the *transition elements* are in the center of the periodic table in Periods 4 through 7. These elements lie between the very active metals of the *s* block and the less reactive metals of the *p*-block elements. In this section we discuss properties shared by all the transition elements.

The transition elements in Period 7 beginning with rutherfordium, element 104, are all radioactive. They have been made synthetically in *very* small amounts, just several atoms in some cases, and, therefore, little is currently known about their properties.

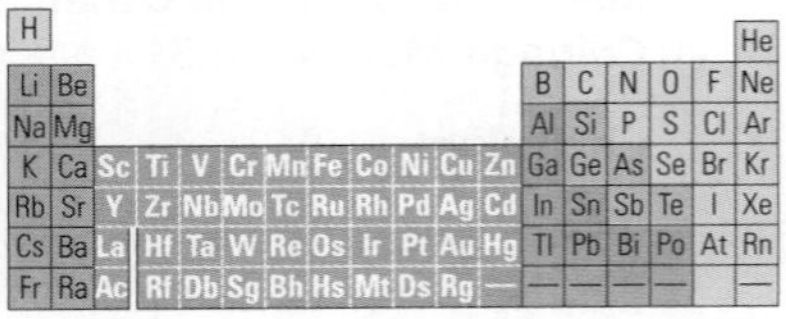

Compared to the *s*-block and *p*-block elements, there is a slow, steady transition in properties from one transition metal to the next, hence the name transition elements. These generalities apply to the transition elements:

- All are metals that conduct electricity well, but to varying degrees.
- Most are ductile (able to be drawn into a wire) and malleable (able to be hammered into thin sheets).
- Except for gold and copper, they are silvery-white or bluish.
- They usually have higher melting and boiling points than the main group elements; tungsten has the highest melting point of any metal, 3410 °C. Mercury is an exception, being the only liquid metal at room temperature.
- They generally have high densities; osmium (22.61 g/mL) and iridium (22.65 g/mL) are the most dense metals, even more dense than gold (19.3 g/mL).
- Many form brightly colored compounds (see figure in the margin).
- Some are paramagnetic; a few are ferromagnetic **(← *p. 305*)**.
- They form complex ions (Section 22.6).
- Most have multiple oxidation states; the scandium (+3) and zinc (+2) groups are exceptions.

Oxidation states (oxidation numbers) are described in Sections 5.4 and 19.1.

The transition elements at the beginning of each series show considerable differences in their chemical behavior from those at the end of each series. Such differences in chemical behavior can be attributed to the number and distribution of *d*-orbital electrons (Table 22.1). At the left side of the series, members of the scandium group of elements (Sc, Y, La) are reactive metals like the alkaline-earth metals, their predecessors in each period. On the other hand, the zinc family members (Zn, Cd, Hg) are not like other transition elements in that the zinc family members have filled $(n-1)\,d$ and ns sublevels. In fact, the elements of the zinc family are sometimes not classified as transition elements.

Many transition metal ions form colored aqueous solutions. Concentrated aqueous solutions of nitrate salts containing *(left to right):* Fe^{3+}, Co^{2+}, Ni^{2+}, and Cu^{2+}.

Electron Configurations

All transition elements have the electron configuration

$$[\text{noble gas}](n-1)d^x ns^y$$

where n is the period number (4 through 7), x is the number of d electrons (1 through 10), and y is the number of s electrons (1 or 2, except in palladium), as summarized in Table 22.1. The number of d electrons increases from left to right across a transition metal series. Elements in the zinc group (Zn, Cd, Hg) at the end of the series have filled $(n-1)\,d^{10}$ sublevels. In the preceding group the elements

Table 22.1 Outermost Electron Configurations of *d*-Block Elements

Deviations (marked in color) occur when a nonstandard configuration is more stable.

Configuration	$(n-1)d\ ns^2$	$(n-1)d^2\ ns^2$	$(n-1)d^3\ ns^2$	$(n-1)d^4\ ns^2$	$(n-1)d^5\ ns^2$	$(n-1)d^6\ ns^2$	$(n-1)d^7\ ns^2$	$(n-1)d^8\ ns^2$	$(n-1)d^9\ ns^2$	$(n-1)d^{10}\ ns^2$
First series:	$_{21}$**Sc** $3d^1 4s^2$	$_{22}$**Ti** $3d^2 4s^2$	$_{23}$**V** $3d^3 4s^2$	$_{24}$**Cr** $3d^5 4s^1$	$_{25}$**Mn** $3d^5 4s^2$	$_{26}$**Fe** $3d^6 4s^2$	$_{27}$**Co** $3d^7 4s^2$	$_{28}$**Ni** $3d^8 4s^2$	$_{29}$**Cu** $3d^{10} 4s^1$	$_{30}$**Zn** $3d^{10} 4s^2$
Second series:	$_{39}$**Y** $4d^1 5s^2$	$_{40}$**Zr** $4d^2 5s^2$	$_{41}$**Nb** $4d^4 5s^1$	$_{42}$**Mo** $4d^5 5s^1$	$_{43}$**Tc** $4d^5 5s^2$	$_{44}$**Ru** $4d^7 5s^1$	$_{45}$**Rh** $4d^8 5s^1$	$_{46}$**Pd** $4d^{10}$	$_{47}$**Ag** $4d^{10} 5s^1$	$_{48}$**Cd** $4d^{10} 5s^2$
Third series:	$_{57}$**La** $5d^1 6s^2$	$_{72}$**Hf** $4f^{14}5d^2 6s^2$	$_{73}$**Ta** $4f^{14}5d^3 6s^2$	$_{74}$**W** $4f^{14}5d^4 6s^2$	$_{75}$**Re** $4f^{14}5d^5 6s^2$	$_{76}$**Os** $4f^{14}5d^6 6s^2$	$_{77}$**Ir** $4f^{14}5d^7 6s^2$	$_{78}$**Pt** $4f^{14}5d^9 6s^1$	$_{79}$**Au** $4f^{14}5d^{10} 6s^1$	$_{80}$**Hg** $4f^{14}5d^{10} 6s^2$

4*f*-elements intervene

Cu, Ag, and Au also have filled $(n - 1)\ d^{10}$ sublevels, along with half-filled ns^1 sublevels, as described below.

The progressive filling of the *d* atomic orbitals from Sc to Zn in the first series is not uniform, as seen in Table 22.2. The first three elements of the series—Sc, Ti, and V—have $[\mathrm{Ar}]3d^1 4s^2$, $[\mathrm{Ar}]3d^2 4s^2$, and $[\mathrm{Ar}]3d^3 4s^2$ electron configurations, respectively. The filling sequence changes at chromium, which has a ground state electron configuration of $[\mathrm{Ar}]3d^5 4s^1$ with *two* half-filled sublevels, which is a lower energy state than the expected $[\mathrm{Ar}]3d^4 4s^2$ configuration. The sequence reverts back to that expected from manganese through nickel, with the pairing of 3*d* electrons. It changes again with copper, which has a filled $3d^{10}$ sublevel and a half-filled $4s^1$ sublevel (Table 22.2) that is more stable than the expected $[\mathrm{Ar}]3d^9 4s^2$ configuration. The first series ends with zinc, which has filled 3*d and* 4*s* sublevels, $[\mathrm{Ar}]3d^{10}4s^2$.

Table 22.2 Atomic Orbital Occupancy of the First Transition Series Elements

Element	Partial Atomic Orbital Diagram 3*d*					4*s*	4*p*			Unpaired Electrons
Sc	↑					↑↓				1
Ti	↑	↑				↑↓				2
V	↑	↑	↑			↑↓				3
Cr*	↑	↑	↑	↑	↑	↑				6
Mn	↑	↑	↑	↑	↑	↑↓				5
Fe	↑↓	↑	↑	↑	↑	↑↓				4
Co	↑↓	↑↓	↑	↑	↑	↑↓				3
Ni	↑↓	↑↓	↑↓	↑	↑	↑↓				2
Cu*	↑↓	↑↓	↑↓	↑↓	↑↓	↑				1
Zn	↑↓	↑↓	↑↓	↑↓	↑↓	↑↓				0

*Exceptions to normal orbital occupancy sequence.

If a substance is diamagnetic (has no unpaired electrons), its apparent mass is slightly reduced when the magnetic field is "on."

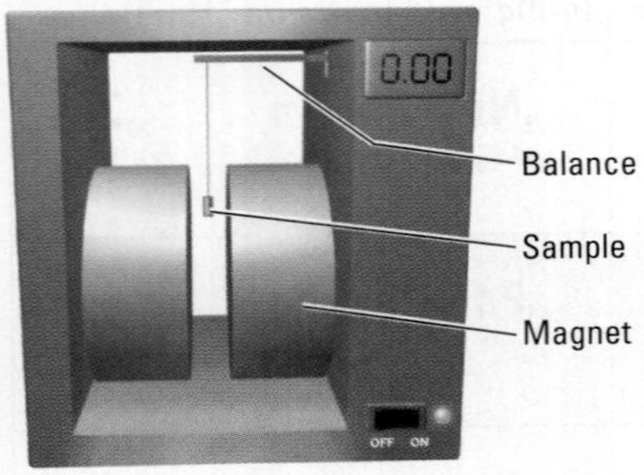

If a substance is paramagnetic (has unpaired electrons), its apparent mass increases when the field is "on" because the balance arm feels an additional force as the sample is attracted by the magnetic field.

Figure 22.1 Measurement of magnetic behavior.

The **PROBLEM-SOLVING STRATEGY** in this book is

- Analyze the problem
- Plan a solution
- Execute the plan
- Check that the result is reasonable

Appendix A.1 explains this in detail.

Transition metal atoms lose electrons to form transition metal cations. ***When transition metal atoms lose electrons to form cations, the*** **ns** ***electrons are lost before the*** **(n − 1)d** ***electrons.*** Thus, both 4*s* electrons are lost when Fe^{2+} and Fe^{3+} form. These ions differ in their number of 3*d* electrons, not 4*s* electrons, that are lost; each ion has no 4*s* electrons.

Fe atom [Ar]$3d^6 4s^2$ ⟶ Fe^{2+} ion + 2 e^- [Ar]$3d^6 4s^0$

3d					4s		3d					4s
↑↓	↑	↑	↑	↑	↑↓		↑↓	↑	↑	↑	↑	

Fe atom [Ar]$3d^6 4s^2$ ⟶ Fe^{3+} ion + 3 e^- [Ar]$3d^5 4s^0$

3d					4s		3d					4s
↑↓	↑	↑	↑	↑	↑↓		↑	↑	↑	↑	↑	

Magnetic susceptibility measurements confirm the electron configurations of the first row and other transition metal ions (Figure 22.1). The *magnetic moment* is a value calculated from the measured paramagnetism of a sample and indicates the number of unpaired electrons in the sample ***(← p. 305).*** As seen from Figure 22.2, the greater the number of unpaired electrons, the greater the magnetic moment of the substance. Figure 22.2 provides experimental evidence confirming that the 4*s* electrons are removed first leaving Fe^{2+} and Fe^{3+} with four and five unpaired electrons, respectively. If the 3*d* electrons were removed first, these ions would have four (Fe^{2+}) and three (Fe^{3+}) unpaired electrons.

PROBLEM-SOLVING EXAMPLE 22.1 Transition Metal Ion Electron Configurations

Use partial atomic orbital box diagrams to explain the number of unpaired electrons shown in Figure 22.2 for (a) Ti^{3+}, (b) Cr^{2+}, and (c) Cu^{2+}.

Answer

(a) [Ar]$3d^1 4s^0$

3d					4s
↑					

(b) [Ar]$3d^4 4s^0$

3d					4s
↑	↑	↑	↑		

(c) [Ar]$3d^9 4s^0$

3d					4s
↑↓	↑↓	↑↓	↑↓	↑	

Strategy and Explanation When an ion is formed from a Period 4 transition metal, 4*s* electrons are removed before 3*d* electrons, giving rise to the number of unpaired electrons in the ion.

(a) Ti atom [Ar]$3d^2 4s^2$ ⟶ Ti^{3+} ion + 3 e^- [Ar]$3d^1 4s^0$

3d					4s		3d					4s
↑	↑				↑↓		↑					

(b) Cr atom [Ar]$3d^5 4s^1$ ⟶ Cr^{2+} ion + 2 e^- [Ar]$3d^4 4s^0$

3d					4s		3d					4s
↑	↑	↑	↑	↑	↑		↑	↑	↑	↑		

(c) Cu atom [Ar]$3d^{10} 4s^1$ ⟶ Cu^{2+} ion + 2 e^- [Ar]$3d^9 4s^0$

3d					4s		3d					4s
↑↓	↑↓	↑↓	↑↓	↑↓	↑		↑↓	↑↓	↑↓	↑↓	↑	

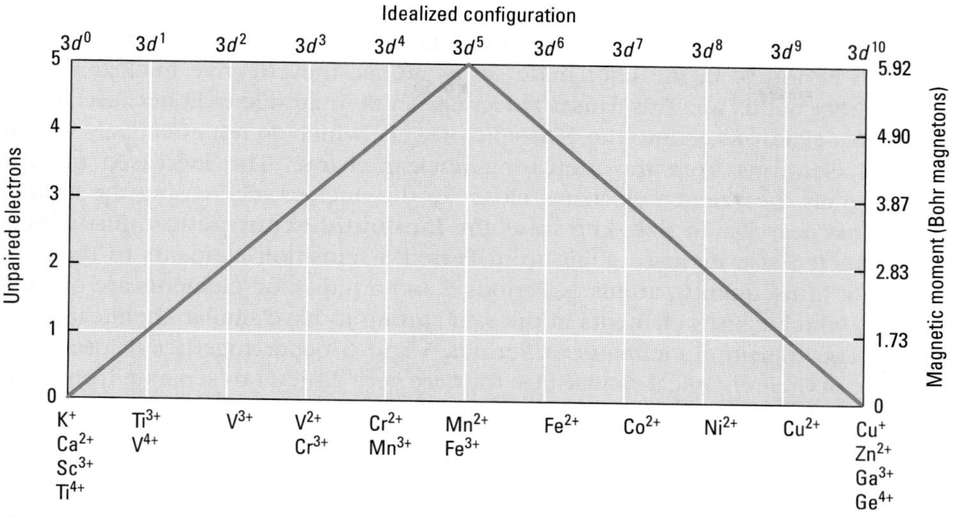

Figure 22.2 The number of unpaired electrons in the first-row transition metal ions and their magnetic moments. (The Bohr magneton is a unit that measures paramagnetism.)

PROBLEM-SOLVING PRACTICE **22.1**

Use partial atomic orbital box diagrams to explain the number of unpaired electrons shown in Figure 22.2 for (a) Mn^{2+} (b) Cr^{3+}.

PROBLEM-SOLVING PRACTICE answers are provided at the back of this book in Appendix K.

Trends in Atomic Radii of Transition Elements

Transition metals have less variation in atomic radii across a period than do main group elements. Across a row of transition elements, the atomic radii decrease steadily and then increase slightly; Mn is an exception (Figure 22.3). The decrease occurs because as each *d* electron is added, there is also an increase of one unit of nuclear charge (a proton). Repulsions among the electrons are not sufficient to counteract the added attraction of the nucleus to the electrons, so the atomic radius decreases. Toward the end of each transition metal series, the radii increase slightly due to several factors, including electron-electron repulsion as electrons are paired in *d* atomic orbitals.

The radii of the Period 5 elements are, as expected, greater than those of the Period 4 transition metals. What is unexpected is what occurs with the atomic radii in going from the Period 5 to Period 6. Instead of Period 6 radii being larger, they are nearly same as those of Period 5. This is a consequence of the lanthanide

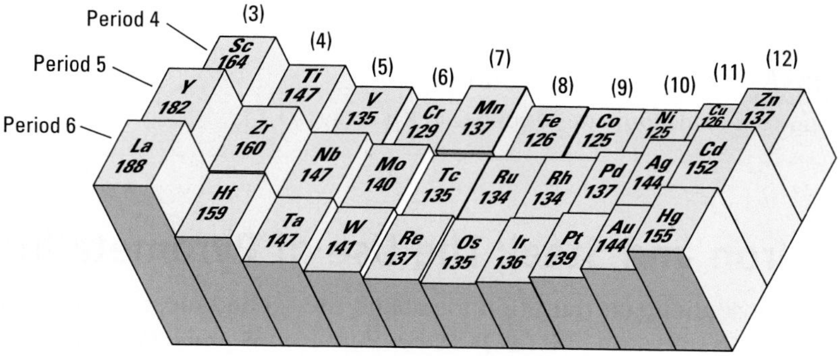

Figure 22.3 Radii of transition metal atoms (in picometers).

series of elements (La, element 57, to Lu, element 71) intervening between barium, the Period 6 alkaline-earth element, and hafnium, the next transition element of Period 6. In the lanthanide series atoms, the effective nuclear charge (← ***p. 309***) builds up. This causes a decrease in their atomic radii because all the additional electrons go into the 4*f* atomic orbitals, which do not effectively screen valence electrons from the increasing nuclear charge. The increased nuclear charge pulls the valence electrons closer to the nucleus, decreasing the atomic radii. This decrease in size, known as the **lanthanide contraction,** just offsets the expected size increase going from Period 5 transition elements to those of Period 6. Consequently, atoms of Periods 5 and 6 transition elements are of similar size, which causes elements in the same group to have similar chemical properties. The transition elements of Periods 5 and 6 occur together in ores, and because of their chemical similarities, they are very difficult to separate from each other.

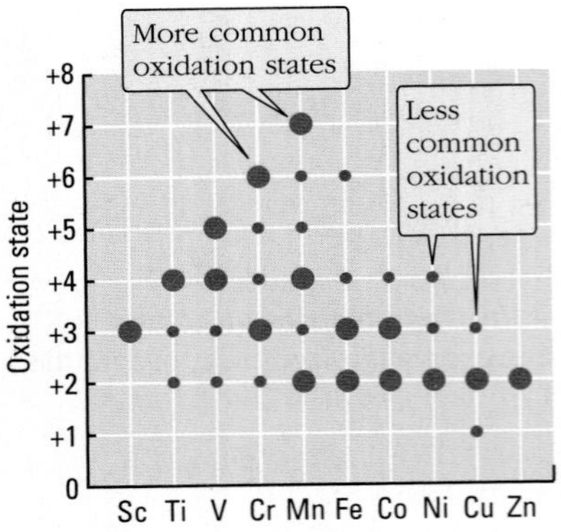

Figure 22.4 Oxidation states of first transition series elements. Note that the greatest number of oxidation states occurs with elements in the middle of the first transition series.

Oxidation States

Except for the scandium group (+3) and zinc group (+2) elements, all transition metals have multiple oxidation states. The oxidation states of the first transition series elements are listed in Figure 22.4. Manganese, for example, has three common oxidation states: +2 in Mn^{2+}, +4 in MnO_2, and +7 in MnO_4^-. Iron has two common oxidation states: +2 in FeO and +3 in Fe_2O_3. Less common oxidation states are also noted in Figure 22.4.

Transition metals that form 2+ ions do so, in general, by losing two *ns* electrons before losing any (*n* − 1) *d* electrons. Higher oxidation states involve losing (*n* − 1) *d* electrons as well. The maximum oxidation state for the first five elements in the first series—Sc through Mn—equals the total of the (*n* − 1) *d* plus *ns* electrons. Thus, the maximum oxidation state of chromium is +6, which is found in CrO_4^{2-} and $Cr_2O_7^{2-}$, and that of manganese is +7, which is found in MnO_4^-. These high oxidation states make $Cr_2O_7^{2-}$ (in acidic solution) and MnO_4^- strong oxidizing agents. In general, compounds in which the transition metal has a low oxidation state tend to be ionic, whereas compounds with transition metals in high oxidation states are relatively covalent. Thus, $MnCl_2$ (mp 650 °C) is an ionic solid containing Mn^{2+} and Cl^- ions. On the other hand, MnO_4^- is a polyatomic ion containing covalent Mn—O bonds.

Cr^0 [Ar]$3d^54s^1$

Mn^0 [Ar]$3d^54s^2$

Ions of transition elements with partially filled *d* atomic orbitals can accept or donate electrons, a property that makes them effective components of catalysts. In iron compounds, for example, iron can be present as Fe^{2+} (reduced form) or Fe^{3+} (oxidized form). Each iron ion acts as an electron shuttle, losing or gaining electrons between the oxidized and reduced forms when catalyzing electron transfer reactions, such as the production of ammonia by the Haber-Bosch process (← ***p. 689***).

Answers to **EXERCISES** are provided at the back of this book in Appendix L.

EXERCISE 22.1 Oxidation States of Transition Metals

Determine the oxidation state of the transition metal(s) in each compound.

(a) $Na_2V_4O_{11}$ (b) $KAgF_4$ (c) $CoTiO_3$ (d) $MnAl_2O_4$

EXERCISE 22.2 Oxidation State

Determine the oxidation state of copper in $K_2Ca[Cu(NO_2)_6]$.

22.2 Iron and Steel: The Use of Pyrometallurgy

Iron is the most abundant transition metal and the second most abundant metallic element in Earth's crust (4.7%) (← ***p. 998***). Pure iron is a silvery-white, rather soft metal. Its great commercial importance comes with the addition of very small amounts of carbon or other alloying elements to it to form steel.

In air, Fe^{2+}(aq) is oxidized to Fe^{3+}(aq).

$$4\,Fe^{2+}(aq) + O_2(g) + 4\,H^+(g) \longrightarrow 4\,Fe^{3+}(aq) + 2\,H_2O(\ell)$$

Aqueous Fe^{3+} reacts with water to form a hydrated oxide known as *rust.*

$$2\,Fe^{3+}(aq) + 4\,H_2O(\ell) \longrightarrow Fe_2O_3 \cdot H_2O(s) + 6\,H^+(aq)$$

Iron reacts with weakly oxidizing acids such as HCl and acetic acid to form the pale green $[Fe(H_2O)_6]^{2+}$ ion.

$$Fe(s) + 2\,H^+(aq) \longrightarrow Fe^{2+}(aq) + H_2(g) \qquad E° = +0.44\text{ V}$$

Fe^{2+}(aq) is the hydrated ion, $[Fe(H_2O)_6]^{2+}$.

When reacted with strongly oxidizing acids such as dilute nitric acid, the metal is oxidized directly to Fe^{3+}.

$$Fe(s) + 4\,H^+(aq) + NO_3^-(aq) \longrightarrow Fe^{3+}(aq) + NO(g) + 2\,H_2O \quad E° = +1.00\text{ V}$$

The principal iron ores are hematite, Fe_2O_3, and magnetite, Fe_3O_4, which are found in large deposits in Minnesota, Russia, Brazil, India, and Australia. Iron production involves steps to concentrate and purify the ores. Iron ions in the oxide ores are reduced to the metal by using carbon in the form of coke as the reducing agent at high temperatures in a blast furnace (Figure 22.5). **Pyrometallurgy** is the extraction of a metal from its ore using chemical reactions carried out at high temperatures.

Iron also occurs in nature as the pyrite, FeS_2. This mineral is not used as an ore because in steel making it is difficult to remove all the sulfur, which makes the steel brittle.

To extract metallic iron, a mixture of iron ore, coke, and limestone, $CaCO_3$, is fed into the top of the blast furnace, and a blast of heated air or oxygen is forced up from the bottom of the furnace. The coke reacts exothermically with the heated air, producing a high temperature that speeds up the iron-forming reactions, which makes the process economical. The iron ore is reduced to metallic iron by these reactions

$$2\,C(s) + O_2(g) \longrightarrow 2\,CO(g) \qquad \text{exothermic}$$

$$Fe_2O_3(s) + 3\,CO(g) \longrightarrow 2\,Fe(s) + 3\,CO_2(g) \qquad \text{exothermic}$$

Limestone is added to remove silica-containing impurities in the ore.

$$CaCO_3(s) \longrightarrow CaO(s) + CO_2(g) \qquad \text{endothermic}$$

$$CaO(s) + SiO_2(s) \longrightarrow CaSiO_3(\ell) \qquad \text{endothermic}$$

Calcium silicate and other metal silicates form *slag*, which is a liquid at the temperature of the blast furnace. The result is the formation of two layers at the bottom of the furnace. The lower, more dense liquid is molten iron that contains a substantial concentration of dissolved carbon and smaller concentrations of other impurities. The upper liquid layer is the slag. Periodically, the blast furnace is tapped from the bottom to draw off the molten iron. The liquid slag is drawn off at a port higher in the furnace (Figure 22.5).

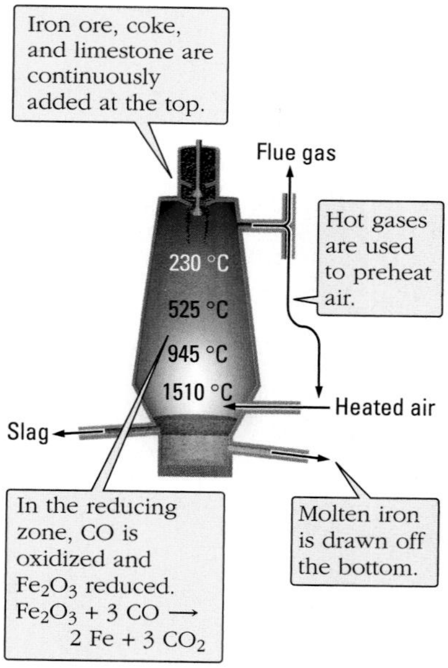
Iron ore, coke, and limestone are continuously added at the top.
Flue gas
Hot gases are used to preheat air.
230 °C
525 °C
945 °C
1510 °C
Heated air
Slag
In the reducing zone, CO is oxidized and Fe_2O_3 reduced. $Fe_2O_3 + 3\,CO \longrightarrow 2\,Fe + 3\,CO_2$
Molten iron is drawn off the bottom.

Figure 22.5 Diagram of a blast furnace. Iron ore is reduced to iron in a blast furnace.

The iron produced by a blast furnace is *pig iron*, a material that is brittle due to impurities of as much as 4.5% carbon, 1.7% manganese, 0.3% phosphorus, 0.04% sulfur, and 1% silicon. The principal embrittling material is cementite, Fe_3C, an iron carbide formed at the temperatures of the blast furnace.

$$3\,Fe(s) + C(s) \longrightarrow Fe_3C(s)$$

Molten pig iron that is poured into molds of a desired shape is called *cast iron.* It can be used directly to make molded automobile engine blocks, brake drums, transmission housings, and the like. Cast iron is too brittle for most structural uses. To convert cast iron or pig iron into **steel,** a much stronger material, the phosphorus, sulfur, and silicon impurities must be removed and the carbon content reduced to about 1.3%.

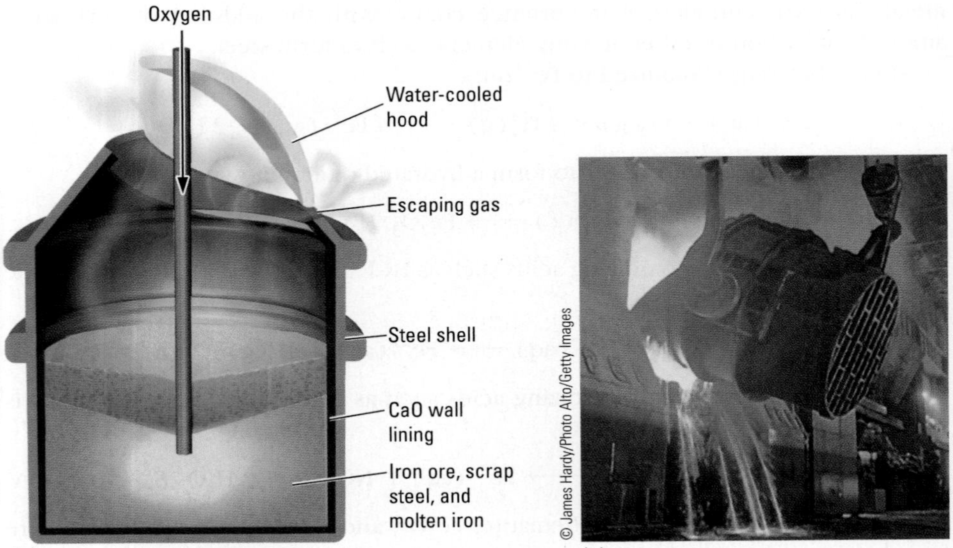

(a) (b)

Figure 22.6 The basic oxygen process for making steel. Much of the steel manufactured today is produced by the basic oxygen process. (a) Schematic diagram of a basic oxygen furnace. (b) Molten steel being poured from a basic oxygen furnace.

Steel

Many iron alloys are known collectively as *steels,* each with its own particular structural properties. One of the most common is carbon steel, an iron alloy containing 0.5 to 1.3% carbon. To convert pig iron to steel, the excess carbon is oxidized away using oxygen. Thus, whereas extracting iron from an ore is a *reduction* process, steel making is an *oxidation* process. One of several techniques used to make steel is the *basic oxygen process* (Figure 22.6), in which pure oxygen is blown through a water-cooled ceramic tube that is pushed below the surface of molten, impure iron. At the high temperatures of the melt (1900 °C), the dissolved carbon reacts rapidly with the oxygen to form gaseous carbon monoxide and carbon dioxide, which are vented. The scale of the basic oxygen process operation is impressive. About 200 tons of molten pig iron, 100 tons of scrap iron, and 20 tons of limestone are loaded into the furnace at a time. The steel is produced within an hour using such a process.

The composition of steel is varied by adding silicon, chromium, manganese, molybdenum, nickel, or other metals to give the steel specific physical, chemical, and mechanical properties. Table 22.3 lists the composition and uses of some com-

Table 22.3 Some Steels and Their Uses

Name	Composition	Properties	Uses
Carbon steel	98.7% Fe, 1.3% C	Hard	Sheet steel, tools
Manganese steel	10–18% Mn, 90–82% Fe, 0.5% C	Hard, resistant to wear	Railroad rails, safes, armor plate
Stainless steel	14–18% Cr, 7–9% Ni, 79–73% Fe, 0.2% C	Resistant to corrosion	Cutlery, instruments
Nickel steel	2–4% Ni, 98–96% Fe, 0.5% C	Hard, elastic, resistant to corrosion	Drive shafts, gears, cables
Duriron	12–15% Si, 88–85% Fe, 0.85% C	Resistant to corrosion by acids	Pipes
High-speed steel	14–20% W, 86–80% Fe, 0.5% C	Retains temper when hot	High-speed cutting tools

mon steel alloys. Magnetic alloys can be permanent magnets, such as those in audio speakers and computer hard drives, or temporary magnets such as those in electric motors, generators, and transformers. Alnico is the general name for a series of popular permanent magnets containing Al, Ni, Co, Fe, and sometimes Cu and Ti. Alnico V, for example, contains 51% Fe, as well as four other elements—14% Ni, 24% Co, 8% Al, and 3% Cu.

Blacksmiths and other steel fabricators have long known that the properties of steel are also affected by its processing temperature, cooling rates, and hammering, rolling, and extrusion. If hot steel is cooled quickly by immersing it in water or oil, the carbon in the steel will remain as cementite, Fe_3C, resulting in hard, but brittle, steel. By rapid cooling, followed by controlled reheating, a process called *tempering,* the cementite-to-graphite ratio is adjusted and the properties of the resultant steel varied further.

An Alnico magnet picking up iron or steel objects.

The artist Michelangelo wrote about using fire to transform substances:

"It is with fire that blacksmiths iron subdue
Unto fair form, the image of their thought . . ."

Sonnet 59

PROBLEM-SOLVING EXAMPLE 22.2 Cementite

Cementite, Fe_3C, is produced as an impurity during iron production in a blast furnace. To decrease the amount of cementite produced, the blast furnace temperature is decreased. Is the formation of cementite exothermic or endothermic? Explain your answer.

Answer Endothermic

Strategy and Explanation We can apply Le Chatelier's principle by noting that lowering the temperature increases the amount of iron and carbon not converted to cementite. This must mean that raising the temperature increases the amount of cementite formed, indicating that cementite formation is an endothermic reaction.

$$3\ Fe(s) + C(s) \rightleftharpoons Fe_3C \quad \text{endothermic} \quad (\Delta H > 0)$$

PROBLEM-SOLVING PRACTICE 22.2

In a blast furnace at approximately 1000 °C, much of the carbon is converted into cementite, Fe_3C, during the production of iron, which is used subsequently to produce steel. If the temperature of the steel is decreased rapidly, cementite is trapped in the iron to produce a steel with a high cementite concentration causing the steel to be brittle. Explain why this is the case. If the steel is cooled slowly, what would the effect be on the properties of the steel?

PROBLEM-SOLVING EXAMPLE 22.3 Iron Production

(a) Calculate how much carbon monoxide is needed to form 1.00×10^3 kg iron from hematite, Fe_2O_3, and from magnetite, Fe_3O_4.

(b) Calculate how much carbon in the form of coke must be used to prepare the total amount of CO needed for the two reductions in part (a).

Answer

(a) 7.52×10^5 g CO for hematite and 6.68×10^5 g CO for magnetite

(b) 6.09×10^5 g C

Strategy and Explanation Write the balanced chemical equation and use its stoichiometric factors to relate the mass of iron to the mass of carbon monoxide needed. The equation for the reduction of hematite to iron is

$$Fe_2O_3(s) + 3\ CO(g) \longrightarrow 2\ Fe(s) + 3\ CO_2(g)$$

(a) Applying the information from the balanced chemical equation, the mass of CO required to form 1.00×10^3 kg Fe from hematite is

$$1.00 \times 10^3 \text{ kg Fe}\left(\frac{10^3 \text{ g Fe}}{1 \text{ kg Fe}}\right)\left(\frac{1 \text{ mol Fe}}{55.85 \text{ g Fe}}\right)\left(\frac{3 \text{ mol CO}}{2 \text{ mol Fe}}\right)\left(\frac{28.00 \text{ g CO}}{1 \text{ mol CO}}\right)$$
$$= 7.52 \times 10^5 \text{ g CO}$$

The CO required to reduce magnetite can be calculated the same way based on the equation

$$Fe_3O_4(s) + 4\,CO(g) \longrightarrow 3\,Fe(s) + 4\,CO_2(g)$$

$$1.00 \times 10^3\text{ kg Fe}\left(\frac{10^3\text{ g Fe}}{1\text{ kg Fe}}\right)\left(\frac{1\text{ mol Fe}}{55.85\text{ g Fe}}\right)\left(\frac{4\text{ mol CO}}{3\text{ mol Fe}}\right)\left(\frac{28.00\text{ g CO}}{1\text{ mol CO}}\right) = 6.68 \times 10^5\text{ g CO}$$

(b) The carbon monoxide is prepared by the reaction

$$2\,C(s) + O_2(g) \longrightarrow 2\,CO(g)$$

From part (a), the total mass of CO required is $(7.52 \times 10^5\text{ g}) + (6.68 \times 10^5\text{ g}) = 1.42 \times 10^6\text{ g}$. The required amount of carbon is

$$1.42 \times 10^6\text{ g CO}\left(\frac{1\text{ mol CO}}{28.00\text{ g CO}}\right)\left(\frac{2\text{ mol C}}{2\text{ mol CO}}\right)\left(\frac{12.01\text{ g C}}{1\text{ mol C}}\right) = 6.09 \times 10^5\text{ g C}$$

☑ **Reasonable Answer Check**

(a) Hematite: 10^3 kg iron is equivalent to about

$$10^6\text{ g Fe} \times \frac{1\text{ mol Fe}}{56\text{ g Fe}} = 1.8 \times 10^4\text{ mol Fe}$$

which requires 1.5 times that number of moles of CO, about 2.7×10^4 mol CO. This is equivalent to approximately 8×10^5 g CO.

Magnetite: Using the same approach, we estimate that magnetite requires 7×10^5 mol CO. The estimated values for CO needed for each ore are close to the calculated values, so the answers are reasonable.

(b) The sum of the estimated masses of CO is about 1.5×10^6 g CO. This is close to 6×10^5 g C, the actual calculated answer, which is reasonable.

PROBLEM-SOLVING PRACTICE 22.3

Calculate the total mass of iron ore needed to produce 1.00×10^3 kg iron by the process described in Problem-Solving Example 22.3.

EXERCISE 22.3 High-Speed Steel

Ultrahigh-speed steel is used in some saw blades. Such a saw blade, weighing 500. g, contains 0.6% C, 4.0% Cr, 18% W, 1.0% Mo, 1.5% V, and 6.0% Co along with iron.

(a) Calculate the mass of W and of Co in the saw blade.

(b) Determine which alloying metal is present in the greatest mole percent, that is,

$$\frac{\text{moles alloying metal}}{\text{moles all alloying metals}} \times 100\%$$

ESTIMATION Steeling Automobiles

Steel is the most recycled consumer product, given that over 96% of steel from automobiles is recycled. Iron and steel make up approximately 65% of the weight of an automobile. The Steel Recycling Institute estimates that 14 million tons of steel scrap from automobiles was recycled in 2008, the equivalent of 13.5 million automobiles. Estimate how much iron ore (tons) containing approximately 2% Fe_3O_4 did not have to be mined that year due to the use of recycled steel. For this estimation problem, assume that the steel is 100% iron.

Approximately 14×10^6 tons of iron (steel) is recycled. Each 10^3 tons iron ore contains 20 tons Fe_3O_4 in which there are 168 tons Fe per 232 tons Fe_3O_4.

$$14 \times 10^6\text{ tons iron} \times \frac{232\text{ tons } Fe_3O_4}{168\text{ tons iron}} \times \frac{10^3\text{ tons iron ore}}{20\text{ tons } Fe_3O_4} = 1 \times 10^9\text{ tons ore}$$

About one billion tons of ore did not have to be mined, transported, and processed because recycled steel was used instead. This saved a substantial quantity of material and energy.

Visit this book's companion website at www.cengage.com/chemistry/moore to work an interactive module based on this material.

22.3 Copper: A Coinage Metal

Copper metal is sometimes found in nature, and evidence suggests that such naturally occurring metallic copper was known and used during the Stone Age. As early as 10,000 years ago, the metal was hammered into useful items such as coins, jewelry, tools, and weapons. Nearly five millennia later, the Bronze Age was ushered in when humans learned how to alloy copper with tin to make bronze.

Native copper.

The Metallurgy of Copper

Native copper, metallic copper that is found in nature uncombined chemically, is not available in sufficient supply to meet current demands for the metal, and so chemical methods have been developed to extract copper metal from its ores. The principal ores are chalcocite, Cu_2S, and chalcopyrite, $CuFeS_2$, which occur in conjunction with two iron sulfides, FeS_2 and FeS.

Modern methods to extract copper from its ores begin with crushing the ore and separating it from rocks. The ore is then heated in air to temperatures high enough to drive off the sulfur as sulfur dioxide, a process called *roasting*.

$$3\,FeS_2(s) + 8\,O_2(g) \xrightarrow{\text{heat}} Fe_3O_4(s) + 6\,SO_2(g)$$

$$2\,CuFeS_2(s) + O_2(s) \xrightarrow{\text{heat}} Cu_2S(s) + 2\,FeS(s) + SO_2(g)$$

An open-pit copper mine near Bagdad, Arizona.

Copper is separated from the iron by melting the Cu_2S and Fe_3O_4 mixture and then combining it with oxygen and SiO_2 to form a liquid iron silicate slag and molten Cu_2S. The slag is less dense than the molten copper(I) sulfide and floats on it, where the slag can be drawn off periodically.

The conversion of copper(I) sulfide to metallic copper takes place in a process similar to the basic oxygen process for steel making. After the iron silicate slag is removed, air is blown through the molten Cu_2S, converting it to Cu_2O, which reacts with the remaining Cu_2S to form copper metal.

$$2\,Cu_2S(\ell) + 3\,O_2(g) \longrightarrow 2\,Cu_2O(\ell) + 2\,SO_2(g)$$

$$2\,Cu_2O(\ell) + Cu_2S(\ell) \longrightarrow 6\,Cu(\ell) + SO_2(g)$$

The resulting copper, called *blister copper,* is about 96% to 99.5% copper; it can be further purified by electrolysis.

The electrorefining of copper is carried out in large electrolytic cells in which anodes of blister copper bars alternate with very thin sheets of pure copper, which are the cathodes (Figure 22.7). A mixture of copper(II) sulfate and dilute sulfuric acid is the electrolyte. Copper metal in the impure mixture is oxidized at the anode to Cu^{2+}, and Cu^{2+} in the electrolyte is reduced to pure metallic copper at the cathode.

Anode:	Cu(s, impure blister copper) $\longrightarrow$ $Cu^{2+}(aq) + 2\,e^-$	
Cathode:	$Cu^{2+}(aq) + 2\,e^-$ $\longrightarrow$ Cu(s, pure)	
Overall Reaction:	Cu(s, impure blister copper) $\longrightarrow$ Cu(s, pure)	

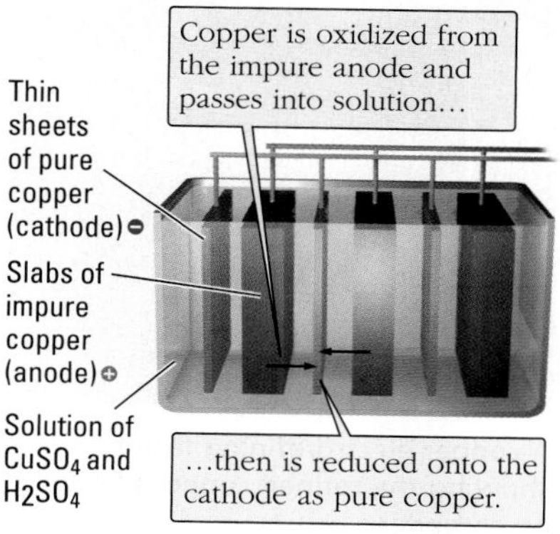

Copper is oxidized from the impure anode and passes into solution...
Thin sheets of pure copper (cathode) −
Slabs of impure copper (anode) +
Solution of CuSO4 and H2SO4
...then is reduced onto the cathode as pure copper.

Figure 22.7 Electrolytic cell for refining copper.

By controlling the voltage, only copper and those impurities (zinc, lead, iron) in the blister copper anode that have a low enough reduction potential (more easily oxidized than Cu) are oxidized and dissolved in the electrolyte. Any less easily oxidized metal impurities, such as metallic gold and silver, are essentially unaffected and drop from the anode as it is consumed during electrolysis, forming an *anode sludge.* The voltage is regulated so that only copper, the least electropositive of the metals in the electrolyte, is plated onto the pure copper cathode. Electrorefined copper is greater than 99.9% pure, a purity required for copper used in electrical applications. Usually, enough gold, silver, and platinum are recovered from the anode sludge to pay for the cost of copper electrorefining.

PROBLEM-SOLVING EXAMPLE 22.4 Copper Electrorefining

Copper is electrorefined by removing impurities such as lead, silver, gold, and zinc. Based on these standard reduction potentials, explain how copper can be separated by electrolysis from these impurities.

	$E°$(V)
$Ag^+(aq) + e^- \longrightarrow Ag(s)$	+0.799
$Au^+(aq) + e^- \longrightarrow Au(s)$	+1.83
$Cu^{2+}(aq) + 2\,e^- \longrightarrow Cu(s)$	+0.340
$Pb^{2+}(aq) + 2\,e^- \longrightarrow Pb(s)$	−0.126
$Zn^{2+}(aq) + 2\,e^- \longrightarrow Zn(s)$	−0.763

Answer At the potential that just oxidizes blister copper from the anode (−0.340 V relative to a standard hydrogen electrode), zinc and lead are also oxidized. Silver and gold are not oxidized and form the anode sludge. This potential is not sufficient to reduce Zn^{2+} and Pb^{2+}, so only copper forms at the cathode.

Strategy and Explanation Consider what reactions could occur at the impure blister copper anode and the pure copper cathode. Adjust the cell potential so that only copper and metals that are more easily oxidized than copper will dissolve. Because the other metals that dissolve are easier to oxidize than copper, their ions will be harder to reduce and therefore copper(II) ions will be reduced at the cathode.

Ag^+ and Au^+ have more positive standard reduction potentials than Cu^{2+}, so Ag^+ and Au^+ are easier to reduce than Cu^{2+}. For the reverse (oxidation) reactions, Cu will be easier to oxidize than Ag or Au. A potential that just oxidizes Cu will not oxidize Ag or Au, which will form the anode sludge.

Pb^{2+} and Zn^{2+} are more difficult to reduce than Cu^{2+}. Therefore, Pb and Zn are easier to oxidize than Cu, and a potential that oxidizes Cu at the anode will also oxidize Zn and Pb. Thus, Cu^{2+}, Zn^{2+}, and Pb^{2+} will be available in solution for reduction at the cathode. Because Pb^{2+} and Zn^{2+} are more difficult to reduce than Cu^{2+}, only Cu^{2+} will react at the cathode, forming pure Cu.

PROBLEM-SOLVING PRACTICE 22.4

Explain how zinc and lead could be separated from each other without plating out copper in the electrolytic cell in Problem-Solving Example 22.4.

Chris Sharp/Photo Researchers, Inc.

A copper electrorefining facility showing the refined copper plated on cathodes.

The mass of pure copper deposited on a cathode by electrorefining can be calculated. For example, a copper electrorefining cell operates at 200. amperes (A) for 24 hours a day for a year. How many kilograms of pure copper are produced, assuming 100% efficiency? This mass can be calculated by determining the number of coulombs of charge used and recognizing that two moles of electrons are needed for each mole of Cu^{2+} reduced to copper metal. From Section 19.12 (← *p. 942*) we know that

$$1 \text{ ampere} = 1 \text{ coulomb/second (C/s) and } 1 \text{ mol e}^- = 9.65 \times 10^4 \text{ C}$$

$$365 \text{ days} \times 200.\text{ A} \times \left(\frac{1 \text{ C/s}}{1 \text{ A}}\right)\left(\frac{24 \text{ h}}{1 \text{ day}}\right)\left(\frac{3600 \text{ s}}{1 \text{ h}}\right) = 6.31 \times 10^9 \text{ C}$$

$$6.31 \times 10^9 \text{ C}\left(\frac{1 \text{ mol e}^-}{9.65 \times 10^4 \text{ C}}\right)\left(\frac{1 \text{ mol Cu}}{2 \text{ mol e}^-}\right) = 3.27 \times 10^4 \text{ mol Cu}$$

$$3.27 \times 10^4 \text{ mol Cu}\left(\frac{63.55 \text{ g Cu}}{1 \text{ mol Cu}}\right)\left(\frac{1 \text{ kg Cu}}{10^3 \text{ g Cu}}\right) = 2.08 \times 10^3 \text{ kg Cu}$$

EXERCISE 22.4 Refining Copper

Calculate the mass of pure copper deposited during electrorefining in a cell operating at 250. A for 12.0 h. Assume the process is 100% efficient.

Bronze and Brass

The Bronze Age began about 3800 BC with the discovery that bronze formed when tin combined with copper. This discovery was likely accidental, brought on by the fact that copper ores and tin ores are often found together. In a stroke of good fortune, some tin ore was present during the reduction of copper ore by the charcoal of a wood fire, and tin ions were also reduced to tin metal. The heated mixture of the resulting copper and tin metals formed bronze. The advantage of bronze is that it is sufficiently hard to keep a cutting edge, something copper cannot do. Bronze usually contains 7 to 10% tin, but bronzes with as much as 20% tin are known. Because of its properties, all early civilizations that produced bronze used it to create weapons as well as exquisite works of art.

An ancient bronze work of art.

On prolonged exposure to moist air, copper or bronze, by a redox reaction, forms a green outer coating (a patina) of copper(II) hydroxycarbonate, $Cu_2(OH)_2CO_3$, often seen on statuary, such as the Statue of Liberty, and on copper-clad roofs.

$$2\,Cu(s) + O_2(g) + CO_2(g) + H_2O(g) \longrightarrow Cu_2(OH)_2CO_3(s)$$

Brass is an alloy made of varying proportions of copper and zinc (20 to 45% Zn), which becomes harder as the percentage of zinc increases. Because brass is easy to forge, cast, and stamp, it is widely used for pipes, valves, and fittings.

Patina on the Statue of Liberty. The copper hydroxycarbonate coating on the statue is responsible for the green color.

Copper is used in all U.S. coins. By 1982, the price of copper had risen to where it was costing the U.S. Treasury Department more than 1 cent to make a penny. Since then, to conserve copper and reduce costs, the penny has been made of 97.5% zinc and 2.5% copper, with the zinc core sandwiched between two thin layers of copper. Silver-colored coins actually contain no silver. The U.S. nickel is made of a copper (75%) and nickel (25%) alloy of uniform composition throughout the coin. The other silver-colored coins are a "sandwich" made of a pure copper core covered with a thin layer of a copper and nickel alloy (91.67% copper). The Sacagawea dollar coin, which looks like gold, contains no gold.

Metallic copper is not attacked by most acids, although it does react with nitric acid. The nitrate ion, NO_3^-, acts as the oxidizing agent. In dilute nitric acid, NO_3^- is reduced to NO; concentrated nitric acid yields NO_2.

$$3\,Cu(s) + 2\,NO_3^-(aq) + 8\,H^+(aq) \longrightarrow 3\,Cu^{2+}(aq) + 2\,NO(g) + 4\,H_2O(\ell)$$

$$Cu(s) + 2\,NO_3^-(aq) + 4\,H^+(aq) \longrightarrow Cu^{2+}(aq) + 2\,NO_2(g) + 2\,H_2O(\ell)$$

Copper(II) sulfate pentahydrate, $CuSO_4 \cdot 5\,H_2O$, is the most widely used copper compound. Commonly called blue vitriol, this compound is used to kill algae and fungi. In the solid it contains the hydrated copper(II) complex ion, $[Cu(H_2O)_4]^{2+}$. The fifth water molecule is bound to the sulfate ion through hydrogen bonding. The water of hydration can be removed by gentle heating or by putting the hydrate into a desiccator. Many aqueous solutions containing copper(II) ions are blue due to the presence of the $[Cu(H_2O)_6]^{2+}$ ion. The color of aqueous copper(II) ions depends on the Lewis base (p. 1067) bonded to the Cu^{2+} ion.

Pre-1982 (left) and post-1982 (right) pennies. A copper-clad penny (*right*) with some of the copper cladding removed to expose the silvery-appearing zinc core.

The reaction of metallic copper with concentrated nitric acid produces brown fumes of NO_2. The solution takes on the green color of the other reaction product, copper(II) nitrate, in concentrated nitric acid.

CHEMISTRY IN THE NEWS — An Apartment with a View

In a soon-to-be-demolished housing complex near London Bridge, English artist Roger Hiorn created SEIZURE, a unique display of art and place that drew more than 25,000 viewers. Hiorn encrusted his entire ground-floor studio apartment from floor to ceiling with blue $CuSO_4 \cdot 5\ H_2O$ crystals. After conducting several smaller-scale experiments, Hiorn and his team scaled up the crystallization to apartment-sized proportions. Funded by the Jerwood Foundation, Hiorn purchased 90 tons of hydrated copper(II) sulfate from an Italian agricultural chemistry company. The compound is used commonly as a fungicide throughout Europe's wine-growing regions. Large stainless steel barrels for preparing a supersaturated solution and crude oil pumps to move it into the apartment were purchased on eBay. After reinforcing the structure and covering the dwelling with plastic sheets, the Hiorn team pumped 90,000 L of a supersaturated copper(II) sulfate solution into his apartment through a hole in the floor of the apartment above.

© Nick Cobbing

Roger Hiorn in his hydrated copper(II) sulfate-encrusted apartment.

After a three-week period during which the supersaturated solution temperature dropped daily, crystallization occurred. Hiorn said, "We found when we drained the apartment that the crystals were far, far bigger than we anticipated." Art critic Adrian Searle commented that, ". . . The walls and ceiling are covered in blue copper sulfate crystals. . . . Every surface is furred and infested; big blue crystals dangle like cubist bats from the light fittings." Because only four people could be in the apartment simultaneously, visitors stood in line for hours waiting to put on rubber boots and then view the cerulean crystalline cavern. Hiorn is considering doing another large-scale crystallization. "Now that we know how to do it, it seems perverse not to do it again," he says.

© Nick Cobbing

$CuSO_4 \cdot 5\ H_2O$ crystals deposited on a mixing tank test stick.

Source: Halford, B. "Concocting a Crystalline Lair." *Chemical & Engineering News*, January 5, 2009; pp. 30–31. **www.cen-online.org**

In a disproportionation reaction, the same substance is oxidized and reduced.

Copper also exists in a +1 oxidation state, Cu^+; however, aqueous copper chemistry involves principally Cu^{2+} because Cu^+ is unstable in aqueous solutions. Cu^{2+} disproportionates (undergoes both oxidation and reduction) into Cu and Cu^{2+}.

$$
\begin{array}{rcll}
Cu^+(aq) + e^- & \longrightarrow & Cu(s) & E^\circ = +0.52\ V \\
Cu^+(aq) & \longrightarrow & Cu^{2+}(aq) + e^- & E^\circ = +0.15\ V \\
\hline
2\ Cu^+(aq) & \longrightarrow & Cu(s) + Cu^{2+}(aq) & E^\circ = +0.37\ V
\end{array}
$$

© Cengage Learning/Charles D. Winters

Hydrated copper(II) sulfate (blue) | Partially anhydrous copper(II) sulfate (white)

Hydrated copper(II) sulfate is converted to anhydrous copper(II) sulfate by heating, which drives off the waters of hydration and causes the color to change.

In basic solution, however, Cu^{2+} is reduced to copper(I) oxide, Cu_2O. This reaction was once used to test for glucose in urine. When heated, a basic solution containing blue aqueous Cu^{2+} and a reducing sugar such as glucose react to form a brick-red precipitate of Cu_2O. The reducing sugar is represented here with its aldehyde functional group, RCHO.

$$\underset{\text{blue}}{2\,Cu^{2+}(aq)} + \underset{\text{reducing sugar}}{RCHO(aq)} + 5\,OH^-(aq) \longrightarrow \underset{\text{brick red}}{Cu_2O(s)} + RCOO^-(aq) + 3\,H_2O(\ell)$$

At elevated temperatures copper reacts with oxygen. Below 1000 °C it forms black copper(II) oxide, CuO. Above 1000 °C copper reacts with oxygen to form red copper(I) oxide, Cu_2O, which is found in the mineral cuprite.

Samples of red copper(I) oxide, Cu_2O, and black copper(II) oxide, CuO.

22.4 Silver and Gold: The Other Coinage Metals

Silver and gold occur in elemental (uncombined) form in nature, although such sources have dwindled. In the past, gold prospectors, such as the forty-niners in the 1849 California gold rush, panned for gold. They simply swirled gold-bearing sand and gravel from streams in a pan. Because of its high density (19.3 g/cm^3), gold settles out from the sand (about 2.5 g/cm^3) and other rocky impurities. At present, the gold content in such deposits is much too low for panning to be commercially effective.

Early civilizations highly prized silver and gold for their luster, corrosion resistance, and workability in making jewelry, art objects, and coins. Silver is the best metallic conductor of heat and electricity. Long synonymous with wealth and power, gold was thought to be a part of the sun, thus its Latin name *aurum,* meaning "bright dawn," from which its chemical symbol is derived. Gold is the most malleable metal, so malleable that it can be rolled thin enough to be transparent. Very thin gold leaf is used to cover domes of churches and capitol buildings.

The lack of reactivity of gold and silver is explained by the standard reduction potentials of their ions, which are much more positive than that of hydrogen (0.00 V).

$$Au^+(aq) + e^- \longrightarrow Au(s) \qquad E° = +1.83\ V$$

$$Ag^+(aq) + e^- \longrightarrow Ag(s) \qquad E° = +0.799\ V$$

Gold can be hammered into ultrathin sheets for decorative purposes.

Both metal ions are fairly strong oxidizing agents. Consequently, neither metal reacts with weak oxidizing acids such as hydrochloric acid, nor do they react readily with oxygen at normal temperatures. When heated in air, gold does not react with oxygen, even at high temperatures. Silver, however, slowly forms silver(I) oxide, Ag_2O, which is unstable and decomposes back to the elements when heated strongly. Silver reacts with oxidizing acids such as nitric acid, HNO_3. However, it takes aqua regia, a 3:1 mixture of concentrated HCl and HNO_3, to dissolve gold.

The ability of aqua regia (Latin, "royal water") to dissolve gold has been known since the 1300s.

$$3\,Ag(s) + 4\,H^+(aq) + NO_3^-(aq) \longrightarrow 3\,Ag^+(aq) + NO(g) + 2\,H_2O(\ell)$$

$$Au(s) + 6\,H^+(aq) + 3\,NO_3^-(aq) + 4\,Cl^-(aq) \longrightarrow AuCl_4^-(aq) + 3\,NO_2(g) + 3\,H_2O(\ell)$$

The oxidation of metallic gold to Au^{3+} is highly unfavorable, but the reaction in aqua regia occurs because as Au^{3+} ions form, they combine with Cl^- ions to form the complex ion, $AuCl_4^-$, in a highly product-favored reaction.

$$Au^{3+}(aq) + 4\,Cl^-(aq) \longrightarrow AuCl_4^-(aq) \qquad K = 1 \times 10^{38}$$

Complex ions such as $AuCl_4^-$ were discussed in Section 17.5 ***(← p. 830)*** and are discussed in detail in Section 22.6 (p. 1055).

EXERCISE 22.5 Putting Silver and Gold into Solution

(a) Identify the oxidizing and reducing agents in the reaction of nitric acid with silver.
(b) Do the same for the reaction of gold with aqua regia.

Gold and silver are both relatively soft metals whose hardness is increased by alloying with other metals. Sterling silver, for example, is 92.5% silver and 7.5% copper. The proportion of gold in its alloys is expressed in carats. Pure gold (24 carats) is too soft to be used in jewelry. It is alloyed with copper or other metals to make the 18-carat and 14-carat gold for jewelry, which are 75% (18/24 × 100%) and 58% (14/24 × 100%) gold, respectively.

As noted earlier (p. 1047), silver and gold are by-products of the electrorefining of copper. They are also obtained from ores. Silver is obtained from its principal ore argentite, Ag_2S, by cyanide extraction. The ore is ground and put into a 0.5% solution of NaCN.

$$Ag_2S(s) + 4\,CN^-(aq) \longrightarrow 2\,Ag(CN)_2^-(aq) + S^{2-}(aq)$$

The formation of the $Ag(CN)_2^-$ complex ion brings the insoluble silver sulfide into solution.

$$\begin{aligned} Ag_2S(s) &\rightleftharpoons 2\,Ag^+(aq) + S^{2-}(aq) \\ 2\,Ag^+(aq) + 4\,CN^-(aq) &\rightleftharpoons 2\,Ag(CN)_2^-(aq) \\ \hline Ag_2S(s) + 4\,CN^-(aq) &\longrightarrow 2\,Ag(CN)_2^-(aq) + S^{2-}(aq) \end{aligned}$$

Powdered zinc, a good reducing agent, is added to the solution to convert Ag^+ to metallic silver.

$$Zn(s) + 2\,Ag(CN)_2^-(aq) \longrightarrow 2\,Ag(s) + Zn(CN)_4^{2-}(aq)$$

The silver can be further purified by electrolytic refining.

Like silver, gold is obtained from ores by cyanide extraction followed by reduction with zinc.

In all of history, the amount of gold that has been mined is relatively small, approximately enough to fill just two Olympic-sized swimming pools.

$$4\,Au(s) + 8\,CN^-(aq) + O_2(g) + 2\,H_2O(\ell) \longrightarrow 4\,Au(CN)_2^-(aq) + 4\,OH^-(aq)$$

$$Zn(s) + 2\,Au(CN)_2^-(aq) \longrightarrow 2\,Au(s) + Zn(CN)_4^{2-}(aq)$$

The cyanide extraction of silver and gold depends on the formation of the $Ag(CN)_2^-$ and $Au(CN)_2^-$ complex ions. Because of the toxicity of CN^-, waste cyanide solutions must be disposed of properly. In places where such has not been the case and the solutions were simply dumped near the processing site, serious environmental damage has occurred.

22.5 Chromium

Chromium is characteristic of the middle transition elements that exhibit multiple oxidation states (p. 1042). In compounds, chromium has oxidation states from +2 to +6 because of its five 3*d* electrons and one 4*s* electron, although +2 and +3 are the most common oxidation states. The +6 oxidation state occurs in CrO_4^{2-} and $Cr_2O_7^{2-}$ ions (Figure 22.8). The +4 and +5 oxidation states are uncommon.

Chromium is a hard, brittle metal that is extremely corrosion resistant due to a chromium oxide surface layer that passivates and protects the metal beneath from further oxidation. In this regard, chromium resembles aluminum ***(← p. 1016).*** Chromium is used in stainless steel and is plated on truck bumpers to give them a bright surface. Its oxides are used in magnetic recording tapes, CrO_2, in abrasives; and as a glass pigment, Cr_2O_3.

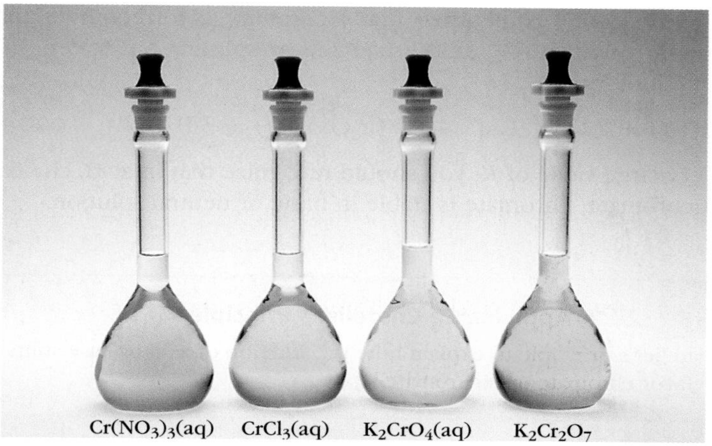

Figure 22.8 Chromium ions in solution. The color changes with the oxidation state of chromium, and the chemical environment of the metal ion. The two flasks on the left contain solutions of chromium ions with oxidation state +3: $Cr(NO_3)_3$ (violet) and $CrCl_3$ (green). The flasks on the right show the colors of the two chromium anions with oxidation state +6: yellow chromate ion, CrO_4^{2-}, and orange dichromate ion, $Cr_2O_7^{2-}$.

The chief chromium ore is chromite, $FeCr_2O_4$, which is treated with lime, CaO, oxygen, and sodium carbonate to form sodium chromate. Water is added to remove the soluble sodium chromate produced.

$$4\ FeCr_2O_4(s) + 8\ CaO(s) + 8\ Na_2CO_3(s) + 7\ O_2(g) \longrightarrow 8\ Na_2CrO_4(s) + 2\ Fe_2O_3(s) + 8\ CaCO_3(s)$$

Sodium chromate is used widely in chrome plating and as an intermediate in the formation of many chromium compounds. One of these compounds is chromium(III) oxide, from which chromium metal is obtained by reduction with aluminum metal.

$$2\ CrO_4^{2-}(aq) + 3\ SO_2(g) + 3\ H_2O(\ell) \longrightarrow Cr_2O_3(s) + 3\ SO_4^{2-} + 2\ H_3O^+(aq)$$

$$Cr_2O_3(s) + 2\ Al(s) \longrightarrow 2\ Cr(s) + Al_2O_3(s)$$

EXERCISE 22.6 Production of Chromium Metal

Use data from Appendix J to calculate the enthalpy change and the Gibbs free energy change for the reduction of chromium(III) oxide by aluminum.

In aqueous solution, chromate ions, CrO_4^{2-}, and dichromate ions, $Cr_2O_7^{2-}$, exist in a highly pH-dependent equilibrium.

$$H^+(aq) + \underset{\text{yellow}}{CrO_4^{2-}(aq)} \rightleftharpoons HCrO_4^-(aq)$$

$$2\ HCrO_4^-(aq) \rightleftharpoons \underset{\text{orange}}{Cr_2O_7^{2-}(aq)} + H_2O(\ell)$$

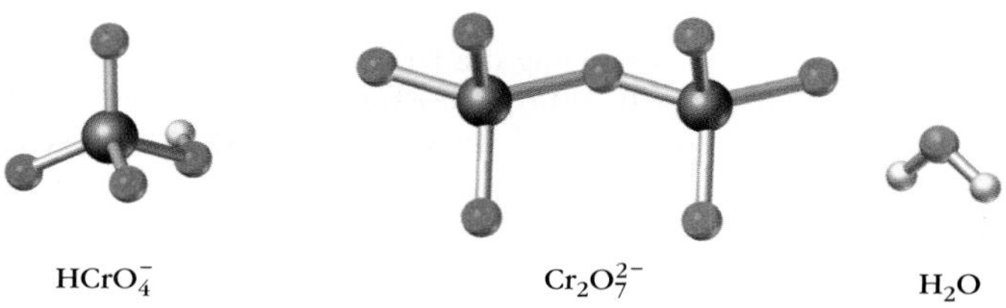

Notice from the second equilibrium that dichromate is formed by a condensation reaction in which two $HCrO_4^-$ units are joined by splitting out water.

The net equilibrium is

$$2\,H_3O^+(aq) + 2\,CrO_4^{2-}(aq) \rightleftharpoons Cr_2O_7^{2-}(aq) + 3\,H_2O(\ell) \qquad K_c = 4 \times 10^{14}$$

From the very large value of K, you should recognize that in acid, chromate is converted to dichromate; chromate is stable in basic or neutral solution.

EXERCISES that are labeled **CONCEPTUAL** are designed to test your understanding of one or more concepts; they usually involve qualitative rather than quantitative thinking.

CONCEPTUAL EXERCISE 22.7 Applying Le Chatelier's Principle

Use Le Chatelier's principle to explain how the addition of acid or base shifts the equilibrium to favor chromate or dichromate.

In highly acidic solution the dichromate ion is a powerful oxidizing agent.

$$Cr_2O_7^{2-}(aq) + 14\,H_3O^+(aq) + 6\,e^- \longrightarrow 2\,Cr^{3+}(aq) + 21\,H_2O(\ell) \qquad E^\circ = +1.33\text{ V}$$

It is sufficiently strong to oxidize alcohols to aldehydes or ketones *(← p. 549)* and aldehydes to carboxylic acids, for example, acetaldehyde to acetic acid.

$$Cr_2O_7^{2-}(aq) + 3\,CH_3CHO(aq) + 8\,H_3O^+(aq) \longrightarrow 2\,Cr^{3+}(aq) + 3\,CH_3COOH(aq) + 12\,H_2O(\ell)$$

The oxidizing strength of dichromate decreases as the pH increases, as shown in Problem-Solving Example 22.5.

PROBLEM-SOLVING EXAMPLE 22.5 Dichromate and pH

Use the Nernst equation *(← p. 926)* to calculate the E_{cell} for the reduction of dichromate ion by iodide ion at a pH of 4.0 and 25 °C with all concentrations other than H_3O^+ equal to 1.0 M. E°_{cell} is 0.795 V.

Answer $E_{cell} = +0.242$ V

Strategy and Explanation We use the balanced equation for the reaction and the Nernst equation to calculate the cell potential at the nonstandard conditions. The balanced equation *(← p. 906)* for the reaction is

$$Cr_2O_7^{2-}(aq) + 6\,I^-(aq) + 14\,H_3O^+(aq) \longrightarrow 2\,Cr^{3+}(aq) + 3\,I_2(aq) + 21\,H_2O(\ell)$$

The Nernst equation is

$$E_{cell} = E^\circ_{cell} - \frac{RT}{nF}\ln Q = 0.795\text{ V} - \frac{0.0257\text{ V}}{n}\ln\frac{[Cr^{3+}]^2[I_2]^3}{[Cr_2O_7^{2-}][I^-]^6[H_3O^+]^{14}}$$

Six moles of electrons are transferred per mole of $Cr_2O_7^{2-}$:

$2\,Cr^{6+} + 6\,e^- \longrightarrow 2\,Cr^{3+}$

$6\,I^- \longrightarrow 3\,I_2 + 6\,e^-$

In the equation, 6 mol e^- are transferred so $n = 6$. All concentrations are 1.0 M except $[H_3O^+]$, which is $10^{-pH} = 10^{-4}$. Therefore, the Nernst equation becomes

$$E_{cell} = 0.795\text{ V} - \frac{0.0257\text{ V}}{6}\ln\frac{1}{(1.0 \times 10^{-4})^{14}}$$

$$= 0.795\text{ V} - \frac{0.0257\text{ V}}{6}\ln(1.0 \times 10^{56})$$

$$= 0.795\text{ V} - 0.553\text{ V} = 0.242\text{ V}$$

☑ **Reasonable Answer Check** Changing the $[H_3O^+]$ from 1.0 M (the standard-state concentration) to 1.0×10^{-4} M (pH = 4.0) decreases the cell potential by 0.553 V. This large change results because of the large coefficient (14) of H_3O^+ in the balanced equation. Dichromate ion is a much weaker oxidizing agent at pH = 4.0 than at standard-state conditions.

PROBLEM-SOLVING PRACTICE 22.5

At what pH does $E_{cell} = 0.00$ V for the reduction of dichromate by iodide ion in acid solution, assuming standard-state concentrations of all species except H_3O^+ ion?

22.6 Coordinate Covalent Bonds: Complex Ions and Coordination Compounds

In Section 8.10, the formation of F_3BNH_3 was described as occurring by the sharing of a lone pair from NH_3 with BF_3. This type of covalent bond, in which one atom contributes *both* electrons for the shared pair, is called a *coordinate covalent bond.* Atoms with lone pairs of electrons, such as nitrogen, phosphorus, and sulfur, can use those lone pairs to form coordinate covalent bonds. For example, the formation of the ammonium ion from ammonia results from formation of a coordinate covalent bond between H^+ and the lone pair of electrons of nitrogen in NH_3.

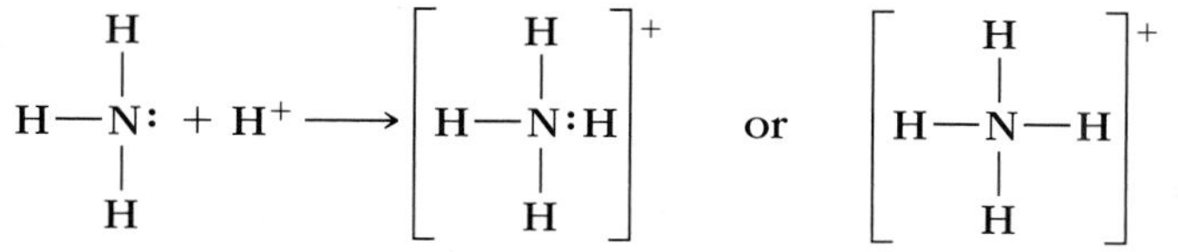

Once the coordinate covalent bond is formed, it is impossible to distinguish which of the N—H bonds it is; all four N—H bonds are equivalent.

Metals and Coordination Compounds

Much of the chemistry of *d*-block transition metals is related to their ability to form coordinate covalent bonds with molecules or ions that have lone pair electrons. Transition metals with vacant *d* atomic orbitals can act as Lewis acids and accept the lone pairs into those orbitals (← *p. 787*).

You have seen that metal ions in aqueous solution are surrounded by water molecules; for example, the Ni^{2+} ion in aqueous solution is surrounded by six water molecules. This type of ion, in which several molecules or ions are connected to a central metal ion or atom by coordinate covalent bonds, is known as a *complex ion.* The molecules or ions bonded to the central metal ion are called **ligands,** from the Latin verb *ligare*, "to bind." Each ligand (a water molecule in this example) has one or more atoms with lone pairs that can form coordinate covalent bonds to the metal ion. To write the formula of a complex ion, the ligand formulas are placed in parentheses following the metal ion. The entire complex ion formula is enclosed by brackets, and the ionic charge, if any, is a superscript outside the brackets. For the nickel(II) complex ion with six water ligands this gives $[Ni(H_2O)_6]^{2+}$.

Ligands act as Lewis bases, electron pair donors; transition metal ions are Lewis acids (electron pair acceptors) (Section 16.9, ← *p. 786*).

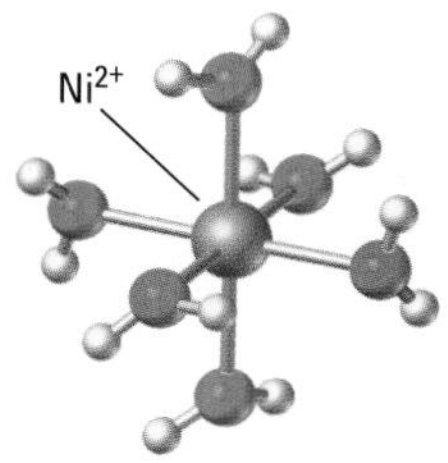

$[Ni(H_2O)_6]^{2+}$

The charge of a complex ion is determined by the charges of the metal ion and the charges of its ligands. In $[Ni(H_2O)_6]^{2+}$ the water ligands have no net charge, so the charge of the complex ion is that of the Ni^{2+} ion. In the complex ion formed by Ni^{2+} with four chloride ions, $[NiCl_4]^{2-}$, the net 2− charge of this complex ion results from the 4− charge of four chloride ions and the 2+ charge of the nickel ion.

Compounds that contain metal ions surrounded by ligands are called **coordination compounds.** Usually, complex ions are combined with oppositely charged ions *(counter ions)* to form neutral compounds. Coordination compounds are generally brightly colored as solids or in solution (Figure 22.8, two left flasks). The complex ion part of a coordination compound's formula is enclosed in brackets; counter ions are outside the brackets, as in the formula $[Ni(H_2O)_6]Cl_2$ of the compound consisting of chloride counter ions with the $[Ni(H_2O)_6]^{2+}$ complex ion. The two Cl^-

When the word *coordinated* is used in chemistry, such as in "the chloride ions in $[NiCl_4]^{2-}$ are coordinated to the nickel ion," it means that coordinate covalent bonds have been formed.

ions compensate for the 2+ charge of the complex ion. $[Ni(H_2O)_6]Cl_2$ is an ionic compound analogous to $CaCl_2$, which also contains a 2+ cation and two Cl^- ions. In some cases, no compensating ions are needed outside the brackets for a coordination compound. For example, the anticancer drug $[Pt(NH_3)_2Cl_2]$ (cisplatin) is a coordination compound containing NH_3 and Cl^- ligands coordinated to a central Pt^{2+} ion. The two Cl^- ions compensate for the charge of the Pt^{2+} ion, resulting in a neutral coordination compound rather than a complex ion.

PROBLEM-SOLVING EXAMPLE 22.6 Coordination Compounds

For the coordination compound $K_3[Fe(CN)_6]$, identify

(a) The central metal ion.

(b) The ligands.

(c) The formula and charge of the complex ion and the charge of the central metal ion.

Answer

(a) Iron (b) Six cyanide ions, CN^- (c) $[Fe(CN)_6]^{3-}$, Fe^{3+}

Strategy and Explanation We apply the guidelines and concepts described previously. In a formula, a complex ion is enclosed in square brackets; within the brackets, ligands coordinated to the central metal ion are enclosed by parentheses.

(a) The iron ion is the central metal ion, as shown by its placement inside the brackets.

(b) Cyanide ions, CN^-, are coordinated to the central iron ion.

(c) The charge on three potassium ions is $(3 \times 1+) = 3+$. Therefore, the compensating charge of the complex ion must be 3−, arising from the 6− charge of six cyanide ions $(6 \times 1- = 6-)$, combined with the 3+ charge of the central iron(III) ion: $(6-) + (3+) = 3-$.

PROBLEM-SOLVING PRACTICE 22.6

For the coordination compound $[Cu(NH_3)_4]SO_4$, identify

(a) The counter ion. (b) The central metal ion.

(c) The ligands. (d) The formula and charge of the complex ion.

CONCEPTUAL EXERCISE 22.8 Coordination Complex Ion

In a complex ion, a central Cr^{3+} ion is bonded to two ammonia molecules, two water molecules, and two hydroxide ions. Give the formula and the net charge of this complex ion.

Naming Complex Ions and Coordination Compounds

Like other compounds, coordination compounds in early times were known by common names, for example, "roseo" salt and Zeise's salt. Since then, a systematic nomenclature has been developed for complex ions and coordination compounds. This nomenclature system indicates the central metal ion and its oxidation state, as well as the number and kinds of ligands. Table 22.4 lists the names and formulas of some common ligands. Although there are extensive rules for such nomenclature, we will consider only some basic aspects of the system by interpreting the names of a neutral coordination compound and two other coordination compounds, one containing a complex cation and the other a complex anion.

Consider the coordination compound $[Co(NH_3)_3(OH)_3]$, which is named triamminetrihydroxocobalt(III). From Table 22.4, we see that the name and formula indicate that three ammonia molecules and three hydroxide ions are bonded to a central cobalt ion. The three hydroxide ions carry a total 3− charge; ammonia mol-

Table 22.4 Names and Formulas of Some Common Ligands

Neutral Ligand	Ligand Name	Anionic Ligand	Ligand Name
NH_3	Ammine	Br^-	Bromo
CO	Carbonyl	CO_3^{2-}	Carbonato
$H_2NCH_2CH_2NH_2$	Ethylenediamine, en	Cl^-	Chloro
H_2O	Aqua	CN^-	Cyano
		F^-	Fluoro
		OH^-	Hydroxo
		$C_2O_4^{2-}$	Oxalato, ox
		NCS^-	Isothiocyanato
		SCN^-	Thiocyanato

ecules have no net charge, and thus cobalt must be Co^{3+} because the compound has no net charge. In naming any coordination compound or complex ion, the *ligands are named in alphabetical order*—in this case ammine for ammonia precedes hydroxo for hydroxide (for anions *-ide* is changed to *o*). *The name and oxidation state (in parentheses) of the metal ion are given last.* Greek prefixes *di, tri, tetra,* and so on are used to *denote the number of times each of these ligands is used.* Such prefixes are ignored when determining the alphabetical order of the ligands.

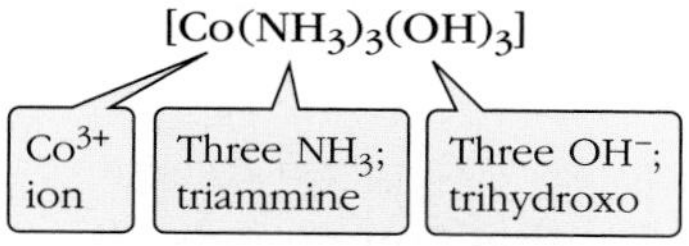

triamminetrihydroxocobalt(III)

Next, consider $[Fe(H_2O)_2(NH_3)_4]Cl_3$, a coordination compound that consists of a complex *cation,* $[Fe(H_2O)_2(NH_3)_4]^{3+}$, and three chloride ions as counter ions. In such cases the *complex cation is always named first, followed by the name of the anionic counter ions.* The compound's name is tetraamminediaquairon(III) chloride. From Table 22.4, we see that the ligands are ammine (NH_3, four of them) and aqua (H_2O, two of them). *For complex cations, the metal ion and its oxidation state follow the names of the ligands.*

Counter ions offset the charge of the complex ion.

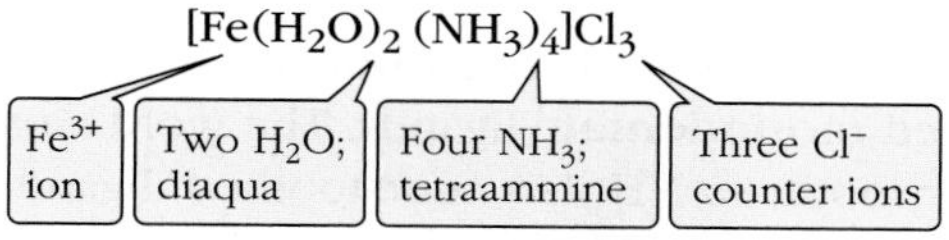

tetraamminediaquairon(III) chloride

The compound $K_2[PtCl_4]$ contains a complex *anion,* $[PtCl_4]^{2-}$, and two K^+ ions as counter ions and is named potassium tetrachloroplatinate(II). As with any ionic compound, the *cation is named first, followed by the anion name.* For complex anions, the central metal ion's name ends in *-ate* followed by its oxidation state in parentheses.

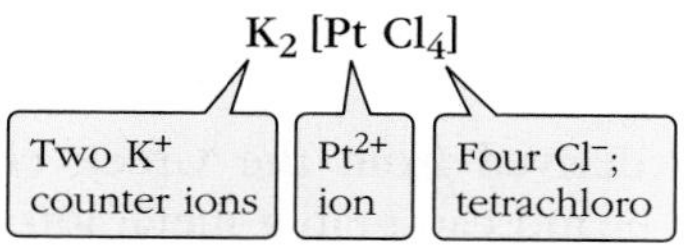

potassium tetrachloroplatinate(II)

PROBLEM-SOLVING EXAMPLE 22.7 Formulas and Names of Coordination Compounds

(a) Write the formula of diamminetriaquahydroxochromium(III) nitrate.
(b) Name $K[Cr(NH_3)_2(C_2O_4)_2]$.

Answer

(a) $[Cr(NH_3)_2(H_2O)_3(OH)](NO_3)_2$
(b) Potassium diamminedioxalatochromate(III)

Strategy and Explanation Use the names and formulas of the ligands in Table 22.4. Compound (a) contains a complex cation, and compound (b) contains a complex anion.

(a)

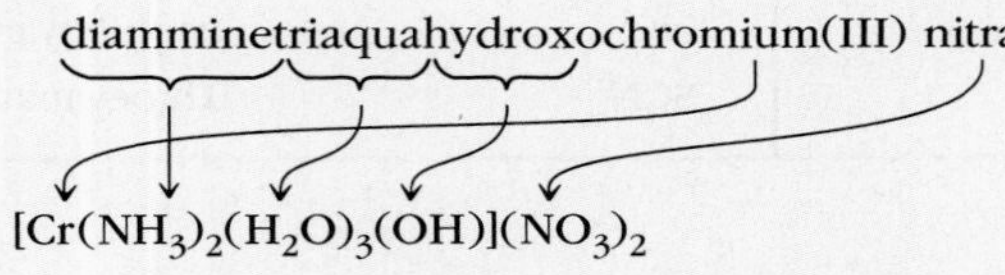

(b)

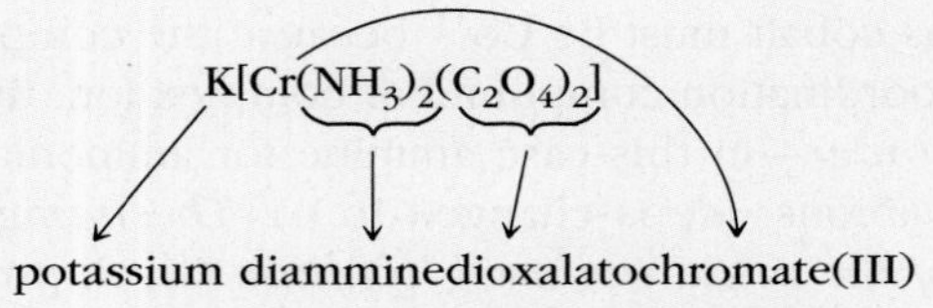

PROBLEM-SOLVING PRACTICE 22.7

(a) Name this coordination compound: $[Ag(NH_3)_2]NO_3$.
(b) Write the formula of pentaaquaisothiocyanatoiron(III) chloride.

CONCEPTUAL EXERCISE 22.9 Coordination Compounds

$CaCl_2$ and $[Ni(H_2O)_6]Cl_2$ have the same formula type, MCl_2. Give the formula of a simple (noncoordination) ionic compound that has a formula analogous to $K_2[NiCl_4]$.

Types of Ligands and Coordination Number

Coordination Number	Examples
2	$[Ag(NH_3)_2]^+$, $[AuCl_2]^-$
4	$[NiCl_4]^{2-}$, $[Pt(NH_3)_4]^{2+}$
6	$[Fe(H_2O)_6]^{2+}$, $[Co(NH_3)_6]^{3+}$

The number of coordinate covalent bonds between the ligands and the central metal ion in a coordination compound is the **coordination number** of the metal ion, usually 2, 4, or 6.

Ligands such as H_2O, NH_3, and Cl^- that form only one coordinate covalent bond to the metal are termed **monodentate** ligands. The word *dentate* derives from the Latin word *dentis,* for tooth, so NH_3 is a "one-toothed" ligand. Common monodentate ligands are shown in Figure 22.9.

Some ligands can form two or more coordinate covalent bonds to the same metal ion because the ligands have two or more atoms with lone pairs separated by several intervening atoms. The general term **polydentate** is used for such ligands. **Bidentate** ligands are those that form *two* coordinate covalent bonds to the central metal ion. A good example is the bidentate ligand 1,2-diaminoethane, $H_2NCH_2CH_2NH_2$, commonly called ethylenediamine and abbreviated *en.* When lone pairs of electrons from both nitrogen atoms in en coordinate to a metal ion, a stable five-membered ring forms (Figure 22.10). Notice that Co^{3+} has a coordination number of 6 in $[Co(en)_3]^{3+}$.

The word "chelating," derived from the Greek *chele*, "claw," describes the pincer-like way in which a ligand can grab a metal ion. Some common **chelating ligands,** those that are polydentate ligands and can bond with the central metal ion using two or more electron pairs, are also shown in Figure 22.9.

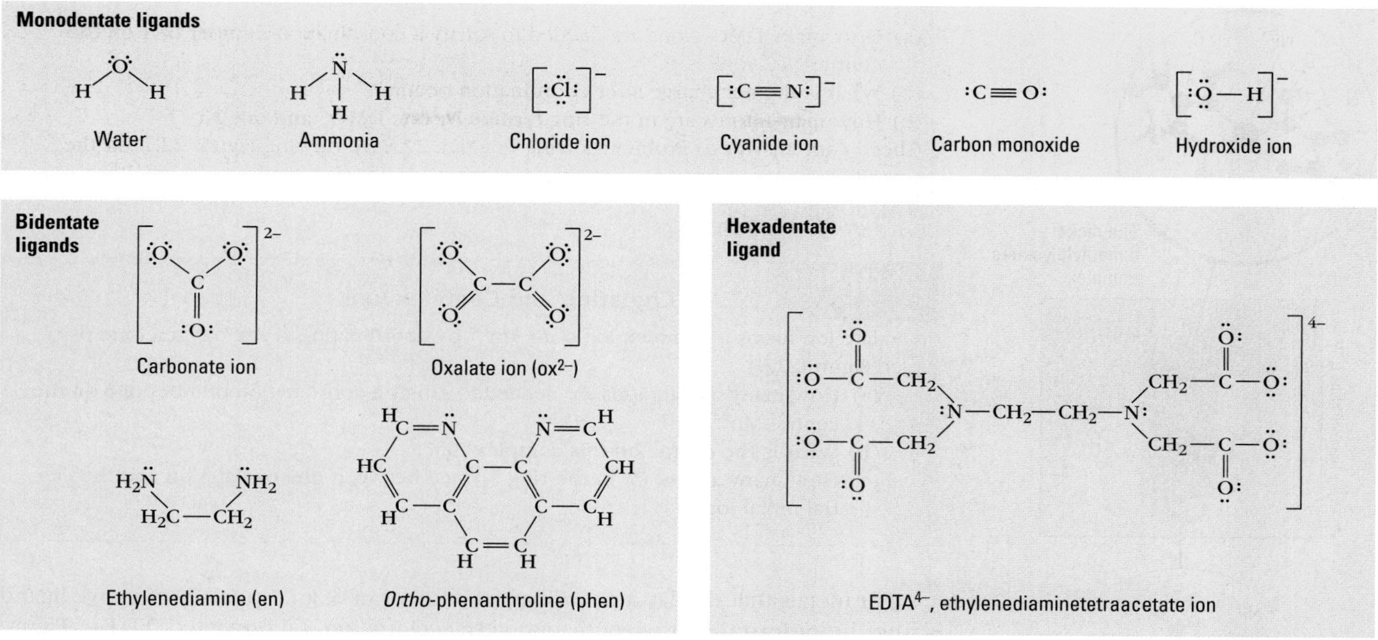
Monodentate ligands
Water
Ammonia
Chloride ion
Cyanide ion
Carbon monoxide
Hydroxide ion
Bidentate ligands
Carbonate ion
Oxalate ion (ox2–)
Ethylenediamine (en)
Ortho-phenanthroline (phen)
Hexadentate ligand
EDTA4–, ethylenediaminetetraacetate ion

Figure 22.9 Monodentate, bidentate, and hexadentate ligands. Chelating ligands have two lone pairs (bidentate) up to six lone pairs (hexadentate) that can be shared with a central metal ion.

PROBLEM-SOLVING EXAMPLE 22.8 Chelating Agents

Two ethylenediamine (en) ligands and two chloride ions form a complex ion with Co^{3+}.
(a) Write the formula of this complex ion.
(b) What is the coordination number of the Co^{3+} ion?
(c) Write the formula of the coordination compound formed by Cl^- counter ions and the Co^{3+} complex ion.

Answer

(a) $[Co(en)_2Cl_2]^+$ (b) 6 (c) $[Co(en)_2Cl_2]Cl$

Strategy and Explanation Recall that ethylenediamine is a bidentate ligand that forms two coordinate covalent bonds per en molecule.

(a) Two en molecules and two chloride ions are bonded to the central cobalt ion, so the formula of the complex ion is $[Co(en)_2Cl_2]^+$. Ethylenediamine is a neutral ligand, each chloride ion is 1−, and cobalt has a 3+ charge. The charge on the complex ion is $2(0) + 2(1-) + (3+) = 1+$.
(b) The coordination number is 6 because there are six coordinate covalent bonds to the central Co^{3+} ion—two from each bidentate ethylenediamine and one from each monodentate chloride ion.
(c) The 1+ charge of the complex ion requires one chloride ion as a counter ion: $[Co(en)_2Cl_2]Cl$.

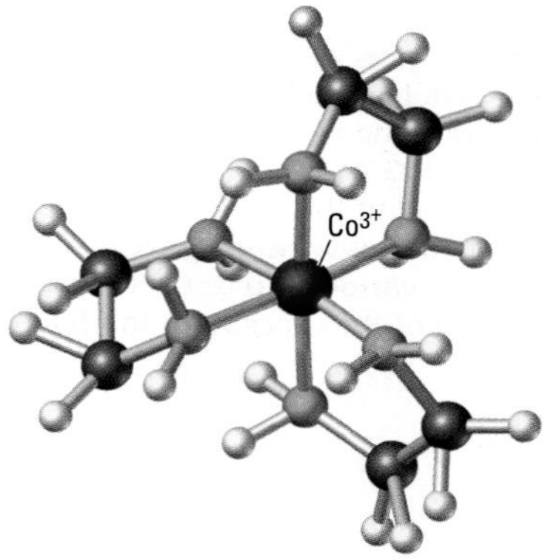
Co3+

Figure 22.10 The $[Co(en)_3]^{3+}$ complex ion. Cobalt ion (Co^{3+}) forms a coordination complex ion with three ethylenediamine ligands.

Note that Cl_2 in the complex ion's formula represents two chloride ions, *not* a diatomic chlorine molecule.

PROBLEM-SOLVING PRACTICE 22.8

The dimethylglyoximate anion (abbreviated DMG^-),

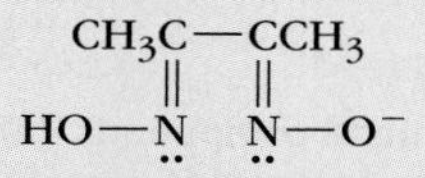

is a bidentate ligand used to test for the presence of nickel. It reacts with Ni^{2+} to form a beautiful red solid in which the Ni^{2+} has a coordination number of 4. DMG^- coordinates to Ni^{2+} by the lone pairs on the nitrogen atoms.

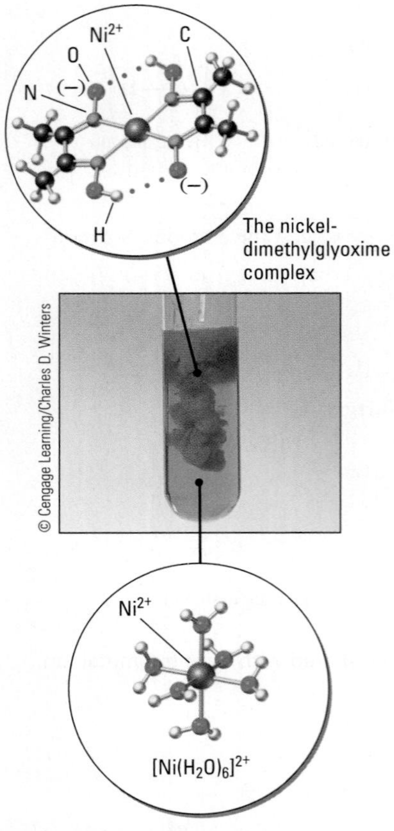

Active Figure 22.11 The nickel-dimethylglyoxime complex. Ni^{2+} ions react with the dimethylglyoximate anion (DMG^-) to form a beautiful red solid. Visit this book's companion website at **www.cengage.com/chemistry/moore** to test your understanding of the concepts in this figure.

(a) How many DMG^- ions are needed to satisfy a coordination number of 4 on the central Ni^{2+} ion?
(b) What is the net charge after coordination occurs?
(c) How many atoms are in the ring formed by one DMG^- and one Ni^{2+}?
Check your answer to Problem-Solving Practice 22.8 by viewing Figure 22.11 at the website.

CONCEPTUAL EXERCISE 22.10 Chelating and Complex Ions

Oxalate ion forms a complex ion with Mn^{2+} by coordinating at the oxygen lone pairs (see Figure 22.9).

(a) How many oxalate ions are needed to satisfy a coordination number of 6 on the central Mn^{2+} ion?
(b) What is the charge on this complex ion?
(c) How many atoms are in the ring formed between one oxalate ion and the central metal ion?

For metals that display a coordination number of 6, an especially effective ligand is the **hexadentate** ethylenediaminetetraacetate ion (abbreviated EDTA; Figure 22.9) that encapsulates and firmly binds metal ions. Hexadentate ligands such as EDTA have six lone pair donor atoms (four O atoms and two N atoms) that can coordinate to a single metal ion, so $EDTA^{4-}$ is an excellent chelating ligand. $EDTA^{4-}$ is often added to commercial salad dressing to remove traces of metal ions from solution, because these metal ions could otherwise accelerate the oxidation of oils in the product and make them rancid.

Another use of $EDTA^{4-}$ is in bathroom cleansers, where it removes hard water deposits of insoluble $CaCO_3$ and $MgCO_3$ by chelating Ca^{2+} or Mg^{2+} ions, allowing them to be rinsed away. EDTA is also used in the treatment of lead and mercury poisoning because it has the ability to chelate these metals and aid in their removal from the body (Figure 22.12).

Coordination compounds of *d*-block transition metals are often colored, and the colors of the complexes of a given transition metal ion depend on both the metal ion and the ligand (Figure 22.13). Many transition metal coordination compounds are used as pigments in paints and dyes. For example, Prussian blue, $Fe_4[Fe(CN)_6]_3$, a deep-blue pigment known for hundreds of years, is the "bluing agent" in engineer-

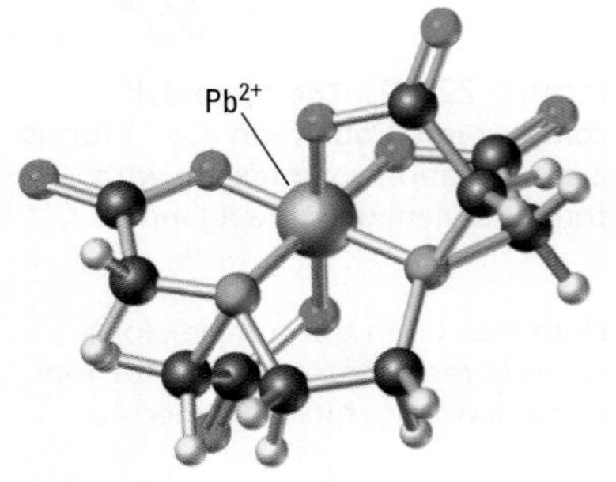

Figure 22.12 A Pb^{2+}-EDTA complex ion. The structure of the chelate formed when the $EDTA^{4-}$ anion forms a complex with Pb^{2+}.

(a) (b)

Figure 22.13 Color of transition metal compounds. (a) Concentrated aqueous solutions of the nitrate salts containing hydrated transition metal ions of (*left to right*) Fe^{3+}, Co^{2+}, Ni^{2+}, Cu^{2+}, and Zn^{2+}. Aqueous Zn^{2+} compounds are colorless. (b) The colors of the complexes of a given transition metal ion depend on the ligand(s). All of the complexes pictured in (b) contain the Ni^{2+} ion. The green solid is $[Ni(H_2O)_6](NO_3)_2$; the purple solid is $[Ni(NH_3)_6]Cl_2$; the red solid is $Ni(dimethylglyoximate)_2$.

CHEMISTRY YOU CAN DO — A Penny for Your Thoughts

You will need the following items to do the experiment:

- Two glasses or plastic cups that will each hold about 50 mL liquid
- About 30 to 40 mL household vinegar
- About 30 to 40 mL household ammonia
- A copper penny

Place the penny in one cup and add 30 to 40 mL vinegar to clean the surface of the penny. Let the penny remain in the vinegar until the surface of the penny is cleaner (reddish-coppery) than it was before (darker copper color). Pour off the vinegar and wash the penny thoroughly in running water.

Next, place the penny in the other cup and add 30 to 40 mL household ammonia. Observe the color of the solution over several hours.

Think about these questions:

1. What did you observe happening to the penny in the ammonia solution?
2. What did you observe happening to the ammonia solution?
3. Interpret what you observed happening to the solution on the nanoscale level, citing observations to support your conclusions.
4. What is necessary to form a complex ion?
5. Are all of these kinds of reactants present in the solution in this experiment? If so, identify them.
6. How do the terms "ligand," "central metal ion," and "coordination complex" apply to your experiment?
7. Try to write a formula for a complex ion that might form in this experiment.

ing blueprints. The origin of colors in coordination compounds will be discussed in Section 22.7 (p. 1065).

Some household products that contain EDTA. Check the label on your shampoo container. It will likely list disodium EDTA as an ingredient. The EDTA in this case has a 2− charge because two of the four organic acid groups have each lost an H^+, but $EDTA^{2-}$ still coordinates to metal ions in the shampoo.

CONCEPTUAL EXERCISE 22.11 Complex Ions

Prussian blue contains two kinds of iron ions. What is the charge of the iron in

(a) The complex ion $[Fe(CN)_6]^{4-}$?

(b) The iron ion not in the complex ion?

Geometry of Coordination Compounds and Complex Ions

The geometry of a complex ion or coordination compound is dictated by the arrangement of the electron donor atoms of the ligands attached to the central metal ion. Although other geometries are possible, we will discuss only the four most common ones, those associated with coordination numbers of 2, 4, and 6. To simplify matters, we will consider only monodentate ligands, L, bonded to a central metal ion, M^{n+}.

Coordination Number = 2, ML_2^{n+} All such complex ions have a *linear* geometry with the two ligands on opposite sides of the central metal ion to give an L—M—L bond angle of 180°, such as that in $[Ag(NH_3)_2]^+$. Other examples are $[CuCl_2]^-$ and $[Au(CN)_2]^-$, the complex ion used to extract gold from ores (p. 1052).

$[Cl—Cu—Cl]^-$

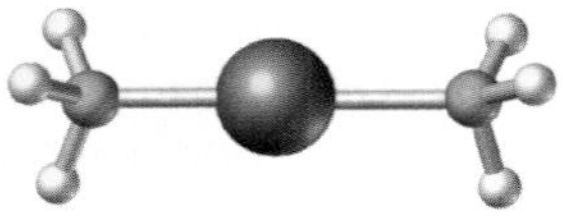

$[H_3N—Ag—NH_3]^+$

Coordination Number = 4, ML_4^{n+} Four-coordinate complex ions have either tetrahedral or square planar geometries. In the *tetrahedral* case, the four monodentate ligands are at the corners of a tetrahedron, such as in $[Zn(NH_3)_4]^{2+}$; L—M—L bond angle is 109.5°. In *square planar* geometry, the ligands lie in a plane at the corners of a square as in $[Ni(CN)_4]^{2-}$ and $[Pt(NH_3)_4]^{2+}$ ions; L—M—L bond angle is 90°.

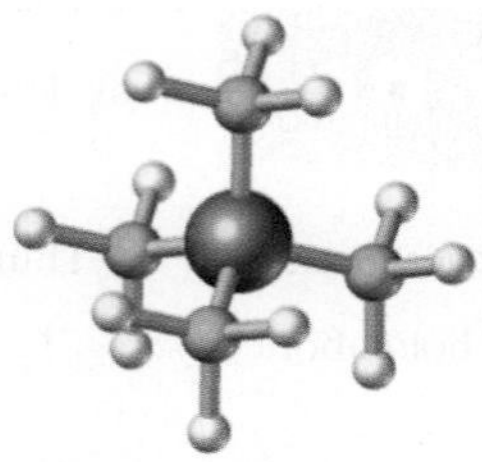

Tetrahedral
$[Zn(NH_3)_4]^{2+}$

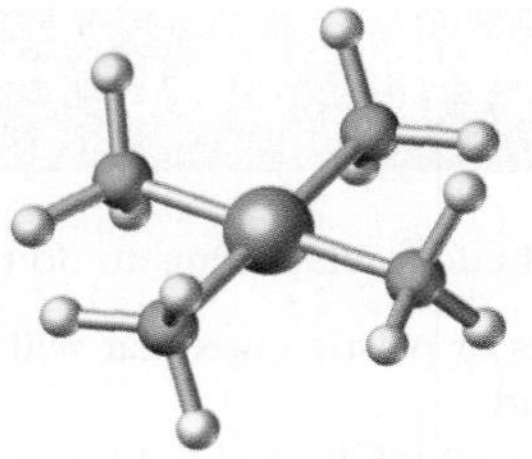

Square planar
$[Pt(NH_3)_4]^{2+}$

Coordination Number = 6, ML_6^{n+} Octahedral geometry is characteristic of this coordination number. The six ligands are at the corners of an octahedron with the central metal ion at its center. Octahedral geometry can be regarded as derived from a square planar geometry by adding two ligands, one above and one below the square plane. Two common octahedral complex ions are $[Co(NH_3)_6]^{3+}$ and $[Fe(CN)_6]^{3-}$, in which the six ligands are equidistant from the central metal ion and all six ligand sites are equivalent; L—M—L bond angle is 90°.

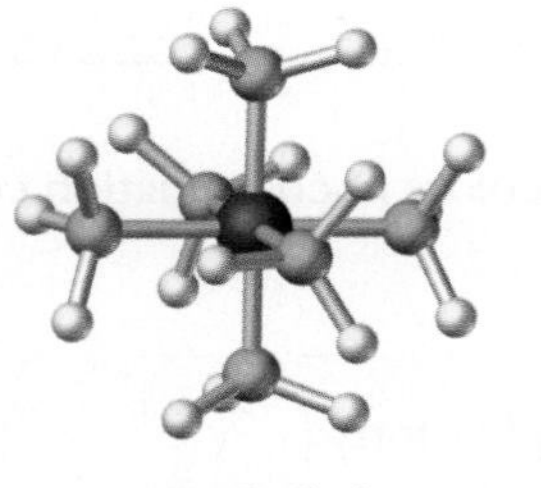

Octahedral
$[Co(NH_3)_6]^{3+}$

Isomerism in Coordination Compounds and Complex Ions

Various types of isomerism have been discussed previously with regard to organic compounds. *Constitutional isomerism* occurs with molecules that have the same molecular formula but differ in the way their atoms are connected together, such as occurs with butane and 2-methylpropane **(← p. 83)**. *Stereoisomerism* is a second general category of isomerism in which the isomers have the same bond connections, but the atoms are arranged differently in space. One type of stereoisomerism is *geometric isomerism,* such as that found in *cis-* and *trans-*1,2-dichloroethene **(← p. 341)**. The other type of stereoisomerism is *optical isomerism,* which occurs when mirror images are nonsuperimposable **(← p. 388)**. Constitutional, geometric, and optical isomers also occur with coordination complex ions and coordination compounds.

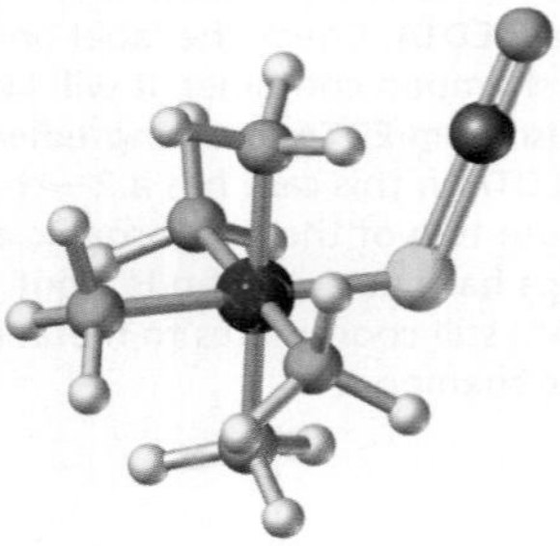

pentaammine-thiocyanatocobalt(III) ion

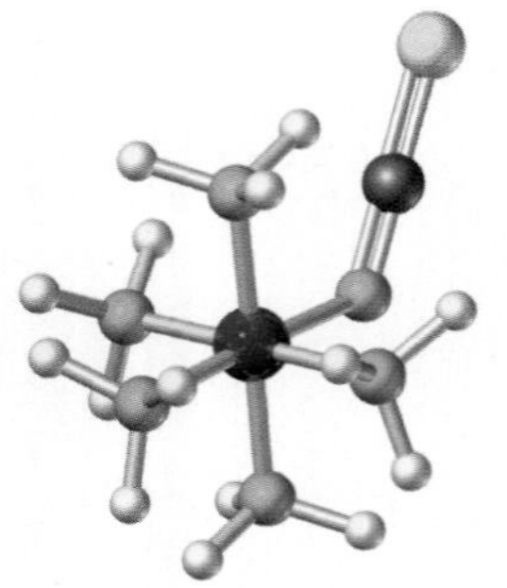

pentaammine-isothiocyanatocobalt(III) ion

Linkage Isomerism, a Type of Constitutional Isomerism Linkage isomerism occurs when a ligand can bond to the central metal using either of two different electron-donating atoms. Thiocyanato $(SCN)^-$ and isothiocyanato $(NCS)^-$ ions are examples of such ligands with coordination to a metal ion by sulfur in the first case and by nitrogen in the second, as illustrated for the Co^{3+} complex ions shown in the margin.

Geometric Isomerism Geometric isomerism does not exist in tetrahedral complex ions because all the corners of a tetrahedron are equivalent. Geometric isomerism, however, does occur with square planar complex ions and compounds of the type Ma_2b_2 or Ma_2bc, where M is the central metal ion and a, b, and c are *different* ligands. The square planar coordination compound $[Pt(NH_3)_2Cl_2]$, an Ma_2b_2 type, occurs in two geometric forms. The *cis*-$[Pt(NH_3)_2Cl_2]$ isomer has the chloride ligands as close as possible at 90° to one another. In *trans*-$[Pt(NH_3)_2Cl_2]$, the chloride ions are as far apart as possible, directly across the square plane of the molecule at 180° from each other.

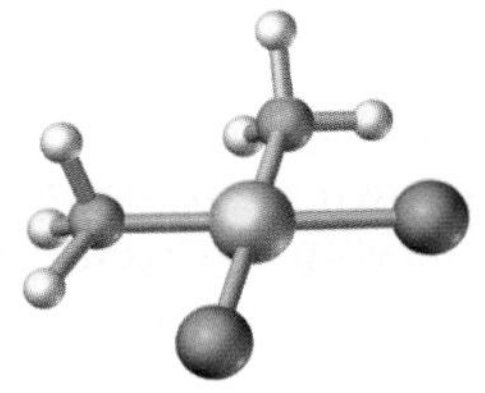

cis-$[Pt(NH_3)_2Cl_2]$

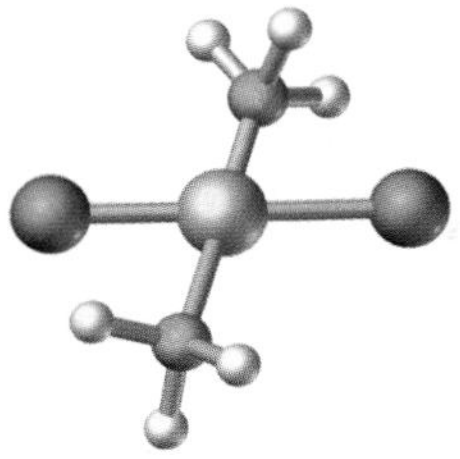

trans-$[Pt(NH_3)_2Cl_2]$

These two isomers differ in water solubility, color, melting point, and chemical reactivity. The *cis* isomer is used in cancer chemotherapy, whereas the *trans* form is not effective against cancer.

Cis-trans isomerism is also possible in octahedral complex ions and compounds, as illustrated with $[Co(NH_3)_4Cl_2]^+$. In this complex ion the *cis* isomer has the chloride ions adjacent to each other; the *trans* isomer has them opposite each other. The differences in properties are striking, particularly the color. The *cis* isomer is violet, whereas the *trans* form is green.

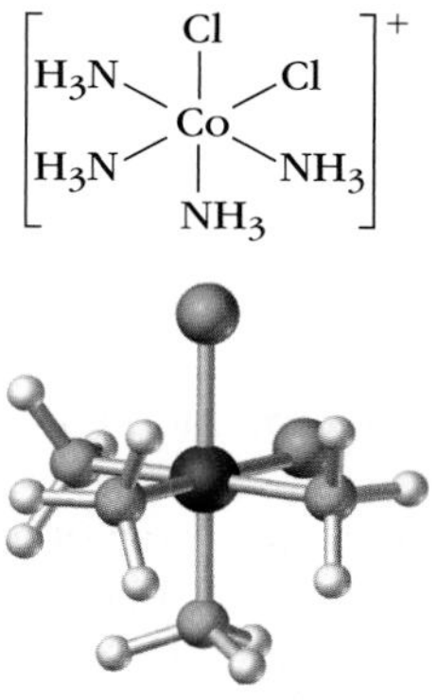

cis-$[Co(NH_3)_4Cl_2]^+$
(violet)

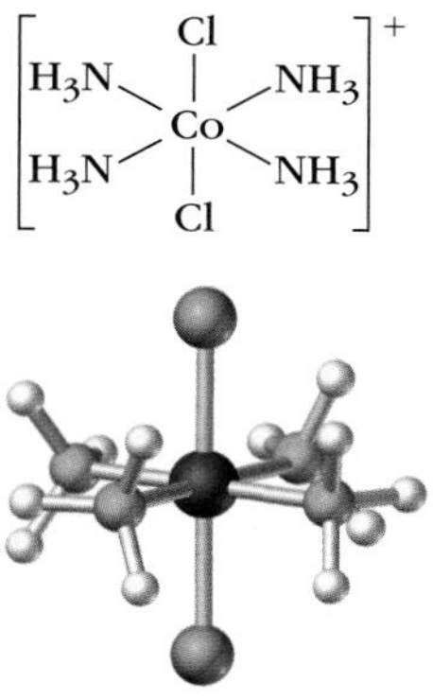

trans-$[Co(NH_3)_4Cl_2]^+$
(green)

Science Photo Library/Photo Researchers, Inc.

Alfred Werner

1866–1919

In 1893, while still a young professor, Alfred Werner published a revolutionary paper about transition metal compounds. He asserted that transition metal ions could exhibit a secondary valence as well as a primary one, such as in $CoCl_3 \cdot 6\ NH_3$ (now written as $[Co(NH_3)_6]Cl_3$). The primary valence was represented by the ionic bonds between Co^{3+} and the chloride ions; the secondary valence was represented by the coordinate covalent bonds between the metal ion and six NH_3 molecules, what we now called the coordination sphere around the central metal ion. Werner also made the inspired proposal that the ammonia molecules were octahedrally coordinated around the Co^{3+} ion, thereby laying the foundation for understanding the geometry of complex ions. For his groundbreaking work, Werner received the 1913 Nobel Prize in Chemistry.

PROBLEM-SOLVING EXAMPLE 22.9 Geometric Isomerism

How many geometric isomers are there for $[Co(en)_2Cl_2]^+$?

Answer Only two geometric isomers are possible, *cis* and *trans.*

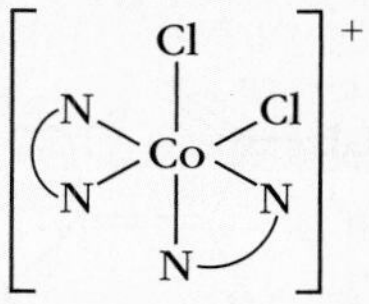

cis

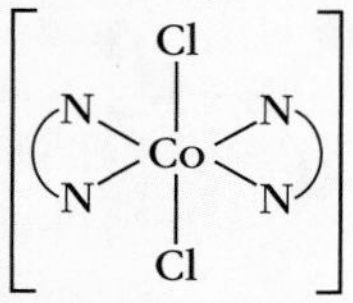

trans

Strategy and Explanation Start by putting the two Cl^- ions in *trans* positions, that is, one at the "top" of the octahedron and the other at the "bottom." The two ethylenediamine ligands (en), represented here as N͡N, occupy the other four sites around the cobalt ion. This is the *trans* isomer. The *cis* isomer has the Cl^- ions in adjacent *(cis)* positions.

PROBLEM-SOLVING PRACTICE 22.9

How many geometric isomers are there for the square planar compound $[Pt(NH_3)_2ClBr]$?

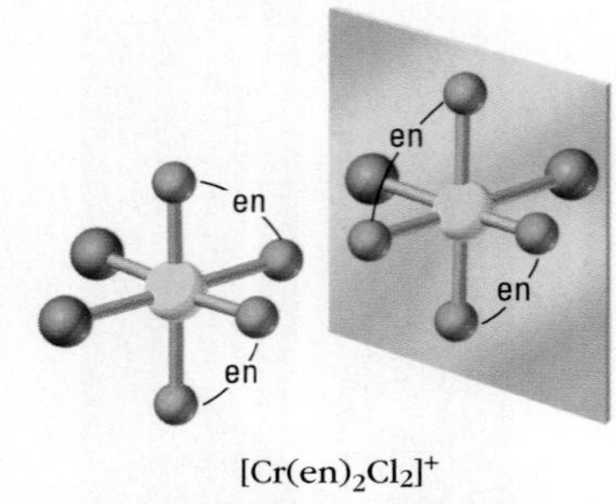

Figure 22.14 Optical isomerism in $[Cr(en)_2Cl_2]^+$. The ion on the *left* cannot be superimposed on its mirror image *(right).*

EXERCISE 22.12 Geometric Isomerism

How many isomers are possible for $[Co(NH_3)_3Cl_3]$? Write the structural formulas of the isomers.

Optical Isomerism Optical isomers are mirror images that are not superimposible. Such nonsuperimposable mirror images are known as *enantiomers* **(← *p. 388*).** An example of a complex ion that has optical isomerism is $[Cr(en)_2Cl_2]^+$. There are two enantiomers, as shown in Figure 22.14. No matter how they are twisted and turned, the two enantiomers are nonsuperimposable.

Optical isomerism is not possible for square planar complexes based on the geometry around the metal ion; the mirror images are superimposable. Although optical isomers of tetrahedral complex ions with four different ligands are theoretically possible, no such stable complexes are known.

Coordination Compounds and Life

Bioinorganic chemistry, the study that applies chemical principles to inorganic ions and compounds in biological systems, is a rapidly growing field centered mainly around coordination compounds. This is because the very existence of living systems depends on many coordination compounds in which metal ions are chelated to the nitrogen and oxygen atoms in proteins and especially in enzymes. Copper-containing proteins, for example, give the blood of crabs, lobsters, and snails its blue color, as well as transport oxygen.

In humans, molecular oxygen, O_2 is transported through the circulatory system by hemoglobin, a very large protein (molecular weight of about 68,000 amu) in red blood cells. Hemoglobin is blue but becomes red when oxygenated. This is why arterial blood is bright red (high O_2 concentration) and blood in veins is bluish (low O_2 concentration).

A hemoglobin molecule carries four O_2 molecules, each of which forms a coordinate covalent bond to one of four Fe^{2+} ions (Figure 22.15). Each Fe^{2+} ion is at the center of one heme, a nonprotein part of the hemoglobin molecule that consists of four linked nitrogen-containing rings. Bound in this way, molecular oxygen is carried to the cells, where it is released as needed by breaking the Fe—O_2 coordinate covalent bond.

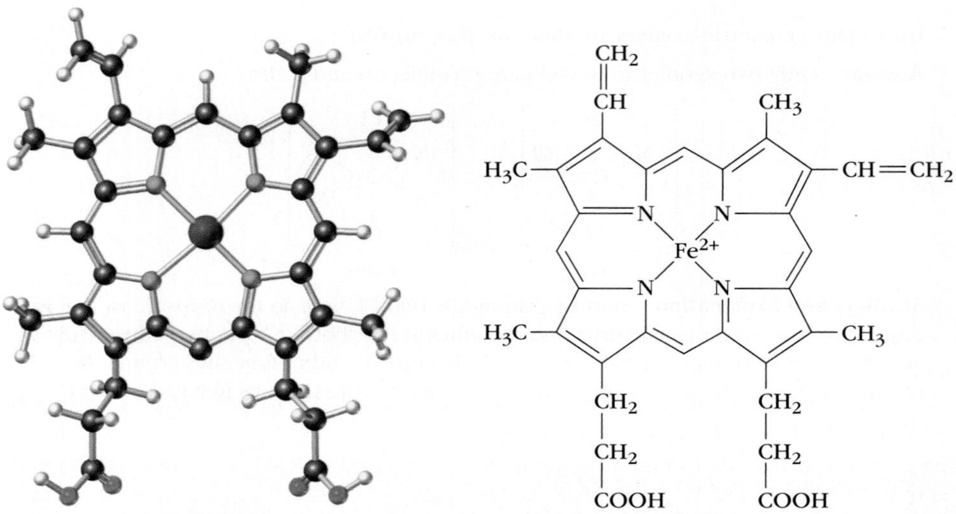

Figure 22.15 Heme, the carrier of Fe^{2+} in hemoglobin. Fe^{2+} is coordinated to four nitrogen atoms in heme. There are four hemes in each hemoglobin molecule.

Other substances that can donate an electron pair can also bond to the Fe^{2+} in heme. Carbon monoxide is such a ligand and forms an exceptionally strong Fe^{2+}—CO bond, nearly 200 times stronger than the O_2—Fe^{2+} bond. Therefore, when a person breathes in CO, it displaces O_2 from hemoglobin and prevents red blood cells from carrying oxygen. The initial effect is drowsiness. But if CO inhalation continues, cells deprived of oxygen can no longer function and the person suffocates.

It is interesting (and fortunate) that N≡N does not behave chemically like C≡O, even though each molecule contains 14 electrons.

Structures similar to the oxygen-carrying unit in hemoglobin are also found in other biologically important compounds, including such diverse ones as myoglobin and vitamin B-12. Myoglobin, like hemoglobin, contains Fe^{2+} and carries and stores molecular oxygen, principally in muscles. At the center of a vitamin B-12 molecule is a Co^{3+} ion bonded to the same type of group that Fe^{2+} bonds to in hemoglobin. Vitamin B-12 is the only known dietary use of cobalt, but it makes cobalt an essential mineral ***(← p. 108)***.

The dietary necessity of zinc for humans has become established only since the 1980s. Zinc, in the form of Zn^{2+} ions, is essential to the functioning of several hundred enzymes, including those that catalyze the breaking of P—O—P bonds in adenosine triphosphate (ATP), an important energy-releasing compound in cells ***(← p. 877)***.

Copper ranks third among biologically important transition metal ions in humans, trailing only iron and zinc. Although we generally excrete any dietary excess of copper, a genetic defect causes Wilson's disease, a condition in which Cu^{2+} accumulates in the liver and brain. Fortunately, Wilson's disease can be treated by administering chelating agents that coordinate excess Cu^{2+} ions, allowing them to be excreted harmlessly.

22.7 Crystal-Field Theory: Color and Magnetism in Coordination Compounds

Bright colors are characteristic of many coordination compounds (Figure 22.13). Any theory about bonding in such compounds needs to address the origin of their colors. One such approach is the crystal-field theory developed by Hans Bethe and J. H. van Vleck. Crystal-field theory explains color as originating from electron transitions between two sets of *d* atomic orbitals in a complex ion, similar to the bright line atomic spectra that originate from electron transitions among atomic orbitals in elements. In many cases of complex ions, the energy difference between the *d* atomic orbitals corresponds with a wavelength within the visible region of the spectrum.

Crystal-Field Theory

The crystal-field theory considers the bonding between ligands and a metal ion to be primarily electrostatic. It assumes that the electron pairs on the ligands create an electrostatic field around the *d* atomic orbitals of the metal ion, such as in the case of the octahedral $[Fe(CN)_6]^{4-}$ complex ion. This electrostatic field alters the relative energies of the various *d* atomic orbitals. Before this interaction occurs, the *d* atomic orbitals in a sublevel of the isolated metal ion, such as the 3*d*, all have the same energy. In the presence of ligands, the *d* atomic orbitals split into two groups of differing energy (Figure 22.16). The higher-energy pair consists of the $d_{x^2-y^2}$ and d_{z^2} atomic orbitals; the lower-energy trio is made up of the d_{xy}, d_{yz}, and d_{xz} atomic orbitals.

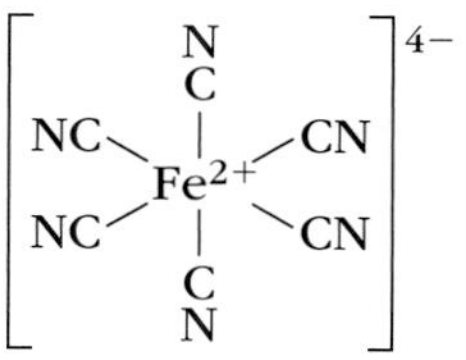

The $[Fe(CN)_6]^{4-}$ complex ion. The six CN^- ions are arranged octahedrally around the central Fe^{2+} ion.

Why does this splitting occur? To understand this, we must first consider the orientations of *d* atomic orbitals on the central metal ion and what happens to the energy of the *d* atomic orbitals when they are approached by the six ligands as an assembly of negative charges. From Figure 22.17 note that in a set of five *d* atomic orbitals, two of the *d* atomic orbitals—$d_{x^2-y^2}$ and the d_{z^2}—have their lobes of maximum electron density directly *along* the *x*-, *y*-, and *z*-axes. In contrast, the other three

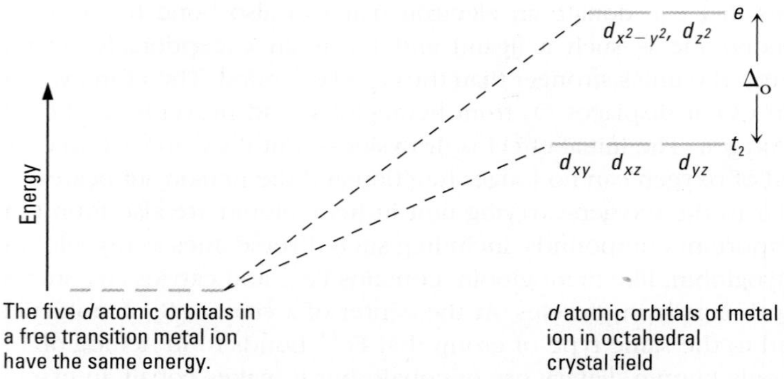

Figure 22.16 The splitting of *d* atomic orbitals in an octahedral field of ligands.

d atomic orbitals—d_{xy}, d_{yz}, and d_{xz}—have their lobes of maximum electron density aligned *between* the *x*-, *y*-, and *z*-axes rather than directly along them. Thus, as the CN^- ligands approach along the *x*-, *y*-, and *z*-axes to form an octahedral field of ligands, the electrostatic field created by the electron pairs of the ligands is felt more strongly by electrons in the $d_{x^2-y^2}$ and d_{z^2} atomic orbitals—those directly along the axes. Electrons in orbitals not directly along the axes—the d_{xy}, d_{yz}, and d_{xz} atomic orbitals—are not disturbed as strongly by the electrostatic field. The overall result is the splitting of the 3*d* atomic orbitals in the complex ion into two sets of differing energy; the higher-energy pair of atomic orbitals, labeled *e*, and the lower-energy trio of atomic orbitals, labeled t_2 (Figure 22.16). The energy difference between the sets of *d* atomic orbitals is known as the **crystal-field splitting energy, Δ.** For an octahedral field of ligands, as in $Fe(CN)_6^{4-}$, a subscript o is added, Δ_o.

Electron Configurations and Magnetic Properties of Coordination Complex Ions

The electron configurations and magnetic properties of transition metal ions can be interpreted by using crystal-field theory. Applying Hund's rule ***(← p. 297)***, electrons occupy the vacant atomic orbitals of lowest energy first. If the vacant atomic orbitals all have the same energy, electrons occupy those orbitals unpaired with their spins parallel. Consequently, in a complex ion where the metal ion has a d^1, d^2, or d^3 electron configuration, the electrons will occupy the lower-energy t_2 set of *d* atomic orbitals (Figure 22.18).

We might expect that the trend of unpaired electrons would continue with the filling of the set of *e* higher-energy *d* atomic orbitals for metal ions with d^4 and d^5

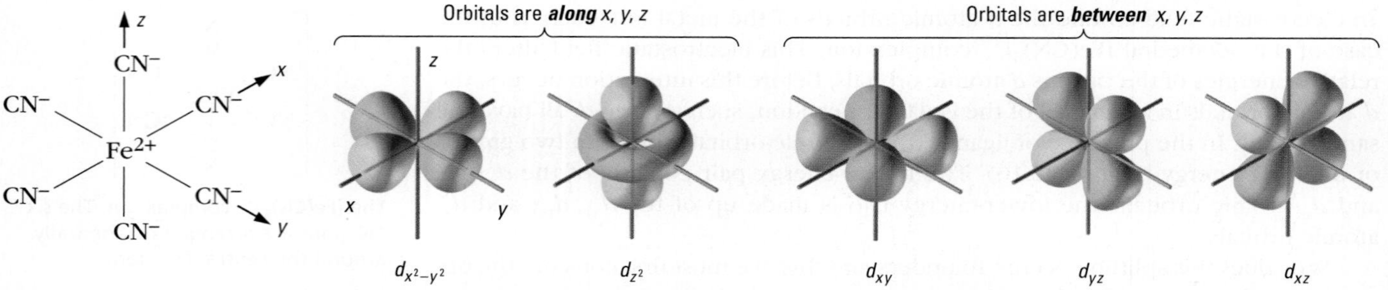

Figure 22.17 The spatial arrangement of *d* atomic orbitals of an Fe^{2+} ion and the octahedral orientation of six CN^- ligands as negative point charges along the *x*-, *y*-, and *z*-axes. The $d_{x^2-y^2}$ and d_{z^2} atomic orbitals of the metal ion are aligned along the axes and oriented toward ligands approaching the six corners of the octahedron. The other three *d* atomic orbitals are oriented between the axes.

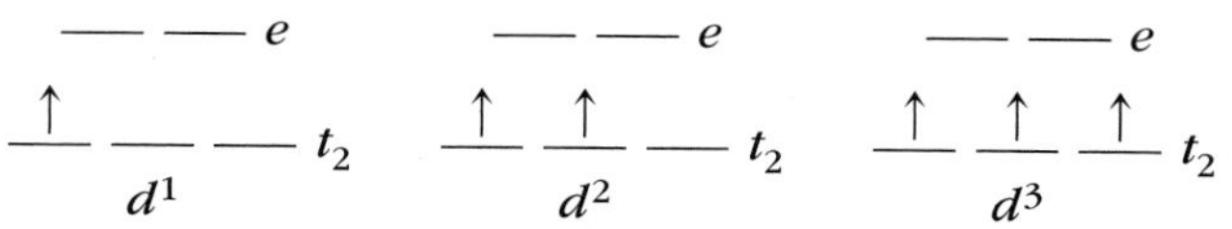

Figure 22.18 Octahedral complex ions with d^1, d^2, and d^3 electron configurations. In each case the electrons are unpaired.

electron configurations. This is true in some, but not all cases. *Whether these electrons are unpaired or paired depends on the magnitude of the crystal-field splitting energy, Δ_o, the energy gap between the metal ion's two sets of* d *atomic orbitals*. That crystal-field splitting energy depends on the metal ion and the ligands. The relative effect of the ligands is given by the **spectrochemical series,** a listing of ligands in the order of their crystal-field splitting energy. The spectrochemical series is determined experimentally. An abbreviated spectrochemical series for octahedral complexes is

$$Cl^- < F^- < H_2O < NCS^- < NH_3 < en < NO_2^- < CN^-$$

weak field ⟶ increasing Δ_o ⟶ strong field

Ligands such as CN^-, NO_2^-, and en that cause a large Δ_o are called *strong-field* ligands; those at the other end of the series, such as Cl^- and F^-, with smaller Δ_o are termed *weak-field* ligands.

Consider two complex ions $[Fe(CN)_6]^{4-}$ and $[Fe(H_2O)_6]^{2+}$, each containing the Fe^{2+} ion with six $3d$ electrons. As seen from the spectrochemical series, cyanide ion, CN^-, is a strong-field ligand with a Δ_o large enough to cause the six d electrons to pair in the three lower-energy t_2 atomic orbitals (Figure 22.19). In $[Fe(H_2O)_6]^{2+}$, water is a weak-field ligand with a Δ_o sufficiently smaller that four of the six d electrons remain unpaired. Using Hund's rule for the aqua complex ion, the first five electrons occupy each of the five d atomic orbitals individually, with pairing occurring when the sixth electron is added into one of the t_2 atomic orbitals. The result is four electrons occupying the t_2 atomic orbitals and the remaining two electrons unpaired in the higher-energy set of d_{z^2} and $d_{x^2-y^2}$ atomic orbitals (Figure 22.19). Thus, $[Fe(H_2O)_6]^{2+}$ is paramagnetic with four unpaired electrons, while $[Fe(CN)_6]^{4-}$ has no unpaired electrons and is diamagnetic (p. 1040).

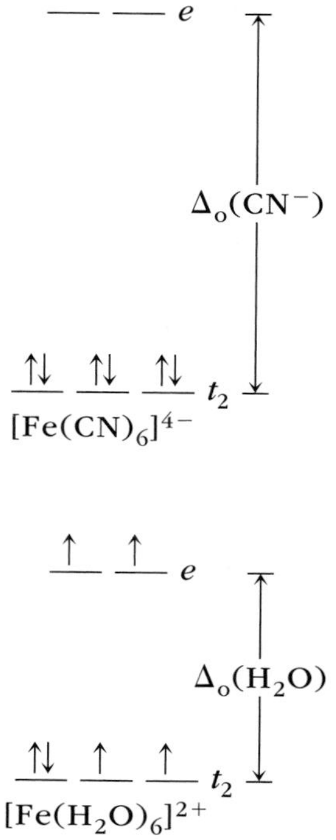

Figure 22.19 Ligands affect the number of unpaired electrons in these two complex ions. Cyanide is a strong-field ligand and the large Δ_o causes maximum electron pairing in Fe^{2+}. Water is a weak-field ligand and the small Δ_o allows the maximum number of unpaired electrons.

How paramagnetic a metal ion is can be measured using a special balance (p. 1040).

High-spin and Low-spin Complexes Pairs of complex ions like this are known for many other transition metal complex ions, those in which the metal ion contains between four and seven inner d electrons. The complex ion with the greater number of unpaired electrons is known as the **high-spin complex;** the **low-spin complex** contains the lesser number of unpaired electrons. High-spin complexes are expected with weak-field ligands where the crystal-field splitting energy Δ_o is small. The opposite applies to low-spin complexes in which strong-field ligands cause maximum pairing of electrons in the set of three t_2 atomic orbitals due to large Δ_o.

High-spin: maximum number of unpaired electrons. $[Fe(H_2O)_6]^{2+}$ is high-spin.

Low-spin: minimum number of unpaired electrons. $[Fe(CN)_6]^{4-}$ is low-spin.

PROBLEM-SOLVING EXAMPLE 22.10 High-spin and Low-spin Complex Ions

Experimental data indicate that $[Co(CN)_6]^{3-}$ is diamagnetic and $[CoF_6]^{3-}$ is paramagnetic.

(a) Use the crystal-field model to illustrate the $3d$ electron configuration of the cobalt ion in each complex.
(b) How many unpaired electrons are in the $3d$ atomic orbitals of the paramagnetic complex?
(c) Which is the low-spin complex?

Answer

(a)

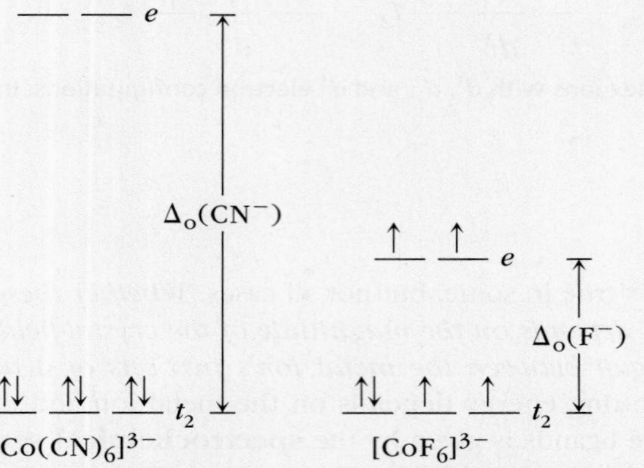

(b) Four (c) $[Co(CN)_6]^{3-}$

Strategy and Explanation We first must determine the number of 3*d* electrons in each case and then use the spectrochemical series to determine the relative crystal field strengths of cyanide and fluoride ions as ligands. Cobalt metal atoms have the $[Ar]4s^2 3d^7$ electron configuration and lose three electrons to form Co^{3+} ions. Recall that in forming ions, transition metals lose *ns* electrons before losing $(n - 1)$ *d* electrons (p. 1040). In this case, two 4*s* electrons are lost plus one 3*d* electron to give Co^{3+} ion an $[Ar]3d^6$ electron configuration.

(a) Cyanide ion is a strong-field ligand that will cause maximum pairing of electrons (low spin). Fluoride ion, a weak-field ligand, is not sufficiently strong to cause maximum *d* electron pairing; instead, the maximum number of unpaired electrons occurs (high spin) as shown in the answer above.

(b) In $[CoF_6]^{3-}$, there are four unpaired 3*d* electrons.

(c) Because $[Co(CN)_6]^{3-}$ is diamagnetic, it contains no unpaired electrons and must be the low-spin complex of these two.

PROBLEM-SOLVING PRACTICE **22.10**

A Cr^{2+} ion contains four 3*d* electrons. How many unpaired electrons are in a high-spin octahedral complex of this metal ion? A low-spin complex of this ion? Is either complex paramagnetic? Explain your answer.

CONCEPTUAL EXERCISE **22.13 High- and Low-spin Complexes**

Explain why Ni^{2+} ions cannot form high- and low-spin complexes.

Tetrahedral and Square Planar Complexes Up to this point we have focused on the application of crystal-field theory to octahedral complexes. It can be applied to tetrahedral and square planar complexes as well, which have different crystal-field splitting patterns (Figure 22.20). The *d* atomic orbital splitting pattern for tetrahedral complexes is the opposite of the octahedral pattern. In tetrahedral complexes the $d_{x^2-y^2}$ and d_{z^2} atomic orbitals are lower in energy by Δ_t than the set of d_{xz}, d_{xy}, and d_{yz} atomic orbitals.

Remember that the d_{z^2} atomic orbital has a donut of electron density in the *x*-*y* plane (Figure 22.17, p. 1066).

In square planar complexes, we assume that ligands are in a plane containing the *x*- and *y*-axes. Because the $d_{x^2-y^2}$ atomic orbital points along these axes, it is the highest in energy, followed by the d_{xy} atomic orbital and then the d_{z^2} atomic orbital. The d_{xz} and d_{yz} atomic orbitals are of equal and lowest energy (Figure 22.20). The crystal-field splitting energy for square planar complexes, Δ_{sp}, is the energy difference between the $d_{x^2-y^2}$ and d_{xy} atomic orbitals. A square planar complex can be considered as forming by removing the two ligands along the *z*-axis from an octahedral complex (Figure 22.21).

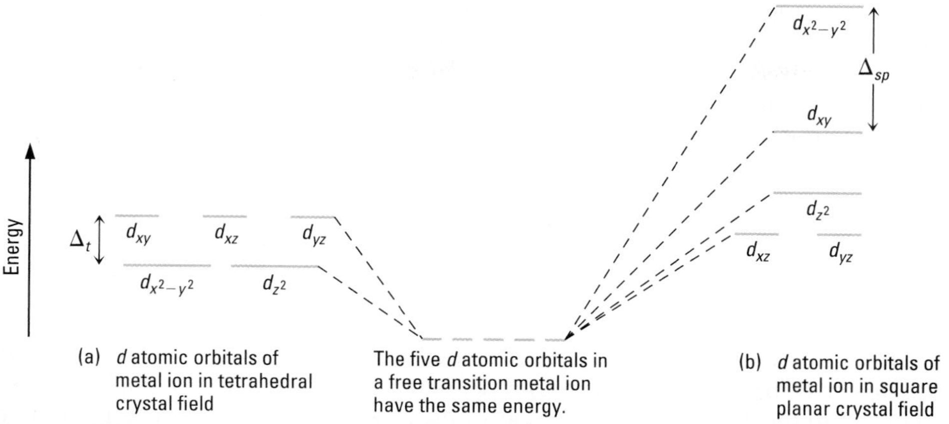

Figure 22.20 Splitting of metal ion *d* atomic orbitals in (a) tetrahedral and (b) square planar complexes.

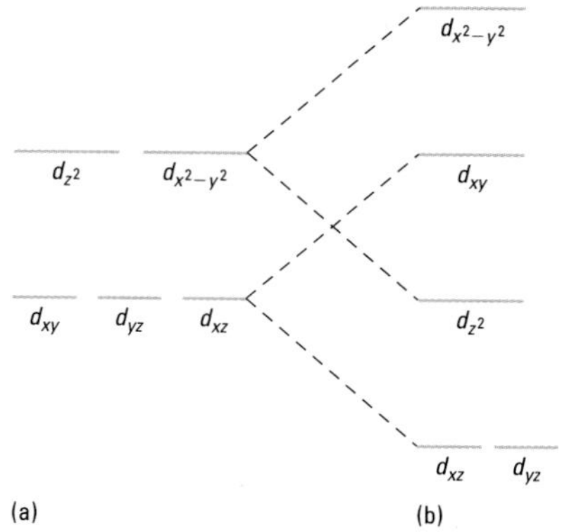

Figure 22.21 The change in relative energy of *d* atomic orbitals in (a) octahedral and (b) square planar complexes. Removing two ligands from the *z*-axis in an octahedral complex and moving the four ligands in the *x-y* plane closer to the metal ion yields a square planar complex.

Color in Coordination Complexes

Most transition metal complexes are brightly colored and explaining their color is a major success of crystal-field theory. The colors arise because of transitions of electrons, called ***d*-to-*d* transitions,** from a lower-energy set of *d* atomic orbitals to a higher-energy *d* atomic orbital of the metal ion. In a coordination complex, the energy difference between these sets of *d* atomic orbitals typically corresponds to photon wavelengths within the visible region of the spectrum (400 nm to 700 nm), thus giving rise to a color when some wavelengths of visible light are absorbed. The color we see is the color of light that is transmitted; *it is the complementary color of the light absorbed.* On the color wheel in Figure 22.22, complementary colors are directly across from each other. Therefore, as seen from Figure 22.22, a complex appears yellow, for example, because it absorbs blue-violet light and transmits yellow light, the complementary color. A $[Ni(H_2O)_6]^{2+}$ solution is green because the complex absorbs red-violet light and transmits green light.

Coordination compounds with metal ions whose *d* atomic orbitals are filled, such as Zn^{2+}, or have no *d* electrons, such as Sc^{3+}, are not colored.

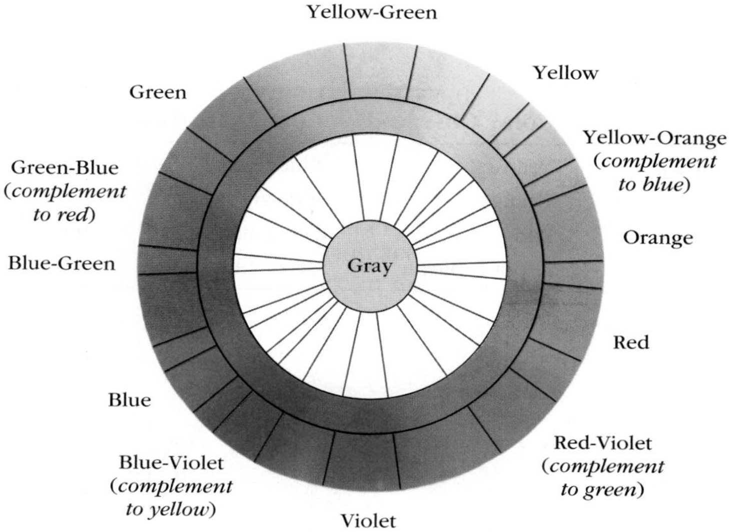

Figure 22.22 The color wheel. The color of a coordination complex is the complementary color to that which is absorbed (directly across from the color on the wheel.) The green color of $[Ni(H_2O)_6]^{2+}$ is the complement of the red-violet light absorbed by the complex ion.

© Cengage Learning/Charles D. Winters

An aqueous solution of Ni^{2+} is green. The $[Ni(H_2O)_6]^{2+}$ complex absorbs red-violet light and transmits green light.

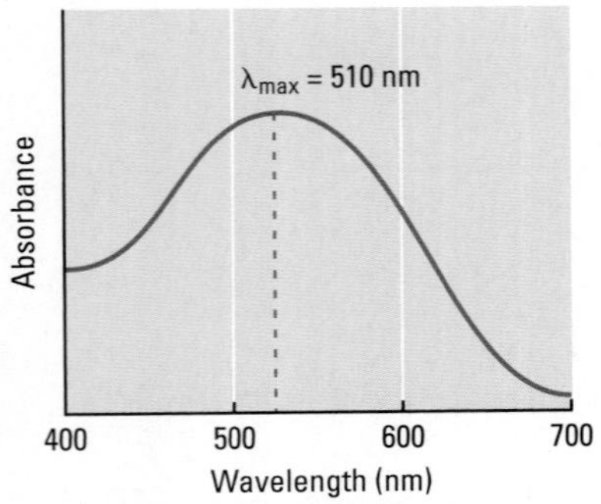

Figure 22.23 The absorption spectrum of $[Ti(H_2O)_6]^{3+}$. The absorption of 510-nm wavelength light causes a *d*-to-*d* transition of the 3*d* electron of $[Ti(H_2O)_6]^{3+}$.

We can apply crystal-field theory to understand the origin of the vivid purple (red-violet) color of the $[Ti(H_2O)_6]^{3+}$ complex ion. This case is particularly simple because Ti^{3+} has only a single 3*d* electron. The complex ion absorbs light at 510 nm, the green region of the visible spectrum, raising the *d* electron from a lower-energy t_2 atomic orbital to a higher-energy *e* atomic orbital (Figure 22.23).

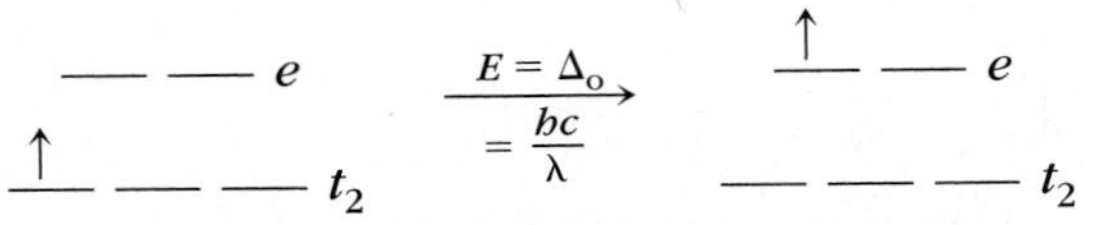

d-to-*d* transition of 3*d* electron of $[Ti(H_2O)_6]^{3+}$.

In this case, the energy difference is a direct measure of Δ_o. We can calculate Δ_o, the crystal-field splitting energy, for the octahedral $[Ti(H_2O)_6]^{3+}$ complex ion by relating it to the wavelength and energy of the absorbed photon. For one photon of 510-nm light,

$$E = \frac{hc}{\lambda} = \frac{(6.626 \times 10^{-34}\ \mathrm{J \cdot s})(2.998 \times 10^{8}\ \mathrm{m/s})}{510 \times 10^{-9}\ \mathrm{m}} = 3.90 \times 10^{-19}\ \mathrm{J}$$

This translates into 235 kJ/mol, which is the splitting energy, Δ_o, the energy separating the t_2 and *e* sets of *d* atomic orbitals.

$$E = 3.90 \times 10^{-19}\ \mathrm{J} \times \frac{1\ \mathrm{kJ}}{10^{3}\ \mathrm{J}} \times \frac{6.022 \times 10^{23}}{1\ \mathrm{mol}} = 2.35 \times 10^{2}\ \mathrm{kJ/mol}$$

Color and the Spectrochemical Series

The color of a coordination complex depends on the splitting energy, which depends on the metal ion and the field strength of its ligands, as given by the spectrochemical series (p. 1067). *Stronger-field ligands cause a larger splitting energy.* Consequently, complexes with such ligands absorb light at shorter wavelengths compared to those with weaker-field ligands, as seen vividly in Figure 22.24 and listed in the table below for a series of Co^{3+} complexes. Note from the table that as weaker-field ligands are substituted for NH_3 in the complex, the wavelength of the absorbed radiation becomes longer, indicating a lower splitting energy.

Complex	Color Observed	Color Absorbed	Approximate Wavelength Absorbed (nm)
(a) $[Co(NH_3)_6]^{3+}$	Yellow	Blue-violet	430
(b) $[Co(NH_3)_5NCS]^{2+}$	Orange	Blue-green	470
(c) $[Co(NH_3)_5H_2O]^{3+}$	Red	Green-blue	500
(d) $[Co(NH_3)_5Cl]^{2+}$	Reddish purple	Green-blue	522

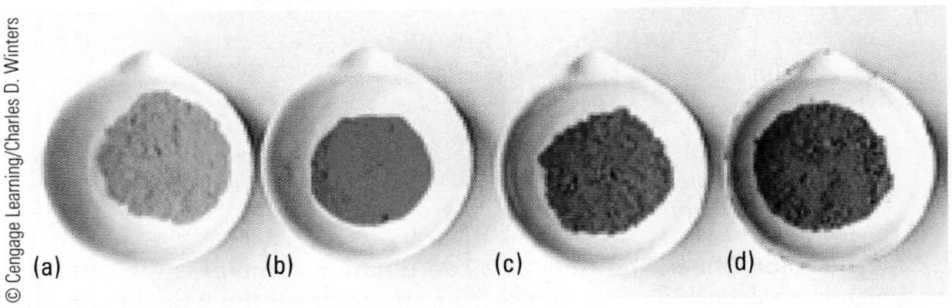

Figure 22.24 The shift in the color of coordination complexes as one of the ligands bonded to Co^{3+} ion is changed. Ligands are: (a) NH_3; (b) NCS^-; (c) H_2O; and (d) Cl^-.

CONCEPTUAL EXERCISE

22.14 Color and Electron Configuration

An aqueous solution of $[Cr(H_2O)_6]^{2+}$ is blue. Predict whether the $3d$ electrons of the central chromium ion are in a low-spin or high-spin configuration.

SUMMARY PROBLEM

Part I

Solid iron(II) sulfide, FeS, is roasted to form either $Fe_2O_3(s)$ or $Fe_3O_4(s)$.

(a) Write balanced equations for the roasting of FeS(s) to $Fe_2O_3(s)$ and to $Fe_3O_4(s)$.

(b) Use thermochemical data from Appendix J to calculate the enthalpy change for these reactions at 25 °C, given $\Delta H_f^\circ = -101.671$ kJ/mol for FeS(s).

(c) Use thermochemical data from Appendix J to calculate the Gibbs free energy change for these reactions at 25 °C, given $S^\circ = 82.81\ \text{J mol}^{-1}\ \text{K}^{-1}$ for FeS(s). Which reaction is more product-favored at this temperature?

(d) Calculate the Gibbs free energy change for the conversion of FeS to Fe_2O_3 and to Fe_3O_4 at 600 °C. Which reaction is more product-favored at this temperature?

Part II

Iron(III) forms a deep red coordination compound, $[Fe(acac)_3]$, with acetylacetonate ions, $acac^-$.

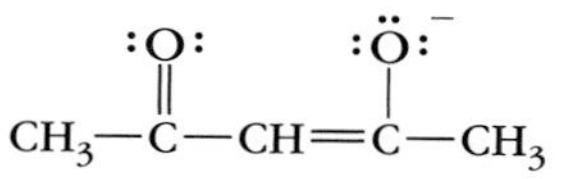

Acetylacetonate ion, $acac^-$

The acetylacetonate anion is a bidentate ligand that furnishes lone pairs from the C═O and C—O oxygens.

(a) Calculate the formal charge ***(← p. 350)*** on each atom in $acac^-$ to verify that one oxygen atom has a negative charge.

(b) Write a structural formula for $[Fe(acac)_3]$.

(c) Give the coordination number of Fe^{3+} in this compound.

(d) Account for the fact that there is no net charge on $[Fe(acac)_3]$.

(e) Why are no counter ions needed for this compound?

IN CLOSING

Having studied this chapter, you should be able to . . .

- Recognize the general properties of transition metals (Section 22.1).
- Write electron configurations and orbital box diagrams for transition metals and their ions (Section 22.1). End-of-chapter questions: 14, 65, 67
- Explain why most transition metals have multiple oxidation states (Section 22.1). Question 16
- Explain trends in sizes of transition metal atomic radii (Section 22.1).
- Describe how iron ore is processed into iron and then into steel (Section 22.2).

OWL and go Chemistry

Sign in at **www.cengage.com/owl** to:

- View tutorials and simulations, develop problem-solving skills, and complete online homework assigned by your professor.
- For quick review and exam prep, download Go Chemistry mini lecture modules from OWL (or purchase them at **www.CengageBrain.com**).

- Discuss how copper is extracted from its ores and purified (Section 22.3). Question 69
- Discuss the chemistry of gold and silver (Section 22.4).
- Discuss the chemistry of chromium (Section 22.5).
- Explain the coordinate covalent bonding of ligands in coordination compounds and complexes (Section 22.6). Questions 28, 54, 60, 62
- Interpret the names and formulas of coordination complex ions and compounds (Section 22.6). Questions 34, 38, 42
- Discuss isomerism in coordination compounds and complex ions (Section 22.6). Question 48
- Give examples of coordination compounds and their uses (Section 22.6).
- Describe crystal-field theory and its applications to interpreting colors and magnetic properties of coordination compounds (Section 22.7).

KEY TERMS

bidentate *(Section 22.6)*
chelating ligands *(22.6)*
coordination compound *(22.6)*
coordination number *(22.6)*
crystal-field splitting energy, Δ *(22.7)*
***d*-to-*d* transitions** *(22.7)*
hexadentate *(22.6)*
high-spin complex *(22.7)*
lanthanide contraction *(22.1)*
ligands *(22.6)*
low-spin complex *(22.7)*
monodentate *(22.6)*
polydentate *(22.6)*
pyrometallurgy *(22.2)*
spectrochemical series *(22.7)*
steel *(22.2)*

QUESTIONS FOR REVIEW AND THOUGHT

OWL Interactive versions of these problems are assignable in OWL.

Blue-numbered questions have short answers at the back of this book in Appendix M and fully worked solutions in the *Student Solutions Manual.*

Review Questions

These questions test vocabulary and simple concepts.

1. What is the primary reducing agent in the production of iron from its ores? Write a balanced chemical equation for this reduction process.
2. Why is lime necessary in the blast furnace reduction of iron ore?
3. What is the difference between pig iron and cast iron?
4. Explain the purpose of each of these materials in the blast-furnace conversion of iron ore to iron.
   (a) Air (b) Limestone (c) Coke
5. Identify what is produced by each of these processes or operations.
   (a) Blast furnace
   (b) Basic oxygen process
   (c) Roasting
6. Identify a common use for Cr, Cu, Fe, Au, and Ag.
7. Name three transition metals that are found "free" in nature.
8. What is the lanthanide contraction? Why does it occur?
9. In general, how do the atomic radii change across the first transition series (Period 4)?
10. What is the distinguishing chemical feature of a ligand?
11. Distinguish between
    (a) A monodentate and a bidentate ligand.
    (b) A *cis* and a *trans* isomer.
    (c) A coordination compound and a coordination complex ion.
    (d) A geometric and an optical isomer.
12. Define these words or phrases and give an example for each.
    (a) Coordination compound (b) Complex ion
    (c) Ligand (d) Chelate
    (e) Bidentate ligand

Topical Questions

These questions are keyed to the major topics in the chapter. Usually a question that is answered at the back of this book is paired with a similar one that is not.

Transition Metals (Section 22.1)

13. Write electron configurations for the 2+ ions of
    (a) Iron. (b) Copper. (c) Chromium.

14. Write electron configurations for the common oxidation states of
    (a) Silver. (b) Gold.
15. Which Period 4 transition metal ions are isoelectronic with
    (a) Zn^{2+} (b) Mn^{2+}
    (c) Cr^{3+} (d) Fe^{3+}
16. Of the two more common oxidation states of chromium in Figure 22.4, which are paramagnetic?
17. Arrange these in order of decreasing strength as oxidizing agents: Mn^{2+}, MnO_4^-, MnO_2. Explain the trend.
18. Arrange these in order of decreasing strength as oxidizing agents: $Cr_2O_7^{2-}$ (in acid), Cr^{2+}, Cr^{3+}. Explain the trend.
19. Write a balanced equation to represent
    (a) The roasting of Cu_2S to copper metal.
    (b) The reduction of Fe_2O_3 with aluminum.
20. Write a balanced equation to represent
    (a) The reduction of Fe_2O_3 with carbon monoxide in a blast furnace.
    (b) The production of hydrogen gas when hydrochloric acid reacts with an iron nail.
21. Balance this redox reaction (in acidic solution).

$$Cu(s) + NO_3^-(aq) \longrightarrow Cu^{2+}(aq) + NO_2(g)$$

22. Balance this redox reaction (in acidic solution).

$$Fe(s) + NO_3^-(aq) \longrightarrow Fe^{3+}(aq) + NO_2(g)$$

23. Determine the oxidation state of the transition metal in each compound.
    (a) V_2O_5 (b) $K_2Cr_2O_7$
    (c) MnO_2 (d) OsO_4
24. Determine the oxidation state of iron in $KFe[Fe(CN)_6]$. Explain your answer.

Coordination Compounds (Section 22.6)

25. In a complex ion, a central ruthenium ion, Ru(III), is bonded to six ammonia molecules.
    (a) Give the formula and net charge for this complex ion.
    (b) How many chloride ions are needed to balance the net charge on this complex ion?
    (c) Write the formula for the compound formed by this complex ion and the chloride ions that are not part of it.
26. In a complex ion, a central Cr^{3+} ion is bonded to two ammonia molecules, three water molecules, and a hydroxide ion.
    (a) Give the formula and charge of the complex ion.
    (b) Identify a single counter ion that could be used with the complex ion to form an uncharged compound.
27. Consider the complex ion $[Cr(NH_3)_2(H_2O)_2Br_2]^+$.
    (a) Identify the ligands and their charges (if any).
    (b) What is the charge on the central metal ion?
    (c) What is the formula of the sulfate salt of this cation?
28. Consider the complex ion $[Co(C_2O_4)_2Cl_2]^{3-}$.
    (a) Identify the ligands and their charges (if any).
    (b) What is the charge on the central metal ion?
    (c) What would be the formula and charge of the complex ion if the $C_2O_4^{2-}$ ions were replaced by NH_3 molecules?
29. Determine the charge of the central metal ion in each case.
    (a) $[Zn(H_2O)_3(OH)]^+$
    (b) $[Pt(NH_3)_3Cl_3]^+$
    (c) $[Cr(CN)_6]^{3-}$
30. For coordination compounds $Na_3[IrCl_6]$ and $[Mo(CO)_4]Br_2$, identify in each case
    (a) The ligands.
    (b) The central metal ion and its charge.
    (c) The formula and charge of the complex ion.
    (d) Any ions not part of the complex ion.
31. Give the coordination number of the central metal ion in
    (a) $[Ni(en)_2(NH_3)_2]^{2+}$
    (b) $[Fe(en)(C_2O_4)Cl_2]^-$
32. Give the coordination number of the central metal ion in
    (a) $[Pt(en)_2]^{2+}$
    (b) $[Cu(C_2O_4)_2]^{2-}$
33. Write a structural formula for the coordination compound $[Cr(en)(NH_3)_2I_2]$, and give the coordination number for the central Cr^{2+} ion.
34. Give the formula of each of these coordination compounds formed with Pt^{2+}.
    (a) Two ammonia molecules and two bromide ions
    (b) One ethylenediamine molecule and two nitrite ions, NO_2^-
    (c) One chloride ion, one bromide ion, and two ammonia molecules
35. Give the charge on the central metal ion in each of these.
    (a) $[VCl_6]^{4-}$ (b) $[Sc(H_2O)_3Cl_3]$
    (c) $[Mn(NO)(CN)_5]^{3-}$ (d) $[Ni(en)_2(NH_3)_2]^{2+}$
36. Identify the coordination number of the metal ion in these coordination complexes.
    (a) $[FeCl_4]^-$ (b) $[PtBr_4]^{2-}$
    (c) $[Mn(en)_3]^{2+}$ (d) $[Cr(NH_3)_5H_2O]^{3+}$
37. Using structural formulas, show how the carbonate ion can be either a monodentate or bidentate ligand to a transition metal cation.
38. Classify each ligand as monodentate, bidentate, and so on.
    (a) $(CH_3)_3P$
    (b) 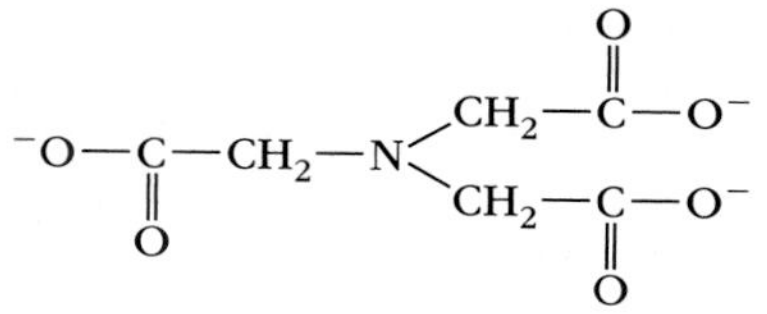
    (c) $H_2N—(CH_2)_2—NH—(CH_2)_2—NH_2$
    (d) H_2O
39. Which of these would be expected to be effective chelating agents? Explain your answer.
    (a) CH_3CH_2OH (b) $H_2N—(CH_2)_3—NH_2$
    (c) 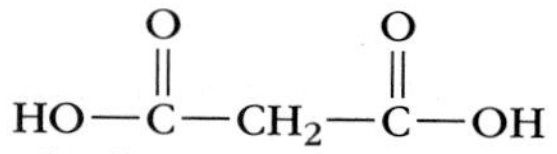 (d) PH_3
40. Give the formula of a simple ionic (noncoordination) compound analogous to $[Rh(en)_3]Cl_3$.

Naming Complex Ions and Coordination Compounds (Section 22.6)

41. Write the formula for
(a) Potassium diaquadioxalatocobaltate(III)
(b) Diamminetriaquahydroxochromium(II) nitrate
(c) Ammonium tetrachlorocuprate(II)
(d) Tetrachloroethylenediaminecobaltate(III)
(e) Triaquatrifluorocobalt(III)

42. Write the name corresponding to each of these formulas.
(a) $[MnCl_4]^{2-}$ (b) $K_3[Fe(C_2O_4)_3]$
(c) $[Pt(NH_3)_2(CN)_2]$ (d) $[Fe(H_2O)_5(OH)]^{2+}$
(e) $[Mn(en)_2Cl_2]$

Geometry of Coordination Complexes (Section 22.6)

43. Sketch the geometry of
(a) *cis*-$[Cu(H_2O)_2Br_4]^{2-}$ (b) *trans*-$[Ni(NH_3)_2(en)_2]^{2+}$

44. Sketch the geometry of
(a) *cis*-$[Pt(H_2O)_2Cl_2]$ (b) *trans*-$[Cr(H_2O)_4Cl_2]^+$

45. The ligand 1,2-diaminocyclohexane

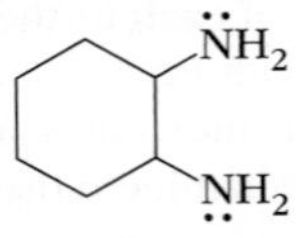

is abbreviated "dech." Sketch the geometry of *cis*-$[Pd(H_2O)_2(dech)]^{2+}$.

46. The acetylacetonate ion (acac)$^-$

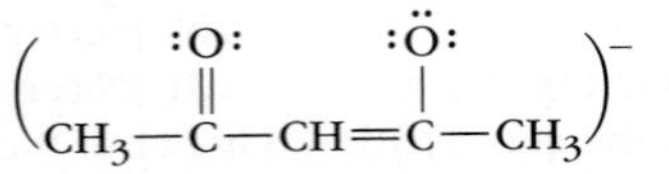

forms a complex with Co^{3+}. Sketch the geometry of $[Co(acac)_3]$.

47. Which of these octahedral coordination complexes can exhibit geometric isomerism?
(a) $[Pt(H_2O)_2Cl_2Br_2]^{2-}$ (b) $[Pt(H_2O)_2Cl_3Br]^-$

48. Which of these octahedral coordination complexes can exhibit geometric isomerism?
(a) $[Cr(H_2O)_3Cl_3]$ (b) $[Cr(H_2O)_4Cl_2]^+$

49. Draw the possible geometric isomers, if any, of
(a) $[Ni(NH_3)_4Cl_2]$
(b) $[Pt(NH_3)_2(SCN)Br]$ (The S in NCS is bonded to Pt^{2+}.)
(c) $[Co(en)Cl_4]^-$

50. Draw the possible geometric isomers, if any.
(a) $[Co(H_2O)_4Cl_2]^+$ (b) $[Pt(NH_3)Cl_3]^-$
(c) $[Co(H_2O)_3Cl_3]$ (d) $[Co(en)_2(NH_3)Br]^{2+}$

Crystal-Field Theory and Magnetic Properties of Complex Ions (Section 22.7)

51. Draw the crystal-field splitting diagrams and put in the *d* electrons for these octahedral complexes. In those cases where they are possible, draw diagrams for both low-spin and high-spin cases.
(a) $[Cr(H_2O)_6]^{2+}$ (b) $[Mn(H_2O)_6]^{2+}$
(c) $[FeF_6]^{3-}$ (d) $[Cr(en)_3]^{3+}$

52. Using the spectrochemical series, predict the actual number of unpaired electrons in each complex in Question 51 for which the possibility of low-spin and high-spin forms exist.

53. Fe^{3+} forms octahedral complexes with NCS^- and with NO_2^- ligands. One complex displays a greater paramagnetism than the other.
(a) Write the formula for each of these complex ions.
(b) Use the spectrochemical series to predict whether the complex ions are high-spin or low-spin.
(c) Identify which complex ion is more paramagnetic.
(d) Draw the crystal-field splitting diagram, including *d* electrons, for each complex ion.

54. Explain why Cr^{2+} forms high-spin and low-spin octahedral complexes, but Cr^{3+} does not.

55. How many unpaired electrons are in the high-spin and low-spin octahedral complexes of Cr^{2+}?

56. Use crystal-field theory to explain why some Co^{3+} octahedral complexes are diamagnetic and others are paramagnetic.

57. Use crystal-field theory to explain why some octahedral Co^{2+} complexes are more paramagnetic than others.

58. Use crystal-field theory to explain why Cu^{2+} does not form high-spin and low-spin octahedral complexes.

Crystal-Field Theory and Color in Complex Ions (Section 22.7)

59. An aqueous solution of $[Ni(NH_3)_6]^{2+}$ is purple. Predict the approximate wavelength and predominant color of light absorbed by the complex.

60. An aqueous solution of $[Rh(C_2O_4)_3]^{3-}$ is yellow. Predict the approximate wavelength and predominant color of light absorbed by the complex.

61. An aqueous solution of $[Rh(C_2O_4)_3]^{3-}$ is yellow; that of $[Rh(en)_3]^{3+}$ is a different color. Oxalate is to the right of en in the spectrochemical series. Predict what the change in color likely will be from $[Rh(C_2O_4)_3]^{3-}$ to $[Rh(en)_3]^{3+}$. In which direction does the absorbed wavelength change?

62. As discussed in Section 22.6, an aqueous solution of $[Ti(H_2O)_6]^{3+}$ is purple (red-violet). Predict how the value of Δ_o would change if all H_2O ligands were replaced by CN^- ligands; by Cl^- ligands. How would the color change in each case?

63. A solution of a complex ion absorbs visible light at a wavelength of 540 nm.
(a) What is the color of the solution?
(b) Calculate the energy of an absorbed photon in joules and in kJ/mol.

General Questions

These questions are not explicitly keyed to chapter topics; many require integration of several concepts.

64. Give the electron configuration of
(a) Ti^{3+} (b) V^{2+} (c) Ni^{3+} (d) Cu^+

65. Give the electron configuration of
(a) Cr^{2+} (b) Zn^{2+} (c) Co^{2+} (d) Mn^{4+}

66. Write a partial atomic orbital box diagram and determine the number of unpaired electrons for each species in Question 64.

67. Write a partial atomic orbital box diagram and determine the number of unpaired electrons for each species in Question 65.

68. Assuming 100% recovery of the metal, which would yield the greater number of grams of copper?
    (a) One kilogram of an ore containing 3.60 mass % azurite, $Cu(OH)_2 \cdot 2\ CuCO_3$
    (b) One kilogram of an ore containing 4.95 mass % chalcopyrite, $CuFeS_2$
69. Calculate what mass of copper could be electroplated from a $CuSO_4$ solution using an electric current of 2.50 A for 5.00 h. Assume 100% efficiency.
70. Copper metal is obtained directly by roasting covelite, CuS.
    (a) Write a balanced equation for this process.
    (b) Assume that the roasting is 90.0% efficient. Calculate how many tons of SO_2 would be released into the air by roasting 500. tons of covelite.
71. Determine what mass of SO_2 is produced when 1.0 ton of chalcocite, Cu_2S, is roasted to Cu_2O.
72. What is the coordination number of the central metal ion in
    (a) $[Pt(NH_3)_2Br_2]$ (b) $[Fe(CN)_6]^{3-}$
    (c) $[Ti(H_2O)Cl_5]^{2-}$ (d) $[Mn(C_2O_4)_3]^{4-}$
73. Determine the coordination number of the central metal ion in
    (a) $[Ni(en)Cl_2]$ (b) $[Mo(CO)_4Br_2]$
    (c) $[Cd(CN)_4]^{2-}$ (d) $[Co(CN)_5(OH)]^{3-}$
74. Draw sketches for all octahedral complexes of Co^{3+} using only ethylenediamine and/or Cl^- as ligands.
75. Draw sketches for as many octahedral complexes as you can for the formula $Co(NH_3)_4Cl_2Br$.
76. In your own words explain why
    (a) $H_2N{-}(CH_2)_3{-}NH_2$ is a bidentate ligand.
    (b) AgCl dissolves in NH_3.
    (c) There are no geometric isomers of tetrahedral complexes.
77. Determine whether each statement is true or false. If it is false, correct the statement.
    (a) The coordination number of the Fe^{3+} ion in $[Fe(H_2O)_4(C_2O_4)]^+$ is five.
    (b) Cu^+ has two unpaired electrons.
    (c) The net charge of a coordination complex of Cr^{3+} with two NH_3, one en, and two H_2O is 2+.
78. Determine whether each statement is true or false. If it is false, correct the statement.
    (a) In $[Pt(NH_3)_4Cl_4]$, platinum has a 4+ charge and a coordination number of six.
    (b) In general, Cu^{2+} is more stable than Cu^+ in aqueous solutions.
79. The metal ion in $[Pt(NH_3)_2(C_2O_4)]$ is surrounded by a square planar array of coordinating atoms.
    (a) Give the oxidation number of the central metal ion.
    (b) Draw the structural formula of this coordination compound.
80. Chromium(III) forms three different compounds with water and chloride ions, all of which have the same composition: 19.51% Cr, 39.92% Cl, and 40.57% water. One of the compounds is violet and dissolves in water to give a complex ion with a 3+ charge plus three chloride ions. All three chloride ions precipitate immediately as AgCl when $AgNO_3$ is added to the solution. Draw the structural formula of this complex ion.

Applying Concepts

These questions test conceptual learning.

81. Iron nails are put into Fe^{2+} aqueous solutions to reduce any Fe^{3+} that forms back to Fe^{2+}. Write a balanced chemical equation for this preventative reaction.
82. Use VSEPR theory to predict the shape and bond angles around chromium in
    (a) Chromate ions (b) Dichromate ions
83. The structure of cyclam is given below.

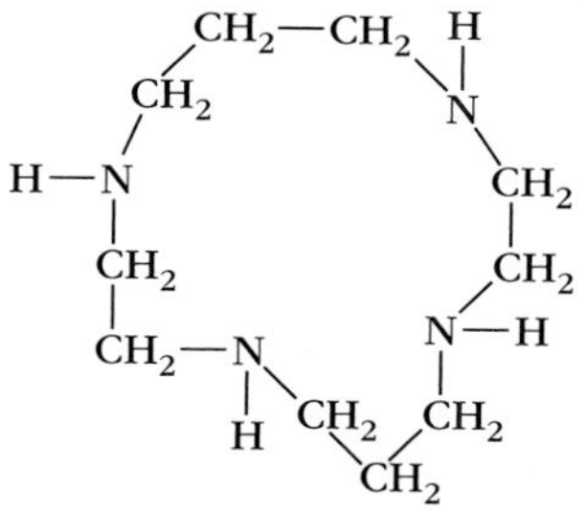

Cyclam can act as a ligand. How many coordinate covalent bonds can one cyclam molecule form with a central metal ion?

84. The compound 1,10-phenanthroline is a chelating agent used in analytical chemistry. Its isomer 4,7-phenanthroline is not. Use these structural formulas to explain this difference.

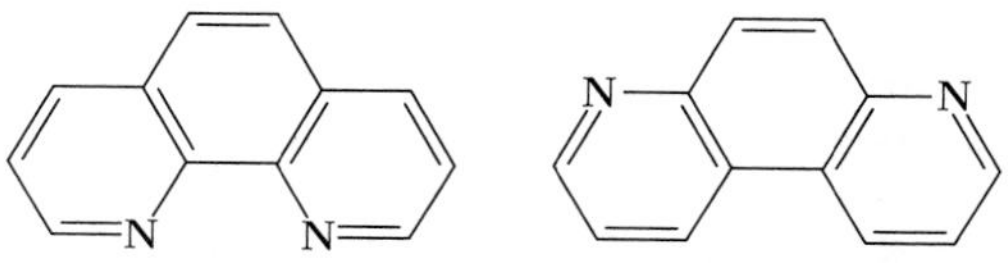

1,10-phenanthroline 4,7-phenanthroline

85. Two different isomers are known with the formula $[Pt(py)_2Cl_2]$, where py represents pyridine, an uncharged monodentate ligand in which an N atom bonds to the metal ion. There is, however, only one structure known for $[Pt(phen)Cl_2]$, where phen represents *ortho*-phenanthroline, an uncharged bidentate ligand (Figure 22.9). Draw the structural formulas of all three molecules and explain why there are isomers in one case, but not the other.
86. An electrochemical cell is made by immersing a strip of chromium into a 1.0-M solution of Cr^{3+} and a strip of gold into a 1.0-M solution of Au^{3+}. The half-cells are connected by a salt bridge. A wire and light bulb complete the circuit.
    (a) Write the balanced chemical equation for the reaction that is product-favored.
    (b) Calculate the cell potential.
    (c) Draw a sketch of the cell and indicate the anode, cathode, and direction of electron flow.
87. Repeat the directions for Question 86 using a cell constructed of a strip of nickel immersed in a 1.0-M Ni^{2+} solution and a strip of silver dipping into a 1.0-M Ag^+ solution.
88. Calculate $\Delta G°$ for the reduction of Fe_2O_3 with CO gas at 25 °C and at 1000 °C. What application does this have to the conversion of iron ore to iron in a blast furnace?

89. To determine the percent iron in an ore, a 1.500-g sample of the ore containing Fe^{2+} is titrated to the equivalence point with 18.6 mL of 0.05012-M $KMnO_4$. The products of the titration are Fe^{3+} and Mn^{2+}. Calculate the weight percent of iron in the ore.

90. Consider the reaction

$$2\ Cu^{+}(aq) \longrightarrow Cu^{2+}(aq) + Cu(s)$$

for which $E^\circ_{cell} = +0.37$ V. Use the Nernst equation to calculate

(a) E when the Cu^{2+} concentration is equal to the Cu^{+} concentration $= 1 \times 10^{-4}$ M.

(b) The concentration of Cu^{+} when the Cu^{2+} concentration $= 1.0$ M and $E = 0.00$ V.

91. Consider the reaction

$$2\ Ag^{+}(aq) \longrightarrow Ag(s) + Ag^{2+}(aq)$$

for which $E^\circ_{cell} = -1.18$ V. Use the Nernst equation to calculate

(a) E when the Ag^{+} concentration $= 1 \times 10^{-4}$ M, which is five times the concentration of Ag^{2+}.

(b) The concentration of Ag^{2+} when the Ag^{+} concentration $= 1.0$ M and $E = 0.00$ V.

92. Use the Nernst equation to calculate E_{cell} for $Cr_2O_7^{2-}$ oxidation of Cl^- in 6.0-M H^+ when the concentration of $Cr_2O_7^{2-}$ = concentration of $Cr^{3+} = 0.10$ M, and all other concentrations = 1.0 M.

93. If 1.00 mol of each of these compounds is dissolved in separate samples of water sufficient to dissolve the compound, how many moles of ions are present in each solution?

(a) $[Pt(en)Cl_2]$
(b) $Na[Cr(en)_2(SO_4)_2]$
(c) $K_3[Au(CN)_4]$
(d) $[Ni(H_2O)_2(NH_3)_4]Cl_2$

94. For each of the compounds in Question 93, state which it would most likely resemble in colligative properties and conductivity: $(NH_2)_2CO$ (urea), KCl, K_2SO_4, or K_3PO_4.

95. In aqueous solution, $[Cr(NH_3)_6]Cl_3$ is yellow, but aqueous $[Cr(NH_3)_5Cl]Cl_2$ is purple. Explain the difference in colors.

96. Early coordination chemists relied on close experimental observation to determine the formulas of coordination compounds. They found, for example, that aqueous $BaCl_2$ did not cause precipitation when added to a solution of a Co^{3+}-containing coordination compound, but precipitation occurred when aqueous silver nitrate was added to a solution of the coordination compound. The coordination compound was known to contain one Co^{3+} ion, one sulfate ion, one chloride ion, and four ammonia molecules. Write the structural formula of the coordination compound that is consistent with the experimental results.

More Challenging Questions

These questions require more thought and integrate several concepts.

97. The bidentate oxalate ion, $C_2O_4^{2-}$, forms octahedral complexes with Fe^{3+} and Ru^{3+} ions.

(a) Write the structural formula for each complex.

(b) The ruthenium complex is low-spin; the iron complex is high-spin. Write the *d*-orbital splitting diagram for each metal ion.

(c) Which complex has the higher Δ_o? Explain your answer.

98. Analysis of a coordination compound gives these results: 22.0% Co, 31.4% N, 6.78% H, and 39.8% Cl. One mole of the compound dissociates in water to form 4 mol ions.

(a) Determine the formula of the compound.

(b) Write the equation for its dissociation in water.

99. A chemist synthesizes two coordination compounds. One compound decomposes at 210 °C, the other at 240 °C. When analyzed, the compounds give the same mass percent data: 52.6% Pt, 7.6% N, 1.63% H, and 38.2% Cl. Both compounds contain a 4+ central metal ion.

(a) Determine the simplest formula of the compounds.

(b) Draw structural formulas for the complexes present.

100. A coordination compound has the simplest formula $PtN_2H_6Cl_2$ with a molar mass of about 600 g/mol. It contains a complex cation and a complex anion. Draw its structural formula.

101. The glycinate ion (gly) is $H_2NCH_2CO_2^-$. It can act as a ligand coordinating through the nitrogen and one of the oxygens. Using $\overset{\frown}{N\ O}$ to represent glycinate ion, draw structural formulas for four stereoisomers of $[Co(gly)_3]$.

102. Five-coordinate coordination complexes are known, including $[CuCl_5]^{3-}$ and $[Ni(CN)_5]^{3-}$. Write the structural formulas and identify a plausible geometry for these complexes.

103. Predict the number of unpaired electrons in a square planar transition metal ion with seven *d* electrons.

104. A coordination compound has the empirical formula $Fe(H_2O)_4(CN)_2$. Its paramagnetism is the equivalent of 2.67 unpaired electrons per Fe ion. Explain how this is possible.

105. Two different compounds are known with the formula $Pd(py)_2Cl_2$, but there is only one compound with the formula $Zn(py)_2Cl_2$. The symbol py is for pyridine, a monodentate ligand. Explain the differences in the Pd and Zn compounds.

106. An octahedral coordination complex ion is formed by the combination of an Fe^{3+} ion and det ligands (det is $H_2NCH_2CH_2NHCH_2CH_2NH_2$). Write a structural formula for the complex ion.

APPENDIX

A Problem Solving and Mathematical Operations

In this book we provide many illustrations of problem solving and many problems for practice. Some are numerical problems that must be solved by mathematical calculations. Others are conceptual problems that must be solved by applying an understanding of the principles of chemistry. Often, it is necessary to use chemical concepts to relate what we know about matter at the nanoscale to the properties of matter at the macroscale. The problems throughout this book are representative of the kinds of problems that chemists and other scientists must regularly solve to pursue their goals, although our problems are often not as difficult as those encountered in the real world.

Problem solving is not a simple skill that can be mastered with a few hours of study or practice. Because there are many different kinds of problems and many different kinds of people who are problem solvers, no hard and fast rules are available that are guaranteed to lead you to solutions. The general guidelines presented in this appendix are, however, helpful in getting you started on any kind of problem and in checking whether your answers are correct. The problem-solving skills you develop in a chemistry course can later be applied to difficult and important problems that may arise in your profession, your personal life, or the society in which you live.

In getting a clear picture of a problem and asking appropriate questions regarding the problem, you need to keep in mind all the principles of chemistry and other subjects that you think may apply. In many real-life problems, not enough information is available for you to arrive at an unambiguous solution; in such cases, try to look up or estimate what is needed and then forge ahead, noting assumptions you have made. Often the hardest part is deciding which principle or idea is most likely to help solve the problem and what information is needed. To some degree this can be a matter of luck or chance. Nevertheless, in the words of Louis Pasteur, "In the field of observation chance only favors those minds which have been prepared." The more practice you have had, and the more principles and facts you can keep in mind, the more likely you are to be able to solve the problems that you face.

A.1 General Problem-Solving Strategies

This book advocates a four-step strategy for approaching and solving problems.

Step 1. **Analyze the problem.** Carefully review the information contained in the problem. What is the problem asking you to find? What key principles are involved? What known information is necessary for solving the problem and what information is there just to place the question in context? Organize the information to see what is necessary and to see the relationships among the known data. Try writing the information in an organized way. If the information is numerical, be sure to include proper units. Can you picture the situation under consideration? Try sketching it and including any relevant dimensions in the sketch.

Step 2. **Plan a solution.** Have you solved a problem of this type before? If you recognize the new problem as similar to ones you know how to solve, you can use the same method that worked before. Try reasoning backward from the units of what is being sought. What data are needed to find an answer in those units?

Can the problem be broken down into smaller pieces, each of which can be solved separately to produce information that can be assembled to solve the entire problem? When a problem can be divided into simpler problems, it often helps to write down a plan that lists the simpler problems and the order in which those problems must be put together to arrive at an overall solution. Many major problems in chemical research have to be solved in this way. In problems in this book we have mostly provided the needed numerical data, but in the laboratory, the first

"The mere formulation of a problem is often far more essential than its solution. To raise new questions, new possibilities, to regard old problems from a new angle, requires creative imagination and marks real advances in science."

—Albert Einstein

aspect of solving a problem is often devising experiments to gather the data or searching databases to find needed information.

If you are still unsure about what to do, do something anyway. It may not be the right thing to do, but as you work on it, the way to solve the problem may become apparent, or you may see what is wrong with your initial approach, thereby making clearer what a good plan would be.

Step 3. **Execute the plan.** Carefully write down each step of a mathematical problem, being sure to keep track of the units. Do the units cancel to give you the answer in the desired units? Don't skip steps. Don't do any except the simplest steps in your head. Once you've written down the steps for a mathematical problem, check what you've written—is it all correct? Students often say they got a problem wrong because they "made a stupid mistake." Teachers—and textbook authors—make mistakes, too. These errors usually arise because they don't take the time to write down the steps of a problem clearly and correctly. In solving a mathematical problem, remember to apply the principles of dimensional analysis and significant figures. Dimensional analysis is introduced in Sections 1.4 ***(← p. 7)*** and 2.3 ***(← p. 45);*** it is reviewed below. Section 2.4 ***(← p. 50)*** and Appendix A.3 (p. A.5) introduce significant figures.

Step 4. ☑ **Check the answer to see whether it is reasonable.** As a final check of your solution to any problem, ask yourself whether the answer is reasonable: Are the units of a numerical answer correct? Is a numerical answer of about the right size? Don't just copy a result from your calculator without thinking about whether it makes sense.

Suppose you have been asked to convert 100. yards to a distance in meters. Using dimensional analysis and some well-known factors for converting from the English system to the metric system, you could write

$$100.\ \text{yd} \times \frac{3\ \text{ft}}{1\ \text{yd}} \times \frac{12\ \text{in.}}{1\ \text{ft}} \times \frac{2.54\ \text{cm}}{1\ \text{in.}} \times \frac{1\ \text{m}}{100\ \text{cm}} = 91.4\ \text{m}$$

To check that a distance of 91.4 m is about right, recall that a yard is a little shorter than a meter. Therefore 100 yd should be a little less than 100 m. If you had mistakenly divided instead of multiplied by 3 ft/yd in the first step, your final answer would have been a little more than 10 m. This is equivalent to only about 30 ft, and you probably know a 100-yd football field is longer than that.

Here is an example that shows how this problem-solving strategy can be applied to a simple problem. The problem has been taken verbatim from Chapter 1, p. 10.

PROBLEM-SOLVING EXAMPLE 1.1 Density

In an old movie thieves are shown running off with pieces of gold bullion that are about a foot long and have a square cross section of about six inches. The volume of each piece of gold is 7000 mL. Calculate the mass of gold and express the result in pounds (lb). Based on your result, is what the movie shows physically possible? (1 lb = 454 g)

Answer 1.4×10^5 g; 300 lb; probably not

Strategy and Explanation A good approach to problem solving is to (1) analyze the problem, (2) plan a solution, (3) execute the plan, and (4) check your result to see whether it is reasonable. (These four steps are described in more detail in Appendix A.1.)

Analyze the problem.

Step 1: *Analyze the problem.* You are asked to calculate the mass of the gold, and you know the volume.

Plan a solution.

Step 2: *Plan a solution.* Density relates mass and volume and is the appropriate proportionality factor, so look up the density in a table. Mass is proportional to volume, so the volume either has to be multiplied by the density or divided by the density. Use the units to decide which.

Execute the plan.

Step 3: *Execute the plan.* According to Table 1.1, the density of gold is 19.32 g/mL. Setting up the calculation so that the unit (milliliter) cancels gives

$$7000\ \cancel{\text{mL}} \times \frac{19.32\ \text{g}}{1\ \cancel{\text{mL}}} = 1.35 \times 10^5\ \text{g}$$

This can be converted to pounds

$$1.35 \times 10^5\ \cancel{\text{g}} \times \frac{1\ \text{lb}}{454\ \cancel{\text{g}}} = 300\ \text{lb}$$

Notice that the result is expressed to one significant figure, because the volume was given to only one significant figure and only multiplications and divisions were done.

☑ **Reasonable Answer Check** Gold is nearly 20 times denser than water. A liter (1000 mL) of water is about a quart and a quart of water (2 pints) weighs about two pounds. Seven liters (7000 mL) of water should weigh 14 lb, and 20 times 14 gives 280 lb, so the answer is reasonable. The movie is not—few people could run while carrying a 300-lb object!

Check that the result is reasonable.

A.2 Numbers, Units, and Quantities

Many scientific problems require you to use mathematics to calculate a result or draw a conclusion. Therefore, knowledge of mathematics and its application to problem solving is important. However, one aspect of scientific calculations is often absent from pure mathematical work: Science deals with *measurements* in which an unknown quantity is compared with a standard or unit of measure. For example, using a balance to determine the mass of an object involves comparing the object's mass with standard masses, usually in multiples or fractions of one gram; the result is reported as some number of grams, say 4.357 g. *Both the number and the unit are important.* If the result had been 123.5 g, this would clearly be different, but a result of 4.357 oz (ounces) would also be different, because the unit "ounce" is different from the unit "gram." A *result that describes the quantitative measurement of a property,* such as 4.357 g, is called a *quantity* (or physical quantity), and it consists of a number and a unit. Chemical problem solving requires calculating with quantities. Notice that whether a quantity is large or small depends on the units as well as the number; the two quantities 123.5 g and 4.357 oz, for example, represent the *same* mass.

A quantity is always treated as though the number and the units are multiplied together; that is, 4.357 g can be handled mathematically as $4.357 \times g$. Using this simple rule, you will see that calculations involving quantities follow the normal rules of algebra and arithmetic: $5\text{ g} + 7\text{ g} = 5 \times \text{g} + 7 \times \text{g} = (5 + 7) \times \text{g} = 12\text{ g}$; or $6\text{ g} \div 2\text{ g} = (6\ \cancel{\text{g}})/(2\ \cancel{\text{g}}) = 3$. (Notice that in the second calculation the unit g appears in both the numerator and the denominator and cancels out, leaving a pure number, 3.) Treating units as algebraic entities has the advantage that *if a calculation is set up correctly, the units will cancel or multiply together so that the final result has appropriate units.* For example, if you measured the size of a sheet of paper and found it to be 8.5 in. by 11 in., the area A of the sheet could be calculated as area = length × width = 11 in. × 8.5 in. = 94 in.^{2}, or 94 square inches. If a calculation is set up incorrectly, the units of the result will be inappropriate. Using units to check whether a calculation has been properly set up is called *dimensional analysis* (← *p. 9*).

This idea of using algebra on units as well as numbers is useful in all kinds of situations. For example, suppose you are having a party for some friends who like pizza. A large pizza consists of 12 slices and costs \$10.75. You expect to need 36 slices of pizza and want to know how much you will have to spend. A strategy for solving the problem is first to figure out how many pizzas you need and then to figure out the cost in dollars. This solution could be diagrammed as

$$\text{Slices} \xrightarrow[\text{step 1}]{\text{slices per pizza}} \text{pizzas} \xrightarrow[\text{step 2}]{\text{dollars per pizza}} \text{dollars}$$

Step 1: Find the number of pizzas required by dividing the number of slices per pizza into the number of slices, thus converting "units" of slices to "units" of pizzas:

$$\text{Number of pizzas} = 36 \text{ slices}\left(\frac{1 \text{ pizza}}{12 \text{ slices}}\right) = 3 \text{ pizzas}$$

If you had multiplied the number of slices times the number of slices per pizza, the result would have been labeled pizzas × slices^2, which does not make sense. In other words, the labels indicate whether multiplication or division is appropriate.

Strictly speaking, slices and pizzas are not units in the same sense that a gram is a unit. Nevertheless, labeling things this way will often help you keep in mind what a number refers to—pizzas, slices, or dollars in this case.

Step 2: Find the total cost by multiplying the cost per pizza by the number of pizzas needed, thus converting "units" of pizzas to "units" of dollars:

$$\text{Total price} = 3 \text{ pizzas}\left(\frac{\$10.75}{1 \text{ pizza}}\right) = \$32.25$$

Notice that in each step you have multiplied by a factor that allowed the initial units to cancel algebraically, giving the answer in the desired units. A factor such as (1 pizza/12 slices) or ($10.75/pizza) is referred to as a *proportionality factor* (**← *p. 10***). This name indicates that it comes from a proportion. For instance, in the pizza problem you could set up the proportion

$$\frac{x \text{ pizzas}}{36 \text{ slices}} = \frac{1 \text{ pizza}}{12 \text{ slices}} \quad \text{or} \quad x \text{ pizzas} = 36 \text{ slices}\left(\frac{1 \text{ pizza}}{12 \text{ slices}}\right) = 3 \text{ pizzas}$$

A proportionality factor such as (1 pizza/12 slices) is also called a *conversion factor*, which indicates that it converts one kind of unit or label to another; in this case the label "slices" is converted to the label "pizzas."

Many everyday scientific problems involve proportionality. For example, the bigger the volume of a solid or liquid substance, the bigger its mass. When the volume is zero, the mass is also zero. These facts indicate that mass, *m*, is directly proportional to volume, *V*, or, symbolically,

$$m \propto V$$

where the symbol $\propto$ means "is proportional to." Whenever a proportion is expressed this way, it can also be expressed as an equality by using a proportionality constant—for example,

$$m = d \times V$$

In this case the proportionality constant, *d*, is called the density of the substance. This equation embodies the definition of density as mass per unit volume, since it can be rearranged algebraically to

$$d = \frac{m}{V}$$

As with any algebraic equation involving three variables, it is possible to calculate any one of the three quantities *m*, *V*, or *d*, provided the other two are known. If density is wanted, simply use the definition of mass per unit volume; if mass or volume is to be calculated, the density can be used as a proportionality factor.

Suppose that you are going to buy a ton of gravel and want to know how big a bin you will need to store it. You know the mass of gravel and want to find the volume of the bin; this implies that density will be useful. If the gravel is primarily limestone, you can assume that its density is about the same as for limestone and look it up. Limestone has the chemical formula $CaCO_3$, and its density is 2.7 kg/L. However, these mass units are different from the units for mass of gravel—namely, tons. Therefore you need to recall or look up the mass of 1 ton (exactly 2000 pounds [lb]) and the fact that there are 2.20 lb per kilogram. This provides enough information to calculate the volume needed. Here is a "roadmap" plan for the calculation:

$$\text{Mass of gravel in tons} \xrightarrow[\text{step1}]{\text{change units}} \text{mass of gravel in kilograms} \xrightarrow[\text{step 2}]{\text{density}} \text{volume of bin}$$

Step 1: Figure out how many kilograms of gravel are in a ton.

$$m_{\text{gravel}} = 1 \text{ ton} = 2000 \text{ lb} = 2000 \text{ lb}\left(\frac{1 \text{ kg}}{2.20 \text{ lb}}\right) = 909 \text{ kg}$$

The fact that there are 2.20 pounds per kilogram implies two proportionality factors: (2.20 lb/1 kg) and (1 kg/2.20 lb). The latter was used because it results in appropriate cancellation of units.

Step 2: Use the density to calculate the volume of 909 kg of gravel.

$$V_{\text{gravel}} = \frac{m_{\text{gravel}}}{d_{\text{gravel}}} = \frac{909 \text{ kg}}{2.7 \text{ kg/L}} = 909 \text{ kg}\left(\frac{1 \text{ L}}{2.7 \text{ kg}}\right) = 340 \text{ L}$$

In this step we used the definition of density, solved algebraically for volume, substituted the two known quantities into the equation, and calculated the result. However, it is quicker simply to remember that mass and volume are related by a proportionality factor called density and to use the units of the quantities to decide whether to multiply or divide by that factor. In this case we divided mass by density because the units kilograms canceled, leaving a result in liters, which is a unit of volume.

Also, it is quicker and more accurate to solve a problem like this one by using a single setup. Then all the calculations can be done at once, and no intermediate results need to be written down. The "roadmap" plan on p. A.4 can serve as a guide to the single-setup solution, which looks like this:

$$V_{\text{gravel}} = 1 \text{ ton}\left(\frac{2000 \text{ lb}}{1 \text{ ton}}\right)\left(\frac{1 \text{ kg}}{2.20 \text{ lb}}\right)\left(\frac{1 \text{ L}}{2.7 \text{ kg}}\right) = 340 \text{ L}$$

To calculate the result, then, you would enter 2000 on your calculator, divide by 2.20, and divide by 2.7. Such a setup makes it easy to see what to multiply and divide by, and the calculation goes more quickly when it can be entered into a calculator all at once.

The liter is not the most convenient volume unit for this problem, however, because it does not relate well to what we want to find out—how big a bin to make. A liter is about the same volume as a quart, but whether you are familiar with liters, quarts, or both, 300 of them is not easy to visualize. Let's convert liters to something we can understand better. A liter is a volume equal to a cube one tenth of a meter (1 dm) on a side; that is, $1\text{L} = 1 \text{ dm}^3$. Consequently,

$$340 \text{ L} = 340 \text{ L}\left(\frac{1 \text{ dm}^3}{1 \text{ L}}\right)\left(\frac{1 \text{ m}}{10 \text{ dm}}\right)^3 = 340 \text{ dm}^3\left(\frac{1 \text{ m}^3}{1000 \text{ dm}^3}\right) = 0.34 \text{ m}^3$$

Thus, the bin would need to have a volume of about one third of a cubic meter; that is, it could be a meter wide, a meter long, and about a third of a meter high and it would hold the ton of gravel.

One more thing should be noted about this example. We don't need to know the volume of the bin very precisely, because being off a bit will make very little difference; it might mean getting a little too much wood to build the bin or not making the bin quite big enough and having a little gravel spill out, but this isn't a big deal. In other cases, such as calculating the quantity of fuel needed to get a space shuttle into orbit, being off by a few percent could be a life-or-death matter. Because it is important to know how precise data are and to be able to evaluate how important precision is, scientific results usually indicate their precision. The simplest way to do so is by means of significant figures.

A.3 Precision, Accuracy, and Significant Figures

The **precision** of a measurement indicates how well several determinations of the same quantity agree. Some devices can make more precise measurements than others. Consider rulers A and B shown in the margin. Ruler A is marked every centimeter and ruler B is marked every millimeter. Clearly you could measure at least to the nearest millimeter with ruler B, and probably you could estimate to the nearest 0.2 mm or so. However, using ruler A you could probably estimate only to the nearest millimeter or so. We say that ruler B allows more precise measurement than ruler A. If several different people used ruler B to measure the length of an iron bar, their measurements would probably agree more closely than if they used ruler A.

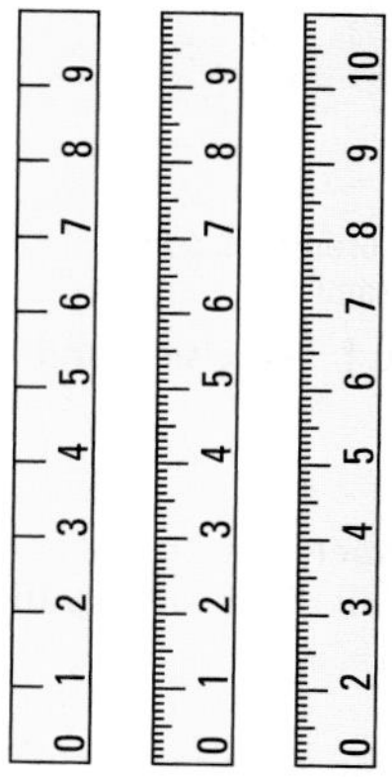

Three rulers.

Precision is also illustrated by the results of throwing darts at a bull's-eye (Figure A.1 on the next page). In part (a), the darts are scattered all over the board; the dart thrower was apparently not very skillful (or threw the darts from a long distance away from the board), and the precision of their placement on the board is low. This is analogous to the results that would be obtained by several people using ruler A. In part (b), the darts are all clustered together, indicating much better reproducibility on the part of the thrower—that is, greater precision. This is analogous to measurements that might be made using ruler B.

Notice also that in Figure A.1b every dart has come very close to the bull's-eye; we describe this result by saying that the thrower has been quite **accurate**—the average of all throws is very close to the accepted position, the bull's-eye. Figure A.1c illustrates that it is possible to be precise without being accurate—the dart thrower has consistently missed the bull's-eye, but all darts are clustered very precisely around the wrong point on the board. This third case is like an experiment with some flaw (either in its design or in a measuring device) that causes all results to differ from the correct value by the same quantity. An example is measuring length with ruler C, on which the scale is incorrectly labeled. The precision (reproducibility) of measurements with ruler C might be quite good, but the accuracy would be poor because all items longer than 1 cm would be off by a centimeter.

In the laboratory we attempt to set up experiments so that the greatest possible accuracy can be obtained. As a further check on accuracy, results may be compared among different laboratories so that any flaw in experimental design or measurement can be detected. For each individual experiment, several measurements are usually made and their precision determined. In most cases, better precision is taken as an indication of better experimental work, and it is necessary to know the precision to compare results among different experimenters. If two different experimenters both had results like those in Figure A.1a, for example, their average values could differ quite a lot before they would say that their results did not agree within experimental error.

In most experiments several different kinds of measurements must be made, and some can be done more precisely than others. It is common sense that *a calculated result can be no more precise than the least precise piece of information that went into the calculation.* This is where the rules for significant figures come in. In the last example in the preceding section, the quantity of gravel was described as "a ton." Usually gravel is measured by weighing an empty truck, putting some gravel in the truck, weighing the truck again, and subtracting the weight of the truck from the weight of the truck plus gravel. The quantity of gravel is not adjusted if there is a bit too much or a bit too little, because that would be a lot of trouble. You might end up with as much as 2200 pounds or as little as 1800 pounds, even though you asked for a ton. In terms of significant figures this would be expressed as 2.0×10^3 lb.

The quantity 2.0×10^3 lb is said to have two significant figures; it designates a quantity in which the 2 is taken to be exactly right but the 0 is not known precisely. (In this case, the number could be as large as 2.2 or as small as 1.8, so the 0 obviously is not exactly right.) In general, in a number that represents a scientific measurement, the last digit on the right is taken to be inexact, but all digits farther to the left are assumed to be exact. When you do calculations using such numbers, you must follow some simple rules so that your results will reflect the precision of all the measurements that go into the calculations. Here are the rules:

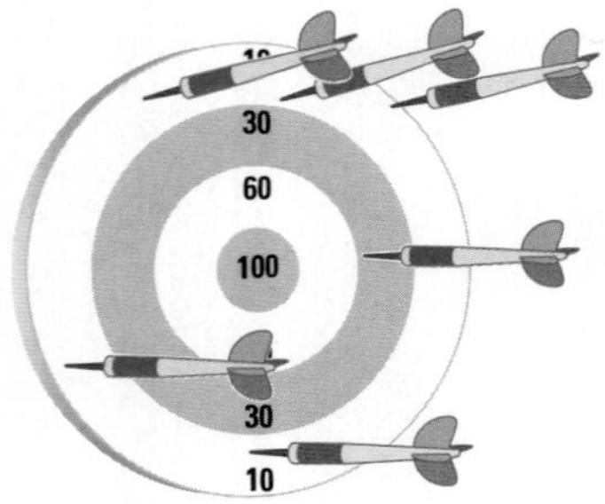

(a) Poor precision

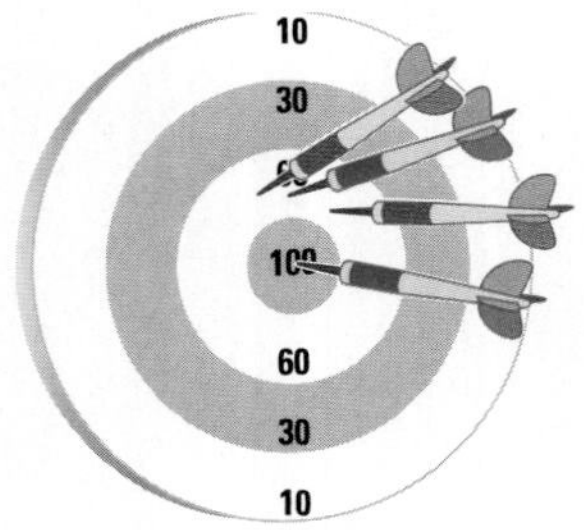

(b) Good precision; good accuracy

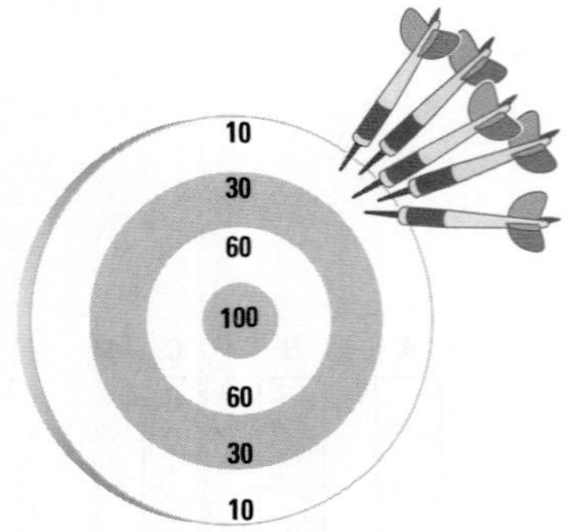

(c) Good precision; poor accuracy

Figure A.1 Precision and accuracy.

For a number written in scientific notation, all digits are significant.

The number π is now known to 1,011,196,691 digits. It is doubtful that you will need that much precision in this course—or ever.

Rule 1: To determine the number of significant figures in a measurement, read the number from left to right and count all digits, starting with the first digit that is *not* zero.

Example	Number of Significant Figures
1.23 g	3
0.00123 g	3; the zeros to the left of the 1 simply locate the decimal point. The number of significant figures is more obvious if you write numbers in scientific notation; thus, $0.00123 = 1.23 \times 10^{-3}$.
2.0 g and 0.020 g	2; both have two significant digits. When a number is greater than 1, *all zeros to the right of the decimal point are significant.* For a number less than 1, only zeros to the right of the first significant digit are significant.
100 g	1; in numbers that do not contain a decimal point, "trailing" zeros may or may not be significant. To eliminate possible confusion, the practice followed in this book is to include a decimal point if the zeros are significant. Thus, 100. has three significant digits, while 100 has only one. Alternatively, we write in scientific notation 1.00×10^2 (three significant digits) or 1×10^2 (one significant digit). For a number written in scientific notation, all digits preceding the 10^x term are significant.
100 cm/m	Infinite number of significant figures, because this is a defined quantity. There are *exactly* 100 centimeters in one meter.
$\pi = 3.1415926\ldots$	The value of π is known to a greater number of significant figures than any data you will ever use in a calculation.

Rule 2: When adding or subtracting, the number of decimal places in the answer should be equal to the number of decimal places in the number with the *fewest* places. Suppose you add three numbers:

0.12	2 significant figures	2 decimal places
1.6	2 significant figures	1 decimal place
+ 10.976	5 significant figures	3 decimal places
12.696		

This sum should be reported as 12.7, a number with one decimal place, because 1.6 has only one decimal place.

Rule 3: In multiplication or division, the number of significant figures in the answer should be the same as that in the quantity with the fewest significant figures.

$$\frac{0.1208}{0.0236} = 0.512 \text{ or, in scientific notation, } 5.12 \times 10^{-1}$$

Since 0.0236 has only three significant figures, while 0.01208 has four, the answer is limited to three significant figures.

Rule 4: When rounding a number to reduce the number of digits, follow these rules:

- The last digit retained is left unchanged if the following digit is less than 5 (4.327 rounded to two significant digits is 4.3).
- The last digit retained is increased by 1 if the following digit is greater than 5 or is a 5 followed by other non-zero digits (4.573 rounded to two significant digits is 4.6).
- If the last digit retained is followed by a single digit 5 only or by a 5 followed by zeros, then the last digit retained is increased by 1 if it is *odd* and remains the same if it is *even* (4.75 rounded to two significant digits is 4.8; 4.850 rounded to two significant digits is 4.8).

The examples in the table below further illustrate these rules.

Full Number	Number Rounded to Three Significant Figures
12.696	12.7
16.249	16.2
18.350	18.4
18.45	18.4
18.351	18.4

One last word regarding significant figures and calculations. In working problems on a pocket calculator, you should do the calculation using all the digits allowed by the calculator and round only at the end of the problem. Rounding in the middle can introduce small errors (called *rounding errors* or *round-off errors*). If your answers do not quite agree with those in the Appendices of this book, rounding errors may be the source of the disagreement.

Now let us consider a problem that is of practical importance and that makes use of all the rules. Suppose you discover that young children are eating chips of paint that flake off a wall in an old house. The paint contains 200. ppm lead by mass (200. milligrams of Pb per kilogram of paint). Suppose that a child eats five such chips. How much lead has the child gotten from the paint?

The ppm unit stands for "parts per million." If a substance is present with a concentration of 1 ppm by mass, there is 1 gram of the substance in 1 million grams of the sample.

As stated, this problem does not include enough information for a solution to be obtained; however, some reasonable assumptions can be made, and they can lead to experiments that could be used to obtain the necessary information. The statement does not say how big the paint chips are. Let's assume that they are 1.0 cm by 1.0 cm so that the area is 1.0 cm^2. Then eating five chips means eating 5.0 cm^2 of paint. (This assumption could be improved by measuring similar chips from the same place.) Since the concentration of lead is reported in units of mass of lead per mass of paint, we need to know the mass of 5.0 cm^2 of paint. This could be determined by measuring the areas of several paint chips and determining the mass of each. Suppose that the results of such measurements were those given in the table.

Mass of Chip (mg)	Area of Chip (cm^2)	Mass per Unit Area (mg/cm^2)
29.6	2.34	12.65
21.9	1.73	12.66
23.6	1.86	12.69

$$\text{Average mass per unit area} = \frac{(12.65 + 12.66 + 12.69)\ \text{mg/cm}^2}{3}$$

$$= 12.67\ \text{mg/cm}^2 = 12.7\ \text{mg/cm}^2$$

The average has been rounded to three significant figures because each experimentally measured mass and area has three significant figures. (Notice that more than three significant figures were kept in the intermediate calculations so as not to lose precision.) Now we can use this information to calculate how much lead the child has consumed.

$$m_{\text{paint}} = 5.0\ \text{cm}^2\ \text{paint}\left(\frac{12.7\ \text{mg paint}}{1\ \text{cm}^2\ \text{paint}}\right)\left(\frac{1\ \text{g}}{1000\ \text{mg}}\right)\left(\frac{1\ \text{kg}}{1000\ \text{g}}\right)$$

$$= 6.35 \times 10^{-5}\ \text{kg paint}$$

$$m_{\text{Pb}} = 6.35 \times 10^{-5}\ \text{kg paint}\left(\frac{200.\ \text{mg Pb}}{1\ \text{kg paint}}\right) = 1.27 \times 10^{-2}\ \text{mg Pb}$$

$$= 1.3 \times 10^{-2}\ \text{mg Pb} = 0.013\ \text{mg Pb}$$

The final result was rounded to two significant figures because there were only two significant figures in the initial 5.0 cm^2 area of the paint chip. This is quite adequate precision, however, for you to determine whether this quantity of lead is likely to harm the child.

The methods of problem solving presented here have been developed over time and represent a good way of keeping track of the precision of results, the units in which those results were obtained, and the correctness of calculations. These methods are not the only way that such goals can be achieved, but they do work well. We recommend that you include units in all calculations and check that they cancel appropriately. It is also important not to overstate the precision of results by keeping too many significant figures. By solving many problems, you should be able to develop your problem-solving skills so that they become second nature and you can use them without thinking about the mechanics. You can then devote all your thought to the logic of a problem solution.

A.4 Electronic Calculators

The directions for calculator use in this section are given for calculators using "algebraic" logic. Such calculators are the most common type used by students in introductory courses. For calculators using RPN logic (such as those made by Hewlett-Packard), the procedure will differ slightly.

The advent of inexpensive electronic calculators has made calculations in introductory chemistry much more straightforward. You are well advised to purchase a calculator that has the capability of performing calculations in scientific notation, has both base 10 and natural logarithms, and is capable of raising any number to any power and of finding any root of any number. In the discussion below, we will point out in general how these functions of your calculator can be used. You should practice using your calculator to carry out arithmetic operations and make certain that you are able to use all of its functions correctly.

Although electronic calculators have greatly simplified calculations, they have also forced us to focus again on significant figures. A calculator easily handles eight or more significant figures, but real laboratory data are rarely known to this precision. Therefore, if you have not already done so, review Appendix A.3 on significant figures, precision, and rounding numbers.

The mathematical skills required to read and study this textbook successfully involve algebra, some geometry, scientific notation, logarithms, and solving quadratic equations. The next three sections review the last three of these topics.

A.5 Exponential or Scientific Notation

In exponential or scientific notation, a number is expressed as a product of two numbers: $N \times 10^n$. The first number, N, is called the digit term and is a number between 1 and 10. The second number, 10^n, the exponential term, is some integer power of 10. For example, 1234 would be written in scientific notation as 1.234×10^3 or 1.234 multiplied by 10 three times.

$$1234 = 1.234 \times 10^1 \times 10^1 \times 10^1 = 1.234 \times 10^3$$

Conversely, a number less than 1, such as 0.01234, would be written as 1.234×10^{-2}. This notation tells us that 1.234 should be divided twice by 10 to obtain 0.01234.

$$0.01234 = \frac{1.234}{10^1 \times 10^1} = 1.234 \times 10^{-1} \times 10^{-1} = 1.234 \times 10^{-2}$$

Some other examples of scientific notation are

$10{,}000 = 1 \times 10^4$	$12{,}345 = 1.2345 \times 10^4$
$1000 = 1 \times 10^3$	$1234 = 1.234 \times 10^3$
$100 = 1 \times 10^2$	$123 = 1.23 \times 10^2$
$10 = 1 \times 10^1$	$12 = 1.2 \times 10^1$
$1 = 1 \times 10^0$	(any number to the zeroth power $= 1$)
$1/10 = 1 \times 10^{-1}$	$0.12 = 1.2 \times 10^{-1}$
$1/100 = 1 \times 10^{-2}$	$0.012 = 1.2 \times 10^{-2}$
$1/1000 = 1 \times 10^{-3}$	$0.0012 = 1.2 \times 10^{-3}$
$1/10{,}000 = 1 \times 10^{-4}$	$0.00012 = 1.2 \times 10^{-4}$

When converting a number to scientific notation, notice that the exponent n is positive if the number is greater than 1 and negative if the number is less than 1. The value of n is the number of places by which the decimal was shifted to obtain the number in scientific notation.

$$1\,2\,3\,4\,5. = 1.2345 \times 10^4$$

Decimal shifted 4 places to the left. Therefore, n is positive and equal to 4.

$$0.0\,0\,1\,2 = 1.2 \times 10^{-3}$$

Decimal shifted 3 places to the right. Therefore, n is negative and equal to 3.

If you wish to convert a number in scientific notation to the usual form, the procedure above is simply reversed.

$$6\,.\,2\,7\,3 \times 10^2 = 627.3$$

Decimal point shifted 2 places to the right, because n is positive and equal to 2.

$$0\,0\,6.273 \times 10^{-3} = 0.006273$$

Decimal point shifted 3 places to the left, because n is negative and equal to 3.

To enter a number in scientific notation into a calculator, first enter the number itself. Then press the EE (Enter Exponent) key followed by n, the power of 10. For example, to enter 6.022×10^{23}, you press these keys in succession:

6 . 0 2 2 EE 2 3 .

(Do not enter the number using the multiplication key, the number 10, and the EE key. This will result in a number that is 10 times bigger than you want. For example, pressing these keys

6 . 0 2 2 × 1 0 EE 2 3

enters the number $6.022 \times 10 \times 10^{23} = 6.022 \times 10^{24}$, because EE 2 3 means 10^{23}.)

There are two final points concerning scientific notation. First, if you are used to working on a computer you may be in the habit of writing a number such as 1.23×10^3 as 1.23E3, or 6.45×10^{-5} as 6.45E-5. Second, some electronic calculators allow you to convert numbers readily to scientific notation. If you have such a calculator, you can change a number shown in the usual form to scientific notation by pressing an appropriate key or keys.

Usually you will handle numbers in scientific notation with a calculator. In case you need to work without a calculator, however, the next few sections describe pencil-and-paper calculations as well as calculator methods.

1. Adding and Subtracting

When adding or subtracting numbers in scientific notation without using a calculator, first convert the numbers to the same powers of 10. Then add or subtract the digit terms.

$$(1.234 \times 10^{-3}) + (5.623 \times 10^{-2}) = (0.1234 \times 10^{-2}) + (5.623 \times 10^{-2})$$
$$= 5.746 \times 10^{-2}$$
$$(6.52 \times 10^2) - (1.56 \times 10^3) = (6.52 \times 10^2) - (15.6 \times 10^2)$$
$$= -9.1 \times 10^2$$

In this calculation, the result has only two significant figures, although each of the original numbers had three. Subtracting two numbers that are nearly the same can reduce the number of significant figures appreciably.

2. Multiplying

Multiply the digit terms in the usual manner, and add the exponents. Express the result with a digit term with only one nonzero digit to the left of the decimal.

$$(1.23 \times 10^3)(7.60 \times 10^2) = (1.23 \times 7.60)(10^3 \times 10^2) = (1.23)(7.60) \times 10^{3+2}$$
$$= 9.35 \times 10^5$$

$$(6.02 \times 10^{23})(2.32 \times 10^{-2}) = (6.02)(2.32) \times 10^{23-2}$$
$$= 13.966 \times 10^{21} = 1.3966 \times 10^{22}$$
$$= 1.40 \times 10^{22} \quad \text{(rounded to 3 significant figures)}$$

3. Dividing

Divide the digit terms in the usual manner, and subtract the exponents. Write the quotient with one nonzero digit to the left of the decimal.

$$\frac{7.60 \times 10^3}{1.23 \times 10^2} = \frac{7.60}{1.23} \times 10^{3-2} = 6.18 \times 10^1$$

$$\frac{6.02 \times 10^{23}}{9.10 \times 10^{-2}} = \frac{6.02}{9.10} \times 10^{(23)-(-2)} = 0.662 \times 10^{25} = 6.62 \times 10^{24}$$

4. Raising Numbers in Scientific Notation to Powers

When raising a number in scientific notation to a power, treat the digit term in the usual manner. Multiply the exponent by the number indicating the power.

$$(1.25 \times 10^3)^2 = (1.25)^2 \times (10^3)^2 = (1.25)^2 \times 10^{3\times 2} = 1.5625 \times 10^6 = 1.56 \times 10^6$$

$$(5.6 \times 10^{-10})^3 = (5.6)^3 \times 10^{(-10)\times 3} = 175.6 \times 10^{-30} = 1.8 \times 10^{-28}$$

Electronic calculators usually have two methods of raising a number to a power. To square a number, enter the number and then press the "x^2" key. To raise a number to any power, use the "y^x" key. For example, to raise 1.42×10^2 to the 4th power, that is, to find $(1.42 \times 10^2)^4$,

(a) Enter 1.42×10^2.
(b) Press "y^x."
(c) Enter 4 (this should appear on the display).
(d) Press "=" and 4.0658689. . . $\times 10^8$ will appear on the display. (The number of digits depends on the calculator.)

As a final step, express the number in the correct number of significant figures (4.07×10^8 in this case).

5. Taking Roots of Numbers in Scientific Notation

Unless you use an electronic calculator, first put the number into a form in which the exponential is exactly divisible by the root. Find the root of the digit term in the usual way, and divide the exponent by the desired root.

$$\sqrt{3.6 \times 10^7} = \sqrt{36 \times 10^6} = \sqrt{36} \times \sqrt{10^6} = 6.0 \times 10^3$$

$$\sqrt[3]{2.1 \times 10^{-7}} = \sqrt[3]{210 \times 10^{-9}} = \sqrt[3]{210} \times \sqrt[3]{10^{-9}} = 5.9 \times 10^{-3}$$

To take a square root on an electronic calculator, enter the number and then press the "$\sqrt{x}$" key. To find a higher root of a number, such as the fourth root of 5.6×10^{-10},

On some calculators, Steps (a) and (c) may be interchanged.

(a) Enter the number, 5.6×10^{-10} in this case.
(b) Press the "$\sqrt[x]{y}$" key. (On most calculators, the sequence you actually use is to press "2ndF" and then "$\sqrt[x]{y}$." Alternatively, you may have to press "INV" and then "y^x".)
(c) Enter the desired root, 4 in this case.
(d) Press "=". The answer here is 4.8646×10^{-3} or 4.9×10^{-3}.

A general procedure for finding any root is to use the "y^x" key. For a square root, x is 0.5 (or $\frac{1}{2}$), whereas it is 0.33 (or $\frac{1}{3}$) for a cube root, 0.25 (or $\frac{1}{4}$) for a fourth root, and so on.

A.6 Logarithms

There are two types of logarithms used in this text: common logarithms (abbreviated log), whose base is 10, and natural logarithms (abbreviated ln), whose base is e (=2.7182818284).

$$\log x = n \qquad \text{where } x = 10^n$$

$$\ln x = m \qquad \text{where } x = e^m$$

Logarithms to the base 10 are needed when dealing with pH.

Most equations in chemistry and physics were developed in natural or base e logarithms, and this practice is followed in this text. The relation between log and ln is

$$\ln x = 2.303 \log x$$

Aside from the different bases of the two logarithms, they are used in the same manner. What follows is largely a description of the use of common logarithms.

The common logarithm of a number is the power to which you must raise 10 to obtain the number. For example, the log of 100 is 2, because you must raise 10 to the power 2 to obtain 100. Other examples are

$$\log 1000 = \log(10^3) = 3$$

$$\log 10 = \log(10^1) = 1$$

$$\log 1 = \log(10^0) = 0$$

$$\log 1/10 = \log(10^{-1}) = -1$$

$$\log 1/10{,}000 = \log(10^{-4}) = -4$$

To obtain the common logarithm of a number other than a simple power of 10, use an electronic calculator. For example,

$$\log 2.10 = 0.3222\text{, which means that } 10^{0.3222} = 2.10$$

$$\log 5.16 = 0.7126\text{, which means that } 10^{0.7126} = 5.16$$

$$\log 3.125 = 0.49485\text{, which means that } 10^{0.49485} = 3.125$$

To check this result on your calculator, enter the number and then press the "log" key.

To obtain the natural logarithm of the numbers above, use a calculator having this function. Enter each number and press "ln".

$$\ln 2.10 = 0.7419\text{, which means that } e^{0.7419} = 2.10$$

$$\ln 5.16 = 1.6409\text{, which means that } e^{1.6409} = 5.16$$

To find the common logarithm of a number greater than 10 or less than 1 with a log table, first express the number in scientific notation. Then find the log of each part of the number and add the logs. For example,

Nomenclature of Logarithms: The number to the left of the decimal in a logarithm is called the *characteristic,* and the number to the right of the decimal is called the *mantissa.*

$$\log 241 = \log(2.41 \times 10^2) = \log 2.4 + \log 10^2$$
$$= 0.382 + 2 = 2.382$$

$$\log 0.00573 = \log(5.73 \times 10^{-3}) = \log 5.73 + \log 10^{-3}$$
$$= 0.758 + (-3) = -2.242$$

Significant Figures and Logarithms

The mantissa (digits to the right of the decimal point in the logarithm) should have as many significant figures as the number whose log was found. (So that you could more clearly see the result obtained with a calculator or a table, this rule was not strictly followed until the last two examples.)

Obtaining Antilogarithms

If you are given the logarithm of a number and need to find the number from it, you need to obtain the "antilogarithm" or "antilog" of the number. There are two common procedures used by electronic calculators to do this:

Procedure A	Procedure B
(a) Enter the log or ln (a number).	(a) Enter the log or ln (a number).
(b) Press 2ndF.	(b) Press INV.
(c) Press 10^x or e^x.	(c) Press log or ln x.

Test one or the other of these procedures with the following examples.

EXERCISE A.1 Find the number whose log is 5.234.

Recall that $\log x = n$, where $x = 10^n$. In this case $n = 5.234$. Enter that number in your calculator and find the value of 10^n, the antilog. In this case,

$$10^{5.234} = 10^{0.234} \times 10^5 = 1.71 \times 10^5$$

Notice that the characteristic (5) sets the decimal point; it is the power of 10 in the exponential form. The mantissa (0.234) gives the value of the number x. Thus, if you use a log table to find x, you need only look up 0.234 in the table and see that it corresponds to 1.71.

EXERCISE A.2 Find the number whose log is −3.456.

$$10^{-3.456} = 10^{0.544} \times 10^{-4} = 3.50 \times 10^{-4}$$

Notice here that −3.456 must be expressed as the sum of −4 and +0.544.

Mathematical Operations Using Logarithms

Because logarithms are exponents, operations involving them follow the same rules as the use of exponents. Thus, multiplying two numbers can be done by adding logarithms.

$$\log xy = \log x + \log y$$

For example, we multiply 563 by 125 by adding their logarithms and finding the antilogarithm of the result.

$$\log 563 = 2.751$$

$$\log 125 = 2.097$$

$$\log (563 \times 125) = 2.751 + 2.097 = 4.848$$

$$563 \times 125 = 10^{4.848} = 10^4 \times 10^{0.848} = 7.05 \times 10^4$$

One number (x) can be divided by another (y) by subtraction of their logarithms.

$$\log \frac{x}{y} = \log x - \log y$$

For example, to divide 125 by 742,

$$\log 125 = 2.097$$

$$\log 742 = 2.870$$

$$\log (125/742) = 2.097 - 2.870 = -0.773$$

$$125/742 = 10^{-0.773} = 10^{0.227} \times 10^{-1} = 1.69 \times 10^{-1}$$

Similarly, powers and roots of numbers can be found using logarithms.

$$\log x^y = y(\log x)$$

$$\log \sqrt[y]{x} = \log x^{1/y} = \frac{1}{y} \log x$$

As an example, find the fourth power of 5.23. First find the log of 5.23 and then multiply it by 4. The result, 2.874, is the log of the answer. Therefore, find the antilog of 2.874.

$$(5.23)^4 = ?$$

$$\log (5.23)^4 = 4 \log 5.23 = 4(0.719) = 2.874$$

$$(5.23)^4 = 10^{2.874} = 748$$

As another example, find the fifth root of 1.89×10^{-9}.

$$\sqrt[5]{1.89 \times 10^{-9}} = (1.89 \times 10^{9})^{1/5} = ?$$

$$\log (1.89 \times 10^{-9})^{1/5} = \frac{1}{5} \log (1.89 \times 10^{-9}) = \frac{1}{5}(-8.724) = -1.745$$

The answer is the antilog of -1.745.

$$(1.89 \times 10^{-9})^{1/5} = 10^{-1.745} = 1.80 \times 10^{-2}$$

A.7 Quadratic Equations

Algebraic equations of the form $ax^2 + bx + c = 0$ are called **quadratic equations.** The coefficients a, b, and c may be either positive or negative. The two roots of the equation may be found using the *quadratic formula.*

$$x = \frac{-b \pm \sqrt{b^2 - 4ac}}{2a}$$

As an example, solve the equation $5x^2 - 3x - 2 = 0$. Here $a = 5$, $b = -3$, and $c = -2$. Therefore,

$$x = \frac{3 \pm \sqrt{(-3^2) - 4(5)(-2)}}{2(5)}$$

$$= \frac{3 \pm \sqrt{9 - (-40)}}{10} = \frac{3 \pm \sqrt{49}}{10} = \frac{3 \pm 7}{10}$$

$$x = 1 \textit{ and } -0.4$$

How do you know which of the two roots is the correct answer? You have to decide in each case which root has physical significance. However, it is usually true in this course that negative values are not significant.

Many calculators have a built-in quadratic-equation solver. If your calculator offers this capability, it will be convenient to use it, and you should study the manual until you know how to use your calculator to obtain the roots of a quadratic equation.

When you have solved a quadratic expression, you should always check your values by substituting them into the original equation. In the example above, we find that $5(1)^2 - 3(1) - 2 = 0$ and that $5(-0.4)^2 - 3(-0.4) - 2 = 0$.

You will encounter quadratic equations in the chapters on chemical equilibria, particularly in Chapters 14, 16, and 17. Here you may be faced with solving an equation such as

$$1.8 \times 10^{-4} = \frac{x^2}{0.0010 - x}$$

This equation can certainly be solved by using the quadratic formula or your calculator (to give $x = 3.4 \times 10^{-4}$). However, you may find the method of successive approximations to be especially convenient. Here you begin by making a reasonable approximation of x. This approximate value is substituted into the original equation, and this expression is solved to give what is hoped to be a more correct value of x. This process is repeated until the answer converges on a particular value of x—that is, until the value of x derived from two successive approximations is the same.

Step 1: Assume that x is so small that $(0.0010 - x) \approx 0.0010$. This means that

$$x^2 = 1.8 \times 10^{-4}(0.0010)$$

$$x = 4.2 \times 10^{-4} \text{ (to 2 significant figures)}$$

Step 2: Substitute the value of x from Step 1 into the denominator (but not the numerator) of the original equation and again solve for x.

$$x^2 = (1.8 \times 10^{-4})(0.0010 - 0.00042)$$
$$x = 3.2 \times 10^{-4}$$

Step 3: Repeat Step 2 using the value of x found in that step.

$$x = \sqrt{1.8 \times 10^{-4}(0.0010 - 0.00032)} = 3.5 \times 10^{-4}$$

Step 4: Continue by repeating the calculation, using the value of x found in the previous step.

Step 5. $x = \sqrt{1.8 \times 10^{-4}(0.0010 - 0.00034)} = 3.4 \times 10^{-4}$

Here we find that iterations after the fourth step give the same value for x, indicating that we have arrived at a valid answer (and the same one obtained from the quadratic formula).

Some final thoughts on using the method of successive approximations: First, in some cases this method does not work. Successive steps may give answers that are random or that diverge from the correct value. For quadratic equations of the form $K = x^2/(C - x)$, the method of approximations will work only as long as $K < 4C$ (assuming one begins with $x = 0$ as the first guess; that is, $K \approx x^2/C$). This will always be true for weak acids and bases.

Second, values of K in the equation $K = x^2/(C - x)$ are usually known only to two significant figures. Therefore, we are justified in carrying out successive steps until two answers are the same to two significant figures.

Finally, if your calculator does not automatically obtain roots of quadratic equations, we highly recommend this method. If your calculator has a memory function, successive approximations can be carried out easily and very rapidly. Even without a memory function, the method of successive approximations is much quicker than solving a quadratic equation by hand.

A.8 Graphing

Mass (g)	Volume (mL)
2.03	0.76
5.27	1.95
9.57	3.54
11.46	4.25
14.96	5.55
18.02	6.68
21.83	8.10
25.17	9.32
30.08	11.14
32.84	12.17
36.27	13.43
36.77	13.62
39.36	14.58

When analyzing experimental data, chemists and other scientists often graph the data to see whether the data agree with a mathematical equation. If the equation does fit the data (often indicated by a linear graph), then the graph can be used to obtain numerical values (parameters) that can be used in the equation to predict information not specifically included in the data set. For example, suppose that you measured the masses and the volumes of several samples of aluminum. Your data set might look like the table in the margin. When these data are plotted with volume along the horizontal (x) axis and mass on the vertical (y) axis, a graph like this one is obtained.

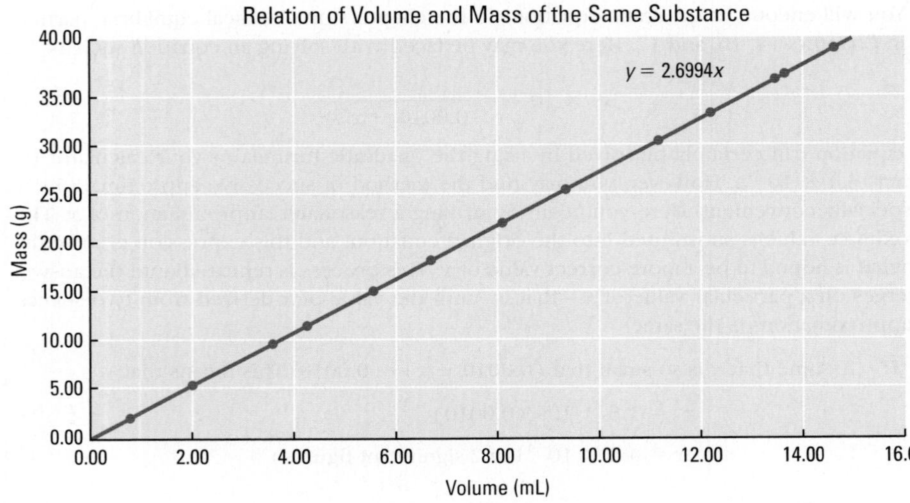

Notice the features of this graph. It has a title that describes its contents, it has a label for each axis that specifies the quantity plotted and the units used, it has equally spaced grid lines in the vertical and horizontal directions, and the numbers associated with those grid lines have the same difference between successive grid lines. (The differences are 2.00 mL on the horizontal axis and 5.00 g on the vertical axis.) The experimental points are indicated by circular markers, and a straight line has been drawn through the points. Because the line passes through all of the points and through the origin (point 0, 0), the data can be represented by the equation $y = 2.70x$, where 2.70 is the slope of the line. (On the graph, the slope is indicated by 2.6994; in the equation, it has been rounded to 2.70 because some of the data have only three significant figures.)

The slope of a graph such as this one can be obtained by choosing two points on the straight line (not two data points). The points should be far apart to obtain the most precise result. The slope is then given by the change in the *y*-axis variable divided by the change in the *x*-axis variable from the first point to the second. An example is shown on the graph below. From the calculation of the slope, it is more obvious that the slope has only three significant figures. For this graph, the slope represents mass divided by volume—that is, density. A good way to measure density is to measure the mass and volume of several samples of the same substance and then plot the data. The resulting graph should be linear and should pass very nearly through the origin. The density can be obtained from its slope. When a graph passes through the origin, we say that the *y*-axis variable is directly proportional to the *x*-axis variable.

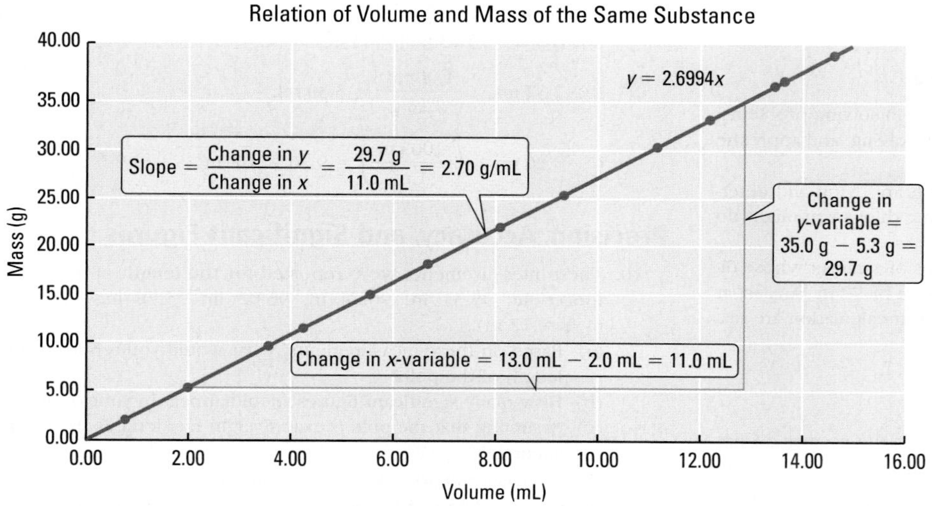

When a graph does not pass through the origin, there is an intercept as well as a slope. The intercept is the value of the *y*-axis variable when the *x*-axis variable equals zero. A good example is the relation between Celsius and Fahrenheit temperatures. Suppose you have measured a series of temperatures using both a Celsius thermometer and a Fahrenheit thermometer. Your data set might look something like the table in the margin.

When graphed, these data look like this.

Temperature (°C)	Temperature (°F)
−5.23	22.586
13.54	56.372
32.96	91.328
48.34	119.012
63.59	146.462
74.89	166.802
88.02	190.436
105.34	221.612

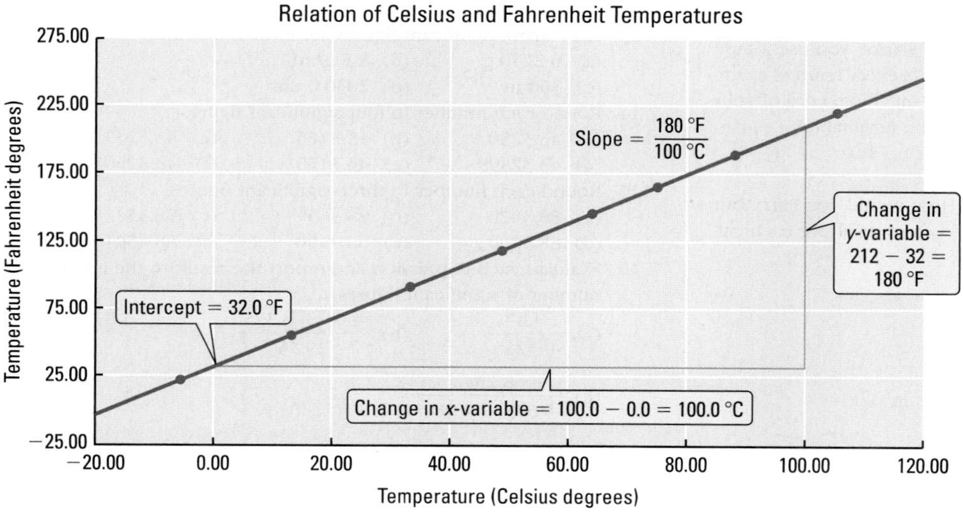

In this case, although the data are on a straight line, that line does not pass through the origin. Instead, it intersects the vertical line corresponding to $x = 0$ (0 °C) at a value of 32 °F. This makes sense, because the normal freezing point of water is 0 °C, which is the same temperature as 32 °F. This value, 32 °F, is the intercept.

If we determine the slope by starting at the intercept value of 32 °F and going to 212° F, the normal boiling point of water, the change in temperature on the Fahrenheit scale is 180 °F. The corresponding change on the Celsius scale is 100 °C, from which we can determine that the slope is 180 °F/100 °C. Thus the equation relating Celsius and Fahrenheit temperatures is

$$y \quad = \quad m \quad x \quad + \quad b$$

$$\text{temperature } °\text{F} = \frac{180\ °\text{F}}{100\ °\text{C}} \times \text{temperature } °\text{C} + 32\ °\text{F}$$

QUESTIONS FOR APPENDIX A

Blue-numbered questions have short answers at the back of this book in Appendix M and fully worked solutions in the *Student Solutions Manual.*

General Problem-Solving Strategies

1. List four steps that can be used for guidance in solving problems. Choose a problem that you are interested in solving, and apply the four steps to that problem.
2. You are asked to study a lake in which fish are dying and determine the cause of their deaths. Suggest three things you might do to define this problem.
3. You have calculated the area in square yards of a carpet whose dimensions were originally given to you as 96 in by 72 in. Suggest at least one way to check that the results of your calculation are reasonable. (12 in = 1 ft; 3 ft = 1 yd)

Numbers, Units, and Quantities

4. When a calculation is done and the units for the answer do not make sense, what can you conclude about the solution to the problem? What would you do if you were faced with a situation like this?
5. The term "quantity" (or "physical quantity") has a specific meaning in science. What is that meaning, and why is it important?
6. To measure the length of a pencil, you use a tape measure calibrated in inches with marks every sixteenth of an inch. Which of these results would be a suitable record of your observation? Why would the other results be unsuitable? (1/16 in = 0.0625 in)
   (a) 8.38 ft (b) 8.38 m (c) 8.38 in
   (d) 8.38 (e) 0.698 ft
7. To measure the inseam length of a pair of slacks, you use a tape measure calibrated in centimeters with marks every tenth of a centimeter. Which of these results would be a suitable record of your observation? Why would the other results be unsuitable?
   (a) 75.0 cm (b) 75.0 m (c) 75.0
   (d) 75.0 in (e) 750 mm
8. What is wrong with each calculation? How would you carry out each calculation correctly? What is the correct result for each calculation?
   (a) $4.32\text{ g} + 5.63\text{ g} = 9.95\text{ g}^2$
   (b) $5.23\text{ g} \times \dfrac{4.87\text{ g}}{1.00\text{ mL}} = 25.5\text{ g}$
   (c) $3.57\text{ cm}^3 \times \left(\dfrac{1\text{ m}}{100\text{ cm}}\right) = 3.57 \times 10^{-2}\text{ m}^3$
9. What is wrong with each calculation? How would you carry out each calculation correctly? What is the correct result for each calculation?
   (a) $7.86\text{ g} - 5.63\text{ g} = 2.23\text{ g}^2$
   (b) $7.37\text{ mL} \times \dfrac{1.00\text{ mL}}{2.23\text{ g}} = 3.30\text{ mL}$
   (c) $9.26\text{ m}^3 \times \left(\dfrac{100\text{ cm}}{1\text{ m}}\right) = 9.26 \times 10^2\text{ m}^3$

Precision, Accuracy, and Significant Figures

10. These measurements were reported for the length of an eight-foot pole: 95.31 in; 96.44 in; 96.02 in; 95.78 in; 95.94 in (1 ft = 12 in).
    (a) Based on these measurements, what would you report as the length of the pole?
    (b) How many significant figures should appear in your result?
    (c) Assuming that the pole is exactly eight feet long, is the result accurate?
11. These measurements were reported for the mass of a sample in the laboratory: 32.54 g; 32.67 g; 31.98 g; 31.76 g; 32.05 g.
    (a) Based on these measurements, what would you report as the mass of the sample?
    (b) How many significant figures should appear in your result?
    (c) The sample is exactly balanced by a weight known to have a mass of 35.0 g. Is the result accurate?
12. How many significant figures are in each quantity?
    (a) 3.274 g (b) 0.0034 L
    (c) 43,000 m (d) 6200. ft
13. How many significant figures are in each quantity?
    (a) 0.2730 g (b) 8.3 g/mL
    (c) 300 m (d) 2030.0 dm^3
14. Round each number to four significant figures.
    (a) 43.3250 (b) 43.3165 (c) 43.3237
    (d) 43.32499 (e) 43.3150 (f) 43.32501
15. Round each number to three significant figures.
    (a) 88.3520 (b) 88.365 (c) 88.45
    (d) 88.5500 (e) 88.2490 (f) 88.4501
16. Evaluate each expression and report the result to the appropriate number of significant figures.
    (a) $\dfrac{4.47}{0.3260}$ (b) $\dfrac{4.03 + 3.325}{29.75}$
    (c) $\dfrac{8.234}{5.673 - 4.987}$

17. Evaluate each expression and report the result to the appropriate number of significant figures.
(a) $\frac{4.47}{0.3260}$ (b) $\frac{4.03 + 3.325}{29.75}$
(c) $\frac{8.234}{5.673 - 4.987}$

Exponential or Scientific Notation

18. Without using a calculator, express each number in scientific notation and with the appropriate number of significant figures.
(a) 76,003 (b) 0.00037 (c) 34,000
19. Without using a calculator, express each number in scientific notation and with the appropriate number of significant figures.
(a) 49,002 (b) 0.0234 (c) 23,400
20. Evaluate each expression using your calculator and report the result to the appropriate number of significant figures with appropriate units.
(a) $\frac{0.7346}{304.2}$
(b) $\frac{(3.45 \times 10^{-3})(1.83 \times 10^{12})}{23.4}$
(c) $3.240 - 4.33 \times 10^{-3}$
(d) $(4.87 \text{ cm})^3$
21. Evaluate each expression using your calculator and report the result to the appropriate number of significant figures with appropriate units.
(a) $\frac{893.0}{0.2032}$
(b) $\frac{(5.4 \times 10^{3})(8.36 \times 10^{-12})}{5.317 \times 10^{-3}}$
(c) $3.240 \times 10^{5} - 8.33 \times 10^{3}$
(d) $(4.87 \text{ cm} + 7.33 \times 10^{-1} \text{ cm})^3$

Logarithms

22. Use your calculator to find the logarithm of each number and report the logarithm to the appropriate number of significant figures.
(a) log(0.7327) (b) ln(34.5)
(c) $\log(6.022 \times 10^{23})$ (d) $\ln(6.022 \times 10^{23})$
(e) $\log\left(\frac{8.34 \times 10^{-5}}{2.38 \times 10^{3}}\right)$
23. Use your calculator to find the logarithm of each number and report the logarithm to the appropriate number of significant figures.
(a) log(54.3) (b) ln(0.0345)
(c) $\log(4.344 \times 10^{-3})$ (d) $\ln(8.64 \times 10^{4})$
(e) $\ln\left(\frac{4.33 \times 10^{24}}{8.32 \times 10^{-2}}\right)$

24. Use your calculator to evaluate each expression and report the result to the appropriate number of significant figures.
(a) antilog(0.7327) (b) antiln(34.5)
(c) $10^{2.043}$ (d) $e^{3.20 \times 10^{-4}}$
(e) exp(4.333/3.275)
25. Use your calculator to evaluate each expression and report the result to the appropriate number of significant figures.
(a) antilog(87.2) (b) antiln(0.0034)
(c) $e^{2.043}$ (d) $10^{(3.20 \times 10^{-4})}$
(e) $\exp(4.3 \times 10^{3}/8.314)$

Quadratic Equations

26. Find the roots of each quadratic equation.
(a) $3.27x^2 + 4.32x - 2.83 = 0$
(b) $x^2 + 4.32 = 4.57x$
27. Find the roots of each quadratic equation.
(a) $8.33x^2 - 2.32x - 7.53 = 0$
(b) $4.3x^2 - 8.37 = -2.22x$

Graphing

28. Graph these data involving mass and volume, label the graph appropriately, and determine whether m is directly proportional to V.

V (mL)	m (g)
0.347	0.756
1.210	2.638
2.443	5.326
7.234	15.76
11.43	24.90

29. Graph these data for heat transfers during a reaction, label the graph appropriately, and determine whether the heat evolved is directly proportional to the amount of reactant consumed.

Amount of Reactant (mol)	Heat Evolved (J)
94.2	43.2
70.7	32.5
65.7	30.1
34.2	15.7
54.3	24.9

APPENDIX

B

Units, Equivalences, and Conversion Factors

B.1 Units of the International System (SI)

The metric system was begun by the French National Assembly in 1790 and has undergone many modifications since its inception. The International System of Units, or *Système International* (SI), which represents an extension of the metric system, was adopted by the 11th General Conference on Weights and Measures in 1960. It is constructed from seven base units, each of which represents a particular physical quantity (Table B.1). More information about the SI is available at **http://physics.nist.gov/cuu/Units/index.html**.

The first five units listed in Table B.1 are particularly useful in chemistry. They are defined as follows:

1. The *meter* is the length of the path traveled by light in a vacuum during a time interval of 1/299,792,458 of a second.
2. The *kilogram* represents the mass of a platinum-iridium block kept at the International Bureau of Weights and Measures in Sevres, France.
3. The *second* is the duration of 9,192,631,770 periods of a certain line in the microwave spectrum of cesium-133.
4. The *kelvin* is 1/273.16 of the temperature interval between absolute zero and the triple point of water (the temperature at which liquid water, ice, and water vapor coexist).
5. The *mole* is the amount of substance that contains as many elementary entities (atoms, molecules, ions, or other particles) as there are atoms in exactly 0.012 kg of carbon-12 (12 g of ^{12}C atoms).

Decimal fractions and multiples of metric and SI units are designated by using the **prefixes** listed in Table B.2. The prefix *kilo-,* for example, means that a unit is multiplied by 10^3.

$$1 \text{ kilogram} = 1 \times 10^3 \text{ grams} = 1000 \text{ grams}$$

The prefix *centi-* means that the unit is multiplied by the factor 10^{-2}.

$$1 \text{ centigram} = 1 \times 10^{-2} \text{ gram} = 0.01 \text{ gram}$$

The prefixes are added to give units of a magnitude appropriate to what is being measured. The distance from New York to London (5.6×10^3 km = 5600 km) is much easier

Table B.1 SI Fundamental Units

Physical Quantity	Name of Unit	Symbol
Length	Meter	m
Mass	Kilogram	kg
Time	Second	s
Temperature	Kelvin	K
Amount of substance	Mole	mol
Electric current	Ampere	A
Luminous intensity	Candela	cd

Table B.2 Prefixes for Metric and SI Units[1]

Factor	Prefix	Symbol	Factor	Prefix	Symbol
10^{18}	exa-	E	10^{-1}	*deci-*	d
10^{15}	peta-	P	10^{-2}	*centi-*	c
10^{12}	tera-	T	10^{-3}	*milli-*	m
10^{9}	giga-	G	10^{-6}	*micro-*	μ
10^{6}	mega-	M	10^{-9}	*nano-*	n
10^{3}	*kilo-*	k	10^{-12}	*pico-*	p
10^{2}	hecto-	h	10^{-15}	femto-	f
10^{1}	deka-	da	10^{-18}	atto-	a

[1]The prefixes most commonly used in chemistry are shown in italics.

to comprehend measured in kilometers than in meters (5.6×10^6 m = 5,600,000 m). Following Table B.2 is a list of units for measuring very small and very large distances.

attometer (am)	0.000000000000000001 meter
femtometer (fm)	0.000000000000001 meter
picometer (pm)	0.000000000001 meter
nanometer (nm)	0.000000001 meter
micrometer (μm)	0.000001 meter
millimeter (mm)	0.001 meter
centimeter (cm)	0.01 meter
decimeter (dm)	0.1 meter
meter (m)	1 meter
dekameter (dam)	10 meters
hectometer (hm)	100 meters
kilometer (km)	1000 meters
megameter (Mm)	1,000,000 meters
gigameter (Gm)	1,000,000,000 meters
terameter (Tm)	1,000,000,000,000 meters
petameter (Pm)	1,000,000,000,000,000 meters
exameter (Em)	1,000,000,000,000,000,000 meters

In the International System of Units, all physical quantities are represented by appropriate combinations of the base units listed in Table B.1. The result is a derived unit for each kind of measured quantity. The most common derived units are listed in Table B.3. It is easy

Table B.3 Derived SI Units

Physical Quantity	Name of Unit	Symbol	Definition	Expressed in Fundamental Units
Area	Square meter	m^2	—	
Volume	Cubic meter	m^3	—	
Density	Kilogram per cubic meter	kg/m^3	—	
Force	Newton	N	$\frac{(\text{kilogram})(\text{meter})}{(\text{second})^2}$	$kg\ m\ s^{-2}$
Pressure	Pascal	Pa	$\frac{(\text{newton})}{(\text{meter})^2}$	$N/m^2 = kg\ m^{-1}\ s^{-2}$
Energy	Joule	J	$\frac{(\text{kilogram})(\text{meter})^2}{(\text{second})^2}$	$kg\ m^2\ s^{-2}$
Electric charge	Coulomb	C	(ampere)(second)	A s
Electric potential difference	Volt	V	$\frac{(\text{joule})}{(\text{ampere})(\text{second})}$	$J\ A^{-1}\ s^{-1} = kg\ m^2\ s^{-3}\ A^{-1}$

to see that the derived unit for area is length × length = meter × meter = square meter, m^2, or that the derived unit for volume is length × length × length = meter × meter × meter = cubic meter, m^3. More complex derived units are arrived at by a similar kind of combination of units. Units such as the joule, which measures energy, have been given simple names that represent the combination of fundamental units by which they are defined.

B.2 Conversion of Units for Physical Quantities

The result of a measurement is a physical quantity, which consists of a number and a unit. Algebraically, a physical quantity can be treated as if the number is multiplied by the unit. To convert a physical quantity from one unit of measure to another requires a conversion factor (proportionality factor) based on equivalences between units of measure such as those given in Table B.4. (See Appendix A.2 for more about physical quantities and proportionality factors.) Each equivalence provides two conversion factors that are the reciprocals of each other. For example, the equivalence between a quart and a liter, 1 quart = 0.9463 liter, gives

$$\frac{1 \text{ quart}}{0.9463 \text{ liter}} \quad \text{There is 1 quart per 0.9463 liter.}$$

$$\frac{0.9463 \text{ liter}}{1 \text{ quart}} \quad \text{There is 0.9463 liter per 1 quart.}$$

The method of canceling units described in Appendix A.2 provides the basis for choosing which conversion factor is needed: It is always the one that allows the unit being converted to be canceled and leaves the new unit uncanceled.

To convert 2 quarts to liters:

$$2 \text{ quarts} \times \frac{0.9463 \text{ liter}}{1 \text{ quart}} = 1.893 \text{ liters}$$

To convert 2 liters to quarts:

$$2 \text{ liters} \times \frac{1 \text{ quart}}{0.9463 \text{ liter}} = 2.113 \text{ quarts}$$

Because of the definitions of Celsius degrees and Fahrenheit degrees, conversions between these temperature scales are a bit more complicated. Both units are based on the properties of water. The Celsius unit is defined by assigning 0 °C as the freezing point of pure water and 100 °C as its boiling point, when the pressure is exactly 1 atm. The size of the Fahrenheit degree is equally arbitrary. Fahrenheit defined 0 °F as the freezing point of a solution in which he had dissolved the maximum quantity of ammonium chloride (because this was the lowest temperature he could reproduce reliably), and he intended 100 °F to be the normal human body temperature (but this value turned out to be 98.6 °F). Today, the reference points are set at exactly 32 °F and 212 °F (the freezing and boiling points of pure water, at 1 atm). The number of units between these two Fahrenheit temperatures is 180 °F. Thus, the Celsius degree is almost twice as large as the Fahrenheit degree; it takes only 5 Celsius degrees to cover the same temperature range as 9 Fahrenheit degrees.

To be entirely correct, we must specify that pure water boils at 100 °C and freezes at 0 °C only when the pressure of the surrounding atmosphere is 1 atm.

$$\frac{100\ °\text{C}}{180\ °\text{F}} = \frac{5\ °\text{C}}{9\ °\text{F}}$$

This relationship is the basis for converting a temperature on one scale to a temperature on the other. If t_C is the numerical value of the temperature in °C and t_F is the numerical value of the temperature in °F, then

$$t_C = \left(\tfrac{5}{9}\right)(t_F - 32)$$
$$t_F = \left(\tfrac{9}{5}\right)t_C + 32$$

For example, to show that your normal body temperature of 98.6 °F corresponds to 37.0 °C, use the first equation.

$$t_C = \left(\tfrac{5}{9}\right)(t_F - 32) = \left(\tfrac{5}{9}\right)(98.6 - 32) = \left(\tfrac{5}{9}\right)(66.6) = 37.0$$

Thus, body temperature in °C = 37.0 °C.

Table B.4 Common Units of Measure

Mass and Weight

1 pound = 453.59 grams = 0.45359 kilogram
1 kilogram = 1000 grams = 2.205 pounds
1 gram = 10 decigrams = 100 centigrams = 1000 milligrams
1 gram = 6.022×10^{23} atomic mass units
1 atomic mass unit = 1.6605×10^{-24} grams
1 short ton = 2000 pounds = 907.2 kilograms
1 long ton = 2240 pounds
1 metric tonne = 1000 kilograms = 2205 pounds

Length

1 inch = 2.54 centimeters (exactly)
1 mile = 5280 feet = 1.609 kilometers
1 yard = 36 inches = 0.9144 meter
1 meter = 100 centimeters = 39.37 inches = 3.281 feet = 1.094 yards
1 kilometer = 1000 meters = 1094 yards = 0.6215 miles
1 Angstrom = 1.0×10^{-8} centimeters = 0.10 nanometers = 100 picometers
= 1.0×10^{-10} meters = 3.937×10^{-9} inches

Volume

1 quart = 0.9463 liters
1 liter = 1.0567 quarts
1 liter = 1 cubic decimeter = 10^3 cubic centimeters = 10^{-3} cubic meters
1 milliliter = 1 cubic centimeters = 0.001 liters = 1.056×10^{-3} quarts
1 cubic foot = 28.316 liters = 29.924 quarts = 7.481 gallons

Force and Pressure

1 atmosphere = 760.0 millimeters of mercury = 1.01325×10^5 pascals
= 14.70 pounds per square inch
1 bar = 10^5 pascals = 0.98692 atmospheres
1 torr = 1 millimeter of mercury
1 pascal = 1 kg m^{-1} s^{-2} = 1 N/m^2

Energy

1 joule = 1×10^7 ergs
1 thermochemical calorie = 4.184 joules = 4.184×10^7 ergs
= 4.129×10^{-2} liter-atmospheres
= 2.612×10^{19} electron volts
1 erg = 1×10^{-7} joules = 2.3901×10^{-8} calories
1 electron volt = 1.6022×10^{-19} joules = 1.6022×10^{-12} ergs = 96.85 kJ/mol[1]
1 liter-atmosphere = 24.217 calories = 101.32 joules = 1.0132×10^9 ergs
1 British thermal unit = 1055.06 joules = 1.05506×10^{10} ergs = 252.2 calories

Temperature

0 K = −273.15 °C
If T_K is the numerical value of the temperature in kelvins, t_C is the numerical value of the temperature in °C, and t_F is the numerical value of the temperature in °F, then

$$T_K = t_C + 273.15$$
$$t_C = \left(\tfrac{5}{9}\right)(t_F - 32)$$
$$t_F = \left(\tfrac{9}{5}\right)t_C + 32$$

[1]The other units in this line are per particle and must be multiplied by 6.022×10^{23} to be strictly comparable.

Laboratory work is almost always done using Celsius units, and we rarely need to make conversions to and from Fahrenheit degrees. It is best to try to calibrate your senses to Celsius units; to help you do so, it is useful to know that water freezes at 0 °C, a comfortable room temperature is about 22 °C, your body temperature is 37 °C, and the hottest water you could leave your hand in for some time is about 60 °C.

QUESTIONS FOR APPENDIX B

Blue-numbered questions have short answers at the back of this book in Appendix M and fully worked solutions in the *Student Solutions Manual.*

Units of the International System

1. Which SI unit accompanied by which prefix would be most convenient for describing each quantity?
   (a) Mass of this book
   (b) Volume of a glass of water
   (c) Thickness of this page
2. Which SI unit accompanied by which prefix would be most convenient for describing each quantity?
   (a) Distance from New York to San Francisco
   (b) Mass of a glass of water
   (c) Area of this page
3. Explain the difference between an SI base (fundamental) unit and a derived SI unit.
4. What is the official SI definition of the mole? Describe in your own words each part of the definition and explain why each part of the definition is important.

Conversion of Units for Physical Quantities

5. Express each quantity in SI base (fundamental) units. Use exponential (scientific) notation whenever it is needed.
   (a) 475 pm (b) 56 Gg (c) 4.28 μA
6. Express each quantity in SI base (fundamental) units. Use exponential (scientific) notation whenever it is needed.
   (a) 32.5 ng (b) 56 Mm (c) 439 pm
7. Express each quantity in SI base (fundamental) units. Use exponential (scientific) notation whenever it is needed.
   (a) 8.7 nm^2 (b) 27.3 aJ (c) 27.3 μN
8. Express each quantity in SI base (fundamental) units. Use exponential (scientific) notation whenever it is needed.
   (a) 56.3 cm^3 (b) 5.62 MJ (c) 33.4 kV
9. Express each quantity in the units indicated. Use scientific notation.
   (a) 1.00 kg in pounds
   (b) 2.45 ton in kilograms
   (c) 1 L in cubic inches (in^3)
   (d) 1 atm in pascals and in bars
10. Express each quantity in the units indicated. Use scientific notation.
   (a) 24.3 amu in grams
   (b) 87.3 mL in cubic feet (ft^3)
   (c) 24.7 dg in ounces
   (d) 1.02 bar in millimeters of mercury (mm Hg) and in torrs
11. Express each temperature in Fahrenheit degrees.
   (a) 37 °C (b) −23.6 °C (c) −40.0 °C
12. Express each temperature in Celsius degrees.
   (a) 180. °F (b) −40.0 °F (c) 28.3 °F

APPENDIX C

Physical Constants[1] and Sources of Data

Quantity	Symbol	SI Units	Traditional Units
Acceleration of gravity	g_n	9.806 65 m/s^2	980.665 cm/s^2
Atomic mass unit ($\frac{1}{12}$ the mass of ^{12}C atom)	amu or u	1.660 538 782(83) $\times$ 10^{-27} kg	1.660 538 782(83) $\times$ 10^{-24} g
Avogadro constant	N_A, L	6.022 141 79(30) $\times$ 10^{23} particles/mol	6.022 141 79(30) $\times$ 10^{23} particles/mol
Bohr radius	a_0	5.291 772 0859(36) $\times$ 10^{-11} m	0.529 177 208 59(36) Å
Boltzmann constant	k	1.380 6504(24) $\times$ 10^{-23} J/K	1.380 6504(24) $\times$ 10^{-16} erg/K
Charge-to-mass ratio of electron	e/m_e	−1.758 820 150(44) $\times$ 10^{11} C/kg	−1.758 820 150(44) $\times$ 10^{8} C/g
Elementary charge (electron or proton charge)	e	1.602 176 487(40) $\times$ 10^{-19} C	1.602 176 487(40) $\times$ 10^{-19} C
Electron rest mass	m_e	9.109 382 15(45) $\times$ 10^{-31} kg	9.109 382 15(45) $\times$ 10^{-28} g
Faraday constant	F	96 485.3399(24) C/mol e$^-$	96 485.3399(24) C/mol e$^-$
		96 485.3399(24) J V^{-1} mol^{-1}	23.060 479 kcal V^{-1} mol^{-1}
Gas constant	R	8.314 472(15) dm^3 Pa mol^{-1} K^{-1}	0.082 057 L atm mol^{-1} K^{-1}
		8.314 472(15) J mol^{-1} K^{-1}	1.9872 cal mol^{-1} K^{-1}
Molar volume (1 atm, 273.15 K)	V_m	22.414 $\times$ 10^{-3} m^3/mol	22.414 L/mol
		22.414 dm^3/mol	
Neutron rest mass	m_n	1.674 927 211(84) $\times$ 10^{-27} kg	1.674 927 211(84) $\times$ 10^{-24} g
		1.008 664 916 amu	1.008 6644 916 amu
Planck's constant	h	6.626 068 96(33) $\times$ 10^{-34} J s	6.626 068 96(33) $\times$ 10^{-27} erg s
Proton rest mass	m_p	1.672 621 637(83) $\times$ 10^{-27} kg	1.672 621 637(83) $\times$ 10^{-24} g
		1.007 276 467 amu	1.007 276 467 amu
Rydberg constant	R_∞	1.097 373 156 8527(73) $\times$ 10^{7} m^{-1}	1.097 303 156 8527(73) $\times$ 10^{5} cm^{-1}
		2.179 871 97(11) $\times$ 10^{-18} J	2.179 871 97(11) $\times$ 10^{-11} erg
Speed of light (in a vacuum)	c, c_0	2.997 924 58 $\times$ 10^{8} m/s	2.997 924 58 $\times$ 10^{10} cm/s
			186 282 mile/s

[1]Data from the National Institute for Standards and Technology reference on constants, units, and uncertainty (in parentheses), **http://physics.nist.gov/cuu/Constants/Index.html.** The notation 1.234 56(12) means that the value of the quantity is 1.234 56 ± 0.000 12; that is, the value can range from 1.234 44 up to 1.234 68. Spaces are used between numbers to make reading easier.

Online Sources

- Finding Chemical Data in Web and Library Sources. University of Adelaide Library
  http://www.library.adelaide.edu.au/guide/sci/Chemistry/propindex.html
- SIRCh: Selected Internet Resources for Chemistry
  http://www.indiana.edu/~cheminfo/cis_ca.html

  SIRCh: Physical Property Information
  http://www.indiana.edu/~cheminfo/ca_ppi.html
- Thermodex. University of Texas at Austin.
  http://thermodex.lib.utexas.edu/
- How Many? A Dictionary of Units of Measurement.
  http://www.unc.edu/~rowlett/units/index.html
- NIST Chemistry Web Book
  http://webbook.nist.gov/chemistry

Print Sources

- Lide, D. R., Ed. *CRC Handbook of Chemistry and Physics,* 89th ed., Boca Raton, FL: CRC Press, 2008–2009.
- Budavari, S., O'Neil, M. J., Smith, A., and Heckelman, P. E., Eds. *The Merck Index:An Encyclopedia of Chemicals, Drugs, and Biologicals,* 14th ed., Rahway, NJ: Merck & Co., 2006.
- Speight, J., Ed. *Lange's Handbook of Chemistry,* 16th ed., New York: McGraw-Hill, 2004.
- Perry, R. H., and Green, D. W., Eds. *Perry's Chemical Engineers' Handbook,* 8th ed., New York: McGraw-Hill, 2007.
- Zwillinger, D., Ed. *CRC Standard Mathematical Tables and Formulae,* 31st ed., Boca Raton, FL: CRC Press, 2003.
- Lewis, R. J., Sr. *Hawley's Condensed Chemical Dictionary* (with CD-ROM), 15th ed., New York: Wiley, 2007.

APPENDIX

D Ground-State Electron Configurations of Atoms

Z	Element	Configuration	Z	Element	Configuration	Z	Element	Configuration
1	H	$1s^1$	40	Zr	[Kr] $4d^2 5s^2$	79	Au	[Xe] $4f^{14} 5d^{10} 6s^1$
2	He	$1s^2$	41	Nb	[Kr] $4d^4 5s^1$	80	Hg	[Xe] $4f^{14} 5d^{10} 6s^2$
3	Li	[He] $2s^1$	42	Mo	[Kr] $4d^5 5s^1$	81	Tl	[Xe] $4f^{14} 5d^{10} 6s^2 6p^1$
4	Be	[He] $2s^2$	43	Tc	[Kr] $4d^5 5s^2$	82	Pb	[Xe] $4f^{14} 5d^{10} 6s^2 6p^2$
5	B	[He] $2s^2 2p^1$	44	Ru	[Kr] $4d^7 5s^1$	83	Bi	[Xe] $4f^{14} 5d^{10} 6s^2 6p^3$
6	C	[He] $2s^2 2p^2$	45	Rh	[Kr] $4d^8 5s^1$	84	Po	[Xe] $4f^{14} 5d^{10} 6s^2 6p^4$
7	N	[He] $2s^2 2p^3$	46	Pd	[Kr] $4d^{10}$	85	At	[Xe] $4f^{14} 5d^{10} 6s^2 6p^5$
8	O	[He] $2s^2 2p^4$	47	Ag	[Kr] $4d^{10} 5s^1$	86	Rn	[Xe] $4f^{14} 5d^{10} 6s^2 6p^6$
9	F	[He] $2s^2 2p^5$	48	Cd	[Kr] $4d^{10} 5s^2$	87	Fr	[Rn] $7s^1$
10	Ne	[He] $2s^2 2p^6$	49	In	[Kr] $4d^{10} 5s^2 5p^1$	88	Ra	[Rn] $7s^2$
11	Na	[Ne] $3s^1$	50	Sn	[Kr] $4d^{10} 5s^2 5p^2$	89	Ac	[Rn] $6d^1 7s^2$
12	Mg	[Ne] $3s^2$	51	Sb	[Kr] $4d^{10} 5s^2 5p^3$	90	Th	[Rn] $6d^2 7s^2$
13	Al	[Ne] $3s^2 3p^1$	52	Te	[Kr] $4d^{10} 5s^2 5p^4$	91	Pa	[Rn] $5f^2 6d^1 7s^2$
14	Si	[Ne]$3s^2 3p^2$	53	I	[Kr] $4d^{10} 5s^2 5p^5$	92	U	[Rn] $5f^3 6d^1 7s^2$
15	P	[Ne] $3s^2 3p^3$	54	Xe	[Kr] $4d^{10} 5s^2 5p^6$	93	Np	[Rn] $5f^4 6d^1 7s^2$
16	S	[Ne] $3s^2 3p^4$	55	Cs	[Xe] $6s^1$	94	Pu	[Rn] $5f^6 7s^2$
17	Cl	[Ne] $3s^2 3p^5$	56	Ba	[Xe] $6s^2$	95	Am	[Rn] $5f^7 7s^2$
18	Ar	[Ne] $3s^2 3p^6$	57	La	[Xe] $5d^1 6s^2$	96	Cm	[Rn] $5f^7 6d^1 7s^2$
19	K	[Ar] $4s^1$	58	Ce	[Xe] $4f^1 5d^1 6s^2$	97	Bk	[Rn] $5f^9 7s^2$
20	Ca	[Ar] $4s^2$	59	Pr	[Xe] $4f^3 6s^2$	98	Cf	[Rn] $5f^{10} 7s^2$
21	Sc	[Ar] $3d^1 4s^2$	60	Nd	[Xe] $4f^4 6s^2$	99	Es	[Rn] $5f^{11} 7s^2$
22	Ti	[Ar] $3d^2 4s^2$	61	Pm	[Xe] $4f^5 6s^2$	100	Fm	[Rn] $5f^{12} 7s^2$
23	V	[Ar] $3d^3 4s^2$	62	Sm	[Xe] $4f^6 6s^2$	101	Md	[Rn] $5f^{13} 7s^2$
24	Cr	[Ar] $3d^5 4s^1$	63	Eu	[Xe] $4f^7 6s^2$	102	No	[Rn] $5f^{14} 7s^2$
25	Mn	[Ar] $3d^5 4s^2$	64	Gd	[Xe] $4f^7 5d^1 6s^2$	103	Lr	[Rn] $5f^{14} 7s^2 7p^1$
26	Fe	[Ar] $3d^6 4s^2$	65	Tb	[Xe] $4f^9 6s^2$	104	Rf	[Rn] $5f^{14} 6d^2 7s^2$
27	Co	[Ar] $3d^7 4s^2$	66	Dy	[Xe] $4f^{10} 6s^2$	105	Db	—
28	Ni	[Ar] $3d^8 4s^2$	67	Ho	[Xe] $4f^{11} 6s^2$	106	Sg	—
29	Cu	[Ar] $3d^{10} 4s^1$	68	Er	[Xe] $4f^{12} 6s^2$	107	Bh	—
30	Zn	[Ar] $3d^{10} 4s^2$	69	Tm	[Xe] $4f^{13} 6s^2$	108	Hs	—
31	Ga	[Ar] $3d^{10} 4s^2 4p^1$	70	Yb	[Xe] $4f^{14} 6s^2$	109	Mt	—
32	Ge	[Ar] $3d^{10} 4s^2 4p^2$	71	Lu	[Xe] $4f^{14} 5d^1 6s^2$	110	Ds	—
33	As	[Ar] $3d^{10} 4s^2 4p^3$	72	Hf	[Xe] $4f^{14} 5d^2 6s^2$	111	Rg	—
34	Se	[Ar] $3d^{10} 4s^2 4p^4$	73	Ta	[Xe] $4f^{14} 5d^3 6s^2$	112	—	—
35	Br	[Ar] $3d^{10} 4s^2 4p^5$	74	W	[Xe] $4f^{14} 5d^4 6s^2$	113	—	—
36	Kr	[Ar] $3d^{10} 4s^2 4p^6$	75	Re	[Xe] $4f^{14} 5d^5 6s^2$	114	—	—
37	Rb	[Kr] $5s^1$	76	Os	[Xe] $4f^{14} 5d^6 6s^2$	115	—	—
38	Sr	[Kr] $5s^2$	77	Ir	[Xe] $4f^{14} 5d^7 6s^2$	116	—	—
39	Y	[Kr] $4d^1 5s^2$	78	Pt	[Xe] $4f^{14} 5d^9 6s^1$	118	—	—

All electron configurations in this table are from experimental determinations except for Lr and Rf, which are predicted theoretically. For elements with atomic number greater than 104 only a few atoms can be produced at a time and it is very difficult to determine electron configurations.

APPENDIX E

Naming Simple Organic Compounds

The systematic nomenclature for organic compounds was proposed by the International Union of Pure and Applied Chemistry (IUPAC). The IUPAC set of rules provides different names for the more than 10 million known organic compounds and allows names to be assigned to new compounds as they are synthesized. Many organic compounds also have *common* names. Usually the common name came first and is widely known. Many consumer products are labeled with the common name, and when only a few isomers are possible, the common name adequately identifies the product for the consumer. However, as illustrated in Section 3.4 ***(← p. 83)***, a system of common names quickly fails when several structural isomers are possible.

E.1 Hydrocarbons

The name of each member of the hydrocarbon classes has two parts. The first part, called the prefix (*meth-*, *eth-*, *prop-*, *but-*, and so on), reflects the number of carbon atoms. When more than four carbons are present, the Greek or Latin number prefixes are used: *pent-*, *hex-*, *hept-*, *oct-*, *non-*, and *dec-*. The second part of the name, called the suffix, tells the class of hydrocarbon. Alkanes have carbon-carbon single bonds, alkenes have carbon-carbon double bonds, and alkynes have carbon-carbon triple bonds.

Unbranched Alkanes and Alkyl Groups

The names of the first 20 unbranched (straight-chain) alkanes are given in Table E.1.

Alkyl groups are named by dropping *-ane* from the parent alkane and adding *-yl* (see Table 3.5 for examples).

Branched-Chain Alkanes

The rules for naming branched-chain alkanes are as follows:

1. *Find the longest continuous chain of carbon atoms; it determines the parent name for the compound.* For example, the following compound has two methyl groups attached to a *heptane* parent; the longest continuous chain contains seven carbon atoms.

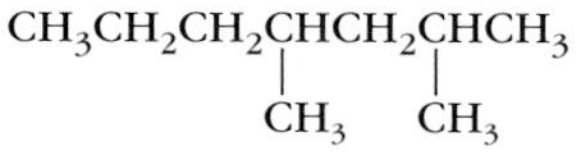
CH3CH2CH2CHCH2CHCH3
CH3
CH3

Table E.1 Names of Unbranched Alkanes

CH_4	Methane	$C_{11}H_{24}$	Undecane
C_2H_6	Ethane	$C_{12}H_{26}$	Dodecane
C_3H_8	Propane	$C_{13}H_{28}$	Tridecane
C_4H_{10}	Butane	$C_{14}H_{30}$	Tetradecane
C_5H_{12}	Pentane	$C_{15}H_{32}$	Pentadecane
C_6H_{14}	Hexane	$C_{16}H_{34}$	Hexadecane
C_7H_{16}	Heptane	$C_{17}H_{36}$	Heptadecane
C_8H_{18}	Octane	$C_{18}H_{38}$	Octadecane
C_9H_{20}	Nonane	$C_{19}H_{40}$	Nonadecane
$C_{10}H_{22}$	Decane	$C_{20}H_{42}$	Eicosane

The longest continuous chain may not be obvious from the way the formula is written, especially for the straight-line format that is commonly used. For example, the longest continuous chain of carbon atoms in the following chain is *eight,* not *four* or *six.*

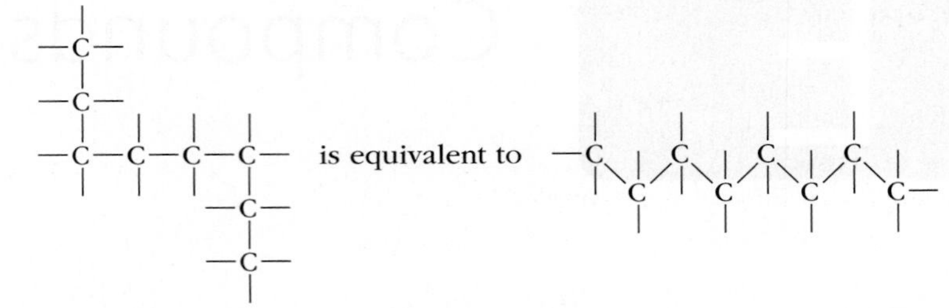
is equivalent to

2. *Number the longest chain beginning with the end of the chain nearest the branching. Use these numbers to designate the location of the attached group. When two or more groups are attached to the parent, give each group a number corresponding to its location on the parent chain.* For example, the name of

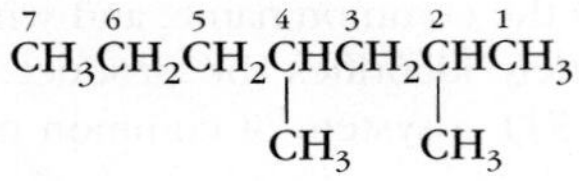

is 2,4-dimethylheptane. The name of the compound below is 3-methylheptane, not 5-methylheptane or 2-ethylhexane.

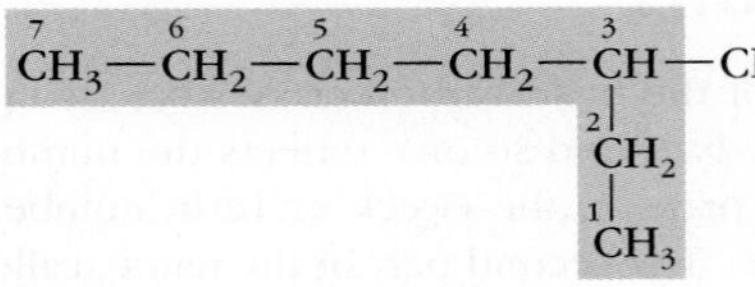

3. *When two or more substituents are identical, indicate this by the use of the prefixes di-, tri-, tetra-, and so on. Positional numbers of the substituents should have the smallest possible sum.*

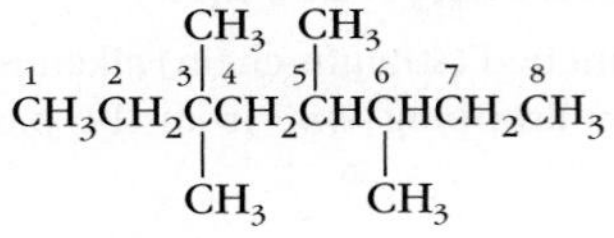

The correct name of this compound is 3,3,5,6-tetramethyloctane.

4. *If there are two or more different groups, the groups are listed alphabetically.*

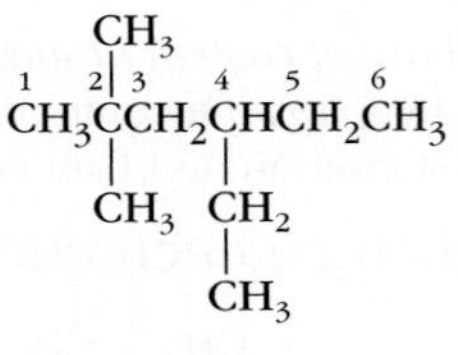

The correct name of this compound is 4-ethyl-2,2-dimethylhexane. Note that the prefix *di-* is ignored in determining alphabetical order.

Alkenes

Alkenes are named by using the prefix to indicate the number of carbon atoms and the suffix *-ene* to indicate one or more double bonds. The systematic names for the first two members of the alkene series are *ethene* and *propene.*

$$CH_2{=}CH_2 \qquad CH_3CH{=}CH_2$$

When groups, such as methyl or ethyl, are attached to carbon atoms in an alkene, the longest hydrocarbon chain is numbered from the end that will give the double bond the lowest number, and then numbers are assigned to the attached groups. For example, the name of

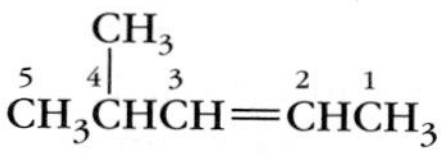

is 4-methyl-2-pentene. See Section 8.5 for a discussion of *cis-trans* isomers of alkenes.

Alkynes

The naming of alkynes is similar to that of alkenes, with the lowest number possible being used to locate the triple bond. For example, the name of

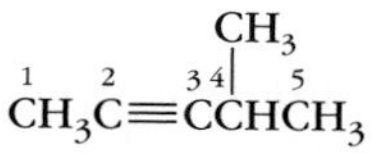

is 4-methyl-2-pentyne.

Benzene Derivatives

Monosubstituted benzene derivatives are named by using a prefix for the substituent. Some examples are

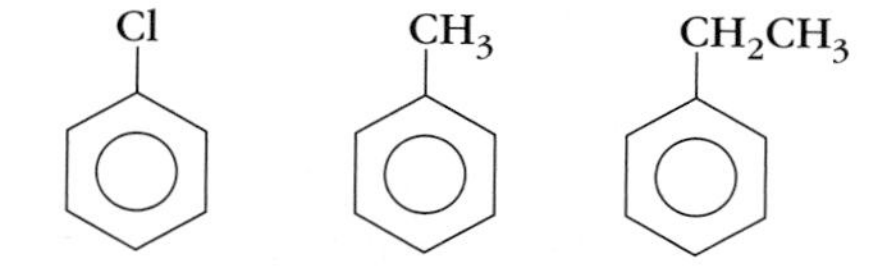

chlorobenzene   methylbenzene (toluene)   ethylbenzene

Three isomers are possible when two groups are substituted for hydrogen atoms on the benzene ring. The relative positions of the substituents are indicated either by the prefixes *ortho-*, *meta-*, and *para-* (abbreviated *o-*, *m-*, and *p-*, respectively) or by numbers. For example,

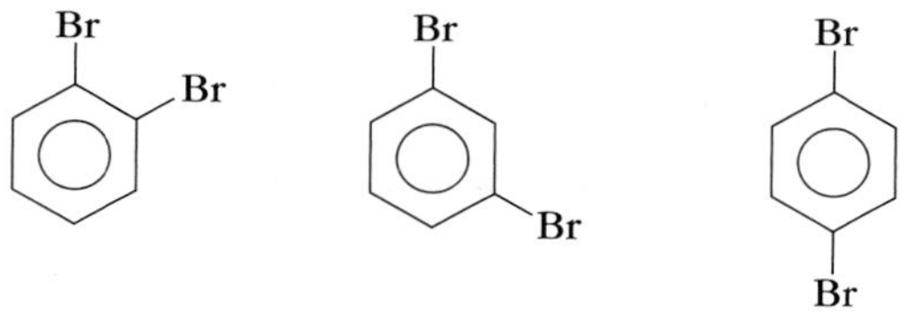

1,2-dibromobenzene (*o*-dibromobenzene)   1,3-dibromobenzene (*m*-dibromobenzene)   1,4-dibromobenzene (*p*-dibromobenzene)

The dimethylbenzenes are called *xylenes.*

If more than two groups are attached to the benzene ring, numbers must be used to identify the positions. The benzene ring is numbered to give the lowest possible numbers to the substituents.

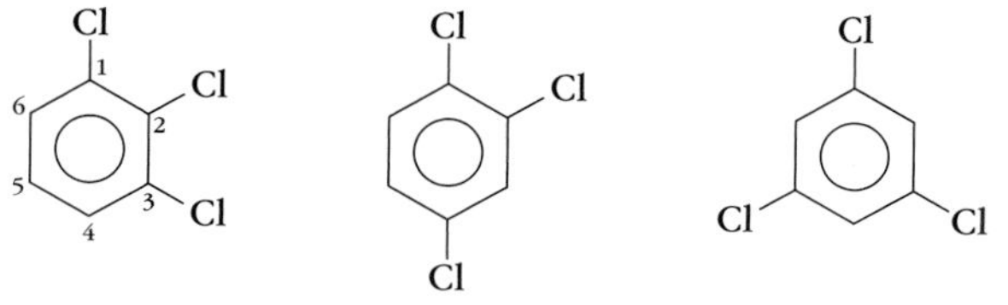

1,2,3-trichlorobenzene   1,2,4-trichlorobenzene   1,3,5-trichlorobenzene

E.2 Functional Groups

An atom or group of atoms that defines the structure of a specific class of organic compounds and determines their properties is called a *functional group* (**← p. 77**). The millions of organic compounds include classes of compounds that are obtained by replacing hydrogen atoms of hydrocarbons with functional groups (Sections 3.1, 12.5, and 12.6). The important functional groups are shown in Table E.2.

The "R" attached to the functional group represents the hydrocarbon framework with one hydrogen removed for each functional group added. The IUPAC system provides a systematic method for naming all members of a given class. For example, alcohols end in *-ol* (methan*ol*); aldehydes end in *-al* (methan*al*); carboxylic acids end in *-oic* (ethan*oic* acid); and ketones end in *-one* (propan*one*).

Alcohols

Isomers are also possible for molecules containing functional groups. For example, three different alcohols are obtained when a hydrogen atom in pentane is replaced by —OH, depending on which hydrogen atom is replaced. The rules for naming the "R" or hydrocarbon framework are the same as those for hydrocarbon compounds.

$CH_3CH_2CH_2CH_2CH_2OH$ 1-pentanol

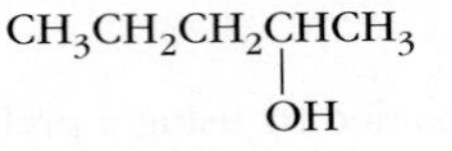

2-pentanol

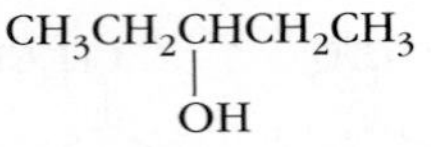

3-pentanol

Compounds with one or more functional groups (Table E.2) and alkyl substituents are named so as to give the functional groups the lowest numbers. For example, the correct name of

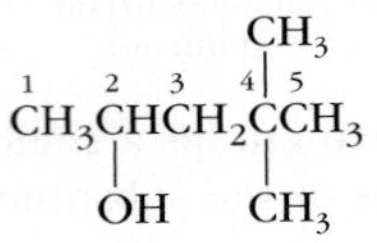

is 4,4-dimethyl-2-pentanol.

Aldehydes and Ketones

The systematic names of the first three aldehydes are methanal, ethanal, and propanal.

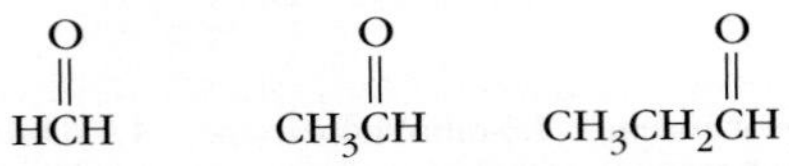

methanal (formaldehyde) ethanal (acetaldehyde) propanal (propionaldehyde)

For ketones, a number is used to designate the position of the carbonyl group, and the chain is numbered in a way that gives the carbonyl carbon the smallest number.

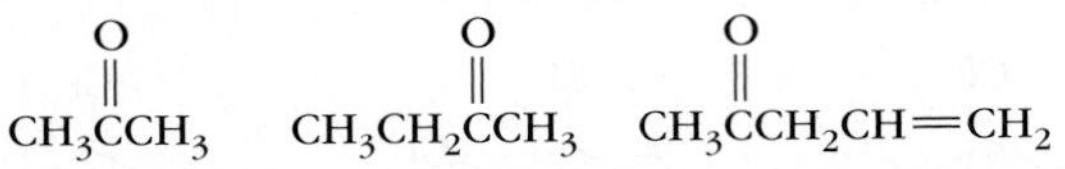

2-propanone (acetone) 2-butanone (methyl ethyl ketone) 4-penten-2-one

Carboxylic Acids

The systematic names of carboxylic acids are obtained by dropping the final *e* of the name of the corresponding alkane and adding *-oic acid.* For example, the name of

$$CH_3CH_2CH_2CH_2CH_2COOH$$

is hexanoic acid. Other examples are

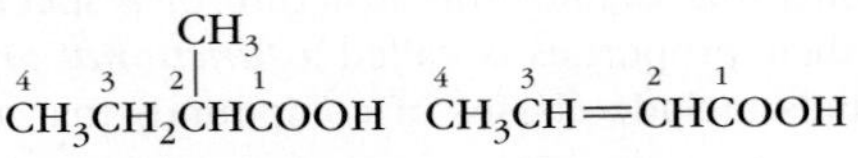

2-methylbutanoic acid 2-butenoic acid

Table E.2 Classes of Organic Compounds Based on Functional Groups[1]

General Formulas of Class Members	Class Name	Typical Compound	Compound Name	Common Use of Sample Compound
$R{-}X$	Halide	$H{-}CH(Cl){-}Cl$	Dichloromethane (methylene chloride)	Solvent
$R{-}OH$	Alcohol	$H{-}CH_2{-}OH$	Methanol (wood alcohol)	Solvent
$R{-}C({=}O){-}H$	Aldehyde	$H{-}C({=}O){-}H$	Methanal (formaldehyde)	Preservative
$R{-}C({=}O){-}OH$	Carboxylic acid	$H{-}CH_2{-}C({=}O){-}OH$	Ethanoic acid (acetic acid)	Vinegar
$R{-}C({=}O){-}R'$	Ketone	$H{-}CH_2{-}C({=}O){-}CH_2{-}H$	Propanone (acetone)	Solvent
$R{-}O{-}R'$	Ether	$C_2H_5{-}O{-}C_2H_5$	Diethyl ether (ethyl ether)	Anesthetic
$R{-}C({=}O){-}O{-}R'$	Ester	$CH_3{-}C({=}O){-}O{-}C_2H_5$	Ethyl ethanoate (ethyl acetate)	Solvent in fingernail polish
$R{-}N(H){-}H$	Amine	$H{-}CH_2{-}N(H){-}H$	Methylamine	Tanning hides (foul odor)
$R{-}C({=}O){-}N(H){-}R'$	Amide	$CH_3{-}C({=}O){-}N(H){-}H$	Acetamide	Plasticizer

[1]R stands for an H or a hydrocarbon group such as $—CH_3$ or $—C_2H_5$. R′ could be a different group from R.

Esters

The systematic names of esters are derived from the names of the alcohol and the acid used to prepare the ester. The general formula for esters is

$$R{-}C({=}O){-}OR'$$

As shown in Section 12.4, the $R{-}C({=}O)$ comes from the acid and the R′O comes from the alcohol. The alcohol part is named first, followed by the name of the acid changed to end in *-ate*. For example,

$$CH_3CH_2C({=}O){-}OCH_3$$

is named methyl propanoate and

$$CH_3C({=}O){-}OCH{=}CH_2$$

is named ethenyl ethanoate.

APPENDIX

F

Ionization Constants for Weak Acids at 25 °C[1]

Acid	Formula and Ionization Equation	K_a
Acetic	$CH_3COOH + H_2O \rightleftharpoons H_3O^+ + CH_3COO^-$	1.8×10^{-5}
Arsenic	$H_3AsO_4 + H_2O \rightleftharpoons H_3O^+ + H_2AsO_4^-$	$K_1 = 6.17 \times 10^{-3}$
	$H_2AsO_4^- + H_2O \rightleftharpoons H_3O^+ + HAsO_4^{2-}$	$K_2 = 1.17 \times 10^{-7}$
	$HAsO_4^{2-} + H_2O \rightleftharpoons H_3O^+ + AsO_4^{3-}$	$K_3 = 3.09 \times 10^{-12}$
Benzoic	$C_6H_5COOH + H_2O \rightleftharpoons H_3O^+ + C_6H_5COO^-$	1.2×10^{-4}
Boric	$B(OH)_3(H_2O) + H_2O \rightleftharpoons H_3O^+ + B(OH)_4^-$	5.8×10^{-10}
Carbonic	$H_2CO_3 + H_2O \rightleftharpoons H_3O^+ + HCO_3^-$	$K_1 = 4.3 \times 10^{-7}$
	$HCO_3^- + H_2O \rightleftharpoons H_3O^+ + CO_3^{2-}$	$K_2 = 4.7 \times 10^{-11}$
Citric	$H_3C_6H_5O_7 + H_2O \rightleftharpoons H_3O^+ + H_2C_6H_5O_7^-$	$K_1 = 1.4 \times 10^{-3}$
	$H_2C_6H_5O_7^- + H_2O \rightleftharpoons H_3O^+ + HC_6H_5O_7^{2-}$	$K_2 = 4.5 \times 10^{-5}$
	$HC_6H_5O_7^{2-} + H_2O \rightleftharpoons H_3O^+ + C_6H_5O_7^{3-}$	$K_3 = 1.5 \times 10^{-6}$
Formic	$HCOOH + H_2O \rightleftharpoons H_3O^+ + HCOO^-$	3.0×10^{-4}
Hydrazoic	$HN_3 + H_2O \rightleftharpoons H_3O^+ + N_3^-$	1.0×10^{-5}
Hydrocyanic	$HCN + H_2O \rightleftharpoons H_3O^+ + CN^-$	3.3×10^{-10}
Hydrofluoric	$HF + H_2O \rightleftharpoons H_3O^+ + F^-$	6.8×10^{-4}
Hydrogen peroxide	$H_2O_2 + H_2O \rightleftharpoons H_3O^+ + HO_2^-$	2.1×10^{-12}
Hydrosulfuric[2]	$H_2S + H_2O \rightleftharpoons H_3O^+ + HS^-$	$K_1 = 1 \times 10^{-7}$
	$HS^- + H_2O \rightleftharpoons H_3O^+ + S^{2-}$	$K_2 = 1 \times 10^{-19}$
Hypochlorous	$HOCl + H_2O \rightleftharpoons H_3O^+ + OCl^-$	6.8×10^{-8}
Oxalic	$H_2C_2O_4 + H_2O \rightleftharpoons H_3O^+ + HC_2O_4^-$	$K_1 = 5.5 \times 10^{-2}$
	$HC_2O_4^- + H_2O \rightleftharpoons H_3O^+ + C_2O_4^{2-}$	$K_2 = 1.4 \times 10^{-4}$
Phenol	$HC_6H_5O + H_2O \rightleftharpoons H_3O^+ + C_6H_5O^-$	1.7×10^{-10}
Phosphoric	$H_3PO_4 + H_2O \rightleftharpoons H_3O^+ + H_2PO_4^-$	$K_1 = 7.2 \times 10^{-3}$
	$H_2PO_4^- + H_2O \rightleftharpoons H_3O^+ + HPO_4^{2-}$	$K_2 = 6.3 \times 10^{-8}$
	$HPO_4^{2-} + H_2O \rightleftharpoons H_3O^+ + PO_4^{3-}$	$K_3 = 4.6 \times 10^{-13}$
Phosphorous	$H_3PO_3 + H_2O \rightleftharpoons H_3O^+ + H_2PO_3^-$	$K_1 = 2.4 \times 10^{-2}$
	$H_2PO_3^- + H_2O \rightleftharpoons H_3O^+ + HPO_3^{2-}$	$K_2 = 2.9 \times 10^{-7}$
Selenic	$H_2SeO_4 + H_2O \rightleftharpoons H_3O^+ + HSeO_4^-$	K_1 = very large
	$HSeO_4^- + H_2O \rightleftharpoons H_3O^+ + SeO_4^{2-}$	$K_2 = 2.2 \times 10^{-2}$
Selenous	$H_2SeO_3 + H_2O \rightleftharpoons H_3O^+ + HSeO_3^-$	$K_1 = 2.5 \times 10^{-3}$
	$HSeO_3^- + H_2O \rightleftharpoons H_3O^+ + SeO_3^{2-}$	$K_2 = 1.6 \times 10^{-9}$
Sulfuric	$H_2SO_4 + H_2O \rightleftharpoons H_3O^+ + HSO_4^-$	K_1 = very large
	$HSO_4^- + H_2O \rightleftharpoons H_3O^+ + SO_4^{2-}$	$K_2 = 1.1 \times 10^{-2}$
Sulfurous	$H_2SO_3 + H_2O \rightleftharpoons H_3O^+ + HSO_3^-$	$K_1 = 1.7 \times 10^{-2}$
	$HSO_3^- + H_2O \rightleftharpoons H_3O^+ + SO_3^{2-}$	$K_2 = 6.3 \times 10^{-8}$
Tellurous	$H_2TeO_3 + H_2O \rightleftharpoons H_3O^+ + HTeO_3^-$	$K_1 = 1.9 \times 10^{-3}$
	$HTeO_3^- + H_2O \rightleftharpoons H_3O^+ + TeO_3^{2-}$	$K_2 = 4.6 \times 10^{-10}$

[1]Taken from Högfeldt, E., and Perrin, D. D. *Stability Constants of Metal-Ion Complexes*, 1st ed. Oxford; New York: Pergamon, 1979–1982. International Union of Pure and Applied Chemistry Commission on Equilibrium Data.

[2]See Myers, R. "The New Low Value for the Second Dissociation Constant for H_2S." *Journal of Chemical Education*, Vol. 63, 1986; pp. 687–690.

APPENDIX G

Ionization Constants for Weak Bases at 25 °C[1]

Base	Formula and Ionization Equation	K_b
Ammonia[3]	$NH_3 + H_2O \rightleftharpoons NH_4^+ + OH^-$	1.77×10^{-5}
Aniline[2]	$C_6H_5NH_2 + H_2O \rightleftharpoons C_6H_5NH_3^+ + OH^-$	3.9×10^{-10}
Dimethylamine[2]	$(CH_3)_2NH + H_2O \rightleftharpoons (CH_3)_2NH_2^+ + OH^-$	5.8×10^{-4}
Ethylenediamine	$(CH_2)_2(NH_2)_2 + H_2O \rightleftharpoons (CH_2)_2(NH_2)_2H^+ + OH^-$	$K_1 = 7.8 \times 10^{-5}$
	$(CH_2)_2(NH_2)_2H^+ + H_2O \rightleftharpoons (CH_2)_2(NH_2)_2H_2^{2+} + OH^-$	$K_2 = 2.1 \times 10^{-8}$
Hydrazine	$N_2H_4 + H_2O \rightleftharpoons N_2H_5^+ + OH^-$	$K_1 = 1.2 \times 10^{-6}$
	$N_2H_5^+ + H_2O \rightleftharpoons N_2H_6^{2+} + OH^-$	$K_2 = 1.3 \times 10^{-15}$
Hydroxylamine	$NH_2OH + H_2O \rightleftharpoons NH_3OH^+ + OH^-$	9.3×10^{-9}
Methylamine	$CH_3NH_2 + H_2O \rightleftharpoons CH_3NH_3^+ + OH^-$	5.0×10^{-4}
Pyridine	$C_5H_5N + H_2O \rightleftharpoons C_5H_5NH^+ + OH^-$	1.6×10^{-9}
Trimethylamine[2]	$(CH_3)_3N + H_2O \rightleftharpoons (CH_3)_3NH^+ + OH^-$	6.2×10^{-5}

[1]Taken from Högfeldt, E., and Perrin, D. D. *Stability Constants of Metal-Ion Complexes,* 1st ed. Oxford; New York: Pergamon, 1979–1982. International Union of Pure and Applied Chemistry Commission on Equilibrium Data.

[2]These values are from Meites, L., Ed. *Handbook of Analytical Chemistry,* 1st ed. New York: McGraw-Hill, 1963.

[3]This value from Read, A. J. *Journal of Solution Chemistry,* Vol. 11, No. 9, 1982, pp. 649–664.

APPENDIX H

Solubility Product Constants for Some Inorganic Compounds at 25 °C[1]

Substance	K_{sp}
Aluminum Compounds	
$AlAsO_4$	1.6×10^{-16}
$Al(OH)_3$ amorphous	1.3×10^{-33}
$AlPO_4$	6.3×10^{-19}
Barium Compounds	
$Ba_3(AsO_4)_2$	8.0×10^{-15}
$BaCO_3$	5.1×10^{-9}
BaC_2O_4	1.6×10^{-7}
$BaCrO_4$	1.2×10^{-10}
BaF_2	1.0×10^{-6}
$Ba(OH)_2$	5×10^{-3}
$Ba_3(PO_4)_2$	3.4×10^{-23}
$BaSeO_4$	3.5×10^{-8}
$BaSO_4$	1.1×10^{-10}
$BaSO_3$	8×10^{-7}
BaS_2O_3	1.6×10^{-5}
Bismuth Compounds	
$BiAsO_4$	4.4×10^{-10}
BiOCl[2]	7.0×10^{-9}
$BiO(OH)$	4×10^{-10}
$Bi(OH)_3$	4×10^{-31}
BiI_3	8.1×10^{-19}
$BiPO_4$	1.3×10^{-23}
Cadmium Compounds	
$Cd_3(AsO_4)_2$	2.2×10^{-33}
$CdCO_3$	5.2×10^{-12}
$Cd(CN)_2$	1.0×10^{-8}
$Cd_2[Fe(CN)_6]$	3.2×10^{-17}
$Cd(OH)_2$ fresh	2.5×10^{-14}
Calcium Compounds	
$Ca_3(AsO_4)_2$	6.8×10^{-19}
$CaCO_3$	2.8×10^{-9}
$CaCrO_4$	7.1×10^{-4}
$CaC_2O_4 \cdot H_2O$[3]	4×10^{-9}
CaF_2	5.3×10^{-9}
$Ca(OH)_2$	5.5×10^{-6}
$CaHPO_4$	1×10^{-7}
$Ca_3(PO_4)_2$	2.0×10^{-29}
$CaSeO_4$	8.1×10^{-4}
$CaSO_4$	9.1×10^{-6}
$CaSO_3$	6.8×10^{-8}

Substance	K_{sp}
Chromium Compounds	
$CrAsO_4$	7.7×10^{-21}
$Cr(OH)_2$	2×10^{-16}
$Cr(OH)_3$	6.3×10^{-31}
$CrPO_4 \cdot 4\,H_2O$ green	2.4×10^{-23}
$CrPO_4 \cdot 4\,H_2O$ violet	1.0×10^{-17}
Cobalt Compounds	
$Co_3(AsO_4)_2$	7.6×10^{-29}
$CoCO_3$	1.4×10^{-13}
$Co(OH)_2$ fresh	1.6×10^{-15}
$Co(OH)_3$	1.6×10^{-44}
$CoHPO_4$	2×10^{-7}
$Co_3(PO_4)_2$	2×10^{-35}
Copper Compounds	
$CuBr$	5.3×10^{-9}
$CuCl$	1.2×10^{-6}
$CuCN$	3.2×10^{-20}
CuI	1.1×10^{-12}
$CuOH$	1×10^{-14}
$CuSCN$	4.8×10^{-15}
$Cu_3(AsO_4)_2$	7.6×10^{-36}
$CuCO_3$	1.4×10^{-10}
$Cu_2[Fe(CN)_6]$	1.3×10^{-16}
$Cu(OH)_2$	2.2×10^{-20}
$Cu_3(PO_4)_2$	1.3×10^{-37}
Gold Compounds	
$AuCl$	2.0×10^{-13}
AuI	1.6×10^{-23}
$AuCl_3$	3.2×10^{-25}
$Au(OH)_3$	5.5×10^{-46}
AuI_3	1×10^{-46}
Iron Compounds	
$FeCO_3$	3.2×10^{-11}
$Fe(OH)_2$	8.0×10^{-16}
$FeC_2O_4 \cdot 2H_2O$[3]	3.2×10^{-7}
$FeAsO_4$	5.7×10^{-21}
$Fe_4[Fe(CN)_6]_3$	3.3×10^{-41}
$Fe(OH)_3$	4×10^{-38}
$FePO_4$	1.3×10^{-22}

Substance	K_{sp}
Lead Compounds	
$Pb_3(AsO_4)_2$	4.0×10^{-36}
$PbBr_2$	4.0×10^{-5}
$PbCO_3$	7.4×10^{-14}
$PbCl_2$	1.6×10^{-5}
$PbCrO_4$	2.8×10^{-13}
PbF_2	2.7×10^{-8}
$Pb(OH)_2$	1.2×10^{-15}
PbI_2	7.1×10^{-9}
PbC_2O_4	4.8×10^{-10}
$PbHPO_4$	1.3×10^{-10}
$Pb_3(PO_4)_2$	8.0×10^{-43}
$PbSeO_4$	1.4×10^{-7}
$PbSO_4$	1.6×10^{-8}
$Pb(SCN)_2$	2.0×10^{-5}
Magnesium Compounds	
$Mg_3(AsO_4)_2$	2.1×10^{-20}
$MgCO_3$	3.5×10^{-8}
$MgCO_3 \cdot 3\ H_2O$[3]	2.1×10^{-5}
$MgC_2O_4 \cdot 2\ H_2O$[3]	1×10^{-8}
MgF_2	6.5×10^{-9}
$Mg(OH)_2$	1.8×10^{-11}
$Mg_3(PO_4)_2$	10^{-23} to 10^{-27}
$MgSeO_3$	1.3×10^{-5}
$MgSO_3$	3.2×10^{-3}
$MgNH_4PO_4$	2.5×10^{-13}
Manganese Compounds	
$Mn_3(AsO_4)_2$	1.9×10^{-29}
$MnCO_3$	1.8×10^{-11}
$Mn_2[Fe(CN)_6]$	8.0×10^{-13}
$Mn(OH)_2$	1.9×10^{-13}
$MnC_2O_4 \cdot 2H_2O$[3]	1.1×10^{-15}
Mercury Compounds	
Hg_2Br_2	5.6×10^{-23}
Hg_2CO_3	8.9×10^{-17}
$Hg_2(CN)_2$	5×10^{-40}
Hg_2Cl_2	1.3×10^{-18}
Hg_2CrO_4	2.0×10^{-9}
$Hg_2(OH)_2$	2.0×10^{-24}
Hg_2I_2	4.5×10^{-29}
Hg_2SO_4	7.4×10^{-7}
Hg_2SO_3	1.0×10^{-27}
$Hg(OH)_2$	3.0×10^{-26}

Substance	K_{sp}
Nickel Compounds	
$Ni_3(AsO_4)_2$	3.1×10^{-26}
$NiCO_3$	6.6×10^{-9}
$2\ Ni(CN)_2 \rightarrow Ni^{2+} + Ni(CN)_4^{2-}$	1.7×10^{-9}
$Ni_2[Fe(CN)_6]$	1.3×10^{-15}
$Ni(OH)_2$ fresh	2.0×10^{-15}
NiC_2O_4	4×10^{-10}
$Ni_3(PO_4)_2$	5×10^{-31}
Silver Compounds	
Ag_3AsO_4	1.0×10^{-22}
$AgBr$	5.0×10^{-13}
Ag_2CO_3	8.1×10^{-12}
$AgCl$	1.8×10^{-10}
Ag_2CrO_4	1.1×10^{-12}
$AgCN$	1.2×10^{-16}
$Ag_2Cr_2O_7$	2.0×10^{-7}
$Ag_4[Fe(CN)_6]$	1.6×10^{-41}
$AgOH$	2.0×10^{-8}
AgI	8.3×10^{-17}
Ag_3PO_4	1.4×10^{-16}
Ag_2SO_4	1.4×10^{-5}
Ag_2SO_3	1.5×10^{-14}
$AgSCN$	1.0×10^{-12}
Strontium Compounds	
$Sr_3(AsO_4)_2$	8.1×10^{-19}
$SrCO_3$	1.1×10^{-10}
$SrCrO_4$	2.2×10^{-5}
$SrC_2O_4 \cdot H_2O$[3]	1.6×10^{-7}
$Sr_3(PO_4)_2$	4.0×10^{-28}
$SrSO_3$	4×10^{-8}
$SrSO_4$	3.2×10^{-7}
Tin Compounds	
$Sn(OH)_2$	1.4×10^{-28}
$Sn(OH)_4$	1×10^{-56}
Zinc Compounds	
$Zn_3(AsO_4)_2$	1.3×10^{-28}
$ZnCO_3$	1.4×10^{-11}
$Zn_2[Fe(CN)_6]$	4.0×10^{-16}
$Zn(OH)_2$	1.2×10^{-17}
ZnC_2O_4	2.7×10^{-8}
$Zn_3(PO_4)_2$	9.0×10^{-33}

[1]Taken from Patnaik, P. *Dean's Analytical Chemistry Handbook,* 2nd ed. New York: McGraw-Hill, 2004, Table 4.2 (published on the web by Knovel, http://www.knovel.com).
[2]Taken from Meites, L., Ed. *Handbook of Analytical Chemistry,* 1st ed. New York: McGraw-Hill, 1963.
[3]Because $[H_2O]$ does not appear in equilibrium constants for equilibria in aqueous solution in general, it does not appear in the K_{sp} expressions for hydrated solids.

No metal sulfides are listed in this table because sulfide ion is such a strong base that the usual solubility product equilibrium equation does not apply. See Myers, R. J. *Journal of Chemical Education,* Vol. 63, 1986; pp. 687–690.

APPENDIX

I

Standard Reduction Potentials in Aqueous Solution at 25 °C[1]

Acidic Solution	Standard Reduction Potential, $E°$ (volts)
$F_2(g) + 2e^- \longrightarrow 2\,F^-(aq)$	2.87
$Co^{3+}(aq) + e^- \longrightarrow Co^{2+}(aq)$	1.92
$Au^+(aq) + e^- \longrightarrow Au(s)$	1.83
$H_2O_2(aq) + 2\,H_3O^+(aq) + 2\,e^- \longrightarrow 4\,H_2O(\ell)$	1.763
$Ce^{4+}(aq) + e^- \longrightarrow Ce^{3+}(aq)$	1.72
$Pb^{4+}(aq) + 2\,e^- \longrightarrow Pb^{2+}(aq)$	1.69
$PbO_2(s) + SO_4^{2-}(aq) + 4\,H_3O^+(aq) + 2\,e^- \longrightarrow PbSO_4(s) + 6\,H_2O(\ell)$	1.690
$NiO_2(s) + 4\,H_3O^+(aq) + 2\,e^- \longrightarrow Ni^{2+}(aq) + 6\,H_2O(\ell)$	1.68
$2\,HClO(aq) + 2\,H_3O^+(aq) + 2\,e^- \longrightarrow Cl_2(g) + 4\,H_2O(\ell)$	1.63
$Au^{3+}(aq) + 3\,e^- \longrightarrow Au(s)$	1.52
$MnO_4^-(aq) + 8\,H_3O^+(aq) + 5\,e^- \longrightarrow Mn^{2+}(aq) + 12\,H_2O(\ell)$	1.51
$BrO_3^-(aq) + 6\,H_3O^+(aq) + 5\,e^- \longrightarrow \frac{1}{2}\,Br_2(aq) + 9\,H_2O(\ell)$	1.478
$2\,ClO_3^-(aq) + 12\,H_3O^+(aq) + 10\,e^- \longrightarrow Cl_2(g) + 18\,H_2O(\ell)$	1.47
$Cr_2O_7^{2-}(aq) + 14\,H_3O^+(aq) + 6\,e^- \longrightarrow 2\,Cr^{3+}(aq) + 21\,H_2O(\ell)$	1.36
$Cl_2(g) + 2\,e^- \longrightarrow 2\,Cl^-(aq)$	1.358
$N_2H_5^+(aq) + 3\,H_3O^+(aq) + 2\,e^- \longrightarrow 2\,NH_4^+(aq) + 3\,H_2O(\ell)$	1.275
$MnO_2(s) + 4\,H_3O^+(aq) + 2\,e^- \longrightarrow Mn^{2+}(aq) + 6\,H_2O(\ell)$	1.23
$O_2(g) + 4\,H_3O^+(aq) + 4\,e^- \longrightarrow 6\,H_2O(\ell)$	1.229
$ClO_4^-(aq) + 2\,H_3O^+(aq) + 2\,e^- \longrightarrow ClO_3^-(aq) + 3\,H_2O(\ell)$	1.201
$IO_3^-(aq) + 6\,H_3O^+(aq) + 5\,e^- \longrightarrow \frac{1}{2}\,I_2(aq) + 9\,H_2O(\ell)$	1.195
$Pt^{2+}(aq) + 2\,e^- \longrightarrow Pt(s)$	1.188
$Br_2(\ell) + 2\,e^- \longrightarrow 2\,Br^-(aq)$	1.066
$AuCl_4^-(aq) + 3\,e^- \longrightarrow Au(s) + 4\,Cl^-(aq)$	1.00
$NO_3^-(aq) + 4\,H_3O^+(aq) + 3\,e^- \longrightarrow NO(g) + 6\,H_2O(\ell)$	0.96
$NO_3^-(aq) + 3\,H_3O^+(aq) + 2\,e^- \longrightarrow HNO_2(aq) + 4\,H_2O(\ell)$	0.94
$Pd^{2+}(aq) + 2\,e^- \longrightarrow Pd(s)$	0.915
$2\,Hg^{2+}(aq) + 2\,e^- \longrightarrow Hg_2^{2+}(aq)$	0.9110
$Hg^{2+}(aq) + 2\,e^- \longrightarrow Hg(\ell)$	0.8535
$SbCl_6^-(aq) + 2e^- \longrightarrow SbCl_4^-(aq) + 2\,Cl^-(aq)$	0.84[2]
$Hg_2^{2+}(aq) + 2\,e^- \longrightarrow 2\,Hg(\ell)$	0.7960
$Ag^+(aq) + e^- \longrightarrow Ag(s)$	0.7991
$Fe^{3+}(aq) + e^- \longrightarrow Fe^{2+}(aq)$	0.771
$[PtCl_4]^{2-}(aq) + 2\,e^- \longrightarrow Pt(s) + 4\,Cl^-(aq)$	0.758
$[PtCl_6]^{2-}(aq) + 2\,e^- \longrightarrow [PtCl_4]^{2-}(aq) + 2\,Cl^-(aq)$	0.726
$O_2(g) + 2\,H_3O^+(aq) + 2\,e^- \longrightarrow H_2O_2(aq) + 2\,H_2O(\ell)$	0.695
$TeO_2(s) + 4\,H_3O^+(aq) + 4\,e^- \longrightarrow Te(s) + 6\,H_2O(\ell)$	0.604
$H_3AsO_4(aq) + 2\,H_3O^+(aq) + 2\,e^- \longrightarrow HAsO_2(aq) + 4\,H_2O(\ell)$	0.560
$I_2(s) + 2\,e^- \longrightarrow 2\,I^-(aq)$	0.535

[1] Taken from Bard, A. J., Parsons, R., and Jordan, J. *Standard Potentials in Aqueous Solution.* New York: Marcel Dekker, 1985. International Union of Pure and Applied Chemistry, Commission on Electrochemistry and Electroanalytical Chemistry.

[2] Taken from Brown, R. A., and Swift, E. H. *Journal of the American Chemical Society,* Vol. 71, 1949; pp. 2719–2723.

Acidic Solution	**Standard Reduction Potential, $E°$ (volts)**
$Cu^+(aq) + e^- \longrightarrow Cu(s)$	0.521
$[RhCl_6]^{3-}(aq) + 3\,e^- \longrightarrow Rh(s) + 6\,Cl^-(aq)$	0.5
$Cu^{2+}(aq) + 2\,e^- \longrightarrow Cu(s)$	0.340
$Hg_2Cl_2(s) + 2\,e^- \longrightarrow 2\,Hg(\ell) + 2\,Cl^-(aq)$	0.27
$AgCl(s) + e^- \longrightarrow Ag(s) + Cl^-(aq)$	0.222
$Cu^{2+}(aq) + e^- \longrightarrow Cu^+(aq)$	0.159
$SO_4^{2-}(aq) + 4\,H_3O^+(aq) + 2\,e^- \longrightarrow H_2SO_3(aq) + 5\,H_2O(\ell)$	0.158
$Sn^{4+}(aq) + 2\,e^- \longrightarrow Sn^{2+}(aq)$	0.15
$S(s) + 2\,H_3O^+(aq) + 2\,e^- \longrightarrow H_2S(aq) + 2\,H_2O(\ell)$	0.144
$AgBr(s) + e^- \longrightarrow Ag(s) + Br^-(aq)$	0.0713
$2\,H_3O^+(aq) + 2\,e^- \longrightarrow H_2(g) + 2\,H_2O(\ell)$ (reference electrode)	0
$N_2O(g) + 6\,H_3O^+(aq) + 4\,e^- \longrightarrow 2\,NH_3OH^+(aq) + 5\,H_2O(\ell)$	−0.05
HgS(s, black) $+ 2\,H_3O^+(aq) + 2\,e^- \longrightarrow Hg(\ell) + H_2S(g) + 2\,H_2O(\ell)$	−0.085
$Se(s) + 2\,H_3O^+(aq) + 2\,e^- \longrightarrow H_2Se(aq) + 2\,H_2O(\ell)$	−0.115
$Pb^{2+}(aq) + 2\,e^- \longrightarrow Pb(s)$	−0.125
$Sn^{2+}(aq) + 2\,e^- \longrightarrow Sn(s)$	−0.1375
$AgI(s) + e^- \longrightarrow Ag(s) + I^-(aq)$	−0.1522
$[SnF_6]^{2-}(aq) + 4\,e^- \longrightarrow Sn(s) + 6\,F^-(aq)$	−0.200
$Ni^{2+}(aq) + 2\,e^- \longrightarrow Ni(s)$	−0.25
$Co^{2+}(aq) + 2\,e^- \longrightarrow Co(s)$	−0.277
$Tl^+(aq) + e^- \longrightarrow Tl(s)$	−0.3363
$PbSO_4(s) + 2\,e^- \longrightarrow Pb(s) + SO_4^{2-}(aq)$	−0.3505
$Cd^{2+}(aq) + 2\,e^- \longrightarrow Cd(s)$	−0.403
$Cr^{3+}(aq) + e^- \longrightarrow Cr^{2+}(aq)$	−0.424
$Fe^{2+}(aq) + 2\,e^- \longrightarrow Fe(s)$	−0.44
$2\,CO_2(g) + 2\,H_3O^+(aq) + 2\,e^- \longrightarrow (COOH)_2(aq) + 2\,H_2O(\ell)$	−0.481
$Ga^{3+}(aq) + 3\,e^- \longrightarrow Ga(s)$	−0.53
$Cr^{3+}(aq) + 3\,e^- \longrightarrow Cr(s)$	−0.74
$Zn^{2+}(aq) + 2\,e^- \longrightarrow Zn(s)$	−0.763
$Cr^{2+}(aq) + 2\,e^- \longrightarrow Cr(s)$	−0.90
$V^{2+}(aq) + 2\,e^- \longrightarrow V(s)$	−1.13
$Mn^{2+}(aq) + 2\,e^- \longrightarrow Mn(s)$	−1.18
$Zr^{4+}(aq) + 4\,e^- \longrightarrow Zr(s)$	−1.55
$Al^{3+}(aq) + 3\,e^- \longrightarrow Al(s)$	−1.676
$H_2(g) + 2\,e^- \longrightarrow 2\,H^-(aq)$	−2.25
$Mg^{2+}(aq) + 2\,e^- \longrightarrow Mg(s)$	−2.356
$Na^+(aq) + e^- \longrightarrow Na(s)$	−2.714
$Ca^{2+}(aq) + 2\,e^- \longrightarrow Ca(s)$	−2.84
$Sr^{2+}(aq) + 2\,e^- \longrightarrow Sr(s)$	−2.89
$Ba^{2+}(aq) + 2\,e^- \longrightarrow Ba(s)$	−2.92
$Rb^+(aq) + e^- \longrightarrow Rb(s)$	−2.925
$K^+(aq) + e^- \longrightarrow K(s)$	−2.925
$Li^+(aq) + e^- \longrightarrow Li(s)$	−3.045

Basic Solution	**Standard Reduction Potential, $E°$ (volts)**
$ClO^-(aq) + H_2O(\ell) + 2\,e^- \longrightarrow Cl^-(aq) + 2\,OH^-(aq)$	0.89
$OOH^-(aq) + H_2O(\ell) + 2\,e^- \longrightarrow 3\,OH^-(aq)$	0.867
$2\,NH_2OH(aq) + 2\,e^- \longrightarrow N_2H_4(aq) + 2\,OH^-(aq)$	0.73
$ClO_3^-(aq) + 3\,H_2O(\ell) + 6\,e^- \longrightarrow Cl^-(aq) + 6\,OH^-(aq)$	0.622
$MnO_4^-(aq) + 2\,H_2O(\ell) + 3\,e^- \longrightarrow MnO_2(s) + 4\,OH^-(aq)$	0.60
$MnO_4^-(aq) + e^- \longrightarrow MnO_4^{2-}(aq)$	0.56
$NiO_2(s) + 2\,H_2O(\ell) + 2\,e^- \longrightarrow Ni(OH)_2(s) + 2\,OH^-(aq)$	0.49
$Ag_2CrO_4(s) + 2\,e^- \longrightarrow 2\,Ag(s) + CrO_4^{2-}(aq)$	0.4491
$O_2(g) + 2\,H_2O(\ell) + 4\,e^- \longrightarrow 4\,OH^-(aq)$	0.401
$ClO_4^-(aq) + H_2O(\ell) + 2\,e^- \longrightarrow ClO_3^-(aq) + 2\,OH^-(aq)$	0.374
$Ag_2O(s) + H_2O(\ell) + 2\,e^- \longrightarrow 2\,Ag(s) + 2\,OH^-(aq)$	0.342
$2\,NO_2^-(aq) + 3\,H_2O(\ell) + 4\,e^- \longrightarrow N_2O(g) + 6\,OH^-(aq)$	0.15
$N_2H_4(aq) + 2\,H_2O(\ell) + 2\,e^- \longrightarrow 2\,NH_3(aq) + 2\,OH^-(aq)$	0.10
$HgO(s) + H_2O(\ell) + 2\,e^- \longrightarrow Hg(\ell) + 2\,OH^-(aq)$	0.0977
$O_2(g) + H_2O(\ell) + 2\,e^- \longrightarrow OOH^-(aq) + OH^-(aq)$	0.0649
$[Co(NH_3)_6]^{3+}(aq) + e^- \longrightarrow [Co(NH_3)_6]^{2+}(aq)$	0.058
$NO_3^-(aq) + H_2O(\ell) + 2\,e^- \longrightarrow NO_2^-(aq) + 2\,OH^-(aq)$	0.01
$MnO_2(s) + 2\,H_2O(\ell) + 2\,e^- \longrightarrow Mn(OH)_2(s) + 2\,OH^-(aq)$	−0.05
$CrO_4^{2-}(aq) + 4\,H_2O(\ell) + 3\,e^- \longrightarrow Cr(OH)_3(s) + 5\,OH^-(aq)$	−0.11
$Cu_2O(s) + H_2O(\ell) + 2\,e^- \longrightarrow 2\,Cu(s) + 2\,OH^-(aq)$	−0.365
$FeO_2^-(aq) + H_2O(\ell) + e^- \longrightarrow HFeO_2^-(aq) + OH^-(aq)$	−0.69
$HFeO_2^-(aq) + 2\,e^- \longrightarrow Fe(s) + 3\,OH^-(aq)$	−0.8
$2\,H_2O(\ell) + 2\,e^- \longrightarrow H_2(g) + 2\,OH^-(aq)$	−0.8277
$2\,NO_3^-(aq) + 2\,H_2O(\ell) + 2\,e^- \longrightarrow N_2O_4(g) + 4\,OH^-(aq)$	−0.86
$SO_4^{2-}(aq) + H_2O(\ell) + 2\,e^- \longrightarrow SO_3^{2-}(aq) + 2\,OH^-(aq)$	−0.936
$N_2(g) + 4\,H_2O(\ell) + 4\,e^- \longrightarrow N_2H_4(aq) + 4\,OH^-(aq)$	−1.16
$Zn(OH)_2(s) + 2\,e^- \longrightarrow Zn(s) + 2\,OH^-(aq)$	−1.246
$[Zn(OH)_4]^{2-}(aq) + 2\,e^- \longrightarrow Zn(s) + 4\,OH^-(aq)$	−1.285
$Cr(OH)_3(s) + 3\,e^- \longrightarrow Cr(s) + 3\,OH^-(aq)$	−1.33
$[Zn(CN)_4]^{2-}(aq) + 2\,e^- \longrightarrow Zn(s) + 4\,CN^-(aq)$	−1.34
$SiO_3^{2-}(aq) + 3\,H_2O(\ell) + 4\,e^- \longrightarrow Si(s) + 6\,OH^-(aq)$	−1.69

APPENDIX J

Selected Thermodynamic Values[1]

Species	ΔH_f° (298.15 K) (kJ/mol)	S° (298.15 K) (J K^{-1} mol^{-1})	ΔG_f° (298.15 K) (kJ/mol)
Aluminum			
Al(s)	0	28.275	0
Al^{3+}(aq)	−531	−321.7	−485
$AlCl_3$(s)	−704.2	110.67	−628.8
Al_2O_3(s, corundum)	−1675.7	50.92	−1582.3
Argon			
Ar(g)	0	154.843	0
Ar(aq)	−12.1	59.4	16.4
Barium			
$BaCl_2$(s)	−858.6	123.68	−810.4
BaO(s)	−553.5	70.42	−525.1
$BaSO_4$(s)	−1473.2	132.2	−1362.2
$BaCO_3$(s)	−1216.3	112.1	85.35
Beryllium			
Be(s)	0	9.5	0
$Be(OH)_2$(s)	−902.5	51.9	−815
Bromine			
Br(g)	111.884	175.022	82.396
$Br_2(\ell)$	0	152.231	0
Br_2(g)	30.907	245.463	3.110
Br_2(aq)	−2.59	130.5	3.93
Br^-(aq)	−121.55	82.4	−103.96
BrCl(g)	14.64	240.10	−0.98
BrF_3(g)	−255.6	292.53	−229.43
HBr(g)	−36.40	198.695	−53.45
Calcium			
Ca(s)	0	41.42	0
Ca(g)	178.2	158.884	144.3
Ca^{2+}(g)	1925.9	—	—
Ca^{2+}(aq)	−542.83	−53.1	−553.58
CaC_2(s)	−59.8	69.96	−64.9
$CaCO_3$(s, calcite)	−1206.92	92.9	−1128.79
$CaCl_2$(s)	−795.8	104.6	−748.1
CaF_2(s)	−1219.6	68.87	−1167.3
CaH_2(s)	−186.2	42	−147.2
CaO(s)	−635.09	39.75	−604.03
CaS(s)	−482.4	56.5	−477.4
$Ca(OH)_2$(s)	−986.09	83.39	−898.49
$Ca(OH)_2$(aq)	−1002.82	−74.5	−868.07
$CaSO_4$(s)	−1434.11	106.7	−1321.79

[1]Taken from Wagman, D. D., Evans, W. H., Parker, V. B., Schumm, R. H., Halow, I., Bailey, S. M., Churney, K. L., and Nuttall, R. "The NBS Tables of Chemical Thermodynamic Properties." *Journal of Physical and Chemical Reference Data,* Vol. 11, Suppl. 2, 1982.

Species	ΔH_f° (298.15 K) (kJ/mol)	S° (298.15 K) (J K^{-1} mol^{-1})	ΔG_f° (298.15 K) (kJ/mol)
Carbon			
C(s, graphite)	0	5.74	0
C(s, diamond)	1.895	2.377	2.9
C(g)	716.682	158.096	671.257
CCl_4(ℓ)	−135.44	216.4	−65.21
CCl_4(g)	−102.9	309.85	−60.59
$CHCl_3$(ℓ)	−134.47	201.7	−73.66
$CHCl_3$(g)	−103.14	295.71	−70.34
CH_4 (g, methane)	−74.81	186.264	−50.72
C_2H_2 (g, ethyne)	226.73	200.94	209.2
C_2H_4 (g, ethene)	52.26	219.56	68.15
C_2H_6 (g, ethane)	−84.68	229.6	−32.82
C_3H_8 (g, propane)	−103.8	269.9	−23.49
C_4H_{10} (g, butane)	−126.148	310.227	−16.985
C_6H_6(ℓ, benzene)	49.03	172.8	124.5
C_6H_{14}(ℓ, hexane)	−198.782	296.018	−4.035
C_8H_{18} (g, octane)	−208.447	466.835	16.718
C_8H_{18}(ℓ, octane)	−249.952	361.205	6.707
CH_3OH(ℓ, methanol)	−238.66	126.8	−166.27
CH_3OH(g, methanol)	−200.66	239.81	−161.96
CH_3OH(aq, methanol)	−245.931	133.1	−175.31
C_2H_5OH(ℓ, ethanol)	−277.69	160.7	−174.78
C_2H_5OH(g, ethanol)	−235.1	282.7	−168.49
C_2H_5OH(aq, ethanol)	−288.3	148.5	−181.64
$C_6H_{12}O_6$ (s, glucose)	−1274.4	235.9	−917.2
CH_3COO^-(aq)	−486.01	86.6	−369.31
CH_3COOH(aq)	−485.76	178.7	−396.46
CH_3COOH(ℓ)	−484.5	159.8	−389.9
CO(g)	−110.525	197.674	−137.168
CO_2(g)	−393.509	213.74	−394.359
H_2CO_3(aq)	−699.65	187.4	−623.08
HCO_3^-(aq)	−691.99	91.2	−586.77
CO_3^{2-}(aq)	−677.14	−56.9	−527.81
$HCOO^-$(aq)	−425.55	92.0	−351.0
HCOOH(aq)	−425.43	163	−372.3
HCOOH(ℓ)	−424.72	128.95	−361.35
CS_2(g)	117.36	237.84	67.12
CS_2(ℓ)	89.70	151.34	65.27
$COCl_2$(g)	−218.8	283.53	−204.6
Cesium			
Cs(s)	0	85.23	0
Cs^+(g)	457.964	—	—
CsCl(s)	−443.04	101.17	−414.53
Chlorine			
Cl(g)	121.679	165.198	105.68
Cl^-(g)	−233.13	—	—
Cl^-(aq)	−167.159	56.5	−131.228
Cl_2(g)	0	223.066	0
Cl_2(aq)	−23.4	121	6.94
HCl(g)	−92.307	186.908	−95.299
HCl(aq)	−167.159	56.5	−131.228

Species	ΔH_f° (298.15 K) (kJ/mol)	S° (298.15 K) (J K^{-1} mol^{-1})	ΔG_f° (298.15 K) (kJ/mol)
$ClO_2(g)$	102.5	256.84	120.5
$Cl_2O(g)$	80.3	266.21	97.9
$ClO^-(aq)$	−107.1	42.0	−36.8
$HClO(aq)$	−120.9	142.	−79.9
$ClF_3(g)$	−163.2	281.61	−123.0
Chromium			
$Cr(s)$	0	23.77	0
$Cr_2O_3(s)$	−1139.7	81.2	−1058.1
$CrCl_3(s)$	−556.5	123	−486.1
Copper			
$Cu(s)$	0	33.15	0
$CuO(s)$	−157.3	42.63	−129.7
$CuCl_2(s)$	−220.1	108.07	−175.7
$CuSO_4(s)$	−771.36	109.	−661.8
Fluorine			
$F_2(g)$	0	202.78	0
$F(g)$	78.99	158.754	61.91
$F^-(g)$	−255.39	—	—
$F^-(aq)$	−332.63	−13.8	−278.79
$HF(g)$	−271.1	173.779	−273.2
HF(aq, un-ionized)	−320.08	88.7	−296.82
HF(aq, ionized)	−332.63	−13.8	−278.79
Hydrogen[2]			
$H_2(g)$	0	130.684	0
$H_2(aq)$	−4.2	57.7	17.6
$HD(g)$	0.318	143.801	−1.464
$D_2(g)$	0	144.960	0
$H(g)$	217.965	114.713	203.247
$H^+(g)$	1536.202	—	—
$H^+(aq)$	0	0	0
$OH^-(aq)$	−229.994	−10.75	−157.244
$H_2O(\ell)$	−285.83	69.91	−237.129
$H_2O(g)$	−241.818	188.825	−228.572
$H_2O_2(\ell)$	−187.78	109.6	−120.35
$H_2O_2(aq)$	−191.17	143.9	−134.03
$HO_2^-(aq)$	−160.33	23.8	−67.3
$HDO(\ell)$	−289.888	79.29	−241.857
$D_2O(\ell)$	−294.600	75.94	−243.439
Iodine			
$I_2(s)$	0	116.135	0
$I_2(g)$	62.438	260.69	19.327
$I_2(aq)$	22.6	137.2	16.40
$I(g)$	106.838	180.791	70.25
$I^-(g)$	−197	—	—
$I^-(aq)$	−55.19	111.3	−51.57
$I_3^-(aq)$	−51.5	239.3	−51.4
$HI(g)$	26.48	206.594	1.70
HI(aq, ionized)	−55.19	111.3	−51.57
$IF(g)$	−95.65	236.17	−118.51

Species	ΔH_f° (298.15 K) (kJ/mol)	S° (298.15 K) ($J\ K^{-1}\ mol^{-1}$)	ΔG_f° (298.15 K) (kJ/mol)
ICl(g)	17.78	247.551	−5.46
ICl_3(s)	−89.5	167.4	−22.29
ICl(ℓ)	−23.89	135.1	−13.58
IBr(g)	40.84	258.773	3.69
Iron			
Fe(s)	0	27.78	0
FeO(s, wustite)	−266.27	57.49	−245.12
Fe_2O_3(s, hematite)	−824.2	87.4	−742.2
Fe_3O_4(s, magnetite)	−1118.4	146.4	−1015.4
$FeCl_2$(s)	−341.79	117.95	−302.3
$FeCl_3$(s)	−399.49	142.3	−344
FeS_2 (s, pyrite)	−178.2	52.93	−166.9
$Fe(CO)_5$(ℓ)	−774	338.1	−705.3
Lead			
Pb(s)	0	64.81	0
$PbCl_2$(s)	−359.41	136	−314.1
PbO(s, yellow)	−217.32	68.7	−187.89
PbS(s)	−100.4	91.2	−98.7
$PbSO_4$(s)	−919.94	148.57	−813.14
$PbCl_4$(ℓ)	−329.3	—	—
PbO_2(s)	−277.4	68.6	−217.33
Lithium			
Li(s)	0	29.12	0
Li^+(g)	685.783	—	—
LiOH(s)	−484.93	42.8	−438.95
LiOH(aq)	−508.48	2.8	−450.58
LiCl(s)	−408.701	59.33	−384.37
Magnesium			
Mg(s)	0	32.68	0
Mg^{2+}(aq)	−466.85	−138.1	−454.8
$MgCl_2$(g)	−400.4	—	—
$MgCl_2$(s)	−641.32	89.62	−591.79
$MgCl_2$(aq)	−801.15	−25.1	−717.1
MgO(s)	−601.70	26.94	−569.43
$Mg(OH)_2$(s)	−924.54	63.18	−833.51
MgS(s)	−346	50.33	−341.8
$MgSO_4$(s)	−1284.9	91.6	−1170.6
$MgCO_3$(s)	−1095.8	65.7	−1012.1
Mercury			
Hg(ℓ)	0	76.02	0
$HgCl_2$(s)	−224.3	146	−178.6
HgO(s, red)	−90.83	70.29	−58.539
HgS(s, red)	−58.2	82.4	−50.6
Nickel			
Ni(s)	0	29.87	0
NiO(s)	−239.7	37.99	−211.7
$NiCl_2$(s)	−305.332	97.65	−259.032

Species	ΔH_f° (298.15 K) (kJ/mol)	S° (298.15 K) (J K^{-1} mol^{-1})	ΔG_f° (298.15 K) (kJ/mol)
Nitrogen			
$N_2(g)$	0	191.61	0
$N_2(aq)$	−10.8	—	—
$N(g)$	472.704	153.298	455.563
$NH_3(g)$	−46.11	192.45	−16.45
$NH_3(aq)$	−80.29	111.3	−26.50
$NH_4^+(aq)$	−132.51	113.4	−79.31
$N_2H_4(\ell)$	50.63	121.21	149.34
$NH_4Cl(s)$	−314.43	94.6	−202.87
$NH_4Cl(aq)$	−299.66	169.9	−210.52
$NH_4NO_3(s)$	−365.56	151.08	−183.87
$NH_4NO_3(aq)$	−339.87	259.8	−190.56
$NO(g)$	90.25	210.761	86.55
$NO_2(g)$	33.18	240.06	51.31
$N_2O(g)$	82.05	219.85	104.20
$N_2O_4(g)$	9.16	304.29	97.89
$N_2O_4(\ell)$	−19.50	209.2	97.54
$NOCl(g)$	51.71	261.69	66.08
$HNO_3(\ell)$	−174.10	155.60	−80.71
$HNO_3(g)$	−135.06	266.38	−74.72
$HNO_3(aq)$	−207.36	146.4	−111.25
$NO_3^-(aq)$	−205.0	146.4	−108.74
$NF_3(g)$	−124.7	260.73	−83.2
Oxygen[2]			
$O_2(g)$	0	205.138	0
$O_2(aq)$	−11.7	110.9	16.4
$O(g)$	249.170	161.055	231.731
$O_3(g)$	142.7	238.93	163.2
$OH^-(aq)$	−229.994	−10.75	−157.244
Phosphorus			
P_4(s, white)	0	164.36	0
P_4(s, red)	−70.4	91.2	−48.4
$P(g)$	314.64	163.193	278.25
$PH_3(g)$	5.4	310.23	13.4
$PCl_3(g)$	−287	311.78	−267.8
$PCl_3(\ell)$	−319.7	217.1	−272.3
$PCl_5(s)$	−443.5	—	—
$P_4O_{10}(s)$	−2984	228.86	−2697.7
$H_3PO_4(s)$	−1279	110.5	−1119.1
Potassium			
$K(s)$	0	64.18	0
$KF(s)$	−567.27	66.57	−537.75
$KCl(s)$	−436.747	82.59	−409.14
$KCl(aq)$	−419.53	159.0	−414.49
$KBr(s)$	−393.798	95.90	−380.66
$KI(s)$	−327.900	106.32	−324.892
$KClO_3(s)$	−397.73	143.1	−296.25
$KOH(s)$	−424.764	78.9	−379.08
$KOH(aq)$	−482.37	91.6	−440.5

Species	ΔH_f° (298.15 K) (kJ/mol)	S° (298.15 K) (J K^{-1} mol^{-1})	ΔG_f° (298.15 K) (kJ/mol)
Silicon			
Si(s)	0	18.83	0
$SiBr_4(\ell)$	−457.3	277.8	−443.8
SiC(s)	−65.3	16.61	−62.8
$SiCl_4(g)$	−657.01	330.73	−616.98
$SiH_4(g)$	34.3	204.62	56.9
$SiF_4(g)$	−1614.94	282.49	−1572.65
SiO_2(s, quartz)	−910.94	41.84	−856.64
Silver			
Ag(s)	0	42.55	0
$Ag^+(aq)$	105.579	72.68	77.107
$Ag_2O(s)$	−31.05	121.3	−11.2
AgCl(s)	−127.068	96.2	−109.789
AgI(s)	−61.84	115.5	−66.19
$AgN_3(s)$	620.60	99.22	591.0
$AgNO_3(s)$	−124.39	140.92	−33.41
$AgNO_3(aq)$	−101.8	219.2	−34.16
Sodium			
Na(s)	0	51.21	0
Na(g)	107.32	153.712	76.761
$Na^+(g)$	609.358	—	—
$Na^+(aq)$	−240.12	59.0	−261.905
NaF(s)	−573.647	51.46	−543.494
NaF(aq)	−572.75	45.2	−540.68
NaCl(s)	−411.153	72.13	−384.138
NaCl(g)	−176.65	229.81	−196.66
NaCl(aq)	−407.27	115.5	−393.133
NaBr(s)	−361.062	86.82	−348.983
NaBr(aq)	−361.665	141.4	−365.849
NaI(s)	−287.78	98.53	−286.06
NaI(aq)	−295.31	170.3	−313.47
NaOH(s)	−425.609	64.455	−379.484
NaOH(aq)	−470.114	48.1	−419.15
$NaClO_3(s)$	−365.774	123.4	−262.259
$NaHCO_3(s)$	−950.81	101.7	−851.0
$Na_2CO_3(s)$	−1130.68	134.98	−1044.44
$Na_2SO_4(s)$	−1387.08	149.58	−1270.16
Sulfur			
S(s, monoclinic)	0.33	—	—
S(s, rhombic)	0	31.80	0
S(g)	278.805	167.821	238.250
$S^{2-}(aq)$	33.1	−14.6	85.8
$S_2Cl_2(g)$	−18.4	331.5	−31.8
$SF_6(g)$	−1209.	291.82	−1105.3
$SF_4(g)$	−774.9	292.03	−731.3
$H_2S(g)$	−20.63	205.79	−33.56
$H_2S(aq)$	−39.7	121	−27.83
$HS^-(aq)$	−17.6	62.8	12.08
$SO_2(g)$	−296.830	248.22	−300.194
$SO_3(g)$	−395.72	256.76	−371.06

Species	ΔH_f° (298.15 K) (kJ/mol)	S° (298.15 K) (J K^{-1} mol^{-1})	ΔG_f° (298.15 K) (kJ/mol)
$SOCl_2(g)$	−212.5	309.77	−198.3
$SO_4^{2-}(aq)$	−909.27	20.1	−744.53
$H_2SO_4(\ell)$	−813.989	156.904	−690.003
$H_2SO_4(aq)$	−909.27	20.1	−744.53
$HSO_4^-(aq)$	−887.34	131.8	−755.91
Tin			
Sn(s, white)	0	51.55	0
Sn(s, gray)	−2.09	44.14	0.13
$SnCl_2(s)$	−325.1	—	—
$SnCl_4(\ell)$	−511.3	258.6	−440.1
$SnCl_4(g)$	−471.5	365.8	−432.2
$SnO_2(s)$	−580.7	52.3	−519.6
Titanium			
Ti(s)	0	30.63	0
$TiCl_4(\ell)$	−804.2	252.34	−737.2
$TiCl_4(g)$	−763.2	354.9	−726.7
$TiO_2(s)$	−939.7	49.92	−884.5
Uranium			
U(s)	0	50.21	0
$UO_2(s)$	−1084.9	77.03	−1031.7
$UO_3(s)$	−1223.8	96.11	−1145.9
$UF_4(s)$	−1914.2	151.67	−1823.3
$UF_6(g)$	−2147.4	377.9	−2063.7
$UF_6(s)$	−2197.0	227.6	−2068.5
Zinc			
Zn(s)	0	41.63	0
$ZnCl_2(s)$	−415.05	111.46	−369.398
ZnO(s)	−348.28	43.64	−318.3
ZnS(s, sphalerite)	−205.98	57.7	−201.29

[2]Many hydrogen-containing and oxygen-containing compounds are listed only under other elements; for example, HNO_3 appears under nitrogen.

APPENDIX K

Answers to Problem-Solving Practice Problems

Chapter 1

1.1 (1) *Analyze the problem:* You are asked to find the volume of the sample, and you know the mass.
(2) *Plan a solution:* Density relates mass and volume and is the appropriate conversion factor, so look up the density in a table. Volume is proportional to mass, so the mass has to be either multiplied by the density or multiplied by the reciprocal of the density. Use the units to decide which.
(3) *Execute the plan:* According to Table 1.1, the density of benzene is 0.880 g/mL. Setting up the calculation so that the unit (grams) cancels gives

$$4.33 \text{ g} \times \frac{1 \text{ mL}}{0.880 \text{ g}} = 4.92 \text{ mL}$$

Notice that the result is expressed to three significant figures, because both the mass and the density had three significant figures.
(4) *Check your answer:* Because the density is a little less than 1.00 g/mL, the volume in milliliters should be a little larger than the mass in grams. The calculated answer, 4.92 mL, is a little larger than the mass, 4.33 g.

1.2 Substance A must be a mixture since some of it dissolves and some, substance B, does not.

Substance C is the soluble portion of substance A. Since all of substance C dissolves in water there is no way to determine how many components it has. Additionally, it is not possible to determine whether the one or more components themselves are elements or compounds. Therefore it is not possible to say whether C is an element, a compound, or a mixture.

The only thing we know about substance B is that it is insoluble in water. We do not know whether it is one insoluble substance, or more than one insoluble substance. Additionally, we do not know whether the substance or substances of B are elements or compounds. Therefore it is not possible to say whether B is an element, a compound, or a mixture.

1.3 Oxygen is O_2; ozone is O_3. Oxygen is a colorless, odorless gas; ozone is a pale blue gas with a pungent odor.

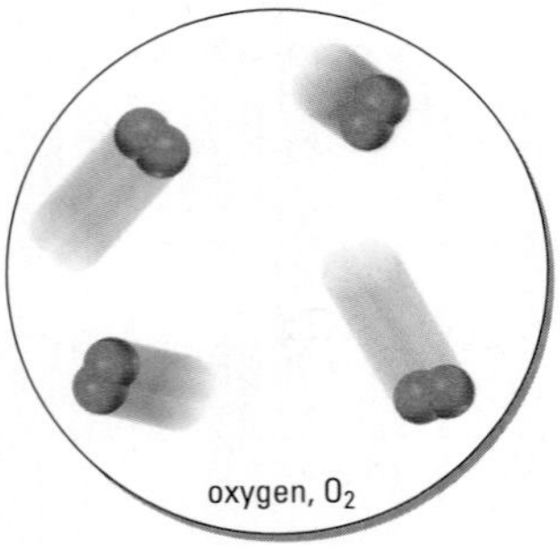
oxygen, O_2

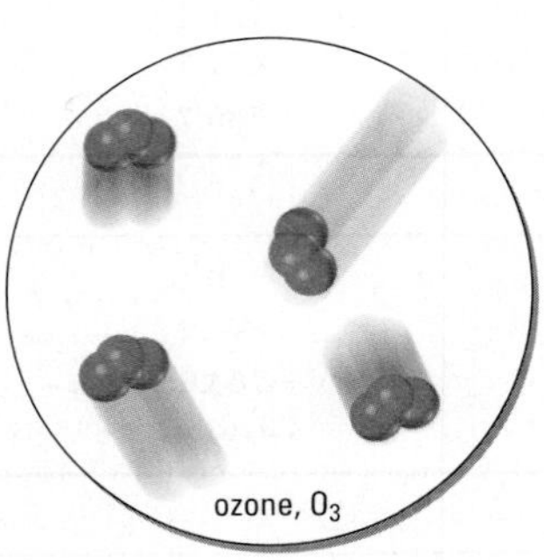
ozone, O_3

Chapter 2

2.1 (a) 10 gal = 40 qt. There are 1.0567 quarts per liter, so

$$40 \text{ qt} \times \frac{1 \text{ L}}{1.0567 \text{ qt}} = 37.9 \text{ L or } 38 \text{ L}$$

(b) $$100 \text{ yds} \times \frac{3 \text{ ft}}{\text{yd}} \times \frac{12 \text{ in}}{\text{ft}} \times \frac{2.54 \text{ cm}}{\text{in}} \times \frac{1 \text{ m}}{100 \text{ cm}} = 91.46 \text{ m}$$

2.2 (a) 1 lb = 453.59 g, so 5 lb = 2268. g
(b) $3 \text{ pt} \times \frac{1 \text{ qt}}{2 \text{ pt}} \times \frac{1 \text{ L}}{1.057 \text{ qt}} \times \frac{1000 \text{ ml}}{1 \text{ L}} = 1420 \text{ ml}$
(c) $\frac{1.420 \text{ L}}{5 \text{ L}} \times 100\% = 28\%$

2.3 Work with the numerator first: 165 mg is 0.165 g. Work with the denominator next: 1 dL = 0.1 L. Therefore, the concentration is

$$\frac{0.165 \text{ g}}{0.1 \text{ L}} = 1.65 \text{ g/L}.$$

2.4 (a) 0.00602 g   3 sf
(b) 22.871 mg   5 sf
(c) 344. °C   3 sf
(d) 100.0 mL   4 sf
(e) 0.00042 m   2 sf
(f) 0.002001 L   4 sf

2.5 (a) 244.2 + 0.1732 = 244.4
(b) 6.19 × 5.2222 = 32.3
(c) $\frac{7.2234 - 11.3851}{4.22} = -0.986$

2.6 (a) A phosphorus atom ($Z = 15$) with 16 neutrons has $A = 31$.
(b) A neon-22 atom has $A = 22$ and $Z = 10$, so the number of electrons must be 10 and the number of neutrons must be $A - Z = 22 - 10 = 12$ neutrons.
(c) The periodic table shows us that the element with 82 protons is lead. The atomic weight of this isotope of lead is 82 + 125 = 207, so the correct symbol is $^{207}_{82}\text{Pb}$.

2.7 The magnesium isotope with 12 neutrons has 12 protons, so $Z = 12$ and the notation is $^{24}_{12}Mg$; the isotope with 13 neutrons has $Z = 12$ and $^{25}_{12}Mg$; and the isotope with 14 neutrons has $Z = 12$ and $^{26}_{12}Mg$.

2.8 75 g wire × (fraction Ni) = g Ni, so 75 g Ni × 0.80 = 60 g Ni. For Cr we have 75 g Cr × 0.20 = 15 g Cr. Or, we could have solved for the mass of Cr from 75 g wire − 60 g Ni = 15 g Cr.

2.9 (a) $4.00 \text{ g Ti} \times \dfrac{1 \text{ mol Ti}}{47.867 \text{ g Ti}} = 8.36 \times 10^{-2} \text{ mol Ti}$

(b) $3.00 \times 10^{-2} \text{ mol Ag} \times \dfrac{107.868 \text{ g Ag}}{1 \text{ mol Ag}} = 3.24 \text{ g Ag}$

Chapter 3

3.1 (a) $C_{10}H_{11}O_{13}N_5P_3$ (b) $C_{18}H_{27}O_3N$ (c) $C_2H_2O_4$

3.2 (a) Sulfur dioxide
(b) Boron trifluoride
(c) Carbon tetrachloride

3.3 Yes, a similar trend would be expected. The absolute values of the boiling points of the chlorine-containing compounds would be different from their alkane parents, but the differences between successive chlorine-containing compounds should follow a similar trend as for the alkanes themselves.

3.4 (a) A Ca^{4+} charge is unlikely because calcium is in Group 2A, the elements of which lose two electrons to form 2+ ions.
(b) Cr^{2+} is possible because chromium is a transition metal ion that forms 2+ and 3+ ions.
(c) Strontium is a Group 2A metal and forms 2+ ions; thus, a Sr^- ion is highly unlikely.

3.5 (a) $CaSO_3$ contains one Ca^{2+} ion and one sulfite ion, SO_3^{2-}.
(b) $Mg_3(PO_4)_2$ contains three Mg^{2+} ions and two phosphate ions, PO_4^{3-}.

3.6 (a) One Mg^{2+} ion and two Br^- ions
(b) Two Li^+ ions and one CO_3^{2-} ion
(c) One NH_4^+ ion and one Cl^- ion
(d) Two Fe^{3+} ions and three SO_4^{2-} ions
(e) CuCl and $CuCl_2$

3.7 (a) KNO_2 is potassium nitrite.
(b) $NaHSO_3$ is sodium hydrogen sulfite or sodium bisulfite.
(c) $Mn(OH)_2$ is manganese(II) hydroxide.
(d) $Mn_2(SO_4)_3$ is manganese(III) sulfate.
(e) Ba_3N_2 is barium nitride.
(f) LiH is lithium hydride.

3.8 (a) KH_2PO_4 (b) CuOH (c) NaClO
(d) NH_4ClO_4 (e) $CrCl_3$ (f) $FeSO_3$

3.9 (a) CH_4 is formed from two nonmetals and is molecular.
(b) $CaBr_2$ is formed from a metal and a nonmetal, so it is ionic.
(c) $MgCl_2$ is formed from a metal and a nonmetal, so it is ionic.
(d) PCl_3 is formed from two nonmetals and is molecular.
(e) KCl is formed from a metal and a nonmetal and is ionic.

3.10 (a) The molar mass of $Ca(NO_3)_2$ is 164.09 g/mol.

$$\frac{12.0 \text{ g}}{164.09 \text{ g/mol}} = 7.31 \times 10^{-2} \text{ mol}$$

(b) The molar mass of $KMnO_4$ is 158.03 g/mol.

$$\frac{12.0 \text{ g}}{158.03 \text{ g/mol}} = 7.59 \times 10^{-2} \text{ mol}$$

(c) The molar mass of Na_2SO_4 is 142.04 g/mol.

$$\frac{12.0 \text{ g}}{142.04 \text{ g/mol}} = 8.45 \times 10^{-2} \text{ mol}$$

3.11 (a) The molar mass of sucrose, $C_{12}H_{22}O_{11}$, is 342.3 g/mol.

$$5.0 \times 10^{-3} \text{ mol sucrose} \times \frac{342.3 \text{ g sucrose}}{1 \text{ mol sucrose}} = 1.7 \text{ g sucrose}$$

(b) $3.0 \times 10^{-6} \text{ mol ACTH} \times \dfrac{4600 \text{ g ACTH}}{1 \text{ mol ACTH}} = 1.4 \times 10^{-2} \text{ g ACTH}$

$$1.4 \times 10^{-2} \text{ g} \times \frac{10^3 \text{ mg}}{1 \text{ g}} = 14 \text{ mg ACTH}$$

3.12 (a) The molar mass of aspirin, $C_9H_8O_4$, is 180.16 g/mol. The molar mass of $Ca_3(PO_4)_2$ is 310.18 g/mol.

$$\frac{12.0 \text{ g}}{180.16 \text{ g/mol}} = 6.66 \times 10^{-2} \text{ mol aspirin}$$

$$\frac{12.0 \text{ g}}{310.18 \text{ g/mol}} = 3.87 \times 10^{-2} \text{ mol } Ca_3(PO_4)_2$$

(b) The molar mass of $Mn_2(SO_4)_3$ is 398.06 g/mol. The molar mass of caffeine, $C_8H_{10}N_4O_2$, is 194.19 g/mol.

$$0.350 \text{ mol } Ca_3(PO_4)_2 \times 398.06 \text{ g/mol} = 139 \text{ g } Ca_3(PO_4)_2$$

$$0.350 \text{ mol caffeine} \times 194.19 \text{ g/mol} = 68.0 \text{ g caffeine}$$

3.13 The mass of Si in 1 mol SiO_2 is 28.0855 g. The mass of O in 1 mol SiO_2 is 31.9988 g.

$$\% \text{ Si in } SiO_2 = \frac{28.0855 \text{ g}}{60.08 \text{ g}} \times 100\% = 46.7\% \text{ Si}$$

$$\% \text{ O in } SiO_2 = \frac{31.9988 \text{ g}}{60.08 \text{ g}} \times 100\% = 53.3\% \text{ O}$$

3.14 The molar mass of hydrated nickel chloride is

$$58.69 \text{ g/mol} + 2(35.45 \text{ g/mol}) + 12(1.008 \text{ g/mol}) + 6(16.00 \text{ g/mol}) = 237.69 \text{ g/mol.}$$

The percentages by weight for each element are found from the ratios of the mass of each element in 1 mole of hydrated nickel chloride to the molar mass of hydrated nickel chloride:

$$\frac{58.69 \text{ g Ni}}{237.69 \text{ g hydrated nickel chloride}} \times 100\% = 24.7\% \text{ Ni}$$

$$\frac{70.90 \text{ g Cl}}{237.69 \text{ g hydrated nickel chloride}} \times 100\% = 29.8\% \text{ Cl}$$

$$\frac{12.096 \text{ g H}}{237.69 \text{ g hydrated nickel chloride}} \times 100\% = 5.09\% \text{ H}$$

$$\frac{96.00 \text{ g O}}{237.69 \text{ g hydrated nickel chloride}} \times 100\% = 40.4\% \text{ O}$$

3.15 $\dfrac{56.34 \text{ g P}}{30.974 \text{ g/mol P}} = 1.82 \text{ mol P}$ $\quad \dfrac{43.66 \text{ g O}}{15.9994 \text{ g/mol}} = 2.73 \text{ mol O}$

$$\frac{2.73 \text{ mol O}}{1.82 \text{ mol P}} = \frac{1.50 \text{ mol O}}{1 \text{ mol P}}$$

Therefore, the mole ratio is 3 O atoms for 2 P atoms. The empirical formula is P_2O_3. The formula weight of this species is 2 mol P × (30.97 g/mol) + 3 mol O × (15.9994 g/mol) = 109.9 g/mol. The given molar mass for the compound is 219.90 g/mol, which is twice as large, so the molecular formula is P_4O_6.

3.16 Find the number of moles of each element in 100.0 g of vitamin C.

$$40.9 \text{ g C} \times \frac{1 \text{ mol C}}{12.011 \text{ g C}} = 3.405 \text{ mol C}$$

$$4.58 \text{ g H} \times \frac{1 \text{ mol H}}{1.0079 \text{ g H}} = 4.544 \text{ mol H}$$

$$54.5 \text{ g O} \times \frac{1 \text{ mol O}}{15.9994 \text{ g O}} = 3.406 \text{ mol O}$$

Find the mole ratios.

$$\frac{4.544 \text{ mol H}}{3.406 \text{ mol O}} = \frac{1.334 \text{ mol H}}{1.000 \text{ mol O}}$$

The same ratio holds for H to C. Using whole numbers, we have $C_3H_4O_3$ for the empirical formula. The empirical formula weight is (3)(12.011) + (4)(1.0079) + (3)(15.9994) = 88.06 g. The molar mass, however, is 176.13 g/mol, so the molecular formula must be twice the empirical formula: $C_6H_8O_6$.

Chapter 4

4.1 (a) N_2; combination (b) O_2; combination
(c) N_2; decomposition

4.2 (a) Decomposition reaction:

$$2\,Al(OH)_3(s) \longrightarrow Al_2O_3(s) + 3\,H_2O(g)$$

(b) Combination reaction:

$$Na_2O(s) + H_2O(\ell) \longrightarrow 2\,NaOH(aq)$$

(c) Combination reaction: $S_8(s) + 24\,F_2(g) \rightarrow 8\,SF_6(g)$

(d) Exchange reaction:

$$3\,NaOH(aq) + H_3PO_4(aq) \longrightarrow Na_3PO_4(aq) + 3\,H_2O(\ell)$$

(e) Displacement:

$$3\,C(s) + Fe_2O_3(s) \longrightarrow 3\,CO(g) + 2\,Fe(\ell)$$

4.3 (a) $2\,Cr(s) + 3\,Cl_2 \rightarrow 2\,CrCl_3(s)$
(b) $As_2O_3(s) + 3\,H_2(g) \rightarrow 2\,As(s) + 3\,H_2O(\ell)$

4.4 (a) $C_2H_5OH + 3\,O_2 \rightarrow 2\,CO_2 + 3\,H_2O$
(b) $C_2H_5OH + 2\,O_2 \rightarrow 2\,CO + 3\,H_2O$

4.5 0.433 mol hematite needs 0.433 × 3 = 1.30 mol CO. Molar mass of CO is 28.01, so 1.30 × 28.01 = 36.4 g CO.

4.6 (a) $0.300 \text{ mol cassiterite} \times \frac{1 \text{ mol Sn}}{1 \text{ mol cassiterite}} \times \frac{118.7 \text{ g Sn}}{1 \text{ mol Sn}} = 35.6 \text{ g Sn}$

(b) $35.6 \text{ g Sn} \times \frac{1 \text{ mol Sn}}{118.7 \text{ g Sn}} \times \frac{2 \text{ mol C}}{1 \text{ mol Sn}} \times \frac{12.01 \text{ g C}}{1 \text{ mol C}} = 7.20 \text{ g C}$

4.7 (a) $57. \text{ g C} \times \frac{1 \text{ mol C}}{12.01 \text{ g C}} \times \frac{1 \text{ mol } O_2}{2 \text{ mol C}} \times \frac{32.0 \text{ g } O_2}{1 \text{ mol } O_2} = 76. \text{ g } O_2$

(b) $57. \text{ g C} \times \frac{1 \text{ mol C}}{12.01 \text{ g C}} \times \frac{2 \text{ mol CO}}{2 \text{ mol C}} \times \frac{28.0 \text{ g CO}}{1 \text{ mol CO}} = 1.3 \times 10^2 \text{ g CO}$

4.8 $6.46 \text{ g } MgCl_2 \times \frac{1 \text{ mol } MgCl_2}{95.2104 \text{ g } MgCl_2} = 6.78 \times 10^{-2} \text{ mol } MgCl_2$

The same number of moles of Mg as $MgCl_2$ are involved, so

$$6.78 \times 10^{-2} \text{ mol Mg} \times \frac{24.3050 \text{ g Mg}}{1 \text{ mol Mg}} = 1.65 \text{ g Mg}$$

$$\frac{1.65 \text{ g Mg}}{1.72\text{-g sample}} \times 100\% = 95.9\% \text{ Mg in sample}$$

4.9 $0.75 \text{ mol } CO_2 \times \frac{1 \text{ mol } (NH_2)_2CO}{1 \text{ mol } CO_2} = 0.75 \text{ mol } (NH_2)_2CO$

4.10 (a) $CS_2(\ell) + 3\,O_2(g) \rightarrow CO_2(g) + 2\,SO_2(g)$
(b) Determine the quantity of CO_2 produced by each reactant; the limiting reactant produces the lesser quantity.

$$3.5 \text{ g } CS_2 \times \frac{1 \text{ mol } CS_2}{76.0 \text{ g } CS_2} \times \frac{1 \text{ mol } CO_2}{1 \text{ mol } CS_2} \times \frac{44.01 \text{ g } CO_2}{1 \text{ mol } CO_2} = 2.0 \text{ g } CO_2$$

$$17.5 \text{ g } O_2 \times \frac{1 \text{ mol } O_2}{31.998 \text{ g } O_2} \times \frac{1 \text{ mol } CO_2}{3 \text{ mol } O_2} \times \frac{44.01 \text{ g } CO_2}{1 \text{ mol } CO_2} = 8.02 \text{ g } CO_2$$

Therefore, CS_2 is the limiting reagent.

(c) The yield of SO_2 must be calculated using the limiting reagent, CS_2.

$$3.5 \text{ g } CS_2 \times \frac{1 \text{ mol } CS_2}{76.0 \text{ g } CS_2} \times \frac{2 \text{ mol } SO_2}{1 \text{ mol } CS_2} \times \frac{64.1 \text{ g } SO_2}{1 \text{ mol } SO_2} = 5.9 \text{ g } SO_2$$

4.11 Find the number of moles of each reactant.

$$100. \text{ g } SiCl_4 \times \frac{1 \text{ mol } SiCl_4}{169.90 \text{ g } SiCl_4} = 0.589 \text{ mol } SiCl_4$$

$$100. \text{ g Mg} \times \frac{1 \text{ mol Mg}}{24.3050 \text{ g Mg}} = 4.11 \text{ mol Mg}$$

Find the mass of Si produced, based on the mass available of each reactant.

$$0.589 \text{ mol } SiCl_4 \times \frac{1 \text{ mol Si}}{1 \text{ mol } SiCl_4} \times \frac{28.0855 \text{ g Si}}{1 \text{ mol Si}} = 16.5 \text{ g Si}$$

$$4.11 \text{ mol Mg} \times \frac{1 \text{ mol Si}}{2 \text{ mol Mg}} \times \frac{28.0855 \text{ g Si}}{1 \text{ mol Si}} = 57.7 \text{ g Si}$$

Thus, $SiCl_4$ is the limiting reactant, and the mass of Si produced is 16.5 g.

4.12 To make 1.0 kg CH_3OH with 85% yield will require using enough reactant to produce 1000/0.85, or 1180 g CH_3OH.

$$1180 \text{ g } CH_3OH \times \frac{1 \text{ mol}}{32.042 \text{ g}} = 36.83 \text{ mol } CH_3OH$$

$$36.83 \text{ mol } CH_3OH \times \frac{2 \text{ mol } H_2}{1 \text{ mol } CH_3OH} = 73.65 \text{ mol } H_2$$

$$73.65 \text{ mol } H_2 \times \frac{2.0158 \text{ g } H_2}{1 \text{ mol } H_2} = 149 \text{ g } H_2$$

4.13 Calculate the mass of Cu_2S you should have produced and compare it with the amount actually produced.

$$2.50 \text{ g Cu} \times \frac{1 \text{ mol Cu}}{63.546 \text{ g Cu}} = 3.93 \times 10^{-2} \text{ mol Cu}$$

$$3.93 \times 10^{-2} \text{ mol Cu} \times \frac{8 \text{ mol } Cu_2S}{16 \text{ mol Cu}} = 1.97 \times 10^{-2} \text{ mol } Cu_2S$$

$$1.97 \times 10^{-2} \text{ mol } Cu_2S \times \frac{159.16 \text{ g } Cu_2S}{1 \text{ mol } Cu_2S} = 3.14 \text{ g } Cu_2S$$

$$\frac{2.53 \text{ g}}{3.14 \text{ g}} \times 100\% = 80.6\% \text{ yield was obtained}$$

Your synthesis met the standard.

4.14 (a) $491 \text{ mg } CO_2 \times \frac{1 \text{ g } CO_2}{10^3 \text{ mg } CO_2} \times \frac{1 \text{ mol } CO_2}{44.01 \text{ g } CO_2} \times \frac{1 \text{ mol C}}{1 \text{ mol } CO_2} = 1.116 \times 10^{-2} \text{ mol C}$

$$1.116 \times 10^{-2} \text{ mol C} \times \frac{12.01 \text{ g C}}{1 \text{ mol C}} = 0.1340 \text{ g C} = 134.0 \text{ mg C}$$

$$100 \text{ mg } H_2O \times \frac{1 \text{ g } H_2O}{10^3 \text{ mg } H_2O} \times \frac{1 \text{ mol } H_2O}{18.02 \text{ g } H_2O} \times \frac{2 \text{ mol H}}{1 \text{ mol } H_2O} = 1.110 \times 10^{-2} \text{ mol H}$$

$$1.110 \times 10^{-2} \text{ mol H} \times \frac{1.008 \text{ g H}}{1 \text{ mol H}} = 1.119 \times 10^{-2} \text{ g H} = 11.2 \text{ mg H}$$

The mass of oxygen in the compound

= total mass − (mass C + mass H)

= 175 mg − (134.0 mg C + 11.2 mg H) = 29.8 mg O

The moles of oxygen are

$$29.8 \text{ mg O} \times \frac{1 \text{ g O}}{10^3 \text{ mg O}} \times \frac{1 \text{ mol O}}{16.00 \text{ g O}} = 1.862 \times 10^{-3} \text{ mol O}$$

The empirical formula can be derived from the mole ratios of the elements.

$$\frac{1.116 \times 10^{-2} \text{ mol C}}{1.862 \times 10^{-3} \text{ mol O}} = 5.993 \text{ mol C/mol O}$$

$$\frac{1.110 \times 10^{-2} \text{ mol H}}{1.862 \times 10^{-3} \text{ mol O}} = 5.961 \text{ mol H/mol O}$$

$$\frac{1.862 \times 10^{-3} \text{ mol O}}{1.862 \times 10^{-3} \text{ mol O}} = 1.000 \text{ mol O}$$

The empirical formula of phenol is C_6H_6O.
(b) The molar mass is needed to determine the molecular formula.

4.15 $0.569 \text{ g Sn} \times \frac{1 \text{ mol Sn}}{118.7 \text{ g Sn}} = 4.794 \times 10^{-3} \text{ mol Sn}$

$$2.434 \text{ g } I_2 \times \frac{1 \text{ mol } I_2}{253.81 \text{ g } I_2} \times \frac{2 \text{ mol I}}{1 \text{ mol } I_2} = 1.918 \times 10^{-2} \text{ mol I}$$

$$\frac{1.918 \times 10^{-2} \text{ mol I}}{4.794 \times 10^{-3} \text{ mol Sn}} = \frac{4.001 \text{ mol I}}{1.000 \text{ mol Sn}}$$

Therefore, the empirical formula is SnI_4.

Chapter 5

5.1 (a) NaF is soluble.
(b) $Ca(CH_3COO)_2$ is soluble.
(c) $SrCl_2$ is soluble.
(d) MgO is not soluble.
(e) $PbCl_2$ is not soluble.
(f) HgS is not soluble.

5.2 (a) This exchange reaction forms insoluble nickel hydroxide and aqueous sodium chloride.

$$NiCl_2(aq) + 2\ NaOH(aq) \longrightarrow Ni(OH)_2(s) + 2\ NaCl(aq)$$

(b) This is an exchange reaction that forms aqueous potassium bromide and a precipitate of calcium carbonate.

$$K_2CO_3(aq) + CaBr_2(aq) \longrightarrow CaCO_3(s) + 2\ KBr(aq)$$

5.3 (a) $BaCl_2(aq) + Na_2SO_4(aq) \rightarrow BaSO_4(s) + 2\ NaCl(aq)$

$$Ba^{2+}(aq) + SO_4^{2-}(aq) \longrightarrow BaSO_4(s)$$

(b) $(NH_4)_2S(aq) + FeCl_2(aq) \rightarrow FeS(s) + 2\ NH_4Cl(aq)$

$$Fe^{2+}(aq) + S^{2-}(aq) \rightarrow FeS(s)$$

5.4 Any of the strong acids in Table 5.2 would also be strong electrolytes. Any of the weak acids or bases in Table 5.2 would be weak electrolytes. Any organic compound that yields no ions on dissolution would be a nonelectrolyte.

5.5 $H_3PO_4(aq) + 3\ NaOH(aq) \rightarrow Na_3PO_4(aq) + 3\ H_2O(\ell)$

5.6 (a) Sulfuric acid and magnesium hydroxide
(b) Carbonic acid and strontium hydroxide

5.7 $2\ HI(aq) + Ca(OH)_2(aq) \longrightarrow CaI_2(aq) + 2\ H_2O(\ell)$

$$2\ HI(aq) + Ca^{2+}(aq) + 2\ OH^-(aq) \longrightarrow Ca^{2+}(aq) + 2\ I^-(aq) + 2\ H_2O(\ell)$$

$$HI(aq) + OH^-(aq) \longrightarrow I^-(aq) + H_2O(\ell)$$

5.8 The oxidation numbers of Fe and Sb are 0 (Rule 1). The oxidation numbers in Sb_2S_3 are +3 for Sb^{3+} and −2 for S^{2-} (Rules 2 and 4). The oxidation numbers in FeS are +2 for Fe^{2+} and −2 for S^{2-} (Rules 2 and 4).

5.9 In the reaction $PbO(s) + CO(g) \rightarrow Pb(s) + CO_2(g)$, Pb^{2+} is reduced to Pb; Pb^{2+} is the oxidizing agent. C^{2+} is oxidized to C^{4+}; C^{2+} is the reducing agent.

5.10 Reactions (a) and (b) will occur. Aluminum is above copper and chromium in Table 5.5; therefore, aluminum will be oxidized and acts as the reducing agent in reactions (a) and (b). In reaction (a), Cu^{2+} is reduced, and Cu^{2+} is the oxidizing agent. Cr^{3+} is the oxidizing agent in reaction (b) and is reduced to Cr metal. Reactions (c) and (d) do not occur because Pt cannot reduce H^+, and Au cannot reduce Ag^+.

5.11 $36.0 \text{ g } Na_2SO_4 \times \frac{1 \text{ mol } Na_2SO_4}{142.0 \text{ g } Na_2SO_4} = 0.254 \text{ mol } Na_2SO_4$

$$\text{Molarity} = \frac{0.254 \text{ mol}}{0.750 \text{ L}} = 0.339 \text{ molar}$$

5.12 $V(\text{conc}) = \frac{0.150 \text{ molar} \times 0.050 \text{ L}}{0.500 \text{ molar}} = 0.015 \text{ L} = 15 \text{ mL}$

5.13 (a) 1.00 L of 0.125-M Na_2CO_3 contains 0.125 mol Na_2CO_3.

$$0.125 \text{ mol} \times \frac{105.99 \text{ g}}{1 \text{ mol}} = 13.2 \text{ g } Na_2CO_3$$

Prepare the solution by adding 13.2 g Na_2CO_3 to a volumetric flask, dissolving it and mixing thoroughly, and adding sufficient water until the solution volume is 1.0 L.
(b) Use water to dilute a specific volume of the 0.125-M solution to 100 mL.

$$V(\text{conc}) = \frac{0.0500 \text{ M} \times 0.100 \text{ L}}{0.125 \text{ M}} = 0.040 \text{ L} = 40.\text{ mL of 0.125-M solution}$$

Therefore, put 40. mL of the more concentrated solution into a container and add water until the solution volume equals 100 mL.
(c) 500 mL of 0.215-M $KMnO_4$ contains 1.70 g $KMnO_4$.

$$0.500 \text{ L} \times \frac{0.0215 \text{ mol } KMnO_4}{1 \text{ L}} = 0.01075 \text{ mol } KMnO_4$$

$$0.01075 \text{ mol } KMnO_4 \times \frac{158.0 \text{ g } KMnO_4}{1 \text{ mol } KMnO_4} = 1.70 \text{ g } KMnO_4$$

Put 1.70 g $KMnO_4$ into a container and add water until the solution volume is 500 mL.
(d) Dilute the more concentrated solution by adding sufficient water to 52.3 mL of 0.0215-M $KMnO_4$ until the solution volume is 250 mL.

$$V(\text{conc}) = \frac{0.00450 \text{ M} \times 0.250 \text{ L}}{0.0215 \text{ M}} = 0.0523 \text{ L} = 52.3 \text{ mL}$$

5.14 $1.2 \times 10^{10} \text{ kg NaOH} \times \frac{1 \text{ mol NaOH}}{0.040 \text{ kg NaOH}} \times \frac{2 \text{ mol NaCl}}{2 \text{ mol NaOH}} \times \frac{58.5 \text{ g NaCl}}{1 \text{ mol NaCl}} \times \frac{1 \text{ L brine}}{360 \text{ g NaCl}} = 4.9 \times 10^{10} \text{ L}$

5.15 We have 50.0 mg or 0.050 g AgBr. The molar mass of AgBr is 187.77 g/mol.

$$0.050 \text{ g AgBr} \times \frac{1 \text{ mol}}{187.77 \text{ g}} = 2.66 \times 10^{-4} \text{ mol AgBr}$$

According to the balanced equation, we need twice as much thiosulfate as AgBr, so we need $2 \times 2.66 \times 10^{-4} = 5.32 \times 10^{-4}$ mol thiosulfate.

$$\frac{5.32 \times 10^{-4} \text{ mol}}{150 \times 10^{-2} \text{ mol/L}} = 0.3549 \text{ L or 35.5 mL of solution}$$

5.16 $H_2SO_4(aq) + 2\ NaOH(aq) \rightarrow Na_2SO_4(aq) + 2\ H_2O(\ell)$

$$\text{Moles NaOH} = (0.0413 \text{ L})(0.100 \text{ M}) = 0.00413 \text{ mol NaOH}$$

$$\text{Moles } H_2SO_4 = \tfrac{1}{2} \times 0.00413 = 0.002065 \text{ mol } H_2SO_4$$

$$\text{Molarity} = \frac{0.002065 \text{ mol}}{0.020 \text{ L}} = 0.103 \text{ M } H_2SO_4$$

Chapter 6

6.1 (a) $160\text{ Cal} \times \frac{1000\text{ cal}}{\text{Cal}} \times \frac{4.184\text{ J}}{\text{cal}} = 6.7 \times 10^5\text{ J}$

(b) $75\text{ W} = 75\text{ J/s}$;
$75\text{ J/s} \times 3.0\text{ h} \times 60\text{ min/h} \times 60\text{ s/min} = 8.1 \times 10^5\text{ J}$

(c) $16\text{ kJ} \times \frac{1\text{ kcal}}{4.184\text{ kJ}} = 3.8\text{ kcal}$

6.2 $\Delta E = -2400\text{ J} = q + w = -1.89\text{ kJ} + w$
$w = -2400\text{ J} + 1.89\text{ kJ} = -2.4\text{ kJ} + 1.89\text{ kJ} = -0.5\text{ kJ}$

6.3 $q = c \times m \times \Delta T = c \times m \times (T_{final} - T_{initial})$

$$T_{final} = T_{initial} + \frac{q}{c \times m} = 5\text{ °C} + \frac{24{,}100\text{ J}}{(0.902\text{ J g}^{-1}\text{ °C}^{-1})(250.\text{ g})}$$
$$= 5\text{ °C} + 106.8\text{ °C} = 112\text{ °C}$$

6.4 Since no work is done, $\Delta E = q$. Assume that the tea has a mass of 250. g and that its specific heat capacity is the same as that of water, namely $4.184\text{ J g}^{-1}\text{ °C}^{-1}$.

$\Delta E = q = (\text{mass}) \times (\text{specific heat capacity}) \times (\text{change in temperature})$

$\Delta E = q = (250.\text{ g}) \times (4.184\text{ J g}^{-1}\text{ °C}^{-1}) \times [(65 - 37)\text{ °C}] = 2.9 \times 10^4\text{ J} = 2.9 \times 10^1\text{ kJ}$

6.5 $q_{water} = -q_{iron}$
$(4.184\text{ J/g °C})(1000.\text{ g})(32.8\text{ °C} - 20.0\text{ °C}) = -(0.451\text{ J/g °C})(400.\text{ g})(32.8\text{ °C} - T_i)$
$T_i = (297 + 32.8)\text{ °C} = 330.\text{ °C}$

6.6 $1.00\text{ g K(s)} \times \frac{1\text{ mL}}{0.86\text{ g}} = 1.16\text{ mL}$;

$1.00\text{ g K}(\ell) \times \frac{1\text{ mL}}{0.82\text{ g}} = 1.22\text{ mL}$

The change in volume is $(1.22 - 1.16)\text{ mL} = 0.06\text{ mL}$.

$w = 0.06\text{ mL} \times \frac{0.10\text{ J}}{1\text{ mL}} = 6 \times 10^{-3}\text{ J}$

$\Delta H = \frac{14.6\text{ cal}}{1\text{ g}} \times 1.00\text{ g} \times \frac{4.184\text{ J}}{1\text{ cal}} = 61.1\text{ J}$

$\Delta E = \Delta H + w = 61.1\text{ J} + (6 \times 10^{-3}\text{ J}) = 61.1\text{ J}$

6.7 (a) $10.0\text{ g } I_2 \times \frac{1\text{ mol } I_2}{253.8\text{ g } I_2} \times \frac{62.4\text{ kJ}}{1\text{ mol } I_2} = 2.46\text{ kJ}$

(b) $3.42\text{ g } I_2 \times \frac{1\text{ mol } I_2}{253.8\text{ g } I_2} \times \frac{-62.4\text{ kJ}}{1\text{ mol } I_2} = -0.841\text{ kJ} = -841\text{ J}$

(c) This process is the reverse of the one in part (a), so $\Delta H°$ is negative. Thus the process is exothermic. The quantity of energy transferred is 841 J.

6.8 The equation as written involves CO_2 as a reactant, but the question asks for CO_2 as a product. Therefore we will have to change the sign of $\Delta H°$. In addition, since the question asks about production of 4 mol CO_2, we must multiply $\Delta H°$ for the endothermic decomposition by 4. So for the production of 4 mol CO_2,

$$\Delta H° = (4) \times (662.8\text{ kJ}) = 2651\text{ kJ}$$

6.9 According to the thermochemical expression, the reaction is endothermic, so 285.8 kJ of energy is transferred into the system per mole of $H_2O(\ell)$ decomposed. Thus,

$$12.6\text{ g } H_2O \times \frac{1\text{ mol } H_2O}{18.02\text{ g } H_2O} \times \frac{285.8\text{ kJ}}{1\text{ mol } H_2O} = 200.\text{ kJ}$$

6.10 $\Delta T = (25.43 - 20.64)\text{ °C} = 4.79\text{ °C}$

$\Delta E_{calorimeter} = \frac{877\text{ J}}{\text{°C}} \times 4.79\text{ °C} = 4.200 \times 10^3\text{ J} = 4.200\text{ kJ}$

$\Delta E_{water} = 832\text{ g} \times \frac{4.184\text{ J}}{\text{g °C}} \times 4.79\text{ °C} = 16.67\text{ kJ}$

$\Delta E_{reaction} = -(q_{calorimeter} + q_{water}) = -(4.200 + 16.67)\text{ kJ} = -20.87\text{ kJ}$

$20.87\text{ kJ} \times \frac{1\text{ kcal}}{4.184\text{ kJ}} \times \frac{1\text{ Cal}}{1\text{ kcal}} = 4.99\text{ Cal}$

Since metabolizing the Fritos chip corresponds to oxidizing it, the result of 4.99 Cal verifies the statement that one chip provides 5 Cal.

6.11 The total volume of the initial solutions is 200. mL, which corresponds to 200. g of solution. The quantities of reactants are 0.10 mol $H^+(aq)$ and 0.050 mol $OH^-(aq)$, so 0.050 mol H_2O is formed.

$$0.050\text{ mol } H_2O \times \frac{-58.6\text{ kJ}}{1\text{ mol } H_2O} = -2.93\text{ kJ}$$

Since $\Delta H°$ is negative, energy is transferred to the water, and its temperature will rise.

$$\Delta T = \frac{q}{c \times m} = \frac{2.93 \times 10^3\text{ J}}{(4.184\text{ J g}^{-1}\text{ °C}^{-1})(200.\text{ g})} = 3.5\text{ °C}$$

The final temperature will be $(20.4 + 3.5)\text{ °C} = 23.9\text{ °C}$.

6.12 The total mass of solution is $270.\text{ mL} \times 1.00\text{ g/mL} + 10.4\text{ g} = 280.4\text{ g}$.

$$q_p = -c \times m \times \Delta T = -(4.184\text{ J g}^{-1}\text{ °C}^{-1})(280.4\text{ g})(8.6\text{ °C}) = -1.009 \times 10^4\text{ J}$$

This quantity of heat transfer does not correspond with the chemical equation as written for 1 mol H_2SO_4. The 1.009×10^4 J is transferred when 10.4 g H_2SO_4 dissolves. This is

$$10.4\text{ g } H_2SO_4 \times \frac{1\text{ mol } H_2SO_4}{98.08\text{ g } H_2SO_4} = 1.060 \times 10^{-1}\text{ mol } H_2SO_4$$

For the balanced chemical equation, in which 1 mol H_2SO_4 dissolves,

$$\frac{\Delta H}{1\text{ mol } H_2SO_4} = \frac{-1.009 \times 10^4\text{ J}}{1.060 \times 10^{-1}\text{ mol } H_2SO_4}$$

$$\Delta H = 1\text{ mol } H_2SO_4 \times \frac{-1.009 \times 10^4\text{ J}}{1.060 \times 10^{-1}\text{ mol } H_2SO_4} = -9.51 \times 10^4\text{ J} = -95.1\text{ kJ}$$

6.13 The balanced chemical equation is

$$2\ Fe_3O_4(s) \rightarrow 6\ FeO(s) + O_2(g)$$

The equations given in the problem can be arranged so that when added they give this balanced equation:

$2 \times [Fe_3O_4(s) \rightarrow 3\ Fe(s) + 2\ O_2(g)]$ $\quad \Delta H° = 2(1118.4\text{ kJ})$

$6 \times [Fe(s) + \frac{1}{2} O_2(g) \rightarrow FeO(s)]$ $\quad \Delta H° = 6(-272.0\text{ kJ})$

So

$$\Delta H° = 2236.8\text{ kJ} - 1632.0\text{ kJ} = 604.8\text{ kJ}$$

6.14 (a) $\frac{1}{2} N_2(g) + \frac{3}{2} H_2(g) \rightarrow NH_3(g)$ $\quad \Delta H_f° = -46.11\text{ kJ/mol}$

(b) $C(\text{graphite}) + \frac{1}{2} O_2(g) \rightarrow CO(g)$ $\quad \Delta H_f° = -110.525\text{ kJ/mol}$

6.15 For the reaction given,

$$\Delta H° = \{6\text{ mol } CO_2(g)\} \times \Delta H_f°\{CO_2(g)\} + \{5\text{ mol } H_2O(g)\} \times \Delta H_f°\{H_2O(g)\} - (2\text{ mol }\{C_3H_5(NO_3)_3(\ell)\} \times \Delta H_f°\{C_3H_5(NO_3)_3(\ell)\}$$
$$= \{6(-393.509) + 5(-241.818) - 2(-364)\}\text{ kJ}$$
$$= -2.84 \times 10^3\text{ kJ}$$

For 10.0 g nitroglycerin (nitro),

$$q = 10.0\text{ g} \times \frac{1\text{ mol nitro}}{227.09\text{ g}} \times \frac{-2.84 \times 10^3\text{ kJ}}{2\text{ mol nitro}} = -62.5\text{ kJ}$$

(The 2 mol nitro in the last factor comes from the coefficient of 2 associated with nitroglycerin in the chemical equation.)

6.16 $SO_2(g) + \frac{1}{2}O_2(g) \rightarrow SO_3(g) \quad \Delta H° = ?$

$$\Delta H° = \Delta H_f°\{SO_3(g)\} - [\Delta H_f°\{SO_2(g)\} + \tfrac{1}{2}\Delta H_f°\{O_2(g)\}]$$
$$= -395.72\text{ kJ} - (-296.830 + 0)\text{ kJ} = -98.89\text{ kJ}$$

6.17 (a) $C_8H_{18}(\ell) + \frac{25}{2}O_2(g) \rightarrow 8\,CO_2(g) + 9\,H_2O(g)$

$$\Delta H° = [8(-393.509) + 9(-241.818) - (-249.952)]\text{ kJ}$$
$$= -5074.48\text{ kJ}$$

$$\frac{5074.48\text{ kJ}}{1\text{ mol }C_8H_{18}} \times \frac{1\text{ mol}}{114.23\text{ g}} = 44.423\text{ kJ/g }C_8H_{18}$$

$$44.423\text{ kJ/g }C_8H_{18} \times 0.699\text{ g/mL} \times \frac{10^3\text{ mL}}{\text{L}} = 3.11 \times 10^4\text{ kJ/L }C_8H_{18}$$

(b) $N_2H_4(\ell) + O_2(g) \rightarrow N_2(g) + 2\,H_2O(g)$

$$\Delta H° = [2(-241.818) - 50.63]\text{ kJ} = -534.26\text{ kJ}$$

$$\frac{534.26\text{ kJ}}{1\text{ mol }N_2H_4} \times \frac{1\text{ mol}}{32.045\text{ g}} = 16.672\text{ kJ/g }N_2H_4$$

$$16.672\text{ kJ/g }N_2H_4 \times 1004\text{ g/L} = 1.67 \times 10^4\text{ kJ/L }N_2H_4$$

(c) $C_6H_{12}O_6(g) + 6\,O_2(g) \rightarrow 6\,CO_2(g) + 6\,H_2O(g)$

$$\Delta H° = [6(-393.509) + 6(-241.818) - (-1274.4)]\text{ kJ}$$
$$= -2537.56\text{ kJ}$$

$$\frac{2537.56\text{ kJ}}{\text{mol }C_6H_{12}O_6} \times \frac{1\text{ mol}}{180.158\text{ g}} = 14.085\text{ kJ/g}$$

$$14.085\text{ kJ/g} \times 1560\text{ g/L} = 2.20 \times 10^4\text{ kJ/L}$$

6.18 $\text{Energy} = 4.2 \times 10^9\text{ t coal} \times \frac{26.4 \times 10^9\text{ J}}{1\text{ t coal}} = 1.1 \times 10^{20}\text{ J}$

$$\text{volume natural gas} = 1.1 \times 10^{20}\text{ J} \times \frac{1\text{ ft}^3\text{ natural gas}}{1.055 \times 10^6\text{ J}}$$
$$= 1.0 \times 10^{14}\text{ ft}^3\text{ natural gas}$$

6.19 (a) $227\text{ g milk} \times \frac{5.0\text{ g carbohydrate}}{100\text{ g milk}} \times \frac{17\text{ kJ}}{1\text{ g carbohydrate}} = 193\text{ kJ}$

$$227\text{ g milk} \times \frac{4.0\text{ g fat}}{100\text{ g milk}} \times \frac{38\text{ kJ}}{1\text{ g fat}} = 345\text{ kJ}$$

$$227\text{ g milk} \times \frac{3.3\text{ g protein}}{100\text{ g milk}} \times \frac{17\text{ kJ}}{1\text{ g protein}} = 127\text{ kJ}$$

Total caloric intake = (193 + 345 + 127) kJ = 665 kJ

Reduce by 10% to digest, absorb, and metabolize, giving 665 kJ − 66.5 kJ = 598 kJ.

(b) Walking requires $2.5 \times \text{BMR} = 2.5 \times \frac{1750\text{ Cal}}{1\text{ day}} \times \frac{4.184\text{ kJ}}{1\text{ Cal}} \times \frac{1\text{ day}}{24\text{ h}} \times \frac{1\text{ h}}{60\text{ min}} = 12.7\text{ kJ/min}$

$$598\text{ kJ} \times \frac{1\text{ min}}{12.7\text{ kJ}} = 47\text{ min}$$

Chapter 7

7.1 (a) $\lambda = \frac{c}{\nu} = \frac{2.998 \times 10^{-8}\text{ m/s}}{9.99 \times 10^{14}\text{ s}^{-1}} = 3.00 \times 10^{-7}\text{ m}$

(b) $3.00 \times 10^{-7}\text{ m} \times \frac{1\text{ nm}}{10^{-9}\text{ m}} = 300.\text{ nm}$

7.2 (a) One photon of ultraviolet radiation has more energy because ν is larger in the UV spectral region than in the microwave region. (b) One photon of blue light has more energy because the blue portion of the visible spectrum has a higher frequency than the green portion of the visible spectrum.

7.3 Any from $n_{hi} > 8$ to $n_{lo} = 2$

7.4 (a) $\nu = \frac{2.179 \times 10^{-18}\text{ J}}{h}\left(\frac{1}{n_i^2} - \frac{1}{n_f^2}\right)$

$$= \left(\frac{2.179 \times 10^{-18}\text{ J}}{6.626 \times 10^{-34}\text{ J}\cdot\text{s}}\right)\left(\frac{1}{6^2} - \frac{1}{4^2}\right)$$
$$= (3.289 \times 10^{15}\text{ s}^{-1})\left(\frac{1}{36} - \frac{1}{16}\right)$$
$$= (3.289 \times 10^{15}\text{ s}^{-1})(-3.47 \times 10^{-2})$$
$$= -1.14 \times 10^{14}\text{ s}^{-1}$$

The negative sign indicates that energy is emitted.

$$\lambda = 2.63 \times 10^{-6}\text{ m}$$

(b) Longer than that of the $n = 7$ to $n = 4$ transition.

7.5 $\lambda = \frac{h}{mv} = \frac{6.626 \times 10^{-34}\text{ J}\cdot\text{s}}{1.67 \times 10^{-27}\text{ kg} \times 2.998 \times 10^7\text{ m/s}}$

$$= 1.32 \times 10^{-14}\text{ m}$$

7.6 (a) 6*d* (b) 5 (c) 2, 1, 0, −1, −2

7.7 (a) $[Ne]\,3s^2 3p^2$ (b) [Ne] 3*s* [↑↓] 3*p* [↑] [↑] [ ]

7.8 The electron configurations for :Se: and :Te: are $[Ar]\,4s^2 3d^{10} 4p^4$ and $[Kr]5s^2 4d^{10} 5p^4$, respectively. Elements in the same main group have similar electron configurations.

7.9 (a) P^{3-} (b) Ca^{2+}

7.10 The ground state Cu atom has a configuration $[Ar]\,4s^1 3d^{10}$. When it loses one electron, it becomes the Cu^+ ion with configuration $[Ar]\,3d^{10}$. There is an added stability for the completely filled set of 3*d* orbitals.

7.11 B < Mg < Na < K

7.12 (a) Cs^+ (b) La^{3+}

7.13 F > N > P > Na

Chapter 8

8.1 (a) 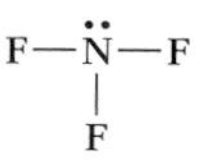 (b) 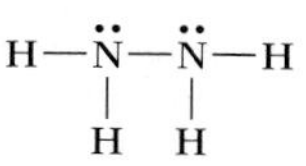

(c) 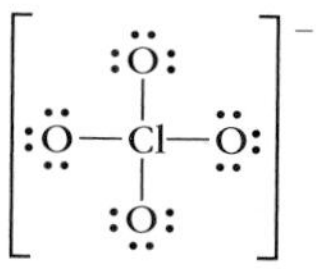

8.2 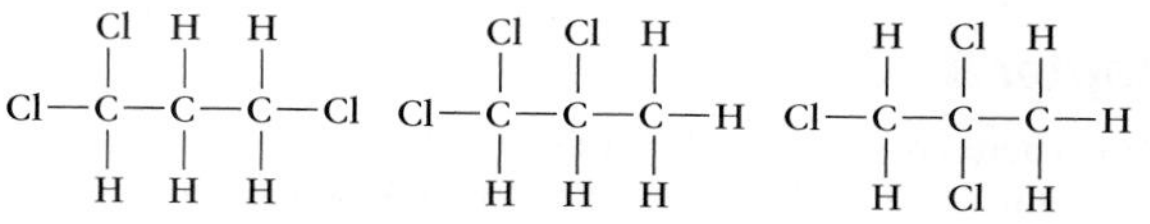

8.3 (a) $[:N\equiv O:]^+$ (b) 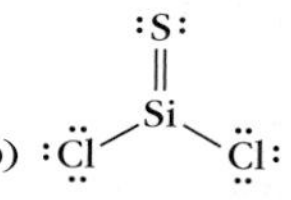

8.4 Only (c) can have geometric isomers. Molecules in (a) and (b) each have the same two groups on one of the double-bonded carbons.

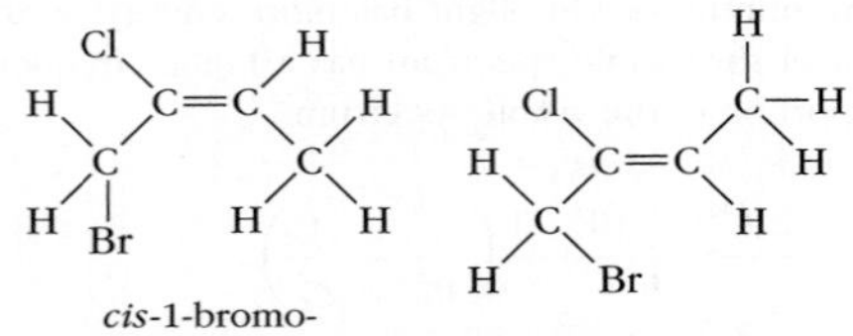

cis-1-bromo-2-chloro-2-butene   *trans*-1-bromo-2-chloro-2-butene

8.5 (a) Si is a larger atom than S. (b) Br is a larger atom than Cl. (c) The greater electron density in the triple bond brings the N≡O atoms closer together than the smaller electron density in the N=O double bond does.

8.6

$$\Delta H = [(4 \text{ mol C—H})(416 \text{ kJ/mol}) + (2 \text{ mol O=O})(498 \text{ kJ/mol})] - [(4 \text{ mol O—H})(467 \text{ kJ/mol}) + (2 \text{ mol C=O})(803 \text{ kJ/mol})] = (2660 \text{ kJ}) - (3474 \text{ kJ}) = -814 \text{ kJ}$$

8.7 (a) B—Cl is more polar; $\overset{\delta+}{\text{B}}—\overset{\delta-}{\text{Cl}}$; O—H is more polar; $\overset{\delta-}{\text{O}}—\overset{\delta+}{\text{H}}$.

8.8 The other Lewis structure is :O≡N—N̈:

	O	N	N
Valence electrons	6	5	5
Lone pair electrons	2	0	6
$\frac{1}{2}$ shared electrons	3	4	1
Formal charge	+1	+1	−2

8.9 The N-to-O bond length in NO_2^- is 124 pm. From Table 8.1, N—O is 136 pm; N=O is 115 pm. Thus, the nature of the bond in NO_2^- is between that of a N—O single bond and a N=O double bond.

8.10 (a) :F̤̈—Be—F̤̈:   (b) :Ö—C̈l—Ö:

(c) 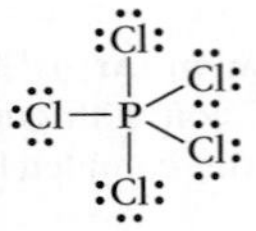   (d) [H—B—H]$^+$

(e) 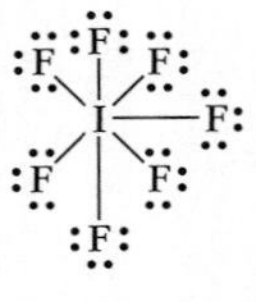

BeF_2—not an octet around the central Be atom; ClO_2—an odd number of electrons around Cl; PCl_5—more than four electron pairs around the central phosphorus atom; BH_2^+—only two electron pairs around the central B atom; IF_7—iodine has seven shared electron pairs.

8.11 Ten valence electrons. The bond order is (8 − 2)/2 = 3. No unpaired electrons. See Table 8.4.

8.12 Bond order = (8 − 5)/2 = 1.5. There is one unpaired electron.

Chapter 9

9.1 There are two Be—F bonds in BeF_2 and no lone pairs on beryllium. Therefore, the electron-pair and the molecular geometry are the same, linear, with 180° Be—F bond angles.

9.2

Central Atom (underlined)	Bond Pairs	Lone Pairs	Electron-Pair Geometry	Molecular Shape
$\underline{Br}O_3^-$	3	1	Tetrahedral	Triangular pyramid
$\underline{Se}F_2$	2	2	Tetrahedral	Angular
$\underline{N}O_2^-$	3	1	Triangular planar	Angular

9.3 (a) ClF_2^-: triangular bipyramidal electron-pair geometry and linear molecular geometry
(b) XeO_3: tetrahedral electron-pair geometry and triangular pyramidal molecular geometry

9.4 The chiral centers are identified by an asterisk in the structural formula. Each of those carbon atoms has four different groups or atoms attached to it.

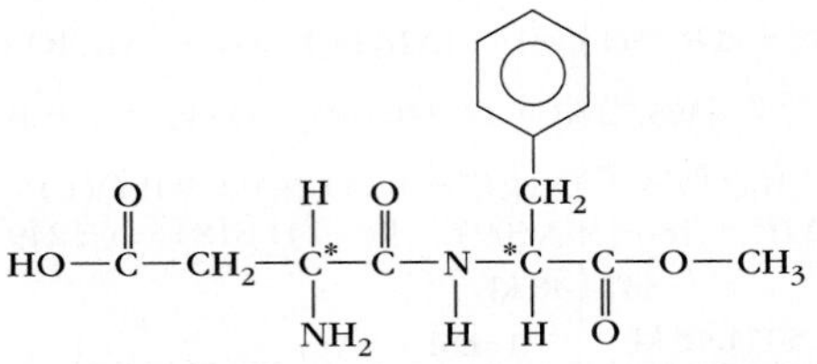

9.5 (a) $BeCl_2$: *sp* hybridization (two bonding electron pairs around the central Be atom), linear geometry

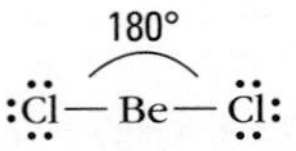

(b) The central N atom has four bonding pairs and no lone pairs in sp^3 hybridized orbitals on N giving tetrahedral electron-pair and molecular geometries.

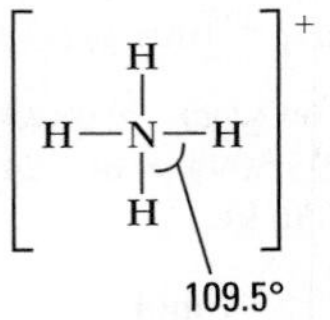

(c) Each carbon is sp^3 hybridized with no lone pairs, so the electron-pair and molecular geometries are both tetrahedral. The sp^3 hybridized oxygen atom has two bonding pairs, each in single bonds to a carbon atom plus two lone pairs giving it a tetrahedral electron-pair geometry and an angular molecular geometry.

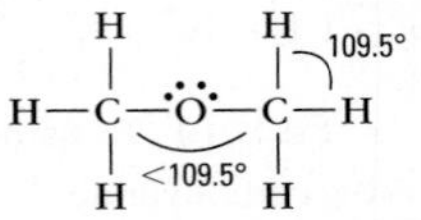

(d) The central boron atom has four single bonds, one to each fluorine atom, and no lone pairs resulting in tetrahedral electron-pair and molecular geometries. The boron is sp^3 hybridized to accommodate the four bonding pairs.

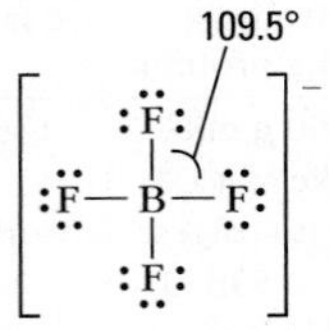

9.6 (a) In HCN, the *sp* hybridized carbon atom is sigma bonded to H and to N, as well as having two pi bonds to N. The sigma and two pi bonds form the C≡N triple bond. The nitrogen is *sp* hybridized with a sigma and two pi bonds to carbon; a lone pair is in the nonbonding *sp* hybrid orbital on N.

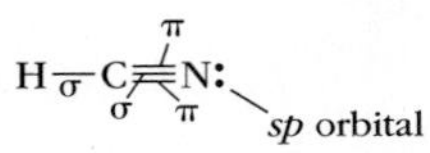

(b) The double-bonded carbon and nitrogen are both sp^2 hybridized. The sp^2 hybrid orbitals on C form sigma bonds to H and to N; the unhybridized *p* orbital on C forms a pi bond with the unhybridized *p* orbital on N. The sp^2 hybrid orbitals on N form sigma bonds to carbon and to H; the N lone pair is in the nonbonding sp^2 hybrid orbital. (See diagram on next page.)

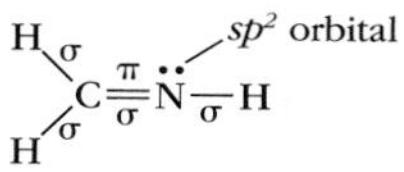

9.7 (a) $BFCl_2$ is a triangular planar molecule with polar B—F and B—Cl bonds. The molecule is polar because the B—F bond is more polar than the B—Cl bonds, resulting in a net dipole. The B—Cl regions are partially positive; the B—F region is partially negative.
(b) NH_2Cl is a triangular pyramidal molecule with polar N—H bonds. (N—Cl is a slightly polar bond; N and Cl have nearly the same electronegativity.) It is a polar molecule because the N—H dipoles do not cancel and produce a net dipole. The partial positive charge is on each H; the partial negative charge is on the N.
(c) SCl_2 is an angular polar molecule. The S—Cl bond dipoles do not cancel each other because they are not symmetrically arranged due to the two lone pairs on S. The partial positive charge resides on the S; the partial negative charges are on each Cl.

9.8 (a) London forces between Kr atoms must be overcome for krypton to melt.
(b) The C—H covalent bonds in propane must be broken to form C and H atoms; the H atoms covalently bond to form H_2.

9.9 (a) London forces occur between N_2 molecules.
(b) CO_2 is nonpolar, and London forces occur between it and polar water molecules. CO_2 can also hydrogen bond to water (H from water and lone pair on the O of CO_2).
(c) London forces occur between the two molecules, but the principal intermolecular forces are the hydrogen bonds between the H on NH_3 with the lone pairs on the OH oxygen, and the hydrogen bonds between the H on the oxygen in CH_3OH and lone pair on nitrogen in NH_3.

Chapter 10

10.1 (a) Pressure in atm =

$$29.5 \text{ in Hg} \times \frac{1 \text{ atm}}{76.0 \text{ cm Hg}} \times \frac{2.54 \text{ cm}}{1 \text{ in}} = 0.986 \text{ atm}$$

(b) Pressure in mm Hg = $29.5 \text{ in Hg} \times \dfrac{25.4 \text{ mm}}{1 \text{ in}} = 749 \text{ mm Hg}$

(c) Pressure in bar =

$$29.5 \text{ in Hg} \times \frac{1.013 \text{ bar}}{760 \text{ mm Hg}} \times \frac{25.4 \text{ mm}}{1 \text{ in}} = 0.999 \text{ bar}$$

(d) Pressure in kPa =

$$29.5 \text{ in Hg} \times \frac{101.3 \text{ kPa}}{760 \text{ mm Hg}} \times \frac{25.4 \text{ mm}}{1 \text{ in}} = 99.9 \text{ kPa}$$

10.2 (a) The temperature remains constant, so the average energy of the gas molecules remains constant. If the volume is decreased, then the gas molecules must hit the walls more frequently, and the pressure is increased.
(b) The temperature remains constant, so the average energy of the gas molecules remains constant. The addition of more molecules within a fixed volume must mean that the molecules hit walls more frequently, so the pressure is increased.

10.3 Volume of NO gas = $1.0 \text{ L } O_2 \times \dfrac{2 \text{ L NO}}{1 \text{ L } O_2} = 2.0 \text{ L NO}$

10.4 $V = \dfrac{nRT}{P}$

$$= \frac{(2.64 \text{ mol})(0.0821 \text{ L atm mol}^{-1} \text{ K}^{-1})(304 \text{ K})}{0.640 \text{ atm}} = 103 \text{ L}$$

10.5 $V_2 = \dfrac{P_1V_1}{P_2} = \dfrac{(1.00 \text{ atm})(400. \text{ mL})}{0.750 \text{ atm}} = 533 \text{ mL}$

10.6 (a) $V_2 = \dfrac{P_1V_1T_2}{P_2T_1}$

$$= \frac{(710 \text{ mm Hg})(21 \text{ mL})(299.6 \text{ K})}{(740 \text{ mm Hg})(295.4 \text{ K})} = 20. \text{ mL}$$

(b) $V_2 = \dfrac{(21 \text{ mL})(299.6 \text{ K})}{(295.4 \text{ K})} = 21 \text{ mL}$

10.7 $\dfrac{10.0 \text{ g } NH_4NO_3}{80.043 \text{ g/mol}} = 0.1249 \text{ mol } NH_4NO_3$

$$0.1249 \text{ mol } NH_4NO_3 \times \frac{7 \text{ mol product gases}}{2 \text{ mol } NH_4NO_3}$$

$$= 0.437 \text{ mol produced}$$

$$V = \frac{(0.437 \text{ mol})(0.0821 \text{ L atm mol}^{-1} \text{ K}^{-1})(298 \text{ K})}{1 \text{ atm}} = 10.7 \text{ L}$$

10.8 $\dfrac{1.0 \text{ g LiOH}}{23.94 \text{ g/mol}} = 0.0418 \text{ mol LiOH}$

$$0.0418 \text{ mol LiOH} \times \frac{1 \text{ mol } CO_2}{2 \text{ mol LiOH}} = 0.0209 \text{ mol } CO_2$$

$$V = \frac{(0.0209 \text{ mol})(0.0821 \text{ L atm mol}^{-1} \text{ K}^{-1})(295 \text{ K})}{1 \text{ atm}}$$

$$= 0.51 \text{ L } CO_2$$

10.9 Amount of gas, $n = \dfrac{PV}{RT}$

$$= \frac{(0.850 \text{ atm})(1.00 \text{ L})}{(0.0821 \text{ L atm mol}^{-1} \text{ K}^{-1})(293 \text{ K})} = 0.0353 \text{ mol}$$

$$\text{Molar mass} = \frac{1.13 \text{ g}}{0.0353 \text{ mol}} = 32.0 \text{ g/mol}$$

The gas is probably oxygen.

10.10 Amount of N_2 = $7.0 \text{ g } N_2 \times \dfrac{1 \text{ mol } N_2}{28.10 \text{ g } N_2} = 0.25 \text{ mol } N_2$

Amount of H_2 = $6.0 \text{ g } H_2 \times \dfrac{1 \text{ mol } H_2}{2.02 \text{ g } H_2} = 3.0 \text{ mol } H_2$

Total number of moles = 3.0 + 0.25 = 3.25 mol

$$P_{total} = \frac{(3.25 \text{ mol})(0.0821 \text{ L atm mol}^{-1} \text{ K}^{-1})(773 \text{ K})}{5.0 \text{ L}} = 41.3 \text{ atm}$$

$$X_{N_2} = \frac{0.25 \text{ mol}}{3.25 \text{ mol}} = 0.077 \qquad X_{H_2} = \frac{3.0 \text{ mol}}{3.25 \text{ mol}} = 0.92$$

$$P_{N_2} = \frac{(0.25 \text{ mol})(0.0821 \text{ L atm mol}^{-1} \text{ K}^{-1})(773 \text{ K})}{5.0 \text{ L}} = 3.2 \text{ atm}$$

$$P_{H_2} = \frac{(3.0 \text{ mol})(0.0821 \text{ L atm mol}^{-1} \text{ K}^{-1})(773 \text{ K})}{5.0 \text{ L}} = 38 \text{ atm}$$

10.11 For NO:

$$n = \frac{(1.0 \text{ atm})(4.0 \text{ L})}{(0.0821 \text{ L atm mol}^{-1} \text{ K}^{-1})(298 \text{ K})} = 0.163 \text{ mol NO}$$

For O_2:

$$n = \frac{(0.40 \text{ atm})(2.0 \text{ L})}{(0.0821 \text{ L atm mol}^{-1} \text{ K}^{-1})(298 \text{ K})} = 0.0327 \text{ mol } O_2$$

All the O_2 is used.

$$0.0327 \text{ mol } O_2 \times \frac{2 \text{ mol NO}}{1 \text{ mol } O_2} = 0.0654 \text{ mol NO used}$$

0.163 mol − 0.0654 mol = 0.0976 mol NO remains

$$0.0327 \text{ mol } O_2 \times \frac{2 \text{ mol } NO_2}{1 \text{ mol } O_2} = 0.0654 \text{ mol } NO_2 \text{ formed}$$

n_{total} = 0.0976 + 0.0654 = 0.163 mol of gas

$$P_{total} = \frac{nRT}{V} = \frac{(0.163 \text{ mol})(0.0821 \text{ L atm mol}^{-1} \text{ K}^{-1})(298 \text{ K})}{6.0 \text{ L}}$$

$$= 0.665 \text{ atm}$$

10.12 $P_{HCl} = P_{total} - P_{water} = 740 \text{ mm Hg} - 21 \text{ mm Hg} = 719 \text{ mm Hg}$

$$n = \frac{PV}{RT} = \frac{(719/760 \text{ atm})(0.260 \text{ L})}{(0.0821 \text{ L atm mol}^{-1} \text{ K}^{-1})(296 \text{ K})}$$

$$= 0.0101 \text{ mol } H_2$$

$$0.0101 \text{ mol} \times 2.0158 \text{ g/mol} = 0.0204 \text{ g} = 20.4 \text{ mg } H_2$$

10.13 $P = \frac{nRT}{V} = \frac{(5.00\text{ mol})(0.0821\text{ L atm mol}^{-1}\text{ K}^{-1})(273\text{ K})}{20.0\text{ L}}$

$= 5.60$ atm

$$P = \left(\frac{nRT}{V - nb}\right) - \left(\frac{n^2a}{V^2}\right)$$

$$P = \left(\frac{(5.00\text{ mol})(0.0821\text{ L atm mol}^{-1}\text{ K}^{-1})(273\text{ K})}{(20.0\text{ L}) - (5.00\text{ mol})(0.0428\text{ L/mol})}\right)$$

$$- \left(\frac{(5.00\text{ mol})^2(2.25\text{ L}^2\text{ atm/mol}^2)}{(20.0\text{ L})^2}\right) = 5.523\text{ atm}$$

$$\text{Percentage difference in pressure} = \frac{5.60\text{ atm} - 5.52\text{ atm}}{5.60\text{ atm}} \times 100\%$$

$= 1.43\%$

Chapter 11

11.1 $\ln\left(\frac{P_2}{143.0\text{ torr}}\right) = \frac{3.21 \times 10^4\text{ J/mol}}{8.31\text{ J mol}^{-1}\text{ K}^{-1}}\left[\frac{1}{303.15\text{ K}} - \frac{1}{333.15\text{ K}}\right]$

$\ln\left(\frac{P_2}{143.0\text{ torr}}\right) = 1.1473$

$\ln P_2 = -1.1473 - \ln(143.0)$

$P_2 = 450.$ torr

11.2 From Table 11.2, $\Delta H_{vap}(Br_2)$ is 29.54 kJ/mol at its normal boiling point.
So 29.54 kJ/mol × 0.500 mol = 14.77 kJ.

11.3 The molar mass of ethanol is 46.05 g/mol.

$$500.\text{ kJ} \times \frac{1\text{ mol}}{38.6\text{ kJ}} \times \frac{46.05\text{ g}}{1\text{ mol}} = 597\text{ g ethanol}$$

11.4 (21.95)/2 = 10.98 kJ

11.5 The gas phase.

11.6 (a) Solid decane is a molecular solid.
(b) Solid $MgCl_2$ is composed of Mg^{2+} and Cl^- ions and is an ionic solid.

11.7 There are two atoms per bcc unit cell. The diagonal of the bcc unit cell is four times the radius of the atoms in the unit cell, so, solving for the edge,

$$\text{Edge} = \frac{4 \times 144\text{ pm}}{\sqrt{3}} = 333\text{ pm}$$

$$\text{Density} = \frac{\text{mass}}{\text{volume}}$$

$$= \frac{(2\text{ Au atoms})(196.97\text{ g Au}/6.022 \times 10^{23}\text{ Au atoms})}{[(333\text{ pm})(1\text{ m}/10^{12}\text{ pm})(10^2\text{ cm/m})]^3}$$

$= 17.7$ g/cm^3

11.8 The edge of the KCl unit cell would be 2 × 152 pm + 2 × 167 pm = 638 pm.
The unit cell of KCl is larger than that of NaCl.

Volume of the unit cell = $(638\text{ pm})^3$

$$= 2.60 \times 10^{18}\text{ pm}^3 \times \left(\frac{10^{-10}\text{ cm}}{\text{pm}}\right)^3$$

$= 2.60 \times 10^{-22}$ cm^3

$$D = \frac{m}{v}$$

$$= \frac{(4\text{ formula units KCl}) \times 74.55\text{ g KCl}/6.022 \times 10^{23}\text{ formula units}}{2.60 \times 10^{-22}\text{ cm}^3}$$

$D = 1.91$ g/cm^3

11.9 Energy transfer required

$$= 1.45\text{ g Al} \times \frac{1\text{ mol Al}}{26.98\text{ g Al}} \times \frac{10.7\text{ kJ}}{\text{mol}} = 0.575\text{ kJ}$$

Chapter 12

12.1 The balanced combustion reaction for methanol vapor is

$$CH_3OH(g) + \tfrac{3}{2}O_2(g) \longrightarrow CO_2(g) + 2\,H_2O(g)$$

Using Hess's law, we see that the heat of combustion of methanol vapor is

$$\Delta H_{comb} = [\Delta H_f^\circ CO_2(g)] + 2[\Delta H_f^\circ H_2O(g)]$$
$$-1[\Delta H_f^\circ CH_3OH(g)]$$
$$= -393.509\text{ kJ/mol} + 2(-241.818\text{ kJ/mol})$$
$$-(-200.66\text{ kJ/mol})$$
$$= -676.49\text{ kJ/mol}$$

For methanol,

$$-676.49\text{ kJ/mol}\left(\frac{1\text{ mol}}{32.04\text{ g}}\right)\left(\frac{0.791\text{ g}}{1\text{ mL}}\right)\left(\frac{1000\text{ mL}}{\text{L}}\right)$$

$= -1.67 \times 10^4$ kJ/L

12.2 $CO(g) + 2\,H_2(g) \rightarrow CH_3OH(g)$

$\Delta H^\circ = \{[(1\text{ mol C}{\equiv}\text{O})(1073\text{ kJ/mol})$
$+ (2\text{ mol H—H})(436\text{ kJ/mol})]\}$
$- \{[(3\text{ mol C—H})(416\text{ kJ/mol})]$
$+ [(1\text{ mol C—O})(336\text{ kJ/mol})] + [(1\text{ mol O—H})(467\text{ kJ/mol})]\}$
$= (1945\text{ kJ}) - (2051\text{ kJ}) = -106\text{ kJ}$

12.3 (a) The first oxidation product of $CH_3CH_2CH_2OH$ is the aldehyde.

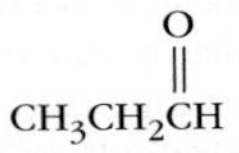

The second oxidation product of $CH_3CH_2CH_2OH$ is the acid.

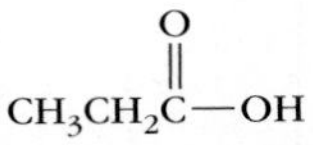

(b) The oxidation product of this secondary alcohol is the ketone.

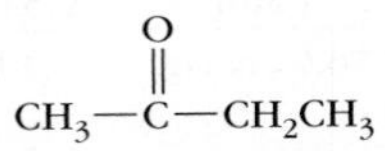

12.4 In this case, stearic acid would be on carbon 2 of glycerol where it would be flanked by oleic acids at carbons 1 and 3 of glycerol. See Problem-Solving Example 12.4 (← *p. 558*) for the structural formulas of stearic and oleic acids.

12.5

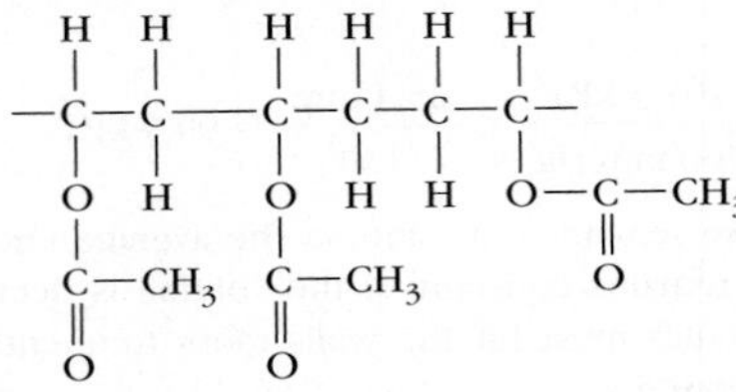

12.6 (a)

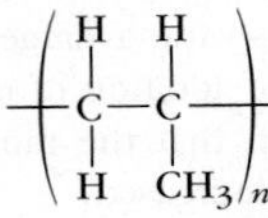

(b)

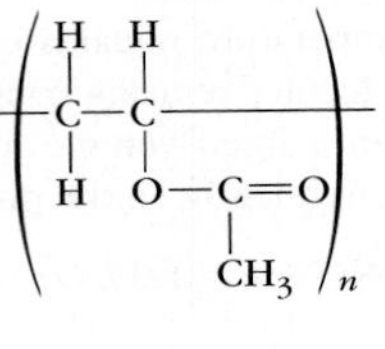

(c)

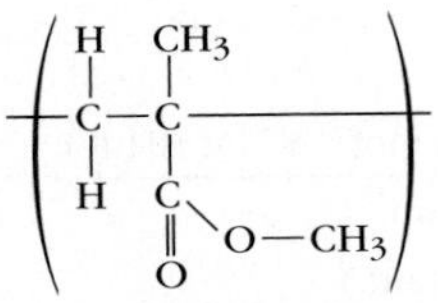

12.7 MM vinyl chloride, C_2H_3Cl, is 62.5 g/mol.

$$\text{Degree of polymerization} = 1.50 \times 10^5\text{ g} \times \frac{1\text{ mol monomer unit}}{62.5\text{ g}}$$

$= 2.40 \times 10^3$ monomer units

12.8 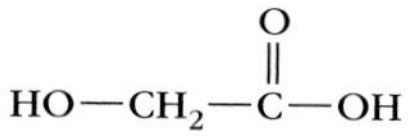

12.9 Twelve moles of water

12.10 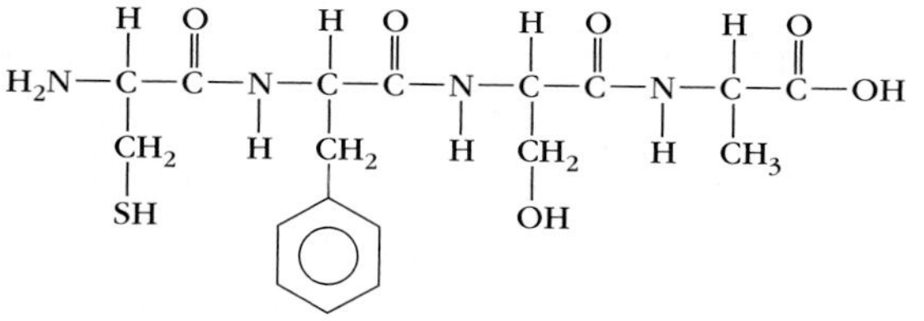

Chapter 13

13.1 (a) Rate $= \dfrac{-\Delta[\mathrm{Cv^+}]}{\Delta t} = 1.27 \times 10^{-6}\ \mathrm{mol\ L^{-1}\ s^{-1}}$

$$\Delta t = \frac{-\Delta[\mathrm{Cv^+}]}{1.27 \times 10^{-6}\ \mathrm{mol\ L^{-1}\ s^{-1}}}$$

$$= \frac{(4.30 \times 10^{-5} - 3.96 \times 10^{-5})\ \mathrm{mol/L}}{1.27 \times 10^{-6}\ \mathrm{mol\ L^{-1}\ s^{-1}}}$$

$$= 2.7\ \mathrm{s}$$

(b) No. The rate of reaction depends on the concentration of Cv^+ and, therefore, becomes slower as the reaction progresses. Therefore, the method used in part (a) works only over a small range of concentrations.

13.2 (a) The balanced chemical equation shows that for every mole of O_2 consumed two moles of N_2O_5 are produced. Therefore, the rate of formation of N_2O_5 is twice the rate of disappearance of O_2.
(b) Four moles of NO_2 are consumed for every mole of O_2 consumed. Therefore, if O_2 is consumed at the rate of 0.0037 mol L^{-1} s^{-1} the rate of disappearance of NO_2 is four times this rate.

$$4 \times (0.0037\ \mathrm{mol\ L^{-1}\ s^{-1}}) = 0.015\ \mathrm{mol\ L^{-1}\ s^{-1}}$$

13.3 (a) The effect of $[OH^-]$ on the rate of reaction cannot be determined, because the $[OH^-]$ is the same in all three experiments.
(b) Rate $= k[\mathrm{Cv^+}]$

(c) $k_1 = \dfrac{\text{rate}}{[\mathrm{Cv^+}]} = \dfrac{1.3 \times 10^{-6}\ \mathrm{mol\ L^{-1}\ s^{-1}}}{4.3 \times 10^{-5}\ \mathrm{mol/L}}$

$= 3.0 \times 10^{-2}\ \mathrm{s^{-1}}$

$k_2 = 3.0 \times 10^{-2}\ \mathrm{s^{-1}}$

$k_3 = 3.0 \times 10^{-2}\ \mathrm{s^{-1}}$

$$k = \frac{k_1 + k_2 + k_3}{3} = 3.0 \times 10^{-2}\ \mathrm{s^{-1}}$$

(d) Rate $= k[\mathrm{Cv^+}]$

$= (3.0 \times 10^{-2}\ \mathrm{s^{-1}})(0.00045\ \mathrm{mol/L})$

$= 1.4 \times 10^{-5}\ \mathrm{mol\ L^{-1}\ s^{-1}}$

(e) Rate $= (3.0 \times 10^{-2}\ \mathrm{s^{-1}})(0.5 \times 0.00045\ \mathrm{mol/L})$

$= 6.8 \times 10^{-6}\ \mathrm{mol\ L^{-1}\ s^{-1}}$

13.4 (a) The order of the reaction with respect to each chemical is the exponent associated with the concentration of that chemical. So the reaction is second-order with respect to NO and it is first-order with respect to Cl_2.
(b) Tripling the concentration of NO will make the reaction go $3^2 = 9$ times faster but decreasing the concentration of Cl_2 by a factor of 8 will make the reaction go at 1/8 the initial rate. If these two changes are made simultaneously the relevant factor will be $9/8 = 1.13$, so the reaction will occur 13% faster than it did under the initial conditions.

13.5 Make three plots of the data.

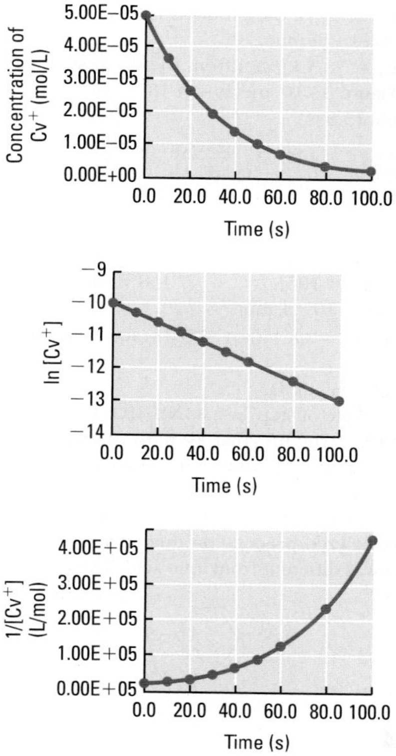

The first-order plot is a straight line and the other plots are curved, so the reaction is first-order. The slope of the first-order plot is $-0.0307\ \mathrm{s^{-1}}$, so $k = -\text{slope} = 0.031\ \mathrm{s^{-1}}$.

13.6 Use the integrated first-order rate law from Table 13.2.

$$\ln[\mathrm{A}]_t = -kt + \ln[\mathrm{A}]_0$$

$$\ln\frac{[\mathrm{A}]_t}{[\mathrm{A}]_0} = -kt$$

$$t = -\frac{1}{k}\ln\frac{[\mathrm{A}]_t}{[\mathrm{A}]_0} = -\left(\frac{1}{3.43 \times 10^{-2}\ \mathrm{d^{-1}}}\right)\ln\left(\frac{0.1}{1.0}\right)$$

$$= -(29.15\ \mathrm{d})(-2.303) = 67.1\ \mathrm{d}$$

13.7 In Figure 13.3 the $[\mathrm{Cv^+}]$ falls from 5.00×10^{-5} M to 2.5×10^{-5} M in 23 s. The $[\mathrm{Cv^+}]$ falls from 2.5×10^{-5} M to 1.25×10^{-5} M between 23 s and 46 s. The two times are equal, so $t_{1/2} = 23$ s.

$$k = \frac{0.693}{t_{1/2}} = \frac{0.693}{23\ \mathrm{s}} = 3.0 \times 10^{-2}\ \mathrm{s^{-1}}$$

13.8 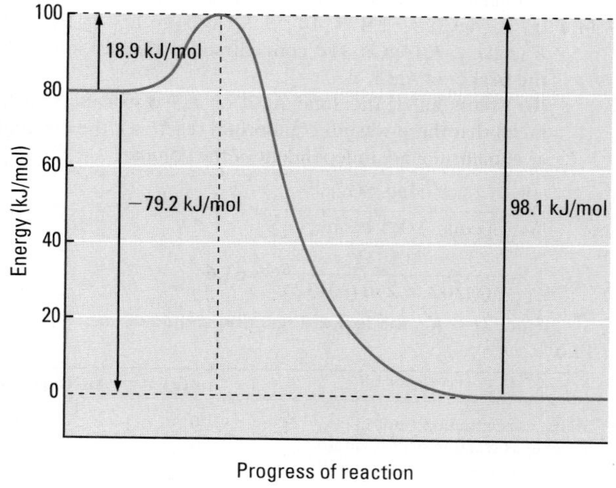

$$Cl_2(g) + 2\ NO(g) \rightarrow 2\ NOCl(g)$$

Since the energy of the products is less than that of reactants, the reaction is exothermic.

13.9 Obtain $E_a = 76.3$ kJ/mol from the discussion and analysis of the data in Figure 13.10 and $k = 4.18 \times 10^{-5}$ L mol^{-1} s^{-1} at 273 K from Table 13.3.

$$\frac{k_1}{k_2} = e^{\left[\frac{E_a}{R}\left(\frac{1}{T_2} - \frac{1}{T_1}\right)\right]} = e^{\left[\frac{76{,}300\ \text{J/mol}}{8.314\ \text{J K}^{-1}\text{mol}^{-1}}\left(\frac{1}{348\ \text{K}} - \frac{1}{273\ \text{K}}\right)\right]}$$

$$= e^{-7.245} = 7.138 \times 10^{-4}$$

$$k_2 = \frac{k_1}{7.138 \times 10^{-4}} = \frac{4.18 \times 10^{-5}\ \text{L mol}^{-1}\ \text{s}^{-1}}{7.138 \times 10^{-4}}$$

$$= 5.86 \times 10^{-2}\ \text{L mol}^{-1}\ \text{s}^{-1}$$

13.10 (a) $2\ NH_3(aq) + OCl^-(aq) \rightarrow N_2H_4(aq) + Cl^-(aq) + H_2O(\ell)$
(b) Step 1
(c) NH_2Cl, OH^-, $N_2H_5^+$
(d) Rate = rate of step 1 = $k\ [NH_3][OCl^-]$

13.11 Choose the structure that is most similar to the structure of *p*-aminobenzoic acid. An enzyme might be inhibited by this molecule, which could fit the active site but not be converted to a product similar to folic acid. Or, the molecule might react in an enzyme-catalyzed process, producing a product whose biological function was different from folic acid. The best choice is

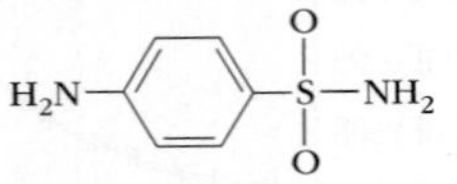

Chapter 14

14.1 (a) $K_c = [CO_2(g)]$ (b) $K_c = \frac{[H_2]}{[HCl]}$
(c) $K_c = \frac{[CO][H_2]^3}{[CH_4][H_2O]}$ (d) $K_c = \frac{[HCN][OH^-]}{[CN^-]}$

14.2 $K_c = K_{c_1} \times K_{c_2} = (4.2 \times 10^{-7})(4.8 \times 10^{-11})$
$= 2.0 \times 10^{-17}$

14.3 $[CH_3COO^-] = \frac{2.96}{100} \times 0.0200\ \text{mol/L} = 5.92 \times 10^{-4}$ M
$[H_3O^+] = 5.92 \times 10^{-4}$ M

$$[CH_3\ COOH] = \frac{100 - 2.96}{100} \times 0.0200\ \text{mol/L}$$
$$= 1.94 \times 10^{-2}\ \text{M}$$

$$K_c = \frac{[H_3O^+][CH_3COO^-]}{[CH_3COOH]} = \frac{(5.92 \times 10^{-4})(5.92 \times 10^{-4})}{1.94 \times 10^{-2}}$$
$$= 1.81 \times 10^{-5}$$

The result agrees with the value in Table 14.1.

14.4 (a) $K_c(AgCl) = 1.8 \times 10^{-10}$; $K_c(AgI) = 1.5 \times 10^{-16}$. Because $K_c(AgI) < K_c(AgCl)$, the concentration of silver ions is larger in the beaker of AgCl.
(b) Unless all of the solid AgCl or AgI dissolves (which would mean that there was no equilibrium reaction), the concentrations at equilibrium are independent of the volume.

14.5 $$Q = \frac{(\text{conc. } SO_3)^2}{(\text{conc. } SO_2)^2(\text{conc. } O_2)}$$
$$= \frac{(0.184)^2}{(0.102 \times 2)^2(0.0132)} = 61.6$$

Since $Q < K_c$, the forward reaction should occur.

14.6

	$AuI(s) \rightleftharpoons$	$Au^+(aq) + I^-(aq)$
Initial concentration (mol/L)	0	0
Change as reaction occurs (mol/L)	$+x$	$+x$
Equilibrium concentration (mol/L)	x	x

$$K_c = 1.6 \times 10^{-23} = [Au^+][I^-] = x^2$$
$$x = \sqrt{1.6 \times 10^{-23}} = 4.0 \times 10^{-12} = [Au^+] = [I^-]$$

14.7 The reaction is the reverse of the one in Table 14.1, so

$$K_c = \frac{1}{1.7 \times 10^2} = 5.9 \times 10^{-3}$$

$$Q = \frac{(\text{conc. } NO_2)^2}{(\text{conc. } N_2O_4)} = \frac{\left(\frac{0.500\ \text{mol}}{4.00\ \text{L}}\right)^2}{\left(\frac{1.00\ \text{mol}}{4.00\ \text{L}}\right)} = 6.25 \times 10^{-2}$$

Because $Q > K_c$, the reaction should go in the reverse direction. Therefore, let x be the change in concentration of N_2O_4, giving the ICE table:

	N_2O_4	$\rightleftharpoons$	$2\ NO_2$
Initial concentration (mol/L)	$\frac{1.00}{4.00} = 0.250$		$\frac{0.500}{4.00} = 0.125$
Change as reaction occurs (mol/L)	x		$-2x$
Equilibrium concentration (mol/L)	$0.250 + x$		$0.125 - 2x$

$$K_c = 5.9 \times 10^{-3} = \frac{(0.125 - 2x)^2}{0.250 + x}$$
$$= \frac{(1.56 \times 10^{-2}) - 0.500x + 4x^2}{0.250 - x}$$
$$(1.48 \times 10^{-3}) + (5.9 \times 10^{-3})x = (1.56 \times 10^{-2}) - 0.500x + 4x^2$$
$$4x^2 - 0.5059x + (1.412 \times 10^{-2}) = 0$$
$$x = \frac{-(-0.5059) \pm \sqrt{(-0.5059)^2 - 4 \times 4 \times 1.412 \times 10^{-2}}}{2 \times 4}$$
$$= \frac{0.5059 \pm \sqrt{3.001 \times 10^{-2}}}{8}$$
$$= \frac{0.5059 \pm 0.1732}{8}$$
$$x = 8.49 \times 10^{-2} \text{ or } x = 4.16 \times 10^{-2}$$

If $x = 8.49 \times 10^{-2}$, then $[N_2O_4] = 0.250 + x = 0.335$ and $[NO_2] = 0.125 - 2x = 0.125 - (2 \times 7.86 \times 10^{-2}) = -0.0448$. A negative concentration is impossible, so x must be 4.16×10^{-2}. Then

$$[N_2O_4] = 0.250 + 0.0416 = 0.292$$
$$[NO_2] = 0.125 - (2 \times 0.0416) = 0.0418$$

As predicted by Q, the reverse reaction has occurred, and the concentration of N_2O_4 has increased.

14.8 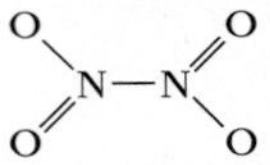 and 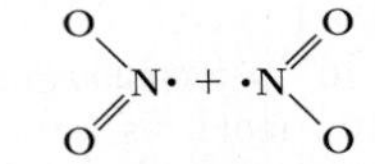

Because a bond is broken and because bond breaking is always endothermic (← p. 238), the reaction must be endothermic. Increasing temperature shifts the equilibrium in the endothermic direction. Figure 14.3 shows that at a higher temperature there is a greater concentration of brown NO_2.

14.9 (a) There are more moles of gas phase reactants than products, so entropy favors the reactants.
(b) Data from Appendix J show that $\Delta H° = -46.11$ kJ. The reaction is exothermic, so the energy effect favors the products.
(c) As T increases the reaction shifts in the endothermic direction, which is toward reactants. The entropy effect also becomes more important at high T, and it favors reactants. There is a greater concentration of NH_3 at low temperature.

Chapter 15

15.1 (a) Ethylene glycol molecules are polar and attracted to each other by dipole-dipole attractions and hydrogen bonding. They will not dissolve in gasoline, a nonpolar substance.
(b) Molecular iodine and carbon tetrachloride are nonpolar; therefore iodine should dissolve readily in carbon tetrachloride.
(c) Motor oil contains a mixture of nonpolar hydrocarbons that will dissolve in carbon tetrachloride, a nonpolar solvent.

15.2 The —OH groups attached to the ring and to the side chain of vitamin C hydrogen-bond to water molecules. The oxygen atoms in and on the ring also form hydrogen bonds to water.

15.3 Use Henry's law, the Henry's law constant, and the fact that air is only 21 mol percent oxygen, a 0.21 mole fraction of oxygen.

$$\text{Pressure of } O_2 = (1.0\text{ atm})\left(\frac{760\text{ mm Hg}}{1\text{ atm}}\right)(0.21) = 160.\text{ mm Hg}$$

$$S_g = k_H P_g = \left(1.66 \times 10^{-6}\frac{\text{mol/L}}{\text{mm Hg}}\right)(160.\text{ mm Hg}) = 2.66 \times 10^{-4}\text{ mol/L}$$

$$(2.66 \times 10^{-4}\text{ mol/L})\left(\frac{32.00\text{ g } O_2}{1\text{ mol } O_2}\right) = 0.0085\text{ g/L or } 8.5\text{ mg/L}$$

15.4 Total mass is 750 + 21.5 = 771.5 g.

$$\text{Weight percent glucose} = \left(\frac{21.5\text{ g}}{771.5\text{ g}}\right) \times 100\% = 2.79\%$$

15.5 $$\left(\frac{30\text{ g Se}}{10^9\text{ g } H_2O}\right)\left(\frac{1\text{ g } H_2O}{1\text{ mL } H_2O}\right)\left(\frac{10^6\ \mu\text{g Se}}{1\text{ g Se}}\right) = 3.0 \times 10^{-2}\ \mu\text{g Se/mL } H_2O$$

Se in 100. mL water

$$= \left(\frac{3.0 \times 10^{-2}\mu\text{g Se}}{1\text{ mL } H_2O}\right)(100.\text{ mL } H_2O) = 3.0\ \mu\text{g Se}$$

15.6 (a) $$(0.250\text{ L})\left(\frac{0.0750\text{ mol NaBr}}{1\text{ L}}\right)\left(\frac{102.9\text{ g NaBr}}{1\text{ mol NaBr}}\right) = 1.93\text{ g NaBr}$$

(b) $$V_c = \frac{M_d \times V_d}{M_c} = \frac{(0.00150\text{ M})(0.500\text{ L})}{0.0750\text{ M}} = 0.0100\text{ L} = 10.0\text{ mL}$$

15.7 (a) $$\left(\frac{0.0556\text{ mol } Mg^{2+}}{1\text{ L soln}}\right)\left(\frac{24.3\text{ g } Mg^{2+}}{1\text{ mol } Mg^{2+}}\right)\left(\frac{1\text{ L soln}}{1.03 \times 10^3\text{ g soln}}\right) = 0.00131\frac{\text{g } Mg^{2+}}{\text{g soln}}$$

(b) 0.00131×10^6 ppm = 1310 ppm

15.8 $$\text{Molarity } H_2O_2 = \frac{\text{moles } H_2O_2}{\text{L solution}}$$

$$= \frac{30.0\text{ g } H_2O_2}{100.\text{ g solution}} \times \frac{1.11\text{ g solution}}{1\text{ mL solution}} \times \frac{10^3\text{ mL solution}}{1\text{ L solution}} \times \frac{1\text{ mol } H_2O_2}{34.0\text{ g } H_2O_2}$$

$$= \frac{3.33 \times 10^4\text{ mol } H_2O_2}{3.40 \times 10^3\text{ L}} = 9.79\text{ M}$$

15.9 $$\text{Molarity, M} = \left(\frac{20\text{ g NaCl}}{100\text{ g solution}}\right)\left(\frac{1.148\text{ g solution}}{1\text{ mL solution}}\right) \times \left(\frac{1\text{ mol NaCl}}{58.5\text{ g NaCl}}\right)\left(\frac{10^3\text{ mL}}{1\text{ L}}\right) = 3.9\text{ mol NaCl/L}$$

$$\text{Molality, } m = \left(\frac{20\text{ g NaCl}}{0.0080\text{ kg } H_2O}\right)\left(\frac{1\text{ mol NaCl}}{58.5\text{ g NaCl}}\right) = 4.3\text{ mol NaCl/kg } H_2O$$

15.10 $$P_{water} = (X_{water})(P^\circ_{water})$$

$$291.2\text{ mm Hg} = (X_{water})(355.1\text{ mm Hg})$$

$$X_{water} = \frac{291.2\text{ mm Hg}}{355.1\text{ mm Hg}} = 0.8201$$

$$X_{urea} = 1.000 - X_{water} = 1.000 - 0.8201 = 0.180$$

15.11 $$\text{Molality of solution} = \left(\frac{1.20\text{ kg ethylene glycol}}{6.50\text{ kg } H_2O}\right)\left(\frac{1000\text{ g}}{\text{kg}}\right)\left(\frac{1\text{ mol ethylene glycol}}{62.068\text{ g ethylene glycol}}\right) = 2.97\text{ mol/kg } H_2O$$

Next, calculate the freezing point depression of a 2.97 mol/kg solution.

$$\Delta T_f = (1.86\ ^\circ\text{C kg mol}^{-1})(2.97\text{ mol/kg}) = 5.52\ ^\circ\text{C}$$

This solution will freeze at −5.52 °C, so this quantity of ethylene glycol will not protect the 6.5 kg water in the tank if the temperature drops to −25 °C.

15.12 F.P. benzene = 5.50 °C; ΔT_f = 5.50 °C − 5.15 °C = 0.35 °C.

$$\text{Molality} = \frac{\Delta T_f}{K_f} = \frac{0.35\ ^\circ\text{C}}{5.10\ ^\circ\text{C kg mol}^{-1}} = 0.0686\text{ mol/kg}$$

$$\frac{0.0686\text{ mol solute}}{1\text{ kg benzene}} \times 0.0500\text{ kg benzene} = 0.00343\text{ mol solute}$$

$$\frac{0.180\text{ g solute}}{0.00343\text{ mol solute}} = 53.\text{ g/mol}$$

15.13 Hemoglobin is a molecular substance so the factor i is 1.

$$c = \frac{\Pi}{RTi} = \frac{1.8 \times 10^{-3}\text{ atm}}{(0.0821\text{ L atm mol}^{-1}\text{ K}^{-1})(298\text{ K})(1)} = 7.36 \times 10^{-5}\text{ mol/L}$$

$$\text{MM} = \frac{5.0\text{ g}}{7.36 \times 10^{-5}\text{ mol}} = 6.8 \times 10^4\text{ g/mol}$$

Chapter 16

16.1

Acid	Its Conjugate Base	Base	Its Conjugate Acid
$H_2PO_4^-$	HPO_4^{2-}	PO_4^{3-}	HPO_4^{2-}
H_2	H^-	NH_2^-	NH_3
HSO_3^-	SO_3^{2-}	ClO_4^-	$HClO_4$
HF	F^-	Br^-	HBr

16.2 $[OH^-] = 3.0 \times 10^{-8}$ M;

$$[H_3O^+] = \frac{1.0 \times 10^{-14}}{3.0 \times 10^{-8}} = 3.3 \times 10^{-7}\text{ M}$$

Therefore, the solution whose $[H_3O^+]$ is 5.0×10^{-4} M is more acidic.

16.3 The 5.0×10^{-9} M OH^- solution is more basic.

16.4 In a 0.040-M solution of NaOH, the $[OH^-]$ is 0.040 M because the NaOH is 100% dissociated; pOH = −log(0.040) = 1.40; pH = 14.00 − 1.40 = 12.60.

16.5 (a) $[H_3O^+] = 10^{-7.90} = 1.3 \times 10^{-8}$ M
(b) A pH of 7.90 is basic.

16.6 (a) $HN_3(aq) \rightleftharpoons H^+(aq) + N_3^-(aq)$ $\quad K_a = \frac{[H^+][N_3^-]}{[HN_3]}$

(b) $HCOOH(aq) \rightleftharpoons H^+(aq) + HCOO^-(aq)$ $\quad K_a = \frac{[H^+][HCOO^-]}{[HCOOH]}$

(c) $HClO_2(aq) \rightleftharpoons H^+(aq) + ClO_2^-(aq)$ $\quad K_a = \frac{[H^+][ClO_2^-]}{[HClO_2]}$

16.7 (a) $CH_3NH_2(aq) + H_2O(\ell) \rightleftharpoons CH_3NH_3^+(aq) + OH^-(aq)$

$$K_b = \frac{[CH_3NH_3^+][OH^-]}{[CH_3NH_2]}$$

(b) $PH_3(aq) + H_2O(\ell) \rightleftharpoons PH_4^+(aq) + OH^-(aq)$

$$K_b = \frac{[PH_4^+][OH^-]}{[PH_3]}$$

(c) $NO_2^-(aq) + H_2O(\ell) \rightleftharpoons HNO_2(aq) + OH^-(aq)$

$$K_b = \frac{[HNO_2][OH^-]}{[NO_2^-]}$$

16.8 Setting up a small ICE table for lactic acid, HLa:

	$HLa + H_2O \rightleftharpoons$	H_3O^+	$+ La^-$
Initial concentration (mol/L)	0.10	10^{-7}	0
Concentration change due to reaction (mol/L)	$-x$	$+x$	$+x$
Equilibrium concentration (mol/L)	$0.10 - x$	x	x

But $x = 10^{-2.43} = 3.7 \times 10^{-3}$ because $x = [H_3O^+]$. Substituting in the K_a expression,

$$K_a = \frac{[H_3O^+][La^-]}{[HLa]} = \frac{(3.7 \times 10^{-3})^2}{0.10 - (3.7 \times 10^{-3})}$$

$$= \frac{1.4 \times 10^{-5}}{0.1} = 1.4 \times 10^{-4}$$

Lactic acid is a stronger acid than propionic acid, $K_a = 1.4 \times 10^{-5}$.

16.9 (a) Using the same methods as shown in the example,

$$\frac{x^2}{0.015} = 1.9 \times 10^{-5}$$

Solving for x, which is $[H_3O^+]$, we get

$$x = \sqrt{(1.9 \times 10^{-5})(0.015)} = 5.3 \times 10^{-4} = [H_3O^+].$$

The pH of this solution is $-\log(5.3 \times 10^{-4}) = 3.27$.

(b) $$\% \text{ ionization} = \frac{[H_3O^+]}{[HN_3]_{\text{initial}}} \times 100\% = \frac{5.4 \times 10^{-4}}{0.015} \times 100\% = 3.6\%$$

16.10 In such cases use the K_b expression and value to calculate $[OH^-]$ and then pOH from $[OH^-]$. Calculate pH from 14 − pOH.

$$C_6H_{11}NH_2(aq) + H_2O(\ell) \rightleftharpoons C_6H_{11}NH_3^+(aq) + OH^-(aq)$$

$$K_b = \frac{[C_6H_{11}NH_3^+][OH^-]}{[C_6H_{11}NH_2]} = 4.6 \times 10^{-4}$$

$$K_b = \frac{[C_6H_{11}NH_3^+][OH^-]}{[C_6H_{11}NH_2]} = \frac{x^2}{0.015 - x} = 4.6 \times 10^{-4}$$

$$x^2 = (0.015 - x)(4.6 \times 10^{-4}) = (6.9 \times 10^{-6}) - (4.6 \times 10^{-4}\,x)$$

Solve the quadratic equation for x.

$$x = \frac{4.8 \times 10^{-3}}{2} = 2.4 \times 10^{-3} = [OH^-]$$

$$\text{pOH} = -\log(2.4 \times 10^{-3}) = 2.62; \quad \text{pH} = 14.00 - 2.62 = 11.38$$

16.11 Using the same methods as those used in the example, letting $x = [OH^-]$ and $[HCO_3^-]$, and using the value of 2.1×10^{-4} for K_b for CO_3^{2-}, we get

$$\frac{x^2}{1.0} = 2.1 \times 10^{-4} \qquad x = \sqrt{2.1 \times 10^{-4}} = 1.45 \times 10^{-2}$$

pOH = 1.84 and pH = 12.16

16.12 NH_4Cl dissolves by dissociating into NH_4^+ and Cl^- ions. The ammonium ions react with water to produce an acidic solution.

$$NH_4Cl(s) \rightleftharpoons NH_4^+(aq) + Cl^-(aq)$$

$$NH_4^+(aq) + H_2O(\ell) \rightleftharpoons H_3O^+(aq) + NH_3(aq)$$

The K_a of $NH_4^+ = \frac{K_w}{K_b} = \frac{1.0 \times 10^{-14}}{1.8 \times 10^{-5}}$;

$$K_a = 5.6 \times 10^{-10} = \frac{[H_3O^+][NH_3]}{[NH_4^+]}$$

NH_4^+	H_2O	H_3O^+	NH_3
Initial	0.10	0	0
Change	$-x$	$+x$	$+x$
Equilibrium	$0.10 - x$	x	x

$$5.6 \times 10^{-10} = \frac{(x)(x)}{(0.10 - x)}$$

Assume $0.10 - x \approx 0.10$ because K_a is so small. Thus, $(5.6 \times 10^{-10})(0.10) = x^2$; $x = [H_3O^+] = 7.48 \times 10^{-6}$ M.

$\text{pH} = -\log[H_3O^+] = -\log(7.48 \times 10^{-6}) = 5.13$

16.13 The molar masses and moles of acid per gram for the five antacids are:

	Formula Weight	Mol Acid/Gram
$Mg(OH)_2$	58.32	1 mol acid/29.16 g antacid
$CaCO_3$	100.10	1 mol acid/50.05 g antacid
$NaHCO_3$	84.00	1 mol acid/84.00 g antacid
$Al(OH)_3$	78.0034	1 mol acid/26.00 g antacid
$NaAl(OH)_2CO_3$	143.99	1 mol acid/36.00 g antacid

Of these antacids, $Al(OH)_3$ neutralizes the most stomach acid per gram of antacid.

Chapter 17

17.1 $$K_a = 4.3 \times 10^{-7} = \frac{[H^+][HCO_3^-]}{[H_2CO_3]} = \frac{H^+(0.025)}{(0.0020)} = [H^+] \times 12.5$$

$$[H^+] = \frac{4.3 \times 10^{-7}}{12.5} = 3.4 \times 10^{-8}$$

$$\text{pH} = -\log(3.4 \times 10^{-8}) = 7.46$$

17.2 $$7.40 = 7.20 + \log(\text{ratio}) = 7.20 + \log\frac{[HPO_4^{2-}]}{[H_2PO_4^-]}$$

$$\log\frac{[HPO_4^{2-}]}{[H_2PO_4^-]} = 7.40 - 7.20 = 0.20$$

$$\frac{[HPO_4^{2-}]}{[H_2PO_4^-]} = 10^{0.20} = 1.6$$

Therefore, $[HPO_4^{2-}] = 1.6 \times [H_2PO_4^-]$.

17.3 (a) Lactic acid-lactate (b) Acetic acid-acetate
(c) Hypochlorous acid-hypochlorite; $H_2PO_4^-/HPO_4^{2-}$
(d) Hydrogen carbonate-carbonate

17.4 Using the Henderson-Hasselbalch equation and the same methods as shown in the example,

(a) Buffer I:

$$\text{pH} = 3.85 + \log\left(\frac{(0.15 - 0.075)}{(0.15 + 0.075)}\right) = 3.85 - 0.48 = 3.37$$

Buffer II:

$$\text{pH} = 3.85 + \log\left(\frac{(0.20 - 0.075)}{(0.15 + 0.075)}\right) = 3.85 - 0.25 = 3.60$$

Buffer III:

$$\text{pH} = 3.85 + \log\left(\frac{(0.15 - 0.075)}{(0.25 + 0.075)}\right) = 3.85 - 0.64 = 3.21$$

(b) Buffer I:

$$\text{pH} = 3.85 + \log\left(\frac{(0.15 + 0.065)}{(0.15 - 0.065)}\right) = 3.85 + 0.64 = 4.25$$

Buffer II:

$$\text{pH} = 3.85 + \log\left(\frac{(0.20 + 0.065)}{(0.15 - 0.065)}\right) = 3.85 + 0.49 = 4.34$$

Buffer III:

$$\text{pH} = 3.85 + \log\left(\frac{(0.15 + 0.065)}{(0.25 - 0.065)}\right) = 3.85 + 0.064 = 3.91$$

17.5 The buffer capacity will be exceeded when just over 0.25 mol KOH is added, which will have reacted with the 0.25 mol $H_2PO_4^-$.

$0.25 \text{ mol } OH^- = 0.25 \text{ mol KOH} \times \frac{56 \text{ g KOH}}{1 \text{ mol KOH}} = 14 \text{ g}$. Thus, slightly more than 14 g KOH will exceed the buffer capacity.

17.6 (a) $[H_3O^+] = \frac{(5.00 \times 10^{-3}) - (1.00 \times 10^{-3})}{0.0500 + 0.0100}$

$= \frac{4.00 \times 10^{-3}}{0.0600} = 6.67 \times 10^{-2}$ M

pH = 1.176 = 1.18

(b) $[H_3O^+] = \frac{(5.00 \times 10^{-3}) - (0.00250)}{0.0500 + 0.0250}$

$= \frac{2.50 \times 10^{-3}}{0.0750} = 3.33 \times 10^{-2}$ M

$pH = -\log(3.33 \times 10^{-2}) = 1.48$

(c) $[H_3O^+] = \frac{(5.00 \times 10^{-3}) - (0.00450)}{0.0500 + 0.0450}$

$= \frac{5.00 \times 10^{-4}}{0.0950} = 5.26 \times 10^{-3}$

$pH = -\log(5.26 \times 10^{-3}) = 2.28$

(d) $[OH^-] = \frac{0.05 \times 10^{-3} \text{ mol } OH^-}{0.0500 \text{ L} + 0.0505 \text{ L}}$

$= 5.0 \times 10^{-4}$ mol/L

$pOH = -\log(5.0 \times 10^{-4}) = 3.30$

pH = 14.00 − 3.30 = 10.70

17.7 (a) Adding 10.0 mL of 0.100-M NaOH is adding (0.100 mol/L) (0.0100 L) = 0.00100 mol OH^-, which neutralizes 0.00100 mol acetic acid, converting it to 0.00100 mol acetate ion.

$pH = pH + \log \frac{[\text{acetate}]}{[\text{acetic acid}]}$

$= 4.74 + \log \frac{(0.00100/0.0600)}{(0.00400/0.0600)}$

$= 4.74 + \log(0.25) = 4.74 + (-0.602) = 4.14$

(b) $pH = 4.74 + \log \frac{(0.00250/0.0750)}{(0.00250/0.0750)}$

$= 4.74 + \log(1) = 4.74$

(c) $pH = 4.74 + \log \frac{(0.00450/0.0950)}{(0.00050/0.0950)}$

$= 4.74 + \log(9) = 4.74 + 0.95 = 5.70$

(d) $[OH^-] = \frac{0.10 \times 10^{-3} \text{ mol } OH^-}{0.0500 \text{ L} + 0.0510 \text{ L}}$

$= 9.9 \times 10^{-4}$ mol/L

$pOH = -\log(9.9 \times 10^{-4})$

= 3.00

pH = 14.00 − 3.00 = 11.00

17.8 (a) $K_{sp} = [Cu^+][Br^-]$ (b) $K_{sp} = [Hg^{2+}][I^-]^2$

(c) $K_{sp} = [Sr^{2+}][SO_4^{2-}]$

17.9 $AgBr(s) \rightleftharpoons Ag^+(aq) + Br^-(aq)$

$K_{sp} = [Ag^+][Br^-] = 5.0 \times 10^{-13} = x^2$; x = solubility

$x = \sqrt{5.0 \times 10^{-13}} = 7.1 \times 10^{-7}$ M

17.10 $Ag_2C_2O_4(s) \rightleftharpoons 2\ Ag^+(aq) + C_2O_4^{2-}(aq)$

$K_{sp} = [Ag^+]^2[C_2O_4^{2-}]$

$K_{sp} = [Ag^+]^2[C_2O_4^{2-}] = (1.4 \times 10^{-4})^2 (6.9 \times 10^{-5})$

$= 1.4 \times 10^{-12}$

17.11

	$PbCl_2 \rightleftharpoons$ Pb^{2+}	+ 2 Cl^-
Initially (mol/L)	0	0.5
Change due to dissolving (mol/L)	+S	0.5 + 2S
Equilibrium (mol/L)	S	0.5 (ignore 2S because S will be small)

$K_{sp} = (S)(0.5)^2 = 1.6 \times 10^{-5}$

$S = \frac{1.6 \times 10^{-5}}{(0.5)^2} = 6.4 \times 10^{-5} \text{ mol/L} = Pb^{2+}$ conc

17.12 (a) $PbCl_2(s) \rightleftharpoons Pb^{2+}(aq) + 2\ Cl^-(aq)$

$K_{sp} = [Pb^{2+}][Cl^-]^2 = 1.6 \times 10^{-5}$. Let S equal the solubility of lead chloride, which equals $[Pb^{2+}]$

$K_{sp} = 1.6 \times 10^{-5} = (S)(2S)^2 = 4S^3$; $S = \sqrt[3]{\frac{1.6 \times 10^{-5}}{4}}$

$= (4.0 \times 10^{-6})^{1/3} = 1.6 \times 10^{-2}$ mol/L

(b) Let solubility of $PbCl_2 = [Pb^{2+}] = S$; $[Cl^-] = 0.20$ M

$[Pb^{2+}] = S = \frac{1.6 \times 10^{-5}}{(0.20)^2} = 4.0 \times 10^{-4}$ mol/L This is less than that in pure water due to the common ion effect of the presence of chloride ion.

17.13 $AgCl(s) \rightleftharpoons Ag^+(aq) + Cl^-(aq)$ $K_{sp} = 1.8 \times 10^{-10}$

$Ag^+(aq) + 2\ S_2O_3^{2-}(aq) \rightleftharpoons Ag(S_2O_3)_2^{3-}(aq)$ $K_f = 2.0 \times 10^{13}$

Net reaction:

$AgCl(s) + 2\ S_2O_3^{2-}(aq) \rightleftharpoons Ag(S_2O_3)_2^{3-}(aq) + Cl^-(aq)$

Therefore, the equilibrium constant for the net reaction is $K_{sp} \times K_f = (1.8 \times 10^{-10})(2.0 \times 10^{13}) = 3.6 \times 10^3$. Because K_{net} is much greater than 1, the net reaction is product-favored, and AgCl is much more soluble in a $Na_2S_2O_3$ solution than it is in water.

17.14 (a) $Q = (1.0 \times 10^{-5})(1.0 \times 10^{-5}) = 1.0 \times 10^{-10} < K_{sp}$; no precipitation.

(b) For precipitation to occur, $Q \geq K_{sp}$; Q = conc Ag^+ × conc Cl^-; $K_{sp} = [Ag^+][Cl^-]$; $1.8 \times 10^{-10} = [Ag^+][Cl^-]$;

$[Cl^-] = \frac{1.8 \times 10^{-10}}{1.0 \times 10^{-5}} = 1.8 \times 10^{-5}$ M

the minimum Cl^- concentration for AgCl precipitation.

17.15 AgCl will precipitate first. $[Cl^-]$ needed to precipitate AgCl:

$[Cl^-] = \frac{1.8 \times 10^{-10}}{1.0 \times 10^{-2}} = 1.8 \times 10^{-8}$ M

$[Cl^-]$ needed to precipitate $PbCl_2$:

$[Cl^-] = \sqrt{\frac{1.6 \times 10^{-5}}{1.0 \times 10^{-1}}} = 1.3 \times 10^{-2}$ M

Chapter 18

18.1 $\Delta S = q_{rev}/T = (30.8 \times 10^3 \text{ J})/(273.15 + 45.3) \text{ K}$

$= (30.8 \times 10^3 \text{ J})/(318.45 \text{ K})$

$\Delta S = 96.7$ J/K

18.2 (a) C(g) has higher $S°$, 158.096 J K^{-1} mol^{-1}, versus 5.740 J $K^{-1}mol^{-1}$ for C (graphite).

(b) Ar(g) has higher $S°$, 154.7 J K^{-1} mol^{-1}, versus 41.42 J $K^{-1}mol^{-1}$ for Ca(s).

(c) KOH(aq) has higher $S°$, 91.6 J K^{-1} mol^{-1}, versus 78.9 J $K^{-1}mol^{-1}$ for KOH(s).

18.3 (a) $\Delta S^\circ = 2\text{ mol CO(g)} \times S^\circ(\text{CO(g)}) + 1\text{ mol O}_2\text{(g)} \times S^\circ(\text{O}_2\text{(g)}) - 2\text{ mol CO}_2\text{(g)} \times S^\circ(\text{CO}_2\text{(g)})$
$= \{2 \times (197.674) + (205.138) - 2 \times (213.74)\}\text{ J/K}$
$= 173.01\text{ J/K}$

(b) $\Delta S^\circ = 1\text{ mol NaCl(aq)} \times S^\circ(\text{NaCl(aq)}) - 1\text{ mol NaCl(s)} \times S^\circ(\text{NaCl(s)})$
$= (115.5 - 72.13)\text{ J/K} = 43.4\text{ J/K}$

(c) $\Delta S^\circ = 1\text{ mol MgO(s)} \times S^\circ(\text{MgO(s)}) + 1\text{ mol CO}_2\text{(g)} \times S^\circ(\text{CO}_2\text{(g)}) - 1\text{ mol MgCO}_3\text{(s)} \times S^\circ(\text{MgCO}_3\text{(s)})$
$= (26.94 + 213.74 - 65.7)\text{ J/K}$
$= 175.0\text{ J/K}$

18.4 $N_2(g) + 3\,H_2(g) \rightarrow 2\,NH_3(g)$

$$\Delta H^\circ = 2\text{ mol NH}_3 \times \Delta H_f^\circ(\text{NH}_3\text{(g)}) = 2(-46.11)\text{ kJ} = -92.22\text{ kJ}$$

$$\Delta S^\circ = 2\text{ mol NH}_3 \times S^\circ(\text{NH}_3\text{(g)}) - 1\text{ mol N}_2\text{(g)} \times S^\circ(\text{N}_2\text{(g)}) - 3\text{ mol H}_2 \times S^\circ(\text{H}_2\text{(g)})$$
$$= 2(192.45)\text{ J/K} - (191.61)\text{ J/K} - 3(130.684)\text{ J/K}$$
$$= -198.76\text{ J/K}$$

$$\Delta S^\circ_{\text{universe}} = \frac{-\Delta H^\circ}{T} + \Delta S^\circ = \frac{92.2\text{ kJ}}{298.15\text{ K}} + (-198.76\text{ J/K})$$
$$= \frac{92{,}200\text{ J}}{298.15\text{ K}} - 198.76\text{ J/K} = 110.5\text{ J/K}$$

The process is product-favored.

18.5 (a) $\Delta H^\circ = \{(-238.66) - (-110.525)\}\text{ kJ} = -128.14\text{ kJ}$
$\Delta S^\circ = \{(126.8) - 197.674 - 2 \times 130.684\}\text{ J/K}$
$= -332.2\text{ J/K}$
$\Delta G^\circ = \Delta H^\circ = T\Delta S^\circ$
$= -128.14 \times 10^3\text{ J} - 298.15\text{ K} \times (-332.2\text{ J/K})$
$= -29.09 \times 10^3\text{ J} = -29.09\text{ kJ}$

(b) $\Delta G^\circ = [-166.27 - (-137.168)]\text{ kJ} = -29.10\text{ kJ}$. The two results agree.

(c) ΔG° is negative. The reaction is product-favored at 298.15 K. Because ΔS° is negative, at very high temperatures the reaction will become reactant-favored.

18.6 (a) $T = \Delta H^\circ/\Delta S^\circ = (-565{,}968\text{ J})/(-173.01\text{ J/K}) = 3271\text{ K}$
(b) The reaction is exothermic and therefore is product-favored at temperatures lower than 3271 K.

18.7 (a) $\Delta G^\circ = \{-553.04 - 527.81 - (-1128.79)\}\text{ kJ}$
$= 47.94\text{ kJ}$

$$K^\circ = e^{-\Delta G^\circ/RT} = e^{-(47.94\text{ kJ/mol})/(8.314\text{ J K}^{-1}\text{ mol}^{-1})(298\text{ K})}$$
$$= e^{-(47{,}940\text{ J/mol})/(8.314\text{ J K}^{-1}\text{ mol}^{-1})(298\text{ K})}$$
$$= e^{-19.35} = 3.9 \times 10^{-9}\text{ (close to } K_c)$$

(b) $K^\circ = e^{-14.68} = 4.2 \times 10^{-7}$ (agrees with K_c)
(c) $K^\circ = e^{-(-1.909)} = 6.75$ (agrees with K_P)
For reactions (a) and (b), $K_c = K^\circ$.

18.8 (a) At 298 K,
$\Delta G^\circ = 2 \times (-16.45)\text{ kJ} = -32.9\text{ kJ}$
$$K^\circ = e^{-(-32{,}900\text{ J})/(8.314\text{ J K}^{-1}\text{ mol}^{-1})(298\text{ K})} = e^{13.28} = 5.8 \times 10^5$$

(b) At 450. K,
$\Delta G^\circ = \Delta H^\circ - T\Delta S^\circ = -92.22\text{ kJ} - (450.)(-0.19876)\text{ kJ}$
$= -2.78\text{ kJ}$
$$K^\circ = e^{-(-2780\text{ J})/(8.314\text{ J K}^{-1})(450.\text{ K})} = 2.10$$

(c) At 800. K,
$\Delta G^\circ = -92.22\text{ kJ} - (800.)(-0.19876)\text{ kJ} = 66.79\text{ kJ}$
$$K^\circ = e^{-(66{,}790\text{ J})/(8.314\text{ J K}^{-1}\text{ mol}^{-1})(800.\text{ K})} = 4.3 \times 10^{-5}$$

18.9 $\Delta G = \Delta G^\circ + RT\ln Q$
ΔG° was calculated in Problem-Solving Practice 18.7 to be 47.94 kJ for the reverse of this reaction. Therefore,
$\Delta G^\circ = -47.94\text{ kJ}$
$$\Delta G = -47.94\text{ kJ/mol} + (8.314\text{ J K}^{-1}\text{ mol}^{-1})(298\text{ K})\ln\left(\frac{1}{(0.023)(0.13)}\right)$$
$$= -47.94\text{ kJ/mol} + 1.44 \times 10^4\text{ J/mol}$$
$$= -47.94\text{ kJ/mol} + 14.4\text{ kJ/mol} = -33.5\text{ kJ/mol}$$

18.10 (a) $\Delta G^\circ = 2(-137.168)\text{ kJ} - 2(-394.359)\text{ kJ} = 514.382\text{ kJ}$. The reaction is reactant-favored, and at least 514.382 kJ of work must be done to make it occur.
(b) $\Delta G^\circ = 2(-742.2)\text{ kJ} = -1484.4\text{ kJ}$. The reaction is product-favored and could do up to 1484.4 kJ of useful work.

18.11 $\Delta G^{\circ\prime} = \Delta G^\circ$ for this reaction because none of the reactants or products requires a standard state different from 1 bar or 1 mol/L.

18.12 (a) The strongest phosphate donor has the most negative ΔG° for its reaction with water to produce dihydrogen phosphate. The $\Delta G^{\circ\prime}$ values are given in Problem-Solving Example 18.12. Creatine phosphate, at -43.1 kJ, has the most negative value and is the strongest phosphate donor.
(b) Glycerol 3-phosphate has the least negative $\Delta G^{\circ\prime}$ at -9.7 kJ and therefore is the weakest phosphate donor.
(c) See parts (a) and (b) for explanation.

18.13 (a) $\Delta G_f^\circ(\text{MgO(s)}) = -569.43\text{ kJ}$, so formation of MgO(s) is product-favored and MgO(s) is thermodynamically stable.
(b) $\Delta G_f^\circ(\text{N}_2\text{H}_4(\ell)) = 149.34\text{ kJ}$; kinetically stable.
(c) $\Delta G_f^\circ(\text{C}_2\text{H}_6\text{(g)}) = -32.82\text{ kJ}$; thermodynamically stable.
(d) $\Delta G_f^\circ(\text{N}_2\text{O(g)}) = 104.20\text{ kJ}$; kinetically stable.

Chapter 19

19.1 Reducing agents are indicated by "red" and oxidizing agents are indicated by "ox." Oxidation numbers are shown above the symbols for the elements.

(a) $\overset{0}{2\,\text{Fe}}(s)_{\text{red}} + \overset{0}{3\,\text{Cl}_2}(g)_{\text{ox}} \rightarrow 2\,\overset{+3}{\text{Fe}}\overset{-1}{\text{Cl}_3}(s)$

(b) $\overset{0}{2\text{H}_2}(g)_{\text{red}} + \overset{0}{\text{O}_2}(g)_{\text{ox}} \rightarrow 2\,\overset{+1}{\text{H}_2}\overset{-2}{\text{O}}(\ell)$

(c) $\overset{0}{\text{Cu}}(S)_{\text{red}} + 2\,\overset{+5}{\text{N}}\overset{-2}{\text{O}_3^-}(aq)_{\text{ox}} + 4\,\overset{+1}{\text{H}_3}\text{O}^+(aq) \rightarrow$
$\overset{+2}{\text{Cu}^{2+}}(aq) + 2\,\overset{+4}{\text{N}}\overset{-2}{\text{O}_2}(g) + 6\,\overset{+1}{\text{H}_2}\overset{-2}{\text{O}}(\ell)$

(d) $\overset{0}{\text{C}}(s)_{\text{red}} + \overset{0}{\text{O}_2}(g)_{\text{ox}} \rightarrow \overset{+4}{\text{C}}\overset{-2}{\text{O}_2}(g)$

(e) $6\,\overset{+2}{\text{Fe}^{2+}}(aq)_{\text{red}} + \overset{+6}{\text{Cr}_2}\overset{-2}{\text{O}_7^{2-}}(aq)_{\text{ox}} + 14\,\overset{+1}{\text{H}_3}\overset{-2}{\text{O}^+}(aq) \rightarrow$
$6\,\overset{+3}{\text{Fe}^{3+}}(aq) + 2\,\overset{+3}{\text{Cr}^{3+}}(aq) + 21\,\overset{+1}{\text{H}_2}\overset{-2}{\text{O}}(\ell)$

19.2 (a) Oxidation half-reaction: $Cd(s) \rightarrow Cd^{2+}(aq) + 2\,e^-$
Reduction half-reaction: $Cu^{2+}(aq) + 2\,e^- \rightarrow Cu(s)$
The sum of the two half-reactions is
$Cd(s) + Cu^{2+}(aq) \rightarrow Cd^{2+}(aq) + Cu(s)$

(b) Oxidation half-reaction: $Mg(s) \rightarrow Mg^{2+}(aq) + 2\,e^-$
Reduction half-reaction: $Cr^{3+}(aq) + 3\,e^- \rightarrow Cr(s)$
The sum of the two half-reactions is
$3\,Mg(s) + 2\,Cr^{3+}(aq) \rightarrow 3\,Mg^{2+}(aq) + 2\,Cr(s)$

(c) Oxidation half-reaction: $Cu^+(aq) \rightarrow Cu^{2+}(aq) + e^-$
Reduction half-reaction: $Fe^{3+}(aq) + e^- \rightarrow Fe^{2+}(aq)$
The sum of the two half-reactions is
$Fe^{3+}(aq) + Cu^+(aq) \rightarrow Fe^{2+}(aq) + Cu^{2+}(aq)$

19.3 ***Step 1.*** This is an oxidation-reduction reaction. It is obvious that Zn is oxidized by its change in oxidation state.
Step 2. The half-reactions are

$Zn(s) \longrightarrow Zn^{2+}(aq)$ (This is the oxidation reaction.)

$Cr_2O_7^{2-}(aq) \longrightarrow 2\ Cr^{3+}(aq)$ (This is the reduction reaction.)

Step 3. Balance the atoms in the half-reactions. The atoms are balanced in the Zn half-reaction. We need to add water and H in the $Cr_2O_7^{2-}$ half-reaction. Fourteen H^+ ions are required on the right to combine with the seven O atoms.

$$Cr_2O_7^{2-}(aq) + 14\ H^+(aq) \longrightarrow 2\ Cr^{3+}(aq) + 7\ H_2O(\ell)$$

Step 4. Balance the half-reactions for charge. Write the Zn half-reaction as

$$Zn(s) \longrightarrow Zn^{2+}(aq) + 2\ e^-$$

and write the $Cr_2O_7^{2-}$ half-reaction as

$$Cr_2O_7^{2-}(aq) + 14\ H^+(aq) + 6\ e^- \longrightarrow 2\ Cr^{3+}(aq) + 7\ H_2O(\ell)$$

Step 5. Multiply the half-reactions by factors to make the number of electrons gained equal to the number lost.

$$3\ [Zn(s) \longrightarrow Zn^{2+}(aq) + 2\ e^-]$$

$$1\ [Cr_2O_7^{2-}(aq) + 14\ H^+(aq) + 6\ e^- \longrightarrow 2\ Cr^{3+}(aq) + 7\ H_2O(\ell)]$$

Step 6. Add the two half-reactions, canceling the electrons.

$$3\ Zn(s) \longrightarrow 3\ Zn^{2+}(aq) + 6\ e^-$$

$$Cr_2O_7^{2-}(aq) + 14\ H^+(aq) + 6\ e^- \longrightarrow 2\ Cr^{3+}(aq) + 7\ H_2O(\ell)$$

$$Cr_2O_7^{2-}(aq) + 3\ Zn(s) + 14\ H^+(aq) \longrightarrow 2\ Cr^{3+}(aq) + 3\ Zn^{2+}(aq) + 7\ H_2O(\ell)$$

Step 7. Everything checks.
Step 8. Water was added in Step 3. The balanced equation is

$$Cr_2O_7^{2-}(aq) + 3\ Zn(s) + 14\ H_3O^+(aq) \longrightarrow 2\ Cr^{3+}(aq) + 3\ Zn^{2+}(aq) + 21\ H_2O(\ell)$$

19.4 ***Step 1.*** This is an oxidation-reduction reaction. The wording of the question says Al reduces NO_3^- ion. Al is oxidized.
Step 2. The half-reactions are:

$Al(s) \longrightarrow Al(OH)_4^-(aq)$ (This is the oxidation reaction.)

$NO_3^-(aq) \longrightarrow NH_3(aq)$ (This is the reduction reaction.)

Step 3. Balance the atoms in the half-reactions. For the Al half-reaction, add four H^+ ions on the right and four water molecules on the left.

$$Al(s) + 4\ H_2O(\ell) \longrightarrow Al(OH)_4^- + 4\ H^+(aq)$$

For the NO_3^- half-reaction,

$$NO_3^-(aq) + 9\ H^+(aq) \longrightarrow NH_3(aq) + 3\ H_2O(\ell)$$

Step 4. Balance the half-reactions for charge. Put $3\ e^-$ on the right in the Al half-reaction

$$Al(s) + 4\ H_2O(\ell) \longrightarrow Al(OH)_4^- + 4\ H^+(aq) + 3\ e^-$$

and put $8\ e^-$ on the left side of the NO_3^- half-reaction.

$$NO_3^-(aq) + 9\ H^+(aq) + 8\ e^- \longrightarrow NH_3(aq) + 3\ H_2O(\ell)$$

Step 5. Multiply the half-reactions by factors to make the electrons gained equal to those lost.

$$8[Al(s) + 4\ H_2O(\ell) \longrightarrow Al(OH)_4^- + 4\ H^+(aq) + 3\ e^-]$$

$$3[NO_3^-(aq) + 9\ H^+(aq) + 8\ e^- \longrightarrow NH_3(aq) + 3\ H_2O(\ell)]$$

Step 6. Remove $H^+(aq)$ ions by adding an appropriate amount of OH^-. For the Al half-reaction, add 32 OH^- ions to get

$$8\ Al(s) + 32\ OH^-(aq) + 32\ H_2O(\ell) \longrightarrow 8\ Al(OH)_4^- + 32\ H_2O + 24\ e^-$$

For the NO_3^- half-reactions, add 27 OH^- ions to get

$$3\ NO_3^-(aq) + 27\ H_2O + 24\ e^- \longrightarrow 3\ NH_3(aq) + 9\ H_2O(\ell) + 27\ OH^-(aq)$$

Step 7. Add both half-reactions and cancel the electrons.

$$8\ Al(s) + 32\ OH^-(aq) + 32\ H_2O(\ell) \longrightarrow 8\ Al(OH)_4^- + 32\ H_2O + 24\ e^-$$

$$3\ NO_3^-(aq) + 27\ H_2O + 24\ e^- \longrightarrow 3\ NH_3(aq) + 9\ H_2O(\ell) + 27\ OH^-(aq)$$

$$3\ NO_3^-(aq) + 8\ Al(s) + 59\ H_2O(\ell) + 32\ OH^-(aq) \longrightarrow 8\ Al(OH)_4^-(aq) + 3\ NH_3(aq) + 27\ OH^-(aq) + 41\ H_2O(\ell)$$

Step 8. Make a final check. Since there are OH^- ions and water molecules on both sides of the equation, cancel them out. This gives the final balanced equation.

$$3\ NO_3^-(aq) + 8\ Al(s) + 18\ H_2O(\ell) + 5\ OH^-(aq) \longrightarrow 8\ Al(OH)_4^-(aq) + 3\ NH_3(aq)$$

(This is a fairly complicated equation to balance. If you balanced this one with a minimum of effort, your understanding of balancing redox equations is rather good. If you had to struggle with one or more of the steps, go back and repeat them.)

19.5 (a) $Ni(s) \rightarrow Ni^{2+}(aq) + 2\ e^-$ (This is the oxidation half-reaction.)

$2\ Ag^+(aq) + 2\ e^- \rightarrow 2\ Ag(s)$ (This is the reduction half-reaction.)

(b) The oxidation of Ni takes place at the anode and the reduction of Ag^+ takes place at the cathode.
(c) Electrons would flow through an external circuit from the anode (where Ni is oxidized) to the cathode (where Ag^+ ions are reduced).
(d) Nitrate ions would flow through the salt bridge to the anode compartment. Potassium ions would flow into the cathode compartment.

19.6 Oxidation half-reaction: $Fe(s) \rightarrow Fe^{2+}(aq, 1\ M) + 2\ e^-$ (anode)

Reduction half-reaction: $Cu^{2+}(aq, 1\ M) + 2\ e^- \rightarrow Cu(s)$ (cathode)

$$E^\circ_{cell} = +0.78\ V = E^\circ_{cathode} - E^\circ_{anode}$$

Since $E^\circ_{cathode} = +0.34$ V, E°_{anode} must be -0.44 V.

19.7 $F_2(g) + 2\ e^- \longrightarrow 2\ F^-(aq)$ $\quad E^\circ_{cathode} = +2.87$ V

$2\ Li(s) \longrightarrow 2\ Li^+(aq) + 2\ e^-$ $\quad E^\circ_{anode} = -3.045$ V

$2\ Li(s) + F_2(g) \longrightarrow 2\ Li^+(aq) + 2\ F^-(aq)$ $\quad E^\circ_{cell} = +5.91$ V

$$E^\circ_{cell} = E^\circ_{cathode} - E^\circ_{anode} = +2.87 - (-3.045) = +5.91\ V$$

19.8 The two half-reactions are

$Hg^{2+}(aq) + 2\ e^- \longrightarrow 2\ Hg(\ell)$ $\quad E^\circ_{cathode} = +0.855$ V

$2\ I^-(aq) \longrightarrow I_2(s) + 2\ e^-$ $\quad E^\circ_{anode} = +0.535$ V

$$E^\circ_{cell} = E^\circ_{cathode} - E^\circ_{anode} = +0.855 - 0.535 = +0.320\ V$$

The reaction is product-favored as written.

19.9 We first need to calculate E°_{cell}, and to do this we break the reaction into two half-reactions.

Ox: $Sn^{2+}(aq) \longrightarrow Sn^{4+}(aq) + 2\ e^-$ $\quad E^\circ_{anode} = +0.15$ V

Red: $I_2(s) + 2\ e^- \longrightarrow 2\ I^-(aq)$ $\quad E^\circ_{cathode} = +0.535$ V

$I_2(s) + Sn^{2+}(aq) \longrightarrow 2\ I^-(aq) + Sn^{4+}(aq)$ $\quad E^\circ_{cell} = +0.385$ V

$$E^\circ_{cell} = E^\circ_{cathode} - E^\circ_{anode} = +0.535 - 0.15 = +0.385\ V$$

E°_{cell} is 0.385 V, and 2 mol of electrons are transferred.

$$\log K = \frac{nE^\circ\ \text{V}}{0.0592\ \text{V}} = \frac{2 \times 0.385\ \text{V}}{0.0592\ \text{V}}$$

$$= 13.00 \text{ and } K = 1 \times 10^{13}$$

The large value of K indicates that the reaction is strongly product-favored as written.

19.10 $E_{cell} = 0.51\ \text{V} - \left(\frac{0.0592\ \text{V}}{2} \times \log \frac{3}{0.010}\right)$

$$= 0.51\ \text{V} - (0.0296\ \text{V} \times \log(300))$$

$$= 0.51\ \text{V} - 0.073\ \text{V} = +0.44\ \text{V}$$

19.11 If the pH = 3.66, then the E_{cell} = 0.217 V.

19.12 (a) The net cell reaction would be

$$2\ Na^+ + 2\ Br^- \longrightarrow 2\ Na + Br_2$$

Sodium ions would be reduced at the cathode and bromide ions would be oxidized at the anode.

(b) H_2 would be produced at the cathode for the same reasons given in Problem-Solving Example 19.8. That reaction is

$$2\ H_2O(\ell) + 2\ e^- \longrightarrow H_2(g) + 2\ OH^-(aq)$$

At the anode, two reactions are possible: the oxidation of water and the oxidation of Br^- ions.

$$6\ H_2O(\ell) \longrightarrow O_2(g) + 4\ H_3O^+(aq) + 4\ e^- \qquad E^\circ = 1.229\ \text{V}$$

$$2\ Br^-(aq) \longrightarrow Br_2(\ell) + 2\ e^- \qquad E^\circ = 1.08\ \text{V}$$

Bromide ions will be oxidized to Br_2 because that potential is smaller. The net cell reaction is

$$2\ H_2O(\ell) + 2\ Br^-(aq) \longrightarrow Br_2(\ell) + H_2(g) + 2\ OH^-(aq)$$

(c) Sn metal will be formed at the cathode because its reduction potential (−0.14 V) is less negative than the potential for the reduction of water. O_2 will form at the anode because the E° value for the oxidation of water is smaller than the E° value for the oxidation of Cl^-. The net cell reaction is

$$2\ Sn^{2+}(aq) + 6\ H_2O(\ell) \longrightarrow 2\ Sn(s) + O_2(g) + 4\ H_3O^+(aq)$$

19.13 First, calculate the quantity of charge:

$$\text{Charge} = (25 \times 10^3\ \text{A})(1\ \text{h})\left(\frac{60\ \text{s}}{1\ \text{min}}\right)\left(\frac{60\ \text{min}}{1\ \text{h}}\right)$$

$$= 9.0 \times 10^7\ \text{A}\cdot\text{s} = 9.0 \times 10^7\ \text{C}$$

Then calculate the mass of Na:

$$\text{Mass of Na} = (9.0 \times 10^7\ \text{C}) \times \left(\frac{1\ \text{mol } e^-}{96{,}500\ \text{C}}\right)\left(\frac{1\ \text{mol Na}}{1\ \text{mol } e^-}\right)\left(\frac{22.99\ \text{g Na}}{1\ \text{mol Na}}\right)$$

$$= 2.1 \times 10^4\ \text{g Na}$$

Chapter 20

20.1 (a) ${}^{237}_{93}Np \rightarrow {}^{4}_{2}He + {}^{233}_{91}Pa$ (b) ${}^{35}_{16}S \rightarrow {}^{0}_{-1}e + {}^{35}_{17}Cl$

20.2 (a) ${}^{11}_{6}C \rightarrow {}^{11}_{5}B + {}^{0}_{1}e$ (b) ${}^{35}_{16}S \rightarrow {}^{35}_{17}Cl + {}^{0}_{-1}e$

(c) ${}^{30}_{15}P \rightarrow {}^{0}_{+1}e + {}^{30}_{14}Si$ (d) ${}^{22}_{11}Na \rightarrow {}^{0}_{-1}e + {}^{22}_{12}Mg$

20.3 (a) ${}^{42}_{19}K \rightarrow {}^{0}_{-1}e + {}^{42}_{20}Ca$ (b) ${}^{234}_{92}U \rightarrow {}^{4}_{2}He + {}^{230}_{90}Th$

(c) ${}^{20}_{9}F \rightarrow {}^{0}_{-1}e + {}^{20}_{10}Ne$

20.4 (a) ${}^{90}_{38}Sr \rightarrow {}^{0}_{-1}e + {}^{90}_{39}Y$

(b) It takes 4 half-lives (4 × 29 y = 116 y) for the activity to decrease to 125 beta particles emitted per minute:

Number of Half-lives	Change of Activity	Total Elapsed Time (y)
1	2000 to 1000	29
2	1000 to 500	58
3	500 to 250	87
4	250 to 125	116

20.5 (a) $t_{1/2} = \dfrac{0.693}{9.3 \times 10^{-3}\ \text{d}^{-1}} = 75\ \text{d}$

(b) $\ln(\text{fraction remaining}) = -k \times t = -(9.3 \times 10^{-3}\ \text{d}^{-1}) \times (100\ \text{d}) = -0.930$

Fraction of iridium-192 remaining = $e^{-0.930}$ = 0.39. Therefore, 39% of the original iridium-192 remains.

20.6 $k = \dfrac{0.693}{1.60 \times 10^3\ \text{y}} = 4.33 \times 10^{-4}\ \text{y}^{-1}$

As of 2004: $\ln(\text{fraction remaining}) = -k \times t = -(4.33 \times 10^{-4}\ \text{y}^{-1}) \times 83\ \text{y} = -3.39 \times 10^{-2}$

Fraction of radium-226 remaining = $e^{-0.0339}$ = 0.965. Therefore, 96.5% of the original radium-226 remains; 0.965 × 1.00 g = 0.965 g.

20.7 $\ln(0.60) = -0.510 = -k \times t$

$$k = \frac{0.693}{t_{1/2}} = \frac{0.693}{12.3\ \text{y}} = 0.0563\ \text{y}^{-1}$$

$$t = \frac{-0.510}{-0.0563\ \text{y}^{-1}} = 9.1\ \text{y}$$

Chapter 21

21.1 $1.5 \times 10^9\ \text{mol Na} \times \dfrac{4\ \text{mol NaCl}}{4\ \text{mol Na}} \times \dfrac{58.5\ \text{g NaCl}}{1\ \text{mol NaCl}} \times \dfrac{1\ \text{lb NaCl}}{454\ \text{g NaCl}} \times \dfrac{1\ \text{ton NaCl}}{2000\ \text{lb NaCl}} = 9.7 \times 10^4\ \text{ton NaCl}$

21.2 $\text{mol of Al} = 1.00\ \text{ton Al} \times \dfrac{2000\ \text{lb Al}}{1\ \text{ton Al}} \times \dfrac{454\ \text{g Al}}{1\ \text{ton Al}} \times \dfrac{1\ \text{mol Al}}{26.98\ \text{g Al}} = 3.37 \times 10^4\ \text{mol Al}$

$Al^{3+} + 3\ e^- \rightarrow Al$; therefore, 3 mol electrons are needed to produce 1 mol Al.

Number of moles of electrons

$$= 3.37 \times 10^4\ \text{mol Al} \times \frac{3\ \text{mol } e^-}{1\ \text{mol Al}}$$

$$= 1.01 \times 10^5\ \text{mol } e^-$$

$$\text{Charge} = 1.01 \times 10^5\ \text{mol } e^- \times \frac{9.65 \times 10^4\ \text{C}}{1\ \text{mol } e^-}$$

$$= 9.75 \times 10^9\ \text{C}$$

$$\text{Time} = 9.75 \times 10^9\ \text{C} \times \frac{1\ \text{s}}{1.00 \times 10^5\ \text{C}} = 9.75 \times 10^4\ \text{s}$$

$$= 1.63 \times 10^3\ \text{min} = 27.0\ \text{h}$$

21.3 I^- is oxidized to I_2; IO_3^- is reduced to I_2. IO_3^- is the oxidizing agent and I^- is the reducing agent.

21.4 (a)

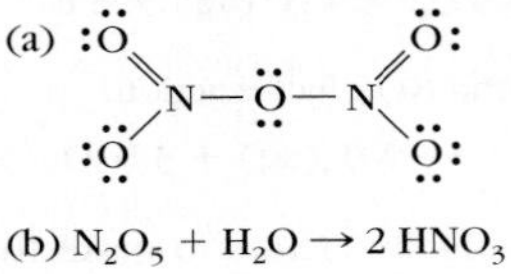

(b) $N_2O_5 + H_2O \rightarrow 2\ HNO_3$

21.5 (a) The formation of NO_2 from NO is exothermic. Thus, lowering the temperature favors the forward reaction (NO_2 formation).
(b) By reacting with water, NO_2 is converted to HNO_3 thereby removing NO_2 from the reaction mixture.

21.6
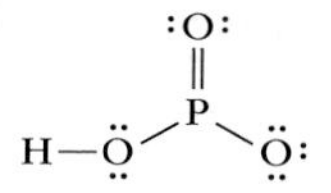

21.7 $S_8(s) + 12\,O_2(g) \longrightarrow 8\,SO_3(g)$

$$1.3 \times 10^{10}\ \text{kg } S_8 \times \frac{1\ \text{mol } S_8}{0.256\ \text{kg } S_8} \times \frac{8\ \text{mol } SO_3}{1\ \text{mol } S_8} \times \frac{0.080\ \text{kg } SO_3}{1\ \text{mol } SO_3} = 3.3 \times 10^{10}\ \text{kg } SO_3$$

21.8 (a) Non-redox reaction; no change in oxidation numbers
(b) Redox reaction; F_2 is the oxidizing agent; S_8 is the reducing agent.

Chapter 22

22.1 (a) Mn [Ar]$3d^5 4s^2$

3d: ↑ ↑ ↑ ↑ ↑ 4s: ↑↓

Mn^{2+} [Ar]$3d^5 4s^0$

3d: ↑ ↑ ↑ ↑ ↑ 4s: (empty)

(b) Cr [Ar]$3d^5 4s^1$

3d: ↑ ↑ ↑ ↑ ↑ 4s: ↑

Cr^{3+} [Ar]$3d^3 4s^0$

3d: ↑ ↑ ↑ (empty) (empty) 4s: (empty)

22.2 Cooling the iron slowly would shift the equilibrium to favor the reverse reaction, the conversion of cementite to iron and carbon (graphite).

22.3 Assume that 50% of the iron comes from each ore.

$$Fe_2O_3\text{: } 5.00 \times 10^2\ \text{kg Fe} \times \frac{159.7\ \text{kg } Fe_2O_3}{111.7\ \text{kg Fe}} = 7.15 \times 10^2\ \text{kg } Fe_2O_3$$

$$Fe_3O_4\text{: } 5.00 \times 10^2\ \text{kg Fe} \times \frac{231.6\ \text{kg } Fe_3O_4}{167.6\ \text{kg Fe}} = 6.91 \times 10^2\ \text{kg } Fe_3O_4$$

Total $= 7.15 \times 10^2$ kg $+ 6.91 \times 10^2$ kg $= 1.41 \times 10^3$ kg

22.4 By controlling the voltage, the zinc could be removed and then the lead.

22.5 $E_{cell} = 0.00\ \text{V} = +0.795\ \text{V} - 0.00985 \log\left(\frac{1}{[H^+]^{14}}\right)$

$$-0.795 = -0.00985 \log \frac{1}{[H^+]^{14}}$$

$$\log \frac{1}{[H^+]^{14}} = \frac{-0.795}{-0.00985} = 80.7$$

$$\frac{1}{[H^+]^{14}} = 10^{80.7} = 5.01 \times 10^{80}$$

$$1 = (5.01 \times 10^{80})[H^+]^{14}$$

$$[H^+]^{14} = \frac{1}{5.01 \times 10^{80}} = 2.00 \times 10^{-81}$$

$$14 \log[H^+] = \log(2.00 \times 10^{-81})$$

$$\log[H^+] = \frac{-80.7}{14} = -5.76$$

$$-\log[H^+] = \text{pH} = 5.76$$

22.6 (a) SO_4^{2-} (b) Cu^{2+}
(c) NH_3 (d) $[Cu(NH_3)_4]^{2+}$

22.7 (a) Diamminesilver(I) nitrate
(b) $[Fe(H_2O)_5(NCS)]Cl_2$

22.8 (a) Two (b) Zero (c) Five

22.9 Two

22.10 High-spin: four unpaired electrons; low-spin: two unpaired electrons; both complexes are paramagnetic due to their unpaired electrons.

APPENDIX

L Answers to Exercises

Chapter 1

1.1 (a) These temperatures can be compared to the boiling point of water, 212 °F or 100 °C. So 110 °C is a higher temperature than 180 °F.
(b) These temperatures can be compared to normal body temperature, 98.6 °F or 37.0 °C. So 36 °C is a lower temperature than 100 °F.
(c) This temperature can be compared to normal body temperature, 37.0 °C. Since body temperature is above the melting point, gallium held in one's hand will melt.

Courtesy of James H. Maynard/ University of Wisconsin

Courtesy of James H. Maynard/ University of Wisconsin

A sample of gallium melts in a gloved hand.

1.2 The figure on page 9 indicates that kerosene is the top layer, vegetable oil is the middle layer, and water is the bottom layer.
(a) Since the least dense liquid will be the top layer and the densest liquid will be the bottom layer, the densities increase in the order kerosene, vegetable oil, water.
(b) If vegetable oil is added to the tube, the top and bottom layers will remain the same, but the middle layer will become larger.
(c) If kerosene is now added to the tube the top layer will grow, but the middle and bottom layers will remain the same. The order of levels will *not change*. Density does not depend on the quantity of material present. So no matter how much of each liquid is present, the densities increase in the order kerosene, vegetable oil, water.

1.3 (a) Properties: blue (qualitative), melts at 99 °C (quantitative)
Change: melting
(b) Properties: white, cubic (both qualitative)
Change: none
(c) Properties: mass of 0.123 g, melts at 327 °C (both quantitative)
Change: melting
(d) Properties: colorless, vaporizes easily (both qualitative), boils at 78 °C, density of 0.789 g/mL (both quantitative)
Changes: vaporizing, boiling

1.4 Physical change: boiling water
Chemical changes: combustion of propane, cooking the egg

1.5 (a) Homogeneous mixture (solution)
(b) Heterogeneous mixture (contains carbon dioxide gas bubbles in a solution of sugar and other substances in water)
(c) Heterogeneous mixture of dirt and oil
(d) Element; diamond is pure carbon.
(e) Modern quarters (since 1965) are composed of a pure copper core (that can be seen when they are viewed side-on) and an outer layer of 75% Cu, 25% Ni alloy, so they are heterogeneous matter. Pre-1965 quarters are fairly pure silver.
(f) Compound; contains carbon, hydrogen, and oxygen

1.6 (a) Energy from the sun warms the ice and the water molecules vibrate more; eventually they break away from their fixed positions in the solid and liquid water forms. As the temperature of the liquid increases, some of the molecules have enough energy to become widely separated from the other molecules, forming water vapor (gas).
(b) Some of the water molecules in the clothes have enough speed and energy to escape from the liquid state and become water vapor; these molecules are carried away from the clothes by breezes or air currents. Eventually nearly every water molecule in the clothes vaporizes, and the clothes become dry.
(c) Water molecules from the air come into contact with the cold glass, and their speeds are decreased, allowing them to become liquid. As more and more molecules enter the liquid state, droplets form on the glass.
(d) Some water molecules escape from the liquid state, forming water vapor. As more and more molecules escape, the ratio of sugar molecules to water molecules becomes larger and larger, and eventually some sugar molecules start to stick together. As more and more sugar molecules stick to each other, a visible crystal forms. Eventually all of the water molecules escape, leaving sugar crystals behind.

1.7 (a) Tellurium, Te, Earth (Latin *tellus* means Earth); uranium, U, for Uranus; neptunium, Np, for Neptune; and plutonium, Pu, for Pluto. (Mercury, like the planet Mercury, is named for a Roman god.)
(b) Californium, Cf
(c) Curium, Cm, for Marie Curie; and meitnerium, Mt, for Lise Meitner
(d) Scandium, Sc, for Scandinavia; gallium, Ga, for France (Latin *Gallia* means France); germanium, Ge, for Germany; ruthenium, Ru, for Russia; europium, Eu, for Europe; polonium, Po, for Poland; francium, Fr, for France; americium, Am, for America; californium, Cf, for California
(e) H, He, C, N, O, F, Ne, P, S, Cl, Ar, Se, Br, Kr, I, Xe, At, Rn

1.8 (a) Elements that consist of diatomic molecules are H, N, O, F, Cl, Br, and I; At is radioactive and there is probably less than 50 mg of naturally occurring At on Earth, but it does form diatomic molecules; H, N, O, plus group 7A.
(b) Metalloids are B, Si, Ge, As, Sb, and Te; along a zig-zag line from B to Te.

1.9 Tin and lead are two different elements; allotropes are two different forms of the same element, so tin and lead are not allotropes.

Chapter 2

2.1 The movement of the comb though your hair removes some electrons, leaving slight charges on your hair and the comb. The charges must sum to zero; therefore, one must be slightly positive and one must be slightly negative, so they attract each other.

2.2 (a) A nucleus is about one ten-thousandth as large as an atom, so $100 \text{ m} \times (1 \times 10^{-4}) = 1 \times 10^{-2} \text{ m} = 1 \text{ cm}$. (b) Many everyday objects are about 1 cm in size—for example, a grape.

2.3 The statement is wrong because two atoms that are isotopes always have the same number of protons. It is the number of neutrons that varies from one isotope of an element to another.

2.4 Atomic weight of lithium
$= (0.07500)(6.015121 \text{ amu}) + (0.9250)(7.016003 \text{ amu})$
$= 0.451134 \text{ amu} + 6.489802 \text{ amu}$
$= 6.940936 \text{ amu}$, or 6.941 amu

2.5 Because the most abundant isotope is magnesium-24 (78.70%), the atomic weight of magnesium is closer to 24 than to 25 or 26, the mass numbers of the other magnesium isotopes, which make up approximately 21% of the remaining mass. The simple arithmetic average is $(24 + 25 + 26)/3 = 25$, which is larger than the atomic weight. In the arithmetic average, the relative abundance of each magnesium isotope is 33%, far less than the actual percent abundance of magnesium-24, and much more than the natural percent abundances of magnesium-25 and magnesium-26.

2.6 There is no reasonable pair of values of the mass numbers for Ga that would have an average value of 69.72.

2.7 Start by calculating the number of moles in 10.00 g of each element.

$$10.00 \text{ g Li} \times \frac{1 \text{ mol Li}}{6.941 \text{ g Li}} = 1.441 \text{ mol Li}$$

$$10.00 \text{ g Ir} \times \frac{1 \text{ mol Ir}}{192.22 \text{ g Ir}} = 0.05202 \text{ mol Ir}$$

Multiply the number of moles of each element by Avogadro's number.

$$1.441 \text{ mol Li} \times 6.022 \times 10^{23} \text{ atoms/mol} = 8.678 \times 10^{23} \text{ atoms Li}$$

$$0.05202 \text{ mol Ir} \times 6.022 \times 10^{23} \text{ atoms/mol} = 3.133 \times 10^{22} \text{ atoms Ir}$$

Find the difference.
$(8.678 \times 10^{23}) - (0.3133 \times 10^{23})$
$= 8.365 \times 10^{23}$ more atoms of Li than Ir

2.8 1. (a) 13 metals: potassium (K), calcium (Ca), scandium (Sc), titanium (Ti), vanadium (V), chromium (Cr), manganese (Mn), iron (Fe), cobalt (Co), nickel (Ni), copper (Cu), zinc (Zn), and gallium (Ga)
(b) Three nonmetals: selenium (Se), bromine (Br), and krypton (Kr)
(c) Two metalloids: germanium (Ge) and arsenic (As)
2. (a) Groups 1A (except hydrogen), 2A, 1B, 2B, 3B, 4B, 5B, 6B, 7B, 8B
(b) Groups 7A and 8 (c) None
3. Period 6

2.9 The question should ask: Which of the following men does not have an *element* named after him? The correct answer to the question, after it is properly posed, is Isaac Newton.

Chapter 3

3.1 Propylene glycol structural formula:

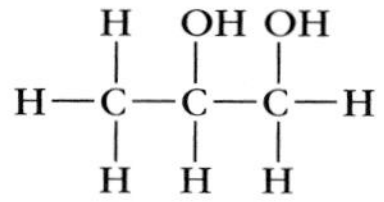

Condensed formula:

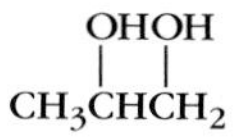

Molecular formula: $C_3H_8O_2$

3.2 (a) CS_2 (b) PCl_3 (c) SBr_2
(d) SeO_2 (e) OF_2 (f) XeO_3

3.3 (a) $C_{16}H_{34}$ and $C_{28}H_{58}$
(b) $C_{14}H_{30}$ has 30 hydrogen atoms.
(c) If n is the number of carbon atoms, then the number of hydrogens should be $2n + 2$. For $C_{16}H_{34}$, $16 \times 2 + 2 = 34$. For $C_{28}H_{58}$, $2 \times 28 + 2 = 58$. For $C_{14}H_{30}$, $2 \times 14 + 2 = 30$.

3.4 The structural and condensed formulas for three constitutional isomers of five-carbon alkanes (pentanes) are

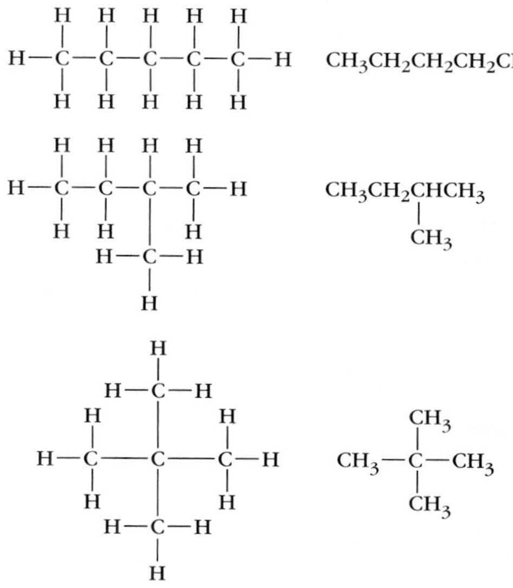

3.5 The compound is most likely ionic because ionic compounds are usually solids at room temperature (as a consequence of strong attractions between oppositely charged ions) and conduct electricity when in solution (as a consequence of movement of ions).

3.6 (a) 174.18 g/mol (b) 386.66 g/mol
(c) 398.07 g/mol (d) 194.19 g/mol

3.7 The statement is true. Because both compounds have the same formula, they have the same molar mass. Thus, 100 g of each compound contains the same number of moles.

3.8 Epsom salt is $MgSO_4 \cdot 7H_2O$, which has a molar mass of 246 g/mol.

$$20 \text{ g} \times \frac{1 \text{ mol}}{246 \text{ g}} = 8.1 \times 10^{-2} \text{ mol Epsom salt}$$

3.9 (a) SF_6 molar mass is 146.06 g/mol; 1.000 mol SF_6 contains 32.07 g S and $18.9984 \times 6 = 113.99$ g F. The mass percents are

$$\frac{32.07 \text{ g S}}{146.06 \text{ g } SF_6} \times 100\% = 21.96\% \text{ S}$$

$$100.0\% - 21.96\% = 78.04\% \text{ F}$$

(b) $C_{12}H_{22}O_{11}$ has a molar mass of 342.3 g/mol; 1.000 mol $C_{12}H_{22}O_{11}$ contains

$$12.011 \times 12 = 144.13 \text{ g C}$$

$$1.0079 \times 22 = 22.174 \text{ g H}$$

$$15.9994 \times 11 = 175.99 \text{ g O}$$

The mass percents of the three elements are

$$\frac{144.13 \text{ g C}}{342.3 \text{ g}} \times 100\% = 42.12\% \text{ C}$$

$$\frac{22.174 \text{ g H}}{342.3 \text{ g}} \times 100\% = 6.478\% \text{ H}$$

$$\frac{175.99 \text{ g O}}{342.3 \text{ g}} \times 100\% = 51.41\% \text{ O}$$

(c) $Al_2(SO_4)_3$ molar mass is 342.15 g/mol; 1.000 mol $Al_2(SO_4)_3$ contains

$$26.9815 \times 2 = 53.96 \text{ g Al}$$

$$32.066 \times 3 = 96.20 \text{ g S}$$

$$15.9994 \times 12 = 192.0 \text{ g O}$$

The mass percents of the three elements are

$$\frac{53.96 \text{ g Al}}{342.15 \text{ g}} \times 100\% = 15.77\% \text{ Al}$$

$$\frac{96.20 \text{ g S}}{342.15 \text{ g}} \times 100\% = 28.12\% \text{ S}$$

$$\frac{192.0 \text{ g O}}{342.15 \text{ g}} \times 100\% = 56.12\% \text{ O}$$

(d) $U(OTeF_5)_6$ molar mass is 1669.6 g/mol; 1.000 mol $U(OTeF_5)_6$ contains 238.0289 g of U and

$$15.9994 \times 6 = 96.00 \text{ g O}$$

$$127.60 \times 6 = 765.6 \text{ g Te}$$

$$18.9984 \times 30 = 570.0 \text{ g F}$$

The mass percents of the four elements are

$$\frac{238.0289 \text{ g U}}{1669.6 \text{ g}} \times 100\% = 14.26\% \text{ U}$$

$$\frac{96.00 \text{ g O}}{1669.6 \text{ g}} \times 100\% = 5.750\% \text{ O}$$

$$\frac{765.6 \text{ g Te}}{1669.6 \text{ g}} \times 100\% = 45.86\% \text{ Te}$$

$$\frac{570.0 \text{ g F}}{1669.6 \text{ g}} \times 100\% = 34.14\% \text{ F}$$

3.10 (a) Carbon, nitrogen, oxygen, phosphorus, hydrogen, selenium, sulfur, Cl, Br, I (b) Calcium and magnesium (c) Chloride, bromide, and iodide (d) Iron, copper, zinc, vanadium (also chromium, manganese, cobalt, nickel, molybdenum, and cadmium)

Chapter 4

4.1 One mol of methane reacts with 2 mol oxygen to produce 1 mol carbon dioxide and 2 mol water.

4.2 (a) The total mass of reactants {4 Fe(s) + 3 O_2(g)} must equal the total mass of products {2 Fe_2O_3(s)}, which is 2.50 g.

(b) The stoichiometric coefficients are 4, 3, and 2.

(c) $1.000 \times 10^4 \text{ O atoms} \times \frac{1 \text{ O}_2 \text{ molecule}}{2 \text{ O atoms}} \times \frac{4 \text{ Fe atoms}}{3 \text{ O}_2 \text{ molecules}} = 6.667 \times 10^3 \text{ Fe atoms}$

4.3 (a) Not balanced; the number of oxygen atoms do not match.

(b) Not balanced; the number of bromine atoms do not match.

(c) Not balanced; the number of sulfur atoms do not match.

4.4 (a) To predict the product of a combination reaction between two elements, we need to know the ion that will be formed by each element when combined. (b) For calcium, Ca^{2+} ions are formed, and for fluorine, F^- ions are formed. (c) The product is CaF_2.

4.5 (a) Magnesium chloride, $MgCl_2$

(b) Magnesium oxide, MgO, and carbon dioxide, CO_2

4.6 $\frac{2 \text{ mol Al}}{3 \text{ mol Br}_2}, \frac{2 \text{ mol Al}}{1 \text{ mol Al}_2\text{Br}_6}, \frac{3 \text{ mol Br}_2}{1 \text{ mol Al}_2\text{Br}_6}$, and their reciprocals

4.7 $0.300 \text{ mol CH}_4 \times \frac{2 \text{ mol H}_2\text{O}}{1 \text{ mol CH}_4} = 0.600 \text{ mol H}_2\text{O}$

$$0.600 \text{ mol H}_2\text{O} \times \frac{18.02 \text{ g H}_2\text{O}}{1 \text{ mol H}_2\text{O}} = 10.8 \text{ g H}_2\text{O}$$

4.8 (a) $300. \text{ g urea} \times \frac{1 \text{ mol urea}}{60.06 \text{ g urea}} \times \frac{2 \text{ mol NH}_3}{1 \text{ mol urea}} \times \frac{17.03 \text{ g NH}_3}{1 \text{ mol NH}_3} = 170. \text{ g NH}_3$

$$100. \text{ g H}_2\text{O} \times \frac{1 \text{ mol H}_2\text{O}}{18.02 \text{ g H}_2\text{O}} \times \frac{2 \text{ mol NH}_3}{1 \text{ mol H}_2\text{O}} \times \frac{17.03 \text{ g NH}_3}{1 \text{ mol NH}_3} = 189. \text{ g NH}_3$$

Therefore, urea is the limiting reactant.

(b) 170. g NH_3

$$300. \text{ g urea} \times \frac{1 \text{ mol urea}}{60.06 \text{ g urea}} \times \frac{1 \text{ mol CO}_2}{1 \text{ mol urea}} \times \frac{44.01 \text{ g CO}_2}{1 \text{ mol CO}_2} = 220. \text{ g CO}_2$$

(c) $300. \text{ g urea} \times \frac{1 \text{ mol urea}}{60.06 \text{ g urea}} \times \frac{1 \text{ mol H}_2\text{O}}{1 \text{ mol urea}} \times \frac{18.02 \text{ g H}_2\text{O}}{1 \text{ mol H}_2\text{O}} = 90.0 \text{ g H}_2\text{O}$

100. g − 90.0 g = 10.0 g H_2O remains

4.9 (1) Impure reactants; (2) Inaccurate weighing of reactants and products

4.10 Assuming that the nicotine is pure, weigh a sample of nicotine and burn the sample. Separately collect and weigh the carbon dioxide and water generated, and calculate the moles and grams of carbon and hydrogen collected. By mass difference, determine the mass of nitrogen in the original sample, then calculate the moles of nitrogen. Calculate the mole ratios of carbon, hydrogen, and nitrogen in nicotine to determine its empirical formula.

Chapter 5

5.1 It is possible for an exchange reaction to form two different precipitates—for example, the reaction between barium hydroxide and iron(II) sulfate:

$$Ba^{2+}(aq) + 2\,OH^-(aq) + Fe^{2+}(aq) + SO_4^{2-}(aq) \rightarrow BaSO_4(s) + Fe(OH)_2(s)$$

5.2 $H_3PO_4(aq) \rightleftharpoons H_2PO_4^-(aq) + H^+(aq)$

$H_2PO_4^-(aq) \rightleftharpoons HPO_4^{2-}(aq) + H^+(aq)$

$HPO_4^{2-}(aq) \rightleftharpoons PO_4^{3-}(aq) + H^+(aq)$

5.3 (a) Hydrogen ions and perchlorate ions:

$$HClO_4(aq) \rightarrow H^+(aq) + ClO_4^-(aq)$$

(b) $Ca(OH)_2(aq) \rightarrow Ca^{2+}(aq) + 2\,OH^-(aq)$

5.4 (a) $H^+(aq) + Cl^-(aq) + K^+(aq) + OH^-(aq) \rightarrow H_2O(\ell) + K^+(aq) + Cl^-(aq)$

$H^+(aq) + OH^-(aq) \rightarrow H_2O(\ell)$

(b) $2\,H^+(aq) + SO_4^{2-}(aq) + Ba^{2+}(aq) + 2\,OH^-(aq) \rightarrow 2\,H_2O(\ell) + BaSO_4(s)$

$H^+(aq) + OH^-(aq) \rightarrow H_2O(\ell)$

$Ba^{2+}(aq) + SO_4^{2-}(aq) \rightarrow BaSO_4(s)$

(c) $2\,CH_3COOH(aq) + Ca^{2+}(aq) + 2\,OH^-(aq) \rightarrow Ca^{2+}(aq) + 2\,CH_3COO^-(aq) + 2\,H_2O(\ell)$

$CH_3COOH(aq) + OH^-(aq) \rightarrow H_2O(\ell) + CH_3COO^-(aq)$

5.5 $Al(OH)_3(s) + 3\,H^+(aq) + 3\,Cl^-(aq) \rightarrow 3\,H_2O(\ell) + Al^{3+}(aq) + 3\,Cl^-(aq)$

$H^+(aq) + OH^-(aq) \rightarrow H_2O(\ell)$

5.6 (a) The products are aqueous sodium sulfate, water, and carbon dioxide gas.

$$Na_2CO_3(aq) + H_2SO_4(aq) \rightarrow Na_2SO_4(aq) + H_2O(\ell) + CO_2(g)$$
$$2\,H^+(aq) + CO_3^{2-}(aq) \rightarrow H_2O(\ell) + CO_2(g)$$

(b) The products are aqueous iron(II) chloride and hydrogen sulfide gas.

$$FeS(s) + 2\,HCl(aq) \rightarrow FeCl_2(aq) + H_2S(g)$$
$$2\,H^+(aq) + S^{2-}(aq) \rightarrow H_2S(g)$$

(c) The products are aqueous potassium chloride, water, and sulfur dioxide gas.

$$K_2SO_3(aq) + 2\,HCl(aq) \rightarrow 2\,KCl(aq) + H_2O(\ell) + SO_2(g)$$
$$2\,H^+(aq) + SO_3^{2-}(aq) \rightarrow H_2O(\ell) + SO_2(g)$$

5.7 (a) Gas-forming reaction; the products are aqueous nickel sulfate, water, and carbon dioxide gas.

$$NiCO_3(s) + H_2SO_4(aq) \rightarrow NiSO_4(aq) + H_2O(\ell) + CO_2(g)$$
$$NiCO_3(s) + 2\,H^+(aq) \rightarrow Ni^{2+}(aq) + H_2O(\ell) + CO_2(g)$$

(b) Acid-base reaction; nitric acid reacts with strontium hydroxide, a base, to produce water and strontium nitrate, a salt.

$$2\,HNO_3(aq) + Sr(OH)_2(s) \rightarrow Sr(NO_3)_2(aq) + 2\,H_2O(\ell)$$
$$Sr(OH)_2(s) + 2\,H^+(aq) \rightarrow Sr^{2+}(aq) + 2\,H_2O(\ell)$$

(c) Precipitation reaction; aqueous sodium chloride and insoluble barium oxalate are produced.

$$BaCl_2(aq) + Na_2C_2O_4(aq) \rightarrow BaC_2O_4(s) + 2\,NaCl(aq)$$
$$Ba^{2+}(aq) + C_2O_4^{2-}(aq) \rightarrow BaC_2O_4(s)$$

(d) Precipitation and gas-forming reaction; lead sulfate precipitates and carbon dioxide gas is released.

$$PbCO_3(s) + H_2SO_4(aq) \rightarrow PbSO_4(s) + H_2O(\ell) + CO_2(g)$$
$$Pb^{2+}(aq) + SO_4^{2-}(aq) \rightarrow PbSO_4(s)$$
$$2\,H^+(aq) + CO_3^{2-}(aq) \rightarrow H_2O(\ell) + CO_2(g)$$

5.8 In the reaction $2\,Ca(s) + O_2(g) \rightarrow 2\,CaO(s)$, Ca loses electrons, is oxidized, and is the reducing agent; O gains electrons, is reduced, and is the oxidizing agent.

5.9 $Cl_2(g) + Ca(s) \rightarrow CaCl_2(s)$. $Cl_2(g)$ is the oxidizing agent.

5.10 (a) This is not a redox reaction. Nitric acid is a strong oxidizing agent, but here it serves as an acid.

(b) In this redox reaction, chromium metal (Cr) is oxidized (loses electrons) to form Cr^{3+} ions in Cr_2O_3; oxygen (O_2) is reduced (gains electrons) to form oxide ions, O^{2-}. Oxygen is the oxidizing agent, and chromium is the reducing agent.

(c) This is an acid-base reaction, but not a redox reaction; there are no strong oxidizing or reducing agents present.

(d) Copper is oxidized and chlorine is reduced in this redox reaction, in which copper is the reducing agent and chlorine is the oxidizing agent. The equations are $Cu \rightarrow Cu^{2+} + 2e^-$ and $Cl_2 + 2e^- \rightarrow 2\,Cl^-$.

5.11 (a) Carbon in oxalate ion, $C_2O_4^{2-}$ (oxidation state $= +3$), is oxidized to oxidation state $+4$ in CO_2.

(b) Carbon is reduced from $+4$ in CCl_2F_2 to 0 in C(s).

5.12 (a) $CH_3CH_2OH(\ell) + 3\,O_2(g) \rightarrow 3\,H_2O(\ell) + 2\,CO_2(g)$; redox

(b) $2\,Fe(s) + 6\,HNO_3(aq) \rightarrow 2\,Fe(NO_3)_3(aq) + 3\,H_2(g)$; redox

(c) $AgNO_3(aq) + KBr(aq) \rightarrow AgBr(s) + KNO_3(aq)$; not redox

5.13 Molar mass of cholesterol $= 386.7$ g/mol

$$240\text{ mg} \times \frac{1\text{ g}}{10^3\text{ mg}} \times \frac{1\text{ mol}}{386.7\text{ g}} = 6.21 \times 10^{-4}\text{ mol cholesterol}$$

$$\frac{6.21 \times 10^{-4}\text{ mol}}{0.100\text{ L}} = 6.2 \times 10^{-3}\text{ M}$$

5.14 (a) $6.37\text{ g Al(NO}_3)_3 \times \dfrac{1\text{ mol Al(NO}_3)_3}{213.0\text{ g Al(NO}_3)_3} = 0.0299\text{ mol Al(NO}_3)_3$;

$$\frac{0.0299\text{ mol}}{0.250\text{ L}} = 0.120\text{ M Al(NO}_3)_3$$

(b) Molarity: $Al^{3+} = 0.120$; $NO_3^- = 3(0.120) = 0.360$

5.15 If the description of solution preparation is always worded in terms of adding enough solvent to make a specific volume of solution, then any possible expansion or contraction has no effect on the molarity of the solution. The denominator of the definition of molarity is liters of *solution*.

5.16 The moles of HCl in the concentrated solution are given by (6.0 mol/L)(0.100 L) = 0.6 mol HCl. The moles of HCl in the dilute solution are given by (1.20 mol/L)(0.500 L) = 0.6 mol HCl.

5.17 The molarity could be increased by evaporating some of the solvent.

5.18 $0.0193\text{ L} \times \dfrac{0.200\text{ mol AgNO}_3}{1\text{ L}} \times \dfrac{1\text{ mol Ag}^+}{1\text{ mol AgNO}_3}$

$$\times \frac{1\text{ mol Cl}^-}{1\text{ mol Ag}^+} \times \frac{1\text{ mol NaCl}}{1\text{ mol Cl}^-} = 3.86 \times 10^{-3}\text{ mol NaCl}$$

$$\frac{3.86 \times 10^{-3}\text{ mol NaCl}}{0.0250\text{ L}} = 0.154\text{ M NaCl}$$

Chapter 6

6.1 You transfer some mechanical energy to the ball to accelerate it upward. The ball's potential energy increases the higher it gets, but its kinetic energy decreases by an equal quantity, and eventually it stops rising and begins to fall. As it falls, some of the ball's potential energy changes to kinetic energy, and the ball goes faster and faster until it hits the floor. When the ball hits the floor, some of its kinetic energy is transferred to the atoms, molecules, or ions that make up the floor, causing them to move faster. Much of the ball's kinetic energy is used to compress the ball, squeezing together and deforming the molecules within the ball. When those molecules move apart and their original shape is restored, the ball bounces up, but not to the full height from which it was dropped. Eventually all of the ball's energy is transferred, and the ball stops moving. The nanoscale particles in the floor and the ball (and some in the air that the ball fell through) are moving faster on average, and the temperature of the floor, the ball, and the air is slightly higher. The energy has spread out over a much larger number of particles.

6.2

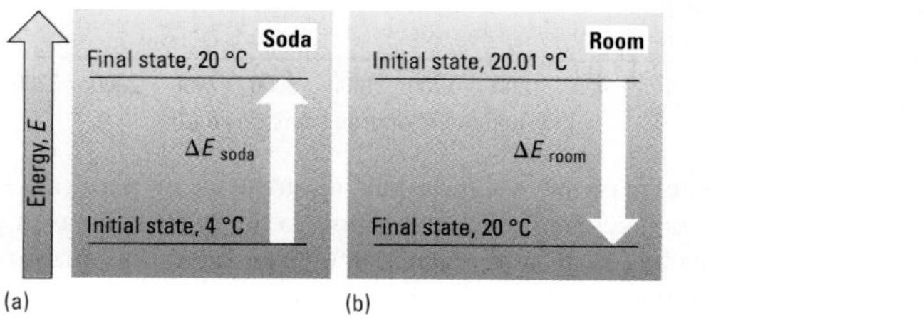

6.3 (a) Heat transfer $= (25.0\text{ °C} - 1.0\text{ °C})\left(\dfrac{1.5\text{ kJ}}{1.0\text{ °C}}\right) = 36\text{ kJ}$

(b) The system is the can and the liquid it contains.

(c) The surroundings includes the air and other materials in contact with the can, or close to the can.

(d) ΔE is negative because the system transferred energy to the surroundings as it cooled; $\Delta E = -36$ kJ.

(e)

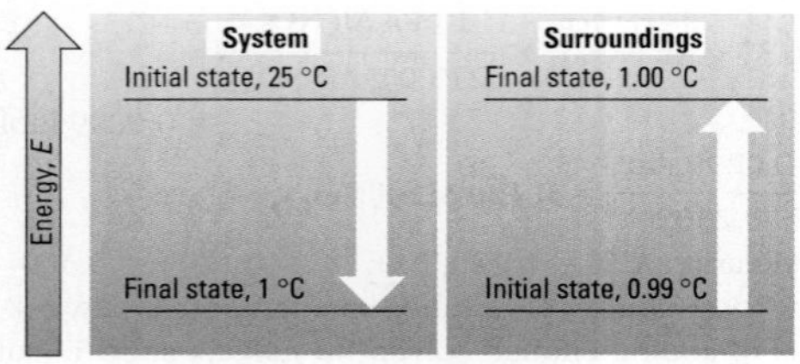

The surroundings are warmed very slightly, say from 0.99 °C to 1.00 °C, by the heat transfer from the can of soda.

6.4 The same quantity of energy is transferred out of each beaker and the mass of each sample is the same. Therefore the sample with the smaller specific heat capacity will cool more. Look up the specific heat capacities in Table 6.1. Because glass has a larger specific heat capacity than carbon, the carbon will cool more and therefore will have the lower temperature.

6.5 The calculation for Al is given as an example.

$$\text{Molar heat capacity} = \frac{0.902\text{ J}}{\text{g °C}} \times \frac{26.98\text{ g}}{1\text{ mol}} = 24.3\text{ J mol}^{-1}\text{ °C}^{-1}$$

Metal	Molar Heat Capacity (J mol^{-1} °C^{-1})	Metal	Molar Heat Capacity (J mol^{-1} °C^{-1})
Al	24.3	Cu	24.5
Fe	25.2	Au	25.2

The molar heat capacities of most metals are about 25 J mol^{-1} °C^{-1}. This rule does not work for ethanol or other compounds listed in Table 6.2.

6.6 (a) Since the heat of vaporization is almost seven times larger than the heat of fusion, the temperature stays constant at 100 °C almost seven times longer than it stays constant at 0 °C. It stays constant at 0 °C for slightly less time than it takes to heat the water from 0 °C to 100 °C (see graph). Because the heating is at a constant rate, time is proportional to quantity of energy transferred.

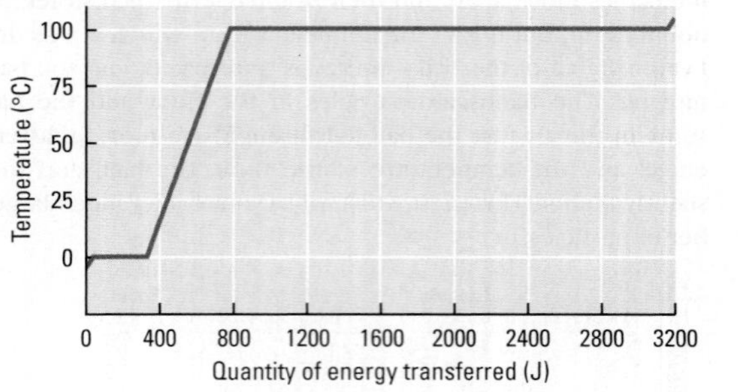

(b) The mass of water is half as great as in part (a), so each process takes half as long. A graph to the same scale as in part (a) begins at 105 °C and reaches −5 °C with half the quantity of energy transferred.

6.7 Heat of fusion: 237 g × 333 J/g = 78.9 kJ
Heating liquid: 237 g × 4.184 J g^{-1} °C^{-1} × 100.0 °C = 99.2 kJ
Heat of vaporization: 237 g × 2260 J/g = 536 kJ
Total heating = (78.9 + 99.2 + 536) kJ = 714 kJ

6.8 The direction of energy transfer is indicated by the sign of the enthalpy change. Transfer to the system corresponds to a positive enthalpy change.

6.9 Because of heats of fusion and heats of vaporization, the enthalpy change is different when a reactant or product is in a different state.

6.10 When 1.0 mol of H_2 reacts (Equation 6.3), $\Delta H = -241.8$ kJ. When half that much H_2 reacts, ΔH is half as great; that is, 0.5 × (−241.8) = −120.9 kJ.

6.11 $\frac{-92.22\text{ kJ}}{1\text{ mol } N_2(g)}$ $\frac{-92.22\text{ kJ}}{3\text{ mol } H_2(g)}$ $\frac{-92.22\text{ kJ}}{2\text{ mol } NH_3(g)}$

$\frac{1\text{ mol } N_2(g)}{-92.22\text{ kJ}}$ $\frac{3\text{ mol } H_2(g)}{-92.22\text{ kJ}}$ $\frac{2\text{ mol } NH_3(g)}{-92.22\text{ kJ}}$

6.12 The reaction used must be exothermic. Because it can be started by opening the package, it probably involves oxygen from the air, and the sealed package prevents the reaction from occurring before it is needed. Many metals can be oxidized easily and exothermically. The reaction of iron with oxygen (as in the *Chemistry You Can Do* experiment, **← p. 235**) is a good candidate.

6.13 Yes, it would violate the first law of thermodynamics. According to the supposition, we could create energy by starting with 2 mol HCl, breaking all the molecules apart, recombining the atoms to form 1 mol H_2 and 1 mol Cl_2, and then reacting the H_2 and Cl_2 to give 2 HCl.

$2\text{ HCl} \longrightarrow \text{atoms} \longrightarrow H_2 + Cl_2$ $\Delta H° = +185$ kJ

$H_2 + Cl_2 \longrightarrow 2\text{ HCl}$ $\Delta H° = -190.$ kJ

The net effect of these two processes is that there is still 2 mol HCl, but 5 kJ of energy has been created. This is impossible according to the first law of thermodynamics.

6.14 (a) In the reaction $2\text{ HF} \rightarrow H_2 + F_2$ there are two bonds in the two reactant molecules and two bonds in the two product molecules. Since the reaction is endothermic, the bonds in the reactant molecules must be stronger than in the products.
(b) For the reaction $2\text{ }H_2O \rightarrow 2\text{ }H_2 + O_2$, there are four bonds in the two reactant molecules but only three bonds in the three product molecules. The reaction is endothermic because more bonds are broken than are formed.

6.15 $C_6H_{12}O_6(s) + 6\text{ }O_2(g) \rightarrow 6\text{ }CO_2(g) + 6\text{ }H_2O(\ell)$
Because the volume of any ideal gas is proportional to the amount (moles) of gas, and because there are 6 mol of gaseous reactant and 6 mol of gaseous product, there will be very little change in volume. Almost no work will be done, and $\Delta H \cong \Delta E$.

6.16 In the Problem-Solving Example 6.11 the reaction produced 0.25 mol NaCl and the heat transfer associated with the reaction caused 500 mL of solution to warm by 7 °C.
(a) Here the reaction produces 0.20 mol NaCl by neutralizing 0.20 mol NaOH and the heat transfer warms 400. mL water. So there is less heat transfer but it will be heating a smaller volume. The quantity of reaction is $\frac{0.20\text{ mol}}{0.25\text{ mol}} = 0.80$ as much, so the heat transfer is 0.80 as much. The quantity of water to be heated is $\frac{400.\text{mL}}{500.\text{mL}} = 0.800$ as much. Therefore the combined effect on temperature is the same as in Problem-Solving Example 6.11, and the temperature change associated with this process will also be 7 °C.
(b) Here the limiting reactant is 0.10 mol NaOH. (Only half of the H_2SO_4 is used up.) The heat transfer from the reaction warms 200. mL water. So there is less heat transfer but it will heat a smaller volume. The quantity of reaction is $\frac{0.10\text{ mol}}{0.25\text{ mol}} = 0.40$ as much, so the heat transfer is 0.40 as much. The quantity of water is $\frac{200.}{400.} = 0.40$ as much. Therefore once again the combined effect on temperature is the same as in Problem-Solving Example 6.11, and the temperature change associated with this process will also be 7 °C.

6.17 $N_2(g) \rightarrow N_2(g)$
(a) The product is the same as the reactant, so there is no change—nothing happens.
(b) Since product and reactant are the same, $\Delta H = 0$.

6.18 (a) It takes 160 kJ to break 1 mol N—N bonds, 4 × 391 kJ to break 4 mol N—H bonds, and 498 kJ to break 1 mol of bonds in O_2.
(b) Forming 1 mol of bonds in N_2 releases 946 kJ, and forming 4 mol O—H bonds releases 4 × 467 kJ.
(c) Therefore, $\Delta H = \{160 + (4 \times 391) + 498\}$ kJ $- \{946 + (4 \times 467)\}$ kJ $= -592$ kJ

$N_2H_4(g) + O_2(g) \rightarrow N_2(g) + 2\text{ }H_2O(g)$ $\Delta H° = -592$ kJ

6.19 In carbon dioxide, carbon is in its highest oxidation state (+4) and cannot be oxidized further; in carbon dioxide, carbon has already combined with the maximum possible amount of oxygen.

6.20 According to the thermochemical equation, 2801.6 kJ is released per mole of glucose.

$$\frac{2801.6 \text{ kJ}}{1 \text{ mol}} \times \frac{1 \text{ mol}}{180.16 \text{ g}} \times \frac{1 \text{ kcal}}{4.184 \text{ kJ}} \times \frac{1 \text{ Cal}}{1 \text{ kcal}} = 3.717 \text{ Cal}$$

This rounds to 4 Cal.

6.21 (a) The average BMR is given in the text as 1750 Cal/day for a 70-kg male at rest. So the BMR is

$$\frac{1750 \text{ Cal}}{\text{d}} \times \frac{1000 \text{ cal}}{\text{Cal}} \times \frac{4.184 \text{ J}}{\text{cal}} = 7.32 \times 10^6 \text{ J/day}$$

$$(7.322 \times 10^6 \text{ J/day}) \times (1 \text{ day}/24 \text{ h}) \times (1 \text{ h}/60 \text{ min}) \times (1 \text{ min}/60 \text{ s}) = 84.75 \text{ W}$$

(b) The average BMR for a 70-kg male playing basketball is 7 times the above or 593.2 W. Typical incandescent light bulbs are in the range of 50–250 W.

Chapter 7

7.1 Wavelength and frequency are inversely related. Therefore, low-frequency radiation has long-wavelength radiation.

7.2 (a) Ultraviolet; (b) 1.20×10^{15} Hz has shorter wavelength than 7.898×10^{14} Hz radiation.

7.3 For 4.275 μm radiation:

$$E = \frac{hc}{\lambda} = \frac{(6.626 \times 10^{-34} \text{ Js})(2.998 \times 10^{-8} \text{ m/s})}{4.257 \text{ μm} \times \left(\frac{1 \text{ μm}}{10^{-6} \text{ m}}\right)} = 4.666 \times 10^{-20} \text{ J}$$

A similar calculation for 15.00 μm indicates 1.324×10^{-20} J. Thus, 4.257 μm radiation has the greater amount of energy.

7.4 In a sample of excited hydrogen gas there are many atoms, and each can exist in one of the excited states possible for hydrogen. The observed spectral lines are a result of all the possible transitions of all these hydrogen atoms.

7.5 (a) Emitted (b) Absorbed
(c) Emitted (d) Emitted

7.6 $(2.179 \times 10^{-18} \text{ J/photon}) \times \left(\frac{1 \text{ kJ}}{10^3 \text{ J}}\right)\left(\frac{6.022 \times 10^{23} \text{ photons}}{1 \text{ mol}}\right)$
$= 1312$ kJ/mol photons

7.7 (a) $5d$ (b) $4f$ (c) $6p$

7.8 (a) The $n = 3$ level can have only three types of sublevels—s, p, and d.
(b) The $n = 2$ level can have only s and p sublevels, not d sublevels ($\ell = 2$).

7.9 3, 0, 0, $+\frac{1}{2}$; 3, 0, 0 $-\frac{1}{2}$

7.10 (a) 3, 1, −1, $+\frac{1}{2}$ and 3, 1, 0, $+\frac{1}{2}$ (b) 3, 1, 1, $+\frac{1}{2}$ and 3, 0, 0, $+\frac{1}{2}$

7.11 One electron is in the $3p_y$ orbital; the other electron is in the $3p_z$ orbital.

7.12 (a) The maximum number of electrons in the $n = 3$ level is 18 (2 electrons per orbital). The orbitals would be designated $3s$, $3p_x$, $3p_y$, $3p_z$, $3d_{z^2}$, $3d_{xy}$, $3d_{yz}$, $3d_{xz}$, and $3d_{x^2-y^2}$.
(b) The maximum number of electrons in the $n = 5$ level is 50. The orbitals would be designated $5s$, $5p_x$, $5p_y$, $5p_z$, $5d_{yz}$, $5d_{xz}$, $5d_{xy}$, $5d_{z^2}$, $5d_{x^2-y^2}$, and the seven $5f$ orbitals and the nine $5g$, which are not designated by name in the text.

7.13 The first shell that could contain g orbitals would be the $n = 5$ shell. There would be nine g orbitals.

7.14 The [Ar]$3d^4 4s^2$ configuration for chromium has four unpaired electrons, and the [Ar]$3d^5 4s^1$ configuration has six unpaired electrons.

7.15 For the chlorine atom, $n = 3$, and there are seven electrons in this highest energy level. The configuration is

$3s$ [↑↓] $3p$ [↑↓] [↑↓] [↑]. For the selenium atom, the highest energy level is $n = 4$, and there are six electrons in the $n = 4$ level. The configuration is $4s$ [↑↓] $4p$ [↑↓] [↑] [↑].

7.16 (a) VF_5 (V^{5+} has no unpaired electrons); (b) VF_2 (V^{2+} has three unpaired electrons); (c) VF_3 (V^{3+} has two unpaired electrons).

Chapter 8

8.1 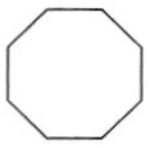 C_8H_{16}

8.2 N_2 has only 10 valence electrons. The Lewis structure shown has 14 valence electrons.

8.3 None of the structures are correct. (a) is incorrect because sulfur does not have an octet of electrons (it has only six); (b) is incorrect because, although it shows the correct number of valence electrons (26), there is a double bond between F and N rather than a single bond with a lone pair on N; (c) is incorrect because the left carbon has five bonds; (d) is incorrect because OCCl should have 17 valence electrons, not 18 as shown.

8.4 (a) C_5H_{10} (b) Two

8.5 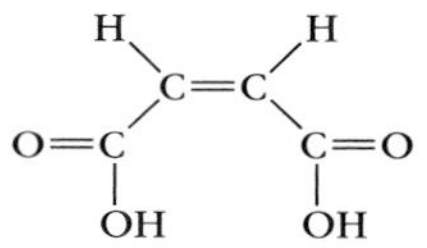
maleic acid
(the *cis* isomer)

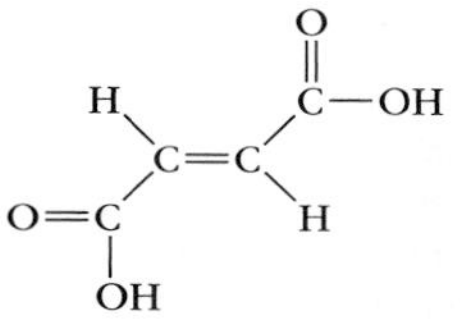
fumaric acid
(the *trans* isomer)

8.6 C—N > C=N > C≡N. The order of decreasing bond energy is the reverse order: C≡N > C=N > C—N. See Tables 8.1 and 8.2.

8.7 (a) The electronegativity difference between sodium and chlorine is 1.8, sufficient to cause electron transfer from sodium to chlorine to form Na^+ and Cl^- ions. Molten NaCl conducts an electric current, indicating the presence of ions.
(b) There is an electronegativity difference of 1.4 in BrF, which is sufficient to form a polar covalent bond, but not great enough to cause electron transfer leading to ion formation.

8.8 The Lewis structure of hydrazine is

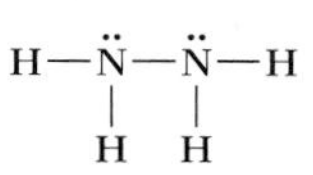

	H	H	N	N	H	H
Valence electrons	1	1	5	5	1	1
Lone pair electrons	0	0	2	2	0	0
$\frac{1}{2}$ shared electrons	1	1	3	3	1	1
Formal charge	0	0	0	0	0	0

8.9 Atoms cannot be rearranged to derive a resonance structure. There is no N-to-O bond in cyanate ion; therefore, such an arrangement cannot be a resonance structure of cyanate ion.

8.10 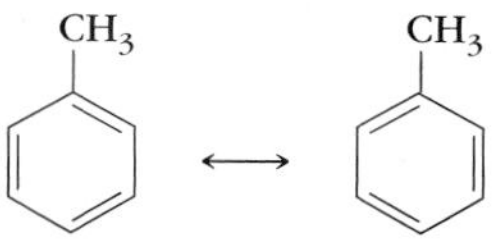

8.11 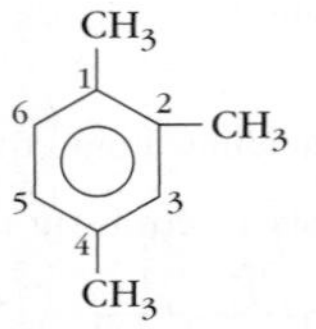

1,2,4-trimethylbenzene

The ring is numbered to give the lowest numbers for the substituents in 1, 2, 3-trimethylbenzene.

8.12 1. (a) 20 carbon atoms and 30 hydrogen atoms
(b) The carbon atom at the top of the six-membered ring
(c) Five C═C double bonds
2. (a) $C_{29}H_{50}O_2$ (b) No C═C double bonds
(c) The H—O bond

Chapter 9

9.1 When the central atom has no lone pairs

9.2 The triangular bipyramidal shape has three of the five pairs situated in equatorial positions 120° apart and the remaining two pairs in axial positions. The square pyramidal shape has four of the atoms bonded to the central atom in a square plane, with the other bonded atom directly above the central atom and equidistant from the other four.

9.3 (a) AX_2E_3, linear; (b) AX_3E_1, triangular pyramid;
(c) AX_2E_3, linear

9.4 Pi bonding is not possible for a carbon atom with sp^3 hybridization because the atom has no unhybridized $2p$ orbitals. All of its $2p$ orbitals have been hybridized.

9.5 (a) Bromine is more electronegative than iodine, and the H—Br bond is more polar than the H—I bond.
(b) Chlorine is more electronegative than the other two halogens; therefore, the C—Cl bond is more polar than the C—Br and C—I bonds.

9.6 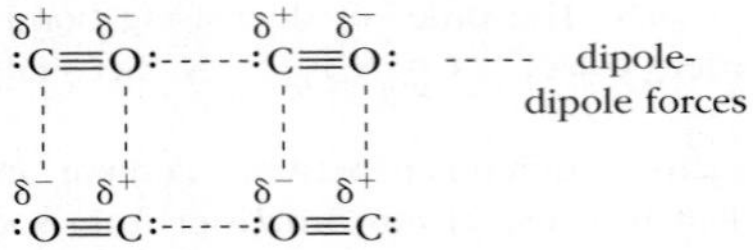

9.7 The F—H · · · F—H hydrogen bond is the strongest because the electronegativity difference between H and F produces a more polar F—H bond than does the lesser electronegativity difference between O and H or N and H in the O—H or N—H bonds.

9.8 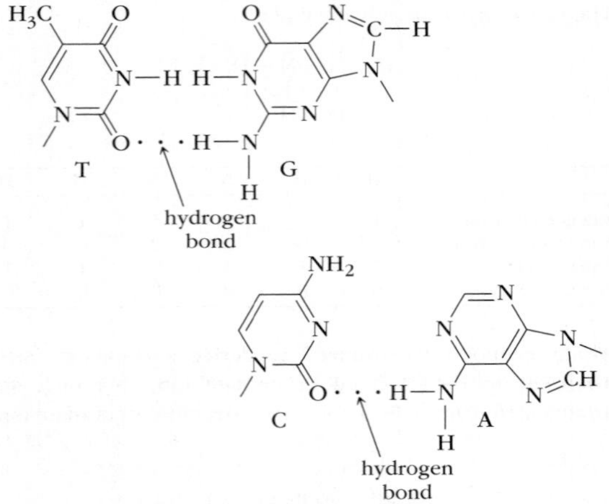

9.9 Replication would be very difficult because strong, covalent bonds would have to be broken. DNA replicates easily because it is weaker hydrogen bonds, not covalent bonds, that occur between the base pairs and hold the two DNA strands together.

Chapter 10

10.1 $(2.7 \times 10^{14} \text{ molecules}) \times \left(\dfrac{64.06 \text{ g } SO_2}{6.02 \times 10^{23} \text{ molecules}}\right) = 2.9 \times 10^{-8} \text{ g } SO_2$

10.2 First, gas molecules are far apart. This allows most light to pass through. Second, molecules are much smaller than the wavelengths of visible light. This means that the waves are not reflected or diffracted by the molecules.

10.3 As more gas molecules are added to a container of fixed volume, there will be more collisions of all of the gas molecules with the container walls. This causes the observed pressure to rise.

10.4 All have the same kinetic energy at the same temperature.

10.5 For a sample of helium, the plot would look like the curve marked He in Figure 10.7. When an equal number of argon molecules, which are heavier, are added to the helium, the distribution of molecular speeds would look like the sum of the curves marked He and O_2 in Figure 10.7, except that the curve for Ar would have its peak a little to the left of the O_2 curve.

10.6 (a) The balloon placed in the freezer will be smaller than the one kept at room temperature because its sample of helium is colder.
(b) Upon warming, the helium balloon that had been in the freezer will be either the same size as the balloon kept at room temperature or perhaps slightly larger because there is a greater chance that He atoms leaked out of the room temperature balloon during the time the other balloon was kept in the freezer. This would be caused by the faster-moving He atoms in the room temperature balloon having more chances to escape from tiny openings in the balloon's walls.

10.7 The gas in the shock absorbers will be more highly compressed. The gas molecules will be closer together. The gas molecules will collide with the walls of the shock absorber more often, and the pressure exerted will be larger.

10.8 Increasing the temperature of a gas causes the gas molecules to move faster, on average. This means that each collision with the container walls involves greater force, because, on average, a molecule is moving faster and hits the wall harder. If the container remained the same (constant volume), there would also be more collisions with the container wall because faster-moving molecules would hit the walls more often. Increasing the volume of the container, on the other hand, requires that the faster-moving molecules must travel a greater distance before they strike the container walls. Increasing the volume enough would just balance the greater numbers of harder collisions caused by increased temperature. To maintain a constant volume requires that the pressure increases to match the greater pressure due to more and harder collisions of gas molecules with the walls.

10.9 Two moles of O_2 gas are required for the combustion of one mole of methane gas. If air were pure O_2, the oxygen delivery tube would need to be twice as large as the delivery tube for methane. Since air is only one-fifth O_2, the air delivery tube would need to be 10 times larger than the methane delivery tube to ensure complete combustion.

10.10 Using the ratio of 100 balloons/26.8 g He, calculate the number of balloons 41.8 g He can fill.

$$\text{Balloons} = (41.8 \text{ g He})\left(\frac{100 \text{ balloons}}{26.8 \text{ g He}}\right) = 155 \text{ balloons}$$

This much He will fill more balloons than needed.

10.11 1. Increase the pressure.
2. Decrease the temperature.
3. Remove some of the gas by reaction to form a nongaseous product.

10.12 Density of Cl_2 at 25 °C and 0.750 atm $= \dfrac{PM}{RT}$

$$= \frac{(0.750 \text{ atm})(70.906 \text{ g/mol})}{(0.0821 \text{ L atm mol}^{-1} \text{ K}^{-1})(298 \text{ K})} = 2.17 \text{ g/L}$$

$$\text{Density of } SO_2 \text{ at 25 °C and 0.750 atm} = \frac{PM}{RT}$$

$$= \frac{(0.750\ \text{atm})(64.06\ \text{g/mol})}{(0.0821\ \text{L atm mol}^{-1}\ \text{K}^{-1})(298\ \text{K})} = 1.96\ \text{g/L}$$

$$\text{Density of } Cl_2 \text{ at 35 °C and 0.750 atm} = \frac{PM}{RT}$$

$$= \frac{(0.750\ \text{atm})(70.906\ \text{g/mol})}{(0.0821\ \text{L atm mol}^{-1}\ \text{K}^{-1})(308\ \text{K})} = 2.10\ \text{g/L}$$

$$\text{Density of } SO_2 \text{ at 25 °C and 2.60 atm} = \frac{PM}{RT}$$

$$= \frac{(2.60\ \text{atm})(64.06\ \text{g/mol})}{(0.0821\ \text{L atm mol}^{-1}\ \text{K}^{-1})(298\ \text{K})} = 6.81\ \text{g/L}$$

10.13 Density of He = 1.66×10^{-4} g/mL
Density of Li = 0.53 g/mL
Since the density of He is so much less than that of Li, the atoms in a sample of He must be much farther apart than the atoms in a sample of Li. This idea is in keeping with the general principle of the kinetic-molecular theory that the particles making up a gas are far from one another.

10.14 A 50-50 mixture of N_2 and O_2 would have less N_2 in it than does air. Since O_2 molecules have greater mass than N_2 molecules, this 50-50 mixture has greater density than air.

10.15 (a) If lowering the temperature causes the volume to decrease, by $PV = nRT$, the pressure can be assumed to be constant. The value of n is unchanged. Since both P and n remain unchanged, the partial pressures of the gases in the mixture remain unchanged.
(b) When the total pressure of a gas mixture increases, the partial pressure of each gas in the mixture increases because the partial pressure of each gas in the mixture is the product of the mole fraction for that gas and the total pressure.

10.16 We can calculate the total number of moles of gas in the flask from the given information.

$$n = \frac{PV}{RT} = \frac{\left(\frac{626}{760}\right)\text{atm}\ (0.355\ \text{L})}{(0.0821\ \text{L atm mol}^{-1}\ \text{K}^{-1})(308\ \text{K})} = 0.01156\ \text{mol gas}$$

The number of moles of Ne is

$$0.146\ \text{g Ne} \times \frac{1\ \text{mol Ne}}{20.18\ \text{g/mol}} = 0.007235\ \text{mol Ne}$$

We find the number of moles of Ar by subtraction.

$$0.01156\ \text{mol gas} - 0.007235\ \text{mol Ne} = 0.004325\ \text{mol Ar}$$

$$0.004325\ \text{mol Ar} \times 39.948\ \text{g/mol} = 0.173\ \text{g Ar}$$

10.17 The value of n depends directly on the measured pressure, P. Intermolecular attractions in a real gas would cause the measured P to be slightly smaller than for an ideal gas. The lower value of P would cause the calculated number of moles to be somewhat smaller. Using this slightly smaller value of n in the denominator would cause the calculated molar mass to be a little larger than it should be.

10.18 Mass of S burned per hour = $(3.06 \times 10^6\ \text{kg})(0.04)$

$$= 1 \times 10^5\ \text{kg}$$

$$\text{Mass of } SO_2 \text{ per hour} = (1 \times 10^5\ \text{kg})\left(\frac{64.06\ \text{kg } SO_2}{32.07\ \text{kg S}}\right) 2 \times 10^5\ \text{kg}$$

$$\text{Mass of } SO_2 \text{ per year} = (2 \times 10^5\ \text{kg/hr})(8760\ \text{hr/yr}) = 2 \times 10^9\ \text{kg/yr}$$

10.19 Vol percent SO_2 = $(5 \text{ parts } SO_2/10^6 \text{ parts air}) \times 100\%$

$$= 5 \times 10^{-4}\ \text{vol\%}$$

10.20 1 metric ton = 1000 kg = 10^6 g = 1 Mg

$$\text{Mass of } HNO_3 = (400\ \text{Mg } N_2)\left(\frac{2\ \text{Mg NO}}{\text{Mg } N_2}\right)$$

$$\times \left(\frac{1\ \text{Mg } NO_2}{1\ \text{Mg NO}}\right)\left(\frac{1\ \text{Mg } HNO_3}{1\ \text{Mg } NO_2}\right) = 800\ \text{Mg } HNO_3$$

10.21 $\cdot NO_2 \xrightarrow{h\nu} NO\cdot + \cdot O\cdot$

$O_3 \xrightarrow{h\nu} O_2 + \cdot O\cdot$

$\cdot O\cdot + O_2 \longrightarrow O_3$

10.22 Among the many possible molecules would be: NO, which comes from automobile combustion; NO_2, which comes from reactions of NO and O_2 in the atmosphere; O_3, which comes from reactions of NO, NO_2, and O_2 in the atmosphere; and hydrocarbons, and oxidation products of hydrocarbon reactions with O_2 and other reactive species.

10.23 Natural sources: animal respiration, forest fires, decay of cellulose materials, partial digestion of carbohydrates, volcanoes. Human sources: burning fossil fuels, burning agricultural wastes and refined cellulose products such as paper, decay of carbon compounds in landfills.

10.24 (a) 4.8%; (b) 7.0%; (c) 8.6%; 1990–2008 showed the greatest percentage increase in CO_2.

10.25 The fluctuations occur because of the seasons. Photosynthesis, which uses CO_2, is greatest during the spring and summer, accounting for lower CO_2 levels.

10.26 For this calculation, we can use the figure of 450×10^9 passenger miles. Using the ratio of 2×10^3 kg CO_2/3000 passenger miles, we can calculate the CO_2 released.

$$\text{Quantity of } CO_2 = 450 \times 10^9\ \text{passenger mi} \times \frac{2 \times 10^3\ \text{kg } CO_2}{3 \times 10^3\ \text{passenger mi}} = 3 \times 10^{11}\ \text{kg } CO_2$$

If a typical automobile gets 20 mi/gal of fuel and about 1.5 passengers are transported for every mile the automobile travels, then an automobile gets 30 passenger miles per gallon. The number of gallons used is

$$\text{Volume of gasoline} = 450 \times 10^9\ \text{passenger mi} \times \frac{1\ \text{gal gasoline}}{30\ \text{passenger mi}} = 2 \times 10^{10}\ \text{gal gasoline}$$

Assume the gasoline produces about the same mass of CO_2 per gallon as does jet fuel, or 2×10^3 kg CO_2/200 gal.

$$\text{Quantity of } CO_2 \text{ from gasoline} = 2 \times 10^{10}\ \text{gal gasoline} \times \frac{2 \times 10^3\ \text{kg } CO_2}{200\ \text{gal gasoline}} = 2 \times 10^{11}\ \text{kg } CO_2$$

So the numbers are about the same for these two modes of transportation.

10.27 He meant the burning coal converted its carbon into carbon dioxide in the air, thus increasing atmospheric CO_2 concentration.

Chapter 11

11.1 The London forces are greater between bromoform molecules than between chloroform molecules because the bromoform molecules have more electrons. This stronger intermolecular attraction causes the $CHBr_3$ molecules to exhibit a greater surface tension. (The dipole in each molecule contributes less than the London forces to the intermolecular attractions.)

11.2 (a) Water and glycerol would have similar surface tensions because each of them can engage in extensive hydrogen bonding.
(b) Octane and decane would have similar surface tensions because both are alkane hydrocarbons and would have very similar intermolecular interactions.

11.3 (a) 62 °C (b) 0 °C (c) 80 °C

11.4 Bubbles form within a boiling liquid when the vapor pressure of the liquid equals the pressure of the surroundings of the liquid sample. The bubbles are actually filled with vapor of the boiling liquid. One way to prove this would be to trap some of these bubbles and allow them to condense. They would condense to form the liquid that had boiled.

11.5 The evaporating water carries with it thermal energy from the water inside the pot. In addition, a large quantity of thermal energy is required to cause the water to evaporate. Much of this thermal energy comes from the water inside the pot.

11.6 (a) Bromine molecules have more electrons than chlorine molecules. Therefore, bromine molecules are held together by stronger intermolecular attractions.
(b) Ammonia molecules are attracted to one another by hydrogen bonds. This causes ammonia to have a higher boiling point than that of methane, which has no hydrogen bonding.

11.7 Two moles of liquid bromine crystallizing liberates 21.59 kJ of energy. One mole of liquid water crystallizing liberates 6.02 kJ of energy.

11.8 High humidity conditions make the evaporation of water or the sublimation of ice less favorable. Under these conditions, the sublimation of ice required to make the frost-free refrigerator work is less favorable, so the defrost cycle is less effective.

11.9 The impurity molecules are less likely to be converted from the solid phase to the vapor phase. This causes them to be left behind as the molecules that sublime go into the gas phase and then condense at some other place. The molecules that condense are almost all of the same kind, so the sublimed sample is much purer than the original.

11.10 The curve is the vapor pressure curve. Upward: condensation. Downward: vaporization. Left to right: vaporization. Right to left: condensation.

11.11 (a) If liquid CO_2 is slowly released from a cylinder of CO_2, gaseous CO_2 is formed. The temperature remains constant (at room temperature) because there is time for energy to be transferred from the surroundings to separate the CO_2 molecules from their intermolecular attractions. This can be seen from the phase diagram as the phase changes from liquid to vapor as the pressure decreases.
(b) If the pressure is suddenly released, the attractive forces between a large number of CO_2 molecules must be overcome, which requires energy. This energy comes from the surroundings as well as from the CO_2 molecules themselves, causing the temperature of both the surroundings and the CO_2 molecules to decrease. On the phase diagram for CO_2, a decrease in both temperature and pressure moves into a region where only solid CO_2 exists.

11.12 It is predicted that a small concentration of gold will be found in the lead and that a small concentration of lead will be found in the gold. This will occur because of the movement of the metal atoms with time, as predicted by the kinetic molecular theory.

11.13 One Po atom belongs to its unit cell. Two Li atoms belong to its unit cell. Four Ca atoms belong to its unit cell.

Primitive cubic

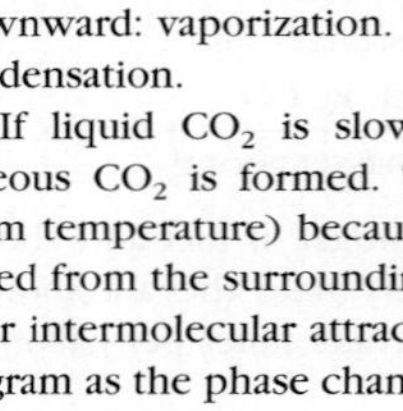

1 Po atom
Each of the 8 atoms
contributes 1/8 to unit cell

Body-centered cubic

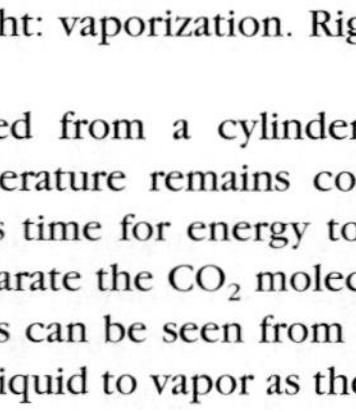

2 Li atoms
8/8 from corner atoms
+ 1 atom at center

Face-centered cubic

4 Ca atoms
8/8 from corner atoms
+ 6/2 from each atom on
the 6 faces contributing 1/2 atom

11.14 Each Cs^+ ion at the center of the cube has eight Cl^- ions as its neighbors. One eighth of each Cl^- ion belongs to that Cs^+ ion. So the formula for this salt must be a 1 : 1 ratio of Cs^+ ions to Cl^- ions, or CsCl.

11.15 Cooling a liquid above its freezing point causes the temperature to decrease. When the liquid begins to solidify, energy is released as atoms, molecules, or ions move closer together to form in the solid crystal lattice. This causes the temperature to remain constant until all the molecules in the liquid have positioned themselves in the lattice. Further cooling then causes the temperature to decrease. The shape of this curve is common to all substances that can exist as liquids.

11.16 Increasing strength of metallic bonding is related to increasing numbers of valence electrons. In the transition metals, the presence of *d*-orbital electrons causes stronger metallic bonding. Beyond a half-filled set of *d*-orbitals, however, extra electrons have the effect of decreasing the strength of metallic bonding.

Chapter 12

12.1 (a) $4\ H_2$
(b) $C_7H_{16} \rightarrow C_7H_8 + 4\ H_2$
(c) Toluene has a much higher octane number and can be used as an octane enhancer.

12.2 $\%\ \text{oxygen} = \left(\dfrac{16.0}{46.0}\right) \times 100\% = 34.8\%$

Ethanol is more highly oxygenated than the hydrocarbons in gasoline.

12.3 (a) $C_2H_5OH + 2\ O_2 \rightarrow 2\ CO + 3\ H_2O$

$$\frac{\text{g CO}}{\text{g ethanol}} = \frac{2(28.0\ \text{g})}{46.0\ \text{g}} = 1.22\ \text{g CO/g ethanol}$$

(b) $C_7H_8 + \frac{11}{2}\ O_2 \rightarrow 7\ CO + 4\ H_2O$

$$\frac{\text{g CO}}{\text{g toluene}} = \frac{7(28.0\ \text{g})}{92.0\ \text{g}} = 2.13\ \text{g CO/g toluene}$$

(c) Ethanol produces less CO per gram than does toluene.

12.4 $CH_4(g) + \frac{3}{2}\ O_2(g) \rightarrow CO(g) + 2\ H_2O(g)$
Thus, 28.0 g CO is produced per 16.0 g CH_4;

$$\frac{28.0\ \text{g CO}}{16.0\ \text{g}\ CH_4} = 1.75\ \text{g CO/g}\ CH_4$$

$C_8H_{18}(\ell) + \frac{17}{2}\ O_2(g) \rightarrow 8\ CO(g) + 9\ H_2O(g)$
114.0 g octane produces 224.0 g CO;

$$\frac{224\ \text{g CO}}{114.0\ \text{g octane}} = 1.96\ \text{g CO/g}\ C_8H_{18}$$

$\frac{1.96}{1.75} = 1.12$, which indicates 12% more CO is produced per gram of octane than per gram of methane.

12.5 Ten or so carbon atoms in an alcohol molecule will make it much less water-soluble than alcohols with fewer numbers of carbon atoms.

12.6 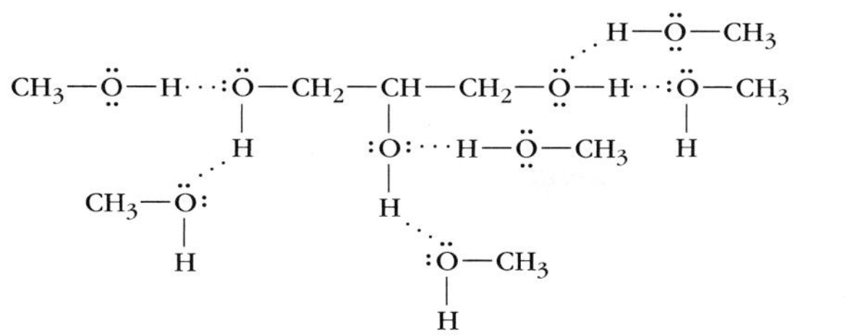

12.7 The acetaldehyde molecule has two fewer hydrogen atoms compared with the ethanol molecule. Loss of hydrogen is oxidation. The acetaldehyde molecule is more oxidized than the ethanol molecule. Comparing the formulas for acetaldehyde and acetic acid, the hydrogen atoms are the same, but the acetic acid molecule has one additional oxygen atom. Gain of oxygen is oxidation. So the acetic acid molecule is more oxidized than the acetaldehyde molecule.

12.8 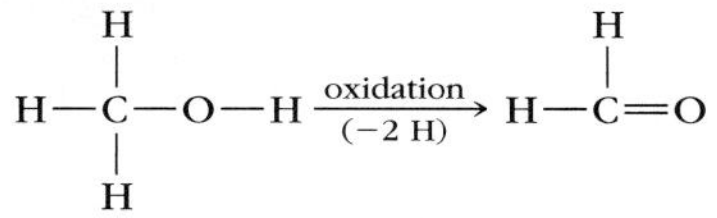

12.9 $CH_3CH_2CH_2OH$, 1-propanol

12.10 (a) Estradiol contains two alcohol groups (—OH).

(b) Secondary alcohol

(c) Oxidation (removal of two hydrogens)

(d) Estradiol: Two alcohol groups, aromatic ring, one $—CH_3$ group

Testosterone: One alcohol group; one ketone group; C=C double bond in ring; two $—CH_3$ groups

12.11 Conversion of a secondary alcohol on the five-membered ring to a ketone by oxidation (removal of 2 H atoms).

12.12 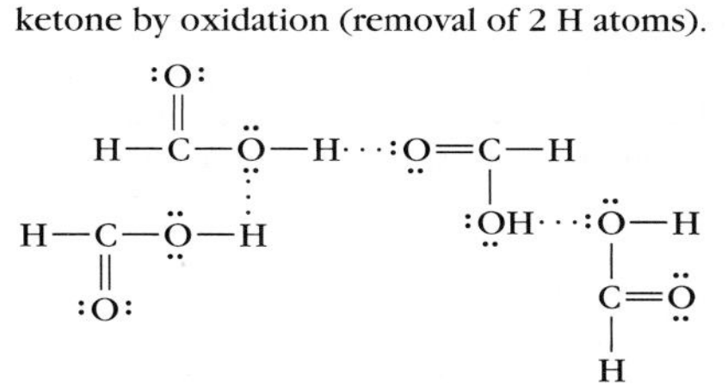

12.13 (a)

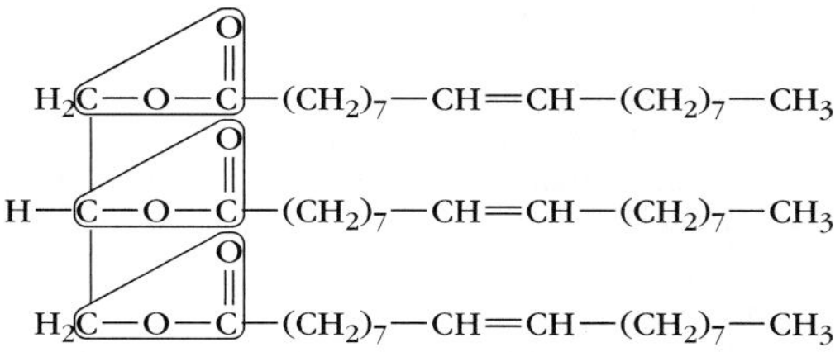

(b)

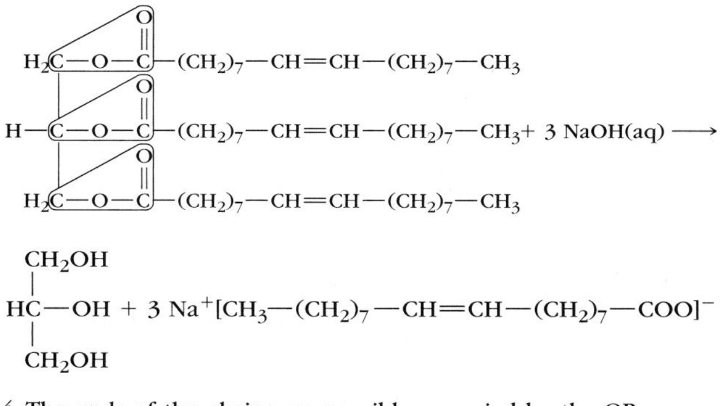

12.14 The ends of the chains are possibly occupied by the OR groups from the initiator molecules.

12.15

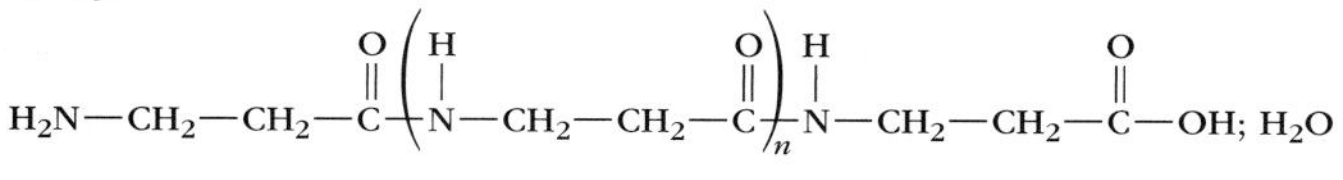

12.16 (1) Amine (2) Carboxylic acid (3) Amide (4) Ester

12.17 The —OH groups in this molecule allow it to be extensively hydrogen-bonded with solvent water molecules.

12.18 Cellulose contains glucose molecules linked together by *trans* 1,4 linkages. Humans lack the enzymes needed to catalyze breaking of *trans* 1, 4 linkages. Ruminant animals have large colonies of bacteria and protozoa that live in the forestomach and digest cellulose.

12.19 If humans could digest cellulose, then common plants that are easy to grow could become food. There might be less reliance upon cultivation of plants for food. In addition, the entire plant could be used for food rather than just certain parts eaten and the other parts wasted. On the other hand, in times of famine, there might not be enough cellulose to go around. Destroying trees and other plants for food might cause enlargements of desert regions and the disappearance of entire species of plants.

12.20 Serine and glutamine can hydrogen-bond to one another if they were close in two adjacent protein chains because they have polar groups containing H atoms in their R groups. Glycine and valine would not because they have no additional polar groups in their R groups.

12.21 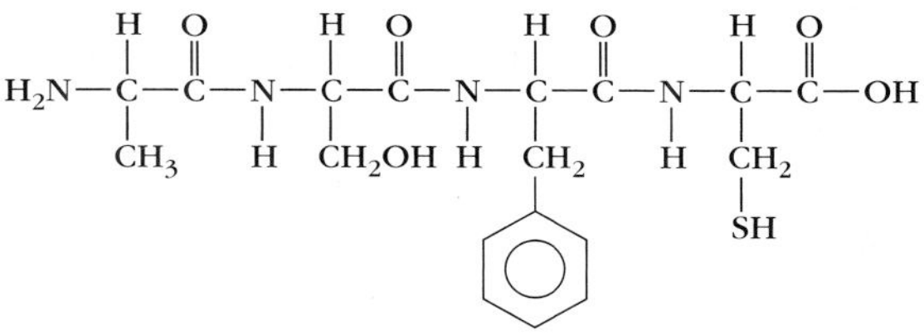

Chapter 13

13.1 (a)

i. $$\text{Rate} = -\frac{\Delta[\text{Cv}^+]}{\Delta t}$$
$$= -\frac{(0.793 \times 10^{-5} - 1.46 \times 10^{-5})\ \text{mol/L}}{(60.0 - 40.0)\ \text{s}}$$
$$= \frac{6.67 \times 10^{-6}\ \text{mol/L}}{20.0\ \text{s}} = 3.3 \times 10^{-7}\ \text{mol L}^{-1}\ \text{s}^{-1}$$

ii. $$\text{Rate} = -\frac{(0.429 \times 10^{-5} - 2.71 \times 10^{-5})\ \text{mol/L}}{(80.0 - 20.0)\ \text{s}}$$
$$= 3.8 \times 10^{-7}\ \text{mol L}^{-1}\ \text{s}^{-1}$$

iii. $$\text{Rate} = -\frac{(0.232 \times 10^{-5} - 5.00 \times 10^{-5})\ \text{mol/L}}{(100.0 - 0.0)\ \text{s}}$$
$$= 4.8 \times 10^{-7}\ \text{mol L}^{-1}\ \text{s}^{-1}$$

(b)

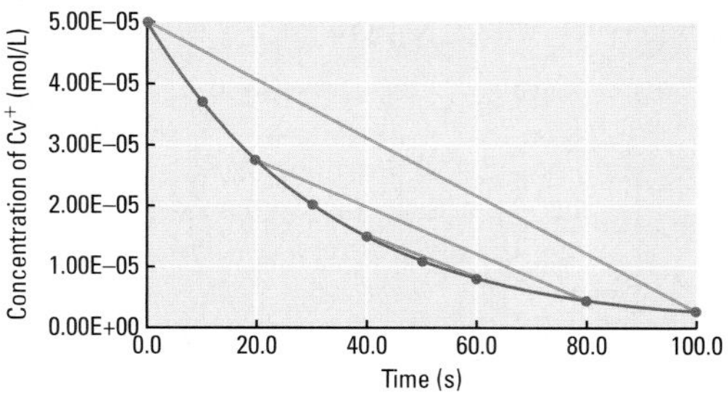

(c) The rate is faster when the concentration of Cv^+ is larger. As the reaction takes place, the rate gets slower. There is a much larger change in concentration from 0 s to 50 s than from 50 s to 100 s. Therefore, the Δ $[Cv^+]$ is more than three times bigger for the time range from 20 s to 80 s than it is for the time range from 40 s to 60 s, even though Δt is exactly three times larger.

13.2 (a) Rate (1) is twice rate (3); rate (2) is twice rate (4); rate (3) is twice rate (5).

(b) In each case, the $[Cv^+]$ is twice as great when the rate is twice as great.

(c) Yes, the rate doubles when $[Cv^+]$ doubles.

13.3 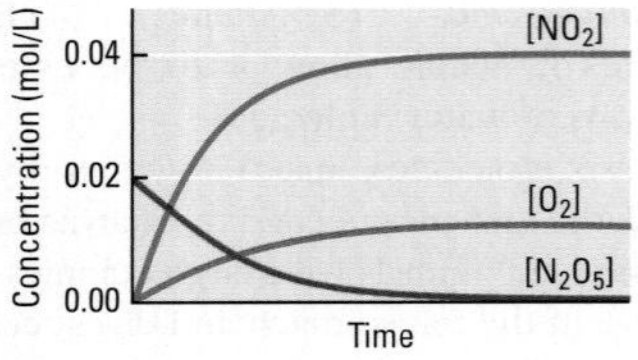

13.4 $\text{Rate}_1 = k[CH_3COOCH_3][OH^-]$
$\text{Rate}_2 = k(2[CH_3COOCH_3])(\frac{1}{2}[OH^-]) = k[CH_3COOCH_3][OH^-]$
The rate is unchanged.

13.5

$$\text{Rate} = k[NO]^x[Cl_2]^y$$

Taking the log of both sides and applying to experiments 4 and 2 gives

$$\log(\text{Rate}_4) = \log(k) + x\log[NO]_4 + y\log[Cl_2]_4$$

$$\log(\text{Rate}_2) = \log(k) + x\log[NO]_2 + y\log[Cl_2]_2$$

Recognizing that $[Cl_2]_4 = [Cl_2]_2$ and then subtracting the second equation from the first gives

$$\log(\text{Rate}_4) - \log(\text{Rate}_2) = x\log[NO]_4 - x\log[NO]_2$$

Using the properties of logarithms gives

$$\log\{\text{Rate}_4/\text{Rate}_2\} = x\{\log[NO]_4/[NO]_2\}$$

So

$$x = \log\{\text{Rate}_4/\text{Rate}_2\}/\{\log[NO]_4/[NO]_2\} \quad \text{QED}$$

Now we can insert the actual numbers from experiments 4 and 2:

$$x = [\log(6.60 \times 10^{-4}/1.65 \times 10^{-4})]/\log(0.04/0.02)$$

$$x = \log 4/\log 2 = 0.602/0.301 = 2$$

13.6 (a) $CH_3NC \rightarrow CH_3CN$ unimolecular
(b) $2\,HI \rightarrow H_2 + I_2$ bimolecular
(c) $NO_2Cl \rightarrow NO_2 + Cl$ unimolecular
(d) $C_4H_8 \rightarrow C_4H_8$ unimolecular
(e) $NO_2Cl + Cl \rightarrow NO_2 + Cl_2$ bimolecular

13.7 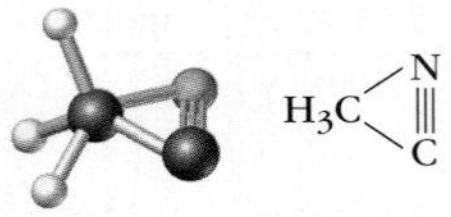

The Lewis structure has five pairs of electrons around one C atom, which would not be stable. However, this is a transition state, which is, by definition, unstable.

13.8 (1) 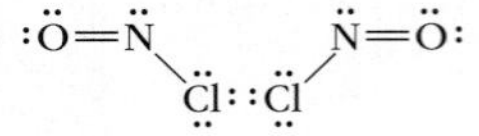

(2) 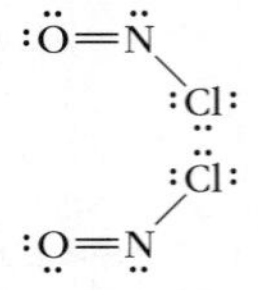

(3) 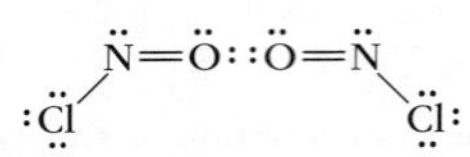

(4) 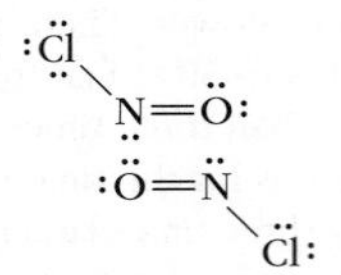

(1) and (2) are much more likely to result in a reaction than are (3) and (4).

13.9 (a) $k = Ae^{-Ea/RT}$

$$= (6.31 \times 10^8\ \text{L mol}^{-1}\ \text{s}^{-1})\, e^{\frac{-10{,}000\ \text{J/mol}}{(8.314\ \text{J K}^{-1}\ \text{mol}^{-1})(370\ \text{K})}}$$

$$k = 2.4 \times 10^7\ \text{L mol}^{-1}\ \text{s}^{-1}$$

(b) $\text{Rate} = k[NO][O_3]$

$$= (2.4 \times 10^7\ \text{L mol}^{-1}\ \text{s}^{-1})(1.0 \times 10^{-3}\ \text{mol/L}) \times (5.0 \times 10^{-4}\ \text{mol/L})$$

$$= 12\ \text{mol L}^{-1}\ \text{s}^{-1}$$

13.10 The reaction does not occur in a single step. If it did, the rate law would be $\text{Rate} = k[NO_2][CO]$.

13.11 $\text{Rate} = k[HOI][I^-]$

13.12 (a) $2\,ICl(g) + H_2(g) \rightarrow 2\,HCl(g) + I_2(g)$
(b) $\text{Rate} = k_2[HI][ICl]$
However, HI is an intermediate. Assume that the concentration of HI reaches a steady state.

Rate of step +1 = rate of step −1 + rate of step 2

Also assume that step 2 is much slower than step +1 and step −1. Then

$$k_1[ICl][H_2] = k_{-1}[HI][HCl]$$

$$[HI] = \frac{k_1[ICl][H_2]}{k_{-1}[HCl]}$$

$$\text{Rate} = k_2[ICl][HI] = k_2[ICl] \times \frac{k_1[ICl][H_2]}{k_{-1}[HCl]}$$

$$= \frac{k_1k_2}{k_{-1}}[ICl]^2[H_2][HCl]^{-1}$$

(c) The rate is inversely proportional to the concentration of the product, HCl.
(d) Because [HCl] increases as the reaction proceeds, the rate of reaction will decrease more quickly over time than it would if $[HCl]^{-1}$ were not in the rate law. However, the rate constant will not change.

13.13 (a) $Ce^{4+} + \cancel{Mn^{2+}} \rightarrow Ce^{3+} + \cancel{Mn^{3+}}$
$Ce^{4+} + \cancel{Mn^{3+}} \rightarrow Ce^{3+} + \cancel{Mn^{4+}}$
$\cancel{Mn^{4+}} + Tl^+ \rightarrow \cancel{Mn^{2+}} + Tl^{3+}$
$2\,Ce^{4+} + Tl^+ \rightarrow 2\,Ce^{3+} + Tl^{3+}$
(b) Intermediates are Mn^{3+} and Mn^{4+}.
(c) The catalyst is Mn^{2+}.
(d) $\text{Rate} = k[Ce^{4+}][Mn^{2+}]$
(e) $\text{Rate} = k[Ce^{4+}]^2[Mn^{2+}][Ce^{3+}]^{-1}$

13.14 (a) The concentration of a homogeneous catalyst *must* appear in the rate law.
(b) A catalyst does not appear in the equation for an overall reaction.
(c) A *homogeneous* catalyst must always be in the same phase as the reactants.

Chapter 14

14.1 (a) $(\text{conc. Al}) = \frac{2.70\ \text{g}}{\text{mL}} \times \frac{1000\ \text{mL}}{1\ \text{L}} \times \frac{1\ \text{mol}}{26.98\ \text{g}}$

$$= 100.\ \text{mol/L}$$

(b) $(\text{conc. benzene}) = \frac{0.880\ \text{g}}{\text{mL}} \times \frac{1000\ \text{mL}}{1\ \text{L}} \times \frac{1\ \text{mol}}{78.11\ \text{g}}$

$$= 11.3\ \text{mol/L}$$

(c) $(\text{conc. water}) = \frac{0.998\ \text{g}}{\text{mL}} \times \frac{1000\ \text{mL}}{1\ \text{L}} \times \frac{1\ \text{mol}}{18.02\ \text{g}}$

$$= 55.4\ \text{mol/L}$$

(d) $(\text{conc. Au}) = \frac{19.32\ \text{g}}{\text{mL}} \times \frac{1000\ \text{mL}}{1\ \text{L}} \times \frac{1\ \text{mol}}{196.97\ \text{g}}$

$$= 98.1\ \text{mol/L}$$

14.2 The mixture is not at equilibrium, but the reaction is so slow that there is no change in concentrations. You could show that the system was not at equilibrium by providing a catalyst or by raising the temperature to speed up the reaction.

14.3 (a) The new mixture is not at equilibrium because the quotient (conc. *trans*)/(conc. *cis*) no longer equals the equilibrium constant. Because (conc. *cis*) was halved, the quotient is twice K_c.
(b) The rate *trans* → *cis* remains the same as before, because (conc. *trans*) did not change. The rate *cis* → *trans* is only half as great, because (conc. *cis*) is half as great as at equilibrium.
(c) At 600 K, K_c is 1.47. Thus, [*trans*] = 1.47[*cis*].
(d) 0.15 mol/L

14.4 (a) If the coefficients of an equation are halved, the numerical value for the new equilibrium constant is the square root of the previous equilibrium constant. So the new equilibrium constant is $(6.25 \times 10^{-58})^{1/2} = 2.50 \times 10^{-29}$.
(b) If a chemical equation is reversed, the value for the new equilibrium constant is the reciprocal of the previous equilibrium constant. So the new equilibrium constant is $1/(6.25 \times 10^{-58}) = 1.60 \times 10^{57}$.

14.5 (a) $K_P = K_c \times (RT)^{\Delta n} = (3.5 \times 10^8\ \text{L}^2\ \text{mol}^{-2})$

$$\times \{(0.082057\ \text{L atm K}^{-1}\ \text{mol}^{-1})(298)\}^{-2}$$
$$= 5.8 \times 10^5\ \text{atm}^{-2}$$

(b) $K_P = (3.2 \times 10^{81}\ \text{L mol}^{-1})$

$$\times \{(0.082057\ \text{L atm mol}^{-1}\ \text{K}^{-1})(298\ \text{K})\}^{-1}$$
$$= 1.3 \times 10^{80}\ \text{atm}^{-1}$$

(c) $K_P = K_c = 1.7 \times 10^{-3}$
(d) $K_P = (1.7 \times 10^2\ \text{L mol}^{-1})$

$$\times \{(0.082057\ \text{L atm mol}^{-1}\ \text{K}^{-1})(298\ \text{K})\}^{-1}$$
$$= 6.9\ \text{atm}^{-1}$$

14.6 (a) Because K_c for the forward reaction is small, K_c' for the reverse reaction is large.
(b) $K_c' = \dfrac{1}{K_c} = \dfrac{1}{1.8 \times 10^{-5}} = 5.6 \times 10^4$
(c) Ammonium ions and hydroxide ions should react, using up nearly all of whichever is the limiting reactant.
(d) $NH_4^+(aq) + OH^-(aq) \rightleftharpoons NH_3(aq) + H_2O(\ell)$
$NH_3(aq) \rightleftharpoons NH_3(g)$
You might detect the odor of $NH_3(g)$ above the solution. A piece of moist red litmus paper above the solution would turn blue.

14.7 Q should have the same mathematical form as K_P, so for the general Equation 14.1

$$Q_P = \frac{P_C^c \times P_D^d}{P_A^a \times P_B^b}$$

The rules for Q_P and K_P are analogous to those for Q and K_c:
If $Q_P > K_P$, then the reverse reaction occurs.
If $Q_P = K_P$, then the system is at equilibrium.
If $Q_P < K_P$, the forward reaction occurs.

14.8 $PCl_5(s) \rightleftharpoons PCl_3(g) + Cl_2(g)$
(a) Adding Cl_2 shifts the equilibrium to the left.
(b) Adding PCl_3 to the container shifts the equilibrium to the left.
(c) Because $PCl_5(s)$ does not appear in the K_c expression, adding some will not affect the equilibrium.

14.9 For the first equilibrium, [*trans*] = 1.2 mmol/L and [*cis*] = 0.80 mmol/L; this gives $K_c = 1.2/0.80 = 1.5$. For the second equilibrium, [*trans*] = 0.96 mmol/L and [*cis*] = 0.64 mmol/L; this gives $K_c = 0.96/0.64 = 1.5$, the same value. K does not change unless the temperature changes.

14.10 $$Q = \frac{(\text{conc. } H_2)(\text{conc. } I_2)}{(\text{conc. HI})^2} = \frac{(0.102 \times 3)(0.0018 \times 3)}{(0.096 \times 3)^2}$$
$$= \frac{(0.102)(0.0018)(9)}{(0.096)^2(9)} = 0.020 = K_c$$

Since $Q = K_c$, the system is at equilibrium under the new conditions. No shift is needed and none occurs.

14.11 From Problem-Solving Practice 14.7, $[NO_2] = 0.0418$ mol/L and $[N_2O_4] = 0.292$ mol/L.
Decreasing the volume from 4.00 to 1.33 L increases the concentrations as

$$(\text{conc. } NO_2) = 0.0418 \times \frac{4.00}{1.33} = 0.1257\ \text{mol/L}$$

$$(\text{conc. } N_2O_4) = 0.292 \times \frac{4.00}{1.33} = 0.8782\ \text{mol/L}$$

$$Q = \frac{(\text{conc. } NO_2)^2}{(\text{conc. } N_2O_4)} = \frac{(0.1257)^2}{0.8782} = 1.80 \times 10^{-2}$$

which is greater than K_c. Therefore, the equilibrium should shift to the left. Let x be the change in concentration of NO_2, (which should be negative).

	N_2O_4	$\rightleftharpoons$	$2\ NO_2$
Initial concentration (mol/L)	0.8782		0.1257
Change as reaction occurs (mol/L)	$-\frac{1}{2}x$		x
New equilibrium concentration (mol/L)	$0.8782 - \frac{1}{2}x$		$0.1257 + x$

$$K_c = 5.9 \times 10^{-3} = \frac{(0.1257 + x)^2}{0.8782 - 0.500x}$$
$$= \frac{(1.580 \times 10^{-2}) + (2.514 \times 10^{-1})x + x^2}{0.8782 - 0.500x}$$

$$5.18 \times 10^{-3} - (2.95 \times 10^{-3})x$$
$$= (1.580 \times 10^{-2}) + (2.514 \times 10^{-1})x + x^2$$

$$x^2 + 0.2544x + (1.062 \times 10^{-2}) = 0$$

$$x = \frac{-0.2544 \pm \sqrt{6.472 \times 10^{-2} - (4 \times 1 \times 1.062 \times 10^{-2})}}{2}$$

$$x = \frac{-0.2544 \pm 0.1491}{2}$$

$$x = -0.0526 \quad \text{or} \quad x = -0.2018$$

The second root is mathematically reasonable, but results in a negative value for the $[NO_2]$. The new equilibrium concentrations are

$$[NO_2] = 0.1257 - 0.0526 = 0.0730\ \text{mol/L}$$

$$[N_2O_4] = 0.8782 - \tfrac{1}{2}(-0.0526) = 0.9045\ \text{mol/L}$$

Compared to the initial equilibrium, the concentrations have changed by

$$NO_2\text{: } \frac{0.0730}{0.0418} = 1.75 \qquad N_2O_4\text{: } \frac{0.945}{0.292} = 3.10$$

The concentration of N_2O_4 did increase by more than a factor of 3. The concentration of NO_2 increased by 1.75, which is less than a factor of 3 increase.

14.12

	How Reaction System Changes	Equilibrium Shifts	Change in K_c?
(a) Add reactant	Some reactants consumed	To right	No
(b) Remove reactant	More reactants formed	To left	No
(c) Add product	More reactants formed	To left	No
(d) Remove product	More products formed	To right	No
(e) Increase *P* by decreasing *V*	Total pressure decreases	Toward fewer gas molecules	No
(f) Decrease *P* by increasing *V*	Total pressure increases	Toward more gas molecules	No
(g) Increase *T*	Heat transfer into system	In endothermic direction	Yes
(h) Decrease *T*	Heat transfer out of system	In exothermic direction	Yes

If a substance is added or removed, the equilibrium is affected only if the substance's concentration appears in the equilibrium constant expression, or if its addition or removal changes concentrations that appear in the equilibrium constant expression. Changing pressure by changing volume affects an equilibrium only for gas phase reactions in which there is a difference in the number of moles of gaseous reactants and products.

14.13 (a) The reaction is exothermic. $\Delta H° = -46.11$ kJ.
(b) The reaction is not favored by entropy.
(c) The reaction produces more products at low temperatures.
(d) If you increase *T* the reaction will go faster, but a smaller amount of products will be produced.

Chapter 15

15.1 The data in Table 15.2 indicate that the solubility of alcohols decreases as the hydrocarbon chain lengthens. Thus, 1-octanol is less soluble in water than 1-heptanol and 1-decanol should be even less soluble than 1-octanol.

15.2 Methanol is more water soluble than is octanol, but octanol is more soluble in gasoline. The octanol molecule is more hydrocarbon-like and this explains its solubility in gasoline. The methanol molecule is more water-like and this explains its greater solubility in water.

15.3 34 g NH_4Cl would crystallize from solution at 20 °C.

15.4 (a) Unsaturated (b) Supersaturated (c) Saturated

15.5 The solubility of CO_2 decreases with increasing temperature, and the beverage loses its carbonation, causing it to go "flat."

15.6 Putting back water that is too warm would decrease the solubility of oxygen in the lake or river, thereby decreasing the oxygen concentration. This could cause a fish kill if the oxygen concentration dropped sufficiently.

15.7 Hot solvent would cause more of the solute to dissolve because Le Chatelier's principle states that at higher temperature an equilibrium will shift in the endothermic direction.

15.8 (a) Mass fraction of $NaHCO_3$

$$= \frac{0.20}{1000 + 6.5 + 0.20 + 0.1 + 0.10} = 2.0 \times 10^{-4}$$

Wt. fraction $= 2.0 \times 10^{-4} \times 100\% = 2.0 \times 10^{-2}$

(b) KCl and $CaCl_2$ each have the lowest mass fraction, 0.015.

15.9 (a) The 20. ppb sample has the higher lead concentration. (The other sample is 3.0 ppb lead.)
(b) 0.015 mg/L is equivalent to 0.015 ppm, which is 15 ppb. The 20-ppb sample exceeds the EPA limit; the 3.0-ppb sample does not.

15.10 $$\frac{280\text{ mg}}{1\text{ d}} \times \frac{0.500\text{ L}}{12\text{ mg}} = \frac{12\text{ L}}{\text{d}} = 12\text{ bottles per day}$$

15.11 $$3.5 \times 10^{20}\text{ gal} \times \frac{3.785\text{ L}}{\text{gal}} \times \frac{1 \times 10^{-3}\text{ mg Au}}{1\text{ L}}$$
$$= 1 \times 10^{18}\text{ mg Au} = 1 \times 10^{15}\text{ g Au}$$
$$1 \times 10^{15}\text{ g Au} \times \frac{1\text{ lb Au}}{454\text{ g Au}} = 2 \times 10^{12}\text{ lb Au}$$

15.12 (a) $355\text{ mL} \times \frac{1.01\text{ g}}{\text{mL}} = 359$ g solution;

$$\frac{46.3 \times 10^{-3}\text{ g caffeine}}{359\text{ g solution}} = \frac{1.29 \times 10^{-4}\text{ g caffeine}}{1\text{ g solution}}$$
$$= \frac{129\text{ g caffeine}}{10^6\text{ g solution}} = 129\text{ ppm}$$

(b) $$\text{M} = \frac{\text{moles solute}}{\text{L solution}}$$
$$= \frac{(46.3 \times 10^{-3}\text{ g caffeine}) \times \left(\frac{1\text{ mol caffeine}}{194.2\text{ g caffeine}}\right)}{0.355\text{ L}}$$
$$= 6.70 \times 10^{-4}\text{ M}$$

(c) $$m = \frac{\text{moles solute}}{\text{kg solution}}$$
$$= \frac{(46.3 \times 10^{-3}\text{ g caffeine}) \times \left(\frac{1\text{ mol caffeine}}{194.2\text{ g caffeine}}\right)}{0.359\text{ kg solution}}$$
$$= 6.63 \times 10^{-4}\ m$$

15.13 (a) Moles of solute and kilograms of solvent
(b) Molar mass of the solute and the density of the solution

15.14 $$50.0\text{ g sucrose} \times \left(\frac{1\text{ mol sucrose}}{342\text{ g sucrose}}\right) = 0.146\text{ mol sucrose}$$

$$100.0\text{ g water} \times \left(\frac{1\text{ mol water}}{18.0\text{ g water}}\right) = 5.56\text{ mol water}$$

$$X_{H_2O} = \frac{5.56}{5.56 + 0.146} = \frac{5.56}{5.706} = 0.974$$

Applying Raoult's law:

$$P_{\text{water}} = (X_{\text{water}})(P^0_{\text{water}}) = (0.974)(71.88\text{ mm Hg}) = 70.0\text{ mm Hg}$$

15.15 $\Delta T_b = (2.53\ °\text{C kg mol}^{-1})(0.10\text{ mol/kg}) = 0.25\ °\text{C}$. The boiling point of the solution is 80.10 °C + 0.25 °C = 80.35 °C.

15.16 First, calculate the required molality of the solution that would have a freezing point of −30. °C.

$$\Delta T_f = -30.\ °\text{C} = (-1.86\ °\text{C}\cdot\text{kg mol}^{-1}) \times m$$

$$m = \frac{-30.\ °\text{C}}{-1.86\ °\text{C kg mol}^{-1}} = 16.1\text{ mol/kg}$$

To protect 4.0 kg of water from this freezing temperature, you would need 4.0 × 16.1 mol of ethylene glycol, or 64 mol.

$$64\text{ mol}\left(\frac{62.1\text{ g}}{\text{mol}}\right)\left(\frac{1\text{ mL}}{1.113\text{ g}}\right)\left(\frac{1\text{ L}}{1000\text{ mL}}\right) = 3.6\text{ L}$$

15.17 $\Delta T_f = K_f \times m \times i$;

$$4.78\ °\text{C} = (1.86\ °\text{C kg mol}^{-1})(2.0\text{ mol/kg})(i)$$

$$i = \frac{4.78\ °\text{C}}{(2.0\text{ mol/kg})(1.86\ °\text{C kg mol}^{-1})} = 1.28$$

Degree of dissociation: If completely dissociated, 1 mol $CaCl_2$ should yield 3 mol ions

$$CaCl_2 \longrightarrow Ca^{2+} + 2\,Cl^-$$

and *i* should be 3. The degree of dissociation in this solution is $1.28/3 = 0.427 \approx 43\%$.

15.18 The 0.02 mol/kg solution of ordinary soap would contain more particles since a soap is a salt of a fatty acid while sucrose is a nonelectrolyte.

15.19 $\frac{3\text{ qt}}{\text{d}} \times \frac{1\text{ L}}{1.06\text{ qt}} \times \frac{0.050\text{ mg Se}}{\text{L}} \times \frac{10^3\ \mu\text{g}}{1\text{ mg}} = \frac{142\ \mu\text{g Se}}{\text{d}}$

15.20 0.0250 ppm Pb means that for every liter (1 kg) of water, there is 0.0250 mg, or 25.0 μg of Pb. Using this factor

$$\text{Volume of water} = 100.0\ \mu\text{g Pb}\left(\frac{1\text{ L}}{25.0\ \mu\text{g Pb}}\right) = 4.00\text{ L}$$

Chapter 16

16.1 (a) Acid (b) Base
(c) Acid (d) Base

16.2 More water molecules are available per NH_3 molecule in a very dilute solution of NH_3.

16.3 (a) $H_3O^+(aq) + CN^-(aq)$
(b) $H_3O^+\ (aq) + Br^-\ (aq)$
(c) $CH_3NH_3^+(aq) + OH^-(aq)$
(d) $(CH_3)_2NH_2^+(aq) + OH^-(aq)$

16.4 (a) $HSO_4^-(aq) + H_2O(\ell) \rightarrow H_2SO_4(aq) + OH^-(aq)$
(b) H_2SO_4 is the conjugate acid of HSO_4^-; water is the conjugate acid of OH^-.
(c) Yes; it can be an H^+ donor or acceptor.

16.5 The reverse reaction is favored because HSO_4^- is a stronger acid than CH_3COOH.

16.6 (a) 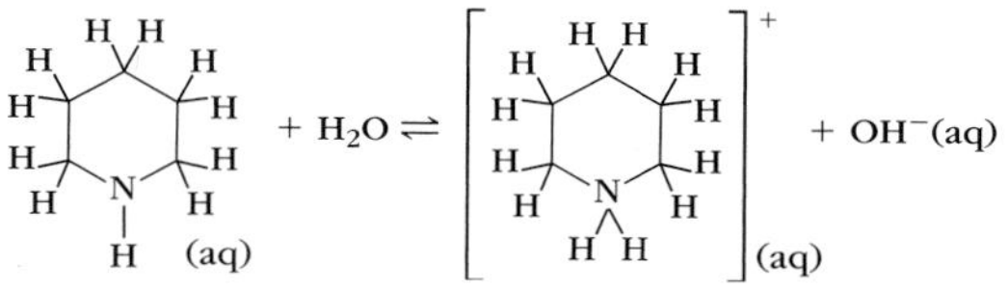

(b) 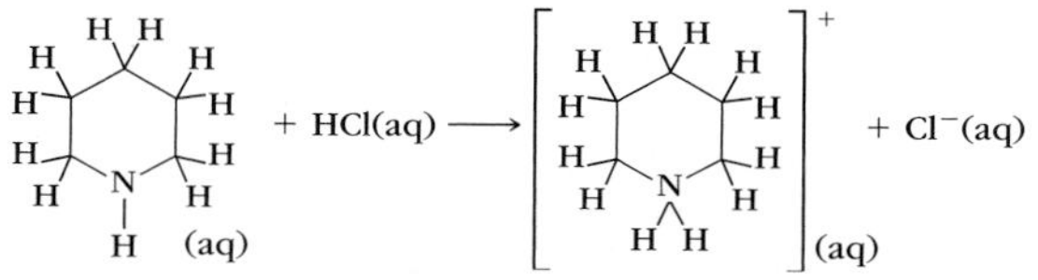

16.7 (a) Amine (b) Neither (e) Amine and acid
(c) Acid (d) Amine

16.8 The pH values of 0.1-M solutions of these two strong acids would be essentially the same since they both are 100% ionized, resulting in $[H_3O^+]$ values that are the same.

16.9 Lime juice is more acidic: pH $= -\log [H_3O^+] = -\log (1.3 \times 10^{-2}) = 1.89$; pH carbonated beverage is 2.10 (less acidic).

16.10 Because pH + pOH = 14.0, both solutions have a pOH of 8.5. The $[H_3O^+] = 10^{-pH} = 10^{-5.5} = 3.16 \times 10^{-6}$ M.

16.11 Pyruvic acid is the stronger acid as indicated by its larger K_a value. Lactic acid's ionization reaction is more reactant-favored (less acid ionizes).

16.12 Being negatively charged, the HSO_4^- ion has a lower tendency to lose a positively charged proton because of the electrostatic attractions of opposite charges.

16.13 (a) Step 1: $HOOC{-}COOH(aq) + H_2O(\ell) \rightleftharpoons H_3O^+(aq) + HOOC{-}COO^-(aq)$

Step 2: $HOOC{-}COO^-(aq) + H_2O(\ell) \rightleftharpoons H_3O^+(aq) + {}^-OOC{-}COO^-(aq)$

(b) Step 1: $C_3H_5(COOH)_3(aq) + H_2O \rightleftharpoons H_3O^+(aq) + C_3H_5(COOH)_2COO^-(aq)$

Step 2: $C_3H_5(COOH)_2COO^-(aq) + H_2O \rightleftharpoons H_3O^+(aq) + C_3H_5(COOH)(COO)_2^{2-}(aq)$

Step 3: $C_3H_5(COOH)(COO)_2^{2-}(aq) + H_2O \rightleftharpoons H_3O^+(aq) + C_3H_5(COO)_3^{3-}(aq)$

16.14 (a) Fluorobenzoic acid (b) Chloroacetic acid
In both cases, the more electronegative halogen atom increases electron withdrawal from the acidic hydrogen, thereby increasing its partial positive charge.

16.15 Oxalic acid is the stronger acid because it has a greater number of oxygens.

16.16 (a) 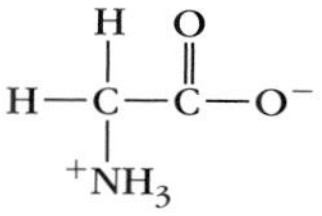

(b) 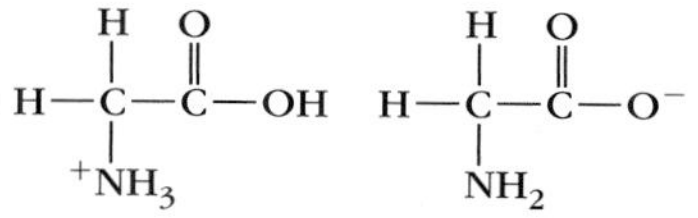

16.17 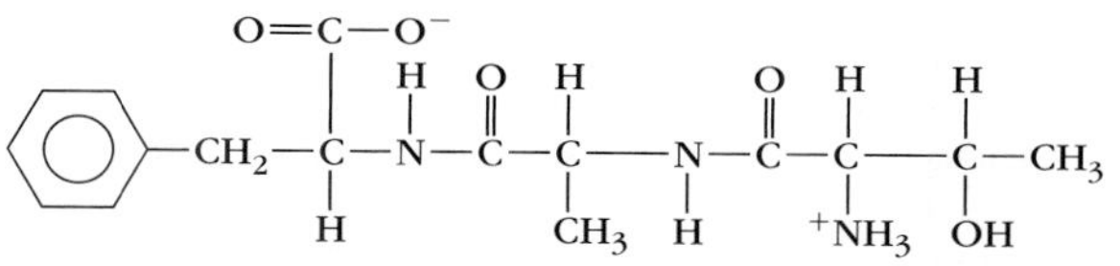

16.18 Set up an ICE table for the dissociation of nitrous acid.

	$HNO_2(aq)$ ⇌	$H^+(aq)$ +	NO_2^-
Initial conc., M	0.495	0	0
Conc. change, M	$-x$	$+x$	$+x$
Equil. conc., M	$0.495 - x$	$+x$	$+x$

Substitute the values into the equilibrium constant expression.

$$K_a = 4.5 \times 10^{-4} = \frac{[H^+][NO_2^-]}{[HNO_2]} = \frac{(x)(x)}{(0.495 - x)} = \frac{x^2}{(0.495 - x)}$$

Assume that x is small compared to 0.495 and the equation simplifies to

$$K_a = 4.5 \times 10^{-4} = \frac{[H^+][NO_2^-]}{[HNO_2]} = \frac{(x)(x)}{(0.495 - x)} \approx \frac{x^2}{(0.495)}$$

$$x^2 \approx (4.5 \times 10^{-4})(0.495) = 2.23 \times 10^{-4}$$

$$x = 1.49 \times 10^{-2}\text{ M} = [H^+]$$

$$\text{pH} = -\log (1.49 \times 10^{-2}) = 1.83$$

The simplifying assumption is reasonable because the percent ionization of 0.495-M nitrous acid, (x/initial concentration) × 100%, is less than 5%.

16.19 $K_b = \frac{1.0 \times 10^{-14}}{K_a} = \frac{1.0 \times 10^{-14}}{1.3 \times 10^{-10}} = 7.7 \times 10^{-5}$; carbonate; by comparing K_b values.

16.20 The pH of soaps is >7 due to the reaction of the conjugate base in the soap with water to form a basic solution.

16.21 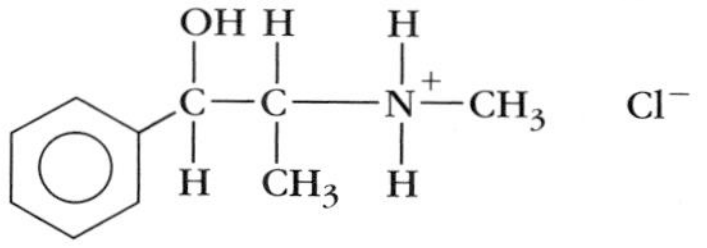

16.22 Ammonium acetate. The pH of a solution of this salt will be 7.

16.23 (a) Lewis base (b) Lewis acid
(c) Lewis acid and base (d) Lewis acid
(e) Lewis acid (f) Lewis base

16.24 The solution is slightly basic. The K_b of acetate ion, 5.6×10^{-10}, is slightly greater than the K_a of $Ni(H_2O)_6^{2+}$, 2.5×10^{-11}.

16.25 A reaction had these conditions:

	$H_2O(\ell)$ + $Ni(H_2O)_6^{2+}(aq)$ ⇌	$Ni(H_2O)_5(OH)^+(aq)$ +	$H_3O^+(aq)$
Initial (mol/L)	0.15	0	10^{-7}
Change (mol/L)	$-x$	$+x$	$+x$
Equilibrium (mol/L)	$0.15 - x$	x	x

Substituting these values in the equilibrium constant expression, and simplifying $0.15 - x$ to be 0.15 because the value of K_a is so small

$$K_a = \frac{[Ni(H_2O)_5(OH)^+][H_3O^+]}{[Ni(H_2O)_6^{2+}]}$$

$$= \frac{(x)(x)}{(0.15 - x)} \approx \frac{x^2}{0.15} = 2.5 \times 10^{-11}$$

Solving for x, which is the $[H_3O^+]$

$$x = \sqrt{(0.15)(2.5 \times 10^{-11})} = 1.9 \times 10^{-6}$$

So, the pH of this solution $= -\log(1.9 \times 10^{-6}) = 5.72$.

16.26 Strong bases would cause damage to tissue.

16.27 Baking powder and baking soda are sources of bicarbonate; buttermilk and baking powder are sources of acid.

16.28 Set up an ICE table for the hydrolysis reaction.

	$H_2O + CO_3^{2-}$	$\rightleftharpoons HCO_3^-$	$+ OH^-$
Initial concentration (mol/L)	5.2	0	10^{-7}
Concentration change due to reaction (mol/L)	$-x$	$+x$	$+x$
Equilibrium concentration (mol/L)	$5.2 - x$	x	x

Using the K_b expression and substituting the values from the table

$$K_b = \frac{K_w}{K_a(HCO_3^-)} = \frac{[HCO_3^-][OH^-]}{[CO_3^{2-}]} = \frac{x^2}{5.2 - x}$$

$$\approx \frac{x^2}{5.2} = 2.1 \times 10^{-4}$$

$$x = [OH^-] = \sqrt{(5.2)(2.1 \times 10^{-4})} = 3.3 \times 10^{-2}$$

$$pOH = -\log(3.3 \times 10^{-2}) = 1.48$$

$$pH = 14.00 - 1.48 = 12.52$$

Chapter 17

17.1 HCl and NaCl: no; has no significant H^+ acceptor (Cl^- is a very poor base).

KOH and KCl: no; has no H^+ donor.

17.2 $pH = 7.20 + \log \frac{(0.0025)}{(0.0015)} = 7.20 + \log(1.67)$

$= 7.20 + 0.22 = 7.42$

17.3 $pH = pK_a + \log \frac{[\text{acetate}]}{[\text{acetic acid}]}$

$$4.68 = 4.74 + \log \frac{[\text{acetate}]}{[\text{acetic acid}]}$$

$$\log \frac{[\text{acetate}]}{[\text{acetic acid}]} = 4.68 - 4.74 = -0.06;$$

$$\frac{[\text{acetate}]}{[\text{acetic acid}]} = 10^{-0.06} = 0.86$$

Therefore, [acetate] = 0.86 × [acetic acid].

17.4 (a) $pH = 6.37 + \log \frac{[0.25]}{[0.10]} = 6.37 + 0.398 = 6.77$

(b) $pH = 7.20 + \log \frac{[HPO_4^{2-}]}{[H_2PO_4^-]}$

$$= 7.20 + \log \frac{(0.25)}{(0.10)} = 7.20 + 0.398 = 7.60$$

17.5 Since CO_2 reacts to form an acid, H_2CO_3, the phosphate ion that is the stronger base, HPO_4^{2-}, will be used to counteract excess CO_2.

17.6

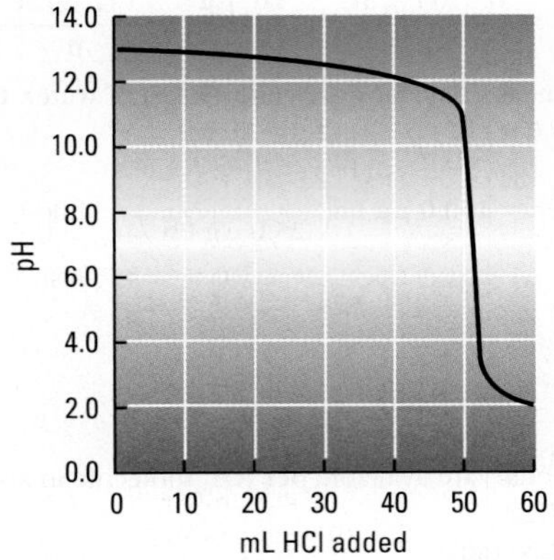

17.7 The addition of 30.0 mL of 0.100-M NaOH neutralizes 30.0 mL of 0.100-M acetic acid, forming 0.00300 mol of acetate ions, which is in 80.0 mL of solution. There is (0.0200 L)(0.100 mol/L) = 0.00200 mol acetic acid that is unreacted.

$$K_a = 1.8 \times 10^{-5} = \frac{[H^+][C_2H_3O_2^-]}{[HC_2H_3O_2]}$$

$$= \frac{[H^+] \times \left(\frac{0.00300 \text{ mol}}{0.0800 \text{ L}}\right)}{\left(\frac{0.00200 \text{ mol}}{0.0800 \text{ L}}\right)}$$

$$1.8 \times 10^{-5} = \frac{[H^+] \times (0.0375)}{(0.025)} = [H^+] \times 1.5$$

$$[H^+] = \frac{1.8 \times 10^{-5}}{1.5} = 1.2 \times 10^{-5}$$

$$pH = -\log(1.2 \times 10^{-5}) = 4.92$$

17.8 As NaOH is added, it reacts with acetic acid to form sodium acetate. After 20.0 mL NaOH has been added, just less than half of the acetic acid has been converted to sodium acetate; when 30.0 mL NaOH has been added, just over half of the acetic acid has been neutralized. Thus, after 20.0 mL and 30.0 mL base have been added, the solution contains approximately equal amounts of acetic acid and acetate ion, its conjugate base, which acts as a buffer.

17.9 Because Reaction (b) occurs to an appreciable extent, CO_3^{2-} is used as it forms by Reaction (a), causing additional $CaCO_3(s)$ to dissolve.

17.10 (a) The excess iodide would create a stress on the equilibria involving $I^-(aq)$ and shift them to the left; some AgI and some PbI_2 would precipitate from solution.

(b) The added SO_4^{2-} would cause the precipitation of $BaSO_4$.

Chapter 18

18.1 (a) $H_2O(\ell) \rightarrow H_2O(g)$ Product-favored

(b) $SiO_2(s) \rightarrow Si(s) + O_2(g)$ Reactant-favored

(c) $(C_6H_{10}O_5)_n(s) + 6n\ O_2(g) \rightarrow$
$6n\ CO_2(g) + 5n\ H_2O(g)$ Product-favored

(d) $C_{12}H_{22}O_{11}(s) \rightarrow C_{12}H_{22}O_{11}(aq)$ Product-favored

18.2 A*** A**B* A*B** B***

If C, D, and E are added, there are many more arrangements in addition to these:

A*B*C*	A*B*D*	A*B*E*	A*C*D*	A*C*E*
A*D*E*	B*C*D*	B*C*E*	B*D*E*	C*D*E*
A**C*	A**D*	A**E*	B**C*	B**D*
B**E*	C**A*	C**B*	C**D*	C**E*
D**A*	D**B*	D**C*	D**E*	E**A*
E**B*	E**C*	E**D*	C***	D***
E***				

There are 35 possible arrangements, but only 4 of them have the energy confined to atoms A and B. The probability that all energy remains with A and B is thus 4/35 = 0.114, or a little more than 11%.

18.3 Using Celsius temperature and $\Delta S = q_{rev}/T$, if the temperature were -10 °C, the value of ΔS would be negative, in disagreement with the fact that transfer of energy to a sample should increase molecular motion and, hence, entropy.

18.4 (a) The reactant is a gas. The products are also gases, but the number of molecules has increased, so entropy is greater for products. (Entropy increases.)
(b) The reactant is a solid. The product is a solution. Mixing sodium and chloride ions among water molecules results in greater entropy for the product. (Entropy increases.)
(c) The reactant is a solid. The products are a solid and a gas. The much larger entropy of the gas results in greater entropy for the products. (Entropy increases.)

18.5 (a) Because $\Delta S_{surroundings} = -\Delta H/T$ at a given temperature, the larger the value of T, the smaller the value of $\Delta S_{surroundings}$.
(b) If ΔS_{system} does not change much with temperature, then $S_{universe}$ must also get smaller. In this case, because ΔS_{system} is negative, $\Delta S_{universe}$ would become negative at a high enough temperature.

18.6 (a) The reaction would have gaseous water as a product.
(b) Both $\Delta H°$ and $\Delta S°$ would change. $\Delta H° = -10{,}232.197$ kJ and $\Delta S° = 756.57$ J/K.
(c) If any of the reactants or products change to a different phase (s, ℓ, or g) over the range of temperature, $\Delta H°$ and $\Delta S°$ will change significantly at the temperature of the phase transition.

18.7

Reaction	ΔH°, 298 K (kJ)	ΔS°, 298 K (J/K)
(a)	−1410.94	−267.67
(b)	467.87	560.32
(c)	−393.509	2.862
(d)	1241.2	−461.50

Reaction (a) is product-favored at low T (room temperature) and reactant-favored at high T.
Reaction (b) is reactant-favored at low T, product-favored at high T.
Reaction (c) is product-favored at all values of T.
Reaction (d) is reactant-favored at all values of T.

18.8 (a)
$$\Delta S°_{system} = 2\text{ mol HCl(g)} \times S°(\text{HCl[g]}) - 1\text{ mol H}_2\text{[g]} \times S°(\text{H}_2\text{[g]}) - 1\text{ mol Cl}_2\text{[g]} \times S°(\text{Cl}_2\text{[g]})$$
$$= (2 \times 186.908 - 130.684 - 223.066)\text{ J/K}$$
$$= 20.055\text{ J/K}$$
$$\Delta S°_{surroundings} = -\Delta H°/T = -\Delta H°/T$$
$$= -[2\text{ mol HCl(g)} \times (-92.307\text{ kJ/mol})]/298.15\text{ K}$$
$$= 619.20\text{ J/K}$$
(b) $\Delta S°_{universe} = (619.20 + 20.066)\text{ J/K} = 639.27\text{ J/K}$ (Product-favored)

18.9

Sign of ΔH°	Sign of ΔS°	Sign of ΔG°	Product-favored?
Negative (exothermic)	Positive	Negative	Yes
Negative (exothermic)	Negative	Depends on T	Yes at low T; no at high T
Positive (endothermic)	Positive	Depends on T	No at low T; yes at high T
Positive (endothermic)	Negative	Positive	No

18.10 (a) At 400 °C the equation is

$$2\text{ HgO(s)} \longrightarrow 2\text{ Hg(g)} + \text{O}_2\text{(g)}$$

Because Hg(g) is a product, instead of Hg(ℓ), both $\Delta H°$ and $\Delta S°$ will have significantly different values above 356 °C from their values below 356 °C. Therefore, the method of estimating $\Delta G°$ would not work above 356 °C.
(b) At 400 °C the entropy change should be more positive, which would make the reaction more product-favored. Because $\Delta H°$ and $\Delta S°$ are both positive, the reaction is product-favored at high temperatures.

18.11 (a) If the extent of reaction is 0.10, then 0.10 mol of NaOH has reacted with 0.10 mol of CO_2 to produce 0.10 mol of $NaHCO_3$.

$$\Delta G°(0.10\text{ extent}) = 0.10\text{ mol} \times \Delta G_f°(\text{NaHCO}_3\text{[s]}) - 0.10\text{ mol} \times \Delta G_f°(\text{NaOH[s]}) - 0.10\text{ mol} \times \Delta G_f°(\text{CO}_2\text{[g]})$$
$$= -0.10 \times 851.0\text{ kJ} + 0.10 \times 379.484\text{ kJ} + 0.10 \times 394.359\text{ kJ}$$
$$= -7.72\text{ kJ}$$

Similarly,

$$\Delta G°(0.40\text{ extent}) = -30.9\text{ kJ } (= 0.40[-72.2\text{ kJ}])$$
$$\Delta G°(0.80\text{ extent}) = -61.8\text{ kJ } (= 0.80[-72.2\text{ kJ}])$$

(b) In each case, $\Delta G°(x\text{ extent}) = x\Delta G°(\text{full extent})$, which verifies the statement.
(c) Since $\Delta G°(x\text{ extent}) = x\Delta G°(\text{full extent})$,

$$y = xm + b$$

where b = 0 and m = $\Delta G°$(full extent) = slope.

18.12 (a)
$$\Delta S°(\text{i}) = \{2 \times (27.78) + \tfrac{3}{2} \times (205.138) - (87.40)\}\text{ J/K} = 275.86\text{ kJ}$$
$$\Delta H°(\text{i}) = -\Delta H_f°(\text{Fe}_2\text{O}_3\text{[s]}) = 824.2\text{ kJ}$$
$$\Delta G°(\text{i}) = -\Delta G_f°(\text{Fe}_2\text{O}_3\text{[s]}) = 742.2\text{ kJ}$$
$$\Delta S°(\text{ii}) = \{50.92 - 2 \times (28.3) - \tfrac{3}{2} \times (205.138)\}\text{ J/K} = -313.4\text{ kJ}$$
$$\Delta H°(\text{ii}) = \Delta H_f°(\text{Al}_2\text{O}_3\text{[s]}) = -1675.7\text{ kJ}$$
$$\Delta G°(\text{ii}) = \Delta G_f°(\text{Al}_2\text{O}_3\text{[s]}) = -1582.3\text{ kJ}$$

Step (i) is reactant-favored. Step (ii) is product-favored.
(b) Net reaction
$Fe_2O_3(s) + 2\ Al(s) \rightarrow 2\ Fe(s) + Al_2O_3(s)$
$\Delta S° = 275.86\text{ J/K} + (-313.4\text{ J/K}) = -37.5\text{ J/K}$
$\Delta H° = 824.2\text{ kJ} + (-1675.7\text{ kJ}) = -851.5\text{ kJ}$
$\Delta G° = 742.2\text{ kJ} + (-1582.3\text{ kJ}) = -840.1\text{ kJ}$
The net reaction has negative $\Delta G°$ and is therefore product-favored. For the *net* reaction, $\Delta S°$, $\Delta H°$, and $\Delta G°$ are all negative.
(c) If the two reactions are coupled, it is possible to obtain iron from iron(III) oxide even though that reaction is not product-favored by itself. The large negative $\Delta G°$ for formation of $Al_2O_3(s)$ makes the overall $\Delta G°$ negative for the coupled reactions.
(d) $Mg(s) + \frac{1}{2} O_2(g) \rightarrow MgO(s)$
$$\Delta G° = \Delta G_f°(\text{MgO[s]}) = -569.43\text{ kJ}$$
Coupling the reactions, we have

$$\text{Fe}_2\text{O}_3\text{(s)} \longrightarrow 2\text{ Fe(s)} + \tfrac{3}{2}\text{O}_2\text{(g)} \qquad \Delta G_1° = 742.2\text{ kJ}$$
$$3 \times (\text{Mg[s]} + \tfrac{1}{2}\text{O}_2\text{[g]} \longrightarrow \text{MgO[s]}) \qquad \Delta G_2° = 3(-569.43)\text{ kJ} = -1708.29\text{ kJ}$$
$$\text{Fe}_2\text{O}_3\text{(s)} + 3\text{ Mg(s)} \longrightarrow 2\text{ Fe(s)} + 3\text{ MgO(s)} \qquad \Delta G_3° = -966.1\text{ kJ}$$

18.13 $\Delta G° = -2870\text{ kJ} + 32 \times (30.5\text{ kJ}) = -1894$ kJ. The 1894 kJ of Gibbs free energy is transformed into thermal energy.

18.14 64,500 g ATP/50 g ATP = 1290 times each ADP must be recycled to ATP on average each day.

Chapter 19

19.1 This is an application of the law of conservation of matter. If the number of electrons gained were different from the number of electrons lost, some electrons must have been created or destroyed.

19.2 Removal of the salt bridge would effectively switch off the flow of electricity from the battery.

19.3 Avogadro's number of electrons is 96,500 coulombs of charge, so it is 96,500 times as large as one coulomb of charge.

19.4 The zinc anode could be weighed before the battery was put into use. After a period of time, the zinc anode could be dried and reweighed. A loss in weight would be interpreted as being caused by the loss of Zn atoms from the surface through oxidation.

19.5 No, because Hg^{2+} ions can oxidize Al metal to Al^{3+} ions. The net cell reaction is

$$2\ Al(s) + 3\ Hg^{2+}(aq) \longrightarrow 2\ Al^{3+}(aq) + 3\ Hg(\ell)$$

$$E_{cell} = +2.51\ V$$

19.6 For this table,
(a) V^{2+} ion is the weakest oxidizing agent.
(b) Cl_2 is the strongest oxidizing agent.
(c) V is the strongest reducing agent.
(d) Cl^- is the weakest reducing agent.
(e) No, E_{cell} for that reaction would be <0.
(f) No, E_{cell} for that reaction would be <0.
(g) Pb can reduce I_2 and Cl_2.

19.7 In Table 19.1, Sb would be above H_2 and Pb would be below H_2. For Sb, the reduction potential would be between 0.00 and +0.337 V, and for Pb the value would be between 0.00 and −0.14 V.

19.8 For Na^+: $E_{ion} = 61.5 \log\left(\frac{150}{18}\right) = 61.5 \log(8.33)$

$$= 61.5 \times 0.921 = 57\ mV$$

19.9 During charging, the reactions at each electrode are reversed. At the electrode that is normally the anode, the charging reaction is

$$Cd(OH)_2(s) + 2\ e^- \longrightarrow Cd(s) + 2\ OH^-(aq)$$

This is reduction, so this electrode is now a cathode.
At the electrode that is normally the cathode, the charging reaction is

$$Ni(OH)_2 + OH^-(aq) \longrightarrow NiO(OH)(s) + H_2O(\ell) + e^-$$

This is oxidation, so this electrode is now an anode.

19.10 Remove the lead cathodes and as much sulfuric acid as you can from the discharged battery. Find some steel and construct a battery with Cl_2 gas flowing across a piece of steel. The two half-reactions would be

$$Cl_2(g) + 2\ e^- \longrightarrow 2\ Cl^-(aq) \qquad +1.36\ V$$

$$Pb(s) + SO_4^{2-}(aq) \longrightarrow PbSO_4(s) + 2\ e^- \qquad +0.356\ V$$

$$E_{cell} = 1.36 + 0.356 = 1.71\ V$$

19.11 Potassium metal was produced at the cathode.
Oxidation reaction: $2\ F^-\ (molten) \rightarrow F_2(g) + 2\ e^-$
Reduction reaction: $2(K^+[molten] + e^- \rightarrow K[\ell])$
Net cell reaction:
$2\ K^+\ (molten) + 2F^-\ (molten) \rightarrow 2\ K(\ell) + F_2(g)$

19.12 Reaction (c) making 2 mol of Cu from Cu^{2+} would require 4 Faradays of electricity. Two F are required for part (b), and 3 F are required for part (a).

19.13 First, calculate how many coulombs of electricity are required to make this much aluminum.

$$(2000.\ ton\ Al)\left(\frac{1000.\ kg\ Al}{1\ ton\ Al}\right)\left(\frac{10^3\ g}{1\ kg}\right)\left(\frac{1\ mol\ Al}{26.982\ g\ Al}\right)$$

$$\times \left(\frac{3\ mol\ e^-}{1\ mol\ Al}\right)\left(\frac{96{,}500\ C}{1\ mol\ e^-}\right) = 2.146 \times 10^{13}\ C$$

Next, using the product of charge and voltage, calculate how many joules are required; then convert to kilowatt-hours.

$$Energy = (2.146 \times 10^{13}\ C)(4.0\ V)\left(\frac{1\ J}{(1\ C)(1\ V)}\right)$$

$$\times \left(\frac{1\ kWh}{3.60 \times 10^6\ J}\right) = 2.4 \times 10^7\ kWh$$

19.14 To calculate how much energy is stored in a battery, you need the voltage and the number of coulombs of charge the battery can provide. The voltage is generally given on the battery label. To determine the number of coulombs available, you would have to disassemble the battery and determine the masses of the chemicals at the cathode and anode.

19.15 $(0.50\ A)(20.\ min)\left(\frac{60\ s}{1\ min}\right)\left(\frac{1\ C}{1\ A\ s}\right)\left(\frac{1\ mol\ e^-}{96{,}500\ C}\right)$

$$\times \left(\frac{1\ mol\ Ag}{1\ mol\ e^-}\right)\left(\frac{107.9\ g\ Ag}{1\ mol\ Ag}\right) = 0.67\ g\ Ag$$

19.16 No, not all metals corrode as easily. Three metals that would corrode about as readily as Fe and Al are Zn, Mg, and Cd. Three metals that do not corrode as readily as Fe and Al are Cu, Ag, and Au. These three metals are used in making coins and jewelry. Metals fall into these two broad groups because of their relative ease of oxidation compared with the oxidation of H_2. In Table 19.1, you can see this breakdown easily.

19.17 (b) > (a) > (d) > (c)
Sand by the seashore, (b), would contain both moisture and salts, which would aid corrosion. Moist clay, (a), would contain water but less dissolved salts. If an iron object were embedded within the clay, its impervious nature might prevent oxygen from getting to the iron, which would also lower the rate of corrosion. Desert sand in Arizona, (d), would be quite dry, and this low-moisture environment would not lead to a rapid rate of corrosion. On the moon, (c), there would be a lack of moisture and oxygen. This would lead to a very low rate of corrosion.

Chapter 20

20.1 ${}^{235}_{92}U \rightarrow {}^{4}_{2}He + {}^{231}_{90}Th \qquad {}^{231}_{90}Th \rightarrow {}^{0}_{-1}e + {}^{231}_{91}Pa$

${}^{231}_{91}Pa \rightarrow {}^{4}_{2}He + {}^{227}_{89}Ac \qquad {}^{227}_{89}Ac \rightarrow {}^{4}_{2}He + {}^{223}_{87}Fr$

${}^{223}_{87}Fr \rightarrow {}^{0}_{-1}e + {}^{223}_{88}Ra$

20.2 ${}^{26}_{13}Al \rightarrow {}^{0}_{+1}e + {}^{26}_{12}Mg \qquad {}^{26}_{13}Al + {}^{0}_{-1}e \rightarrow {}^{26}_{12}Mg$

20.3 Mass difference = $\Delta m = -0.03438$ g/mol
$\Delta E = (-3.438 \times 10^{-5}\ kg/mol)(2.998 \times 10^8\ m/s)^2$
$= -3.090 \times 10^{12}\ J/mol = -3.090 \times 10^9\ kJ/mol$
E_b per nucleon = 5.150×10^8 kJ/nucleon
E_b for ${}^6Li$ is smaller than E_b for ${}^4He$; therefore, helium-4 is more stable than lithium-6.

20.4 From the graph it can be seen that the binding energy per nucleon increases more sharply for the fusion of lighter elements than it does for heavy elements undergoing fission. Therefore, fusion is more exothermic per gram than fission.

20.5 $(\frac{1}{2})^{10} = 9.8 \times 10^{-4}$; this is equivalent to 0.098% of the radioisotope remaining.

20.6 All the lead came from the decay of ${}^{238}U$; therefore, at the time the rock was dated, N = 100 and N_0 = 209. The decay constant, k, can be determined:

$$k = \frac{0.693}{4.46 \times 10^9\ y} = 1.55 \times 10^{-10}\ y^{-1}$$

The age of the rock (t) can be calculated using Equation 20.3:

$$\ln \frac{100}{209} = -(1.55 \times 10^{-10}\ y^{-1}) \times t$$

$$t = 4.76 \times 10^9\ y$$

20.7 Ethylene is derived from petroleum, which was formed millennia ago. The half-life of ^{14}C is 5730 y, and thus much of ethylene's ^{14}C would have decayed and would be much less than that of the ^{14}C alcohol produced by fermentation.

20.8 (a) $^{13}_{6}C + ^{1}_{0}n \rightarrow ^{4}_{2}He + ^{10}_{4}Be$

(b) $^{14}_{7}N + ^{4}_{2}He \rightarrow ^{1}_{0}n + ^{17}_{9}F$

(c) $^{253}_{99}Es + ^{4}_{2}He \rightarrow ^{1}_{0}n + ^{256}_{101}Md$

20.9 $^{208}_{82}Pb + ^{70}_{30}Zn \rightarrow ^{277}_{112}E + ^{1}_{0}n$

20.10 Burning a metric ton of coal produces 2.8×10^7 kJ of energy.

$$\left(\frac{2.8 \times 10^4 \text{ kJ}}{1.0 \text{ kg}}\right)\left(\frac{10^3 \text{ kg}}{\text{metric ton}}\right) = 2.8 \times 10^7 \text{ kJ of energy}$$

The fission of 1.0 kg of ^{235}U produces

$$\frac{2.1 \times 10^{10} \text{ kJ}}{0.235 \text{ kg } ^{235}U} = 8.93 \times 10^{10} \text{ kJ}$$

It would require burning 3.2×10^3 metric tons of coal to equal the amount of energy from 1.0 kg of ^{235}U:

$$8.93 \times 10^{10} \text{ kJ from } ^{235}U \times \frac{1 \text{ metric ton coal}}{2.8 \times 10^7 \text{ kJ}}$$

$$= 3.2 \times 10^3 \text{metric tons}$$

20.11 $k = \frac{0.693}{29.1 \text{ y}} = 2.38 \times 10^{-2} \text{ y}^{-1}$

$\ln(\text{fraction}) = -(2.38 \times 10^{-2} \text{ y}^{-1})$ (18 y, as of 2004) $= -0.4284$

$\text{fraction} = e^{-0.4284} = 0.652 = 65.2\%$

20.12 $k = \frac{0.693}{30.2 \text{ y}} = 2.29 \times 10^{-2} \text{ y}^{-1}$

(a) 60% drop in activity; 40% activity remaining

$\ln(0.40) = -0.916 = -(2.29 \times 10^{-2} \text{ y}^{-1}) \times t$

$$t = \frac{-0.916}{-2.29 \times 10^{-2} \text{ y}^{-1}} = 40 \text{ y}$$

(b) 90% drop in activity, 10% remains

$\ln(0.10) = -2.30 = -(2.29 \times 10^{-2} \text{ y}^{-1}) \times t$

$$t = \frac{-2.30}{-2.29 \times 10^{-2} \text{ y}^{-1}} = 100 \text{ y}$$

20.13 (a) $^{7}_{3}Li + ^{1}_{1}H \rightarrow ^{1}_{0}n + ^{7}_{4}Be$

(b) $^{2}_{1}H + ^{3}_{2}He \rightarrow ^{4}_{2}He + ^{1}_{1}H$

20.14 $k = \frac{0.693}{3.82 \text{ d}} = 0.181 \text{ d}^{-1}$

(a) The drop from 8 to 4 pCi represents one half-life, 3.82 days.

(b) $\ln\left(\frac{1.5}{8}\right) = -1.67 = -(0.181 \text{ d}^{-1}) \times t$

$$t = \frac{-1.67}{-0.181 \text{ d}^{-1}} = 9.25 \text{ d}$$

20.15 $k = \frac{0.693}{78.2 \text{ h}} = 8.86 \times 10^{-3} \text{ h}^{-1}$

$\ln(0.10) = -2.30 = -(8.86 \times 10^{-3} \text{ h}^{-1}) \times t$

$t = 260$ h

20.16 Iron-59 $k = \frac{0.693}{44.5 \text{ d}} = 1.557 \times 10^{-2} \text{ d}^{-1}$

Chromium-51 $k = \frac{0.693}{27.7 \text{ d}} = 0.0250 \text{ d}^{-1}$

Fractions remaining:

^{59}Fe: $\ln(\text{fraction}) = -(1.557 \times 10^{-2} \text{ d}^{-1}) \times 90$ d

$\text{fraction} = e^{-1.40} = 0.246$; 80 mg $\times$ 0.246 = 19.7 mg left

^{51}Cr: $\ln(\text{fraction}) = -(0.0250 \text{ d}^{-1}) \times 90$ d; 10.5 mg left

Alternatively, consider the fact that 90 days is approximately two half-lives of ^{59}Fe. Therefore, approximately $\frac{3}{4}$ of it (about 60 mg) has decayed after 90 days, and about 20 mg remains. In that same time, ^{51}Cr has undergone more than three half-lives, so that less than $\frac{1}{8}$ remains (less than 12.5 mg).

Chapter 21

21.1 $^{12}_{6}C + ^{4}_{2}He \rightarrow ^{16}_{8}O$

$^{16}_{8}O + ^{4}_{2}He \rightarrow ^{20}_{10}Ne$

$^{20}_{10}Ne + ^{4}_{2}He \rightarrow ^{24}_{12}Mg$

21.2 $^{130}_{48}Cd \rightarrow ^{0}_{-1}e + ^{130}_{49}In$

$^{130}_{49}In \rightarrow ^{0}_{-1}e + ^{130}_{50}Sn$

$^{130}_{50}Sn \rightarrow ^{0}_{-1}e + ^{130}_{51}Sb$

$^{130}_{51}Sb \rightarrow ^{0}_{-1}e + ^{130}_{52}Te$

21.3 Two of the four oxygens in an SiO_4 unit are shared with other SiO_4 tetrahedra. Therefore, for each SiO_4 unit,

1 Si + two oxygen not shared + 2 oxygen shared $= SiO_{2+1}$

$= SiO_3^{2-}$

21.4 $5.0 \text{ L} \times \frac{1.4 \text{ g}}{\text{mL}} \times \frac{10^3 \text{ mL}}{1 \text{ L}} = 7.0 \times 10^3 \text{ g } O_2$

$7.0 \times 10^3 \text{ g} \times \frac{1 \text{ mol}}{32.0 \text{ g}} = 2.2 \times 10^2$ mol; $PV = nRT$

$$V = \frac{(2.2 \times 10^2 \text{ mol})(0.0821 \text{ L·atm·mol}^{-1}\text{·K}^{-1})(273 \text{ K})}{1 \text{ atm}}$$

$= 4.9 \times 10^3$ L

21.5 $2 \text{ NaCl}(\ell) \rightarrow 2 \text{ Na}(\ell) + Cl_2(g)$

Coulombs (C) = (A)(s)

$$= 2.0 \times 10^4 \text{ A} \times 24 \text{ h} \times \frac{3600 \text{ s}}{\text{h}} = 1.7 \times 10^9 \text{ C}$$

$$1.7 \times 10^9 \text{ C} \times \frac{1 \text{ mol e}^-}{9.65 \times 10^4 \text{ C}} = 1.8 \times 10^4 \text{ mol e}^-$$

$$1.8 \times 10^4 \text{ mol e}^- \times \frac{1 \text{ mol Na}}{1 \text{ mol e}^-} \times \frac{23.0 \text{ g Na}}{1 \text{ mol Na}} \times \frac{1 \text{ lb}}{454 \text{ g}} \times \frac{1 \text{ ton}}{2000 \text{ lb}} = 0.46 \text{ tons Na}$$

$$1.8 \times 10^4 \text{ mol e}^- \times \frac{1 \text{ mol } Cl_2}{2 \text{ mol e}^-} \times \frac{70.9 \text{ g } Cl_2}{1 \text{ mol } Cl_2} \times \frac{1 \text{ lb}}{454 \text{ g}} \times \frac{1 \text{ ton}}{2000 \text{ lb}} = 0.70 \text{ tons } Cl_2$$

21.6 $C = 2.00 \times 10^4 \text{ A} \times 100. \text{ h} \times \frac{3600 \text{ s}}{\text{h}} = 7.20 \times 10^9 \text{ C}$

$$7.20 \times 10^9 \text{ C} \times \frac{1 \text{ mol e}^-}{9.65 \times 10^4 \text{ C}} \times \frac{1 \text{ mol NaOH}}{1 \text{ mol e}^-} \times \frac{40.00 \text{ g NaOH}}{1 \text{ mol NaOH}} \times \frac{1 \text{ lb}}{454 \text{ g}} \times \frac{1 \text{ ton}}{2000 \text{ lb}}$$

$= 3.29$ tons NaOH

21.7 Phosphorus in $Ca_3(PO_4)_2$ has an oxidation state of +5; it is reduced to zero in P_4. Carbon is oxidized to CO (oxidation state changes from zero to −2).

21.8 $\frac{93 \text{ g}}{502 \text{ g}} \times 100\% = 19\%$ P

21.9 Br^- (reducing agent) is oxidized to Br_2; Cl_2 (oxidizing agent) is reduced to Cl^-.

21.10 $I_2(s) + 2\,e^- \rightarrow 2\,I^-(aq)$ $E° = +0.535$ V
$Br_2(\ell) + 2\,e^- \rightarrow 2\,Br^-(aq)$ $E° = +1.08$ V

$$I_2(s) + 2\,e^- \rightarrow 2\,I^-(aq) \quad E°_{cathode} = +0.535\text{ V}$$
$$2\,Br^-(aq) \rightarrow Br_2(\ell) + 2\,e^- + \quad E°_{anode} = -1.08\text{ V}$$
$$I_2(s) + 2\,Br^-(aq) \rightarrow 2\,I^-(aq) + Br_2(\ell) \quad E°_{cell} = -0.545\text{ V}$$

$I_2(s)$ is below $Br^-(aq)$ so iodine will not oxidize bromide ion to bromine, as seen by the negative cell voltage. Rather, bromine will oxidize iodide to iodine.

21.11 Oxide, O^{2-}, $[:\ddot{\underset{..}{O}}:]^{2-}$; peroxide, O_2^{2-}, $[:\ddot{\underset{..}{O}}:\ddot{\underset{..}{O}}:]^{2-}$;

superoxide, O_2^-, $[:\ddot{\underset{..}{O}}:\ddot{\underset{..}{O}}\cdot]^-$
MO diagrams:

Peroxide ion, $O_2^{2-}: \sigma_{2s}(\uparrow\downarrow)\,\sigma^*_{2s}(\uparrow\downarrow)\,\pi_{2p}(\uparrow\downarrow)\,\pi_{2p}(\uparrow\downarrow)$
$\sigma_{2p}(\uparrow\downarrow)\,\pi^*_{2p}(\uparrow\downarrow)\,\pi^*_{2p}(\uparrow\downarrow)$

Superoxide ion, $O_2^-: \sigma_{2s}(\uparrow\downarrow)\,\sigma^*_{2s}(\uparrow\downarrow)\,\pi_{2p}(\uparrow\downarrow)\,\pi_{2p}(\uparrow\downarrow)$
$\sigma_{2p}(\uparrow\downarrow)\,\pi^*_{2p}(\uparrow\downarrow)\,\pi^*_{2p}(\uparrow)$

21.12 (a) CaO (b) BaO_2 (c) Sr_3N_2 (d) CaC_2
21.13 (a) Mg_3N_2 (b) $3\,Mg(s) + N_2(g) \rightarrow Mg_3N_2(s)$

21.14 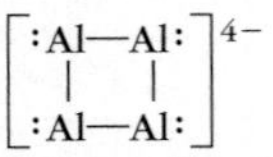

21.15 NO_3^-; oxidation number of oxygen is −2; oxidation number of N is +5.
NH_4^+; oxidation number of hydrogen is +1; oxidation number of nitrogen is −3.

21.16 $$150.\text{ g NaN}_3 \times \frac{1\text{ mol NaN}_3}{65.0\text{ g NaN}_3} \times \frac{3\text{ mol N}_2}{2\text{ mol NaN}_3} = 3.46\text{ mol N}_2$$

$$V = \frac{nRT}{P} = \frac{(3.46)(0.0821)(273)}{(1)} = 77.6\text{ L}$$

Chapter 22

22.1 (a) V^{+5} (b) Ag^{+3} (c) Co^{+2} (d) Mn^{+2}

22.2 Cu^{+2}

22.3 (a) $$500\text{ g} \times \frac{18\text{ g W}}{100\text{ g}} = 90\text{ g W}$$

$$500\text{ g} \times \frac{6.0\text{ g Co}}{100\text{ g}} = 30\text{ g Co}$$

(b) Iron is present in the greatest mole percent.

$$500\text{ g} \times \frac{68.9\text{ g Fe}}{100\text{ g}} = 344\text{ g Fe} = 6.2\text{ mol}$$

$$\frac{6.2}{8.0} \times 100\% = 77.5\%\text{ Fe}$$

22.4 $Cu^{2+}(aq) + 2\,e^- \rightarrow Cu(s)$

$$12.0\text{ h}\left(\frac{3600\text{ s}}{\text{h}}\right)\left(\frac{250\text{ C}}{\text{s}}\right)\left(\frac{1\text{ mol e}^-}{9.65 \times 10^4\text{ C}}\right)$$
$$\times \left(\frac{1\text{ mol Cu}}{2\text{ mol e}^-}\right)\left(\frac{63.55\text{ g Cu}}{1\text{ mol Cu}}\right) = 3.56 \times 10^3\text{ g Cu}$$

22.5 (a) NO_3^- is the oxidizing agent; silver metal is the reducing agent.
(b) NO_3^- is the oxidizing agent; gold is the reducing agent.

22.6 $2\,Al(s) + Cr_2O_3(s) \rightarrow Al_2O_3(s) + 2\,Cr(s)$

$$\Delta H_f° = \Delta H_f°(Al_2O_3(s)) - \Delta H_f°(Cr_2O_3(s))$$
$$= (-1675.7\text{ kJ}) - (-1139.7\text{ kJ})$$
$$= -536.0\text{ kJ/mol Cr}_2\text{O}_3$$

$$\Delta G° = \Delta G_f°(Al_2O_3(s)) - \Delta G_f°(Cr_2O_3(s))$$
$$= (-1582.3\text{ kJ}) - (-1058.1\text{ kJ})$$
$$-524.2\text{ kJ/mol Cr}_2\text{O}_3$$

22.7 The addition of acid shifts the equilibrium to the right, converting CrO_4^{2-} to $Cr_2O_7^{2-}$. Added base reacts with H^+ to form water, causing the equilibrium to shift to the left, converting $Cr_2O_7^{2-}$ to CrO_4^{2-}.

22.8 $[Cr(H_2O)_2(NH_3)_2(OH)_2]^+$

22.9 Should be $M_2^+X^{2-}$; such as K_2SO_4

22.10 (a) Three (b) 4− (c) Five

22.11 (a) Fe^{2+} (b) Fe^{3+}

22.12 Two isomers

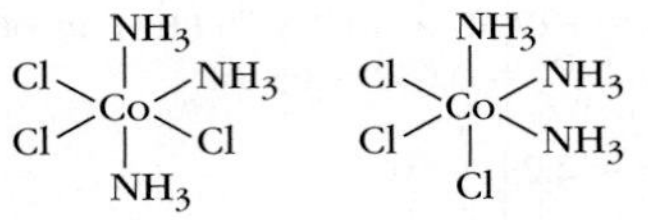

isomer 1 isomer 2

22.13 Ni^{2+} ions have a $3d^8$ configuration, which has three pairs of electrons in the t_2 orbitals and an unpaired electron in each of the two e orbitals. No other electron configuration is possible, so high- and low-spin Ni^{2+} complex ions are not formed.

22.14 Water is a weaker-field ligand (spectrochemical series), as indicated by the blue color of the complex ion due to absorption of longer-wavelength light. Therefore, the d orbital electrons would have the maximum number of unpaired electrons and the complex ion would be high-spin.

APPENDIX M

Answers to Selected Questions for Review and Thought

Chapter 1

11. (a) Quantitative (b) Qualitative (c) Qualitative
(d) Quantitative and qualitative (e) Qualitative

13. (a) Qualitative (b) Quantitative
(c) Quantitative and qualitative (d) Qualitative

15. Sulfur is a pale yellow, powdery solid. Bromine is a dark, red-brown liquid and a red-brown gas that fills the upper part of the flask. Both the melting point and the boiling point of sulfur must be above room temperature. The boiling point, but not the melting point, of bromine must be above room temperature. Both substances are colored. Most of their other properties appear to be different.

17. The liquid will boil because your body temperature of 37 °C is above the boiling point of 20 °C.

19. (a) 20 °C (b) 100 °C
(c) 60 °C (d) 20 °F

21. Copper

23. Aluminum

25. 3.9×10^3 g

27. (a) Physical (b) Chemical
(c) Chemical (d) Physical

29. (a) Chemical (b) Chemical (c) Physical

31. (a) An outside source of energy is forcing a chemical reaction to occur.
(b) A chemical reaction is releasing energy and causing work to be done.
(c) A chemical reaction is releasing energy and causing work to be done.
(d) An outside source of energy is forcing a chemical reaction to occur.

33. Heterogeneous; use a magnet.

35. (a) Homogeneous (b) Heterogeneous
(c) Heterogeneous (d) Heterogeneous

37. (a) Evaporate the water.
(b) Use a magnet.
(c) Dissolve the sugar in water, separate solution from solid, then evaporate the water.

39. (a) A compound decomposed
(b) A compound decomposed

41. (a) Heterogeneous mixture (b) Pure compound
(c) Heterogeneous mixture (d) Homogeneous mixture

43. (a) Heterogeneous mixture (b) Pure compound
(c) Element (d) Homogeneous mixture

45. (a) No (b) Maybe

47. The macroscopic world; a parallelepiped shape; the atom crystal arrangement is a parallelepiped shape.

49. Microscale world

51. Nanoscale

53. Carbon dioxide molecules are crowded in the unopened can. When opened, the molecules quickly escape through the hole.

55. When sucrose is heated, the motion of the atoms increases. Only when that motion is extreme enough, will the bonds in the sucrose break, allowing for the formation of new bonds to produce the "caramelization" products.

57. (a) 3.275×10^4 m (b) 3.42×10^4 nm (c) 1.21×10^{-3} μm

59. Because atoms in the starting materials must all be accounted for in the substances produced, and because the mass of each atom does not change, there would be no change in the mass.

61. (Remember, you are instructed to use your own words to answer this question.) Consider two compounds that both contain the same two elements. In each compound, the proportion of these two elements is a whole-number integer ratio. Because they are different compounds, these ratios must be different. If you pick a sample of each of these compounds such that both samples contain the same number of atoms of the first element, and then you count the number of atoms of the second type, you will find that a small integer relationship exists between the number of atoms of the second type in the first compound and the number of atoms of the second type in the second compound.

63. If two compounds contain the same elements and samples of those two compounds both contain the same mass of one element, then the ratio of the masses of the other elements will be small whole numbers.

65. Many responses are equally valid here. Common examples given here: (a) iron, Fe; gold, Au (b) carbon, C; hydrogen, H (c) boron, B; silicon, Si (d) nitrogen, N_2; oxygen, O_2.

67.

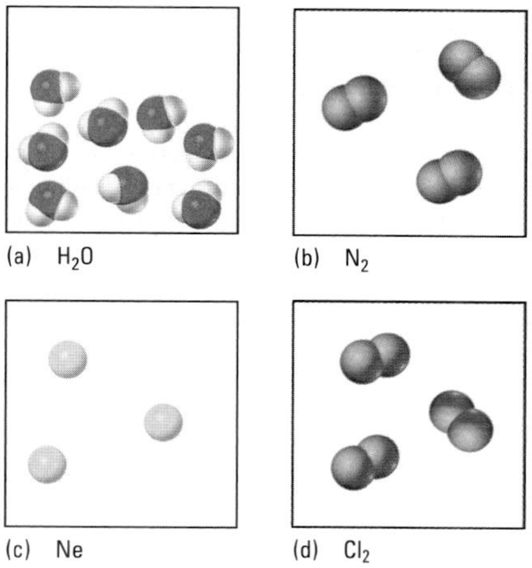

(a) H_2O
(b) N_2
(c) Ne
(d) Cl_2

69. $2\ H_2(g) + O_2(g) \longrightarrow 2\ H_2O(g)$

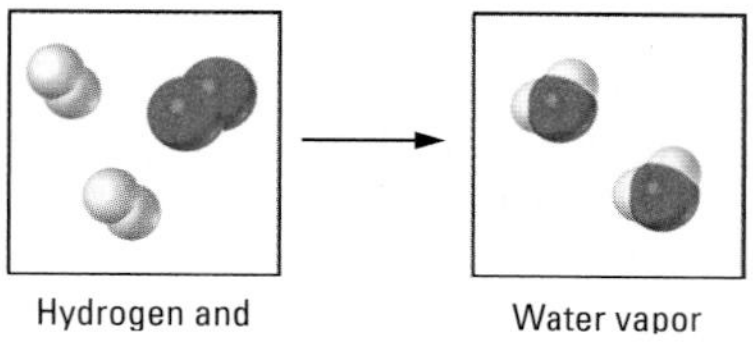

Hydrogen and
Water vapor

71. 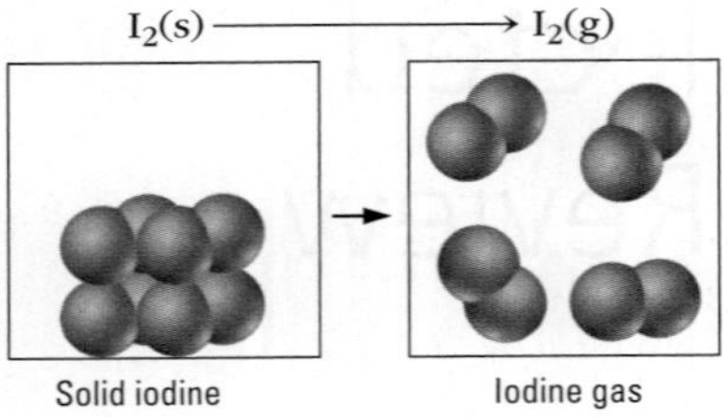
$I_2(s)$
$I_2(g)$
Solid iodine
Iodine gas

73. (a) Mass is quantitative and related to a physical property. Colors are qualitative and related to physical properties. Reaction is qualitative and related to a chemical property.
(b) Mass is quantitative and related to a physical property. The fact that a chemical reaction occurs between substances is qualitative information and related to a chemical property.
75. In solid calcium, smaller radius atoms are more closely packed, making a smaller volume. In solid potassium, larger radius atoms are less closely packed, making a larger volume.
77. 508 m
79. The Teflon piece will be on the bottom in liquid C_6F_{14}, with water floating on it. The HDPE will be floating on the water in the liquid C_6H_{14}.
81. They are different by how the atoms are organized and bonded together.
83. (a) Bromobenzene (b) Gold (c) Lead
85. (a) 2.7×10^2 mL ice
(b) Bulging, cracking, deformed, or broken
87. Gold
89. (a) Water layer on top of bromobenzene layer.
(b) If it is poured slowly and carefully, ethanol will float on top of the water and slowly dissolve in the water. Both ethanol and water will float on the bromobenzene.
(c) Stirring will speed up dissolving of ethanol and water in each other. After stirring only two layers will remain.
91. Drawing (b)
93. 6.02×10^{-29} m^3
95. (a) Lavender (b) Gray and blue (c) Orange and lavender
97. S and Se; they are in the same group.
99. It is difficult to prove that something cannot be broken down (see Section 1.3).
101. (a) Nickel, lead, and magnesium (b) Titanium
103. (a) 32.1 g S (b) 29.8 g ZnS
105. No, samples contain variable percentages of iron.
107. 3.1 L
109. Obtain four or more lemons. Keep one unaltered to be the "control" case. Perform designated tasks to others, including applying both tasks to the same lemon; then juice all of them, recording results, such as juice volume and ease of task. Repeat with more lemons to achieve better reliability. Hypothesis: Disrupting the "juice sacks" inside the pulp helps to release the juice more easily.

Chapter 2

7. 40,000 cm
9. 614 cm, 242 cm, 20.1 ft
11. 76.2 kg
13. 2.0×10^3 cm^3, 2.0 L
15. 1550 in^2
17. 2.8×10^9 m^3
22. (a) 4 (b) 3 (c) 4
(d) 4 (e) 3
24. (a) 4.33×10^{-4} (b) 4.47×10^1
(c) 2.25×10^1 (d) 8.84×10^{-3}
26. (a) 1.9 g/mL (b) 218.4 cm^3
(c) 0.0217 (d) 5.21×10^{-5}
28. Ions
30. Using a magnetic field
35. Number of neutrons
37. 27 protons, 27 electrons, and 33 neutrons
39. 78.92 amu/atom
41. (a) 9 (b) 48 (c) 70
43. (a) ${}^{23}_{11}Na$ (b) ${}^{39}_{18}Ar$ (c) ${}^{69}_{31}Ga$
45. (a) 20 e^-, 20 p^+, 20 n^0 (b) 50 e^-, 50 p^+, 69 n^0
(c) 94 e^-, 94 p^+, 150 n^0
47.

Z	A	Number of Neutrons	Element
35	81	46	Br
46	108	62	Pd
77	192	115	Ir
63	151	88	Eu

49. ${}^{18}_{9}X$, ${}^{20}_{9}X$ and ${}^{15}_{9}X$
51. 80.1% silver, 19.9% copper
53. 245 g sulfuric acid
55. 0.93% Na
57. 6.941 amu/atom
59. 60.12% ${}^{69}Ga$, 39.87% ${}^{71}Ga$
61. ${}^{7}Li$
63. Pair (2), dozen (12), gross (144), million (1,000,000) (Other answers are possible.)
66. (a) 27 g B (b) 0.48 g O_2
(c) 6.98×10^{-2} g Fe (d) 2.61×10^3 g H
68. (a) 1.9998 mol Cu (b) 0.499 mol Ca
(c) 0.6208 mol Al (d) 3.1×10^{-4} mol K
(e) 2.1×10^{-5} mol Am
70. 2.19 mol Na
72. 9.42×10^{-5} mol Kr
74. 4.131×10^{23} Cr atoms
76. 1.055×10^{-22} g Cu
78. In a group, elements share the same vertical column. In a period, elements share the same horizontal row.
83. Transition elements: iron, copper, chromium
Halogens: fluorine and chlorine
Alkali metal: sodium
(Other answers are possible.)
85. Five; nonmetal: carbon (C), metalloids: silicon (Si) and germanium (Ge), and metals: tin (Sn) and lead (Pb).
87. (a) I (b) In (c) Ir (d) Fe
89. Sixth period
91. (a) Mg (b) Na (c) C (d) S
(e) I (f) Mg (g) Kr (h) S
(i) Ge [Other answers are possible for (a), (b), and (i).]
93. (a) Iron or magnesium (b) Hydrogen (c) Silicon
(d) Iron (e) Chlorine
97. (a) 0.197 nm (b) 197 pm
99. (a) 0.178 nm^3 (b) 1.78×10^{-22} cm^3
102. (a) 1.67×10^{21} molecules (b) 1.67×10^8 km
104. 89 tons/yr
106. ${}^{39}K$
109. (a) Ti; 22; 47.88 (b) Group 4B; Period 4; zirconium, hafnium, rutherfordium (c) light-weight and strong (d) strong, low-density, highly corrosion resistant, occurs widely, light weight, high-temperature stability (Other answers are possible.)
111. 0.038 mol
113. $5,700
115. 3.4 mol, 2.0×10^{24} atoms
116. (a) Not possible (b) Possible (c) Not possible
(d) Not possible (e) Possible (f) Not possible
118. (a) Same (b) Second (c) Same
(d) Same (e) Same (f) Second
(g) Same (h) First (i) Second
(j) First
121. (a) K (b) Ar (c) Cu (d) Ge
(e) H (f) Al (g) O (h) Ca
(i) Br (j) P

123. (a) Se (b) ^{39}K (c) ^{79}Br (d) ^{20}Ne
124. (a) 270 mL (b) No
127. (a) $^{79}Br—^{79}Br$, $^{79}Br—^{81}Br$, $^{81}Br—^{81}Br$
(b) 78.918 g/mol, 80.196 g/mol
(c) 79.90 g/mol (d) 51.1% ^{79}Br, 48.9% ^{81}Br

Chapter 3

9. (a) BrF_3 (b) XeF_2 (c) P_2F_4 (d) $C_{15}H_{32}$ (e) N_2H_4
11. Butanol, $C_4H_{10}O$, $CH_3CH_2CH_2CH_2OH$

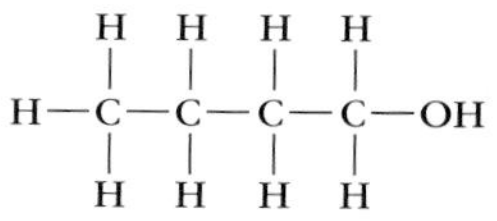

Pentanol, $C_5H_{12}O$, $CH_3CH_2CH_2CH_2CH_2OH$

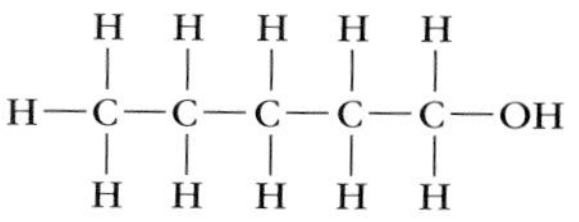

14. (a) C_6H_6 (b) $C_6H_8O_6$
16. (a) 1 calcium atom, 2 carbon atoms, and 4 oxygen atoms
(b) 8 carbon atoms and 8 hydrogen atoms
(c) 2 nitrogen atoms, 8 hydrogen atoms, 1 sulfur atom, and 4 oxygen atoms
(d) 1 platinum atom, 2 nitrogen atoms, 6 hydrogen atoms, and 2 chlorine atoms
(e) 4 potassium atoms, 1 iron atom, 6 carbon atoms, and 6 nitrogen atoms
18. (a) Sulfur dioxide
(b) Carbon tetrachloride
(c) Tetraphosphorus decasulfide
(d) Sulfur tetrofluoride
20. (a) Hydrogen bromide
(b) Chlorine trifluoride
(c) Dichlorine heptaoxide
(d) Boron triiodide
22. Three
24. 16
26. (a) Same number of atoms of each kind
(b) Different bonding arrangements
28. (a) Li^+ (b) Sr^{2+} (c) Al^{3+} (d) Ca^{2+}
(e) Zn^{2+}
30. Ba: 2+, Br: −1
32. (a) +2 (b) +2 (c) +2 or +3 (d) +3
34. CoO, Co_2O_3
36. (c) and (d) are correct formulas. (a) $AlCl_3$ (b) NaF
38. Mn
40. (a) 1 Pb^{2+} and 2 NO_3^- (b) 1 Ni^{2+} and 1 CO_3^{2-}
(c) 3 NH_4^+ and 1 PO_4^{3-} (d) 2 K^+ and 1 SO_4^{2-}
42. $BaSO_4$, barium ion, 2+, sulfate, 2−; $Mg(NO_3)_2$, magnesium ion, 2+, nitrate, 1−; $NaCH_3CO_2$, sodium ion, 1+, acetate, 1−
44. (a) $Ni(NO_3)_2$ (b) $NaHCO_3$ (c) LiClO
(d) $Mg(ClO_3)_2$ (e) $CaSO_3$
46. (b), (c), and (e) are ionic.
48. Only (e) is ionic, with metal and nonmetal combined. (a)-(d) are composed of only nonmetals.
50. (a) $(NH_4)_2CO_3$ (b) CaI_2 (c) $CuBr_2$ (d) $AlPO_4$
52. (a) Potassium sulfide (b) Nickel(II) sulfate (c) Ammonium phosphate (d) Aluminum hydroxide (e) Cobalt(III) sulfate
54. MgO; MgO has higher ionic charges and smaller ion sizes than NaCl.
56. Conducts electricity in water; check electrical conductivity; examples: NaCl and $C_{12}H_{22}O_{11}$. (Other answers are possible.)
58. Molecular compounds are generally not ionic compounds and, therefore, would not ionize in water.
60. (a) K^+ and OH^- (b) K^+ and SO_4^{2-}
(c) Na^+ and NO_3^- (d) NH_4^+ and Cl^-
62. (a) Box 2 (b) Box 3 (c) Box 1
64. (a) and (d)
66.

	CH_3OH	Carbon
No. of moles	One	One
No. of molecules or atoms	6.022×10^{23} molecules	6.022×10^{23} atoms
Molar mass	32.0417 g/mol	12.0107 g/mol

	Hydrogen	Oxygen
No. of moles	Four	One
No. of molecules or atoms	2.409×10^{24} atoms	6.022×10^{23} atoms
Molar mass	4.0316 g/mol	15.9994 g/mol

68. (a) 159.688 g/mol (b) 67.806 g/mol (c) 44.0128 g/mol
(d) 197.905 g/mol (e) 176.1238 g/mol
70. (a) 0.0312 mol (b) 0.0101 mol (c) 0.0125 mol
(d) 0.00406 mol (e) 0.00599 mol
72. (a) 179.855 g/mol (b) 36.0 g (c) 0.0259 mol
74. (a) 151.1622 g/mol (b) 0.0352 mol (c) 25.1 g
76. 1.80×10^{-3} mol; 1.09×10^{21} molecules
78. (a) 0.400 mol (b) 0.250 mol (c) 0.628 mol
80. 2.7×10^{23} atoms
82. 1.2×10^{24} molecules
84. (a) 0.250 mol CF_3CH_2F
(b) 6.02×10^{23} F atoms
87. (a) 239.3 g/mol PbS, 86.60% Pb, 13.40% S
(b) 30.0688 g/mol C_2H_6, 79.8881% C, 20.1119% H
(c) 60.0518 g/mol CH_3CO_2H, 40.0011% C, 6.7135% H, 53.2854% O
(d) 80.0432 g/mol NH_4NO_3, 34.9979% C, 5.0368% H, 59.9654% O
89. 58.0% M in MO
91. 245.745 g/mol, 25.858% Cu, 22.7992% N, 5.74197% H, 13.048% S, 32.5528% O
93. One
95. (a) −2 (b) Al_2X_3 (c) Se
98. An empirical formula shows the simplest whole-number ratio of the elements, e.g., CH_3; a molecular formula shows the actual number of atoms of each element, e.g., C_2H_6.
100. $C_4H_4O_4$
102. $C_2H_3OF_3$
104. KNO_3
106. B_5H_7
108. (a) C_3H_4 (b) C_9H_{12}
110. (a) C_5H_7N (b) $C_{10}H_{14}N_2$
112. $C_5H_{14}N_2$
114. Five
122. (a) $C_7H_5N_3O_6$ (b) $C_3H_7NO_3$
125. (a) (i) Chlorine tribromide (ii) Nitrogen trichloride
(iii) Calcium sulfate (iv) Heptane (v) Xenon tetrafluoride
(vi) Oxygen difluoride (vii) Sodium iodide
(viii) Aluminum sulfide (ix) Phosphorus pentachloride
(x) Potassium phosphate
(b) (iii), (vii), (viii), and (x)
127. (a) 2.39 mol (b) 21.7 cm^3
129. (a) CO_2F_2 (b) CO_2F_2
131. $MnC_9H_7O_3$
133. (a) 1.66×10^{-3} mol (b) 0.346 g
135. 2.14×10^5 g

138.

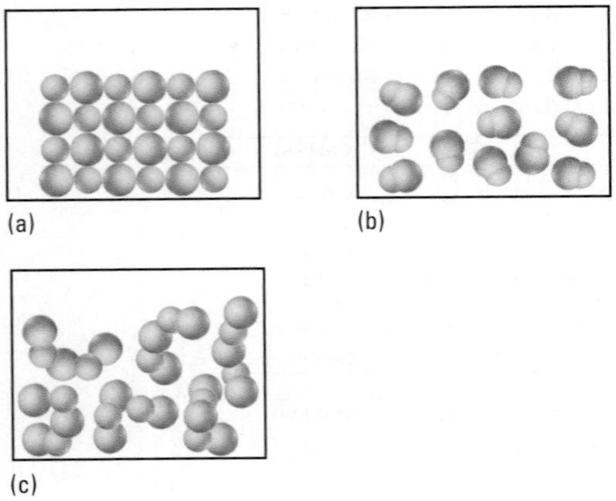

140. (a) Three (b) Three pairs are identical:

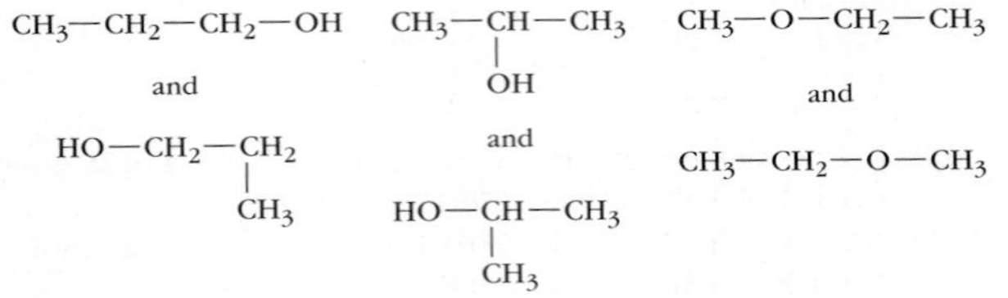

$CH_3—CH_2—CH_2—OH$ and $HO—CH_2—CH_2—CH_3$ (the final CH_3 printed below the second CH_2)

$CH_3—CH(OH)—CH_3$ and $HO—CH(CH_3)—CH_3$

$CH_3—O—CH_2—CH_3$ and $CH_3—CH_2—O—CH_3$

142. Tl_2CO_3, Tl_2SO_4
144. (a) Perbromate, bromate, bromite, hypobromite
(b) Selenate, selenite
147. CO_2
149. $MgSO_3$, magnesium sulfite
150. (a) 0.0130 mol Ni (b) NiF_2
(c) Nickel(II) fluoride
153. 5.0 lb N, 4.4 lb P, 4.2 lb K
155. 75.0% sulfate

Chapter 4

10.

	NH_3	O_2
No. molecules	4 × Avogadro's number	5 × Avogadro's number
No. atoms	16	10
No. moles of molecules	4	5
Mass	68.1216 g	159.9940 g
Total mass of reactants	228.1156 g	
Total mass of products		

	NO	H_2O
No. molecules	4 × Avogadro's number	6 × Avogadro's number
No. atoms	8	18
No. moles of molecules	4	6
Mass	120.0244 g	108.0912 g
Total mass of reactants		
Total mass of products	228.1156 g	

12. (a) 1.00 g (b) 2 for Mg, 1 for O_2, and 2 for MgO
(c) 50 atoms of Mg
14. $4\ Fe(s) + 3\ O_2(g) \rightarrow 2\ Fe_2O_3(s)$
16. Equation (b)
18.

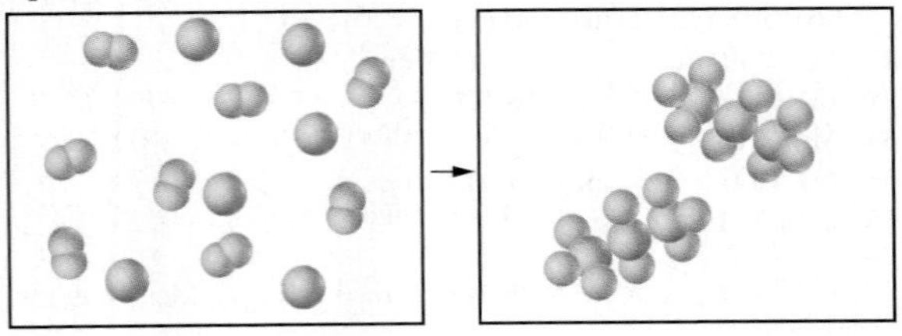

20. (a) Box (a) (b) Box (c)
22. $X_2 + Y_2 \rightarrow XY_2 + X$
24. (a) Combination (b) Decomposition (c) Exchange
(d) Displacement
26. (a) Decomposition (b) Displacement (c) Combination
(d) Exchange
28. (a) $2\ C_4H_{10}(g) + 13\ O_2(g) \rightarrow 8\ CO_2(g) + 10\ H_2O(g)$
(b) $C_6H_{12}O_6(s) + 6\ O_2(g) \rightarrow 6\ CO_2(g) + 6\ H_2O(g)$
(c) $2\ C_4H_8O(\ell) + 11\ O_2(g) \rightarrow 8\ CO_2(g) + 8\ H_2O(g)$
30. (a) $2\ Mg(s) + O_2(g) \rightarrow 2\ MgO(s)$, magnesium oxide
(b) $2\ Ca(s) + O_2(g) \rightarrow 2\ CaO(s)$, calcium oxide
(c) $4\ In(s) + 3\ O_2(g) \rightarrow 2\ In_2O_3(s)$, indium oxide
32. (a) $2\ K(s) + Cl_2(g) \rightarrow 2\ KCl(s)$, potassium chloride
(b) $Mg(s) + Br_2(\ell) \rightarrow MgBr_2(s)$, magnesium bromide
(c) $2\ Al(s) + 3\ F_2(g) \rightarrow 2\ AlF_3(s)$, aluminum fluoride
34. (a) $4\ A1(s) + 3\ O_2(g) \rightarrow 2\ A1_2O_3(s)$
(b) $N_2(g) + 3\ H_2(g) \rightarrow 2\ NH_3(g)$
(c) $2\ C_6H_6(\ell) + 15\ O_2(g) \rightarrow 6\ H_2O(\ell) + 12\ CO_2(g)$
36. (a) $UO_2(s) + 4\ HF(\ell) \rightarrow UF_4(s) + 2\ H_2O(\ell)$
(b) $B_2O_3(s) + 6\ HF(\ell) \rightarrow 2\ BF_3(s) + 3\ H_2O(\ell)$
(c) $BF_3(g) + 3\ H_2O(\ell) \rightarrow 3\ HF(\ell) + H_3BO_3(s)$
38. (a) $H_2NCl(aq) + 2\ NH_3(g) \rightarrow NH_4Cl(aq) + N_2H_4(aq)$
(b) $(CH_3)_2N_2H_2(\ell) + 2\ N_2O_4(g)$
$\rightarrow 3\ N_2(g) + 4\ H_2O(g) + 2\ CO_2(g)$
(c) $CaC_2(s) + 2\ H_2O(\ell) \rightarrow Ca(OH)_2(s) + C_2H_2(g)$
40. (a) $C_6H_{12}O_6 + 6\ O_2 \rightarrow 6\ CO_2 + 6\ H_2O$
(b) $C_5H_{12} + 8\ O_2 \rightarrow 5\ CO_2 + 6\ H_2O$
(c) $2\ C_7H_{14}O_2 + 19\ O_2 \rightarrow 14\ CO_2 + 14\ H_2O$
(d) $C_2H_4O_2 + 2\ O_2 \rightarrow 2\ CO_2 + 2\ H_2O$
42. 50.0 mol HCl
44. 12.8 g
46. 1.1 mol O_2, 35 g O_2, 1.0×10^2 g NO_2
48. (a) 12.7 g Cl_2 (b) 0.179 mol $FeCl_2$, 22.7 g $FeCl_2$ expected
50.

$(NH_4)_2PtCl_6$	Pt	HCl
12.35 g	5.428 g	5.410 g
0.02782 mol	0.02782 mol	0.1484 mol

52. (a) 0.148 mol H_2O (b) 5.89 g TiO_2, 10.8 g HCl
54. 2.0 mol, 36.0304 g
56. 0.699 g Ga and 0.751 g As
58. (a) $4\ Fe(s) + 3\ O_2(g) \rightarrow 2\ Fe_2O_3(s)$
(b) 7.98 g (c) 2.40 g
60. (a) $CCl_2F_2 + 2\ Na_2C_2O_4 \rightarrow C + 4\ CO_2 + 2\ NaCl + 2\ NaF$
(b) 170. g $Na_2C_2O_4$ (b) 112 g CO_2
62. (a) 699 g (b) 526 g
64. $BaCl_2$, 1.12081 g $BaSO_4$
66. (a) Cl_2 is limiting. (b) 5.08 g Al_2Cl_6
(c) 1.67 g Al unreacted
68. (a) CO (b) 1.3 g H_2 (c) 85.2 g
70. 0 mol CaO, 0.19 mol NH_4Cl, 2.00 mol H_2O, 4.00 mol NH_3,
2.00 mol $CaCl_2$
72. 1.40 kg Fe
74. Graph (a)
75. (a) CH_4 (b) 188 g (c) 739 g
76. 699 g, 93.5%
77. 86.3 g
79. 56.0%
81. 8.8%
83. 5.3 g SCl_2
86. SO_3
88. CH
90. $C_3H_6O_2$
94. (a) $C_9H_{11}NO_4$ (b) $C_9H_{11}NO_4$
95. 21.6 g N_2
97. Element (b)
99. 12.5 g $Pt(NH_3)_2Cl_2$
101. SiH_4
103. KOH, KOH

105. Two butane molecules react with 13 diatomic oxygen molecules to produce 8 carbon dioxide molecules and 10 water molecules. Two moles of gaseous butane molecules react with 13 moles of gaseous diatomic oxygen molecules to produce 8 moles of gaseous carbon dioxide molecules and 10 moles of liquid water molecules.
107. A_3B
109. Ag^+, Cu^{2+}, and NO_3^-
112. Equation (b)
114. $A_2 + 4\ BC_2 \rightarrow 4\ C_2 + 2\ AB_2$
116. When the metal mass is less than 1.2 g, the metal is the limiting reactant. When the metal mass is greater than 1.2 g, the bromine is the limiting reactant.
118. $H_2(g) + 3\ Fe_2O_3(s) \rightarrow H_2O(\ell) + 2\ Fe_3O_4(s)$
120. (a) 2/1 (b) C
122. 44.9 amu
124. 0 g $AgNO_3$, 9.82 g Na_2CO_3, 6.79 g Ag_2CO_3, 4.19 g $NaNO_3$
126. 99.7% CH_3OH, 0.3% C_2H_5OH

Chapter 5

11. All soluble (a) Fe^{2+} and ClO_4^- (b) Na^+ and SO_4^{2-} (c) K^+ and Br^- (d) Na^+ and CO_3^{2-}
13. All soluble (a) K^+ and HPO_4^{2-} (b) Na^+ and ClO^- (c) Mg^{2+} and Cl^- (d) Ca^{2+} and OH^- (e) Al^{3+} and Br^-
15. Box (b)
17. Box (b)
19. $2\ HNO_3(aq) + Ca(OH)_2(aq) \rightarrow 2\ H_2O(\ell) + Ca(NO_3)_2(aq)$
21. (a) $MnCl_2(aq) + Na_2S(aq) \rightarrow MnS(s) + 2\ NaCl(aq)$
(b) No precipitate
(c) No precipitate
(d) $Hg(NO_3)_2(aq) + Na_2S(aq) \rightarrow HgS(s) + 2\ NaNO_3(aq)$
(e) $Pb(NO_3)_2(aq) + 2\ HCl(aq) \rightarrow PbCl_2(s) + 2\ HNO_3(aq)$
(f) $BaCl_2(aq) + H_2SO_4(aq) \rightarrow BaSO_4(s) + 2\ HCl(aq)$
23. (a) CuS insoluble;
$Cu^{2+} + H_2S(aq) \rightarrow CuS(s) + 2\ H^+$;
spectator ion is Cl^-.
(b) $CaCO_3$ insoluble;
$Ca^{2+} + CO_3^{2-} \rightarrow CaCO_3(s)$; spectator ions are K^+ and Cl^-.
(c) AgI insoluble;
$Ag^+ + I^- \rightarrow AgI(s)$; spectator ions are Na^+ and NO_3^-.
25. Complete ionic equation:
$2\ K^+ + CO_3^{2-} + Cu^{2+} + 2\ NO_3^- \rightarrow CuCO_3(s) + 2\ K^+ + 2\ NO_3^-$
Net ionic equation:
$CO_3^{2-} + Cu^{2+} \rightarrow CuCO_3(s)$
Precipitate is copper(II) carbonate.
27. (a) $Zn(s) + 2\ HCl(aq) \rightarrow H_2(g) + ZnCl_2(aq)$
$Zn(s) + 2\ H^+(aq) + 2\ Cl^-(aq) \rightarrow H_2(g) + Zn^{2+}(aq) + 2\ Cl^-(aq)$
$Zn(s) + 2\ H^+(aq) \rightarrow H_2(g) + Zn^{2+}(aq)$
(b) $Mg(OH)_2(s) + 2\ HCl(aq) \rightarrow MgCl_2(aq) + 2\ H_2O(\ell)$
$Mg(OH)_2(s) + 2\ H^+(aq) + 2\ Cl^-(aq) \rightarrow Mg^{2+}(aq) + 2\ Cl^-(aq) + 2\ H_2O(\ell)$
$Mg(OH)_2(s) + 2\ H^+(aq) \rightarrow Mg^{2+}(aq) + 2\ H_2O(\ell)$
(c) $2\ HNO_3(aq) + CaCO_3(s) \rightarrow Ca(NO_3)_2(aq) + H_2O(\ell) + CO_2(g)$
$2\ H^+(aq) + 2\ NO_3^-(aq) + CaCO_3(s) \rightarrow Ca^{2+}(aq) + 2\ NO_3^-(aq) + H_2O(\ell) + CO_2(g)$
$2\ H^+(aq) + CaCO_3(s) \rightarrow Ca^{2+}(aq) + H_2O(\ell) + CO_2(g)$
(d) $4\ HCl(aq) + MnO_2(s) \rightarrow MnCl_2(aq) + Cl_2(g) + 2\ H_2O(\ell)$
$4\ H^+(aq) + 4\ Cl^-(aq) + MnO_2(s) \rightarrow Mn^{2+}(aq) + 2\ Cl^-(aq) + Cl_2(g) + 2\ H_2O(\ell)$
$4\ H^+(aq) + 2\ Cl^-(aq) + MnO_2(s) \rightarrow Mn^{2+}(aq) + Cl_2(g) + 2\ H_2O(\ell)$
29. (a) $Ca(OH)_2(s) + 2\ HNO_3(aq) \rightarrow Ca(NO_3)_2(aq) + 2\ H_2O(\ell)$
$Ca(OH)_2(s) + 2\ H^+(aq) + 2\ NO_3^-(aq) \rightarrow Ca^{2+}(aq) + 2\ NO_3^-(aq) + 2\ H_2O(\ell)$
$Ca(OH)_2(s) + 2\ H^+(aq) \rightarrow Ca^{2+}(aq) + 2\ H_2O(\ell)$
(b) $BaCl_2(aq) + Na_2CO_3(aq) \rightarrow BaCO_3(s) + 2\ NaCl(aq)$
$Ba^{2+}(aq) + 2\ Cl^-(aq) + 2\ Na^+(aq) + CO_3^{2-}(aq) \rightarrow BaCO_3(s) + 2\ Na^+(aq) + 2\ Cl^-(aq)$
$Ba^{2+}(aq) + CO_3^{2-}(aq) \rightarrow BaCO_3(s)$
(c) $2\ Na_3PO_4(aq) + 3\ Ni(NO_3)_2(aq) \rightarrow Ni_3(PO_4)_2(s) + 6\ NaNO_3(aq)$
$6\ Na^+(aq) + 2\ PO_4^{3-}(aq) + 3\ Ni^{2+}(aq) + 6\ NO_3^-(aq) \rightarrow Ni_3(PO_4)_2(s) + 6\ Na^+(aq) + 6\ NO_3^-(aq)$
$2\ PO_4^{3-}(aq) + 3\ Ni^{2+}(aq) \rightarrow Ni_3(PO_4)_2(s)$
31. $Ba(OH)_2(aq) + 2\ HNO_3(aq) \longrightarrow Ba(NO_3)_2(aq) + 2\ H_2O(\ell)$
33. $CdCl_2(aq) + 2\ NaOH(aq) \longrightarrow Cd(OH)_2(s) + 2\ NaCl(aq)$
$Cd^{2+}(aq) + 2\ Cl^-(aq) + 2\ Na^+(aq) + 2\ OH^-(aq) \longrightarrow Cd(OH)_2(s) + 2\ Na^+(aq) + 2\ Cl^-(aq)$
$Cd^{2+}(aq) + 2\ OH^-(aq) \longrightarrow Cd(OH)_2(s)$
35. $Pb(NO_3)_2(aq) + 2\ KCl(aq) \rightarrow PbCl_2(s) + 2\ KNO_3(aq)$
Reactants: lead(II) nitrate and potassium chloride
Products: lead(II) chloride and potassium nitrate
38. (a) Base strong, K^+ and OH^-
(b) Base strong, Mg^{2+} and OH^-
(c) Acid weak, small amounts of H^+ and ClO^-
(d) Acid strong, H^+ and Br^-
(e) Base strong, Li^+ and OH^-
(f) Acid weak, small amounts of H^+, HSO_3^-, and SO_3^{2-}
40. (a) Acid: HNO_2, base: NaOH
complete ionic form:
$HNO_2(aq) + Na^+ + OH^- \longrightarrow H_2O(\ell) + Na^+ + NO_2^-$
net ionic form:
$HNO_2(aq) + OH^- \longrightarrow H_2O(\ell) + NO_2^-$
(b) Acid: H_2SO_4 base: $Ca(OH)_2$
complete ionic and net ionic forms:
$H^+ + HSO_4^- + Ca(OH)_2(s) \longrightarrow 2\ H_2O(\ell) + CaSO_4(s)$
(c) Acid: HI, base: NaOH
complete ionic form:
$H^+ + I^- + Na^+ + OH^- \longrightarrow H_2O(\ell) + Na^+ + I^-$
net ionic form: $H^+ + OH^- \longrightarrow H_2O(\ell)$
(d) Acid: H_3PO_4, base: $Mg(OH)_2$
complete ionic and net ionic forms:
$2\ H_3PO_4(aq) + 3\ Mg(OH)_2(s) \longrightarrow 6\ H_2O(\ell) + Mg_3(PO_4)_2(s)$
42. (a) Precipitation reaction; products are NaCl and MnS;
$MnCl_2(aq) + Na_2S(aq) \rightarrow 2\ NaCl(aq) + MnS(s)$ (b) Precipitation reaction; products are NaCl and $ZnCO_3$;
$Na_2CO_3(aq) + ZnCl_2(aq) \rightarrow 2\ NaCl(aq) + ZnCO_3(s)$
(c) Gas-forming reaction; products are $KClO_4$, H_2O, and CO_2;
$K_2CO_3(aq) + 2\ HClO_4(aq) \rightarrow 2\ KClO_4(aq) + H_2O(\ell) + CO_2(g)$
44. (a) Ox. # O = −2, Ox. # S = +6
(b) Ox. # O = −2, Ox. # H = +1, Ox. # N = +5
(c) Ox. # K = +1, Ox. # O = −2, Ox. # Mn = +7
(d) Ox. # O = −2, Ox. # H = +1
(e) Ox. # Li = +1, Ox. # O = −2, Ox. # H = +1
(f) Ox. # Cl = −1, Ox. # H = +1, Ox. # C = 0
46. (a) S: +6, O: −2 (b) N: +5, O: −2
(c) Mn: +7, O: −2 (d) Cr: +3, O: −2, H: +1
(e) P: +5, O: −2, H: +1 (f) S: +2, O: −2
48. (a) −1 (b) +1 (c) +3 (d) +5 (e) +7
50. (a) −2 (b) 0 (c) +2 (d) +4 (e) +6
52. Only reaction (b) is an oxidation-reduction reaction; oxidation numbers change. Reaction (a) is precipitation and reaction (c) is acid-base neutralization.
55. Substances (b), (c), and (d)
57. (a) $CO_2(g)$ or $CO(g)$ (b) $PCl_3(g)$ or $PCl_5(g)$
(c) $TiCl_2(s)$ or $TiCl_4(s)$ (d) $Mg_3N_2(s)$
(e) $Fe_2O_3(s)$ (f) $NO_2(g)$
59. F_2 is the best oxidizing agent; I^- anion is the best reducing agent.
61. (b) Br_2 is reduced. NaI is oxidized. Br_2 is oxidizing agent. NaI is reducing agent.
(c) F_2 is reduced. NaCl is oxidized. F_2 is oxidizing agent. NaCl is reducing agent.
(d) Cl_2 is reduced. NaBr is oxidized. Cl_2 is oxidizing agent. NaBr is reducing agent.

64. This is just an example:
$Fe(s) + 2\ HCl(aq) \longrightarrow FeCl_2(aq) + H_2(g)$
(a) Fe is oxidized. (b) HCl is reduced.
(c) The reducing agent is Fe(s).
(d) The oxidizing agent is HCl(aq).
(Other answers are possible.)
66. (a) N.R. (b) N.R. (c) N.R.
(d) $Au^{3+}(aq) + 3\ Ag(s) \longrightarrow 3\ Ag^+(aq) + Au(s)$
68. 0.12-M Ba^{2+}, 0.24-M Cl^-
70. (a) 0.254-M Na_2CO_3 (b) 0.508-M Na^+, 0.254-M CO_3^{2-}
72. 0.494 g $KMnO_4$
74. 5.08×10^3 mL
76. 0.0150-M $CuSO_4$
78. Method (b)
80. 39.4 g $NiSO_4 \cdot 6H_2O$
83. 0.205 g Na_2CO_3
85. 121 mL HNO_3 solution
87. 22.9 mL NaOH solution
89. 0.18 g AgCl, NaCl, 0.0080-M NaCl
91. (a) Step (ii) is wrong. Steps (iii) and (iv) are wrong because Step (ii) is wrong.
(b) 0.00394 g citric acid
93. 1.192-M HCl
95. 96.8% pure
96. 16.1%
97. 104 g/mol
99. Chloride is the spectator ion.
$CaCO_3(s) + 2\ H^+(aq) \longrightarrow CO_2(g) + Ca^{2+}(aq) + H_2O(\ell)$
gas-forming exchange reaction
101. (a) $(NH_4)_2S(aq) + Hg(NO_3)_2(aq) \longrightarrow HgS(s) + 2\ NH_4NO_3(aq)$
(b) Reactants: ammonium sulfide, mercury(II) nitrate; products: mercury(II) sulfide, ammonium nitrate
(c) $S^{2-}(aq) + Hg^{2+}(aq) \longrightarrow HgS(s)$
(d) Precipitation reaction
103. (a) Combination reaction; product: $H_2SO_4(aq)$
(b) Combination reaction; product: $SrH_2(s)$
(c) Displacement reaction; products: $MgSO_4(aq)$ and $H_2(g)$
(d) Exchange (precipitation) reaction; products: $Ag_3PO_4(s)$ and $NaNO_3(aq)$
(e) Decomposition and gas-forming reaction; products: CaO(s), $H_2O(\ell)$, and $CO_2(g)$
(f) Oxidation-reduction reaction; products: $Fe^{2+}(aq)$ and $Sn^{4+}(aq)$
105. (a) $NH_3(aq)$, H_2O (b) $CH_3CO_2H(aq)$, H_2O
(c) Na^+, OH^-, H_2O (d) H^+, Br^-, H_2O
107. (c) is a redox reaction. The oxidizing agent is Ti. The reducing agent is Mg.
109. (a) $CaF_2(s) + H_2SO_4(aq) \longrightarrow 2\ HF(g) + CaSO_4(s)$
Reactant: calcium fluoride and sulfuric acid
Products: hydrogen fluoride and calcium sulfate
(b) Precipitation reaction.
(c) Carbon tetrachloride, antimony(V) chloride, hydrogen chloride
(d) CCl_3F
111. 6.28% impurity
112. Case 1 (a) Before: clear colorless solution; after: solid at the bottom of beaker with colorless solution above it
(b)

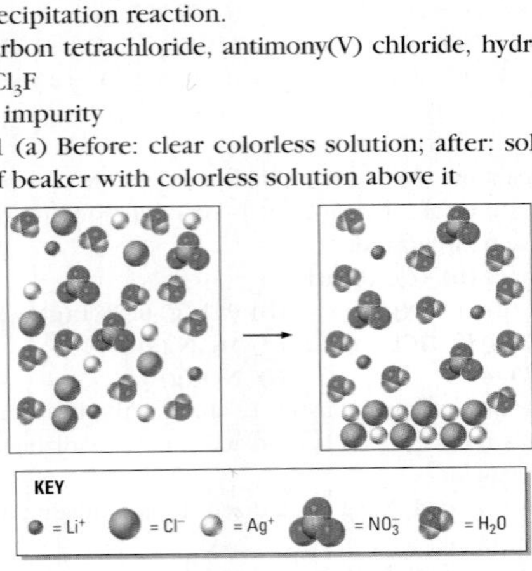
KEY
= Li+
= Cl−
= Ag+
= NO3−
= H2O

(c) $Li^+ + Cl^- + Ag^+ + NO_3^- \rightarrow Li^+ + AgCl(s) + NO_3^-$

Case 2 (a) Before and after: clear colorless solutions
(b)

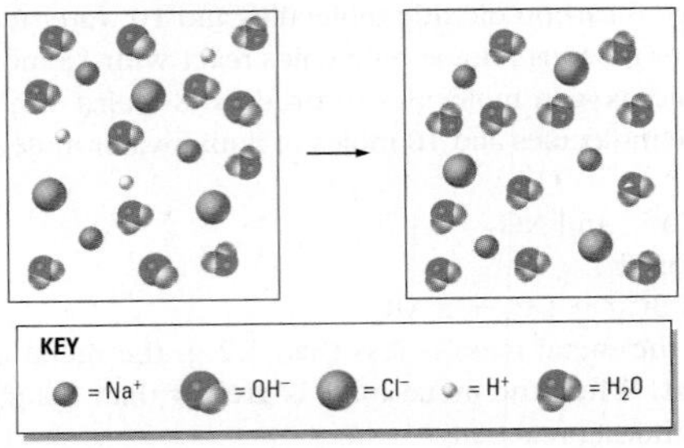
KEY
= Na+
= OH−
= Cl−
= H+
= H2O

(c) $Na^+ + OH^- + H^+ + Cl^- \rightarrow Na^+ + H_2O(\ell) + Cl^-$
114. (a) Combine $Ba(OH)_2(aq)$ and $H_2SO_4(aq)$
(b) Combine $Na_2SO_4(aq)$ and $Ba(NO_3)_2(aq)$
(c) Combine $BaCO_3(s)$ and $H_2SO_4(aq)$
116. HCl(aq); insoluble lead(II) chloride would precipitate if Pb^{2+} was present, but no precipitate would be seen if only Ba^{2+} was present.
118. The products are the result of something being oxidized and something being reduced. Only the reactants are oxidized and reduced.
120. (d)
122. (a) and (d) are correct.
124. (a) $MgBr_2$, magnesium bromide; $CaBr_2$, calcium bromide; $SrBr_2$, strontium bromide
(b) $Mg(s) + Br_2(\ell) \rightarrow MgBr_2(s)$
$Ca(s) + Br_2(\ell) \rightarrow CaBr_2(s)$
$Sr(s) + Br_2(\ell) \rightarrow SrBr_2(s)$
(c) Oxidation-reduction
(d) The point where increase stops gives: $\frac{\text{g metal}}{\text{g product}}$. The different metals have different molar masses, so the ratios will be different. Use grams to find moles and then set up a ratio.
126. (a) Groups C and D: $Ag^+ + Cl^- \rightarrow AgCl(s)$, Groups A and B: $Ag^+ + Br^- \rightarrow AgBr(s)$
(b) Different silver halide produced in Group C and D than in Groups A and B
(c) Bromide is heavier than chloride.
130. 0.0154-M $CaSO_4$; 0.341 g $CaSO_4$ undissolved
132. 184 mL
135. Three

Chapter 6

11. (a) 399 Cal (b) 5.0×10^6 J/day
13. 1.12×10^4 J
15. 3.60×10^6 J, $0.03
17. the first product
19. (a) The chemical potential energy of the atoms in the match, fuse, and fuel are converted to thermal energy (due to the combustion reaction), potential energy (as the rocket's altitude increases), and light energy (colorful sparkles). (b) The chemical potential energy of the atoms in the fuel is converted to thermal energy (due to the combustion reaction), some of which is converted to kinetic energy (for the movement of the vehicle).
21. (a) The system: the plant (stem, leaves, roots, etc.); the surroundings: anything not the plant (air, soil, water, sun, etc.) (b) To study the plant growing, we must isolate it and see how it interacts with its surroundings. (c) Light energy and carbon dioxide are absorbed by the leaves and are converted to other molecules storing the energy as chemical energy that is used to increase the size of the plant. Minerals and water are absorbed through the soil. CO_2 is absorbed from the air. The plant expels oxygen and other waste materials into the surroundings.
23. (a) The system: NH_4Cl; the surroundings: anything not NH_4Cl, including the water. (b) To study the release of energy during the phase change of this chemical, we must isolate it and see how it interacts with the surroundings. (c) The system's interaction with the surroundings causes heat to be transferred into the system and from the surroundings. There is no material transfer in this

process, but there is a change in the specific interaction between the water and system. (d) Endothermic

25. $\Delta E = +32$ J
27.

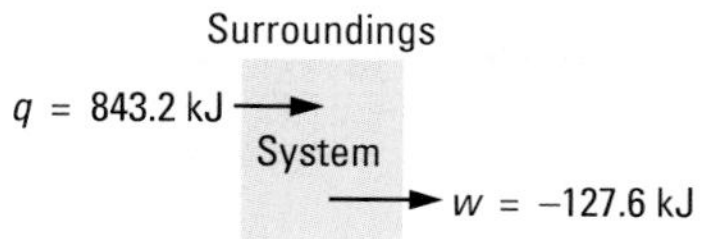

Surroundings
q = 843.2 kJ
System
w = −127.6 kJ

$\Delta E = 715.6$ J

29. Process (a) requires greater transfer of energy than (b).
31. It takes less time to raise the Cu sample to body temperature.
33. 1.0×10^2 kJ
35. 136 J mol^{-1} K^{-1}
37. 0.270 J/g·°C
39. More energy (1.48×10^6 J) is absorbed by the water than by the ethylene glycol (9.56×10^5 J).
41. 330. °C
43. (a) 0.45 J g^{-1} $°C^{-1}$, 25 J mol^{-1} $°C^{-1}$
45. Gold
47. 160 °C
49. Positive; Negative
51. 330. kJ
53. 10.3 kJ
55. 273 J
57.

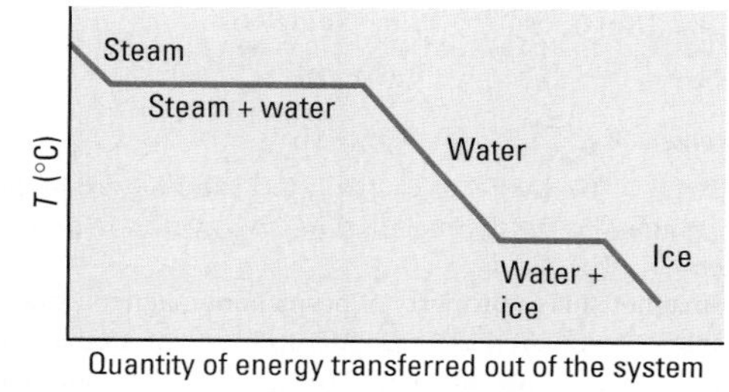

Steam
Steam + water
Water
Water + ice
Ice
T (°C)
Quantity of energy transferred out of the system

59. (a) X(s) (b) Enthalpy of fusion
    (c) Enthalpy of vaporization is positive.
61. (a) Negative (b) Positive
63. 150 J
65. 271 J; endothermic
67. Endothermic
69. $CaO(s) + H_2O(\ell) \rightarrow Ca^{2+}(aq) + 2\ OH^-(aq)$; exothermic
71. 49.3 kJ
73. 6.0 kJ of energy is used to convert one mole of ice into liquid water.
75. (a) −210 kJ (b) −33 kJ
77. (a) $\frac{1}{2}\ C_8H_{18}(\ell) + \frac{25}{4}\ O_2(g) \rightarrow 4\ CO_2(g) + \frac{9}{2}\ H_2O(\ell)$
    $\Delta H° = -2748.0$ kJ
    (b) $100\ C_8H_{18}(\ell) + 1250\ O_2(g) \rightarrow 800\ CO_2(g) + 900\ H_2O(\ell)$
    $\Delta H° = -5.4960 \times 10^5$ kJ
    (c) $C_8H_{18}(\ell) + \frac{25}{2}\ O_2(g) \rightarrow 8\ CO_2(g) + 9\ H_2O(\ell)$
    $\Delta H° = -5496.0$ kJ
79. $\frac{464.8\ \text{kJ}}{1\ \text{mol CaO}}, \frac{464.8\ \text{kJ}}{3\ \text{mol C}}, \frac{464.8\ \text{kJ}}{1\ \text{mol CaC}_2}, \frac{464.8\ \text{kJ}}{1\ \text{mol CO}}, \frac{1\ \text{mol CaO}}{464.8\ \text{kJ}},$
    $\frac{3\ \text{mol C}}{464.8\ \text{kJ}}, \frac{1\ \text{mol CaC}_2}{464.8\ \text{kJ}}, \frac{1\ \text{mol CO}}{464.8\ \text{kJ}}$
81. -3.3×10^4 kJ
83. $\Delta H = -1450$ kJ/mol
85. 6.74×10^3 kJ
87. 6×10^4 kJ released
89. HF
91. For $H_2 + F_2 \longrightarrow 2\ HF$ reaction: (a) 594 kJ (b) −1132 kJ
    (c) −538 kJ
    For $H_2 + Cl_2 \longrightarrow 2\ HCl$ reaction: (a) 678 kJ (b) −862 kJ
    (c) −184 kJ (d) ΔH of HF is most exothermic.
93. Endothermic
94. 0.522 J/g·°C
96. 23.9 °C
98. (a) 1.4×10^4 J (b) −42 kJ
100. 2.20×10^3 kJ
102. 394 kJ/mol evolved
104. −43.6 kJ
106. −320. kJ
108. $\Delta H_f°(PbO) = -217.3$ kJ/mol, 2.6×10^2 kJ evolved
110. (a) $C_2H_5OH(\ell) \rightarrow 2\ C(s) + 3\ H_2(g) + \frac{1}{2}\ O_2(g)$
     $\Delta H° = +277.69$ kJ
     (b) $NaF(s) \rightarrow Na(s) + \frac{1}{2}\ F_2(g)$ $\Delta H° = +573.647$ kJ
     (c) $MgSO_4(s) \rightarrow Mg(s) + S(s) + 2\ O_2(g)$ $\Delta H° = +1284.9$ kJ
     (d) $NH_4NO_3(s) \rightarrow N_2(g) + 2\ H_2(g) + \frac{3}{2}\ O_2(g)$
     $\Delta H° = +365.56$ kJ
112. (a) $2\ Al(s) + \frac{3}{2}\ O_2(g) \rightarrow Al_2O_3(s)$ $\Delta H° = -1675.7$ kJ
     (b) $Ti(s) + 2\ Cl_2(g) \rightarrow TiCl_4(\ell)$ $\Delta H° = -804.2$ kJ
     (c) $N_2(g) + 2\ H_2(g) + \frac{3}{2}\ O_2(g) \rightarrow NH_4NO_3(s)$
     $\Delta H° = -365.56$ kJ
114. (a) −1675.7 kJ/mol (b) −205.4 kJ
116. −584 kJ/mol
118. $\Delta H° = -98.89$ kJ
120. (a) 178.32 kJ (b) −595.2 kJ (c) 44 kJ
122. (a) −0.50 kJ (b) +18 kJ (c) −6.50 kJ (d) −4.13 kJ
124. 103.6 kJ/mol
126. 41.2 kJ evolved
128. 1.1×10^2 kg
130. 44.422 kJ/g octane > 19.927 kJ/g methanol
132. 720 kJ
134. 2.2 hours walking
136. Gold reaches 100 °C first.
138. 4.82×10^4 J
140. 75.4 g
142. $\Delta H_f°(B_2H_6) = 36$ kJ/mol
144. −96.4 kJ
146. (a) −69.14 kJ (b) 933 kJ evolved
148. $\Delta H_f°(CH_3OH) = -200.660$ kJ/mol
150. 0.745 m/s
151. (a) 36.03 kJ evolved (b) 1.18×10^4 kJ evolved
153. Step 1: −137.23 kJ, Step 2: 275.341 kJ, Step 3: 103.71 kJ
     $H_2O\ (g) \longrightarrow H_2(g) + \frac{1}{2}\ O_2(g)$
     $\Delta H° = 241.82$ kJ, endothermic.
155. Melting is endothermic. Freezing is exothermic.
157. Substance A
159. Greater; a larger mass of water will contain a larger quantity of thermal energy at a given temperature.
161. The given reaction produces 2 mol SO_3. Formation enthalpy from Table 6.2 is for the production of 1 mol SO_3.
163. $\Delta E = 310$ J; $w = 0$ J
165. $\Delta H_f°(OF_2) = 18$ kJ/mol
167. CH_4, 50.014 kJ/g; C_2H_6, 47.484 kJ/g; C_3H_8, 46.354 kJ/g; C_4H_{10}, 45.7140 kJ/g; $CH_4 > C_2H_6 > C_3H_8 > C_4H_{10}$
169. Measure the heat of formation of the calcium and sulfur oxides, CaO and SO_3. Use those enthalpies, the given $\Delta H°$ for CaO and SO_3 and Hess's law to calculate the $\Delta H_f°$ for $CaSO_4$.

$\cancel{CaO} + \cancel{SO_3} \longrightarrow CaSO_4$		$\Delta H° = -403.3$ kJ
$Ca + \frac{1}{2}\ O_2 \longrightarrow \cancel{CaO}$		$\Delta H° = -635.09$ kJ
+ $S + \frac{3}{2}\ O_2 \longrightarrow \cancel{SO_3}$		$\Delta H° = -395.72$ kJ
$Ca + S + 2\ O_2 \longrightarrow CaSO_4$		$\Delta H_f° = -1434.1$ kJ

170. 0.15 kJ; 0.060 g apples
172. (a) 26.6 °C (Above $C_6H_8O_6$ masses of 8.81 g, NaOH is the limiting reactant.)
     (b) $C_6H_8O_6$ limits in Experiments 1–3 and NaOH limits in Experiments 4 and 5.
     (c) Ascorbic acid has one hydrogen ion; equal quantities of reactant in Experiment 3 at stoichiometric equivalence point.
174. Note: Great variability in answers may result from specific assumptions made.
     (a) Approx. 1.0×10^{10} kJ
     (b) Approx. 2.1 metric kilotons TNT

(c) 15 kilotons for Hiroshima bomb, 20 kilotons for Nagasaki; max ever = 50,000 kilotons; approx. 14% Hiroshima bomb or 10% Nagasaki bomb
(d) Approx. 20 hours of continuous hurricane damage

Chapter 7

10. Short wavelength and high frequency
12. (a) Radio waves have less energy than infrared.
(b) Microwaves are higher frequency than radio waves.
14. (a) 3.00×10^{-3} m (b) 6.63×10^{-23} J/photon
(c) 39.9 J/mol
16. (d) < (c) < (a) < (b)
18. 6.06×10^{14} Hz
20. 4.4×10^{-19} J/photon
22. Many types of electromagnetic radiation, including visible, ultraviolet, infrared, microwaves
24. 1.1×10^{15} Hz, 7.4×10^{-19} J/photon
26. X-ray (8.42×10^{-17} J/photon) energy is larger than that of orange light (3.18×10^{-19} J/photon).
28. 1.20×10^{8} J/mol
30. Photons of this light are too low in energy. Increasing the intensity only increases the number of photons, not their individual energy.
32. No; insufficient energy
34. Line emission spectra are mostly dark, with discrete bands of light. Sunlight is a continuous rainbow of colors.
36. The higher-energy state to the lower-energy state; difference
38. (a) Absorbed (b) Emitted (c) Absorbed (d) Emitted
40. (a), (b), (d)
42. $\Delta E = -6.18 \times 10^{-19}$ J; $\lambda = 320.5$ nm; ultraviolet
44. $\Delta E = -1.839 \times 10^{-17}$ J; 10.80 nm; ultraviolet
46. 4.576×10^{-19} J absorbed, 434.0 nm
48. 0.05 nm
50. (a) First electron: $n = 1$, $\ell = 0$, $m_\ell = 0$, $m_s = +\frac{1}{2}$; second electron: $n = 1$, $\ell = 0$, $m_\ell = 0$, $m_s = -\frac{1}{2}$; third electron: $n = 2$, $\ell = 0$, $m_\ell = 0$, $m_s = +\frac{1}{2}$; fourth electron: $n = 2$, $\ell = 0$, $m_\ell = 0$, $m_s = -\frac{1}{2}$; fifth electron: $n = 2$, $\ell = 1$, $m_\ell = 1$, $m_s = +\frac{1}{2}$ (The fifth electron could have different m_ℓ and m_s values.)
(b) $n = 3$, $\ell = 0$, $m_\ell = 0$, $m_s = +\frac{1}{2}$ and $n = 3$, $\ell = 0$, $m_\ell = 0$, $m_s = -\frac{1}{2}$
(c) $n = 3$, $\ell = 2$, $m_\ell = 2$, $m_s = +\frac{1}{2}$ (Different m_ℓ and m_s values are also possible.)
52. (a) Cannot occur, m_ℓ too large (b) Can occur
(c) Cannot occur, m_s cannot be 1, here.
(d) Cannot occur, ℓ must be less than n. (e) Can occur
54. (a) $n = 4$, $\ell = 0$, $m_\ell = 0$, $m_s = +\frac{1}{2}$
(b) $n = 3$, $\ell = 1$, $m_\ell = 1$, $m_s = -\frac{1}{2}$
(c) $n = 3$, $\ell = 2$, $m_\ell = 0$, $m_s = +\frac{1}{2}$
56. Four subshells
58. Electrons do not follow simple paths as planets do.
60. Orbits have predetermined paths—position and momentum are both exactly known at all times. Heisenberg's uncertainty principle says that we cannot know both simultaneously.
62. d, p, and s orbitals; nine orbitals total
63. $_{12}$Mg: $1s^22s^22p^63s^2$; $_{17}$Cl: $1s^22s^22p^63s^23p^5$
64. $_{13}$Al: $1s^22s^22p^63s^23p^1$; $_{16}$S: $1s^22s^22p^63s^23p^4$
66. $_{32}$Ge: $1s^22s^22p^63s^23p^63d^{10}4s^24p^2$
68. Oxygen; Group 6A has 6 valence electrons with valence electron configuration of ns^2np^4.
69. (a) Cs^+, Se^{2-} (b) Cs^+, Se^{2-}
70. (a) $4s$ orbital must be full.
(b) Orbital labels must be 3, not 2; electrons in $3p$ subshell should be in separate orbitals with parallel spin.
(c) $4d$ orbitals must be completely filled before $5p$ orbitals start filling.

72. (a) V

3d 4s
[Ar] [↑][↑][↑][ ][ ] [↑↓]

(b) V^{2+}

3d
[Ar] [↑][↑][↑][ ][ ]

(c) V^{4+}

3d
[Ar] [↑][ ][ ][ ][ ]

74. 18 elements, all possible orbital electron combinations are already used
76. Mn: $1s^22s^22p^63s^23p^63d^54s^2$; it has 5 unpaired electrons:

3d 4s
[Ar] [↑][↑][↑][↑][↑] [↑↓]

Mn^{2+}: $1s^22s^22p^63s^23p^63d^5$; it has 5 unpaired electrons:

3d
[Ar] [↑][↑][↑][↑][↑]

Mn^{3+}: $1s^22s^22p^63s^23p^63d^4$; it has 4 unpaired electrons:

3d
[Ar] [↑][↑][↑][↑][ ]

78. [Ne]$3s^23p^4$
80. (a) $_{63}$Eu: [Xe]$4f^76s^2$ (b) $_{70}$Eu: [Xe]$4f^{14}6s^2$
82. (a) ·Sr· (b) :B̈r̤·
(c) ·Ġa· (d) ·S̈ḅ·
84. (a) [Ar] (b) [Ar] (c) [Ne]; Ca^{2+} and K^+ are isoelectronic.
86. $_{50}$Sn: [Kr]$4d^{10}5s^25p^2$; $_{50}Sn^{2+}$: [Kr]$4d^{10}5s^2$; $_{50}Sn^{4+}$: [Kr]$4d^{10}$
87. (a) Zn (b) Cr
88. Ferromagnetism is a property of permanent magnets. It occurs when the spins of unpaired electrons in a cluster of atoms (called a domain) in the solid are all aligned in the same direction. Only metals in the Fe, Co, Ni subgroup (Group 8B) exhibit this property.
90. In both paramagnetic and ferromagnetic substances, atoms have unpaired spins and thus are attracted to magnets. Ferromagnetic substances retain their aligned spins after an external magnetic field has been removed, so they can function as magnets. Paramagnetic substances lose their aligned spins after a time and, therefore, cannot be used as permanent magnets.
92. (a) TiO_2; it is diagmagnetic (b) TiO; it is paramagnetic
94. P < Ge < Ca < Sr < Rb
96. (a) Rb smaller (b) O smaller (c) Br smaller
(d) Ba^{2+} smaller (e) Ca^{2+} smaller
98. Al < Mg < P < F
100. (c)
102. Na; it must have one valence electron.
104. (a) Al (b) Al (c) Al < B < C
106. (a) H^- (b) N^{3-} (c) F^-
108. Adding a negative electron to a negatively charged ion requires additional energy to overcome the coulombic charge repulsion.
110. −862 kJ
112. Red < Yellow < Green < Violet
114. Seven pairs
116. (a) He (b) Sc (c) Na
118. (a) F < O < S (b) S
120. (a) Sulfur (b) Radium (c) Nitrogen
(d) Ruthenium (e) Copper
122. (a) Z (b) Z
124. In^{4+}, Fe^{6+}, and Sn^{5+}; very high successive ionization energies
126. (a) False; shorter, not longer (b) True
(c) True (d) False; inversely, not directly
128. (a) B, D, I; (b) E; (c) D; (d) A, G; (e) H, C

130. (a) Directly related, not inversely related
(b) Inversely proportional to *the square of* the principle quantum number, not inversely proportional to the principle quantum number
(c) Before, not as soon as
(d) Wavelength, not frequency
132. Ultraviolet; 91 nm or shorter wavelength is needed to ionize hydrogen.
134. (a) $[Rn]5f^{14}6d^{10}7s^{2}7p^{6}8s^{2}$
(b) Magnesium (c) EtO, $EtCl_2$
136. $1s^{2}2s^{2}2p^{4}$
138. XCl
140. (a) Ground state (b) Could be ground state or excited state
(c) Excited state (d) Impossible
(e) Excited state (f) Excited state
142. $[Rn]5f^{14}5g^{18}6d^{10}6f^{14}7s^{2}7p^{6}7d^{10}8s^{2}8p^{2}$, Group 4A
143. (a) Increase, decrease
(b) Helium
(c) 5 and 13
(d) He has only two electrons; there are no electrons left.
(e) The first electron is a valence electron, but the second electron is a core electron.
(f) $Mg^{2+}(g) \rightarrow Mg^{3+}(g) + e^{-}$
146. (a) 4.34×10^{-19} J (b) 6.54×10^{14} Hz
148. 2.18×10^{-18} J
150. To remove an electron from a N atom requires disrupting a half-filled subshell (p^3), which is relatively stable, so the ionization of O requires less energy than the ionization of N.
152. (a) Ir (b) $[Rn]7s^{2}5f^{14}6d^{7}$
154. (a) Transition metals (b) $[Rn]7s^{2}5f^{14}6d^{10}$

Chapter 8

14. (a) 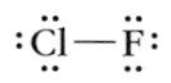 (b) 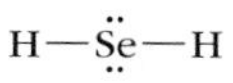

(c) 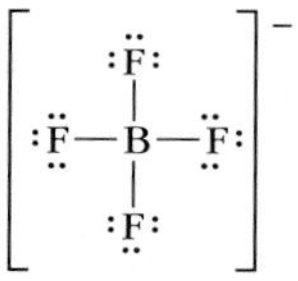 (d) 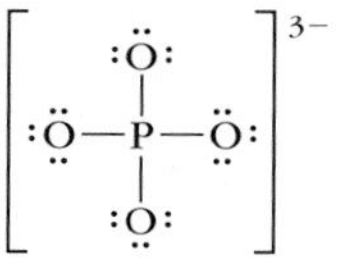

16. (a) 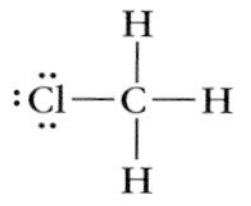 (b) 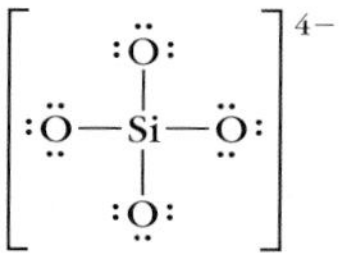

(c) 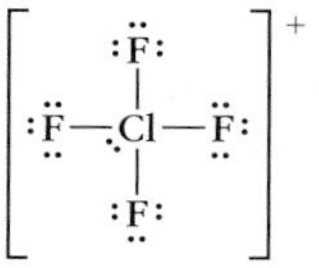 (d) 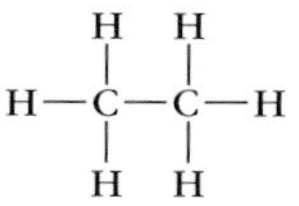

18. (a) 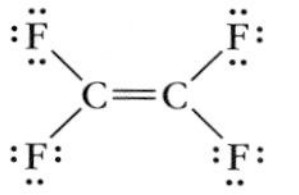 (b) 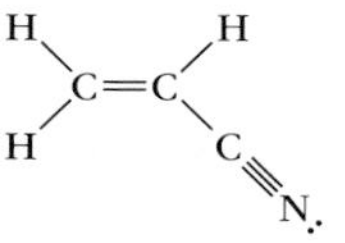

22. (a) Incorrect; the F atoms are missing electrons.
(b) Incorrect; the structure has 10 electrons, but needs 12 electrons.
(c) Incorrect. The structure has three too many electrons; Carbon has nine electrons. The single electron should be deleted. Oxygen has ten electrons; delete one pair of electrons.
(d) Incorrect; one hydrogen atom has more than two electrons. One hydrogen atom is completely missing.
(e) Incorrect; the structure has 16 electrons, but needs 18 electrons. The N atom doesn't follow the octet rule. It needs another pair of electrons in the form of a lone pair.

24. Four branched-chain compounds

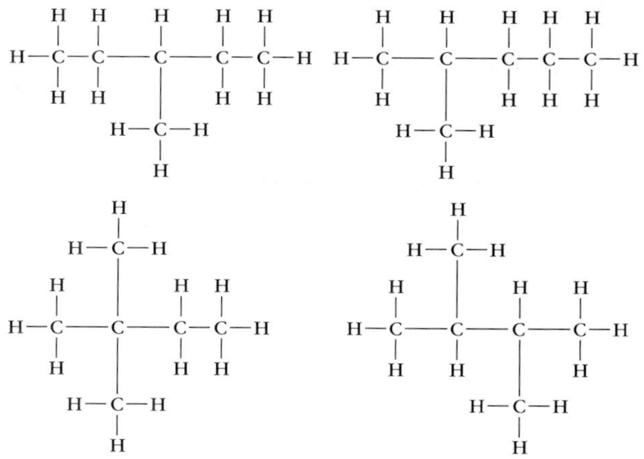

26. (a) Alkyne (b) Alkane (c) Alkene
28. *cis*-2-pentene *trans*-2-pentene

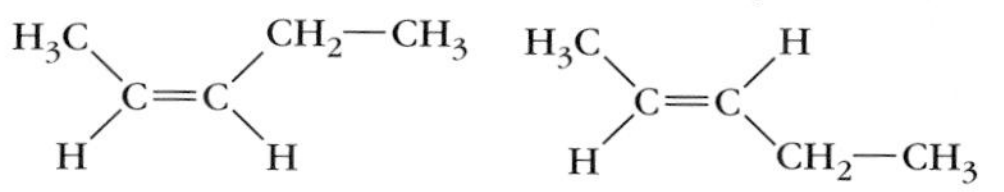

30. (a) No
(b) Yes *cis* *trans*

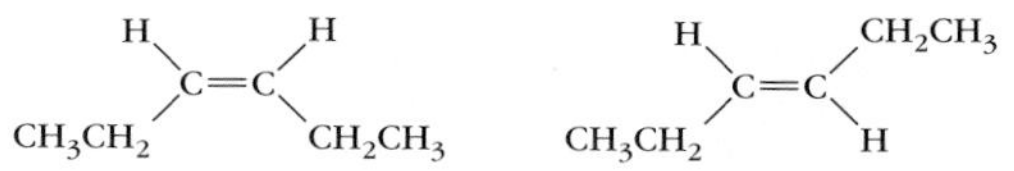

(c) Yes *cis* *trans*

(d) No
32. No; free rotation about the single C—C bond prevents it.
34. (a) B—Cl (b) C—O (c) P—O (d) C=O
36. (a)
38. CO
40. CO_3^{2-} has longer C—O bonds.
42. −92 kJ; the reaction is exothermic.
44. HF has the strongest bond; −538. kJ; −184 kJ; −103 kJ; −11 kJ. The reaction of H_2 with F_2 is most exothermic.
46. (a) N, C, Br, O (b) S—O is the most polar.
48. (a) All the bonds in urea are somewhat polar.
(b) The most polar bond is C=O; the O end is partially negative.
50. The total formal charge on a molecule is always zero. The total formal charge on an ion is the ionic charge.
52. (a) 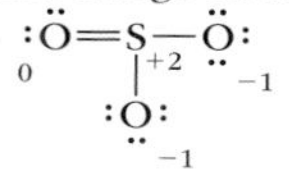 (b) 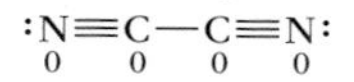

(c) 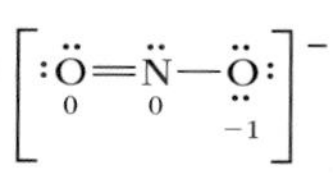

54. (a) 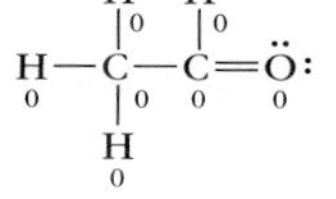 (b) 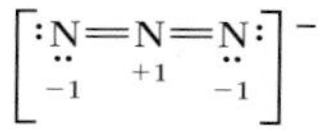

(c) 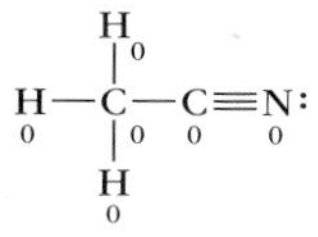

56. 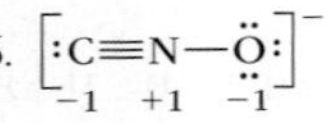

58. 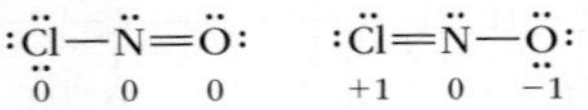

60. (a)

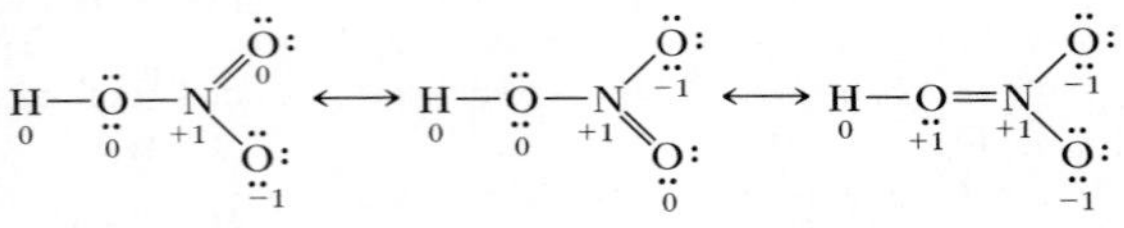

(b)

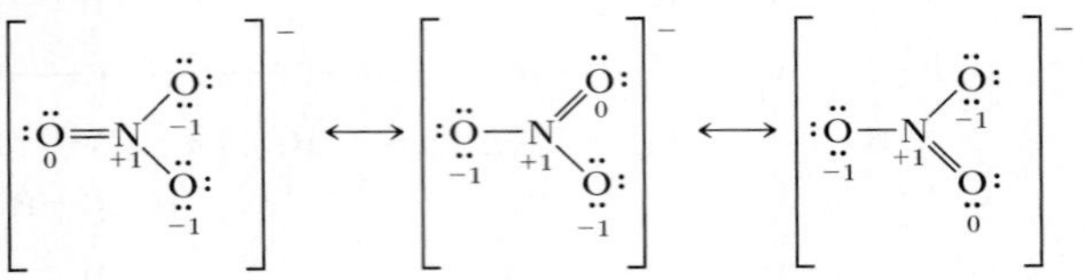

62. 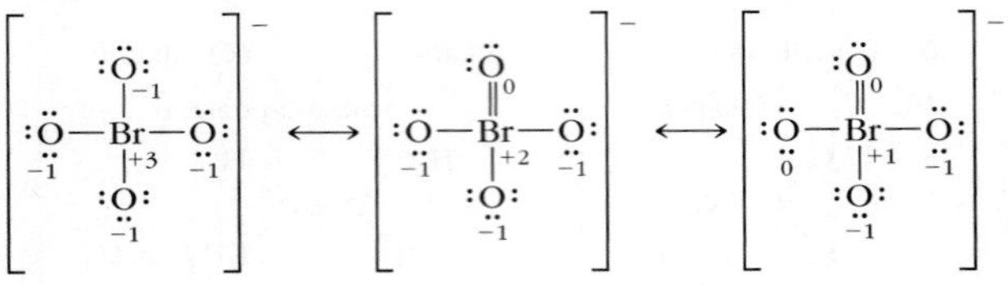

Most plausible 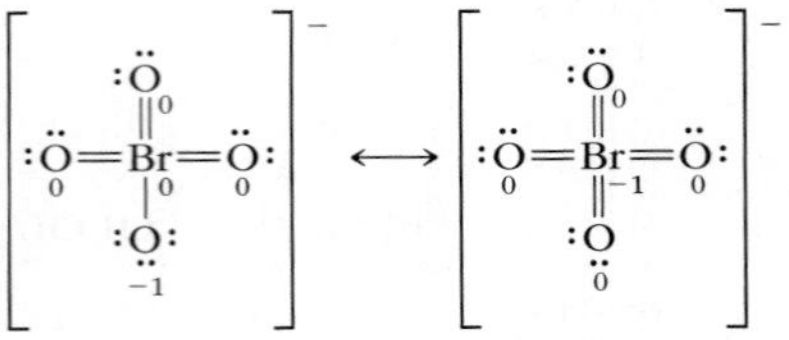

65. 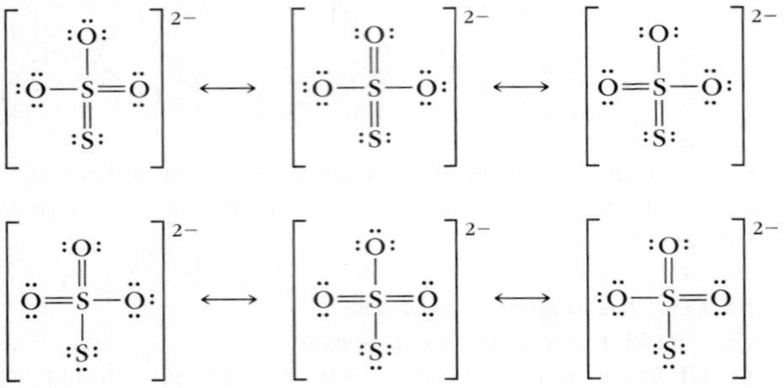

67. (a) 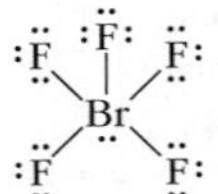 (b) 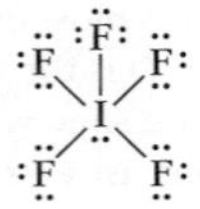 (c) 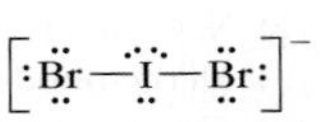

69. P, Cl, Se, Sn

71. Against; with C═C and C—C, the bond lengths would be different.

73. $C_{14}H_{10}$

75.

	σ_{2s}	σ^*_{2s}	π_{2p}	π_{2p}	σ_{2p}	π^*_{2p}	π^*_{2p}	σ^*_{2p}
NO^+	(↑↓)	(↑↓)	(↑↓)	(↑↓)	(↑↓)	( )	( )	( )

NO^+ has three bonds and no unpaired electrons.

78. Si—F; Si is in the third period and less electronegative than C, S, or O.

80. Yes; "close" means similar electronegativities, therefore covalent bonds. "Far apart" means different electronegativities, therefore ionic bonds.

82. (a) The C═C is shorter. (b) The C═C is stronger.

(c) $\overset{\delta^+}{C}\equiv\overset{\delta^-}{N}$

84. 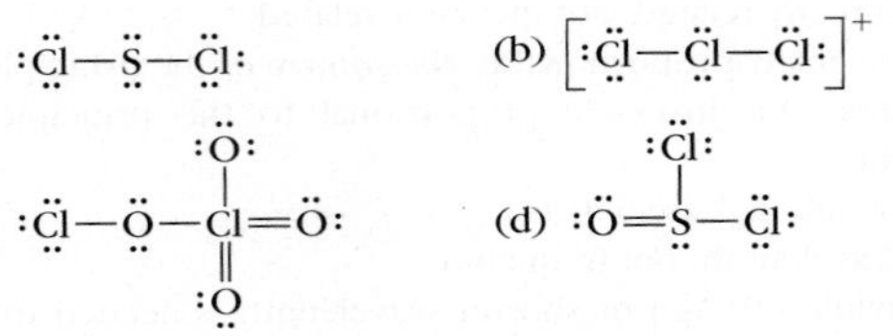

86. O—O < Cl—O < O—H < O═O < O═C

88. (a) Aromatic (b) Aromatic (c) Neither category
(d) Alkane (e) Alkane (f) Neither category

90. Forgot to subtract one electron for the positive charge.

94. Atoms are not bonded to the same atoms.

96. 

98. Cl: 2.7, S: 2.3, Br: 2.6, Se: 2.4, As: 2.1

100. (a) Box 7 (b) Box 1 (c) Boxes 4 and 9 (d) Box 10 (e) Box 12 (f) Box 2

102. NF_5; you cannot expand the octet of N.

104. 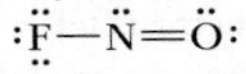 (a) N—F
(b) N═O (c) F—N

106. (a) C—O (b) C≡N (c) C—O

108. (a) 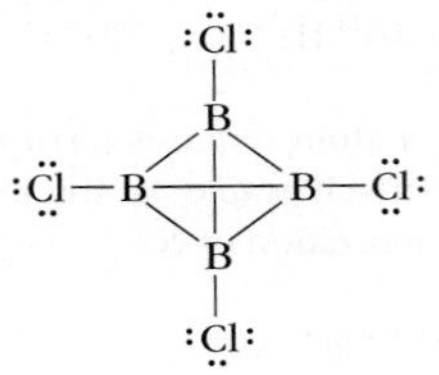

(b) 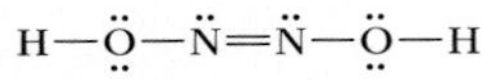

110. (a) 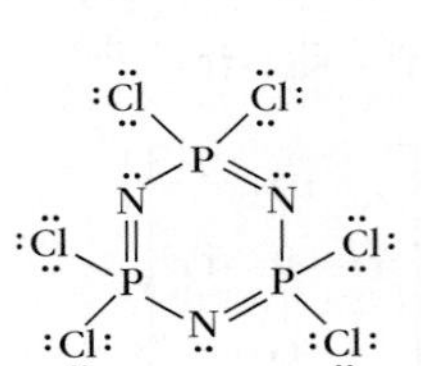 (b) 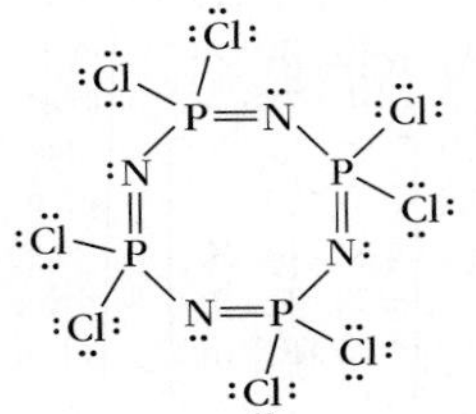

112. 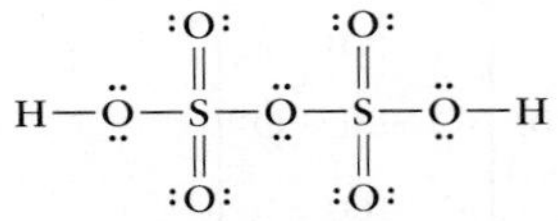

114. The NON structure has higher formal charges and positive formal charge on the higher-EN O atom.

117. (a) 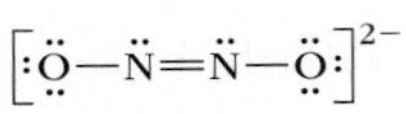 (b) 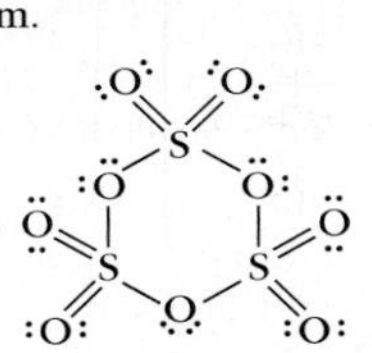

119. 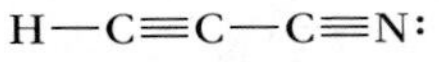

121. 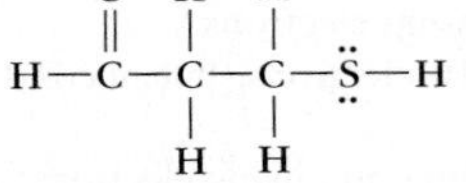

123. 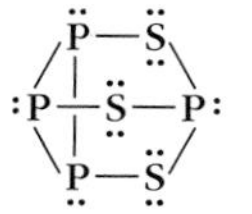

125. 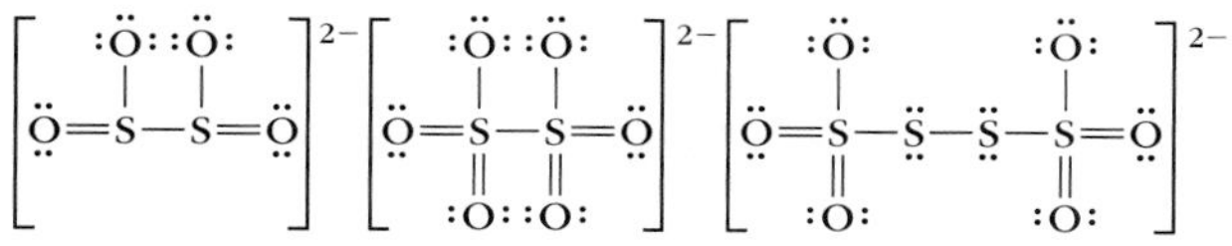

128. (a) 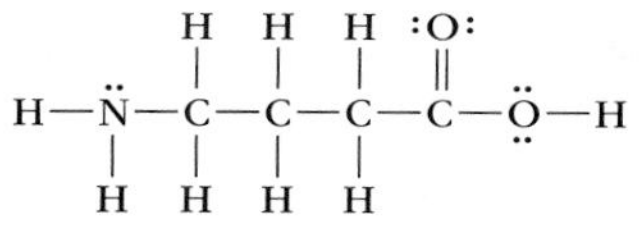

(b) C—C (c) C═O (d) C═O

Chapter 9

13. (a) H—Be—H
Linear

(b) 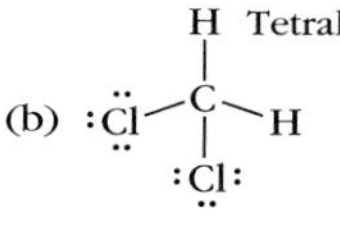 Tetrahedral

(c) 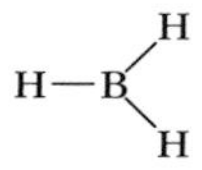 Triangular planar

(d) 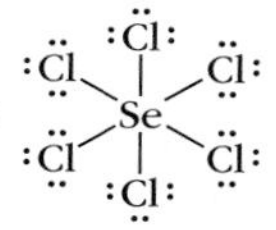 Octahedral

(e) 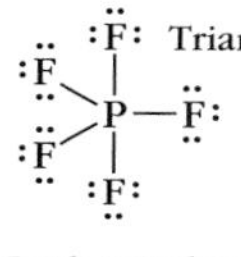 Triangular bipyramidal

15. (a) Both tetrahedral (b) Tetrahedral, angular (109.5°)
(c) Both triangular planar (d) Both triangular planar
17. (a) Both triangular planar (b) Both triangular planar
(c) Tetrahedral, triangular pyramidal
(d) Tetrahedral, triangular pyramidal
Three atoms bonded to the central atom gives triangular shape; however, structures with 26 electrons have the shape of a triangular pyramidal and structures with 24 electrons have the shape of a triangular planar.
19. (a) Both octahedral
(b) Triangular bipyramidal and see-saw
(c) Both triangular bipyramidal
(d) Octahedral and square planar
21. (a) 120° (b) 120°
(c) H—C—H angle is 120°, C—C—N angle is 180°
(d) H—O—N angle = 109.5°, O—N—O angle = 120°
23. (a) 90°, 120°, and 180° (b) 120° and 90° (c) 90° and 120°
25. The O—N—O angle in NO_2^+ is larger than in NO_2.

27. (a) 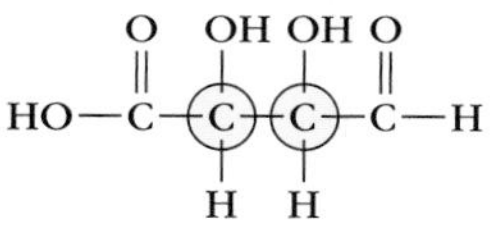 (b) No chiral centers

(c) 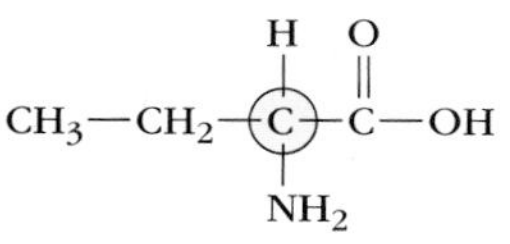

29. (a) No chiral centers (b) No chiral centers

(c) 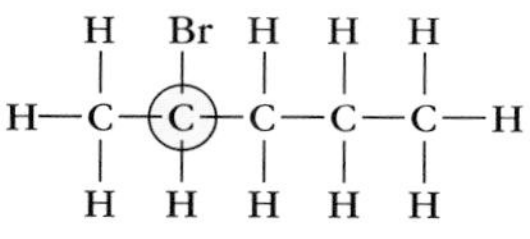 (d) No chiral centers

31. If four different things are bonded to one atom in the structural formula, then the compound can exist as a pair of enantiomers.
33. (a) sp^3 (b) sp (c) sp^2
35. Tetrahedral, sp^3 hybridized carbon atom
37. The central S atom in SCl_2 has sp^3 hybridization; the central S atom in OCS has sp hybridization.
38. PCl_4^+ is tetrahedral, 109.5° bond angles; PCl_5 is triangular bipyramidal, 120° and 90° bond angles; PCl_6^- is octahedral with 90° bond angles.
42. The N atom is sp^3 hybridized with 109.5° angles. The first two carbons are sp^3 hybridized with 109.5° angles. The third carbon is sp^2 hybridized with 120° angles. The single-bonded oxygen is sp^3 hybridized with 109.5° angles. The double-bonded O is sp^2 hybridized with 120° angles.
44. (a) The first two carbon atoms are sp^3 hybridized with 109.5° angles. The third and fourth carbon atoms are sp hybridized with 180° angles.
(b) C≡C (c) C≡C

46. (a) 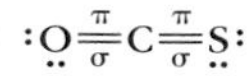 (b) 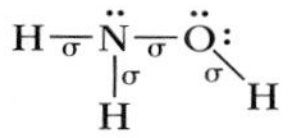

(c) 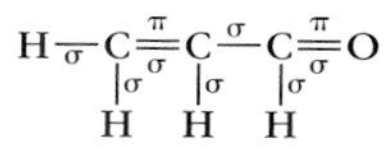 (d) 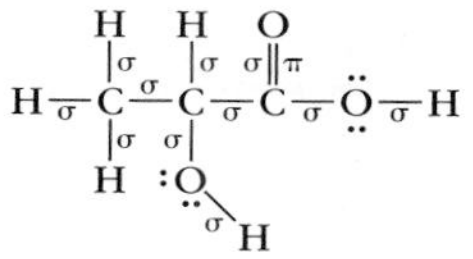

48. (a) Six (b) Three (c) sp (d) sp (e) Both sp^2
50. (a) H_2O (b) CO_2 and CCl_4 (c) F
52. (a) Nonpolar
(b) Polar; the H-side of the molecule is more positive.
(c) Polar; the H-side of the molecule is more positive.
(d) Nonpolar
54. (a) The Br—F bond has a larger electronegativity difference.
(b) The H—O bond has a larger electronegativity difference.

56. (a) 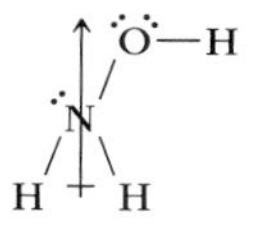 (b) 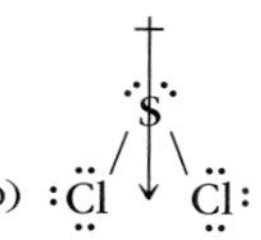

58.

Interaction	Distance	Example
Ion-ion	Longest range	Na^+ interaction with Cl^-
Ion-dipole	Long range	Na^+ ions in H_2O
Dipole-dipole	Medium range	H_2O interaction with H_2O
Dipole-induced dipole	Short range	H_2O interaction with Br_2
Induced dipole-induced dipole	Shortest range	Br_2 interaction with Br_2

60. Wax molecules interact using London forces. Water molecules interact using hydrogen bonding. The water molecules interact much more strongly with other water molecules; hence, beads form as the water tries to avoid contact with the surface of the wax.
62. (c), (d), and (e)
64. Vitamin C is capable of forming hydrogen bonds with water.
66. (a) London forces (b) London forces
(c) Intramolecular (covalent) forces
(d) Dipole-dipole force (e) Hydrogen bonding
69. (a) C—C stretch (b) O—H stretch
70. (a) UV > IR (b) UV < IR (c) UV > IR
71. 427 nm, 483 nm; 427 nm is the more energetic transition.

73.

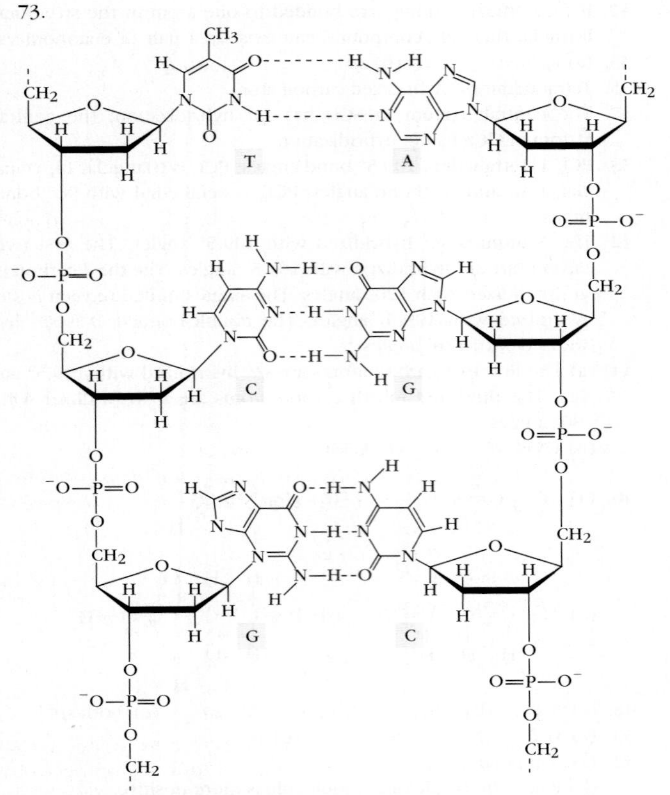

75. (a) (1) NCC = angle 180°, (2) HCH angle = 109.5°,
(3) HCO angle = 109.5°
(b) C=O (c) C=O

77. (a) $\ddot{O}=C=C=C=\ddot{O}$ (b) 180° (c) 180°

79. It has an extended conjugated pi-system, so yes, it should absorb visible light.

81. (a)

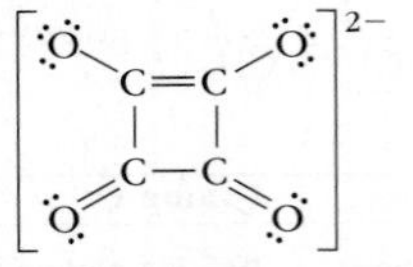

(b) Each C is sp^2 hybridized.

83. HCl has $\delta^+ = 2.69 \times 10^{-20}$ C and HF has $\delta^+ = 6.95 \times 10^{-20}$ C; therefore, F is more electronegative than Cl.

85. KF has $\delta^+ = 1.32 \times 10^{-19}$ C, which is 81.6% of the charge of the electron, so KF is not completely ionic.

87. If the polarity of the bonds exactly cancel each other, a molecule will be nonpolar.

89. (a) B, D, and H; E and I; C and F
(b) D and H
(c) A and G
(d) E and I
(e) B, C, D, E, F, H, and I
(f) A
(g) F
(h) H
(i) A
(j) F

91. (a)

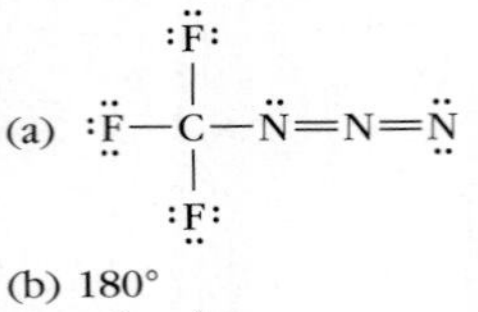

(b) 180°
(c) sp^2, sp^3

(d)

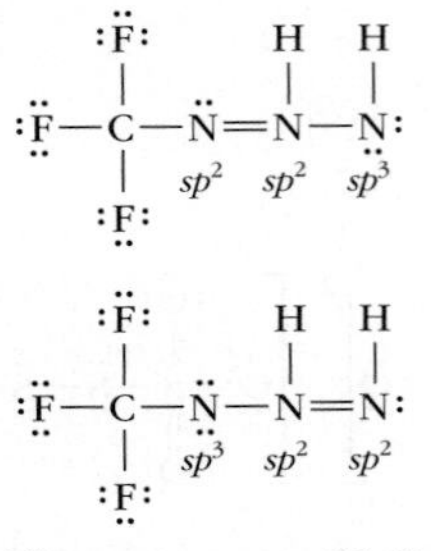

95. (a) Nitrogen (b) Boron (c) Phosphorus
(d) Sulfur (There are other right answers.)

97. Five

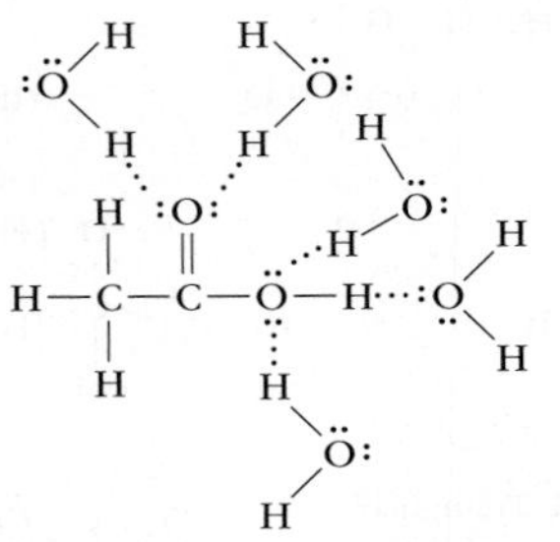

99. Diagram (d) is correct.

101. 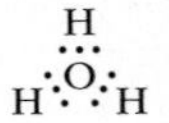 Triangular planar; Triangular planar 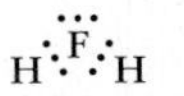 Triangular planar; bent

103. (a)

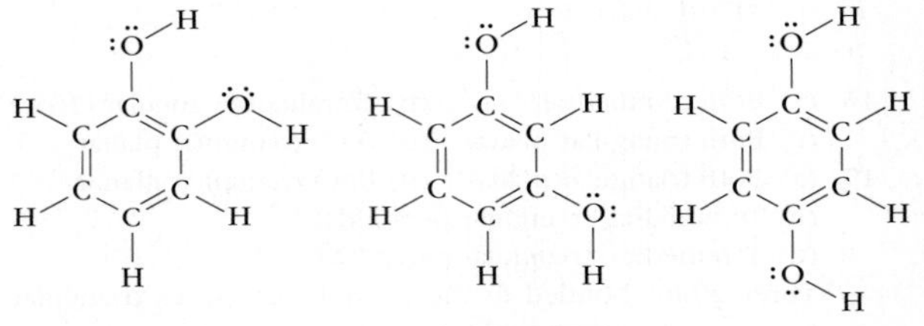

(i) (ii) (iii)

(b) (i) < (ii) < (iii)

105. (a)

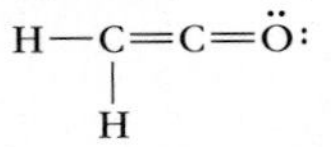

(b) Left C is triangular planar, H—C—C and H—C—H bond angles both 120°; right C is linear, H—C—C angle is 120°, and C—C—O angle is 180°.
(c) sp^2, sp, sp (d) Polar

107. (a) :N≡N—N̈—H 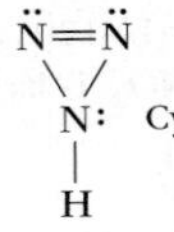 Cyclotriazene
Hydrogen azide

(b) sp, sp, sp^2 (c) sp^2, sp^2, sp^3
(d) 3, 4 (e) 2, 1
(f) Hydrogen azide 120°, 180°; cyclotriazene 60°, 60°, 109.5°

109. (a) Angle 1: 120°, angle 2: 120°, angle 3: 109.5° (b) sp^3
(c) For O with two single bonds, sp^3; for O with one double bond, sp^2

111. (a)

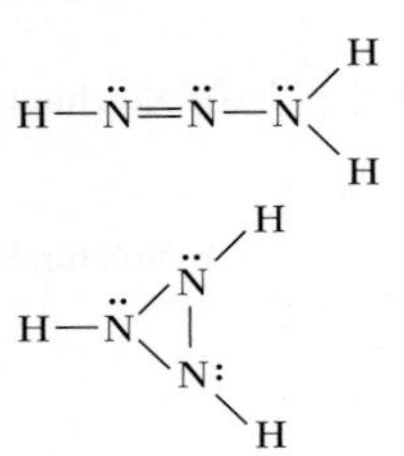

(b) 120°, 60°

113. (a) sp^3 (b) sp^2 (c) sp^2 (d) 8σ, 2π

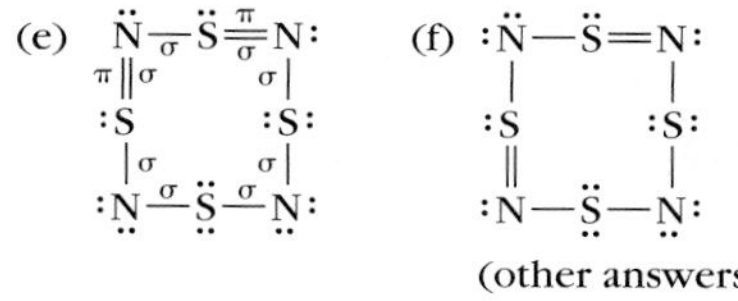

(other answers are possible)

Chapter 10

11. Nitrogen serves to moderate the reactiveness of oxygen by diluting it. Oxygen sustains animal life as a reactant in the conversion of food to energy. Oxygen is produced by plants in the process of photosynthesis.

13.

Molecule	ppm	ppb	
N_2	780,840	780,840,000	↑
O_2	209,480	209,480,000	
Ar	9,340	9,340,000	
CO_2	330	330,000	
Ne	18.2	18,200	>1 ppm
H_2	10.	10,000	
He	5.2	5,200	
CH_4	2	2,000	↓
Kr	1	1,000	↑
CO	0.1	100	between
Xe	0.08	80	1 ppm
O_3	0.02	20	and
NH_3	0.01	10	1 ppb
NO_2	0.001	1	↓
SO_2	0.0002	0.2	<1 ppb

15. 1.2×10^8 metric tons SO_2, 2×10^6 metric tons SO_2
16. (a) 0.947 atm (b) 950. mm Hg (c) 542 torr (d) 98.7 kPa (e) 6.91 atm
18. 14 m
20. With a perfect vacuum at the top of the well, atmospheric pressure can only push water up to about 34 feet. So, the well cannot be deeper than that, not even using a high-quality vacuum pump.
22. I. A gas is composed of molecules whose size is much smaller than the distances between them.
    II. Gas molecules move randomly at various speeds and in every possible direction.
    III. Except when molecules collide, forces of attraction and repulsion between them are negligible.
    IV. When collisions occur, they are elastic.
    V. The average kinetic energy of gas molecules is proportional to the absolute temperature.

    Postulate I will become false at very high pressures. Postulates III and IV will become false at very high pressures or very low temperatures. Postulate II is most likely to always be correct.
24. Slowest $CH_2Cl_2 < Kr < N_2 < CH_4$ fastest
26. Ne will arrive first.
29. 4.2×10^{-5} mol CO
30. 25.5 L
32. 62.5 mm Hg
35. 172 mm Hg
37. 26.5 mL
39. −96 °C
41. 4.00 atm
43. 501 mL
45. 0.507 atm
47. Largest number in (d); smallest number in (c)
49. 6.0 L H_2
51. 0.67 L CO_2; about 15% of the loaf is CO_2.
53. 10.4 L O_2, 10.4 L H_2O
55. 21 mm Hg
57. 10.0 L Br_2
59. 1.44 g $Ni(CO)_4$
61. (a) $2\ CH_3OH(\ell) + 3\ O_2(g) \rightarrow 2\ CO_2(g) + 4\ H_2O(g)$
    (b) 1.1×10^3 L CO_2
63. 130. g/mol
65. 2.7×10^3 mL
67. P_{He} is 7.000 times greater than P_{N_2}.
69. 3.7×10^{-4} g/L
71. $P_{tot} = 4.51$ atm
73. (a) 154 mm Hg
    (b) $X_{N_2} = 0.777$, $X_{O_2} = 0.208$, $X_{Ar} = 0.0093$, $X_{H_2O} = 0.0053$, $X_{CO_2} = 0.0003$
    (c) 77.7% N_2, 20.8% O_2, 0.93% Ar, 0.03% CO_2, 0.54% H_2O. This sample is wet. Table 1.3 gives percentages for dry air, so the percentages are slightly different due to the water in this sample.
75. (a) $P_{tot} = 1.98$ atm
    (b) $P_{O_2} = 0.438$ atm, $P_{N_2} = 0.182$ atm, $P_{Ar} = 1.36$ atm
77. Membrane irritation: 1×10^{-4} atm; Fatal narcosis: 0.02 atm
79. 0.0041; 3.1 mm Hg; the mean partial pressure includes humid and dry air, summer and winter, worldwide.
82. 18. mL $H_2O(\ell)$, 22.4 L $H_2O(g)$; No, because the vapor pressure of water at 0 °C is 4.6 mm Hg; we cannot achieve 1 atm pressure for this gas at this temperature.
84. Molecular attractions become larger at higher pressures. Molecules hitting the walls hit them with somewhat less force due to the opposing attractions of other molecules.
86. N_2 is more like an ideal gas at high P than CO_2.
88. These reactions can be found in Section 10.11.
    (a) $CF_3Cl \xrightarrow{h\nu} \cdot CF_3 + \cdot Cl$ (b) $\cdot Cl + \cdot O\cdot \rightarrow ClO\cdot$
    (c) $ClO\cdot + \cdot O\cdot \rightarrow \cdot Cl + O_2$
90. CH_3F has no C—Cl bonds, which in CH_3Cl are readily broken when exposed to UV light to produce ozone-depleting radical halogen, ·Cl.
92. CFCs are not toxic. Refrigerants used before CFCs were very dangerous. One example is NH_3, a strong-smelling, reactive chemical. Use the keywords "CFCs" or "refrigerants" and "toxicity" in any web browser.
95. Primary pollutants (e.g., particle pollutants, including aerosols and particulates, sulfur dioxide, nitrogen oxides, hydrocarbons); secondary pollutants (e.g., ozone); (see Section 10.10 for details).
97. Adsorption means firmly attaching to a surface. Absorption means drawing into the bulk of a solid or liquid.
99. 1.6×10^9 metric tons, 2.3×10^6 hours
101. $O_2 \xrightarrow{h\nu} \cdot O\cdot + \cdot O\cdot$, No, the reaction will not occur if the light is visible light.
103. SO_2, coal and oil, $2\ SO_2 + O_2 \longrightarrow 2\ SO_3$
105. $N_2 + O_2 \xrightarrow{heat} 2\ NO$; reaction takes elemental nitrogen and makes a compound of nitrogen.
107. (a) $P_i = 1.33 \times 10^{-4}$ mm Hg
     (b) $X_{SO_2} = 1.75 \times 10^{-7}$
     (c) 500. μg SO_2
108. Greenhouse effect = trapping of heat by atmospheric gases. Global warming = increase of the average global temperature. Global warming is caused by an increase in the amount of greenhouse gases in the atmosphere.
110. Examples of CO_2 sources: animal respiration, burning fossil fuels and other plant materials, decomposition of organic matter, etc. Examples of CO_2 removal: photosynthesis in plants, being dissolved in rain water, incorporation into carbonate and bicarbonate compounds in the oceans, etc. Currently, CO_2 production exceeds CO_2 removal.
112. (a) Before: $P_{H_2} = 3.7$ atm; $P_{Cl_2} = 4.9$ atm
     (b) Before: $P_{tot} = 8.6$ atm (c) After: $P_{tot} = 8.6$ atm
     (d) Cl_2; 0.5 moles remain (e) $P_{HCl} = 7.4$ atm; $P_{Cl_2} = 1.2$ atm
     (f) $P = 8.9$ atm
114. Statements (a), (b), (c), and (d) are true.

116.

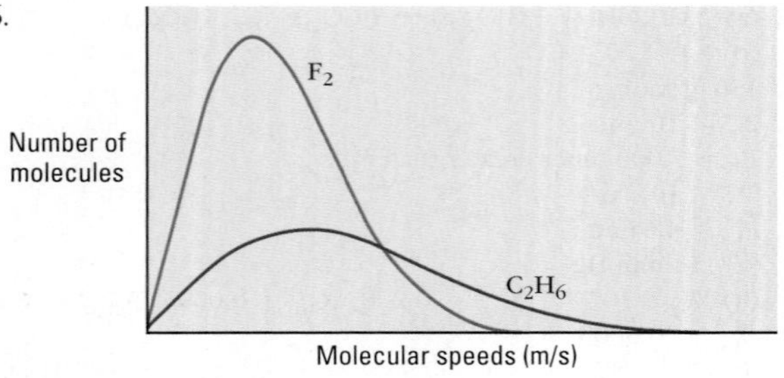

119. (a) Pressure: Box (c) = Box (a) > Box (b) = Box (d)
(b) Density: Box (c) > Box (d) > Box (a) > Box (b)
(c) Average kinetic energy: Box (a) = Box (b) = Box (c) = Box (d)
(d) Average molecular speed: Box (a) = Box (b) > Box (c) = Box (d)

121. For reference, the initial state looks like this:

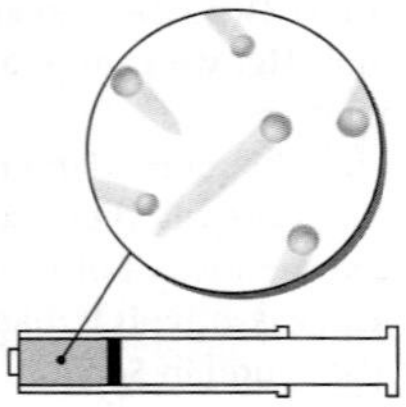

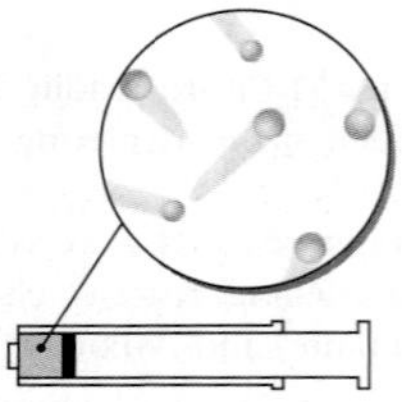

(a)

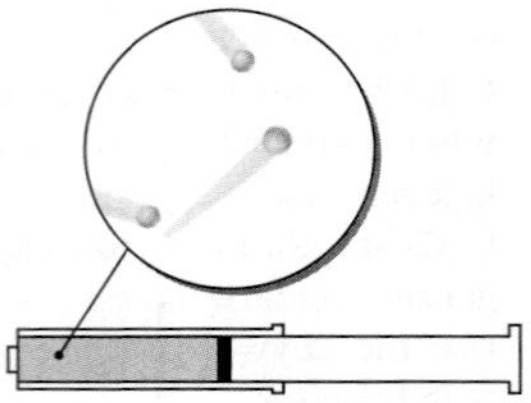

(b)

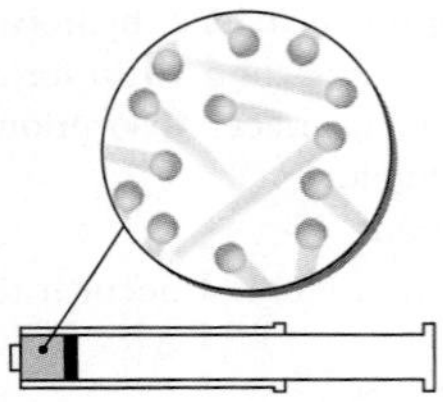

(c)

123. Box (b). The initial-to-final volume ratio is 2:1, so, for every two molecules of gas reactants, there must be one molecule of gas products. 6 reactant molecules must produce 3 product molecules.

125. (a) 64.1 g/mol
(b) Empirical formula = CHF; molecular formula = $C_2H_2F_2$
(c) If the F atoms are on different C atoms

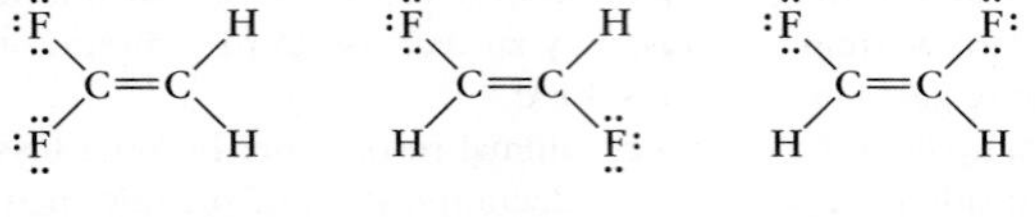

127. (a) NH_3; Ox. # N = −3 (b) NH_4^+; Ox. # N = −3
(c) N_2O; Ox. # N = +1 (d) N_2; Ox. # N = 0
(e) HNO_3; Ox. # N = +5 (f) HNO_2; Ox # N = +3
(g) NO_2; Ox. # N = +4

129. (a) $\cdot O \cdot + H_2O \rightarrow \cdot OH + \cdot OH$
(b) $\cdot NO_2 + \cdot OH \rightarrow HNO_3$
(c) $RH + \cdot OH \rightarrow \cdot R + H_2O$

130. $P_{tot} = 0.88$ atm

132. $\frac{m_{Ne}}{m_{Ar}} = 1$

134. (a) 29.1 mol CO_2; 14.6 mol N_2; 2.43 mol O_2 (b) 1.1×10^3 L
(c) $P_{N_2} = 0.317$ atm; $P_{CO_2} = 0.631$ atm; $P_{O_2} = 0.0527$ atm

136. 458 torr

138. 4.5 mm^3

140. (a) More significant, because of more collisions
(b) More significant, because of more collisions
(c) Less significant, because the molecules will move faster

Chapter 11

13. At higher temperatures, the molecules move around more. The increased random motion disrupts the intermolecular interactions responsible for surface tension.

15. Reduce the pressure

17. The molecules of water in your sweat have a wide distribution of molecular speeds. The fastest of these molecules are more likely to escape the liquid state into the gas phase. The low-speed molecules left behind will have a lower average speed and, therefore, a lower average kinetic energy that, according to kinetic molecular theory, implies that the temperature lowers.

19. 1.5×10^6 kJ

21. 233 kJ

23. 2.00 kJ required for Hg is more than 1.13 kJ required for H_2O sample. Hg has a much greater mass.

25. NH_3 has a relatively large boiling point because the molecules interact using relatively strong hydrogen-bonding intermolecular forces. The increase in the boiling points of the series PH_3, AsH_3, and SbH_3 is related to the increasing London dispersion intermolecular forces experienced, due to the larger central atom in the molecule. (Size: P < As < Sb)

27. Methanol molecules are capable of hydrogen bonding, whereas formaldehyde molecules use dipole-dipole forces to interact. Molecules experiencing stronger intermolecular forces (such as methanol here) will have higher boiling points and lower vapor pressures compared to molecules experiencing weaker intermolecular forces (such as formaldehyde here).

28. Substance D; it has lower vapor pressure than A, B, or C at any given temperature.

30. 0.21 atm, approximately 57 °C

32. 1600 mm Hg

34. $\Delta H_{vap} = 70.$ kJ/mol

36. The interparticle forces in the solid are very strong.

38. The intermolecular forces between the molecules of H_2O in the solid (hydrogen bonding) are stronger than the intermolecular forces between the molecules of H_2S (dipole-dipole).

40. 27 kJ

42. 51.9 g CCl_2F_2

44. LiF, because the ions are smaller, making the charges more localized and closer together, causing a higher coulombic interaction between the ions.

46. The highest melting point is (a). The extended network of covalent bonds in SiC are the strongest interparticle forces of the listed choices. The lowest melting point is (d). Both I_2 and $CH_3CH_2CH_2CH_3$ interact using the weakest intermolecular forces—London dispersion forces—but large I_2 has many more electrons, so its London forces are relatively stronger.

48. The freezer compartment of a frost-free refrigerator keeps the air so cold and dry that ice inside the freezer compartment sublimes [(s) → (g)]. The hail stones would eventually disappear.

49. (a) Region A: solid; region B: liquid; region C: gas. (b) Point 1: solid and gas; point 2: solid, liquid, and gas; point 3: liquid and gas; point 5: solid and liquid.

51. (a) Gas phase (b) Liquid phase (c) Solid phase

53. Water molecules interact using relatively strong hydrogen-bonding intermolecular forces, so greater kinetic energy is needed for molecules to escape into the gas phase.

55. The crystal structure in ice maximizes the hydrogen-bonding capacity and leaves considerable open spaces compared to the liquid. See discussion in Section 11.4 for more details.

59. (a) Molecular (b) Metallic (c) Network (d) Ionic

61. (a) Amorphous; it decomposes before melting and does not conduct electricity. (b) Molecular; low melting point and nonconducting (c) Ionic; high melting point and only liquid (molten ions) conduct electricity. (d) Metallic; both solid and liquid conduct electricity.
63. (a) Molecular (b) Ionic
(c) Metallic or network (d) Amorphous
65. See Figure 11.21 and its description
67. 220 pm
70. Diagonal is 696 pm, side length is 401 pm.
72. No; the ratio of ions in the unit cell must reflect the empirical formula of the compound.
74. (a) 152 pm
(b) Radius of I^- = 212 pm and radius of Li^+ = 88.0 pm
(c) It is reasonable that the atom is larger than the cation. The assumption that anions touch anions seems unreasonable; there would be some repulsion and probably a small gap or distortion.
75. Carbon atoms in diamond are sp^3 hybridized and are tetrahedrally bonded to four other carbon atoms. Carbon atoms in pure graphite are sp^2 hybridized and bonded with a triangle planar shape to other carbon atoms. These bonds are partially double bonded so they are shorter than the single bonds in diamond. However, the planar sheets of sp^2 hybridized carbon atoms are only weakly attracted by intermolecular forces to adjacent layers, so these interplanar distances are much longer than the C—C single bonds. Therefore, graphite is less dense than diamond.
76. Diamond and graphite are examples of network solids. Neither is soluble in water or most other solvents. Atoms in a covalent network are held together by covalent bonds, which are much stronger than intermolecular forces among the atoms and molecules of a solvent. Therefore, dissolving would be highly endothermic and is not likely to occur.
77. Diamond is an electrical insulator because all the electrons are in single bonds shared between two specific atoms and cannot move around. However, graphite is a good conductor of electricity because its electrons are delocalized in conjugated double bonds that allow the electrons to move easily through the graphite sheets.
79. $\nu = 5.30 \times 10^{17} s^{-1}$; (a) 3.51×10^{-16} J/photon
(b) 2.11×10^8 J/mol, X ray
81. 361 pm
84. In a conductor, the valence band is only partially filled, whereas, in an insulator, the valence band is completely full, the conduction band is empty, and there is a wide energy gap between the two. In a semiconductor, the gap between the valence band and the conduction band is very small so that electrons are easily excited into the conduction band.
86. Substance (c), Ag, has the greatest electrical conductivity because it is a metal. Substance (d), P_4, has the smallest electrical conductivity because it is a nonmetal. (The other two are metalloids.)
88. A superconductor is a substance that is able to conduct electricity with no resistance. Two examples are $YBa_2Cu_3O_7$ and $Hg_{0.8}Tl_{0.2}Ba_2Ca_2Cu_3O_{8.23}$.
90. $SiO_2(s) + 2\ C(s) \rightarrow Si(s) + 2\ CO(g)$; Si is being reduced and C is being oxidized. $SiCl_2(\ell) + 2\ Mg(s) \rightarrow Si(s) + 2\ MgCl_2(s)$; Si is being reduced and Mg is being oxidized.
92. Doping is the intentional addition of small amounts of specific impurities into very pure silicon. Group III elements are used because they have one less electron per atom than the Group IV silicon. Group V elements are used because they have one more electron per atom.
95. The amorphous solids known as glasses are different from NaCl because they lack symmetry or long-range order, whereas ionic solids such as NaCl are extremely symmetrical. NaCl must be heated to melting temperatures and then cooled very slowly to make a glass.
97. (a) Two examples of oxide ceramics: Al_2O_3 and MgO
(b) Two examples of nonoxide ceramics: Si_3N_2 and SiC.
(other answers are possible)
99. 780 kJ, 1.6×10^4 kJ
101. (a) Dipole-dipole forces and London forces
(b) $CH_4 < NH_3 < SO_2 < H_2O$
102. (a) Approx. 80 mm Hg (b) Approx. 18 °C
(c) Approx. 640 mm Hg (d) Diethyl ether and ethanol
(e) Diethyl ether evaporates immediately. Ethanol and water remain liquid.
(f) Water
104. The butane in the lighter is under great enough pressure so that the vapor pressure of butane at room temperature is less than the pressure inside the lighter. Hence, it exists as a liquid.
106. 1 and C, 2 and E, 3 and B, 4 and F, 5 and G, 6 and H, 7 and A
108. Each has the same fraction of filled space. The fraction of spaces filled by the closest packed, equal-sized spheres is the same, no matter what the size of the spheres.
109. Vapor-phase water condenses on contact with skin and condensation is exothermic, which imparts more energy to the skin.
112. (a) Condensation, freezing (b) Triple point
(c) Melting point curve
114. (a) Approx. 560 mm Hg (b) Benzene (c) 73 °C
(d) Methyl ethyl ether, approx. 7 °C; carbon disulfide, approx. 47 °C; benzene, approx. 81 °C
116. 22.2 °C
118. 0.533 g/cm^3

Chapter 12

22. (a) 20–200 °C
(b) The octane rating of the straight-run gasoline fraction is 55.
(c) No; because the octane rating is far lower than regular gasoline we buy at the pump (86–94), that means it would cause far more pre-ignition than we expect from the gasoline. It would need to be reformulated to make it an acceptable motor fuel.
24. Gasolines contain molecules in the liquid phase that, at ambient temperatures, can easily overcome their intermolecular forces and escape into the vapor phase. That means all gasolines evaporate easily.
27. (a) $CH_3—CH_2—CH_2—OH$

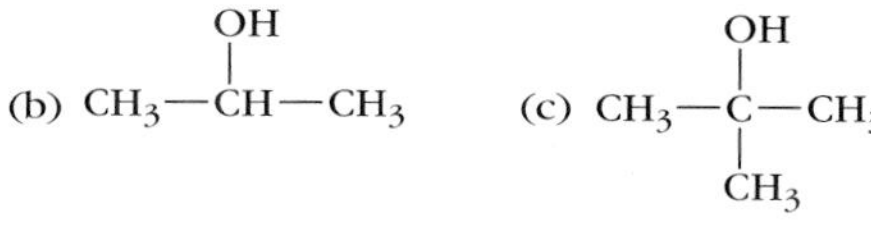

(There are other correct answers)

29. (a) Tertiary (b) Primary (c) Secondary
(d) Secondary (e) Tertiary (f) Secondary

31. 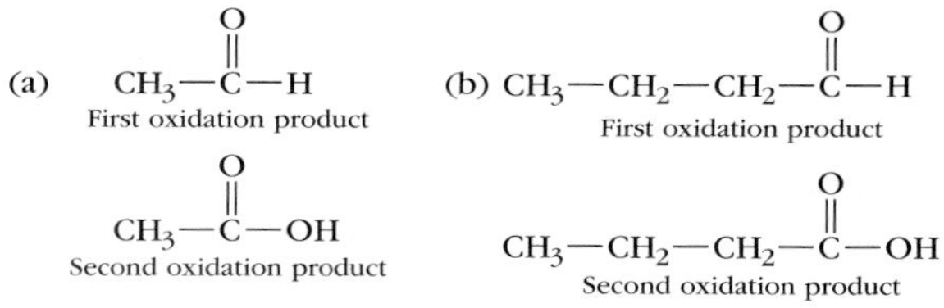
First oxidation product
First oxidation product
Second oxidation product
Second oxidation product

33. 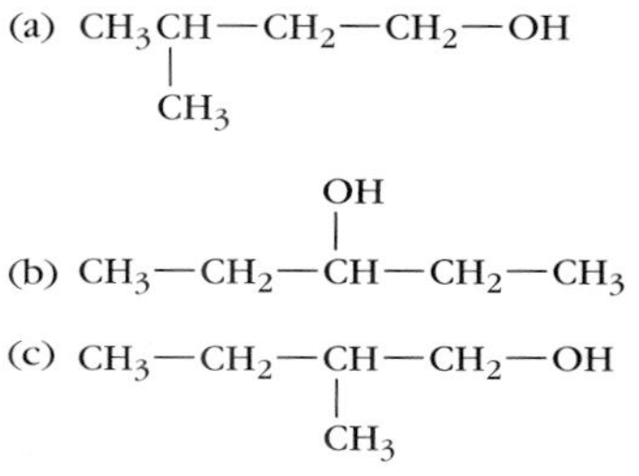

35. Wood alcohol (methanol) is made by heating hardwoods such as beech, hickory, maple, or birch. Grain alcohol (ethanol) is made from the fermentation of plant materials such as grains.
37. —OH groups are a common site of hydrogen-bonding intermolecular forces. Their presence would increase the solubility of the biological molecule in water and create specific interactions with other biological molecules.
38. Radiowaves
41. (a) $CH_3COOCH_2CH_3$ (b) $CH_3CH_2COOCH_2CH_2CH_3$
(c) $CH_3CH_2COOCH_3$
43. (a) $CH_3CH_2COOH + CH_3OH$
(b) $HCOOH + CH_3CH_2OH$
(c) $CH_3COOH + CH_3CH_2OH$
45. Examples of thermoplastics are milk jugs (polyethylene), sunglasses and toys (polystyrene), and CD audio discs (polycarbonates). Thermoplastics soften and flow when heated.

47. (a) (b)

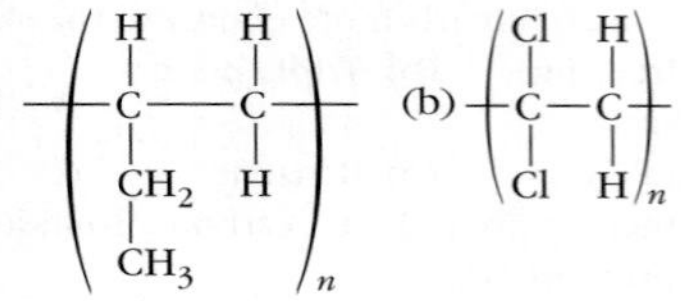

(c)

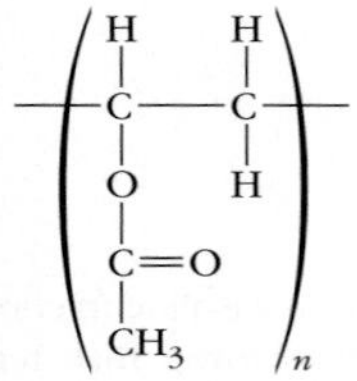

49. (a)

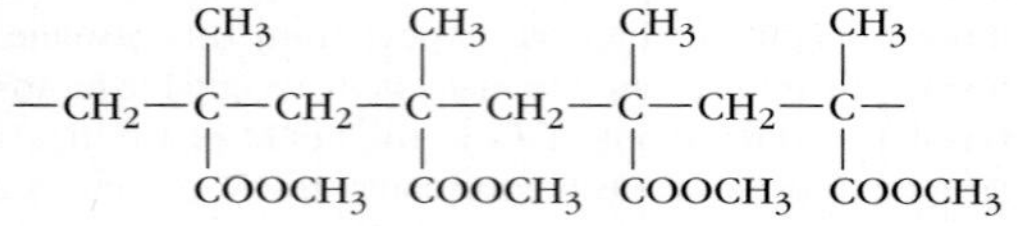

(b)

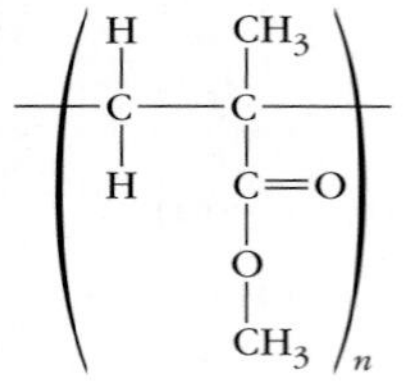

51. (a) $5\ H_2C{=}CCl_2 + 2\ \cdot OR \longrightarrow$

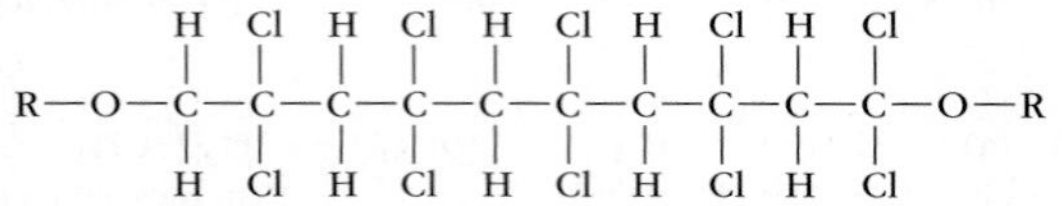

(b) $5\ F_2C{=}CF_2 + 2\ \cdot OR \longrightarrow$

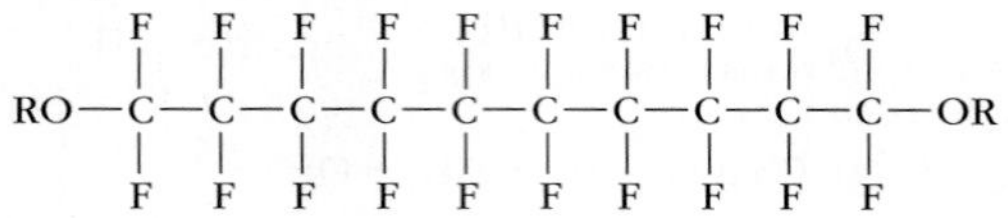

52. (a) (b) (c) (d) (e)

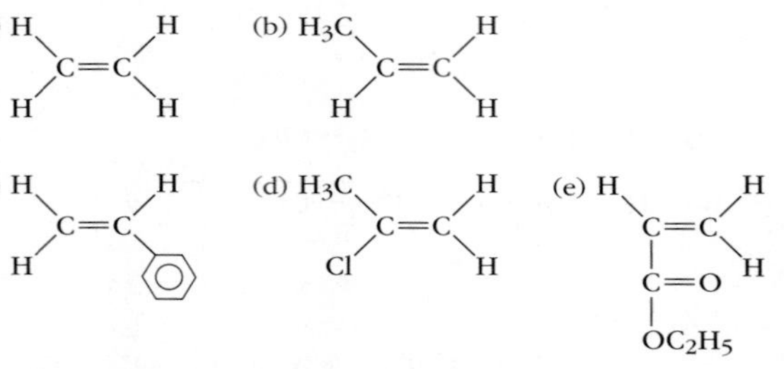

53. Isoprene, 2-methyl-1,3-butadiene; *cis*-isomer

55.

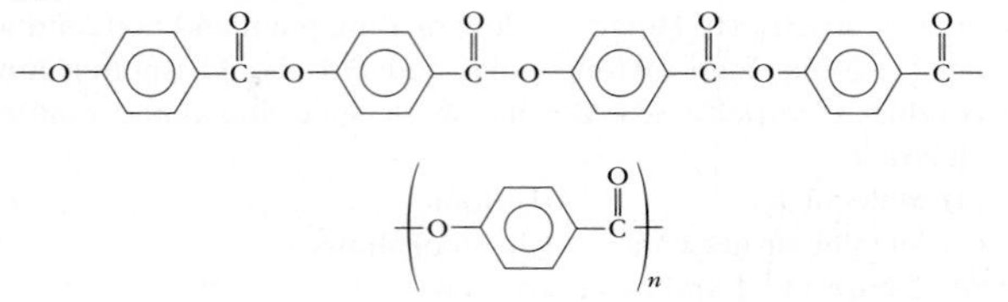

57. Carboxylic acid and alcohol
59. Carboxylic acid and amine; nylon
61. One major difference is that the protein polymer's monomers are not all alike. Different side chains change the properties of the protein.
63.

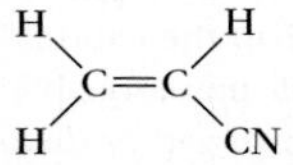

65. 16,000 monomer units
67. (a)

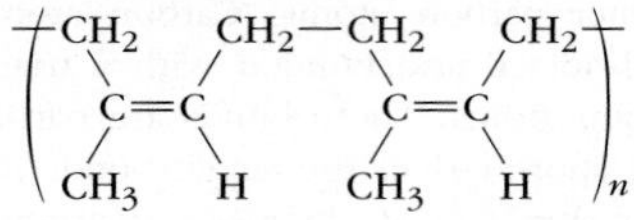

(b)

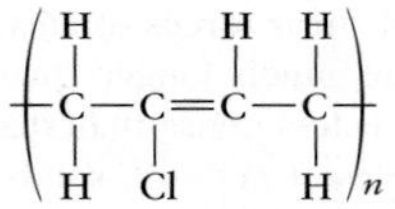

(c)

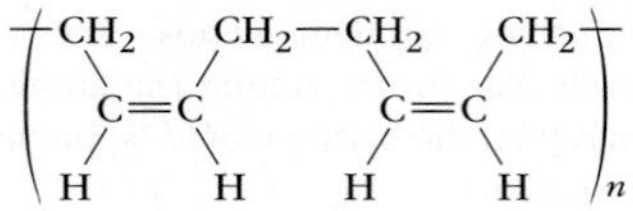

68. Major end uses for recycled PET include fiberfill for ski jackets and sleeping bags, carpet fibers, and tennis balls. HDPE is converted into a fiber used for sportswear, insulating wrap for new buildings, and very durable shipping containers.
70. Proteins, DNA, RNA
71. Amine, amide, alcohol, and carboxylic acid
72.

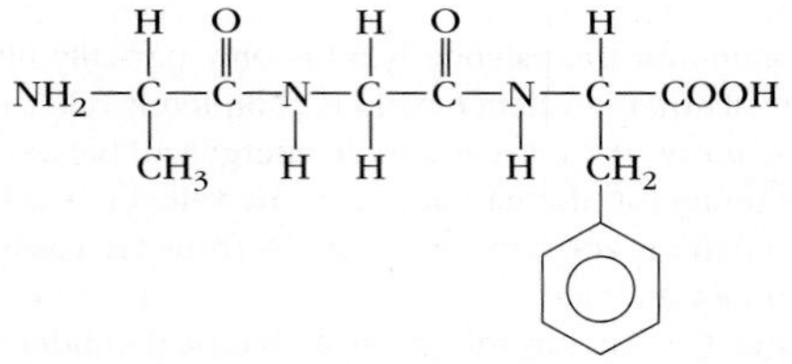

74. (a) A monosaccharide is a molecule composed of one simple sugar molecule, while disaccharides are molecules composed of two simple sugar molecules.
(b) Disaccharides have only two simple sugar molecules, whereas polysaccharides have many.
76. Starch and cellulose
78. (a) Glycogen contains glucose linked with the glycosidic linkages in "*cis*-positions," and cellulose contains glucose with the glycosidic linkages in "*trans*-positions." Humans do not have the enzymes required to break the *trans*-linkage in cellulose.
(b) Cows have the enzymes for breaking the *trans*-linkage of cellulose.
80. It will show up earlier on the chromatograph. Polar molecules would not be attracted to the stationary phase, so they would exit the chamber more quickly.
82. $CH_3CH_2CH_2CH_2CH_2CH_2CH_2CH_2CH_2CH_2OH$ is a larger molecule than CH_3CH_2OH. The polar alcohol group will interact well with the water; however, the nonpolar end of the molecule will not. The longer nonpolar end of the decanol will not be miscible in water, lowering the solubility compared to smaller, more polar ethanol.
84. Vulcanized rubber has short chains of sulfur atoms that bond together the polymer chains of natural rubber.

85. 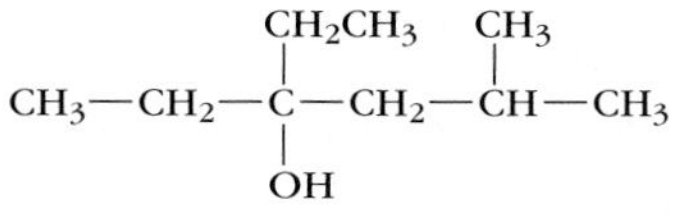

A tertiary alcohol

88. (a) $2\ C_8H_{18}(\ell) + 25\ O_2(g) \rightarrow 16\ CO_2(g) + 18\ H_2O(g)$
(b) 1.4×10^3 L CO_2

90. Glycogen has the glycosidic linkages in "*cis*-positions," whereas cellulose has glycosidic linkages in "*trans*-positions."

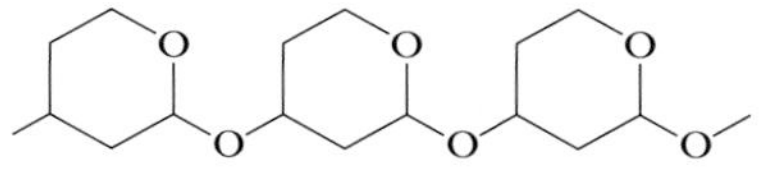

Glycogen

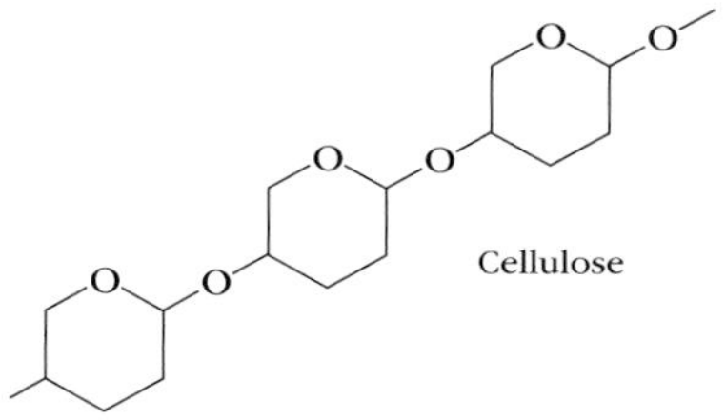

Cellulose

92. 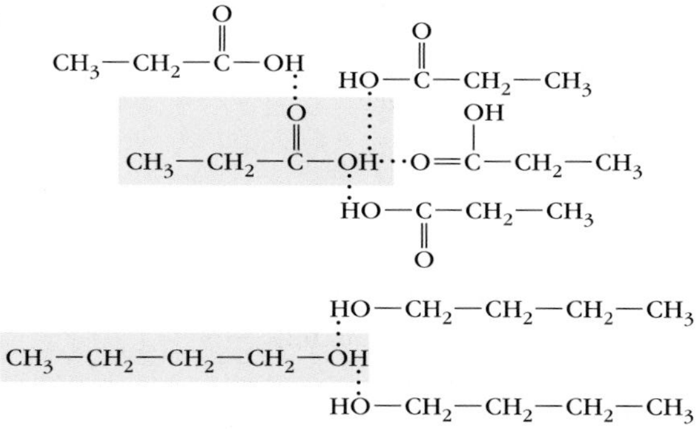

Propanoic acid boils at higher *T*, due to more H-bonding.

94. $CH_3—C{\equiv}C—H$

96. 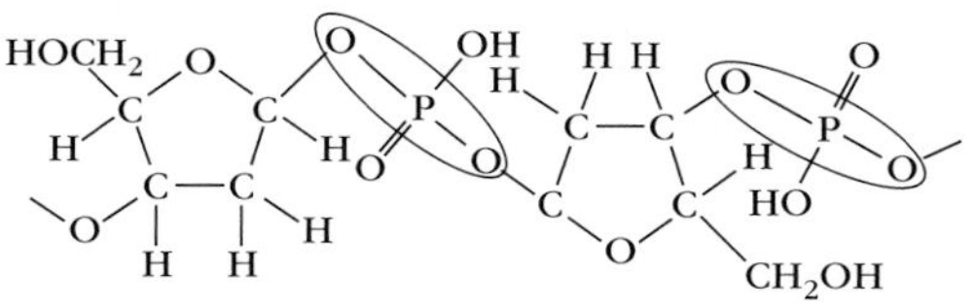

98. Some data that you would need to know are sources and amounts of CO_2 generated over time to determine additional CO_2, photosynthesis rate of depletion per tree per year, average number of trees per acre, the number of acres of land in Australia that could support trees, and the allowable tree density. Other data besides these may need to be contemplated, so consider it a challenge to think of other things you would need to know.

100. (a) Box C (b) Box G (c) Box G (d) Box F (e) Box D (f) Box B (g) Box A (h) Box C (i) Box I (j) Box E

103. (a) X ethanol; Y acetic acid; Z acetaldehyde
(b)

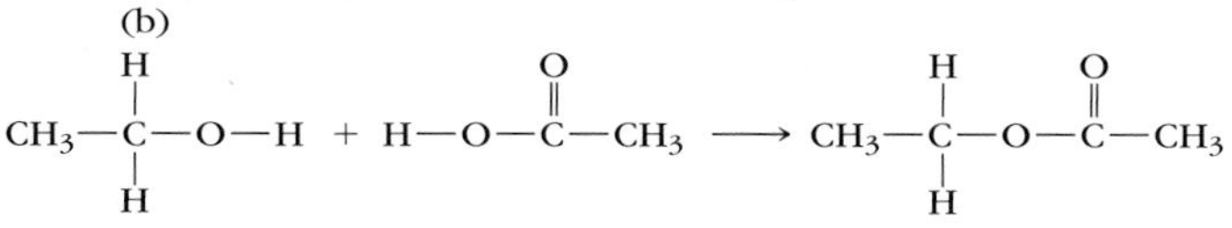

$$CH_3C(=O)—O—H + H_2O \rightleftharpoons H_3O^+ + CH_3C(=O)—O^-$$

(c) Ethanol is a primary alcohol that is initially oxidized to acetaldehyde; further oxidation of acetaldehyde produced acetic acid. When acetaldehyde is the initial reactant, it is directly oxidized to acetic acid.

105. (a) CH_2 (b) Alkene
(c) 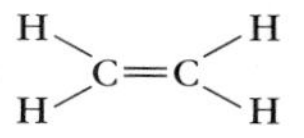

(Other structures also work.)

107. 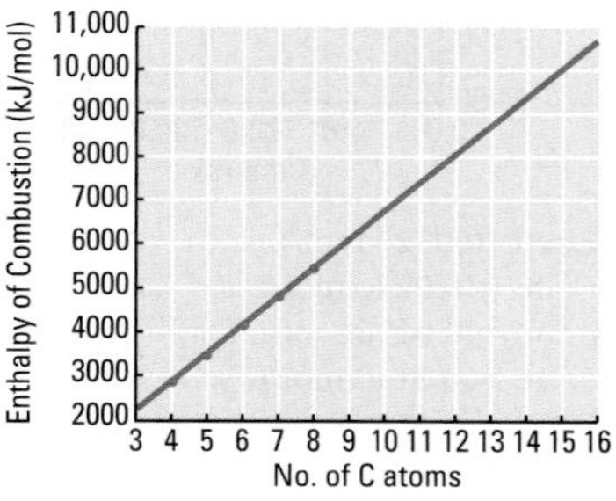

Estimate: C_3, 2200 kJ/mol; C_9, 6100 kJ/mol; C_{16}, 10,700 kJ/mol

109. 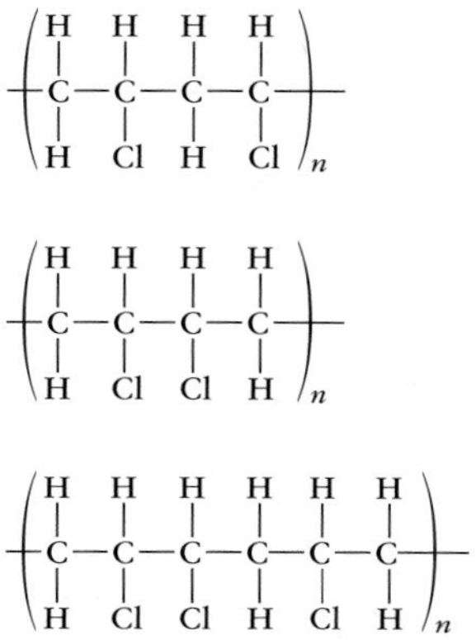

111. (a) 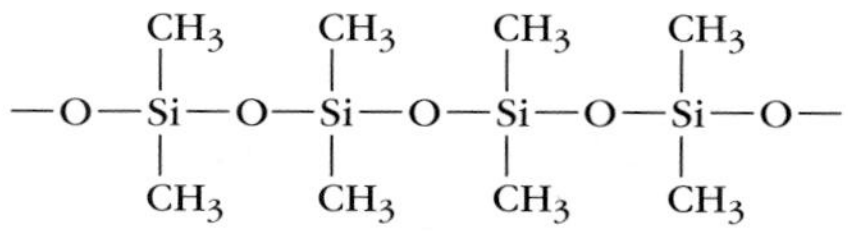

(b) Condensation polymerization

113. 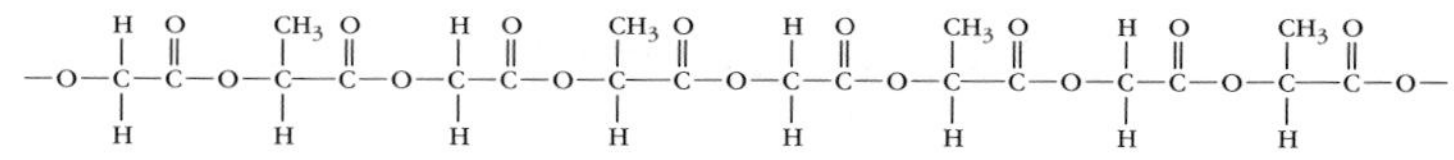

Chapter 13

9. (d) dissolves fastest; (a) dissolves slowest. Rate of dissolving is larger when the grains of sugar are smaller, because there is more surface contact with the solvent.

11. (a)

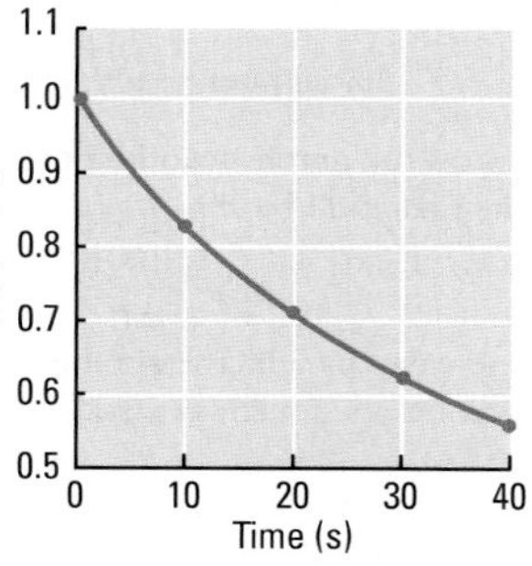

0.0167 M/s, 0.0119 M/s, 0.0089 M/s, 0.0070 M/s
The concentration of the reactant is decreasing.
(b) The rate of change of [B] is twice as fast as the rate of change of [A].
(c) 0.0238 M/s

13. (a)

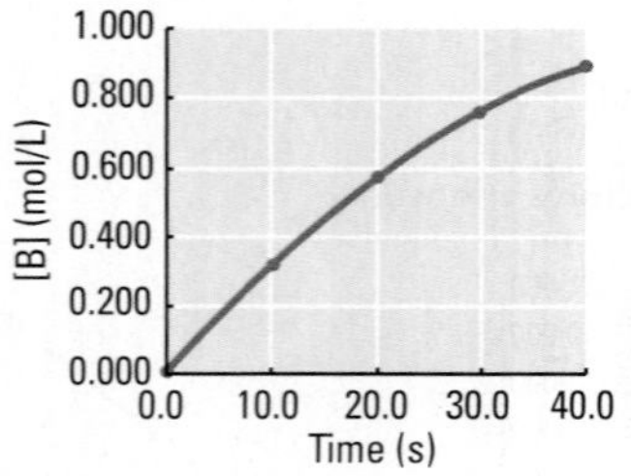

0–10 s, rate = 0.0326 mol/L·s;
10–20 s, rate = 0.0246 mol/L·s;
20–30 s, rate = 0.0178 mol/L·s;
30–40 s, rate = 0.0140 mol/L·s;
Yes

(b) Half as fast; 0.0123 mol/L·s
(c) 0.0160 mol/L·s

15. (a) 0.23 mol/L·h (b) 0.20 mol/L·h (c) 0.161 mol/L·h
(d) 0.12 mol/L·h (e) 0.090 mol/L·h (f) 0.066 mol/L·h

17. (a) Calculate the average concentration for each time interval and plot average concentration vs. average rate:

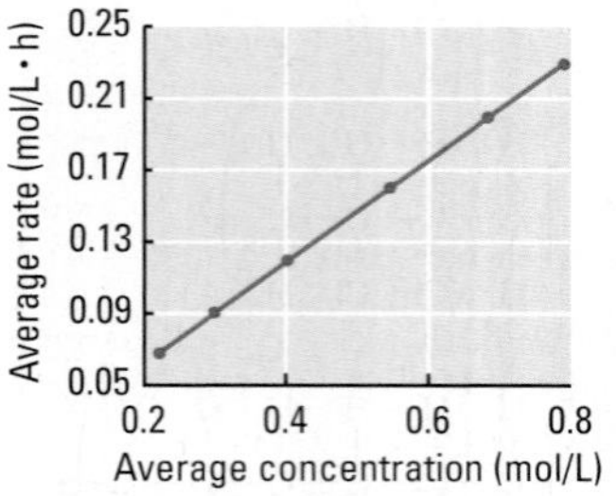

The linear relationship shows: Rate = $k[N_2O_5]$.
(b) $k = 0.29\ h^{-1}$

19.

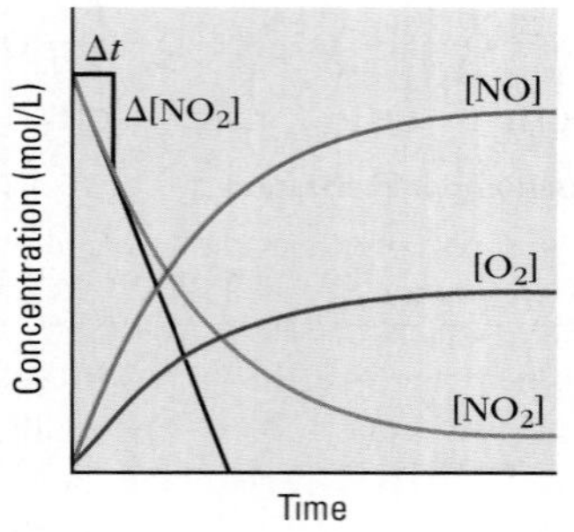

(a) To calculate initial rate, obtain values for Δt and Δ(concentration) near initial time, where the curve can still be approximated by a straight line (the dashed line, above) then divide Δ(concentration) by Δt and change the sign to make it positive, e.g.,

$$\text{Initial rate} = \frac{-\Delta[NO_2]}{\Delta t}$$

(b) The curves on the graph become horizontal after a very long time, so the final rate will be zero.

21. (a) $\dfrac{-\Delta[N_2O_4]}{\Delta t}$ (b) $\dfrac{\Delta[N_2]}{\Delta t}$

23. (a) The rate increases by a factor of nine.
(b) The rate will be one fourth as fast.

25. (a) Rate = $k[NO_2]^2$
(b) The rate will be one fourth as fast.
(c) The rate is unchanged.

27. (a) (i) 9.0×10^{-4} M/h, (ii) 1.8×10^{-3} M/h, (iii) 3.6×10^{-3} M/h

(b) If the initial concentration of $Pt(NH_3)_2Cl_2$ is high, the rate of disappearance of $Pt(NH_3)_2Cl_2$ is high. If the initial concentration of $Pt(NH_3)_2Cl_2$ is low, the rate of disappearance of $Pt(NH_3)_2Cl_2$ is low. The rate of disappearance of $Pt(NH_3)_2Cl_2$ is directly proportional to $[Pt(NH_3)_2Cl_2]$.
(c) The rate law shows direct proportionality between rate and $[Pt(NH_3)_2Cl_2]$.
(d) When the initial $[Pt(NH_3)_2Cl_2]$ is high, the rate of appearance of Cl^- is high. When the initial concentration is low, the rate of appearance of Cl^- is low. The rate of appearance of Cl^- is directly proportional to $[Pt(NH_3)_2Cl_2]$.

29. $\dfrac{\Delta[C]}{\Delta t} = (3.0 \times 10^{-3}\ min^{-1})[A]$

30. (a) Rate = k[I][II]
(b) $k = 1.04 \dfrac{L}{mol \cdot s}$

32. (a) The order with respect to NO is one and with respect to H_2 is one.
(b) The reaction is second-order.
(c) Rate = $k[NO][H_2]$
(d) $k = 2.4 \times 10^2\ L\ mol^{-1}\ s^{-1}$
(e) Initial rate = $1.5 \times 10^{-2}\ mol\ L^{-1}\ s^{-1}$

34. (a) First-order in A and third-order in B, fourth-order overall
(b) First-order in A and first-order in B, second-order overall
(c) First-order in A and zero-order in B, first-order overall
(d) Third-order in A and first-order in B, fourth-order overall

36. Equation (a) cannot be right

38. (a) $\dfrac{-\Delta[NH_2]}{\Delta t} = k$
(b) $2.5 \times 10^{-4}\ mol\ L^{-1}\ min^{-1}$
(c) Every concentration above zero.

39. (a) Rate = k[phenyl acetate]
(b) First-order in phenyl acetate
(c) $k = 1.259\ s^{-1}$ (d) $0.13 \dfrac{mol}{L \cdot s}$

40. $k = 0.0450 s^{-1}$; first order

42. (a) Rate = $k[NO_2][CO]$
(b) First-order in both NO_2 and CO
(c) $4.2 \times 10^8 \dfrac{L}{mol \cdot h}$

44. (a) Rate = $k[CH_3COCH_3][H_3O^+]$; first-order in both H_3O^+ and CH_3COCH_3, and zero-order in Br_2
(b) $4 \times 10^{-3} \dfrac{L}{mol \cdot s}$ (c) $2 \times 10^{-5} \dfrac{mol}{L \cdot s}$

46.

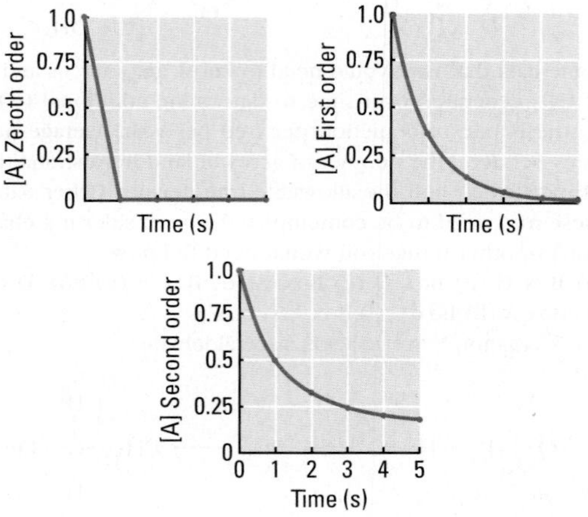

Only the first graph constructed here can be compared directly with one of the graphs in Figure 13.5—in particular, Figure 13.5(c). They both show a downward linear functional dependence of [A] versus time, which is characteristic of zero-order reactions.

48. 4.3 mg

50. 11.4 days
51. (a) 0.16 mol/L (b) 90. s (c) 120 s
53. $t = 5.49 \times 10^4$ s
55. (a) 0.02 M (b) 14 y (c) 5.55 y
56. (a) Not elementary (b) Bimolecular and elementary
(c) Not elementary (d) Unimolecular and elementary
58. $NO + O_3$, NO is an asymmetric molecule and Cl is a symmetric atom.
60. $E_a = 19$ kJ/mol, Ratio = 1.8
62. 10.7 times faster.
64. 3×10^2 kJ
65. (a) $E_a = 120.$ kJ/mol, $A = 1.22 \times 10^{14}\ s^{-1}$
(b) $k = 1.7 \times 10^{-3}\ s^{-1}$
67. (a) $E_a = 22.2$ kJ/mol, $A = 6.66 \times 10^7 \frac{L^2}{mol^2 \cdot s}$
(b) $k = 8.39 \times 10^4 \frac{L^2}{mol^2 \cdot s}$
69. (a) 3×10^{-20} (b) 4×10^{-16}
(c) 4×10^{-10} (d) 1.9×10^{-6}
71. (a) $1 \times 10^{-5} \frac{mol}{L \cdot s}$ (b) $2 \times 10^{1} \frac{mol}{L \cdot s}$
73. 3×10^2 kJ/mol
75. (a) rate = $k[Cl][ICl]$
(b) rate = $k[O][O_3]$
(c) rate = $k[NO_2]^2$
77. (a)

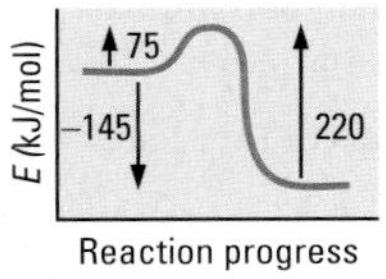

(a)

(b)

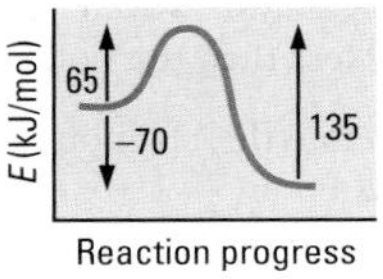

(b)

(c)

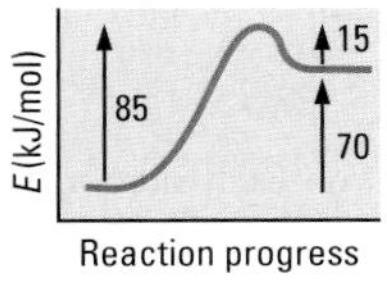

(c)

79. (a) Reaction (b) (b) Reaction (c)
81. (a) Reaction (c) (b) Reaction (a)
83. (a) Rate = $k[NO][NO_3]$ (b) Rate = $k[O][O_3]$
(c) Rate = $k[(CH_3)_3CBr]$ (d) Rate = $k[HI]^2$
85. (a) $NO_2 + F_2 \longrightarrow FNO_2 + \cancel{F}$
$+ NO_2 + \cancel{F} \longrightarrow FNO_2$
$2\ NO_2 + F_2 \longrightarrow 2\ FNO_2$
(b) The first step is rate-determining.
87. (a) $NO_2 + \cancel{NO_2} \longrightarrow NO + \cancel{NO_3}$
$+ \cancel{NO_3} + CO \longrightarrow \cancel{NO_2} + CO_2$
$NO_2 + CO \longrightarrow NO + CO_2$
(b) Bimolecular
(c) Rate = $k[NO_2]^2$
(d) NO_3
89. (a) Rate = $k'[NO]^2[Cl_2]$
(b) $NO + NO \rightleftharpoons N_2O_2$ fast
$N_2O_2 + Cl_2 \rightarrow 2\ NOCl$ slow
(c) $NO + Cl_2 \rightarrow NOCl + Cl$ slow
$NO + Cl \rightarrow NOCl$ fast
There can be other answers for (b) and (c).
91. (a) $CH_3COOH + H_2O \rightleftharpoons CH_3COOH + CH_3OH$
(b) rate = $k'[CH_3COOH][H_3O^+]$
(c) Catalyst: H_3O^+
(d) Intermediates: $H_3C(OH)OCH_3^+$, $H_3C(H_2O)(OH)OCH_3^+$, $H_3C(OH)_2OHCH_3^+$, and H_2O.
93. Only mechanism (a) is compatible with the observed rate law.
95. (a) $CH_3OH + Br^- + H^+ \rightarrow CH_3Br + H_2O$
(b)

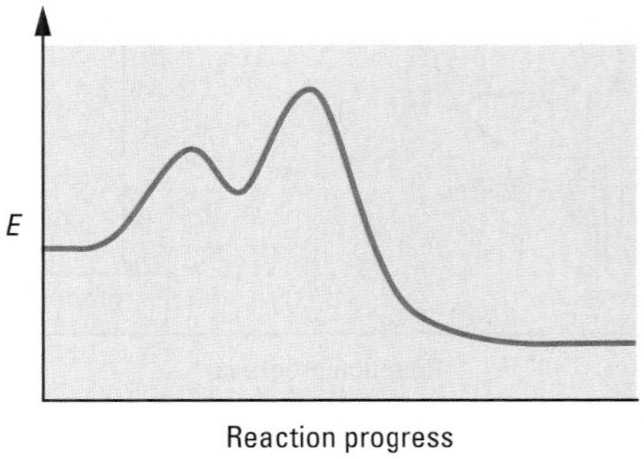

(c) Rate = $k_2\ [CH_3OH_2^+][Br^-]$ slow
Fast equilibrium, forward rate = reverse rate
$k_1\ [CH_3OH][H^+] = k_{-1}\ [CH_3OH_2^+]$
Substitute for intermediate
Rate = $k_2\left(\frac{k_1}{k_{-1}} = [CH_3OH][H^+]\right)[Br^-]$
Rate = $k\ [CH_3OH][H^+][Br^-]$
97. (a) is true.
(b) A catalyst is never consumed in a reaction.
(c) A catalyst need not be the same phase as the reactants.
(d) A catalyst can change the course of a reaction, but the same products are always formed.
99. (a) and (c); (a) homogeneous (c) heterogeneous
101. 38 times
102. Enzyme: A protein that catalyzes a biological reaction.
Cofactor: A small molecule that interacts with an enzyme to allow it to catalyze a biological reaction.
Polypeptides: Molecules made by polymerizing several amino acids.
Monomer: The small molecule reactant in the formation of a polymer.
Polysaccharide: An ether-linked chain of monomer sugars.
Lysozyme: An enzyme that speeds the hydrolysis of certain polysaccharides.
Substrate: The reactant of a reaction catalyzed by an enzyme.
Active site: That part of an enzyme where the substrate binds to the enzyme in preparation for conversion into products.
Proteins: Large polypeptide molecules that serve as structural and functional molecules in living organisms.
Inhibition: When a molecule that is not the substrate enters and occupies the active site of an enzyme preventing substrate binding and reaction.
104. 30. times faster
106. The active site on the enzyme (where the substrate binds to form an intermediate complex) is designed to accommodate the substrate's four terminal O atoms. Malonate and oxalate also have four terminal O atoms. When either of these two ions occupy the active site of an enzyme, they prevent the substrate from binding. With a smaller number of active sites free for reaction, the rate of the succinate dehydrogenation reaction decreases.
108. Rate = $k \frac{[X][HA]}{[A^-]}$
First-order with respect to HA
Doubling [HA] doubles the rate of the reaction
109. Catalysts make possible the production of vital products. Without them, many necessities and luxuries could not be made efficiently, if at all.
111. (a) To maximize the surface area and increase contact with the heterogeneous catalyst
(b) Catalysis happens only at the surface of the metal; strips or rods with less surface area would be less efficient and more costly.
113. Exothermic

114.

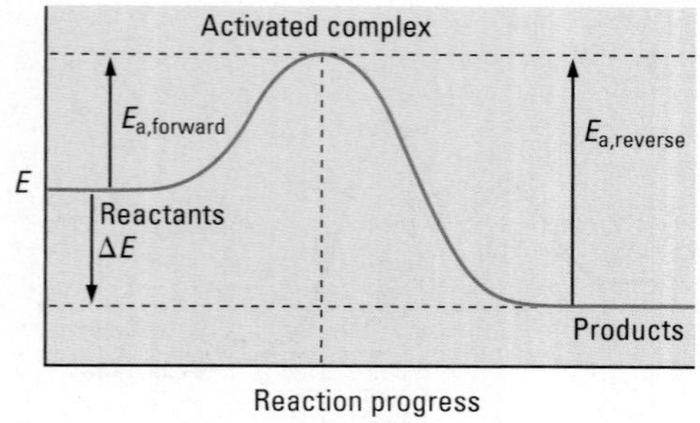

$\Delta E = E_{a,\ forward} - E_{a,\ reverse}$

116. (a)

Time	$[C_6H_{12}]$	$[C_{12}H_{10}]$	$[H_2]$
1.00	—	—	0.144
2.00	—	0.034	0.238
3.00	0.088	—	0.308

(b) Rate = 0.0369

118. (a) True
(b) False. "The reaction rate decreases as a first-order reaction proceeds at a constant temperature."
(c) True
(d) False. "As a second-order reaction proceeds at a constant temperature, the rate constant does not change."
(Other corrections for (b) and (d) are also possible.)

120. Rate $= k[H_2][NO]^2$

122. (a) NO is second-order, O_2 is first-order.
(b) Rate $= k[NO]^2[O_2]$ (c) $25 \dfrac{L^2}{mol^2 \cdot s}$
(d) $7.8 \times 10^{-4} \dfrac{mol}{L \cdot s}$ (e) $-\dfrac{\Delta[NO]}{\Delta t} = 2.0 \times 10^{-4} \dfrac{mol}{L \cdot s}$, $+\dfrac{\Delta[NO_2]}{\Delta t} = 2.0 \times 10^{-4} \dfrac{mol}{L \cdot s}$

124. (a) First-order in $HCrO_4^-$, first-order in H_2O_2, and first-order in H_3O^+
(b) Cancel intermediates, H_2CrO_4 and $H_2CrO(O_2)_2$, and add the three reactions.
(c) Second step

126. ***Mechanism 1:*** $rate_1 = rate_{-1}$
$k_1[NO][O_2] = k_{-1}[NO_3]$.
slow $rate_2 = k_2[NO_3][NO]$
$= k_2\left(\dfrac{k_1}{k_{-1}}[NO][O_2]\right)[NO]$
so, rate $= k[NO]^2[O_2]$
Mechanism 2: $rate_1 = rate_{-1}$
$k_1[NO]^2 = k_{-1}[N_2O_2]$.
slow $rate_2 = k_2[N_2O_2][O_2]$
$= k_2\left(\dfrac{k_1}{k_{-1}}[NO]^2\right)[O_2]$
so, rate $= k[NO]^2[O_2]$

128. The catalytic role of the chlorine atom (produced from the light decomposition of chlorofluorocarbons) in the mechanism for the destruction of ozone indicates that even small amounts of CFCs released into the atmosphere pose a serious risk.

130. Curve A represents $[H_2O(g)]$ increase with time, Curve B represents $[O_2(g)]$ increase with time, and Curve C represents $[H_2O_2(g)]$ decrease with time.

132. Snapshot (b)

134. Rate $= k[A]^3[B][C]^2$

136. (a) Three (b) Exothermic

137. E_a is very, very small—approximately zero.

139. The Pb must permanently bind with the Pt atoms on the surface of the metal occupying the active sites of the catalyst.

140. 94.7 s

142. (a) Rate $= 2.4 \times 10^{-7}$ mol L^{-1} s^{-1} $+ k[BSC][F^-]$
(b) 0.3 L mol^{-1} s^{-1}

144. 0.127 s^{-1}, 54.6 s

146. (a) 2.8×10^3 s (b) 1.4×10^4 s (c) 2.0×10^4 s

148. (a) $E_a = 3 \times 10^2$ kJ (b) $t = 2 \times 10^3$ s

150. Note: there are other correct answers to this question; the following are examples:

(a)	$CH_3CO_2CH_3 + H_3O^+ \longrightarrow CH_3COHOCH_3 + H_2O$	slow
	$CH_3COHOCH_3 + H_2O \longrightarrow CH_3COH(OH_2)OCH_3$	fast
	$CH_3COH(OH_2)OCH_3 \longrightarrow CH_3C(OH)_2 + CH_3OH$	fast
	$CH_3C(OH)_2 + H_2O \longrightarrow CH_3COOH + H_3O^+$	fast
(b)	$H_2 + I_2 \longrightarrow 2\ HI$	slow
(c)	$H_2 + Pt(s) \longrightarrow PtH_2$	fast
	$PtH_2 + I_2 \longrightarrow PtH_2I_2$	slow
	$PtH_2I_2 + O_2 \longrightarrow PtI_2O + H_2O$	fast
	$PtI_2O + H_2 \longrightarrow Pt(s) + I_2 + H_2O$	fast
(d)	$H_2 \longrightarrow 2\ H$	fast
	$H + CO \longrightarrow HCO$	slow
	$HCO + H \longrightarrow H_2CO$	fast

152. Approximately 26 times faster

154. Estimated $E_a = 402 \times 10^{-21}$ J/molecule; it is the same to one significant figure as the E_a given in Figure 13.7.

156. The speed of the reaction is related to the energy of the production of a photon that can break a bond in iodine and make a free radical catalyst. Wavelengths less than 800 nm have photons with energy less than the bond enthalpy of I_2.

158. (i) Define the reaction rate in terms of [A]: Rate $= -\dfrac{\Delta[A]}{\Delta t}$
(ii) Write the rate law using [A], k and t: $-\dfrac{\Delta[A]}{\Delta t} = k[A]$.
Calculus uses "d" instead of "Δ": $-\dfrac{d[A]}{dt} = k[A]$.
(iii) Separate the variables: $\dfrac{d[A]}{[A]} = -kdt$
(iv) Do a separate definite integral on each side of the equation:
$$\int_0^t \frac{d[A]}{[A]} = -k\int_0^t dt$$
$\ln[A]_t - \ln[A]_0 = -k(t_t - t_0)$
(v) With $t = t_t - t_0$:
$\ln[A]_t - \ln[A]_0 = -kt$
$\ln[A]_t = -kt + \ln[A]_0$
The above equation is in the proper form, with $y = \ln[A]_t$, $m = -k$, $x = t$, and $b = \ln[A]_0$.
(vi) $t = t_{1/2}$, when $[A]_t = \dfrac{1}{2}[A]_0$
$\ln\dfrac{1}{2}[A]_0 - \ln[A]_0 = -kt_{1/2}$
$\ln[A]_0 - \ln 2 - \ln[A]_0 = -kt_{1/2}$
$-\ln 2 = -kt_{1/2}$
$t_{1/2} = \dfrac{-\ln 2}{-k} = \dfrac{\ln 2}{k}$

Chapter 14

8. There are many answers to this question; this is an example: Prepare a sample of N_2O_4 in which the N atoms are the heavier isotopes ^{15}N. Introduce the heavy isotope of N_2O_4 into an equilibrium mixture of N_2O_4 and NO_2. Use spectroscopic methods, such as infrared spectroscopy to observe the distribution of the radioisotope among the reactants and products.

10. (a) 0 °C (b) Dynamic equilibrium; Molecules at the interface between the water and the ice may detach from the ice and enter the liquid phase or may attach to the solid phase leaving the liquid phase.

12.

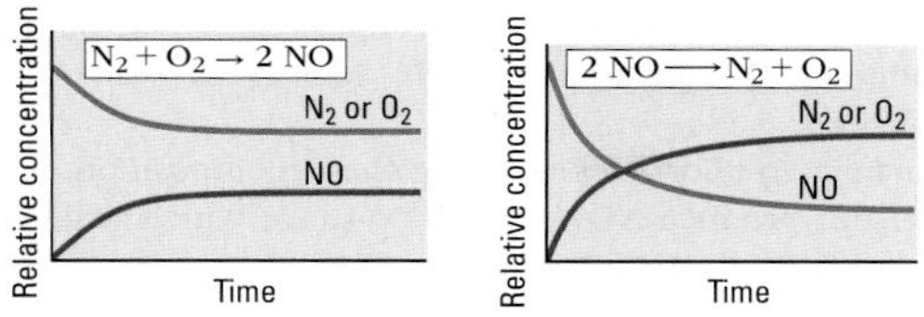

14. (a) $K = \frac{[O_3]^2}{[O_2]^3}$ (b) $K = \frac{[Fe(CO)_5]}{[CO]^5}$

(c) $K = [NH_3]^2[CO_2][H_2O]$

(d) $K = [Ag^+][SO_4^{2-}]$

16. (a) $K_c = \frac{[CO][H_2O]}{[CO_2][H_2]}$ (b) $K_c = [CO_2]$

(c) $K_c = \frac{[CH_4]^2[Cl_2]^3}{[CHCl_3]^2[H_2]^3}$ (d) $K_c = \frac{[HI]^2}{[H_2][I_2]}$

(e) $K_c = \frac{[NO]^2[Cl_2]}{[NOCl]^2}$

17. (a) $K_c = \frac{[H_2O]^2[O_2]}{[H_2O_2]^2}$ (b) $K_c = \frac{[PCl_5]}{[PCl_3][Cl_2]}$

(c) $K_c = [CO]^2$ (d) $K_c = \frac{[H_2S]}{[H_2]}$

19. (a) $K = \frac{[Ni(CO)_4]}{[CO]^4}$ (b) $K = \frac{[H_3O^+][F^-]}{[HF]}$

(c) $K = \frac{[Cl^-]^2}{[Cl_2][Br^-]^2}$

21. (a) $K_c = [PCl_3][Cl_2]$ (b) $K_c = \frac{[CoCl_4^{2-}]}{[Co(H_2O)_6^{2+}][Cl^-]^4}$

(c) $K_c = \frac{[CH_3COO^-][H_3O^+]}{[CH_3COOH]}$

(d) $K_c = \frac{[OF_2][HF]^2}{[F_2]^2[H_2O]}$

23. (a) $K_c = [H_2O]^2$ (b) $K_c = \frac{[HF]^4}{[SiF_4][H_2O]^2}$

(c) $K_c = \frac{[HCl]^2}{[H_2O]}$

25. (a) $2\ SO_2(g) + 2\ H_2O(g) \longrightarrow 2\ H_2S(g) + 3\ O_2(g)$

(b) $\frac{1}{2}\ F_2(g) + \frac{1}{2}\ I_2(g) \longrightarrow IF(g)$

(c) $Cl_2(g) + 2\ Br^-(aq) \longrightarrow Br_2(\ell) + 2\ Cl^-(aq)$

27. Equation (e)
29. (a) 0.87 (b) 1.3
31. 48
33. (a) $K_P = \frac{P^2_{H_2O}P_{O_2}}{P^2_{H_2O_2}}$ (b) $K_P = \frac{P_{PCl_5}}{P_{PCl_3}P_{Cl_2}}$

(c) $K_P = P^2_{CO}$ (d) $K_P = \frac{P_{H_2S}}{P_{H_2}}$

35. 3.0×10^{20}
37. $K_P = 0.0161$
39. 1.4×10^2
41. 2.6×10^{-9}
43. $K_P = 1.6$
45. (a) $K_c = 1.0$ (b) $K_c = 1.0$ (c) $K_c = 1$
47. $K_P = 0.108$
48. $K_c = 0.075$
50. $K_c = 75$
52. 3.9×10^{-4}
54. 1.4×10^3
55. Reactions (b) and (c) are product-favored. Most reactant-favored is (a), then (b), and then (c).
57. (d)
58. (a) $Ag_2SO_4(s) \rightleftharpoons 2\ Ag^+(aq) + SO_4^{2-}(aq)$

$K = [Ag^+]^2[SO_4^{2-}] = 1.7 \times 10^{-5}$

$Ag_2S(s) \rightleftharpoons 2\ Ag^+(aq) + S^{2-}(aq)$

$K = [Ag^+]^2[S^{2-}] = 6 \times 10^{-30}$

(b) $Ag_2SO_4(s)$ (c) $Ag_2S(s)$

60. (a)

	butane $\rightleftharpoons$	2-methylpropane
Conc. initial	0.100 mol/L	0.100 mol/L
Change conc.	$-x$	$+x$
Equilibrium conc.	$0.100 - x$	$0.100 + x$

(b) $K_c = \frac{[\text{2-methylpropane}]}{[\text{butane}]} = 2.5 = \frac{0.100 + x}{0.100 - x}$ $x = 0.043$

(c) [2-methylpropane] = 0.024 mol/L, [butane] = 0.010 mol/L

62. 3.39 g $C_6H_{12}(g)$
64. $[HI] = 4 \times 10^{-2}$ M; $[H_2] = [I_2] = 5 \times 10^{-3}$ M
66. $[Br_2] = [F_2] = 0.0466$ mol/L; $[BrF] = 0.347$ mol/L
67. (a) 1.94 mol HI (b) 1.92 mol HI (c) 1.98 mol HI
69. (a) $[CO] = [H_2O] = 0.95$ mol/L, $[CO_2] = [H_2] = 0.0489$ mol/L

(b) $[CO] = [H_2O] = 1.90$ mol/L, $[CO_2] = [H_2] = 0.0977$ mol/L

71. (a) $[COBr_2] = 0.107$ mol/L; $[CO] = [Br_2] = 0.143$ mol/L

(b) 57.1%

72. (a) No (b) Proceed toward products
73. No; increases
76. (a) No (b) Proceed toward reactants

(c) $[N_2] = [O_2] = 0.025$ M; $[NO] = 0.0010$ M

78. (a) No (b) Toward reactants
80. (a) Left (b) Left (c) Left (d) Right
82. Choice (b); because heat is a reactant, increasing the temperature drives the reaction toward the products.
84. (a) Left (b) Right (c) No shift
85. (a) Reverse (b) Forward (c) Forward
87.

Change	$[Br_2]$	[HBr]	K_c	K_P
Some H_2 is added to the container.	Decrease	Increase	No change	No change
Temperature of the gases in the container is increased.	Increase	Decrease	Decrease	Decrease
The pressure of HBr is increased.	Increase	Increase	No change	No change

89. (a) No change (b) Left (c) Left
91. (a) Right (b) Left (c) Right
93. (a) (i) (b) (ii) (c) (i) (d) (iii) (e) (iii)
95. (a) (1) Increases (2) Increases (3) No change (4) Increases (5) Increases

(b) (5) increases K.

97. $[PCl_5] = 0.0198$ M; $[PCl_3] = 0.0231$ M; $[Cl_2] = 0.0403$ M
99. (a) Energy effect (b) Entropy effect (c) Neither
101. (a) Insufficient information is available.

(b) Greater than 1, products favored

(c) Less than 1, reactants favored

103. A reaction will only go significantly toward products if it is product-favored. If a reaction is accompanied by an increase in entropy (a favorable entropy effect) and the products are lower in energy (a favorable enthalpy effect), then products are favored. However, if the reaction is favored only by an entropy effect, then it needs high temperatures to assist the endothermic process. If the reaction is favored only by an enthalpy effect, then it needs low temperatures to keep randomness at a minimum.
105. (a) First reaction: $\Delta H = -296.830$ kJ, second reaction: $\Delta H = -197.78$ kJ, third reaction: $\Delta H = -132.44$ kJ

(b) All three are exothermic, none are endothermic.

(c) None of the reactions have entropy increase, the second and third reaction have entropy decrease, and the first reaction entropy is about the same.

(d) Low temperature favors all three reactions.

107. $K_c = [H^+][OH^-]$, $K_c = \dfrac{[CH_3COO^-][H^+]}{[CH_3COOH]}$,

$K_c = \dfrac{[NH_3]^2}{[N_2][H_2]^3}$, $K_c = \dfrac{[CO_2][H_2]}{[CO][H_2O]}$

First reaction

(a) $(1.0 + x)(1.0 + x) = 1.0 \times 10^{-14}$

(b) Quadratic (c) N/A

Second reaction

(a) $\dfrac{(1.0 + x)(1.0 + x)}{(1.0 - x)} = 1.8 \times 10^{-5}$

(b) Quadratic (c) N/A

Third reaction

(a) $\dfrac{(1.0 + 2x)^2}{(1.0 - x)(1.0 - 3x)^3} = 3.5 \times 10^8$

(b) Not quadratic

(c) Use approximation techniques

Fourth reaction

(a) $\dfrac{(1.0 + x)(1.0 + x)}{(1.00 - x)(1.00 - x)} = 4.00$

(b) This equation is quadratic.

(c) Not applicable

109. (a) negative

(b) 4.0×10^2

(c) 0.25%

110. (a)

Species	Br_2	Cl_2	F_2	H_2	N_2	O_2
[*E*] (mol/L)	0.28	0.057	1.44	1.76×10^{-5}	4×10^{-14}	4.0×10^{-6}

(b) F_2; (At this temperature, the lowest bond energy is predicted from the reaction that gives the most products.) Compared to Table 8.2, the product production decreases as the bond energy increases: 156 kJ F_2, 193 kJ Br_2, 242 kJ Cl_2, 436 kJ H_2, 498 kJ O_2, 946 kJ N_2. Lewis structures of F_2, Br_2, Cl_2, H_2 have a single bond and more products are produced than O_2 with double bond and N_2 with triple bonds.

111. $P_{N_2O_4} = 1.1$ atm; $P_{NO_2} = 0.397$ atm

114. 2.3 atm

116. (a) 0.927 atm (b) 0.0420 mol

118. Cases a and b will have decreased [HI] at equilibrium, and Cases c and d will have increased [HI].

120. (a) 0.18

(b) $P_{CO} = 1.02$ atm; $P_{Br_2} = 0.52$ atm; $P_{COBr_2} = 0.10$ atm

123. It is at equilibrium at 600. K. No more experiments are needed.

125. Le Chatlier's principle says the equilibrium will shift to compensate for changes made. The pressure causes the H_2O molecules to be pushed together. Because water is more dense than ice, some of the ice will melt to the liquid state, reducing the pressure.

127. (a) (iii) (b) (i)

129. In the warmer sample, the molecules would be moving faster and more NO_2 molecules would be seen. In the cooler sample, the molecules would be moving somewhat slower and fewer NO_2 molecules would be seen. In both samples, the molecules are moving very fast. The average speed of gas molecules is commonly hundreds of miles per hour. In both samples, one would see a dynamic equilibrium with some N_2O_4 molecules decomposing and some NO_2 molecules reacting with each other, at equal rates.

131. Diagrams (b), (c), and (d)

133. The following is an example of a diagram. (Other answers are possible.)

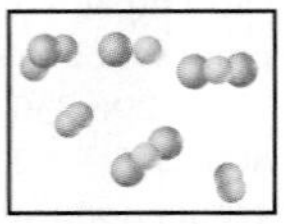

135. B

137. Dynamic equilibria with small values of K introduce a small amount of D^+ ions in place of H^+ ions in the place of the acidic hydrogen.

139. Dynamic equilibria representing the decomposition of the dimer $N_2O_4(g)$ produce $NO_2(g)$ and $^*NO_2(g)$, which will occasionally recombine into the mixed dimer, $O_2N^*{-}NO_2(g)$

$O_2N{-}NO_2(g) \rightleftharpoons 2\ NO_2(g)$

$O_2N^*{-}^*NO_2(g) \rightleftharpoons 2\ ^*NO_2(g)$

$O_2N^*{-}NO_2(g) \rightleftharpoons {}^*NO_2(g) + NO_2(g)$

141. 1.3 atm

143. (a) 1.00×10^{-6} (b) 0.089 mol

145. (a) $[PCl_3] = [Cl_2] = 0.14$ mol/L, $[PCl_5] = 0.005$ mol/L

(b) $[PCl_3] = 0.29$ mol/L, $[Cl_2] = 0.14$ mol/L, $[PCl_5] = 0.01$ mol/L

147. (a) 56

(b) Before, $P_{TOT} = 5.2$ atm; after $P_{TOT} = 5.2$ atm

(c) $P_{H_2} = P_{I_2} = 0.55$ atm; $P_{HI} = 4.1$ atm

149. $K_c = \dfrac{A_f e^{(-E_{a,r}/RT)}}{A_r e^{(-E_{a,r}/RT)}} = \dfrac{A_f}{A_r} e^{[(E_{a,r} - E_{a,f})/RT]}$

When a catalyst is added, the activation energies of both the forward and reverse reactions get smaller:

$K_{c,\ cat} = \dfrac{A_{f,cat}}{A_{r,cat}} e^{[(E_{a,r,cat} - E_{a,f,cat})/RT]}$

The frequency factors did not change and, because the reaction has to conserve energy, the E_a values must each be reduced by the same amount of energy X:

$A_{f,cat} = A_f$ $E_{a,f,cat} = E_{a,f} - X$

$A_{r,cat} = A_r$ $E_{a,r,cat} = E_{a,r} - X$

Notice that

$E_{a,f,cat} - E_{a,r,cat} = (E_{a,f} - X) - (E_{a,r} - X)$
$= E_{a,f} - X - E_{a,r} + X = E_{a,f} - E_{a,r}$

Substituting into the $K_{c,cat}$ equation gives

$K_{c,cat} = \dfrac{A_f}{A_r} e^{[(E_{a,r} - E_{a,f})/RT]} = K_c$

Therefore, the equilibrium state, as described quantitatively by the equilibrium constant, does not change with the addition of a catalyst.

150. (a) $2\ NO_2(g) + Br_2(g) \longrightarrow 2\ NOBr(g)$

(b) Rate forward, step 1 = rate reverse, step 1

$k_1[NO][Br_2] = k_{-1}[NOBr_2]$

$K_{c,step1} = \dfrac{k_1}{k_{-1}} = \dfrac{[NOBr_2]}{[NO][Br_2]}$

Rate forward, step 2 = rate reverse, step 2

$k_2[NOBr_2][NO] = k_{-2}[NOBr]^2$

$K_{c,step2} = \dfrac{k_2}{k_{-2}} = \dfrac{[NOBr]^2}{[NOBr_2][NO]}$

(c) $K_{c,overall} = \dfrac{[NOBr]^2}{[NO]^2[Br_2]} = \left(\dfrac{[NOBr_2]}{[NO][Br_2]}\right)\left(\dfrac{[NOBr]^2}{[NO][NOBr_2]}\right)$

$= \left(\dfrac{k_1}{k_{-1}}\right)\left(\dfrac{k_2}{k_{-2}}\right) = \dfrac{k_1k_2}{k_{-1}k_{-2}}$

(d) Yes

151. (a) Not at equilibrium; right

(b) Decrease

153. The $PbCl_2$ beaker

155. (a) $K_p = 0.03126$; $K_c = 0.0128$

(b) $K_P = 1$

(c) $K_c = 1/(RT_{bp})$

157. 1.15%

Chapter 15

25. If the solid interacts with the solvent using similar (or stronger) intermolecular forces, it will dissolve readily. If the solute interacts with the solvent using different intermolecular forces than those experienced in the solvent, it will be almost insoluble. For example, consider dissolving an ionic solid in water and oil. The interactions between the ions in the solid and water are very strong, since ions would be attracted to the highly polar water molecule; hence, the solid would have a high solubility. However, the ions in the solid interact with each other much more strongly than the London dispersion forces experienced between the non-polar hydrocarbons in the oil; hence, the solid would have a low solubility. (Other examples exist.)
27. Beaker (c)
29. The dissolving process was endothermic, so the temperature dropped as more solute was added. The solubility of the solid at the lower temperature is lower, so some of the solid did not dissolve. As the solution warmed, however, the solubility increased again. What remained of the solid dissolved. The solution was saturated at the lower temperature, but is no longer saturated at the current temperature.
30. Exothermic
31. When an organic acid has a large (non-polar) piece, it interacts primarily using London dispersion intermolecular forces. Since water interacts via hydrogen-bonding intermolecular forces, it would rather interact with itself than with the acid. Hence, the solubility of the large organic acids drops, and some are completely insoluble.
33. The positive H end of the very polar water molecule interacts with the negative ions. The negative O end of the very polar water molecule interacts with the positive ions.

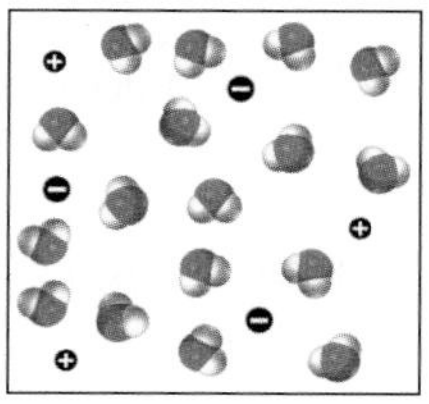

35. 1×10^{-3} M
38. 0.00732% by weight
40. $1 \text{ ppb} = \dfrac{1 \text{ g part}}{10^9 \text{ g whole}} \times \dfrac{1 \text{ μg part}}{10^{-6} \text{ g part}} \times \dfrac{1000 \text{ g whole}}{1 \text{ kg whole}} = \dfrac{1 \text{ μg part}}{1 \text{ kg whole}}$
42. 90. g ethanol
45. 1.6×10^{-6} g Pb
47. (a) 160. g NH_4 Cl (b) 83.9 g KCl (c) 7.46 g Na_2SO_4
48. (a) 0.762 M (b) 0.174 M (c) 0.0126 M (d) 0.167 M
51. 96% H_2SO_4
53. 0.1-M NaCl
55. 59 g
57. 1.2×10^{-7} M
60. (a) 108 ppm
(b) 5.64×10^{-4} mol/L
(c) 5.59×10^{-4} mol/kg
63. Curve (a). The addition of solute lowers vapor pressure of the solvent.
65. 100.26 °C
67. (a) < (d) < (b) < (c)
69. $T_f = -1.65$ °C, $T_b = 100.46$ °C
71. $X_{H_2O} = 0.79999$, 712 g sucrose
73. 1.9×10^2 g/mol
75. 3.6×10^2 g/mol; $C_{20}H_{16}Fe_2$
77. 1.8×10^2 g/mol, $C_{14}H_{10}$
79. (a) 2.5 kg (b) 104.2 °C
81. 29 atm
84. 5×10^{-4} g As
86. No
88. The lime-soda process relies on the precipitation of insoluble compounds to remove the "hard water" ions. The ion-exchange process relies on the high charge of the "hard water" ions to attract them to an ion-exchange resin, thereby removing them from the water.
90. Water is sprayed into the air to oxidize organic substances dissolved in it.
92. Fish take in oxygen by extracting it from the water, while plants take in carbon dioxide by extracting it from the water. The concentrations of these gases in calm water drop unless they are replenished. The concentration of dissolved gases in the water is replenished by bubbling air through the water in the aquarium.
94. 4%
96. Water in the cells of the wood leaked out, since the osmotic pressure inside the cells was less than that of the seawater in which the wood was sitting.
98. 28% NH_3
100. 0.982 *m*, 10.2%
102. (a) = (c) < (d) < (b)
104. 1.77 g
106. (a) $Na_2SO_4(aq) + Ba(NO_3)_2(aq) \rightarrow 2\ NaNO_3(aq) + BaSO_4(s)$
(b) 15.0 mL (c) 12.0 mL
108. (a)

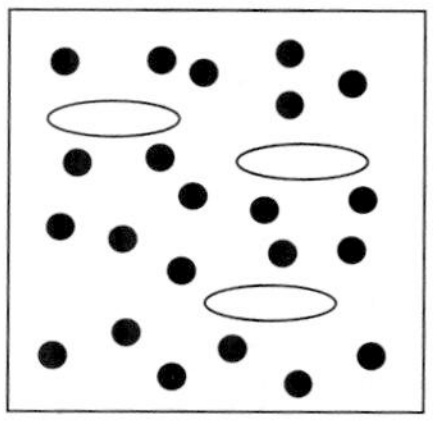

(b)

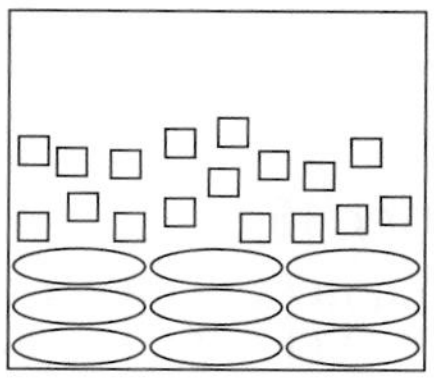

110. (a) Unsaturated (b) Supersaturated
(c) Supersaturated (d) Unsaturated
112.

Compound	Mass of Compound	Mass of Water	Mass Fraction of Solute	Weight Percent of Solute	ppm of Solute
Lye	75.0 g	125 g	0.375	37.5%	3.75×10^5
Glycerol	33 g	200. g	0.14	14%	1.4×10^5
Acetylene	0.0015	2×10^2 g	0.000009	0.0009%	9

114. Molecules slow down. The reduced motion prevents them from randomly translocating as they had in the liquid state. As a result, the intermolecular forces between one molecule and the next begin to organize them into a crystal form. The presence of a nonvolatile solute disrupts the formation of the crystal. Its size and shape will be different from that of the solute. Intermolecular forces between the solute and solvent are also different from those of solvent molecules with each other. To form the crystalline solid, the solute has to be excluded. If the ice in an iceberg is in regular crystalline form, the water will be pure. Only if the ice is crushed or dirty will other particles be included. So, melting an iceberg will produce relatively pure water.
116. (a) Seawater contains more dissolved solutes than fresh water. The presence of a solute lowers the freezing point. That means a lower temperature is required to freeze the seawater than to freeze fresh water.
(b) Salt added to a mixture of ice and water will lower the freezing point of the water. If the ice cream is mixed at a lower temperature, its temperature will drop faster; hence, it will freeze faster.
119. The empirical and molecular formulas are both $C_{18}H_{24}Cr$.
121. (a) No (b) 108.9°
123. 28 m
125. 0.30 M

127. (a)

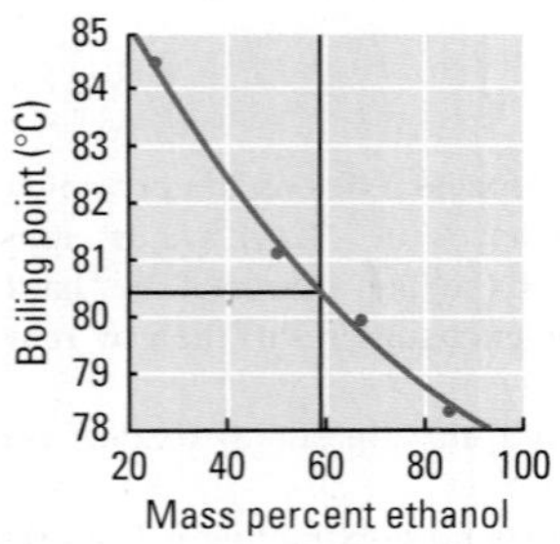

Approx. 80.4 °C
(b) 12.3 M; 44.1 mol/kg

129. (a) Freezes at −3.36°, when the solute concentration is 0.50 *m*.

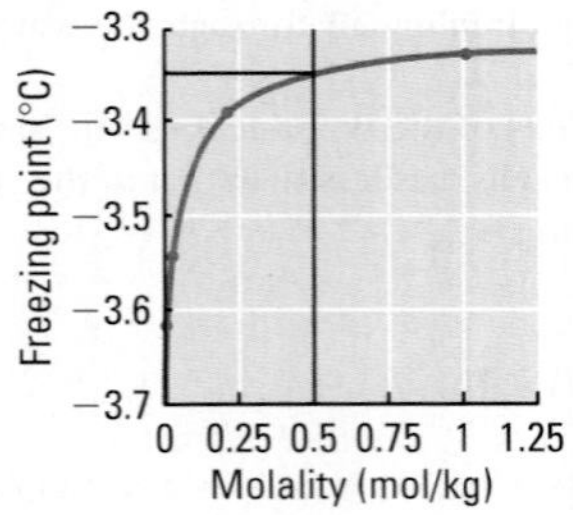

(b) 39,000%, 9,400%, 810%, 80%

131. (a) 6300 ppm, 6300000 ppb
(b) 0.040 M (c) 4.99×10^5 bottles

Chapter 16

12. (a) $HCO_3^- + H_2O \rightleftharpoons CO_3^{2-} + H_3O^+$
(b) $HCl + H_2O \rightleftharpoons Cl^- + H_3O^+$
(c) $CH_3COOH + H_2O \rightleftharpoons CH_3COO^- + H_3O^+$
(d) $HCN + H_2O \rightleftharpoons CN^- + H_3O^+$

14. (a) $HIO + H_2O \rightleftharpoons IO^- + H_3O^+$
(b) $CH_3(CH_2)_4COOH + H_2O \rightleftharpoons CH_3(CH_2)_4COO^- + H_3O^+$
(c) $HOOCCOOH + H_2O \rightleftharpoons HOOCCOO^- + H_3O^+$
$HOOCCOO^- + H_2O \rightleftharpoons {}^-OOCCOO^- + H_3O^+$
(d) $CH_3NH_3^+ + H_2O \rightleftharpoons CH_3NH_2 + H_3O^+$

16. (a) $HSO_4^- + H_2O \rightleftharpoons H_2SO_4 + OH^-$
(b) $CH_3NH_2 + H_2O \rightleftharpoons CH_3NH_3^+ + OH^-$
(c) $I^- + H_2O \rightleftharpoons HI + OH^-$
(d) $H_2PO_4^- + H_2O \rightleftharpoons H_3PO_4 + OH^-$

20. (a) H_2SO_4 (b) HNO_3 (c) $HClO_4$
(d) $HClO_3$ (e) H_2SO_4

22. (a) I^-, iodide, conjugate base
(b) HNO_3, nitric acid, conjugate acid
(c) HCO_3^-, hydrogen carbonate ion, conjugate acid
(d) HCO_3^-, hydrogen carbonate ion, conjugate base
(e) SO_4^{2-}, sulfate ion, conjugate base, H_2SO_4, sulfuric acid, conjugate acid
(f) HSO_3^-, hydrogen sulfite ion, conjugate acid

24. Pairs (b), (c), and (d)

26. (a) Reactant acid = H_2O, reactant base = HS^-, product conjugate acid = H_2S, product conjugate base = OH^-
(b) Reactant acid = NH_4^+, reactant base = S^{2-}, product conjugate acid = HS^-, product conjugate base = NH_3
(c) Reactant acid = HSO_4^-, reactant base = HCO_3^-, product conjugate acid = H_2CO_3, product conjugate base = SO_4^{2-}
(d) Reactant acid = NH_3, reactant base = NH_2^-, product conjugate base = NH_2^-, product conjugate acid = NH_3

28. (a) Reactant acid = CH_3COOH, reactant base = CN^-, product conjugate acid = HCN, product conjugate base = CH_3COO^-
(b) Reactant acid = H_2O, reactant base = O^{2-}, product conjugate acid = OH^-, product conjugate base = OH^-
(c) Reactant acid = H_2O, reactant base = HCO_2^-, product conjugate acid = HCOOH, product conjugate base = OH^-

31. (a) $CO_3^{2-} + H_2O \rightleftharpoons HCO_3^- + OH^-$
$HCO_3^- + H_2O \rightleftharpoons H_2CO_3 + OH^-$
(b) $H_3AsO_4 + H_2O \rightleftharpoons H_2AsO_4^- + H_3O^+$
$H_2AsO_4^- + H_2O \rightleftharpoons HAsO_4^{2-} + H_3O^+$
$HAsO_4^{2-} + H_2O \rightleftharpoons AsO_4^{3-} + H_3O^+$
(c) $NH_2CH_2COO^- + H_2O \rightleftharpoons {}^+NH_3CH_2COO^- + OH^-$
${}^+NH_3CH_2COO^- + H_2O \rightleftharpoons {}^+NH_3CH_2COOH + OH^-$

33. 3×10^{-11} M, basic

35. 3.6×10^{-3} M, acidic

37. pH = 12.40, pOH = 1.60

39. pOH = 12.51

41. 2 g HCl

43.

pH	$[H_3O^+]$ (M)	$[OH^-]$ (M)	Acidic or Basic
(a) 6.21	6.1×10^{-7}	1.6×10^{-8}	Acidic
(b) 5.34	4.5×10^{-6}	2.2×10^{-9}	Acidic
(c) 4.67	2.1×10^{-5}	4.7×10^{-10}	Acidic
(d) 1.60	2.5×10^{-2}	4.0×10^{-13}	Acidic
(e) 9.12	7.6×10^{-10}	1.3×10^{-5}	Basic

45. (a) 100 times more acidic (b) 10^4 times more basic
(c) 16 times more basic (d) 2×10^4 times more acidic

47. (a) 5.0×10^{-9} M (b) Basic

49. (a) $F^-(aq) + H_2O(\ell) \rightleftharpoons HF(aq) + OH^-(aq)$ $K = \frac{[HF][OH^-]}{[F^-]}$
(b) $NH_3(aq) + H_2O(\ell) \rightleftharpoons NH_4^+(aq) + OH^-(aq)$
$K = \frac{[NH_4^+][OH^-]}{[NH_3]}$
(c) $H_2CO_3(aq) + H_2O(\ell) \rightleftharpoons HCO_3^-(aq) + H_3O^+(aq)$
$K = \frac{[HCO_3^-][H_3O^+]}{[H_2CO_3]}$
(d) $H_3PO_4(aq) + H_2O(\ell) \rightleftharpoons H_2PO_4^-(aq) + H_3O^+(aq)$
$K = \frac{[H_2PO_4^-][H_3O^+]}{[H_3PO_4]}$
(e) $CH_3COO^-(aq) + H_2O(\ell) \rightleftharpoons CH_3COOH(aq) + OH^-(aq)$
$K = \frac{[CH_3COOH][OH^-]}{[CH_3COO^-]}$
(f) $S^{2-}(aq) + H_2O(\ell) \rightleftharpoons HS^-(aq) + OH^-(aq)$
$K = \frac{[HS^-][OH^-]}{[S^{2-}]}$

51. (a) NH_3 (b) K_2S (c) $NaCH_3COO$ (d) KCN

54. (a) 2.19 (b) 5.13 (c) 8.09 (d) 9.02
(e) 13.30 (f) 1.55 (g) 9.68 (h) 7.00

56. $K_a = 1.6 \times 10^{-5}$

58. $[H_3O^+] = [A^-] = 1.3 \times 10^{-5}$ M, [HA] = 0.040 M

60. $K_a = 1.4 \times 10^{-5}$

62. 8.83

64. $K_b = 1.3 \times 10^{-3}$

66. (a) $C_{10}H_{15}NH_2 + H_2O \rightleftharpoons C_{10}H_{15}NH_3^+ + OH^-$
(b) 10.48

68. 3.28

72. (a) CN^-, product-favored (b) HS^-, reactant-favored
(c) $H_2(g)$, reactant-favored

74. (a) $NH_4^+ + HPO_4^{2-} \rightarrow NH_3 + H_2PO_4^-$, reactant-favored
(b) $CH_3COOH + OH^- \rightarrow CH_3COO^- + H_2O$, product-favored
(c) $HSO_4^- + H_2PO_4^- \rightarrow H_3PO_4 + SO_4^{2-}$, product-favored
(d) $CH_3COOH + F^- \rightarrow CH_3COO^- + HF$, reactant-favored

76. (a) pH < 7 (b) pH > 7 (c) pH = 7

78. (a) pH > 7 (b) pH > 7 (c) pH > 7

80. (a), (b), (c), (d), and (e)

82. All three are Lewis bases; CO_2 is a Lewis acid.

84. Cr^{3+} and SO_3 are Lewis acids. CH_3NH_2 is a Lewis base.

86. (a) I_2 is a Lewis acid and I^- is a Lewis base.
(b) BF_3 is a Lewis acid and SO_2 is a Lewis base.
(c) Au^+ is a Lewis acid and CN^- is a Lewis base.
(d) CO_2 is a Lewis acid and H_2O is a Lewis base.

88.

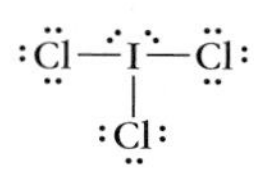

T-shaped

It functions as a Lewis acid.

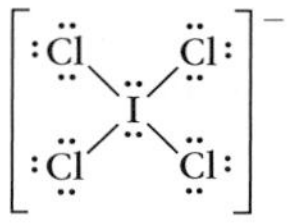

Square planar

90. $[Zn(H_2O)(OH)_3]^-$
92. $Na_2CO_3 + 2\ CH_3(CH_2)_{16}COOH \rightarrow$
$2\ CH_3(CH_2)_{16}COONa + H_2O + CO_2$
94. Dishwasher detergent is very basic and should not be used to wash anything by hand, including a car. If it gets into the engine area, it can also dissolve automobile grease and oil, which could prevent the engine from running correctly.
96. Lemon juice is acidic and neutralizes the basic amines. The acid formed from the neutralized base is an ion and not volatile.
98. (a) Weak acid (b) Strong base (c) Strong acid
(d) Weak base (e) Strong base (f) Amphiprotic
100. (a) Less than 7 (b) Equal to 7 (c) Greater than 7
102. 2.85
104. (a) Increase (b) Stays the same (c) Stays the same
106. 9.59
108. At 10 °C, pH = 7.27; at 25 °C, pH = 6.998; and at 50 °C, pH = 6.631. At 10 °C, at 25 °C, and at 50 °C the solutions are neutral, since $[H_3O^+] = [OH^-]$.
110. (a) $H_2O > Cl^- = H_3O^+ >> OH^-$
(b) $H_2O > Na^+ = ClO_4^- >> H_3O^+ = OH^-$
(c) $H_2O > HNO_2 > H_3O^+ = NO_2^- >> OH^-$
(d) $H_2O > Na^+ \cong ClO^- > OH^- = HClO >> H_3O^+$
(e) $H_2O > NH_4^+ \cong Cl^- > H_3O^+ = NH_3 >> OH^-$
(f) $H_2O > Na^+ = OH^- >> H_3O^+$
112. Conjugates must differ by just one H^+.
114. Arrhenius theory: Electron pairs on the solvent water molecules (Lewis base) form a bond with the hydrogen ion (Lewis acid) producing aqueous H_3O^+ ions. Electron pairs on the OH^- ions (Lewis base) form a bond with the hydrogen ion (Lewis acid) in the solvent water molecule, producing aqueous OH^- ions.
Brønsted-Lowry theory: The H^+ ion from the Brønsted-Lowry acid is bonded to a Brønsted-Lowry base using an electron pair on the base. The electron-pair acceptor, the H1 ion, is the Lewis acid and the electron-pair donor is the Lewis base.
116. (a) 2.01 (b) 23 times
119. (a) HY; the calculated K_a is largest.
(b) Z^-; the smallest K_a acid has the strongest conjugate base.
122. (a) Boxes C, F, and G (b) Box I (c) Boxes A and H
(d) Boxes D and E (e) Boxes B and E
(f) Box I (g) Boxes D and E
124. Lactic acid sample
125. 0.76 L; No
126. Yes, higher pH
128. Br has a higher electronegativity than H. $ClNH_2$ is weaker, because Cl has a higher electronegativity than Br.
130. Strongest HM > HQ > HZ weakest; $K_{a,HZ} = 1 \times 10^{-5}$, $K_{a,HQ} = 1 \times 10^{-3}$, $K_{a,HM} = 1 \times 10^{-1}$ or larger
132. (a) Weak (b) Weak
(c) Acid/base conjugates (d) 6.29
134. Dilute solution is 9%, saturated solution is 2%.
136. (a) $NH_2CH_2CH_2CH_2CH_2CH_2NH_2$
(b) $NH_3CH_2CH_2CH_2CH_2CH_2NH_3^{2+}$ (c) 9.13
138. The carboxylic acid substituent is an acid (related to $pK_a = 2.18$). The two NH_2 groups are bases (related to $pK_a = 8.95$ and 10.53).

Chapter 17

16. (c)
18. (a) pH = 2.1 (b) pH = 7.20 (c) pH = 12.33
20. (a) Lactic acid and lactate ion
(b) Acetic acid and acetate ion
(c) HClO and ClO^-
(d) HCO_3^- and CO_3^{2-}
22. 4.8 g
24. pH = 8.62
26. 9.55; 9.51
28. Sample (b). Only (b) results in a solution containing a conjugate acid/base pair.
30. (a) ΔpH = 0.1 (b) Δ pH = 3.8 (c) Δ pH = 7.25
32. (a) pH = 5.02 (b) pH = 4.99 (c) pH = 4.06
33. (a) Acid 2; higher initial pH
(b) Acid 1 pH ≈ 8.0; Acid 2 pH ≈ 10.0
(c) The weaker acid produces a more basic salt.
(d) Acid 1 is more highly ionized
(e) Bromthymol blue for Acid 1. Phenolphthalein for Acid 2.
34. At the equivalence point, the solution contains the conjugate base of the weak acid. Weaker acids have stronger conjugate bases. Stronger bases have a higher pH, so the titration of a weaker acid will have a more basic equivalence point.
36. (a) Bromthymol blue
(b) Phenolphthalein
(c) Methyl red
(d) Bromthymol blue. Suitable pH color changes
38. 0.0253-M HCl
40. 93.6%
42. (a) 29.2 mL (b) 600. mL (c) 1.20 L (d) 2.7 mL
44. (a) pH = 3.62 (b) pH = 8.31 (c) pH = 12.15
46. (a) pH = 0.824 (b) pH = 1.30 (c) pH = 3.8
(d) pH = 7.000 (e) pH = 10.2 (f) 12.48

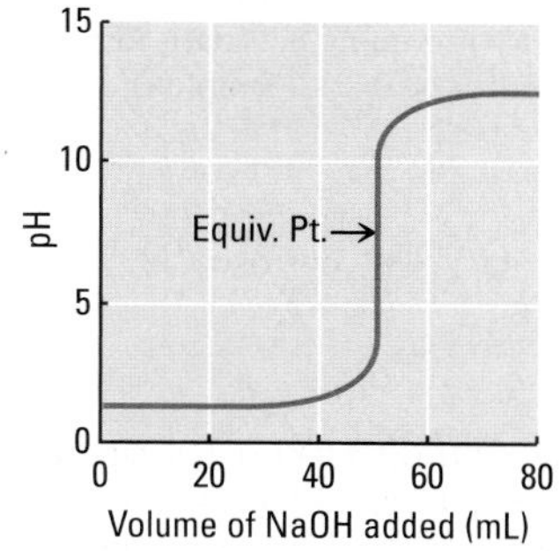

48. NO_2 reaction: $2\ NO_2(g) + H_2O(g) \rightarrow HNO_3(g) + HNO_2(g)$
SO_3 reactions: $2\ SO_2(g) + O_2(g) \rightarrow 2\ SO_3(g)$
$SO_3(g) + H_2O(g) \rightarrow H_2SO_4(g)$
50. $CaCO_3(s) + 2\ H^+(aq) \rightarrow Ca^{2+}(aq) + CO_2(g) + H_2O(\ell)$
54. (a) $FeCO_3(s) \rightleftharpoons Fe^{2+}(aq) + CO_3^{2-}(aq)$ $K_{sp} = [Fe^{2+}][CO_3^{2-}]$
(b) $Ag_2SO_4(s) \rightleftharpoons 2\ Ag^+(aq) + SO_4^{2-}(aq)$ $K_{sp} = [Ag^+]^2[SO_4^{2-}]$
(c) $Ca_3(PO_4)_2(s) \rightleftharpoons 3\ Ca^{2+}(aq) + 2\ PO_4^{3-}(aq)$
$K_{sp} = [Ca^{2+}]^3[PO_4^{3-}]^2$
(d) $Mn(OH)_2(s) \rightleftharpoons Mn^{2+}(aq) + 2\ OH^-(aq)$
$K_{sp} = [Mn^{2+}][OH^-]^2$
55. $K_{sp} = 3.0 \times 10^{-18}$
56. $K_{sp} = 2.22 \times 10^{-4}$
58. $K_{sp} = 2.2 \times 10^{-12}$
60. $K_{sp} = 1.7 \times 10^{-5}$
62. 6.2×10^{-11}-M Zn^{2+}
63. (a) 3.7×10^{-6} M;
(b) 2.8×10^{-10} M;
(c) 2.8×10^{-10} M
64. 3.2×10^{-5} mol/L
66. (a) $K_{sp} = 1 \times 10^{-11}$ (b) $[OH^-]$ must be 0.008 M or higher.
68. 1.2×10^{-9} M or lower
70. pH = 9.16

72. (a) $Ag^+(aq) + 2\ CN^-(aq) \rightleftharpoons [Ag(CN)_2]^-(aq)$

$$K_f = \frac{[[Ag(CN)_2]^-]}{[Ag^+][CN^-]^2}$$

(b) $Cd^{2+}(aq) + 4\ NH_3(aq) \rightleftharpoons [Cd(NH_3)_3]^{2+}(aq)$

$$K_f = \frac{[[Cd(NH_3)_4]^{2+}]}{[Cd^{2+}][NH_3]^4}$$

74. 0.0063 mol or more
76. (a) $Zn(OH)_2(s) + 2\ H_3O^+(aq) \rightarrow Zn^{2+}(aq) + 4\ H_2O(\ell)$
$Zn(OH)_2(s) + 2\ OH^-(aq) \rightarrow [Zn(OH)_4]^{2-}(aq)$
(b) $Sb(OH)_3(s) + 3\ H_3O^+(aq) \rightarrow Sb^{3+}(aq) + 6\ H_2O(\ell)$
$Sb(OH)_3(s) + OH^-(aq) \rightarrow [Sb(OH)_4]^-(aq)$
78. (a) H_2O, CH_3COO^-, Na^+, CH_3COOH, H_3O^+, OH^-
(b) pH = 4.95
(c) pH = 5.05
(d) $CH_3COOH(aq) + H_2O(\ell) \rightleftharpoons CH_3COO^-(aq) + H_3O^+(aq)$
80. Ratio = 1.6
82. 0.020 mol added to 1 L
84. (a) pH = 2.78 (b) pH = 5.39
86. $K_a = 3.5 \times 10^{-6}$
88. No change
90. $K_{sp} = 2.2 \times 10^{-12}$
92. The tiny amount of base (CH_3COO^-) present is insufficient to prevent the pH from changing dramatically if a strong acid is introduced into the solution.
94. $K_a = 2.3 \times 10^{-4}$
96. Blood pH decreases; acidosis
98. (a) Adding Ca^{2+} drives the reaction more toward reactants, making more apatite.
(b) Acid reacts with OH^-, removing a product and driving the reaction toward the products, causing apatite to decompose.
100. Acidosis; increase
102. (a) Box C (b) Boxes C and D (c) Box E (d) Box D
(e) Box A (f) Box B (g) Box E
104. (a) Box C (b) Box D (c) Box B (d) Box A
106. Sample A: $NaHCO_3$; Sample B: NaOH; Sample C: Mixture of NaOH and Na_2CO_3 and/or $NaHCO_3$; Sample D: Na_2CO_3
108. 5.64
110. 3.22
112. 3.2 g
114. (a) 1.1×10^{-9} M (b) 0.010 M

Chapter 18

15. (a) $2\ H_2O(\ell) \rightarrow 2\ H_2(g) + O_2(g)$, reactant-favored
(b) $C_8H_{18}(\ell) \rightarrow C_8H_{18}(g)$, product-favored
(c) $C_{12}H_{22}O_{11}(s) \rightarrow C_{12}H_{22}O_{11}(aq)$, product-favored
17. (a) $\frac{1}{2}$ (b) $\frac{1}{2}$ (c) 50 of each
19. (a) Probability of $\frac{1}{2}$ in flask A; probability of $\frac{1}{2}$ in flask B
(b) 50 in flask A and 50 in flask B
21. (a) Negative (b) Positive (c) Positive
23. (a) Positive (b) Positive (c) Positive
25. (a) Item 2 (b) Item 2 (c) Item 2
27. (a) NaCl(s) (b) $P_4(g)$ (c) $NH_4NO_3(aq)$
29. (a) $Ga(\ell)$ (b) $AsH_3(g)$ (c) NaF(s)
31. (a) Negative (b) Positive
(c) Negative (d) Positive
33. (a) Negative (b) Negative (c) Positive
35. 112 J K^{-1} mol^{-1}
37. (a) 2.63 J/K (b) 1000 J/K (c) 2.45 J/K
39. (a) +113.0 J/K (b) +38.17 J/K
41. (a) −120.64 J/K (b) 156.9 J/K
(c) −198.76 J/K (d) 160.6 J/K
43. (a) −173.01 J K^{-1} (b) −326.69 J K^{-1} (c) 137.55 J K^{-1}
45. $\Delta S° = -247.7$ J/K; Cannot tell without $\Delta H°$ also, since that is needed to calculate $\Delta G°$. $\Delta H° = -88.13$ kJ and $\Delta G° = -14.3$ kJ, so it is product-favored.
47. Product-favored at low temperatures; the exothermicity is sufficient to favor products if the temperature is low enough to overcome the decrease in entropy.
49. Exothermic reactions with an increase in disorder, exhibited by a larger number of gas-phase products than gas-phase reactants, never need help from the surroundings to favor products.
51. (a) Enthalpy change (ΔH) is negative; entropy change (ΔS) is positive.
(b) Using enthalpy of formations and Hess's law:
$\Delta H° = -184.28$ kJ; $\Delta S° = -7.7$ J K^{-1}
The enthalpy change is negative, as predicted in (a). But the entropy change is negative, not as predicted in (a). The aqueous solute must have sufficient order to compensate for the high disorder of the gas. The value of −7.7 J K^{-1} is small.
53. Entropy increase is insufficient to drive this highly endothermic reaction to form products without assistance from the surroundings at this temperature.
55. (a) $\Delta S° = -37.52$ J/K, $\Delta H° = -851.5$ kJ, product-favored at low temperature
(b) $\Delta S° = -21.77$ J/K, $\Delta H° = 66.36$ kJ, never product-favored
57. (a) $\Delta S_{univ} = 4.92 \times 10^3$ J/K (b) -1.47×10^3 kJ
(c) Yes; ethane is used as a fuel; hence, we might expect that its combustion reaction is product-favored.
59. In the first line, $\Delta G°_{system}$ is negative. In the last line, $\Delta G°_{system}$ is positive.
61. $\Delta G° = \Delta H° - T\Delta S°$ Here $\Delta H°$ is negative, and $\Delta S°$ is positive, so $\Delta G° = -|\Delta H°| - |T\Delta S°| = -(|\Delta H°| + |T\Delta S°|) < 0$.
63. $\Delta G° = 28.63$ kJ; reactant-favored
65. $\Delta G° = 462.28$ kJ; uncatalyzed; it would not be a good way to make Si.
67. (a) −141.05 kJ (b) 141.73 kJ (c) −959.43 kJ
Reactions (a) and (c) are product-favored.
69. (a) 385.7 K (b) 835.1 K
71. (a) 49.7 kJ (b) −178.2 kJ (c) −1267.5 kJ
73. (a) $\Delta H° = 178.32$ kJ, $\Delta S° = 160.6$ J/K, $\Delta G° = 130.5$ kJ
(b) Reactant-favored
(c) No; it is only product-favored at high temperatures.
(d) 1110. K
75. $\Delta G_f°(Ca(OH)_2) = -867.8$ kJ
77. (a) $K = 4 \times 10^{-34}$ (b) $K = 5 \times 10^{-31}$
79. (a) $\Delta G° = -100.97$ kJ, product-favored
(b) $K = 5 \times 10^{17}$. When $\Delta G°$ is negative, K is larger than 1.
81. (a) $K°_{298} = 2 \times 10^{12}$, product-favored
$K°_{1000} = 2.0 \times 10^{-2}$, reactant-favored
(b) $K°_{298} = 10^{135}$, product-favored
$K°_{1000} = 3 \times 10^{33}$, product-favored
83. (a) $K°_{800} = 8.7 \times 10^{26}$, product-favored
(b) $K°_{500} = 7 \times 10^{10}$, product-favored
(c) $K°_{2000} = 1.3 \times 10^{-15}$, reactant-favored
85. (a) $\Delta G° = -106$ kJ/mol (b) $\Delta G° = 8.55$ kJ/mol
(c) $\Delta G° = -33.8$ kJ/mol
87. (a) can be harnessed; (b) and (c) require work to be done.
89. Reaction 87(b), 5.068 g graphite oxidized; reaction 87(c), 26.94 g graphite oxidized
91. (a) $2\ C(s) + 2\ Cl_2(g) + TiO_2(s) \rightarrow TiCl_4(g) + 2\ CO(g)$
(b) $\Delta H°$ −44.6 kJ; $\Delta S° = 242.7$ J K^{-1}; $\Delta G° = -116.5$ kJ
(c) Product-favored (d) more reactant-favored
93. CuO, Ag_2O, HgO, and PbO
95. (a) Five O—H bonds, seven C—O bonds, seven C—H bonds, five C—C bonds, and six O=O bonds are broken. Twelve C=O bonds and 12 O—H bonds are formed. $\Delta H° \cong -2873$ kJ.
(b) Interactive forces in condensed phases (solid glucose and liquid water) are neglected in this calculation.
97. (a) 6.46 mol ATP per mol of glucose
(b) $\Delta G° = -106$ kJ (c) Product-favored
99. The combustions of coal, petroleum, and natural gas are the most common sources used to supply free energy. We also use solar and nuclear energy as well as the kinetic energy of wind and water. (There may be other answers.)

101. (a) $\Delta G° = -86.5$ kJ, product-favored
(b) $\Delta G° = -873.1$ kJ (c) No (d) Yes

103. For $CH_4(g)$, $\Delta G° = -817.90$ kJ
For $C_6H_6(\ell)$, $\Delta G° = -3202.0$ kJ
For $CH_3OH(\ell)$, $\Delta G° = -702.34$ kJ
Organic compounds are complex molecular systems that require significant rearrangement of atoms and bonds to undergo combustion. This makes them likely candidates for being kinetically stable.

105. (a) 5.5×10^{18} J/yr (b) 1.5×10^{16} J/day
(c) 1.7×10^{11} J/s (d) 1.7×10^{11} W
(e) 6×10^2 W/person

107. (a) Reaction 2 (b) Reactions 1 and 5
(c) Reaction 2 (d) Reactions 2 and 3
(e) None of them

109. (a) $\Delta G° = 141.73$ kJ (b) No (c) Yes
(d) $K_P = 1 \times 10^4$ (e) $K_c = 7 \times 10^1$

111. (a) $\Delta G° = 31.8$ kJ (b) $K_P = P_{Hg(g)}$
(c) $K° = 2.7 \times 10^{-6}$ (d) $P_{Hg(g)} = 2.7 \times 10^{-6}$ atm
(e) $T = 450$ K

113. Scrambled is a very disordered state for an egg. The second law of thermodynamics says that the more disordered state is the more probable state. Putting the delicate tissues and fluids back where they were before the scrambling occurred would take a great deal of energy. Humpty Dumpty is a fictional character who was also an egg. He fell off a wall. A very probable result of that fall is for an egg to become scrambled. The story goes on to tell that all the energy of the king's horses and men was not sufficient to put Humpty together again.

114. Absolute entropies can be determined because the minimum value of $S°$ is zero at $T = 0$ K. It is not possible to define conditions for a specific minimum value for internal energy, enthalpy, or Gibbs free energy of a substance, so relative quantities must be used.

116. Many of the oxides have negative enthalpies of reaction, which means their oxidations are exothermic. These are probably product-favored reactions.

117. NaCl, in an orderly crystal structure, and pure water, with only O—H hydrogen bonding interactions in the liquid state, are far more ordered than the dispersed hydrated sodium and chloride ions interacting with the water molecules.

119. (a) False (b) False (c) True
(d) True (e) True

121. The energy obtained from nutrients is stored as ATP. The source of the energy needed to synthesize the sugars was sunlight used by plants to produce the sugars and other carbohydrates.

123. $\Delta G = 0$ means products are favored; however, the equilibrium state will always have some reactants present, too. To get all the reactants to go away requires the removal of the products from the reactants, so the reaction continues forward.

125. (a) C (b) A

127. (a) 58.78 J/K (b) −53.29 J/K (c) −173.93 J/K
Adding more hydrogen makes the $\Delta S°$ more negative.

129. Product favored; $\Delta H°$ is negative. Assuming the $S°$ of C_8H_{16} and C_8H_{18} are approximately the same value, $\Delta S°$ is negative, but not large enough to give $\Delta G°$ a different sign; hence, $\Delta G°$ is negative, also.

131. (a) 331.51 K (b) 371. K

133. (a) $K_P = 1.5 \times 10^7$ (b) Product-favored (c) $K_c = 3.7 \times 10^8$

135. (a) $7\ C(s) + 6\ H_2O(g) \rightarrow 2\ C_2H_6(g) + 3\ CO_2(g)$
$5\ C(s) + 4\ H_2O(g) \rightarrow C_3H_8(g) + 2\ CO_2(g)$
$3\ C(s) + 4\ H_2O(g) \rightarrow 2\ CH_3OH(\ell) + CO_2(g)$
(b) For C_2H_6, $\Delta H° = 101.02$ kJ, $\Delta G° = 122.72$ kJ,
$\Delta S° = -72.71$ J/K
For C_3H_8, $\Delta H° = 76.5$ kJ, $\Delta G° = 102.08$ kJ, $\Delta S° = -86.6$ J/K
For CH_3OH, $\Delta H° = 96.44$ kJ, $\Delta G° = 187.39$ kJ,
$\Delta S° = -305.2$ J/K
None of these are feasible. $\Delta G°$ is positive. In addition, $\Delta H°$ is positive and $\Delta S°$ is negative, suggesting that there is no temperature at which the products would be favored.

137. $\Delta G° = -RT \ln K = \Delta H° - T\Delta S°$

$$\ln K = -\frac{\Delta H°}{RT} + \frac{\Delta S°}{R}$$

If $\ln K$ is plotted against $1/T$, the slope would be $-\Delta H°/R$, and the y-intercept would be $\Delta S°/R$. The linear graph shows that $\Delta S°$ and $\Delta H°$ are independent of temperature.

139. (a) (ii) (b) (i) (c) (ii)

141. 205.4 J/mol·K

Chapter 19

10. (a) $Zn(s) \rightarrow Zn^{2+}(aq) + 2\ e^-$
(b) $2\ H_3O^+(aq) + 2\ e^- \rightarrow 2\ H_2O(\ell) + H_2(g)$
(c) $Sn^{4+}(aq) + 2\ e^- \rightarrow Sn^{2+}(aq)$
(d) $Cl_2(g) + 2\ e^- \rightarrow 2\ Cl^-(aq)$
(e) $6\ H_2O(\ell) + SO_2(g) \rightarrow SO_4^{2-}(aq) + 4\ H_3O^+(aq) + 2\ e^-$

12. (a) $Al(s) \rightarrow Al^{3+}(aq) + 3\ e^-$
$Cl_2(g) + 2\ e^- \rightarrow 2\ Cl^-(aq)$
(b) $Fe^{2+}(aq) \rightarrow Fe^{3+}(aq) + e^-$
$MnO_4^-(aq) + 8\ H_3O^+(aq) + 5\ e^- \rightarrow Mn^{2+}(aq) + 12\ H_2O(\ell)$
(c) $FeS(s) + 12\ H_2O(\ell) \rightarrow$
$Fe^{3+}(aq) + SO_4^{2-}(aq) + 8\ H_2O(\ell) + 9\ e^-$
$NO_3^-(aq) + 4\ H_3O^+(aq) + 3\ e^- \rightarrow NO(g) + 6\ H_2O(\ell)$

14. $4\ Zn(s) + 7\ OH^-(aq) + NO_3^-(aq) + 6\ H_2O(\ell)$
$\rightarrow 4\ Zn(OH)_4^{2-} + NH_3(aq)$

16. (a) $3\ CO(g) + O_3(g) \rightarrow 3\ CO_2(g)$
O_3 is the oxidizing agent; CO is the reducing agent.
(b) $H_2(g) + Cl_2(g) \rightarrow 2\ HCl(g)$
Cl_2 is the oxidizing agent; H_2 is the reducing agent.
(c) $H_2O_2(aq) + Ti^{2+}(aq) + 2\ H_3O^+(aq) \rightarrow 4\ H_2O(\ell) + Ti^{4+}(aq)$
H_2O_2 is the oxidizing agent; Ti^{2+} is the reducing agent.
(d) $2\ MnO_4^-(aq) + 6\ Cl^-(aq) + 8\ H_3O^+(aq) \rightarrow$
$2\ MnO_2(s) + 3\ Cl_2(g) + 12\ H_2O(\ell)$
MnO_4^- is the oxidizing agent; Cl^- is the reducing agent.
(e) $4\ FeS_2(s) + 11\ O_2(g) \rightarrow 2\ Fe_2O_3(s) + 8\ SO_2(g)$
O_2 is the oxidizing agent; FeS_2 is the reducing agent.
(f) $O_3(g) + NO(g) \rightarrow O_2(g) + NO_2(g)$
O_3 is the oxidizing agent; NO is the reducing agent.
(g) $Zn(Hg)(amalgam) + HgO(s) \rightarrow ZnO(s) + 2\ Hg(\ell)$
HgO is the oxidizing agent; Zn(Hg) is the reducing agent.

18. The generation of electricity occurs when electrons are transmitted through a wire from the metal to the cation. Here, the transfer of electrons would occur directly from the metal to the cation and the electrons would not flow through any wire.

20. In chemistry, they are conventionally written as reduction reactions.

22. (a) $Zn(s) + Pb^{2+}(aq) \rightarrow Zn^{2+}(aq) + Pb(s)$
(b) Oxidation of zinc occurs at the anode. The reduction of lead occurs at the cathode. The anode is metallic zinc. The cathode is metallic lead.
(c)

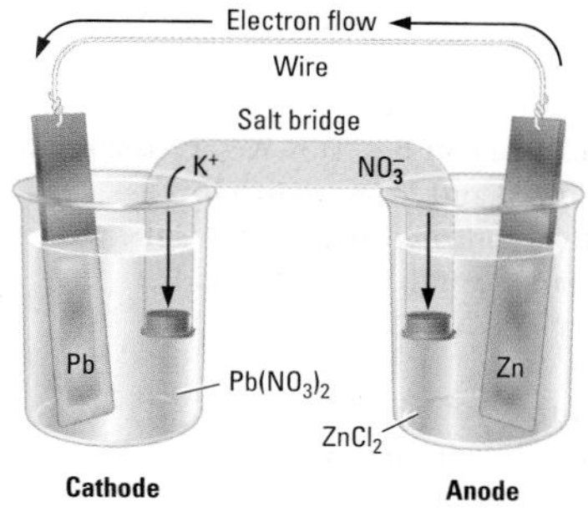

24. 7.5×10^3 C

26. (a) $Cu(s) \rightarrow Cu^{2+}(aq) + 2\ e^-$
$Ag^+(aq) + e^- \rightarrow Ag(s)$
(b) The copper half-reaction is oxidation and it occurs in the anode compartment. The silver half-reaction is reduction and it occurs in the cathode compartment.

28. Li is the strongest reducing agent and Li^+ is the weakest oxidizing agent. F_2 is the strongest oxidizing agent and F^- is the weakest reducing agent.
30. Worst oxidizing agent (d) $<$ (c) $<$ (a) $<$ (b) best oxidizing agent
32. (a) 2.89 V (b) −0.028 V (c) 0.65 V (d) 1.164 V
    Reactions (a), (c), and (d) are product-favored.
34. (a) Al^{3+} (b) Ce^{4+} (c) Al (d) Ce^{3+} (e) Yes (f) No
    (g) Mercury(I) ion, silver ion, and cerium(IV) ion
    (h) Hg, Sn, Ni, Al
36. Greater; less
38. (a) 1.55 V (b) −1196 kJ, 1.55 V
40. −409 kJ
42. 1×10^{-18}, 102 kJ
46. $K° = 4 \times 10^1$
48. (a) 1.20 V (b) 1.16 V (c) 3 M
50. −0.378 V
52. (conc. Pb^{2+})/(conc. Sn^{2+}) = 0.38
54. (a) $Ni^{2+}(aq) + Cd(s) \rightarrow Ni(s) + Cd^{2+}(aq)$
    (b) Cd is oxidized, Ni^{2+} is reduced, Ni^{2+} is the oxidizing agent, and Cd is the reducing agent.
    (c) Metallic Cd is the anode and metallic Ni is the cathode.
    (d) 0.15 V
    (e) Electrons flow from the Cd electrode to the Ni electrode.
    (f) Toward the anode compartment
56. A fuel cell has a continuous supply of reactants and will be useable for as long as the reactants are supplied. A battery contains all the reactants of the reaction. Once the reactants are gone, the battery is no longer useable.
58. (a) The N_2H_4 half-reaction occurs at the anode and the O_2 half-reaction occurs at the cathode.
    (b) $N_2H_4(g) + O_2(g) \rightarrow N_2(g) + 2\ H_2O(\ell)$
    (c) 7.5 g N_2H_4 (d) 7.5 g O_2
59. $O_2(g)$ produced at anode; $H_2(g)$ produced at cathode; 2 mol H_2 per mol O_2
61. Au^{3+}, Hg^{2+}, Ag^+, Hg_2^{2+}, Fe^{3+}, Cu^{2+}, Sn^{4+}, Sn^{2+}, Ni^{2+}, Cd^{2+}, Fe^{2+}, Zn^{2+}
63. H_2, Br_2, and OH^- are produced. After the reaction is complete, the solution contains Na^+, OH^-, a small amount of dissolved Br_2 (though it has low solubility in water), and a very small amount of H_3O^+. H_2 is formed at the cathode. Br_2 is formed at the anode.
65. 0.16 g Ag
67. 5.93 g Cu
69. 2.7×10^5 g Al
71. 1.9×10^2 g Pb
73. 0.10 g Zn
75. 0.043 g Li; 0.64 g Pb
77. 6.85 min
80. Ions increase the electrolytic capacity of the solution.
82. Chromium is highly resistant to corrosion and protects iron in steel from oxidizing.
85. 0.00689 g Cu, 0.00195 g Al
87. Worst reducing agent B $<$ D $<$ A $<$ C best reducing agent
89. (a) B is oxidized and A^{2+} is reduced.
    (b) A^{2+} is the oxidizing agent and B is the reducing agent.
    (c) B is the anode and A is the cathode.
    (d) $A^{2+} + 2e^- \rightarrow A$
    $B \rightarrow B^{2+} + 2e^-$
    (e) A gains mass.
    (f) Electrons flow from B to A.
    (g) K^+ ions will migrate toward the A^{2+} solution.
91. (a) The reaction with water is spontaneous:
    $$4\ Co^{3+}(aq) + 6\ H_2O(\ell) \rightarrow 4\ Co^{2+}(aq) + O_2(aq) + 4\ H_3O^+(aq) \qquad E°_{cell} = 0.591\ V$$
    (b) The reaction with oxygen in air is spontaneous:
    $$4\ Fe^{2+}(aq) + O_2(aq) + 4\ H_3O^+(aq) \rightarrow 4\ Fe^{3+}(aq) + 6\ H_2O(\ell) \qquad E°_{cell} = 0.458\ V$$
93. $E_{cell} = 0.379$ V
95. (a) 9.5×10^6 g HF (b) 1.7×10^3 kWh
97. $Cl_2(g) + 2\ Br^-(aq) \rightarrow Br_2(\ell) + 2\ Cl^-(aq)$
99. 4+
101. 75 s
103. 4×10^{-6}-M Cu^{2+}

Chapter 20

11. (a) α emission (b) β emission (c) Electron capture or positron emission (d) β emission
13. (a) $^{238}_{92}U$ (b) $^{32}_{15}P$ (c) $^{10}_{5}B$ (d) $^{0}_{-1}e$
    (e) $^{15}_{7}N$
15. (a) $^{28}_{12}Mg \rightarrow ^{28}_{13}Al + ^{0}_{-1}e$
    (b) $^{238}_{92}U + ^{12}_{6}C \rightarrow 4^{1}_{0}n + ^{246}_{98}Cf$
    (c) $^{2}_{1}H + ^{3}_{2}He \rightarrow ^{4}_{2}He + ^{1}_{1}H$
    (d) $^{38}_{19}K \rightarrow ^{38}_{18}Ar + ^{0}_{+1}e$
    (e) $^{175}_{78}Pt \rightarrow ^{4}_{2}He + ^{171}_{76}Os$
17. $^{231}_{90}Th$, $^{231}_{91}Pa$, $^{227}_{89}Ac$, $^{227}_{90}Th$, $^{223}_{88}Ra$
19. $^{222}_{86}Rn \rightarrow ^{4}_{2}He + ^{218}_{84}Po$
    $^{218}_{84}Po \rightarrow ^{4}_{2}He + ^{214}_{82}Pb$
    $^{214}_{82}Pb \rightarrow ^{0}_{-1}e + ^{214}_{83}Bi$
    $^{214}_{83}Bi \rightarrow ^{0}_{-1}e + ^{214}_{84}Po$
    $^{214}_{84}Po \rightarrow ^{4}_{2}He + ^{210}_{82}Pb$
    $^{210}_{82}Pb \rightarrow ^{0}_{-1}e + ^{210}_{83}Bi$
    $^{210}_{83}Bi \rightarrow ^{0}_{-1}e + ^{210}_{84}Po$
    $^{210}_{84}Po \rightarrow ^{4}_{2}He + ^{206}_{82}Pb$
20. (a) $^{19}_{10}Ne \rightarrow ^{19}_{9}F + ^{0}_{+1}e$ (b) $^{230}_{90}Th \rightarrow ^{0}_{-1}e + ^{230}_{91}Pa$
    (c) $^{82}_{35}Br \rightarrow ^{0}_{-1}e + ^{82}_{36}Kr$ (d) $^{212}_{84}Po \rightarrow ^{4}_{2}He + ^{208}_{82}Pb$
22. The binding energy per nucleon of ^{10}B is 6.252×10^8 kJ/mol nucleon. The binding energy per nucleon of ^{11}B is 6.688×10^8 kJ/mol nucleon. ^{11}B is more stable than ^{10}B because its binding energy is larger.
24. -1.7394×10^{11} kJ/mol ^{238}U; 7.3086×10^8 kJ/mol nucleon
27. 5 mg
29. (a) $^{131}_{53}I \rightarrow ^{131}_{54}Xe + ^{0}_{-1}e$ (b) 1.56 mg
31. 4.0 y
33. 34.8 d
35. 2.58×10^3 y; Approx. 570 B.C.
37. 1.6×10^5 y
39. $^{239}_{94}Pu + 2^{1}_{0}n \rightarrow ^{0}_{-1}e + ^{241}_{95}Am$
41. $^{12}_{6}C$
43. Cadmium rods (a neutron absorber to control the rate of the fission reaction), uranium rods (a source of fuel, since uranium is a reactant in the nuclear equation), and water (used for cooling by removing excess heat and used in steam/water cycle for the production of turning torque for the generator)
45. (a) $^{140}_{54}Xe$ (b) $^{101}_{41}Nb$ (c) $^{92}_{36}Kr$
47. 6.9×10^3 barrels
49. A rad is a measure of the amount of radiation absorbed. A rem includes a quality factor that better describes the biological impact of a radiation dose. The unit rem would be more appropriate when talking about the effects of an atomic bomb on humans. The unit gray (Gy) is 100 rad.
51. Since most elements have some proportion of unstable isotopes that decay and we are composed of these elements (e.g., ^{14}C), our bodies emit radiation particles.
53. The gamma ray is a high-energy photon. Its interaction with matter is most likely just to impart large quantities of energy. The alpha and beta particles are charged particles of matter, which could interact, and possibly react, with the matter composing the food.
55. 0.13 L
57. (a) $^{0}_{-1}e$ (b) $^{4}_{2}He$ (c) $^{87}_{35}Br$ (d) $^{216}_{284}Po$
    (e) $^{68}_{31}Ga$
59. 2 mg

61. 3.6×10^9 y
63. (a) $^{247}_{299}Es$ (b) $^{16}_{8}O$ (c) $^{4}_{2}He$ (d) $^{12}_{6}C$
(e) $^{10}_{15}B$
65. Alpha and beta radiation decay particles are charged ($^{4}_{2}He^{21}$ and $^{0}_{-1}e^{-}$), so they are better able to interact with and ionize tissues, disrupting the function of the cancer cells. Gamma radiation, like X rays, goes through soft tissue without much of it being absorbed. This is less likely to interfere with the cancerous cells.
67. ^{20}Ne is stable. The ^{17}Ne is likely to decay by positron emission, increasing the ratio of neutrons to protons. The ^{23}Ne is likely to decay by beta emission, decreasing the ratio of neutrons to protons.
69. A nuclear reaction occurred, making products. Therefore, some of the lost mass is found in the decay particles, if the decay is alpha or beta decay, and almost all of the rest is found in the element produced by the reaction.
71. They are spontaneous reactions due to the nature of the nucleus. No other species are involved.
73. 2.8×10^4 kg
75. 75.1 y
77. 3.92×10^3 y

Chapter 21

20. (a) Cathode (b) Chlorine gas (c) 2.027×10^9 C
(d) 1.6×10^3 kJ/mol
22. 7×10^3 kWh
24. 67.1 g Al
26. Answers are in **bold.**

Formula	Name	Oxidation State of Phosphorus
$\mathbf{P_4}$	Phosphorus	**0**
$(NH_4)_2HPO_4$	**Ammonium hydrogen phosphate**	+5
$\mathbf{H_3PO_4}$	Phosphoric acid	+5
$\mathbf{P_4O_{10}}$	Tetraphosphorus decaoxide	+5
$Ca_3(PO_4)_2$	**Calcium phosphate**	+5
$\mathbf{Ca(H_2PO_4)_2}$	Calcium dihydrogen phosphate	+5

28. 1.7×10^7 tons
30.

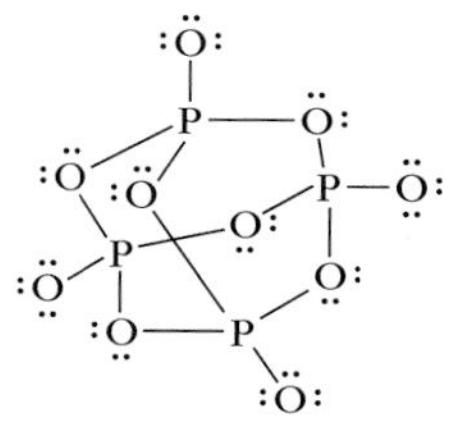

32. 962 K
34. Raw materials: sulfur, water, oxygen, and catalyst (Pt or VO_5)
$S_8(s) + 8\ O_2\ (g) \rightarrow 8\ SO_2\ (g)$
$2\ SO_2\ (g) + O_2\ (g) \rightarrow 2\ SO_3\ (g)$
$SO_3\ (g) + H_2SO_4(\ell) \rightarrow H_2S_2O_7(\ell)$
$H_2S_2O_7\ (\ell) + H_2O(\ell) \rightarrow 2\ H_2SO_4\ (aq)$
36.

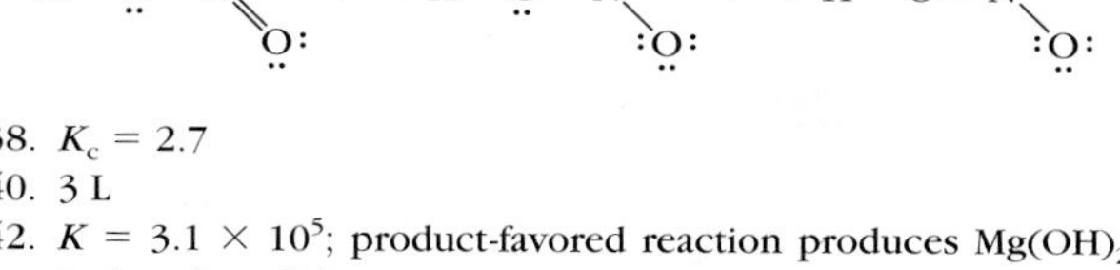

38. $K_c = 2.7$
40. 3 L
42. $K = 3.1 \times 10^5$; product-favored reaction produces $Mg(OH)_2$ (s). Isolate the solid.
Putting seawater in the presence of $Ca(OH)_2$ will cause a precipitate of $Mg(OH)_2$ to form. The solid can be isolated after it settles.
44. $\ddot{N}{=}N{=}\ddot{N}{-}\ddot{N}{=}\ddot{O}$ (Other structures are possible.)
46. (a) $H{-}\ddot{N}{=}\ddot{N}{=}\ddot{N} \leftrightarrow H{-}\ddot{N}{-}N{\equiv}N{:}$
(b) +140. kJ with single and triple bond; +410. kJ with two double bonds
48. (a) $NO_2(g) + NO(g) \rightarrow N_2O_3(g)$
(b) 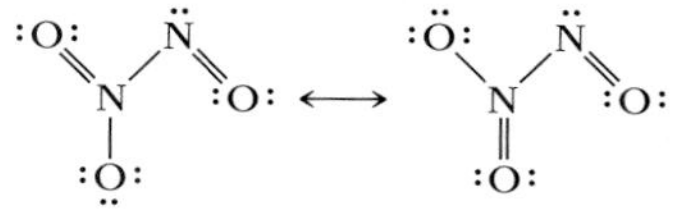
(c) The O—N—O angle is 120° and the N—N—O angle is slightly less than 120°.
51. (a) 7×10^2 kJ
(b) 3 V must be used to overcome the cell potential.
53. (a) One; 120 °C, 1×10^{-4} atm (b) (i) Solid rhombic (ii) Liquid (iii) Solid monoclinic (iv) Vapor
55. (a) $4\ H_3O^+\ (aq) + MnO(s) + 2\ Br^-(aq) \rightarrow$
$Mn^{2+}(aq) + 6\ H_2O(\ell) + Br_2(\ell)$
(b) 0.0813 mol Br^- (c) 3.54 g MnO_2
63. 6.3×10^5 J/mol
66. (a) $P_4(s) + 5\ O_2(g) \rightarrow P_4O_{10}(s)$ (b) pH = 2.25
(c) 25.0 g $Ca_3(PO_4)_2$ (d) 5.42 L
68. (a) Cl is reduced; Al and N are oxidized. (b) −2674 kJ
70. −2708. kJ/mol
72. $K_p = 1.985$
74. 2961 kJ, Mg^{2+} is a much smaller ion with higher charge density than Sr^{2+}, so the $Mg^{2+} + F^-$ ions are closer together and the ionic bond is much stronger.
76. B_6H_{12}

Chapter 22

14. (a) Ag [Kr] $4d^{10}5s^1$; Ag^+ [Kr] $4d^{10}$
(b) Au [Xe] $4f^{14}5d^{10}6s^1$; Au^+ [Xe]$4f^{14}\ 5d^{10}$; Au^{3+} [Xe] $4f^{14}\ 5d^8$
16. Cr^{2+} and Cr^{3+}
18. $Cr_2O_7^{2-}$ (in acid) > Cr^{3+} > Cr^{2+}; the species with a more positive oxidation state of Cr has greater tendency to be reduced.
E°_{red}, respectively, is +1.33 V > −0.74 V > −0.91 V, so the more positive the oxidation state, the better the oxidizing agent.
20. (a) $Fe_2O_3(s) + 3\ CO(g) \rightarrow 2\ Fe(s) + 3\ CO_2(g)$
(b) $Fe(s) + 2\ H_3O^+(aq) \rightarrow Fe^{2+}\ (aq) + 2\ H_2O(\ell) + H_2(g)$ or $2\ Fe(s) + 6\ H_3O^+\ (aq) \rightarrow 2\ Fe^{3+}(aq) + 6\ H_2O(\ell) + 3\ H_2(g)$
22. $6\ H_3O^+(aq) + 3\ NO_3^-(aq) + Fe(s) \rightarrow$
$Fe^{3+}(aq) + 3\ NO_2(g) + 9\ H_2O(\ell)$
26. (a) $[Cr(NH_3)_2(H_2O)_3(OH)]^{2+}$; +2
(b) The counter ion would be an anion, such as Cl^-.
28. (a) $C_2O_4^{2-}$ with 2− charge and Cl− with 1− charge
(b) 3+ charge
(c) $[Co(NH_3)_4Cl_2]^+$, with 1+ charge
30. For $Na_3[IrCl_6]$: (a) Six Cl^- (b) Ir with 3+ charge
(c) $[IrCl_6]^{3-}$ with 3− charge (d) Na^+
For $[Mo(CO)_4Br_2]$: (a) Four CO (b) Mo with 2+ charge
(c) $[Mo(CO)_4]^{2+}$ with 2+ charge (d) Br^-
32. (a) Four (b) Four
34. (a) $[Pt(NH_3)_2Br_2]$ (b) $[Pt(en)(NO_2)_2]$
(c) $[Pt(NH_3)_2BrCl]$
36. (a) Four (b) Four
(c) Six (d) Six
38. (a) Monodentate (b) Tetradentate
(c) Tridentate (d) Monodentate
40. For example, $FeCl_3$ (Other answers are possible.)
42. (a) Tetrachloromanganate (II)
(b) Potassium trioxalatoferrate (III)
(c) Diamminedicyanoplatinum (II)
(d) Pentaquahydroxoiron (III)
(e) Diethylenediaminedichloromanganese (II)

44. (a) 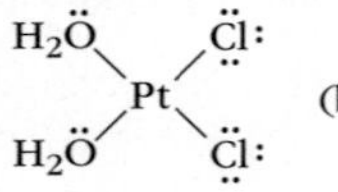 (b) 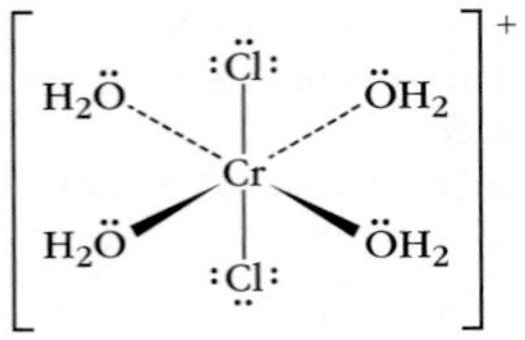

46. 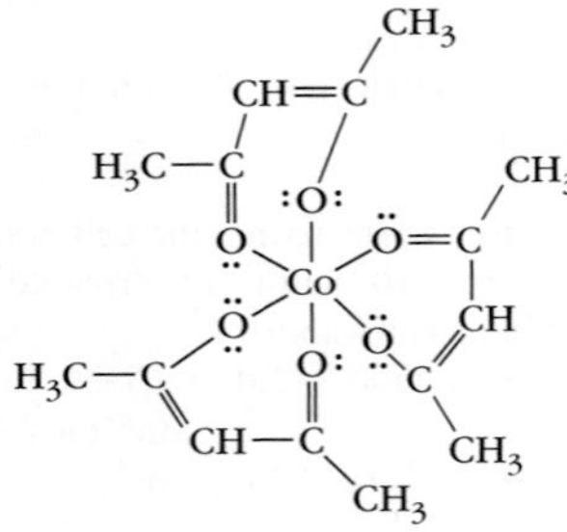

48. (a) and (b)

50. (a), (c), and (d)

52. (a) 4 (b) 5 (c) 5 (d) 3

54. Cr^{2+} has four electrons. Cr^{3+} has three electrons. Since there are three low-energy t_2 orbitals, the first three electrons can fill one into each t_2 orbital. The fourth electron (only found in Cr^{2+}) is required to pair up with one of the other electrons (low spin) or to span the Δ_o gap to go into an *e* orbital (high spin).

56. If Δ_o is large, electrons fill the t_2 orbitals and end up all paired. If Δ_o is small, all five orbitals are half-filled before any pairing begins, resulting in four unpaired electrons. So, high-spin Co^{3+} complexes are paramagnetic and low-spin Co^{3+} complexes are diamagnetic.

58. Cu^{2+} has nine *d*-electrons. All t_2 orbitals are filled, so high- and low-spin complexes do not form.

60. violet light; Approximately 400 nm

62. With CN^-: Δ_o increases and the observed color shifts closer to yellow-orange.
With Cl^-: Δ_o decreases and the observed color shifts closer to blue.

65. (a) [Ar] $3d^4$ (b) [Ar] $3d^{10}$ (c) [Ar] $3d^7$ (d) [Ar] $3d^3$

67.

(a) 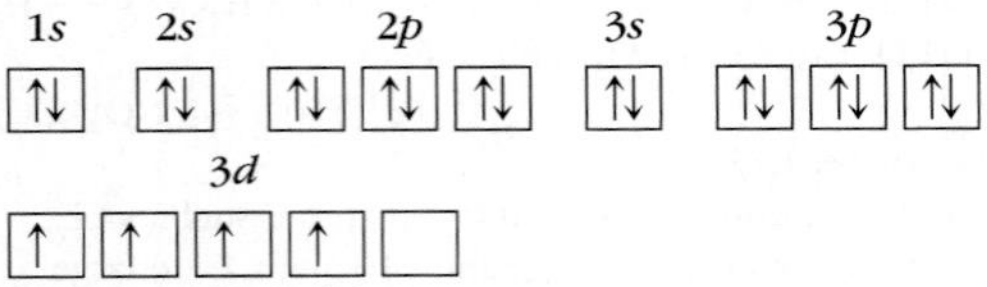

Four unpaired electrons

(b) 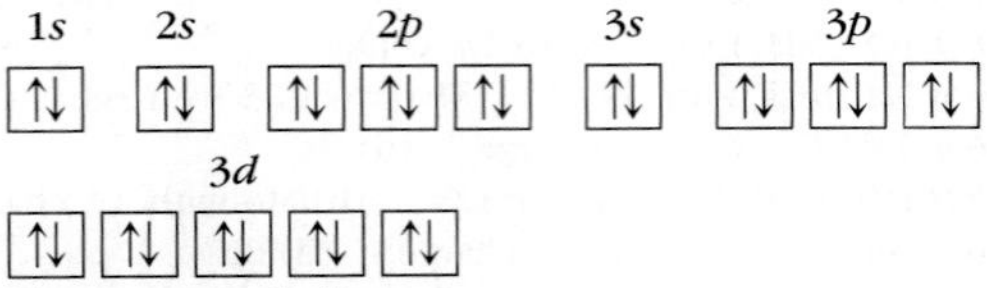

Zero unpaired electrons

(c) 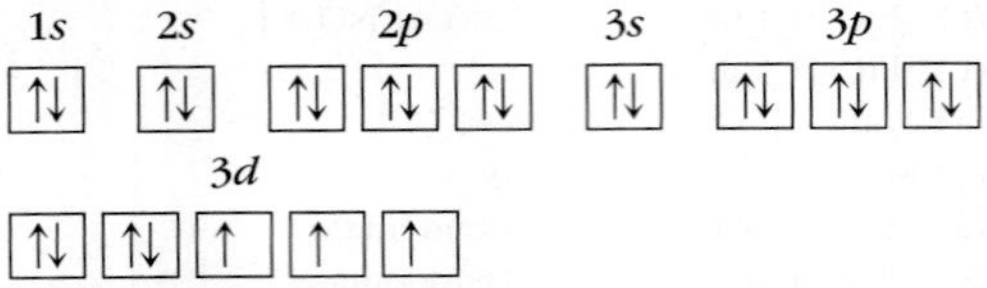

Three unpaired electrons

(d) 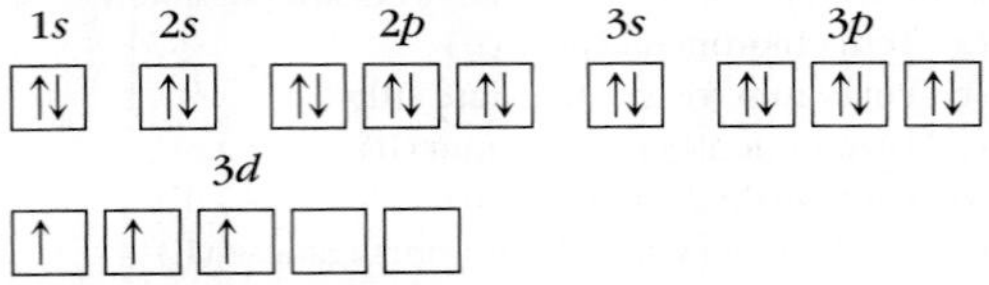

Three unpaired electrons

69. 14.8 g Cu

71. 0.40 ton SO_2

73. (a) Four (b) Six (c) Four (d) Six

75. 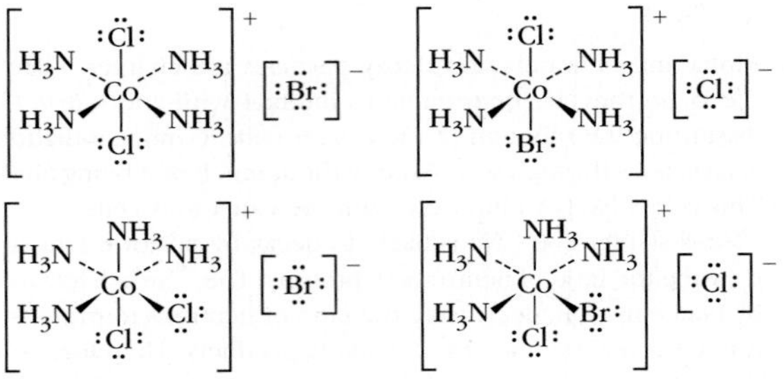

77. (a) False; the coordination number is six.
(b) False; Cu^+ has no unpaired electrons.
(c) False; the net charge is 3+.

78. (a) False; "In[$Pt(NH_3)_4$]Cl_4 the Pt has a 4+ charge and a coordination number of four."
(b) True

79. (a) +2
(b) 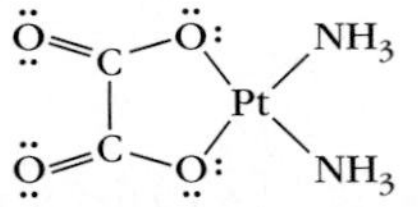

81. $Fe(s) + 2\ Fe^{3+}\ (aq) \rightarrow 3\ Fe^{2+}\ (aq)$

83. Four

85. 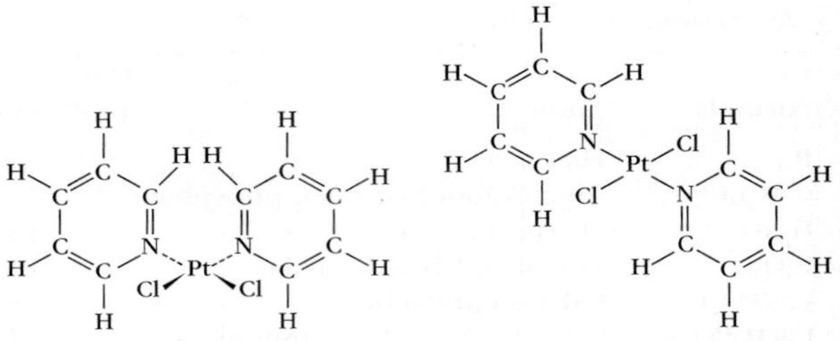

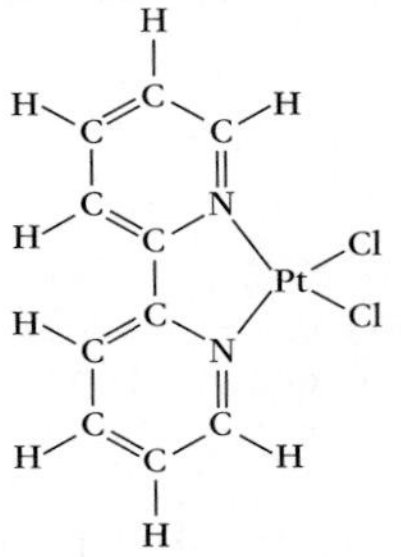

The bidentate ortho-phenanthroline is not able to form the *trans*-isomer.

87. (a) $2\ Ag^+ + Ni \rightarrow Ni^{2+} + 2\ Ag$
(b) +1.05V

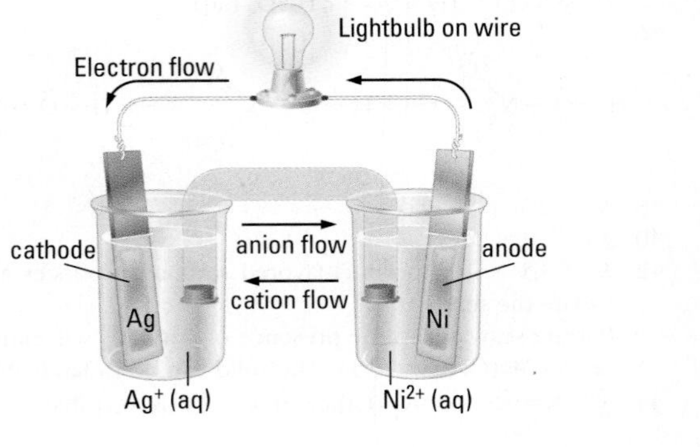

89. 17.4% Fe

91. (a) −1.38 V (b) 1×10^{-20}-M Ag^{2+}

93. (a) No ions (b) 2.00 mol ions (c) 4.00 mol ions (d) 3.00 mol ions

96.

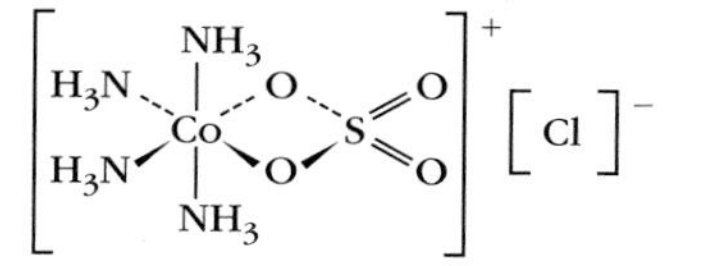

98. (a) $[Co(NH_3)_6]Cl_3$
(b) $[Co(NH_3)_6]Cl_3 \rightarrow [Co(NH_3)_6]^{3+}(aq) + 3\ Cl^-(aq)$

100.

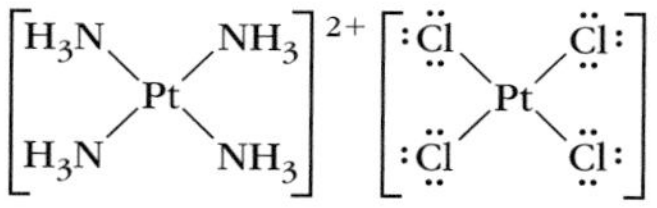

102.

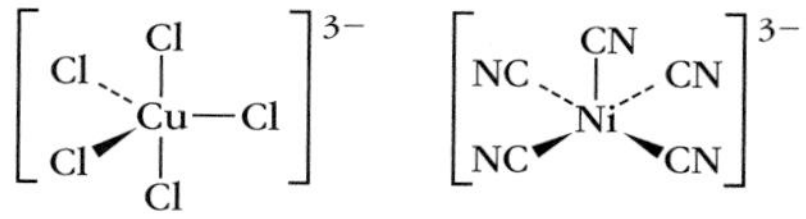

Triangular bipyramidal or square pyramidal

104. A mixture of high spin and low spin complexes, in which the high spin species is favored by a 2:1 ratio, would give the observed 2.67 unpaired electrons per iron ion. High spin iron(II) has four unpaired electrons, while low spin iron(II) has zero unpaired electrons. This translates to 8 unpaired electrons for every three iron ions or 8/3 = 2.67 unpaired electrons per iron ion.

106.

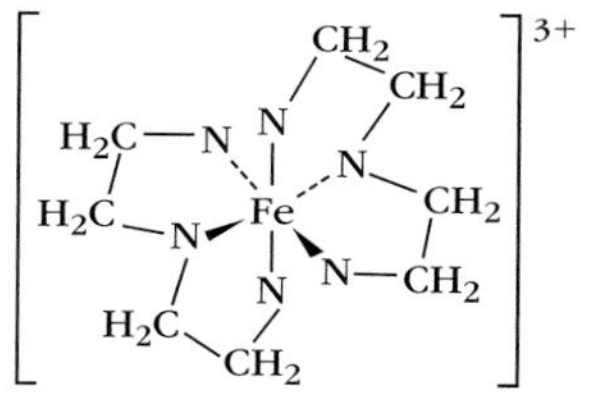

(One other structure is possible, too.)

Appendix A

4. The calculation is incomplete or incorrect. Check for incomplete unit conversions (e.g., cm^3 to L, g to kg, mm Hg to atm, etc.) and check to see that the numerator and denominator of unit factors are placed such that the unwanted units can cancel (e.g., $\frac{g}{mL}$ or $\frac{mL}{g}$).

6. Answer (c) gives the properly reported observation. Answers (a) and (b) do not have the proper units. Answer (d) is incomplete, with no units. Answer (e) shows the conversion of the observed measurement to new units, making (e) the result of a calculation, not the observed measurement.

8. (a) The units should not be squared. To do this correctly, add values with common units (g) and give the answer the same units (g). The result is 9.95 g.
(b) The unit factor is not set up to cancel unwanted units (g); the units reported (g) are not the units resulting from this calculation $(\frac{g^2}{mL})$. To do this correctly, $5.23\ g \times \frac{1.00\ mL}{4.87\ g} = 1.07\ mL$.
(c) The unit factor must be cubed to cancel all unwanted units; the units reported (m^3) are not the units resulting from this calculation ($m \cdot cm^2$). To do this correctly, $3.57\ cm^3 \times (\frac{1\ m}{100\ cm})^3 = 3.57 \times 10^{-6}\ m^3$

10. (a) 7.994 ft ± 0.043 ft
(b) Three, because uncertainty is in the hundredths place
(c) The result is accurate (i.e., the true answer, 8.000 ft, is within the range of the uncertainty described by these values, 7.951 ft–8.037 ft), though not very precise.

12. (a) 4 (b) 2 (c) 2 (d) 4

14. (a) 43.32 (b) 43.32 (c) 43.32 (d) 43.32 (e) 43.32 (f) 43.33

16. (a) 13.7 (b) 0.247 (c) 12.0

18. (a) 7.6003×10^4 (b) 3.7×10^{-4} (c) 3.4000×10^4

20. (a) 2.415×10^{-3} (b) 2.70×10^8 (c) 3.236 (d) 116 cm^3

22. (a) −0.1351 (b) 3.541 (c) 23.7797 (d) 54.7549 (e) −7.455

24. (a) 5.404 (b) 1×10^{15} (c) 110. (d) 1.000320 (e) 1.68

26. (a) 0.480, −1.80 (b) 5.37, −0.804

28.

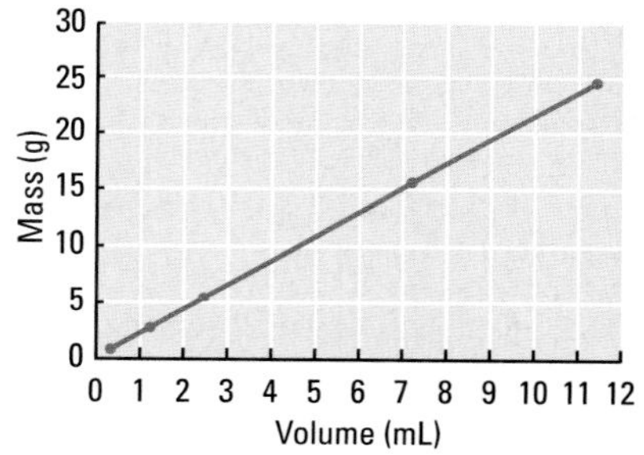

Yes, y is proportional to x.

Appendix B

1. (a) Kilogram, no additional prefix (b) Cubic meters, nano- (c) Meters, milli-

3. SI base (fundamental) units are set by convention; derived units are based on the fundamental units.

5. (a) 4.75×10^{-10} m (b) 5.6×10^7 kg (c) 4.28×10^{-6} A

7. (a) $8.7 \times 10^{-18}\ m^2$ (b) 2.73×10^{-17} J (c) 2.73×10^{-5} N

9. (a) 2.20 lb (b) 2.22×10^3 kg (c) 60 in^3 (d) 1×10^5 Pa and 1 bar

11. (a) 99 °F (b) −10.5 °F (c) −40.0 °F

Glossary

absolute temperature scale (Kelvin temperature scale) A temperature scale on which the zero is the lowest possible temperature and the degree is the same size as the Celsius degree.

absolute zero Lowest possible temperature, 0 K.

absorb To draw a substance into the bulk of a liquid or a solid (compare with adsorb).

achiral Describes an object that can be superimposed on its mirror image; opposite of chiral.

acid (Arrhenius) A substance that increases the concentration of hydronium ions, H_3O^+, in aqueous solution. (See also **Brønsted-Lowry acid, Lewis acid.**)

acid-base reaction A reaction in which one molecule or ion (an acid) transfers a proton, H^+, to another molecule or ion (a base).

acid ionization constant (K_a) The equilibrium constant for the reaction of a weak acid with water to produce hydronium ions and the conjugate base of the weak acid.

acid ionization constant expression Mathematical expression in which the product of the equilibrium concentrations of hydronium ion and conjugate base is divided by the equilibrium concentration of the un-ionized conjugate acid.

acid rain Rain (or other precipitation) with a pH below about 5.6 (the pH of unpolluted rain water).

acidic solution An aqueous solution in which the concentration of hydronium ion exceeds the concentration of hydroxide ion.

actinides The elements after actinium in the seventh period; in actinides the 5*f* subshell is being filled.

activated complex A molecular structure corresponding to the maximum of a plot of energy versus reaction progress; also known as the transition state.

activation energy (E_a) The potential energy difference between reactants and activated complex; the minimum energy that reactant molecules must have to be converted to product molecules.

active site The part of an enzyme molecule that binds the substrate to help it to react.

activity (*A*) A measure of the rate of nuclear decay, given as disintegrations per unit time.

actual yield The quantity of a reaction product obtained experimentally; less than the theoretical yield.

addition polymer A polymer made when monomer molecules join directly with one another, with no other products formed in the reaction.

adsorb To attract and hold a substance on a surface (compare with absorb).

aerosols Small particles (1 nm to about 10,000 nm in diameter) that remain suspended indefinitely in air.

air pollutant A substance that degrades air quality.

alcohol An organic compound containing a hydroxyl group (—OH) covalently bonded to a saturated carbon atom.

aldehyde An organic compound characterized by a carbonyl group in which the carbon atom is bonded to a hydrogen atom; a molecule containing the —CHO functional group.

alkali metals The Group 1A elements in the periodic table (except hydrogen).

alkaline-earth metals The elements in Group 2A of the periodic table.

alkane Any of a class of hydrocarbons characterized by the presence of only single carbon-carbon and carbon-hydrogen bonds.

alkene Any of a class of hydrocarbons characterized by the presence of a carbon-carbon double bond.

alkyl group A fragment of an alkane structure that results from the removal of a hydrogen atom from the alkane.

alkyne Any of a class of hydrocarbons characterized by the presence of a carbon-carbon triple bond.

allotropes Different forms of the same element that exist in the same physical state under the same conditions of temperature and pressure.

alpha carbon The carbon adjacent to the acid group (—COOH) in an amino acid.

alpha (α) particles Positively charged (2+) particles ejected at high speeds from certain radioactive nuclei; the nuclei of helium atoms.

alpha radiation Radiation composed of alpha particles (helium nuclei).

amide An organic compound characterized by the presence of a carbonyl group in which the carbon atom is bonded to a nitrogen atom (—$CONH_2$, —CONHR, —$CONR_2$); the product of the reaction of an amine with a carboxylic acid.

amide linkage A linkage consisting of a carbonyl group bonded to a nitrogen atom that connects monomers in a polymer.

amine An organic compound containing an —NH_2, —NHR, or —NR_2 functional group.

amino acids Organic molecules containing a carboxyl group with an R group and an amine group on the alpha carbon; building block monomers of proteins.

amorphous solid A solid whose constituent nanoscale particles have no long-range repeating structure.

amount of substance (amount) A measure of the number of elementary entities (such as atoms, ions, molecules) in a sample of matter compared with the number of elementary entities in exactly 0.012 kg pure ^{12}C. Also called molar amount.

ampere The SI unit of electrical current; involves the flow of one coulomb of charge per second.

amphiprotic Describes a molecule or ion that can either donate or accept a proton, H^+, in acid-base reactions; an amphiprotic species can serve as either an acid or a base depending on the circumstances.

amphoteric Refers to a substance that can act as either an acid or a base.

anion An ion with a negative electrical charge.

anode The electrode of an electrochemical cell at which oxidation occurs.

anodic inhibition The prevention of oxidation of an active metal by painting it, coating it with grease or oil, or allowing a thin film of metal oxide to form.

antibonding molecular orbital A higher-energy molecular orbital that, if occupied by electrons, does not result in attraction between the atoms.

antioxidant Reducing agent that converts free radicals and other reactive oxygen species into less reactive substances.

aqueous solution A solution in which water is the solvent.

aromatic compound Any of a class of organic compounds characterized by the presence of one or more benzene rings or benzene-like rings.

Arrhenius equation Mathematical relation that gives the temperature dependence of the reaction rate constant; $k = Ae^{-E_a/RT}$.

asymmetric Describes a molecule or object that is not symmetrical. Refers to chiral atoms and structures that can have a molecular dipole.

atom The smallest particle of an element that can be involved in chemical combination with another element.

atom economy The fraction of atoms of starting materials incorporated into the desired final product in a chemical reaction.

atomic force microscope Analytical instrument that produces images of individual atoms or molecules on a surface.

atomic mass unit (amu) The unit of a scale of relative atomic masses of the elements; 1 amu = 1/12 the mass of an atom of ^{12}C.

atomic number The number of protons in the nucleus of an atom of an element.

atomic radius One-half the distance between the nuclei centers of two like atoms in a molecule.

atomic structure The identity and arrangement of subatomic particles in an atom.

atomic theory Dalton's theory that each element consists of atoms that are the smallest particles that embody the chemical properties of the element and that are different from atoms of every other element.

atomic weight The average mass of an atom in a representative sample of atoms of an element.

autoionization The equilibrium reaction in which water molecules react with each other to form hydronium ions and hydroxide ions.

average reaction rate A reaction rate calculated from a change in concentration divided by a change in time.

Avogadro's law The volume of a gas, at a given temperature and pressure, is directly proportional to the amount of gas.

Avogadro's number The number of particles in a mole of any substance, 6.022×10^{23}.

axial position(s) Positions above and below the equatorial plane in a triangular bipyramidal structure.

background radiation Radiation from natural and synthetic radioactive sources to which all members of a population are exposed.

balanced chemical equation A chemical equation that shows equal numbers of atoms of each kind in the products and the reactants.

bar A pressure unit equal to 100,000 Pa.

barometer A device for measuring atmospheric pressure.

basal metabolic rate The energy required to maintain an organism that is awake, at rest, and not digesting or metabolizing food.

base (Arrhenius) Substance that increases concentration of hydroxide ions, OH^-, in aqueous solution. (See also **Brønsted-Lowry base, Lewis base**.)

base ionization constant (K_b) The equilibrium constant for the reaction of a weak base with water to produce hydroxide ions and the conjugate acid of the weak base.

base ionization constant expression Mathematical expression in which the product of the equilibrium concentrations of hydroxide ion and conjugate acid is divided by the equilibrium concentration of the conjugate base.

basic solution An aqueous solution in which the concentration of hydroxide ion is greater than the concentration of hydronium ion.

battery (voltaic cell) An electrochemical cell (or group of cells) in which a product-favored oxidation-reduction reaction is used to produce an electric current.

becquerel A unit of radioactivity equal to 1 nuclear disintegration per second.

beta particles Electrons ejected from certain radioactive nuclei.

beta radiation Radiation composed of electrons.

bidentate ligand A ligand that has two atoms with lone pairs that can form coordinate covalent bonds to the same metal ion.

bimolecular reaction An elementary reaction in which two particles must collide for products to be formed.

binary molecular compound A molecular compound whose molecules contain atoms of only two elements.

binding energy The energy required to separate all nucleons in an atomic nucleus.

binding energy per nucleon The energy per nucleon required to separate all nucleons in an atomic nucleus.

biodegradable Capable of being decomposed by biological means, especially by bacterial action.

boiling The process whereby a liquid vaporizes throughout when its vapor pressure equals atmospheric pressure.

boiling point The temperature at which the equilibrium vapor pressure of a liquid equals the external pressure on the liquid.

boiling-point elevation A colligative property; the difference between the higher boiling point of a solution containing a non-volatile solute and the normal boiling point of the pure solvent.

bond Attractive force between two atoms holding them together, for example, as part of a molecule.

bond angle The angle between the bonds to two atoms that are bonded to the same third atom.

bond enthalpy (bond energy) The change in enthalpy when a mole of chemical bonds of a given type is broken, separating the bonded atoms; the atoms and molecules must be in the gas phase.

bond length The distance between the centers of the nuclei of two bonded atoms.

bonding electrons Electron pairs shared in covalent bonds.

bonding molecular orbital A lower-energy molecular orbital that can be occupied by bonding electrons.

bonding pair A pair of valence electrons that are shared between two atoms.

Born-Haber cycle A stepwise thermochemical cycle in which the constituent elements are converted to ions and combined to form an ionic compound.

boundary surface A surface within which there is a specified probability (often 90%) that an electron will be found.

Boyle's law The volume of a confined ideal gas varies inversely with the applied pressure, at constant temperature and amount of gas.

Brønsted-Lowry acid A hydrogen ion donor.

Brønsted-Lowry acid-base reaction A reaction in which an acid donates a hydrogen ion and a base accepts the hydrogen ion.

Brønsted-Lowry base A hydrogen ion acceptor.

buckyball Buckminsterfullerene; an allotrope of carbon consisting of molecules in which 60 carbon atoms are arranged in a cage-like structure consisting of five-membered rings sharing edges with six-membered rings.

buffer See **buffer solution.**

buffer capacity The quantity of acid or base a buffer can accommodate without a significant pH change (more than one pH unit).

buffer solution A solution that resists changes in pH when limited amounts of acids or bases are added; it contains a weak acid and its conjugate base, or a weak base and its conjugate acid.

caloric value The energy of complete combustion of a stated size sample of a food, usually reported in Calories (kilocalories).

calorie (cal) A unit of energy equal to 4.184 J. Approximately 1 cal is required to raise the temperature of 1 g of liquid water by 1 °C.

Calorie (Cal) A unit of energy equal to 4.184 kJ = 1 kcal. (See also **kilocalorie.**)

calorimeter A device for measuring the quantity of thermal energy transferred during a chemical reaction or some other process.

capillary action The process whereby a liquid rises in a small-diameter tube due to noncovalent interactions between the liquid and the tube's material.

carbohydrates Biochemical compounds with the general formula $C_x(H_2O)_y$, in which x and y are whole numbers.

carbonyl group An organic functional group consisting of carbon bonded to two other atoms and double bonded to oxygen; $>C{=}O$.

carboxylic acid An organic compound characterized by the presence of the carboxyl group (—COOH).

catalyst A substance that increases the rate of a reaction but is not consumed in the overall reaction.

catalytic cracking A petroleum refining process using a catalyst, heat, and pressure to break long-chain hydrocarbons into shorter-chain hydrocarbons, including both alkanes and alkenes suitable for gasoline.

catalytic reforming A petroleum refining process in which straight-chain hydrocarbons are converted to branched-chain hydrocarbons and aromatics for use in gasoline and the manufacture of other organic compounds.

catenation Formation of chains and rings by bonds between atoms of the same element.

cathode The electrode of an electrochemical cell at which reduction occurs.

cathodic protection A process of protecting a metal from corrosion whereby it is made the cathode by connecting it electrically to a more reactive metal or to the negative pole of a DC voltage supply.

cation An ion with a positive electrical charge.

cell voltage The electromotive force of an electrochemical cell; the quantity of work in joules a cell can produce per coulomb of charge that the chemical reaction produces.

Celsius temperature scale A scale defined by the freezing (0 °C) and boiling (100 °C) points of pure water, at 1 atm.

cement A solid consisting of microscopic particles containing compounds of calcium, iron, aluminum, silicon, and oxygen in varying proportions and tightly bound to one another.

ceramics Materials fashioned from clay or other natural materials at room temperature and then hardened by heat.

CFCs See **chlorofluorocarbons.**

change of state A physical process in which one state of matter is changed into another (such as melting a solid to form a liquid).

Charles's law The volume of an ideal gas at constant pressure and amount of gas varies directly with its absolute temperature.

chelating ligand A ligand that uses more than one atom to bind to the same metal ion in a coordination complex.

chemical bond Attractive force between two atoms holding them together, for example, as part of a molecule.

chemical change (chemical reaction) A process in which substances (reactants) change into other substances (products) by rearrangement, combination, or separation of atoms.

chemical compound (compound) A pure substance (e.g., sucrose or water) that can be decomposed into two or more different pure substances; homogeneous, constant-composition matter that consists of two or more chemically combined elements.

chemical element (element) A substance (e.g., carbon, hydrogen, or oxygen) that cannot be decomposed into two or more new substances by chemical or physical means.

chemical equilibrium A state in which the concentrations of reactants and products remain constant because the rates of forward and reverse reactions are equal.

chemical formula (formula) A notation combining element symbols and numerical subscripts that shows the relative numbers of each kind of atom in a molecule or formula unit of a substance.

chemical fuel A substance that reacts exothermically with atmospheric oxygen and is available at reasonable cost and in reasonable quantity.

chemical kinetics The study of the speeds of chemical reactions and the nanoscale pathways or rearrangements by which atoms, ions, and molecules are converted from reactants to products.

chemical periodicity, law of Law stating that the properties of the elements are periodic functions of atomic number.

chemical property Describes the kinds of chemical reactions that chemical elements or compounds can undergo.

chemical reaction (chemical change) A process in which substances (reactants) change into other substances (products) by rearrangements, combination, or separation of atoms.

chemistry The study of matter and the changes it can undergo.

chemotrophs (See also **phototrophs.**) Organisms that must depend on phototrophs to create the chemical substances from which they obtain Gibbs free energy.

chiral Describes a molecule or object that is not superimposable on its mirror image.

chlor-alkali process Electrolysis process for producing chlorine and sodium hydroxide from aqueous sodium chloride.

chlorination Addition of chlorine or a chlorine compound to kill bacteria in municipal water supplies; HOCl formed in water is the antibacterial agent.

chlorofluorocarbons (CFCs) Compounds of carbon, fluorine, and chlorine. CFCs have been implicated in stratospheric ozone depletion.

***cis* isomer** The isomer in which two like substituents are on the same side of a carbon-carbon double bond, the same side of a ring of carbon atoms, or the same side of a coordination complex.

***cis-trans* isomerism** A form of stereoisomerism in which the isomers have the same molecular formula and the same atom-to-atom bonding sequence, but differ in the location of pairs of substituents on the same side or on opposite sides of a molecule or ion.

Clausius-Clapeyron equation Equation that gives the relationship between vapor pressure and temperature.

closest packing Arranging atoms so that they are packed into the minimum volume.

coagulation The process in which the protective charge layer on colloidal particles is overcome, causing them to aggregate into a soft, semisolid, or solid mass.

coefficients (stoichiometric coefficients) The multiplying numbers assigned to the formulas in a chemical equation in order to balance the equation.

cofactor An inorganic or organic molecule or ion required by an enzyme to carry out its catalytic function.

colligative properties Properties of solutions that depend only on the concentration of solute particles in the solution, not on the nature of the solute particles.

colloid A state intermediate between a solution and a suspension, in which solute particles are large enough to scatter light, but too small to settle out; found in gas, liquid, and solid states.

combination reaction A reaction in which two reactants combine to give a single product.

combined gas law A form of the ideal gas law that relates the *P, V, T* of a given amount of gas before and after a change: $P_1V_1/T_1 = P_2V_2/T_2$.

combining volumes, law of At constant temperature and pressure, the volumes of reacting gases are always in the ratios of small whole numbers.

combustion analysis A quantitative method to obtain percent composition data for compounds that can burn in oxygen.

combustion reaction A reaction in which an element or compound burns in air or oxygen.

common ion effect Shift in equilibrium position that results from addition of an ion identical to one in the equilibrium.

complementary base pair Bases, each in a different DNA strand, that hydrogen-bond to each other: guanine with cytosine and adenine with thymine.

complex ion An ion with several molecules or ions connected to a central metal ion by coordinate covalent bonds.

composites Materials with components that may be metals, polymers, and ceramics.

compound See **chemical compound.**

compressibility The property of a gas that allows it to be compacted into a smaller volume by application of pressure.

concentration The relative quantities of solute and solvent in a solution.

concentration cell An electrochemical cell in which the voltage is generated because of a difference in concentrations of the same chemical species.

concrete A mixture of cement, sand, and aggregate (crushed stone or pebbles) in varying proportions that reacts with water and carbon dioxide to form a rock-hard solid.

condensation Process whereby a molecule in the gas phase enters the liquid phase.

condensation polymer A polymer made from the condensation reaction of monomer molecules that contain two or more functional groups.

condensation reaction A chemical reaction in which two (or more) molecules combine to form a larger molecule, simultaneously producing a small molecule such as water.

condensed formula A chemical formula of an organic compound indicating how atoms are grouped together in a molecule.

conduction band In a solid, an energy band that contains electrons of higher energy than those in the valence band, the electrons are free to move about the solid.

conductor A material that conducts electric current; has an overlapping valence band and conduction band.

conjugate acid-base pair A pair of molecules or ions related to one another by the loss and gain of a single hydrogen ion.

conjugated Refers to a system of alternating single and double bonds in a molecule.

conservation of energy, law of (first law of thermodynamics) Law stating that energy can be neither created nor destroyed—the total energy of the universe is constant.

conservation of mass, law of Law stating that there is no detectable change in mass during an ordinary chemical reaction.

constant composition, law of Law stating that a chemical compound always contains the same elements in the same proportions by mass.

constitutional isomers (structural isomers) Compounds with the same molecular formula that differ in the order in which their atoms are bonded together.

continuous phase The solvent-like dispersing medium in a colloid.

continuous spectrum A spectrum consisting of all possible wavelengths.

conversion factor (proportionality factor) A relationship between two measurement units derived from the proportionality of one quantity to another (e.g., density is the conversion factor between mass and volume).

coordinate covalent bond A chemical bond in which both of the two electrons forming the bond were originally associated with the same one of the two bonded atoms.

coordination complex An ion or uncharged molecule in which several molecules or ions are connected to a central metal ion by coordinate covalent bonds.

coordination compound A compound in which complex ions are combined with oppositely charged ions to form an uncharged compound; an uncharged coordination complex.

coordination number The number of coordinate covalent bonds between ligands and a central metal ion in a coordination complex.

copolymer A polymer formed by combining two different types of monomers.

core electrons The electrons in the filled inner shells of an atom.

corrosion Oxidation of a metal exposed to the environment.

coulomb The unit of electrical charge equal to the quantity of charge that passes a fixed point in an electrical circuit when a current of one ampere flows for one second.

Coulomb's law Law that represents the force of attraction between two charged particles; $F = k(q_1q_2/d^2)$.

covalent bond Interatomic attraction resulting from the sharing of electrons between two atoms.

critical mass The minimum quantity of fissionable material needed to support a self-sustaining chain reaction.

critical point The temperature and pressure above which there is no longer any distinction between liquid and gas phases; the end of the liquid-gas equilibrium line in a phase diagram.

critical pressure The vapor pressure of a liquid at its critical temperature.

critical temperature The temperature above which there is no distinction between liquid and vapor phases.

cryogens Liquefied gases that have temperatures below -150 °C.

crystal-field splitting energy Energy difference between sets of *d* orbitals on the central metal ion in a coordination compound.

crystal-field theory Theory that predicts spectra and magnetism of coordination compounds based on electrostatic bonding between ligands and a metal ion.

crystal lattice The ordered, repeating arrangement of ions, molecules, or atoms in a crystalline solid.

crystalline solids Solids with an ordered arrangement of atoms, molecules, or ions that results in planar faces and sharp angles of the crystals.

crystallization The process in which mobile atoms, molecules, or ions in a liquid or solution convert into a crystalline solid.

cubic close packing The three-dimensional structure that results when atoms or ions are closest packed in the *abcabc* arrangement.

cubic unit cell A unit cell with equal-length edges that meet at 90° angles.

curie (Ci) A unit of radioactivity equal to 3.7×10^{10} disintegrations per second.

***d*-to-*d* transition** Excitation of an electron from a lower-energy *d*-orbital to a higher-energy *d*-orbital in a transition-metal ion.

Dalton's law of partial pressures The total pressure exerted by a mixture of gases is the sum of the partial pressures of the individual gases in the mixture.

decomposition reaction A reaction in which a compound breaks down chemically to form two or more simpler substances.

degree of polymerization Number of repeating units in a polymer chain.

delocalized electrons Electrons, such as in benzene, that are spread over several atoms in a molecule or polyatomic ion.

denaturation Disruption in protein secondary and tertiary structure brought on by high temperature, heavy metals, and other substances.

density The ratio of the mass of an object to its volume.

deoxyribonucleic acid (DNA) A double-stranded polymer of nucleotides that stores genetic information.

deposition The process of a gas converting directly to a solid.

detergent(s) Molecules whose structure contains a long hydrocarbon portion that is hydrophobic and a polar end that is hydrophilic.

dew point Temperature at which the actual partial pressure of water vapor equals the equilibrium vapor pressure.

diamagnetic Describes atoms or ions in which all the electrons are paired in filled shells so their magnetic fields effectively cancel each other.

diatomic molecule A molecule that contains two atoms.

dietary minerals Essential elements that are not carbon, hydrogen, oxygen, or nitrogen.

diffusion Spread of gas molecules of one type through those of another type.

dilution Reduction of the concentration of solute(s) in a solution by adding solvent (water in the case of an aqueous solution).

dimensional analysis A method of using units in calculations to check for correctness.

dimer A molecule made from two smaller units.

dipole-dipole attraction The noncovalent force of attraction between any two polar molecules or polar regions in the same large molecule.

dipole moment The product of the magnitude of the partial charges ($\delta+$ and $\delta-$) of a molecule times the distance of separation between the charges.

disaccharides Carbohydrates such as sucrose consisting of two monosaccharide units.

dispersed phase The larger-than-molecule-sized particles that are distributed uniformly throughout a colloid.

displacement reaction A reaction in which one element reacts with a compound to form a new compound and release a different element.

dissociation The separation of cations from anions when an ionic solid dissolves in water.

doping Adding a tiny concentration of one substance (a *dopant*) to improve the semiconducting properties of another.

double bond A bond formed by sharing two pairs of electrons between the same two atoms.

dynamic equilibrium A balance between opposing reactions occurring at equal rates.

effective nuclear charge The nuclear positive charge experienced by outer-shell electrons in a many-electron atom.

effusion Escape of gas molecules from a container through a tiny hole into a vacuum.

electrochemical cell A combination of anode, cathode, and other materials arranged so that a product-favored oxidation-reduction reaction can cause a current to flow or an electric current can cause a reactant-favored redox reaction to occur.

electrochemistry The study of the relationship between electron flow and oxidation-reduction reactions.

electrode A device such as a metal plate or wire that conducts electrons into and out of a system.

electrolysis The use of electrical energy to produce a chemical change.

electrolyte A substance that ionizes or dissociates when dissolved in water to form an electrically conducting solution.

electromagnetic radiation Radiation that consists of perpendicular, oscillating electric and magnetic fields that travel through space at the same rate (the speed of light: 186,000 miles/s or 10^8 m/s in a vacuum).

electromotive force (emf) The difference in electrical potential energy between the two electrodes in an electrochemical cell, measured in volts.

electron A negatively charged subatomic particle that occupies most of the volume of an atom.

electron affinity The energy change when a mole of electrons is added to a mole of atoms in the gas phase.

electron capture A radioactive decay process in which one of an atom's inner-shell electrons is captured by the nucleus, which decreases the atomic number by 1.

electron configuration The complete description of the orbitals occupied by all the electrons in an atom or ion.

electron density The probability of finding an electron within a tiny volume in an atom; determined by the square of the wave function.

electron-pair geometry The geometry around a central atom including the spatial positions of bonding and lone electron pairs.

electronegativity A measure of the ability of an atom in a molecule to attract bonding electrons to itself.

electronically excited molecule A molecule whose potential energy is greater than the minimum (ground-state) energy because of a change in its electronic structure.

element (chemical element) A substance (e.g., carbon, hydrogen, and oxygen) that cannot be decomposed into two or more new substances by chemical or physical means.

elementary reaction A nanoscale reaction whose equation indicates exactly which atoms, ions, or molecules collide or change as the reaction occurs.

empirical formula A formula showing the simplest possible ratio of atoms of elements in a compound.

emulsion A colloid consisting of a liquid dispersed in a second liquid; formed by the presence of an *emulsifier* that coats and stabilizes dispersed-phase particles.

enantiomers A pair of molecules consisting of a chiral molecule and its mirror-image isomer.

end point The point at which the indicator changes color during a titration.

endergonic Refers to a reaction that requires input of Gibbs free energy; applies to biochemical reactions that are reactant-favored at body temperature.

endothermic (process) A process in which thermal energy must be transferred into a thermodynamic system in order to maintain constant temperature.

energy The capacity to do work.

energy band In a solid, a large group of orbitals whose energies are closely spaced; in an atomic solid the average energy of a band equals the energy of the corresponding orbital in an individual atom.

energy conservation The conservation of useful energy, that is, of Gibbs free energy.

energy density The quantity of energy released per unit volume of a fuel.

enthalpy change (ΔH) The quantity of thermal energy transferred when a process takes place at constant temperature and pressure.

enthalpy of fusion The enthalpy change when a substance melts; the quantity of energy that must be transferred when a substance melts at constant temperature and pressure.

enthalpy of solution The quantity of thermal energy transferred when a solution is formed at constant T and P.

enthalpy of sublimation The enthalpy change when a solid sublimes; the quantity of energy, at constant pressure, that must be transferred to cause a solid to vaporize.

enthalpy of vaporization The enthalpy change when a substance vaporizes: the quantity of energy that must be transferred when a liquid vaporizes at constant temperature and pressure.

entropy A measure of the number of ways energy can be distributed in a system; a measure of the dispersal of energy in a system.

enzyme A highly efficient biochemical catalyst for one or more reactions in a living system.

enzyme-substrate complex The combination formed by the binding of an enzyme with a substrate through noncovalent forces.

equatorial position(s) Position(s) lying on the equator of an imaginary sphere around a triangular bipyramidal molecular or ionic structure.

equilibrium concentration The concentration of a substance (usually expressed as molarity) in a system that has reached the equilibrium state.

equilibrium constant (K) A quotient of equilibrium concentrations of product and reactant substances that has a constant value for a given reaction at a given temperature.

equilibrium constant expression The mathematical expression associated with an equilibrium constant.

equilibrium vapor pressure Pressure of the vapor of a substance in equilibrium with its liquid or solid in a closed container.

equivalence point The point in a titration at which a stoichiometrically equivalent amount of one substance has been added to another substance.

ester An organic compound formed by the reaction of an alcohol and a carboxylic acid

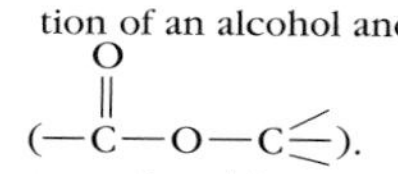

evaporation The process of conversion of a liquid to a gas.

exchange reaction A reaction in which cations and anions that were partners in the reactants are interchanged in the products.

excited state The unstable state of an atom or molecule in which at least one electron does not have its lowest possible energy.

exergonic Refers to a reaction that releases Gibbs free energy; applies to biochemical reactions that are product-favored at body temperature.

exothermic Refers to a process in which thermal energy must be transferred out of a thermodynamic system in order to maintain constant temperature.

extent of reaction The fraction of reactants that has been converted to products.

Faraday constant (F) The quantity of electric charge on one mole of electrons, 9.6485×10^4 C/mol.

fat A solid triester of fatty acids with glycerol.

ferromagnetic A substance that contains clusters of atoms with unpaired electrons whose magnetic spins become aligned, causing permanent magnetism.

first law of thermodynamics (law of conservation of energy) Energy can neither be created nor destroyed—the total energy of the universe is constant.

formal charge The charge a bonded atom would have if its electrons were shared equally.

formation constant (K_f) The equilibrium constant for the formation of a complex ion.

formula (chemical formula) A notation combining element symbols and numerical subscripts that shows the relative numbers of each kind of atom in a molecule or formula unit of a substance.

formula unit The simplest cation-anion grouping represented by the formula of an ionic compound; also the collection of atoms represented by any formula.

formula weight The sum of the atomic weights in amu of all the atoms in a compound's formula.

fractional distillation The process of refining petroleum (or another mixture) by distillation to separate it into groups (fractions) of compounds having distinctive boiling point ranges.

Frasch process Process for recovering sulfur from underground deposits by melting the sulfur with superheated water.

free radical An atom, ion, or molecule that contains one or more unpaired electrons; usually highly reactive.

freezing point Temperature at which a pure liquid freezes—same as the melting point.

freezing-point lowering A colligative property; the difference between the freezing point of a pure solvent and the freezing point of a solution in which a nonvolatile nonelectrolyte solute is dissolved in the solvent.

frequency (ν) The number of complete traveling waves passing a point in a given period of time (cycles per second).

frequency factor The factor (A) in the Arrhenius equation that depends on how often molecules collide when all concentrations are 1 mol/L and on whether the molecules are properly oriented to react when they collide.

fuel cell An electrochemical cell that converts the chemical energy of fuels directly into electricity.

fuel value The quantity of energy released when 1 g of a fuel is burned.

fullerenes Allotropic forms of carbon that consist of many five- and six-membered rings of carbon atoms sharing edges.

functional group An atom or group of atoms that imparts characteristic properties and defines a given class of organic compounds (e.g., the —OH group is present in all alcohols).

galvanized Has a thin coating of zinc metal that forms an oxide coating impervious to oxygen, thereby protecting a less active metal, such as iron, from corrosion.

gamma radiation Radiation composed of highly energetic photons.

gas A phase or state of matter in which a substance has no definite shape and has a volume determined by the volume of its container.

gasohol A blended motor fuel consisting of 90% gasoline and 10% ethanol by mass.

gene The unique sequence of nitrogen bases in DNA that codes for the synthesis of a specific protein; carrier of a genetic trait.

Gibbs free energy A thermodynamic function that decreases for any product-favored system. For a process at constant temperature and pressure, $\Delta G = \Delta H - T\Delta S$.

glass Amorphous, clear solids formed from silicates and other oxides.

global warming Increase in temperature at Earth's surface as a result of the greenhouse effect amplified by increasing concentrations of carbon dioxide and other greenhouse gases.

glycogen A highly branched, high-molar-mass polymer of glucose found in animals.

glycosidic linkage The C—O—C bond that connects monosaccharides in disaccharides and polysaccharides; forms between carbons 1 and 4 or 1 and 6 of linked monosaccharides.

gram The basic unit of mass in the metric system; equal to 1×10^{-3} kg.

gray The SI unit of absorbed radiation dose equal to the absorption of 1 joule per kilogram of material.

greenhouse effect Atmospheric warming caused when atmospheric carbon dioxide, water vapor, methane, ozone, and other greenhouse gases absorb infrared radiation reradiated from Earth.

ground state The state of an atom or molecule in which all of the electrons are in their lowest possible energy levels.

groups The vertical columns of the periodic table of the elements.

Haber-Bosch process The process developed by Fritz Haber and Carl Bosch for the direct synthesis of ammonia from its elements.

half-cell One half of an electrochemical cell in which only the anode or cathode is located.

half-life, $t_{1/2}$ The time required for the concentration of one reactant to reach half its original value; radioactivity—the time required for the activity of a radioactive sample to reach half of its original value.

half-reaction A reaction that represents either an oxidation or a reduction process.

halide ion A monatomic ion (1−) of a halogen.

halogens The elements in Group 7A of the periodic table.

heat (heating) The energy-transfer process between two samples of matter at different temperatures.

heat capacity The quantity of energy that must be transferred to an object to raise its temperature by 1 °C.

heat of See **enthalpy of.**

heating curve A plot of the temperature of a substance versus the quantity of energy transferred to it by heating.

helium burning The fusion of helium nuclei to form beryllium-8, as it occurs in stars.

Henderson-Hasselbalch equation The equation describing the relationships among the pH of a buffer solution, the pK_a of the acid, and the concentrations of the acid and its conjugate base.

Henry's law A mathematical expression for the relationship of gas pressure and solubility; $S_g = k_H P_g$.

Hess's law If two or more chemical equations can be combined to give another equation, the enthalpy change for that equation will be the sum of the enthalpy changes for the equations that were combined. A similar situation applies to entropy and Gibbs free energy.

heterogeneous catalyst A catalyst that is in a different phase from that of the reaction mixture.

heterogeneous mixture A mixture in which components remain separate and can be observed as individual substances or phases.

heterogeneous reaction A reaction that takes place at an interface between two phases, solid and gas, for example.

hexadentate A ligand in which each of six different atoms donates an electron pair to a coordinated central metal ion.

hexagonal close packing The three-dimensional structure that results when layers of atoms in a solid are closest packed in the *ababab* arrangement.

high-spin complex A complex ion that has the maximum possible number of unpaired electrons.

homogeneous catalyst A catalyst that is in the same phase as that of the reaction mixture.

homogeneous mixture A mixture of two or more substances in a single phase that is uniform throughout; also called a solution.

homogeneous reaction A reaction in which the reactants and products are all in the same phase.

Hund's rule Electrons pair only after each orbital in a subshell is occupied by a single electron.

hybrid orbitals Orbitals formed on a single atom by combining atomic orbitals of appropriate energy and orientation.

hybridized Refers to atomic orbitals of proper energy and orientation that have combined to form hybrid orbitals.

hydrate A solid compound that has a stoichiometric amount of water molecules bonded to metal ions or trapped within its crystal lattice.

hydration The binding of one or more water molecules to an ion or molecule within a solution or within a crystal lattice.

hydrocarbon An organic compound composed only of carbon and hydrogen.

hydrogen bond Noncovalent interaction between a hydrogen atom and a very electronegative atom to produce an unusually strong dipole-dipole force.

hydrogen burning The fusion of hydrogen nuclei (protons) to form helium, as it occurs in stars.

hydrogenation An addition reaction in which hydrogen adds to the double bond of an alkene; the catalyzed reaction of H_2 with a liquid triglyceride to produce saturated fatty acid chains, which convert the triglyceride into a semisolid or solid.

hydrolysis A reaction in which a bond is broken by reaction with a water molecule and the —H and —OH of the water add to the atoms of the broken bond.

hydronium ion H_3O^+; the simplest proton-water complex; responsible for acidity.

hydrophilic "Water-loving," a term describing a polar molecule or part of a molecule that is strongly attracted to water molecules.

hydrophobic "Water-fearing," a term describing a nonpolar molecule or part of a molecule that is not attracted to water molecules.

hydroxide ion OH^- ion; bases increase the concentration of hydroxide ions in solution.

hypertonic Refers to a solution having a higher concentration of nanoscale particles and therefore a higher osmotic pressure than another solution.

hypothesis A tentative explanation for an observation and a basis for experimentation.

hypotonic Refers to a solution having a lower solute concentration of nanoscale particles and therefore a lower osmotic pressure than another solution.

ideal gas A gas that behaves exactly as described by the ideal gas law, and by Boyle's, Charles's, and Avogadro's laws.

ideal gas constant The proportionality constant, *R*, in the equation $PV = nRT$; $R = 0.0821\ \text{L atm mol}^{-1}\ \text{K}^{-1} = 8.314\ \text{J K}^{-1}\ \text{mol}^{-1}$.

ideal gas law A law that relates pressure, volume, amount (moles), and temperature for an ideal gas; the relationship expressed by the equation $PV = nRT$.

immiscible Describes two liquids that form two separate phases when mixed because each is only slightly soluble in the other.

induced dipole A temporary dipole created by a momentary uneven distribution of electrons in a molecule or atom.

induced fit The change in the shape of an enzyme, its substrate, or both when they bind.

inhibitor A molecule or ion other than the substrate that causes a decrease in catalytic activity of an enzyme.

initial rate The instantaneous rate of a reaction determined at the very beginning of the reaction.

initiation The breaking of a carbon-carbon double bond in a polymerization reaction to produce a molecule with highly reactive sites that react with other molecules to produce a polymer; first step in a chain reaction.

inorganic compound A chemical compound that is not an organic compound; usually of mineral or nonbiological origin.

insoluble Describes a solute, almost none of which dissolves in a solvent.

instantaneous reaction rate The rate at a particular time after a reaction has begun.

insulator A material that has a large energy gap between fully occupied and empty energy bands, and does not conduct electricity.

intermediate An atom, molecule, or ion that is produced in one step of a reaction mechanism and used up in a later step of the mechanism.

intermolecular forces Noncovalent attractions between separate molecules.

internal energy The sum of the individual energies (kinetic and potential) of all of the nanoscale particles (atoms, molecules, or ions) in a sample of matter.

ion An atom or group of atoms that has lost or gained one or more electrons so that it is no longer electrically neutral.

ion product (*Q*) A value found from an expression with the same mathematical form as the solubility product expression (K_{sp}) but using the actual concentrations rather than equilibrium concentrations of the species involved. (See **reaction quotient.**)

ionic bonding Forces of attraction between cations and anions in the crystal lattice of an ionic compound.

ionic compound A compound that consists of positive and negative ions (cations and anions).

ionic hydrate Ionic compounds that incorporate water molecules in the ionic crystal lattice.

ionic radius Radius of an anion or cation in an ionic compound.

ionization constant for water (K_w) The equilibrium constant that is the mathematical product of the hydronium ion concentration and the concentration of hydroxide ion in any aqueous solution; $K_w = 1 \times 10^{-14}$ at 25 °C.

ionization energy The energy needed to remove a mole of electrons from a mole of atoms in the gas phase.

isoelectronic Refers to atoms and ions that have identical electron configurations.

isomers Compounds that have the same molecular formula but different arrangements of atoms.

isotonic Refers to a solution having the same concentration of nanoscale particles and therefore the same osmotic pressure as another solution.

isotopes Forms of an element composed of atoms with the same atomic number but different mass numbers owing to a difference in the number of neutrons.

joule (J) A unit of energy equal to 1 kg m^2/s^2. The kinetic energy of a 2-kg object traveling at a speed of 1 m/s.

Kelvin temperature scale (See also **absolute temperature scale.**) A temperature scale on which the zero is the lowest possible temperature and the degree is the same size as a Celsius degree.

ketone An organic compound characterized by the presence of a carbonyl group in which the carbon atom is bonded to two other carbon atoms ($R_2C{=}O$).

kilocalorie (kcal or Cal) (See also **calorie.**) A unit of energy equal to 4.184 kJ. Approximately 1 kcal (1 Cal) is required to raise the temperature of 1 kg of liquid water by 1 °C. The food Calorie.

kinetic energy Energy that an object has because of its motion. Equal to $\frac{1}{2}mv^2$, where m is the object's mass and v is its velocity.

kinetic-molecular theory The theory that matter consists of nanoscale particles that are in constant, random motion.

lanthanide contraction The decrease in atomic radii across the fourth-period and the fifth-period *f*-block elements (lanthanides and actinides) due to the lack of electron screening by electrons in the 4*f* and 5*f* orbitals.

lanthanides The elements after lanthanum in the sixth period in which the 4*f* subshell is being filled.

lattice energy Enthalpy of formation of 1 mol of an ionic solid from its separated gaseous ions.

law A statement that summarizes a wide range of experimental results and has not been contradicted by experiments.

law of chemical periodicity Law stating that the properties of the elements are periodic functions of atomic number.

law of combining volumes At constant temperature and pressure, the volumes of reacting gases are always in the ratios of small whole numbers.

law of conservation of energy (first law of thermodynamics) Law stating that energy can be neither created nor destroyed—the total energy of the universe is constant.

law of conservation of mass Law stating that there is no detectable change in mass in an ordinary chemical reaction.

law of constant composition Law stating that a chemical compound always contains the same elements in the same proportions by mass.

law of multiple proportions When two elements A and B can combine in two or more ways, the mass ratio A : B in one compound is a small-whole-number multiple of the mass ratio A : B in the other compound.

Le Chatelier's principle If a system is at equilibrium and the conditions are changed so that it is no longer at equilibrium, the system will react to give a new equilibrium in a way that partially counteracts the change.

Lewis acid A molecule or ion that can accept an electron pair from another atom, molecule, or ion to form a new bond.

Lewis base A molecule or ion that can donate an electron pair to another atom, molecule, or ion to form a new bond.

Lewis dot symbol An atomic symbol with dots representing valence electrons.

Lewis structure Structural formula for a molecule that shows all valence electrons as dots or as lines that represent covalent bonds.

ligands Atoms, molecules, or ions bonded to a central atom, such as the central metal ion in a coordination complex.

limiting reactant The reactant present in limited supply that controls the amount of product formed in a reaction.

line emission spectrum A spectrum produced by excited atoms and consisting of discrete wavelengths of light.

linear Molecular geometry in which there is a 180° angle between bonded atoms.

lipid bilayer The structure of cell membranes that are composed of two aligned layers of phospholipids with their hydrophobic regions within the bilayer.

liquid A phase of matter in which a substance has no definite shape but a definite volume.

London forces Forces resulting from the attraction between positive and negative regions of momentary (induced) dipoles in neighboring molecules.

lone-pair electrons Paired valence electrons unused in bond formation; also called nonbonding or unshared pairs.

low-spin complex A coordination complex that has the minimum possible number of unpaired electrons.

macromolecule A very large polymer molecule made by chemically joining many small molecules (monomers).

macroscale Refers to samples of matter that can be observed by the unaided human senses; samples of matter large enough to be seen, measured, and handled.

main-group elements Elements in the eight A groups to the left and right of the transition elements in the periodic table; the *s*- and *p*-block elements.

major minerals Dietary minerals present in humans in quantities greater than 100 mg per kg of body weight.

mass A measure of an object's resistance to acceleration.

mass fraction The ratio of the mass of one component to the total mass of a sample.

mass number The number of protons plus neutrons in the nucleus of an atom of an element.

mass percent The mass fraction multiplied by 100%.

mass spectrometer An analytical instrument used to measure atomic and molecular masses directly.

mass spectrum A plot of ion abundance versus the mass of the ions; produced by a mass spectrometer.

materials science The science of the relationships between the structure and the chemical and physical properties of materials.

matter Anything that has mass and occupies space.

melting point The temperature at which the structure of a pure solid collapses and the solid changes to a liquid.

meniscus A concave or convex surface that forms on a liquid as a result of the balance of noncovalent forces in a narrow container.

metabolism All of the chemical reactions that occur as an organism converts food nutrients into constituents of living cells, to stored Gibbs free energy, or to thermal energy.

metal An element that is malleable, ductile, forms alloys, and conducts an electric current.

metal activity series A ranking of relative reactivity of metals in displacement and other kinds of reactions.

metallic bonding In solid metals, the nondirectional attraction between positive metal ions and the surrounding sea of negatively charged electrons.

metalloid An element that has some typically metallic properties and other properties that are more characteristic of nonmetals.

methyl group A $-CH_3$ group.

metric system A decimalized measurement system.

micelles Colloid-sized particles built up from many surfactant molecules; micelles can transport various materials within them.

microscale Refers to samples of matter so small that they have to be viewed with a microscope.

millimeters of mercury (mm Hg) A unit of pressure related to the height of a column of mercury in a mercury barometer (760 mm Hg = 1 atm = 101.3 kPa).

mineral A naturally occurring inorganic compound with a characteristic composition and crystal structure.

miscible Describes two liquids that will dissolve in each other in any proportion.

MO (see molecular orbital)

model A mechanical or mathematical way to make a theory more concrete, such as a molecular model.

molality (*m*) A concentration term equal to the molar amount of solute per kilogram of solvent.

molar amount See **amount.**

molar enthalpy of fusion The energy transfer required to melt 1 mol of a pure solid.

molar heat capacity The quantity of energy that must be transferred to 1 mol of a substance to increase its temperature by 1 °C.

molar mass The mass in grams of 1 mol of atoms, molecules, or formula units of one kind, numerically equal to the atomic or molecular weight in amu.

molar solubility The solubility of a solute in a solvent, expressed in moles per liter.

molarity (M) Solute concentration expressed as the molar amount of solute per liter of solution.

mole (mol) The amount of substance that contains as many elementary particles as there are atoms in exactly 0.012 kg of carbon-12 isotope.

mole fraction (X) The ratio of number of moles of one component to the total number of moles in a mixture of substances.

mole ratio (stoichiometric factor) A mole-to-mole ratio relating the molar amount of a reactant or product to the molar amount of another reactant or product.

molecular compound A compound composed of atoms of two or more elements chemically combined in molecules.

molecular formula A formula that expresses the number of atoms of each type within one molecule of a substance.

molecular geometry The three-dimensional arrangement of atoms in a molecule.

molecular orbitals (MOs) Orbitals generated by combining atomic orbitals; MOs extend over an entire molecule.

molecular weight The sum of the atomic weights of all the atoms in a substance's formula.

molecule The smallest particle of an element or compound that exists independently, and retains the chemical properties of that element or compound.

momentum The product of the mass (m) times the velocity (v) of an object in motion.

monatomic ion An ion consisting of one atom bearing an electrical charge.

monodentate ligand A ligand that donates one electron pair to a coordinated metal ion.

monomer The small repeating unit from which a polymer is formed.

monoprotic acid An acid that can donate a single hydrogen ion per molecule or polyatomic ion.

monosaccharides The simplest carbohydrates, composed of one saccharide unit.

monounsaturated fatty acid Refers to fatty acids, such as oleic acid, that contain only one carbon-carbon double bond per molecule.

mortar A mixture of cement, sand, and lime that reacts with water and carbon dioxide to form a rock-hard solid.

multiple covalent bonds Double or triple covalent bonds.

multiple proportions, law of When two elements A and B can combine in two or more ways, the mass ratio A : B in one compound is a small-whole-number multiple of the mass ratio A : B in the other compound.

***n*-type semiconductor** A material made by doping a semiconductor with an impurity that leaves extra valence electrons.

nanoscale Refers to samples of matter (e.g., atoms and molecules) whose normal dimensions are in the 1 to 100 nanometer range.

nanotubes Members of the family of fullerenes in which graphite-like layers of carbon atoms form cylindrical shapes.

Nernst equation The equation relating the potential of an electrochemical cell to the concentrations of the chemical species involved in the oxidation-reduction reactions occurring in the cell.

net ionic equation A chemical equation in which only those molecules or ions undergoing chemical changes in the course of the reaction are represented.

network solid A solid consisting of one huge molecule in which all atoms are connected via a network of covalent bonds.

neurons Specialized cells that are part of animals' nervous systems and that function according to electrochemical principles.

neutral solution A solution containing equal concentrations of H_3O^+ and OH^-; a solution that is neither acidic nor basic.

neutron An electrically neutral subatomic particle found in the nucleus.

newton (N) The SI unit of force; equal to 1 kg times an acceleration of 1 m/s^2; 1 kg m s^{-2}.

nitrogen cycle The natural cycle of chemical transformations involving nitrogen and its compounds.

nitrogen fixation The conversion of atmospheric nitrogen, N_2, to nitrogen compounds utilizable by plants or industry.

noble gas notation An abbreviated electron configuration of an element in which filled inner shells are represented by the symbol of the preceding noble gas in brackets. For Al, this would be [Ne]$3s^2 3p^1$.

noble gases Gaseous elements in Group 8A; the least reactive elements.

nonbiodegradable Not capable of being decomposed by microorganisms.

noncovalent interactions All forces of attraction other than covalent, ionic, or metallic bonding.

nonelectrolyte A substance that dissolves in water to form a solution that does not conduct electricity.

nonmetal Element that does not have the chemical and physical properties of a metal.

nonpolar covalent bond A bond in which the electron pair is shared equally by the bonded atoms.

nonpolar molecule A molecule that is not polar either because it has no polar bonds or because its polar bonds are oriented symmetrically so that they cancel each other.

NO_x Oxides of nitrogen.

normal boiling point The temperature at which the vapor pressure of a liquid equals 1 atm.

nuclear burning The nuclear fusion reactions by which elements are formed in stars.

nuclear decay Spontaneous emission of radioactivity by an unstable nucleus that is converted into a more stable nucleus.

nuclear fission The highly exothermic process by which very heavy fissionable nuclei split to form lighter nuclei.

nuclear fusion The highly exothermic process by which very light nuclei combine to form heavier nuclei.

nuclear magnetic resonance The process in which the nuclear spins of atoms align in a magnetic field and absorb radio frequency photons to become excited. These excited atoms then return to a lower energy state when they emit the absorbed radiofrequency photons.

nuclear medicine The use of radioisotopes in medical diagnosis and therapy.

nuclear reaction A process in which one or more atomic nuclei change into one or more different nuclei.

nuclear reactor A container in which a controlled nuclear reaction takes place.

nucleon A nuclear particle, either a neutron or a proton.

nucleotide Repeating unit in DNA, composed of one sugar unit, one phosphate group, and one cyclic nitrogen base.

nucleus (atomic) The tiny central core of an atom; contains protons and neutrons. (There are no neutrons in hydrogen-1.)

nutrients The chemical raw materials, eaten as food, that are needed for survival of an organism.

octahedral Molecular geometry of six groups around a central atom in which all groups are at angles of 90° to other groups.

octane number A measure of the ability of a gasoline to burn smoothly in an internal-combustion engine.

octet rule In forming bonds, many main group elements gain, lose, or share electrons to achieve a stable electron configuration characterized by eight valence electrons.

optical fiber A fiber made of glass constructed so that light can pass through it with little loss of intensity; used for transmission of information.

orbital A region of an atom or molecule within which there is a significant probability that an electron will be found.

orbital shape The shape of an electron density distribution determined by an orbital.

order of reaction The reaction rate dependency on the concentration of a reactant or product, expressed as an exponent of a concentration term in the rate equation.

ores Minerals containing a sufficiently high concentration of an element to make its extraction profitable.

organic compound A compound of carbon with hydrogen, possibly also oxygen, nitrogen, sulfur, phosphorus, or other elements.

osmosis The movement of a solvent (such as water) through a semipermeable membrane from a region of lower solute concentration to a region of higher solute concentration.

osmotic pressure (Π) The pressure that must be applied to a solution to stop osmosis from a sample of pure solvent.

overall reaction order The sum of the exponents for all concentration terms in the rate equation.

oxidation The loss of electrons by an atom, ion, or molecule, leading to an increase in oxidation number.

oxidation number (oxidation state) The hypothetical charge an atom would have if all bonds to that atom were completely ionic.

oxidation-reduction reaction (redox reaction) A reaction involving the transfer of one or more electrons from one species to another so that oxidation numbers change.

oxides Compounds of oxygen combined with another element.

oxidized The result when an atom, molecule, or ion loses one or more electrons.

oxidizing agent The substance that accepts electron(s) and is reduced in an oxidation-reduction reaction.

oxoacids Acids in which the acidic hydrogen is bonded directly to an oxygen atom.

oxoanion A polyatomic anion that contains oxygen.

oxygenated gasolines Blends of gasoline with oxygen-containing organic compounds such as methanol, ethanol, and *tertiary*-butyl alcohol.

ozone hole Regions of ozone depletion in the stratosphere centered on Earth's poles, most significantly the South Pole.

ozone layer Region of maximum relative ozone concentration in the stratosphere.

***p*-block elements** Main-group elements in Groups 3A through 8A whose valence electrons consist of outermost *s* and *p* electrons.

***p-n* junction** An interface between *p*-type and *n*-type semiconductors that produces a rectifier that allows current to flow in only one direction.

***p*-type semiconductor** A material made by doping a semiconductor with an impurity that leaves a deficiency of valence electrons.

paramagnetic Refers to atoms, molecules, or ions that are attracted to a magnetic field because they have unpaired electrons in incompletely filled electron subshells.

partial hydrogenation Addition of hydrogen to some of the carbon-carbon double bonds in a triglyceride (a fat or oil).

partial pressure The pressure that one gas in a mixture of gases would exert if it occupied the same volume at the same temperature as the mixture.

particulate Atmospheric solid particles, generally larger than 10,000 nm in diameter.

parts per billion (ppb) One part in one billion (10^9) parts.

parts per million (ppm) One part in one million (10^6) parts.

parts per trillion (ppt) One part in one trillion (10^{12}) parts.

pascal (Pa) The SI unit of pressure; $1\ Pa = 1\ N/m^2$.

Pauli exclusion principle An atomic principle that states that, at most, two electrons can be assigned to the same orbital in the same atom or molecule, and these two electrons must have opposite spins.

peptide linkage The amide linkage between two amino acid molecules; found in proteins.

percent abundance The percentage of atoms of a particular isotope in a natural sample of a pure element.

percent composition by mass The percentage of the mass of a compound represented by each of its constituent elements.

percent yield The ratio of actual yield to theoretical yield, multiplied by 100%.

periodic table A table of elements arranged in order of increasing atomic number so that those with similar chemical and physical properties fall in the same vertical groups.

periods The horizontal rows of the periodic table of the elements.

petroleum fractions The mixtures of hundreds of hydrocarbons in the same boiling point range obtained from the fractional distillation of petroleum.

pH The negative logarithm of the hydronium ion concentration ($-\log [H_3O^+]$).

pH meter An instrument for measuring pH of solutions using electrochemical principles.

phase Any of the three states of matter: gas, liquid, solid. Also, one of two or more solid-state structures of the same substance, such as iron in a body-centered cubic or face-centered cubic structure.

phase change A physical process in which one state or phase of matter is changed into another (such as melting a solid to form a liquid).

phase diagram A diagram showing the relationships among the phases of a substance (solid, liquid, and gas), at different temperatures and pressures.

phospholipid Glycerol derivative with two long, nonpolar fatty-acid chains and a polar phosphate group; present in cell membranes.

photochemical reactions Chemical reactions that take place as a result of absorption of photons by reactant molecules.

photochemical smog Smog produced by strong oxidizing agents, such as ozone and oxides of nitrogen, NO_x, that undergo light-initiated reactions with hydrocarbons.

photodissociation Splitting of a molecule into two free radicals by absorption of an ultraviolet photon.

photoelectric effect The emission of electrons by some metals when illuminated by light of certain wavelengths.

photon A massless particle of light whose energy is given by $h\nu$, where ν is the frequency of the light and h is Planck's constant.

photosynthesis A series of reactions in a green plant that combines carbon dioxide with water to form carbohydrate and oxygen.

phototrophs (See also **chemotrophs.**) Organisms that can carry out photosynthesis and therefore can use sunlight to supply their Gibbs free energy needs.

physical changes Changes in the physical properties of a substance, such as the transformation of a solid to a liquid.

physical properties Properties (e.g., melting point or density) that can be observed and measured without changing the composition of a substance.

pi (π) bond A bond formed by the sideways overlap of parallel *p* orbitals.

Planck's constant The proportionality constant, h, that relates energy of a photon to its frequency. The value of h is 6.626×10^{-34} J s.

plasma A state of matter consisting of unbound nuclei and electrons.

plastic A polymeric material that has a soft or liquid state in which it can be molded or otherwise shaped. (See also **thermoplastic** and **thermosetting plastic**.)

polar covalent bond A covalent bond between atoms with different electronegativities; bonding electrons are shared unequally between the atoms.

polar molecule A molecule that is polar because it has polar bonds arranged so that electron density is concentrated at one end of the molecule.

polarization The induction of a temporary dipole in a molecule or atom by shifting of electron distribution.

polluted water Water that is unsuitable for an intended use, such as drinking, washing, irrigation, or industrial use.

polyamides Polymers in which the monomer units are connected by amide bonds.

polyatomic ion An ion consisting of more than one atom.

polydentate ligand Refers to ligands that can form two or more coordinate covalent bonds to the same metal ion.

polyester A polymer in which the monomer units are connected by ester bonds.

polymer A large molecule composed of many smaller repeating units, usually arranged in a chain-like structure.

polypeptide A polymer composed of 20 to 50 amino acid residues joined by peptide linkages (amide linkages).

polyprotic acids Acids that can donate more than one hydrogen ion per molecule.

polysaccharides Carbohydrates that consist of many monosaccharide units.

polyunsaturated acid A carboxylic acid containing two or more carbon-carbon double bonds; commonly refers to a fatty acid.

positron A nuclear particle having the same mass as an electron, but a positive charge.

potential energy Energy that an object has because of its position.

precipitate An insoluble solid product of an exchange reaction in aqueous solution.

pressure The force exerted on an object divided by the area over which the force is exerted.

primary battery A voltaic cell (or battery of cells) in which the oxidation and reduction half-reactions cannot easily be reversed to restore the cell to its original state.

primary pollutants Pollutants that enter the environment directly from their sources.

primary structure of proteins The sequence of amino acids along the polymer chain in a protein molecule.

principal energy level An energy level containing orbitals with the same quantum number ($n = 1, 2, 3 \ldots$).

principal quantum number An integer assigned to each of the allowed main electron energy levels in an atom.

product A substance formed as a result of a chemical reaction.

product-favored system A system in which, when a reaction appears to be over, products predominate over reactants.

proportionality factor (conversion factor) A relationship between two measurement units derived from the proportionality of one quantity to another (e.g., density is the conversion factor between mass and volume).

proton A positively charged subatomic particle found in the nucleus.

pyrometallurgy The extraction of a metal from its ore using chemical reactions carried out at high temperatures.

qualitative In observations, nonnumerical experimental information, such as a description of color or texture.

quantitative Numerical information, such as the mass or volume of a substance, expressed in appropriate units.

quantum The smallest possible unit of a distinct quantity; for example, the smallest possible unit of energy for electromagnetic radiation of a given frequency.

quantum theory The theory that energy comes in very small packets (quanta); this is analogous to matter occurring in very small particles—atoms.

racemic mixture A mixture of equal amounts of enantiomers of a chiral compound.

rad A unit of radioactivity; a measure of the energy of radiation absorbed by a substance, 1.00×10^{-2} J/kg.

radial distribution plot A graph showing the probability of finding an electron as a function of distance from the nucleus of an atom.

radioactive series A series of nuclear reactions in which a radioactive isotope undergoes successive nuclear transformations resulting ultimately in a stable, nonradioactive isotope.

radioactivity The spontaneous emission of energy and/or subatomic particles by unstable atomic nuclei; the energy or particles so emitted.

Raoult's law A mathematical expression for the vapor pressure of the solvent in a solution; $P_1 = X_1P_1^0$.

rate The change in some measurable quantity per unit time.

rate constant (*k*) A proportionality constant relating reaction rate and concentrations of reactants and other species that affect the rate of a specific reaction.

rate law (rate equation) A mathematical equation that summarizes the relationship between concentrations and reaction rate.

rate-limiting step The slowest step in a reaction mechanism.

reactant A substance that is initially present and undergoes change in a chemical reaction.

reactant-favored system A system in which, when a reaction appears to be over, reactants predominate over products.

reaction intermediate An atom, molecule, or ion produced in one step and used in a later step in a reaction mechanism; does not appear in the equation for the overall reaction.

reaction mechanism A sequence of unimolecular and bimolecular elementary reactions by which an overall reaction may occur.

reaction quotient (*Q*) A value found from an expression with the same mathematical form as the equilibrium constant expression but with the actual concentrations in a mixture not at equilibrium.

reaction rate The change in concentration of a reactant or product per unit time.

redox reaction (oxidation-reduction reaction) A reaction involving the transfer of one or more electrons from one species to another so that oxidation numbers change.

reduced The result when an atom, molecule, or ion gains one or more electrons.

reducing agent The atom, molecule, or ion that donates electron(s) and is oxidized in an oxidation-reduction reaction.

reduction The gain of electrons by an atom, ion, or molecule, leading to a decrease in its oxidation number.

reformulated gasolines Oxygenated gasolines with lower volatility and containing a lower percentage of aromatic hydrocarbons than regular gasoline.

relative humidity In the atmosphere, the ratio of actual partial pressure to equilibrium vapor pressure of water at the prevailing temperature.

rem A unit of radioactivity; 1 rem has the physiological effect of 1 roentgen of radiation.

replication The copying of DNA during regular cell division.

resonance hybrid The actual structure of a molecule that can be represented by more than one Lewis structure.

resonance structures The possible structures of a molecule for which more than one Lewis structure can be written, differing by the arrangement of electrons but having the same arrangement of atomic nuclei.

reverse osmosis Application of pressure greater than the osmotic pressure to cause solvent to flow through a semipermeable membrane from a concentrated solution to a solution of lower solute concentration.

reversible process A process for which a very small change in conditions will cause a reversal in direction.

roentgen (R) A unit of radioactivity; 1 R corresponds to deposition of 93.3×10^{-7} J per gram of tissue.

***s*-block elements** Main-group elements in Groups 1A and 2A whose valence electrons are *s* electrons.

salt An ionic compound whose cation comes from a base and whose anion comes from an acid.

salt bridge A device for maintaining balance of ion charges in the compartments of an electrochemical cell.

saponification The hydrolysis of a triglyceride (a fat or oil) by reaction with NaOH to give sodium salts that are soaps.

saturated fats Fats (or oils) that contain only carbon-carbon single bonds in their hydrocarbon chains.

saturated hydrocarbon Hydrocarbon in which carbon atoms are bonded to the maximum number of hydrogen atoms.

saturated solution A solution in which the concentration of solute is the concentration that would be in equilibrium with undissolved solute at a given temperature.

scanning tunneling microscope An analytical instrument that produces images of individual atoms or molecules on a surface.

screening effect Reduction of the effective attraction between nucleus and valence electrons as a result of repulsion of the outer valence electrons by electrons in inner shells.

second law of thermodynamics The total entropy of the universe (the system and surroundings) is continually increasing. In any product-favored system, the entropy of the universe is greater after a reaction than it was before.

secondary battery A voltaic cell (or battery of cells) in which the oxidation and reduction half-reactions can be reversed to restore the cell to its original state after discharge.

secondary pollutants Pollutants that are formed by chemical reactions of primary pollutants.

secondary structure of proteins Regular repeating patterns of molecular structure in proteins, such as α-helix, β-sheet.

semiconductor Material with a narrow energy gap between the valence band and the conduction band; a conductor when an electric field or a higher temperature is applied.

semipermeable membrane A thin layer of material through which only certain kinds of molecules can pass.

shell A collection of orbitals with the same value of the principal quantum number, *n*.

shifting an equilibrium Changing the conditions of an equilibrium system so that the system is no longer at equilibrium and there is a net reaction in either the forward or reverse direction until equilibrium is reestablished.

SI units The International System of Units (Système International), the officially recognized measurement system used by scientists throughout the world.

sievert The SI unit of effective dose of absorbed radiation, 1 Sv = 100 rem.

sigma (σ) bond A bond formed by head-to-head orbital overlap along the bond axis.

significant figures (digits) All digits in an experimentally determined numerical value that correspond with the precision of the experimental measurement; if a measurement is precise to between 1 part in 100 and 1 part in 1000, then the result has three significant figures, such as 3.25×10^2.

simple sugars Monosaccharides and disaccharides.

single covalent bond A bond formed by sharing one pair of electrons between the same two atoms.

smog A mixture of smoke (particulate matter), fog (an aerosol), and other substances that degrade air quality.

solar cell A device that converts solar photons into electricity; based on doped silicon.

solid A state of matter in which a substance has a definite shape and volume.

solubility The maximum amount of solute that will dissolve in a given volume of solvent at a given temperature when pure solute is in equilibrium with the solution.

solubility product constant (K_{sp}) An equilibrium constant that is the product of concentrations of ions in a solution in equilibrium with a solid ionic compound.

solubility product expression Molar concentrations of a cation and anion, each raised to a power equal to its coefficient in the balanced chemical equation for the solubility equilibrium.

solute The material dissolved in a solution.

solution A homogeneous mixture of two or more substances in a single phase.

solvent The medium in which a solute is dissolved to form a solution.

***sp* hybrid orbitals** Orbitals of the same atom formed by the combination of one *s* orbital and one *p* orbital.

sp^2 hybrid orbitals Orbitals of the same atom formed by the combination of one *s* orbital and two *p* orbitals.

sp^3 hybrid orbitals Orbitals of the same atom formed by the combination of one *s* orbital and three *p* orbitals.

specific heat capacity The quantity of energy that must be transferred to 1 g of a substance to increase its temperature by 1 °C.

spectator ion An ion that is present in a solution in which a reaction takes place, but is not involved in the net process.

spectrochemical series A list of ligands in the order of their crystal-field splitting energy.

spectroscopy Use of electromagnetic radiation to study the nature of matter.

spectrum A plot of the intensity of light (photons per unit time) as a function of the wavelength or frequency of light.

standard atmosphere (atm) A unit of pressure; 1 atm = 101.325 kPa = 1.01325 bar = 760 mm Hg exactly.

standard-state conditions These are 1 bar pressure for all gases, 1-M concentration for all solutes, at a specified temperature.

standard enthalpy change The enthalpy change when a process occurs with reactants and products all in their standard states.

standard equilibrium constant ($K°$) An equilibrium constant in which each concentration (or pressure) is divided by the standard-state concentration (or pressure); if concentrations are expressed in moles per liter (or pressures in bars) then the concentration (or pressure) equilibrium constant equals the standard equilibrium constant $\Delta G° = -RT \ln K°$.

standard hydrogen electrode The electrode against which standard reduction potentials are measured, consisting of a platinum electrode at which 1-M hydronium ion is reduced to hydrogen gas at 1 bar.

standard molar enthalpy of formation The standard enthalpy change for forming 1 mol of a compound from its elements, with all substances in their standard states.

standard molar volume The volume occupied by exactly 1 mol of an ideal gas at standard temperature (0 °C) and pressure (1 atm), equal to 22.414 L.

standard reduction potential ($E°$) The potential of an electrochemical cell when a given electrode as the cathode is paired with a standard hydrogen electrode as the anode under standard conditions.

standard state The most stable form of a substance in the physical state in which it exists at 1 bar and a specified temperature.

standard solution A solution whose concentration is known accurately.

standard temperature and pressure (STP) A temperature of 0 °C and a pressure of 1 atm.

standard voltages Electrochemical cell voltages measured under standard conditions.

state function A property whose value is invariably the same if a system is in the same state.

steel A material made from iron with most P, S, and Si impurities removed, a low carbon content, and possibly other alloying metals.

steric factor A factor in the expression for rate of reaction that reflects the fact that some three-dimensional orientations of colliding molecules are more likely to result in reaction than others.

stoichiometric coefficients The multiplying numbers assigned to the species in a chemical equation in order to balance the equation.

stoichiometric factor (mole ratio) A factor relating number of moles of a reactant or product to number of moles of another reactant or product.

stoichiometry The study of the quantitative relations between amounts of reactants and products in chemical reactions.

stratosphere The region of the atmosphere approximately 12 to 50 km above sea level.

stratospheric ozone layer Region of maximum relative concentration of ozone in the stratosphere.

strong acid An acid that ionizes completely in aqueous solution.

strong base A base that ionizes completely in aqueous solution.

strong electrolyte An electrolyte that consists solely of ions in aqueous solution.

structural formulas Formulas written to show how atoms in a molecule or polyatomic ion are connected to each other.

structural isomers (constitutional isomers) Compounds with the same molecular formula that differ in the order in which their atoms are bonded.

sublimation Conversion of a solid directly to a gas with no formation of liquid.

subshell A group of atomic orbitals with the same n and ℓ quantum numbers.

substance Matter of a particular kind; each substance, when pure, has a well-defined composition and a set of characteristic properties that differ from the properties of any other substance.

substrate A molecule or molecules whose reaction is catalyzed by an enzyme.

superconductor A substance that, below some temperature, offers no resistance to the flow of electric current.

supercritical fluid A substance above its critical temperature and pressure; has density characteristic of a liquid, but flow properties of a gas.

supersaturated solution A solution that temporarily contains more solute per unit volume than a saturated solution at a given temperature.

surface tension The energy required to overcome the attractive forces between molecules at the surface of a liquid.

surfactant(s) Compounds consisting of molecules that have both a hydrophobic part and a hydrophilic part.

surroundings Everything that can exchange energy with a thermodynamic system.

system In thermodynamics, that part of the universe that is singled out for observation and analysis. The region of primary concern.

temperature The physical property of matter that determines whether one object can heat another.

tertiary structure of proteins The overall three-dimensional folding of protein molecules.

tetrahedral Molecular geometry of four atoms or groups of atoms around a central atom with bond angles of 109.5°.

theoretical yield The maximum quantity of product theoretically obtainable from a given quantity of reactant in a chemical reaction.

theory A unifying principle that explains a body of facts and the laws based on them.

thermal equilibrium The condition of equal temperatures achieved between two samples of matter that are in contact.

thermochemical expression A balanced chemical equation, including specification of the states of matter of reactants and products, together with the corresponding value of the enthalpy change.

thermodynamics The science of heat, work, and the transformations of each into the other.

thermoplastic A plastic that can be repeatedly softened by heating and hardened by cooling.

thermosetting plastic A polymer that melts upon initial heating and forms cross-links so that it cannot be melted again without decomposition.

third law of thermodynamics A perfect crystal of any substance at 0 K has the lowest possible entropy.

titrant The solution being added from a buret to another solution during a titration.

titration A procedure whereby a substance in a standard solution reacts with a known stoichiometry with a substance whose concentration is to be determined.

titration curve A plot of the progress of a titration as a function of the volume of titrant added.

torr A unit of pressure equivalent to 1 mm Hg.

trace elements (See also **major minerals.**) The dietary minerals that are present in smaller concentrations than the major minerals, sometimes far smaller concentrations.

tracer A radioisotope used to track the pathway of a chemical reaction, industrial process, or medical procedure.

trans **isomer** The isomer in which two like substituents are on opposite sides of a carbon-carbon double bond, a ring of carbon atoms, or a coordination complex.

transition elements Elements that lie in rows 4 through 7 of the periodic table in which *d* or *f* subshells are being filled; comprising scandium through zinc, yttrium through cadmium, lanthanum through mercury, and actinium and elements of higher atomic number.

transition state A molecular structure corresponding to the maximum of a plot of energy versus reaction progress; also known as the activated complex.

triangular bipyramidal Molecular geometry of five groups around a central atom in which three groups are in equatorial positions and two are in axial positions.

triangular planar Molecular geometry of three groups at the corners of an equilateral triangle around a central atom at the center of the triangle.

triglycerides Esters in which a glycerol molecule is joined with three fatty acid molecules.

triple bond A bond formed by sharing three pairs of electrons between two atoms.

triple point The point on a temperature/pressure phase diagram of a substance where solid, liquid, and gas phases are all in equilibrium.

troposphere The lowest region of the atmosphere, extending from Earth's surface to an altitude of about 12 km.

Tyndall effect Scattering of visible light by a colloid.

uncertainty principle The statement that it is impossible to determine simultaneously the exact position and the exact momentum of an electron.

unimolecular reaction A reaction in which the rearrangement of the structure of a single molecule produces the product molecule or molecules.

unit cell A small portion of a crystal lattice that can be replicated in each of three directions to generate the entire lattice.

unsaturated fats Fats (or oils) that contain one or more carbon-carbon double bonds in their hydrocarbon chains.

unsaturated hydrocarbon A hydrocarbon containing double or triple carbon-carbon bonds.

unsaturated solution A solution that contains a smaller concentration of solute than the concentration of a saturated solution at a given temperature.

valence band In a solid, an energy band (group of closely spaced orbitals) that contains valence electrons.

valence bond model A theoretical model that describes a covalent bond as resulting from an overlap of one orbital on each of the bonded atoms.

valence electrons Electrons in an atom's highest occupied principal shell and in partially filled subshells of lower principal shells; electrons available to participate in bonding.

valence-shell electron-pair repulsion model (VSEPR) A simple model used to predict the shapes of molecules and polyatomic ions based on repulsions between bonding pairs and lone pairs around a central atom.

van der Waals equation An equation of state for gases that takes into account the volume occupied by molecules and noncovalent attractions between molecules:

$$\left[P + a\left(\frac{n}{V}\right)^2\right][V - bn] = nRT.$$

vaporization The change of a substance from the liquid to the gas phase.

vapor pressure The pressure of the vapor of a substance in equilibrium with its liquid or solid in a sealed container.

viscosity The resistance of a liquid to flow.

volatility The tendency of a liquid to vaporize.

volt (V) Electrical potential energy difference defined so that 1 joule (work) is performed when 1 coulomb (charge) moves through 1 volt (potential difference).

voltaic cell An electrochemical cell in which a product-favored oxidation-reduction reaction is used to produce an electric current.

water of hydration The water molecules trapped within the crystal lattice of an ionic hydrate or coordinated to a metal ion in a crystal lattice or in solution.

wave functions Solutions to the Schrödinger wave equation that describe the behavior of an electron in an atom.

wavelength (λ) The distance between adjacent crests (or troughs) in a wave.

weak acid An acid that is only partially ionized in aqueous solution.

weak base A base that is only partially ionized in aqueous solution.

weak electrolyte An electrolyte that is only partially ionized in aqueous solution.

weight percent A mass fraction expressed as a percent by multiplying by 100%; used for elemental composition of a compound, composition of a solute in solution.

work (working) A mechanical process that transfers energy to or from an object.

X-ray crystallography The science of determining nanoscale crystal structure by measuring the diffraction of X-rays by a crystal.

zone refining A purification process in which a molten zone is moved through a solid sample, causing the impurities to concentrate in the liquefied portion.

zwitterion A structure containing both a positive charge and a negative charge, commonly due to loss and gain of a hydrogen ion within the same molecule.

Index

Page numbers in **bold face** refer to key term definitions.
Page numbers in *italics* refer to images or diagrams.
Page numbers followed by a "t" indicate a table.
ES-2 and ES-3 refer to inside front cover.
ES-6 and ES-7 refer to inside back cover.
A.1 indicates appendix page 1.

A

E

F

G

H

I

M

Useful Conversion Factors and Relationships

Length

SI unit: meter (m)

1 kilometer = 1000 meters
= 0.62137 mile
1 meter = 100 centimeters
1 centimeter = 10 millimeters
1 nanometer = 1×10^{-9} meter
1 picometer = 1×10^{-12} meter
1 inch = 2.54 centimeter (exactly)
1 Ångström = 1×10^{-10} meter

Mass

SI unit: kilogram (kg)

1 kilogram = 1000 grams
1 gram = 1000 milligrams
1 pound = 453.59237 grams = 16 ounces
1 ton = 2000 pounds

Volume

SI unit: cubic meter (m^3)

1 liter (L) = 1×10^{-3} m^3
= 1 dm^3 = 1000 cm^3
= 1.056710 quarts
1 gallon = 4 quarts

Energy

SI unit: joule (J)

1 joule = 1 kg m^2/s^2
= 0.23901 calorie
= 1 C × 1 V
1 calorie = 4.184 joules

Note: Numbers that reflect definitions, such as 1000 meters = 1 kilometer or 1 ton = 2000 pounds are exact, that is, they have an infinite number of significant figures

Pressure

SI unit: pascal (Pa)

1 pascal = 1 N/m^2
= 1 kg m^{-1} s^{-1}
1 atmosphere = 101.325 kilopascals
= 760. mm Hg = 760. torr
= 14.70 lb/in.^{2}
1 bar = 1×10^5 Pa

Temperature

SI unit: kelvin (K)

If T_K is the numeric value of the temperature in kelvins, t_C the numeric value of the temperature in °C, and t_F the numeric value of the temperature in °F, then

$$T_K = t_C + 273.15$$

$$t_C = \left(\frac{5}{9}\right)(t_F - 32)$$

$$t_F = \left(\frac{9}{5}\right)t_C + 32$$

Location of Useful Tables and Figures

Atomic, Ionic, and Molecular Properties

Atomic radii	Figures 7.22, 7.23
Bond energies	Table 8.2
Bond lengths	Table 8.1
Electron affinities	Table 7.8
Electron configurations	Table 7.5, Figure 7.17, Appendix D
Electronegativity	Figure 8.6
Ionic radii	Figure 7.24
Ionization energies	Table 7.7, Figures 7.25, 7.26
Molecular geometries	Figures 9.4, 9.5, 9.6
Properties of molecular and ionic compounds	Table 3.9
Properties of solids	Table 11.5

Thermodynamic Properties

Specific heat capacities	Table 6.1
Standard molar enthalpies of formation	Table 6.2, App. J
Composition and caloric values of some foods	Table 6.4
Energy equivalencies of fuels	Table 6.3
Standard molar entropies	Table 18.1, App. J
Enthalpy, Gibbs free energy, entropy	Appendix J
Equilibrium constants	Table 14.1 (Also see Appendices F, G, and H)

Acids, Bases, and Salts

Acid-base properties of ions in solution	Table 16.5
Common acids and bases	Table 5.2
Common polyatomic ions	Table 3.7
Solubility guidelines for ionic compounds	Table 5.1
Common monatomic ions	Figure 3.2
Amino acids	Table 12.8
Buffer systems	Table 17.1
Acid and base ionization constants	Table 16.2 (See also Appendices F, G)
Solubility product constants	Appendix H

Other Useful Tables and Figures

Common oxidizing and reducing agents	Table 5.3
Common ligands	Table 22.4, Figure 22.9
Electromagnetic spectrum	Figure 7.1
Composition of dry air	Table 10.1
Standard reduction potentials	Table 19.1 (See also Appendix I)
Density values	Table 1.1